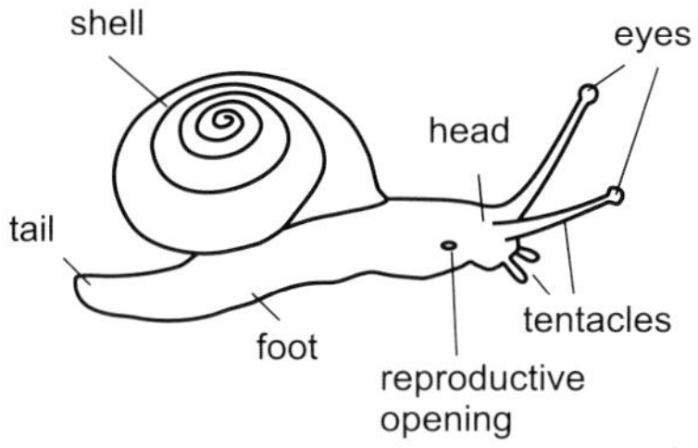

External appearence of a pulmonate snail

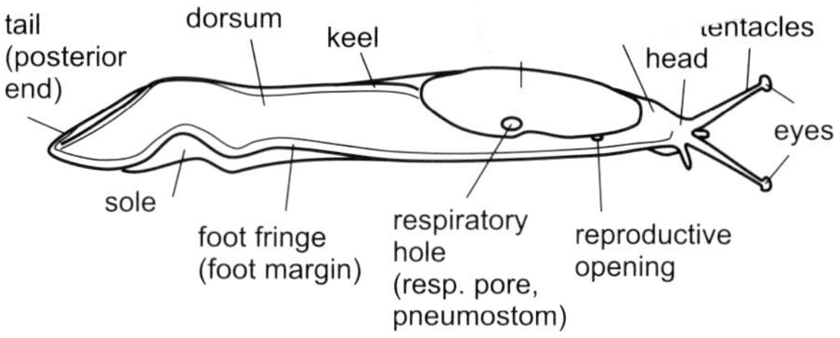

External appearence of a terrestrial slug

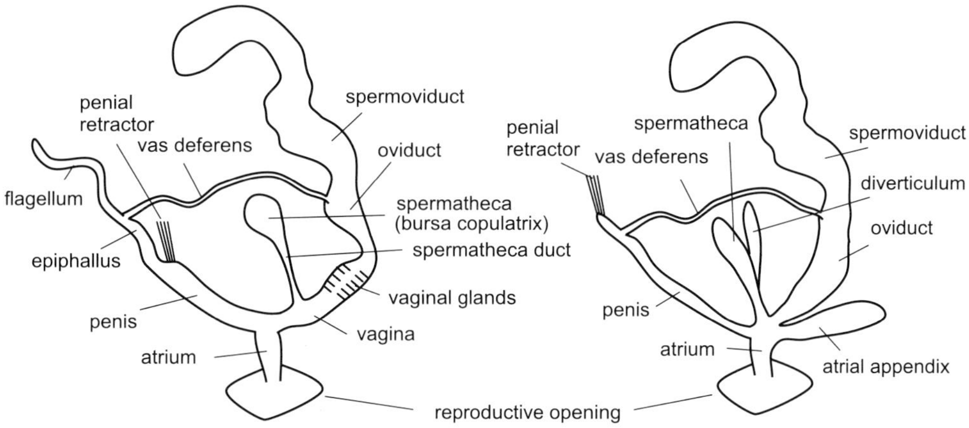

Copulatory organs after dissection

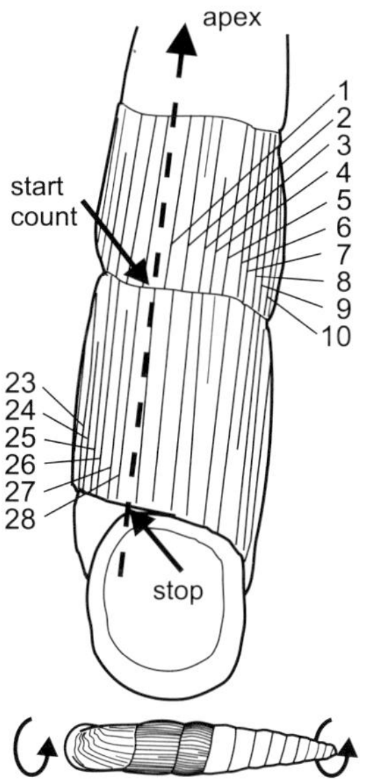

Method of counting the ribs in a clausiliid shell
(Giusti et al. 1995: 332, Welter-Schultes 2010: 149)

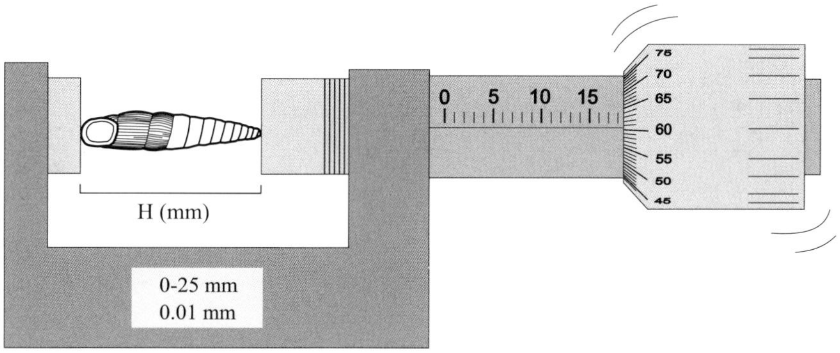

Shell height

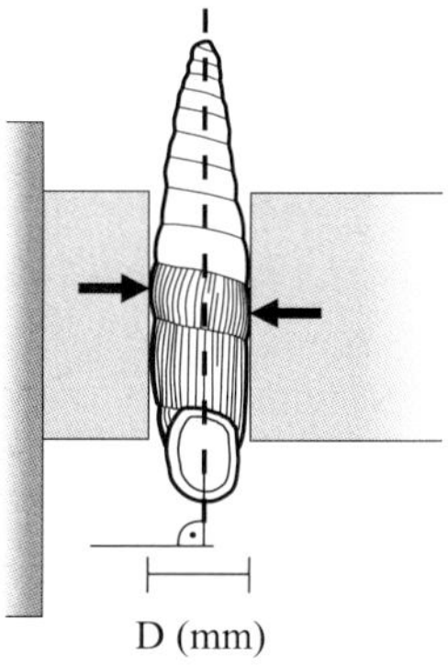

Shell diameter

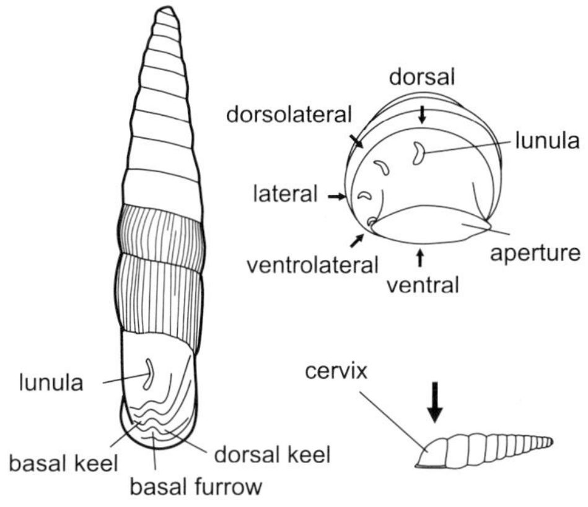

Cervix in a clausiliid shell

A3

Iceland
Sweden
Finland
Norway
Estonia
Latvia
Denmark
Lithuania
Russia
Netherlands
Ireland
Great Britain
Poland
Belarus
Germany
Belgium
Czech Republic
Ukraine
Slovakia
France
Austria
Moldova
Switzerland
Hungary
Slovenia
Romania
Croatia
Bosnia
Serbia
Crimea
Italy
Bulgaria
Corsica
Albania
Portugal
Spain
Balearic Islands
Sardinia
Greece
Turkey
Sicily
Algeria
Malta
Crete
Cyprus
Syria
Morocco
Tunisia

F. W. Welter-Schultes

European non-marine molluscs,
a guide for species identification

Bestimmungsbuch für europäische Land- und Süsswassermollusken

This book intends to provide a guide to identify European mollusc species. If we like to protect nature, we must know how the animals look like.

Europe in the sense of this book has its eastern limit in the middle of Turkey and the Moscow region. The Azores, Madeira and Canary Islands are not contained.
Europe hosts about 2140 currently known species of non-marine molluscs (53 bivalves and almost 2090 gastropods). Currently known means that we are more or less able to distinguish them. The fringes are very blurry, many species are not well known, some are not much more than candidate names. Others are known to contain several species which need to be studied. Again others may not be true species and need to be lumped with others if a bioscientific concept based on reproductive isolation is applied. Not included are several hundred species of the Hydrobioidea and related groups of very small local freshwater snails, which only the experts are able to identify.

This guide was compiled thoroughly and with all care. Like always in such a work, much published information must have been overlooked. There is certainly a good chance to find mistakes. Hopefully not too many. It is always disappointing to look for information in a textbook and to realise that the published information is incorrect or was not up to date at the time.
Well. It is better to finish and publish such a work after 5 years with some errors and unresearched or outdated sources than to try doing a job perfectly and in the end never finish it at all.
The most important point in a project of this size is to give the references used to compile the information, so that others can verify the statements and sources to understand possible errors. In www.animalbase.org the online links to the original descriptions are given. Taxonomists should use these links and verify the sources themselves. It is no option to assume that authors of apparently well-researched books like this one never made mistakes.

The main content of this book was compiled in the museum **Haus der Natur Cismar** in northern Germany, which has its focus on malacology, with a large shell collection and a library displaying most malacological journals.
The basic work was compiled in the Zoological Institute of Göttingen University.

The malacological museum Haus der Natur Cismar

I have not been able to verify all original descriptions of names, but I have indicated those that were not verified.
I screened the past 20-30 years of the **malacological journals** with European content for the biological information. Most journals are present in Haus der Natur. I had however no access for screening the *Journal of Conchology*. I checked most malacological journals, but I must have missed content published in general bioscientific journals. Screening these would have taken much more time.

If not otherwise indicated, almost all **photos** of shells were taken with Vollrath Wiese's excellent equipment (4 flashes) in Haus der Natur, in numerous sessions between 2006 and 2012. Documenting and processing these photos required the largest portion of time. Some 10,000 photos were taken in Cismar, of which 1500 are shown here. All **millimeter scale** sheets were always photographed together with the shells in the same photo sequence. Images were processed with software tools, but the scale sheets were not copied in from pictures taken at other occasions.

Some 10,000 shells have been photographed. To save time I rarely **cleaned** shells when I took the pictures. I tried to take care that the dust did not cover decisive parts for the identification. I did not care to clean the usually black background in the pictures, the background is not relevant for identifying species and would have contributed to more delay.

The pictures are copyright free, they can be downloaded from www.animalbase.org. As said above, if we want to protect nature, we need to know how to identify the animals. Molluscs are usually identified visually, mostly by their shells. We have more than 2000 mollusc species in Europe and we need to know how they look like. This means that we need **natural history museums** which allow us to work with their collections, and to take reliable photos of the species that we need to identify and to protect. Haus der Natur Cismar provided this opportunity. Some European museums prefer hiding their treasures and would not allow to provide images for copyright free usage by a broad public. The result of this behaviour is that young students are confronted with difficulties in learning more about Europe's fauna.
The museums in Budapest, Innsbruck and the Academy of Natural Sciences Philadelphia also provided copyright free pictures of their collection material. Philadelphia offered me to use their lots of several hundred additional species not shown here, which were not present in Cismar. Unfortunately this would have caused another year of delay. Maybe somewhen in the future I find the time to do this and upload these pictures to AnimalBase.

I usually gave references from where I took information on species, except in the most well-known species. I have translated the diagnoses from the German (often from Westerlund's 1884-1890 work), French, Italian and other languages, occasionally also Latin, into a relatively simple standard English which I hope will be understood. Doing this I used a finely tuned vocabulary. This vocabulary is perhaps best understood by comparing the descriptions of those species which have figures.

This book's textual content is largely identical with that of the species homepages of www.animalbase.org.
AnimalBase is a service hosted by the University of Göttingen, online since 2004, relatively stable. It is an open source project, experts can obtain a login account and modify or upload content (also images). Reported mistakes can quickly be corrected, new information can be added.
Please feel invited to report suggestions for corrections, together we can contribute to improving the knowledge on the molluscs of Europe.

The Zoological Institute, University of Göttingen. AnimalBase was developed in this 130 year-old building.

One focus of this work is **nomenclature**.
More young students should feel invited to study Europe's natural heritage, including its molluscs, and we should make it easy for them. I have tried to give references for the nomenclatural background at every occasion where questions may arise. The most frequent questions will probably be if I have perhaps made a mistake somewhere or if I have perhaps not consulted the appropriate literature source.

The stability of names is important, which is an argument to use always the same name for a species. The ICZN Code was established for this purpose. In the character sensitive electronic age it is good to

use always exactly the same letters for genus, species, author and year.
A logical consequence of this thought is that authorships in taxonomic names should have **no initials** - simply because those who do not know the initials cannot add them. The only solution for obtaining exactly the same letter combinations in independently compiled data files is to omit them everywhere.

In some cases it may be useful to use an incorrect spelling for a species, for example for a very well-known and widely used name that has almost always been spelled incorrectly. Art. 33.3.1 of the ICZN Code states that if an incorrect spelling is in **"prevailing usage"** it is deemed to be correct, and gives as example the well-known protozoan hemoflagellate *Trypanosoma brucei* (original spelling: *brucii*).

This example indicates that the "prevailing usage" regulations were not thought to be applied to rarely mentioned species or subspecies, or be abused by taxonomists who made a mistake and then start arguing that the incorrect name is now in "prevailing usage". Most molluscan names are not so frequently used. I did not apply the provisions for "prevailing usage" to those names. In cases of doubt we should always use the original spelling.
Only in few cases it might be useful to conserve incorrect nomenclature. If the malacological names are correct and have a solid nomenclatural base, they are best equipped for the future.

Now, have fun with this guide, and I wish you a successful identification of the molluscs.

Eine kurze Einleitung auf deutsch

Dies ist ein Bestimmungsbuch für die europäischen Molluskenarten. Wenn wir die Natur schützen wollen, müssen wir wissen, wie die Tiere aussehen.

Das Buch ist in Deutschland entstanden. Bewusst wurde für diesen Zweck ein einfaches Englisch gewählt, das für Interessierte in allen europäischen Ländern leicht verständlich sein sollte.

Europa im Sinne dieses Buches schliesst die halbe Türkei ein, geht im Osten bis in die Moskauer Region, und enthält nicht die Azoren, Madeira und die Kanaren.
Europa beherbergt etwa 2140 derzeit bekannte Arten von Binnenmollusken (53 Muscheln und knapp 2090 Schnecken). Derzeit bekannt heisst, wir sind mehr oder weniger in der Lage, sie zu unterscheiden. Die Randbereiche sind dabei ziemlich verschwommen. Viele Arten sind kaum bekannt, einige kaum mehr als provisorische Namen. Bei anderen Artnamen wissen wir, dass sie mehrere Arten beinhalten, die noch untersucht werden müssen. Noch andere wiederum dürften keine echten Arten sein und müssten mit anderen zusammengefasst werden, wenn ein biowissenschaftliches Artkonzept angewandt wird, welches auf reproduktiver Isolation beruht.
Nicht enthalten sind mehrere hundert Arten von Hydrobiiden und verwandten Gruppen, das sind sehr kleine Süsswasserschnecken, die quasi nur die Experten kennen.

Dieses Bestimmungsbuch wurde gründlich und mit einiger Sorgfalt erstellt, aber wie immer bei Arbeiten dieser Grösse dürfte eine Menge übersehen worden sein und die Chancen sind gut, Fehler zu finden. Hoffentlich nicht allzu viele. Natürlich ist es immer eine Enttäuschung, in einem Lehrbuch nach Informationen zu suchen und dann zu erkennen, dass eine Information falsch ist oder zum Zeitpunkt der Publikation schon nicht mehr aktuell war.
Es ist aber besser, eine solche Arbeit nach 5 Jahren überhaupt fertig zu stellen, als zu versuchen, sie perfekt zu machen und sie dann am Ende nie zu veröffentlichen.

Das wichigste dabei ist, die Quellen anzugeben. Nur dann können andere die Aussagen überprüfen und mögliche Fehler verstehen. In www.animalbase.org werden die Online-Links zu den Original-Beschreibungen angegeben. Taxonomen sollten diese Links nutzen und im Zweifel immer die Quellen selbst überprüfen. Niemand sollte sich darauf verlassen, dass Autoren vermeintlich gut recherchierter Bücher nie einen Fehler gemacht haben.

Der Inhalt dieses Buches wurde zum grossen Teil im Museum **Haus der Natur Cismar** zusammengestellt, das auf die wissenschaftliche Weichtierforschung spezialisiert ist. Es besitzt eine umfassende Molluskensammlung und eine gut ausgestattete Fachbibliothek.
Die Basisarbeit wurde im Zoologischen Institut der Universität Göttingen erledigt.

Das malakologische Museum Haus der Natur Cismar

Ich konnte nicht alle Originalbeschreibungen überprüfen, habe aber angegeben, welche nicht überprüft wurden.
Für die biologischen Informationen wurden die letzten 20-30 Jahre in den europäischen **Fachzeitschriften** durchgegangen. Die meisten sind Haus der Natur zugänglich. Nur das *Journal of Conchology* konnte ich nicht Seite für Seite durchsehen.
Mit dieser Methode konnten allerdings Artikel allgemeinen biologischen Zeitschriften nicht erfasst werden. Das hätte deutlich mehr Aufwand bedeutet.

Soweit nicht anders angegeben, wurden fast alle **Fotos** in Vollrath Wieses ausgezeichnetem Fotostudio (mit 4 Blitzen) im Haus der Natur Cismar aufgenommen, an zahlreichen Terminen zwischen 2006 und 2012. Die meiste Zeit wurde dabei für die Dokumentation und Bildbearbeitung verwendet. Von über 10.000 entstandenen Fotos werden hier etwa 1500 gezeigt.
Die **Millimeterpapiere** wurden immer direkt neben den Gehäusen fotografiert. Die Bilder wurden nachbearbeitet, aber die Millimeterpapiere stammen immer aus derselben Fotosequenz. Diese wurden nicht nachträglich von woanders in das Bild kopiert.
Etwa 10.000 Gehäuse wurden fotografiert. Es wäre zeitlich kaum möglich gewesen, jedes dieser Gehäuse gründlich zu reinigen. In der Natur sind sie in der Regel auch nicht sauber. Ich habe darauf geachtet, dass der Schmutz nicht entscheidende Bestimmungsmerkmale verdeckt. Aus demselben Grund der Zeitersparnis habe ich auch den schwarzen Hintergrund nicht gereinigt. Auch der ist nicht wichtig für die Artbestimmung.

Diese Bilder sind frei, sie können auch von www.animalbase.org heruntergeladen werden. Wenn wir die Natur schützen wollen, müssen wir wissen, wie die Tiere aussehen. Die meisten Weichtiere werden anhand ihrer Gehäuse bestimmt. Wir haben mehr als 2000 Molluskenarten in Europa und müssen wissen, wie wir sie bestimmen können. Das bedeutet, wir brauchen **Naturkundemuseen**, die uns erlauben, mit ihren Sammlungen zu arbeiten und zuverlässige Fotos von den Arten zu machen, die wir bestimmen und schützen sollen. Das Haus der Natur Cismar hat das ermöglicht. So etwas ist nicht selbstverständlich. Einige europäische Museen behalten ihre Schätze lieber in den Schubladen und erlauben keine freien Bilder ihres Sammlundsmaterials. Junge Leute haben oft Schwierigkeiten, wenn sie mehr über Europas Tierwelt erfahren wollen.

Die Museen in Budapest, Innsbruck und die Academy of Natural Sciences in Philadelphia haben ebenfalls freie Bilder von Material ihrer Sammlungen zur Verfügung gestellt. Philadelphia hat sogar angeboten, einige hundert weitere Arten zur Verfügung zu stellen, die in Cismar nicht vorhanden waren, vor allem aus der Hemmen-Sammlung. Leider hätte das eine Verzögerung von noch einem weiteren Jahr bewirkt.

In der Regel habe ich die Quellen angegeben, wo die Informationen über die Arten herstammen. Ausser bei den bekanntesten Arten. Dazu habe ich die Beschreibungen aus den deutschen (oft von Westerlund 1884-1890), französischen, italienischen und anderen Quellen (Latein eher selten) in ein relativ einfaches Standard-Englisch übersetzt. Dabei habe ich versucht, einen fein abgestimmten Wortschatz zu verwenden, der sich möglichst oft wiederholt und vielleicht am besten durch den Vergleich mit den Fotos verständlich wird.

Der Inhalt dieses Buches stimmt weitgehend mit dem der species homepages von www.animalbase.org überein.

AnimalBase ist ein Service der Universität Göttingen, online und relativ stabil seit 2004. Es ist ein offenes Projekt, Interessierte und Experten können sich einloggen, Inhalte ändern und Bilder hochladen. Neue Informationen können hinzugefügt und inhaltliche Fehler relativ schnell korrigiert werden.

Das Zoologische Institut der Universität Göttingen. AnimalBase ist hier entstanden.

Alle Interessierten sind gerne eingeladen, sich an diesem Projekt zu beteiligen. Gemeinsam können wir mehr Kenntnisse über die Weichtiere Europas zusammentragen.

Ein Schwerpunkt dieser Arbeit ist die **Nomenklatur**.

Es braucht mehr junge Leute, um Europas Natur zu erforschen, auch die Mollusken, und wir als Fachleute sollten es ihnen nicht unnötig schwer machen. Bei der Nomenklatur habe ich versucht, möglichst überall, wo Fragen auftauchen können, immer den Hintergrund zu erklären und die Quellen anzugeben. Die häufigsten Fragen stellen sich immer dann, wenn mögliche Fehler auftauchen.

Die Stabilität der Namen ist wichtig, für eine Art sollte möglichst überall immer der gleiche Name verwendet werden. Dafür gibt es den ICZN Code. Im elektronischen Zeitalter kommt es sogar auf jedes einzelne Zeichen an. Gattung, Art, Autor und Jahr sollten daher überall ganz genau übereinstimmen.
Die logische Konsequenz davon ist, dass die Autoren in Tiernamen **keine Vornameninitialen** haben sollten - einfach deswegen, weil diejenigen, die die Vornamen nicht kennen, sie auch nicht hinzufügen können. Um die genau gleichen Buchstabenkombinationen in unabhängig erstellten Dateien zu erhalten, ist die einzige Lösung, die Initialen überall wegzulassen.

Manchmal macht es Sinn, eine falsche Schreibweise für eine Art zu verwenden, beispielsweise wenn ein sehr weit verbreiteter Name immer schon falsch geschrieben wurde. Der Ausdruck in Artikel 33.3.1 im ICZN-Code, der das erlaubt, heisst "**prevailing usage**", überwiegender Gebrauch. Als Beispiel wird der weltbekannte Einzeller *Trypanosoma brucei* angeführt (ursprüngliche Schreibweise: *brucii*).
Dieses angeführte Beispiel zeigt, dass die "prevailing usage"-Regel nicht für selten erwähnte Arten oder Unterarten gedacht war, und auch nicht für Taxonomen, die einen Fehler gemacht haben und dann anfangen zu argumentieren, der falsche Name sei jetzt in "prevailing usage". Die meisten Weichtiernamen sind nicht derart weltbekannt. Für diese habe ich die Regel nicht angewandt. Im Zweifelsfall sollte immer die Originalschreibweise genommen werden.
Nur in wenigen Fällen kann es sinnvoll sein, eine falsche Nomenklatur beizubehalten. Wenn die Namen der Weichtiere auf einer soliden und korrekten Grundlage basieren, sind sie für die Zukunft am besten aufgestellt.

So, und jetzt viel Spass bei der Arbeit mit diesem Nachschlagewerk, und viel Erfolg bei der Bestimmung der Mollusken.

Methods and sources used to compile this guide

Names and nomenclature

Black type: names that are regarded as correct and which are used as names here in our classification ("valid" in the sense of the 4th edition of the ICZN Code). Note that some species can have more than one name in the classification used here (*Esperiana esperi* and *Fagotia esperi*, both names are used here and both are regarded as correct).
Grey type: Names which are not regarded as correct here. Names printed in grey type in the species presentation shall not be regarded as having been used as names for taxa (in the sense of Art. 23.9.2), even if they might appear in black type elsewhere in the book (quick-id, index etc.).
Names "without gender agreement" are also given, which could be used should gender agreement and mandatory change once be removed from the ICZN Code.

Statistics and gender agreement

This guide presents 2090 species of non-marine gastropods and 53 freshwater bivalves (classified in some 350 genera, this number is blurry because it depends from personal views).

In Lepidoptera the endings of specific names are usually not changed to agree in gender with the genus. This has historical reasons and has to do with the special situation that the gender of *Papilio* has been disputed since the 1600s and Linné never used a declinable adjective for any of his hundreds of *Papilio* butterfly names. Many lepidopterists never learned to apply gender agreement and ended up by simply not doing it.
In malacology the situation is not much better, especially if neuter gender is involved.
The totel number of specific names mentioned in this work is 2480. 48 % of these were undeclinable words, 42 % were declinable adjectives. In 10 % of the names I did not know (and not research) what it was.
34 % of the species were originally in the same genus as today, 47 % in another genus with the same gender, and 19 % in another genus with different gender. Of the latter, the endings of 197 adjectival names were correctly changed (= 9 % of the European species).
In 51 of 1179 adjectival names used here the ending was originally incorrect (an example is *Unio manca*). Many such names have been used in the incorrect forms until recently (*Mastus violacea*, *Deroceras keaensis*). In 37 cases such incorrect endings had to be subsequently changed.
So in 9 % of the species the adjectival declination change was correctly done, but in 2 % not. This means that the correct declination according to the gender agreement has worked correctly in only 80 % of the cases.
These figures reflect that modern zoologists are not skilled experts in Latin language, and that the malacological science could take advantage from removing the rules for gender agreement from the Code (or from substantially simplifying them).
To get an idea about the result I have given the "without gender agreement" forms of the concerned names.

Original descriptions

98 % of the original sources were verified. I usually gave the **page of description** where the name was made available by meeting the conditions of Art. 11 and 12 of the Code. For publications prior to 1900 I often gave the **plate number** in addition to the page number (bot only if the name was mentioned on the plate), although this is usually not necessary. Names were in most cases made available in the text passages, where a reference to the plate was given. But in some cases plates with printed names were issued prior to the text passages, and I might have missed information on true dates of publication for plates. So I gave the plate numbers, this might be helpful should plates turn out to have been published at a different date.
In 87.5 % of the specific and generic names the authority and year of the taxonomic name corresponded to those of the work. The authority was the same in 97 %, the year in 89 % of the cases. In 1.6 % of the names both authority and year differed from the bibliographical data of the work.

Authorship for taxonomic names

I addressed the issue of authorship in a previous paper (Welter-Schultes 2011). The main points were:
- No initials are used here in the taxonomic names (Müller, Pfeiffer, Boettger; not O. F. Müller, L. Pfeiffer, O. Boettger), only in the literature references, and only if this fascilitates finding the author in the alphabetically sorted list of references (so in O. F. Müller's case the initials are not necessary).
- The author is here consistently spelled as in the original source (*Melanopsis praemorsa* (Linnæus, 1758), *Melanopsis cariosa* (Linné, 1767)).
- Art. 50.1.1 is strictly followed, as demanded by Bank 2011. My previous proposal (Welter-Schultes 2011) is not used here. Art. 50.1 is also strictly followed. If the author of the name differed from that of the work, then we have two options:
  - i - If the author for the name was among the authors of the work ("A & B in A, B & C"), then these are the authors of the name and it is not necessary to look at the responsibility for the description (Art. 50.1).
  - ii - If not ("D in A," or "A & D in A, B & C"), then the authorships for the name and for the description must be the same. If this was not so, then the author of the work must be the author of the name (Art. 50.1.1).

This is the pure application of the rules. It is not exactly in line with Bank 2011 who allowed several self-made exceptions that were not in line with the Code and Sabrosky 1974 who provided a guide with examples for the interpretation of these rules. The main differences:
- The attribution of the authorship for the description must be explicit (or "clearcut") in the work itself, external or implicit evidence must not be considered (except only in anonymously published works and in early Swedish dissertations).
- Every single part of a description has nomenclaturally the same value. If the work was published by A, the name was attributed to D, part 1 of the description to D and part 2 to A or to nobody, then A is the author of the name. No matter how important or necessary part 2 was.
- If the work was by A, the name attributed to D and the description to E, then A is the author of the name (disregarding the fact that A was not responsible for anything in such a case).
- If the work was published by A & B, the name was attributed to B and the description to A, then B is the author of the name.
- If previous authors including myself have attributed a name to an author that was not in line with these rules, this is not repeated (there is no "stability of authorships" and no rule that mistakes must be perpetuated).
- Names with incorrect authorships in the Official List must be corrected (Art. 80.6.2).

Art. 11.6.1 (names first established as synonyms)

The ICZN Code is not a binding law, zoologists follow it voluntarily because of the chaos that would result in the absence of rules. Art. 11.6.1 is an example for the chaos that comes if rules are not worked out thoroughly.
Art. 11 d of the 3rd edition of the ICZN Code (1985-1999) ruled that a name published in synonymy was not made available. In the 4th edition (2000) this rule was broken by adding Art. 11.6.1, an Article that allows to skip basic rules of zoological nomenclature:

> **11.6.1.** *However, if such a name published as a junior synonym had been treated before 1961 as an available name and either adopted as the name of a taxon or treated as a senior homonym, it is made available thereby but dates from its first publication as a synonym (for type species if a genus-group name see Article 67.12; for name-bearing type if a species-group name see Article 72.4.3; for authorship see Article 50.7).*

The given example was about an insect genus. The problems arise when it comes to mollusc species.

Some points are clear in the Code, but can be overlooked.
1 - The synonym must have been mentioned as a **synonym of an available name** (Art. 11.6). A "sensu name" is not an available name (Art. 49). This was overlooked in the *Cecilioides veneta* case.
2 - The **types** must follow Art. 72.4.3. If no specimens were cited with the synonym, then the types are those from the validating author. This was overlooked in the *Balea heydeni / lucifuga* case. "Cited" means cited, so more information than only a name and an author must be given. The Code's guide for this provision was the insect example (Meigen 1818: 82) in which the presence of physical specimens was verbally expressed.

Many early authors (Rossmässler, Pfeiffer and others) published long lists of synonyms, totalling thousands of names which have never been recorded by nomenclators or indexers. These were often names taken from labels provided by shell dealers (Ziegler, Parreyss) for undescribed and badly known material. Years later someone would provide a description, not necessarily in the same sense as the previously mentioned synonyms.

The present work contains many expressions of the form "for those who interprete the Code in this way, the name must be X, for the others Y". This expresses the chaos produced by this new Article.

The main problems:

- Like every new name, such a synonym must (really??) meet the conditions of Art. 12, so it must basically be equipped with a **description**.
- The validating author must (really??) have had the **same taxon** in mind than the author who mentioned the synonym.
- The validating author must not (really not??) have had expressed **doubts** that both belonged to the same taxon.
- The validating author must (really??) have shown awareness of the previously published synonym (when "adopting" the name), for example by providing a **bibliographical reference** to it.
- If a synonym was later validated, but another name for the same species is also available from the same original source - was the original author not the First Reviser to select **precedence** between both names?
- Which name takes precedence if name 1 was used for a subspecies, and name 2 was initially mentioned as a synonym of a species, but the validating author used it for a subspecies?
- Another conflict arises when the name was first mentioned as a **synonym of a variety** which had no name, but only a letter. Under Art. 11.6 such a synonym cannot be validated because it was not used as a synonym "of a name then treated as valid", but Kadolsky 2012: 78 proposed to neglect this. I have preliminarily accepted this proposal and have used the corresponding names and authors here. But it would be better to give up this practice, and to look for solid names for the involved few species, names which were regularly established in the usual form.

Only for a few names of European species the validation of such synonyms is useful or necessary. Mostly they create only problems. An increasing number of species (currently more than 50 names, 2 % of the molluscs presented here) are suffering from these unclear provisions in the current version of Art. 11.6.1, and no end is visible. Every year new old synonyms are discovered in old sources and continuously new problems are created. The produced damage is serious and outweights by far the consequences if this Article would be removed from the Code (or only accepted for genera), so that perhaps 5 or 10 rarely mentioned species would have to obtain a new name, and a few others a different authorship.

Type designations "by group"

Many names of subgenera were established to substitute a term used for a group of species, for which a term was used in the form of "groupf of...", "Gruppe der ..." or "groupe de..".

Examples:

Boettger 1887 established *Montenegrina* as an infrasubgeneric name in the subgenus *Clausilia (Delima)* to substitute the term "Gruppe der *[Clausilia] cattaroensis*".

Locard 1890 established a subgeneric name *Anodonta (Collobiana)*, to substitute a term "Groupe de l'*A[nodonta]. colloba*".

Malacologists have often recognised these constructions as type fixations "by original designation" (Nordsieck 1972: 26, Schileyko 2000: 652, Graf 2010: 9). This view is not tenable under Art. 67.5, following which "*the term 'designation' in relation to fixation of a type species [Arts. 68, 69] must be rigidly construed*" (French Code "*doit être pris dans une acception étroite*"). Art. 67.5.1 rules clearly that "*mention of a species as an example of a genus or subgenus*" is not a designation under the Code. Art. 68.2 demands that a type species must be "*explicitly designated*".

This problems concerns several names used as genera today. I have added the remark that the "Gruppe der..." construction does not meet the requirements of Art. 67.5 and 68.2.

"**Systematisches Conchylien-Cabinet von Martini und Chemnitz**" (SCC)

In this multivolume work plates and texts were usually issued at different dates. Names of taxa were not printed on the plates, but the plates were issued with printed wrappers containing the names of the figured species. These wrappers were not printed for the purpose of providing a permanent scientific record, they were clearly issued for the purpose of helping the subscribers to compile the volumes and were thought to be destroyed after binding. They did not represent published work in the sense of the Code (Art. 8.1.1).

If each wrapper belonged to one particular volume, as in the ornithological multivolume work by Temminck & Laugier de Chartrouse 1821-1838, the wrappers are considered as published work because the publisher provided an opportunity for the subscribers to bind the wrapper inside the corresponding volumes and thus, to preserve the information (Dickinson 2011).

In SCC the subscribers were not equipped by the publisher with this opportunity. Each Lieferung contained a single wrapper and material belonging to several volumes (usually two or three, Welter-Schultes 1999). By consequence most SCC wrappers of the early years have nowhere been preserved and are entirely lost (for example in the clausiliid volume SCC 1 (14) by Küster 1847). Even the preserved wrappers cannot be considered as published work, because under Art. 8.1.1 the initial scope is decisive.

In Nordsieck's publications on clausiliids (and in some other malacological sources, but not all) the names were taken from the wrappers (without consulting them, because they were lost).

In entomology, vertebrate zoology and other fields such wrappers are not regarded as published work. In malacology this should not be done either. For the dates of publication and spellings of names I regard only the SCC text pages as relevant, not the wrappers.

Morphology

The counting of the shell whorls (0.5 whorls accuracy) follows Kerney & Cameron 1979: 13. This is not the method used by Westerlund (Fauna pal. Binnenconch.) and Gittenberger, Ba. & R. 1970: 17 (Fig. 4b) which counts 0.5 whorls more. The morphological methods and the names of shell structures are shown on the figures of the cover inside A1-A2 and on p. 678.

The shell diameter (D) corresponds always to the maximum diameter, the shell height (H) to the perpendicular axis.

Red Lists

Many regional and national Red Lists have been published in Europe. I often used these for local species, but had to take into account that the criteria for the categories of threat were not standardised. I also used the European Red List (Cuttelod et al. 2011, www.iucnredlist.org). In this Red List *all* species of certain groups were listed (contradicting its headline "Red List of *Threatened* Species"), but some groups of terrestrial gastropods (clausiliids, zonitids, slugs) were not included.

This method explains that species like *Dreissena polymorpha* and *Helix aspersa* appeared on such Red Lists, but extremely threatened local clausiliids or zonitids not. This must neither be misunderstood as that the local endemics would generally not be at risk, nor that the term "on the Red List" would imply any potential need of protection.

The currently applied criteria of risk evaluation in the EU Red List are occasionally questionable and do not always adequately reflect the risk of extinction. Some local endemics (for example *Cochlostoma*) may obtain a status as Near Threatened or Vulnerable if they have small restricted ranges, just because they meet the area requirements, and not because any of their populations would in any realistic sense be at risk.

It is obvious that various members of the EU team did not align their criteria (very well visible in the vertiginids or the genus *Agardhiella*), so that species without any known threats and known from 4 locations can get a Least Concern status, species of the same genus (*Agardhiella*) known from 9 locations a Near Threatened status, and species from a different genus (*Vertigo*) and 4 known locations can get the Endangered status. Occasionally a strictly buerocratic application of criteria and exaggerated concentration on endemisms instead of true risks had the effect that extremely threatened but widespread freshwater species (*Anisus*, *Myxas glutinosa*) gained the same Least Concern status as *Dreissena polymorpha* and other pests.

I added a comment when citing a Red List status alone would not adequately reflect real threats - in both directions (exaggerated and underestimated threats).

General sources used to compile this book

Published books

The last published comprehensive work on European non-marine molluscs had been **Westerlund**'s impressive multivolume work "*Fauna der in der paläarctischen Region (...) lebenden Binnenconchylien*" (1884-1890), which included also northern Africa and large parts of Asia.

In the past decades **Kerney & Cameron**'s "*Field guide to the landsnails of Britain and North-west Europe*" (1979, updated 1983 in a German version, often translated and reprinted) became a standard equipment to the study of terrestrial gastropods in the entire continent. Two decades later this work was not available any more. Neither were its copyrighted pictures, a situation which gave rise to the decision taken in Haus der Natur Cismar in 2006 to create the present work, and not to sell the rights for its content or pictures to anyone in the future.

In 1990 appeared **Falkner**'s "*Binnenmollusken*" (*Steinbachs Naturführer 10*). This was the first work after 100 years in which not only the northern and central, but also the most important southern and eastern European species were included. This book contained excellent pictures of living animals which had never been shown before.

Fauna Europaea

I used Fauna Europaea (www.faunaeur.org, since 2004) as a very helpful database with a lot of precious information on species numbers, synonyms, systematics and distribution. The data for the non-marine molluscs were compiled by R. Bank in 2004 (Gastropoda) and R. Araujo slightly later (Bivalvia). It is a pure list of names, without biological information, but with many synonyms and countries. The main shortcomings:

- For the names of gastropods no published references were given, neither to the original sources, nor to the sources used for the current classification, nor to those used for the countries of occurrence.
- There have been updates, but irregularly and without saving access to the previously given information (the meaning of names of the *Anisus* species changed probably around 2010).
- Unpublished information was incorporated, species appeared which had not been mentioned in the recent literature. Species were classified in genera in which they had never been classified before (*Chilostoma* and others).
- No information from Russian and Ukrainian sources was incorporated.
- Technical bugs can obviously not be fixed. The ability to show correctly the diacritic marks in authorships was lost around 2009.

CLECOM lists

Fauna Europaea was based on CLECOM lists. The list of names proposed in the present work was compiled independently. The CLECOM lists were consulted, but not adopted.

The taxonomic names lists of the CLECOM project are not mandatory or something like a nomenclaturally official malacological data resource. Every researcher has to verify the original sources for the names personally and must not rely on others assuming that they have already done the job correctly and under the present-day standards.

For any debatable statement in a scientific publication, a published reference to a scientific study should be given. CLECOM did not coinsider this point. They provided long lists of names, but did not publish the sources and references on which all these names were based. They did not explain the choice of doubtful names, spellings or dates of publication. It is not possible to verify or falsify their results and conclusions by using the same methods.

By neglecting the need to provide published references the CLECOM group made a fundamental mistake in a very basic state of their project. Their published lists cannot really be used as a reliable source. This is really a pity, because it must have taken long time by excellently skilled experts to compile these lists.

AnimalBase

AnimalBase (www.animalbase.org, since 2004) was created by myself as a general list of animal species, and in the following years I "abused" it to provide detailed biological information and pictures on European molluscs. AnimalBase is an open source project, any researcher who likes to contribute can obtain a login and submit information and pictures. Various European malacologists have done this. The person and date is recorded when content is submitted. Previously published information is saved in accessible history files, and can be restored. Any given name can be traced to the original description, to which an online link is provided. The nomenclatural information is usually commented, especially if it deviates from other sources.
The main shortcomings:
- Like most other online databases the information is not regularly updated. All information is based on manual input.
- When I gave biological information I did not add the bibliographic reference to each single statement. Instead I generally listed all references in the "Comments" section. So the references were given, but it is not easy to see which statement was based on which literature reference (it is necessary to analyse the history section for this purpose).
- It is not possible to improve the program, there is no personal staff paid for fixing program errors. Errors leading to several blackouts in a week from a few minutes until 12 hours at weekends cannot be fixed.
- The web service has been stable since 2004, but nobody can say how long the University of Göttingen can run it properly. AnimalBase content is not connected to other services in biodiversity informatics.

AnimalBase was used here to store most of the information published here, including most pictures. Most descriptions of mophology and habitats were copied from the corresponding AnimalBase pages. If mistakes in this book are reported they will be corrected at the corresponding AnimalBase genus, species or taxon summary pages ("taxon" is the term used for the originally established name, "species" and "genus" for the biological units).

Geographic ranges

In the genus sections I tried to indicate if genera lived also outside Europe. If there is no such indication this should usually mean that the genus is restricted to the study area and consists only of the listed species.
The maps of some countries on pp. 675 and 677 might be useful to get oriented in the localities of the figured specimens.

Range maps

Compiling the range maps was difficult. I tried to show where the species approximately live. The entire range of a widespread species is often not completely and accurately known. Light colour corresponds to rare occurrences or regions where the species can be expected and from where I did not have data. Dark colour indicates that the species is not extremely rare, but does not necessarily mean that the information is based on published records. A species like *Pisidium casertanum* that occurs all over Europe can be expected to live in Belarus, although I found no reliable published records for this country.

How the range maps were compiled

Range maps for species with smaller ranges were easier to compile because more complete information was available, widespread species were more problematic. Land snails were generally easier than freshwater species (Kerney et al. 1983 covered central and N Europe, the S European species are usually less widespread). Many groups of S European land snails were accurately revised in the past years, range maps for these were usually taken from there.
For the others (mainly the widespread freshwater species) I used various sources with varying degrees of reliability. Reliability depended from the difficulty to identify a particular species and to distinguish it from others.

Well elaborated atlases or lists with **detailed distributional data** have been published for several countries, regions or federal states. In most cases highly reliable. The first faunistic works in such an excellently accurate standard were published by Pavlović 1912 for Serbia and by Nobre 1913 for Portugal. Many others followed, mainly after the 1970s. Almost every year a new atlas for a country or region is currently added to Europe's molluscan treasure.
Coarse distributional data were available for some regions (Italy, Andalusia, Bulgaria, Romania). Schütt (2001) gave coarse distributional data for Turkey.
I had no accurate sources for France, Spain, Finland, N Sweden, SE Germany (Bayern, Sachsen), Bosnia-Croatia, Greece, Belarus, Ukraine and Russia. The maps published by Kerney et al. 1983 included as a secondary source some of these regions. Internet resources were checked for France (http://inpn.mnhn.fr/isb) and Sweden (Artportalen småkryp, www.artportalen.se/bugs).

I also checked Haus der Natur Cismar collection material (electronic file) and unpublished sources for Crete (my own collections, Vardinoyannis 1994) and Kiklādes islands (Mylonas 1982). Occurrences in small countries were verified by using Fauna Europaea (www.faunaeur.org) (Benelux, Baltic countries, former Yugoslavian countries etc., however this resource largely skipped Russia, Ukraine, Belarus and Moldova). The least reliable source were the GBIF maps (http://data.gbif.org/species), which were good for Benelux and S Germany, but they had usually no records for S and E Europe.
The main gaps which remained were Greece, Spain, Ukraine, Belarus and Russia. I checked many Spanish papers with local faunas, but this did not really help and I must have missed many. In Russian sources I found general statements, only in some cases detailed data on local occurrences of species. In Ukraine it was slightly better.
This means that range limits indicated in the maps for Spain, mainland Greece, Belarus and Russia are often estimated, or, more directly said, fantasy. If too exaggerated, I indicated this in a comment. The regions with the lowest levels of published faunistic research in Europe were certainly Belarus and mainland Greece.

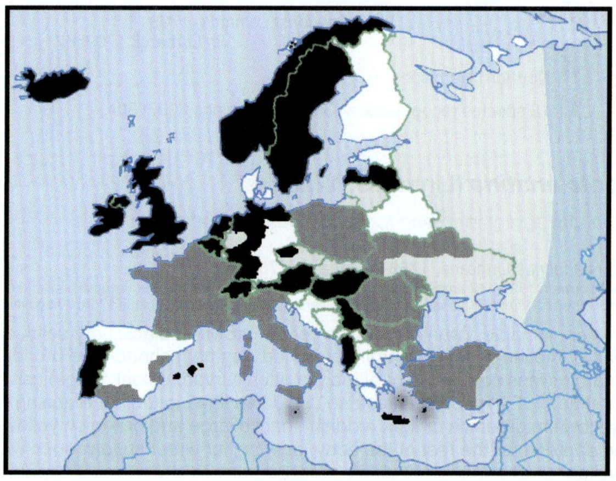

Accuracy of the mollusc record by countries. Black, detailed distributional data. Grey, coarse distributional data. Light grey, general statements. White, no local faunas.

The following main sources for countries and regions were used (in alphabetic order by countries):

Dhora & Welter-Schultes 1996 (Albania), Bertrand 2000 (Andorra), Klemm 1973; Reischütz 1986 (Austria), De Wilde et al. 1986 (Belgium), Damjanov & Likharev 1975 (Bulgaria), Flasar 1998 (Czech Republic: NW Bohemia), Germain 1930, http://inpn.mnhn.fr/isb/servlet (2010) (France), Bürk & Jungbluth 1982 (Germany: Baden-Württemberg), Jungbluth 1978 (Germany: Hessen), Zettler et al. 2006 (Germany: Mecklenburg-Vorpommern), Jungbluth et al. 1989 (Germany: Niedersachsen), Vogt et al. 1994 (Germany: Rheinland-Pfalz), Wiese 1991 (Germany: Schleswig-Holstein), Kerney 1999 (Great Britain), Vardinoyannis 1994 (Greece: Crete), Mylonas 1982 (Greece: Cyclades), Pintér & Suara 2004 (Hungary), Mandahl-Barth 1938 (Iceland), Kerney 1999 (Ireland), Alzona 1971, Cossignani & Cossignani 1995 (Italy), Boato et al. 1984 (Italy: Alpi Liguri), Rudzīte et al. 2010 (Latvia), Beckmann 1987, Giusti, Mang. & S. 1995, Kolouch 2003 (Malta), Coadă & Welter-Schultes 2011 (Moldova), Gittenberger et al. 1984 (Netherlands), Økland 1925, Økland 1990 (Norway), Wiktor 2004 (Poland), Nobre 1913, Nobre 1941 (Portugal), Grossu 1981, 1983, 1986, 1987, Cioboiu 2006 (Romania), Egorov 2008, Sysoev & Shileyko 2009 (Russia), Tappert 2009 (Russia: Moscow region), Pavlović 1912, Tomić 1959 (Serbia), Ruiz Ruiz et al. 2006 (Spain: Andalusia), Beckmann 2007 (Spain: Balearic Islands), Martínez-Ortí & Robles 2003 (Spain: Comunidad Valenciana), Waldén 2007, www.artportalen.se/bugs (2010) (Sweden), Turner et al. 1998 (Switzerland), Schütt 1993, 1996, 2001, 2005 (Turkey), Tappert & Korniushin 2001, Kornuishin et al. 2002, Sverlova 2006, Egorov 2008, Sysoev & Shileyko 2009, Balashov & Gural-Sverlova 2012 (Ukraine).

Acknowledgements

Without Vollrath Wiese and the Haus der Natur Cismar team's help this book could never have been compiled in this form. He provided the equipment for taking the pictures. Most pictures of shells were taken by myself in the laboratory of Haus der Natur Cismar HNC from HNC material (except some shells obtained in loan from M. Sosso, J. Grego, C. Audibert, C. R. Altaba and Z. Fehér/Museum Budapest). Large parts of this book were written in Cismar, most modern malacological literature was consulted there.
I am grateful to colleagues who provided helpful comments in nomenclatural problems: Cristian Ruiz Altaba, Rafael Araujo, Ruud Bank, Edi Gittenberger, Dietrich Kadolsky, Fabio Liberto and Gary Rosenberg, as well as numerous members of the [iczn-list] and [Taxacom] mailing lists and the naturamediterraneo web community in Italy. I took advantage from an unpublished manuscript by J.-P. Rocroi & P. Bouchet on the generic names. Numerous malacologists helped with literature and identified species.
I thank all those who allowed me to use their pictures of shells and live animals: Claus Ableiter, Roy Anderson, Cédric Audibert, Ettore Balocchi, Peter Barfield, Jean Béguinot, Harald Berger, Joseph Berger, Cesare Bogi, Clemens M. Brandstetter, Guillaume Brocker, Luis Miguel Bugallo Sánchez, Vera Buhl, Agneta Chaniecka, Karolina Chaniecka, David Cilia, József Dózsai, Zoltán Fehér, Paolo Fontanesi, Vittorio Formenti, Emilio Garcia, Charles Geerts, Dilian Georgiev, Ana Gonçalves, David González, Jozef Grego, Burçin A. Gümüş, Alessandro Hallgass, Stefan Heim, Panagiotis Helmis, Raymond Huet, Atanas Irikov, Valter Jacinto, Adrienne Jochum, Georgios Konstantinou, Sergey Leonov, Fabio Liberto, José Liétor Gallego, Emidio Machado, Michal Maňas, Francesco Mascia, Carlo Morelli, Jiří Novák, Aydın Örstan, Panayotis Ovalis, Paolo Paolini, Giuseppe Pocaterra, Massimo Prodan, Agatino Reitano, Walter Renda, Robert Reuselaars, Ira Richling, Gonçalo M. Rosa, Florian Schwabe, Eric Simon, Maurizio Sosso, Jan Steger, Jivko Stoilov, Anna Sulikowska-Drozd, Pascal Van Herck, Angelo Vannozzi, Rafael Vázquez Romero, Alexander M. Weigand, Vollrath Wiese, Paolo Zaccaria, Enrico Zallot, Alex Zelenko.
Z. Fehér contributed with diagnoses of some species. A. Jochum checked the chapters on *Carychium* and *Zospeum*. A. Örstan checked the Quick ID plates for Turkish species. P. Subai corrected the chapters on Aciculidae, *Argna*, *Agardhiella*, *Poiretia*, *Lindholmiola* and Ariantinae.

Finally I like to thank Christine for her patience and for having supported me so much during this work.

Bivalvia

Genera sorted alphabetically:
1 - Unionoidea: *Anodonta, Margaritifera, Microcondylaea, Potomida, Pseudanodonta, Sinanodonta, Unio*
2 - *Corbicula*
3 - Dreissenidae: *Congeria, Dreissena, Mytilopsis*
4 - Sphaeriidae: *Musculium, Pisidium, Sphaerium*

Anodonta

Anodonta Lamarck, 1799
in Lamarck 1799: 87

Type species *Mytilus cygneus* Linnæus, 1758
(monotypy)

Northern Hemisphere

13 species recognized by Graf & Cummings 2007: 304.

Anodonta anatina (Linnæus, 1758)

Without gender agreement: *Anodonta anatinus* (Linnæus, 1758)

Mytilus anatinus Linnæus, 1758 in Linnæus 1758: 706
Mordan & Woodward 1990: 112 proposed to designate a neotype from Wales, after unsuccessful attempts to do so in the 1960s. The original types were from Sweden (Linnæus' material) and Britain (Lister's material, to which Linnæus 1758: 706 gave a bibliographical reference).

Shell very variable, usually greenish yellow, brown or black, very solid, often twice as heavy as *A. cygnea*, frontal part of the inside lower margin thickened, ligament shorter and broader than in *A. cygnea*, dorsal and ventral margins less parallel, ventral margin often curved, dorsal margin can be straighter. Embryonal shell smoother than in *A. cygnea*, with coarse undulated lines, often not reaching the margin.
Animal yellowish, greenish or light reddish, mantle large and brownish yellow, gills greyish and like gaze, foot large and darker than the rest of the body, opening for water supply much wider than in *A. cygnea*, and with short papillae.

Calm rivers, river bays, lakes, creeks, in muddy or sandy substrate, usually 2-3 m deep, not below 20 m, tolerates slightly faster water currents than *A. cygnea*. In Switzerland in up to 1800 m altitude, usually below 1200 m, higher than the other freshwater mussels. Eutrophic conditions are better tolerated than by *Unio pictorum*, but less than by *A. cygnea*.

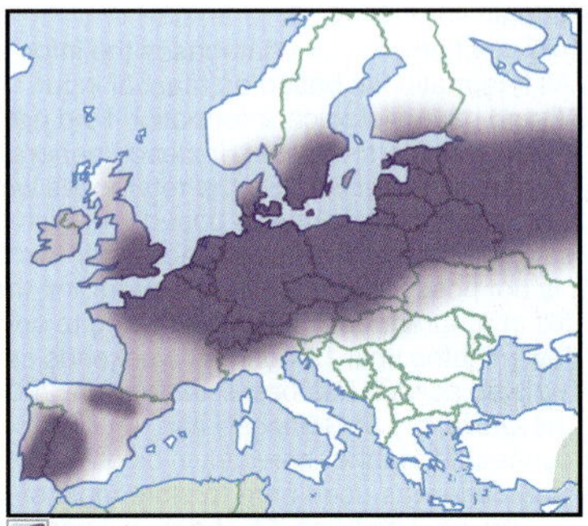

Europe and N Asia

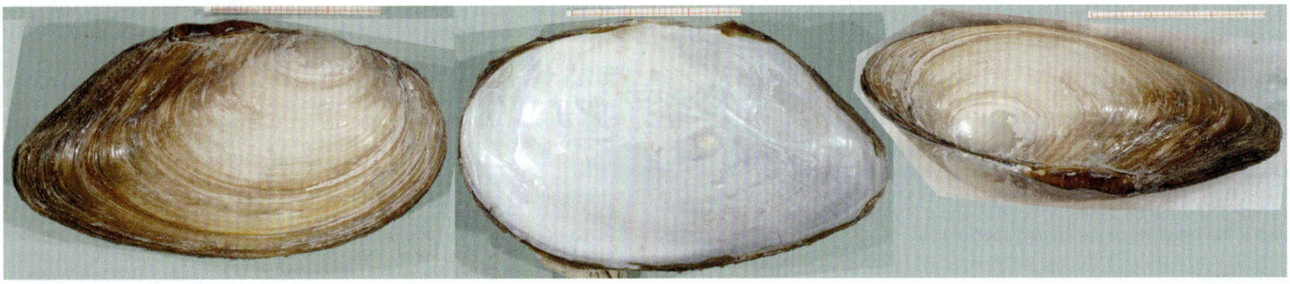

80-150 x 50-60 x 25-35 (height) mm

Germany: Schleswig-Holstein, Grosser Eutiner See

Glochidia (0.35 x 0.35 mm) are mature in autumn and are released in spring. Size and age of maturity depends on water temperatures, animals in warm waters reach maturity after 2 years, in cold waters they need up to 5 years and the shells stay smaller. Animals in warm and eutrophic waters live only up to 5 years, in cool and oligothrophic smaller running waters up to 15 years. Hermaphroditic animals may occasionally occur. Host fishes are among others *Perca fluviatilis*, *Gobio gobio* and various local species, in N Germany also *Salmo trutta*, *Gasterosteus aculeatus* and *Leuciscus leuciscus.*

Generally threatened by continuous destructions of habitats, water pollutions and agricultural fertilization. Local forms are increasingly threatened by uniform forms introduced with gamefish.

Anodonta cygnea (Linnæus, 1758)

Without gender agreement: *Anodonta cygneus* (Linnæus, 1758)

Mytilus cygneus Linnæus, 1758 in Linnæus 1758: 706

Shell yellowish or greenish brown, thin, not very solid, ligament long and narrow, without teeth, lower side of frontal interior margin not thickened. Embryonal shell finely striated along the growth lines, the lines reach the margin.
Differs from *A. anatina* in its larger shell with dorsal and ventral margins being straighter and more parallel, the growth lines of the embryonal shell are finer and less undulated, and reach the margin. Animal like *A. anatina*.

Lakes, old river arms, artificial lakes, needs silent waters, only rarely in running waters, usually in lowlands. Prefers muddy substrate bare of vegetation. Artificial lakes are usually reached when infected fishes are brought in, young mussels are rarely dispersed by water birds. Tolerates eutrophic conditions, but not in the last toxic stadium. Up to 1500 m in Switzerland.

Animals are hermaphrodites. Host fishes for glochidia are *Salmo trutta*, *Perca fluviatilis*, *Leuciscus leuciscus*, *Sander lucioperca* and *Gasterosteus aculeatus*, but not *Rhodus sericeus*. Glochidia are mature in autumn, are released in spring (in England and in Italy), their size (0.35 x 0.35 mm) does not increase much in the rays of host fishes, until they sink down to the bottom to found new populations.

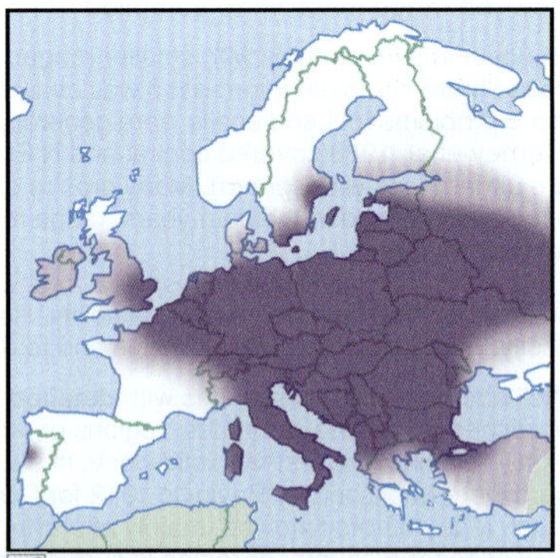

Europe to Caucasus

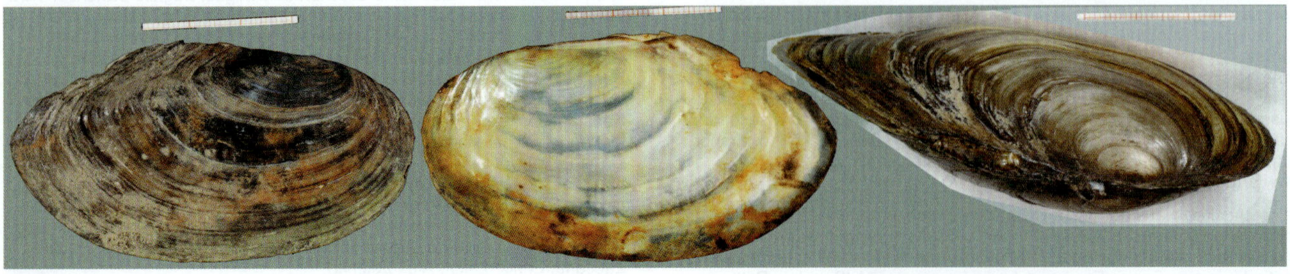

60-120 x 120-200 x 30-60 (height) mm

(1-2) Austria: Niederösterreich, Rosskopfarm near Stopfenreuth/Donau,
(3) Germany: Schleswig-Holstein, Kremper Au near Krummbek

EU Red List: Near Threatened NT. Threatened by continuous destruction of habitats, water current regulations, exaggerated pollution by fertilizers, also by natural drying out of waters.
In Niederösterreich (Austria) all bivalves were defined in 2002 by law as fishes and by this way taken outside any form of protection (Reischütz 2009). In the Fischereiverordnung 2002 *A. cygnea* was defined as an artificially introduced species, there has never been any scientific support for such a view.

Margaritifera

Margaritifera Schumacher, 1815 in Schumacher [1815]: 7 (*Margartifera*)

Type species *Margaritifera fluviatilis* Schumacher, 1815 (monotypy)

ICZN Opinion 495: *Margartifera* ruled to be an incorrect original spelling of *Margaritifera*. Type species is not *Mya margaritifera* Linnæus, 1758. Schumacher 1815 included one nominal species (*Mar. fluviatilis*) and cited *Mya margaritifera* as a synonym. Monotypy is superior to absolute tautonymy (Art. 68.1), so type by monotypy (Art. 68.3).

Pseudunio Haas, 1910 in Haas 1910: 182, type species *Unio sinuata* Lamarck, 1819 [in Lamarck 1819: 70] (monotypy) (= currently *Margaritifera auricularia* (Spengler, 1793))

Larval stage (glochidium) is parasitic on the epithelium of fishes (grasps gill filaments). Species are strictly host specific.

Margaritifera auricularia (Spengler, 1793)

Without gender agreement: *Margaritifera auricularius* (Spengler, 1793)

Unio auricularius Spengler, 1793 in Spengler 1793: 54

Shell black or dark brown, very solid (up to 0.6 kg), embryonal shell corroded, ligament strong, lower margin of the shell concave.

Large rivers, usually in the deepest and medial zones of the water current, buried half inside the substrate.

Known hosts are the Spanish endemic fish *Salaria fluviatilis* (not in France), *Acipenser sturio* (only these two are native in Europe), and the exotic fishes *Gambusia hollbrooki, Acipenser baerii, Acipenser naccarrii.* Needs several years to reach maturity. Age 50-80 years, may reach more than 100 years.

EU Red List: Critically Endangered CR. Functionally extinct. Strictly protected by law in France, including its habitats. No juveniles are known to exist in any of the Spanish populations, which are entirely disconnected from their host fishes and moreover threatened by habitat destruction. Under these conditions the species cannot survive.

López & Altaba 2005, López et al. 2007.

Was recently (Falkner 1990: 254, Falkner et al. 2001: 66) classified as *Pseudunio auricularius.* This classification was not confirmed by Huff et al. 2004: 385 and Araujo et al. 2009 who recommended not to use *Pseudunio* as a genus or subgenus.

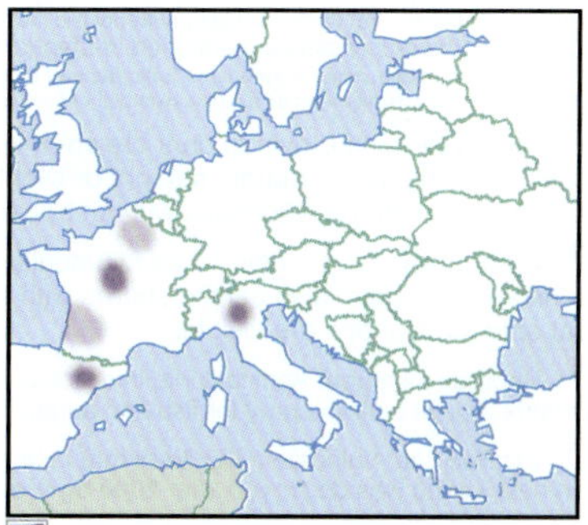

In the 1800s in most rivers of France, until 1990 extinct except perhaps in the middle Loire section, perhaps in N Italy, NE Spain

70-90 x 110-180 x 30-50 (height) mm

Spain: Zaragoza, Ebro near Sastago

Margaritifera margaritifera (Linnæus, 1758)

Mya margaritifera Linnæus, 1758 in Linnæus 1758: 671

Shell dark brown to almost black, often with white scratches, oval elongate, embryonal whorls usually corroded, teeth strong.

Upper and middle sections of running oligotrophic waters, in clean and cool water free from mud and suspended matter, usually on crystalline water and sandy substrate (only in Ireland also on calcareous substrate), often burrowed between pebbles and boulders. In Portugal often in the deepest sections of rivers.

Females produce several million glochidia (0.05 mm) which must be swallowed by a fish of the genus *Salmo* (*Salmo salar, Salmo trutta*), where they grow as gill parasites to a size of 0.5 mm. Then they leave the host and settle in the bottom of the water. They appear at the bottom surface when they have reached a size of 2 cm and an age of 4-5 years. Average age up to 80-100 years, in Spain average 35 (maximum 65 years), in cold northern regions up to 100-190 years, maximum 210 years, belongs to the longest lived animals in the world.
One animal of 2700 contains a pearl.

EU Red List: Critically Endangered CR. In many regions extinct. Strictly protected by law in France, including its habitats. Threatened by pollution, excessive fertilization in farmlands, habitat destruction (including soft changes such as increase of turbidity downstream by dredging), lack of hosts and destruction by pearl fishers. Only 2-3 % of the former populations are still present, often for decades without offspring.

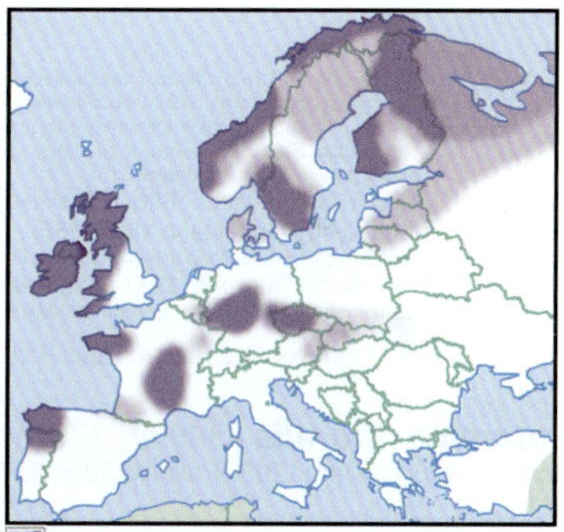

Europe to N Russia, E Siberia to Kamchatka and N Japan, E North America (Newfoundland to Pennsylvania)

45-60 x 100-130 x 30-40 (height) mm, maximum up to 160 mm

Finland: Pykajöki Kärsamäki Lahmkoski

Microcondylaea

Microcondylaea Vest, 1866
in Vest 1866: 197

Type species *Unio bonellii* Férussac, 1827
(monotypy)

Contained originally *Unio bonellii* and with doubts *Anodonta lata* Rafinesque, 1820 (monotypy, Art. 68.3).

Only one species recognized (Falkner et al. 2001: 66).

Microcondylaea bonellii (Férussac, 1827)

Unio bonellii Férussac, 1827 [not verified]
Original combination not verified but likely. I did not find the publication.
Alasmodonta compressa as mentioned in Menke 1830: 106 (nomen nudum)
No description added, only the synonyms "*Unio Bonelli* Fér., *Unio depressus* Mhlfld. [= *Unio depressus* Pfeiffer, 1825 in C. Pfeiffer 1825: 32]", no bibliographic reference. Not treated as a junior synonym of an available name (Art. 11.6), the available name *depressus* was treated as a synonym.

Shell yellowish brown to dark brown or with greenish hue, elliptical, very flat, vortex finely corrugated, ligament very long, lower margin straight or slightly concave.
Foot long and strong. Glochidia without hooks.

Clean moving or running waters, usually in shallow zones with sandy substrate, occasionally overgrown by higher water plants, also at lake shores with some wave movement, often deeply buried inside the substrate so that only the siphons are visible.

Reproduction from April to June, infectious larvae are released as worm-like objects with mature glochidia embedded in a matrix. These look like prey for fish and attract possible host fishes. Host fishes unknown. Age up to 15 years.

EU Red List: Vulnerable VU. Very rare in Italy, the species was regarded as extinct until 1984. A large Albanian population near Lezhë was destroyed after 2000 by river regulations.

Nagel et al. 2007, Reischütz et al. 2008.

Formerly known under the name *Microcondylaea compressa* (Menke, 1830).

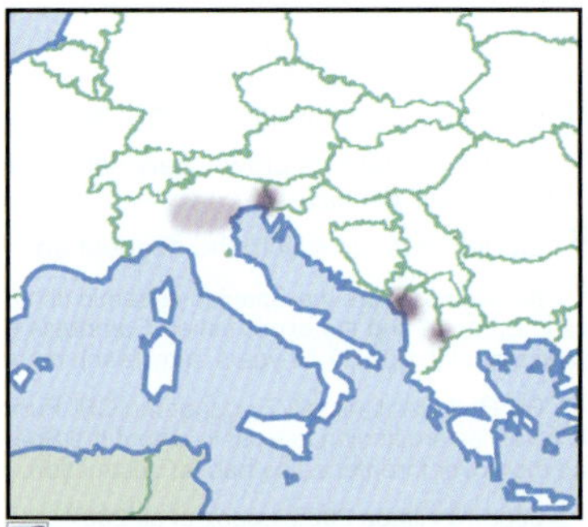

N Italy (Po plain, recent only in Isonzo river, in the past reaching S Ticino in Switzerland), Montenegro to Albania

25-35 x 50-70 x 17-24 (height) mm, maximum 100 mm

N Italy: Larghetto near Piadena, W Mantua

Potomida

Potomida Swainson, 1840
in Swainson 1840: 268 *Unio (Potomida)*

Type species *Unio corrugata* Swainson, 1840
(original designation, Swainson 1840: 281)
(currently = *Unio littoralis* Cuvier, 1798)

Established on p. 268 as a subgenus of *Unio*, on p. 281 as genus, on p. 379 apparently classified as a subgenus of *Mysca*.
Contained originally (p. 379) *Unio sinuata* Lamarck, 1819, *Unio corrugata* Swainson, 1840 (non Philippson 1788) and *Unio littoralis* Cuvier, 1798. Swainson 1840 used the term "the typical species" on p. 281, but probably not in the sense of a unique name-bearing type, so the type designation can or cannot be recognized under Art. 67.5. For those who do not accept Swainson's 1840 type designation (Altaba 2007: 148), the next available type is *Unio sinuata* Lamarck, 1819 designated by Gray 1847: 196 currently = *Margaritifera auricularia* (Spengler, 1793), in this view *Potomida* is a synonym of *Margaritifera* Schumacher, 1815.
Psilunio Stefanescu, 1896 in Stefanescu 1896: 44, type species *Unio craiovensis* Tournouër, 1880 (in Tournouër 1880: 97 *Unio Craiovensis*) (subsequent designation by Cossmann 1897: 156)
Contained originally *Unio craiovensis* and *Psilunio recurvus* Stefanescu, 1896, fossil species from Romania.
Rhombunio Germain, 1911 in Germain 1911: 67, type species *Unio rothi* Bourguignat, 1865 (subsequent designation by Haas 1969: 271)
Contained originally 3 species from the Middle East, not *Unio littoralis* (given as type by Ellis 1946) which was mentioned but not originally included, Art. 67.5.3.

Araujo et al. 2009: 28 used *Potomida*. I would recommend to use this name.

Potomida littoralis (Cuvier, 1798)
Unio littoralis Cuvier, 1798, *Psilunio littoralis* (Cuvier, 1798) (3 names are in use)

Unio littoralis Cuvier, 1798 in Cuvier 1798: 425
The generic name depends on the personal view of a more or less restrictive interpretation of Art. 67.5. *Unio littoralis* is currently the mostl widely used name, but not in accordance with the phylogenetic classification. All three names are used as valid here.
Google hits 07-2009: "unio-littoralis" 1900, "potomida-littoralis" 800, "psilunio-littoralis" 200, "rhombunio-littoralis" 20.

Shell greenish black, interior with slightly violet hue, shape variable but usually short and high, embryonal shell usually white without periostracum, teeth very strong. 1-2 cm sized juveniles are light brown with undulated surface structures.
Animal yellowish or greenish grey, mantle yellow, gills dark reddish brown, foot yellowish.

In rivers, usually in middle and lower river sections or in river canals, up to 6-10 m depth, prefers slowly moving waters, occasionally also in river affluents, canals and lakes. Usually between gravels and stones, sometimes also in sandy substrate. Needs clean waters, but does not occur in oligotrophic mountain waters.

Maturity is reached after 4-5 years (size 4 cm). In Spain pregnant females were observed in July and August, juveniles hatched in September, glochidia (0.21 mm) were observed from March to October. Known host fishes are *Barbus* species, perhaps also *Chondrostoma*.

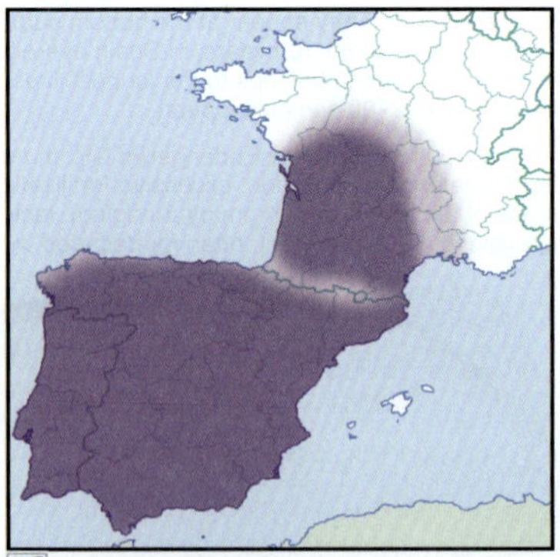

SW Europe, N Africa (Tunisia to Morocco), Middle East, perhaps also Greece

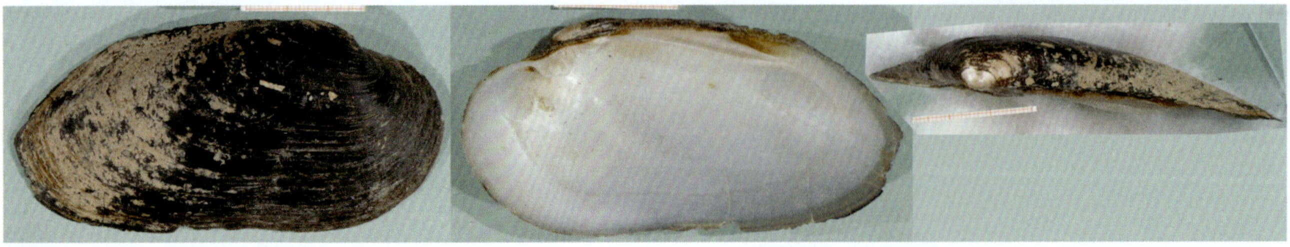

40-50 x 50-70 x 25-30 (height) mm

S Spain: Guadalquivir near Córdoba

EU Red List: Near Threatened NT. Vulnerable in Spain (Verdú & Galante 2006). In Spain rapidly declining, very problematic in terms of conservation, threatened by habitat destruction and competing introduced species. Had been common in Portugal before 1940, today only few populations are left. Nothing known in Greece.

Nobre 1941: 232, Pérez-Quintero et al. 2004, Graf & Cummings 2007: 309, Araujo et al. 2009: 28.

Pseudanodonta — *Pseudanodonta* Bourguignat, 1877 in Bourguignat 1877: 99 *Pseudodon (Pseudanodonta)*

Type species *Anodonta complanata* Rossmässler, 1835 (subsequent designation by Westerlund 1902: 137)

Pseudanodonta complanata (Rossmässler, 1835)

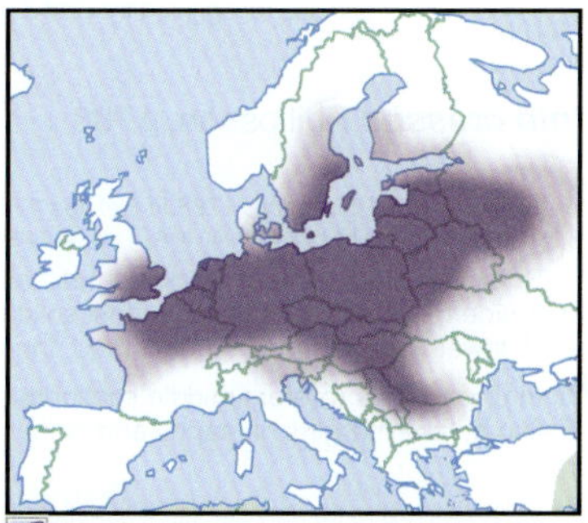

Anodonta complanata Rossmässler, 1835 in Rossmässler 1835-1837 (1): 112, Taf. 3

Shell flat, oval with short frontal part (right side in > shape in Fig. 1), posterior part slightly pointed and directed downwards, ligament consisting of indistinct tubercles, embryonal shell not corrugated but with small tubercles. Shape not very variable, its form is typical. Males are larger than females, there are no hermaphrodites.

In lowland rivers in sections of slow water movement, old river arms, margins of lakes, prefers muddy soils, up to 11 m depth, extremely sensitive against water pollution and eutrophic conditions, particularly juveniles.

Non-gravid period in May, glochidia formation in June to August, glochidia are mature in September, they are released in the following April. Females brood 5,000-50,000 glochidia, depending on mussel length. Sex ratio in populations 2.5 females : 1 male. Known host fishes are *Salmo trutta*, *Perca fluviatilis*, *Gasterosteus aculeatus*, *Gymnocephalus cernuus*, *Pungitius pungitius*, *Oncorhynchus mykiss* and *Stizostedion lucioperca*.

EU Red List: Near Threatened NT. Scattered and rare, has become locally extinct due to habitat destruction and pollution. Some of the largest populations live in East Anglia, Britain. Critically Endangered in Germany.

Great Britain (central and S England and E Wales) and from France to Ukraine and E Baltic Sea region

Buttstedt 2007, McIvor & Alridge 2007.

35-50 x 50-80 x 20-30 (height) mm, rarely up to 100 mm

NW Russia: Kavgolovskoye Ozero N of St. Peterburg

Current consensus is that *P. complanata* is regarded as a single species (Falkner et al. 2001, Anderson 2005). Kantor & Sysoev (2005) regarded Falkner et al.'s 2001 subspecies as separate species (*klettii*, *elongata*, *nordenskioldi*).

Sinanodonta — *Sinanodonta* Modell, 1945 in Modell 1945: 62

Type species *Symphynota magnifica* Lea, 1834 (original designation by Fischer 1886: 1003)

Expressly established as a new replacement name for *Pteranodon* Fischer, 1886 in Fischer et al. 1887: 1003, type species *Symphynota magnifica* Lea, 1834 (original designation) (not *Pteranodon* March, 1876 (Sauropsida).
Modell 1945: 96 gave as type *Symphynota woodiana* Lea, 1834, this was incorrect (Art. 67.8).

Central to E Asia, Japan

Two species recognized by Graf & Cummings 2007: 304.

Sinanodonta woodiana (Lea, 1834)

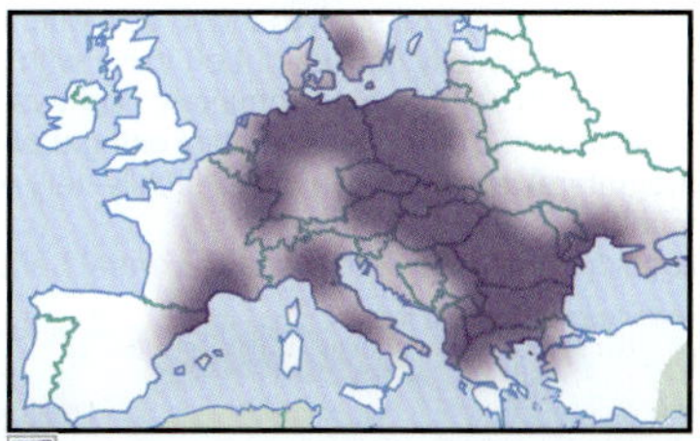

Symphynota woodiana Lea, 1834 in Lea 1834: 42, Pl. 5 (*Symphynota Woodiana*)

Shell brown or blackish, sometimes with greenish hue, colour very variable but usually more intensively coloured at the periphery, elliptical to almost spherical, lower margin strongly convex.

From slowly running rivers to eutrophic ponds. Introduced to Europe with fishes originally imported to control algae growth in eutrophic silent waters.

Kraszewski 2007, De Vico et al. 2007, Proschwitz 2008.

In Romania and Hungary in stable populations since 1979, still spreading, reached Sweden after 2000 with infected carp and goldfish.

E Asia from Amur (Heilongjiang) to Kampuchea, dispersed elsewhere

100-170 x 120-200 x 30-45 (height) mm

China: Guangzhou

Unio

Unio Philipsson, 1788
in Retzius 1788: 16

Type species *Mya pictorum* Linnæus, 1758
(fixed by ICZN under the plenary powers)

ICZN Opinion 495 (1957): gender masculine. Author of the work was Philipsson, in an early Swedish dissertation. The new name was not attributed to a particular author.

Europe to Middle East and Nile river

7 species recognized by Graf & Cummings 2007: 304. The Iberian populations were revised by Araujo et al. 2009.

Unio crassus Philipsson, 1788

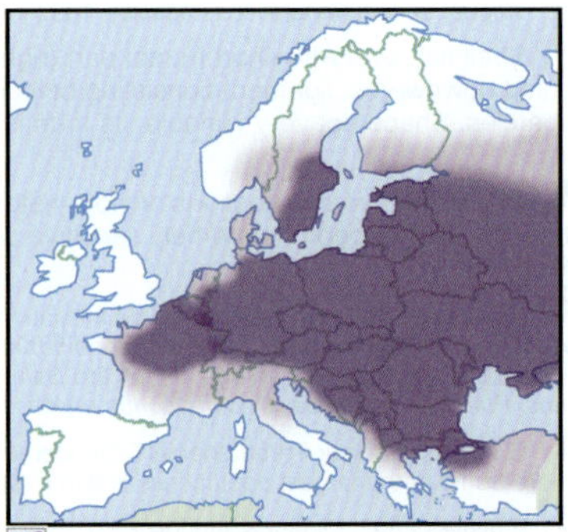

Europe except British Isles to Black Sea region and Iraq

Unio crassus Philipsson, 1788 in Retzius 1788: 17 (*Unio Crassus*)
Author of the work was Philipsson, in an early Swedish dissertation. The new name was not attributed to a particular author.

Shell very dark, almost black, sometimes with greenish streaks, short and elliptical (posterior margin on upper side rounded, not straight), most solid parts located behind the embryonic shell, which is usually corroded, lower margin straight, not curved, the main tooth is steeply conical and pointed.

In clean rivers (prefers middle sections) and smaller running waters in lowlands, not above 600 m, in the SE also at lake margins. In sandy and stony substrate, usually entirely inside. Not on contaminated muddy bottoms.

Juveniles are extremely sensitive against any kind of water pollution and need high oxygen contents, adults cannot reproduce in running waters with nitrate contents exceeding 10 mg/l. Life span 20-30 years. Sexes are separate, high population densities are needed for reproduction.
Hosts are *Cottus gobio*, *Phoxinus phoxinus*, *Leuciscus cephalus*, *Scardinius erythrophthalmus*, *Gymnocephalus cernuus* and *Perca fluviatilis*, in N Germany also *Gasterosteus aculeatus*. Different populations may have different hosts, sometimes only some local populations of fishes can serve as hosts.

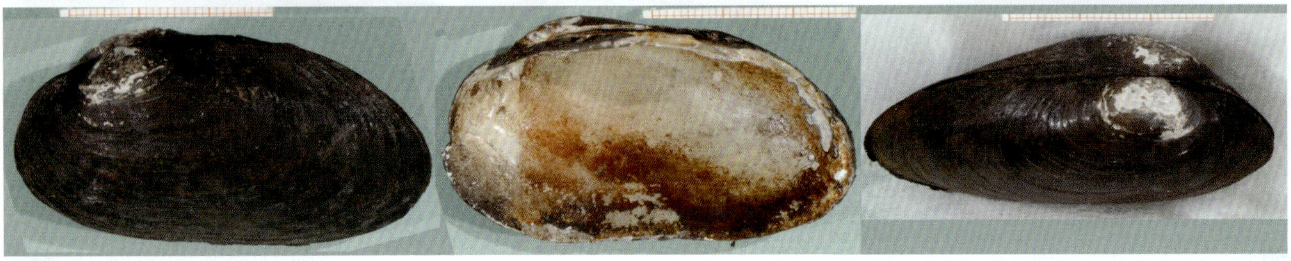

25-45 x 40-70 x 20-28 (height) mm

N Germany: Mecklenburg-Vorpommern, Sude near Radelübbe

EU Red List: Vulnerable VU. Critically Endangered in Germany, Switzerland (already vanished from most Cantons) and Austria. Strictly protected by law in France, including its habitats. Until the 1950s the most frequent freshwater bivalve of Europe. Threatened by water pollution, drainage, regulations of small running waters and disappearing of host fish species. In some regions living populations of *Unio crassus* have the power to stop industrial projects.

Populations from Spain and Portugal were recently classified as *Unio tumidiformis* based on molecular studies (Araujo et al. 2009: 49).

Unio delphinus Spengler, 1793

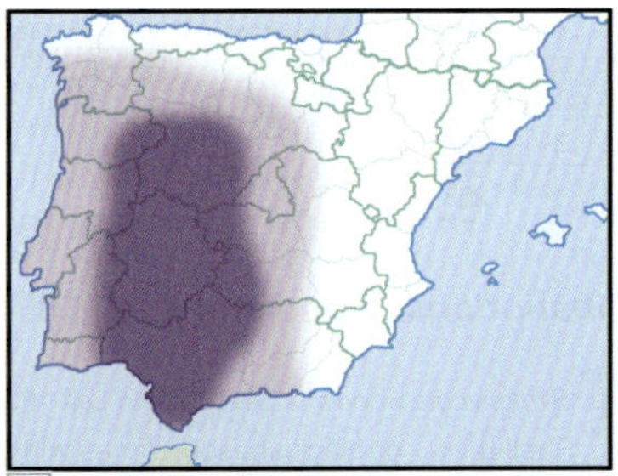

E Iberian peninsula, Morocco

Unio delphinus Spengler, 1793 in Spengler 1793: 63

Shell greenish yellow to dark brown, solid, oval and usually elongated, occasionally more rounded or nearly rectangular. In hard or moderately hard waters south of Tajo river the dorsal margin descends abruptly towards the posterior end, in populations of soft waters the dorsal margin is higher and the posterior end more rounded. Embryonal shell rounded and with tubercles parallel to the growth lines, not always well visible, in acid waters usually corroded. Left valve with 2 pseudocardinal teeth, joint or separated, and 2 lateral teeth, the inferior tooth being more prominent. Right valve with one pseudocardinal and almost always crenulated tooth, a characteristic supracardinal thickening (between dorsal shell margin and pseudocardinal tooth), and one posterior lateral laminar tooth. Shell locally very variable, looks very much like *Unio pictorum*, shells can also be confounded with *Unio mancus*.
Glochidia triangular with a strong ventral band of spines, size 0.23 x 0.21 x 0.15 mm.

Up to 100 mm long

Spain: Guadalquivir near Sevilla

In all kinds of rivers and creeks with permanent water supply, most frequent under the shade of trees on sandy substrate, but also between stones and in habitats exposed to the sun, rarely in lakes and reservoirs. In waters affected by temporary droughts animals can survive in locally remaining water basins together with fishes.

Reproduction in summer. Pregnant females were observed in February in S Spain, embryos and glochidia occur between May and July (Morocco to N Spain). Host fishes unknown, but *Barbus*, *Chondrostoma*, *Gasterosteus aculeatus* and *Squalinus* are likely candidates.

EU Red List: Near Threatened NT. Threatened by habitat destruction, exhaustive water management and pollution. Still common in Spain and Portugal but steadily declining, populations from many localities have already been lost.

Araujo et al. 2009: 45.

Was considered until recently as a synonym of *U. pictorum*, but is disconnected from the main range of that species.

Unio gibbus Spengler, 1793

Unio gibbus Spengler, 1793 in Spengler 1793: 64

Shell thin, light, green to brown, slightly inflated, oval or rounded, equivalve, inequilateral, usually elevated at dorsal posterior end. Anterior end rounded in a circular arc, posterior end rounded but dorsally elevated, forming a marked keel in ligamental area. Umbones rounded, prominent, never eroded, clearer in colour that rest of shell. Left valve with two laminar pseudo-cardinal teeth, sometimes crenulated, not always well separated, posterior tooth beaked and more elevated. 2 laminar posterior lateral teeth variable in length, the lower tooth always more elevated. Right valve with pseudo-cardinal tooth generally beaked, sometimes flat, and very tall laminar lateral posterior tooth.

Spain: Arroyo del Salado, Morón near Sevilla (habitat and populations today destroyed) Picture: Drouet 1893: Pl. 1 Fig. 4

Shell length up to 85 mm

S Iberian peninsula (recent only near Cádiz), Morocco, Algeria, Tunisia

Glochidia rounded triangular without ventral hook, size 0.21 x 0.21 x 0.068 mm.

In rivers, on sandy or stony substrate, preferably under the shade of trees, tolerates also faster water currents, as well as interruptions of the river water current, but needs pools that do not dry out during the dry season.

Females in Spain are pregnant at the end of February and carry embryos in spring (Spain) or early summer (Morocco). Host fishes unknown.

EU Red List: Critically Endangered CR. The population near Cádiz has been seriously declining due to recent habitat destructions (the site at Barbate river completely dried out in 2007). Populations reported by Drouet 1893: 204 from near Sevilla have entirely disappeared due to habitat destructions. Currently the most seriously threatened unionoid species on European territory, but could still survive in N Africa where the rivers are also seriously under human pressure.

Araujo et al. 2009: 40.

Unio mancus Lamarck, 1819

Without gender agreement: *Unio manca* Lamarck, 1819

Unio manca Lamarck, 1819 in Lamarck 1819: 80
Unio elongatula Pfeiffer, 1825 in C. Pfeiffer 1825: 35
Lamarck 1819 and Pfeiffer 1825 treated *Unio* as if it was of feminine gender.
Unio moquinianus Dupuy, 1843 in Dupuy 1843: 80 [not verified]
Unio aleronii Companyo & Massot, 1845 in Companyo & Massot 1845: 234
Unio valentinus Rossmässler, 1854 in Rossmässler 1854-1859 (13/14): 37, Taf. 69

Shell yellowish or blackish brown, often with greenish zones, oval, thin and finely sculptured or coarse and solid, inequilateral with anterior part rounded and posterior part truncated. Interior white and shiny with well marked pallial line between scars of both anterior and posterior muscles. Sipho with typical conical papillas, sometimes branched. Embryonal shell not very prominent in the adult shell, sculpture simple or undulated, lower margin straight or slightly convex. Very variable and shells can be confounded with *Unio delphinus* in Spain.

Mediterranean countries from NE Spain to W Iran, upper Rhine valley

25-40 x 50-80 x 12-23 (height) mm, occasionally larger but rarely exceeding 90 mm

Croatia: Istria, Zudetici near Motovun

Rivers, old river arms, lakes, freshwater channels, needs standing clean water on sandy bottoms, often between roots of trees, half buried in the substrate, not frequent in muddy soils and on stony bottoms. Seems to require hard water exceeding 100 mg/l calcium. Little pollution is tolerated, but not too much. In Switzerland up to 500 m.

Possible host fishes in Spain are *Barbus graellsii* (the only fish known to act as host in nature), *Barbus haasi*, *Squalinus pyrenaicus*, *Squalinus cephalus*, *Chondrostoma miegii*, *Phoxinus phoxinus* and *Salaria fluviatilis*. Pregnant females were observed in July and August in Spain, glochidia were present between March and October.

EU Red List: Near Threatened NT. Threatened by water pollution and habitat destructions, many populations have disappeared since the late 1800s, often as a direct consequence of ingressions of waste waters in densely populated regions.

Araujo et al. 2009: 32.

Considered by Haas 1969 as a subspecies of *U. elongatulus* (not in accordance with the Principle of Priority), this name is still used in some Red Lists. Recent authors have shifted to the older name. Haas' classification is still basically accepted (except *U. e. requienii* Michaud), but subspecies are not recognized by many recent authors. We are still waiting for someone to revise the whole *U. mancus* complex in Europe. Based on unpublished results of mitochondrial DNA studies (Araujo et al. 2009: 33) *U. aleronii* from SW France was considered a synonym of *U. mancus*. Seems to have reached the Rhine valley via the Rhône-Rhine Canal (Glöer & Zettler 2005: 20).

Unio pictorum (Linnæus, 1758)

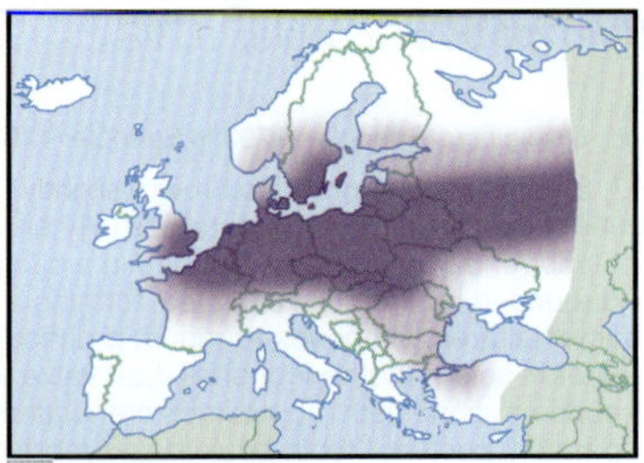

Europe to Ural region

Mya pictorum Linnæus, 1758 in Linnæus 1758: 671

Shell variable in colour and shape, often greenish yellow or brownish without radial streaks, long oval, lower margin not as curved as in *Unio tumidus*, embryonal shell not very prominent and usually corroded, close to the anterior end, posterior end curved, teeth not very prominent.
Animal yellow with reddish grey hue, mantle brown, foot reddish and long, gills greyish.

In rivers, old river arms, sometimes in lakes, reservoirs and canals, mainly in lowlands. Lives inside the sandy or silty substrate, usually in deeper zones, up to 6 m deep. Avoids muddy and stony bottoms. In Switzerland up to 700 m.

Host fishes in N Germany *Perca fluviatilis*, *Gymnocephalus cernuus*, *Rutilus rutilus* and *Scardinius erythrophthalmus*.

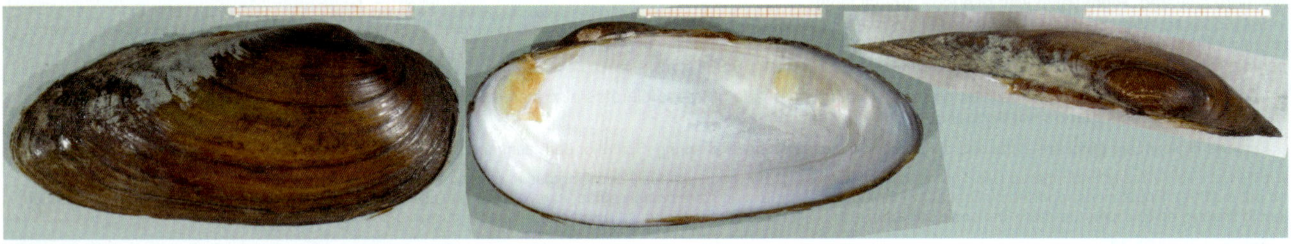

30-40 x 70-100 x 23-28 (height) mm, rarely up to 140 mm long

Latvia: Gauja river near Murjani, near Riga

Still frequent in the 1990s but generally declining, characteristic local forms are continuously replaced by global forms introduced with gamefish (Falkner 1990). Threatened by water pollution and increasingly eutrophic conditions, and by a change in behaviour of host fishes for glochidia (the fishes tend to feed more on plankton than at the bottom, so that the rate of infection with glochidia has decreased). On the other hand *Unio pictorum* has invaded river reservoirs where it has replaced the critically endangered *Unio crassus*.

Unio ravoisieri Deshayes, 1847

N Africa (Tunisia and Algeria E of Moulouya river), isolated in NE Spain (Bañolas lake and Fluviá river basin)

Unio ravoisieri Deshayes, 1847 in Deshayes 1848: Pl. 108 (*Unio Ravoisieri*)
Unio penchinatianus Bourguignat, 1865 in Bourguignat 1863-1865: 141

Shell brown, often greenish or yellowish (in lake Bañolas yellowish brown with white encrusted posterior part), dorsal and ventral margins nearly parallel, anterior dorsal margin characteristically rounded, forming a conspicuous arc (this occurs only rarely in *Unio mancus*). Embryonal shells smooth and not very prominent, usually with two well marked rows of tubercles (but may be absent in not eroded shells).

Shells from rivers up to 60 mm long, exceptionally 95 mm, in Bañolas lake up to 105 mm

In middle and lower sections of small and not very deep rivers, usually hidden between roots of trees, under the shade of trees. In lake Bañolas deeply buried in the sediment at the bottom of the lake. In the Spanish locations syntopically with *Unio mancus*.

Pregnant females in NE Spain occur in June and July, in Tunisia larvas are released in March. Host fishes unknown, but *Barbus meridionalis* and *Leuciscus cephalus* are possible candidates.

Endangered in Spain (Araujo et al. 2009) due to the rare occurrence and small size of the populations. In lake Bañolas threatened by artificial introduction of exotic fishes replacing the native host species.

Araujo et al. 2009: 53.

Classification was very weakly justified and based on unpublished results of molecular studies, in 2009 no evidence was provided that this was an independent species. The conclusions of the molecular studies seem to be substantiated and tenable (R. Araujo, e-mail, 12-2009), preliminary publication of the status as an independent species was also necessary for the protection of the species. It is expected that this species will be confirmed in future studies.

Unio tumidiformis Da Silva e Castro, 1885

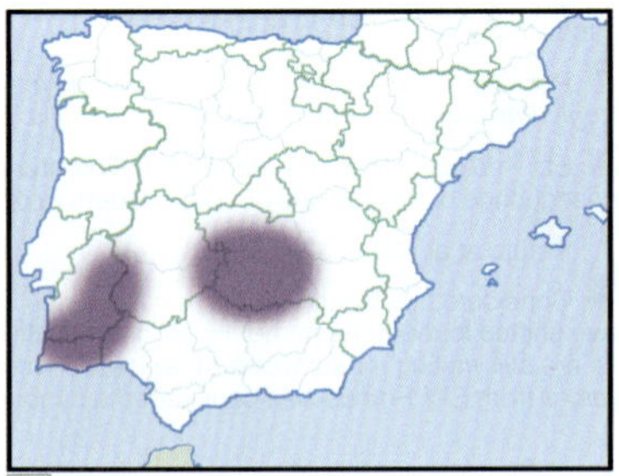

Spain, Portugal

Unio tumidiformis Da Silva e Castro, 1885 in Da Silva e Castro 1885: 283

Shell small, greenish yellow to dark brown, usually with yellow, green or reddish radial streaks, oval, elongate (but less than in *Unio delphinus*) and broad (rarely more than 2.5 x longer than broad), surface usually not very corroded, lower margin straight or weakly curved, anterior part very short and rounded, posterior part long and high. Embryonal shell rounded and prominent, with parallel W-like structures (hardly visible in eroded shells).
Glochidia triangular with a spiny ventral band, size 0.20 x 0.16 x 0.14 mm.

Shell length up to 50 mm, rarely longer

Small and moderate streams, relatively deeply buried in fine sediment near the banks, under the shade of trees, occasionally in lakes. During temporary periods of drought the species survives in local remains of water, together with fishes.

Reproduction in summer. Females with glochidia can be found between March and July, glochidia are released as a loose conglutinate. Five fish species in the genus *Squalius* are good hosts for the species. Metamorphosis takes place over 10 days at an average temperature of 22° C.

EU Red List: Vulnerable VU. Very rare, very restricted distribution, many populations are isolated and consist only of few individuals, lives in a highly sensitive habitat. One population had thousands of individuals including juveniles prior to 2005 and disappeared almost entirely after 2005.

Araujo et al. 2009: 49.

Had been classified as *Unio crassus* before 2009.

Unio tumidus Philipsson, 1788

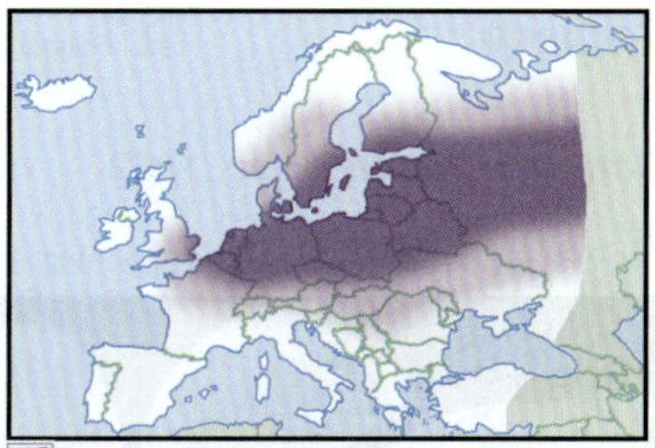

Europe to W Ural region

Unio tumidus Philipsson, 1788 in Retzius 1788: 17

Shell greenish or brownish, often with yellowish or greenish radial streaks from the centre to the periphery, inflated, solid, lower margin more curved than in *Unio pictorum*.

Slowly running rivers, old river arms and lakes, also in artificial lakes and fish ponds, prefers sandy substrate and weakly moving waters, up to 9 m deep. Avoids stony substrate and deep soft mud, seems to require cleaner and better oxygenated waters than the other *Unio* species. Mostly in lowlands, in Switzerland up to 1000 m.

Host fishes for glochidia are *Perca fluviatilis*, *Rutilus rutilus* and *Scardinius erythrophthalmus*, in N Germany also *Gymnocephalus cernuus*, *Gasterosteus aculeatus* and *Blicca bjoerkna*.

25-40 x 50-80 x 23-35 (height) mm, rarely up to 120 mm

Latvia: Gaja river near Adazi, near Riga

Threatened by continuous water pollution and habitat destruction. Critically Endangered in Austria, Endangered in Germany and Spain, Vulnerable in Switzerland, Slovakia and the Czech Republic. Has also declined in Britain due to draining and change in character of canal habitats.

Corbicula

Corbicula Megerle von Mühlfeld, 1811 in Megerle von Mühlfeld 1811: 56

Type species *Tellina fluminalis* Müller, 1774 (monotypy)

Corbicula fluminalis (Müller, 1774)

Tellina fluminalis Müller, 1774 in Müller 1774: 205

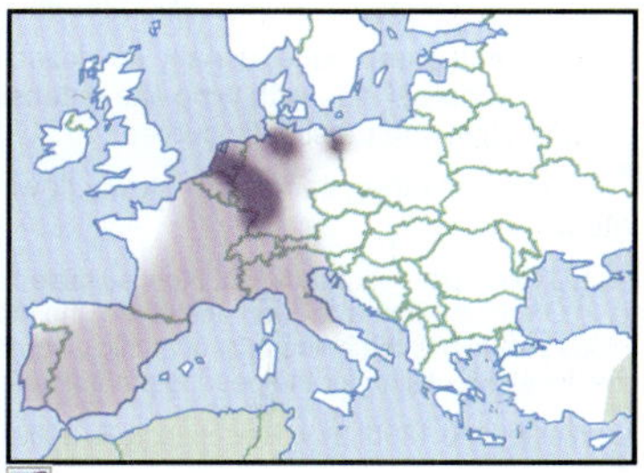

Originally E Asia, introduced to North America and Europe

Up to 40 x 40 mm

Germany: Rhein near Düsseldorf

Shell yellowish green (also fully grown shells), slightly asymmetrical or (shaped, usually wider than long, vortex inflated, concentrical riblets narrowly spaced, usually less than 1 mm, shell inside pale bluish to pale violet.

In Europe at river margins, rapidly extending its range, much rarer than *Corbicula fluminea*, about 10 % of *Corbicula* individuals belong to this species. Seems to prefer brackish water, and to tolerate chemically polluted water.

Maturity is reached after 3 years.

Taxonomic status seems to be unclear.

Corbicula fluminea (Müller, 1774)

Tellina fluminea Müller, 1774 in Müller 1774: 206

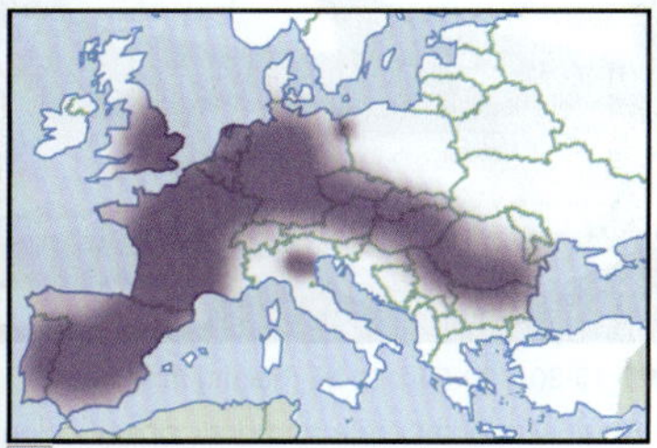

Originally E Asia, introduced to North America and Europe
Reached in 1980 Portugal and S France, 1994 Netherlands, 2000 Romania.

Up to 40 x 40 mm

Germany: Rhein near Düsseldorf

Shell yellowish green (juveniles) to dark brownish(adults), slightly asymmetrical, usually longer than wide, vortex inflated, concentrical riblets widely spaced, usually more than 1 mm, shell inside whitish to pale bluish.

In Europe at river margins, rapidly extending its range, much more abundant than *Corbicula fluminalis*, about 90 % of *Corbicula* individuals belong to this species. In some regions in Europe extremely frequent where present, causing serious problems to indigenous species.

Maturity is reached after 3 years.

Congeria

Congeria Partsch, 1835 in Partsch 1835: 97

Type species *Congeria subglobosa* Partsch, 1835 (subsequent designation by Pilsbry 1911: 95)

Congeria kusceri Bole, 1962

Congeria kusceri Bole, 1962 in Bole 1962: 57 (*Congeria kuščeri*)

16-22 x 7-14 x 11-16 mm

S Croatia: near Vrgorac

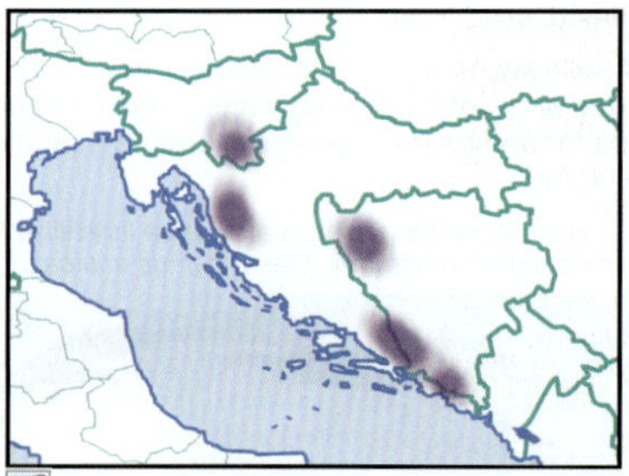

Slovenia to SW Bosnia

Shell light brown, thin but solid and with thin periostracum, rounded and without keel, surface prominently and irregularly ribbed, finely striated between the ribs. Byssus with 40-80 horny threads. Shell shape variable.

In subterranean waters of the Dinaric karst region, at temperatures between 6 and 19℃, usually at 9℃, attached with short and thick byssus threads to the walls of either permanently or periodically inundated karst caves. Their tiny foot does not allow significant locomotion after attachment.

EU Red List: Vulnerable VU. Extrelemy rare, known only from a limited number of caves. Sensitive against pollution of habitat, by domestic sources and agricultural run-off. Protected under Fauna und Flora Habitats Directive FFH 2 and 4.

Jalžić 1998 (8 populations in Croatia), Morton et al. 1998 (biology and anatomy), Jalžić 2001 (distribution), Stepien et al. 2001 (molecular diversity).

C. kusceri is the world's only known subterranean bivalve mollusc.

Dreissena

Dreissena Van Beneden, 1835 in Van Beneden 1835: 25 (*Driessena*)

Type species *Mytilus polymorphus* Pallas, 1771 (monotypy)

Date was Feb 1835. Established on pp. 25-26 as *Driessena*, without description and with one species *Mytilus polymorphus* included. The name of the person was given as M. Driessens on p. 26, so no inadvertent error (Art. 32.5). ICZN Opinion 351: name to be spelled *Dreissena*.

Dreissena blanci Westerlund, 1890

Dreissena bourguignati var. *blanci* Westerlund, 1890 in Westerlund 1890: 319

Shell with sharp edge, usually no ziczac colour pattern, dorsal side with wing-like extension, ventral margin can be slightly curved.

Lakes and other standing or slowly flowing waters.

EU Red List: Vulnerable VU. Most suitable habitats in Greek waters have been destroyed, the most serious problem is the entire lack of public awareness of the local population for the need to protect natural water bodies and their fauna and flora (Albrecht, Lo. et al. 2006).

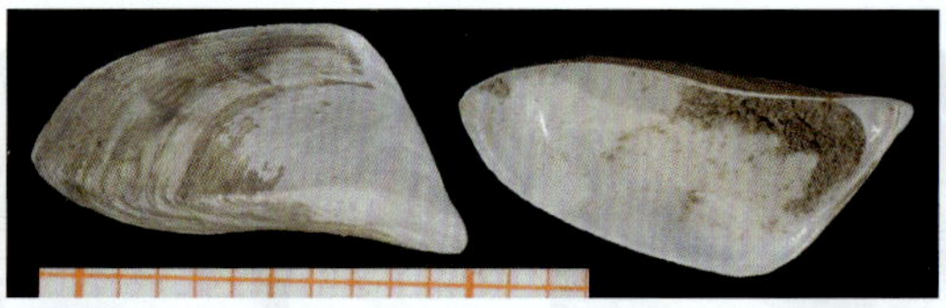

9 x 25 x 10 (height) mm

Albania: Lake Shkodra, near Shkodër

Albania, Greece, W Turkey

Westerlund 1890: 319, Reischütz & R. 2003: 29 (Etoloakarnania, lakes Trichonia and Lisimachia), Falkner 1990: 268, Dhora & W. 1996: 172, Dhora 2004: 77 (ecology).

Tentative and uncertain classification, unclear in which sense this name has been used. It is necessary to study the *Dreissena* populations of Lake Shkodra to verify the relationships with populations from elsewhere (they were not studied by Gelembiuk et al. 2006).

Dreissena polymorpha (Pallas, 1771)

Without gender agreement: *Dreissena polymorphus* (Pallas, 1771)

Mytilus polymorphus Pallas, 1771 in Pallas 1771: 478

15-30 x 20-50 x 10-25 (height) mm

S Germany: Bodensee, Sipplingen

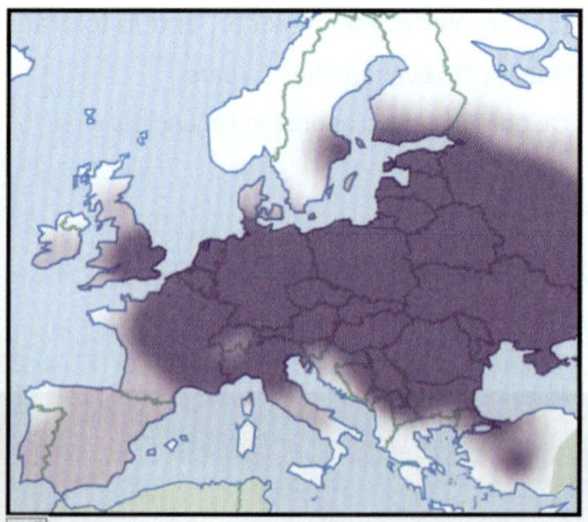

Originally SE Europe to Caspian Sea, in the 1800s artificially spread with ships over central and W Europe, later all over the world

Shell with blunt edge and dark ziczac colour pattern.

In rivers and lakes, shells attach to wood logs and other submerged objects, also to ships, with which the species has been spread over large distances. Tolerates slightly brackish water. Requires water temperatures above 10℃ to grow and above 14℃ to reproduce. This excludes deeper zones of some alpine lakes. In Switzerland in up to 900 m altitude.

6-9 years after having been introduced to new waters substantial increases of abundance are observed under favourable conditons. In this phase predators get adapted to the new source of prey (diving birds, fishes, hirudines and plathelminths), so that *D. polymorpha* abundance declines up to 4 m depth.

First records in Britain in the 1820s, introduced with timber from the Baltic, by 1850 its distribution approximated that of today. Around Manchester the species has recently declined, other disappearances in England are due to the abandonment of canals. Serious problems were observed in Bulgaria when reservoirs for the water supply were invaded in the early 2000s.

Spawning in Germany takes place in restricted months in the season depending on local conditions, at water temperatures above 12℃. Eggs are expulsed in small mucus agglomerations, free swimming veliger larvae transform to juvenile mussels and attach themselves with their byssus to substrate of almost any kind. Maturity is reached at 8-10 mm size, shells may grow up to 5 cm in 2 years, and then produce 1 million eggs annually. Average life span 5 years.

Borcherding 1991, Stepien et al. 2002, Gelembiuk et al. 2006.

Dreissena rostriformis (Deshayes, 1838)

Dreissena bugensis Andrusov, 1897

Mytilus rostriformis Deshayes, 1838 in Deshayes 1838: 61
Dreissena bugensis Andrusov, 1897 in Andrusov 1897: 285 [not verified]

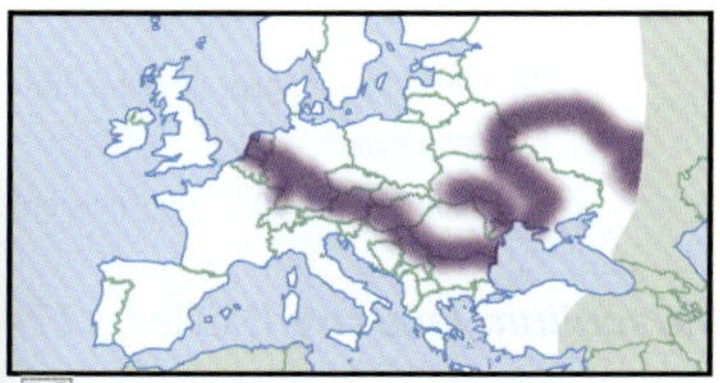

Originally Ukraine (Dnjepr river drainage, Bug river), introduced to other parts of Europe (Dnepr, Dnestr, Volga, Caspian Sea, Danube to Rhine), since 1988/1989 in N America, since 2004 in Romania, since 2006 in the Netherlands

Up to 40 mm — USA: New York state, Rochester/Lake Ontario — Academy of Natural Sciences Philadelphia ANSP

Shell evenly black, cream or whitish, no zebra pattern, byssal groove small, at ventral side near hinge, midventral line of closed shell not straight but curved (valves asymmetrical if seen from ventral or frontal side), no ridge but rounded between ventral and dorsal surfaces. Ventral side convex, can sometimes be distinguished by placing shells on the ventral side; *D. rostriformis* will topple over, *D. polymorpha* will not.

In rivers and lakes, same habitats and requirements as *D. polymorpha.*

Prodigious water filterer, removing substantial amounts of phytoplankton and suspended particulate from the water. Impacts are similar to those of the *D. polymorpha.* By removing the phytoplankton, the species in turn decreases the food source for zooplankton, therefore altering the food web.
Impacts associated with the filtration of water include increases in water transparency, decreases in mean chlorophyll a concentrations, and accumulation of pseudofeces. *D. rostriformis bugensis* is often regarded as a subspecies, sometimes as an independent species.

Therriault et. al 2004, Schonenberg & Gittenberger 2008.

Occasionally called *Dreissena bugensis.*

Dreissena stankovici Lvova & Starobogatov, 1982

Dreissena stankovici Lvova & Starobogatov, 1982 in Lvova & Starobogatov 1982: 1749

Differs only slightly from *Dreissena polymorpha*, the main differences showed up in the molecular analyses.

At the shore of lakes.

EU Red List: Near Threatened NT. Threatened by artificial changes of water levels due to over-extraction of water.

Gelembiuk et al. 2006, Reischütz et al. 2008: 3.

Could be conspecific with *Dreissena presbensis* Kobelt, 1915. If true it would take this name.

SE Europe: Lake Ohrid and Lake Prespa, possibly some other lakes

15-30 x 20-50 x 10-25 (height) mm — Albania: Lake Ohrid, 4 km S Pishkurat

Mytilopsis

Mytilopsis Conrad, 1858 in Conrad 1858: 167

Type species *Mytilus leucophaeatus* Conrad, 1831 (subsequent designation by Dall 1898: 808)

Gender feminine (Art. 30.1.2).

Mytilopsis leucophaeata (Conrad, 1831)

Without gender agreement: *Mytilopsis leucophaeatus* (Conrad, 1831)

Mytilus leucophaeatus Conrad, 1831 in Conrad 1831: 263 (*Mytilus leucophæatus*)

6-11 x 12-20 x 7-12 (height) mm — Germany: Nord-Ostsee-Kanal, 20 km E Hochdonn

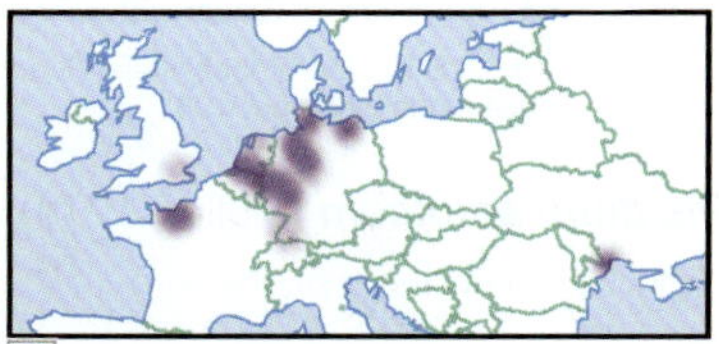

Originally North America, introduced since 1835 to European waterways

Shell rounded on upper side without edge, periostracum scaly.

In brackish waters, often in low abundance among competing species, and more weakly attached to substrate. Populations can grow rapidly under favourable conditions.

Egg diameter 0.06 mm, umbos in many larvae present after 4 days, metamorphosed juveniles settle after 9 days, size 1 mm after 30-50 days under laboratory conditions at 23 °C.

Therriault et al. 2004, Kennedy 2011.

Musculium

Musculium Link, 1807
in Link 1807: 152

Type species *Tellina lacustris* Müller, 1774
(monotypy)

Shell thin, fragile, umbone narrow, hinge somewhat reduced. High number of embryos, nephridium with long and narrow funnel and small excretory sac.

Worldwide

Korniushin 2001: 108 (revision), Korniushin & Glaubrecht 2002 (phylogeny).

Musculium lacustre (Müller, 1774)

Without gender agreement: *Musculium lacustris* (Müller, 1774)

Tellina lacustris Müller, 1774 in Müller, 1774: 204

5-8 x 7-15 x 8-10 (height) mm

Germany: Schleswig-Holstein, Lauenburg

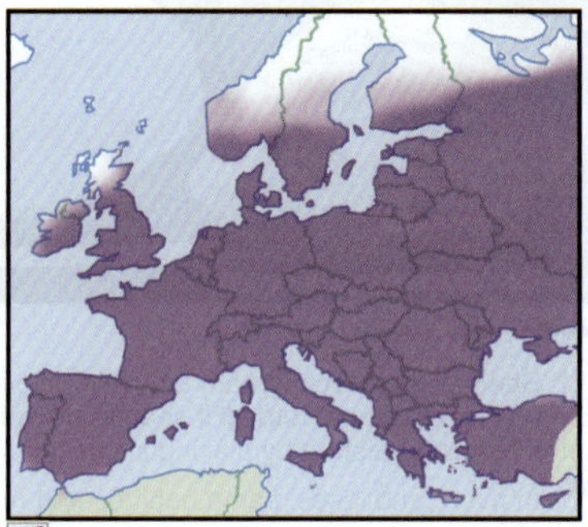

Worldwide with gaps

Shell horny yellowish to greyish, very thin and almost transparent, embryonal shell appears more pointed and outside the main shell than in other species of this size.
Animal very light yellow with reddish hue, slightly transparent, foot up to twice as long as the shell, siphons long.

Small and temporary water bodies, swamps, not very deep, inside fine sediments, is able to crawl on plants and tolerates anaerobic bottoms substrates, in SW Spain up to 400 m altitude.

In Portugal, Germany and England not rare, less frequent in Wales, rare in Scotland, Ireland and SW Spain. Has probably declined to some extent in Britain because of agricultural drainage and removal of field ponds, more seriously in Ireland and Wales, disappeared since 1900 entirely from the region around Dublin.

Musculium transversum (Say, 1829)

Without gender agreement: *Musculium transversa* (Say, 1829)

Cyclas transversa Say, 1829 in Say 1829: 356 [not verified]

11-15 x 9-11 x 7-8 (height) mm

USA: Pennsylvania, near Philadelphia
Academy of Natural Sciences Philadelphia ANSP

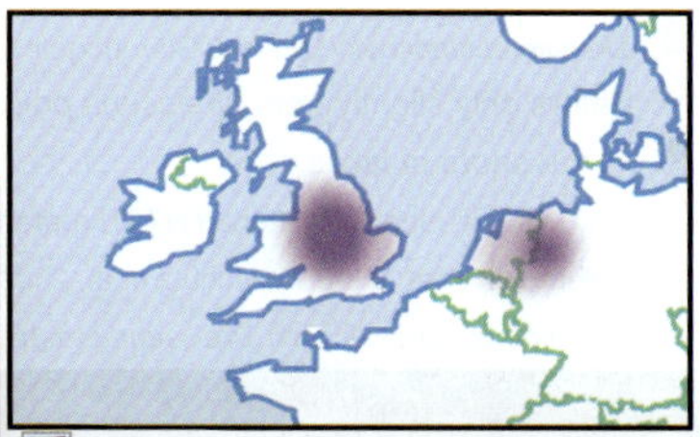

North America (Florida and Mexico to S NW Territories and Quebec), accidentally introduced (1850s) to England, the Netherlands and W Germany

Shell pale geenish yellow, prominently and very irregularly striated, inflated, anterior part short, posterior part long and blunt, vortex wide and inflated, not much protruding, ligament relatively long and strong, shell inside bluish white, 1 strong cardinal tooth at the left valve, two cardinal teeth at the right valve, the posterior tooth very strong, the anterior tooth higher and thinner, between both a broad triangular space for the opposite tooth.

In Britain only in canals and canalized rivers, in muddy substrate, tolerates anaerobic bottom conditions in urban canals.

Introduced to the English canal system in the 1850s, has spread until it was widely distributed and has declined after 1900 for unknown reasons to become a rarity. Still survives at some scattered localities. Has probably vanished in the Netherlands.

Pisidium

Pisidium Pfeiffer, 1821
in C. Pfeiffer 1821: 17, 123

Type species *Tellina amnica* Müller, 1774
(fixed under the plenary powers)

ICZN Opinion 336: type species designated by Gray 1847: 185. This is not tenable. Anton 1839: 14 gave as type *Cyclas obtusalis* Lamarck, 1818. Gray 1847 (09-11-1847): 184-185 did not designate *Tellina amnica* as type for *Pisidium*, but for *Pisum* Megerle von Mühlfeld, 1811, of which Gray regarded *Pisidium* to be a synonym. So the type species should be regarded as fixed under the plenary powers. Herrmannsen 1847-1849 [07-12-1847]: 275 gave as type *Tellina amnica*.

Euglesa Jenyns, 1832 in Jenyns 1832: 302 (*Pisidium* synonym *Euglesa*)

Mentioned as *Euglesa Henslowiana* "Leach MSS. Brit. Mus." as a synonym for *Pisidium pusillum*. The name was not used for a taxon by Turton & Gray 1840: 282, but was mentioned there as a synonym *Euglesia* for *Pisidium*. Used for a genus *Euglesa* by Leach 1852: 291, containing one species "*Euglesa Henslowiana*", from near Cambridge, with description, classified as *Pisidium henslowanum* by mussel-project.ua.edu (11.2011). The circumstances are quite chaotic and I have not been able to come to a definite conclusion which one would be the type species in this case.

Korniushin & Glaubrecht 2002 (phylogeny).

In S Europe probably also *P. annandalei* Prashad, 1925, known from India and SE Asia. I have no recently published records for this species to confirm the statements by Kuiper 1981.

Pisidium amnicum (Müller, 1774)

Without gender agreement: *Pisidium amnica* (Müller, 1774)

Tellina amnica Müller, 1774 in Müller 1774: 205

4-8 x 7-11 mm

N Germany: Elbe-Lübeck-Kanal near Lauenburg

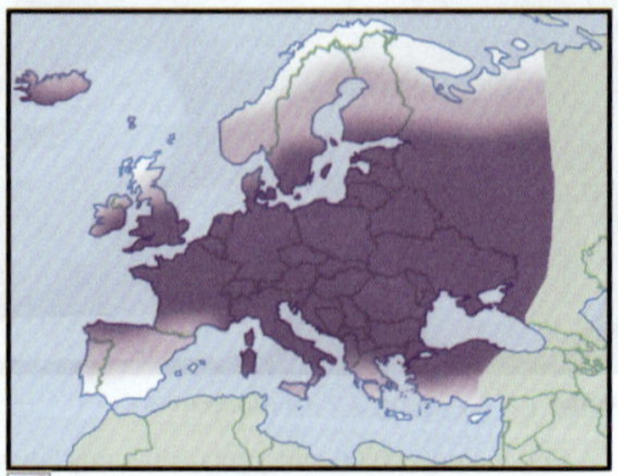

Palearctic and E North America

Shell thin, larger than the other *Pisidium* species, brownish with some concentric dark bands, irregularly ribbed.
Animal whitish or greyish, slightly transparent, foot not long, siphon short.

Larger creeks, rivers, canals, also in litoral zones of large lakes, generally in clear and clean waters with some water movement, in the substrate near water plants. Usually not below 10 m depth, maximum 35 m. In S in up to 700 m, in BG 1000 m.

In Switzerland Vulnerable, threatened by habitat destructions and water pollution. Rare in most parts of Scandinavia. Endangered in Spain, Germany and Austria.

Pisidium casertanum (Poli, 1791)

Cardium casertanum Poli, 1791 in Poli 1791: 65, Tab. 16 (*Cardivm casertanvm*)
Pisidium casertanum var. *ponderosa* Stelfox, 1918 in Stelfox 1918: 294

Shell rounded and inflated, in rivers oblique and solid.
Animal whitish or greyish, slightly transparent, foot not very long, siphon subconical.

Small and temporary water bodies, springs, temporary waters, creeks, rivers, with little water current, prefers clear water. Depth seems to depend from other competing *Pisidium* species, sometimes only above and sometimes only below 40 m. In Switzerland in up to 2700 m, in Bulgaria 2400 m, Scotland 1000 m.

Clarke 1981: 394.

Perhaps the world's most widely distributed non-marine mollusc, frequent in Europe.
Piechocki 1989: 291 gave a summary on previous studies concerning the identity of *ponderosum*, usually regarded as a form of *P. casertanum*, which could not be firmly established as a separate species in previous studies. Glöer & Zettler 2005: 21 reported without detailed data that the two forms occurred syntopically and differed in their ecological preferences. This was not in contrast to Piechocki's 1989 argument that the observed differences were probably ecologically induced. The presence of two species with independent gene pools should be demonstrated in a more detailed study.

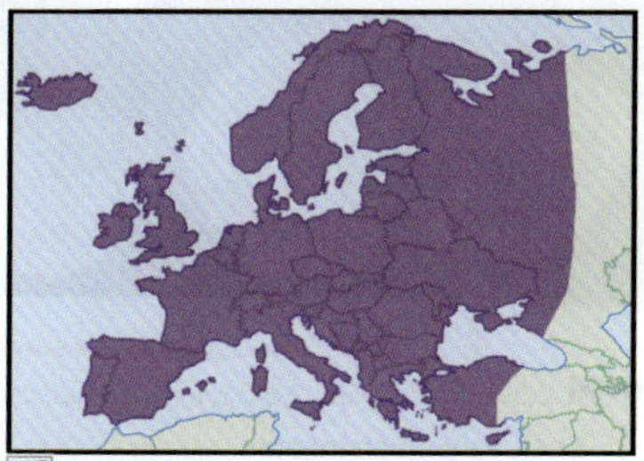

Worldwide, in Canada to 70° N

3-4 x 4-5 mm

Germany: Mecklenburg-Vorpommern, near Schaalsee

Pisidium conventus Clessin, 1877

Pisidium conventus Clessin, 1877 in Clessin 1877: 181, Taf. 3 (*Pisidium Conventus*)

Shell light horny yellowish, very gently rounded, not much inflated, weakly shiny, anterior part narrow and weakly pointed, posterior part broadly truncated, vortex broad, blunt and not very prominent, left valve with veery fine cardinal teeth, lateral teeth high and prominent, frontal tooth short and with a long hook.

2.2-2.8 x 1.1-1.5 x 2-2.4 (height) mm

In N Europe in cold standing waters of mountains, often deeply shaded, in Britain up to 750 m altitude and down to 300 m depth (deeper is not possible in Britain). In the Alps in deep waters of alpine lakes below 15 m. In high alpine lakes also at the lake margins. In Switzerland in up to 2600 m altitude.

Threatened by water pollution, populations of many large Swiss lakes have disappeared since the early 1900s. Perhaps less threatened in N Europe where it is underrecorded and lives in remote waters far from urban centres.

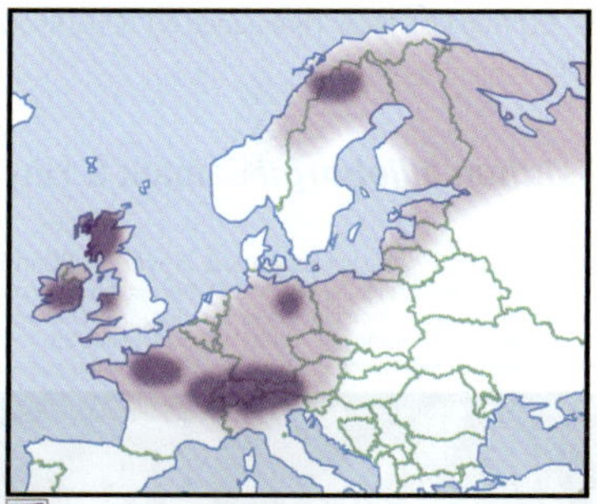

Europe to N Russia,
northern USA to Newfoundland and Alaska

Pisidium globulare Westerlund, 1873

Pisidium globulare Westerlund, 1873 in Westerlund 1873: 532
Name was attributed to Clessin, main description was written by Westerlund, but below (p. 533) Westerlund cited a text Anm[ärkningar] attributed to Clessin where the new species was compared with *Pisidium amnicum*. Under Sabrosky's guide (the *Ceriodes durani* example, there was no clearcut attribution of Clessin's authorship for the main description, just having provided the information does not suffice) and Art. 50.1.1 (both name and description would have to be attributed to the same different author) the name must be attributed to Westerlund alone, the author of the work.

Shell horny yellowish, inflated, thin, very finely and regularly striated, upper margin rounded, anterior margin very weakly pointed, posterior margin rounded and slightly truncated, lower margin rounded, vortex broad and protruding, left valve with short outer cardinal tooth, inner cardinal tooth bent, frontal lateral tooth very high and pointed. Reliable differences to *P. casertanum* in the anatomy.

3.7-5.5 x 3 x 4 (height) mm

In muddy substrate of small and temporary water bodies.

Westerlund 1890: 21, Korniushin & Glaubrecht 2002 (anatomical differences to *P. casertanum*), Horsák & Neumanová 2004, Glöer 2006, Heber & Patzner 2009.

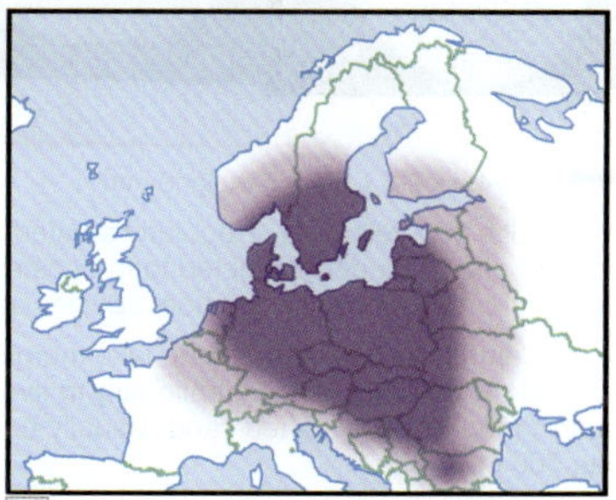

Europe (?)

Pisidium henslowanum (Sheppard, 1823)

Without gender agreement: *Pisidium henslowana* (Sheppard, 1823)

Tellina henslowana Sheppard, 1823 in Sheppard 1823: 150 (*Tellina Henslowana*)

2-3.5 x 3-5 x 2-3.5 (height) mm

Germany: Schleswig-Holstein, Alster near Poppenbüttler Schleuse

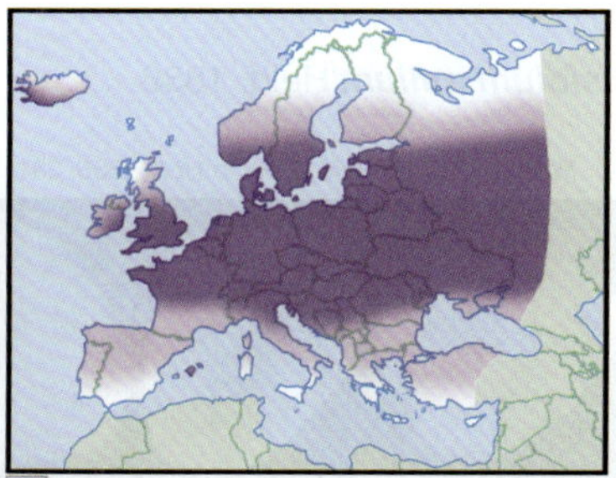

Eurasia, not known from N Africa, introduced to E North America (Great Lakes to S Quebec)

Shell light brownish, densely striated, opaque and weakly shiny, oval, with oblique vortex fold.
Animal whitish, with short foot and siphon.

Creeks and other small running freshwaters, marshland drains, canals, also in lowland rivers and lakes, between roots of water plants. Almost never in closed standing waters. Needs eutrophic conditions with sandy or solid substrate rich in organic compounds, on calcareous base. Usually up to 15-25 m deep, maximum 40 m. In Switzerland in up to 900 m altitude.

Juveniles with 0.75 mm shell diameter remain some time inside the adult shell, 7-15 juveniles were observed per individual.

Vulnerable in Germany.

Pisidium hibernicum Westerlund, 1894

Pisidium hibernicum Westerlund, 1894 in Westerlund 1894: 205

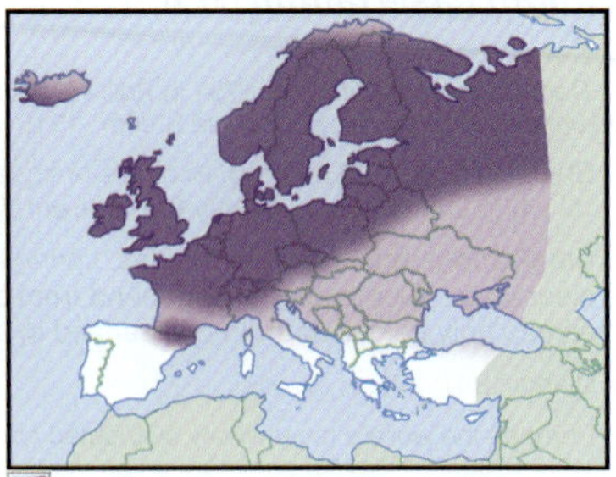

2.5-3.5 x 2.5-3.5 x 2.5-3.5 mm, occasionally > 5 mm

Sweden: Torne-Lappmark near Björkstugan

Palearctic

Shell very much inflated.

Lakes and smaller standing waters, slowly moving rivers and channels, usually in clean and clear water, avoids water with periodical droughts in temperate climate zones. Prefers substrate with rich organic compounds. Usually not below 10 m depth, maximum 30 m. In Spain in clean standing waters of high mountains. In NW Italy in up to 2800 m, Pyrenees 2200 m, Austria 2100 m, Britain 700 m.

Number of juveniles seems to be substantially lower than in other *Pisidium* species.

Endangered in Germany and Austria, Vulnerable in Spain and S Switzerland.

Pisidium hinzi Kuiper, 1975

Pisidium hinzi Kuiper, 1975 in Kuiper 1975: 27

Shell greyish white, solid, not translucent, anterior and posterior parts evenly rounded, vortex slightly flattened above, located clearly at posterior side, growth lines flat, irregular and indistinct. Looks like *P. obtusale*, but posterior tooth P3 without any callus at the proximal side.

Up to 3.5 x 2.6 x 3.1 (height) mm

In small lakes, 2-10 m deep (not in swampy habitats like *P. obtusale*).

EU Red List: Near Threatened NT. Only known from few Scandinavian localities.

Norway to N Russia and Kamchatka

Pisidium lilljeborgii Esmark & Hoyer, 1886

Pisidium lilljeborgii Esmark & Hoyer, 1886 in Esmark & Hoyer 1886: 119 (*Pisidium Lilljeborgii*)

Name was originally attibuted to Clessin, but not the description, so the authors of the work must be authors of the new name (Art. 50.1.1 and Sabrosky 1974, the *Ceriodes durani* example, there was no clearcut attribution of Clessin's authorship for the description, just having with some probability provided the information does not suffice).

The name ends in *-ii* and not in *-i* (in Welter-Schultes 2011: 38 I stated that the ending was *-i*, with a claim to have consulted on the original source. I did consult the original source, but missed to copy the correct ending into the relevant file at that occasion. Sorry for the confusion).

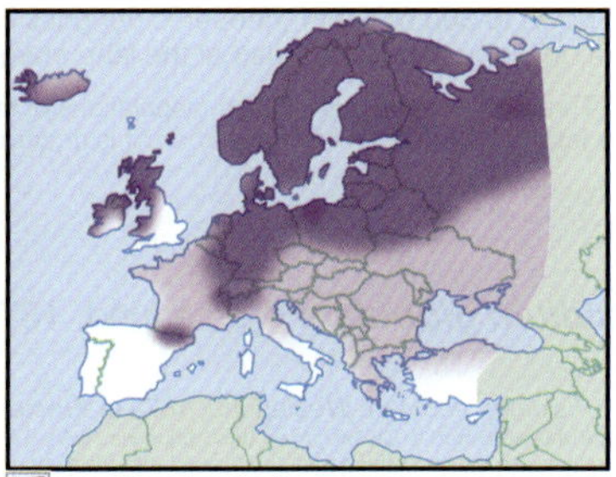

3.0-4.5 x 2-3 x 2.5-3.5 (height) mm

Sweden: Torne-Lappmark near Björksugan

Northern Hemisphere

Shell horny yellowish, irregularly striated, strongly inflated and high, anterior part roundly pointed, posterior part truncated, vortex strongly inflated, rounded and protruding, directed to the posterior side.
Smaller than *P. waldeni* when both species occur together. In *P. lilljeborgii* the upper margin is shorter.

In central Europe restricted to lake margins with water movement on fine sandy substrate, in Britain to lakes in the highland zone. In colder climates also small lakes, but only very exceptionally in rivers or canals, no other running waters. Usually in soft waters, tolerates a wide range of bottom conditions from organic mud to well-washed sand. Usually up to 10 m deep, exceptionally up to 40 m. In the Pyrenees in up to 2300 m, Switzerland 2100 m, in Britain more than 700 m.

Number of juveniles seems to be substantially lower than in other *Pisidium* species.

In W Ireland, W Wales, NW England and Scotland restricted to highland zones where the species is not rare and appears to be secure. Endangered in Austria and Bayern, Vulnerable in Spain and S Switzerland.

Pisidium milium Held, 1836

Pisidium milium Held, 1836 in Held 1836: 281

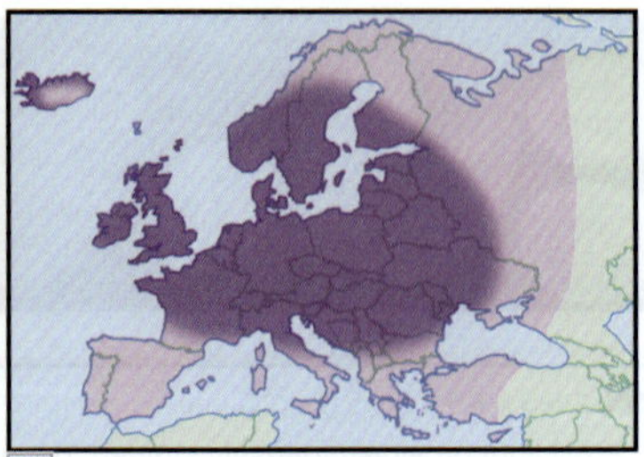

2-2.5 x 2.5-3 x 2-2.5 (height) mm

Sweden: Torne-Lappmark near Tornehamn

Northern Hemisphere

Shell light yellowish to brownish, shiny, obliquely trapezoid, strongly inflated.

Standing und running waters with muddy substrate and usually good aquatic flora, tolerates soft water and moderately eutrophic conditions, but disappears from too eutrophic and too much organically or chemically polluted waters. Avoids intensive water movement. Survives short periods of drought, but tends to avoid waters liable to dry up. Usually in shallow waters up to 10 m depth, maximum not below 25 m. In Switzerland in up to 1300 m altitude.

Pisidium moitessierianum Paladilhe, 1866

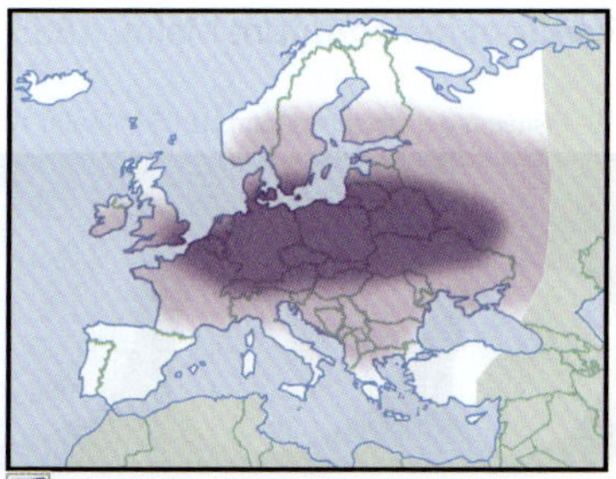

Europe (or Palearctic?)

Pisidium moitessierianum Paladilhe, 1866 in Paladilhe 1866: 172, Pl. 13 (*Pis. Moitessierianum*)

The very small size is characteristic.

In slowly running waters, in Switzerland in large lakes with moderately moving water and solid sandy substrate, in Britain restricted to canals and lowland rivers. Requires slowly moving, well-oxygenated hard water over clean, unpolluted substrates varying from fine mud to sand. In Switzerland in up to 450 m altitude. Usually associated with *Pisidium supinum* and other species.

In Finland (water temperature 1-20°C) one generation per year, minimum mature size is 1.0 mm, parents have around 7 larvae, life span 2 years.

Usually in low abundance where present. Extinct in Ireland, Endangered in Austria, Vulnerable in Germany. Should at least be considered as Critically Endangered in Wales (last records from the 1940s).

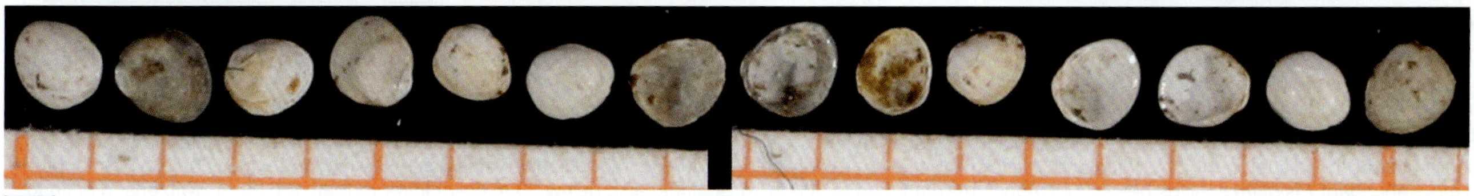

1.5-2.0 mm

Germany: Hellbach near Drüsensee

Pisidium nitidum Jenyns, 1832

Pisidium nitidum Jenyns, 1832 in Jenyns 1832: 304, Tab. 20
Pisidium nitidum var. *crassa* Stelfox, 1918 in Stelfox 1918: 295
Pisidium edlaueri Kuiper, 1960 in Kuiper 1960: [not verified]
Pisidium maasseni Kuiper, 1987 in Kuiper 1987: 163

1.5-2.5 x 2.5-3.5 x 2-3 (height) mm

Slovakia: Zohorsky Potok

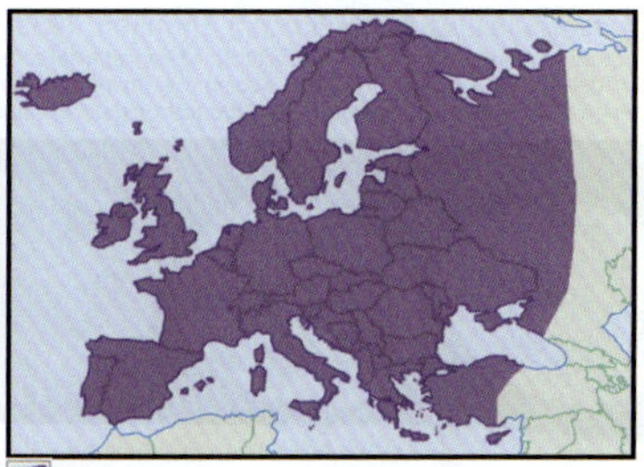

Northern Hemisphere

Shell pale yellowish or whitish, smooth and very shiny, 3-5 grooves around vortex. Animal whitish, foot moderately long, siphon short.

Margins of lakes and lowland rivers with water plants, in sandy and muddy substrate, prefers clean and unpolluted bottoms with good oxygenation. Not frequent in isolated small water bodies, avoids swampy and anaerobic places and those liable to dry up. Tolerates some water movement and cold temperatures, was very frequent in postglacial lakes. Usually up to 20 m deep, rarely below 50 m. In SE France in up to 2500 m altitude, in Britain up to 500 m and mainly in lowlands.

Piechocki 1989: 285 gave some arguments that *crassa* (usually listed among the forms of *P. nitidum*) should be regarded as a separate species. This view is not followed here, the presence of an independent species should be firmly established in a more detailed study. Schultheiß et al. 2008 found no molecular constraints for the presence of separate species in the lakes of Ohrid (*edlaueri*) and Prespa (*maasseni*).

Pisidium obtusale (Lamarck, 1818)

Without gender agreement: *Pisidium obtusalis* (Lamarck, 1818)

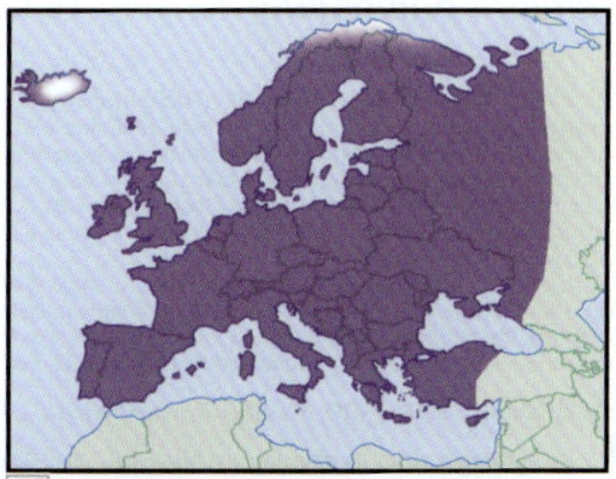

Northern Hemisphere

Cyclas obtusalis Lamarck, 1818 in Lamarck 1818: 559

Shell spherically inflated, horny yellowish, or with reddish or greyish hue, semi-transparent, weakly striated, vortex widely rounded, upper margin extremely convex.
Animal greyish, sometimes with yellowish or reddish hue, foot longer than shell, siphon short.

Swamps, standing water bodies tending to become land. The only *Pisidium* species that requires high standing water plants (Cyperaceae, Juncaceae). Tolerates very soft water. In Switzerland in up to 1800 m altitude, in Scotland up to more than 700 m.

Has certainly declined in many parts of Europe.

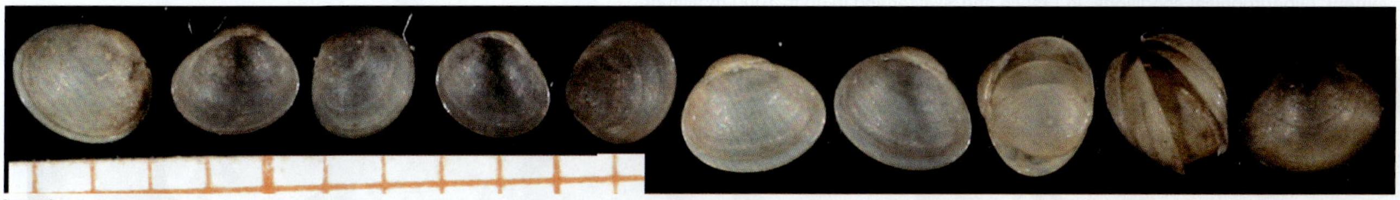

1.5-3 x 2-3.5 x 2-3.5 (height) mm

Germany: Mecklenburg-Vorpommern near Schaalsee

Pisidium personatum Malm, 1855

Pisidium personatum Malm, 1855 in Malm 1855: 107

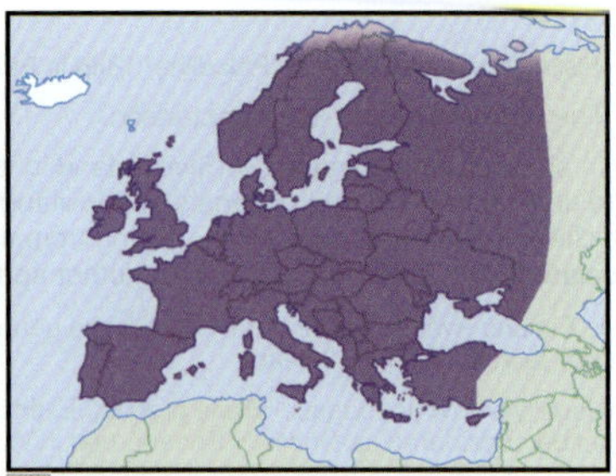

3-3.5 x 3.5-4 x 2-2.5 (height) mm

N Germany: Flensburger Förde

Europe, N Africa, Near East

Shell like *Pisidium obtusale*, smaller and less inflated, anterior part longer and narrower than posterior section, vortex very little protruding, valves form an angle of 70-80° at the lower side.
Animal with sipho short, finely folded.

In various poorest aquatic habitats from small standing waters to springs and lakes, grassy ponds, roadside ditches, also temporary waters and occasionally under ground litter in marshy woodland. Artificial habitats are rapidly colonized. Also in deep zones of lakes, in Switzerland usually below 15 m depth, the most abundant *Pisidium* species below 30 m depth, maximum up to 300 m, in Scotland also below 30 m down to more than 100 m. In Switzerland in up to 2800 m altitude, in the Pyrenees below 1000 m, in Britain up to 900 m. In Britain rarely associated with any other bivalve, sometimes with *P. casertanum*, in deep lakes with *P. conventus*.

Pisidium pseudosphaerium Ehrmann, 1933

Pisidium milium v. *pseudo-sphærium* as mentioned by Favre 1927: 366

Favre used the term "pseudo-sphaerium" to describe a certain *Sphaerium*-like form of shells as such, as a technical term suggested by it being repeatedly used for various species ("pseudo-sphaerium" for 5 species, "ponderosum" for 3 species). These expressions should be regarded as technical terms and not as taxonomic names (Piechocki 1989: 278). Favre 1927 could also have used the term to describe a certain form in artificially created objects (ceramic objects etc.). Adler 1993: 54 regarded it as a scientific name for a taxon (of infrasubspecific rank), this view is not accepted here.

Pisidium milium f. *pseudosphærium* Ehrmann, 1933 in Ehrmann 1933: 241

Mentioned with short description and without bibliographical reference, used as a subspecific name under Art. 45.6.4, eventually making Favre's name available under Art. 45.6.4.1 for those who accept that Favre used the term as a taxonomic name. Ehrmann 1933 did not give a bibliographically unambiguous reference and Favre 1927 had employed the term *pseudosphaerium* at various occasions, Ehrmann did not specify which name was adopted.

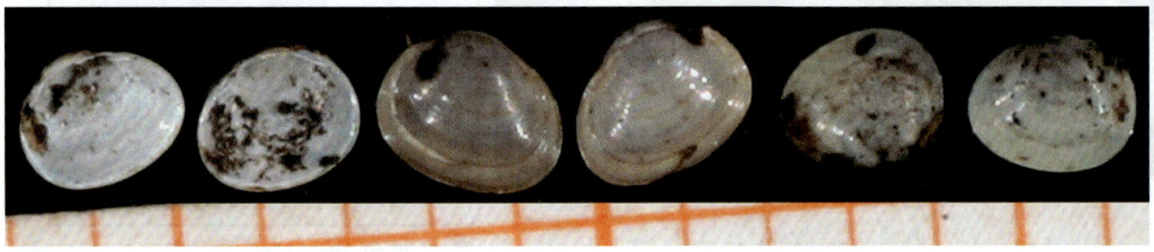

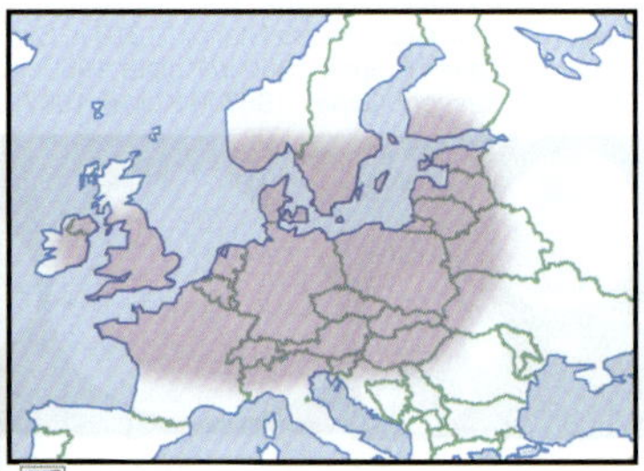

E Ireland, England and France to Hungary, S Sweden and Estonia
Exact distribution little known.

2.5-3.2 x 2.0-2.6 mm

Germany: Schleswig-Holstein, Middelburger See

Shell like *Pisidium milium*, but smaller, more inflated, valves very regularly oval, cardinal plateau narrower, lateral teeth smaller and weaker, ligament longer and occupies the whole plateau.

Lives in swamps, dried lakes and ditches, mud grooves filled with water, with abundant vegetation and connection to larger water bodies, requires clear and clean water and calcareous substrate. Usually not in isolated small water bodies. In Ireland in abandoned canals, in England in lowland marsh drains and ponds. The shell is usually covered with algae. Does not tolerate any pollution, artificial disturbance, drought and sinking of ground water levels. Disappears from habitats with unstable swamp conditions. Known from up to 500 m altitude. Occurs either alone or associated with *P. obtusale* and *P. milium*, rarely with other species.

Scattered and local everywhere. Critically Endangered in Austria and Germany, Endangered in Switzerland, rare in England and Ireland.

Adler 1993, Groh et al. 2009.

Pisidium pulchellum Jenyns, 1832

Pisidium pulchellum Jenyns, 1832 in Jenyns 1832: 306

Jenyns mentioned a synonym *Pera pulchella* attributed to Leach for a variety ß (beta), with reference to types deposited in the British Museum. Jenyns used the name *Pisidium pulchellum* for the main form which had a different description (and different types) as var. ß. *Pera pulchella* "Leach MSS." was previously mentioned by Alder 1830: 41 as a synonym of *Cyclas fontinalis* Draparnaud, 1801, with a description for *C. fontinalis* (indirectly, by comparison of the following species *Cyclas gibba* with *C. fontinalis*), and with reference to types deposited in the British Museum. Jenyns 1832 did not use the name *Pisidium pulchellum* in the same sense as Alder 1830 had mentioned it as a synonym (this is clear because that synonym was classified with var. ß by Jenyns 1832).

For those who require a bibliographic reference for making available a synonym under Art. 11.6.1, the name must be attributed to Jenyns 1832. For those who require that the author who adopts a name must have used it for subjectively the same taxon as the author who had mentioned it as a synonym, the name must be attributed to Jenyns 1832. For those who require that the synonym must be equipped with a proper description when first mentioned, the name must be attributed to Jenyns 1832. For the others, to Alder 1830, and the identity of the name would eventually change.

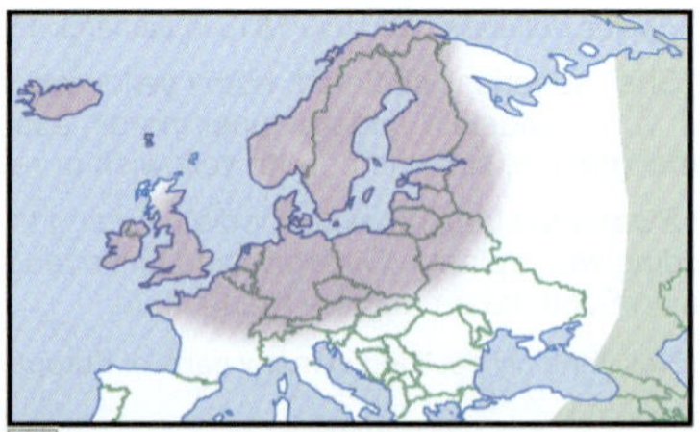

N and central Europe, in the SE to Ukraine (not Hebrides, not Orkneys)

Shell smaller than *Pisidium casertanum*, finely and regularly striated, shiny.

In springs and likewise small and cool waters with some movement, in France mostly in highlands, in Britain mainly in lowlands, usually up to 10 m deep. Does not tolerate periodical droughts, absence of water movement, hypertrophical conditions. In France and Switzerland in up to 1000 m altitude.

Extinct in Bayern, Critically Endangered in Germany, scattered and local in the British Isles.

1.5-2.5 x 2-3 mm

Germany: Hamburg, Curslack near Bergedorf

Pisidium subtruncatum Malm, 1855

Pisidium subtruncatum Malm, 1855 in Malm 1855: 92

Shell with very short distal end, vortex narrow.
Foot longer than shell.

Usually in silent running water bodies of any size with muddy ground, litoral of lakes, not in periodically drying small water bodies. Not restricted to calcareous substrate, tolerates mild pollution. Usually up to 5-30 m deep, maximum 140 m. In Switzerland in up to 1800 m altitude, usually below 1000 m, in Britain rare above 300 m.

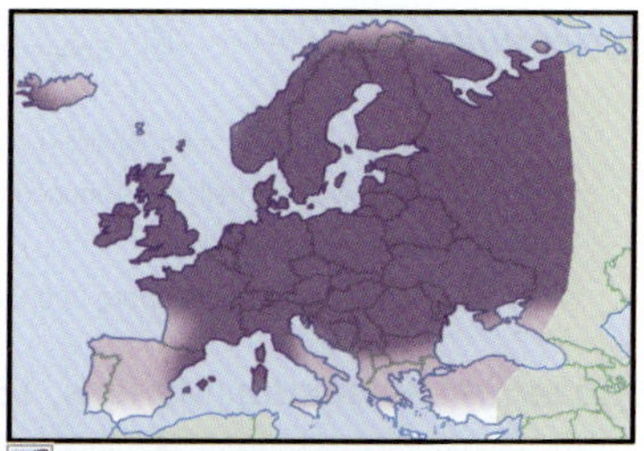

Northern Hemisphere

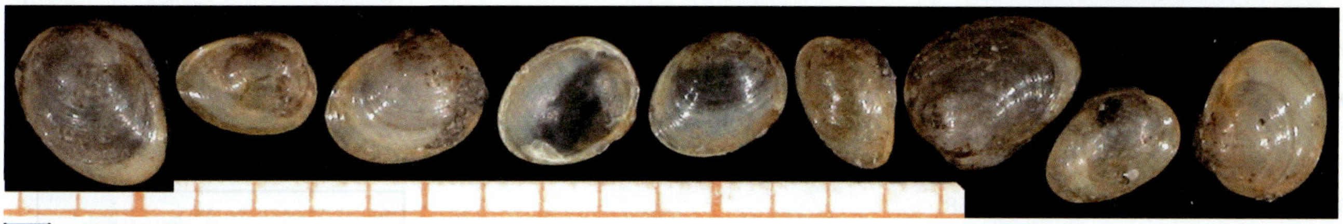

2-3 x 3-4 x 2-2.5 (height) mm

Germany: Mecklenburg-Vorpommern, Schaalsee

Pisidium supinum Schmidt, 1851

Pisidium supinum Schmidt, 1851 in Schmidt 1851: 119

Shell solid, usually yellowish, finely ribbed, obliquely triangular, vortex fold oblique.

Only in large lowland rivers and canals, with clean, slowly moving well-oxygenated hard water, normally in clean mud or silt, sometimes also on sandy substrates. In France in up to 400 m altitude. Frequent in England.

Threatened by bottom pollution. Extinct in Ireland, Endangered in Austria, Vulnerable in Germany.

Could be a hypertrophied form of *P. henslowanum*, but mixed occurrences are known (Kerney 1999: 227).

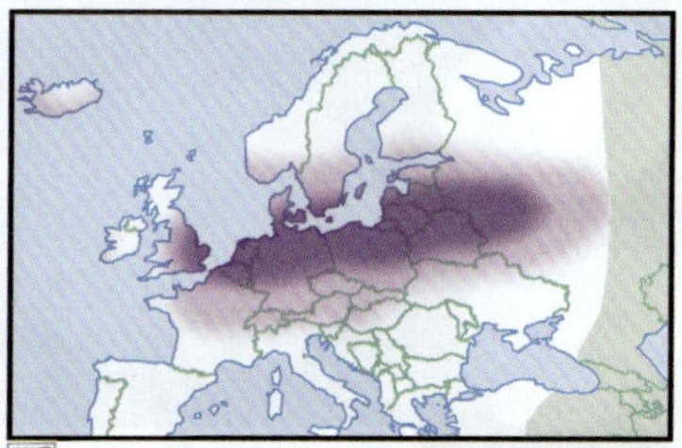

Palearctic and NE North America

3-5 x 3-5 mm

Germany: Main near Gemünden

Pisidium tenuilineatum Stelfox, 1918

Pisidium tenuilineatum Stelfox, 1918 in Stelfox 1918: 296, Pl. 8

Shell very regularly striated, shiny, vortex towards posterior end, anterior margin bluntly pointed, posterior margin somewhat truncated.
Animal whitish with large and short siphon.

Lowland rivers and canals, also at margins of lakes with rich calcium contents, very exceptionally in large ponds and limestone spring pools. Prefers silty and muddy substrates in clean, hard and unpolluted waters. In Switzerland in up to 700 m altitude.

In France (water temperature 2-23 °C) 2-3 generations per year emerge from May to September, minimum mature size is 1.1 mm, minimum parent size with shelled larvae 1.3 mm, gravid individuals can be found throughout the year. Parents carry 5-10 shelled larvae, occasionally 20. Populations are usually small. Life span about 1-2 years.

Extremely local everwhere, and certainly declining. Endangered in Austria and Germany, Vulnerable in Albania, rare in England (vanished from large parts of its original range, it seems that all populations north of London are extinct, presumably due to water pollution).

Mouthon 2008 (life cycle).

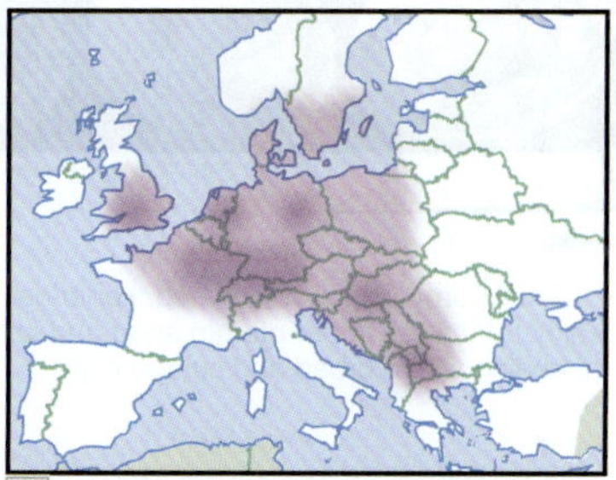

Central and E Europe, Balkan peninsula, in Sweden up to 60° N, England

1.5-2.4 x 1-2 x 0.8-1.4 (height) mm

N Germany: Briese near Brinkenwerder

Pisidium waldeni Kuiper, 1975

Pisidium waldeni Kuiper, 1975 in Kuiper 1975: 31

Shell greyish yellow, not translucent, with concentric and flat riblets, moderatley inflated, anterior and posterior parts almost evenly rounded, vortex directed to posterior side, cardinal teeth in left valve very short.
Shell quite variable, also within a population, lateral teeth can be reduced or long.

Up to 4.7 x 4.3 x 3.2 (height) mm

In lakes, 2-25 m deep.

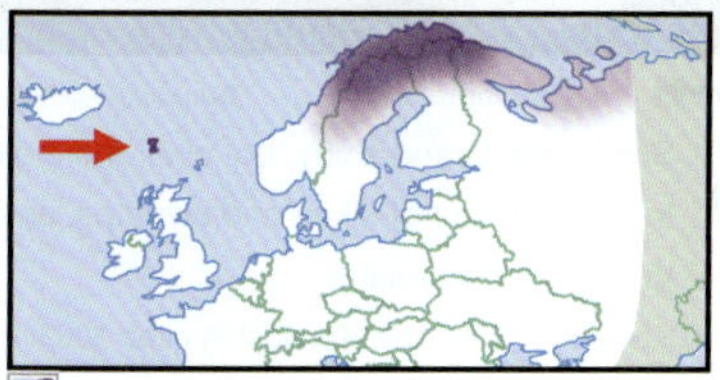

N Europe, Canada

Sphaerium

Sphaerium Scopoli, 1777
in Scopoli 1777: 397 *Sphaerivm*

Type species *Tellina cornea* Linnæus, 1758
(monotypy)

Type not by subsequent designation by Gray 1847: 184 (as stated in ICZN Opinion 94, Direction 72), corrected to "monotypy" in Direction 117 (1985).

Thick-walled shell, non-reduced hinge, low or moderate number of embryos per brood pouch, nephridium with broad funnel and large excretory sac.

Northern Hemisphere

Korniushin & Hackenberg 2000, Korniushin 2001: 80 (revision), Korniushin & Glaubrecht 2002 (phylogeny).

Sphaerium asiaticum (Martens, 1864)

Without gender agreement: *Sphaerium asiatica* (Martens, 1864)

Cyclas asiatica Martens, 1864 in Martens 1864: 349 (*Cyclas Asiatica*)

Shell slightly broader and more elongated in posterior than in anterior section, finely striated, lower margin curved, vortex low, not protruding, lateral teeth protruded and doubled.

7.5-9 x 6 x 6-7 (height) mm

NW Russia to S Kamchatka

Sphaerium corneum (Linnæus, 1758)

Without gender agreement: *Sphaerium cornea* (Linnæus, 1758)

Tellina cornea Linnæus, 1758 in Linnæus 1758: 678

6-11 x 8-14 x 7-13 (height) mm

N Germany: Hagener Au, Passader See

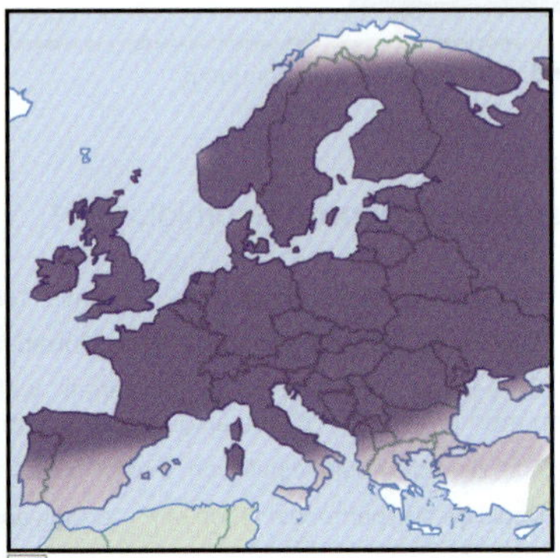

Eurasia, introduced to E North America (St. Lawrence river system)

Shell yellowish brown, interior side light bluish, muscular impression not well visible, thin, almost smooth, often ball-like inflated, embryonal shell almost central, in larger rivers flatter, trapezoid and pale yellow. Teeth c2 and c4 slightly bent and about the same length, pores rare and randomly placed. Animal greyish or yellowish, with long grey siphons with reddish ends, foot slightly longer than shell.

Any water bodies, tolerant against moderate pollution, but not against desiccation, up to 40 m deep. *Sphaerium* is able to leave the substrate and crawl on aquatic vegetation, so that it can escape toxic conditions at the bottom. In Portugal in silent waters on sandy or muddy substrate, locally abundant. In Switzerland in up to 700 m altitude, not in higher alpine regions.

Reproduction almost year-round in England, usually two annual generations at a time, some individuals may survive the annual cycle (Dussart 1979).

Sphaerium nitidum Clessin, 1877

Sphaerium nitidum Clessin, 1877 in Clessin 1877: 98, Taf. 12

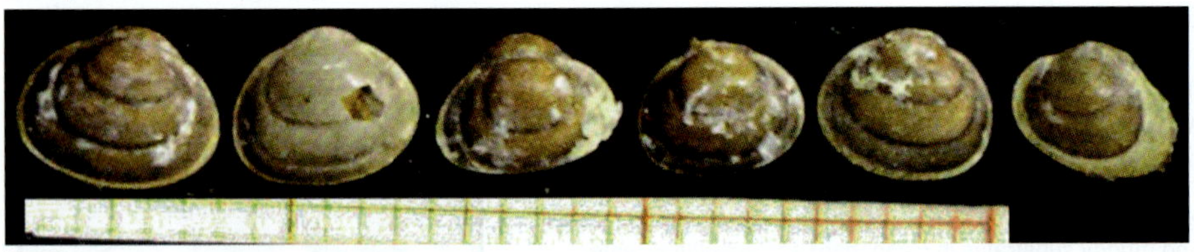

6-7 x 4 x 5-6 (height) mm

Canada: Nueltin Lake, Keewatin

Northern regions of Northern Hemisphere

Academy of Natural Sciences Philadelphia ANSP

Shell horny yellowish, inflated, spherical, finely striated, also on vortex, annual growth lines prominent, very shiny, anterior section rounded, posterior part weakly truncated, upper margin short and almost straight, lower margin almost straight with blunt edges at lateral margins, vortex broad and tube-like protruding, ending with a broad and curved cap.

Sphaerium nucleus (Studer, 1820)

Cyclas nucleus Studer, 1820 in Studer 1820: 93

8 x 8 x 8 mm

Belgium: Liège, cold water basins of Botanical Garden (scale bar 8 mm)

Eric Simon

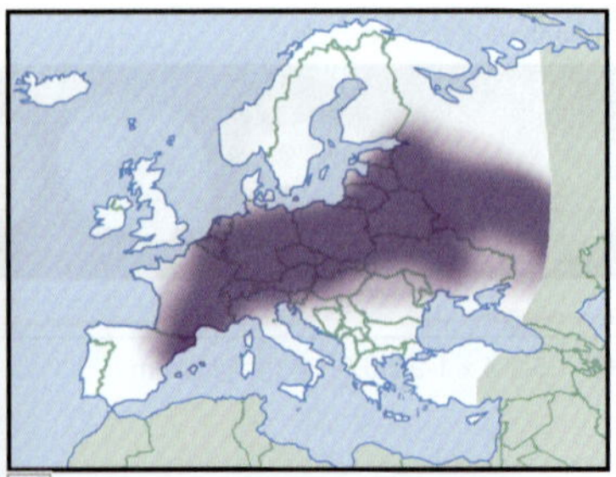

Europe to central Asia (Kyrgyzstan)
Range insufficiently known, was not distinguished from *Sph. corneum* before 2001.

Shell smooth and shiny, not oval, more globulous than *Sphaerium corneum*. Differs from other species in the characteristic shape of its nephridium.

In silent waters of lakes and swamps, and other standing water bodies.

Vulnerable in Germany.

Korniushin 2001: 90, Korniushin et al. 2002: 469, Glöer 2006, Jungbluth & Knorre 2009: 12.

Sphaerium ovale (Férussac, 1807)

Without gender agreement: *Sphaerium ovalis* (Férussac, 1807)

Cyclas ovalis Férussac, 1807 in J. Férussac 1807: 128, 136

I am giving Férussac's single authorship here, as this has been traditionally done. I have not researched which one of the two Férussac's was responsible for this name (or both). It seems to have been overlooked that both Férussacs were co-authors of the 1807 work which should better be cited as "Férussac & Férussac 1807", and that the responsibilities were explained in the introduction.

Used by Férussac 1807: 128 without description, indicated as a new name, next to a name *Cyclas lacustris* as used by Draparnaud 1805. Later *C. ovalis* was used as a name for a species, for example by Westerlund 1890: 111 (with bibliographical reference to Férussac 1807).

Férussac 1807: 128-129 presented name usages for various species by various authors Daudebard (= Férussac 1807), Draparnaud 1805, Müller 1774 and Gmelin 1791. For species No. 176 Férussac used the name *Cyclas lacustris* and stated that Müller and Gmelin used *Tellina lacustris* for it, while *Cyclas caliculata* was used by Draparnaud. Species No. 175 was called *ovalis* "nob." by Férussac, the only given synonym was *lacustris* as used by Draparnaud. Draparnaud 1805: 130 used Müller's name (misspelled *lacutris*) and gave a description.

Falkner 2000: 32 interpreted Férussac's name as a new replacement name for *Cyclas lacustris* sensu Draparnaud, not *Tellina lacustris* Müller, 1774. This was not tenable because (1) the name was not expressly established as a new replacement name (Art. 72.7), (2) a new replacement name can only be proposed for an available name (Art. 12.2.3), Draparnaud just used Müller's name, and this name was not thought to be replaced.

For a regular new name a bibliographic reference to a description is needed (Art. 12.2.1, 72.4.4). "Draparnaud" (on p. 129) is here regarded as a bibliographic reference ("Draparnaud" referred exclusively to the 1805 work, see p. 85 etc., p. 98, p. 114), *lacustris* was mentioned only once in that work, on p. 130, with a description. Férussac 1807: 99 said that he also incorporated corrections of mistakes in Draparnaud's work, taken from unpublished correspondence, so the term "*lacustris*" on p. 129 referred quite clearly to Draparnaud's 1805: 130 description.

Sphaerium levinodis var. *radiatum* Westerlund, 1897 in Westerlund 1897: 131

There are many other candidate names, *Sphaerium radiatum*seems to be one of them (Korniushin 2001: 92), but there seem to exist earlier names.

Shell yellowish grey, weakly rhomboid, not much inflated, very finely striated, with silky gloss, anterior part slightly pointed, posterior part blunt and broadened, vortex small and pointed, cardinal teeth long and bent, pores dense and regularly placed, muscle impressions separated.

7-10 x 5.5-8 x 4-5.5 (height) mm

In silent waters, small lakes and old river arms.

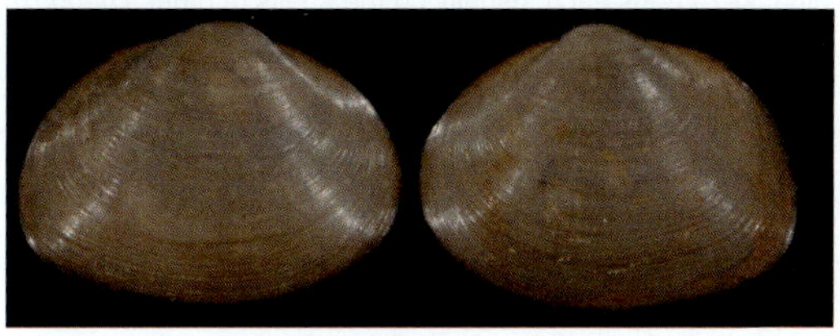

Germany: Schleswig-Holstein, near Lübeck

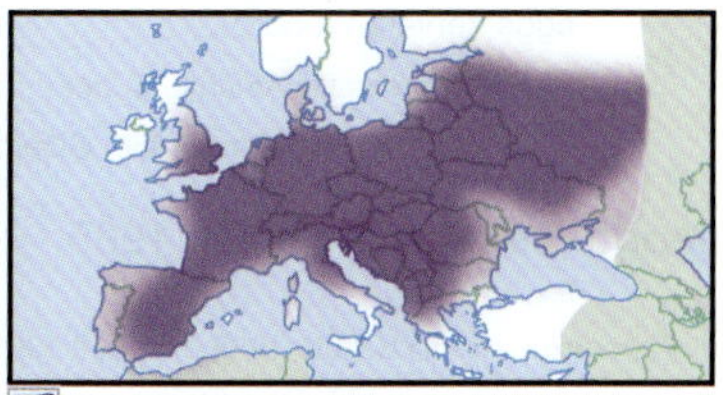

Europe, W Siberia

Sphaerium rivicola (Lamarck, 1818)

Cyclas rivicola Lamarck, 1818 in Lamarck 1818: 558

Germany: Bayern, Main near Gemünden

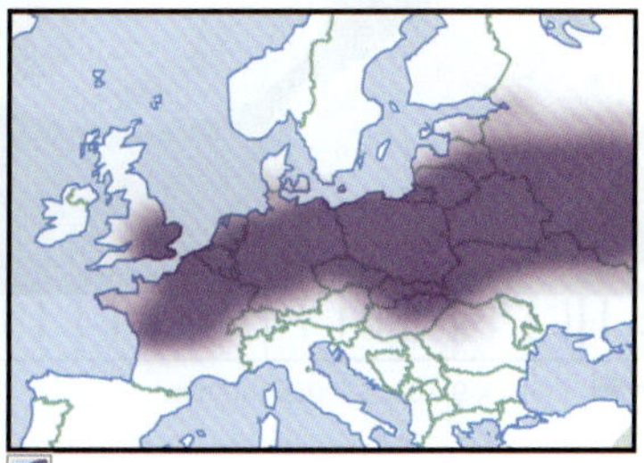

Central and E Europe (not Alps), England

10-16 x 15-25 x 17-20 (height) mm

Shell solid and finely ribbed, olive greenish or brownish, with some pale yellowish zones, sometimes with slightly darker transversal streaks, weakly shiny, almost opaque. Juveniles slightly more elongated and evenly yellow.
Animal yellowish grey or brownish, gills brownish or reddish, siphons whitish and almost equally short. External gills 6 x 1.5 mm.

Restricted to large rivers and canals, prefers silent zones and well-oxygenated hard-water. In Britain nearly always associated with *Viviparus viviparus*, in relatively deep water.

Juveniles or eggs can sometimes be seen at the exterior gills. After hatch the juveniles have a diameter of 4 mm, they are relatively flat with a large foot, the mother makes them leave quickly.

Threatened by pollution and habitat change, some populations in England were lost. Critically Endangered in Germany.

Sphaerium solidum (Normand, 1844)

Without gender agreement: *Sphaerium solida* (Normand, 1844)

Cyclas solida Normand, 1844 in Normand 1844: 6

Germany: Hamburg, Neßsand

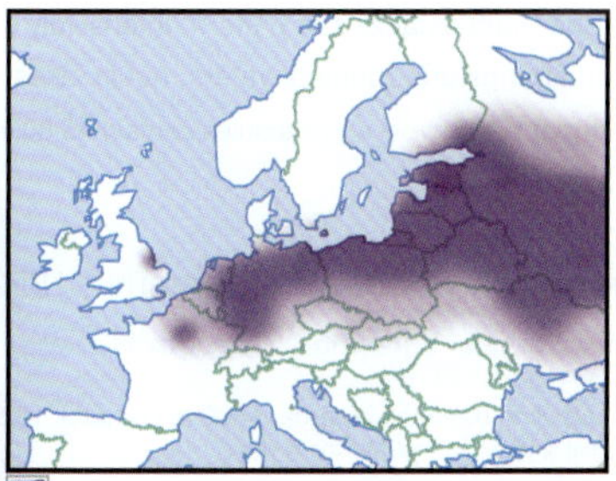

Central and E Europe to central Siberia
Lives possibly in Hungary, but I have not seen any recently published reliable records.

4-6 x 6-10 x 5-7.5 (height) mm

Shell solid with coarse concentric ribs, evenly pale yellow or pale greyish brown. Juveniles greyish.
Animal white with short siphons.

In fine sands of large rivers and reservoirs, in relatively deep waters.

EU Red List: Near Threatened NT, assessment probably too weak. This species seems to be almost extinct in the European Union. Extinct in the Czech Republic, Critically Endangered in Germany, still present in a few populations and protected by law in Poland. In England Endangered, only along a 15 km transect of the artificially deepened and canalized river Witham in Lincolnshire, treatened by pollution. Situation in central Russia probably better, but no data exist on the matter.

Freshwater gastropods

Genera sorted taxonomically and by shell shape.

Theodoxus

Theodoxus Montfort, 1810
in Montfort, 1810: 350

Type species *Theodoxus lutetianus* Montfort, 1810
(original designation)
(= currently *Theodoxus fluviatilis* Linnæus, 1758)

On p. 350 spelled *Theodoxis*, on p. 351 *Theodoxus*. First Reviser not researched.

Shell semiovular with flat apertural plain, no umbilicus. Spire and inner whorls are dissolved.

Animals are attached to stones, often also under stones, in up to 5-6 m depth, feeding on algae covers. *Theodoxus* needs rough surfaces to be able to digest its food, stony substrate is necessary. Green algae are not consumed, *Theodoxus* has no cellulase.

Animals lay egg capsules containing 30-70 eggs each, usually on the shells of other *Theodoxus* animals. Only 1 juvenile grows, the other eggs serve as food.

Europe and N Africa to S Iran

Schütt 1965, Grossu 1986, Glöer 2002: 48, Glöer & Meier-Brook 2003: 28, Bunje & Lindberg 2007, Shubrat 2007.

Theodoxus altenai Schütt, 1965

Theodoxus altenai Schütt, 1965 in Schütt 1965: 46

Shell exceptionally thin, light horny yellowish, translucent and shiny, whorls flat and very rapidly increasing, with brown quadrangular colour pattern.
Operculum reddish, without apophysis.

In lakes.

S Turkey (Vilayet Antalya)

11-12 x 9-10 x 6-7 (height) mm

Turkey: near Kırkgöz (?) N of Antalya

Theodoxus anatolicus (Récluz, 1841)

Without gender agreement: *Theodoxus anatolica* (Récluz, 1841)

Nerita anatolica Récluz, 1841 in Récluz 1841: 342 (*Nerita Anatolica*)

8-13 x 6-9 mm

S Turkey: İçel, Yenikaş near Aydıncık

SW Turkey and adjacent islands

Shell with axial ziczac pattern, intensively shiny. Operculum greyish with weak rib and short and pointed apophysis.

In springs and running waters, not in lakes.

1 mm sized ovoid egg capsules contain 45-60 eggs.

Theodoxus baeticus (Lamarck, 1822)

Without gender agreement: *Theodoxus baetica* (Lamarck, 1822)

Neritina baetica Lamarck, 1822 in Lamarck 1822: 188 (*Neritina Bætica*)
Original spelling was not *Neritina boetica*.

Shell half globulous, greyish blue or brownish, with dark ziczac pattern or bands, smooth, shiny, apex eroded, 3 whorls. Operculum grey with reddish margin and strong rib, apophysis short and indented.

In running waters (springs, creeks, rivers) on hard substrate (rocks, stones, pebbles). Feeds on diatoms and other algae, and detritus.

EU Red List: Critically Endangered CR. Threatened by water pollution, agricultural contamination, habitat destruction and water withdrawal, also by shell collectors. Populations are increasingly isolated from each other.

6-7 x 4-5.5 mm

Spain: Huelva, Cortelazor
Rafael Vázquez Romero

Spain

Falkner 1990: 114, Gómez Moliner et al. 2001: 90, Martínez-Ortí & Robles 2003: 179, Martínez-Ortí et al. in Verdú & Galante 2006: 334, Talaván Gómez & T. 2006: 35, Altaba 2007: 25.

Taxonomic status of the populations unclear, could be a complex composed of several different species.

Theodoxus brauneri (Lindholm, 1908)

Ukraine

Neritina brauneri Lindholm, 1908 in Lindholm 1908: 217

Shell evenly black or with white spots or stripes, thin but solid, translucent, finely striated, apex blunt or eroded, 3 whorls with deep suture, last whorl predominating, apertural margin thin, columellar area whitish or light grey and with sharp margin, slightly wider on upper than on lower side. Operculum horny yellowish with red margin.

6-7x 4.5-5 mm

Theodoxus callosus (Deshayes, 1835)

Theodoxus varius (Menke, 1828)

Without gender agreement: *Theodoxus callosa* (Deshayes, 1835)

W Greece, S Albania

Neritina varia as mentioned in Menke 1828: 27 (nomen nudum)
Menke 1828 mentioned *Neritina varia* without description or bibliographical reference (also in Menke 1830: 49), without synonyms. *Neritina varia* cannot be used. *Neritina callosa* Deshayes, 1835 is a candidate.

Neritina rivalis De Cristofori & Jan, 1832 in De Cristofori & Jan 1832: [Conchylia] 8 (nomen nudum)
Mentioned without description or reference, attributed to Ziegler. www.faunaeur.org (2009, 2012) gave "*rivalis* Jan 1830", www.iucnredlist.org (2012) gave "*rivalis* Jan, 1833". I did not find where this name was made available, it was not listed by Sherborn 1922.

Neritina callosa Deshayes, 1835 in Deshayes [1835]: 156, Pl. 19

Shell small, shiny, brownish yellow, reddish or greenish brown with white spots, columellar area whitish. Operculum with prominent apophysis at the inner side of the operculum begins at the lower part of the end of the relatively short rib, margin orange.

In springs. Possibly also in running waters such as small mountain creeks.

Vulnerable in Albania (Dhora 2004: 141)

4-7 x 3-5 mm

Greece: Fokís, Delfi, Kirra near Itea

Theodoxus danubialis (Pfeiffer, 1828)

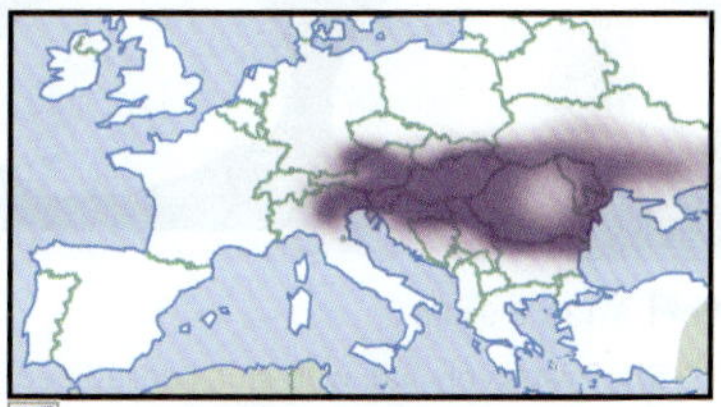

N Italy to Slovakia, Danube river region from SE Germany to Ukraine

Nerita danubialis Pfeiffer, 1828 in C. Pfeiffer 1828: 48

Shell whitish or yellowish with dark zigzag pattern, in the east also uniformly black, aperture almost circular. Operculum pale yellow with light brownish margin, apophysis flat and in a groove.

In rivers and lakes, prefers clean and well-oxygenated water currents with stony substrate. Sensitive to water pollution.

Critically Endangered in Germany and Austria, Endangered in Czech Republic. Has dramatically declined in Austria, most populations in the Danube river are already lost.

Bunje 2007, Fehér et al. 2009, Stadnychenko & Tarasova-Shubrat 2012 (reports for Ukraine outside the lower Danube region are probably not reliable).

8-9 x 6-7 mm

Germany: W of Kelheim

Theodoxus elongatulus (Morelet, 1845)

Without gender agreement: *Theodoxus elongatula* (Morelet, 1845)

Neritina elongatula Morelet, 1845 in Morelet 1845: 96, Pl. 9

Shell very convex, weakly striated, whitish (rarely yellowish) with dark reddish ziczac lines which can merge to a reticulate pattern.

7.5-8.5 x 4-5 mm

Westerlund 1886: 154.

Portugal: Rio Nabão near Tomar

Spain, Portugal

Ana Gonçalves

Theodoxus euxinus (Clessin, 1886)

Without gender agreement: *Theodoxus euxina* (Clessin, 1886)

Neritina euxina Clessin, 1886 in Clessin 1886: 55

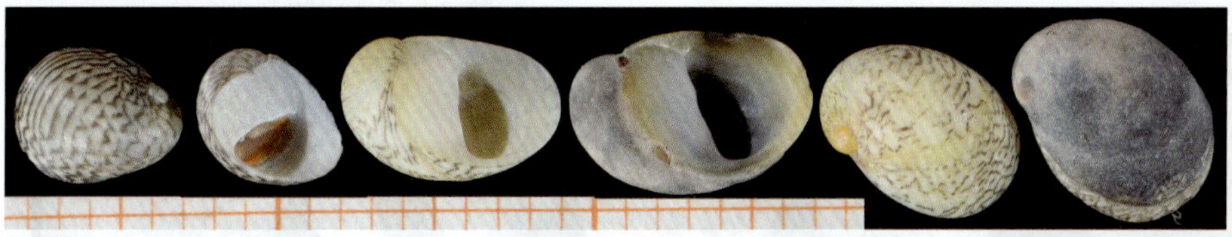

6-8 x 3-5 mm

Romania: Tulcea, Danube delta near Sulina

W Black Sea region

Shell shiny, smooth, with coarse white ziczac pattern.

In woodlands of river deltas, on rocks, sandy or muddy bottom.

Westerlund 1886: 156, Grossu 1986: 161.

Theodoxus fluviatilis (Linnæus, 1758)

Nerita fluviatilis Linnæus, 1758 in Linnæus 1758: 777
Theodoxus velox Anistratenko, 1999 in Anistratenko et al. 1999 [not verified]

Shell basically whitish or yellowish with dark reddish or violetish reticulate pattern, very variably arranged, sometimes partly in bands, occcasionally evenly dark, in the Mediterranean region slightly different forms. Operculum light reddish with red margin, with broad rib and without apophysis. Animal light yellow, head black, tentacles greyish and long, eyes big and black, foot whitish.

Central and lower parts of rivers, also in brackish water, requires stony substrate. Tolerates mild organic pollution, low oxygen contents (down to below 2 mg/liter) but not too long periods of droughts or ice. Up to 13 m deep. Feeds on algae cover on stones.

45-65 eggs (diameter 0.8-0.9 mm) are laid from mid-April to October, they are fixed on stones or shells of bivalves or other *Theodoxus*, 0.5-1 mm sized juveniles hatch after 30 days (25°C) (65 days with 20°C), life span 2-3 years.

Mainly threatened by water current regulations and pollution in densely populated regions. Extinct in the Czech Republic, Critically Endangered in Switzerland, Endangered in Germany.

Gassies 1849: 185, Moquin-Tandon 1849: 549, Germain 1931: 681, Bunje & Lindberg 2007.

Th. velox from the Black Sea region is probably only a form of *Th. fluviatilis*.

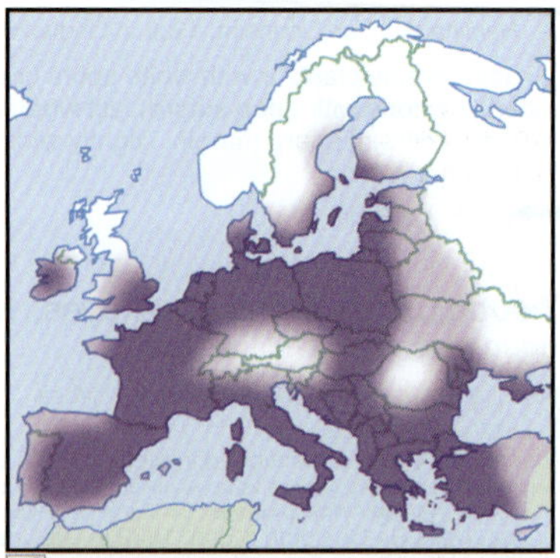

Europe except Alps and regions immediately north of the Alps

5-9 x 4-6.5 mm, up to 13 mm

Germany: Schleswig-Holstein, Wittensee

Theodoxus heldreichi (Martens, 1879)

Neritina heldreichi Martens, 1879 in Martens 1879: 225

8-13 x 6-9 mm

Turkey: Vilayet Konya, Beyşehir Gölü

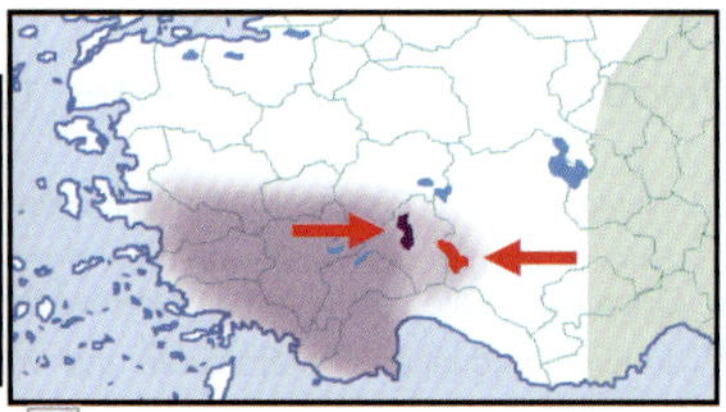

Turkey: Eğirdir Gölü, Beyşehir Gölü

Shell relatively flat, with small coiled section, often with violet reticulated colour pattern on white background, like in some forms of *T. fluviatilis*. Operculum usually without apophysis.

Litoral of central Anatolian lakes, lives between stones, often found washed ashore at the lake margins. Another form of this species lives in running waters

Seriously threatened in Beyşehir Gölü due to increasing salt concentrations (perhaps already extinct, 2010), population in Eğirdir Gölü seems to be stable.

Schütt 1965: 46, Yıldırım 1999, G. Hartz (unpublished, Jun 2010).

Theodoxus hispalensis (Martens, 1879)

Neritina hispalensis Martens, 1879 in Martens 1879: 230, Taf. 22 (*Neritina Hispalensis*)

6-8 x 5.5-6.5 mm

Spain: Guadalquivir, near Lora del Río

Spain (Guadalquivir)
Reported from near Sevilla.

Shell finely corrugated, not shiny, pale yellow, with numerous perpendicular dark streaks.

Westerlund 1886: 154.

Theodoxus meridionalis (Philippi, 1836)

Nerita meridionalis Philippi, 1836 in Philippi 1836: 159, Tab. 9

Shell cone-shaped to half-egg-shaped, pale yellowish with dark red to blackish ziczac streaks, almost smooth, weakly shiny.

In cold springs and small running waters in immediate vicinity of springs. Dispersed by birds.

Known from 14 sites in Sicily, but threatened by continuous habitat destructions in the course of channeling water sources and cementation of the beds of running waters. Rare in Tunisia.

Italy (Sicily), NW Tunisia

7-8 x 4-5 mm

Italy: Sicily, Anapo river S of Siracusa

Theodoxus pallasi Lindholm, 1924

Theodoxus pallasi Lindholm, 1924 in Lindholm 1924: 34
Expressly established as a new replacement name for
Neritina liturata Eichwald, 1838 in Eichwald, 1838: 156 (non Schultze, 1826)

3-5 x 2-4 mm — Caspian Sea (territory of former Soviet Union)

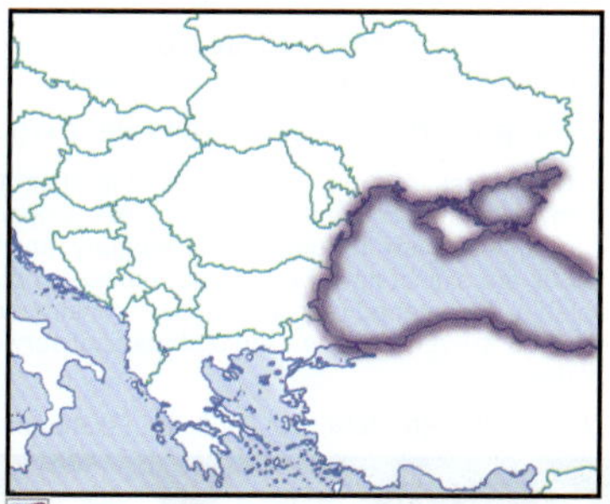

Caspian Sea, Black Sea, Azov Sea, Aral Lake

Shell usually light grey with dark and coarse zigzag patterns, or yellowish with brown zigzag patterns, zigzag streaks are sometimes interrupted, colour pattern varies locally, 1.5 whorls.
Differs from *Theodoxus fluviatilis* in its smaller size and proportionally lower shell height, *T. fluviatilis* has finer zigzag lines.

In brackish water with 1 % salt content.

Eggs are laid in clutches of 30-40 eggs between algae, which form some kind of a nest.

Eichwald 1838: 156, Dybowski 1888: 56, Grossu 1986: 162.

Theodoxus prevostianus (Pfeiffer, 1828)

Without gender agreement: *Theodoxus prevostiana* (Pfeiffer, 1828)

Nerita prevostiana Pfeiffer, 1828 in C. Pfeiffer 1828: 49

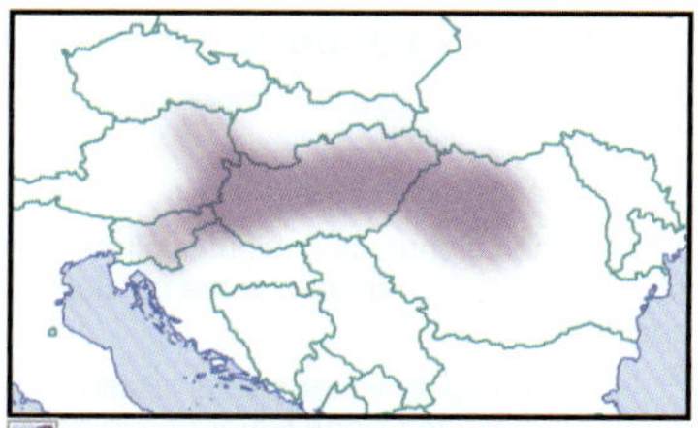

E Austria and Slovenia to Romania

Shell usually evenly black or dark violet, finely striated, columella white and opaque, last whorl distinctly descending. Operculum dark grey with orange margin, apophysis flat and in a groove.

In thermal springs (15-20 °C) and the initial sections of their outflows.

EU Red List: Endangered EN. Extremely rare and has declined considerably in the past 50 years, only 4 populations remained. Threatened by construction works, drops of ground water levels and capturing of springs. Extinct in Romania, Critically Endangered in Austria and Hungary. The only surviving Hungarian population at Kács is at serious risk by planned construction works. It was planned in 2010 to reintroduce the species at 3 suitable Hungarian sites where the native populations were recently lost.

Bunje 2007, Fehér et al. 2009 (suggested that *T. danubialis* and *prevostianus* were different species, apparently a case of paraphyly).

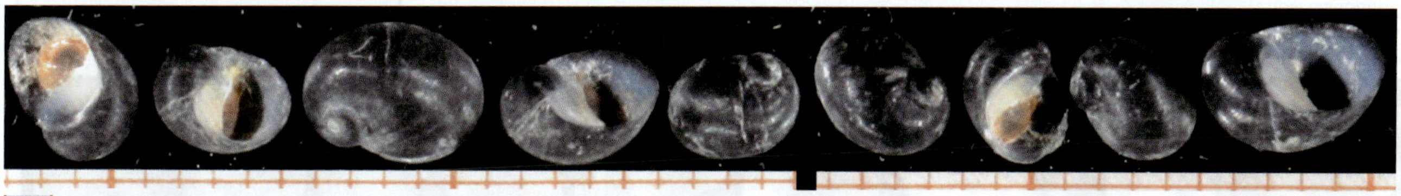

4-6 x 3-4 mm — Austria: Niederösterreich, Vösslau near Wien

Theodoxus saulcyi (Bourguignat, 1852)

Neritina saulcyi Bourguignat, 1852 in Bourguignat 1852: 25

3-5 x 2-4 mm — Greece: E Crete, near Agios Nikólaos

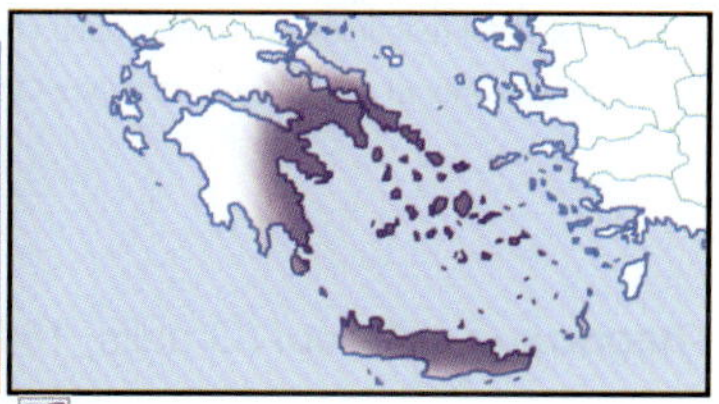

Greece

Shell rather thin, blackish violet, shiny, with fine black lines.

Habitats in Crete include standing and running waters (Kournas lake, springs at Spíli, river estuary at Georgioúpoli).

Westerlund 1886: 155, Schütt 1986 (the only one species in Crete), Frank 1988: 82, Bunje & Lindberg 2007 (probably a form of *Th. fluviatilis*).

Theodoxus subterrelictus Schütt, 1963

Theodoxus subterrelictus Schütt, 1963 in Schütt 1963: 212

4-6 x 3-5 x 2-3 (height) mm — Bosnia-Herzegovina: Trebinje — Zoltán Fehér

Dalmatia, Bosnia-Herzegovina

Shell not pigmented, grey, opaque, with very weak transversally arranged dashes, shell relatively thin, apertural margin sharp and slightly broadened towards outside, last whorl with irregular riblets. Operculum thick, calcareous, grey with light reddish margin and a hollow rib at the inner side, apophysis slender and running half-way parallel to the rib.

In subterranean small running waters of karstic areas.

EU Red List: Endangered EN. Only found in a small area and in sensitive habitats.

Theodoxus transversalis (Pfeiffer, 1828)

Nerita transversalis Pfeiffer, 1828 in C. Pfeiffer 1828: 48

Shell yellow to pale lead grey, often with indistinct darker spiral colour zones. Operculum reddish with red margin, apophysis prominent and screwed.

In large rivers, prefers clean and well oxygenated water. Often below stones.

EU Red List: Endangered EN. Extinct in Austria and Romania, probably also in Slovakia, Critically Endangered in Germany. Used to occur in Danube river from Donauwörth downwards, today only two scattered populations in Germany, none in Austria, a few in Bulgaria and Hungary, and in Tisa and Dnestr rivers. The last populations in Romania reported in the 1960s are extinct now. Most of the remnant populations can be found in smaller tributaries where water quality is still acceptable. Threatened by pollution, dam constructions and the invasive species *Th. fluviatilis*, which is the major problem in the Danube river sections. Can probably not survive in the Danube, only isolated populations could survive in some tributaries.

Glöer 2002: 50, Reischütz 2009: 29, Fischer et al. 2009: 45 (Lake Dojran, Rep. Makedonija).

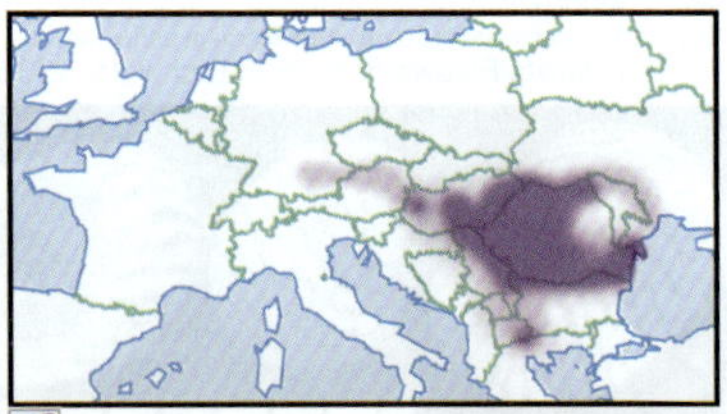

SE Europe

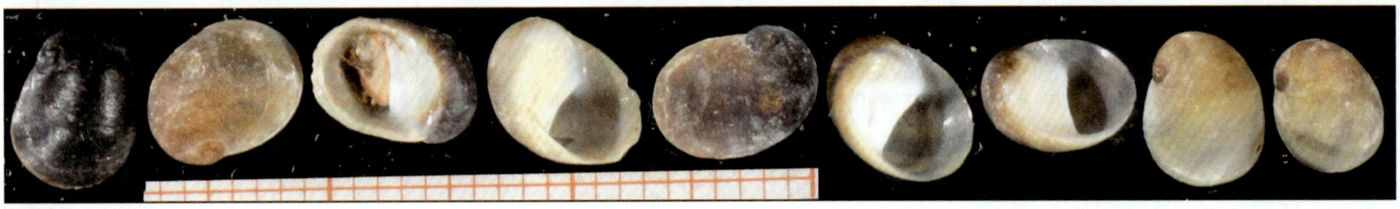

6-8 x 4-6 mm

Austria: Donau near Kuchelau, near Wien (population today extinct)

Theodoxus valentinus (Graells, 1846)

Without gender agreement: *Theodoxus valentina* (Graells, 1846)

Neritina valentina Graells, 1846 in Graells 1846: 21

Shell half globulous to highly spired, dark or red with ziczac pattern, very solid, 4 convex whorls, last whorl with a characteristic groove at the palatal side. Operculum whitish with narrow rib and short entirely connected apophysis.

In upwellings of water called "ullals", creeks and rivers, with very constant water temperature between 16 and 18 °C. Requires clean water rich in calcium content and hard substrate to which the animals attach strongly. Animals feed on diatoms and algae, the most important predator (of the egg capsules) is the ostracod *Cypridopsis vidua*.

Eggs are laid throughout the year in oval capsules (0.8-1 mm), fixed on shells or stones, only one egg will develop, life span 18-22 months.

Spain (València)

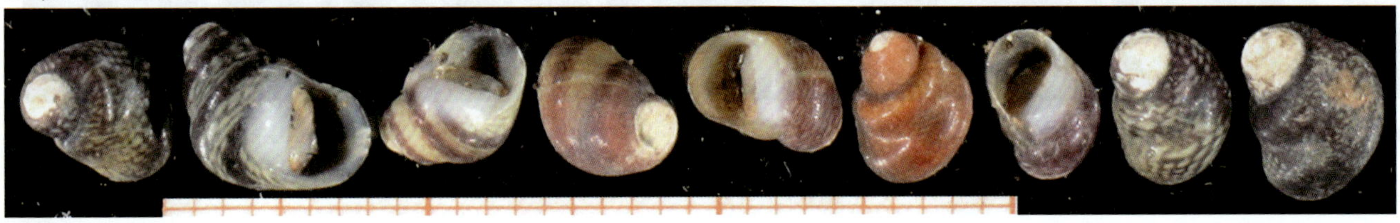

5-9 x 5-7 mm

Spain: Albufera S València

EU Red List: Critically Endangered CR. Endemic species. The original habitat in Río de los Santos at Venta del Conde was destroyed in the 1980s (by drought, the species had been abundant there), obviously the species was recently introduced to a protected area of riu Verd near Xàtiva. Threatened by water level changes due to exploitation of water resources, by contamination by fertilizers and pesticides, also by shell collectors.

Gómez Moliner et al. 2001: 94, Martínez-Ortí & Robles 2003: 182, Martínez-Ortí et al. in Verdú & Galante 2006: 335.

Theodoxus velascoi (Graells, 1846)

Neritina velascoi Graells, 1846 in Graells 1846: 20 (*Neritina Velascoi*)

Shell relatively highly spired, solid, 4 rapidly increasing whorls, last whorl inflated and regularly rounded.
Differs from *Th. valentinus* in its less highly spired shell and the absence of a groove at the palatal side.

In upwellings of water (ullals in local term), creeks and rivers, with very constant water temperature between 16 and 18 °C. Animals are locally extremely abundant and feed on diatoms and algae, the most important predator (of the egg capsules) is the ostracod *Cypridopsis vidua*.

Eggs are laid throughout the year in oval capsules (0.8-1 mm), fixed on shells or stones, only one egg will develop, life span 18-22 months.

Spain: Valencia, Río Verde

5-7 x 8-11 mm

José Liétor Gallego

Spain (València)

Critically Endangered. Endemic species, protected and rapidly declining, disappeared entirely from Albaida river and the only population left is in riu Verd. Threatened by water pollution and habitat destruction, also by shell collectors.

Gómez Moliner et al. 2001: 94, Martínez-Ortí & Robles 2003: 185, Martínez-Ortí et al. in Verdú & Galante 2006: 337.

Viviparus *Viviparus* Montfort, 1810
in Montfort 1810: 246

Type species *Viviparus fluviorum* Montfort, 1810
(original designation)
(= currently *Viviparus viviparus* (Linnæus, 1758))

ICZN Opinion 573: type species *Helix vivipara* Linnæus, 1758 "by original designation" as interpreted under the plenary powers (ICZN 1987: 185). This interpretation was incorrect because Montfort 1810 designated *V. fluviorum* as type and listed *Helix vivipara* as a synonym. Montfort referred to a specimen from Netherlands with pointed apex (possibly *V. contectus*). It is convenient to accept *V. fluviorum* as type by original designation, and to interprete this name as a junior synonym of *Helix vivipara*.

Shells smooth cone-shaped with convex whorls, shell height 20-60 mm, with concentrically growing horny operculum, sex dimorphism (females usually more inflated and bigger). First whorls in juvenile shells have hairs, these are mostly eroded in adults.

Grazing is typical, but the snails can also just sit there for weeks and filter objects out of the water current.

Baršiene et al. 2000, Glöer 2002, Soes et al. 2009.

Viviparus acerosus Bourguignat, 1862

Without gender agreement: *Viviparus acerosa* Bourguignat, 1862

Viviparus acerosa Bourguignat, 1862 in Bourguignat 1862: 115, Pl. 5 (*Vivipara acerosa*)
Vivipara was an incorrect subsequent spelling for *Viviparus* Montfort, 1810.

Shell usually greenish or dirty greyish yellow with three reddish brown colour bands, apex minute, pointed and white, 6-7 initially not very convex whorls, lower whorls with deeper suture, umbilicus open in females, almost closed in males, males with less convex shells.

Large rivers with silent water currents in the Danube system.

75 % of the animals are females.

Critically Endangered in Bayern and Austria, Endangered in Germany, Vulnerable in Czech Republic and Slovakia. Also the populations in Donauauen National Park have seriously declined, 80 % of the last populations in Austria were lost until 2009.

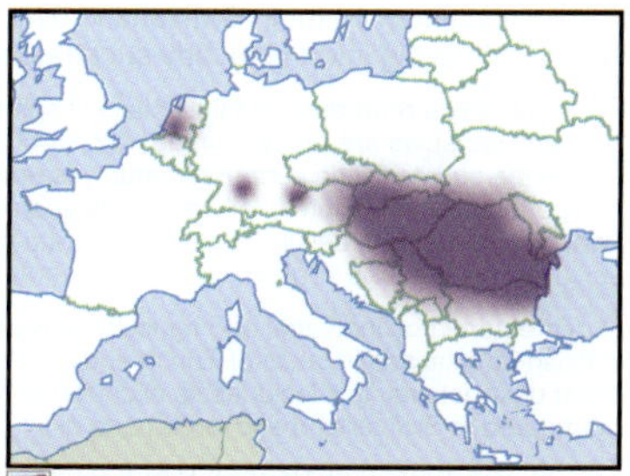

Danube and large adjacent rivers from Wien downwards, isolated in Germany near Passau and Geislingen

30-50 x 25-35 mm, females are more slender than males

Slovakia: Botany Vychodosl Niz

Soes et al. 2009.

Commercially sold in central Europe for use in garden ponds and aquaria. Escaped snails are able to reproduce fast, for example in the Netherlands. They are easily overlooked in the wild, the shells differ only weakly from *V. contectus* and *V. viviparus*.

Viviparus ater (De Cristofori & Jan, 1832)

Without gender agreement: *Viviparus atra* (De Cristofori & Jan, 1832)

Paludina atra De Cristofori & Jan, 1832 in De Cristofori & Jan 1832: 3

Shell olive brown to black, sometimes with colour bands, apex pointed, very variable, 4.5-5.5 not very convex whorls, suture less deep than in the other *Viviparus* species, umbilicus narrow but open.
First 2 whorls increase more rapidly than in *V. acerosus*. Last whorl in juvenile shells rather angulated than rounded, no apical hairs. 9 chromosomes.

In rivers and lakes, shallow waters. In Switzerland up to 500 m altitude. In the warm season in shallower water, in the winter in deeper zones, animals dig inside the substrate and remain inactive.
Animals feed on algae and bacterial covers, organic parts of muddy substrate, gills are able to filter particles out of the water. Predated by rats and water birds.

Ovoviviparous. Females reach maturity after 2 years, they are gravid 2-8 months, can also be gravid in the winter season. Age 8-10 years.

Vulnerable in Switzerland, but successfully introduced to several lakes.

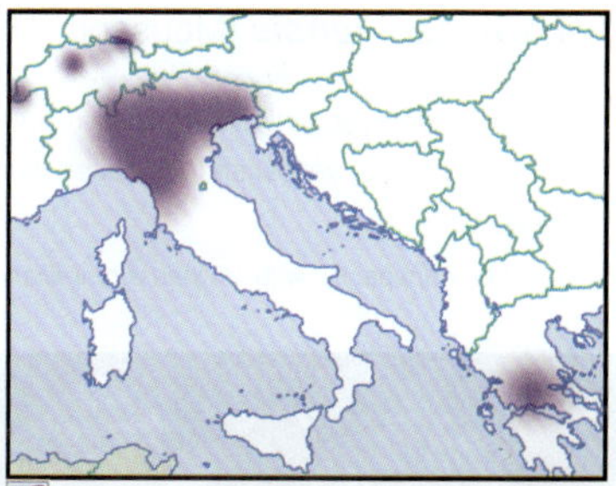

Originally N Italy, isolated in Central Greece, introduced to lakes of Geneva, Zürich and Bodensee

35-45 x 25-35 mm

N Italy: Lago di Garda

Porter & Ribi 1994, Ribi & Porter 1995, Bodon et al. 1995: 21, Katoh & Ribi 1996, Katoh & Ribi 1997, Trüb & Ribi 1997, Ribi & Oertli 2000, Glöer 2002: 65.

The species is able to hybridize with *V. contectus* where they co-occur, interspecific matings are frequent, but such matings seem to produce only 1 % offspring. The survival rates of hybrids seem to be the same as those of the parental species. Gene introgression is low and seems to be limited by the different chromosome numbers.

Viviparus contectus (Millet, 1813)

Without gender agreement: *Viviparus contectum* (Millet, 1813)

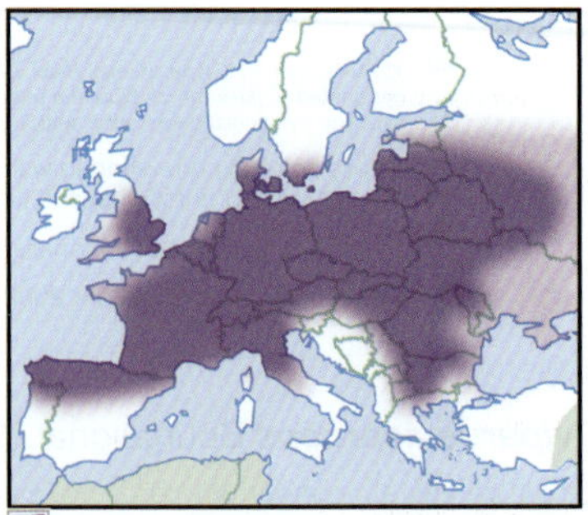

N and central Europe to Slovakia, N Spain and Portugal, mainland Italy, Balkans to Greece (not Norway, not Finland, in Sweden S of 57°N)

Cyclostoma contectum Millet, 1813 in Millet 1813: 5

Shell evenly greenish brown or with blurry colour bands, often with hammer pattern, apex strongly pointed, 5.5-6 strongly convex whorls with deep suture, umbilicus narrow but open. Differs from *V. viviparus* in its pointed apex.
Animal blackish or greenish, with many yellow dots, tentacles long, in males the right tentacle is shorter, eyes big, circular and black, located on two small tubercles at the base of the tentacles. 7 chromosomes.

Standing small and large waters with rich vegetation, swamps and temporary water bodies. Tolerates acid water and up to 0.4 % salt. Artificially created new habitats are rarely accepted in Austria. In N Italy up to 4-10 m deep, maximum 15 m. In Switzerland up to 400 m, in Germany 800 m. Feeds grazing on mud, or by filtrating water. Animals dig into the muddy substrate to hibernate.

In N Greece reproduction takes place in spring and autumn, sex ratio is 1:1 (also in E Poland), in Greece and France 12-30 egg capsules are present in the uterus of females (in Poland 5-15, in autumn less), the 2.5-10 mm sized juveniles hatch in the course of the summer, maturity is reached after 3 months for the spring generation and 8 months for the autumn generation, at 17 mm shell height, life span up to 13 years, in Greece around 5 years. Females may retain embryos at water temperatures below 15°C, they migrate from 0.5 m to deper water (1-2 m) in October and hibernate in the mud, to give birth in the next season.

Threatened by industrial and agricultural water pollution, closing of small water bodies due to agricultural land use, drainage, sinking of ground water levels. Attempts to establish populations in lakes of Zurich and Geneva were unsuccessful. Clearly declining in N England and around London (1999). Has seriously declined in Austria due to habitat destructions, this includes its occurrence in Donauauen National Park.
Endangered in Switzerland and large parts of W Germany (Baden-Württemberg, Hessen, Nordrhein-Westfalen), Vulnerable in Germany and Austria, rare in Portugal.

Germain 1931: 600, Porter & Ribi 1994, Ribi & Porter 1995, Eleutheriadis & Lazaridou-Dimitriadou 1995 (2 papers), Katoh & Ribi 1996, Katoh & Ribi 1997, Trüb & Ribi 1997, Ribi & Oertli 2000, Jakubik 2009.

The names *V. viviparus* and *V. contectus* were not consistently used until 1959, previous reports of *V. viviparus* partly included *V. contectus* (and sources which worked with information published before 1959).
The species is able to hybridize with *V. ater* where they co-occur, interspecific matings are frequent, but such matings seem to produce only 1 % offspring. The survival rates of hybrids seem to be the same as those of the parental species. Gene introgression is low and seems to be limited by the different chromosome numbers.

25-55 x 20-40 mm, very variable

Germany: Schleswig-Holstein, near Grömitz

Viviparus janinensis (Mousson, 1859)

NW Greece: Ipiros

Paludina inflata var. *janinensis* Mousson, 1859 in Mousson 1859: 281

Shell very large, yellowish to blackish brown, apex blunt, 6-7 very convex whorls with deep suture.

In lakes.

Could be conspecific with *Viviparus mamillatus*.

60 x 40 mm

NW Greece: Ipiros, Lake Ioannina

Viviparus mamillatus (Küster, 1852)

Without gender agreement: *Viviparus mamillata* (Küster, 1852)

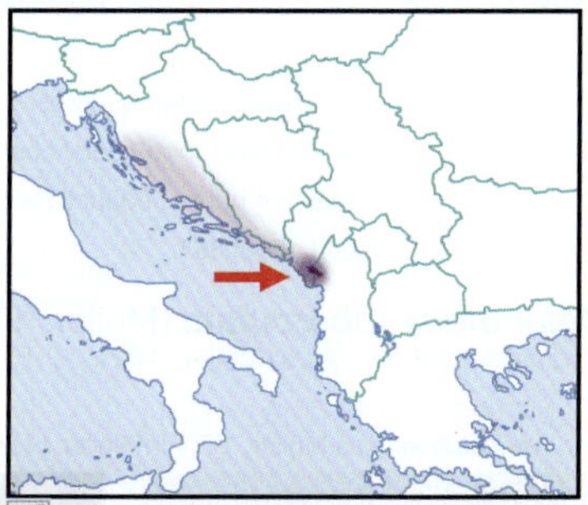

Paludina mamillata Küster, 1852 in Küster 1852: 20, Taf. 4

Shell usually evenly olive brown, prominently striated, sometimes with hammer pattern, apex blunt, 5-6 convex whorls, males more slender than females.

In lakes, often associated with submerged macrophytes. In Shkodra Lake in the warm season found in swallow coastal environments, in the winter (when the water level rises in the lake) in deeper zones.

Viviparous, in August females carry 10-30 juveniles in different dimensions, in procreate cubs about 10 x 10 mm size. The procreation period seems to last relatively long time. Maximum age is estimated to about 4 years (Dh. Dhora, unpublished).

Vulnerable in Albania (Dhora 2004: 141).

Possibly conspecific with *V. janinensis* from Greece. The taxonomic identities of SE European forms and populations remain to be understood.

Croatia to N Albania (Shkodra Lake)

40-60 x 30-40 mm

Albania: Lake Shkodër, W Shkodër

Viviparus viviparus (Linnæus, 1758)

Without gender agreement: *Viviparus vivipara* (Linnæus, 1758)

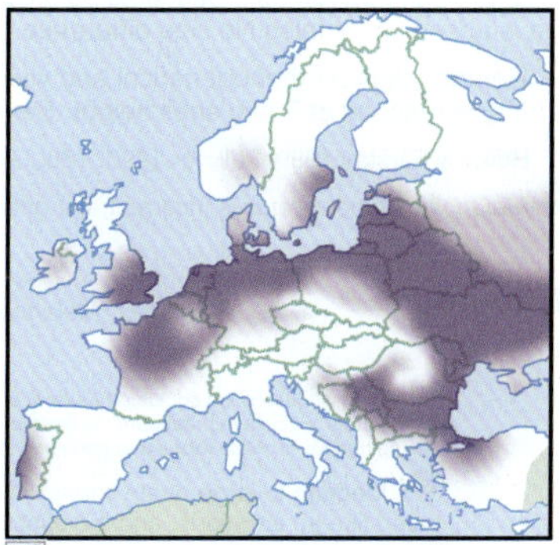

Helix vivipara Linnæus, 1758 in Linnæus 1758: 772
Viviparus fluviorum Montfort, 1810 in Montfort 1810: 247

Shell dark greenish brown or greyish yellow with three reddish brown spiral bands, striated but not with hammer pattern, apex blunt, 5.5-6 weakly convex whorls, last whorl relatively large when compared to the other *Viviparus* species, umbilicus narrow.
Differs from the other species in its blunt apex. 9 chromosomes.

In slowly moving waters, lowland rivers, canals and in the litoral of big lakes, on muddy substrate, usually not in small isolated standing waters. Requires high oxygen contents, pH tolerance 6.8-8.6, up to 0.3 % salt, dimensions are smaller in brackish water. Suspension feeder, lies for long periods in the mud with the mouth uppermost (diet 85 % detritus, 10 % higher plants and 5 % green algae).

Snails in Poland are active between May and October, egg size 3-7 mm, adult females in Poland contain 1-4 embryos, in France up to 70 were observed, juveniles have 3-4 mm diameter at hatch, maturity is reached at 20 mm shell height after 2 years. Maximum age up to 5-10 years.

Not in serious decline, but many populations are threatened by destruction of habitats and water pollution. Endangered in Germany (populations in Neckar river are lost), rare in Poland, Sweden, Norway and Denmark.

Germain 1931: 598, Jakubik 2006, Jakubik & Lewandowski 2007.

Lowlands from England to Ural, in the north up to S Sweden and S Norway (60° N), not Finland, not in highlands and in the upper and central Danube regions (not Switzerland, Austria, Hungary, Slovakia)

Does not live in Austria (true distributional gap), but *V. contectus* has frequently been misidentified as *V. viviparus* and the incorrect reports are frequently cited in recent publications. The names *V. viviparus* and *V. contectus* were not consistently used until 1959, previous reports of *V. viviparus* partly included *V. contectus* (and sources which worked with information published before 1959).

25-35 x 20-26 mm

Poland: Warszawa, near Wisla bridge

Melanoides

Melanoides Olivier, 1804
in Olivier 1804: 40

Type species *Melanoides fasciolata* Olivier, 1804
(monotypy)

Genus was treated as feminine by Olivier 1804 in combination with the adjectival name *fasciolata* (gender feminine, Art. 30.1.4.4).

Shells with high spire, radial ribs often nodulose when crossed by spiral sculptural elements. Operculum with few whorls and basal nucleus.

Females are parthenogenetic with a brood pouch that is separated from the oviduct.

Neubert 1998: 351.

Melanoides tuberculata (Müller, 1774)

Melanoides tuberculatus (Müller, 1774)

Nerita tuberculata Müller, 1774 in Müller 1774: 191

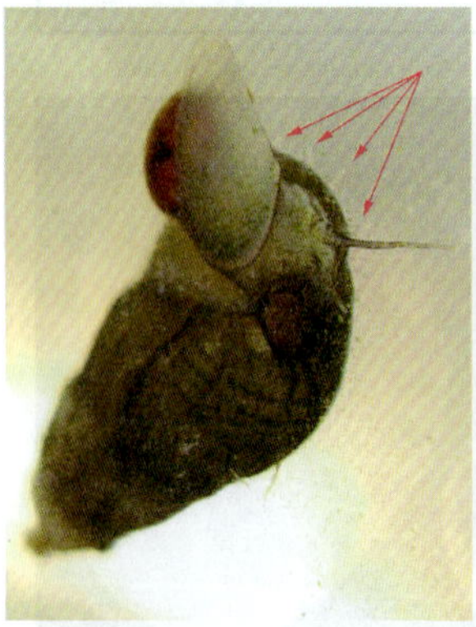

Czech Republic: Olomouc

Michal Maňas

Arrows are pointing to the pallial tentacles for breathing.

20-50 x 7-16 mm

Austria: Kärnten, Villach

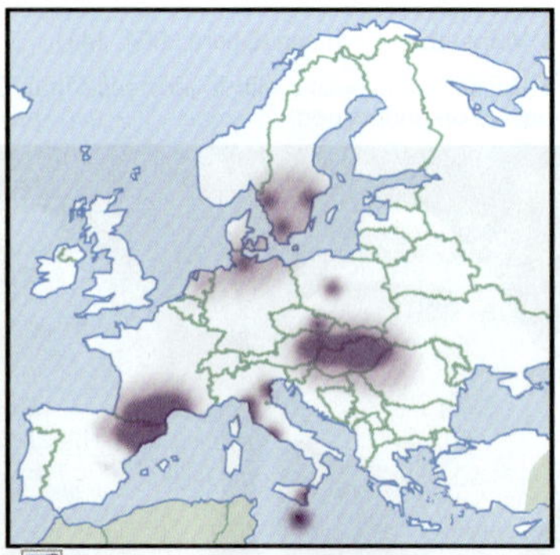

Africa to SE Asia, introduced to Europe and Azores

Tropical and subtropical freshwaters, frequent in botanical gardens and aquaria, also in artificially warmed up waters. Lives at the bottom of standing of moderately moving warm waters, for example in thermal springs.

Shell high, light horny with brown streaks or spots, spirally coarsely striated, 10-15 slightly convex whorls.
Empty shells cannot be reliably sex-classified. Gonads of females are cream, yellow or pale orange. Live males have a red-coloured testis, large and subequal in size to hepatopancreas, noticeable as a dark area in the upper whorls (whorls 5-6 from the apex), visible by holding the shell against a strong light, the dark area is absent in females.No size difference between males and females, but males have generally narrower shells.

Reproduction parthenogenetical and viviparous. In the Far East nearly exclusively females with few sterile males, bisexual populations were found in Israel. Maturity is reached at 7 mm shell height. Sex-reversal with age does not seem to occur.

Heller & Farstay 1989, Falkner 1990: 128, Glöer 2002: 74.

Melanoides tuberculatus is incorrect, the genus is feminine.

Melanopsis

Melanopsis Férussac, 1807
in J Férussac 1807: 70

Type species *Melania costata* Olivier, 1804
(in Olivier 1804: 294, Pl. 31)
(subsequent designation by Gray 1847: 153)

Original description contained *Melania buccinoidea* Olivier, *Bulimus antidiluvianus* Poiret, *Melania costata* (possibly a junior subjective synonym of *Buccinum praemorsum*Linnæus, 1758), and beyond doubts "*Buccinum prærosum*" as listed by Gmelin [1791]: 3489 (an incorrect subsequent spelling for *Buccinum praemorsum* Linnæus, 1758) (not eligible under Art. 67.2.5 because of the doubts). The authorship for *Melanopsis* was attributed to A. E. J. P. J. F. d'Audebard de Férussac alone, this was precisely explained on p. xij in the original source.

Medium to large ovate shells which may be smooth to coarsely ribbed. Aperture acute-oval and usually with a broad distinct parietal callus, often concentrated like an angularis, and a distinct siphonal canal. The operculum is paucispiral with a basal nucleus.

Glaubrecht 1993 (proposed to regard all circummediterranean populations as *M. praemorsa*), Neubert 1998: 352 (rejected Glaubrecht's proposal but came basically to the same taxonomic conclusion in Arabia), Bába 2004 (recognized several species with partly overlapping ranges), Heller et al. 2005 (various species in the Levant).

Neubert 1998 argued that a superspecies should be monophyletic, and that Glaubrecht 1993 should have provided evidence for a monophyletic origin. This argument seems weak. A widely distributed species can easily become paraphyletic or polyphyletic. The definite event to stop gene flow within a species is the onset of reproductive isolation and not a branching event of an isolated subpopulation that makes the parent species paraphyletic (Gofas 2001: 1335).
Heller et al. 2005 recognised various different species, but admitted that the classification based on shell morphology was subjective.
A conclusive study to understand the specific identities of the forms of this species complex would be desirable.

Melanopsis cariosa (Linné, 1767)

Without gender agreement: *Melanopsis cariosus* (Linné, 1767)

Murex cariosus Linné, 1767 in Linné 1767: 1220

18-22 x 10-12 mm

Spain: Guadalquivir near Córdoba

Morocco, Spain (coastal rivers of lower Andalucía), Portugal (Sado river near Setúbal)

Shell horny yellowish to light brown, ribs always present, sometimes with nodules near suture, 3-4 whorls, last whorl predominating, accounting for 80-90 % of the shell height, apertural margin inside and at parietal side white.

Coastal rivers and small running waters with clean waters, also in standing waters. Prefers hard substrate.

EU Red List: Near Threatened NT. Vulnerable in Spain. Threatened by high human pressure, contamination and habitat destruction.

Nobre 1941: 225, Falkner 1990: 128, Martínez-Ortí et al. in Verdú & Galante 2006: 330.

Preliminarily classified as an independent species here, more detailed studies are needed to firmly establish the taxonomic status of this form.

Melanopsis praemorsa (Linnæus, 1758)

Without gender agreement: *Melanopsis praemorsum* (Linnæus, 1758)

Buccinum praemorsum Linnæus, 1758 in Linnæus 1758: 740 (*Buccinum præmorsum*)
Melania buccinoidea Olivier, 1801 in Olivier 1801: [atlas] p. vij, Pl. 17 (*Melania buccinoïdea*)
Melania parreyssii Philippi, 1847 in Philippi 1847: 176, Tab. 4 *Melania (s. Melanopsis) Parreyssii*
Melanopsis lorcana Guirao, 1854 in Guirao 1854: 32 (*Melanopsis Lorcana*)
Melanopsis lorcana var. *etrusca* Westerlund, 1886 in Westerlund 1886: 128

Name mentioned with description and with bibliographical reference to Brot 1862.
Name first mentioned by Brot 1862: 63 as "*etrusca* Villa (v. min)", with locality "Toscane", without description or bibliographical reference, under the name *Melanopsis* "*Dufourii* Fér.", for which a bibliographical reference (to "Rossm. Icon. 835-839") and locality "Espagne" was given. Since *etrusca* was not equipped with a description, and the reference to Rossmässler did not cover the taxon *etrusca*, the synonym *etrusca* was not equipped with a description or indication (Art. 12.1).
Mentioned by Brot 1874: 434 (date 1876) as a synonym for a variety β of *Melanopsis dufourii* Férussac, 1822, without description or bibliographical reference, but with reference to Fig. 3 [Taf. 47] of the same work, which was published in 1879. The name used by Brot for the subordinate was β, which was not an available name as required by Art. 11.6, and thus *etrusca* cannot have been made available under Art. 11.6.1.
For those who do not think that a new name must have a description or indication (and that Art. 12.1 does not apply to names first mentioned in synonymy), the name can be attributed to Brot 1862, validated by Westerlund 1886. For the others, the name must be attributed to Westerlund 1886.

Melanopsis penchinati Bourguignat, 1868 in Bourguignat 1863-1868: 293, Pl. 40 (*Melanopsis Penchinati*)

10-20 x 6-10 mm — Greece: E Crete, 20 km SW Sitía

Spain and N Africa to Iraq

Shell very variable, also within a population, dark brownish with orange or violet hue, whorls not convex, shell cone-shaped and pointed, sometimes with colour bands, smooth or ribbed, aperture cut out at lower side, columella merges abruptly into basal margin of aperture.

In freshwaters of any kind, in Morocco also in subterranean canals. Usually not on muddy substrate. Needs calcareous freshwater with permanently high oxygen contents. Sensitive against organic pollution. In N Africa the species tolerates 13-25°C, salt up to 3.5 % and pH 7-8.3. Feeds mainly on diatoms and algae on hard substrate, but consumes also detritus and macrophytes, in aquaria also remains of fish food etc.
Important intermediate host for larval stages (cercaria) of various trematodes, especially in the Near and Middle East.

Sex ration m:f = 2:1. Self-fertilization is possible, some populations tend to be clonal. 10-25 eggs (diameter 0.23-0.46 mm) are laid in brownish clutches (2-5 x 1 mm) at free undersides of stones, at softer substrate eggs are laid separately and not in clutches (Mouahid et al. 1996), juveniles (0.3 mm, 1.3 whorls) hatch in spring, life span probably 5 years (Elkarmi & Ismail 2006).

EU Red List: several species are listed independently.
Critically Endangered CR: *Melanopsis parreyssii*, known only from one spring in Hungary and in serious decline, threatened by ongoing pollution and habitat disturbance.
Critically Endangered CR: *Melanopsis penchinati*, known only from one thermal spring in Spain near Zaragoza, threatened by habitat disturbance and competing introduced *Melanoides tuberculata*. .
Endangered EN: *Melanopsis etrusca*, known only from Italy (Toscana) in thermal springs, several populations have been lost in the recent past due to habitat destructions.
Near Threatened NT: *Melanopsis lorcana*, from S Spain, threatened by over-exploitation of water resources.

Polytypic or monotypic species with variable shells, many names have been established, many forms have been called subspecies or species, possibly not justified (Neubert 1998: 372). Ribbed populations could be united with smooth populations, as done by Glaubrecht 1993 and Neubert 1998, but they seem to live sympatrically with each other (for example near Iskenderun, S Turkey). But these could be clonal populations of potentially the same species. A more detailed knowledge on the distribution of ribbed and smooth forms would be desirable and an important precondition for studies on the nature and degree of gene flow (and possible barriers) in conchologically distinct populations. Ribbed populations in Spain are almost extinct by now (*M. penchinati*), so there might not remain time to study them any more.
It is certainly useful and justified to list the endangered forms as separate species in the Red Lists, and to protect their last habitats.

Esperiana

Esperiana Bourguignat, 1877
in Bourguignat 1877: 88 *Melanopsis (Esperiana)*

Type species *Melanopsis esperi* Férussac, 1823
(original designation)

Name established in agreement with the Code. There is an internal agreement among malacologists not to use many names established by the "Nouvelle École " group (Bourguignat, Caziot and others) between 1860 and 1915, because their behaviour is considered as a case of abuse of zoological nomenclature. By this convention all generic names ending in *-iana* are blacklisted. This procedure is not covered by the Code, but applying this convention is important because otherwise many well established names would be threatened (Hausdorf 1996: 13). Here I would like to propose to exclude the 3 names *Esperiana* Bourguignat, 1877, *Holandriana* Bourguignat, 1884 and *Hohenwarthiana* Bourguignat, 1864 from the blacklist (because in contrast to the other *-iana* names these names have actually been used by many authors in the recent past), and to amend the convention: *-iana* names established by the Nouvelle École and not used for taxa between 1950 and 2000 should not be used.
For those who would not like to accept this proposal, *Fagotia* Bourguignat, 1884 seems to be the next available name.

Fagotia Bourguignat, 1884 in Bourguignat 1884: 30, type species *Melanopsis esperi* Férussac, 1823 (subsequent designation by Westerlund 1902: 131)

Esperiana esperi (Férussac, 1823)

Fagotia esperi (Férussac, 1823)

Melanopsis esperi Férussac, 1823 in A. Férussac 1823: 160 (separate p. 31) (*Melanopsis Esperi*)

15-25 x 6-11 mm — Serbia: Gornjak gorge

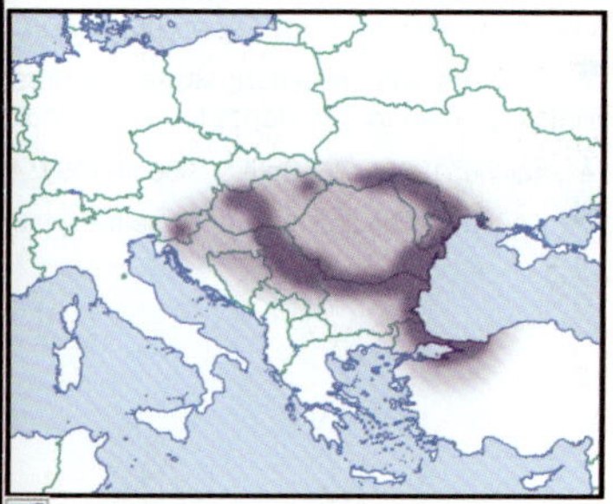

SE Europe (W Black Sea region to Hungary), NW Anatolia

Shell horny coloured or pale olive greenish with irregular reddish brown spots, occasionally evenly black, smooth or finely striated, upper whorls often eroded, 6-8 almost flat whorls, basal margin rounded, columella merges slowly and constantly into basal margin.
Apex less pointed than in *Melanopsis praemorsa*.

In rivers and lakes, at stones, submerged roots or on mud at river margins, pH 5-8, oxygen level not below 12 mg O2/dm3, optimum depth 80 cm, they tend to migrate vertically down to 2 m depth to escape cold surface temperatures. In Ukraine usually in lower densities than *Microcolpia daudebartii*.

In Ukraine threatened by river regulation and industrial pollution.

Bole & Slapnik 1997: 143 (S Slovenia), Glöer 2002: 77, Stelmashchuk 2012 (Ukraine).

Holandriana

Holandriana Bourguignat, 1884 in Bourguignat 1884: 7 *Melanella (Holandriana)*

Type species *Melania holandrii* Pfeiffer, 1828 (subsequent designation, if not earlier then here)

Contained originally 10 species, name established in agreement with the Code. See the comment under *Esperiana*, where I propose to use the name *Holandriana*. For those who would not like to accept this proposal, *Amphimelania* Fischer, 1885 seems to be the next available name.

Melanella Swainson, 1840 in Swainson 1840: 199, 341 *Melania (Melanella)*, not *Melanella* Bowdich, 1822 (Gastropoda: Eulimidae)
Established with description, but without species included. First author to have included species was not researched.

Amphimelania Fischer, 1885 in Fischer et al. 1887: 701 *Melania (Amphimelania)*, type species *Melania holandrii* Pfeiffer, 1828 (designation not researched)
Expressly established as a new replacement name for *Melanella* Swainson, 1840, not *Melanella* Bowdich, 1822. Fischer included only *M. holandrii*.

Whorls pointed, columella merges constantly into basal margin of aperture, shell pale yellow, sometimes with 3-4 more or less distinct colour bands, columella and aperture brown, all characters very variable.

Rivers and creeks slightly below water surface at medium to high water currents.

Holandriana holandrii (Pfeiffer, 1828)
Amphimelania holandrii (Pfeiffer, 1828)

Melania holandrii Pfeiffer, 1828 in C. Pfeiffer 1828: 47 (*Melania Holandrii*)

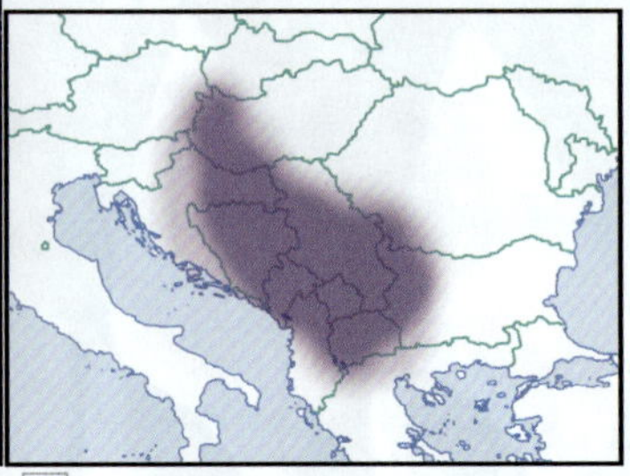

14-16 x 9-10 mm

Serbia: Gornjak gorge

W Hungary to Albania and Bulgaria

Shell yellowish or with green hue, sometimes with 3-4 colour bands, often only visible from inside the shell, apex pointed, 4-6 convex whorls, aperture higher than broad, inside brownish, parietal margin brownish, columella merges constantly into rounded basal margin. All characters very variable.

Rivers and creeks slightly below water surface, at stones or wood, at medium to high water currents.

Extinct in Austria after habitat destruction, Endangered in Albania.

Microcolpia

Microcolpia Bourguignat, 1884 in Bourguignat, 1884: 49

Type species *Melanopsis acicularis* Férussac, 1823 (subsequent designation by Wenz 1938: 691, or earlier)
(= currently *Microcolpia daudebartii* (Prevost, 1821))

Microcolpia daudebartii (Prevost, 1821)
Fagotia daudebartii (Prevost, 1821)

Melanopsis daudebartii Prevost, 1821 in Prevost 1821: 136
Melanopsis acicularis Férussac, 1823 in A. Férussac, 1823: 160 (*Mélanopsis acicularis*)

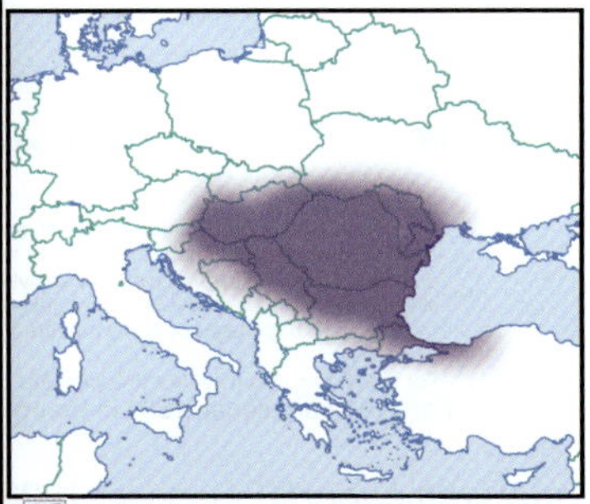

10-25 x 4-8 mm

Slovakia: Kravany nad Dunajom, Podunajska niz.

W Black Sea region to Slovakia and E Austria

Shell dark reddish brown to violet black, shiny, very finely striated, 7-8 almost flat whorls, basal margin rounded, columella merges abruptly into basal margin of aperture, in most populations with weak basal furrow at the columellar side.
Shell more slender than *Esperiana esperi*.

In rivers and creeks, at stones, submerged roots or on mud, pH 5-8, oxygen level not below 12 mg O2/dm3, optimum depth 80 cm, they tend to migrate vertically down to 2 m depth to escape cold surface temperatures. In Ukraine usually in higher densities than *E. esperi*.

Falkner 1990: 128, Bole & Slapnik 1997: 143 (S Slovenia), Glöer 2002: 78, Stelmashchuk 2012 (Ukraine).

It seems to be convenient to use the generic name *Microcolpia* for this species, just to avoid the *Fagotia/Esperiana* problem (they have the same type species).

Rissooidea and **Hydrobioidea**

Most species are not listed here, they are very local, taxonomy and nomenclature are too difficult to understand for the non-expert.

Writing this book took me 4 years, and I would have needed at least one more year, had I decided to include this group. Hundreds of species are recognised and classified in more than 100 genera. Almost every time a species is mentioned in the literature it has another genus. This has made it extremely sophisticated to learn more about these animals and aggregate information on species from various sources. The game becomes particularly difficult if the same specific names like *albanica*, *bourguignati*, *bosniaca*, *falkneri*, *macedonica* or *stankovici* are over and over again established in closely related different genera in the family Hydrobiidae, and even the authors and the years of such names do not differ or were cited incorrectly. In botany it would not be allowed to have the same specific name used in different genera of the same family. It would make sense to apply this rule also here.

Only a few widespread and important species are listed here. Also these species have suffered these problems, some have been classified in not less than 5 different genera in recent publications.

Bithynia

Bithynia Leach, 1818 in Abel 1818: 362

Type species *Helix tentaculata* Linnæus, 1758 (subsequent designation by Herrmannsen 1846 [01 Dec 1846]: 114)

ICZN Opinion 475: type species *Helix tentaculata* Linnæus, 1758 by original designation. This was incorrect. Leach in Abel (1818: 362) gave both *Helix tentaculata* and *Cyclostoma impurum* Draparnaud, 1801 (: p. 41) as type species.
Bythinia = incorrect subsequent spelling for *Bithynia*.
Author and year of names of species which were originally combined with this misspelling are not set in parentheses (Art. 51.3.1).

Shell small to medium-sized, operculum calcareous and concentrically ridged.

Contains possibly 3 widespread species and a number of local forms, which have been regarded as separate species by some authors. Only *B. graeca* is listed here. Its identity is probably not understood either, but its life cycle was studied in detail.
The published record only contains descriptions of many new species, including very accurately described species until recently (Glöer & Pešić 2006, Glöer & Pešić 2007, Glöer & Maassen 2009, Glöer, F. & P. 2010). But the relationships between these forms, their distributional ranges and particularly the nature of morphological and anatomical variation at spatial and population level has not been studied in detail. This would be a prerequisite for justifying the presence of separate species.
Studying accurately and in detail the local forms is important and necessary, but establishing new specific names for every single local form is a very questionable procedure that is probably not going to facilitate future studies on this group. More field studies are needed, especially in SE Europe, in a much denser coverage of area, to come to reliable conclusions concerning the classification.
Some important names which have occasionally been used for more than only extremely local forms:

Paludina boissieri Küster, 1852 in Küster 1852: 35, Taf. 7 (*P. Boissieri*) : Corsica, S Italy (?)
Bithynia schwabii Frauenfeld, 1865 in Frauenfeld 1865: 528, Taf. 10 (*Bythinia Schwabii*) : Republic of Makedonija, N Greece (?)
Bithynia walderdorffii Frauenfeld, 1865 in Frauenfeld 1865: 528, Taf. 9 (*Bythinia Walderdorffii*) : S Croatia (Dalmatia)
Bithynia leachii var. *italica* Paulucci, 1879 in Paulucci 1879: 197, Tav. 9 (*Bythinia leachi* var. *Italica*) : S Italy, Corsica
Bithynia badiella var. *candiota* Westerlund, 1886 in Westerlund 1886: 22 (*Bythinia badiella* var. *candiota*) : S Greece

Bithynia graeca Westerlund, 1879

N Greece (Ipiros, Macedonia)

Bithynia graeca Westerlund, 1879 in Westerlund & Blanc 1879: 137 (*Bythinia græca*) Type locality Ioánnina Lake.
Bythinia = incorrect subsequent spelling for *Bithynia*, so author and year not in parentheses (Art. 51.3.1). Name and description (p. 139) in original source were both attributed to Westerlund alone.

Shell densely striated, 5-6 convex whorls, aperture oval, umbilicus open.

9-11 x 6-7 mm

Lakes and marchlands, on exposed surfaces of the undersides of rocks, facing the bottom but not touching it. Snails can remain without food and outside the water for 6-7 months.

Reproductive period begins in April, population increases 10-fold until May-June (this is the period when two cohorts exist in the field), then decreases again until August. Snails hibernate between August and March, overwinter mortality is lower than mortality after breeding. Most adults die after egg-laying, maximum age little more than 1 year.

Eleutheriadis & Lazaridou-Dimitriadou 1995, 2001.

Bithynia leachii (Sheppard, 1823)

Turbo leachii Sheppard, 1823 in Sheppard 1823: 152

Germany: Schleswig-Holstein, Kellersee near Eutin (identification uncertain)

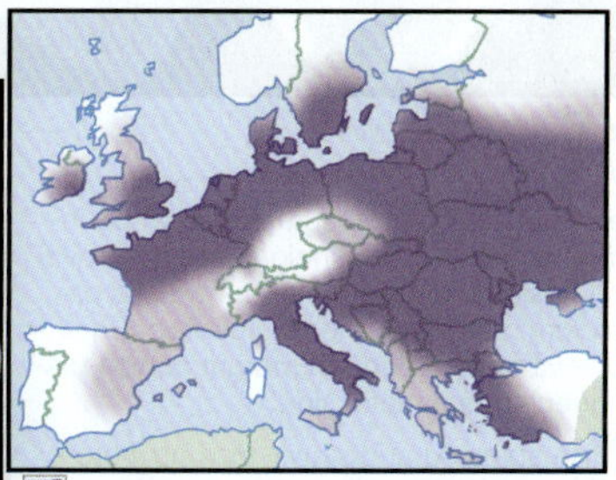

N Africa and Europe to E Siberia, not Alps and S Germany

4-8 x 3-7 mm

Shell brownish or greyish, 4-4.5 strongly convex whorls with very deep suture, aperture and operculum upside rounded. Juveniles look like small adults.

Silent slow-moving waters with rich vegetation, in central Europe and Britain in similar habitats to *B. tentaculata*, up to 3 m deep. Tolerates brackish water with up to 0.5 % salt. In Switzerland in up to 400 m altitude. Seems to be much more sensitive against water contamination.

Eggs are laid mid-May to end-July, usually only juveniles hibernate, exceptionally also adults (reaching up to 9 mm shell height).

Some populations in Switzerland seem to have increased in the past decades (Zurich lake), some decreased or became extinct. In Britain locally declining, habitats in Ireland and Scotland are potentially threatened. Critically Endangered in Bayern, Nordrhein-Westfalen, Sachsen/Thüringen, Czech Republic and Austria, Endangered in Germany, Vulnerable in Switzerland and Poland.

Hungary: Lake Balaton, Káptalanfüred (determined by P. Glöer)

Zoltán Fehér

Bithynia tentaculata (Linnæus, 1758)

Helix tentaculata Linnæus, 1758 in Linnæus 1758: 774

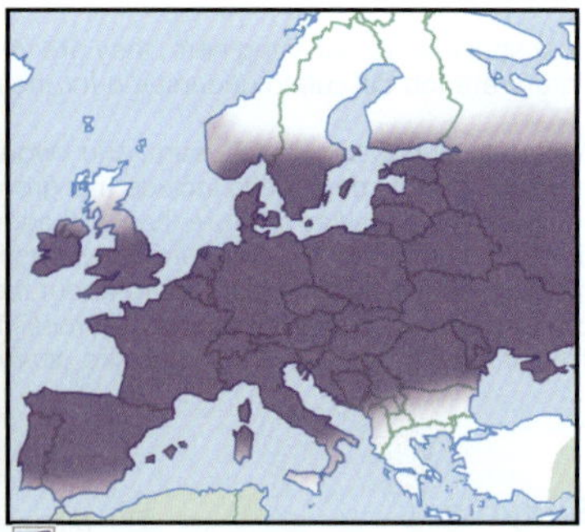

7-13 x 5-9 mm

Germany: Schleswig-Holstein, Alster near Wurksfelder Mühle

Europe to W Siberia and Kashmir

Shell horny yellowish, often covered with a brownish or blackish surface layer, 5 not very convex whorls with deep suture, aperture and operculum upside pointed.
Animal blackish or dark brown with small rounded yellow spots, head small, tentacles long with big black eyes.

Moving and standing waters, also in temporarily drying water bodies, on muddy bottoms and plants, usually very abundant where present. Seems to prefer abundant and rich vegetation, muddy substrate, high oxygen contents and hard waters. In running waters usually only at the lower side of stones. In Portugal not on non-calcareous granitic substrate. Feeds grazing on periphyton, but can also use its ctenidia to filter phytoplankton. Up to 45 m deep, but rarely below 15 m. In Switzerland in up to 1400 m altitude. Importand diet of winter-visiting diving ducks in Northern Ireland.

Eggs are mostly laid in May and June, in France May to August, in 2-3 rows of 12-30 mm long bands containing 10-70 eggs, attached to aquatic plants, juveniles hatch after 14-25 days, offspring per female in N Germany between 120 and 170, in standing waters more, maturity is reached in the second year, life span 1-2 years, exceptionally 3 years.

Has seriously declined in Moldova (Dniester river) down to 0.5 % of its previous population density in the 1980s (Coadă & Welter-Schultes 2011: 42).

Not in N Scandinavia and Greece, introduced to North America (S Quebec to Great Lakes and mid-Atlantic USA).

Bouchard-Chantereaux 1838: 82 (reproduction), Germain 1931: 604,Prince 1967, Clarke 1981: 78, Falkner 1990: 126, Glöer 2002: 84.

Bithynia transsilvanica (Bielz, 1853)

Bithynia troschelii (Paasch, 1842)

Paludina troschelii Paasch, 1842 in Paasch 1842: 301, Taf. 6 (*Paludina Troschelii*)
The lectotype designation by Glöer 2002 (SMF 4063/1) is invalid under Art. 74.2 because Falkner 2003: 30 demonstrated with reasonable arguments that this was not a syntype under Art. 72.4.1.1. Glöer 2004's additional arguments were based on speculations and not on evidence as required by Art. 72.4.1.1 (evidence that the specimens must have been recognised as the new taxon when the nominal species was established), Falkner's 2003 arguments were not proven as incorrect. Lectotype was fixed by Falkner 2003: 32, specimen no longer exists, but correctly designated under Art. 74.4, synonym of *B. leachii*. Name can currently not be used for this species.
Paludina transsilvanica Bielz, 1853 in E. A. Bielz 1853: 164

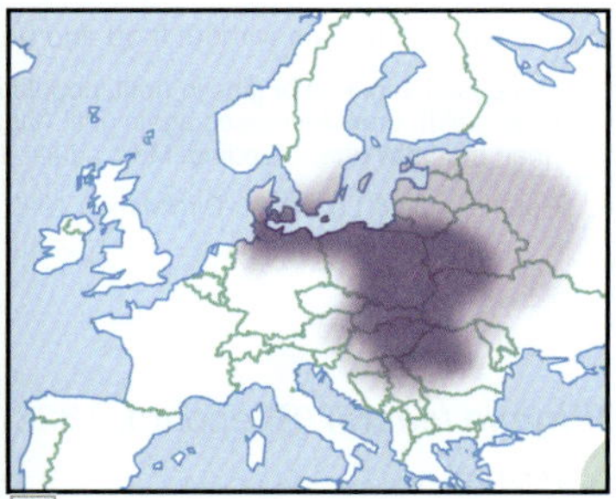

Central and E Europe, from Sweden and Denmark to Austria, Romania and Russia (not S Germany)

8-12 x 5-6.5 mm

Germany: Mecklenburg-Vorpommern

Shell horny brown, 5-5.5 strongly convex whorls with deep suture, last whorl inflated and 1/3 of shell height, aperture oval and upside bluntly angular, as well as operculum, umbilicus slightly covered. Shell quite variable, apertural margin can be disconnected at parietal side.

In standing and slowly moving waters rich in vegetation, often at wood or in reeds at lake margins.

Endangered in Poland.

Hungary: Békéscsaba, Élővíz-csatorna (determined by P. Glöer) Zoltán Fehér

A very difficult species. Distribution in E Europe not very well known.
Glöer 2002 regarded *troschelii* as a subspecies of *B. leachii* because hybrids occurred frequently. Glöer & Fehér 2004 suggested presence of 2 distinct species in Hungary, Glöer & Sîrbu 2005 found it in Romania. Glöer, F. & S. 2006 tried to find differences between *leachii* and the present species and only found the penis morphology as a stable distinctive mark to support reproductive isolation, if specimens with intermediate characters are classified as hybrids between the two species. Glöer & Diercking 2010: 68 reported both species from Hamburg, from largely the same sites. So current state of knowledge is that we have a second species next to *leachii*, but its name is not *troschelii*. Molecular studies would be desirable to understand the genetic background.
I would not to use the name *troschelii* any more and to use *transsilvanica* instead. Previous reports for *troschelii* included misidentifications, so it is convenient to use a clean name.

Heleobia

Heleobia Stimpson, 1865
in Stimpson 1865: 47

Type species *Paludestrina culminea* D'Orbigny, 1840
(subsequent designation by Pilsbry 1911: 550)
(= currently *Heleobia culminea* (D'Orbigny, 1840) from South America)

Genus established estblished conditionally, "for *P. culminea* and its allies". These allies were listed on p. 46, *P. cuminginana* d'Orbigny, 1840 and *P. parchappii* d'Orbigny, 1840.

Heleobia stagnorum (Gmelin, 1791)

Helix stagnorum Gmelin, 1791 in Gmelin [1791]: 3653

3-4 x 1.8-2 mm

Netherlands: Zuriksee

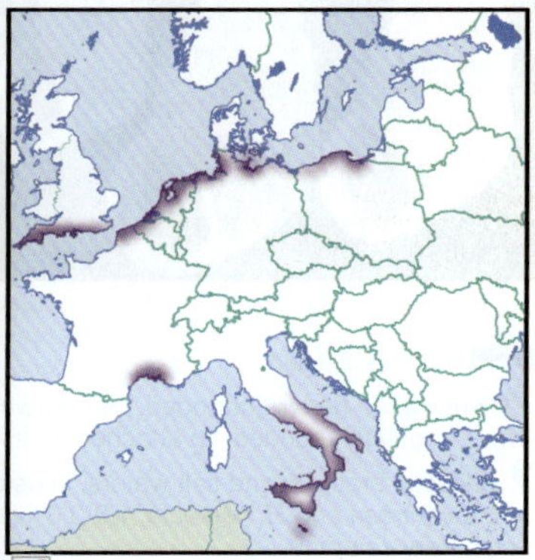

Europe
Possibly also in Greece and NW Turkey.
Northern and southern European populations could also belong to different species.

Shell smooth, 4-5 very convex whorls, apertural margin very slightly detached, upside slightly pointed. The animal is dark grey with light grey borders at the foot, terminal end of snout light grey, tentacles evenly dark with whitish longitudinal lines and a whitish zone at the base around the eyes. The colour seems to be quite constant and characteristic for this species in Italy.
Very much like *Hydrobia ventrosa*, last whorl slightly less predominating.

In the Netherlands in a few non-tidal lagoons and drainage ditches near the sea, in slightly brackish water, 0.2-0-8 % salinity. In Italy only in waters with very low salinity.

Populations in the Netherlands have been seriously declining, the few remaining sites are threatened by urban pollution and by changes in salinity caused by coastal control.

This species has been classified in many genera after 1970 (*Hydrobia, Heleobia, Semisalsa, Ventrosia, Ecrobia*), and it seems that no researcher likes to work on this problem again. So *stagnorum* has remained untouched for some decades in the South American genus *Heleobia*, where it had been placed by Giusti & Pezzoli 1984. This seems to have become the preferred genus for this species.
Google general search (02.2011): *Hydrobia* 174 : *Heleobia* 296 : *Semisalsa* 48 : *Ventrosia* 22 : *Ecrobia* 0
Google Scholar (02.2011): *Hydrobia* 58 (10 since 2004) : *Heleobia* 21 (10 since 2004) : *Semisalsa* 15 : *Ventrosia* 5 : *Ecrobia* 0

Hydrobia

Hydrobia Hartmann, 1821
in Hartmann 1821 (Neue Alpina 1): 258

Type species *Cyclostoma acutum* Draparnaud, 1805
(monotypy)
(= currently *Hydrobia acuta* (Draparnaud, 1805), a marine species)

Contained originally *acuta, thermarum* and *diaphana* in Neue Alpina, and *acuta, Cyclostoma vitreum* Draparnaud, 1801 and *minuta* in Sturm 1821 (which appreared later). The names *thermarum* and *diaphana* were nomina nuda. If taken from Sturm, then Gray 1847: 151 would have designated *C. acutum* as type.
Ecrobia Stimpson, 1865 in Stimpson 1865: 42, type species *Turbo minutus* Totten, 1834 from New England (original designation)
Peringia Paladilhe, 1874 in Paladilhe 1874: 6, type species *Turbo ulvae* Pennant, 1777 (subsequent designation by Clessin 1880: 184)
Obrovia Radoman, 1974 in Radoman 1974: 286, type species *Obrovia salaria* Radoman, 1974 (original designation)
Ventrosia Radoman, 1977 in Radoman 1977: 208, type species *Turbo ventrosus* Montagu, 1803 (ICZN decision)
Radoman 1977 had originally designated as type species *Helix stagnorum* Gmelin, 1971, this type species was misinterpreted for *Turbo ventrosus* Montagu, 1803 (Giusti & Pezzoli 1984). ICZN Opinion 2034 (30 Jun 2003): type species *Turbo ventrosus* fixed under the plenary powers. Today such a case would be solved under Art. 70.3.

Hydrobia neglecta Muus, 1963

Hydrobia neglecta Muus, 1963 in Muus 1963: 132

Shell much like *H. ventrosa*, shell light yellow brown or greyish, thin, very finely striated, 5-7 whorls, rounded but more flattened, apex bluntly elevated, aperture angled above, peristome continuous.
Penis smaller than in *Peringia ulvae*, stout (not pointed as in *H. ventrosa*), its rounded tip bent to the right of penis axis.
Experienced researchers might be able to separate 95 % of the shells between *H. neglecta* and *H. ventrosa*, presence or absence of a pointed tip at the penis is always a definite character for identification. Newly hatched snails have no tubercles.

3-4 x 1.5-2 mm

In fjords, in brackish water, 1.0-3.3 % salinity, usually around 1.8 %, sheltered lagoons and marsh drains, no tides, habitats are permanently submerged.

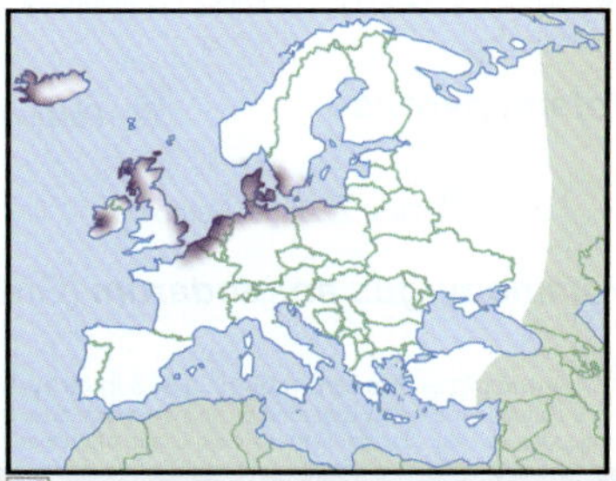

NW Europe, Iceland

Eggs (diameter 0.13-0.16 mm) are deposited twice in the year, in spring and late summer, or only once, depending on the climatic conditions. Juveniles hatched in spring and early summer grow from 0.29 mm (larger than *H. ventrosa*) to 2.3 mm until late autumn and survive the winter. Juveniles from late summer grow to a size of 1.3 mm in September when their growth is stopped. The 2.3 mm large snails start growing again in May, they begin reproducing at the age of 1 year and reach 3.4 mm in November, some snails can survive the winter (mortality 70 % until February), approximate life span in Denmark 18 months.

Much more local than *H. ventrosa*, to that extent vulnerable in the small nuber of known sites.

Fish & Fish 1981, Siegismund 1982, Kerney 1999: 31.

Occasionally classified in *Obrovia* (but I found no hits for *Obrovia* in Google Scholar, 02.2011)

Hydrobia ulvae (Pennant, 1777)

Turbo ulvae Pennant, 1777 in Pennant 1777: 132, Pl. 86 (*Turbo ulvæ*)

4-5.5 x 2.5-3 mm

Germany: Schleswig-Holstein, St. Peter

Europe

Shell often heavily corroded, usually whitish with brown peristome present on the last whorl, 5-7 very weakly convex whorls, regularly increasing but not always regularly rounded, lip attached to last whorl.

Coastal brackish and salt waters, in estuaries, intertidal habitats and saltmarshes, seems to prefer more exposed localities with less vegetation than *Hydrobia ventrosa* and *H. neglecta*, 1.0-3.3 % salinity. Most common within the upper half of the intertidal zone. Feeds on detritus, also consumes seaweeds directly. Very common.

Mostly classified in *Hydrobia*, often in *Peringia*. Google Scholar (02.2011): *Hydrobia* 3294 : *Peringia* 240.

Hydrobia ventrosa (Montagu, 1803)

Without gender agreement: *Hydrobia ventrosus* (Montagu, 1803)

Turbo ventrosus Montagu, 1803 in Montagu 1803: 317

Expressly established as a new replacement name for *Turbo eburneus* Jacobs, 1798 (in Adams & Kanmacher 1798: 637 [not verified]) (considered by Montagu 1803 as inappropriate, *T. eburneus* = nomen oblitum under Art. 23.9.1 by Kadolsky 2012: 69 who provided the 25 references).

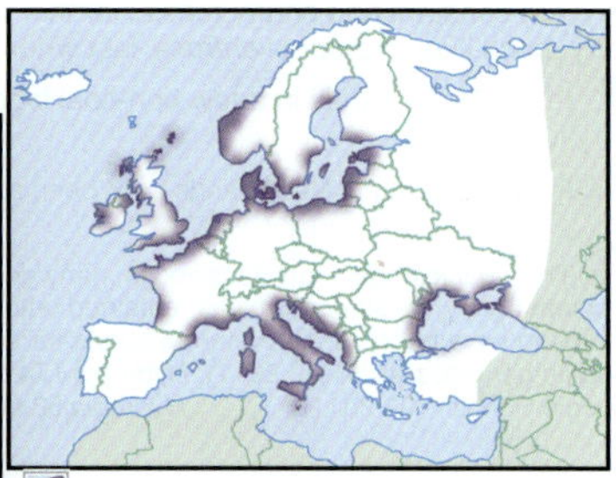

Europe, Black Sea

Knowledge on distribution scattered.

Habitats are more vulnerable than those of *Peringia ulvae*. Not seriously threatened in England.

3-4 x 1.5-2 mm

Netherlands: near Zuriksee

Shell very finely striated, 5-7 convex whorls, slightly more convex than in *H. neglecta*, smaller shells with 5 whorls are slightly less slender than *H. neglecta*, suture deep, aperture upside rounded or only slightly pointed, attached to last whorl, lip weakly developed. Penis with a pointed tip, not blunt as in *H. neglecta*. Newly hatched snails have small tubercles (diameter 0.04-0.08 µm).

In fjords and coastal waters, quiet estuaries, drainage ditches in coastal marshes, in Denmark about 0.6-2.0 % salinity, in Britain 0.5-2.5 %. Lives more or less permanently submerged, on mud or in aquatic vegetation. Lives often sympatrically with *H. neglecta*, predominating below 1.5 % salinity.

Eggs (diameter 0.13-0.16 mm) are deposited twice in the year, in spring and late summer, or only once, depending on the climatic conditions. Juveniles hatched in spring grow from 0.24 mm (smaller than *H. neglecta*) to 2.5 mm until late autumn and survive the winter. Juveniles from late summer grow to a size of 1.5 mm in September when their growth is stopped. Snails start growing from April, they begin reproducing at the age of 1 year, they reach a size of 3.6 mm in November and do not survive the next winter, life span in Denmark 18 months.

Fish & Fish 1981, Siegismund 1982, Kerney 1999: 31, Kadolsky 2012: 69.

Usually classified in *Hydrobia*, occasionally in *Ecrobia* or *Ventrosia*. Google Scholar (02.2011): *Hydrobia* 932 : *Ventrosia* 35 : *Ecrobia* 4.

Potamopyrgus

Potamopyrgus Stimpson, 1865 in Stimpson 1865: 49

Type species *Melania corolla* Gould, 1848 (original designation)
(= currently *Potamopyrgus corolla* (Gould, 1848) from New Zealand)

New Zealand, Europe

Potamopyrgus antipodarum (Gray, 1843)

Amnicola antipodarum Gray, 1843 in Gray 1843: 241 (*Amnicola antipodanum*)

A. antipodanum is widely regarded as an inadvertent error for *antipodarum*. Gray did not explain the name, but on p. 247 a species was named *Bulimus antipodarum*. I would suggest to maintain usage of *antipodarum*.

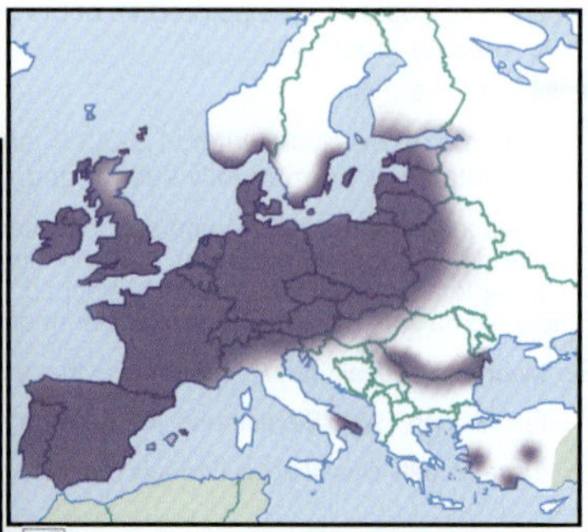

New Zealand, Europe

Originally from New Zealand, introduced around 1850 to England, since then expanded its range all over Europe, in Switzerland since 1972.

3-4.5 x 1.5-2.2 mm

Germany: Schleswig-Holstein, Dobersdorfer See

Parthenogenetic reproduction is possible.

Shell with 5 not very convex whorls, suture quite deep, aperture upside pointed, not attached to last whorl.

Brackish and freshwater (0-1.7 % salinity) in flowing waters of any kind and lakes, often mass occurences (up to 100,000 animals per square meter). Up to 10 m deep. Range expansion is favoured by nitrate pollution of non-marine waters, dispersal by birds and boats is easy. Tolerates mild pollution. In Switzerland in up to 650 m altitude. Feeds on detritus. Very common in Europe, on of the most frequent freshwater molluscs in the British Isles, very common in S Spain.

Pyrgula

Pyrgula De Cristofori & Jan, 1832
in De Cristofori & Jan 1832: 4

Type species *Turbo annulatus* Linnæus, 1758
(monotypy)

Contains a few species.

Pyrgula annulata (Linnæus, 1758)

Without gender agreement: *Pyrgula annulatus* (Linnæus, 1758)

Turbo annulatus Linnæus, 1758 in Linnæus 1758: 767

Shell with 2 sharp characteristic keels, one at the periphery and one below, also at the apertural margin.

In N Italy and N Albania in lakes, in Dalmatia in springs.

Schütt 1968, Falkner 1990: 122, Bodon et al. 1995: 28, Dhora & Welter-Schultes 1996: 106.

4-10 x 2.5-4 mm

N Italy: Lago di Garda

N Italy to N Albania

Lithoglyphus

Lithoglyphus Menke, 1830
in Menke 1830: 42 *Paludina (Lithoglyphus)*

Type species *Paludina naticoides* Pfeiffe
(subsequent designation by Herrmannsen 1846 [25-5-1847

Established by Menke 1830 as a subgenus *Lithoglyphus* "Zgl. [= Ziegler]" (*Paludina* section D) without bibliographical reference, with a very short description and 2 species included *Paludina fusca* Pfeiffer, 1828 and *Paludina naticoides* Pfeiffer, 1828.
First mentioned as *Lithoglyp. eburneus* "Megerle ab Mühlfeld" by Hartmann 1821: 57 (unavailable, nomen nudum and abbreviation was ambiguous). C. Pfeiffer 1828: 47 mentioned *Lithoglyphus fuscus* "Ziegler" as a synonym for *Paludina fusca* Pfeiffer, 1828 from Slovenia. No author from before 1830 referred to Pfeiffer 1828 bibliographically and adopted this name.
For those who require a bibliographical reference for making available a synonym under Art. 11.6.1, the name must be attributed to Menke 1830, and the type species must be *Paludina naticoides*. For the others (Falkner et al. 2001: 23), to Pfeiffer 1828, and the type species must be *Paludina fusca*.
Fitzinger 1833: 116 established *Lithoglypter* (type species *Paludina naticoides* Pfeiffer, 1828 by monotypy), and mentioned *Lithoglypus* "Megerle von Mühlfeld" as a synonym. Krynicki 1837: 57 established *Lithoclypus* "Meg[erle von Mühlfeld]" (type species *Paludina naticoides* Pfeiffer, 1828 by monotypy).
Google hits (Apr 2009): Lithoglyphus 3800, Lithoclyphus 2, Lithoglypus 9, Lithoclypus 9, Lithoglyptus 4, Lithoclyptus 6, Lithoglypter 2.
It is convenient to accept Menke's 1830 authorship for the genus. If Menke's 1830 name is regarded as not having been used as a subgeneric name in Menke's classification (because the other sections A-C and likewise sections in other genera did not always have unambiguous names), the next available generic name for *Paludina naticoides* is *Lithoglypter* Fitzinger, 1833, but this name has never been used.

Lithoglyphus naticoides (Pfeiffer, 1828)

Paludina naticoides Pfeiffer, 1828 in C. Pfeiffer 1828: 45

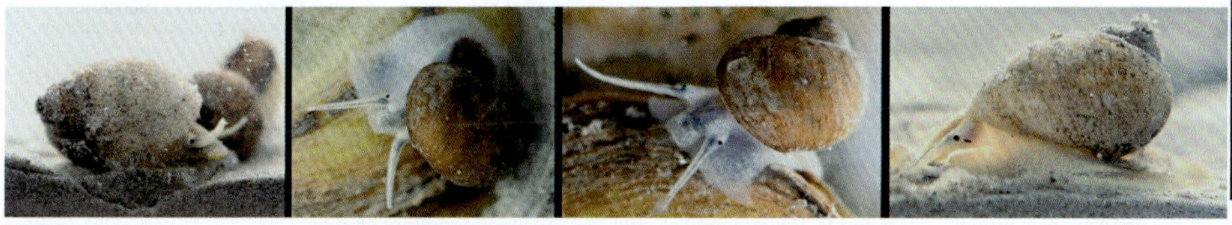

7-10 x 7-10 mm, males often smaller than females

Austria: Oberösterreich, Donau near Plesching — Jan Steger

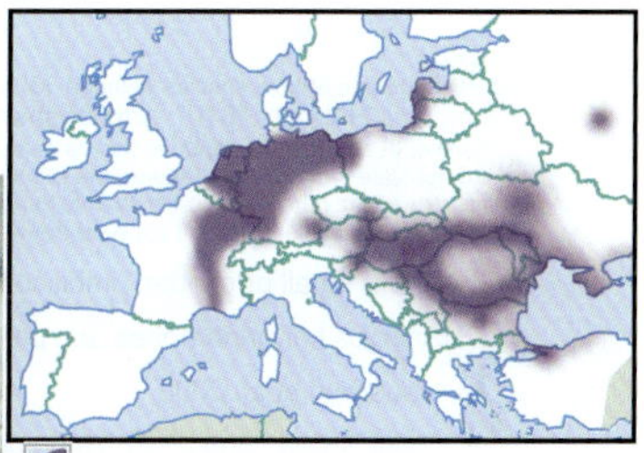

Europe
Originally only in Black Sea rivers, Danube up to Regensburg, after 1800 introduced to Elbe and Rhine regions by artificial canals, since after 1900 in France, very slowly expanding its range there.

Shell globulous, light grey to greenish yellow or dark, last whorl predominating. Animal with broad foot.

In rivers at sites with low water currents, on solid soils (rocky bottoms or stones), requires high oxygen and calcium carbonate contents. Sensitive against too high water temperatures.

Annual life cycles were observed in France and the Netherlands. In France eggs are laid from March to June in a capsule on other *Lithoglyphus* shells (occasionally also on *Valvata piscinalis*), juveniles (0.8-1 mm) hatch after some weeks, they grow rapidly and reach a size of 6 mm im October. They survive only one winter and die after egg laying, latest in June. Maximum population densities are observed in June.

After 1960 in central Europe almost extinct due to water pollution, also affected by global warming. Endangered in Germany and Poland.

Falkner 1990: 126, Mouthon 2007 (life cycle).

Marstoniopsis

Marstoniopsis Van Regteren Altena, 1936
in Van Regteren Altena 1936: 68

Type species *Hydrobia steinii* Martens, 1858
(original designation)
(currently = *Marstoniopsis insubrica* (Küster, 1853))

Falniowski & Wilke 2001.

Marstoniopsis insubrica (Küster, 1853)

Marstoniopsis scholtzi (Schmidt, 1856)

Paludina insubrica Küster, 1853 in Küster 1852: 77
Hydrobia scholtzi Schmidt, 1856 in Schmidt 1856: 158 (*Hydrobia Scholtzi*) Type locality in Poland.

Shell transparent, horny greyish to brownish, finely striated, 4-5 slowly and regularly increasing convex whorls, apertural height less than 1/2 of shell height, margin conneted at parietal side and angulated on upper side, often with white lip inside, columellar margin weakly reflected, umbilicus narrow.
More globulous than *Bythinella*, suture deeper, whorls more convex. Animals with yellow spots under the eyes. Differs from *Potamopyrgus antipodarum* in its more regularly rounded aperture, more convex whorls and broader apex.

3 x 2 mm

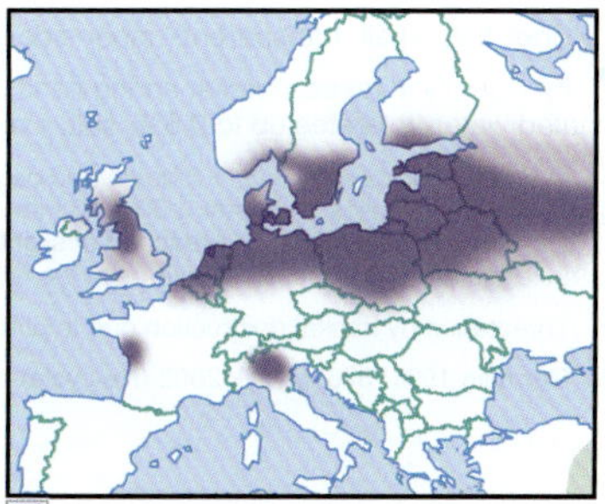

Central and N Europe, Britain

Lakes with rich vegetation, silent bays of lowland rivers, slowly moving waters in canals. Occurs near Berlin exclusively on hard substrate, such as submerged water plants, wood, stones or even metal waste, usually on the sides that are not exposed to sunlight. In Hamburg on muddy ground, often associated with submerged dead wood logs, crawling at the lower side of submerged wood and leaves. Appears to tolerate mild industrial pollution near Manchester and Berlin, but not too much. Tolerates wave movement, seems to prefer habitats with wave action in S Sweden.

Annual life cycle, reproduction in May/June, adults die few days later, so that only juveniles can be found in late summer and autumn.

Threatened by continuous alterations and destructions of habitats. Endangered or vulnerable in Germany (but populations seem to have recovered in the past years), rare in England.

Westerlund 1886: 38, 73, Falkner 1990: 124, Kerney 1999: 37, R. Müller 2008 (not very rare near Berlin), Glöer & Diercking 2010: 73 (not extremely rare in Hamburg).

M. scholtzi has occasionally been classified as a separate species. Falniowski & Wilke 2001 found no morphological differences and low genetic divergence. Glöer & Zettler 2005: 14 argued that the populations were geographically isolated and regarded both taxa as separate species. In the biological species concept (Mayr 1942) reproductive isolation must be caused by intrinsic factors.

Truncatella

Truncatella Risso, 1826
in Risso 1826: 124

Type species *Truncatella costulata* Risso, 1826
(subsequent designation by Lowe 1855: 217)
(= subjective synonym of *Helix subcylindrica* Linné, 1767)

ICZN Opinions 344 and 1664: type species designated by Lowe 1855: 217. This was correct, Lowe 1855 gave *Cyclostoma truncatulum* Draparnaud, 1801 as type species and listed *costulata* as a synonym. So the type species was *T. costulata* fixed under Art. 69.2.2.

Truncatella subcylindrica (Linné, 1767)

Helix subcylindrica Linné, 1767 in Linné 1767: 1248

Shell decollate, pale yellow to light brown, sometimes ribbed, aperture oblique, no umbilicus. Juveniles cylindrical in their first 3-4 whorls. Adults slightly smaller and strikingly different, cone-shaped with small apex, like *Melanopsis* with very small and oblique aperture, regular ribs from 3rd whorl onwards. Animal relatively large, more or less dark grey, upper tentacles not very long, lower tentacles very short.

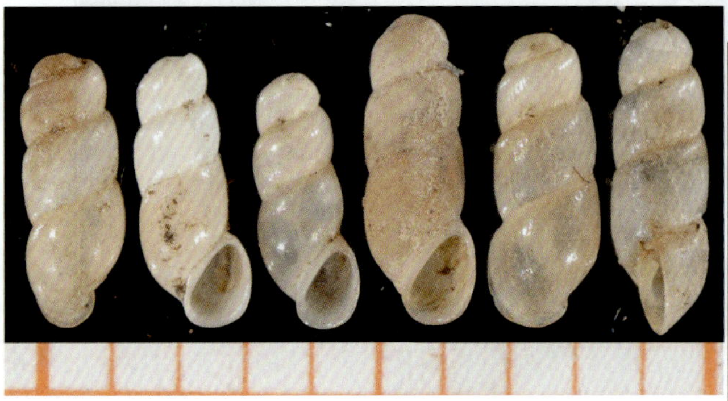

Spain: Illes Balears, Eivissa/Ibiza (only juveniles)

5 x 1.5-2 mm (adults up to 5 mm, juveniles more than 5 mm)

European and N African coasts

Marine supralitoral, usually outside the water, in large numbers under algae and organic litoral deposits, up to 15 cm inside the soil, more rarely under rocks, also in freshwater of river estuaries. Rarely found alive. Prefers fresh air but can survive long periods submerged.

Nobre 1941: 165, Alzona 1971: 39, Bodon et al. 1995: 28, Kerney 1999: 38, Schütt 2001: 11, Martínez-Ortí & Robles 2005.

Valvata

Valvata Müller, 1773
in Müller 1773: [25]

Type species *Valvata cristata* Müller, 1774
(subsequent monotypy by Müller 1774: 198)

Appeared without species included. ICZN Opinion 335: type species by monotypy. This must be subsequent monotypy.

"*Cincinna* Hübner, 1810": unpublished, not made available.

Falkner et al. 2002: 93 regarded this name as available, without providing evidence that (1) Hübner's letter was published work under Art. 8 and 9 (both Menke 1848: 55 and Férussac [1821-1822]: 83 did not state that Hübner's letters were printed and published for providing a permanent record), and (2) that the name was made available by meeting the conditions in Art. 11 and 12 (Menke mentioned the name, but no constraints to assume that Hübner had combined it with a description or indication). The assumption that Hübner might have employed it for *Valvata piscinalis* is arbitrary and unsubstantiated. I do not consider the view expressed by Falkner et al. 2002 as tenable. Menke 1848 did not make the name available either.

Shells small or minute, umbilicus open, whorls rounded, aperture circular, operculum with central nucleus.

Animals hermaphrodites and partly viviparous, partly ovoviviparous.

Hauswald et al. 2008.

Glöer & Diercking 2010: 79 recognised another species *V. ambigua*, this was not based on published evidence. The same applies to *V. alpestris* mentioned by Glöer & Zettler 2005: 15, also other (partly new) species recently attributed to *Valvata* require confirmation in independent studies.

Valvata cristata Müller, 1774

Valvata cristata Müller, 1774 in Müller 1774: 198

0.6-1.5 x 2-3.5 mm

Germany: Schleswig-Holstein, Plöner See

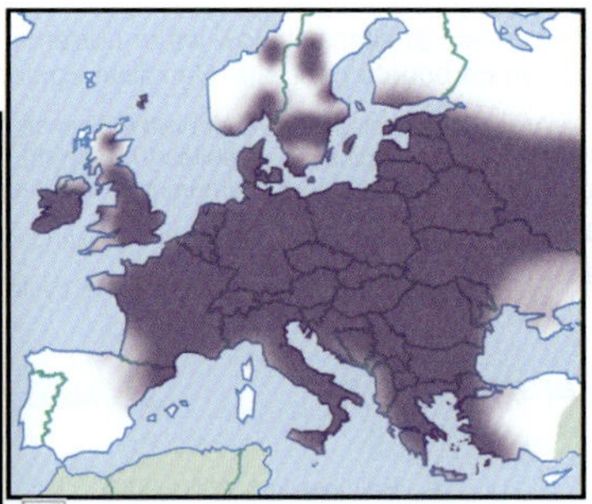

Europe
Distribution in Spain and Russia uncertain.

Shell flat, small, transparent, upper side slightly depressed, 3-3.5 whorls, aperture circular, umbilicus wide and open, more than 1/3 of shell diameter.

Frequently in lakes, creeks, springs and ponds, rarely in temporary waters. Prefers eutrophical habitats with rich vegetation, muddy substrate, well oxygenated water, tolerates up to 0.5 % salt, exceptionally also in subterranean waters. In Switzerland up to 1500 m.

In Poland copulation takes 1-2 hours and begins in spring, it is repeated various times, eggs (0.2-0.3 mm) are deposited in cocoons containing 1-4 eggs, attached to aquatic plants, juveniles (diameter 0.32-0.43 mm, 0.5-0.8 whorls) hatch after 8-11 days at 24-20 °C (25 days at 12 °C), 2 mm size and 2 whorls are reached after 180 days in the laboratory, female maturity is reached at slightly more than 2 mm diameter, at 2.9 mm the gonads of almost all snails contain oocytes, in lifetime some 50 cocoons containing 150 eggs are laid, life span 1-3 years in the laboratory.

Threatened by habitat destruction due to pollution and drainage. In some regions considered as Vulnerable or Endangered.

Germain 1931: 676, Myzyk 2002 (life cycle).

Valvata hirsutecostata Poliński, 1929

Valvata hirsutecostata Poliński, 1929 in Poliński 1929: 138

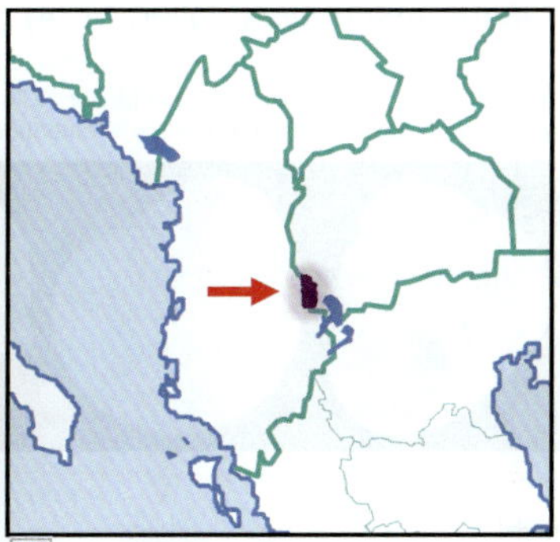

5-7 x 7-9 mm Rep. Makedonija: Lake Ohrid, near Ohrid Museum Budapest HNHM Lake Ohrid

Shell light, with fine and sharp ribs, rapidly increasing whorls, first whorls flat, 2.5-3 whorls with deep suture, last whorls descending, umbilicus wide, 1/4-1/3 of shell diameter.

Sublitoral zone, 20-30 m depth.

EU Red List: Vulnerable VU. Not very abundant where present. Potentially threatened by eventual changes of water quality (caused by deforestation and urban waste waters) or temperature.

Valvata klemmi Schütt, 1962

Valvata klemmi Schütt, 1962 in Schütt 1962: 158

Shell yellowish brown, translucent, with four strong keels, of which usually two are well visible at the periphery, and about 20 prominent and regularly arranged radial ribs on the penultimate whorl, with 5-10 weaker but also regularly arranged riblets in between, apertural margin sharp, umbilicus narrow and deep. Operculum thin, yellowish brown, with 7-9 whorls and a central nucleus, margin with edges to fit the edges of the aperture provoked by the keels.

9-10 x 9-10 mm

In the littoral zones of large lakes, around 5 m below lake surface, on hard substrate, sand and mud.

EU Red List: Endangered EN. Very rare, shortly before extinction, threatened by habitat destruction. Was thought to be extinct in 2006, but found alive in 2007.

Reischütz & R. 2003: 29, Albrecht, Ha. et al. 2009 (Lake Trichonída).

W Central Greece

Valvata macrostoma Mörch, 1864

Valvata macrostoma Mörch, 1864 in Mörch 1864: 321

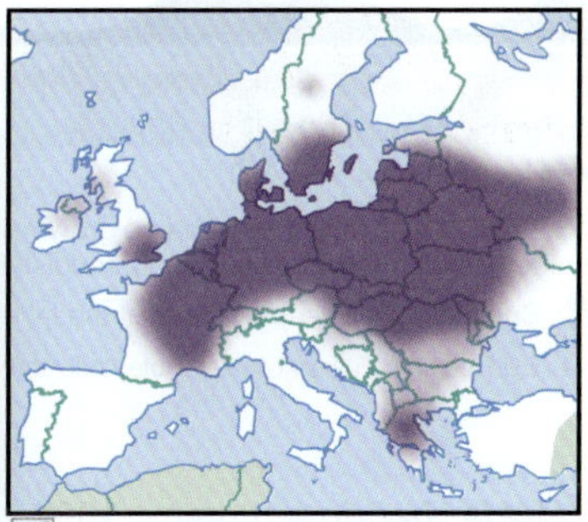

2-3 x 3.5-5 mm N Germany: Ratzeburger See near Rothenhusen Europe

Shell with fine and sharp ribs, rapidly increasing whorls, first whorls flat, last whorls descending, umbilicus wide, 1/4-1/3 of shell diameter.

Temporary water bodies in river floodplains, in the north also at swampy lake margins and drainage ditches in marshland, well vegetated habitats. Usually frequent if present. In Switzerland up to 1000 m.

Reproduction sexual, eggs are laid in cocoons (7-12 eggs per cocoon, maximum 35), female maturity is reached after 2-4 months (at 2.5-3 mm diameter and 2.5-3 whorls), life span 1-2 years, in the laboratory longer at unfavourable life conditions. Snails produce up to 150 cocoons during their lifetime.

Threatened due to its rare occurrence combined with increased destruction and eutrophication of moist habitats all over central Europe and Britain. Critically Endangered in Switzerland, Czech Republic and Germany, Endangered in Austria and the Netherlands, Vulnerable in England and Poland.

Falkner 1990: 120, Vogt et al. 1994: 53, Turner et al. 1998: 93, Kerney 1999: 28, Tappert & Korniushin 2001: 16 (*pulchella*), Glöer 2002: 187, Myzyk 2004, Pintér & Suara 2004: 41 (*pulchella*), Faltýnkova et al. 2008 (central Finland, determination requires confirmation, probably this was *piscinalis*), Jungbluth & Knorre 2009: 10.

Recognized as an independent species since the 1980s, often called *Valvata pulchella*.

Valvata piscinalis (Müller, 1774)

Nerita piscinalis Müller, 1774 in Müller 1774: 172
Valvata ambigua Westerlund, 1873 in Westerlund 1873: 439

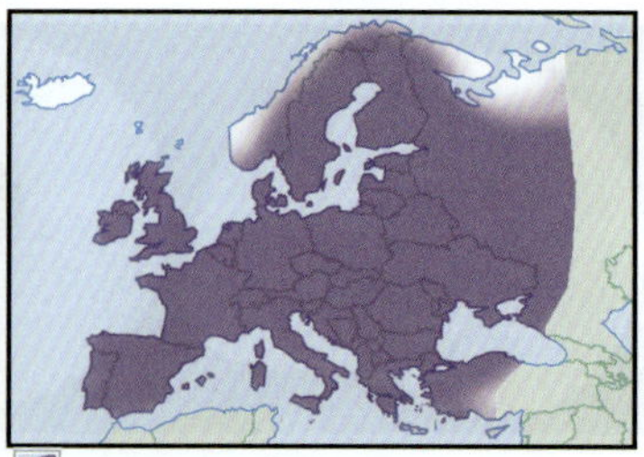

3.5-5 x 4-6 mm, locally variable — Germany: Schleswig-Holstein, Passader See — W Eurasia or Palearctic

Shell whitish yellow to greenish grey or brown, helicoid with narrow umbilicus, shell height variable, aperture almost circular, slightly angulated, umbilicus slightly wider in the Alps, partly covered by the reflected columellar margin.
Animal dirty yellow and almost transparent, tentacles long, branchial apparatus at the right side of the head, eyes big and located at the internal side of the bases of the tentacles.

Clear standing and slowly moving waters, up to 80 m deep, commonly 3-10 m, on muddy or silty substrate. In Switzerland up to 1000 m. Frequently burrowed in mud so that only feeler and tentacles are visible. Requires high oxygen content, intolerant against humine substances, but indicates eutrophic conditions in waters of good quality. Tolerates soft water, maximum salt concentration 0.4 %. Feeds on detritus. Important diet of fishes and anatid water birds.

Reproduction sexual, in Poland from April to July, elsewhere until September. Eggs are laid in cocoons (15-30, maximum 75 eggs per cocoon), juveniles hatch after 15-30 days (12-16 days in France) and appear from July to October, female maturity is reached after 45-65 days (at 4 mm diameter, 3.5 whorls), in less favourable food conditions slightly later at smaller shell size. Age between 5 and 21 months, longer life was observed under less favourable food conditions in the laboratory. Snails produce up to 100 cocoons during their lifetime.

A common species, but in many parts declined or extinct due to water pollution. Vulnerable in Austria, decreasing in Bayern. In 2010 population density in Moldova has dropped down to 0.5 % of its maximum values in the 1980s (Coadă & Welter-Schultes 2011: 42).

Germain 1931: 670, Prince 1967, Clarke 1981: 44, Falkner 1990: 120, Vogt et al. 1994: 54, Turner et al. 1998: 95, Kerney 1999: 29, Glöer 2002: 190, Myzyk 2007. Introduced to North America (Lake Erie to S Quebec).

Glöer & Diercking 2010: 79 claimed that they were able to distinguish two species in Hamburg without intermediate forms (*piscinalis* and *ambigua*), but they did not substatiate their statements with published results of taxonomic studies.

Valvata relicta (Poliński, 1929)

Without gender agreement: *Valvata relictus* (Poliński, 1929)

Gyraulus relictus Poliński, 1929 in Poliński 1929: 164

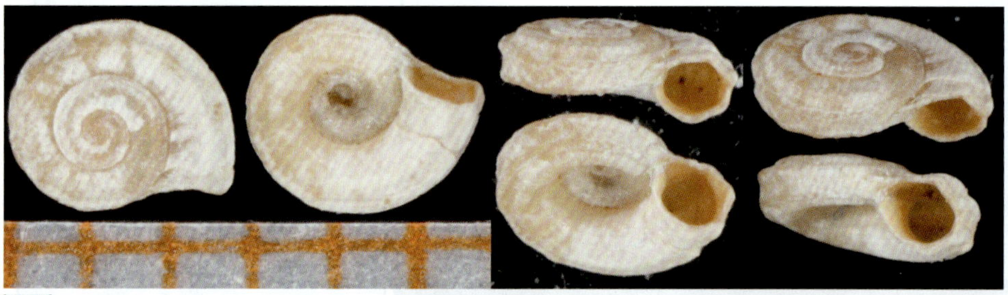

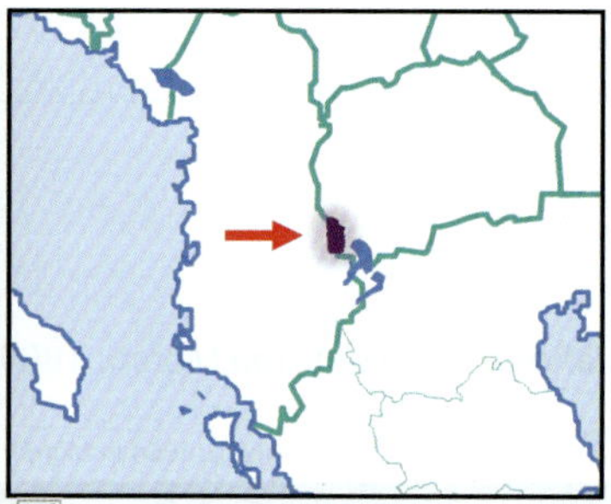

1.2-1.5 x 3-4 mm — Albania: Lake Ohrid, Tushemisht near Pogradec — Museum Budapest HNHM — Lake Ohrid

Shell horny whitish with coarse growth lines, solid and with four keels in regular distance between upper and lower side, 2.5-3 convex whorls with well marked suture, aperture inside with reddish hue, umbilicus wide, 1/3 of diameter.

Sublittoral and deeper zones of the lake, 35-100 m depth, on various kinds of hard substrate like *Dreissena* shells and stones.

EU Red List: Vulnerable VU. Potentially threatened by eventual changes of water quality (caused by deforestation and urban waste waters).

Valvata rhabdota Sturany, 1894

Valvata rhabdota Sturany, 1894 in Sturany 1894: 381, Taf. 19

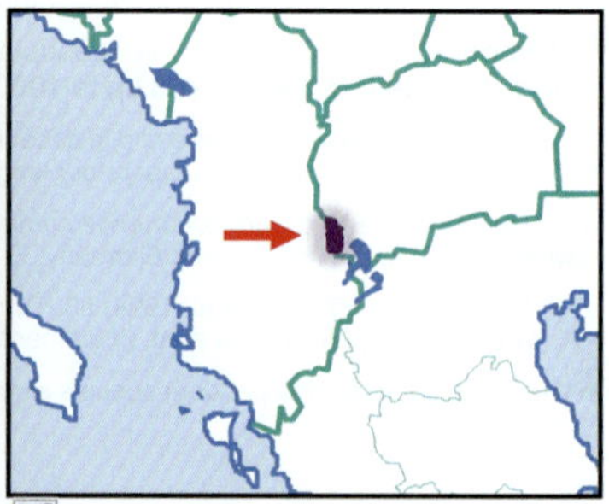

4.5-7 x 6-8.5 mm — Rep. Makedonija: Lake Ohrid, near Ohrid — Lake Ohrid

Shell pale yellow or white, densely and regularly ribbed, 2.5-3 rounded whorls with deep suture, last whorl broader and descending near aperture, aperture nearly circular, margin sharp and connected at parietal side, umbilicus wide.

Shallow waters of the lake in the sublitoral zone, 5-30 m depth. Usually abundant where present.

EU Red List: Near Threatened NT. The species is larger than *V. relicta* and can cope better with artificial sediment loads, so slightly less threatened.

Poliński 1929: 139, Hauswald et al. 2008.

Valvata saulcyi Bourguignat, 1853

Valvata saulcyi Bourguignat, 1853 in Bourguignat 1853: 68

Shell horny brown, almost smooth, apex blunt, 3-4 convex whorls with deep suture, last whorl rounded and large, apertural margins approach each other closely, nearly connected, umbilicus wide.

Westerlund 1886: 137 (reports for Syria and Egypt), Bodon et al. 1995: 44 (Sicily).

1.7-2.3 x 2.5-3.5 mm

Italy: Sicily, Anapo river deposits near Siracusa

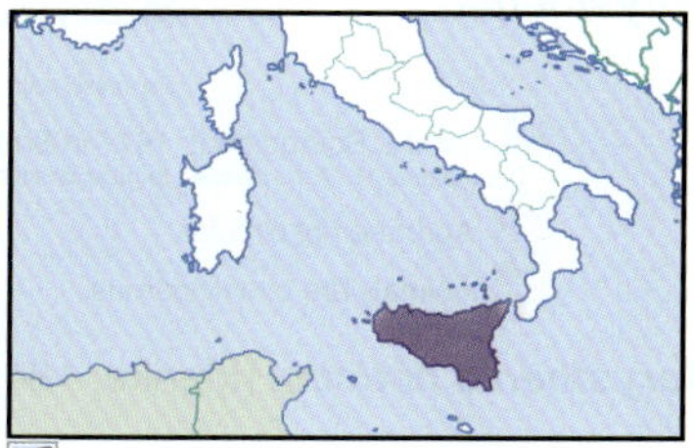

Syria to Egypt (?), Italy (Sicily)

Massimo Prodan

Valvata sibirica Middendorff, 1851

Valvata cristata var. *sibirica* Middendorff, 1851 in Middendorff 1851: 299 (*Valvata cristata* var. *Sibirica*)

Shell horny greenish, usually shiny, densely and finely ribbed, upper side flat, 3-3.5 convex whorls, last whorl inflated, aperture large and circular, umbilicus wide, 1/3 of shell diameter. Shell not very variable.
Related to *V. macrostoma*, but shell always flat, periostracum usually lighter, umbilicus wider.

In standing and slowly running tundra waters, at stones or at the bottom. Prefers in Norway pH around 7.5-9.5.

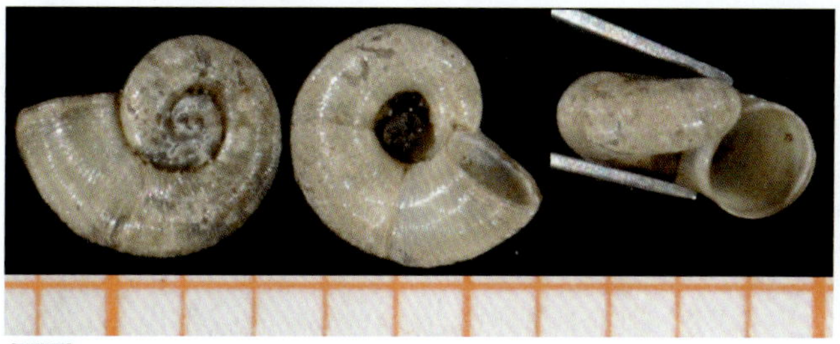

2-2.5 x 3-4 mm

N Russia: Tajmyr, nearKhatanga

N Siberia to central Sweden

Glöer 2002: 188.

Valvata stenotrema Poliński, 1929

Valvata strenotrema Poliński, 1929 in Poliński 1929: 135

4.5-6 x 4-5.5 mm

Rep. Makedonija: Lake Prespa, Otesevo (determination uncertain)

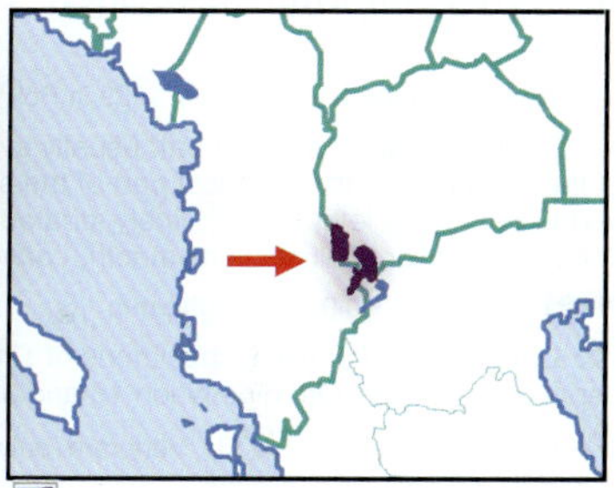

Lake Ohrid and Lake Prespa

Shell yellow to almost white, finely striated, 4-5 whorls, aperture nearly circular, height approximates shell height, margin connected at parietal side and apically slightly angular, umbilicus narrow and deep, slightly covered by the reflected columellar margin.

Shallow to sublitoral zones of Lake Ohrid, 5-50 m depth, also at sandy beaches.

EU Red List: Near Threatened NT, only Lake Ohrid. Endemic species of either Lake Ohrid alone, or both Lake Ohrid and Lake Prespa.

Sturany 1894: 382, Polinski 1929: 135, Dhora 2004: 141, Hauswald et al. 2008 (only Lake Ohrid), Fehér & Erőss 2009.

Valvata studeri Boeters & Falkner, 1998

Valvata studeri Boeters & Falkner, 1998 in Boeters & Falkner 1998: 115, Taf. 14, Taf. 16
Valvata pulchella Studer, 1789 in Coxe 1789: 381 (not this species)
Established without description, but with a reference to "Geofr. Ner. 4", junior synonym of *Valvata piscinalis* Müller, 1774. Studer 1820: 91 had used the name *Valvata pulchella* for the present species and until 1998 the name *Valvata pulchella* was commonly used in the 1820 sense. A name *Valvata pulchella* Studer, 1820 was not established, Studer 1820 just used the 1789 name for a misidentified species.

Shell flat to cone-shaped, dark brown and usually very shiny, smooth, last whorl weakly striated, 3-4 whorls with deep suture, umbilicus moderate, 1/7 of diameter.
Body of the animal brown, often bluish iridescent.

2.3-3.5 x 2.8-5.2 mm

Flat fenlands of northern and western Alps promontories. Eutrophic old river waters, swampy lake margins. Intolerant against drought, requires high oxygen content. In Switzerland in 350-750 m altitude.

Alps

Scattered and rare. Threatened by continuous destruction of moist habitats. Critically Endangered in Switzerland, Bayern and Tirol (Austria), Endangered in Austria.

Falkner 1990: 120, Turner et al. 1998: 94 (*pulchella*), Glöer 2002: 195.

Little is known about this species and its distribution. Prior to 1998 known under the name *Valvata pulchella*, this name was also used for *V. macrostoma*.

Borysthenia

Borysthenia Lindholm, 1914
in Lindholm 1914 ["1913"]: 167
Expressly established as a new replacement name for

Type species *Valvata jelskii* Crosse, 1863
(original designation by Bourguignat 1877)
(= currently *Borysthenia naticina* (Menke, 1845))

Jelskia Bourguignat, 1877 in Bourguignat, 1877: 92 not *Jelskia* Taczanowski, 1871 (Araneae)
Bourguignat 1877: 92 claimed to have described *Jelskia* in 1867, but there is no evidence.

Aperture not circular.

Animals are ovoviviparous.

Borysthenia naticina (Menke, 1845)

Valvata naticina Menke, 1845 in Menke 1845: 129

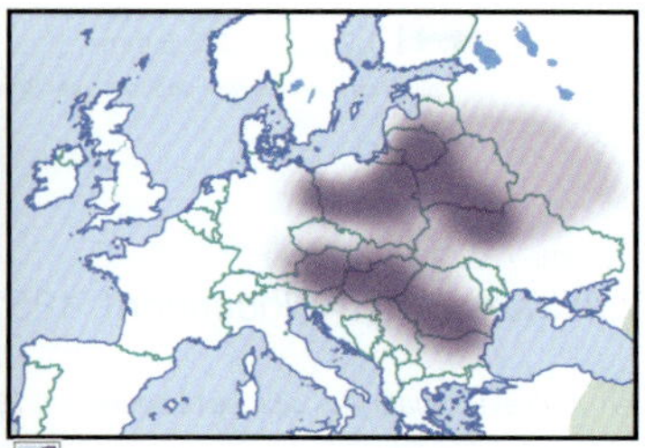

2.5-4.5 x 3.5-5.5 mm

Poland: near Gorzów Wielkopolski

Central to E Europe

Shell horny yellowish, solid and shiny, finely striated, 3.5-4 moderately convex whorls, last whorl strongly inflated, aperture very oblique, pointed at apical side, basal margin straight, columellar margin not much reflected, umbilicus open, narrow and partly covered by the oblique columellar margin.

At the margins of large rivers and lakes, on fine sandy substrate.

Extinct in Austria, Critically Endangered in Germany and Poland.

Westerlund 1886: 136, Grossu 1986: 189, Glöer 2002: 196.

Acroloxus

Acroloxus Beck, 1838
in Beck 1837: 124 *Ancylus (Acroloxus)*

Type species *Patella lacustris* Linnæus, 1758
(subsequent designation by Herrmannsen 1846 [01-09-1846]: 16)

ICZN Opinion 636: type species designated by Herrmannsen 1846.

Acroloxus improvisus Poliński, 1929

Acroloxus improvisus Poliński, 1929 in Poliński 1929: 166

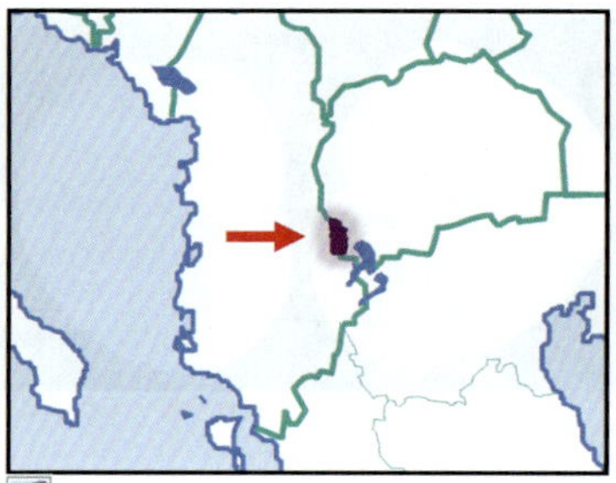

Lake Ohrid

Shell colourless, anterior part usually symmetrical, posterior part not, its posterior tip is bent towards the right of the median line, the broadest part of the shell tends to be located anterior to its middle. Apical region pointed and turned to the left (30-45°), finely striated with 30 longitudinal lines, connected to the rest of the shell by a slight concavity. Lower shell section smooth, convex, periostracum follows all around the aperture.

2-2.5 x 3-4 x 0.9-1.4 (height) mm

Lives on dead and live specimens of *Dreissena polymorpha*, which predominate the bottom material in the depth zone of Lake Ohrid between 18 and 35 m.

EU Red List: Vulnerable VU. Very rare, only at few localities. Potentially threatened by changes in water quality and if a competing species is introduced.

Hubendick 1960.

Acroloxus lacustris (Linnæus, 1758)

Patella lacustris Linnæus, 1758 in Linnæus 1758: 783
ICZN Opinion 363: as interpreted by Müller 1774: 199-200.

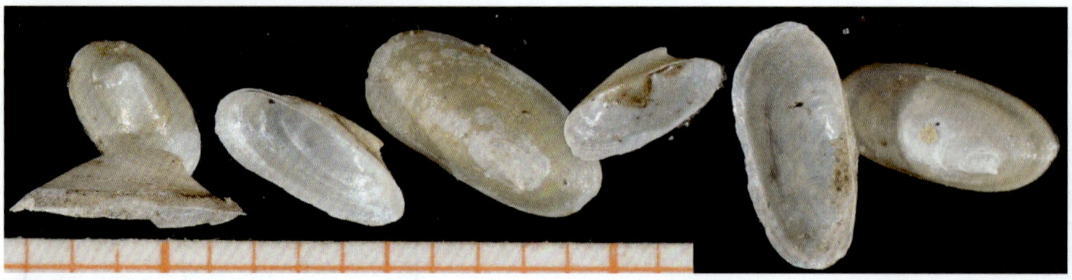

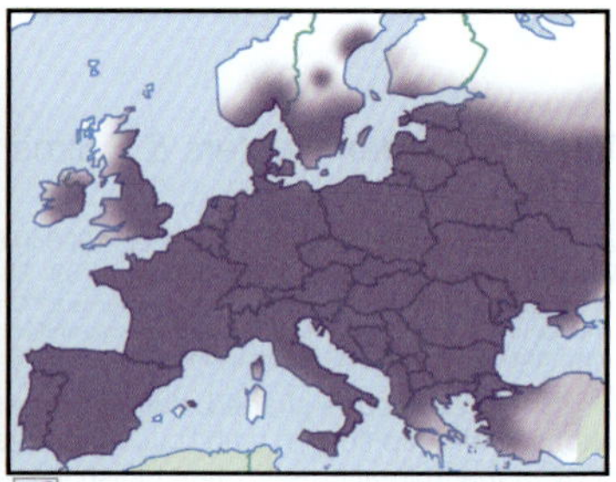

Up to 3-3.5 x 4-7 x 1.2-2 (height) mm

Germany: Schleswig-Holstein, Lensahn

Europe

Shell yellowish grey to light brown, thin, translucent, flat, peak sharp and bent to the left and posterior side.

Standing or (more rarely) slowly running clean waters with rich vegetation and rich oxygen content. Usually found at aquatic plants, submerged timber and lower sides of swimming leaves, also on bivalve shells. Feeds on algae cover and microorganisms. Calciphile, but tolerates soft water, pH tolerance 6-8.9. Sensitive against water pollution. Mainly in lowlands, in Switzerland up to 1000 m.

Transparent eggs (diameter 1 mm) are laid in 2 mm long ovoid capsules (difficult to count, usually 4-12 per capsule), juveniles hatch after 20-25 days in France.

Widespread in Europe, but declining due to continuous destruction of moist habitats and water pollution. Vulnerable in Switzerland, Bayern, Baden-Württemberg and Austria.

Moquin-Tandon 1855: 488, Germain 1931: 552, Glöer 2002: 198.

Acroloxus macedonicus Hadžišče, 1959

Acroloxus macedonicus Hadžišče, 1959 in Hadžišče 1956: 99 Date was not 1956 (Radoman 1962: 69)

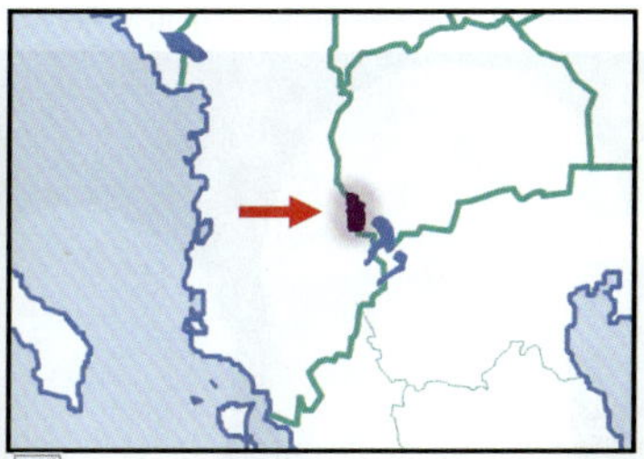

Lake Ohrid

Up to 1.4-1.8 x 2-2.5 x 0.6 (height) mm

Rep. Makedonija: Lake Ohrid, near Sv. Kaneo

Museum Budapest HNHM

Shell variable in different populations, aperture usually obliquely elliptic (anterior end to the left and posterior end to the right of the median line), apex much blunter than in *A. improvisus*, points slightly to the left, distance from the margin is 1/5 of the shell length or less, apical region with a fine sculpture of 30-35 longitudinal ridges partly crossed by extremely fine transverse ridges, both can also be present on the lower section of the shell. In some populations many shells have a strong radiating macrosculpture on the lower shell section, consisting of down to 15-5 ridges extending to the margin which then becomes wavy, like in coarse marine pectinid or cardiid shells. Other populations have mostly smooth shells and if present, the ridges do not affect the apertural margin.

Restricted to habitats providing limestone rocks and boulders 30-50 cm below lake surface. The animals are usually found below these rocks. Such habitats are only present at certain sections of the lake shore, so that populations are isolated from each other.

EU Red List: Critically Endangered CR. Endemic species of Lake Ohrid, restricted to very limited stretches of the shore zone, where the animals can be abundant. Threatened by changes in the sediment composition of the lake shore in response to erosion of the adjacent slopes, caused by fires.

Hubendick 1960.

Acroloxus tetensi (Kuščer, 1932)

Ancylus fluviatilis tetensi Kuščer, 1932 in Kuščer 1932: 61

Shell small, thin and fragile, apex pointed and strongly bent to the left side.

Lives in cave waters with high calcium content.

EU Red List: Vulnerable VU. Only known from two localities, potentially threatened by water pollution.

2-3 x 3-4 mm

Kuščer 1932 gave no size. Estimated after Massimo Prodan's specimen.

Slovenia: Rakov Škocjan near Postojna

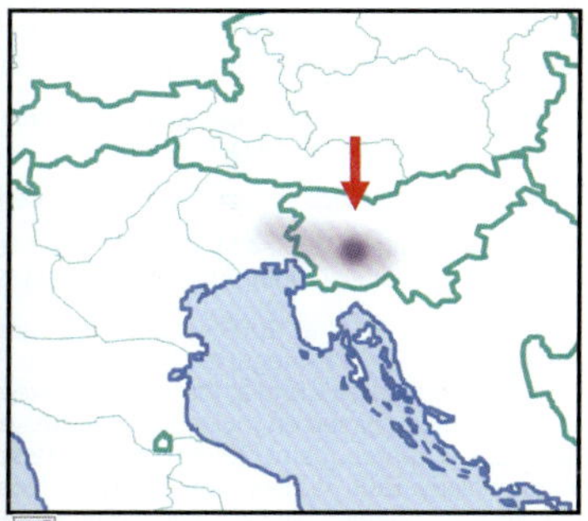

Slovenia

Massimo Prodan

Ancylus

Ancylus Müller, 1773 in Müller 1773: [25]

Type species *Ancylus fluviatilis* Müller, 1774 (fixed under the plenary powers)

Appeared without species. Müller 1774: 199 added *A. lacustris* and *A. fluviatilis*. ICZN Opinion 363 [sic]: type species designated under the plenary powers. Year was given as 1774, this was incorrect, must be 1773. Anton 1839: 26 gave as type *A. fluviatilis*.

Small limpet-like shells with a blunt apex. The apex and often the shell show coarse radial ridges.

Ancylus fluviatilis Müller, 1774

Ancylus fluviatilis Müller, 1774 in Müller 1774: 201

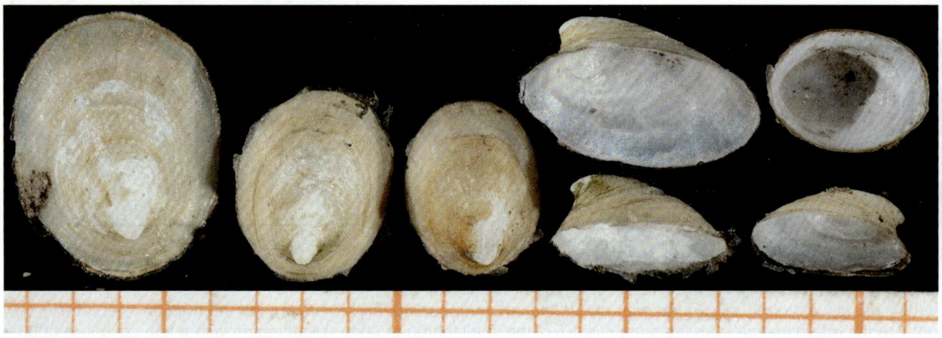

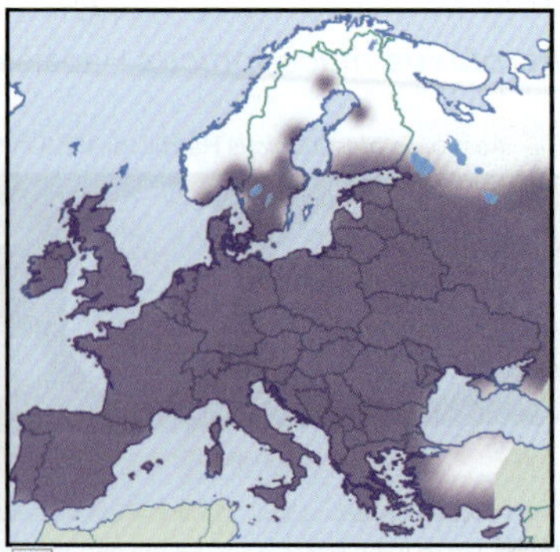

N Africa, Europe, SW Asia

3.5-8 x 2-3.5 mm

Germany: Rheinland-Pfalz, Nahe river near Staudernheim

Portugal: Algarve, near Castro Marim
Valter Jacinto

Austria: Oberösterreich, near Zwettl an der Rodl
Jan Steger

Shell light yellowish to reddish brown, thin, translucent, with reticulate sculpture, higher than *Acroloxus* and *Ferrissia*, apex blunt when seen from above, more pointed from the side, bent backwards and very weakly to the right side.
Animal grey with black dots near the head, almost entirely covered by the shell, tentacles triangular with eyes at their base, genital pore and pneumostome very small and located at the right side.

On and under stones and plant leaves in oxygen rich and clear quick-flowing waters, from small creeks to rivers, also in springs and at lake margins with wave movement. Feeds on algae cover. Not in rivers which modify their beds significantly after floods, not in seasonal mountain creeks. Disappears from waters where too much organic material is deposited on the substrate, or where silt and sand covers the stones where the species feeds on. The species is able to recolonise lost territory, and is a good indicator of the pollution status of waters. Preferred pH in Norway 6.6-7.8. Lives also in higher altitudes, up to 800 m in E Germany, 1000 m in Switzerland, 2300 m in Bulgaria.

Eggs are laid in 3 mm sized and amber coloured capsules containing 9-12 eggs each, 7-10 capsules per individual, fixed to stones or other submerged solid surfaces, juveniles hatch after 24-27 days in France.

Water pollution has reduced its range in central Europe, locally it also affected populations in Britain.

Moquin-Tandon 1855: 484, Germain 1931: 548, Hubendick 1960: 525, Schütt 1982, Økland 1990, Falkner 1990: 130, 132, Pujante Mora et al. 1998, Pfenninger, St. et al. 2003, Ruiz Ruiz et al. [2006], Albrecht, Tr. et al. 2006.

Pfenninger, St. et al. 2003 suggested a complex of 4 reproductively isolated species (Clade 1 = N Portugal to N Europe, 2 = Portugal, 3 = Canary Islands to Turkey, 4 = S Portugal to Croatia; the region between Bosnia, Greece and W Turkey was not studied), based on molecular analyses and shell morphology.
Soldatenko & Starobogatov 2004 proposed 12 species in W Eurasia based on genital morphology. Both studies were literally independent, the researchers used different names for presumably the same S European species and did not cite each other.
Soldatenko 2009 and Albrecht, Tr. et al. 2006 did this and showed at least awareness that other research teams were working on the same genus.
In a study that did not focus on the European ranges Albrecht, Tr. et al. 2006 used once again different names (clade 1 = *fluviatilis*, 2 = A, 3 = B, 4 = C) and included a few additional localities.
Soldatenko 2009 gave detailed information on the shell morphology of two Mediterranean species but was naturally not able to compare these results (and align the names) with those of Pfenninger et al. 2003 who had missed to present the results of their morphological analyses in a form that enabled other research teams to reproduce their studies and confirm or falsify their conclusions.

It seems likely that the *A. fluviatilis* complex contains various reproductively separated lineages. Currently too little is known, the Russian and central European research teams are still largely disconnected from each other and have presented incomplete studies in terms of range analyses. The Russian researchers presented their results certainly in a more useful form, but seemed to have based their results on very little material from the Mediterranean region, while the central European research teams seemed to have failed to present reproducible results and detailed data in their morphological analyses, which would enable others to continue these studies and to determine collected material.
Ancylus fluviatilis is a very frequent species in southern Europe, it is the most frequent freshwater mollusc in large parts of the Iberian peninsula, so studying the nature of the European diversity more in detail should not be limited by the availability of the animals.

Ancylus lapicidus Hubendick, 1960

Ancylus lapicidus Hubendick, 1960 in Hubendick 1960: 512, Pl. 2-3

Shell with colourless and transparent periostracum, apical region almost cornet-shaped with 20 distinct longitudinal ridges, there are also weak traces of transversal sculpture, adult shell looks scalarid with a varying number of concentric eaves-like outgrowths between apical region and aperture, like in *A. scalariformis*, aperture almost recularly elliptical.
The main difference to *A. scalariformis* is the prominently sculptured apical region. When the eaves-like outgrowths are disregarded, the general shape of the shell is even, not step-shaped.

1.8-1.9 x 2.3-2.6 x 0.9-1 (height) mm

At the W shore of Lake Ohrid under loose limestone rocks, 30-50 cm below water surface.

EU Red List: Endangered EN. Not abundant, threatened by increasingly eutrophic conditions and pollution.

Hubendick 1960, Dhora & Welter-Schultes 1996: 173. Not mentioned by Soldatenko & Starobogatov 2004.

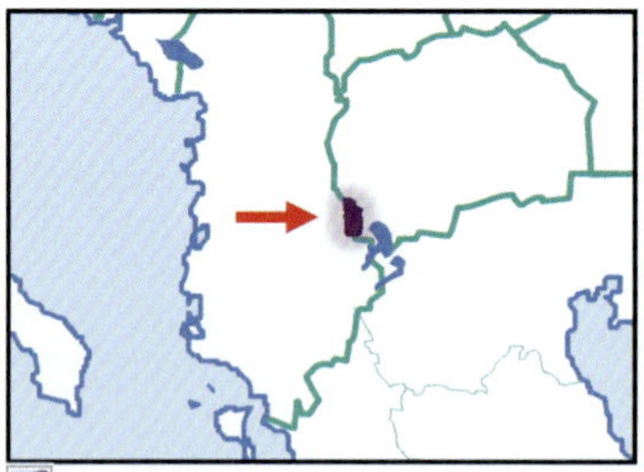

Lake Ohrid

Ancylus scalariformis Stanković & Radoman, 1953

Ancylus scalariformis Stanković & Radoman, 1953 in Stanković & Radoman 1953: 118
Original spelling was not *scalariniformis*.

Periostracum pale brown, apical region cornet-shaped, points posteriorly, bluntly pointed, with 20 weak longitudinal lines which are usually worn and only weakly visible in adult shells. Between apical region and aperture there are a varying number of eaves-like concentrical outgrowths making the shell looking scalarid, their width increases with distance from apex, they are narrower than in *A. lapicidus*, rarely exceeding 10 % of the shell. Immediately behind each scale the shell surface often slopes steeply ventrally, and then continues at a less steep angle, producing a more or less step-like shell shape.
Differs from *A. lapicidus* in its less sculptured apical region, without prominent longitudinal ridges and without any transversal sculpture. When the eaves-like outgrowths are disregarded, the general shape of the shell is step-shaped, not even.

2.3-4.5 x 3.5-6 x 3 (height) mm

Lives on dead and live specimens of *Dreissena polymorpha*, which predominate the bottom material in the depth zone of Lake Ohrid between 18 and 35 m.

EU Red List: Vulnerable VU. Potentially threatened by increasing pollution and changes of water temperature.

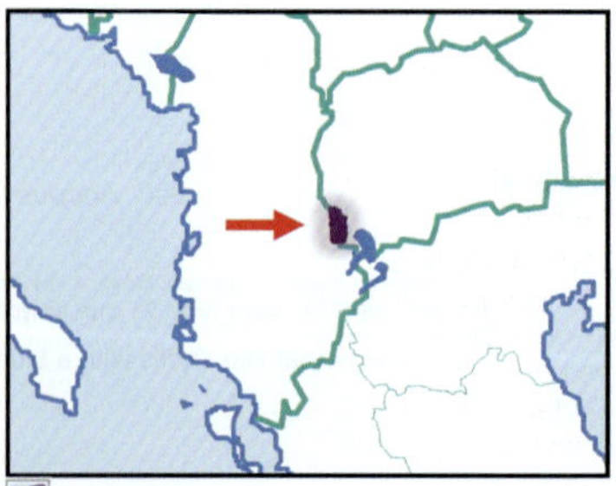

Lake Ohrid

Hubendick 1960, Dhora & Welter-Schultes 1996: 173.

Ancylus tapirulus Poliński, 1929

Ancylus tapirulus Poliński, 1929 in Poliński 1929: 165

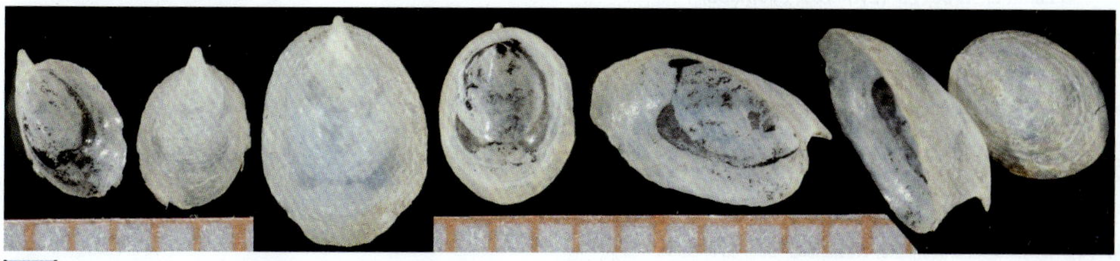

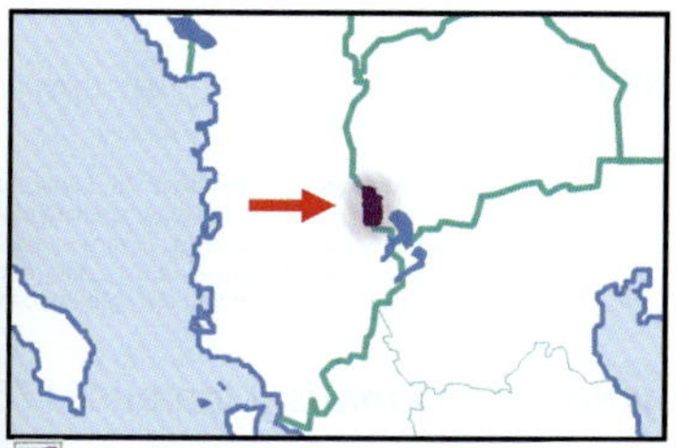

Lake Ohrid

2-4.5 x 3-6 x 1.5-3 (height) mm

Rep. Makedonija: Lake Ohrid, Sv. Naum — Museum Budapest HNHM

Hubendick 1960, Dhora & Welter-Schultes 1996: 173.

Periostracum yellowish brown, cracks off when the shell falls dry, apical region cornet-shaped and slender, like a tapir snout, longitudinally striated with 20-25 lines, usually worn and difficult to trace in adults, apex blunt and smooth. Apical region is usually delimited from the rest of the shell by a furrow or constriction, shell almost smooth with very weak concentric growth lines, sometimes with a prominent scalarid-like growth line between apical region and aperture.

Lives on dead and live specimens of *Dreissena polymorpha*, the main bottom material in the depth zone of Lake Ohrid between 18 and 35 m.

EU Red List: Endangered EN. A rare species, threatened by pollution and increasingly eutrophic conditions caused by intensive agriculture contaminating Lake Prespa.

Ferrissia

Ferrissia Walker, 1903 in Walker 1903: 15 *Ancylus (Ferrissia)*

Type species *Ancylus rivularis* Say, 1817 (original designation)
(= currently *Ferrissia rivularis* (Say, 1817) from USA and Canada)

N and central America, Europe, E Asia, Africa, Australia

Ferrissia fragilis (Tryon, 1863)

Ferrissia isseli (Bourguignat, 1866), *Ferrissia clessiniana* (Jickeli, 1882), *Ferrissia wautieri* (Mirolli, 1960)

Ancylus fragilis Tryon, 1863 in Tryon 1863: 149 Type locality in California, USA.
Ancylus isseli Bourguignat, 1866 in Bourguignat 1863-1868: 214 Type locality in N Egypt.
Ancylus clessinianus Jickeli, 1882 in Jickeli 1882: 366 (*Ancylus Clessinianus*) Type locality in N Egypt.
Watsonula wautieri Mirolli, 1960 in Mirolli 1960: 154 Type locality in France.

Germany: Nordrhein-Westfalen, Ascheberg-Herbern

3-4 x 1.4-1.7 x 0.8-1.2 (height) mm, up to 6 mm long

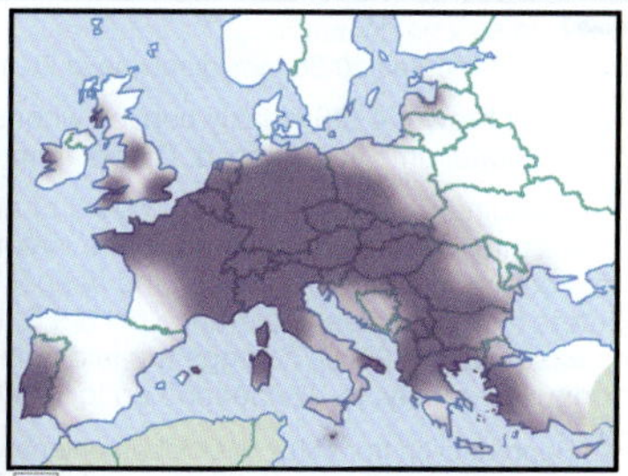

Originally N America (USA, S Ontario, S Quebec, SW British Columbia), introduced to Europe and E Asia

Clarke 1981: 222, Richardot 1979, Spyra 2008.

Shell oval, transparent, apex blunt, broad and turned to the right side slightly behind the centre in the posterior section.
Differs from *Acroloxus lacustris* by its apex being bent to the right side.

Wide ecological amplitude, usually in well vegetated stagnant or slowly moving waters.
Lives at submerged water plants in silent shallow waters, tolerates hypertrophic conditions, perhaps also seasonal desiccation, up to 1 m deep. In Switzerland up to 600 m.

In the summer juveniles need only 11 days from hatching to first egg deposition (at 1.7 mm size), in the winter this takes longer, growth need 80 days and eggs are deposited when the size is 2.9 mm.
Shell formation: juveniles and adults of permanent freshwater habitats have an ancyloid shell. During waterless periods in temporary water bodies a calcareous septum covers partially the aperture in submature animals (septifer form), so they can survive periods of drought. When the conditions are favourable again, a new shell is formed around the aperture, and then the animals reproduce. Such forms are rarely observed in central Europe.

Cryptic invader of European and East Asian freshwater ecosystems, was until 2006 known in Europe under various names. Current classification after Walther et al. 2006 (2 papers), Beran & Horsák 2007, Cilia 2009 (Malta).

Galba

Galba Schrank, 1803 in Schrank 1803: 262, 285

Type species *Buccinum truncatulum* Müller, 1774 (fixed under the plenary powers)

ICZN Op.1896 (1998): type species designated under the plenary powers. Genus contained originally only *Galba pusilla* Schrank, 1803 (in Schrank 1803: 285), types of *Galba pusilla* are lost (Starobogatov 1997 in Bulletin of Zoological Nomenclature 54: 19-21). It was not necessary to refer this case to the Commission, Starobogatov could simply have designated a neotype for *Galba pusilla*.

Shell small and turreted with a spire of nearly the same height as the aperture, protoconch conical and smooth, suture deeply impressed, whorls well rounded, columella straight and reflected over the umbilicus, umbilicus often open and slit-like.

Lymnaeidae. Here the currently established European genus-species combinations are used. General comments see under *Stagnicola*.

Galba truncatula (Müller, 1774)

Without gender agreement: *Galba truncatulum* (Müller, 1774)

Buccinum truncatulum Müller, 1774 in Müller 1774: 130

Germany: Schleswig-Holstein, Kronsbek

5-9 x 2-4.5 mm, dwarfs 3 mm high, giants up to 15 mm

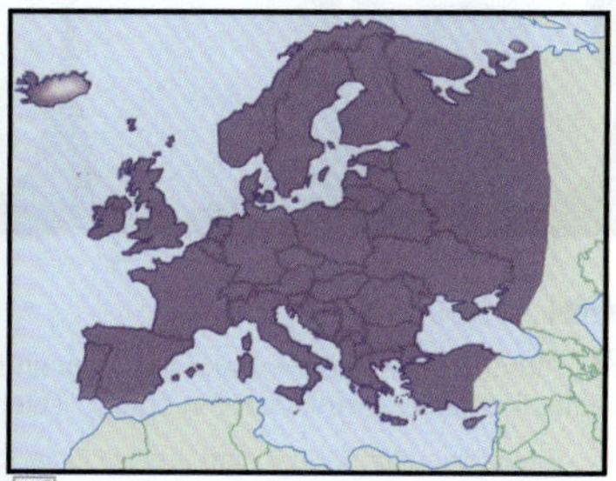

Africa, Europe, W and N Asia, Alaska, dispersed elsewhere

Moquin-Tandon 1855: 473, Germain 1931: 501, Falkner 1990: 136, Glöer 2002: 202.

Shell light, yellow or horny brownish, densely and regularly striated, 5-6 whorls characteristically step-like separated, aperture oval, height less than 1/2 of shell height, margin broadly reflected at columellar side, columellar edge straight, umbilicus open and deep. Genitalia with short phallotheca, provagina very broad.

Semiterrestrial. Typical habitats are in swampy forests and on moist meadows. On water plants and stones of small and very small water bodies, in containers of artificial water supply, in larger waters only at flat margins, often outside the water in humid habitats. Tolerates disturbed and poorly vegetated places, pH tolerance up tp 9.6. In Switzerland up to 2800 m, in Bulgaria 2000 m, Norway 1300 m. Frequent.
In Europe temporary host for the parasite *Fasciola hepatica* and other trematodes.

12-20 eggs (diameter 0.5 mm) are laid in rounded clutches (diameter 5-6 mm) and occasionally fixed to solid substrate. Up to 3 generations per year, depends on climate, life span up to 1 year.

Lymnaea

Lymnaea Lamarck, 1799
in Lamarck 1799: 75 (*Lymnœa*, *Lymnaea*)

Type species *Helix stagnalis* Linnæus, 1758
(monotypy)

Lymnœa was written on p. 75, *Lymnaea* on the Tableau. The printer did obviously not have the æ ligature and had the choice between using separate letters *ae* and the œ ligature. ICZN Opinion 495: *Lymnoea* = incorrect original spelling of *Lymnaea*.
Incorrect subsequent spellings: *Lymnaeus*, *Lymneus* , *Lymnoeus*, *Lymnea*, *Lymnoea*, *Lymnula*, *Lymnus*, *Limnaea*, *Limnaeus*, *Limneus*, *Limnoeus*, *Limnoea*, *Limneus*, *Limnea* - but not *Limnaea* Poli, 1791 (Bivalvia) which was used for a different group of species.
ICZN Opinion 495 stated that the type species was designated by Fleming 1818: 313. Montfort 1810: 259 was earlier. Lamarck 1799: 75 included only one species *Helix stagnalis*.

Lymnaea stagnalis (Linnæus, 1758)

Helix stagnalis Linnæus, 1758 in Linnæus 1758: 774
Helix fragilis Linnæus, 1758 in Linnæus 1758: 774 First Reviser unrecorded, either Westerlund 1885: 25 or earlier.

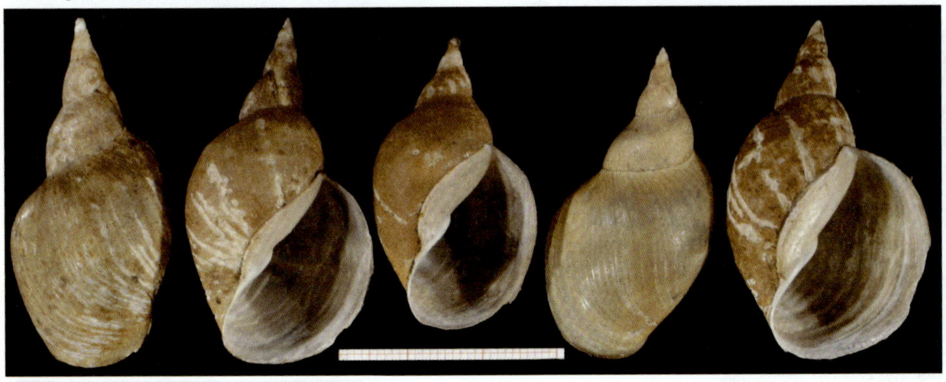

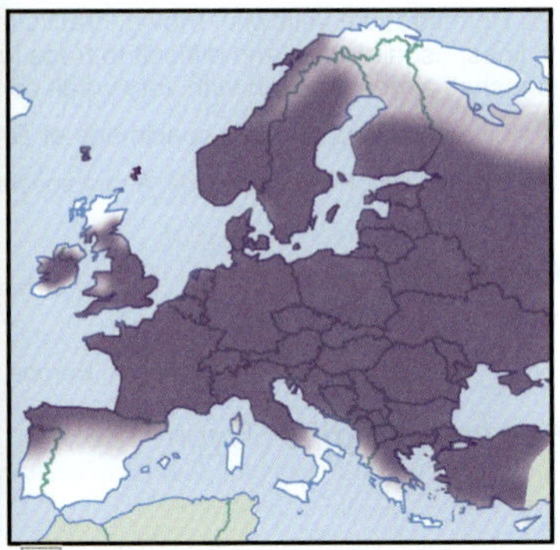

40-50 x 22-30 mm
Dwarfs in Baltic Sea 20-30 mm, giants up to 70 mm

Germany: Schleswig-Holstein, Lauenburg

Northern Hemisphere

Shell brown, 4.5-6 weakly convex whorls, upper whorls pointed, last whorl suddenly inflated, its diameter is more than a continuous increase of that of the upper whorls, umbilicus closed. Differs from other species in its large size.

Standing and slowly moving waters of any size with rich vegetation, in river plains also in temporary waters. Prefers shallow places, rarely below 20 m, juveniles may often be found outside the water on plants. Often successfully introduced into garden ponds, but in nature rarely in small closed standing waters. More common in larger habitats and deeper waters than other lymnaeids in N Ireland.
Prefers high pH values, tolerance pH 6-9.5, temporary droughts and freezing of ponds are tolerated, up to 0.7 % salt, temperature optimum 19-24 °C, mortality increases above 26 °C, no eggs are laid below 13 °C. When water temperatures decrease below 5 °C the animal does not return any more to the water surface for oxygen supply. In Switzerland up to 1700 m, in Norway 800 m.

Reproduction from May to October, one new generation per year. Oval eggs (1.5 x 2.2 mm) are laid in clutches of 30-40 mm size arranged in rows and containing 40-140 eggs, juveniles hatch after 15-30 days and have 1.3 mm size. Maturity is reached after 2 years, life span not much more than 2 years.

Moquin-Tandon 1855: 471, Germain 1931: 483, Falkner 1990: 138, Glöer 2002: 222.

Lymnaea fragilis was regarded as a separate species by Kruglov & Starobogatov 1993: 67, 71 who gave a range from entire Europe to Siberia, and found that *fragilis* was reproductively isolated from *stagnalis* by "the method of experimental hybridization". This laboratoy method alone would not provide evidence for reproductively isolated units in nature, especially if the two forms are claimed to live sympatrically in broad regions.
Glöer & Diercking 2010: 88 reported *L. fragilis* from various localities in Hamburg and argued without support of published results of scientific studies that *fragilis* did not differ anatomically, but was slightly more slender than *L. stagnalis* (photos suggested diameters of 15.0 mm (*fragilis*) and 15.4 mm (*stagnalis*) at 30 mm shell height, certainly not a statistically significant difference) and had more whorls (photos showed no differences).
Helix fragilis was considered as a synonym of *L. stagnalis* by Hubendick 1951, this classification seems to represent the current status of published research.

Myxas

Lymnaea section § 1 synonym *Myxas* Sowerby, 1822
in Sowerby 1821-1834 (part 7): p. [3] *Limnea (Myxas)*

Type species *Buccinum glutinosum* Müller, 1774
(monotypy)

Name *Myxas* appeared only in the text, not on the plate. Mentioned as a synonym of a subgroup of *Lymnaea*, which was names § 1, and had no available name. So it was not mentioned as a junior synonym of an available name (as demanded by Art. 11.6). For those who neglect this (as proposed by Kadolsky 2012: 78, this is fully accepted here because this is a generic and not a specific name), the name was made available under Art. 11.6.1 by a subsequent source which I have not researched. The rules of the Code (Art. 11.6) should be modified to cover such cases.

Myxas glutinosa (Müller, 1774)

Buccinum glutinosum Müller, 1774 in Müller 1774: 129

Shell brownish or greenish, extremely thin, very transparent and shiny, upper whorls almost flat, last whorl inflated and predominating, aperture more than 90 % of shell height, umbilicus closed. Mantle covers entirely the upper shell surface.

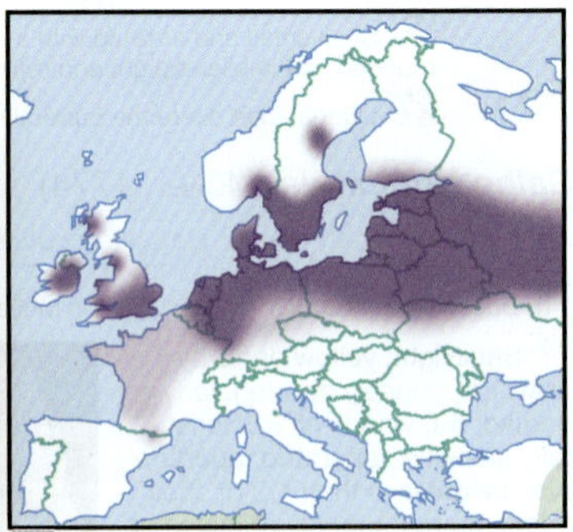

7-15 x 6-13 mm

Germany: Niedersachsen, Hohnstorf (from 1951, habitat today destroyed)

Europe

Silent creeks, old river arms, canals and lakes, restricted to lowlands, requires very clean water, usually on solid substrate, often at stones, up to 6 m depth, in shallow waters only in spring. Tolerates up to 0.3 % salt, pH tolerance 6.0-7.0.

Eggs (diameter 0.7-0.8 mm) are laid in flat clutches of 4-5 x 15-20 mm size containing 30-40 eggs (Moquin-Tandon 1855), from April to November. Snails hatched in spring lay eggs in fall and die, snails hatched in fall hibernate, lay eggs in spring and may live up to 1 year (Glöer 2002, this scenario differs slightly from Calsson 2001). Animals reach maturity in the first year at 8 mm shell height, in the following spring they grow to 13-15 mm shell height, breeding takes part in the sunny shallow water zones. In SW Finland in spring, individuals live on rocks where they feed on epilithic algae and propagate, then die. In late summer the new generation hatches and is mainly found on leaves of water lilies. During this time they grow, feeding either on epiphytic algae or on decaying leaves. In late autumn, having reached their adult size, they descend to deeper water to hibernate (Carlsson 2001).

EU Red List: Least Concern LC, certainly too weak, the study by Carlsson 2001 was apparently not consulted. Extinct in Austria, Czech Republic, most parts of Germany, last record from N Ireland from 1900. Critically Endangered in Germany and Poland, Endangered in England and Wales, Vulnerable in Ireland. In Germany most habitats have been destroyed, only very few populations survived in Mecklenburg-Vorpommern and Schleswig-Holstein. Has declined dramatically in Britain, threatened by intensive farming. In Ireland in the Royal Canal, which has been abandoned and degenerating since the 1990s, the species is expected to decline seriously if the habitat is not protected. 90 % decline in the Netherlands since 1960. 3 stations known in Norway, 7 in S Sweden, possibly still present in Denmark (last records 1949), in Finland in 21 lakes, rare in Poland, possibly still present in Estonia, Latvia and Lithuania (last records from 1942). Situation in Russia and Belarus unclear.

Moquin-Tandon 1855: 461, Germain 1931: 506, Anderson 1996: 5, Kerney 1999: 57, Carlsson 2001, Glöer 2002: 220.

Omphiscola

Omphiscola Rafinesque, 1819
in Rafinesque 1819: 423

Type species *Buccinum glabrum* Müller, 1774
(subsequent monotypy by Beck 1837 [1838]: 110)

Appeared without species, first species included by Beck 1837 [1838]. Name mentioned by Blainville 1825 (Dict. 36: 111 and Manuel de malacologie: 449, only French name Omphiscole), Deshayes 1832 (Enc. Meth. 3: 663, French name and no species included) and Menke 1830: 38 (no species included).
Gender disputable and meaning of the name unclear. If considered as a compound name it would be masculine under Art. 30.1.4.2. If not considered as a compound name it would be feminine under Art. 30.2.4. I see no evidence for a compound name. Falkner et al. 2002: 79 speculated that the name was derived from the Greek words omfalos (= umbilicus) and skolios (= torsion), transferred to a latinized feminine form. But this would have given *Omphaloscolia*, not *Omphiscola*. Beck 1837 [1838] treated it as feminine.

Omphiscola glabra (Müller, 1774)

Omphiscola glaber (Müller, 1774)
Without gender agreement: *Omphiscola glabrum* (Müller, 1774)

Buccinum glabrum Müller, 1774 in Müller 1774: 135
Lymnaea glabra var. *reticulata* Gassies, 1867 in Gassies 1867: 128, Pl. 1 (*Limnaea glabra* var. D *reticulata*)

Shell strongly cylindrical, light horny coloured, often with brownish or blackish cover, apex blunt, 7-8 moderately convex whorls, last whorl twice as high as the narrow aperture, aperture often with white lip.

Small waters rich in vegetation, swamps and standing forest waters with leaf litter, often in water with organic iron contents and low calcium contents. In Britain in small standing waters low in nutrients, with poor aquatic flora, often in waters drying out periodically.
Usually not in habitats with high molluscan diversity, usually on uncultivated land. Calciphile, pH tolerance 5.4-8.8.

9-12 x 3-4 mm, height up to 20 mm — N Germany: near Lauenburg/Elbe

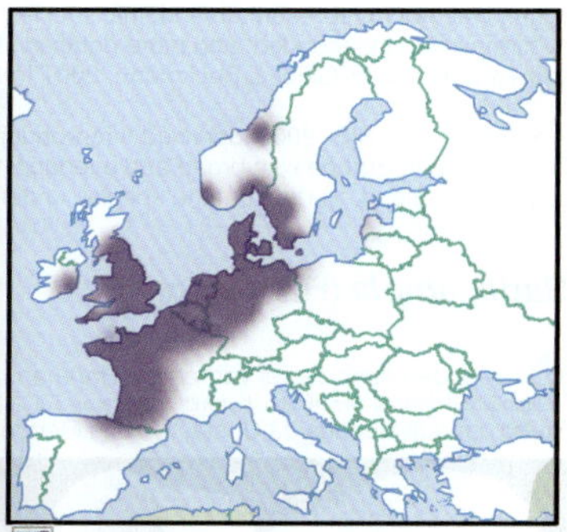

Pyrenees to Scandinavia, Britain, perhaps Latvia

Moquin-Tandon 1855: 478, Germain 1931: 504, Falkner 1990: 136, Glöer 2002: 210, Falkner et al. 2002.

Reproduction begins in May, 15-30 eggs (diameter 0.6-0.9 mm) are laid in 10-15 mm long clutches fixed to solid substrate, juveniles hatch after 15-25 days. 2 generations per year.

EU Red List: Near Threatened NT. Very scattered and rare, at many places extinct. Seriously threatened by continuous habitat destruction (drainage, intensive farming), the probably last population in Ireland was destroyed by drainage in 1980 in Wexford. Acriculturally induced eutrophication has certainly destroyed populations. Disappeared widely from urbanized areas such as London. 30 % of the habitats in the Netherlands were destroyed since 1965. Extinct in Bayern, Critically Endangered in W Germany (Rheinland-Pfalz, Saarland, Nordrhein-Westfalen, Hessen), Endangered in Germany, Vulnerable in Great Britain.

Pseudosuccinea

Pseudosuccinea Baker, 1908
in F. C. Baker 1908: 943

Type species *Lymnaea columella* Say, 1817
(original designation)

Original description contained only name and type species.

North and South America, dispersed elsewhere

Pseudosuccinea columella (Say, 1817)

Lymnaea columella Say, 1817 in Say 1817: 14 (*Lymnæa columella*)

15-20 x 8-13 mm — Germany: Berlin-Dahlem

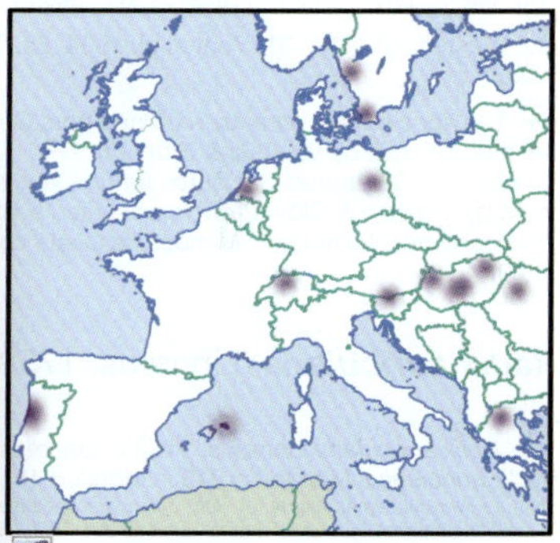

SE Canada and S Manitoba to S America, worldwide dispersed

Clarke 1981: 118, Turner et al. 1998: 417, Glöer 2002: 224, Glöer & Sîrbu 2006, Beckmann 2007: 47, Reischütz & Reischütz 2004: 9, De Oliveira et al. 2010 (Portugal).

Shell horny brown, thin, translucent, fragile, very finely striated, upper whorls pointed, 3.5-4 weakly convex whorls with shallow suture, last whorl predominating, aperture ovate, upper palatal margin descending steeply, columellar margin reflected only at its upper section, lower columellar margin sharp and straight.
Animal dusky with whitish spots, eyes small and black, located at the inner base of the tentacles.

In N America in stagnant waters, at the edges of lakes, ponds, muddy and sluggish streams, among lily pads and reeds on sticks and mud. In Europe predominantly in greenhouses, sometimes also outside (Austria, Hungary, Greece), needs warm waters and does not survive central European winter temperatures. Also found outside the water on floating leaves of aquatic plants. In N Greece in a spring near a road. Feeds on aquatic plants and algae.

Radix

Radix Montfort, 1810
in Montfort 1810: 266

Type species *Radix auriculatus* Montfort, 1810
(original designation)
(= currently *Radix auricularia* (Linnæus, 1758))

Montfort treated the name *Radix* masculine by combining it with the specific epithet *auriculatus*. *Helix auricularia* was given as a synonym.

Spire of the shell short, last whorl considerably enlarged.

Pfenninger, Col. & S. 2006 and Correa et al. 2010 suggested that *Radix* is monophyletic and a natural taxon. Remigio 2002 suggested paraphyly.

The names of important species of *Radix* were changed after a molecular study by Bargues et al. 2001: *Helix balthica* Linnæus, 1758, *Buccinum lagotis* Schrank, 1803, *Lymnaea ovata* Draparnaud, 1805, *Limnaeus pereger* var. *labiatus* Rossmässler, 1835 and *Buccinum peregrum* Müller, 1774. The names *Radix ampla* and *Radix auricularia* were not changed.

Buccinum peregrum Müller, 1774 was suggested to belong to the species for which *Radix ovata* had been in use until 2001 (although its until-2001-considered subjective synonym *Buccinum lagotis* was described earlier). Since 2001 *lagotis* has been considered a separate species, and the three names *ovata*, *peregra* and *balthica* have been considered subjective synonyms of each other.

So if the conclusions by Bargues et al. 2001 are accepted, it goes this way:
true *peregra* = until 2001 *ovata* = since 2001 *balthica*
true *balthica* = not used outside Russia until 2001 = since 2001 *balthica*
true *ovata* = until 2001 *ovata* = since 2001 *balthica*
true *lagotis* = until 2001 *ovata* = since 2001 *lagotis*
true *labiata* = until 2001 *peregra* = since 2001 *labiata*.
Both names *ovata* and *peregra* were not used any more under this system. All old *peregra* records in western publications referred to *labiata*, the old *ovata* records referred partly to *balthica* and to *lagotis*, usually to *balthica*.
Russian and Ukrainian authors have classified *Radix* as a subgenus of *Lymnaea* until recently (Kruglov & Starobogatov 1993, Prozorova 2009), and used *balthica* since the 1980s (Kruglov & Starobogatov 1983) for a species regarded as different from *ovata*. Kruglov & Starobogatov 1993 recognised many different *Radix* species for northern Eurasia, several of these had type localities in central and W Europe but have not recently been regarded as true species in those countries.
Glöer & Pešić 2008 established a new name *Radix skutaris* for a form from Montenegro, claimed to represent an endemic species from Skadar Lake. This requires confirmation in independent and more detailed studies at a larger scale. The same applies to *Radix lilli* Glöer & Beckmann, 2007 from Mallorca.

Pfenninger, Col. & S. 2006 identified 6 reproductively isolated clades within the *Radix* group in Europe with broad and overlapping geographical ranges. Very difficult to understand. Sorting out the species under these circumstances seems to require studying thousands of populations at molecular level. Populations at one site seem to belong to a single clade, but in close vicinity the next population may belong to another clade. Molecular distances seem to suggest that the clades are genetically separated since a million years or so. Their morphology does not allow to associate shells reliably with clades.
It seems that morphological and associated anatomical differences are mainly environmentally induced. Narrow shells are found in temporal and stagnant waters with soft substrate, shells with broad apertures are found in deeper waters and running waters with hard substrate.

In summary, the traditional taxonomic separation into species as presented here, is perhaps not in agreement with the genetic background, but at the current state of research it seems we don't have anything better with which we could work. In the present list the molecular/taxonomic approach proposed by Bargues et al. 2001 is used, which has the shortcoming that the conclusions were probably too premature and not confirmed in subsequent studies such as those by Pfenninger et al. 2006. So in this sense the published conclusions by Bargues et al. 2001 may serve as a bad example that coming up too quickly with taxonomic implications of such a molecular study can unnecessarily distort an approved taxonomic system.
The alternative solution would be to classify all *Radix* species as one single species and wait with the subdivision at the specific level until we know more (in this case we would select one of the Linnean names *balthica* or *auricularia* for this species complex).

Radix ampla (Hartmann, 1821)

Lymnaea auricularius var. *ampla* Hartmann, 1821 in Hartmann 1821 (N. Alpina): 250 (*Limneus auricularius* var. *ampla*)
Established without description, but with a figure on Taf. 2 (Fig. 29).
Turbo patulus Mendes da Costa, 1778 in Mendes da Costa 1778: 95 (Pl. 5 Fig. 17, this species?)

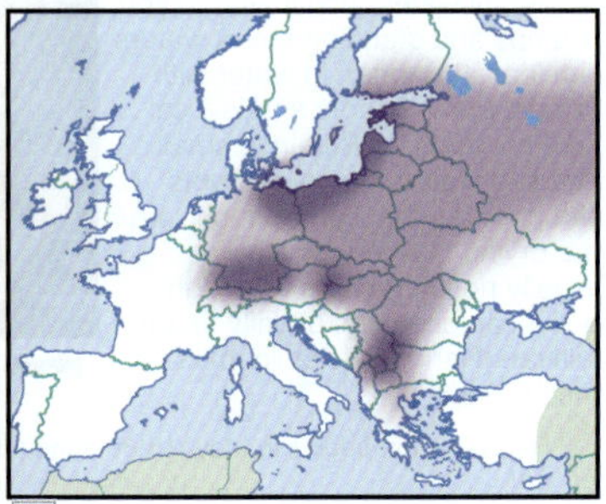

20-40 x 18-38 mm

Germany: Bayern, near Ansbach, Neuses/Rezat

Central and E Europe to Siberia

Shell solid, first whorls only 1-2 mm high, with very deep suture depressed in adult shell and ascending at last whorl, last whorl and aperture extremely inflated, margin usually reflected, upper section of palatal side ascending, columellaris usually weak, straight and often parallel to the straight outline of the columellar apertural margin, umbilicus open and partly covered by the columellar margin. Looks very much like *R. auricularia*, but with distinctly larger aperture, almost absent columellar fold, juveniles look like the adults, spermatheca duct short.

Large standing or slowly moving waters in lowlands and small lakes in subalpine zones, on muddy substrates. Never found swimming, never feeding on fresh plants, always at the bottom, or crawling on stones. In Switzerland up to 1500 m.

Adult females lay eggs at the lake margin at the water line, and fix the eggs at the shells of other female *Radix ampla*.

EU Red List: Least Concern LC, threat certainly underestimated. Declined so much that survival is questionable. Some scattered occurrences were reported for Switzerland, but despite intensive research the species could nowhere be found recently. Critically Endangered in Germany, Endangered in Switzerland and parts of Austria.

Westerlund 1885: 31, Vogt et al. 1994: 65, Turner et al. 1998: 108, Glöer 2002: 215, Glöer & Sîrbu 2006, Vinarski & Glöer 2007, Jungbluth & Knorre 2009: 10, Prozorova 2009.

The species was recently recorded from Germany (Bayern, Mecklenburg-Vorpommern) and Poland. Indicated range in E and SE Europe uncertain. Identity is based on the interpretation by Vinarski & Glöer 2007.
Prozorova 2009 argued that *Turbo patulus* was not a synonym of *Helix balthica* Linnæus, 1758 but of *Lymnaea ampla* (and that the figure of the *Turbo patulus* lectotype designated by Vinarski & Glöer 2007 represented a young form of *Radix ampla*). She argued that the form of the shell of the figured lectotype was in agreement with *R. ampla*, but it remained unexplained how Mendes da Costa could have obtained shells of *R. ampla* from Britain if the species did not live there (as reasonably argued by Vinarski & Glöer 2007).

Radix auricularia (Linnæus, 1758)

Helix auricularia Linnæus, 1758 in Linnæus 1758: 774 (*Helix Auricularia*)
Radix auriculatus Montfort, 1810 in Montfort 1810: 267
Not a new replacement name for *Helix auricularia* because this name was only given as a synonym, without expressed statement that and why this name would be replaced for nomenclatural reasons (Art. 72.7, Glossary). Montfort gave only a description for the genus *Radix*, for which *R. auriculatus* was indicated as type species, but added a figure and bibliographic references that made the specific name available.

Spire of the shell short, last whorl considerably enlarged.

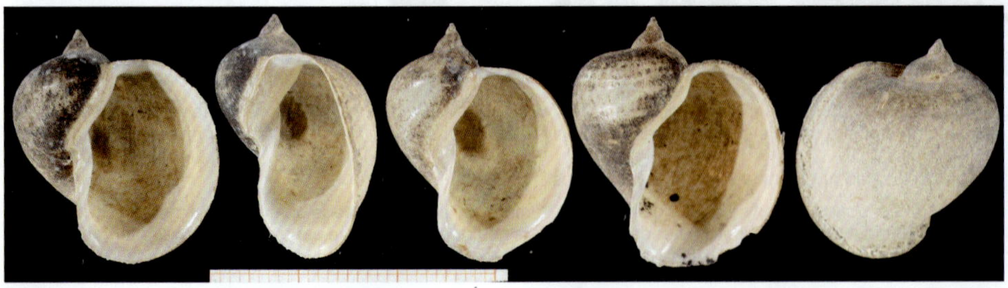

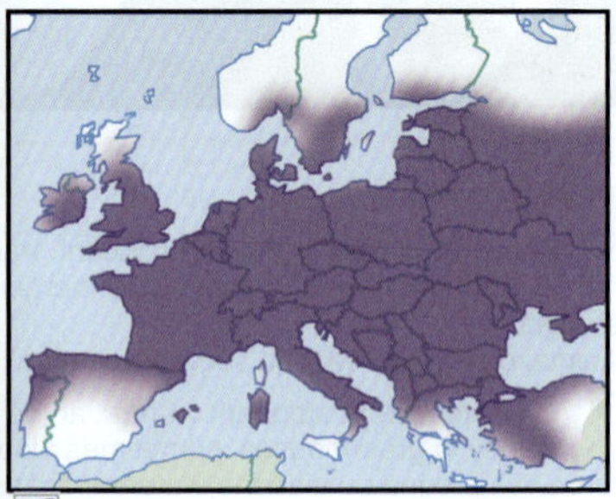

10-35 x 7-27 mm, usually 14-24 x 12-18 mm

Germany: Schleswig-Holstein, Plön

Europe, SW and N Asia, introduced to North America

Shell with upper whorls small and pointed, last whorl inflated, provoking a concave outline of the upper section of the shell, apertural height slightly larger than apertural diameter, margin on upper palatal side almost horizontally, columella twisted at the aperture with strong blunt edge, umbilicus usually closed. Juveniles are strongly pointed and high-spired.
Species can be identified by its height/diameter ratio, also juveniles. Differs from *R. balthica* in the concave outline of the upper part, upper palatal apertural margin is not horizontal in *R. balthica*. Differs from *R. ampla* in its very long spermatheca duct.

Lakes with rich vegetation, old river arms and silent stream bays. Up to 25 m deep. Tolerates up to 0.6 % salt, pH tolerance 7-9.9. Usually not in small standing waters, but sometimes successfully introduced into ornamental ponds in gardens and parks. In Switzerland up to 1800 m, usually in lowlands, also in Norway and Britain. Feeds on overgrown soft substrates, consumes mainly green algae and protozoans.

30-80 oval eggs (1.5 x 1.8 mm) are laid in clutches of 15-30 mm between July and September, fixed at stones and aquatic plants, juveniles hatch after 2-3 weeks, life span 1 year.

Critically Endangered in Tirol (Austria), Vulnerable in parts of N Germany, not very frequent in Spain. Seems to be sensitive against water pollution, may disappear locally in Britain, but no general decline was observed until 1999.

Gassies 1849: 163 (*Lymnaea trencaleonis*), Moquin-Tandon 1855: 463, Germain 1931: 487, Glöer 2002: 213.

Radix balthica (Linnæus, 1758)

Helix balthica Linnæus, 1758 in Linnæus 1758: 775

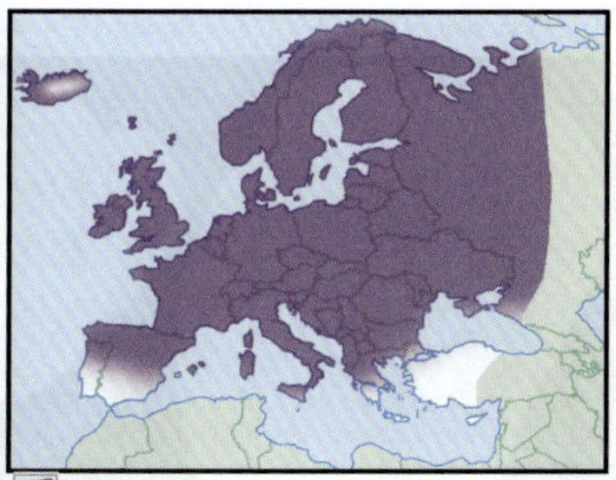

10-20 x 8-14 mm — Estonia: near Harku — N and W Eurasia

Shell light horny coloured, translucent, not very solid, egg-shaped, like a half-grown *R. auricularia*, 4.5-5 moderately convex whorls, first whorls with convex outline, last whorl inflated and deviating from the convex outline, aperture can have a weak lip inside, upper palatal margin usually immediately descending. Shell very variable.
Mantle basically black with light, mostly circular spots, anterior section of mantle light with dark pattern.

Creeks, rivers, old river arms and lakes, on plants, stony and muddy substrate, prefers calcareous waters. Highly adaptive, tolerates up to 1.4 % salt, pH tolerance 5.8-9.9, temperature optimum 15-25℃. Prefers 0.2-2 m depth, goes down to 10 m, exceptionally deeper. In Switzerland in up to 2500 m altitude, only rarely above the timberline. Feeds on algae and bacterial cover on hard substrate, also on detritus on soft bottoms, consumes in some habitats mainly green algae and protozoans, in others mainly detritus. Usually very frequent, the most commonly recorded freshwater mollusc in the British Isles.
Sensitive against presence of competing species, death rates increase with presence of *Lymnaea stagnalis*, does not survive next to *Radix auricularia*.

Usually 1 generation per year in Germany, copulation and egg-laying in March, life span 1 year (Glöer 2002: 217). In Zurich lake 2 generations per year, the summer generation lives 3-4 months, another generation hibernates and lives 9-11 months (Walter 1978, Turner et al. 1998: 105). In France between June and September 40-60 oval eggs (1-1.5 mm long) are laid, juveniles hatch after 14-16 days (Moquin-Tandon 1855: 465, Germain 1931: 488).

Until 2001 known under the name *Radix ovata*, that name is not used any more under the system proposed by Bargues et al. 2001.

Radix labiata (Rossmässler, 1835)

Without gender agreement: *Radix labiatus* (Rossmässler, 1835)

Lymnaea pereger var. *labiatus* Rossmässler, 1835 in Rossmässler 1835-1837 (1): 98 (*Limnaeus pereger* var. *labiatus*)
Described in the text under the chapter of *Limnaeus pereger*. Rossmässler observed two forms, one with and one without lip, and then said that there were two forms *Limnaeus pereger* var. [alpha]. *major* and *Limnaeus pereger* var. [beta]. *labiatus*. He did not explicitly say which one is which, the selected names seemed to be clear enough. Taf. 2 Fig. 54 did not explicitly refer to one of the forms.

12-20 x 7-13 mm — Germany: Bayern, near Tittling — Europe, N Africa

Shell dark horny to brown, often with brownish black cover, apex pointed, outline of upper portion of shell straight, 4.5-5 moderately convex whorls with deep and almost horizontal suture, last whorl weakly convex (on left side in apertural view), sometimes straight, aperture without lip, 2/3 of shell height, upper palatal margin sharply descending.
Animal greenish yellow with blackish and golden dots, foot light or grey, anterior 1/4 of mantle light with dark pattern, tentacles triangular and not very long, eyes at their base and very small.
Differs anatomically from *R. balthica* in the presence of a spermatheca duct.

Standing or slowly running small water bodies, on plants, stones and sandy bottoms, survives drought periods better than the other *Radix* species. In Switzerland this species prefers muddy substrate with vegetation (*R. balthica* shows no preferences among muddy and stony substrates), up to 2700 m altitude. Also in high alpine waters which are free of ice during only 3-4 months of the year. In Scotland and Wales up to 700 m, in Bulgaria 2300 m. Frequent in E Spain and Portugal. In Switzerland nowhere really frequent. Feeds on filamentous green algae.

In France between July and September 20-80 eggs (diameter 1 mm) are laid per clutch, one individual can lay more than 25 clutches with few days between each deposit, together more than 1000 eggs per individual in a season. Juveniles hatch after 20-25 days. In Switzerland two generations per year, one in summer with a life span of 3-4 months, one hibernating generation with a maximum age of 9-11 months. Shell height increases rapidly within 2 weeks from 5 to 10-12 mm. Adults die shortly after egg-laying.

Moquin-Tandon 1855: 468, Germain 1931: 494, Walter 1978, Turner et al. 1998: 104, Falkner et al. 2002: 94, Glöer 2002: 216.

The name *Radix peregra* (Müller, 1774) was used until 2001, and is not in use any more under the system proposed by Bargues et al. 2001. All *peregra* records refer to *labiata*, before and after 2001. After 2001 *peregra* is only used in outdated classifications in the meaning of *labiata*.

Radix lagotis (Schrank, 1803)

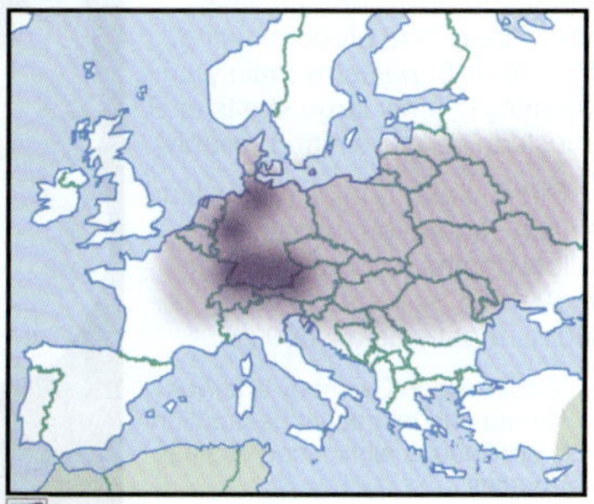

Europe

Buccinum lagotis Schrank, 1803 in Schrank 1803: 290 (*Buccinum Lagotis*)

Shell horny coloured, thin, finely striated, shiny, first whorls 4-5 mm high, pointed, almost scalarid, with straight outline, 4-5 convex whorls with deep suture, last whorl moderately inflated, suddenly ascending, aperture 15-21 x 9-12 mm, simple, margin not reflected, upper palatal margin nearly horizontal and running straight for a while until the curve is initiated, columellaris usually sharp and curved, with a weak fold or none, umbilicus very narrow.

15-25 x 10-21 mm

In standing waters in meadows, creeks, ponds and old river arms.

Regarded as Critically Endangered in Germany by Jungbluth & Knorre 2009, but this was based on insufficient knowledge about this species (Glöer & Diercking 2010: 94).

Westerlund 1885: 33 (*lagotis*), Glöer 2002: 219, Glöer & Diercking 2010: 94.

Distribution not well known, reliable records require molecular analyses. Previous records for *R. ovata* (and also some for *R. peregra*?) may refer to *R. lagotis*, present records for *R. balthica* may contain *R. lagotis* if not determined by molecular analyses. Glöer & Diercking 2010 stated that *R. lagotis* from Hamburg was determined anatomically without explaining how this was done. No anatomical differences between *lagotis* and *balthica* are known.

Radix pinteri Schütt, 1974

Radix pinteri Schütt, 1974 in Schütt 1974: 471, Taf. 1

15 x 14 mm — Rep. Makedonija: Lake Prespa, near Perovo — Zoltán Fehér

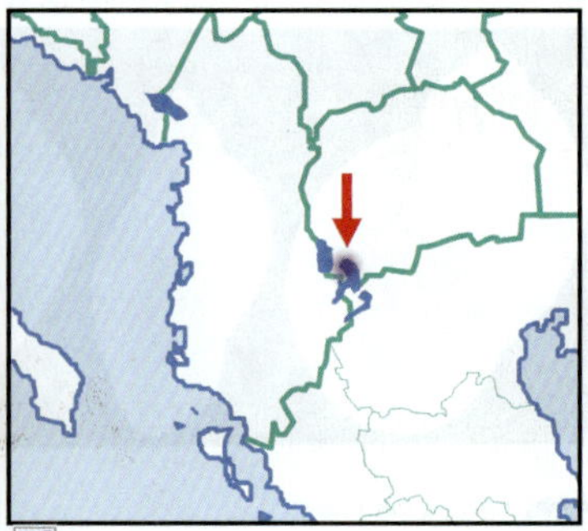

Lake Prespa

Albrecht et al. 2008.

Shell small, weakly striated, apex blunt and rounded, not pointed, aperture very large, elliptical, columellar margin completely detached from the penultimate whorl.
Animal whith some blackish brown dashes on the upper side and colourless on its lower side.

Lives on fine sandy and muddy ground behind the sea shore vegetational zone in the sublitoral zone of the NW part of Lake Prespa. Animals graze on mud and submerged parts of water plants, pulmonal respiration is reduced. Feeds on diatoms.

EU Red List: Endangered EN. Lives in a very restricted and specialized habitat only in Lake Prespa. Threatened by water pollution (eutrophication, herbicides).

Radix relicta Poliński, 1929

Radix relicta Poliński, 1929 in Poliński 1929: 158

Shell light brownish, striated, 3-3.5 whorls, upper whorls weakly rounded and rather flat, only the apex is pointed, last whorl very much inflated, aperture 85 % of shell height.

Litoral and sublitoral zones of Lake Ohrid, 0-30 m depth.

Albrecht et al. 2008.

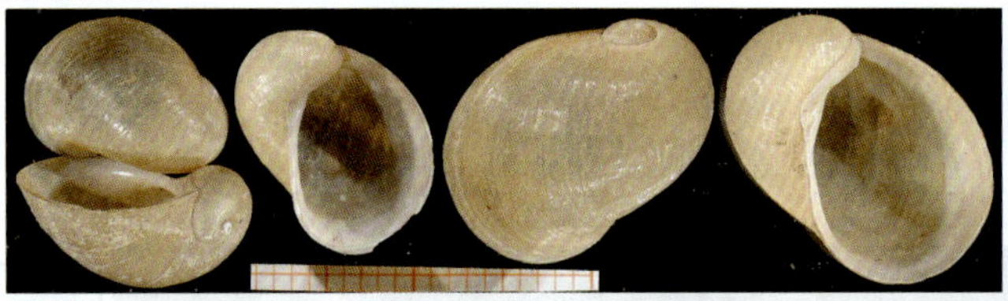

13-17 x 11-14 mm — Albania: Drilonë near Pogradec

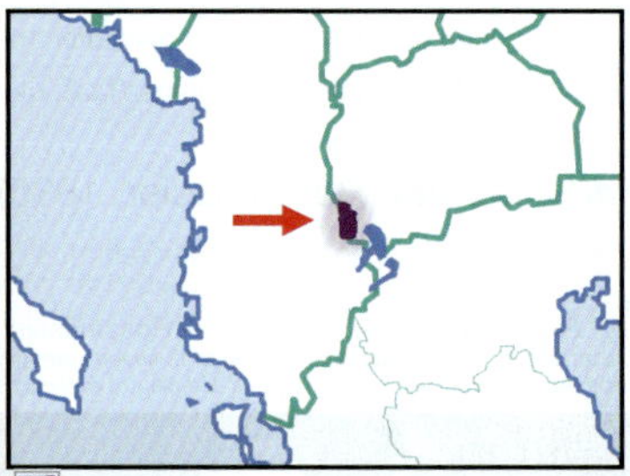

Lake Ohrid

Stagnicola

Stagnicola Jeffreys, 1830 in Jeffreys 1830: 376 *Limneus* synonym *Stagnicola*

Type species *Lymnaea communis* Jeffreys, 1830 (monotypy) (= currently *Stagnicola palustris* (Müller, 1774))

Date was 27 May 1830. Mentioned as "*Stagnicola communis* Leach MSS", as a synonym of *Limneus* (= *Lymnaea*) *communis*. *Stagnicola* was frequently used as a name for a genus before 1961 (Art. 11.6.1).
The name was attributed to Leach, but Leach did not contribute to making the name available, so in a strict application of Art. 50.1.1 the name in such a case must be attributed to the author of the work. Not *Stagnicola* Brehm, 1830 (Aves) from Oct 1830, type species *Fulica chloropus* Linnæus, 1758 by monotypy.

Catascopia Meier-Brook & Bargues, 2002 in Meier-Brook & Bargues 2002: 83, type species *Lymnaea catascopium* Say, 1817 (original designation) (= currently *Stagnicola catascopium* (Say, 1817) from North America, often classified in *Lymnaea*)
Name exclusively based on differential DNA, but available because the number of base pairs in a certain sequence was described.

Highspired shells with a spire of the same height or much longer than the aperture.

Palearctic, North America

The genus seems to be badly known, reliable records are rare, distributional ranges certainly not fully understood. A revision with a clear guide how to distinguish species, their full ranges and local variations would be useful.
Catascopia was recently established with a North American type species, its generic validity has not been accepted by North American authors who have continued using *Stagnicola* following Remigio 2002 and Correa et al. 2010: [7]. *Catascopia* is not accepted here either.
Correa et al. 2010 suggested that the group containing the North American species (*elodes*, *catascopium* etc.) and *occulta* (= *terebra*) composed a monophyletic group (this would be the *Catascopia* group), but also that the taxonomy of all lymnaeid species would generally have to be revised. If their phylogenetic tree corresponds with reality, we could be dealing with four monophyletic groups (1) *Radix*, (2) *Austropeplea* and others, (3) *Lymnaea* (containing *L. stagnalis*, *Omphiscola*, and *Stagnicola* except *terebra*), (4) *Galba* and many others including all other European groups. They decided to keep preliminarily the widely used genus-species combinations and to accept that the generic names did not reflect their taxonomic conclusions.
This is also done here. At the present state of research it makes no sense to distort the established genus-species combinations.
A new taxonomic system is needed. Hubendick's 1951 approach to use *Lymnaea* for all lymnaeid species would more appropriately reflect the current state of research, and would probably be superior to a system comprising many different generic names. A new taxonomic system at the generic level should also take into account that it should enable taxonomists to determine the appropriate genus for fossil species.
Recently several new species were described (Glöer & Yıldırım 2006, Glöer & Pešić 2009), their identities still require confirmation in independent studies.

Stagnicola corvus (Gmelin, 1791)

Helix corvus Gmelin, 1791 in Gmelin [1791]: 3665 (*Helix Corvus*)

Shell finely striated, larger than *S. palustris*, often with black cover, 6-8 weakly convex whorls, last whorl inflated and predominating. Differs from *S. palustris* mainly anatomically (only large shells of *S. corvus* can be determined reliably).

Silent water bodies of any kind with rich vegetation, usually in shallow depths, exceptionally below 50 m. In Switzerland up to 1800 m, mostly below 600 m.

Vulnerable in Germany and Austria. Threatened by habitat destruction due to continuous elimination of small water bodies.

13-34 x 6-17 mm — Germany: Wedeler Marsch near Hamburg

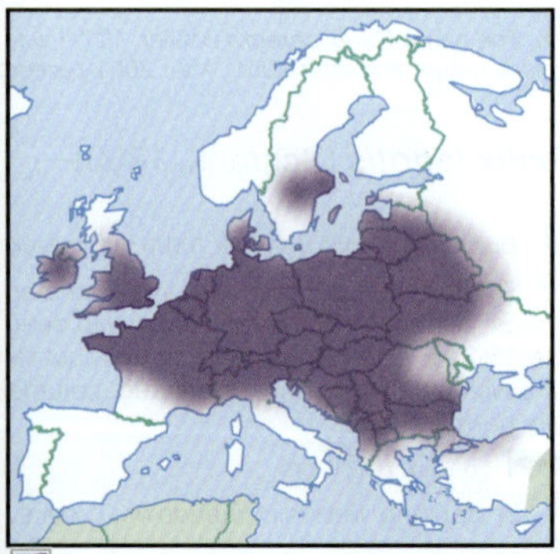

Europe
Reliable records are rare, usually not distinguished from *S. palustris* in Britain or Netherlands. Distribution in E Europe badly known.

Stagnicola fuscus (Pfeiffer, 1821)

Lymnaea fuscus Pfeiffer, 1821 in C. Pfeiffer 1821: 92 (*Limnaeus fuscus*)

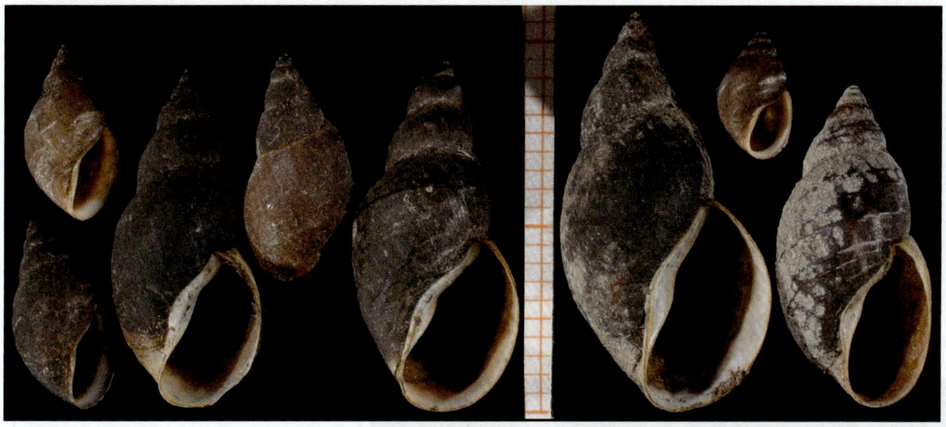

Germany: Mecklenburg, near Ratzeburger See

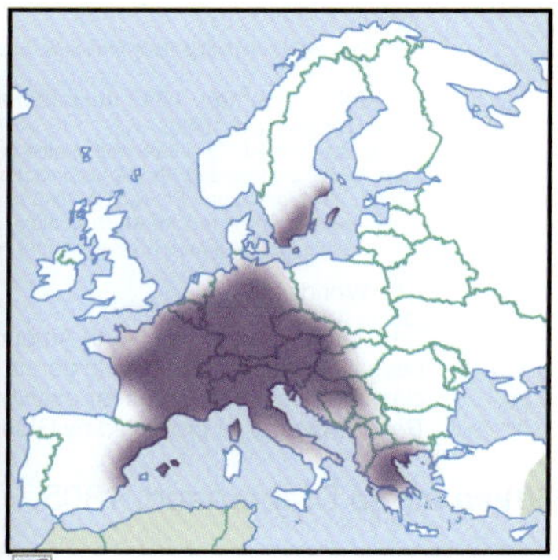

Europe

Threatened by continuous destruction of small water bodies. Vulnerable in Switzerland and Germany, decreasing in Bayern.

10-25 x 6-12 mm

Glöer 2002: 207, Glöer & Yıldırım 2006.

Shell slender, brown, whorls often not very convex, irregularly striated, apertural height about 50 % of shell height. Differs anatomically from the other species, prostata with 2 folds, praeputium : phallotheca = 3:1.

At well vegetated margins of moderately large standing waters, shallow grooves, swampy small standing waters. In Switzerland up to 500 m.

Stagnicola palustris (Müller, 1774)

Without gender agreement: *Stagnicola palustre* (Müller, 1774)

Buccinum palustre Müller, 1774 in Müller 1774: 131
Lymnaea turricula Held, 1836 in Held 1836: 278 (*Limnaea turricula*)

Germany: Mecklenburg-Vorpommern, Barther Oie

Europe

Moquin-Tandon 1855: 475, Germain 1931: 497, Falkner 1990: 136, Glöer 2002: 205.

10-18 x 6-10 mm
Height rarely exceeding 20 mm, shape very variable

Shell brown, greyish brown or with a yellowish or violet hue, finely striated, 4.5-6 convex whorls with well marked suture, last whorl only little inflated, apertural height slightly less than 50 % of shell height.
Differences to *S. corvus* mainly anatomical. Prostata with one fold, praeputium length equals phallotheca.
Animal dark, almost black with violet hue, covered with minute black and golden spots, tentacles black and triangular, eyes situated on small tubercles at their base.

Mostly in permanent and silent waters of any kind with rich vegetation, mainly in lowlands, preferred pH around 7. Swamps, shallow drains, also in periodically desiccating waters. More rarely in open waters. Tolerates slightly brackish water. Can also be found above the water surface. In Norway up to 400 m.
The *turricula* form was reported from temporary water bodies in periodically flooded habitats and well vegetated slow-moving waters.

Oval eggs (diameter 0.7-1.5 mm) are laid in 15-20 mm long clutches of 50-90 eggs.

Threatened by agricultural drainage. Vulnerable in Austria and parts of Germany (Nordrhein-Westfalen/Hessen), declined in England, not very frequent in Spain.

Reliable information is rare because small specimens can only be determined anatomically. *Stagnicola turricula* was considered as a separate species until 2005. Bargues et al. 2005 saw no genetical difference in molecular analyses. Reproductive isolation was not tested.
Its shell is more slender, whorls less convex than the main form of *S. palustris*. Shell horny brown (not violet), 6-7 slowly increasing whorls with deep suture, last whorl not much inflated, with strong columellaris, apertural height distinctly smaller than shell height, aperture inside often with a strong reddish brown callus. *S. turricula* replaces *S. palustris* in the Danube region, it lives often in temporary water bodies in periodically flooded areas.

Stagnicola terebra (Westerlund, 1885)

Lymnaea palustris var. *terebra* Westerlund, 1885 in Westerlund 1885: 46 (*Limnæa (Limnophysa) palustris* var. *terebra*)
Galba occulta Jackiewicz, 1959 in Jackiewicz 1959: 39

Shell light horny yellowish, striated, 5-6 usually convex whorls with deep suture, last whorl only slightly inflated, apertural height considerably less than 50 % of shell height, columellar margin can be white in adults.

9-17 x 5-7 mm

Small temporary standing waters in forests, needs desintegrating leaves and wood inside the water. Tolerates relatively long periods of drought. Very common in Asian Russia, in some regions in Siberia very abundant and in almost every suitable habitat.

Vulnerable in Poland.

Glöer 2002: 206, Vinarski & Glöer 2008.

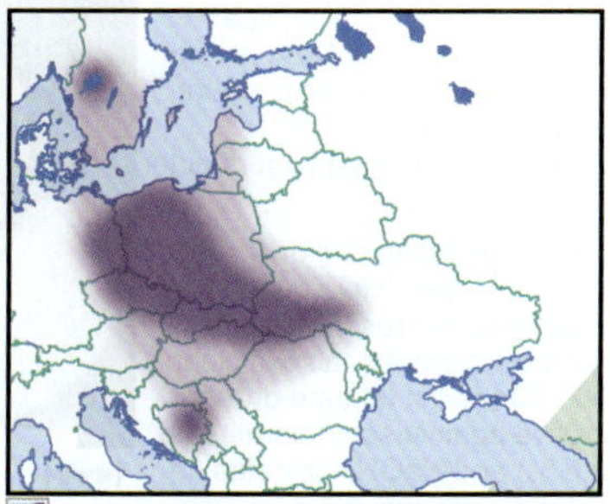

E Europe and central to E Siberia
Distribution insufficiently known, obviously disjunct with a large gap in European Russia.

Classified by Meier-Brook & Bargues 2002 and Bargues et al. 2005 in the genus *Catascopia*, this approach is rejected here. Correa et al. 2010 showed that the currently used genus-species combinations in recent Lymnaeidae do probably not reflect the phylogenetic relationships, and that Hubendick's 1951 proposal to use *Lymnaea* for all lymnaeid species would probably be most appropriate. They decided not to distort the currently widely used genus-species combinations and to wait until the taxonomic relationships are more firmly established and confirmed in independent studies. This is also done here.

Physa

Physa Draparnaud, 1801
in Draparnaud 1801: 31, 52

Type species *Bulla fontinalis* Linnæus, 1758
(subsequent designation by Children 1823 [Quart. J. Sci. Lit. Arts 15]: 243
[not verified])

Contained originally *Bulla fontinalis*, *hypnorum* and *scaturiginum*. Turton 1831: 10 gave as type *Bulla hypnorum* Linnæus, 1758, currently *Aplexa hypnorum*. Anton 1839: 48 gave as type *Bulla fontinalis*.

Physella Haldeman, 1843 in Haldeman 1842-1847: 14, type species *Physa globosa* Haldeman, 1841 (subsequent designation by Dall 1870: 355) (= currently *Physa acuta* (Draparnaud, 1805))

Date was Jan 1843 (this date was printed on the subscription wrapper of part 6). Haldeman (1843: 14) referred to plate [*Physa*] 5, where 4 species were figured and listed: *Physa distorta*, *Ph. hypnorum*, *Ph. globosa* and *Ph. microstoma*. Falkner et al. 2002: 96 and Bank et al. 2007: 46 argued that following Kennard & Woodward 1926: 97 the name *Physella* was printed on the cover of the 4th part in 1842, and dated *Physella* to 1842 (also in Falkner et al. 2001: 27 and Bank et al. 2001: 84). This is probably not tenable because such covers were printed for the binders and not for providing a permanent scientific record, which would qualify them as published work under Art. 8.1.1. Falkner et al. 2002 did not provide evidence that the name *Physella* had actually been made available on the cover. On the other subscription wrappers only the generic names were printed, without names of species, so this would have been a nomen nudum.

Worldwide

It seems that most North American sources classify *acuta*, *gyrina* and *heterostropha* in *Physa* and not in *Physella*. In recent European sources *Physella* has been classified as a separate genus (Neubert 1998: 359, Glöer & Diercking 2010: 100). It makes no sense that Europeans and North Americans use different classifications and names for exactly the same species. The broader proposal is *Physa*, so I would recommend to use this genus worldwide.
Physa adversa as proposed by Glöer & Diercking 2010: 99 is not accepted here, its specific identity was not based on published evidence.

Physa acuta Draparnaud, 1805

Physa acuta Draparnaud, 1805 in Draparnaud 1805: 55, Pl. 3
Lymnaea heterostropha Say, 1817 in Say 1817: B-7 No. 2, Pl. 2

8-16 x 4.5-9 mm

Germany: Elbe near Lauenburg

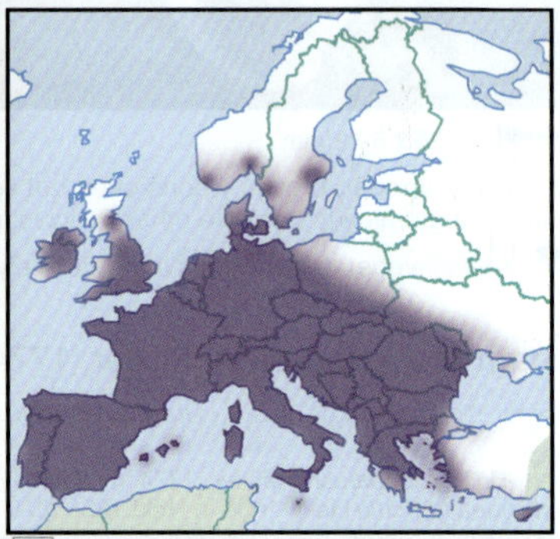

Worldwide, originally from either SW Europe or E North America

Slovakia: Vinné lake near Michalovce
Jozef Grego

Slovakia: Vinné lake near Michalovce
Jozef Grego

Shell light horny yellowish, sometimes slightly transparent, pale shiny, apex pointed, 4-6 not very convex whorls without deep suture, aperture 50-75 % of shell height, columella and apertural lip in fully grown shells white, occasionally reddish. Animal blackish or dark grey with violet hue, mantle with golden spots which are visible below the shell, tentacles greyish and almost transparent with small blackish eyes, foot usually lighter with fine black pigment spots.

In N America very common, in all kinds of perennial water bodies and temporarily flooded habitats, usually among vegetation. In Europe in warm standing and slowly moving waters, at margins of lakes, silent river sections, tolerates short periods of drought and eutrophic conditions. Feeds mainly on detritus and green algae.
In W Portugal frequently in waters supplying rice crops. In Switzerland up to 800 m, but not frequent except in warm springs, in Germany in lowlands. In many regions in Spain the most frequent freshwater mollusc.

Slightly protandrous. Oval eggs (0.8-1 x 0.5 mm) are laid in elongate packages (4-20 x 2.5-5 mm) containing 20-400 eggs (usually below 50), juveniles hatch after 20 days in temperate France (15 days in S Europe), initial size 0.5-0.7 mm, maturity is reached after 17-18 months. Two generations per year.

Moquin-Tandon 1855: 452, Germain 1931: 510, Nobre 1941: 207, Clarke 1981: 156 (Newfoundland, New Brunswick to New Jersey and Pennsylvania), Falkner 1990: 130, Manganelli et al. 1995: 10, 35, Pujante Mora et al. 1998, Turner et al. 1998: 113, Dillon et al. 2002, Glöer 2002: 238, Pérez-Quintero et al. 2004, Anderson 2005, Glöer & Diercking 2010: 101.

Very important species.
Dillon et al. 2002 found that *Ph. heterostropha* from E USA and *Ph. acuta* from Ireland and France produced fertile offspring under laboratory conditions, and regarded both as conspecific. Anderson 2003 found that both taxa did not differ morphologically in Ireland and followed Dillon et al.'s 2002 view. Glöer & Zettler 2005: 17 and Glöer & Diercking 2010 argued that Anderson 1996 had separated two species anatomically and maintained the previous classification as two independent species. This view is not followed here.
Google (02.2011, Germany) general search: *Physa* 37300, *Physella* 13300; Google Scholar: *Physa* 1700, *Physella* 435 (likewise proportions for *heterostropha*).

Physa fontinalis (Linnæus, 1758)

Bulla fontinalis Linnæus, 1758 in Linnæus 1758: 727

7-12 x 4-7 mm

Germany: Schleswig-Holstein, Hagener Au

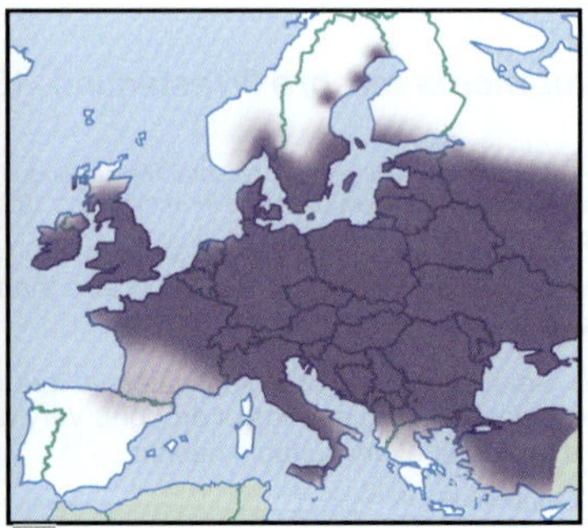

Holarctic

Shell sinistral, oval, light horny coloured, thin, very shiny, apex blunt and rounded, 3-4 whorls, upper whorls weakly convex, suture often weakly whitish.
Mantle touches in long fingers outer surface of the shell.

Clear standing or slowly moving water bodies rich in vegetation. Usually in lakes, reservoirs and slowly moving rivers, also in clean creeks. Does not tolerate drought. Sensitive against salt (tolerates up to 0.6 %) and NaOH, indicates unpolluted conditions, pH tolerance 5.4-9.6. In Switzerland in up to 1000 m. Feeds on submerged plants (*Potamogeton*), or detritus.

Reproduction from May to autumn, egg packages with 5-22 eggs (diameter 0.5 mm) are fixed at submerged plants, 1-3 times in a year, juveniles hatch after 18-20 days in France. Animals which hatched in autumn hibernate and live up to 11-14 months, those which hatched in spring live only 3-4 months.

Threatened by fertilization and water pollution in densely populated regions. Endangered in Bayern, Switzerland and Austria, Vulnerable in Germany.

Moquin-Tandon 1855; 451, Germain 1931: 509, Najim 1961, Glöer 2002: 235.

Physa gyrina Say, 1821

Physa gyrina Say, 1821 in Say 1821: 171

Shell light horny brownish, upper whorls with reddish hue, smooth, translucent, apex pointed, 3-5 convex whorls with slightly impressed suture, last whorl predominating, aperture 55-75 % of shell height, with broad white callus at parietal side, palatal margin sharp, basal margin white and weakly reflected, columellar margin thickened. Shell shape and apertural height variable.

Standing and slowly running waters in North America, abundant, tolerates organic water pollution. Where no other species occurs this species indicates eutrophic conditions.
Escaped in Great Britain and Ireland from aquaria into canals and similar habitats, spreading since the 1990s.

Reproduction principally in spring.

Canada: Québec, St. Basile le Grand

7-24 x 3.5-16 mm

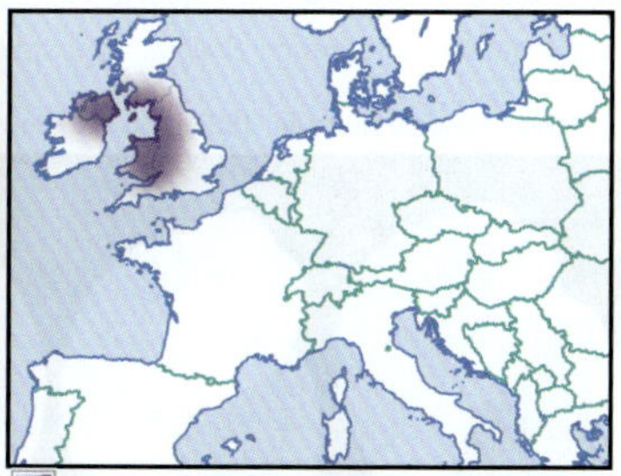

N America, introduced to Europe
Possibly also in Spain (www.faunaeur.org, 2007/2011, without published reference).

Clarke 1981: 152 (Alaska to S Quebec, California and Gulf of Mexico), Anderson 1996: 4, Glöer 2002: 239, http://data.gbif.org/species (02.2011).
Google (02.2011, Germany) general search: classified in *Physa* 6100, *Physella* 2700; Google Scholar: *Physa* 1000, *Physella* 168.

Aplexa

Aplexa Fleming, 1820
in Fleming 1830: 617

Type species *Bulla hypnorum* Linnæus, 1758
(subsequent designation by Herrmannsen 1846: 65)

Established as a genus, contained originally *Bulla hypnorum* and *Bulla rivalis* Maton & Rackett, 1807. ICZN Opinion 335 (1955) (and Bank, F. & P. 2007: 45) indicated monotypy, this was incorrect.

Aplexa hypnorum (Linnæus, 1758)

Bulla hypnorum Linnæus, 1758 in Linnæus 1758: 727 (*Bulla Hypnorum*)

9-15 x 4-6 mm

Germany: Schleswig-Holstein, Kieler Bucht near Laboe

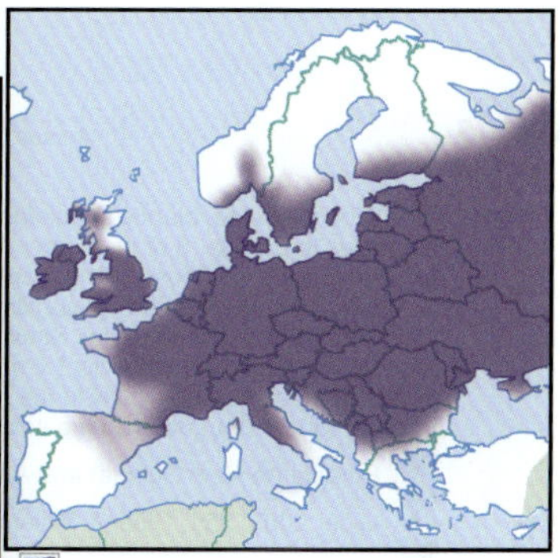

Holarctic

Shell sinistral, slender, yellowish to reddish brown, thin, shiny, 4-6 weakly convex whorls with light suture, aperture narrow. Animal black, mantle not touching outer surface of the shell.

Small and temporary water bodies rich in vegetation, in lowlands, often on mud bottoms. Also at swampy and forested margins of lakes, in swampy deciduous forests and very moist meadows. Tolerates temporary droughts, pH tolerance 5.8-8.5, feeds on detritus and rotting leaves fallen down from terrestrial plants. In Switzerland up to 1000 m, in Norway 300 m.
The animal moves very quickly, highest velocity of all freshwater snails, often outside the water in the vegetation at the margin.

Oval eggs (0.9-1 mm long) are laid in elongate packages (6-10 x 3-4 mm) containing 7-25 eggs, loose between rotting leaves, juveniles hatch after 14-16 days, initial size 1 mm, they hibernate and grow when temperatures exceed 7 °C. Maturity is either reached after 2 months (Glöer 2002 who reported one generation per year), or at the end of the second year after 20-22 months (Moquin-Tandon 1855, Germain 1931).

In Europe threatened by continuous destruction of moist habitats. In Britain many habitats were lost after 1900 by infilling of farm ponds and ditches. Extinct in Sweden, Endangered in parts of Germany, Switzerland and Austria, Vulnerable in Poland, Germany, Switzerland and Austria. Scattered and declining in Ireland.

Moquin-Tandon 1855: 455, Germain 1931: 512, Clarke 1981: 172, Falkner 1990: 130, Vogt et al. 1994: 68, Anderson 1996: 4, Turner et al. 1998: 111, Kerney 1999: 48, Glöer 2002: 234.

Bulinus

Bulinus Müller, 1781
in Müller 1781: 6

Type species *Bulinus senegalensis* Müller, 1781
(Linnaean tautonymy)
(= currently *Bulinus senegalensis* Müller, 1781 from W Africa)

Included originally 4 species, three from Müller 1774 and *B. senegalensis* with a reference to Adanson 1757: 5. Müller 1781: 6 explained that he adopted Adanson's name *Bulinus* for the new genus. This was in accordance with his citation under *B. senegalensis*, so that this should be regaded as type fixation by Linnean tautonymy.

Shell sinistral with gently curved whorls, angular or carinate. The penis does not project freely into the penis sheath. It is a long, coiled, eversible tube attaching at the upper and lower end of the sheath.

Neubert 1998: 358.

Bulinus truncatus (Audouin, 1827)

Without gender agreement: *Bulinus truncata* (Audouin, 1827)

Physa truncata Audouin, 1827 (I do not know where originally published)

9-13 x 6-8 mm

Italy: SE Sicily, near Siracusa

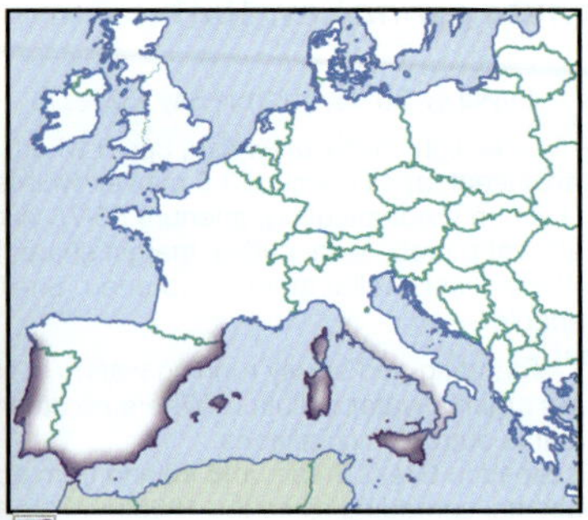

SE Asia, Africa, S Europe

Germain 1931: 514, Nobre 1941: 209, Manganelli et al. 1995: 11, Pérez-Quintero et al. 2004, Beckmann 2007: 49 (rare or extinct in Mallorca and Menorca).

Shell oval, brownish yellow or whitish, opaque or slightly transparent, 3-4 slightly angled and scalariniform whorls, last whorls looking step-like. Apertural margin with an angle between columellar and parietal sides.
Animal olive green, with yellow dots in anterior part, red inside the shell, foot reddish, tentacles long and cylindrical, eyes in the anterior base of the tentacles.

In S Spain reported from localities in coastal vicinity, always associated with algae films or macrophytes. In Portugal only in coastal regions, always very abundant where present, on muddy substrate and on water plants in rivers and smaller slowly running waters.

Eggs are laid in elongate and 4 mm sized packages containing 8-20 eggs, fixed to stones, aquatic plants or mollusc shells, juveniles hatch after 12 days, maturity is reached after 4-5 months.

Used to be abundant near Porto (Portugal) before 1940, until it was entirely replaced by *Physa acuta*, which had perhaps been introduced by migrating water birds.

Anisus

Anisus Studer, 1820 in Studer 1820: 91

Type species *Helix spirorbis* Linnæus, 1758 (subsequent designation by Gray 1847: 180)

Whorls keeled, bluntly angulated or rounded, shell flat.
5 species in Europe, 3 of them very difficult to distinguish. Glöer & Meier-Brook 2008 gave a guide, but did not present statistical analyses and avoided statements on infraspecific variations of the mentioned shell characters.
A. vortex is sharply keeled, *A. vorticulus* bluntly (and the species is smaller), both are easy to recognise. *A. septemgyratus* is slightly angulated, the periphery of the other 2 species is rounded. Similarly sized shells of these 3 species differ in their shell height (diameter 4 mm at beginning of last whorl: *septemgyratus* and *leucostoma* 0.85 mm, *spirorbis* 1.2 mm; at the end of the last whorl *septemgyratus* 0.9 mm, *leucostoma* 1.0 mm, *spirorbis* 1.4 mm). The three species can reliably be distinguished anatomically in the number of segments or transverse folds (diverticles) of the prostata gland (> 30 in *septemgyratus*, 20-25 in *leucostoma*, 10-15 in *spirorbis*), without overlapping values and obviously not influenced by region or age.

Eurasia (to New Guinea) and Africa

Glöer 2002: 254, Glöer & Meier-Brook 2008.

Kantor et al. 2010: 104 recognised numerous species in Europe in excess of what has recently been recognised by western researchers. Here the western system is adopted, mainly because the Russian references did not provide scientific arguments and did not explain how the so-called species (some of which had their type localities in central Europe) should be recognised and differentiated from other species. This applied also to *strauchianus* (from Ukraine).

Without a sound and detailed published study Falkner et al. 2001 and Falkner et al. 2002 confused the *Anisus* concept for Europe and rejected the traditional use of several names. *A. leucostoma* was called *septemgyratus*, *A. septemgyratus* was called *calculiformis*, *A. spirorbis* was maintained and included *leucostoma* as a synonym. This classification was immediately accepted for the Fauna Europaea database (Bank 2004 in www.faunaeur.org).
Falkner et al.'s 2002 statements were not based on unambiguosly fixed type specimens, and they were only partly accepted by Glöer 2002: 254 who accepted to use *calculiformis* and argued that the name *septemgyratus* should not be used any more. Glöer & Meier-Brook 2008 demonstrated that Falkner et al. 2002 had based their preliminary conclusions on misinterpreted material, they finally rejected Falkner et al.'s 2002 entire concept, fixed a neotype for *leucostoma* and returned to use *septemgyratus*. Around 2010 www.faunaeur.org returned to the traditionally used names, but many internet sources had meanwhile copied the information, classified taxa under the untenable concept and associated information on distributional ranges to names which were not corrected later.
Anisus species are threatened and need protection, and a stable taxonomy with stable usages of names is a prerequisite to save the last surviving populations of these animals. Distorting their taxonomy will certainly not help them at all.

The range maps presented here are tentative and I had to neglect some recent reports for *leucostoma*, *spirorbis* and *septemgyratus* because it was unclear to which taxon which name referred in the literature.

Anisus leucostoma (Millet, 1813)

Anisus leucostomus (Millet, 1813) (Google hits 2009: 35 %, 2011: 18 %, Google Scholar 2011: 29 %)

Planorbis leucostoma Millet, 1813 in Millet 1813: 16
Planorbis is masculine and Millet wrote *leucostoma*, name is to be treated as a noun in apposition under Art. 31.2.2. Falkner et al. 2002: 99 considered this taxon as a synonym of *Helix spirorbis* Linnæus, 1758, but did not fix a type for it. Glöer & Meier-Brook 2008: 92 validly fixed a neotype, so that the traditional usage of the name can be maintained.

1-1.5 x 6-9 mm

Germany: Schleswig-Holstein near Schrevenborn

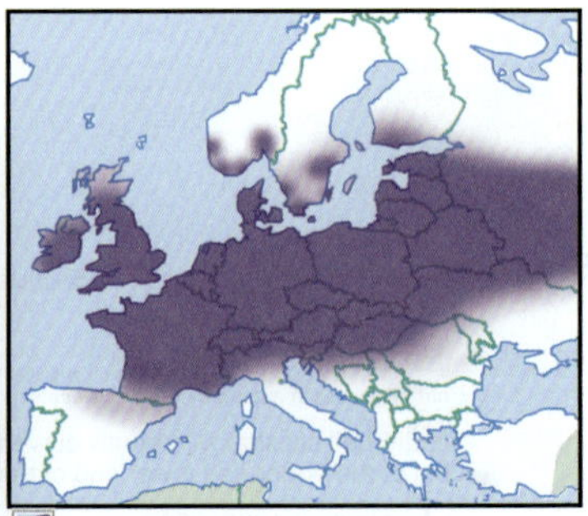

Europe to central Siberia

Shell usually brown, very shiny, usually upper side almost flat and lower side depressed, or the contrary, 6-6.5 slightly convex whorls, bluntly keeled, whorls almost with a second edge, last whorl not much wider than penultimate whorl. Whorls more narrowly coiled and last whorl narrower than in *A. spirorbis*. Prostata gland with 20-25 segments or transverse folds (diverticles).

In various habitats, like *A. spirorbis*, but also in mountain regions. Typical habitats are periodical and very small water bodies, moist habitats at periodically flooded river and lake margins, in Britain often in waters drying out in the summer, mainly in lowlands. Longer periods of drought are tolerated. Prefers pH 6.2-7.8 in Norway. In Switzerland up to 1800 m, in Germany 600 m.

Reproduction between April and September.

Many habitats were destroyed by drainage and fertilization. Critically Endangered in Tirol, Vulnerable in Austria, Switzerland and Bayern.

Falkner et al. 2001: 28 used the name *septemgyratus* for this species, this was adopted by Bank in 2004 for www.faunaeur.org (removed in 2010). Hubendick 1951 and others considered this form as conspecific with *Anisus spirorbis*.

Anisus septemgyratus (Rossmässler, 1835)

Planorbis septemgyratus Rossmässler, 1835 in Rossmässler 1835-1837 (1): 106, Taf. 2
Established for a species from near Bolzano/Bozen, N Italy. No types seem to exist for this taxon, most recent authors have used it for this species here. Falkner et al. 2002 argued that material collected at Bolzano belonged to *A. leucostoma*. Glöer & Meier-Brook 2008 demonstrated that Falkner et al.'s conclusions were based on insufficient knowledge of the group and misinterpretations of their collected material which did not belong to *A. leucostoma*.
Planorbis calculiformis Sandberger, 1874 in Sandberger 1870-1875: 779, Taf. 33

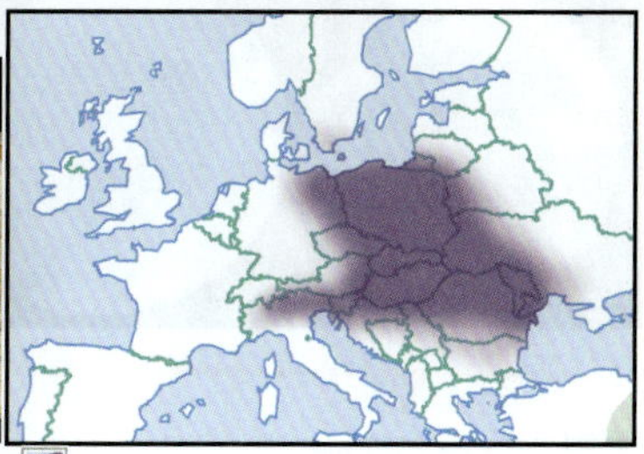

1.0-1.2 x 7-8 mm — E Slovakia: Východoslovenská nížina, Malica — Europe

Shell horny brown, thin, with dense growth lines, 7-8.5 regularly increasing whorls with well-marked suture, slightly angulated at functional lower side, aperture oval, rounded, small and oblique.
Aperture slightly smaller than in *A. leucostoma*, height of the last whorl at 4 mm shell diameter ranges at 0.9 mm, slightly lower than in *A. leucostoma* (1.0 mm). Prostata gland with 30 to more than 50 segments or transverse folds (diverticles).

Small standing waters, ponds and swamps, tolerates temporary droughts.

Extinct in Austria, Critically Endangered in Czech Republic, Vulnerable in Poland and Germany.

Falkner et al. 2001: 28 used the name *calculiformis* for this species, this was accepted by Glöer 2002: 254 and adopted by Bank in 2004 for www.faunaeur.org. Falkner et al. 2001 used *septemgyratus* for *A. leucostoma*, this was not accepted by subsequent authors, but adopted in 2004 for www.faunaeur.org. As explained above, this classification was based on unpublished studies and misidentified species. Glöer & Meier-Brook 2008 provided a solid base for using *septemgyratus* in the traditionally used sense, this classification was adopted in www.faunaeur.org around 2010. The www.faunaeur.org records "*septemgyratus*" between 2004 and 2010 referred to *A. leucostoma*. Many sources after 2002 worked with this incorrect classification, so that many records for *A. septemgyratus* in France and N Italy (and probably also elsewhere) after 2002 did not refer to this species, but to *A. leucostoma*.

Anisus spirorbis (Linnæus, 1758)

Helix spirorbis Linnæus, 1758 in Linnæus 1758: 770 Type locality in Sweden.

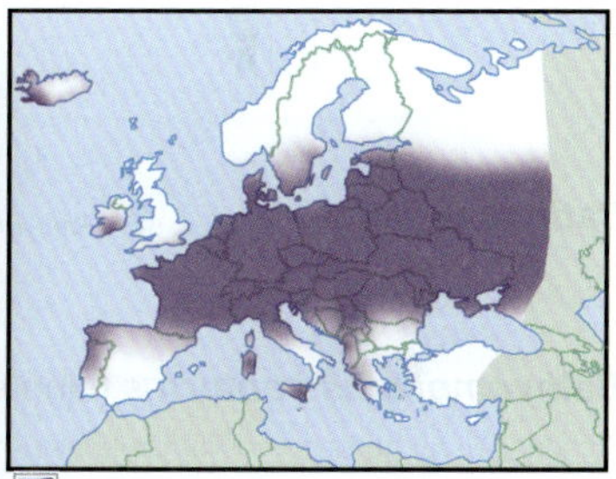

1.4-1.5 x 5-6 mm — Germany: Schleswig-Holstein near Dinghorst — Europe to Siberia
Range insufficiently known.

Shell light horny yellowish, finely striated, translucent, 4-5 slightly convex and rapidly increasing whorls, last whorl width 2 x of penultimate whorl. Last whorl broader than in *A. leucostoma*, from which it also differs in its larger shell height.
Animal very small and short, brownish red or blackish grey, tentacles transparent and not very long, with small black eyes.

Temporary natural water bodies in lowlands and river plains, on water plants, also in periodic swamps and moist meadows. Usually abundant if present. In Switzerland up to 1500 m.

Endangered in Tirol and Germany, Vulnerable in Switzerland.

Occurrence in Sweden uncertain following Glöer & Meier-Brook 2008: 92 who reported that Proschwitz (unpublished data) did not find the species, but did not consider the numerous published reports for Sweden by Waldén 2007. They might have referred to *A. leucostoma*. Reports for W Greece (Rähle 1980: 206) and Portugal (Nobre 1941: 202) remained uncommented by Glöer & Meier-Brook 2008 who gave range limits between France and Siberia excluding the Balkan peninsula.

Anisus vortex (Linnæus, 1758)

Helix vortex Linnæus, 1758 in Linnæus 1758: 770 (*Helix Vortex*)

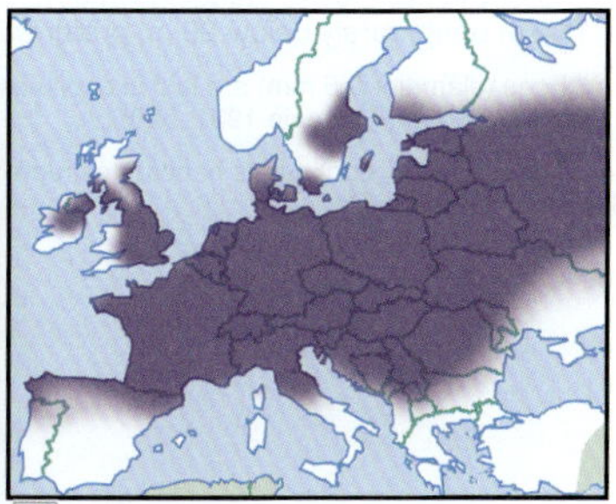

0.8-1.5 x 4-9 mm — Germany: Schleswig-Holstein, near Diekendorn — Europe and W Siberia

Shell with very sharp keel, flattened at functional lower side, concave at upper side, 6-7 whorls, flat on upper side and strongly convex with edge and deep suture at lower side.

Standing and slowly running waters with rich vegetation, also temporary water bodies but needs moist substrate remaining in drought periods (desiccation is not tolerated). In Switzerland up to 1000 m, in Britain mainly in lowlands.

Eggs (diameter 0.5 mm) are laid in ovoid capsules (diameter 4 mm) containing 10-12 eggs each, juveniles hatch after 10-12 days.

Rare in the Alps. Vulnerable in Germany, Switzerland and Austria, decreasing in Bayern.

Moquin-Tandon 1855: 433, Germain 1931: 528, Falkner 1990: 132, Glöer 2002: 261.

Anisus vorticulus (Troschel, 1834)

Planorbis vorticulus Troschel, 1834 in Troschel 1834: 51

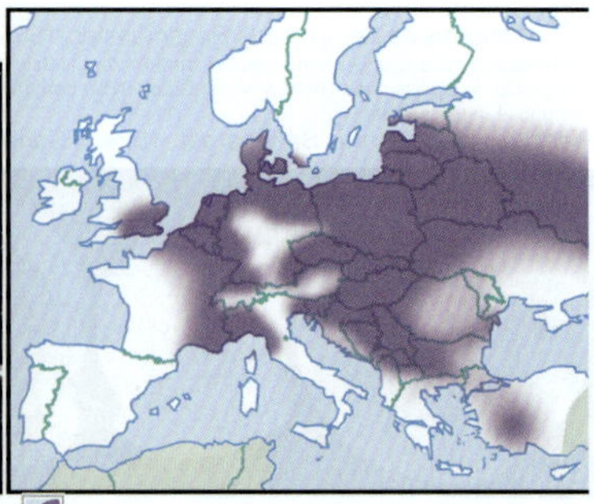

0.6-1.2 x 2.5-7 mm Germany: Schleswig-Holstein, Hansdorfer See Europe to Caspian Sea

Shell horny brown, upper side slightly concave, lower side almost flat, 5-5.5 convex whorls with distinct suture, last whorl with blunt edge in the centre of the periphery, aperture elliptical and usually oblique.
Like *A. vortex*, usually smaller, keel blunt, whorls rounded at both sides.

Standing or slowly floating clean waters usually rich in vegetation (*Lemna trisulca*, *Hydrocharis*, *Ceratophyllum*), oxygen and calcium contents, no visible sediment preference. Often floating on the surface among *Lemna*. Feeds on diatoms or algae which need sunlight (cutting the bankside vegetation improves the conditions for *A. vorticulus*). Not sensitive against careful weed removal and against measures to remove 1 m thick anaerobe mud layers. Tolerates frozen waters in winter and dried-out waters in sumer. Does not tolerate hypertrophic conditions, sensitive against too fast water currents and against destruction of water banks by cattle. In Switzerland up to 500 m. Seems to be dispersed by birds and can increase population densities rapidly under appropriate conditions.

Snails in NW Poland begin copulating at 3 mm shell diameter. Together up to 500 eggs (0.55 x 0.45 mm) per individual are laid between May and August in cocoons containing 1-6 eggs, juveniles hatch after 10 days (24°C), diameter reaches 4 mm after 3 months, no growth between November and March, maximum size is reached in June. Life span up to 18 months, snails usually die in the second winter.

EU Red List: Near Threatened NT. Critically Endangered in Germany, Czech Republic and Austria, Endangered in Italy and Switzerland, Vulnerable in England and Poland. Scattered and rare, threatened by continuous habitat destructions, drainage, overfrequent dredging, fertilization in agricultural zones and introduction of nutrients to water bodies. All surviving populations in England are threatened by these causes. Populations in Austria have dramatically declined since the 1990s, including in Donauauen National Park, where the species became extinct between 1991 and 2006 by mismanagement of the National Park administration. The last surviving populations in Austria (near Bad Deutsch-Altenburg in an old river arm with low water current) are seriously threatened by changes in river regulation (2006/2009) expected to cause another dramatic decline and possibly local extinction.

Glöer 2002: 262, Terrier et al. 2006, Glöer & Groh 2007 (measures for conservation), Myzyk 2008 (life cycle), Reischütz 2009 (situation in Austria).

Bathyomphalus

Bathyomphalus Charpentier, 1837 in Charpentier 1837: 20 *Planorbis (Bathyomphalus)*

Type species *Helix contorta* Linnæus, 1758 (monotypy)

Bathyomphalus contortus (Linnæus, 1758)

Without gender agreement: *Bathyomphalus contorta* (Linnæus, 1758)

Helix contorta Linnæus, 1758 in Linnæus 1758: 770

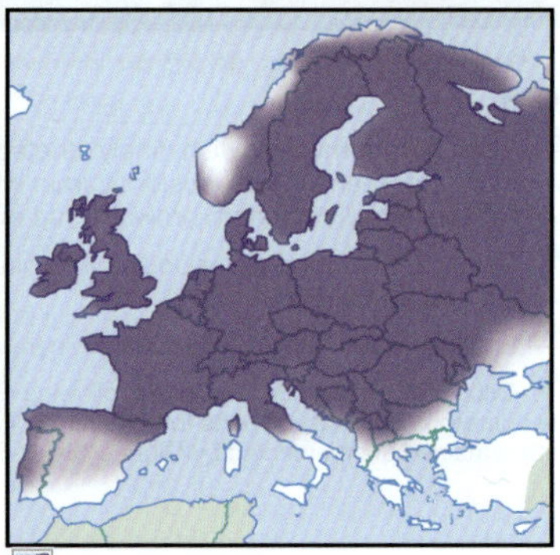

1-2 x 3-6 mm Germany: Schleswig-Holstein, Lindewitt near Schleswig

Europe to E Siberia

Shell reddish horny brown, often with black or brown encrustations, finely striated, up to 7-8 densely coiled and rounded whorls with deep suture, whorls higher than wide, lower side almost flat, upper side with large umbilicus, more than 1/3 of shell diameter, aperture narrow. Shell very characteristical, cannot be confounded.
Animal blackish dark red, tentacles very long, eyes small and black.

Mostly in permanent standing water bodies, with rich vegetation, tolerates humine substances, at lake margins, also in larger bays of large rivers. Usually up to 2 m deep, rarely below 20 m, preferred pH 5.2-9.6. In Switzerland up to 1800 m.

Eggs (diameter 0.5 mm) are laid in capsules of 3 mm size containing 6-8 eggs each, juveniles hatch after 10-12 days, maturity is reached after 22 months (Moquin-Tandon 1855: 443, Germain 1931: 520).

Threatened by continuous habitat destruction, construction projects in densely populated regions, and agricultural fertilization. Vulnerable in Switzerland, Austria and Bayern, very rare in Portugal.

Gyraulus

Gyraulus Charpentier, 1837
in Charpentier 1837: 21 *Planorbis (Gyraulus)*

Type species *Planorbis albus* Müller, 1774
(subsequent designation by Dall 1870: 351)

Established without description, with 4 species included. Name in original source attributed to Agassiz, but the composition of the species was Charpentier's responsibility. Contained originally *Planorbis hispidus* Draparnaud, 1805 (with synonym "*Pl. albus* Stud.") *subcarinatus, corneus* Studer, 1820, *imbricatus* and *cristatus*.

Shell small, disciform, smooth or covered by a fine spiral to reticulate sculpture, aperture wide, apertural diameter exceeds apertural height, whorls rapidly increasing in width. Penis with subterminal opening and sclerotised stylet.

Northern Hemisphere and Africa, possibly Australia

Meier-Brook 1983, Neubert 1998: 356.

Recently a number of *Gyraulus* species from a few isolated spots were briefly described as new (Glöer & Pešić 2007, Glöer & Rähle 2009 and others), without visibly contributing to a more profound understanding of the diversity of this group. These so-called new species are not listed here, their status as separate species needs to be confirmed in independent studies at larger scales. It makes however sense to set them on Red Lists, if this helps to protect their habitats. The study by Glöer & Vinarski 2009 on *Gyraulus stromi* was based on accurate and detailed field data.

Gyraulus acronicus (Férussac, 1807)

Planorbis acronicus Férussac, 1807 in J. Férussac 1807: 105
I am giving Férussac's single authorship here, as this has been traditionally done. I have not researched which one of the two Férussac's was responsible for this name (or both). It seems to have been overlooked that both Férussacs were co-authors of the 1807 work which should better be cited as "Férussac & Férussac 1807", and that the responsibilities were explained in the introduction.

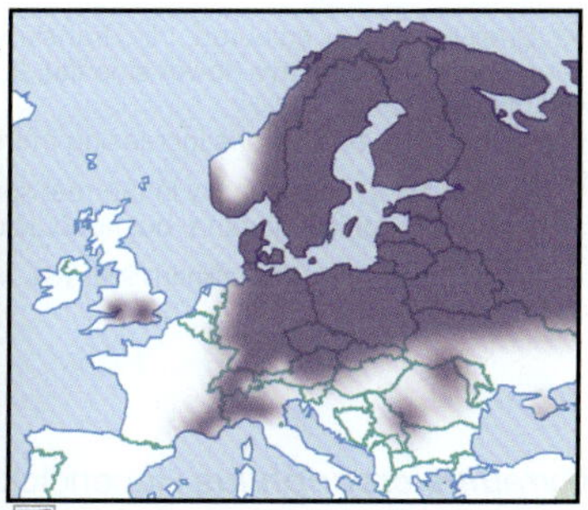

1-1.3 x 5-7 mm

Sweden: Torne-Lappmark

Europe to Lena river basin

Shell yellowish to horny brownish, finely and regularly striated, not shiny, 4-5 convex whorls with deep suture, upper side concave, lower side flat with impressed first whorls, last whorl rounded or angulated at periphery, not keeled.
Shells with rounded whorls can be confounded with *G. laevis* and *G. parvus*, which have shiny surfaces and anatomical differences. *G. stromi* is larger and has keeled whorls.

In England in clean, slow-flowing calcareous waters with densely vegetated margins, appears to be restricted to short sections of small rivers near their confluences with the Thames river. In Germany mainly in standing waters of various sizes, in Switzerland in shallow waters rich in vegetation with connections to large or small lakes. Tolerant against cold temperatures and humine acids, preferred pH in Norway 5.2-9.6. In Switzerland up to 700 m.

In Switzerland most habitats have been destroyed since 1900, small lakes were contaminated by fertilization, only few populations survived. In Norway present in 10 % of the lakes. Very restricted in its distribution in England. Critically Endangered in Switzerland and Germany, Endangered in Poland, Czech Republic, Slovakia and Tirol, Vulnerable in England and Austria.

Killeen & McFarland 2004, Glöer & Vinarski 2009.

Gyraulus albidus Radoman, 1953

Gyraulus albidus Radoman, 1953 in Radoman 1953: 55

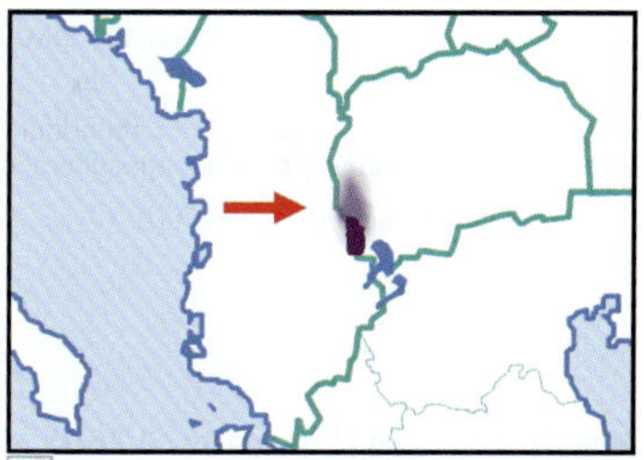

Shell whitish or very pale brown, almost transparent, in adults with weak yellowish hue, regularly striated, often also with very fine spiral lines, 3-4 whorls, strictly discoidal, equally convex at both sides, smoothly rounded at the periphery, first whorls depressed (concave), last whorl descending, aperture very oblique, umbilicus large but not very deep, so that upper and lower sides appear almost symmetrically.

1.0-1.2 x 3.2-4.7 mm

Strictly limited to the lake margin of Lake Ohrid, 0.5-10 m depth, on soft bottom and sandy beaches, occasionally on gravel and stones at the shore.

EU Red List: Vulnerable VU. Potentially threatened by habitat degradation due to denser human occupation of the sandy bays of the lake, and agricultural contamination resulting in enhanced reed belt growth.

Hubendick & Radoman 1959: 239.

Rep. Makedonija, Albania

Gyraulus albus (Müller, 1774)

Planorbis albus Müller, 1774 in Müller 1774: 164

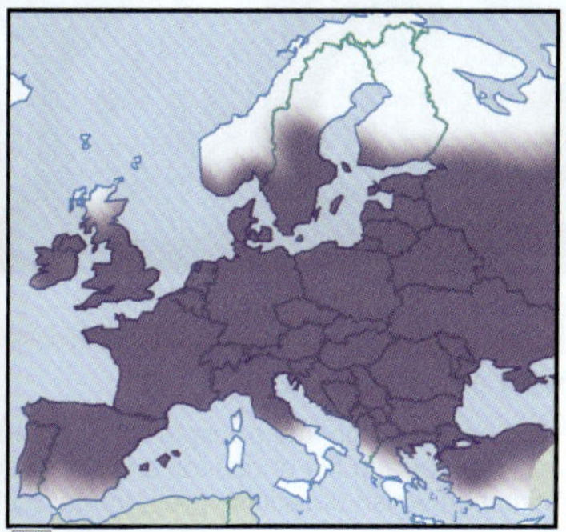

1.5-2.5 x 4-7 mm

Germany: Niedersachsen, Franzhausen

Eurasia

Shell yellowish or greenish white or reddish, transparent, radially and spirally finely striated, 3-4 regularly rounded whorls, no keel, lower flat side with depressed first whorls, upper side concave with large umbilicus, more than 1/3 of shell diameter, aperture bent obliquely to the upper side. The reticulate pattern is characteristic.
Animal dark or yellowish brown with numerous small black dots, tentacles very long and transparent, eyes small, at the base of the tentacles.

Standing and slowly running waters, at margins of lakes, in shallow depths on mud and water plants. Wide ecological amplitude, tolerates mild pollution, but not seasonal drought. In Switzerland in up to 1500 m altitude.

Reproduction between March and September, eggs (diameter 0.4 mm) are laid in 6-8 ovoid capsules (3-4 mm diameter) containing 4-10 eggs each, juveniles hatch after 10-14 days, maturity is reached after 20-23 months.

Moquin-Tandon 1855: 440, Germain 1931: 535, Falkner 1990: 134, Glöer 2002: 268.

Gyraulus chinensis (Dunker, 1848)

Planorbis chinensis Dunker, 1848 in Dunker 1848: 41

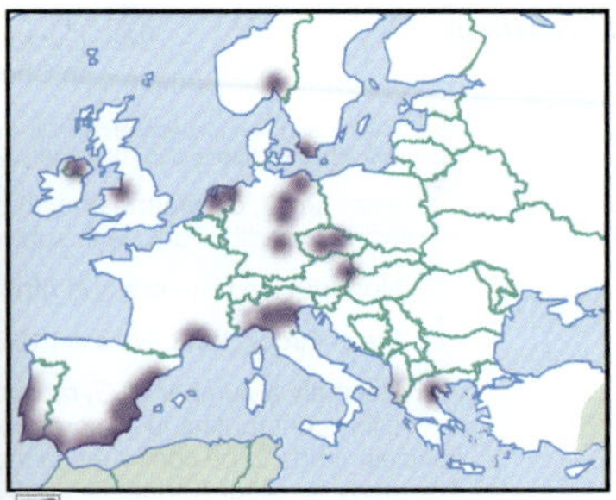

E and S Asia (Iran to India, Korea, S Japan and Neuguinea), introduced elsewhere
Map based on published sources, probably more frequent in Europe.

0.8-1.5 x 3-5 mm — Spain: Ebro delta

Shell light horny coloured, thin, translucent, shiny, striated and sometimes with fine spiral lines forming a fine reticulate pattern and spirally arranged short hairs, upper side concave, lower side flat, 3.5-4 whorls, last whorl broadened, aperture slightly bent to the upper side. The animal is light grey, head and foot with blackish spots.
Flatter than *G. albus*, whorls increasing more slowly, mantle with large pigment patches visible from outside the shell.

In rice crops, occasionally in thermal springs. Intermediate host of various trematodes, also in Spain. Survives in central and N European greenhouses, fish tanks and tropical aquaria, but not outside.

Meier-Brook 1983: 97, Falkner 1990: 134, Manganelli et al. 1995: 11, Brown et al. 1998, Glöer 2002: 271, Anderson 2005, Beran & Glöer 2006.

Gyraulus crenophilus Hubendick & Radoman, 1959

Gyraulus crenophilus Hubendick & Radoman, 1959 in Hubendick & Radoman 1959: 223

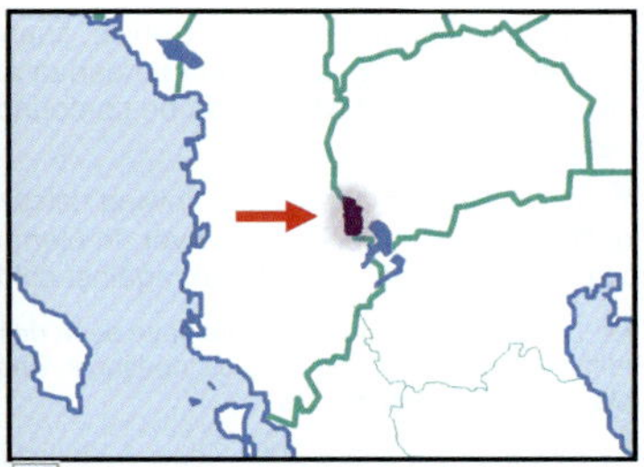

Up to 7 x 10 mm — Rep. Makedonija: Lake Ohrid, near Ohrid — Museum Budapest HNHM — Lake Ohrid

Shell pseudodextral, coarsely striated and with fine spiral lines and sometimes with small periostracal hairs in juveniles, 3.5-4 whorls, first 2 whorls on pseudo-upper side almost flat, last whorls usually smoothly rounded, aperture connected at pseudo-lower side, margin sharp, umbilicus moderate and slightly covered by the reflected columellar margin.
Shells often covered by algae. Animal greyish, sometimes blackish, tentacles usually densely pigmented.

Restricted to some springs near the lake shore of Lake Ohrid, lives in small creeks formed by the spring water, on stones and rocks in constant water temperatures of 12-13 °C.

EU Red List: Endangered EN. Not abundant where present, threatened by contamination of the springs by domestic waste and aquaculture, sediments from new constructions, deforestation and erosion, agricultural contamination near Lake Prespa resulting in increased nutrient load in Lake Ohrid, and touristic pressure.

Dhora & Welter-Schultes 1996: 110.

Gyraulus crista (Linnæus, 1758)

Nautilus crista Linnæus, 1758 in Linnæus 1758: 709 (*Nautilus Crista*)

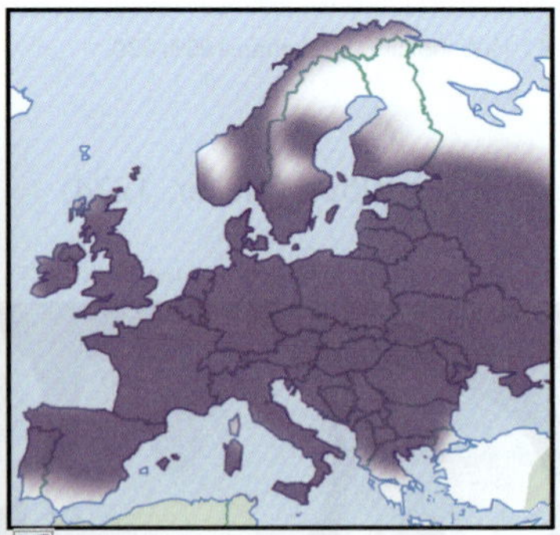

0.7-1.1 x 2-3 mm — Germany: Schleswig-Holstein, Selenter See — Northern Hemisphere

Shell very thin and transparent with dark periostracum, smooth or finely striated, sometimes with periostracum riblets, 3 whorls, angulated or keeled, convex at upper and flat or convex on lower side, aperture amost circular, located on the upper side of the periphery.
Animal brownish or reddish grey, frontal part blackish, with many black dots, tentacles long and cylindrical, eyes black.

Permanent and silent small water bodies with rich vegetation, on leaves of water plants, also at vegetated lake margins. Tolerates soft water, preferred pH 5.7-9.6. Has good powers of passive dispersal. In Switzerland in up to 1400 m altitude, in Britain in lowlands.

Eggs (diameter 0.25 mm) are laid in 1.5 mm sized capsules containing 3-6 eggs each, juveniles hatch after 10-12 days.

In Europe in habitats that are continuously destroyed by landscape clearing and construction projects.

Moquin-Tandon 1855: 438, Germain 1931: 540, Clarke 1981: 184, Falkner 1990: 134, Glöer 2002: 277.

Indicated range in Finland and NW Russia uncertain, rare above 62° N. In Canada S Ontario to S Alberta, rare above 55° N.

Gyraulus fontinalis Hubendick & Radoman, 1959

Gyraulus fontinalis Hubendick & Radoman, 1959 in Hubendick & Radoman 1959: 231

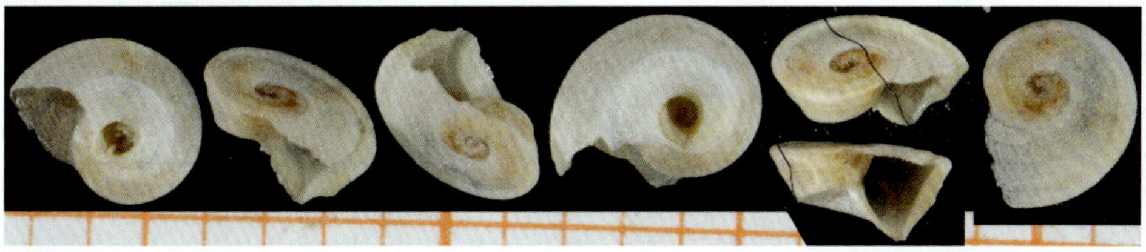

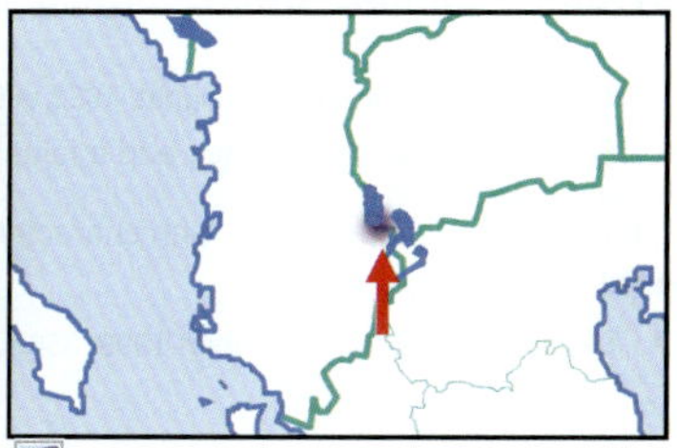

Up to 5 x 11 mm

Rep. Makedonija: Sveti Naum near Lake Ohrid

Lake Ohrid

Shell pseudodextral, coarsely striated and with a finer and more regular (not very prominent) spiral sculpture, pseudo-upper side slightly convex, 3-4 whorls, first whorls flat on pseudo-upper side, keeled (later angulated) at periphery and keeled again basally at the pseudo-lower side, aperture almost triangular in juveniles, more rounded in larger shells, margin sharp and slightly reflected at the columellar side, umbilicus wide.
Differs from *G. trapezoides* in the lack of a predominant spiral sculpture and a narrower umbilicus.
The animal is basically greyish, with reddish tentacles, head and sides of the foot.

On the bottom of a small pond-like lake, water temperature constantly 10-11 °C.

EU Red List: Endangered EN. Extremely restricted range, very sensitive on chemistry and temperature, threatened by increasing pollution and aquaculture.

Gyraulus laevis (Alder, 1838)

Planorbis laevis Alder, 1838 in Alder 1838: 337 (*Planorbis lævis*)

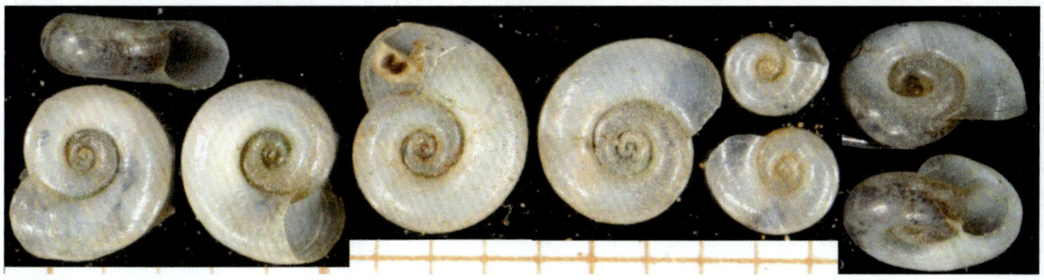

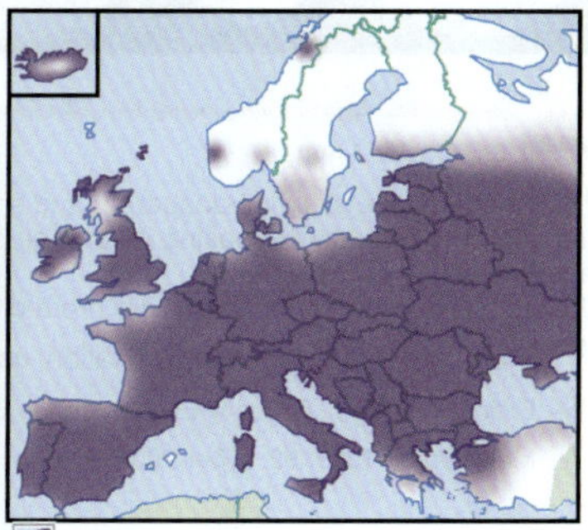

1-2.5 x 3.0-4.5 mm

Germany: Schleswig-Holstein, near Diekenhorn

Europe

Shell horny brownish, shiny, irregularly striated, no spiral sculpture, 3-4 strongly convex whorls with deep suture, at lower side in one regular plain, never keeled, upper side concave.
Usually slightly larger than *G. parvus*, penultimate whorl does not deviate from the plain.

Sunny shallow water zones of clean and silent waters with moderately rich vegetation. Often in undisturbed waters with eutrophic conditions but with rich oxygen contents, acts also as a pioneer species (in lake margins that are in the process of becoming land). If the other conditions are optimal the species tolerates high nitrate contents. Tolerates slightly brackish water, pH values in Norway 9-9.9. In lowland England often noted in newly formed reservoirs and ornamental lakes in public parts. In Switzerland in up to 1500 m altitude.

Very scattered and rare, always local in Europe, at many places extinct or shortly before. In Britain it was among the first species after the retreat of the glacial ice, it seems to have good powers of passive dispersal. Its present distribution in Britain is partly relict, and partly due to its ability to colonize new artificial habitats, so no general decline has been observed. Extinct in Sweden, Critically Endangered in Germany and Switzerland, Endangered in the Netherlands, Ireland, Poland and Tirol, Vulnerable in Austria and Norway (only at 2 locations), Rare in the Czech Republic, Slovakia and Denmark.

Gyraulus lychnidicus Hesse, 1928

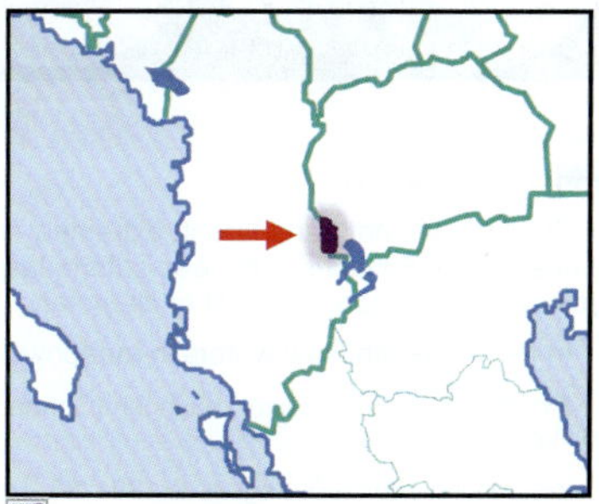

Lake Ohrid

Gyraulus lychnidicus Hesse, 1928 in Hesse 1928: 6, Taf. 1

Shell pseudodextral, whitish yellow, first whorls smooth and not depressed, adult shell with prominent spiral lines, also at the pseudo-lower side and inside the umbilicus, 3-4 whorls with deep suture, step-like separated from each other and distinctly keeled (scalarid), with a second keel running along the pseudo-lower side, keels become less prominent near the aperture, apertural margins approach each other closely, connected by a white layer at the parietal side, umbilicus very wide.
Animal can be entirely pigmented or practically unpigmented.

3-8 x 5-10 mm, usually 3.5-4.5 x 4-6 mm

Littoral zone of Lake Ohrid, above *Chara* meadows, usually on stones, also on gravel.

EU Red List: Near Threatened NT. Abundant, but potentially threatened by declining water quality.

Hubendick & Radoman 1959: 234, Dhora & Welter-Schultes 1996: 173.

Gyraulus parvus (Say, 1817)

Planorbis parvus Say, 1817 in Say 1817: B-6, Pl. 1

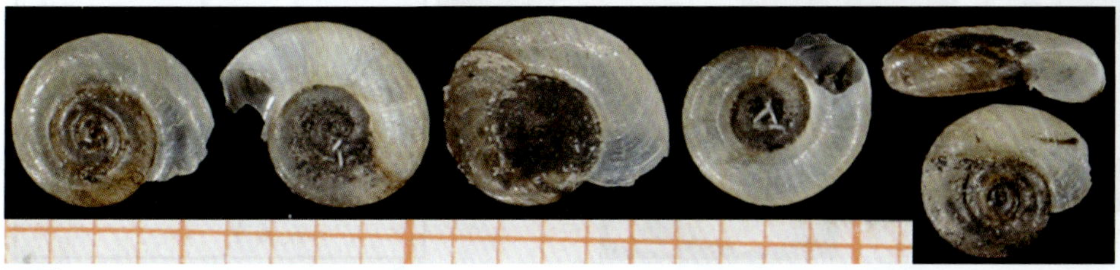

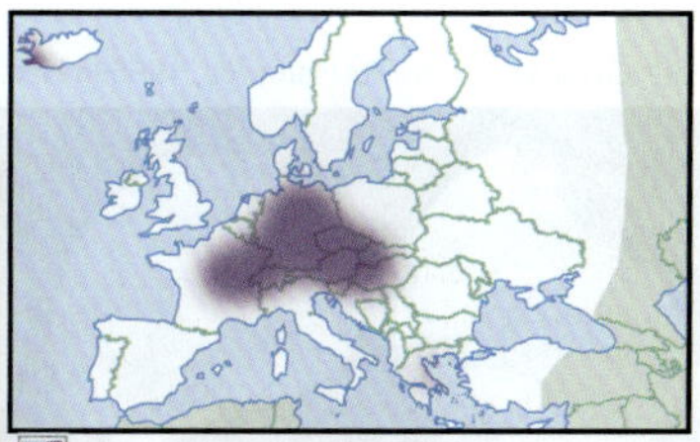

N America (Alaska to S Hudson Bay and Labrador, N USA), S Greenland, Iceland. Introduced to Europe, rapidly expanding its range eastwards. In Germany since 1973, Switzerland 1994, Czech Republic 2001.

0.7-1.3 x 2.5-5 mm

Germany: Schleswig-Holstein, Beltingharder Koog

Shell usually whitish-grey or light horny coloured, transparent, moderately shiny, thin, irregularly striated, 4-5 rounded whorls without keel, upper side concave, lower side flat with obliquely distorted/deviating penultimate whorl, aperture can be oblique towards to upper side, umbilicus more than 1/3 of shell diameter. Animal not much pigmented.

In North America in all kinds of permanent or temporary water-filled habitats that support submersed aquatic vegetation, mostly on muddy bottom, abundant. In Europe in shallow standing waters of various sizes rich in vegetation, mostly in artificial water bodies in gardens or reservoirs, with few competing species. In Switzerland in up to 500 m altitude.

Clarke 1981: 180, Falkner 1990: 134, Turner et al. 1998: 128, Glöer 2002: 273.

Gyraulus piscinarum (Bourguignat, 1852)

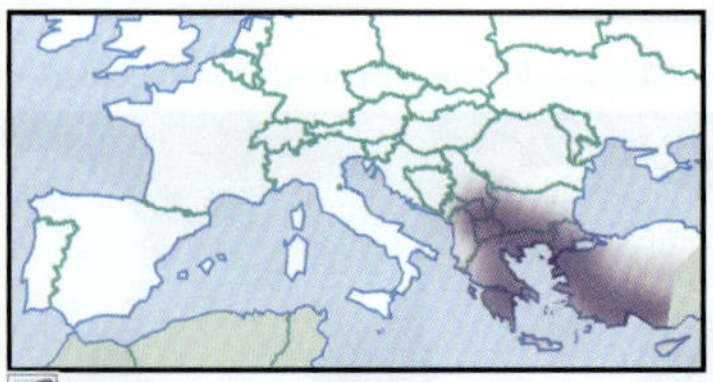

SE Europe to Oman
Indicated range uncertain.

Planorbis piscinarum Bourguignat, 1852 in Bourguignat 1852: 22 (*Planorbis Piscinarum*) Type locality in Egypt.

Shell with axial growth lines and a faint reticulate pattern, 3-4 rounded whorls, or slightly angulate at periphery, subsurface convex.
Spermatheca of tadpole-type, with characteristic diffuse mantle pigmentation.

Diameter up to 6 mm

Meier-Brook 1983: 52, Neubert 1998: 356.

Gyraulus riparius (Westerlund, 1865)

Planorbis riparius Westerlund, 1865 in Westerlund 1865: 106

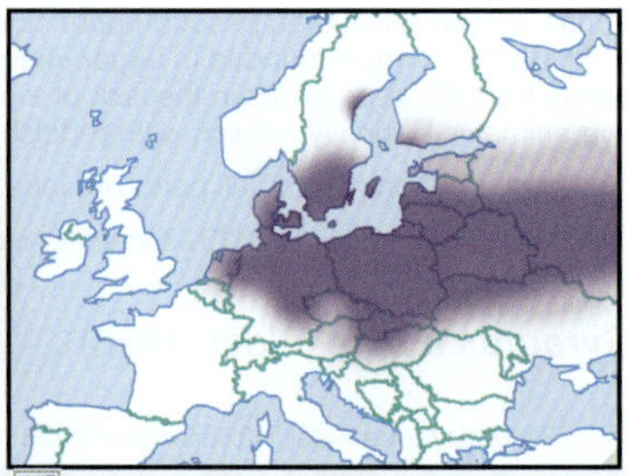

0.5-0.8 x 2-3 mm — Germany: Schleswig-Hostein, Kalksee near Preetz

Europe to Ural and Caspian Sea

Shell whitish, slightly transparent and shiny, weakly striated, 3-3.5 rounded whorls, with a blunt edge at the periphery, upper side concave and lower side flat, last whorl rapidly increasing before the aperture, aperture slightly bent upwards and elliptical, umbilicus on upper side slightly excentrical, 30 % of shell diameter.
Juvenile *Hippeutis complanatus* are more distinctly angulated, the aperture is heart-shaped.

Standing waters with rich vegetation, usually in the litoral zone of lakes. In many lakes in Finland.

Extinct in Bayern, Critically Endangered in Germany, Vulnerable in Poland.

Meier-Brook 1983: 92, Glöer 2002: 274.

Gyraulus rossmaessleri (Schmidt, 1852)

Planorbis rossmaessleri Schmidt, 1852 in Schmidt 1852: 179 (*Planorbis Rossmaessleri*)
Name was attributed to Auerswald, description was partly formatted in quotation marks (probably indicating Auerswald's authorship), partly expressly by Schmidt, so under Art. 50.1.1 the authorship for the new name must be attributed to Schmidt alone.

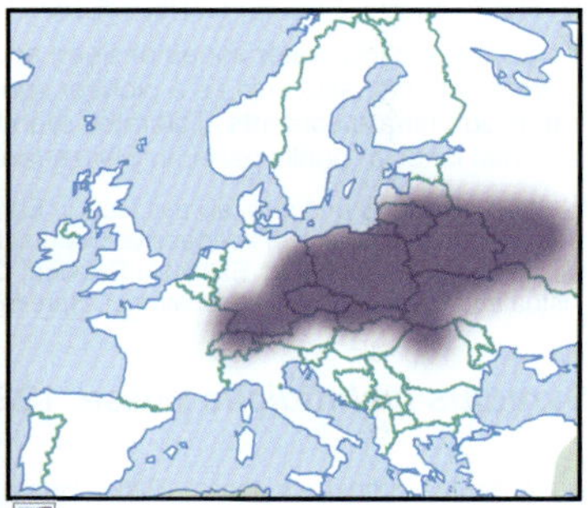

1.3-1.5 x 4-6 mm — Poland: Lubrzanka river — Museum Budapest HNHM

Europe
Reliable records are rare.

Shell brown, not transparent, extremely finely and regularly striated, with fine spiral lines, flat with first whorls slightly depressed on lower side, concave on upper side, 4-4.5 whorls with deep suture, last whorl rounded, not ascending near aperture, aperture not oblique, inside usually with milky white lip, umbilicus at upper side wide, more than 1/3 of diameter.

In temporary shallow waters in meadows, also in swamps. Found particularly in spring.

Extinct in Switzerland and Thüringen (Germany), Critically Endangered in Germany and Czech Republic. Populations are rapidly decreasing, the few habitats are continuously being destroyed.

Westerlund 1885: 80, Vogt et al. 1994: 83, Turner et al. 1998: 129, Glöer 2002: 276, Glöer & Sîrbu 2006.

Gyraulus stankovici Hadžišče, 1955

Gyraulus stankovici Hadžišče, 1955 in Hadžišče 1955: 175 (*G. (Carinogyraulus) stankovići*)

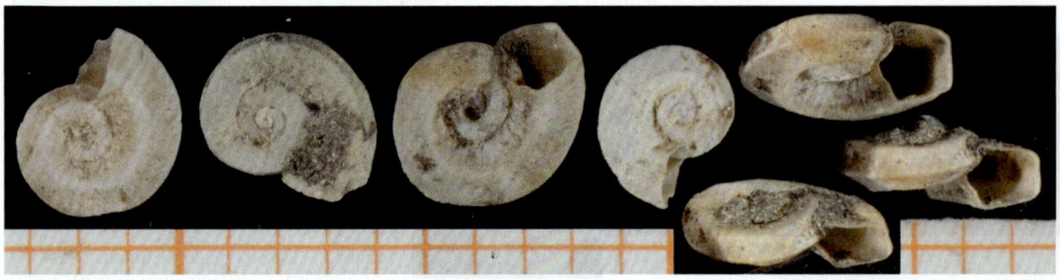

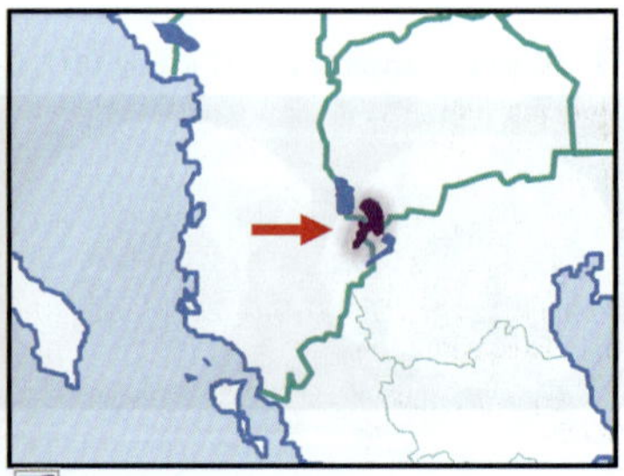

1.8-2.1 x 4-5 mm — Rep. Makedonija: Lake Prespa, Otesevo

Lake Prespa

Shell dirty whitish, finely striated, upper side depressed, 3.5 whorls with marked suture and 3 edges (2 at periphery and one at lower side), aperture pentagonal, connected at parietal side by a distinct layer, umbilicus wide, 1/3 of shell diameter.

Rocky habitats in the littoral zone.

EU Red List: Endangered EN. Not abundant. Seriously threatened by abruptly changing water levels, agricultural contamination, introduced fish and increased pollution.

Gyraulus stromi (Westerlund, 1881)

Planorbis stromi Westerlund, 1881 in Westerlund 1881 (Öfversigt 4): 63 (*Pl. (Gyraulus) strömi*)
The name *strömi* was not of German origin (probably dedicated to the Norvegian naturalist Hans Strøm (1726–1797)), so the name must be converted to *stromi*, not *stroemi* (Art. 32.5.2.1).

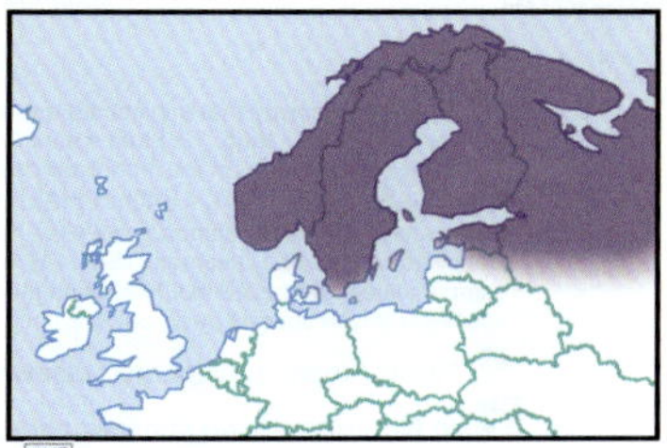

Norway to E Russia

Shell finely striated, whorls keeled, regularly rounded at both sides, first whorls deep on functional upper side, lower side slightly umbilicated. Animal light grey.
G. acronicus has rounded or angled whorls and is smaller.

1-1.5 x 5-9 mm

Tends to be more frequent than *G. acronicus* in the northern regions.

Glöer & Vinarski 2009.

This species was overlooked in Scandinavia after 1900, but was well known in Russia since the mid-1900s. Often incorrectly spelled *G. stroemi*.

Gyraulus trapezoides Poliński, 1929

Gyraulus trapezoides Poliński, 1929 in Poliński 1929: 162

Rep. Makedonija: Orce Nikolov near Lake Ohrid

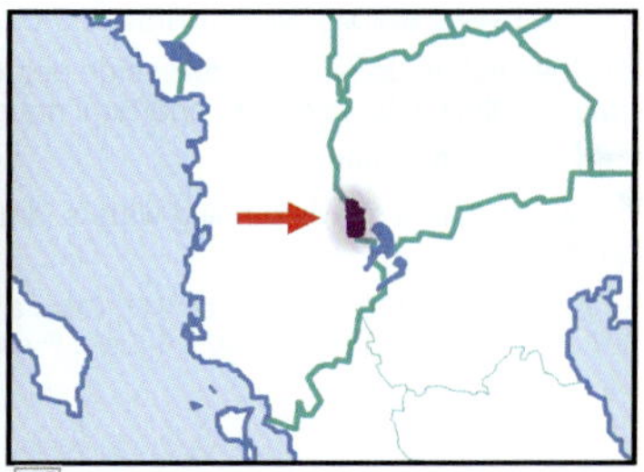

Lake Ohrid

2.5-3 x 7-9 mm

Shell beige to pale brown, with a rough spiral sculpture and rows of periostracal hairs, pseudo-upper side usually flat, 3-3.5 whorls, with two keels, one at the periphery at the pseudo-upper side, one half-way at the pseudo-lower side, opening of the umbilicus also sharply angulated, aperture often with a pronounced trapezoid profile, or triangular with a rounded section at the columellar/parietal side, umbilicus widely open. Shell shape variable, unusually highly spired pseudo-dextral shells are also recorded.
Differs from *G. fontinalis* in its more prominent spiral sculpture and wider umbilicus.

Litoral and sublitoral zone of the lake, 5-25 m depth, seems to prefer the bottom, on *Chara* meadows and in the shell zone.

EU Red List: Endangered EN. The rarest *Gyraulus* species of Lake Ohrid, threatened by increasing pollution and sediment load impacting the algae on the hard substrate.

Hubendick & Radoman 1959: 237, Dhora & Welter-Schultes 1996: 173, Dhora 2004: 143, Fehér & Erőss 2009.

Hippeutis

Hippeutis Charpentier, 1837
in Charpentier 1837: 22 *Planorbis (Hippeutis)*

Type species *Helix complanata* Linnæus, 1758
(subsequent designation by ???)

Established without description, with 2 species included (*complanatus* with synonym *lenticularis*, and *Planorbis nitidus* Müller, 1774). Name in original source attributed to Agassiz, but the composition of the species was Charpentier's responsibility. Küster, Dunker & Clessin 1886 [1878]: 34 gave as type *Helix fontana* Lightfoot, 1786, which was not originally contained. Westerlund 1902: 121 gave as type *Pl. nitidus*. Wenz & Zilch 1959-1960: 113 gave as type *Helix complanata*.

Hippeutis complanatus (Linnæus, 1758)

Without gender agreement: *Hippeutis complanata* (Linnæus, 1758)

Helix complanata Linnæus, 1758 in Linnæus 1758: 769
Planorbis stossichi Clessin, 1878 in Clessin 1878: 126, Taf. 5 (*Planorbis Stossichi*)

Germany: Schleswig-Holstein, Plöner See

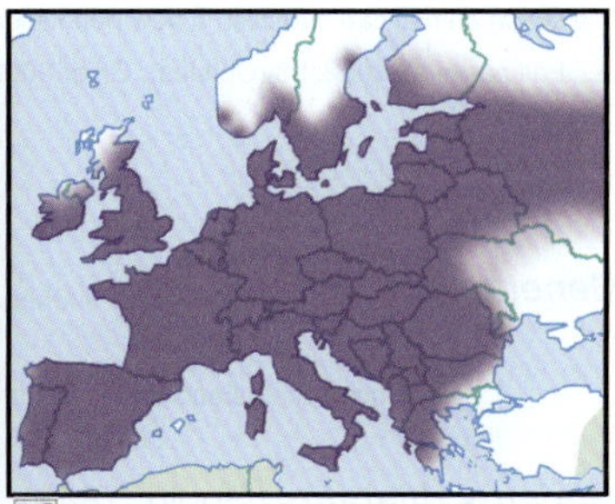

NW Africa and Europe to W Siberia

0.7-1.2 x 3-5 mm

Shell usually whitish grey or light yellow, pale silky shiny, finely striated, translucent, lenticular with 3-4 strongly keeled whorls, keel in the centre of the periphery, without inner compartments visible from outside, lower side almost flat, upper side with umbilicus 1/6-1/4 of shell diameter. Looks much like *Segmentina nitida*, but aperture not so much bent to the upper side.
Animal small, adults blackish or dark reddish brown, juveniles yellowish grey, slightly transparent, with many small black dots, tentacles very long and transparent, eyes big and black.

Standing or slowly running shallow waters rich in vegetation, on muddy or loamy bottoms or water plants, also temporary water bodies, in Britain often in small closed ponds. Also at vegetated margins of lakes. Calciphile, does not tolerate desiccation, preferred pH in Norway 6.1-9.6. In Switzerland in up to 1100 m, in Bulgaria 700 m.

Eggs (diameter 0.2-0.25 mm) are deposited in clutches or agglomerations of 3-5 eggs on *Hippeutis* shells, juveniles hatch after 10-15 days, maturity is reached after 23 months.

Habitats are threatened by drainage, intensive agriculture and fertilization, sinking of ground water levels. Endangered in Tirol, Vulnerable in Austria, Switzerland, large parts of Germany, in SW Spain not frequent, in Portugal only known from few localities.

Moquin-Tandon 1855: 426, Germain 1931: 541, Westerlund 1885: 78 (*stossichi*), Falkner 1990: 134, Turner et al. 1998: 131, Kerney 1999: 68, Glöer 2002: 279, Pérez-Quintero et al. 2004, Hubenov 2007.

I found no constraints in the literature to regard *stossichi* from Dalmatia as a different species.

Helisoma

Helisoma Swainson, 1840
in Swainson 1840: 337 *Planorbis (Helisoma)*

Type species *Planorbis bicarinatus* Say, 1819
(monotypy, fixed under Art. 70.3 here)
(= currently *Helisoma anceps* (Menke, 1830))

Type by monotypy (not original designation, some genera had more than one species listed), cited by Swainson as "*H. bicarinata* Sow. Gen. f. 4". It is likely that Sowerby 1821-1834 misidentified the species (Binney 1865: 124) and meant *Planorbis campanulatus* Say, 1821. Under Art. 70.3 of the Code shall now be fixed *Planorbis bicarinatus* as type species for *Helisoma*, misidentified as *Planorbis campanulatus* in the monotypic designation by Swainson 1840: 337.

Planorbella Haldeman, 1843 in Haldeman 1842-1847: 14, type species *Planorbis campanulatus* Say, 1821 (subsequent designation by Dall 1870: 351 or earlier)
(= currently *Planorbella campanulata* (Say, 1821))
Date was Jan. 1843. Haldeman (1843: 14) referred to plate [*Planorbis*] 1, where two species were figured: *Planorbis bicarinatus* and *Pl. campanulatus*. The verbal description on p. 14 ("aperture campanulate") suggested that Haldeman meant *Pl. campanulatus* and possibly liked to exclude *Pl. bicarinatus*. But Haldeman did not explicitly refer to *Pl. campanulatus* as the one and only included species. So type species designation is not possible by monotypy, only subsequently in a later work.

North America, dispersed elsewhere

In North American sources *anceps* and *duryi* are usually classified in *Helisoma*, in some European sources in *Planorbella*. It may be convenient to use the North American generic classification also in the other continents to which North American species were artificially dispersed.
Google Scholar 12.2011: "helisoma anceps" 704, "planorbella anceps" 23 (only European sources)
Google Scholar 12.2011: "helisoma duryi" 793, "planorbella duryi" 120

Helisoma anceps (Menke, 1830)

Planorbis anceps Menke, 1830 in Menke 1830: 36
Established without description but with a reference to "Lister Conch. tab. 139 fig. 44" = Lister et al. 1686: p. [13].

Shell flat on upper side, lower side depressed, upper side with edge, 4-4.5 whorls, last whorl inflated and with straight outline in last section, apertural margin auricular.

6-12 x 10-21 mm

Standing and slowly running waters, among aquatic plants, avoids temporary waters. Widely distributed in aquaria.

Clarke 1981: 198, Glöer 2002: 247, Glöer & Zettler 2005: 17, Glöer & Sîrbu 2006 (not in W Romania), Reitano et al. 2007: 313 (Italy), Erőss et al. 2005: 6 (Lake Prespa).

North America (Mexico to Hudson Bay), occasionally introduced to European waters

Helisoma duryi (Wetherby, 1879)

Planorbis duryi Wetherby, 1879 in Wetherby 1879: 99
Original spelling was not *durii* (incorrectly cited in Zoological Record 1880: Moll. 92).

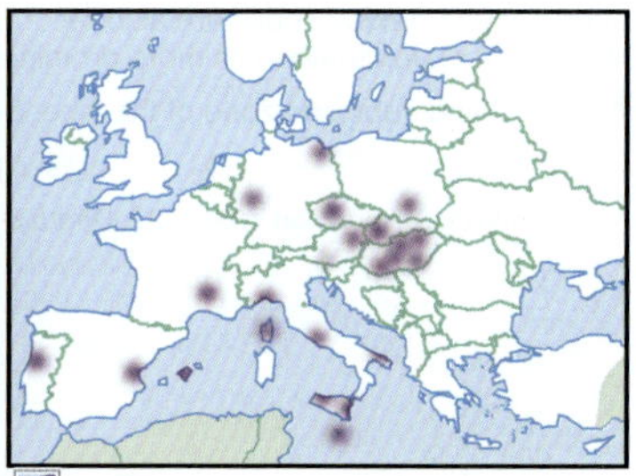

Up to 10 x 25 mm

(1-2) Slovakia: thermal pond in Piešťany spa, (3-9) Austria: Villach (1-2) Jozef Grego

S USA (Florida), introduced to South America, Hawaii, S and E Africa, E Atlantic islands and Europe

Shell light orange or yellowish, smaller than *Planorbarius corneus*, lower side flat with blunt edge at the peripery, upper side with deep umbilicus, 4-4.5 whorls, last whorl not too much inflated, apertural margin not reflected. Animal reddish or pink, olive or blue, with pigment spots.

In Europe in outdoor habitats restricted to artificially warmed waters, in temperate zones there are probably no permanent populations. In Sicily restricted to small artificial ponds and fountains.
The species is sold for aquaria in Europe. Snails feed on desintegrating plants, detritus, dead aquaria animals, algae, fish food and vegetables. Animals can survive short periods of drought, they withdraw deeply inside the shell.

13-17 eggs are fixed in a transparent envelope at hard substrate, juveniles hatch (24-26 °C) after 10 days. Self-fertilization is possible, at a lower hatch rate of only 5 %. One single animal can establish a new population.

Falkner et al. 2002: 98 (Corsica), Glöer 2002, Alexandrowicz 2003 (Italy: Lago Albano; Poland: Kraków), Beckmann 2007: 49 (Mallorca), Reitano et al. 2007: 313 (Italy).

Menetus

Menetus Adams & Adams, 1855
in Adams & Adams 1858: 262 *Planorbis (Menetus)*

Type species *Planorbis opercularis* Gould, 1847
(subsequent designation by Dall 1870: 351)
(= currently *Menetus dilatatus* (Gould, 1841))

United States and British Columbia

Menetus dilatatus (Gould, 1841)

Planorbis dilatatus Gould, 1841 in Gould 1841: 210
Not *Planorbis dilatatus* Pfeiffer, 1841 (in L. Pfeiffer 1841: 225) from Hungary which was hopefully published later.

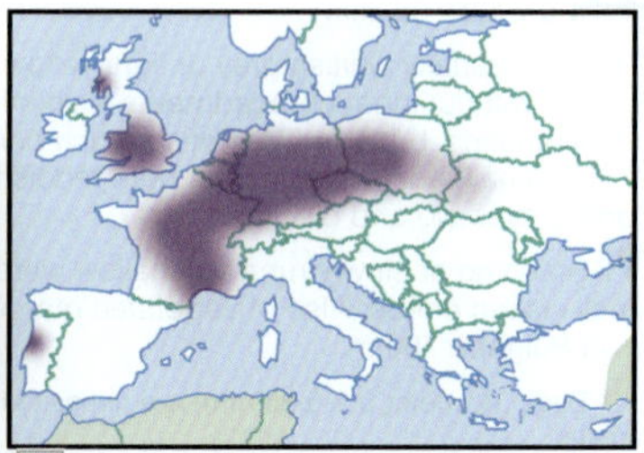

0.8-1 x 2-4 mm

Germany: Baden-Württemberg, Freiburg

North America (New England to Florida and Texas), introduced to Europe

Shell yellowish green, weakly striated and with reticulate pattern, lower side flat, upper side rounded, 2-3 whorls, angulated at the lower side of the periphery, last whorl inflated, rounded in its last section, aperture rounded, margin of fully grown shells reflected, umbilicus on upper side, narrow and deep.

In North America in small ponds and humid habitats on moss. In Europe in silent zones of slowly running waters. In Britain found in canals and lakes, was often associated with discharge of heated water (discharges of condenser water of steam engines near mills, cooling water from a nuclear power plant), but survives also freezing temperatures. Seems to tolerate anaerobic bottoms.

Kerney 1999: 71, Glöer 2002: 248, De Oliveira et al. 2010 (Portugal).

Planorbarius

Planorbarius Duméril, 1806
in Duméril 1806: 164 (*planorbarius*)

Type species *Helix cornea* Linnæus, 1758
(subsequent monotypy by Froriep 1806: 165)

Appeared without species.

Planorbarius corneus (Linnæus, 1758)

Without gender agreement: *Planorbarius cornea* (Linnæus, 1758)

Helix cornea Linnæus, 1758 in Linnæus 1758: 770

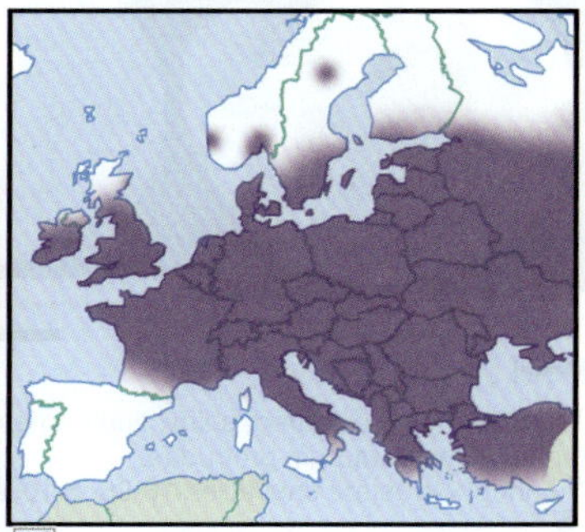

10-17 x 22-40 mm

Germany: Schleswig-Holstein, Worth near Geesthacht

W Eurasia

Germany: Rheinland-Pfalz, Speyer
Claus Ableiter

Shell large, light yellowish with brown, reddish or greenish periostracum, radially and spirally weakly striated, upper side weakly and lower side deeply depressed, 3-4.5 rounded whorls, last whorl predominating. Animal brown or reddish.

Silent waters in lowlands with rich vegetation, standing or slowly moving, tolerates periodical droughts and anaerobic bottoms, temperature optimum 19-20 °C, perfers pH around 6-9. Poor powers of passive dispersal, but frequently introduced to artificial ponds. Feeds mainly or exclusively on detritus. Often released from aquaria, but does not survive permanently in highlands. In Switzerland up to 800 m, in Austria 500 m, Erzgebirge (Germany/Czech Rep.) 750 m.

Reproduction in spring and autumn at water temperatures above 15 °C, eggs (diameter 1.2-1.7 mm) are laid in mostly elongate capsules of 8-15 mm width, each strain containing 12-40 eggs, fixed to aquatic plants, embryos are reddish with transparent shells, juveniles hatch after 14-16 days, life span up to 3 years. Self-fertilization is possible, one single released animal can establish a stable population, but only 5 % of the juveniles in self-fertilized eggs will hatch.

Moquin-Tandon 1855: 445, Germain 1931: 518, Falkner 1990: 132, Glöer 2002: 244.

Planorbarius metidjensis (Forbes, 1838)

Planorbis metidjensis Forbes, 1838 in Forbes 1838: 254, Pl. 12 (*Planorbis Metidjensis, Planorbis Metidgensis*)
Spelled *Planorbis Metidgensis* on p. 254 in the text, and *Planorbis Metidjensis* on Pl. 12. In the text the name of the plain of Metidja was mentioned several times. The spelling is taken from Pl. 12. The *g* in the name on p. 254 was probably an inadvertent error (Art. 32.5).
Planorbis dufourii Graells, 1846 in Graells 1846: 11

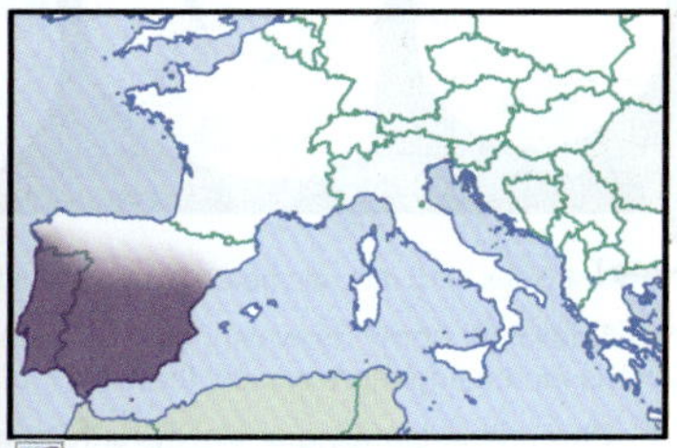

8 x 16-18 mm — Juveniles (5 mm) — Portugal: Algarve, Corte do Gago near Castro Marim — Valter Jacinto

NW Africa, S Portugal, S Spain
Indicated range in central and N Spain uncertain.

Shell light yellowish with brown, reddish or greenish periostracum, wrinkled and spirally weakly striated, 3-4 rounded whorls, last whorl predominating, upper side with deeper umbilicus, lower side flatter. Animal almost black with greyish foot and tentacles, exes are located at the internal sides of the tentacles' base.

In standing slowly moving water bodies, also in containers for artificial water supply, on mud, stones and water plants. Not rare in S Portugal and SW Spain.

20-30 eggs are laid in an envelope few days after copulation and fixed at shells or submerged water plants, juveniles hatch after 4 weeks.

Graells 1846: 11, Nobre 1941: 192, Talaván Gómez & T. 2006: 35 (Almería), Pérez-Quintero 2009. Includes *Planorbis dufourii* described from central Spain, a junior synonym.

Presence of a true distributional gap between *Pl. medidjensis* in NE Spain and *Pl. corneus* in S France needs to be verified.

Portugal: Algarve, Corte do Gago near Castro Marim
Valter Jacinto (determined by Daniel Rondelaud)

Planorbis

Planorbis Müller, 1773
in Müller 1773: [25]

Type species *Helix planorbis* Linnæus, 1758
(subsequent absolute tautonymy)

Appeared without species. First (14) species were included by Müller 1774: 152. On p. 157 under *Pl. carinatus* was cited the Linnean description of *Helix planorbis* (No. 662) without mentioning this name. On p. 161 in a discussion under *Pl. umbilicatus* the name *Helix planorbis* was mentioned. Kennard & Woodward 1924: 10 accepted *Helix planorbis* as type by absolute tautonymy. All previous authors had given as type *Helix cornea* Linnæus, 1758, which was not originally contained.
For those who do not accept subsequent absolute tautonymy (the Code has a gap and allows only subsequent monotypy, but does not talk about subsequent absolute tautonymy), then the type was fixed by subsequent designation by Kennard & Woodward 1924: 10. Cowie 1998: 46 gave as type *Pl. carinatus*, but overlooked that Müller had mentioned the name *Helix planorbis* on p. 161 (Cowie cited Kennard & Woodward 1924: 9, but seems to have overlooked their relevant statement on p. 10, Cowie 1998 argued that *Helix planorbis* had not been mentioned by Müller 1774).
ICZN Opinion 335: type species by absolute tautonymy, but this is not possible because no species were originally included in 1773. Should be subsequent absolute tautonymy.

Shell medium-sized, disciform, periphery bluntly angulated or keeled. Penis sheath short, penis without sclerotisation or stylet.

Soldatenko & Starobogatov 2000 recognised 12 species in this genus, of which 10 were reported to live in Europe (*Pl. planorbis*, *umbilicatus* Müller, 1774, *carinatus*, *kubanicus*, *philippianus* Monterosato, 1902 (in Caziot 1902: 262), *intermixtus*, *rhombeus*, *dubius*, *macedonicus*, *presbensis*). This system has not been accepted by western authors.

Glöer & Pešić 2010 proposed a system based on the number of prostate diverticles as the main feature to distinguish species, but they also demonstrated that this character was locally very variable, obviously providing a striking contrast to their conclusions.

They recognised 7 species in SE Europe (including a new taxon known from only one locality in Montenegro, which was not compared with other species), but did not study the degree of local variation at small scales. *Pl. planorbis* was subdivided into a northern and southern group, the southern group was called *Pl. intermixtus*, but no arguments were given to justify the arbitrary separation along a line Albania-Istanbul-N Caucasus margin-SE Caspian Sea. In the contrary, the few data provided for the prostate diverticles suggested that this character varied gradually over area. The suggested classification is not accepted here because the results were based on too few data (only 1 studied population of *Pl. carinatus* was included). It would be useful to study the degree of morphological variation within populations and at regional level, or to test hypotheses based on morphological and anatomical observations with molecular studies.

Planorbis atticus Bourguignat, 1852

Planorbis atticus Bourguignat, 1852 in Bourguignat 1852: 22

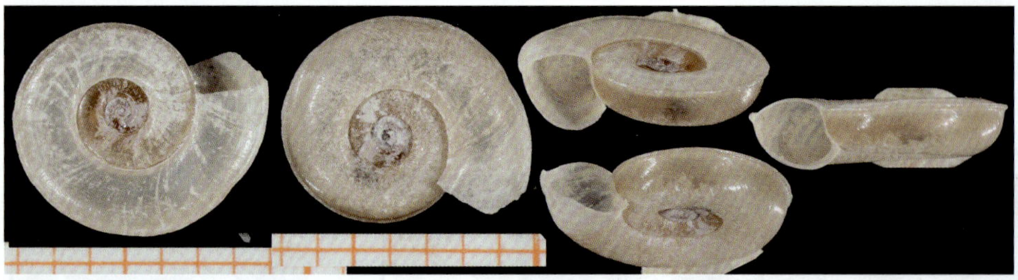

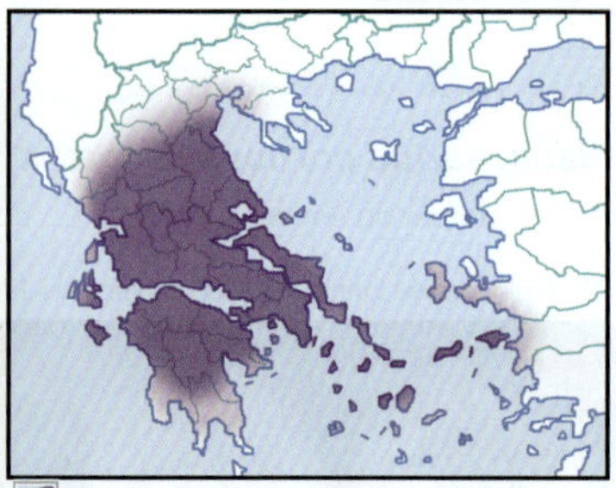

2.2-2.8 x 6-9 mm — Greece: Evia, northern part — Greece

Shell horny coloured, shiny, finely striated, equally convex at both sides, 4.5 whorls, last whorl width 2 x of preceding whorl, rounded or finely keeled, aperture oval, outer margin protruded.
Number of prostate diverticles 16-26, 5-10 less than local populations of *Pl. planorbis*.

In slowly running freshwaters.

Westerlund 1885: 69, Meier-Brook 1976, Rähle 1980 (Kefaloniá, Zákinthos), Bank & Maassen 1998: 47 (Sámos), Glöer & Zettler 2009 (perhaps also Crete), Glöer & Pešić 2010.

Soldatenko & Starobogatov 2000: 23 considered this as a synonym of either *Pl. planorbis* or *Pl. philippianus*.

Planorbis carinatus Müller, 1774

Planorbis carinatus Müller, 1774 in Müller 1774: 157
Possibly established to substitute *Helix planorbis* Linnæus, 1758 to avoid tautonymy (the description was taken from Linné 1767: 1242 No. 662 which was cited, but the references to figures were not exactly the same). Not a new replacement name for *Helix planorbis* Linnæus, 1758 because there was no expressed statement (Art. 72.7).
Planorbis kubanicus Soldatenko & Starobogatov, 1998 in Soldatenko & St. 1998: 59
Planorbis vitojensis Glöer & Pešić, 2010 in Glöer & Pešić 2010: 252

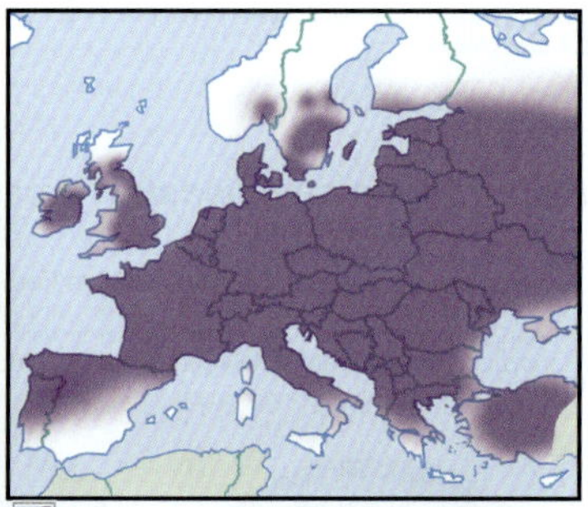

1.5-3 x 6-15 mm, diameter 10 mm at 4 whorls — Germany: Schleswig-Holstein, Dobersdorfer See — Europe to W Siberia

Shell light horny brownish, often translucent, 4.5-5 whorls with weak suture on both sides and a very sharp keel, usually near the centre of the periphery, whorls increasing more rapidly than in *Pl. planorbis*, last whorl rounded at both sides, functional lower side of adult shell not flat.

Permanent standing or slowly running water bodies, on muddy substrate, with rich vegetation, richer in oxygen than for *Pl. planorbis*. Avoids marshy places which may dry out periodically, up to 18 m depth, pH preference 7.4-7.9, may feed exclusively on detritus. In Switzerland up to 1000 m, in Germany 230 m.

Oval eggs (diameter 0.6-0.8 mm) are laid in capsules in single strings containing 10-20 eggs each, between April and August, juveniles hatch after 10-15 days.

Disappears with increasingly eutrophic conditions and pollution more rapidly than *Pl. planorbis*. Critically Endangered in Tirol, Endangered in Germany and Albania, Vulnerable in Switzerland and Czech Republic, Rare in Portugal.

Moquin-Tandon 1855: 431, Germain 1931: 525, Falkner 1990: 132, Glöer 2002: 253.

The separate specific identity of two local forms of Montenegro (*vitojensis*) and Krasnodar/S Russia (*kubanicus*) was not substantiated by Glöer & Pešić 2010 or previous studies, this would have required more detailed knowledge on the regional variation of *Pl. carinatus*.

Planorbis macedonicus Sturany, 1894

Planorbis macedonicus Sturany, 1894 in Sturany 1894: 379, Taf. 18 (*Planorbis Macedonicus*)

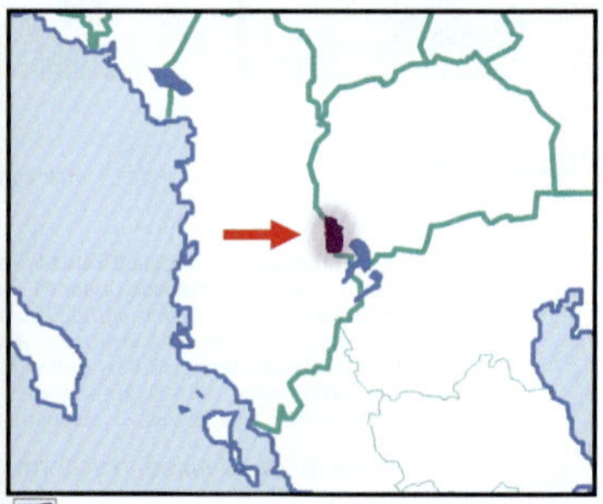

2.2-2.7 x 6-8 mm — Rep. Makedonija: Lake Ohrid, near Pestani — Lake Ohrid

Shell yellowish, striated, weakly shiny and translucent, first whorls slightly depressed, lower side concave, 3-4 whorls with deep suture, last whorl width more than 2 x of penultimate whorl, sometimes with keel at lower side, slightly descending near aperture, umbilicus very wide, open and perspectival.
Penis without stylet and with terminal pore (so not to be classified in *Gyraulus*).

At the shore of Lake Ohrid, apparently not rare in certain seasons, on hard substrate like rocks or stones.

EU Red List: Endangered EN. Threatened by increasing pollution and agricultural contamination impacting the algae growing on the hard substrate.

Hubendick & Radoman 1959: 243, Dhora 2004: 143.

Planorbis moquini Requien, 1848

Planorbis moquini Requien, 1848 in Requien 1848: 50 Type locality in Corsica.
Description given in footnote 23.

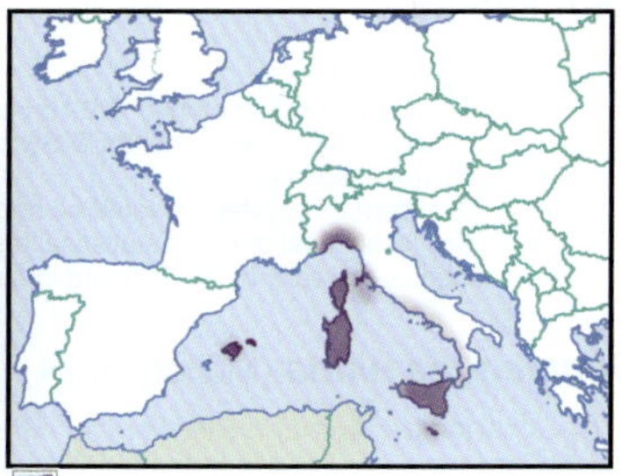

Shell horny brownish, transparent and shiny, striated, upper side concave, 3.5 regularly increasing whorls, not keeled, aperture slightly bent downwards, umbilicus 1/3 of diameter. Shells in Sardinia slightly flatter than in Corsica.
Shell does not differ much from *Gyraulus laevis*, the penultimate whorl seen from the lower side appears slightly flatter in *Pl. moquini* than in *G. laevis* where it seems more rounded.
Differs from *G. laevis* anatomically in the absence of a sclerotised stylet in the penis.
Animal light grey with light mantle pigmentation, in Sardinia mantle pigmentation with dark spots.

1.5-2.5 x 3-4.5 mm

Italy: Sardinia, Nuoro, Laconi

S Europe

In springs and small creeks.

Giusti 1976, Sparacio 1992, Manganelli et al. 1995: 11, 36, Beckmann 2007: 51, Glöer & Bouzid 2008, Glöer & Zettler 2009.

Sparacio 1992 did not find anatomical differences between Sicilian individuals and those of the Toscana archipelago. Glöer & Zettler 2009 argued that the populations outside Corsica should be classified as different species, but did not provide arguments to substatiate this statement.

Planorbis planorbis (Linnæus, 1758)

Helix planorbis Linnæus, 1758 in Linnæus 1758: 769 (*Helix Planorbis*)
Planorbis intermixtus Mousson, 1874 in Mousson 1874: 45

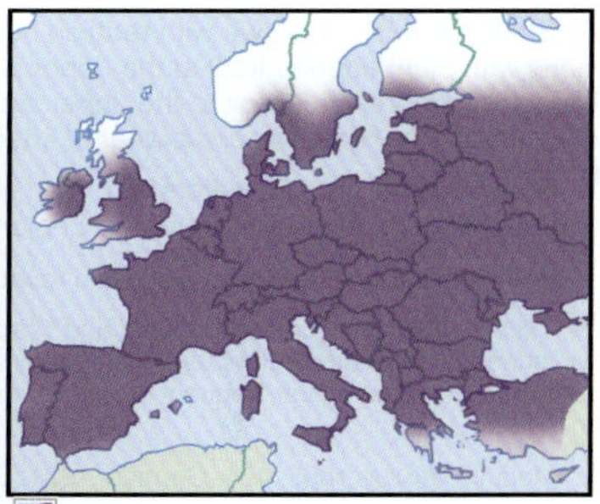

3-5 x 11-15 mm

Austria: Burgenland, Neusiedler See near Illmitz

W Eurasia

Shell horny yellowish to reddish, 4-5 whorls increasing slowly and regularly, keeled, functional lower side almost flat, last whorl with fathom-like keel near functional lower side.
Animal blackish to almost black with reddish hue, tentacles long and conical, eyes small and black.

Shallow standing and slowly running freshwaters, on muddy substrate, also muddy ponds and temporarily drying flood grooves, up to 1 m depth. Does not tolerate intensive water movements. Relatively tolerant against eutrophic conditions, salt up to 0.4 %, pH preferred 6.6-6.9, temperature optimum 19 °C, temporary droughts are survived. In Switzerland up to 1900 m, in Britain mainly in lowlands.

Reproduction in France between April and July, 6 days after copulation 8-10 egg capsules (width 3-5 mm) with 3-30 (usually 13-16) oval eggs (diameter 0.5-0.8 mm) each are laid, juveniles hatch after 11-14 days (20 °C), for France 14-16 days were reported, life span 3-11 years. Self-fertilization does not seem to produce offspring.

Threatened by continuous habitat destructions. Not very common in Portugal. Has obviously declined in Mallorca and Ibiza.

Germain 1931: 522, Falkner 1990: 132, Glöer 2002: 251.

Planorbis presbensis Sturany, 1894

Planorbis presbensis Sturany, 1894 in Sturany 1894: 380, Taf. 18 (*Planorbis Presbensis*)

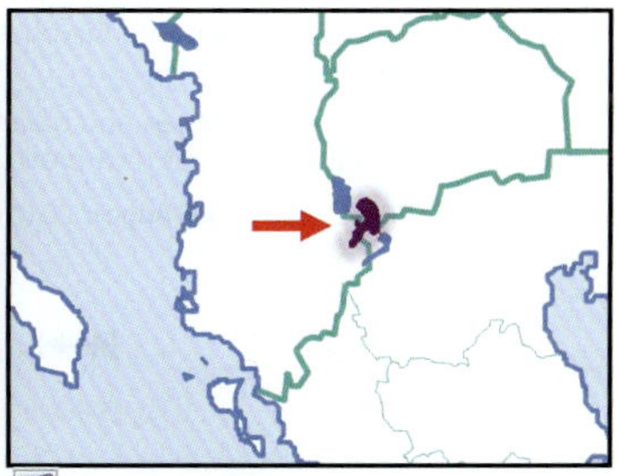

4-6 x 7-9 mm

Rep. Makedonija: Lake Prespa, Otesevo

Lake Prespa

Shell yellowish green to brownish with whitish surface layer, finely striated, upper side flat, lower side concave, 3-4 rapidly increasing whorls, last whorl width 2 x of preceding whorl, rounded, descending near aperture, apertural margin sharp, connected at parietal side by a thin layer, umbilicus wide and perspecivical. Shell height variable, depends on how deep the last whorl descends in its last section.

In the littoral zone of Lake Prespa, in rocky habitats.

EU Red List: Vulnerable VU. Threatened by abruptly changing water levels, agricultural contamination and introduced fish.

Meier-Brook 1976: 115, Dhora 2004: 143, Fehér & Erőss 2009 (Albanian parts).

Segmentina

Segmentina Fleming, 1818
in Fleming 1818: 309 [not verified]

Type species *Nautilus lacustris* Lightfoot, 1786
(monotypy)
(= currently *Segmentina nitida* (Müller, 1774))

Kruglov & Soldatenko 1997 proposed to subdivide this group into 3 subgenera and 13 species, most from central western European countries. Their classification was adopted by Kantor & Sysoev 2005: 222, but obviously not outside Russia except for the European Red List (www.iucnredlist.org, 06.2012).
It is not used here.
To accept this system it would be necessary to provide substantiated arguments for the presence of so many distinct and genetically isolated species, including their local variations and statistical tests if the system was based on differential shell morphology, their exact areas of distribution, and a guide how to determine a species correctly in the field and in museum collections.

Segmentina nitida (Müller, 1774)

Without gender agreement: *Segmentina nitidus* (Müller, 1774)

Planorbis nitidus Müller, 1774 in Müller 1774: 163
Nautilus lacustris Lightfoot, 1786 in Lightfoot 1786 (Phil. Trans.): 163, Tab. I

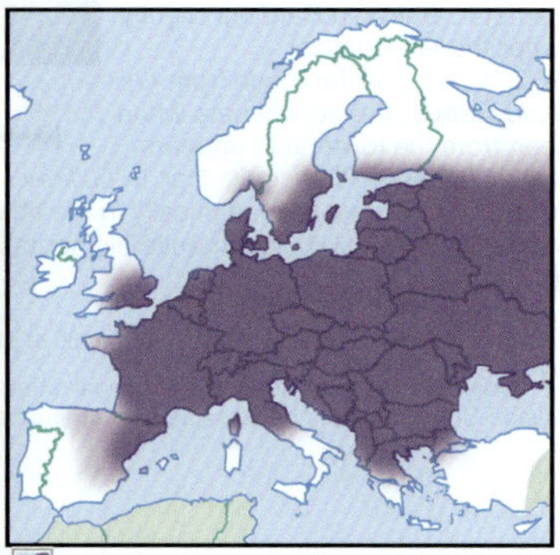

Europe to central Siberia (Lena)

1-1.5 x 4-6 mm (1-3) Czech Republic: Moravia, near Olomouc, (4-7) N Germany: Ostholstein, near Malente (1-3): Jiří Novák, www.biolib.cz

Shell yellowish or brown with reddish hue, translucent, usually convex on upper side and flat or concave (rarely weakly convex) on lower side, 4-5 convex whorls with blunt edge or keel at the upper side of the periphery, last whorl with 2-4 white radial streaks as if separating various chambers, aperture very oblique and depressed, bent to the upper side, umbilicus on upper narrow and side deep, on lower side usually very wide. Shell variable in several characters. Looks much like *Hippeutis complanatus*, which has a more lenticular shape and no white radial streaks.

Silent and permanent waters with high humine contents (blackish water) and *Lemna trisulca*, shallow waters, standing water bodies on meadows, periodical drought is sometimes tolerated. In England in drainage ditches in marsh levels. In Switzerland up to 650 m.

Eggs (diameter 0.3-0.5 mm) are laid in rounded capsules (size 1.2-1.5 mm) containing 3-8 eggs each, in August, juveniles hatch after 10-12 days in France.

Extinct in Sweden, central England and Wales (all populations of central England and Wales disappeared until the 1950s), Critically Endangered in Austria, Endangered in Switzerland and Bayern, Vulnerable in Germany and England, rare in Czech Republic. Only known from one station in SE Norway. Has declined dramatically in England since the early 1800s due to over-frequent dredging, lowering of water levels, pollution and drainage. The few surviving populations are in marginal areas with low levels of fertilization.

Moquin-Tandon 1855: 424, Germain 1931: 542, Westerlund 1885: 86, Falkner 1990: 134, Glöer 2002: 280.

Otina

Otina Gray, 1847
in Gray 1847: 156 *Velutina* synonym *Otina*

Type species *Helix otis* Turton, 1819
(original designation)
(= currently *Otina ovata* (Brown, 1827))

ICZN Opinion 335: type species by monotypy, this was incorrect, type was originally designated. Name was mentioned as a subjective synonym of *Velutina*, so it must have been validated later under Art.11.6.1. Monterosato 1906: 125 used *Otina* as a genus, with bibliographical reference to Gray 1847: 156. Whitfield 1865: 265 used *Otina* as a subgenus for *Velutina*, in a different sense than Gray had done this, and did not give a bibliographic reference to Gray 1847.

Otina ovata (Brown, 1827)

Without gender agreement: *Otina ovatum* (Brown, 1827)

Galericulum ovatum Brown, 1827 in Brown 1827: 23, Pl. 38 Fig. 27-28 [not verified]

Shell very small, 1.5 whorls, last whorl inflated, aperture oval and very large, like in succineids, columellar and parietal sides of the margin thickened.
Animal cannot withdraw entirely into the shell. No tentacles, eyes look like in comic snails.

Intertidal zone on rocks, algae and balanoids. Often in empty bivalve shells and moist rock crevices.

Glöer 2002: 289, Sosso & Dell'Angelo 2010 (Sardinia).

2-5 x 1.5-3 mm Northern Ireland: Down County Roy Anderson

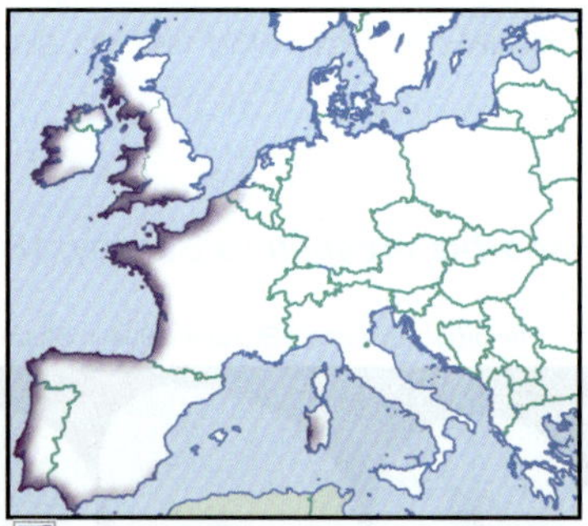

W and SW Europe
Not previously reported from Portugal and Spain, information based on Museum Rotterdam records.

Onchidella

Onchidella Gray, 1850
in M. E. Gray 1850: 117

Type species *Onchidium nigricans* Quoy & Gaimard, 1832
(subsequent designation by Fischer & Crosse 1878: 687)
(= currently *Onchidella nigricans* (Quoy & Gaimard, 1832) from New Zealand)

Dark and inconspicuous sea slugs without shell, body covered with coarse and evenly spaced tubercles, head with two thick tentacles with eyes at their tip. Pulmonate slug, mantle cavity acts as a lung and is situated near the anus.

Slugs lay eggs, a veliger larval stage is completed within the egg.

Worldwide except Arctic and Antarctic

Onchidella celtica (Cuvier, 1816)

Without gender agreement: *Onchidella celticum* (Cuvier, 1816)

Onchidium celticum Cuvier, 1816 in Cuvier 1817: 411, footnote 4
True date was before 2 Dec 1816 (Roux 1976), 1817 is incorrect. Original description short: "petite espèce".
Onchidium nanum Philippi, 1844 in Philippi 1844: 101, Tab. 20

Slug greyish-green with whitish head, pale grey foot and black tentacles, round to oval with evenly spaced tubercles, body elongate when crawling. When the animal is motionless head and foot are not visible.

Animals live amphibiously in the eulitoral zone in wet crevices and on rocks. They avoid direct sunlight and prefer to leave their hideouts when the sky is overcast. Animals follow the receding tide down the shore crawling on algae and return up the shore as the tide rises. In Britain mainly active between April and November, not rare in Devon and Cornwall.

Up to 15 x 8 mm (in Britain 13 x 6 mm) England: Cornwall, Tregardock Beach Peter Barfield

Britain to Italy and Azores

Weiss & Wägele 1998: 70 (anatomy, morphology), Bedulli et al. 1995: 22 (Italy), Dayrat 2009 (regarded *Onchidella nana* as a synonym).

Ovatella

Ovatella Bivona Bernardi, 1832
in Bivona Bernardi 1832: 58

Type species *Ovatella punctata* Bivona Bernardi, 1832
(subsequent designation by Monterosato 1906: 126)
(= currently *Ovatella firminii* Payraudeau, 1827)

Contained originally *Ovatella punctata* Bivona Bernardi, 1832 and *Ovatella polita*. Gray 1847: 159, 179 gave as type *Voluta bidentata* Montagu, 1808 which was not originally contained.

Ovatella firminii (Payraudeau, 1827)

Auricula firminii Payraudeau, 1827 in Payraudeau 1826: 105
Ovatella punctata Bivona Bernardi, 1832 in Bivona Bernardi 1832: 58
Both an Italian name Ovatella puntata and the Latin name *Ovatella punctata* were given.

Shell of juveniles with short reddish yellow hairs, their scars are visible in adult shells forming a regular and reticulate pattern (with radial and spiral lines connecting the hair scars), about 8 weakly convex whorls, aperture with very strong and long lower parietalis, strong middle parietalis, sometimes a weak upper parietalis, strong horizontal columellaris, palatal callus with 1-2 tubercles, no umbilicus. Larger than *Myosotella myosotis*, which has no reticulate surface pattern.

Between sea algae, found in marine litter deposits.

7-11 x 4-7 mm Italy: Sardinia, Oristano, Mandriola Maurizio Sosso

Mediterranean coasts

Myosotella

Myosotella Monterosato, 1906
in Monterosato 1906: 126

Type species *Auricula myosotis* Draparnaud, 1801
(subsequent designation by Wenz 1923: 1139)

Frias Martins 1996: 192 classified *Myosotella* as a separate genus. It seems that this has been widely accepted.
The taxon *denticulata* is commonly considered as an open-coast form of *M. myosotis*, not recognized as independent by all authors. Falkner et al. 2002: 100 argued that the two forms live in different habitats in the Netherlands. This alone was certainly not sufficient for regarding them as separate species.
A special study would be needed to provide evidence for the presence of two separated gene pools. Their life cycles were not found to differ.

Myosotella myosotis (Draparnaud, 1801)

Auricula myosotis Draparnaud, 1801 in Draparnaud 1801: 53
Voluta denticulata Montagu, 1803 in Montagu 1803: 234
Auricula myosotis var. *payraudeaui* Shuttleworth, 1843 in Shuttleworth 1843: 18 (*Auricula myosotis* var. *Payraudeaui*)

5-10 x 2.5-3.5 mm

Germany: Nordfriesland, near Schlüttsiel

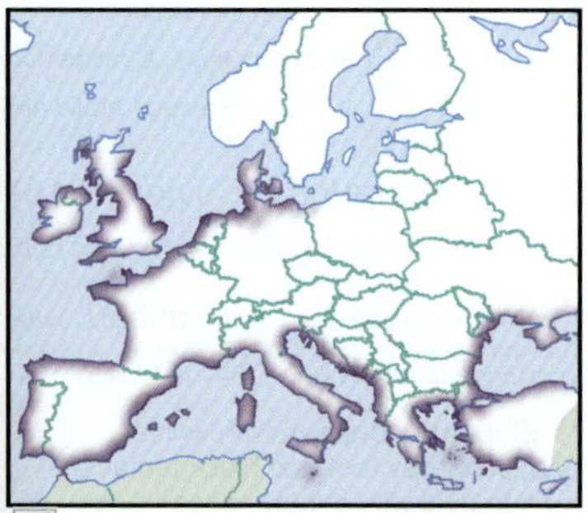

Black Sea to North Atlantic (USA)

Not frequent in SW Spain. Critically Endangered in Germany. Threatened by human impact on coastal environments.

W Ireland: Galway County — Roy Anderson

Shell brownish to yellowish, smooth or finely striated, shiny, 7-8 weakly convex whorls, apertural margin white and usually with white layer at parietal side, parietalis very strong and horizonal, often a small second parietalis is present, in open coast forms (*denticulata*) 3-6 short and thin folds near the palatal margin at the inner lip (they look like drops), columellaris strong, umbilicus usually covered. Animal light grey, upper side and tentacles slightly darker, two tentacles, eyes located at the base of the tentacles, at the inner side facing each other, not at the outer side like in other prosobranchs.

Always close to beaches without strong wave movement, among sea algae washed ashore, but not marine, under stones or wood, on muddy substrate, often together with freshwater molluscs, also in meadows close to the sea shore. Usually brackish water habitats, rarely at non-marine brackish waters. More terrestrial than aquatic, lives mostly outside the water. Feeds on silicate algae, detritus and rotting water plants.
Optimal salt concentration 1.8 %, maximum tolerance 0.9-9.9 %.

Hermaphrodites. Copulation in NW Germany during the entire year, but most frequently in April/May and August/September, few weeks later 15-80 eggs (diameter 0.5-0.6 mm) in an envelope are deposited in soil cavities, usually in clusters by various animals. Juveniles in N Germany hatch after 3-7 weeks (12-15 days in France). At 2.5 whorls size the inner parts of the upper whorls are entirely reduced, male maturity is reached shortly before the shell is fully grown after 1 year, female and full maturity after 1.5 years, then they can copulate both as females and males, life span 3-4 years. Animals occasionally crawl inside soil crevices shortly before they die.

Moquin-Tandon 1855: 415, Germain 1931: 560, Meyer 1955, Frias Martins 1996: 194, Schütt 2001: 16, Glöer 2002: 292, Wiese & Richling 2008.

Leucophytia

Leucophytia Winckworth, 1949 in Winckworth 1949: 38
Expressly established as a new replacement name for

Type species *Voluta bidentata* Montagu, 1808
(subsequent designation by Herrmannsen 1846: 586)

Leuconia Gray, 1840 in Turton & Gray 1840: 227 not *Leuconia* Grant, 1833 (Porifera)
Auriculinella Tausch, 1886 in Tausch 1886: 16, type species *Auriculinella whitei* Tausch, 1886 (monotypy)
(= currently *Auriculinella whitei*, Upper Cretaceous from Hungary)

Differs from *Ovatella* by a furrow in the anterior third of the sole.

Occasionally classified in the genus *Auriculinella*, which was established for a fossil (Cretaceous) species from Hungary with shouldered whorls and very prominent ribs (Bandel & Riedel 1994: 25). Conchologically slightly different fossil species are classified in different genera, for example *Leopoldium* Bandel & Riedel, 1994. A very close relationship of *L. bidentata* with *A. whitei* is not obvious.

Leucophytia bidentata (Montagu, 1808)

Voluta bidentata Montagu, 1808 in Montagu 1808: 100
Not *Voluta bidentata* Schröter, 1804, which is a nomen oblitum under Art. 23.9 (fixed by Falkner et al. 2002: 101).

Shell with horny to reddish brown periostracum, apex pointed, 6-7 very weakly convex whorls, last whorl 75 % of shell height, aperture with white layer at parietal side, parietalis usually strong, columellaris moderate or indistinct, no umbilicus.

A semi-marine species, often on open shores, around high-tide mark, under stones and in marine litter deposits, tolerates higher salt concentrations than *O. myosotis*. Less frequently in estuaries or tidal lagoons. Lives mostly outside the water, feeds on detritus.

Reproduction between June and September, 12-30 eggs (diameter 0.6-0.7 mm) are laid in agglomerations fixed to solid surfaces, juveniles hatch after 14-16 days, maturity is reached at the end of the second year.

Moquin-Tandon 1855: 417, Germain 1931: 563, Manganelli et al. 1995: 10, Kerney 1999: 47, Schütt 2001: 18, Glöer 2002: 292.

5-7 x 2-4 mm

NW France: Finistère, Île Callot near Roscoff (H = 4.0 mm)

Mediterranean to North Sea

Eric Simon

Pseudomelampus

Pseudomelampus Pallary, 1900
in Pallary 1900: 240 *Alexia (Pseudomelampus)*

Type species *Melampus exiguus* Lowe, 1832
(subsequent designation by Westerlund 1902: 117)

Contained originally *M. exiguus*, *biscayensis* Fischer, *Alexia kochi* Pallary, 1900, and *Alexia jolyi* Pallary, 1900.

Pseudomelampus exiguus (Lowe, 1832)

Melampus exiguus Lowe, 1832 in Lowe 1832: 291, Pl. 13

Shell ovoid, solid, pinkish brown with numerous spiral threads on the surface, spire short, last whorl not more than 90 % of shell height, aperture with several prominent teeth.

Shell height up to around 6 mm

Rolán 2005: 254.

Similar shells were found in N Morocco and Málaga, either this species or *Pseudomelampus kochi* (Pallary, 1900) from Algeria.

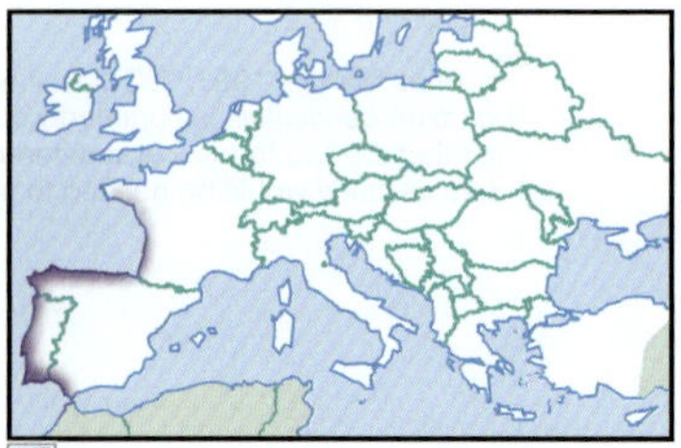

France to Angola

Terrestrial prosobranch gastropods

Genera sorted by size: *Hydrocena*, *Pholeoteras*, Aciculidae, *Carychium* and *Zospeum*, *Cochlostoma* and *Striolata*, Pomatiasidae

Hydrocena

Hydrocena Küster, 1844
in Küster, 1844: 646 *Paludina (Hydrocena)*

Type species *Paludina sirkii* Küster, 1844
(monotypy)
(currently = *Hydrocena cattaroensis* (Pfeiffer, 1841))

Hydrocaena is an incorrect subsequent spelling. Author was not Pfeiffer 1847 (Z. Malak.: 112) who was too late.

Hydrocena cattaroensis (Pfeiffer, 1841)

Without gender agreement: *Hydrocena cattaroense* (Pfeiffer, 1841)

Cyclostoma cattaroense Pfeiffer, 1841 in L. Pfeiffer 1841: 225

2-3 x 2-2.5 mm — Montenegro: Ljuta spring near Donji Orahovac N Kotor — Museum Budapest HNHM

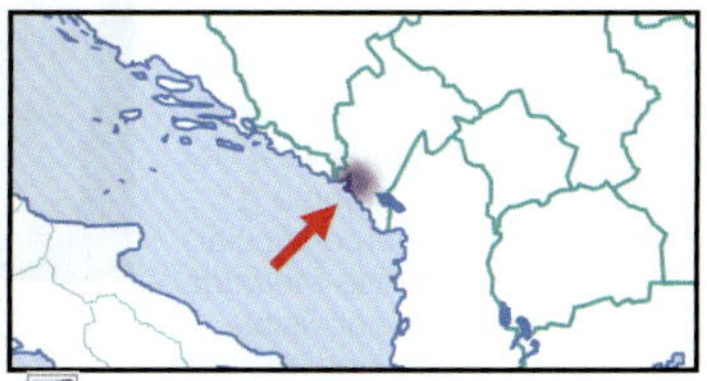

W Montenegro

Shell orange-red, 3-4 convex whorls, usually covered with camouflage.

At rocks and in humid canyons.

Pholeoteras

Pholeoteras Sturany, 1904
in Sturany 1904 (Anz. Wien): 118

Type species *Pholeoteras euthrix* Sturany, 1904
(monopypy)

Also published by Sturany in Nachrichtsblatt DMG (July 1904) (p. 106 there), I do not know which one was published earlier.

The shells of this small subterranean genus are rarely found and are morphologically slightly variable, inviting taxonomists to describe a new species at every instance when a new population or shell is discovered. It would be necessary to study more closely the local variation of the characters, and find constraints for inherent reproductive isolation in such differences.

Gittenberger 1977 confirmed that material from Kérkira agreed well with shells from Hercegovina. *Ph. zilchi* from Ipiros was claimed to differ slightly in the shape of the aperture and in being more slender. But these characters seem to be variable in Hercegovina. If a separate species is based on morphological differences, the spatial variations should be analysed appropriately and taken into account.
Ph. olympios from the E coast in Greece can certainly not be accepted as a true species. The characters claimed to differ from the others (more slender shape, more convex whorls and slightly narrower shell sculpture) were not tested for being statistically significant. The size was given as 5.2 x 2.4 mm (suggesting that only one shell was measured), either an error in the calibration of the microscope, or overlooked as a remarkable difference to any other known specimens in this group. 10 ribs per mm were given for the last whorl, clearly contradicting the statement that the ribs were narrower in this species than in the others.

Pholeoteras euthrix Sturany, 1904

Pholeoteras euthrix Sturany, 1904 in Sturany 1904: 118
Pholeoteras zilchi Subai, 1993 in Subai 1993: 63
Pholeoteras olympios Reischütz, Reischütz & Fischer, 2008 in Reischütz et al. 2008: 5

Shell yellowish white to brownish, not shiny, with 6 whorls, reticulate surface pattern composed of radial riblets with minute hairs (only in fresh shells) crossed by strong spiral lines (55-75 ribs on the last whorl, = 20 ribs/mm), suture deep, aperture about 1/6 of shell height, almost circular without teeth, umbilicus very narrow. Shells in the south appear to be more slender, but the picture shows that also populations in Hercegovina display some variation in shell shape.

2.2-3 x 0.9-1.3 mm — Croatia or Hercegovina: Ombla near Dubrovnik

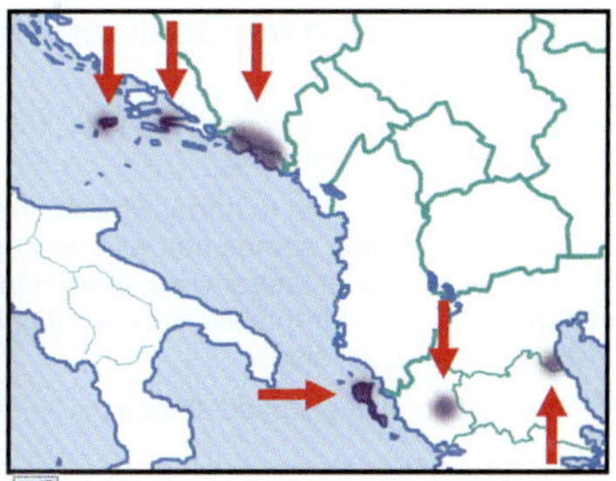

S Dalmatia to N Greece

Subterranean, also in caves, washed to the surface during rain, with subterranean water currents.

Kobelt 1906: 23, Gittenberger 1977: 47, Subai 1993: 63, Reischütz & R. 2003, Reischütz et al. 2008.

Aciculidae: *Acicula*, *Menkia*, *Platyla*, *Renea*

The presentation of these 4 genera was kindly checked by P. Subai.

Acicula

Acicula Hartmann, 1821
in Hartmann 1821: 205. 215

Type species *Bulimus lineatus* Draparnaud, 1801
(monotypy)

Genus was erected by Hartmann 1821: 205, but in a footnote of the same page he removed this name and classified this species in the genus *Acmaea*. ICZN Opinion 344: validated under the plenary powers (ICZN 1987: 40).

Shell cylindrical, 1.3-6.5 x 0.5-2 mm, shiny, with numerous radlial lines or grooves, suture with or without edge, cervix in some species with a distinct transversal callus.

Deep in litter layers, under dead woodlogs, carnivorous, species usually feed on eggs of other snails.

Europe, Caucasian region, N Africa

Boeters et al. 1989 (revision).

One species from S France (Pyrenees) needs confirmation, could be a fossil form, classified in *Acicula* by Gittenberger 2004:
Renea douctouyrensis Bertrand, 2004 in Bertrand 2004: 41

Acicula beneckei (Andreae, 1883)

Acme beneckei Andreae, 1883 in Andreae 1883: 137 (*Acme Beneckei*)

Shell reddish brown, striated (12-18 lines at penultimate whorl), last whorl slightly ascending, aperture with distinct angularis, parietal callus well developed, margin in lateral view with a distinct sinulus on the upper side (difference to *A. lineata* and *A. lineolata*), margin sharp without lip, umbilicus covered by a weak callus. The lower margin of the sinulus goes parallel with the suture, in lateral view like a Greek upper-case gamma (Γ, though not entirely perpendicular), very much like in *Renea moutonii*. In frontal view this is not visible and looks like *lineata* and *lineolata*. *A. lineolata* has a much weaker sinulus and a smaller size.

3.9-4.9 x 1.3-1.6 mm

Italy: Bergamo, Val Taleggio, Olda

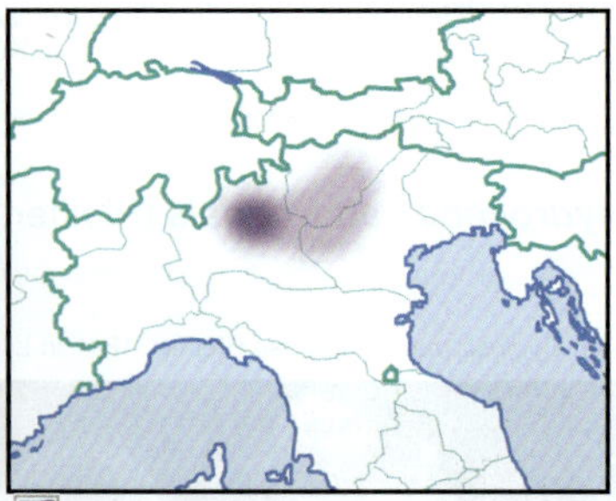

N Italy

EU Red List: Near Threatened NT. Restricted to a small area, potentially threatened by road constructions in the narrow valleys.

Boeters et al. 1989: 28.

Acicula benoiti (Bourguignat, 1864)

Acme benoiti Bourguignat, 1864 in Bourguignat 1864: 218 (*Acme Benoiti*)

Shell pale brownish, striated (15-26 lines at penultimate whorl), 6.5-7.5 almost flat whorls, no sinulus in a lateral view of the aperture, aperture inside thickened, callus covers umbilicus weakly, angularis weak, parietal callus well limited in frontal side but not thickened there, a sometimes slightly lighter cervical callus runs outside along the aperture from the suture to the umbilicus. Usually larger than *A. lallemanti* from N Africa and more widely striated, not many differences to *A. algerensis* from Morocco, larger than *A. szigethyannae*, apex blunter, more widely striated and apertural shape different.

Subterranean habitats in humid zones.

2.9-4.7 x 1.3-1.45 mm

Italy: Sicily, near Siracusa

Italy (Sicily)

EU Red List: Vulnerable VU.

Acicula corcyrensis (Boettger, 1883)

Acme lineata var. *corcyrensis* Boettger, 1883 in O. Boettger 1883: 319 (*Acme lineata* var. *Corcyrensis*)

Shell reddish brown to horny whitish, irregularly striated (20-30 lines at penultimate whorl), sometimes with an edge below the suture, aperture with a small sinulus at lateral view, margin inside thickened, parietal callus sharply limited at frontal side, cervical callus present and often stronger at the basal side.

Was found at S exposed slopes.

2.6-4.5 x 1.1-1.6 mm

Greece: Kérkira, Paleokastritsa

Massimo Prodan

W Greece

Acicula disjuncta Boeters, Gittenberger & Subai, 1989

Acicula disjuncta Boeters, Gittenberger & Subai, 1989 in Boeters et al. 1989: 36

Shell large, brownish, widely striated (13-20 lines at penultimate whorl), aperture without distinct sinulus at lateral view, weak callus covers umbilicus, cervical callus weak.

Was found at S exposed slopes, subterranean habitats of rocky habitats, mostly limestone.

EU Red List: Near Threatened NT. Only known from 4 localities.

4.6-6.2 x 1.7-1.95 mm

Italy: Calabria, Catanzaro, Tiriolo
H = 6 mm

Cesare Bogi

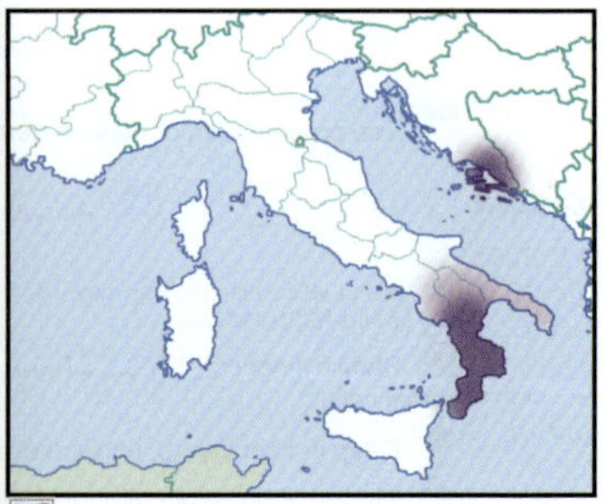

Croatia (Dalmatia), S Italy

Acicula fusca (Montagu, 1803)

Without gender agreement: *Acicula fuscus* (Montagu, 1803)

Turbo fuscus Montagu, 1803 in Montagu 1803: 330
not *Turbo fuscus* Müller, 1776 in Müller 1776: 242 - nomen oblitum under Art. 23.9, fixed here:
In accordance with Art. 23.9 *Turbo fuscus* Montagu, 1803 shall be used as "valid" (name in a current classification), *Turbo fuscus* Müller, 1776 has not been used for a taxon since 1899, *Turbo fuscus* Montagu, 1803 was used as a "valid" name in the following 27 papers:
Sevo 1974: 8, Devriese 1976, Kerney 1976: 75, Zilch 1976: 126, Fretter & G. 1978: 121, Seidl 1978: 96, Kerney & Cameron 1979: 54, Kerney et al. 1980, Kerney et al. 1983: 69, Ross 1984, Holyoak & S. 1985: 69, De Wilde et al. 1986, Prieto et al. 1987: 261, Van Goethem 1988, Boeters et al. 1989: 43, Korn 1990, Wiese 1991: 34, Vogt et al. 1994: 50, Anderson 1996: 3, Kerney 1999: 43, Falkner et al. 2001, Roper 2003: 28, Moorkens & K. 2004, Anderson 2005, Kobialka et al. 2006, Jungbluth & K. 2008, Kobialka et al. 2009: 16.

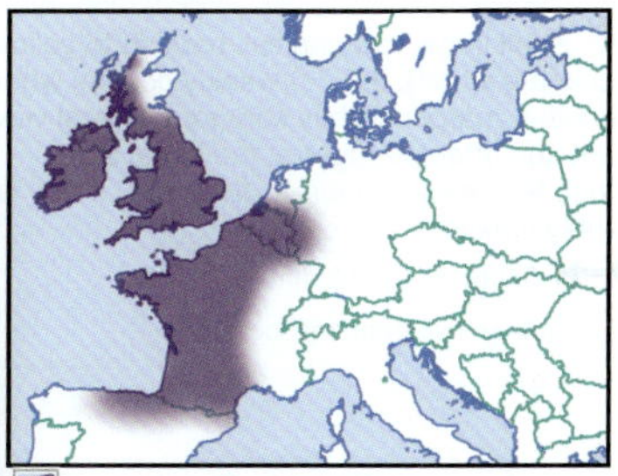

W Europe

Shell small, reddish brown, striated (15-23 lines at penultimate whorl), last whorl slightly ascending, apertural margin not thickened inside, callus does not cover the umbilicus, angularis very weak, parietal callus weak or absent, no cervical callus present.

Hygrophilous habitats, ground litter in old and undisturbed deciduous woods, there often on outcropping chalk, basalt or limestone, in moss near springs. Does not live in forest habitats affected by summer droughts, and does not tolerate human disturbance. In Britain and Ireland also in exposed habitats such as sea cliffs and mossy roadside banks.

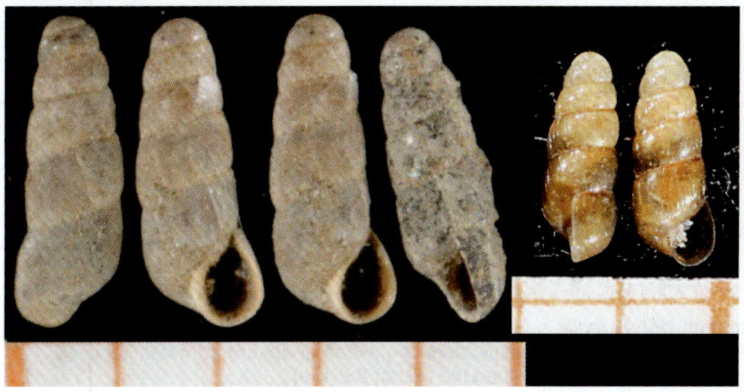

(1-4) France: near Dijon, (5-6) Spain: Guipuzzeva near San Sebastián

2.0-2.95 x 0.9-1.0 mm

Threatened by habitat destructions, intensive farming and modern forest management. Populations have declined substantially in Britain and Ireland. Vulnerable in Germany and Ireland. EU Red List: Least Concern LC, but obviously much more threatened than some local endemic species for which higher categories of threat were given.

Acicula hausdorfi Boeters, Gittenberger & Subai, 1989

Acicula hausdorfi Boeters, Gittenberger & Subai, 1989 in Boeters et al. 1989: 49

Shell horny brownish, densely and irregularly striated, 5-6 weakly convex whorls, last whorl distinctly ascending, apertural margin broad, with a small sinulus in frontal view, but not in lateral view, cervical callus strong and lighter than the shell, very strong at basal side.
Differs from *A. corcyrensis* mainly by its dense radial riblets. Not as dense as in the slightly larger *A. multilineata.*

2.9-4.2 x 1.15-1.5 mm

Leaf litter in shady and humid environments such as woodlands.

EU Red List: Near Threatened NT. Only known from few localities, woodlands in Greece are generally at high risk.

Boeters et al. 1989: 49.

W Greece

Acicula lineata (Draparnaud, 1801)

Without gender agreement: *Acicula lineatus* (Draparnaud, 1801)

Bulimus lineatus Draparnaud, 1801 in Draparnaud 1801: 67
Not *Bulimus lineatus* Bruguière, 1789 in Bruguière 1792: 323, nomen oblitum proposed under Art. 23.9 by Falkner et al. 2002: 69. This action was taken in error, the name is used for a frequent species *Macroceramus lineatus* (Bruguière, 1789) (Urocoptidae) from Haiti. The Commission has been asked to protect both names (Welter-Schultes 2011: 250), meanwhile usage should be maintained under Art. 23.9.5 and 23.10.
Acme sublineata Andreae, 1883 in Andreae 1883: 138

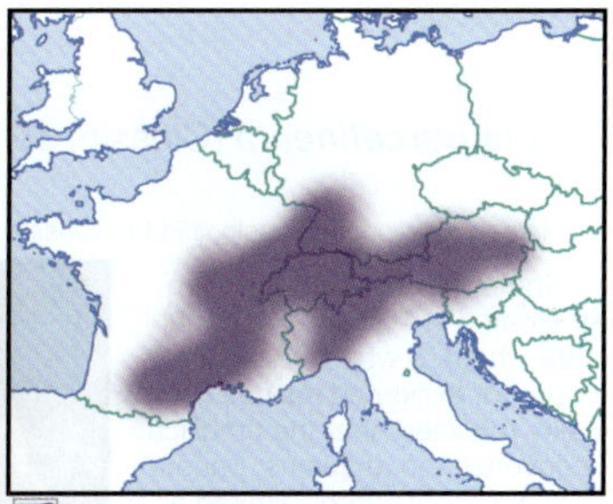

Alps and adjacent highlands

Shell reddish brown, irregularly but dinstinctly striated (20-30 lines at penultimate whorl), 5-6.5 weakly convex and from whorl 2 onwards only very slowly increasing whorls, often with edge below the suture, last whorl ascending near aperture, parietal callus at frontal side well delimited, margin not very broad, (shaped in lateral view, cervical callus present, its colour can be slightly different, umbilicus closed.

2.8-3.7 x 1.0-1.3 mm

in N Italy and S Switzerland 2.6-3.2 x 0.95-1.15 mm

Switzerland: Gorges de Mautier

Differs from *A. lineolata* in its smaller size and less slender and more cylindrical shape, from whorl 2 onwards the whorls increase more slowly in diameter. *A. fusca* is smaller and its umbilicus is open. *Platyla polita* has no radial lines.

Ground litter in open woods, rock rubble. In Switzerland in up to 1800 m.

The slightly smaller shells in N Italy and S Switzerland were classified by Boeters et al. 1989 as a subspecies *A. l. sublineata*, based on shell size as the only distinguishing character.

Acicula lineolata (Pini, 1885)

Acme lineata var. *lineolata* Pini, 1885 in Pini 1885: 241
Acme lineata mut. *villae* Stabile, 1859 (in Stabile 1859: 166, 182) is an older name which was never used after 1899, Boeters et al. 1989: 67 argued that *villae* might be a senior synonym of *lineolata*. The description of *villae* ("Testa majore. - Alt. 4; Diam. 1 millim.") gave no constraints that *lineolata* was meant.
Acicula lineolata banki Boeters, Gittenberger & Subai, 1989 in Boeters, Gittenberger & Subai 1989: 69

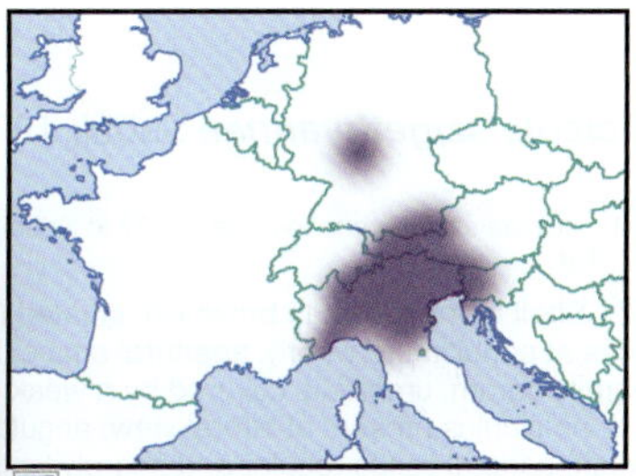

N Italy to Germany

Shell reddish brown, striated with 25-40 lines on the penultimate whorl, 5.5-7 weakly convex and from whorl 2 onwards moderately slowly and steadily increasing whorls, last whorl ascending near aperture, apertural margin inside not thickened, weak callus near umbilicus present, angularis weak, parietal callus at frontal side well delimited, margin not very broad, (shaped in lateral view, most populations have a weak sinulus seen from lateral view (no sinulus in many populations west of Lago di Garda and near Brixen), umbilicus closed.

3.4-4.25 x 1.2-1.4 mm

Italy: Laveno near Varese (Lago Maggiore)

Differs from *A. lineata* in its larger size, more slender and more cylindrical shape, the lower whorls increase more rapidly in diameter. *A. beneckei* is larger and has a much more pronounced sinulus in lateral view, like a Greek upper-case gamma Γ, though not entirely perpendicular.

Open woods, at rocks and rock rubble, in leaf litter. Up to 1000 m in Switzerland.

Boeters et al. 1989 recognised two subspecies *A. l. lineolata* and *A. l. banki*, representing two conchological forms, *banki* (with sinulus) being more widespread, *l. lineolata* (without sinulus, only west of Lago di Garda and near Brixen) may have evolved convergently in various regions. It is also possible that the form *banki* with sinulus belongs to *A. beneckei*.

Acicula miaphene Subai, 2009

Acicula miaphene Subai, 2009 in Subai 2009: 104

Shell large, low conical, reddish to yellowish brown, densely and irregularly striated (55-75 grooves at penultimate whorl), 6.5-7.5 weakly convex whorls with shallow suture, last whorl 1/2 of shell height, slightly ascending, margin slightly convex in lateral view, no sinulus, parietal callus well developed, cervical callus present and often stronger at the basal side.
Much less slender than *A. disjuncta*, which is much more densely striated. Other *Acicula* species are considerably smaller.

5.3-6.1 x 1.9-2.2 mm

Only known from one locality.

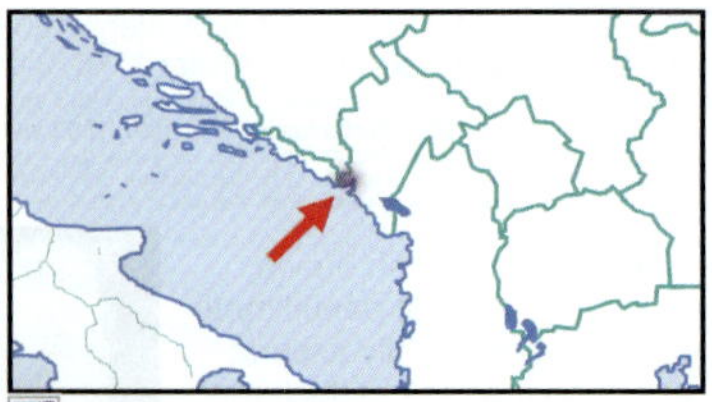

Montenegro

Acicula multilineata Boeters, Gittenberger & Subai, 1989

Acicula multilineata Boeters, Gittenberger & Subai, 1989 in Boeters et al. 1989: 74

Shell large, light horny coloured, with extremely dense radial sculpture. Aperture with weak sinulus at lateral view, margin inside strongly thickened, umbilicus covered by a callus, angularis weak or absent, parietal callus sharply limited at frontal side, cervical callus strong and lighter than the rest of the shell.

3.8-4.4 x 1.5-1.6 mm

Rocky habitats.

EU Red List: Vulnerable VU. Only known from 3 localities, potentially threatened by habitat destruction.

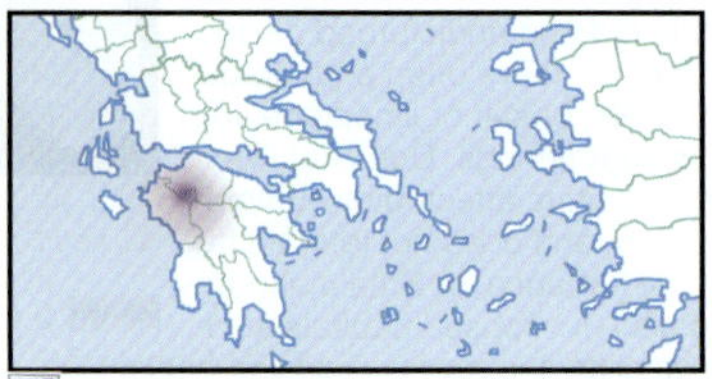

Greece (Pelopónisos)

Acicula norrisi Gittenberger & Boeters, 1977

Acicula norrisi Gittenberger & Boeters, 1977 in Gittenberger & Boeters 1977: 126

Shell colourless or horny brown, widely striated with 11-18 lines on the penultimate whorl, riblets more widely spaced than in *A. lallemanti*, apertural margin covers umbilicus, angularis weak, parietal callus strong and sharply limited at frontal side, cervical callus weak or absent.

2.8-3.5 x 1.1-1.15 mm

Humid habitats in the leaf litter layer under tree trunks and stones, also in rock crevices of scarcely vegetated slopes exposed to sunshine, not necessarily on calcareous substrate.

EU Red List: Vulnerable VU. Only known from 5 localities in Gibraltar and S Spain, one locality in Gibraltar being threatened by construction plans which could result in habitat destruction (www.iucnredlist.org, 07.2012).

Boeters et al. 1989: 75, Talaván Gómez & T. 2006: 37, Gavira Romero et al. 2009.

Gibraltar, S Spain

Acicula parcelineata (Clessin, 1911)

Acme parcelineata Clessin, 1911 in Clessin 1911: 165

Shell whitish, small and slender, striated (14-23 lines at penultimate whorl), 5 whorls, apertural margin not thickened inside, weakly reflected near the umbilicus but covers it, no angularis, parietal callus weak but sharply limited at frontal side, no cervical callus present.
Much like *Acicula isseli*, but no edge under the suture, growth lines slightly less dense.

2.0-2.3 x 0.75-0.85 mm

(1, 5-7) Karolina Chaniecka
(2-4) Museum Budapest HNHM

(1, 5-7) Poland: Pieniny mountains, (2-4) Poland: near Dukla

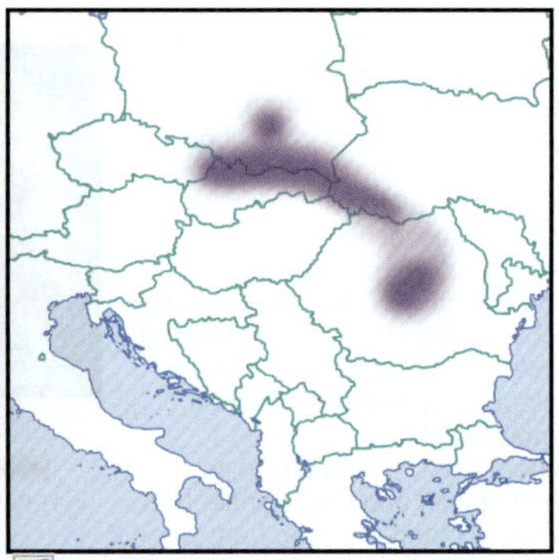

Carpathians (Moravia to Romania)

Acicula szigethyannae Subai, 1977

Acicula szigethyannae Subai, 1977 in Subai 1977: 38, Taf. 1

Shell horny brown to brownish, striated (19-27 lines at penultimate whorl), apertural shape like a parallelogram, umbilicus covered by a weak callus, no sinulus present at lateral view, angularis weak, parietal callus regular and sharply limited at frontal side, weak cervical callus present, can be lighter than the rest of the shell. Smaller than *A. disjuncta*, relatively more whorls than *A. lallemanti*, apertural shape differs from *A. lineata*.

Eikenboom 1998.

(1-2) Italy: Abruzzo, Val d'Arano, Ovindoli, (3-5) Italy: Lucca, Vianova di Careggine

2.7-3.8 x 0.9-1.1 mm

(1-2) Angelo Vannozzi

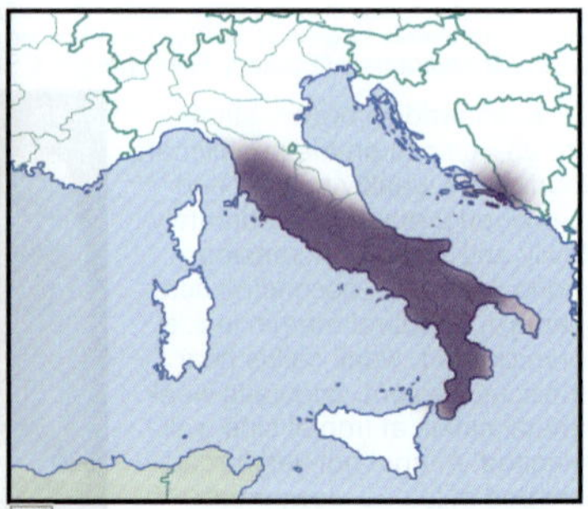

Italy, Croatia (Dalmatia)

Acicula vezzanii Bodon, 1994

Acicula vezzanii Bodon, 1994 in Bodon 1994: 212.
Dedicated to Mrs. Gabriella Vezzani, name not to be corrected to *vezzaniae* (Art. 32.5). Art. 31.1 has no effective force.

Shell light yellowish brown and transparent, with a reddish brown line below the suture, finely and regularly striated, 5-6 weakly convex whorls with well-marked but not deep suture, apertural margin thickened, cervical callus prominent and rounded when seen from lateral view, well delimited at the backside, no apertural sinus at lateral view, umbilicus closed.
Larger than *A. hausdorfi*, cervical callus slightly better delimited at the backside. *A. multilineata* is smoother, more conical, cervical callus merges more gradually down at the backside, like in *A. hausdorfi*.

Limestone substrate, 30-50 cm deep, also in small caves, usually close to small water currents running in open and rocky terrain. Not found in forests.

3.7-4.7 x 1.3-1.5 mm

Italy: Genova, Santa Margherita Ligure

N Italy

EU Red List: Near Threatened NT, possibly exaggerated.

Menkia

Menkia Boeters, Gittenberger & Subai, 1985 in Boeters et al. 1985: 59

Type species *Menkia horsti* Boeters, Gittenberger & Subai, 1985 (original designation)

Shell cylindrical, 1.8-2.8 x 0.6-1.1 mm, with radial and spiral grooves, no edge at the suture, no angularis, no cervical callus.

Boeters et al. 1989.

Menkia dewinteri Gittenberger, 1991

Menkia dewinteri Gittenberger, 1991 in Gittenberger 1991: 252

Shell weakly striated with faintly visible spiral striae, apex blunt, broader than in *M. horsti*, 5-6 moderately convex whorls. Generally broader and more cylindrical/less pointed than *M. horsti*.

2.6 x 1 mm

Obviously rare, only known from one locality.

N Spain (Cantabria)

Menkia horsti Boeters, Gittenberger & Subai, 1985

Menkia horsti Boeters, Gittenberger & Subai, 1985 in Boeters et al. 1985: 60

Shell light to dark brownish, radial growth lines 10-18 at the penultimate whorl, spiral sculpture with 11-18 spiral lines at penultimate whorl, apertual margin without sinulus at lateral view, margin sharp and not thickened, no angularis, parietal callus very thin, no cervical callus.

Humid soil habitats in calcareous rocky landscapes, from near the coast to above 1500 m altitude.

2.2-2.8 x 0.9 mm

Spain: Astrurias, near Ribadesella

N Spain (Asturias)

EU Red List: Near Threatened NT. Known from six localities.

Menkia rolani Gittenberger, 1991

Menkia rolani Gittenberger, 1991 in Gittenberger 1991: 254

Shell light yellowish brown, translucent, irregularly striated and with some faint spiral striae, almost cylindrical with blunt apex, whorls moderately and regularly convex, aperture small, slightly broadened at basal margin (best seen in lateral view).

1.9-2.3 x 0.7 mm

Found in a sandy deposit in a fountain, true habitat unknown, lives sympatrically with *M. horsti* and *Ac. fusca*.

Endangered (Gómez Moliner et al. 2001). Only known from one locality.

N Spain (Asturias)

Platyla

Platyla Moquin-Tandon, 1856 in Moquin-Tandon 1855: 509 *Acme (Platyla)*

Type species *Acme dupuyi* Paladilhe, 1868 (subsequent designation, fixed under Art. 70.3, here)

Originally included were 2 species: *Turbo fuscus* Montagu, 1803 (with *Truncatella polita* Hartmann, 1840 among its synonyms) and *Paludina simoniana* Saint-Simon, 1848 (: 38). Westerlund 1902: 122 selected as type *Truncatella polita*. Moquin-Tandon's (1855: 510) detailed description of *Acme (Platyla) fusca* suggests that *Acme dupuyi* Paladilhe, 1868 and not *Turbo fuscus* Montagu, 1803 was meant, and that the type species was misidentified by Moquin-Tandon 1855. Boeters et al. 1989: 91 gave as type *Acme dupuyi*. This was ineffective in 1989. Under Art. 70.3 of the Code *Acme dupuyi* Paladilhe, 1868 shall now be designated and fixed as type species for *Platyla*, misidentified as *Turbo fuscus* Montagu, 1803 (for which *Truncatella polita* Hartmann, 1821 was cited as a synonym) by Moquin-Tandon 1855: 509.

Shells smooth and shiny, usually no radial grooves, shell size 1.15-5.6 x 0.4-1.9 mm. Suture in some species with edge, cervical callus in many species present.

Deep in litter layers, under dead woodlogs, carnivorous, species usually feed on eggs of other snails.

W Europe to Turkey and W Russia

Boeters et al. 1989 (revision), Bodon & Cianfanelli 2008 (Italy).

Platyla banatica (Rossmässler, 1842)

Pupula lineata var. *banatica* Rossmässler, 1842 in Rossmässler 1838-1844 (11): 12, Taf. 53
(Page number 2 in error for 12)

Shell yellowish brown, aperture ascending at penultimate whorl, last whorl about 1/2 of shell height, parietal callus well developed, angularis weak, umbilicus sometimes covered by a weak callus, cervical callus usually strongly developed, quite sharply limited at its backside.

Subterranean habitats in rocky environments, usually limestone.

3.95-5.4 x 1-35-1.65 mm

Hungary: Mecsek mountains, Nagy-mély-völgy
Zoltán Fehér

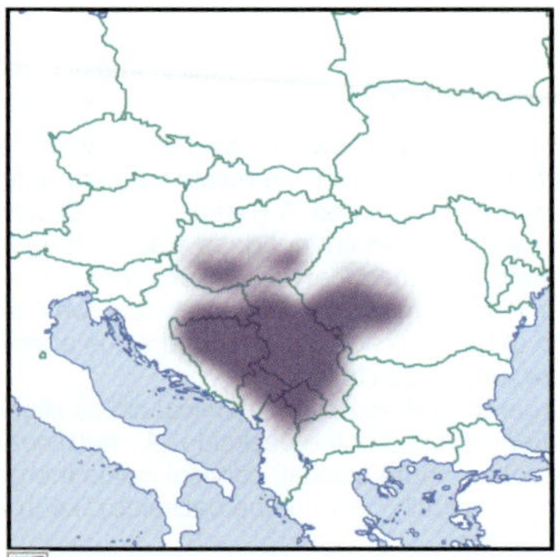

SE Europe

Platyla callostoma (Clessin, 1911)

Acme callostoma Clessin, 1911 in Clessin 1911: 166

Shell yellowish brown, 5-6 whorls, last whorl 1/2 of shell height, apertual margin broad, angularis present, parietal callus well developed, umbilical callus thin, cervical callus prominent and well limited at backside. Western populations smaller with a more slender cervical callus.

2.8-3.6 x 1.0-1.3 mm

Spain: Caranceja near Villanueva de la Pena

Central N Spain and Pyrenees

Leaf litter in moist deciduous forests, mostly under ivy.

Platyla corpulenta Subai, 2009

Platyla corpulenta Subai, 2009 in Subai 2009: 104

Shell whitish or translucent, slightly conical, 5.5-6 very weakly convex whorls, last whorl 55-60 % of shell height, slightly ascending, aperture with thin parietal callus, weak or no angularis, no sinulus, margin reflected and slightly (shaped in lateral view, no separate umbilical callus, no cervical callus.
Larger than *P. procax* and without cervical callus.

3.7-4.3 x 1.7-1.9 mm

Subterranean.

Only known from 3 localities.

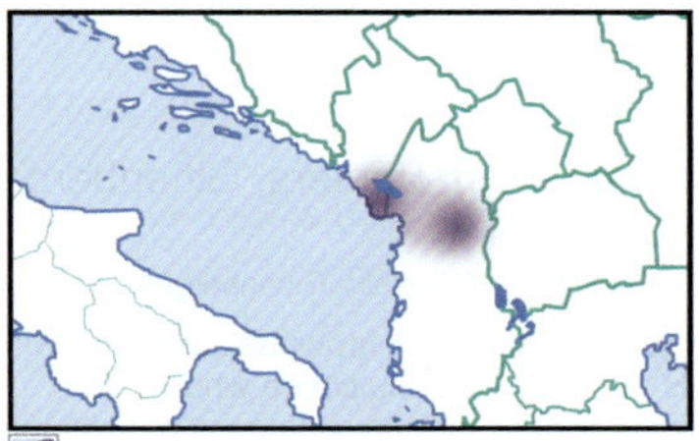

S Montenegro, N Albania

Platyla cryptomena (Folin & Bérillon, 1877)

Acme cryptomena Folin & Bérillon, 1877 in Folin & Bérillon 1877: 445
The name of the author Folin follows the conventions for French titles of nobility, where the "de" is not added to the surname (examples Lamarck, Buffon).

Shell reddish to yellowish brown, apertural margin with a minute sinulus at lateral view, parietal callus well developed, angularis distinct, umbilical callus narrow, cervical callus reddish brown and very broad, flattened but well limited at backside. Shell smaller and more slender than *Pl. callostoma*, the wide and flat cervical callus is characteristic.

2.5-3.5 x 0.8-1.0 mm

Spain: (1-5) Guipuzzeva near San Sebastián, (6-8) Asturias, near Ribadesella

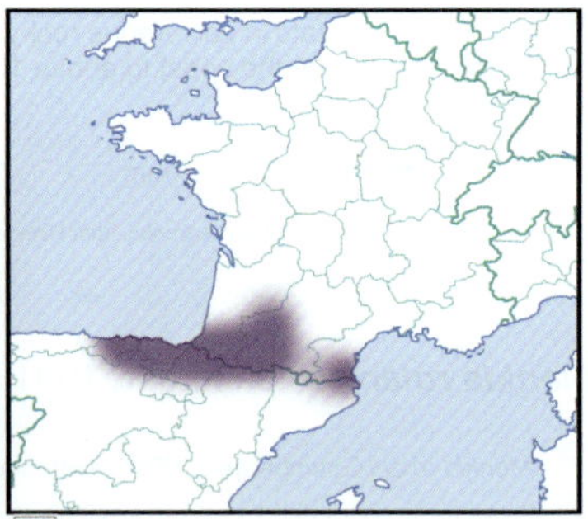

N Spain, S France

Ground litter in moist deciduous forests.

Platyla curtii (Wagner, 1912)

Acme curtii Wagner, 1912 in A. J. Wagner 1912: 259

Shell reddish to yellowish brown, last whorl exceeds 1/2 of shell height, slightly ascending, aperture with a minute sinulus at lateral view, apertural margin thickened, parietal callus well developed, angularis present, umbilical callus thin but covers umbilicus, cervical callus weak und indistinct, not well limited at its backside, differs markedly from those in *Pl. polita* and *Pl. gracilis*. No edge below suture.

Subterranean habitats among rocks, mostly limestone.

Bodon & Cianfanelli 2008.

2.55-3.1 x 1.0-1.15 mm

Croatia: Istria, Veprinac

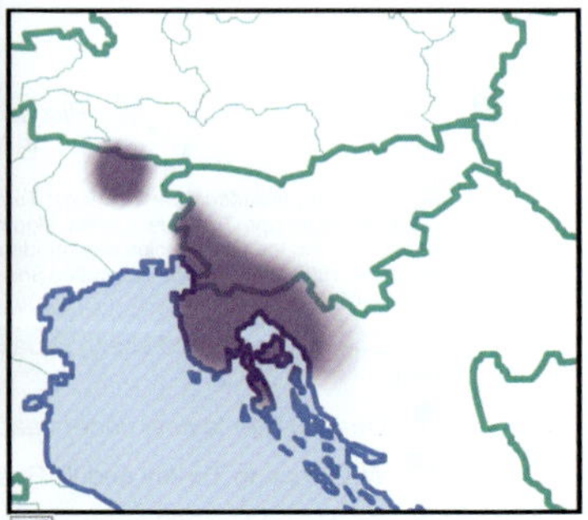

NE Italy to Croatian coastlands

Platyla dupuyi (Paladilhe, 1868)

France: Haute-Marne, near Arc en Barrois
3.0-3.6 x 1.15-1.25 mm
Jean Béguinot

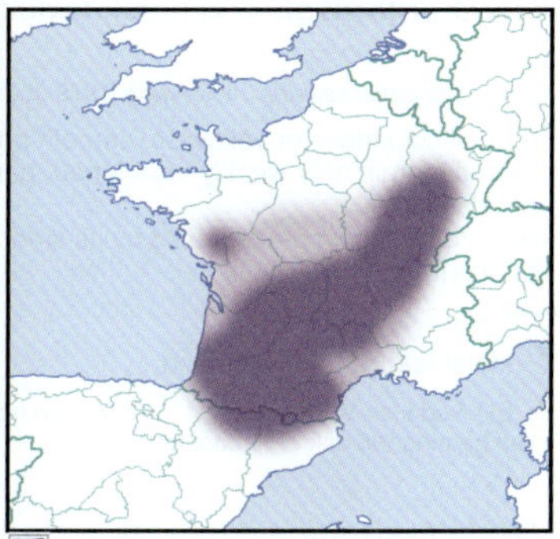

N Spain, France

Acme dupuyi Paladilhe, 1868 in Paladilhe 1868: 274 [= 81], pl. 13 [= 4] (*Acme Dupuyi*)

Shell yellowish brown, 5.5-6 weakly convex whorls, last whorl ascending, no edge below suture, aperture without sinulus, parietal callus well developed, angularis present, umbilical callus very weak, practically no cervical callus present.
Differs from *Pl. curtii* in its larger size and the absence of a clear cervical callus.

Deciduous woods and well-vegetated stony mountain slopes.

Kerney & Cameron 1979: 56, Boeters et al. 1989: 108.

Platyla elisabethae (Pintér & Szigethy, 1973)

Acicula elisabethae Pintér & Szigethy, 1973 in Pintér & Szigethy 1973: 97

Shell yellowish brown, shiny, 6-7 weakly convex whorls, last whorl slightly ascending, apertural margin broad, parietal callus strong, cervical callus characteristic, narrow but very strong, increases its height backwards, between apertural margin and cervical callus a small groove is visible in frontal view.

In rock crevices.

3.9-4.6 x 1.4-1.6 mm

Croatia: Dalmatia, Peljesac, near Ston

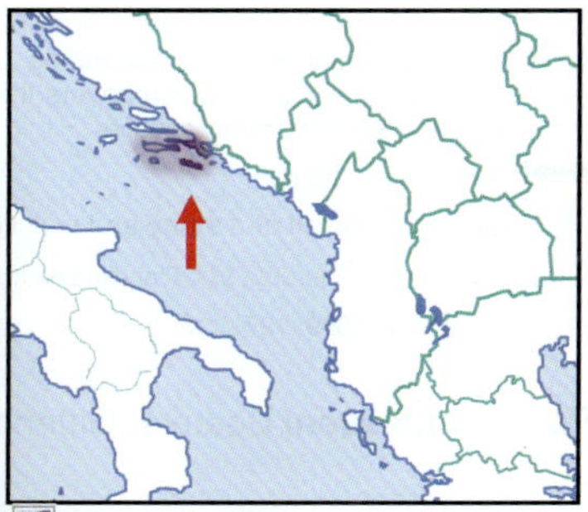

Croatia (S Dalmatia)

EU Red List: Near Threatened NT. Restricted to a small range.

Platyla feheri Subai, 2009

Platyla feheri Subai, 2009 in Subai 2009: 105

Shell slender and conical, whitish or translucent, 6.5-7 convex whorls with deep suture, last whorl 45-50 % of shell height, slightly ascending, margin slightly (shaped in lateral view, parietal callus very thin, no angularis, no sinulus, cervical callus present but not very prominent.
P. polita is reddish brown, smaller, less slender, not conical, with one whorl more and stronger cervical callus. *P. banatica* is larger with much stronger cervical callus. *P. wilhelmi* is brownish, has a sinulus and is more slender at equal shell height.

3.4-4 x 1.3-1.4 mm

Subterranean.

Only known from one locality.

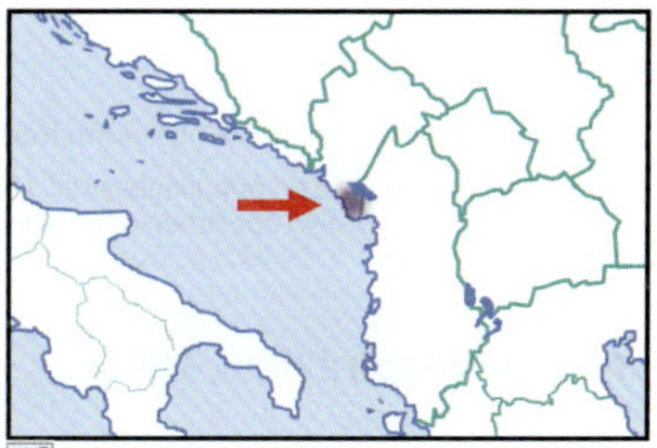

S Montenegro

Platyla foliniana (Nevill, 1879)

Acme foliniana Nevill, 1879 in Nevill 1879: 341 (*Acme Foliniana*)

Shell yellowish brown or colourless, smooth, 6-6.5 moderately convex whorls, last whorl slightly ascending, parietal callus relatively thin, angularis weak, umbilical callus thin, cervical callus broad and high, well delimited backwards, like a double lip.
Differs from other species mainly in its large size.

EU Red List: Vulnerable VU. Restricted to a small area.

Boeters et al. 1989: 116, Bodon & Cianfanelli 2008.

3.9-5.6 x 1.55-1.9 mm

Italy: Imperia, Vallone del Passo
Determined in 2001 by Maurizio Sosso

SE France, NW Italy

Platyla gracilis (Clessin, 1877)

Acme gracilis Clessin, 1877 in Clessin 1877: 42

Shell reddish to yellowish brown, 5-6 weakly convex whorls, with edge below suture, last whorl slightly ascending, aperture with hardly visible sinulus, margin reflected, parietal callus well developed, angularis weak, columellar callus narrow, cervical callus narrow and high, well delimited at backside, concave between backside and apertural margin. Shell locally variable.
Smaller and more slender than *Pl. polita*, pale shiny and more transparent.

2.05-3.45 x 0.8-1.25 mm
Austria: Kärnten, Loibltal, Tscheppaschlucht

Deciduous forests, rock rubble, often in isolated habitats, under rocks, in litter and in light soil. Needs calcareous substrate.

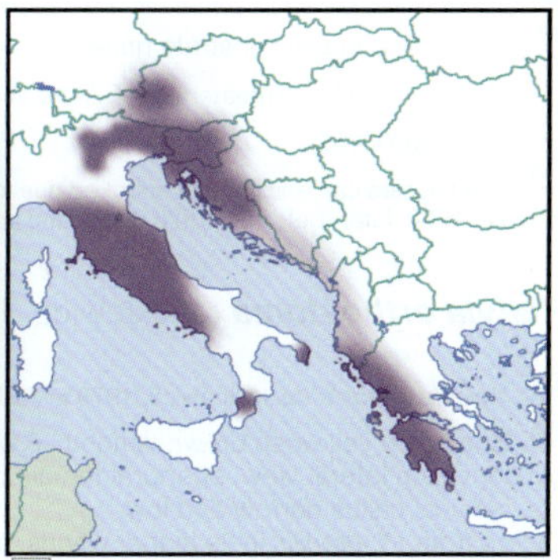

Austria to Italy and Greece

Bole & Sl. 1997: 145, Bodon & Ci. 2008.

Platyla jankowskiana (Jackiewicz, 1979)

Acicula jankowskiana Jackiewicz, 1979 in Jackiewicz 1979: 96

Shell colourless or yellowish, smooth and shiny, 5-6 weakly convex whorls, last whorl slightly ascending, apertural margin only weakly broadened, parietal callus thin, no angularis, no separate columellar callus, no cervical callus.
Differs from *Acicula parcelineata* by its smooth shell. *Platyla perpusilla* is smaller and more slender, with deeper suture.

2.4-2.6 x 0.85-9 mm

Moist leaf litter in woodland.

EU Red List: Vulnerable VU. Only known from one locality in a National Park, there threatened by recreational activities bearing the risk of habitat destruction.

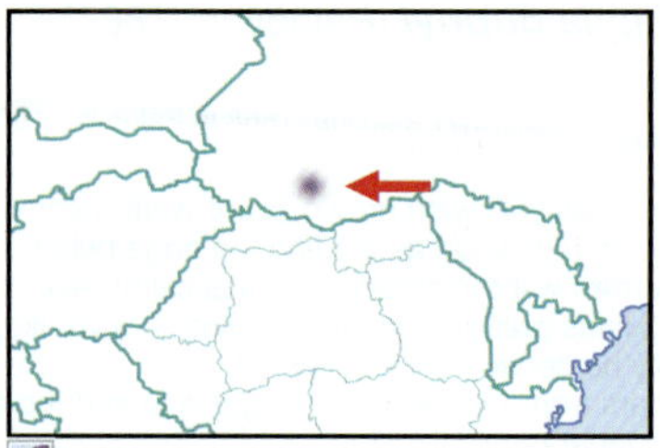

SW Ukraine

Boeters et al. 1989: 125.

Platyla lusitanica (Holyoak & Seddon, 1985)

Acicula lusitanica Holyoak & Seddon, 1985 in Holyoak & Seddon 1985: 63

Shell conical, looks like juvenile, light yellowish brown, with some growth lines, 5.5-6.5 convex whorls, last whorl with largest diameter, not ascending, aperture small, margin blunt, parietal callus extremely thin, connected to columellar callus, which covers the umbilicus, no angularis, no separate umbilical callus, no cervical callus. Easy to tell from other species.

2.6-3.5 x 0.9-1.0 mm

Reported from soil of a stony and well vegetated place near a spring in Caldas de Monchique.

EU Red List: Vulnerable VU. Only known from two localities.

Boeters et al. 1989: 127.

SW Portugal

Platyla maasseni Boeters, Gittenberger & Subai, 1989

Platyla maasseni Boeters, Gittenberger & Subai, 1989 in Boeters et al. 1989: 128

Shell whitish to brownish, 3.5 moderately convex whorls, last whorl > 1/2 of shell height, ascending near aperture, apertural margin broadened, slightly convex in lateral view, parietal callus moderate, angularis weak, no umbilicus, umbilical callus as broad as columella, cervical callus broad but not high, well delimited at both sides, broadening downwards to the base.
Smaller than *Pl. gracilis*, cervical callus flatter and broader. *Pl. similis* has no angularis, a less broadened apertural margin and a flatter cervical callus. *Pl. microspira* is more slender whitout angularis and with a flatter cervical callus.

1.8-1.9 x 0.8 mm

EU Red List: Vulnerable VU. Only known from one locality.

Description was based on one single shell, more has not been found. The authors believed that the shell did not represent an unusually grown specimen of another species. The presence of a stable population of this Croatian species needs to be confirmed.

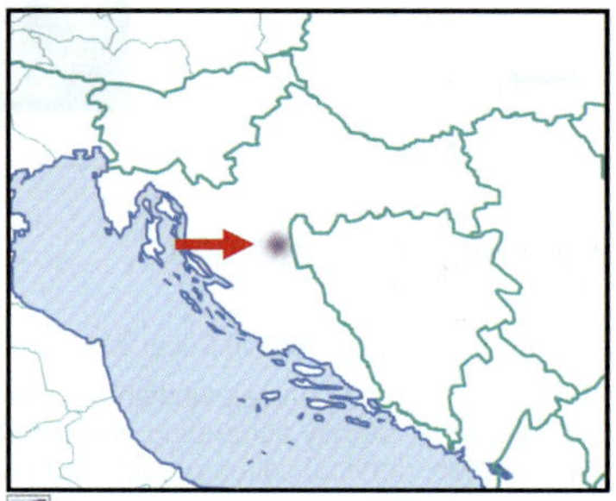

Croatia

Platyla microspira (Pini, 1885)

Acme microspira Pini, 1885 in Pini 1885: 239, Tav. 12
Date was Feb 1885 (indicated on the wrapper preserved in SUB Göttingen), not 1884.

Shell colourless or with brownish hue, 4-5 weakly convex whorls, last whorl weakly ascending before aperture, apertural margin broadened, (shaped in lateral view, parietal callus only present near columellar side, no angularis, no separate umbilical callus, cervical callus flat and indistinct, well delimited at both sides, broader at basal side.
Differs from *P. perpusilla* in its flat cervical callus and less cylindrical shape. *P. maasseni* has a larger aperture with angularis and a higher cervical callus. *P. stussineri* is smaller with more cylindical shape and has a sharp apertural margin without cervical callus, \ shaped in lateral view.

Italy: Bergamo, Bracca
Maurizio Sosso

Boeters et al. 1989: 130, Bodon & Cianfanelli 2008, Lengyel & Páll-G. 2010: 92.

Cianfanelli et al. 2000 reported a species from Sardinia which is more common there than *Pl. sardoa* and should preliminarily be identified with *P. microspira* based on the morphological record.

1.75-2.15 x 0.65-0.75 mm

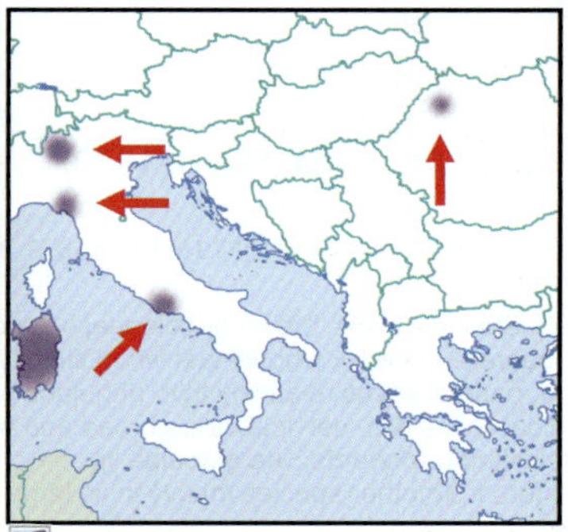

N Italy, NW Romania

Platyla minutissima Boeters, Gittenberger & Subai, 1989

Platyla minutissima Boeters, Gittenberger & Subai, 1989 in Boeters et al. 1989: 132

Shell very small, colourless or light brownish, transparent, regularly cylindrical, 4-4.5 weakly convex whorls, last whorl not ascending at aperture, aperture simple, no separate umbilical callus, no cervical callus.

1.1-1.25 x 0.4 mm. One of the smallest terrestrial gastropods in Europe.

Was found in soil substrate under shrubs in a dry valley.

EU Red List: Near Threatened NT. In Kérkira potentially threatened by habitat destruction.

Subai 2009: 109.

The species can easily be overlooked due to its small size. The population in Kérkira seems to be stable, it was first found in 1972, later (before 1989) it was found again.

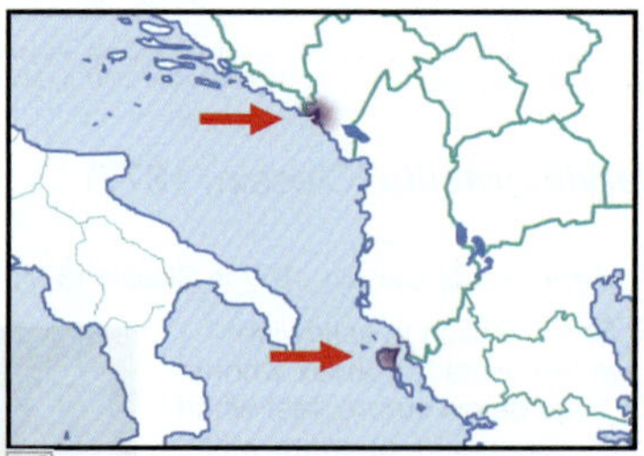

Greece (Kérkira), W Montenegro

Platyla orthostoma (Jackiewicz, 1979)

Acicula orthostoma Jackiewicz, 1979 in Jackiewicz 1979: 135, Tab. 1

Shell pale yellowish brown, smooth, 6 weakly convex whorls, last whorl slightly ascending, apertural margin convex in lateral view, apertural margin connected at parietal side by a broad callus, no angularis, no sinulus, cervical callus narrow and low.
Differs from *Pl. gracilis* in its higher whorls, a shallower suture and flatter cervical callus.

3.4 x 1.0 mm

EU Red List: Near Threatened NT. Only known from few localities in a small and restricted area.

Boeters et al. 1989: 135.

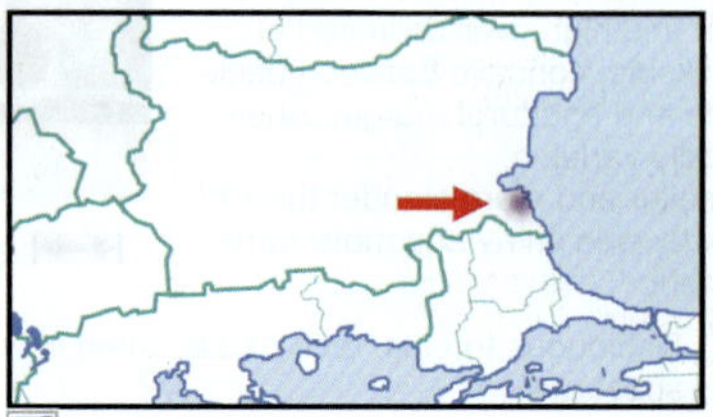

SE Bulgaria

Platyla peloponnesica Boeters, Gittenberger & Subai, 1989

Greece (S Pelopónisos)

Platyla peloponnesica Boeters, Gittenberger & Subai, 1989 in Boeters et al. 1989: 137

Shell yellowish to reddish brown, slightly conical, 5-5.5 moderately convex whorls with a narrow band below the suture, last/penultimate whorl height more than 50 % of shell height, last whorl slightly ascending, aperture with sinulus, no angularis but with a weak angular callus, strong parietal callus, columellar callus as broad as columella, cervix with very weak cervical callus.
Between *Pl. similis* and *Pl. wilhelmi*. *Pl. wilhelmi* is lighter with larger aperture, *Pl. similis* is smaller and has a weaker parietal callus.

2.7-3.0 x 1.0-1.15 mm

EU Red List: Vulnerable VU. Only known from one locality.

Platyla perpusilla (Reinhardt, 1880)

Acme perpusilla Reinhardt, 1880 in Reinhardt 1880: 46

Shell colourless, transparent, almost cylindrical, 4.5-5.5 weakly convex whorls, last whorl very weakly ascending, apertural margin slightly convex in lateral view, parietal callus strong, no angularis, umbilical callus short and weak, cervix with a very weak thickening, no callus.
Small size, but larger than *Pl. stussineri* and *Pl. minutissima*.

Boeters et al. 1989: 139.

In E Hungary found in river deposits, origin NW Romania or SE Ukraine.

1.3-2.0 x 0.7 mm

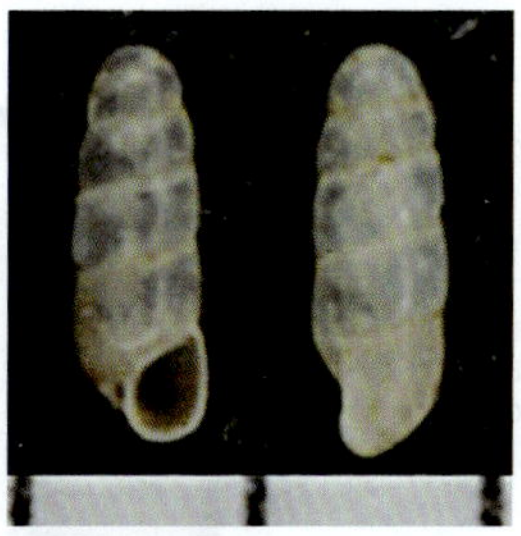

Hungary: Tiszatelek, Tisza river deposits
Zoltán Fehér

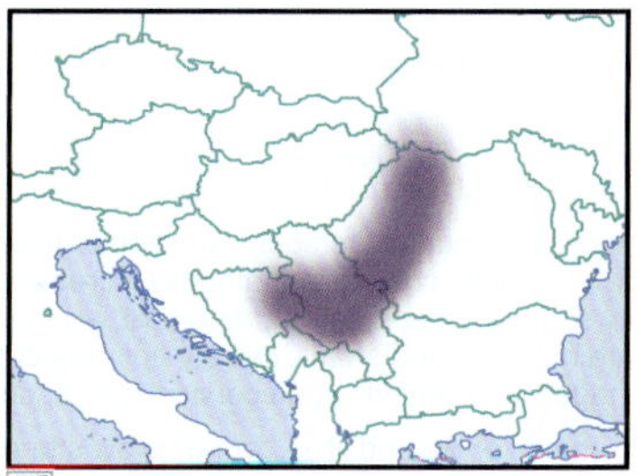

SE Europe

Platyla pezzolii Boeters, Gittenberger & Subai, 1989

Platyla pezzolii Boeters, Gittenberger & Subai, 1989 in Boeters et al. 1989: 141

Shell reddish to yellowish brown, 6-5 moderately convex whorls, last whorl ascending near aperture, aperture with angularis, palatal margin with very small sinulus, separate umbilical callus present, cervical callus very prominent and characteristic, moderately broad, in lateral view convex between backside and margin, very high at basal side, visible as a separate structure in frontal view at columellar and palatal side.
In P. *gracilis* the callus is weaker and in lateral view concave between backside and margin, not convex or semicircular.

EU Red List: Near Threatened NT. Only known from few localities in a small restricted area.

Bodon & Cianfanelli 2008.

2.85-3.35 x 1.05-1.15 mm

Italy: Treviso, near Vittorio Veneto

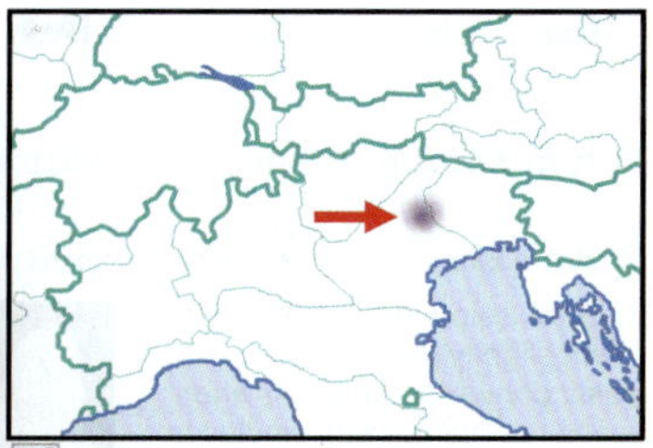

NE Italy

Leaf litter and soil in rocky habitats.

Massimo Prodan

Platyla pinteri (Subai, 1976)

Acicula pinteri Subai, 1976 in Subai 1976: 119

Shell colourless to weakly yellowish brown, smooth, very slender and slightly conical, 5.5-6.5 weakly convex whorls, apertural margin with weak sinulus (but not in lateral view), no angularis, cervical callus narrow and flat, well delimited at backside.
Pl. banatica is less slender, darker, cervical callus higher and less well delimited at backside.

In soil litter under shrubs on limestone substrate, up to 1100 m.

3.5-5 x 1.2-1.4 mm

Albania: Gjirokastër

W Greece, S Albania

Platyla polita (Hartmann, 1840)

Truncatella acicularis polita Hartmann, 1840 in Hartmann 1844: 5, Tab. 2
Name originally *Truncatella (Pupula) polita* on Tab. 2, *Truncatella (Pupula) acicularis polita* on p. 5. Hartmann (1844) frequently wrote "*Pupula polita*" in the text, but on p. 4 he made clear that *Pupula* was meant as a subgenus ("Untergattung") of *Truncatella*. Type locality Nyon (Switzerland) and Schöneberg (23 km NNW Kassel, Germany); Boeters et al. 1989: 145 intended to "restrict" the type locality to Schöneberg, but this was ineffective since no name-bearing type was selected (Art. 76).
Acme oedogyra Paladilhe, 1868 in Paladilhe 1868: 236

Shell reddish brown, smooth, 5.5-6 moderately convex whorls, outline almost straight, last whorl slightly ascending, margin internally thickened, angularis weak, cervical callus fmoderately broad, sharply delimited at posterior side, usually also at frontal side.
Shell larger and less slender than *Pl. gracilis*, with broader cervical callus. *Pl. subdiaphana* is larger and more slender. *Pl. similis* has a sinulus and no cervical callus.

2.4-3.85 x 1.05-1.35 mm

Germany: Schleswig-Holstein, near Eckernförde

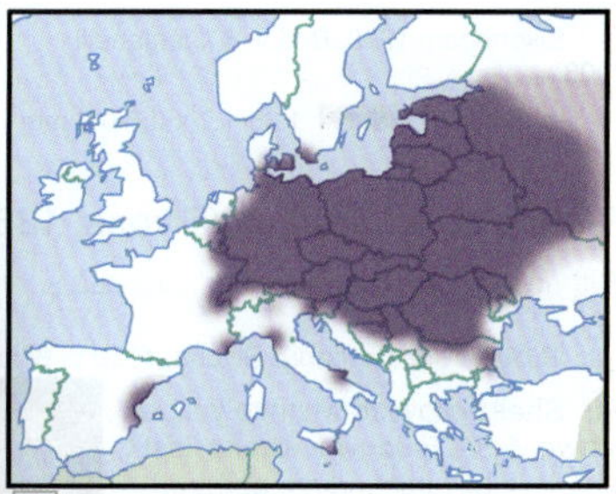

Europe

In moderately humid forests in soil litter, under dead wood logs and stones. The species is usually not found alive, but appears when large quantities of soil are sieved, in Spain for example from bases of rocks.

The system proposed by Boeters et al. 1998 is used here, where *Pl. polita* was regarded as a morphologically variable species. In total three taxonomic systems are currently used:
(1) Boeters et al. 1989: 145, 153 (*oedogyra* "clearly belonging to" *Platyla polita*, *similis* an independent species, *perpusilla* an independent species).
(2) Anistratenko 1998, Sysoev & Schileyko 2009: 13 (*oedogyra* an independent species, *similis* a synonym of *oedogyra*, *perpusilla* an independent species).
(3) Egorov & Greke 2003: 20, Kantor & Sysoev 2005: 41, Egorov 2008: 20 (*oedogyra* an independent species, *similis* and *perpusilla* synonyms of it).
Egorov & Greke 2003 argued that *oedogyra* from W Ukraine (Lviv, Zhitomir) had a smaller size and shell height, more convex whorls and differed from *polita* in its apertural margin, they argued that both forms occurred syntopically. But they did not provide data to substantiate the arguments. A detailed study would be useful, particularly on the nature of morphological variations within the *Pl. polita* group, or statistical evidence that we are clearly dealing with two separate units at some localities.

Platyla procax Boeters, Gittenberger & Subai, 1989

Platyla procax Boeters, Gittenberger & Subai, 1989 in Boeters et al. 1989: 157

Shell whitish to yellowish brown, with some coarse radial lines, slightly conical, looking like a juvenile shell with 4-5.5 moderately convex whorls, last whorl ascending near aperture, aperture with thin parietal callus, no angularis, margin **S** shaped in lateral view, no separate umbilical callus, cervical callus narrow and weak, not well delimited at both sides.
Shell broader than usual in Aciculidae, last whorl with largest diameter. Differs from *P. corpulenta* in its smaller size and presence of a cervical callus.

3.25-3.6 x 1.5-1.55 mm

Was found at or in a cave at 1300 m above sea level.

EU Red List: Vulnerable VU. Only known from one cave between Virpazar and Petrovac na Moru (P. Subai).

Montenegro

Platyla sardoa Cianfanelli, Talenti, Bodon & Manganelli, 2000

Platyla sardoa Cianfanelli, Talenti, Bodon & Manganelli, 2000 in Cianfanelli et al. 2000: 63

Shell colourless, smooth with some growth lines, transparent, 5.5-7.5 slightly convex and steadily increasing whorls, last whorl very slightly ascending near aperture, aperture relatively small, margin simple and sharp, usually not connected at parietal side, \ shaped in lateral view, no cervical callus, umbilicus closed.

In soil or ground litter, found below stones, from sea level up to 800 m.

EU Red List: Near Threatened NT. Restricted range, but locally frequent.

Bodon & Cianfanelli 2008.

2.7-3.8 x 0.9-1.1 mm

Italy (E Sardinia)

Italy: Sardinia, Nuoro, Cala Gonone, Cala Fujli

Platyla similis (Reinhardt, 1880)

Acme similis Reinhardt, 1880 in Reinhardt 1880: 46

Shell colourless to yellowish brown, slightly conical, 4.5-6 weakly convex whorls, last whorl ascending, margin with sinulus in frontal and lateral view, **(** shaped in lateral view, no cervical callus, only a weak thickening.
Pl. jankowskiana is more cylindrical, aperture without sinulus. *Pl. polita* is larger, has a moderately broad and at the backside well delimited cervical callus and no sinulus.

Bodon & Cianfanelli 2008.

2.0-2.8 x 0.7-1.1 mm

Italy: Lucca, near Pizzorno

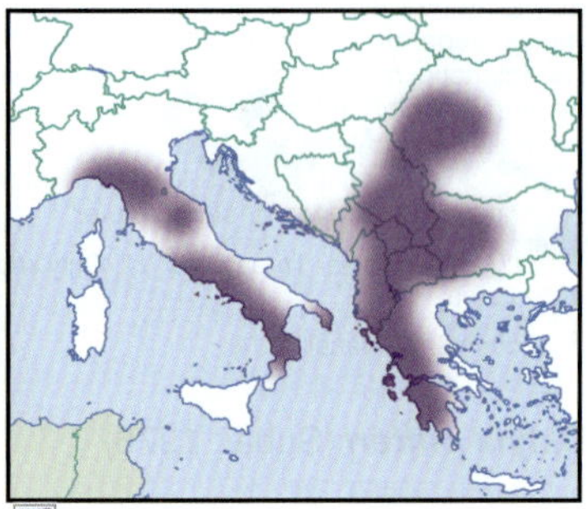

SE Europe, Italy
Pliocene: from England to Czech Republic.

Platyla stussineri (Boettger, 1884)

Acme stussineri Boettger, 1884 in O. Boettger 1884: 185 (*Acme Stussineri*)

Shell very small, colourless with brownish yellow hue, very small and cylindrical, 4-5 weakly convex whorls, last whorl not ascending, parietal callus very weak, no cervical callus.
Animal with strongly pigmented eyes at the bases of the tentacles.

Not a cave snail, but needs humidity.

Eikenboom 1998, Bodon & Cianfanelli 2008.

1.35-1.7 x 0.5-0.6 mm

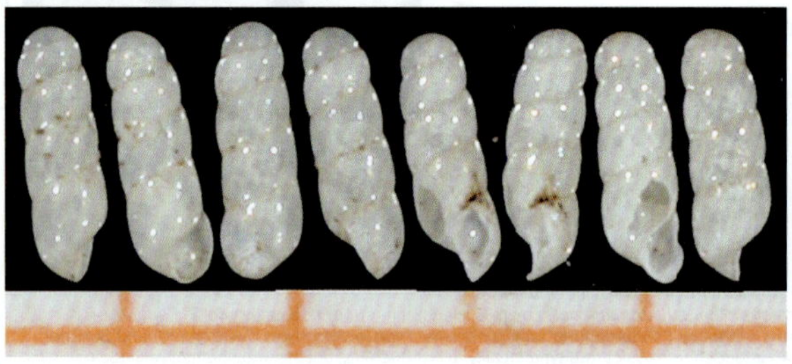

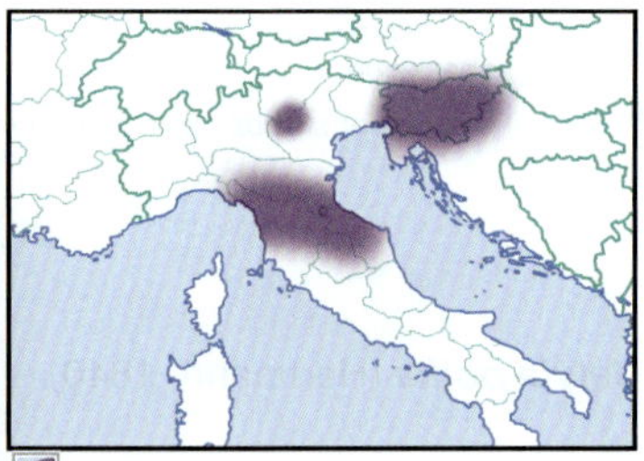

Slovenia to central Italy

Italy: Lucca, Pizzorno (river deposits)

Platyla subdiaphana (Bivona, 1839)

Without gender agreement: *Platyla subdiaphanus* (Bivona, 1839)

Bulimus subdiaphanus Bivona, 1839 in Bivona 1839: 20 [not verified]

Shell yellowish white to yellow brown, 5.5-6.5 convex whorls, last whorl slightly ascending, aperture with thin layer at parietal side and a weak sinulus (only in frontal view), angularis weak, umbilical callus thin and narrow, cervical callus strong, often white, moderately narrow and distinctly delimited at backside.
Suture deeper than in *Pl. pinteri*. *Pl. gracilis* is smaller, has an edge near the suture and it cervical callus usually a concave structure between backside and apertural margin. *Pl. polita* is slightly smaller, less slender, with an edge below the suture. It also seems that the diameter of the aperture in *Pl. subdiaphana* is slightly larger than in other species.

3.5-4.5 x 1.2-1.5 mm

Italy: Sicily, near Valderice

Museum Budapest HNHM

Boeters et al. 1989: 168, Bodon & Cianfanelli 2008.

Italy (NW Sicily)
Specimens from S Sicily could belong to this species or to *Pl. polita*.

Found in soil litter in limestone areas, lives probably in subterranean habitats.

EU Red List: Near Threatened NT. Only known from 2 localities.

Platyla talentii Bodon & Cianfanelli, 2008

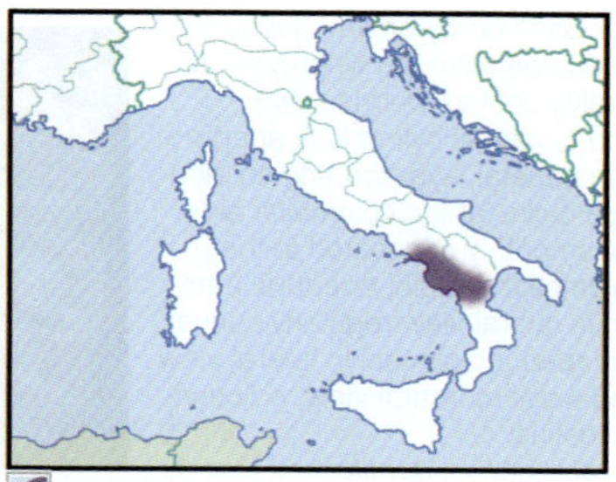

S Italy

Platyla talentii Bodon & Cianfanelli, 2008 in Bodon & Cianfanelli 2008: 28

Shell brown yellowish, slightly trapezoidal, apertural margin slightly reflected, cervical callus strong and broad (1/5 of shell diameter), with a shallow groove at its anterior part near the aperture and getting more prominent towards its posterior part, where it ends abruptly.
Differs from *P. elisabethae* in its slightly broader cervical callus (slightly more protruding posteriorly), and the posterior part being less prominently delimited (more C-like and not L-like).

3-4.5 x 1.2-1.5 mm

In leaf litter of deciduous forests, under limestone rocks and stones.

EU Red List: Near Threatened NT.

Bodon & Cianfanelli 2008 published a very accurate study, but unfortunately missed to analyse the degree of variation in the thickness of the cervical callus in S Italy and Dalmatia. It seems that some Croatian specimens of *P. elisabethae* have a broader cervical callus, like the Italian population.

Platyla turcica Boeters, Gittenberger & Subai, 1989

NW Turkey (Vilayet Bolu)

Platyla turcica Boeters, Gittenberger & Subai, 1989 in Boeters et al. 1989: 171

Shell colourless, smooth, very small and cylindrical, 4 almost flat whorls, first whorl as broad as the others, last whorl not ascending, aperture simple without sinulus, no cervical callus.

1.5-1.6 x 0.5 mm

Only known from one locality (Schütt 2001: 13).

Platyla wilhelmi (Wagner, 1910)

Acme wilhelmi Wagner, 1910 in A. J. Wagner 1910: 187

Shell brownish to yellowish, 5.5-6 weakly convex whorls, last whorl relatively large and ascending, looks somehow like a *Cecilioides* species with low aperture, apertural margin slightly thickened, some shells have a very weak cervical callus.

Found in organic litter at rocks, subterranean habitats.

Boeters et al. 1989: 172, Dhora & Welter-Schultes 1996: 107, Deli & Subai 2011: 46.

3.15-3.9 x 1.15-1.4 mm

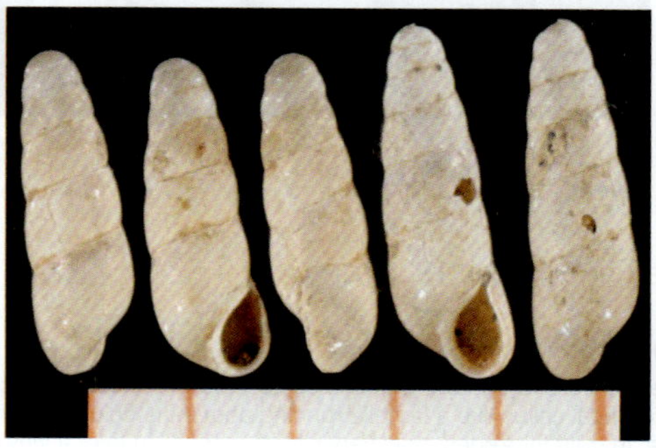

Albania: near Shkodër

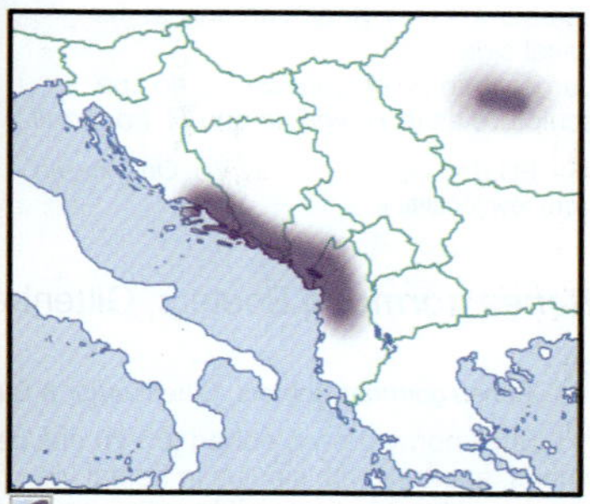

S Croatia to Albania, W Romania

Renea

Renea Nevill, 1880 in Nevill 1880: 137

Type species *Renea bourguignatiana* Nevill, 1880 (original designation)

Shell cylindrical, 1.8-6.3 x 0.6-2 mm, densely and regularly ribbed (30-80 ribs/penultimate whorl), no edge at the suture, last whorl slightly ascending near aperture, cervical callus absent or very weak.
In SE France we observe a succession of various species displaying a west-east directed gradient from a very long and prominent angular sinulus towards a regular aperture without sinulus (*moutonii-paillona-gormonti-bourguignatiana-gentilei / elegantissima*).

Recent from Albania to French Alps, fossil (Oligocene/Miocene) also in central Europe

Boeters et al. 1989 (revision).

Renea bourguignatiana Nevill, 1880

Renea bourguignatiana Nevill, 1880 in Nevill 1880: 138, Pl. 14

Shell reddish or horny brown, densely and regularly ribbed (24-30 ribs/mm, 50-65 ribs at penultimate whorl), slender, 6.5-8 weakly convex whorls, last 3 whorls equal in diameter, apertural margin **C** or **(** shaped in lateral view and some shells with sinulus, parietal margin strongly protruded, no cervical callus.

Lives in a little basin surrounded by woodland, in damp ground litter.

Bodon & Boato 1987, Boeters et al. 1989: 180, Bodon et al. 1989: 21, www.iucnredlist.org (07.2012).

3.2-3.8 x 1.0-1.1 mm

NW Italy: Imperia, Ventimiglia, Ciaixe

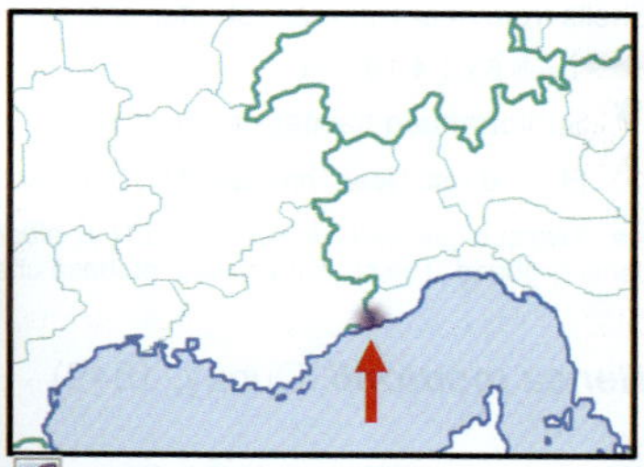

NW Italy, SE France
Fossils were found in SE France (Menton).

EU Red List: Critically Endangered CR. A living fossil restricted to a small locality, threatened by the ongoing urbanization of the region and habitat destruction (parking lots and garages have been built into the gorge where the species had been found in the 1980s) and invasive plants (*Impatiens*, probably the most serious threat). Protected by French law, but habitat destruction is apparently not forbidden.

Renea elegantissima (Pini, 1886)

Acme elegantissima Pini, 1886 in Pini 1886: 521

Shell horny brown, ribbed (55-65 ribs/penultimate whorl), 6-7.5 weakly convex whorls, lower whorls almost cylindrical, aperture with parietal layer forming an elongate callus at the columellar side, not smoothly connected to the columellar margin, apertural margin in lateral view regularly convex (shaped, no sinulus in lateral view and weak in frontal view, no cervical callus.
Larger than *R. gentilei* and less cylindrical.

3.1-3.6 x 1.1-1.2 mm

Italy: Savona, Finali Ligure, Rio Ponci river deposit

A relatively widespread but obviously rare species.

NW Italy, SE France

Renea gentilei (Pollonera, 1889)

Acme gentilei Pollonera, 1889 in Pollonera 1889: 53, Tav. 2 (*Acme Gentilei*)

Shell horny brown, regularly ribbed (50-60 ribs/penultimate whorl), 6-7.5 moderately to weakly convex whorls, aperture with callus at parietal side, not merging smoothly into the columellar margin but running towards to the umbilicus, apertural margin slightly thickened, in lateral view **I** shaped or only weakly convex, no sinulus in lateral view, usually no cervical callus, sometimes very weak and only at the basal side.
Larger than *R. elegantissima*, no fine sculpture as in *R. veneta* and *R. gormonti*.

3.8-4.2 x 1.25-1.35 mm

Italy: Piemonte, Cuneo, Bossea

EU Red List: Vulnerable VU. Only known from few localities.

NW Italy

Renea gormonti Boeters, Gittenberger & Subai, 1989

Renea gormonti Boeters, Gittenberger & Subai, 1989 in Boeters et al. 1989: 185

Shell horny brown, ribbed (55-70 ribs/penultimate whorl) and with fine spiral lines, 6-7 moderately convex whorls, aperture with sinulus in frontal view, parietal callus like a lamella running above the columellar margin, angularis-like structure visible in the aperture, margin thickened, only very slightly convex in lateral view, cervical callus narrow, partly lighter than the rest of the shell.

3.5-4.1 x 1.2-1.4 mm

Found in shady habitats in thick layers of leaves and rock rubble, and in a more open grassy slope with some rocks.

EU Red List: Vulnerable VU. A rare species, known from only 3 localities, potentially threatened by urbanization.

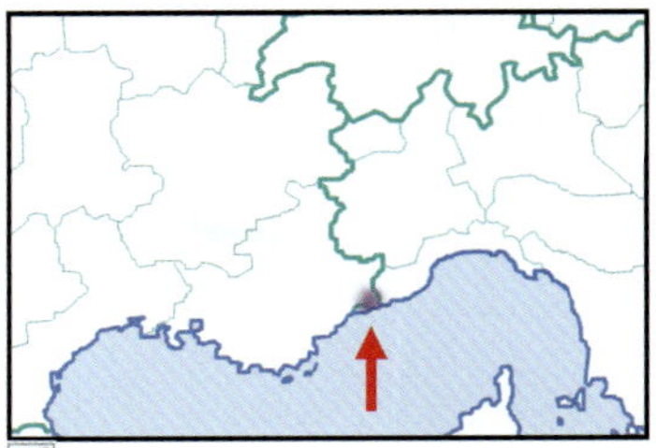

SE France (Alpes-Maritimes)

Renea kobelti (Wagner, 1910)

Acme kobelti Wagner, 1910 in A. J. Wagner 1910: 187
Renea kobelti albanica Boeters, Gittenberger & Subai, 1989 in Boeters et al. 1989: 189

Shell pale horny brown, transparent, ribbed (35-60 ribs/penultimate whorl), 6-7.5 moderately convex whorls, aperture with angularis and without cervical callus, margin (shaped in lateral view, no sinulus in lateral view, no cervical callus.
Shells from near Kotor are shorter and have less ribs than between Virpazar and central Albania.

4-5.6 x 1.4-1.7 mm

Subterranean habitats in rocky areas, mostly limestone.

EU Red List: Near Threatened NT. Only known from few localities in a broad area.

Two subspecies *R. k. kobelti* and *R. k. albanica* have been recognized. *R. k. albanica* is slightly higher and more densely ribbed. The spatial variation of these characters remains to be analysed, more field material would be necessary.

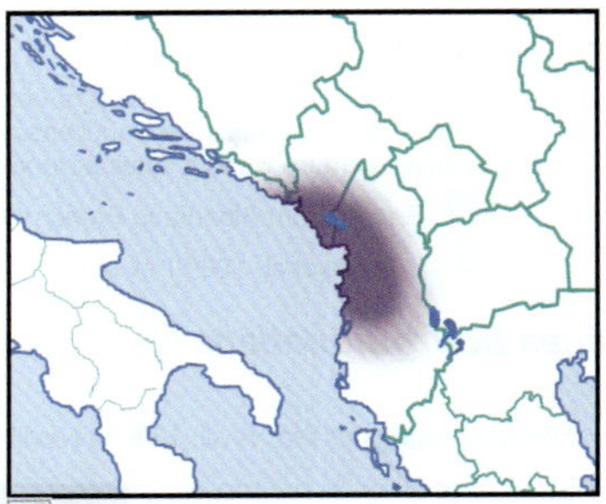

Albania, Montenegro

Renea moutonii (Dupuy, 1849)

Acme moutonii Dupuy, 1849 in Dupuy 1849: [3] (*Acme Moutonii*)
Caziotia singularis Pollonera, 1905 in Pollonera 1905: 2, Tav. [1]

Shell light horny brown, finely ribbed (50-75 ribs at penultimate whorl), 6-7.5 moderately convex whorls, last 3 whorls almost equal in diameter, aperture with prominent sinulus, no angularis, margin very obliquely (shaped in lateral view, in some populations with an exaggerated sinulus (a **P**-like opening at the suture in the last quarter of the last whorl), no cervical callus, but the aperture is slightly reflected at the basal side. Diameter larger than in *R. bourguignatiana* and *R. paillona*.

3.1-4.1 x 1.1-1.3 mm

France: Alpes-Maritimes, near Pont du Loup

Found in soil litter in valleys, moist habitats.

SE France (Alpes-Maritimes)

EU Red List: Near Threatened NT. Only known from 10 or so localities.

Falkner et al. 2002: 68.

Renea paillona Boeters, Gittenberger & Subai, 1989

Renea paillona Boeters, Gittenberger & Subai, 1989 in Boeters et al. 1989: 198

Shell light horny brown, regularly ribbed (55-65 ribs at penultimate whorl), 6.5-7.5 weakly convex whorls, last whorl ascending, aperture with parietal callus becoming gradually stronger towards umbilicus, it passes over the columellar margin like a lamella, margin (shaped in lateral view with strong sinulus, also in frontal view, no cervical callus.
Differs from *R. bourguignatiana* in its broader aperture with less thickened margin and narrower sinulus.

EU Red List: Vulnerable VU. Only known from few localities, obviously a rare species. Protected under the French law.

3.1-3.8 x 1.1-1.15 mm

Italy: Imperia, Vallone del Passo (determined by M. Sosso)

SE France, NW Italy

Renea spectabilis (Rossmässler, 1839)

Without gender agreement: *Renea spectabile* (Rossmässler, 1839)

Carychium spectabile Rossmässler, 1839 in Rossmässler 1838-1844 (9-10): 36, Taf. 49

Shell horny brown, ribbed (55-70 ribs at penultimate whorl), slightly conical, 6.5-8 moderately convex whorls, last whorl slightly ascending, aperture with layer at parietal side and a small angular callus, cervical callus very strong and usually lighter, rounded at backside, (shaped in lateral view with a small sinulus, umbilicus closed by the prolonged parietal callus.
The only *Renea* species with a very strong cervical callus.

Deciduous forests under rocks, in litter and in light soil, also in caves.

5.0-6.3 x 1.9-2.0 mm

NE Italy: Alpi Giulie, Gorizia

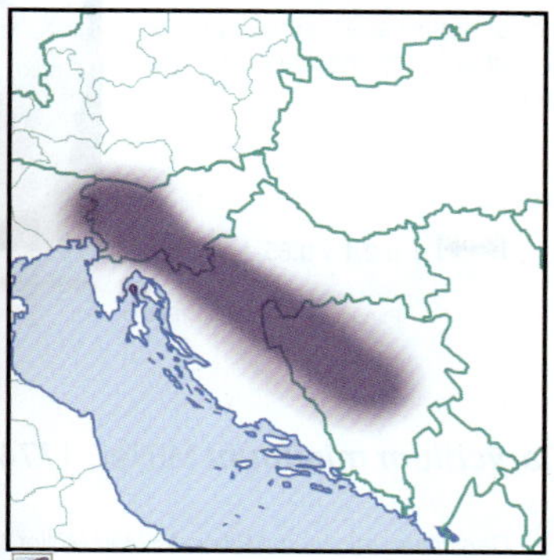

Slovenia to Bosnia

Štamol & Velkovrh 1995 (N Cres), Bole & Slapnik 1997: 145.

Renea veneta (Pirona, 1865)

Acicula spectabilis var. *veneta* Pirona, 1865 in Pirona 1865: 697

Shell brown, regularly ribbed (45-60 ribs at penultmate whorl), conical, outline almost straight-sided, 6.5-7.5 weakly convex whorls, aperture with angular tooth and deep sinulus in frontal view, margin inside thickened, in lateral view slightly (shaped, no sinulus in lateral view, cervical callus narrow, most prominent at basal side, umbilicus closed by parietal callus.

Limestone rocks, but also in mountains forests in the soil and on dead wood logs.

3.7-4.5 x 1.3-1.5 mm

Austria: Oberösterreich, Hallstadt

Austria (Salzburg), N Italy, Slovenia

Carychium

Carychium Müller, 1773 in Müller 1773: [24]

Type species *Carychium minimum* Müller, 1774 (subsequent monotypy by Müller 1774: 125)

ICZN Opinion 335: type fixed by monotypy, should be subsequent monotypy. Date was given as 1774, must be 1773 (Art. 80.6.2).

Hermaphrodites with functional males and females, females have penis reduced. Large eggs 0.4 mm in diameter. Up to 4 generations per year.

This genus presentation was kindly checked by A. Jochum.
Weigand, Gö. & J. 2012 suggested that *Carychium minimum* displayed locally some morphological variation in terms of shell sculpture and height/width ratio, which has previously been associated with the presence of separate species.
The results of DNA barcoding brought no visible difference between *mariae* from N Italy and *minimum*, *mariae* having been regarded as a separate species that was characterised by a striated shell. The authors also reported that *minimum* and *tridentatum* possessed striated shells in N Italy, so this character could be ecologically induced.
The identities of *biondii*, *hellenicum* and *paganettii* should also be revised, their original descriptions seem to point to a similar situation as in *mariae.*
DNA studies by Weigand, Jo. et al. 2011 suggested that *minimum*, *tridentatum*, *ibazoricum* and an undescribed form from Bulgaria were separate species. More research is necessary.

Carychium biondii Paulucci, 1882

Carychium biondii Paulucci, 1882 in Paulucci 1882: 305 (*Carychium Biondii*)

Shell usually not as slender as *C. tridentatum* (H/D about 2), weakly striated, last whorl perhaps slightly more predominating.

Identity should be revised.

1.8-2.2 x 1 mm

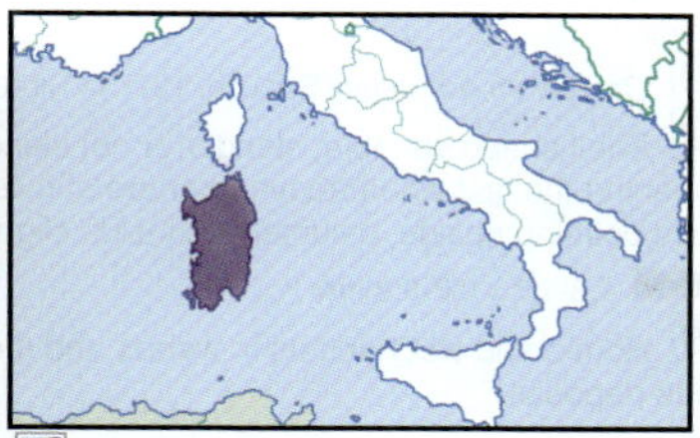

Italy (Sardinia)

Italy: Sardinia, Nuoro, Laconi

Carychium hellenicum Bank & Gittenberger, 1985

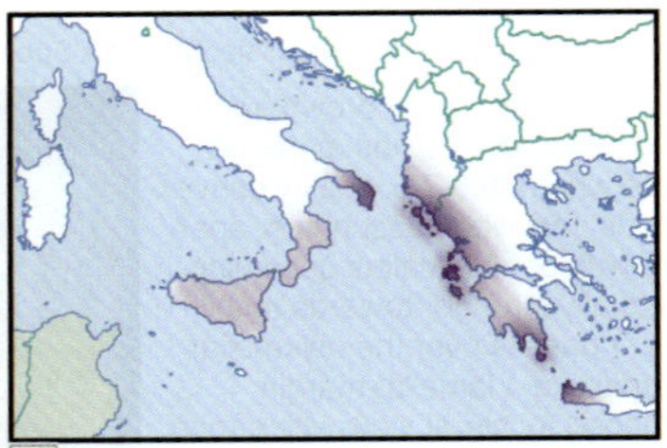

W Greece, S Italy

Carychium hellenicum Bank & Gittenberger, 1985 in Bank & Gittenberger 1985: 95

Shell 1.6-2.0 times as high as broad, sculpture relatively vague transverse, internal part of parietalis strongly contorted with prominent curve upward, abrupt decline of parietalis already in last whorl.

1.5-2.0 x 0.8-1.1 mm

In S Italy found in soil samples of swamps and humid habitats near lakes and lagoons, lives probably in the leaf litter.

Ferreri et al. 2005: 46 (S Italy).

The description suggests that this form could be conspecific with *C. minimum*.

Carychium ibazoricum Bank & Gittenberger, 1985

Carychium ibazoricum Bank & Gittenberger, 1985 in Bank & Gittenberger 1985: 85
Expressly established as a new replacement name for
Auricula gracilis Morelet, 1845 (in Morelet 1845: 76, Pl. 7) non Grateloup, 1828

Rather big shell, slender (height/diameter ratio = 2.0-2.2), with prominent (sharp and transverse) riblets, granular microsculpture on entire shell, sinuous edge of (rather long) internal part of parietalis.

1.8-2.1 x 0.85-1.1 mm

Portugal: Litoral, Bosco Bucao

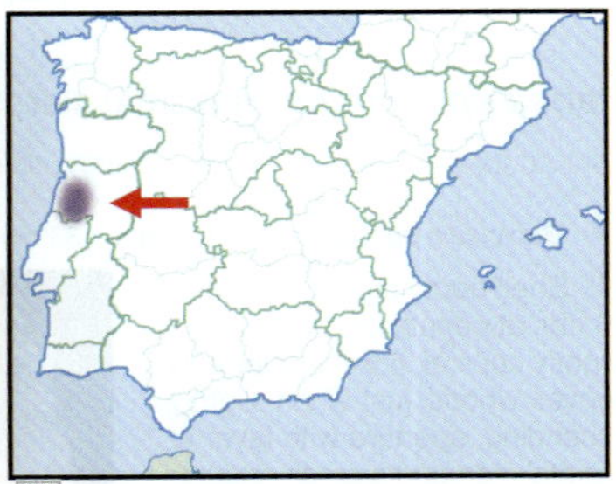

Portugal, Azores

Carychium minimum Müller, 1774

Carychium minimum Müller, 1774 in Müller 1774: 125
Carychium mariae Paulucci, 1878 in Paulucci 1878: 41 (*Carychium Mariæ*)

Less slender than *C. tridentatum* (H/D < 2). When the last whorl is opened above the aperture, the parietalis descends in a simple way downwards.
Animal whitish with weak yellowish hue, tentacles thick and conical, black eyes at their base.

Slovakia: Low Tatra National Park
Jozef Grego

1.6-2.2 x 0.9-1.1 mm

N Germany: Schleswig-Holstein, near Westensee

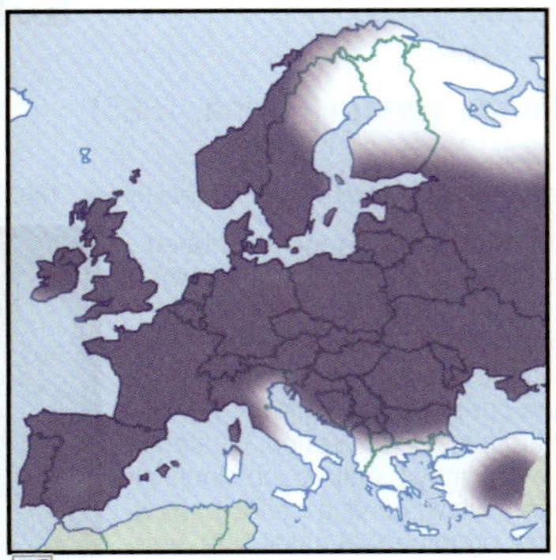

Eurasia, introduced to North America

Permanently wet habitats, swamps, river plain woods, wet meadows, usually near silent water bodies. At stones, plants or submerged objects such as pieces of wood or desintegrating tree branches. Does not tolerate drought, sensitive against sinking of ground water levels and drainage. Virtually amphibious, can survive long winter flooding. In Switzerland up to 1800 m.

Declining in Bayern. Common in Britain and Ireland.

Weigand, Gö. & J. 2012 (regarded *mariae* as a synonym).

Might also contain *biondii*, *hellenicum* and *paganettii*.

Carychium paganettii Zimmermann, 1925

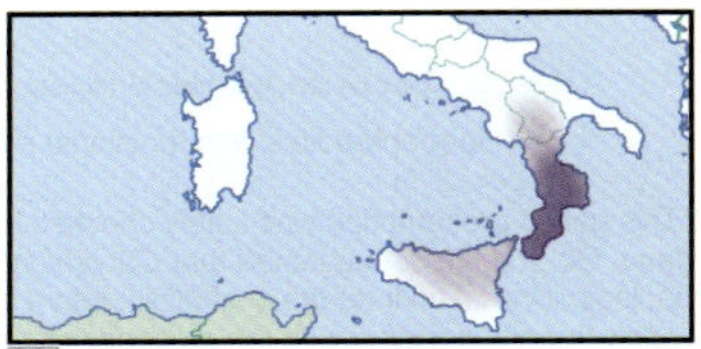

Italy (Calabria, Sicily)

Carychium paganettii Zimmermann, 1925 in Zimmermann 1925: 328, 332, Taf. 4 (*C. Paganettii*)

Shell like *C. minimum* but larger, apertural lip slightly weaker, shell almost smooth.

1.8-2.3 x 0.9-1.1 mm

Zimmermann 1925 (Calabria and Sicily), Manganelli et al. 1995: 10, Ferreri et al. 2005: 126 (only Calabria and Basilicata).

Classification uncertain, the differences to *C. minimum* pointed out by Zimmermann 1925 seem to be too weak to provide evidence that this is an independent species.

Carychium schlickumi Strauch, 1977

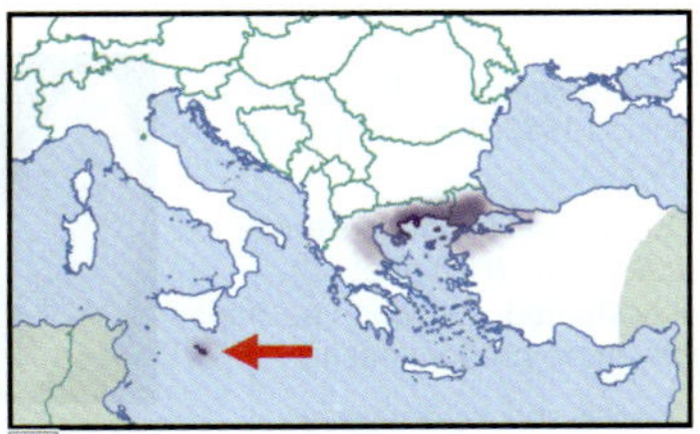

Malta, N Greece, NW Turkey

Carychium schlickumi Strauch, 1977 in Strauch 1977: 168

Shell shape variable, slender or not slender, exterior can resemble both *C. minimum* and *C. tridentatum*. Opening the last whorl above aperture shows parietalis descending in a single fold which has a prominent broad lobe pointing upside, columellaris inside slightly weaker than in the other two species.

1.6-1.7 x 0.7-0.8 mm

Humid and wet habitats in woods and meadows, under leaves and decaying wood, frequently found in riverside deposits.

Bank & Gittenberger 1985: 94, Giusti et al. 1995, Schütt 2001: 15.

Fossil present in a wide range between France and Caucasus.

Carychium tridentatum (Risso, 1826)

Without gender agreement: *Carychium tridentata* (Risso, 1826)

Saraphia tridentata Risso, 1826 in Risso 1826: 84

Shell more slender than *C. minimum* (H/D > 2). Opening the last whorl above aperture shows parietalis descending in a characteristic double curve.

1.8-2.3 x 0.8-0.9 mm

Germany: Bayern, Berchtesgadener Alpen

Europe to Caucasus, N Africa

Wide ecological amplitude, often in highlands, requires permanent humid conditions, but less humid than *C. minimum.* In moderately humid forests and in humid meadows, there in deep litter layers and between stones. Leaves have to be permanently humid and partly covered with black worm faeces. Indicates continuously forested areas in Ireland, does not occur in recently established or coniferous plantations. In Switzerland up to 2200 m.

Protandric hermaphrodite, penis and vas deferens only present in July-August. 5-6 eggs are laid often at distinct places, egg diameter 0.3 mm, shell height 0.4 mm after hatch in spring, growth until November to 1.2-1.5 mm, juveniles have high survival rates (60 % in the second year), populations consist of 30 % adults, age presumably up to 3 years.

Doll 1982, Morton 1954, Anderson 1996: 4, Weigand, Gö. & J. 2012.

Alexander M. Weigand

Slovenia: near Gornji Ig village

Zospeum

Zospeum Bourguignat, 1856 in Bourguignat 1856: 504

Type species *Carychium spelaeum* Rossmässler, 1839 (subsequent designation by Westerlund 1902: 116)

Shell small (1-2.5 x 0.8-1.5 mm), smooth or striated, aperture with parietalis and sometimes a columellar fold visible. For species determination it is usually necessary to open the last whorl of the shell, next to the aperture, and to look at the folds around the columella.

Subterranean, usually found in caves, but also in karstic rock crevices, from near sea level to 1850 m. Inside caves live on walls, on the bottom or on rotten wood, and feed probably on detritus.

Life cycle unknown. They move distances of 1-15 cm per week (average 7 mm per day) at 8 °C temperature and 97-100 % humidity.

Bole 1974 (revision), Slapnik 1994 (distribution E Slovenia), Slapnik 2001 (activity and movements in *Z. isselianum*), Slapnik & Ozimec 2004 (distribution Croatia), Weigand et al. 2011 (DNA barcoding).

This genus presentation was kindly checked by A. Jochum.
Weigand et al. 2011 found that 3 *Z. isselianum* populations from 3 different caves clustered so remotely from each other that they would normally be classified as different species, if we rely on usual values of molecular distances. Only 3 caves were incorporated to the analyses. The species is known from more than 300 caves in Slovenia - so will they host 300 different species? Slovenia and Croatia have more than 6000 caves in which *Zospeum* could live - so does this mean all these caves would potentially host 6000 separate species?

Zospeum alpestre (Freyer, 1855)

Carychium alpestre Freyer, 1855 in Freyer 1855: 19, Taf.

Shell smooth, parietalis higher than in *Z. isselianum*, forming an irregularly curved line, spiralis well developed at dorsal side of columella.

1.2-1.8 x 1.0-1.2 mm

(1) Massimo Prodan, (2-3) Adrienne Jochum
(1) could also be *Z. isselianum*

(1) N Italy: Udine, Cave of Villanova, (2-3) Slovenia: Lozekarjeva Zijalka Olseva

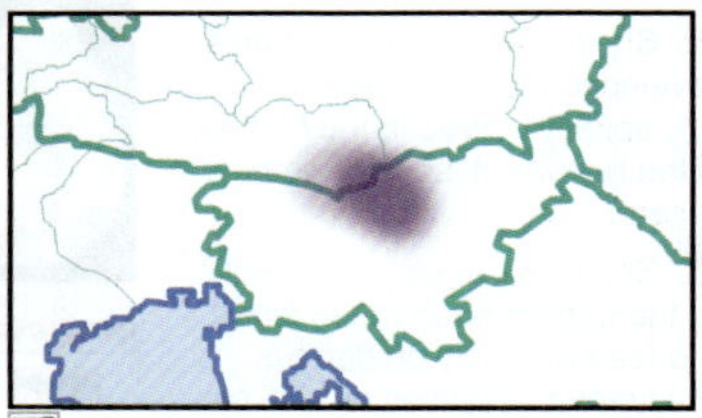

Central Slovenia, S Austria

Zospeum amoenum (Frauenfeld, 1856)

Carychium amoenum Frauenfeld, 1856 in Frauenfeld 1856: 82

Shell smooth, colourless and transparent, aperture without folds, rarely with a weak spiralis, no parietalis.

Occurs sympatrically with *Z. isselianum*.

1.0-1.7 x 0.8-1.1 mm

Slovenia: Marko cave, NE Smrna Gora

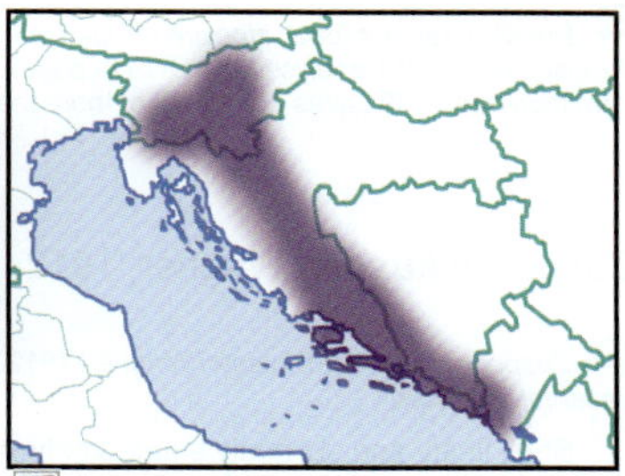

Slovenia to Montenegro

Zospeum bellesi Gittenberger, 1973

Zospeum bellesi Gittenberger, 1973 in Gittenberger 1973: 137

Shell egg-shaped, sides slightly convex, 6-7 evenly rounded whorls, aperture without teeth, apertural margin almost circular.

1.3-1.9 x 0.9-1.15 mm, locally variable

Found in small caves in canyons and valleys.

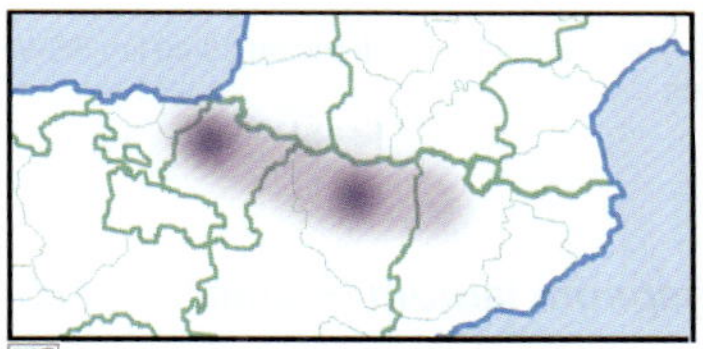

NE Spain (Huesca, Navarra)

Zospeum biscaiense Gómez & Prieto, 1983

Zospeum biscaiense Gómez & Prieto, 1983 in Gómez & Prieto 1983: 8

Shell strongly conical, sides straight or even slightly concave in the upper half, 6-6.5 convex whorls, aperture oblique and elongate, curved when seen from lateral view and also when seen from below, with 4 teeth (2 palatal and 2 parietal), columella without lamella.

1.15-1.3 x 1.0-1.1 mm

Lives only inside caves, at the calcareous wall faces inside.

EU Red List: Vulnerable VU. Restricted to only few localities (Verdú & Galante 2006: 353).

N Spain (Bizkaia)

Zospeum exiguum Kuščer, 1932

Zospeum obesum exiguum Kuščer, 1932 in Kuščer 1932: 60

Shell less slender than other *Zospeum* species, smooth or weakly striated, aperture relatively large, columella twisted, columellaris prominent and ascending very steeply, over more than one whorl.

1.2-1.8 x 1.1-1.3 mm

In flowing water inside caves. This is an aquatic snail, it does not live on the dry parts of caves, in contrast to all other *Zospeum* species.

EU Red List: Vulnerable VU. Only known from few localities. Threatened by pollution and changing water levels inside the caves caused by hydroelectric usage of rivers.

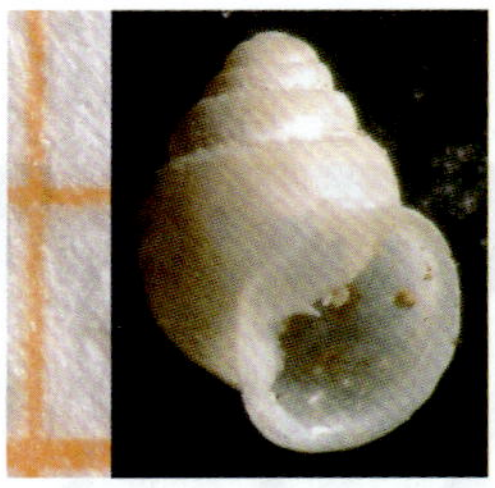

Slovenia: Notranjsko Kračka, Postojna, Lipsenj near Cerknica

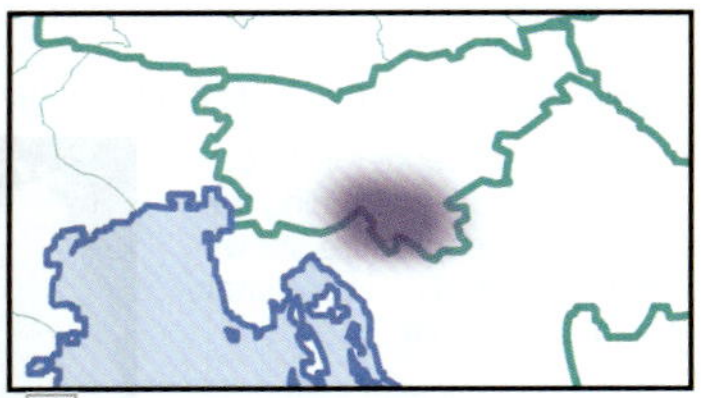

S Slovenia, NW Croatia

Massimo Prodan

Zospeum frauenfeldii (Freyer, 1855)

Carychium frauenfeldii Freyer, 1855 in Freyer 1855: 19 (*Carychium Frauenfeldii*)
Original spelling was not *frauenfeldi*.

Shell prominently and densely striated, parietalis high and sharp, regularly twisted, goes deep inside, columellaris relatively low, goes from dorsal side of columella further inside.

1.5-2.4 x 1.1-1.6 mm, locally variable

A frequent species S of Ljubljana.

Bole 1974: 285.

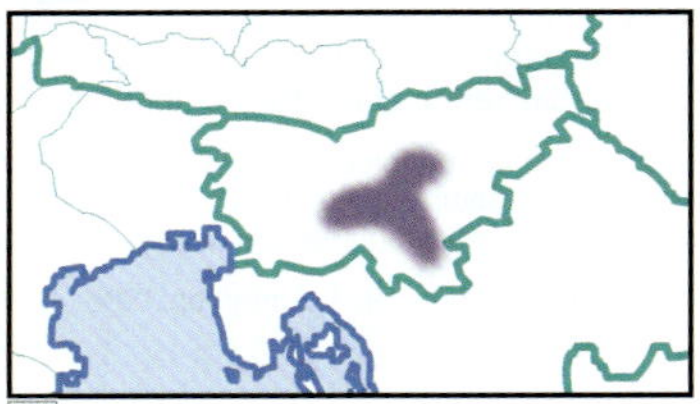

Slovenia

Zospeum globosum Kuščer, 1928

Zospeum globosum Kuščer, 1928 in Kuščer 1928: 185

Shell almost smooth or weakly and regularly striated, fresh shell transparent, 5-6 whorls, aperture rounded, margin thickened and reflected, parietalis prominent and well visible in the aperture, but in some populations weaker, columellar fold deep.

Restricted to cave habitats.

1.3-1.8 x 0.9-1.3 mm

Italy: Lombardia, Brescia, Funtanà di Nalmese

Massimo Prodan

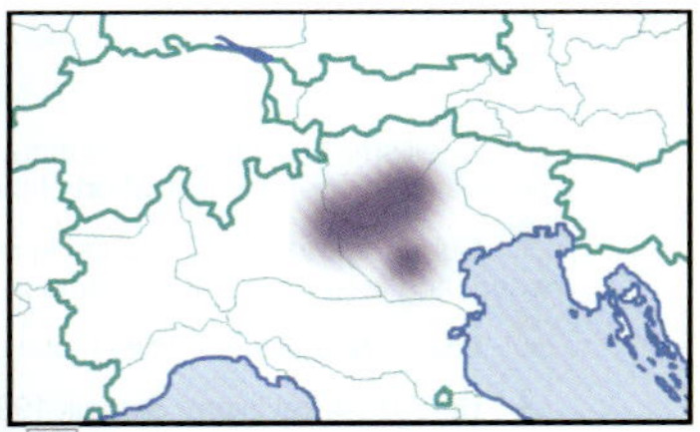

N Italy (Brescia to Brenta valley)

Zospeum isselianum Pollonera, 1887

Zospeum isselianum Pollonera, 1887 in Pollonera 1887: 205, Tav. 6 (*Zospeum Isselianum*)
Date was 15 Sep 1887.

Shell smooth, parietalis less developed than in *Z. alpestre*, regularly spirally curved in the first half of the last whorl, spiralis weak or absent.

Snails live in caves, on walls, on the bottom or on rotten wood, and feed probably on detritus. They move distances of 1-15 cm per week (average 7 mm per day) at 8 °C and 97-100 % humidity.

1.1-1.8 x 0.8-1.2 mm

Slovenia: Turkova jama near Zaplani

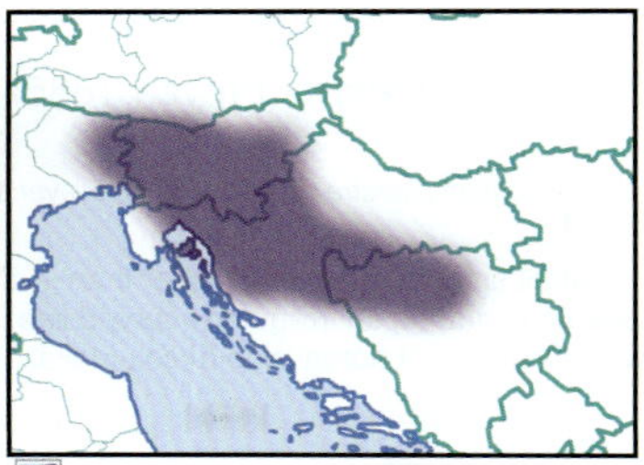

NE Italy and Austria to Bosnia

Bole 1974: 284, Manganelli et al. 1995: 10, Slapnik 2001 (movements).

Previous reports for *Z. alpestre* for Italy referred to this species.
Weigand et al. 2011 analysed the DNA of populations from 3 caves and found that they were as remotely related to each other as were different species. *Zospeum* is known from more than 300 caves - of a total of more than 6000 caves in Slovenia and Croatia. If they all host different species...

Zospeum kusceri Wagner, 1912

Zospeum frauenfeldi kusceri Wagner, 1912 in A. J. Wagner 1912: 257 (*Zospeum frauenfeldi Kusceri*)

Shell finely and obliquely striated only immediately below the suture, above the suture the shell is smooth, columellaris long and prominent, an additional fold is present at the inner wall of the penultimate whorl (visible when the last whorl of the shell is opened).

The species is most frequent 20-40 km SSW of Ljubljana, less frequent in other regions.

1.2-1.8 x 1.0-1.1 mm

Massimo Prodan

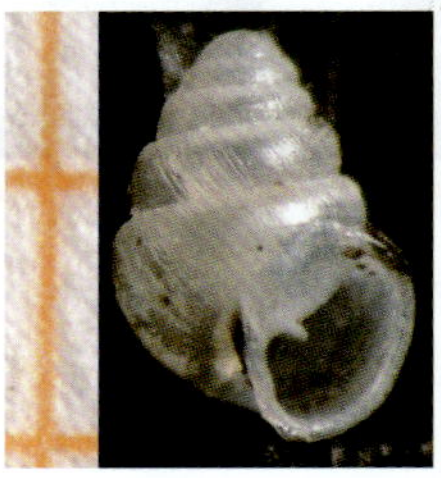

Slovenia: Obalno-kračka, Hrpelje-Kozina, cave near Bać

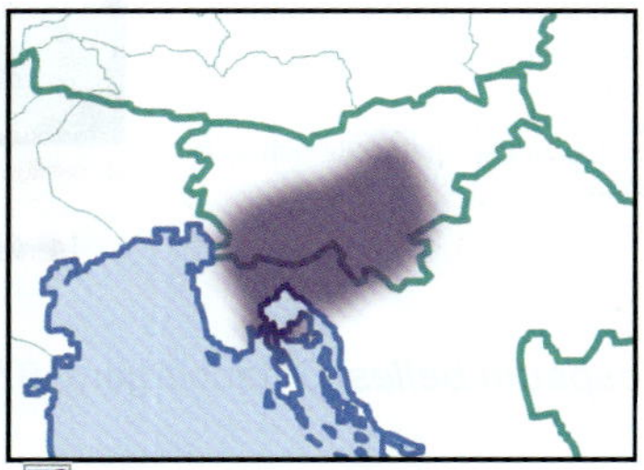

Slovenia, NW Croatia

Zospeum lautum (Frauenfeld, 1854)

Carychium lautum Frauenfeld, 1854 in Frauenfeld 1854: 33

Shell smooth, with 3 well visible folds in the aperture, columellaris and two parietal folds. The lower parietal fold is very prominent.

Bole 1974: 265, Manganelli et al. 1995: 10.

Most finds are from the northern region around Ljubljana, in other regions the species is relatively rare.

1.5-2 x 1.1-1.3 mm

Slovenia: Antra

Museum Budapest HNHM

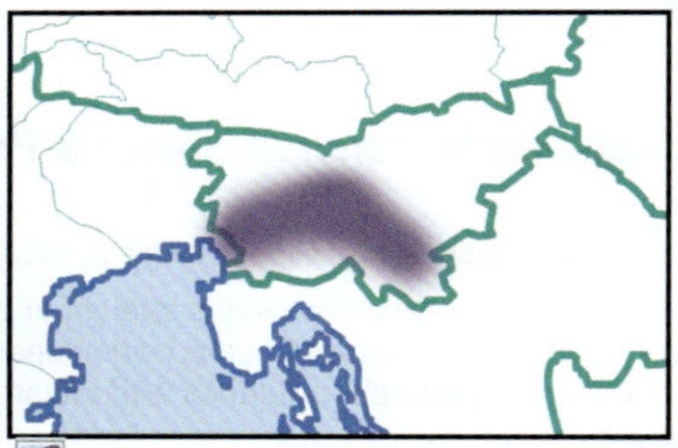

Slovenia, NE Italy (near Trieste)

Zospeum likanum Bole, 1960

Zospeum likanum Bole, 1960 in Bole 1960: 61

Shell translucent, almost smooth, 4-5 moderately convex whorls, last whorl and aperture large, apertural margin reflected, presence of a prominent fold next to the columella, columellaris weak, with one weak bulge. Differs from *Z. alpestre* in its slightly larger size, more oblique aperture, and a less prominent parietalis.

1.5-1.8 x 1.0-1.4 mm

Scattered and nowhere in large numbers.

Bole 1974: 37 (classified it as a subspecies of *Z. amoenum*), Slapnik & Ozimec 2004: 120 (classified it as a separate species, they had collected more material).

NW and W Croatia

Zospeum obesum (Frauenfeld, 1854)

Carychium obesum Frauenfeld, 1854 in Frauenfeld 1854: 34

Shell larger than other *Zospeum* species, smooth or striated, mostly weakly striated, parietalis high and sharp, regularly curved and goes deep inside, columellaris absent.

Only in some caves of Krka and Kolpa river systems. Several reports for *Z. obesum* in the literature before Bole 1974 referred to other species.

1.8-2.3 x 1.4-1.9 mm

Slovenia: (1) Osrednjeslovenska, spring Poltarica near Krka, (2) Gradiček, Krška jama

(1) Massimo Prodan, (2) Adrienne Jochum

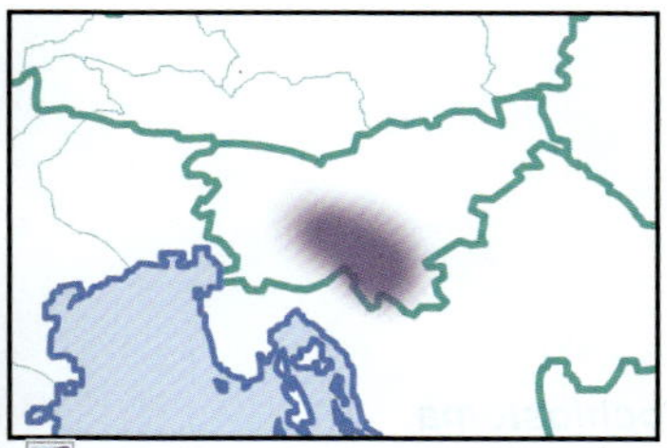

S Slovenia

Zospeum pretneri Bole, 1960

Zospeum pretneri Bole, 1960 in Bole 1960: 61

Shell small with very small aperture, smooth, apertural margin thickened and connected by a thick layer at the parietal side, parietalis low and not visible in the aperture, begins deeper inside at the dorsal side of the columella, no other folds present.

Extremely rare, only known from 5 caves in Croatia.

Bole 1974: 286, Slapnik & Ozimec 2004.

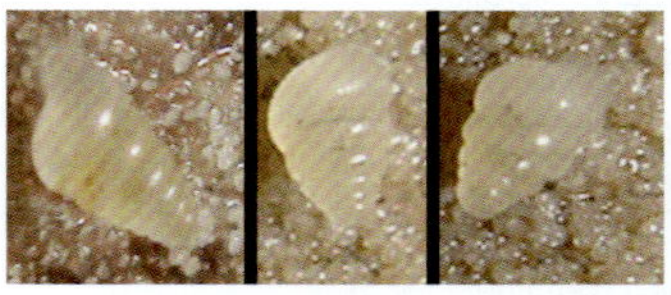

Croatia: Gračac, Kesići, lower Cerovačka cave

Alexander M. Weigand

1.0-1.4 x 0.6-0.9 mm

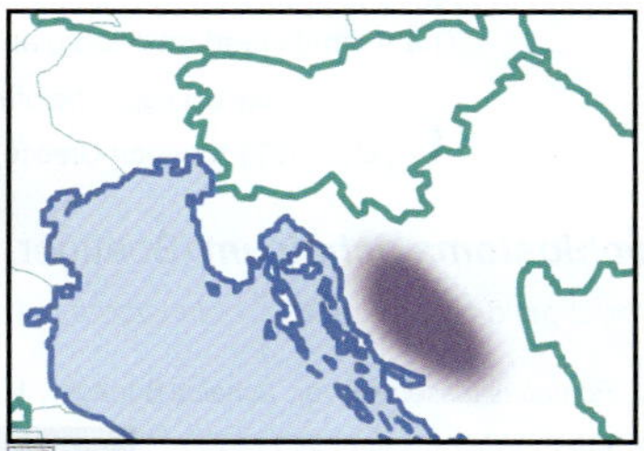

NW Croatia

Zospeum schaufussi Frauenfeld, 1862

Zospeum schaufussi Frauenfeld, 1862 in Frauenfeld 1862: 971 (*Zospeum Schaufussi*)

Shell conical with almost straight sides, 5-6 evenly rounded whorls, aperture with strong parietal callus which has a 90° angle at the columellar side and another angle at the angular (upper) side, making the aperture looking non-circular, sometimes with a small tooth at the columellar-parietal corner indicating a weak parietalis. Differs from the smaller *Z. suarezi* in its non-circular aperture, evenly rounded whorls and a much weaker or absent parietalis.

1.2-1.45 x 0.9-1.0 mm

Gittenberger 1980: 205.

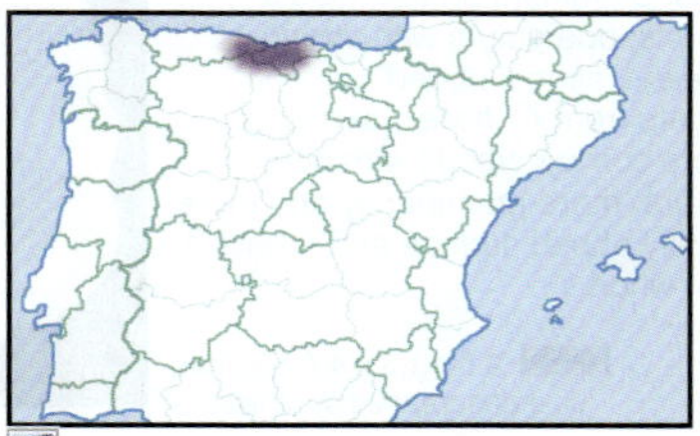

N Spain

Zospeum spelaeum (Rossmässler, 1839)

Carychium spelaeum Rossmässler, 1839 in Rossmässler 1838-1844 (9-10): 36, Taf. 49

Shell very variable, smooth, striated or ribbed, parietalis prominent and regularly curved, a second parietalis can be present, columellaris low, sometimes also palatalis and infrapalatalis present. Surface sculpture varies locally: ribbed W and SW of Ljubljana, ribs only below the suture E and S of Ljubljana, around Trieste and S of Trieste to N Croatian Istria, smoother surface in populations between Trieste and 30 km SW of Ljubljana.

1.4-2.3 x 0.8-1.3 mm

Slovenia: Turkova jama near Zaplani

In caves and probably likewise subterranean habitats, occasionally found in river deposits.

Bole 1974: 286, Slapnik & Ozimec 2004.

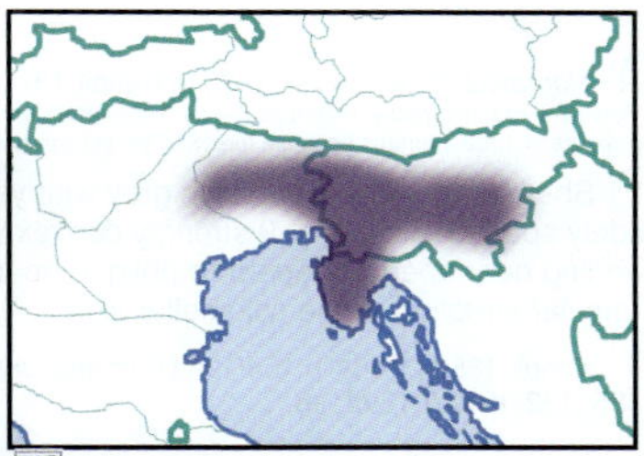

NE Italy, Slovenia and Istria

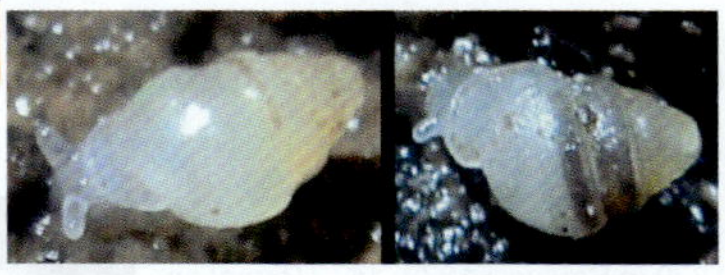

Slovenia: near Gornji Ig

Adrienne Jochum

Zospeum suarezi Gittenberger, 1980

Zospeum suarezi Gittenberger, 1980 in Gittenberger 1980: 204

Shell conical with almost straight sides, colourless, translucent and shiny, with 5-6 somewhat shouldered whorls, apertural margin almost circular incorporating a distinct parietal callus, parietalis sometimes visible in an oblique view from outside and gets more prominent deeper inside at 3/4 whorl, two additional folds can be present inside. Differs from the larger *Z. schaufussi* in its more circular aperture, shouldered whorls and its parietalis.

0.95-1.45 x 0.7-1.0 mm

Spain, Castilla y León, Monte Santiago, Cueva de los Paúles

Adrienne Jochum

N Spain (Bizkaia)

Zospeum subobesum Bole, 1974

Zospeum subobesum Bole, 1974 in Bole 1974: 278

Shell smaller and less slender than *Z. obesum*, smooth or weakly spirally and radially striated in a reticulate pattern, parietalis weakly developed, located closer to the columella than in other species, columella smooth and straight.

1.4-1.7 x 1.2-1.4 mm

(1) Slovenia: Spodnjeposavska, Dvorce near Brezice, (2) Croatia: Tounj near Ogulin
(1) Massimo Prodan, (2) Alexander M. Weigand

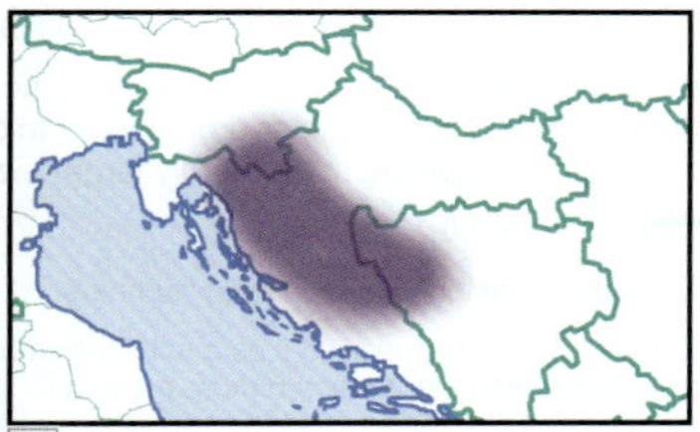

S Slovenia to NW Bosnia

Cochlostoma

Cochlostoma Jan, 1830 in Jan 1830: 6 *Cyclostoma (Cochlostoma)*

Type species *Cyclostoma maculatum* Draparnaud, 1805 (subsequent designation by Wenz 1923 (23): 1773)

Mantle caves work as pulmones, operculum horny with several whorls, penis above right tentacle, sex dimorphism often recognisable at the shell.

The species live in rocky habitats, usually limestone, during rain at the face of the rocks, they hide in crevices or at the base of the rocks. The animals feed on dead plants and microflora.

Females lay big eggs, the offspring hatches after 3-6 weeks.

Schütt 1977 (revision Greece), Bank 1988 (revision NE Italy), Bole 1994 (Slovenia), Gofas 2001 (Spain, Pyrenees), Fehér 2004 (subgenus *Titanopoma*).

Cochlostoma achaicum (Boettger, 1885)

Without gender agreement: *Cochlostoma achaica* (Boettger, 1885)

Pomatias tesselatus var. *achaica* Boettger, 1885 in O. Boettger 1885: 123 (*Pomatias tesselatus* var. *Achaica*)

Shell grey or bownish, rarely with 3 rows of blurry reddish brown spots on last whorl, widely ribbed, most prominent on the middle whorls, operculum with strong streaks, almost ribbed.
Differs from *C. hellenicum* in its more convex whorls with deeper suture, more prominently and more widely ribbed surface and rounded last whorl.

9.5-13.5 x 4.5-5.5 mm, size descreases southwards

Greece (Pelopónisos)

Westerlund 1890: 92, Schütt 1977: 29.

Greece: Pelopónisos, near Megalópoli

Cochlostoma affine (Benoit, 1875)

Pomatias affine Benoit, 1876 in Benoit 1875 (Bull. SMI 1): 155
Pomatias was incorrectly treated as neuter. Description extremely short ("small, elegant"). Date often given as of 1882, perhaps because the 1876 description is not regarded as sufficient?

Shell horny brown, or bluish grey with yellowish first whorls, ribbed, ribs widely spaced and white, 9 strongly convex whorls, last whorl strongly ascending near aperture, aperture oblique, margin broad and white, with auricular structure at the columellar side.

Benoit 1882 ("1881"): 154/194 (*Pomatias affinis*, with diagnosis), Westerlund 1885: 132, Kobelt 1892: 96.

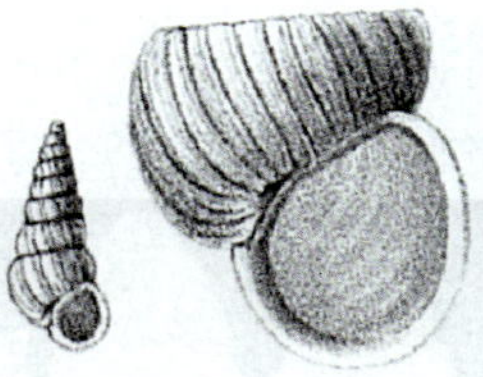

Picture: Kobelt 1892: Icon. 5 Fig. 916

7-9.5 x 2.7-3.7 mm

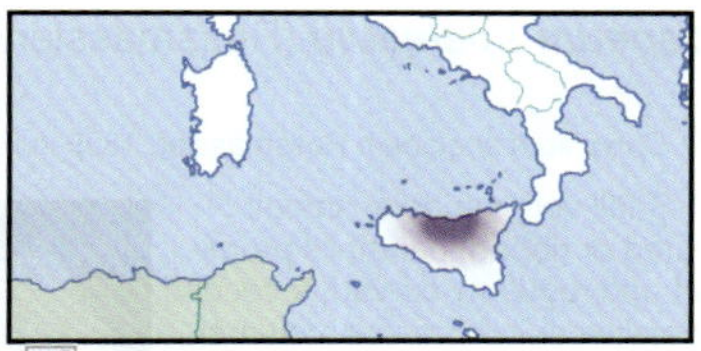

Italy (Sicily)

EU Red List: Vulnerable VU. Only known from one locality.

Cochlostoma alleryanum (Paulucci, 1879)

Without gender agreement: *Cochlostoma alleryanus* (Paulucci, 1879)

Pomatias alleryanus Paulucci, 1879 in Paulucci 1879: 16 (*Pomatias Alleryanus*)

Shell grey with light brownish rows of spots, densely and prominently striated, lower whorls weakly convex, aperture inside white.

6-8 x 2.5-3 mm

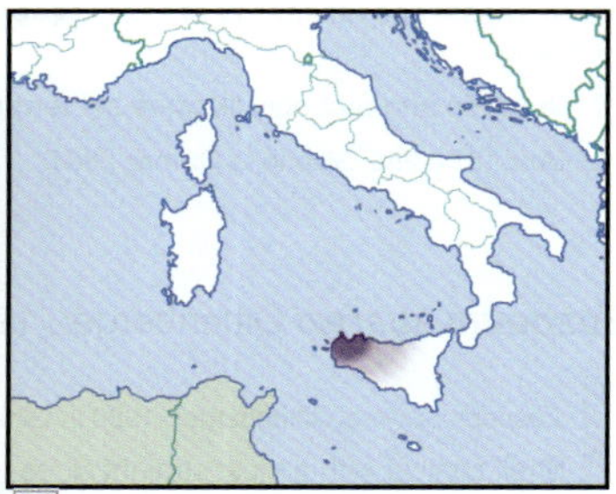

Italy (Sicily)

Italy: Sicily, Trapani, Monte Cofano

Cochlostoma anomphale Boeckel, 1939

Cochlostoma septemspirale anomphale Boeckel, 1939 in Boeckel 1939: 45

Shell reddish brown with milky greyish blue hue in living snails, two colour bands of light brown blurry spots, on last whorl three bands, finely, densely and regularly ribbed, last whorl with a charcteristic inflation 1.5 mm behind the aperture.
Smaller and more slender than *Coc. septemspirale*.

Forest habitats, rocks and rock rubble.

5.9-7.6 x 2.8-3.6 mm

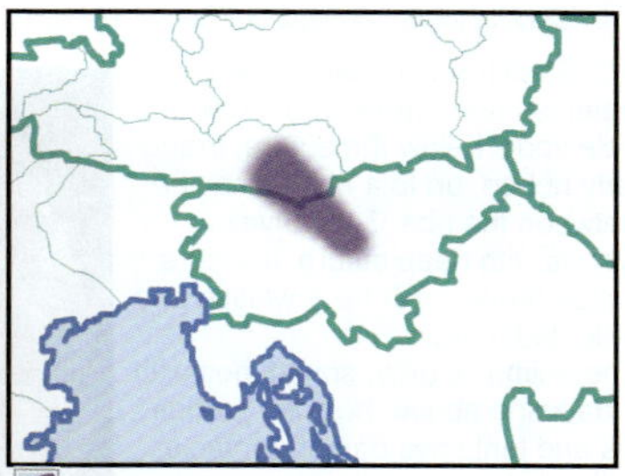

S Austria, N Slovenia

Kerney et al. 1983: 67, Bole 1994.

Slovenia: Julian Alps, Bohinj

Cochlostoma apricum (Mousson, 1847)

Cyclostoma apricum Mousson, 1847 in Mousson 1847: 47

Shell thin, pale yellowish brown with many reddish brown spots which can merge into two colour bands on lower whorls, slightly transparent, 9-10 fine and very regular ribs/mm, 8 weakly convex whorls with shallow suture, apertural margin thin and slightly reflected, broad on columellar and narrow on palatal side.

Rocks, rock rubble, in forests, mostly montane, 400-900 m.

Kerney et al. 1983: 65, Falkner 1990: 116, Raven 1990: 36.

7-10 x 3-4.5 mm

SE France: Savoie, 10 km SSW Chambéry

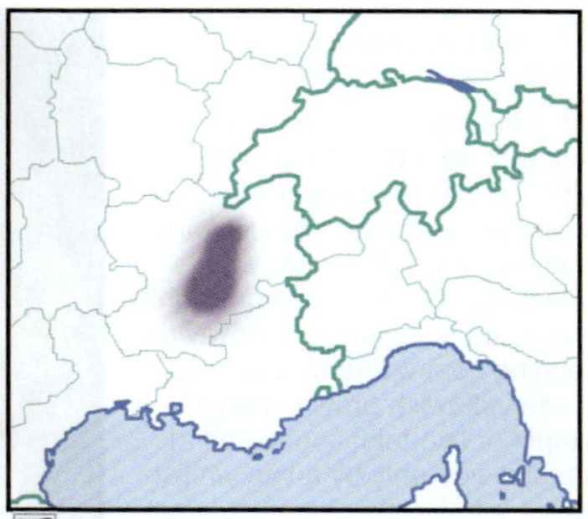

SW Alps

Cochlostoma asturicum (Raven, 1990)

Obscurella asturicum Raven, 1990 in Raven 1990: 52
Obscurella was treated as neuter by Raven.

Shell violet brown with white ribs, giving it a grey appearance, 7-8 convex whorls with deep suture, last whorl suddenly and distinctly ascending near aperture, apertural margin broad, white and strongly reflected, passing via yellow into reddish brown inside the aperture.
The animal is dark grey, tentacles black, near the base with a broad white ring, sole yellowish to pale grey.

Limestone rocks, on rock faces and below the blocks, 400-1500 m.

Gofas 2001: 1322.

8.8-10.6 x 4.2-4.9 mm

Spain: Asturias, Lago de la Ercina

N Spain

EU Red List: Near Threatened NT. This is only because it has a small range and meets the EU area requirements. The species is known from 7 localities in stable populations and not at any risk.

Cochlostoma auritum (Rossmässler, 1837)

Cyclostoma auritum Rossmässler, 1837 in Rossmässler 1835-1837 (5-6): 50, Taf. 28

9-13 x 4.5-6 mm

(1) Montenegro: near Herceg-Novi, (2-6) Albania: Zogaj near Shkodër

(1) Cédric Audibert

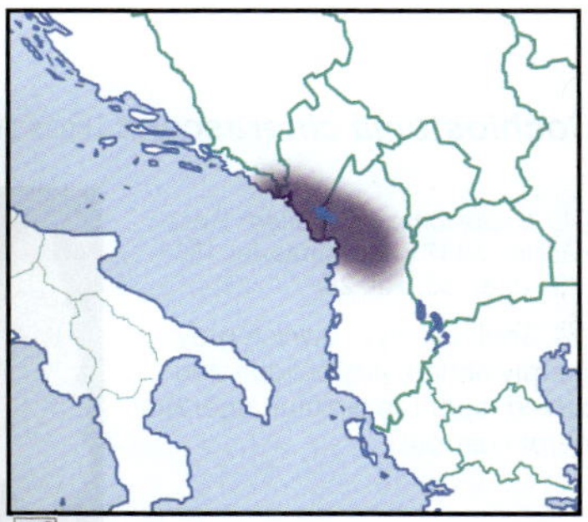

W Montenegro, N Albania
Reports for S Dalmatia referred possibly only to the vicinity of Kotor, today in Montenegro.

Shell large with wide irregular and partly prominent ribs, often decollated, operculum thick and with calcaeous layer.

Limestone habitats in mountains.

Falkner 1990: 116, Dhora & Welter-Schultes 1996, Fehér 2004, Fehér & Erőss 2009 (not central and S Albania).

Cochlostoma bicostulatum Gofas, 1989

Cochlostoma bicostulatum Gofas, 1989 in Gofas 1989: 45

Shell horny brown, often with greenish hue, rarely with a row of pale spots below the suture, irregularly ribbed, up to 4 riblets may be between the ribs, 7-9 convex whorls with deep suture, aperture inside brown, with thick whitish and reflected margin.
The animal is grey, sometimes with a dark line above the sole, proboscis and tentacles darker, tentacles with a light ring at their base, sole white.

Humid and shady habitats, overgrown limestone rocks, rarely on bare rock faces, between the coast and 700 m.

7-13 x 3.9-5.1 mm

Spain: near Llanes W of Santander

N Spain (Asturias to Bizkaia)

Raven 1990: 40, Gofas 2001: 1322.

Cochlostoma braueri (Wagner, 1897)

Pomatias braueri Wagner, 1897 in A. J. Wagner 1897:574, Taf. 1

Shell light horny brownish to yellowish with bluish hue, often with two indistinct rows of light brown colour bands, densely and finely ribbed, 7-8 convex whorls with deep suture, last whorl slightly angulated at lower side, slowly ascending near aperture, aperture inside yellowish brown, margin simple or two-fold, connected at parietal side only by a thin whitish layer, not narrower at the columellar side, no auricular structure present.

Relatively dry limestone rocks with poor vegetation, 800-1200 m.

7-8 x 3.5-4 mm

Croatia: Velebit mountains, N of Visočica peak

W Croatia

Zoltán Fehér

Cochlostoma canestrinii (Adami, 1876)

Pomatias canestrinii Adami, 1876 in Adami 1876: 79, Tav. 1 (*Pomatias Canestrinii*)

Shell dark horny brown, finely ribbed, weakly shiny, 9-11 rather flat whorls, slightly angular below periphery, last whorl slightly ascending and getting enamel white near aperture. First 1-4 whorls are often broken off.
Shell larger than other species and well distinguished by its enamel white aperture.

Limestone rocks in mountains, above 1500 m.

EU Red List: Vulnerable VU. This is only because its small range in Monte Presolana (Bergamo: Valle di Scalve, Valle Seriana) meets the EU area requirements. No currently visible threats to habitat or populations.

11-16 x 4.6-5.8 mm

Italy: Bergamo, Colere, Monte Presolana

NW Italy (Lombardia)

Raven 1990: 55, Bodon et al. 1995: 19.

Cochlostoma cinerascens (Rossmässler, 1837)

Cyclostoma cinerascens Rossmässler, 1837 in Rossmässler 1835-1837 (5-6): 53, Taf. 28

Shell evenly brownish grey, sharply ribbed, ribs oblique and curved, aperture circular, apertural margin narrow.

6.5-7.5 x 3.5 mm

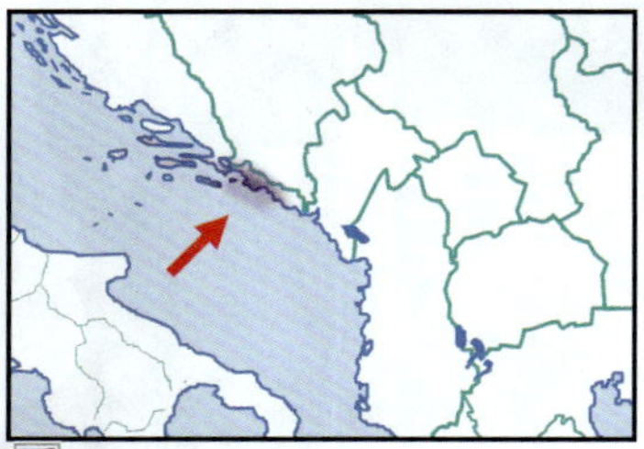

Croatia (S Dalmatia)

Croatia: S Dalmatia, Lopud island

Cochlostoma conicum ([Vallot], 1801)

Without gender agreement: *Cochlostoma conicus* ([Vallot], 1801)

Turbo conicus [Vallot], 1801 in [Vallot] 1801: 6
[Vallot] 1801 was an internal school script, not published work in the sense of Art. 8.1.1.

Shell conical, striated, 8 whorls, apertural lip white and reflected, umbilicus weak.

Falkner et al. 2002: 66.
Not in Moquin-Tandon 1855, Pfeiffer 1852, Pfeiffer 1876, Locard 1882, Westerlund 1885.

10-12 x 4.5-5.5 mm

France: Haute-Marne, near Arc en Barrois

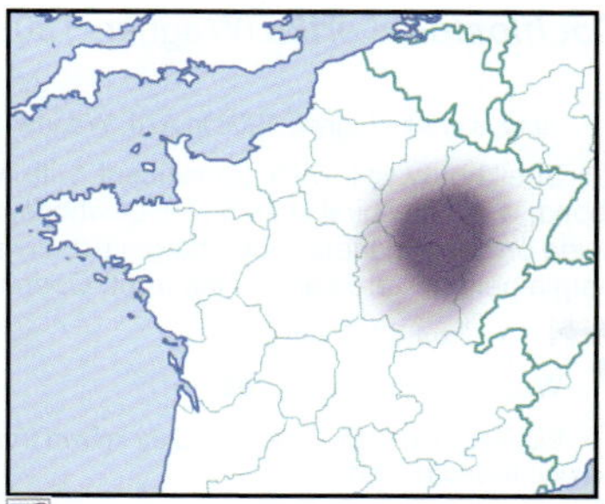

France

Jean Béguinot

Description taken from [Vallot] 1801, no size given there. Size taken from photos by J. Béguinot.
Germain 1931: 576 and Gofas 2001: 1291 regarded it as a synonym of *C. obscurum*, Falkner et al. 2002: 66 argued that it was a distinct species. More research is necessary.

Cochlostoma cretense (Maltzan, 1887)

Without gender agreement: *Cochlostoma cretensis* (Maltzan, 1887)

Pomatias cretensis Maltzan, 1887 in Maltzan 1887: 119

Shell horny greyish with three very weak rows of brownish colour spots, finely ribbed, suture deep, umbilicus very narrow.

Rocky limestone habitats.

Schütt 1977: 18, Vardinoyannis (unpublished thesis, 1994), HNC data.

The only *Cochlostoma* species in Crete.

6-8 x 3-3.5 mm

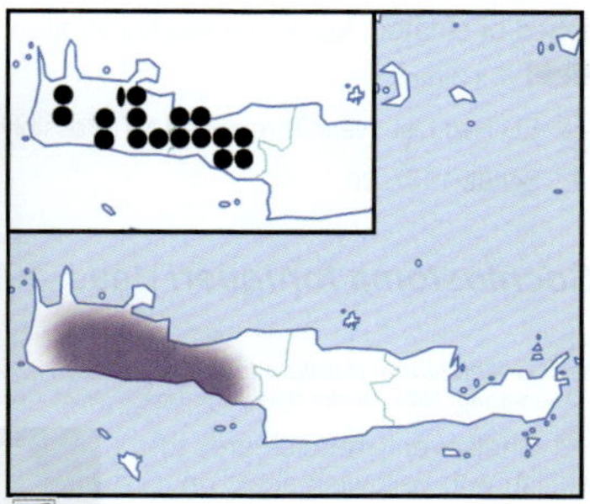

Greece (W Crete)

Greece: W Crete, Samaria gorge

Cochlostoma crosseanum (Paulucci, 1879)

Without gender agreement: *Cochlostoma crosseanus* (Paulucci, 1879)

Pomatias crosseanus Paulucci, 1879 in Paulucci 1879: 15 (*Pomatias Crosseanus*) (not *Pomatias Partioti* var. *Crosseana* Saint-Simon, 1867 in Saint-Simon 1867: 11 from SW France)
Pomatias pinianum De Stefani, 1879 is a junior synonym.

Shell light horny whitish, apex yellowish, ribs widely spaced and not prominent, 7-9 moderately convex whorls, aperture inside yellowish, margin white, not detached but usually connected by a callus, columellar margin weakly ear-shaped, umbilicus closed.

Rocks and rock rubble in mountains, often very abundant.

8-10 x 3-4 mm

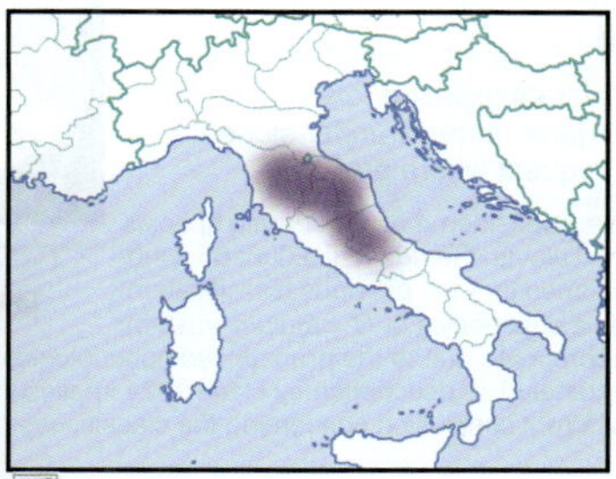

Italy (Toscana to Abruzzo)

Westerlund 1885: 116, 129, Zilch 1958: 67, Alzona 1971: 21, Bodon et al. 1995: 19, 49, Nordsieck 2011: 17.

Italy: Marche, Monti Sibillini, Tofe near Talenge

Cochlostoma dalmatinum (Pfeiffer, 1863)

Without gender agreement: *Cochlostoma dalmatinus* (Pfeiffer, 1863)

Pomatias dalmatinus Pfeiffer, 1863 in L. Pfeiffer 1863: 136

Shell yellowish with violet hue, white and thin ribs prominent and widely spaced, apertural margin simple or weakly reflected.

12 x 5 mm

EU Red List: Vulnerable VU. This is only because the small range near Herceg-Novi meets the EU area requirements.

Westerlund 1885: 120, Gittenberger 1976: 280, Štamol 2010: 5.

Montenegro (near Kotor)

Cochlostoma elegans (Clessin, 1879)

Pomatias elegans Clessin, 1879 in Clessin 1879 (NachrBl. DMG): 122

Shell slender, yellowish grey with weak reddish brown spots arranged in two rows on the upper and three rows on the lower whorls, irregularly and very obliquely ribbed (this is characteristic), more finely ribbed on the last whorl, aperture circular, columellar margin not ear-shaped

Has nothing to do with *Pomatias elegans* (Müller, 1774).
Schütt 1977: 19 reported the species also from Ipiros. The identity of this form should be confirmed.

5-7 x 2.5-3.3 mm

Croatia: Velebit mountains near Mali Halani

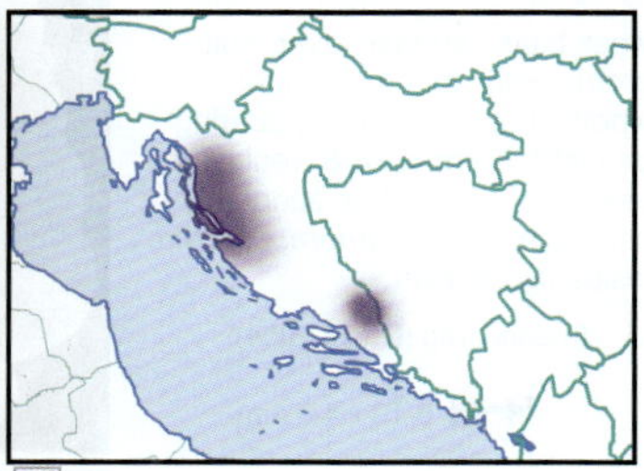

W Croatia

Štamol & Mužinić 1998.

Cochlostoma erika (Wagner, 1906)

Auritus erika Wagner, 1906 in A. J. Wagner 1906: 132, Taf. 4

Shell horny yellowish, translucent, with white spots near the umbilicus, regularly, finely and sharply ribbed, with fine spiral lines on lower whorls, 9 convex whorls, last whorl ascending near aperture, aperture inside brown at parietal side, margin usually simple, broadened and sharp, auricular structure at columellar margin bent upwards and to the front. Operculum thin, transparent, with 4 indistinct whorls.

10-11 x 4.7-5.3 mm

Was only found in caves.

EU Red List: Vulnerable VU. Only known from 3 localities. Potentially threatened by habitat destruction associated with increased tourism.

Gittenberger 1976: 281, www.iucnredlist.org (07.2012).

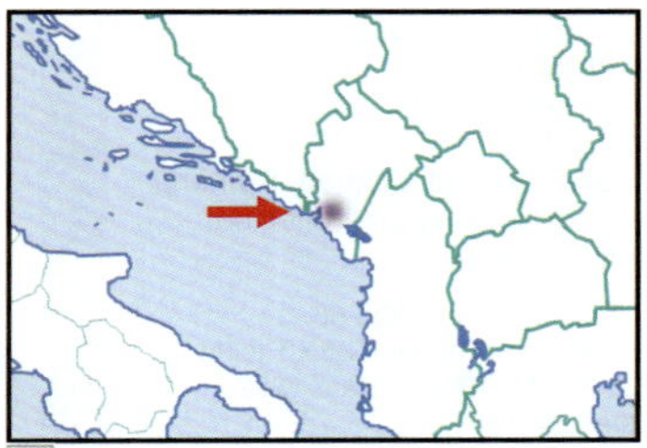

Montenegro

Cochlostoma euboicum (Westerlund, 1885)

Without gender agreement: *Cochlostoma euboicus* (Westerlund, 1885)

Pomatias banaticus var. *euboicus* Westerlund, 1885 in Westerlund 1885: 125

Shell horny coloured with distinct upper row of colour spots, coarsley ribbed, antepenultimate whorl finely ribbed or striated, last whorl weakly striated.

7 x 3 mm

EU Red List: Near Threatened NT. Possibly restricted to a small area. No real threats known.

Schütt 1977: 20.

Greece (Evia)

Cochlostoma fontqueri Haas, 1924

Cochlostoma patulum fontqueri Haas, 1924 in Haas 1924: 148, Taf. 8
Date was Apr 1924. On the plate caption in the genus *Cyclostoma*, but in the text p. 149 Haas gave a statement that this was not deliberate.

Shell evenly reddish grey to whitish with yellowish apex, no spots, 7-10 regular and prominent ribs/mm, not weaker on last whorl, umbilicus closed.
The animal is black.
Ribs perhaps more regularly spaced than in *C. patulum.*

Scarcely vegetated rocky habitats exposed to sunshine, in organic material under stones. Frequent where present.

Considered as a separate species by Talaván Gómez & T. 2006: 65, who claimed that the previous classification as a subspecies of *C. patulum* was not appropriate due to the pronounced conchological differences and the far distance between the ranges. But they did not give any detail how the forms actually differed from one another. The description by Haas 1924 implies that he did not know that the ranges of the two forms were so much separated, otherwise he would possibly have come to a different conclusion concerning the classification.

6.5-7 x 3.2-3.5 mm

Spain: Tarragona, near Tortosa

NE Spain

Cochlostoma fuchsi Fehér, 2004

Cochlostoma fuchsi Fehér, 2004 in Fehér 2004: 42

Shell with 3 bands of reddish brown spots, evenly ribbed, 7-7.5 convex whorls with deep suture, apertural margin broad with prominent columellar and parietal auriculations.
Operculum with rudimentary upper plate, its top absent, septa reduced to ribs on the perpendicular branch of the basal plate.
Ribs more widely spaced and more prominent than in *Coc. tesselatum excisum*, the species are best distinguished by the structure of their opercula.

7-10 x 3.9-4.5 mm

Albania: Elbasan, Shushicë

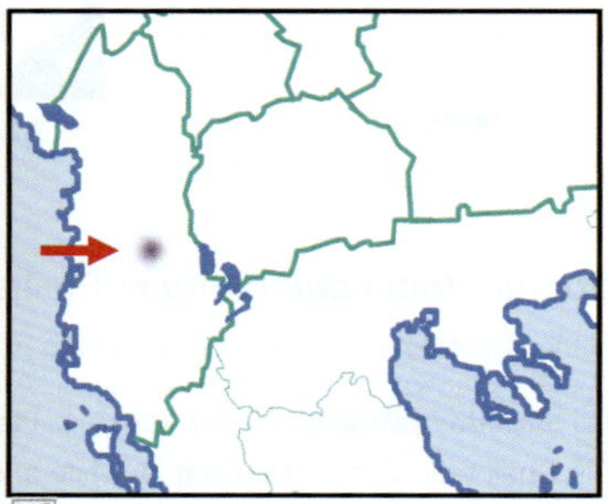

Central Albania

EU Red List: Vulnerable VU. Only known from one locality, but might also live elsewhere. No real threats known.

Cochlostoma georgi (Wagner, 1906)

Auritus georgi Wagner, 1906 in A. J. Wagner 1906: 138, Taf. 4

Shell evenly brownish, with a weak bluish surface layer, quite prominently ribbed on upper whorls, ribs almost disappearing on last 1-2 whorls, 6-8 weakly convex whorls, last whorl suddenly ascending near aperture, aperture inside light brown.

Endangered (Dhora 2004).

8-10 x 4-5 mm

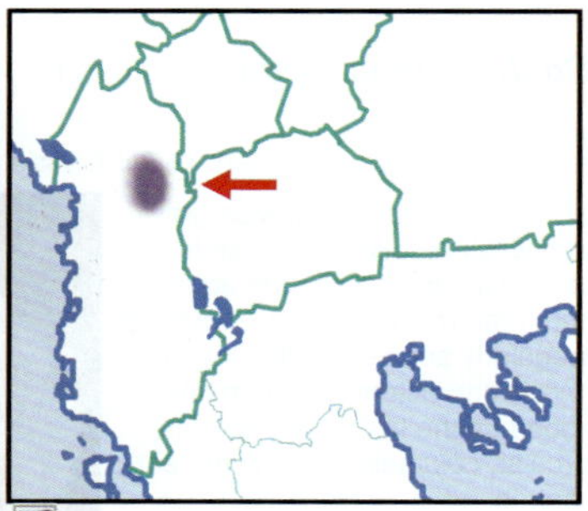

N Albania

NE Albania: near Oroshi

Cochlostoma gigas Gofas & Backeljau, 1994

Cochlostoma gigas Gofas & Backeljau, 1994 in Gofas & Backeljau 1994: 49

Shell horny brown with whitish spots or dashes and white apex, densely ribbed, 8-9 convex whorls, at the last whorl occasionally slightly angulated above the suture, aperture with white broad and reflected margin.

On limestone rock faces and walls, not too much exposed.

Gofas 2001:1306.

15-18 x 7-8 mm (females), 13.5-16.5 x 6-7.5 mm (males)

Spain: Aragón, Añisco valley

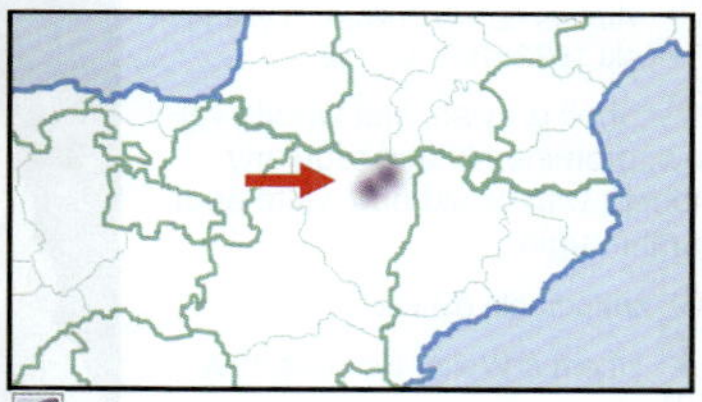

NE Spain (Huesca)

Charles Geerts

Cochlostoma gracile (Pfeiffer, 1849)

Cyclostoma gracile Pfeiffer, 1849 in L. Pfeiffer 1846: 191, Taf. 26

Pomatias gracilis was previously mentioned as a nomen nudum by Pfeiffer 1847: 110 (and attributed to Küster).

Shell very small, evenly grey or horny brown, transparent with thin and sharp ribs, on last whorl weaker, aperture auricular.

Kerney et al. 1983: 67, Falkner 1990: 116, Bole 1994.

5-7.5 x 2.8-3.2 mm

Croatia: Apatišanska Duliba near Senj

SE Austria to NW Greece

Cochlostoma hellenicum (Saint-Simon, 1869)

Without gender agreement: *Cochlostoma hellenicus* (Saint-Simon, 1869)

Pomatias hellenicus Saint-Simon, 1869 in Saint-Simon 1869: 8 (*Pomatias Hellenicus*)

Shell brownish red without any spots, with densely and regularly arranged ribs, 6-7.5 weakly convex whorls with flat suture, apertural margin with two lips, columellar wall thickened, last whorl occasionally angulated.
Differs from other species in its compact form, weakly convex whorls with flat suture (less pronounced in the east), last whorl is narrower at its base.

9-11 x 3.5-5.5 mm

Greece: Viotía, near Eptálofos

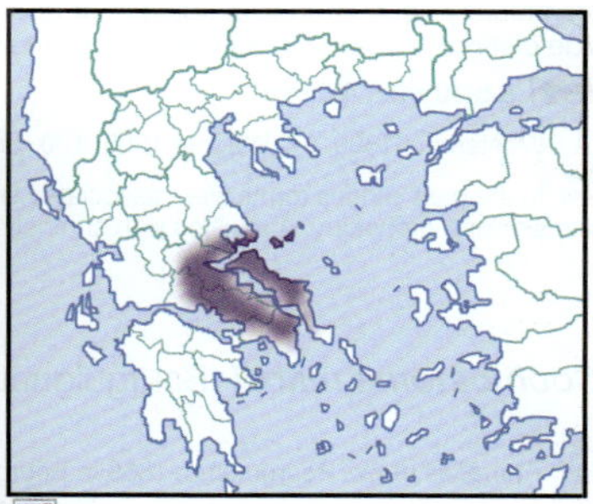

Greece

Pfeiffer 1876: 205, Westerlund 1885: 122, Schütt 1977: 31, Liebegott 1986: 10.

Cochlostoma henricae (Strobel, 1851)

Pomatias henricae Strobel, 1851 in Strobel 1851: 18

Shell greyish brown with reddish hue, sometimes bluish, no spots or colour bands, 9-11 whorls, lower whorls almost smooth, aperture inside brown, at least a little, umbilicus open.
Females with more convex whorls (but less than in *C. septemspirale*).
Differs from *C. villae* in its usually open umbilicus, columellar lobe not being fold backwards and at least some brown colour inside the aperture (this seems to be a quite constant character).

At limestone rocks, up to 2000 m.

7.7-11.5 x 3.5-4.8 mm

Italy: Udine, near Pontebba pass

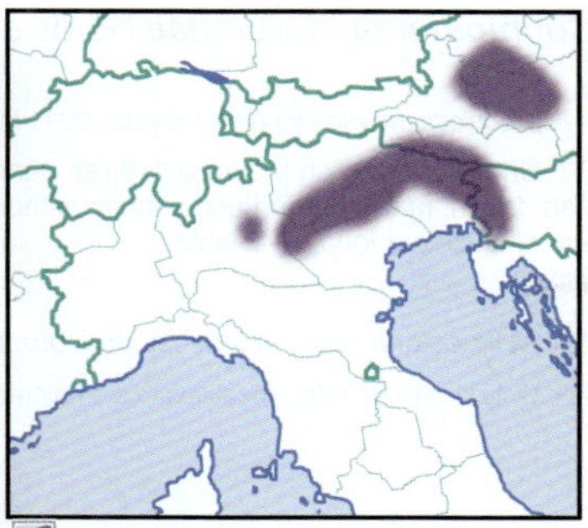

E and S Alps

Kerney et al. 1983: 64, Falkner 1990: 116, Bank 1988: 152, Bodon et al. 1995: 19.

Cochlostoma hidalgoi (Crosse, 1864)

Pomatias hidalgoi Crosse, 1864 in Crosse 1864: 24, Pl. 2 (*Pomatias Hidalgoi*)

Shell yellowish brown or dark violet, apex yellow, ribbed, lower whorls weakly striated, 7-9 not very convex whorls, suture shallow and white, aperture oblique, margin broad and white. Shell very variable in size, rib numbers, whorl convexity and colour.

Limestone rocky habitats of various kinds, shady und unshaded, or nearby in the vegetation.

8-15 x 4.2-6.8 mm

NE Spain: Lleida

N Spain

Pfeiffer 1876: 167, Westerlund 1885: 116, Raven 1990: 43, Gofas 2001: 1313.

Cochlostoma hoyeri (Poliński, 1922)

Auritus hoyeri Poliński, 1922 in Poliński 1922: 163, Taf. 12

Shell greyish, first whorls yellow, prominently and regularly ribbed, with broad and white apertural margin.

Vulnerable (Dhora 2004).

Dhora & Welter-Schultes 1996: 101.

12-13 x 6-7 mm

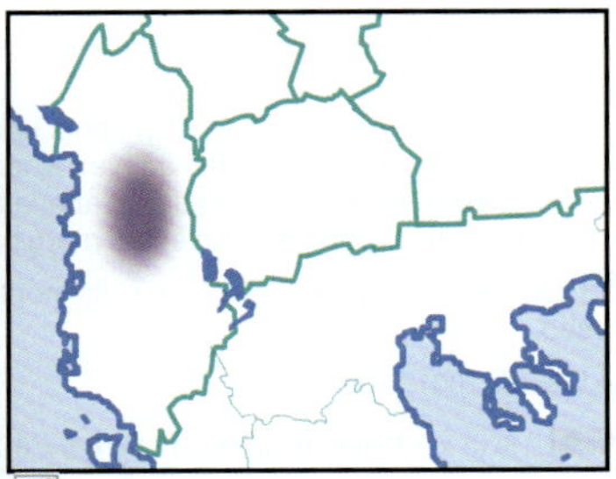

Central Albania

Albania: Mali i Krujës, 900 m

Cochlostoma hueti (Kobelt, 1882)

Pomatias hueti Kobelt, 1882 in Kobelt 1882: 121 (*Pomatias Hueti*)

Shell greyish, regularly striated.

Was found at tree trunks.

Only known from a collection of 1853, never found again. In the region of Istanbul there are hardly any tree trunks left. If the species ever lived there, I would expect it to be extinct now.

15 x 4 mm

Turkey: near Istanbul
Picture: Kobelt 1892: Taf. 151 Fig. 891

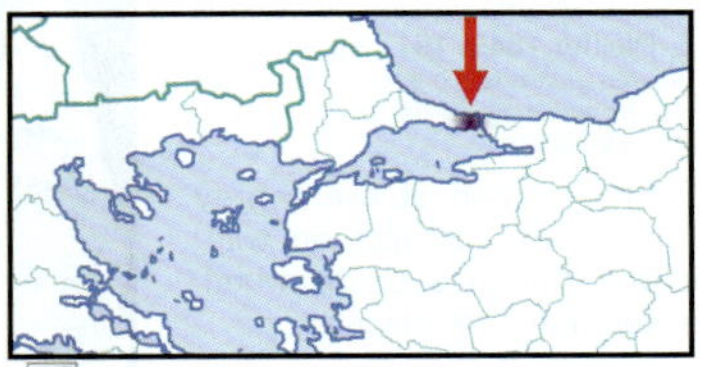

Turkey (near Istanbul)

Cochlostoma kleciaki (Braun, 1887)

Pomatias kleciaki Braun, 1887 in Braun 1887: 110 (*Pomatias Kleciaki*)

Shell evenly horny yellowish or reddish without spots, shiny, with white ribs, 7-9 whorls, last whorl less prominently ribbed and ascending near aperture, apertural margin two-fold, auricular structure at columellar side distinctly broader, bent upwards and to the front.

6.5-10 x 2.5-4.5 mm

Westerlund 1890: 97, Wagner 1906: 130, Schütt 1977: 20, Dhora & Welter-Schultes 1996: 101.

In the past various forms were associated with this species, but it seems, not all forms between Croatia and Greece with similar morphology. A revision with an analysis of the local variations of these forms is needed.

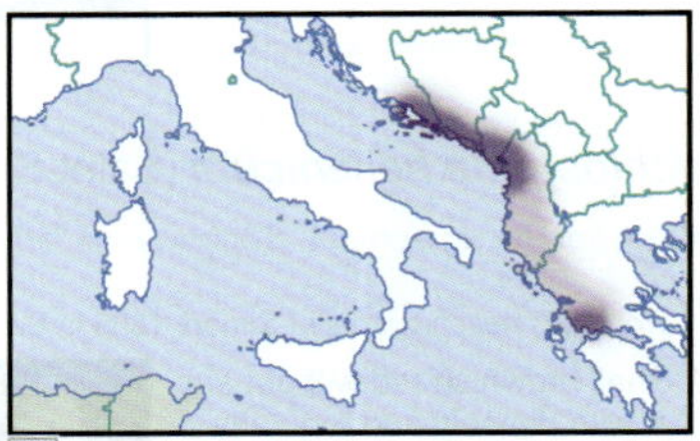

Croatia (Dalmatia) to Greece

Cochlostoma macei (Bourguignat, 1869)

Pomatias macei Bourguignat, 1869 in Bourguignat 1869: 56 (*Pomatias Macei*)

Shell thin, horny greyish, strongly ribbed, apex pointed, whorls very convex, ribs denser on the last whorl.

10 x 4 mm

Westerlund 1885: 126, Germain 1931: 585.

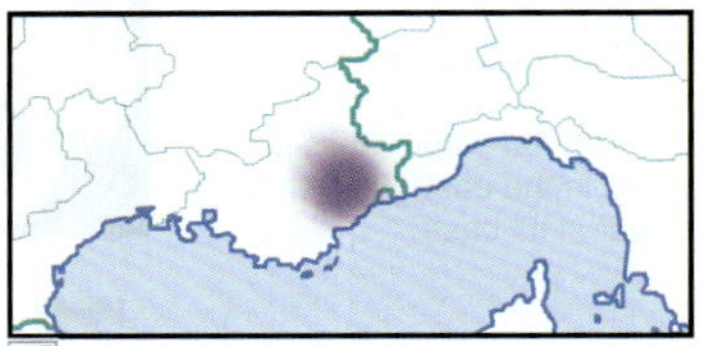

SE France

Cochlostoma mariannae Nordsieck, 2011

Cochlostoma mariannae Nordsieck, 2011 in Nordsieck 2011: 15

Shell light greyish brown with three bands of brown spots, moderately densely ribbed, aperture inside reddish brown, margin broad and circular without ear (difference to *C. crosseanum*), umbilicus closed. Rib density and shell colour locally variable.

7.3-9.5 x 3.3-4.2 mm

At limestone rocks in a deciduous forest.

Only known from few localities in a small restricted area east of L'Aquila in Abruzzo.

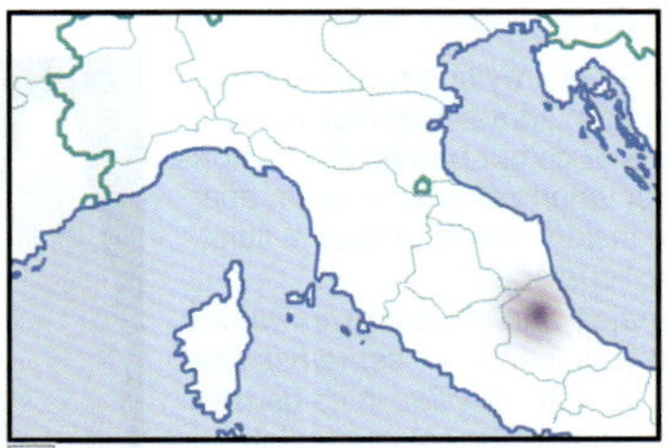

Central Italy (Abruzzo)

Cochlostoma martorelli (Servain, 1880)

Pomatias martorelli Servain, 1880 in Servain 1880: 144 (*Pomatias Martorelli*)

Shell blackish yellow, regularly and finely ribbed, suture quite deep, aperture almost circular, margin two-fold, umbilicus narrow.

In mountains above 800 m, at walls and rocks close to water sources, and at high altitudes above 1500 m with high humidity (clouds).

Westerlund 1885: 117, Gofas 2001: 1306, Martínez-Ortí & Robles 2003: 57.

10-13.5 x 4-5 mm

Spain: Barcelona, Montserrat, Pas de Francés

E Pyrenees to Tarragona

Cochlostoma mienisi Schütt, 1978

Cochlostoma mienisi Schütt, 1978 in Schütt 1978: 64

Shell evenly greyish, finely and very regularly ribbed, 7-8 moderately convex whorls, apertural margin broad.

Only known from Kekova island.

8-11 x 4-5 mm

Turkey (Kekova island)

Turkey: Vilayet Antalya, island of Kekova

Cochlostoma montanum (Issel, 1866)

Without gender agreement: *Cochlostoma montanus* (Issel, 1866)

Pomatias patulum var. *montanus* Issel, 1866 in Issel 1866*:* 29
Pomatias cassiniacus Paulucci, 1878 in Paulucci 1878*:* 45 (*Pomatias Cassiniacus*)
Paulucci 1878 attributed the name to Saint-Simon and explained on p. 45 that the diagnosis was provided by Saint-Simon. The text suggests that this referred only to the Latin diagnosis. It is clear from the context that the French part of the description (comparisons with other species) was written by Paulucci. So Paulucci provided parts of the description in the original source. In a strict application of Art. 50.1.1 the name in such a case must be attributed to the author of the work (because Saint-Simon was not alone responsible for the description).
Pomatias adamii Paulucci, 1879 in Paulucci 1879 [Feb 1879]: 17 (*Pomatias Adamii*)

Shell yellowish grey, apex yellow, rarely with red or yellowish red spots or dashes, densely and regularly striated, less densely ribbed in the south, apertural margin simple, margin broad, getting abruptly narrower at columellar side (moderately or weakly ear-shaped).

Westerlund 1885: 130, Boato et al. 1984: Tab. 1, Bodon et al. 1995: 19, Nordsieck 2011: 17.

Bodon et al. 1995 classified *montanus* and *cassiniacus* in one species and *adamii* as a second species. Nordsieck 2011 classified S Italian forms from Lazio, Abruzzo, Campania and Calabria as *C. montanum cassiniacum*, a senior synonym of *adamii*.

5.5-7 x 2.5-3 mm, in the south 7-9 x 2.5-3 mm

Italy: Lucca, Pizzorno

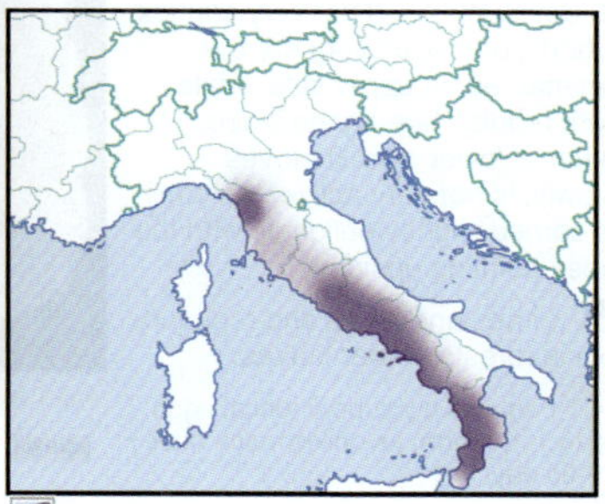

Italy

Cochlostoma mostarense (Wagner, 1906)

Without gender agreement: *Cochlostoma mostarensis* (Wagner, 1906)

Pomatias mostarensis Wagner, 1906 in A. J. Wagner 1906: 99, Taf. 3

Shell brownish with greyish hue and a light zone near the aperture, often with 3 indistinct rows of spots on the lower whorls, finely and densely ribbed, last whorl more densely striated, 7-9 whorls, last whorl slowly but distinctly ascending near aperture, apertural margin sharp, simple or weakly two-fold, not or only weakly connected by a deep whitish layer at the parietal side, umbilicus narrow.

EU Red List: Near Threatened NT. Only known from a small range, meets the EU area requirements but no true threats are known.

4.5-6.5 x 2-3 mm

Bosnia: Podvelez near Mostar

Herzegovina (near Mostar)

Rocky habitats on hills.

Cochlostoma nanum (Westerlund, 1879)

Without gender agreement: *Cochlostoma nanus* (Westerlund, 1879)

Pomatias nanus Westerlund, 1879 in Westerlund 1879: 167

Shell evenly horny reddish or with reddish brown colour bands, sometimes with greyish blue hue, ribs thin and low, on last whorl fine and dense.

Westerlund 1885: 132, Kerney et al. 1983: 66, Bole 1994.

6.5-8.1 x 3.3-3.8 mm

Austria to Croatia

Croatia: Risnjak mountain, 30 km NW Rijeka

Cochlostoma nouleti (Dupuy, 1851)

Pomatias nouleti Dupuy, 1851 in Dupuy 1847-1852: 510, 513, Pl. 25 (*Pomatias Nouleti*)
Date for p. 510 was May 1851.

Shell horny to dark greyish brown, often with white spots near suture, ribbed (6 ribs/mm), 7-10 slightly convex whorls, can be angled below the periphery, apertural margin white and reflected, not broad, umbilicus narrow.
The animal is grey with grey head and tentacles, and a white ring at the base of the tentacles, sole usually white.

Rocks and rock rubble, montane, mostly in river valleys, 400-1600 m. Not rare.

8-12 x 4-5.3 mm

France: Hautes-Pyrénées, Lourdes

NE Pyrenees

Moquin-Tandon 1855: 500, Kerney et al. 1983: 65, Raven 1990: 37, Gofas 2001: 1326.

Cochlostoma obscurum (Draparnaud, 1805)

Cyclostoma obscurum Draparnaud, 1805 in Draparnaud 1805: 39

Shell large, greyish white or brown, often with two rows of blurry dark spots, regularly ribbed (8-9 ribs/mm), 6-8 not very convex whorls, umbilicus very narrow.
Animal red on upper side, white with reddish hue on the sides, paler at lower side. It moves slowly, is not very active, the shell is carried slightly obliquely with the apex pointing upside.

Limestone rocks and rock rubble in forests, under leaves.

Females lay subcylindrical eggs (size 1-1.5 mm), juveniles hatch after 20-25 days.

10-14 x 4-5 mm

France: Basses-Pyrenées, Mousserolles near Bayonne

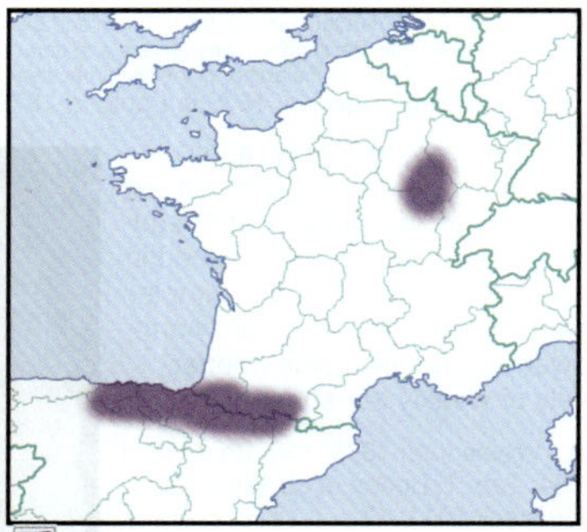

N central Spain to central Pyrenees, N central France

Moquin-Tandon 1855: 499, Germain 1931: 576, Kerney et al. 1983: 64, Raven 1990: 30, Gofas 2001.

EU Red List: Near Threatened NT. The species has been declining in France, various populations from type localities of closely related forms have been lost.

Cochlostoma oscitans Gofas, 1989

Cochlostoma oscitans Gofas, 1989 in Gofas 1989: 46

Shell violetish grey with whitish apex and ribs, very finely and regularly ribbed, 7-8 moderately convex whorls, aperture inside weakly reddish, margin broad, not connected at parietal side, columellar margin not auriculated.

9-12 x 5-6.5 mm, males slightly smaller (height 8-10 mm)

Exposed habitats at very steep or perpendicular limestone rocks. Animals live sympatrically with *Coc. bicostulatum*, *C. hidalgoi* and *C. obscurum.* They intend to copulate, but obviously unsuccessfully.

Vulnerable (Gómez et al. in Verdú & Galante 2006: 315). Only known from few localities.

N Spain

Cochlostoma pageti Klemm, 1962

Cochlostoma pageti Klemm, 1962 in Klemm 1962: 213, Taf. 1

Shell regularly ribbed, greyish yellow, rarely with brown spots below suture, aperture inside weakly brown, circular with sharp margins, columellar margin with a distinct appendix at its end, characteristically twisted against the apertural plain in about 90 degrees, connected to the shell. Operculum fragile and horny.

6.3-7.6 x 2.5-3 mm

Rocky habitats in high altitudes.

Schütt 1977: 20.

NW Greece (Ipiros)

Cochlostoma paladilhianum (Saint-Simon, 1869)

Without gender agreement: *Cochlostoma paladilhianus* (Saint-Simon, 1869)

Pomatias paladilhianus Saint-Simon, 1869 in Saint-Simon 1869: 5 (*Pomatias Paladilhianus*)

Shell evenly yellowish grey, finely ribbed, 6-8 slightly convex whorls, apertural margin simple or two-fold, slightly reflected, auricular structure weak.

EU Red List: Vulnerable VU. Populations near Palermo are potentially threatened by habitat destruction.

9-12 x 3.5-4 mm

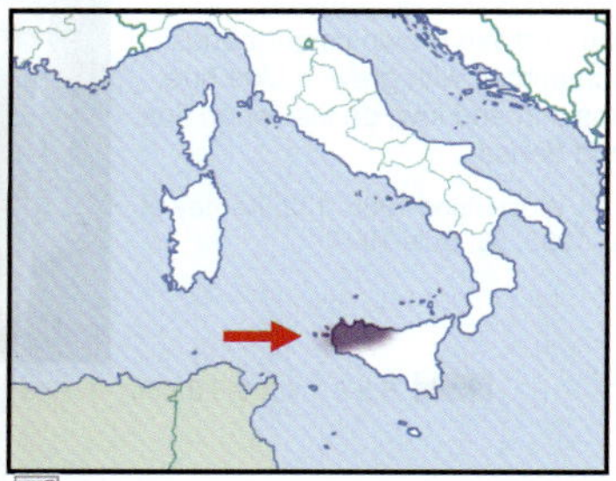

Italy (NW Sicily and Favignana)

Italy: Sicily, Monte Pellegrino

Cochlostoma parnonis Schütt, 1981

Cochlostoma achaicum parnonis Schütt, 1981 in Schütt 1981: 43
The word *parnonis* can be regarded as a noun in the genitive singular (Normand David, e-mail 01-2011), so it is not changed to *parnone* under Art. 31.2.2.

Shell light brown with bluish hue, with 3 spiral rows of dark brown spots, finely ribbed (80 ribs on last whorl), 7-8 convex whorls with deep suture, apertural margin broadened and simple, not detached, with a thin callus at parietal side, auricular structure at columella deeply situated, operculum typical with 3.5 whorls.
Differs from *Coc. achaicum* in its smaller size.

7.5-8.5 x 4-4.5 mm

At limestone rocks and conglomerate, also at rocks at road margins, and at old walls.

EU Red List: Near Threatened NT. Only known from few localities, but may have a wider range.

Greece (Pelopónisos: Arkadia)

Cochlostoma partioti (Saint-Simon, 1848)

Cyclostoma partioti Saint-Simon, 1848 in Saint-Simon 1848: 36 (*Cyclostoma Partioti*)
Name in original description attributed to Moquin-Tandon, but description was by Saint-Simon.

Shell grey or brownish violet, sometimes with 1-2 dark colour bands, 13-17 fine and sometimes irregular ribs/mm visible under microscope, whorls convex (but less than in *Coc. septemspirale*).
Animal greyish white to reddish grey, with fine and irregular black dots, paler at lower side. Crawling animal slow, shy, carrying its shell obliquely.

9-10 x 3.5-4.5 mm

Limestone rocks and rock rubble, usually in mountains.

Moquin-Tandon 1855: 501, Germain 1931: 577, Kerney et al. 1983: 65.

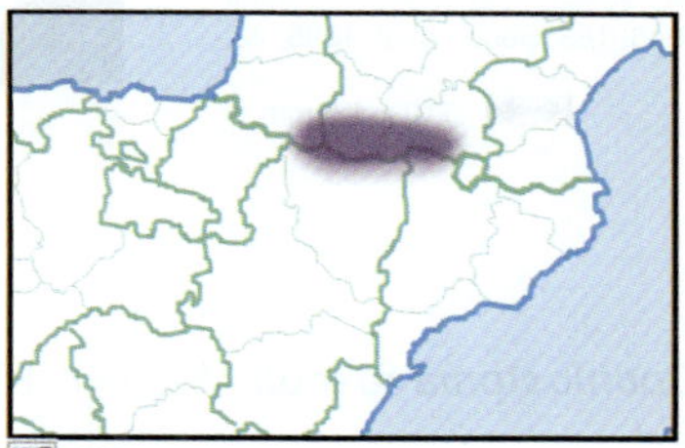

Pyrenees

Cochlostoma patulum (Draparnaud, 1801)

Cyclostoma patulum Draparnaud, 1801 in Draparnaud 1801: 39

Shell small, evenly reddish grey to whitish, no spots, 7-10 weak and irregular ribs/mm, umbilicus closed.
Animal small (4 mm long), reddish grey with fine blackish dots.

At rocks, in rock crevices, in dry meadows and more rarely at old walls.

Moquin-Tandon 1855: 505, Germain 1931: 583, Kerney et al. 1983: 65, Falkner 1990: 16, Bank 1988: 169, Taleván Gómez & T. 2006: 65 (classified *fontqueri* as a separate species).

5-8 x 2-3 mm

S France, NW Italy

France: Vaucluse, Forêt de Venasque

Cochlostoma philippianum (Gredler, 1853)

Pomatias philippianum Gredler, 1853 in Gredler 1853: 51 (*Pomatias Philippianum*)
Gredler treated *Pomatias* as neuter.

Shell slender, with or without weak colour bands, usually ribbed, often alternating a weak rib and a prominent rib, in both populations also non-ribbed forms occur, 8.8-10.5 whorls, columellar apertural margin with a lobe, umbilicus completely covered.
In ribbed populations the weak ribs between prominent ribs are characteristic.
Differs from *C. villae* in its columellar lobe not being directed backwards but rather flat and attached to the penultimate whorl.

Kobelt 1902 [Das Tierreich]: 525, Bank 1988: 160, Falkner 1990: 116.

7-10 x 3.3-3.9 mm

NE Italy: (1-3) Belluno, Casere Bortot (H = 9.1 mm), (4-5) Udine/Pordenone, Monte Zuccul (H = 7.5 mm)

NE Italy

Enrico Zallot

Cochlostoma pinteri Fehér, 2004

Cochlostoma pinteri Fehér, 2004 in Fehér 2004: 39

Shell with 3 bands of brown spots, evenly ribbed, more densely on the last whorl, 7-10 convex whorls with deep suture, aperture often with brown spots visible inside, margin broad and often twofold, slightly protruding below the small parietal auriculation, columellar auriculation far distant from the last whorl.

EU Red List: Near Threatened NT. Only known from few localities, which are not under real threats.

9-12 x 4.5-5.5 mm

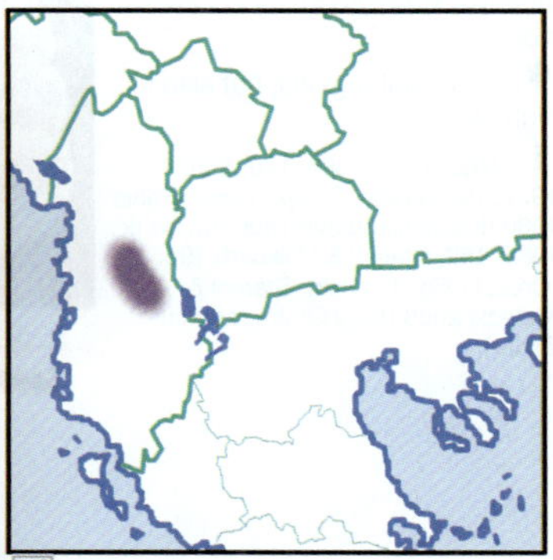

Central Albania

Albania: 3 km N Shtyllës pass near Tiranë

Cochlostoma porro (Strobel, 1850)

Pomatias porro Strobel, 1850 in Strobel 1850: 123

Original spelling was not *porroi*, not to be corrected under Art. 32.5. The name is not frequently used, so it is convenient to use the name in its original spelling.

Shell dark reddish brown with rows of red spots, red colour band at lower side, with reddish or white ribs, last whorl almost smooth, shiny, 7-8 whorls, umbilicus closed.

Westerlund 1885: 133, Bank 1988: 163, Bodon et al. 1995: 20.

7-10 x 4-5 mm

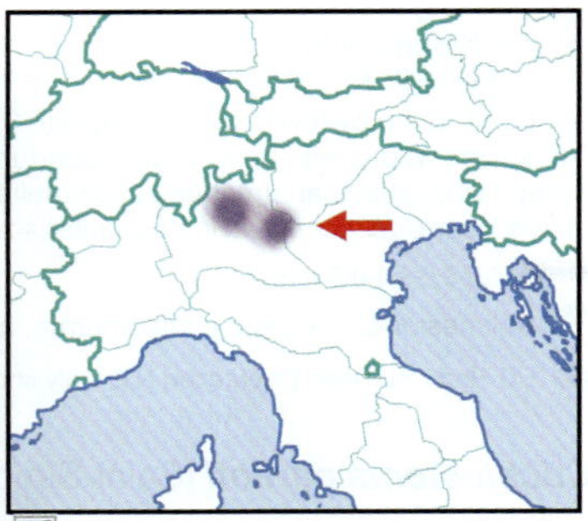

NE Italy

Italy: Trento, Val Ampola near Tiarno di Sopra

Cochlostoma roseoli (Wagner, 1901)

Pomatias roseoli Wagner, 1901 in A. J. Wagner 1901: 64

Shell large, often decollated, with regular and prominent white ribs, apertural margin very wide, operculum with thin calcareous layer.

9-12 x 4.5-6 mm

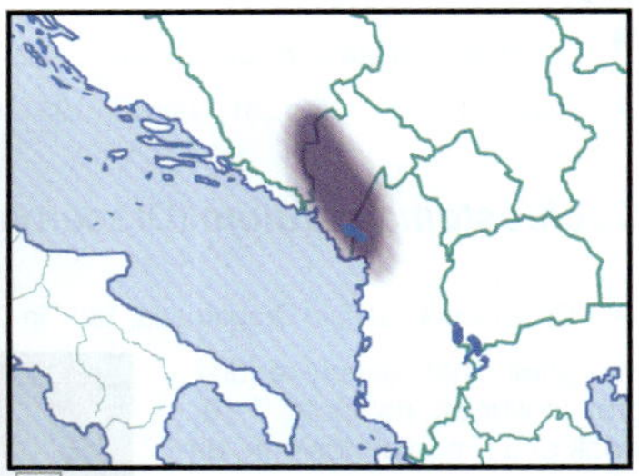

Herzegovina to N Albania

Albania: Malësia, near Bogë

Cochlostoma sardoum (Westerlund, 1890)

Without gender agreement: *Cochlostoma sardous* (Westerlund, 1890)

Pomatias sardous Westerlund, 1890 in Westerlund 1890: 96

Original spelling was not *sardosus*.

Shell grey or light horny grey, last whorl whitish, with rows of dark reddish spots, near the umbilicus a reddish colour band, irregularly ribbed, whorls quite convex, apertural margin relatively thin, not circular, columellar margin usually with lobe.

Limestone rocks with little vegetation.

Bodon et al. 1995: 20.

The only one species in Sardinia, could be conspecific with a mainland form.

7-9 x 3.5-4 mm

Italy: Sardinia, Nuoro, Cala Fujli near Dorgali

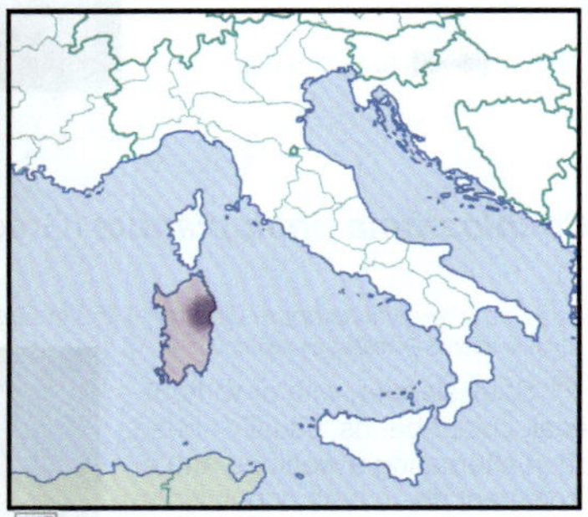

Italy (Sardinia)

Cochlostoma scalarinum (Villa, 1841)

Pomatias scalarinum Villa, 1841 in Villa & Villa 1841: 58

Name was attributed to "Villa" in original source. The authorship for the description was not explained. Villa was one of the two co-authors of the work (A. Villa & G. ["J."] B. Villa). Under Art. 50.1. the name for the taxon must be attributed to Villa alone. Villa & Villa treated *Pomatias* as neuter.

Shell horny greyish, with indistinct rows of spots, regularly and widely ribbed, whorls convex, apertural margin broad, with a / shaped lobe-like structure at the columellar side.

In coastal regions, but also in high altitudes.

Westerlund 1886: 110, Schütt 1977: 19 (Kérkira, Pelopónisos), Rähle 1980 (Kefaloniá above 1400 m), Bank 1988: 157, Štamol & Velkovrh 1995 (Cres, Lošinj, Plavnik), Štamol & Kletečki 2005 (Dugi Otok island off Zadar).

7.5-8 x 3-3.5 mm

Croatia: Krk island, Aleksandrovo

NE Italy to Greece

Cochlostoma septemspirale (Razoumowsky, 1789)

Without gender agreement: *Cochlostoma septemspiralis* (Razoumowsky, 1789)

Helix septemspiralis Razoumowsky, 1789 in Razoumowsky 1789: 278 (*Helix Septem Spiralis*)
Pomatias intermedius Pini, 1885 in Pini 1885: 370, Tav. 12

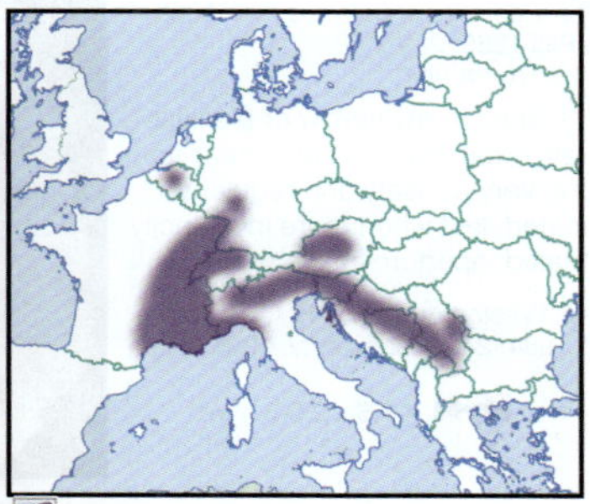

Shell light greyish brown with 3 rows of reddish brown spots, 6-8 regular ribs/mm, 7-10 convex whorls, aperture inside whitish, margin broad, at columellar side slightly broader and C shaped without distinct lobe, umbilicus open. No significant sex dimorphism.
Animal medium-sized (5-6.5 mm long), greyish with weak brownish hue.

6.7-10.2 x 3.2-4.4 mm

Italy: Lago di Garda, Gardone

S Europe

Moquin-Tandon 1855: 503, Prince 1967, Kerney et al. 1983: 63, Bank 1985: 166, Bank 1988: 157, Štamol & V. 1995, Turner et al. 1998: 70.

Forest habitats, rocks, rock rubble, walls, grassy slopes, up to 2100 m. Feeds on desintegrating plant substrate, sometimes also on algae cover of limestones. The most widely distributed *Cochlostoma* species.
Animal slow, not very active, very shy, comes outside the shell when put onto a cold object. Active only at wet weather, operculum is closed when the soil is dry. May climb trees up to 2 m at very wet weather conditions. Hibernates between stones, under leaves and grasses, activity seems to cease when temperatures go below 6-7 °C.

Eggs (diameter 1.0-1.1 mm, occasionally down to 0.6 mm) are laid in clutches of up to 10 eggs from April to October, mainly May to June, about 1 cm below the suface. Eggs are covered by the female with faeces and mucus. Juveniles hatch after 45-60 days, full shell size is reached after roughly 1 year.

Alzona 1971 regarded *intermedius* from N Italy as a synonym of *septemspirale*. The form has never been found again, the original description does not provide constraints for the presence of a separate species. The original material is lost.

Cochlostoma simrothi (Caziot, 1908)

Pomatias simrothi Caziot, 1908 in Caziot 1908: 460 (*Pomatias (Auritus (Patuliana)) Simrothi*)
Caziot 1908 attributed the name to Pollonera and gave a statement "il [Pollonera] a décrit cette forme", probably referring to the Latin diagnosis. But it is clear from the context that also Caziot contributed to the description of the species. Under Art. 50.1.1 the authorship must be attributed to the author of the work.

Shell densely and regularly striated, 8-9 whorls, last whorl rounded (not subangulate at the base) and suddenly ascending near aperture, apertural margin broad, white, reflected and sometimes two-fold, columellar margin auriculated, margins connected by a thin callus at parietal side, no umbilicus.

Cool habitats at rocky limestone walls, 350-1850 m.

Alzona 1971: 21, Boato et al. 1984: 245, Tab. 1.

Possibly conspecific with *Cochlostoma patulum*.

7-9 x 2.5-3.8 mm

France: Val Roya, Gorges de Saorge

SE France, NW Italy

Maurizio Sosso

Cochlostoma stossichi (Hirc, 1881)

Pomatias stossichi Hirc, 1881 in Hirc 1881: 522 (*Pomatias Stossichi*)

Shell pale bluish grey, finely and densely striated, on last whorl indistinctly, 6-7 very convex whorls, apex yellow and shiny, aperture inside reddish, apertural margin white.

EU Red List: Near Threatened NT. Only few populations are known, some are located near the city of Rijeka.

Westerlund 1885: 134, Bole 1994.

4-5.5 x 3-3.5 mm

NW Croatia: Ponikve near Rijeka

S Slovenia, NW Croatia

Cochlostoma sturanii (Wagner, 1897)

Pomatias gracilis var. *sturanii* Wagner, 1897 in A. J. Wagner 1897: 609, Taf. 8
Original spelling was not *sturanyi*.

Shell light horny yellowish with 2-3 weak and interrupted rows of brownish spots, moderately and slightly irregularly ribbed, last whorl striated, 7-8 convex whorls with deep suture, last whorl ascending, aperture inside light yellowish brown, margin broad and two-fold, columellar margin with angulated auricular structure.

6-7 x 3-3.5 mm

Croatia to N Albania

Croatia: Plitvicka jezera National Park

Cochlostoma subalpinum (Pini, 1885)

Without gender agreement: *Cochlostoma subalpinus* (Pini, 1885)

Pomatias subalpinus Pini, 1885 in Pini 1885: 233
Date was not 1884.

Shell horny yellow or blackish grey, often with red spots near suture, very densely and regularly striated, last whorl more indistinctly striated, aperture inside whitish.

Westerlund 1885: 127, Boato et al. 1984: 246, Bodon et al. 1995: 20.

7-7.5 x 2.5-3 mm

NW Italy, SE France

Italy: Cuneo, Viozene, Le Vene

Cochlostoma tergestinum (Westerlund, 1878)

Without gender agreement: *Cochlostoma tergestinus* (Westerlund, 1878)

Pomatias tergestinus Westerlund, 1878 in Westerlund 1878: 109

Shell horny grey with reddish brown or light brown bands of spots, distinctly ribbed, ribs disappear slowly on last whorl, apex yellow and shiny, apertural margin two-fold, ear-shaped at columellar side.

Limestone rock faces and rock rubble at scarcely vegetated and sunny slopes, in Austria up to 1100 m. Feeds on lichen attached to stones and rocks.

Westerlund 1885: 127, Bole 1994, Mildner 1997, Fehér and Erőss 2009. Not in Albania.

6.5-8 x 3.5-4 mm

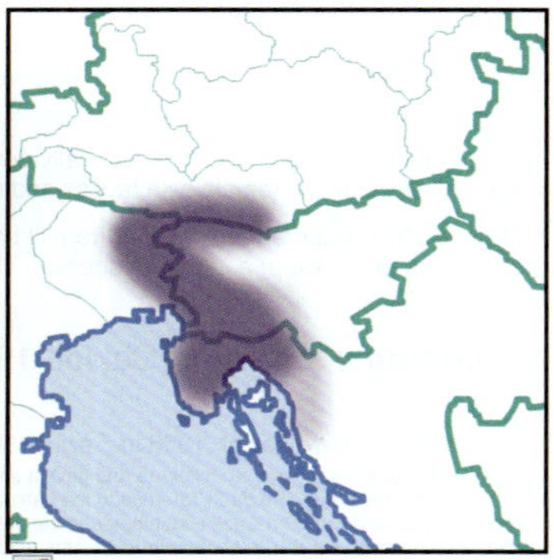

SE Austria to Croatia

Croatia: Istria, Dvograd

Cochlostoma tesselatum (Rossmässler, 1837)

Cyclostoma tesselatum Rossmässler, 1837 in Rossmässler 1835-1837 (5-6): 53, Taf. 28

Shell greyish with reddish brown spots arranged in rows, distinctly ribbed, ribs appear whitish, rib density varies locally.

Westerlund 1885: 121, Schütt 1977: 22, Rähle 1980, Dhora & Welter-Schultes 1996: 102, Fehér & Erőss 2009.

Morphologically variable species, several subspecies have been recognised by some authors.

8-11 x 4-5 mm

(1) Greece: Kérkira, Pantokrator mountain, (2-5) Albania: Mali Gradecit near Çorovodë

W Greece, S Albania

Cochlostoma villae (Strobel, 1851)

Pomatias maculatum var. *villae* Strobel, 1851 in Strobel 1851: 17 (*Pomatias maculatum* var. *Villae*)
Name dedicated to Antonio Villa or Giovanni Battista Villa or to both, the ending *-ae* is not corrected to *-i* or *-orum* (Art. 32.5). Art. 31.1 has no effective force.

Shell horny yellow with reddish brown spots, upper whorls prominently and widely ribbed, lower whorls often more finely and densely ribbed, 20-25 % of the shells in a population can be smooth, 9-11 whorls, aperture inside shiny with broad white layer, apertural margin two-fold (but not always), with lobe at columellar side directed backwards, not touching the lower side of the penultimate whorl, but closing the umbilicus.
Differs from *C. henricae* in its columellar lobe being fold backwards and never touching the penultimate whorl, closed umbilicus (in *C. henricae* usually open), and white colour inside the aperture (in *C. henricae* brown, this seems to be a very constant character). *C. septemspirale* has no distinct columellar lobe seen from frontal view, and always an open umbilicus. *C. philippianum* has a larger protoconch, the columellar lobe is flat and touches the last whorl.

7.6-10.5 x 3.4-4.0 mm

Italy: (1-2) Vicenza, Lago di Fimon, (3-5) Veneto, Monti Berici
(1-2) Enrico Zallot

N Italy

Humid and shady habitats at limestone rocks, also at tree trunks. Not found on volcanic rocks.

Westerlund 1885: 127, Bank 1988: 166, Bodon et al. 1995: 20.

Cochlostoma waldemari (Wagner, 1897)

Pomatias waldemari Wagner, 1897 in A. J. Wagner 1897: 616, Taf. 8-9

Shell evenly horny reddish to yellowish, finely and densely ribbed on upper whorls, ribs weaker on last two whorls, apex narrow and pointed, 7-8 convex whorls, cervix white.

At and near rocks, also in meadows near rocks and under stones.

Boeckel 1939 (subspecies of *Coc. tergestinum*), Kerney et al. 1983: 66, Bole 1994.

5.7-7.8 x 2.7-3.6 mm

NW Slovenia: near Bled

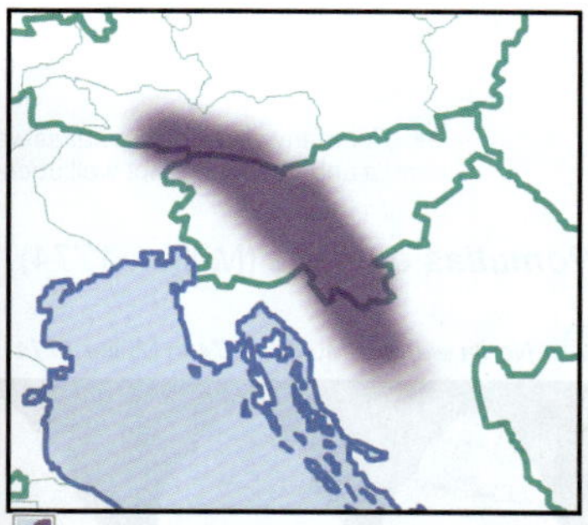

Austria, W Slovenia, Croatia

Cochlostoma westerlundi (Paulucci, 1879)

Pomatias westerlundi Paulucci, 1879 in Paulucci 1879: 20 (*Pomatias Westerlundi*)

Shell whitish yellow with rows of red spots, upper whorls ribbed, lower whorls less distinctly and densely ribbed, last whorl indistinctly striated, lower whorls with 5-8 prominent parallel spiral lines, columellar apertural margin with lobe, umbilicus closed or very narrow.

Limestone rocks, walls and stones on calcareous substrate, in Sicily in humid and shady habitats.

Westerlund 1885: 122, Bodon et al. 1995: 20, Ferreri et al. 2005: 37, Reitano et al. 2009: 182.

Species needs a revision.

11-11.5 x 4 mm

Italy: Sicily, Siracusa

S Italy

Cochlostoma zawinkanum (Wagner, 1906)

Without gender agreement: *Cochlostoma zawinkanus* (Wagner, 1906)

Auritus sturanii zawinkanus Wagner, 1906 in A. J. Wagner 1906: 127

Shell yellow to light reddish brown without bands, shiny, weakly and widely striated, less on last whorl, last whorl slightly inflated and weakly ascending near aperture, apertural margin conspicuously thickened.

7.5-8.5 x 3.4-3.8 mm

Endemic species in SW Slovenia, only known from few localities.

Was classified with *C. sturanii* until Bole 1994 observed that both forms occurred sympatrically.

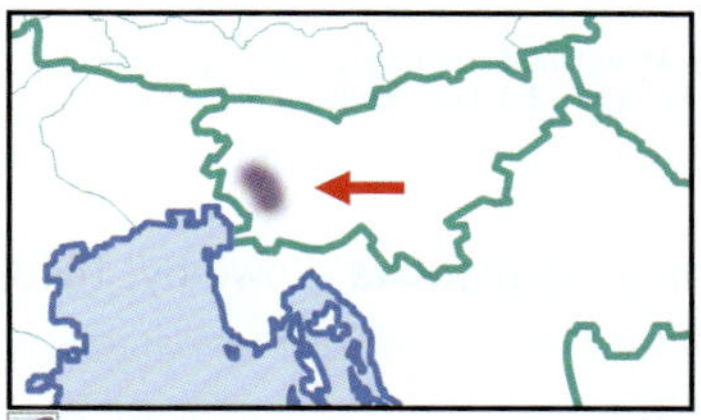

SW Slovenia

Striolata

Striolata Wagner, 1897 in A. J. Wagner 1897: 590 *Pomatias (Rhabdotakra (Striolata))*

Type species *Pomatias striolatum* Porro, 1840 (monotypy)

Established as an infrasubgeneric name, available (Art. 10.4).

Toffolettia Giusti, 1971 in Giusti 1971: 367, type species *Pomatias striolatum* Porro, 1840 (original designation)

Junior objective synonym of *Striolata*.

Shell cone-shaped, apertural margin not auricular, umbilicus narrow. Male genitalia: penis short, vas deferens open (not inside penis as in *Cochlostoma*). Female genitalia with a short receptaculum seminis (absent in *Cochlostoma*).

Giusti 1971.

Striolata striolata (Porro, 1840)

Without gender agreement: *Striolata striolatum* (Porro, 1840)

Pomatias striolatum Porro, 1840 in Porro 1840: 106

Porro treated *Pomatias* as neuter.

8-9 x 3.5-4.5 mm

Italy: near Genova, (1-2) San Rocco di Camogli, (3-7) Nervi (1-2) Giuseppe Pocaterra

NW Italy, SE France (?)

Shell shape like *Cochlostoma*, light horny brownish, often with three indistinct colour bands, with relatively dense white ribs, very oblique and strongly in S form, 7-8 only slightly convex whorls, apertural margins not connected at parietal side, not auriculated at columellar side, umbilicus narrow. Animal greyish, tentacles conical at their base.

In cool habitats in gorges at limestone rocks. In E Liguria a common species, in the west less frequent.

Westerlund 1885: 112, Germain 1931: 581, Giusti 1971: 371, Boato et al. 1984: 247.

Pomatias

Pomatias Studer, 1789
in Coxe 1789: 388

Type species *Nerita elegans* Müller, 1774 (monotypy)

Author was Studer (Coxe 1789: 334). Contained originally *Nerita elegans* and *Pomatias variegatus* (numen nudum). *Pomatias* is a masculine Greek word and has to be treated as masculine (Art. 30.1.2).

For many decades *Pomatias* was used for the group of species currently known as *Cochlostoma*. The phylogenetic relationships between *Pomatias*, *Tudorella* and *Leonia* are not well understood (Pfenninger et al. 2010).

Pomatias elegans (Müller, 1774)

Nerita elegans Müller, 1774 in Müller 1774: 177

13-18 x 9-12 mm

Albania: Skrapar, 20 km SE Berat

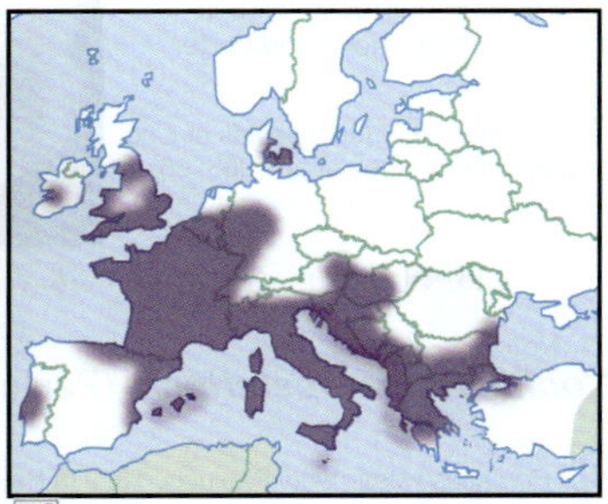

W Europe to NW Turkey, N Africa

Turkey: Istanbul Aydın Örstan

Shell moderately slender, light reddish with a colour pattern by whitish flames, radially and spirally prominently striated provoking a reticulate pattern, apex pointed, 4-5 very convex whorls, aperture and calcareous operculum upside pointed.
Males usually slightly smaller, more slender and with smaller aperture than females.
Animal large, brownish grey, tentacles long and almost cylindrical, transversally striated.

Open woods, shrublands, rock rubble, maritime grasslands, with smooth and loose rubbly soils allowing to dig at least 10 cm deep into the substrate, to estivate or hibernate. Restricted to calcareous substrate. Northern limit connected with the 2 °C January isotherm line. In Spain up to more than 1300 m, in Switzerland up to 1000 m, Bulgaria 900 m.
Mucus is slightly acid and reacts with calcareous substrate. Snails crawl at temperatures down to 8 °C, hibernating snails need at least 10-11 °C to become active. Hibernating snails suvive frosts down to at least -6 °C for several days, lethal temperatures must be lower. Drought periods lasting several months are no problem for survival, but snails remain inactive during this period. Estivating animals start crawling 30 min after getting in contact with water. Water is taken by skin and intestine, not directly by the mouth. For opening the operculum a minimum relative humidity of 95 % is needed, snails cannot get active when being submerged under water. Although high humidity is needed for crawling and feeding, *P. elegans* cannot suvive more than 4 weeks on permanently wet soils.
Snails feed on dead leaves and desintegrating wood, dry leaves are preferred, about 2 mg per day (average weight of adults 450-500 mg). Also pure cellulosis is digested. They do not feed below soil surface.

Copulation in central Europe from spring to autumn, 50-60 eggs (size 2 mm) are laid only in autumn, singly one by one into the soil, not in clutches, egg surfaces are covered with slime, so that soil particles surround the egg for camouflage. Juveniles hatch after 20 days or hibernate (in central Europe) and the juveniles hatch after 3 months, they have 2 whorls (2 mm shell height), 7 mm shell height are reached after 2 months, maturity after 18 months. Maximum age about 4-5 years.

Threatened in central Europe by intensive farming, also in vineyards, by habitat destruction, fertilizers, insectizides, herbizides and fungizides. Has seriously declined in central and E England. Endangered in Switzerland and Ireland, Vulnerable in Germany.

Gassies 1849: 73, Kilian 1951, Frömming 1954: 371, Prince 1967, Damjanov & Likharev 1975: 87, Kerney 1976, Turner et al. 1998: 75.

Pomatias glaucus (Sowerby, 1843)

Without gender agreement: *Pomatias glaucum* (Sowerby, 1843)

Cyclostoma glaucum Sowerby, 1843 in Sowerby 1842-1887: 100, Pl. 24

15-17 x 15-17 mm

Turkey: Vilayet Antalya, Alanya (determination uncertain)

S Turkey, Cyprus

Shell brownish violet, indistinctly reticulated, spiral lines flat and regular, radial lines weak, 4.5 whorls, aperture not or only little detached, rounded egg-shaped, upside bluntly angulate, inside brownish.

Not in Schütt 2001, occurrence in S Turkey uncertain.

Pomatias rivularis (Eichwald, 1829)

Cyclostoma rivularis Eichwald, 1829 in Eichwald 1829: 302 Eichwald treated *Cyclostoma* as feminine or masculine.

11.5-16 x 11-17 mm

Turkey: Vilayet Antalya, Demre

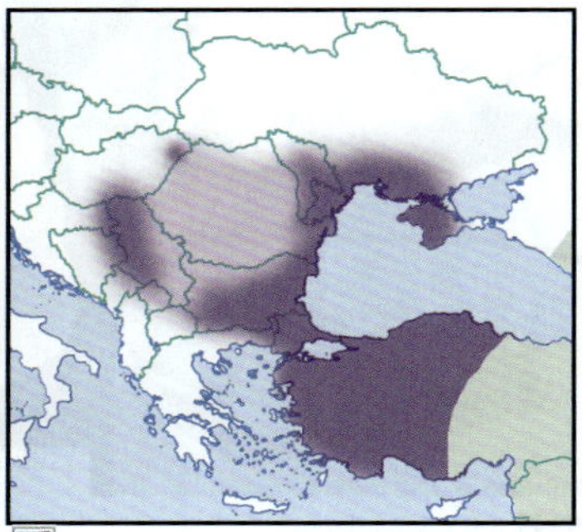

SE Europe, Turkey, N Caucasus

Shell diameter relatively larger than in *Pom. elegans*, with prominent spiral lines but radial lines very weak and reticulate pattern less prominent, 4-5 convex whorls, aperture and operculum upside rounded, only very slightly pointed.

In leaf litter near shrubs and in forests, often on loamy soils, in Bulgaria up to 1200 m.

Schütt 2001: 9, Sysoev & Schileyko 2009: 14, Coadă & Welter-Schultes 2011: 42.

Pomatias rivulare is incorrect.

Tudorella

Tudorella Fischer, 1885
in Fischer, Œhlert & Woodward 1887: 747 *Cyclostoma (Tudorella)*

Type species *Cyclostoma ferruginea* Lamarck, 1822 (monotypy)

Most of the time burrowed in soil and under limestone rocks near coasts, active only at moist weather conditions.

W Mediterranean region

The phylogenetic relationships between *Pomatias*, *Tudorella* and *Leonia* are not well understood. In a combined morphological, anatomical and molecular study Martínez-Ortí, Ar. et al. 2008 distinguished 3 species *mauretanica*, *sulcata* and *ferruginea*.
Pfenninger et al. 2010 studied some more locations (not Portugal, Libya and others), confirmed basically the previous results and were able to subdivide *sulcata* into 6 separate lineages which were defined as species. The conclusions concerning the species identities in the *T. sulcata* complex (subspecies in 2008, species in 2010) seemed to be either arbitrary or premature, they require confirmation by other methods.
More research seems to be necessary. It would also be interesting to analyse the phylogenetical relationships among the fossil species. *Tudorella* is found in S Europe since the early Miocene (Véla et al. 2008), various fossils seem to be closely related to *T. sulcata* and *T. ferruginea*.

Tudorella ferruginea (Lamarck, 1822)

Cyclostoma ferruginea Lamarck, 1822 in Lamarck 1822: 147
Lamarck treated *Cyclostoma* as feminine or masculine.

18-22 x 10-11 mm sexual dimorphism (females larger)

Spain: Illes Balears, Palma de Mallorca

Spain: Balearic islands (Mallorca, Menorca, Cabrera, Dragonera)
In Ibiza only fossils.

Shell reddish brown with spots, very slender, apex blunt, spiral sculpture moderately prominent, radial lines weak and very dense, 4-5 weakly convex whorls with shallow suture, last whorl 60-65 % of shell height.

Gasull 1964: 67, Falkner 1990: 122, Beckmann 2007: 35, Martínez-Ortí, Ar. et al. 2008.

Tudorella mauretanica (Pallary, 1899)

Without gender agreement: *Tudorella mauretanicum* (Pallary, 1899)

Cyclostoma mauretanicum Pallary, 1899 in Pallary 1898: 131
True date was 11 Feb 1899 (Winckworth 1936).

Shell light reddish brown, with moderately prominent spiral sculpture, radial lines weak, 3.5 rapidly increasing convex whorls with deep suture, first whorls smooth, last whorl height 75 % of shell height and 2 x of preceding whorl, from which it is detached and separarted by a narrow gap, aperture circular in its lower part, angulated in the upper part, inside orange.

Portugal: Algarve, Alvor near Portimão

16-22 x 11-16 mm, sexual dimorphism (females larger)

NW Africa, S Iberian peninsula

Nobre 1941: 222 (S Portugal), Martínez-Ortí & Robles 2005, Arrébola et al. in Verdú & Galante 2006: 333, Talaván Gómez & T. 2006: 35, Martínez-Ortí, Ar. et al. 2008, Pfenninger et al. 2010 (did not study specimens from Portugal).

In Spanish coastlands threatened by rapidly growing cities and fire. In some localities of Comunidad Valenciana measures were untertaken to transfer snails from habitats shortly prior to destruction to more protected places, possibly successfully.

Tudorella sulcata (Draparnaud, 1805)

Without gender agreement: *Tudorella sulcatum* (Draparnaud, 1805)

Cyclostoma sulcatum Draparnaud, 1805 in Draparnaud 1805: 33, Pl. 13
Cycostoma melitense Sowerby, 1843 in Sowerby 1842-1887 (Vol. 1): 105

13-17 x 10-11 mm, sexual dimorphism (females are larger)

(1) France: near Marseille, (2) Italy: Sardinia, Alghero, (3) Italy: Sicily, near Palermo, (4) Italy: Sicily, Agrigento, (5-6): Malta: Gozo

SW Europe, N Africa

Shell slender, whitish to orange or reddish with dark colour bands, fine and irregular radial striae, very prominent spiral ridges except on first 2 whorls, apex pointed, 5 very convex whorls, aperture inside orange.
Animal large, foot dark brownish, tentacles slightly conical, eyes located at exterior side at the base of the tentacles.

In coastal vicinity below 100 m in limestone rock rubble, covered with open pine woods or shrubs.

Vulnerable. In France extinct in Alpes-Maritimes, extremely rare in Var and still abundant only in Bouches du Rhône, threatened by rapid coastal urbanisation.

Often listed as *Pomatias sulcatus*. Martínez-Ortí, Ar. et al. 2008 proposed to recognise two subspecies, *melitensis* from Malta and *sulcata* from Italy, France and N Africa. Two years later the same team (Pfenninger et al. 2010) proposed to subdivide *T. sulcata* into 6 species, one was *melitensis*, based on molecular studies which require confirmation. The form of Malta differs slightly morphologically, but not the others.

Leonia

Leonia Gray, 1850
in J. E. Gray 1850: 34

Type species *Cyclostoma mammillaris* Lamarck, 1822
(monotypy)

Author often given as Baird, but I see no evidence for this author in the original work. A note on p. [3] suggested that Gray was the author. Author was not Gray 1840.

The shell is smoother than in *Tudorella*, the whorls less convex and the aperture less detached from the penultimate whorl.

The phylogenetic relationships between *Pomatias*, *Tudorella* and *Leonia* are not well understood (Pfenninger et al. 2010).

Leonia mammillaris (Lamarck, 1822)

Leonia mamillaris (Lamarck, 1822)

Cyclostoma mammillaris Lamarck, 1822 in Lamarck 1822: 147
Not spelled *mamillaris*, no reference to previous works. Lamarck treated *Cyclostoma* as feminine or masculine.

17-20 x 11-12 mm (females), 16-17 x 9-10 mm (males)

S Spain: N of Alicante

Algeria to Morocco, S Spain

Shell smoother than *Pomatias* and *Tudorella*, white to light reddish brown, often with dark spots below suture, operculum convex towards the outer side.

Under stones and in halva grass spots, in Spain on stony calcareous substrate in Mediterranean pine wood and shrub vegetation, up to 600 m.

Falkner 1990: 122, Martínez-Ortí & Robles 2005, Talaván Gómez & T. 2006: 35.

Pupilloidea and other small groups

Genera sorted by shell shape and size (within families alphabetically):
1 - High spire, few whorls, no umbilicus, apex pointed: Ferussaciidae, Cochlicopidae, *Hypnophila*, Succineidae, *Poiretia*
2 - High spire, many whorls, narrow umbilicus, apex blunt: Vertiginidae, *Gulella*, *Pupilla*, Lauriidae, Argnidae, Orculidae, *Spelaeoconcha*, Chondrinidae, Enidae, *Rumina*
3 - Rounded or flat shell: *Punctum*, *Paralaoma*, *Lucilla*, Valloniidae, *Zoogenetes*, *Euconulus*, *Pyramidula*, *Klemmia*, *Virpazaria*, *Spelaeodiscus*, *Discus*, *Pleurodiscus*

Calaxis

Calaxis Letourneux & Bourguignat, 1887
in Letourneux & Bourguignat 1887: 114

Type species *Tornatellina hierosolymarum* Roth, 1855
(subsequent designation by Westerlund 1902: 114)

Author was not Bourguignat alone ("les genres que nous comprenons dans cette famille [Ferussaciidae] sont: [*Ferussacia*, *Hypnophila*, *Calaxis*, *Hohenwarthia*, *Coelostele*]"). Contained originally *hierosolymarum*, *rothi*, *saulcyi* and *moussonianus*.

Parietal tooth usually present. Very occasionally parietal teeth can also be present in *Cecilioides* specimens of S Europe.

E Mediterranean region

Calaxis hierosolymarum (Roth, 1855)

Tornatellina hierosolymarum Roth, 1855 in Roth 1855: 39, Tab. 1 (*Tornatellina Hierosolymarum*)
Calaxis rothi var. *cypria* Kobelt, 1896 in Rolle & Kobelt 1895-1897: 59, Taf. 20

4-7 x 2.5 mm

Cyprus: Kato Pafos

Palestine/Israel, Syria, Cyprus

Shell with parietalis and columellaris visible in the aperture.

In subterranean habitats.

The taxon *cypria* from Cyprus should be considered as a synonym. In my studies on Ferussaciidae I have found no evidence to support the presence of an independent *Calaxis* species in Cyprus.

Cecilioides

Cecilioides Férussac, 1814
in A. Férussac 1814: 48

Type species *Buccinum acicula* Müller, 1774
(fixed by ICZN under the plenary powers)

ICZN Opinion 335 (1955): available, gender was stated to be feminine (should be accepted as by definition under the plenary powers, otherwise it would be masculine under Art. 30.1.4.4 because it was not treated as feminine by Férussac 1814). Type species was said by monotypy. I think that Férussac 1814 was non-binominal. Férussac 1814 mixed up French and Latin scientific names and did not apply consistently the Linnean system of Latin names for his scientific names of the animals. For one animal he used a French scientific name, for the next animal a Latin scientific name. Almost all were binominal. Sometimes the genus was Latin and the species French. It is however nowhere in the Code said that French scientific names were not scientific names, so the work was consistently binominal under Art. 11.4 and we also could immediately start looking for ten thousands of scientific names established by Buffon and co-workers. I would not recommend this because this would totally distort Linnean nomenclature.
If we demand that the work of an author who used French binominal names as scientific names for animals, either consistently or partly mixed with Latin names, must be rejected for the purposes of zoological nomenclature because of not consistently having applied the Linnean system, then Férussac's 1814 work must be rejected for nomenclature. I would clearly favour this interpretation.
The alternative solution would always imply that we would not have a clear argument where to set the limit between a work based on the French system and one on the Linnean system. Not possessing such a clear limit would open the option to define any French name as a scientific name in the Linnean system. Some would perhaps only count the names which have similar orthography, others would ask why not to take all French names.
The name *Cecilioides* was a name in the French orthography (in Latin it would be spelled *Caecilioides*). I would favour an interpretation by which the Commission made this name available in 1955 under the plenary powers (despite of having been published in a "non-binominal" work, here meant in the sense that the Linnean system was not applied consistently), and fixed its type *Buccinum acicula* also under the plenary powers (and not by monotypy, since no name in Férussac's 1814 work can have any status in nomenclature).

Shells small, transparent, fragile, columella concave, more or less distinctly obliquely truncate at base, usually somewhat calloused. Body colour unpigmented, animal blind without eyes.
Radula: marginalia cusps serrated, marginalia shape broad, marginalia number 10 (*C. acicula* and *C. genezarethensis*), laterals number 5-9, one ovotestis lobule, penis simple, receptacle base not muscular.

In subterranean soil habitats.

Oviparous (*C. acicula* and *C. genezarethensis* were studied).

Africa to central Europe and Near East

Pilsbry 1907-1908, Pilsbry 1909-1910: 1, Wächtler 1929, Heller et al. 1991: 289-293.

In extended studies on *Cecilioides* which involved measuring thousands of shells mainly from Greece and Turkey I came to the conclusion that in S Europe we seem to have two forms which I would regard as species, in some regions perhaps three: *C. acicula* from Turkey to Spain and N Europe, the second species is known as the larger species and occurs only in the south between Turkey and Spain.

In **Italy** this larger form is known as *C. janii*, the oldest name for it.
Manganelli et al. 1995 recognized two forms in Italy including Sicily, *acicula* and *janii*. This was in agreement with the results of my studies and seems to reflect what we currently know about these forms in Italy.
Beckmann & Falkner 2003: 38 recognized at least five species in **Sicily** (*stephaniana*, *actoniana*, *acicula*, *petitiana*, *rizzeana*, perhaps also *bourguignatiana*), but I think this study was too weak to support these conclusions. To propose a solid classification in *Cecilioides* it is certainly not sufficient to present figures of a low number of quite different shells, without studying the local and spatial variation, and particularly, without analysing many shells to obtain statistically confident results.

In **Greece** we have a perhaps different form *C. tumulorum*, my studied material from former Yugoslavia was too scarce for reliable statements on its range. In Greece and Turkey the third species could be called *C. michoniana*, but it is likely that it is conspecific with *C. janii*.

It is possible that **Sicily** and **Crete** host more *Cecilioides* species than only two, but this needs to be firmly established. It seems more likely that we are dealing with a highly variable species complex (*acicula* / *tumulorum*) with some gene exchange between various morphs. The *tumulorum* form seems to hybridize with *acicula* in many regions, while in others they seem to live sympatrically, and again in other regions (S Turkey) the *tumulorum* forms seem to hybridize with *janii*. Recognizing three species is useful, *acicula* and *janii* are relatively well supported, *tumulorum* as a third species is less well supported in some regions but generally it seems to make sense to separate it.

In **Romania** *C. acicula* seems to be widespread. Grossu 1955: 1579 reported a rare form as *Hohenwarthia orghidani*, the values of which correspond with *C. janii*.

In **Russia, Ukraine and Caucasus** Sysoev & Schileyko 2009: 111 used two names *acicula* and *raddei*. They distinguished both forms by the relative apertural height (their statements contrasted the figures 53F and 53G, in the text *acicula* and *raddei* were probably mixed up, and suggested that the authors did not understand that the relative apertural height is not static but decreases with increasing shell height). The values for *raddei* corresponded well with those for *janii*.

In **Spain** we also seem to have three species, *C. connollyi* could be a name for a third form. I have not seen enough material from the Iberian peninsula.

Cecilioides acicula (Müller, 1774)

Buccinum acicula Müller, 1774 in Müller 1774: 150

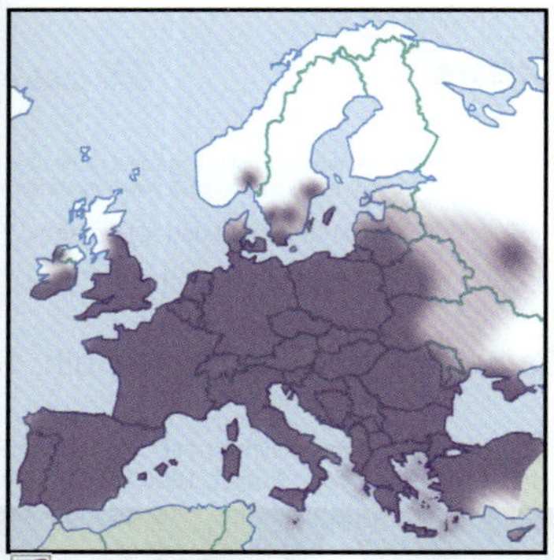

Up to 4 x 1.1 mm, rarely larger

Germany: near Xanten

Europe to central Asia, N Africa

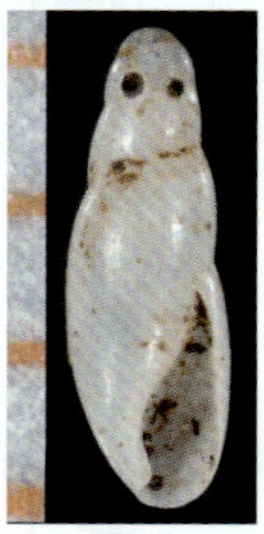

S Turkey: Konya, near Beyşehir Gölü

Shell white transparent, slender, columella obliquely truncate at the base, height of aperture relatively small in relation to shell height, but the proportion differs with the stage of growth and must be seen in relation to shell height.
Apertural height = AH, taken at the right margin, Last whorl height = LH, taken at the left margin in the same position, Shell height = H.
AH/H in pure (undisturbed, typical) *C. acicula* 0.5-0.65 (for H = 2 mm), 0.4-0.5 (H = 3 mm), 0.34-0.44 (4 mm), 0.31-0.41 (H = 5 mm).
LH/H in pure (undisturbed, typical) *C. acicula* 0.63-0.78 (for H = 2 mm), 0.56-0.7 (H = 3 mm), 0.48-0.62 (4 mm), 0.43-0.54 (H = 5 mm).
Animal whitish, almost transparent, upper tentacles without pigments, the animal is blind.

Subterranean in soil, living well (usually 20-40 cm, locally up to 2 m or more) below the soil, in organic litter of rock crevices, among plant roots, often found in organic river deposits, or in ant hills and in the cavities of ancient buried bones. Rarely found alive. In Switzerland up to 1800 m, rare in the central alpine regions.

Reproduction in E Germany between May and July, copulation lasts about one hour, eggs (size 0.75 mm) are laid singly 10-20 days later, in up to 30 cm depth, animals die probably after having laid some 2-3 eggs, perhaps more, but it seems that the animal cannot produce more than 11-13 eggs. Growth, maturity and life span unknown.

Wächtler 1929. The AH/H and LH/H values were obtained in my own studies (Welter-Schultes 1998: 49).

Cecilioides connollyi Tomlin, 1943

Cecilioides connollyi Tomlin, 1943 in Tomlin 1943: 23

Shell large, 8 whorls, last whorl height 70 % of shell height, apertural height 50 % (all sizes referring to 10 mm shell height).

10 x 2.2-2.3 mm

Currently not more than a candidate name for a third form of the S Iberian peninsula. More research is necessary.

S Iberian peninsula

Cecilioides janii (De Betta & Martinati, 1855)

Achatina janii De Betta & Martinati, 1855 in De Betta & Martinati 1855 [28 Feb]: 59 Type locality in N Italy.
Appeared a few days after Strobel's mention of the name *veneta*. Name proposed conditionally, without description but with reference to De Betta 1852: Tav. 1 Fig. 2.
Achatina aciculoides synonym *veneta* as mentioned in Strobel, 1855 [c. 15 Feb]: 137 (synonym and nomen nudum, no description or indication added)
In the original source not used as a synonym of a name treated as valid, but without description or indication as a synonym for "*aciculoides* sensu auct." and "*acicula* sensu De Betta", both were not available names in the sense of Art. 11.6 and Art. 49). ICZN 2004 (Opinion 2080): name placed on Official List, thus available under Art. 80.6.1. The Commission had not been informed that the name was not equipped with a description or indication in 1855, and that it was not mentioned as a synonym of an available name.
Strobel 1857: 248 listed a taxon *Glandina acicula* var. *veneta* "Charp.", without bibliographical reference and without description.
Cionella (Acicula) veneta Albers, 1860 in Albers & Martens 1860: 260 Type locality in N Italy.
First occasion at which the name *veneta* was made available. Established without description, with bibliographic reference to Philippi 1836: Tav. 8 Fig. 26, and De Betta 1852: pl. Fig. 2. Not *Acicula spectabilis* var. *veneta* Pirona, 1865. The original genus, authorship and year for the name of the taxon as placed on the Official List must be replaced by "*Cionella veneta* Albers, 1860" (G. Rosenberg. pers. comm., 2011).
Bank, Fa. & G. 2000 designated a lectotype collected near Trento, from the De Betta collection in Verona. This was in line with Albers' 1860 indication, so *Cionella veneta* Albers, 1860 is an objective junior synonym of *Achatina janii*.
Ferussacia michoniana Bourguignat, 1864 in Bourguignat, 1863-1868 [01 May 1864]: 115, Pl. 19 Type locality in Israel/Palestine, near Jerusalem.
Cochlicopa raddei Boettger, 1879 in O. Boettger 1879: 25, Taf. 1 (*Cochlicopa (Hohenwartiana) Raddei*) Type locality in Georgia.
Hohenwarthia orghidani Grossu, 1955 in Grossu 1955: 1579 Type locality in S Romania.

3-7 x 1-2 mm

Montenegro: Kronistar near Duravci

S Europe, Turkey

Shell grows larger than *C. acicula*, aperture higher in relation to shell height (in shells exceeding 3.5 mm shell height the AH/H and LH/H values are all without overlap larger than in *C. acicula*), shell less slender, apex perhaps narrower, columella perhaps more clearly truncate at the base. In "pure" *C. janii* forms the apertural height also exceeds the values for *C. tumulorum* in Greek and Turkish populations.
The animal is blind like *C. acicula*.

Subterranean, lives inside calcareous stony soils, at the base of old walls composed of natural stones, at rocks and in rock crevices, also under flowerpots in gardens. In Switzerland up to 500 m.

Vulnerable in Switzerland. It is possible that this species has a wide distribution from S Europe to the Near East. It is also possible that the relatively similar form of Turkey and Greece is a different species, *C. michoniana*, but unless evidence is provided for this assumption I prefer classifying *janii* and *michoniana* as conspecific.

Giusti 1976, Kerney & Cameron 1979: 150, Falkner 1990: 168, Giusti et al. 1995: 299, Manganelli et al. 1995: 22, Turner et al. 1998: 238, Bank, Fa. & G. 2000.

Cecilioides tumulorum (Bourguignat, 1856)

Caecilianella tumulorum Bourguignat, 1856 in Bourguignat 1856 (Rev. (2) 8): 424, Pl. 12 Type locality near Athens, Greece.

4-8 x 1.3-2 mm

Greece: E Crete, island of Agios Nikólaos near Móchlos

SE Europe, Turkey

Shell generally larger than *C. acicula*, but not necessarily with larger shell diameter, height of the aperture larger than in *C. acicula* shells of the same shell height. Shell parameters vary locally.

Subterranean, often sympatric with *C. acicula*.

This name can be used for Greek forms. It seems to interbreed with *C. acicula* in some regions resulting in mixed hybrid populations, and to live sympatrically at other localities. Could also be classified as a subspecies of *C. acicula*.

Coilostele

Coilostele Benson, 1864 in Benson 1864: 136
Gender feminine because derived from Greek σπηλη.

Type species *Coilostele scalaris* Benson, 1864 (monotypy)
(= currently *Coilostele scalaris* Benson, 1864 from India)

Shell turreted, cylindrical, translucent and thin with a blunt protoconch of usually 3 whorls. The teleoconch whorls with a straight outline, smooth to ribbed with a deep suture. In the upper teleoconch whorls, the inner parietal walls and the columella are completely resorbed. In the last two whorls, these structures are still present, the columella is indented with an oblique lamella.

S Spain and N Africa to S Arabia and India

Currently (Neubert 1998: 411) classified in Ferussaciidae, but could also belong to Ellobiidae, no anatomical information available.

Coilostele acus (Pfeiffer, 1853)

Bulimus acus Pfeiffer, 1853 in L. Pfeiffer 1853 (Proc. Zool. Soc. 20): 60

Shell whitish translucent, weakly striated, 6-8 moderately convex whorls, looks like a small enid, just with more whorls, last whorl less than 1/2 of shell height, LH/H around 0.4, aperture rounded like in an enid shell, less than 1/4 of shell height, no tooth in the aperture.

Known only from Guadalquivir river deposits near Sevilla, habitat unknown.

3.5-4 x 0.7-0.8 mm

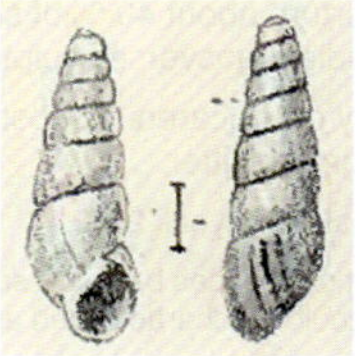

Picture: Küster & Pfeiffer 1845-1855: Taf. 69 Fig. 12-14

Spain: near Sevilla

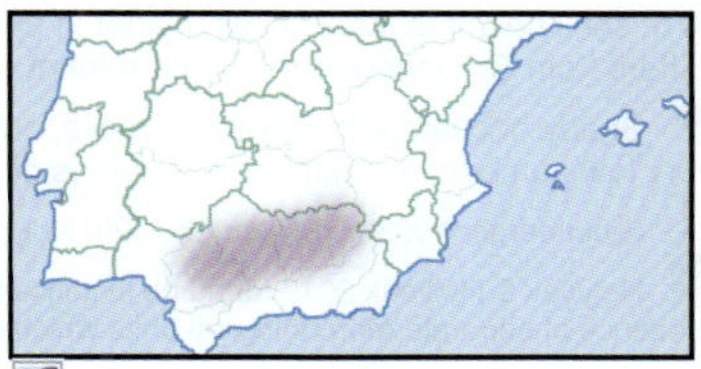

S Spain

Cryptazeca

Cryptazeca Folin & Bérillon, 1877 in Folin & Bérillon 1877: 441

Type species *Azeca monodonta* Folin & Bérillon, 1877 (monotypy)

Contained originally *A. monodonta*, and two subspecies var. *hyalina* and var. *subcylindrica*, type by monotypy after Art. 68.3.
The name of the author Folin follows the conventions for French titles of nobility, where the "de" is not added to the surname (examples Lamarck, Buffon).

Adults can be distinguished from immatures by a thickened edge of the parietal margin of the aperture.

Gómez 1990 (2 papers), Holyoak & Holyoak 2012.

Madeira et al. 2010 suggested in a molecular study that *Cryptazeca* was closely related to *Azeca* and *Hypnophila*, not to *Ferussacia* or *Cochlicopa*.

Cryptazeca elongata Gómez, 1990

Cryptazeca elongata Gómez, 1990 in Gómez 1990: 366

Shell colourless, very slender, 7-7.5 weakly convex whorls, last whorl little less than half shell height, aperture with a thin callus or layer at the parietal side, columellaris with a small tooth, aperture in lateral view upper third concave, lower part convex, umbilicus closed.
Differs from *C. spelaea* in its more slender shell (height/diameter ratio exceeds 2.8).

6.2-7 x 2-2.35 mm

Found in caves on the soil, in zones of high humididy and absence of light.

EU Red List: Vulnerable VU. Only known from one single cave.

N Spain

Cryptazeca monodonta (Folin & Bérillon, 1877)

Azeca monodonta Folin & Bérillon, 1877 in Folin & Bérillon 1877: 199
The name of the author Folin follows the conventions for French titles of nobility, where the "de" is not added to the surname (examples Lamarck, Buffon).
Cryptazeca kobelti Gittenberger, 1983 in Gittenberger 1983: 313

Shell horny brown, not slender, smooth, shiny, 4-5 whorls, last whorl height 2/3 of shell height in adult shells, parietal callus well developed, at the columellar base a small denticle is present, in side view the upper half of the apertural margin is slightly convex and inclined backward, lower half more straight but similarly, symmetrically, inclined backward.
Differs from *C. subcylindrica* in its much smoother and less slender shell (height/diameter ratio exceeds 2.35), the last whorl being more predominant and the general shape more ovoid.

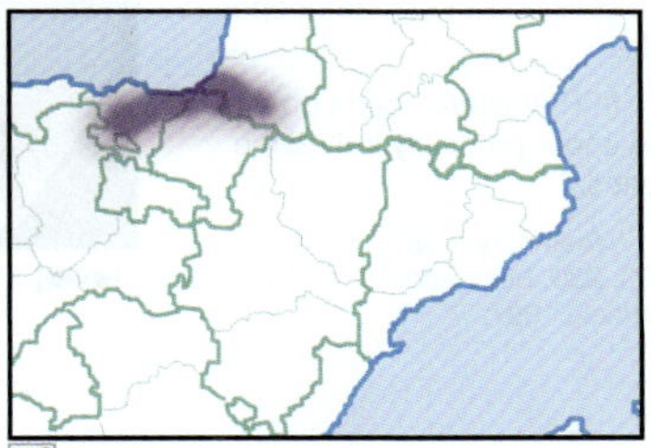

Basque Country

3.2-4 x 1.5-1.8 mm

Leaf litter habitats in moist deciduous woodlands, obviously confined to limestone substrate but not always associated with rocky outcrops.

Occurrences are rare and always very local. The species occurs at sites which are rich in land snail species, always in very low abundance.

Gittenberger 1983: 315, Gómez 1990, Holyoak & Holyoak 2012: 93.

C. monodonta and *kobelti* were originally distinguished by shell shape. This was initially based on very few specimens and not confirmed in studies by Holyoak & Holyoak 2012 who analysed more shells and observed that the values overlapped.

Cryptazeca spelaea Gómez, 1990

Cryptazeca spelaea Gómez, 1990 in Gómez 1990: 220

Shell moderately slender, 7-7.5 whorls, last whorl not as dominant as in other species, apertural lip slightly curved inward except in the upper part, denticle present at the columellar bottom, in side view the upper half of the apertural margin is strongly convex ")", the lower half concave "(", umbilicus closed.
Differs from *C. elongata* in its less slender shell (height/diameter ratio less than 2.8).

7.2-8.2 x 2.6-2.9 mm

In caves on guano, organic litter, crawling on stalagmites, habitats with total absence of light.

EU Red List: Vulnerable VU. Only known from two caves.

Gómez Moliner et al. 2001: 120.

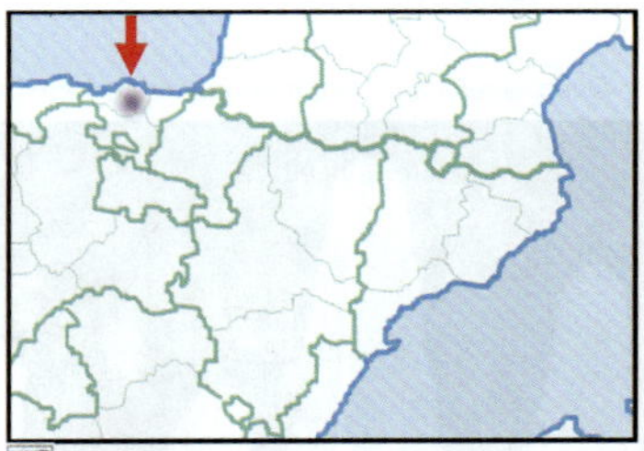

N Spain (Bizkaia)

Cryptazeca subcylindrica Folin & Bérillon, 1877

Cryptazeca monodonta var. *subcylindrica* Folin & Bérillon, 1877 in Folin & Bérillon 1877: 443, Pl. 3 (*Cryptazeca monodonta* var. *Subcylindrica*)
Name was established as intentionally new, not *Helix subcylindrica* Linné, 1767, currently *Truncatella subcylindrica*.
The name of the author Folin follows the conventions for French titles of nobility, where the "de" is not added to the surname (examples Lamarck, Buffon).
Ferussacia vasconica Kobelt, 1894 in Kobelt 1894: 37, Taf. 188

2.9-4.7 x 1.1-1.7 mm

Spain: Santander, Liendo near Laredo

N Spain to W Pyrenees

A very rare species, always found in low numbers.

Shell colourless or horny brown, shiny, with a relatively prominent radial/transverse sculpture, most prominent below the suture, 5-5.5 whorls, last whorl 60 % of shell height in adult shells, height of aperture about 40 % of shell height, parietal callus well developed. Shell size and shape variable.
Differs from *C. monodonta* in its prominent radial/transverse sculpture and more slender shell (height/diameter ratio exceeds 2.35).

Shady habitats with limestone rocks, only on calcareous substrate, occasionally under closed cover of deciduous woodland, also in caves. Usually found in very low numbers in voluminous sievings of ground litter.

Gittenberger 1983: 316, Gómez 1990, Holyoak & Holyoak 2012.

Holyoak & Holyoak 2012 found no significant differences between *subcylindrica* in France and *vasconica* in Spain, their analyses did not provide constraints to support the presence of separate morphological units which could be characterised as species.

Ferussacia

Ferussacia Risso, 1826
in Risso 1826: 80

Type species *Ferussacia gronoviana* Risso, 1826
(subsequent designation by Pilsbry 1908: 216)
(= currently *Ferussacia folliculum* (Schröter, 1784))

Originally spelled *Ferrussacia*, French name on p. 80 was given as Férussacie. On p. 434 both names were spelled with rr (*Ferrussacia*, Ferrusacie). In the text "Féruss." was frequently cited in bibliographic contexts, on p. 80 at two occasions. This suggests that *Ferrussacia* was an inadvertent error, to be corrected under Art. 32.5.
Contained originally *F. gronoviana* and *F. sayea* Risso, 1826. Herrmannsen 1846: 441, Gray 1847: 178, Pfeiffer 1848: 543, Lowe 1855: 200 and Albers & Martens 1860 gave as type *Helix folliculum*.

Canary Islands, Madeira, Mediterranean region, N Africa, Mauritius

Ferussacia carnea (Risso, 1826)

Pegea carnea Risso, 1826 in Risso 1826: 88
Description based on shells brought to Nice, France, with cereals from N Africa.

Shell yellowish or brown with orange-red hue, shiny, 5-6 flattened whorls, margin more regularly rounded than in *F. folliculus*, so that both apex and base are more pointed, aperture higher and more pointed at both sides, parietalis present but can be deeper inside.

In Pantelleria found on the soil and under partically buried big stones.

Westerlund 1887: 161, Giusti 1973: 230, Manganelli et al. 1995: 22, Sparacio 1997.

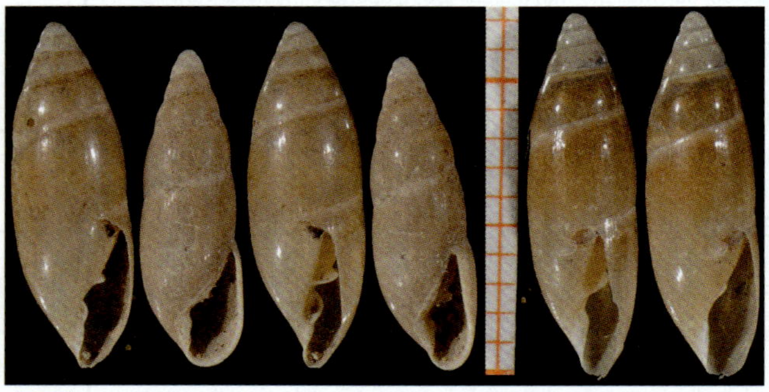

11-13 x 3.5-5 mm (1-4) Tunisia: near Tūnis, (5-6) Algeria: near Qusanṭīnah/Constantine

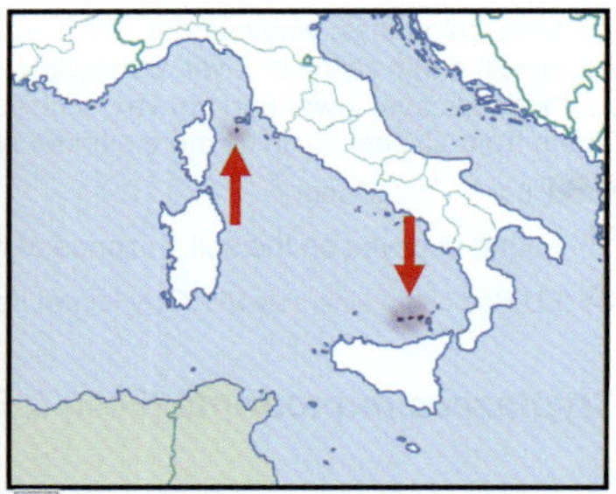

N Africa, Italy (islands of Pianosa and Eolian islands)
In Italy considered as introduced, the identity there was confirmed anatomically, but the northern African populations require more detailed studies.

Ferussacia folliculum (Schröter, 1784)

Ferussacia folliculus (Schröter, 1784), *Ferussacia follicula* (Schröter, 1784)

Helix folliculum Schröter, 1784 in Schröter 1784: 237, 699 (*Helix Folliculum*)
Listed by Gmelin [1791]: 3654 as *Helix Folliculus*, with reference to the same figure (by Gronovius 1781) that Schröter 1784 also had given.
Currently known as *F. folliculus* (usage about 70 %) or *F. follicula* (usage about 30 %), correct is *F. folliculum* (usage about 1-2 %, probably increasing).
There are 3 words in Latin, all are nouns: folliculus (small bag), follicula (the bag) and folliculum (a small bag). Subsequent users thought that *folliculum* was incorrect, or was an adjective, so they called the species *folliculus* and *follicula*. Under Art. 32.5 the original spelling *folliculum* must be used. I would suggest to use *folliculum* (though it has not been used until 2010), it makes no sense to select one of the two currently used incorrect spellings. Alba et al. 2011: 48 selected this spelling in their checklist of Catalonia, so it is probably going to become accepted.

6-9 x 2.5-3.5 mm

Spain: Cau Tunis near Barcelona

W Mediterranean region

Portugal: Algarve, near Faro

Charles Geerts

Shell horny yellowish or pale brown, smooth and very shiny, 5-6 almost flat whorls, last whorl more than 2/3 of shell height, aperture in adults 40 % of shell height (in juveniles more than 50 %), columellar callus present, better visible in an oblique view, apertural margin connected by a transparent parietal callus, no parietalis. Shell very variable.
Animal long, yellow or yellowish with greenish hue, darker on dorsum and head, upper tentacles short, lower tentacles only in the form of short tubercles.

On plants, under desintegrating leaves and stones, mostly in coastal vicinity and lowlands, in natural and disturbed habitats, usually in small and concentrated populations.

Ovoviviparous. Parent animals incubate several embryos of various sizes, juveniles near hatch have 1.5 whorls of an oval transparent shell, greenish foot and black eyes.

Moquin-Tandon 1855: 306 (incubation), Germain 1930: 322, Nobre 1941: 164, Falkner 1990: 168, Reitano et al. 2009: 190.

Hohenwartiana

Hohenwartiana Bourguignat, 1864
in Bourguignat 1864: 117 *Ferussacia (Euferussacia (Hohenwartiana))*

Type species *Achatina hohenwarti* Rossmässler, 1839 (original designation)

Date was 01 May 1864. Also published in Revue et Magasin de Zoologie (2) 11: 201 (31 July 1864), with the same explicitly mentioned type species. Name established in agreement with the Code. See the comment under *Esperiana*. I propose to use the name *Hohenwartiana*.
Hohenwarthia as mentioned by Letourneux & Bourguignat 1887: 115, incorrect subsequent spelling for *Hohenwartiana*
For those who would not like to accept using *Hohenwartiana*, *Hohenwarthia* was argued (Bank et al. 2001) to be the next available name. But *Hohenwarthia* referred bibliographically to Bourguignat's 1864 description of *Hohenwartiana*, and was only an incorrect subsequent spelling of that name. What should be the type species of an incorrect subsequent spelling of a blacklisted genus?

Shell fragile, translucent, 5-7 nearly flat whorls, the last tapering to the base, aperture without teeth, columella straight, tapering or a little excised below, but not really truncate, not reaching the base. Differs from *Cecilioides* in the absence of a columellar truncation.

Mediterranean region

Pilsbry 1907-1908: 327, Wenz & Zilch 1959-1960: 337, Giusti 1973: 222.

Hohenwarthia orghidani Grossu 1955 (in Grossu 1955: 1579) does not belong to *Hohenwartiana*, but apparently to *Cecilioides* (columellar base truncated, not straight), shell size and apertural height given by Grossu 1981: 241 suggest that it is conspecific with *Cecillioides janii*.

Hohenwartiana hohenwarti (Rossmässler, 1839)

Achatina hohenwarti Rossmässler, 1839 in Rossmässler 1838-1844 (9-10): 34, Taf. 49 (*Achatina Hohenwarti*)
Misspelled *Hohenwardti* on Taf. 49 and in the index p. 45.
Achatina aradasiana Benoit, 1862 in Benoit 1875 (Ill. Sist.): 24 (*Achatina Aradasiana*)

4-7.5 x 1.7-2.5 mm

Spain: Valencia, Acequia del Rey

S Europe

Pilsbry 1907-1908: 335, Westerlund 1887: 174, Giusti 1973: 222, Giusti et al. 1995: 303, Manganelli et al. 1995: 22.

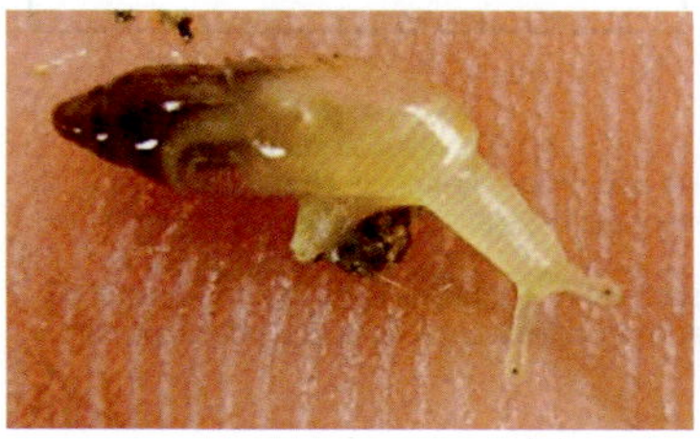

Malta: Gnajn Tuffiena

David Cilia

Shell yellowish, shiny, 4-5 weakly and irregularly convex whorls, aperture simple, outer margin convex, columella short and straight, goes in a smooth movement downwards, not reaching the base (in *Cecilioides* abruptly/truncated).
The animal is greyish with moderately dark upper tentacles, or evenly light yellowish, but always with dark pigments at its fully functional eyes.
Differs from *Ferussacia folliculum* in its smaller and more slender shell.

In microcavities inside the soil under rocks, walls and boulders. Snails occasionally appear at the surface after heavy rainfalls. In Malta rare and restricted to few sites in natural habitats under stones embedded in soil.

I see no constraints in the descriptions of the shell that the taxon *aradasiana* from Sicily represents a separate species. Giusti 1973 analysed genitalia and radula of two specimens from N Italy and Lipari island (N of Sicily) and found only weak differences. He argued that two species *aradasiana* and *hohenwartiana* were present in Sicily, but did not explain how to distinguish them. The specimen from Lipari was assigned to *aradasiana*, but it remained unclear why.

Cochlicopa

Cochlicopa Férussac, 1821
in A. Férussac [1821-1822]: 24 (folio 28) *Helix (Cochlicopa)*

Type species *Helix lubrica* Müller, 1774
(subsequent designation by Westerlund 1902: 113)

Established with short description (ouverture étroite, coquille ovoïde ou turriculée) and without species included, 06 Apr 1821 (livraison 9), made available as a regular new name at this occasion. In the synonymy Férussac mentioned a name *Polyphemus* Montfort, 1810 and a name *Styloides* (a nomen nudum).
First species included in folio p. 54 (26-05-1821) (quarto p. 50, 13-07-1821): *Helix priamus*, *rosea*, *mulleri*, *poireti*, *glans*, and many others, also *lubrica*. Doubtlessly this was a subsequent use of *Cochlicopa* as established a few pages before in the same work.
Type *Helix lubrica* designated by Westerlund 1902: 113. Waldén & Gittenberger in Gittenberger 1983: 308 intended to designate as type once again *Helix lubrica*, they argued that previous authors had not mentioned that the genus was established in April 1821. But this was not necessary under Art. 67.7.
Kennard 1942: 113 argued that *Cochlicopa* was a substitute name for *Polyphemus* Montfort, 1810 (non Müller, 1776, Crustacea) (type species *Bulimus glans* Bruguière, 1792 from the Caribbean, = currently *Oleacina voluta* (Gmelin, 1791)). Regarding *Cochlicopa* as a new replacement name is not tenable today because Férussac 1821 gave no expressed statement as demanded in Art. 72.7 to qualify for a new replacement name. For those who do not require an expressed statement for a new replacement name, *Cochlicopa* must be considered as a junior synonym of *Oleacina* [Röding], 1798, and cannot be used here.

Self-fertilising is the main breeding strategy.

Eurasia, northern Africa, Iceland, North America

Armbruster & Schlegel 1994 (allozyme study), Armbruster 1995 (differential shell morphology), Starobogatov 1996 (taxonomic revision resulting in 27 species in Eurasia, based on comparative shell morphology), Armbruster 1997 (3 species *lubrica*, *lubricella* and nitens in allozyme studies, *lubricella* seems to represent a paraphyletic arrangement), Armbruster & Bernhard 2000 (previous taxonomy was confirmed in ribosomal RNA studies).

The results of molecular studies suggest a high degree of self-fertilisation and only a weak correlation between morphological differences and phylogenetic units. Only *lubrica* and *nitens* were confirmed as independent units or species, *lubricella* was weakly supported as an independent unit, other such units were not found.

Starobogatov 1996 recognised 10 species in Europe (*collina*, *curta*, *lubrica*, *lubricella*, *lubricoides*, *major*, *minima*, *nitens*, *pfeifferi*, *repentina*). These have been accepted by subsequent Russian authors until recently (Kantor & Sysoev 2005: 230, Egorov 2008: 30 except *repentina*, Sysoev & Schileyko 2009: 23), but not in central and western Europe. In the light of the results of molecular studies Starobogatov's 1996 approach seems too weakly supported.

Cochlicopa lubrica (Müller, 1774)

Helix lubrica Müller, 1774 in Müller 1774: 104
Cochlicopa repentina Hudec, 1960 in Hudec 1960: 287

Falkner et al. 2002 intended to protect this name against its senior subjective synonyms *Zua crassula* Fagot, 1879 and *Zua locardi* Pollonera, 1885 under Art. 23.9.1 and provided the 25 references. This was ineffective, *Cochlicopa lubrica locardi* was used by Germain 1930: 457 and Alzona 1971: 61.

Shell horny yellowish to reddish brown, shiny, shape like a small enid, 4-5 moderately convex whorls, juveniles not keeled, aperture inside slightly thickened, no teeth, no umbilicus.
Animal greyish or blackish, tentacles dark.
Slightly larger than *C. lubricella*, less slender, whorls slightly more convex. Single shells can be very difficult to tell apart from *C. lubricella*.

5-7.5 x 2.4-2.9 mm

Germany: Schleswig-Holstein, near Plön

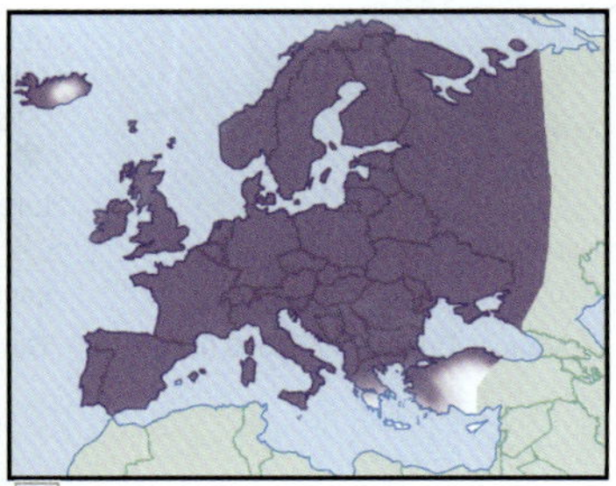

Holarctic

Gassies 1849: 122, Frömming 1954: 29, Umiński & Focht 1979, Armbruster & Schlegel 1994, Armbruster 1995, Turner et al. 1998: 139, Schütt 2001: 22.

Slovakia: S Rožňavská kotlina basin

Wide ecological amplitude, usually in moderately humid habitats, valley meadows and forests, under stones and wood, in plains and highlands. Tolerates non-calcareous soils, in Norway most abundant at pH 6.0. Feeds on live and dead plants and small mushrooms. In Switzerland in up to 2600 m altitude, Bulgaria 1300 m.

In Poland breeding is possible at any time, final size and maturity is reached at 21-24 months, life span more than 3 years. In France deposition of some 20 eggs per individual on humid soil below plants was observed in April and May, juveniles hatched after 2 weeks, adults were observed at the end of the summer. In Germany juveniles hatched after 3 weeks, 2 whorls at 1.5 mm shell height was reached after 2 weeks, 3 whorls (1.8 mm) after 5-6 weeks.

Includes the form known until the 1990s as *Cochlicopa repentina* Hudec, 1960. It seems that animals with *repentina*-like reproduction organs can be bread from pure *Cochlicopa lubrica*, conchological and molecular (isozyme) characters did not provide evidence for a separate species *C. repentina*.

Jozef Grego

Cochlicopa lubricella (Porro, 1838)

Bulimus lubricus var. *lubricella* Porro, 1838 in Porro 1838: 53 (with short description)
Name mentioned by Rossmässler 1834 (Heft 2): 6, as *Achatina lubricella* from Austria in the synonymy of *Achatina lubrica*, nomen nudum without description or indication, and by Rossmässler 1835-1837 (1): 88 as a variety of *Achatina lubrica* without description (nomen nudum). Name not available under Art. 11.6.1 from Rossmässler 1834 because Rossmässler did not give a description or indication.

Shell light horny coloured, smaller and more slender than *C. lubrica*, whorls slightly less convex.

Prefers meadows at S exposed dry slopes, usually dryer habitats than *C. lubrica*, sometimes both species occur together. Seems to be more confined to limestone substrate, in Scandinavia very sensitive against acid rainfalls. In Switzerland rarely in forests, up to 2800 m, in Bulgaria 1800 m.

Armbruster & Schlegel 1994 (allozyme studies) and Armbruster 1995 (shell morphology) suggested presence of a separate true species, in molecular studies by Armbruster & Bernhard 2000 this was not confirmed.

4.0-5.7 x 1.8-2.3 mm

Germany: near Göttingen

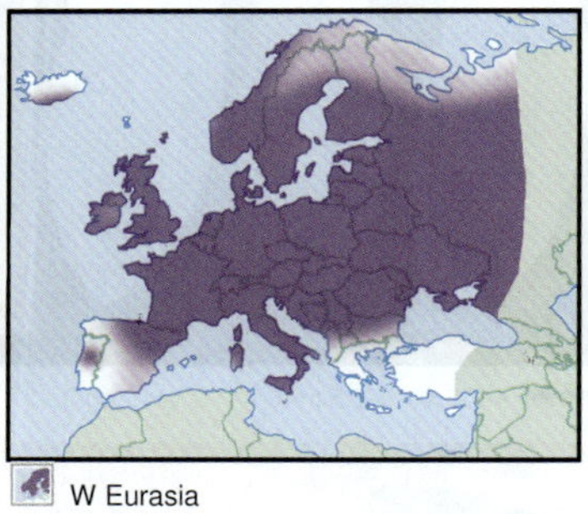

W Eurasia

The taxonomic status as a separate species has repeatedly been disputed (Schütt 1996: 21 = Schütt 2010: 22, Ruiz Ruiz et al. [2006]: 214). The main arguments were that previous studies did not rule out the hypothesis that environmental, seasonal and parasitological factors could influence shell size and shape. Armbruster 1995 did not find sympatric populations of *lubrica* and *lubricella*. De Oliveira 2009: 57 found that at one locality in Portugal both morphs were present but not statistically significant. If true, then this could rule out environmental factors to determine shell shape of individuals.
Seasonal factors could be tested, but the results obtained by Umiński & Focht 1979: 184 in Poland suggest that breeding is possible at any time and adults do not die immediately after egg laying. If the season had influence on the shell shape, this would result in the presence of all classes of shell shapes in individuals collected at one single site. Armbruster's 1995 results seemed to contrast such a scenario.

Cochlicopa nitens (Gallenstein, 1848)

Bulimus nitens Gallenstein, 1848 in Gallenstein 1848: 10

Shell colour dark horny brown to coffee brown, intensively shiny, larger than the other species.

In periodically flooded swamps, irrigation canals, very moist meadows, also at water margins, on calcareous substrate. In Switzerland up to 1200 m.

Critically Endangered in Germany and Austria, Endangered in Switzerland. Rare everywhere, threatened by intensive land use and habitat destruction.

Recognized as an independent species (from *C. lubrica*) since 1956, previous reports for *C. lubrica* may include this species.

Bulgaria: Beli Lom river deposit
Dilian Georgiev

Slovakia: Slovenský Kras National Park
Jozef Grego

6.2-7.5 x 2.8-3.4 mm

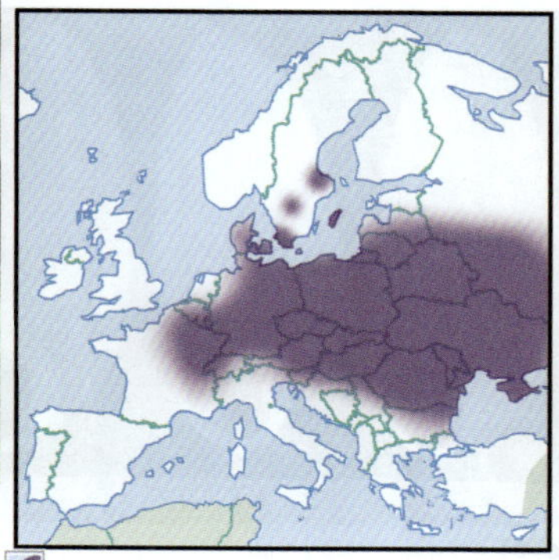

Central and E Europe

Azeca

Azeca Fleming, 1828 in Fleming 1828: 269

Type species *Azeca tridens* Fleming, 1828 (monotypy)
(= currently *Azeca goodalli* (Férussac, 1821))

ICZN Opinion 335: type species *Turbo tridens* Pulteney, 1799 by monotypy. Author and date were incorrect under Art. 11.10 (Pulteney 1799: 46 had misidentified a species as *Helix tridens* Müller, 1774, Fleming 1828 deliberately employed this misidentified name for a species that served as type for *Azeca* by monotypy), must be corrected to *Azeca tridens* Fleming, 1828.

Azeca goodalli (Férussac, 1821)

Helix goodalli Férussac, 1821 in A. Férussac [1821-1822]: 71 (folio p. 75) (*Helix (Cochlodonta) goodalli*)
Azeca tridens Fleming, 1828 in Fleming 1828: 269
ICZN Opinion 335: *Turbo tridens* Pulteney, 1799 suppressed (for priority) against *Helix goodalli* Férussac, 1821. This must be corrected because under Art. 11.10 the name must be *Azeca tridens* Fleming, 1828.

Shell reddish, yellowish brown or white, shape perfectly ovoid including the aperture, aperture with 1-2 parietal, 1-2 columellar and 1-2 palatal folds, no umbilicus.

Deciduous forests with lower vegetation, at humid and moderately open places, in moss and ground litter, usually on calcareous soils. Colonies are usually small.

Vulnerable in Germany.

Moquin-Tandon 1855: 302, Kerney et al. 1983: 83, Gittenberger 1983 (Spain), Falkner 1990: 140, Vogt et al. 1994: 98, Kerney 1999: 80.

5-8 x 2.4-3 mm

Germany: Schleswig-Holstein, Dieksee

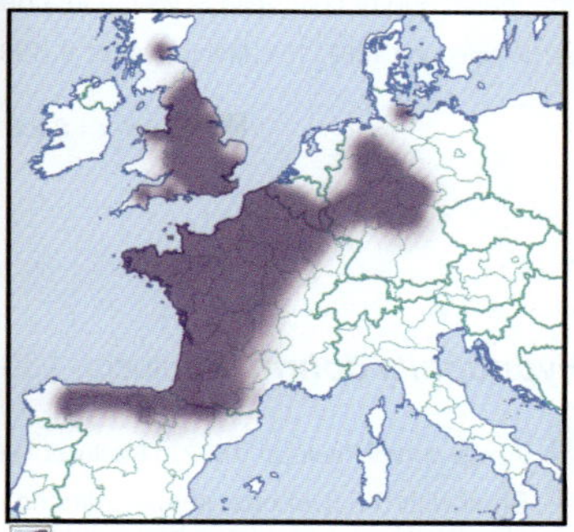

N Spain to Germany, Great Britain

Hypnophila

Hypnophila Bourguignat, 1858 in Bourguignat 1860: 102 *Azeca (Hypnophila)*

Type species *Bulimus pupaeformis* Cantraine, 1835 (subsequent designation by Letourneux & Bourguignat 1887: 114)

(1) NW Africa (Algeria to Morocco), (2) Iberian peninsula to possibly S France, (3) Toscano archipelago, M. Argentario peninsula on Italian mainland, Corsica, Sardinia, Lipari islands, Sicily, (4) Balkans from Croatia to Greece

Hypnophila bisacchii Giusti, 1970

Hypnophila bisacchii Giusti, 1970 in Giusti 1970: 68

Shell yellowish brown, 6 almost flat whorls with weak suture, last whorl height about half shell height, aperture small, margin slightly reflected, basal part less rounded and more pointed than in other species, columella inside widened (can only be seen in broken shells). Shell shape cylindrical-ovoid, the main difference to other Italian *Hypnophila* species, like *H. pupaeformis* but smaller.

6-6.2 x 2.3-2.4 mm

Italy (central E Sardinia)

Italy: Sardinia, Nuoro, Cala Gonone, Cala Fuili

Hypnophila boissii (Dupuy, 1851)

Zua boissii Dupuy, 1851 in Dupuy 1847-1852: 332, Pl. 15 (*Zua Boissii*)
Date was 1851. Spelled *Zua Boissii* on p. 332 and on the plate explanations, *Zua Boyssii* on p. 333 and p. 334, expressly dedicated to M. de Boissy.

Shell pale brownish, almost smooth, shiny, with a weak spiral sculpture, 4.5-5 whorls, aperture with weak parietal callus, columellar base truncated (points to lower section of last whorl), palatal wall of aperture slightly thickened.

Westerlund 1887: 150, Gittenberger 1983: 306.

4-5 x 1.7-1.9 mm

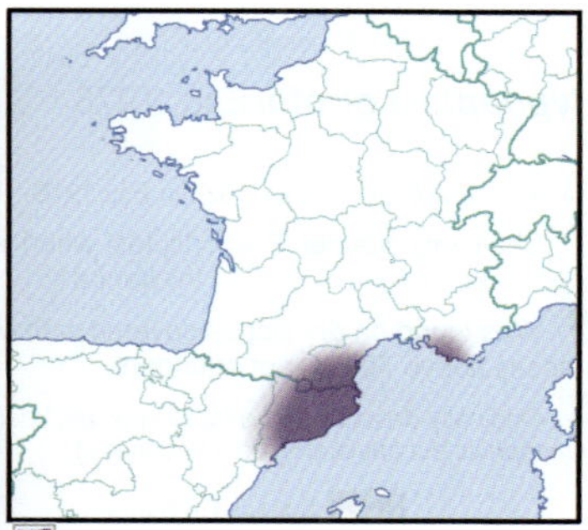

NE Spain, S France

Spain: Cataluña, Girona, near Lloret del Mar

Hypnophila cyclothyra (Boettger, 1885)

Cochlicopa cyclothyra Boettger, 1885 in O. Boettger 1885: 121

5.5-6.6 x 2.5-2.8 mm — Greece: Akarnania, near Agrinio — Museum Budapest HNHM — Greece

Aperture more rounded than in the other species, callus ring-like and more clearly protruding, lowest point of the shell situated behind the apertural lip.

Gittenberger 1993: 523.

Hypnophila cylindracea (Calcara, 1840)

Without gender agreement: *Hypnophila cylindraceus* (Calcara, 1840)

Bulimus cylindraceus Calcara, 1840 in Calcara 1840: 33

Shell nearly cylindrical, whitish, translucent, 5-6 whorls, apertural margin simple and sharp, connected at parietal side by a callus ending in an elongated angular knot, palatal margin with narrow sinulus at angular side, basal end of columella is visible in a perpendicular view.

Westerlund 1887: 151, Manganelli et al. 1995: 12.

6-7 x 1.8-2.6 mm

S Italy: Marettimo island W Sicily

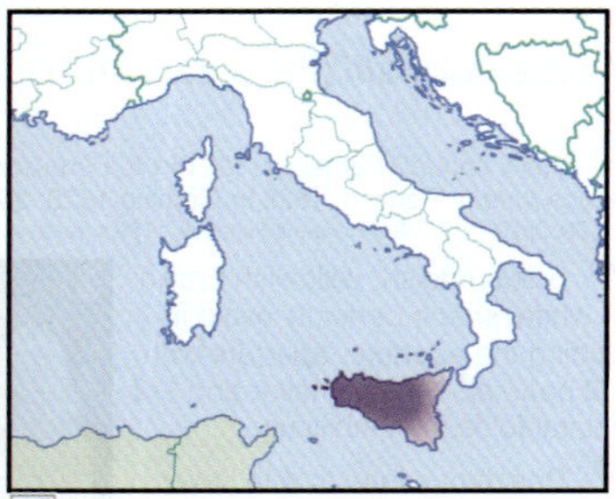

Italy (Sicily)

Hypnophila dohrni (Paulucci, 1882)

Azeca dohrni Paulucci, 1882 in Paulucci 1882: 271 (*Azeca Dohrni*)

Shell with 5-5.5 whorls, shape cylindrical, last whorl not much wider than penultimate whorl, aperture in basal section rounded, angular callus present.

Under stones and in cavities in the soil, on calcareous substrate.

Westerlund 1887: 150, Giusti 1970: 71, Manganelli et al. 1995: 12.

5.5-6.5 x 1.8 mm

Italy: Toscana, Argentario (Grosseto), Porto Ercole
Cesare Bogi

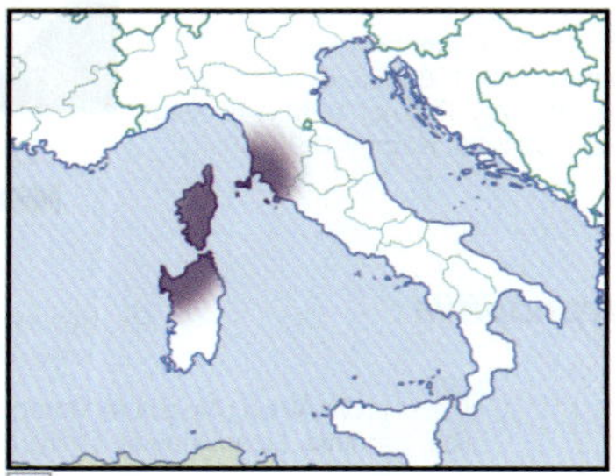

Italy/France: Toscana to Sardinia

Hypnophila emiliana (Bourguignat, 1858)

Azeca Emiliana Bourguignat, 1858 in Bourguignat 1860: 105, Pl. 14 (*Azeca (Hypnophila) Emiliana*)

Shell cylindrical with 6 whorls, whorls very high, last whorl diameter not much wider penultimate whorl, aperture with angular callus, rounded in basal section. Shell shape like *H. incerta* but larger.

Was found under stones in ancient ruins.

7-8 x 2.6-2.8 mm

Italy: Marettimo island, Trapani
Cesare Bogi

Italy: island of Marettimo W Sicily

Hypnophila girottii Esu, 1978

Hypnophila girottii Esu, 1978 in Esu 1978: 16

Shell very slender, 8 whorls, last whorl less than 50 % of shell height, palatal wall of apertural margin very slightly thickened.

Was found on dry slopes covered with Mediterranean shrub vegetation, also among grasses above 700 m, not abundant.

Originally described as a fossil species, later found alive. Anatomy unknown, classified in *Hypnophila*, but could also be a ferussaciid.

5 x 1.5-1.6 mm

Italy: Sardinia, Nuoro, Cala Gonone, Cala Fuili (H = 5 mm)
Cesare Bogi

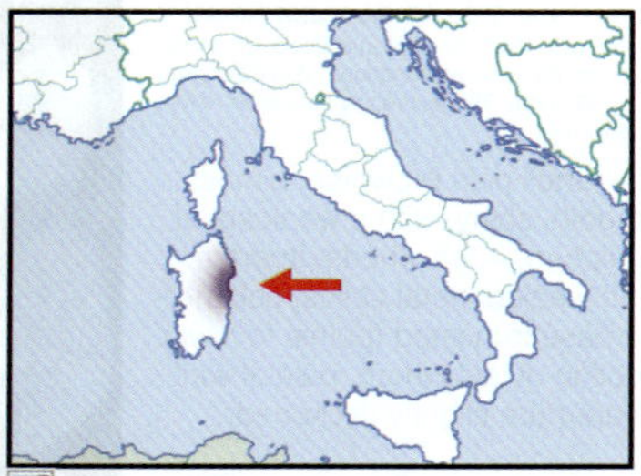

Italy (E Sardinia)

Hypnophila incerta (Bourguignat, 1858)

Azeca incerta Bourguignat, 1858 in Bourguignat 1860: 108 Pl. 14

Shell with 5-5.5 whorls, shape like *Cochlicopa*, last whorl not much wider than penultimate whorl, aperture in basal section rounded (not pointed near the columella), angular callus present.
Differs from *H. dohrni* from Sardinia in its basal apertural margin being slightly less pointed, in all other aspects both species are very similar. Smaller than *H. emiliana*.

5-6 x 1.9-2.1 mm

Almost everywhere, also in extremely difficult habitats such as at the crater margin of Stromboli, under lava layers, dry olive tree leaves, and below stones hidden under dense vergetation.

Giusti 1970: 71, Giusti 1973: 128.

Fabio Liberto

S Italy: Lipari island

Italy (Isole Eolie)

Hypnophila malagana Gittenberger & Menkhorst, 1983

Hypnophila malagana Gittenberger & Menkhorst, 1983 in Gittenberger 1983: 304
Gittenberger & Menkhorst contributed both to the description, as explained in the original source.

Shell slender, 6-7.5 whorls, aperture smaller in relation to shell height than on the other *Hypnophila* species, with prominent parietal callus, rising into a marked angularis, columella rounded and not truncated at the base, so not attaching or reclining to the lower part of the last whorl. Umbilicus closed. In juveniles of up to 5 whorls a prominent columellar callus is present, columellar base is truncate.
Differs from *H. boissii* in its angularis and the columellar callus not being attached to the last whorl.

Dry limestone rocks and rock crevices of dry slopes.

Not very frequent, had been overlooked for a long time.

4.8-6.2 x 1.9-2.3 mm

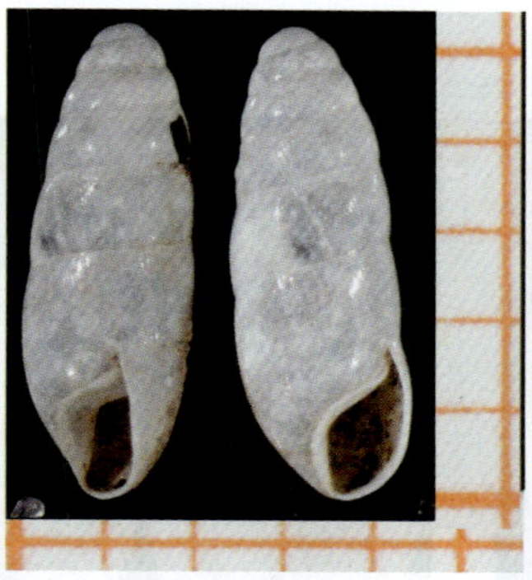

S Spain (Cádiz to Castellón)

Spain: Málaga, near Ojén

Hypnophila polita (Porro, 1838)

Without gender agreement: *Hypnophila politus* (Porro, 1838)

Bulimus politus Porro, 1838 in Porro 1838: 4, Pl. 2

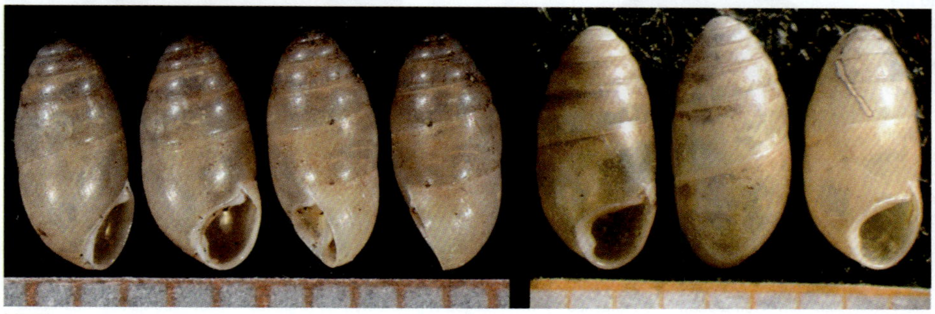

NE Greece, probably S Albania

5.6-6.3 x 2.5-2.7 mm

Greece: Kérkira, (1-4) Nisáki, (5-7) Dasia

(5-7) Massimo Prodan
(1-4) Museum Budapest HNHM

Shell spindle-shaped, tapering nearly equally strong above and below the periphery, columella with prominent denticle, more acute than in the other S Balkan species, aperture somewhat narrowed near the basis.

Gittenberger 1993: 519-521.

Hypnophila pupaeformis (Cantraine, 1835)

Bulimus pupaeformis Cantraine, 1835 in Cantraine 1835: 380 (*Bulimus pupæformis*)

Shell ovoid with 6-7 whorls, largest diameter in lower third, aperture small, usually with a weak angular callus, rounded in basal section.

Gittenberger 1993: 518, Dhora & Welter-Schultes 1996: 115.

6.5-7.6 x 2.6-3.3 mm

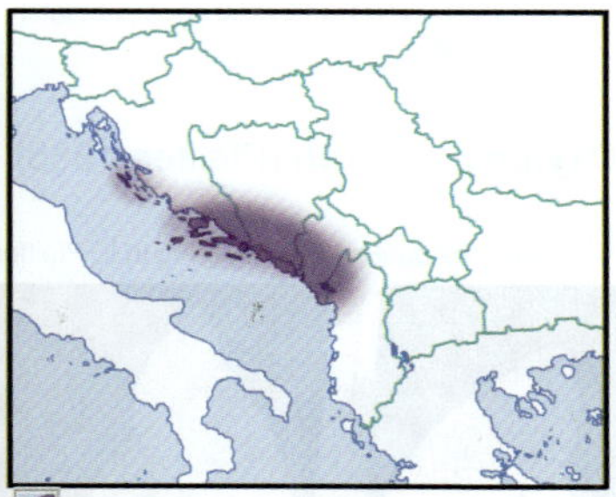

Croatia to Albania

N Albania: Malësia, near Tamarë

Hypnophila remyi (Boettger, 1949)

Azeca remyi Boettger, 1949 in C. R. Boettger 1949: 167

Shell horny coloured, shiny, high, 6-7 very weakly convex whorls, aperture simple without angular callus, apertural margin slightly thickened at the outer side.

7.5-8.5 x 2.5 mm

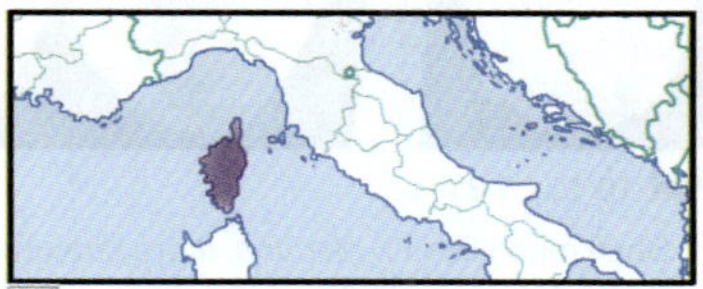

France (Corsica)

Hypnophila zacynthia (Roth, 1855)

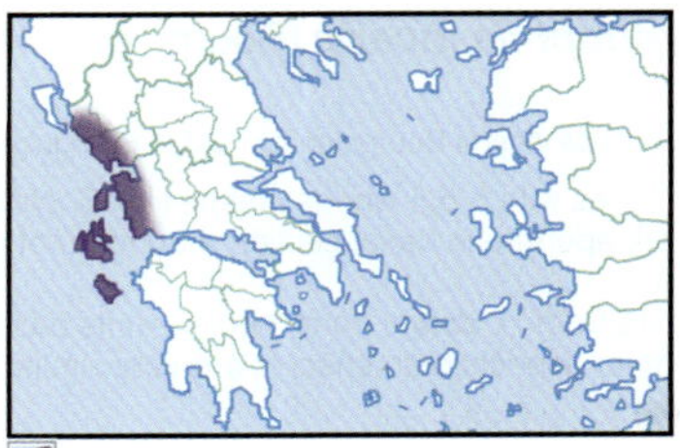

W Greece

Azeca zacynthia Roth, 1855 in Roth 1855: 39, Tab. 1

Penultimate and last whorl nearly equally wide, strongly flattened, columella with no or nearly obsolete denticle, aperture broadly rounded below.

5.2-6.6 x 2.4-2.6 mm

Various kinds of lowland habitats, at the soil, at rocks or at walls of old ruins, frequent and up to 700 m in Kefaloniá.

Rähle 1980, Gittenberger 1993: 521.

Succinea

Succinea Draparnaud, 1801 in Draparnaud 1801: 32, 55

Type species *Helix succinea* Müller, 1774 (absolute tautonymy) (= currently *Succinea putris* Linnæus, 1758)

Contained originally *Succinea amphibia* Draparnaud, 1801 and *Succinea oblonga* Draparnaud, 1801. Among the cited synonyms of *S. amphibia* the name *Helix succinea* Müller, 1774 (in Müller, 1774: 97) was listed, so type species fixed by absolute tautonymy (Art. 68.4).

Succinea putris (Linnæus, 1758)

Helix putris Linnæus, 1758 in Linnæus 1758: 774

Germany: near Kiel

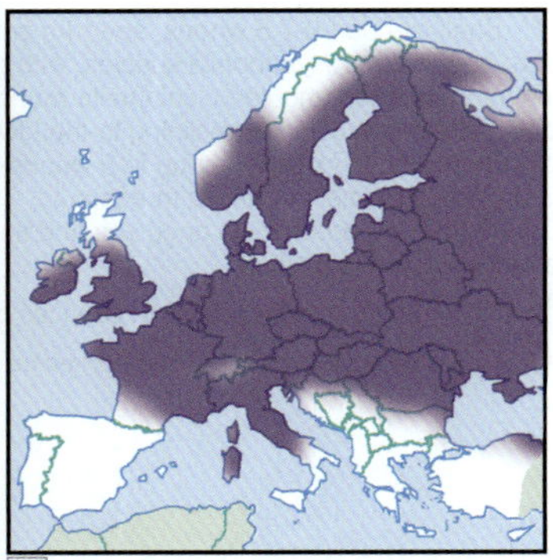

Europe, W Asia and N Asia

10-17 x 6-8 mm, maximum up to 27 mm

Shell amber yellowish with 3-4 whorls, usually slightly larger than *Oxyloma* species. Animal yellow with reddish hue, getting lighter downwards to the sides, can also be dark grey, but not covered with dark spots. Differs from *Oxyloma* in a simple, freely exposed epiphallus without penis attachment. Animal moves slowly and does not withdraw easily when touched.

Germany: N Bayern, near Bad Brückenau

Czech Republic: Olomouc — Michal Maňas

On reeds and other vegetation at water margins, on permanently humid meadows, in river plain woodlands and swampy forests. In Switzerland up to 1800 m, but generally rare in alpine zones. Often climbs erect plants. Can hibernate at considerable distances from its normal feeding grounds.

Reproduction in Poland, Netherlands and France from May to September, spermatozoa are reciprocally exchanged, elongate eggs (1 x 1.5 mm) in clutches of 20-30 (up to 90) eggs are fixed in yellowish envelopes (8 x 4 mm) to stones, aquatic plants or to the soil, individuals may lay more than 10 clutches, juveniles hatch after 10-15 days in France and Germany (at colder temperatures up to 22 days), maturity is reached after 1 year.
Intermediate host for the bird parasite *Leucochloridium macrostomum* (Trematoda). Often attacked by larval stages of predatory flies (Diptera). The semiparasitical *Pherbellia schoenherri* deposits its eggs at the suture of the snail, which defends itself partly successfully by encapsulating the larva at the columellar margin.

Gassies 1849: 69, Moquin-Tandon 1855: 55, Germain 1930: 461, Frömming 1954: 15, Jackiewicz 1980, Grossu 1987: 179, Kerney et al. 1983: 79, Turner et al. 1998: 233, Kerney 1999: 77.

Oxyloma

Oxyloma Westerlund, 1885 in Westerlund 1885: 1, 7 *Succinea (Oxyloma)*

Type species *Succinea dunkeri* Pfeiffer, 1865 (subsequent designation by Westerlund 1902: 116)

Originally included species: *dunkeri, hungarica, bourguignati, megalonyxia, servaini*. Lindholm 1927: 330 and others gave as type *Succinea hungarica* Hazay, 1880 by monotypy.

Lays up to 150 eggs per clutch, in gallert masses, like frogs.

Eurasia, N Africa

Oxyloma dunkeri (Pfeiffer, 1865)

Succinea dunkeri Pfeiffer, 1865 in L. Pfeiffer 1865: 101 (*Succinea Dunkeri*) Type locality Danube delta swamps, Romania.

S Russia: Volga delta near Mumra

Hungary to Caspian Sea

10-15 x 5-8 mm

Shell much like *O. elegans*, more straightly stretched, whorls flatter.
Distal end of penial appendix spherical. Not easy to distinguish this species from *O. elegans* anatomically.

Closely related to *O. elegans*. The typical *dunkeri* form lives in the Danube delta.

On muddy soils in periodically flooded habitats.

Van Regteren Altena 1958, Falkner 1990: 166, Sysoev & Schileyko 2009: 21.

Oxyloma elegans (Risso, 1826)

Succinea elegans Risso, 1826 in Risso 1826: 59
Succinea pfeifferi Rossmässler, 1835 in Rossmässler 1835-1837 (1): 92 (*Succinea Pfeifferi*)

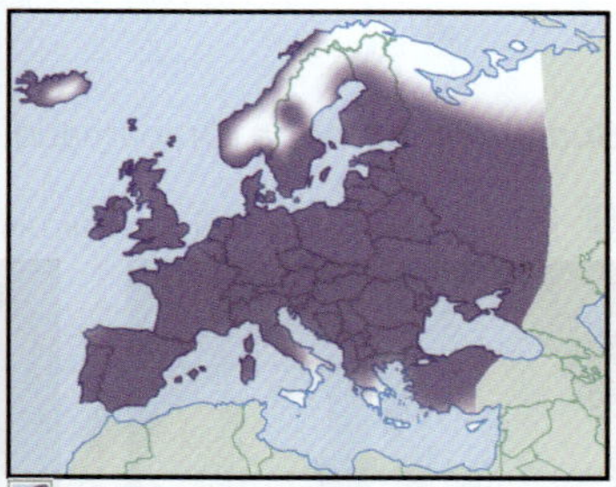

9-17 x 6-8 mm, up to 20 mm

N Germany: Fehmarn island, near Wallnau

NW Africa and Europe to N Asia

Shell yellowish amber, usually less than 3 whorls. Animal usually light with dark pigments on upper side (*Succinea* has no spots).
Epiphallus slightly curved and inside a short penis prolongation (difference to *Succinea*), distance between penis and spermatheca duct very short (difference to *O. sarsii*).

Slovakia: Slovak Karst mountains, Plešivec
Jozef Grego

Wetland of all kinds except in woods, more at the soil than *Succinea putris*, often on plants standing in the water. Tolerates periodical flooding and sparse vegetation. Feeds on fading parts of plants. Also in gardens and flower containers. In Bulgaria up to 800 m.
Intermediate host for the bird parasite *Leucochloridium macrostomum*.

Reproduction in France and Germany between May and September, clutches of 10-150 eggs (0.7-1.0 mm) are fixed at stones or aquatic plants, juveniles hatch after 8-16 days and have initially very short tentacles. Maturity is reached in the second year and clutches of 15-20 eggs are laid. In the following year the individuals reach full size in July and may lay more than 100 eggs per clutch, after that the animals die in late summer or autumn at the age of 2 years.

Moquin-Tandon 1855: 59, Germain 1930: 466 (life cycles, *S. pfeifferi*), Frömming 1954: 22.

Oxyloma pinteri Grossu, 1987

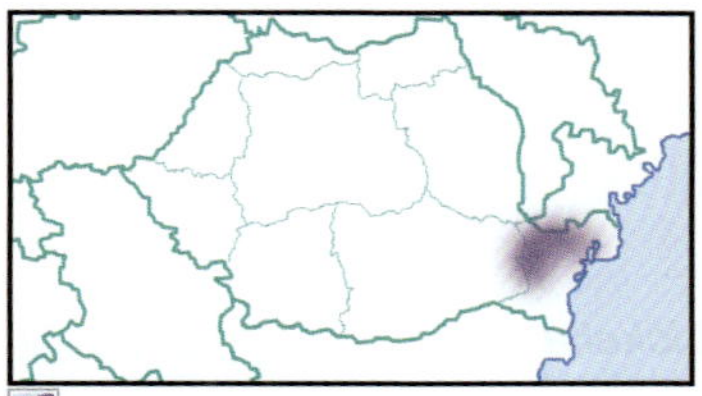

Romania (Dobrogea)

Oxyloma pinteri Grossu, 1987 in Grossu 1987: 200

Shell pale yellowish, striated, relatively slender, with 2.5 whorls, apex small and pointed, last whorl inflated, 75 % of shell height.

15-17 x 7-8 mm

In the Danube river basin in immediate vicinity of the river.

Oxyloma sarsii (Esmark, 1886)

Succinea putris var. *sarsii* Esmark, 1886 in Esmark & Hoyer 1886: 108 (*Succinea putris* var. *Sarsii*)

The name was attributed to Esmark, the description was not attributed to a particular person, so both authors were responsible for the description. Bank (2011: 10, 15) attributed the name to Esmark alone. This was in line with a strict interpretation of Art. 50.1.

10-15 x 6-9 mm, maximum up to 20 mm

Germany: Schleswig-Holstein, Aschau river

Europe to N Siberia and Yakutsk

Shell and animal like *O. elegans*, animal with dark pigments. Anatomical difference see under *O. elegans*.

Water body margins without much vegetation at rivers and lakes, particularly in periodically flooded areas. Follows water line, never far from standing water, usually crawling on water plants.

Vulnerable in England and Germany. Threatened by continuous habitat destruction and pollution. In SE England the few surviving populations in two restricted regions are all at risk from pollution, dredging or river-management schemes.

Kerney et al. 1983: 83, Falkner 1990: 166, Manganelli et al. 1995: 12, Kerney 1999: 79, Sysoev & Schileyko 2009: 22.

Reports for central Norway, W France, N Italy, Hungary, Crimea and other southern regions should be anatomically confirmed.

Succinella

Succinella Mabille, 1871
in Mabille [1871]: 82 *Succinea (Succinella)*

Type species *Succinea oblonga* Draparnaud, 1801
(subsequent designation by C. R. Boettger 1947: 190)

Contained originally *Succinea lutetiana, oblonga, humilis, arenaria* and *baudonii.*

Lays single eggs, only few eggs per clutch.

Eurasia

Succinella oblonga (Draparnaud, 1801)

Succinea oblonga Draparnaud, 1801 in Draparnaud 1801: 56

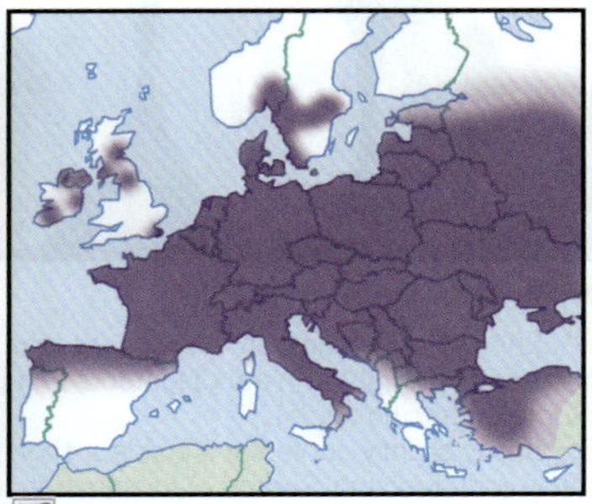

5-8 x 3.0-4.5 mm

Germany: Schleswig-Holstein, Grömitz

Europe and N Asia

Shell at marine coasts intensively amber coloured, inside the land usually pale greyish yellow, but can also be reddish amber. Smaller than *Succinea* and *Oxyloma*, 3-3.5 whorls, whorls more convex. Can be very similar to *Quickella arenaria*, embryonal whorls look more symmetric when seen from above, whorls increase less rapidly, *S. oblonga* shells of equal shell height have more whorls, usually 3 or more.
Shell is frequently covered with faeces to protect against drying out and for camouflage.
Differences to *Q. arenaria* in Switzerland mainly anatomical: long penis with epiphallus, long vas deferens and no black spot at the insertion of retractor at penis.

At the soil, often in humid and open habitats, sparsely vegetated, frequent in periodically flooded areas, damp cattle pastures, in N Portugal at a river margin, but also in relatively dry and sunny habitats. In Switzerland up to 2100 m, in Bulgaria 1300 m. Feeds on green algae and rotting parts of plants. Juveniles are easily dispersed by birds.

Eggs (diameter 0.5 mm) are laid between May and September in France and Germany, juveniles hatch after 2 weeks, maturity is reached at the end of the same year, later hatched juveniles hibernate and reach maturity in the next year, age 1-2 years.

Vulnerable in Bayern, rare in Britain. Disappears quickly after a change of the environmental conditions, but can reach new habitats relatively easy through bird dispersal. Was common in lowland England in the glacial periods, declined with forest growth, increased again with prehistoric and Roman deforestation, and declined again with increased modern habitat destruction.

Moquin-Tandon 1855: 61 (life cycle), Germain 1930: 471, Frömming 1954: 26.

Quickella

Quickella Boettger, 1939
in C. R. Boettger 1939: 50

Type species *Succinea arenaria* Potiez & Michaud, 1838
(original designation)

Epiphallus absent, vas deferens inserts directly into penis, penis without envelope.

Occasionally classified as a subgenus of *Catinella* from Hawaii and Samoa. Only one species.

Quickella arenaria (Potiez & Michaud, 1838)

Succinea arenaria Potiez & Michaud, 1838 in Potiez & Michaud 1838: 67

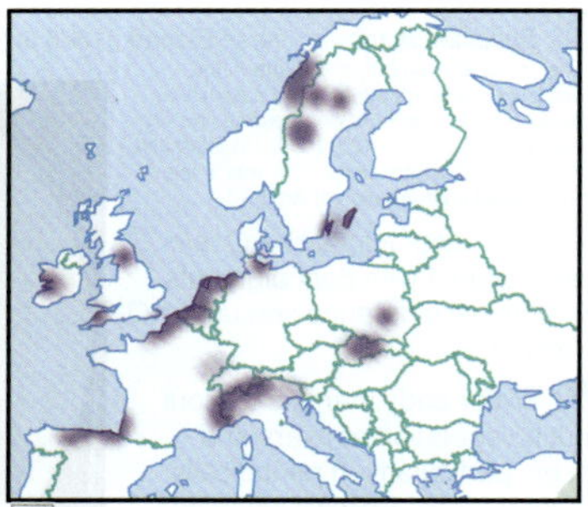

5.5-7 x 4-4.5 mm

Sweden: Öland, near Kastlösa

N Spain to Sweden and Poland

Shell can be very similar to *Succinella oblonga*, yellowish to reddish amber coloured. Embryonal whorls look more asymmetric when seen from above, they increase more rapidly than in *S. oblonga*, shells of equal shell height have less whorls. Animal evenly black on upper side, blackish grey on lower side. Penis short with a black spot at vas deferens and retractor insertion, vas deferens inserts directly into penis, epiphallus absent, penis without envelope (*S. oblonga* has a long penis with epiphallus and a long vas deferens).

In moist grassy hollows in dunes of marine coasts, or in wet and swampy habitats with small running waters (this is obligate, the water must be unaffected by deep flooding), sparsely vegetated ground, on sand or mud, calciphile (pH 7.5-8.5). In Switzerland up to 2200 m. Shell usually encrusted, animals can be found crawling directly at water current margins on humid soil or mosses, sometimes climbing short sections on grasses and herbs. At several sites in Britain the open habitat is maintained by grazing.

EU Red List: Near Threatened NT. Critically Endangered in Germany, Endangered in Switzerland, N England, Cornwall and Ireland. Scattered and rare due to its relictual distribution. Initially it was widespread at the end of the glacial period, declined after forest growth. In Switzerland and Ireland threatened by habitat destruction (drainage).

Forcart 1970: 109, Wüthrich 1984, Holyoak & Holyoak 2009.

Poiretia

Poiretia Fischer, 1883 in Fischer et al. 1887: 452 *Glandina (Poiretia)*

Type species *Bulimus algirus* Bruguière, 1792 (monotypy)
(= currently *Poiretia algira* (Bruguière, 1792) from Algeria)

Shell white to dark yellowish, slender, striated, 5-7 whorls, no umbilicus, aperture upside pointed, about 1/2 of shell height. Animal slender, brown to yellowish, sole and foot margins lighter, surface granulated, upper tentacles 2 x as long as lower tentacles. Genital organs simple and relatively uniform in all species.

Carnivorous, predators of earthworms and snails. The animal consumes also moderately large terrestrial snails such as clausiliids and *Pomatias*. *Poiretia* crawls on the shell and remains motionless on it for about two days in which an acid is secreted by a gland at the foot. The acid slowly dissolves the shell. The radula is not used for the attack.

Algeria and Italy to Caucasus

Wächtler 1927 (mode of consuming earthworms, Subai 1980 (revision, 6 species), F. Liberto in www.naturamediterraneo.com (2007, photo documentation of the mode of consuming *Pomatias*).

For a long time the name of the N African species *P. algira* was used for European populations. This genus presentation was kindly checked by P. Subai.

Poiretia compressa (Mousson, 1859)

Glandina compressa Mousson, 1859 in Mousson 1859: 21

25-38 x 8-12 mm

(1) Greece: Zákinthos, (2-9) Albania: Gjirokastër

(1) Panagiotis Helmis

W Greece, S Albania
In Libya probably introduced.

Subai 1980: 164, Rähle 1980: 216, Dhora & Welter-Schultes 1996: 139.

Shell white with yellowish hue, the most slender of all *Poiretia* species, finely striated, first whorls more prominently ribbed than in *P. cornea*, up to 5.5 whorls, not convex, suture without fathom-like margin, aperture narrow (width 5-7 mm), columella almost straight, not curved.

In all kinds of lowland habitats, in Kefaloniá not above 500 m.

Poiretia cornea (Brumati, 1838)

Achatina cornea Brumati, 1838 in Brumati 1838: 35

Croatia: Dalmatia, Brač island

Jozef Grego

Open habitats with shrubs, and forests.

32-47 x 10-17 mm

Albania: Shkodër

NE Italy to S Albania and W Serbia; Gargano peninsula (Italy), Libya

Shell light dirty yellowish, thin, finely striated, more slender than *P. dilatata*, 5.5-6.5 whorls, not very convex (less than *P. dilatata*), suture with fathom-like margin, aperture narrow (width 5.5-9.5 mm), columella not much curved.

Pavlović 1912: 18, Tomić 1959: 8, Subai 1980: 158, Falkner 1990: 168, Manganelli et al. 1995: 23, Štamol & V. 1995 (Croatia: Cres, Lošinj), Dhora & W. 1996: 139.

Poiretia delesserti (Bourguignat, 1852)

Bulimus delesserti Bourguignat, 1852 in Bourguignat 1852: 17 (*Bulimus Delesserti*)

Greece: Ipiros, Ioánnina

Cédric Audibert

34-38 x 13-15 mm

Albania: near Tepëlenë

NW Greece to central Albania

Shell light dirty yellow, solid, not slender, very finely ribbed, apex narrower than in Italian *P. ditalata*, up to more than 6 convex whorls, suture with fathom-like margin, aperture regularly wide (6.5-8 mm), columella strongly curved at the base.

Open habitats with shrubs, and forests.

Poiretia dilatata (Philippi, 1836)

Without gender agreement: *Poiretia dilatatus* (Philippi, 1836)

Achatina algira var. ß synonym *Polyphemus dilatatus* Philippi, 1836 in Philippi 1836: 141
Mentioned with description. Not mentioned as a junior synonym of an available name (as demanded by Art. 11.6), but as a synonym for a subordinate variant for which the Greek letter ß was used. For those who neglect this (as proposed by Kadolsky 2012: 78, this is preliminarily accepted here), the name was made available under Art. 11.6.1 by Beck 1837: 78 who used the name as *Achatina (Glandina) dilatata*, with a nearly bibliographical reference "Polyphemus dilatatus Zigl. Chr et J! - Achatina algira var. ß Philp.". For the others, the name must be attributed to Beck 1837.

Shell yellowish with greenish or brownish hue, 6-7 whorls, more convex than in *P. compressa*, suture with fathom-like margin, apertural width regular (5.5-9 mm).

In a broad variety of habitats, rocky limestone outcrops with scarce vegetation in mountains, but also in orange plantations and densely vegetated aeas in flatlands.

Alzona 1971: 115, Subai 1980, Maassen 1991: 35, Vardinoyannis (1994, unpublished thesis), unpublished data (HNC).

(1-3) Greece: W central Crete, near Sélia, (4-7) Italy: Sicily, near Segesta

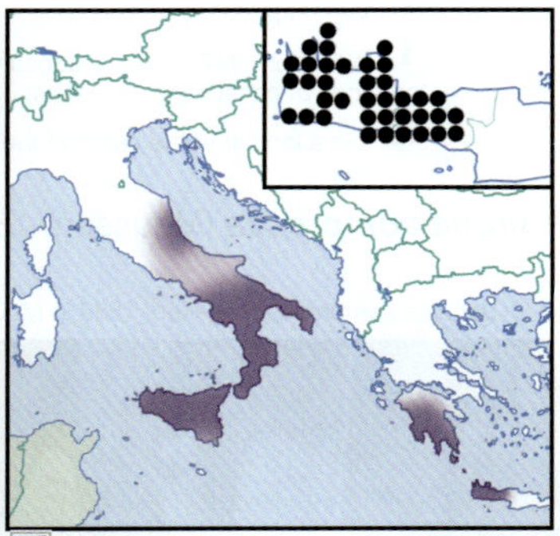

30-46 x 10-17 mm

Italy, SW Greece

The classification by Subai 1980 was based on the morphological record. The populations from Italy and Greece could belong to separate species.

Columella

Columella Westerlund, 1878
in Westerlund 1876-1878: 193 *Pupa (Columella)*

Type species *Pupa inornata* Michaud, 1831 (monotypy)
(= currently *Columella edentula* (Draparnaud, 1805))

Shell dextral, up to 7.5 whorls. Shape cylindrical, rarely conical. Aperture simple, margin sharp. Shells up to 4 mm high, diameter up to 2.3 mm. Spermatheca duct several times longer than the spermatheca, penial retractor inserted on the spermiduct or below the penis tip.

Pokryszko 1990.

Columella aspera Waldén, 1966

Columella aspera Waldén, 1966 in Waldén 1966: 53

Germany: Mecklenburg-Vorpommern, Kampenwerder near Schaalsee

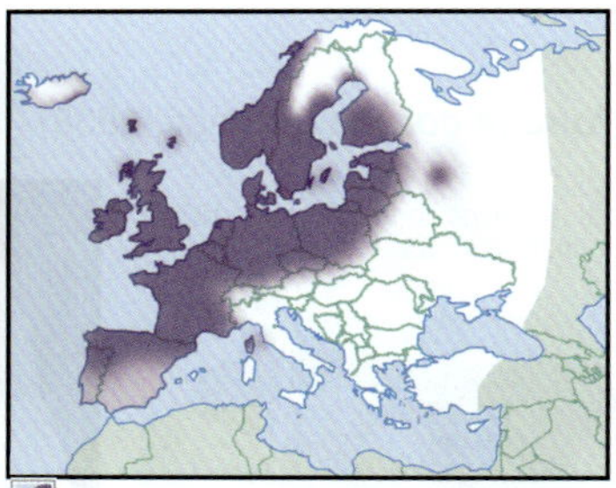

2.0-2.5 x 1.3-1.4 mm

Palearctic

Shell dark greyish brown or horny brown, not shiny, finely and regularly striated, 4-5 whorls, aperture simple without teeth. Often slightly darker than *C. edentula*.

Forests and poor uncultivated grasslands, often montane, relatively dry habitats. Avoids calcareous habitats, can survive on acid soils covered by ferns, where usually no other shelled molluscs can survive. In N Poland only in forests or in shady habitats near forest margins. May co-occur in some overlapping zones with *C. edentula*, which occupies the lower habitat, closer to the water body.

Reproduction in Poland between June and September, very few single oval and uncalcified eggs (0.7-0.85 mm long) are laid at rainy days, on the concave upper side of moss leaves which attach to the stems in periods of lower humidity, juveniles hatch after 20-30 days (22-17 °C), they grow slowly and most juveniles hibernate to reach maturity in the next year, growth continues more slowly till death, maximum age up to 3 years.

Myzyk 2011: 77.

Columella columella ([Martens], 1830)

Pupa columella [Martens], 1830 in [Martens] 1830: 171 (*Pupa Columella*)
Name originally attributed to Benz, author's name not given in the journal volume.

N Sweden: Torne-Lappmark, near Kopparåsen

2.65-4.0 x 1.25-1.5 mm

N Eurasia

Shell brown, shiny with weak growth lines, up to 7.5 whorls, almost cylindrical, penultimate and antepenultimate whorls of equal size, last whorl often significantly broader. Larger than *C. edentula*, apex flatter.

In marshy grasslands and wet subarctic woodlands, grey willow thickets, alpine meadows, on calcareous substrate. In Switzerland between 1000 and 2900 m.

Critically Endangered in Bayern, Endangered in Vorarlberg and Tirol. No reliable recent records for Baden-Württemberg, possibly extinct there.

Columella edentula (Draparnaud, 1805)

Pupa edentula Draparnaud, 1805 in Draparnaud 1805: 59, Pl. 3

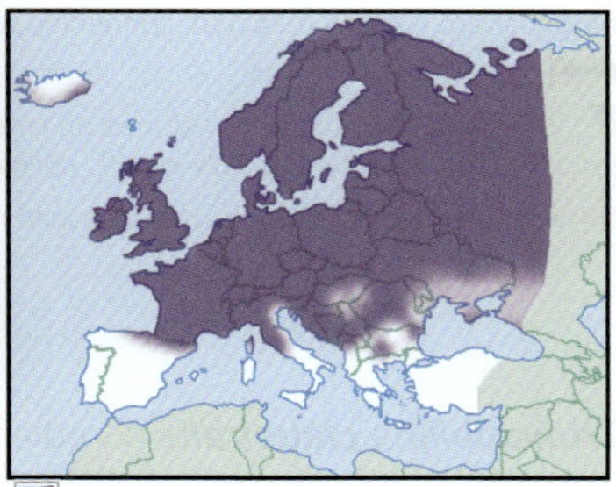

2.5-3 x 1.3-1.5 mm

Germany: Mecklenburg-Vorpommern, Steinort near Schaalsee

Holarctic

Shell yellowish brown, shiny, finely striated, cylindrical with conical apex, 4.5-6 moderately convex whorls, last whorl with largest diameter, aperture fragile without teeth. Apex more pointed than in *C. columella*. Usually lighter than *C. aspera*.

Very moist to wet habitats on calcareous substrate. Forests, alder swamps and open habitats, most abundant in relatively humid, calcareous places in lowlands. In the summer found on plants, later in leaf and plant litter. Does not tolerate short periods of drought, very sensitive against lowering of groundwater levels (for Britain this has not been reported, seems to live on meadows and in forests there). Snails can also feed at 10 °C daytime temperature in the winter. May co-occur in some overlapping zones with *C. aspera* and occupies the lower habitat, closer to the water body. In Switzerland up to 2300 m, in Bulgaria 1500 m. Difficult to rear in captivity.

Reproduction in Poland between June and September, few single oval and uncalcified eggs (0.7-0.8 mm long) are laid, juveniles hatch after some 20 days, they grow slowly and most hibernate with 2-5 whorls to reach maturity in the next year, growth continues till death, maximum age perhaps 3 years.

Myzyk 2011: 79.

Truncatellina

Truncatellina Lowe, 1852 in Lowe 1852: 275 *Pupa (Truncatellina)*

Type species *Pupa linearis* Lowe, 1852 (monotypy)
(= currently *Truncatellina linearis* (Lowe, 1852) from Madeira)

Lowe 1855: 207 gave *Pupa minutissima* as type, incorrectly.

Africa, Europe to central Asia

Pokryszko 1990.

Truncatellina arcyensis Klemm, 1943

Truncatellina arcyensis Klemm, 1943 in Klemm 1943: 100

Shell yellowish-brown, not shiny, no ribs (only irregular growth lines), no teeth in the aperture.

Meadows on stony and relatively dry limestone slopes of hills.

EU Red List: Endangered EN. Only known from 3 localities. Threatened by changes in agricultural practices, provoking changes in the vegetation.

2.2-2.5 x 0.9 mm

Central France (Yonne)

France: Arcy sur Cure, N Avalon

The identity of this form as a separate species should be verified by molecular methods.

Truncatellina callicratis (Scacchi, 1833)

Turbo callicratis Scacchi, 1833 in Scacchi 1833: 11 (*Turbo Callicratis*)
Truncatellina beckmanni Quintana Cardona, 2010 in Quintana Cardona 2010: 153

Algeria, S Europe to central Asia
Probably with large range gaps, not reported from Turkey and Ukraine. Present in the Caucasus region and central Asian mountains.

1.6-2.2 x 0.7-0.9 mm

Italy: Lago di Garda, near Torbole

Shell ribs weaker und less regular than in *T. cylindrica*, sometimes the shell is even smooth, aperture with 3 teeth (but can be reduced, generally weaker in British specimens), palatal tooth in the aperture well visible.

Very dry habitats on calcareous meadows, open and rocky hill slopes. In S England also on cliffs and among old quarries. In Switzerland up to 2200 m.

A rare species. Critically Endangered in Germany, rare and extremely scattered in Austria, rare in S England, Vulnerable in N Switzerland.

T. beckmanni from Menorca was claimed to differ from *T. callicratis* in its shell height (populations in Menorca ø 1.40 mm, Ibiza ø 1.70, Iberian peninsula ø 1.77). Besides of the problematic procedure to measure various different populations from these regions and lump their values, it was not considered that Mediterranean *Truncatellina* populations are known to vary considerably in their mean shell height over very short distances. In two populations of *T. rothi* collected at 1 km distance and at the same date on the small Greek island of Gávdos I obtained mean shell height values of 1.40 mm and 1.56 mm (Welter-Schultes 1998: 29). It would be necessary to test if these differences could be environmentally induced. This seems likely because in a shell without predetermined end of growth the final size can be determined by external factors.

Truncatellina cameroni Triantis & Pokryszko, 2004

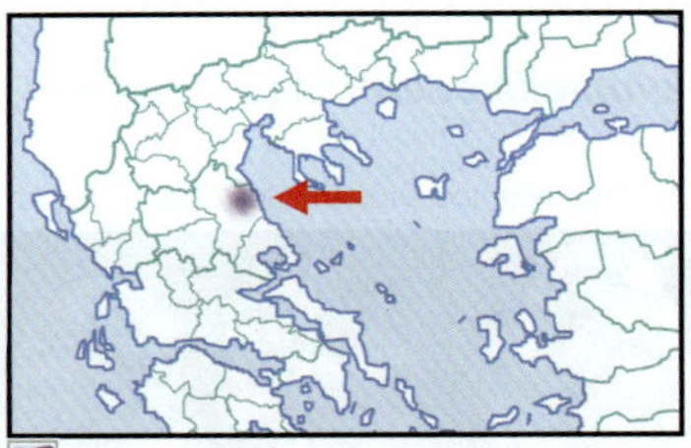

Greece (Mt. Ossa)

Truncatellina cameroni Triantis & Pokryszko, 2004 in Triantis, Pokryszko, Vardinoyannis & Mylonas 2004: 393
Name originally attributed to Triantis & Pokryszko, co-authors of the work and thus authors of the name under Art. 50.1.

Shell finely and densely striated but not ribbed, with 5.5-6 moderately convex whorls, aperture without teeth, margin sharp, slightly reflected, connected at parietal side by a thin whitish layer. Largest species of *Truncatellina*.

2.8-3.3 x 1.6-1.75 mm

In ground litter and under stones on limestone substrate, in open forests and alpine shrubland, between 700 and 2000 m. Lives syntopically with *T. callicratis* and *T. cylindrica*.

Known as a species with a small restricted range, but no threats are known. Probably more widespread.

Truncatellina claustralis (Gredler, 1856)

Pupa claustralis Gredler, 1856 in Gredler 1856: 116, Taf. 2
Pupa opisthodon Reinhardt, 1879 in Reinhardt 1879: 138

1.5-1.8 x 0.75 mm

Hungary: Bükk mountains, Szalajka valley

S Europe to Caucasus
Very scattered and rare in the Alps.

Shell smaller than the other species, very light horny yellowish, very finely ribbed, characteristically shiny, last whorl slightly flattened at the periphery, 3 teeth, in a perpendicular view only parietal and columellar tooth visible, not palatal, palatal tooth is deeper in the aperture and visible from outside as a white spot.

Dry habitats on limestone hills and among rocks and rock rubble, characteristic in *Artemisia* and *Helianthemum* stocks. In the Alps also on S exposed meadows on rock rubble. Up to 1000 m in the Alps, Spain and Bulgaria.

Vulnerable in Austria and Switzerland, rare in E Spain.

Wagner 1928: 44, Damjanov & Likharev 1975: 126, Kerney et al. 1983: 89, Grossu 1987: 314, Manganelli et al. 1995: 13, Turner et al. 1998: 183, Schütt 2001: 77, Martínez-Ortí & Robles 2003: 84, Ferreri et al. 2005: 51, Štamol & Kletečki 2005, Hubenov 2007, Sysoev & Schileyko 2009: 54.

The taxon *opisthodon* from SW Romania and Bulgaria had almost the same diagnosis and size as *claustralis*. Wagner 1927: 44 and Grossu 1987 recognised it as a separate species without explaining why this form falls outside the known variability of *T. claustralis*. Such a statement would be necessary to have an indication why this would represent a separate species. It rather seems that we are dealing with marginal populations of the widespread *T. claustralis* which does not seem to occur in central Romania.

Truncatellina costulata (Nilsson, 1823)

Pupa costulata Nilsson, 1823 in Nilsson 1822: 51

1.6-1.95 x 0.5-0.6 mm

Sweden: Öland, Hulterstad

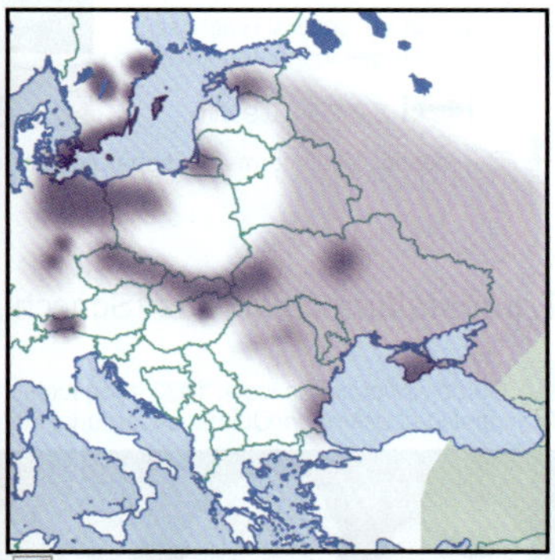

Europe to Caucasus

Shell ribs more widely spaced that in *T. cylindrica* and *T. callicratis*, adult shell with white and prominent apertural margin, parietal tooth well visible, palatal and columellar teeth present but not well visible.

Dry habitats: dry grasslands, sand dunes, open woodlands, roadside sunny tree stands.

Endangered in Germany.

Damjanov & Likharev 1975: 129, Kerney et al. 1983: 89, Jungbluth et al. 1989: 95, Pokryszko 1990: 216, Jungbluth & Knorre 2009: 11, Sysoev & Schileyko 2009: 54 (gave a general statament that the species occurs between Estonia and Ulyanovsk), Balashov & Gural-Sverlova 2012: 95.

Range insufficiently known. Partly the large range gaps seem to be real (E Poland and Latvia should have some records), partly they are due to the weak collection record (no reliable data published for Russia and Belarus).

Truncatellina cylindrica (Férussac, 1807)

Vertigo cylindrica Férussac, 1807 in J. Férussac 1807: 52

I am giving Férussac's single authorship here, as this has been traditionally done. I have not researched which one of the two Férussac's was responsible for this name (or both). It seems to have been overlooked that both Férussacs were co-authors of the 1807 work which should better be cited as "Férussac & Férussac 1807", and that the responsibilities were explained in the introduction.

Italy: Sicily, Latornia del Paradiso near Siracusa

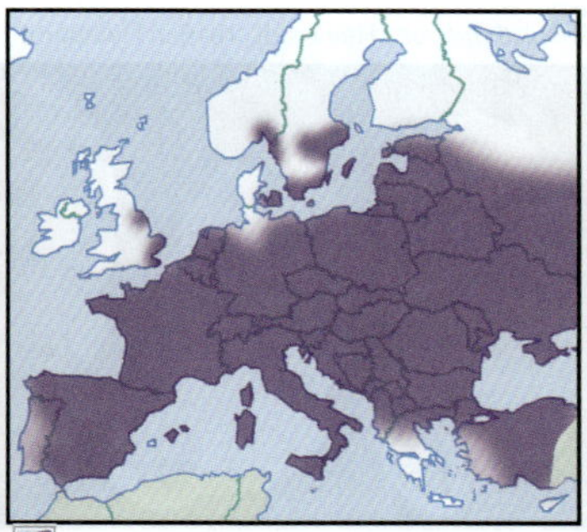

Europe to central Russia, Caucasus and NW Africa

1.6-2 x 0.82-0.90 mm

Myzyk 2005: 174, Myzyk 2011: 76.

Shell golden horny brownish, not shiny, ribbed, spaces between the ribs approximately equal in width than the ribs, shell appears slightly conical, 5-5.5 whorls, last whorl with largest diameter, penultimate and antepenultimate whorls only 97-98 % of the diameter of the last whorl, aperture without teeth, lip not much reflected. In Poland winter growth interruptions are visible on most shells as light streaks and growth disturbances.

Sunny dry and moderately dry hills on calcareous substrate, grassland on sandy and stony ground, plant covers on limestone rocks, in the south also in light forests. In central Europe and England typically under *Sedum* and *Artemisia*. In Switzerland often at S exposed dry and warm limestone rocks, in up to 2800 m altitude, in Bulgaria up to 1900 m.

Reproduction in Poland between June and September, 4-11 single spherical or lenticular eggs (0.5-0.6 mm, with opaque calcified envelope) are laid per individual in a season, the calcified envelope helps embryos to survive droughts, juveniles hatch after 14-20 days (24-20°C), 30 % reach maturity in the same year but most juveniles hibernate with 1-4 whorls and reach maturity in the next year, adults usually die after reproduction at the age of 1 year, maximum age up to 2 years.

Extinct in Ireland and probably in Scotland, Endangered in Vorarlberg and Niedersachsen, Vulnerable in England and Germany. Threatened by continuous habitat destruction, most populations in Britain have disappeared, attempts to refind old sites failed frequently.

Truncatellina laeviuscula (Küster, 1850)

Pupa minutissima var. A synonym *Pupa laeviuscula* Küster, 1850 in Küster 1852: 101, Taf. 14

Mentioned with description and figure. Not mentioned as a junior synonym of an available name (as demanded by Art. 11.6), but as a synonym for a subordinate variant for which the letter A was used. For those who neglect this (as proposed by Kadolsky 2012: 78, this is preliminarily accepted here), the name was made available under Art. 11.6.1 by Westerlund 1887: 129 who used the name as *Pupa minutissima* forma *læviuscula*, with bibliographical reference to Küster 1852: Taf. 14 Fig. 6-8. For the others, the name must be attributed to Westerlund 1887.

NE Italy (Servola near Trieste)

Shell horny yellowish, weakly striated or almost smooth, shiny, 5-6 moderately convex whorls, apertural margin sharp, white, no teeth visible in the aperture, umbilicus narrow.

2 x 0.7 mm

Was found at a grassy mountain slope.

More research seems to be necessary about this form.

Truncatellina lussinensis Štamol, 1995

Truncatellina lussinensis Štamol, 1995 in Štamol 1995: 99

Shell with variably but generally widely spaced ribs, 6-7.5 moderately convex whorls, aperture usually without teeth, rarely with a deep palatal lamella (not visible in the aperture).
Differs from sympatric *T. callicratis* in the lack of teeth in the aperture (*T. callicratis* has 3 teeth).

Found in pine woods (*Pinus halepensis*), sympatrically with *T. callicratis* but in much smaller numbers (2-3 specimens of *T. lussinensis* are found with 30-40 specimens of *T. callicratis*).

Štamol 1995 did neither provide evidence that *callicratis* and *lussinensis* were separate species or mophological units, nor that the morphological background of the involved forms was fully understood.
Massimo Prodan studied *Truncatellina* on various islands and observed that only on Lošinj *lussinensis*-like forms without teeth occurred, and that the morphological parameters of *lussinensis* and local *callicratis* did not overlap (e-mail, Sep 2009). This observation supports the presence of a separate species, but a scientific study would be desirable.

Croatia: Lošinj island, Veli Losinj

1.9-2.8 x 0.8-1.0 mm

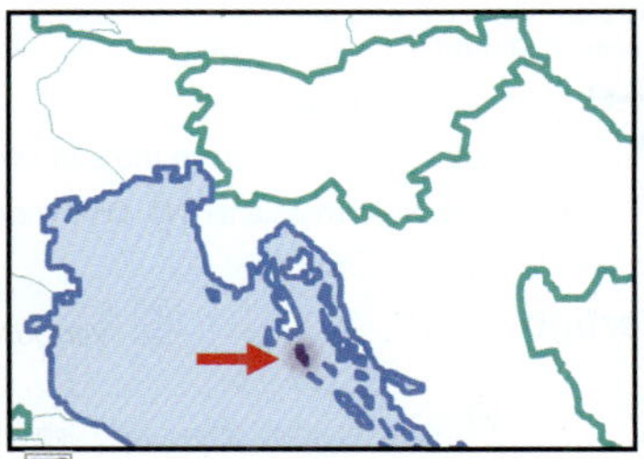

Croatia (island of Lošinj)

EU Red List: Vulnerable VU. Rare, but status possibly exaggerated. Potentially threatened by tourism. Not found on other islands in equivalent habitats.

Massimo Prodan

Truncatellina monodon (Held, 1837)

Vertigo monodon Held, 1837 in Held 1837: 304

Austria: Kärnten, Kupitzklamm near Eisenkappl

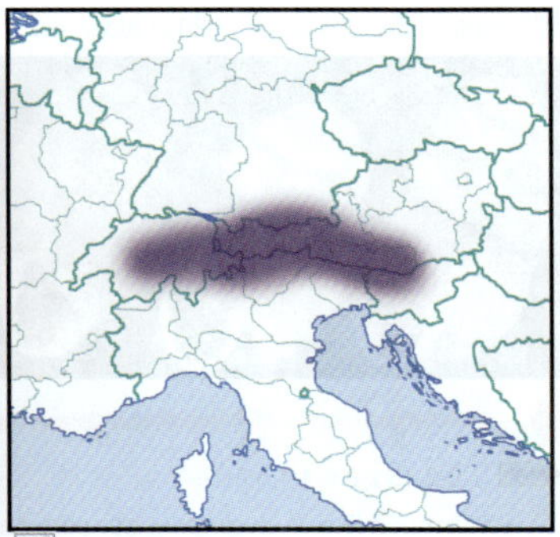

Alps

2.2-2.3 x 0.9 mm

Kerney et al. 1983: 90.

Shell brown, no ribs, only fine, dense and irregular growth lines, 3 teeth, of which only the parietal tooth is visible.

Meadows on limestone and rock rubble, dry mountain slopes, in valley slopes covered by vegetation, also at limestone rocks in humid forests. In the Alps with wider ecological amplitude than the other *Truncatellinas*. In Switzerland up to 2500 m.

Vulnerable in Switzerland, rare in Bayern and Vorarlberg.

Truncatellina rothi (Reinhardt, 1916)

Isthmia rothi Reinhardt, 1916 in Reinhardt 1916: 164 (*Isthmia Rothi*)

Greece: Lefkáda island

Albania to S Greece and Turkey

Kirchner et al. 1997, Welter-Schultes 1998: 28.

1.2-1.8 x 0.8-0.9 mm

Shell like *T. cylindrica*, usually more coarsely ribbed, spaces between ribs wider than the ribs, whorls usually more convex, diameter of last 3 whorls equal (last whorl not larger). Shell height varies locally and within a population, the diameter varies at a much lower amplitude.

Locally frequent, but often overlooked. In Gávdos island, which was systematically screened for minute species, it was present all over the island, usually in organic compounds on the soil under pine and *Juniperus phoenicea* trees and lower shrubs, on limestone or sandy substrate. Rare under *Juniperus macrocarpa* or *Pistacia lentiscus*, which have bigger leaves. Not frequent in more shady habitats, seems to require high temperatures, frequent at southern slopes. Can be blown by the wind over surprisingly long distances.

Range not well known, more research on local shell variations would be useful.

Truncatellina uniarmata (Küster, 1856)

Pupa uniarmata Küster, 1856 in Küster 1856: 77

Shell horny yellow, sharply ribbed, whorls strongly convex, one palatal tooth in the aperture, columella thickened, almost with umbilicus.

2 x 0.7-0.8 mm

Pfeiffer 1877: 619, Westerlund 1887: 128.

More research seems to be necessary. Its original locality in or near Trieste was so inaccurately recorded that it has not been found again in the past 100 years. It is not even possible to say in which country it lives.

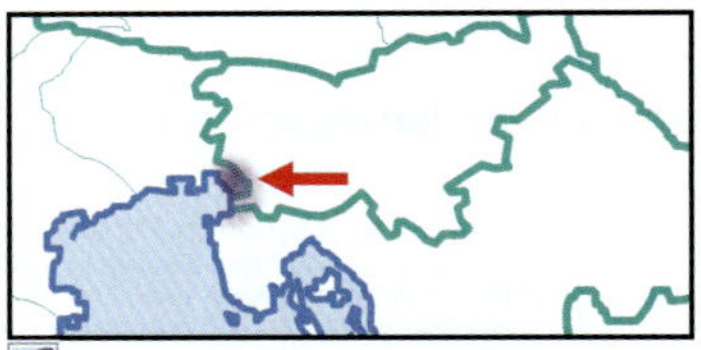

Italy or Slovenia or Croatia (near Trieste)

Truncatellina velkovrhi Štamol, 1995

Truncatellina velkovrhi Štamol, 1995 in Štamol 1995: 97

Shell finely striated, shiny, 5-6 slightly convex whorls, no parietalis, columellaris present but not always visivle from the aperture, palatal tooth visible in a perpendicular view of the aperture. Differs from sympatric *T. callicratis* in its smaller size and shiny shell surface, *T. callicratis* is strongly ribbed. *T. monodon* is considerably larger (diameter 0.9 mm).

1.3-1.75 x 0.65-0.75 mm

Found in a vertical cliff above the sea shore, partly overgrown with grassy vegetation.

EU Red List: Near Threatened NT. Only known from two very small islands. No threats currently known.

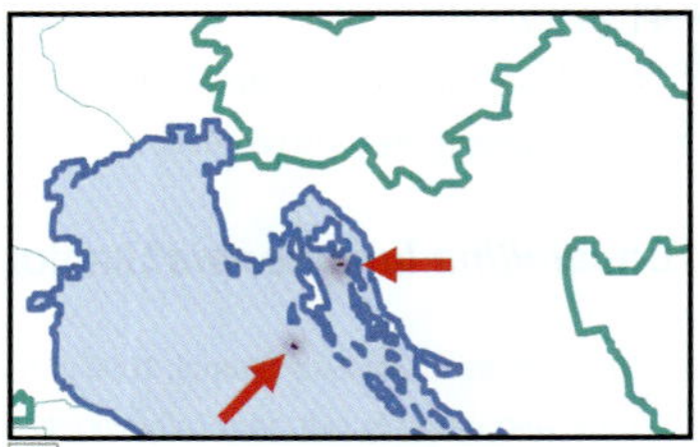

Croatia (islands of Sv. Grgur and Susak)

Vertigo

Vertigo Müller, 1773 in Müller 1773: [25]

Type species *Vertigo pusilla* Müller, 1774 (subsequent monotypy by Müller 1774: 124)

Appeared without species. ICZN Opinion 335: type species by monotypy. This should be subsequent monotypy.

Reproduction was impressively well studied in many species (Myzyk 2011 and previous works). As a general comment, the average number of eggs per season and the breeding periods show considerable variation. Some snails may lay their last egg in the season in May, others of the same population in October. In most species egg-laying takes place every 1-3 days, but individuals can interrupt this for up to several weeks. The numbers of eggs laid per individual can vary between 10 and 120 in a population.

Northern Hemisphere

Pokryszko 1990 (Poland), Pokryszko 2003, Proschwitz 2003 (Scandinavia), Myzyk 2005 (egg sizes and structure).

Vertigo alpestris Alder, 1838

Vertigo alpestris Alder, 1838 in Alder 1838: 340

Slovakia: Slovenský Raj, Vely Sokol

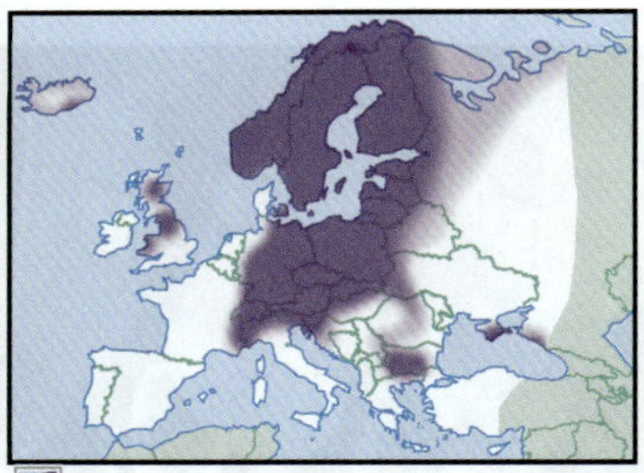

1.6-2.15 x 0.9-1.1 mm

Europe, N Asia, islands off Newfoundland

Shell much like *V. pygmaea*, but cervical crest absent.

Lowlands and lower mountain forests on calcareous substrate, often at mountain slopes covered by woods with loose boulders and rich accumulations of leaf litter, also among moss. In high mountains in sunny and open rocky habitats, open dry forests and overgrown rock rubble below limestone rocks. Occurs also above the timberline (in Poland up to 2000 m, Switzerland 2500 m, Bulgaria 2300 m). In England and Wales almost exclusively at old stone walls, in Scotland among mosses in limestone rock rubble.

Critically Endangered in Germany, Vulnerable in Vorarlberg.

Proschwitz 2003: 29, Proschwitz 2004.

Vertigo angustior Jeffreys, 1830

Vertigo angustior Jeffreys, 1830 in Jeffreys 1830: 361

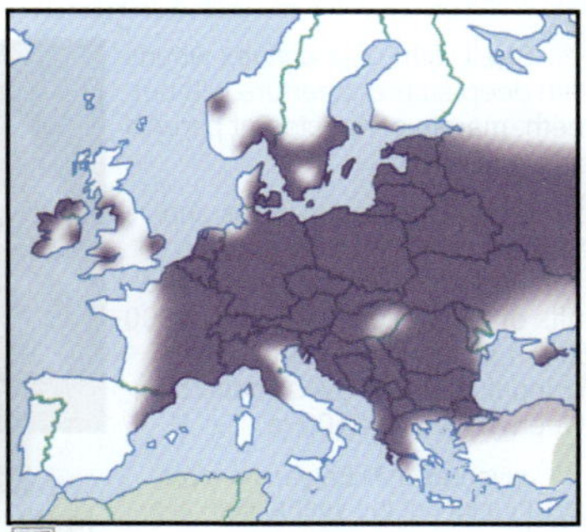

1.5-1.9 x 0.9-1.0 mm

Germany: Schleswig-Holstein, Aassee near Eckernförde

W Eurasia to N Iran

Shell sinistral, smaller than *V. pusilla*, with strongly developed cervical crest.

Calciphile, habitat preferences vary with climate. In central Europe in habitats with constantly high humidity, in Scandinavia in open dry habitats and sandy dunes in coastlands with low vegetation cover, in Britain restricted to permanently moist open habitats with short grasses, herbs or mosses quickly warmed by the sun. Found in litter, in summer at bases of stems and leaves. Usually up to 1150 m in Switzerland, occasionally 1900 m.

Juveniles are present at all times between March and October in Wales, reproductive productivity depends on weather conditions (delayed at dry weather), egg size 0.55-0.65 mm, egg-to-egg generation time 60-70 days. In Poland one egg (0.5-0.65 mm) per day is laid on wet grasses and leaves on grass between May and June, at longer intervals until September, occasionally also in winter, 20-70 eggs per individual in a season. Juveniles hatch after 11-16 days (23-19°C), full size with complete aperture is reached after 40-55 days of activity (no activity at low air humidity), maturity is usually reached in June-August, life span up to 2 years, snails die about 10 days after having laid the last egg.

EU Red List: Near Threatened NT. Critically Endangered in Spain (recently known from only two localities, one of which was destroyed). Endangered in S Switzerland, Bayern, Niedersachsen and Great Britain, Vulnerable in Ireland, Germany and Switzerland.
A rare and declining species, everywhere very local. Threatened by drainage, lowering of groundwater levels, construction projects or other changes in land use, and fertilization. Many populations were destroyed in the last years.

Pokryszko 1990: 211, Cameron 2003: 98, Martínez-Ortí & Robles 2003: 90, Myzyk 2005, Boschi 2011: 303, Myzyk 2011: 57.

Vertigo antivertigo (Draparnaud, 1801)

Pupa antivertigo Draparnaud, 1801 in Draparnaud 1801: 57 (*Pupa anti-vertigo*)

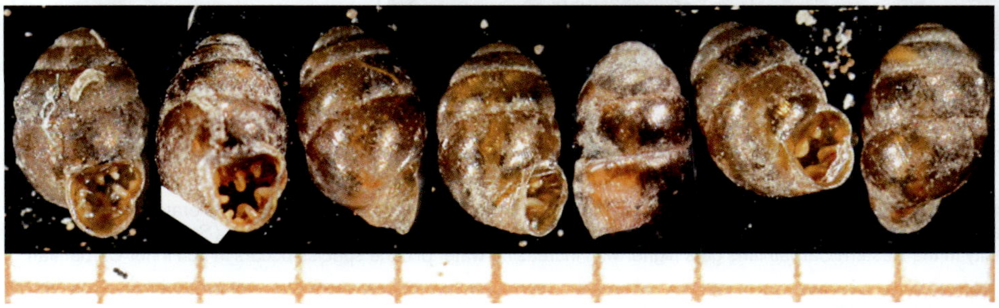

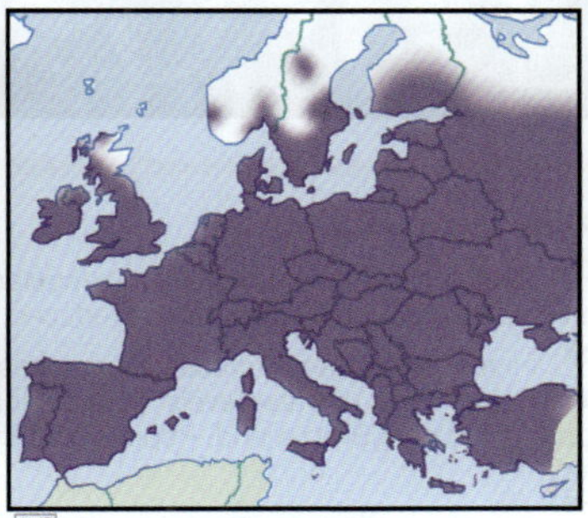

1.7-2.3 x 1.2-1.4 mm

N Germany: Ostfriesland, Leda

W Europe to Transbaical region

Shell ovate, ovate-cylindrical, exceptionally cylindrical, aperture with 1-11 teeth, exceptionally simple. Columellar tooth in shape of a fold perpendicular to the columella or knob-shaped. Differs from *V. moulinsiana* in the heart-shaped aperture with its basal margin much more narrowly reflexed, smaller size, and larger and stronger teeth.

Hygrophile, in swampy meadows, river and lake margins, found even in regularly flooded areas. Never in places drying up for a part of the year. Mostly among rotting plant matter. In Switzerland up to 1500 m, in Poland 900 m, Bulgaria 1000 m. Penetrates tundra zone. Hibernates in Poland in litter and moss, sometimes also among dry glass blades hanging above the frozen water surface.

Reproduction in Poland between May and September, single eggs (0.6-0.7 mm) are laid at 0.5-1 day intervals, up to 45 per month and 150 per season, juveniles hatch after 10-20 days (25-18°C) (70 days at 14°C), if hatched underwater they crawl to the surface, full size is reached after 30-40 days, most snails reach maturity in the same year and start egg laying 5 days later, life span up to 3 years.

Endangered in Bayern and Austria, Vulnerable in Germany and S Switzerland. In Poland locally common but does not reach high densities. Many habitats have been destroyed by drainage, river management schemes and other kinds of habitat destruction.

Myzyk 2005, Myzyk 2011: 68.

Vertigo extima (Westerlund, 1877)

Pupa arctica var. *extima* Westerlund, 1877 in Westerlund 1877: 42

Shell larger than *V. modesta*, horny brownish or light yellowish brown, transparent, smooth, with umbilicus, aperture usually without teeth, sometimes 2-3 teeth deep inside and of the same colour like the shell, outer margin with a thin lip inside.

2.7-3.0 x 1.5-1.8 mm

Subarctic and tundra zones, humid valleys with rich vegetation, also swamp margins with short *Salix* trees, needs humid soils, 300-700 m in Scandinavia, pH 5.5-7.2 (usually above 6.0). Not in wide tundra regions without humid soils and with poor vegetation.

Westerlund 1887: 132, Waldén 1986, Proschwitz 2003: 31, Sysoev & Schileyko 2009: 48.

N Scandinavia, N Siberia, W Alaska

Vertigo genesii (Gredler, 1856)

Pupa genesii Gredler, 1856 in Gredler 1856: 122, Taf. 2 (*Pupa Genesii*)

1.6-2.0 x 1.0-1.2 mm

Sweden: Lappland, Tärna, Stenträsk (habitat today destroyed)

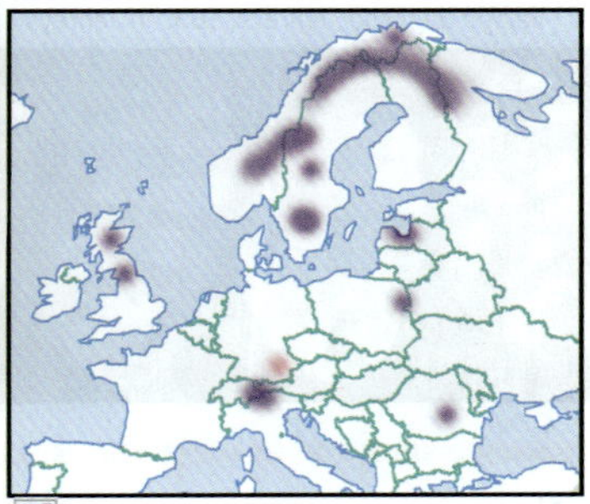

N Europe, Alps

Shell with 3.5-4 convex whorls with deep suture, aperture without teeth, margin connected at parietal side by a callus.

Open calcareous mires and swamps, mostly on soligenous type, found among plant litter and tufts of vegetation. In Britain in 300-500 m altitude in calcareous N-facing flushes. In Switzerland between 1500 and 2100 m.

Juveniles in N England are present throughout the year and on average they comprise 55% of the population, their mortality is high. The reproductive cycle is not strongly correlated with weather conditions. Abundance seems to increase between August and November.

Extinct in Bayern (habitat of the last population near Gröbenzell was recently destroyed by drainage), Endangered in Switzerland and Britain. The habitats in Britain (known since 1980) are relictual, very small and vulnerable to accidental damage.

Kerney et al. 1983: 94, Pokryszko 1990: 205, Manganelli et al. 1995: 13, Turner et al. 1998: 193, Kerney 1999: 98, Cameron 2003: 105, Proschwitz 2003: 32, Killeen 2005.

Vertigo geyeri Lindholm, 1925

Vertigo genesii geyeri Lindholm, 1925 in Lindholm 1925: 241

1.6-1.95 x 1.05-1.2 mm

Austria: Kärnten, Langsee near St. Georgen

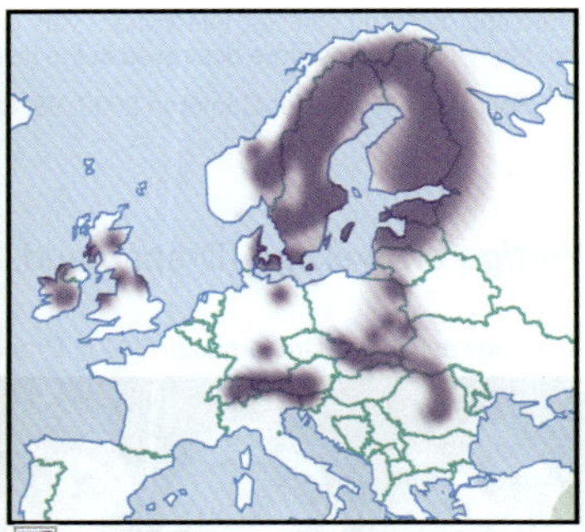

N and central Europe

Shell reddish brown, weakly and regularly striated and shiny, 4 convex whorls, aperture with a weak parietal callus and usually 4 teeth (1 parietal, 1 columellar, 2 palatal), but some can be reduced.

In very wet, swampy places, in marshes of stable water level, found among plant litter and tufts of vegetation, calciphile. Does not tolerate periodic drying, flooding or drops of water levels. In Scandinavia restricted to calcareous swamps and alluvial sands with narrow-leaved swamp plants (*Schoenus ferrugineus*). Up to 1400 m in Switzerland. Population density in the western Carpathians gets higher with increasing water pH, the species occurs in rich inner Carpathian fens.

Juveniles in Wales are present throughout the year, there is no clear annual life cycle. Eggs are laid singly, one individual may produce 1-10 eggs in lifetime.

Critically Endangered in Switzerland and Germany, Endangered in Austria, Wales, Ireland, N England and Scotland. Isolated and scattered, range seems to be generally declining, fossil postglacial finds are more frequent. All populations in Britain are threatened by habitat loss, lowering of water levels and drainage, several populations in Britain and Ireland were destroyed by drainage after the 1930s. Populations in Poland and Slovakia are well known and need protection.

Pokryszko 1990: 208, Kerney 1999: 99, Cameron 2003: 105, Proschwitz 2003: 34, Horsák & Hájek 2005, Boschi 2011: 299, Schenková et al. 2012.

Vertigo heldi (Clessin, 1877)

Pupa heldi Clessin, 1877 in Clessin 1877: 49 (*Pupa (Vertigo) Heldi*) Type locality unknown, Donau river deposits near Regensburg.

2.4-2.7 x 1.1-1.25 mm

Germany: Neckar river deposit near Neckarsteinach

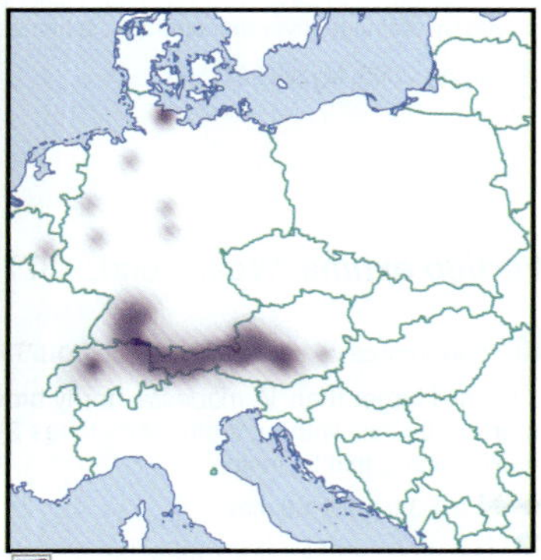

Central Europe

Shell brown, finely striated and shiny, cylindrical with 5-6 whorls, last 3 whorls of almost equal diameter, apertural margin sharp, without teeth or sometimes with weak parietal and palatal teeth.

Swampy meadows and damp grassy habitats, mainly in highlands at low altitude levels, in Switzerland found at 300-400 m.

EU Red List: Endangered EN. Critically Endangered in Switzerland, Germany and Austria. Extremely rare, only few isolated populations are known. Theatened by habitat destruction (converting wet meadows into arable land destroys completely the habitat).

Kerney et al. 1983: 92, Jungbluth et al. 1989: 100, Turner et al. 1998: 190, Jungbluth & Knorre 2009: 10.

Classification as a species uncertain, needs to be verified. Perhaps restricted to N Alps and Schwäbische Alb mountains, but occasionally reported from other parts of central Europe.

Vertigo lilljeborgi (Westerlund, 1871)

Pupa lilljeborgi Westerlund, 1871 in Westerlund 1871: 90
Westerlund gave a description and referred to *Vertigo lilljeborgi* from Coll. Typ. Moll. Su. 1868 n:o 60 (= Westerlund 1868, presumably an unpublished source).

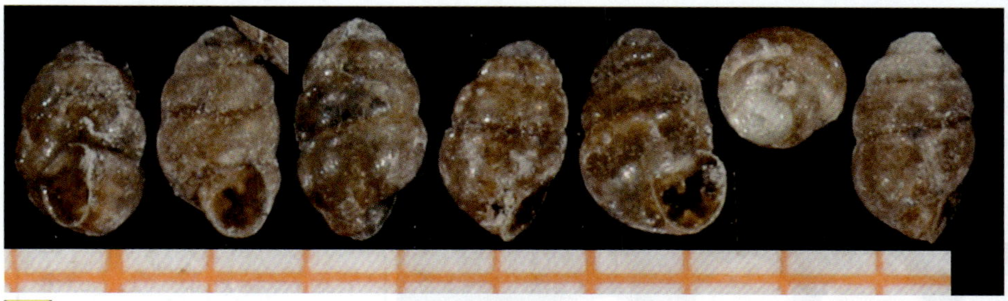

1.7-2.05 x 1.15-1.35 mm

N Sweden: Torne-Lappmark, S Tornehamn

N Europe, Pyrenees

Shell egg-shaped, brown, translucent, shiny and with faint regular growth lines, 3.5-4.5 convex whorls, last whorl very large, aperture with 4 narrow folds (1 parietal, 1 columellar, 2 palatal), margin sharp, slightly reflected and usually connected by a weak parietal callus. Differs from *V. moulinsiana* in the structure of its reproductive system.

Open marches and fens, *Carex* and *Juncus* swamps, at lake and river margins, periodically flooded habitats, also oligotrophic habitats of pH 5, but avoiding decidedly poor conditions.

EU Red List: Near Threatened NT. Endangered in Spain, rare in Great Britain and Ireland. Threatened by agricultural drainage and water current management, several populations in Great Britain and Ireland have been destroyed by this way since the 1930s.

Kerney et al. 1983: 94 (Scandinavia including entire Finland), Pokryszko 1990: 188, Kerney 1999: 97, Gómez Moliner et al. 2001, Proschwitz 2003: 35, Proschwitz 2004.

Mentioned by Sysoev & Schileyko 2009: 50 as a synonym of *V. moulinsiana*, "to Lithuania in the north" (obviously they had no records from NW Russia).

Vertigo modesta (Say, 1824)
Vertigo arctica (Wallenberg, 1858)

Pupa modesta Say, 1824 in Keating 1824: 259, Pl. 15
For clarifying the identity of this name, it is necessary to study material from Northwest Territories, Canada (Keating 1824) (www.mollbase.de, 11-2006).
Pupa hoppii Möller, 1842 in Möller 1842: 4 (*Pupa Hoppii*)
Must not be corrected to *hoppei* under Art. 32.5.
Pupa arctica Wallenberg, 1858 in Wallenberg 1858 (Malak. Bl. 5): 32
Pupa tirolensis Gredler, 1869 in Gredler 1869: 912
Pupa krauseana Reinhardt, 1883 in Reinhardt 1883: 38 (*Pupa (Vertigo) Krauseana*)

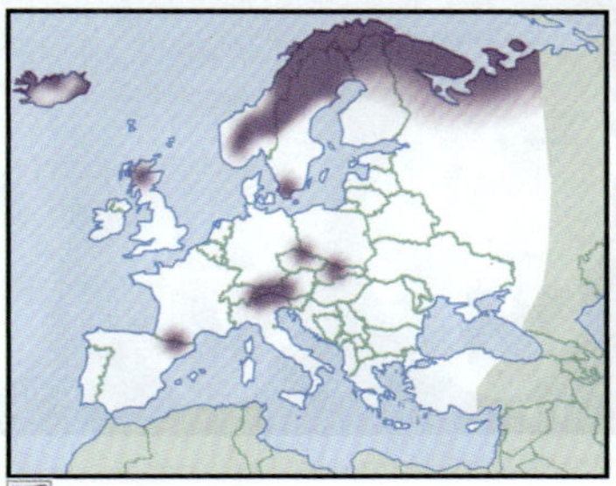

Shell pale yellowish brown, regularly and finely striated, silky shiny, with 4-5 moderately convex whorls, apertural margin sharp and connected by a callus at parietal side, 3 teeth (parietal, columellar, palatal).

2.05-2.6 x 1.25-1.55 mm

Shady calcareous mountain habitats, in Scandinavia in subarctic woodlands, open or partly open valley slopes, fens, in Scotland in similar habitats on N exposed slopes in 800-1000 m altitude, on calcareous substrate. Found in leaf-litter or tufts of vegetation. In the Alps on overgrown rock rubble at creek margins, and on swampy alpine meadows, not restricted to calcareous substrate. In Switzerland not above the timberline, between 1600 and 2100 m.

Eurasia, or Eurasia and Greenland, or Northern Hemisphere

EU Red List: Near Threatened NT. Critically Endangered in Switzerland and Germany, Endangered in Scotland and Austria.In the Alps very scattered and rare. In Scotland considered as glacial relicts.

Waldén 1986: 40, Pokryszko 1990: 196 (classified *arctica* as an independent species), Turner et al. 1998: 192, Kerney 1999: 96, Proschwitz 2003: 36, Proschwitz 2004 (classified the European form as *modesta*), Sysoev & Schileyko 2009: 50 (classified W and E Siberian forms as *V. modesta*).

In Europe *arctica*, in Greenland *hoppii*, in E Siberia *krauseana*, in the Alps *tirolensis*. If (only) the populations of Greenland and Europe are regarded conspecific (this seems to be likely, www.mollbase.de, 11.2006), the name must be *V. hoppii*. If not, the European species must be *V. arctica*. If European and Canadian populations are regarded as conspecific, the name must be *V. modesta*.

Vertigo moulinsiana (Dupuy, 1849)

Pupa moulinsiana Dupuy, 1849 in Dupuy 1849: [4] (*Pupa Moulinsiana*)
Vertigo laevigata Gallenstein, 1848 in Gallenstein 1848: 11 (nomen oblitum under Art. 23.9, fixed here)
Not *Pupa laevigata* as cited by Falkner et al. 2002: 110.
Google hits 05.2009: "vertigo-moulinsiana" 8900 hits, "vertigo-laevigata" no hits except 2 citations of the original description.
In accordance with Art. 23.9 *Pupa moulinsiana* Dupuy, 1849 shall be used for the taxon, *Vertigo laevigata* Gallenstein, 1848 has not been used as a name for a taxon since 1899 and shall be regarded as nomen oblitum, *Pupa moulinsiana* Dupuy, 1849 was used as a name for our taxon in the following 32 papers: Klemm [1973], Damjanov & L. 1975: 123, Jungbluth 1978, Kerney & C. 1979: 72, Kerney et al. 1983, Shilejko 1984, De Wilde et al. 1986, Van Goethem 1988, Wiese 1991, Pokryszko 1990: 183, Falkner 1990: 142, Vogt et al. 1994: 124, Manganelli et al. 1995: 13, Schütt 1996: 71, Turner et al. 1998: 191, Flasar 1998, Kerney 1999: 95, Gómez Moliner et al. 2001: 123, Manganelli et al. 2001, Schütt 2001: 72, Falkner et al. 2002: 110, Proschwitz 2003: 36, Jueg 2004, Pintér & Suara 2004, Quintana Cardona 2006: 25, Zettler et al. 2006, Beckmann 2007: 67, Jungbluth & K. 2008, Kobialka et al. 2009: 18, Książkiewicz 2009, Sysoev & Schileyko 2009: 50, Balashov & Gural-Sverlova 2012: 95.

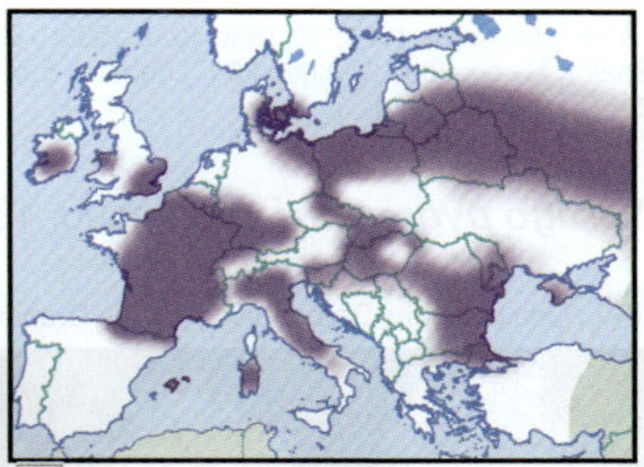

Europe to Transcaucasia

2.25-2.75 x 1.3-1.65 mm

Germany: Schulensee near Kiel

France: Isère, Châbon, Grand Lemps

Cédric Audibert

Shell egg-shaped, yellowish or reddish brown, translucent, shiny and almost smooth, 3.5-4.5 convex whorls, last whorl very large, accounting for 2/3 of the total height, aperture with 4 narrow folds (1 parietal, 1 columellar, 2 palatal, the palatal teeth arising from a pale rib-like transversal callus), margin sharp, slightly reflected and usually connected by a weak parietal callus. Winter growth interruptions are visible in form of a whitish streak in 60 % of the shells. Differs from *V. lilljeborgi* anatomically and in its rib-like transversal palatal callus. The callus is also visible from outside.

Restricted to old calcareous wetlands, marshes and fens, beds of reed, on lake shores and river banks. Climbs up plants at water body margins to stay 30-50 cm above ground or water surface in Britain, on the ground only to pass the winter. In Poland the youngest snails stay near the surface, larger juveniles climb higher, only adults may climb up to 1 m. Hibernating snails are usually found attached to grass blades and dried sedges hanging about 20-50 cm above the ground or water table, and inside dense grasses, rarely in litter and moss on the ground.

Does not tolerate any human disturbance, even less than *V. antivertigo*, never found in habitats regularly mowed or grazed. Animal feeds on fungi growing on marsh plants. Usually in lowlands, in Switzerland at 400-600 m, sensitive against low winter temperatures, but at night temperatures above 0 °C the animals may still be active until November in Poland.

Reproduction from April to September, some 20 single eggs (0.65-0.85 mm) are laid in 2-3 day or longer intervals, at mosses or dense grasses near the water edge, juveniles hatch after 10-22 days (24-17 °C) (67 days at 13 °C), shell increases from 1.2 to 3 whorls in the first month, followed by a temporary slow-growth period of another month, full shell size is reached after 3 months of activity, maturity 0-2 weeks after shell completion. Life span usually below 2 years, maximum age more than 3 years.

EU Red List: Vulnerable VU. Critically Endangered in Bayern, Rheinland-Pfalz, Austria and S Switzerland, Endangered in Germany and N Switzerland, Endangered and very rare in Italy. The populations in Wales should be considered at least as Endangered in this scheme.
In Germany threatened by habitat destructions, in Poland most historically confirmed populations are extinct because their habitats were destroyed, but the species was able to colonise new habitats and is more frequent than formerly believed. Only very few records from Wales and central England.

Schütt 2001: 72, Proschwitz 2003: 36 (Denmark), Myzyk 2005 (egg size), Książkiewicz 2009 (Poland), Sysoev & Schileyko 2009: 50, Myzyk 2011: 62.

Vertigo parcedentata (Braun, 1847)

Pupa parcedentata Braun, 1847 in Braun 1847: 51

Differs from *V. pygmaea* in its larger size, shell thinner, teeth very weak.

Open calcareous highland fens, 1000-1200 m.

1.7-2.3 x 1.15-1.25 mm

France: Alsace, near Münchhausen

Norway
Elsewhere in Europe known as fossils, associated with cold climatic phases in central Europe (loess fauna).

Proschwitz 2003: 37.

EU Red List: Vulnerable VU. Possibly exaggerated. The only endemic mollusc species of Norway, restricted to a limited and remote area in Dovrefjell. No threats known.

Vertigo pusilla Müller, 1774

Vertigo pusilla Müller, 1774 in Müller 1774: 124

1.6-2.2 x 1.0-1.2 mm

Germany: Schleswig-Holstein, Kellersee near Eutin

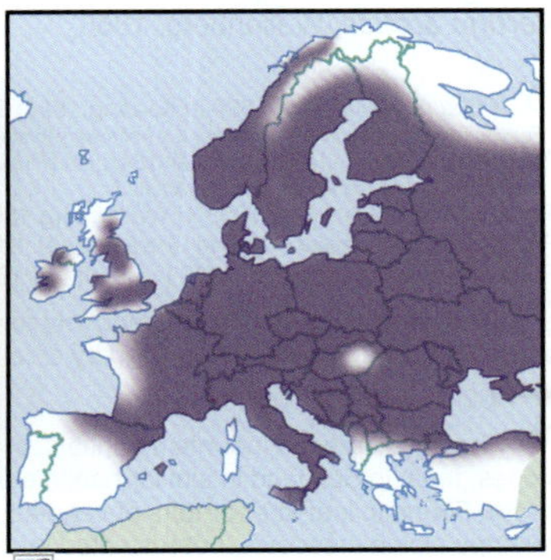

Europe to Caucasus and Altai

Shell sinistral, yellowish brown, almost smooth with irregular weak growth lines, cervical crest weakly developed, aperture with at least 6 teeth (2 parietal, 2 columellar and 2 palatal), sometimes up to 9, margin rounded, heart-shaped with a convex section at the palatal side, connected at parietal side by a white callus. Winter growth interruptions are almost always visible as light streaks followed by a constriction of the whorl. Larger than *V. angustior*.

In woodlands and open forests, prefers rocks, occasionally also in sand-dunes and rocky alpine meadows. In humid spots of otherwise relatively dry habitats. Found in leaf-litter, tree hollows, also stone walls. Climbs hazel trunks up to 2 m high. In Switzerland up to 1900 m, Bulgaria 1600 m.

5-30 eggs (size 0.57-0.67 mm) are laid by one individual during lifetime, after self-fertilization or copulation. Reproduction in Switzerland between March and October (in Poland May-September), eggs are laid singly in small humid holes in substrate that serves as food for the offspring, juveniles hatch after 11-20 days (at 24-18 °C in Poland), maturity is reached after 30-40 days under favourable conditions, in Poland usually in the next summer, but 3-4 generations per year are possible, maximum age up to 3 years.

Widespread but local in the British Isles, certainly declining in England. Vulnerable in Spain and Bayern, rare in Ireland.

Kerney et al. 1983: 90, Pokryszko 1990: 169, Falkner 1990: 142, Pokryszko 1992, Kerney 1999: 91, Schütt 2001: 70, Martínez-Ortí & Robles 2003: 87 (6 localities in Spain), Myzyk 2005, Waldén 2007, Fehér & Erőss 2009 (N Albania, not Greece, not Rep. Makedonija), Sysoev & Schileyko 2009: 51, Örstan 2010 (N Turkey), Boschi 2011: 285, Myzyk 2011: 70.

Vertigo pygmaea (Draparnaud, 1801)

Pupa pygmaea Draparnaud, 1801 in Draparnaud 1801: 57 (*Pupa pygmæa*)

1.4-2.1 x 0.95-1.2 mm

Germany: Bayern, Günzburg

Holarctic

Shell yellowish to dark brown, very weakly striated, 4.3-4.7 whorls, cervical crest present (difference to *V. alpestris*), aperture with 1 parietal and 3-6 other folds, palatal teeth usually linked by a callus round their bases, margin connected at parietal side by a weak callus in fully grown shells. Winter growth interruptions are nearly always visible in form of a whitish streak.

Open habitats of very variable humidity, from dry sunny slopes with little vegetation to humid or marshy meadows with lush plant cover, also in habitats under strong human-based pressure, such as gardens, pastures, railway embankments. In Portugal in mosses and under stones, same habitats as *Leiostyla anglica*. In Britain frequent at the roots of grass. Does not tolerate shade. In Switzerland up to 1800 m, in Bulgaria 1500 m.

Reproduction in Poland between June and September, usually in mid-summer, single eggs (0.5-0.6 mm) are laid in a humid environment (along main nerves of plant leaves, often covered with detritus) at 2-3 day intervals (but eggs may be retained for a few days and laid in series, probably to escape dry weather conditions), up to 70 per season, juveniles hatch after 11-15 days (24-21 °C), most juveniles hibernate with 2-3 whorls, maturity is reached in July-August of the next year, life span up to 3 years.

Pokryszko 1990: 178, Myzyk 2005, Fehér & Erőss 2009 (N Albania), Sysoev & Schileyko 2009: 51, Boschi 2011: 291, Myzyk 2011: 67.

Vertigo ronnebyensis (Westerlund, 1871)

Pupa ronnebyensis Westerlund, 1871 in Westerlund 1871: 94

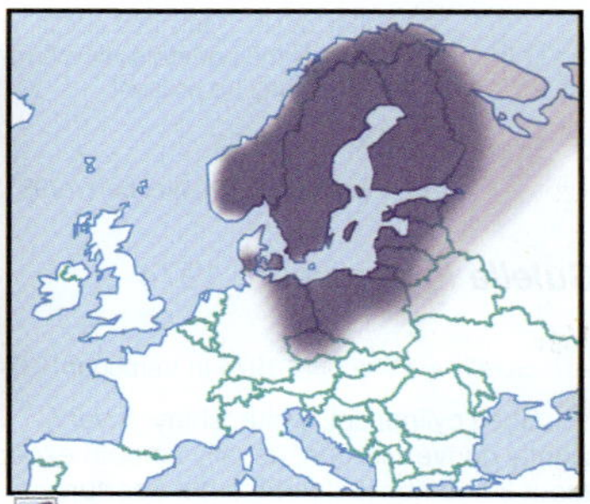

2.0-2.4 x 1.15-1.35 mm

Sweden: Värmland, Fjornshöjden

Central and N Europe

Shell yellowish brown, translucent, silky shiny with very fine regular growth lines, 4-5.5 weakly convex whorls, last whorl getting narrower basally, penultimate whorl has equally large diameter, aperture usually with 3 small denticles (1 parietal, 1 columellar, 1 palatal), margin sharp, not white, parietal callus weak. Winter growth interruptions are nearly always visible in form of a whitish streak.

Woodlands, forests of all kind, in N Europe typically in coniferous forests, in dry places, under ground litter and moss, also on poor, non-calcareous soil. Abundance depends on humidity and vegetational cover, clearfelling and growth of young trees have strong effects. In dry years the snails are only found in moist places, in humid summers also on the forest floor.

Reproduction in Poland between June and September, single eggs (0.6-0.75 mm) are laid at 1-2 day intervals, about 15-40 per season, juveniles hatch after 14-22 days (22-18 °C), most juveniles hibernate with 2-3 whorls, maturity is often reached in July-August of the next year, some individuals hibernate possibly twice as juveniles, life span up to 3 years.

Very rare outside Scandinavia. Endangered in Germany.

Westerlund 1871 (Fauna moll.): 263, Pokryszko 1990: 201, Proschwitz 1990 (variation), Proschwitz 2003: 39, Myzyk 2005, Myzyk 2011: 73.

Vertigo substriata (Jeffreys, 1833)

Alaea substriata Jeffreys, 1833 in Jeffreys 1833: 515 (*Alæa substriata*)

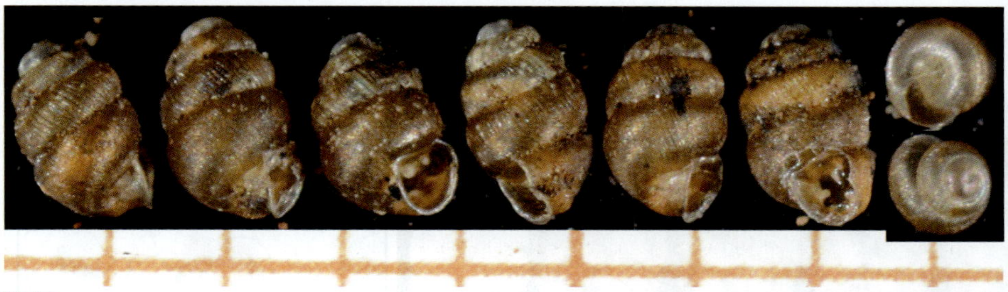

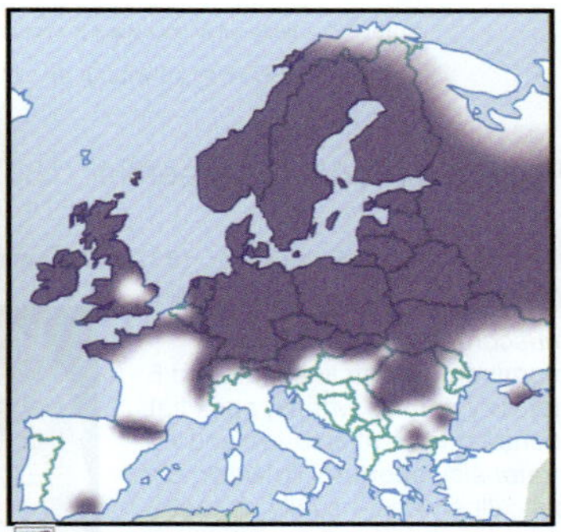

1.5-2.0 x 0.95-1.2 mm

Germany: Schleswig-Holstein, Lottsee

W Eurasia, isolated in Altai

Shell brown, iridescent, characteristic surface sculpture consisting of very fine and regular ribs, cervical crest present but not very strong, 4-4.5 strongly convex whorls with deep suture, aperture with 2 parietal, 1-2 columellar and 2 palatal folds, margin thin but not sharp, slightly reflected, connected by a parietal callus. Only 50 % of the shells show winter growth interruptions in the form of white streaks.

In humid places in mixed, deciduous (N Europe: coniferous) forests, sometimes also in humid open or half-open habitats (stream banks with rich herb vegetation, meadows, marshy grasslands, fens), sometimes also stone walls. In British highlands not uncommon in wet areas of rough grazing. Tolerates acid soils, but not human disturbance. In Poland up to 1000 m, in Switzerland up to 2000 m.

Reproduction in Poland between June and September, single eggs (0.6-0.7 mm) are laid at 1-2 day intervals, about 30-50 per season, juveniles hatch after 12-22 days (24-18 °C), most juveniles hibernate with 2-3 whorls, maturity is usually reached in July-August of the next year, life span up to 3 years.

Endangered in Austria and parts of Germany, Vulnerable in Switzerland and Germany. Scattered distribution with large gaps between populations, never very abundant, but except in Spain and Bulgaria usually not rare. Has seriously declined around London and in the British midlands, surviving populations are at risk from intensive farming.

Steusloff 1949, Proschwitz 2003: 40, Proschwitz 2004, Myzyk 2005, Myzyk 2011: 74.

Vertigo ultimathule Proschwitz, 2007

Vertigo ultimathule Proschwitz, 2007 in Proschwitz 2007: 73

Often given as *Vertigo ultimathule* von Proschwitz, 2007. In continental Europe the German conventions are applied, by which the title of nobility "von" is not added when the author's surname is cited alone. This applies also in Sweden, for example in the case of the founder of zoological nomenclature, Linné (citation of the full name: Carl von Linné, citation in alphabetical lists: Linné, Carl von).

N Scandinavia

Shell yellowish brown, almost cylindrical (not ovoid), coarsely and irregularly striated, 4.5 convex whorls with deep suture, last whorl not narrowing markedly at the base, aperture without teeth or with a small parietal tooth deeply inside, apertural margin sharp.
V. ronnebyensis has 3-4 teeth and a darker brown shell, more finely and more regularly striated (the relationship in this character resembles that between *Columella aspera* and *Col. edentula*). Smaller than *V. extima*.

2.1-2.2 x 1.3-1.4 mm

Creek valleys and mountain slopes with shrubs and Betula trees, occasionally with rich herbal vegetation, 200-600 m.

EU Red List: Near Threatened NT. A rare species, only few populations and a few dozen specimens were found. Some populations live close to roads and could be at risk from construction works.

Proschwitz 2010: 15.

The original description of *V. ultimathule* was based on only 4 specimens, without evidence provided in a sound study that these belonged to a separate species, for example by statistical evaluations of those shell characters claimed to differ in the involved species. Alternative hypotheses were not tested. Proschwitz 2010 added a more accurate description and a much better guide how to distinguish this species from other closely related species, based on the study of considerably more specimens from more localities.

Gulella

Gulella Pfeiffer, 1856
in L. Pfeiffer 1855-1856: 173 *Ennea (Gulella)*

Type species *Pupa menkeana* Pfeiffer, 1853
(subsequent designation by Albers & Martens 1860: 298)
(= currently *Gulella menkeana* (Pfeiffer, 1853) from South Africa)

Established without description, contained originally 7 species.

Small ovoid to turreted shells, smooth to coarsely ribbed. The aperture is rounded to nearly quadrate with a flared peristome. The almost heavy dentition is composed of the parietalis, which may be connected to an angularis. The aperture is often constricted palatally, displaying a strong palatalis. One to two columellar lamellae may be present.

Tropical Africa

Verdcourt 1974, Neubert 1998: 379, Oke & Chokor 2010.

Gulella io Verdcourt, 1974

Gulella io Verdcourt, 1974 in Verdcourt 1974: 121

Shell cylindrical, white, shiny, 5 very slightly convex whorls, usually smooth except about 8 fine ribs just behind the aperture, sometimes with some more ribs on the whorls, suture slightly margined, aperture rounded, about 0.9-1.0 mm in diameter, apertural margin thickened with 4 teeth: parietal very strong (1/3 of apertural diameter), columellar, basal and palatal teeth regular. Like a short *Acicula* species but with 4 teeth in the aperture.

One of the most frequent species of S Nigeria. In Europe only in some greenhouses.

Oke & Alohan 2006, Oke & Chokor 2009, Oke & Chokor 2010.

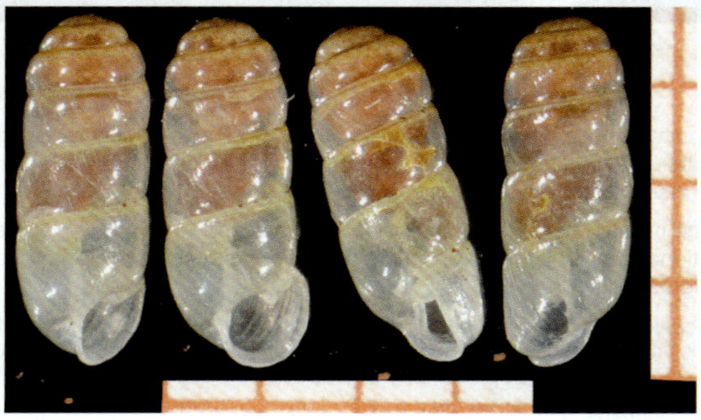

3.5-4.0 x 1.2-1.4 mm

Germany: Hamburg, greenhouse Alter Botanischer Garten

Originally from tropical Africa, introduced to European greenhouses

Pupilla

Pupilla Fleming, 1828
in Fleming 1828: 268 *Pupa* synonym *Pupilla*

Type species *Pupa marginata* Draparnaud, 1801
(monotypy)
(= currently *Pupilla muscorum* (Linnæus, 1758))

First mentioned as a synonym, "This species was sent me many years ago, by Dr Leach, from Battersea, under the title *Pupilla marginata*.". ICZN Opinion 335: name was made available by Fleming 1828 (this makes it unnecessary to verify the presence of conditions under Art. 11.6.1).

Jueg 1997, Proschwitz et al. 2009.

Pupilla alpicola (Charpentier, 1837)

Pupa alpicola Charpentier, 1837 in Charpentier 1837: 16, Tab. 2

Shell slightly larger than *P. muscorum*, with slightly more prominent growth lines, around 6 not very convex whorls, apex flat (contrast to *P. muscorum* and *P. pratensis*), apertural lip weak, cervical callus weak, usually no teeth, occasionally a weak parietal tooth. Proschwitz et al. 2009 reported the presence of a characteristic groove in the external surface of the last whorl, at about 1/4 of the distance from its base (in oblique light visible in the aperture as a narrow, delicate ridge).

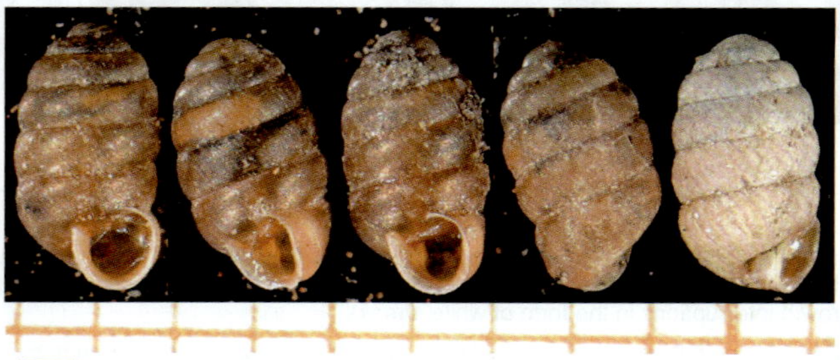

2.8-3.3 x 1.75-1.8 mm

Austria: Tirol, near Grins

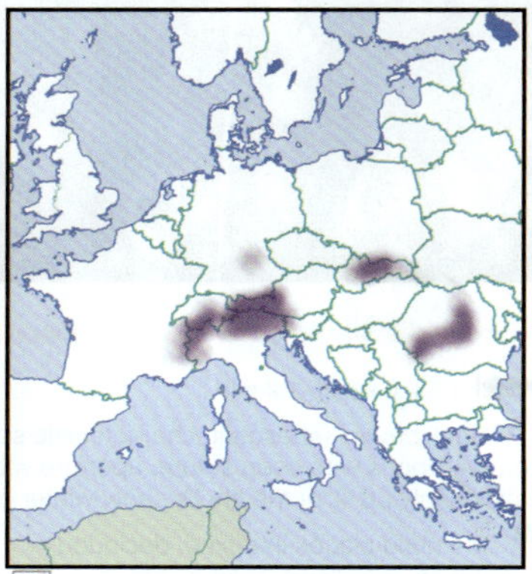

Alps, Carpathians, Altai

In moss of wet meadows in high alpine regions, mostly in calcareous fens with sparce herbal vegetation, requires extremely high calcium carbonate precipitation (tufa) and low nutrients. In Switzerland between 900 and 2600 m altitude, in Romania up to 2400 m.

Scattered distribution, rare, most habitats are close to alpine cattle pastures, this implies a danger. Glacial relict, occurs not more distant than 40 km from known Holocene refugia in the W Carpathians. Locally threatened by spontaneous succession towards shrubby and forest habitats after cessation of traditional land management. Critically Endangered in Germany and Poland, Vulnerable in Switzerland, potentially threatened in Austria. Populations in NW Bavaria were lost by habitat destruction. Most populations in Poland have been lost.

Kerney et al. 1983: 119, Grossu 1987: 278, Falkner 1990: 148, Turner et al. 1998: 164, Zajac 2005: 14, Proschwitz et al. 2009: 318, Boschi 2011: 247, Horsák et al. 2011 (previous records for Albania referred to *P. sterrii*), Horsák et al. 2012.

Pupilla bigranata (Rossmässler, 1839)

Pupa bigranata Rossmässler, 1839 in Rossmässler 1838-1844 (9-10): 27, Taf. 49 No type locality given, original types lost.

Smaller than *P. muscorum*, weakly striated, whorls slightly more rounded, cervical callus strong, lip reflected, two teeth palatal and parietal, palatal tooth relatively strong.

Dry and warm grassy slopes on calcareous substrate, in Switzerland in up to 2000 m altitude.

Extinct in Bayern, Critically Endangered in Switzerland, Endangered in Germany, rare in Austria.
Very scattered in Switzerland, only 3 of 20 formerly known populations seem to have survived.

2.5-3.5 x 1.5 mm (diameter not very variable)

Switzerland: Gorges de Moutier

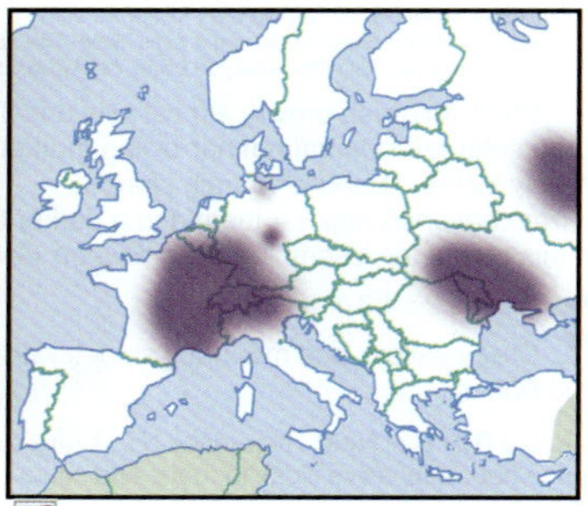

Europe to central Asia (?)

Falkner 1990: 148, Vogt et al. 1994: 104, Turner et al. 1998: 164, Bößneck 2009, Sysoev & Schileyko 2009: 43, Boschi 2011: 243.

A revision is needed. It seems that this name has not been used by all research groups in the same sense. Manganelli et al. 1995: 15 considered it as a synonym of *P. muscorum*. Sysoev & Schileyko 2009 and Balashov & Gural-Sverlova 2012: 94 used the name for a species distributed from W Ukraine and W Russia to W Siberia, Caucasus and central Asia.

Pupilla muscorum (Linnæus, 1758)

Turbo muscorum Linnæus, 1758 in Linnæus 1758: 767

Shell usually light brown, varies from reddish brown to horny grey, weakly striated or almost smooth, 5-6.5 weakly convex whorls, suture not very deep, aperture usually with with well developed lip, cervical callus strongly developed, like a dam, parietal tooth usually present, palatal tooth sometimes too.
Differs from *P. pratensis* in its thicker, smaller and more slender shell, lighter and more variable colour and stronger apertural lip.
Animal small, elliptical, dark with lighter sides and foot, upper tentacles not very long, lower tentacles very short.

E France: Jura, Pontarlier
Musée des Confluences Lyon

3.0-4.0 x 1.65-1.75 mm (diameter not very variable)

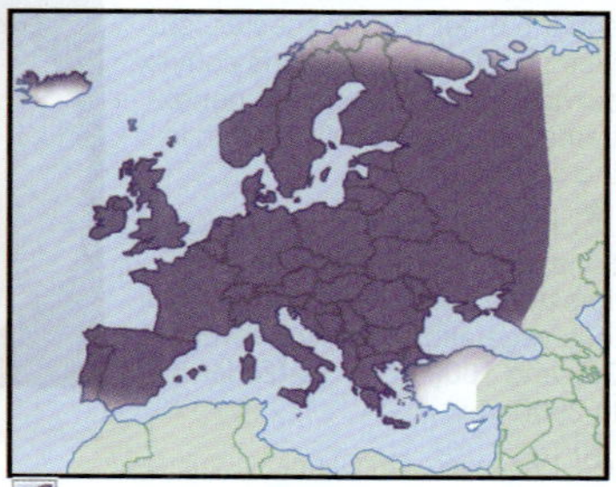

Northern Hemisphere

On dry meadows, sand dunes, open and sunny habitats. Calciphile. In Portugal found under stones, dead leaves and in mosses. In Britain frequent in sheep-grazed calcareous grasslands. In the Alps up to 2400 m, in Bulgaria 1200 m. Feeds mainly on dead plant remains on the soil, occasionally also green leaves.

Ovoviviparous. Reproduction in France mainly between July and September, maximum release period in Poland is June-August, the species is able to hibernate with eggs (diameter 1-2 mm) and can release eggs with partly grown embryos (size 1 mm, 1.5 whorls) in favourable seasons. Individuals incubate 1-8 (mostly 3-7) embryos thoughout the year, activity in Poland is from March to October. Maturity is probably reached in the second or third season. Gravid individuals are not able to copulate, 0-20 % of the individuals at a given time are not gravid.

Declining in Bayern. Threatened by increasingly intensive farming on old calcareous grasslands in Britain.

Moquin-Tandon 1853: 225, Moquin-Tandon 1855: 392, Germain 1930: 423, Pokryszko 2001, Proschwitz et al. 2009.

Pupilla pratensis (Clessin, 1871)

Pupa muscorum var. *pratensis* Clessin, 1871 in Clessin 1871: 101

3.5-4.5 x 1.85-2.05 mm

Germany: Schleswig-Holstein, Kitzeberg near Kieler Förde

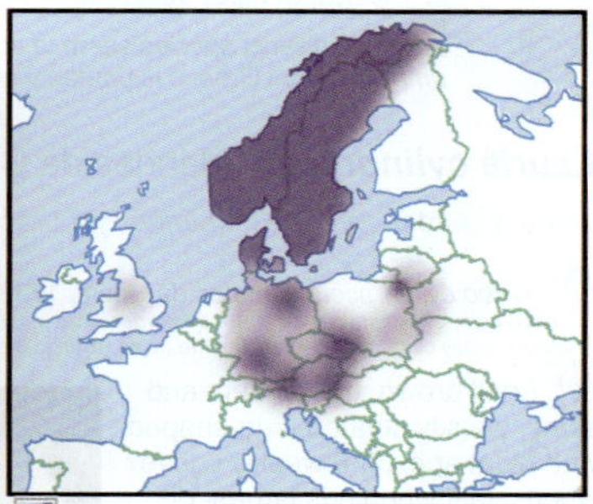

Central Europe to N Scandinavia, possibly Britain

Shell dark to brown, translucent and thin, with fine and distinct growth lines, 6-7.5 whorls with deep suture, outline elongate and broad, aperture rounded with weakly developed lip, usually toothless, or with weak parietal and occasionally a palatal tooth, sometimes with weak cervical callus.
Differs from *P. muscorum*, with which it lives sympatrically, in its thinner, larger and broader shell, darker and less variable colour and weaker apertural lip.

Open moist and wetland habitats, often on calcareous substrate, in Scandinavia and Poland in calcareous fens and meadows.

Endangered in Czech Republic, problematic situation also in Poland, some populations were destroyed by drainage.

Jueg 1997, Proschwitz et al. 2009, Horsák et al. 2010, Horsák et al. 2012.

Was misidentified with *P. muscorum* before 2009. It is still necessary to verify if the description and the differences to *P. muscorum* are invariable and stable in all European populations.

Pupilla sterrii (Forster, 1840)

Pupa sterrii Forster, 1840 in Forster 1840: 469 (*Pupa Sterrii*)

The name was attributed to Voith, the description contained two diagnoses, one by Forster and one by Voith. Bank 2011: 16 attributed the name to Voith alone and argued that Forster's description was only a summary of the information given in Voith's description and "has no effect on the authorship of the taxon", in other words, it was unnecessary or not important enough.
A description is either present or not. Forster did add a description, his text alone would have made the name available, so in a strict application of Art. 50.1.1 the name must be attributed to the author of the work.

Shell light horny coloured or reddish brown, weathering grey, finely ribbed, whorls strongly convex with deep suture, aperture with 2 teeth, columellar and parietal, cervical callus moderate and white.

In grass of very dry and sunny limestone rocks, between stones, on dry meadows on calcareous substrate. In N Italy up to 2800 m.

Ovoviviparous.

Endangered in Germany. Scattered distribution. In lower altitudes threatened by continuous habitat destructions.

2.8-3.5 x 1.5-1.7 mm

Germany: Rhine river deposit near Rheinhausen

The figured specimens are not typical, their whorls are not very convex.

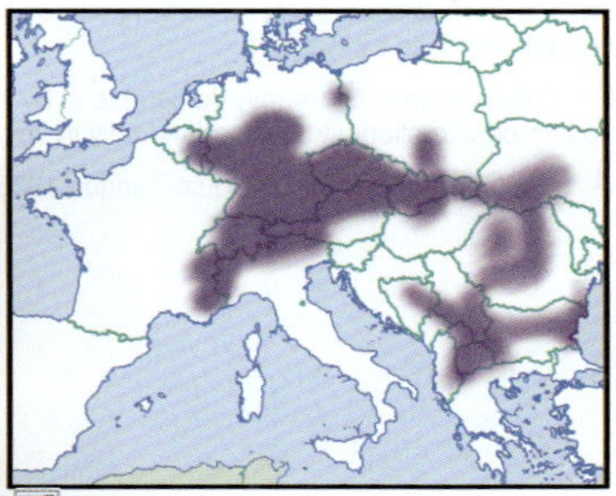

Alps and Carpathians to Turkey and NW China

Kerney et al. 1983: 120, Grossu 1987: 279, Falkner 1990: 148, Sysoev & Schileyko 2009: 45, Boschi 2011: 245, Horsák et al. 2011 (previous records for *P. alpicola* from Albania referred to *P. sterrii*).

Pupilla triplicata (Studer, 1820)

Glischrus triplicata Studer, 1820 in Studer 1820: 89 (*Glischrus (Pupa) triplicata*)

Much smaller than *P. sterrii* and the other *Pupilla* species, surface structure finer, whorls convex, cervical callus present but not extremely strong, usually 3 teeth in the aperture.

In grass at limestone rocks, dry and sunny habitats, often at limestone rock rubble with xerophilous vegetation. Mainly between 300 and 1000 m, in Switzerland up to 2500 m.

Endangered in Germany, Vulnerable in Austria. Scattered distribution, populations in lowlands are threatened by habitat destruction.

Austria: Niederösterreich, Scheiblingskirchen

2.2-4 x 1.3-1.4 mm (diameter should not exceed much 1.4 mm)

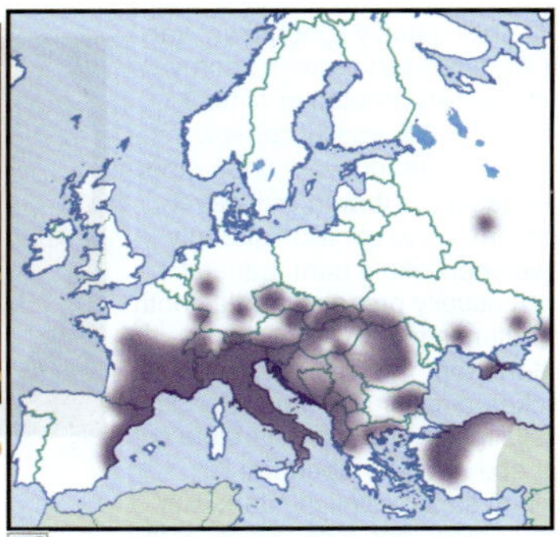

S Europe to central Asia

Kerney et al. 1983: 119, Grossu 1987: 281, Bech 1988: 109, Falkner 1990: 148, Turner et al. 1998: 166, Schütt 2001: 67, Proschwitz et al. 2009: 321 (argued that *P. triplicata* was "certainly" a species complex), Sysoev & Schileyko 2009: 46, Boschi 2011: 241, Balashov & Gural-Sverlova 2012: 94.

Lauria

Lauria Gray, 1840
in Turton & Gray 1840: 193 *Pupa (Lauria)*

Type species *Pupa umbilicata* Draparnaud, 1801
(subsequent designation by Herrmannsen 1846 [25 May 1847]: 578)
(= currently *Lauria cylindracea* (Mendes da Costa, 1778))

Shell small oval to subcylindrical, weakly to strongly ribbed, aperture often with angularis, parietalis and columellaris, but the folds may be reduced.

Ovoviviparous.

Africa, Europe, Caucasus, Arabia

Two distinct species seem to occur in Europe, but their morphology and local variations are not understood and require more detailed research. A third taxon *umbilicus* from Greece is probably not a separate species, but it is unclear to which species it belongs.

Lauria cylindracea (Mendes da Costa, 1778)

Without gender agreement: *Lauria cylindraceus* (Mendes da Costa, 1778)

Turbo cylindraceus Mendes da Costa, 1778 in Mendes da Costa 1778: 89, Tab. 5
Spelled once *cylindraceous* on p. 89, *cylindraceus* on p. 89 and index (page V), *cylindraceous* corrected to *cylindraceus* in errata.
Author usually cited as Da Costa, but the name given in the original source was Mendes da Costa.

Shell brown, transparent and shiny, weakly striated, egg-shaped with blunt apex, 5-6 weakly convex whorls, last whorl has largest diameter, aperture with parietalis and with or without angular tooth, margin white, sharp and reflected in fully grown specimens, usually with whitish parietal callus, umbilicus open and narrow.
Juveniles with additional folds visible from outside the shell.
Animal small, dark with lighter sides and foot, upper tentacles short, lower tentacles very short. The animal crawls with the shell in a high and almost straight position.

3-4.5 x 1.8 mm

Portugal: Lisboa, Castelo São Jorge

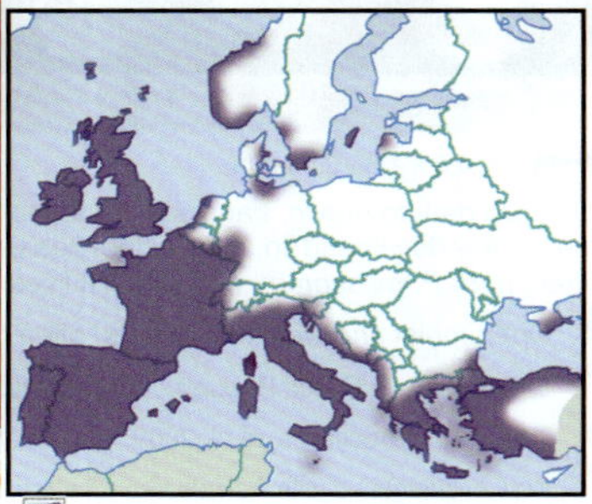

Europe to Caucasus and Arabia

Forests, rocks, meadows, not too humid habitats. Often in high numbers under stone wall cover of *Hedera helix*. In S Crimea specimen numbers are 10 times higher in dry than in humid habitats, but in Portugal it prefers humid and shady habitats where it is very abundant on mosses and under stones and desintegrating leaves, also on the bark of trees. In Britain one of the most frequent species, very common on stone walls at the top under *Hedera helix*. In S Europe often abundant, the second most common land snail of Portugal after *Helix aspersa*, very common at old walls in Albania. In Switzerland up to 800 m.

Ovoviviparous, never lays eggs, reproduction in July and August, parent individuals incubate 3-5 embryos. Juveniles have 1.5-2.5 whorls (diameter 1 mm), maturity is reached after 2.5 years, life span 4 years or more. No embryos are observed between November and January. Gravid individuals are probably not able to copulate.
Gassies 1849 observed that 15-20 very small eggs were laid between April and June, under dead leaves and stony particles. These might have belonged to a different species, misidentified by Gassies.

Endangered in Germany, Vulnerable in Switzerland and Albania. Has declined in the English midlands, threatened by removal of old stone walls and replacing them by modern constructions. In Switzerland only in very warm and dry habitats, most habitats have already been destroyed, only very few populations have survived.

Moquin-Tandon 1853, Moquin-Tandon 1855: 391, Germain 1930: 428, Heller & Sivan 1997.

Lauria sempronii (Charpentier, 1837)

Pupa sempronii Charpentier, 1837 in Charpentier 1837: 15, Tab. 2 (*Pupa (Sphyradium) Sempronii*)
Lauria reischuetzi Falkner, 1985 in Falkner 1985: 45

Shell weakly reddish brown, less whorls than *L. cylindracea* (only 5-6), weaker folds in the aperture, angular tooth less often present and never connected to the outer lip, also weak in juveniles, margin less reflected, columellar fold invisible in the aperture, base more rounded, zone around the umbilicus less keeled or angulated.

2.5-3.2 x 1.5 mm, occasionally smaller

Italy: Toscana, (1) Lucca, Camporgiano, (2-6) Arezzo, Chiusi della Verna

(2-6) Angelo Vannozzi

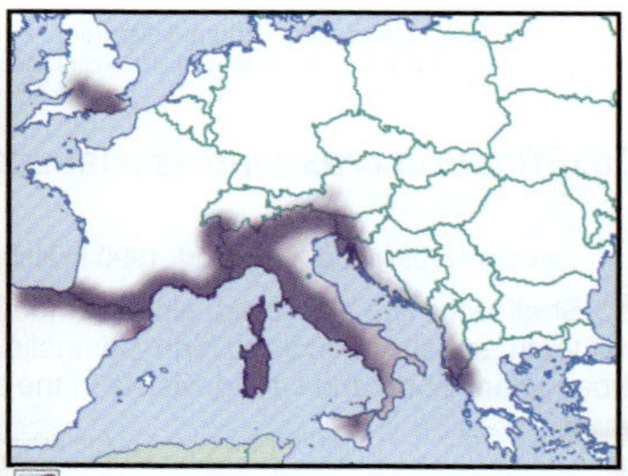

S Europe, S England

Croatia: Cres, Porozina

Massimo Prodan

Forests, rocks, old walls covered with moss vegetation, exceptionally also at tree trunks covered with moss, usually in mountains, but rare above 1000 m. Prefers limestone, but not restricted to calcareous substrate. In Britain found on a calcareous stone wall in a moist location on the side of a valley.
Was not found sympatrically with *L. cylindracea* in Italy, but it was observed that the habitats of the two species did not differ there.

Possibly extinct in Austria, Endangered in England, Vulnerable in Switzerland.

Pilsbry 1922-1926: 55, Kerney et al. 1983: 121, Turner et al. 1998: 145, Kerney 1999: 106, Larraz Azcárate & Rolán 2003, Fehér & Erőss 2009.

The original description of *L. reischuetzi* was based on a single shell from unknown origin in Istria, found in a rich river deposit. The main difference was its slightly smaller size (2.4 x 1.4 mm). The statements on differences to *L. sempronii* would have required a statistical significance test and more knowledge about the Croatian populations of *Lauria*. Falkner 1985 did not provide evidence that the morphological background of the involved taxa was fully understood, and that the unusually small size could not be explained otherwise.
Štamol & Velkovrh 1995: 234 reported *L. reischuetzi* from 3 stations in the north of Cres, the authors found 40 specimens but did not explain how they distinguished them from *L. sempronii*. I see currently no published evidence for the presence of two separate species *L. sempronii* and *L. reischuetzi*.
Reitano et al. 2007: 315 reported *L. sempronii* from 1500 m altitude in Sicily, determination was based on shell height and columellar folds. Shell height is often meaningless in pupilloids, shell diameter is much more important, apertural folds display local variations. The figured shell had 1.7 mm diameter, slightly smaller than usual in *L. cylindracea* but still closer to its usual values than to 1.5 mm of *L. sempronii*.

Lauria umbilicus (Roth, 1839)

Pupa umbilicus Roth, 1839 in Roth 1839: 20, Tab. 2
Not *Lauria umbilica*, name is to be treated as a noun under Art. 31.2.2.

Shell finely striated, not very shiny, 6 convex whorls, last whorl slightly ascending near aperture, sharply impressed at its base, aperture almost triangular, parietalis long, connected to margin, umbilicus wide.

2.3-3.5 x 1.4-1.9 mm

Westerlund 1887: 81.

The identity of this taxon needs to be confirmed. Mylonas (unpublished Ph. D. thesis, 1982) classified the form of Síros as *Lauria cylindracea*. The description of *umbilicus* deviated considerably from *cylindracea*, so it is preliminarily kept here.

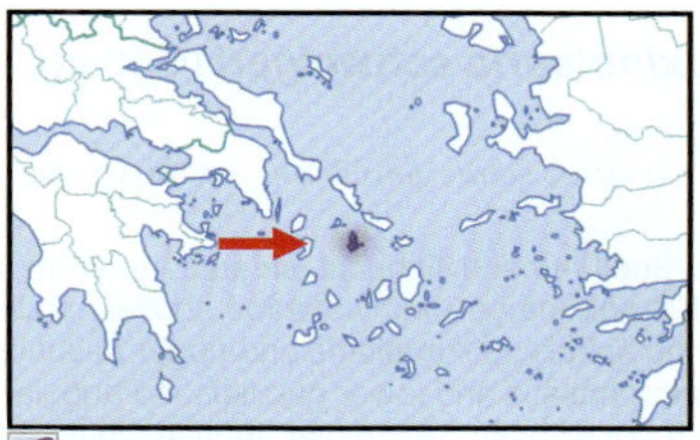

Greece (island of Síros)

Leiostyla

Leiostyla Lowe, 1852
in Lowe 1852: 276 *Pupa (Leiostyla)*

Type species *Pupa vincta* Lowe, 1852
(subsequent designation by Pilsbry 1922-1926 [1922]: 67)
(= currently *Leiostyla vincta* (Lowe, 1852) from Madeira)

Contained originally 6 species, *vincta, irrigua, laurinia, laevigata, recta, macilenta*. Lowe 1855: 208 gave as type *Vertigo anglica* Férussac, 1821, which was not originally contained.

Azores, Madeira, Canary Islands, W Europe

Leiostyla anglica (Férussac, 1821)

Vertigo anglica Férussac, 1821 in A. Férussac [1821-1822]: 64
Férussac's name preceded Wood 1828, the entry in the Official List (from ICZN Opinion 336) *Turbo anglicus* Wood, 1828 is incorrect.

Shell egg-shaped, reddish brown, almost smooth or weakly striated, 5-6.5 very weakly convex whorls, apertural margin rounded, U-shaped with pronounced circular angular sinus and strong angularis, strong parietalis deeper inside, strong columellaris, 2 weaker and deeper palatal folds.
Animal light, almost transparent except greyish head and tentacles, upper tentacles long, lower tentacles very short, like tubercles.

3.1-3.7 x 1.7-2.1 mm

British Isles: Isle of Man, Glennhage, Sykes

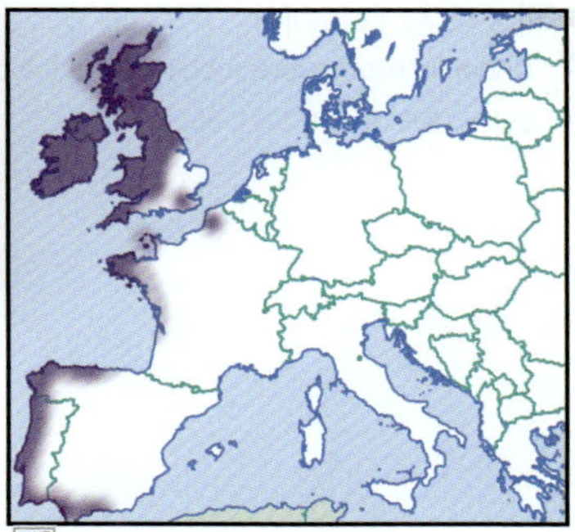

W Europe, Madeira, Algeria

In the west of Ireland and Scotland in humid habitats of all kinds, shaded forests and open pasture, more eastwards mostly in undisturbed old woodland. Under desintegrating leaves, under stones and on humid mosses, seems to tolerate high temperatures. In central and S England it is a useful indicator of ancient woodland that has never been clear felled. In NW Spain in very humid and shady habitats under stones and fallen tree trunks.

Vulnerable in Ireland. Seriously declining in central and S England, threatened by woodland clearances, has disappeared almost entirely from SE England since prehistorical times. Should be considered as Critically Endangered in SE England, and at least as Vulnerable in central England. More frequent in Cornwall, Wales, N England and Scotland.

Nobre 1941: 148, Rolán & Otero-Schmitt 1988: 112, Seddon et al. 1993, Anderson 1996: 8, Kerney 1999: 104, Rolán 2003, Irikov & Erőss 2008 (previous Bulgarian reports referred to *Euxinolauria schweigeri*), Torres 2010: 38.

Euxinolauria

Euxinolauria Lindholm, 1924
in Lindholm, 1924: 66
Expressly established as a new replacement name for

Type species *Pupa caucasica* Pfeiffer, 1857
(absolute tautonymy by Caziot & Margier 1909)
(= currently *Euxinolauria caucasica* (Pfeiffer, 1857) from Caucasus)

Caucasica Caziot & Margier, 1909 in Caziot & Margier 1909: 141 *Lauria (Caucasica)* not *Caucasica* Boettger, 1877 (Clausiliidae)
Caucasica Caziot & Margier, 1909 contained originally 4 species, among which was *Pupa caucasica* Pfeiffer, 1857.

Euxinolauria crassilabris (Hausdorf, 1990)

Leiostyla crassilabris Hausdorf, 1990 in Hausdorf 1990: 53

Shell looks like an orculid, almost smooth, whorls weakly convex, aperture with broad white margin and several teeth: spiralis strong and high, parietalis lower, columellaris strong, basalis short, lower palatalis strong, subangularis short and strong, visible in the aperture.

3.3-3.7 x 1.7-1.8 mm

Shady limestone rock rubble habitats.

Only known from one locality.

Schütt 2001: 51.

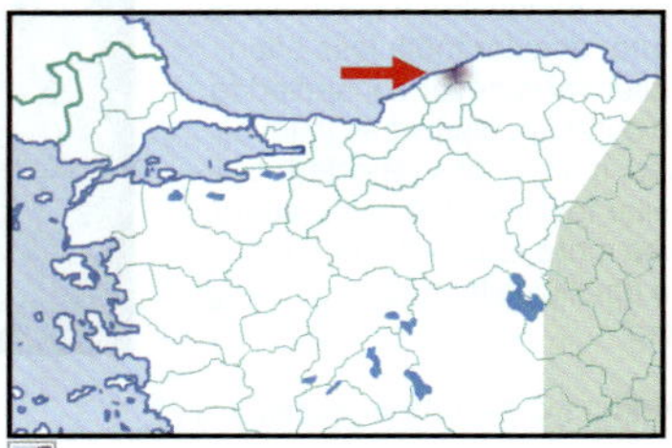

NW Turkey

Euxinolauria paphlagonica (Hausdorf, 1990)

Leiostyla paphlagonica Hausdorf, 1990 in Hausdorf 1990: 52

Shell brown and transparent, very finely striated, almost cylindrical, whorls convex with deep suture, last whorl with largest diameter, apertural lip weakly reflected, spiralis very prominent, parietalis less prominent, columellaris strong, lower palatalis strong, subangularis short and strong, no basalis. In some populations a basal keel is present, with insections at the inner and outer side.

3.5-4.2 x 1.7-2.0 mm

At exposed limestone rocks.

Possibly rare, only known from few localities.

Schütt 2001: 51.

NW Turkey

Euxinolauria schweigeri (Götting, 1963)

Leiostyla schweigeri Götting, 1963 in Götting 1963: 31

Shell rather cylindrical, striated, aperture large with broad margin, with insertion above spiralis, strong spiralis, columellaris and upper palatalis, weaker basalis and lower palatalis, no angular tooth but with almost circular angular sinus. Shell less egg-shaped than *Leiostyla anglica*, whorls more convex, basalis and lower palatalis stronger.

Shady forest habitats in leaf litter and rock rubble.

3.5-4 x 2.0-2.4 mm S Bulgaria: Strandza mountains, Kiten village Dilian Georgiev

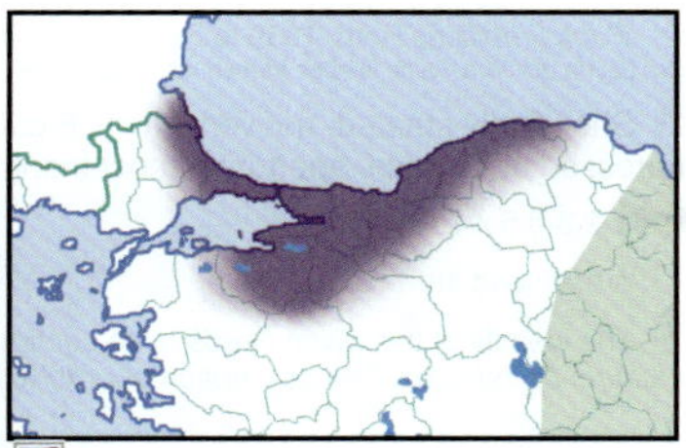

SE Bulgaria, N Turkey

Damjanov & Likharev 1975: 104 ("*Leiostyla anglica*", also from "Silistar" in NE Bulgaria, needs comfirmation), Hausdorf 1990: 49, Schütt 2001: 54, Irikov & Erőss 2008.

Euxinolauria superba (Hausdorf, 1990)

Leiostyla superba Hausdorf, 1990 in Hausdorf 1990: 49

Shell horny coloured, more or less densely ribbed, slightly conical, aperture circular with some 9 folds inside (2 strong parietal, 2 weak angular at both sides, 2-3 columellar, 2-3 palatal), margin not very broad, umbilicus very narrow. Size variable, but not much within a population, shells in the south are smaller and less densely ribbed.

2.4-3.4 x 1.9-2.6 mm

Under limestone rock rubble and at the base of tree trunks in forests.

Subai 1993, Schütt 2001: 56.

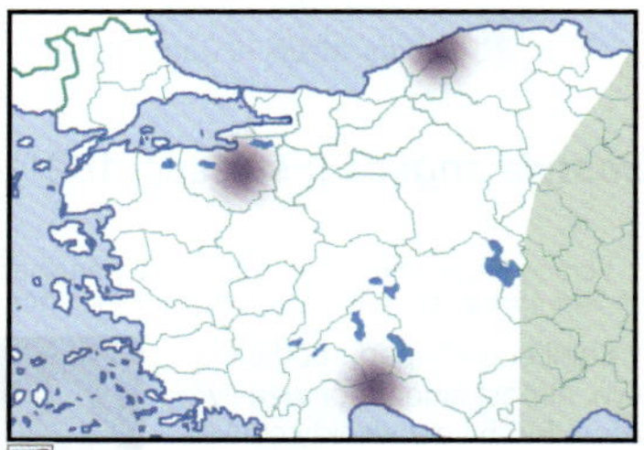

NW and SW Turkey

Agardhiella

Agardhiella Hesse, 1923
in Hesse 1923: 195 *Agardhia (Agardhiella)*

Type species *Pupa truncatella* Pfeiffer, 1841 (monotypy)

Established for "die Sippe der *A[gardhia]. truncatella*" sensu Wagner 1914 (in Sturany & Wagner 1914). Only *truncatella* was cited by name, the others were not originally contained under Art. 67.2.3.

Subai 2008 (revision of species of the Adriatic Sea bordering countries), Subai 2011 (revision E Balkan)

Agardhiella armata (Clessin, 1887)

Sphyradium parreyssii var. *armata* Clessin, 1887 in Clessin 1887: 247 (*Sphyradium Parreyssii* var. *armata*)

Shell densely ribbed (50-80 ribs on last whorl) with minute riblets between main ribs, 5.5-7.5 whorls, aperture U shaped, with weak parietalis, elongate angularis, often a short infraparietalis, sometimes a faint columellaris, slightly impressed in centre of palatal wall, sometimes with weak palatalis. Prominence of apertural folds locally variable, weaker in Bulgaria, stronger near Skopje, very weak in parts of Serbia, locally very strong in Romania.
Differs from *Ag. caesa* in the presence of a parietalis. *Ag. lamellata* has stronger columellaris, parietalis and palatalis, and also a basalis (always absent in *Ag. armata*). *Ag. parreyssii* has no angularis and a stronger palatalis.

3.5-6.5 x 1.2-2.4 mm, locally very variable

Rep. Makedonija: near Skopje

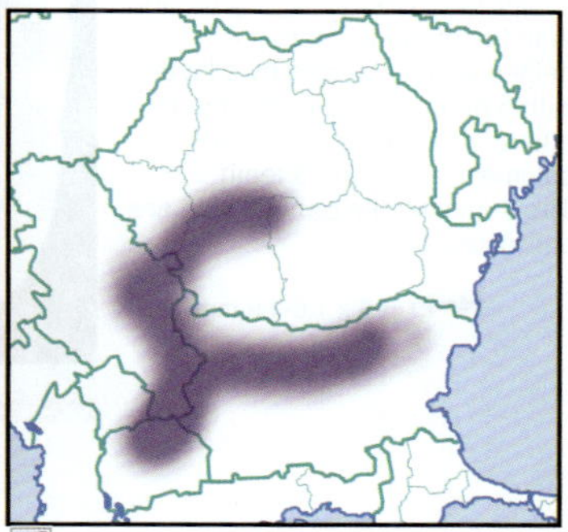

Romania and Bulgaria to Rep. Makedonija

Subai 2011: 81.

Agardhiella banatica (Zilch, 1958)

Argna banatica Zilch, 1958 in Zilch 1958: 151, Taf. 11-12

Shell whitish opaque, very finely ribbed (70-90 ribs on last whorl), 6-7 weakly convex whorls, aperture rhomboid (difference to all other species except *Ag. tunde*) and with a basal sinulus downwards below the columella, palatal margin almost straight, slightly reflected at lower palatal and basal sides, basal margin concave, margins connected at parietal side by a callus, infraparietalis strong, separated from the columellar margin by a deep groove, columella ends basally in a fold, visible in a perpendicular view, umbilicus narrow.
Differs from *Ag. tunde* in the absence of a strong parietalis and an elongate palatalis.

4-5 x 1.4-1.8 mm

Romania: Caraş-Severin, Băile Herculane

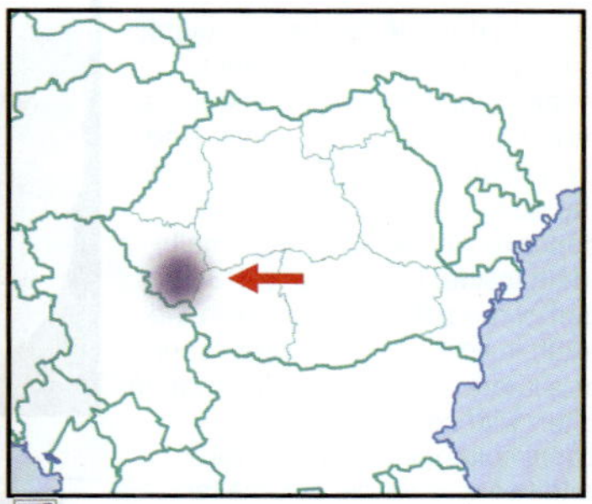

Romania (Banat)

Subterranean, seems to be a cave species, usually found in river deposits.

EU Red List: Near Threatened NT. Only known from 5 localities.

Grossu 1987: 245, Subai 2011: 84.

Agardhiella biarmata (Boettger, 1880)

Pupa truncatella var. *biarmata* Boettger, 1880 in O. Boettger 1880 (Ber. Offenb. Ver.): 109

Shell ribbed (35-55 ribs on last whorl), 6-7 whorls, last whorl more than 1/2 of shell height, aperture narrow with weak infraparietalis, well developed parietalis and weak angularis, strongly developed palatal tooth, sometimes with columellar teeth, umbilicus narrow. Aperture narrower than in *A. truncatella*, with a parietalis and a columellar tooth.

Subai 2008: 84.

3.2-3.9 x 1.4-1.8 mm

Croatia: Korčula island, Pupnatska Luka
Massimo Prodan

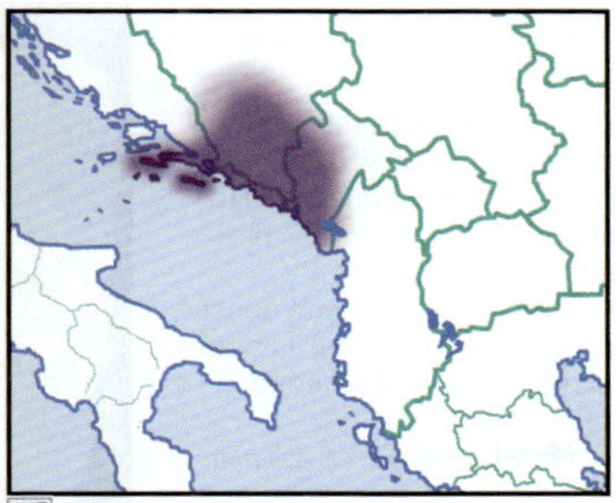

S Dalmatia to Montenegro

Agardhiella caesa (Westerlund, 1878)

Pupa parreyssi var. *caesa* Westerlund, 1878 in Westerlund 1876-1878: 188 (*Pupa (Sphyradium) parreyssi* var. *cæsa)*

Shell densely ribbed (55-90 ribs on last whorl), aperture U shaped, getting slightly broader in its lower part, with strong infraparietalis, weaker angularis, no other folds, a deep canal is present between infraparietalis and columellar wall pointing at right angles inside the aperture.
Differs from *Ag. truncatella* in its denser ribs and wider aperture in the lower section. *Ag. armata* has a parietalis, *Ag. domokosi* has no teeth and wider ribs.

4-6 x 1.4-2.2 mm

Romania: Gorj, Runcu

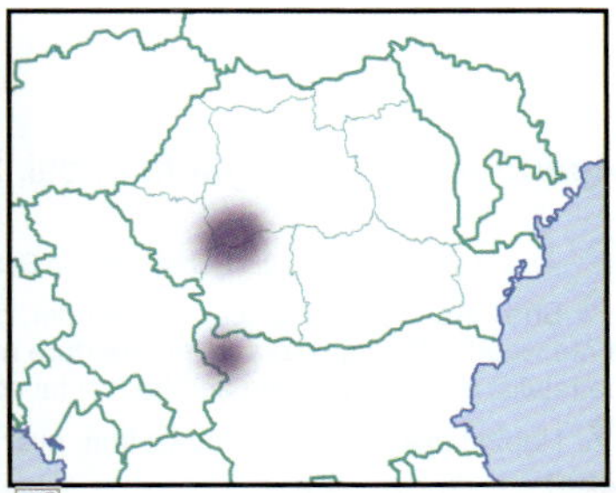

SW Romania, NW Bulgaria

Subai 2011: 85.

Ag. caesa, *Ag. truncatella* and *Ag. armata* seem to be closely related, but were found sympatrically at various locations without intermediate forms.

Agardhiella crassilabris (Grossu & Negrea, 1968)

Argna crassilabris Grossu & Negrea, 1968 in Grossu & Negrea 1968: 729

Shell opaque, regularly striated (60-80 ribs at last whorl), 7-8 whorls, apertural margin sharp at the uppermost palatal margin, broad from middle palatal to columellar side, interior slightly broader than the margins, margin connected at parietal side by a layer with 5 or more equally prominent and relatively long parallel folds, columellar side with 3 teeth.
The 5 folds at the parietal side are characteristic.

5-6 x 1.6-1.9 mm

Romania: NW Gorj, Motru Sec

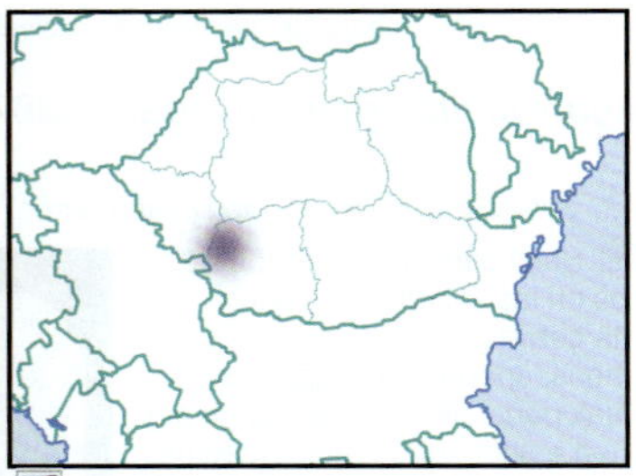

Romania (SW Carpathians)

Very humid habitats.

EU Red List: Near Threatened NT. Only known from 4 localities. No real threats known.

Grossu 1987: 247, Subai 2011: 87.

Agardhiella dabovici Gittenberger, 1975

Agardhiella dabovici Gittenberger, 1975 in Gittenberger 1975: 280

Shell sharply ribbed (35-50 ribs/last whorl), 6.5-7.5 moderately convex whorls with deep suture, last whorl 51-56 % of shell height, aperture with strong parietalis, weak infraparietalis and angularis, margin connected at parietal side by a strong callus, often repeating the rib structure, palatal margin concave with two merged small teeth inside, the concave 1/3 section of the palatal margin with the teeth inside protrudes when seen in a lateral view, columellar margin often with some knots, basal/columellar section inside with two well visible columellar folds, umbilicus narrow.
A. zoltanorum has a higher and narrower aperture, and a long columellaris.
The animal is blind, eyes have no pigments.

3.6-5.7 x 1.7-2.2 mm

Montenegro: near Boljevići S of Virpazar

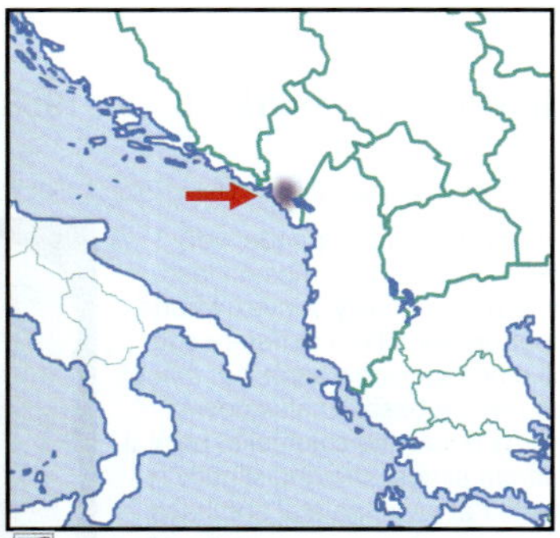

W Montenegro

Known from caves.

EU Red List: Near Threatened NT. Only known from 4 localities. No real threats known.

Subai 2008: 93.

Agardhiella domokosi Subai, 2011

Agardhiella domokosi Subai, 2011 in Subai 2011: 89

Shell with moderately widely spaced ribs (50-60 ribs on last whorl), 7-8 whorls, aperture U shaped without any folds, thickened at all sides.
Ag. incerta is smaller with denser and less prominent ribs, its apertural margin is not thickened.

4.5-5.5 x 1.6-1.9 mm

W Romania: Caraş-Severin, near Anina

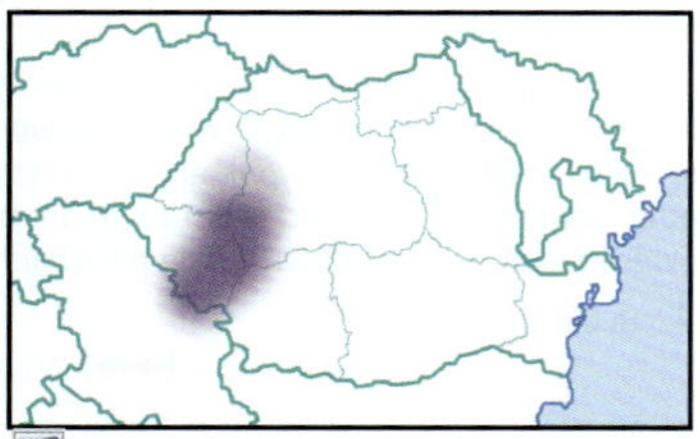

SW Romania

Agardhiella extravaganta Subai, 2008

Agardhiella extravaganta Subai, 2008 in Subai 2008: 95

Shell broad and biconical, not slender, ribbed (40-60 ribs on the last whorl), 6.5-7 convex whorls, penultimate whorl with largest diameter, aperture nearly triangular with angularis and infraparietalis, extremely strong parietalis, extremely strong and horizontal columellaris, palatal margin slightly convex, umbilicus open.

Habitat unknown, only known from river deposits.

3.6-4.5 x 2.2-2.4 mm

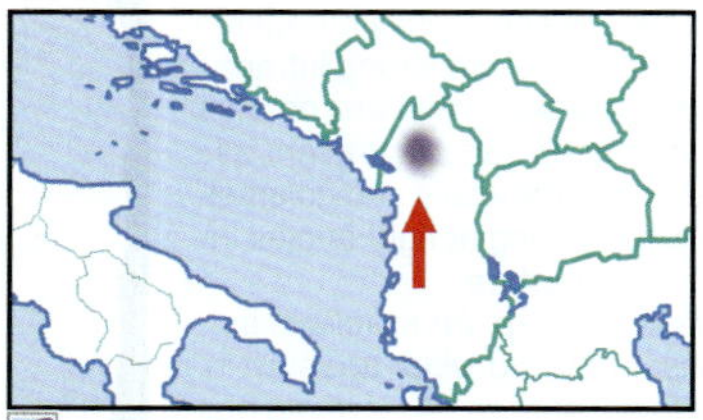

N Albania

Agardhiella formosa (Westerlund, 1887)

Pupa truncatella var. *formosa* Westerlund, 1887 in Westerlund 1887: 88

Name was first mentioned as *Pupa truncatella* var. ß synonym *Pupa formosa* Pfeiffer, 1848 in L. Pfeiffer 1848: 304, with a description, specimens were not cited. Perhaps the first to have used this name for a taxon was Westerlund 1887: 88 who gave a short description and a reference to Küster 1852: Taf. 4 Fig. 22-23, not to Pfeiffer 1848.

Pfeiffer 1848 gave localities from Slovenia to Turkey. Westerlund 1887 used the name for a form from "Dalmatia". Gittenberger 1975: 284 attributed the name to Pfeiffer 1848 and restricted the name to a form from S Dalmatia, Bosnia and Montenegro. Subai 2008: 91 used the name for a microendemic species from Montenegro, and designated a neotype since Pfeiffer's specimens were lost, but did not consider that Westerlund's and Küster's specimens were the types that had to be traced under Art. 72.4.3. So the neotype is probably invalid (Art. 75.3.4) or at least threatened if syntypes are discovered which would set aside the neotype (Art. 75.8).

For those who require a bibliographic reference for making available a synonym under Art. 11.6.1, the name must be attributed to Westerlund 1887. For the others, to Pfeiffer 1848. The types would remain the same.

Shell densely ribbed (35-50 ribs on the last whorl), 6-7 whorls, aperture with strong parietalis, much weaker infraparietalis and angularis, palatal margin of the aperture slightly concave with a weak tooth-like callus corresponding inside, sometimes with weak folds or teeth at the basal or columellar sides.
Differs from *A. truncatella* in its strong parietalis and concavely indented palatal apertural margin. *A. biarmata* is smaller with denser and weaker ribs, and has a smaller aperture.

3.4-5 x 1.5-2 mm

Subterranean, in caves and rock crevices.

Gittenberger 1975: 284, Subai 2008: 91.

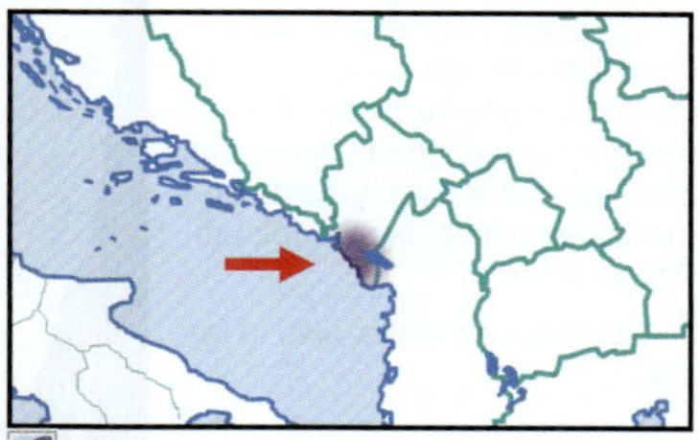

W Montenegro

Agardhiella grossui (Zilch, 1958)

Argna grossui Zilch, 1958 in Zilch 1958: 150, Taf. 11-12

Shell white, translucent, densely ribbed (60-75 ribs on last whorl), more sharply on last whorl, 6-7 weakly convex whorls, aperture very narrow, oblique, palatal margin extremely concave, upper palatal margin producing a sinulus running from parietalis to the angular edge, lower palatal, basal and columellar margins U-shaped, broadly reflected with a sharp and smoothly running edge inside, margins connected at parietal side by a strong callus, parietalis S shaped at the margin, 3 small infraparietal teeth present on the callus between columellar margin and parietalis, on the angular side another tooth is present, columellar fold visible inside the aperture, umbilicus narrow.
In *Ag. stenostoma* and *Ag. reinhardti* the aperture is less narrowed. *Ag. zoltanorum* has a much broader shell at similar height and a wider umbilicus.

4-5 x 1.4-1.7 mm

Romania: Ponoarele near Baia de Aramă

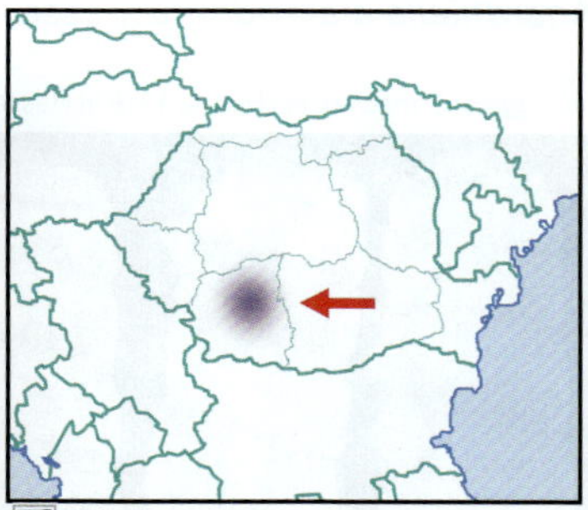

Romania (Oltenia)

Was found in caves.

EU Red List: Near Threatened NT. Only known from 3 localities, obviously rare.

Grossu 1987: 248, Subai 2011: 92.

Agardhiella incerta Grossu, 1987

Agardhiella incerta Grossu, 1987 in Grossu 1987: 253

Shell looks like a small and juvenile *Agardhiella*, very finely ribbed (60-90 ribs on last whorl), aperture simple without teeth, margin not thickened or reflected, not connected by a callus or white layer at the parietal side (difference to otherwise similar species), palatal margin straight or weakly convex, umbilicus narrow and rounded.

3.7-4.7 x 1.3-1.6 mm

On limestone rocks in valleys, also in forests.

Subai 2011: 94.

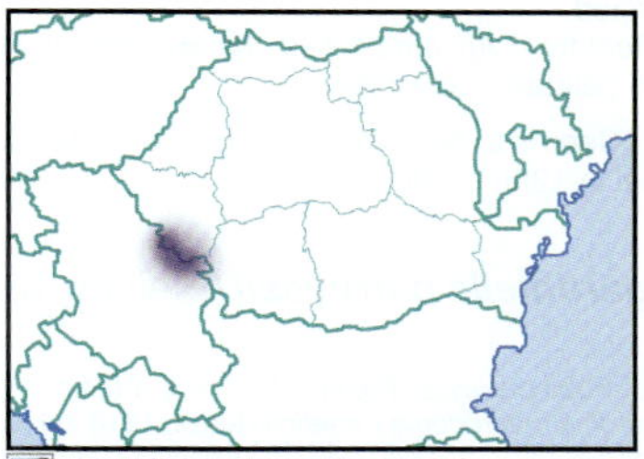

SW Romania, NE Serbia

Agardhiella lamellata (Clessin, 1887)

Sphyradium parreyssii var. *lamellata* Clessin, 1887 in Clessin 1887: 248 (*Sphyradium Parreyssii* var. *lamellata*)
Agardhiella prahovensis Grossu, 1986 in Grossu 1986: 8

Shell white and densely ribbed (60-75 ribs/last whorl), aperture narrow and bold-X shaped with infraparietalis, parietalis, angularis and occasionally some more small folds at the parietal side, palatal and columellar teeth prominent (the strong columellar tooth is characteristic), of similar size and sitting opposite to each other, basalis prominent and sitting opposite to parietalis, umbilicus closed. Apertural characters quite constant.
Aperture larger and broader than in *Ag. macrodonta* (the palatal tooth in *Ag. lamellata* is smaller). *Ag. armata* and *Ag. macrodonta* have no basalis.

4-5.5 x 1.3-1.8 mm, diameter variable, also within a population

Romania: Vâlcea, Vânturarița mountains, 1073 m

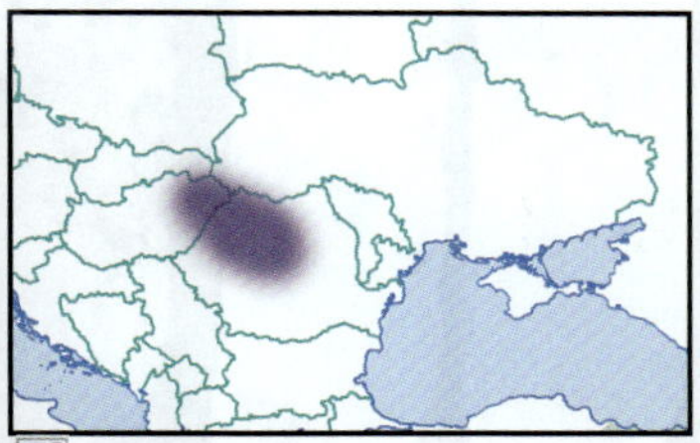

W Romania
Also found in river deposits in E Hungary and S Romania, and fossil in S Slovakia.

In mountains at humid decaying wood, under stones and leaf litter, also in rock crevices, not restricted to limestone habitats, occasionally also 20 cm deep in sand.

Kerney et al. 1983: 124, Grossu 1987: 243, Subai 2011: 96.

Agardhiella langaleta Subai, 2011

Agardhiella langaleta Subai, 2011 in Subai 2011: 99

Shell sharply and moderately densely ribbed (40-55 ribs/last whorl), 6-8 whorls, aperture narrow and (shaped, palatal tooth very strong, palatal side concave, parietalis and columellar tooth weak, parietal margin detached (this is characteristic), more deeply inside another parietal tooth is present, umbilicus narrow.

At limestone rocks, also in forests.

4.5-5.5 x 1.5-1.8 mm

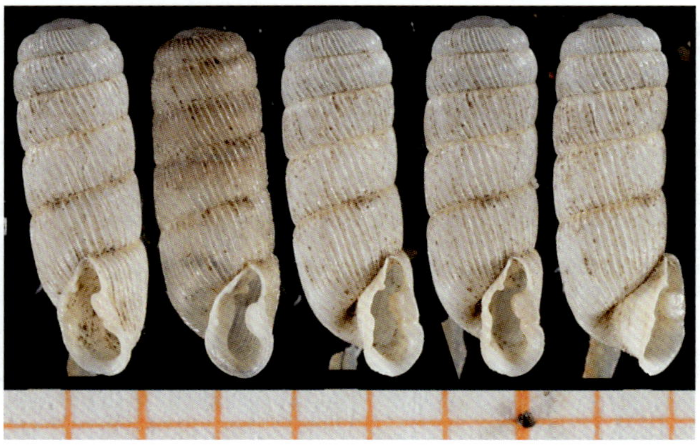

W Bulgaria (Iskar valley)

EU Red List: Near Threatened NT. Only known from a small restricted range. No real threats known.

NW Bulgaria: Iskar valley, Ritlite

Agardhiella macrodonta (Hesse, 1916)

Agardhia macrodonta Hesse, 1916 in Hesse 1916: 116

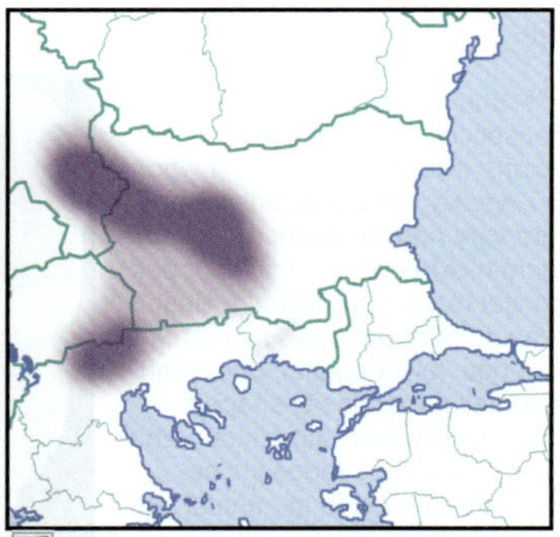

E Serbia, W Bulgaria, N Greece

A rare species, should be protected.

Damjanov & Likharev 1975: 105, Irikov & Mollov 2006: 764, Irikov & Georgiev 2008: 136, Subai 2011: 100.

3.5-4.4 x 1.3-1.7 mm

Bulgaria: near Panagyurishte

Sometimes misspelled *A. macrostoma*.

Shell densely ribbed (40-65 ribs/last whorl), apex blunt, 6-7 whorls, aperture not very narrow, approximately r shaped, lower section U shaped, parietalis well developed, occasionally with other weak parietal folds, palatal tooth strong, sometimes basally elongated, palatal side slightly concave, columellaris usually present and narrowing the aperture more deeply inside, columellar side slightly concave, margin reflected and sharp, umbilicus narrow and comma shaped.

Differs from *Ag. lamellata* in the absence of a basal tooth. *Ag. parreyssii* has usually no columellar tooth and a straight or convex columellar margin, parietalis and palatalis are weaker.

Seems to be confined to limestone, was found in rock crevices, among herbal plant roots, in old forests and city park forests with broad leaves. From sea level to 1200 m altitude.

Agardhiella parreyssii (Pfeiffer, 1848)

Pupa parreyssii Pfeiffer, 1848 in L. Pfeiffer 1848: 311 (*Pupa Parreyssii*)
Agardhia macrodonta rumelica Hesse, 1916 in Hesse 1916: 117
Argna buresi Urbański, 1960 in Urbański 1960: 65 (*Argna (Agardhiella) bureši*)
Agardhiella licherdopoli Grossu, 1986 in Grossu 1986: 9
Argna densicostata Grossu & Negrea, 1968 in Grossu & Negrea 1968: 731
Argna nana Grossu & Negrea, 1968 in Grossu & Negrea 1968: 731

Bulgaria, N Greece

Among rocks and under leaf litter, or near cave entrances, in forests, open terrain or near marshes. Usually on limestone substrate, occasionaly also on marble and crystalline rocks. Difficult to find in their habitats, often found in river deposits.

3.3-5.7 x 1.2-1.7 mm

(1-4) S Bulgaria, near Peshtera, (5-7) N Bulgaria, Sredna Gora, Ostra Mogila

Subai 2011: 103.

Shell whitish and transparent, usually finely ribbed (40-80 ribs/last whorl), 6-8 whorls, aperture moderately broad, no angularis, oblique parietalis and palatal fold in the upper part of the palatal wall usually present, margin with depression at palatalis, sometimes with infraparietalis, rarely with columellaris and infracolumellaris (but no prominent tooth deep in the aperture), umbilicus narrow.

Very variable in terms of shell size and prominence of apertural folds. In the centre of the range of the species the teeth can be entirely absent. Individual shells with well developed parietalis and strong palatal tooth have been called *rumelica*; these characters are variable within populations.

A. macrodonta has stronger parietalis and palatalis, and a well developed tooth deep inside the aperture, its columellar side is slightly concave (curved inwards). *A. domokosi* has a larger diameter, larger rib interspaces and whorls more slowly increasing, at equal height *A. domokosi* has one more whorl. Toothless forms of *A. armata* are larger and have a larger diameter, a larger aperture with an angularis and weaker palatalis and parietalis.

Reports prior to 2011 for Serbia, Romania and Hungary referred to other species, often to *A. domokosi*. Reports for *rumelica* referred to *parreyssii*.

Agardhiella pirotana Subai, 2011

Agardhiella pirotana Subai, 2011 in Subai 2011: 107

Shell densely ribbed (50-70 ribs/last whorl), 5.5-6.5 whorls, aperture U shaped, broadening to columellar side, no folds except a shallow elongate angularis, palatal wall straight or slightly convex, umbilicus narrow.
Differs from *Ag. caesa* and *Ag. truncatella* in the absence of a strong infraparietalis, the shell of *Ag. domokosi* is broader and has more prominent ribs, toothless forms of *Ag. parreyssii* have no angularis, Serbian populations of *Ag. armata* with weak folds are larger and the folds are evenly weak, their angularis is not stronger.

3.5-4.5 x 1.3-1.6 mm

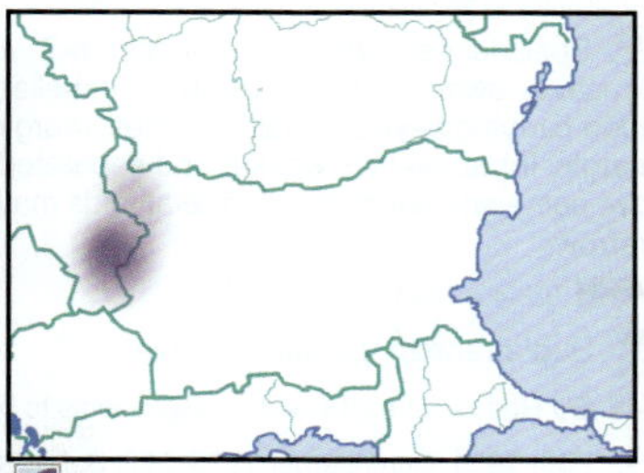

SE Serbia, NW Bulgaria

Rocky limestone habitats.

EU Red List: Near Threatened NT. Only known from 4 localities. No real threats known.

Differs only very slightly from similar taxa.

Serbia: 25 km S Bela Palanka

Agardhiella reinhardti (Zilch, 1958)

Argna reinhardti Zilch, 1958 in Zilch 1958: 151, Taf. 11-12
Argna angustistoma Grossu & Negrea, 1968 in Grossu & Negrea 1968: 728

Shell whitish opaque, very finely ribbed (65-70 ribs/last whorl), 6.5-7 weakly convex whorls, aperture moderately narrow, palatal margin concave and with weak denticle inside, lower palatal, basal and columellar margin U shaped and slightly reflected, parietal callus present but not very prominent, infraparietalis short and prominent, parietalis moderate, angularis weak, columellar fold inside weakly visible, columellar margin undulating in lateral view and not straight (main difference to similar other species), umbilicus very narrow.
Ag. grossui has a much narrower aperture. *Ag. serbica* is very similar, its umbilicus is slightly wider. *Ag. stenostoma* has more coarsely arranged ribs and always a smaller shell. *Ag. zoltanorum* is more coarsely ribbed, has a much broader shell at similar height and a wider umbilicus.

4.5-5.5 x 1.6-1.9 mm

Rocky limestone habitats.

EU Red List: Near Threatened NT. Only known from few localities, apparently a rare species. No real threats known.

Grossu 1987: 244, Subai 2011: 108, www.iucnredlist.org (07.2012).

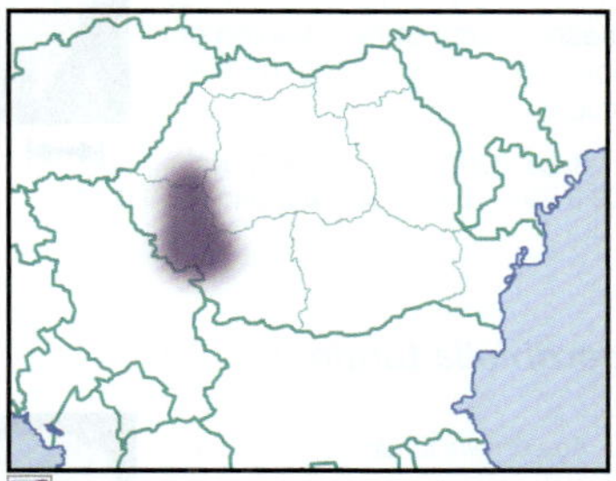

W Romania

Agardhiella serbica Subai, 2011

Agardhiella serbica Subai, 2011 in Subai 2011: 110

Shell finely ribbed (60-80 ribs/last whorl), 5.5-6.5 whorls, apertural U shaped in lower section, margin moderately broad, infraparietalis short, parietalis well developed, rarely with more small folds at the parietal side, columellar side slightly concave, palatal side slightly concave and impressed, palatal margin in the lower part broad, in the upper third sharp, umbilicus moderately wide.
Differs from *Ag. reinhardti* in its broader aperture and wider umbilicus, in *Ag. grossui* the aperture is even more compressed. *Ag. stenostoma* and *Ag. grossui* have a much narrower umbilicus, *Ag. zoltanorum* is more coarsely ribbed, has a broader shell and a wider umbilicus.

3.9-4.8 x 1.45-1.75 mm

Serbia: Stol mountain near Luka

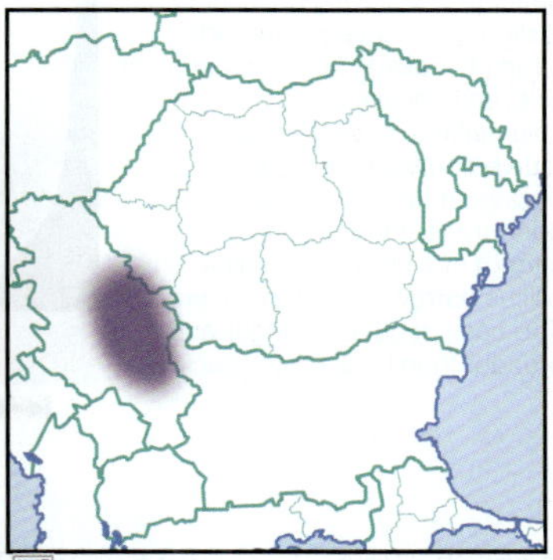

SE Serbia

On limestone rocks, often in forests.

EU Red List: Near Threatened NT. Possibly exaggerated. Not known from many localities. No real threats known.

Agardhiella skipetarica (Wagner, 1914)

Agardhia truncatella skipetarica Wagner, 1914 in Sturany & Wagner 1915: 65, Taf. 18

Shell weakly but densely ribbed (55-65 ribs on the last whorl), 5.5-6.5 whorls, aperture without teeth, with weak lip inside, umbilicus narrow. Differs from the other species in its slender shell, weak and dense ribs, and no teeth in the aperture.

Critically Endangered in Albania (Dhora 2004: 144), possibly exaggerated.

Subai 2008: 89 (distribution in Greece obviously restricted to Kérkira), Fehér & Erőss 2009 (N Albania, range overlaps with *A. truncatella*).

3.4-4.3 x 1.45-1.65 mm

Greece: Kérkira, Dasia — Massimo Prodan

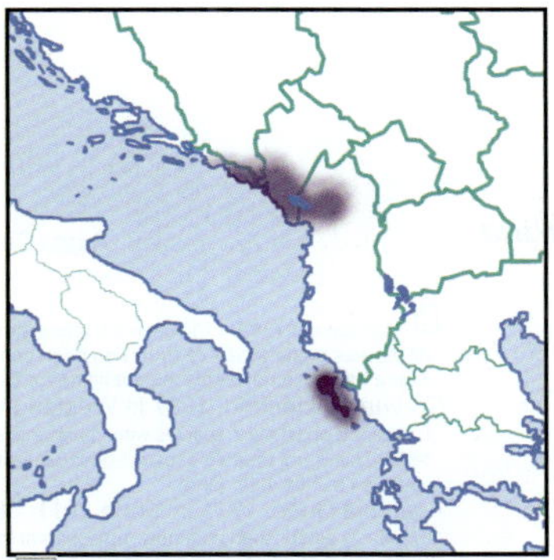

S Dalmatia to N Albania, Kérkira

Agardhiella stenostoma (Flach, 1890)

Pupa stenostoma Flach, 1890 in Flach 1890: 53, footnote 2 (*Pupa (Coryna) stenostoma*)
Original genus was *Pupa* (p. 50), *Coryna* is to be seen as a subgenus according to the style of the work.

Shell densely ribbed, translucent, 6-7 weakly convex whorls, aperture narrow and oblique, palatal margin concave, lower palatal, basal and columellar margin U-shaped and moderately broadly reflected, inside with edge carrying several small denticles, margin connected at parietal side by a callus, parietalis S-shaped at the margin, infraparietalis weaker and separated from the nearby columellar margin by a narrow groove, angularis low, some additional small parietal folds may be present, columella inside with a prominent fold, umbilicus very narrow.

3.2-4 x 1.3-1.55 mm

Subterranean, perhaps in caves.

EU Red List: Least Concern LC. Seems to be a rare species, only known from 6 localities.

Zilch 1958, Gittenberger 1975: 285, Subai 2008: 87.

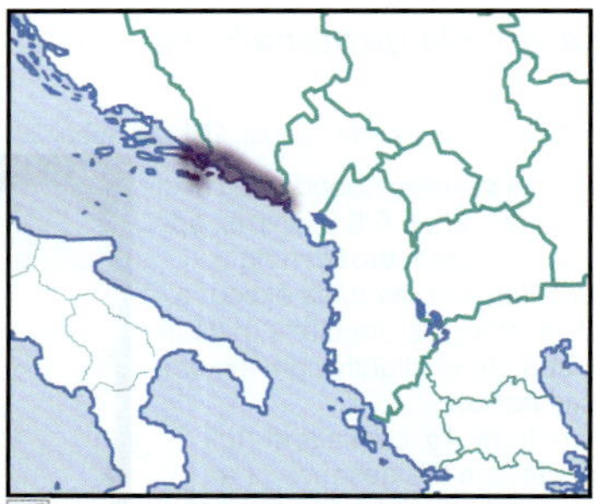

S Dalmatia to N Montenegro

Agardhiella truncatella (Pfeiffer, 1841)

Pupa truncatella Pfeiffer, 1841 L. Pfeiffer 1841: 46 (*Pupa Truncatella*)

Shell pale yellow, ribbed (45 ribs on last whorl), last whorl 50 % of shell height, aperture 25 %, narrow without folds and structures at parietal side.

Under leaves in shrublands and forests, under rotting wood and stones on humid forests soils, at bases of tree trunks, also on dry slopes and on rocks, 300-1000 m altitude.

Subai 2008: 76. Fehér & Erőss 2009 (N and central Albania), Subai 2011: 112.

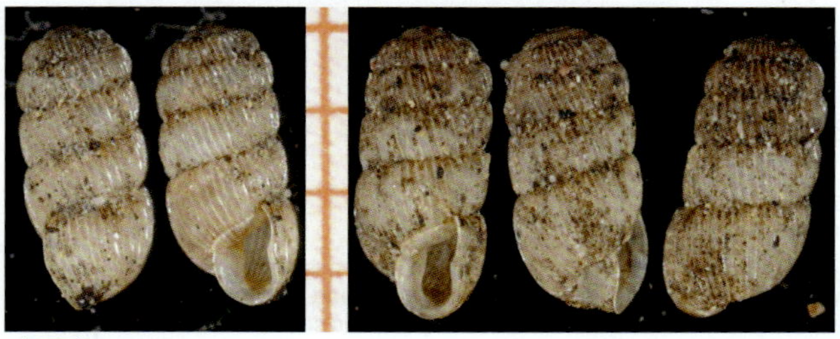

3.0-4.5 x 1.3-1.8 mm

Austria: Kärnten, Tscheppaschlucht

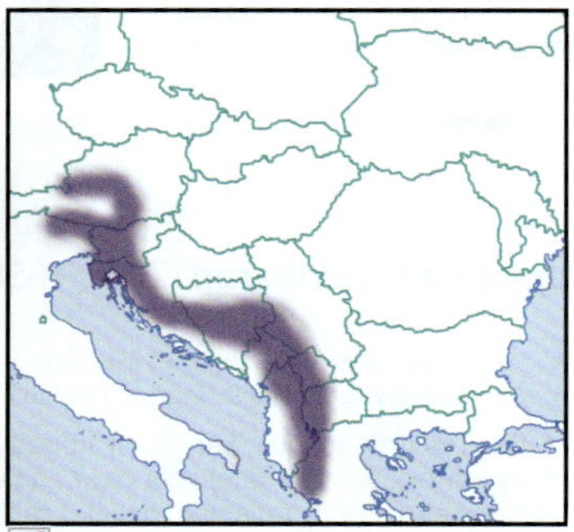

Austria to Greece

Agardhiella tunde Deli, 2010

Agardhiella tunde Deli, 2010 in Deli 2010: 315

Shell very finely ribbed (70-110 ribs/last whorl), 6.5-8 whorls, aperture narrow, with infraparietalis, parietalis and angularis and often some more shallow folds at parietal side, strong palatal tooth at a slightly concave palatal side, strong basalis at an almost straight or even concave basal margin, columellar side very concave (curved outwards at its base), producing an oblique x shape of the aperture (but the parietalis is not so strong) in a rhomboid outline, umbilicus narrow and often covered. The basal tooth is stronger and broader than in all other species.

4-5.5 x 1.45-1.7 mm

Romania: Gorj, Pârvuleşti near Târgu Jiu

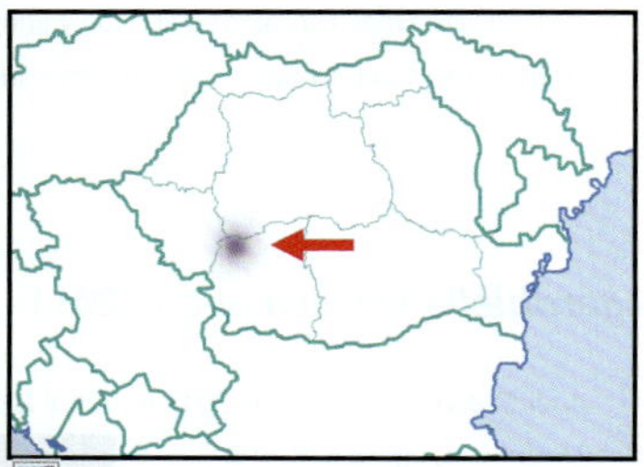

SW Romania (Gorj)

Limestone habitats, found in gorges and river deposits, probably mainly a subterranean species.

EU Red List: Vulnerable VU. Only known from 3 localities. No real threats known.

Subai 2011: 112.

Agardhiella zoltanorum Subai, 2008

Agardhiella zoltanorum Subai, 2008 in Subai 2008: 97

Shell cylindrical, moderately strongly ribbed (45-65 ribs/last whorl), 6-7 whorls, aperture with strong parietalis and moderate infraparietalis, vertical columellaris, palatal margin strongly concave (producing a very narrow aperture), upper palatal margin thin, lower margin broad, umbilicus moderately wide.
Less slender and less densely ribbed than *Ag. reinhardti* and *Ag. stenostoma*, shell broader and umbilicus wider than in *Ag. grossui*.

3.5-4.7 x 1.7-2.3 mm

Subai 2011: 93.

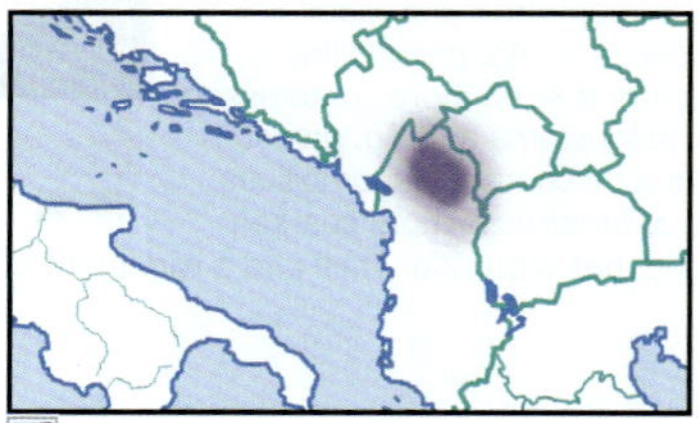

N Albania

Argna

Argna Cossmann, 1889
in Cossmann 1889: 1104 *Pupa (Argna)*

Type species *Pupa proexcessiva* Sacco, 1888
(monotypy)
(= currently *Argna biplicata* (Michaud, 1831))

Cossmann 1889: 1104 gave a summary of Sacco 1888. The name *Argna* was just mentioned as "*Pupa (Argna) proexcessiva*" in a list of species cited from Sacco's work, without any description. *Argna* was made available by having been combined with a specific name. Sacco 1888: 76 described *Pupa (Coryna) proexcessiva*, so Cossmann just replaced *Coryna* by *Argna* without any comment. *Argna* was not a new replacement name for *Coryna* Westerlund, 1887 non Bosc, 1802 (Cnidaria) because Cossmann 1889 did not give an expressed statement (Art. 67.8).

Coryna Westerlund, 1887 in Westerlund 1887: 78, 87, type species *Pupa biplicata* Michaud, 1831 (subsequent designation by Caziot & Margier 1909: 139)
Contained originally 7 species with variants and synonyms. Pilsbry 1922-1926: 129 regarded *Coryna* Westerlund, 1887 as a new replacement name for *Sphyradium* Hartmann, 1841 non Charpentier, 1837. This is not tenable because because Hartmann 1840: 53 had just used Charpentier's name and did not establish a new name, Westerlund 1887: 87 agreed with this view and did not intend to establish a substitute name.

Agardhia Gude, 1911 in Gude 1911: 361
Expressly established as a new replacement for *Coryna* Westerlund, 1887 non Bosc 1802. Gude 1911 gave as type *Pupa ferrari* Porro, 1838, this was incorrect because *Coryna* had as type already *Pupa biplicata* designated by Caziot & Margier 1909. *Agardhia* is a junior subjective synonym for *Argna*.

This genus presentation was kindly checked by P. Subai.

Argna bielzi (Rossmässler, 1859)

Pupa bielzi Rossmässler, 1859 in Rossmässler 1854-1859 (17-18): 109, Taf. 84 (*Pupa Bielzi*)

Carpathians

Under leaves and litter in highland forests.

Kerney et al. 1983: 123, Sysoev & Schileyko 2009: 36.

4.5-6.7 x 1.7-2.0 mm

(1-5) Romania: Argeş, Piatra Craiului, (6-7) Slovakia: Slovenský Raj, Velky Sokol

Shell horny yellow and transparent, cylindrical with blunt apex, aperture not very narrow with 5 quite variable folds: parietalis usually strong, columellaris strong, middle palatalis strong, upper and lower palatalis smaller.

Argna biplicata (Michaud, 1831)

Pupa biplicata Michaud, 1831 in Michaud 1831: 62, Pl. 15

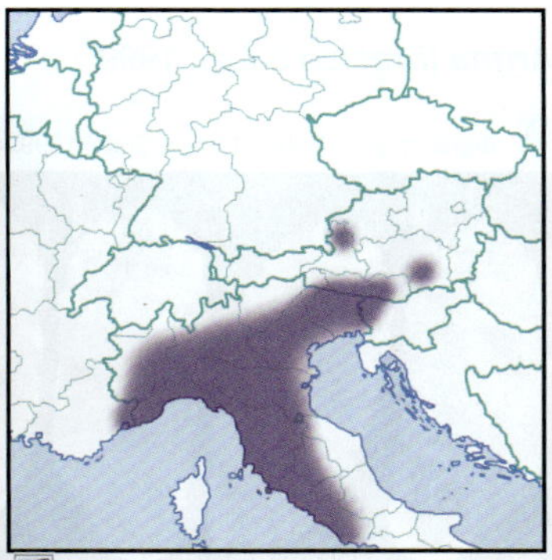

S Alps, Apennines
Isolated in Valencia (Spain) and NW Greece

In humid and sheltered habitats, in soil substrate and moss, at old wood covered by rock rubble, often deep inside rock crevices, rarely found alive.

4-4.5 x 1.5-1.7 mm

Austria: Kärnten, Gailtal near Mauthen

Shell pale yellow, shiny and transparent, almost smooth, very finely striated, cylindrical but very slightly conical, aperture with 4 folds (parietalis, columellaris, strong and long lower palatalis, weak upper palatalis), umbilicus narrow. Shell more slender than *A. ferrari*.

Seems to be locally variable, various subspecies have been recognised. Their alignment with criteria usually applied in biodiversity research (intrincic partial barriers to gene flow) needs to be verified.

Argna bourguignatiana (Nevill, 1880)

Pupa bourguignatiana Nevill, 1880 in Nevill 1880: 127, Pl. 13

NW Italy, SE France

Leaf litter in limestone zones.

EU Red List: Vulnerable VU. Only known from 3 localities, has never been found alive.

4-4.6 x 1.4-1.6 mm

Italy: Imperia, Vallone del Passo

(1-6) Vollrath Wiese

Shell white, shiny and transparent, very finely striated, cylindrical, 8-9 whorls, aperture with some 4 folds (strong parietalis, prominent columellaris, lower palatalis, upper palatalis), umbilicus very narrow.

Differences to *A. biplicata* and systematic position unclear. Obviously not found sympatrically with *A. biplicata*, perhaps only an extreme form at the margin of its range. Nevill 1880 argued that the apertural characters differed clearly, but did not explain how.

Argna ferrari (Porro, 1838)

Pupa ferrari Porro, 1838 in Porro 1838: 57

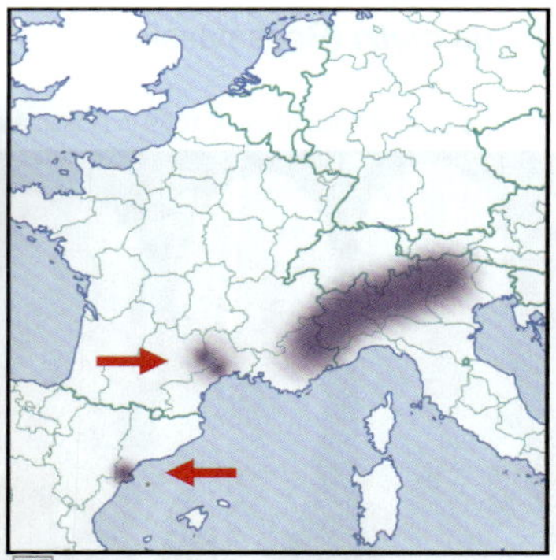

S Alps, central S France, N Spain

Vulnerable in Switzerland.

Falkner 1990: 146, Kerney et al. 1983: 122, Manganelli et al. 1995: 15, Turner et al. 1998: 146, Bertrand 2000: 41.

3.5-5.5 x 1.8-2.0 mm

N Italy: Baveno, near Lago Maggiore

Shell horny brown, densely and regularly ribbed, sometimes evenly cylindrical, sometimes with largest diameter in upper third, apex not very blunt, 7-9 not very convex whorls, aperture narrow with 4 folds not touching apertural margin (1 parietal, 1 columellar, 2 palatal deeper inside), palatal margin slightly convex with an additional palatal tooth at the margin, umbilicus open and deep.

In forests under leaves and between rocks, also in rock crevices, humid habitats. Not restricted to calcareous substrate, but prefers limestone. In Switzerland up to 1700 m.

Argna thracica Subai, 1999

Argna thracica Subai, 1999 in Subai 1999: 43

3.5-4.2 x 1.7-1.8 mm

NE Greece: near Xánthi

NE Greece

Shell brownish, densely ribbed, cylindrical, apex not very blunt, 8-9 moderately convex whorls, apertural margin white and reflected, aperture with 3 folds visible inside (parietal, columellar and palatal), palatal margin slightly convex (weaker than in *A. ferrari*) and with an additional palatal tooth touching the margin, umbilicus open.

Argna valsabina (Spinelli, 1851)

Pupa valsabina Spinelli, 1851 in Spinelli 1851: 138 (*Pupa Valsabina*)

Shell white, smooth, 8 almost straight whorls, aperture narrow and remarkably protruded, margin reflected and rounded, concave at palatal side.

In caves and deep limestone rock crevices, mainly subterranean.

EU Red List: Vulnerable VU. Threatened by habitat destruction.

Gittenberger 1974: 9, Eikenboom 1993.

4 x 1.2-1.3 mm

Italy: Brescia, Marmentino, river deposits (H = 4 mm)

Cesare Bogi

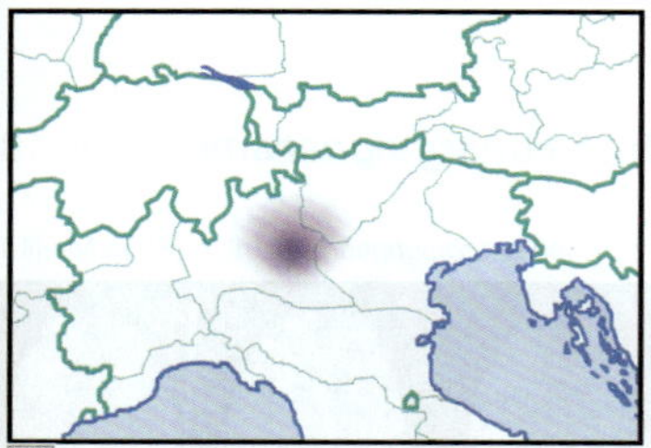

N Italy (S Alps)

Orculidae: *Alvariella, Orcula, Orculella, Pagodulina, Pilorcula, Schileykula, Sphyradium, Odontocyclas, Walklea, Speleodentorcula*

Alvariella

Alvariella Hausdorf, 1996 in Hausdorf 1996: 13, 53

Type species *Alvariella multiplicata* Hausdorf, 1996 (original designation)

Palatal folds and parietalis extending inside (not restricted to the apertural margin), no angularis, subangularis and spiralis. Proximal penis section very slender, penis appendix like in *Walklea, Pilorcula* and *Orculella*, retractor without branch, inserts at epiphallus (not at penis).

Hausdorf 1996: 7.

Alvariella multiplicata Hausdorf, 1996

Alvariella multiplicata Hausdorf, 1996 in Hausdorf 1996: 53

Last 2 whorls narrower than antepenultimate whorl and less prominently ribbed, aperture large, columellar folds visible deep inside aperture, parietalis strong, 3-4 palatal folds in the last whorl, one of which is also visible in the aperture.

6.4-7.7 x 2.9-3.2 mm

Under leaf litter near limestone rocks.

Schütt 2001: 48.

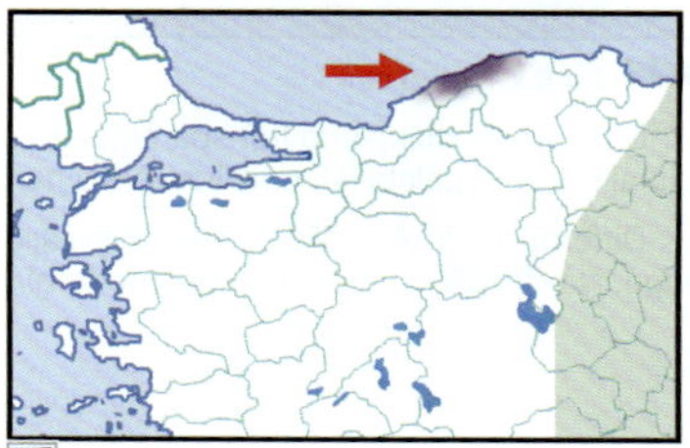

N Turkey

Orcula

Orcula Held, 1838
in Held 1838: 919

Type species *Pupa dolium* Draparnaud, 1801
(subsequent designation by Herrmannsen 1847-1849 [08 Sep 1847]: 158)

Shell cylindrical, apex more or less depressed conical, first whorls without any distinct sculpture, apertural lip and palatal callus do not form a uniform entity, palatal callus can be reduced, parietalis extending far inside (not restricted to the aperturtal margin), no angularis, subangularis and spiralis, palatal folds restricted to the last 1/4 of the last whorl.
No penial appendix present, retractor inserts at the base of the flagellum or more proximal at the penis.

Zimmermann 1932, Klemm 1967, Hausdorf 1996: 7.

Orcula austriaca Zimmermann, 1932

Orcula spoliata austriaca Zimmermann, 1932 in Zimmermann 1932: 37

6-7 x 2.6-2.8 mm

Austria: Niederösterreich, Rossbachklamm N of Höllental

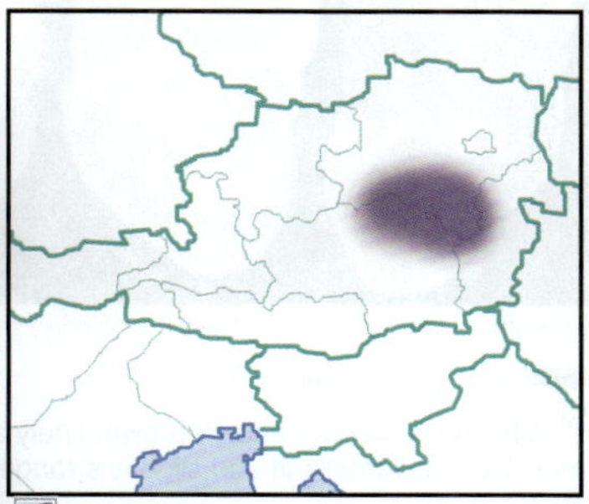

Austria (E Alps)

Shell like *Orcula gularis*, strong palatal callus turned inside.

Humid and dark gorges to S exposed dry grassy or rocky slopes, at old wood, in organic litter and at rocks.

Klemm [1973], Kerney et al. 1983: 101.

Orcula conica (Rossmässler, 1837)

Pupa conica Rossmässler, 1837 in Rossmässler 1835-1837 (5-6): 17, Taf. 23

5.5-7.0 x 3.0-3.5 mm

Austria: Kärnten, Gailtaler Alpen, near Weissensee

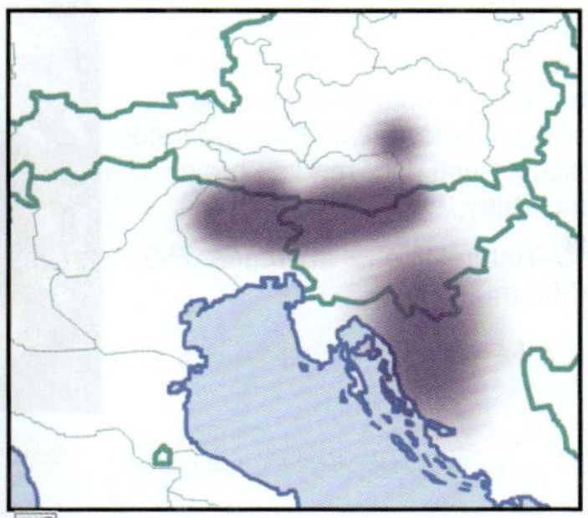

Austria to NE Italy and Croatia

Humid limestone rock rubble slopes, under leaf litter and rotting wood logs.

Shell egg-like cone-shaped (conical except last 2-3 whorls), 1 parietal and 2 columellar folds, aperture slightly oblique.

Orcula dobrogica (Grossu, 1986)

Without gender agreement: *Orcula dobrogicum* (Grossu, 1986)

Sphyradium dobrogicum Grossu, 1986 in Grossu 1986 (Trav. Mus. Hist. Nat. 28): 11

Shell ovoid, yellowish, regularly ribbed, 35-40 ribs on penultimate whorl, with spaces between ribs 3-4 x wider than ribs, 7 whorls, last whorl distinctly narrower than penultimate whorl and ascending near aperture, aperture with parietalis and visible columellaris, margin not much reflected, inside whitish and relatively sharp.

4.2 x 2.5 mm

Leaf litter at the bottom with some herbal vegetation in forests, in relatively dry habitats.

Must be rare, only 1 specimen has ever been found.

Was classified by R. Bank in www.faunaeur.org (07.2008) in the genus *Orcula* (possibly without any published reference behind). More research seems to be necessary, but a classification in *Orcula* seems justified, Grossu's classification in *Sphyradium* was unsubstantiated.

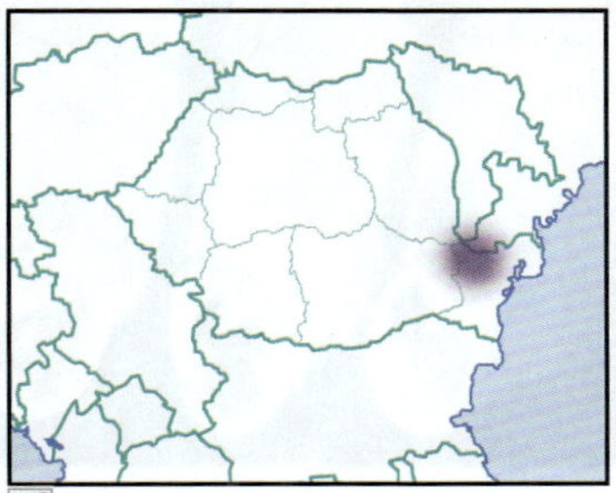

Romania (Tulcea)

Orcula dolium (Draparnaud, 1801)

Pupa dolium Draparnaud, 1801 in Draparnaud 1801: 58

6.7-9 x 3-3.6 mm

Germany: Baden, Kaiserstuhl

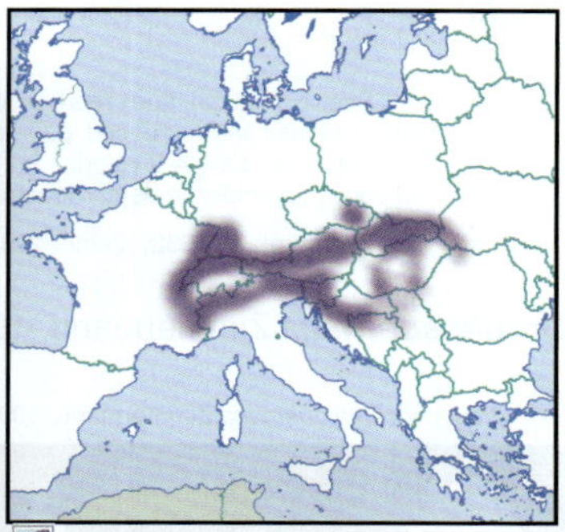

Alps (not in the high mountains inside) and some adjacent highlands, Carpathians, range in Croatia and Bosnia uncertain

Kerney et al. 1983: 99, Falkner 1990: 146, Turner et al. 1998: 147.

Shell yellowish to reddish brown, finely and irregularly sriated, apex conical, short and pointed, 8-10.5 whorls, last whorl slightly ascending, aperture with weak upper columellaris and slightly stronger lower columellaris, parietalis strong. Locally very variable. Juveniles with camouflage.

Limestone rocks, litter in stony forest slopes, with high humidity tolerance, occurring in forest humid habitats as well as on dry and sunny slopes, up to 2100 m. A frequent species, occurs for example on practically every mountain slope in Salzburger Kalkalpen. Scattered distribution in Switzerland and N Italy.

Various subspecies have been recognised. Their alignment with criteria usually applied in biodiversity research (intrincic partial barriers to gene flow) needs to be verified.

Orcula fuchsi Zimmermann, 1931

Orcula fuchsi Zimmermann, 1931 in Zimmermann 1931: 44, Taf. 4

Shell very slender, apex pointed conical, last whorl relatively larger than in other *Orcula* species, no columellar lamellas visible, aperture with strong palatal callus. Shell dark reddish brown and shiny.

Habitat like *O. austriaca*, 750-1500 m.

6.6-7.0 x 2.3-2.4 mm

Austria: Niederösterreich, near Kernhof

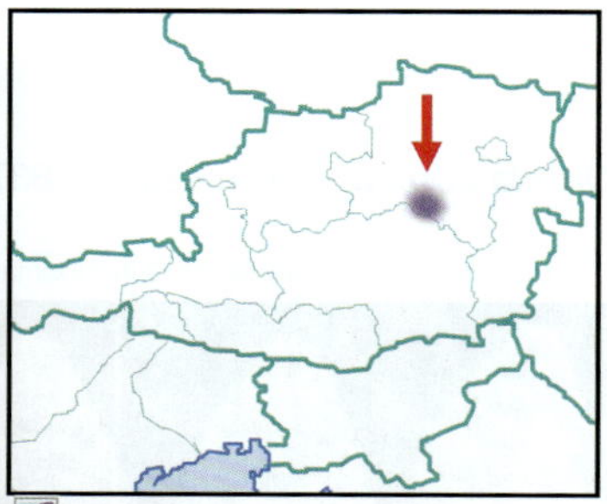

Austria (E Alps)

EU Red List: Critically Endangered CR. Only from few localities, threatened by modern forest management.

Orcula gularis (Rossmässler, 1837)

Pupa gularis Rossmässler, 1837 in Rossmässler 1835-1837 (5-6): 17, Taf. 23

6.5-7 x 2.8 mm

Austria: Kärnten, Tscheppaschlucht

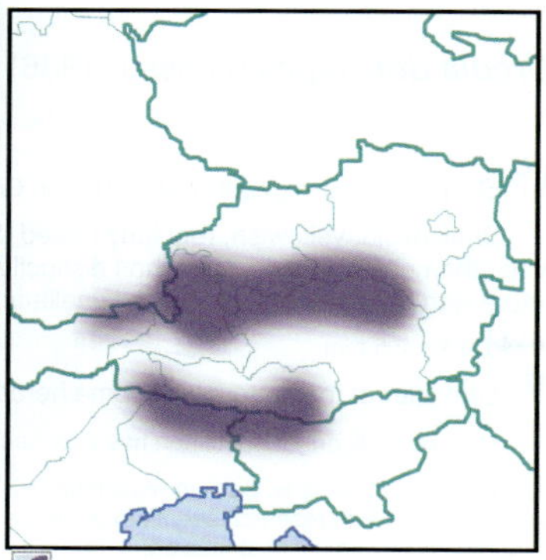

Central Alps

Smaller and more slender than *O. dolium*, apex rounded, aperture with strong palatal tooth, columellar folds deeper and weaker than in *O. dolium*.
The animal is greyish brown, getting lighter down to the sides.

Rock rubble on cool and humid limestone slopes. More calciphile than *O. dolium*, requires more humidity.

Klemm [1973], Kerney et al. 1983: 99, Falkner 1990: 146, Manganelli et al. 1995: 14.

Austria: Oberösterreich, Kalkalpen, Hoher Nock near Feichtauhütte
Jan Steger

Orcula jetschini (Kimakowicz, 1883)

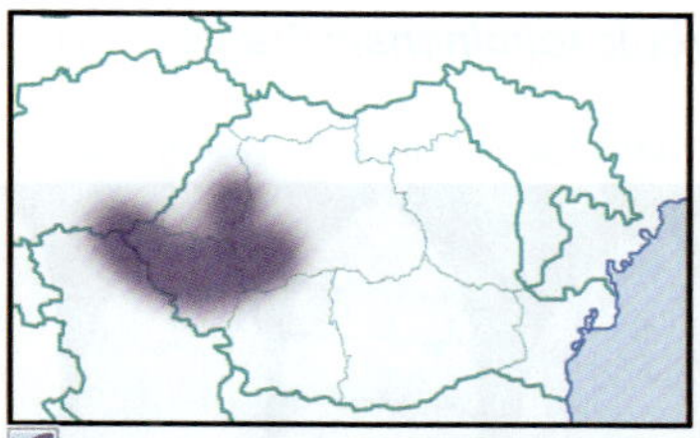

W Romania and adjacent regions

Pupa jetschini Kimakowicz, 1883 in Kimakowicz 1883: 44 (*Pupa (Orcula) Jetschini*)
Original genus was not *Orcula*.

Shell much less slender than the other *Orcula* species, yellowish brown, often with a red hue, with oblique and wide ribs, apex rounded, parietalis oblique, 2-3 columellar folds, umbilicus always present.

5.8-7.2 x 3.5-3.7 mm

Bielz 1867: 95 (*O. dolium* with rounded apex), Kerney et al. 1983: 103, Grossu 1987: 223, Lengyel & Páll-Gergely 2010: 93.

Orcula pseudodolium Wagner, 1912

Orcula gularis pseudodolium Wagner, 1912 in A. J. Wagner 1912: 252

Shell like *O. gularis*, but larger, apex rounded, last 4 whorls cylindrical, aperture with strong palatal callus, 2 columellar lamellas visible.

Under rock rubble covered by vegetation.

EU Red List: Near Threatened NT. Known from some 40 localities, seems to have been declining. Threatened by tourist activities and afforestation with *Picea* trees.

Kerney et al. 1983: 100, www.iucnredlist.org (07.2012).

7 x 3.0-3.5 mm

Austria: Oberösterreich, Gaisberg, Molln

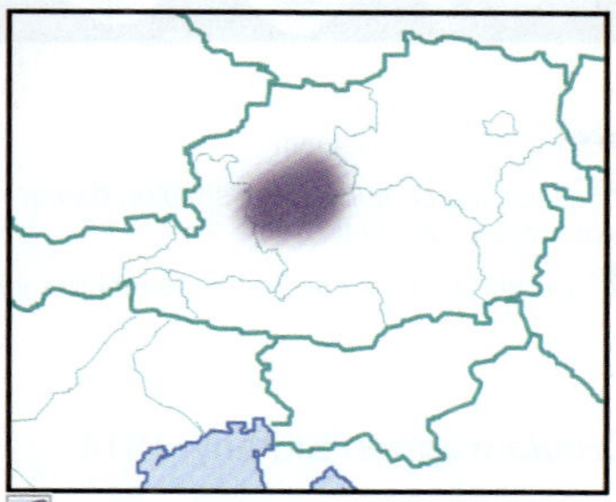

Austria

Orcula restituta (Westerlund, 1887)

Pupa gularis forma *restituta* Westerlund, 1887 in Westerlund 1887: 85

Very much like *O. gularis* but without palatalis, slightly smaller, palatal callus sometimes present, 3 lamellas in the aperture equally well developed.

Rock rubble on cool and humid limestone slopes, also at the base of rocks.

Kerney et al. 1983: 102.

6.0-6.5 x 2.8 mm

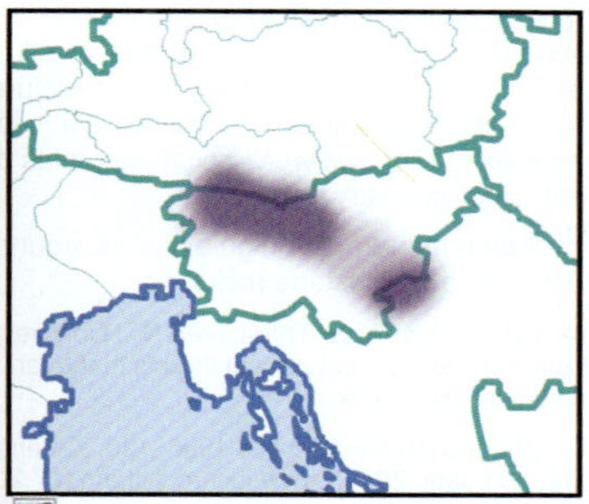

SE Alps (to N Croatia?)

Croatia: Žumberak mountains near Jastrebarsko (determination uncertain)

Orcula schmidtii (Küster, 1843)

Pupa schmidtii Küster, 1843 in Küster 1852: 26, Taf. 3 (*Pupa Schmidtii*)

5-7 x 2.9-3.2 mm

N Albania: Malësia, near Bogë

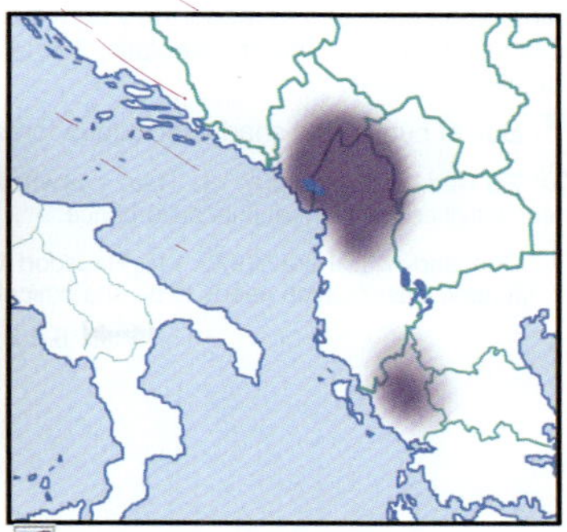

Montenegro, Albania, NW Greece

Shell horny yellowish to horny reddish, irregularly and prominently striated, 8-9 relatively convex whorls, suture deep, apex short and pointed, umbilicus open, aperture with strong parietalis, columellaris and infracolumellaris, usually no white layer at parietal side. Shells tend to be higher in the south.

Rocky shrublands or forests, under stones or at the base of rocks.

Hausdorf 1987: 51 (previous reports for Rep. Makedonija referred to other species), Reischütz & Sattmann 1990, Dhora & Welter-Schultes 1996: 120.

Orcula spoliata (Rossmässler, 1837)

N Italy, Austria

Pupa gularis var. *spoliata* Rossmässler, 1837 in Rossmässler 1835-1837 (5-6): 18, Taf. 23 Type locality "Tiroler Alpen".

Like *O. gularis*, aperture without palatal tooth but with callus, last whorl narrower than preceding two whorls.

6.2 x 2.9 mm

Rock rubble and vegetated boulder slopes, also at rocky outcrops among grass.

Seems to be rare, only known from dead shells.

Klemm [1973], Kerney et al. 1983: 101, Manganelli et al. 1995: 14.

Identity uncertain and needs to be confirmed, possibly only a local form of *O. gularis*.

Orcula tolminensis Wagner, 1912

Orcula gularis tolminensis Wagner, 1912 in A. J. Wagner 1912: 253

4.8-5.5 x 2.2-2.4 mm

Austria: Kärnten, Kupitzklamm near Eisenkappl

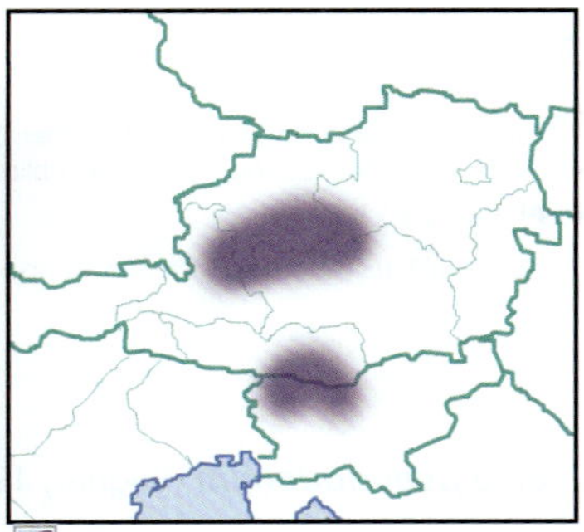

Austria, Slovenia (E Alps)

Klemm [1973], Kerney et al. 1983: 100.

Shell like *O. gularis*, but smaller, darker reddish brown, apex rounded, last 4 whorls cylindrical, aperture with strong palatalis and palatal callus, columellar folds weak or absent.

Under plant litter at limestone rocks covered with vegetation, dead grass packs, in humid and dry habitats, even on S-exposed slopes without vegetation.

Orcula wagneri Sturany, 1914

Orcula wagneri Sturany, 1914 in Sturany & Wagner 1915: 63, Taf. 15

Shell horny yellowish to reddish, fresh shells with periostracum riblets producing a silky sheen, 7-8 very weakly convex whorls, last whorl slightly ascending, aperture with strong palatal callus, also visible from outside as a yellow structure, margin connected at parietal side by a thin layer, parietalis strong, two weak columellar folds present, umbilicus very narrow.

Alpine meadows in high altitudes around 2000 m under limestone rocks.

EU Red List: Near Threatened NT. Only very few populations known, but no real threats in the remote and undisturbed sites.

Dhora & Welter-Schultes 1996: 120, Z. Fehér (unpublished data, 2011), www.iucnredlist.org (07.2012).

5.3-6.5 x 2.9-3.3 mm

Albania: Mali i Tomorrit near Berat, 2250 m

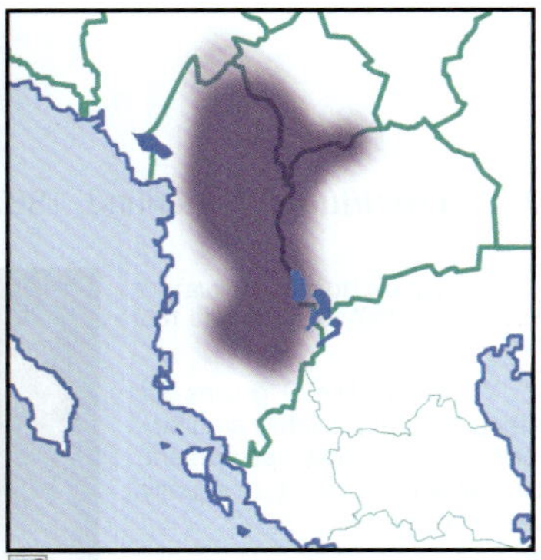

Albania, Rep. Makedonija, Kosovo

Zoltán Fehér

Orcula zilchi Urbański, 1960

Orcula zilchi Urbański, 1960 in Urbański 1960: 57

Shell reddish brown, irregularly and obliquely striated, slightly translucent and shiny, apex conical, 7-9 moderately convex whorls, last whorl narrower than preceding whorl, aperture with strong parietalis, 2 small columellar folds, no palatal callus. Shell usually encrusted.

Soil of humid and shady deciduous forests, under leaf litter and rotting wood.

EU Red List: Vulnerable VU. Rare everwhere where it lives. Threatened by forest management activities and recreational disturbance.

Damjanov & Likharev 1975: 115, Hausdorf 1996: 14, Schütt 2001: 24, Páll-Gergely 2010: 91. Generic classification needs to be anatomically confirmed.

6-8 x 2.5-3.5 mm

SE Bulgaria: Strandza mountains, Kosti village

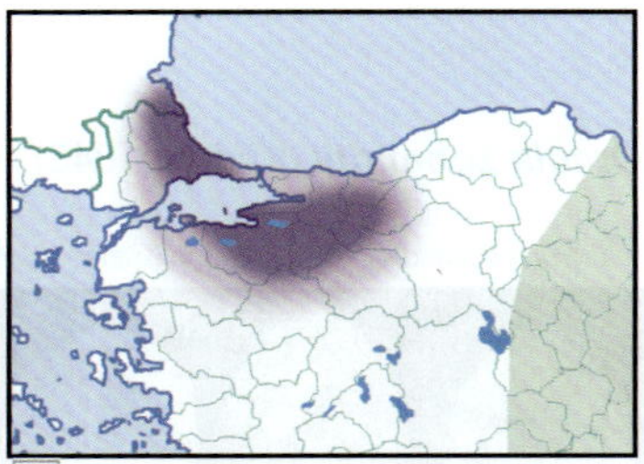

SE Bulgaria, NW Turkey

Atanas Irikov

Orculella

Orculella Steenberg, 1925 in Steenberg, 1925: 176

Type species *Pupa libanotica* Tristram, 1865 (original designation, fixed under Art. 70.3 here) (= currently *Orculella sirianocoriensis* (Mousson, 1854))

Steenberg 1925: 176 designated originally *Pupa orientalis* Pfeiffer, 1861 (in L. Pfeiffer 1861: 168, = currently *Orculella orientalis* (Pfeiffer, 1861) from NW Syria and Hatay) and referred to the species examined by Wiegmann 1901, so he must have meant *Pupa libanotica* Tristram, 1865 (in Tristram 1865: 538) (Hausdorf 1996: 11). Hausdorf 1996 argued that *Pupa libanotica* should be regarded as type species for *Orculella*. Under Art. 70.3 of the Code shall now be fixed *Pupa libanotica* as type species for *Orculella*, misidentified as *Pupa orientalis* in the original designation by Steenberg 1925: 176.

Shell cylindrical, apex more domed than in *Orcula*, first whorls with spiral lines, parietalis extending far inside (not restricted to the aprtural margin), no angularis, subangularis and spiralis, palatal folds inside the shell. No conchologically separation from *Schileykula* and *Sphyradium* is possible, except that *Sphyradium* has no palatal folds.
Penial appendix has some separate sections, retractor inserts at the base of flagellum or more proximal at penis.

Spain and N Africa to Azerbaijan and Iran

Hausdorf 1996, Gittenberger & Hausdorf 2004.

Orculella astirakiensis Gittenberger & Hausdorf, 2004

Orculella astirakiensis Gittenberger & Hausdorf, 2004 in Gittenberger & H. 2004: 104

Like *O. cretimaxima*, but less slender, smaller, last whorl narrowed, columellar apertural wall not clearly concave, apetrture with spiralis and only two columellar folds.

Rocky limestone habitats, in plant litter at the base of stones and rocks, also in crevices.

7.4-9.3 x 3.8-4.0 mm

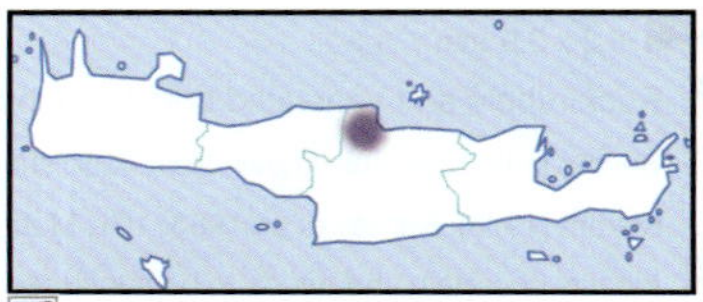

Greece (N central Crete)

Orculella bulgarica (Hesse, 1915)

Orcula bulgarica Hesse, 1915 in Hesse 1915: 62

Apex rounded, broad apertural margin, reduced apertural folds: parietalis very short, columellar folds almost absent.

Sandy highland habitats, often near lake sediments. In Spain in permanently humid and almost undisturbed habitats such as small marshes.

Critically Endangered in Spain, only few populations are known to have survived since the Plio-Pleistocene, several of which have disappeared in the decade before 2005 (in 2006 only one habitat remained). Rare in Bulgaria.

Damjanov & Likharev 1975: 114, Schütt 1996: 24, Hausdorf 1996: 14, Garrido et al. 2005, Arrébola in Verdú & Galante 2006: 394, Irikov & Erőss 2008, Robles & Martínez-Ortí 2009.

5-8 x 3.0-3.2 mm

NE Bulgaria: near Beloslav village

E Spain, Bulgaria to Armenia

Atanas Irikov

Orculella creantirudis Gittenberger & Hausdorf, 2004

Orculella creantirudis Gittenberger & Hausdorf, 2004 in Gittenberger & H. 2004: 98

5.7-7.8 x 2.6-3.5 mm

Greece: Crete, Díkti mountains, near Psichró

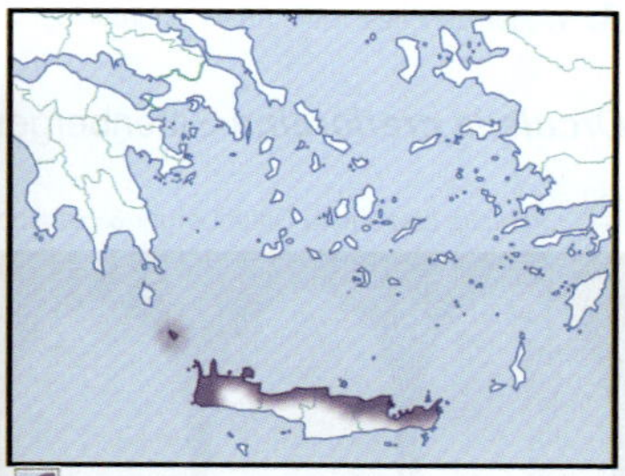

Greece (Andikíthira, Crete)

Shell medium-sized, cylindrical, whorls flattened, with prominent radial sculpture, palatal apertural lip strongly protruding.

Rocky limestone habitats.

Orculella creticostata Gittenberger & Hausdorf, 2004

Orculella creticostata Gittenberger & Hausdorf, 2004 in Gittenberger & H. 2004: 102

Shell with convex whorls, prominent and regularly spaced radial ribs, columellar lamellae obsolete, parietalis low. Can easily be recognized because of the very constant sculpture.

Rocky limestone habitats.

4.2-5.9 x 2.4-2.7 mm

Greece (W Crete)

Orculella cretilasithi Gittenberger & Hausdorf, 2004

Orculella cretilasithi Gittenberger & Hausdorf, 2004 in Gittenberger & H. 2004: 112

Shell broadly cylindrical to elongated ovoid, regular dense riblets on teleoconch whorls only, palatal apertual lip not clearly protruding.

5.7-7.8 x 3.0-3.7 mm

Rocky limestone habitats.

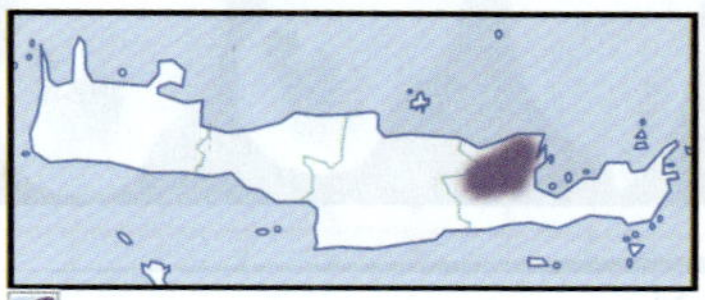

Greece (E Crete)

Orculella cretimaxima Gittenberger & Hausdorf, 2004

Orculella cretimaxima Gittenberger & Hausdorf, 2004 in Gittenberger & H. 2004: 103

Shell very large, slender cylindrical, whorls flattened and with growth lines only, columellar apertural wall concave, palatal lip not protruding. Aperture with spiralis and sometimes a third columellar lamella inside.

Rocky limestone habitats.

9.1-12.7 x 4.0-4.2 mm

Greece: Crete, near Achláda

Greece (N central Crete)

Orculella cretiminuta Gittenberger & Hausdorf, 2004

Orculella cretiminuta Gittenberger & Hausdorf, 2004 in Gittenberger & H. 2004: 106

3.6-5.3 x 2.0-2.5 mm

Greece: S central Crete, near Pitsídia

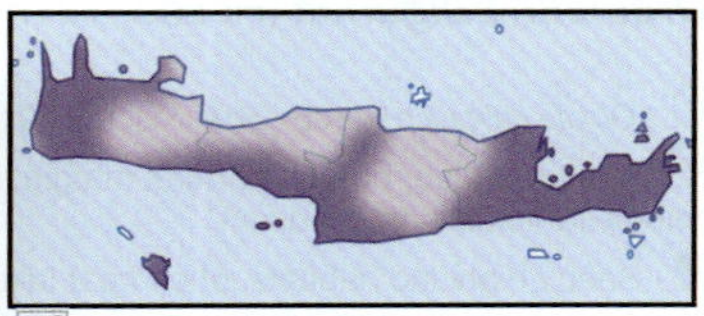

Greece (Crete, Gávdos)

Diverse habitats, seems to require shadow, in soil litter in open or dense forests on Neogene substrate. Frequent between rocks and in soil litter next to rocks. Rare on Mesozoic substrate. Also in sandy habitats, frequent under *Juniperus* trees. Rare in rocky areas without much vegetation.

Shell small, broad, nearly ovoid, 6-8 whorls, with prominent radial scuplture only on the initial teleoconch whorls, no sutural canal, parietalis high, two long columellar folds.

Welter-Schultes 1998: 38-41 (morphological analyses of the Gávdos population, not considered by Gittenberger & Hausdorf 2004).

Orculella cretioreina Gittenberger & Hausdorf, 2004

Orculella cretioreina Gittenberger & Hausdorf, 2004 in Gittenberger & H. 2004: 113

6.4-9.4 x 2.8-3.5 mm

Greece: central Crete, Idi mountains S Anógia, 1500 m

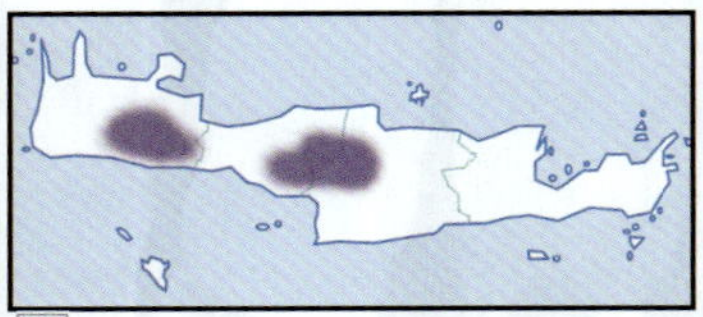

Greece (Crete)

Shell light yellowish brown, rather glossy, whorls flattened, irregular riblets on initial teleoconch whorls only, sutural canal unclear, palatal apertural lip not clearly protruding.

Very frequent in high mountain zones of 1000-2250 m.

Orculella critica (Pfeiffer, 1856)

Pupa critica Pfeiffer, 1856 in L. Pfeiffer 1856 (Malak. Bl. 3): 177

5.0-6.8 x 2.4-2.6 mm

Greece: Attikí, Athína, Argiroúpoli

Greece, W Turkey

Various limestone habitats. In Turkey much rarer than *O. ignorata*.

Shell cylindrical, apex rounded, umbilicus open, parietalis short, two weak columellar folds deep in the aperture, no palatal folds. Tends to exhibit local forms.

Differences to *O. ignorata*: shell more regularly cylindrical (in *O. ignorata* the upper half of the shell may be slightly broadened), upper teleoconch whorls prominently ribbed, lower whorls usually only striated, parietalis often not clearly extending into the penultimate whorl (in *O. ignorata* the parietalis is still high at the ventral side of the shell), palatal lip less prominently protruding. Penis with relatively complex appendix (not in *O. ignorata*).

Orculella diensis Gittenberger & Hausdorf, 2004

Orculella diensis Gittenberger & Hausdorf, 2004 in Gittenberger & H. 2004: 115

4.8-5.9 x 2.0-2.3 mm

Greece: E parts of Día island N of Crete

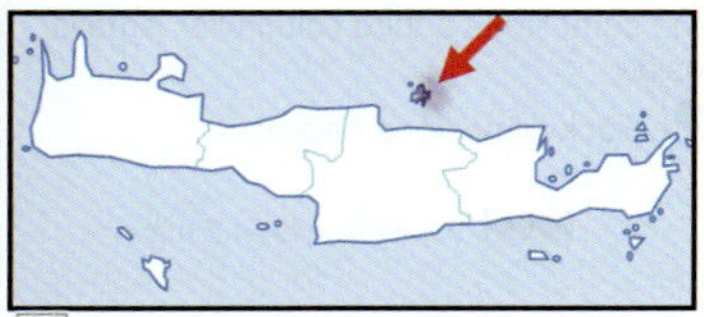

Greece (Día island N of Crete)

Shell very slender cylindrical, with irregular and indistinct ribs, strongly inflated upper whorls, distinct sutural canal.

Rocky limestone habitats, often in rock crevices. Moderately frequent.

Schultes & Wiese 1990: 25 ("glatt").

Orculella exaggerata (Fuchs & Käufel, 1936)

Greece (Kárpathos, Kásos, Sariá)

Orcula doliolum exaggerata Fuchs & Käufel, 1936 in Fuchs & K. 1936: 544

Shell prominently ribbed, ribs accentuated by lamelliform periostracum, conspicuous sutural canal, strongly protruding palatal lip.

5.8-7.6 x 2.6-2.8 mm

Gittenberger & Hausdorf 2004: 118.

Orculella fodela Gittenberger & Hausdorf, 2004

Greece (N central Crete)

Orculella fodela Gittenberger & Hausdorf, 2004 in Gittenberger & H. 2004: 105

Smaller than *O. cretimaxima*, with widely spaced ribs on the initial teleoconch whorls, shell less broad and more slender than *O. astirakiensis*.

6.8-8.4 x 3.0-3.3 mm

Rocky limestone habitats.

Orculella franciscoi Gittenberger & Hausdorf, 2004

Orculella franciscoi Gittenberger & Hausdorf, 2004 in Gittenberger & H. 2004: 117

Greece: Día island N of Crete

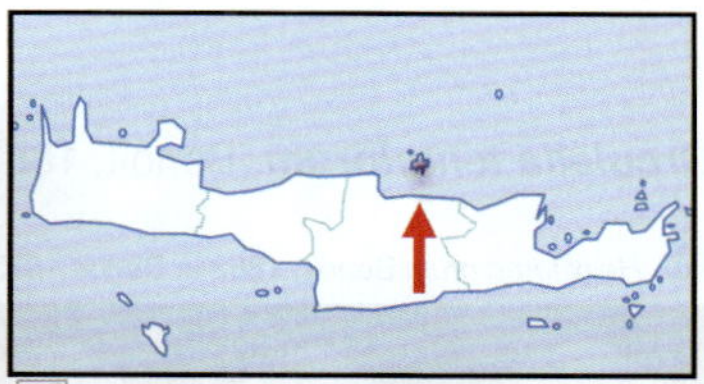

Greece (Día island N of Crete)

Seems to be restricted to habitats in rock crevices at the southern coast of Día.

Apparently a very rare species, restricted to very few localities, where it is not abundant.

Schultes & Wiese 1990: 25 ("glatt").

4.6-5.1 x 2.2-2.3 mm

Shell small, moderately slender cylindrical, upper whorls convex, lower whorls increasingly more flattened, penultimate and preceding whorl with irregular, rather obsolete ribs. Shell looks like *O. diensis*, only less slender because of its lower number of whorls.

Orculella ignorata Hausdorf, 1996

Orculella ignorata Hausdorf, 1996 in Hausdorf 1996: 19 Type locality Fethiye, S Turkey.
The holotype was collected in 1894 - when the species was described 102 years later the type locality might already have been destroyed.

Greece: island of Psérimos

Greece, SW Turkey, Cyprus
In Cyprus rare, possibly introduced (specimens collected in Akrotíri by G. Konstantinou, 2012).

Limestone habitats of any kind. Frequent in SW Turkey.

4.8-8.1 x 2.4-3.4 mm

Shell with upper whorls prominently ribbed, lower ones only striated and often slightly narrowed, parietalis long and prominent, clearly reaching into the penultimate whorl (difference to *O. critica*). Relatively simple penial appendix, unlike in *O. critica*. In *O. critica* the shell is more regularly cylindrical.

Orculella scalaris Gittenberger & Hausdorf, 2004

Orculella scalaris Gittenberger & Hausdorf, 2004 in Gittenberger & H. 2004: 116

Greece: Día island N of Crete

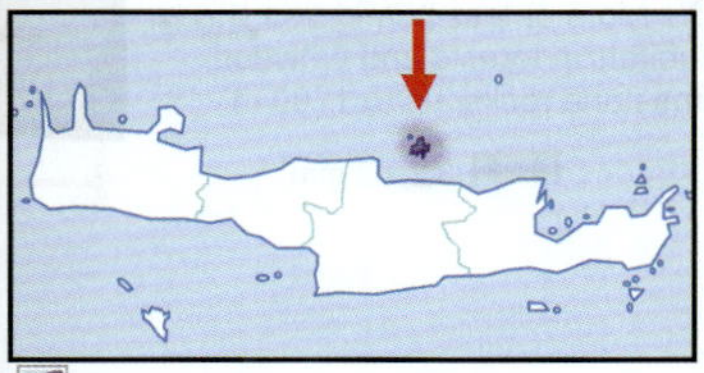

Greece (Día island N of Crete)

Rocky limestone habitats. The most frequent *Orculella* species of Día.

Schultes & Wiese 1990: 25 ("gerippt").

4.0-6.1 x 2.4-2.7 mm

Shell not very slender, apex flattened like in no other species, whorls strongly inflated, shell is formed like a winding staircase, with prominent regular ribs, no distinct sutural canal.

Orculella sirianocoriensis (Mousson, 1854)

Pupa dolium var. *sirianocoriensis* Mousson, 1854 in Mousson 1854: 386 [sic, not 368]

Cyprus and S Turkey to Iran

Limestone mountain slopes.

Hausdorf 1996: 33, Schütt 2001: 36.

6-12 x 3-6 mm, size locally variable S Turkey: Tokmar Kale near Silifke Museum Budapest HNHM

Shell horny coloured, finely striated, apex weakly pointed, 7-11 flattened whorls, often with a white line at suture, aperture with distinct parietal callus, parietalis strong, subangularis present, columellaris usually as a lamella in the last whorl, supracolumellaris flatter.

Orculella templorum (Benoit, 1862)

Helix templorum Benoit, 1862 in Benoit 1875 (Ill. Sist.): 194

Sicily and W Libya

EU Red List: Near Threatened NT. Threatened by exaggerated agricultural land use, not very abundant.

Hausdorf 1988, Reitano et al. 2009: 185.

The only orculid of Sicily, *Sph. doliolum* does not live there.

5.5-7 x 2.8-3.1 mm Italy: Sicily, Cava d'Ispica Museum Budapest HNHM

Shell horny brownish, translucent, distinctly striated, 7.5-8 whorls, diameter largest in the upper half, parietalis prominent, angular callus weak, aperture rounded and not narrowed at the base (in contrast to *Sphyradium doliolum*), margin white and reflected. Juveniles with 3 whorls have a parietalis and a columellaris.

In litter, earthworm burrows and under stones, humid and arid limestone habitats distant from the coasts, also in areas affected by fire.

Pagodulina

Pagodulina Clessin, 1876 in Clessin 1876: 198 *Pupa (Pagodulina)*

Type species *Pupa pagodula* Des Moulins, 1830 (monotypy)

Shell cylindrical, relatively small, apertural margin connected at parietal side, parietalis extending far inside, palatal folds inside the last whorl, no angularis, subangularis and spiralis. No conchological difference to *Pilorcula*, but usually smaller.
Penis without appendix, retractor inserts at epiphallus.

W Europe to Caucasus

Bole 1995, Hausdorf 1996: 7.

The genus seems to need a revision, for which it is necessary to understand the degree of local morphological variation of various populations in SE Europe. Various forms, especially in Greece, have until recently been regarded as separate species by various authors, but in the published record I see currently no evidence to support the presence of many independent species.

Pagodulina austeniana (Nevill, 1880)

Pupa austeniana Nevill, 1880 in Nevill 1880: 130, Pl. 13

Like *P. pagodula*, less densely ribbed, more oval. Below a long palatalis a second, shorter fold is always present, sometimes a third fold.

More or less humid and shady habitats in forests and between limestone rocks, up to 1700 m.

3.2 x 2 mm

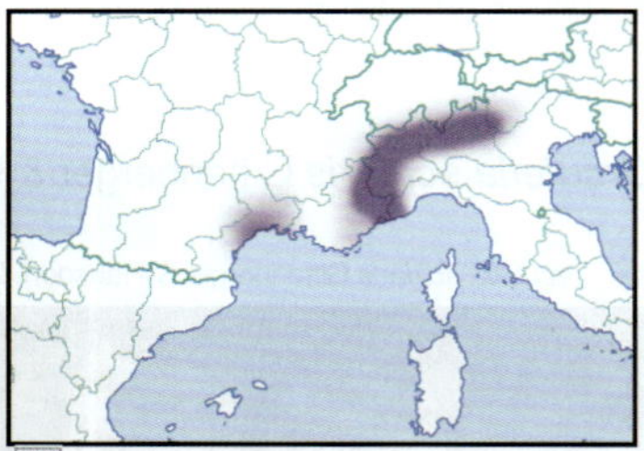

SW Alps

Italy: Lago di Como, near Cernobbio

Pagodulina epirotes Klemm, 1939

Pagodulina epirotes Klemm, 1939 in Klemm 1939: 255, Taf. 1

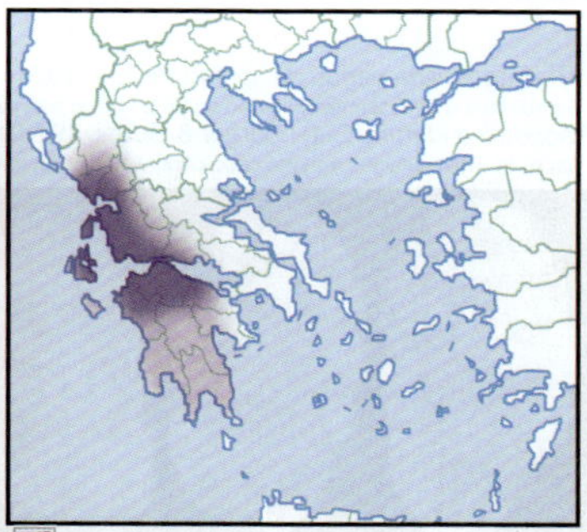

Greece (Ipiros to Pelopónisos)

In humid limestone rock rubble covered by mosses.

3-4.5 x 1.4-1.8 mm

Greece: Pelopónisos, near Káto Loutró

Shell irregularly ribbed, on last whorl not more densely (5-9 ribs/mm), 5.5-6.5 moderately convex whorls, upper section of the last whorl bluntly keeled, apertural margin white and slightly reflected, palatal margin concave with a tooth, palatalis strong, columellar side with a weak tooth, no spiralis, umbilicus open. Differs from *P. subdola* in its blunt keel at the last whorl.

Pagodulina hauseri Gittenberger, 1978

Pagodulina hauseri Gittenberger, 1978 in Gittenberger 1978: 40

Shell densely and regularly ribbed (11-17 ribs/mm on the penultimate whorl), 6-8 strongly convex whorls, columellaris strong and nearly vertical, only one palatal fold, no spiralis, umbilicus entirely closed.
Differs from *P. subdola* in its smaller size, only one instead of 2 palatal folds and in the absence of a spiralis.

2.5-3 x 1.7-1.8 mm

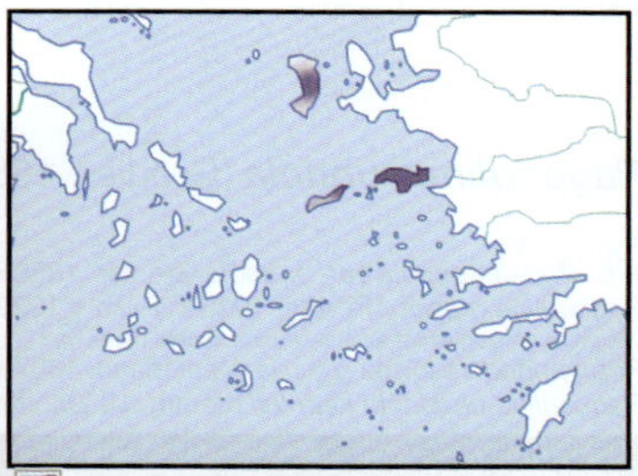

Greece (Ikaría, Sámos, Chíos)

Greece: Sámos, near Pláka

Pagodulina pagodula (Des Moulins, 1830)

Pupa pagodula Des Moulins, 1830 in Des Moulins 1830: 161
Pupa tschapecki Gredler, 1877 in Gredler 1877: 4 (*Pupa Tschapecki*)

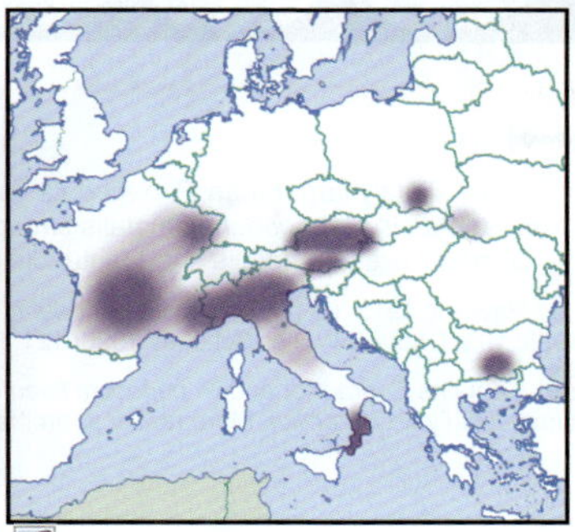

Europe
Range not well known.

2.8-3.3 x 1.8-2.2 mm

Austria: Niederösterreich, Schwarzau

Shell egg-shaped, fine and sharp ribs producing a silky shine, last whorl strongly ascending, only one palatal fold relativley deep in the last whorl, not visible from the aperture, umbilicus closed.

In forests under dead leaf litter, dark and shady habitats.

Klemm 1939: 253, Klemm [1973], Damjanov & Likharev 1975: 117 (anatomy), Kerney et al. 1983: 103, 105, Falkner 1990: 146, Manganelli et al. 1995: 14, 38.

The taxon *tschapecki* is very rarely found, it has been considered by several authors as an independent species, but seems to represent only parasitically attacked *P. pagodula* specimens. It is higher (4-4.1 x 2-2.1 mm) and has 9-10 whorls.

Pagodulina pisidica Schütt, 1993

Pagodulina pisidica Schütt, 1993 in Schütt 1993: 52

S Turkey

Humid and shady limestone rock habitats.

3.5-4.5 x 2.0-2.5 mm

S Turkey: Side

Museum Budapest HNHM

Shell usually widely ribbed, parietalis short, palatalis long and visible in an oblique view from the aperture.

Schütt 1996: 56, Hausdorf 1996: 38, 65.

Pagodulina sparsa Pilsbry, 1924

Pagodulina pagodula sparsa Pilsbry, 1924 in Pilsbry 1922-1926 [16 July 1924]: 171
Established with description, figured on Pl. 20 which was published later.
Pagodulina klemmi Gittenberger & Subai, 1978 in Gittenberger 1978: 41
On p. 41 was explained that the description was by Gittenberger & Subai.

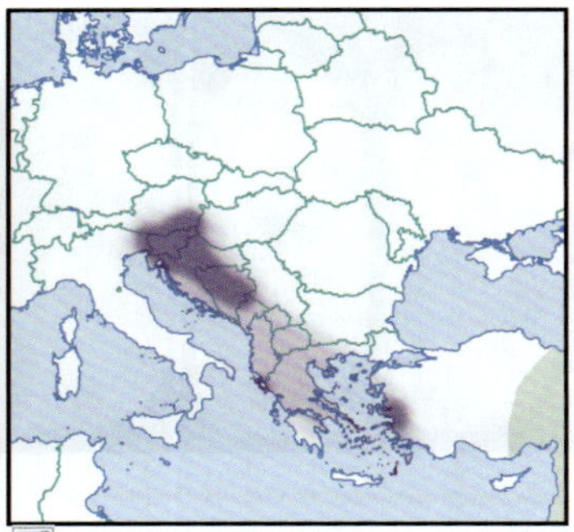

2.3-3 x 1.2-1.9 mm

Greece: Kárpathos, near Othos

SE Europe, W Turkey

Apertural margin strongly ascending, can reach the suture of the penultimate whorl, no parietalis. In S Greece smaller than between Alps and Bosnia, irregularly ribbed, most coarsely at the penultimate whorl (7 ribs/mm), 5.5-6 strongly convex whorls, last whorls flattened, apertural margin broadly reflected, columellaris strong, two palatal folds present.

Organic forest litter, humid habitats, in calcareous karst regions very frequent, between sea level and 1800 m.

Bole 1995: 104, Hausdorf 1996: 67 (SE Alps to Bosnia, isolated in Kérkira, Andros, Kárpathos).

The degree of variation in S Greece has not been analysed or understood, there is currently no evidence to support the presence of several independent species.

Pagodulina subdola (Gredler, 1856)

Pupa pagodula var. *subdola* Gredler, 1856 in Gredler 1856: 108
Pagodulina gracilior Pilsbry, 1926 in Pilsbry 1922-1926: 320
Expressly established as a new replacement name for
Pupa pagodula forma *gracilis* Westerlund, 1887 in Westerlund 1887: 91 (not Sowerby, Issel, Rossmässler and others)
Pagodulina pagodula kaeufeli Klemm, 1939 in Klemm 1939: 248, Taf. 1

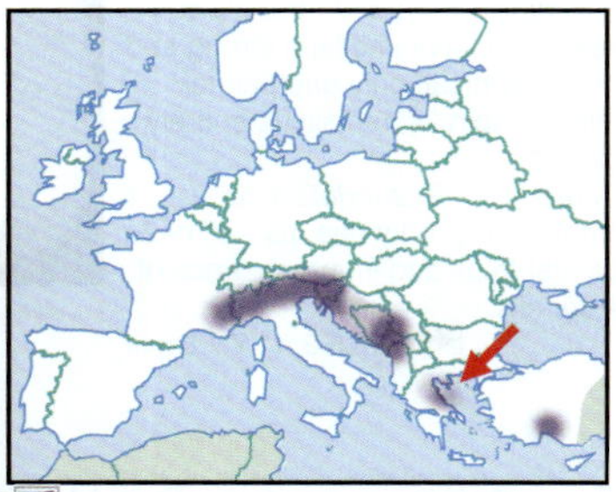

S France to S central Turkey

Humid and wet habitats, forests under dead leaves, between rocks, often near water sources, from sea level to 1600 m.

3-4 x 1.7-2.1 mm

Austria: Kärnten, Karnische Alpen, Plöckenstrasse

Vulnerable in Albania.

Shell slightly larger than *P. pagodula*, more cylindrical, ribs on last whorl often denser, 6-9 weakly convex whorls, last whorl slightly ascending, columellaris stronger than in *P. pagodula*, palatalis strong, below 1-2 additional folds can be present, invisible from the aperture but often visible from outside the shell near the umbilical area, parietalis short, suturalis present, umbilicus closed.

Pavlović 1912: 65 (W Serbia, as *pagodula*), Klemm 1939: 235, Tomić 1959: 32, Gittenberger 1978: 40, Liebegott 1986, Falkner 1990: 146, Kerney et al. 1983: 104, Bole 1995: 105, Dhora & Welter-Schultes 1996: 121, Hausdorf 1996: 67, Schütt 2001: 59.

Contains *P. gracilior* and *P. pagodula kaeufeli* according to Hausdorf in Dhora & Welter-Schultes 1996. Bole 1995: 104 classified it as a separate species because it lived sympatrically with *gracilior*. More research on these forms seems to be necessary.

Pilorcula

Pilorcula Germain, 1912
in Germain 1912: 448

Type species *Pupa raymondi* Bourguignat, 1863
(monotypy)
(= currently *Pilorcula raymondi* (Bourguignat, 1863) from Lebanon/Palestine)

Shell cylindrical, apertual margin connected at parietal side, no conchological difference to *Pagodulina*, but usually larger. Penial appendix like in *Orculella*, *Walklea* and *Alvariella*, retractor inserts at epiphallus (not at penis).

Caucasus and Turkey to Palestine, scattered and relictual distribution.

Hausdorf 1996: 7, 13.

Pilorcula trifilaris (Mousson, 1856)

Pupa trifilaris Mousson, 1856 in Mousson 1856: 398

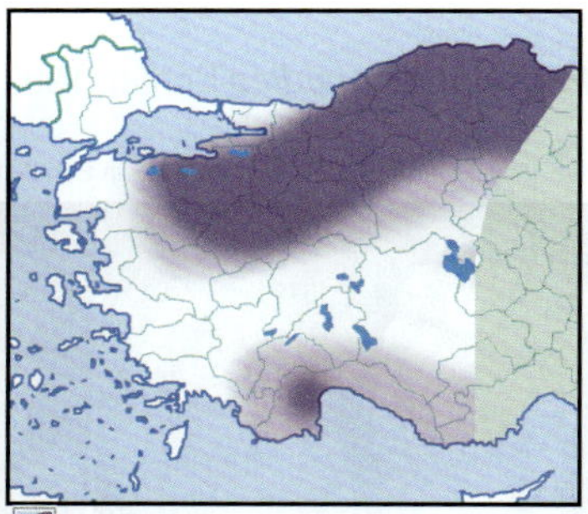

Turkey to Georgia and S Russia (coastal region near Black Sea)

3.2-7 x 2.1-3.1 mm

S Turkey: Kozan

Museum Budapest HNHM

Shell widely ribbed, fresh shells with two rows of spines, 6.5-9 whorls, weak spiral sculpture, last whorl slightly narrower than penultimate whorl, apertural margin broad and white, lip not reflected, parietalis strong, strong parietal callus present, two columellar folds visible in the aperture.

Leaf litter in deciduous forests, mountain river valleys with dense grass vegetation, also rocky slopes and rock rubble.

Hausdorf 1996: 58, Schütt 2001: 46, Egorov 2008: 24, Sysoev & Schileyko 2009: 30.

Schileykula

Schileykula Gittenberger, 1983 in Gittenberger 1983: 327

Type species *Schileykula batumensis* Gittenberger, 1983 (original designation)

Gittenberger 1983: 327 designated originally *Pupa doliolum* var. *batumensis* Retowski, 1889 (in Retowski 1889: 254, = currently *Schileykula batumensis* (Retowski, 1889) from NE Turkey to Adsharya) and referred to the species examined by Hesse 1914, so he must have meant *Pupa scyphus* Pfeiffer, 1848 (see Neubert 1993: 55, Hausdorf 1996: 13) and used the name *batumensis* as a deliberate misidentification of the original name. Under Art. 67.13.1. Gittenberger established a new specific name *Schileykula batumensis* Gittenberger, 1983 and fixed this as type species.

Shell like *Orculella*, palatal fold deep inside the last whorl, similar to *Pagodulina* and *Pilorcula raymondi*, no angularis, subangularis and spiralis. Penis without appendix, retractor inserts directly at epiphallus distal of the flagellum.

Gittenberger 1983: 327, Hausdorf 1996: 7, 13.

Schileykula nordsiecki Hausdorf, 1996

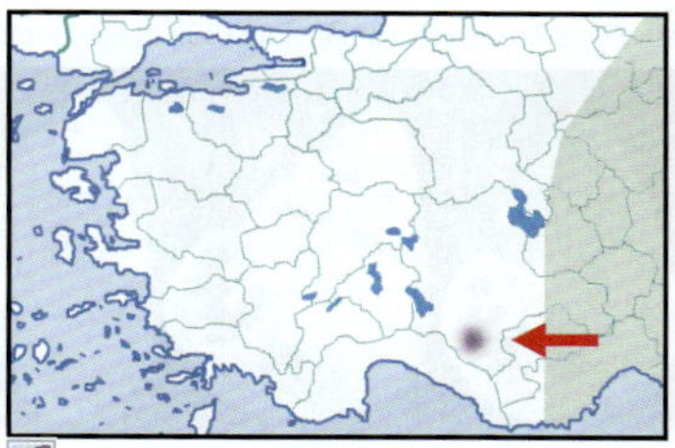

Schileykula nordsiecki Hausdorf, 1996 in Hausdorf 1996: 40

Shell cylindrical, apertural margin strong and inforced, columellaris usually prominent, apex slightly conical and pointed. Shell does not differ from local populations of *Sch. scyphus*, differences were observed in the genitalia.

6.8-7.4 x 2-7-2.9 mm

Only known from one locality.

S Turkey (Vilayet Konya)

Schileykula scyphus (Pfeiffer, 1848)

Pupa scyphus Pfeiffer, 1848 in L. Pfeiffer 1848: 7

5.5-8.5 x 3-3.5 mm

Turkey: E Antalya, Serik Belek

NW Turkey to NW Iran

Shell cylindrical, apertural margin strong and inforced, columellaris usually prominent, apex slightly conical and pointed.

Limestone habitats.

Sphyradium

Sphyradium Charpentier, 1837 in Charpentier 1837: 15 *Pupa (Sphyradium)*

Type species *Bulimus doliolum* Bruguière, 1792 (subsequent designation by Albers & Martens 1860: XVIII, 295)

Shell like *Orculella*, *Orcula* and *Schileykula*, but no palatal folds, no angularis, subangularis and spiralis. Proximal part of penis very slender, no penial appendix present, retractor inserts at a blind container proximal of the short flagellum.

Pyrenees to N Iran and central Asia

Hausdorf 1996: 7.

Only one species.

Sphyradium doliolum (Bruguière, 1792)

Bulimus doliolum Bruguière, 1792 in Bruguière 1792: 351

Pyrenees to N Iran and central Asia

Endangered in Germany and N Switzerland, Vulnerable in S Switzerland and Bayern.

4.5-6 x 2-2.5 mm

Austria: Osttirol, Iselsberg

Shell light horny coloured, widely ribbed, last 2-3 whorls narrower than those in the upper third, two columellar folds not separated. Juveniles with wide umbilicus and fresh shells often with small spines at the periphery.

In forests under leaves and rock rubble, at dry walls, often near water springs, needs warm temperatures, not confined to limestone. In SW Switzerland in warm and dry habitats, up to 1600 m, in Bulgaria 1900 m.

Damjanov & L. 1975: 116, Kerney et al. 1983: 103, Hausdorf 1988, Falkner 1990: 146, Manganelli et al. 1995: 14, Turner et al. 1998: 148, Sysoev & Sch. 2009: 31.

Odontocyclas

Odontocyclas Schlüter, 1838
in Schlüter 1838: 10

Type species *Pupa kokeilii* Rossmässler, 1837
(monotypy)

Shell conical, parietalis and palatal folds restricted to apertural margin, angularis and short subangularis present, no spiralis. No conchological difference to *Walklea*.
Penis without appendix, retractor inserts at penis, no diverticulum.

Hausdorf 1996: 7, 11.

Odontocyclas kokeilii (Rossmässler, 1837)

Pupa kokeilii Rossmässler, 1837 in Rossmässler 1835-1837 (5-6): 18, Taf. 23 (*Pupa Kokeilii*)

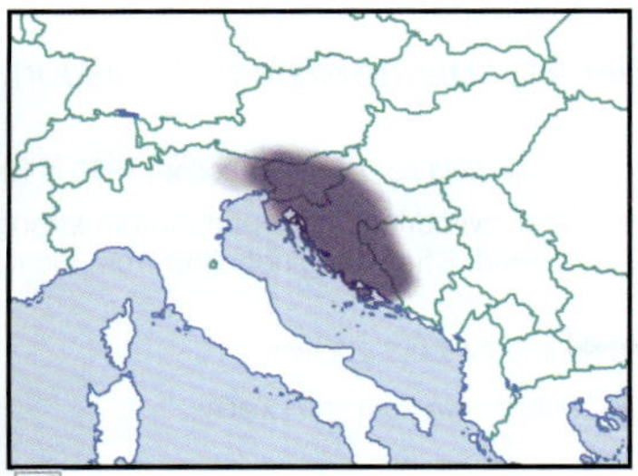

3-4 x 2-3 mm

Croatia: Dalmatia, Clissa near Split

SE Alps to Dalmatia

Shell characteristically cone-shaped, umbilicus open, aperture circular with many folds, nearly always covered with soil or faeces particles. Juveniles without apertural folds. Shell cannot be confounded, smaller than *Walklea rossmaessleri*, last whorl less predominant and outline straighter.

In moss near limestones, under leaves in deciduous forests and on walls, also under stones. Humid and shady habitats, restricted to calcareous substrate.

Walklea

Walklea Gittenberger, 1978
in Gittenberger 1978: 11

Type species *Pupa rossmaessleri* Rossmässler, 1838
(original designation)

Shell conical, parietalis and palatal folds restricted to apertural margin, angularis and short subangularis present, no spiralis. No conchological difference to *Odontocyclas*.
Penis with appendix, retractor inserts at penis, diverticulum present (in contrast to *Odontocyclas*).

Gittenberger 1978: 11, Hausdorf 1996: 7, 11.

Walklea rossmaessleri (Rossmässler, 1838)

Pupa rossmaessleri Rossmässler, 1838 in Rossmässler 1838-1844 (7-8): 39 (footnote) (*Pupa Rossmaessleri*)

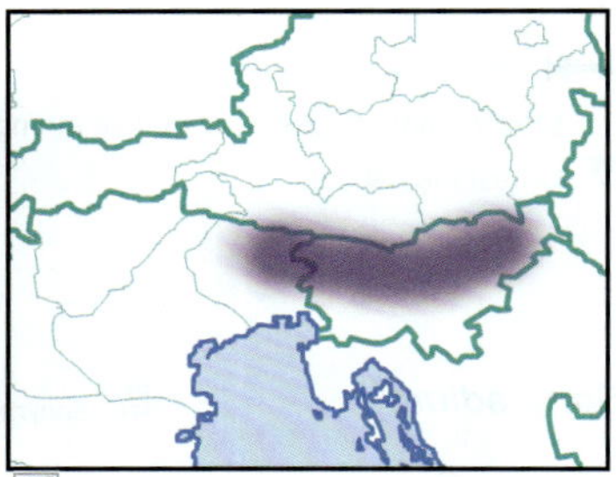

5-6 x 4-4.5 mm

NE Italy: Friuli, Monte Sabotino near Gorizia

SE Alps

Shell cone-shaped, light horny brown, finely ribbed, umbilicus closed. Juveniles without apertural folds. Larger than *Odontocyclas kokeilii*, last whorl more predominant and outline more rounded.

Under leaf litter and rocks.

Speleodentorcula

Speleodentorcula Gittenberger, 1985
in Gittenberger 1985: 222

Type species *Speleodentorcula beroni* Gittenberger, 1985
(original designation)

Shell small, cylindrical, striated, with parietal, palatal and columellar teeth in the aperture. Spermatheca simple, penis with flagellum and a complicated appendix with its own retractor muscle.

Hausdorf 1996: 10.

Last orculid in this list. Could also belong to Argnidae or Lauriidae.

Speleodentorcula beroni Gittenberger, 1985

Speleodentorcula beroni Gittenberger, 1985 in Gittenberger 1985: 222

Shell colourless, finely and regularly striated, shape like *Lauria cylindracea*, aperture like in *Pagodulina*, 7-8 relatively convex whorls, last whorl suddenly ascending near aperture, aperture with 5 teeth (strong parietalis, 2 columellar and 2 palatal), umbilicus narrow but open. Eyes pigmented.

4.2-5.2 x 2.4-2.7 mm

A cave species, found in absolutely dark sections at humid limestones.

EU Red List: Vulnerable VU. Only known from one locality, not abundant. Potentially threatened by any habitat alterations in the cave.

Greece (Evia)

Spelaeoconcha

Spelaeoconcha Sturany, 1901 in Sturany 1901: 761

Type species *Spelaeoconcha paganettii* Sturany, 1901 (monotypy)

Shell small, whitish, without teeth in the aperture, apertural margin curved seen from a side view, umbilicus open.

Systematic position unclear and gliding. Maassen 1989 (Enidae), Bank et al. 2001 (Pupilloidea incertae sedis), www.faunaeur.org (04.2011) (Vertiginidae: Nesopupinae). I do not know on which published evidence the various classifications were based.

Spelaeoconcha paganettii Sturany, 1901

Spelaeoconcha paganettii Sturany, 1901 in Sturany 1901: 761

Shell whitish or pale yellowish, almost smooth, very shiny, 7 not very convex whorls, last whorl suddenly ascending near aperture, aperture without teeth, margins thickened except on upper side, and with a wavy contour seen from the side, umbilicus open but very narrow, only partly covered by columellar margin of the aperture.

Subterranean habitats in karstic caves and fissures inside rocks.

Maassen 1989 (recognised 3 subspecies).

5.4-6.1 x 2.5-2.6 mm

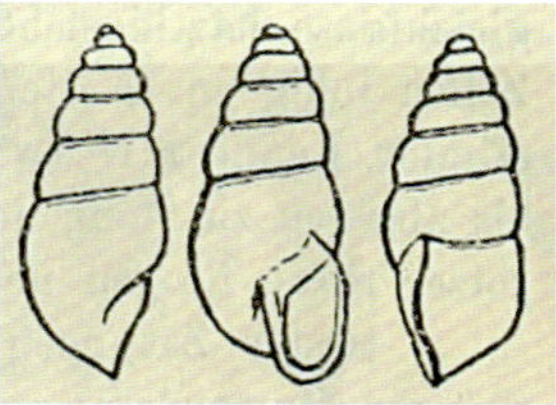

Croatia: Dalmatia, island of Korčula
Picture: Sturany 1901

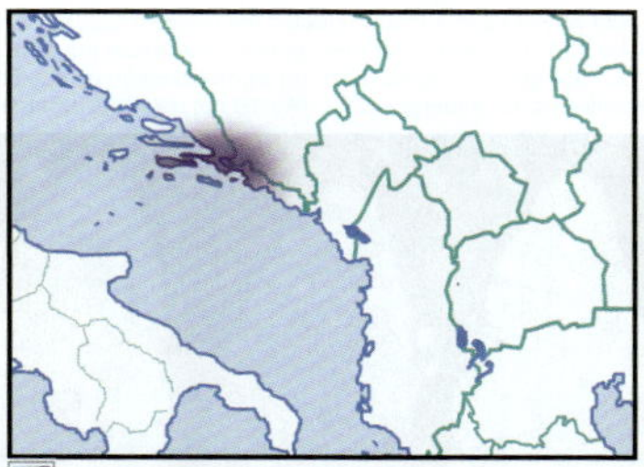

Croatia (Dalmatia), Herzegovina

Chondrinidae: *Abida*, *Chondrina*, *Granaria*, *Granopupa*, *Rupestrella*, *Solatopupa*

Abida

Abida Turton, 1831 in Turton 1831: 101 *Vertigo* synonym *Abida*

Type species *Pupa secale* Draparnaud, 1801 (monotypy)

ICZN Opinion 335: available (this makes it unnecessary to verify the presence of conditions under Art. 11.6.1). Author was not Leach (Art. 50.7). Type species by monotypy (Art. 67.12).

Shell weakly coloured to dark brown, with relatively well expressed ribs on all whorls, upper and lower palatal teeth equally strong developed.

Ground dwelling snails at limestone rocks.

Gittenberger 1973: 82, Gittenberger 1984: 62, Kokshoorn [2009]: 123, Kokshoorn & Gittenberger 2010.

9 species, only *A. secale* occurs outside the close Pyrenees vicinity. Up to 4 species may occur sympatrically.

Abida ateni Gittenberger, 1973

Abida secale ateni Gittenberger, 1973 in Gittenberger 1973: 122

Shell looks like *Chondrina*, weakly ribbed, 7-8 weakly convex whorls, aperture connected at parietal side by a weak callus, angularis connected with spiralis, subangularis weak, infraparietalis often present, infracolumellaris slightly less prominent than columellaris, both not reaching the margin, basalis often present, 3 palatal folds, all reaching the margin, suprapalatalis present, umbilicus covered.
Does not differ much from *A. secale*, has less whorls and its lower palatalis and infrapalatalis are neither indented nor interrupted.

5.8-6.9 x 2.2-2.6 mm

Valley habitats at 300-400 m altitude.

EU Red List: Vulnerable VU. Only 3 populations known.

Kokshoorn & Gittenberger 2010: 23.

The molecular record suggests that it is more or less closely related to *A. partioti*.

S France (Pyrenees)

Abida attenuata (Fagot, 1886)

Pupa attenuata Fagot, 1886 in Fagot 1886: 202

Shell very slender, finely and regularly striated, 8.5-11 moderately convex whorls, last whorl keeled at the base, aperture connected at parietal side, with angularis and subangularis, 1-2 parietal, 2 columellar and 3 palatal folds, umbilicus open.
Differs from the other *Abida* species in the presence of a subangularis.

Kokshoorn & Gittenberger 2010: 24.

The disjunct distribution was confirmed in molecular studies. Has apparently hybridised with *A. secale* in S France, resulting in introgression of its mitochondrial genome.

6.5-9 x 2-2.4 mm

France: Aude, Ussous-lès-Bains

France (Aude), N Spain (Burgos)

Abida bigerrensis (Moquin-Tandon, 1856)

Abida bigoriensis (Rossmässler, 1837)

Pupa ringens Michaud, 1831 in Michaud 1831: 64

Not *Pupa ringens* Jeffreys, 1830 in Jeffreys 1830: 356 (never used after 1899, synonym of *Leiostyla anglica* (Férussac, 1821))

Pupa bigoriensis Rossmässler, 1837 in Rossmässler 1835-1837 (5-6): 14, Taf. 23

Junior homonym (Art. 58.7) or subsequent use and validation under Art. 11.6.1 of *Pupa megacheilos* var. *pusilla* synonym *Pupa Bigorriensis* Des Moulins, 1835 (= currently *Chondrina pusilla* (Des Moulins, 1835))

Pupa ringens var. *bigerrensis* Moquin-Tandon, 1856 in Moquin-Tandon 1855: 362 (unjustified emendation for *Pupa bigoriensis*)

Pupa ringens var. *bigerrensis* was an unjustified emendation for *Pupa bigoriensis* Rossmässler, 1837 which was explicitly cited (Art. 33.2.1), but only (1) if Art. 33.2.1 is literally interpreted because no intention was visible why Moquin-Tandon selected the different spelling, and (2) for those who require a bibliographic reference and expressed use for the same taxon for validating a name under Art. 11.6.1 (only in this interpretation of Art. 11.6.1 Rossmässler's use of the name would not have validated Des Moulins's 1835 name and could be used for a different taxon). If both conditions are accepted (these interpretations of Art. 33.2.1 and Art. 11.6.1 are generally favoured here) *bigerrensis* can be used as a substitute name for *Pupa bigoriensis* Rossmässler, 1837 (Art. 33.2.3) because in contrast to Rossmässler's name, *bigerrensis* is not a homonym of *bigorriensis*. Gittenberger 1973: 146 did not use *ringens* because of the primary homonymy. This was a correct decision in 1973. Today *ringens* could be used under Art. 23.9 if it was currently used.

5.0-6.6 x 2.2-2.6 mm

France: Sarrance near Oloron

N Spain to central Pyrenees

Shell smaller than the other *Abida* species, last whorl with smaller diameter than penultimate whorl, with a shoulder near the suture, aperture with a thin white layer at parietal side, 9 or more teeth, one single infraparietal and 3 strong palatal.

Limestone rocks and rock rubble, up to 1400 m. Usually found in smaller numbers than *A. secale*, *A. occidentalis* and *A. pyrenaearia*.

Gittenberger 1973: 146, Kerney & Cameron 1979: 86, Kokshoorn & Gittenberger 2010: 25.

Must not be confounded with *Chondrina bigorriensis* (Des Moulins, 1835) (= Ch. pusilla).

Abida cylindrica (Michaud, 1829)

Pupa cylindrica Michaud, 1829 in Michaud 1829: 268, pl. [5]

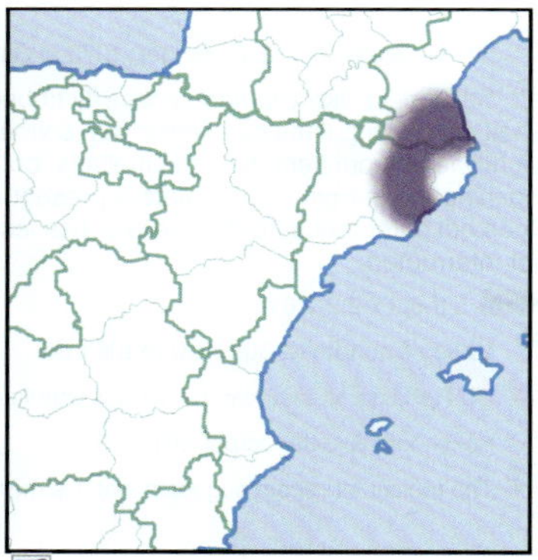

6.2-8.6 x 2.7-3.0 mm

Spain: Sant Estare de Llemena

E Pyrenees

Shell finely striated, shape cylindrical (this is characteristic) with conical apical part, 11-12 whorls, aperture with angular tooth, parietalis, 2 weak columellar teeth, 4 palatal teeth, 3 of which are long and reach the margin, margin not continuously connected at parietal side.
Shape like in *A. gittenbergeri*, apertural folds like in *A. partioti*.

Under stones and dead leaves.

Gittenberger 1973: 138, Kokshoorn [2009]: 142, Kokshoorn & Gittenberger 2010: 24.

Abida gittenbergeri Bößneck, 2000

Abida gittenbergeri Bößneck, 2000 in Bößneck 2000: 5

Shell light horny brown, densely ribbed, shape cylindrical (like in *A. cylindrica*) with conical apical part, 10-13 moderately convex whorls, apertural margin white and reflected, usually not connected at parietal side, or by a thin layer, palatal side weakly concave, 3 moderate palatal folds (very even and parallel, also visible from outside), 2 strong columellar not connected to margin, 3 parietal, umbilicus open.
Shape like in *A. cylindrica*, apertural folds like in *A. secale*.
Animal light grey with brownish anterior part and darker bluish head and tentacles.

7.5-10 x 2.9-3.3 mm

At limestone rocks under leaf litter, under rocks and in crevices of walls, also in mixed forests, 140-1100 m.

Life span possibly several years.

EU Red List: Near Threatened NT. Very local occurrence, but was found at many localities.

Tarruella Ruestes 2006, Kokshoorn [2009]: 142, Kokshoorn & Gittenberger 2010: 25.

E Pyrenees

Abida occidentalis (Fagot, 1888)

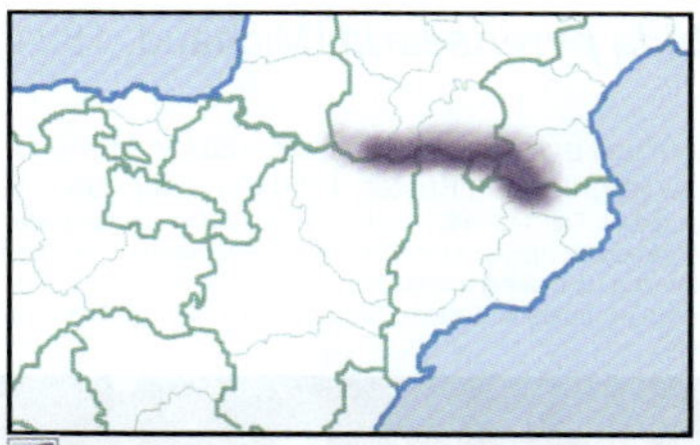

N Pyrenees

Pupa occidentalis Fagot, 1888 in Fagot 1887-1888: 194 [not verified]

Shell like *A. secale*, more slender, apertural margin continuously connected at parietal side by a prominent white layer, teeth almost like in *A. pyrenaearia*, palatal folds shorter, their distal end is usually visible in the aperture.

5.7-8.1 x 2.0-2.3 mm

In mountains, limestone rocks and rock rubble, up to 1400 m.

Gittenberger 1973: 137, Kerney et al. 1983: 110, Kokshoorn [2009]: 142, Kokshoorn & Gittenberger 2010: 24.

Abida partioti (Saint-Simon, 1848)

Pupa partioti Saint-Simon, 1848 in Saint-Simon 1848: 28 (*Pupa Partioti*)
Name originally attributed to Moquin-Tandon, but description was by Saint-Simon alone.
Abida escudiei Geniez & Bertrand, 2001 in Geniez & Bertrand 2001: 27

NE Spain: Lleida, near Senterada

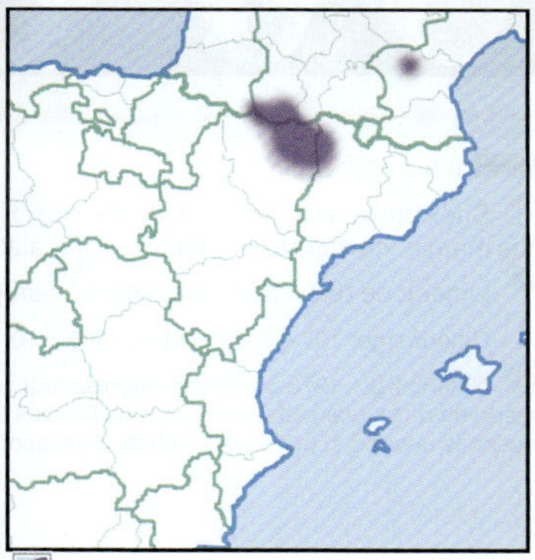

Pyrenees

6.5-8 x 2.2-2.7 mm

Shell brown, very finely and regularly ribbed, 7-10 weakly convex whorls, in parietal region only a thin white layer (apertural margin not continuous), aperture with 9 or more teeth, infracolumellar a spiral-like curved fold, characteristical in the aperture, umbilicus very narrow but usually open.

Limestone rocks and rock rubble, in mountains.

Gittenberger 2006: 20, Kokshoorn [2009]: 143, Kokshoorn & Gittenberger 2010: 25.

Abida polyodon (Draparnaud, 1801)

Pupa polyodon Draparnaud, 1801 in Draparnaud 1801: 60

Spain: E Pyrenees

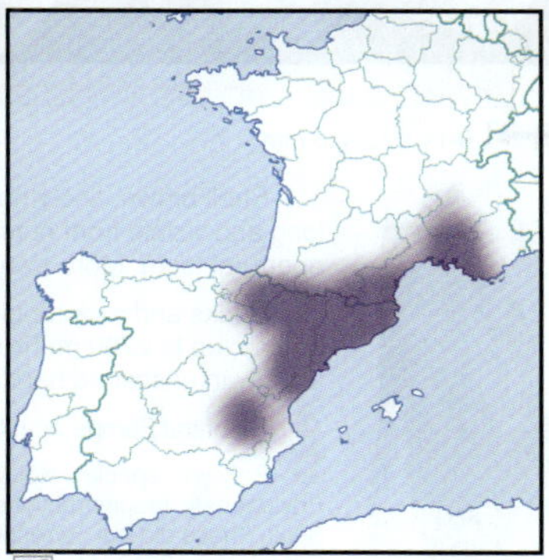

E Spain, S France, Algeria

6.3-15.3 x 2.5-4.5 mm

Shell very regularly and densely ribbed, with largest diameter at penultimate whorl, apertural margin with very thin white layer at parietal side, many teeth in the aperture, with many additional folds between the main folds.

Usually on dry limestone meadows or in shrubland, less frequent in rocky habitats. Mainly in lowlands, up to 1300 m.

Gittenberger 1973: 153.

Abida pyrenaearia (Michaud, 1831)

Pupa pyrenaearia Michaud, 1831 in Michaud 1831: 66, Pl. 15 (*Pupa pyrenæaria*)
Pupa vergniesiana Küster, 1850 in Küster 1852: 103 (*Pupa Vergniesiana*) Type locality in Ariège, France, at some distance N of Andorra.
Figured on Taf. 14 [1847], name was not mentioned on plate, only perhaps on the wrapper (spelling unknown), which was not published work under Art. 8.1.1. So the date was not 1847.
Mentioned by Pfeiffer 1848: 342 (*Pupa Vergnesiana* [sic]) as a synonym of *Pupa pyrenarearia*, with short description and reference to Küster 1852 [1847]: Taf. 14 Fig. 13-16. For those who require a bibliographical reference for making available a name under Art. 11.6.1, the name must be attributed to Küster 1850 and be spelled *vergniesiana*. For the others, to Pfeiffer 1848 and be spelled *vergnesiana* (those who regard the printed wrappers in Systematisches Conchylien-Cabinet as published work under Art. 8.1.1 have a problem with the spelling because no wrapper was preserved and they cannot verify how the name was spelled there).

6-8 x 2.2-2.5 mm — Andorra: La Massana — N and central Pyrenees

Shell brown with fine and regular ribs, like *A. secale*, but more slender and more cylindrical, apertural margin connected on parietal side by a thick continuous margin, no subangular tooth, suprapalatal tooth only weakly developed.

Limestone rocks and rock rubble, mainly above 1000 m.

Gittenberger 1973, Kerney et al. 1983: 110, Bertrand 2000: 25, Kokshoorn [2009]: 141, Kokshoorn & Gittenberger 2010: 24.

Gittenberger 1973 classified *vergniesiana* as a subspecies of *pyrenaearia*. Kokshoorn [2009] classified *vergniesiana* from two localities in Andorra as a separate species, based on molecular distance. This result needs to be confirmed, the molecular and shell records seem to contradict each other (Kokshoorn & Gittenberger 2010). More information is needed to understand which populations are reproductively isolated from which ones.

Abida secale (Draparnaud, 1801)

Pupa secale Draparnaud, 1801 in Draparnaud 1801: 59

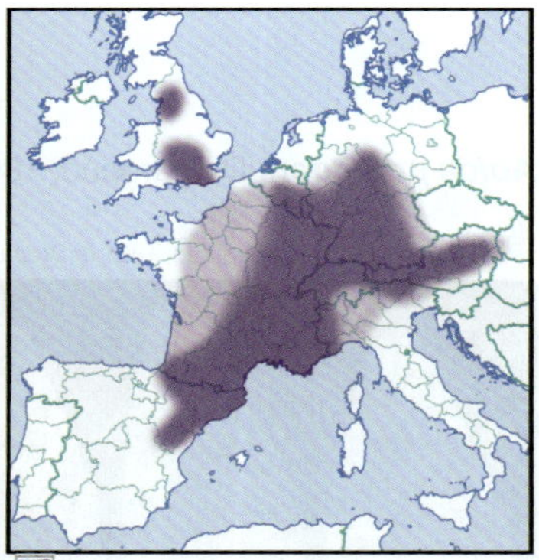

6-11 x 2.3-2.8 mm — Germany: Höxter — Spain to W Slovakia and England

Austria: Tirol, Brandstatt near Reutte

Shell brown, finely ribbed, aperture with thin parietal callus and some 9 teeth: angularis stronger than parietalis, principal palatal folds long and visible from outside the shell, umbilicus open. Shell very variable displaying local variations in rib prominence, size parameters and apertural teeth. More slender than *Granaria frumentum*, no cervical callus.

Rocks and rock rubble in forests, not too dark, on calcareous substrate, meadows with bare rocks showing through, also above the timberline, up to 2700 m. Frequently found at trees, in France also in dense populations in dry and sunny meadows devoid of rocks. In Britain usually in unshaded habitats.

In central Europe threatened by habitat destruction in forest management, in Britain by changes of land use.

Polytypic species, Kokshoorn & Gittenberger 2010: 11 recognised numerous subspecies, mostly in the Pyrenees and Spain.
Phylogenetic reconstructions by Kokshoorn & Gittenberger 2012 on a molecular base produced various clades and lineages within this species in NE Spain, these did not correspond well with traditional subspecies based on shell morphology. The identities of the subspecies recognised by Kokshoorn & Gittenberger 2010: 11 and previous authors did not involve partial barriers to reproduction along hybrid zones, so they seem only weakly justified.
The phylogeny has not been fully understood and various hypotheses have been proposed to explain the observed patterns.

Abida vasconica (Kobelt, 1882)

Pupa vasconica Kobelt, 1882 in Kobelt 1882: 72

N Spain

In 200-2000 m altitude.

Gittenberger 1973: 150.

5.4-7.1 x 2.0-2.4 mm

N Spain: Picos de Europa, Parador Fuende De, 1200 m

Shell horny reddish brown, finely ribbed, 8-9 convex whorls with deep suture, last whorl ascending and not wider than penultimate whorl, apertural margins weakly connected by a thin callus at parietal side, angularis prominent, parietalis weaker and deeper inside, columellaris weak and deep inside, 3 palatal folds of which 2 reach apertural margin.

Chondrina

Chondrina Reichenbach, 1828 in Reichenbach 1828: 93 (or 6) [not verified]

Type species *Bulimus avenaceus* Bruguière, 1792 (subsequent designation or monotypy by Reichenbach 1836: 152 [not verified])

Species live on not too humid calcareous rock faces, densest populations are found on S and SW exposed rocks with weak vegetational covers, population densities are lower at W exposed rocks, and again lower on E and N exposed rocks. At the margins of limestone regions also non-calcareous substrate is exceptionally inhabited. Not more than 2 *Chondrina* species are found sympatrically, morphology is only weakly influenced by altitude. All species in the S Alps occur up to 1800 m, sometimes they occur also above 2000 m.

Central and S Europe, N Africa

Gittenberger 1973: 158, Gittenberger 1984: 62, Kokshoorn [2009]: 144 (phylogeny, checklist of species and subspecies, preliminary revision).

Chondrina altimirai Gittenberger, 1973

Chondrina altimirai Gittenberger, 1973 in Gittenberger 1973: 200

Shell light horny brown, finely ribbed or striated, 6-8 moderately convex whorls, aperture with two weak palatal folds, usually no folds at the parietal side, sometimes with angular and columellar folds, margin slightly thickened and reflected, at the base slightly angulated (between columellar and basal sides), producing a weak keel outside.

In mountain habitats between 500 and 1500 m.

EU Red List: Near Threatened NT. Only known from 4 localities. No real threats known.

6.4-10.2 x 2.6-3.3 mm

Spain: Girona, near Olot

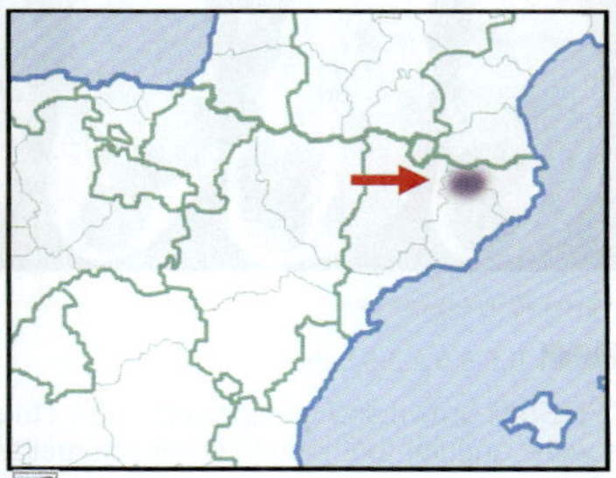

NE Spain (E Pyrenees)

Paolo Zaccaria

Tiroler Landesmuseen Innsbruck

Chondrina arcadica (Reinhardt, 1881)

Torquilla avenacea var. *arcadica* Reinhardt, 1881 in Reinhardt 1881: 137
Pupa avenacea var. *clienta* Westerlund, 1883 in Westerlund 1883: 60

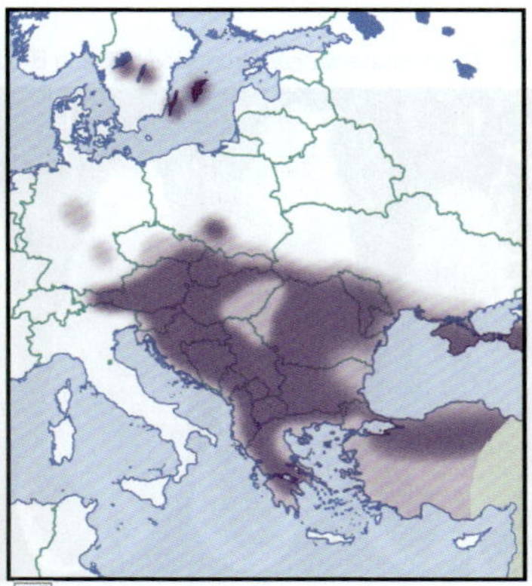

5.5-7.5 x 2.5 mm

Austria: Kärnten, near Zell-Freibach

Central Europe to Caucasus

Shell dark grey or reddish brown, not shiny, distinctly and irregularly striated, apertural margin sharp and little reflected, with a thin white layer at parietal side, strong angular tooth, weaker and deeper parietalis, 2 columellar folds, usually 3 palatal folds, of which only 2 are well developed. Shell smaller than *Ch. avenacea*, and more conical. Differs from *Abida secale* by the large relative size of the last whorl. Slightly larger than *Rupestrella philippii*.

Dry and open habitats on calcareous substrate or rock rubble, usually in mountains. In the Alps restricted to rocky surfaces and walls directly exposed to the sun, usually at S exposed slopes. Mean rate of dispersal is around 7-10 cm per day, 1 m per year, 2.5 m per year is much. Exceptionally, for example during dry weather periods, also beech tree trunks up to 10 m distant from rocks can be reached by the snails. In Bulgaria up to 1900 m. Feeds on lichen.

Klemm 1951, Nordsieck 1970: 261, Gittenberger 1973: 209, Klemm [1973]: 156, Gittenberger 1984: 62, Baur & Baur 1995, Balashov & Gural-Sverlova 2012: 94.

Also known as *Chondrina clienta*. *Glischrus hordeum* Studer, 1820 is not a synonym (Turner et al. 1998: 158).

Chondrina arigonis (Rossmässler, 1859)

Pupa arigonis Rossmässler, 1859 in Rossmässler 1854-1859 (17-18): 105, Taf. 84 (*Pupa Arigonis*)

5-8.5 x 2.2-3.4 mm — Spain: Valencia, Jeresa, Los Congles — E Spain (Alicante to València)

Shell violet brownish, weakly striated, last whorl rounded at the base, suture deep, aperture with strong and oblique angularis, strong parietalis, strong columellaris (deeper inside), weaker infracolumellaris, 2 short but distinct palatal folds.

Limestone rocks, from sea level to up to 1300 m.

EU Red List: Near Threatened NT. Threatened by fire, expanding cities and road construction. Known from only around 10 localities.

Gittenberger 1973: 225, Martínez-Ortí & Robles 2003: 72, Kokshoorn [2009]: 147, 153.

Chondrina ascendens (Westerlund, 1878)

Pupa jumillensis var. *ascendens* Westerlund, 1878 in Westerlund 1876-1878: 179
Chondrina falkneri Gittenberger, 2002 in Gittenberger 2002: 41, Colour-Pl. 8b
Chondrina gerhardi Gittenberger, 2002 in Gittenberger 2002: 42

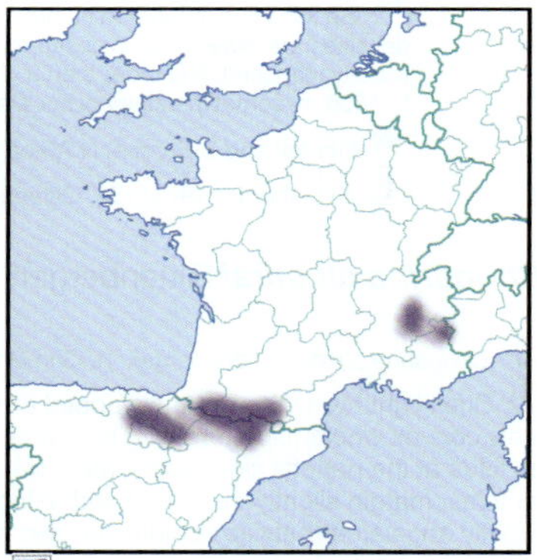

6.5-8.5 x 2.5-3 mm — Spain: Barcelona, Begas (perhaps misidentified) — Pyrenees, N Spain

Shell brown and transparent, finely striated, apertural margin sharp and not reflected, not connected by a white layer at parietal side, aperture usually with a weak angular tooth and a weak columellar tooth, no other teeth.

Open rocky habitats in mountains.

The taxa *falkneri* and *gerhardi* (from W Alps, only known from few localities there, and only for this reason considered as Vulnerable in the EU Red List) are considered as synonyms of *Ch. ascendens* here, following Bourguignat 1864 (Malac. Grande-Chartreuse). No difference was given to distinguish the two taxa *falkneri* and *gerhardi*, and no constraint was provided that they cannot be classified within *Ch. ascendens*. The classification by Bourguignat 1864 should be maintained until new evidence is found in a detailed study.

Chondrina avenacea (Bruguière, 1792)

Without gender agreement: *Chondrina avenaceus* (Bruguière, 1792)

Bulimus avenaceus Bruguière, 1792 in Bruguière 1792: 355

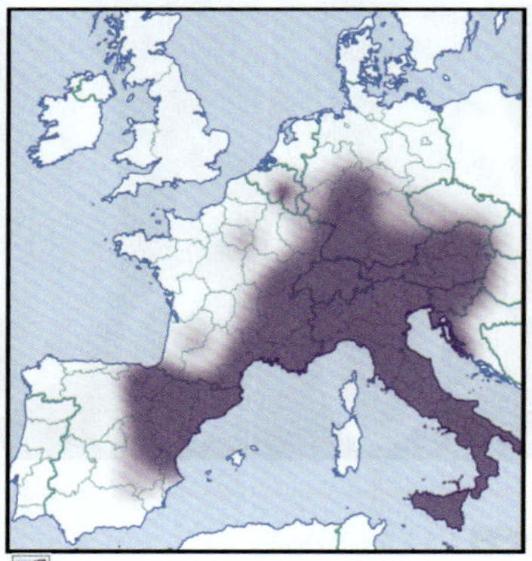

6-8 x 2.3-2.5 mm — Italy: Marche, San Felipe near Pertizano — Italy to Germany and Spain

Shell dark reddish brown, irregularly striated, not shiny, 7 slightly convex whorls with deep suture, last whorl 1/3 of total shell, aperture without parietal callus, some 7 folds: strong angularis, 3 palatal folds short (tooth-like), upper palatalis relatively upside, opposite to the parietalis and reaching the lip. Shell often covered with stone dust or lichens. Locally variable.

Dry open habitats, characteristically at limestone rock faces and walls, rare above 2000 m, maximum up to 2700 m, in Bulgaria 1800 m. In the Alps restricted to rocks directly exposed to sunshine, usually S exposed rocks, at N exposed walls only at W and E exposed places. Only active on wet surfaces or at high humidity. Feeds on endolithic lichens and algae.

Between 1 and 90 % of the individuals in a population are aphallic. Shell growth is completed after 3-5 years, maximum age 10 years.

Chondrina bergomensis (Küster, 1850)

Pupa bergomensis Küster, 1850 in Küster 1852: 114

N Italy (Val Brembana to Sogno)

5-7 x 2.1-2.6 mm

Italy: Bergamo, Val Brembana

Limestone rocks.

Shell light horny greyish brown, more prominently striated than other *Chondrina* species, cervix with a keel-like structure, aperture with narrow and sharp white margin, long and prominent angularis weakly connected to spiralis, strong parietalis and columellaris, weak infracolumellaris, 3 short palatal folds prominent upper, weaker lower and again weaker infrapalatalis, palatal folds slightly more apically located than opposite columellar/parietal folds, umbilicus open.

EU Red List: Near Threatened NT. Only known from a small restricted range near Bergamo. No real threats known.

Chondrina calpica (Westerlund, 1872)

Pupa calpica Westerlund, 1872 in Westerlund 1872: 27

N Morocco, S Spain

5.4-7.9 x 2.3-3.1 mm

Morocco: Tetuan

In Spain from sea level to 1000 m.

Shell brown, coarsely ribbed, ribs can be white, 6-8 moderately convex whorls, apertural margin weakly thickened or straight, usually not reflected, parietalis, columellaris and 2 palatal folds are always present, infracolumellaris and other weak folds can occur, palatal margin can be impressed forming a weak groove which can extend for 0.5 whorls.

Chondrina centralis (Fagot, 1891)

Pupa centralis Fagot, 1891 in Fagot 1891: 132

Shell weakly brown and transparent, prominently and less densely ribbed than the other *Chondrina* species, aperture usually with 6 teeth.

5.5-6.5 x 2.5-2.7 mm

Open rocky limestone habitats between 1000 and 1700 m.

EU Red List: Vulnerable VU. Only known from 4 localities. No real threats known.

Kerney & Cameron 1979: 89, Kerney et al. 1983: 117.

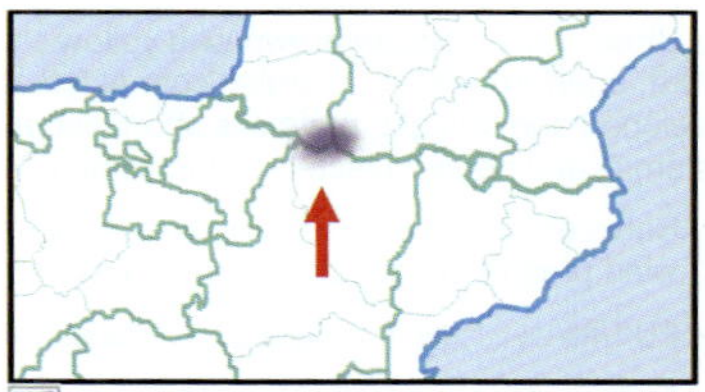

Central Pyrenees

Chondrina farinesii (Des Moulins, 1835)

Pupa farinesii Des Moulins, 1835 in Des Moulins 1835: 156, Pl. 2 (*Pupa Farinesii*)

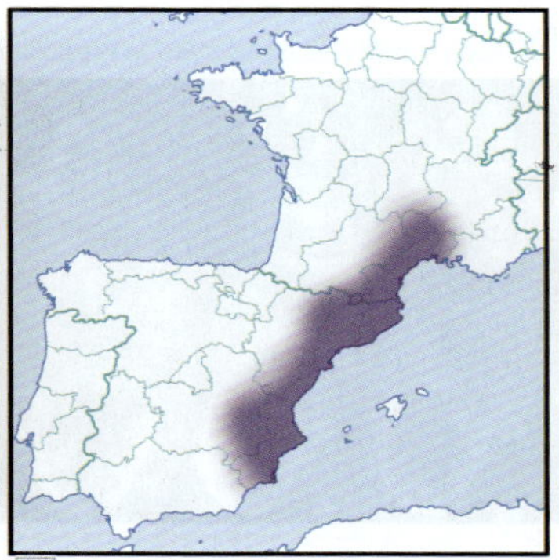

4-6.5 x 2.0-2.5 mm

Andorra: near La Vella, 1070 m

E Spain to S France

Shell brown, sometimes with light bluish hue, smooth or finely striated, 6-8 convex whorls, apertural margin sharp and thin, not reflected, not connected at parietal side by a white layer but columellar and palatal margins approach each other closely, aperture either without teeth, or with moderately developed angular tooth, deeper inside a weak parietal and a weak columellar tooth, umbilicus very narrow.

Westerlund 1887: 100, Gittenberger 1973, Kokshoorn [2009]: 146, 151 (paraphyletic species).

Chondrina gasulli Gittenberger, 1973

Chondrina gasulli Gittenberger, 1973 in Gittenberger 1973: 246

5.7-7.6 x 2.6-2.9 mm

Spain: Valencia, Oliva, Duques de Gandia

E Spain (Alicante, Valencia)

EU Red List: Vulnerable VU. Only known from few localities, threatened by fire and habitat destruction, has been declining.

Gittenberger 1973: 246, Martínez-Ortí & Robles 2003: 75, Kokshoorn [2009]: 158.

Shell light horny brown, regularly ribbed or on last whorl striated, 6.5-7 moderately convex whorls, apertural margin thickened and white, angularis usually weak, parietalis and columellaris strong, infracolumellaris weaker, no palatal folds. Differs from *Ch. farinesii* in its coarser ribs and weaker apertural folds.

Kokshoorn [2009] classified 4 additional populations from nearby locations in close relationship with this species and regarded the current knowledge on intraspecific variation as insufficient.

Chondrina generosensis Nordsieck, 1962

Chondrina megacheilos generosensis Nordsieck, 1962 in Nordsieck 1962: 11, Taf. 1

Shell light horny yellow, irregularly striated, 6-5.5-6.5 convex whorls, cervix rounded, aperture with weak folds inside (only 3 palatal folds, in opposition to the corresponding columellar/parietal folds, spiralis weak or wanting, no suprapalatalis, no suturalis), angularis present, margin white and weakly reflected, not connected at parietal side.

Limestone rocks and walls in open habitats as well as in shady forests, near water falls and also on mountain peaks. Prefers in lower altitudes cool and N exposed places. Between 700 and 1700 m.

4.6-7.0 x 2.1-2.9 mm

Switzerland: Ticino, Lago di Lugano

Paolo Zaccaria

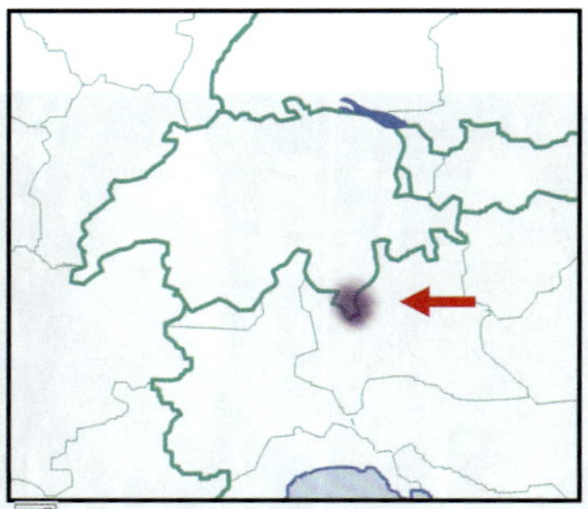

S Alps (S Ticino and near Como)

Turner et al. 1998: 161.

Tiroler Landesmuseen Innsbruck

Chondrina granatensis Alonso, 1974

Chondrina farinesii granatensis Alonso, 1974 in Alonso 1974: 87

Shell horny violetish brown with yellowish apex, striated, 7-8 whorls with moderate suture, aperture whitish, margins connected at parietal side by a callus, angularis well developed and straight, parietalis short, columellaris strong, 2 long palatal folds, umbilicus small and partly reflected by the columellar margin.

Between or at the base of limestone rocks, found up to 10 cm deep in soil, sunny and wind protected habitats, in mountains at 800-1760 m.

EU Red List: Near Threatened NT. Probably exaggerated, no real threats known.

Arrébola & Gómez 1998, Martínez-Ortí et al. 2004 (with figure).

5.5-6.5 x 2-2.5 mm

Spain: near Jaén

S Spain (Jaén to Albacete)

José Liétor Gallego

Chondrina kobelti (Westerlund, 1887)

Pupa kobelti Westerlund, 1887 in Westerlund 1887: 102

3.7-6.7 x 1.8-2.6 mm

Spain: Picos de Europa, Parador Fuende De, 1200 m

N Spain

Rocky limestone habitats in karstic aeas, syntopically with *Ch. avenacea*, in up to 1800 m.

Shell dark violetish brown, coarsely ribbed, less prominently on the last whorl, 5-7 usually strongly convex whorls last whorl not wider than penultimate whorl, aperture V-shaped, margin broad but not reflected, angularis low and connected to spiralis if a spiralis is present, parietalis strong, columellaris very strong, infracolumellaris weaker, 4 palatal folds (the upper and lower folds weaker) usually exactly opposed to the parietal/columellar folds. Differs from syntopically occurring *Ch. kobeltoides* in its smaller size, and from *Ch. avenacea* and others in its V-shaped aperture with straight palatal margin.

Chondrina kobeltoides Gittenberger, 1973

Chondrina kobeltoides Gittenberger, 1973 in Gittenberger 1973: 175

Spain: León, Oseja de Sajambre

N Spain (Oviedo, Santander)

Seems to prefer slightly less sunny habitats than *Ch. kobelti*, with which it occurs syntopically at most places, altitudes 250-1100 m.

5.6-8.5 x 2.3-3.1 mm

Shell violetish brown, finely striated, 6-8 strongly convex whorls, apertural margin not reflected, basal side slightly pointed, angularis connected to spiralis, parietalis strong, columellaris strong, infracolumellaris much weaker, 4 palatal folds exactly opposing the parietal/columellar system. Larger than *Ch. kobelti* at places where both species occur syntopically.

Chondrina lusitanica (Pfeiffer, 1848)

Pupa lusitanica Pfeiffer, 1848 in L. Pfeiffer 1848: 347

Portugal: Obidos, 70 km NE Lisboa

Central and S Portugal

At old walls and limestone rocks.

Nobre 1941: 140.

7-9 x 2.5-3 mm

Shell evenly dark brown, prominently striated, not shiny, 7 not very convex whorls with deep suture, last whorl 1/3 of total shell, aperture without parietal callus, some 7 folds: strong angularis, weak parietalis more inside, 3 strong palatal folds reaching the lip, upside another weak palatal denticle, strong upper and weaker lower columellaris.

Animal small and dark with lighter sides, upper tentacles not very long, lower tentacles very short.

Chondrina maginensis Arrébola & Gómez, 1998

Chondrina maginensis Arrébola & Gómez, 1998 in Arrébola & Gómez 1998: 110

Shell conical, not fusiform, first whorls with smaller diameter than usual in *Chondrina*, shell generally more slender (H/D = 3.0-3.5, shells with less than 9 whorls have less than 2 mm diameter), shell finely striated, aperture not rounded at the palatal side, 5 teeth: 2 palatal, 1 parietal, 2 columellar. Shells of live animals covered with dust and lichens, possibly to protect against insolation.

5.6-7.1 x 2.4-2.7 mm

Calcareous rocks and walls, often in rock crevices, well adapted to dry mountain habitats, 600-800 m. Population density much lower than in other *Chondrina* populations.

EU Red List: Vulnerable VU. Only known from few localities. Threatened by fire and increasingly dry conditions.

Classified in *Chondrina* by Kokshoorn [2009]: 63, 145.

S Spain (Sierra Mágina)

Chondrina megacheilos (De Cristofori & Jan, 1832)

Chondrus megacheilos De Cristofori & Jan, 1832 in De Cristofori & Jan 1832: 3

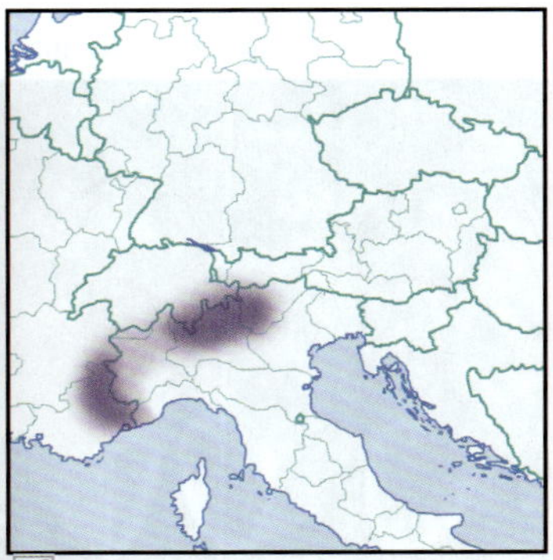

5-13 x 2.5-5 mm

Italy: Monte Baldo near Malcesine

S Alps

Kerney et al. 1983: 115, Gittenberger 1984: 62, Manganelli et al. 1995: 15, Turner et al. 1998: 160.

Shell brown, often with whitish surface layers, irregularly striated, 6-7 weakly convex whorls, cervix often with a keel-like structure, aperture with broad lip running from columellar margin to mid-palatal side, inside with many folds (strong angularis connected to usually strong spiralis, parietalis, columellaris, weaker infracolumellaris, up to 5 palatal folds, weak or absent suturalis, suprapalatalis, strong upper palatalis, shorter lower palatalis, again shorter infrapalatalis, no basalis), margin often white and reflected, umbilicus open.

Larger than *Ch. avenacea*, shape more conical, aperture with 4 palatal folds. The animal is dark.

Dry and open habitats on calcareous substrate, between rocks, prefers S and SW exposed uncovered rocky faces. Sometimes also on non-calcareous rocks. Mainly above 1000 m, up to 1700 m in Switzerland.

Critically Endangered in Switzerland. Threatened by rock quarrying.

Some authors proposed to subdivide this species into various subspecies. In www.iucnredlist.org 07.2012 a division into 4 subspecies was proposed, one of which undescribed and no description was given for it in the Red List. The genetic relationships between those populations remain to be studied. The result should also be aligned with usual standards for subspecies in commonly applied biodiversity concepts (which are based on partial barriers to reproduction).

Chondrina multidentata (Strobel, 1851)

Pupa avena var. *multidentata* Strobel, 1851 in Strobel 1851: 35

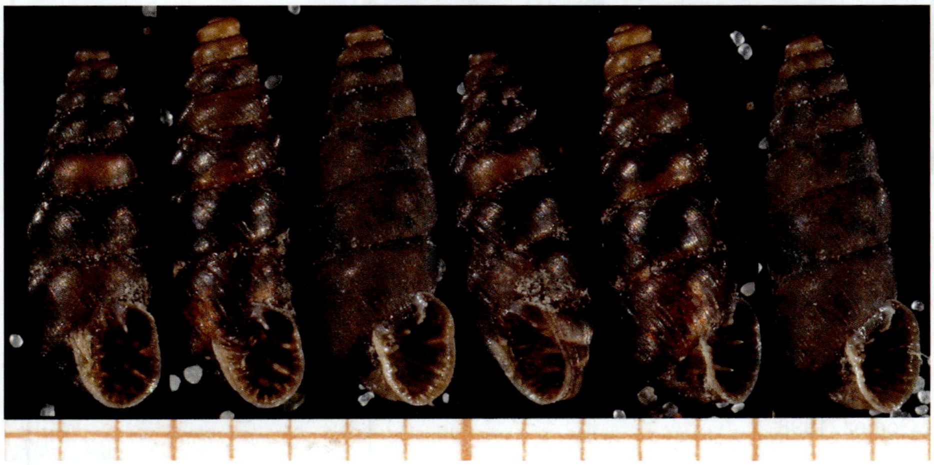

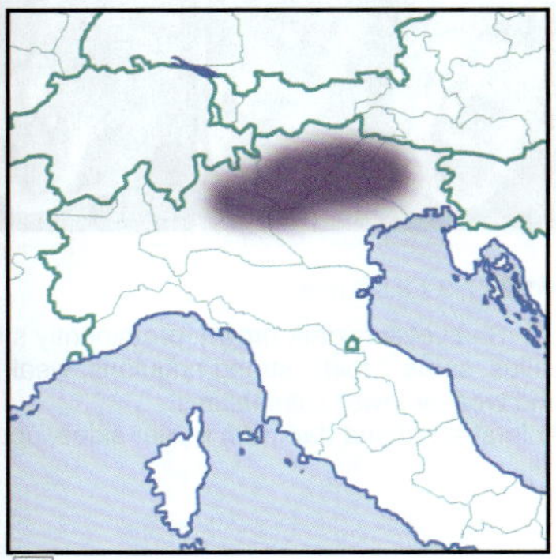

4.5-9 x 2.0-3.7 mm

Italy: Alto Adige/Südtirol, Valfonda near Tobiacco

S Alps (Lago di Como to Friuli)

Limestone rocks in mountains, up to 1700 m.

Nordsieck 1962: 12.

Shell dark violetish brown, 6.5-7.5 weakly convex whorls, last whorl weakly inflated below suture, cervix with keel, and a second keel-like fold running parallel, aperture with angularis and spiralis connected to each other, parietalis, columellaris, infracolumellaris, 3-4 palatal folds, margin whitish and weakly reflected with up to 20 small marginal teeth, umbilicus widely open. Differs from the other species in its marginal teeth.

Chondrina oligodonta (Del Prete, 1879)

Pupa avenacea var. *oligodonta* Del Prete, 1879 in Del Prete 1879: 80, Tav. 1

5.5-7 x 2.5-3.0 mm

Italy: Massa Carrara, Alpi Apuane, Monte Altissimo

N Italy: Alpi Apuane

Outcrops of natural calcareous rocks.

Shell dark reddish brown with light bluish hue, very finely striated, apertural margin white, sharp and not reflected, often connected at parietal side by a thin white layer, no teeth inside.

EU Red List: Vulnerable VU. Threatened by quarrying of the Carrara marbles which can lead to total destruction of the habitats of the species.

Chondrina pusilla (Des Moulins, 1835)

Chondrina bigorriensis (Des Moulins, 1835)

Pupa megacheilos var. *pusilla* Des Moulins, 1835 in Des Moulins 1835: 163, Pl. 2
Pupa megacheilos var. *pusilla* synonym *Pupa bigorriensis* Des Moulins, 1835 in Des Moulins 1835: 159, 163 (*Pupa Bigorriensis*)

Nomenclatural situation very weak and problematic. Frequently used as a name for a species *Chondrina bigorriensis* by subsequent authors until recently, but many have used the spelling *bigoriensis* or other spellings This has led to very much confusion and it is very difficult to understand which taxon was meant at which occasion.
Rossmässler 1835-1837 (5-6) [1837]: 14 used a name *Pupa bigoriensis*, attributed to Michaud, without bibliographic reference to Des Moulins 1835, for a species currently known as *Abida bigerrensis* (Moquin-Tandon, 1856). It is possible that initially the name *bigoriensis* had been used in unpublished sources for both species, which were either not distinguished or misidentified. Both species were from the same locality (Bagnères-de-Bigorre) and look very much alike.
Rossmässler's 1837 name *bigoriensis* was a homonym of *bigorriensis* (Art. 58.7), so for those who do not require a bibliographic reference for making a name available under Art. 11.6.1, and for those who do not require that the author who validated the name must have used it for the same species as the synonym, the name *bigorriensis* was validated at this occasion by Rossmässler in 1837 and must be attributed to Des Moulins 1835. The name-bearing type under Art. 72.4.3 would be the shell figured by Des Moulins 1835 Pl. 2 Fig. D. This would imply that Rossmässler's name could not be used for the *Abida* species. It would also imply that the name *bigerrensis* Moquin-Tandon, 1856 could not be used for the *Abida* species either, because also this usage was based on a misidentification of the taxon that must be attributed to Des Moulins, 1835.
For the others, the name *bigorriensis* was made available at another occasion (at least by Pilsbry & Cooke 1918-1920: 29 who referred bibliographically to Des Moulins 1835). This interpretation of Art. 11.6.1 is generally favoured here.
Pilsbry & Cooke 1918-1920: 29 asked why the name *pusilla* could not be used for this species. I am asking the same question here. Using *pusilla* would be much more convenient. The names *pusilla* and *bigorriensis* were both established by Des Moulins 1835: 163.
First Reviser was Des Moulins 1835 who selected *pusilla*.
Vertigo pusilla Müller, 1774 had been classified very occasionally and only for a few years in the 1840s (Pfeiffer 1848: 364) in the genus *Pupa*, so *pusilla* Des Moulins, 1835 was regarded by some authors as permanently invalid (Art. 59.3). I believe that this is why *pusilla* has not been used. Art. 59.3.1 rules that in case of confusion the Commission must be asked for a ruling. This is certainly a clear case of confusion (two closely related species were named and subsequently misspelled *Pupa bigorriensis*, *Pupa bigoriensis*, *Pupa bigorrensis* and *Pupa bigerrensis*), many confound and misspell these names. I would suggest solving the problem by simply not using *Chondrina bigorriensis*.

7-9 x 2.5-2.8 mm

France: Hautes-Pyrenées, near Grezian

N Pyrenees

Shell more conical than *Ch. avenacea*, last whorl with larger diameter than penultimate whorl, not equal in size, aperture with 3 palatal teeth reaching the margin, a 4th (supra)palatal tooth is small but usually present.

Open rocky habitats in mountains, up to 2000 m.

Chondrina ripkeni Gittenberger, 1973

Chondrina ripkeni Gittenberger, 1973 in Gittenberger 1973: 203

Shell dark horny brown, usually coarsely ribbed, 6.5-7 moderately convex whorls, apertural margin thickened and reflected, basal side slightly pointed or angular producing a weak keel outside, angularis connected to spiralis, parietalis and columellaris moderate, infracolumellaris weaker and shorter, 3 palatal folds, sometimes also a weak suprapalatalis, palatal system usually shifted slightly to the basal side.
More coarsely ribbed and with less whorls than *Ch. pusilla*.

6.1-7.0 x 2.5-2.7 mm

Rocky habitats in mountains at 800-1500 m.

EU Red List: Near Threatened NT. Only known from few localities, range fragmented, but no real threats known.

N Spain (Basque mountains)

Chondrina spelta (Beck, 1837)

Torquilla spelta Beck, 1837 in Beck 1837: 86

4-8 x 2.1-3 mm

Albania: Shkodër

Central Albania to Croatia

Shell reddish brown, striated, conical, apertural margin with lip, infrapalatalis usually not present, suprapalatalis almost always absent. Smaller than *Ch. arcadica*, teeth less well developed, slightly larger than *Rupestrella*. Locally variable.

Rocky limestone habitats, a relatively frequent species.

Nordsieck 1970, Gittenberger 1973: 162, Gittenberger 1984: 62, Dhora & Welter-Schultes 1996: 118.

The local variations in shell morphology have invited taxonomists to recognise many local subspecies. It remains to be verified if such subspecies are tenable in the light of commonly applied concepts in biodiversity research, where partial barriers to reproduction should be responsible for partial gene flow, and morphological differences at both sides of hybrid zones a result of such a genetically induced background. The alternative hypothesis, presence of continuous character variations over area, could also explain the variation and would not justify a subspecies model. The second option seems much more likely.

Chondrina tatrica Ložek, 1948

Chondrina tatrica Ložek, 1948 in Ložek 1948: 85

6-7.5 x 2.2-2.8 mm

Slovakia: Karst, Zadielska dol.

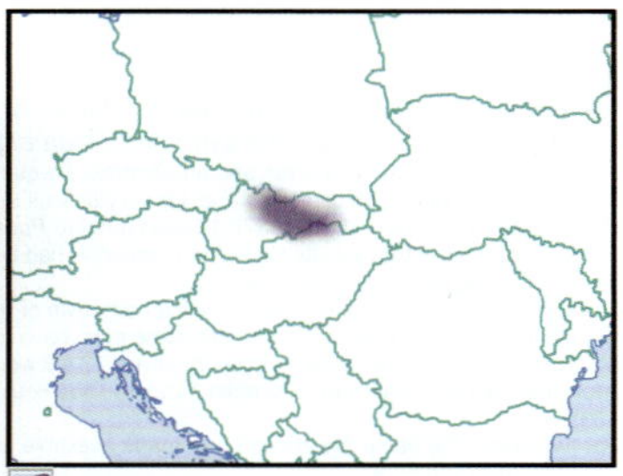

Slovakia

Slovakia: Slovenský Kras, Zadiel gorge

Shell dark violet brown, often with light blue hue, with fine and indistinct streaks, sometimes smooth, aperture more or less like *Ch. avenacea* but with weaker folds, margin thin, sharp and almost not reflected, not connected by a white layer at parietal side, palatal margin inside often thickened.

At limestone rocks not covered by vegetation, preferably at large rocks.

Jozef Grego

Chondrina tenuimarginata (Des Moulins, 1835)

Pupa megacheilos var. *tenuimarginata* Des Moulins, 1835 in Des Moulins 1835: 163, Pl. 2

8-14 x 3-4 mm

Spain: Cataluña, Girona, Vidra

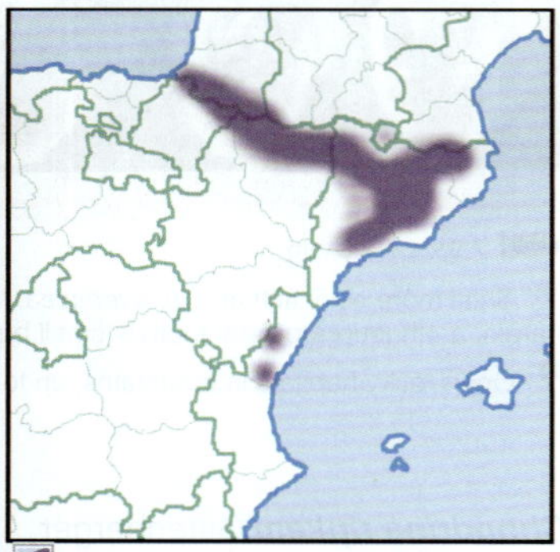

S Pyrenees, E Spain (Castellón)

Open rocky habitats in mountains, up to 1700 m.

Gittenberger 1973, Kerney et al. 1983: 116, Martínez-Ortí & Robles 2003: 78.

Shell larger than *Ch. pusilla*, with weaker reddish hue in the brown colour, apertural lip flat, presence of an additional sutural tooth between angularis and palatal side.

Granaria

Granaria Held, 1838 in Held 1838: 918

Type species *Pupa frumentum* Draparnaud, 1801 (Subsequent designation by Herrmannsen 1846 [18-04-1847]: 488)

Shell small to medium-sized, cylindrical, aperture with 4 palatal, 1 parietal and 2 columellar lamellae. A penial caecum is present, retractor inserts in a medium position on the epiphallus.

Near limestone rocks at the ground, not on open rock faces.

Central and S Europe to Pakistan and Yemen

Gittenberger 1973, Kokshoorn [2009]: 120 (checklist of 7 species).

Granaria braunii (Rossmässler, 1842)

Pupa braunii Rossmässler, 1842 in Rossmässler 1838-1844 (11): 10, Taf. 53 (*Pupa Braunii*)

5.8-9.2 x 2.2-2.8 mm

Spain: Pyrenees, Guypasena

E Spain and Pyrenees

Germain 1930: 397, Gittenberger 1973: 57, Kerney et al. 1983: 109, Gittenberger & Ripken 1993, Martínez-Ortí & Robles 2003: 81, Uribe et al. 2007.

Whorls weakly convex, apertural margin thick and white, also in the parietal region, thickened on the palatal side, usually 6 teeth: parietalis, strong angularis, only one columellar fold (the other *Granaria*s have two), two palatal folds.

Between rocks and rock rubble, in 400-1600 m. Shell size and shape are not correlated with altitude in the Pyrenees.

Life cycle unknown, oval eggs (length 1 mm) were observed.

Threatened in E Spain (Alicante-València) by fire and growing urbanized areas in combination with the small number of known populations.

Granaria frumentum (Draparnaud, 1801)

Pupa frumentum Draparnaud, 1801 in Draparnaud 1801: 59
Pupa frumentum var. *illyrica* Rossmässler, 1835 in Rossmässler 1835-1837 (1): 82

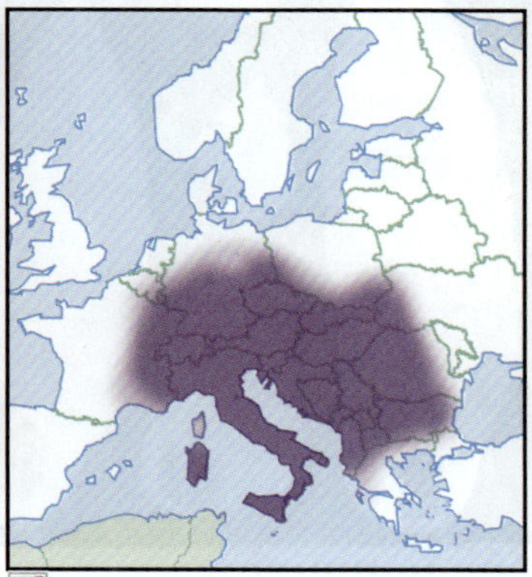

frumentum form 6.5-8.0 x 2.7-3.0 mm or larger — Austria: Tirol, near Zams — Central Europe to SE Bulgaria

Shell weakly brownish, finely and regularly ribbed, aperture usually with 8 folds: 4 palatal lamellas also visible from outside, angularis stronger than parietalis, cervix with strong callus. Shell diameter larger than in *Abida secale* and *Chondrina avenacea*. Shells from S exposed slopes are larger than from N exposed slopes.
The *illyrica* form is larger than *frumentum*, cervical callus weak or absent, shell less prominently ribbed on the last 2-3 whorls.

Dry open calcareous habitats: very characteristically on sunny slopes with low grasses, grasslands, rock rubble, walls. Sensitive against modification of habitats. Rarely above 1000 m, in Switzerland *frumentum* not above 800 m (*illyrica* up to 1300 m), in Bulgaria up to 1500 m.

illyrica form 9-11.5 x 3.2-4 mm — Italy: Teodo near Padova

Critically Endangered in Vorarlberg, Endangered in Germany, Vulnerable in Switzerland.

Classification follows Gittenberger 1973: 57 and Kokshoorn [2009]: 121 who found no evidence that the two forms *frumentum* and *illyrica* may occur sympatrically and without being interconnected by intermediate populations in a hybrid zone. The *illyrica* form is conchologically variable, and often classified as an independent species (Kerney et al. 1983: 108, Manganelli et al. 1995: 14, Turner et al. 1998: 152), in a range from Italy and S Switzerland to SE Austria and Balkans.
Some authors have recognised many more subspecies based on the locally variable shell morphology, but their status has not been verified in terms of commonly applied subspecies concepts in biodiversity research, where partial barriers to reproduction should be decisive.

Italy: Lago di Garda, Gargnano

Granaria stabili (Martens, 1865)

Granaria stabilei (Martens, 1865)

Pupa mortilleti Stabile, 1864 (in Stabile 1864: 96, *Pupa Mortilleti*) not *Pupa (Torquilla) Mortilleti* Martens, 1860 (in Albers & Martens 1860: 287, = currently *Solatopupa psarolena* (Bourguignat, 1858)).
Pupa stabili Martens, 1865 in Martens 1865: 233
Expressly established as a new replacement name for *Pupa mortilleti* Stabile, 1864 non Martens, 1860.
The name *stabili* was correctly formed from the Latinized name Stabilus of the Italian name Stabile, not to be corrected to *stabilei* under Art. 32.5.
Google general search 04.2009: stabili 5, stabilei 107
Google general search 09.2010: stabili 14, stabilei 61
Google general search 07.2012: stabili 62, stabilei 117
Google Scholar 12.2011: stabili 1, stabilei 7.
These figures indicate a very rarely used name. In such a case the original spelling should be used.

5.6-9.3 x 2.0-2.6 mm — Italy: Cuneo, Prato Nevoso, Rifugio La Balma — SW Alps

Shell more slender than the other *Granaria*s and more conical (last whorl with largest diamater), lower whorls nearly smooth, aperture with thin and fragile margin, no parietal callus, 7 or so teeth: lower palatalis long, prominent and ~ like curved.

Dry habitats among rocks and rock rubble, 500-2700 m.

Granaria variabilis (Draparnaud, 1801)

Pupa variabilis Draparnaud, 1801 in Draparnaud 1801: 60

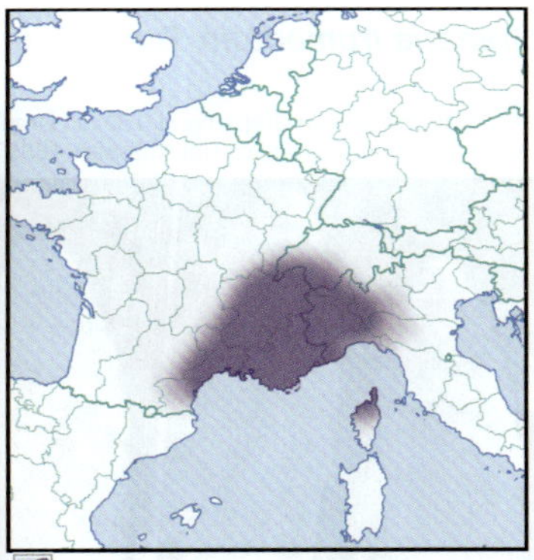

S France to N central Italy

Dry and warm habitats, calcareous meadows, rocks and rock rubble, dry walls. Usually in lowlands, in the Alps up to 1600 m (Switzerland 1100 m).

Populations declined in the region of lake Geneva.

6.7-19 x 2.6-4.5 mm

France: Provence, near Montélimar

Shell pale brown, shape very variable, aperture less variable. Usually more slender and larger than *G. frumentum*, rather smooth on the last whorls, whorls not convex, no cervical callus, aperture with strong white lip and some 7 teeth, lower palatalis remarkably strong and connected to the lip, no parietal callus.

Granopupa

Granopupa Boettger, 1889
in O. Boettger 1889: 249

Type species *Pupa granum* Draparnaud, 1801
(monotypy)

Established without description, with one species included.

Shell small, with fine evenly spaced ribs, aperture with 4 palatal, 1 parietal and 2 columellar folds. Male genital system without penial caecum. Internally with coarse papillae in the distal epiphallus lumen.

Canary Islands and Morocco to Tunisia, Portugal to Arabia and Afghanistan, Ascension island (Atlantic)

Gittenberger 1984: 57, Neubert 1998: 367, Kokshoorn [2009]: 120.

Only one species.

Granopupa granum (Draparnaud, 1801)

Pupa granum Draparnaud, 1801 in Draparnaud 1801: 59

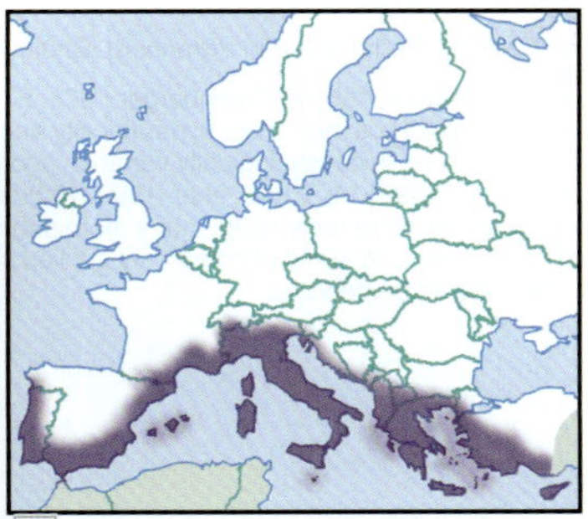

Canary Islands to Afghanistan

Nobre 1941: 142, Gittenberger 1973: 36, Kerney et al. 1983: 107, Manganelli et al. 1995: 14, Welter-Schultes 1998: 34, Turner et al. 1998: 150, Schütt 2001: 78, Kokshoorn [2009], Sysoev & Schileyko 2009: 55.

3.5-6 x 1.4-1.8 mm

Greece: Kárpathos, near bay of Pigadia

Fresh shell light horny brown, very finely striated, 6-8 strongly convex whorls, last 2-3 whorls with same diameter, aperture with 7 or so teeth (strong parietalis, strong lower palatalis, usually no angularis). Apex (first whorl) diameter 0.55-0.6 mm, smaller than *Rupestrella philippii* (0.7 mm).
Animal small, grey or brown with lighter sides and foot, upper tentacles relatively long, lower tentacles conical and very short.

Dry and sunny habitats, often very abundant, at the ground below limestone rocks and in crevices. In S Switzerland at S exposed slopes, up to 1400 m.

Rupestrella

Rupestrella Monterosato, 1894
in Monterosato 1894: 170 *Pupa (Rupestrella)*

Type species *Bulimus rupestris* Philippi, 1836
(original designation)

Rupestrella dupotetii (Terver, 1839)

Vertigo dupotetii Terver, 1839 in Terver 1839: 32, Pl. 4 (*Vertigo Dupotelii*)
Misspelled *Dupotelii* on p. 32, but expressly dedicated to Dupotet and spelled *Dupotetii* on p. [40] (explication des planches), inadvertent error must be corrected (Art. 32.5).

Shell dark brown with bluish hue, finely striated, whorls rounded with deep suture, no teeth in the aperture.

Westerlund 1887: 105, Arrébola & Gittenberger 1993, Kokshoorn [2009]: 179, Torres 2010: 39.

4-5 x 2.0-2.3 mm

Spain: Málaga, El Palo
José Liétor Gallego

NW Africa, S Spain

Rupestrella homala (Westerlund, 1892)

Pupa homala Westerlund, 1892 in Westerlund 1892: 192

Shell reddish brown, weakly striated, 5-6 convex whorls with deep suture, last whorl slightly larger than penultimate whorl and slightly ascending, aperture without teeth.
Differs from *R. rupestris* in its less slender shell, much less convex whorls and shallower suture.

4-6 x 2.1-2.4 mm

EU Red List: Vulnerable VU. Restricted to few localities, no real threats known.

Kokshoorn & Gittenberger 2010: 40. Not listed by Manganelli et al. 1995: 14.

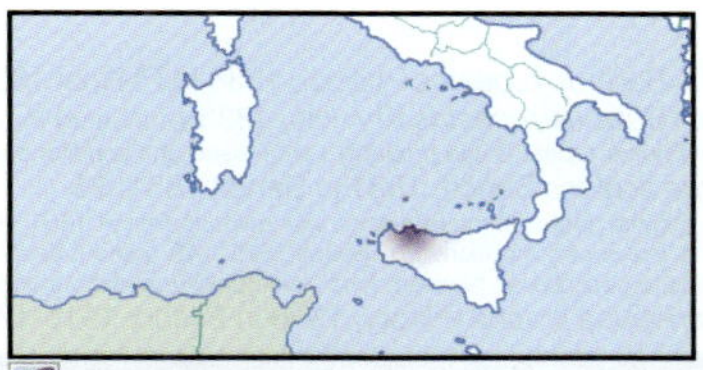

Italy (Sicily)

Rupestrella occulta (Rossmässler, 1839)

Pupa occulta Rossmässler, 1839 in Rossmässler 1838-1844 (9-10): 25, Taf. 49

Shell light horny brown to brown, not shiny, weakly and finely striated, 5-6 convex whorls, last whorl relatively large but not wider than penultimate whorl, aperture with 3 folds (parietal, columellar and palatal), sometimes possibly with angularis and a second palatal fold.

Westerlund 1887: 105, Pilsbry 1916-1918: 353, Manganelli et al. 1995: 14 ("entità da rivedere"), Kokshoorn & Gittenberger 2010: 42 (only names without reference to a scientific study to support the proposed classifications).

Species needs a revision.

4.5-6 x 1.7-2.1 mm

Italy: Sicily

Italy: NE Sicily

Picture: Rossmässler 1839 (Icon.): Taf. 49 Fig. 638

Rupestrella philippii (Cantraine, 1840)

Pupa philippii Cantraine, 1840 in Cantraine 1840: 140
Rupestrella jaeckeli Beckmann, 2002 in Beckmann 2002: 72

3.5-6 x 1.4-2 mm

Greece: Kálimnos, Kálimnos town

S Europe, N Libya

Shell finely ribbed or striated, or nearly smooth, 5.5-6 weakly convex whorls with shallow suture, apex diameter broad (0.65-0.75 mm; in *R. rhodia* 0.55-0.6 mm), initial whorls look like truncated, last whorl with largest diameter, aperture with some 5 teeth or less, strong upper palatalis, relatively strong parietalis and columellaris, usually a well developed angularis, columellaris usually the strongest.

Attached to limestone habitats, not too open and not too sunny, snails hide in crevices. Predated by *Drilus* beetles in Sicily. On Gávdos island (Greece) in valleys with pine trees.

Manganelli et al. 1995: 14, Štamol & Velkovrh 1995, Welter-Schultes 1998: 36, Schütt 2001: 80, Örstan et al. 2005, Štamol & Kletečki 2005, Renda et al. 2011.

R. jaeckeli was not compared with *R. philippii* in its original description to justify its status as an independent species. Renda et al. 2011 came to the conclusion that *jaeckeli* falls within the usual range of variation of local populations of *R. philippii*.

Paratype *jaeckeli* Italy: Sicily, Monte Biaggio S of Agrigento

Rupestrella rhodia (Roth, 1839)

Pupa rhodia Roth, 1839 in Roth 1839: 19, Tab. 2

3.5-6 x 1.4-1.8 mm

Greece: central Crete, Mt. Kédros

SE Europe, W Turkey

Apex (first whorl) diameter 0.5-0.6 mm, smaller than in *R. philippii* (0.7 mm), first whorls conical and pointed, last whorl not as predominant as in *R. philippii*, aperture with 6-7 folds: strong upper and lower palatalis, strong columellaris, no well developed angularis.

Limestone habitats, more shady than *Granopupa granum*.

Westerlund 1887: 106, Štamol & Velkovrh 1995 (Cres, Lošinj), Welter-Schultes 1998: 36, Schütt 2001: 79, Štamol & Kletečki 2005, Balashov & Gural-Sverlova 2012: 94.

Rupestrella rupestris (Philippi, 1836)

Bulimus rupestris Philippi, 1836 in Philippi 1836: 141 (figured on Tab. 8 Fig. 18 [sic])
not *Bulimus rupestris* Krynicki, 1833 (= currently *Peristoma rupestre* (Krynicki, 1833) (Enidae))
The name *rupestris* can probably not be used, the conditons of Art. 23.9.1 are not met. Perhaps under Art. 23.9.5.
Pupa scalaris Benoit, 1881 in Benoit 1881: 96
In original publication a reference was given to Benoit 1857-1862: Tav. 5 Fig. 37, but in that publication the name was not mentioned on the plate or in the text. Previously mentioned by Benoit 1875: 149, without description or indication, only with a locality, name was not made available there. Was probably used to describe an aberrant form, eventually unavailable under Art. 1.3.2.
Granopupa coloba Pilsbry, 1918 in Pilsbry 1916-1918: 352 (*Granopupa (Rupestrella) rupestris coloba*)

3.5-4.5 x 1.9-2.5 mm

Italy: Sicily, Monte Pellegrino W Palermo

Italy (Sicily, Capri)

Westerlund 1887: 104 (*rupestris*), 105 (*scalaris*).

Shell horny yellowish or brown, not shiny, coarsely striated, 4-5 convex whorls, regularly increasing, also the last whorl is wider than the penultimate whorl so that the margin is not convex but straight and the shell looks like a juvenile *Rupestrella*, aperture without folds or teeth, umbilicus open.

EU Red List: Near Threatened NT. Probably exaggerated. It was only because some authors had recognised populations from single localities as subspecies.

The presence of subspecies as recognised by Beckmann 2002 (some of which seem to represent locally extreme forms and were restricted to one locality) should be verified and aligned with commonly applied subspecies concepts based on partial barriers to gene flow.

Solatopupa

Solatopupa Pilsbry, 1917
in Pilsbry 1916-1918: 234 *Chondrina (Solatopupa)*

Type species *Bulimus similis* Bruguière, 1792 (original designation)

Gittenberger 1973: 62, Boato 1988, Boato 1991, Ketmaier et al. 2006.

Solatopupa cianensis (Caziot, 1910)

Pupa cianensis Caziot, 1910 in Caziot 1910: 318

8.9-11.3 x 3.0-3.6 mm

France: Alpes-Maritimes, Gorges de Cians

SE France (Cians, Daluis)

Only on red Permian sandstone rocks, 800-1000 m.

EU Red List: Vulnerable VU. Only known from few localities which are not protected.

Shell horny brown with reddish hue due to dust particles, fragile, 7-8 whorls, apertural margin sharp, not thickened, slightly reflected, usually no folds in the aperture, sometimes with weak collumellaris and 2 palatal folds, umbilicus open and visible in an oblique view.

Solatopupa guidoni (Caziot, 1904)

Pupa similis var. *guidoni* Caziot, 1904 in Caziot 1904: 38 (*Pupa similis* var. *Guidoni*)

9-14 x 3-4 mm

Italy: Sardinia, Sassari, Alghero, Capo Caccia

N Corsica, NW Sardinia, Elba

Open habitats at limestone rocks in highlands covered by Mediterranean shrubs, in Sardinia not below 300 m.

Rare and scattered, only 13 populations are known, 2 in Sardinia and 2 in Elba, the others in Corsica.

Giusti 1970 (Atti Toscana B 77): 71, Gittenberger 1973: 68, Ketmaier et al. 2010.

Shell cannot be distinguished well from *S. similis*. Both species differ in its anatomy (structure of penis). The animal is light brown with pale grey tentacles.

Ketmaier et al. 2010 found that the populations at Elba and Sardinia were closely related to northern Corsican populations, and that the central inland Corsican populations formed a second group.

Solatopupa juliana (Issel, 1866)

Pupa quinquedentata var. *juliana* Issel, 1866 in Issel 1866: 21 (*Pupa quinquedentata* var. *Juliana*)

N Italy

Gittenberger 1973: 63 (synonym of *S. similis*), Boato 1991, Cossignani & C. 1995, Manganelli et al. 1995: 14, 38.

11-15 x 3-4 mm

Italy: Lucca, Pizzorno

Shell whitish with reddish or violet hue and violetish brown first whorls, 9-10 weakly convex whorls, aperture whitish or yellowish with shiny white lip and 5-6 teeth, angularis strong, parietalis strong, upper columellaris weak, lower columellaris very weak, two strong palatal folds.

Diagnosis uncertain and based on material collected by M. Sosso in Lucca. Issel 1866 gave no description and just stated that the shell was more reddish than *quinquedentata* and that this could be environmentally induced. Not in Westerlund 1887.

Solatopupa pallida (Rossmässler, 1842)

Pupa pallida Rossmässler, 1842 in Rossmässler 1838-1844 (11): 11, Taf. 53

NW Italy

Usually in rocky habitats, in coastal regions often at outcrops facing the sea, occasionally also in conglomerates.

Alzona 1971: 76, Gittenberger 1973: 72, Boato et al. 1984: 277, Cossignani & C. 1995, Manganelli et al. 1995: 14, 38.

8.8-11.6 x 3.1-3.6 mm

N Italy: La Spezia

Shell pale white to bluish with horny brown first whorls and indistinct light horny brown spots, 7.5-8.5 whorls, apertural margin slightly thickened, rounded at basal side, weak angularis, parietalis and columellaris can be present, no palatal folds, umbilicus narrow or closed.

Solatopupa psarolena (Bourguignat, 1858)

Without gender agreement: *Solatopupa psarolenus* (Bourguignat, 1859)

Bulimus cinereus Mortillet, 1851 in Mortillet 1851: 96 (*Bulimus cinerens*) not *Pupa cinerea* Draparnaud, 1801 [in Draparnaud 1801: 61] (= currently *Solatopupa similis*), not *Bulimus cinereus* Reeve, 1848 from Bolivia

Bulimus cinerens was an inadvertent error for *cinereus* and must be corrected (Art. 32.5). Not *Pupa cinerea* Draparnaud, 1801 (with which it was compared by Mortillet 1851) (= currently *Solatopupa similis*), but a secondary homonym if both are classified in the same genus.

Bulimus psarolenus Bourguignat, 1858 in Bourguignat 1858-1859: Pl. 15

Established as a regular new name on Pl. 15 of vol. 10 (dated to 1858), only a name and one figured specimen (Fig. 1-2). On p. 51 of vol. 11 (1859) expressly presented as a replacement name for *Bulimus cinereus* Mortillet, 1851 non *Bulimus cinereus* Reeve, 1848 from Bolivia. Not a new replacement name in the sense of the Code because the 1858 publication is alone relevant.

6.5-8.2 x 2.8-3.2 mm

Italy: Imperia, Monte Pietravecchia

SW Alps, near the coast

Shell whitish or bluish and horny brown marbled, cone-shaped, 6-7 strongly convex whorls, aperture without folds, margin not thickened and only reflected at columellar side, umbilicus narrow and open, only visible in an oblique view.

In mountain habitats at 500-1100 m, locally very restricted localities, usually abundant where present.

EU Red List: Vulnerable VU. One of its few localities (gorge de Saorge) was destroyed by railroad construction in 1907, Caziot 1907 anticipated that this population was endangered, no animals were found there after the 1950s. The case demonstrated how traffic construction can lead to the local extinction of populations. Today a major threat are road constructins and road enlargements. Only 4 known localities are left.

Caziot 1907, Gittenberger 1973: 78, Manganelli et al. 1995: 14, www.iucnredlist.org (07.2012).

Solatopupa similis (Bruguière, 1792)

Bulimus similis Bruguière, 1792 in Bruguière 1792: 355

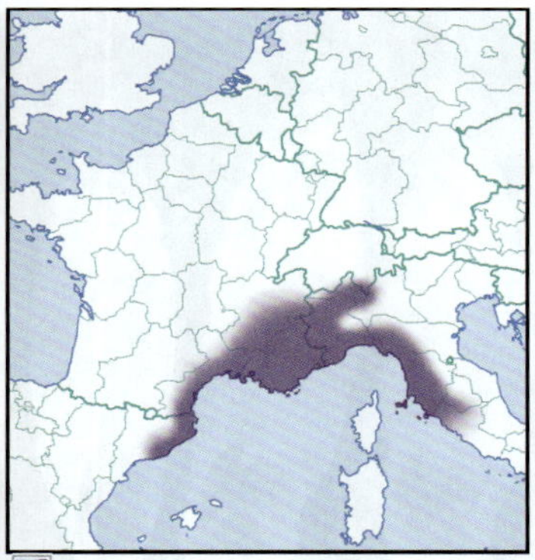

9.0-14.2 x 3.0-4.2 mm

Italy: Liguria, near Andora

NE Spain to Italy

Shell pale white or bluish, horny brown marbled, aperture with 2 equally strong palatal folds, also visible from outside.

At sunny limestone rocks in coastal regions, in the S Alps at S exposed slopes. In Switzerland in up to 700 m altitude. Feeds on lichen covers of stones. This species is also able to colonise newly and artificially created rocky habitats in disturbed areas such as margins of newly built roads.

France: Marseille, Calanques de Marseille
Cédric Audibert

Enidae

Gümüş & Neubert 2012: 29 presented a consistent system to classify enid genera. This system is used here with exception of *Eubrephulus* (here classified in *Chondrula*).
The presented generic characters for the enids (derived from a personal communication by R. Bank) were not equipped with sufficiently distinctive diagnoses that would allow to understand the entire system. It would also be useful to align the proposed system with results of molecular studies.

Brephulopsis

Brephulopsis Lindholm, 1925 in Lindholm 1925: 28 *Zebrina (Zebrina (Brephulopsis))*

Type species *Chondrus attenuatus* Krynicki, 1833 (original designation)

Feminine under Art. 30.1.2. Not a new replacement name for a name *Brephulus* sensu Westerlund 1902 (not Beck 1837) because this is not possible under Art. 49. Westerlund 1902 only used the name *Brephulus* and attributed it to Beck 1837 (and gave *attenuatus* as type, which was not originally contained).

Ramusculus Lindholm, 1925 in Lindholm 1925: 28 *Zebrina (Zebrina (Ramusculus))*, type species *Bulimus subulatus* Rossmässler, 1837 (original designation)

Penis with appendix, caecum at central position of the epiphallus.

Páll-Gergely 2009, Gümüş & Neubert 2012: 29.

Gümüş & Neubert 2012 gave no arguments why *subulata* and the other two species were classified in the same genus, contrasting Páll-Gergely 2009 who did not compare *Ramusculus subulatus* with *Brephulopsis*. This should be done somewhere because it is not easy to understand the proposed classification.

Brephulopsis bidens (Krynicki, 1833)

Chondrus bidens Krynicki, 1833 in Krynicki 1833: 401 (figured on Tab. 7 Fig. 3 [sic])

9-19 x 3-5.5 mm

Ukraine: Crimea, Jev Pastoria

S Ukraine (Crimea)

In crevices of limestone rocks and under stones and boulders.

Balashov & Gural-Sverlova 2012: 96.

Shell white or with dark streaks, shiny, 7.5-12 whorls, in the aperture on the palatal side a long white fold is present, parallel to the suture, columella callose and prominent, twisted, apertural lip weak.

Brephulopsis fusiformis (Menke, 1828)

Brephulopsis cylindrica (Menke, 1828)
Without gender agreement: *Brephulopsis cylindricus* (Menke, 1828)

Bulimus fusiformis Menke, 1828 in Menke 1828: 77
Established with description.
Bulimus cylindricus Menke, 1828 in Menke 1828: 77
Established with description. Not *Bulimus cylindricus* Gray, 1825 in Gray 1825: 414.
I did not find 25 references to meet the conditions of Art. 23.9, to protect Menke's name *cylindricus*. If this cannot be done, *fusiformis* should be used.

Shell bluish white or white, sometimes with dark streaks, opaque, 9-12 not very convex whorls, aperture inside yellow-brownish, with a weak white lip, columella slightly thickened.

22-35 x 7.5-9 mm

Rocky habitats in mountains and lowlands, grasslands and urbanised areas.

Pfeiffer & Clessin 1881: 272 (under *4 lineatus*), 295, Westerlund 1887: 7, Likharev & R. 1952: 184.

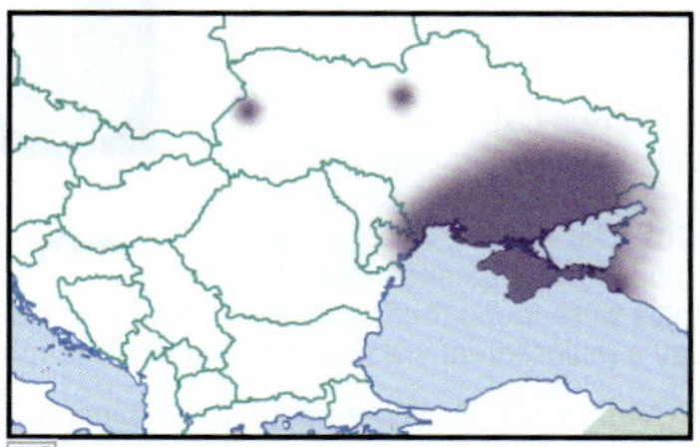

S Ukraine, Moldova, SW Russia

Brephulopsis subulata (Rossmässler, 1837)

Without gender agreement: *Brephulopsis subulatus* (Rossmässler, 1837)

Bulimus subulatus Rossmässler, 1837 in Rossmässler 1835-1837 (5-6): 48, Taf. 28

Shell grey with brown streaks, almost smooth, shiny, thin, not very translucent, 10-11 whorls, aperture simple, 1/3 of shell height, not reflected, umbilicus very narrow.

Rocky habitats in mountains, in rock crevices and under stones.

Westerlund 1887: 26, Likharev & Rammelmejer 1952: 185, Sysoev & Schileyko 2009: 84, Balashov & Gural-Sverlova 2012: 96 (*Ramusculus subulatus*).

10-14 x 2.5-3 mm

Ukraine: Crimea, Kuchuk Yanyshar mountains

Ukraine (Crimea mountains)

Buliminus

Buliminus Beck, 1837 in Beck 1837: 68 *Bulimus (Buliminus)*

Type species *Bulimus labrosus* Olivier, 1804 (monotypy by Ehrenberg 1831: [Gasterpoda p. 10]) (= currently *Buliminus labrosus* (Olivier, 1804) from Near East)

Beck cited the name as "*Buliminus* (Ehrbg.) B[eck].", and among the synonyms the name "*Bulimina* Ehrbg.". This indicated that Beck modified deliberately *Bulimina* Ehrenberg, 1831, so this is to be interpreted as an unjustified emendation. It can be used as a substitute name because *Bulimina* Ehernberg cannot be used.

Bulimina Ehrenberg, 1831 in Ehrenberg 1831: [Gasterpoda p. 10], type species *Bulimus labrosus* Olivier, 1804 (monotypy) (not *Bulimina* d'Orbigny, 1826 (Foraminifera)

Buliminus carneus (Pfeiffer, 1846)

Bulimus carneus Pfeiffer, 1846 in Philippi 1847: 114, Tab. 4

Name was attributed to Pfeiffer, authorship of the entire description was also attributed to Pfeiffer.

Shell brownish, shiny with irregular growth lines, 6 whorls, last 2 whorls equal in diameter, aperture very broad, about 1/3 of shell height, with white rounded and reflected margin, at parietal side a thin white layer, palatal side deviating outside the cylindrical shell shape, no folds in the aperture but columellar margin relatively thick, umbilicus narrow.

Stony and xerothermous habitats with low vegetation, frequent near Antalya.

Bank & Hovestadt 1991: 12, Schütt 2001: 112.

15-25 x 7-9 mm

S Turkey: Perge 2 km ENE Antalya (H = 18.3 mm)

S Turkey

In Cyprus only at one locality (Ormidia), possibly introduced.

Emilio Garcia

Buliminus lycicus (Pfeiffer, 1846)

Bulimus lycicus Pfeiffer, 1846 in Philippi 1847: 113, Tab. 4

Name was attributed to Pfeiffer, authorship of the entire description was also attributed to Pfeiffer.

Shell horny yellowish, 5 not very convex whorls, last whorl predominating, last whorls and aperture very large, aperture with white margin, connected by a thin white layer at parietal side, no teeth or lamellas, columella oblique and visible in the aperture, umbilicus usually covered.

22-24 x 11.5-12.5 mm

Shrubby limestone habitats in mountains.

Schütt 2001: 115.

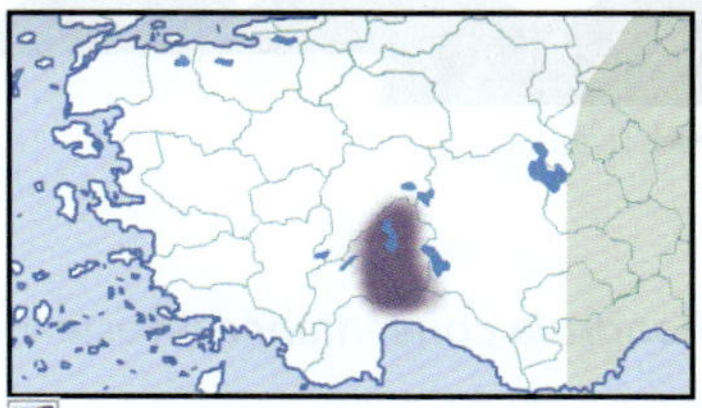

SW Turkey

Chondrula

Chondrula Beck, 1837 in Beck 1837: 87 *Pupa (Chondrula)*

Type species *Helix tridens* Müller, 1774 (subsequent designation by Herrmannsen 1846 [01-12-1846]: 231)

Spaniodonta Kobelt & Möllendorff, 1902 in Kobelt 1902: 1022 *Chondrula (Spaniodonta)*, type species *Buliminus diodon* Retowski, 1883 (original designation)

Authorship of Kobelt & Möllendorff was explained on p. 1021. No description, but a type species was given.

Eubrephulus Wagner, 1928 in A. J. Wagner 1928: 310, type species *Bulimus bicallosus* Pfeiffer, 1847 (subsequent designation by Forcart 1940: 250)

Europe and N Africa to Iran

Gümüş & Neubert 2012: 29 classified *Eubrephulus* as a separate genus containing 2 species *bicallosus* and *orientalis*, but gave no arguments how to distinguish *Eubrephulus* from *Chondrula*.

Chondrula beieri Klemm, 1962

Chondrula beieri Klemm, 1962 in Klemm 1962: 233, Taf. 2

Shell slender, parietalis well developed, palatal tooth present, columellaris absent.

8-10.5 x 3.2-3.6 mm

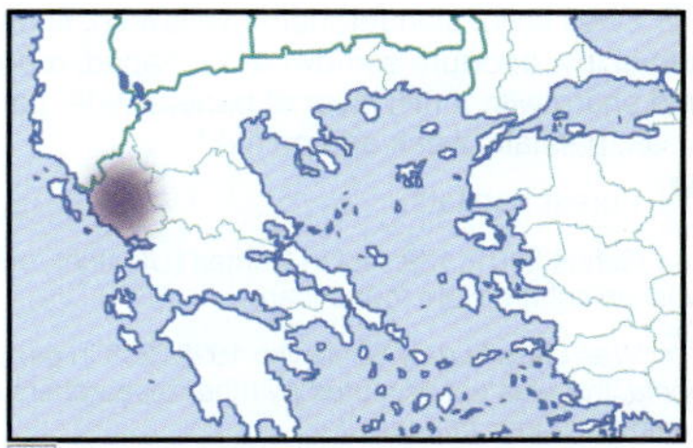

Greece (Ipiros)

Chondrula bergeri (Roth, 1839)

Pupa bergeri Roth, 1839 in Roth 1839: 19, Tab. 2

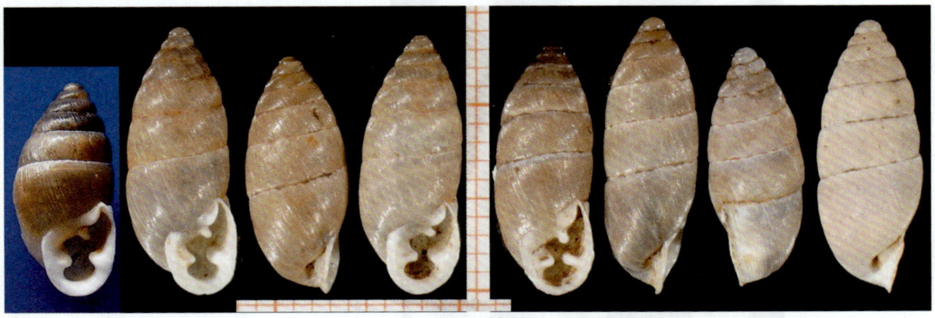

Greece

13-16 x 5-6 mm (1) Greece: Attiki, near Thiva, (2-8) Greece: Attikí, W Marathon (1) Panayotis Ovalis

Heller 1976: 381.

Shell dextral, horny brown, regularly striated, aperture with white lip inside and 4 prominent teeth: angular, parietal, palatal and columellar, margin not always connected by a thin white layer between angular knot and columellar side, umbilicus open. Shell very much like *Ch. tridens*, but with very prominent angular tooth.

The species has not recently been studied in detail. Its relationships with other *Chondrula* species are unclear, could be a regional form of *Chondrula tridens*.

Chondrula bicallosa (Pfeiffer, 1847)

Without gender agreement: *Chondrula bicallosus* (Pfeiffer, 1847)

Bulimus bicallosus Pfeiffer, 1847 in L. Pfeiffer 1847: 192

16-23 x 6-7 mm

Bulgaria: Sakar mountains near Sladin

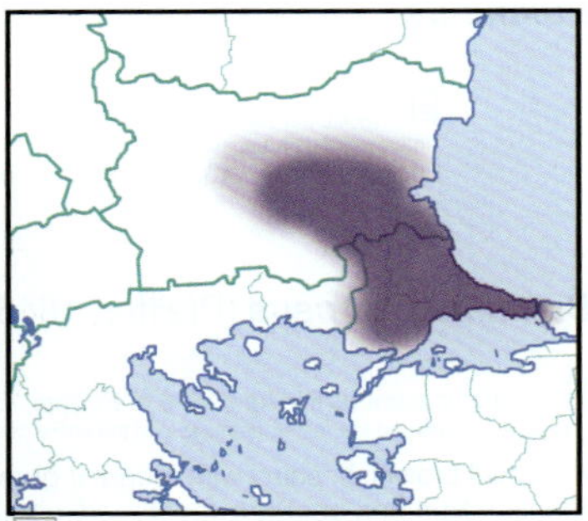

SE Bulgaria, European Turkey

Shell slender, horny brown, weakly shiny, apex pointed, aperture with broad white and reflected margin, curved at columellar side, usually connected by a thin white layer at parietal side with knots at both sides, no parietalis, columella forming a twisted lamella ending at the columellar margin of the aperture, palatal tooth present, umbilicus very narrow due to the curved columellar margin of the aperture.

Humid habitats in valleys and river plains.

Gittenberger 1984: 332, Schütt 2001: 140, Gümüş & N. 2012: 29 (*Eubrephulus*).

Chondrula consentanea (Westerlund, 1887)

Without gender agreement: *Chondrula consentaneus* (Westerlund, 1887)

Buliminus consentaneus Westerlund, 1887 in Westerlund 1887: 38

Westerlund 1887 gave a reference to Küster 1870: 87 where *Bulimus (Napaeus) consentaneus* was mentioned without description (nomen nudum).

Shell horny reddish, finely striated, lower whorls not very convex, suture whitish, aperture with a weak parietalis and a weak callus at the outer margin, parietal side with a thin whitish layer between the apertural margins, lip thin, columella weakly callose.

10 x 4 mm

Possibly only a synonym of *Chondrula tridens*, a revision is needed.

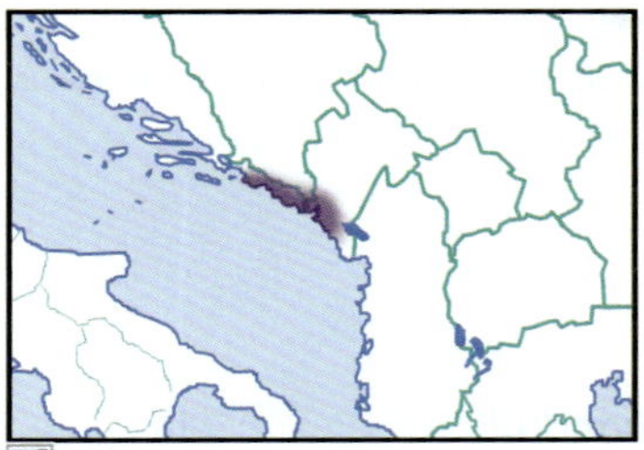

Montenegro, Dalmatia

Chondrula diodon (Retowski, 1883)

Buliminus diodon Retowski, 1883 in Retowski 1883: 55

Shell solid, slender, horny yellowish, finely striated and shiny, aperture somewhat V-shaped, margin white and sharp with a thin layer at parietal side, parietalis very weak, palatalis weak and pointed.

Forest habitats.

Schütt 2001: 139. Not in Crimea (Ukraine), only shells washed ashore were found there.

Was classified by Schileyko 1998: 208 in genus *Spaniodonta*, this was not accepted by other researchers (see also comment under *Imparietula*).

6.5-10 x 2.5-3.5 mm, locally variable

N Turkey: Vilayet Kastamonu, 25 km N Ilgaz

N Turkey

Chondrula jaczewskii (Wagner, 1928)

Brephulus jaczewskii Wagner, 1928 in A. J. Wagner 1928: 312

NE Greece, European Turkey

Limestones and canyons.

Schütt 2001: 142.

Needs a revision, might belong to the genus *Mastus*.

14-20 x 4.5-5 mm

Greece: Makedonia, near Philippi

Shell yellowish to reddish horny, transparent, not shiny, lower whorls with very fine and irregular growth lines, shell slender, apex pointed, whorls not convex, aperture without teeth, margin thick and white, connected by a very prominent and thick callus at parietal side with angular knot being a part of it, and which merges without visible interruption with columellar margin, margin between angular knot and thick palatal wall thin, forming a narrow sinulus pointing to the apex.

Chondrula lugorensis Wagner, 1914

Chondrula lugorensis Wagner, 1914 in Sturany & Wagner 1915: 62, Taf. 14

Shell dirty yellowish grey, translucent, yellowish opaque near the aperture, coarsely and irregularly striated, 8 weakly convex whorls, last whorl not descending, aperture simple without teeth, margin straight and thin, with a white lip inside, sometimes forming some teeth-like structures, margins connected at parietal side by a thin layer, but do not approach each other much, umbilicus very narrow.

Under stones or at the base of limestone rocks in mountains, up to 1900 m.

EU Red List: Near Threatened NT. Only known from few localities in remote areas, probably more widespread. No real threats known.

Fehér & Erőss 2009: 10.

8-11 x 3.5-4 mm

Albania: Vlorë, Mali i Çikës, 1850 m

SW Albania

Zoltán Fehér

Chondrula lycaonica Sturany, 1904

Chondrula lycaonica Sturany, 1904 in Sturany 1904: 118

Shell light yellowish, weakly shiny, coarsely and irregularly striated, aperture with a white margin except at parietal side, strong angular tooth, deep and weak parietalis, strong horizontal columellaris, often also with a basal tooth, strong palatal tooth, umbilicus very narrow.

Humid habitats in calcareous mountains. Locally a frequent species.

Schütt 2001: 135, Gümüş & Neubert 2012: 29 (classified the species in the genus *Chondrus*, but without reference to a published study to justify this classification).

8-11 x 3.0-3.5 mm

Turkey: Denizli

Aydın Örstan

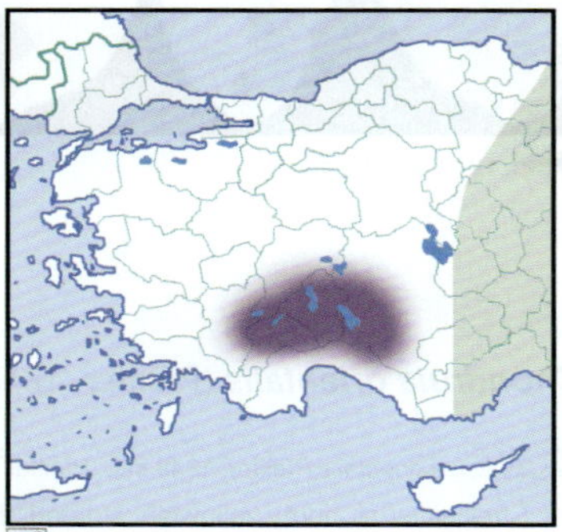

S Turkey

Chondrula macedonica Wagner, 1914

Chondrula macedonica Wagner, 1914 in Sturany & Wagner 1915: 62, Taf. 14

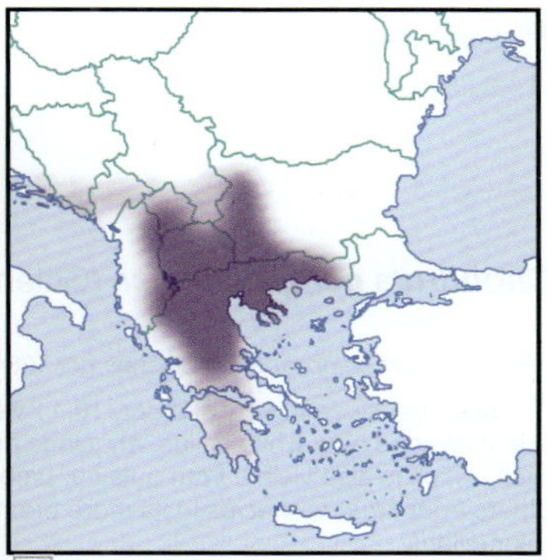

SE Europe

Moderately humid to dry habitats in mountains, calciphile, in Bulgaria up to 1000 m.

Damjanov & L. 1975: 156, Dhora & W. 1996: 124, Hubenov 2007: 136.

7-13 x 3-4 mm

N Greece: Makedonía, near Asprovălta

Shell much like *Chondrula tridens*, parietalis stronger than angular knot, columellar and palatal teeth less prominent, apertural margin connected at parietal side by a thick white layer.

Chondrula microtragus (Rossmässler, 1839)

Pupa microtragus Rossmässler, 1839 in Rossmässler 1838-1844 (9-10): 30, Taf. 49
Gender is unchangeable, derived from tragus which is not an adjective, name is not *Chondrula microtraga*.

8-11 x 4.0-4.5 mm

Greece: Makedonia, Philippi near Kavála

Shell dextral, light horny brown to whitish grey, finely striated, aperture with 4 teeth (angular tubercle, 1 parietal, 1 columellar, 1 palatal) and basal a swollen convex contour (in *C. tricuspidata* the basal part is thin and concave), margin very broad, usually connected at parietal side by at least a thin white layer. Shell very variable. Juveniles with camouflage.

Damjanov & L. 1975: 154, Heller 1976: 380, Liebegott 1986, Grossu 1987: 372, Falkner 1990: 152, Dhora & W. 1996: 124, Schütt 2001: 136, Sysoev & Sch. 2009: 81.

SE Europe, NW Turkey

Dry habitats of all kinds in highlands, calciphile, in Bulgaria up to 1000 m. In Romania mainly in sunny coastland habitats under shrubs or on the soil or sand, almost always associated with *Zebrina varnensis* and *Ch. tridens.*

Chondrula munita (Westerlund, 1894)

Without gender agreement: *Chondrula munitus* (Westerlund, 1894)

Buliminus munitus Westerlund, 1894 in Westerlund 1894: 170
Original spelling was *munitus*, the context of the publication gave no constraints for an inadvertent error for *minutus*.

12-16 x 3-3.5 mm

Greece: S Pelopónisos, Máni peninsula, marine deposits (determination uncertain, diameter not 3-3.5 mm)

Greece (Peloponisos)

Shell horny brown, weakly and densely striated, 10-12 whorls with shallow suture, apertural margin white, with 3 teeth parietal, columellar and palatal.

Shells from S Pelopónisos possess an angular tooth (like *Ch. bergeri*) and a white parietal layer (like *Ch. microtragus* and *Ch. macedonica*), palatal and columellar teeth are not situated opposite to each other (columellar fold is more basal, like in *Ch. microtragus*, not like in *Ch. bergeri*). More research on the local variation of the apertural folds of *Chondrula* forms in S Greece is necessary.

Chondrula orientalis (Pfeiffer, 1848)

Bulimus orientalis Pfeiffer, 1848 in L. Pfeiffer 1848: 6

Shell sinistral, horny yellowish, striated, thin, apex pointed, suture with white hue, aperture with light lip inside, margin connected at parietal side by a weak white layer, angular knot present, no parietalis, columella forming a strong twisted columellaris like in clausiliids, upper palatalis and sometimes a lower palatalis also present, umbilicus open and perpendicular to shell axis due to the twisted columella.

14.5-18.5 x 5 mm

Forest habitats.

Gittenberger 1984: 333, Schütt 2001: 141, Gümüş & Neubert 2012: 29 (classified this species in *Eubrephulus*).

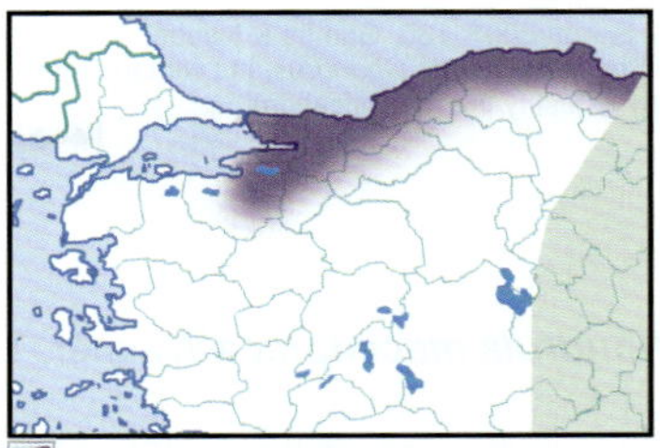

NW Turkey to Samsun

Chondrula peloponnesica (Gittenberger, 1984)

Chondrula peloponnesica Gittenberger 1984 in Gittenberger 1984: 330

Shell translucent, very slender, 10-11 weakly convex whorls, apertural lip reflected, inside thickened by a white callus, angularis present, columellaris extends one whorl into the shell, which is also visible from outside, umbilicus very narrow.
Differs from *Mastus* species in its more prominent angular-parietal callus and its well visible columellaris.

Found at a rocky slope in coastal vicinity.

14-16 x 3.7-4.3 mm

Greece: Pelopónisos, "Moni Loukos" (?) (H =12 mm)
Robert Reuselaars

Greece (Pelopónisos: Arkadía)

Chondrula pindica (Westerlund, 1894)

Without gender agreement: *Chondrula pindicus* (Westerlund, 1894)

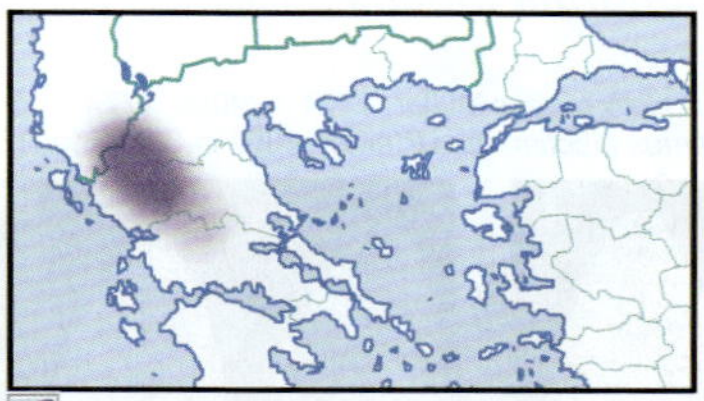

Buliminus pindicus Westerlund, 1894 in Westerlund 1894: 170

Shell whitish with weak brownish streaks, solid, weakly striated, 7-8 whorls, subangulated at suture, last whorl ascending, aperture rounded at basal side, with parietal and palatal teeth, margin white and reflected.

7-8 x 2.5-3 mm

Rocky mountain slopes, between 600 and 2100 m.

Reischütz & Sattmann 1990: 257, Sattmann & Reischütz 1994: 45.

NW Greece

Chondrula quinquedentata (Rossmässler, 1837)

Pupa quinquedentata Rossmässler, 1837 in Rossmässler 1835-1837 (5-6): 9, Taf. 23

Not *Pupa pygmaea* var. *quinquedentata* Hartmann, 1821 (Vertiginidae), previously mentioned as *Vertigo quinquedentata* without description by Studer in Coxe 1789: 388 (nomen nudum). If this vertiginid name was not made available before 1821, Rossmässler's name would be a junior homonym and should be protected under Art. 23.9.1.

Shell brownish yellow, 6-8 almost flattened whorls, aperture whitish or yellowish with shiny white lip and 5-6 teeth, angularis weak or moderate, parietalis strong, upper columellaris weak, lower columellaris strong, palatalis strong.

Among leaf litter, under stones and at rocks.

12-15 x 5-6 mm

Croatia: Istria, near Rovinj

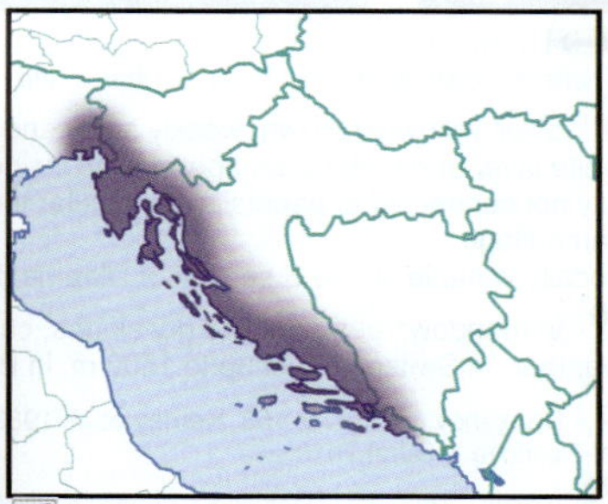

NE Italy to Dalmatia and Bosnia

Westerlund 1887: 41, Alzona 1971: 81, Manganelli et al. 1995: 16, Štamol & Velkovrh 1995 (Cres, only at one of 80 stations).

Chondrula sturmii (Küster, 1852)

Pupa sturmii Küster, 1852 in Küster 1852: 132, Taf. 17 (*Pupa Sturmii*)

Shell small, slender and conical, with 9-10 whorls looking more like those of a pointed clausiliid, last whorl with largest diameter, cervix lower than basal apertural margin, apertural margin white, connected at parietal side by a strong white and rounded layer, with an angular knot, parietalis and palatal tooth present.

Forest habitats in highlands.

Schütt 2001: 143.

8 x 2.5 mm

José Liétor Gallego

Turkey: near Bursa

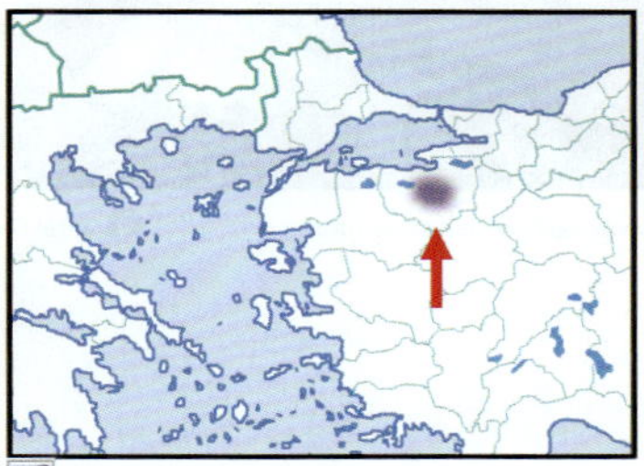

NW Turkey (Vilayet Bursa)

Chondrula tricuspidata (Küster, 1843)

Pupa tricuspidata Küster, 1843 in Küster 1852: 62, Taf. 8

8.5-11.5 x 4-5 mm

Greece: Makedonía, Pangéo mountains, 1800 m

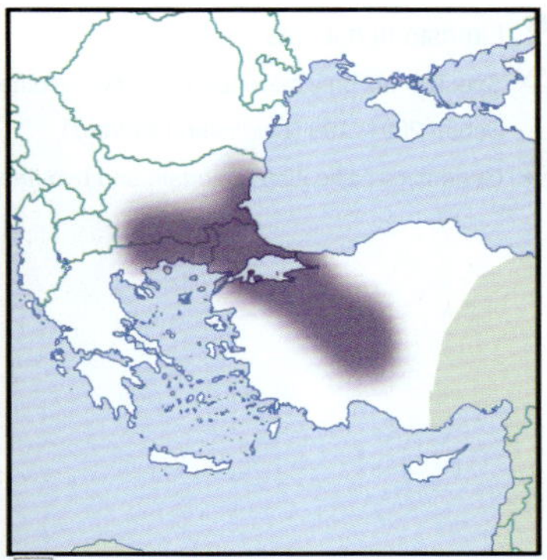

SE Europe, W Turkey

Limestone rocks and rock rubble.

Damjanov & Likharev 1975: 155, Heller 1976: 380, Schütt 2001: 137.

Perhaps only a form of *Chondrula microtragus.*

Shell like *Ch. microtragus*, basal part of the aperture thin and concave (not thickened and convex), and parietal tooth united with the angular tubercle.

Chondrula tridens (Müller, 1774)

Helix tridens Müller, 1774 in Müller 1774: 106
Bulimus albolimbatus Pfeiffer, 1859 in L. Pfeiffer 1859: 129

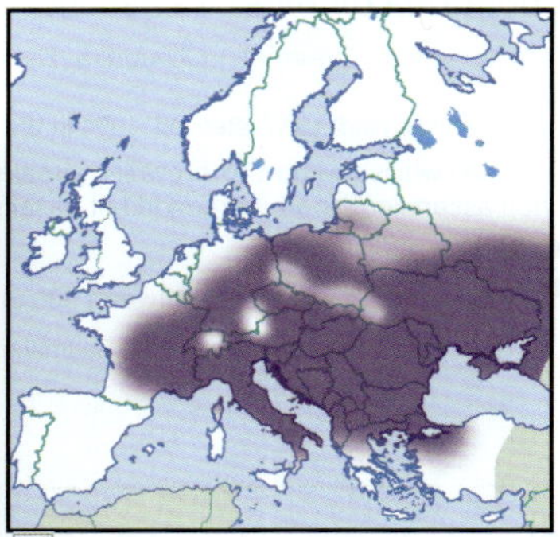

Albania: Mali i Kakariqit at road Shkodër-Lezhë

Europe to S Ural and N Iran

7-16 x 4-5 mm
Shells in the southeast can be larger than in the north.

Critically Endangered in Germany and Switzerland.

Shell yellowish brown, weakly shiny and finely striated, 7-8 almost straight whorls, aperture with white, sharp and slightly reflected margin, connected by a white layer at parietal side, apertural lip inside strong, also visible from outside, parietalis strong and not reaching margin, angular knot often present and usually not connected to parietalis, columellar tooth moderate, lower palatal tooth strong, a weak upper palatal tooth is sometimes also present. Juveniles with camouflage.
Locally variable, in the E section of Ukraine the aperture is narrower, especially the space between angularis and palatal side.

In meadows of sunny and dry slopes, characteristic grassland snail, usually in the soil, near rocks, appears at the surface only at warm and humid weather. In Switzerland in up to 1500 m, in Bulgaria 1900 m. Feeds on dead parts of plants.

Damjanov & L. 1975: 153, Kerney et al. 1983: 133, Falkner 1990: 152, Manganelli et al. 1995: 16, Turner et al. 1998: 198, Schütt 2001: 138, Sysoev & Sch. 2009: 81, Gural-Sverlova & Gural 2010.

Grossu 1987: 369 classified *Chondrula albolimbata* as an independent species (penis much more inflated in its distal section). This taxon was considered as a synonym of *Chondrula tridens* by Egorov 2008: 45 and Sysoev & Schileyko 2009: 81.

Slovakia: Spišský hrad castle ruins — Jozef Grego

Chondrula tridens — Russia: Volga river, Parepta

Slovakia: Slovenský Kras National Park
Jozef Grego

Chondrula tridens

Chondrula werneri (Sturany, 1902)

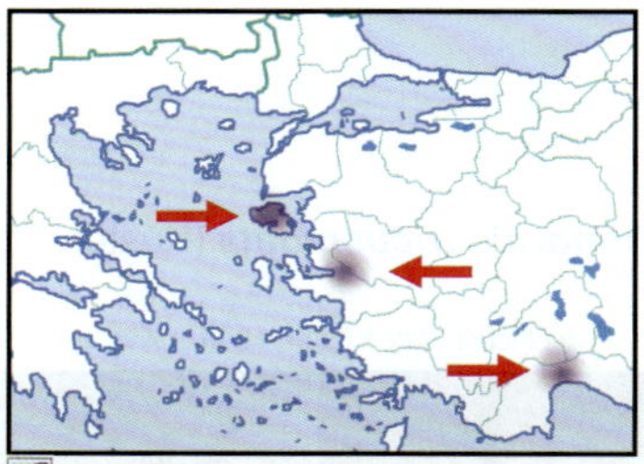

Buliminus werneri Sturany, 1902 in Sturany 1902: 134, Taf. 2

Shell solid, horny brown with irregular blurry growth lines, aperture with 4 prominent folds (parietalis curved in J form, columellaris long, basalis, palatal), umbilicus very narrow.

9.7-10.3 x 3.6-4.0 mm

Limestone habitats.

Species has only been found 4 times at different sites, seems to be extremely rare.

Schütt 2001: 108 (*Ljudmilena werneri*).

Generic classification uncertain and hypothetical, needs to be verified anatomically.

Greece (Lésvos), W Turkey

Chondrus

Chondrus Cuvier, 1816
in Cuvier 1817: 408

Type species *Bulimus zebra* Olivier, 1801
(subsequent designation by Gray 1847: 175)
(= currently *Chondrus zebrula* (Férussac, 1821))

Cuvier's original description contained 2 species, *Bulimus zebra* and *Bulimus avenaceus* Bruguière, 1792.

Chondrus tournefortianus (Férussac, 1821)

Without gender agreement: *Chondrus tournefortiana* (Férussac, 1821)

Helix tournefortia Férussac, 1821 in A. Férussac [1821-1822]: 58 (folio p. 62)
Established without description but with reference to Tournefort 1717 (vol. 3): lett. 21 p. 308.
Helix tournefortiana Férussac, 1821 in A. Férussac [1821-1822]: 70 (folio p. 74)
Corrected original spelling for *Helix tournefortia*. Date of both pages was 13-07-1821.

13-29 x 5-8.5 mm

Turkey: Vilayet Amasya, Amasya

Central Turkey to SE Bulgaria

Shell sinistral, white, rarely with some dark radial streaks, 9-12 slightly convex whorls, aperture of very variable size, inside brownish yellow, margin solid and slightly reflected, connected at parietal side with by a prominent white layer, parietalis strong and slightly deeper inside than the other two folds, columellaris strong and touching apertural margin, palatalis strong.

Humid and wet habitats with herbal vegetation, in lowlands, river plains.

Mating was observed in October. Sperm exchange probably unilateral, not simultaneously reciprocal.

Columna 1616: 16 (description p. 17), Tournefort 1717: 308 (Kütahya), Saint-Simon 1856: 44 (anatomy), Damjanov & Likharev 1975: 148; Schütt 2001: 129., Örstan 2009.

Chondrus zebrula (Férussac, 1821)

Bulimus zebra Olivier, 1801 in Olivier 1801: 225, Pl. 17 (not *Bulimus zebra* Bruguière, 1792 in Bruguière 1792: 357 from Madagascar)
Helix zebriola Férussac, 1821 in A. Férussac [1821-1822]: 58 (folio p. 62)
Established expressly as new ("nobis") without description but with bibliographical reference to Olivier 1801: Pl. 17 Fig. 10. Not a new replacement name because no expressed statement was given (Art. 72.7).
Helix zebrula Férussac, 1821 in A. Férussac [1821-1822]: 70 (folio p. 74)
Corrected original spelling for *Helix zebriola*. Date of both pages was 13-07-1821.

9-19 x 3.5-5 mm

(1-8): N Turkey: 25 km N Ilgaz, 1500 m

(9-10): Greece: Ermioni near Náfplio

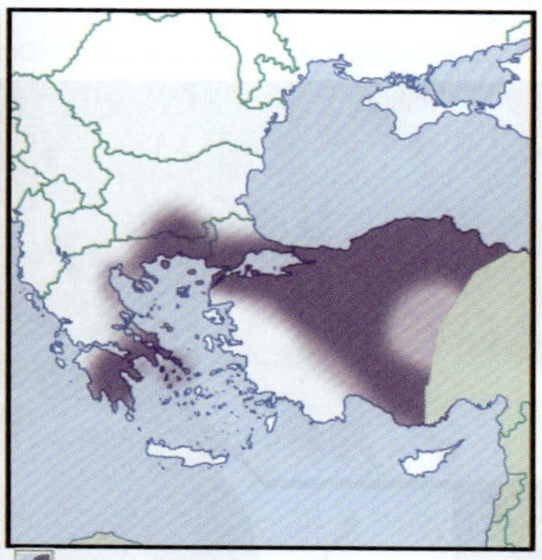

SE Europe to central Turkey

Shell solid and very slender, white with dark radial streaks, with faint striations, clear and deep spiral ornamentation, forming a reticulate granular pattern, 8-9 very slightly convex whorls, from 6th whorl onwards with equal diameter, aperture white with slightly rounded and reflected margin, connected at parietal side by a thick white layer, usually with an angular knot, usually 3 teeth in the aperture (parietal, columellar and palatal, not touching margin, parietalis slightly deeper inside). The teeth can be reduced in some individuals within a population.

Open habitats or under Mediterranean shrubs, among limestones or on calcareous substrate.

Damjanov & Likharev 1975: 146, Heller 1976: 377, Schütt 2001: 130, Irikov & Mollov 2006: 772, Sysoev & Schileyko 2009: 82.

This species was formerly known under the incorrect name *Chondrus zebra* (Olivier, 1801).

Ena Turton, 1831
in Turton 1831: 80 *Bulimus* synonym *Ena*

Type species *Bulimus montanus* Draparnaud, 1801
(subsequent designation by Herrmannsen 1847 [18-04-1847]: 421)

Author was not Leach because Turton 1831 introduced the name as a synonym of *Bulimus* in the combination *Ena montana* (synonym of *Bulimus montanus*, with a reference to "Leach, Mollusc. p. 112", also on p. 81 *Ena obscura* for *Bulimus obscurus*, reference to Leach "p. 113"), and the author of a work is alone responsible for introducing synonyms (Art. 50.7), here Turton. *Ena* was used as a name by Forcart 1940: 126 (with bibliographical reference to Turton 1831). Name made available under Art. 11.6.1.

Shell large, brownish, elongated oval without apertural teeth, apertural margins not approaching each other closely at parietal side. Penial papilla usually well-developed and imperforate, penial appendix present, no penial caecum, caecum at the middle section of the epiphallus present, as well as a more or less short flagellum, and a diverticulum to the spermatheca. The only difference to *Merdigera* (imperforate penial papilla) has not been studied in all species assigned to *Ena*.

Bank & Menkhorst 1992, Hausdorf 2001 (revision Turkey).

Ena concolor (Westerlund, 1887)

Buliminus cefalonicus forma *concolor* Westerlund, 1887 in Westerlund 1887: 28

Shell like *Napaeopsis cefalonica* but larger and colour evenly whitish, 6-7 weakly convex whorls, aperture more rounded, white layer at parietal side present, margin slightly more rounded at upper palatal and columellar sides, umbilicus more open.

Dhora & W. 1996: 125, Hausdorf 2001: 1631.

17-19 x 6-7 mm

Albania: Mal i Krujë, 750 m

Bosnia to Rep. Makedonija

Ena elongata (Kobelt, 1877)

Without gender agreement: *Ena elongatus* (Kobelt, 1877)

Buliminus graecus var. *elongatus* Kobelt, 1877 in Kobelt 1877: 67, Taf. 137

Higher than *Ena montana*, 9 whorls, base broader.

18-20 x 7.5 mm

Westerlund 1887: 30, Bank & Menkhorst 1992, Hausdorf 2001: 1636.

Classification uncertain, penial papilla not studied.

S Greece (Pelopónisos)

Ena montana (Draparnaud, 1801)

Without gender agreement: *Ena montanus* (Draparnaud, 1801)

Bulimus montanus Draparnaud, 1801 in Draparnaud 1801: 65

14-17 x 6-7 mm

S Germany: Bayern, near Walchensee

Shell brown, not very shiny, radially striated and also very finely and irregularly spirally striated, producing a characteristic granular surface pattern, aperture reflected and without folds, margin connected at parietal side by a thin white layer, umbilicus open. Animal light with violet hue and dark spots on head and dorsum. Juveniles often with camouflage.

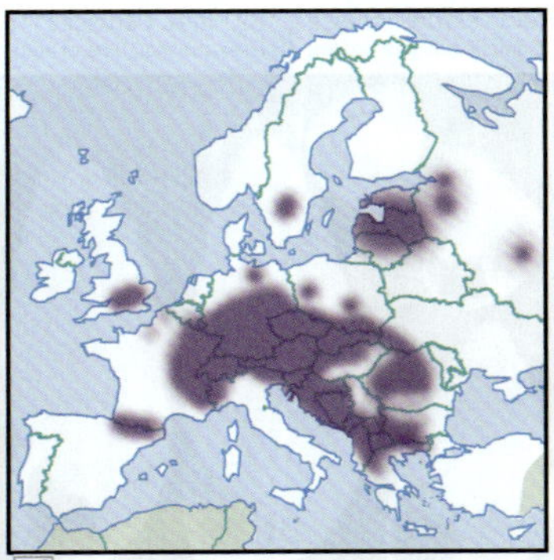

Europe, scattered in Russia

Humid deciduous forests in hills, highlands and mountains, in Britain on well-drained calcareous soils, in central Europe a characteristic snail in river plain woods. In Switzerland up to 2500 m, but very rare above the timberline, in Bulgaria 2000 m. Feeds on faded and dead plant material. Climbs rocks and smooth tree trunks during rainfall.

Rare in S England.

Ena monticola (Roth, 1856)

Bulimus monticola Roth, 1856 in Roth 1856: 3

15-21 x 6-7 mm

Greece: Thessalía, Tríkala, Pili

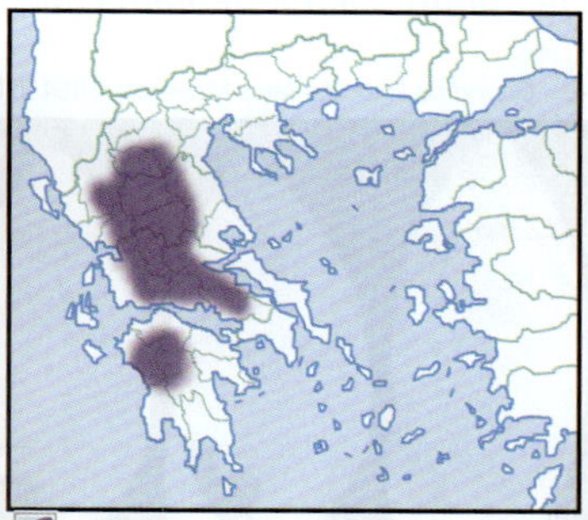

Greece

Shell horny brown, irregularly and finely striated, whorls almost flat, suture white, aperture oval and slightly quadrangular, margin strongly bent backwards, lip white.

Westerlund 1887: 27, Bank & Menkhorst 1992.

Ena nogellii (Roth, 1850)

Bulimus nogellii Roth, 1850 in M. Wagner 1850: 341 (*Bulimus Nogellii*)

Shell horny brownish, more slender than *Ena menkhorsti*, with weak growth lines, 7-8 slightly inflated whorls, last whorl not ascending near aperture, aperture without teeth and with moderate lip inside, columella visible in the aperture and curved, margin reflected and slightly expanded, umbilicus narrow and partly covered.

Forests in moderate and high altitudes. A widespread species.

13-22 x 5-7 mm

Turkey; Vilayet Samsum, S of Taflan

N central Turkey

Ena subtilis (Rossmässler, 1837)

Bulimus subtilis Rossmässler, 1837 in Rossmässler 1835-1837 (5-6): 47, Taf. 28 Type locality not recorded.

14-16 x 4.5-5 mm

Albania: Lepushë

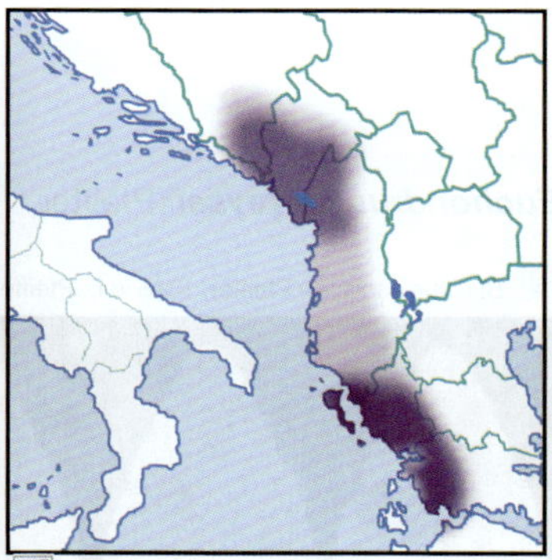

Bosnia to Greece

Westerlund 1887: 26, Bank & Menkhorst 1992, Dhora & Welter-Schultes 1996: 125.

Shell horny yellowish to brown, shiny, weakly striated, 7-8 very weakly convex whorls, suture white, aperture with broad white margin and prominent lip, reflected, connected at parietal side by a thin whitish layer, umbilicus almost covered by columellar margin of the aperture.

Euchondrus

Euchondrus Böttger, 1883 in O. Böttger 1883 (Ber. Offenb. 22/23): 173 *Buliminus (Euchondrus)*

Type species *Pupa chondriformis* Mousson, 1861 (monotypy)

Established without description, but with one species contained.

Forcart 1940: 199, Wenz & Zilch 1959-1960: 182, Bank & Hovestadt 1991 (revision Cyprus).

Euchondrus ledereri (Pfeiffer, 1868)

Pupa ledereri Pfeiffer, 1868 in L. Pfeiffer 1868: 316 (*Pupa Ledereri*)

Shell horny brown, finely striated, aperture with white margin, sometimes connected at parietal side by a thin white layer, folds in the aperture: parietalis, angularis weak or absent, angular tubercle, columellaris perpendicular to the columellar margin, upper and lower palatalis. Differs from *Eu. borealis* in the presence of only one parietal and no basal tooth.

7-9 x 2.6-3.1 mm

Open habitats near the Mediterranean Sea.

Heller 1976: 380, Bank & Hovestadt 1991: 2, Schütt 2001: 151.

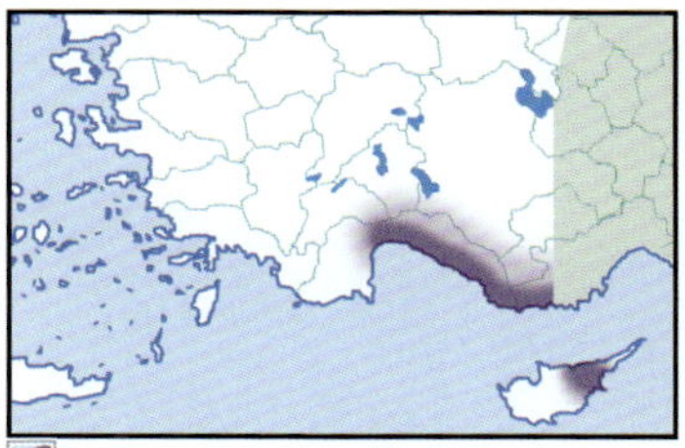

Turkey (Antalya) to Israel, Cyprus

Euchondrus limbodentatus (Mousson, 1854)

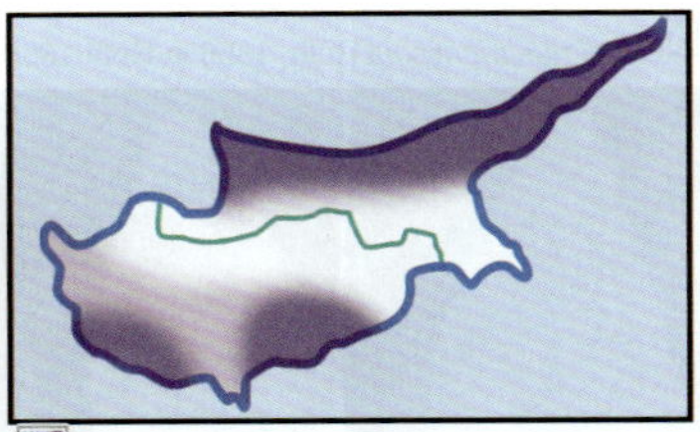

Chondrus limbodentatus Mousson, 1854 in Mousson 1854: 385

9-13 x 4.5-6 mm

N Cyprus: Bellapais

Cyprus

Seems to be restricted to habitats in coastal vicinity.

EU Red List: Near Threatened NT. Threatened by fire, overgrazing, and increasing land use and habitat destruction in coastal vicinities associated with tourism.

Bank & Hovestadt 1991: 2.

Shell ovoid with pointed apex, horny brown with white radial streaks, 5-6 not convex whorls, apertural margin broad, white, not connected by a white layer at parietal side, with angular tooth, parietalis, columellar tooth, weak basal tooth, moderate lower palatal tooth, and a weak upper palatal tubercle.

Euchondrus nucifragus (Pfeiffer, 1848)

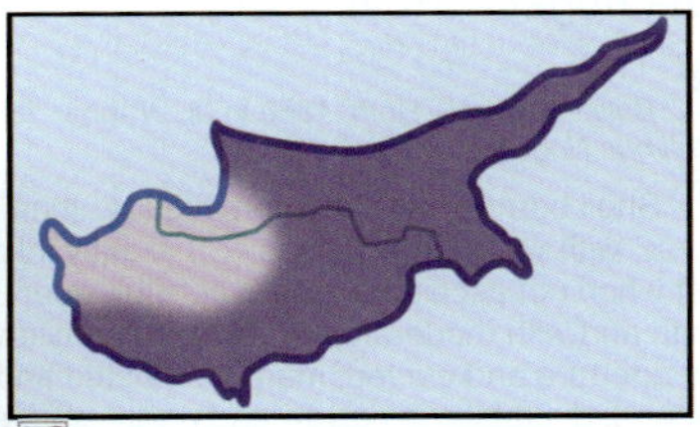

Bulimus nucifragus Pfeiffer, 1848 in L. Pfeiffer 1848: 135

Shell elliptical with pointed apex, light yellowish, shiny, weakly striated, 5-6 whorls, apertural margin broad and white, aperture more than 1/3 of shell height, with angular tooth, parietalis, columellaris perpendicular, 3 palatal teeth (can be weak or absent in the SW). Diameter smaller than in *Eu. limbodentatus*, lower whorls usually less prominently striated.

Westerlund 1887: 47, Bank & Hovestadt 1991: 4.

6.5-10 x 2.5-3.6 mm

S Cyprus: Lárnaka
Paolo Zaccaria
Tiroler Landesmuseen Innsbruck

Cyprus

S Cyprus
Georgios Konstantinou

Euchondrus parreyssi (Pfeiffer, 1846)

Bulimus parreyssi Pfeiffer, 1846 in L. Pfeiffer 1846 (Symbolae): 86

6.5-13 x 3.4-5.2 mm, size extremely variable, also within a population

W Cyprus: Latsí

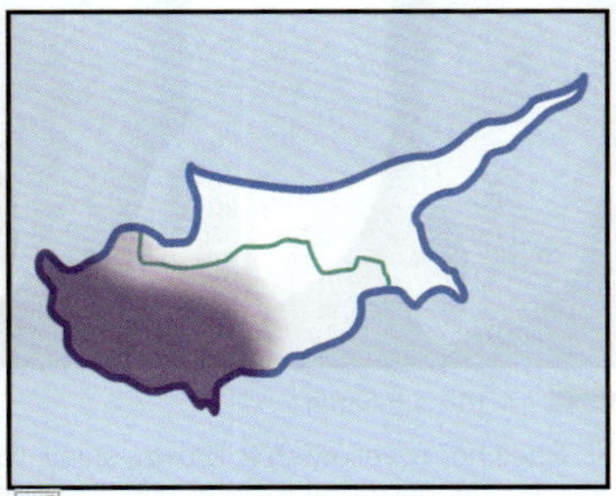

W Cyprus

At the soil among shrubs in open habitats. A frequent species.

Westerlund 1887: 43, Bank & Hovestadt 1991: 6.

Shell whitish and light horny brown marbled, usually coarsely striated, 5-7 whorls, upper whorls convex and lower whorls almost flattened, aperture with angular tubercle and usually a prominent white lip, a small second tooth may be deeply inside, umbilicus very narrow.
The only species in Cyprus with weak teeth at parietal and columellar and no teeth at the palatal sides. However, some populations of *Eu. nucifragus* in SW Cyprus can have very weak palatal teeth. This looks like a recult of some kind of restricted gene ingressions, since both species occur sympatrically.

Imparietula

Imparietula Lindholm, 1925
in Lindholm 1925: 30 *Jaminia (Imparietula)*

Type species *Bulimus leucodon* Pfeiffer, 1846
(original designation)
(= currently *Imparietula leucodon* (Pfeiffer, 1846) from NE Turkey)

Pseudochondrula Hesse, 1933 in Hesse 1933: 167, type species *Buliminus florenskii* Rosen, 1914 (original designation)

SE Europe to Transcaucasia

Taxonomists produced some chaos in the generic classification. *Pseudochondrula* and *Imparietula* seem to be closely related, but Schileyko 1998: 208 did not know the anatomy of *leucodon*, which is necessary to give a definite statement on the genus to be used for *seductilis*.
Forcart 1940 and Bank 1988 classified *seductilis* and *leucodon* in *Imparietula*. Shilejko 1984 and Schileyko 1998 classified *seductilis* in *Pseudochondrula*, *leucodon* in *Spaniodonta* Kobelt & Möllendorff, 1902 (its type species *Buliminus diodon* Retowski, 1883, is here classified as *Chondrula diodon*). Schütt 2001 and Sysoev & Schileyko 2009 classified *leucodon* in *Imparietula*, *seductilis* in *Pseudochondrula*.
The best solution might be to use the classification from before 1984 and to leave all species within *Imparietula*, and *diodon* in *Chondrula*.
Currently *Imparietula* and *Pseudochondrula* are used for *seductilis*. *Imparietula* is the older name.

Imparietula seductilis (Rossmässler, 1837)

Pupa seductilis Rossmässler, 1837 in Rossmässler 1835-1837 (5-6): 10, Taf. 23

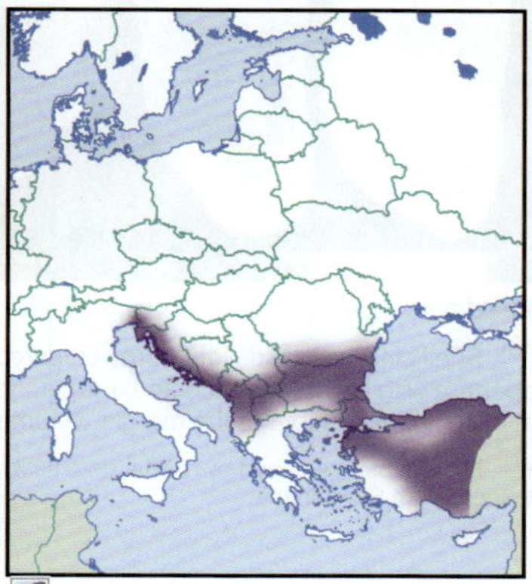

8-13 x 4-5 mm

Bulgaria: Varna

Slovenia to Turkey and Armenia

Shell sinistral, dark horny brown with fine growth lines, weakly shiny, 8-9 weakly convex whorls, last 3 whorls equal in diameter, aperture with parietalis, sometimes angular callus, usually columellaris (ending in a knot), often a palatal tooth.

Limestone habitats, not in forests.

Tomić 1959: 30, Damjanov & L. 1975: 150 (anatomy), Dhora & W. 1996: 126, Schütt 2001: 95, De Mattia 2004, Sysoev & Schileyko 2009: 63.

Classified by some authors in the genus *Pseudochondrula*.

Jaminia

Jaminia Risso, 1826
in Risso 1826: 88

Type species *Jaminia heterostropha* Risso, 1826
(subsequent designation by Gray 1847: 176)
(= currently *Jaminia quadridens* (Müller, 1774))

Central and S Europe to Caucasus and Near East

The differences between *J. loweii*, *J. thiesseana* and *J. quadridens* do not seem to be well pronounced, and that the morphological differences described by different authors contradict each other. Perhaps the three forms are conspecific, and there is only some local variation.

Jaminia loewii (Philippi, 1844)

Bulimus loewii Philippi, 1844 in Philippi 1844: 106 footnote (*Bulimus Loewii*)

10-16 x 3-4.5 mm, shells from inland Turkey are smaller

Greece: island of Psérimos

SW Turkey, Greece (Dodekánisa)
Occurrence in Cyprus requires confirmation.

Rocky limestone habitats of any kind. Species occurs in colonies, often with large gaps in between, tens of kilometers wide even if suitable habitats are present.

Heller 1976: 378, 284, Bank & Hovestadt 1991: 9, Bank & Maassen 1998: 54, Schütt 2001: 109, Örstan & Yıldırım 2003, Örstan et al. 2005, Triantis et al. 2008: 478.

Shell sinistral, horny brown, weakly and regularly striated, sometimes almost smooth, shiny, 8-10 whorls, last 4 whorls with equal diameter, apertural margin white, connected by a white callus at parietal side, which can be thick, angular knot can be present, parietal tooth slightly more inside but prominent, columellar tooth at the margin, columellaris visible inside the aperture like in a clausiliid, palatal tooth also reaching apertural margin. Coastal colonies have larger shells than inland colonies, but no differences were found in the genitalia.

Possibly conspecific with *J. thiesseana* and *J. quadridens*, more research is necessary.

Jaminia quadridens (Müller, 1774)

Helix quadridens Müller, 1774 in Müller 1774: 107

7-15 x 3.5-4 mm

Albania: near Pogradec

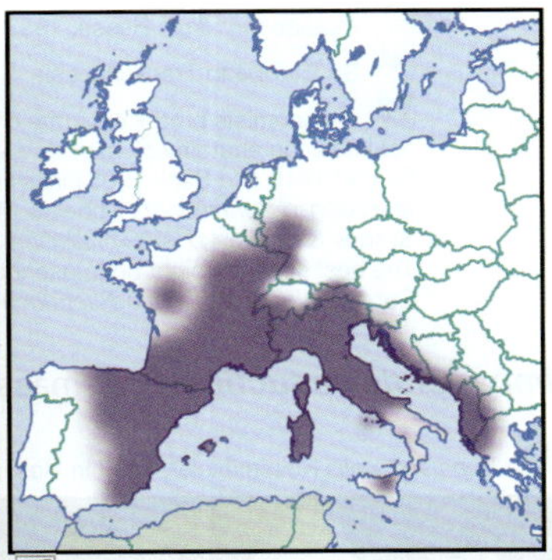

Spain to Germany and N Greece
Grossu 1987: 344 reported the species from E Romania. This requires confirmation.

Dry and warm habitats on meadow slopes between stones. Prefers calcareous substrate. Tolerates extensive grazing and is found on sheep and goat pastures. In the Alps up to 2400 m.

Shell sinistral, solid, light horny coloured, weakly shiny with irregular fine growth lines, 7.5-9 very weakly convex whorls, last 3-4 whorls equal in diameter, aperture with broad and reflected margin, no thick white callus at parietal side, sometimes with a thin white layer, deep parietalis, weak angular knot if present at all, columellar tooth at the margin, columellaris visible deeply inside the aperture, strong palatal tooth. Juveniles with distinct edge at the base of the whorls.

In the Alps threatened by habitat destruction like all species specialized on dry habitats. Critically Endangered in parts of Austria (Vorarlberg, Tirol, Osttirol) and Germany, Vulnerable in Switzerland.

Jaminia thiesseana (Westerlund & Blanc, 1879)

Without gender agreement: *Jaminia thiesseanus* (Westerlund, 1879)

Buliminus thiesseanus Westerlund & Blanc, 1879 in Westerlund & Blanc 1879: 92
Name was attributed to Mousson, description was attributed to Westerlund. In a strict application of Art. 50.1.1 the name in such a case must be attributed to the author of the work.

Shell sinistral, pale horny brown, weakly striated, apertural margin white, slightly reflected, sometimes with a very thick layer at parietal side, 4 teeth in the aperture: 1 parietal (sometimes fused with angular callus), 1 palatal, 2 columellar (might be weak or partly fused). Differs from *J. quadridens* and *J. loewii* in its more oblique (directed to the left side) aperture, possibly also by the stronger and more lamelliform folds.

On rocks and under stones.

11-14.5 x 2.5-4 mm

Westerlund 1887: 53, Heller 1976: 378, Liebegott 1986.

Sattmann & Reischütz 1994: 45 reported the species from NW Greece, but the degrees of local variation of shell characters and their taxonomic significance have never been studied in detail. Is possible that this form is conspecific with both *J. quadridens* and *J. loewii*.

Greece: Fokída, Sikiá
(H = 15 mm)
Robert Reuselaars

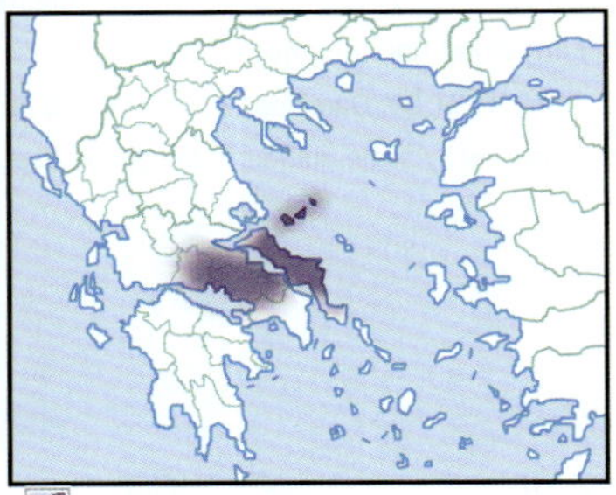

Greece (Evia, Viotia)

Leucomastus

Leucomastus Wagner, 1928
in A. J. Wagner 1928 ["1927"]: 313

Type species *Leucomastus buresi* Wagner, 1928 (monotypy)
(= currently *Leucomastus kindermanni*)

Shell not very elongate, surface without riblets, aperture wide, no parietal callus. Penis with appendix, caecum at central position of the epiphallus (in *Zebrina* subterminal).

Open habitats, snails live at the bottom vegetation.

Gümüş & Neubert 2012: 29 separated *Zebrina* into *Rhabdoena*, *Leucomastus* and *Zebrina*, and presented taxonomic arguments.

Leucomastus dardanus (Philippi, 1844)

Bulimus dardanus Philippi, 1844 in Philippi 1845: 158

22-30 x 8-11 mm, in the Bosporus region slightly smaller

Turkey: Vilayet Bursa, İsnik

(1-3) Panayotis Ovalis

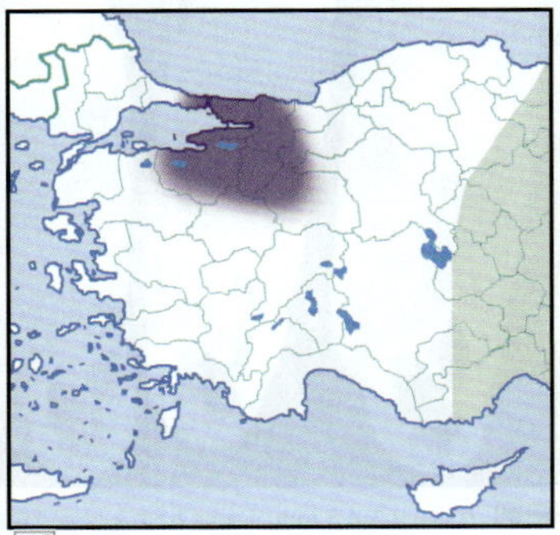

N Turkey

Open grasslands, partly with small shrubs, animals climb on grasses and shrubs to avoid bottom heat.

Falkner 1990: 152, Schütt 1996: 133, Schütt 2001: 124.

Shell slender, dirty white with unclear dark streaks, apex pointed, outline of shell at first whorls straight, aperture relatively high in relation to shell height, margin connected at parietal side by a thin white layer, umbilicus narrow.

Leucomastus eburneus (Pfeiffer, 1842)

Bulimus eburneus Pfeiffer, 1842 in L. Pfeiffer 1842: 44

S Turkey

Open habitats, grasslands and rocky areas. Climbs up grasses to avoid bottom heat.

This species does not live in Ródos and Kárpathos (K. Vardinoyannis, unpublished Ph. D. thesis, 1994).

19-28 x 6.5-8 mm

Turkey: Karatepe

Shell very slender, white with few brownish streaks, apex blunt or rounded, whorls more convex than in other *Zebrina* species, aperture inside light brownish, apertural height only 7-8 mm, umbilicus very narrow. Shell narrower than *Zebrina fasciolata*, more whorls, smaller aperture. Differs from *L. dardanus* in its smaller apertural height in relation to shell height, more slender.

Leucomastus kindermanni (Pfeiffer, 1853)

Bulimus kindermanni Pfeiffer, 1853 in L. Pfeiffer 1853 (Mon. Helic.): 433 (*Bulimus Kindermanni*)

Pfeiffer 1853: 434 referred correctly to Küster & Pfeiffer 1845-1855: Taf. 36 Fig. 16-17, unpublished at the time, this plate was published in 1854. Date was not 1850.

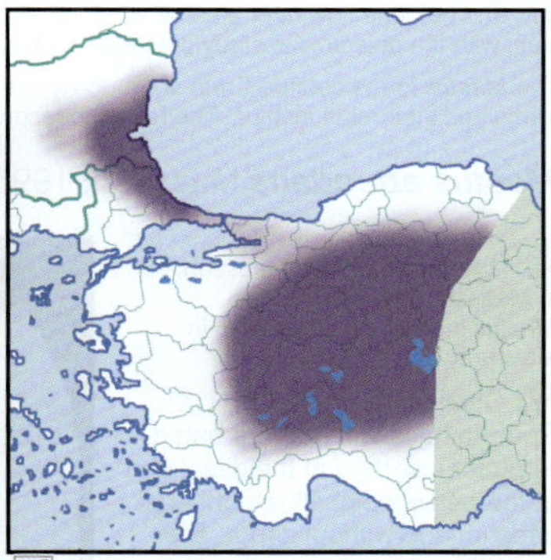

13-17 x 6-9 mm

S Bulgaria: near Savino

Shell not slender, dirty white to reddish, with fine growth lines, apex pointed, often with reddish hue, umbilicus almost covered.

S Turkey to SE Bulgaria

Among shrubby vegetation and limestone rocks in dry steppe habitats, also in highlands.

Leucomastus varnensis (Pfeiffer, 1847)

Bulimus varnensis Pfeiffer, 1847 in L. Pfeiffer 1847 (Z. f. Malak. 4): 192
Leucomastus buresi Wagner, 1928 in A. J. Wagner 1928: 313 (*Leucomastus bureši*)

19-26 x 6-7 mm

Bulgaria: Kavarna

Shell very slender, white with weak brownish streaks, 8-9 whorls, aperture with an angle and an overlapping fold between columellar and parietal side, umbilicus almost covered.

E Romania, E Bulgaria

Shrubs in highlands. Restricted to open habitats on hills or flatlands covered with grasses, herbs or small shrubs, often on sandy bottom in coastal vicinity, locally very common, occasionally also near lakes in the interior.

Damjanov & Likharev 1975: 145, Grossu 1987: 358.

Mastus

Mastus Beck, 1837
in Beck 1837: 73 *Bulimus (Mastus)*

Type species *Helix pupa* Linnæus, 1758
(subsequent designation by Herrmannsen 1847-1849 [17-07-1847]: 24

Shells horny brownish with weakly convex whorls, not very deep suture, weakly shiny with weak growth lines, aperture simple with white and sharp margin, usually with weak white layer at parietal side and often with an angular knot, no other teeth. Shell sizes and shapes are locally variable within certain limits. *Mastus* species differ anatomically from each other, as has been shown by Maassen 1995 for Crete, but it has not been clarified the correlation between the observed anatomical differences and reproductive isolation.

Greece to Romania and Turkey

Maassen 1995, Welter-Schultes 2000, Parmakelis et al. 2005.

Mastus shows a high diversity in the island of Crete, which is currently not well understood. The animals occur constantly over area with up to three conchologically distinct forms living sympatrically. Various approaches have been undertaken to provide names for the forms.
Heller 1976 used two names (*olivaceus* and *cretensis*), based on shell size. Vardinoyannis (unpublished thesis, 1994) used four names (*cretensis*, *olivaceus*, *turgidus* and *etuberculatus*) without arguments.
Based on anatomical studies, Maassen 1995 claimed that there were more than 15 parapatric species in Crete. Populations in S central Crete and large forms in central to eastern Crete (*cretensis*) were not studied.
Based on shell morphology, Welter-Schultes 2000 (Folia Malacologica) argued that there were probably not more than 10 species in Crete. I did not study the western Cretan forms, and had also some gaps in important regions in central Crete. Widespread populations with small shells in S central Crete (Iráklio lowlands to Asteroúsia mountains) had no name in Maassen's system.
Parmakelis et al. 2003 came to the conclusion that the taxonomic concept proposed by Maassen 1995 was not well established. A very high degree of molecular variation obtained on the small island of Koufonísi was difficult to explain.
Parmakelis et al. 2005 analysed all important forms, but did not present conclusions on the specific classification except that two forms were classified as new species without names (they were called *Mastus* n. sp. [with space, from Kritsá] and *Mastus* n.sp. [without space, from Thriptí]). The 2005 work suggested that some species can probably be united, although not always the same as suggested by Welter-Schultes 2000, and in total perhaps less. Parmakelis et al. 2005 did not study the widely distributed small *Mastus* forms in S central Crete.
What we can do now, is to extract some conclusions based on the degree of more or less close molecular relationships as provided by Parmakelis et al. 2005, and to work with a preliminary system of all populations known to occur in Crete.

1 - *sitiensis* and *itanosensis* seem to be very closely related, they are parapatrically distributed and are classified as one species here, *M. sitiensis*.
2 - *hemmeni*, *sphakiota* and *procax* seem to be closely related, they are parapatrically distributed and are classified as one species here, *M. sphakiota*.
3 - these two groups seem to be moderately closely related, together with *ierapetrana*. *M. ierapetrana* is classified as an independent species here.
4 - *cretensis* seems to be a homogenous group distributed between Iráklio and Pachiá Ammos/Thriptí, it seems to be more closely related to *abundans-subaii* than to *hemmeni* from the Réthimno lowlands. *M. cretensis* is classified as an independent species here.
5 - *subaii* and *abundans* seem to be closely related, they are parapatrically distributed and are classified as one species here, *M. abundans*.
6 - these two groups seem to be moderately closely related.
7 - *riedeli* and Kritsá (= *Mastus* n. sp. with space) seem to be closely related, they are parapatrically distributed and are classified as one species here, *M. riedeli*. The widespread small species in the Iráklio lowlands and Asteroúsia mountains (Welter-Schultes 2000: 155, we could call it *Mastus* n. sp. with two spaces) is preliminarily classified with this form here. The conchological record suggests that this is currently the most likely option.
8 - *olivaceus*, *violacea* and *gittenbergeri* seem to represent relatively isolated groups, they are classified as independent species here.
9 - another unknown species from Thriptí (*Mastus* n.sp. without space) also seems to represent an isolated population, perhaps a mountain relict with affinities to *olivaceus*. More research is necessary, the species should be compared with southern populations of *M. gittenbergeri*.
10 - *butoti* also seems to represent a relatively isolated population and is classified as an independent species here.

The correlation between the observed anatomical differences and reproductive isolation is not yet understood. The question is where to set the limit between continuous gene flow and no gene flow over area between parapatrically distributed populations. A closer look with analyses of populations living "between" the parapatrically distributed species, with the objective to study presence and nature of gradients between so-called species, would be useful to understand the nature of the variation in this group.

The *Mastus* forms outside Crete are much less well known, often names were only tentatively used. Much more research is necessary to understand the identities of these forms and their relationships. The faunistic record in mainland Greece is so weak that we do not even know where exactly this genus is distributed at all.

Mastus abundans Maassen, 1995

Mastus abundans Maassen, 1995 in Maassen 1995: 55
Mastus subaii Maassen, 1995 in Maassen 1995: 57

Shell fairly small. Spermatheca without true bursa, retractor inserting at the distal end of the penis, epiphallus with large caecum and without septa, spermatophore with long and sculptured spine.

Species is slightly more abundant than the other *Mastus* species.

11.3-14.5 x 4.5-5.5 mm

Greece: E Crete, near Vrouchás N of Eloúnda

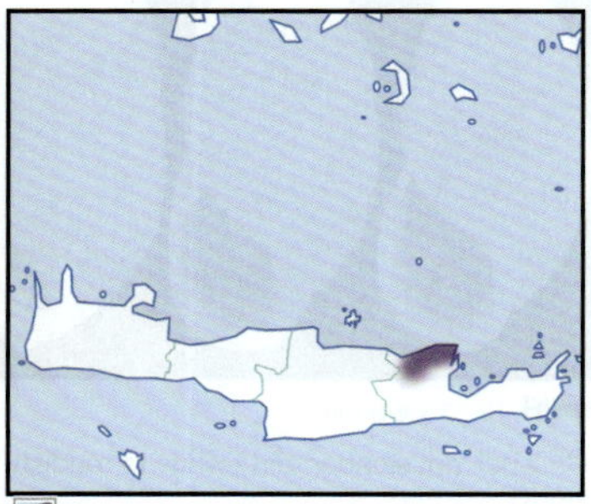

Greece (NE central Crete)

M. subaii and *M. abundans* clustered very closely in molecular analyses by Parmakelis et al. 2005.

Mastus amenazada Welter-Schultes, 1999

Mastus amenazada Welter-Schultes, 1999 in Welter-Schultes 1999: 101
Name to be used as a noun.

Shell larger than *Mastus claudia*, usually no angular tubercle.

Seems to require a minimum of vegetational cover.

14-17 x 6-7 mm

Greece: island of Gavdopoúla S of Crete

Greece (island of Gavdopoúla)

EU Red List: Vulnerable VU. Perhaps underestimated. The species with the large shell is threatened by overgrazing by goats.
In the late 1990s it was planned to destroy the entire 2.6 km^2 island and to convert it into a gigantic container port. These plans were made public by Greek environmental activists and were fortunately stopped in 1999.

Mastus athensis (Pfeiffer, 1847)

Bulimus athensis Pfeiffer, 1847 in L. Pfeiffer 1847: 191

Shell dark horny brown, finely striated, last whorl 1/3 of shell height, aperture with white margin, connected at parietal side by a prominent white layer at the parietal side, with an angular tooth.

20-24 x 6-7 mm

Westerlund 1887: 15.

Identity should be revised, could be conspecific with *Chondrula jaczewskii* (Wagner, 1928) (*Chondrula* or *Mastus*).

N Greece (Athos mountain)

Mastus bielzi (Kimakowicz, 1890)

Dentistomus bielzi Kimakowicz, 1890 in Kimakowicz 1890: 88 (*Dentistomus (Amphitrorsus) Bielzi*)
Established as a new name with a reference to Bielz 1853: 119 *Dentistomus (Chondrulus) eximius* or *Buliminus grandis* (it is unclear to me which name was meant, Kimakowicz may have intended to establish a substitute name, but there was no expressed statement as required by Art. 72.7).

Shell horny brown with reddish hue, densely striated, apertural margin connected at parietal side by a thin white layer, angular tubercle present.

In the leaf litter layer in forests, under stones or wood logs, appears at the surface mainly after warm rains.

Grossu 1987: 375, Sysoev & Schileyko 2009: 81.

16-22 x 6-10 mm

Romania: Transilvania

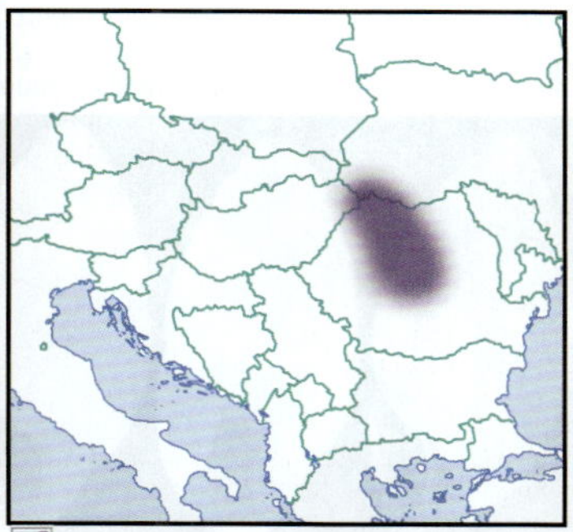

Romania (Transilvania), W Ukraine

Mastus butoti Maassen, 1995

Mastus butoti Maassen, 1995 in Maassen 1995: 37

Shell small, aperture small. Spermatheca with bursa and diverticulum, penis retractor inserting at the distal end of the penis, epiphallus short.

9.2-12.4 x 4.1-4.8 mm

Lives sympatrically with the more common *M. subaii* at 2 km W of Limin Chersonísou. Classified by Welter-Schultes 2000 as *M. sitiensis*.

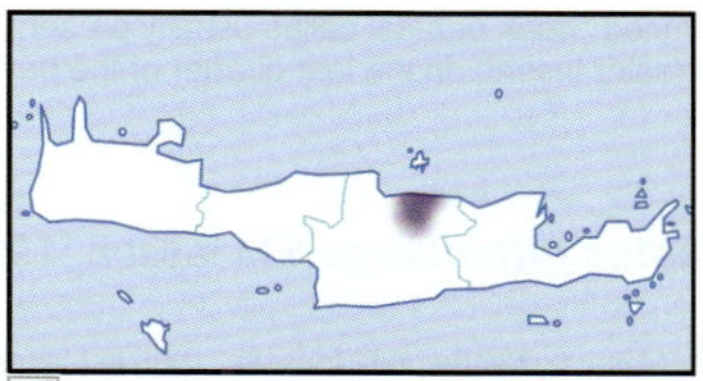

Greece (N central Crete)

Mastus claudia Maassen & Welter-Schultes, 1998

Mastus claudia Maassen & Welter-Schultes, 1998 in Welter-Schultes 1998: 41
Name and description were attributed to both co-authors.

12-15 x 4.5-5 mm

Greece: island of Gávdos S of Crete

Greece (island of Gávdos)

Shell brownish yellow, coarsely striated, upper whorls moderately convex, lower whorls less, angular tubercle present.

Under shrubs and pine trees, rare in open habitats, prefers Neogene substrate and does not live in areas of Mesozoic limestone.

EU Red List: Vulnerable VU. Certainly exaggerated. Abundant and widespread, potentially threatened by fire. The island was inhabited by 8000 people in the Roman epoch 2000 years ago and the species survived that pressure (today < 50 persons).

Mastus cretensis (Pfeiffer, 1846)

Bulimus cretensis Pfeiffer, 1846 in Philippi 1847: 124
Name was attributed to Pfeiffer, authorship of the entire description was also attributed to Pfeiffer.

Shell large. Animal light with dark spots.

At the soil in shrubland vegetation, usually under shrubs.

Maassen 1995, Welter-Schultes 2000: 155.

Greece: Dragonáda island N of E Crete
Vollrath Wiese

16-22 x 6.2-7.4 mm, varies locally

Greece: N Crete, near Goúves E of Iráklio

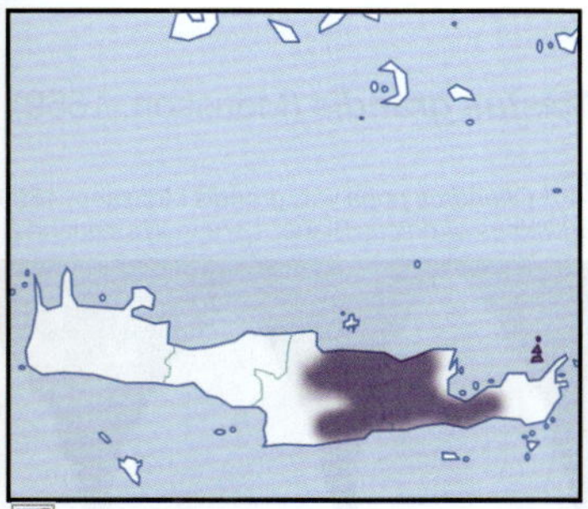

Greece (Crete)

Mastus emarginatus (Deshayes, 1835)

Bulimus emarginatus Deshayes, 1835 in Deshayes [1835]: 165, Pl. 19 Type locality in Pelopónisos.
Bulimus ehrenbergi Pfeiffer, 1847 in L. Pfeiffer 1847: 113 (*Bulimus Ehrenbergi*) Type locality Andikíthira.

Shell horny brown, apertural margin connected by a weak layer at parietal side, angular tubercle present.

15-23 x 7.5-10 mm

Heller 1976: 388, Parmakelis et al. 2005 (only Andikíthira, *ehrenbergi*).

Could include *ehrenbergi* from Andikíthira. More research is necessary. The EU Red List (www.iucnredlist.org, 2011) gave a status "Vulnerable" for the population of Andikíthira (only because of the small island size, no real threats known), which was regarded as a separate species (island endemic) and incorrectly called *emarginatus*.

S Greece

Mastus etuberculatus (Frauenfeld, 1867)

Without gender agreement: *Mastus etuberculata* (Frauenfeld, 1867)

Ena etuberculata Frauenfeld, 1867 in Frauenfeld 1867: 777

11-14 x 6-7 mm

Turkey: Dalyan, near Kaunos

Shell horny brown to olive green, regularly striated, inflated (not very slender), angular tubercle often well pronounced, weak parietal callus usually present, umbilicus open.
Retractor inserts on the last quarter of the penis, caecum is situated in the middle of the epiphallus.

Greece and Bulgaria to S Turkey

Limestone rocks and rock rubble.

Heller 1976: 381, Liebegott 1986, Reischütz 1988, Schütt 2001: 144, Örstan et al. 2005, Irikov & Erőss 2008 ("various parts of Bulgaria").

Comprises probably various species.

Mastus gittenbergeri Maassen, 1995

Mastus gittenbergeri Maassen, 1995 in Maassen 1995: 36

11.7-15.3 x 5-6 mm

Greece: E Crete, near Eloúnda

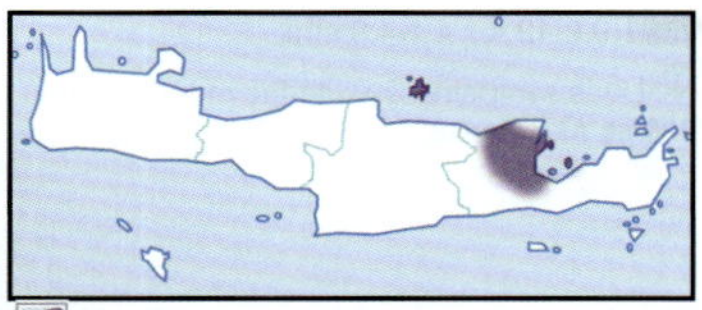

Greece (E central Crete)

Shell slender, apical whorls convex, the others more flattened. Penis papilla very flat.

Greece: island of Día — Vollrath Wiese

This *Mastus* species separated quite well in the conchological analysis (Welter-Schultes 2000), it occurs syntopically with *M. abundans*, *M. cretensis* and populations of the *M. sitiensis* complex. Also the anatomical differences (Maassen 1995) were relatively well pronounced.
The populations of Día are tentatively classified with this species, based on the conchological similarity. They might represent a separate endemic species.

Mastus giuricus (Westerlund, 1902)

Buliminus giuricus Westerlund, 1902 in Westerlund 1902: 40

Shell light brown, smooth or very weakly striated, 6-7 whorls, apertural margin white, with angular tubercle, columella very thin. Differs from *M. grandis* at least in its very small size.

10-12 x 5-5.5 mm

Parmakelis et al. 2005: 995, Triantis et al. 2008 (Skíros), www.iucnredlist.org (08.2012).

There are two forms on Skíros, more research is needed to understand their identities. The second form was called *M. carneolus* by Parmakelis et al. 2005, probably without analysing the type material from Istanbul (see *M. rossmaessleri*).

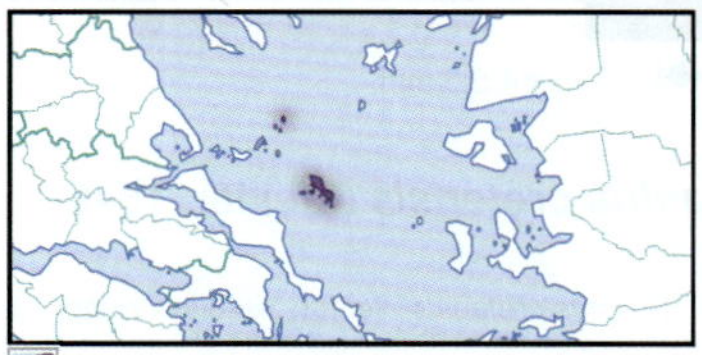

Greece (N Sporades, Skíros)

Mastus grandis (Mousson, 1859)

Chondrus pupa var. *grandis* Mousson, 1859 in Mousson 1859: 32
Buliminus dirphicus Blanc, 1879 in Westerlund & Blanc 1879: 90

16-19 x 7-8 mm

Greece: Parnasós, Agoriani

NW Greece to central Albania

Shell horny yellowish to brownish, shiny, finely and irregularly striated, apertural margin white and shiny, connected at parietal side by a thick white layer, angular tooth strong and broad.

Rähle 1980: 207, Dhora & W. 1996: 126.

Has often been misidentified as *M. pupa*. The populations from Evia (*dirphicus*) are classified with *M. grandis* here. Parmakelis et al. 2005 used *dirphicus* for the form of Evia, but did not study its relationships to the more western forms. More research is necessary to see if there are various different large-sized *Mastus* forms between Albania and Evia, where exactly they are distributed and if there are true gaps between their areas, for example between Evia and the mainland. The current state of published research does not allow more than to classify all forms with one single species.

Mastus ierapetranus Maassen, 1995

Without gender agreement: *Mastus ierapetrana* Maassen, 1995

Mastus ierapetrana Maassen, 1995 in Maassen 1995: 43
Should have been *ierapetranus*, *Mastus* is masculine and *ierapetrana* was meant as an adjective.

Shell like *M. sitiensis*, but smaller and in most regions slender.

20-24 x 6-7 mm

Clustered close to *M. procax* in molecular analyses (Parmakelis et al. 2005), but not in the ITS2 analysis (where it came close to *sitiensis* and *itanosensis*). Distribution suggested from the conchological record, the populations from near Kritsá could also belong to *A. riedeli*.

Greece (E Crete)

Mastus olivaceus (Pfeiffer, 1846)

Bulimus olivaceus Pfeiffer, 1846 in Philippi 1847: 124, Tab. 5
Name was attributed to Pfeiffer, authorship of the entire description was also attributed to Pfeiffer.

Shell slender, average size.

Appears to be confined to the high mountains.

Has been used in a much broader sense in the past, when this name was referred to all small *Mastus* species in Crete. Seems to be restricted to the Samariá gorge, perhaps also the nearby mountains.

12.6-14.5 x 5.0-5.4 mm

Greece: W Crete, Samariá gorge

Greece (W Crete)

Mastus ponticus (Retowski, 1887)

Buliminus ponticus Retowski, 1887 in Retowski 1887: 28, Taf. 1

Shell slender, horny yellow and shiny, aperture with white lip and usually an angular knot, umbilicus narrow.

Forest slopes.

14-17 x 5-6 mm

NE Turkey: near Erzurum (determination uncertain or locality confounded, the species is known to live between Sinop and Samsun)

N Turkey (Sinop to Samsun)

Mastus pupa (Linnæus, 1758)

Helix pupa Linnæus, 1758 in Linnæus 1758: 773 (*Helix Pupa*)

11-17 x 4-7 mm, locally variable

Italy: Sicily, Agrigento

NW Africa, S Europe

Shell whitish to dirty yellowish or deep brown, 6-8 regularly increasing whorls, finely striated or with prominently marked growth lines, apex pointed, last whorl 1/2 of shell height, aperture with white lip (but less prominent than in *M. grandis*), angular tooth usually present and prominent, narrow umbilicus present.

Open shrubland habitats, under stones, also frequently in disturbed habitats.

Beckmann 1990: 81 (size variation), Manganelli et al. 1995: 16, Giusti, Ma. & S. 1995: 238, Irikov & Erőss 2008 (not in Bulgaria), Abbes et al. 2009 (NW Africa), Štamol 2010: 18 (probably not in Croatia).

Previous reports for Albania and N Greece referred to *M. grandis* or other species, reports for Dalmatia also to specimens washed ashore at the coasts.

Mastus pusio (Broderip, 1836)

Bulinus pusio Broderip, 1836 in Broderip 1836: 45
Original genus was not *Bulimus*.
Buliminus unius Boettger, 1885 in O. Boettger 1885: 23

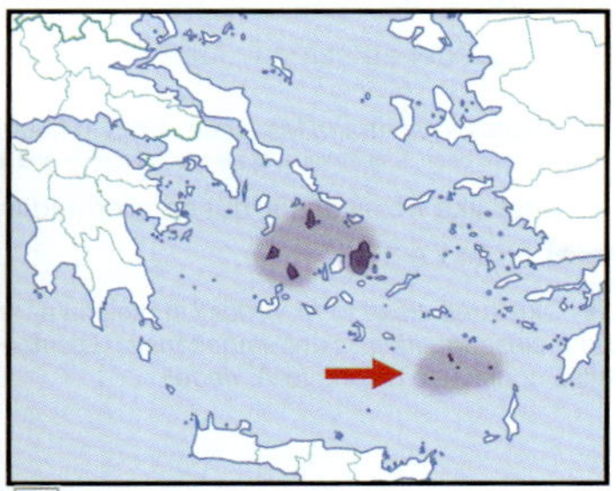

Greece (Kikládes)

10-16 x 6.5-8 mm

Greece: island of Náxos (1-5), Greece: island of Síros (6-7)

Heller 1976: 387, Parmakelis et al. 2005.

Shell much inflated (diameter much larger than in *M. etuberculatus*), olive-horny brown, distinctly striated, and with very short spiral lines, apex blunt, whorls slightly convex, rapidly increasing, aperture with white lip inside, apertural margins faintly connected at parietal side, no angular tooth, columella thickened, apertural height larger than in *M. etuberculatus* in relation to shell height.

Heller 1976 regarded *unius* as a synonym of *M. pusio*. In molecular studies (Parmakelis et al. 2005) *M. pusio* clustered surprisingly inside all the species from Crete.

Mastus riedeli Maassen, 1995

Mastus riedeli Maassen, 1995 in Maassen 1995: 53

Shell small, whorls moderately convex, aperture with angular tubercle.
Spermatheca without true bursa, retractor inserting proximal of the distal end of the penis, proximal epiphallus near caecum with some prominent septa, spermatophore distinctly sculptured.

11.5-14.5 x 5.0-5.4 mm

Considered by Welter-Schultes 2000 as a synoynm of *M. sitiensis*. Parmakelis et al. 2005 found it relatively isolated.

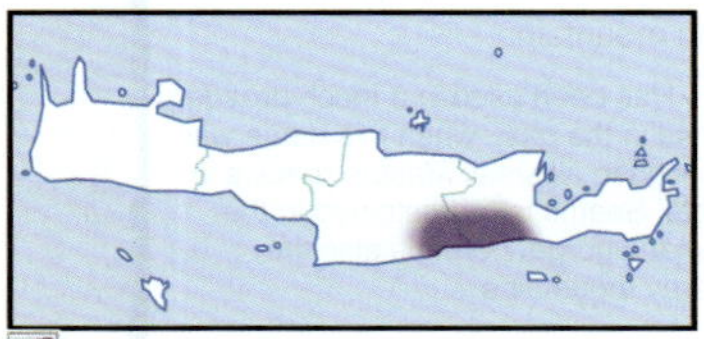

Greece (SE Crete)

Mastus rossmaessleri (Pfeiffer, 1847)

Bulimus rossmaessleri Pfeiffer, 1847 in L. Pfeiffer 1847: 113 (*Bulimus Rossmaessleri*) Type locality unknown.
Chondrus carneolus Mousson, 1863 in Mousson 1863: 285 Type locality near Istanbul.

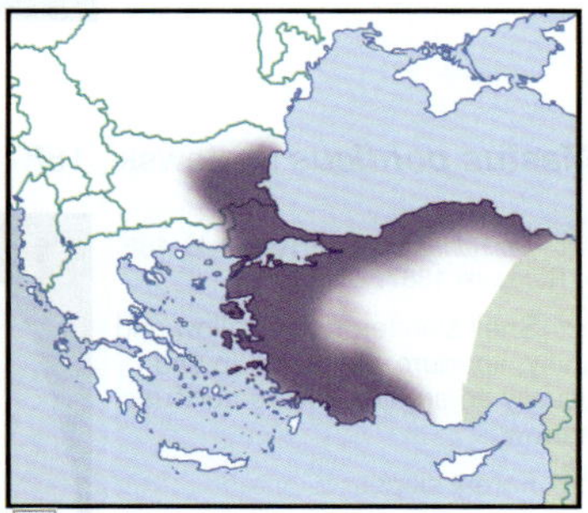

W Turkey and E Aegean islands to Bulgaria

Shell yellowish to dark horny brown, not very shiny, aperture usually with white lip and angular knot.

Open habitats under stones or among limestone rocks, with little or dense vegetation, often in the soil. Not very abundant.

9-17 x 4-6 mm

(1-5) Bulgaria: Sakar mountains, near Sladun, (6-9) SW Turkey: Vilayet Antalya, near Kaş

Hudec 1965 regarded *rossmaessleri*, *carneolus* and some other taxa as conspecific. This view was only partly followed (Bar & Butot 1986), obviously many malacologists did not know this publication. Schütt 1996: 154 classified *M. carneolus* as an independent species, Schütt 2001 as a subspecies of *M. rossmaessleri*, with reference to Hudec 1965. Recently Bulgarian authors (Georgiev 2005, Georgiev & Stoycheva 2009) classified Bulgarian populations as *M. rossmaessleri*. More research is necessary to understand the regional variation of the shell characters. Particularly the size seems to differ locally.
Parmakelis et al. 2005 used *carneolus* for a form from the W Aegean island of Giúra, this was a tentative classification and probably not based on comparative analyses of populations from near Istanbul. In the same study E Aegean populations were classified as *M. rossmaessleri*. The same applies to a form reported from Skíros reported by Triantis et al. 2008. More research is needed to clarify the relationships between these forms.

Mastus sitiensis Maassen, 1995

Mastus sitiensis Maassen, 1995 in Maassen 1995: 39
Mastus itanosensis Maassen, 1995 in Maassen 1995: 41

Greece (Crete)

Shell of average size, aperture without angular tubercle.
Spermatheca with bursa and diverticulum, retractor inserting distally on the penis (proximal of the distal end in *itanosensis*), epiphallus long (in *itanosensis* with a long proximal and short distal part), spermatophore with a row of little, flattened spines.

10-14 x 5-6 mm

Greece: E Crete, near Sitía

Studying shell morphology Welter-Schultes 2000 considered *itanosensis*, *ierapetrana*, *riedeli*, *subaii* and *butoti* as synonyms of *M. sitiensis*. In molecular studies (Parmakelis et al. 2005) *M. itanosensis* clustered very close to *M. sitiensis*, but the other taxa did not cluster closely with *sitiensis*.
East of a line Tourlotí-Koutsourás the shells are larger and less slender (Welter-Schultes 2000: 155). So this line seems to correspond with a break in the gene flow over area. The next taxon in the west seems to be *M. ierapetrana*, characterized by smaller and more slender shells, the next one *M. riedeli* with larger and less slender shells, like *M. sitiensis*. It is more difficult to detect conchologically closely related local populations in the Iráklio lowlands.

Mastus sphakiotus Maassen, 1995

Without gender agreement: *Mastus sphakiota* Maassen, 1995

Mastus sphakiota Maassen, 1995 in Maassen 1995: 45
Maassen's original name did not agree in gender with *Mastus*.
Mastus hemmeni Maassen, 1995 in Maassen 1995: 48
Mastus procax Maassen, 1995 in Maassen 1995: 46

Shell small, very slender, spindle-shaped.
Gentalia slender, spermatheca with bursa and diverticulum, retractor inserting proximal of the distal end of the penis, lumen of distal epiphallus with several septa.

10-18 x 4-6.5 mm

Greece: W Crete, near Ammoudári

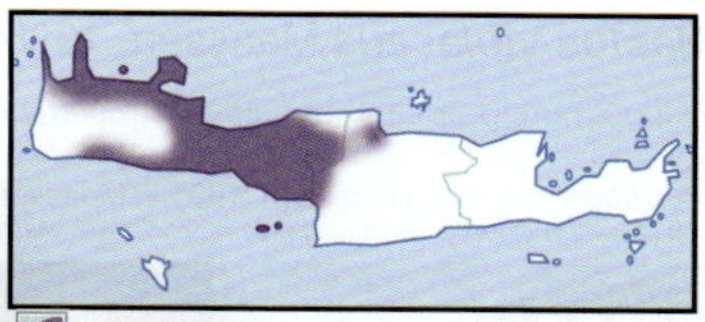

Greece (central and W Crete)

Welter-Schultes 2000 suspected a close relationship of *hemmeni* and *cretensis*. This was not supported in molecular studies (Parmakelis et al. 2005). Their results suggested that the 3 taxa *sphakiotus*, *hemmeni* and *procax* were closely related, *hemmeni* perhaps paraphyletic. This could have been an artifact because 3 *hemmeni* populations were analysed, of the other species only 1 or 2. Such a study could be more enlighting if 10 distant populations of each taxon were analysed, to see if they really produce closed groups.

Mastus transsylvanicus (Kimakowicz, 1883)

Buliminus transsylvanicus Kimakowicz, 1883 in Kimakowicz 1883: 40
Mastus was given as subgenus (or section), the generic name *Buliminus* was mentioned on p. 38.

9-15 x 4-6 mm, locally very variable

Romania: Transilvania, Bucegi mountains

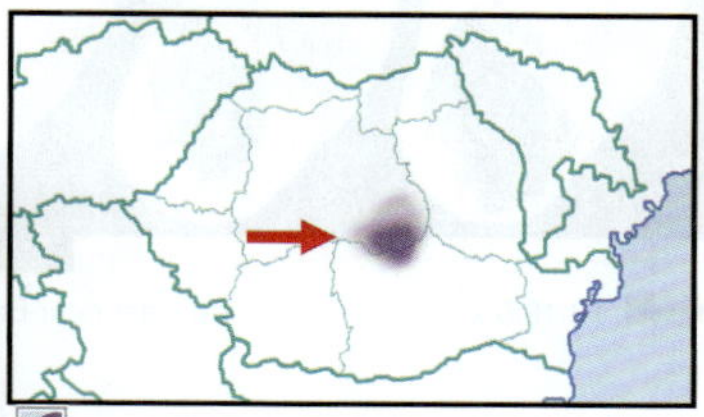

Romania (SE Transilvania)

Shell light horny brownish, upper whorls darker, apex pointed, apertural margin usually connected at parietal side with a thin white layer, no angular tubercle.

Humid and shadow habitats in mountain forests, in the leaf litter layer or among herbs and mushrooms.

Westerlund 1887: 18, Kimakowicz 1890: 89, Grossu 1987: 376.

Mastus turgidus (Kobelt, 1877)

Buliminus turgidus Kobelt, 1877 in Kobelt 1877: 69, Taf. 137

9-11.5 x 4.8-5.3 mm

Greece: Ródos

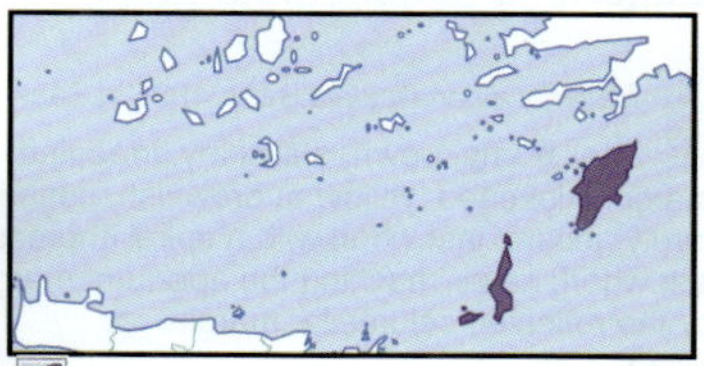

SE Greece

Westerlund 1887: 18.

Could be conspecific with *M. emarginatus*. Parmakelis et al. 2005 regarded it as an independent species. It clustered closely with *M. ehrenbergi* from Andikíthira, not closely with the species from Crete. *M. emarginatus* from Pelopónisos was not analysed.

Shell brownish, weakly and finely striated, aperture with white margin, connected by a thin whitish layer at parietal side, angular tubercle usually present.

Mastus venerabilis (Pfeiffer, 1853)

Bulimus venerabilis Pfeiffer, 1853 in L. Pfeiffer 1853 (Mon. Helic.): 352
Dentistomus venerabilis var. *alpicola* Kimakowicz, 1890 in Kimakowicz 1890: 91

17-22 x 7-9 mm

Romania: SE Transilvania, Postăvarul mountains

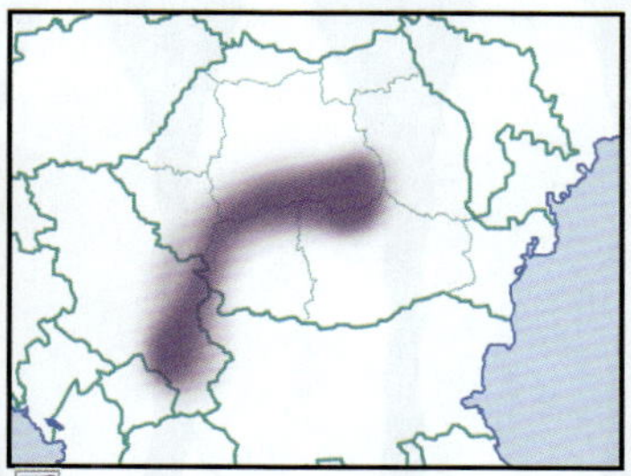

Romania, SE Serbia

Shell sinistral, olive horny brown, first 5 whorls 1/4 of shell height, last whorl 1/3 of shell height, angular tooth present.

Humid and shady habitats in forests at the soil, under stones or herbs, on the soil among roots, climbs occasionally on rocks or tree trunks. In Romania in up to 2300 m.

Westerlund 1887: 21, Pavlović 1912: 61, Tomić 1959: 30, Grossu 1987: 379.

The name *Mastus alpicolus* is incorrect, should be *Mastus alpicola*. This is not regarded as a separate species here. Grossu 1987 listed anatomical differences, but these should be confirmed and the local variation of anatomical characters should be better understood.

Mastus violaceus Maassen, 1995

Without gender agreement: *Mastus violacea* Maassen, 1995

Mastus violacea Maassen, 1995 in Maassen 1995: 52
Maassen's original name did not agree in gender with *Mastus.*

16-18.5 x 5.6-6.2 mm (in the west near Balí 4.9-6.0 mm)

Greece: N central Crete, (shells) near Balí, (animal) near Síses

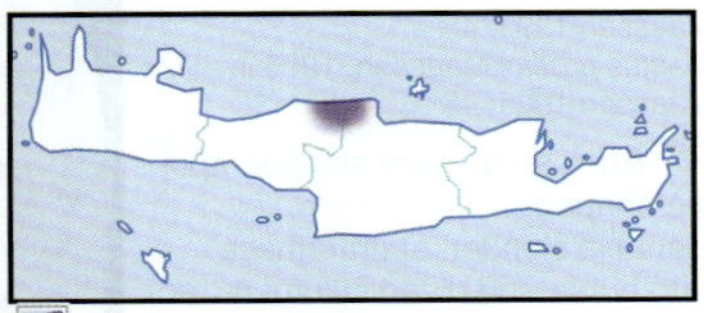

Greece (N central Crete)

Large-sized *Mastus* fraction, the shells of the Síses population are usually of violet colour, aperture with a prominent angular tubercle. Spermatheca with bursa and diverticulum, retractor inserting proximal of the distal end of the penis, epiphallus robust with septa in the distal part, spermatophore with a row of small spines. Animal light greyish brown covered with violetish brown spots on dorsum, upper side of head, tentacles and sides.

Clustered relatively close to Aegean island species in molecular analyses (Parmakelis et al. 2005).

Meijeriella

Meijeriella Bank, 1985 in Bank 1985: 41

Type species *Meijeriella canaliculata* Bank, 1985 (original designation)

Meijeriella differs from *Meyeriella* Krausse, 1917 (in Krausse 1917: 95, Hymenoptera) in more than one letter (Art. 56.2).

Borlumastus Örstan & Yıldırım, 2004 in Örstan & Yıldırım 2004: 126, type species *Ena yildirimi* Schütt, 1995 (original designation)

Gümüş & Neubert 2012.

Gümüş & Neubert 2012: 25 did not regard the conchological and anatomical characters of *Borlumastus* as sufficient to qualify for a separate genus.

Meijeriella canaliculata Bank, 1985

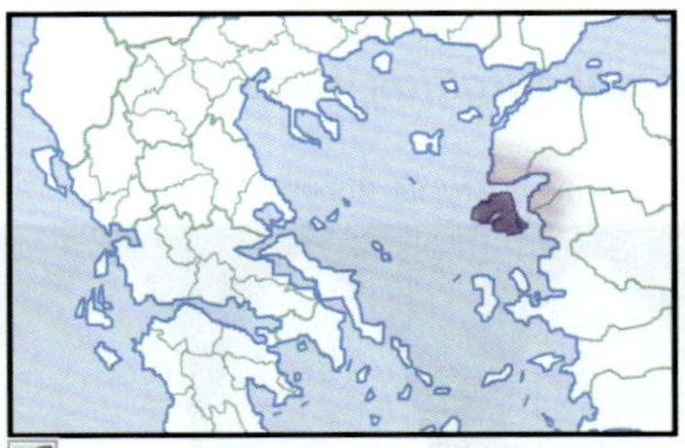

Greece (Lésvos)

Meijeriella canaliculata Bank, 1985 in Bank 1985: 42

Shell horny brown, dull shiny, irregularly striated, 7.5-9.5 whorls, aperture about 25 % of shell height, margin thickened, yellow or brownish, connected by a callus at the parietal side, angularis usually prominent, rarely absent, last whorl with a spiral impression running from near the angularis for not more than 3/4 of the last whorl, never reaching the apertural margin at its posterior end, umbilicus very narrow and partly covered by the reflected columellar margin.

11-15 x 3.2-3.3 mm

Open habitats at rocks, also olive orchards, relatively frequent.

Meijeriella frivaldskyi (Pfeiffer, 1847)

Bulimus frivaldskyi Pfeiffer, 1847 in L. Pfeiffer 1847: 191 (*Bulimus Frivaldskyi*)

11.5-16 x 4-6 mm

Turkey: Troya

NW Turkey

Shady habitats, under stones.

Bank 1985: 42, Hausdorf 2001: 1628, Gümüş & Neubert 2012: 27.

Shell horny brown, weakly shiny, with weak growth lines and no granular surface structure, whorls not convex, aperture relatively small, umbilicus partly covered.

Shell higher and more slender than *Ena montana*, aperture relatively smaller. Differs from *M. canaliculata* in its rounded and not canaliculated last whorl, and in its thickened palatal margin of the aperture. No anatomical differences.

Meijeriella yildirimi (Schütt, 1995)

Ena yildirimi Schütt, 1995 in Schütt 1995: 164, Taf. 1

Shell relatively high, almost smooth, 7.5-8 almost flat whorls, penultimate whorl with largest diameter, aperture always without angular tooth, palatal tooth or knot usually present, umbilicus covered.

At limestone rocks and below in rock rubble.

Gümüş & Neubert 2012: 25.

12-18 x 4-5 mm

Turkey: Vilayet Isparta
Aydın Örstan

SW Turkey (Isparta)

Merdigera

Merdigera Held, 1838 in Held 1838: 917

Type species *Helix obscura* Müller, 1774 (subsequent designation by Herrmannsen 1847-1849 [17-07-1847]: 39)

Perforate penial papilla, the only difference to *Ena*.

The only difference to *Ena* (perforate penial papilla) has not been studied in all species assigned to *Ena*. Manganelli et al. 1995: 39 argued that the difference was of a minor degree and did not justify separating the genera.

Merdigera obscura (Müller, 1774)

Helix obscura Müller, 1774 in Müller 1774: 103

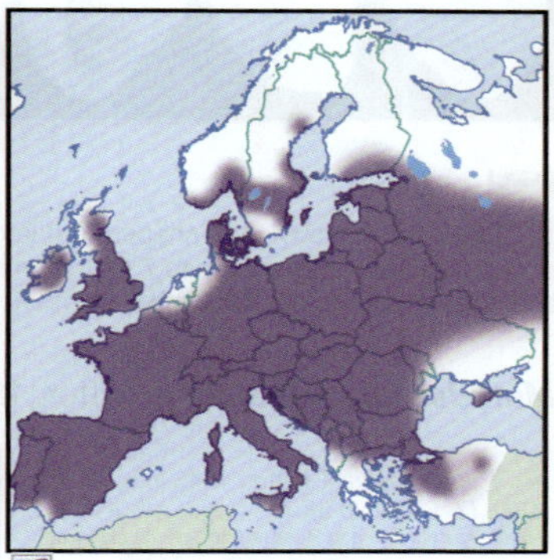

8.5-10.5 x 3-4 mm

Germany: Schleswig-Holstein, Kellersee near Eutin

NW Africa, Europe to Uzbekistan

Slovakia: Banská Bystrica, Podlavice
Jozef Grego

Shell evenly brown, densely radially striated, 6 slightly convex whorls with deep suture, apertural margin reflected and white, not connected by a callus or white layer at parietal side. Animal dark brown, lighter at sides and foot, upper tentacles long, lower tentacles length 1/4 of upper tentacles. The animal crawls with the shell in a very high position and not much bent to the right side. Juveniles and adults often with camouflage.

Shady and rocky habitats in deciduous forests at tree trunks and under leaf litter, also outside forests, under stones, at walls, vegetational litter and on mosses, requires less humidity than *Ena montana*. Juveniles climb vertical surfaces and tree trunks in wet weather. In Switzerland up to 2200 m, in Bulgaria 2000 m. Usually not very abundant, but a frequent species.

12-20 oval eggs (1.3 x 1.5 mm) are laid between May and October, juveniles hatch after 2 weeks, maturity is reached in the first months of the second year.

Gassies 1849: 112, Moquin-Tandon 1855: 291, Germain 1930: 372 (life cycles).

Also known as *Ena obscura*. It seems that the majority of users has recently used *Merdigera*.
Google (03.2011): ena 37 %, merdigera 63 %;
Google Scholar (03.2011): ena : merdigera (since 1992) 47 : 53 %, (since 2000): 40 : 60 %, (since 2005) 33 : 67 %.

Improvisa

Improvisa Schileyko, 1978 in Shilejko 1978: 846

Type species *Chondrus pupoides* Krynicki, 1833 (original designation)

Name was spelled "*Improvisa* Schileyko" in Latin script in the original work, on title page Shilejko in Russian script.
Bollingeria Forcart, 1940 in Forcart 1940: 194, type species *Bulimus lamelliferus* Rossmässler, 1858 (fixed under Art. 70.3, here)
Originally designated type species *Chondrus pupoides* Krynicki, 1833. Forcart 1940 misidentified the type species (Bank & Hovestadt 1991: 9) and meant *Bulimus lamelliferus* Rossmässler, 1858. Under Art. 70.3 of the Code shall now be fixed *Bulimus lamelliferus* as type species for *Bollingeria*, misidentified as *Chondrus pupoides* in the original designation by Forcart 1940: 194.

Improvisa pupoides (Krynicki, 1833)

(Not in study area)

Chondrus pupoides Krynicki, 1833 in Krynicki 1833: 410

N Caucasus, Armenia, Georgia

5.5-6.5 x 3.0-3.1 mm

S Russia: Pyatigorsk, Mashuk

Hudec 1972: 211, Shilejko 1984: 272, Bank & Hovestadt 1991: 9, Schütt 1996: 163, Schütt 2001: 152.

Shell like *Multidentula lamellifera*, differs only anatomically. Only figured here to show how *Multidentula lamellifera* looks like.

Multidentula

Multidentula Lindholm, 1925
in Lindholm 1925: 30 *Jaminia (Imparietula (Multidentula))*

Type species *Bulimus ovularis* Olivier, 1801
(original designation)

Multidentula lamellifera (Rossmässler, 1858)

Without gender agreement: *Multidentula lamelliferus* (Rossmässler, 1858)

Bulimus lamelliferus Rossmässler, 1858 in Rossmässler 1854-1859 (17-18): 95, Taf. 83

Shell looks like *Improvisa pupoides* from N Caucasus, Armenia and Georgia, horny brown with 5 whorls, aperture with a thick white margin and 3 strong folds in the aperture: angular callus present, parietalis strong, columellaris relatively strong, basalis weak, lower palatalis strong, upper palatalis weak, suprapalatalis very weak. Umbilicus open.
Differs from *I. pupoides* in the presence of a diverticle at the receptaculum seminis.

4-6.5 x 2.6-3.2 mm

In forests and relictual sections of previous forests. Frequent in Amasya and Tokat, elsewhere rare.

See the insert of *Improvisa pupoides* above, the shells look similar.

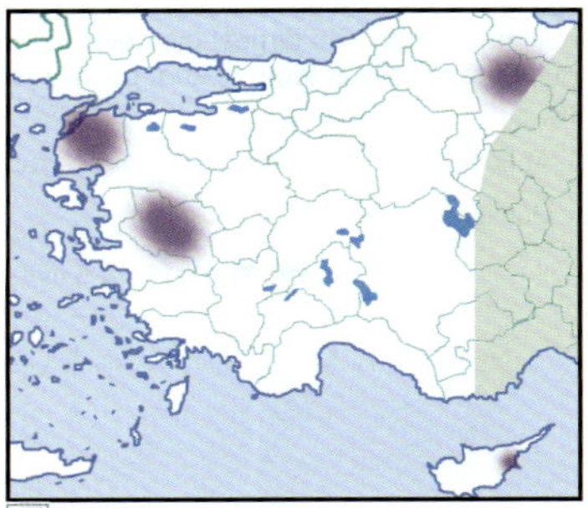

Scattered in Turkey, Cyprus

Multidentula ovularis (Olivier, 1801)

Bulimus ovularis Olivier, 1801 in Olivier 1801: 225, Pl. 18

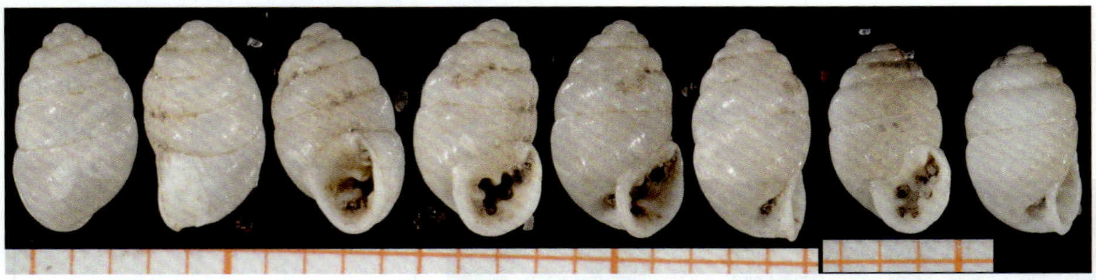

4.6-6.2 x 2.9-3.3 mm

Bulgaria: Albena

Shell horny brown, egg-shaped with largest diameter in lower third, 5.5 convex whorls, aperture usually with a strong angular callus and up to 7 teeth: strong angularis, parietalis, columellaris, basalis, upper and lower palatalis, and a weak suprapalatalis. Umbilicus open.

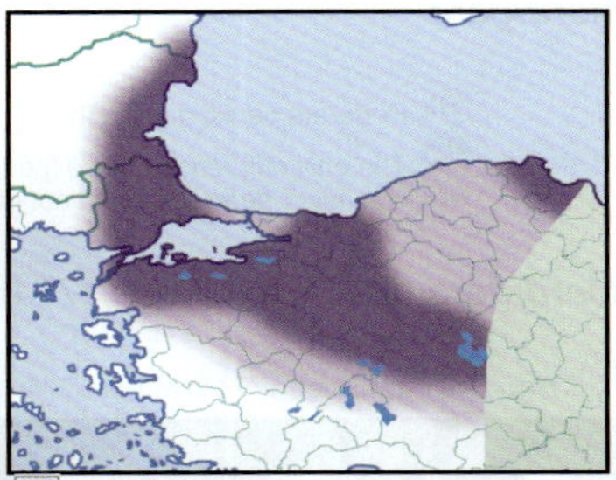

Israel to Turkey and SE Romania

At and near limestones, in SE Romania in dry and sunny habitats on hills among grass.

Damjanov & Likharev 1975: 149, Grossu 1987: 387, Schütt 2001: 146.

Multidentula squalina (Pfeiffer, 1848)

Pupa squalina Pfeiffer, 1848 in L. Pfeiffer 1848 (Z. f. Malak. 5): 7
Name was attributed to Rossmässler, description was not attributed to a particular person. In a strict application of Art. 50.1.1 the name in such a case must be attributed to the author of the work.

Shell like a chondrinid, dark horny brown with 6-7 convex whorls, suture deep, apex looks somehow rounded, largest diameter at last whorl, aperture with white lip and strong teeth: weak angular callus, very strong parietalis, strong columellaris, regular basalis, strong lower palatalis, regular upper palatalis, often a very weak suprapalatalis, umbilicus open.

At the bottom of limestone rocks.

Damjanov & L.1975: 150, Schütt 2001: 148, Urbański 1960, Fehér & E. 2009: 10.

Albania: Skrapar, Mali i Ostrovicës near Backë, 1750 m
Zoltán Fehér

6.0-7.2 x 2.2-2.3 mm

Turkey, isolated in Albania
In Bulgaria only seashore deposits.

Multidentula stylus (Pfeiffer, 1848)

Bulimus stylus Pfeiffer, 1848 in L. Pfeiffer 1848 (Mon. Helic.): 135

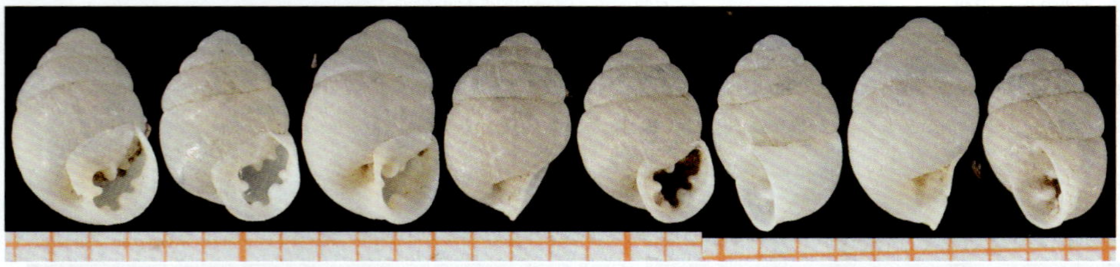

4.5-6 x 3.2-3.7 mm

Cyprus: 2.5 km from Agios Georgios to Lana bay

Shell horny yellowish, aperture 1/2 of shell height, with 6 teeth deeply inside: 1 large and 1 small parietal, 1 large and 1 small columellar, 2 closely together on a narrow palatal callus.

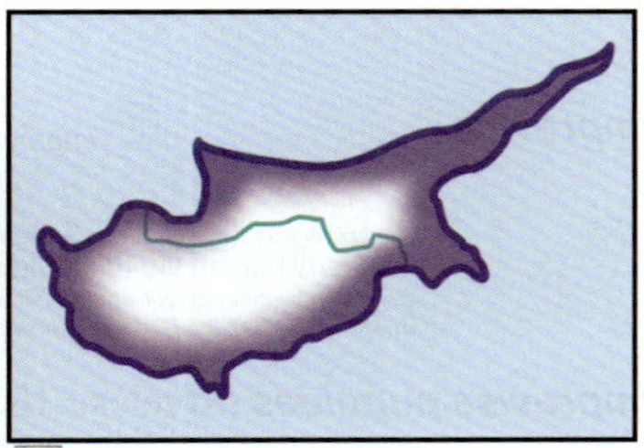

Cyprus

Seems to be restricted to coastal habitats. Most records are from the south.

Westerlund 1887: 47, Bank & Hovestadt 1991: 7 (*Euchondrus stylus*).

Multidentula substurmii (Retowski, 1887)

Buliminus sturmii var. *substurmii* Retowski, 1887 in Retowski 1887: 34 (*Bul. (Euchondrus) Sturmii* var. *substurmii*)
Imparietula microdon Schütt, 1995 in Schütt 1995: 164

Shell sinistral, light horny brown, translucent and shiny, 8-9 convex whorls with deep suture, aperture with a white lip except at parietal side, apertural margin sharp, parietalis weak and deep inside, columellaris hardly visible from aperture, palatal callus present, no other teeth, no umbilicus.

7-8 x 1.9-2.1 mm

Westerlund 1887: 56, Schütt 2001: 149.

Imparietula microdon seems to be a junior synonym (R. Bank, unpublished).

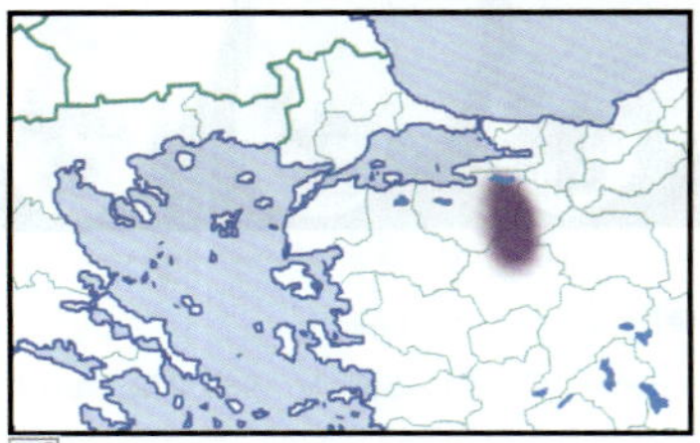

W Turkey

Napaeopsis

Napaeopsis Sturany & Wagner, 1914 in Sturany & Wagner 1914 (Anz.): 192 *Napaeus (Napaeopsis)*

Type species *Buliminus merditanus* Sturany, 1907 (fixed under Art. 70.3, here)

Gender feminine (Art. 30.1.2). I do not know the precedence between the two publications by Sturany & Wagner 1915 (preprint Denkschr. Wien from 1914: 40, 41 and 95) and Sturany & Wagner 1914 (Anz. Wien): 192. If Anz. Wien has precedence, we have the following situation:
Sturany & Wagner 1914 (Anz. Wien): 192 established *Napaeopsis* with a monotypic type species *Bulimus cefalonicus* Mousson, 1859, authorship must be attributed to Sturany & Wagner.
If preprint Denkschr. Wien has precedence, the authorship must be attributed to Wagner alone (specified on p. 95), and the subgenus contained 3 species *cefalonicus* (with subspecies *latifianus*), *zebianus* and *winneguthi*. Hesse 1933: 153 designated as type *Buliminus cefalonicus* Mousson, 1859.
In both cases the type species was misidentified and should have referred to *Buliminus merditanus* Sturany, 1907, currently *Napaeopsis merditana* (Sturany, 1907) (see Bank & Menkhorst 1992: 139). The specific status of this species needs to be confirmed, it could also be a junior synonym of *B. cefalonicus*. Gittenberger 2008: 271 regarded this as the type species of *Napaeopsis*, without taking action under Art. 70.3. Under Art. 70.3 of the Code shall now be fixed *Buliminus merditanus* as type species for *Napaeopsis*, misidentified as *Bulimus cefalonicus* in the monotypic designation by Sturany & Wagner 1914: 192.

Differs from *Zebrina* by not having neither a diverticle at the receptaculum seminis, nor a caecum at the epiphallus. Shell usually smaller, thinner, brownish, last whorls have a pattern of spots and streaks.

Most *Napaeopsis* species live at or close to limestone rocks, often in high altitudes and always protected from direct sunshine.

SE Europe to Asia Minor

Bank & Menkhorst 1992: 139.

Napaeopsis cefalonica (Mousson, 1859)

Without gender agreement: *Napaeopsis cefalonicus* (Mousson, 1859)

Bulimus cefalonicus Mousson, 1859 in Mousson 1859: 31
Napaeopsis mennoi Gittenberger, 2008 in Gittenberger 2008: 272

Taxonomic delimitation unclear, more research on the local variation is necessary. Bank & Menkhorst 1992 regarded it as a microendemic species from Kefaloniá, but it is also possible to classify Albanian populations with this species (Dhora & Welter-Schultes 1996).
The taxon *mennoi* from N Pelopónisos is classified with *cefalonicus* here, in its original description no constraints were given to classify it as a separate species. It was claimed to differ mainly in colour patterns, size and a very slightly more reflected apertural lip, but the local variations of these characters in the adjacent regions in the north were not studied and such characters are known to vary over area.

14-15 x 5-6 mm

Albania: 20 km SE Berat

NW Greece, S Albania

Shell not very solid, striated, 6-7 moderately convex whorls, apertural margin sharp, connected at parietal side by a thin layer, no teeth, umbilicus narrow.

Under the bark of wood, at limestone rocks, in humid rock rubble covered by mosses. In Kefaloniá only above 1200 m.

Napaeopsis merditana (Sturany, 1907)

Without gender agreement: *Napaeopsis merditanus* (Sturany, 1907)

Buliminus merditanus Sturany, 1907 in Sturany 1907: 231

11-14 x 4-5.5 mm

Albania: Kukës, Bicaj

Could be considered as a synonym of a polytypic or locally variable species *N. cefalonica*.

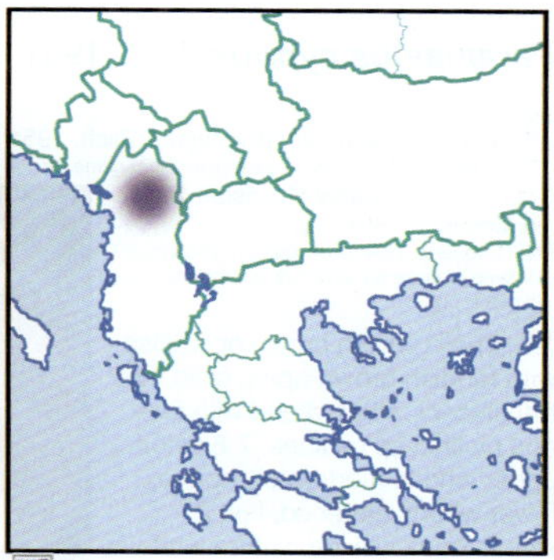

N Albania

Shell whitish with horny brown radial streaks and light yellowish first 2-3 whorls, coarsely and irregularly striated, 6-7.5 moderately convex whorls with well-marked dark suture, aperture inside light brown with a weak whitish lip, margin sharp, not connected at parietal side, umbilicus very narrow.

Napaeopsis minima Bank & Menkhorst, 1992

Without gender agreement: *Napaeopsis minimus* Bank & Menkhorst, 1992

Napaeopsis minimus Bank & Menkhorst, 1992 in Bank & Menkhorst 1992: 141

Shell horny brown with weak whitish streaks or dashes, usually not translucent, apex relatively pointed, 6-8 convex whorls with deep suture, aperture inside grey, streaks and dashes visible from inside, margin connected at parietal side by a thin whitish and translucent layer, not reflected, sharp, with a white lip inside.

9-11 x 4.3-4.6 mm

Differs from *N. cefalonicus* only in its smaller size. Anatomy unknown.

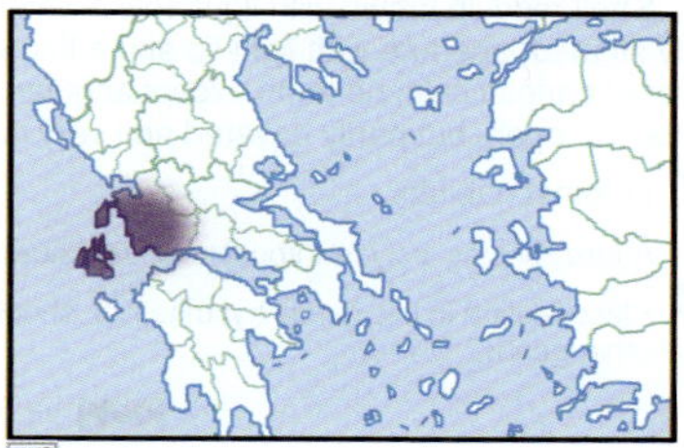

W Greece

Napaeopsis ossica (Boettger, 1885)

Without gender agreement: *Napaeopsis ossicus* (Boettger, 1885)

Buliminus ossicus Boettger 1885 in O. Boettger 1885: 180, Taf. 4 (*Buliminus (Ena) Ossicus*)

11-13 x 4-5 mm

Greece: Thessalia, 30 km NNE Lárisa

N Greece

Shell whitish with indistinct horny streaks or dashes, whorls convex, irregularly striated, last whorl 1/3 of shell height, aperture with white margin and often a broad white lip, columella slightly concave.

Paramastus

Paramastus Hesse, 1933 in Hesse 1933: 152, 181

Type species *Paramastus cyprius* Zilch, 1951 (fixed under Art. 70.3, here)

Hesse 1933 designated originally as type species *Bulimus episomus* Bourguignat, 1857, after examination of specimens from Cyprus which currently belong to *Paramastus cyprius* Zilch, 1951, so the type species was misidentified (Bank & Hovestadt 1991: 11). *B. episomus* lives in the Near East, not in Cyprus, and belongs to the same genus.
Under Art. 70.3 of the Code shall now be fixed *Paramastus cyprius* as type species for *Paramastus*, misidentified as *Bulimus episomus* in the original designation by Hesse 1933.

Shell subcylindrical to broadly fusiform. Protoconch mammillate, the first whorls are conical. The surface is characterised by a rugose axial sculpture with spiral incisions. Peristome broadly flared, internally with a thickened lip, a small angularis is present. The male copulatory organs with or without a penial appendix, a penial caecum and a flagellum is always present.

Libya and Turkey to Iran

Paramastus cyprius Zilch, 1951

Paramastus attenuatus cyprius Zilch, 1951 in Zilch 1951: 39
Expressly established as a new replacement name for *Chondrus attenuatus* Mousson, 1854 in Mousson 1854: 282
Not *Chondrus attenuatus* Krynicki, 1833 from Ukraine (= currently *Brephulopsis bidens* (Krynicki, 1833))

Shell greyish brown or whitish with reddish brown apex, solid, not translucent, striated and with fine and distinct spiral lines, 7-8 whorls, upper whorls moderately convex, lower whorls flattened, last 2 whorls equally wide, suture whitish, aperture brown, margin hardly reflected, connected by a distinct white callus at parietal side, umbilicus very narrow.

16-21 x 7.5-10 mm

Cyprus

Georgios Konstantinou

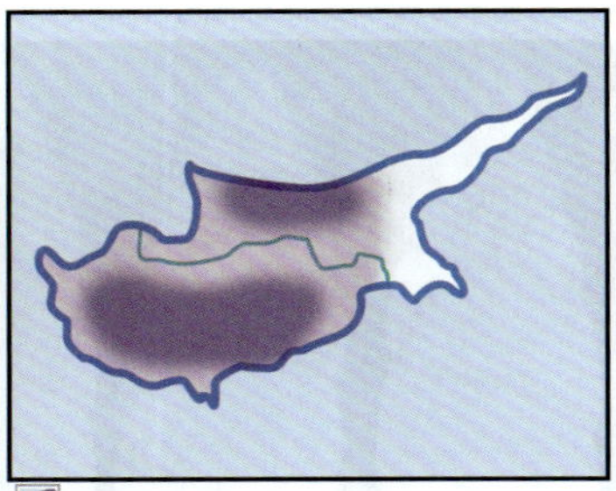

Cyprus

Tends to prefer habitats not too far from the coast.

Bank & Hovestadt 1991: 10.

Paramastus goettingi Forcart, 1961

Paramastus goettingi Forcart, 1961 in Forcart 1961: 177 (*Paramastus? goettingi*)

Shell reddish brown with light apical whorls, irregularly striated, aperture with light lip and a thin whitish layer at parietal side, no teeth except angular tooth, umbilicus open, but partly covered and very narrow.

In forest habitats.

A rare species, only few specimens are known.

Classification in this genus hypothetical, should belong to another genus.

24-27 x 11 mm

N Turkey: near Isirganlik N of Kürre

NW Turkey

Paramastus spratti (Pfeiffer, 1846)

Bulimus spratti Pfeiffer, 1846 in Philippi 1847: 114, Tab. 4 (*Bulimus Spratti*)
Name was attributed to Pfeiffer, the authorship of the entire description was also attributed to Pfeiffer.

Shell egg-shaped or conical with very large diameter, periostracum greenish yellow with darker brownish radial streaks, shell below periostracum white and striated, 4.5-5 not very convex whorls, last whorl predominating, aperture broad with white and relatively sharp margin, connected at parietal side by a thin whitish callus, columella oblique and visible inside aperture, umbilicus open.

At rocks with vegetational cover.

17-22 x 10.5-12 mm

SW Turkey: Vilayet Antalya, Sillyon near Antalya

SW Turkey

Only known from few localities.

Peristoma

Peristoma Krynicki, 1833 in Krynicki 1833: 420

Type species *Peristoma merdueniana* Krynicki, 1833 (monotypy)

Was treated as feminine by Krynicki 1833, but should be neuter.

Peristoma merduenianum Krynicki, 1833

Without gender agreement: *Peristoma merdueniana* Krynicki, 1833

Peristoma merdueniana Krynicki, 1833 in Krynicki 1833: 421, pl. 9

Shell thin, almost smooth, horny yellow, whorls convex, aperture oval or elliptical, connected at parietal side by a white layer, umbilicus open.

13.5-15 x 5.5-6.5 mm

Humid and shady habitats in mountains.

EU Red List: Near Threatened NT. Threatened by increasing temperatures associated with climate change.

Westerlund 1887: 27, Sysoev & Schileyko 2009: 84.

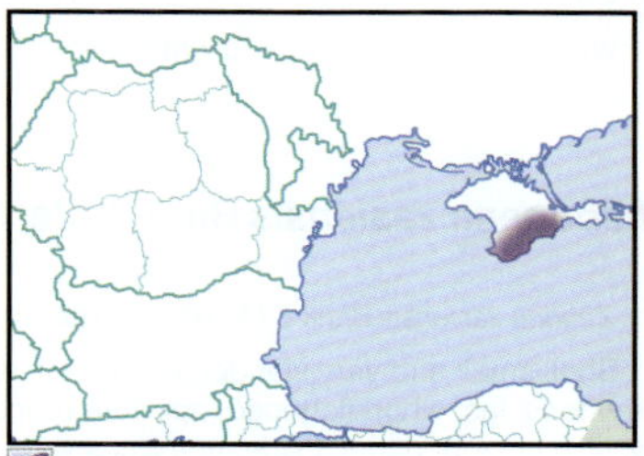

Ukraine (Crimea mountains)

Peristoma rupestre (Krynicki, 1833)

Without gender agreement: *Peristoma rupestris* (Krynicki, 1833)

Bulimus rupestris Krynicki, 1833 in Krynicki 1833: 413 (Tab. 8 Fig. 5 [sic])
Not *Bulimus rupestris* Philippi, 1836 (Chondrinidae) (= *Rupestrella rupestris*)

Shell yellowish horny brown, thin, not very shiny, weakly striated, 7-8 whorls, last whorl inflated, apertural margin slightly reflected, with weak lip inside, umbilicus narrow.

Humid and shady habitats at mountain water currents or in gorges.

EU Red List: Near Threatened NT. Threatened by increasing temperatures associated with climate change.

17-20 x 6-7 mm

Paolo Zaccaria

Ukraine: S Crimea, near Alushta
Tiroler Landesmuseen Innsbruck

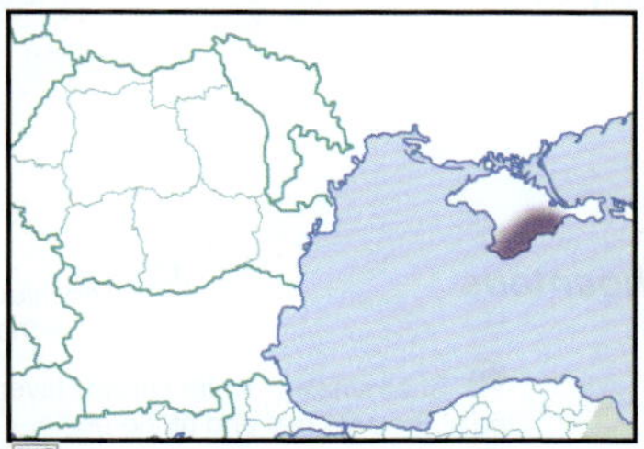

Ukraine (Crimea mountains)

Westerlund 1887: 27, Sysoev & Schileyko 2009: 84.

Rhabdoena

Rhabdoena Kobelt & Möllendorff, 1902 in Kobelt 1902: 1021

Type species *Buliminus caesius* Boettger, 1885 (original designation)
(= currently *Rhabdoena cosensis*)

Shell elongate with rounded aperture and callus at parietal side, surface with fine riblets. Penis with appendix, caecum at central position of the epiphallus (in *Zebrina* subterminal).

Obligate rock dwellers.

Gümüş & Neubert 2012: 29 separated *Zebrina* into *Rhabdoena*, *Leucomastus* and *Zebrina*, and presented taxonomic arguments.

Rhabdoena cosensis (Reeve, 1849)

Bulimus cosensis Reeve, 1849 in Reeve 1849: Pl. 67 No. 474 (*Bulimus Cosensis*)
Buliminus caesius Boettger, 1885 in O. Boettger 1885: 24 (*Buliminus (Zebrina) cæsius*)

Shell medium slender, not as solid as the other *Zebrina* species, apex pointed, first whorls brownish, lower whorls dirty white with blue hue and brownish streaks and dashes, apertural margin slightly reflected, umbilicus partly covered.

Limestones in lowlands and moderate highlands.

Bar & Butot 1986, Bank & Neuteboom 1988: 49, Schütt 2001: 123.

13-23 x 5.5-8.5 mm

(1-4) Greece: Kos, near Platánio,
(5) W Turkey: Vilayet Manisa, Spil mountain

W Turkey, E Greek islands

(5) Aydın Örstan

Rhabdoena mirifica Bank & Menkhorst, 1992

Zebrina mirifica Bank & Menkhorst, 1992 in Bank & Menkhorst 1992: 136

Shell slender, usually whitish or with bluish hue, apex can be horny yellowish, with some streaks, not very shiny, striated, 8-9 moderately convex whorls, apertural margin shortly reflected, inside thickened by a white callus, palatal and columellar margin approaching each other considerably at parietal side, connected by a whitish layer.

14-17 x 4.6-5.3 mm

Open forests and grasslands among shrubs, low vegetation and rocky outcrops in high altitudes.

EU Red List: Near Threatened NT. Only known from a few remote ocalities in Thásos. No real threats known.

www.iucnredlist.org (07.2012, K. Triantis).

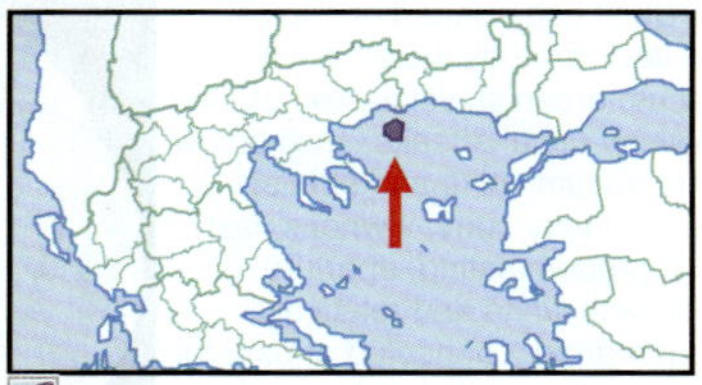

Greece (Thasos)

Rhabdoena stokesi (Boettger, 1885)

Buliminus stokesi Boettger, 1885 in O. Boettger 1885: 25

Shell solid, horny whitish with yellowish streaks, little shiny, whorls almost flat, last whorl 40 % of shell height, aperture inside relatively dark, with a white thin lip.

17-18 x 5.5-6 mm

Rocky habitats, hides in crevices.

EU Red List: Near Threatened NT. Lives only on Amorgós. No real threats known.

Westerlund 1887: 28, www.iucnredlist.org (07.2012, K. Triantis).

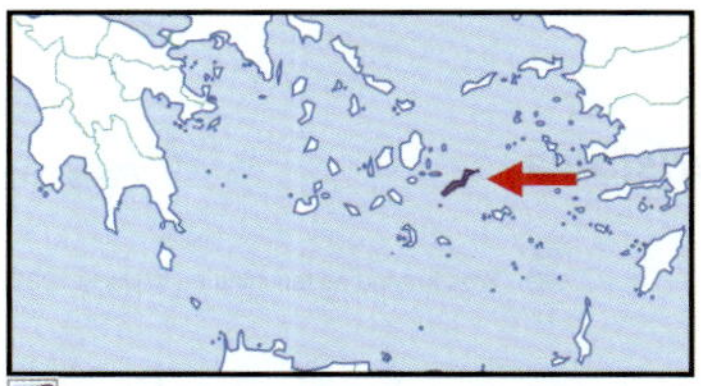

Greece (Amorgós)

Rhabdoena zasiensis Bank & Menkhorst, 1992

Zebrina zasiensis Bank & Menkhorst, 1992 in Bank & Menkhorst 1992: 134

Shell small and very slender, horny yellowish with white surface layer, quite regularly striated, 8-9 very convex whorls, last whorl flattened, aperture rounded, inside whitish or horny yellowish, margin sharp, connected by a whitish layer at parietal side, umbilicus wide.
Looks almost like a slender *Cochlostoma* with its convex whorls and rounded aperture.

11-13 x 3-4 mm

Rocky habitats, hides in crevices.

EU Red List: Near Threatened NT. Range restricted to a small area at Zas mountain, the species is rare and consists of only few mature individuals (but its life cycle is unknown).

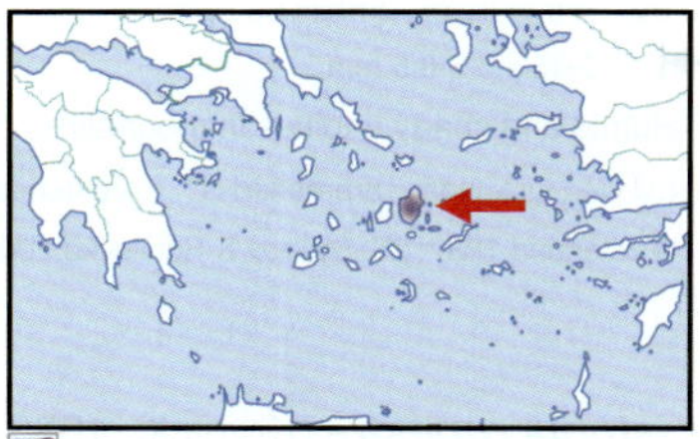

Greece (Naxos)

Thoanteus

Thoanteus Lindholm, 1925
in Lindholm 1925: 28 *Zebrina (Zebrina (Thoanteus))*

Type species *Bulimus gibber* Krynicki, 1833 (original designation)

Shell sinistral, white surface layer present but often weak. Penis with appendix, caecum at subterminal position of the epiphallus, presence of a knotty swelling at the proximal end of the penis.

Snails live on exposed calcareous rocks, not on the ground.

Ukraine (Crimea), N Turkey

Hausdorf 1993, Hausdorf 1994., Schileyko 1998: 227, Gümüş & Neubert 2012: 29 (presented a consistent system to classify enid genera).

The disjunct distribution is explained to have originated by passive dispersal from Crimean populations to N Turkey (perhaps by birds), rather than as a remnant of a formerly continuous distribution around the Black Sea.

Thoanteus corneus Hausdorf, 1993

Thoanteus corneus Hausdorf, 1993 in Hausdorf 1993: 91

Shell sinistral, fragile, horny brownish with some spots of a weak white surface layer, with irregular growth lines and partly with some spiral lines, 6.2-7.5 whorls, aperture without teeth, columellar and palatal sides of the aperture closer together than in the other *Thoanteus* species, umbilicus open.
Differs from the other *Thoanteus* species by it smaller size and its reduced appendix section A1.

12.4-15.8 x 6.2-7.4 mm

Reported from rocky habitats near a village.

Only known from a small area.

Schütt 2001: 134.

N Turkey (Vilayet Kastamonu)

Thoanteus ferrarii Hausdorf, 1994

Thoanteus ferrarii Hausdorf, 1994 in Hausdorf 1994: 354

Shell sinistral, brownish with uneven white surface layer, 7-7.5 weakly convex whorls, aperture without teeth, apertural margin little reflected and slightly thickened inside, very thin white layer at parietal side, umbilicus open.
Appendix retractor inserts laterally on the appendix section A1 (not subterminally like in the other *Thoanteus* species).

17.5-19.1 x 7.7-8.1 mm

On limestone rocks covered with lichens.

EU Red List: Least Concern LC. Only known from one locality, the population seems to be stable, no threats known.

Ukraine (Crimea mountains)

Thoanteus gibber (Krynicki, 1833)

Bulimus gibber Krynicki, 1833 in Krynicki 1833: 416, Tab. 8 (*Bulimus gibler*, *Bulimus gibber*)
Original spelling on p. 416 *gibler*, but spelled *gibber* on p. 389 and Tab. 8. The spelling *gibler* was an inadvertent error in typesetting, *gibber* is a Latin word and *gibler* gives no sense.

Shell sinistral, with a marked white surface layer, apex pointed, 6-7 weakly convex whorls, aperture without teeth, margin sharp and slightly reflected, connected at parietal side by a thin white layer, umbilicus open and partly covered.
Presence of a distinct coecum growing out of the bulge that originates from the insertion of epiphallus at proximal end of penis.

On limestone rocks, widespread and abundant.

19-21 x 8-9 mm

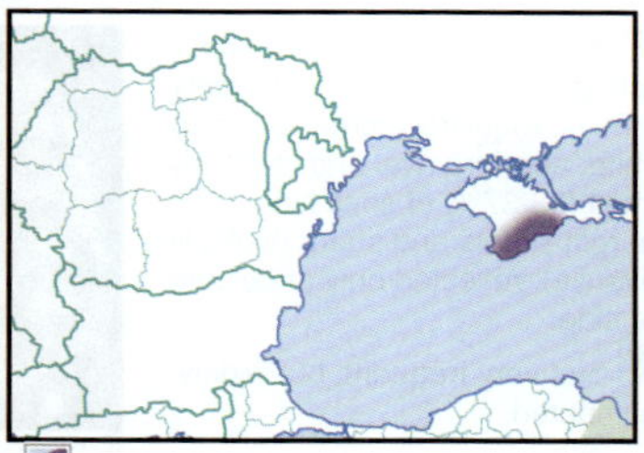

Ukraine (Crimea mountains)

Ukraine: Crimea, Chatyr Dag

Thoanteus zilchi Hausdorf, 1993

Thoanteus zilchi Hausdorf, 1993 in Hausdorf 1993: 89

Shell sinistral, irregularly striated, bluish-horny coloured with an irregular opaque surface layer, no teeth in the aperture, umbilicus open.
Differs from *Th. gibber* by its weaker and more irregular white surface layer, weaker spiral striae, its smaller relative shell height and lower number of whorls (6.5-7.3 in *zilchi*, 7-7.5 in *gibber*), anatomically by its terminal insertion of appendix section A3 in A2.

17.1-21.2 x 8.9-10.2 mm

Rocky limestone habitats.

Only known from a small restricted area.

Schütt 2001: 134.

N Turkey (Vilayet Kastamonu)

Turanena

Turanena Lindholm, 1922 in Lindholm 1922: 275

Type species *Buliminus herzi* Boettger, 1889 (original designation)
(= currently *Turanena herzi* (Boettger, 1889) from Iran)

Intended as a new replacement name for *Pseudonapaeus* sensu Kobelt & Möllendorff, 1902 non Westerlund, 1887. *Turanena* must be interpreted as a regular new name, established without description but with one species included, the originally designated type species. Not a new replacement name because Kobelt & Möllendorff 1902: 1021 just used Westerlund's 1887 name *Pseudonapaeus* and gave an incorrect type species.

Pseudonapaeus Westerlund, 1887 in Westerlund 1887: 66 *Buliminus (Pseudonapæus)*, type species *Buliminus asiaticus* Martens, 1880 [in Martens 1880 (Conch. Mitth. 1880): 29, Taf. 69, = currently not in genus *Turanena*] (subsequent designation by Lindholm 1922: 275)
Lindholm 1922 thought that the first species in a list must be its type species, this was *B. asiaticus*, and at this occasion Lindholm fixed *B. asiaticus* as type. Westerlund's 1887 list contained originally 8 species, but I did not see not a name *Buliminus herzi* Boettger, 1889, which was given as type by Kobelt & Möllendorff 1902: 1021 and Bank & Menkhorst 1992: 147.

(1) N Iran to Greece and (2) central Asia

Bank & Menkhorst 1992: 147, Gittenberger & Menkhorst 1993, Schileyko & Moisseeva 1995.

Turanena albrechti Rähle, 1988

Turanena albrechti Rähle, 1988 in Rähle 1988: 176

Shell brownish, partly covered by a white surface layer, whorls strongly convex (like *Cochlostoma*), aperture small (32-38 % of shell height), inside brownish, almost circular, outer margin sharp, columellar margin reflected, umbilicus narrow.
Differs from neighbouring *Turanena* species in its strongly convex whorls.

5.0-5.5 x 2.6-3.1 mm

Was found in 1800 m altitude, at low sun-exposed limestone rocks in a small valley with rich vegetation.

Gittenberger & Menkhorst 1993: 73, Schütt 2001: 86.

Could be conspecific with *T. forcartiana* Schnell, 1979 from E central Turkey.

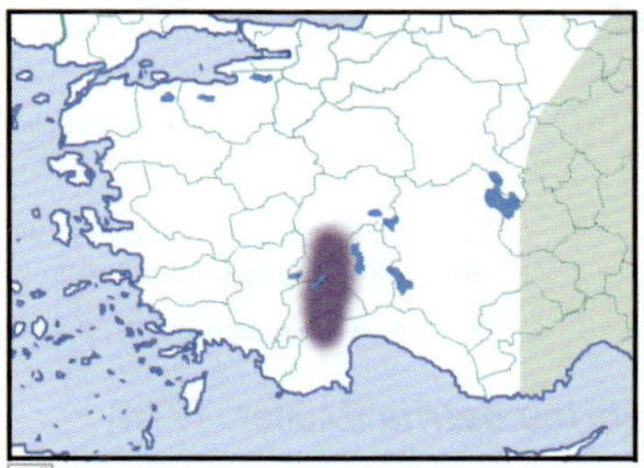

SW Turkey

Turanena carpathia (Boettger, 1885)

Without gender agreement: *Turanena carpathius* (Boettger, 1885)

Buliminus carpathius Boettger, 1885 in O. Boettger 1885: 26

12-13 x 4.5-5 mm

Greece: Karpathos, near Aperi

Shell horny brown, high, conical, 5-7 convex whorls, distinctly striated, regularly increasing, aperture elongate.

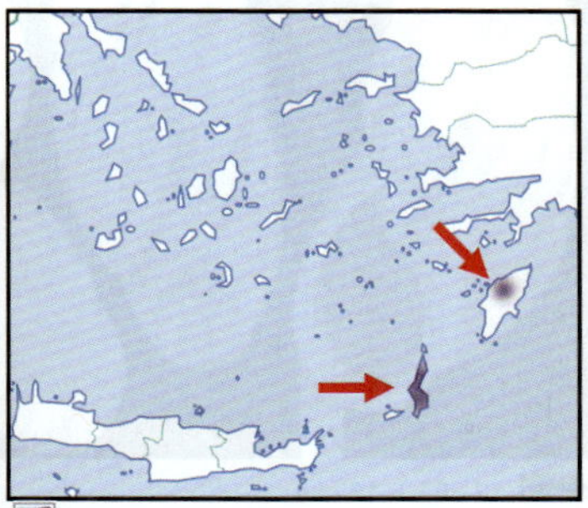

Greece (Kárpathos, Ródos)

Limestone rock crevices in high altitudes.

EU Red List: Near Threatened NT. A rare species, no major threats known.

Heller 1976: 376, Maassen 1981, Pieper 1984: 51, Frank 1997: 8.

Turanena hemmeni Bank & Butot, 1990

Turanena hemmeni Bank & Butot, 1990 in Bank & Butot 1990: 197

Fresh shell light brown and transparent, whorls not extremely convex, regularly increasing, suture white, aperture oval and relatively small (33-37 % of shell height), margin simple and sharp (looks like juvenile), reflected only at columellar side.

Not very frequent, but widely distributed.

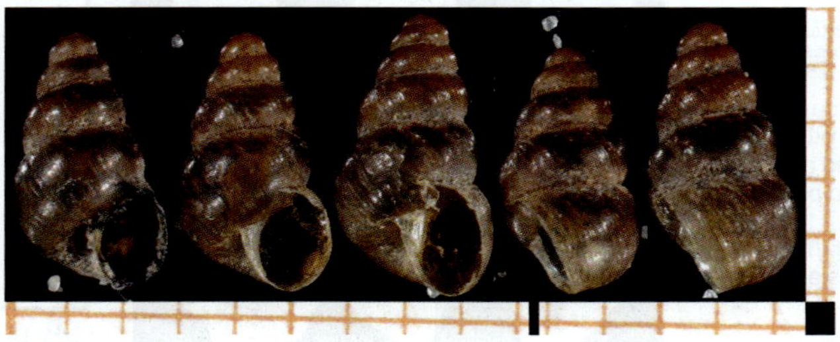

6.8-8.7 x 3.1-3.7 mm — Turkey: Vilayet Içel, Gülek Bogazı

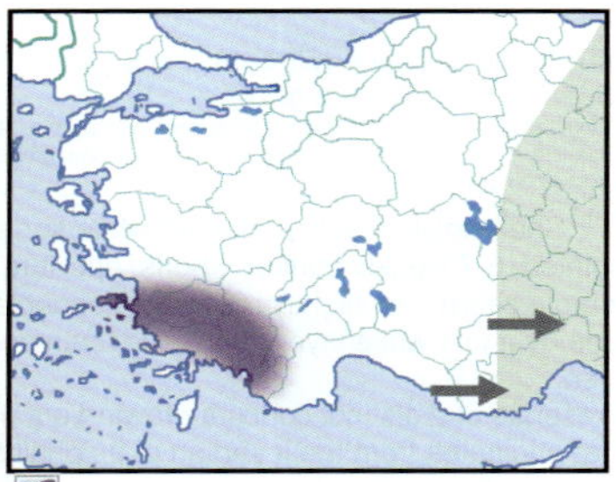

W and S central Turkey, Sámos

Schütt 2001: 87, Örstan et al. 2007: 333.

Turanena katerinae Gittenberger, 1996

Turanena katerinae Gittenberger, 1996 in Gittenberger 1996: 9

Shell with olive brown periostracum, irregularly striated, 4-5 relatively convex whorls, aperture about 40 % of shell height, margin sharp, thin, not reflected, umbilicus narrow.

5-7 x 3-4 mm

Was found in high mountains at 1600 m.

EU Red List: Near Threatened NT. Only known from two very remote localities, no real threats known.

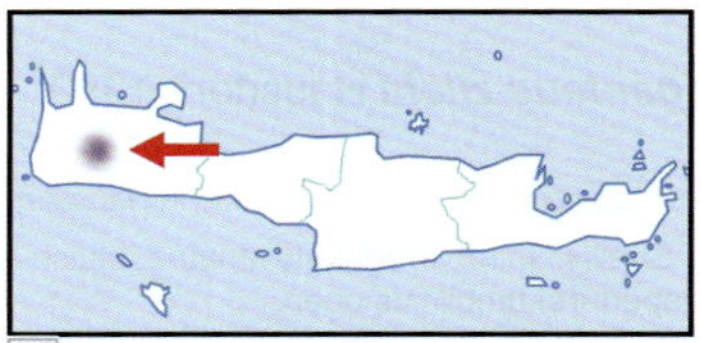

Greece (W Crete)

Turanena tuccari Gittenberger, 1986

Turanena tuccari Gittenberger, 1986 in Gittenberger 1986: 212

7.9-11.3 x 3.7-6.2 mm — Turkey: Vilayet Antalya, near Selge

Shell brownish and mostly covered by a bluish white surface layer, apex horny brown, whorls weakly convex, regularly increasing, aperture inside horny brown, margin white, broad and reflected, umbilicus very narrow. Shells W of Antalya are more slender than in the northeast.

SW Turkey

Rocky habitats, often together with clausiliids, up to 1100 m. A common species in the environs of Antalya.

Gittenberger & Menkhorst 1993: 80, Schütt 2001: 88.

Zebrina

Zebrina Held, 1838 in Held 1838: 917

Type species *Bulimus fasciolatus* Olivier, 1801 (subsequent designation by Herrmannsen 1847-1849 [Mar 1849]: 714)

Contained originally 4 species: *Bulimus fasciolatus*, "*B. dealbatus* Say, *B. radiatus* Bruguière and *B. illibatus* Ziegler". Gray 1847: 175 gave *Helix detrita* Müller, 1774 as type for *Bulimulus* Leach, 1814, *Zebrina* was mentioned as a synonym of *Bulimulus*. Bank & Menkhorst 1992: 118 gave *Bulimus radiatus* Bruguière, 1789 as type designated by Gray 1847: 175, I did not find *B. radiatus* mentioned by Gray 1847.

Shell large, usually whitish with darker radial colour streaks, aperture wide, without lamellas, no callus at parietal side, inside coloured. Penis with appendix, caecum at subterminal position of the epiphallus (in *Rhabdoena* central).

Gümüş & Neubert 2012: 29 separated *Zebrina* into *Rhabdoena*, *Leucomastus* and *Zebrina*, and presented taxonomic arguments.

Zebrina detrita (Müller, 1774)

Helix detrita Müller, 1774 in Müller 1774: 101

12-30 x 8-12 mm — S Hungary: Szársomlyó

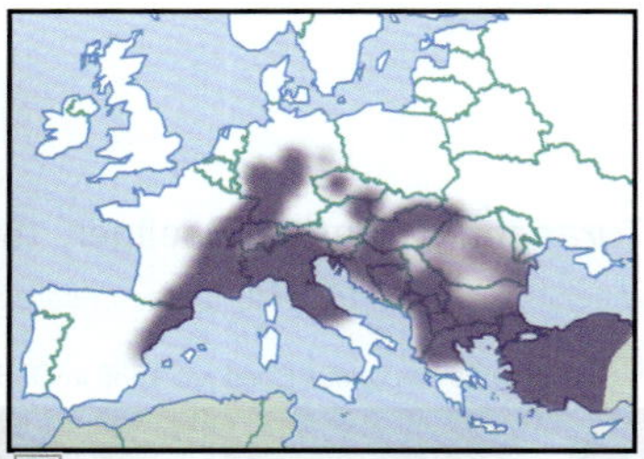

Europe

Shell creamy white with irregular and blurry reddish brown radial streaks, apex blunt, aperture light brownish, umbilicus open and partly covered.

Endangered in Austria and Germany, Vulnerable in Switzerland. Threatened by agricultural activities, especially by cutting of grasses in the summer season.

Meadows of warm and dry slopes, rocky grasslands. In central Europe the animals must be able to hibernate in deeper zones in the soil. In the summer the animals escape bottom heat (up to 70 °C) by climbing up grasses. In Switzerland up to 1800 m, in Bulgaria 2500 m. Main intermediate host of *Dicrocoelium lanceolatum*.

Frömming 1954: 45, Prince 1967, Damjanov & Likharev 1975: 142, Kerney et al. 1983: 135, Falkner 1990: 150, Fehér & Erőss 2009.

Reproduction in Germany in two periods April/May and August/September, copulation only at rainy days, takes about 1 hour in which both partners exchange spermatophores, snails can copulate various times, 10-20 days after copulation the eggs (2.0-2.4 mm) are laid in clutches of 15-70 eggs, juveniles hatch after 27-36 days, full size is reached after 2 years. Maximum age 4-5 years.

At the periphery of the range of this species in central Albania, isolated from the main populations, Fehér & Erőss 2009 found a consistently sinistral population and classified it as a new subspecies, *Z. detrita sallake*.
Sinistral chirality is known to provide obstacles to copulation, but this alone would not be decisive. For a subspecies under broadly accepted concepts in biodiversity research (Mallet 2001) partial barriers to gene flow should be present. This would be the case if there was gene flow between both populations at a contact zone, and the genes for dextral chirality would not pass through the barrier. In the absence of a contact zone other constraints should be provided to show that such a scenario would be likely.

Zebrina fasciolata (Olivier, 1801)

Without gender agreement: *Zebrina fasciolatus* (Olivier, 1801)

Bulimus fasciolatus Olivier, 1801 in Olivier 1801: 415, Pl. 17

SE Greece to Palestine
In Kárpathos only marine deposits.

At stones in lowlands, prefers limestone, usually in dry shrublands.

15-25 x 6-8 mm   Greece: Ródos   (7-8) Panayotis Ovalis

Shell not very slender, dirty whitish with irregular brownish streaks, apex blunt, aperture inside dark brownish, umbilicus covered. Smaller and narrower than *Z. detrita*, aperture darker brownish inside and smaller in relation to shell height, but larger and with less whorls than in *Leucomastus eburneus*.

Rumina

Rumina Risso, 1826
in Risso 1826: 79

Type species *Helix decollata* Linnæus, 1758
(monotypy)

Shell with high spire, decollated with some 5 whorls remaining. Genital system primitive, vagina and penis of similar length, no dart sacs or mucus glands, spermatheca duct short.

Cross-fertilisation prevails, but self-fertilisation is also possible, at least in *R. decollata*.

N Africa, S Europe

Selander & Kaufman 1973, Dundee 1986, Bank & Gittenberger 1993, Carr 2002.

Prior to 1993 only one species was believed to exist.

Rumina decollata (Linnæus, 1758)

Helix decollata Linnæus, 1758 in Linnæus 1758: 773

25-40 x 10-14 mm   Portugal: Algarve, Tavira

N Africa, S Europe

Portugal: Algarve, Faro   Valter Jacinto

Shell in living snails brownish and shiny with white apertural lip, empty shells white, diameter larger than in *R. saharica*, whorls slightly more convex. Shell decollate, initial whorls are disconnected by deliberate truncation, the apical opening is closed by a plate.
Animal relatively short, whitish with dark pigments on the body except in the grooves, upper tentacles long, grey and slightly transparent, lower tentacles very short.

In Europe often in dry and open habitats, prefers shady habitats in Portugal, between plants and under stones, or hidden in the soil. In N Africa in all regions covered by cultivated areas, grasslands and semideserts, up to 5-10 km distant from the coast, exceptionally up to 80 km or more, usually below 800m, with more isolated distribution patterns in the south, not in true deserts and on salt substrate. Not very frequent on sandy and loamy substrate. Mostly nocturnal, in France and N Africa.

Life cycle and decollation in France were studied by Brisson in 1753/1754 and by Gassies in 1842/1843 under laboratory conditions (Dundee 1986 obtained the same results in Louisiana, USA).
30-40 eggs (diameter 2.5 mm) are laid inside the soil between May and October, adult animals estivate inside the soil 1-3 months covering their eggs, hatched juveniles have initially 3 whorls, 6 whorls after 2 weeks, 9-10 whorls after 6 weeks. Decollation of the first whorls begins in the second month, snails hibernate mostly inactively inside the soil, then growth and decollation continues. Initially 3 whorls are lost (remaining 6 whorls), then 2.5 (remaining 4.5), later one whorl at a time (total number of lost whorls is 8), after 5 decollations full shell size and maturity is reached at the age of 11-13 months. Copulation of fully grown individuals takes place in the first active days in spring (also observed in N Africa), offspring from the previous year copulates in June shortly after the last decollation.
In the process of decollation, a first step consists in perforating the shell at a particular line from inside, then the animal withdraws itself from the upper whorls and builds a new tap, and finally the upper whorls are lost by external influences.
In N Africa snails estivate up to 10 cm deep inside the soil, under stones, rarely in roots systems of shrubs.

Brisson 1759, Gassies 1847, Moquin-Tandon 1855: 311, Germain 1930: 320, Kaltenbach 1951.

Reports for *R. decollata* should be checked whether the author was aware of the difference between *R. decollata* and *R. saharica*.

Rumina saharica Pallary, 1901

Rumina decollata var. *saharica* Pallary, 1901 in Pallary 1901: 141, Pl. 3

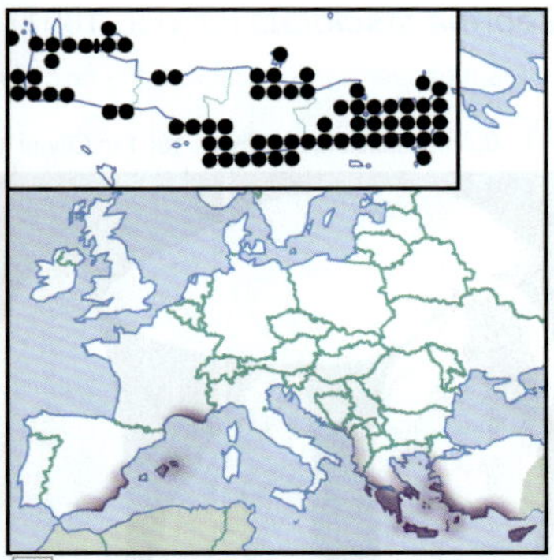

20-35 x 7-9 mm

(1-4) Greece: W Crete, Akrotíri, Moní Ag. Triáda; (5-8) Greece: island of Psérimos

N Africa, S Europe, Turkey

Greece: Síros
Panayotis Ovalis

W Turkey: near Bodrum

Shell small to medium-sized, subcylindrical, whorls flatter and shell diameter smaller than in *R. decollata*. Animal white with yellowish hue, or with dark pigments, tentacles long and transparent.

In Greece usually in cultivated lowlands, also in disturbed habitats, near road margins, in and around villages, not in high mountains. In N Africa at elevations of more than 1400 m.

Carr 2002, Mienis 2008.

Was not separated from *R. decollata* prior to 1993. Range uncertain. Data for Crete (insert map): Vardinoyannis (1994, unpublished thesis), unpublished data (HNC).

Punctum

Punctum Morse, 1864
in Morse 1864: 5, 27

Type species *Helix minutissima* Lea, 1841
(monotypy)
(= currently *Punctum pygmaeum* (Draparnaud, 1801))

Usually very small species.

Oviparous, eggs are relatively large sized and flattened like pills.

Worldwide

Punctum pygmaeum (Draparnaud, 1801)

Without gender agreement: *Punctum pygmaea* (Draparnaud, 1801)

Helix pygmaea Draparnaud, 1801 in Draparnaud 1801: 93 (*Helix pygmæa*)

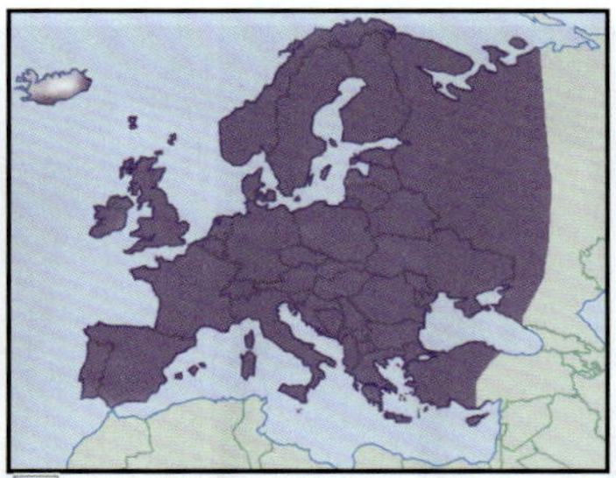

0.6-0.8 x 1.2-1.6 mm

Germany: Bayern, W Allgäu between Trögern and Weiler

Europe, W and N Asia

Ø 0.4 mm at 1 whorl, 0.65-0.7 mm (2 whorls), 1.2-1.3 mm (3 whorls)

Shell light horny brown, almost flat, extremely densely and regularly striated, silky shiny, 3-3.5 moderately convex whorls, aperture rounded, margin thin, not reflected and without lip, umbilicus wide (25 % of shell diameter).
Could be confounded with juvenile Pupilloidea, but they usually have a smaller umbilicus.

Mostly in litter layers of forests, characteristical for beech forests on acid soils, but usually more abundant on calcareous substrate, sometimes also in open habitats, wide ecological amplitude between wet habitats and warm and dry places. Not in heavily disturbed habitats. In Portugal under dead leaves and bark, under mosses and lichens on bark, also at old walls. In Switzerland up to 2500 m, in Bulgaria 1500 m. A frequent species, common nearly thoughout.

Individuals are hermaphrodites, but reproduction is also possible without partner. 1-16 (average 6) eggs (diameter 0.4-0.5 mm) are laid singly during a period of several days between May and September, juveniles hatch after 15-20 days, shell diameter of hatchlings 0.49 mm, snails grow quickly until reaching a shell diameter of 1.3-1.4 mm (sexual maturity is reached), less quickly afterwards but they never stop growing. Average age 200 days.

Gassies 1849: 105 (life cycle), Baur 1989 (life cycle).

Paralaoma

Paralaoma Iredale, 1913
in Iredale 1913: 380

Type species *Paralaoma raoulensis* Iredale, 1913
(subsequent designation by Wenz & Zilch 1959-1960: 204)
(= currently *Paralaoma servilis* (Shuttleworth, 1852))

Contained originally *Paralaoma raoulensis* Iredale, 1913, and *Paralaoma ambigua* Iredale, 1913.

Paralaoma servilis (Shuttleworth, 1852)

Helix servilis Shuttleworth, 1852 in Shuttleworth 1852 [May 1852]: 140
Helix caputspinulae Reeve, 1852 in Reeve 1854 [Oct 1852]: Pl. 133 No. 818 (*Helix caput-spinulæ*)

Spain: Aracena

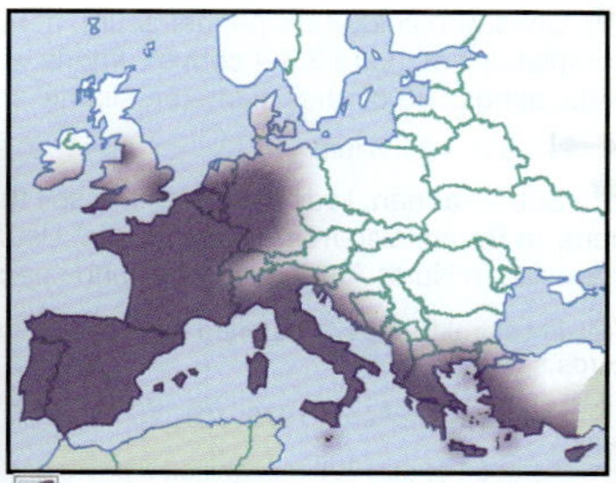

Possibly worldwide, or Australasian and introduced to Europe and Atlantic islands

1 x 1.5-2.0 mm

1.5-1.6 mm at 3 whorls, 1.9-2.0 mm at 3.5 whorls

Shell golden brownish, quite flat, ribs widely spaced and regular, 3.5 whorls, aperture almost circular, umbilicus wide (20-25 % of shell diameter). *Gittenbergia sororcula* is larger at 3-3.5 whorls.

In soil litter, relatively humid and shady habitats. In Britain in disturbed habitats such as at the side of roads and gardens.

Formerly known under names of *Pleuropunctum micropleuros* (Paget 1854), *Toltecia pusilla* (Lowe, 1831) and *Paralaoma caputspinulae* (Reeve, 1852). In Germany introduced to bamboo plantations (Schmid 2002).

Lucilla

Lucilla Lowe, 1852
in Lowe 1852: 115 *Helix (Lucilla)*

Type species *Helix scintilla* Lowe, 1852
(monotypy)

Helicodiscus Morse, 1864 in Morse 1864: 5, 24, 25, type species *Helix lineata* Say, 1817 (monotypy) (= currently *Helicodiscus parallelus* (Say, 1821) from North America)
Hebetodiscus Baker, 1929 in H. B. Baker 1929: 86, type species *Helicodiscus singleyanus inermis* Baker, 1929 (monotypy) (= ? currently *Lucilla singleyana*)

Shell almost exactly like that of *Hawaiia minuscula*, but glossy and with almost smooth shell surface, occasionally with very fine spiral lines. Retractor only at penis and not at penis and epiphallus (as in *Hawaiia*). Marginal teeth of radula are serrated (blade-like in *Hawaiia*). The last difference may be the best method to distinguish *Lucilla* and *Hawaiia*.

North America and Europe

Turgeon et al. 1998, Örstan (unpublished 2007, differences to *Hawaiia*), Horsák et al. 2009 (gave a guide to distinguish both European species).

Lucilla has precedence. There are many species classified in this group (*Helicodiscus*) in North America, two in Europe. One or both European species are thought to have been introduced from North America long time ago and could be conspecific with some North American species (the type locality of *singleyana* is in Texas, of *inermis* in Tennessee, of *scintilla* in Madeira).
The differences between the two forms found in Europe were not well understood prior to Horsák et al. 2009, but various previous authors had distinguished two species since the 1960s (called *L. singleyana* and *L. scintilla* by Falkner et al. 2002: 118). Some differences were found in anatomical studies in North America and Europe with preliminary and possibly premature conclusions on the generic (North America) or specific level (Giusti 1976: 309, new species in Italy). It seems that the forms of both continents have not been analysed anatomically in a combined study. Some species are known from Pleistocene deposits in Europe, but it remains unknown if these represented the origin of one or both recent European species (or if their ancestors were later introduced from North America).
Current consensus is that molecular studies are needed to sort out the relationships within this group and that classifying all species (20 or so in North Ameica) in more than one genus *Lucilla* would not be firmly established or justified at the current state of scientific research (Hotopp et al. 2010: 28).

Lucilla scintilla (Lowe, 1852)

Helix scintilla Lowe, 1852 in Lowe 1852: 115 Type locality Madeira.
Helicodiscus singleyanus inermis Baker, 1929 in H. B. Baker 1929: 86 Type locality in Tennessee, USA.

Germany: Nordrhein-Westfalen, near Opladen

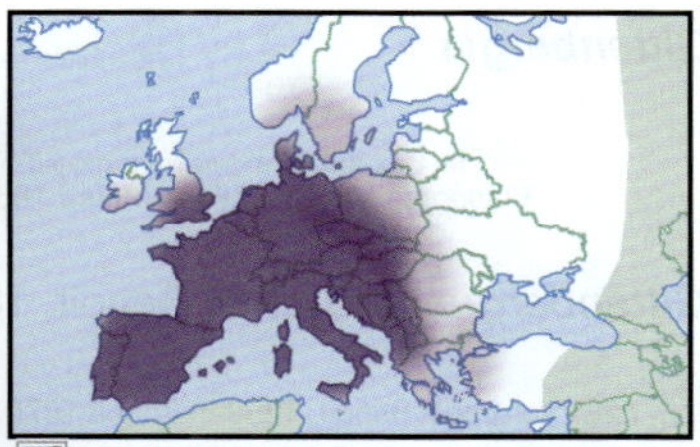

North America, in Madeira and Europe possibly introduced

Up to 1.1 x 2.2 mm

1.5 mm at 3 whorls, 2.0 mm at 3.5 whorls

Shell with distinctly yellowish periostracum layer, not as transparent as *Vitrea*, shiny, finely and irregularly striated, shape slightly conical, up to 3.5-4 convex whorls with relatively deep suture, aperture simple, margin straight, umbilicus wide, almost 1/3 of shell diameter, slightly deeper than in *L. singleyana*. The animal is blind.

Subterranean, lives deeper in the soil than other land snails. Usually only found when large quantities of soil are analysed, or in river deposits. In North America in highly disturbed anthropogenic habitats, in gardens and lawns, at roadsides.

Distribution in Europe weakly known. Reliable records only after 2009.

Lucilla singleyana (Pilsbry, 1889)

Without gender agreement: *Lucilla singleyanus* (Pilsbry, 1889)

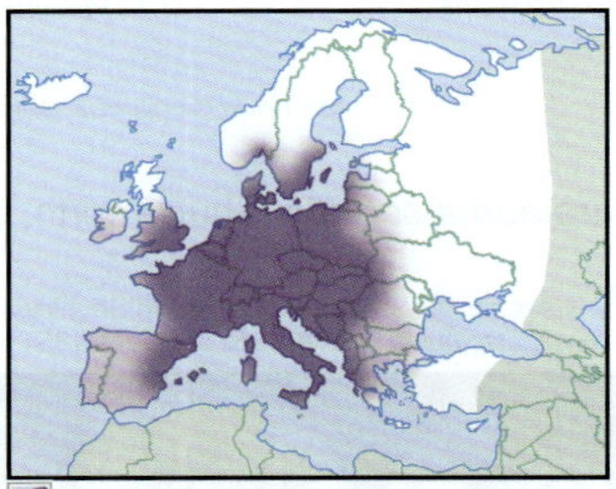

Zonites singleyanus Pilsbry, 1889 in Pilsbry 1889: 84 Type locality in Texas, USA.
Date of publication was 14 May 1889 (given on p. [6] of the journal volume, one page behind the title page, indicating that at this date a copy was presented to the meeting of the Academy, that this was a copy of the print run, and that from this moment on the copies were sent to the subscribers), previous statements for "1890" were incorrect.

Shell with colourless periostracum, not as transparent as *Vitrea*, shiny, finely and irregularly striated, shape completely flat, up to 3.5-4 convex whorls with relatively deep suture, aperture simple, margin straight, umbilicus wide, almost 1/3 of shell diameter, slightly shallower than in *L. scintilla*. The animal is blind.

Up to 1.2 x 3 mm

Subterranean, lives deeper in the soil than other land snails. In Switzerland known from vineyards and gardens, in Britain also from greenhouses. Usually only found when large quantities of soil are analysed, or in river deposits. In North America a grassland species, particularly in rocky grasslands.

North America, in Europe introduced (?)

In central Europe threatened by intensive land use and contamination of vineyard soils. In Britain known since 1975, only few scattered records since then, no fossil records.

Kerney & Cameron 1979: 101, Turner et al. 1998: 242, Kerney 1999: 116, Sverlova 2006: 126, Horsák et al. 2009.

Turgeon et al. 1998 recognized *inermis* and *singleyanus* (they should be in *Lucilla* if they are congeneric with *scintilla*) as two independent North American species. European reports from before 2009 are expected to include *L. scintilla*.

Acanthinula

Acanthinula Beck, 1847
in Beck 1847: 122

Type species *Helix aculeata* Müller, 1774
(subsequent designation by Albers & Martens 1860: XIII, 100)

Established without description, but with 2 species included (*Helix aculeata*, *Helix lamellata* Jeffreys, 1830).

Acanthinula aculeata (Müller, 1774)

Helix aculeata Müller, 1774 in Müller 1774: 81

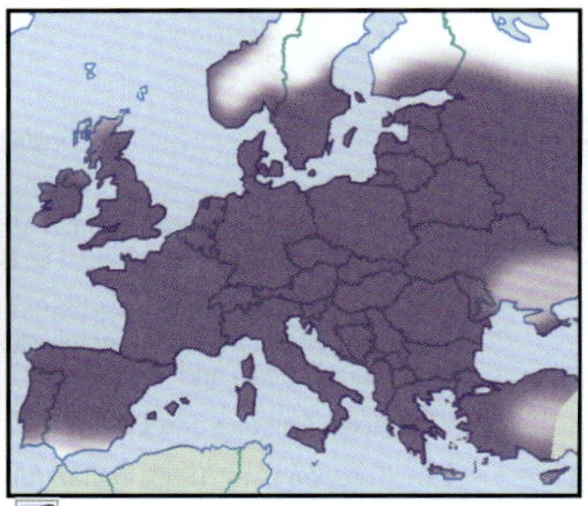

Up to 2 x 2 mm

Portugal: below Castro Mouro near Sintra

N Africa and Europe to central Russia and Transcaucasia

Periostracum ribs with characteristic spines.

In deciduous forests and under shrubs, in leaf litter and under fallen timber in shady places, often on calcareous substrate, also directly on overgrown limestone rocks in forests. Requires a minimum of humidity in the south. In Liechtenstein up to 1900 m, rare above 1200 m, in Bulgaria up to 1500 m.

Common in some regions and rare in others. In Britain slightly declining due to habitat loss by intensive farming, in Austria and Bayern populations have decreased due to habitat destructions by forest management.

Gittenbergia

Gittenbergia Giusti, Castagnolo & Manganelli, 1985
in Giusti et al. 1985: 84, 88

Type species *Helix sororcula* Benoit, 1859
(monotypy)

No type species originally designated, no statement that the name was new.
Planogyra Morse, 1864 in Morse 1864: 5, 24, type species *Planogyra astericus* (Morse) from N America (monotypy)

Gittenbergia sororcula (Benoit, 1859)

Helix sororcula Benoit, 1859 in Benoit 1875 (Ill. Sist.): 148-149

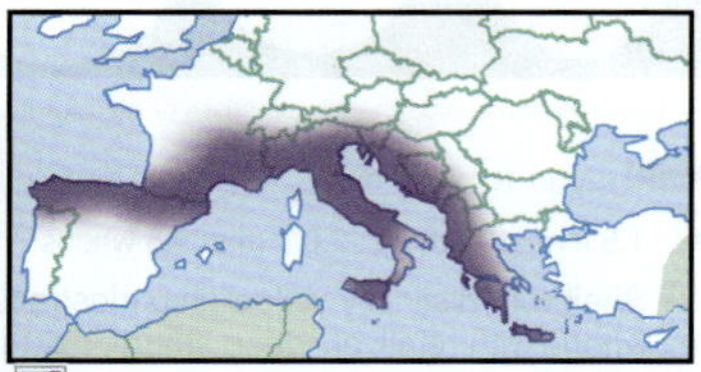

1.6-1.9 x 2.3-2.7 mm

Italy: (1-3) Abruzzo, L'Aquila, Prati di Tivo, (4-7) Brescia, Lago di Garda, Gargnano (1-3) Angelo Vannozzi

S Europe, Algeria
Scattered distribution with many gaps, probably an old fauna element.

Ø 1.7-2.0 mm at 3 whorls, 2.4-2.7 mm at 3.5 whorls

Shell weakly yellow, with fine and widely spaced ribs, very fine and regular intercostal riblets, spiral lines visible under high magnification, 4 convex whorls, umbilicus wide and deep, 1/4 of shell diameter. Differs from *Vallonia costata* in its usually darker colour, well visible spiral shell sculpture, slighly more densely coiled whorls and absence of *Vallonia*'s strongly reflected margin. *Paralaoma servilis* is smaller at 3-3.5 whorls.

Between and below rocks and rock rubble, well hidden, mainly in mountains, seems to prefer calcareous substrate. In NW Spain in moderately humid habitats in forests between rocks and soil with dry leaves or branches of shrubs. In the Apennines not rare above 1500 m. The ribs help the snail not to get washed down the leaves when it rains. In NW Spain abundant where it is present.

Kerney et al. 1983: 132, Giusti & Manganelli 1984: 86, Rolán & Otero-Schmitt 1988: 113 (NW Spain), Gittenberger 1993: 322 (Greece), Manganelli et al. 1995: 16 (Italy), Fehér & Erőss 2009 (Albania).

Prior to 1985 classified in *Planogyra* or *Vallonia*.

Plagyrona — *Plagyrona* Gittenberger, 1977 in Gittenberger 1977: 297

Type species *Helix debeauxiana* Bourguignat, 1863 (original designation)
(= currently *Plagyrona placida* (Shuttleworth, 1852))

Plagyrona placida (Shuttleworth, 1852)

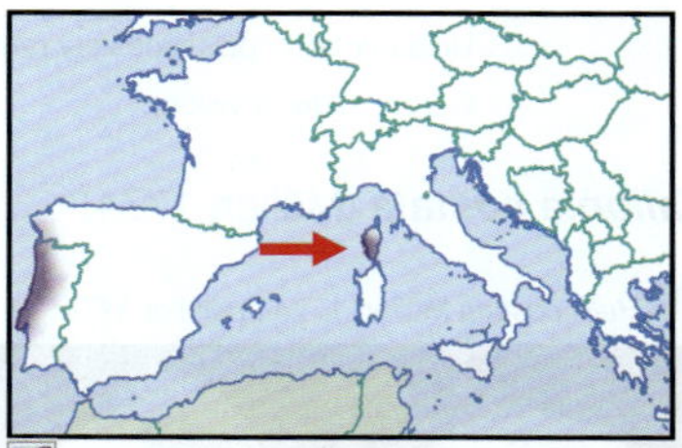

S Europe, NW Africa

Helix placida Shuttleworth, 1852 in Shuttleworth 1852: 140

Shell translucent, with very fine spiral lines on the inital whorls and widely spaced radial riblets on the last whorls (but no major ribs), 3.5 whorls, last whorl not much descending, apertural margin sharp and not thickened, umbilicus wide, 1/4 of diameter.

1.2-1.5 x 1.8-2.2 mm

In forests.

Gittenberger 1977: 297, Falkner et al. 2002: 107.

Spermodea — *Spermodea* Westerlund, 1902 in Westerlund 1902: 90 *Aulaca (Spermodea)*

Type species *Helix lamellata* Jeffreys, 1830 (subsequent designation by Pilsbry 1922-1926 [Mar 1926]: 186)

Contained originally *Helix lamellata* and *Helix spermatia* Castro, which Westerlund indicated both as types.

Spermodea lamellata (Jeffreys, 1830)

Helix lamellata Jeffreys, 1830 in Jeffreys 1830: 333

Germany: Schleswig-Holstein, Schwabstedter Geest

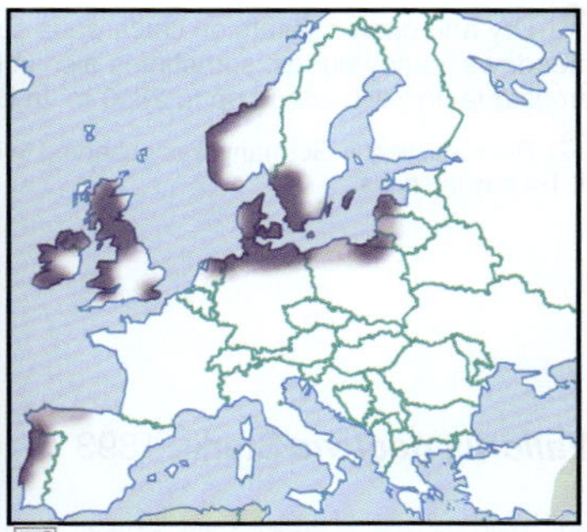

N and W Europe

Up to 2 x 2 mm

Shell yellow, with caracteristic semispherical shape like a bee basket, with fine, white and very regular ribs, silky shiny, ribs not much weaker on lower side, apertural margin simple and breaks easily, no folds inside, umbilicus narrow. Juveniles more whitish with weaker ribs at lower side, umbilicus relatively wider.
Falls immediately down when percepting vibrations.

Old deciduous woodlands, humid habitats in leaf litter above humid or wet spots, under rottning tree trunks and branches. Tolerates non-calcareous substrate. In Britain a useful indicator of old woodland that has never been clear felled, but can also live in partially replanted or managed forests. In Ireland frequent in old semi-natural forests, mostly found in deep leaf litter under beech trees. In NW Spain in leaf litter layers of humid habitats in mixed oak and *Eucalyptus* forests.

EU Red List: Near Threatened NT. Endangered in Ireland and Niedersachsen. Should at least be considered as Endangered in central and S England. Many populations in N Britain are potentially threatened. Many previously known populations in central and S England have been destroyed since 1880 by clearance and forest management. In Ireland 70 % of the populations were lost in the past 30 years. In Sweden over 50 % were lost in 50 years. In NW Spain rare, but abundant where present.

Kerney et al. 1983: 131, Rolán & Otero-Schmitt 1988: 112, Jungbluth et al. 1989: 114, Falkner 1990: 148, Anderson 1996: 8, Kerney 1999: 111, De Oliveiro 2007: 29, Sysoev & Schileyko 2009: 37.

Vallonia

Vallonia Risso, 1826
in Risso 1826: 101

Type species *Vallonia rosalia* Risso, 1826 (monotypy)
(= currently *Vallonia pulchella* Müller, 1774)

Shell small to minute (diameter 1.7-4.0 mm), remarkably wider than high (height 0.84-2.16 mm), suture deep and distinct, umbilicus always open and wide, aperture oblique, usually with well-expressed lip and never with teeth or folds. Colour always light, from white to greyish yellow or greyish brown.

Usually open habitats from wet to relatively dry, at the soil, calciphile, usually not in forests. Since Eocene.

Uniparental, copulations very rare.

Gerber 1996 (revision).

Vallonia costata (Müller, 1774)

Helix costata Müller, 1774 in Müller 1774: 31

Sweden: Öland

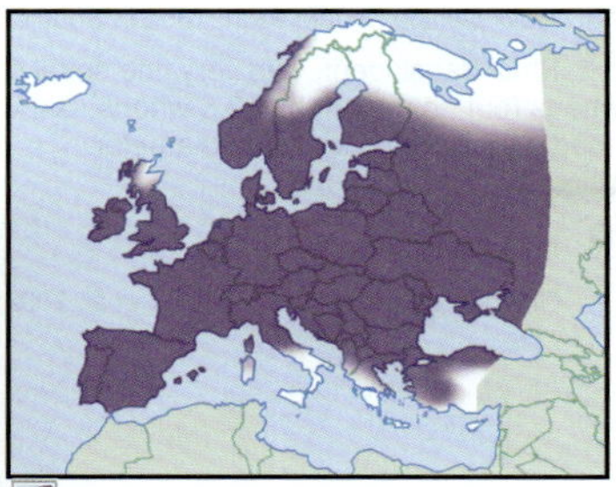

N Africa and Europe to central Asia

Slovakia: Slovenská Ľupča, Príboj — Jozef Grego

1.1-1.6 x 2.2-2.7 mm

Ø 2.0-2.2 mm at 3.0 whorls

Shell flat with regular and widely spaced conchin/periostracum riblets (difficult to recognize in weathered shells - inside the umbilicus the ribs are usually better preserved), striated between the ribs, 3.2-3.3 whorls, last whorl descending, aperture oblique and in adult specimens broadly reflected. No spiral sculpture as in *Gittenbergia sororcula*.

Dry and open habitats on calcareous subground, grassy and sunny slopes, rock rubble, stone walls, in short meadows, sandy dunes, sometimes also in light forests, tolerates light shading, rare in swamps. Wide ecological amplitude. In Switzerland up to 2800 m, in Bulgaria 1300 m, in N Britain mainly costal and not in the highlands.

Reproduction in Germany was observed in March, copulation seems to be rare, flattened eggs (0.8 mm) are laid singly at 1-4 day intervals.

Frömming 1954: 41, Falkner 1990: 150, Kerney et al. 1983: 125, Gerber 1996: 156.

Vallonia declivis Sterki, 1893

Vallonia declivis Sterki, 1893 in Sterki 1893: 251, Pl. 32

Germany: Neckar river deposit near Neckarsteinach

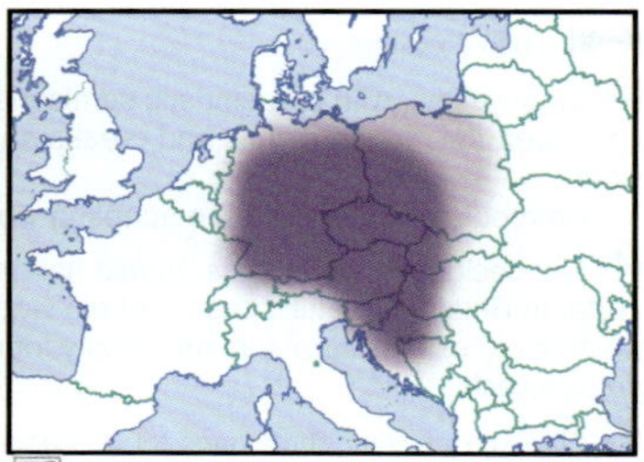

Central Europe

1.3-1.8 x 2.6-3.0 mm

Shell larger than other European *Vallonia* species, lip not distinctly visible from above.

Open and humid meadows, often in river plains, in Switzerland up to 1400 m.

EU Red List: Near Threatened NT. Extremely rare, very scattered distribution. Critically Endangered in Switzerland, Austria and Germany.

Vallonia enniensis (Gredler, 1856)

Helix pulchella var. *enniensis* Gredler, 1856 in Gredler 1856: 56

Belarus: Kletsk SW of Minsk

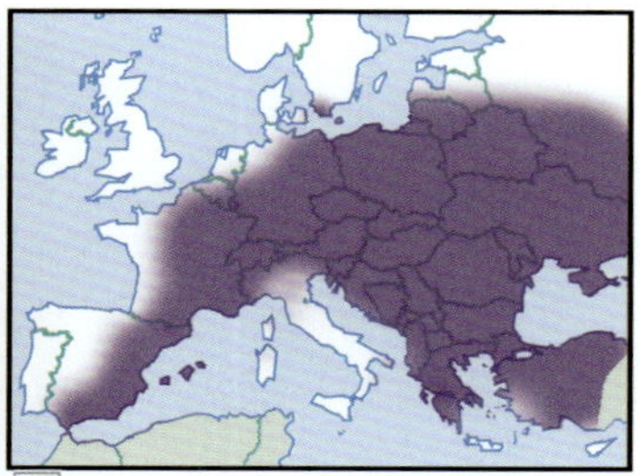

S to E Europe, Turkey and Egypt

1-1.5 x 2.2-2.5 mm

Shell very much like *V. pulchella*, but with very faint and fine ribs (45-60 on last whorl, much more densely ribbed than *V. costata*).

Humid and wet habitats in lowlands, on humid and uncultivated meadows and in calcareous swamps, sunny hills with springs in highlands. Frequently found as fossils. In Switzerland up to 1000 m.

Critically Endangered in Switzerland and Germany, Endangered in Austria. Threatened by habitat destruction, the scattered populations are usually isolated, extirpation of populations is usually definite because the species cannot repopulate the area from neighbouring areas.

Vallonia excentrica Sterki, 1893

Vallonia excentrica Sterki, 1893 in Sterki 1893: 249, Pl. 32

0.9-1.4 x 2.0-2.2 mm

Germany: Bayern, Neustadt an der Aisch

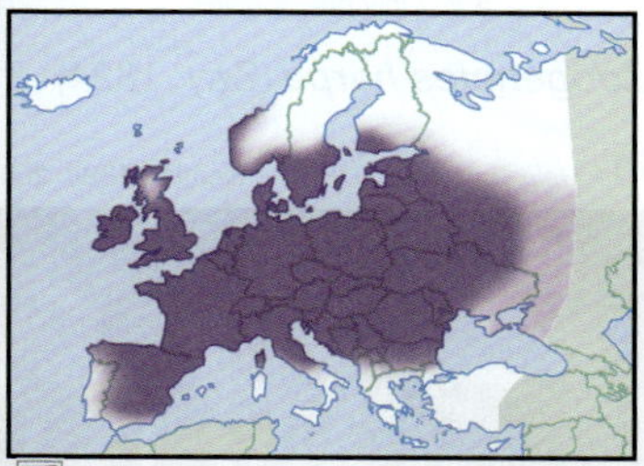

E North America (Mississippi to Newfoundland), Europe

Shell much like *V. pulchella*, usually smaller (3 whorls only), weakly coloured and transparent, last quarter of last whorl remarkably increased giving the shell an elliptical shape, lip not so distinctly visible from above as in *V. pulchella*.

Open and usually dry habitats, short meadows, rock rubble, sandy dunes, not in forests and swamps. Tolerates less calcareous soils than *V. costata*. In Switzerland up to 1500 m, in Bulgaria 1200 m, rare in Scottish highlands. In Britain more frequent than *V. pulchella* and more tolerant against human disturbance than *V. costata*.

Copulations are rare and take place only between euphallic animals. Eggs are laid one by one every few days, not in clutches, at the soil between soil particles or in small crevices. Juveniles hatch after 1-2 weeks, depends on temperature, full shell size is reached after 7 weeks (25 °C). 2 generations per year in New Zealand, hibernating snails reach maturity in spring, snails hatched in spring reach maturity in autumn.

Vallonia pulchella (Müller, 1774)

Helix pulchella Müller, 1774 in Müller 1774: 30

1-1.5 x 2.0-2.5 mm

Turkey: Antalya, Belek near Serik

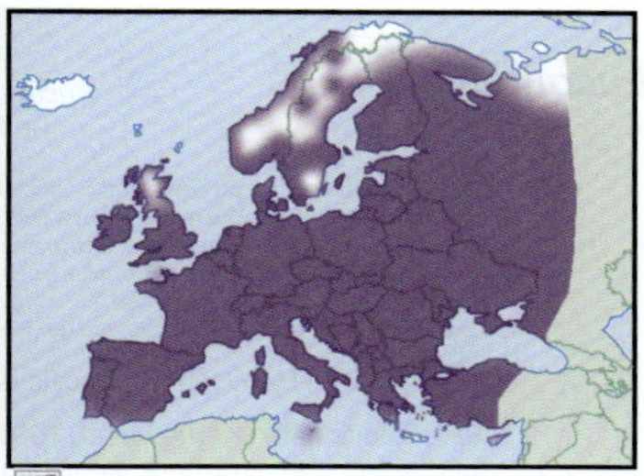

Northern Hemisphere

Czech Republic: Olomouc
Michal Maňas

2.0-2.2 mm at 3.0 whorls

Shell nearly circular, white, finely and irregularly striated, 3.2-3.3 whorls, slightly more elevated whorls than *V. costata*, last whorl wider immediately near aperture, not much descending, aperture not much oblique, lip weaker than in *V. costata*.
Animal milky white, tentacles short, foot rounded at posterior end.

In meadows of open habitats on calcareous subground, prefers swampy meadows, generally on more humid soils than *V. costata* and *V. enniensis*, sand dunes, sometimes also on dry meadows and rock rubble, not in forests. In Turkey slightly more tolerant agaist heat than *V. costata*. In Switzerland up to 2000 m, rarely above the timberline, in Bulgaria 1200 m. Rarely in British and Scandinavian highlands, in N England and Scotland only at few spots in coastal zones.

Copulations seem to be rare, some 20 single eggs (0.7-0.8 mm) are laid at the soil among soil particles or in small crevices (not in clutches), juveniles hatch after 15-20 days in Austria and France (12 days at 21-22 °C), in nature 15-20 days were observed, full shell size is reached after 2 months (21-22 °C), egg laying is initiated one week later, under optimal conditions up to one egg per day, during several months. Snails die after having finished the egg laying period. There may be 2 generations per year, snails seem to be able to hibernate. In France egg laying was observed between July and September and maturity in the beginning of the second year, possibly this was the second generation.

Gassies 1849: 102 (life cycle), Germain 1930: 382, Frömming 1954: 39, Falkner 1990: 150, Kerney et al. 1983: 127, Gerber 1996: 48.

Vallonia suevica Geyer, 1908

Vallonia suevica Geyer, 1908 in Geyer 1908: 324, Taf. 3-4

1-1.5 x 2.2-2.5 mm

Germany: Neckar river deposits near Neckarsteinach

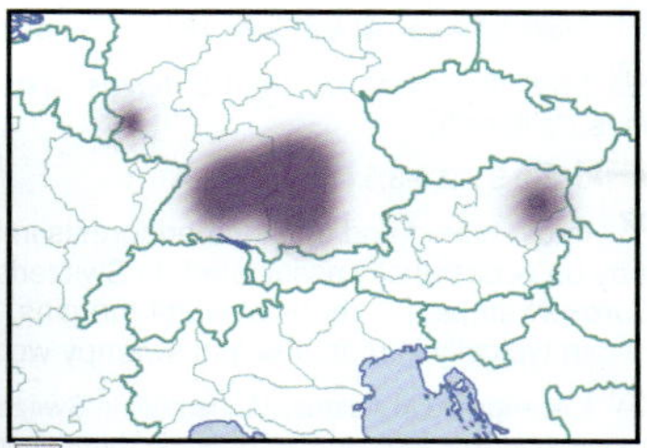

S Germany, Austria
Neckar valley and some other valleys of Württemberg (Ammer, Tauber, Kocher, Nagold, upper Donau), rare and scattered occurrences in other regions. Probably the only endemic mollusc species in Garmany.

Shell much like *V. costata*, but without ribs, only very fine streaks, weakly shiny.

Moderately dry and fresh meadows (Arrhenateretum salvietosum, with *Sanguisorba minor* and *Centaurea scabiosa*), not intensively cultivated, mostly in river plain lowlands, lives together with *V. costata* and *V. excentrica*.

EU Red List: Endangered EN. Critically Endangered in Bayern. Rapidly declining due to habitat destruction, threatened by modern riverside management (flood preventions) and occupation of river plain flatlands for industrialised agricultural use. Most populations have been lost. Does probably not live in Austria any more.

Zoogenetes

Zoogenetes Morse, 1864 in Morse 1864: 5, 32 *Zoögenetes*

Type species *Helix harpa* Say, 1824 (monotypy)

Northern Hemisphere

Only one species.

Zoogenetes harpa (Say, 1824)

Helix harpa Say, 1824 in Keating 1824: 256, Pl. 15

Norway: Lillehammer

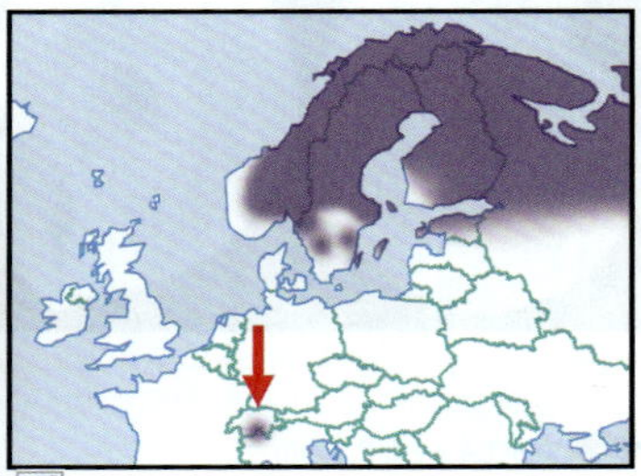

Northern polar regions and high alpine mountains (isolated in Rocky Mountains, Alps, Caucasus)

3-4 x 2.4-3 mm

Shell conical, like a juvenile enid, quite brown, only weakly calcified, periostracum builds fine and almost regular riblets, 3.2-4 whorls, aperture simple.

Coniferous forests, often on non-calcareous subground with *Vaccinium*, in Switzerland on soils with pH 3.5-5.5. Up to 2200 m in the Alps. May be dispersed by birds.

Kerney et al. 1983: 133, Turner et al. 1998: 177, Sysoev & Schileyko 2009: 37.

Euconulus

Euconulus Reinhardt, 1883 in Reinhardt 1883: 86 *Conulus (Euconulus)*

Type species *Helix fulva* Müller, 1774 (subsequent designation by Baker 1922: 50)

Original description contained 5 species *fulvus*, *praticola*, *pupula*, *pustulinus* and *phyllophilus*; *incertus* was included with doubts. Established as a subgenus of *Conulus* Fitzinger, 1833 (not *Conulus* Leske, 1778 (Echinodermata)), not a new replacement name because there was no expressed statement referring to the act of replacing a name, and because *Conulus* was the used generic name for the species of this subgenus and two other subgenera (strongly suggesting that *Euconulus* was not meant to replace *Conulus*, otherwise the generic names for the other two subgenera should have been *Euconulus* (= *E.*) and not *Conulus* (= *C.*)).
Euconulus is potentially threatened if an author before 1883 mentioned both spellings *Petasina* Beck, 1847 and *Petasia* Beck, 1837 (non Stephens, 1828, Lepidoptera), so that the emended name *Petasina* would become available and get *Helix trochiformis* Montagu, 1803 (currently in *Euconulus*) as its type under Art. 33.2.3, designated by Gray 1847.

Northern Hemisphere, Pacific islands

Traditionally (Kerney et al. 1983: 200) two species were recognized in Britain and parts of Europe, *E. fulvus* and *E. alderi*. Both seem to be easily distinguishable in Britain where *alderi* has been used for a darker and smaller woodland species. Outside Britain the situation seems to be more difficult, *alderi* seems to have been used for a larger species from open wetland habitats (Colling et al. 2008: 114).
The classification of *Euconulus* was distorted by Falkner, Ripken & Falkner 2002: 120 who listed 4 species (*fulvus*, *trochiformis*, *praticola* and *callopisticus*) in a new classification based on preliminary results of a detailed study that was claimed of being prepared at that time (Proschwitz, Falkner & Ripken). They used the following names:

Helix trochiformis Montagu, 1803 in Montagu 1803: 427 Type locality Wiltshire and Devonshire, S England.
Montagu 1803 gave a figure (Pl. 11 Fig. 9) and cited as references *Trochus terrestris* Pennant, 1777 (as used and described by Mendes da Costa 1778: 35), Lister 1678 (p. 123), "Phil. Trans. No. 105 f. 9", and *Helix trochulus* Müller, 1774 (= juvenile *Helix obscura* Müller, 1774, currently *Merdigera obscura*).
Zonites callopisticus Bourguignat, 1880 in Servain 1880: 30 Type locality near Lyon, France.
Servain (1880: 29) mentioned the name and attributed it to "Bourguignat, in Sched., 1875", and then on pp. 30-33 quoted a text written by Bourguignat, who described in detail 7 species, and where the name *callopisticus* was also connected with the 1875 date. The name had most probably not been published in 1875.
Hyalina praticola Reinhardt, 1883 in Reinhardt 1883: 40 Type locality Brandenburg, N Germany.

In the years that followed such a detailed study was not published. Bank, Falkner & Proschwitz 2007: 49 simply repeated the unsubstantiated arguments which were not based on a sound and published scientific study.
The logical consequence was that subsequent authors did not accept the proposed new classification. Those who tried to test it came to different results. Anderson 2005 was not able to confirm the validity of the new classification in Britain. Audibert 2010 (unpublished) observed that the given criteria were not applicable for the two forms observed in France.
Wiktor 2004: 217 maintained *fulvus* and *alderi* for the two distinguishable forms in Poland, Anderson 2005 for Britain and Ireland, Waldén 2007: 122 for Sweden, Egorov 2008: 67 for Latvia. In all major countries except Germany the new system failed to gain acceptance.
We have the same situation as before 2002: the genus needs a revision.

Euconulus alderi (Gray, 1840)

Helix fulva var. *alderi* Gray, 1840 in Turton & Gray 1840: 148 (*Helix fulva* var. 2 *Alderi*) Type locality in Great Britain.

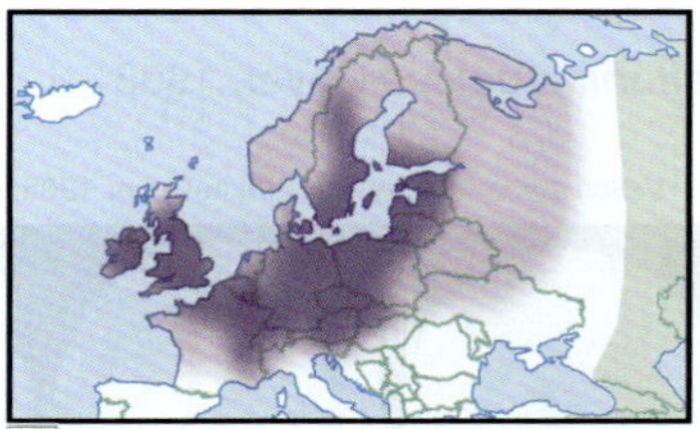

Europe or Holarctic

Shell slightly darker than *E. fulvus*, spire more elevated, lower whorls not very shiny, with well marked spiral lines ("sillons").

2-2.5 x 2.8-3.5 mm

In old humid and undisturbed forests in the leaf litter layer, generally wetter than for *E. fulvus*, with which it may be occasionally associated. In Switzerland restricted to wet habitats, characteristically in semi-europhic *Carex* swamps and reeds at water margins, up to 2500 m. Lives typically together with *Vertigo antivertigo*. In Britain typically in marshes and swampy woodland.

Endangered in Austria, Vulnerable in Switzerland, Niedersachsen, Rheinland-Pfalz and Bayern. Obviously rare and local, in France only known from 3 localities, more frequent but scattered in Switzerland. Threatened by continuous destructions of moist habitats, drainage and fertilization. In Britain obviously widespread but probably also declining.

The identity of this species has been recognized since the 1970s.
Falkner et al. 2002 intended to classify *alderi* as a synonym of *trochiformis*, provoking a change of the name *alderi* to *trochiformis*. Most authors (Manganelli et al. 1995: 22, Schütt 2001: 290, Waldén 2007: 124) have regarded *trochiformis* as a synonym of *fulvus*.
In the light of the situation that *alderi* had previously been used for various taxa, it would be a good idea to use a different name for a new and well established concept, not *trochiformis*. But what Falkner et al. 2002 actually did was introducing a new usage of a previosuly unused name *trochiformis* without a sound concept behind. If *trochiformis* is currently used, it is not used for a well established taxon.
I would currently not recommend to use *trochiformis*, so I tend to agree with Anderson 2005 and Egorov 2008 that the status of the taxa can only be established after a detailed revision of the genus *Euconulus*.

Euconulus fulvus (Müller, 1774)

Without gender agreement: *Euconulus fulva* (Müller, 1774)

Helix fulva Müller, 1774 in Müller 1774: 56 Type locality Frederiksdal near København, Denmark.

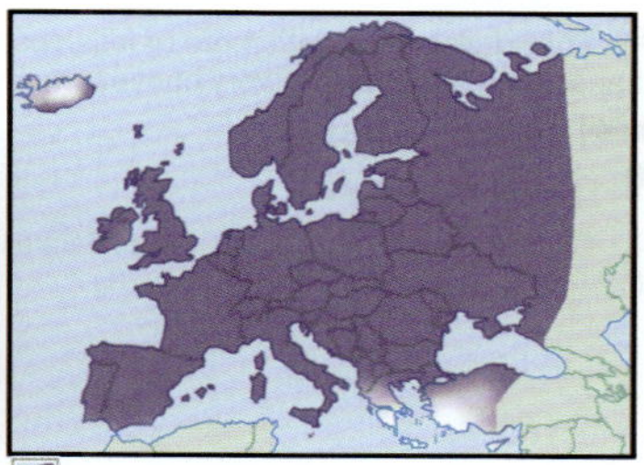

2.0-2.5 x 2.8-3.5 mm — Germany: Schleswig-Holstein, near Schleswig

Northern Hemisphere

Shell broader than high, yellowish brown, shell surface on upper side rather pale, lower side rather silky and smooth, no umbilicus (difference to *Pyramidula*), last whorl with a trace of a keel.

Usually in forests of all kind, prefers cool and humid soils with dead wood in montane and subalpine coniferous forests. Also in humid or dry meadows and grassy alpine habitats. Tolerates non-calcareous soils. In Switzerland up to 2900 m, very rare above 2400 m, in Bulgaria up to 1500 m.
One of the most frequent land snails in Switzerland, next to *Arianta arbustorum* and *Discus rotundatus*.

In S France 20-30 eggs are laid in April and May under leaves, juveniles hatch after 2 weeks (Gassies 1849: 75).

Records from before the 1970s are not reliable and included other species. Only one species lives at the type locality (Falkner et al. 2002: 120).

Pyramidula

Pyramidula Fitzinger, 1833 in Fitzinger 1833: 95

Type species *Helix rupestris* Draparnaud, 1801 (monotypy)

At limestone rock surfaces.

W Europe to Japan

Gittenberger & Bank 1996 (revision Europe), Martínez-Ortí et al. 2007 (revision Spain), Balashov & Gural-Sverlova 2011 (E Europe).

It seems that Gittenberger & Bank's 1996 results were not sufficiently clearly presented. They gave only coarse distributional data and blurry guides in the morphological diagnoses of the species, with overlapping character expressions. It was difficult to understand why presenting more detailed distributional data, for example in the form of maps, should have produced "an increasing confusion in the literature". Such maps would have been very useful to understand the concepts and would have helped to identify collection material.
Subsequent authors could not successfully determine collected or deposited material in S Europe. Until recently it has been difficult to identify species in *Pyramidula* records in local faunas. It seems that Martínez-Ortí et al. 2007 did not apply the same criteria and came to slightly different conclusions.

I analysed many specimens from the Haus der Natur Cismar (HNC) collection and material provided by Cédric Audibert, and tried to align my results with those presented by Gittenberger & Bank 1996 and Martínez-Ortí et al. 2007.
This did not always produce the same results. R. Bank in www.faunaeur-org (2004) reported *P. pusilla* and *P. rupestris* for Andorra, while Martínez-Ortí et al. 2007 suggested that the two species occurring in Andorra were *P. umbilicata* and *P. pusilla*. One sample in the HNC collection from Andorra contained *P. umbilicata* and *P. pusilla*, as interpreted by Martínez-Ortí et al. 2007.
Martínez-Ortí et al. 2007 identified samples from Els Ports (Tarragona) as *P. pusilla* and *P. rupestris*. I had one lot from there and identified it as *P. umbilicata*, I saw no differences between this and the samples from England and Andorra.
A more detailed study on the regional shell variation would be useful.

Pyramidula cephalonica (Westerlund, 1898)

Patula rupestris var. *cephalonica* Westerlund, 1898 in Westerlund 1898: 156

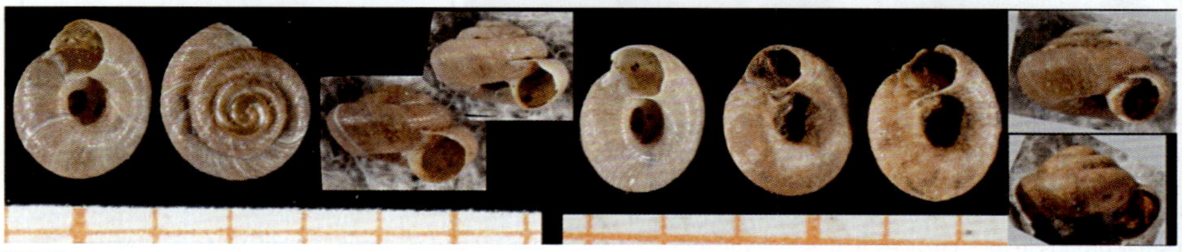

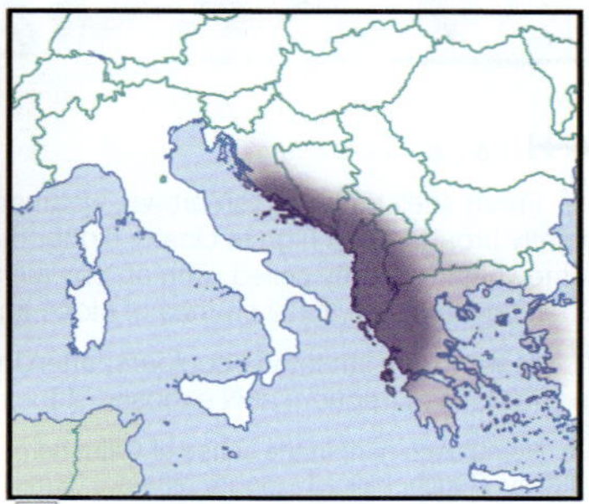

1.4-1.9 x 1.8-2.6 mm — (1-4) Montenegro: Lovcen pass, (5-9) Albania: Tushemisht near Pogradec

Shell conical, yellowish brown, whorls regularly increasing, with very prominent growth lines or riblets, broader than high, aperture relatively small, umbilicus wide, 1/3 of diameter or broader.

Gittenberger & Bank 1996: 72.

N Dalmatia to Greece and Turkey

Pyramidula chorismenostoma (Westerlund & Blanc, 1879)

Helix rupestris chorismenostoma Westerlund & Blanc, 1879 in Westerlund & Blanc 1879: 32, Pl. 1

2-3 x 1.7-2.5 mm — W Turkey: near Bodrum

Shell scalarid, last whorl separated from penultimate whorl, light to dark brown, with very prominent growth lines. Cannot be confounded, the only shell in Europe with such a detached last whorl.

At limestone rocks and among rock rubble, usually in open habitats, up to 700 m in Crete.

Central and S Greece, W Turkey

Vardinoyannis 1994 (unpublished thesis), Gittenberger & Bank 1996: 72, Örstan et al. 2005, Triantis et al. 2008: 474, Örstan & Welter-Schultes (unpublished, Dilek peninsula).

Pyramidula jaenensis (Clessin, 1882)

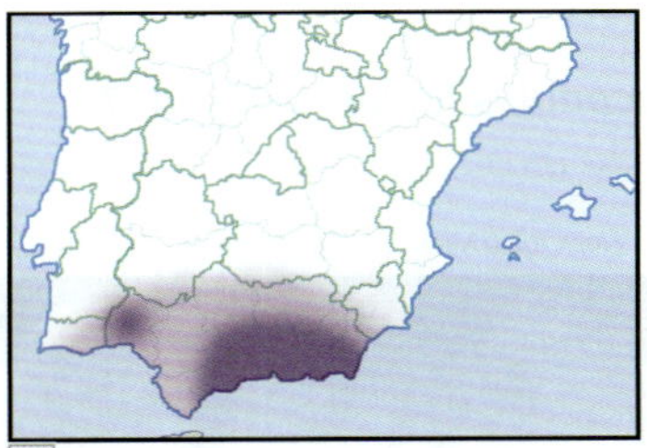

S Spain

Helix jaenensis Clessin, 1882 in Clessin 1882: 187 (*Helix (Patula) Jaenensis*)

Shell dark brown, with more or less obsolete growth lines, high conical, height exceeds diameter, more egg-shaped than the others, 4 very convex whorls, umbilicus narrow, less than 1/4 of diameter.

2.2-3.2 x 1.8-2.6 mm

On limestone rocks.

Ortiz de Zárate Rocandio & Ortiz de Zárate López 1961: 189, Gittenberger & Bank 1996: 72, Martínez-Ortí et al. 2007: 78.

Pyramidula pusilla Gittenberger & Bank, 1996

Helix pusilla mentioned by [Vallot] 1801: 5 (not available, not correctly published under Art. 8.1.1 and 8.1.2)
Pyramidula pusilla Gittenberger & Bank, 1996 in Gittenberger & Bank 1996: 74
Gittenberger & Bank 1996: 74 attributed the name to [Vallot] 1801. Martínez-Ortí et al. 2007: 78 attributed the name to Gittenberger & Bank 1996 and did not regard Vallot's name as available (their view that [Vallot] 1801 cannot be considered as published work is regarded as tenable here, in accordance with Reitano et al. 2007: 314). The name was cited before 1960 in synonymy lists under *Pyramidula rupestris*, for example by Drouet 1867: 41, Germain 1930: 385, but probably not used for a taxon before 1960.

1.4-2.5 x 1.8-2.8 mm — SE France: Ain, Belmont-Luthézieu, Champdossin

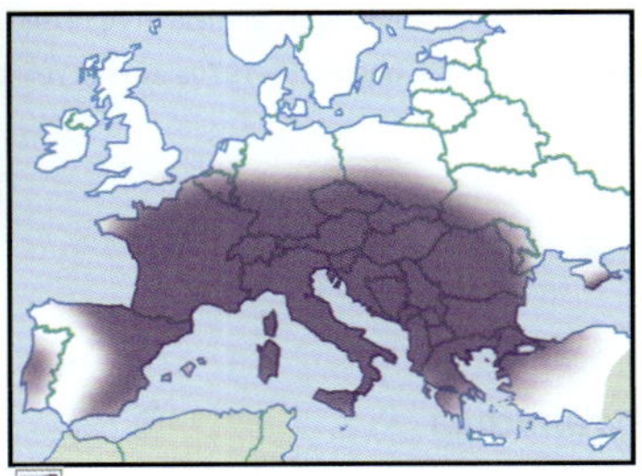

Mediterranean region, Europe

Shell dark brown, low conical, clearly broader than high, growth lines fine and narrow, sometimes weak, apical whorls regularly increasing, umbilicus 1/4 of diameter, in Spain locally 1/3.

Steep limestone rocks or rocks with calcareous compounds, up to 3000 m. Feeds on endolithic lichens, mainly dispersed by birds.

Ovoviviparous. Reproduction in summer. Parent individuals incubate 3-7 (usually 4-6) embryos of up to 0.8 mm size and 1.5 whorls. Juveniles are released singly after reaching this size.

Moquin-Tandon 1853: 226, Moquin-Tandon 1855: 192, Gittenberger & Bank 1996: 74, Turner et al. 1998: 167, Martínez-Ortí et al. 2007: 78, Reitano et al. 2007: 314 (Sicily, Boschi 2011: 249, Balashov & Gural-Sverlova 2011.

There are possibly older synonyms (Bank 2011: 16), this should be verified in a taxonomic study.

Pyramidula rupestris (Draparnaud, 1801)

Helix rupestris Draparnaud, 1801 in Draparnaud 1801: 71

1.8-2.8 x 1.8-2.8 mm — S France: near Nîmes

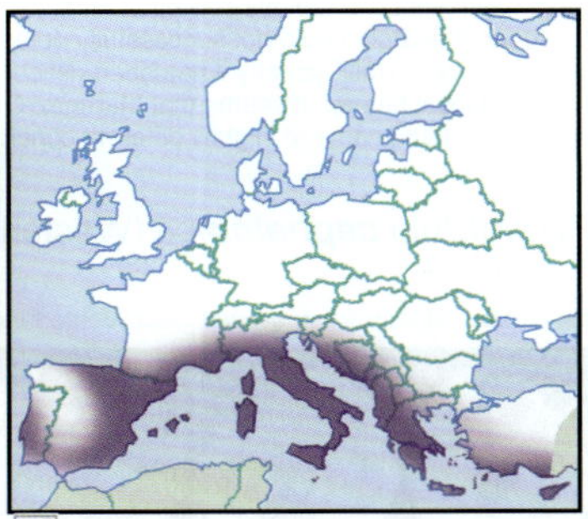

S Europe to Syria, Israel and possibly central Asia, Armenia, N Africa and Canary Islands

Fresh shell dark reddish brown, weathered greyish white, conical with nearly straight sides, varying from slightly broader than high to clearly higher than broad, with more or less obsolete irregular growth lines, first whorls somewhat irregularly coiled, with an obliquely bulging part, umbilicus narrow, 1/4 or less of shell diameter. Often covered with small crumbs of stone and remains of lichens. Animal black.

At warm-dry limestone boulders, often in small groups in rock crevices and very small rock grooves. At S exposed slopes direct insolation is tolerated, the species is less frequent at N exposed slopes. Only exceptionally distant from rocks.

Name used here in the sense of Gittenberger & Bank 1996: 76. Prior to 1996 this name was used for several species, but also after 1996 the name has been used inconsistently. Conchologically slightly different forms were reported by Balashov & Gural-Sverlova 2011 from Armenia and Kyrgyzstan.

Pyramidula umbilicata (Montagu, 1803)

Bulla umbilicata Montagu, 1803 in Montagu 1803: 222
ICZN Opinion 549: validated under the plenary powers as *Bulla umbilicata* from p. 222 (not *Bulla umbilicata* [Röding], 1798 which was suppressed).

1.2-1.7 x 2-3 mm — Ireland: Aran islands, Kilronan

Great Britain, Ireland, N Spain and Pyrenees

Shell greyish brown, low conical, much broader than high, with very fine growth lines, umbilicus wide, 1/3 of shell diameter.

Limestone rocks and walls built with lime mortar, in depressions and crevices. Active at night and at wet weather, feeds on lichens and algae. The species has good powers of passive dispersal.

Ovoviviparous.

Habitats at walls are threatened by removal or repair with Portland cement, food plants are affected by pollution and acid rain in E England.

Name used here in the sense of Gittenberger & Bank 1996: 76 and Martínez-Ortí et al. 2007: 79. Kerney 1999: 84 used the name *P. rupestris* for this form.

Klemmia

Klemmia Gittenberger, 1969 in Gittenberger 1969: 304

Type species *Klemmia sinistrorsa* Gittenberger, 1969 (original designation)

Shell sinistral, colourless, aperture simple and moon-shaped, no teeth.

Subterranean, usually found in caves.

Klemmia magnicosta Gittenberger, 1975

Klemmia magnicosta Gittenberger, 1975 in Gittenberger 1975: 275

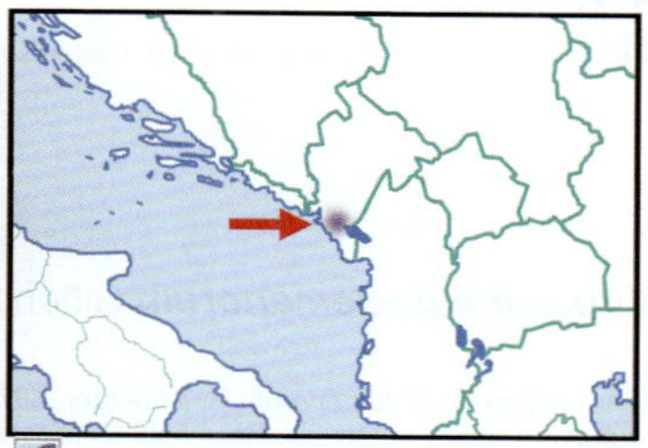

2.1 x 3.3-3.5 mm — Montenegro: Komarno — Museum Budapest HNHM — S Montenegro

Shell transparent, colourless, ribs more widely spaced than in *K. sinistrorsa*, umbilicus narrower, 1/10-1/11 of diameter. Living animals were found to be covered with soil substrate.

Klemmia sinistrorsa Gittenberger, 1969

Klemmia sinistrorsa Gittenberger, 1969 in Gittenberger 1969: 304

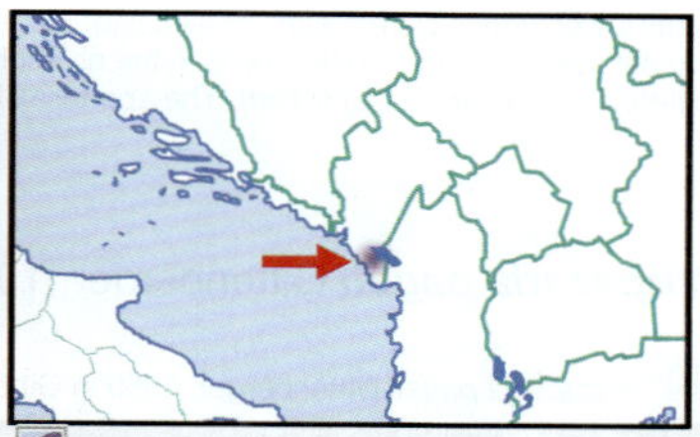

1.7-2.5 x 3.0-4.3 mm — Montenegro: Seoca village near Virpazar — S Montenegro

Shell, transparent, colourless, first whorl radially sculptured, then smoother, last whorls densely ribbed with riblets on and between the ribs (55-80 ribs on last whorl), aperture simple, moon-shaped, umbilicus regular, 1/7-1/9 of diameter.

Cave snail, shell of living animals covered with dust particles.

Virpazaria

Virpazaria Gittenberger, 1969 in Gittenberger 1969: 297

Type species *Virpazaria adrianae* Gittenberger, 1969 (original designation)

Shell dextral, colourless, aperture moon-shaped, margin connected at parietal side, forming a distinct wall.

In caves and rock crevices.

This genus seems to need a revision and a better understanding, not every new population that is discovered represents a new species. Some forms (*deelemanorum*, *adrianae*) seem to be more widespread and show relatively little local variation over area in size, rib density, umbilicus width, apertural teeth and prominence of the apertural margin. Others were only found at one locality and are either microendemics or not recent (*aspec.*, *ripkeni*, *backhuysi*). *V. pageti* was repeatedly found in river deposits near Herceg-Novi and could either be extinct or live somewhere in the nearby mountains.

Virpazaria adrianae Gittenberger, 1969

Virpazaria adrianae Gittenberger, 1969 in Gittenberger 1969: 298

Shell densely ribbed (40-50 ribs on last whorl), aperture moon-shaped, parietal apertural margin not regularly rounded, lower half of the margin not covered by parietal margin in a frontal view of the aperture, lip divided into 3 parts, umbilicus very wide (almost 1/3 of shell diameter).

Only known from caves.

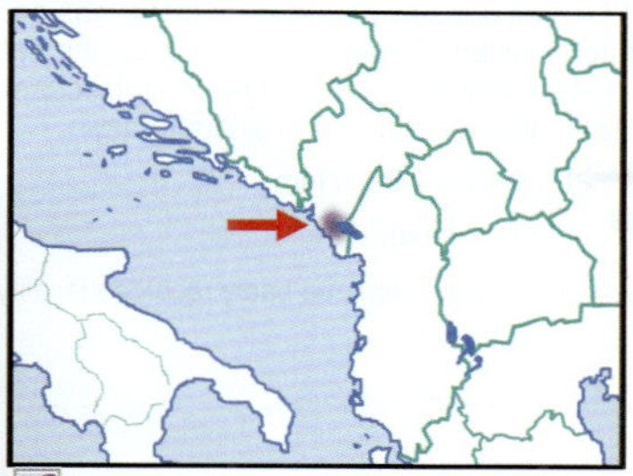

S Montenegro
This species was found at various localities.

1.6-1.9 x 3.0-3.5 mm — Montenegro: Začir — Museum Budapest HNHM

Virpazaria aspectulabeatidis Reischütz, Reischütz & Subai, 2009

Virpazaria aspectulabeatidis Reischütz, Reischütz & Subai, 2009 in Reischütz & R. 2009: 56
Subai's co-authorship was explicitly explained (p. 56). Appendix B 5 of the Code was probably violated (name is certainly not easily memorable).

Shell light yellow, translucent, ribbed (50 ribs/last whorl), 5-5.5 whorls, upper side moderately convex, lower side strongly convex, periphery bluntly angled in the upper section, aperture narrow, margin reflected and parallel at parietal and basal sides, flattened on upper side, umbilicus very narrow, 1/11 of diameter, 2/3 covered by the reflected columellar margin. From the lower side this looks like if the parallel apertural margins run in a = form from the periphery directly into the umbilicus. This is a unique character.

2.2 x 3 mm

Not found in caves, but in limestone rock crevices and holes at S exposed mountain slopes, up to 500 m.

Subai 2009: 109: Fig. 9 [sic, not 10].

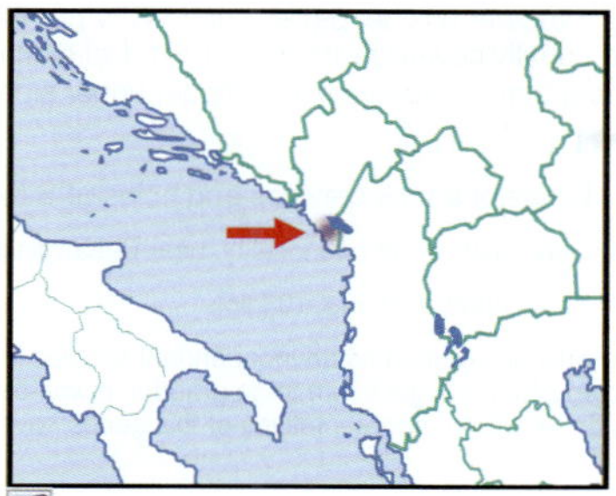

S Montenegro

Virpazaria backhuysi Gittenberger, 1969

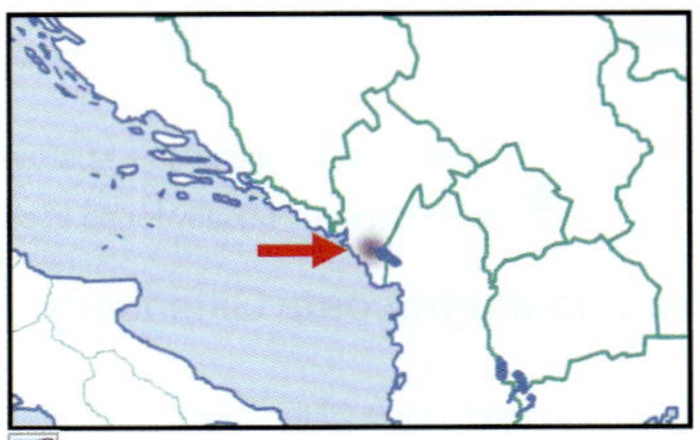

S Montenegro

Virpazaria backhuysi Gittenberger, 1969 in Gittenberger 1969: 299

Shell ribbed (40 ribs on last whorl), parietal apertural margin prominent and rounded, covers entirely the lower apertural margin in a frontal view of the aperture, apertural lip parallel and with a callus near the outer margin, opposite on the parietal side a small tooth, umbilicus wide, 2/7 of diameter and partly covered by apertural margin.

1.5 x 2.5 mm

Only known from one single shell. Perhaps not recent.

Virpazaria deelemanorum Gittenberger, 1975

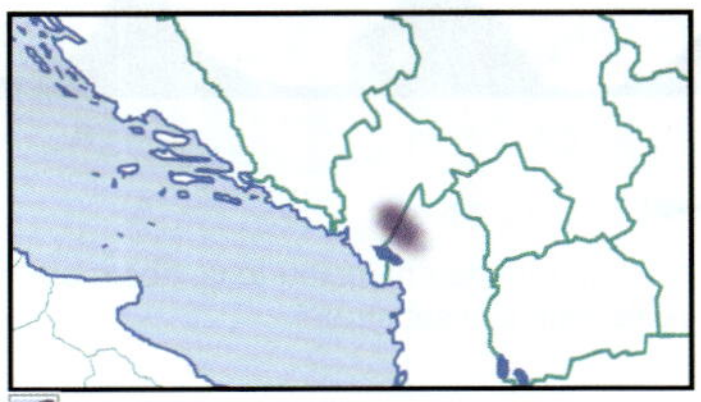

S Montenegro, N Albania

Virpazaria deelemanorum Gittenberger, 1975 in Gittenberger 1975: 268
Virpazaria dhorai Reischütz, Reischütz & Reischütz, 2010 in Reischütz et al. 2010: 29

Shell larger than the other *Virpazaria* species, with dense and high ribs (45-50 on last whorl, higher than in *V. adrianae*), parietal margin of the aperture narrow, palatal margin with only two knots, not covered by palatal margin in a frontal view, umbilicus 2/7 of shell diameter.

1.8-2.3 x 3.5-4.2 mm, perhaps slightly smaller in Albania.

Subterranean, in caves and deep rock crevices.

Seems to be relatively widely distributed at the northeastern mountains of the Skutari Lake plain, like an equivalent of *V. adrianae* in the southeast. Could possibly have a larger range along the edge of this mountain chain.
The taxon *dhorai* from Albania is classified with this species here. The claimed differences were smaller size (3.5 instead of 3.7-4.2 mm in Montenegro), higher whorls (perhaps lower whorls were meant, shell height 1.8 mm instead of 2.1-2.3 mm in Montenegro) and very weak differences in the prominence of the ribs which were not shown. I cannot see the claimed differences in the rib sculpture, the figures suggest perfect agreement with Gittenberger's 1975 figure. The sizes were not given accurately and the differences certainly insignificant. The aperture of the Albanian population fits very well with those from near Podgorica.

Virpazaria pageti Gittenberger, 1969

Virpazaria pageti Gittenberger, 1969 in Gittenberger 1969: 301

Montenegro: W of Herceg-Novi, river deposit near Sutorina

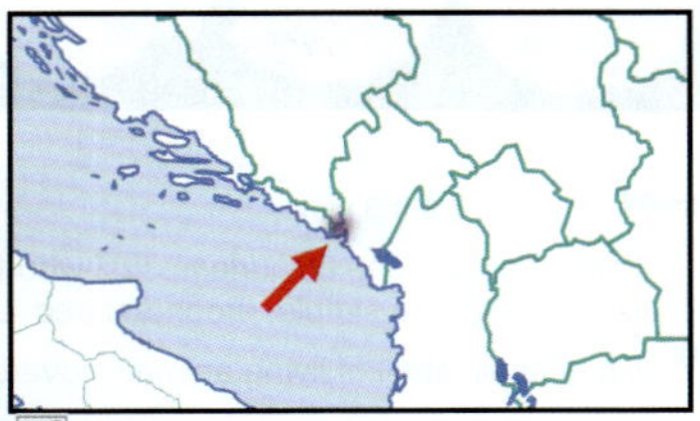

W Montenegro

Only known from river deposits. Possibly not recent.

1.3-1.5 x 2.5-3 mm

Shell on the upper side flat, ribbed (50 ribs on last whorl), apertural margins thick, parietal margins covers lower margin in a frontal view entirely, lip without callus-like structures, umbilicus wide (almost 1/3 of shell diameter), partly covered by last quarter of last whorl.

Gittenberger 1988.

Virpazaria ripkeni Gittenberger, 1969

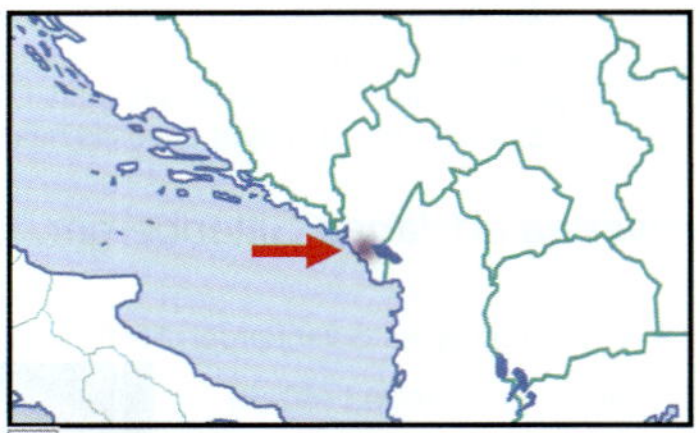

S Montenegro

Virpazaria ripkeni Gittenberger, 1969 in Gittenberger 1969: 302

Shell densely ribbed (around 80 ribs on last whorl), upper side not flat, parietal margin of the aperture not much thickened, regularly curved, on the palatal side 2 folds and one basal callus present, no lip, inside of the aperture visible in a frontal view, umbilicus wide (2/7 of shell diameter). Differs from the other species in its dense ribs and a thinner parietal margin.

1.9-2.2 x 3.5-3.9 mm

Inside a cave.

Only known from one badly recorded locality, not found again in 1975.

Virpazaria stojaspali Reischütz, Reischütz & Subai, 2009

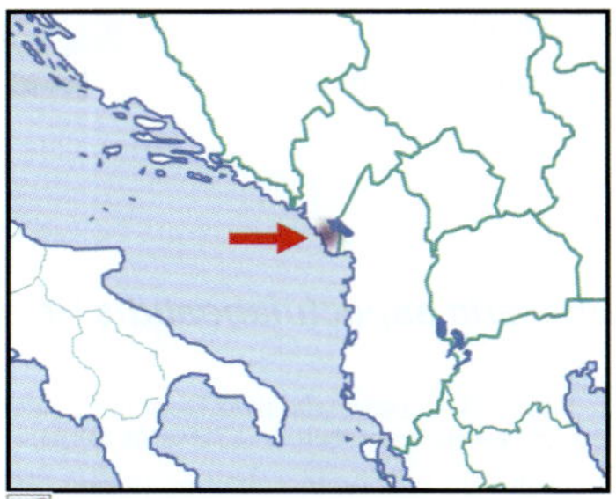

S Montenegro

Virpazaria stojaspali Reischütz, Reischütz & Subai, 2009 in Reischütz & R. 2009: 58
Subai's co-authorship was explicitly explained (p. 56).

Shell white, sharply ribbed (30 ribs/last whorl), equally ribbed on lower side, 5 moderately convex whorls, upper side slightly concave, periphery weakly angled on upper and lower side and moderately convex in (form in between, lower side rounded, aperture very narrow, parietal and basal margins not exactly parallel, aperture getting narrower upside, parietal ridge very prominent and covers aperture when seen in frontal view, umbilicus wide and usually covered with dust, 1/5-1/4 of diameter.
Differs from *V. adrianae* in its flatter upper side.

1.8 x 2.8 mm

Limestone rock crevices and holes at S exposed mountain slopes, up to 500 m.

Only known from one locality, near Besa village in Montenegro.

Subai 2009: 109: Fig. 10 [sic].

This is regarded as an independent species here, but could also belong to *adrianae*. Size, rib density and umbilicus width seem to fit, the differences in shell periphery and apertural margin seem to be gradual. It would be interesting to know if *Virpazaria* populations exist in the 10 km extending area between Virpazar and Besa village, a region where no representatives of this genus have been reported.

Spelaeodiscus

Spelaeodiscus Brusina, 1886 in Brusina 1886: 37 *Patula (Spelaeodiscus)*

Type species *Helix hauffeni* Schmidt, 1855 (monotypy)

Aspasita Westerlund, 1889 in Westerlund 1889: 18, 26, type species *Helix triaria* Rossmässler, 1839 (subsequent designation by Pilsbry 1922-1926 [1926]: 180)
Contained originally *Helix triaria, trinodis* and *triadis.*

Shell dextral, yellowish or transparent, apertural margin not distinctly connected at parietal side.

Schileyko 1998 classified *Aspasita* as a separate genus. But it seems that this proposal was not accepted by all authors. The three spelaeodiscine genera *Klemmia* (sinistral shell), *Virpazaria* (dextral shell and parietal wall at aperture) and *Spelaeodiscus* (dextral shell and no parietal wall at aperture) are well delimited. There is no such clear definition for *Aspasita*. Considering the low number of species I see no need to subdivide *Spelaeodiscus* and to classify *Aspasita* as a separate genus.

Spelaeodiscus albanicus (Wagner, 1914)

Without gender agreement: *Spelaeodiscus albanica* (Wagner, 1914)

Aspasita albanica Wagner, 1914 in Sturany & Wagner 1915: 67, Taf. 2

Shell yellowish white, translucent and slightly opaque, very densely ribbed, more than 70 ribs on the last whorl, apex smooth, 3-4 whorls with deep suture, last whorl not descending, aperture with 3 cavities, margin connected by a thin layer at parietal side, with weak callus-like teeth at both the basal and the palatal sides, umbilicus very wide and perspectival.
The species has less whorls than *S. triarius*, a wider umbilicus and is more densely ribbed, its last whorl is not descending. Differs from *S. hauffeni* in its denser ribs, last whorl not descending, and its teeth in the aperture.

3-4 x 1.9-2.5 mm

Probably subterranean or a true cave snail.

Gittenberger 1969: 294.

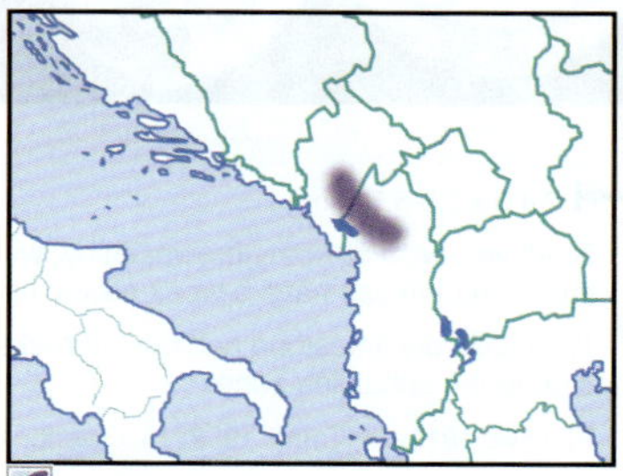

Montenegro, N Albania

Spelaeodiscus bulgaricus (Subai & Dedov, 2008)

Without gender agreement: *Spelaeodiscus bulgarica* (Subai & Dedov, 2008)

Aspasita bulgarica Subai & Dedov, 2008 in Subai & Dedov 2008: 116

3-3.7 x 4.5-5.3 mm Bulgaria: Vraca, near Ledenika cave Museum Budapest HNHM

W Bulgaria

Only known from two locations in W Bulgaria near Vraca.

Shell yellowish brown, densely ribbed (60-70 ribs on last whorl), 5.5-6.5 convex whorls, apertural margin thin, sharp and slightly reflected, without palatal tooth, no parietal callus, umbilicus wide and perspectival, at last whorl slightly excentrical.

Spelaeodiscus dejongi Gittenberger, 1969

Spelaeodiscus dejongi Gittenberger, 1969 in Gittenberger 1969: 293

Shell transparent, smaller than *S. hauffeni*, very densely ribbed, 80 ribs on the last whorl, aperture without teeth, umbilicus wide, 1/4 of diameter.

1.5-1.7 x 2.6-2.9 mm

Only known from caves.

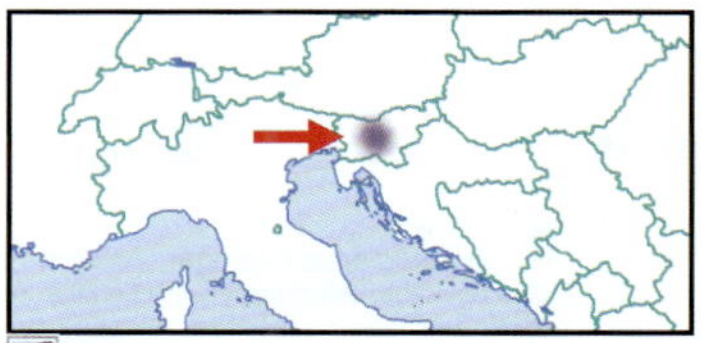

Slovenia

Spelaeodiscus hauffeni (Schmidt, 1855)

Helix hauffeni Schmidt, 1855 in Schmidt 1855: 3

Shell yellowish white, translucent and slightly opaque, regularly ribbed, 60-70 ribs on the last whorl, apex smooth, 3-4 whorls with deep suture, last whorl descending, aperture simple, with a weak lip, margin weakly thickened, connected by a thin layer at parietal side, umbilicus wide and perspectival. Shell larger than *S. dejongi*.
Animal white, almost transparent, with 4 tentacles, no eyes.

3-4 x 1.9-2.3 mm

In caves.

Restricted to a small area in Slovenia.

Sturany & Wagner 1915: 67, Gittenberger 1969: 292.

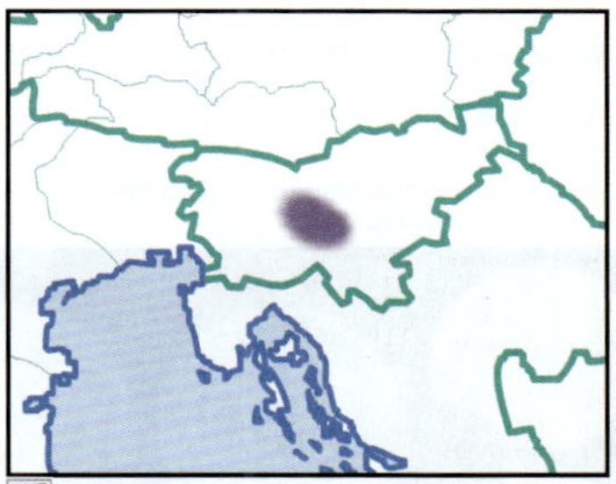

Central Slovenia

Spelaeodiscus obodensis Bole, 1965

Spelaeodiscus obodensis Bole, 1965 in Bole 1965: 350

Shell with 40-50 sharp ribs on the last whorl, aperture less oblique and narrower than in *S. hauffeni*, shape differs also from *S. dejongi*.

In caves.

Gittenberger 1969: 296.

1.5 x 2.6 mm Montenegro: near Zacir, near Virpazar

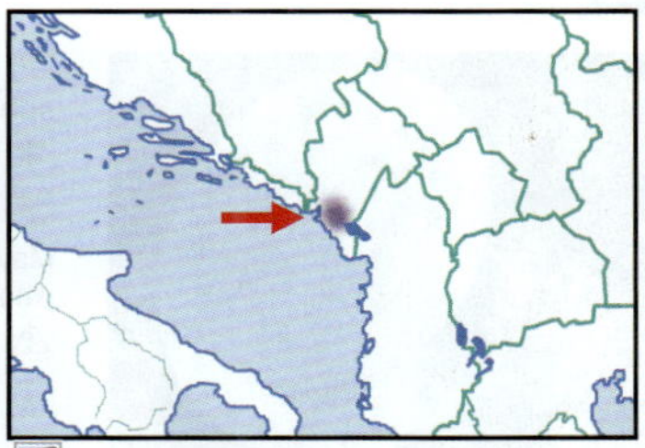

S Montenegro

Spelaeodiscus triarius (Rossmässler, 1839)

Without gender agreement: *Spelaeodiscus triaria* (Rossmässler, 1839)

Helix triaria Rossmässler, 1839 in Rossmässler 1838-1844 (9-10):13, Taf. 47

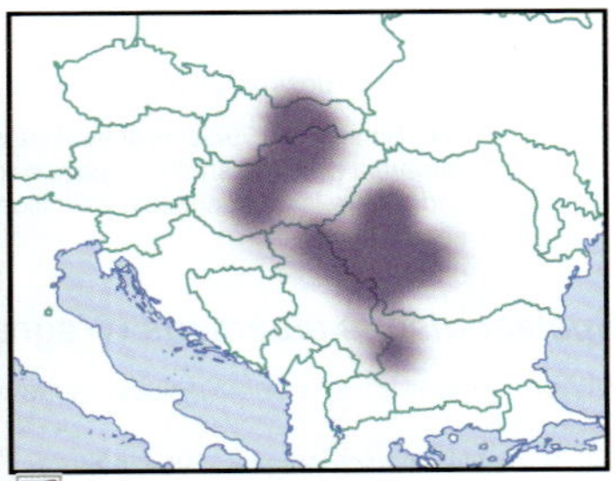

Slovakia to Bulgaria

2.4-3 x 3.5-5.5 mm

Slovakia: High Tatra, Kammene Vrata

Rare in Bulgaria.

Shell yellowish or horny brown, shiny, widely ribbed, 50-55 ribs on last whorl, ribs on lower side weaker, aperture with a lip, sometimes with a palatal tooth, columellar margin can have a weak callus, umbilicus wide and perspectival.

Relatively dry limestone habitats, at rocks, between stones, in rock crevices, deeply hidden in grass or between leaves, at the bottom near rocks. In highlands up to the subalpine zone.

Pavlović 1912: 34, Tomić 1959: 18, Gittenberger 1969: 289, Damjanov & Likharev 1975: 137, Kerney et al. 1983: 125, Grossu 1987: 268.

Spelaeodiscus unidentatus Bole, 1961

Spelaeodiscus unidentatus Bole, 1961 in Bole 1961: 206 [not verified]

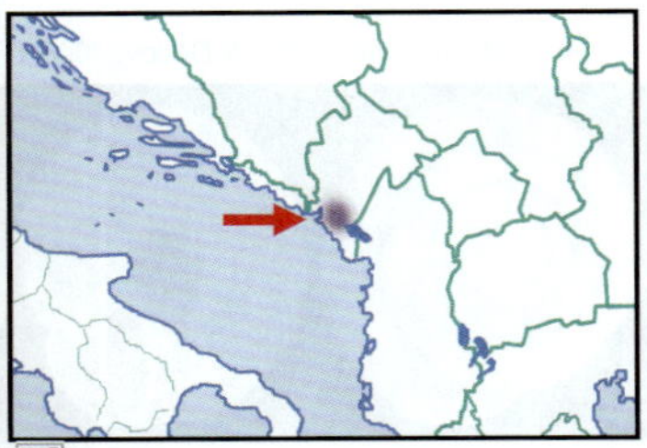

S Montenegro

1.4-1.5 x 2.8-3 mm

Montenegro: Zacir near Virpazar

Only known from caves.

Shell densely ribbed (35-45 ribs/penultimate whorl), whorls moderately convex, upper side relatively flat, last whorl not much descending near aperture, aperture with a very strong ands horizontal palatal tooth at the margin, palatal margin concave, and a weaker tooth at the basal side, producing a sinulus between the two teeth.

Discus

Discus Fitzinger, 1833 in Fitzinger 1833: 99

Type species *Helix ruderata* Hartmann, 1821 (subsequent designation by Gray 1847: 174)

Gonyodiscus Fitzinger, 1833 in Fitzinger 1833: 98, type species *Helix perspectiva* Megerle von Mühlfeld, 1816 (monotypy)
First Reviser to select precedence between *Discus* and *Gonyodiscus* was (were?) Adams & Adams 1858 [1855]: 116.

Small to medium sized, umbilicus wide and perspectival, with ribs.

Europe, northern Asia, North America

A subspecific classification based on shells and genitalia (*perspectivus* and *rotundatus* in a subgenus *Gonyodiscus*) was not supported by biological data obtained by Kuźnik-Kowalska 2006: 35.

Discus perspectivus (Megerle von Mühlfeld, 1816)

Without gender agreement: *Discus perspectiva* (Megerle von Mühlfeld, 1816)

Helix perspectiva Megerle von Mühlfeld, 1816 in Megerle von Mühlfeld 1816: 11, Tab. 2
Year of publication was not 1818.

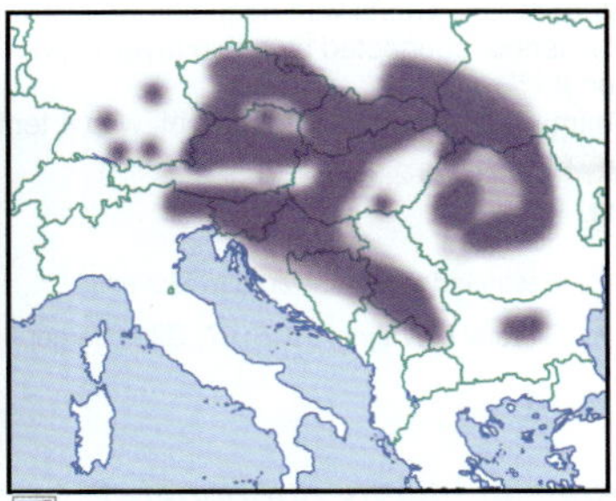

1.6-1.9 x 5.5-6.5 mm

Austria: Kärnten, Hochosterwitz

E Alps to central Bulgaria

Shell very flat with sharp keel, finely ribbed, weakly spotted, 5.5-6.5 whorls, umbilicus extremely wide and perspectival, almost 50 % of shell diameter, somehow like *Caracollina lenticula*.

Under litter in humid mountain slope forests, calciphile, in Bulgaria in up to 1500 m.

Reproduction is probably not uniparental. 15-25 eggs (1 mm diameter) per individual are laid in batches of 2-5 eggs between May and August, juveniles (initially with 1.8-2.3 whorls) hatch nearly synchronously after 26-33 days, they increase at a rate of 1 whorl in 4 months, maturity is reached in the second season at slightly more than 5 whorls, life span 2-3 years.

Vulnerable in Germany, rare in Bulgaria.

Kuźnik-Kowalska 2005.

Slovakia: Slovenská Ľupča
Jozef Grego

Discus rotundatus (Müller, 1774)

Without gender agreement: *Discus rotundata* (Müller, 1774)

Helix rotundata Müller, 1774 in Müller 1774: 29

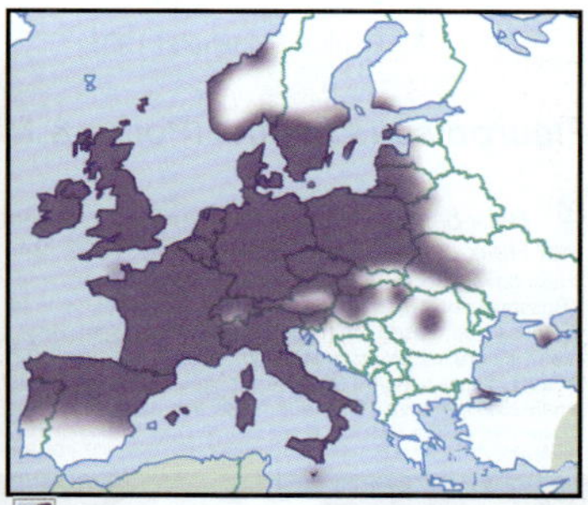

2-4 x 5.5-7 mm

Germany: Nordrhein-Westfalen, near Höxter

Central Atlantic islands, Europe

Germany: Göttingen

Shell horny coloured with reddish brown spots, quite flat, coarsely ribbed, weakly keeled at periphery, umbilicus more 1/3 of shell diameter or more. Animal in central Europe bluish on upper side, greyish white on foot below inferior tentacles, in S Europe bluish black.

In central and N Europe in forests at dead wood logs, also at humid shady places in open habitats. In Britain in nearly all kinds of moderately moist and shady places, tolerates also non-calcareous substrate. In the south under stones and desintegrating leaves in humid and shady habitats, on humus and in soil litter at old walls, often in colonies of numerous individuals. Also in cultivated regions, in gardens and open buildings. In Switzerland in up to 2100 m. Competes in higher altitudes with *D. ruderatus*, the two species can rarely coexist. In Ireland and Great Britain probably the most frequent land snail.

Uniparental reproduction is regular. The reproductive season from May to October, 20-50 eggs (0.7-1.5 mm diameter, larger parents lay larger eggs) per individual are laid between May and September, 2-3 weeks after copulation, in batches of 2-15 eggs below decaying leaves under trees, juveniles (initially with 1.5-2.3 whorls) hatch after 10-30 days, they increase at a rate of 0.5 whorls per month, maturity is reached in the second season at 5.5 whorls, life span 2.5-3.5 years. Life cycles in France and Poland only with minor differences.

Gassies 1849: 105, Moquin-Tandon 1855: 107, Germain 1930: 167, Kuźnik-Kow. 1999, Örstan 2003: 27.

Discus ruderatus (Hartmann, 1821)

Without gender agreement: *Discus ruderata* (Hartmann, 1821)

Helix ruderata Hartmann, 1821 in Hartmann 1821: 231

Name not to be taken from Férussac 1821 (incorrect on Official List) because Férussac (quarto p. 40, 26 May 1821) mentioned the name as a nomen nudum without description or indication. Same with *Glischrus ruderata* mentioned by Studer 1820.

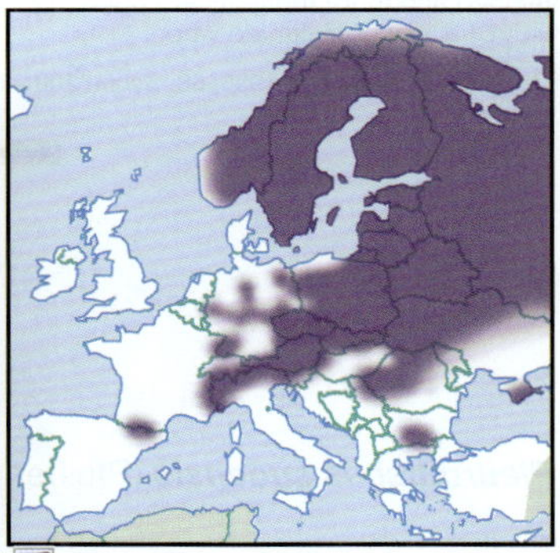

2.0-3.5 x 5.5-7.0 mm

Austria: Tirol, Steinach

N Eurasia, W Canada

Slovakia: Slovenské Rudohorie mountains, 1000 m

Jozef Grego

Shell usually evenly horny to reddish brown, rounded at periphery, not as flat as *D. rotundatus*, more finely ribbed, aperture almost circular, umbilicus 1/3 of shell diameter.

At and under rotting wood and stones, often in coniferous forests. In leaf litter layers of montane and subalpine forests, in higher altitudes also in open habitats. In Switzerland in up to 2800 m, most frequent between 1000 and 2000 m. Can be abundant above the timberline if sufficient stones are present for the snails to hide.

Uniparental reproduction is very rare. The reproductive season in Poland lasts from April to October, 6-15 eggs (1.2 x 1.3 mm, larger parents do not lay larger eggs) per individual are laid, in batches of 2-4 eggs, juveniles (initially with 1.5-2.5 whorls) hatch nearly synchronously after 17-34 days, success rate 50 %, they increase at a rate of 0.5 whorls per month, maturity is reached in the second season at slightly more than 4 whorls after 4 months, growth continues slowly at a rate of less than 1 whorl per year, life span 2-3 years.

Extinct in Great Britain since 8000 years ago. Endangered in Germany, rare in Bulgaria.

Kuźnik-Kowalska 2006: 35.

Pleurodiscus

Pleurodiscus Wenz, 1919
in Wenz 1919: 78 *Gonyodiscus (Pleurodiscus)*

Type species *Helix balmei* Potiez & Michaud, 1838
(original designation)

S Europe, N Africa

Bank & Menkhorst 1991.

Pleurodiscus balmei (Potiez & Michaud, 1838)

Helix balmei Potiez & Michaud, 1838 in Potiez & M. 1838: 120 (*Helix Balmei*)
not *Helix balmei* Beck, 1837 in Beck 1837: 24 (*Helix (Circinaria) Balmei*) (nomen oblitum under Art. 23.9, fixed here)

Helix balmei Beck, 1837 was established as a new name without description but with a reference to Férussac & Deshayes 1819-1851 [1832]: Pl. 82 Fig. 1 (showing a specimen of *Balcanodiscus frivaldskyanus* (Rossmässler, 1842)), not as a new replacement name for *Helix convexa* Férussac, 1832 non Link, 1807 (because (1) no expressed statement was given, Art. 72.7, (2) *Helix convexa* was mentioned as a synonym by Férussac, who used the name *Helix frivaldskyana*). Beck 1837 attributed the name to Michaud, but Potiez & Michaud described a different species. So Beck's name was probably new. The name *Helix balmei* Beck, 1837 is here suppressed under Art. 23.9:
In accordance with Art. 23.9 *Helix balmei* Potiez & Michaud, 1838 shall be used for the taxon, *Helix balmei* Beck, 1837 has not been used as a name for a taxon since 1899 and shall be regarded as nomen oblitum, *Helix balmei* Potiez & Michaud, 1838 was used as a name for our taxon in the following 26 papers:
Forcart 1965: 89, Zilch 1969: 231, Alzona 1971: 80, Giusti 1973: 137, Paget 1976: 710, Kerney & Cameron 1979: 206, Maassen 1981: 32, Shea 1982, Beckmann 1987, Bank & N. 1988: 49, Falkner 1990: 148, Bank & Menkhorst 1991: 63, Schütt 1993: 69, Cossignani & C. 1995, Giusti et al. 1995: 234, Manganelli et al. 1995: 16, Schütt 1996: 63, Frank 1997: 166, Bank & Maassen 1998: 52, Schileyko 1998: 121, Schütt 2001: 64, Kolouch 2003, Örstan, P. et al. 2005: 184, Örstan, Y. et al. 2005: 6, Hausdorf 2006: 376, Beckmann 2007: 131.

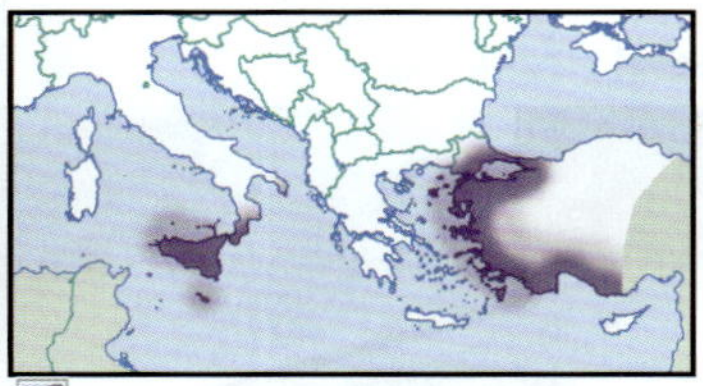

5-8 x 7-11 mm

Italy: Sicily, Taormina

S Europe, Near East and N Africa

1.0-1.2 mm at 1 whorl, 2.4-2.6 mm (2 whorls), 3.8-4.2 mm (3), 6.0-6.5 mm (4), 8.5-9.5 mm (5 whorls)

Shell pale golden-brown, sharply ribbed, less regularly than *P. cyprius*, last whorl width 1.25 x of penultimate whorl, apertural margin simple and sharp, umbilicus wide and deep.

Mainly in open and ruderal habitats, under stones, among rocks, in soil crevices and caves, also in wooded areas, requires some shadow in Malta. Not restricted to calcareous substrate.

Pleurodiscus cyprius (Kobelt, 1896)

Without gender agreement: *Pleurodiscus cypria* (Kobelt, 1896)

Patula sudensis var. *cypria* Kobelt, 1896 in Rolle & Kobelt 1896: 56, Taf. 22

Differs from *P. balmei* in its larger shell, ribs more sharply developed and more regularly arranged, 5.5-5.8 whorls, last whorl width 1.4 x of penultimate whorl.

Bank & Menkhorst 1991: 68, Schütt 2001: 64.

6.0-6.9 x 11.0-14.5 mm

N Cyprus: 10 km W Kyrenia

Stefan Heim

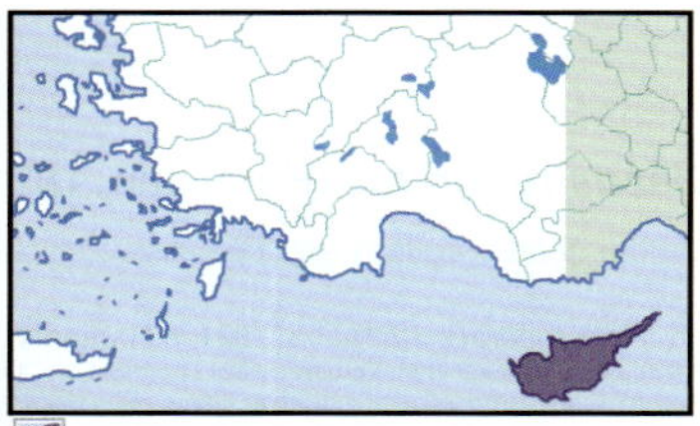

Cyprus

Tiroler Landesmuseen Innsbruck

Pleurodiscus sudensis (Pfeiffer, 1846)

Helix sudensis Pfeiffer, 1846 in L. Pfeiffer 1846 (Symbolae): 67 (*Helix Sudensis*)

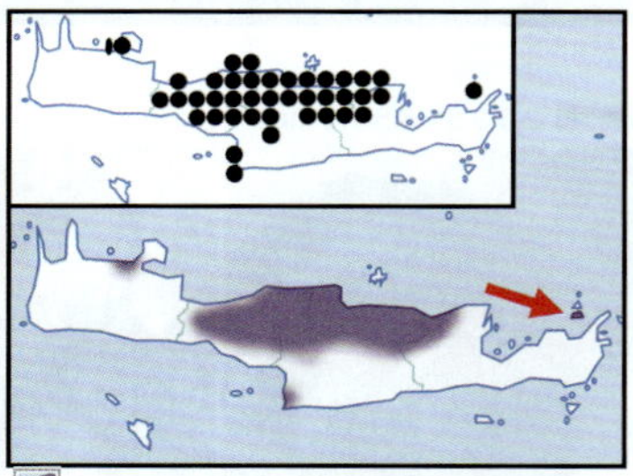

7.5-9.5 x 10-15 mm

Greece: N central Crete, near Goniés

Greece (central Crete)

Shell often reddish brown due to dust particles, very regularly and finely ribbed, also at lower side, 4.5-5 convex whorls, umbilicus deep, 1/7 of shell diameter.

Bank & Menkhorst 1991: 69, Schultes & Wiese 1991, Vardinoyannis 1994 (unpublished thesis), unpublished data (HNC collection).

Clausiliidae

Genera sorted by taxonomic subgroup and alphabetically:
1 - *Serrulina* subgroup
2 - *Alopia* subgroup (Romania-Yugoslavia-Albania)
3 - *Albinaria* subgroup (Greece-Turkey)
4 - *Idyla-Elia* subgroup (Greece-Turkey)
5 - all-others subgroup (not Turkey)
6 - *Balea* subgroup

European Clausiliidae are often locally extremely variable in their shell morphology and many species have been subdivided into numerous (= many many many many) subspecies following a concept established by H. Nordsieck. In this concept (a comprehensive summary was given by Nordsieck 2007), species were defined or recognized on the base of so-called essential characters, and subspecies by other essential characters regarded to be inferior.
The problem of this concept was that species and subspecies depended on the personal opinion of the researcher who took the subjective decision on the importance of the so-called essential characters.
In this concept the morphological characters have usually been verbally described using a finely tuned language, but not in a reproducible form that would have allowed repeating a study and coming to the same result.
Such a method does generally not allow falsifying results in detail. Even the same researcher applying exactly the same concept can come to different conclusions and classifications if a group is studied in different years. Nordsieck 2004 classified *Albinaria xanthostoma loosjesi* as a subspecies. Nordsieck 2007 classified it as an independent species *Albinaria loosjesi*, based on exactly the same material and methods.
I am not convinced that the currently applied traditional classification of European clausiliid species is the best one. In a detailed case study in Crete (Welter-Schultes 2010) I transferred shell characters to numerical values and obtained without major problems reproducible and falsifiable results. Several widespread species displayed complex spatial gradients in various characters, which usually varied gradually in several independent directions over area. This concerned also characters which had been considered as "essential" and taxonomically important for the identification of species and subspecies (Nordsieck 2004). Narrow hybrid zones between clearly distinct populations at both sides - distinct in more than one independent shell character - were found only exceptionally (*Albinaria hippolyti* was such an example). In Crete it was possible and useful to apply the concepts for species and subspecies as established by Mayr 1942 and Wilson & Brown 1953, currently used in many other fields of biodiversity research (Mallet 2001), in combination with recent knowledge on hybrid zones (Barton & Hewitt 1985). This reduced the number of species in Crete from 37 to 31, and that of the additional name-bearing subspecies from more than 40 (most forms had no name, certainly more than 100 could be described in traditional concepts) down to 8.
Under these interdisciplinarily applied concepts the populations should be recognized as different taxonomic units only if they differ clearly and in more than one morphological character from each other, given that no other evidence is provided for the presence of reproductive isolation of taxa. The idea behind is that hybrid zones provide partial barriers to reproduction, so that some genes do not pass from one population to the other.

I expect that applying a similar concept for other groups, particularly those of *Montenegrina*, *Bulgarica* and *Alopia*, would yield likewise results: a slightly lower number of species, and a considerable reduction down to few or no tenable subspecies.

The traditional concept would result in a steadily ongoing continuous inflation of subspecific taxa. Such a system would create a lot of work in establishing new names, but I would not expect this to be associated with a similar increase of understanding of the nature behind the diversity.
It sounds generally more meaningful to give statements like "populations in the west tend to be smaller and have a more prominent subcolumellaris" than to say "this species has two subspecies".

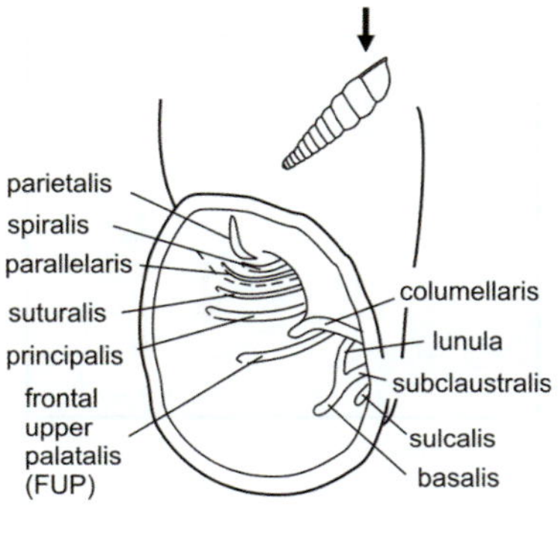

Apertural folds

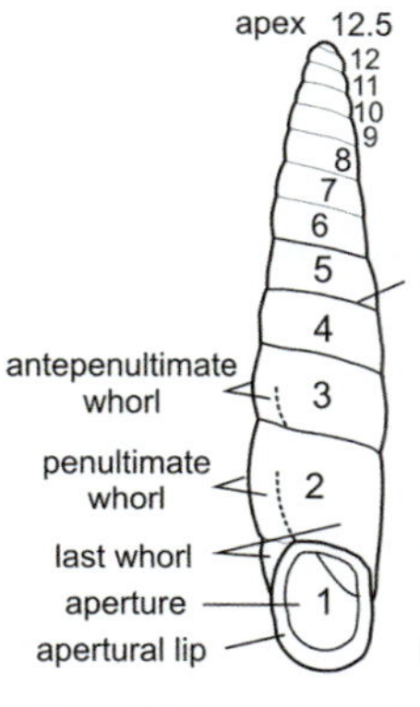

Simplified whorl count

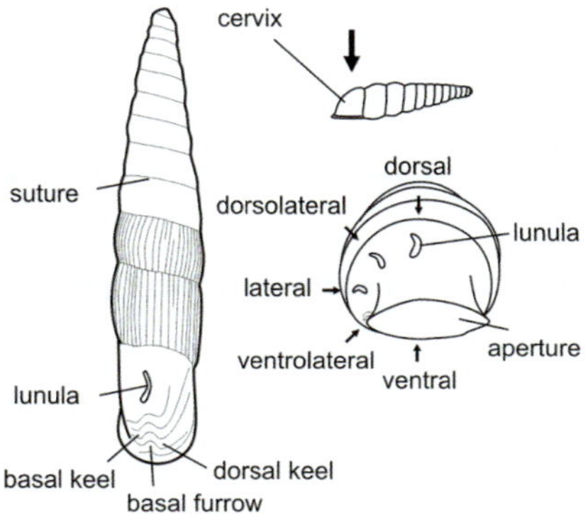

Cervix

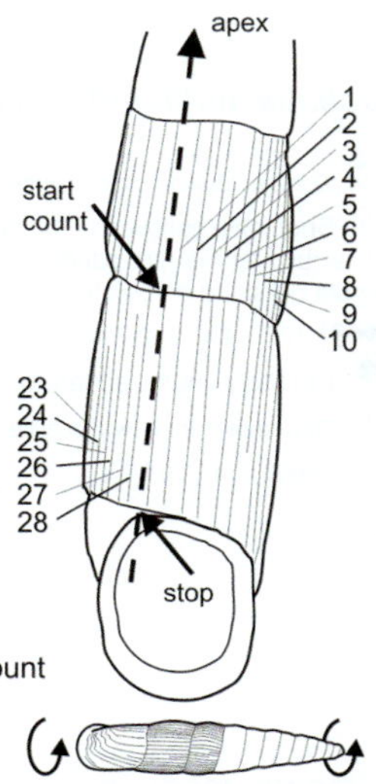

Rib count

1 - *Serrulina* subgroup

Bofilliella, Dobatia, Graecophaedusa, Neniatlanta, Sciocochlea, Serrulina, Tsoukatosia

Bofilliella

Bofilliella Ehrmann, 1927
in Ehrmann 1927: 56 *Laminifera (Bofilliella)*

Type species *Nenia subarcuata* Bofill y Poch, 1897 (monotypy)

Bofilliella subarcuata (Bofill y Poch, 1897)

Nenia subarcuata Bofill y Poch, 1897 in Bofill y Poch 1897: 366

Shell usually decollated with 5-6 whorls remaining, light greyish, transparent and siny, regularly and finely ribbed or striated, whorls with deep suture, cervix rounded, aperture distinctly detached and protruded, with uncommonly shaped sinulus and a 90 degree angle at the parietalis, several denticles between parietalis and columellar side, columellaris not prominent and ascending vertically, no palatal folds, no visible subcolumellaris.

Humid habitats, usually in forests on calcareous substrate, in leaf litter and below tree trunks and stones, and in the soil below the organic surface layer, also in the vegetation at creek margins and in river valleys.
Obviously mainly subterranean in the soil in depths depending on seasonal humidity gradients, animals feed on dead vegetable material.

9-12 x 1.9-2.5 mm

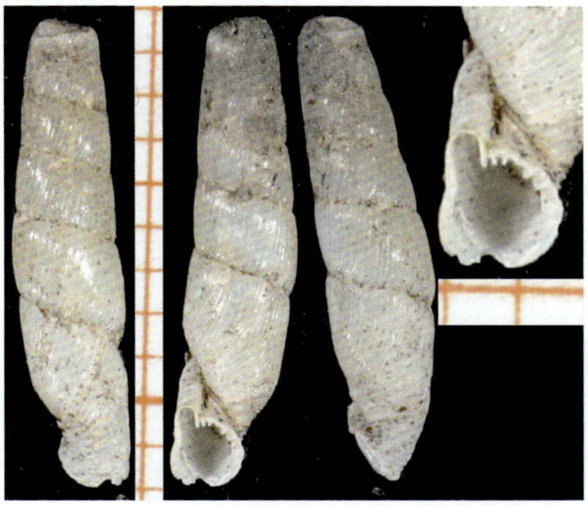
Spain: Barcelona, Cueva Vora Fosca Naullen

E Pyrenees

IUCN: Vulnerable VU D2. It is unclear how rare the species is; its habitat is not rare, but it is poorly accessible.

Gittenberger, Me. & Ra. 1980: 14, Gittenberger 2007: 149.

Dobatia

Dobatia Nordsieck, 1973
in Nordsieck 1973: 80

Type species *Serrulina goettingi* Brandt, 1961
(original designation)

Dobatia goettingi (Brandt, 1961)

Serrulina goettingi Brandt, 1961 in Brandt 1961: 18, Taf. 2

Turkey: near Lake Abant

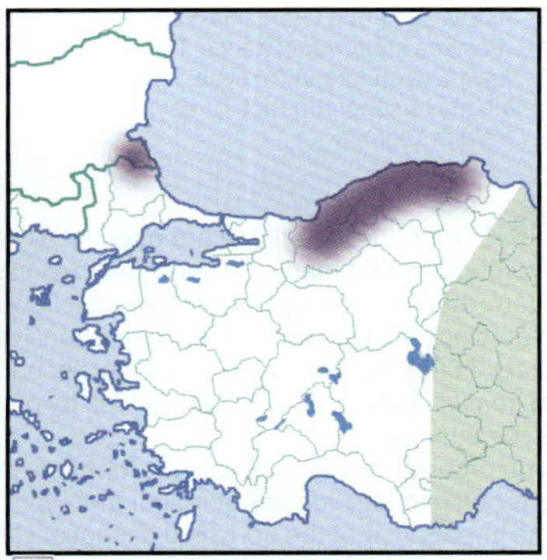

N Turkey, SE Bulgaria

Shell slender, horny greenish, finely ribbed, apex blunt, cervix rounded without keels, aperture with excentric sinulus, several teeth at apertural margin between parietalis and columellaris, 2-3 folds between columellaris and subcolumellaris, principalis short but visible in the aperture and from outside the shell.

11-13 x 2.5-3 mm

Forests, in rotting wood, under bark and in forest soil litter.

Dobatia multidentifera Neubert, 1992

Dobatia multidentifera Neubert, 1992 in Neubert 1992: 66

Shell pale yellowish, 10 ribbed whorls, aperture more or less protruded, aperture with densely arranged teeth all around its margin, parietalis, columellaris and subcolumellaris visible in an obblique view of the aperture, lunula dorsal to dorsolateral.

12.5-14 x 2.3-2.7 mm

Probably subterranean, was found in woody ground litter of a steep slope.

Only known from two localities.

Schütt 2001: 304.

S Turkey

Graecophaedusa

Graecophaedusa Rähle, 1982
in Rähle 1982: 45

Type species *Graecophaedusa sperrlei* Rähle, 1982
(original designation)

Graecophaedusa sperrlei Rähle, 1982

Graecophaedusa sperrlei Rähle, 1982 in Rähle 1982: 46

Apex bold, 9 whorls, first 3 whorls smooth, the others densely ribbed, cervix rounded, aperture pierced between parietalis and below subcolumellaris with numerous (11-15) teeth rolled around the apertural margin, apertural margin at palatal and basal sides smooth, lunella ventrolateral.

11-12 x 2.3 mm

Probably subterranean, found in deep rock crevices.

Considered as a relictual species, the group was much more widely distributed in the Tertiary.

N Greece (Pangéo mountains)

Neniatlanta

Neniatlanta Bourguignat, 1876
in Bourguignat 1876: 20 *Nenia (Neniatlanta)*

Type species *Clausilia pauli* Mabille, 1865
(subsequent designation by Lindholm 1924: 68)

Contained originally *Nenia pauli* and *Nenia mabilli*.

Neniatlanta pauli (Mabille, 1865)

Clausilia pauli Mabille, 1865 in Mabille 1865: 259, Pl. 14 (*Clausilia Pauli*)

13-15.5 x 2.7-3 mm

Spain: Navarra, Roncesvalles

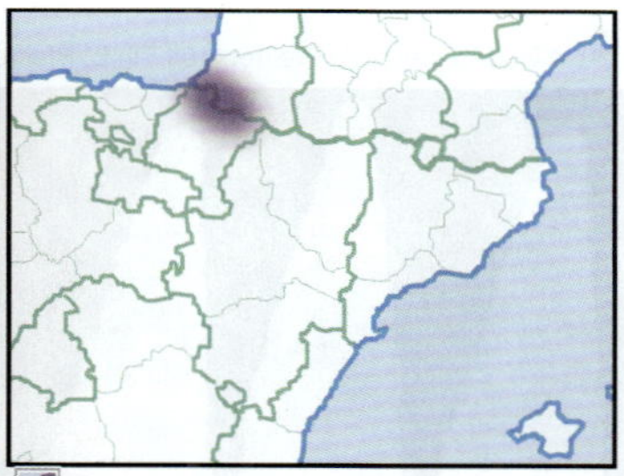

Pyrenees

Shell horny reddish brown, finely ribbed, 11-13 whorls with deep suture, cervix more coarsely ribbed, last whorl detached and protruded, aperture inside orange, with a small triangular area at the suture above the aperture, columellar margin of the aperture with 4-5 folds, short frontal lower palatalis connected to lunula.

Shady habitats in woods and among rocks, in and outside limestone areas.

IUCN 1996: Vulnerable VU D2. Known only from few places, threatened by deforestation, not rare where it occurs. Protected by law in France, eggs and animals must not be collected (but habitat destruction is allowed).

Gitttenberger 2007: 148 (classified in genus *Laminifera*), Larraz et al. 2007.

Sciocochlea

Sciocochlea Boettger, 1935
in C. R. Boettger 1935: 53 (footnote) *Serrulina (Sciocochlea)*
Expressly established as a new replacement name for

Type species *Serrulina collasi* Sturany, 1904
(monotypy by Wagner 1914: 34, 38)

Phygas Wagner, 1914 in A. J. Wagner 1914 (Sitz.-Br. Akad Wiss. I 123): 34, 38 not *Phygas* Treitschke, 1833 (Lepidoptera)
Wagner mentioned *Phygas* uncommented without description, just in combination with the name "*colasi* Stur.".

Shell small, apex blunt, apertural margin with intralamellar folds between columellaris and subcolumellaris.

Gittenberger 2000: 81, Reischütz & Reischütz 2004: 5.

Only one species recognized here.
The situation as currently known suggests that we have several conchologically slightly different forms from a relatively small number of localities in Kérkira, Ipiros and SW Albania. There is some variation among these isolated populations. Initially these populations were described as new species, later with more finds between the stations after 2000, also as subspecies.
Reischütz & Reischütz 2009: 34 found that the variation within a cave can be considerable, this affected shell size and possibly also rib prominence. It was also considered as possible that some shells were very old and shell characters could vary over time.
At the current state of research it seems questionable to subdivide the genus into different species. It would be necessary to study morphological variations within populations and look for possible gradients over area.

Currently I see no constraints that the morphological differences between the distant and isolated populations are associated with barriers to reproductive isolation, as demanded for biological species. Geographical isolation alone is probably not sufficient.
We seem to have likewise problems as in other cave snail groups, like in *Zospeum* (where 6000 caves could host 6000 species) or *Virpazaria*.
Some characters which were reported to differ among *Sciocochlea* species ("columellaris slightly steeper ascending", shell size, rib prominence and rib density) are not self explaining in this context, these characters are known to vary over area in other clausiliid groups in Greece. For some other characters (prominence of folds at the columellar margin, certain aspects of shell shape) some statistical evaluations on their local degrees of variation would be useful for a better understanding of their taxonomical value.

Sciocochlea collasi (Sturany, 1904)

Serrulina collasi Sturany, 1904 in Sturany 1904 (Anz. Wien): 118
Also published by Sturany in Nachrichtsblatt DMG (July 1904) (p. 105 there), I do not know which one was published earlier.
Sciocochlea nordsiecki Subai, 1993 in Subai 1993: 65
Sciocochlea cryptica Subai & Szekeres, 1999 in Subai & Szekeres 1999: 9
Sciocochlea llogaraensis Reischütz & Reischütz, 2009 in Reischütz & Reischütz 2009: 47

8-11 x 2-2.9 mm

Greece: (1-4) Ipiros, 13 km S Mavronéri, (5-9) Kérkira, Katsouri cave near Barbati

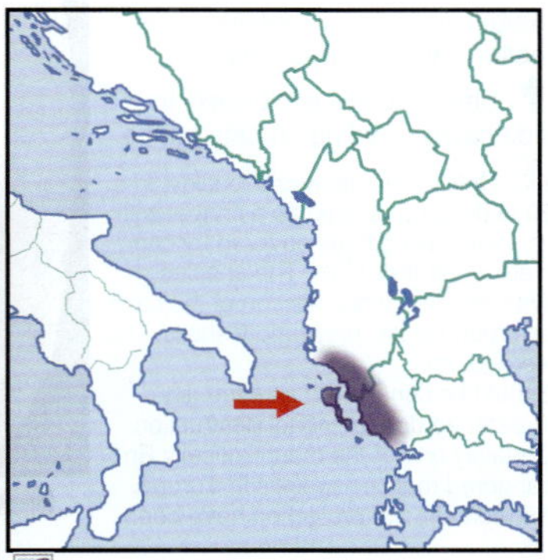

NW Greece, SW Albania

Shell yellowish white or whitish, transparent and shiny, almost smooth or ribbed, 7-8 whorls, penultimate or last whorl with largest diameter, last whorl often bent to the left side, parietalis strong, columellaris usually not very strong, often hidden inside, usually with some intralamellar folds at the apertural margin between columellaris and the shortly visible subcolumellaris. Locally variable in terms of ribs, shell size, diameter of the last whorl and apertural folds.

In caves, subterranean, probably associated with deep-reaching tree root systems, obviously feeding on symbiotic fungi. Animals seem to require relatively high winter temperatures and high humidity. Only rarely found at the surface.

Restricted to caves, rare and scattered, in Greece threatened by abuse of habitats as garbage dumps (Reischütz & R. 2009: 34).

Reischütz & Reischütz 2004: 5, Reischütz & R. 2009: 33, Reischütz, Re. & R. 2010: 38.

Serrulina

Serrulina Mousson, 1873 in Mousson 1873: 214

Type species *Clausilia sieversi* Pfeiffer, 1871 (subsequent designation by Lindholm 1924: 63)

Contained originally only *Cl. sieversi* and *Clausilia semilamellata*. Martens 1880: 70 and Westerlund 1902: 113 gave *Cl. serrulata* as type species, which was not originally contained.

Serrulina serrulata (Pfeiffer, 1847)

Clausilia serrulata Pfeiffer, 1847 in L. Pfeiffer 1847: 71

11-15 x 2.6-3.2 mm

Bulgaria: Strandza mountains, near Malko Târnovo

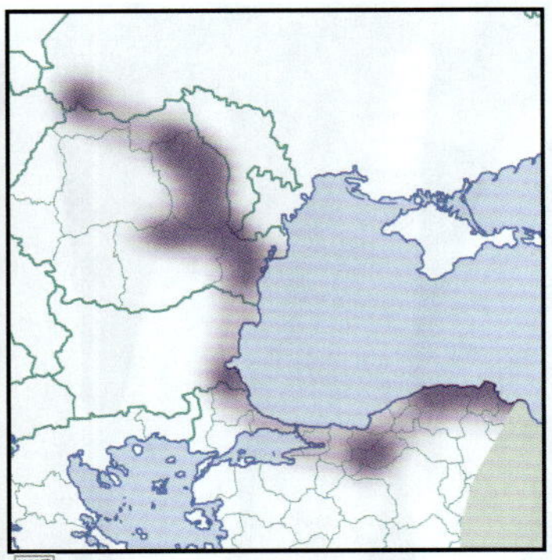

SE Europe to central Caucasus

Shell translucent, finely ribbed, apertural margin serrated with many teeth except at the angular sinulus, columellaris almost entirely hidden, principalis long, parietalis prominent.

Only in undisturbed deciduous forests, under rotting bark and inside the fibres of humid desintegrating wood, the species needs large decaying tree logs. Social animals.

Damjanov & Likharev 1975: 169, Grossu 1981: 20, Schütt 2001: 302, Páll-Gergely & Roibu 2011.

Tsoukatosia

Tsoukatosia Gittenberger, 2000 in Gittenberger 2000: 85

Type species *Tsoukatosia liae* Gittenberger, 2000 (original designation)

Shell dextral, ribbed, apex broad, spiralis and columellaris reaching equally far inside, lunula laterally situated, upper palatalis present.

In rock crevices of Mesozoic limestone rocks, or truly subterranean.

Taxonomic situation slightly different from *Sciocochlea*. Less populations are known, from a slightly broader area, suggesting a higher level of geographic isolation. The populations differ morphologically at a relatively low degree (especially *christinae* and *subaii*), but seem to have different ecological requirements. This might justify to maintain them preliminarily as independent species. More taxonomic research is needed.

Tsoukatosia christinae Reischütz & Reischütz, 2003

Tsoukatosia christinae Reischütz & Reischütz, 2003 in Reischütz & R. 2003: 1

Shell colourless and transparent, old shells with reddish hue from the cave walls, ribbed, 8-9 flattened whorls with deep suture, cervix with a transversal furrow, apertural margin detached and with up to 30 folds all over, going round the margin, including parietalis, columellaris and subcollumellaris.

Seems to live exclusively inside caves, feeding on bats guano.

Only known from one locality, in a 20 m deep cave used as a chapel with its walls having been covered by concrete in the first 15 m. The species lives only in the last section of the cave on natural walls, only empty shells were found.
Should be considered as seriously threatened due to habitat destruction. Certainly one of the most Critically Endangered mollusc species in Europe. No measures of protection have been taken.

12.5-14 x 2.6-3.0 mm

Greece: Pelopónisos, 12 km S Dirráchi

Greece (Pelopónisos: Messinia)

Reischütz & Reischütz 2004: 5.

Tsoukatosia liae Gittenberger, 2000

Tsoukatosia liae Gittenberger, 2000 in Gittenberger 2000: 85

Shell regularly ribbed, broad apex and second whorl width 60 % of maximum shell diameter, 8-8.5 whorls, last two whorls bent to the right side, apertural lip thickened and reflected, parietal wall curved yielding a broadly rounded sinus above the parietalis, columellaris invisible in frontal view, aperture with 4-6 folds at the columellar side surpassing the apertural margin (columellaris, subcolumellaris and a few additional folds). Upper palatalis present but not visible in a perpendicular view of the aperture.

10.5-11.5 x 1.0-1.1 mm

Mesozoic limestone rock crevices with reddish soil, very difficult to find.

A very rare species.

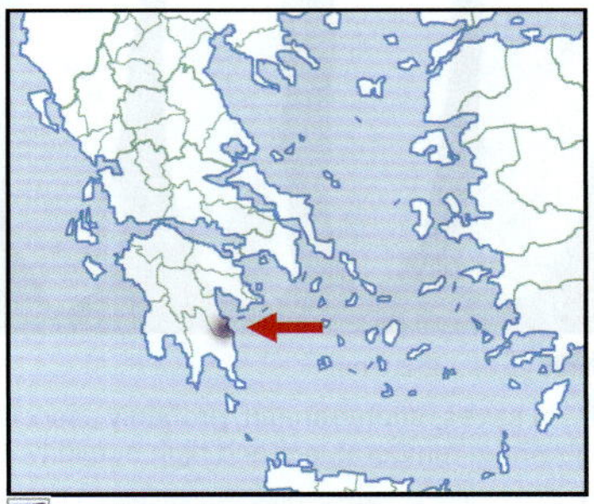

Greece (Pelopónisos: Arkadia)

Tsoukatosia subaii Hunyadi & Szekeres, 2009

Greece (Central Greece: Fokís)

Tsoukatosia subaii Hunyadi & Szekeres, 2009 in Hunyadi & Szekeres 2009: 2

Shell colourless, translucent, 8-9 whorls, finely striated, slightly more prominent on last whorl, cervix with crest-like basal keel, rounded at doral side, with a distinct furrow in between, apertural large, margin serrated with denticles, parietalis slightly more prominent, columellaris and subcolumellaris hardly visible in a perpendicular view, lunula lateral.
Differs from *T. liae* in its serrated parietal apertural margin, from *T. christinae* in its smaller size, finer sculpture and less prominent basal furrow.

10-11 x 1.9-2.5 mm

Known from soil samples with access to subterranean habitats.

A rare species.

Reischütz, Re. & R. 2010: 37.

2 - *Alopia* subgroup (Romania-Yugoslavia-Albania)

Agathylla, *Alopia*, *Herilla*, *Medora*, *Triloba*

Agathylla

Agathylla Adams & Adams, 1855
in Adams & Adams 1858: 184 *Clausilia (Agathylla)*

Type species *Clausilia exarata* Rossmässler, 1835
(subsequent designation by Albers & Martens 1860: XVII, 275)

Agathylla abrupta (Küster, 1847)

Clausilia abrupta Küster, 1847 in Küster 1847: 15 (figured on Taf. 1 Fig. 13-16)
Date was not 1845 because text is relevant. Taf. 1 did not contain names, wrapper issued with plate (in 1845) did not constitute published work (Art. 8.1.1), and was probably nowhere preserved.

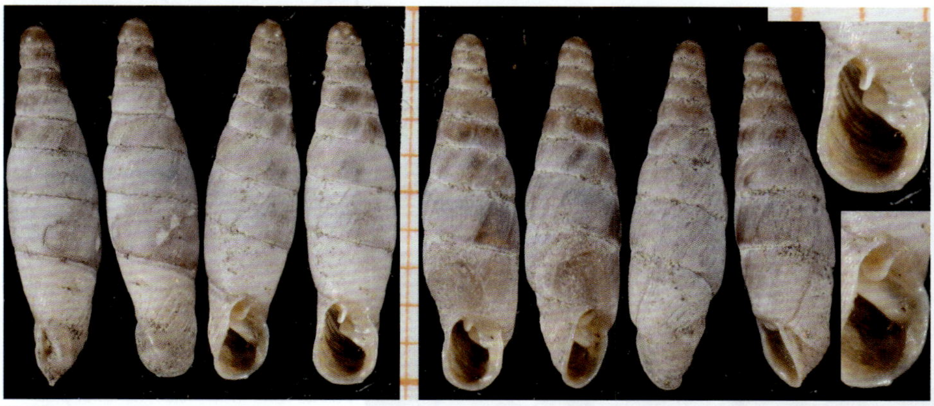

9-10 x 2.5 mm

Croatia: Dalmatia, Pelješac

Montenegro, S Dalmatia

Shell grey, smooth, cervix ribbed, 8-9 whorls, last whorl narrower, columellaris strong, horizontal, upper palatalis present, lower palatalis very short and connected to lunula.

Westerlund 1884: 107.

Agathylla biloba (Wagner, 1914)

Alopia biloba Wagner, 1914 in Sturany & Wagner 1915: 69, Taf. 15

9-12 x 2-2.5 mm

Albania: Shkodër

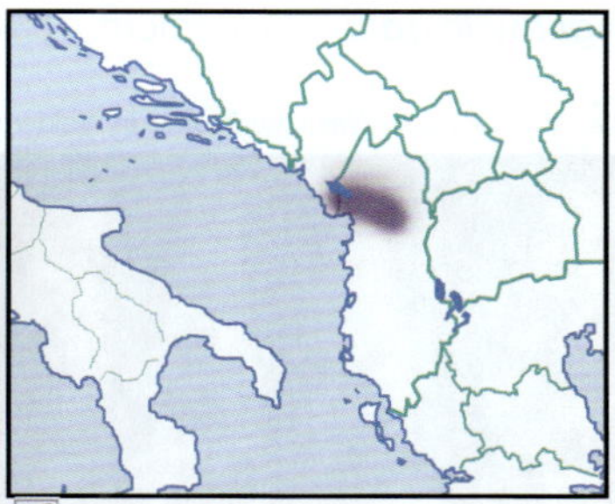

N Albania, SE Montenegro

Shell reddish with yellowish hue, ribbed, apex yellowish brown, 8-10 whorls, cervix with rounded basal and weak dorsal keel, aperture detached, margin thin, parietalis reaches margin and passes spiralis, columellaris prominent and nearly horizontal, frontal upper and lower palatalis usually very short, lunula absent, subcolumellaris only weakly visible in the aperture. Clausilium two-lobed, S-shaped, with groove, columellar lobe long and pointed, outer lobe rounded.

Nordsieck 1996: 7.

Agathylla exarata (Rossmässler, 1835)

Clausilia exarata Rossmässler, 1835 in Rossmässler 1835-1837 (2): 13, Taf. 7

17-23 x 2.8-3.5 mm

Croatia: Dalmatia, near Nova Sela

S Dalmatia, Herzegovina

Shell very slender, violetish grey with brown hue and fine white ribs, fragile, not shiny, 11-13 whorls, cervix with two keels twisted among each other, the dorsal keel being very prominent, and a narrow furrow in between, aperture inside orange, with frontal upper palatalis and lunula, margin detached and protruded.

Westerlund 1884: 111, Falkner 1990: 154.

Agathylla formosa (Rossmässler, 1835)

Clausilia formosa Rossmässler, 1835 in Rossmässler 1835-1837 (2): 14, Taf. 7

7-10 x 2.3-2.7 mm

Montenegro: near Herceg-Novi

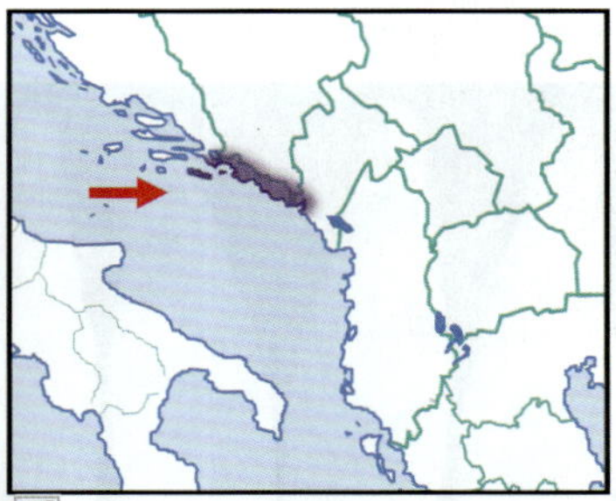

Croatia (S Dalmatia), Montenegro

Shell fragile, grey with yellowish red hue, not shiny, very finely and densely ribbed or striated, cervix more coarsely ribbed, suture with white margin, upper palatalis present, rarely with a second palatal tooth deep inside, no lower palatalis.

Westerlund 1884: 108.

Agathylla goldi (Walderdorff, 1864)

Clausilia goldi Walderdorff, 1864 in Walderdorff 1864: 508 (*Clausilia Goldi*)

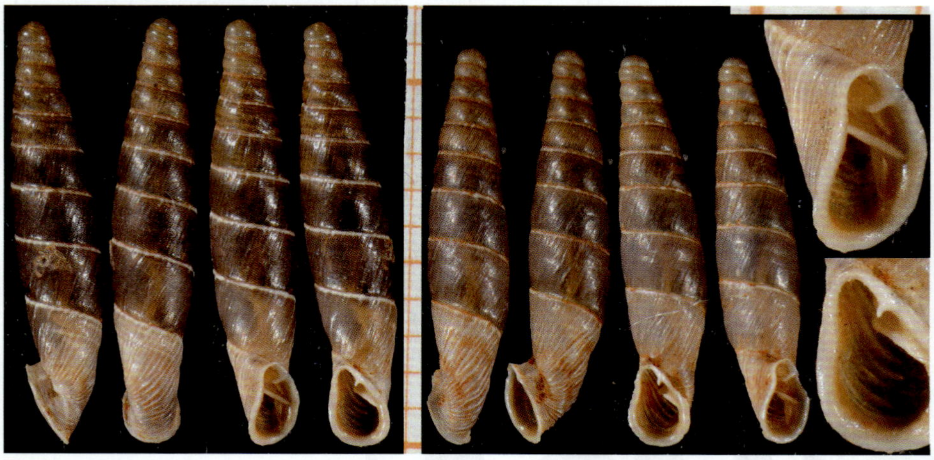

11-14 x 2.2-2.6 mm

Montenegro: Lovćen pass

Montenegro

Shell violet horny coloured or whitish, shiny, almost smooth, with corrugated ribs at cervix, columellaris slightly oblique, no lower palatalis, subcolumellaris only visible in an oblique view of the aperture.

Westerlund 1884: 110.

Agathylla lamellosa (Wagner, 1829)

Clausilia lamellosa Wagner, 1829 in Schubert & Wagner 1829: 190

11-13 x 2.0-2.5 mm

Croatia: Dalmatia, Mlini S of Dubrovnik

Croatia (S Dalmatia), Montenegro

Shell yellowish brown with white ribs, cervix simple without basal furrow, no frontal upper palatalis. Smaller than *Ag. exarata*, less slender, cervix without additional keel.

Westerlund 1884: 109, Falkner 1990: 154.

Agathylla merditana (Wagner, 1914)

Alopia biloba merditana Wagner, 1914 in Sturany & Wagner 1915: 70, Taf. 15

In figure caption of Taf. 15 spelled "*A. (A.) biloba merditan*", an inadvertent error.

8-14 x 1.8-2.7 mm

Albania: (1-3) 6 km SW Leskovik; (4-7) Mali Gradecit E Çorovodë

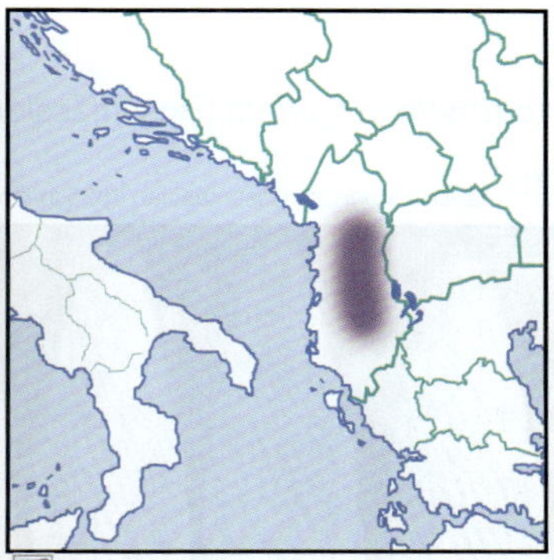

Albania

Shell yellowish or reddish brown, nearly smooth except sometimes near the suture and in the last section of the cervix, slightly larger and less slender than *A. biloba*.

Nordsieck 1996: 7.

Agathylla narentana (Schmidt, 1868)

Clausilia narentana Schmidt, 1868 in Schmidt 1868: 99

15-17 x 2.7-3.3 mm

Croatia: Dalmatia, Gradac near Metković

Croatia (S Dalmatia)

Shell slender, finely ribbed, coarsely striated between the ribs, subcolumellaris and clausilium hardly visible in an oblique view of the aperture, lunula lateral.

Westerlund 1884: 109.

Agathylla neutra (Westerlund, 1898)

Clausilia neutra Westerlund, 1898 in Westerlund 1898: 170

Shell brown, smooth, 8-9 whorls, apertural margin connected at parietal side, parietalis long, columellaris straight and almost horizontal, 2 palatal folds, no lunula, subcolumellaris visible in the aperture.

8-12 x 1.8-2.4 mm

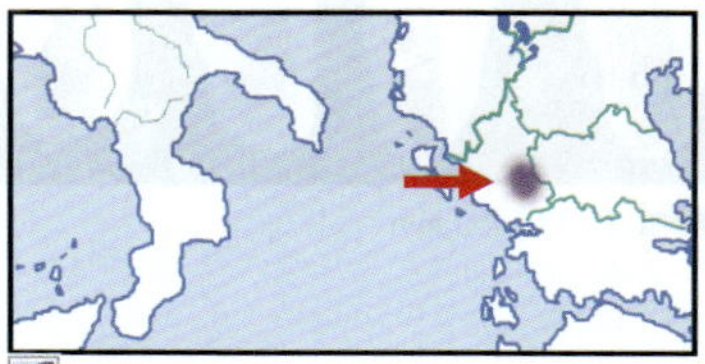

NW Greece

Agathylla regularis (Pfeiffer, 1861)

Clausilia regularis Pfeiffer, 1861 in L. Pfeiffer 1861: 170, Taf. 3

15-17 x 2.5-3 mm

Montenegro: Risan

Montenegro

Shell slender, horny bluish with white ribs, 11 whorls, aperture oblique, pear-shaped, apertural margin detached, dark, parietalis reaching the margin, lunula not visible.

Westerlund 1884: 109.

Agathylla strigillata (Rossmässler, 1835)

Clausilia strigillata Rossmässler, 1835 in Rossmässler 1835-1837 (2): 13, Taf. 7

9-10 x 2.5-2.9 mm

Croatia: Dalmatia, Peljesac

Croatia (Dalmatia)

Shell light greyish or with brown hue, with white delicate ribs, aperture rounded, margin reflected, upper palatalis present, rarely a second palatal tooth deeper inside, no lower palatalis, principalis shorter than in *Ag. formosa*.

Westerlund 1884: 108.

Agathylla sulcosa (Wagner, 1829)

Clausilia sulcosa Wagner, 1829 in Schubert & Wagner 1829: 190

12-18 x 2-3 mm

Croatia: Dalmatia, Komolac cemetery near Dubrovnik

Croatia (Dalmatia)

Shell grey with violet hue, densely and sharply ribbed, ribs white, or almost smooh, 10-12 whorls, aperture semioval pear-shaped, columellar margin very short, palatal margin much longer, columellaris oblique and long, one long and one shorter upper palatal fold present, lower palatalis long and very oblique, subcolumellaris visible in a perpendicular view.

Westerlund 1884: 108.

Agathylla viperina (Westerlund, 1901)

Clausilia viperina Westerlund, 1901 in Westerlund 1901: 94

9-11 x 2.3-2.7 mm

Croatia: Dalmatia, Orebić, Sveti Ilija mountain

Croatia (Dalmatia: Orebić)

Shell reddish brown with horny brown apex and whitish surface layers, weakly and coarsely ribbed, 9-10 whorls, aperture inside brownish, margin rounded and slightly reflected, parietalis and columellaris strong, columellaris almost horizontal, subcolumellaris visible in an oblique view.

Alopia

Alopia Adams & Adams, 1855 in Adams & Adams 1858: 181 *Clausilia (Alopia)*

Type species *Clausilia bielzii* Pfeiffer, 1849 (subsequent designation by Westerlund 1902: 108)

Contained originally *Cl. bielzii*, *Cl. lischkeana* and *Cl. pruinosa*. Albers & Martens 1860: XVII, 273 gave as type *Clausilia livida* Menke, 1828. Vest 1867: 162 gave as type *Cl. plumbea*. Both were not originally contained.

Kimakowiczia Szekeres, 1969 in Szekeres 1969: 27 *Alopia (Kimakowiczia)*, type species *Balea glauca* Bielz, 1853 (original designation)

Grossu 1981 (20 species, revision with detailed descriptions and range maps), Nordsieck 2007 (23 species, only a list), Nordsieck 2008 (22 species with 53 additional subspecies, dry revision without reproducible results and range maps).

Grossu 1981 and Nordsieck 2008 subdivided the genus into two subgenera *Alopia* and *Kimakowiczia*.

Species in *Alopia (Kimakowiczia)*: *A. glauca*, *A. maciana*, *A. cyclostoma* (= *pomatias*) and *A. soosiana*. All others in *Alopia (Alopia)*:

Grossu 1981 (17 species): *bielzii*, *bogatensis*, *canescens*, *fussi*, *glorifica*, *helenae*, *hildegardae*, *lischkeana*, *livida*, *meschendorferi*, *nefasta*, *nixa*, *plumbea*, *straminicollis*, *subcosticollis* (with *occulta*), *valachiensis* (with *zagani*), *vranceana*.

Nordsieck 2007/2008 (18-19 species): *bielzii*, *bogatensis*, *canescens*, *fussi*, *glorifica* (with *valachiensis*), *helenae*, *hildegardae*, *lischkeana*, *livida*, *meschendorferi*, *nefasta*, *nixa*, *occulta* (2008 again in *subcosticollis*), *petrensis* Nordsieck 1995, *plumbea*, *straminicollis*, *subcosticollis*, *vranceana*, *zagani*.

Nordsieck 1979: 57 reported that *Alopia livida* can locally interbreed with *straminicollis*, *nixa* and *glorifica*, producing so-called hybrid populations. It is possible that we are dealing with one single species complex as proposed by Szekeres 1976, and that there are less true species in *Alopia* than recognized under Nordsieck's 2008 system. The term "hybrid populations" was also used by Nordsieck 2004 for the *Albinaria cretensis* complex in Crete, where the field record suggests a highly variable species continuously distributed over area, displaying gradual character transitions and no hybrids as such (Welter-Schultes 2010). The situation in the *Alopia livida* complex suggests a similar situation. More research, involving detailed studies on local character variations, seems to be necessary to understand the nature and diversity of these populations. All this should be based on reproducible results. I see currently no evidence in the published record to justify the presence of several true species in the *Alopia livida* complex.

14 species are recognized in *Alopia (Alopia)* here: *bielzii*, *bogatensis*, *canescens*, *fussi*, *helenae*, *hildegardae*, *lischkeana*, *livida* (with *glorifica*, *nixa*, *straminicollis*, *valachiensis*, *zagani*), *meschendorferi*, *nefasta*, *petrensis*, *plumbea*, *subcosticollis* (with *occulta*), *vranceana*.

Alopia bielzii (Pfeiffer, 1849)

Clausilia bielzii Pfeiffer, 1849 in L. Pfeiffer 1849: 121 (*Clausilia Bielzii*)

9-20 x 2.8-4.7 mm, locally variable, usually 14-18 x 3.7-4.5 mm

Romania: Hunedoara castle

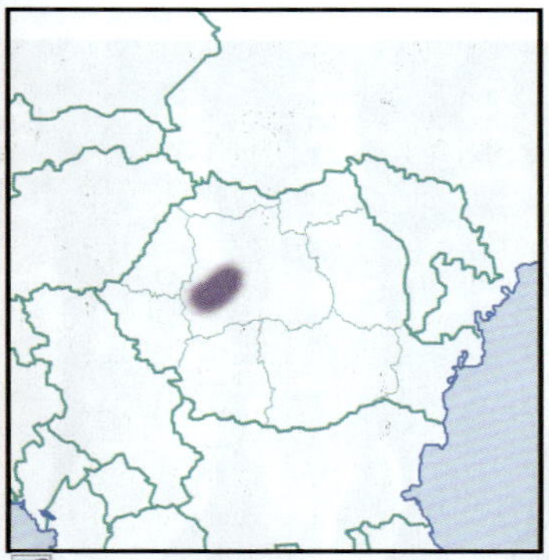

Romania (SW Transilvania)

Shell pale greyish horny brown, irregularly and widely ribbed, aperture without lip, columellaris strong, 4 prominent palatal folds, 3 of them long and one short.

Limestone outcrops and conglomerates. Locally frequent.

Westerlund 1884: 12, Grossu 1981: 36.

Slovakia: Slovenský Kras National Park
Determination uncertain, the species has not previously been reported from Slovakia.
Jozef Grego

Alopia bogatensis (Bielz, 1856)

Clausilia bogatensis Bielz, 1856 in E. A. Bielz 1856: 222

Shell thin, horny yellowish, with fine and widely spaced corrugated ribs, sometimes ribs only near the suture, suture white, cervix weakly convex and more prominently ribbed, columellaris very prominent, almost horizontal, 4 palatal folds, subcolumellaris well visible.

Shady and humid limestone habitats in mountains, 400-1800 m.

Westerlund 1884: 16, Grossu 1981: 78.

13-19 x 3.2-4.2 mm, seems to be locally variable

Romania: Perşani mountains, Bogatr valley

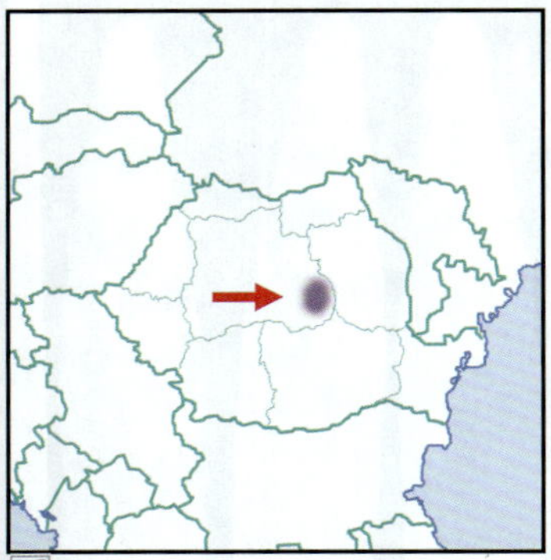

Romania (Perşani mountains)

Alopia canescens (Charpentier, 1852)

Clausilia canescens Charpentier, 1852 in Charpentier 1852: 364

Romania: Ciucaş mountains near peak Ciucaş

Romania (Ciukaş mountains)

12-17 x 3.7-4.7 mm, locally variable and larger

Westerlund 1884: 14, Grossu 1981: 113.

Shell sinistral, usually reddish violet with light blue surface layers, striated or almost smooth, suture white and often papillated, cervix inflated, light with white ribs, aperture inside brownish, parietalis moderately prominent, columellaris weak and relatively deep inside, not visible in the aperture, 2-3 usually small palatal folds, sometimes the upper fold stronger, subcolumellaris weakly visible, clausilium very slender.

Alopia cyclostoma (Bielz, 1858)

Balea cyclostoma Bielz, 1858 in E. A. Bielz 1858: 147

Not *Clausilia cyclostoma* Pfeiffer, 1850 (= currently *Cyclonenia cyclostoma* (Pfeiffer, 1850) from Korea). *Balea cyclostoma* Bielz, 1858 is not available if both are classified in the same genus (next available name for *Balea cyclostoma* would then be *Clausilia pomatias* Westerlund, 1884).

Balea cyclostoma synonym *Clausilia Pomatias* Pfeiffer, 1865 in L. Pfeiffer 1860-1866: 265

Balea cyclostoma synonym *Clausilia Pomatias* Pfeiffer, 1868 in L. Pfeiffer 1868: 367

At both occasions mentioned as a synonym, with descriptions. First used as a name for a taxon by Kobelt 1879: 54 with a bibliographical reference to Pfeiffer 1868: 397 (Kobelt cited *Cl. pomatias* from there), in the intention to establish a new replacement name for *Balea cyclostoma* Bielz, 1858 not *Clausilia cyclostoma* Pfeiffer, 1850 (the Code does not regulate which ones should be the types in such a case) (Kobelt 1879 gave also a bibliographical reference to Pfeiffer 1860-1866: 265, but did not cite the name *Cl. pomatias* from this source). For those who demand a biliographical reference for making available a synonym under Art. 11.6.1, the name must be attributed to Pfeiffer 1865. For the others, to Pfeiffer 1868.

For those who regard Art. 59.3 as useful and intelligent, the substitute name *Clausilia pomatias* Pfeiffer, 1868 (or 1865, depends on the interpretation of Art. 11.6.1) must be used by those who think that that name "is" in use (the name was used as *Alopia pomatias* by Nordsieck 1996: 91, Nordsieck 2007, Nordsieck 2008). I do not use it here, in line with Grossu 1981: 139.

I think Art. 59.3 should be removed from the Code or substantially modified. The Article was thought to protect usage of widely used names, not for fishing out unused old replacement names for rarely mentioned species. For defining widespread use of a name the criteria of Art. 23.9.2 could serve as a guide.

13-17 x 3.6-4.4 mm

Romania: W part of Bucegi mountains

Romania (Bucegi mountains)

Shell brown violet with light greyish blue surface layers, irregularly striated, 9-10 whorls with white suture, sometimes papillated, aperture large and almost circular, margin very broad, parietalis smaller than columellaris.

Westerlund 1884: 13, Grossu 1981: 139 (*Alopia cyclostoma*).

Alopia fussi (Kimakowicz, 1894)

Clausilia fussi Kimakowicz, 1894 in Kimakowicz 1894: 52 (*Clausilia (Alopia) Fussi*)

Shell dextral, reddish brown with a layer of light violetish hue on the lower whorls, whitish near aperture, finely striated, also on cervix, 8-9 whorls, aperture detached, inside with light brownish hue, margin slightly reflected, parietalis and columellaris moderate, no other folds present.

Restricted to high mountains, 1900-2500 m.

Grossu 1981: 104.

12-16 x 3.8-4.2 mm

Romania: Bucegi mountains, Vurful Doamnele, 1850 m

Romania (Bucegi mountains)

Alopia glauca (Bielz, 1853)

Balea glauca Bielz, 1853 in E. A. Bielz 1853: 120

Shell reddish violet with whitish or bluish white surface layers, or entirely white, smooth or ribbed, 8-9 whorls with white suture, suture sometimes almost papillated, cervix well rounded and in the last section white, apertural margin detached, parietalis small, columellaris even smaller, often almost absent, no palatal folds, no subcolumellaris and no clausilium present. Several shell characters seem to display spatial variations, or are variable within a population.

Westerlund 1884: 13, Grossu 1981: 135, Bădărău & Szekeres 2001.

11-18 x 3.6-4.7 mm, very variable, also within a population

Romania: Transilvania, "Dongökö"

Romania, Ukraine (E Carpathians)

Alopia helenae Kimakowicz, 1928

Alopia nefasta helenae Kimakowicz, 1928 in Kimakowicz 1928: 119, Taf. 3 (*Alopia nefasta-helenae*)

Shell dextral, reddish brown with light violetish surface layers on lower whorls, with widely spaced low ribs or weakly and irregularly striated, cervix on last section coarsely ribbed, 8-10 whorls, suture sometimes white, aperture inside yellowish or light brownish, margin continuous at parietal side and usually detached, thickened inside, principalis and upper frontal palatalis visible also from outside, basalis indistinct, subcolumellaris often visible in a perpendicular view.

Limestone rocks and conglomerates in mountains, 1000-1900 m. Frequent where present.

Grossu 1981: 124.

14-20 x 3-4.5 mm

Romania (SE Ciucaş mountains)

Romania: Ciucaş mountains, Piriul Alba, 1150 m

Alopia hildegardae Kimakowicz, 1931

Alopia hildegardae Kimakowicz, 1931 in Kimakowicz 1931: 41, Taf. 2

13-23 x 3.5-5 mm, locally variable

Romania: Transilvania, jud. Gorj, Polovrani

Romania (SW Carpathians)

Shell dextral, yellowish brown or light horny yellowish, lighter near the aperture, weakly striated or almost smooth, also on cervix, 8-12 whorls (locally variable), suture distinct and white, apertural margin detached, broad and white, parietalis and columellaris well developed, spiralis and frontal upper palatalis well visible in the aperture, subcolumellaris visible.

Very humid limestone rock habitats.

Grossu 1981: 54.

Alopia lischkeana (Charpentier, 1852)

Clausilia lischkeana Charpentier, 1852 in Charpentier 1852: 361, Pl. 11 (*Clausilia Lischkeana*)

13-22 x 3.5-5.5 mm, locally variable

Romania: Zărneşti Riu valley

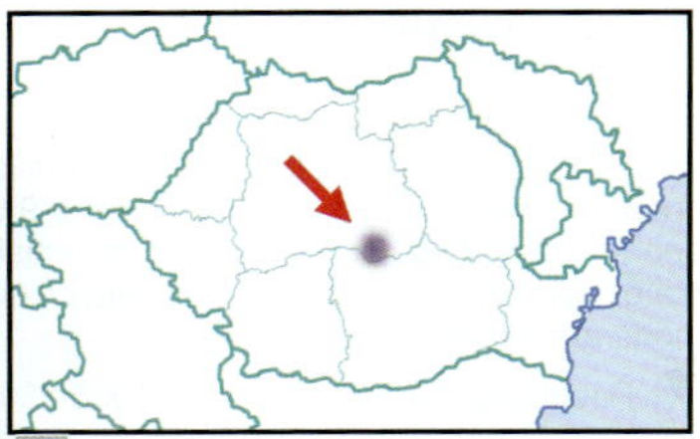

Romania (Piatra Craiului)

Shell dextral, brown violet or light horny brown, strongly ribbed, striated or almost smooth, suture and beginning of ribs white, cervix light or white, cervical ribs more prominent, apertural margin very broad, lip white or brownish, parietalis prominent and long, columellaris prominent and horizontal in frontal section, 4 palatal folds of which 3 are very small. Several shell characters seem to be locally variable.

Westerlund 1884: 11, Grossu 1981: 69.

Alopia livida (Menke, 1828)

Clausilia livida Menke, 1828 in Menke 1828: 77 (Established with description)
Clausilia glorifica Charpentier, 1852 in Charpentier 1852: 364
Clausilia straminicollis Charpentier, 1852 in Charpentier 1852: 364, Pl. 11
Clausilia regalis var. *valachiensis* Boettger, 1879 in Kobelt 1879: 58, Taf. 168
Spelling was not *valachensis*.
Clausilia nixa Kimakowicz, 1894 in Kimakowicz 1894: 39
Alopia valachensis zagani Szekeres, 1969 in Szekeres 1969: 31

12-22 x 3.5-5 mm, locally variable

(1-4, *livida*) Romania: Bucegi, near Sinaja, (5-6, *zagani*): Romania: Ciucaş mountains, Gropsoaerele valley (7-8, *straminicollis*) Romania: Bucegi

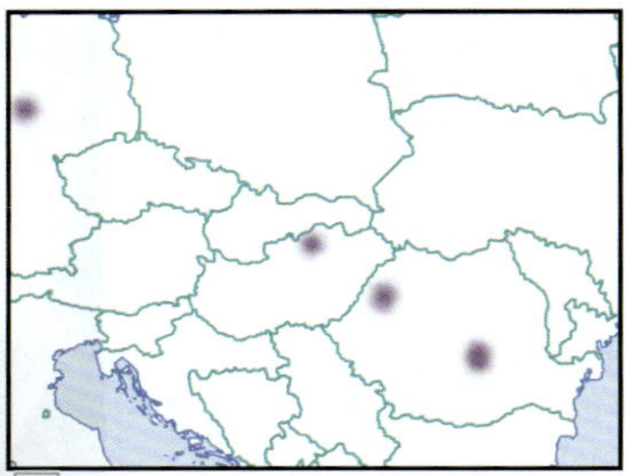

Romania (Bucegi and Bihor mountains); introduced to Hungary (Bükk mountains), and N Germany (Harz)

Shell dextral or sinistral, violet or brownish with or without greyish surface layers, or with white layers, smooth or weakly ribbed, shiny, suture and cervix white, 8-11 whorls with white or sometimes papillated suture, cervix finely striated or ribbed, aperture broad and oval, not always detached, often with a white lip inside, parietalis weak or strong, columellaris usually weak and deep inside, oblique and often S-shaped, but locally more prominent, usually no palatal folds (sometimes 2-3), no spiralis, subcolumellaris absent or very weak, clausilium absent or small. Shell characters display spatial variations.

The taxa *straminicollis*, *valachiensis*, *zagani*, *nixa* and *glorifica* are sinistral, *livida* is dextral.

Nordsieck 1979 reported that *livida* can locally interbreed with *straminicollis*, *nixa* and *glorifica*, it seems likely that we are dealing with one single polytypic species (as proposed by Szekeres 1976). More detailed studies are needed to verify the identity of this species complex (Grossu 1981: 109), and also of the taxa *valachiensis* and *zagani* which seem to belong to this complex as well.

Alopia maciana Bădărău & Szekeres, 2001

Alopia maciana Bădărău & Szekeres, 2001 in Bădărău & Szekeres 2001: 65

13.8-17.1 x 3.4-3.8 mm

Romania: jud. Cluj, Mt. Beliora, 1300 m

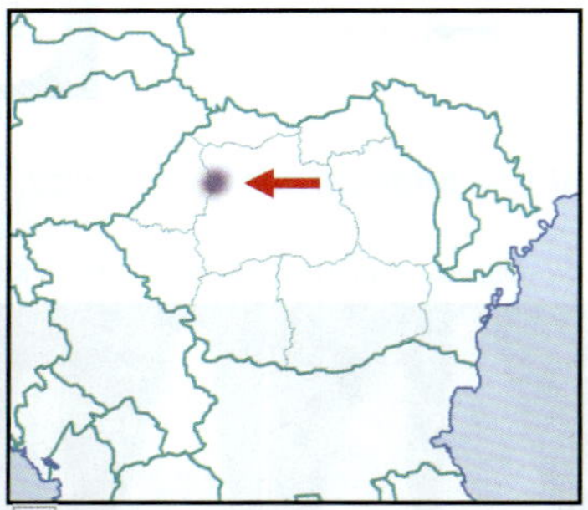

Romania (Gilău-Muntele Mare)

Shell dark brown with violet hue, ribbed, 8-10 moderately convex whorls with white suture, cervix rounded, aperture detached, margin broad and light yellowish, parietalis weak, reaching short spiralis inside, columellaris strong, principalis very short, upper palatalis weak and very short if present at all, no lunula, sometimes a very weak basalis, subcolumellaris not visible in a perpendicular view.

Only known from one locality.

Alopia meschendorferi (Bielz, 1858)

Clausilia meschendorferi Bielz, 1858 in E. A. Bielz 1858: 149 (*Clausilia Meschendorferi*)

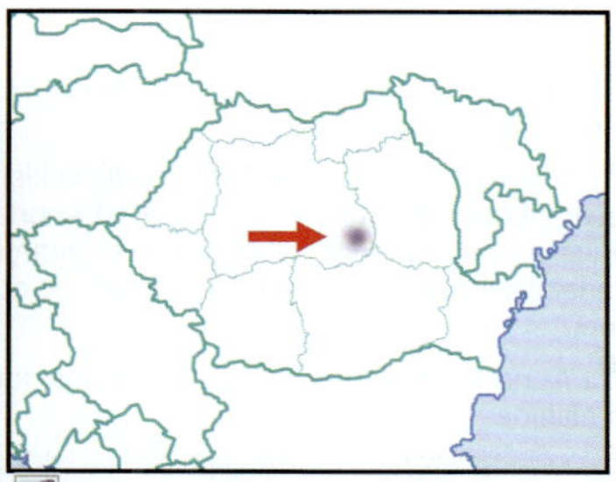

16-23 x 4.3-5.2 mm

Romania: Transilvania, Magura Codlei

Romania (SW Perşani mountains)

Shell dextral, horny yellowish, smooth, very shiny, white suture, cervix weakly corrugated, aperture brownish, margin reflected with distinct white lip and 4 palatal folds, 2 of which are visible in the aperture. Animal yellowish grey, on dorsum darker.

Eastern mountain slopes at limestone rocks in shady and humid habitats.

Grossu 1981: 75.

This is an isolated population or species, could also be classified as *A. lischkeana*.

Alopia nefasta (Kimakowicz, 1894)

Clausilia nefasta Kimakowicz, 1894 in Kimakowicz 1894: 51

Shell dextral brown-violet, usually with a whitish layer, weakly striated or almost smooth, also at cervix, 8-10 whorls, aperture inside white or slightly brownish, margin strongly reflected, with lip inside, parietalis weak, columellaris reaches apertural lip, no other folds present.
Differs anatomically from *A. livida*.

Humid habitats at limestone rocks in mountains, 1500-1600 m.

Grossu 1981: 109.

14-19 x 4.5-5 mm

Romania: Ciucaş mountains, Piriul Berei valley

Romania (Ciucaş mountains)

Alopia petrensis Nordsieck, 1996

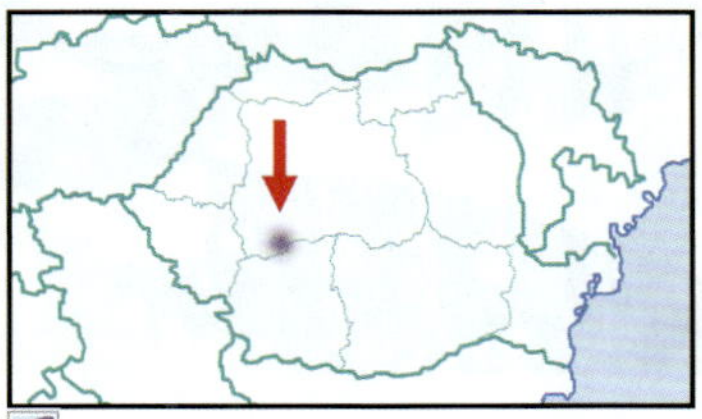

Clausilia straminicollis forma *occidentalis* Boettger, 1879 in Kobelt 1879: 58 (not *Clausilia laminata* var. *occidentalis* Bourguignat, 1877 in Bourguignat 1877: 18)
Alopia petrensis Nordsieck, 1996 in Nordsieck 1996: 91
Date was not 1995. Expressly established as a new replacement name for *Clausilia straminicollis* form *occidentalis* Boettger, 1879 non Bourguignat, 1876.

Differs from *A. straminicollis* mainly in its yellowish horny brown shell colour.

19-23 x 4.7-5.2 mm

Status as an independent species questionable, needs further research. Grossu 1981: 45 classified *occidentalis* Boettger as a synonym of *Alopia straminicollis straminicollis*.

Romania (SW Transilvania: Cheile Taia)

Alopia plumbea (Rossmässler, 1839)

Clausilia plumbea Rossmässler, 1839 in Rossmässler 1838-1844 (9-10): 16, Taf. 48

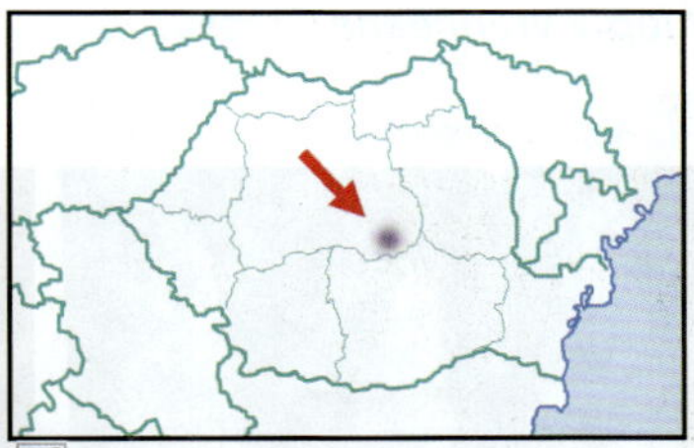

14-26 x 3-5.2 mm, locally variable

Romania: Bran castle

Romania (mountains S of Braşov)

Shell brown, often with bluish surface layers, or intensively blue violet, or almost white, or yellowish horny grey, almost smooth, also at the cervix, 9-11 whorls with white suture, cervix regularly rounded, bottom side of cervix white, columellaris strong and rounded, 4 palatal folds of which only one is very short, often visible from outside, subcolumellaris visible in a perpendicular view, clausilium very broad. Shell colour seems to vary locally.

Westerlund 1884: 15, Grossu 1981: 81.

Alopia soosiana Agócsy & Pócs, 1961

Alopia soosiana Agócsy & Pócs, 1961 in Agócsy & Pócs 1961: 533 (*Alopia soósiana*)

Shell sinistral, reddish brown with whitish surface layer, ribbed, ribs widely spaced, 9-10 whorls usually with white suture, aperture inside brownish, margin reflected, parietalis moderate, columellaris weak, no other folds present.

At northern slopes of high mountains, found in 1700 m altitude.

Must be rare and restricted to few localities, difficult to find.

Grossu 1981: 137, Bădărău & Szekeres 2001.

13-15 x 2.9-3.5 mm

Romania: Făgăraş mountains, Arpasu Mare valley, 1700 m

Romania (Făgăraş mountains)

Alopia subcosticollis (Schmidt, 1868)

Clausilia straminicollis var. *subcosticollis* Schmidt, 1868 in Schmidt 1868: 28
Alopia occulta Kimakowicz, 1931 in Kimakowicz 1931: 41, Taf. 2

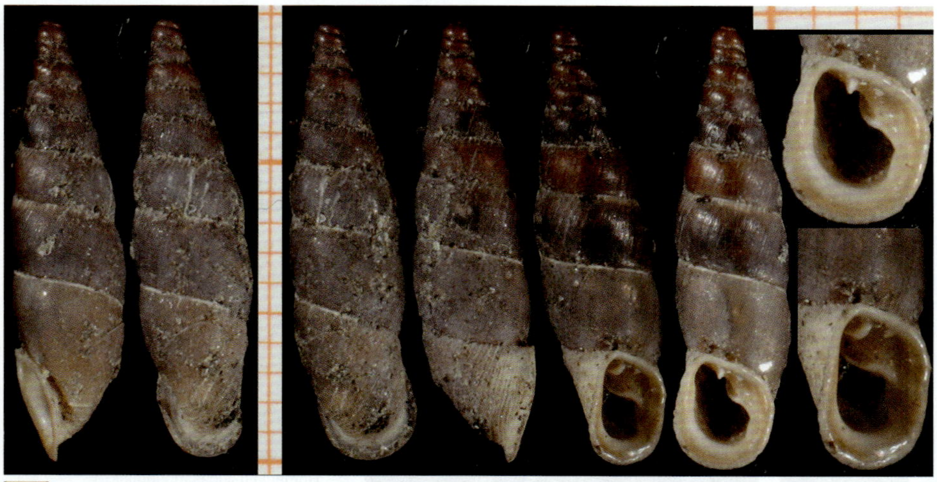

15-21 x 3.7-5 mm

Romania: 2 km N Baile Olanesti

Romania (SW Carpathians)

Shell sinistral, reddish violet, often with milky surface layers, almost smooth, suture white, 9-11 relatively convex whorls, cervix with strong and usually short white corrugated ribs coming from the keel region, columellaris prominent and curved, 3 palatal folds.

Westerlund 1884: 15, Grossu 1981: 41 (*occulta* in *subcosticollis*), Nordsieck 2007 (*occulta* as a separate species), Nordsieck 2008 (*occulta* in *subcosticollis*).

Romania, Retezat mountains, Cheile Butii
Jozef Grego

Alopia vranceana Grossu, 1967

Alopia vranceana Grossu, 1967 in Grossu 1967: 9

15-21 x 4.5-5 mm

Romania: jud. Vrancea, Cheile Tişiţa

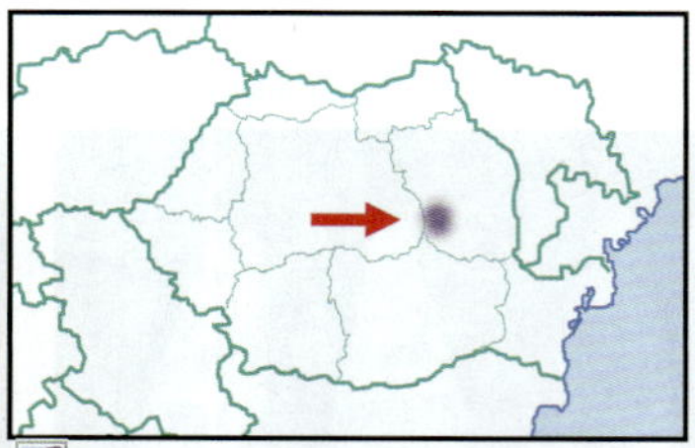

Romania (Vrancea mountains)

Shell sinistral, reddish brown with white surface layers, almost smooth or weakly striated, lighter and ribbed in the last section of cervix, shiny, 9-11 whorls with white suture, apertural margin reflected and with lip, parietalis long and well developed, columellaris reaches lip, principalis, 2 palatal folds and basalis present, subcolumellaris visible in an oblique view.

At steep limestone rocks, sometimes also at conglomerates, was found in 500-600 m.

Grossu 1981: 129.

Herilla

Herilla Adams & Adams, 1855
in Adams & Adams 1858: 181 *Clausilia (Herilla)*

Type species *Clausilia dacica* Pfeiffer, 1848
(subsequent designation by Albers & Martens 1860: 280)
(= currently *Herilla ziegleri* (Küster, 1847))

Shell usually smooth, yellowish to reddish brown with white suture, sometimes with weak white surface layer, cervix usually rounded, aperture detached or not, lunula dorsal or dorsolateral, subcolumellaris visible in a perpendicular view of the aperture or not. Size very variable, 12-37 x 3.5-8.5 mm. There are no definite morphological differences between *Alopia* and *Herilla*.

Nordsieck 1971.

Herilla bosniensis (Pfeiffer, 1868)

Clausilia bosniensis Pfeiffer, 1868 in L. Pfeiffer 1868: 475 (*Clausilia Bosniensis*)

20-27 x 5-7 mm, perhaps locally variable

Bosnia: Vrbas gorge near Banja Luka

Croatia to N Albania

Shell cherry or violet brown, almost smooth, 9-11 whorls, aperture oval or with weak sinulus, margin detached or not, columellaris prominent and nearly horizontal, pincipalis short, upper palatalis long, often passing over and behind lunula, often there are more palatal folds not connected to lunula, suturalis sometimes present, lunula not very long, subcolumellaris and lower palatalis equally long in their frontal sections, subcolumellaris usually not visible in a perpendicular view. Clausilium with two lobes, but lobes much smaller than in *H. ziegleri*. Some shell characters display considerable spatial variations.

Nordsieck 1971 suggested that *bosniensis* and *illyrica* are interconnected by intermediate forms (*hannae* and others), and thus may represent one single biological species. Introduced to Austria near Wien.

Herilla durmitoris (Boettger, 1909)

Clausilia durmitoris Boettger, 1909 in Wohlberedt 1909: 669, Taf. 54
Description and name were exclusively attributed to Boettger.

14-20 x 3.7-5 mm

Montenegro: Durmitor mountains, Bobotov Kok

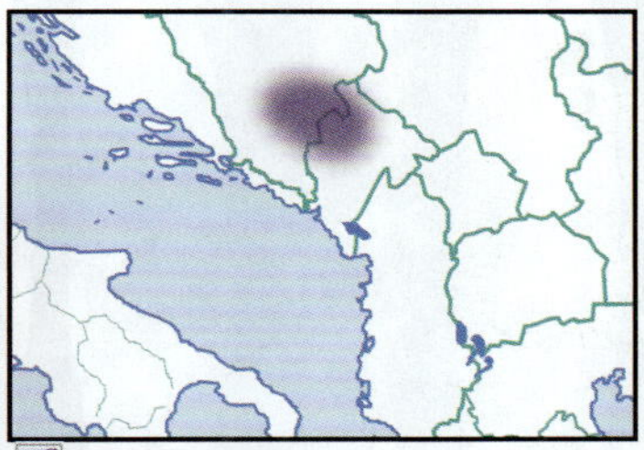

Montenegro, Bosnia

Shell weakly striated, 9-11 whorls, cervix rounded, apertural margin not detached, parietalis not reaching spiralis, columellaris not prominent, lunula weak and usually without middle section, subcolumellaris usually not visible in a perpendicular view. Clausilium plate with two lobes, outer lobe protruded, not closing perfectly.

Nordsieck 1971: 63.

Herilla illyrica (Möllendorff, 1899)

Clausilia illyrica Möllendorff, 1899 in Möllendorff 1899: 152

Shell light violetish with brownish first whorls, almost smooth, 11-13 whorls, aperture not detached but margin connected, parietalis variably long, columellaris variable, lunula weak in its upper section, frontal upper palatalis usually present and often connected to distal upper palatalis, basalis short and straight, subcolumellaris not visible in a perpendicular view. Several shell characters display local variations.

Nordsieck 1971: 74.

Could be connected to *H. bosniensis* by intermediate forms and thus represent one single species.

20-31 x 5-8.5 mm

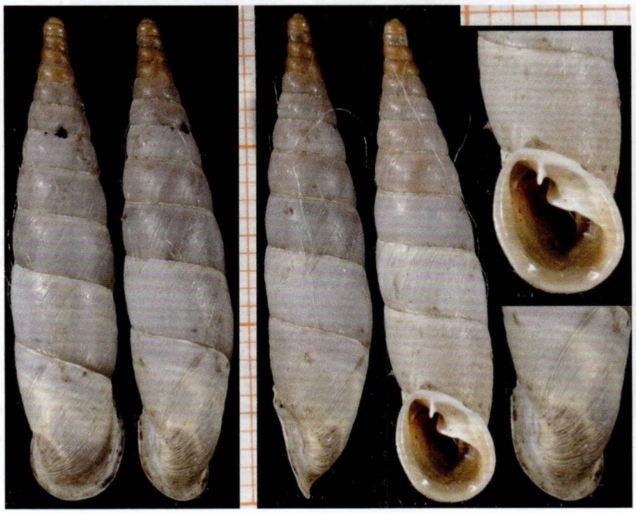

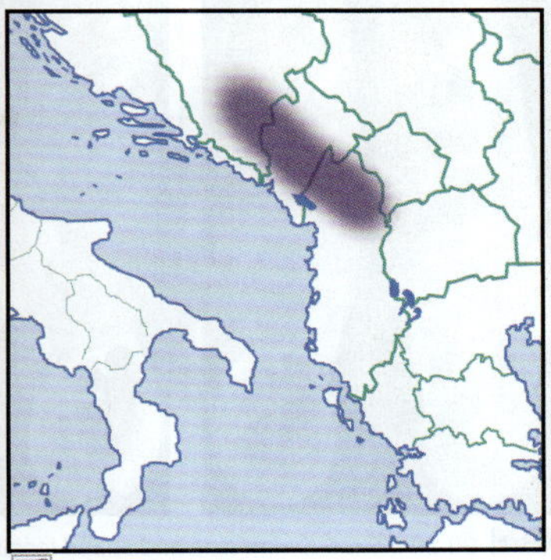

Bosnia to Albania

Albania: Periferi Kukës, Bicaj

Herilla jabucica (Böttger, 1907)

Clausilia jabucica Böttger, 1907 in Wohlberedt 1907: 550 (*Clausilia (Herilla) iabucıca*)
Originally spelled *iabucıca*, inadvertent error for *jabucica*. There must have been dirt particles on the printing plate. Name attributed to Böttger, description also entirely attributed to Böttger.

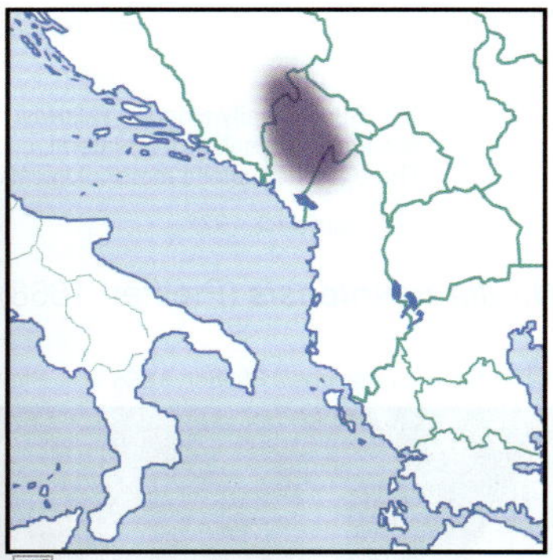

20-26 x 5-6.5 mm

Albania: Malësia, near Vermosh

Museum Budapest HNHM

Nordsieck 1971: 59, Dhora 2004: 145.

Bosnia to N Albania

Shell almost smooth, 10-12 whorls, apertural margin not detached, parietalis and columellaris variable, lunula often weak or absent in its middle section, frontal upper palatalis usually present and connected to distal upper palatalis, basalis present, subcolumellaris sometimes visible in a perpendicular view. Clausilium plate like in *H. ziegleri* but outer lobe never pointed.

Herilla pavlovici (Wagner, 1914)

Alopia pavlovici Wagner, 1914 in A. J. Wagner 1914 (Icon. 21): Taf. 590
Plate was published Feb 1914, name was printed on the plate, text p. 38 was published in Jan 1915 and is irrelevant.

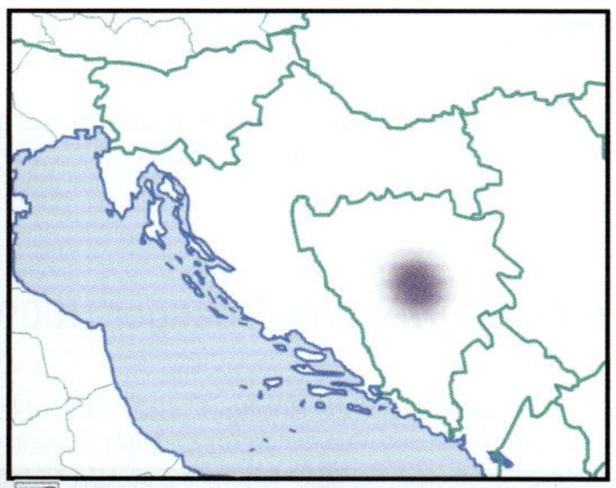

13.5-16.5 x 3.5-3.8 mm

Bosnia: near Fojnica

Museum Budapest HNHM

Nordsieck 1971: 64.

The morphological differences between *durmitoris* and *pavlovici* seem to be very weak.

Bosnia

Shell yellowish brown with green hue, suture white and papillated, irregularly striated, last whorl ribbed, 9-10 whorls, aperture inside horny yellowish, parietalis low and short, columellaris low, 2 small palatal folds present, lunula and basalis largely reduced or absent, subcolumellaris rarely visible in a perpendicular view. Differs from *H. durmitoris* in its more thickened apertural margin, subcolumellaris more often visible, lunula more reduced, palatal folds often equally strong.

Was found in caves.

Herilla ziegleri (Küster, 1847)

Clausilia ziegleri Küster, 1847 in Küster 1847: 16 (*Clausilia Ziegleri*)
Figured on Taf. 1 Fig. 17-18 [1845], name not printed on plate, only perhaps on the wrapper, which was not published work (Art. 8.1.1).

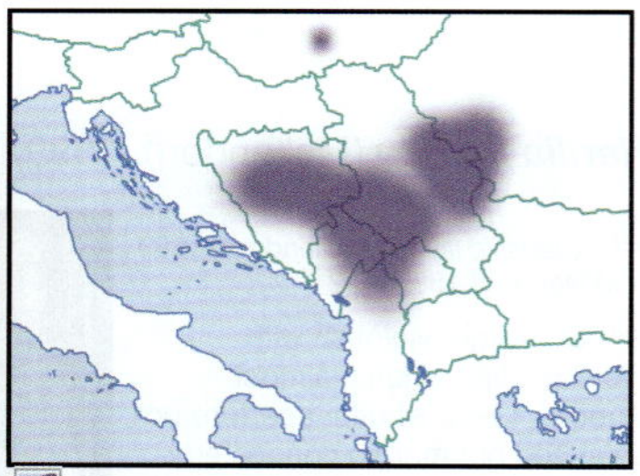

20-29 x 5.5-6.5 mm, size and shape locally variable

Serbia: Rudo/Lin

Bosnia to Romania (Banat)

Shell cherry brown red or light reddish brown, often with milky surface layers, finely striated or smooth, 9-11 whorls, suture white and papillated on upper whorls, apertural margin not always detached, 2 upper palatal folds, the lower one short and connected with lunula, basalis prominent, sometimes extremely prominent, lunula dorsal, well visible from outside. Clausilium with two lobes, right lobe short and pointed, left lobe longer and curved.

Westerlund 1884: 49.

Medora

Medora Adams & Adams, 1855 in Adams & Adams 1858: 183 *Clausilia (Medora)*

Type species *Clausilia macarana* Rossmässler, 1834 (subsequent designation by Albers & Martens 1860: XVII, 276) (= currently *Medora macascarensis* (Sowerby, 1828))

Nordsieck 1970, Giusti, Gr. et al. 1986.

The classical subdivision of this group into species and subspecies followed Nordsieck 1970 and complementary studies. In a detailed study Giusti, Gr. et al. 1986: 314 suggested that this method would seem subjective and depend on specific weights granted to certain morphological characters (see detailed comment under *M. macascarensis*).

Medora agnata (Pfeiffer, 1848)

Clausilia agnata Pfeiffer, 1848 in L. Pfeiffer 1848: 435
Name mentioned by Küster 1847: 38 [1850], Taf. 4 Fig. 7-9 [1847], name not printed on plate, only perhaps on the wrapper, which was not published work (Art. 8.1.1).

14-22 x 3-4 mm

Croatia: Mali Alan

NW Croatia

Shell horny or opal yellow whitish, almost smooth, shiny, cervix finely striated with indistinct basal keel, aperture nearly rectangular with curved lower side, margin usually not connected at parietal side, or only by a very thin whitish layer, lip weak, columellaris oblique and thickened in frontal section, lower palatalis often short but can be longer, very close to subcolumellaris, lunula broad at lower side, upper side curved, well visible from outside.

Westerlund 1884: 105.

Medora almissana (Pfeiffer, 1848)

Clausilia almissana Pfeiffer, 1848 in L. Pfeiffer 1848: 432
Name mentioned by Küster 1847: 33 [1850], Taf. 3 Fig. 24-31 [1847], name not printed on plate, only perhaps on the wrapper, which was not published work (Art. 8.1.1).

18-32 x 4.5-6.5 mm, must be locally variable

Croatia: Dalmatia, Almissa

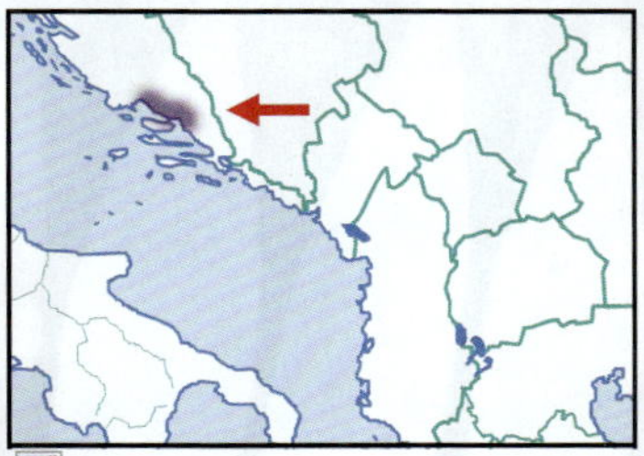

Croatia (Dalmatia)
Introduced to Germany at Weltenburg/Donau.

Shell milky bluish violet with horny brown apex, almost smooth, 9-11 whorls, cervix ribbed and nearly rounded with basal keel, aperture very broad and rounded, inside brownish, margin detached, columellaris prominent and concave in its lower frontal section, spiralis strongly curved outwards, suturalis strong, upper palatalis present, lower palatalis (not basalis) short and broad, connected to lunula, lunula dorsal and curved in its upper section.

Westerlund 1884: 102, Falkner 1990: 154.

Medora armata (Küster, 1847)

Clausilia armata Küster, 1847 in Küster 1847: 22
Figured on Taf. 2 Fig. 8-10 [1844], name not printed on plate, only perhaps on the wrapper, which was not published work (Art. 8.1.1).

15-19 x 3.2-4 mm

Croatia: Dalmatia, Brač island, Supetar

Museum Budapest HNHM

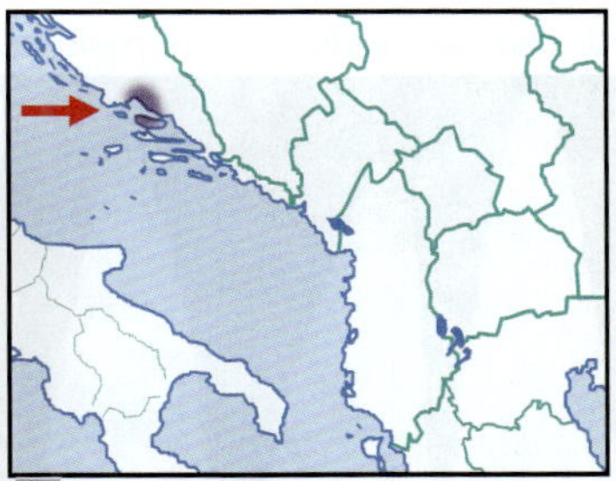

Croatia (Dalmatia)

Shell solid, prominently ribbed (or percostate) on a striated background, aperture large, lower palatalis strong, subcolumellaris and clausilium well visible in an oblique view, lunular almost dorsal. Smaller and more slender than *M. almissana*.

Very small area of distribution, which should be seriously protected.

Westerlund 1884: 109, Falkner 1990: 154.

Medora contracta (Rossmässler, 1842)

Clausilia dalmatina var. *contracta* Rossmässler, 1842 in Rossmässler 1838-1844 (11): 4, Taf. 52

17-25 x 5.3-6.2 mm

Montenegro: Morača valley near Podgorica

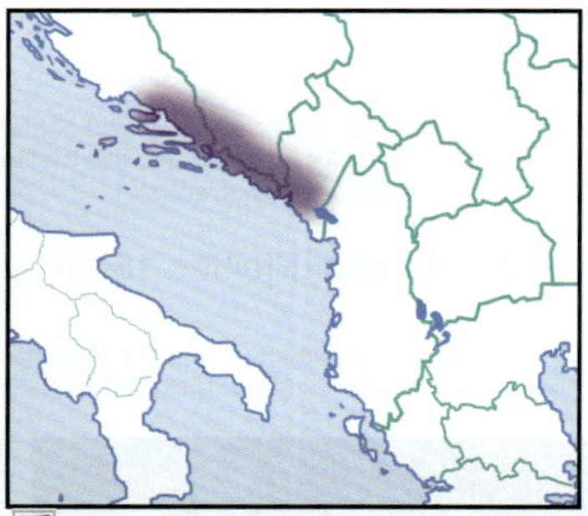

Croatia (Dalmatia) to Montenegro

Shell not very high, 8-9 whorls, cervix strongly corrugated, aperture inside light yellowish brown, 2 upper palatal folds, the upper one long and parallel to the weak suturalis, the lower one short, deeply inside and connected to the lunula, subcolumellaris visible in an oblique view.

Westerlund 1884: 99.

Medora dalmatina (Rossmässler, 1834)

Clausilia marmorata Rossmässler, 1834 in Rossmässler [1834] (2): 7
Clausilia marmorata synonym *Clausilia dalmatina* Rossmässler, 1834 in Rossmässler [1834] (2): 7

Rossmässler [1834] mentioned both names and selected *marmorata*. Correct name should be *M. marmorata* (Rossmässler, 1834). This name *marmorata* was not used after 1834, another name *marmorata* does probably not interfere.

Google general search (11.2010): 86 "medora-dalmatina" 86 hits, "medora-marmorata" 0. There are probably not 25 citations for *Medora dalmatina* to suppress *marmorata* under Art. 23.9. I would not like to be the first one to use *Medora marmorata*, at least not without consulting other opinions. But if anyone starts using *marmorata* I would not use *dalmatina* any more.

Clausilia dalmatina Rossmässler, 1835 in Rossmässler 1835-1837 (2): 7, Taf. 7 (First use of this name for a taxon).

20-32 x 4.8-6.7 mm, locally variable

Dalmatia: Vrgorac, Biokovo mountains

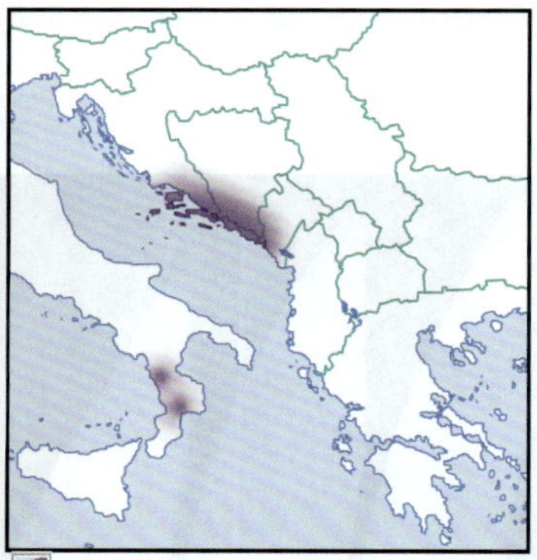

Croatia to Montenegro, S Italy

Westerlund 1884: 100, Nordsieck 1970, Alzona 1971: 83, Manganelli et al. 1995: 23.

Shell greyish milky white with horny brown apex, blue or violetish grey marbled, smooth or finely striated, 9-11 whorls, cervix coarsely or densely ribbed without prominent keels, some local forms have a weak basal keel, aperture large, inside yellowish brown, margin usually detached, columellaris often forked inside, suturalis very delicate, 2 upper palatal folds, lower palatalis slightly oblique or perpendicular and well visible. Last whorls and cervix locally variable.

The ribbed *Medora hiltrudae* could also belong to this group. The Italian populations need special attention, they could also belong to *M. macascarensis*.

Medora eris (Pfeiffer, 1866)

Clausilia eris Pfeiffer, 1866 in L. Pfeiffer 1866: 150 (*Clausilia Eris*)

20-28 x 5.3-7.3 mm, perhaps locally variable

Croatia: Dalmatia, near Makarska

Museum Budapest HNHM

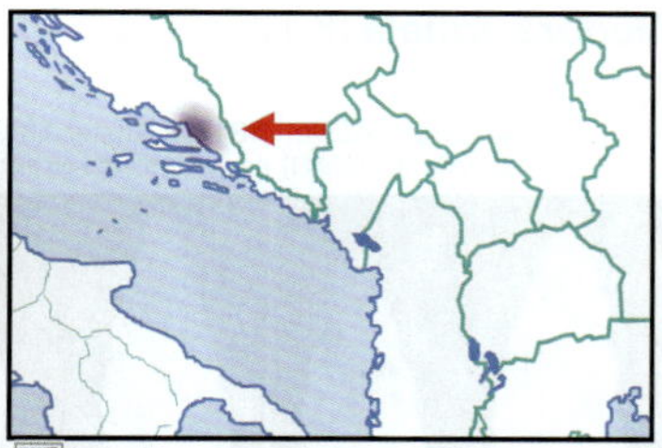

Croatia (Dalmatia)

Shell whitish grey or bluish with blackish and horny brownish spots, almost smooth, shiny, aperture oval, yellowish white, cervix can be ribbed, apertural margin attached at parietal side, columellaris very oblique and low, lower palatalis short and shiny white, well visible, hardly or not connected to lunula, subcolumellaris strong, visible from the aperture and not reaching deeply inside, lunula almost dorsal and broad. Colour possibly locally variable.

Westerlund 1884: 103.

Medora hiltrudae Nordsieck, 1970

Medora hiltrudae Nordsieck, 1970 in Nordsieck 1970: 38

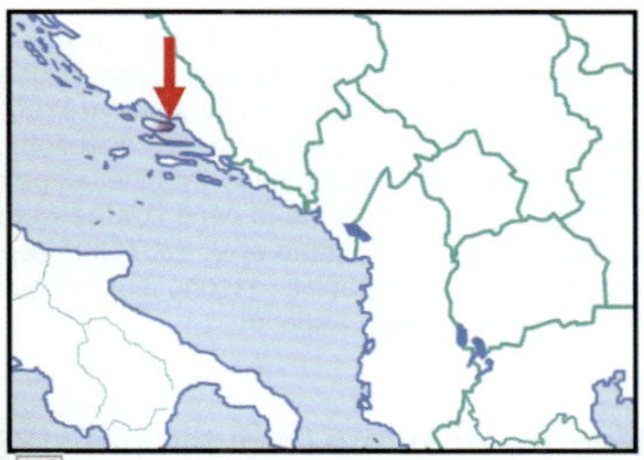

Croatia (Dalmatia)

Shell moderately ribbed (18-25 ribs/penultimate whorl), on cervix more widely spaced, apertural margin detached and slightly protruded, columellaris low and very oblique, distal end of spiralis simple, lunula dorsolateral, frontal upper palatalis absent or weak and disconnected from lunula and distal upper palatalis, basalis visible in a perpendicular view, subcolumellaris not visible in the aperture. Differs from *M. dalmatina* in its surface sculpture, weak or absent frontal upper palatalis and simple distal end of spiralis.

17-20 x 4.5-5.1 mm — Croatia: Dalmatia, Brač island, Sumartin

Very restricted range only around Sumartin bay on Brač island, Dalmatia.

It is likely that this form belongs to *Medora dalmatina*, a species that needs to be better understood in its local variations.

Medora lesinensis (Pfeiffer, 1848)

Clausilia lesinensis Pfeiffer, 1848 in L. Pfeiffer 1848: 434

Name mentioned by Küster 1847: 37 [1850], Taf. 4 Fig. 1-3 [1847], name not printed on plate, only perhaps on the wrapper, which was not published work (Art. 8.1.1).

Croatia (Dalmatia)

Westerlund 1884: 105.

22-24 x 4.6-5.4 mm — Montenegro: near Kameno — Museum Budapest HNHM

Shell slender, violetish grey with darker spots on the middle whorls, obliquely and finely striated, 12-13 whorls with white suture, cervix finely and densely ribbed, aperture egg-shaped and yellowish brown, margin not connected at parietal side, parietalis reaching the margin, columellaris oblique, not very high, shiny white, lunula broad, straight and almost dorsal, lower palatalis very short and connected to lunula.

Medora macascarensis (Sowerby, 1828)

Clausilia macascarensis Sowerby, 1828 in Sowerby 1821-1834 (part 30): p. [2], pl. [168] or [30] (*Cl. Macascarensis*)

Original spelling was not *macarscarensis*.

Clausilia albescens Menke, 1830 in Menke 1830: 130 (established with description)

Clausilia dalmatina var. *italiana* Küster, 1847 in Küster 1847: 32 (*Clausilia dalmatina* var. *Italiana*)

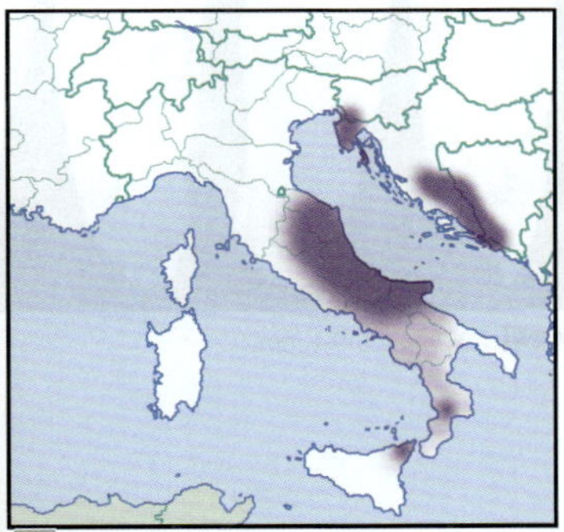

Slovenia to Herzegovina, Italy

Shell light violetish grey with horny brown apex, 9-10 whorls, cervix coarsely ribbed with basal keel, aperture very large and rounded, inside brownish, margin light brown, always detached, columellaris prominent, inside thinner and weakly curved, upper palatalis often slightly shorter than half length of the principalis, lower palatalis short, hardly visible in a perpendicular view, subcolumellaris strong and visible in a slightly oblique view.

22-31 x 5-6.5 mm, locally variable — Croatia: Dalmatia, Versic cave near Makarska

Westerlund 1884: 103, Alzona 1971: 83, Giusti, Grapelli et al. 1986, Manganelli et al. 1995: 23 (*M. albescens* in Italy), Nordsieck 1995: 92 (considered *albescens* as a synonym of *macascarensis*), Štamol & Velkovrh 1995 (N Croatia: Cres, only at steep slopes and cliffs), Fehér 2000 (gave 12 subspecies, 2 in SW Slovenia and Istria, 10 in central to S Dalmatia and W Bosnia, Italy not mentioned), Nordsieck 2007: 43 (*M. macascarensis* with 13 subspecies, *M. italiana* with 5 subspecies), Reitano et al. 2007: 322 (one more subspecies for *M. italiana* from Sicily).

Clausilia italiana Küster was considered by Nordsieck 1970 as a separate species. In a combined conchological, anatomical and electrophoretical study dedicated to this complex Giusti, Gr. et al. 1986: 316 came to the conclusion that it was neither possible to separate *italiana* from *albescens*, nor to establish the presence of subspecies within this whole group. This argument has not been weakened afterwards. Until present researchers failed to provide evidence for the presence of well defined uniform subspecies against the much more likely hypothesis of continuous character variations running over area in various directions, a situation in which establishing subspecies would not be aligned with the usage of this term in biodiversity concepts (existence of partial barriers to gene flow between populations).
Nordsieck 2007 and Reitano et al. 2007: 325 separated *italiana* and the Balcanic forms mainly for geographic reasons, but it should be noted that also the Balcanic forms occur in disjunct regions.

For weakening or specifying the conclusions of Giusti, Gr. et al. 1986 another specially designed and detailed study would be needed. Characters of genitalia and radula provided little diagnostic information, but attempts to transfer conchological characters into numerical values were successful. This is a good precondition to study regional trends at a much denser scale, since only some 10 shells per locality need to be analysed (Welter-Schultes 2010). It would be necessary to select independent characters. The effect of shell height H on diameter D needs to be discounted, to consider only H/D, the same with apertural sizes. Relative whorl number could be added, a previously overlooked character. For some characters (subcolumellaris etc.) numerical values at finer scales could be obtained.
Conclusions from such morphological studies could afterwards be tested with molecular methods (it would probably be inefficient to study hundreds of lots with molecular methods without knowledge on the morphological background of these populations).

Medora proxima (Walderdorff, 1864)

Clausilia proxima Walderdorff, 1864 in Walderdorff 1864: 508

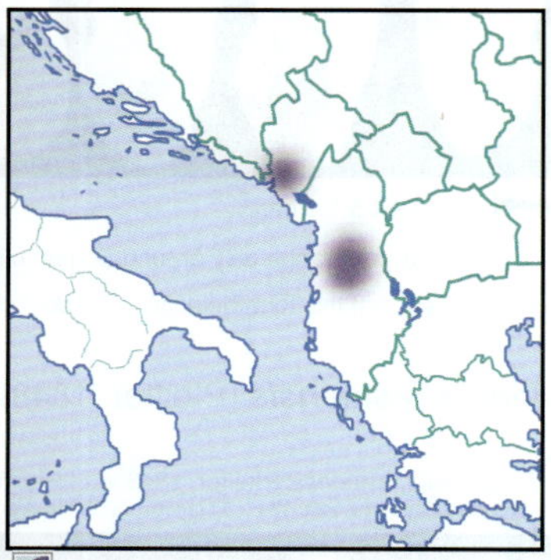

Montenegro to central Albania

19-26 x 4.7-5.8 mm, shape probably locally variable

N Albania: Mali Krujë

Westerlund 1884: 106.

Shell bluish white with some violet spots and horny brown apex, almost smooth, 8-10 whorls, cervix sometimes ribbed, aperture large, rounded, inside yellow brownish or whitish, margin yellowish or brownish, connected at parietal side, sometimes slightly detached, parietalis inside curved, columellaris often with a thin branch on its upper side leading inwards, upper palatalis short, strong and deep inside, lower palatalis almost absent, spiralis and columellaris inside far distant from each other, but parallel and almost equally long.

Medora stenostoma (Rossmässler, 1839)

Clausilia macarana var. *stenostoma* Rossmässler, 1839 in Rossmässler 1838-1844 (9-10): 21, Taf. 48

Croatia (Dalmatia)

Shell like *M. macascarensis* but smaller and more slender, aperture smaller and narrower, apertural folds weaker, cervix finely ribbed, subcolumellaris weaker.

20-24 x 4.6-5.4 mm

Croatia: Biokovo, Sv. Jure

Museum Budapest HNHM

Westerlund 1884: 104.

Triloba

Triloba Vest, 1867
in Vest 1867: 165 *Clausilia (Triloba)*

Type species *Clausilia sandrii* Küster, 1847
(original designation)

Triloba sandrii (Küster, 1847)

Clausilia sandrii Küster, 1847 in Küster 1847: 28 (*Clausilia Sandrii*)
Figured on Taf. 2 Fig. 20-23 [1844], name not printed on plate, only perhaps on the wrapper, which was not published work (Art. 8.1.1).

22-27 x 5.5-7 mm, perhaps locally variable

Albania: Mali i Dajtit

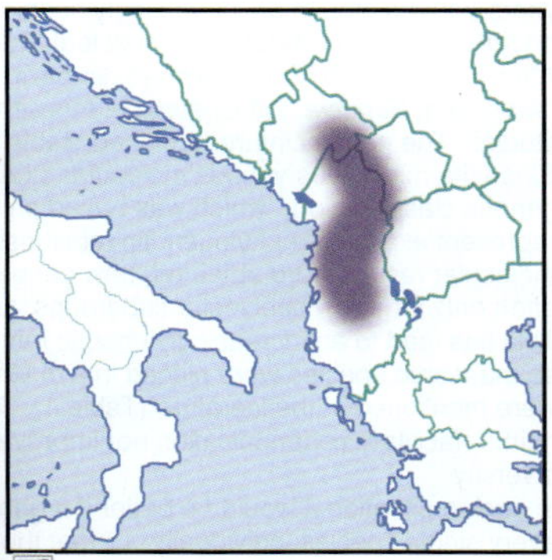

Montenegro, Albania

Shell reddish brown, densely striated on upper whorls, smooth on lower whorls, apertural margin connected by a callus on parietal side, suturalis delicate, 4 palatal folds (long, short, short, long), clausilium subdivided in two parts with its inner lobe.

Westerlund 1884: 16.

Triloba thaumasia (Sturany, 1907)

Clausilia thaumasia Sturany, 1907 in Sturany 1907: 233

21-24 x 5.4-6.2 mm

Rep. Makedonija, Bistra mountains, Sveti Jovan Bigorsk monastery

Deciduous forests in herbal and shrub vegetation at the base of limestone rocks.

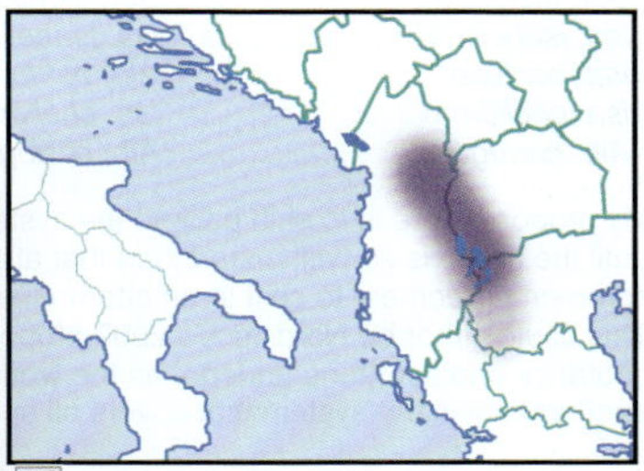

Rep. Makedonija, Greece, Albania

Shell dark reddish brown, striated, 10-11 whorls with shallow and whitish suture, aperture inside light reddish, margin slightly thickened, attached and connected by a callus at parietal side, parietalis almost perpendicular, separated from spiralis, columellaris obliquely horizontal, no lunula, frontal upper palatalis and basalis prominent, subcolumellaris not visible in the aperture. Clausilium unevenly two-lobed, sometimes with a weak 3rd lobe in between.

Dedov & Neubert 2002: 202, Dedov & Neubert 2009: 90.

3 - *Albinaria* subgroup (Greece-Turkey)

Albinaria, Isabellaria, Sericata, Carinigera, Cristataria

The generic classification of the members of this group in the past 10 years has resulted in an increasingly sophisticated system produced by taxonomists, for non-insiders this must look like the absolute chaos.
After Nordsieck 1977 had finished his basic revisions of the genera *Albinaria*, *Isabellaria*, *Sericata* and *Cristataria*, this served as a solid base for further studies. The nomenclature remained stable for 20 years into the 1990s.
Since the mid-1990s various molecular studies suggested that Nordsieck's generic classification - which was based on very few shell characters - did not represent exactly the phylogenetic relationships. Very quickly after the first molecular results were obtained, new generic classifications were proposed, often only for a few species or subgroups, and lasting only for a few years. This has lead to an increasingly chaotic mix of generic names in this group, so that some species were placed in a different genus almost every once they were mentioned in the literature (Table 1). Such a procedure does certainly neither facilitate communication nor improve our understanding of molluscan diversity.
In such a situation it would be better if a generic name was proposed for every single species individually, so that the names of the species would not change with every new study. But for such a model we would not need a binominal nomenclature.
At the current state of research it is not possible to present a consistent and accepted taxonomical system for all species involved.
The last proposal by Nordsieck 2007 was to classify all species of this group in one single genus, for which *Albinaria* was selected. Certainly an intelligent proposal, worth to be considered for a future classification.
Prior to the 1900s all clausiliids were classified in one single genus (*Clausilia*) until the system was better understood. In the past 15 years we have seen that the groups in Greece and Turkey are not well understood at all, so using one single generic name for them would be a logical consequence. And also useful, because it would enable the non-specialist at any time of the year to use the correct generic name for a species in this group. It would also enable researchers to combine newly discovered species with a correct generic name.
A shortcoming of this proposal is that this would result in once again new genus-species combinations for several species of *Cristataria*, *Carinigera* and *Isabellaria*, that had never been combined with the generic name *Albinaria*.
Associated with this problem is another problem that Nordsieck 2007 did not solve when he classified all species in *Albinaria*: the resulting secondary junior homonyms must obtain substitute names, including at least the species currently known as *Albinaria schuetti*, *Carinigera pellucida* and *Albinaria zilchi*, possibly more. This would unnecessarily distort nomenclature, particularly if not all researchers would accept this proposal.

Table 1. Some major works of the past 30 years and their divergent classifications of various species (*colbeauiana* and *hedenborgi* from the Near East do not live in Europe, abbreviations refer to *Albinaria, Carinigera, Cristataria, Inchoata, Isabellaria, Sericata*).

Nordsieck 1977	Gittenberger 1998	Nordsieck 1999	Gitttenberger & Uit de Weerd 2006	Nordsieck 2007
Cri. colbeauiana	*Cri. colbeauiana*	*Cri. colbeauiana*	*Cri. colbeauiana*	*Alb. colbeauiana*
Alb. hedenborgi	*Alb. hedenborgi*	*Alb. hedenborgi*		*Alb. hedenborgi*
Car. eximia	*Car. eximia*	*Car. eximia*	*Car. eximia*	*Alb. eximia*
Car. megdova	*Car. megdova*	*Car. megdova*	*Inc. megdova*	*Alb. megdova*
Ser. sericata	*Ser. sericata*	*Alb. sericata*	*Isa. sericata*	*Alb. sericata*
Ser. dextrorsa	*Ser. dextrorsa*	*Alb. dextrorsa*	*Car. dextrorsa*	*Alb. dextrorsa*
Ser. inchoata	*Ser. inchoata*	*Alb. inchoata*	*Inc. inchoata*	*Alb. inchoata*
Isa. isabellina	*Isa. isabellina*	*Isa. isabellina*	*Isa. isabellina*	*Alb. isabellina*
Isa. confusa	*Alb. confusa*	*Isa. confusa*	*Alb. confusa*	*Alb. confusa*
Isa. chelidromia	*Isa. chelidromia*	*Isa. chelidromia*	*Car. chelidromia*	*Alb. chelidromia*
Alb. corrugata	*Alb. corrugata*	*Alb. corrugata*	*Alb. corrugata*	*Alb. corrugata*

My proposal here is to shift back to the system of Nordsieck 1977, and wait until the group is so well understood that all involved research groups obtain common agreement to shift to an alternative and consistent classification. This could either be Nordsieck's 2007 proposal, if this obtains common acceptance and someone undertakes the work to dissolve all homonymies, or another consistent system that covers all species.
I would also strongly recommend not to establish new specfic names in this group that have already been established in one of the genera involved here. It can generally not be recommended to establish over and over again the same specific names in closely related genera. In botany it is not allowed to establish the same specific name twice in the same family. This idea should serve as a guide here, too.

Albinaria

Albinaria Vest, 1867
in Vest 1867: 170 *Clausilia (Albinaria)*

Type species *Clausilia caerulea* Deshayes, 1835
(original designation)

Rocky limestone outcrops, either atached to the rock faces or more hidden under boulders and in rock crevices.
The most significant predator is a beetle (Drilidae), 10-50 % of the shells show characteristic oval boreholes, there seem to be local and periodical oscillations of predation pressure in the course of several years. Still a mystery is an unknown organism perforating circular boreholes in *Albinaria* shells. Other predators of minor importance are rodents, birds, *Poiretia* snails in some regions, carabid beetles and *Eopolita* snails.
Dispersal ability is relatively low at 0.6-2.5 m yr^{-1}, levels of gene flow are extremely low. The degree of inbreeding is low, values for the panmictic unit are small, only 50 m^2.

Active season from October to April, copulation occurs immediately and synchronously after awakening at the first rainfalls in autumn, this period lasts about one week. 20-30 days after copulation 5-8 eggs are laid in November, juveniles hatch after 2-3 weeks. Snails begin estivating in April, independently from the end of the rainfall season. Shell growth is completed after 2 years, maturity is reached shortly after, maximum age about 7 years.

S Albania, Greece, Cyprus, W and S Turkey, Lebanon

Revisions: Wagner 1924, Nordsieck 1977. Lists of species: Nordsieck 1999, Welter-Schultes 2000: 171, Giokas & Mylonas 2002 (life cycle), Nordsieck 2007 (classified all *Sericata*, *Isabellaria*, *Carinigera* and *Cristataria* species within *Albinaria*), Welter-Schultes 2010 (revision Crete, with references for the biological information).

Albinaria adrianae Gittenberger, 1979

Albinaria adrianae Gittenberger, 1979 in Gittenberger 1979: 83

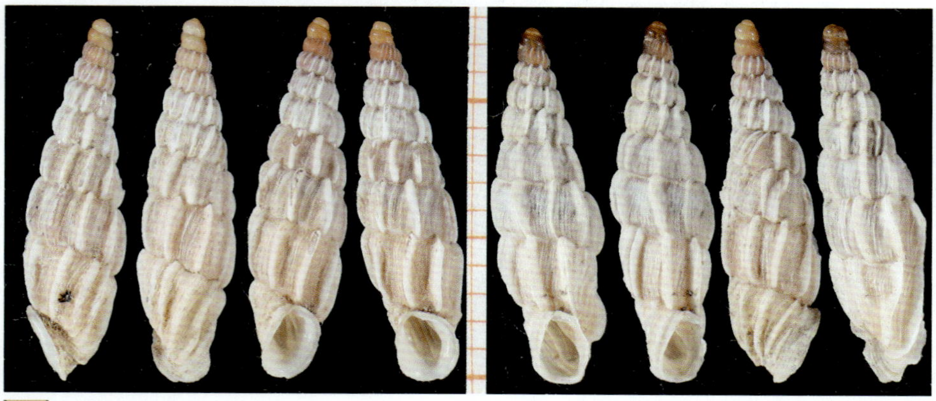

Greece: Kefaloniá, 21 km E Argostoli

10-17 x 2.8-4.3 mm

Greece (S Kefaloniá)

Open rocky habitats, on the rock faces.

Kemperman 1992: 200.

Shell whitish with dark apex, strongly ribbed or with percostate ribs, locally variable, between 6 and 22 ribs/penultimate whorl, outline of upper whorls straight or slightly convex, 9-10 whorls, aperture detached, margin sharp, parietalis weak, no parallelaris, columellaris low and oblique, no lunula, frontal upper palatalis very short and thin, subcolumellaris usually visible in an oblique view.

Albinaria adriani (Gittenberger, 1987)

Isabellaria adriani Gittenberger, 1987 in Gittenberger 1987: 79

11-15 x 3.1-3.8 mm

Greece: Pelopónisos, near Astros

Greece (E Pelopónisos)

Shell whitish with light horny brown spots, weakly striated, with 8-10 whorls, cervix weakly and irregularly ribbed, with strong basal keel, rounded dorsal keel (callus-like) and a furrow in between, apertural margin usually very broad and reflected, parietalis not very strong, spiralis very weak, columellaris weak, lunula dorsal, subcolumellaris sometimes invisible in the aperture but clearly passes lunula. Differs from *A. butoti* and *A. edmundi* in its dorsal keel, and from *A. linnei* in its smoother shell, less prominent columellaris and the subcolumellaris not being blocked by the lunula.

Open rocky habitats, on the rock faces.

A. linnei and *A. adriani* are certainly closely related. Sufficient specimens were collected for the shell record to suggest reproductive isolation, hybrids were not found.

Albinaria alajana (Boettger, 1896)

Clausilia alajana Boettger, 1896 in O. Boettger 1896: 124

14-18 x 3.2-3.6 mm

Turkey: Alanya

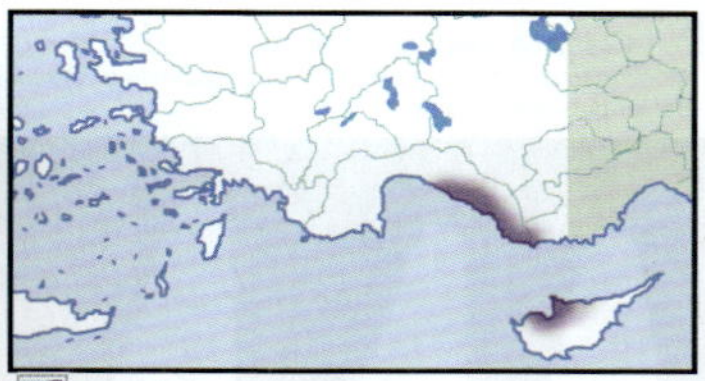

S Turkey, N Cyprus

Shell grey, often covered by red dust, otherwise with bluish hue, ribbed, dorsal keel weak, basal keel stronger, columellaris strong, subcolumellaris only visible in an oblique view of the aperture, principalis long, lunella strong, dorsal to dorsolateral, no frontal upper palatalis, basalis possible. Shells near Alanya weakly ribbed, between Manavgat and Anamur and in Cyprus more strongly ribbed, no basalis in Cyprus.

Limestones in coastal vicinity.

Nordsieck 1993: 7, 18, 23, Schütt 2001: 314.

Three subspecies (*alajana*, *dobati* and *cypria*), but probably not really justified since they are mainly based on rib prominence and presence of a basalis.

Albinaria amalthea (Westerlund, 1878)

Clausilia amalthea Westerlund, 1878 in Westerlund 1878: 245
Clausilia xanthostoma Böttger, 1883 in O. Böttger 1883 (Nachr.-Bl. DMG 15): 108
Albinaria xanthostoma loosjesi Nordsieck, 1977 in Nordsieck 1977: 80

15-24 x 3.3-4.7 mm

Greece: W Crete, near Koufos

Greece (NW Crete)

Shell white or cream, sometimes with some spots, sometimes with a violet hue, smooth or finely ribbed (35-60 ribs/penultimate whorl), 11-13 whorls, both cervical keels relatively weak, apertural lip reflected, parietalis strong, columellaris usually weak, slightly more prominently developed in the northeast, subcolumellaris weak in the east, strong near and west of Manoliópoulo, basalis moderately developed, frontal upper palatalis strong in the east and usually absent near and west of Manoliópoulo. The population on Agria Gramvoúsa is densely ribbed and has a weak columellaris, a relatively weak subcolumellaris, a very weak basalis (if at all), and a moderately developed frontal upper palatalis.
Differs from *A. tenuicostata* in the west in its well visible subcolumellaris.

At limestone rocks, hides in rock crevices.

Welter-Schultes 2010: 224.

Albinaria anatolica (Roth, 1839)

Clausilia anatolica Roth, 1839 in Roth 1839: 21, Tab. 2

14-22 x 3.5-6.5 mm, locally extremely variable

Turkey: Vilayet Antalya, near Kaş

SW Turkey

Shell greyish with red hue from dust, finely ribbed, cervix with strong basal keel and weak dorsal keel, parietalis long and usually weak, columellaris strong, subcolumellaris usually well visible, lunula dorsolateral, no frontal upper palatalis.

Limestone rocks.

Nordsieck 1993: 6, Neubert et al. 2000, Schütt 2001: 315.

Albinaria arcadica (Boettger, 1878)

Clausilia maculosa subsp. *arcadica* Boettger, 1878 in O. Boettger 1878: 139

Mentioned by Pfeiffer (1868: 467) as a synonym of *Clausilia maculosa*, without description. Boettger 1878 described it as new without reference to an earlier published source.

16-20 x 3.5-4.6 mm

Greece: Iti mountains SW Lamia

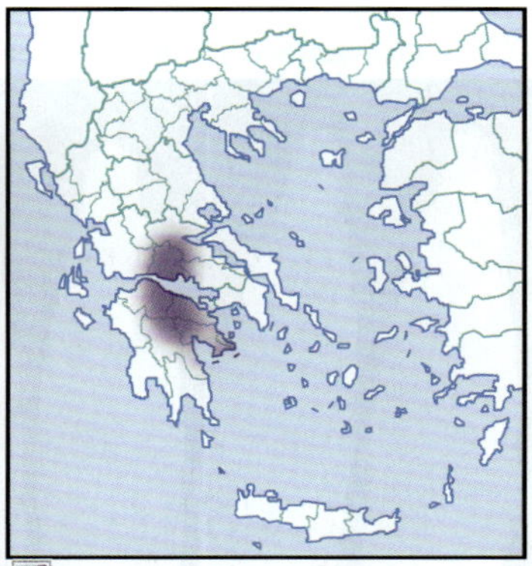

Central Greece and Pelopónisos

Shell yellowish white with distinct dark streaks, or brownish-white marbled, almost smooth or finely striated, 10-12 whorls, cervix ribbed, aperture small, oval, inside white, margin with a very weak lip, parietalis often small, subcolumellaris hardly visible.

Westerlund 1884: 139.

Albinaria argynnis (Westerlund, 1898)

Clausilia argynnis Westerlund, 1898 in Westerlund 1898: 168

16-20 x 4.5-5.5 mm

Greece: Lakonía, Liménas Iérakos

Greece (E Pelopónisos, Attikí)

Shell dirty whitish, with bluish hue or light horny brownish streaks, smooth or weakly striated, apex horny yellowish, last whorl finely ribbed, 9-11 whorls, aperture detached, parietalis reaches margin, columellaris almost horizontal, strong but does not reach margin, weakly forked inside, parallelaris distinct, lunula dorsal, frontal upper palatalis connected to lunula, basalis present, subcolumellaris usually visible.

Albinaria ariadne Schilthuizen & Gittenberger, 1991

Albinaria ariadne Schilthuizen & Gittenberger, 1991 in Schilthuizen & Gittenberger 1991: 111

14-18 x 3.7-4.3 mm

Greece: N central Crete, near Orthés

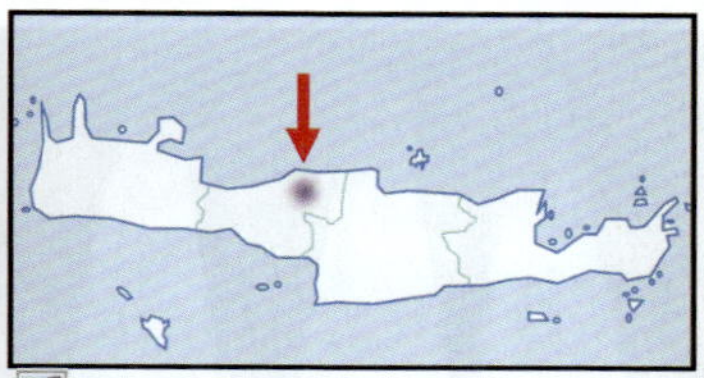

Greece (N central Crete)

Shell ribbed, last whorl less densely ribbed, 11-13 whorls, weak dorsal and basal keels, aperture with strong parietalis, moderate columellaris and basalis. Differs from *A. idaea* in the presence of a basalis and in a weaker and less oblique basal keel.

Limestone rocks, rock crevices.

Welter-Schultes 2010: 203.

It is possible that *A. ariadne* and *A. idaea* are connected in an unexplored 3 km extending transect, and that both are conspecific.

Albinaria arthuriana (Boettger, 1878)

Clausilia arthuriana Boettger, 1878 in O. Boettger 1878: 72 (*Clausilia (Albinaria) Arthuriana*)

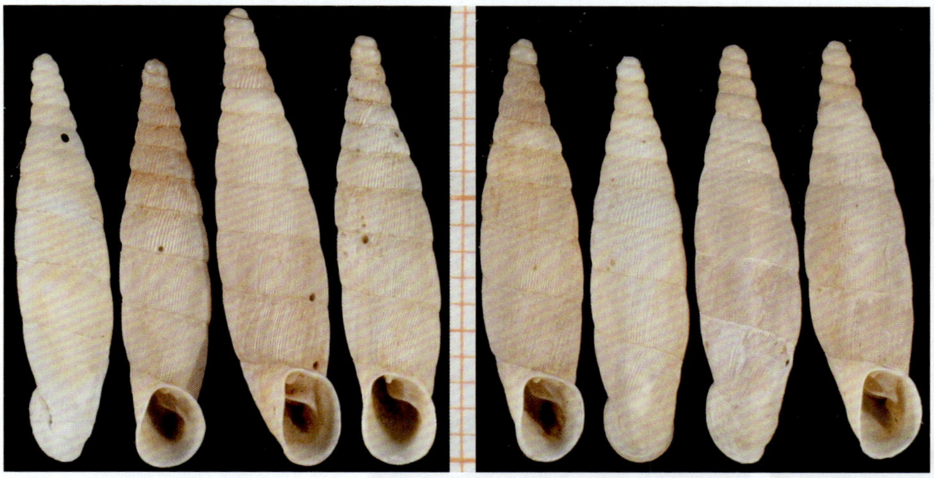

17-21 x 3.6-4.8 mm

Greece: E central Crete, near Sísi

Westerlund 1884: 120, Welter-Schultes 2000 (Schr. Malak. 16), Schilthuizen et al. 2004, Welter-Schultes 2010: 174.

Was classified as a subspecies of *A. hippolyti* by Nordsieck 1977 and Welter-Schultes 2000, this was inappropriate.

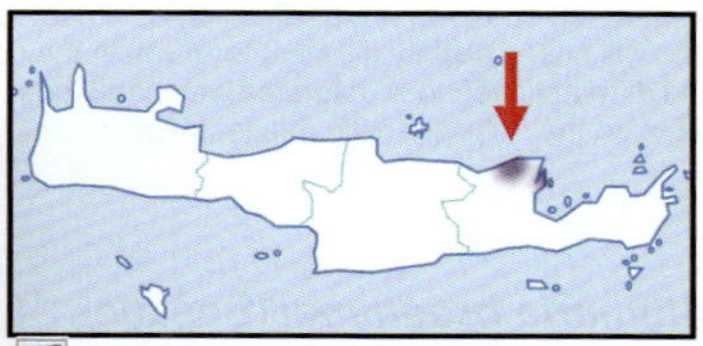

Greece (E central Crete)

Shell evenly cream or with violetish or brownish hue, finely and densely ribbed (8-16 ribs/mm), 10-12 whorls, basal keel usually prominent, dorsal keel usually weak, apertural lip often rounded and reflected, aperture with subclaustralis, usually a suturalis, usually no parallelaris, usually frontal upper palatalis, sometimes a second palatal fold between FUP and principalis is present, lunula dorsal, basalis present and relatively distant from subcolumellaris, subclaustralis present, subcolumellaris usually well developed. Differs from *A. hippolyti* by the presence of a subclaustralis and the relatively large distance between basalis and subcolumellaris.

Limestone rocks, usually in rock crevices of large outcrops.

Albinaria basalifera Neubert, 1992

Albinaria basalifera Neubert, 1992 in Neubert 1992: 68

Shell whitish, striated, cervix finely ribbed, with extremely prominent dorsal and basal keels, aperture as a consequence with deep basal channel, apertural margin broad and white, columellaris plate-like, subcolumellaris short and deep inside, subclaustralis present, basalis strong and connected to lunula.

20-22 x 4.5-5 mm

Rocky habitats.

Schütt 2001: 316.

SW Turkey

Albinaria brevicollis (Pfeiffer, 1850)

Clausilia brevicollis Pfeiffer, 1850 in L. Pfeiffer 1850 (Z. f. Malak. 6): 107

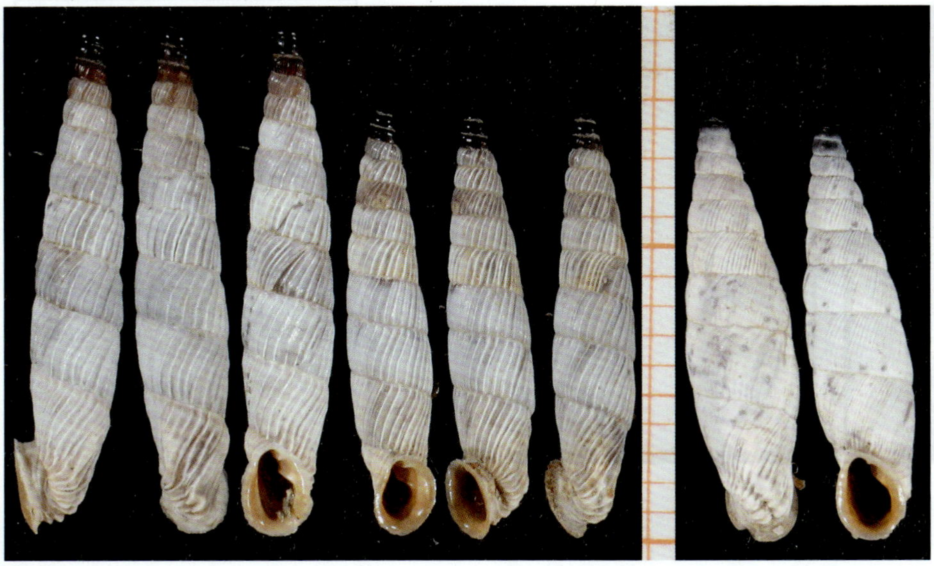

13-18 x 3.5-4 mm

(1-6) Greece: Ródos, near Charáki, (7-8) Turkey: Bodrum, castle

SE Greece (not Crete), W Turkey

Shell usually almost smooth or weakly striated (in Ródos some populations are ribbed, more slender and with less solid shell), white with bluish hue, often with dark spots, apex usually black, cervix and upper whorls ribbed, with short dorsal and basal keels, aperture inside orange, oval, parietalis weak and reaches margin, principalis and subcolumellaris not prominent in the aperture, lunella dorsolateral.

Limestone rocks and walls.

Pfeiffer 1955 (detailed map of forms in Rodos and figures of all forms), Zilch 1977: 325 (19 subspecies), Schütt 2001: 317, Örstan et al. 2005, Douris et al. 2007.

Introduced to the castles of Bodrum and probably also to Kos, but probably native in Iasos N of Bodrum peninsula.
I regard the subspecific classification as outdated. If the range of a widespread and locally variable species is flooded with only few spots remaining, it is loglical that every island will host a slightly different form within the usual variation of the species. It is not useful to use a trinominal system in such a situation. The classification at species level and differentiation to *A. caerulea* was partly doubted in molecular studies (Douris et al. 2007), suggesting that the morphological record may have provoked misleading conclusions on their phylogenetic relationships.

Albinaria broemmei (Boettger, 1892)

Clausilia broemmei Boettger, 1892 in O. Boettger 1892: 64

Shell whitish or with bluish hue, smooth or weakly striated, with few ribs near aperture, 11-12 whorls, last 2-3 whorls extremely inflated, outline in upper section extremely concave, aperture small, width 1/2 of penultimate whorl diameter, inside with light reddish or orange hue, parietalis very weak, columellaris and other folds invisible in a perpendicular view.

13-16 x 5-6 mm

Greece (N Pelopónisos)

Greece: N Pelopónisos, Chelmós mountains

Albinaria butoti (Nordsieck, 1984)

Isabellaria butoti Nordsieck, 1984 in Nordsieck 1984: 195

13.5-18 x 3.3-4.5 mm

Greece: Pelopónisos, Goritsá near Spárti

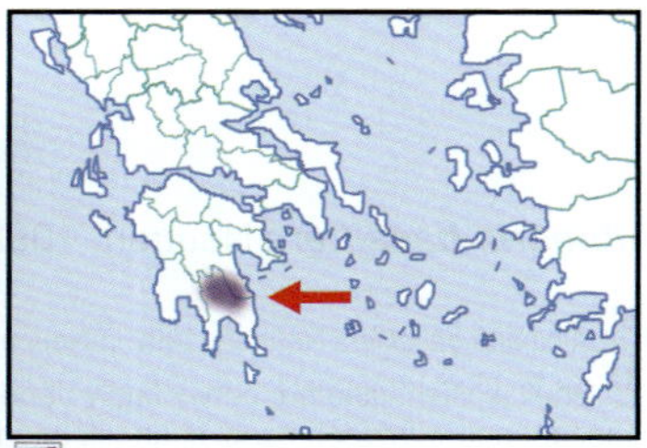

Greece (S Pelopónisos)

Shell usually evenly white, weakly or more prominently striated, cervix densely and very regularly ribbed, 9-11 whorls, cervix with strong basal and weak dorsal keel, aperture nearly oval, margin detached and slightly reflected, parietalis strong, parallelaris strong, spiralis weak and usually not reaching parietalis, lunula dorsal or dorsolateral, basalis short or absent, no frontal upper palatalis, subcolumellaris usually not visible in a perpendicular view.
Morphologically intermediate between *A. campylauchen* and *A. confusa*. *A. campylauchen* has no spiralis, in *A. confusa* the spiralis is stronger than the parallelaris.

Was found at and under low rocks in 100 m altitude.

Nordsieck 1984: 195, Gittenberger 1987: 80.

It is likely that this species belongs to a polytypic and locally very variable group containing *A. campylauchen*, *A. grisea* and *A. confusa*, with smooth morphological gradients connecting all these forms. Nordsieck (1984: 196) interpreted these as a formation of bastards. More research on the degree of local variations in these forms seems to be necessary. If the situation is like in the *A. cretensis* complex from Crete, the classification should be modified (1 instead of 4 species, *A. grisea*).

Albinaria caerulea (Deshayes, 1835)

Clausilia caerulea Deshayes, 1835 in Deshayes [1835]: 166, Pl. 19 (*Clausilia cærulea*)
Original spelling was not *coerulea*. *Cl. coerulea* as used by Rossmässler, 1835 (some months later) was not a new name, but a subsequent use. Synonyms and history:
Clausilia caerulea Deshayes, 1833 (also used in this sense by Boettger 1878) = *coerulea* as used by Rossmässler, 1835 (also used in this sense by Charpentier 1852, Nordsieck 1977) = *bifurcata* Deshayes, 1851.
Clausilia olivieri Roth, 1839 (also used in this sense by Boettger 1878) = *caerulea* as used by Deshayes 1851 = *corrugata* as used by Charpentier 1852.

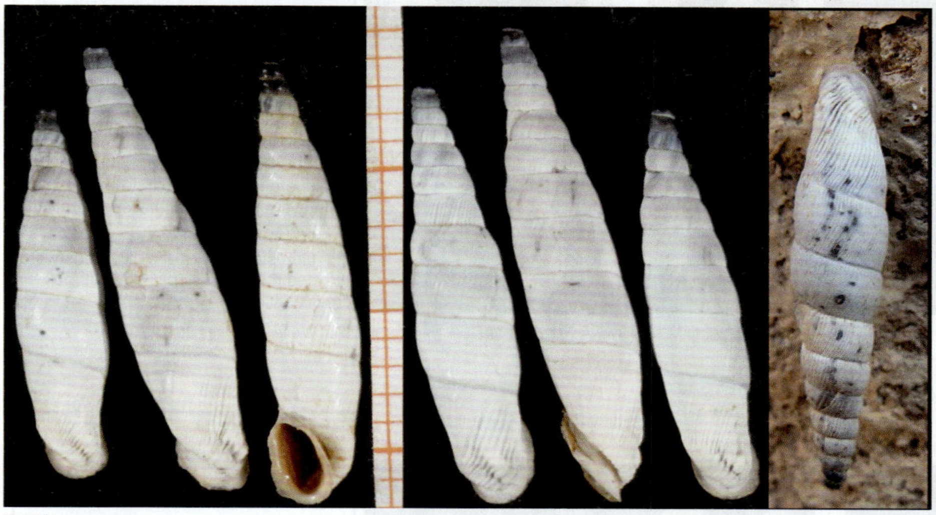

10-18 x 3-4 mm (1-6) Greece: Folégandros, Kardiotissa, (7) Turkey: Vilayet Izmir, Efes (7) Cédric Audibert

Greece (Kikládes), W Turkey

Shell variable, usually unribbed with strong white surface layer, often with dark or brown spots and dashes, cervix ribbed, dorsal keel strong, basal keel weak but long, extending to parietal side of the aperture (difference to *A. brevicollis*), columellaris weak, aperture inside orange, parietalis usually prominent, lunella dorsolateral and strongly curved, no basalis or frontal upper palatalis.

Limestone rocks and walls.

Bar & Butot 1986: 80, Nordsieck 1996: 94, Schütt 2001: 318, Nordsieck 2002, Giokas & Mylonas 2002, Douris et al. 2007, Giokas et al. 2007.

Several subspecies were recognized until recently, but like in most *Albinaria* species with spatial shell variations such a subclassification would be too weakly justified (see *A. brevicollis*). The classification at species level and differentiation to *A. brevicollis* was partly doubted in molecular studies (Douris et al. 2007), suggesting that the morphological record may have provoked misleading conclusions on their phylogenetic relationships.

Albinaria campylauchen (Boettger, 1883)

Clausilia campylauchen Boettger, 1883 in O. Boettger 1883 (Proc. London): 340

Shell whitish with horny brownish streaks and dark apex, 8-10 whorls, basal keel moderate, dorsal keel strong, aperture rounded, margin broad and protruded, lip very weak, columellaris high, parallelaris distinct and inside high, principalis distinct, ends in a callus in the aperture.

12-16 x 3.2-4 mm

Westerlund 1884: 167, Gittenberger 1987: 80, Giokas et al. 2000.

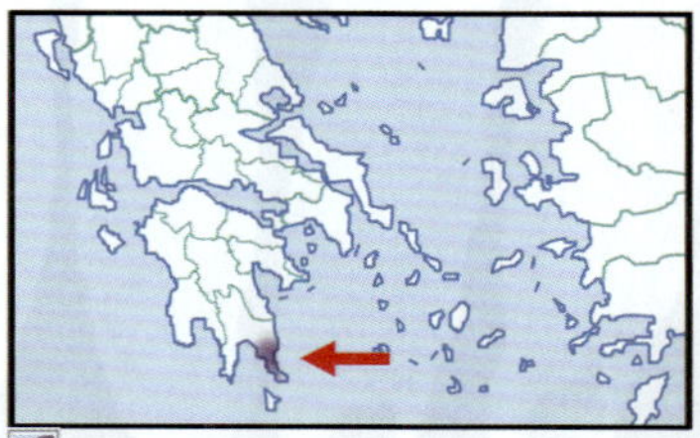

Greece (S Pelopónisos)

Albinaria candida (Pfeiffer, 1850)

Clausilia candida Pfeiffer, 1850 in L. Pfeiffer 1850: 135
Clausilia homaloraphe Pfeiffer, 1850 in L. Pfeiffer 1850: 137
Date was not 1849. First Reviser Nordsieck 1977: 300 selected *candida* if both are regarded as conspecific.

14-22 x 3.3-4.6 mm Greece: W Crete, near Chaniá

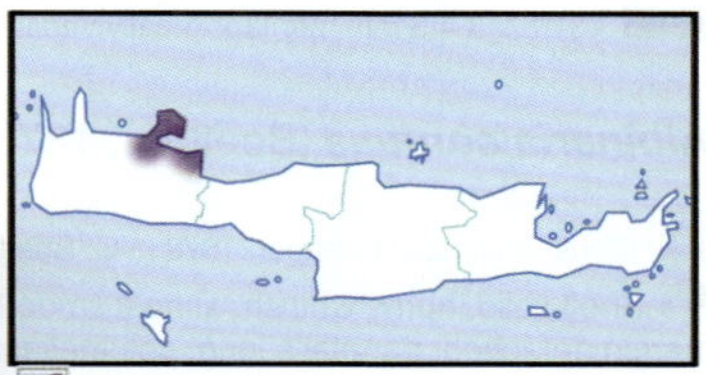

Greece (NW Crete)

Shell whitish, often with dense spots, smooth or densely ribbed, 10-11 whorls, cervix less densely ribbed and with two distinct keels and a broad basal furrow, apertural margin rounded and reflected, dark yellowish, columellaris very prominently developed, occupies really much space in the aperture, subcolumellaris moderate or strong, visible in a perpendicular view in the south, but not always on Akrotíri, basalis strong, frontal upper palatalis moderate or strong.

At limestone rocks, hides in rock crevices.

Westerlund 1884: 119, Welter-Schultes 2000 (Schr. Malak. 16): 42, Nordsieck 2004: 54, Welter-Schultes 2010: 225.

Could be conspecific with *A. amalthea*, this depends on the nature of a possible *hippolyti*-like hybrid zone near Nerokoúros, which needs to be studied in detail.

Albinaria cerigottana (Boettger, 1894)

Clausilia cerigottana Boettger, 1894 in O. Boettger 1894: 9

Shell white with dark apex, sometimes with irregular brownish streaks or dashes, densely striated, 12-14 whorls, cervix with weak basal and dorsal keels and a basal furrow, aperture 3.7-4.3 mm high and 2.2-2.8 mm wide, inside brownish.

17-21 x 3.7-4.5 mm

A frequent species in rocky habitats.

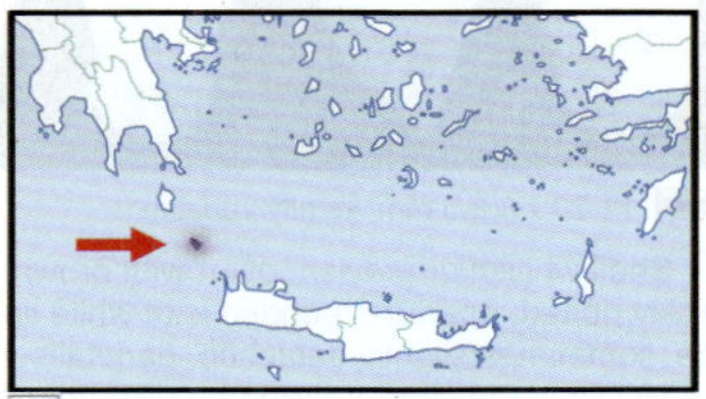

Greece (island of Andikíthira)

Albinaria christae Wiese, 1989

Albinaria christae Wiese, 1989 in Wiese 1989: 56

11-16 x 3-4.6 mm

Greece: E Crete, between Pláka and Vrouchás

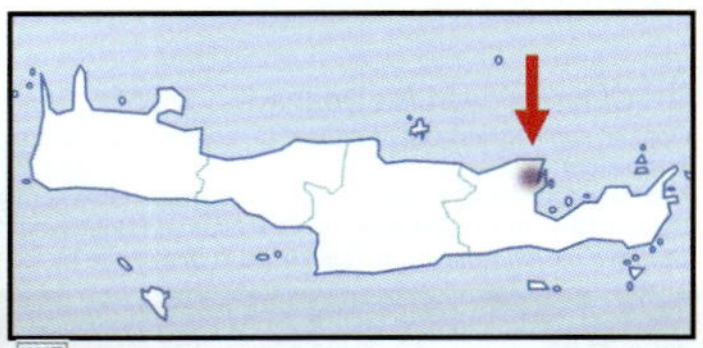

Greece (E Crete)

Shell decollate with 4-7 whorls remaining in adult shells, whitish cream with horny brown apex, finely ribbed, more coarsely on the cervix, basal and dorsal keel present but not very prominent, apertural margin not rounded, detached and slightly protruding, columellaris hardly in the aperture.

Limestone rocks of a rocky eastwards directed mountain slope, hides in rock crevices.

Welter-Schultes 2010: 204.

Albinaria compressa (Pfeiffer, 1850)

Clausilia compressa Pfeiffer, 1850 in L. Pfeiffer 1850 (Proc. London): 136

13-18 x 3-4.3 mm, perhaps locally variable

Greece: Kíthira, Keramotó

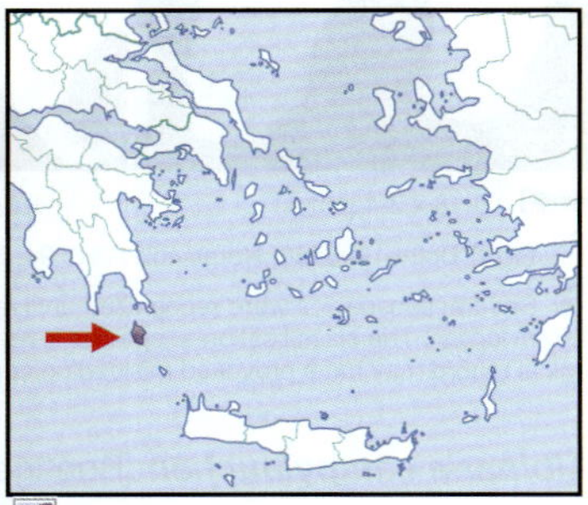

Greece (Kíthira)

Shell evenly white or bluish white with horny brown spots, striated or ribbed, 9-12 whorls, cervix with distinct basal and dorsal keels, V-like connected, aperture inside dark (brown), margin detached, white and thin, columellaris inside forked, palatal fold small, subcolumellaris and lunula deep inside and not visible from the aperture. Colour and ribs seem to display spatial variations.

Westerlund 1884: 126.

Albinaria confusa (Boettger, 1878)

Clausilia confusa Boettger, 1878 in O. Boettger 1878 (Jahrb. DMG 5): 304, Taf. 10

Shell solid, horny whitish, almost smooth on middle whorls, cervix weakly ribbed, 7-8.5 whorls, apertural margin detached, parietalis long, columellaris deep inside and S-shaped, palatal folds absent or very short, subcolumellaris deep inside and only visible in an oblique view.

11-15 x 3.3-4.2 mm

Westerlund 1884: 170, Gittenberger 1987: 80.

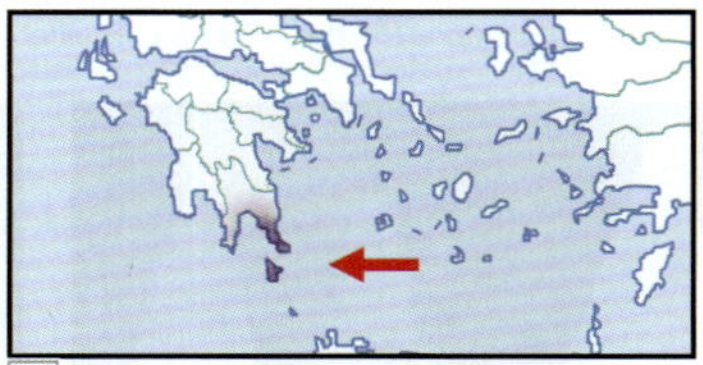

Greece (S Pelopónisos, Kíthira)

Albinaria contaminata (Rossmässler, 1835)

Clausilia contaminata Rossmässler, 1835 in Rossmässler 1835-1837 (2): 11, Taf. 7

11-20 x 3.8-5 mm, locally variable

Greece: Zákinthos, Kastelli

Shell whitish or greyish, often with brownish spots or streaks, almost smooth or striated, 9-11.5 whorls, cervix weakly ribbed, aperture rounded with white lip, inside often brownish, margin sometimes connected at parietal side, sometimes slightly detached, parietalis very weak or absent, far distant from spiralis, columellaris deep inside (but visible in the aperture) and far distant from parietalis, lunula almost dorsal, broad, slightly curved. Colour and apertural shape locally variable.

Westerlund 1884: 140, Kemperman 1992: 176.

NW Greece

Greece: Zákinthos   Panagiotis Helmis

Albinaria corrugata (Bruguière, 1792)

Without gender agreement: *Albinaria corrugatus* (Bruguière, 1792)

Bulimus corrugatus Bruguière, 1792 in Bruguière 1792: 354

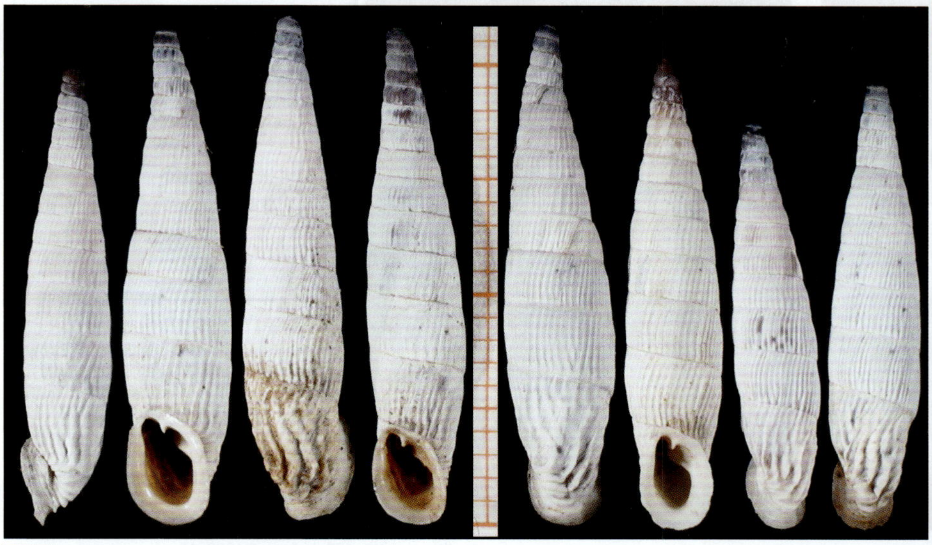

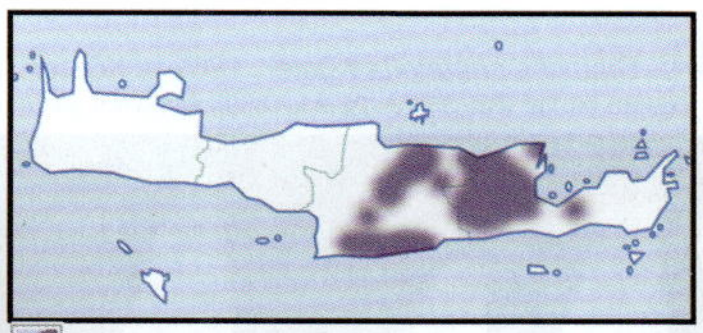

Greece (central Crete)

Greece: central Crete, near Emparos — Vollrath Wiese

14-27 x 3.2-5 mm, locally variable — Greece: central Crete, near Chersónisos

Shells are smaller (diameter 3.2-3.8 mm) at the southern margin of the island, larger (4.5-5 mm) in the Díkti mountains.

Shell locally extremely variable, smooth and white (often with bluish hue) in many mountain populations, or ribbed (mainly in lowlands), or percostate (near Embaros), apex smooth, first ribs appear usually after 2.0 whorls, exceptionally after 1.7 whorls, in smooth shells ribs vanish 2-3 whorls later, 11.5-14.5 whorls, cervix in most regions characteristically corrugated without a distinct dorsal keel, basal keel present but in most populations not very prominent, apertural margin in most regions sharp, no frontal upper palatalis, in most regions no basalis.

A. co. leonisorum: shell densely ribbed with a characteristically inflated basal cervical keel (bulge), cervix less corrugated, apertural margin reflected.

A. co. fulvula: shell brownish with white ribs, cervix slightly less corrugated, basalis usually present but not strong.

Differs from *A. praeclara* in a smoother apex (ribs start at whorl 1.7-2.0, in *A. praeclara* at 1.5 or earlier), this is the only reliable way to distinguish both species. *A. praeclara* has a slightly more prominently ribbed shell surface, the cervix has less ribs and is less corrugated, and the basal keel is less prominent.

Limestone rocks, usually at the face of the rocks, but some ribbed populations hide in rock crevices.

Welter-Schultes 2000 (Schr. Malak. 16): 43, Welter-Schultes 2010: 187.

Albinaria cretensis (Rossmässler, 1836)

Clausilia cretensis Rossmässler, 1836 in Rossmässler 1835-1837 (4): 11, Taf. 18

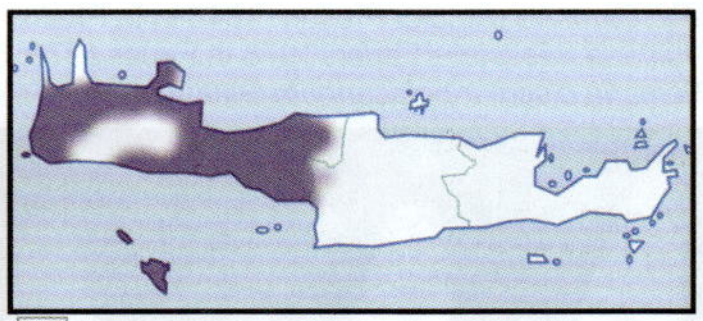

Greece (W to central Crete)

15-22 x 3.5-4.3 mm, locally very variable — Greece: NW Crete, (1-6) Akrotíri peninsula near Stavrós, (7-8) Episkopí near Réthimno (UTM KV 5812, incorrectly labelled as COF in Welter-Schultes 2010: 231, Fig. 38, on this plate *A. corrugata fulvula* was incorrectly labelled as CR)

Shell locally very variable, whitish or cream, finely ribbed or smooth, 10-13 whorls, cervix usually with two keels which are not extremely prominent, apertural margin rounded in many populations, aperture inside usually whitish or cream. Shells locally intensively spotted (N and W of the Idi mountains to Réthimno, also near Chaniá), ribs are more prominent between Plakiás and Agía Galíni, N of the Idi mountains and at the western coast and near Paleochóra, dorsal keel more prominent between Agía Galíni and Plakiás and near Paleochóra, very weak round the Lefká Ori mountains and on Akrotíri. Aperture without basalis and rontal upper palatalis, columellaris not very prominent, populations between Chaniá and Soúda have a well developed parallelaris, also near Kournás, Réthimno and Chóra Sfakíon.

Limestone rocks, often at the face of the rocks, but some populations hide in rock crevices.

Nordsieck 2004, Welter-Schultes 2010: 206.

Nordsieck 2004 subdivided the complex into 4 independent species (*virginea*, *troglodytes*, *byzantina*, *cretensis*). Populations of *virginea* and *troglodytes* turned out to be connected by smooth gradients (Welter-Schultes 2010), *troglodytes* and *byzantina* were also connected by smooth gradients, although they may live sympatrically at some restricted sites with rich cultural history, possibly a result of artificial dispersal in historical times.

The taxa *sphakiota* and *tenuicostata* seem to be reproductively isolated from the *cretensis* complex.

Albinaria cristatella (Küster, 1861)

Clausilia cristatella Küster, 1861 in Küster 1847: 293
Figured on Taf. Taf. 33 Fig. 22-24 [1860], name not printed on plate, only perhaps on the wrapper, which was not published work (Art. 8.1.1).

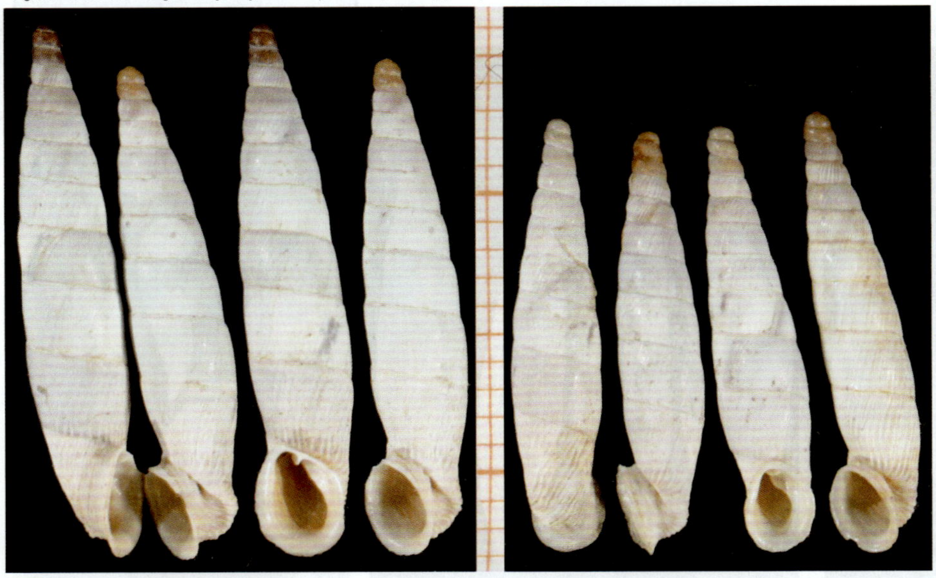

13-20 x 3-4 mm

Greece: island of Skíros

Greece (Vórii Sporádes)

Shell slender, white with horny brown apex, sometimes with grey spots, indistinctly and irregularly striated, often with weak spiral lines, 9-11 whorls, cervix densely ribbed and with long and crest-like basal keel, aperture rounded, yellowish, margin detached, slightly protruded and not much broadened.

Westerlund 1884: 134, Nordsieck 1984: 193, Liebegott 1986: 18.

Albinaria cytherae (Boettger, 1894)

Clausilia cytherae Boettger, 1894 in O. Boettger 1894: 4

Shell white with brownish apex, irregularly striated or almost smooth, 10-12 whorls with deep suture, aperture inside brownish, 3 mm high and 2.1 mm wide, parietalis moderate, columellaris weak and inside forked, lunula dorsolateral, no parallelaris, subcolumellaris weak.
Smaller than *A. compressa*, with smooth surface.

12-14 x 2.5-3 mm

Possibly a rare species.

Relationships with *A. compressa* still need to be studied. No transitions between both forms seem to have been found, but the last study seems to date from 1894.

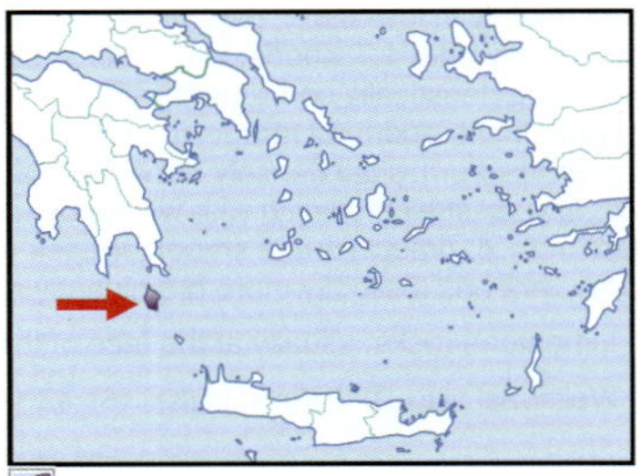

Greece (Kíthira)

Albinaria discolor (Pfeiffer, 1846)

Clausilia discolor Pfeiffer, 1846 in L. Pfeiffer 1846: 93
Albinaria haessleini Fauer, 1978 in Fauer 1978: 57
Albinaria discolor pelocarinata Gittenberger, 1994 in Gittenberger 1994: 60

13-18 x 3-4.5 mm, locally variable

Greece: island of Égina

Greece

Shell evenly white or with grey spots, or dark marbled, apex horny brown, usually weakly ribbed or striated, 9-12 whorls, cervix densely ribbed, with short, prominent and very oblique (to dorsal side) basal keel, aperture inside white or yellowish brown, sometimes with strong white lip, columellaris deep inside, principalis strong, lunula dorsolateral or almost lateral, short and well visible.
Locally variable, near Molái (*haessleini*) ribbed and with weak or without principalis, near Demonía (*pelocarinata*) ribbed and with a stronger basal keel and dorsal lunula.
In *A. edmundi* the basal keel is less oblique.

Westerlund 1884: 136, Gittenberger 1994: 61 (classified *haessleini* and *pelocarinata* as subspecies of *discolor*), Nordsieck 1999: 10, Giokas et al. 2000, Giokas & Mylonas 2002.

Nordsieck 1999 regarded *discolor*, *pelocarinata* and *haessleini* as separate species, and rejected Gittenberger's 1994: 60 and Gittenberger & Schilthuizen's 1996: 297 classification. But at the same occasion Nordsieck gave good arguments that the forms seem to be well interconnected, strengthening Gittenberger's 1994 conclusion to regard the forms as belonging to one species.
Nordsieck's 1999 main arguments to separate the species (white surface layer, ribs, length of the parietalis, prominence of the basal keel and the "tendency to develop a basalis" in *haessleini* and *pelocarinata*) were extremely weak, such characters are known to vary over area. Moreover they were not accompanied by detailed field data which would show the spatial distribution of the character states.

Greece: S Pelopónisos, near Molái

Albinaria eburnea (Pfeiffer, 1854)

Clausilia eburnea Pfeiffer, 1854 in L. Pfeiffer 1854: 296

17-24 x 3.5-5 mm

Greece: W Crete, 11 km SW Láki

Greece (W Crete)

Shell white, sometimes with bluish hue, apex often dark, smooth, cervix finely ribbed, some populations with cervical keels, aperture inside usually orange or brownish, appears to be shorter attached to the last whorl than in *A. cretensis* and *A. tenuicostata*, columellaris not very prominent, no basalis and frontal upper palatalis, some populations with a parallelaris.

Limestone rocks in the Lefká Ori mountains, in the south above 1000 m, in the north above 800 m.

Greece: W Crete, 3 km SE Epanohóri David González

Nordsieck 2004, Welter-Schultes 2010: 223.

Nordsieck 2004 subdivided this species into 4 subspecies, however without detailed knowledge on their distributional ranges. The nominate subspecies was collected by Spratt in the 1840s in some high mountain region at 2000 m altitude and has never been found again. It is likely that Nordsieck's so-called subspecies represent extreme forms of continuous gradients of a widely distributed species.

Albinaria edmundi (Gittenberger, 1987)

Isabellaria edmundi Gittenberger, 1987 in Gittenberger 1987: 81

12-17 x 2.9-3.7 mm

Greece: Pelopónisos, 12 km W Leonídio

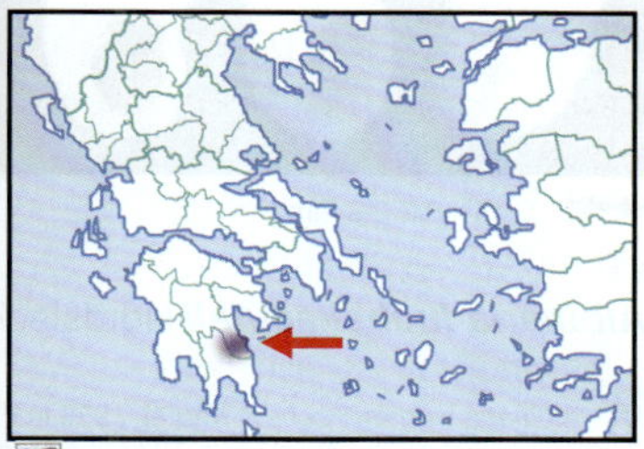

Greece (E Pelopónisos)

Shell solid, white or with bluish hue, with horny brown spots, almost smooth, 8-10.5 whorls, cervix regularly ribbed, with prominent basal keel and weak dorsal keel, aperture oval, margin broad and reflected, detached, parietalis weak, spiralis very weak, columellaris low (or weak), lunula dorsal, subcolumellaris invisible in a perpendicular view.
Differs from *A. adriani* and *A. butoti* in its more slender shell with oval aperture, *A. adriani* had a dorsal callus.

Restricted to a small area 3-12 km W of Leonidio.

Classified in *Albinaria* by Gittenberger 1998.

Greece: Pelopónisos, 12 km W Leonídio
(7-8) Stefan Heim
(7-8) Tiroler Landesmuseen Innsbruck

Albinaria evelynae Gittenberger, 1998

Albinaria evelynae Gittenberger, 1998 in Gittenberger 1998: 265

Shell light greyish cream with whitish ribs and light horny brown apex, aperture inside whitish, parietalis weak, columellaris low and rather straight in the aperture, no suturalis, lunula prominent and dorsal, no palatalis, no basalis, subcolumellaris only visible in an oblique view.

11.5-15 x 3.2-3.7 mm

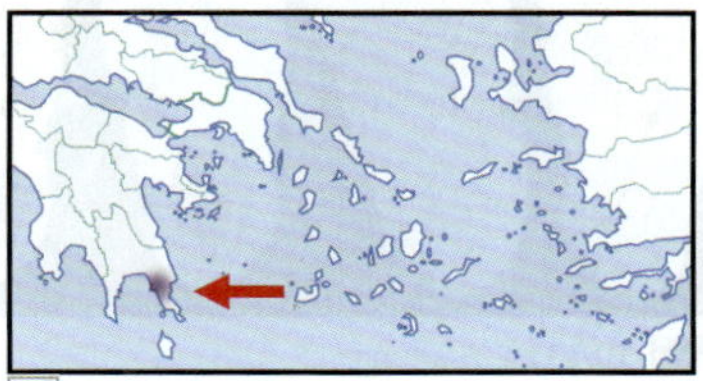

Greece (S Pelopónisos)

Albinaria forbesiana (Pfeiffer, 1846)

Clausilia forbesiana Pfeiffer, 1846 in L. Pfeiffer 1846: 93 (*Clausilia Forbesiana*)

Shell usually finely ribbed, aperture oval (almost circular), cervix with dorsal and basal keels, parietalis narrow, columellaris stronger, principalis strong, lunellar dorsal to dorsolateral, basalis and frontal upper palatalis sometimes present.

Limestone rocks and walls.

Nordsieck 1993: 5, Schütt 2001: 319.

It is possible that various local forms (with variable rib density, prominence of dorsal keel, presence of basalis and frontal upper palatalis) represent subspecies (*f. forbesiana*, *f. dunkeri* and *f. boettgeri* after Nordsieck 1993) connected by hybrid zones like in *A. hippolyti*, but until this is not established, no subspecies should be recognized. Possibly conspecific with *Albinaria inauris* (Boettger, 1896). Another name in this system could be *Clausilia luschani* Martens, 1889 (in Martens 1889 (Sitz.-Ber. Berlin): 182).

14-22 x 2.5-3.5 mm

Turkey: Vilyet Muğla, Fethiye
Burçin A. Gümüş

SW Turkey

Albinaria freytagi (Boettger, 1889)

Clausilia freytagi Boettger, 1889 in O. Boettger 1889 (Abh. Senck. 16): 39, Taf. (*Clausilia Freytagi*)

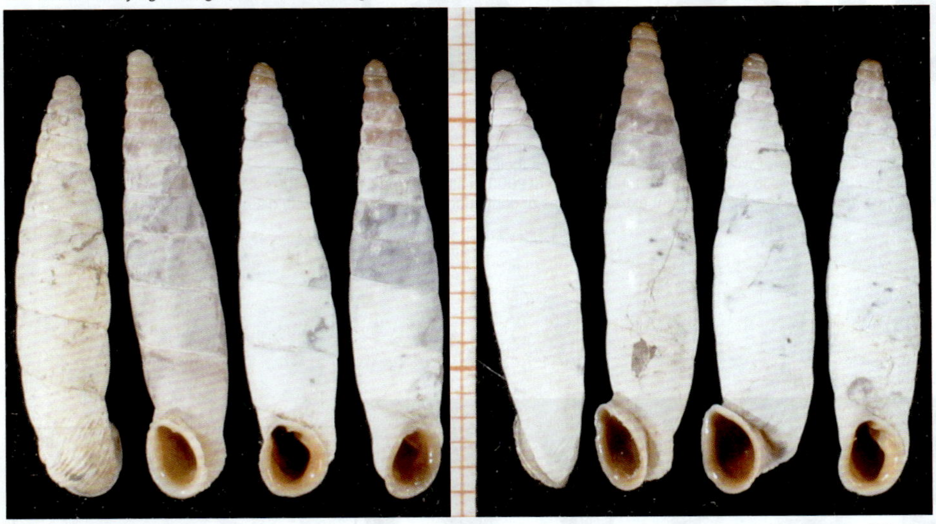

13-16 x 3.2-3.8 mm

Greece: W Sámos, Marathókambos

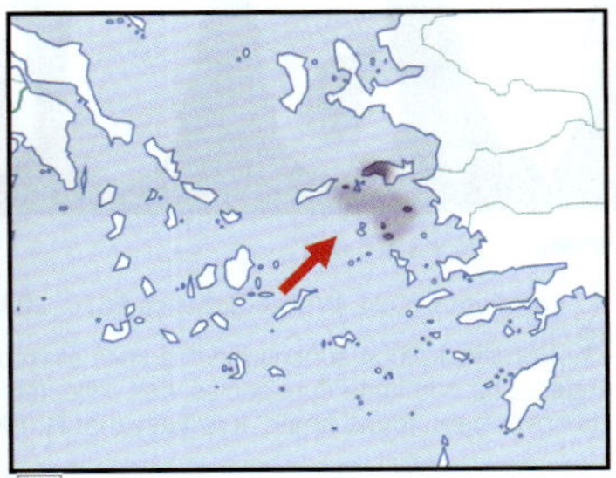

Greece (E Aegean islands)

Shell bluish white with some spots and horny brown apex, smooth, last section of cervix finely ribbed, 11-13 whorls, cervix with very prominent bulge-shaped basal keel, aperture detached, small, nearly circular, inside reddish brown, parietalis weak, columellaris weak, subcolumellaris not visible in the aperture.

Westerlund 1890: 61, Zilch 1977: 324 (subspecies of *A. caerulea*), Nordsieck 1999: 14 (independent species, based on comparisons of the degree of differences in related species).

Albinaria fuchskaeufeli Nordsieck, 1977

Albinaria werneri Fuchs & Käufel, 1936 in Fuchs & Käufel 1936: 599
Albinaria fuchskaeufeli Nordsieck, 1977 in Nordsieck 1977: 297
Expressly established as a new replacement name for *Albinaria werneri* Fuchs & Käufel, 1936 (not *Clausilia heterochroa* f. *werneri* Sturany, 1902, currently *Albinaria grisea* (Deshayes, 1835))

Shell slender, dirty yellowish white, prominently ribbed (20-25 ribs/penultimate whorl), cervix not ribbed (only at the suture and at the basal keel) but very finely striated, 10-11 whorls, basal keel crest-like and ribbed, aperture small, margin sharp and not much reflected, no parietalis, columellaris weak and hardly visible in the aperture, principalis strong, subcolumellaris not visible in the aperture.

12.3-14.3 x 2.5-3 mm

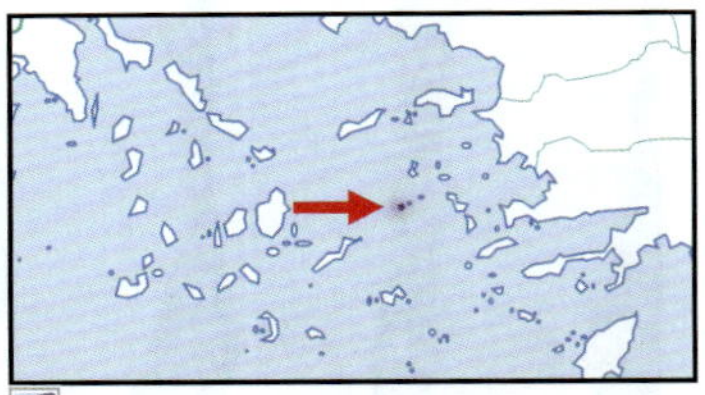

Greece (Kínaros island)

Albinaria gerolimena Nordsieck, 1974

Albinaria voithii gerolimena Nordsieck, 1974 in Nordsieck 1974: 139

10-13 x 2.7-3.4 mm

Greece: Lakonía, Máni, Gerolímenas

Greece (S Pelopónisos)

Shell dextral, whitish with indistinct brown spots, ribbed, ribs more widely spaced on cervix, 8-10 whorls, cervical keel prominent, often two keels are developed, aperture usually detached, parietalis varies locally but does usually not reach spiralis, lunula dorsolateral, clausilium plate visible in an oblique view.

Nordsieck 1999: 8 (revision of the *A. voithii* complex).

Was formerly classified as a subspecies of *A. voithii*, from which it differs in its smaller size and dorsolateral lunula. This alone does not seem to be really much, a more detailed study should clarify the relationships in terms of reproductive isolation and spatial morphological variations of the involved forms.

Albinaria grayana (Pfeiffer, 1846)

Clausilia grayana Pfeiffer, 1846 in L. Pfeiffer 1846: 94 (*Clausilia Grayana*)

Shell thin, greyish cream with some white spots, finely ribbed, 11-12 convex whorls with deep suture, aperture small and inside brown or yellow, margin detached, columellaris very oblique, almost concave at the base, principalis long, lunula almost dorsal and hardly visible.

Westerlund 1884: 137.

14-19 x 3-4 mm

Greece: Kíthira, Kapsáli

Greece (Kíthira, Elafónisos)

Albinaria greeni Tomlin, 1935

Albinaria greeni Tomlin, 1935 in Tomlin 1935: 164

11-15 x 3.5-4.5 mm (decollated)

Cyprus: near Pissoúri bay

Gittenberger & Neuteboom 1989.

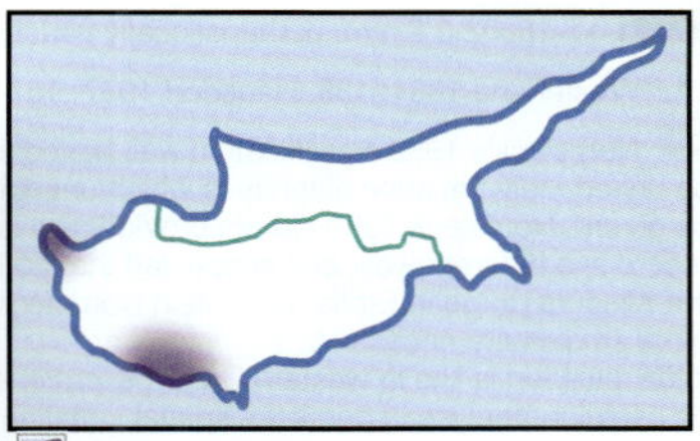

SW Cyprus

Shell decollated with 5-7 whorls remaining, whitish with brownish hue, ribbed (40-50 ribs/penultimate whorl), ribs coarser on last whorl, cervix with strong basal and dorsal keels and a basal furrow in between, aperture inside brownish or orange, detached, margin white, very broad and reflected, parietalis weak, columellaris weak and steep columellaris, subcolumellaris visible in an oblique view.

Restricted to a few localities in SW Cyprus.

Albinaria grisea (Deshayes, 1835)

Clausilia grisea Deshayes, 1835 in Deshayes [1835]: 168, Pl. 19

Shell whitish cream or grey with horny brown spots or streaks, striated or weakly ribbed, 8-10 whorls, the last 3 whorls almost equal in diameter with a straight outline, cervix finely ribbed and with moderate basal and dorsal keels and a basal furrow, aperture oval, inside whitish, margin shortly detached, parietalis passes spiralis at far distance, columellaris prominent, nearly transversal, narrow in frontal section, ascending S-like, subcolumellaris visible in a perpendicular view, lunula dorsal and strong, almost straight, only with a hook at its upper end.

Westerlund 1884: 138, Nordsieck 1995: 93, Neubert 1998: 133.

13-19 x 3-4.5 mm

Greece: Pelopónisos, Náfplio

Greece (Evia to S Pelopónisos)

Name used correctly since 1995. Roth 1856 and Boettger 1878 used the name *Clausilia maculosa* for this species, Nordsieck 1977 *Albinaria grisea thiesseana*. The name *grisea* was misinterpreted by Roth 1839, Roth 1865, Boettger 1878 and Nordsieck 1977, and referred to *Clausilia petrosa* Pfeiffer, 1849.

Albinaria hians (Boettger, 1878)

Clausilia hians Boettger, 1878 in O. Boettger 1878 (Novit. Conch. 5): 125

11-15 x 2-3.2 mm, perhaps locally variable

Greece: Pelopónisos, Pátras

W Greece

Shell thin, whitish grey or milky white, apex horny brown, usually densely ribbed, 8-11 whorls, last whorl only very little narrower, cervix strongly and slightly less densely ribbed, basal keel blunt and curved, aperture small, oval, inside yellow or white, margin thin, slightly detached, parietalis almost reaching margin, far distant from spiralis, principalis extremely weak, no lunula.

Westerlund 1884: 136, Hausdorf 1987: 167 (in *A. scopulosa*), Nordsieck 1999: 9 (independent species).

Nordsieck's 1999 classification was based on the argument "species should not be united, only because diagnoses which unequivocally separate the species cannot be given". Nordsieck 1999 did not provide new data, but interpreted Hausdorf's 1987 record in a different way, and suspected that both species may occur syntopically in the Klisoúra gorge in Etolía. I maintain Nordsieck's 1999 classification here, but more studies are necessary.
If the situation is like in western Crete (*A. cretensis*), which it seems to be, it would be useful to analyse in detail the spatial variation of the shell characters of the involved so-called subspecies between S Albania and Pelopónisos, and see if and how the forms are interconnected by gradients. Hausdorf 1987 suggested that one form (*porrecta*) provided a transition between *scopulosa* and *hians*. Nordsieck 1999 argued that most characters of this form were not transitional, and did not present results of a detailed study to substantiate his arguments. He admitted that the shell sculpture of *porrecta* was transitional. This could suggest uninterrupted gene flow. Both authors concorded that (with one possible exception mentioned by Nordsieck 1999) all forms were distributed parapatrically. A high degree of morphological difference between distant populations would certainly not qualify for the presence of independent species.

Albinaria hippolyti (Boettger, 1878)

Clausilia hippolyti Boettger, 1878 in O. Boettger 1878 (Novit. Conch. 5): 73 (*Clausilia (Albinaria) Hippolyti*)

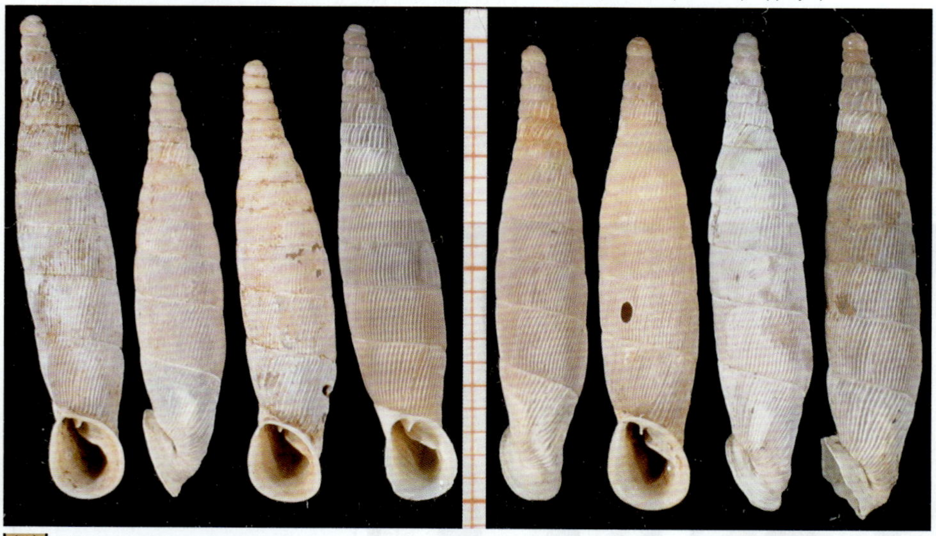

14-22 x 3.4-4.5 mm, locally variable

Greece: central Crete, near Káto Asítes

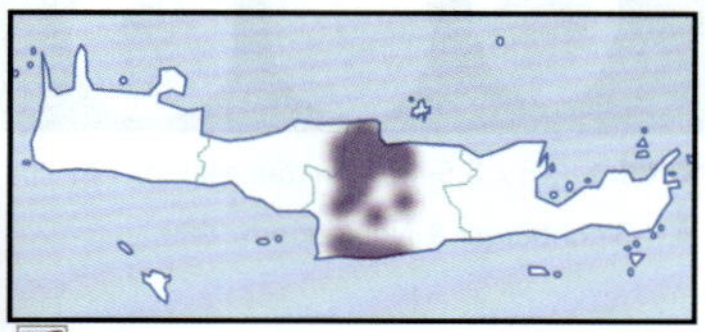

Greece (central Crete)

Shell usually brownish, finely ribbed, apertural margin rounded and often reflected, distance between basalis and subcolumellaris relatively short, no subclaustralis.
A. hi. hippolyti (between Fódele and Iráklio): ribs narrow (4-9 ribs/mm), lunula dorsolateral, usually no dorsal keel and no basalis, no frontal upper palatalis.
A. hi. aphrodite (between Doxaró and Profítis Ilías): ribs usually sharp and wide (3-6 ribs/mm), frontal upper palatalis usually present, basalis present, lunula doral or dorsolateral, subcolumellaris prominent, usually visible in a perpendicaular view.
A. hi. harmonia (near Tílisos): ribs sharp and wide (3-6 ribs/mm), apertural lip strongly reflected, basal and dorsal keels equally prominent, no frontal upper palatalis, basalis usually present, lunula dorsal or dorsolateral, subcolumellaris weak and invisible in a perpendicular view.
A. hi. holtzi (Idi mountains): usually densely ribbed (5-10 ribs/mm), parietalis always long and high, frontal upper palatalis absent, basalis usually present, lunula lateral, subcolumellaris weak and usually invisible in a perpendicular view.
A. hi. asterousea (Asteroúsia mountains): ribs very dense (7-12 ribs/mm), no frontal upper palatalis, no basalis, lunula lateral or ventrolateral, subcolumellaris weak and usualy insible in a perpendicular view, columellaris twisted around the columella very narrowly, its left margin pointing towards the (absent) frontal upper palatalis instead of to the principalis.

Greece: central Crete, near Tílisos

Vollrath Wiese

In crevices of limestone rocks, *A. hi. asterousea* lives very cryptically.

Schilthuizen et al. 1993 (revision), Schilthuizen 1994, Welter-Schultes 2000 (Schr. Malak. 16): 80, Schilthuizen et al. 2004, Welter-Schultes 2010: 171.

Model species for locally well defined subspecies separated by characteristic hybrid zones. *A. arthuriana* is a separate species (Nordsieck 1999, Schilthuizen et al. 2004).

Albinaria idaea (Pfeiffer, 1850)

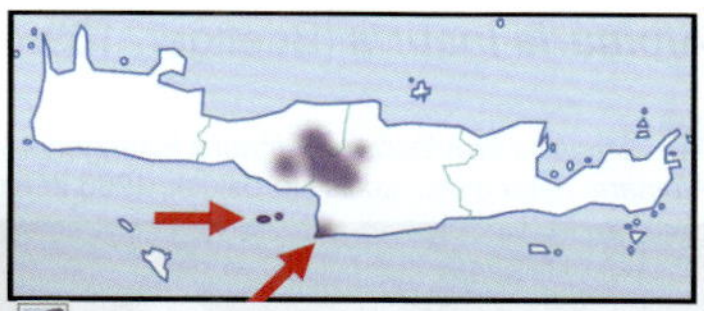

Clausilia idaea Pfeiffer, 1850 in L. Pfeiffer 1850: 140 (*Clausilia idæa*)

16.5-23 x 3.4-4.6 mm, locally variable — Greece: central Crete, near Orthés

Greece (central Crete, Paximádia)

Shell light brownish, or with reddish hue, smooth, ribbed or percostate, most populations densely ribbed (35-50 ribs at penultimate whorl below 1000 m, up to 90 above 1000 m, rib density is altitude dependent), with more (west) or less (east) prominent riblets between widely spaced ribs, ribs on the second whorl usually prominent, cervix in ribbed populations much less densely ribbed with additional ribs in a small band below the suture, 11-15 whorls, cervix with well developed basal and weak dorsal keel, apertural margin usually not strongly reflected, parietalis locally variable, columellaris not very prominent, lunula dorsolateral, no other folds, subcolumellaris weak and usually invisible in the aperture.

Most shell characters display considerable spatial variations. Some populations may resemble *A. praeclara* which has never very dense ribs or a smooth shell. In percostate shells of *A. praeclara* the first 3 whorls are often equal in diameter, in *A. idaea* the diameter increases earlier.

At the base of large pre-Neogene limestone rocks, boulders and walls, seems to require some vegetational cover, needs larger boulders than other species and is usually less abundant, hides in rock crevices. Most populations live in the mountain slope forests, the species tends to live cryptically and does not occur in high numbers. Basically a mountain species, between 400 and 2000 m, some isolated populations live close to the sea (Paximádia islands, Kefáli).

Welter-Schultes 2000 (Biol. J. Linn. Soc.), Welter-Schultes 2000 (Schr. Malak. 16): 50, 110, 176, Welter-Schultes 2010: 167 (no subspecies recognised).

Albinaria idyllica (Gittenberger, 1987)

Isabellaria idyllica Gittenberger, 1987 in Gittenberger 1987: 83

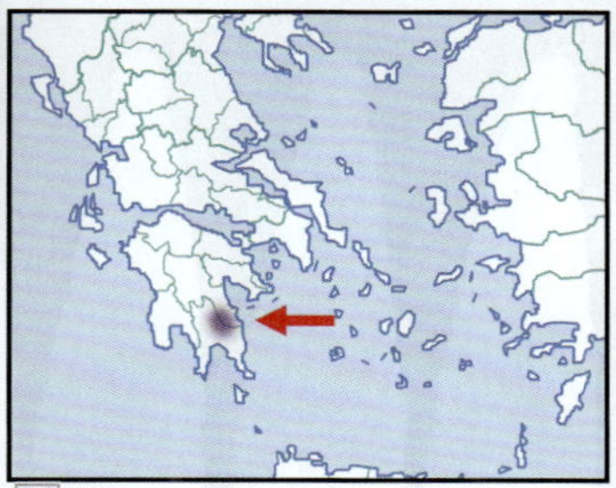

16.5-21 x 4-5 mm — Greece: Pelopónisos, Párnonas mountains near Agios Pétros

Greece (E Pelopónisos)

Shell white, often with bluish hue, with light horny brown spots, weakly striated, 8-10 whorls, cervix striated, with prominent basal keel, no dorsal keel, apertural margin detached and not very broadly reflected, parietalis strong, columellaris moderate, lunula dorsal, no suturalis, subcolumellaris invisible in a perpendicular view.

Was found in a moist valley with abundant vegetation and large deciduous trees, at partly overgrown rocks, in 1100 m altitude at the NW slopes of the Parnon mountains.

Albinaria ietswaarti Gittenberger & Menkhorst, 1992

Albinaria ietswaarti Gittenberger & Menkhorst, 1992 in Gittenberger & Menkhorst 1992: 193

17.5-20 (decollated 2-3 mm shorter) x 4.5-5.5 mm — Turkey: Vilayet Antalya, Dimçay valley E of Alanya

S Turkey (Vilayet Antalya)

Shell (often? usually but not always?) decollated, smooth except coarsely ribbed cervix, white with bluish hue and characteristic horny brown radial streaks, aperture inside orange, columellaris strong but not reaching the apertural margin, subcolumellaris almost invisibly deep inside, lunellar dorsal, principalis weakly visible from outside.

Rocky walls in the Dimçay valley gorge.

Only known from few localities.

Nordsieck 1993: 5, Schütt 2001: 320.

This species shows a parapatric distributional pattern with *A. inauris*, no intermediates were observed, both could eventually be conspecific.

Albinaria inauris (Boettger, 1896)

Clausilia brevicollis var. *inauris* Boettger, 1896 in O. Boettger 1896: 126
Albinaria inauris costicollis Nordsieck, 1993 in Nordsieck 1993: 9

15-22 x 3.5-5 mm

Turkey: SE of Antalya

S Turkey (Vilayet Antalya)

Shell smooth or weakly ribbed, usually with white surface layer, cervix less finely ribbed with short dorsal and basal keels (like in *A. forbesiana*), aperture oval, inside with orange hue, principalis long, lunula dorsal to dorsolateral, no frontal upper palatalis.

At limestone rocks.

Nordsieck 1993: 5, Schütt 1996: 285.

"Subspecies" *A. i. costicollis* in the Karpuz valley differs in rib and cervix prominence and parietalis prominence (Nordsieck 1993). These characters show continuous spatial gradients in *A. idaea* from Crete (Welter-Schultes 2000) and would certainly not justify a subspecific classification of *A. inauris*. The species is closely related to *A. forbesiana*, with which it might be conspecific. A further candidate to belong to this species complex is *A. ietswaarti*, which is usually decollated. So this might be a true species.

Albinaria ithomensis Nordsieck, 1984

Albinaria messenica ithomensis Nordsieck, 1984 in Nordsieck 1984: 202

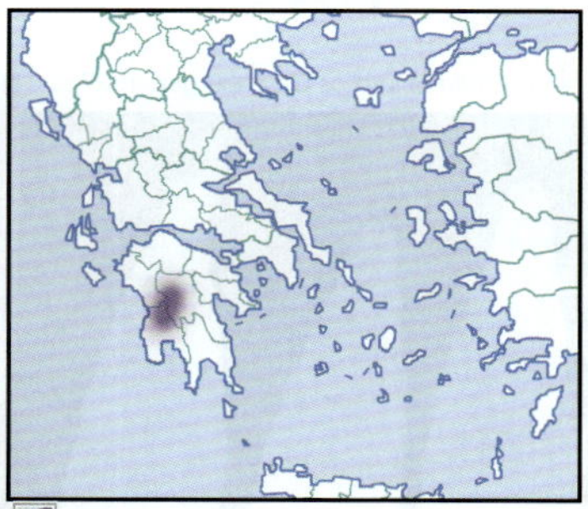

14-19 x 3-4 mm

Greece: Messinía, Mavrommáti

Greece (W Pelopónisos)

Shell like *A. messenica*, more slender, weakly and irregularly ribbed, basalis usually very weak, clausilium plate narrower and always visible in an oblique view. Differs from *A. scopulosa* in the presence of a weak basalis.

Nordsieck 1999: 9 classified it as an independent species with close relationship with *A. scopulosa*. More field research seems to be necessary to clarify the relationships of the involved forms, particularly the local variations of the characters used to distinguish the species.

Albinaria jaeckeli Wiese, 1989

Albinaria jaeckeli Wiese, 1989 in Wiese 1989: 44

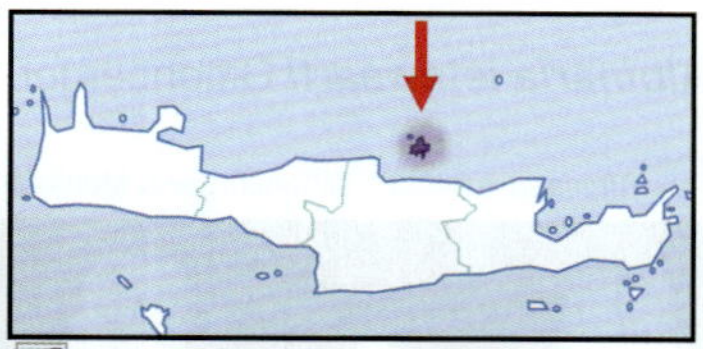

11-13 x 3.5-4.5 mm

Greece: island of Dia N of Crete

Greece (island of Día N of Crete)

Shell decollate with 5-6 whorls remaining in the adult shell, light horny brownish with translucent horny yellowish apex, almost smooth, also at the cervix, complete shells have 13-14 whorls, cervix rounded, aperture not detached, margin not reflected and not connected at parietal side, columellaris not visible in the aperture, columella bulbous in the last whorl producing a bulge on the right side of the shell, parietalis moderate, lunula ventral and well visible from outside.

Lives very cryptically in rock crevices of steep limestones, at N and E exposed slopes, not found at S exposed slopes, scattered distribution, not continuously distributed over area.

At one locality the number of dead shells collected after 3 years was the same as before (around 50), suggesting that dead shells are entirely replaced in such a time span and that the number of individuals is not low.

Only known from 5 localities. Not threatened. The island has a problem of overgrazing by protected wild goats, but this species lives in deep crevices and is not affected.

Schultes & Wiese 1990, Mesher & Welter-Schultes 2008, Welter-Schultes 2010: 205.

Albinaria janicollis Schultes & Wiese, 1991

Albinaria janicollis Schultes & Wiese, 1991 in Schultes & Wiese 1991: 81

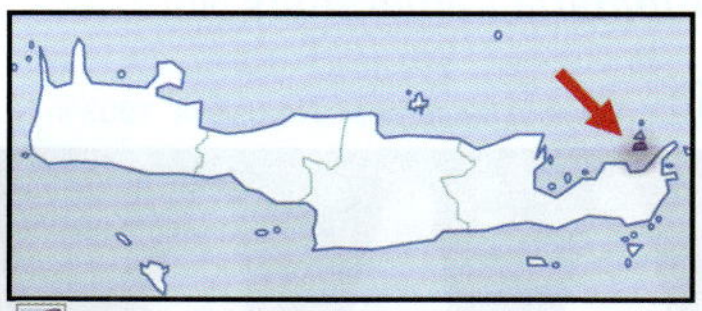

Greece (Gianisáda N of E Crete)

Shell unusually small, white but usually covered with red dust particles, prominently ribbed, ribs wider than the deep spaces in between, 13-15 whorls, extremely regularly incresing with characteristically straight outline, last whorl wider than penultimate whorl and in the last 1/4 section detached and protruding, cervix covered by extremely prominent and high ribs, aperture very small, margin thin and fragile, reflected, parietalis weak, columellaris visible in an oblique view, lunula dorsal or dorsolateral, subcolumellaris invisible in the aperture. Cannot be confounded with any other clausiliid from this region.

13-16 x 2.7-3.1 mm

Greece: island of Gianisáda N of E Crete

In Mesozoic limestone rock crevices.

Welter-Schultes 2010: 205 (spatial variation of shell characters).

Restricted to Gianisáda island, lives in various parts of it and is not threatened.

Albinaria janisadana Loosjes, 1955

Albinaria vermiculata janisadana Loosjes, 1955 in Loosjes 1955: 875
Loosjes 1955 established 3 names, also *A. v. dragonarana* and *A. v. paximadiana*. First Reviser: Schultes & Wiese 1991: 80 (selected *janisadana*).

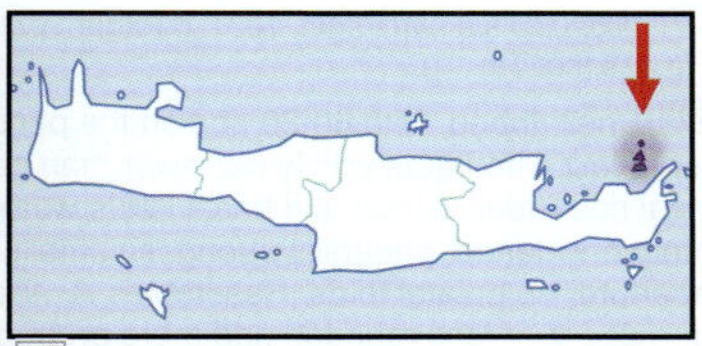

Greece (Dionisádes islands)

Shell dirty whitish and ribbed in most populations (rib numbers decrease from SW to NE), or white with bluish hue and almost smooth, 12-15 whorls, basal keel strong, dorsal keel moderate, with a distinct and unusually narrow basal furrow in between, sometimes forming a steep V, apertural margin sharp, columellaris deep inside and weakly visible in a perpendicular view.

Limestone rocks and between stones.

16-23 x 3.7-4.7 mm

Greece: Dionisádes islands N of E Crete, Prasonisi

Schultes & Wiese 1991, Welter-Schultes 2010: 205 (no subspecies recognised).

The rib density varies gradually from high values in the W of the island group to low values in the W of the island group. Non-ribbed populations live in the extreme west and are connected to ribbed populations by hybrid zones. With its well developed basal keel and furrow, *A. janisadana* has affinities with *A. sturanyi*, *A. rebeli* and *A. moreletiana*.

Greece: island of Paximada N of E Crete

Vollrath Wiese

Albinaria jonica (Pfeiffer, 1866)

Clausilia jonica Pfeiffer, 1866 in L. Pfeiffer 1866: 147

Greece (NW Kefaloniá)

Shell milky white with corrugated and fine ribs, 9-10 almost convex whorls, cervix with two weak keels, aperture large and oval, yellow, margin detached but not very broad, parietalis almost reaching the margin, inside almost reaching spiralis, columellaris strong, almost horizontal, principalis deep inside, subcolumellaris visible in an oblique view, lunula dorsal and arrow-shaped.

12-14 x 3-3.8 mm

Greece: Kefaloniá, 20 km NNE Argostoli

Westerlund 1884: 135, Rähle 1980: 213, Kemperman 1992: 162.

Prefers low rocks and ascends sometimes also shrubs.

Albinaria kemerensis Nordsieck, 1993

Albinaria kemerensis Nordsieck, 1993 in Nordsieck 1993: 17

18-23 x 4.5-5.5 mm

Turkey: Vilayet Antalya, Kesme canyon near Kemer

SW Turkey

Shell bluish white with horny brown dashes or spots, weakly striated or smooth, last whorl ribbed, no dorsal keel, basal keel flat, aperture egg-shaped with broad margin, inside yellowish brown, columellaris not very prominent, lunella dorsal to dorsolateral, no frontal upper palatalis, clausilium plate visible in an oblique view of the aperture, subcolumellaris not visible in a perpendicular view of the aperture.

Limestone rocks in the gorge.

Schütt 2001: 322.

Evolved obviously from an isolated population of *A. lycica*, with which it shows a parapatric distribution. It is possible that the two forms could be connected by a hybrid zone or spatial gradients and thus, be conspecific. More field research in this region is needed.

Albinaria klemmi Paget, 1971

Albinaria klemmi Paget, 1971 in Paget 1971: 462 (figured on Taf. 2 Fig. 1-5 [not Taf. 1])

Shell ribbed, with 60-80 ribs on the penultimate whorl, diameter of last whorl gets visibly narrower than penultimate whorl, cervical ribs wider, dorsal and basal keels weak, aperture protruded, margin widened, parietalis strong, columellaris prominent in the aperture, no palatal folds. Resembles *A. anatolica*, but more slender, aperture less rounded and more protruded, ribs denser, columellaris more prominent.

Does not live at limestone rocks, which is uncommon for *Albinaria* species, but exclusively under stones, in small soil niches and caves, sometimes at the feet of hanging rocks.

Closely related to *A. anatolica*, also anatomically.

18-21 x 4.1-4.5 mm

Greece: Ródos, Péfki

Greece (E Ródos)

Albinaria krueperi (Pfeiffer, 1866)

Clausilia krueperi Pfeiffer, 1866 in L. Pfeiffer 1866: 152 (*Clausilia Krüperi*)

Shell whitish or horny grey with dark transparent streaks and spots, apex horny brown, almost smooth, regularly striated or finely ribbed, 8-11 whorls, cervix densely ribbed, with weak basal and dorsal keels and a weak basal furrow, aperture rounded, inside whitish, margin not necessarily connected at parietal side, if yes then slightly detached, parietalis strong, passing spiralis at far distance, columellaris oblique and with a callus in frontal section, principalis not much longer than lunula, lunula almost lateral, subcolumellaris visible in the aperture.

Westerlund 1884: 138, Neubert 1998: 151 (map).

Hybrid zones were found with *A. grisea*.

12-16 x 3-4 mm, perhaps locally variable

W Greece

Greece: Pelopónisos, Katákolon

Albinaria latelamellaris Neubert, Örstan & Welter-Schultes, 2000

Albinaria latelamellaris Neubert, Örstan & Welter-Schultes, 2000 in Neubert et al. 2000: 117

22-29 x 6-7 mm

Turkey: Vilayet Antalya, 7.5 km W Yavı near Demre

SW Turkey

Shell substantially larger than Albinaria anatolica, aperture fragile and very small in relation to maximum shell diameter at antepenultimate whorl, basalis strong, columellaris very prominent, subcolumellaris short, frontal upper palatalis present.

Limestone rocks.

Schütt 2001: 322.

Albinaria lerosiensis (Pfeiffer, 1841)

Clausilia lerosiensis Pfeiffer, 1841 in L. Pfeiffer 1841: 47

14-18 x 3.5-4 mm

Greece: island of Psérimos

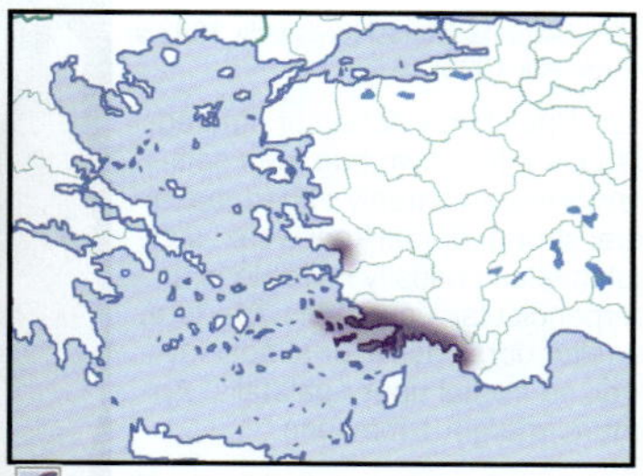

SW Turkey, E Aegean islands

Shell ribbed, grey with brown hue from soil, usually 20-35 ribs/mm and up to 3-5 lines between the ribs, apex pointed, no dorsal keel, basal keel weak, aperture almost egg-shaped, parietalis and columellais inconspicuous, usually no basalis and frontal upper palatalis, lunellar dorsal, clausilium visible in an oblique view of the aperture. Shell characters exhibit spatial gradients.

Limestone rocks.

Schütt 2001: 323, Örstan 2001: 42.

Albinaria li Welter-Schultes, 1999

Albinaria li Welter-Schultes, 1999 in Welter-Schultes 1999: 68

14-18 x 3.2-3.8 mm

Greece: S central Crete, near Kalí Liménes

Welter-Schultes 2000: 101, Welter-Schultes 2010: 203.

With its regular ribs this species was used as a model species to analyse shell parameter relationships.

Greece (S central Crete)

Shell white with brownish apex, finely and uncommonly regularly ribbed, 12-14 whorls, largest shell diameter often located closer to the aperture than in other species, basal keel stronger than dorsal keel and distinctly ribbed. Differs from the sympatric *A. corrugata* in the structure of the basal keel. The animal is black.

Limestone rock faces.

Very narrow restricted range near Kalí Liménes, S central Crete.

Greece: S central Crete, near Kalí Liménes

Vollrath Wiese

Albinaria linnei Gittenberger, 2008

Albinaria linnei Gittenberger, 2008 in Gittenberger 2008: 65

Shell whitish with some spots and light yellowish brown first whorls, finely and regularly ribbed, 8-9 whorls, basal keel weak, dorsal keel prominent, aperture slightly (0.5 mm)detached, margin broad, translucent and fragile, parietalis short, no spiralis, fulcrans prominent and running between parallelaris and columellaris, columellaris prominent and lobe-like in the aperture, suturalis short, frontal upper palatalis ends in a palatal callus, lunula dorsal with high basal part that blocks off the subcolumellaris, no basalis. Clausilium plate 2.5 x longer than broad, with pointed tip.
Differs from sympatrical Albinaria adriani in its more distinctly ribbed shell, the more prominent columellaris and the subcolumellaris being blocked by the lunula.

12.5-15.5 x 3.7-4.5 mm

On low limestone rocks.

A. linnei and *A. adriani* are certainly closely related, but enough specimens were collected to provide evidence in the shell record for reproductive isolation, hybrids were not found.

Greece (E Pelopónisos)

Albinaria lycica Nordsieck, 1993

Albinaria lycica Nordsieck, 1993 in Nordsieck 1993: 16

Shell finely ribbed or striated, usually with brown hue from soil, cervix more coarsely ribbed, basal keel strong, dorsal keel weak, subcolumellaris usually invisible in a perpendicular view of the aperture, lunellar dorsolateral or almost lateral, no frontal upper palatalis. Apertural margin 1 mm long.

"Subspecies" *phaselis* differs in rib prominence, shell shape, prominence of the principalis, relationship parietalis/spiralis, and structure of the columellaris - it is not very likely that we are dealing with uniform entities connected by hybrid zones, which would justify a subspecific classification. The nature of the relationship to *A. kemerensis* should be verified in the field, they might belong to the same species.

15-21 x 4.5-5.5 mm

Turkey: Vilayet Antalya, NE of Finike

SW Turkey

Limestone rocks.

Nordsieck 1993: 6, 17, Schütt 2001: 324.

Albinaria maculosa (Deshayes, 1835)

Clausilia maculosa Deshayes, 1835 in Deshayes 1835: 167, Pl. 19
Clausilia schuchii Rossmässler, 1836 in Rossmässler 1835-1837 (4): 13, Taf. 18

13-19 x 3.5-4.8 mm

Greece: S Pelopónisos, Máni peninsula

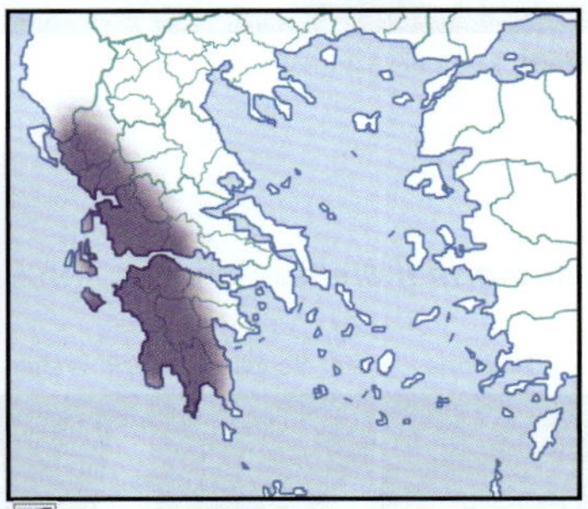

S Albania, W and S Greece

Shell greyish white or yellowish cream with darker transparent streaks, weakly striated or almost smooth, 9-11 whorls, cervix finely ribbed, weakly keeled, aperture inside whitish, margin slightly detached, with white lip, parietalis inside passing the spiralis, columellaris thickened in its frontal part, lunula almost lateral, curved, upper end reaching far inside, subcolumellaris visible in a perpendicular view, occasionally reaching the margin.

Westerlund 1884: 139, Nordsieck 1977: 290 (identity of types clarified), Nordsieck 1996 (Heldia 2): 93, Neubert 1998: 127.

Following Neubert 1998: 128 the name *maculosa* was entirely or partly correctly used by Pfeiffer 1848, Charpentier 1852, Bourguignat 1853, Martens 1872 and Westerlund 1884, incorrectly (referring to *Cl. grisea* Deshayes, 1835, or to *Cl. petrosa* Pfeiffer, 1849) by Küster 1852, Roth 1856, Boettger 1878 and Boettger 1889. Nordsieck 1996 proposed to consider the name *Cl. maculosa* as a nomen oblitum, and to use *Cl. schuchii* instead. Nordsieck 1996 and Neubert 1998: 127 announced to present the case to the Commission, but no action has been taken.
The true identity of *Cl. maculosa* is known since 1977, the identity of *Cl. grisea* is known since 1995. Both names were incorrectly used before.
The names are too rarely mentioned in the literature to qualify for a solution under Art. 23.9.2. Nomenclature can only be stable if it follows the Code. Since no case has been brought to the Commission, *Cl. maculosa* should be used for the species, *Cl. schuchii* is a junior synonym.

Albinaria maltzani (Böttger, 1883)

Clausilia maltzani Böttger, 1883 in O. Böttger 1883 (Nachr.Bl. DMG 15): 111 (*Clausilia (Albinaria) Maltzani*)

15-20 x 3.9-4.8 mm

Greece: E Crete, near Kouroúnes

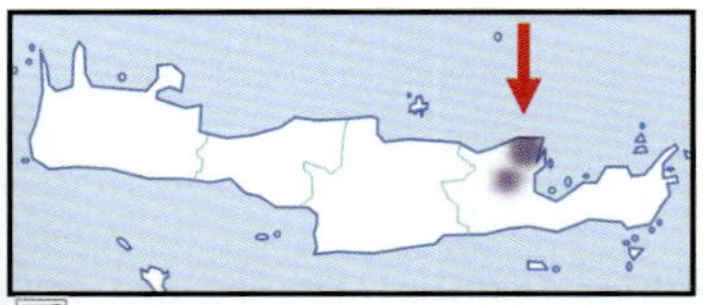

Greece (E Crete)

Shell usually brownish due to dust particles, usually prominently ribbed (21-45 ribs with weak local gradients), cervix less densely ribbed, first ribs visible from whorl 1.3-1.8 onwards (gradient from 1.3 in the south to 1.8 in the north), the first ribs look like faintly visible growth lines but they are equally distant from each other, 10.5-13 whorls, basal keel very prominent and crest-like in most populations, but with a gradient towards a normally developed basal keel in the northeast near Dílakos, dorsal keel weak, apertural margin sharp and broad.
Differs from neighbouring *A. corrugata* and *A. praeclara* in the sculpture of the first whorls (*A. praeclara* from whorl 0.5-0.8, in *A. corrugata* from whorl 2.1 or later).
Populations with a crest-like basal keel cannot be confounded with any other Cretan species. This keel was formerly thought to be characteristical, but the only reliable tool for identification are the ribs at the first whorls.

Limestone rocks, usually at outcrops, also at single bourlders, but boulders must not be too small.

Welter-Schultes 2010: 196.

A. maltzani clustered with *A. praeclara* and *A. lerosiensis* from Dodekánisa in mtDNA analyses (Douris et al. 1998). A low degree of genetical interchange with *A. arthuriana* (at Mílatos cave) is indicated by a prominent basal keel of this form (Welter-Schultes 2000: 89).
Crest-like basal keels exist also in other *Albinaria* species, which could be compared with *A. maltzani*.

Albinaria manselli (Boettger, 1883)

Clausilia manselli Boettger, 1883 in O. Boettger 1883 (Proc. London): 332

14-18 x 3.0-3.9 mm

Greece: E Crete, (1-4) Kaváli SW island, (5-8) near Xerókambos

Greece (E Crete)

Shell relatively small, prominently ribbed (20-28 ribs on the penultimate whorl), on Kaváli SW and NW with percostate ribs (7-8 ribs), rib density not much different on the cervix, apex ribbed. Differs markedly from local *A. teres* populations, but not necessarily from other far distant *A. teres* populations which might have similar character expressions.

Limestone rocks, hides in rock crevices. Was not found syntopically with *A. teres*, but it is possible that at some localities both species live together.

Occupies a very small restricted range in SE Crete near Xerókambos.

Classification as a species has been disputed. Nordsieck (1977, 1999) classified it as a subspecies of *A. teres*, Wiese 1990 and Welter-Schultes 2010: 181 as an independent species. The morphology cannot be distinguished from some distant *A. teres* populations, but locally *manselli* seems to be reproductively isolated from *A. teres*. More research is needed, it is possible that there might be a hybrid zone. It is also possible that Nordsieck's classification is more convenient.

Albinaria mavromoustakisi Brandt, 1961

Albinaria mavromoustakisi Brandt, 1961 in Brandt 1961: 14

Shell decollate with 6-7 whorls remaining, dirty whitish, ribbed (35-40 ribs/penultimate whorl), basal keel moderate and curved, dorsal section of cervix rounded, aperture detached and slightly protruded, margin weakly reflected, parietalis short, columellaris weak and oblique, lunula weak and dorsal or dorsolateral, frontal upper palatalis short, no basalis, subcolumellaris visible in an oblique view.

11.5-14 x 2.9-3.6 mm (decollated)

Was found at some small isolated rocks at the feet of higher mountains.

Only known from a small area 18-20 km W Kerínia/Girne.

Gittenberger & Neuteboom 1989.

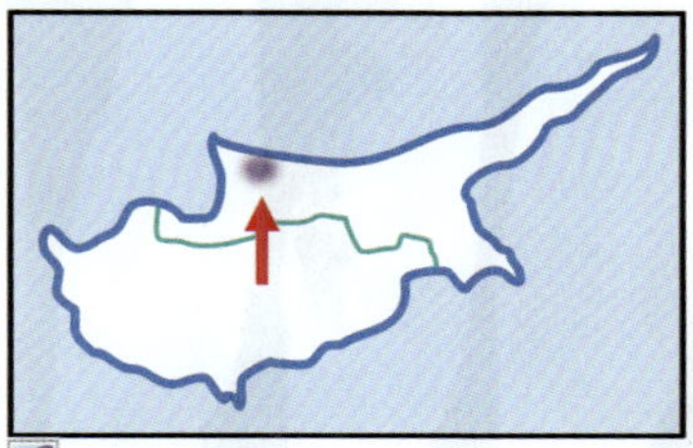

N central Cyprus

Albinaria menelaus (Martens, 1873)

Clausilia menelaus Martens, 1873 in Martens 1873: 41, Taf. 2 (*Clausilia Menelaus*)

12-20 x 3-4.5 mm, size locally variable

Greece: Pelopónisos, Taígetos, Anavrití

Westerlund 1884: 144, Nordsieck 1999: 8 (revision of the *A. voithii* complex).

Greece (S Pelopónisos)
Also in the south of the Máni peninsula (recent finds).

Shell dextral, smooth (white with bluish hue, apex light brownish, cervix weakly corrugated) or ribbed, 9-12 whorls, cervix nearly rounded with weak basal keel, aperture not or only weakly detached, parietalis long, columellaris weak, lunula dorsolateral or lateral.

Albinaria monocristata Neubert, 1992

Albinaria monocristata Neubert, 1992 in Neubert 1992: 67

16-18 x 3.5-4.5 mm

Turkey: Vilayet Içel, near Akdere

S Turkey

Shell ribbed, apex slender and pointed, cervix with characteristical very sharp basal keel, like in *A. schuetti*, provoking a deep basal channel in the aperture, frontal upper palatalis usually present.

Limestone rocks.

Nordsieck 1993: 7, Schütt 2001: 325.

Could be conspecific with *A. schuetti*, some populations of which have also a very prominent basal keel, and with *A. supercarinata* and *A. pellucida*.

Not in the study area covered by this book, but, well.

Albinaria moreletiana (Boettger, 1878)

Clausilia moreletiana Boettger, 1878 in O. Boettger 1878 (Novit. Conch. 5): 96 (*Clausilia (Albinaria) Moreletiana*)

17-25 x 4.1-4.8 mm

Greece: E Crete, Spinalónga

Lives sympatrically with *A. corrugata*, no hybrids were found.

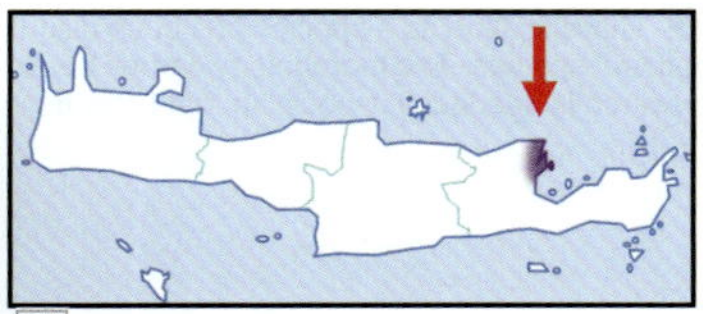

Greece (E Crete)

Shell white, sometimes with bluish hue, smooth or weakly ribbed (35-45 ribs/penultimate whorl where visible), 11-13 whorls, strong basal and dorsal keels, with a basal furrow in between, cervix ribbed but less densely than in *A. corrugata*, number of whorls lower than in neighbouring *A. corrugata* shells of the same height, cervix not corrugated, apertural margin in some populations very broad.

19 mm high *A. moreletiana* have 11.5 whorls, local *A. corrugata* of the same size have 14 whorls, variation +- 1 whorl, no overlap between the two species.

Limestone rocks, usually at rock faces.

Welter-Schultes 2010: 198.

Albinaria munda (Rossmässler, 1836)

Clausilia munda Rossmässler, 1836 in Rossmässler 1835-1837 (4): 11, Taf. 18

Shell greyish or bluish horny coloured, ribbed or striated, smoother shells with dark spots or dashes, cervix with long dorsal and basal keels, basal keel can be crest-like, aperture inside orange, without characteristic features, no basalis or prontal upper palatalis.

Mesozoic limestone and marble rocks.

Nordsieck 1993: 6, Örstan, Y. et al. 2005, Schütt 2001: 326.

Shells are locally variable in terms of rib prominence, subspecies are certainly not justified. Might be conspecific with *A. puella*.

13-17 x 3-4 mm

Greece: Kos, Díkeo mountain

E Greece, W Turkey

Albinaria myrensis Nordsieck, 1993

Albinaria myrensis Nordsieck, 1993 in Nordsieck 1993: 19

Shell brown with white suture, ribs blunt except on cervix, dorsal keel weak, basal keel prominent, lunula dorsolateral or lateral, basalis present but not always visible in a perpendicular view of the aperture, no frontal upper palatalis, subcolumellaris not visible in a perpendicular view.

Limestone rocks around the Demre valley. Perhaps restricted to higher altitudes.

Scattered distribution in a few restricted areas.

Neubert et al. 2000: 115, Schütt 2001: 327.

Could be conspecific with *A. alajana*, differs mainly in the structure of the lunula and is geographically isolated.

13-16 x 3.5-4 mm

SW Turkey

Turkey: Vilayet Antalya, Side

Burçin A. Gümüş

Albinaria nivea (Pfeiffer, 1854)

Clausilia nivea Pfeiffer, 1854 in L. Pfeiffer 1854: 296

Shell solid, white with brown apex and dark or brownish spots or streaks, almost smooth, 9-11 whorls, cervix weakly ribbed, basal keel present, aperture inside yellowish or white, margin broad and shortly detached, parietalis small, columellaris low and oblique, one palatal fold present, lunula and subcolumellaris not visible in the aperture.

Westerlund 1884: 145.

13-16 x 2.7-3.5 mm

Greece (S Pelopónisos)

Greece: Pelopónisos, Kalamáta

Albinaria olivieri (Roth, 1839)

Clausilia olivieri Roth, 1839 in Roth 1839: 21, Tab. 2

14-21 x 3.5-4.7 mm

Greece: (1-4) Kásos, (5-8) Kárpathos

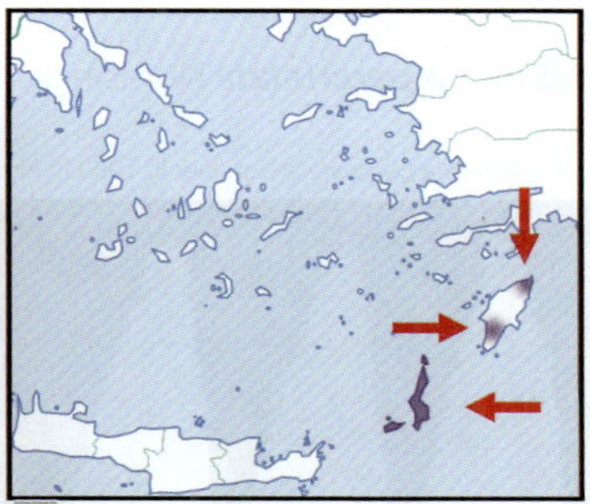

Greece (Kárpathos group, Ródos)

Shell evenly dirty white or with horny brown spots, apex brown or black, striated to very prominently ribbed, 11-14 whorls, cervix ribbed and often corrugated, aperture with usually brownish lip, margin slightly detached, parietalis long and low, usually reaching margin, columellaris low, oblique and visible in the aperture. Rib density and prominence locally variable, percostate populations are known from Kásos.

Westerlund 1884: 132, Paget 1976: 766, Nordsieck 1996: 94.

Name *Cl. olivieri* was used correctly by Boettger 1878, = *Cl. caerulea* sensu Deshayes 1851, = *Cl. corrugata* sensu Charpentier 1852.

Albinaria papillifera Nordsieck, 1993

Turkey (Vilayet Içel)

Albinaria papillifera Nordsieck, 1993 in Nordsieck 1993: 12

Turkey: Vilayet Icel, Anamur, Mamure kalesi

Shell greyish brown, some populations densely ribbed, others smooth, upper whorls with papilla-like structures at the suture provoked by rib thickening or up to 3 ribs merging together, dorsal keel usually stronger than basal keel, subcolumellaris and basalis usually visible in a perpendicular view of the aperture, frontal upper palatalis usually present, clausilium plate also visible.

17-23 x 4-5 mm

Rocky habitats.

Schütt 2001: 328.

Several subspecies were described on differences in shell size, rib prominence, dorsal keel prominence, frontal upper palatalis and basalis, location of the lunellar. It is possible that uniform local entities are separated by each others by hybrid zones, but this needs to be established against the alternative model of continuous shell variations over area, before subspecies can be recognized.

Albinaria pellucida Nordsieck, 1993

Albinaria pellucida Nordsieck, 1993 in Nordsieck 1993: 21

Shell transparent, brownish, finely ribbed, basal keel very prominent and crest-like, longer than in *A. schuetti*, no dorsal keel, subcolumellaris invisible in a perpendicular view of the aperture, basalis short or absent, no frontal upper palatalis.

16-19 x 4-5 mm

Only known from one locality.

Schütt 2001: 329.

Could be conspecific with *A. schuetti*, like *A. monocristata*, *A. pellucida* and *A. supercarinata*, more research is needed to firmly establish an independent species.

Turkey (Vilayet Antalya)

Albinaria percristata Nordsieck, 1993

Turkey (Vilayet Antalya)

Albinaria percristata Nordsieck, 1993 in Nordsieck 1993: 10

Turkey: Vilayet Antalya, Iotape NW of Gazipaşa

Shell greyish brown, weakly ribbed, cervix more prominently ribbed, dorsal keel extremely prominent, basal keel less prominent, often curves in an S-like form, apertural margin very broad, aperture detached from penultimate whorl, frontal upper palatalis and basalis usually present and in some populations connected at proximal end, clausilium plate visible in the aperture.

17-21 x 4-5 mm

Rocky habitats.

Schütt 2001: 330.

4 subspecies were recognised, based on differences in shell size, rib prominence on whorls and cervix, separated proximal ends of subcolumellaris and basalis, presence/absence of a frontal upper palatalis. Only the last two features could indicate the presence of uniform entities within a region, connected to neighbouring groups by hybrid zones, which would justify a subspecific classification. The other characters show continuous spatial gradients in almost every *Albinaria* species. Unless hybrid zones between populations are observed, no subspecies can be seriously recognized here.

Albinaria petrosa (Pfeiffer, 1849)

Clausilia petrosa Pfeiffer, 1849 in L. Pfeiffer 1849 (Z. f. Malak. 5): 122
Clausilia profuga Charpentier, 1852 in Charpentier 1852: 370, Pl. 11
Clausilia grisea var. *bigibbosula* Westerlund, 1878 in Westerlund 1876-1878: 241
Albinaria hohorsti Nordsieck, 1984 in Nordsieck 1984: 199
Albinaria mixta Nordsieck, 1984 in Nordsieck 1984: 200
Albinaria litoraria Neubert, 1998 in Neubert 1998: 142

Paget 1971 designated a neotype for *petrosa*. Nordsieck 1977: 296 disputed the validity of this neotype - unsubstantiated because all requirements of Art. 75.3 were met. A serious problem for the research is that the neotype designated by Paget 1971 came from a very old collection without known locality.
Nordsieck 1977: 296 argued that the "true" *petrosa* was conspecific with *A. grisea*.
Neubert 1998: 141 recognised the neotype as valid and found that it was conchologically similar to forms from near Achladokambos. He identified differences to *A. grisea* in the mode of ribbing and lunula position.
Nordsieck 1999: 7 suggested that *A. petrosa* could be conspecific with *A. mixta*.
Nordsieck 2007 used *petrosa* in Neubert's 1998 sense, and described (p. 163) a new subspecies of *A. petrosa* which was argued to be conspecific with *A. bigibbosula*, a transitional "hybrid" subspecies between *A. petrosa* and *A. mixta*, and a transitional "hybrid" subspecies between the closely related *litoraria* ("still regarded as a species") and *profuga*.
This suggests that at the current state of research the entire complex *bigibbosula-petrosa-mixta-hohorsti-profuga-litoraria* is locally interconnected by transitional forms, like in the *A. cretensis* complex in Crete. The oldest name is *petrosa*.

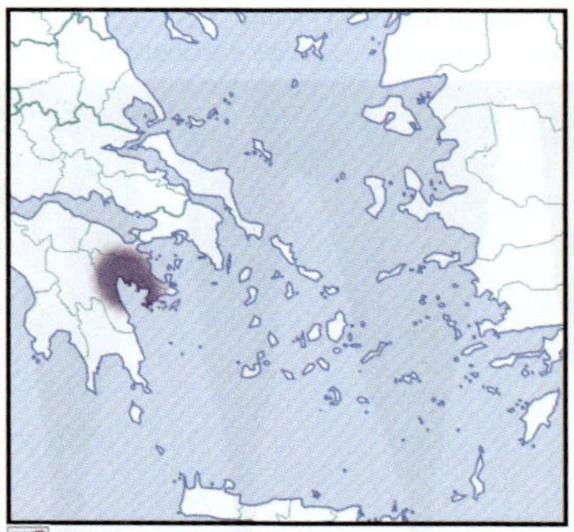

Greece (E Pelopónisos)

11.5-19 x 3-5 mm, locally variable

Greece: Pelopónisos, (1-6) Arkadía, Paralía Astros (*litoraria*), (7-9) Kándia near Náfplio (*hohorsti*), (10-11) Argos (*profuga*)
(10-11) Museum Budapest HNHM

The descriptions vary among the local forms and do partly not differ from *A. grisea*.
petrosa: Shell thin, light horny brown, silky shiny, finely and densely ribbed or striated, 10-12 convex whorls, aperture oval, inside yellow, columellaris in a flattened S shape, lunula dorsal with a blunt angle on its upper side, no parallelaris, subcolumellaris visible in an oblique view.
bigibbosula: Shell conical, whitish or greyish with horny brown spots, 9-11 whorls, weakly striated or ribbed, aperture inside white, margin shortly detached, lunula dorsal and strong.
hohorsti: Shell whitish with horny brown flames, upper whorls horny brown, weakly striated, cervix densely ribbed, 8-10 whorls, cervix with strong dorsal keel, aperture detached, parietalis passing spiralis, columellaris prominent, horizontal and in the upper section of the aperture, lunula dorsal, frontal upper palatalis usually present, subcolumellaris visible in a perpendicular view.
litoraria: Shell white to light horny brown, densely ribbed, denser on cervix, basal keel weak, dorsal keel very weak, apertural margin detached, reflected but not very broad, parietalis prominent and unusually long, there is an additional lamella parallel to the spiralis between spiralis and columella, columellaris slightly truncate in the aperture, lunula dorsal to dorsolateral, small basalis often present.
mixta: Shell evenly whitish or flamed, densely ribbed (35-45 ribs/penultimate whorl), 8-9 whorls, dorsal keel usually present, apertural margin usually detached, reflected and rounded, columellaris weak, lunula dorsal, clausilium plate visible in an oblique view, basalis short or absent, no frontal upper palatalis, palatal wall usually present, subcolumellaris visible in a perpendicular view. Outer margin of clausilium strongly bent, exceeds lunula.
profuga: Shell greyish, sharply ribbed, ribs white, 9-11 whorls, cervix with prominent basal and dorsal keels somewhat twisted around each other (dorsal keel oblique and running to the basal side), aperture broad, margin detached and slightly protruded, columellaris concave in its lower section, almost horizontal and located much upside in the aperture.

At limestone rocks.

Westerlund 1884: 114 (*petrosa*), 121 (profuga), 138 (*grisea*), Nordsieck 1995: 93, Neubert 1998: 140, 141, Nordsieck 1999: 7.

It is possible that *A. hohorsti* is connected to *A. grisea* by either a hybrid zone or a smooth morphological gradient (near Tolon), more local research seems to be necessary.
The form *mixta* could be the result of a hybridisation of a *hohorsti*-like form with *Sericata lutracana*.

Albinaria pondika Welter-Schultes, 2010

Albinaria pondika Welter-Schultes, 2010 in Welter-Schultes 2010: 224

13-14 x 2.9-3.2 mm

Greece: island of Pondikonísi W of Crete (H = 13.5-13.9 mm)

Vollrath Wiese

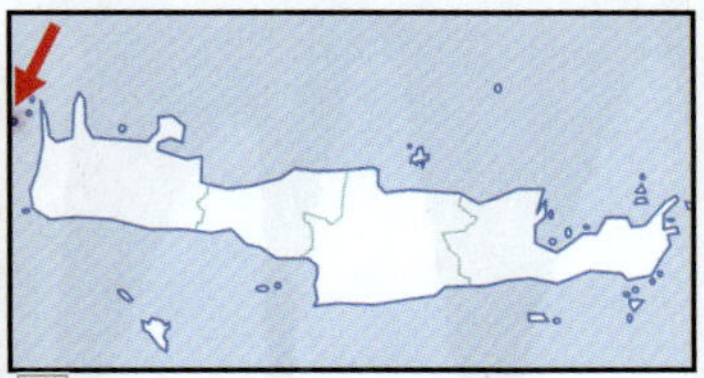

Greece (Pondikonísi W of Crete)

Shell light violetish or brownish, unusually small, densely ribbed, columellaris, frontal upper palatalis and basalis very strong, frontal upper palatalis with a hook at its frontal end, clausilium visible in a perpendicular view of the aperture.

Classification as a species depends on the relationship between *A. amalthea* and *A. candida*. It would be necessary to study possible zones of contact of both taxa near Nerokoúros in W Crete - if there is a hybrid zone like observed in the subspecies of *A. hippolyti*, the species from Pondikonísi should be classified as a subspecies of *A. candida*.

Albinaria praeclara (Pfeiffer, 1853)

Clausilia praeclara Pfeiffer, 1853 in L. Pfeiffer 1853: 150
Clausilia extensa var. *parallelifera* Boettger, 1878 in O. Boettger 1878: 103

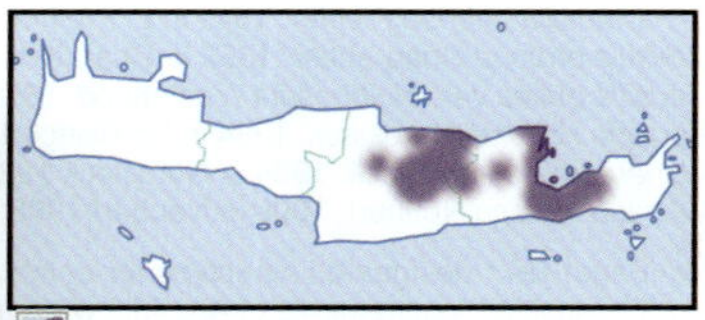

Greece (central and E Crete)

Shell usually brownish or reddish brown due to dust particles, relatively densely ribbed or with percostate ribs, with more or less prominent riblets between the ribs (with gradients, generally more prominent in the east but values overlap, most prominent and scale-like near Agios Nikólaos), cervix usually less densely ribbed, apex ribbed in the east (ribs start from whorl 0.5-0.8), ribs start later in the west (1.2-1.5, still earlier than in *A. corrugata*), apertural margin sharp and fragile, no frontal upper palatalis or basalis.
Two disjunctly distributed subspecies are recognized here, *A. pr. praeclara* east of Díkti and A. *pr. parallelifera* west of Díkti, distinguished by the apex ribs. *A. praeclara* is the only species in Crete where percostate forms occupy larger ranges.

15-25 x 3.2-5.0 mm, locally variable.

Greece: E Crete, (1-4) near Kaló Chorió, (5-8) Páno Loúmas

Shells are small E of Páchia Ammos and E of Iráklio, and large near Kastéli Pediádas, Ierápetra and between Agios Nikólaos and W of Eloúnda.

Limestone rocks, shells hide in rock crevices, occasionally also in small depressions in the rocks, percostate snails are often found estivating in groups of various snails.

Welter-Schultes 2010: 183 recognised two subspecies, distinguished in the apical ribs without overlap (ribs initiate at whorl position 0.5-0.8 in *A. pr. praeclara*, east of Mt. Díkti, and at whorl position 1.2-1.5 in *A. pr. parallelifera*, west of Mt. Díkti). They could also be classified as separate species, this is certainly a subjective decision. I decided to accept Nordsieck's 1977 and 1984 classification. Molecular studies could shed more light on their relationships.

Albinaria proteus (Boettger, 1889)

Clausilia proteus Boettger, 1889 in O. Boettger 1889 (Abh. Senck. 16): 49 (*Clausilia (Albinaria) Proteus*)
Not *Clausilia (Euxina) corpulenta* var. *proteus* Retowski, 1889 in Retowski 1889: 259 (= currently *Elia corpulenta* (Pfeiffer, 1848) from N Turkey), I don't know which name was published earlier. If both were from 31 Dec 1889, Boettger's name for a species would have precedence (Art. 24.1, 57.7).
Clausilia unicolor Boettger, 1878 in O. Boettger 1878: 118
Not this species, but misidentified between 1883 and 1999. *C. unicolor* is a synonym of *A. cretensis* from Crete.

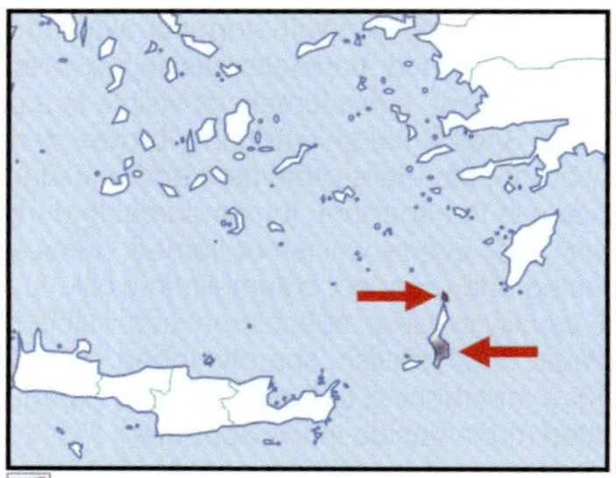

Shell white with brown apex, often with spots, smooth and solid, or weakly striated, cervix with prominent and slightly corrugated ribs, 10-12 whorls, basal keel strong, dorsal keel slightly weaker, aperture nearly circular, inside yellowish brown, parietalis and columellaris not very prominent, subcolumellaris only visible in a very oblique view.

16-23 x 3.4-5 mm

Boettger 1883: 336, Westerlund 1890: 63, Pieper 1969, Nordsieck 1977: 296, Zilch 1977: 332, Nordsieck 1999: 15.

Boettger 1883 and subsequent authors used the name *unicolor* for populations of Kárpathos, but Nordsieck 1999 discovered that the types belonged to *A. cretensis* from W Crete.
Pieper 1969 considered *unicolor* and *proteus* as subspecies of *A. caerulea*, introduced to Kárpathos. Nordsieck 1999 hypothesized that the two forms of Kárpathos could be two different species, especially if they occurred sympatrically without transitions, but had no new data to substantiate this hypothesis.
More research on Kárpathos is necessary.

Greece (Kárpathos, Saría)

Albinaria puella (Pfeiffer, 1850)

Clausilia puella Pfeiffer, 1850 in L. Pfeiffer 1850 (Proc. London 17): 136

SW Turkey, E Aegean islands

Shell smooth or finely streaked, white or with brown spots or dashes, cervix ribbed, dorsal keel weak, basal keel not very prominent, apertural margin narrow, parietalis and columellaris weak, lunella dorsal and also weak, two short palatal folds present.

Limestone rocks and walls in coastal vicinity.

12.5-15.5 x 3-3.5 mm

Greece: island of Ikaría

Bar & Butot 1986: 79, Bank & Maassen 1998: 66, Schütt 2001: 331.

Separating subspecies by rib prominence, as has been done in this species, is certainly not justified.

Albinaria rebeli Wagner, 1924

Albinaria rebeli Wagner, 1924 in A. J. Wagner 1924: 16

14-15 x 3.5-4.3 mm

Greece: E Crete, near Kavoúsi

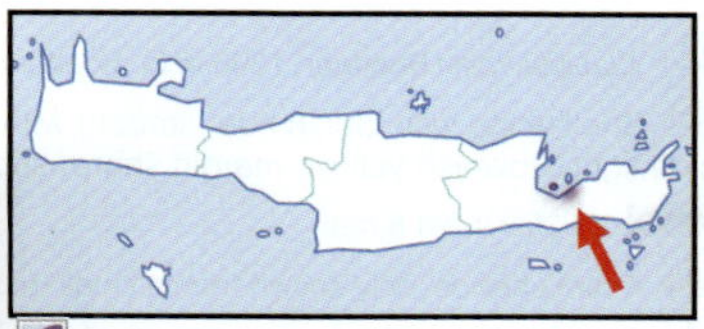

Greece (E Crete)

Shell decollate with 6-9 whorls remaining in the adult shell, white with light horny yellowish apex (5-10 % of the snails do not lose the apex during lifetime), almost smooth, cervix weakly ribbed, with strong basal and dorsal keels arranged in a V pattern, aperture inside reddish, margin sharp.

Limestone rock faces, prefers steep rocky slopes.

Restricted to a gorge NW of Kavoúsi, to the adjacent summits in the northeast, and to the island of Kónida. The species is frequent and currently not threatened.

Welter-Schultes 2010: 204.

A. rebeli was discovered in 1904 by R. Sturany, but the locality was not given accurately and later forgotten. A copy of Sturany's expedition report (Seidl 1978, Mitt. Braunau) led to the rediscovery of the species in 1987. Gittenberger 1991 (Biol. J. Linn. Soc. 43) studied the behaviour of the three sympatrically living species *A. teres*, *A. praeclara* and *A. rebeli* in the gorge, and found that competitive exclusion must be responsible for the selection of particular habitats. *A. rebeli* was also the subject of a study on shell decollation (Gittenberger & Povel 1995).

Albinaria rechingeri Paget, 1971

Albinaria rechingeri Paget, 1971 in Paget 1971: 465 (figured on Taf. 1 Fig. 1-4 [sic, not Taf. 2])

Shell weakly striated or almost smooth, first whorls dark brown, the others milky bluish due to a strong white surface layer, with some dark spots or dashes, cervix ribbed, basal and dorsal keels weak, aperture protruded, parietalis low, columellaris relatively deep inside, no palatal folds, perhaps a palatal callus.
Differs from local *A. brevicollis* populations by its brownish apex and its more inflated and less slender shape.

16-19 x 3.7-4.3 mm

Lives exclusively at limestone rocks, syntopically with *A. brevicollis*.

Occurs only in a small restricted area near Lindos.

Had been regarded as a subspecies of *A. brevicollis* by Nordsieck 1977: 295, but lives syntopically with *A. brevicollis* near Lindos.

Greece (E Ródos)

Albinaria retusa (Olivier, 1801)

Without gender agreement: *Albinaria retusus* (Olivier, 1801)

Bulimus retusus Olivier, 1801 in Olivier 1801: 416, Pl. 17

13-15 x 3.5-4.5 mm, locally variable

Greece: island of Día N of Crete, northern slopes

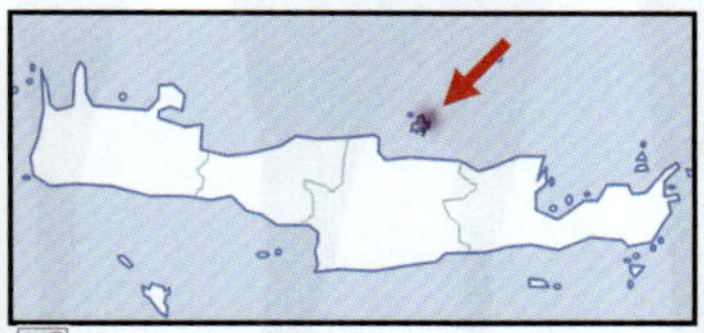

Greece (island of Día N of Crete)

Shell decollate with 5-6 whorls remaining in the adult shell (1-2 % of the shells are not decollate), reddish to violetish brown, more or less prominently ribbed (ribs white, 20-45 ribs/penultimate whorl, varies locally, one population has smooth shells and marbled colour pattern), cervix ribbed but less densely than upper whorls, complete shells have 13-15 whorls, much more regularly increasing than *A. torticollis*, apertural margin sharp, no basalis or frontal upper palatalis, parietalis and columellaris weak, subcolumellaris not visible in the aperture.

Steep limestone rocks and their upper margins, hides in rock crevices. The species is frequent where it occurs.

Schultes & Wiese 1990, Welter-Schultes 1992, Welter-Schultes 2000 (Schr. Malak. 16), Mesher & Welter-Schultes 2008, Welter-Schultes 2010: 204.

An area with hybrids of *A. retusa* and *A. torticollis* was discovered in the south, but the hybrids did not seem to be fertile.

Albinaria rollei (Boettger, 1896)

Clausilia rollei Boettger, 1896 in O. Boettger 1896: 126

Shell white with bluish hue, smooth with weakly striated cervix, 11-13 whorls with deep suture, aperture inside light brownish yellow, margin sharp, subcolumellaris distinct and well visible in an oblique view.

21-25 x 4.2-4.8 mm

A rare species, only known from a small restricted range.

Gittenberger & Neuteboom 1989.

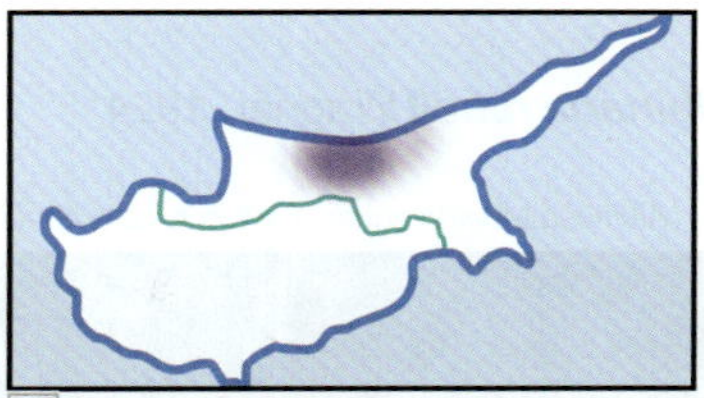

N central Cyprus

Albinaria saxatilis (Pfeiffer, 1846)

Clausilia saxatilis Pfeiffer, 1846 in L. Pfeiffer 1846 (Symb.): 94
Clausilia avia Charpentier, 1852 in Charpentier 1852: 376

12-18 x 2.8-4.5 mm

W Cyprus: near Agios Geórgios W Pégia

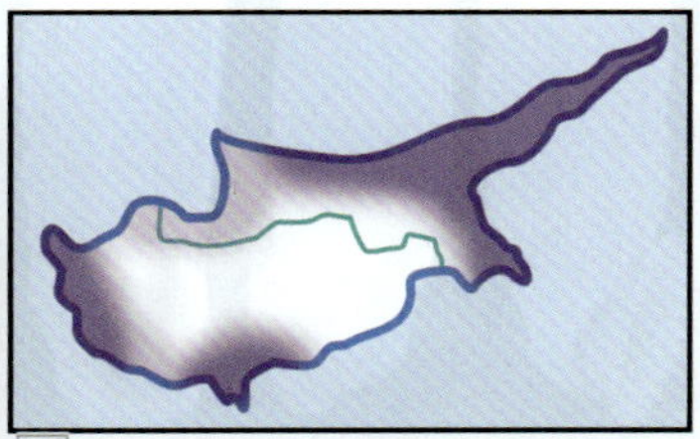

Cyprus

Shell grey with reddish hue, densely and prominently ribbed, 9-12 whorls, outline slightly concave at whorls 5-7 giving a pointed apex, basal keel prominent, no dorsal keel, aperture small, margin detached and slightly protruded, lunula almost dorsal. Southern populations (*avia*) with dorsal keel perhaps slightly more pronounced, cervix less impressed, columellaris steeper, lunula less deep in the aperture.

Westerlund 1884: 129, Falkner 1990: 156, Gittenberger & Neuteboom 1989: 128, 133, Nordsieck 1999: 13.

Classification follows Gittenberger & Neuteboom 1989 who regarded the differences between *avia* and *saxatilis* only as gradual. The spatial variation of shell characters remain to be studied in more detail to see if a classification as two independent species would be appropriate, as suggested by Nordsieck 1999 without detailed data. More field research is necessary.

Albinaria schuetti Nordsieck, 1984

Albinaria schuetti Nordsieck, 1984 in Nordsieck 1984: 197

13-19 x 3.5-5 mm

Turkey: Vilayet Antalya, Manavgat

Nordsieck 1993: 7, 24, Schütt 2001: 332.

SW Turkey

Shell smooth or ribbed, sometimes with dark spots, no dorsal keel, basal keel prominent (or even crest-like, particularly in smooth shells), parietalis weaker than spiralis, subcolumellaris usually invisible in a perpendicular view of the aperture, lunellar dorsal to lateral, basalis weak or absent, no frontal upper palatalis, basal groove provoked by prominent basal keel.

Limestone rocks.

4 subspecies were recognized, mainly based on shell size, rib prominence, prominence of the basal keel and prominence of the basalis. It has not yet been verified if we are not dealing with a widely distributed species with spatial character variations running in various directions, a situation in which it would be misleading to classify certain populations as different subspecies.
Could be conspecific with *A. monocristata*, *A. pellucida*, and *A. supercarinata*. All these taxa have a very prominent basal keel, but are geographically isolated with small ranges, while *A. schuetti* seems to occupy a large range of 50 x 60 km.

Albinaria scopulosa (Charpentier, 1852)

Clausilia scopulosa Charpentier, 1852 in Charpentier 1852: 374

12-17 x 2.8-4 mm

Albania: near Përmet

W Greece, S Albania

Shell white with bluish hue, often with horny brown apex, weakly striated, 9-10 convex whorls, cervix finely ribbed with short dorsal and basal keels and a distinct furrow in between, aperture inside dark yellow, margin shortly detached, not very broad, parietalis very weak, upper palatalis almost dorsal, short and oblique, lunula very weak.

Westerlund 1884: 135, Rähle 1980: 213 (Zákinthos), Dhora & Welter-Schultes 1996 (Albania).

See the discussion under *A. hians*, which could be conspecific with *A. scopulosa*.

Greece: Ipiros, Ditikó Zagóri, Víkos village

Cédric Audibert

Albinaria senilis (Rossmässler, 1836)

Clausilia senilis Rossmässler, 1836 in Rossmässler 1833-1837 (4): 11, Taf. 18

Previously mentioned by Rossmässler 1834 (Heft 2): 7 in the chapter of *Clausilia marmorata*, without description and without clear classification.

13-19 x 3-5.5 mm, very variable

Greece: Kefaloniá, Argostóli

Westerlund 1884: 142, Falkner 1990: 156, Kemperman 1992: 194, Fehér & Erőss 2009 (S Albania: Delvinë).

W Greece, S Albania

Shell solid, yellowish or greyish white with horny brown apex, sometimes with dark spots, ribbed, 11-12 whorls, cervix with one oblique central keel, apertural margin detached and protruded, parietalis weak, columellaris extremely weak, not visible in a perpendicular view, the other folds also weak, subcolumellaris not visible in the aperture.

Albinaria solicola Neubert, 1998

Albinaria solicola Neubert, 1998 in Neubert 1998 (Basteria 62): 146

Shell white with darker dashes or streaks, almost smooth, cervix finely striated, basal keel moderate, dorsal keel weak, with a furrow in between, apertural margin reflected and usually blunt, parietalis strong, overlaps with spiralis, columellaris truncated and thickened at its lower end, lunula dorsolateral or lateral, basalis usually present, sometimes as a callus.

13-17 x 3.6-4.5 mm

Greece (E Pelopónisos)

Albinaria sphakiota (Maltzan, 1887)

Clausilia cretensis var. *sphakiota* Maltzan, 1887 in Maltzan 1887: 119

16-22 x 3.7-4.3 mm Greece: SW Crete, Arádena gorge

Greece (SW Crete)

Shell whitish with horny brown apex, smooth or weakly striated, upper whorls and last whorl coarsely ribbed, 11-13 whorls, basal keel moderate, dorsal keel slightly more prominent, aperture inside brownish, parietalis closer near columellaris than in sympatrical *A. cretensis* forms, columellaris shorter and lower at Imbros, equally strong or slightly higher at Arádena, does not reach the apertural margin at Imbros (at Arádena yes), lunula lateral to ventrolateral.
Differs from sympatric *A. cretensis* by smoother shells and a deeper situated lunula, locally there are some additional differences. At Imbros *A. cretensis* has a weakly ribbed shell, a weaker dorsal keel, in the middle part of the gorge the shells are smaller.

Limestone rocks in steep mountain gorges (Arádena gorge, Imbros gorge, Asfendos gorge).

The species occurs in very restricted and disjunct areas in gorges. It is known from 3 gorges and currently not threatened.

Nordsieck 1999, A. Gittenberger et al. 2001 (molecular evidence for reproductive isolation with *A. cretensis*), Nordsieck 2004: 60, Welter-Schultes 2010: 219.

Albinaria spratti (Pfeiffer, 1846)

Clausilia spratti Pfeiffer, 1846 in L. Pfeiffer 1846 (Symb.): 94 (*Clausilia Spratti*)

24-30 x 4.6-5.7 mm Greece: N central Crete, near Pérama

Greece (N central Crete)

Shell uncommonly large, brownish, densely to very densely ribbed, cervix much less densely ribbed and slightly corrugated, 13-15 whorls, often decollate with 9-11 whorls remaining in the adult shell, cervix with weak basal and no dorsal keel, apertural margin sharp, no basalis or frontal upper palatalis, subcolumellaris invisible in the aperture.

Limestone rocks, hides in rock crevices.

Welter-Schultes 2010: 203.

Albinaria sturanyi Wagner, 1924

Albinaria sturanyi Wagner, 1924 in A. J. Wagner 1924: 18

18-23 x 3.7-4.5 mm Greece: E Crete, Aféndis Kavoúsi mountains, near Thriptí, 1320 m

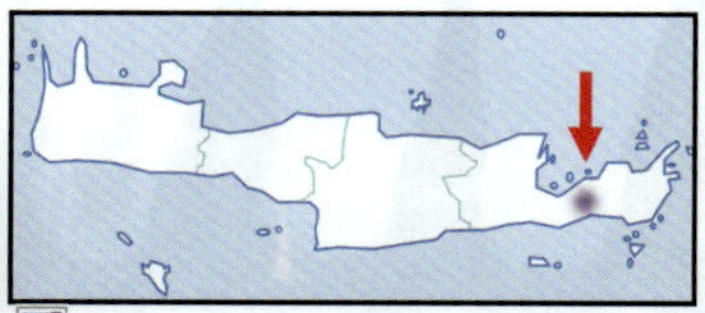

Greece (E Crete)

Shell white, often with bluish hue, smooth or very weakly ribbed (30-40 ribs on the penultimate whorl), cervix weakly ribbed, cervical ribs not corrugated, with strong basal and dorsal keel, and a basal furrow in between, apex usually blackish, aperture more rounded than in *A. teres*, margin less sharp.

Limestone rocks in high altitudes, at rock faces.

Nordsieck 1999, Welter-Schultes 2010: 181.

In a strictly biological concept based on reproductive isolation this taxon could be considered as a subspecies of *A. teres* because it lives parapatrically with *A. teres* and is connected by a narrow hybrid zone with only partial barriers to gene flow.

Albinaria subaii (Nordsieck, 1984)

Sericata subaii Nordsieck, 1984 in Nordsieck 1984: 193

10.5-13.5 x 2.7-3.2 mm

Nordsieck 1999: 4.

Greece: (1-5) Máni peninsula, marine deposits, (6-8) Lakonía, Dirón

(6-8) Paolo Zaccaria

(6-8) Tiroler Landesmuseen Innsbruck

Greece (S Pelopónisos)

Shell white with brownish flames, densely ribbed, 8-10 whorls, cervix with weak dorsal keel, apertural margin usually detached, parietalis usually not reaching spiralis, parallelaris thickened at its proximal end, columellaris low and S-shaped, lunula dorsal to dorsolateral, clausilium plate well visible in an oblique view, basalis strong and visible in a perpendicular view, no frontal upper palatalis, subcolumellaris usually not visible in a perpendicular view.

Albinaria sublamellosa (Boettger, 1883)

Clausilia sublamellosa Boettger, 1883 in O. Boettger 1883 (Proc. London): 327

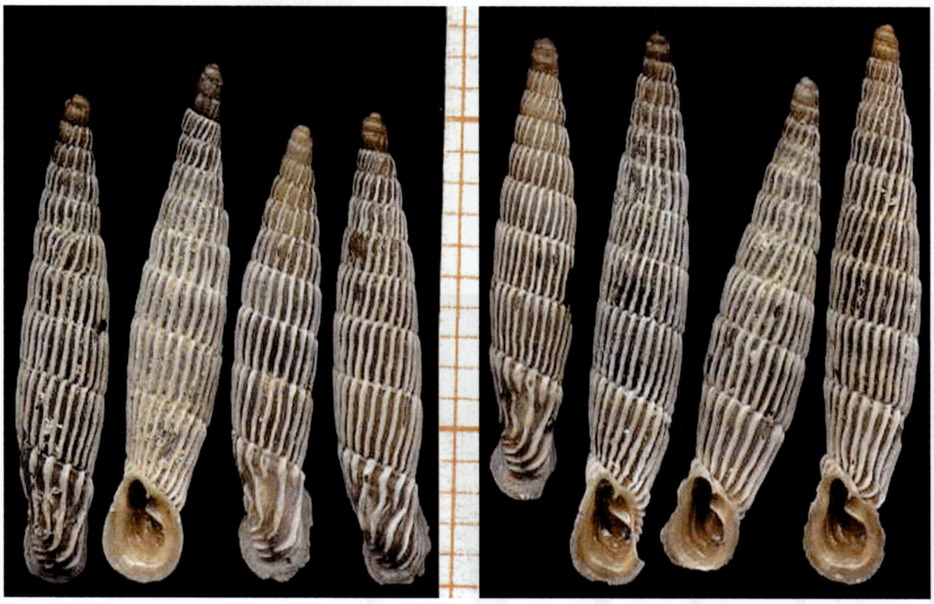

13-18 x 2.7-3.1 mm

Greece: W Crete, Samariá gorge

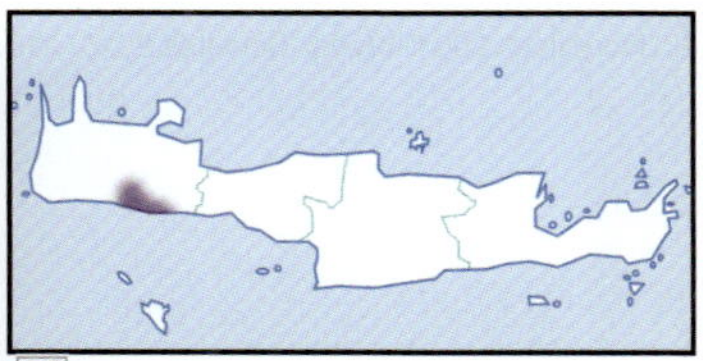

Greece (SW Crete)

Only a small portion if the presumed range of this species is known.

Shell small, usually reddish brown, ribbed (18-30 white ribs on the penultimate whorl, 2.2-3 ribs/mm), in one population with percostate ribs (6-8 ribs), 11-13 whorls, dorsal keel stronger near Chóra Sfakíon than in the west near Agía Roúmeli, aperture inside reddish brown, columellaris strongly developed and often slightly twisted upwards in its frontal part, principalis and spiralis strong and pointing exactly to one another, parallelaris sometimes present, subcolumellaris usually well visible, but not always in a perpendicular view.

Limestone rocks at southern mountain slopes and in gorges.

Welter-Schultes 2000, Nordsieck 2004: 61, Welter-Schultes 2010: 220.

Albinaria supercarinata Gittenberger & Menkhorst, 1992

Albinaria supercarinata Gittenberger & Menkhorst, 1992 in Gittenberger & M. 1992: 194

14-20 x 4-5 mm

Turkey: Vilayet Antalya, near Ob, E of Alanya

Nordsieck 1993: 7, Schütt 2001: 333.

SW Turkey

Shell finely ribbed, basal keel extremely prominent, crest-like and curved, and finely ribbed, longer than in *A. schuetti*, subcolumellaris hardly visible at all.

In steep valleys.

This species lives only in a small area.

It is also possible that this form belongs to *A. schuetti*, as well as *A. monocristata*. In *A. maltzani* in Crete a likewise high local variation of the cerical keel was observed, with smooth gradients between populations.

Albinaria tenuicostata (Pfeiffer, 1864)

Clausilia tenuicostata Pfeiffer, 1864 in L. Pfeiffer 1864: 604

16-21 x 3.5-4.5 mm

Greece: SW Crete, Soúgia

Limestone rocks, at rock faces, not on non-calcareous substrate.

Nordsieck 2004, Welter-Schultes 2010: 216.

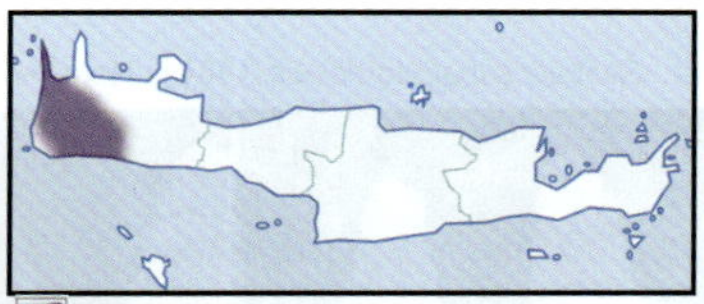

Greece (W Crete)

Shell usually whitish, usually weakly ribbed or smooth, (ribs prominent near Strovlés, Enneachoria, and between Rodováni and east of Soúgia), rib numbers around 30-55 with local gradients, dorsal keel usually weak, aperture without frontal upper palatalis and usually without basalis (only 1/10 of the shells in some populations have a weak basalis), parallelaris absent or very weak, subcolumellaris deep inside, usually only visible in an oblique view, columellaris regular but deeper inside east of Soúgia, clausilium plate usually visible in the aperture in the south, less frequently in the north.
Differs from *A. amalthea* south of Rodopós in its much deeper and less well visible subcolumellaris, and in the absence of a basalis. The subcolumellaris of sympatric *A. cretensis* forms is usually substantially deeper inside than in *A. tenuicostata*, except in the region of Paleochóra where both species are very similar and where we have not yet learned to distinguish both species.

Albinaria terebra (Pfeiffer, 1853)

Clausilia terebra Pfeiffer, 1853 in L. Pfeiffer 1853 (Z. f. Malak. 10): 151

3.1-3.7 x 15-18 mm, locally not very variable

Greece: S central Crete, near Pitsídia

Lowland hills around the Mesará plain, usually on Neogene substrate, in the north up to 600 m.

Greece (S central Crete)

Shell white with light horny brown apex, often brownish due to dust particles, ribbed (25-35 ribs/penultimate whorl, in the south down to 16 ribs), 12-14 whorls, outline concave in upper section at whorls 5-7 with a pointed apex, cervix with moderate basal and dorsal keels and a characteristically distinct basal furrow in between, aperture inside yellowish.
A. terebra is characterized by its well developed and clearly distinct basal furrow on the cervix, an unequivocal feature of this species.

Welter-Schultes 2001 (Paleobiology 27), Welter-Schultes 2010: 199.

Large parts of its present-day range must have been colonized by the species after late Neogene periods of tectonical uplift, when the formerly submerged land became hospitable for *Albinaria* land snails. It is possible that range expansions are documented in the morphological record.

Albinaria teres (Olivier, 1801)

Bulimus teres Olivier, 1801 in Olivier 1801: 417, Pl. 17

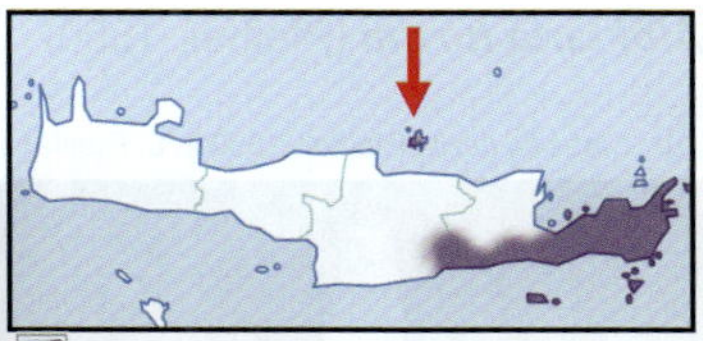

Greece: E Crete, near Péfki

17-25 x 3.3-4.9 mm, locally variable.

Shells near the southern coast are generally smaller (diameter 3.1-3.7 mm), shells near Sitía unusually slender and high (up to 30 mm).

Welter-Schultes 2010: 174 (spatial variation of shell morphology).

Greece (E Crete)

Shell usually whitish, ribbed or smooth (smooth in the mountains near Orinó and at the southern coast west of Goúdouras, almost smooth at Akra Síderos and on Día island, percostate at one site near Moní Toploú and west of Tsoútsouros, weakly ribbed SW of Zákros and in a 10-20 km W-E extending zone near Istro), rib numbers usually 25-35, (with local gradients, 15-25 between Tsoútsouros and Péfkos, 40-55 W and N of Zákros), cervix ribbed, usually more coarsely than the upper whorls, and very slightly corrugated (not as much as in *A. corrugata*), basal keel well visible, dorsal keel in most populations well visible but weaker, aperture oval, margin usually sharp, without basalis and frontal upper palatalis.

Usually at limestone rocks, but also on non-calcareous substrate, usually on the rock faces.

The morphological record suggests that the populations of the Thripti mountain summits (*sturanyi*) have an independent phylogenetic history, *sturanyi* interbreeds with "typical" *A. teres* in nature producing a hybrid zone and is classified as a separate species here. *A. manselli* was classified as an independent species by Wiese 1990, this should be maintained until new results are available.

Albinaria thiesseae (Boettger, 1880)

Clausilia thiesseae Boettger, 1880 in O. Boettger 1880: 48 (*Clausilia (Albinaria) Thiesseae*)

Greece: Akarnanía, Etolikó

9-13 x 2.5-3.3 mm

W Greece

Shell dirty white with indistinct grey spots, opaque, finely and densely ribbed, 8-10 unusually convex whorls, cervix rounded, indistinctly keeled, apertural margin widely detached and much broadened, columellaris oblique and S-shaped, parallelaris very prominent and long.

Westerlund 1884: 122.

Albinaria torticollis (Olivier, 1801)

Bulimus torticollis Olivier, 1801 in Olivier 1801: 416, Pl. 17

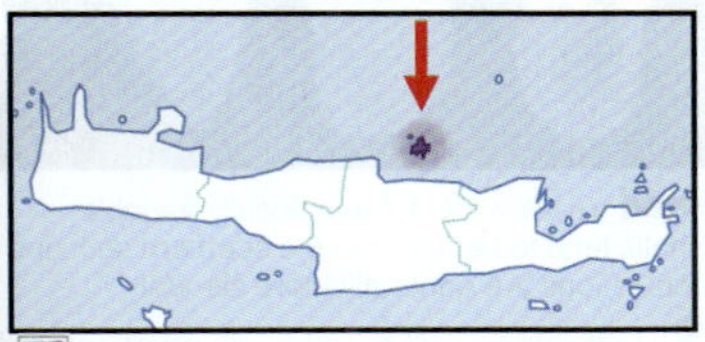

Greece: island of Día N of Crete

10-12 x 3.5-4.5 mm

Greece (Island of Día N of Crete)

Shell decollate with 4-6 whorls remaining in the adult shell (less than 1 % of the individuals are not decollate), usually reddish brown due to dust particles, extremely densely ribbed (70-130 ribs/penultimate whorl), complete shells have 16-17 whorls, first 10 whorls almost equal in diameter, then suddenly increasing to maximum diameter, last 5 whorls are almost equal in diameter and with straight outline, last whorl completely detached and characteristically protruding, cervix with very long and strong basal and dorsal keels.

Among limestone rocks, estivates in crevices and under stones. Very frequent.

Schultes & Wiese 1990, Welter-Schultes 1992, Nordsieck 1999, Welter-Schultes 2000, Mesher & Welter-Schultes 2008, Welter-Schultes 2010: 204.

An area with hybrids of *A. retusa* and *A. torticollis* was discovered in the south, but the hybrids did not seem to be fertile.

Albinaria turrita (Pfeiffer, 1850)

Clausilia turrita Pfeiffer, 1850 in L. Pfeiffer 1850 (Proc. London 17): 135

18-26 x 3.5-5 mm

Greece: Mílos

Greece (W Kikládes)
Makrónisos, Kéa, Sífnos, Mílos.

Shell white with dark dots, smooth or ribbed, 11-14 whorls, cervix rounded without prominent keels, also in smooth forms ribbed in its last section, aperture rounded, margin detached, lip yellowish, parietalis very small. Many characters seem to be locally variable.

Boettger 1878: 113, Westerlund 1884: 132, Nordsieck 1977: 294, Giokas & Mylonas 2002, Douris et al. 2007.

The identity of the populations and their classification as a species (in agreement with the morphological record) was confirmed in molecular studies (Douris et al. 2007).

Albinaria violacea Schilthuizen & Gittenberger, 1990

Albinaria violacea Schilthuizen & Gittenberger, 1990 in Schilthuizen & Gittenberger 1990: 138
Albinaria ulrikae Schilthuizen & Gittenberger, 1990 in Schilthuizen & Gittenberger 1990: 137
First Reviser: Welter-Schultes 2000: 98.

16-24 x 3.4-4.7 mm, locally variable.
Shells tend to be larger in the southern sections of Síses and Achláda, the Achláda populations are generally more slender.

Greece: N central Crete, near Balí

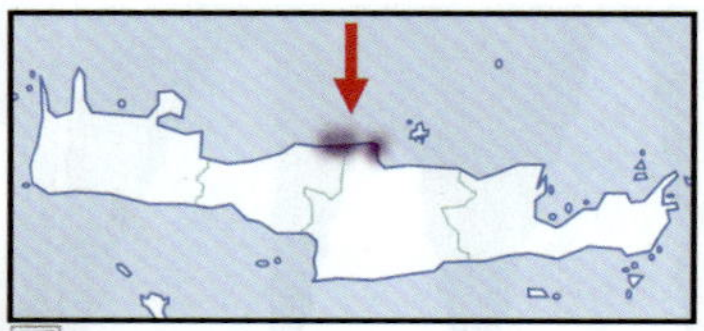

Greece (N central Crete)

Shell light brown or violet, finely and densely ribbed (60-100 ribs/penultimate whorl), 11-15 whorls, largest diameter at antepenultimate whorl, apertural margin rounded, lip reflected, basalis strong, frontal upper palatalis (FUP) and subclaustralis present, lunula dorsal to lateral (locally variable), columella twisted to the right side (pointing between FUP and basalis) at Achláda and E Síses, it points more or less exactly to the FUP (which is usual in Cretan *Albinaria*) in S Síses and W Balí, and between FUP and principalis in W Síses and E Balí, subcolumellaris invisible in a perpendicular view.

Greece: N central Crete, near Balí — Vollrath Wiese

Limestone rocks, relatively cryptical, hides in rock crevices.

Schilthuizen 1994 (both), Welter-Schultes 2000 (Schr. Malak. 16), Welter-Schultes 2010: 199.

A. violacea has 3 disjunct ranges with 2-3 km distance in between, includes *ulrikae* here. Complete reproductive isolation between *A. hippolyti* and *A. violacea* was found by Schilthuizen 1994 (Biol. J. Linn. Soc. 52).

Albinaria virgo (Mousson, 1854)

Clausilia virgo Mousson, 1854 in Mousson 1854: 387

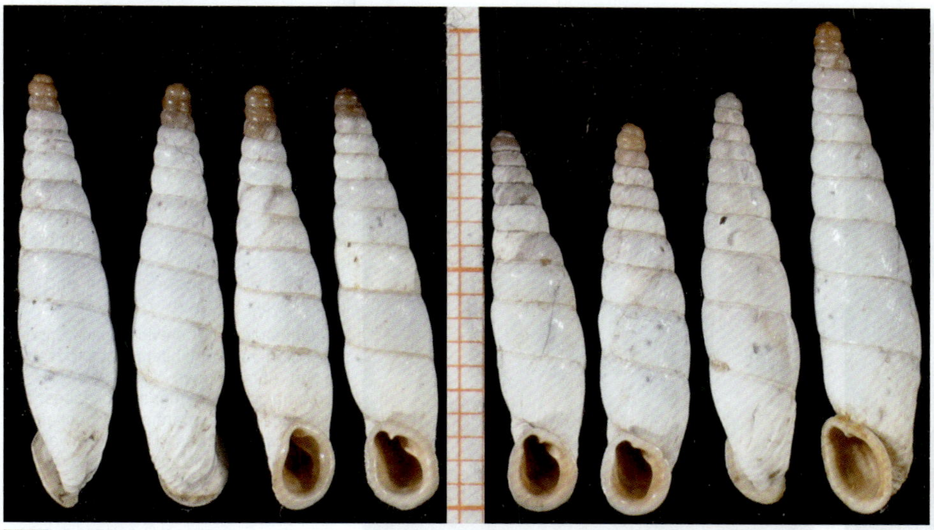

15-19 x 3.3-4.3 mm

N Cyprus: Sina monastery near Lápithos/Lapta

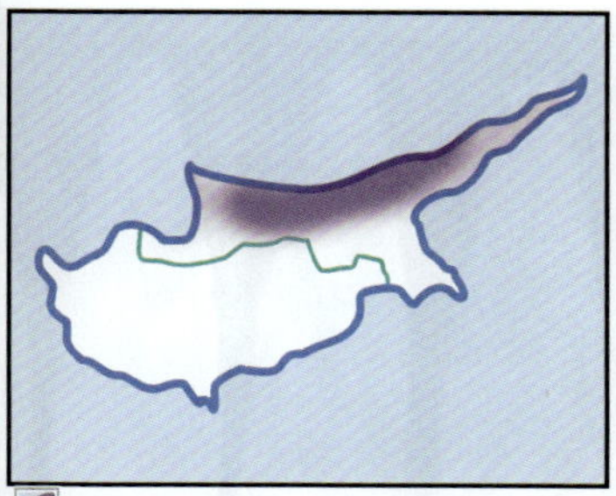

N Cyprus

Shell milky white with some dark dots, shiny, indistinctly striated, 10-12 whorls, apex blunt and often horny brown, cervix ribbed, slightly corrugated with weak dorsal and often slightly stronger basal keel, aperture small, inside brown yellowish, margin distinctly detached, broadly reflected, with white lip.

Westerlund 1884: 134, Gittenberger & Neuteboom 1989.

Albinaria voithii (Rossmässler, 1836)

Clausilia voithii Rossmässler, 1836 in Rossmässler 1835-1837 (4): 11, Taf. 18 (*Clausilia Voithii*)

12-18 x 3.2-4.2 mm, perhaps locally variable

Greece: Pelopónisos, Lakonía, Palea Mistra near Spárti

Westerlund 1884: 144, Nordsieck 1999: 8 (revision of *A. voithii*), Giokas & Mylonas 2002.

Relationships to *A. gerolimena* and *A. menelaus* remain to be studied.

Greece (S Pelopónisos)

Shell dextral, white or bluish white, often with some dark spots, apex often light brownish, main whorls weakly ribbed or smooth, 9-12 whorls, cervix ribbed, with one single central keel (running to the basal side), in some populations very prominently developed (like in *A. maltzani*), apertural margin detached, parietalis weak or absent, columellaris weak, parallelaris often present, lunella short and dorsolateral or lateral.

Albinaria wettsteini Fuchs & Käufel, 1936

Albinaria wettsteini Fuchs & Käufel, 1936 in Fuchs & Käufel 1936: 600

Shell dirty yellowish, very finely and densely ribbed, 9-10 whorls, aperture detached and protruded, last whorl like with two very prominent basal keels, one coming from dorsal and leading to the parietal margin of the aperture, the inner basal keel is sharp and ends dorsally, parietalis weak, columellaris weak and not visible in a perpendicular view, lunula dorsal, broadly connected to principalis, subcolumellaris not visible in the aperture.

11.6-15.2 x 2.7-3.3 mm

Greece: island of Ofidoúsa

Greece (Ofidoúsa island W of Astipálea)

Albinaria wiesei Gittenberger, 1988

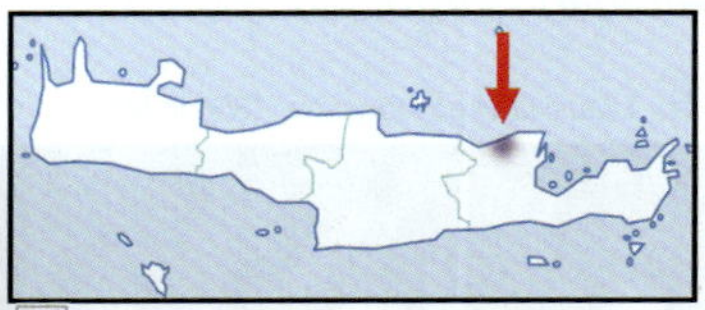

Albinaria wiesei Gittenberger, 1988 in Gittenberger et al. 1988: 111
Gittenberger's authorship for the description was explained in a note on p. 111.

3.2-3.4 x 14-15 mm

Greece: E central Crete, Selinári monastery near Neápoli

Gittenberger 1991 (Biol. J. Linn. Soc.), Welter-Schultes 2010: 204.

Greece (E central Crete)

Shell unusually small, brownish with violet hue, darker than in other *Albinaria* species, often marbled by whitish surface layers, densely striated on upper whorls and cervix, lower whorls smooth, 10-12 whorls, cervix with weak basal and dorsal keels, aperture inside reddish, margin rounded, columellaris very prominent, lunula dorsal.

Limestone rock crevices with herbal vegetation, humid habitats, not exposed to the sun. At rainy days *A. wiesei* and *A. corrugata* crawl together at rock faces, but at sunshine they separate their habitats, *A. corrugata* attaches to the rocky walls and *A. wiesei* crawls into the rock crevices.

Microendemic species, occupies a very small range. Currently not threatened despite a nearby major road.

Albinaria zilchi Fauer, 1993

Albinaria zilchi Fauer, 1993 in Fauer 1993: 50

Shell dirty whitish, ribbed, basal keel strong, dorsal keel weak, subcolumellaris invisible from the aperture, lunella dorsal or dorsolateral, parallelaris longer than spiralis at both sides.

15.5-19.5 x 3.8-5.0 mm

At limestone rocks in karstic habitats with poor vegetation, less frequent than sympatrically occuring *A. grisea*.

Greece (E Pelopónisos)

Isabellaria

Isabellaria Vest, 1867
in Vest 1867: 188 *Clausilia (Isabellaria)*

Type species *Clausilia isabellina* Pfeiffer, 1842 (original designation)

Species composition according to Nordsieck 1999 with exceptions. Gittenberger 1998 proposed to classify several species with *Albinaria*, this concerned at least the taxa *adriani*, *butoti*, *campylauchen* (inclusing *sikeensis*), *confusa*, *edmundi*, *idyllica* and *vrondamasa*. Nordsieck 1999: 6 and Nordsieck 2007: 48 classified them in *Isabellaria*. They are classified in *Albinaria* here.

Isabellaria almae (Boettger, 1889)

Clausilia almae Boettger, 1889 in O. Boettger 1889: 54 (*Clausilia (Papillifera) Almae*)

17-21 x 4.5-5 mm

Greece: Fokís, SW Amfissa, near Agía Efthimía

Greece (S central Stereá Elláda)

Shell light horny brownish, finely striated, 9-10 whorls, outline of whorls 3-5 slightly concave, aperture detached, parietalis prominent, columellaris moderate and inside forked, lunula dorsal and with a callus at its basal end.

Westerlund 1890: 68, Uit de Weerd 2004: 55.

Isabellaria chelidromia (Boettger, 1889)

Clausilia chelidromia Boettger, 1889 in O. Boettger 1889 (Abh. Senck. 16): 55, (*Clausilia (Papillifera) Chelidromia*)

Shell dark violet with weak bluish cover, weakly striated and shiny, cervix finely ribbed, 9-10 slightly convex whorls with white suture, aperture nearly circular and shortly detached, margin rounded, parietalis low and long, columellaris not very prominent, lunula lateral, no palatal folds, no principalis and suturalis, subcolumellaris strong and visible in a perpendicular view, almost reaching margin.

Westerlund 1890: 69, Nordsieck 1984: 193, Liebegott 1986, Uit de Weerd 2004: 55.

15-19 x 4-4.8 mm

Greece: Vórii Sporádes, Alónisos

Greece (Vórii Sporádes)

Isabellaria clandestina (Rossmässler, 1857)

Clausilia clandestina Rossmässler, 1857 in Rossmässler 1857: 40

Shell light brown, weakly striated, 8-10 whorls with dark suture, aperture with white lip, margin connected at parietal side, parietalis curved, reaching the margin, columellaris very oblique and narrowly forked, principalis weak or absent, lower palatalis strong, subcolumellaris visible in an oblique view, lunula dorsolateral and curved in its upper section.

Westerlund 1884: 171, Uit de Weerd 2004: 55.

13-20 x 3-4 mm, perhaps locally variable

Greece: S of Vólos, near Alos ruins

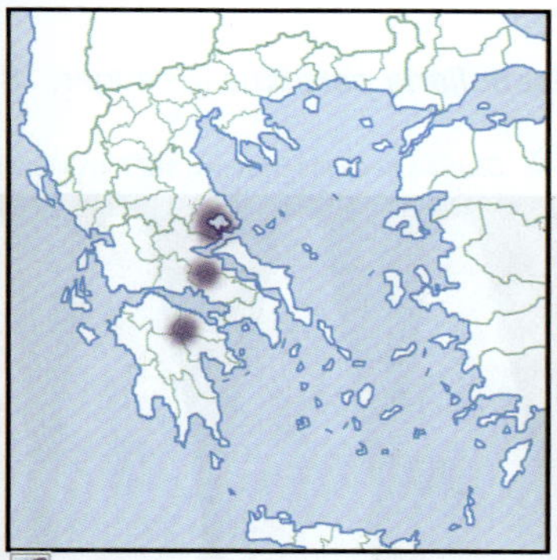

Greece (Thessalía to Pelopónisos)

Isabellaria isabellina (Pfeiffer, 1842)

Clausilia isabellina Pfeiffer, 1842 in L. Pfeiffer 1842: 60

13-16 x 3-4 mm

Greece: Kórinthos

Greece mainland, Salamína, Egina

Shell whitish, finely ribbed, 8-10 whorls, cervix with blunt and short keel, aperture rounded, sometimes margin very broad, parietalis curved and low, columellaris very oblique, low and in inner section often forked, 3 sutural folds, principalis strong, lunula strong, curved, often connected to strong subcolumellaris.

Westerlund 1884: 166.

Isabellaria leucorhaphe (Boettger, 1878)

Clausilia leucorhaphe Boettger, 1878 in O. Boettger 1878 (Jahrb. DMG 5): 295, Taf. 10

Shell dark horny brown or horny yellow, smooth, shiny, 9-10 whorls, cervix striated, aperture almost circular with weak dark lip, margin broadened, columellaris inside forked, 2 fine and sometimes very weak sutural folds, principalis very small, clausilium well visible.

17-20 x 3.8-4.7 mm

Westerlund 1884: 168, Liebegott 1986: 16, Uit de Weerd 2004: 55.

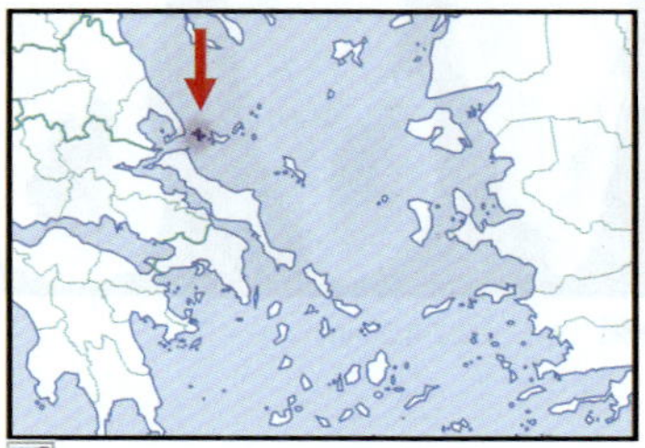

Greece (island of Skíathos)

Isabellaria lophauchena (Sturany, 1894)

Clausuilia lophauchena Sturany, 1894 in Sturany 1894: 374, Taf. 18

14-17 x 3-4 mm

Rep. Makedonija: Vardar riverside near Demir Kapija

N Greece, S Rep. Makedonija

Shell brown, almost smooth, 8-10 whorls, cervix with moderate basal and strong and oblique dorsal keel, basal furrow also visible from inside, aperture with lip, margin detached, parietalis and columellaris prominent, one palatal fold above lunula, lunula dorsolateral.

On the surface of limestone rocks, under stones and rock crevices.

Falkner 1990: 154, Uit de Weerd 2004: 55, Dedov & Neubert 2009: 93.

Isabellaria perplana (Boettger, 1877)

Clausilia perplana Boettger, 1877 in O. Boettger 1877: 51 (footnote), Taf. 4

17-19 x 3.9-4.6 mm

Greece: Fokís, near Aráchova

Greece (S central Stereá Elláda)

Shell horny brownish, almost smooth, 9-10 whorls, cervix flattened, aperture with white lip, margin thickened, columellaris strong, S shaped and very close to the parietal margin, one distinct sutural fold present, connected to the strong principalis.

Westerlund 1884: 168, Uit de Weerd 2004: 55.

Isabellaria praecipua Sajó, 1968

Isabellaria praecipua Sajó, 1968 in Sajó 1968: 58

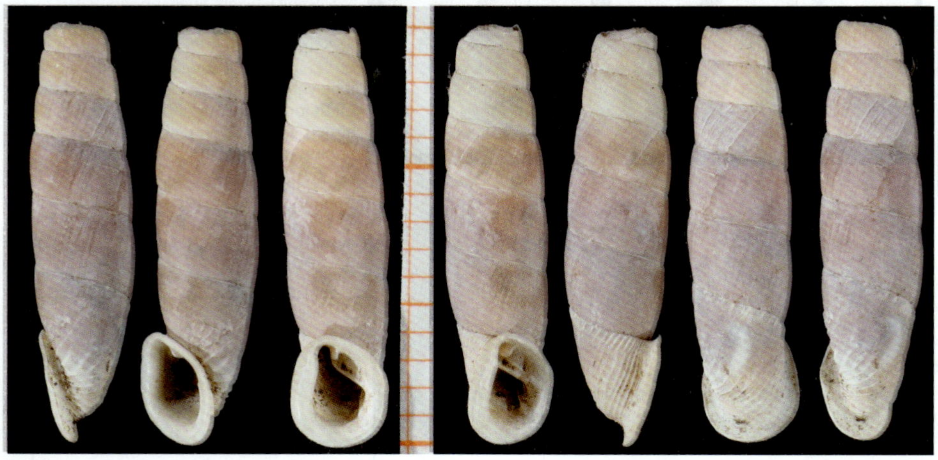

14-17 x 3.7-4.3 mm

Greece: Makedonía, Káto Vérmio

N Greece (W Makedonía)

Shell decollate with 5-7 whorls remaining, violetish brown with white suture, ribbed (30-50 ribs/penultimate whorl), cervix with short curved basal keel, aperture detached, margin broadly reflected at basal side and sharper at parietal side, parietalis short, columellaris not very prominent and nearly horizontal, lunula dorsal, no palatal folds, basal furrow present, subcolumellaris well visible in a perpendicular view.

Uit de Weerd 2004: 55.

Isabellaria praestans (Westerlund, 1893)

Clausilia praestans Westerlund, 1893 in Westerlund 1893: 127

19-23 x 4.6-5.4 mm

Greece: Vórii Sporádes, Skópelos

Museum Budapest HNHM

Greece (Vórii Sporádes)
Only on Alónisos, Peristéra, Kirá Panagía, Skópelos.

Shell white with bluish hue, densely striated, 10-11 whorls with papillated suture, aperture rounded, spiralis visible, frontal upper palatalis present, lunula dorsal, subcolumellaris not very distant from columellaris.

Nordsieck 1984: 193, Liebegott 1986: 16, Uit de Weerd 2004: 55.

Isabellaria riedeli Brandt, 1961

Isabellaria riedeli Brandt, 1961 in Brandt 1961: 12

Shell dark horny brown, striated, cervix densely ribbed, 8-12 whorls with weak whitish suture, umbilicus open and nearly perpendicular, cervix with more or less prominent basal and dorsal keel, aperture inside brown, margin detached or not, parietalis reaches margin, columellaris seems to be variable, lunula lateral to ventrolateral, no principalis, no palatal folds, subcolumellaris not visible in the aperture.
Clausilium plate strongly bent, groove-shaped and with edge at the terminal end.

Uit de Weerd 2004: 55.

15-20 x 3.5-4.5 mm

Greece: N Evia, Prokópi

Greece (N central Evia)

Isabellaria saxicola (Pfeiffer, 1848)

Clausilia saxicola Pfeiffer, 1848 in L. Pfeiffer 1848 (Mon. Helic. 2): 443

13-18 x 3-4.2 mm, locally variable

Westerlund 1884: 169, Uit de Weerd 2004: 55.

Greece: Attikí, Athína

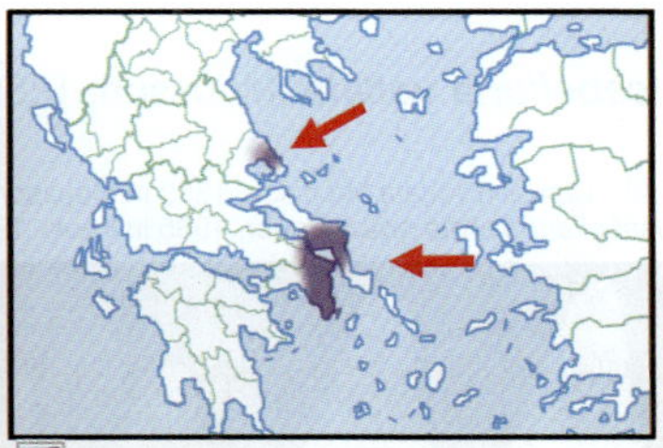

Greece (Evia, Attikí)

Shell horny violet or brown, sometimes reddish, finely striated or smooth, 7-9 whorls, suture densely papillated or white, aperture large and rounded, with weak lip, margin not always connected at parietal side, parietalis variable, columellaris curved and often horizontal, 1-3 indistinct sutural folds, up to two small palatal folds, principalis short or absent, subcolumellaris usually short, often far distant from the columella, sometimes connected to lunula, lunula dorsal, clausilium well visible in the aperture. Several characters seem to display spatial variations.

Isabellaria thermopylarum (Pfeiffer, 1850)

Clausilia thermopylarum Pfeiffer, 1850 in L. Pfeiffer 1850 (Proc. London 17): 139 (*Cl. Thermopylarum*)

17-21 x 3.7-5 mm

Greece: Fokída, Parnasós

Westerlund 1884: 168, Falkner 1990: 154, Uit de Weerd 2004: 55.

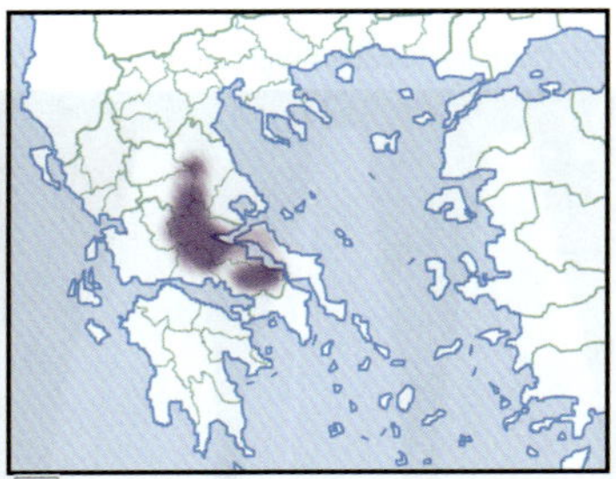

Greece (Thessalía, Stereá Elláda)

Shell whitish to horny brownish or greyish brown, almost smooth, 7-10 whorls, cervix densely ribbed, rounded with very weak basal and dorsal keels, aperture usually large with white lip, margin connected at parietal side, columellaris curved, above lunula 3 long palatal folds instead of one single principalis, often visible from outside, lunula dorsal to dorsolateral, curved and with a hook at its upper end, at lower side with a very short basalis, subcolumellaris ends near lunula.

Isabellaria thessalonica Nordsieck, 1972

Isabellaria thessalonica Nordsieck, 1972 in Nordsieck 1972: 18

14-17 x 3.2-3.7 mm

Greece: W Chalkidikí, near Eleochória

Greece (E Makedonía)

Shell brown with whitish ribs (30-50 ribs/penultimate whorl), less densely ribbed on cervix, 8-11 whorls, aperture detached, margin broadly reflected, parietalis strong, columellaris prominent, lunula dorsal, no palatal folds, usually 3 sutural folds, basalis very short or absent, subcolumellaris usually not visible in a perpendicular view.

Isabellaria vallata (Mousson, 1859)

Clausilia vallata Mousson, 1859 in Mousson 1859: 274
Isabellaria vallata errata Fauer, 1985 in Fauer 1985: 301

15-21 x 3.5-4.5 mm

Greece: Ipiros, Protópappas

Westerlund 1884: 172, Welter-Schultes 2000 (Schr. Malak. 16): 184, Uit de Weerd 2004: 55.

Ranges of *I. vallata vallata* (Ipiros) and *I. v. errata* (Central Greece) are separated by a large gap.

N Greece

Shell light horny brownish or with violet or yellow hue, indistinctly striated, 8-10 whorls, suture at upper whorls finely papillated, cervix finely ribbed, with prominent basal keel, aperture inside very narrow with a broad white lip, margin weakly connected at parietal side, not always detached, parietalis variable, columellaris short and oblique, almost straight, subcolumellaris often visible in a perpendicular view.

Sericata

Sericata Böttger, 1878
in O. Böttger 1878 (Ber. Offenb. 17/18): 42 *Clausilia (Sericata)*

Type species *Clausilia sericata* Pfeiffer, 1850 (monotypy)

See the comments above (*Albinaria*). Regarding *Sericata* as an independent genus is certainly not justified, neither morphologically nor genetically, but due to the chaos produced in this group in the past years it seems currently most convenient to shift back to Nordsieck's 1977 classification and wait a few decades.

Sericata abyssoclista (Boettger, 1883)

Clausilia abyssoclista Boettger, 1883 in O. Boettger 1883 (Proc. London): 340

Shell dark horny brown, densely striated, 10-12 whorls, aperture with white lip, principalis long, close to suturalis and inside passing the lunula, lunula ventrolateral, short and broad, no palatal folds.

15-18 x 3.3-4.2 mm

Westerlund 1884: 58, Uit de Weerd 2004: 56.

Greece: Pelopónisos, Argolídos, Paleá Epídavros
Paolo Zaccaria
Tiroler Landesmuseen Innsbruck

Greece (NE Pelopónisos)

Sericata albicosta (Boettger, 1877)

Clausilia albicosta Boettger, 1877 in O. Boettger 1877 (Nachr.Bl. DMG 9): 67

17-21 x 4-5 mm

Greece: near Mt. Olimbos

Greece (S Makedonía)

Shell brownish or with violet hue, with dense white ribs, cervix rounded without keels, equally densely ribbed, aperture oval rounded, columellaris deep inside, principalis small, upper palatalis very short, both visible as white structures from outside, lunula very small, clausilium narrow and pointed at its frontal end.

Westerlund 1884: 109, Uit de Weerd 2004: 56.

Sericata bathyclista (Boettger, 1879)

Clausilia bathyclista Boettger, 1879 in O. Boettger 1879 (Jahrb. DMG 6): 122, Taf. 3

13-18 x 3.3-4.2 mm, perhaps locally variable

Greece: Evia, Límni   Museum Budapest HNHM

Westerlund 1884: 57, Uit de Weerd 2004: 56.

Greece (NW Evia)

Shell dark horny brown, weakly striated or almost smooth, 9-11 whorls, cervix weakly ribbed, rounded at lower side with basal keel and long and curved basal furrow, parietalis passing end of spiralis, columellaris almost straight, ascending obliquely, suturalis long and distinct, principalis long but not passing lunula, upper palatalis curved and not touching lunula, lower palatalis connected to lunula, subcolumellaris not visible in the aperture.

Sericata calabacensis (Westerlund, 1892)

Clausilia calabacensis Westerlund, 1892 in Westerlund 1892: 39

14-18 x 2.6-3.2 mm

Greece: Evritanía, Píndos, Mt. Veloúchi (determination uncertain)

Greece (W Thessalia)

Shell whitish with horny brown apex, coarsely but not prominently ribbed, 8-11 whorls, cervix rounded without dorsal keel, aperture oval without sinulus, distinctly detached, parietalis and columellaris moderate, lunula dorsal, upper palatalis present, subcolumelaris not visible in a perpendicular view.

Uit de Weerd 2004: 56.

Sericata dextrorsa (Boettger, 1877)

Clausilia dextrorsa Boettger, 1877 in O. Boettger 1877 (Nachr.Bl. DMG 9): 68

17-20 x 3.5-4.7 mm

Greece: Makedonía, Véria

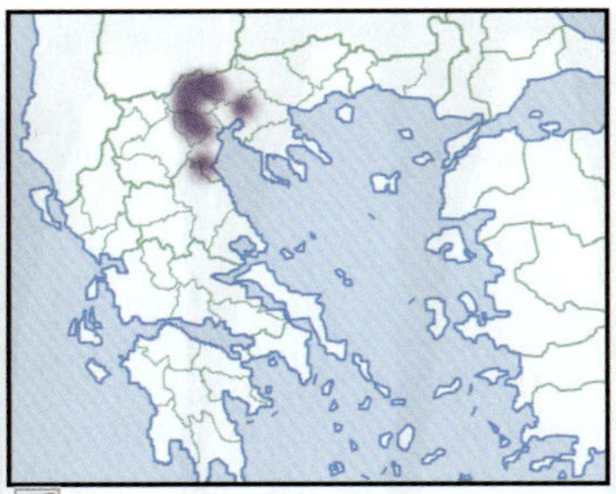

Greece (Makedonía)

Shell dextral, brown with white suture, almost smooth, shiny, 9-11 whorls, cervix with weak basal and dorsal keels, apertural margin usually slightly detached, columellaris forked backwards, lunula broad and often S-shaped, connected to principalis, lower palatalis in form of an indistinct callus.

Westerlund 1884: 155, Uit de Weerd 2004: 56.

Sericata inchoata (Boettger, 1889)

Clausilia inchoata Boettger, 1889 in O. Boettger 1889: 33

Shell horny brown, weakly striated, shiny, 9-12 whorls with white suture, often papillated, cervix rounded, aperture U-shaped, inside yellowish brown, margin white and not connected at parietal side, parietalis strong, columellaris with nodules and not very prominent, lunula lateral and]-shaped, no basalis, subcolumellaris visible in the aperture.

Uit de Weerd 2004: 35.

10-12 x 2.3-2.8 mm

Greece: Ipiros, Platanoúsa, 650 m

Greece (Ipiros)

Sericata liebegottae Nordsieck, 1984

Sericata liebegottae Nordsieck, 1984 in Nordsieck 1984: 192

Shell reddish to greyish brown, smooth or weakly striated, cervix very finely striated, 9-10 whorls with weak and fine white suture, basal keel strong, no dorsal keel, apertural margin usually detached, parallelaris present, columellaris moderate or low, lunula dorsolateral to lateral, clausilium plate visible in an oblique view, basalis short and strong, frontal upper palatalis only present in Adelfopoúla and Adelfi, sometimes also a second frontal palatal fold can be present, subcolumellaris usually invisible in a perpendicular view.

At low limestone rocks and under limestones.

Uit de Weerd 2004: 56.

13.5-20 x 3.3-4.3 mm

Greece: Vórii Sporádes, Prásos

Greece (Vórii Sporádes: small islands Adélfi, Adelfópoulo, Prásos, Kórakas, Skandíli)

Sericata lutracana Nordsieck, 1977

Sericata lutracana Nordsieck, 1977 in Nordsieck 1977: 78

14-16 x 3.4-3.8 mm

Greece: Loutráki

Museum Budapest HNHM

Greece (Kórinthos region)

Shell yellowish brown with whitish suture, very finely ribbed, cervix with weak basal and dorsal keels, aperture detached, parietalis inside running parallel with initiating spiralis, columellaris weak and in upper section of aperture, frontal upper palatalis variably strong and separated from lunula, lunula dorsal, basalis very short, subcolumellaris visible in a perpendicular view, clausilium plate well visible in an oblique view.

Uit de Weerd 2004: 56.

Sericata parnassia (Boettger, 1888)

Clausilia parnassia Boettger, 1888 in O. Boettger 1888: 54 (*Clausilia (Papillifera) Parnassia*)

Shell dark violet, upper and main whorls nearly smooth, shiny, striated near aperture, 9-12 whorls, suture often white, cervix rounded, apertural margin shortly detached and broad, parietalis reaching margin, columellaris moderate, no palatal folds, lunula dorsolateral, subcolumellaris not very well visible.

Westerlund 1890: 56, Uit de Weerd 2004: 56.

20-25 x 4.5-5.5 mm

Greece: Delfi

Paolo Zaccaria

Tiroler Landesmuseen Innsbruck

Greece (Parnassos mountains)

Sericata regina Nordsieck, 1972

Sericata regina Nordsieck, 1972 in Nordsieck 1972: 15

13-17 x 3.1-3.6 mm

Greece: Ipiros, Kerasóna

Greece (S Ipiros)

Shell greyish brown with white suture, whitish surface layer, weakly striated, cervix finely and densely ribbed, 10-12 whorls, cervix with moderate basal keel, dorsal section rounded, aperture detached, parietalis passing spiralis, columellaris not very prominent, lunula lateral, basalis very weak if present, subcolumellaris always visible in a perpendicular view.

Uit de Weerd 2004: 35.

Sericata sericata (Pfeiffer, 1850)

Clausilia sericata Pfeiffer, 1850 in L. Pfeiffer 1850 (Proc. London 17): 140

15-24 x 4-5 mm, perhaps locally variable

Greece: N Evia, near Vasiliká

Greece (Evia)

Shell brown or dark horny brownish, almost smooth or ribbed, 9-12 whorls, suture often white, cervix rounded, apertural margin shortly detached and broad, parietalis reaching margin, columellaris moderate, some irregular short palatal folds sometimes present, lunula dorsal. Ribs and apertural folds seem to display spatial variations.

Westerlund 1884: 121, Uit de Weerd 2004: 56.

Sericata stussineri (Boettger, 1885)

Clausilia stussineri Boettger, 1885 in O. Boettger 1885: 189, Taf. 4 (*Clausilia (Papillifera) Stussineri*)

16-23 x 3-4.5 mm, locally variable

Greece: NE Thessalía, Témbi valley, near Agía Paraskeví

Westerlund 1884: [2. Beilage] p. 5, Uit de Weerd 2004: 56.

Greece (Thessalía, S Makedonía)

Shell horny brown, almost smooth, shiny, last whorl weakly striated, 9-11 whorls, suture light, cervix with strong dorsal and weak basal keel, aperture sitting slightly lower than usual in clausiliids, inside dark, margin shortly detached, parietalis almost reaching the margin, not connected to spiralis which runs a short section by its side, columellaris prominent and curved, lunula lateral and S-shaped, connected to principalis, upper palatalis sometimes present, lower palatalis deeply inside and short, subcolumellaris distinct.

Sericata tantilla (Brandt, 1962)

Albinaria tantilla Brandt, 1962 in Brandt 1962: 146

11-16 x 2.5-3 mm

Greece: Thessalía, near Tírnavos

Greece (Thessalía)

Shell light horny brown, ribbed (30-40 whorls/penultimate whorl), more densely ribbed at cervix, 9-11 whorls, cervix with weak basal and sharper dorsal keel, apertural margin white, slightly reflected, usually detached, columellaris prominent and slightly oblique, lunula dorsolateral, basalis short, subclaustralis fine and running parallel to subcolumellaris, no suturalis, no parallelaris, no frontal upper palatalis, subcolumellaris visible in an oblique view. Clausilium plate narow, slightly curved and with rounded terminal end.

Sericata torifera (Boettger, 1885)

Clausilia torifera Boettger, 1885 in O. Boettger 1885: 191

16-19 x 4.5-5 mm

Greece: Thessalía, near Theópetra

Greece (Thessalía)

Shell evenly brown or dark brown, distinctly striated, suture white, papillated on upper whorls, 9-10 whorls, aperture relatively large, margin broad and reflected with a thickened white lip, suturalis and parallelaris prominent and long, lunula long and S-shaped.

Westerlund 1884: [2. Beilage] p. 5, Uit de Weerd 2004: 56.

Carinigera

Carinigera Möllendorff, 1873 in Möllendorff 1873: 141

Type species *Clausilia eximia* Möllendorff, 1873 (monotypy)

Classification traditionally based on very few anatomical characters (presence of a relatively muscular vaginal retractor and a penial papilla), not confirmed by molecular reseaches. See comments above, under *Albinaria*.

Carinigera buresi (Wagner, 1928)

Delima buresi Wagner, 1928 in A. J. Wagner 1928: 330 (*Delima (Carinigera) bureši*)

13-19 x 3-4 mm

Greece: Makedonía, near Kallífitos NE Dráma

NE Greece, S Bulgaria

Shell not decollated, brown with white suture, usually smooth, 9-11 whorls, suture often weakly papillated on upper whorls, cervix densely ribbed, apertural margin not always detached, columellaris prominent, often horizontal, frontal upper palatalis sometimes present.

Extremely calciphile and probably very drought-resistant.

Damjanov & Likharev 1975: 180, Falkner 1990: 158, Dedov 1998, Uit de Weerd 2004: 56, Irikov & Mollov 2006: 775.

Carinigera drenovoensis (Brandt, 1961)

Delima drenovoensis Brandt, 1961 in Brandt 1961: 5

Shell light horny brown, densely ribbed (45-65 ribs/penultimate whorl), 10-11 whorls, cervix rounded with weak basal keel, apertural margin rounded and shortly detached, parietalis short and prominent, columellaris nearly horizontal, lunula short and dorsal, no prominent palatal folds, basalis extremely short if present at all, subcolumellaris only visible in an oblique view. Clausilium plate strongly screwed and groove-shaped with rounded anterior end.

Uit de Weerd 2004: 25, 56.

15.5-18 x 3.2-3.5 mm

Rep. Makedonija, Kavadarci Raec gorge, road to Prilep

S Rep. Makedonija

Carinigera eximia (Möllendorff, 1873)

Clausilia eximia Möllendorff, 1873 in Möllendorff 1873: 140, Taf. 4

Shell brown, almost smooth, 9-10 whorls, cervix striated with keel, apertural margin white and detached, parietalis straight, columellaris prominent and curved.

Westerlund 1884: 152, Pavlović 1912: 108, Tomić 1959: 46, Uit de Weerd 2004: 25, 56.

12.5-15 x 2.7-3.3 mm

Serbia: Nisava canyon, near Ostravica

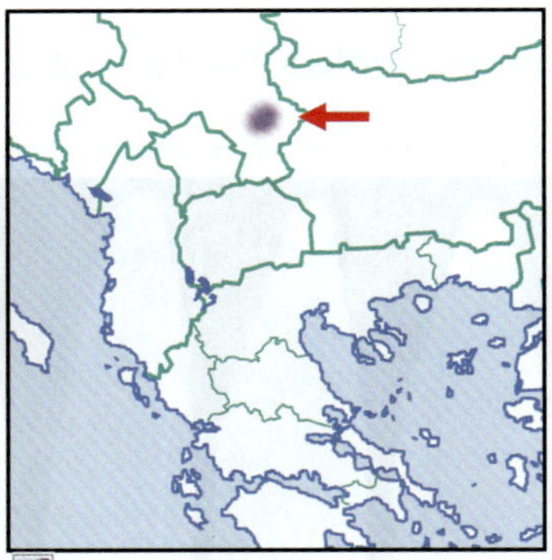

E Serbia

Carinigera haussknechti (Boettger, 1886)

Clausilia haussknechti Boettger, 1886 in O. Boettger 1886: 61, Taf. 2 (*Clausilia (Albinaria) Haussknechti*)

13-17 x 3-3.5 mm

Greece: Thessalía, Píli near Tríkala

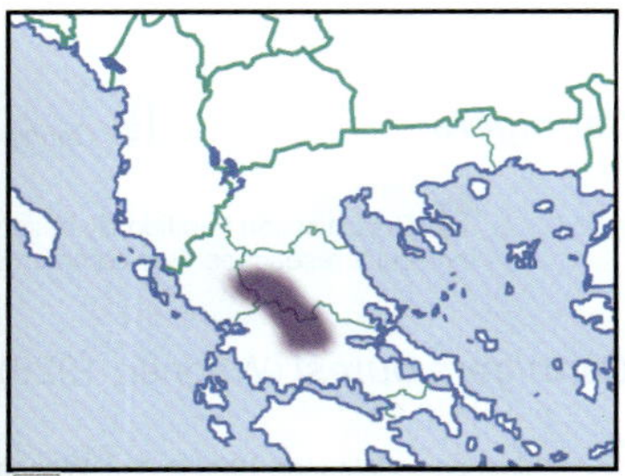

Greece (Ipiros to Stereá Elláda)

Shell dark horny brownish, very finely striated, slightly coarser at the cervix, 9-11 whorls, cervix with strong basal and dorsal keels and a basal forrow in between, also inside the aperture, columellaris not very prominent, frontal upper palatalis short, basalis short, subcolumellaris visible in an oblique view.

Westerlund 1890: 65, Nordsieck 1974: 148, Uit de Weerd 2004: 35.

The morphological differences between the forms *haussknechti*, *pharsalica* and *megdova* are relatively weak and it is possible that the populations are interconnected by intermediate forms, so that they form only one species. It would be necessary to quantify the (very few) differentiating morphological characters and to see their spatial distribution and possible gradients.

Carinigera megdova Nordsieck, 1974

Carinigera megdova Nordsieck, 1974 in Nordsieck 1974: 147

Greece (Ipiros to Stereá Elláda)

Shell horny brown, weakly striated, cervix more prominently striated, 11-13 whorls with white suture, cervix with weak basal keel, dorsal keel very weak if present, aperture not detached, parietalis and columellaris prominent, lunula dorsolateral, basalis very short, no palatal callus, no frontal upper palatalis, subcolumellaris visible in a perpendicular view. Clausilium plate usually with edge at the outer margin.

18-23 x 3.7-4.5 mm

Greece: W Central Greece, near Ditikí Frangísta

Differs from *C. haussknechti* in the presence of a short basalis, absence of a sulcalis, clausilium plate with edge at the outer margin. *C. pharsalica* has a less slender shell, the cervical keels are weaker, the suture is even less papillated, a dorsal lunula and no frontal upper palatalis, the outer edge of the clausilium is weaker.

Uit de Weerd 2004: 35.

Possibly conspecific with *C. haussknechti* and *C. pharsalica*, see comment under *C. haussknechti*.

Carinigera octava Brandt, 1962

Carinigera octava Brandt, 1962 in Brandt 1962: 139

Shell horny greyish brown, smooth with some growth lines, cervix striated, 10-12 almost straight whorls with white suture, cervix with short basal and weak dorsal keel, apertural margin weakly reflected, usually connected at parietal side, parietalis strong, parallelaris usually distinct, columellaris moderately prominent and horizontal, lunula dorsal, connected to subclaustralis which does not meet the subcolumellaris, basalis very short and rudimentary, no fromtal upper palatalis, subcolumellaris always visible in an oblique view.

Uit de Weerd 2004: 25, 56.

18-21 x 3.4-3.9 mm

Rep. Makedonija: Veles

Central Rep. Makedonija

Carinigera pellucida Dedov & Neubert, 2002

Carinigera pellucida Dedov & Neubert, 2002 in Dedov & Neubert 2002: 204

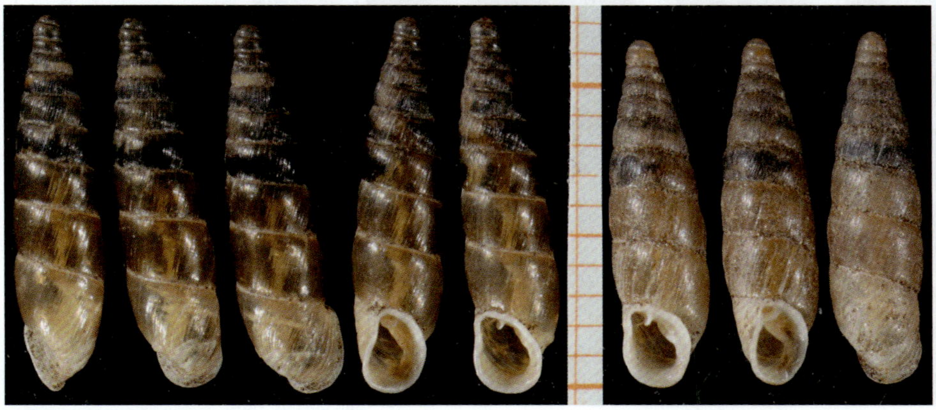

NW Rep. Makedonija

Shell light brown to greyish, striated, cervix densely ribbed, 9-11 whorls with white suture, aperture not detached, margin whitish and thickened, parietalis short and separated from spiralis, spiralis visible in the aperture, columellaris weak and steep, lunula dorsal and weak or absent, 2 small palatal folds present, subcolumellaris visible in an oblique view.

11-15 x 3-3.5 mm

Montenegro: Sar Planina, Titov Vrch, 2740 m

Open habitats, meadows, river valleys, alpine zones in 1600-2500 m.

Uit de Weerd 2004: 56.

Carinigera pharsalica Nordsieck, 1974

Carinigera pharsalica Nordsieck, 1974 in Nordsieck 1974: 146

16-22 x 3.9-4.9 mm

Greece: Thessalía, near Fársala

Museum Budapest HNHM

Greece (Thessalia)

Shell horny brown, weakly striated, cervix more prominently striated, 9-11 whorls with white suture, cervix with weak basal and dorsal keel, apertural margin usually attached, parietalis and columellaris prominent, lunula dorsal, basalis very short, palatal callus present, connected to a weak frontal upper palatalis, clausilium plate visible in an oblique view, subcolumellaris visible in a perpendicular view.

At N-exposed low limestone rocks and walls.

Uit de Weerd 2004: 25, 56.

Possibly conspecific with *C. haussknechti* and *C. megdova*, see comment under *C. haussknechti*.

Carinigera schuetti Brandt, 1962

Carinigera schuetti Brandt, 1962 in Brandt 1962: 140

Shell light brownish, usually decollate with 8-10 whorls remaining, finely and regularly ribbed, cervix with basal keel, aperture slightly detached, lunula lateral, frontal upper palatalis present.

Damjanov & Likharev 1975: 180, Falkner 1990: 158, Uit de Weerd 2004: 25, 56.

14-22 x 3.5-4.4 mm

SW Bulgaria: Slavyanka moutnains, Ali Botush summit

Atanas Irikov

SE Rep. Makedonija to NE Greece

Carinigera septima Brandt, 1962

Carinigera septima Brandt, 1962 in Brandt 1962: 138

15-17 x 3-3.6 mm

Rep. Makedonija: Negotino, Demir Kapija

S Rep. Makedonija

Shell horny greyish, last whorl lighter, obliquely striated, cervix more prominently striated, 9-10 weakly convex whorls with white suture, suture papillated, more weakly at lower whorls, cervix with very weak basal and dorsal keels, apertural margin narrow, connected at parietal side, not always detached, columellaris prominent but not horizontal, lunula dorsal, subclaustralis very weak and not connected to subcolumellaris, subcolumellaris visible in an oblique view.

On the surface of limestone rocks, under stones and rock crevices.

Uit de Weerd 2004: 25, Dedov & Neubert 2009: 93.

Carinigera superba Nordsieck, 1977

Carinigera superba Nordsieck, 1977 in Nordsieck 1977: 83

16-20 x 3.8-4.6 mm

Greece: Makedonía, near Kokkinógia NW Dráma

NE Greece (E Makedonía)

Shell not decollated, brownish with widely spaced white ribs, almost percostate, 9-10 whorls, suture white, cervix rounded, no frontal upper palatalis.

Falkner 1990: 158, Uit de Weerd 2004: 56.

4 - *Idyla-Elia* subgroup (Greece-Turkey)

Elia, *Euxina*, *Galeata*, *Idyla*, *Olympicola*, *Phrygica*, *Sprattia*, *Strumosa*

Elia

Elia Adams & Adams, 1855
in Adams & Adams 1858: 180 *Clausilia (Elia)*

Type species *Clausilia moesta* Rossmässler, 1839
(subsequent designation by Lindholm 1924: 65)

Type designation by Lindholm 1924: 65 "*Cl. mæsta* Fér.". Contained originally "*Cl. Mœsta* Férus.", *biformis*, *corpulenta*, *plicata*, *serrulata* and *thessalonica*. *Helix maesta* as mentioned by Férussac [1821-1822]: 63 is a nomen nudum, the name of the type species must be *Clausilia moesta* Rossmässler, 1839.

Turkey to Near East and Caucasus. Only the W and central Turkish species are listed here.

Elia corpulenta (Pfeiffer, 1848)

Clausilia corpulenta Pfeiffer, 1848 in L. Pfeiffer 1848 (Z. f. Malak. 5): 7
Clausilia corpulenta var. *proteus* Retowski, 1889 in Retowski 1889: 259
Not *Clausilia (Albinaria) proteus* Boettger, 1889 from Kárpathos, currently *Albinaria proteus* (Boettger, 1889). I do not know which name has precedence. If both were from 31 Dec 1889, Boettger's name for a species would have precedence (Art. 24.1, 57.7).

11.5-14.5 x 3.8-4.8 mm

Turkey: Vilayet Bursa, Bursa

Schütt 2001: 375.

Subspecies classification very questionable, mainly based on shell size, rib density and rib prominence.

NW and N central Turkey

Shell quite variable, reddish brown, ribbed or weakly striated, basal keel slightly stronger than dorsal keel, apertural margin with many short lamellas, parietalis strong, columellaris flat and long, principalis and upper palatalis very long, three more palatal folds present. Ribs dense near Bursa, weak or prominent near Samsun and Sinop.

Mountain forests, also at lower altitudes, in soil litter and under stones.

Elia huebneri (Pfeiffer, 1848)

Clausilia huebneri Pfeiffer, 1848 in L. Pfeiffer 1848 (Mon. Helic. 2): 473 (*Clausilia Hübneri*)

Shell horny brown, ribbed, 11-12 whorls, suture papillated, basal keel strong, columellaris deep inside and oblique, forked backwards, 4 palatal folds of which 3 are long, lunula distinct, subcolumellaris not visible in the aperture.

13-16 x 3.3-3.7 mm

Forests and shrubs at limestones.

Westerlund 1884: 34, Schütt 2001: 378.

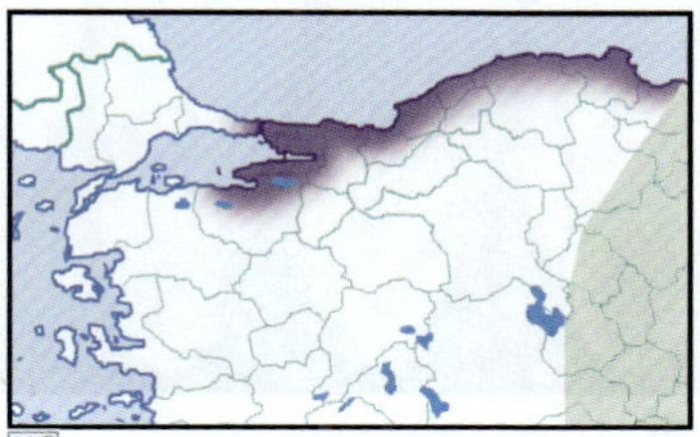

NW and N central Turkey

Elia moesta (Rossmässler, 1839)

Clausilia moesta Rossmässler, 1839 in Rossmässler 1838-1844 (9-10): 23, Taf. 48

12-17 x 3.6-4.4 mm

Turkey: Vilayet Hatay, Cevlik, Titus Tüneli

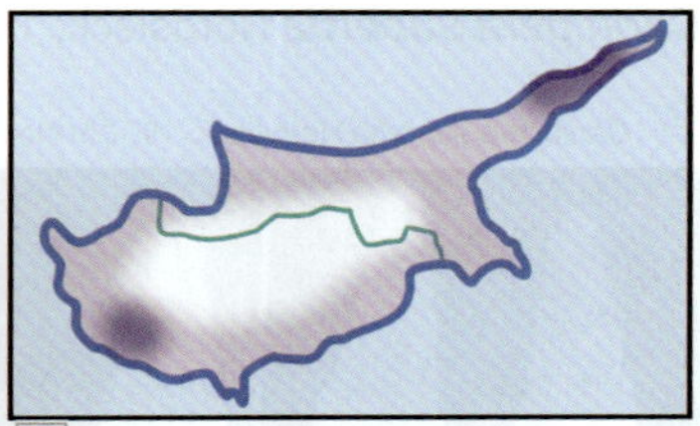

Lebanon to S Turkey, Cyprus

Shell brown, very finely striated, apertural margin with many short folds, parietalis reaching deep inside, ccolumellaris deep inside and very weak, lunula dorsal and distict, 5 palatal folds, all thickened near aperture and very well visible, subcolumellaris curved in last section and well visible in a perpendicular view.

Relatively dry forest soils in mountains.

Westerlund 1884: 29, 36, Bank & Menkhorst 1994: 100, Schütt 2001: 376.

Elia retowskii Nordsieck, 1984

Elia retowskii Nordsieck, 1984 in Nordsieck 1984: 203

18-22 x 4.2-4.9 mm

Turkey: Vilayet Kastamonu/Bartın, near Kapısuyu

Limestone rocks and walls.

Schütt 2001: 379.

N central Turkey

Shell yellowish white, very densely and finely striated, more coarsely at last whorl, 10-12 whorls, basal keel oblique and very prominent, no dorsal keel, apertural margin detached and reflected, columellaris weak, hardly visible in a perpendicular view, lunula dorsolateral, clausilium plate hardly or not visible in an oblique view, frontal and distal palatal folds connected, basalis short and connected to lunula, no midle palatal fold, subcolumellaris invisible in a perpendicular view.

Shell larger than *E. huebneri*, cervix without dorsal keel, columellaris weaker, subcolumellaris shorter, lunula deeper inside, basalis shorter and middle palatalis absent.

Euxina

Euxina Boettger, 1877 in O. Boettger 1877 (Palaeontogr.): 83 *Clausilia (Euxina)*

Type species *Clausilia hetaera* Pfeiffer, 1848 (subsequent designation by Westerlund 1902: 109)

Contained originally many species classified in 9 infrasubgeneric groups.

SE Bulgaria, N Turkey

Euxina circumdata (Pfeiffer, 1848)

Clausilia circumdata Pfeiffer, 1848 in L. Pfeiffer 1848 (Z. f. Malak. 5): 9

12.5-13.5 x 3.0-3.3 mm

Turkey: Istanbul, Yedikale

Under limestone rocks and shrubs, in forest soil litter.

Damjanov & Likharev 1975: 183, Schütt 2001: 371.

SE Bulgaria, N Turkey

Shell brown, densely ribbed, also on cervix, no dorsal keel, basal a keel-like structure, aperture inside brown, parietalis strong, columellaris deep inside, subcolumellaris not very prominent, lunella broad, principalis very long, palatal callus present but weak.

Euxina hetaera (Pfeiffer, 1848)

Clausilia hetaera Pfeiffer, 1848 in L. Pfeiffer 1848 (Z. f. Malak. 5): 10

Shell light brownish, silk-like shiny, densely striated, no dorsal keel, basal with keel-like structure, aperture inside brown, lip light, columellaris deep inside but prominent, visible in a perpendicular view, strong palatal callus between basal furrow and suture.

Deciduous forests and shrubs, under rotting leaves and at the base of limestones.

Schütt 2001: 372.

11-14 x 2.5-3.5 mm

Turkey: Istanbul
Burçin A. Gümüş

NW Turkey

Euxina persica (Boettger, 1879)

Clausilia persica Boettger, 1879 in O. Boettger 1879: 117, Taf. 3

16.5-21 x 5.0-5.5 mm

Bulgaria: Strandza mountains, (1-6) Marina Reka, (7-8) near Mladejko
(7-8) Atanas Irikov

Damjanov & Likharev 1975: 184, Schütt 2001: 373, Irikov & Mollov 2006: 780.

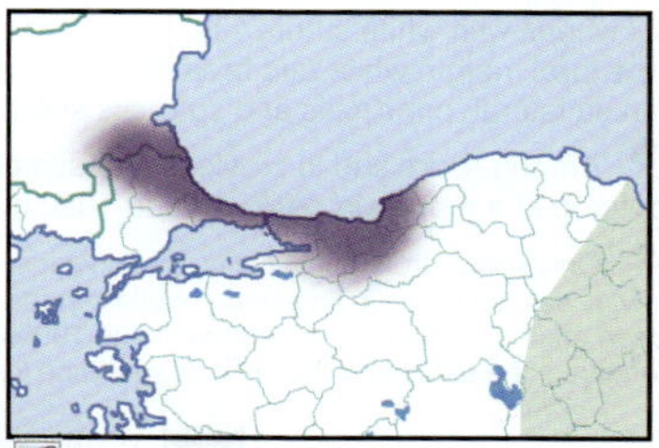

Bulgaria and NW Turkey (not Iran)

Shell brown, densely ribbed or striated, apex blunt, dorsal and basal keels weak, aperture inside brown, lip light, parietalis strong, columellaris deep inside, subcolumellaris hardly visible, lunella absent, short curved palatalis often present, principalis strong, clausilium plate with two scars at anterior end.

Deciduous forests on limestone, at wood and under leaves.

Euxina promta (Schmidt, 1868)

Euxina pontica (Retowski, 1887)

Clausilia promta Schmidt, 1868 in Schmidt 1868: 151
Clausilia pontica Retowski, 1887 in Retowski 1887: 39, Taf. 1
Nordsieck 1993: 36 proposed to ask the Commission to suppress *Clausilia promta* Schmidt, 1868, but no action has been taken in the past 19 years. *Euxina pontica* was not used in 25 recent publications to qualify for Art. 23.9. The older name has precedence.

Shell dark horny brown to reddish, sometimes with white streaks, finely ribbed, with silky shine, basal keel very strong, dorsal keel weak, aperture with white margin and basal furrow, parietalis short but prominent, columellaris deep inside and hardly visible in a perpendicular view, lunella absent or weak.

18.5-21 x 4.5-5.2 mm

Under bark of rotting tree trunks.

Schütt 2001: 374, Irikov & Mollov 2006: 780 (*E. pontica*, not in Bulgaria, previous records were misidentifications).

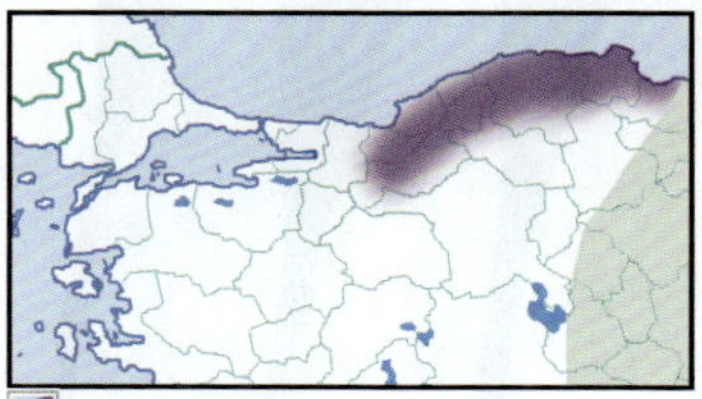

N Turkey

Galeata

Galeata Boettger, 1877
in O. Boettger 1877 (Palaeontogr.): 84 *Clausilia (Euxina (Galeata))*

Type species *Clausilia galeata* Rossmässler, 1839
(absolute tautonymy)
(= currently *Galeata galeata* (Rossmässler, 1839) from S central Turkey)

Established for "Gruppe der *Schwerzenbachi*" and contained originally *Cl. schwerzenbachii* Pfeiffer, 1848 and *Cl. galeata*. Following Art. 67.5.1 original designation cannot be accepted (Boettger 1877 just used this species as an example for the sub-subgenus, he did not apply a type species concept and did not designate type species for those genera which were established at subgeneric level). Previous authors have regarded *Cl. schwerzenbachii* as type, this was incorrect.

Galeata schwerzenbachii (Pfeiffer, 1848)

Clausilia biplicata var. ζ synonym *Clausilia Schwerzenbachii* Pfeiffer, 1848 in L. Pfeiffer 1848 (Mon. Helic. 2): 470
Mentioned with description. Not mentioned as a junior synonym of an available name (as demanded by Art. 11.6), but as a synonym for a subordinate variant for which the letter ζ was used. For those who neglect this (as proposed by Kadolsky 2012: 78, this is preliminarily accepted here), the name was made available under Art. 11.6.1 by Charpentier 1852: 396 who used the name as *Clausilia schwerzenbachii*, with bibliographical reference to Pfeiffer 1848 ("*biplicata*, var. ζ. Pfr. l. c."). For the others, the name must be attributed to Charpentier 1852.

13.5-15.0 x 3.1-3.3 mm

Bulgaria: Strandza mountains, near Malko Târnovo

E Bulgaria, N Turkey

Shell horny or dark brown, densely ribbed or striated, dorsal keel present, basal keel very prominent, aperture with light lip and basal furrow, columellaris and subcolumellaris deep inside, principalis long, upper palatalis curved, lunula absent.

Mountain forests.

Damjanov & Likharev 1975: 188, Bank & Menkhorst 1994: 104, Schütt 2001: 389.

Idyla

Idyla Adams & Adams, 1855
in Adams & Adams 1858: 180 *Clausilia (Idyla)*

Type species *Clausilia bicristata* Rossmässler, 1839
(subsequent designation by Albers & Martens 1860: XVII, 284)

Shell medium-sized, relatively dark, smooth or ribbed, basal and dorsal keels usually present, lunula almost absent or only the lowest part left as a callus, columellaris and other apertural folds weak.

Neubert 1995: 104 (Genitalia), Nordsieck 1994, Örstan et al. 2007.

A revision of the species in this group would be useful. It seems that they display local variations, and that the criteria to separate specific and subspecific taxa have not been applied consistently.

Idyla aydinensis Örstan, 2007

Idyla aydinensis Örstan, 2007 in Örstan, Yıldırım, Ceylan, Pearce & Welter-Schultes 2007: 330
Name was attributed to Örstan alone. Under Art. 50.1 the responsibility for the description does not need to be explained if the author of the name is a co-author of the work.

Shell brown, densely and regularly ribbed, more coarsely on last whorl, 11-13 whorls with white suture line, basal and dorsal keels moderately strong, apertural margin white and broadly reflected, parietalis and columellaris weak, lunula lateral, principalis and suturalis short, visible from outside.

Among rocks and in crevices of a scarcely vegetated marble limestone outcrop at a mountain peak, around 1030 m.

Only known from one locality. The species is rare, individual numbers appear to be low. It was not found in limestone habitats in the same area below 1000 m. There had been attempts to exploit the marble in the 1990s, this was abandoned. If the marble is removed, the habitat for the species would be lost and the species would not survive.

15-19 x 3.1-3.4 mm
W Turkey: Aydın mountains, Gökkiriş near Köşk
(H = 17.7 mm)

W Turkey (Aydın mountains)

Aydın Örstan

Idyla bicristata (Rossmässler, 1839)

Clausilia bicristata Rossmässler, 1839 in Rossmässler 1838-1844 (9-10): 16, Taf. 48

15-18 x 3.5-4 mm

Greece: 16 km W Thíva

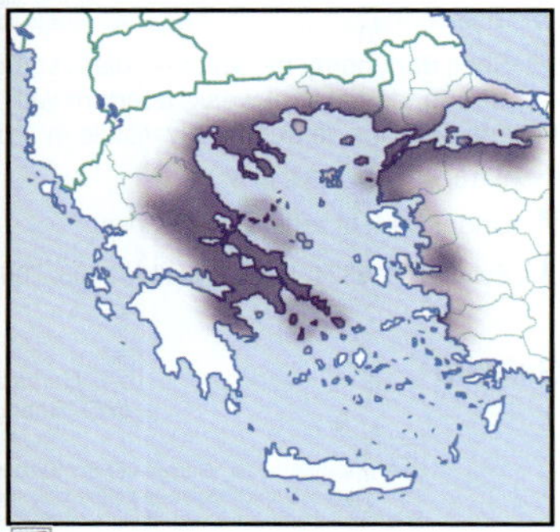

Greece, Turkey

Shell weakly violet brown, shiny, smooth or weakly ribbed, cervix ribbed, dorsal keel very prominent and crest-like, basal keel strong, apertural margin usually quite broad, parietalis weak and short, spiralis very short, subcolumellaris invisible in the aperture. 3 sutural folds present. Differs from other *Idyla* species in its strong dorsal keel.

Limestone rocks.

Zilch 1976: 209, Nordsieck 1984: 193, Schütt 2001: 337, Şeşen & Schütt 2005.

Greece: Chalkidikí, Sithonía, Koufós
Jozef Grego

Idyla castalia (Roth, 1856)

Clausilia castalia Roth, 1856 in Roth 1856: 5 (*Clausilia Castalia*)

12-19 x 2.7-4.5 mm, locally variable

N Greece: Mt. Olimbos, 2000 m

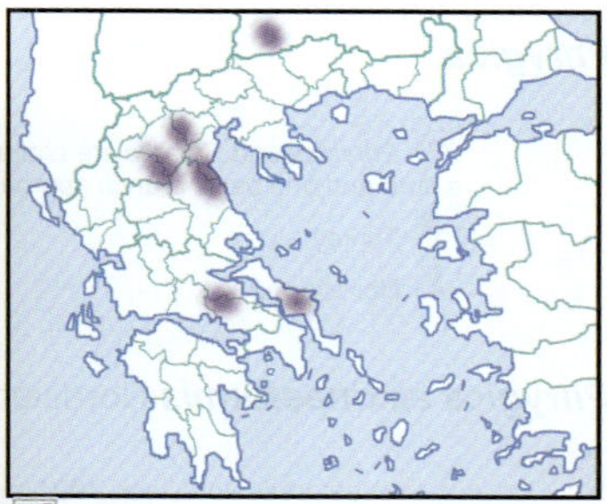

Bulgaria, Greece

Shell horny brownish, finely ribbed, 9-13 often relatively convex whorls, last whorl with largest diameter, cervix with basal keel, dorsal keel very weak or absent, aperture small, margin detached, columellaris deep inside and close to parietalis, almost invisible in the aperture, principalis very weak or absent, upper section of lunula indistinct or absent, or replaced by two very short rudimentary folds of the upper palatalis, palatal callus like a tubercle on the left side.

Moderately humid to dry limestone habitats, not below 900 m, in Bulgaria up to 2600 m.

Westerlund 1884: 103, Damjanov & Likharev 1975: 189, Gittenberger 2002: 135, Hubenov 2007.

Idyla liebegottae Nordsieck, 1994

Idyla liebegottae Nordsieck, 1994 in Nordsieck 1994: 7, 15

Shell reddish brown, bluntly and regularly ribbed, ribs more widely spaced at the cervix, cervix with basal keel, aperture detached, parietalis strong, often with small folds between parietalis and columellaris, columellaris weak, hardly visible in a perpendicular view, lunula dorsal to dorsolateral, palatal structures weak, subcolumellaris ends deep inside near lunula, clausilium plate visible in an oblique view.
Penis longer than in the other *Idyla* species, retractor strong.

16-20 x 4.4-5.1 mm

Oviparous.

Nordsieck 1994 claimed that the apex was blunter (normal, not pointed) than in the other *Idyla* species. I did not understand what was meant, I saw no remarkable differences.

Greece (Thessalía)

Idyla pelobsoleta Gittenberger, 1993

Idyla pelobsoleta Gittenberger, 1993 in Gittenberger 1993: 325

Shell dark, shiny, almost smooth, basal and dorsal keels not very prominent, aperture oval, margin reflected but not flattened, columellaris weak, subcolumellaris invisible from outside.

12.5-18 x 3.3-4.2 mm

High mountain habitats.

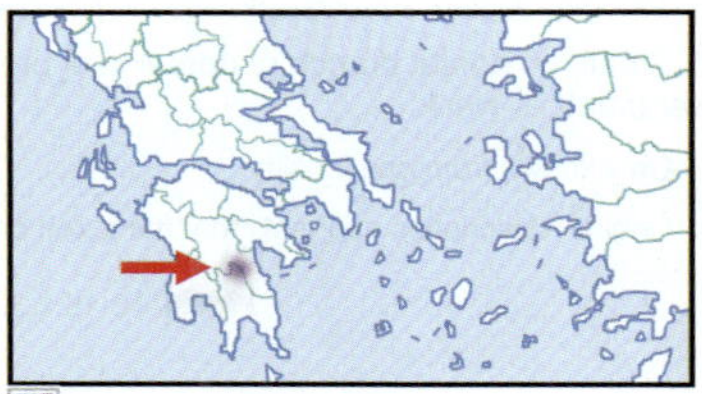

Greece (Pelopónisos)

Olympicola

Olympicola Hesse, 1916
in Hesse 1916: 124
Expressly established as a new replacement name for

Olympia Vest, 1867 in Vest 1867: 191 *Clausilia (Olympia)* not *Olympia* Risso, 1826 (Crustacea)
Gender of *Olympicola* is masculine by default because Hesse 1916 did not combine it with a specific name and did not talk about its gender.

Type species *Clausilia olympica* Pfeiffer, 1848 (original designation by Vest 1867)

Olympicola olympicus (Pfeiffer, 1848)

Without gender agreement: *Olympicola olympica* (Pfeiffer, 1848)

Clausilia olympica Pfeiffer, 1848 in L. Pfeiffer 1848 (Z. f. Malak. 5): 8

9-11 x 2.5-3 mm

Turkey: Bilecik, near Bozuyuk

NW Turkey

Shell small, ribbed, reddish brown, aperture widely detached from penultimate whorl (protruded), columellaris uncommonly weak, almost invisible in the aperture.

Limestones in forests.

Bank & Menkhorst 1994: 106, Schütt 2001: 335.

Phrygica

Phrygica Nordsieck, 1994 in Nordsieck 1994: 9

Type species *Phrygica riedeli* Nordsieck, 1994 (original designation)

Phrygica and *Sprattia* are characterized by a prominent feature of the dorsal keel, which branches off in two parts, one dorsal branch pointing towards the suture and one basal branch approaching the lower edge of the last whorl. Clausiliar apparatus of normal type, in *Sprattia* of G type.

Turkey

Nordsieck 2004.

Phrygica euxinaeformis Nordsieck, 1994

Phrygica euxinaeformis Nordsieck, 1994 in Nordsieck 1994: 21

14-15 x 3.2-4 mm

Turkey: Vilayet Isparta, near Akşehir

Central Turkey

Shell yellowish to reddish brown, densely ribbbed, less densely on the cervix, basal keel strong, dorsal keel weaker, aperture slightly detached and protruded, margin white, lunula dorsal or dorsolateral, upper palatalis present.

Was found at a cave entry.

Only known from one locality.

Schütt 2001: 365.

Phrygica ilegiensis Nordsieck, 2004

Phrygica ilegiensis Nordsieck, 2004 in Nordsieck 2004: 8

Small, no dorsal branch of dorsal keel present, columellaris low, lunella and lower palatalis weak.

Only known from one locality.

Original description suggests that the taxon could be conspecific with *P. riedeli*.

13-16 x 3.1-3.6 mm

Turkey: Vilayet Isparta, Ileğ mountain
Burçin A. Gümüş

SW Turkey

Phrygica jelskii Nordsieck, 1994

Phrygica jelskii Nordsieck, 1994 in Nordsieck 1994: 21

15-19 x 3.6-4.4 mm

Turkey: Bilecik, near Şeyh Edebali

NW Turkey

Shell yellowish to reddish brown, very slender, almost smooth, cervix distinctly and densely ribbed, basal and dorsal keels present, basalis rarely present.

Gorges in limestone mountains.

Schütt 2001: 364.

Phrygica raehlei Nordsieck, 1994

Phrygica raehlei Nordsieck, 1994 in Nordsieck 1994: 21

11-12 x 2.7-3 mm

Turkey: Vilayet Antalya, W of Saklıkent

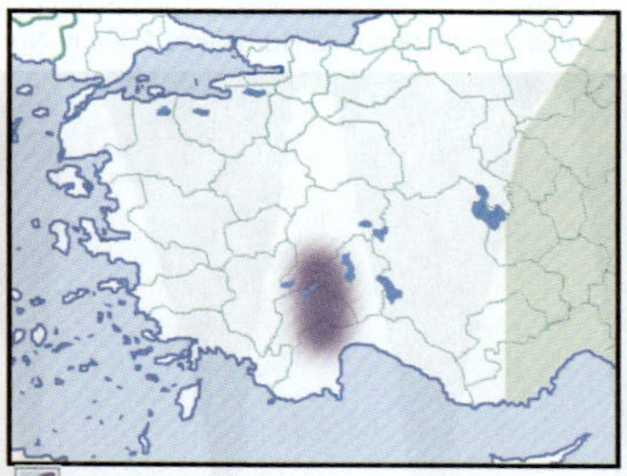

SW Turkey

Shell almost cone-shaped, only 9.5-10.5 whorls, ribbed, dorsal and basal keels present, frontal upper palatalis strong, lunula distinct.

Limestone habitats.

Only known from one locality near Saklıkent.

Schütt 2001: 365.

Phrygica riedeli Nordsieck, 1994

Phrygica riedeli Nordsieck, 1994 in Nordsieck 1994: 21

18-27 x 4.1-5.4 mm

Turkey: Vilayet Burdur, 30 km SW Burdur

SW Turkey

Shell greyish brown, densely ribbed, basal and dorsal keels present, aperture tetragonal rhomboid, folds relatively deep inside.

Limestone rocks.

Schütt 2001: 363.

Sprattia

Sprattia Boettger, 1883
in O. Boettger 1883 (Proc. London): 341 *Clausilia (Oligoptychia (Sprattia))*

Type species *Clausilia sowerbyana* Pfeiffer, 1850 (monotypy)

Phrygica and *Sprattia* are characterized by a prominent feature of the dorsal keel, which branches off in two parts, one dorsal branch pointing towards the suture and one basal branch approaching the lower edge of the last whorl (difficult to photograph, perhaps best seen in *Sprattia sowerbyana* No. 7 and 8). Clausiliar apparatus of G type, in *Phrygica* of N (normal) type.

S Turkey

Yıldırım 1997, Nordsieck 2004: 9.

Sprattia was revised by Nordsieck 2004 who recognized 8 species (*A. aksoylari, aksuensis, beycola, bicarinata, blissi, pseudophrygica, sillyonensis* and *sowerbyana*) and 9 additional subspecies. The problems in a morphological differentiation between species were obvious. Nordsieck 2004: 10 identified 7 characters with two or three states each, to be used for separating species, but in most cases the species displayed both character states at once, so that several selected combinations of 7 character states would fit for various species (0-0-1-0-0-0-1 for *sowerbyana, blissi, aksuensis* and *bicarinata*; 0-0-1-0-0-1-1 for *beycola* and *pseudophrygica*). Nordsieck used additional features, such as a white surface layer, small size, prominent ribs or a fine surface sculpture, for differentiating species. The result of this procedure was that species differed from each other in one single character.

Yıldırım 1997 classified *aksoylari* with *sowerbyana* and suggested that we were dealing with locally variable species in this group.

I follow Yıldırım's 1997 view here and suggest to revise the genus again and more critically, and to present more information about local morphological gradients. It is possible that we are dealing with very few morphologically variable species.

Sprattia beycola Nordsieck, 1994

Sprattia beycola Nordsieck, 1994 in Nordsieck 1994: 26
Sprattia pseudophrygica Nordsieck, 2004 in Nordsieck 2004: 26

S Turkey

Shell greyish brown, ribbed or finely striated, occasionally with white surface layer, dorsal keel slightly stronger than basal keel, parietalis strong, no frontal upper palatalis, basalis short or absent. Differs from the other species in the absence of palatal folds.

Bases of limestone rocks.

18-22 x 4.5-5 mm

Turkey: Antalya, E of Saklıkent, 1760 m

Schütt 2001: 366.

The taxon *pseudophrygica* is classified with *S. beycola* here, the morphological differences given for *pseudophrygica* (small size, prominent ribs and no white surface layer) are certainly not sufficient to provide evidence for the presence of a separate species.

Sprattia blissi (Boettger, 1899)

Clausilia blissi Boettger, 1899 in O. Boettger 1899: 168
Sprattia aksuensis Nordsieck, 2004 in Nordsieck 2004: 21

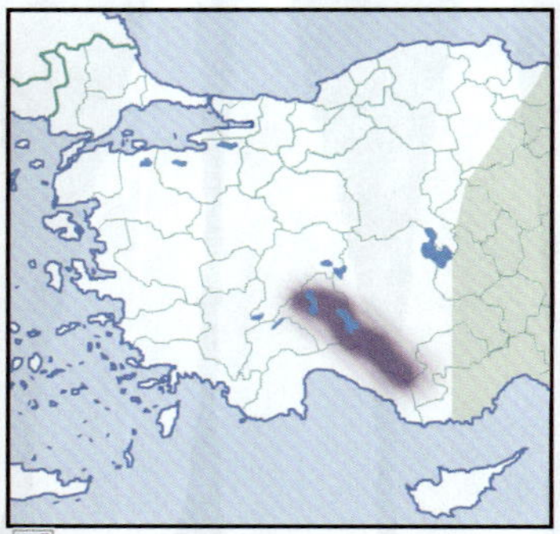

S Turkey

Shell dark brownish, weakly ribbed, 13-16 whorls, suture white, basal and dorsal keels strong, parietalis small, distant from both spiralis and parallelaris, columellaris deep inside, lunula strong, visible from outside, basalis very short, subcolumellaris invisible in the aperture.

20-28 x 3.7-4.4 mm

Turkey: Konya, Beyşehir Gölü

Schütt 2001: 367.

The taxon *askuensis* occurs in immediate vicinity at the northern range margin of the quite variable *S. blissi*, Nordsieck 2004 did not give differences to justify a classification as an independent species.

Sprattia sillyonensis Nordsieck, 1994

Sprattia sillyonensis Nordsieck, 1994 in Nordsieck 1994: 25

S Turkey

Shell reddish brown, shiny, only exceptionally decollate, weakly striated, basal and dorsal keels strong, parietalis long, parallelaris and frontal upper palatalis present, columellaris not very prominent, basalis short or absent, sulcalis present. Differs from *S. sowerbyana* in its weaker columellaris.

Limestone rocks and walls, lives on an isolated mountain.

18-25 x 5-6 mm

Turkey: Vilayet Antalya, Sillyon NE Antalya

Schütt 2001: 369.

Nordsieck's 2004 data suggest a close relationship with both *S. sowerbyana* and *S. blissi*.

Sprattia sowerbyana (Pfeiffer, 1850)

Clausilia sowerbyana Pfeiffer, 1850 in L. Pfeiffer 1850 (Proc. London 17): 141 (*Clausilia Sowerbyana*)
Date was not 1849.
Sprattia sowerbyana aksoylari Yıldırım, 1997 in Yıldırım 1997: 219 [not verified]

17-27 (undecollate) x 4.5-5.5 mm

Turkey: Vilayet Antalya, Seleucia

S Turkey

Shell brownish, in some regions decollate, densely ribbed, basal and dorsal keels very strong, aperture inside brownish, margin broad, columellaris prominent, lunula broad and strong, with short upper palatalis and two aditional short folds above, can be visible from outside, subcolumellaris hardly visible in the aperture. Differs from *S. sillyonensis* in its usually more prominent columellaris.

Limestone and conglomerate rocks.

Schütt 2001: 368, Nordsieck 2004.

Nordsieck's 2004 data suggest a close relationship with both *S. sowerbyana* and *S. blissi.* The prominent basal branch of the dorsal keel of *aksoylari* was taken to justify a classification as a different species by Nordsieck 2004. I follow Yıldırım's 1997 classification and see no constraints to classify this form in the northern section of the sowerbyana range as a separate species based on one single character alone. It rather seems to me that this character is also variable in the other *sowerbyana* populations. A more detailed geographical analysis, like in *Albinaria maltzani* from Crete (Welter-Schultes 2010), would be useful to understand better the local variations of the dorsal keel.

Strumosa

Strumosa Boettger, 1877
in O. Boettger 1877 (Palaeontogr.): 84 *Clausilia (Euxina (Strumosa))*

Type species *Clausilia strumosa* Pfeiffer, 1848
(absolute tautonymy)

Turkey

Strumosa abanti (Brandt, 1961)

Euxina abanti Brandt, 1961 in Brandt 1961: 17

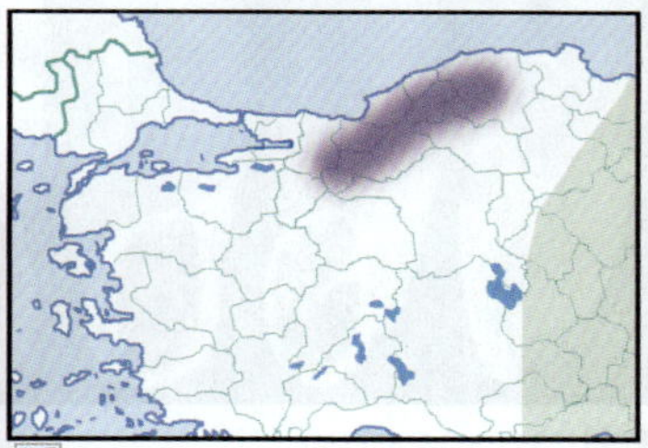

16-18 x 3.4-4 mm

Turkey: 24 km NE Ilgaz

N Turkey

Shell cherry brown, densely ribbed, 10-13 whorls with white suture, aperture inside brown, margin weakly detached, parietalis at the margin, columellaris steep, almost perpendicular, hardly visible in the aperture. Locally variable, some forms have no spiralis, palatal folds and clausilium.

Deciduous and coniferous forests.

Schütt 2001: 338.

Strumosa strumosa (Pfeiffer, 1848)

Clausilia strumosa Pfeiffer, 1848 in L. Pfeiffer 1848 (Z. f. Malak. 5): 8

11-16 x 3-4 mm

W Turkey: Uludağ, 1350 m

W Turkey

Shell brown to horny olive green, almost smooth or weakly striated, cervix also almost smooth, with basal keel, no dorsal keel, aperture with whitish margin, columellaris weak, does not reach apertural margin, no lunula, 2-4 deep palatal folds.

Under stones, at tree roots, at creek margins.

Schütt 2001: 339.

5 - all-others subgroup (not Turkey)

Charpentieria, Clausilia, Cochlodina, Delima, Dilataria, Erjavecia, Euxinella, Fusulus, Graciliaria, Julica, Lampedusa, Leucostigma, Macedonica, Macrogastra, Mentissa, Montenegrina, Muticaria, Neostyriaca, Papillifera, Protoherilla, Pseudofusulus, Ruthenica, Siciliaria, Strigilodelima

Charpentieria

Charpentieria Stabile, 1864 in Stabile 1864: 80 *Clausilia (Charpentieria)*

Type species *Glischrus dyodon* Studer, 1820 (subsequent designation by Lindholm 1924: 60)

Type not designated by Boettger 1877: 53. Boettger just wrote "Gruppe der *diodon*" and gave the name *Charpentieria* as a synonym, the term "type" did not appear.

Nordsieck 1969, Nordsieck 2002.

The modern classification of the corresponding group of clausiliids (Delimini) was established by Nordsieck 1969. Based on a different personal view of the morphological record Nordsieck 2002 (Mitt. DMG 67) classified several species of the *Siciliaria* group in *Charpentieria*. These species had been classified in *Siciliaria* since 1969. The morphological background underlying the new classification did not differ substantially from that of 1969 and the new classification was not a result of recently obtained new constraints (and if so, they were not published). Changing well established names just on the base of a different personal view of the same morphological record ("because there are no differences between these groups which are valid for all species", Nordsieck 2002: 33) is arbitrary and probably not useful.

It seems that Nordsieck 2002 tried to argue that in the meantime he came to the conclusion that the characters used in 1969 to distinguish *Siciliaria* and *Charpentieria* did not provide a solid base for the separation at the generic level (but obviously yes at the subgeneric level).
Names are an important tool for communication, especially in electronic environments. A genus-species combination of an animal's name should only be changed if this is absolutely necessary. Since Nordsieck 2002 did not present basically new data and did not provide evidence that the Nordsieck 1969 study was based on incorrect data, *Siciliaria* should be left as an independent genus and the well established system of Nordsieck 1969 should be maintained.

Charpentieria clavata (Rossmässler, 1836)

Clausilia clavata Rossmässler, 1836 in Rossmässler 1835-1837 (4): 12, Taf. 18

13-20 x 3.4-4.5 mm

N Italy: Bergamo, Val Taleggio

Westerlund 1884: 67, Manganelli et al. 1995: 24.

N Italy

Shell horny brown, densely and weakly or prominently ribbed (white ribs on brown shell, 30-45 ribs/penultimate whorl), 11-12 convex whorls, suture not deep, often with white commas below, columellaris deep inside, small and almost horizontal, lunula more or less strongly curved, subcolumellaris hardly visible. Size and rib prominence locally variable. A gradient towards smaller shells was observed with higher elevation.

Charpentieria dyodon (Studer, 1820)

Glischrus dyodon Studer, 1820 in Studer 1820: 89 (*Glischrus (Clausilia) dyodon*)
Clausilia thomasiana Küster, 1850 in Küster 1847: 48, Taf. 5 (*Clausilia Thomasiana*)

10-13 x 2.6-3 mm

Italy: Piemonte, Netro near Ivrea

Endemic species with a small range, which has partly been destroyed recently by military construction projects, massive road construction and constructions for avalanche protection. Vulnerable in Switzerland.

NW Italy, S Switzerland

Shell brown or olive green, smooth or striated, columellaris not very strong, cervix nearly rounded, usually no lunula, no frontal palatal folds, subcolumellaris well visible in a perpendicular view.
Like *Ch. itala*, but usually smaller, scarcely papillated, lunula usually absent.

Humid habitats among steep to perpendicular silicate rocks, in mountains. Between 850 and 1150 m.

Westerlund 1884: 97, Nordsieck 1979: 259 (regarded *dyodon* and *thomasiana* as separate species), Kerney & Cameron 1979: 158, Falkner 1990: 160, Manganelli et al. 1995: 24 (regarded *dyodon* and *thomasiana* as separate species), Turner et al. 1998: 210, Nordsieck 2002: 36 (regarded *thomasiana* and *dyodon* as conspecific because intermediate populations were found), Nordsieck 2006: 10.

Charpentieria itala (Martens, 1824)

Clausilia itala Martens, 1824 in Martens 1824: 442

14-23 x 3-5 mm, locally variable

Italy: Veneto, Monti Berici, Barbarano

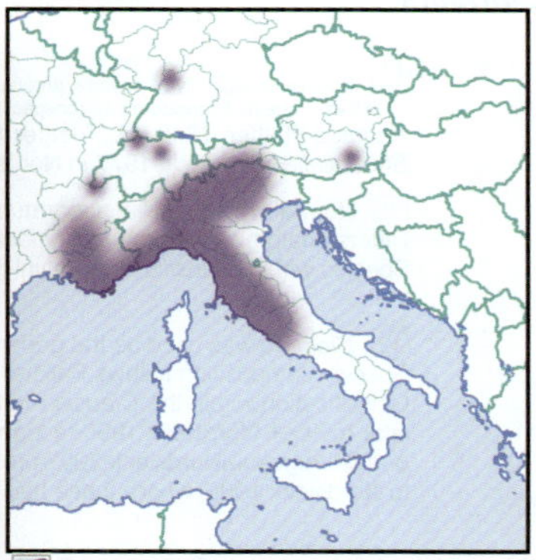

Italy, S Switzerland, SE France
Introduced (probably with vines) to various sites in central Europe.

Shell brown, smooth, striated or moderately ribbed, suture often weakly white and usually papillated, columellaris more prominent than in *Ch. clavata*, cervix nearly rounded, lunula strongly curved, no frontal palatal folds, subcolumellaris well visible in a perpendicular view. Animal brownish to black, locally variable.

At rocks and old walls, often inside human settlements. Feeds on algae and lichens on hard substrate. Less frequent in forests at tree trunks and in leaf litter. In Switzerland up to 1700 m.

Life cycle unknown, oval eggs (2 x 2.5 mm) were observed.

Moquin-Tandon 1855: 326, Westerlund 1884: 64, Kerney & Cameron 1979: 157, Falkner 1990: 160, Manganelli et al. 1995: 24 (listed 9 subspecies, but regarded this classification as doubtful on p. 47), Turner et al. 1998: 212.

Subspecies have traditionally been based on few morphological characters, the nature of their local variations remain to be studied in detail and potential narrow hybrid zones between populations separated by substantially reduced gene flow remain to be identified.

Charpentieria ornata (Rossmässler, 1836)

Clausilia ornata Rossmässler, 1836 in Rossmässler 1835-1837 (3): 9, Taf. 12

9-17 x 3-4 mm, locally variable

Croatia: Zagreb

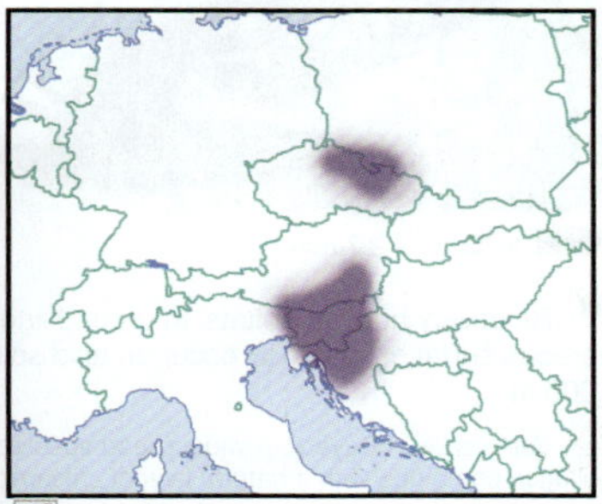

Sudetes, and SE Alps to Croatia

Like *Ch. itala*, but apertural margin never detached, palatal groove usually prominent, lunula evenly rounded in a)-shape (seen from outside).
Animal dark grey, dorsal side, head and tentacles nearly black.

At rocks (also on basaltic rocks) and old walls.

Westerlund 1884: 64, Falkner 1990: 160, Manganelli et al. 1995: 24.

Charpentieria stenzii (Rossmässler, 1836)

Clausilia stenzii Rossmässler, 1836 in Rossmässler 1835-1837 (3): 19, Taf. 12 (*C. Stenzii*)

14-18 x 3.3-4.5 mm

Slovenia: Julian Alps, Krainska Gora

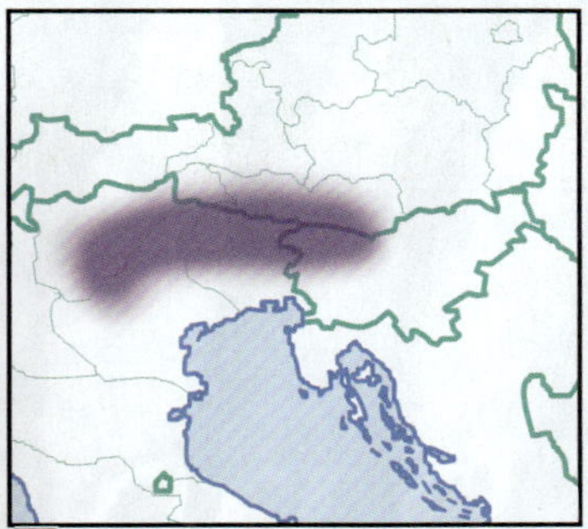

S and SE Alps

Shell dark brown, smooth, finely striated or ribbed, shiny, 10-11 whorls, suture white and irregularly papillated, cervix often more coarsely striated, parietalis very short and not reaching the margin, lower palatalis absent, columellaris weak, lunula dorsal or dorsolateral,]-shaped and slightly oblique (seen from outside), subcolumellaris visible in an oblique view. Locally variable, mainly in terms of rib density and prominence.

Westerlund 1884: 68, Bank 1987, Fauer 1991, Manganelli et al. 1995: 24.

Like in *Ch. itala* many local forms have been classified as subspecies, they remain to be aligned with generally accepted subspecies concepts in biodiversity research.

Clausilia

Clausilia Draparnaud, 1805 in Draparnaud 1805: 24

Type species *Pupa rugosa* Draparnaud, 1801 (fixed under the plenary powers in ICZN Opinion 1455 (1987))

Contained originally 11 species, among these were *Pupa rugosa* and *Turbo bidens* Linnaeus, 1758. Turton 1831: 6 gave *Turbo bidens* "Montagu" as type, having in mind *Turbo bidentatus* Strøm, 1765 which he had considered as conspecific with *Pupa rugosa*. Today such a case would be solved under Art. 70.3.
Andraea Pfeiffer, 1848 in L. Pfeiffer 1848: 538 (*Clausilia (Andraea)*), type species *Clausilia dubia* Draparnaud, 1805 (original designation)
Strobeliella Nordsieck, 1977 in Nordsieck 1977: 98 (*Clausilia (Strobeliella)*), type species *Clausilia whateliana* Strobel, 1850 (original designation)

Normal type clausilial apparatus, with regular lunula and clausilium plate, *Neostyriaca* has a G type clausilial apparatus with weak lunula and short principalis. Differs anatomically from *Neostyriaca* in the proximal section of the epiphallus (or parepiphallus) being slender (in *Neostyriaca* partly or entirely thickened). Species within *Clausilia* differ mainly in their columellaris, subcolumellaris and clausilium plate. Shell shape, sculpture, apertural folds and margin are less important.

Clausilia was used as the generic name for almost all clausiliid species between 1805 and the 1920s.
Species composition follows Falkner et al. 2001: 42, determination of widespread European species see Nordsieck 2002.
Classification within the *Clausilia* complex seems to be arbitrary. Nordsieck 1966 argued that *Neostyriaca* must be classified as a subgenus of *Clausilia*. This was later revised. Nordsieck 1990 recognised subgenera *Strobeliella* (containing *whateliana*), *Andraea* (containing *dubia*) and *Clausilia* (containing *bidentata*, *cruciata*, *pumila*, *rugosa*). Nordsieck 2007 presented a pure list of species and recognised subgenera *Strobeliella* (containing *whateliana* and 3 other species with reference to Nordsieck 1966, where these had been classified as subspecies of *whateliana*), and *Clausilia* (containing the other 5 species including *dubia*).

Clausilia bidentata (Strøm, 1765)

Without gender agreement: *Clausilia bidentatus* (Strøm, 1765)

Turbo bidentatus Strøm, 1765 in Strøm 1765: 436 (*Turbo Bidentatus*)

9-12 x 2.3-2.7 mm

(1-4) Italy: Genova, Acquasanta, (5-8) Monaco
(1-4) Ettore Balocchi

Spain to Finland

Usually larger than *C. rugosa*, fresh shells dark brown or cherry red, silky shiny, or greyish hue, usually with narrow ribs (50-70 on penultimate whorl, 10-11 per mm), first whorls increase continuously, cervix with keel, columellaris less s-like (in *C. rugosa* more s-like) and with triangular or indistinct/weak double fold. Clausilium plate at lower side without edge at margin (terminal lobe indistinct or absent).

Relatively humid habitats, in ground litter, herbage, under fallen timber, rocks and old walls. Climbs up vertical surfaces in moist weather. Usually on calcareous surfaces, but may occur on acid soils at mortared stone walls. Feeds on lichens and other epiphytes on limestone rocks and trees. Rarely found above 1000 m.

Vulnerable in Bayern. A widespread species, but has declined in some regions of central and eastern England (midlands and around London) due to a combination of air pollution combined with a natural lack of calcareous substrate.

Kerney et al. 1983: 228, Nordsieck 1990 (history of misidentifications), Falkner 1990: 162, Manganelli et al. 1995: 26, Turner et al. 1998: 222, Kerney 1999: 171, Nordsieck 2002: 36 (SW range limit in S Portugal to N Spain), Martínez-Ortí & Robles 2003: 96 (Castellón, southernmost occurrence in Spain).

Clausilia cruciata (Studer, 1820)

Glischrus cruciata Studer, 1820 in Studer 1820: 89 (*Glischrus (Clausilia) cruciata*)

9-12 x 2.3-2.6 mm

Germany: Bayern, Jettingen near Augsburg

Europe

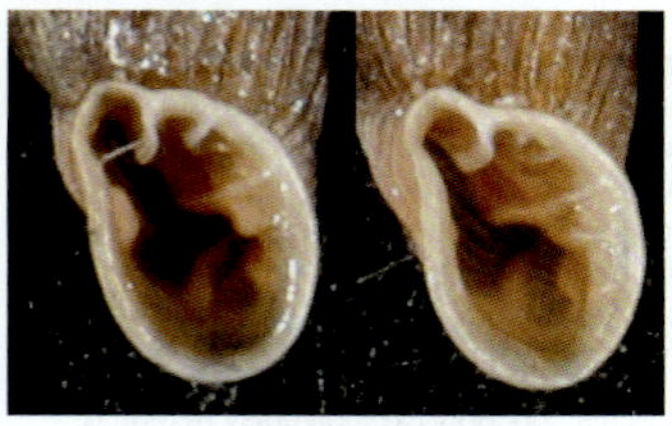

Shell like *C. bidentata* and *C. pumila*, more widely ribbed (40-55 ribs on the penultimate whorl), first whorls not increasing continuously but diameter of third whorl is almost that of the second whorl, palatal callus usually strong, columellaris with step-like bifurcation, subcolumellaris more strongly curved near aperture than in *C. pumila*. Clausilium plate at lower side with distinct edge at margin (terminal lobe).

Usually in humid and dark forests, under soil litter or at tree trunks, above the timberline in open rocky habitats. Mainly in mountains, in Switzerland up to 2100 m.

Vulnerable in Germany.

Kerney & Cameron 1979: 167, Manganelli et al. 1995: 26, Nordsieck 2002, Sulikowska-Drozd 2005: 74.

Clausilia dubia Draparnaud, 1805

Clausilia dubia Draparnaud, 1805 in Draparnaud 1805: 70

11-14 x 2.7-3.2 mm

Germany: Bayern, Altmühljura near Riedenburg

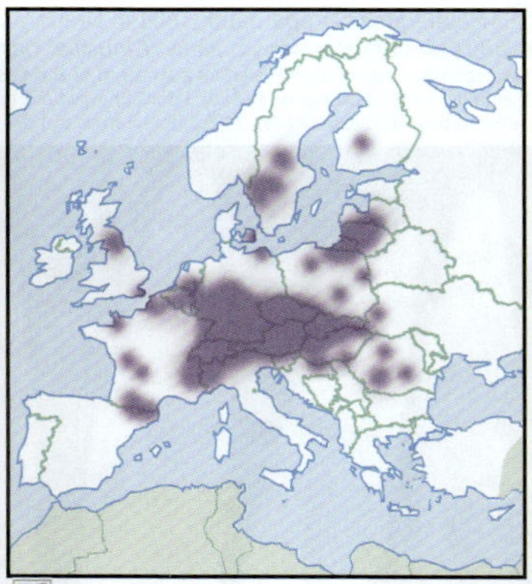

Europe

Usually larger than *C. bidentata*, usually densely ribbed, some populations more widely ribbed, no folds between parietalis and columellaris, columellaris with double fold or not, frontal upper palatalis present in a callus-like form, basal groove in the aperture present, subcolumellaris in frontal section only little curved, not S-like near the aperture and running along the basal groove. Basal cervical keel more prominent than in *Macrogastra.* Shells are highly variable, also within populations.

Characteristically on humid shady rocks, at trees and old walls, usually in exposed highland regions, more rarely in forests, calciphile. In Switzerland up to 2400 m, in N England mostly between 250 and 600 m. Hides in crevices and ground litter, climbs up vertical surfaces in moist weather. Feeds on epiphytic lichens and algae.

Vulnerable in Germany. In SE Poland not very abundant. Can be rare in lower altitudes in Switzerland. In N England threatened by air pollution and destruction of habitats, some populations have been lost since the 1800s.

Kerney et al. 1983: 228, Manganelli et al. 1995: 26, Edlinger 1997 (morphological variation, populations did not match preconceived expectations of well established races or subspecies occurring in well delimited areas), Kerney 1999: 172, Nordsieck 2002, Sulikowska-Drozd 2005: 73.

Clausilia pumila Pfeiffer, 1828

Clausilia pumila Pfeiffer, 1828 in C. Pfeiffer 1828: 41

11-14 x 2.9-3.5 mm

Germany: Bayrischer Wald, near Erlau

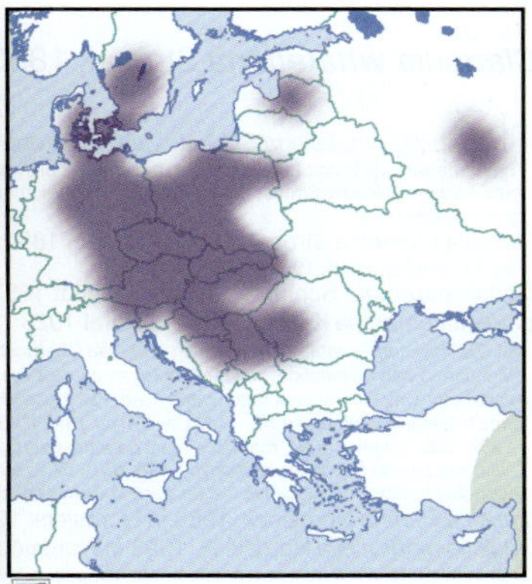

Europe

Shell like *C. bidentata*, more widely ribbed, diameter of third whorl small like second whorl (in the other *Clausilia* species except *C. cruciata* the upper whorls increase continuously), columellaris with step-like double fold (in *C. bidentata* triangular or indistinct), palatal callus weak or absent (unlike *C. cruciata*), basal cervical keel relatively prominent. Terminal lobe of clausilium present.

Humid forests, usually in soil litter or under fallen trees. Not significantly ascending trees. In Bulgaria up to 1900 m.

Endangered in Germany. Not very abundant in SE Poland.

Kerney & Cameron 1979: 168, Falkner 1990: 162, Manganelli et al. 1995: 26, Nordsieck 2002, Sulikowska-Drozd 2005: 75, Hubenov 2007.

Slovakia: Slovenský Raj National Park

Jozef Grego

Clausilia rugosa (Draparnaud, 1801)

Pupa rugosa Draparnaud, 1801 in Draparnaud 1801: 63
Bulimus perversus synonym *Helix (Clausilia) parvula* as mentioned in J. Férussac 1807: 122, 133 (synonym and nomen nudum)
No description added, name listed as a synonym of *Bulimus perversus* as used by Poiret 1789 (either as a new name, but much more probably a misuse of *Turbo perversus* Linnæus, 1758 = currently *Balea perversa* (Linnæus, 1758), I did not verify that). I don't know where this name was made available by a later author who should have referred to Férussac 1807, but even if that was done, it remains unclear if the name can have been made available under Art. 11.6.1 from Férussac 1807 who had not given a description.

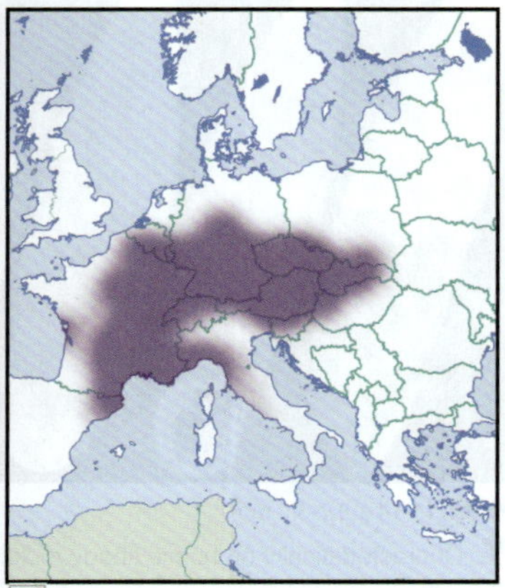

7-12 x 1.9-2.5 mm

Germany: Hessen, Fulda

Central Europe

(1) Germany: Bayern, near Walchensee, (2) France: Isère, Sablons
(2) Cédric Audibert

Shell brown, with ribs or growth lines or smooth, first whorls increase more or less continuously, columellaris usually without double fold, clausilium with or without edge at margin (terminal lobe present or not). Shell variable, many local forms with differences in ribs and clausilium plate structure. Usually smaller than *C. bidentata*. Locally variable in terms of ribs, lunula and clausilium plate.

Relatively humid habitats between rocks and rock rubble, at old walls, preferably in forests. Calciphile, in Switzerland up to 2400 m.

15-20 eggs (diameter 1.5 mm) are laid between August and October under stones or in crevices of rocks and old walls, juveniles hatch after 2 weeks.

Gassies 1849: 129, Moquin-Tandon 1855: 332 (*Cl. perversa*), Germain 1930: 353, Nobre 1941: 156, Gittenberger 1982, Kerney et al. 1983: 227 (*C. parvula*), Falkner 1990: 162, Turner et al. 1998: 220, Nordsieck 1990 (history of misidentifications), Nordsieck 2002.

This species has also been called *Clausilia parvula*. It has long been misidentified and not correctly distinguished by most authors from *bidentata* before the 1980s.
10 recent so-called subspecies have been recognized (Nordsieck 2007), mainly based on rib prominence, rib density and vaguely defined characters states of clausilium plate and lunula (Nordsieck 1990). The descriptions suggest spatially variable characters in gradients running independently over area. For example the subspecies *belonidea* from SE France was defined "rib prominence like *parvula*, lunula and clausilium plate like *rugosa*" and was considered by Nordsieck 1990: 148 as a transitional form between the two. In these definitions the so-called subspecies do not represent genetically partially separated entities with hybrid zones in between.

Clausilia whateliana Strobel, 1850

Clausilia whateliana Strobel, 1850 in Strobel 1850: 120
The name was attributed to Charpentier, a diagnosis provided by Charpentier was cited in quotes in a footnote, another diagnosis and additional descriptive remarks were given by Strobel. Under Art. 50.1.1 the authorship must be attributed to the author of the work, Strobel (Chapentier was not alone responsible for the description).
Clausilia brembina Strobel, 1850 in Strobel 1850: 120
First Reviser Nordsieck 1966 selected *whateliana*.
Clausilia exoptata Schmidt, 1856 in Schmidt 1857: 62
Iphigenia umbrosa Käufel, 1928 in Käufel 1928: 100 (*Iphigenia whatelyana umbrosa*)
Not *Clausilia commutata* var. γ synonym *umbrosa* Charpentier, 1852 in Charpentier 1852: 366 (= currently in *Cochlodina*).
Mentioned with description. Not mentioned as a junior synonym of an available name (as demanded by Art. 11.6), but as a synonym for a subordinate variant for which the Greek letter γ was used. For those who neglect this (as proposed by Kadolsky 2012: 78, this is preliminarily accepted here), the name was made available under Art. 11.6.1 by Küster 1847 [1853]: 121 who used the name as *Clausilia umbrosa*, with bibliographical reference to Charpentier 1852: 366. For the others, the name must be attributed to Küster 1853.
This is not a case of primary homonymy (Art. 53.3, 57.2, because Käufel established the name in *Iphigenia*, Charpentier/Küster in *Clausilia*), and not of secondary homonymy (because the taxa are currently not considered congeneric (Art. 59.1)). The name *umbrosa* Käufel, 1928 can be used.
Clausilia klemmi Nordsieck, 1966 in Nordsieck 1966: 30 (*Clausilia whateliana klemmi*)
Clausilia umbrosella Nordsieck, 1993 in Nordsieck 1993: 39 (*Clausilia whateliana umbrosella*)
Expressly established as a new replacement name for *Iphigenia whateliana umbrosa* Käufel, 1928, not *Clausilia umbrosa* Charpentier, 1852, in the erroneous belief that Käufel's name could not be used.

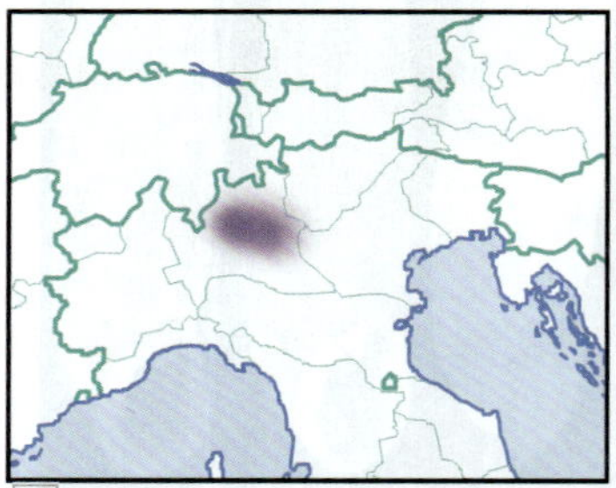

N Italy

9-11 x 1.7-2.5 mm

Italy: Lombardia, Bergamo, Valle Seriana

Shell reddish brown, whorls 2 and 3 of equal diameter, finely ribbed or striated (30-65 ribs/penultimate whorl), can be weaker on lowest whorls, 11-15 whorls, cervix slightly more prominently ribbed, dorsal and basal sides of cervix inflated, aperture very small, margin white and detached, principalis long, palatal callus strong, runs like a fold parallel to the margin and has a denticle on its upper and lower sides, columellaris ends abruptly in some distance from the margin, it reaches the margin in some populations, but not in a smooth curve (difference to *Neostyriaca strobel*), lunula dorsal, subcolumellaris curved, clausilium usually pointed at lower side.
Displays local variations in the rib prominence, expression of the columellaris and terminal lobe of the clausilium. Anatomically the forms were found to be uniform. *N. strobel* has more prominent cervical keels and a G type clausilial apparatus with weak lunula and very short principalis.

Restricted to calcareous substrate, usually in shady and humid habitats, N and NE exposed limestone rocks or rocks without slope, also in mixed and deciduous forests, but never at long distance from rocks, up to 1800 m. Most records have been from valleys, it is unclear if this was because the species is restricted to valley habitats.

Westerlund 1884: 182, Nordsieck 1966 (5 subspecies), Manganelli et al. 1995: 26 (5 subspecies), Nordsieck 2007: 60 (4 species, one with 2 subspecies, with reference to Nordsieck 1966).

The morphological variations have been associated with different taxa (subspecies in Nordsieck 1966, species in Nordsieck 2007 - Nordsieck's 2007 classification contradicted the 1966 work which was cited as reference) from W to E: (Lago di Como) *brembina*, (Brembo) *brembina* and *whateliana* mixed, also with intermedi-

ates, and *exoptata* in a side valley Val Serina, (Serio) *whateliana* and *klemmi*, (Lago d'Idro) *umbrosa*.
The exact distribution of character expressions and their variations over area could shed more light on the nature of diversity of this complex. Names of more or less vaguely defined subspecies alone are not always a helpful tool.
We could be dealing with partially or totally separated genetical entities, or with a single gene pool and several characters varying independently over area. The mixed occurrences of *brembina* and *whateliana* in the Brembo valley could either point to the presence of two different species without gene exchange living there for a long time, or to a single species with convergent local variations of dependent characters, to recent human-based dispersal events, or to past dispersal events from high mountain forms having been transported by giant floods to deeper zones in the valley. If populations were dispersed it would be interesting to study closely the contact zones with putatively local populations (also in the Serio valley between *whateliana* and *klemmi*).
Molecular research could help to understand this species complex, if this is combined with a more profound knowledge of forms living the higher mountain areas outside valleys at slopes at which it can be excluded that populations originated in dispersal events caused by giant floods.

Cochlodina

Cochlodina Férussac, 1821
in A. Férussac [1821-1822] [06 Apr 1821]: 24 (folio p. 28) *Helix (Cochlodina)*

Type species *Helix derugata* Férussac, 1821
(subsequent designation by Pilsbry 1922: 31)
(= currently *Cochlodina laminata* (Montagu, 1803))

Established on p. 24 (folio p. 28) with description, without species included. First species included by Férussac [1821-1822] [26 May 1821]: 61 (folio p. 65). Under his new species *Helix (Cochlodina) derugata* was listed (p. 63, folio p. 67) among others "*Clausilia bidens*, Draparnaud, Hist., Pl. IV, Fig. 5 à 7".
Draparnaud 1805: 68 gave a description (he had *Cochlodina laminata* in mind) and cited among the bibliographical references *Helix bidens* "Müll. 315." (= *Helix bidens* as used by Müller 1774: 116).
Müller 1774 gave a short description referring to a *Cochlodina* species and cited among various references the Linnean description of *Turbo bidens*. So Draparnaud used the Linnean name.
Férussac [1821-1822]: 62 (folio p. 66) listed *Turbo bidens* Linnæus, 1758 as a synonym of *Helix (Cochlodina) papillaris*, which he attributed to Müller, so Férussac showed awareness of Draparnaud's 1805 misidentification of the name *bidens*. Pilsbry 1922 (Nautilus 36): 31 designated "*Clausilia bidens* Draparnaud" as type species for *Cochlodina*. Under Art. 69.2.4 this subsequent designation converts *Clausilia bidens* sensu Draparnaud 1805 into one of the original included nominal species (Art. 72.2.1, 67.13.2), but "the species so designated is the nominal species denoted by the name of the taxonomic species actually involved (and not the nominal species cited)" (Art. 69.2.4). Férussac cited *Clausilia bidens* sensu Draparnaud 1805 as a synonym of *Helix derugata* Férussac, 1821, which must be regarded as the nominal species actually involved in the sense of Art. 69.2.4.
Kadolsky 2009: 24 argued that *Helix derugata* Férussac, 1821 was an "unnecessary synonym" of *Turbo laminatus* Montagu, 1803, and that the type species of *Cochlodina* must be "*Turbo bidens* sensu Draparnaud, 1805, non Linnaeus, 1758 (currently *Cochlodina laminata* (Montagu, 1803) [*Turbo*])". *T. bidens* sensu Draparnaud cannot be the type species because this is not a "nominal species", as demanded by Art. 69.2.4. A nominal species must be "denoted by an available name" (Glossary, nominal taxon). The question is, which one of the various subjective and objective synonyms of "the taxonomic species actually involved" is to be taken as the type species, and where is this nominal species selected?
The first person in this case to have selected a name to be used for the involved taxonomic species was Férussac 1821 who used *Helix derugata* for it. *T. laminatus* "des Anglois" was among the originally included nominal species, cited in the synonymy of *H. derugata*. The argument that in a personal view expressed in 2009 this name was an unnecessary synonym of *T. laminatus* is irrelevant. *H. derugata* is an available name with its own types, its identity is defined by the rules of the Code.
Currently the name *T. laminatus* is used for this species, but tomorrow this can be another name. This depends on taxonomic judgements. A type species must remain fixed.

Europe and N Africa

Nordsieck 1969 (two papers on *Cochlodina*), Nordsieck 2007: 52 (list of species, adopted here except that *inaequalis* is classified in *C. laminata* following Gittenberger 2007: 79).

Cochlodina cerata (Rossmässler, 1836)

Clausilia cerata Rossmässler, 1836 in Rossmässler 1835-1837 (4): 15, Taf. 18

15-21 x 3-4 mm

Slovakia: Licvik Ihrik, Slovensky Raj

Westerlund 1884: 18, Grossu 1981: 156, Kerney et al. 1983: 215.

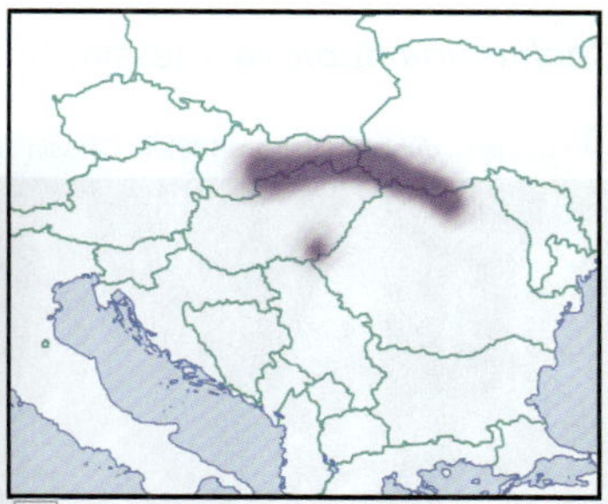

Slovakia to N Romania

Shell yellowish with reddish to olive brown hue, cervix at lower side yellowish, 10-12 whorls, apertural margin with broad and often reddish lip, parietalis short, not reaching margin, columellaris not very prominent, up to 3 small palatal folds, subcolumellaris curved and visible in a perpendicular view.

In forests, mainly in moist habitats at the botton, under rocks and stones, rarely found on the rocks or climbing on tree trunks and wood logs. From highlands to subalpine stage.

Rare in Romania.

Cochlodina comensis (Pfeiffer, 1850)

Clausilia comensis Pfeiffer, 1850 in L. Pfeiffer 1850: 108

11-15 x 2.8-3.6 mm, locally variable

N Italy: Nago near Lago di Garda

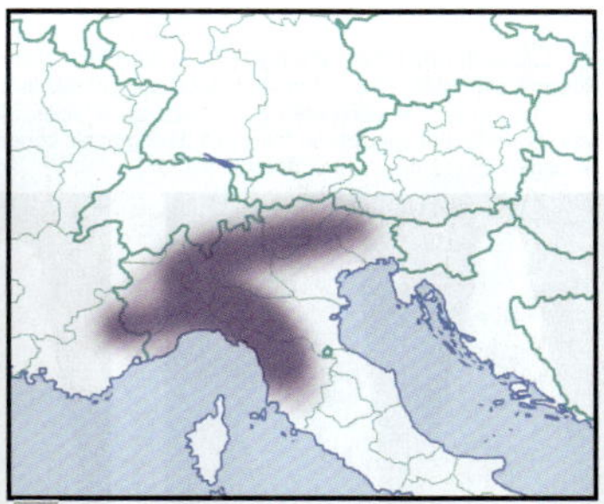

N Italy, S Switzerland

Shell horny yellowish, finely striated, 10-11 whorls, cervix in last section finely ribbed, apertural margin not detached, columellaris very prominent, palatal wall strong and white, 4 palatal folds. Locally variable in terms of size and prominence of the palatal folds, in some populations the two lower palatal folds are connected to each other forming a Y.

At rocks and old walls, under stones, also in the leaf litter layer of deciduous forests. In Switzerland up to 800 m.

Scattered in Switzerland, nowhere abundant.

Westerlund 1884: 26, Kerney & Cameron 1979: 157, Turner et al. 1998: 208.

Cochlodina costata (Pfeiffer, 1828)

Clausilia costata Pfeiffer, 1828 in C. Pfeiffer 1828: 42
Clausilia commutata Rossmässler, 1836 in Rossmässler 1835-1837 (4): 19, Taf. 18
Clausilia curta Rossmässler, 1836 in Rossmässler 1835-1837 (4): 19, Taf. 18

10-18 x 3-4 mm, locally variable

Italy: Veneto, near Bassano del Grappa

Westerlund 1884: 22, Kerney & Cameron 1979: 156, Falkner 1990: 158, Manganelli et al. 1995: 23, Nordsieck 2007: 52.

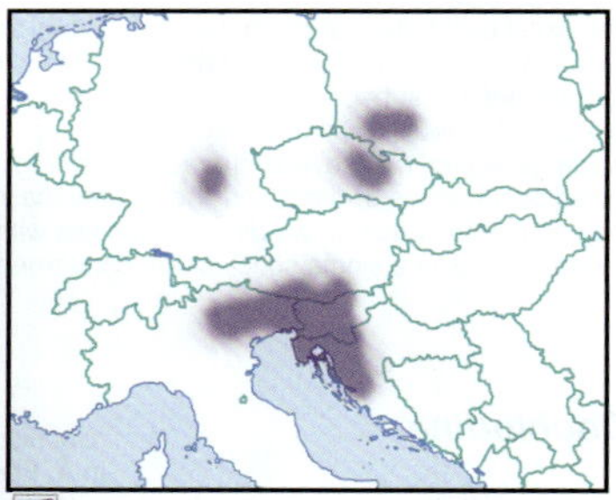

Europe

Shell horny brown, weakly striated to ribbed, 10-12 whorls, cervix more coarsely ribbed, with weak basal and dorsal keels, aperture usually with prominent palatal callus, margin not always connected at parietal side, up to 4 palatal folds, frontal lower palatalis breaking through palatal callus, upper palatalis often connected with a middle palatal fold, subcolumellaris well visible. Colour, apertural shape and palatal folds locally variable.
More slender than *C. laminata*, palatal callus stronger, palatal folds different.

In mountains, lives in humid and shady habitats in woods and among rocks.

Cochlodina dubiosa (Clessin, 1882)

Clausilia dubiosa Clessin, 1882 in Clessin 1882: 137

13-17 x 3.3-3.8 mm

S Austria: Kärnten, Tscheppaschlucht

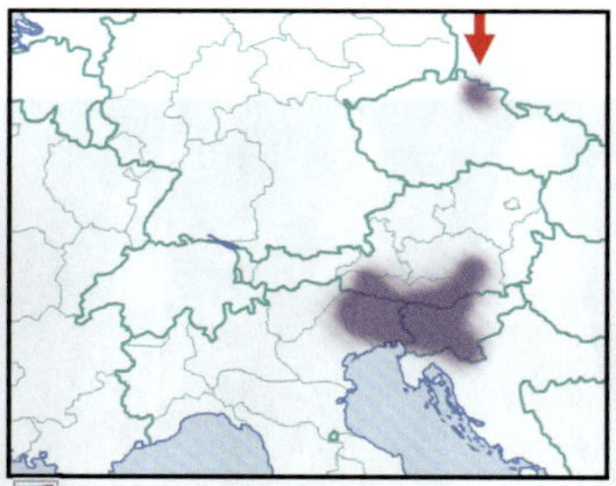

SE Alps, N Czech Republic

Shell smaller and more slender than *C. laminata*, 9-12 weakly convex whorls, aperture pear-shaped, parietalis reaches spiralis, principalis slightly bent downwards at its frontal end, palatal callus strong, not crossed by basalis, subcolumellaris visible in a perpendicular view.

At trees, rotting wood, humid habitats of montane stages up to subalpine coniferous forests.

Kerney et al. 1983: 214, Manganelli et al. 1995: 23, Nordsieck 2007 (Mitt. DMG): 25.

Cochlodina fimbriata (Rossmässler, 1835)

Clausilia bidens synonym *Clausilia fimbriata* Rossmässler, 1835 in Rossmässler 1835-1837 (1): 77
Rossmässler 1835: (1) p. 77 (a short description of this form was given there) did not recognize *Cl. fimbriata* as a separate species. Rossmässler 1835: (2) p. 11 Taf. 7 Fig. 106 recognized it as a separate species, albeit with some doubts. In the index of Rossmässler 1837: (5-6) p. 61 both (1) p. 77 and (2) p. 11 were given for the species, so there is a bibliographic connection to the first name, which can be regarded as having been adopted under Art. 11.6.1 by Rossmässler (1835: (2) 11 or at least in 1837).

14-16 x 3.6-4.4 mm

S Austria: Kärnten, Singerberg near St. Johann

Endangered in Germany, Vulnerable in parts of Austria.

Kerney & Cameron 1979: 155, Falkner 1990: 158, Manganelli et al. 1995: 23, Bole & Slapnik 1997: 147, Turner et al. 1998: 206.

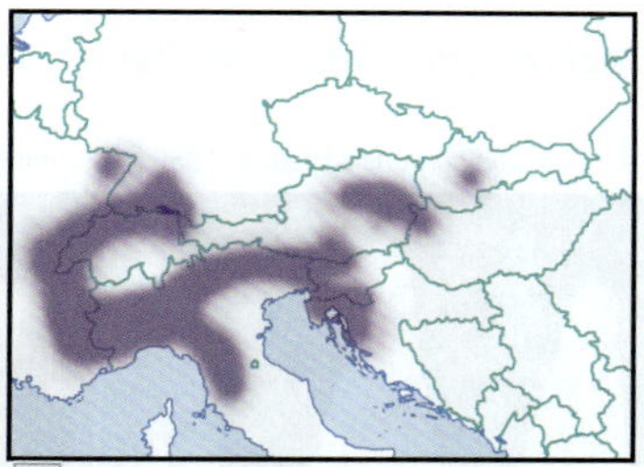

S central Europe

Like *C. laminata*, but with more prominent riblets, colour usually lighter, strong palatal callus present, lower palatalis weak and does not reach the callus (contrast to *C. laminata* where the callus is weak and the lower palatalis stronger).

Humid and shady habitats in woods and among limestone rocks, restricted to mountains. More confined to calcareous substrate than *C. laminata*, and requires more humidity, prefers rocky deciduous forests, not in coniferous forests. In Switzerland up to 1600 m. In S Slovenia the most common clausiliid species, on tree trunks and rocks, under bark and leaves, climbs high into trees in wet weather.

Cochlodina incisa (Küster, 1876)

Clausilia incisa Küster, 1876 in Küster 1875: 20

Usually given as Küster 1876, but could be 1875, there is no evidence for 1876 in the original work.
Kadolsky 2009: 27 selected a neotype for *Turbo bidens* Linnæus, 1758, conspecific with *C. incisa*. The Linnean name had never been used for this species before. The identity of the name *Turbo bidens* was clarified in Opinion 2176 in 2007 (senior synonym of *Helix papillaris* Müller, 1774) and there was no exceptional need to select a neotype. A Case is currently pending at the Commission (Case 3581) to set aside Kadolsky's 2009 neotype fixation, and to maintain the usage of *Clausilia incisa* for the present species (Welter-Schultes 2012: 85).

16-19 x 4-5 mm

Italy: Abruzzo, near Villetta

Italy

Shell brown, finely striated, 10-11 whorls, cervix rounded with very weak basal keel, palatal wall of aperture weakly reddish, parietalis long, columellaris often with two knots at frontal side. Colour lighter than in Italian *C. laminata*, palatal folds obviously more prominent.

In leaf litter and rotting wood, generally at lower altitudes than *C. laminata*.

Westerlund 1884: 21, Manganelli et al. 1995: 23.

Cochlodina kuesteri (Rossmässler, 1836)

Clausilia kuesteri Rossmässler, 1836 in Rossmässler 1835-1837 (3): 16 (*Clausilia Küsteri*)

Described in the text, without figure.

13-20 x 3-4.5 mm, very variable, also within populations

Italy: Sardinia, Nuoro, Sadali

Italy: Sardinia and Gorgona

Shell horny yellowish or brownish with violet hue, finely striated or almost smooth, 11-13 whorls, cervix rounded, palatal wall in the aperture absent or white and weak, apertural margin slightly detached, folds moderate, parietalis almost reaching margin, subcolumellaris visible in an oblique view. One form is reported to possess 7-9 small teeth at the apertural margin.

Westerlund 1884: 25, Manganelli et al. 1995: 23. Not in Corsica.

Cochlodina laminata (Montagu, 1803)

Without gender agreement: *Cochlodina laminatus* (Montagu, 1803)

Turbo laminatus Montagu, 1803 in Montagu 1803: 359
Clausilia melanostoma var. *inaequalis* Schmidt, 1868 in Schmidt 1868: 32
Not *Clausilia bidens* var. *inaequalis* Gallenstein, 1852 which has not been used after 1899, but the conditions of Art. 23.9.2 are not met, the name was only mentioned a few times by two authors in the past 50 years. Gittenberger 2007: 79 announced to bring the case to the Commission (Art. 23.9.5 does not apply because the 1852 name has not been in use recently, 23.9.1 and 23.9.3 apply).

13-18 x 3.5-4.4 mm, locally variable

Latvia: Daugavpils, Naujene

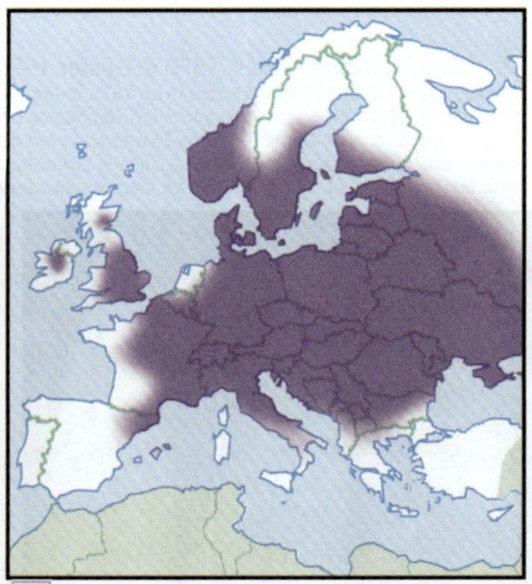

Europe to Caucasus

Shell brown to cherry red, nearly smooth, shiny, parietalis small, columellaris more prominent, palatal callus weak, subcolumellaris visible in an oblique view. Differences to other *Cochlodina* species in their frontal lower palatalis (FLP) and palatal callus: in *C. costata* the callus is strong with FLP breaking through, in *C. fimbriata* FLP does not reach the strong callus, in *C. orthostoma* FLP does not reach the callus, in *C. laminata* the callus is usually weak, FLP reaches the callus and sometimes breaks through.

In deciduous forests at the soil or under shrubs, leaf litter or among moss, at humid weather conditions also ascending vertical surfaces. Tolerates some degree of disturbance, also found in modern beech plantations. In S Bulgaria most abundant on shady rocks covered with moss, prefers high humidity. In Switzerland and Bulgaria up to 2000 m. In Italy generally in higher altitudes than *C. incisa*. Feeds inside moist and fungous leaf litter and desintegrating wood.

10-20 eggs (occasionally up to 150, diameter 2 mm) are laid in clutches between August and October, juveniles hatch after 3 weeks, maturity is reached at the end of the second year.

Vulnerable in some parts of Britain, threatened by habitat loss. Rare in Ireland, certain isolated populations are at risk.

Gassies 1849: 126, Moquin-Tandon 1855: 318, Germain 1930: 338, Gittenberger 1967 (Arch. Moll.), Nordsieck 1993: 34, Nordsieck 2005, Nordsieck 2007 (Mitt. DMG), Gittenberger 2007 (classified *inaequalis* and *laminata* as conspecific).

Gittenberger 1967 classified *inaequalis* as conspecific with *laminata* because the form was interconnected by gradual transitions to *C. laminata grossa*. Nordsieck 2007 classified *inaequalis* as an independent species and did neither consider nor weaken Gittenberger's 1967 argument, and consequently, Gittenberger 2007 maintained the 1967 classification which was based on presence of these gradual transitions.

Cochlodina marisi (Pfeiffer, 1868)

Clausilia transsilvanica synonym *Clausilia Marisi* Pfeiffer, 1868 in L. Pfeiffer 1868 (Mon. Helic. 6): 406
Mentioned with description and with a reference to Schmidt "Clausil." p. 28 (?, not Schmidt 1857: 28), and to Bielz 1863. Previously mentioned as *Cl. transsilvanica* synonym *Cl. Marisi* by Bielz 1863: 127 (with description) and corrected to *Cl. Marusii* in the index on p. V. First used as a name as *Clausilia (Clausiliastra (Marpessa)) Marisi* by Pfeiffer & Clessin 1881: 369 with bibliographical reference to Pfeiffer 1868: 406, making Pfeiffer's 1868 synonym *marisi* available under Art. 11.6.1. For those who require a bibliographical reference for making available a synonym under Art. 11.6.1, the name must be *Cl. marisi* Pfeiffer, 1868. For the others, *Cl. marusii* Bielz, 1863.

14-19 x 3-4.2 mm

Romania: Transilvania, Cluj

W Romania

Shell horny yellowish to horny brown, thin, almost smooth, very shiny, 11-12 whorls, cerxix rounded, cervical callus strong and reddish yellow, apertural margin connected at parietal side but not detached, white, parietalis oblique and reaching margin, columellaris moderate, 3-4 palatal folds.

Restricted to forests, humid and shady habitats, on tree trunks and rotting wood, often also on rocky walls and limestone rocks like *Alopia* species.

Westerlund 1884: 18, Grossu 1981: 155.

Cochlodina meisneriana (Shuttleworth, 1843)

Clausilia meisneriana Shuttleworth, 1843 in Shuttleworth 1843: 18 (*Clausilia Meisneriana*)

15-21 x 3.2-3.8 mm

France: Corsica, Vezzani

France (Corsica)

Shell pale horny brown, very finely ribbed, 12-14 whorls, aperture almost rectangular, margin detached, palatal wall deep inside and oblique, visible from outside as a yellowish structure, parietalis not reaching margin, subcolumellaris slightly hidden.

Westerlund 1884: 25, Manganelli et al. 1995: 23.

Cochlodina orthostoma (Menke, 1828)

Clausilia orthostoma Menke, 1828 in Menke 1828: 78
Established with description.

11-14 x 2.7-3.3 mm

Germany: Bayern, Oberpfalz, Ruine Wolfstein near Neumarkt

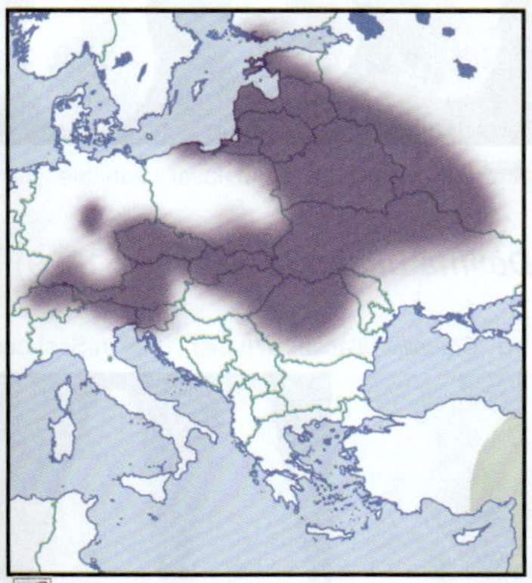

Europe

Shell small, slender, ribbed, cervix nearly rounded, no palatal folds visible in the aperture, palatal callus usually weak.

Deciduous woods in humid and shady rocky habitats, often at tree trunks. In Switzerland up to 1400 m.

Ovoviviparous.

Relictual distribution, scatttered with large distances between populations. Endangered in Germany and Austria.

Kerney & Cameron 1979: 157, Falkner 1990: 158, Turner et al. 1998: 209.

Slovakia: Slovenský Kras National Park
Jozef Grego

Cochlodina triloba (Boettger, 1878)

Clausilia laminata var. *triloba* Boettger, 1878 in O. Boettger 1878 (Jahrb. DMG 5): 34

12-15 x 3.2-3.8 mm

Slovenia: Lokev, Sežana

Museum Budapest HNHM

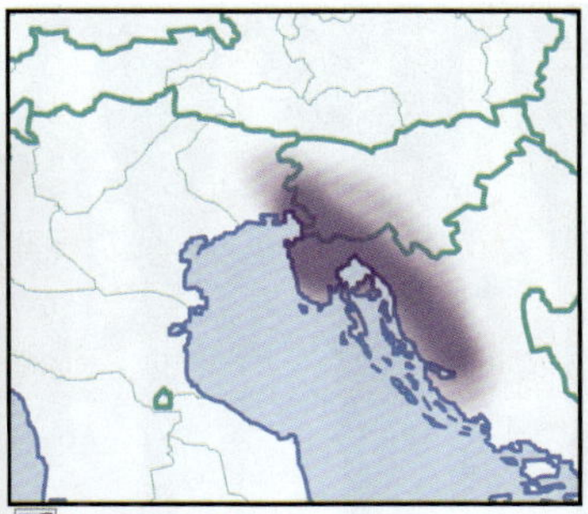

NE Italy to NW Croatia

Shell horny yellowish, striated, not very slender, columellaris in frontal part emerging as a weak nodule, palatal wall strong and white, clausilium broad and weakly bulged, and split into two parts with columellar lobe.

Westerlund 1884: 20, Manganelli et al. 1995: 23.

Delima

Delima Hartmann, 1842
in Hartmann 1844: 156 *Clausilia (Delima)*

Type species *Clausilia laevissima* Rossmässler, 1834
(subsequent designation by Pfeiffer 1848 (Mon. Helic. 2): 542)

Originally without description and without species included, but with a reference to "Abtheil. 8 Pfeiff. ohne *albescens*, *decipiens* und *gastrolepta*". This should be understood as a bibliographical reference to Pfeiffer 1841 (Symb. 1): 32, the 8th section (containing *albescens*, *cattaroensis*, *laevissima*, *decorata*, *decipiens*, *gastrolepta*). The name *Delima* was thus made available under Art. 12.2.5 because *cattaroensis*, *laevissima* and *decorata* were included by bibliographic reference. However *Delima* had no originally included nominal species (Art. 67.2.3, see also the comment under *Papillifera*). Pfeiffer 1848: 542 was the first author to have included a nominal species to *Delima* and designated this one species as type, *Clausilia laevissima*.

Nordsieck 1969 (revision), Zilch et al. 2002 (species and subspecies classification), Nordsieck 2007: 99 (proposed a concept of the *D. blanda* group with so-called hybrid populations).

This genus needs a revision. It would be necessary to know the exact ranges of the shell character states used for classification, the exact locations where populations of different taxa live sympatrically without interbreeding, and to study the nature of the so-called hybrid populations.

Delima albocincta (Pfeiffer, 1841)

Clausilia albocincta Pfeiffer, 1841 in L. Pfeiffer 1841 (Symb. 1): 48

16-28 x 3.8-5.2 mm, locally variable — Croatia: Dalmatia, Krka valley near Drniš

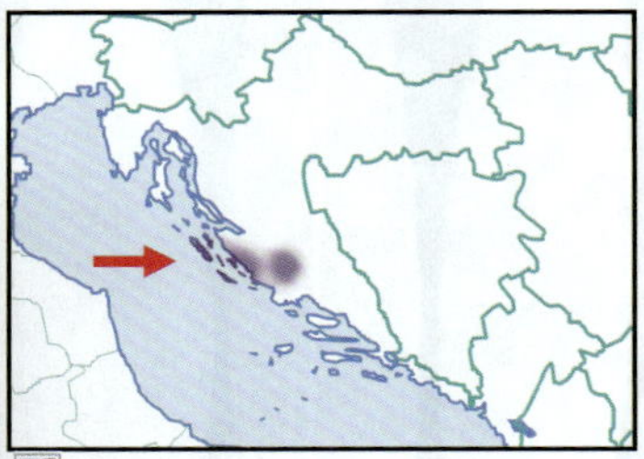

Croatia (N Dalmatia)

Shell light or horny to dark brown, almost smooth, suture white, 10-12 whorls, cervix finely striated with strong basal and dorsal keels and a deep basal furrow in between, aperture with weak yellow brownish lip, margin often detached, columellaris strong and very oblique, palatal wall translucent, suturalis and principalis strong, lower palatalis almost situated in keel furrow, lunula almost straight, subcolumellaris invisible.

Westerlund 1884: 86, Nordsieck 1969: 272, Nordsieck 2007: 99.

D. blanda group.

Delima bilabiata (Wagner, 1829)

Clausilia bilabiata Wagner, 1829 in Schubert & Wagner 1829: 189

14-20 x 3.2-4.3 mm — Croatia: (1) Rab island, (2-8) Dalmatia, Pelješac peninsula — (1) Jozef Grego

NE Italy to N Albania

Shell usually strongly ribbed, at least at upper whorls and cervix, no cervical callus, aperture usually not detached, lunula dorsal, subclaustralis present and usually separated from lunula, basalis short or absent.

I have not seen strongly ribbed specimens.

Westerlund 1884: 80, Nordsieck 1969: 273, Nordsieck 1993: 35, Manganelli et al. 1995: 24, Štamol & V. 1995 (N Croatia: Cres, Lošinj, Trstenik island E Lošinj), Dhora & W. 1996: 135.

Delima binotata (Rossmässler, 1836)

Clausilia binotata Rossmässler, 1836 in Rossmässler 1835-1837 (3): 15, Taf. 12

13-20 x 3-4.2 mm — Bosnia: Zenica

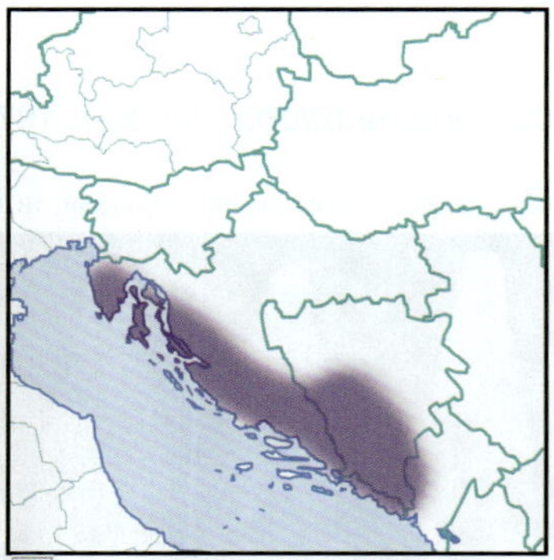

Croatia (Istria) to Montenegro

Shell horny yellow to dark brown, almost smooth or striated, cervix ribbed, apertural margin connected at parietal side, not detached, parietalis reaching margin, columellaris strong, almost horizontal and S-shaped, lower palatalis strong, lunula dorsal, (-shaped, subcolumellaris usually long.

Westerlund 1884: 91, Nordsieck 1969: 271.

Delima blanda (Rossmässler, 1836)

Clausilia blanda Rossmässler, 1836 in Rossmässler 1835-1837 (3). 13, Taf. 12
Clausilia conspurcata Rossmässler, 1836 in Rossmässler 1835-1837 (4): 18, Taf. 18

12-19 x 3.2-4.2 mm, size and shape locally variable

Croatia: Dalmatia, (1-5) Zvennac island off Zadar (*blanda*), (6-8) near Split (*conspurcata*)

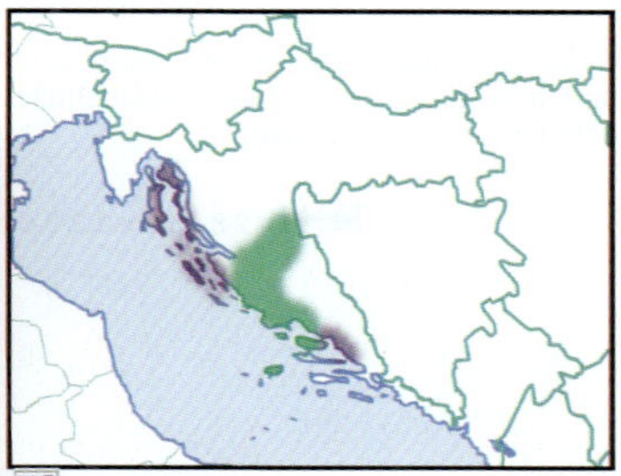

W Croatia
Green: *D. blanda* "*conspurcata*"

Shell horny or amber yellow, almost smooth, translucent and shiny, 9-11 whorls, suture in the central populations slightly papillated on upper whorls, cervix finely striated, apertural margin not or weakly connected at parietal side, columellaris thickened in frontal section, lower palatalis usually very weak, sometimes strong, lunula dorsolateral, usually straight and curved on upper side, touching principalis, subcolumellaris visible in a perpendicular view. Apertural shape variable.

Westerlund 1884: 81, 90, Nordsieck 1969: 271 (*D. conspurcata blanda*), Falkner 1990: 158, Zilch et al. 2002: 205, Nordsieck 2007: 99.

Contains probably *conspurcata* (regarded as a subspecies of *blanda* by Nordsieck 2007 and Zilch et al. 2002: 205), *albocincta*, *latilabris*, *pachystoma* (local intermediates were found by Nordsieck 2007: 99) and *helenae* (closely related to *D. latilabris*). More research and a better knowledge on the spatial variation of characters would be necessary to understand the nature of the diversity in this group. The map presented by Nordsieck 2007: 101 was unclear.

Delima edmibrani Štamol & Slapnik, 2002

Delima edmibrani Štamol & Slapnik, 2002 in Štamol & Slapnik 2002: 240

Shell reddish brown with white suture, weakly striated, 9-11 whorls, cervix with strong basal and weak dorsal keel, aperture often weakly detached, margin white and narrow, parietalis strong, columellaris prominent, principalis sometimes crazy and twisting around, lunula dorsolateral, often very weak, basalis strong and very long, frontal upper palatalis strong, more small folds can be present, subcolumellaris usually visible in a perpendicular view, often reaching margin.

14.6-18.0 x 3.2-3.7 mm

At vertical high cliffs at the sea, and ruins of stone houses.

Croatia (Dugi Otok island)

Delima giselae Wagner, 1914

Delima giselae Wagner, 1914 in Sturany & Wagner 1915: 75, Taf. 16

Shell yellowish to light reddish brown, translucent and shiny, weakly ribbed or striated, cervix more prominently ribbed, 8-10 whorls, aperture with strong white palatal callus, also visible from outside, separated by a reddish brown zone from the margin, margin yellowish and broad, parietalis reaches margin and passes spiralis inside, no frontal upper palatalis, lunula strong, subcolumellaris well visible in the aperture. Clausilium S-shaped with rounded tip, one-lobed.

13-17 x 3.2-3.8 mm

Obviously a forest species.

Nordsieck 1969: 274.

Only recorded once, from "Dinara Kolmut", I could not find this locality.

Montenegro/Bosnia/Croatia (?)

Delima helenae (Küster, 1876)

Clausilia helenae Küster, 1876 in Küster 1875: 118 (*Clausilia Helenae*)
Usually given as Küster 1876, but could be 1875, there is no evidence for 1876 in the original work.

Shell intensively cherry brown, very finely striated, suture white, with white papillae at some distance, cervix densely ribbed, aperture broadly egg-shaped, margin connected at parietal side, palatal wall brown-red, columellaris white, lower palatalis indistinct, lunula almost dorsal, subcolumellaris appearing in a curve at lower side.

18-20 x 4-4.7 mm

Westerlund 1884: 83, Nordsieck 1969: 272, Nordsieck 2007: 99.

D. blanda group.

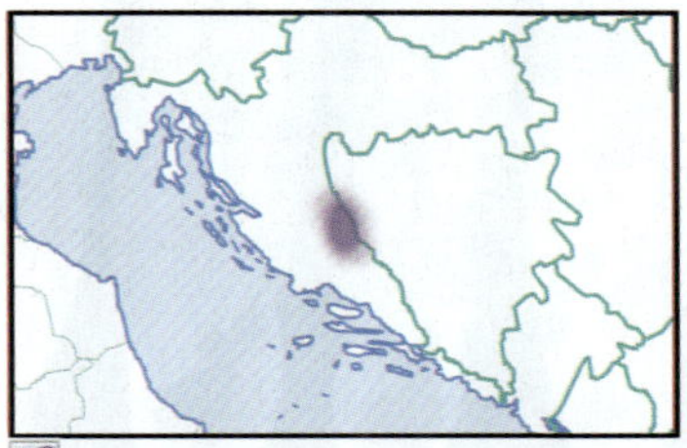

Croatia (central Dalmatia)

Delima hiltrudis Nordsieck, 1969

Delima hiltrudis Nordsieck, 1969 in Nordsieck 1969: 274, 283

Weak sculpture, no cervical callus, apertural margin simple and attached at parietal side, subclaustralis distinct, usually connected to lunella, no basalis.

12.8-15.3 x 3.2-3.6 mm

Croatia: Dalmatia, island of Brač
Massimo Prodan

Croatia (island of Brač)

Delima laevissima (Rossmässler, 1834)

Clausilia laevissima Rossmässler, 1834 in Rossmässler 1834 (1): 9

14-28 x 3-5 mm, locally variable

Croatia: Dalmatia, Dubrovnik

Croatia (S Dalmatia)

Shell yellowish brown, smooth, suture whitish and not papillated, cervix weakly striated and with moderate or prominent basal keel, apertural margin connected at parietal side but not detached, lunula dorsolateral.

Westerlund 1884: 72, Nordsieck 1969: 270, Falkner 1990: 158, Manganelli et al. 1995: 24.

Delima latilabris (Wagner, 1829)

Clausilia latilabris Wagner, 1829 in Schubert & Wagner 1829: 191

17-23 x 4-5 mm

Croatia: Dalmatia, Obrovac near Zadar

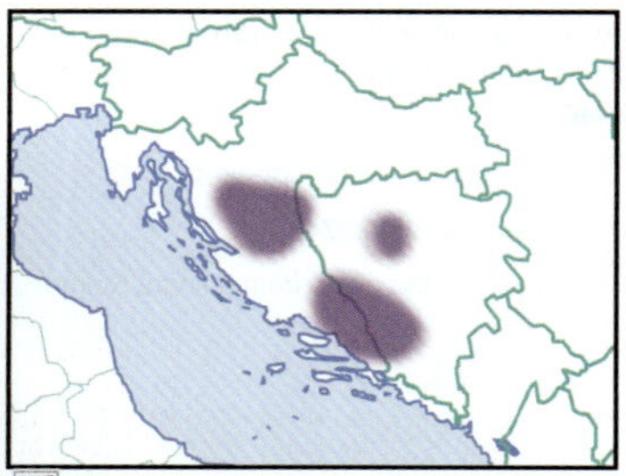

Croatia, Bosnia-Herzegovina

Shell reddish with horny yellow or brownish hue, very finely striated, 9-13 whorls, suture usually white, apertural margin connected at parietal side, not detached, columellaris oblique in frontal section, palatal wall inside like a fold, not visible from outside, lower palatalis short or absent, lunula lateral,)-shaped (seen from outside), connected to principalis, subcolumellaris usually not well visible.

Westerlund 1884: 86, Nordsieck 1969: 272 (*D. decipiens*), Nordsieck 1993: 34, Nordsieck 2007: 99.

D. blanda group.

Delima montenegrina (Pfeiffer, 1848)

Clausilia montenegrina Pfeiffer, 1848 in L. Pfeiffer 1848: 451

9-15 x 2.6-3.7 mm, locally variable

Albania: Shkodër

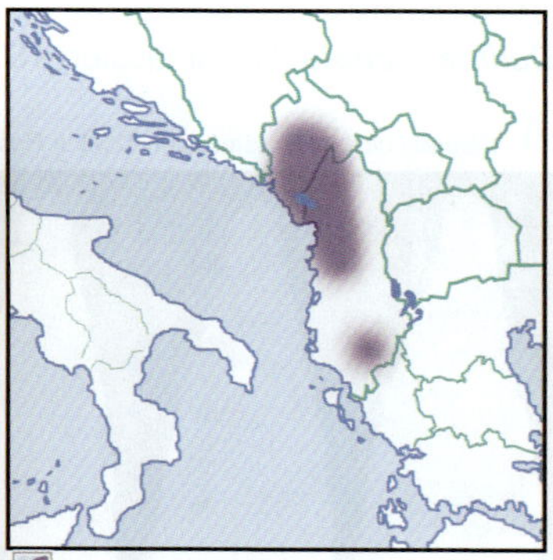

Montenegro to Albania

Shell with papillated suture, lunula dorsal to dorsolateral, upper palatalis long at proximal end (in contrast to *D. laevissima* and *D. binotata*) (but obviously not in Albania), basalis sometimes present.

In N Albania frequent, rocky habitats.

Westerlund 1884: 75, Nordsieck 1969: 270, Dhora & Welter-Schultes 1996: 135.

Delima pachystoma (Pfeiffer, 1848)

Clausilia pachystoma Pfeiffer, 1848 in L. Pfeiffer 1848: 439

Shell yellowish to horny brown, or with reddish hue, very finely striated, 11-13 whorls, suture often finely papillated, cervix finely ribbed with narrow basal keel, no basal furrow, aperture oblique with a strong white callus, lower palatalis directed against basal tubercle, lunula lateral, subcolumellaris visible in the aperture.

15-21 x 3.5-4.8 mm, locally variable

Westerlund 1884: 85, Nordsieck 1969: 272, Nordsieck 2007: 99.

D. blanda group.

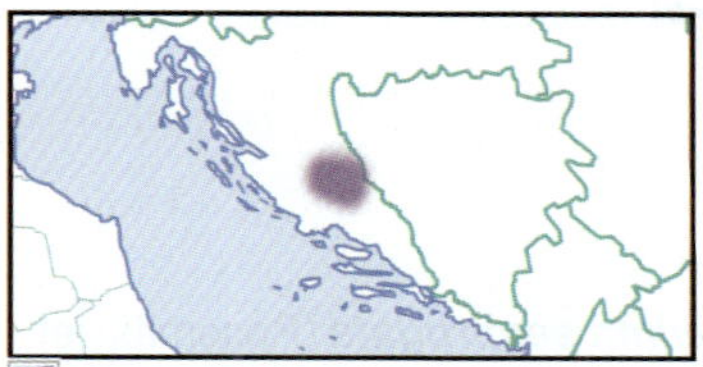

Croatia (central Dalmatia)

Delima pellucida (Pfeiffer, 1848)

Clausilia pellucida Pfeiffer, 1848 in L. Pfeiffer 1848: 456

11-16 x 2.5-4 mm

Croatia: Dalmatia, Kastel Gomilic NW Split

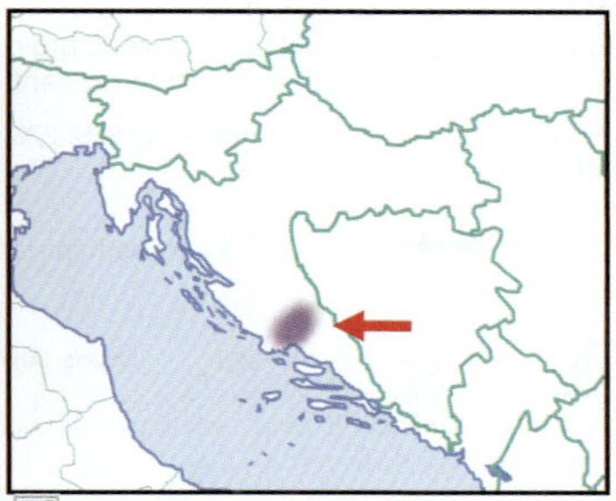

Croatia (central Dalmatia)

Shell horny yellowish, striated or finely ribbed, 9-10 whorls, suture usually papillated, aperture with weak lip, columellaris not very close to parietalis, lower palatalis short but often visible in a perpendicular view, subcolumellaris deep inside.

Westerlund 1884: 72, Nordsieck 1969: 271.

Delima pfeifferi (Küster, 1850)

Clausilia pfeifferi Küster, 1850 in Küster 1847: 59 (figured on Taf. 6 Fig. 14-17) (*Clausilia Pfeifferi*)
Text from 1850, plate from 1853.

17-20 x 3.8-4.7 mm

Croatia: Vrlika

Croatia (central Dalmatia)

Shell yellow brown, often slightly violet, almost smooth, suture white, cervix ribbed, with two keels, aperture broad, margin shortly detached, parietalis not reaching margin, columellaris almost horizontal, lunula dorsal,)-shaped (seen from outside), subcolumellaris visible in an oblique view.

Westerlund 1884: 90, Nordsieck 1969: 271.

Delima piceata (Rossmässler, 1836)

Clausilia piceata Rossmässler, 1836 in Rossmässler 1835-1837 (4): 18, Taf. 18

12-20 x 3.8-4.5 mm, perhaps locally variable

Italy: Abruzzo, Molise, N Termoli

Central Italy

Shell horny yellowish, smooth, apertural margin connected at parietal side but not detached, lunula straight, visible from outside and close to the principalis, subcolumellaris visible in a perpendicular view.
Differs from *Siciliaria paestana* in a subclaustralis and a strong subcolumellaris.

In Mediterranean macchia vegetation.

Westerlund 1884: 63, Manganelli et al. 1995: 24.

Delima semirugata (Rossmässler, 1836)

Clausilia semirugata Rossmässler, 1836 in Rossmässler 1835-1837 (4): 21, Taf. 18

Shell much less slender than *D. conspurcata*, shell height only 2/3 and shell diameter slightly larger, more ribbed and more papillae on upper whorls, cervix ribbed, aperture with palatal groove, margin usually detached, no frontal lower palatalis present.

Westerlund 1884: 77, Nordsieck 1969: 273, Falkner 1990: 158.

10-17 x 2.7-4 mm, perhaps locally variable

Croatia: Dalmatia, Makarska

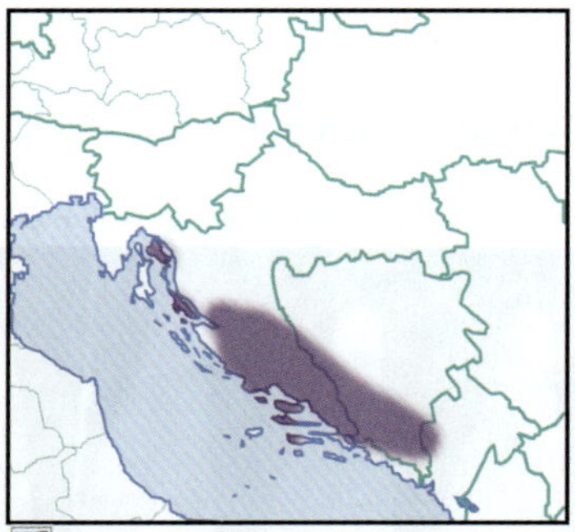

Croatia to NW Montenegro

Delima subcylindrica (Rossmässler, 1836)

Clausilia subcylindrica Rossmässler, 1836 in Rossmässler 1835-1837 (3): 14, Taf. 12

Shell pale reddish horny yellow, almost smooth or weakly striated, 9-10 whorls, apex blunt, apertural margin slightly detached, columellaris prominent, broad and horizontal, principalis not superceeding lunula, lower palatalis often connected with lunula, lunula straight, subcolumellaris visible in the aperture.

Westerlund 1884: 70, Nordsieck 1969: 273.

9-13 x 2.2-2.7 mm

Croatia: Dalmatia, Korčula

Croatia (Dalmatia)

Delima substricta (Pfeiffer, 1866)

Delima amoena (Schmidt, 1868)

Clausilia latilabris var. γ synonym *Clausilia amoena* Pfeiffer, 1848 in L. Pfeiffer 1848: 448

Mentioned with description. Not mentioned as a junior synonym of an available name (as demanded by Art. 11.6), but as a synonym for a subordinate variant for which the Greek letter γ was used. For those who neglect this (as proposed by Kadolsky 2012: 78, this is preliminarily accepted here), the name was eventually made available under Art. 11.6.1 by Schmidt 1868: 60, 62 who used the name as *Clausilia amoena*, but without any bibliographical reference to a previous source and without clear evidence that the same taxon was meant. For the others, the name must be attributed to Schmidt 1868.

Clausilia amoena Schmidt 1868 in Schmidt 1868: 60, 62

Mentioned as "*Cl. amoena* Küst. (teste Parr.)", with description by Schmidt, and no reference to a previous publication.

For those who demand a bibliographical reference for making available a synonym under Art. 11.6.1, the name must be attributed to Schmidt 1868. This applies also to those who do not like to accept Kadolsky's 2012: 78 proposal. For the others, to Pfeiffer 1848, but eventually only if we are definitely dealing with the same taxon. Küster 1875: 77 gave a description and reference to Schmidt 1868: 62.

If the name is attributed to Schmidt, 1868, the name *Cl. substricta* interferes:

Clausilia subcylindrica var. γ synonym *substricta* Charpentier, 1852 in Charpentier 1852: 383

Mentioned with description and attributed to "Parr[eyss]". Not mentioned as a junior synonym of an available name (as demanded by Art. 11.6), but as a synonym for a subordinate variant for which the Greek letter γ was used. For those who neglect this (as proposed by Kadolsky 2012: 78, this is preliminarily accepted here), the name was eventually made available under Art. 11.6.1 by Pfeiffer 1866: 153 who used the name as *Clausilia substricta*, but without bibliographical reference to Charpentier 1852. For the others, the name must be attributed to Pfeiffer 1866.

Clausilia substricta Pfeiffer, 1866 in L. Pfeiffer 1866: 153

Mentioned as a new name, attributed to "Parreyss", with description by Pfeiffer and without reference to Charpentier 1852. Both names may refer to different taxa. For those who demand a bibliographical reference for making available a synonym under Art. 11.6.1, the name must be attributed to Pfeiffer, 1866. This applies also to those who do not like to accept Kadolsky's 2012: 78 proposal. For the others, to Charpentier, 1852, but eventually only if we are definitely dealing with the same taxon.

The problems with the name of this species are created by unclear provisions in Art. 11.6.1 of the Code which is not very helpful in this case.

Croatia: Dalmatia, island of Mljet, near Pola

Massimo Prodan

13-15 x 2.7-3.3 mm

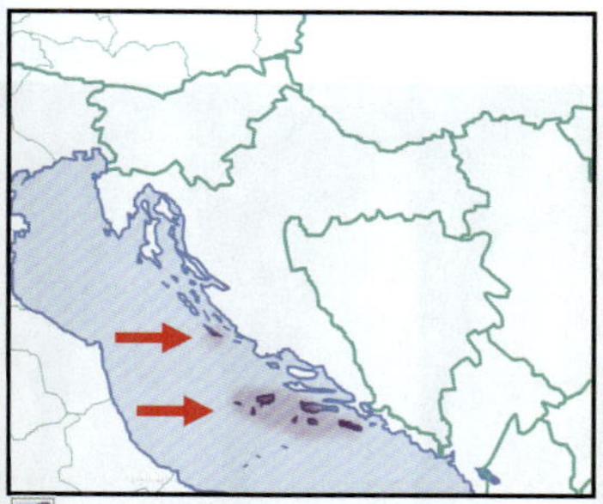

Croatia (Dalmatian islands)

Shell finely ribbed or striated, lunula more or less complete, upper palatalis and subclaustralis more or less separated from lunula, no parallelaris.

Westerlund 1884: 71, Nordsieck 1969: 272 (*Delima substricta*), Nordsieck 1993: 35, Nordsieck 2007: 52 (*D. amoena*).

Delima vidovichii (Pfeiffer, 1846)

Clausilia vidovichii Pfeiffer, 1846 in L. Pfeiffer 1846 (Symb. 3): 95 (*Clausilia Vidovichii*)

12-17 x 3.3-4.3 mm

Croatia: Dalmatia, Brodarica near Šibenik

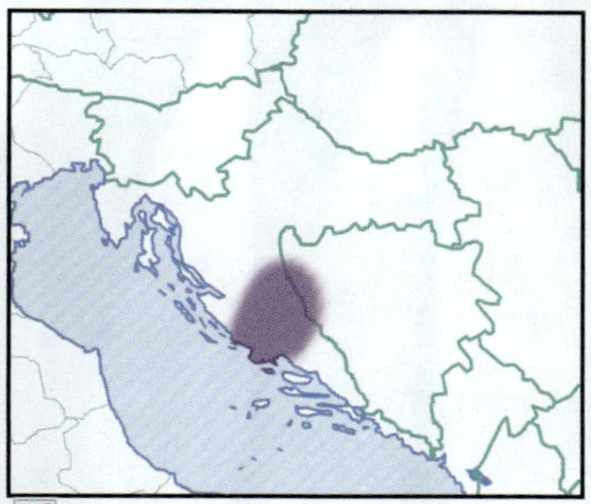

Croatia (Dalmatia), NW Bosnia

Shell very weakly striated, cervix finely ribbed, suture papillated, apertural margin usually not connected at parietal side, columellaris deep inside and oblique, lower palatalis absent, lunula dorsal, oblique and almost straight, subcolumellaris usually weak.

Westerlund 1884: 76, Nordsieck 1969: 274.

Dilataria

Dilataria Vest, 1867 in Vest 1867: 188 *Clausilia (Dilataria)*

Type species *Clausilia succineata* Rossmässler, 1836 (original designation)

A revision would be useful. The species do not differ much from each other, except in their shell size and shape. The classification of *D. marcki* and *D. boettgeriana* as different species is arbitrary.

Dilataria boettgeriana (Paulucci, 1878)

Clausilia boettgeriana Paulucci, 1878 in Paulucci 1878: 40 (*Clausilia Boettgeriana*)

10-16 x 2.7-3.7 mm

Italy: Monti del Matese

Westerlund 1884: 95, Manganelli et al. 1995: 24.

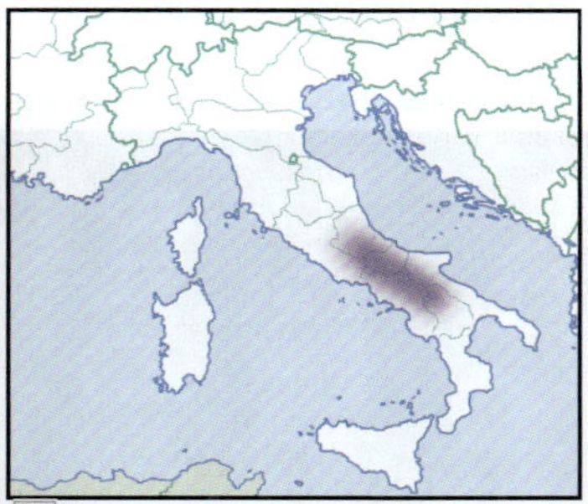

S Italy

Shell olive yellowish, striated, 8-10 whorls, cervix in last section slightly more prominently striated, rounded and often yellowish, aperture with strong white lip, margin connected at parietal side, often slightly detached, principalis and upper palatalis present, subcolumellaris almost invisible in the aperture, clausilium rounded and not broadened in frontal part.

Mountain habitats in beech forests and above the timberline.

Dilataria marcki (Pfeiffer, 1868)

Clausilia marcki Pfeiffer, 1868 in L. Pfeiffer 1868: 408 (*Clausilia Marcki*)

13-17 x 3.3-4.1 mm

Croatia: Plješevica mountains S of Plitvička lakes

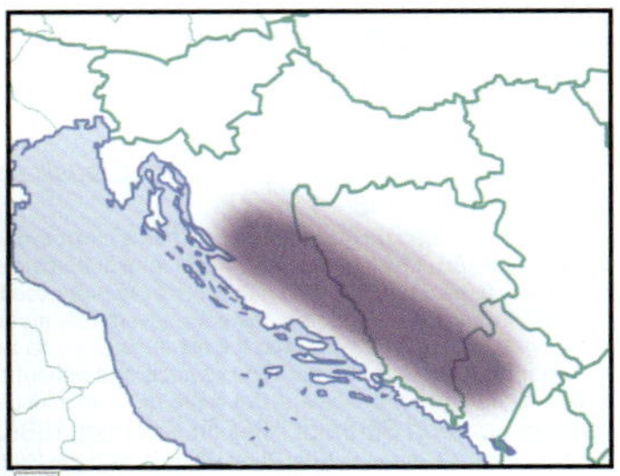

Croatia to Montenegro

Shell reddish horny brown, weakly ribbed or almost smooth, 9-10 whorls, cervix in last section slightly more prominently ribbed, aperture large with broad white lip, margin connected at parietal side but not detached, principalis and at some distance two upper palatal folds (the upper one long and the lower short and often connected to lunula), clausilium rounded and not broadened in frontal part.

Westerlund 1884: 95.

Dilataria succineata (Rossmässler, 1836)

Clausilia succineata Rossmässler, 1836 in Rossmässler 1835-1837 (3): 21, Taf. 12

11-14 x 2.3-2.8 mm, perhaps locally variable

Slovenia: Julian Alps, near Bled

In high altitudes under stones, was found above the timberline.

Westerlund 1884: 95, Manganelli et al. 1995: 24.

Austria and NE Italy to S Croatia

Shell brown yellow to dark horny brown, finely ribbed to almost smooth, 10-12 whorls, apertural margin thickened and connected at parietal side, sometimes with strong white lip, one long and one short palatal fold (the short one below the inner end of the principalis), clausilium broadened and rounded in its frontal section. Several characters seem to display spatial variations.
Animal small and blackish grey.

Erjavecia

Erjavecia Brusina, 1870
in Brusina 1870: 24
Expressly established as a new replacement name for

Type species *Clausilia bergeri* Rossmässler, 1836
(original designation by Vest 1867: 193)

Trigonostoma Vest, 1867 in Vest 1867: 193 *Clausilia (Trigonostoma)* not *Trigonostoma* Blainville, 1825 (Gastropoda: Cancellariidae) and various other authors

Erjavecia bergeri (Rossmässler, 1836)

Clausilia bergeri Rossmässler, 1836 in Rossmässler 1835-1837 (3): 18, Taf. 12 (*Clausilia Bergeri*)

10-12 x 2.3-3 mm

Austria: Oberösterreich, Vorderer Gosausee

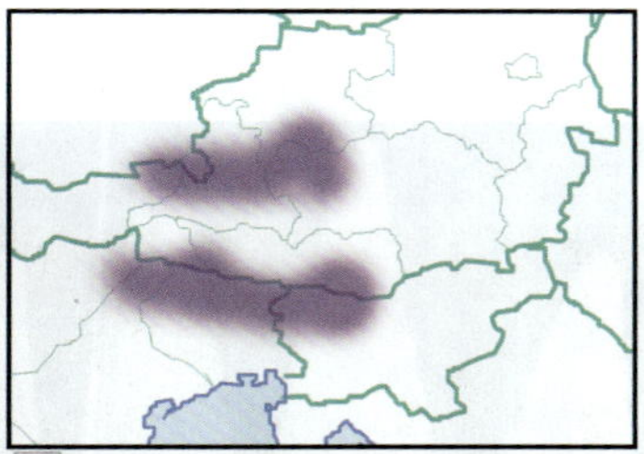

E Alps (Tirol to Slovenia)

Shell violet brown, almost smooth, cervix with prominent basal keel and basal furrow, and a distant and oblique dorsal keel, aperture nearly triangulous with sinulus next to parietalis and pointed basal side ending in a deep basal groove.

Ovoviviparous.

Westerlund 1884: 181, Klemm 1951, Kerney & Cameron 1979: 159, Falkner 1990: 162, Manganelli et al. 1995: 26.

Restricted to rocks exposed to direct sunshine, where the species can stick in the bright sun at steep rocky walls. Adult animals cluster in groups, often in horizontal rock crevices where snails hang down from the top. In large numbers only at S exposed walls, but sometimes also found in small numbers in other habitats, also at N exposed slopes, but only very exceptionally not in rocky habitats. The snails cannot easily cross grasslands or areas covered by herbs to reach other rocky habitats. In mountains between 400 and 2500 m, best conditions between 1200 and 1400 m.

Euxinella

Euxinella Nordsieck, 1973 in Nordsieck 1973: 182

Type species *Euxinella radikae* Nordsieck, 1973 (original designation)

Euxinella radikae Nordsieck, 1973

Euxinella radikae Nordsieck, 1973 in Nordsieck 1973: 183

8-11 x 2.1-2.3 mm

Rep. Makedonija: 3 km from Mavrovo Hanovi to Debar

Dedov & Neubert 2009: 94.

W Rep. Makedonija

Shell yellowish brown, ribbed, 9-10 whorls, last whorl bent to the left side, basal keel strong, dorsal section rounded, aperture detached and protruded, with a narrow umbilicus above the parietal margin, parietalis long and connected to spiralis, columellaris weak, basal furrow usually present, lunula dorsal, but very weak as a fold, no palatal folds, subcolumellaris ends near lunula.

Open and rocky habitats between 800 m and the high alpine zone at 2000 m, usually but not necessarily on limestone.

Euxinella subaii Dedov & Neubert, 2009

Euxinella subaii Dedov & Neubert, 2009 in Dedov & Neubert 2009: 94

Shell brownish with greenish hue, finely ribbed, shiny, 8-10 whorls, basal keel strong, parietalis connected to spiralis with a weak depression in the contact zone, strong palatal callus well visible in the aperture and connected to lower palatalis, lunula dorsal to dorsolateral, subcolumellaris long and running parallel to the basal groove.
Differs from *E. radikae* in its less slender shell, slightly less prominent ribs, its strong palatal callus and a very long subcolumellaris.

9-10 x 2.3-2.5 mm

Found under leaf litter and on wood logs in Fagus forest in 1000 and 1700 m altitude.

Only known from 2 localities, where 2 empty shells were collected.

Not *Albinaria subaii* (Nordsieck, 1984), not *Tsoukatosia subaii* Hunyadi & Szekeres, 2009, both from Greece. This was the third clausiliid species with the name *subaii*. In botany this would not be allowed.

S Rep. Makedonija

Fusulus

Fusulus Fitzinger, 1833 in Fitzinger 1833: 105

Type species *Clausilia interrupta* Pfeiffer, 1828 (subsequent designation by Albers & Martens 1860: XVII)

Contained originally *C. interrupta* and *Pupa fragilis* Draparnaud, 1801 (= currently *Balea perversa* (Linnæus, 1758)).

Fusulus approximans (Schmidt, 1856)

Clausilia approximans Schmidt, 1856 in Schmidt 1857: 35
Clausilia subgracilis Schmidt, 1851 in Schmidt 1851: 183

The name *Clausilia subgracilis* should have priority, but although the synonymy was known (Schmidt 1857: 35, Westerlund 1884: 183, Pfeiffer & Clessin 1881: 402, Nordsieck 1993: 39-40), the junior synonym *approximans* has always been used for this species.
I would try to suppress *Cl. subgracilis* under Art. 23.9. I do not think *Cl. subgracilis* had been used after 1899, but I do not have sufficient references citing *approximans*. The following sources are known to me: Klemm [1973]: 309, Kerney et al. 1983: 231, Frank 1992: 464, Nordsieck 1993: 39, Mildner & Rathmayr 1999: 652, Schileyko 2000: 701, Nordsieck 2001: 21, Kofler & Mildner 2004: 130. I expect more in the Austrian literature. Pure lists of names of species (questionable under Art. 23.9.6): Falkner et al. 2001: 44, Nordsieck 2007: 61.

9-11 x 1.8-2.2 mm

Austria: Kärnten, Hochstuhl

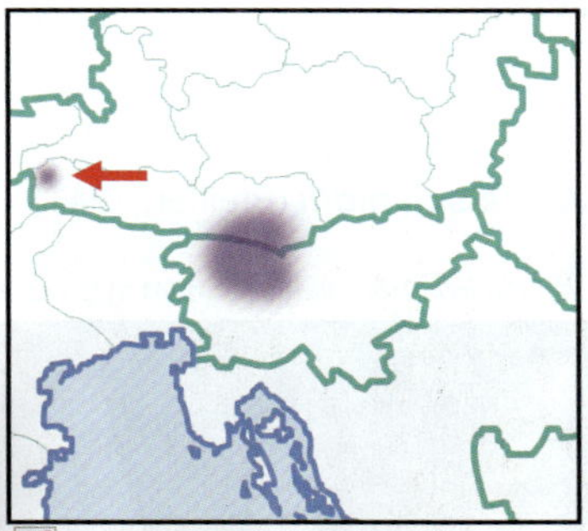

S Austria, NW Slovenia

Shell cherry brown, finely ribbed, 10-13 whorls, cervix more coarsely ribbed, with strong basal keel and deep basal furrow, apertural margin detached, columellaris deep inside, palatal wall strong and parallel to margin, less strong in its middle section, lunula almost straight, subcolumellaris curved and forked in frontal section, visible in an oblique view, clausilium pointed at lower side.

In Austria considered as rare.

Westerlund 1884: 183, Kerney et al. 1983: 231.

Fusulus interruptus (Pfeiffer, 1828)

Without gender agreement: *Fusulus interrupta* (Pfeiffer, 1828)

Clausilia interrupta Pfeiffer, 1828 in C. Pfeiffer 1828: 39

9.5-10.5 x 2.5 mm

Austria: Steiermark, Tarmeben near Peggau

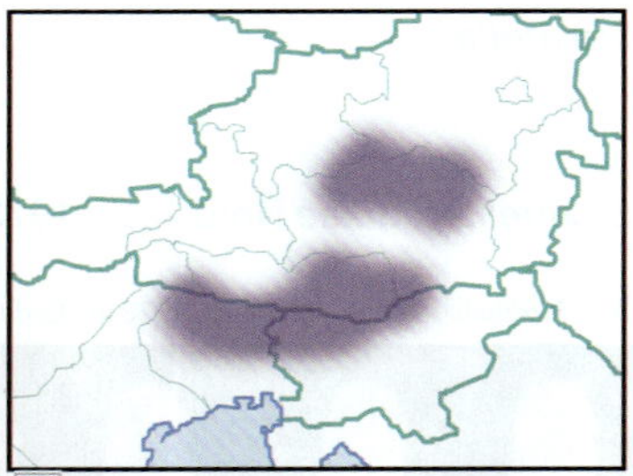

SE Alps

Shell ribbed, aperture attached to the last whorl, margin interrupted, with ribs between parietalis and columellaris.

Under stones, rotten wood and dead leaves in forests, also at rocks and trees. In mountains.

Kerney et al. 1983: 231, Manganelli et al. 1995: 27.

Graciliaria

Graciliaria Bielz, 1867
in E. A. Bielz 1867: 150 *Clausilia (Graciliaria)*

Type species *Clausilia concilians* Bielz, 1853 (subsequent designation by Vest 1867: 192)
(= currently *Graciliaria inserta* (Porro, 1841))

Contained originally *Cl. concilians* and *Cl. filograna*.

Graciliaria inserta (Porro, 1841)

Clausilia inserta Porro, 1841 in Villa & Villa 1841: 58
Name was attributed to Porro in original source. Author of the description was Porro, not Villa & Villa (see footnote p. 54).

11-16 x 2.7-3.5 mm

Romania: Transilvania, Paiana Lunei

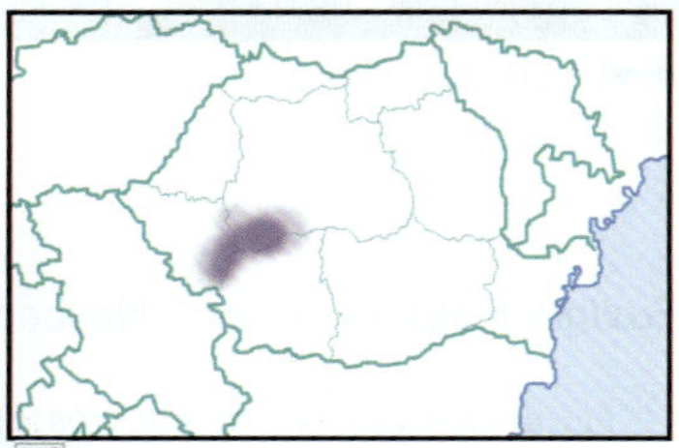

Romania (SW Carpathians)

Shell reddish brown, striated or ribbed, 9-12 whorls, cervix with basal keel and basal furrow, apertural margin detached, with sinulus between parietalis and palatal side, columellaris weak and close to parietalis, no suturalis, principalis short and deeply inside behind clausilium, palatal callus weak, no lower palatalis, lunula dorsal, subcolumellaris hardly visible in the aperture, clausilium broad and deep inside.

Mountain habitats at trees, under wood logs and stones, rarely on the bark.

Westerlund 1884: 178, Grossu 1981: 159 (*Gr. concilians*).

Julica

Julica Nordsieck, 1963
in Nordsieck 1963: 102 *Iphigenia (Julica)*

Type species *Clausilia schmidtii* Pfeiffer, 1841 (original designation)

Julica schmidtii (Pfeiffer, 1841)

Clausilia schmidtii Pfeiffer, 1841 in L. Pfeiffer 1841: 49 (*Clausilia Schmidtii*)

10-14 x 2.4-3 mm

NE Italy: NW of Uccea

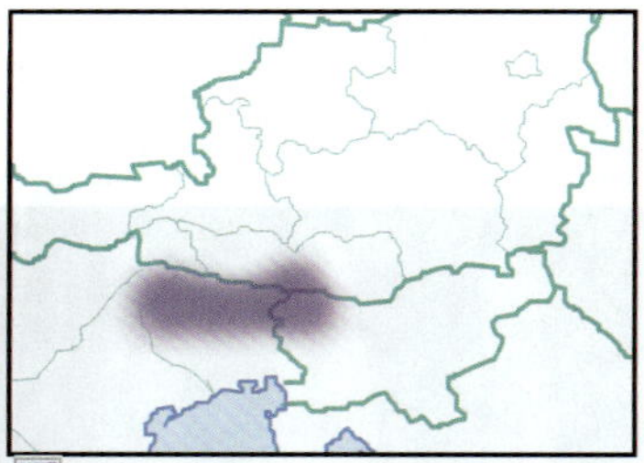

SE Alps

Shell violet brown, smooth or weakly striated, cervix with moderate but pronounced and light basal keel and weak basal impression, aperture with sinulus next to parietalis, palatal wall concave, basal side rounded without groove, two frontal palatal folds. The light basal keel contrasting a dark and nearly smooth shell is characteristic. Local *Clausilia* shells are ribbed.

Rocky habitats.

Westerlund 1884: 183, Falkner 1990: 162, Manganelli et al. 1995: 26.

Lampedusa

Lampedusa Boettger, 1877
in O. Boettger 1877 (Palaeontogr. Sup. 3): 50 *Clausilia (Papillifera (Lampedusa))*

Type species *Clausilia lopadusae* Calcara, 1847
(monotypy)

Contained originally 1 species cited as "*Lampedusae* Calc.". Type not by original designation ("Gruppe der..." construction does not meet the requirements of Art. 67.5 and 68.2).

Holyoak 1986 (proposed to include *Muticaria*), Giusti, Manganelli & Schembri 1995.

Lampedusa imitatrix (Boettger, 1879)

Clausilia imitatrix Boettger, 1879 in O. Boettger 1879: 120, Taf. 3

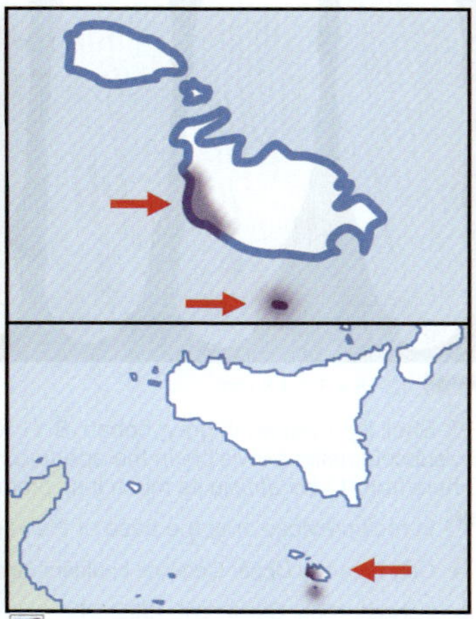

13-19 x 3.2-4.3 mm

Malta: Ħad-Dingli, Kuncizzjoni church

Maltese islands (SW Malta, Filfla)

Shell whitish cream, finely ribbed, 8-10 whorls, cervix with two indistinct keels, basal side of cervix occasionally inflated, apertural margin detached and slightly protruded, parietalis not very prominent and far distant from spiralis, columellaris visible, no suturalis, principalis high and often visible in a perpendicular view, lunula lateral and very oblique, basalis deep inside and not or weakly visible in an oblique view, subcolumellaris not visible in the aperture, clausilium plate long.
Differs from the usually decollated *Muticaria macrostoma* in its weaker dorsal keel, much weaker subcolumellaris, lateral lunula (less folds visible at the basal side) and a strong principalis. *L. melitensis* has a weaker columellaris, no basalis and short clausilium plate.

On bare limestone rocks, often in crevices and at the bases of grasses growing in shallow accumulations of soil on rocks. Occurs sympatrically with *Mut. macrostoma*. On Filfla island under stones or in rock rubble.

Westerlund 1884: 57, Holyoak 1986, Giusti, Manganelli & Schembri 1995: 330.

Lampedusa lopadusae (Calcara, 1846)

Clausilia lopadusae Calcara, 1846 in Calcara 1846: 3 (*Clausilia Lopadusae*)

14-20 x 3.5-4.3 mm

Italy: island of Lampedusa, Guitgia

Italy (Lampedusa and Lampione)

Shell greyish with reddish hue due to dust particles, densely ribbed, ribs white, 9-10 whorls, cervix more coarsely ribbed with moderate basal and dorsal keels, aperture rounded and inside narrow with a very broad margin and lip, parietalis moderate, columellaris not very prominent, lunula small and straight, almost dorsal.

Westerlund 1884: 122, Manganelli et al. 1995: 23, 46.

Lampedusa melitensis (Caruana Gatto, 1892)

Clausilia melitensis Caruana Gatto, 1892 in Caruana Gatto 1892: 149

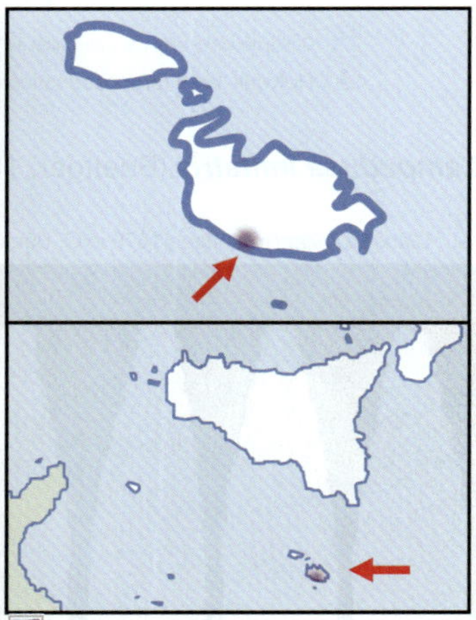

12-18 x 3.2-4.8 mm

Malta: Dingli Cliffs

SW Malta

Shell light yellowish grey, ribbed, 8-11 whorls, aperture detached and slightly protruded, columellaris not visible as a fold, lunula very deep inside, no basalis, subcolumellaris invisible in the aperture, clausilium plate short.
Differs from *L. imitatrix* in its much less prominent columellaris, absence of a basalis and short clausilium plate.

In rock crevices or at the base of the vegetation, also resting on the clay slopes at the foot of the rdum.

Only on a few Upper Coralline boulders detached from the rdum edges in the Rdum il-Maddalena (Rdum Madliena, Dingli Cliffs) area.

Holyoak 1986, Giusti et al. 1995: 343.

Leucostigma

Leucostigma Wagner, 1919
in A. J. Wagner 1919: 145 *Papillifera (Leucostigma)*

Type species *Clausilia leucostigma* Rossmässler, 1836
(absolute tautonymy)
(= currently *Leucostigma candidescens* (Rossmässler, 1835))

Aperture without or with weak spiralis, subcolumellaris strong, lunula dorsal, no frontal upper palatalis (difference to *Muticaria*).

Nordsieck 2011 (revision).

Nordsieck 2011 distinguished about 40 geographic forms at 220 distinct localities which were grouped in six sinistral and three dextral subspecies with exact distributional data (the dextral taxa being counterparts of local sinistral forms). Besides coiling four characters were used to define subspecies: white surface layer, papillated suture, aperture detached or not, and the columellaris. The delimination of the subspecies was argued to have been somewhat arbitrary, local forms were connected to each other by gradients and transitional forms could not be classified unequivocally.
However at some stations sharp transitional gradients between morphologically distinct forms were found. These sites were not analysed for the presence of hybrid zones in the sense of Barton & Hewitt 1985 or Schilthuizen's 1994/1995 studies in *Albinaria hippolyti*. It is possible that we are dealing with partial barriers to reproduction at such zones, and with true subspecies, but this remains to be studied.

Leucostigma candidescens (Rossmässler, 1835)

Clausilia candidescens Rossmässler, 1835 in Rossmässler 1835-1837 (2): 10, Taf. 7

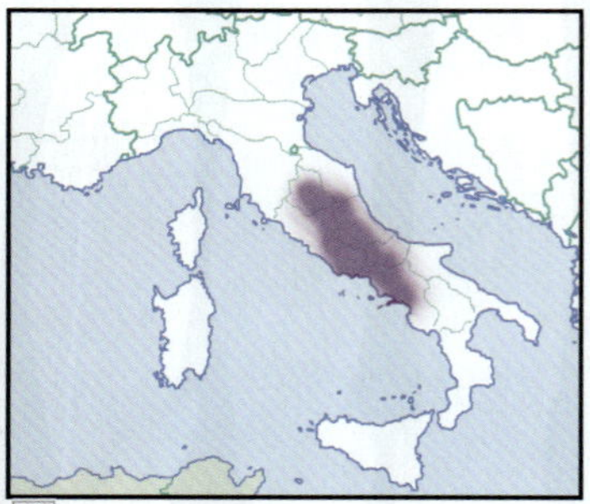

12-22 x 3.2-5 mm, size and shape locally variable

Italy: (1-4) Lazio, between Genatano and Subiaco
(5-8) Perugia

Italy

Shell milky white with violet hue, or brown, apex often brownish, almost smooth, 8-10 whorls, suture white, often papillated, cervix striated and nearly rounded, aperture rounded with white, yellowish white or brownish lip, margin not always connected at parietal side, parietalis moderate, columellaris usually not very prominent, oblique, inside weakly curved, lunula dorsal, subcolumellaris usually strong. Colour, suture and aperture display spatial variations.

Limestone rocks in mountains.

Westerlund 1884: 172, 173, Alzona 1971: 96, Falkner 1990: 156, Manganelli et al. 1995: 23.

Macedonica

Macedonica Boettger, 1877
in O. Boettger 1877 (Palaeontogr.): 30 *Clausilia (Triloba (Macedonica))*

Type species *Clausilia macedonica* Rossmässler, 1839 (monotypy)

Type species not originally designated (Gruppe der... construction does not meet the requirements of Art. 67.5 and 68.2). First Reviser to select between the simultaneously established names *Clausilia (Marpessa (Serbica))* and *Clausilia (Triloba (Macedonica))* was Thiele 1931: 545.

This genus needs a revision. The differences between the *Macedonica* species given by Nordsieck 1974 and 1977 are not really convincing. Several units classified as species differ from each other only gradually. Morphological gradients between local populations were not researched, but it seems likely that they exist. Nordsieck's classification is kept here, but it is probably not tenable.

Macedonica brabeneci Nordsieck, 1977

Macedonica brabeneci Nordsieck, 1977 in Nordsieck 1977: 88

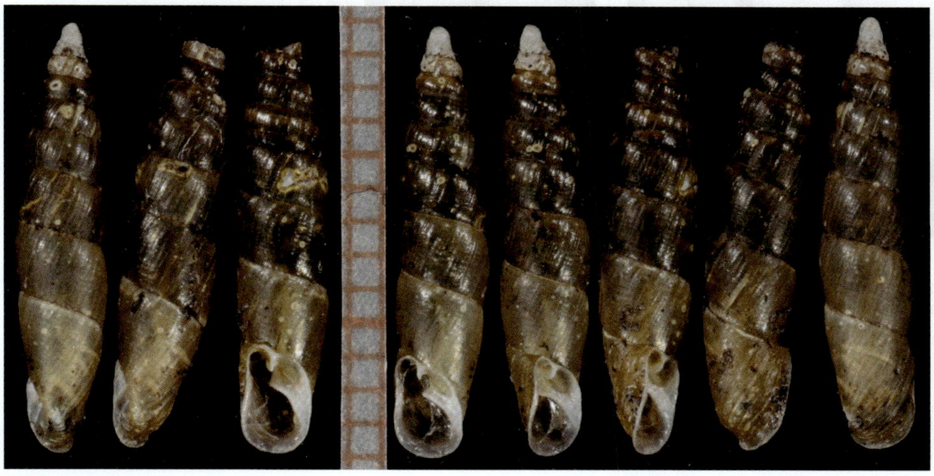

13-16.5 x 2.8-3.4 mm — Bulgaria: Smoljan — Museum Budapest HNHM

Irikov 2006: 81, Irikov & Mollov 2006: 778.

Differences to *M. marginata* are only gradual and of a minor degree, Nordsieck 1977 did not provide evidence for the presence of independent species, the same applies to *M. thasia* and *M. ypsilon*.

S Bulgaria

Shell olive green, often decollated, weakly striated, cervix very finely striated, cervix with weak basal keel, parietalis strong and short, usually not reaching spiralis, columellaris prominent and nearly horizontal, lunula dorsolateral, upper and lower palatalis present, no middle palatalis, no lunula, subcolumellaris sometimes visible in a perpendicular view. Differs from *Mac. marginata* mainly in shell shape (more slender), colour, a less prominent rib sculpture, simple apertural margin and slightly weaker clausiliar apparatus. All these differences are gradual.

Macedonica frauenfeldi (Rossmässler, 1856)

Clausilia frauenfeldi Rossmässler, 1856 in Rossmässler 1854-1859 (15-16): 57, Taf. 75 (*Clausilia Frauenfeldi*)
Date was Apr 1856. Name also mentioned by Pfeiffer 1856 (Malak. Bl. 3, Oct 1856): 180.

17-24 x 4-5.5 mm — Bulgaria: W Stara Planina mountains, near Vraca — Atanas Irikov

Moderately humid habitats, calciphile, in Bulgaria up to 800 m.

Westerlund 1884: 48, Pavlović 1912: 90, Tomić 1959: 43, Damjanov & Likharev 1975: 177, Hubenov 2007.

Bulgaria, SE Serbia, Rep. of Makedonija

Shell cherry brown, often with bluish hue, almost smooth or weakly striated, shiny, 10-11 whorls, cervix striated with weak basal keel and more prominent and oblique dorsal keel, apertural margin not always connected at parietal side, 3 palatal folds, all almost entirely visible in the aperture, subcolumellaris visible in an oblique view.

Macedonica guicciardii (Roth, 1856)

Clausilia guicciardii Roth, 1856 in Roth 1856: 3 (*Clausilia Guicciardii*)

Shell violetish grey, weakly striated, shiny, suture white, cervix ribbed, 2 short palatal folds.

21-25 x 4.3-5.2 mm

Westerlund 1884: 10.

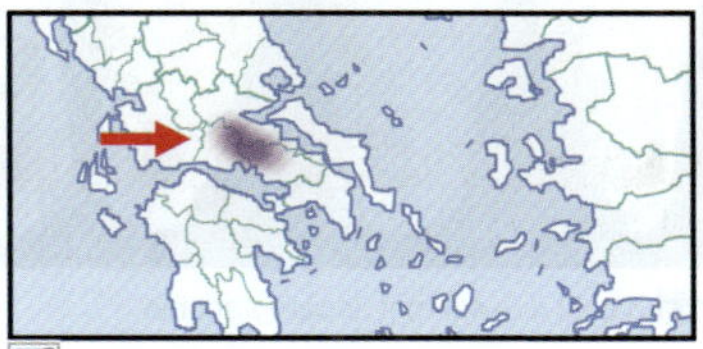

Central Greece (Parnasós)

Macedonica hartmuti Irikov, 2003

Macedonica hartmuti Irikov, 2003 in Irikov 2003: 143

11-13 x 2.9-3.3 mm — Bulgaria: W Rodopi mountains, Dobrostan ridge — (7-8) Atanas Irikov

S Bulgaria (W Rodopi mountains)

Shell horny to violetish brown with yellowish to reddish brown first whorls, often with whitish surface layer, coarsely ribbed, ribs broad, not prominent and very widely spaced (8-12 ribs on penultimate whorl), 9-10 whorls, cervix with distinct basal keel, dorsal section rounded, aperture detached, margin narrow and slightly reflected, parietalis reaching margin, columellaris very prominent and horizontal, no lunula, principalis strong, upper palatalis curved and visible in the aperture, subcolumellaris indistinct. Clausilium plate broad, two-lobed with a narrow cut in between.

The ribs are characteristical and look like those of hybrids in narrow (less than 200 m) transitional zones separating compact ranges of smooth and very prominently ribbed (percostate) forms, for example in *Albinaria* in Crete.

On S exposed, smooth and nearly vertical limestones with scarce herbaceous vegetation and without much snow in the winter. Sometimes active during the winter, 1300 m.

Only known from one locality.

Irikov & Mollov 2006: 778.

Macedonica macedonica (Rossmässler, 1839)

Clausilia macedonica Rossmässler, 1839 in Rossmässler 1838-1844 (9-10): 18, Taf. 48
Macedonica macedonica pirinensis Jaeckel, 1954 in Jaeckel 1954: 63

Shell violet grey or violetish brown, finely ribbed, apex often brownish yellow, cervix densely ribbed, suturalis long, frontal upper, middle and lower palatalis present, lower palatalis often visible in a perpendicular view, subcolumellaris often visible in a perpendicular view.

19-23 x 3.7-4.8 mm

Westerlund 1884: 17, Nordsieck 1974: 141, Damjanov & Likharev 1975: 176.

May contain *M. ypsilon* and *M. thasia* (see comments there).
The taxon *pirinensis* Jaeckel is sometimes recognized as an independent species. It differs from *M. macedonica* only gradually and in characters such as cervical rib prominence and density, shell colour, and width of the apertural margin. The subcolumellaris is often visible in *M. macedonica*, not in *pirinensis*. It is possible that authors misunderstood the names used for the palatal folds (Jaeckel 1954 thought that there might be a third palatal fold above the basalis, but it is possible that previous authors called the basalis the lowest palatal fold).

SE Rep. Makedonija, SW Bulgaria, N Greece

Macedonica marginata (Rossmässler, 1835)

Clausilia marginata Rossmässler, 1835 in Rossmässler 1835-1837 (2): 12, Taf. 7
Rossmässler 1835: 12 did not refer to a previously mentioned name.

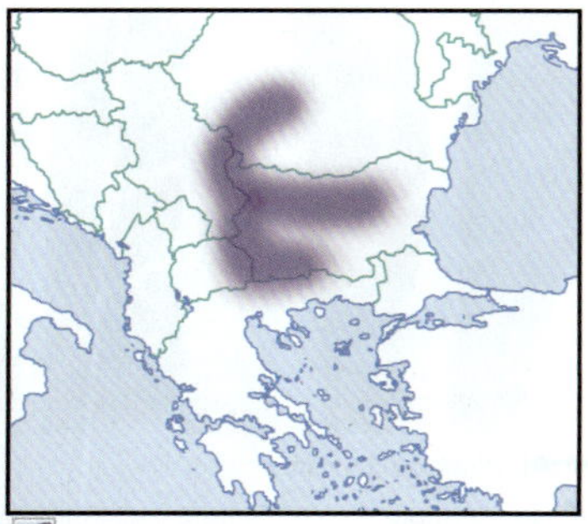

12-23 x 3-4.5 mm, locally variable — (1-4) Romania: Transilvania, Obu Moecin, (5-8) Bulgaria: W Rhodopi, Zabardo village — (5-8) Atanas Irikov

Romania: Retezat mountains, Cheile Butii — Jozef Grego

SE Europe

Shell horny to dark brown, almost smooth or weakly striated, weakly shiny, 9-12 whorls, cervix striated and almost rounded, aperture often with yellowish brown or white lip, margin usually detached, parietalis prominent and usually reaches margin, columellaris prominent, almost horizontal, upper and lower palatalis present and often visible in a perpendicular view, no lunula, subcolumellaris visible in a perpendicular view. Several characters seem to be locally variable, in S Bulgaria populations with a dorsal lunula were found.

In Bulgaria usually in rocky habitats, but can also be found in leaf litter of beech forests, more rarely in oak forests, calciphile, 1000-2000 m.

Westerlund 1884: 19, Damjanov & Likharev 1975: 174, Falkner 1990: 158, Irikov & Mollov 2006: 776, Hubenov 2007.

May contain *brabeneci*, *thasia* and *ypsilon*.

Macedonica martae Sajó, 1968

Macedonica marginata martae Sajó, 1968 in Sajó 1968: 449

S Bulgaria (Pirin mountains)

Shell pale violetish brown, with regular and widely spaced but indistinct ribs, 10-11 whorls, cervix rounded, aperture not detached, margin light brownish, rounded and connected at parietal side, parietalis weak, columellaris very weak, principalis, frontal upper palatalis and basalis very weak and deep inside, subcolumellaris not visible in the aperture.

13.8-16.8 x 3.9-4.1 mm

Bulgaria: Pirin mountains, near summit Vichren, 2800 m

Atanas Irikov

Limestones in the alpine zone of Pirin mountains in 2400-2800 m altitude.

A rare species known only from a small range.

Damjanov & Likharev 1975: 178, Hubenov 2007.

Macedonica pindica Gittenberger, 2002

Macedonica pindica Gittenberger, 2002 in Gittenberger 2002: 130

N Greece

Shell bluish-grey or brownish with patchy whitish surface layer, coarsely and irregularly striated or with coarse white ribs, 10-13 whorls, cervix rounded without keels, apertural margin detached and reflected, parietalis moderate, columellaris prominent, upper palatalis present, principalis prominent, subcolumellaris hardly visible in a very oblique view.

20-26 x 4.7-5.5 mm

Greece: Makedonía, Spilea Portitsa Faragi

Macedonica pinteri Sajó, 1968

Macedonica marginata pinteri Sajó, 1968 in Sajó 1968: 447

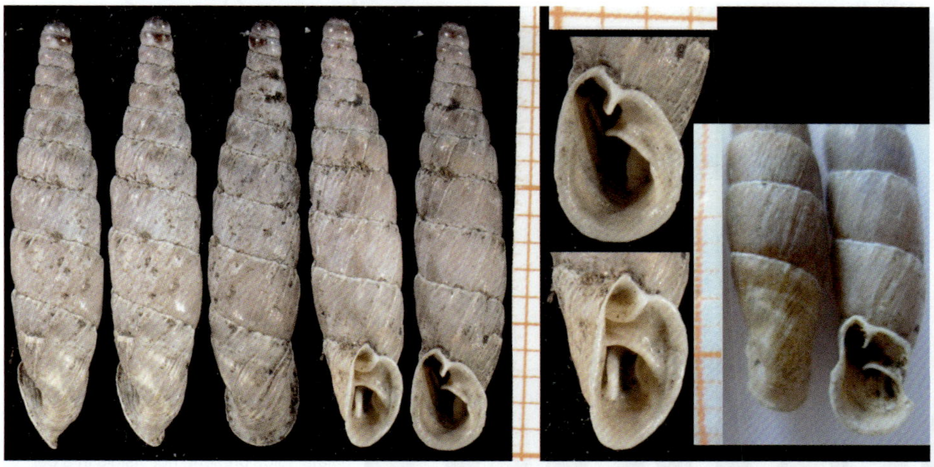

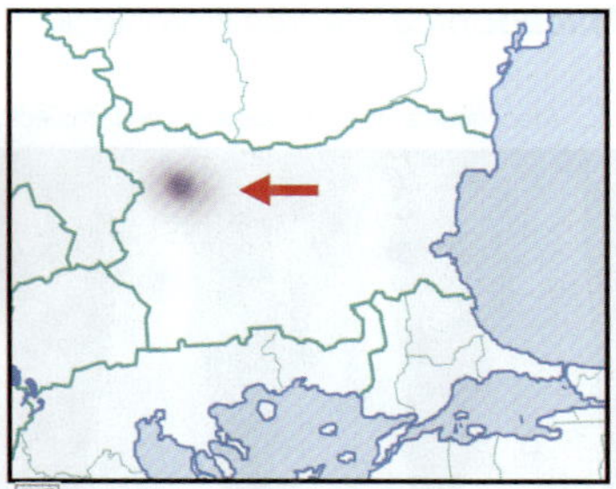

NW Bulgaria

Shell pale yellowish brown, irregularly and very coarsely ribbed, 10-12 whorls, cervix rounded without keels, aperture not always detached but margin connected at parietal side, parietalis strong, ends apically next to spiralis, columellaris very prominent and horizontal, long principalis, extremely strong basalis and subcolumellaris visible in a perpendicular view. Clausilium plate two-lobed with incision, tilted and bent backwards.

15-18 x 3.4-3.9 mm

(1-7) Bulgaria: W Stara Planina, Vratzata, (8-9) Bulgaria: W Stara Planina, near Vratza

(8-9) Atanas Irikov

Damjanov & Likharev 1975: 178.

Limestone rocks in mountains.

Macedonica pirotana (Pavlović, 1912)

Clausiliastra pirotana Pavlović, 1912 in Pavlović 1912: 82

15-19 x 2.8-3.2 mm

Serbia: Pirot

Museum Budapest HNHM

SE Serbia

Shell weakly ribbed or striated, 11-12 whorls, aperture with strong parietalis and columellaris, 4 folds on the palatal side (principalis, frontal upper and lower palatalis and basalis) all visible from outside, basal and palatal margins broad, clausilium two-lobed, ending in a hook at the inner (columellar) lobe, the other lobe ends more broadly.

Macedonica schatzmayri (Wagner, 1914)

Serbica schatzmayri Wagner, 1914 in Sturany & Wagner 1915: 71, Taf. 16

15-25 x 4-4.7 mm

Greece: Makedonía, Pangéo mountains, 1900 m

Was found in mountains, above 1700 m.

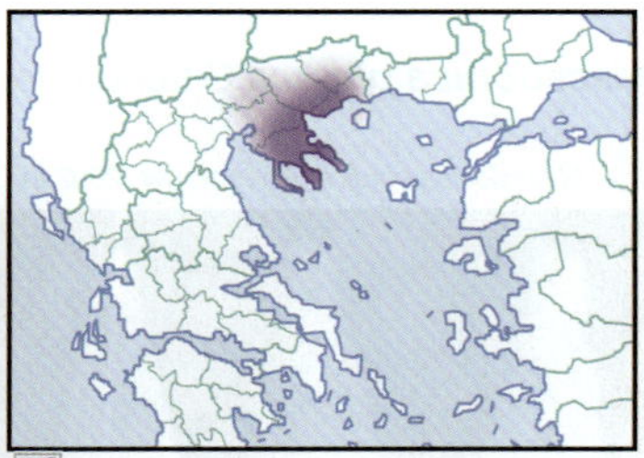

N Greece (Makedonia)

Shell dark reddish brown, or with light violetish surface layer, weakly striated, last whorl more densely and more prominently striated, 9-15 whorls with white suture, cervix rounded or with weak basal keel, aperture inside brown, margin yellowish white and slightly reflected, connected at parietal side and usually detached, sinulus slightly directed upwards, parietalis long, low and separated from the end of the deep spiralis, columellaris strongly curved and horizontal, no lunula, principalis, upper and lower palatalis present and often visible from outside, subcolumellaris visible in an oblique view. Clausilium slightly S-shaped, plate broad and two-lobed, columellar lobe bluntly pointed, outer lobe shorter and broader.

Macedonica slavica Nordsieck, 1974

Macedonica macedonica slavica Nordsieck, 1974 in Nordsieck 1974: 142

18-25 x 4-4.8 mm

Rep. Makedonija: between Prilep and Trojaci

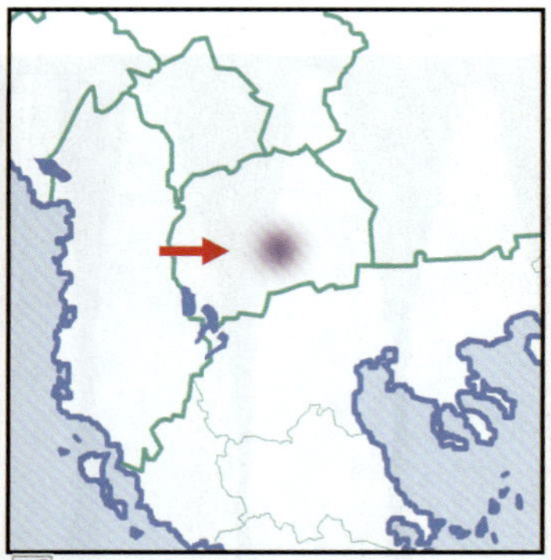

Rep. Makedonija

Shell irregularly striated, cervix more prominently striated, 12-14 whorls, cervix rounded, parietalis weak, usually not reaching spiralis, columellaris prominent, 3 palatal folds present, not visible in a perpendicular view, subcolumellaris often visible in a perpendicular view.

At humid conglomerate rocks in a dry valley.

Initially described as a subspecies of *M. macedonica*, from which it seems to differ only in very few characters, could belong to this species.

Macedonica teodorae Irikov, 2006

Macedonica teodorae Irikov, 2006 in Irikov 2006: 82

15-20 x 4-4.5 mm

Only known from one locality.

Irikov & Mollov 2006: 779.

S Bulgaria: Rodopi mountains, near Zabardo village, 1600 m
(1-3) Jivko Stoilov, (4-7) Atanas Irikov

S Bulgaria

Shell horny brown, coarsely and very regularly ribbed (15-24 ribs/penultimate whorl), 10-12 weakly convex whorls with light suture, aperture slightly detached, margin with lip and moderately reflected, parietalis short and not reaching spiralis, columellaris prominent and often horizontal, no lunula, lower and upper palatalis strong, middle palatalis may occur, subcolumellaris visible in a perpendicular view.

On S exposed limestones in an isolated outcrop, in shady parts of the rocks covered by *Geranium* and moss, 1600 m.

Macedonica thasia Nordsieck, 1977

Macedonica thasia Nordsieck, 1977 in Nordsieck 1977: 86

16-21 x 3.7-4.6 mm

Greece: Thásos, Panagía

Museum Budapest HNHM

Greece (Thásos)

Shell violetish brown with white suture, weakly striated, last whorl finely ribbed, aperture detached, margin broadened, parietalis strong and short, usually not reaching spiralis, columellaris prominent and nearly horizontal, lunula dorsolateral to lateral, lower palatalis longer than upper palatalis, often a short middle palatalis is also present, subcolumellaris usually not visible in a perpendicular view.

At limestone walls in mountains.

Status as an independent species very weak. Nordsieck 1977 gave no substantial differences to *M. macedonica* (both were supposed to differ in absence/weak presence of a whitish surface layer, rib prominence at the initial whorls and by the sometimes-presence of a middle palatalis in *M. thasia*), no differences to *M. ypsilon* were given, which had almost the identical description.

Macedonica ypsilon Nordsieck, 1977

Macedonica ypsilon Nordsieck, 1977 in Nordsieck 1977: 87

14-19 x 3.4-4.1 mm

Greece: Makedonía, Loutrá Eleftherón SW Orfanó

Museum Budapest HNHM

Greece (Pangéo mountains)

Shell horny brown, very weakly striated, cervix finely ribbed, aperture detached, parietalis strong and short, usually not reaching spiralis, columellaris prominent and nearly horizontal, lunula dorsolateral to lateral, lower palatalis longer than upper palatalis, often a short middle palatalis is also present, subcolumellaris usually visible in a perpendicular view.

Status as an independent species very weak. Nordsieck 1977 gave no substantial differences to *M. macedonica* (both were supposed to differ in absence/weak presence of a whitish surface layer and by the sometimes-presence of a middle palatalis in *M. ypsilon*), no differences to *M. thasia* were given, which had almost the identical description. Was claimed to differ from *M. marginata* in its simple apertural margin, presence of a middle palatalis and a clausilium plate like in *M. macedonica*.

Macedonica zilchi Urbański, 1972

Macedonica zilchi Urbański, 1972 in Urbański 1972: 251

19-25 x 5-6.2 mm

Bulgaria: Rodopi, (1-7) near Dyavolsko Gardo cave, (8-9) Trigrad village
(8-9) Jivko Stoilov

Prefers humid and shady rocks, also at steep outcrops, feeds on few species of lichens and fungus with local distribution.

A rare species, restricted to Trigradski Karst area.

Irikov 2006: 81, Irikov & Mollov 2006: 779, Irikov & Erőss 2008.

S Bulgaria

Shell light reddish brown with light bluish opaque surface layer, translucent and weakly shiny, prominently and irregularly striated, cervix coarsely ribbed, 10-13 whorls with whitish suture, cervix with weak basal keel, aperture detached, inside yellowish brown, margin fragile and weakly reflected, parietalis strong and distant from spiralis, columellaris prominent, nearly horizontal, principalis well developed, frontal upper palatalis and basalis deeper inside, no lunula, subcolumellaris not visible in a perpendicular view. Clausilium S-shaped, two-lobed, outer lobe broader and rounded.

Macrogastra

Macrogastra Hartmann, 1841
in Hartmann 1844: 48

Type species *Pupa ventricosa* Draparnaud, 1801 (monotypy)

Hartmann just mentioned a name *Macrogastra ventricosa*, uncommented, but since he meant explicitly a clausiliid species the generic name is unambiguous.

Species composition follows Nordsieck 2006: 56. It is not easy to understand the diagnoses given there.

Macrogastra asphaltina (Rossmässler, 1836)

Clausilia ventricosa var. synonym *Clausilia asphaltina* Rossmässler, 1836 in Rossmässler 1835-1837 (4): 21, 26, Taf. 18

Mentioned without description, but with figure. Caption to Fig. 276 reads "*Cl. ventr.* var. (*Cl. asphaltina* Z.)".
Not mentioned as a junior synonym of an available name (as demanded by Art. 11.6), but as a synonym for a subordinate variant for which no name was used. For those who neglect this (as proposed by Kadolsky 2012: 78, this is preliminarily accepted here), the name was eventually made available under Art. 11.6.1 by Strobel 1844: 330 who used the name as *Clausilia ventricosa* var. *asphaltina*, with bibliographical reference to Rossmässler 1836. For the others, the name must be attributed to Strobel 1844.
Strobel 1844: 330 listed (without description) a form "*Clausilia ventricosa* var. *asphaltina* Ziegl. - Rossm. Ic.? Innsbruck", it is unclear where the "?" should refer to, probably to Innsbruck and not to Rossm. Ic., so that this would have been a name used as valid with bibliographical reference to Rossmässler 1835-1837. If so, then Strobel 1844 made Rossmässler's 1836 synonym available under Art. 11.6.1.
If not, then the name must possibly be referred to Gredler 1856: 139 who used *Clausilia asphaltina* with description.

17-19 x 4-4.3 mm

Austria: Kärnten, Karawanken, Rosenbach

Strobel 1855: 164, 171, Falkner 1990: 160, Manganelli et al. 1995: 26.

S Alps

Shell large, moderately densely ribbed, cervix flattened without basal keel, aperture at lower side well rounded, columellaris in the aperture branched like a horizontal K (not always well visible), subcolumellaris visible in a perpendicular or slightly oblique view of the aperture, various small folds between parietalis and columellaris (not in all but at least in some individuals in a population), no palatal folds.
Differs from *M. ventricosa* in its basal keel being cut off, its aperture at lower side being slightly more rounded (palatal margin slightly more concave), stronger palatal callus, and the presence of small (interlamellar) folds between parietalis and columellaris in many specimens in a population.

In coniferous mountain forests.

Macrogastra badia (Pfeiffer, 1828)

Clausilia badia Pfeiffer, 1828 in C. Pfeiffer 1828: 41

12-16 x 2.8-4 mm Austria: Steiermark, Alpe Ursula

Central Europe

Shell reddish brown, ribbed or prominently striated, shiny, 9-11 whorls, cervix equally densely striated, with two keels and a weak basal furrow, columellaris oblique, principalis passing lunula distally, palatal wall almost parallel to the margin, thickened by subcolumellaris, subcolumellaris hardly visible in the aperture.

Humid forests, under soil litter and on tree trunks and mossy rocks. In mountains, sensitive against higher temperatures.

Oviparous, batches of 1-3 eggs (1.4 x 1.5 mm) are laid in May and June, in dense moss or pieces of rotting timber, juveniles hatch asynchronously (1-2 days difference) after 16-19 days, adult size is reached after 7-8 months in the laboratory, in the wild juveniles reach 6-8 whorls before hibernating, growth is completed in the summer of the next year, reproduction starts in the third year of life, life span exceeds 3 years.

Threatened by habitat destruction and forest management. Critically Endangered in Poland, Endangered in Czech Republic.

Westerlund 1884: 204, Kerney & Cameron 1979: 164, Manganelli et al. 1995: 26, Nordsieck 2003, Nordsieck 2006: 57, Maltz & Pokryszko 2009.

Macrogastra basileensis (Rossmässler, 1835)
Macrogastra attenuata (Rossmässler, 1835)
Macrogastra lineolata (Held, 1836)

Clausilia ventricosa var. *basileensis* Rossmässler, 1835 in Rossmässler 1835-1837 (1): 78, 80; (2): 10

On p. 78 classified as a variety with a short description, one rank below the species. On p. 80 another taxon *dedecora* was classified one rank below *basileensis*, two ranks below the species and thus, infrasubspecific. On p. 10 of Heft 2 (same date, Aug. 1835) mentioned in a form as if it were treated as a species. In Heft 4 (1836) on p. 22 *Cl. basileensis* mentioned as a synonym of a polynominal name (!) Clausilia ventricosa var. minor interlamellari pliculato, on p. 26 (plate figure captions) used as a name of the taxon of Fig. 279. The name was made available at the first occasion in 1835 on p. 78. Nordsieck 1993 expressed the view that the name was originally mentioned as a synonym of *Cl. ventricosa*, and that it was validated later under Art. 11.6.1. My interpretation is slightly different, but both views come to the conclusion that the name *Clausilia basileensis* Rossmässler, 1835 is available.

Clausilia plicatula synonym *Clausilia attenuata* Rossmässler, 1835 in Rossmässler 1835-1837 (1): 80

Mentioned with short description. Also mentioned as a synonym by Rossmässler 1838-1844 (7-8) [1838]: 18 Taf. 34 (Fig. 474), from "Risuta in Illyrien". Westerlund 1884: 203 referred bibliographically to this 1838 source, but used *attenuata* at infrasubspecific rank. I do not know a source from before 1961 to have used *attenuata* and referred to Rossmässler 1835. If regarded as available and regarded as conspecific with *basileensis*, the name *basileensis* should take precedence under Art. 24.1 because *attenuata* was not "proposed" at all by Rossmässler 1835, it was only cited as a synonym. The first usage was later.

Clausilia lineolata Held, 1836 in Held 1836: 275

Clausilia lineolata var. *attenuata* Schmidt, 1856 in Schmidt 1857: 16, Taf. 9

Mentioned without bibliographical references to previous publications. Name also mentioned on plate.
For those who demand a bibliographical reference for making available a synonym under Art. 11.6.1, *C. lineolata* Held, 1836 has precedence over *C. attenuata* Schmidt, 1856. In any case *basileensis* has precedence over *lineolata*, independent from any interpretation of Art. 11.6.1.
For those who regard *attenuata* Rossmässler, 1835 as having been validated under Art. 11.6.1 by Schmidt in 1856, both names *basileensis* and *attenuata* were first used at subspecific rank. The First Reviser to select precedence between both names was probably Westerlund 1884: 203 who selected *basileensis*.

Shell smaller than *M. ventricosa*, 1-2 weak folds between parietalis and columellaris, columellaris in the aperture branched like a horizontal K, aperture at lower part well curved, no basal furrow, palatal callus often present and developed like a white indistinct line, subcolumellaris visible in a perpendicular view of the aperture.

Humid forests, under leaf litter, boulders and stones, and between rocks covered by mosses, predominantly in mountains, in Switzerland up to 1400 m.

Reproduction probably between August and October, 1-11 eggs (size 1-2 mm) are laid in clutches below bark or in soil crevices, juveniles hatch after 20-30 days, they grow at a rate of approximately 1 whorl per month and reach 4 mm size after 3 months.

13-16 x 3.3-3.7 mm Italy: Lombardia, Lago di Garda, near Novazza

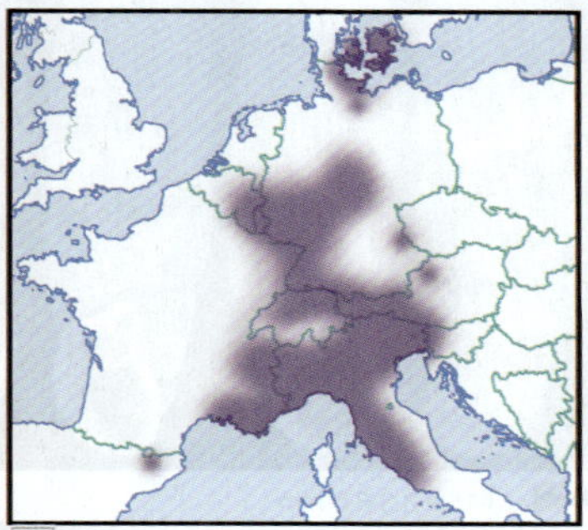

Europe

Loosjes 1941 (life cycle), Kerney et al. 1983: 220, Nordsieck 1993: 37, Manganelli et al. 1995: 26, Turner et al. 1998: 216, Taruella et al. 2009.

Sometimes called *M. lineolata*, sometimes *M. attenuata*, sometimes *M. lineolata attenuata*, sometimes *M. attenuata lineolata*, sometimes in combinations using the name *basileensis*.
This looks quite chaotic.
Prior to 1993 this species was known as *M. lineolata*.
Nordsieck 1993 proposed to use the name *attenuata* for the species, *lineolata* for its northern so-called subspecies and announced to bring up a case to the Commission to suppress *basileensis* - which has never happened. This would probably not have been successful since the stability of names was distorted anyway: *lineolata* was not used for the species any more by Nordsieck 1993 (and 60 % of the subsequent authors), only for a subspecies, and *attenuata* had not been used for the species until 1993. *Cl. basileensis* was used in various sources after 1899 (Klemm 1973: 333, Kofler 1977: 68, Stummer 1983: 125, Frank 1992: 458), so it cannot be suppressed under Art. 23.9. It remained unexplained why Nordsieck 1993 proposed to use *attenuata* and not *basileensis*.
A revision of this species with a helpful and reproducible presentation of the regional distribution of character states (Kerney et al. 1983 did not report a frontal upper palatalis, in Italy the taxon seems to possess one) would be useful, as well as aligning the name with the commonly accepted rules of naming animals (for which purpose the ICZN Code could be used). The name *basileensis* seems to have precedence and I am asking myself why this name has not been used since long. Nordsieck's 1993 proposal did not bring a solution, neither in the understanding of the taxonomic background nor in the name.

Macrogastra densestriata (Rossmässler, 1836)

Clausilia densestriata Rossmässler, 1836 in Rossmässler 1835-1837 (4): 21, Taf. 18 Type locality near Klagenfurt, Kärnten.

12-16 x 3-3.6 mm

Austria: Steiermark, Leutschach

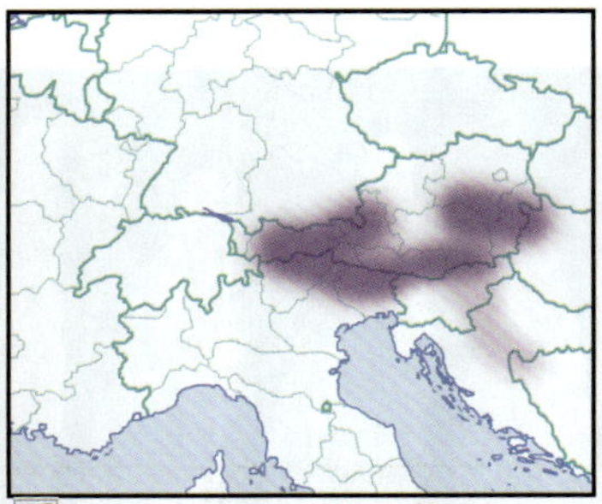

E Alps

Like *M. attenuata*, but aperture rounded, palatal callus stronger, folds between lamellas more pronounced, frontal lower palatalis present, shell in the southern populations usually densely ribbed, but in the N Alps the ribs are wider spaced.

Humid and shady rocky habitats, usually in mountain woodlands.

Kerney & Cameron 1979: 163, Falkner 1990: 161, Manganelli et al. 1995: 26.

There are reports for NE Italy, but not for Slovenia, some sources suggest Croatia and W Bosnia... more detailed research on the local variations of shell characters in *Macrogastra* species would be useful to understand better the situation and to develop methods for reliable species identification in all regions where the genus occurs.

Macrogastra latestriata (Schmidt, 1856)
Macrogastra borealis (Boettger, 1878)

Clausilia latestriata Schmidt, 1856 in Schmidt 1857: 27

Not *Clausilia alboguttulata* var. B synonym *Clausilia latestriata* as mentioned by Küster 1850 ["1847"]: 70 (= currently *Charpentieria itala* (Martens, 1824))

Küster mentioned two synonyms of his var. B, both attributed to Charpentier (*Clausilia ornata* var. *latestriata* and *Clausilia latestriata*) (with a description given for var. B, and a figured specimen).

Not mentioned as a junior synonym of an available name (as demanded by Art. 11.6), but as a synonym for a subordinate variant for which the letter B was used. For those who neglect this (as proposed by Kadolsky 2012: 78, this is preliminarily accepted here), the name was eventually made available under Art. 11.6.1 by Pfeiffer 1868: 470 who used the name as *Clausilia ornata* var. *latestriata*, attributed to Charpentier, but without bibliographical reference to a previous source. For the others, the name must be attributed to Pfeiffer 1868.

Pfeiffer 1868: 470, 495, 508 used the name *latestriata* at 3 occasions for 3 different taxa, without bibliographical reference and description for Küster's name (the one on p. 470 was attributed to Charpentier). The types for Küster's taxon under Art. 72.4.3 would probably be those of Pfeiffer, because Küster's figured specimens were cited with the name var. B and not with one of the two names *latestriata* in the published synonymy.

I did not find a citation where a bibliographical reference to Küster 1847 [true date 1850] was given, which would have made this name available for those who require a bibliographical reference for validating a synonym under Art. 11.6.1. Küster's name has been used after 1899 (Nordsieck 2007: 54, *Charpentieria itala latestriata*).

Clausilia latestriata var. *borealis* Boettger, 1878 in O. Boettger 1878 (Nachr.Bl. DMG 10): 136

Maybe there are other names that could interfere, such as *Clausilia schwabi* or *Clausilia rugulosa*, often mentioned as synonyms and occasionally used for varieties.

For those who require a bibliographical reference for validating a synonym under Art. 11.6.1, the name must be *Macrogastra latestriata* (Schmidt, 1856). For the others, the name could possibly be *Macrogastra borealis* (Boettger, 1878), but the job is not finished and they might be successful in finding older synonyms.

13-15 x 3.3-3.6 mm

Poland: Beskid Niski mountains, Pulawy Dolne

Karolina Chaniecka

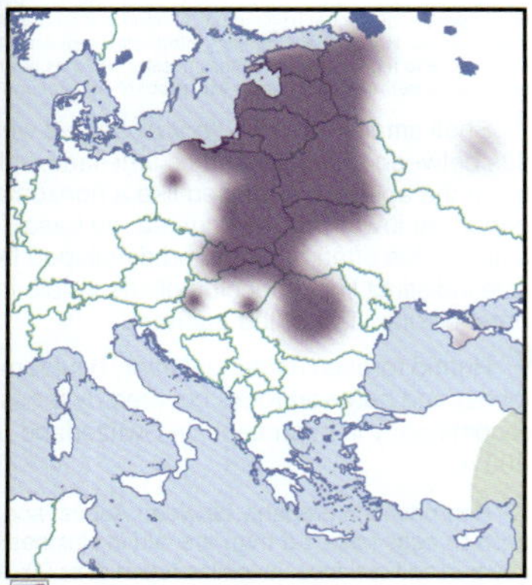

Central and E Europe

Columellaris deeper and weaker than in *M. densestriata*, and less densely ribbed than other *Macrogastra* species (35-45 ribs in the penultimate whorl), palatal callus very weak, aperture relatively circular.

Up to 1200 m, lower forest zone, upland forests. One of the most common clausiliids in SE Poland.

Westerlund 1884: 207 (*latestriata*), Kerney et al. 1983: 223 (*latestriata*), Sulikowska-Drozd 2005: 70 (*borealis*).

Slovakia: Poloniny National Park

Jozef Grego

Macrogastra mellae (Stabile, 1864)

Clausilia mellae Stabile, 1864 in Stabile 1864: 89 (*Clausilia (Iphigenia) Mellæ*)

10-13 x 2.7-3.6 mm

N Italy: Novara near Lago Maggiore

SW Alps

Shell cherry brown, densely ribbed, on the last two whorls often crossed by fine spiral lines forming a reticulate pattern, 10-13 whorls, cervix with basal keel, aperture broad and rounded, columellaris white, weakly forked inwards, principalis widely passing lunula, palatal wall parallel to margin, subcolumellaris visible in an oblique view.

Westerlund 1884: 203, Manganelli et al. 1995: 26 (subspecies of *Mac. attenuata*), Nordsieck 2003, Nordsieck 2006: 10.

Macrogastra plicatula (Draparnaud, 1801)

Pupa plicatula Draparnaud, 1801 in Draparnaud 1801: 64

11-14 x 2.8-3.2 mm

Italy: Piemont, Santuario di Oropa near Biella

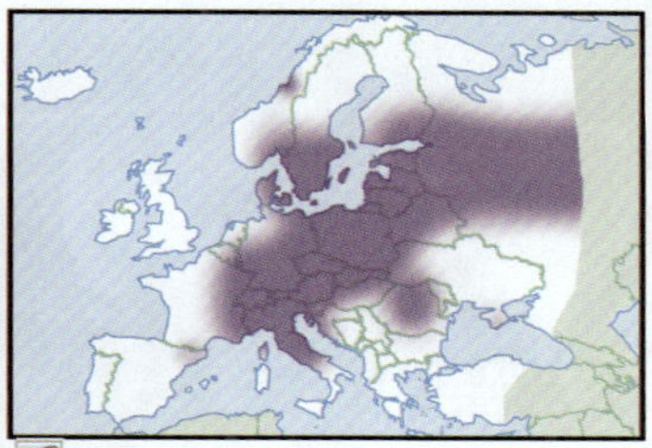

Europe to central Russia

Shell reddish brown, aperture elliptical, smaller and more slender than *M. attenuata*, basal keel not very prominent (unlike in *Clausilia* species), folds present between parietalis and columellaris, no frontal lower palatalis, subcolumellaris visible in a perpendicular view.

Deciduous and coniferous forests, at rocks and trees, preferably in humid habitats, also at rocks and walls outside forests. Prefers calcareous substrate. In Switzerland up to 2300 m.

Austria: Tirol, near Brandstatt

30-40 eggs (diameter 1.5 mm) are laid from May to June and August to October, juveniles hatch after 18-20 days, full shell size and maturity are reached at the end of the second year. Gassies 1849 reported eggs in May and June, Moquin-Tandon 1855 and Germain 1930 between August and September/October, suggesting that two periods of reproduction could be present in this species. It is possible that adult specimens lay eggs in spring, while offspring of the last year which reaches maturity in the course of the summer season, begins reproducing in late summer.

Gassies 1849: 129, Moquin-Tandon 1855: 340, Germain 1930: 365, Kerney & Cameron 1979: 164, Falkner 1990: 160, Manganelli et al. 1995: 26, Turner et al. 1998: 217, Sulikowska-Drozd 2005: 69.

Macrogastra rolphii (Turton, 1826)

Clausilia Rolphii Turton, 1826 in Turton 1826: 565

Regarded by Turton 1826 as conspecific with *Clausilia plicatula*, equipped with a description, literally taken the name was made available as a regular new name because Turton did not expressly specify which name he used for the species. Turton 1831: 71 used *plicatula* for the species and classified *rolphii* as a synonym. ICZN Opinion 336: available from the 1831 publication. This must be corrected to 1826 (Falkner et al. 2002: 113, Bank 2011).

Clausilia portensis Luso da Silva, 1872 in Luso da Silva 1872: 260

11-15 x 3.2-3.9 mm

W Germany: Rheinland, Siebengebirge, Löwenburg

W Europe

Shell reddish brown, smaller than *M. ventricosa*, more densely ribbed (50-70 ribs on penultimate whorl, in Portugal denser), 1-3 small folds present between parietalis and columellaris, no palatal callus and no palatal folds.

Some 10 eggs are laid in May and June under dead leaves.

Local in Britain, some isolated colonies are at risk, threatened by habitat destruction.

Humid deciduous forests, in mossy ground litter, among rocks and under fallen timber, not ascending trees, also in more open habitats among grass. Avoids coniferous plantations. Usually on calcareous substrate, but tolerates acid soils. In Portugal in dark and humid places, more or less deeply hidden among decaying leaves, under stones and tree trunks, also buried below soil surface.

Gassies 1849: 128, Nobre 1941: 158, Gittenberger & Ripken 1998, Norsieck 2006: 56 (included *portensis*), De Oliveira 2007.

Macrogastra tumida (Rossmässler, 1835)

Clausilia ventricosa synonym *Clausilia tumida* Rossmässler, 1835 in Rossmässler 1835-1837 (1): 78
Mentioned with a description for *Cl. ventricosa*.
Clausilia tumida Rossmässler, 1836 in Rossmässler 1835-1837 (3): 21
Name difficult to interpret. Rossmässler 1836: 21 gave no direct statement that he did not regard *Cl. tumida* as a valid name, implying that he indirectly treated it as a name for a species, and referred to Rossmässler 1835: 78 bibliographically. Rossmässler 1836 thus made the name available from Rossmässler 1835 under Art. 11.6.1.

13-17 x 3.5-4.4 mm, locally variable

SW Poland ("Silesia")

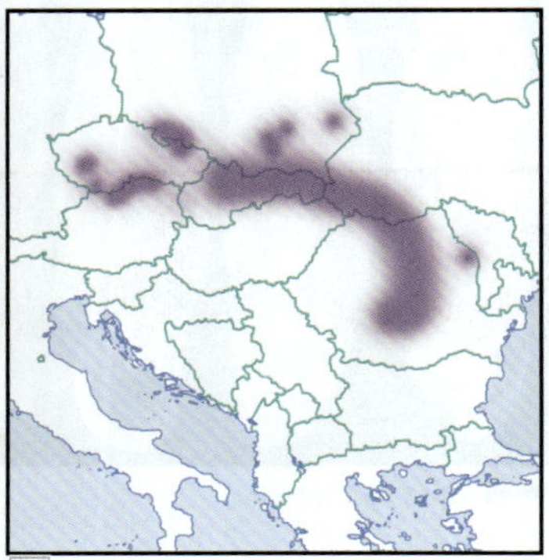

Carpathians

Shell brown, regularly ribbed, silky shiny, 9-12 whorls, cervix with usually weak dorsal and basal keels and a basal furrow, columellaris prominent and curved, often with a weak branch inwards and a short and indistinct branch towards the margin, often with a small fold between parietalis and columellaris, principalis widely passing lunula, palatal callus parallel to margin, thickened on upper and lower side, often absent, subcolumellaris curved in its frontal section and running almost horizontally, visible in an oblique view.

Mountain forests in leaf litter, under stones and rotting timber, up to 2100 m.

Westerlund 1884: 201, Grossu 1981: 187, Falkner et al. 2001: 41, Sulikowska-Drozd 2005: 70.

Slovakia: Slovenská Ľupča

Slovakia: Slovenská Ľupča
Jozef Grego

Macrogastra ventricosa (Draparnaud, 1801)

Pupa ventricosa Draparnaud, 1801 in Draparnaud 1801: 62

17-19 x 3.8-4.4 mm

Germany: Schleswig-Holstein, near Dänisch-Nienhof

Humid forests, under leaf litter, boulders, stones, and between rocks covered by mosses, usually in mountains. In Switzerland up to 1800 m. Climbs up tree trunks and rocks during wet weather.

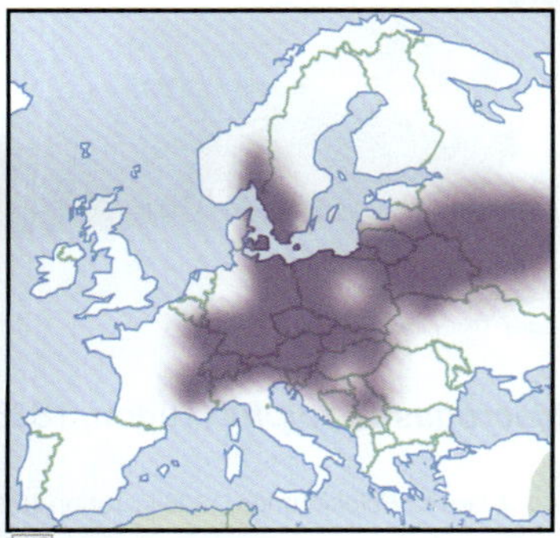

Europe to central Russia

Shell large, moderately densely ribbed, cervix with basal keel, aperture at lower side smoothly curved, subcolumellaris visible in a slightly oblique (nearly perpendicular) view of the aperture, no folds between parietalis and columellaris, no palatal folds.
Differs from *Balea biplicata* in not having a basal furrow in the aperture. *M. asphaltina* has a flatter cervix, a moderate palatal callus and many shells have small folds between parietalis and columellaris.

Mentissa

Mentissa Adams & Adams, 1855 in Adams & Adams 1858: 185 *Clausilia (Mentissa)*

Type species *Clausilia canalifera* Rossmässler, 1836 (subsequent designation by Albers & Martens 1860: XVII, 285)

Contained originally 6 species.

Bajdashnikov 1990 (2 papers).

Mentissa canalifera (Rossmässler, 1836)

Clausilia canalifera Rossmässler, 1836 in Rossmässler 1835-1837 (3): 17, Taf. 12

12-20 x 3-4.2 mm

Ukraine: Crimea, Karabel

Ukraine (Crimea)

Shell brown, finely striated or ribbed, 11-13 whorls, cervix slightly more prominently ribbed, with moderate basal and weak dorsal keel, parietalis moderate, columellaris weak, not reaching margin and often not well visible in a perpendicular view, more inside higher, 3 palatal folds, no lunula, subcolumellaris weak.

Westerlund 1884: 27, Bajdashnikov 1990 (Zool. Zh. 69 (1)): 28.

Mentissa gracilicosta (Rossmässler, 1836)

Clausilia gracilicosta Rossmässler, 1836 in Rossmässler 1835-1837 (3): 18, Taf. 12

15-23 x 3.1-4.5 mm

Ukraine: Crimea, Sevastopol

Ukraine (S Crimea)

Shell variable displaying local variants, more or less densely ribbed, usually smaller than *M. velutina*.

Baidashnikov 1990 (both papers).

Taxonomic status unclear, there seem to exist intermediates between *M. gracilicosta* and *M. velutina*.

Mentissa velutina Baidashnikov, 1990

Mentissa velutina Baidashnikov, 1990 in Baidashnikov 1990 (Zool. Zh. 69 (1)): 21

Author was spelled Baidashnikov in the Latin transcription in the original work.

Shell slightly larger than *M. gracilicosta*, weakly striated or almost smooth, aperture with prominent white lip.

17-22 x 3.7-4.5 mm

Lives parapatrically with *M. gracilicosta*, there seem to exist intermediate forms. Taxonomic situation should be verified.

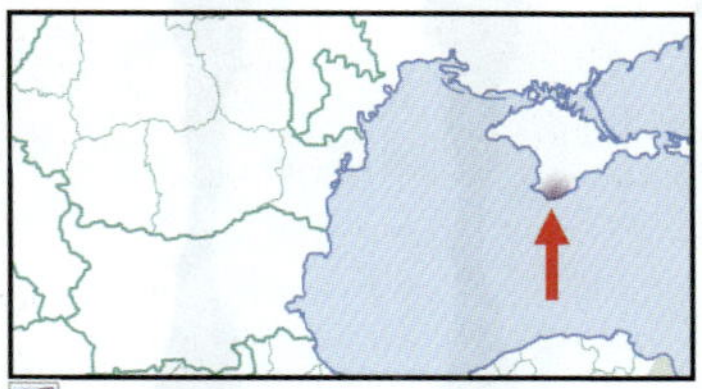

Ukraine (S Crimea)

Micridyla

Micridyla Nordsieck, 1973
in Nordsieck 1973: 181 *Idyla (Micridyla)*

Type species *Idyla pinteri* Nordsieck, 1973
(original designation)

Micridyla pinteri (Nordsieck, 1973)

Idyla pinteri Nordsieck, 1973 in Nordsieck 1973: 181

6-13 x 2-3.2 mm, seems to be locally very variable

Bulgaria: W Rodopi mountains, near Ravnogor

Petrophilic and calcereous species. Prefers humid and shady limestones. Between 600 and 1700 m.

Only known from a small range in the W Rodopi mountains.

Irikov & Mollov 2006: 780.

S Bulgaria (W Rodopi mountains)

Shell horny brown, finely ribbed, 8-11 weakly convex whorls, cervix with weak basal keel and a weak basal furrow, dorsal section rounded, aperture detached, parietalis usually short, no spiralis, columellaris not very prominent, lunula dorsolateral, subcolumellaris visible in the aperture. Clausilium plate rhomboid and pointed with weak edge at the outer side, distal part with weak edge, inside slightly thickened.

Montenegrina

Montenegrina Boettger, 1877
in O. Boettger 1877 (Pal. Sup. 3): 37 *Clausilia (Delima (Montenegrina))*

Type species *Clausilia cattaroensis* Rossmässler, 1835
(subsequent designation by Lindholm 1924: 57)

Contained originally 6 species. Type not by original designation (Gruppe der... construction does not meet the requirements of Art. 67.5 and 68.2).

Montenegro to W Rep. Makedonija, W Greek Makedonia and Ipiros.

Nordsieck 1972: 1 (revision), Nordsieck 1974 (revision), Uit de Weerd 2004: 21 (molecular phylogeny), Nordsieck 2009 (revision).

Nordsieck 2009 proposed to subdivide the genus and established various species groups based on assumed relationships as inferred from taxonomically relevant shell characters and distribution. The previous classification was only slightly modified. In 7 groups he recognized 22 species and 58 additional subspecies. The author listed accurately the characters to distinguish the taxa, but then he gave only lists of names of species and subspecies, without explaining how exactly they differed from each other, how they could be identified, where exactly they were distributed and why they were classified at these ranks. Characters to distinguish forms were only given for the 7 groups.
Those who are interested in identifying snails in this group need to consult at least 80 descriptions of species and subspecies and to check numerous sources and hundreds of localities to get an idea about the distributional ranges of certain forms or character expressions - a high obstacle that makes it almost impossible to understand the diversity.

A comprehensive review of this group with maps to understand local distributions and variations of the taxonomically important characters, which can all be transferred into numerical values, and gaps in the ranges as well as in the collection record, would be highly appreciated and a prerequisite to understand the diversity in this genus.
Dedov & Neubert 2009: 90 pointed out that no information was available to see whether we might be dealing with networks of intergradating populations, and I agree with them in that Nordsieck's 2009 revision did not solve the problem. Perhaps we have only 7 species with high degrees of spatial character variations? Who knows.
I took the descriptions of the species from their nominate forms, occasionally from Westerlund's 1884 diagnoses, and did not check all the many many additional descriptions of taxa for details on the local variations of the involved forms. I regard 3 of Nordsieck's species as too weakly justified, so this account has 19 species.

Montenegrina apfelbecki (Sturany, 1907)

Clausilia apfelbecki Sturany, 1907 in Sturany 1907: 233

12-16 x 3-3.7 mm

N Albania: 11 km from Bogë to Theth, 1800 m

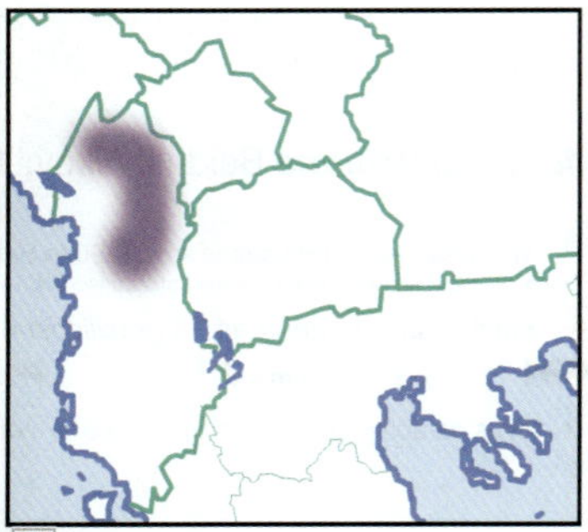

N Albania

Dhora & Welter-Schultes 1996: 131, Dhora 2004: 145, Erőss et al. 2006.

Shell greenish-yellow to yellowish brown with light brown apex, diffuse shiny, almost smooth, last whorl finely striated, 8-11 strongly convex whorls with deep white suture, last whorl inflated between palatal fold and suture, basal keel crest-like with narrowly open umbilicus, aperture detached, with thick white margin, parietalis low and separated from spiralis, columellaris not prominent, lunula thickened and usually straight, frontal upper palatalis long, almost reaching margin, subcolumellaris weakly visible in the aperture.

Montenegrina cattaroensis (Rossmässler, 1835)

Clausilia cattaroensis Rossmässler, 1835 in Rossmässler 1835-1837 (2): 8, Taf. 7

The name was attributed to Ziegler, the main text of the description was witten by Rossmässler in his style. On p. 9 after the subtitles "Thier" and "Aufenthalt" Rossmässler quoted a text written by Ziegler who compared the species with *Clausilia laevissima*. This passage belonged to the original description. In a strict application of Art. 50.1.1 the name in such a case must be attributed to the author of the work.

15-25 x 3-4.7 mm

Montenegro: Lovćen pass

Montenegro

Shell light yellowish brown, upper whorls darker, almost smooth, cervix striated, nearly rounded or with weak basal and dorsal keels, apertural margin sharp and not connected at parietal side, frontal upper palatalis present, lower palatalis long, lunula dorsolateral, principalis usually visible in a perpendicular view of the aperture.

Westerlund 1884: 53, Falkner 1990: 156.

Montenegrina dennisi Gittenberger, 2002

Montenegrina dennisi Gittenberger, 2002 in Gittenberger 2002: 134

15-21 x 4.3-5.2 mm

Greece: Grevená, near Zákas

Uit de Weerd 2004: 25.

N Greece

Shell yellowish brown, last whorl lighter, weakly striated or almost smooth, 9-11 whorls, cervix striated shortly near aperture, rounded without keel, apertural margin broad and reflected, in some populations slightly detached, in others not connected at parietal side, columellaris weak, usually no upper palatalis, lunula dorsal, basalis low and short, solcalis short and prominent, subcolumellaris clearly visible in an oblique view.

Montenegrina dofleini (Wagner, 1928)

Delima dofleini Wagner, 1928 in Hesse 1928: 19

Name and description were both attributed to Wagner in the original source.

Montenegrina kastoriae Nordsieck, 1972 in Nordsieck 1972: 34

16-19 x 4-4.5 mm

Greece: Makedonia, near Kastoria

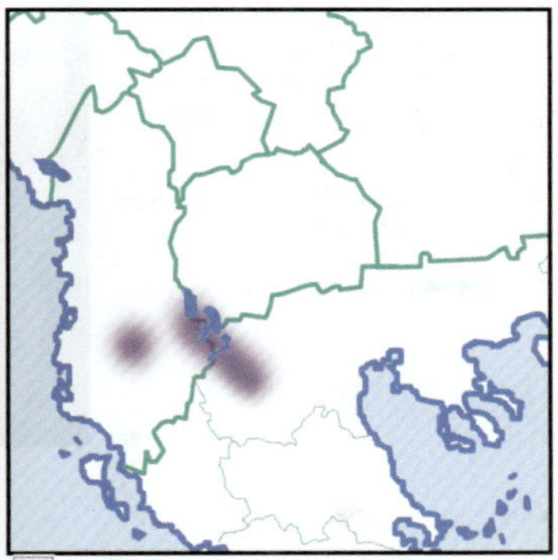

Albania, NW Greece

Shell light horny brown, shiny, finely and indistinctly striated, slightly more prominently at the cervix, cervix usually rounded, aperture not detached, connected at parietal side by a thin callus, margin distinctly reflected, parietalis prominent, columellaris not very prominent, ascends steeply.

Nordsieck 1972, 2007 and 2009 classified *kastoriae* as a species, Nordsieck 1974 and 1988 as a subspecies of *M. dofleini*. Páll-Gergely 2010: 152 did not see any characters that would justify to regard both as separate species.

Montenegrina drimmeri Fehér & Szekeres, 2006

Montenegrina drimmeri Fehér & Szekeres, 2006 in Erőss, Fehér & Szekeres 2006: 182

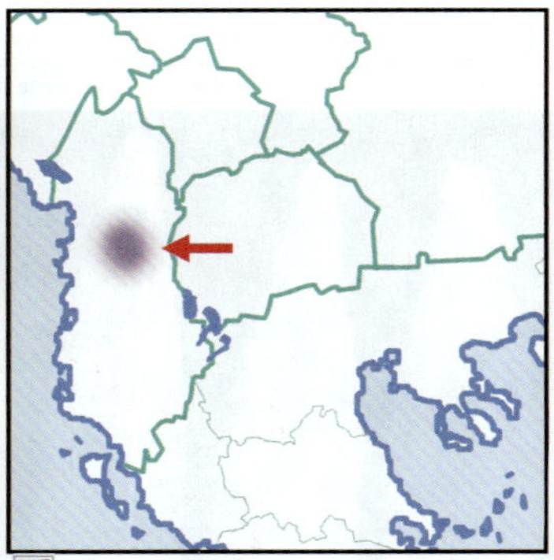

11-16 x 3.8-4.7 mm

Albania: Lumi i Varoshit near Peshkopi

Central Albania

Shell white with light horny brownish first whorls, translucent, finely ribbed, 8-10 convex whorls, lower 3 whorls much inflated and predominating, last whorl with spiral furrow, inflated between upper palatalis and suture, basal keel strong, umbilicus open, aperture not detached, margin broadly reflected and connected by a callus at parietal side, parietalis strong, columellaris deep inside and not visible in a perpendicular view, lunula dorsolateral, basalis present, frontal upper palatalis very strong, subcolumellaris only visible in a very oblique view, clausilium plate not visible in the aperture.

Montenegrina fuchsi Brandt, 1961

Montenegrina fuchsi Brandt, 1961 in Brandt 1961: 2

Shell light brown, weakly and irregularly striated, cervix more prominently striated, 8-10 whorls, aperture detached and very strongly protruded, upper margin of last section with edge, aperture inside brown, margin sharp and weakly reflected, parietalis short, columellaris located in the uppermost section in the aperture, lunula dorsal to dorsolateral, no frontal palatal fold, no basalis, subcolumellaris visible in an oblique view. Clausilium plate narrow, groove-shaped and rounded at its terminal end.

11-15 x 2.7-3.4 mm

Albania: Vejusa bridge, road Tepelenë-Kelcyrë

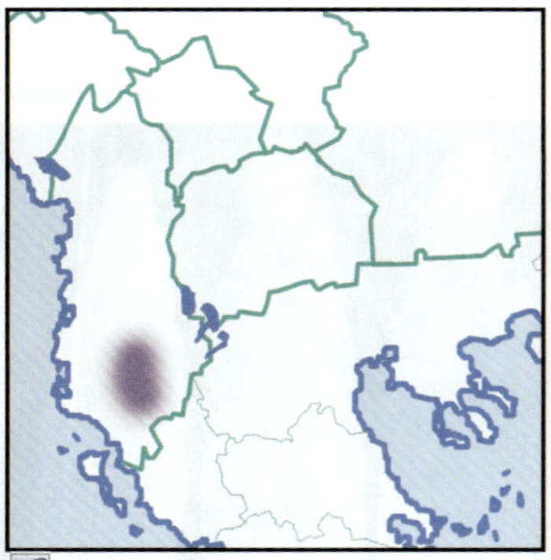

S Albania

Montenegrina helvola (Küster, 1860)

Clausilia helvola Küster, 1860 in Küster 1847: 176 (figured on Taf. 19 Fig. 15-18)

Taf. 19 was from 1853 but without name printed on plate (wrapper was not preserved and is not published work), text from 1860.

Shell light reddish brown, upper whorls darker, indistinctly striated, 9-12 whorls, cervix more prominently striated, dorsal keel moderate, basal keel strong, apertural margin detached and protruded, columellaris deep inside and oblique, visible only in an oblique view of the aperture, two palatal folds, lunula and lower palatalis S-shaped, subcolumellaris very weak.

Westerlund 1884: 40, Dhora & Welter-Schultes 1996: 132.

15-18 x 3.2-3.8 mm

Albania: Krujë

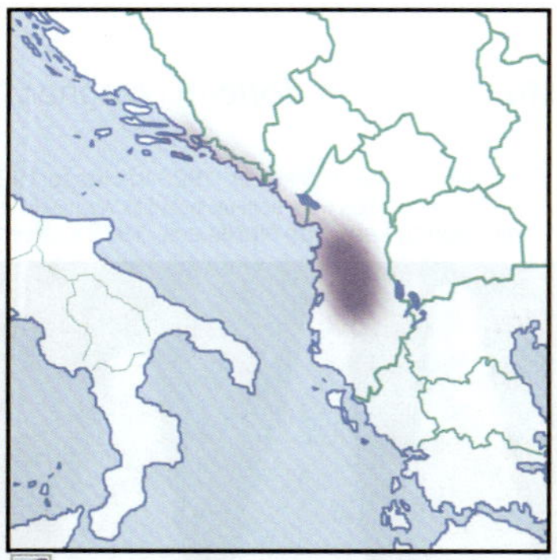

Albania, possibly also Dalmatia

Montenegrina hiltrudae Nordsieck, 1972

Montenegrina hiltrudae Nordsieck, 1972 in Nordsieck 1972: 33

Shell brownish, ribbed, 9-11 whorls, cervix rounded with basal keel, aperture usually detached, inside narrow, margin broadly reflected, frontal upper palatalis present, lunula dorsal, basalis weak and usually disconnected from lunula, subcolumellaris not visible in a perpendicular view.
Differs from *M. rugilabris* in its ribbed shell, detached aperture and invisible subcolumellaris.

14-17.5 x 4-5 mm

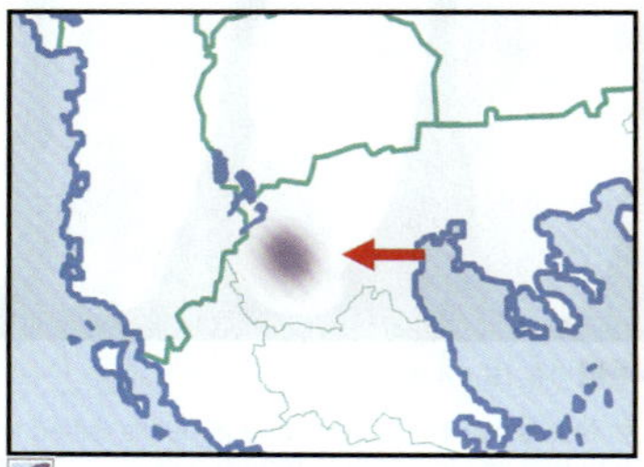

N Greece (W Makedonía)

Montenegrina janinensis (Mousson, 1859)

Clausilia janinensis Mousson, 1859 in Mousson 1859: 276

12-18 x 2.7-3.5 mm

Greece: Ipiros, near Lake Ioánnina

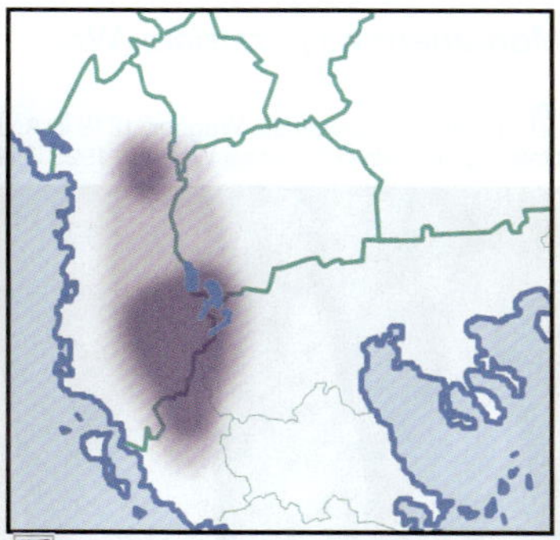

Albania to NW Greece

Shell light brownish, thin, almost smooth, 9-10 whorls, suture white, aperture with thin lip, margin connected at parietal side only by a thin layer, two upper palatal folds, lower palatalis very short, lunula dorsolateral, broad and short.

Westerlund 1884: 54, Nordsieck 1996: 10.

Montenegrina laxa (Küster, 1861)

Clausilia laxa Küster, 1861 in Küster 1847: 276 (figured on Taf. 31 Fig. 14-16)
Taf. 31 was from 1860 but without name printed on plate (wrapper was not preserved and is not published work), text from 1861.
Montenegrina laxa chiasma Nordsieck, 1972 in Nordsieck 1972: 30
Montenegrina dedovi Nordsieck, 2009 in Nordsieck 2009: 75

13-23 x 3.9-5.4 mm

Albania: near Librazhd

Montenegro to NW Greece

Shell light brownish, thin, smooth, 9-11 whorls, cervix rounded with umbilicus at basal side, apertural margin not connected at parietal side, broadly reflected, lower palatalis short, only the extreme ends of the two upper palatal folds are visible in a perpendicular view, but not the lower palatal fold and subcolumellaris, lunula lateral.

Westerlund 1884: 54, Nordsieck 1972: 30, Nordsieck 1996: 8, Nordsieck 2009: 75.

Nordsieck 2009 classified ***chiasma*** as a separate species, which had previously between classified as a subspecies of *M. laxa*. The morphological record suggests that *chiasma* is an intermediate form between *laxa* and *perstriata*, in that it was reported to have weaker folds including a low columellaris (like *laxa*), dorsolateral lunula (like *perstriata*) and an intermediate expression of the subcolumellaris (weak in *laxa*, strong in *perstriata*). The easiest explanation for such a situation is that both *perstriata* and *laxa* must be conspecific, connected by the intermediate form *chiasma* (a detailed analysis of the ranges would be helpful). Simply classifying the intermediate form as a separate species is probably not an appropriate solution of the problem.

*M. **dedovi*** is classified with *M. laxa* here. Nordsieck's 2009 arguments to classify this taxon as a separate species are too weak (smaller average size 13-15 x 3.8-4.1 mm, trend towards a reduced frontal upper palatalis, the fact that it lives in a higher altitude and that there exists an additional distant and isolated population in NW Greece).
Clausiliids are known to be distributed along large altitudinal gradients (*Albinaria idaea* and *A. corrugata* in Crete live between sea level and 2000 m), this is usually associated with some local variation in shell parameters. The observed difference in shell size is within locally observed shell size differences in Crete, and do not fall outside the size ranges of other *Montenegrina* species.

Montenegrina minuscula Erőss & Szekeres, 2006

Montenegrina minuscula Erőss & Szekeres, 2006 in Erőss, Fehér & Szekeres 2006: 184

9-11.5 x 2.3-2.9 mm

Albania: near Qafa e Murrës, road Peshkopi-Burrel, 980 m

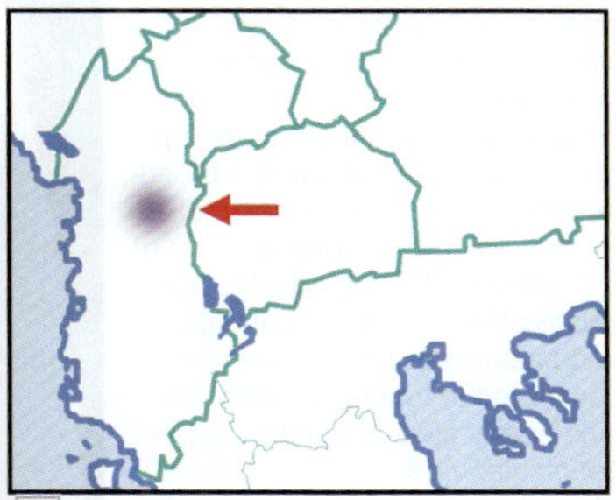

Albania

Shell horny yellowish or brown, smooth, shiny, 8.5-10 convex whorls, last whorl with a spiral furrow at the palatal side, giving the last whorl a concave outine, aperture detached, parietalis long, overlapping with parallel running end of spiralis, columellaris not very prominent, lunula lateral and short, basalis separated from lunula (if present at all), clausilium not visible in the aperture, subcolumellaris only visible in a very oblique view. Smaller than most other *Montenegrina* species, lunula more lateral than in *M. fuchsi*, lunula dorsolateral and weaker in *M. apfelbecki*.

Montenegrina perstriata (Wagner, 1919)

Delima laxa perstriata Wagner, 1919 in A. J. Wagner 1919: 71
Delima perstriata ochridensis Wagner, 1925 in A. J. Wagner 1925: 62, Taf. 14

16-24 x 4-5 mm

(1-4) E Albania: Periferi Kukës, near Bushtricë, (5-8) Rep. Makedonija: Lake Ohrid, near Ohrid
(5-8) Museum Budapest HNHM

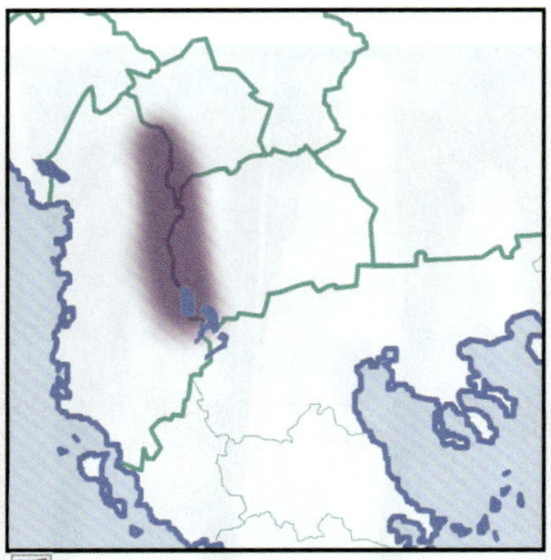

E Albania and vicinity

Shell brown, weakly striated, 9-11 whorls, suture white at last 1-2 whorls, cervix rounded with umbilicus at basal side, apertural margin not or weakly connected at parietal side by a callus, slightly or broadly reflected, parietalis strong, columellaris descending steeply, basalis short, lunula lateral and irregular or dorsolateral, subcolumellaris only visible in an oblique view.

Wagner 1925: 61, Nordsieck 1988: 201, Nordsieck 2009, Fehér & Erőss 2009: 10.

The taxon ***ochridensis*** from near Ohrid was classified as a separate species between 1972 and 2009. Nordsieck 1988 argued that *ochridensis* was closer related to *M. rugilabris* from NW Greece and others than to *M. perstriata* from E Albania and should be regarded as an independent species, key characters were sculpture, cervix, apertural margin and palatal folds.
Wagner 1925 and Nordsieck 2009 regarded *ochridensis* and *perstriata* as conspecific.

Montenegrina rugilabris (Mousson, 1859)

Clausilia rugilabris Mousson, 1859 in Mousson 1859: 275

17-21 x 3.8-5 mm

Greece: Ipiros, Lake Ioánnina

NW Greece (Ipiros)

Shell brownish with violet hue, almost smooth, 10-11 whorls with white suture, cervix rounded with basal keel, aperture inside narrow, margin weakly connected at parietal side and not detached, broadly reflected, upper palatalis short, basalis short and connected to dorsolateral lunula, subcolumellaris often visible in a perpendicular view.

Westerlund 1884: 53, Nordsieck 1972: 33.

Montenegrina sattmanni Nordsieck, 1988

Montenegrina sattmanni Nordsieck, 1988 in Nordsieck 1988: 200

Shell yellowish brown, upper whorls densely ribbed, lower whorls smoother, cervix very densely ribbed, 12-13 whorls, cervix with an indented structure, aperture not detached, margin broadened, parietalis strong, columellaris weaker and in the upper section of the aperture, lunula dorsal to dorsolateral, frontal upper palatalis and basalis prominent, and often connected to lunula, subcolumellaris nearly always visible in a perpendicular view.
Clausilium with strongly bent outer margin, looking compressed, especially in its lower section.
Differs from *M. irmengardis* in its colour, thickened apertural margin and compressed clausilium plate, from *M. rugilabris* and *M. ochridensis* in its higher located and less steep columellaris, basalis often connected to lunula, frontal upper palatalis often connected to distal upper palatalis, and compressed clausilium plate.

17-21.5 x 4.6-5.5 mm

Albania: near Zvezdë, 1030 m

NW Greece, SE Albania (Prespa Lake region)

Montenegrina skipetarica (Soós, 1924)

Clausilia skipetarica Soós, 1924 in Soós 1924: 181
Montenegrina irmengardis Klemm, 1962 in Klemm 1962: 242
Montenegrina ersekensis Nordsieck, 1996 in Nordsieck 1996: 9

16-26 x 4.5-5.6 mm

(1-4) Greece: Ipiros, Platanoúsa, (5-9) Albania: near Ersekë, 1150 m

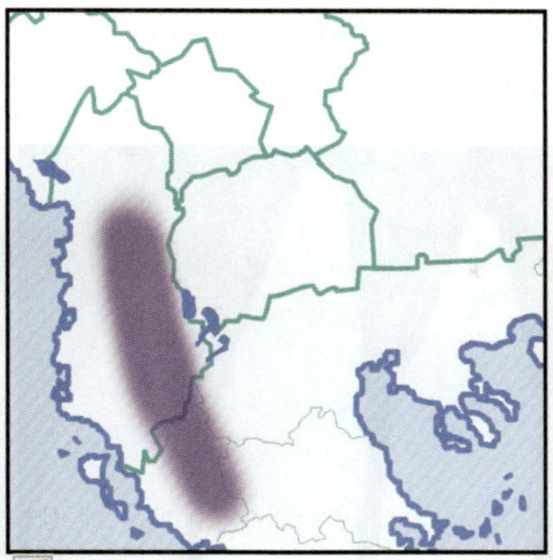

NW Greece, Albania

Shell violet brown with weak white line along suture, 11-12 whorls, parietalis very prominent, columellaris weaker and oblique, lunula dorsal or dorsolateral, connected to frontal lower palatalis, subcolumellaris visible in a perpendicular view.

Greece: Ipiros, near Monodéndri — Cédric Audibert

Erőss et al. 2006: 203, Nordsieck 2009: 74.

Erőss et al. 2006 and Nordsieck 2009 regarded *skipetarica* and ***irmengardis*** as conspecific, *skipetarica* had previously been classified with *perstriata* and most populations of the present species had been known as *M. irmengardis*.
The taxon ***ersekensis*** differs from *irmengardis* only in a lighter shell colour and prolongated sulcalis (= lowest palatal fold). Nordsieck 2009 classified it as a subspecies of *skipetarica*. This seems tenable.

Montenegrina stankovici (Urbański, 1960)

Delima stankovici Urbański, 1960 in Urbański 1960: 51 (*Delima (Delima (Heteroptycha)) stankovići*)

Shell light horny brown, sometimes with bluish hue, weakly striated, 8-10 whorls, cervix rounded, aperture large, whitish, margin weakly connected by a thin callus at the parietal side, parietalis strong and short, separated from spiralis, collumellaris not very prominent and almost straight, principalis visible in a perpendicular view, lunula very weak, subcolumellaris visible in an oblique view. Clausilium plate bluntly pointed.

Shady limestone rocks, in rock crevices and under mosses, 700 m altitude.

15-17 x 3.6-4.4 mm

Rep. Makedonija: Lake Ohrid, Sveti Jovan

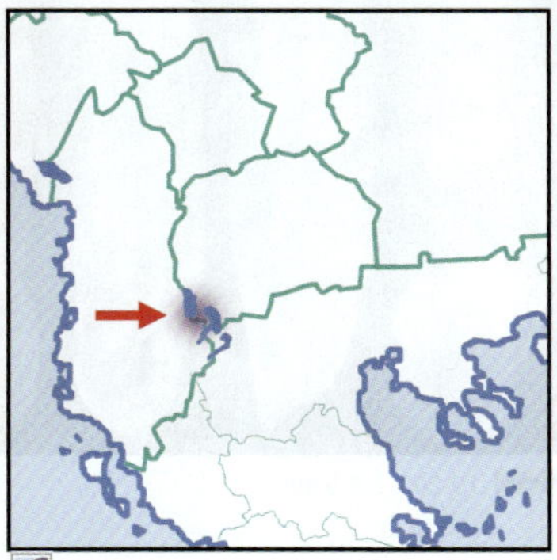

Rep. Makedonija/Albania (S Lake Ohrid region)

Montenegrina subcristata (Pfeiffer, 1848)

Clausilia subcristata Pfeiffer, 1848 in L. Pfeiffer 1848 (Mon. Helic. 2): 438
Author was not Küster. Küster 1847: Taf. 4 was from 1847 but without name printed on plate (wrapper was not preserved and is not published work), text from 1850.

16-24 x 4-5 mm

Albania: Shkodër

Küster 1847: 39, Taf. 4 Fig. 11-13, Westerlund 1884: 54.

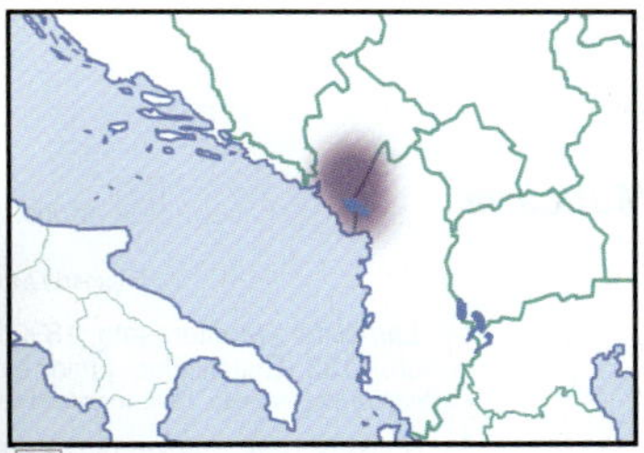

N Albania, Montenegro

Shell light brownish, slightly transparent, almost smooth, 9-11 whorls, aperture with lip, margin not connected at parietal side, slightly broadened, principalis upper palatalis well visible in a perpendicular view, lower palatalis long but usually not visible in a perpendicular view, no fold between lower palatalis and subcolumellaris, lunula lateral, upper side broad, lower side oblique and narrower.

Albania: Shkodër — Vollrath Wiese

Montenegrina umbilicata (Boettger, 1879)

Clausilia umbilicata Boettger, 1879 in O. Boettger 1879: 102, Taf. 2

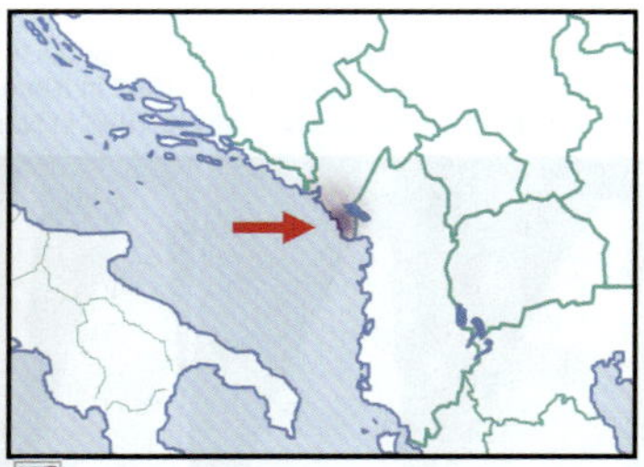

14-24 x 4-5 mm

Montenegro: Stari Bar

Westerlund 1884: 53.

Shell light, slightly transparent, almost smooth, 8-11 whorls, the last two whorls often comprising half shell height, suture often white, cervix weakly striated and flattened with conspicuous umbilicus, apertural margin sharp and not connected at parietal side, slightly broadened, lower palatalis short and visible in a perpendicular view, principalis and upper palatalis also well visible, lunula dorsolateral, subcolumellaris visible in a perpendicular view.

Montenegrina zilchi Nordsieck, 1974

Montenegrina zilchi Nordsieck, 1974 in Nordsieck 1974: 156

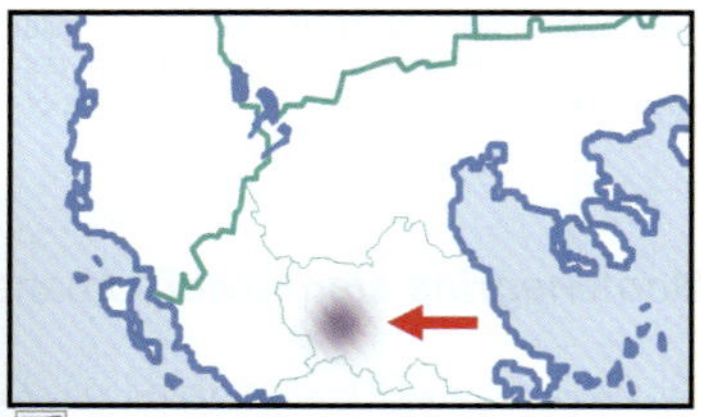

Greece (Thessalía)

19-26 x 5-6.5 mm

Greece: Thessalía, near Píli

Museum Budapest HNHM

Shell horny brown, almost smooth, cervix finely striated, 9-11 whorls, cervix rounded with weak basal keel, apertural margin not detached and not connected at parietal side, parietalis strong, columellaris weak, lunula dorsal, basalis strong and connected to lunula, frontal upper palatalis strong, lower palatal fold usually present, subcolumellaris visible in a perpendicular view. Clausilium with strongly bent and protruded outer margin.
Differs from *M. irmengardis* in its strong basalis connected to the lunula, two palatal folds and its clausilium with curved outer margin.

Muticaria

Muticaria Lindholm, 1925 in Lindholm 1925: 262
Expressly established as a new replacement name for

Type species *Clausilia scalaris* Pfeiffer, 1850 (subsequent designation by Lindholm 1925: 262)
(= currently *Muticaria macrostoma* (Cantraine, 1835))

Lamellifera Monterosato, 1892 in Monterosato 1892: 30 *Clausilia (Lamellifera)* not *Lamellifera* Westerlund, 1890 in Westerlund 1890 (Katal. eur. Binnenconch.): 133 (Clausiliidae, junior synonym of *Neniatlanta* Bourguignat, 1876)
Lamellifera Monterosato, 1892 contained originally *Clausilia scalaris* and *Clausilia syracusana* Philippi, 1836.

Holyoak 1986, Beckmann 1990, Giusti, Manganelli & Schembri 1995, Colomba et al. 2010.

Holyoak 1986: 217 argued that it was not necessary to classify the few closely related species in different genera *Lampedusa* and *Muticaria*, as proposed by Nordsieck 1979. Gittenberger & Beckmann 1987 followed this view, Beckmann 1990 rejected it because other similar clausiliid G and N type genus pairs should also be regarded as subgenera. Giusti et al. 1995: 340 rejected it because they saw more differences between *Muticaria* and *Lampedusa* species than within *Lampedusa* species (they argued that if the *Lampedusa* species are subdivided into two subgenera, the *Muticaria* species must be separated at a higher taxonomic level).
I can agree with Holyoak's 1986 reasoning, but the two genera have been used by most subsequent authors, so I would not unnecessarily touch common usage of the involved genus-species combinations.

Thake 1985 suggested that an ancestor of *Muticaria* / *Lampedusa* invaded the Sicily-Malta-Lampedusa landmass some 5 M yr ago and evolved into different species after marine ingression, *Muticaria* in Sicily. Later during repeated land connections Sicilian snails reached Malta and occupied large parts of the Maltese islands. The morphological record suggests that some following invaders hybridised partly with local populations.
Beckmann 1990 took the presence of *neuteboomi* in Sicily as evidence against this hybridisation theory in Malta.

Giusti et al. 1995: 368 also rejected Thake's hypothesis, but for different reasons. In unpublished allozyme polymorphism studies in Malta and Sicily they found that changes in shell morphology, particular in the ribbing on which Thake's 1985 theory was largely based, must have occurred repeatedly and would not indicate phylogenetic relationships. Phylogenetic relationships seemed to reflect geographical distance more than morphological similarity. They suggested that the Maltese *Muticaria* population evolved from a single common ancestor and the visible morphological variation was due to local events without taxonomic significance. For the same reason they rejected divisions into various subspecies.
Based on results of molecular DNA studies of 10 specimens from two localities in Sicily Colomba et al. 2010 argued that *neuteboomi* was an independent species. These conclusions were partly based on results of unpublished studies.

The situation remains unclear, after more than 30 years of intensive studies by various teams.
Thake and Beckmann based their conclusions on the morphological record, but its taxonomic significance was probably overestimated. The Giusti et al. team published only conclusions of allozyme studies without presenting reproducible results. The Colomba et al. team did partly the same with DNA studies; they have not presented sufficient results to extract a firmly established classification.

Muticaria macrostoma (Cantraine, 1835)

Clausilia macrosoma Cantraine, 1835 in Cantraine 1835: 381 (not *Cl. macrostoma*)

Emended by Cantraine 1840: 153 to *Clausilia macrostoma* - this was an unjustified emendation (because the error was not visible in the work itself, *macrosoma* is also a Latin word) although Cantraine 1840 made the printer responsible for the misspelling in the origfinal work ("J'ai nommé macrostome (et non macrosome comme il est imprimé dans le Bulletin de l'académie)"). Beckmann & Gittenberger 1987: 335 argued that *macrosoma* has no meaning in malacology because it was not used for a gastropod species in the period 1800-1850, in contrast to *macrostoma*, which had been used by many authors for shells with large aperture. This however, is a reference to external evidence, which is not allowed in the Code's regulation.
Since the name has ever and exclusively been used as *macrostoma*, it might be better to continue using the unjustified emendation in combination with the 1835 date (Art. 33.3.1). Should one author in the future use the spelling *macrosoma*, I would not use *macrostoma* any more. The name is only rarely used in the literature.

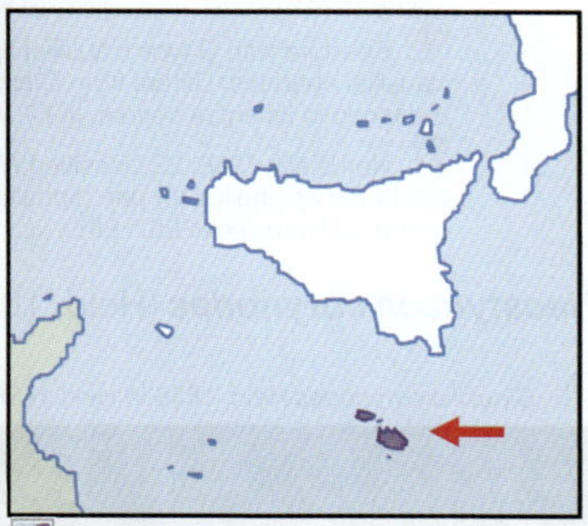

12-19 (decollate down to 8 mm) x 3.5-6 mm

Malta: (1-4) Malta, Dingli Cliffs, (5-8) Gozo, Munxar

Maltese islands

Malta: W coast of St. Paul's Bay

Shell usually decollated with 4-6 whorls remaining in adult shells (sometimes not decollated and possessing only 7-8 whorls), white or grey with brownish hue, usually densely ribbed, dorsal keel stronger than basal keel, often with a furrow in between, apertural margin detached and slightly protruded, columellaris weak, running obliquely in the uppermost section of the aperture, 3 relatively long palatal folds present between suture and lunula/frontal upper palatalis, instead of a principalis, parallelaris present, lunula dorsal, strong and with a very conspicuous callus at its frontal end, often visible in a perpendicular view, not connected to subcolumellaris, subcolumellaris usually visible at least in an oblique view, sometimes very strong, running obliquely inside looking as if it was a second columellaris, the clausilium is well visible with its pointed end in the aperture between subcolumellaris and lunula.
Shell shape and ribs locally variable, can also be more coarsely ribbed and scalarid (near St. Paul's Bay) or inflated (Gozo).
Differs from *Lampedusa imitatrix* in its stronger dorsal keel, the 3 folds between suture and frontal upper palatalis, and more folds visible at the basal side of the aperture (strong subcolumellaris and strong dorsal lunula, tip of clausilium).

On limestone rocks and rdum. Usually in crevices or in vegetation near rocks, sometimes also under stones, in rock rubble and in rubble walls surrounding cultivated and abandoned fields. Widespread and common.

Westerlund 1884: 166 (*Cl. syracusana*), Thake 1985, Holyoak 1985 (synonym of *Lampedusa syracusana*), Giusti et al. 1995: 348.

Thake's 1985 and Holyoak's 1986 detailed observations suggested the presence of morphological gradients over area with intermediate populations and sometimes narrow hybrid zones. Giusti et al. 1995 did not find that the morphological variation was reflected in the molecular record.

Muticaria syracusana (Philippi, 1836)

Clausilia syracusana Philippi, 1836 in Philippi 1836: 139, Tab. 8
Muticaria neuteboomi Beckmann, 1990 in Beckmann 1990: 78

11-16 x 3-5.5 mm

Italy: Sicily, (1-5) Siracusa, (6-9) Cava d'Ispica (1-5 *syracusana*, 6-9 *neuteboomi*)

Italy (SE Sicily)

Shell usually decollated with 5-7 whorls remaining in adult shells, grey with brownish hue and white ribs, cervix with prominent basal keel, apertural margin detached and slightly protruded, columellaris weak, oblique and S-shaped, 3 or 4 palatal folds, parallelaris prominent, lunula strong, dorsal and not connected to subcolumellaris.
Shells near Siracusa are more slender, less densely ribbed and differ from inland populations in the number of palatal folds. Some authors associate the palatal folds with separated species: *syracusana* with 3 short palatal folds between suture and lunula, instead of a principalis, the lowest one connected to the frontal upper palatalis, *neuteboomi* with 4 separate palatal folds, in other words with a small gap between frontal upper palatalis and lowest principalis, the uppermost fold (= principalis) being longer.

In rocky habitats.

Westerlund 1884: 165, Holyoak 1986, Falkner 1990: 156, Giusti et al. 1995: 358, Manganelli et al. 1995: 23, Reitano et al. 2009: 191 (seem to have misidentified shells north of Siracusa as *M. syracusana*, the figures suggested a *neuteboomi* form), Colomba et al. 2010.

Giusti et al. 1995: 365 came to the conclusion that in Sicily *neuteboomi* must be a local form of *syracusana*. The differences in shell shape and ribs mentioned by Beckmann 1990 fall in the usual variability of a S European clausiliid species, the folds above the lunula should also reflect some local variation.
Colomba et al. 2010 found that the ranges of both forms did not overlap: *syracusana* was restricted to the coastlands near Siracusa, *neuteboomi* in a broader area and also in higher altitudes. In a molecular study they analysed 10 animals of two populations (Siracusa and Cava d'Ispica, distance 50 km) and found that the 5 animals of the one population clustered separately from the other population. Combined with the morphological difference they explained the result by the presence of two separate species.
The results presented by Colomba et al. 2010 did not sufficiently support their conclusions, but results of additional unpublished studies were also considered. The published observation that individuals of two distant populations yield two separate clusters is a predictable result and can be explained by spatial distance and low gene flow over area. For justifying separate species it would be necessary to have a closer look on possible transitional zones near Siracusa. The obtained degree of molecular divergence between the two populations would have to be compared with values obtained in forms living in close vicinity near Siracusa, and with spatially very distant *neuteboomi* forms collected near Caltanissetta and Catania. It would also be interesting to compare these with values obtained from Malta.
At the current state of scientific research I would not regard the classification of *M. neuteboomi* as an independent species as justified.

Neostyriaca

Type species ? *Clausilia corynodes* Held, 1836 (? monotypy)

Neostyriaca Wagner, 1920 (??? or *Neostyria*?) in A. J. Wagner 1920: 107 (??? or 72?)

Neostyria Wagner, 1920 in Wagner 1920 : 72 *Clausilia (Neostyria)*, type species *Clausilia corynodes* Held, 1836 (monotypy)
Established on p. 72 [Apr 1920] as *Neostyria*, with description and one species included, *Clausilia corynodes*. Wagner 1920: 107 [July 1920] wrote "*Neostyriaca*" in the headline, and "*Neostyria*" in the footline, without a statement which one would be the correct spelling to be used, and included two species *Cl. corynodes* and *Clausilia styriaca* Schmidt, 1856. The most likely explanation inside the original source is that of a misspelling in the headline of p. 107. On p. IV of the journal volume (list of new names established in the entire journal volume, Date Oct 1920), the name was cited as *Neostyriaca* (*Neostyria* was not mentioned). Kennard & Woodward 1923: 304 gave *Clausilia styriaca* as type for *Neostyriaca*.
The name is currently used as *Neostyriaca*. I do not know why, since the footnote on p. 107 does not allow the conclusion that Wagner intended to correct the name, in the contrary. The footnote reinforced the use of the original spelling *Neostyria*. If nothing else interferes the name should be spelled *Neostyria*, type species *Cl. corynodes* by monotypy. I only use the spelling *Neostyriaca* here because it is possible that I have overlooked some details.

Aperture with G type clausilial apparatus (weak or no spiralis, principalis short, two parallel folds on palatal side, lunula weak), *Clausilia* has a normal type clausilial aparatus. Differs from *Clausilia whateliana* in its epiphallus (parepiphallus) being at least partly thickened (in *corynodes* only in its distal section, in *strobel* thickened in its entire course, in *Cl. whateliana* not thickened).

Nordsieck 1966: 22 classified *Neostyriaca* as a subgenus in *Clausilia* because *Clausilia* species were found to differ only in the thickness of the proximal section of the epiphallus (or parepiphallus), and because the G type clausilial apparatus was not regarded as sufficiently significant for a genus. Later this view was changed (Nordsieck 2007: 60).

Neostyriaca corynodes (Held, 1836)

Clausilia corynodes Held, 1836 in Held 1836: 276

8-13 x 2-2.5 mm — Austria: Oberösterreich near Gmunden

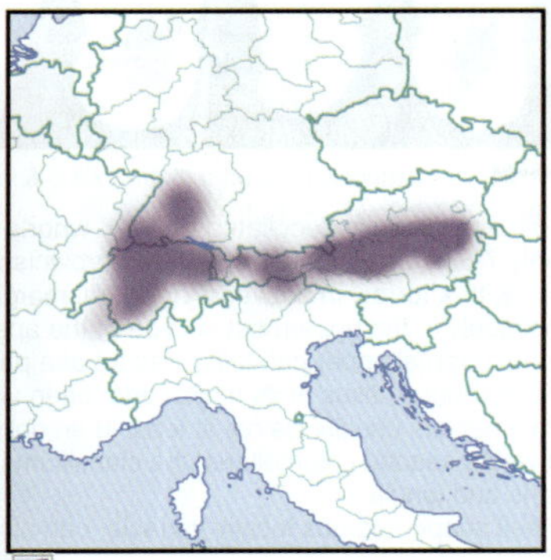

N Alps and adjacent highlands

Shell very weakly ribbed or almost smooth, basal keel strong, cervical groove weakly developed, dorsal side of the cervix rounded, aperture small, columellaris without double folds, its lower margin forming a smooth curve, instead of principalis there are two palatal folds, palatal callus usually white and prominent, runs parallel to the margin and ends suddenly at the basal side, basal groove present but does not reach margin, no lunula.

Differs from *Julica schmidtii* in its basal groove in the lower part of the aperture (basal side rounded and no groove in *J. schmidtii*). *N. strobel* has a more prominently and entirely ribbed shell, more pronounced cervical keels and a deeper basal groove. *Clausilia rugosa* has a weaker palatal callus and a strong lower palatalis.

Limestone rocks in forests, on trees and shady slopes covered with mosses, never at long distances from rocks. Rarely outside forests at overgrown limestone rocks. Restricted to mountains, rare above 1200 m, in Switzerland up to 2300 m.

Westerlund 1884: 178, Klemm 1951, Nordsieck 1966, Klemm 1969, Kerney & Cameron 1979: 168, Falkner 1990: 162, Turner et al. 1998: 224.

No reliable reports from France (where forms of *Clausilia rugosa* were confounded with this species). See also the comment under *N. strobel.*

Neostyriaca strobel (Strobel, 1850)

Clausilia strobel Strobel, 1850 in Strobel 1850: 121
Original spelling was not *strobeli*, not to be corrected under Art. 32.5. The name is not very frequently used, so it is convenient to use the name in its original spelling.

9-13 x 2.4-3 mm — Italy: Alto Adige, Val Sattaro

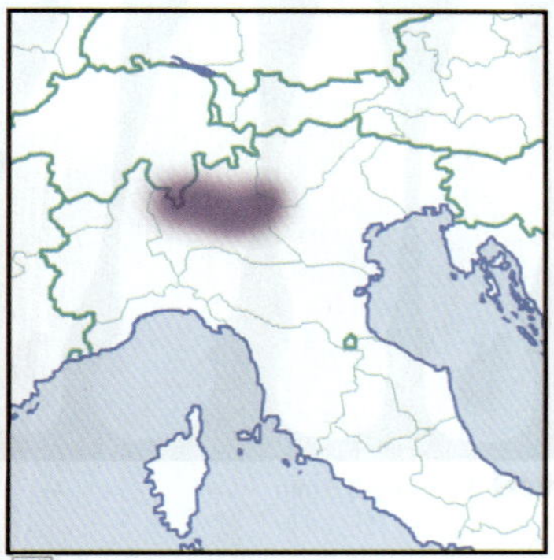

N Italy, S Switzerland

Shell brown, entirely ribbed, not slender, whorls 2 and 3 equal in diameter, the next whorls increase quickly, cervix with strong basal and moderate dorsal keel with a groove in between, aperture very small compared with the penultimate whorl, with prominent lower palatalis, palatal callus often divided into two parts, basal groove present and reaches margin.

Differs from *N. corynodes* in its entirely and more prominently ribbed shell, more pronounced cervical keels and a deeper basal groove. *Clausilia* species have a weaker palatal callus and usually a lower palatalis. *Cl. whateliana* has weaker cervical keels. In *N. strobel* the columellaris runs in a smooth curve to the margin, in *Cl. whateliana* the columellaris ends more abruptly and does not reach the margin in a smooth curve.

Humid habitats in forests among rocks and ground litter, in shady habitats also outside forests, not restricted to rocks (difference to *Clausilia whateliana*). In S Switzerland in up to 1700 m altitude.

Westerlund 1884: 179 (*strobeli*), Klemm 1969 (*strobeli*), Kerney & Cameron 1979: 169 (*strobeli*), Nordsieck 1993: 39 (*strobel*), Manganelli et al. 1995: 27 (*strobeli*), Cossignani & C. 1995: 84 (*strobel*), Turner et al. 1998: 225 (*strobel*), Falkner et al. 2001: 43 (*strobel*), Nordsieck 2007: 61 (*strobel*), Boschi 2011: 347 (*strobel*).

The differences to *N. corynodes* could also be gradual and the forms could display local variations within their ranges. It seems that *N. corynodes* and *N. strobel* do not occur sympatrically, but it might be interesting to see if the character expressions of the cervix, shell shape and ribs approach each other with closer vicinity.

Papillifera

Papillifera Hartmann, 1842 in Hartmann 1844: 156 *Clausilia (Papillifera)*

Type species *Helix papillaris* Müller, 1774 (subsequent designation by Albers & Martens 1860: 278)

Established without description, contained originally 7 species, listed by Pfeiffer 1841 (Symb. 1): 33 in the 9th section to which Hartmann referred without citing any species by name (*blanda* (with synonym *exilis*), *subcylindrica*, *succineata*, *satura*, *diodon*, *solida*, *interrupta*). This reference made *Papillifera* available under Art. 12.2.5. Art. 67.2.3 stipulates that those were not originally included species. This was possibly an unintended gap in the Code (it makes no sense to regard a genus as available because species were included, and in the next Article say they were not originally included), but Kadolsky 2009: 25 also accepted the rules in this form. The first authors to include nominal species were Adams & Adams 1858 [1855]: 180 who included 7 species, among these were *Helix papillaris* Müller, 1774 and *solida*. Albers & Martens 1860: 278 gave as type *Helix papillaris*. Pfeiffer 1848: 550 gave as type "*Claus. bidentis* L.", = *Turbo bidens* Linnæus, 1758.

Papillifera bidens (Linnæus, 1758)

Papillifera papillaris (Müller, 1774)

Turbo bidens Linnæus, 1758 in Linnæus 1758: 767
Established with a brief diagnosis and reference to Gualtieri 1742: Tab. 4 Fig. C, most probably in error for Fig. D or E because the description did not fit Fig. C. Linné 1767: 1240 gave an additional reference to Bonanno 1684: Cl. 3 Fig. 41. This figure was in agreement with the description. Schröter 1784: 55 was the first to have suspected that Linnæus made a mistake and cited the incorrect figure. The resulting dispute has not been solved in the past 230 years despite various efforts to find types in the Linnean collection (Hanley 1855: 351 associated one specimen with *T. bidens*, it was conspecific with *Helix papillaris* Müller, 1774) and to designate neotypes (Falkner et al. 2002: 113, Kadolsky 2009). A new Case is currently pending (Welter-Schultes 2012: 85).

Helix papillaris Müller, 1774 in Müller 1774: 120
Giusti & Manganelli 2005 asked the Commission to suppress *Turbo bidens* against *Helix papillaris*, which they did not regard as conspecific; they argued that *T. bidens* was conspecific with *Cochlodina incisa* (Küster, 1876). The Commission rejected this request in ICZN Opinion 2176 (ICZN 2007) and accepted the view that *T. bidens* was a senior synonym of *Helix papillaris* Müller, 1774. This implied that *T. bidens* should be used for this species (Alba et al. 2011: 48 followed this view), but the Commission did not fix a neotype for *T. bidens*.

11-17 x 2.5-4.5 mm, locally variable

(1-4) Malta: Gozo, (5) Italy: Roma, (6-8) San Marino

Italian mainland, Sicily, introduced elsewhere. In Turkey also in Istanbul, Iznik, Izmir and Antalya.

Shell whitish to greyish brown, smooth or ribbed, suture dark and strongly papillated, apertural margin connected at parietal side, not always detached, lunula runs from dorsolateral to (upper part) dorsal, above the lunula there are two connected palatal folds instead of a principalis, no lower palatal fold, no basalis. Animal light brownish with light tentacles. Differs from *Siciliaria gibbula* in the absence of a lower palatal fold (*S. gibbula* has a strong fold, well visible in a perpendicular view) and in the colour between the dots at the suture being darker than the rest of the shell.

In Malta in open rocky habitats mainly below small shrubs. In Italy very common and in many regions at almost every limestone outcrop, most common clausiliid species in Sicily.

Had been known under the name *Papillifera papillaris* (Müller, 1774), but *Turbo bidens* Linnæus, 1758 has precedence.
Introduced in 1896 from Rome (Italy) to Cliveden (Buckinghamshire, England), and in the 1880s probably from Greece to Brownsea island (Dorset, England).

Papillifera solida (Draparnaud, 1805)

Clausilia solida Draparnaud, 1805 in Draparnaud 1805: 69
Clausilia deburghiae Paulucci, 1878 in Paulucci 1878: 39 (*Clausilia Deburghiæ*)

10-17 x 3-4 mm, locally variable

(1-6) France: Bouches du Rhône, La Ciotat (*solida*), (7-8) S Italy: Cristallino di Calabria, Palizzi superiore (*deburghiae*)
(7-8) Alessandro Hallgass

S Italy to SE France

Shell horny whitish or greyish yellow, sometimes violet, sometimes thin and transparent, finely striated (in S Calabria ribbed), 9-13 whorls, suture often papillated, cervix finely ribbed with prominent and oblique basal keel, and a deep furrow between basal keel and apertural margin, aperture with weak lip, margin not or weakly connected at parietal side, in S Calabria detached, parietalis very long, reaching the margin, parallelaris distinct, no spiralis, columellaris deep inside, almost horizontal and curved, palatal callus present, sometimes white and prominent, subcolumellaris deep inside, visible in an oblique view.
Several characters display spatial variations.

Westerlund 1884: 174, Manganelli et al. 1995: 25, Nordsieck 2002: 36

Nordsieck 2002 regarded *deburghiae* from S Calabria and *solida* as conspecific because of the absence of a spiralis and presence of a distinct parallelaris).

Protoherilla

Protoherilla Wagner, 1921
in A. J. Wagner 1921: 42

Type species *Protoherilla janickii* Wagner, 1921
(subsequent designation by Haas 1923 [Apr 1923]: 126)
(= currently *Protoherilla baleiformis* (Boettger, 1909))

Contained originally *P. janickii* Wagner, 1921 and *Clausilia baleiformis* Boettger, 1909. Kennard & Woodward 1923 [Oct 1923]: 307 gave as type *Cl. baleiformis*, too late.

Protoherilla baleiformis (Boettger, 1909)

Clausilia baleiformis Boettger, 1909 in Wohlberedt 1909: 668, Taf. 54
Description and name were exclusively attributed to Boettger.

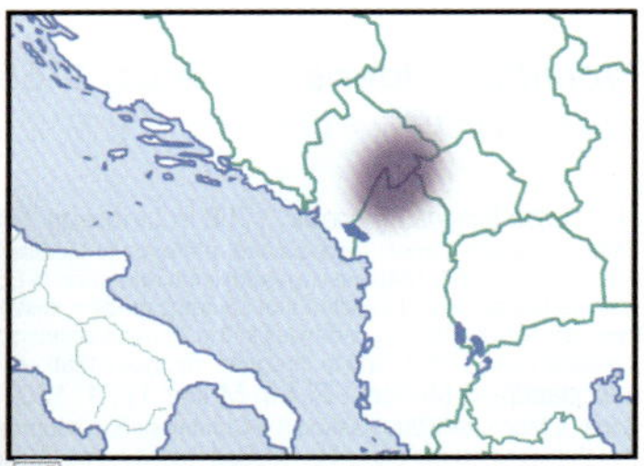

9-11 x 3-3.5 mm

Montenegro: Planinica, 1500 m

Museum Budapest HNHM

SE Montenegro, N Albania

Shell horny olive greenish, weakly striated, 8-10 moderately convex whorls, apertural margin slightly reflected, parietalis short and not reaching margin, in some populations extremely weak, no spiralis, columellaris weak, no palatal folds, no lunula and basalis, subcolumellaris visible in the aperture.

Nordsieck 1972: 37.

Protoherilla mirabilis (Nordsieck, 1972)

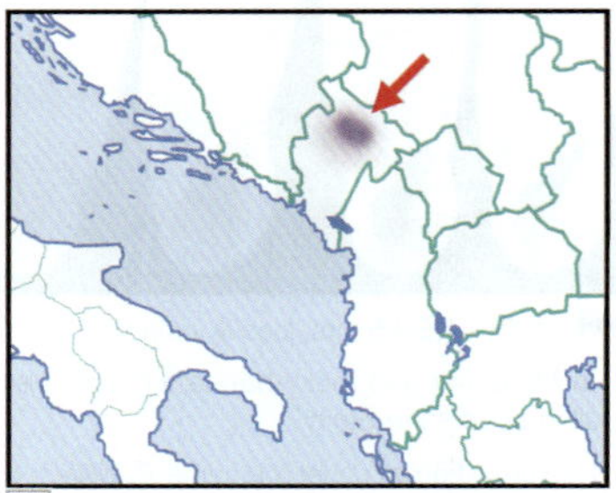

Montenegro

Taraclausilia mirabilis Nordsieck, 1972 in Nordsieck 1972: 36

Shell weakly striated, with spiral sculpture, 9-11 whorls, cervix rounded with weak basal keel, aperture not detached, margin narrow, weakly reflected, not connected at parietal side, palatal margin slightly concave and with sinulus, columellaris moderately steep, lunula dorsal, frontal upper palatalis weak, clausilium plate ueally entirely visible in the aperture, subcolumellaris usually visible in a perpendicular view.

10-12 x 2.5-2.8 mm

Was found at S exposed limestone rocks in a beech forest.

Possibly conspecific with *P. pseudofallax*, see comment there.

Protoherilla pseudofallax (Nordsieck, 1972)

Taraclausilia pseudofallax Nordsieck, 1972 in Nordsieck 1972: 36

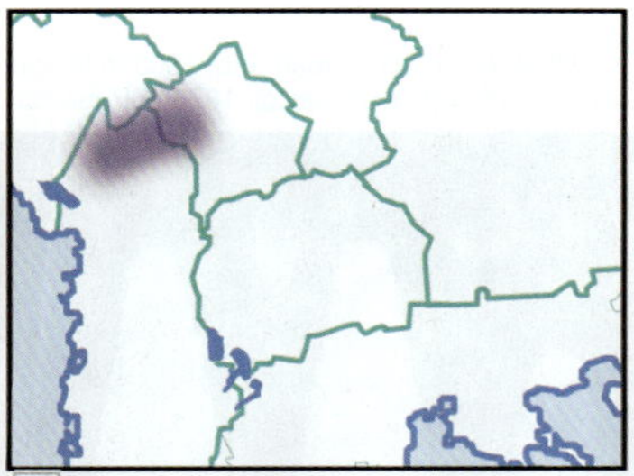

Kosovo, N Albania

Shell horny brown, weakly ribbed, 8-10 whorls, cervix rounded with weak basal keel, aperture not detached, margin narrow, weakly reflected, not connected at parietal side, which is very long, palatal margin slightly concave and with sinulus, parietalis short, not reaching spiralis, columellaris straight and moderately steep, lunula dorsal and weak, frontal upper palatalis very weak, subcolumellaris visible in a perpendicular view.

11-14.5 x 3.5-4.5 mm

N Albania: Malësia, Lepushë

Differs from *P. mirabilis* in a less slender shell shape, slightly coarser ribs, wider parietal section of the aperture, shorter parietalis and principalis, clausilium plate always entirely visible.

In Kosovo found at tree trunks under rotting wood, in 1600 m altitude. In N Albania also in natural mountain forest habitats above 1000 m.

The differences between the taxa *mirabilis* and *pseudofallax* seem to be extremely weak to justify a separation at the level of species. It seems to me that the initially discovered populations from Kosovo and Montenegro represented two extremes of a gradient, and that the N Albanian specimens collected by Dh. Dhora from Malësia (Lepushë, figured here) represented intermediates suggesting that there is a more or less continuous gradient between both forms. Only little more research (evidence for some more intermediate forms between the two extremes) is necessary to finally reject Nordsieck's 1972 classification of the two taxa, and to understand better the spatial variation of the characters.

Pseudofusulus

Pseudofusulus Nordsieck, 1977
in Nordsieck 1977: 96

Type species *Clausilia varians* Pfeiffer, 1828
(original designation)

Pseudofusulus varians (Pfeiffer, 1828)

Clausilia varians Pfeiffer, 1828 in C. Pfeiffer 1828: 40

9-12 x 2.2-3 mm

Austria: Kärnten, Ruine Kraig

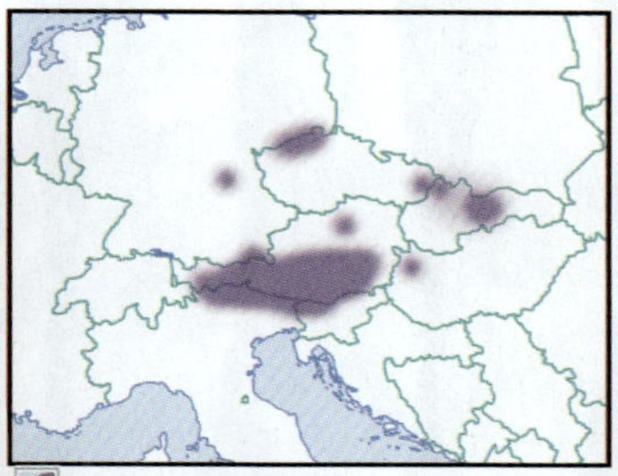

Alps, N Carpathians

Shell horny or greenish brown, finely ribbed, 9-10 whorls, apertural margin detached, columellaris deep inside and forked, hardly visible in a perpendicular view, palatal wall white and prominent, no lunula, subcolumellaris very close to basal furrow, visible in an oblique view. Clausilium not very concave.

Humid and shady habitats in woods, under ground litter and stones, in mountains. In the Czech Republic strictly associated with trees, restricted to undisturbed and hardly accessible natural beech forests.

One of the most endangered species in the Czech Republic, restricted to hidden and undisturbed natural deciduous forests. Many habitats have recently been destroyed by human impact, mainly due to removal of dead wood and clear cutting.

Westerlund 1884: 181, Kerney & Cameron 1979: 159, Manganelli et al. 1995: 26, Lacina 2011.

Ruthenica

Ruthenica Lindholm, 1924
in Lindholm 1924: 68 *Graciliaria (Ruthenica)*

Type species *Clausilia filograna* Rossmässler, 1836
(original designation)

Ruthenica filograna (Rossmässler, 1836)

Clausilia filograna Rossmässler, 1836 in Rossmässler 1835-1837 (4): 17, Taf. 18

7.5-9.0 x 2.0-2.2 mm

Germany: Bayern, Hersbruck

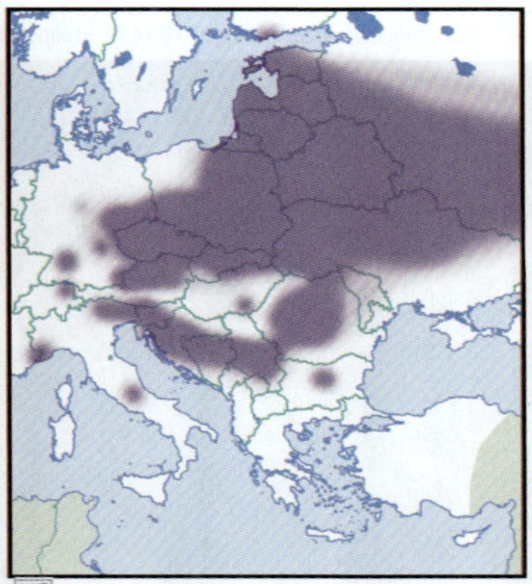

Central and E Europe

Shell much like *Clausilia*, pale yellowish brown, usually sharply ribbed (relatively widely spaced, 30-40 ribs in the penultimate whorl), diameter relatively large at penultimate whorl, apex conical, ending in a sharp point, aperture at lower part not forming a groove, parietalis and columellaris weak, palatal callus present, no lunula. Apertural characters variable, also within a population, without a visible geographical structure.

In ground litter, not confined to rocks, lives also in regions without rocks, at stones and dead wood. Snails hide inside the soil and leave the soil only at favourable weather conditions (warm temperatures and high humidity). Mainly in mountains, very frequent in Romania but restricted to forest habitats below 1500 m.

Klemm 1951, Kerney & Cameron 1979: 158, Grossu 1981: 163, Szybiak & Leśniewska 2008 (morphological variation).

Slovakia: Slovenský Kras National Park
Jozef Grego

Ruthenica gallinae (Bielz, 1861)

Clausilia filograna var. *gallinae* Bielz, 1861 in E. A. Bielz 1861: 216

8-10 x 2.3-2.7 mm

Romania: jud. Alba, Valea Mare

Romania (SW Apuseni mountains)

Shell horny brown, often with white surface layers, weakly ribbed, apex wide, 9-10 whorls, cervix rounded without keel, aperture larger than in *R. filograna*, margin less detached, columellaris weak and close to parietalis, principalis short, not so deep inside but still behind lunula, lunula lateral, clausilium less deep inside.

Westerlund 1884: 180, Grossu 1981: 166.

Siciliaria

Siciliaria Vest, 1867 in Vest 1867: 166 *Clausilia (Siciliaria)*

Type species *Clausilia grohmanniana* Rossmässler, 1836 (original designation)

For the reasons explained above under *Charpentieria* the genus *Siciliaria* should be maintained. I do not follow Nordsieck 2002 and Beckmann 2004 who classified *Siciliaria* as a subgenus of *Charpentieria* and unnecessarily forced a change of the genus-species combinations of more than 20 species.

The specific classification within the genus *Siciliaria* needs a revision. Sicilian groups proposed by Nordsieck 2002:
1a - *crassicostata, eminens, nobilis, spezialensis,*
1b - *calcarae, leucophryna, tiberii,* possibly *riberothi* and *ferrox*
2 - *septemplicata, grohmanniana, scarificata.*
These 12 species were also recognized by Beckmann 2004. Several parapatric so-called species were reported to produce "hybrid populations": *calcarae* / *tiberii, crassicostata* / *eminens, tiberii* / *leucophryna, tiberii* / *septemplicata, septemplicata* / *grohmanniana.*
Nordsieck 2004 used the expression "hybrid population" for a variable species in Crete (*Albinaria cretensis*) of which the extreme populations at ends of continuous morphological gradients over area differed in such a high degree from each other that one could assume these were different species. If the expression "hybrid population" was used in the same sense here, which is likely, Nordsieck's so-called *Siciliaria* species are not tenable in a biological species concept based on the absence of continuous gene flow between populations.
This would reduce the number of true species down to 7: *crassicostata* (including *eminens*), *ferrox, grohmanniana* (including *calcarae, leucophryna, septemplicata* and *tiberii*), *nobilis, riberothi, scarificata, spezialensis.* In the *grohmanniana* complex we have decollated (*grohmanniana, leucophryne* and *tiberii*) and non-decollated populations (*septemplicata, calcarae*).
The morphological gradients of the involved *Siciliaria* populations and the nature of the so-called hybrid populations need to be studied more in detail. It makes no sense to present a revision here, so I copy Nordsieck's 2002 and Beckmann's 2004 classification based on 12 Sicilian species. In a future revision I would expect the number of true species to be substantially reduced.
It would also be useful to have a more detailed knowledge about the ranges of the species. The Sicilian range maps presented here are not very accurate.

Siciliaria calcarae (Philippi, 1844)

Clausilia calcarae Philippi, 1844 in Philippi 1844 (Z. f. Malak 1): 107 (*Clausilia Calcarae*)

18-24 x 4-5 mm

Italy: Sicily, Calatafimi

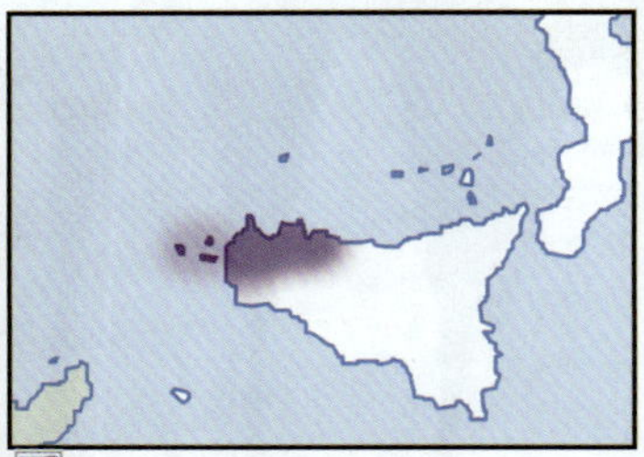

Italy (NW Sicily, Egadi islands)

Shell light horny yellowish, or darker, finely striated, 9-12 whorls, suture papillated, cervix with weak basal keel, columellaris straight or weakly S-shaped, two upper and one lower palatal folds, lower palatalis visible in a perpendicular view, lunula dorsal to dorsolateral, subcolumellaris long but not well visible in the aperture.

Westerlund 1884: 47, Manganelli et al. 1995: 25, Beckmann 2004.

Siciliaria crassicostata (Pfeiffer, 1856)

Clausilia crassicostata Pfeiffer, 1856 in L. Pfeiffer 1856: 184, Taf. 2
Clausilia crassicostata var. *eminens* Schmidt, 1868 in Schmidt 1868: 43 Type locality unknown.

Shell often decollated with 5-9 whorls remaining, whitish cream, coarsely or densely ribbed, cervix with strong basal and weaker dorsal keel, apertural margin shortly detached, columellaris ascending obliquely, 2 upper palatal and 1 lower palatal folds, lunula dorsal and only weakly visible from outside, subcolumellaris long.

Endemic species of Monte Cofano and Favignana island.

Westerlund 1884: 45, Manganelli et al. 1995: 25, Beckmann 2004.

The type locality of *eminens* was not recorded, differs only in more densely arranged ribs, not other differences were observed.

19-23 (decollated) x 5-5.8 mm

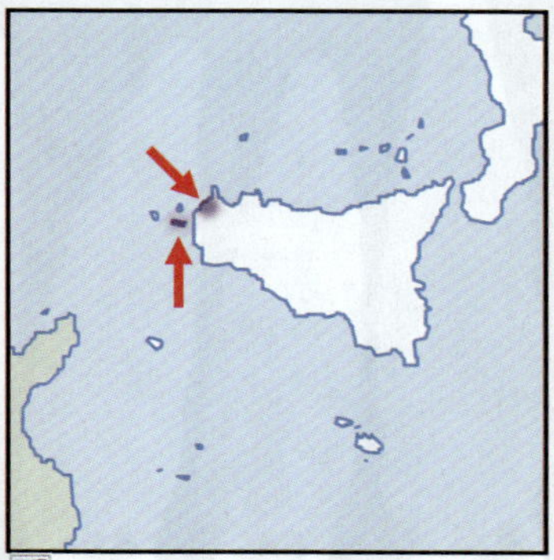

Italy (NW Sicily and Favignana)

Italy: Sicily, Monte Cofano

Siciliaria ernae Fauer, 1978

Siciliaria ernae Fauer, 1978 in Fauer 1978: 263

17-23 x 3.7-4.3 mm

S Italy: Passo Sentinella near Salerno

S Italy (Campania)

Shell violetish brown with white and papillated suture and whitish dashes provoked by opaque surface layers, weakly striated, on cervix finely striated, 11-13 whorls, cervix with strong basal and weak dorsal keel, aperture detached, margin white and slightly reflected, parietalis strong and reaching spiralis, columellaris moderate, lunula dorsal to dorsolateral, subclaustralis sometimes separated from lunula, no frontal upper palatalis, no basalis, subcolumellaris deep inside, not always visible in the aperture. Differs from *S. paestana* in its detached apertural margin and more elongate shell.

Was found in rock crevices of a S exposed mountain slope.

Manganelli et al. 1995: 25.

More research seems to be necessary, it is possible that the form is conspecific with *S. paestana* and the differences in the shell mainly provoked by different habitats. In Monti Alburni specimens were found which displayed intermediate characters and which need to be better understood.

Siciliaria ferrox Brandt, 1961

Siciliaria ferrox Brandt, 1961 in Brandt 1961: 6

Shell dirty yellowish, greyish or brown, ribbed (40-60 ribs/penultimate whorl), cervix equally desely ribbed, 9-12 whorls, suture papillated in the upper whorls, cervix with moderate basal and dorsal keels, aperture inside whitish, margin slightly detached and broadly reflected, parietalis short, columellaris weak and oblique, subcolumellaris visible, lunula dorsal, two small frontal upper palatal folds end in a palatal callus, basalis present, but only sometimes visible in a perpendicular view. Clausilium plate with deep groove, terminal end curved.
Differs from *Papillifera bidens* in the presence of a basalis.

Limestone rocks, together with *Papillifera bidens*.

Manganelli et al. 1995: 25, Beckmann 2004, Reitano et al. 2007: 326.

S Italy: NW Sicily
José Liétor Gallego

14-20 x 3.7-4.6 mm

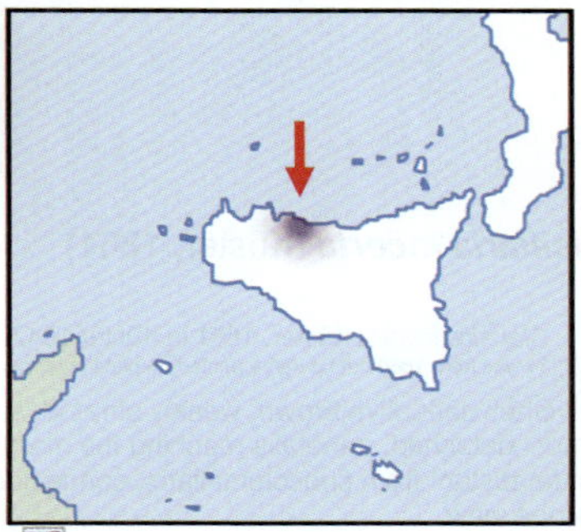

Italy (NW Sicily)

Siciliaria gibbula (Rossmässler, 1836)

Clausilia gibbula Rossmässler, 1836 in Rossmässler 1835-1837 (3): 12, Taf. 12

10-16 x 2.6-3.5 mm

Italy: Veneto, Colli Euganei near Padova, Theodo

Italy to central Dalmatia

Shell horny yellow to horny brown, smooth, finely striated or densely ribbed, shiny and translucent, 8-11 whorls, suture papillated, with cervical callus also visible from inside the aperture, margin usually connected at parietal side but not detached, columellaris S-shaped and almost horizontal, upper palatal fold passing lunula distally, middle palatal fold short and connected to lunula, lower palatalis strong, very well visible in a perpendicular view, connected to lunula and distant from subcolumellaris, subclaustralis present, lunula dorsal and short, subcolumellaris long and visible in the aperture.

Differs from *Papillifera bidens* in the presence of a strong lower palatal fold and in the colour between the dots at the suture being not so very much darker than the rest of the shell. The lunula in *P. bidens* is more oblique.

The animal is brown with light grey foot and lower tentacles.

Westerlund 1884: 55, Falkner 1990: 160, Manganelli et al. 1995: 25, Štamol & Velkovrh 1995 (N Croatia: Cres island, at 2 of 80 stations).

Siciliaria grohmanniana (Rossmässler, 1836)

Clausilia grohmanniana Rossmässler, 1836 in Rossmässler 1835-1837 (3): 7, Taf. 12 (*Clausilia Grohmanniana*)

15-25 (decollated) x 4-6 mm

Italy: NW Sicily, Monte Pellegrino

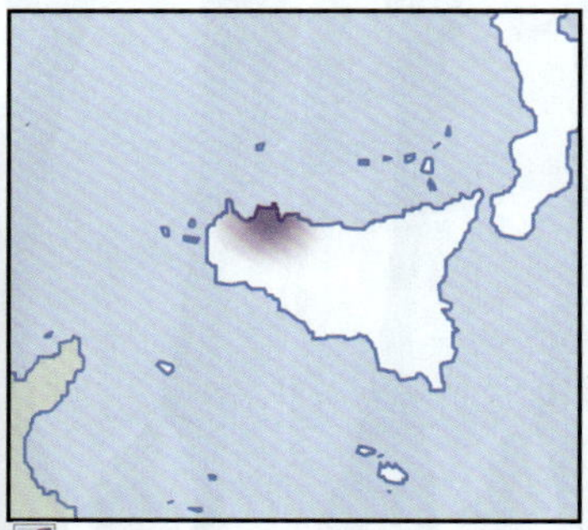

Italy (NW Sicily)

Shell usually decollated with 6-8 whorls remaining, whitish to yellowish brown, ribbed, cervix with prominent basal keel, apertural margin detached, columellaris prominent and curved, horizontal in the aperture, 3 frontal palatal folds, of which 2 are short, the three are connected by a callus at their frontal end, lower palatalis prominent, white, subcolumellaris long, between subcolumellaris and basalis there is an accessory fold for the clausilium (subclaustralis?).

Westerlund 1884: 46, Falkner 1990: 160, Manganelli et al. 1995: 25, Beckmann 2004.

Siciliaria incerta (Küster, 1861)

Clausilia incerta Küster, 1861 in Küster 1847: 299, Taf. 34

Taf. 34 was from 1860 but no name printed on plate (wrapper was not preserved and does not constitute published work under Art. 8.1.1), text from 1861.

Shell dark olive brown, weakly striated, suture papillated, cervix with very weak basal keel, apertural margin almost detached, parietalis reaching the margin, columellaris strong and S-shaped, lower palatalis prominent, not far distant from subcolumellaris, connected to lunula, lunula narrow and curved, subcolumellaris visible in an oblique view.

16-19 x 3.5-4.5 mm

Westerlund 1884: 60, Manganelli et al. 1995: 25.

Exact range unknown. Forms corresponding to this taxon were found in the northern mountain chains between near Palermo and Monti Nebrodi, also near Siracusa, perhaps artificially introduced there.

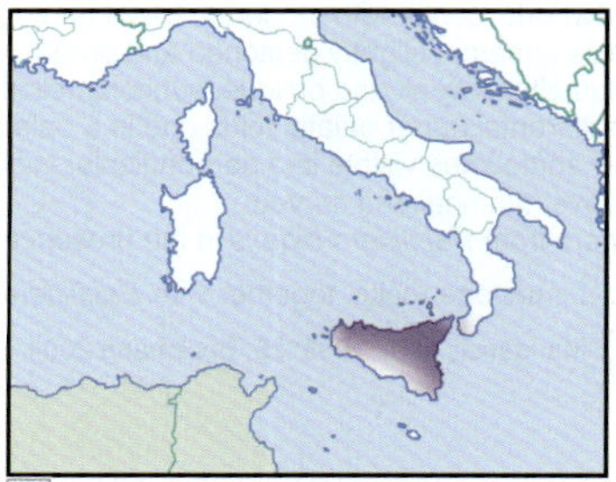

Italy (Sicily)

Siciliaria kobeltiana (Küster, 1876)

Clausilia kobeltiana Küster, 1876 in Küster 1875: 43 (*Clausilia Kobeltiana*)
Date usually given as 1876, but could be 1875, I saw no evidence for 1876 in the original work and don't know a published source that suggested 1876.

13-18 x 3.2-3.9 mm

S Italy: Calabria, Catanzaro, Monte Tiriolo

Alessandro Hallgass

S Italy (Lucania, Calabria)

Shell olive or horny brown, finely ribbed, cervix slightly more coarsely ribbed, suture with dark margin and papillated, cervix with very weak basal keel, columellaris S-shaped and very oblique, upper palatalis sometimes present, lower palatalis short and hardly visible, lunula dorsal and S-shaped, subcolumellaris not very long.

Westerlund 1884: 58, Manganelli et al. 1995: 25.

Siciliaria lamellata (Rossmässler, 1836)

Clausilia lamellata Rossmässler, 1836 in Rossmässler 1835-1837 (4): 14, Taf. 18

Lunula usually more deeply situated than in *S. stigmatica*, frontal upper palatalis if present then connected with upper palatalis.

Westerlund 1884: 58, Nordsieck 1996: 10, Dhora & Welter-Schultes 1996: 135.

13-15 x 2.7-3.3 mm

Greece: Kerkira, Paleokastritsa
Massimo Prodan

SW Albania, NW Greece (Kérkira)

Siciliaria leucophryna (Pfeiffer, 1862)

Clausilia leucophryna Pfeiffer, 1862 in L. Pfeiffer 1862: 204
Original spelling was not *leucophryne*.

Shell decollated with 7-10 whorls remaining, dirty whitish, densely ribbed, cervix slightly more prominently ribbed, with narrow basal keel, aperture inside white, margin strongly detached, columellaris ascending obliquely, two upper and one lower palatal folds, lunula visible as a white structure from outside, subcolumellaris long.

19-22 (decollated) x 4.5-6 mm

Westerlund 1884: 45, Manganelli et al. 1995: 25, Beckmann 2004.

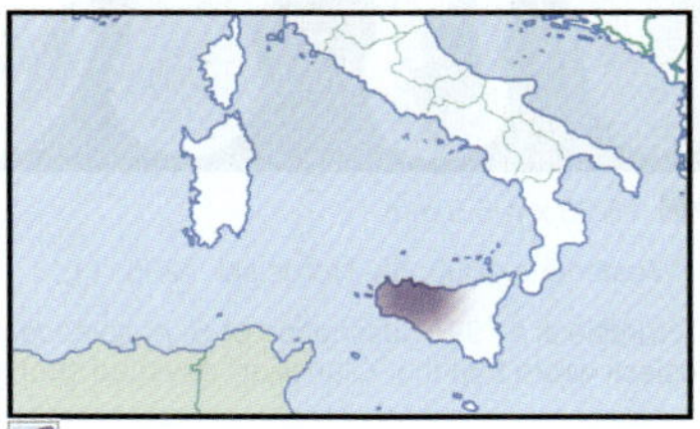

Italy (NW Sicily)

Siciliaria nobilis (Pfeiffer, 1848)

Clausilia nobilis Pfeiffer, 1848 in L. Pfeiffer 1848: 434

22-30 x 5-6 mm

Italy: W Sicily, (1) Monte Cofano, (2-3) Castelluzzo, (4-7) Monte Monaco

(1-3) Giuseppe Pocaterra

Italy (NW Sicily)

Shell decollated with 5-10 whorls remaining, light with some bluish grey spots, densely ribbed or striated, cervix with short basal keel, apertural margin shortly detached, two upper and one lower palatal folds, lunula strong and curved, subcolumellaris long, visible in an oblique view.

Limestone habitats at walls, under stones and in crevices, also in coastal habitats at few meters from the sea shore.

Westerlund 1884: 45, Manganelli et al. 1995: 25, Beckmann 2004.

Siciliaria paestana (Philippi, 1836)

Clausilia paestana Philippi, 1836 in Philippi 1836: 138, 145

Shell brown with horny, olive or reddish hue, almost smooth, suture papillated, 9-12 whorls, cervix ribbed, aperture with prominent white lip, margin connected at parietal side but not detached, columellaris moderate, no lower palatalis, lunula strongly curved inside in its upper section, broadened in its lower section, subcolumellaris visible. Differs from *Delima piceata* in its more reddish colour, a weaker or absent subclaustralis and a weaker subcolumellaris.

Reported from under coniferous trees.

Westerlund 1884: 61, Manganelli et al. 1995: 25.

12-19 x 3-4.5 mm, perhaps locally variable

Italy: Amalfi

Central Italy

Siciliaria pantocratoris (Boettger, 1889)

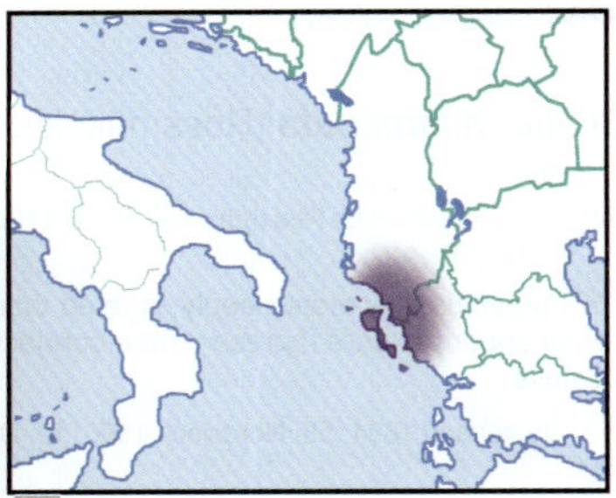

S Albania, NW Greece

Clausilia pantocratoris Boettger, 1889 in O. Boettger 1889: 135
Siciliaria pantocratoris splendens Nordsieck, 1996 in Nordsieck 1996 (Schr. Malak. 9): 12

(1-5) S Albania: Gjirokastër, (6-8) Greece: Kerkira, Mt. Pantocrator
(6-8) Massimo Prodan

15-23 x 4-5.5 mm

Westerlund 1890: 56, Nordsieck 1996: 11.

Shell dark olive or reddish brown, weakly striated or almost smooth, shiny, 10-12 whorls, suture white and irregularly papillated, cervix rounded with strong basal keel, aperture not detached, columellaris prominent, principalis long, lunula dorsal, completely present (nowhere interrupted) and connected to upper palatalis, sulcalis always connected to lunula, subcolumellaris visible in a perpendicular view. Clausilium plate rounded, often broad, not two-lobed, sometimes with a weak outer edge.
Differs from *S. stigmatica* mainly in its long and complete lunula which is connected to the upper palatalis and sulcalis.

Nordsieck's 1996 description of *S. p. splendens* did not explain what exactly were the differences to *S. pantocratoris*, with which *splendens* seems to live parapatrically. Nordsieck wrote that the "Clausiliar" deviated strongly, but did not explain by which visible characters and how they were expressed in both forms.

Siciliaria riberothi Brandt, 1961

Siciliaria leucophryne riberothi Brandt, 1961 in Brandt 1961: 8

Shell decollated with 7-10 whorls remaining, dark brown, irregularly ribbed (30-45 ribs/penultimate whorl), cervix slightly more prominently ribbed, with moderate basal and prominent dorsal keel, aperture inside white, margin strongly detached, columellaris moderately prominent and ascending obliquely, two upper and one lower palatal folds, lunula visible as a white structure from outside, subcolumellaris long.
Differs from *S. leucophryna* in its darker shell, coarser ribs, more prominent cervical keels, more prominent columellaris and in its clausilium, of which the lower margin is not tooth-like protruded.

18-22 x 4.6-5.4 mm

Was initially classified with *S. leucophryna*, from which it seems to differ only weakly.

Italy (S Sicily)

Siciliaria scarificata (Pfeiffer, 1856)

Clausilia scarificata Pfeiffer, 1856 in L. Pfeiffer 1856: 185, Taf. 2

Shell decollated with 7-10 whorls remaining, brown, striated or ribbed, cervix with moderate or weak basal keel, apertural margin light brownish, columellaris ascending in a double S curve, two upper and one lower palatal folds, lunula narrow and S-shaped (seen from outside), subclaustralis present, subcolumellaris long.

Limestone habitats under rocks.

Westerlund 1884: 46, Manganelli et al. 1995: 25, Beckmann 2004.

15-17 x 4-5 mm

S Italy: island of Marettimo W of Sicily
José Liétor Gallego

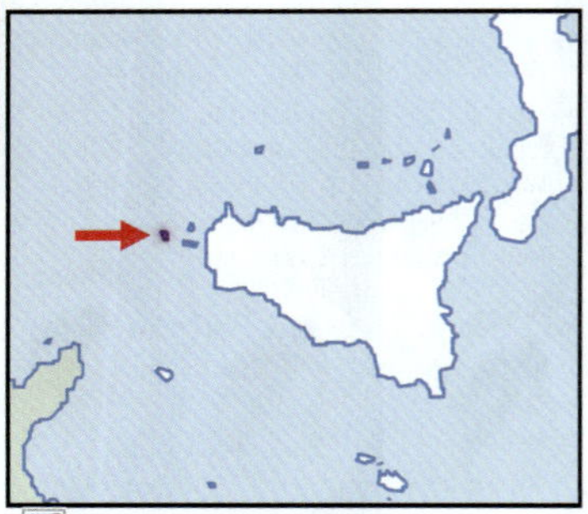

S Italy (Marettimo island)

Siciliaria septemplicata (Philippi, 1836)

Clausilia septemplicata Philippi, 1836 in Philippi 1836: 139, Tab. 8

Italy: Sicily, Palermo

16-22 x 3.5-5.2 mm, perhaps locally variable

Westerlund 1884: 47, Giusti et al. 1995: 327, Manganelli et al. 1995: 25, Beckmann 2004.

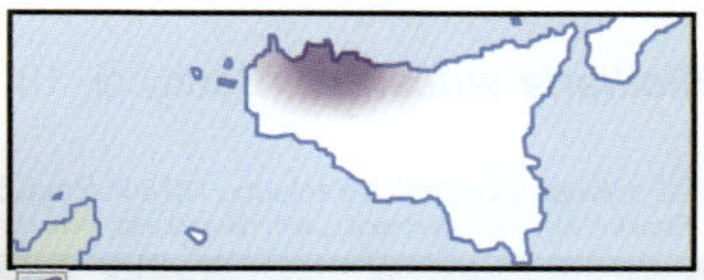

Italy (NW Sicily)

Shell horny or yellowish brown, finely striated, 8-12 whorls, cervix ribbed, with weak basal keel, columellaris prominent and strongly curved, almost horizontal, 3 upper palatal folds (the uppermost one passing the lunula distally), the two other upper palatal folds are shorter and connected by a callus at their frontal end, lower palatalis prominent, subclaustralis present, subcolumellaris long and visible in an oblique view.
Locally variable in terms of size, apertural margin, prominence and curvation of the columellaris, and presence and prominence of palatal folds. The detailed spatial variation is not known, but some extreme variants have been classified as subspecies.

Rocky habitats, in crevices and under stones, also in litter under trees and grasses.

Siciliaria spezialensis Nordsieck, 1984

Siciliaria spezialensis Nordsieck, 1984 in Nordsieck 1984: 202

Shell usually decollated with 6 whorls remaining, whitish, ribbed (35-45 ribs/penultimate whorl), denser and slightly corrugated on cervix, cervix with strong basal and very weak dorsal keel, apertural margin broad, reflected and usually detached, columellaris oblique and not very prominent, lunula dorsal, basalis very weak, no frontal upper palatalis, subcolumellaris usually invisible in a perpendicular view.
Differs from *Sic. nobilis* in the absence of a frontal upper palatalis and absent or much weaker basalis.

20-25 x 5.2-6 mm

Mountain habitats, at the foot of Monte Speziale and in higher altitudes.

Occupies a very small restricted range between Tarpani and San Vito.

Nordsieck 1984: 202, Manganelli et al. 1995: 25, Beckmann 2004.

More research is necessary to clarify the status of this form.

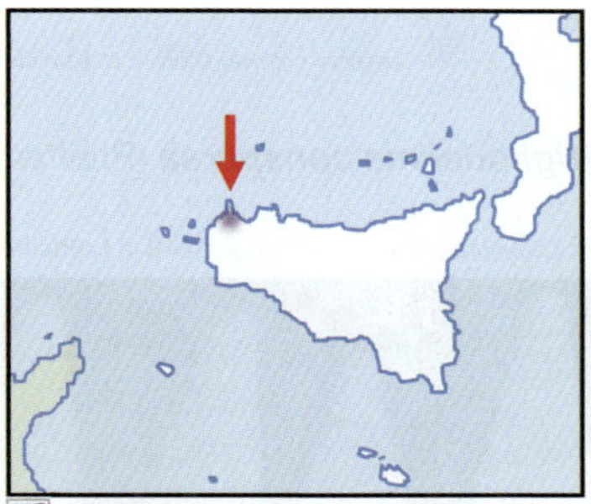

Italy (NW Sicily)

Siciliaria stigmatica (Rossmässler, 1836)

Clausilia stigmatica Rossmässler, 1836 in Rossmässler 1835-1837 (3): 9, Taf. 12

Albania: near Librazhd

13-20 x 3-4 mm, locally variable

Croatia to Greece, SE Italy

Shell cherry brown or horny reddish, striated, suture papillated, cervix with weak basal keel and shallow basal furrow, aperture with very weak lip, margin not or only weakly connected at parietal side, parietalis reaching margin, columellaris runs inside almost without curve, lower palatalis straight below the centre of the columellaris, connected to lunula, lunula dorsal, in its upper section curved, subcolumellaris visible in an oblique view.

Westerlund 1884: 59, Rähle 1980: 216 (Kefaloniá, Zákinthos), Manganelli et al. 1995: 25, Dhora & Welter-Schultes 1996: 136.

Siciliaria tiberii (Schmidt, 1868)

Clausilia tiberii Schmidt, 1868 in Schmidt 1868: 42

Shell light brown to whitish, densely ribbed, 10-12 whorls, cervix with two keels, aperture nearly with a double basal furrow, margin distinctly detached, broadened, columellaris moderate, ascending obliquely, two upper and one lower palatal folds, lower palatalis prominent, subcolumellaris long and visible in an oblique view.

Limestone habitats under rocks.

Westerlund 1884: 46, Manganelli et al. 1995: 25, Beckmann 2004.

20-23 x 4.3-5.1 mm

Italy: Sicily, E of Mondello near Palermo

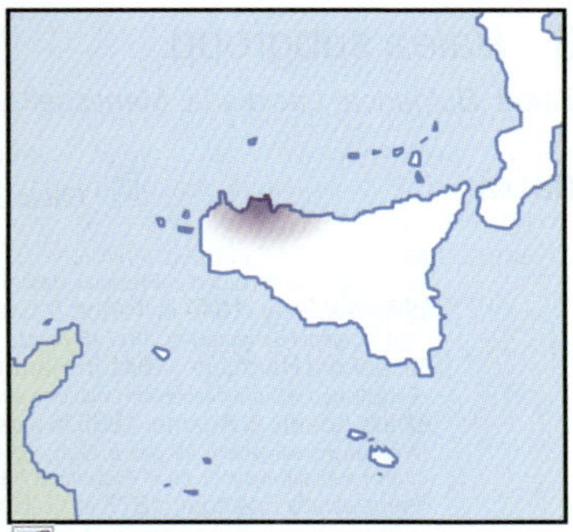

Italy (NW Sicily)

Siciliaria vulcanica (Paulucci, 1878)

Clausilia vulcanica Paulucci, 1878 in Paulucci 1878: 12
Established without description, with reference to "Ben. Moll. Sicil., pl. 6., fig. 18" (= Benoit 1857-1862). The name was not mentioned in Benoit's work.

12-15 x 3.2-3.7 mm

Italy: E Sicily, Acireale

Italy (E Sicily)
Range uncertain.

Shell horny yellow or horny brown, very weakly striated, suture not much papillated, cervix without keel, in last section more prominently striated, apertural margin connected at parietal side, often very slightly detached, columellaris long and horizontal, lower palatalis strong and connected to lunula, lunula dorsal and indistinct, subclaustralis short, subcolumellaris often visible in a perpendicular view.

Westerlund 1884: 55, Westerlund 1901: 58, Alzona 1971: 90.

Strigilodelima

Strigilodelima Wagner, 1924 in A. J. Wagner 1924: 125 *Delima (Strigilodelima)*

Type species *Clausilia platystoma* Küster, 1850 (subsequent designation by Wenz & Zilch 1959-1960: 430) (= currently *Strigilodelima conspersa* (Pfeiffer, 1848))

Contained originally *Clausilia platystoma* Küster, 1850 and *Delima pentheri* Wagner, 1919.

Strigilodelima conspersa (Pfeiffer, 1848)

Clausilia conspersa Pfeiffer, 1848 in L. Pfeiffer 1848 (Mon. Helic. 2): 456

15-20 x 3.3-4.5 mm

Albania: Mali Gradecit E of Çorovodë

In Rep. Makedonija between low altitudes and 2000 m, mainly on limestone in deciduous forests, less frequently in open habitats on alpine limestone rocks. Frequent in Albania.

Westerlund 1884: 67, Dhora & Welter-Schultes 1996: 130, Dedov & Neubert 2009: 92.

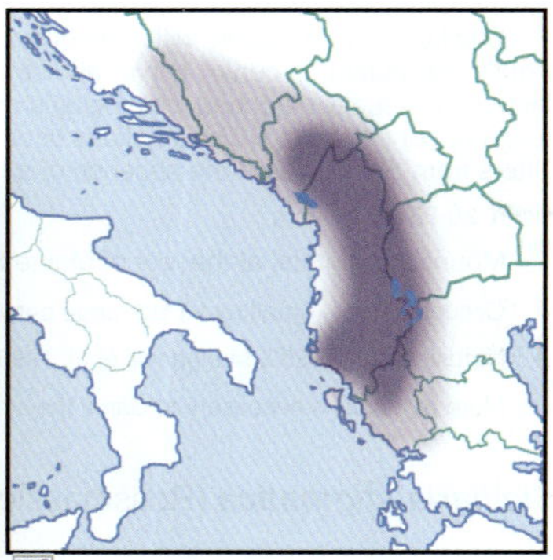

Croatia and Bosnia to N Greece

Shell brown, light brown or with violet hue, densely striated or ribbed, 9-11 whorls, cervix finely ribbed with curved ribs, aperture rounded with strong white lip, margin connected at parietal side by a sometimes very thin white layer, usually not detached, broad, parietalis not reaching margin, columellaris prominent, almost horizontal and far distant from parietalis, lunula dorsolateral, strongly curved, subcolumellaris visible in the aperture and usually reaches apertural base.

Strigilodelima pentheri (Wagner, 1919)

Delima pentheri Wagner, 1919 in A. J. Wagner 1919: 70

Shell reddish brown, often with greyish layers, slightly translucent, regularly and densely ribbed, 9-11 whorls, cervix regularly rounded, aperture inside and at margin yellowish brown, margin detached and broadly reflected, parietalis not reaching margin, passing spiralis inside, columellaris oblique and does not reach margin, basalis present, subcolumellaris visible in a perpendicular view. Clausilium plate forming a groove, pointed in anterior part.

14-18 x 3.3-3.7 mm

Critically Endangered (Dhora 2004: 145). No reasons were given for this assessment.

Dhora & Welter-Schultes 1996: 130.

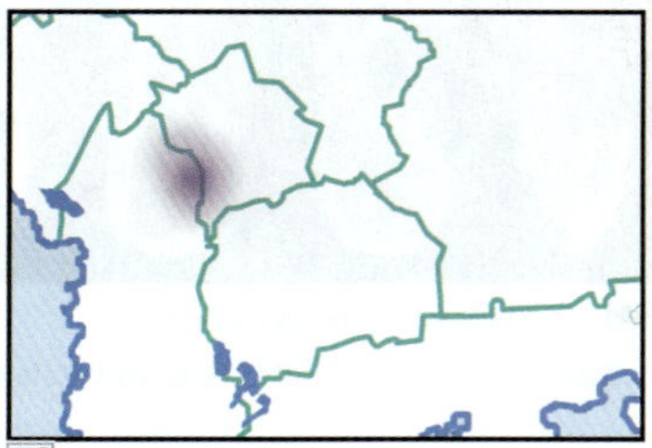

NE Albania

6 - *Balea* subgroup

Balea, *Bulgarica*, *Laciniaria*, *Mentissella*, *Vestia*

Balea

Balea Gray, 1824 in Gray 1824: 61 *Clausilia (Balea)*

Type species *Turbo perversus* Linnæus, 1758 (subsequent designation by Turton 1831: 7)

Originally attributed to Prideaux, description was by Gray. Contained originally *Pupa fragilis* Draparnaud, 1801 with *T. perversus* as one of its synonyms, *B. tristensis* and *B. ventricosa* Gray, 1824.

Iphigenia Gray, 1840 in Turton & Gray 1840: 214 *Clausilia (Iphigenia)*, type *Turbo biplicatus* Montagu, 1803 (subsequent designation by Gray 1847: 177)
Not *Iphigenia* Schumacher, 1817 (Bivalvia).

Plicaphora Hartmann, 1841 in Hartmann 1844: 48, type *Clausilia similis* Hartmann, 1821 (monotypy) (= currently *Balea biplicata* (Montagu, 1803))
Established without description, with one species included. Used for a taxon by Nordsieck 2000: 11 in the belief that it had another type species. Cannot be suppressed under Art. 23.9.1.

Alinda Adams & Adams, 1855 in Adams & Adams 1858: 182 *Clausilia (Alinda)*, type *Turbo biplicatus* Montagu, 1803 (subsequent designation by Gray 1847: 177)
A synonym was given "*Iphigenia*, Gray, not Schum.", this indicated that the name was meant as a new replacement name for *Iphigenia* Gray, 1840, not *Iphigenia* Schumacher, 1817 (Bivalvia). Junior subjective synonym of *Plicaphora* Hartmann, 1841. *Alinda* cannot be used as a name for a genus or subgenus.

Pseudalinda Boettger, 1877 in O. Boettger 1877: 76 *Clausilia (Pseudalinda)*, type *Clausilia montana* Pfeiffer, 1847 (subsequent designation by Westerlund 1902: 111) (= currently *Balea viridana* (Rossmässler, 1836))
Schileyko 2000: 721 gave as type *Balea fallax* by original designation, which was incorrect. Contained originally 11 nominal species.

Contains 3 subgroups, one containing *Balea perversa* and *lucifuga*, one with *Balea biplicata* and related aspecies, and one with *Balea viridana*, *B. fallax* and others. These groups can be classified as genera or subgenera.
Nordsieck 1977: 98 proposed to classify all groups as subgenera in *Balea*. This was accepted by Grossu 1981, Kerney et al. 1983, Falkner 1990 and Falkner et al. 2001, but not by Schileyko 2000 and Nordsieck 2007 (Mitt. DMG) who classified *perversa* in one genus *Balea* and *biplicata* and *viridana* in a second genus for which they used *Alinda*. Nordsieck 2007 justified the separation into two genera only with gradual differences and a suspected independent evolutionary lineage suggested by the molecular record and presence of endemic species of one of the subgroups in Bulgaria.
To escape the problem with *Alinda* and *Plicaphora*, a name which has never been used for this group but which cannot be skipped because Nordsieck 2000 used it, the genus *Balea* is maintained here for the three subgroups, in which the subgroups are classified as subgenera.
Otherwise *Clausilia biplicata* will receive a 4th genus-species combination in less than 50 years (*Laciniaria* was used by Kerney & Cameron 1979 and many previous authors, after 1977 *Balea*, after 2000 *Alinda* and then *Plicaphora* because *Alinda* cannot be used).

Balea biplicata (Montagu, 1803)

Without gender agreement: *Balea biplicatus* (Montagu, 1803)

Turbo biplicatus Montagu, 1803 in Montagu 1803: 361
Clausilia similis Hartmann, 1821 in Hartmann 1821 (Neue Alpina): 217

15-18 x 3.8-4.5 mm

Germany. Bayern, near Miltenberg

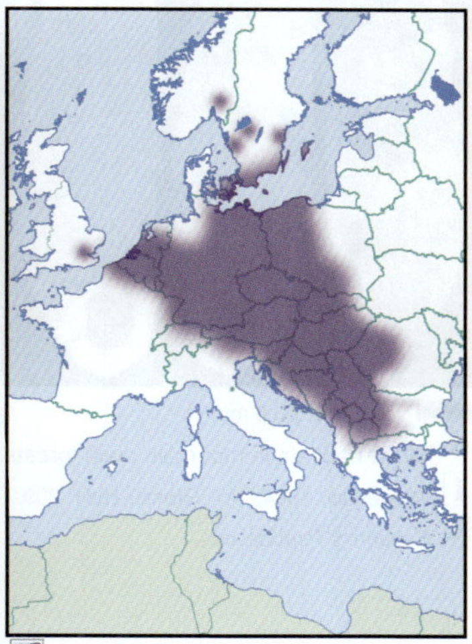

Europe

Germany: Göttingen

Shell brown, moderately widely ribbed, frontal upper palatalis present (and more twisted than in *Laciniaria plicata*), aperture with basal furrow (difference to *Macrogastra ventricosa*), no folds between parietalis and columellaris, subcolumellaris not visible in the aperture. Shell like *L. plicata*, but ribs slightly wider and no small folds at palatal side of aperture.

Shady habitats in forests, under herbal vegetation, leaf litter or between rocks, less frequent in dry and exposed habitats. Most common clausiliid species of Europe. Also in S Bulgaria one of the most common clausiliid species. In Austria and Bulgaria up to 2300 m. In S England in ground litter at the banks of the river Thames, commonly associated with human rubbish, has also been successfully introduced to gardens.

Usually ovoviviparous, but under favourable conditions lays eggs with well-developed embryos, single or in clutches of up to 11 eggs. Reproduction always sexual, in SW Poland from June to August (Netherlands: April to October), juveniles grow at a rate of 1 whorl/month, they hibernate at a size of 3-6 whorls, in the second season they reach 8-10 whorls, maturity is reached when the aperture is completed in the third or fourth season, life span more than 4 years.
Under laboratory conditions maturity of German populations can be reached after 7-16 months, full size earlier (after 4 months), offspring some 40-50 juveniles per parent and season, periods of hibernation also under continuous climate artificial conditions, life span up to 3.5 years.

Apparently introduced to S England near London several 100 years ago, reached many places between the London region and E Cornwall, but the habitats have largely been destroyed by riverside development, only very few populations near London survived, obviously still declining.

Loosjes 1941: 32, Degner 1952, Frömming 1954: 60, Kuźnik-Kowalska 1998 (life cycle), Kerney 1999: 173, Sulikowska-Drozd 2005: 77, Irikov & Mollov 2006: 785.

Subgenus *Plicaphora*.

Balea fallax (Rossmässler, 1836)

Clausilia fallax Rossmässler, 1836 in Rossmässler 1835-1837 (4): 16, Taf. 18

14-22 x 4-5.5 mm, obviously locally variable

Romania: Ciucaş mountains, Kabana Muntele

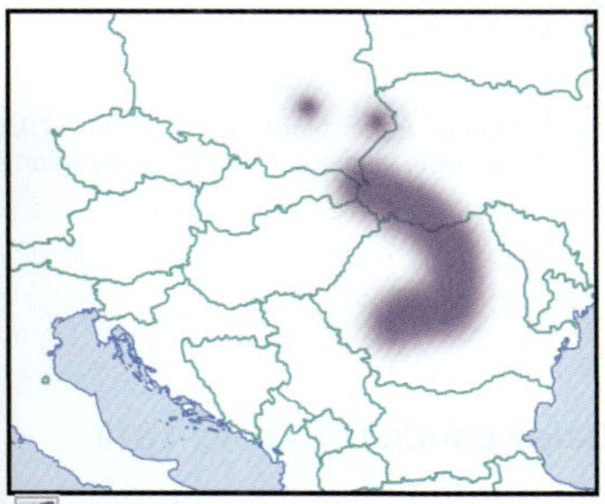

Carpathians (Poland to Romania)

Shell brown, solid, finely ribbed, 10-13 weakly convex whorls, parietalis long and not connected to spiralis, no folds between parietalis and columelaris, columellaris deep inside, slightly concave at lower side and forked inside, no lower palatalis, lunula dorsal and incomplete, subcolumellaris visible in a perpendicular view, next to it a basal furrow is present.

Very rare in Poland.

At trees, in forests, frequently under dead leaf litter and rotting wood, up to 1800 m. In Ukraine in the submontane and lower forest zones.

Reproduction probably only sexual, of Polish populations under laboratory conditions from March to October, 4-11 slightly oval eggs (size 1.7-2 mm) are laid singly or in batches mainly in spring and early autumn, 1-4 batches per animal and year, juveniles hatch after 8-10 days at 18-25 °C, initially with 2.5 whorls, ultimate size is reached after at least 6 months, the first eggs can be laid 5 months after shell growth completion. Between May and August animals retain 10-17 eggs in the uterus, where the embryos grow up to a size of 1 whorl. This reduces the time the eggs need to develop outside.

Westerlund 1884: 146, Kerney et al. 1983: 233, Sulikowska-Drozd 2005: 78, Sulikowska-Drozd & Maltz 2012, Sulikowska-Drozd et al. 2012.

Subgenus *Pseudalinda*. This species has often been confused with *Balea stabilis*.

Balea jugularis (Vest, 1859)

Clausilia critica jugularis Vest, 1859 in Vest 1859: 266 (*Clausilia critica* Varietät b *Clausilia jugularis*)
Was not explicitly (Vest 1859: 263) treated as a name for a species. It is implicite from the context that "*Clausilia jugularis*" was meant as a name for a variety and as such, treated as a name of subspecific rank.

12-15 x 2.5-3.6 mm — Romania: Prahova, Bucegi mountains

Romania (southern Carpathians)

Shell yellowish brown, usually transparent, ribbed, 10-13 weakly convex whorls, parietalis very long, close to spiralis, often 1-2 folds between parietalis and columellaris, palatal callus strong, columellaris deep inside, oblique and forked inside, upper palatalis short and dorsal, no lunula, no lower palatalis, subcolumellaris present.

Moist habitats in mountain plain forests, under rotting wood and bark, sones, among mushrooms and lichens.

Westerlund 1884: 147, Grossu 1981: 209.

Subgenus *Pseudalinda*.

Balea kaeufeli (Brandt, 1962)

Laciniaria kaeufeli Brandt, 1962 in Brandt 1962: 128
Balea kaeufeli eninskoensis Irikov, 2006 in Irikov 2006: 88

9-12 x 2.2-2.7 mm — Bulgaria: Stara Planina, Eninsko gorge E Shipka

Bulgaria

Shell light olive brown, shiny, slightly translucent, striated with spots or densely ribbed, 8-10 moderately convex whorls, last whorl with largest diameter, looks like a juvenile clausiliid, cervix rounded and slightly inflated, apertural margin sharp, detached, white and slightly reflected, parietalis low and sometimes long, no clausilium, no other folds in the aperture, except a very weak and almost perpendicular columellaris visible in an oblique view.

In mountains.

Irikov 2006, Irikov & Erőss 2008.

Subgenus *Balea*. Nordsieck 2007 (Mitt. DMG): 28 treated *B. eninskoensis* as a separate species, based on very weak arguments. Both taxa are from the same region near the town of Shipka, but were collected in different altitudes.

Balea pancici Pavlović, 1912

Balea perversa var. *pancici* Pavlović, 1912 in Pavlović 1912: 79, Tab. I sl. 1-3 (*Balaea perversa* var. *pančići*)

Shell like *B. perversa*, smooth, 7-8 whorls, last whorls and aperture do not have the appearence of a juvenile shell.

8-9 x 2.3-2.7 mm

Subgenus *Balea*. This species needs certainly a revision.

Serbia (Stara Planina)

Balea perversa (Linnæus, 1758)

Without gender agreement: *Balea perversus* (Linnæus, 1758)

Turbo perversus Linnæus, 1758 in Linnæus 1758: 767
Neotype Museum Leiden RMNH 103468 designated by Gittenberger, Preece & Ripken 2006: 146, who explained that they did not find types in Linnean collections, but missed to explain the measures taken to find Lister's (1678) types.
Pupa fragilis Draparnaud, 1801 in Draparnaud 1801: 64

7-10 x 2.5-2.7 mm

Sweden: Öland, between Vickleby and Resmo

Europe

Slovakia: Štiavnické vrchy mountains
Jozef Grego

Shell small, looks like juvenile, pale brownish, often silky shiny, often with distinct riblets, apical whorls cyclindrical, 8-9 whorls, last whorl with largest diameter, densely ribbed, cervix almost without keel, apart from a rudimentary parietalis no folds in the aperture. No clausilium.
Differs from *B. lucifuga* in its less slender and brownish rather than yellowish shell, first whorls increase in diameter less rapidly, sculpture more prominently striated (with usually distinct riblets rather than coarse growth lines). The outline at whorls 2-4 is very slightly concave (in *B. lucifuga* convex).

On mosses and at the bark of trees, also near roads, at walls and rocky slopes, at rocks covered by mosses, less commonly in ground litter. Often on surfaces encrusted with lichens and other epiphytes. Prefers trees with rough bark. May tolerate non-calcareous soils. In Switzerland up to 2000 m. Easily dispersed by birds. Feeds on mosses, algae, lichens and cyanobacteria.

Ovoviviparous, self-fertilization predominates, even in laboratories when snails are kept in pairs. Reproduction in France between August and October, 12-15 eggs (diameter 1.2-1.3 mm) are laid, juveniles hatch after 15-20 days, maturity is reached at the end of the year or at the beginning of the next year. Animals can reach maturity after 3-4 months under favourable conditions, one adult can give birth to 10-20 juveniles per year. Animals can also be active in mild winters.

Critically Endangered in Poland (Zajac 2005), Endangered in Bavaria, Vulnerable in Austria and Germany. Locally threatened by too thorough and too frequent restaurations of old buildings, acid rains, air pollution and cutting of old trees. Has largely disppeared from inside cities. Remains frequent in Ireland, but many colonies in lowland England have certainly disappeared. Extinct around London since the 1920s.

Moquin-Tandon 1855: 349 (life cycle), Germain 1930: 346, Damjanov & Likharev 1975: 219, Kerney & Cameron 1979: 172, Bech 1988: 107, Baur 1990, Wirth et al. 1997, Turner et al. 1998: 228, Kerney 1999: 174, Irikov & Mollov 2006: 786, Gittenberger et al. 2006: 146, Martínez-Ortí 2006: 37, Hubenov 2007.

Subgenus *Balea*.

Balea sarsii Pfeiffer, 1847

Balea lucifuga Bourguignat, 1857
Balea heydeni Maltzan, 1881

Balea fragilis synonym *Balea lucifuga* Gray, 1824 in Gray 1824: 61
Mentioned with description by Gray and attributed to "Leach Mss.", did not refer to this taxon here, but to *Balea perversa*. Name unavailable (Art. 11.6).
Balea sarsii Pfeiffer, 1847 in L. Pfeiffer 1847: 84 (*Balea Sarsii*) Type locality in Norway.
Name originally attributed to Philippi, description was by Pfeiffer (see "Inhalt").
Balea lucifuga Bourguignat, 1857 in Bourguignat 1857: 557 (*Balia lucifuga*)
Established with description, a name "*Balaea lucifuga* Leach, Mss." was mentioned among the synonyms (without bibliographical reference and expressly not in the sense of Gray's 1824 interpretation of the synonym). *Balia* = incorrect subsequent spelling for *Balea*, no parentheses under Art. 51.3.1. Among the synonyms of another species *B. perversa*, Bourguignat 1857: 551 cited *Balea fragilis* as used by Gray 1824 (with bibliographical reference). Bourguignat used *lucifuga* clearly in an intentionally different sense as Gray 1824. This means that when using *lucifuga*, Bourguignat did not adopt Gray's name.
Gittenberger et al.' 2006 intended to "designate" under Art. 24 both *lucifuga* Gray and *Pupa fragilis* Draparnaud, 1801 as synonyms of *Turbo perversus*. This was not covered by the Code. Gittenberger 2010: 110 argued that Bourguignat 1857 made Gray's name available under Art. 11.6.1 by adopting it for a taxon, and designated a neotype for *Balea lucifuga* "Gray, 1824", because Gray's types were lost. Gray 1824 did not cite types, so the types would have to be taken from Bourguignat (Art. 72.4.3), and the neotype selection would be invalid (Art. 75.3.4, 75.3.5). A neotype must not belong to *B. perversa* (Art. 75.3.5).
It is implicit in the Code that Art. 11.6.1 can only be applied if Bourguignat would have used Gray's synonym for the same taxon, not for an expressly different taxon.
The name *lucifuga* must be attributed to Bourguignat 1857, not to Gray 1824.
Balea heydeni Maltzan, 1881 in Maltzan 1881: 162 (*Balea Heydeni*)

6-7 x 2.2-2.5 mm

Portugal: Sintra

W Europe, Azores, Porto Santo

Shell yellowish brown, finely striated, up to 6 convex whorls, apical whorls conical, last whorl with largest diameter, looks like a juvenile clausiliid, no parietalis, no other folds in the aperture.
Differs from *B. perversa* in its more slender and yellowish rather than brownish shell, first whorls increase in diameter more rapidly, sculpture more weakly striated (with coarse growth lines rather than distinct riblets). The outline at whorls 2-4 is very slightly convex (in *B. perversa* concave).

Like *B. perversa*, it is not yet clear whether the two species exhibit differences in their ecological behaviour. On mosses and at the bark of trees, also near roads, at walls and rocky slopes, at rocks, less commonly in ground litter. Often on surfaces encrusted with lichens and other epiphytes. Prefers trees with rough bark, in Portugal shady habitats. May tolerate non-calcareous soils. Easily dispersed by birds. Feeds on mosses, algae, lichens and cyanobacteria.
In France, Belgium and Netherlands mainly coastal (and sympytrically with *B. perversa*), in Portugal, Great Britain and Ireland also in inland localities. In Britain more frequent than *B. perversa*.

Subgenus *Balea*. Gittenberger, Preece & Ripken 2006 suggested the presence of two species in Europe *B. perversa* and *B. heydeni*, but did not analyse material from Norway. Proschwitz 2010 suggested that *B. sarsii* from Norway was conspecific with the present taxon, after having analysed material from its type locality.

Balea serbiana n. nom.

Balea serbica (Möllendorff, 1873)

Balea serbiana n. nom., new replacement name for
Clausilia fallax var. *serbica* Möllendorff, 1873 in Möllendorff 1873: 146 (not *Clausilia (Idyla) serbica* Möllendorff, 1873 (currently in genus *Bulgarica*) from the same paper p. 144)
Nordsieck 1972: 54 proposed to use the name *serbica* for the *Balea* species, but under Art. 57.7 the name *serbica* must be taken for the *Bulgarica* species, and the *Balea* species must obtain a new name. Derivation of name: from Serbia. To be used as a noun in apposition. The name will not provoke a change of the species' position in alphabetical sortings of *Balea* species.

17-19 x 4.5-5 mm — E Montenegro: Čakor

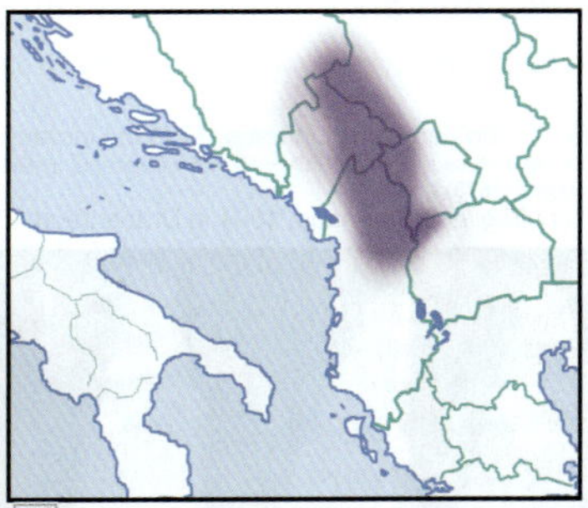

Albania to Serbia

Like *B. fallax* but less slender, yellowish or greenish brown, with shorter parietalis.

Open terrain (not alpine meadows), river valleys and coniferous forests, up to 2200 m.

Westerlund 1884: 146, Damjanov & Likharev 1975: 407, Dhora & W. 1996: 137, Dedov & Neubert 2002: 207.

Subgenus *Pseudalinda*.

Balea stabilis (Pfeiffer, 1847)

Clausilia stabilis Pfeiffer, 1847 in L. Pfeiffer 1847: 70
Balea nordsiecki Dedov & Neubert, 2002 in Dedov & Neubert 2002: 206

11-16 x 3.2-4.0 mm — (1-5) Rep. Makedonija: Gostivar province, near Gorno Jelovce, (6-8) Romania: Transilvania
(1-5) Museum Budapest HNHM

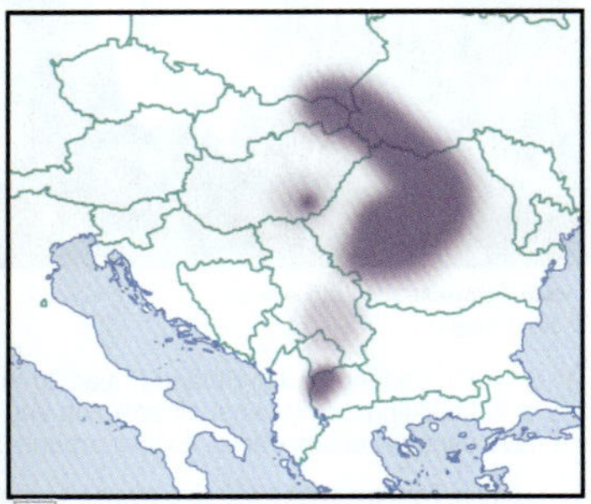

SE Poland to Romania, possibly also Serbia and Makedonija

Shell horny brown or yellowish brown, ribbed, parietalis long, columellaris moderate, upper palatalis long and visible in a perpendicular view, no lunula visible inside and outside, basal groove present, subcolumellaris invisible in the aperture. Shell smaller than *B. fallax*, with which it has often been confounded.

At tree trunks and under rotting wood in humid forests, lowlands and highlands.

Subgenus *Pseudalinda*. Dedov & Neubert 2002 classified populations from Rep. Makedonija as a new species, but did not study local variations of the characters claimed to distinguish *B. nordsiecki* from *B. stabilis* and *B. viridana*. They just referred to a personal communication to H. Nordsieck who "confirmed" their view that the columellaris of *B. nordsiecki* was steeper and its principalis longer than in *B. stabilis*. To meet contemporary scientific standards it would be necessary to analyse the columellaris in a reproducible form, measure the length of the principalis and study other characters in various populations from sufficiently many localities between Poland and Rep. Makedonija, to see and map their local variations, and then eventually come to conclusions based on such results.
The presence of a true range gap (in Kosovo and Serbia) should also be verified. Presence, distribution and identity of *Balea* populations in Kosovo have never been studied. Pavlović 1912 (and Tomić 1959) classified all populations in Serbia with *B. biplicata* and *Laciniaria plicata*. Their classification should be interpreted as a preliminary approach.

Balea viridana (Rossmässler, 1836)

Clausilia viridana Rossmässler, 1836 in Rossmässler 1835-1837 (4): 16, Taf. 18
Clausilia montana Pfeiffer, 1847 in L. Pfeiffer 1847 (Z. f. Malak. 4): 70 t

13-15 x 3.5-4.1 mm — Romania: "Griemaja Dusch"

Westerlund 1884: 147, Grossu 1981: 207, Dedov & Neubert 2002: 206.

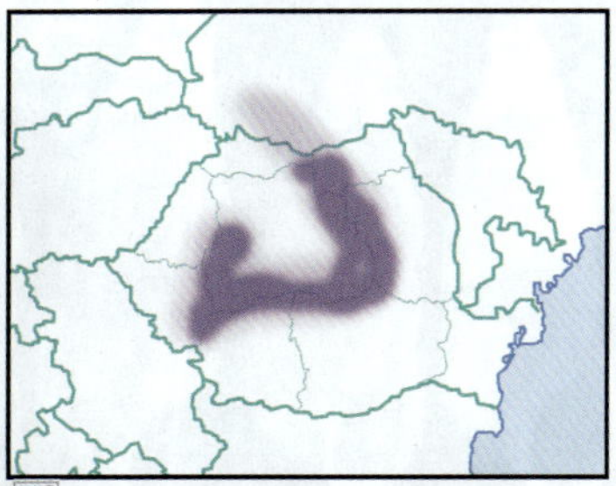

Romania (Carpathians)

Shell greenish or yellowish brown, striated or almost smooth, 10-13 whorls, parietalis long and close to spiralis, columellaris deep inside, oblique and forked inside, palatal callus weak, upper palatalis short, no lower palatalis, lunula dorsal and not developed as a fold, subcolumellaris weak.

Restricted to mountain forests, under leaf litter and rotting wood and bark, among stones, mushrooms and lichens, up to 2000 m altitude.

Subgenus *Pseudalinda*.

Balea vratzatica (Likharev, 1972)

Laciniaria vratzatica Likharev, 1972 in Likharev 1972: 117

10-12 x 2.4-2.6 mm

Bulgaria: Karlukovo cave

Bulgaria

Shell brown, densely ribbed, ribs can be more widely spaced on last whorl, aperture with distinct basal furrow, parietalis reaches apertural margin, columellaris located very high in the aperture, in the upper 1/5 section, very close to the parietalis.

Damjanov & L. 1975: 203, Dedov 1998.

Subgenus *Plicaphora*.

Balea wagneri (Wagner, 1911)

Pseudalinda wagneri Wagner, 1911 ?in Wohlberedt 1911 (??)
I do not know know where this name was established.

15-22 x 4.5-5.5 mm

Bulgaria: (1-7) Stara Planina, near Hristo Danovo, (8-9) Sredna Gora

(8-9) Atanas Irikov

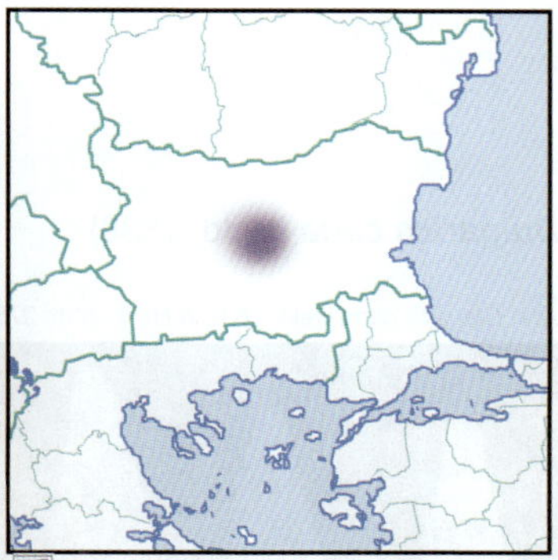

Bulgaria

Nordsieck 2007: 62.

Subgenus *Plicaphora*. I did not find a reference to the original description. Material from D. Georgiev and A. Irikov.

Bulgarica

Bulgarica Boettger, 1877
in O. Boettger 1877 (Palaeontogr.): 93 *Clausilia (Idyla (Bulgarica))*

Type species *Clausilia varnensis* Pfeiffer, 1848 (subsequent designation by Lindholm 1924: 64)

Contained originally 7 species, among these was not *Clausilia bulgarica* Küster, 1860.

Strigillaria Vest, 1867 in Vest 1867: 195, type species *Clausilia vetusta* Rossmässler, 1836 (original designation) (not *Strigilaria*, a nomen nudum mentioned by Rafinesque 1815: 146, Bivalvia)
Vest's name can probably be used under the Code, it was not a homonym of Rafinesque's name because (1) Rafinesque 1815 did not give a description or indication, and (2) spelled the name *Strigilaria*, so the two are not homonyms (Art. 56.2, one-letter difference). It is possible that the genus has to change its name, *Bulgarica* is a junior synonym of *Strigillaria*.

Strigilecula Kennard & Woodward, 1923 in Kennard & Woodward 1923: 304, type species *Clausilia vetusta* Rossmässler, 1836 (designated by Vest 1867: 195)
Expressly established as a new replacement name for "*Strigiliaria*" Vest, 1867, not Rafinesque 1815. The replacement was not necessary, Rafinesque' name unavailable.

Pavlovicia Nordsieck, 1973 in Nordsieck 1973: 192, type species *Bulgarica pavlovici* Nordsieck, 1972 (original designation)

Nordsieck 1973, Nordsieck 1974.

The genus *Bulgarica* needs a critical revision. Possibly the name *Strigillaria* must be used for it.
It seems that we are dealing with a low number of locally variable species, and that the subdivision of various widespread forms into separate species has been as arbitrary as the separation of *Cochlodina inaequalis / C. laminata* or that of various local forms of the *Albinaria cretensis* complex in W Crete. The degree of morphological variation between selected single populations from distant localities should not necessarily be decisive in recognising species. Local morphological gradients which suggest continuous gene flow between such populations should also be taken into account.
Until recently 17 species have been recognized (Nordsieck 2007: 63) (or 19, with *mystica* and *thessalonica*). Their identities within the biological species concept seem only partly justified. Some of these species seem so weakly justified that presenting diagnoses for them would not sound very serious. It is too well visible that the current classification is not aligned with commonly accepted species concepts in biodiversity research.

B. bulgarica / bulgariensis, *B. cana*, *B. denticulata*, *B. fritillaria*, *B. iniucunda*, *B. rugicollis* and *B. vetusta* seem to be tenable, 7 species. I have not seen detailed data for *pygmaea / pavlovici*.
B. denticulata is here regarded as a widespread form that contains *erberi*, *mystica* and *thessalonica*.
B. fritillaria seems to contain *fraudigera*, *hiltrudae*, *pseudofraudigera*, *urbanskii* and *varnensis*.
B. vetusta seems to contain *moellendorffi / serbica*, *pindica* and *stolii*.
I consider the currently published record as insufficient for justifying the presence of more than 8 species. This might be called a dry revision, which should generally be avoided, but it is too obvious that our knowledge does not allow to recognise more. Nordsieck 2007 subdivided the genus into 3 subgenera, without explaining why. *Strigilecula* contained *cana*, *pindica* and *vetusta*, *Pavlovicia* contained *pavlovici*, and *Bulgarica* contained the others.
A taxonomic revision should involve many local populations, detailed analyses of their morphology, spatial distributions of shell characters, and type material.

Bulgarica bulgarica (Küster, 1860)

Bulgarica bulgariensis (Pfeiffer, 1848)

Clausilia pagana γ synonym *Clausilia bulgariensis* Pfeiffer, 1848 in L. Pfeiffer 1848 (Mon. Helic. 2): 469

Mentioned with a short description by Pfeiffer, name attributed to "Friv.". Used as a name by Schmidt 1868: 137, without bibliographical reference to Pfeiffer 1848. For those who demand a bibliographical reference for making available a synonym, the name must be attributed to Schmidt, 1868.

Clausilia bulgarica Küster, 1860 in Küster 1847: 199 (figured on Taf. 21 Fig. 32-35)

Plate from 1855, text from 1860, name not mentioned on plate, only perhaps on a wrapper which was not published work.

11-15 x 2.5-3.5 mm — Bulgaria: Tarnovo, Abanasi

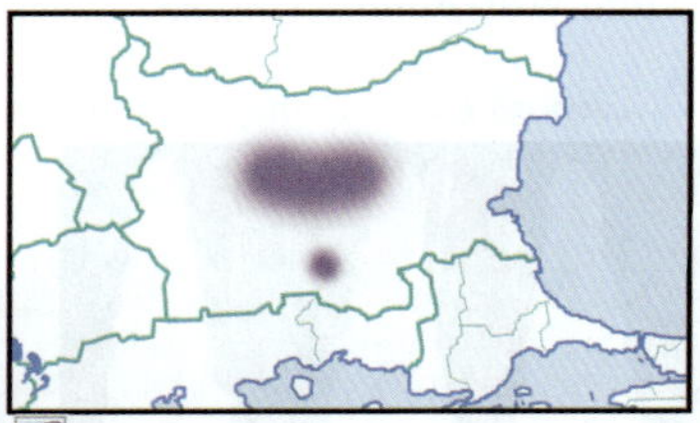

Bulgaria

Shell horny greyish or brownish, finely and densely ribbed, 10-12 whorls, lower palatalis strong, basal groove present, cervix with strong and rounded basal and dorsal keels, cervix unusually narrow at palatal side.
Like *B. rugicollis*, but frontal upper palatalis sometimes connected to distal upper palatalis, and basalis sometimes connected to lunula, clausilium plate with outer edge.

S-exposed limestones with xerothermic conditions. During the summer and the winter the animals use the roots of herbaceous vegetation as shelter. Sometimes animals are active during the winter at rocks not covered by snow (Irikov 2003).

Westerlund 1884: 43 (*Cl. pagana* var. *bulgarica*), Nordsieck 1973: 194 (*Bulgarica intricata*), Irikov & Mollov 2006: 789.

Bulgarian endemic species with disjunct distribution: central Stara Planina mountains and one locality in W Rodopi mountains.

Bulgarica cana (Held, 1836)

Clausilia cana Held, 1836 in Held 1836: 275

16-19 x 3.3-4 mm — S Germany: Bayern, Kaisheim near Donauwörth

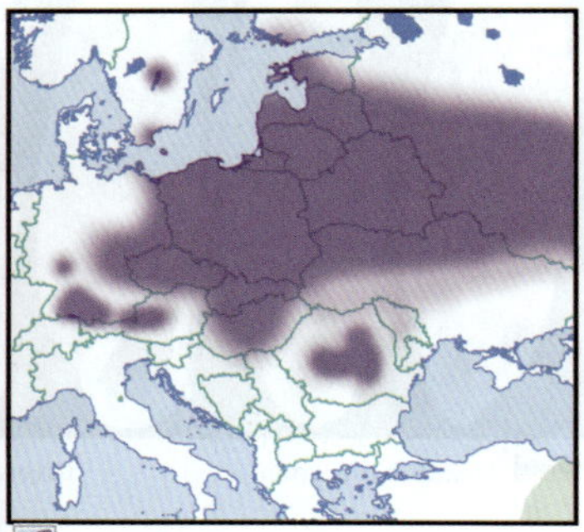

Central and E Europe

Shell brown, solid, ribbed, 11-14 whorls, cervix with basal keel, aperture large with weak palatal callus, columellaris strong and usually brownish red, no frontal upper palatalis, weak frontal lower palatalis connected to lunula, lunula distinct. Clausilium plate narrow, not twisted at lower side.

In Ukrainian Carpathians only in submontane and lower forest zones up to 1300 m. Lives on tree trunks, in moderately humid deciduous and mixed forests. Does not leave the tree trunks in drier weather but hides in bark crevices and under lichens. Tolerates low calcium level in soil.

Critically Endangered in Austria, Endangered in Germany, Vulnerable in Switzerland.

Westerlund 1884: 150, Jungbluth et al. 1989: 186, Falkner 1990: 164, Turner et al. 1998: 230, Sulikowska-Drozd 2005: 82.

Bulgarica denticulata (Olivier, 1801)

Without gender agreement: *Bulgarica denticulatus* (Olivier, 1801)

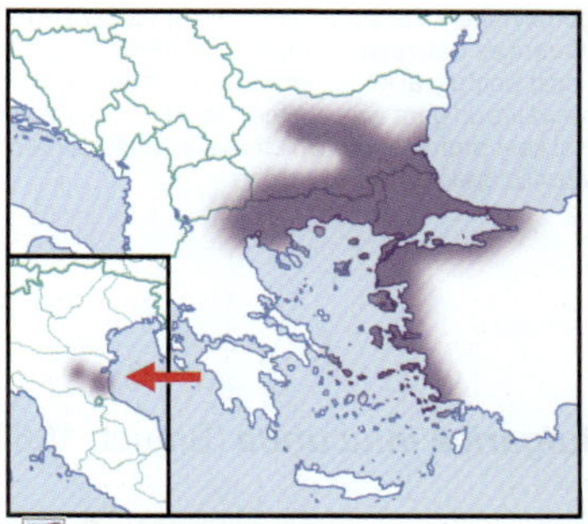

SE Europe, W Turkey

Bulimus denticulatus Olivier, 1801 in Olivier 1801 [atlas]: vij, Pl. 17
Clausilia thessalonica Rossmässler, 1839 in Rossmässler 1838-1844 (9-10): 22, Taf. 48
Laciniaria erberi Frauenfeld, 1867 in Frauenfeld 1867: 775 (*Laciniaria Erberi*)
Clausilia mystica Westerlund, 1893 in Westerlund 1893: 125

Shell brown, densely ribbed, basal keel present, dorsal keel weak, apertural margin broad and white, with many short teeth, parietalis weak, columellaris deep inside, principalis long, basalis prominent and long, subclaustralis present, lunula short and dorsolateral.
The form known as *erberi* has no basalis.
The form known as *mystica* is occasionally smooth, has a smoother apertural margin without small folds at the palatal side, frontal upper palatalis and basalis are sometimes present. This is like an intermediate form between both others.
Differs from sympatric *Laciniaria plicata* in its shorter and more dorsally situated lunula, and presence of a basalis.

Forest soils at the base of limestone rocks, also under stones in open habitats. In Bulgaria at limestones and at silicate rocks, in leaf litter of deciduous forests without rocks, also in pinetree forests down to pH 4.5, up to 1600 m. Highly drought-resistant, broadly adapted to high and low temperatures and humidity. In Ravenna (Italy) surprisingly well adapted to typical city habitats at houses, stone walls, concrete walls, pine logs and litter, also nearby in local pine forests.

denticulata s. str.: NE Greece (Makedonía, Thráki), Rep. of Makedonija, S Bulgaria and European part of Turkey to Istanbul, introduced to Italy (near Ravenna),
mystica: Samothráki,
erberi: W Turkey and adjacent Aegean islands.

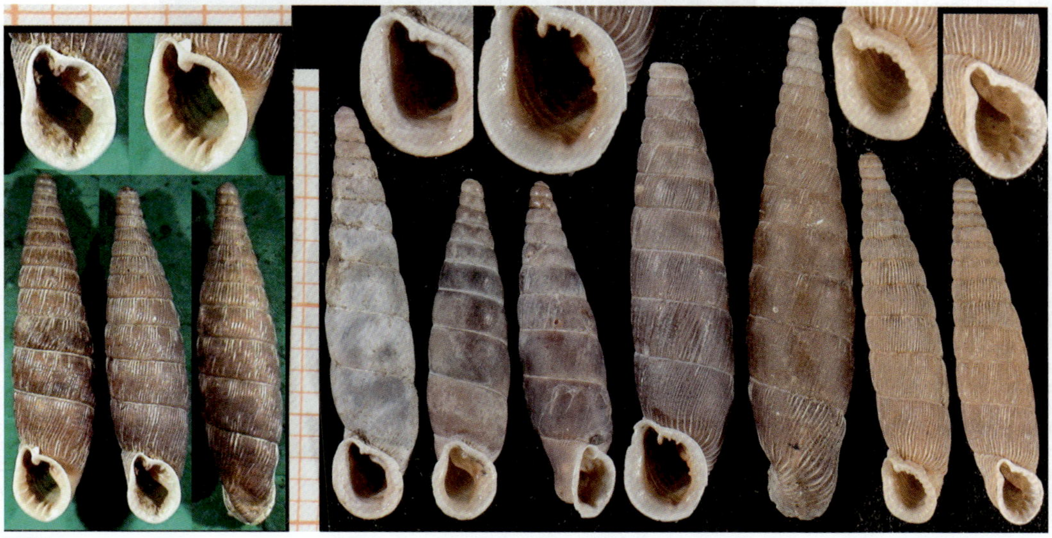

12-19 x 3-4.2 mm

(1-3, *denticulata*): Italy: Ravenna, Casal Borsetti, (4-8, *mystica*): Greece: Samothraki, (9-10, *erberi*): Greece: Psérimos
(1-3) Alessandro Hallgass

Clausilia thessalonica is a synonym of *denticulata* s. str., the name had been used in the belief that *denticulata* referred to *erberi*. The name *denticulata* had been misidentified prior to Nordsieck 2001: 21; reports for *denticulata* from before 2001 referred to *erberi*.
Nordsieck 1973: 197 claimed that *mystica* differed from both *denticulata* and *erberi* by a smooth apertural margin without small denticles. It seems that the shells on which this diagnosis was based were either eroded or that the population of Samothráki is variable, some shells collected recently on Samothráki (D. Georgiev) do have small folds at the parietal and columellar sides.
Nordsieck 2007: 63 recognized *B. erberi* and *B. denticulata* (with additional subspecies *B. d. boettgeri*, *B. d. mystica* and *B. d. thessalonica*), without any published study or arguments to justify this classification.
The *mystica* population seems to exhibit intermediate character expressions (basalis), and is located between both ranges, suggesting that *denticulata*, *erberi* and *mystica* belong to the same gene pool. More local research on the spatial variation of this species complex is necessary.

Bulgarica fritillaria (Frivaldszky, 1835)

Clausilia fritillaria Frivaldszky, 1835 in Frivaldszky 1835: 275, T. 7 k. 12
Clausilia fraudigera Rossmässler, 1839 in Rossmässler 1838-1844 (9-10): 18, Taf. 48
Clausilia varnensis Pfeiffer, 1848 in L. Pfeiffer 1848 (Z. f. Malak. 5): 8
Bulgarica urbanskii Nordsieck, 1973 in Nordsieck 1973: 200
Bulgarica varnensis pseudofraudigera Nordsieck, 1973 in Nordsieck 1973: 199
Bulgarica hiltrudae Nordsieck, 1974 in Nordsieck 1974: 156

13-17 x 3.3-3.9 mm

Bulgaria: Rodopi, S Plovdiv, near Asenova-Krepost (*fraudigera*)

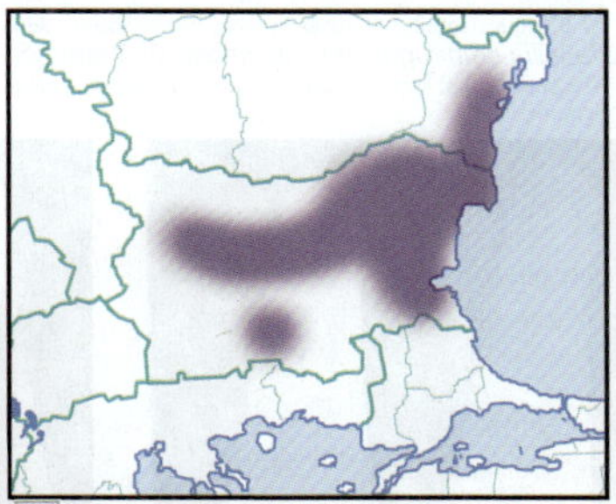

Bulgaria, SE Romania
fritillaria: Rodopi mountains,
fraudigera: Rodopi and Stara Planina,
hiltrudae: NW to central Bulgaria,
urbanskii: E Stara Planina near Sliven (allopatrically with *varnensis* in the same habitat),
varnensis: central Bulgaria to S Romania

Shell horny brown, finely ribbed, 11-15 weakly convex whorls, cervix slightly more coarsely ribbed, with two weak keels, aperture with weak lip, columellaris deep inside, short and horizontal, 2 palatal folds present (upper fold moderate, lower fold very short), subcolumellaris weak and almost hidden behind columella.
The form known as *fraudigera* has almost entirely the same description, 3 well visible palatal folds. The form known as *hiltrudae* is weakly striated or nearly smooth, has more prominent keels and small folds at the columellar margin. The form known as *urbanskii* is more yellowish brown and has no dorsal keel, a long basalis and a weak frontal upper palatalis. The form known as *varnensis* has a moderate basalis and a long frontal upper palatalis. The form known as *pseudofraudigera* does not seem to differ from *fritillaria* and *fraudigera*, I see no differences in the descriptions except perhaps slightly denser ribs.

Open limestone rocks and shady forest habitats, moderately humid, up to 1400 m.

Westerlund 1884: 41, 42, Nordsieck 1974: 158, Damjanov & Likharev 1975: 208, Irikov & Mollov 2006: 787, 788.

Until recently *Bulgarica varnensis*, *fraudigera*, *hiltrudae* and *urbanskii* were recognized as independent species, but there are only very few morphological differences between all these forms (and in H. Nordsieck's concept there are so-called subspecies connecting the so-called pure forms). The forms *varnensis* and *urbanskii* were regarded as separate species only because *fraudigera* and *fritillaria* differ in approximately the same extent from *varnensis* as *urbanskii* does. The morphological differences between *fraudigera* and *fritillaria* are only very weak, the descriptions are almost entirely alike, all forms seem to be locally variable. The form *pseudofraudigera* was claimed to differ from *varnensis* mainly in rib density and cervical keel prominence, which alone would certainly not qualify for an independent clausiliid species in SE Europe.

All this suggests that we are not dealing with true separate species in this complex. More detailed studies on spatial variations of shell characters (constant expressions of characters in wide ranges with abrupt transitions or smooth and gradual changes over wide areas) are needed to establish more firmly the existence of isolated biological species.

Bulgarica iniucunda (Brandt, 1962)

Laciniaria iniucunda Brandt, 1962 in Brandt 1962: 130

Shell horny to brown to cherry red, translucent, sharply or blurry ribbed, can be partly smooth, 11-13 whorls, outline of first 6-7 whorls concave, cervix with prominent basal and strong dorsal keels, separated by a furrow, cervix has a depressed section above the keels, aperture inside brownish, with a basal forrow corresponding to the basal keel, margin sharp and slightly reflected, palatal margin inside occasionally with 2-5 short folds, columellaris very oblique and not very prominent, can be forked inside, 2-5 short marginal folds can be present between columellaris and parietalis, 1-3 between columellaris and basal furrow, lunula dorsolateral and short, upper and lower palatalis and basalis present, sometimes also 1-2 additional palatal folds, subcolumellaris ends very deep inside.

10-14 x 2.5-3.1 mm

Greece: Thraki, Kalifito NE Drama

N Greece (E Makedonía)

Bulgarica pygmaea (Möllendorff, 1873)
Bulgarica pavlovici Nordsieck, 1972

Bulgarica pavlovici Nordsieck, 1972 in Nordsieck 1972: 54
Expressly established as a new replacement name for
Clausilia pygmaea Möllendorff, 1873 in Möllendorff 1873: 142, Taf. 4 (not *Clausilia semirugata* synonym *Clausilia pygmaea* Rossmässler, 1836 in Rossmässler 1835-1837 (4): 21, Taf. 18 Fig. 273 (= currently *Delima bilabiata* (Wagner, 1829))
Established as a new name, no reference to previous papers. For those who demand a bibliographical reference for making available a synonym under Art. 11.6.1, this name is not available from Rossmässler 1836, neither from another author prior to 1873, but from Westerlund 1884:
Clausilia semirugata var. *pygmaea* Westerlund, 1884 in Westerlund 1884: 77 (*Clausilia semirugata* var. *pygmæa*)
This means that for those who demand a bibliographical reference in Art. 11.6.1, *Clausilia pygmaea* Möllendorff, 1873 can be used for the *Bulgarica* species.

8-11 x 1.8-2.2 mm

SE Serbia: 20 km E Niš, Sićevačka Klisura

SE Serbia

Shell finely ribbed or striated, apex inflated (whorls 1-2 wider than whorls 3-4), last whorl more coarsely ribbed, basal and dorsal keels weak, aperture detached, parietalis connected to spiralis, columellaris very deep and hardly visible in the aperture, lunula and basalis weak or absent. Distal end of clausilium plate narrow and digitiform.

Nordsieck 1973: 192.

The relationships of this form with the other *Bulgarica* species remain to be clarified.

Bulgarica rugicollis (Rossmässler, 1836)

Clausilia rugicollis Rossmässler, 1836 in Rossmässler 1835-1837 (3): 18, Taf. 12

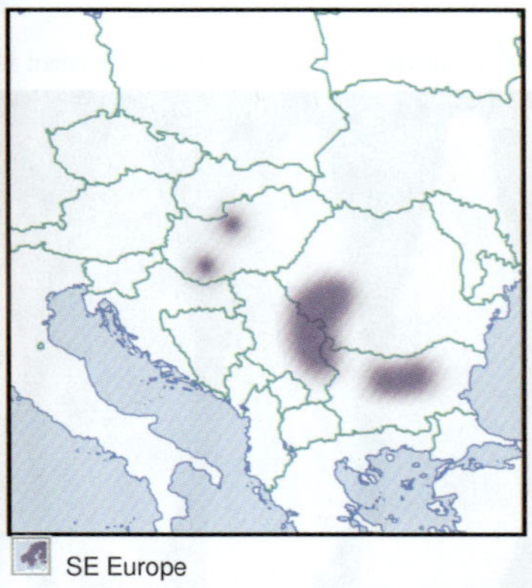

10-20 x 2.9-4 mm, locally variable, in Stara Planina 11-13 x 3-3.5 mm

Romania: Banat, Mehadia

SE Europe

Romania: Aninei mountains, Cheile Carasului

Jozef Grego

Shell ribbed, finely striated or almost smooth, 10-14 whorls, cervix with two very prominent keels and usually ribbed, aperture with basal furrow, columellaris deep inside, short and horizontal, hardly visible in a perpendicular view, frontal upper palatalis and basalis usually weak but variable, subcolumellaris short and deep, clausilium plate with a weak edge at the outer margin. Several characters display spatial variations.

In moderately humid lowland habitats, calciphile, in Bulgaria 250-400 m. In Romania reported exclusively from montains, at the face of rocks, more rarely among stones and in leaf litter under trees.

Westerlund 1884: 43, Nordsieck 1973: 193, Damjanov & Likharev 1975: 207 (*pagana*), 209, Grossu 1981: 228, Hubenov 2007.

Bulgarica vetusta (Rossmässler, 1836)

Clausilia vetusta Rossmässler, 1836 in Rossmässler 1835-1837 (4): 16, Taf. 18

Had previously been mentioned by Rossmässler 1835: 78, without description, nomen nudum there.

Clausilia rugicollis γ synonym var. *Stolii* Pfeiffer, 1859 in L. Pfeiffer 1859: 760

Mentioned as a synonym with a short description by Pfeiffer, and attributed to Zelebor. When Pfeiffer listed subordinate variants by Greek letters, the names behind were usually treated as synonyms. The name was used by Schmidt 1868: 138 as *Clausilia rugicollis* var. *Stollii* [sic] without any description or reference. For those who demand a bibliographical reference for making available a synonym, the name was not made available by Schmidt 1868. For the others, yes.

Clausilia serbica Möllendorff, 1873 in Möllendorff 1873: 144 (figured on Taf. 4 Fig. 7, not Fig. 8) (*Clausilia (Idyla) serbica*)

Not *Clausilia fallax* var. *serbica* Möllendorff, 1873 from p. 146 in the same paper (= currently *Balea serbiana* n. nom.). The name established for the species takes precedence (Art. 24.1, 57.7).

Clausilia stolensis Pfeiffer, 1877 in L. Pfeiffer 1877: 519

Möllendorff 1873: 144 used a name *Clausilia Stolensis* for a taxon without description, and argued that *Stolii* and *Stollii* (without references) were incorrect. If *stolii* was not available, this was a nomen nudum, no new replacement name and no emendation.

Pfeiffer 1877: 519 used a name *Clausilia ?Stolensis* "Möllendorff", without description, and with bibliographical reference to Möllendorff 1873, and mentioned *Cl. Stolii* as a synonym with reference to Pfeiffer 1859: 760. It was the first occasion where a description was provided (by reference to Pfeiffer 1859: 760 for an equivalent synonym) for a name used for a taxon (*Clausilia stolensis* Pfeiffer, 1877).

Bulgarica moellendorffi Nordsieck, 1972 in Nordsieck 1972: 54

Expressly established as a new replacement name for *Clausilia (Idyla) serbica* Möllendorff, 1873 (unnecessary because precedence was clear, Art. 24.1 and 57.7).

Bulgarica vetusta pindica Nordsieck, 1974 in Nordsieck 1974: 158

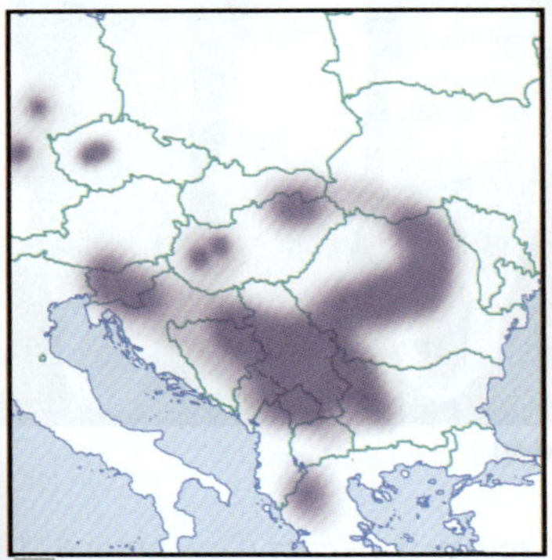

8-16 x 2.2-3.6 mm

W Serbia: Užice

SE Europe

pindica: Pindos mountains in NW Greece,
serbica/moellendorffi: NE Serbia,
stolensis/stolii: NE Serbia,
vetusta: N Albania, Bulgaria and W Carpathians to E Austria, isolated in central and E Germany

Shell brown, ribbed, 11-13 whorls, apex often wider than following whorl, cervix with two keels, aperture relatively small, with basal furrow, columellaris weak but visible in a perpendicular view, palatal callus present, frontal lower palatalis connected to palatal callus and usually not connected to lunula, lunula distinct. Clausilium narrow, forming a furrow, thickened at its pointed end.

More slender than *B. cana*. The form known as *pindica* has a steeper columellaris, no palatal folds and a weak lunula. The form known as *serbica/moellendoffi* a basalis and sometimes smooth shells. The form known as *stolensis/stolii* has weakly stiated or nearly smooth shells.

Endangered in Germany, rare in Bulgaria.

Westerlund 1884: 44.

Nordsieck 1973: 193 considered *stolensis / stolii* as an intermediate form between *serbica / moellendorffi* and *vetusta*. Nordsieck 1974: 158 considered *pindica* and *vetusta* as conspecific. It is likely that we are dealing with a variable species complex. Nordsieck 2007: 63 classified *vetusta* and *pindica* in the subgenus *Strigilecula*, *stolii* and *moellendorffi* in the subgenus *Bulgarica*. In the applied concept this would not mean that such species could not be conspecific.

More research is necessary to firmly establish the presence of independent species. The classification could be based on local studies of spatial variations of the prominence of the apertural folds (presence of sharp transitions or - currently more likely - smooth spatial gradients connecting the populations). Continuous but slow gene flow over area could account for the local variation. This could also become visible in molecular studies, in which it would however be more difficult to study local variations.

Laciniaria

Laciniaria Hartmann, 1842
in Hartmann 1844: Tab. 60 *Clausilia (Laciniaria)*

Type species *Pupa plicata* Draparnaud, 1801 (monotypy)

Clausilia was used as the generic name on Hartmann's plate, *Laciniaria* was meant as a subgenus following the author's style. On the plate combined with the name *plicata*. Mentioned in the text pp. 173, 177 (from 1843).

Laciniaria bajula (Schmidt, 1868)

Clausilia bajula Schmidt, 1868 in Schmidt 1868: 156

10-14 x 3-3.7 mm

Bulgaria: Rodopi, Cudnite Mostove near Zabardo

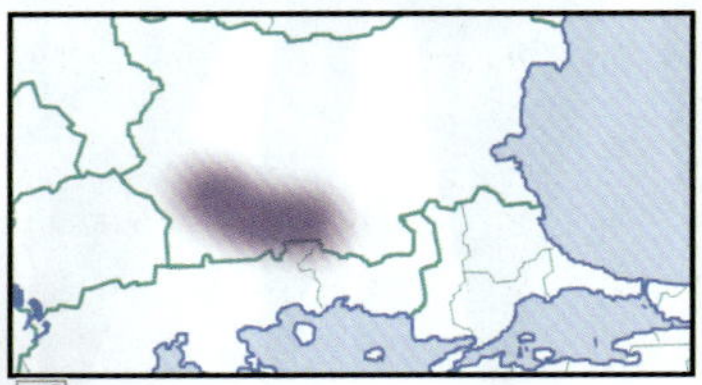

S Bulgaria

Shell horny yellowish, densely striated, 9-12 weakly convex whorls with deep suture, cervix with weak basal keel, aperture with palatal callus, parietalis very short and far distant from spiralis, columellaris deep inside and distant from parietalis, very oblique and forked inside, hardly visible in a perpendicular view, no lower palatalis, lunula almost absent, subcolumellaris weak and often visible in a perpendicular view.

Westerlund 1884: 147, Damjanov & Likharev 1975: 216, Irikov 2006: 81, Irikov & Mollov 2006: 782, Hubenov 2007.

Not necessarily petrophile: often more frequent in leaf litter of humid broad-leaved forests than near limestones but may also live at humid and shady limestones. Up to 1300 m in W Rodopi, 1500 m in Rila mountains.

Laciniaria macilenta (Rossmässler, 1842)

Clausilia macilenta Rossmässler, 1842 in Rossmässler 1838-1844 (11): 6, Taf. 52

9-11 x 1.8-2.5 mm

Bulgaria: Rodopi, Bachkovski monastir near Asenovgrad

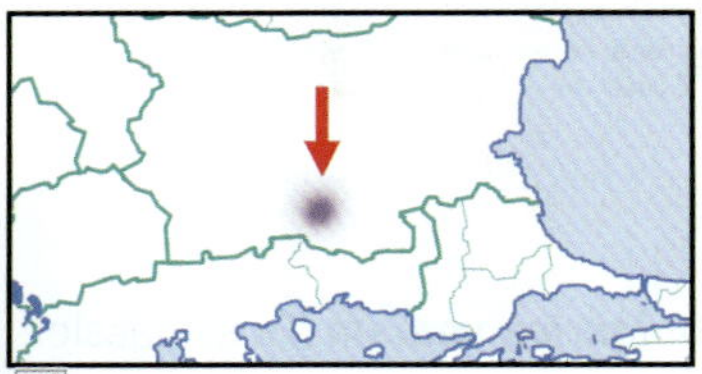

S Bulgaria

Shell brown or horny brown, finely ribbed, 11-14 whorls, cervix with distinct basal keel, apertural margin detached, broadly reflected and with many folds, two upper palatal folds, lunula lateral, deep inside and straight, clausilium forming a furrow.

Petrophilic and calcareous species, which inhabits rock habitats with xerothermic conditions as well as places with high humidity and shade, up to 1600 m.

Only known from few localities in the W Rodopi mountains.

Westerlund 1884: 37, Damjanov & Likharev 1975: 202, Irikov & Mollov 2006: 782.

Laciniaria plicata (Draparnaud, 1801)

Pupa plicata Draparnaud, 1801 in Draparnaud 1801: 63

15-18 x 3.3-3.6 mm

Bulgaria: near Grabovo

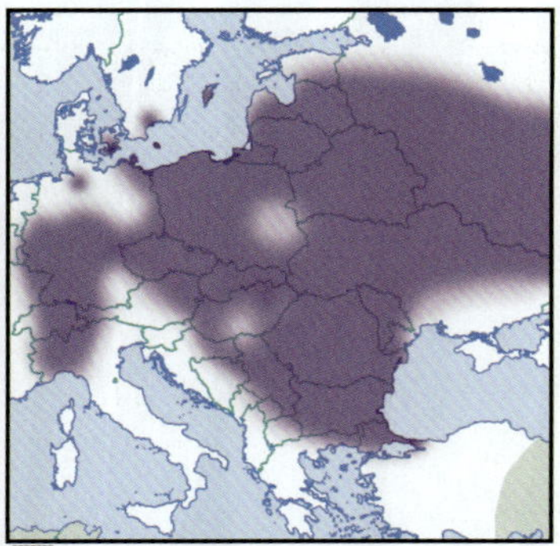

Central Europe to central Russia

Shell brown, ribbed, cervix with distinct basal keel, aperture at lower part with groove, often several small folds at palatal side of aperture, principalis long, frontal upper palatalis present and often well visible in a perpendicular view, no basalis, lunula lateral.
Differs from sympatric *Bulgarica denticulata* in its more laterally situated lunula and absence of a basalis.

Usually humid rocks and old walls in open habitats, less frequent in forests and higher mountains. Not restricted to calcareous substrate. In Bulgaria between rocks and in leaf litter layers of forests. In Switzerland up to 2300 m, but rare above 1200 m, in Bulgaria up to 1500 m. One of the most common clausiliids in lowland and uplands of Poland, also very common in Bulgaria.

Damjanov & Likharev 1975: 194, Kerney et al. 1983: 232, Manganelli et al. 1995: 27, Sulikowska-Drozd 2005: 76, Irikov & Mollov 2006: 780.

Laciniaria potochensis Dedov & Neubert, 2006

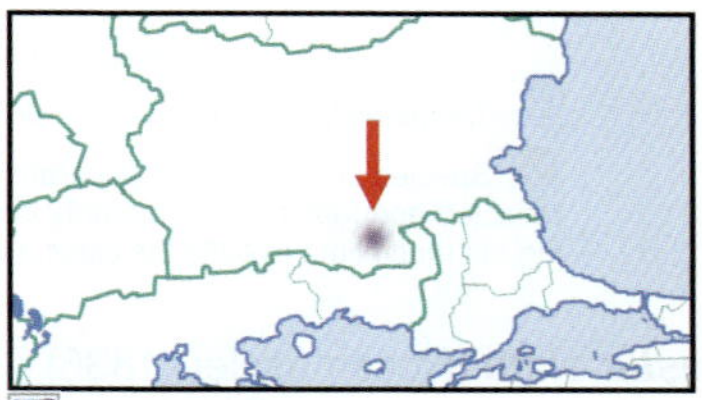

S Bulgaria

Laciniaria potochensis Dedov & Neubert, 2006 in Dedov & Neubert 2006: 74

Small, coarsely ribbed, 3-6 folds in the interlamellar area (apertural margin between parietalis and columellaris).

11-15 x 3.0-3.5 mm

In humid and shady rock crevices, at rain also crawling on the face of the rocks.

Only known from one locality.

Laciniaria pseudostabilis (Westerlund, 1901)

Clausilia plicata var. *pseudostabilis* Westerlund, 1901 in Westerlund 1901: 35

Romania: Latorita mountains, Piatra Fetii W Ciunget

13-15 x 3-3.7 mm

More information on this species would be desirable. Listed by Nordsieck 2007: 62, but not by Grossu 1981.

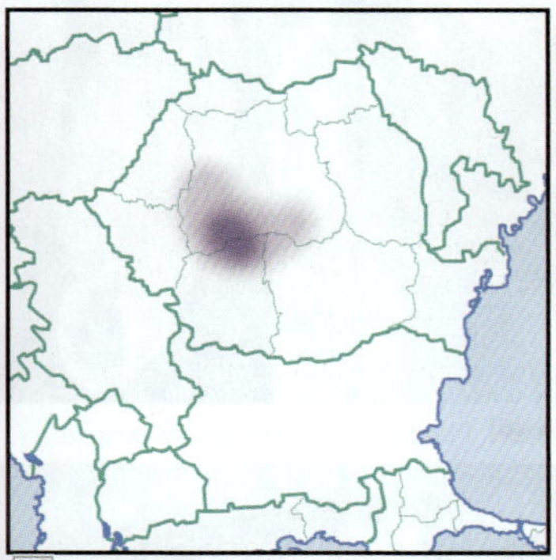

Romania
Range uncertain.

Shell horny brown, regularly ribbed, often denser on last 2 whorls, more widely spaced near aperture, 11-13 whorls, aperture detached, parietalis reaching margin, columellaris straight, oblique and not reaching margin, 1-2 small folds between columellaris and parietalis.

Mentissella

Mentissella Nordsieck, 1973 in Nordsieck 1973: 190

Type species *Clausilia rebeli* Sturany, 1897 (original designation)

Mentissella rebeli (Sturany, 1897)

Clausilia rebeli Sturany, 1897 in Sturany 1897: 117, Taf. 3

Bulgaria: Stara Planina, W Kotel

11-16 x 2.7-4.0 mm

E Bulgaria

Shell slender, brown, densely ribbed with some whitish spots, 10-12 whorls with deep suture, cervix with prominent basal and moderate dorsal keel, aperture distinctly detached, with basal groove, columellaris oblique and not very prominent, 3 palatal folds visible in an oblique view, no lunula, subcolumellaris strong but not visible in a perpendicular view.

Damjanov & Likharev 1975: 210.

Vestia Hesse, 1916
in Hesse 1916: 124
Expressly established as a new replacement name for

Type species *Clausilia elata* Rossmässler, 1836
(original designation by Vest 1867: 190)

Uncinaria Vest, 1867 in Vest 1867: 190 *Clausilia (Uncinaria)* not *Uncinaria* Frölich, 1789 (Nematoda)

Species are nearly ovoviviparous and lay eggs shortly before hatch, growth is completed after 4 months with the formation of the closing apparatus and the reflected apertural margin, maturity is reached shortly after that, age up to 4 years. Delayed maturation in this genus was suggested as a strategy to avoid spending too much time to build the closing apparatus which reinforced the shell (Maltz & Sulikowska-Drozd 2011).

Vestia elata (Rossmässler, 1836)

Clausilia elata Rossmässler, 1836 in Rossmässler 1835-1837 (3): 20, Taf. 12

11-20 x 3.7-4.7 mm, locally variable

Romania: Transilvania

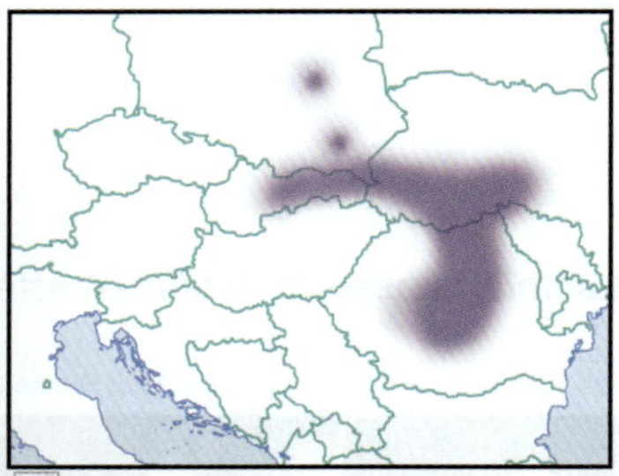

Carpathians

Shell brownish yellow, striated or ribbed, 9-12 whorls, apex blunt, aperture with lip inside, parietalis close to spiralis, columellaris oblique, running in the uppermost part of the aperture, horizontal in its frontal section, no lower palatalis, lunula dorsal and curved, subcolumellaris prominent, clausilium narrowed and pointed at its lower end with a hook.

Humid mountain forests, under rotting logs deep in leaf litter. In Poland in hornbream and beech forests mainly in river valleys. May form very dense populations.

Reproduction in Poland largely restricted to two periods, April-May and August-September, but this can deviate and depends from temperature. Not properly ovoviviparous, eggs (2.0-2.2 x 1.7-1.9 mm) are kept long in the uterus and are laid 2-4 days before hatch, 2-7 eggs per clutch, juveniles consume their own egg envelopes, grow 2 whorls per month and reach full size after 5-6 months in laboratory, in nature growth is slower, full size is reached after 2 years. Life span up to at least 6 years from maturity.

Critically Endangered in SE Poland, only a few isolated populations are known.

Westerlund 1884: 150, Piechocki 1982, Kerney et al. 1983: 235, Sulikowska-Drozd 2005: 79, Sulikowska-Drozd 2008, Sulikowska-Drozd 2009.

Slovakia: Slovenský Raj National Park
Jozef Grego

Vestia gulo (Bielz, 1859)

Clausilia procera var. *gulo* Bielz, 1859 in E. A. Bielz 1859: 222
Name attributed to Rossmässler, but the descriptive diagnosis was given by Bielz.

16-20 x 3.9-4.7 mm

Poland: Krynica

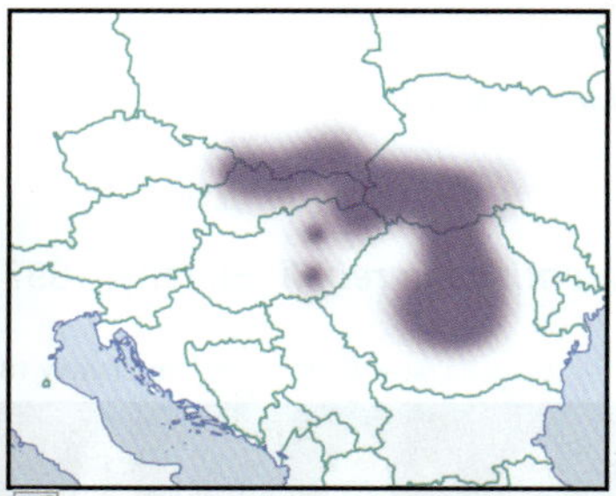

Carpathians

Shell yellowish to reddish brown, ribbed, 10-12 moderately convex whorls, parietalis prominent and close to spiralis, columellaris almost straight, running in the uppermost part of the aperture, usually horizontal and shortly forked in its frontal section, no lower palatalis, lunula dorsal and curved, subcolumellaris prominent, clausilium narrowed and with a hook at its pointed end.

Slovakia: Slanské Vrchy mountains
Jozef Grego

Mountain forests, very humid places, on the banks of brooks, under stones, among rotting sticks and leaves. Avoids big streams and rivers.

Reproduction in Poland from April to August, mainly May-June. Not properly ovoviviparous, eggs are kept long in the uterus and are laid 7-10 days before hatch, 6-15 eggs (very slightly ovoid, 1.7-2 x 1.55-1.8 mm) per clutch, juveniles consume their own egg envelopes and occasionally also those of neighbouring embryos. Life span 3-4 years under laboratory conditions.

Westerlund 1884: 150, Kerney et al. 1983: 236, Sulikowska-Drozd 2005: 80, Sulikowska-Drozd 2009, Balashov & Gural-Sverlova 2012: 97.

Vestia ranojevici (Pavlović, 1912)

Uncinaria turgida var. *ranojevici* Pavlović, 1912 in Pavlović 1912: 105 (*U. turgida* var. *ranojevići*)

13-16 x 3.7-4.5 mm

Bulgaria: Rila mountains, Rilski monastir

Seems to live in moderately humid habitats, in Bulgaria in 500-1800 m.

Irikov & Mollov 2006: 787, Hubenov 2007.

Serbia, Bulgaria, NE Moravia

Shell brown, densely ribbed, cervix more coarsely ribbed, 9-11 whorls, cervix with strong basal and weak dorsal keel, and a weak basal furrow, apertural margin detached and rounded, columellaris not very prominent and descends in a curved line, parietalis connected to margin, subcolumellaris visible in an oblique view.

Vestia roschitzi (Brancsik, 1890)

Clausilia roschitzi Brancsik, 1890 in Brancsik 1890: 74, Tab. 2 (*Clausilia (Strigillaria) Roschitzi*)
Genus name *Clausilia* mentioned on p. 72, it is not clear if this was meant as the genus name for all clausiliid species in the list, which were listed only in the combinations subgenus and species. Established without description but with a figure.

10-13 x 2.9-3.6 mm

Serbia: 9 km E Surdulica

Tomić 1959: 44 (*Unciniaria trigonostoma*, *U. minima*), Dhora & W. 199: 137 (N Albania), Nordsieck 2007: 63.

The diagnosis describes the figured shells from Serbia. Variation and full range of the species remain to be studied.

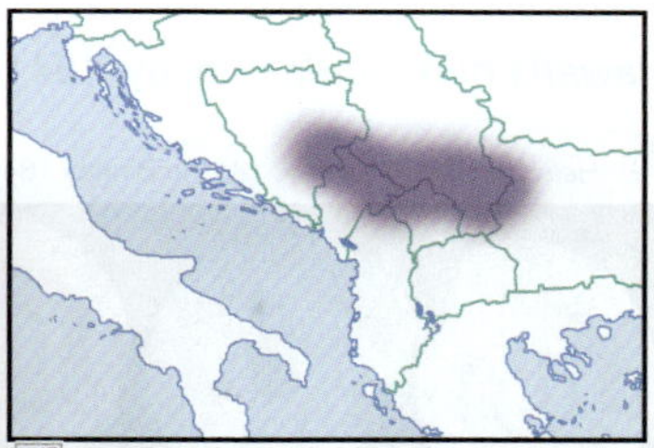

Bosnia to S Serbia

Shell horny brown, densely ribbed, translucent, 11-12 whorls, cervic with strong basal and weak dorsal keel, aperture detached and with sinulus, parietalis oblique and reaches margin, columellaris weak, nearly horizontal, very much upside in the aperture and reached margin, subcolumellaris reaches margin just below, frontal lower palatalis strong and visible in a perpendicular view, also from outside, lunula dorsal, palatal callus can look like denticles, basal furrow present, basalis appears at a position where other species have their subcolumellaris.

Vestia turgida (Rossmässler, 1836)

Clausilia turgida Rossmässler, 1836 in Rossmässler 1835-1837 (3): 20, Taf. 12

11-17 x 3-5 mm, locally variable

Germany: Bayrischer Wald, Buchwalder Leite

Slovakia: Low Tatra National Park

Jozef Grego

Mountains and submontane zones, up to 2100 m, humid forests and scrub, forest marshes, also in drier habitats. Under logs and ground litter.

Reproduction in Poland from April to September, mainly May to July. Ovoviviparous, eggs (very elongated and flattened, 1.9-2.2 x 1.4-1.6 mm) are jelly-like and glued together with mucus, they are kept long in the uterus and hatch either in the parent's body or shortly after, 1-11 eggs per batch, egg cannibalism is not known.

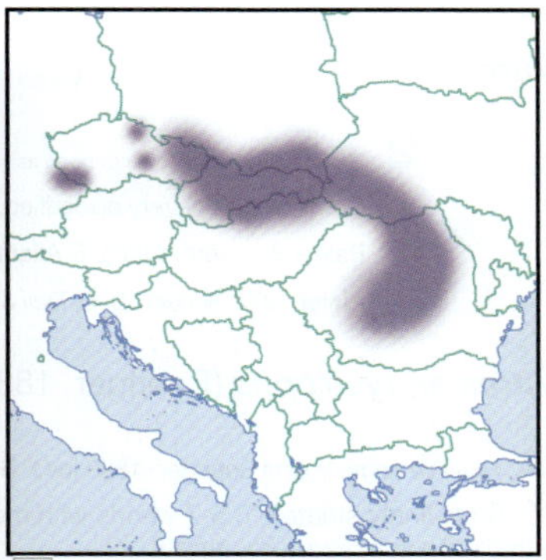

Carpathians

Shell yellowish brown, usually finely ribbed, 9-10 moderately convex whorls, cervix with very weak dorsal keel, aperture relatively small and without basal furrow, parietalis prominent and close to spiralis, columellaris located in the uppermost part of the aperture, straight or slightly curved upside, no lower palatalis, lunula dorsal and curved, weaker in its central part, subcolumellaris prominent, visible in an oblique view. Clausilium narrowed, not twisted, curved and with a hook at its pointed end.

Westerlund 1884: 149, Kerney & Cameron 1979: 172, Falkner 1990: 164, Sulikowska-Drozd 2005: 81, Sulikowska-Drozd 2009.

Zonitids and friends, semislugs and slugs

The traditionally used family-level classification in this group was not consistent with the phylogenetic relationships obtained in modern studies, so some authors changed the taxonomic classification and meanings of widely used names around 2000. They did not wait until molecular results were confirmed in independent studies. Subsequent studies did not always confirm the first and perhaps premature conclusions (Wade et al. 2006).
The result is that the main function of a family name for the purpose of communication was lost, because a name alone (Gastrodontoidea, Zonitidae, *Oxychilus*) does not stand for a broadly understood concept any more. It is necessary to know in which sense such a name is used. This means that in the new systems the taxonomic names have had only preliminary meanings, and that we cannot expect their meanings to remain stable for a few years.
To avoid confusion I prefer using only those family names for which changes of their meanings have not recently been proposed.

The traditional classification has been used in minor variations for many years, and also in modern publications (Alzona 1971, Kerney et al. 1983: 52, Falkner 1990: 275, Manganelli et al. 1995, Egorov 2008, Sysoev & Schileyko 2009, Horsák et al. 2010):
Vitrinidae, Zonitidae (*Vitrea*, *Aegopinella*, *Nesovitrea*, *Oxychilus*, *Zonites*), Daudebardiidae, Gastrodontidae (*Zonitoides*), *Boettgerilla*, Milacidae, Limacidae, Agriolimacidae
Euconulus, Arionidae, *Papilloderma* and *Testacella* were usually outside these groups.

One proposed new classification has been the following (Falkner et al. 2001, Bank et al. 2001, Bouchet & Rocroi 2005):
Gastrodontoidea (*Vitrea*, *Gyralina*, *Euconulus*, *Zonitoides*, Daudebardiinae, *Oxychilus*, *Eopolita*, *Aegopinella*, *Retinella*, *Nesovitrea*)
Parmacelloidea (*Parmacella*, Milacidae)
Zonitoidea (*Zonites*, *-aegopis* group, *Balcanodiscus*)
Limacoidea (Vitrinidae, *Boettgerilla*, Limacidae, Agriolimacidae)
Arionidae, *Papilloderma* and *Testacella* were outside these groups.

In the molecular study by Wade et al. 2006 the groups represented by *Oxychilus*, *Vitrea*, *Tandonia*, *Vitrina*, *Deroceras* and *Euconulus* produced one single monophyletic group Limacoidea, in which *Vitrea* was more closely related to Milacidae than to *Oxychilus* and *Euconulus*. Arionidae and *Testacella* clustered outside.

In the present guide the genera are simply sorted by shell shape and size:
1 - Regularly rounded shells: *Hawaiia*, *Vitrea*, *Troglovitrea*, *Lindbergia*, *Gollumia*, *Gyralina*, *Oxychilus*, *Aegopinella*, *Retinella*, *Nesovitrea*, *Zonitoides*, *-aegopis* group, *Balcanodiscus*, *Meledella*, *Zonites*, *Turcozonites*
2 - Semislugs: *Parmacella*, *Testacella*, Daudebardiidae, Vitrinidae
3 - Slugs: *Papilloderma*, Milacidae, *Boettgerilla*, Limacidae, Agriolimacidae, Arionidae
Euconulus is presented outside this group, at the end of section 3 near *Pyramidula* and *Discus*.

Hawaiia

Hawaiia Gude, 1911
in Gude 1911: 272

Type species *Helix kawaiensis* Reeve, 1854
(original designation)
(= currently *Hawaiia minuscula* (Binney, 1840))

Hawaiia minuscula (Binney, 1841)

Helix minuscula Binney, 1841 in Binney 1841: 435, Pl. 22

USA: Georgia, Wilkes County, Washington near Athens

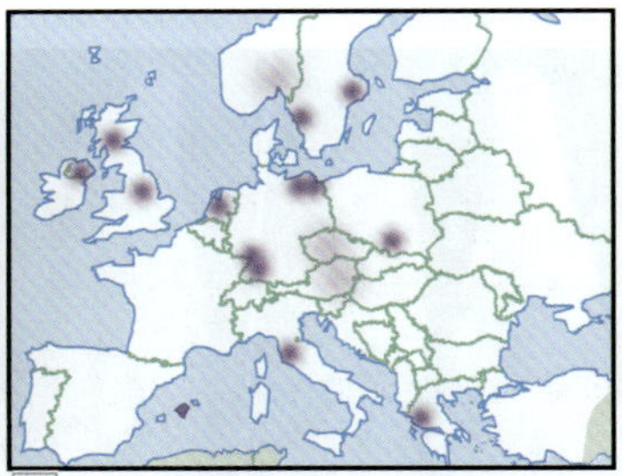

North America and E Asia, introduced elsewhere

0.7-1.2 x 1.7-2.5 mm,

1.45-1.5 mm at 3 whorls, 1.9 mm at 3.5 whorls

Shell whitish, translucent, finely and nearly regularly striated, radial lines crossed by prominent spiral lines, dull silky shiny, surface not glossy, 3.5-4.5 whorls with deep suture, aperture simple, umbilicus wide and deep, 1/4-1/3 of diameter, all whorls visible inside. Shell difficult to distinguish from European *Lucilla* shells, which are smoother and have a glossy surface, their shell is thinner and their suture is shallower.
Retractor at penis and epiphallus (not only at penis as in *Lucilla*). Marginal teeth of radula are blade-like (and not serrated as in *Lucilla*). The last difference may be the best method to distinguish *Lucilla* and *Hawaiia*.

In its native range in a broad variety of habitats, leaf litter of deciduous and mixed forests, shrublands, among stones on grassy slopes, also in gardens. In Europe in greenhouses, in SW Germany occasionally in outdoor bamboo pots.

Kerney & Cameron 1979: 210, Schmid 2002: 392, Schileyko 2003: 1389, www.mollbase.org (01/2007), Örstan (unpublished, 2007), Beckmann 2007: 77 (Mallorca), Kaszuba & Stworzewicz 2008, Horsák et al. 2009, Reischütz & R. 2009: 41.

Vitrea

Vitrea Fitzinger, 1833
in Fitzinger 1833: 99

Type species *Glischrus diaphana* Studer, 1820
(monotypy)

Type species was incorrectly given as *Helix diaphana* Studer, 1820 in Opinion 335 (1955). Studer 1820 classified *Helix* as a subgenus of *Glischrus*.

Conchologically very diversified, but anatomically very homogenous. Only the internal structure of the penis shows reasonable differences.

Palearctic: North Africa, E Atlantic islands, Europe and Asia Minor to Caucasus region and Near East

Pintér 1972, Riedel 1992, Deli & Subai 2011 (Revision S Romania).

Vitrea angystropha (Boettger, 1880)

Hyalinia angystropha Boettger, 1880 in O. Boettger 1880: 380

Shell finely striated, 6.5-7 whorls with moderate suture, upper side flattened, lower side rounded, last whorl with blunt edge, aperture narrow.

1.5-1.7 x 2.7-3.3 mm

Westerlund 1886: 31, Riedel 1992: 12, Riedel 1995: 9.

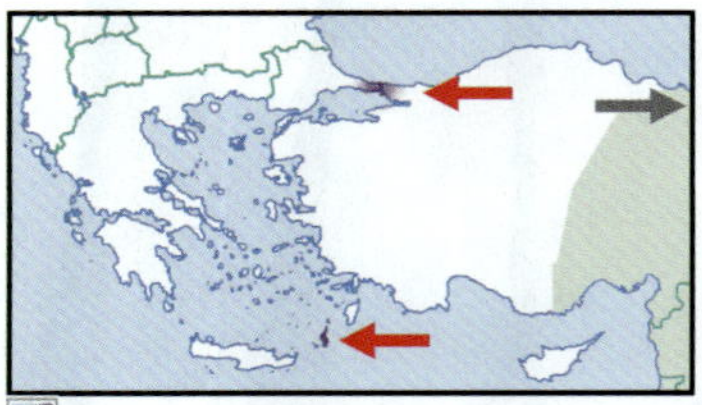

N Iran to Turkey, Kárpathos

Vitrea argolica Riedel, 1962

Vitrea argolica Riedel, 1962 in Riedel 1962: 315

Shell conical, whitish, weakly translucent, shiny and smooth with irregular growth lines, 6-7 whorls, more than most other species, last whorl not much wider than the others, umbilicus narrow but whorls are visible inside.
Can hardly be confounded with any other species.

2.3-2.7 x 3.8-4.0 mm

The species is rather rare. In large numbers only known from Argos, Pelopónisos.

Pintér 1972: 269, Riedel 1992: 12.

Greece (Pelopónisos, Zákinthos)

Vitrea binderi Pintér, 1972

Vitrea binderi Pintér, 1972 in Pintér 1972: 224

Shell conical, yellowish, translucent and shiny with irregular and indistinct growth lines, 6 whorls or more, last whorl not wider than penultimate whorl, weakly keeled, umbilicus very narrow and half covered by columellar margin of the aperture.
Shell cannot be confounded with any other species in this region.

1.5-1.8 x 3.5-4 mm

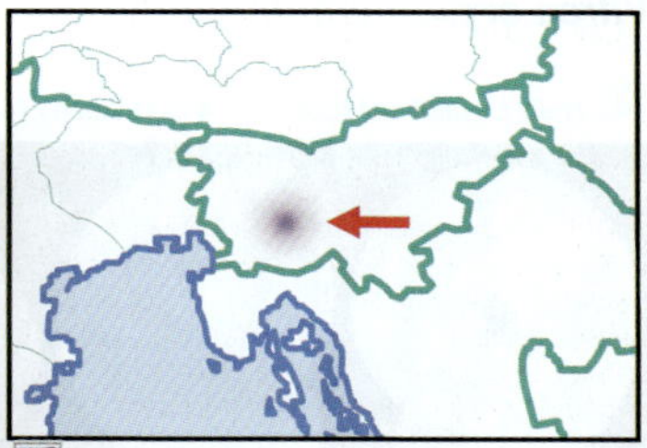

Slovenia

Italy: Slovenia: Planina, Laska Kukava
Massimo Prodan

Vitrea botterii (Pfeiffer, 1853)

Helix crystallina var. synonym *Helix Botterii* Pfeiffer, 1853 in L. Pfeiffer 1853 (Mon. Helic. 3): 66
Mentioned with a very short description: "Obs. Var. paulo apertius umbilicatam ex insula Lesina sub nomine *H. Botterii* Parr. in collectione Philippiana vidi.". Not mentioned as a junior synonym of an available name (as demanded by Art. 11.6), but as a synonym for a subordinate variant for which no name was used. For those who neglect this (as proposed by Kadolsky 2012: 78, this is preliminarily accepted here), the name was made available under Art. 11.6.1 by Bourguignat 1856 (Amén. Malac. 1): 194 who used the name as *Helix Botterii*, with bibliographical reference to Pfeiffer 1853: 66. For the others, the name must be attributed to Bourguignat 1856.

1.3-2 x 3-4 mm

Albania: Mali Gradecit near Çorovodë

Ø 0.4-0.5 mm at 1 whorl, 0.8-0.9 mm (2 wh.), 1.4-1.5 mm (3), 2.1-2.3 mm (4 whorls)

In rocky valleys, not restricted to calcareous substrate but clearly more frequent there. Frequent from S Dalmatia to N Albania, rare in Greece, much less frequent than *V. subrimata* in N Croatia.

Bourguignat 1856: 194, Pintér 1972: 241, Riedel 1992: 13, Štamol & V. 1995, Bole & S. 1997: 148, Deli & S. 2011: 34.

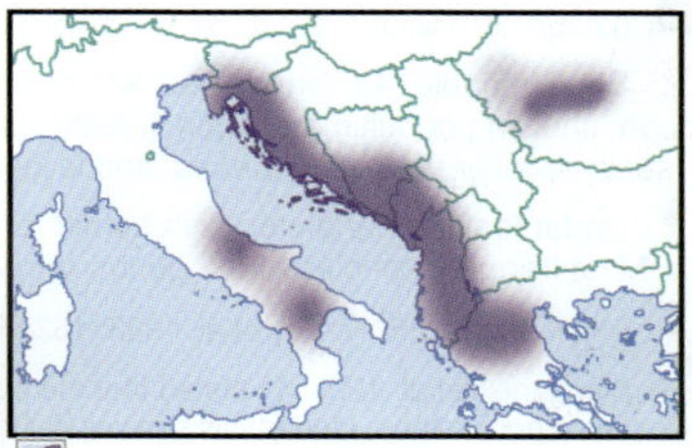

Italy, Slovenia to Greece, Romania

Shell colourless, translucent and shiny, with prominent growth lines, shape variable, periphery rounded and not depressed, 4.5-5 whorls (0.5 whorls more than same-sized *V. crystallina*), last whorl width seen from above 1.4-1.7 x of penultimate whorl (2 x if pen. wh. is taken from suture to suture), umbilicus wide and perspectival, showing all whorls inside.
V. contracta is clearly smaller, *V. illyrica* larger and less densely coiled. In *V. jetschini* the last whorl is inflated.

Vitrea clessini (Hesse, 1882)

Hyalinia clessini Hesse, 1882 in Hesse 1882: 318, Taf. 12 (*Hyalinia Clessini*)

0.9-1.1 x 2-2.5 mm

Greece: E Paximádia island S of central Crete

Ø 0.4-0.45 mm at 1 whorl, 0.85-0.9 mm (2), 1.25-1.3 mm (3), 1.65-1.7 mm (3.5 whorls), 2.0-2.1 mm (4 whorls)

In Gávdos in red soil more frequent than in soil with many organic compounds. Seems to prefer Mesozoic substrate.

Pintér 1972: 276, Riedel 1992: 12, Welter-Schultes 1998: 55.

Greece (Kikládes, Crete)

Shell very similar to *V. contracta*, slightly more than 4 whorls, whorls perhaps more convex, umbilicus wider, showing all whorls inside.

Vitrea contracta (Westerlund, 1871)

Zonites crystallina contracta Westerlund, 1871 in Westerlund 1871: 56

1.0-1.4 x 1.8-3.0 (usually around 2) mm

Greece: S central Crete, near Léndas

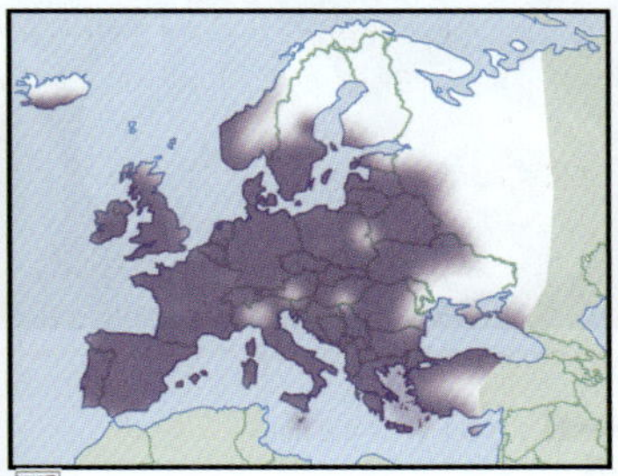

Iceland, Madeira, Azores, N Africa, Europe to Asia Minor and Caucasus

Ø 0.45-0.5 mm at 1 whorl, 0.85-1.0 mm (2 wh.), 1.35-1.5 mm (3), 1.7-1.8 mm (3.5), 2.2-2.3 mm (4 wh.), up to 3.4 mm (5 whorls)

Shell colourless translucent, shiny, almost smooth, 4-5 whorls, last whorl width seen from above 1.4-1.6 x of penultimate whorl, umbilicus moderately wide, getting wider at last whorl. Conchologically very variable, especially in shell size.
The smallest form *zakynthia* Hesse is sometimes regarded as a subspecies or even as a distinct species (Bank 1988, Bank & Neuteboom 1988). But also these shells are variable (for example in their umbilicus) and often almost indistinguishable from juvenile *contracta contracta* (see Riedel 1992).

Warm and dry habitats on calcareous substrate, dry rocky meadows, rock rubble, open forests with rocks. Widespread and common in Greece, in Gávdos on Neogene substrate below shrubs in soil mixed with organic compounds. In central Spain in more humid habitats and locally rare. Tolerant to a broad variety of temperatures and humidity in S Bulgaria, up to 1600 m. In Switzerland up to 1400 m.

Endangered in S Switzerland and Bavaria, Vulnerable in Austria, Switzerland and Rheinland-Pfalz. Scattered and rare in Switzerland, rare in mountains near Salamanca. Threatened by acid rainfalls in Scandinavia, abundance decreased substantially.

Pintér 1972: 272, Damjanov & L. 1975: 263, Riedel 1992, Turner et al. 1998: 264, .Talaván Gómez & T. 2006: 115, Deli & S. 2011: 27, Balashov & G. 2012: 99.

Vitrea crystallina (Müller, 1774)

Helix crystallina Müller, 1774 in Müller 1774: 23

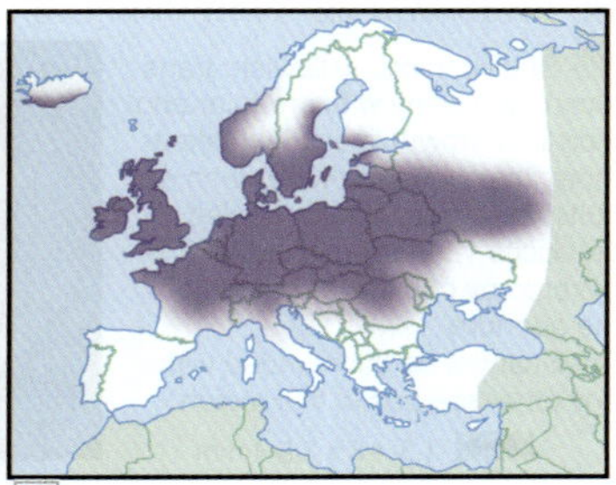

1.4-2.1 x 3-4 mm

Germany: Niedersachsen, Leine river deposits

Europe except in the south

0.6 mm at 1 whorl, 1.2-1.4 mm (2), 2.1-2.3 mm (3 wh.), 3.4-3.7 mm (4 whorls)

Shell variable, 4-5 whorls, last whorl width seen from above 1.5-2 x of penultimate whorl, umbilicus narrow, much narrower than in *V. botterii*, but clearly open, showing penultimate whorl, initially narrower, widened only at last whorl, looks sometimes slightly excentric.
Usually smaller than *V. diaphana*, and larger than *V. contracta* which has more densely coiled whorls.

Under leaf litter and stones in humid forests, in mosses, characterstical for riverside woods, also humid meadows and swamps. In Switzerland up to 2300 m. Most frequent species of the genus *Vitrea*.

Pintér 1972: 235, Kerney et al. 1983: 162, Falkner 1990: 174, Turner et al. 1998: 262, Kerney 1999: 137, Sysoev & Schileyko 2009: 116, Deli & Subai 2011: 30.

Manganelli et al. 1995: 41 stated that the species was probably not present in Italy and previous records based on misidentifications, and assumed artificial dispersal for two Italian populations. I found this species in Gargnano (Lago di Garda).

Vitrea cyprina Westerlund, 1902

Vitrea cyprina Westerlund, 1902 in Westerlund 1902: 22

Shell slightly reddish, shiny and densely striated, 4.5 very slowly increasing whorls, last whorls only slightly wider than penultimate whorl, umbilicus relatively narrow.

Riedel 1995: 12.

1.2-1.4 x 2.2-2.7 mm

Cyprus

Cyprus, S Turkey

Georgios Konstantinou

Vitrea demiobasensis Pintér, 1972

Vitrea demiobasensis Pintér, 1972 in Pintér 1972: 251

Shell slightly brownish, pale shiny and almost smooth, surface very finely granulated, 5.5 whorls, last whorl almost twice as wide as penultimate whorl, umbilicus relatively narrow.

1.3-1.5 x 3.2-4.0 mm

Various habitats, occasionally very abundant.

Riedel 1992: 21.

Greece (S Pelopónisos, Kíthira)

Vitrea diaphana (Studer, 1820)

Glischrus diaphana Studer, 1820 in Studer 1820: 86 (*Glischrus (Helix) diaphana*)

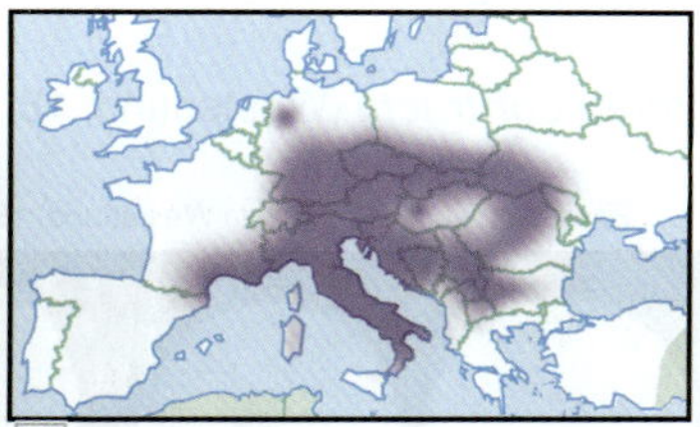

Central to SE Europe

1.8-2.1 x 3.7-4.8 mm

Germany: Bayern, near Hersbruck

2.5-2.7 mm at 3 whorls, 3.3-3.6 mm at 4 whorls

Shell surface smooth and shiny, 5-6 whorls, spire almost flat, last whorl width seen from above 1.7-2.3 x of penultimate whorl, aperture seen from above straight, no umbilicus.
Differs from *V. transsylvanica* in its narrower last whorl and ist apertural margin not being *Nautilus*-like protruded seen from above. *V. erjaveci* has more densely coiled whorls.

Relatively humid montane habitats, characteristically between rocks and rock rubble on wooded slopes. In Bulgaria not in caves, up to 1300 m.

Pintér 1972: 213, Damjanov & Likharev 1975: 258, Kerney et al. 1983: 158, Riedel 1992: 22, Manganelli et al. 1995: 18, Hubenov 2007, Deli & Subai 2011: 8.

Slovakia: Starohorské vrchy mountains near Banská Bystrica
Jozef Grego

Vitrea ephesina Pintér, 1972

Vitrea klemmi ephesina Pintér, 1972 in Pintér 1972: 255

Shell very flat, 5-5.5 whorls, last whorl only slightly wider than penultimate whorl, umbilicus narrow but locally variable.

1.5-2 x 3.8-4.8 mm, in Chíos larger than in Turkey

Rocky limestone and rock rubble habitats, for example in pine forests, inside the soil and in rock crevices.

Riedel 1981: 233, Riedel 1995: 12.

W Turkey, E Greece

Vitrea erjaveci (Brusina, 1870)

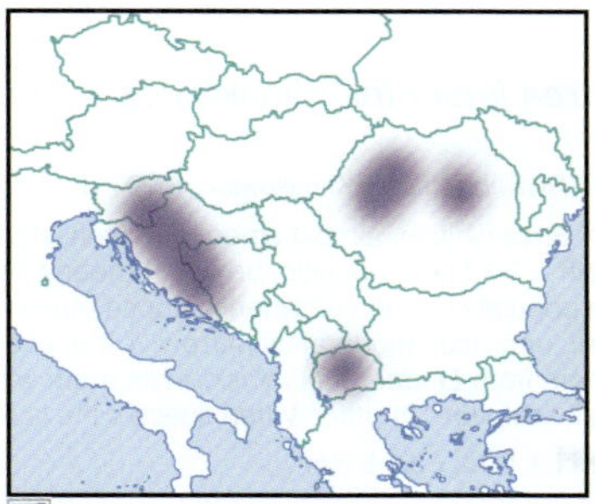

Helicella erjaveci Brusina, 1870 in Brusina 1870 (Contrib. Croat.): 18 (*Helicella Erjaveci*)
Hyalinia densegyrata Kimakowicz, 1890 in Kimakowicz 1890: 36
Original spelling was neither *densigyrata* nor *densigirata*.

Shell pale yellowish white and transparent, not very shiny, upper side densely striated and almost flat, lower side smoother, periphery sometimes slightly keeled, 6-6.5 very slightly convex whorls, last whorl width 1.4-1.6 of penultimate whorl (seen from above), no umbilicus. Shells in Romania are slightly flatter than in the west. Differs from *V. diaphana* and *V. transsylvanica* in its more densely coiled shell and narrower last whorl.

1.4-2.0 x 3-4 mm

Mountain habitats in deciduous forests, not restricted to calcareous substrate.

Kimakowicz 1890: 36 (*densegyrata*), Grossu 1983: 100 (*densigirata*), Deli & Subai 2011: 12 (classified *densegyrata* as a synonym of *erjaveci*).

SE Europe

Vitrea etrusca (Paulucci, 1878)

Hyalina etrusca Paulucci, 1878 in Paulucci 1878: 25 (*Hyalina Etrusca*)

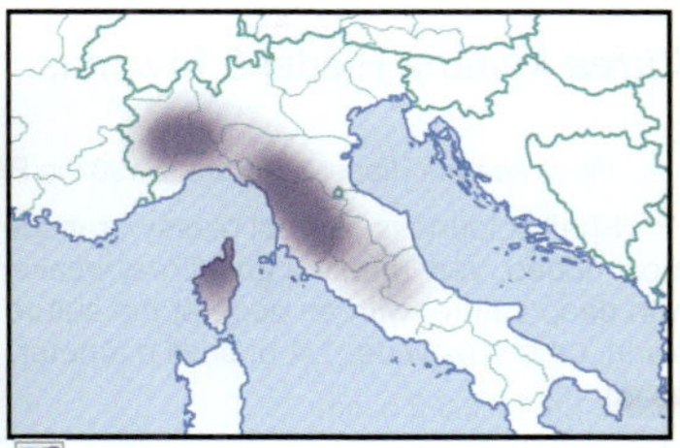

0.9-1.1 x 1.7-2.3 mm (1-3) Italy: Toscana, Chiusi della Verna, (4-5) Italy: Piemonte, Alessandria (1-3) Angelo Vannozzi

0.4-0.45 mm at 1 whorl, 0.85-0.95 mm (2 wh.), 1.4-1.55 mm (3), 1.75-1.9 mm (3.5 whorls)

Reported from under coniferous trees.

Westerlund 1886: 35, Alzona 1971: 119.

Italy, Corsica

Shell smooth, translucent, 4-4.5 slightly convex whorls, regularly increasing, suture impressed.
Differs from *V. botterii* in its wider umbilicus and smaller size.

Vitrea gasulli Riedel & Paul, 1978

Vitrea gasulli Riedel & Paul, 1978 in Riedel & Paul 1978: 54

Shell whitish, transparent and shiny, finely striated on upper side, upper side flat, lower side much more rounded, 4.5-5.5 slowly increasing whorls with shallow suture, umbilicus very narrow and often narrowed by the collumellar margin of the aperture, often covered with dust or mucus.

1.4-1.6 x 3-4 mm

Beckmann 2007: 79 (Ibiza and near Sòller, Mallorca).

V. crystallina does not live on the Balearic islands, reports should refer to *V. gasulli.*

Spain (Illes Balears: Eivissa/Ibiza)

Vitrea ilgazdaglariensis Neubert & Riedel, 1993

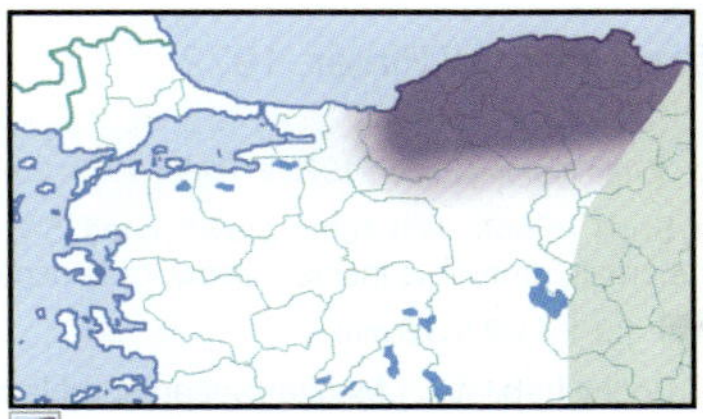

Vitrea ilgazdaglariensis Neubert & Riedel, 1993 in Neubert & Riedel 1993: 156

Shell smooth or weakly striated, translucent and shiny, 4.5-5.5 very densely coiled convex whorls with deep suture, suture with white margin, last whorl not much wider than penultimate whorl, umbilicus narrow, 1/10 of diameter.

1.2-1.3 x 2.4-2.6 mm

Margins of forests in low mountains, under stones and in desintegrating coniferous tree leaves.

Riedel 1995: 14, Schütt 2001: 177.

N Turkey

Vitrea illyrica (Wagner, 1907)

Without gender agreement: *Vitrea illyricus* (Wagner, 1907)

Crystallus illyricus Wagner, 1907 in A. J. Wagner 1907: 103

Shell relatively large, pale whitish translucent, very shiny, densely striated, about 5 whorls, last whorl 1.5 x as wide as penultimate whorl, umbilicus moderately wide and perspectival.
V. botterii is more densely coiled, *V. crystallina* less densely. Juveniles are difficult to tell from *V. botterii* (lower side is perhaps more convex in *V. illyrica*). *V. subaii* has a flatter lower side of the shell, wider and deep umbilicus, and a *Nautilus*-like aperture.

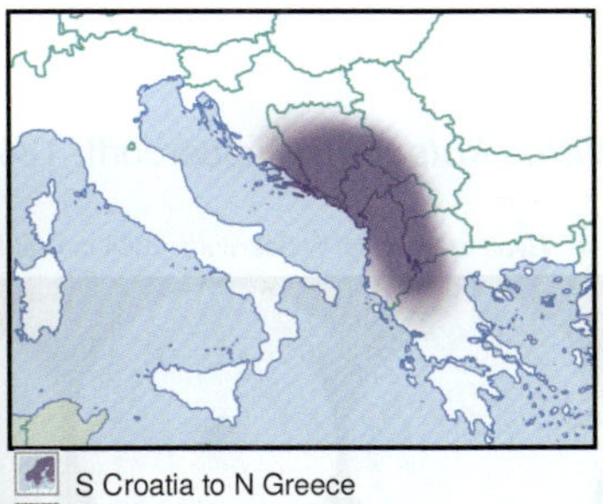

1.6-1.8 x 3.9-4.4 mm Croatia: Dalmatia, island of Korčula Massimo Prodan

0.5-0.6 mm at 1 whorl, 1.1-1.2 mm (2), 1.7-1.8 mm (3), 2.8-3.0 mm (4 whorls)

S Croatia to N Greece

Pavlović 1912: 26, Tomić 1959: 13, Pintér 1972: 247, Riedel 1992: 23.

More than one species may be involved. See also *V. subaii*, this could be a local form of *V. illyrica*.

Vitrea inae De Winter & Ripken, 1991

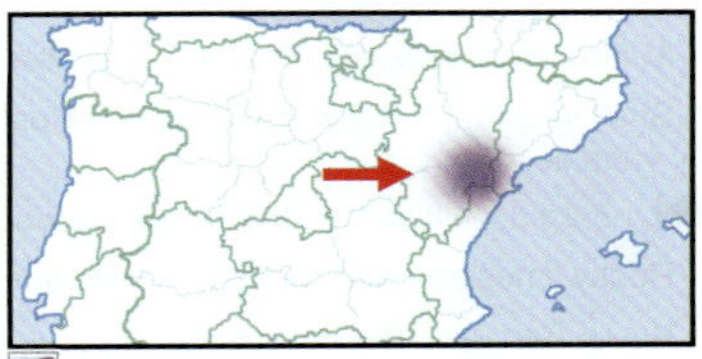

E Spain (Teruel)

Vitrea inae De Winter & Ripken, 1991 in De Winter & Ripken 1991: 5

Shell whitish, partly transparent, very finely in irregularly striated, 3 whorls, last whorl wider, umbilicus wide and shallow, all whorls well visible inside, 1/4-1/3 of diameter.
Whorls increase perhaps slightly more rapidly than in *V. pygmaea*, especially the last whorl is wider.

0.8 x 1.8-1.9 mm

Only known from river deposits.

Vitrea jetschini (Kimakowicz, 1890)

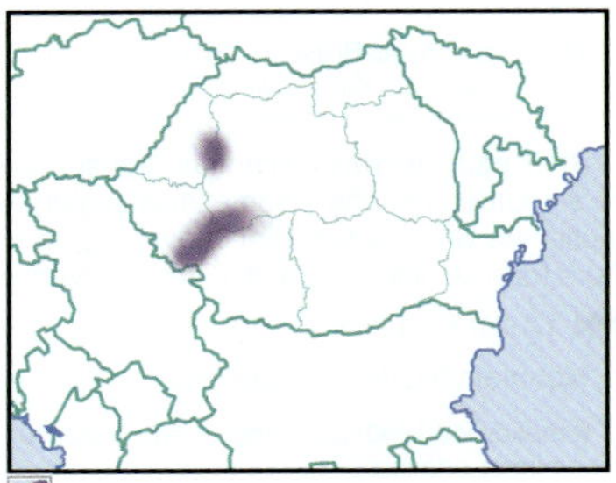

W Romania

Hyalinia jetschini Kimakowicz, 1890 in Kimakowicz 1890: 37 (*Hyalinia (Vitrea) Jetschini*)

Shell whitish, almost smooth or finely striated, 5-6 whorls, first whorls densely coiled, last whorl inflated, width 1.7-2.5 x of penultimate whorl (nearly 3 x if pen. wh. is taken from suture to suture) (however in the northern populations the last whorl is not so much inflated), apertural margin seen from above slightly protruded in (form, umbilicus moderately narrow, 1/8 of shell diameter. Shell shape locally variable.
Differs from *Lindbergia uminskii* in its more whitish and more translucent shell and in the absence of a spiral shell sculpture. In other *Vitrea* species the last whorl is not so much inflated.

1.4-2.2 x 3.5-5 mm

In karstic regions, frequent in small rock crevices and narrow spaces inside rock rubble.

Riedel 1960: 340, Grossu 1983: 106, Deli & Subai 2011: 38.

Vitrea keaana Riedel & Mylonas, 1981

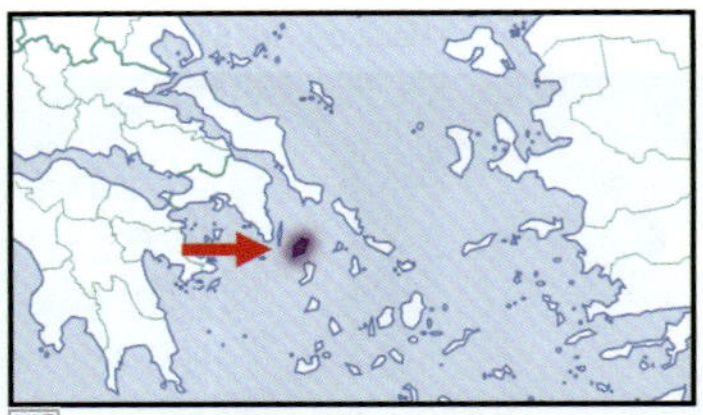

Greece (Kikládes: island of Kea)

Vitrea keaana Riedel & Mylonas, 1981 in Riedel & Mylonas 1981: 117

Shell colourless, almost glassy, transparent and shiny, prominently, densely and regularly striated, or almost ribbed, very unusual for *Vitrea*, weaker at lower side, 5-5.5 strongly convex and slowly increasing whorls with deep suture, with an edge at the suture at the last 2-3 whorls, last whorl not much wider than penultimate whorl, umbilicus wide, 1/4 of shell diameter.

1.4-1.6 x 3-3.5 mm

Riedel 1992: 24.

Vitrea kiliasi Pintér, 1972

Vitrea kiliasi Pintér, 1972 in Pintér 1972: 265

Kosovo: near Pejë/Peć

1.3-1.5 x 3.3-3.5 mm

Obviously a rare species.

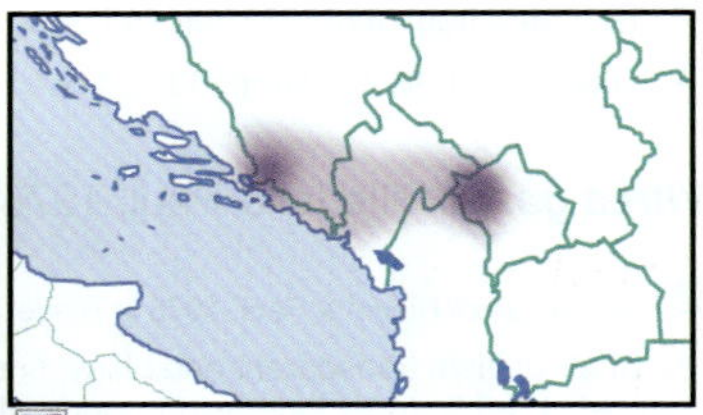

Herzegovina to Kosovo

Shell pale whitish, transparent, almost smooth, usually more than 5 whorls, last whorl almost twice as wide as penultimate whorl, umbilicus very wide, perspectivical, all whorls well visible inside.

Vitrea klemmi Pintér, 1972

Greece (island of Ikaria)

Vitrea klemmi Pintér, 1972 in Pintér 1972: 252

Shell smooth with some growth lines, 5-5.5 whorls, last whorl 1.5 x as wide as penultimate whorl, aperture quite small, umbilicus inside narrow, gets occasionally larger at the last whorl.

1.3-1.7 x 2.5-3.8 mm

Rocky habitats in Mediterranean shrubland vegetation, under stones, in the soil, in rock crevices or under moss, also in caves, usually associated with *V. contracta*. A common species on Ikaria.

Riedel 1981: 231, Riedel 1992: 24.

Vitrea kutschigi (Walderdorff, 1864)

Hyalina kutschigi Walderdorff, 1864 in Walderdorff 1864: 504 (*Hyalina Kutschigi*)

Herzegovina: near Zavala

1.9-2.3 x 4.2-5.3 mm

Westerlund 1886: 36, Pintér 1972: 259.

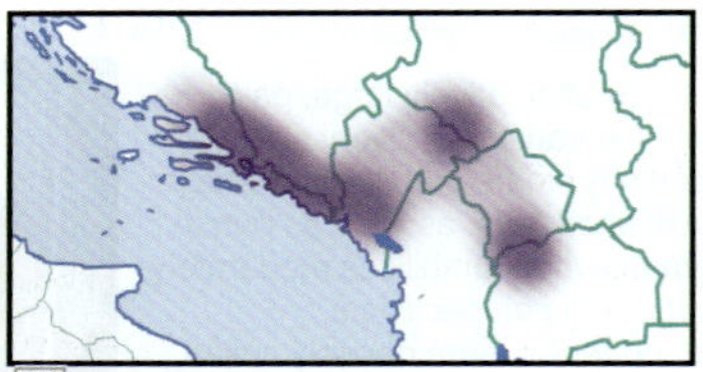

S Dalmatia to Rep. Makedonija

Shell relatively large, depressed, usually flat on upper side, weakly pale yellowish translucent and shiny, 5-6 whorls (not 8), last whorl almost twice as wide as penultimate whorl, umbilicus variable.
Somehow like *Bathyomphalus contortus*.

Vitrea lodosi Riedel, 1984

NW Turkey

Vitrea lodosi Riedel, 1984 in Riedel 1984: 165

Shell colourless, transparent, smooth and shiny, 5.5-6 weakly convex whorls with shallow suture, upper side flat, lower side also relatively flat, periphery regularly rounded, first whorls slowly increasing, last whorl width when seen from upside more than 2 x of penultimate whorl, aperture in its upper section Nautilus-like protruded, no umbilicus.

2.3-3 x 5-6.5 mm

Semisubterranean, in the soil deep under rock rubble, in rock crevices and in caves.

Rare or at least difficult to find, known only from few localities in Vilayet Bursa.

Riedel 1995: 14, Schütt 2001: 178.

Vitrea matsakisi Riedel & Mylonas, 1980

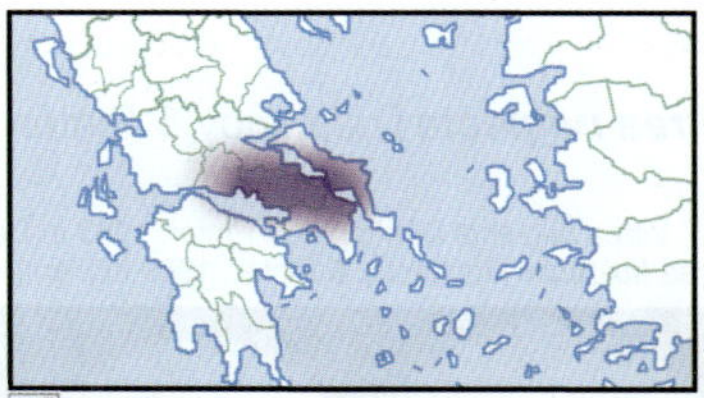

Greece (Evia, Stereá Elláda)

Vitrea matsakisi Riedel & Mylonas, 1980 in Riedel & Mylonas 1980: 135

Shell colourless, smooth, transparent and very shiny, 5.5-6 slowly and regularly increasing whorls with deep suture, last whorl width less than 2 x of penultimate whorl, regularly rounded at periphery, umbilicus wide, 1/5 of shell diameter.

1.4-1.8 x 2.8-3.5 mm

Scarcely vegetated habitats under stones and in rock crevices.

Riedel 1992: 24.

Vitrea meijeri Maassen, 1998

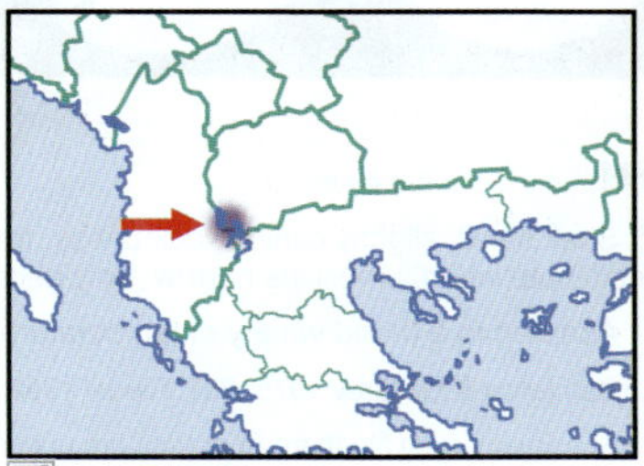

Rep. Makedonija

Vitrea meijeri Maassen, 1998 in Maassen 1998: 215

Shell almost flat, greyish white, transparent, densely and regularly striated, more prominently near the suture, 5-5.5 whorls, suture not deep, last whorl width slightly less than 2 x of preceding whorl, umbilicus wide, 1/7 of diameter.

V. sturanyi is smaller and has a deeper suture, *V. illyrica* is less flat on the upper side and has a narrower umbilicus, *V. siveci* has a broader last whorl and deeper suture, *V. kutschigi* is flatter with flatter lower side, its suture is shallower and its last whorl narrower.

1.7-2.3 x 4.5-5.6 mm

Found in a decidious forest under leaves and stones, 1000 m altitude.

Rare, not much known on this species.

Vitrea mikuskai Pintér, 1977

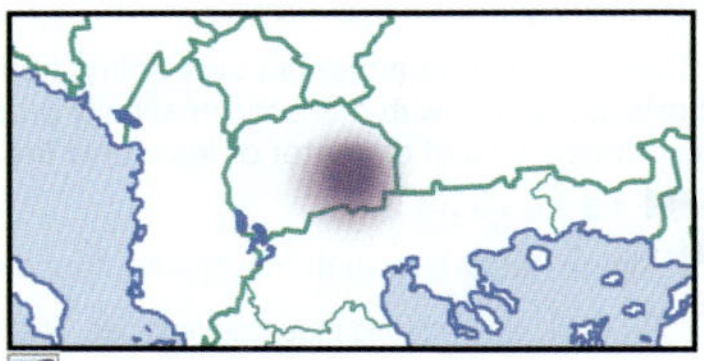

Rep. Makedonija

Vitrea mikuskai Pintér, 1977 in Pintér 1977: 183

Shell yellowish, translucent, almost smooth with fine growth lines and extremely fine spiral sculpture, 4-5 whorls with shallow suture, last whorl width 2 x of penultimate whorl, umbilicus wide, 1/4-1/3 of diameter, all whorls visible inside, getting broader at last whorl.

1.3-1.5 x 2.5-3.2 mm

Found at the base of rocks.

Vitrea minellii Pintér & Giusti, 1983

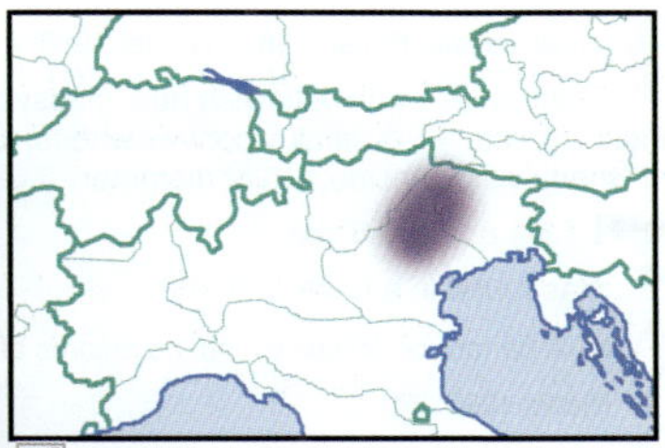

NW Italy

Vitrea minellii Pintér & Giusti, 1983 in Pintér & Giusti 1983: 5

Shell white yellowish, not very transparent and shiny, with few irregular growth lines and some spiral lines at lower side, 5.5 whorls, last whorl almost not wider than penultimate whorl, umbilicus very narrow but open.

Deciduous (beech) and mixed forests.

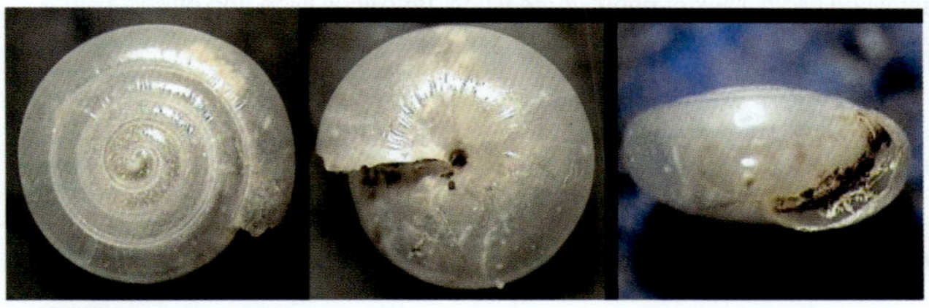

1.7-1.8 x 3.6-3.8 mm

Italy: Treviso, Cansiglio wood   Maurizio Sosso

Vitrea nadejdae Lindholm, 1926

Ukraine (Crimea near Yalta)

Vitrea nadejdae Lindholm, 1926 in Lindholm 1926: 169

Shell white, translucent, weakly striated, 5-5.5 very slowly increasing flat whorls with shallow suture, last whorl slightly wider than penultimate whorl, slightly angulated in upper section, umbilicus moderately narrow, penultimate whorl visible inside.

Shell more depressed than in *V. sorella*, whorls flatter and suture less deep, umbilicus wider.

1.5-1.8 x 3.5-3.8 mm

Sysoev & Schileyko 2009: 117.

Vitrea narbonensis (Clessin, 1877)

Hyalina narbonensis Clessin, 1877 in Clessin 1877: 129, Taf. 1 (*Hyalina Narbonensis*)

Spain: Valencia, Jeresa

1.1-1.4 x 2.5-3.5 mm

Westerlund 1886: 33, Germain 1930: 161, Riedel 1970, 1972: 119. Range insufficiently known.

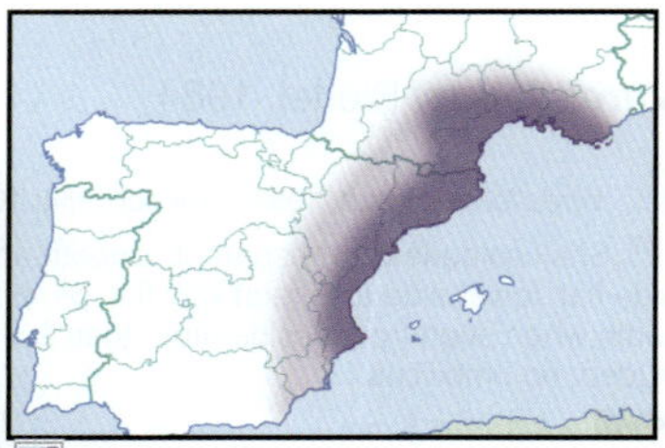

E Spain, S France

Shell very flat, finely and irregularly striated, 5-6 slowly increasing and almost flat whorls, umbilicus very narrow. Flatter than *V. crystallina*, umbilicus narrower.

Vitrea neglecta Damjanov & Pintér, 1969

Vitrea neglecta Damjanov & Pintér, 1969 in Damjanov & Pintér 1969: 35
Vitrea bulgarica Damjanov & Pintér, 1969 in Damjanov & Pintér 1969: 36 (First Reviser: Irikov 2001)

Bulgaria: W Rodopi mountains, Stara Reka, near Peshtera

1.4-1.6 x 2.9-3.4 mm

SE Europe

Shell small, slightly conical, colourless, fresh shells translucent and shiny, surface with fine and distinct radial growth lines, last whorl not much wider than penultimate whorl, umbilicus narrow, only penultimate whorl visible inside.

Adapted to a broad variety of temperatures and humidity, drought-resistant. In Bulgaria up to 1900 m.

Damjanov & Likharev 1975: 261, Riedel 1992: 14, 25, Irikov 2001, Irikov & Mollov 2006: 794, Hubenov 2007.

Specimens from Northern Sporades are uncertain. *Vitrea bulgarica* is a synonym of *neglecta* (Irikov 2001).

Vitrea olympica Riedel & Velkovrh, 1976

Vitrea olympica Riedel & Velkovrh, 1976 in Riedel & Velkovrh 1976: 224

Shell whitish opaque, not very shiny, densely striated, no spiral sculpture, 5.5-6 very slowly increasing whorls, periphery with a characteristically prominent blunt edge in the upper section, also near aperture, umbilicus narrow, 1/10 of diameter or less, only the penultimate whorl visible inside.

1.5-1.6 x 3.0-3.2 mm

Alpine habitats in high mountains, found in 2700 m.

Probably an endemic mountain species, restricted to a small range.

Riedel 1992: 26.

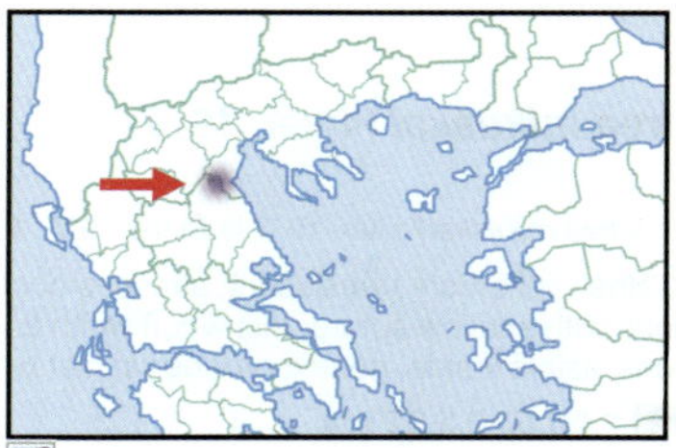

Greece (Mt. Olimbos)

Vitrea ossaea Pintér, 1983

Vitrea ossaea Pintér, 1983 in Pintér 1983: 219

Shell whitish with yellowish hue, glassy translucent and shiny if fresh, older specimens more opaque, thin, finely striated, 4.5-5 weakly convex whorls, last whorl width 1.5 x of penultimate whorl, umbilicus wide and deep, all whorls visible inside, 1/6 of diameter.

1.6-1.8 x 3.4-3.8 mm

Was found in a cave.

Must be rare or difficult to find, no records after 1983.

Riedel 1992: 26.

Could be a form of *V. neglecta* with slightly wider umbilicus. More research is necessary.

Greece (Ossa mountains)

Vitrea pageti Pintér, 1978

Vitrea pageti Pintér, 1978 in Pintér 1978: 170

Shell yellowish white, glassy transparent and shiny, weakly granulated, 5-5.5 moderately convex whorls, last whorl width 2 x of penultimate whorl, apertural margin upside Nautilus-like protruded, umbilicus inside narrow, slowly increasing, 1/6-1/5 of shell diameter at last whorl, all whorls visible inside.

1.8-2.0 x 3.8-4.2 mm

In organic soil litter at limestone rocks.

Obviously rare.

Riedel 1992: 27.

Greece (Ródos)

Vitrea pieperiana Pintér, 1977

Greece (Kárpathos)

Vitrea pieperiana Pintér, 1977 in Pintér 1977: 184

Shell whitish, shape and colour like *Oxychilus hydatinus*, almost smooth and very finely granulated with indistinct growth lines, 4-4.5 whorls with deep suture, spire slightly elevated, regularly rounded at periphery, columellar apertural margin almost perpendicular, umbilicus moderately narrow, 1/6-1/5 of diameter, all whorls visible inside.
Umbilicus wider than in *Oxychilus hydatinus*.

1.9 x 4.1 mm (only 1 shell known)

Must be extremely rare, only one single shell is known, not found again by Riedel 1992.

Riedel 1992: 27.

Vitrea pinteri Riedel & Subai, 1991

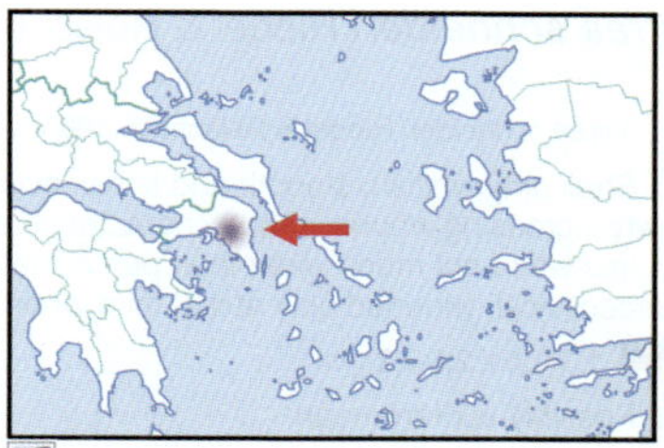

Greece (Attikí: Imittos mountain)

Vitrea pinteri Riedel & Subai, 1991 in Riedel & Subai 1991: 111

Shell dirty whitish, smooth, with weak growth lines, translucent, weakly shiny, 5-5.5 weakly convex whorls with shallow suture, regularly increasing, upper side more flattened and depressed than lower side, last whorl width nearly 2 x of penultimate whorl, umbilicus wide, 1/7-1/6 of shell diameter, showing all whorls inside.
The largest Greek *Vitrea* species.

2.0-2.2 x 4.3-5 mm

Was found at N exposed mountain slopes, in 800 m altitude, rocky habitats.

Obviously extreme local occurrence. Only known from one locality.

Riedel 1992: 27.

Vitrea pseudotrolli Pintér, 1983

Vitrea pseudotrolli Pintér, 1983 in Pintér 1983: 220

Italy: Liguria, Savona, near Albenga

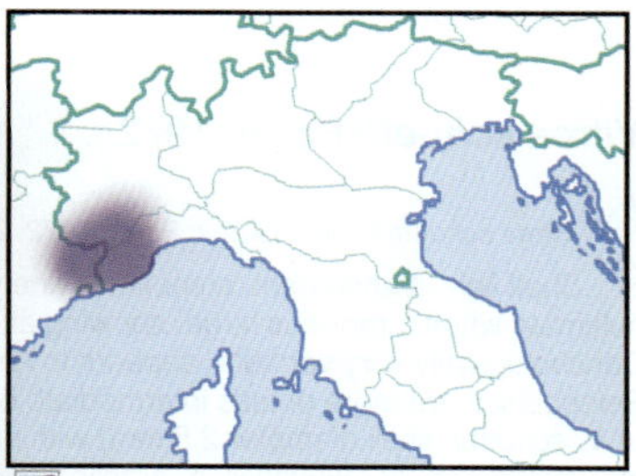

NW Italy, SE France

Manganelli et al. 1995: 18.

1.2-1.4 x 2.5-3.0 mm

Shell yellowish white, translucent and shiny, greyish white if weathered, regularly covered with low riblets, which can be weathered in older shells, 5.5-6 weakly convex and very slowly increasing whorls with shallow suture, last whorl width 1.5 x of penultimate whorl, bluntly keeled above the periphery, umbilicus extremely narrow, better visible in an oblique view.

Vitrea pygmaea (Boettger, 1880)

Hyalinia pygmaea Boettger, 1880 in O. Boettger 1880: 118, Taf. 4

Bulgaria: near Stara Zagora

SE Europe to Turkmenistan

Shell very small, colourless, sometimes slightly pale greenish yellow, 3.5-4 whorls, last whorl twice as wide as penultimate whorl, or even more, umbilicus very wide and perspectivical, all whorls well visible inside. Can usually not be confounded with any other species.

0.7-0.8 x 1.4-2.1 mm

Deciduous oak/hornbeam and pinetree forests with pH 4.5, can survive continuous periods of drought. In Bulgaria between 300 and 1400 m

Rare in Greece.

Pintér 1972: 277, Damjanov & Likharev 1975: 265, Riedel 1992: 28, Riedel 1995: 18, Irikov & Mollov 2006: 794, Fehér & Erőss 2009, Sysoev & Schileyko 2009: 118.

Vitrea riedeli Damjanov & Pintér, 1969

Vitrea riedeli Damjanov & Pintér, 1969 in Damjanov & Pintér 1969: 38

(1-6) W Turkey: near Bodrum, (7-8) Bulgaria: Strandza mountains, Veleka river area
(7-8) Dilian Georgiev

SE Bulgaria, E Greece, W Turkey

Shell colourless, shiny but often pale, 5 or slightly more whorls, last whorl not much wider than penultimate whorl, umbilicus wide, all whorls well visible inside, second widest umbilicus in Bulgaria after *V. vereae*.

1.3-1.4 x 2.6-3.0 mm

0.5 mm at 1 whorl, 0.8-1.0 mm (2), 1.25-1.5 mm (3), 1.6-1.9 mm (3.5), 1.9-2.3 mm (4), 2.15-2.5 mm (4.5), 2.5-2.9 mm (5)
Some populations of *V. riedeli* have an umbilicus narrower than usual, like in *V. schuetti*, or even like in *V. neglecta*.

Damjanov & Likharev 1975: 264, Riedel 1992: 28, Riedel 1995: 18, Irikov & Mollov 2006: 794.

Vitrea riedeliana Paget, 1976

Vitrea riedeliana Paget, 1976 in Paget 1976: 726

Shell white, transparent, lower whorls densely and irregularly ribbed, ribs weaker on lower side, 5-5.5 whorls with deep suture, upper side slightly convex, lower side flatter, umbilicus narrow and perspectivical, 1/11-1/10 of diameter.

1.7-2.0 x 2.8-3.2 mm

Various habitats, common all over the island of Ródos.

Riedel 1992: 29, Riedel 1995: 19.

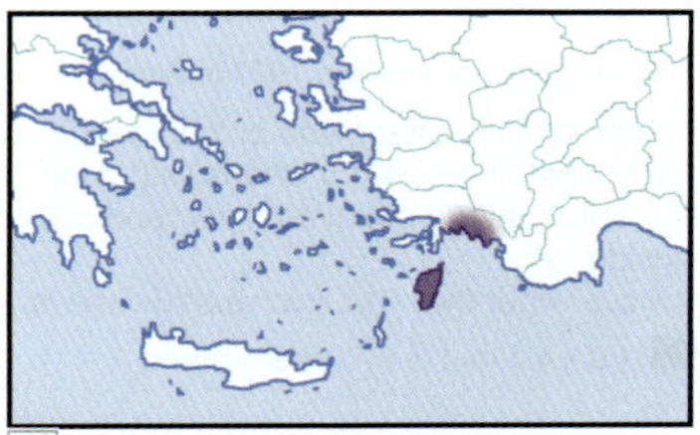

Greece (Ródos), Turkey (Muğla)

Vitrea schneideri Riedel & Reischütz, 1988

Vitrea schneideri Riedel & Reischütz, 1988 in Riedel & Reischütz 1988: 149

Shell dirty whitish, smooth and shiny, occasionally with fine spiral lines or reticulate micropattern, 4.5-5 very slowly increasing whorls, shape characteristically lentiform, also in juvenile shells, with blunt edge at the periphery, last whorl not much wider than penultimate whorl, aperture also with edge, umbilicus moderately wide, not perspectivical, widened only at last third whorl.

1.3-1.5 x 2.6-3.2 mm

Possibly subterranean.

A rare species, difficult to find, only a few specimens are known.

Riedel 1992: 29.

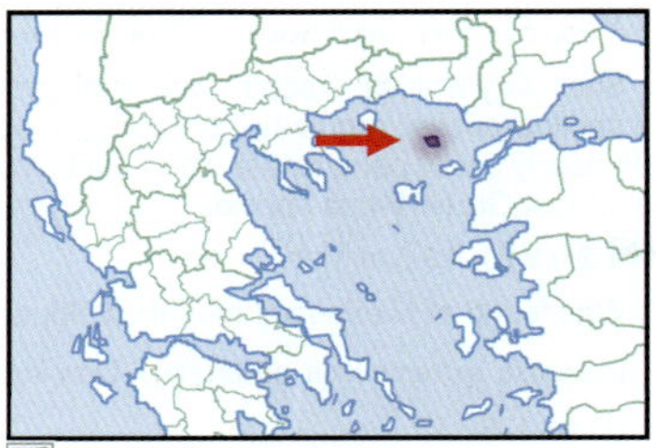

Greece (Samothráki)

Vitrea schuetti Pintér, 1972

Vitrea schuetti Pintér, 1972 in Pintér 1972: 264

Shell light, transparent, sometimes finely and irregularly striated, 5-6 whorls, last whorl twice as wide as penultimate whorl, umbilicus large, showing all whorls inside.
Conchologically very variable, appears to form intermediates with *V. argolica* and *V. matsakisi*. Populations from Peloponisos are more or less intermediate between *V. schuetti* and *V. argolica*. Apart from those, a smaller form of *V. schuetti* (shell diameter 2.5 mm) with narrower umbilicus occurs on the northeastern promontory of Parnon mountains, similar to juvenile *V. argolica* with wider umbilicus - possibly a geographical subspecies.

1.7-2.0 x 3.6-3.9 mm

Riedel 1992: 30, Riedel 1995: 20.

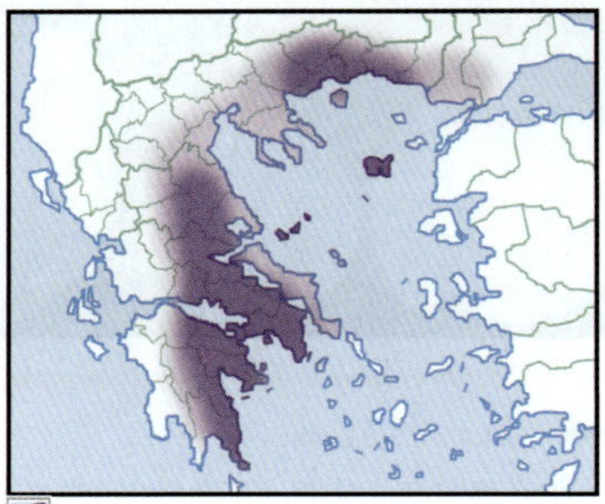

Greece

Vitrea selecta Pintér, 1972

Vitrea zilchi selecta Pintér, 1972 in Pintér 1972: 257

Shell weakly convex on upper side and not extremely flat on lower side, almost smooth, 4.5-5 whorls, last whorl almost twice as wide as penultimate whorl, umbilicus narrow, showing all whorls inside.
Specimens from Monodendri are larger with wider umbilicus. Some populations from Thessalia and Makedonia resemble *V. botterii*. Differs from *V. zilchi* by its more conical shape, smaller size and less whorls.

Around 1.4 x 3.3 mm

Riedel 1992: 31.

Specific separation between *V. selecta* and *V. botterii* is somewhat doubtful, *V. botterii* is highly variable.

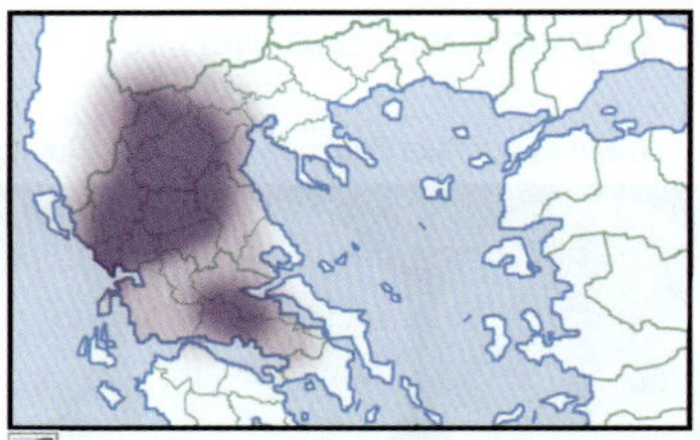

Greece

Vitrea siveci Riedel & Velkovrh, 1976

Vitrea siveci Riedel & Velkovrh, 1976 in Riedel & Velkovrh 1976: 220

Shell whitish or cream, qpaque, shiny and weakly translucent, upper side prominently striated and with a weak and very fine spiral lines, 5-5.5 weakly convex whorls with shallow suture, last whorl width 1.5 x of penultimate whorl, periphery regularly rounded, umbilicus moderately wide and perspectivical, 1/5 of diameter.
One of the largest *Vitrea* species, somehow like *Oxychilus hydatinus* and *O. inpopinatus*, but with more whorls and a much wider umbilicus.

2.2-2.6 x 4.7-5.5 mm

Life specimens were found in soil samples in a cave.

Riedel 1997.

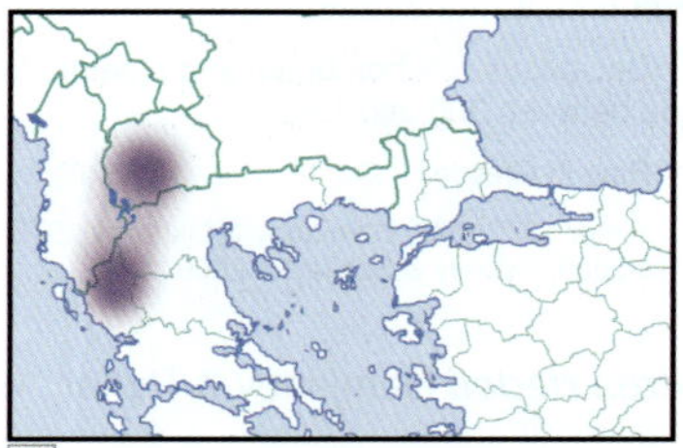

Rep. Makedonija and NW Greece

Vitrea sorella (Mousson, 1863)

Zonites sorella Mousson, 1863 in Mousson 1863: 299

Shell quite flat on upper side, more than 6 whorls, last whorl not much wider than penultimate whorl, umbilicus extremely narrow.
Differs from other *Vitrea* species in its densely coiled whorls and narrow umbilicus.

1.8-2.3 x 3.6-4.3 mm

Moist and shady habitats under limestone rock rubble.

Pintér 1972: 268, Riedel 1995: 20, Hausdorf & Riedel 1996.

N Turkey (Kastamonu, Zonguldak)

Vitrea sossellai Pintér, 1978

Vitrea sossellai Pintér, 1978 in Pintér 1978: 172

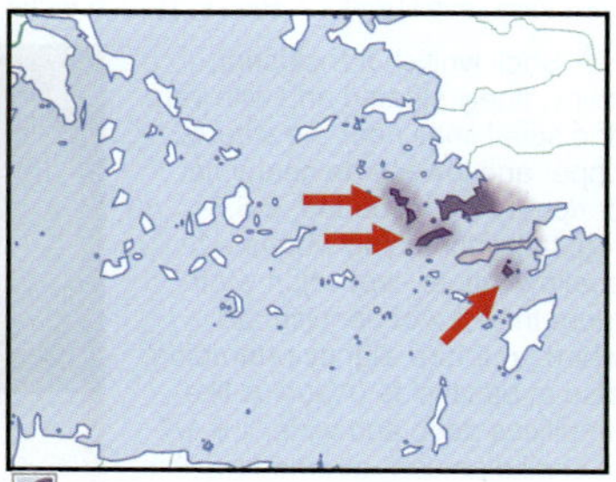

1.7-1.9 x 3.5-4.0 mm

W Turkey: near Bodrum

E Greece, W Turkey (Muğla)

Shell white with yellowish hue, glassy transparent and shiny, weakly granulated, 4.5-5 slowly increasing whorls with shallow suture, apertural margin sometimes with concave curve near insertion, umbilicus narrow, slightly widened at last whorl, 1/9-1/8 of shell diameter.

Riedel 1992: 32 (Greece: Léros, Kálimnos, Kos, Sími), Örstan, Yıldırım et al. 2005 (Turkey: Bodrum peninsula).

Vitrea spelaea (Wagner, 1914)

Without gender agreement: *Vitrea spelaeus* (Wagner, 1914)

Crystallus spelaeus Wagner, 1914 in A. J. Wagner 1914 (Sitz.-Ber. Wien 123): 42

Shell large, solid, whitish translucent, shiny, with blurry growth lines, 4.5-5.5 whorls, last whorl twice as wide as penultimate whorl, umbilicus narrow, showing only the penultimate whorl inside. Cannot be confounded with any other species.

2.0-2.4 x 4.4-5.1 mm

Only known from caves.

Pintér 1972: 266.

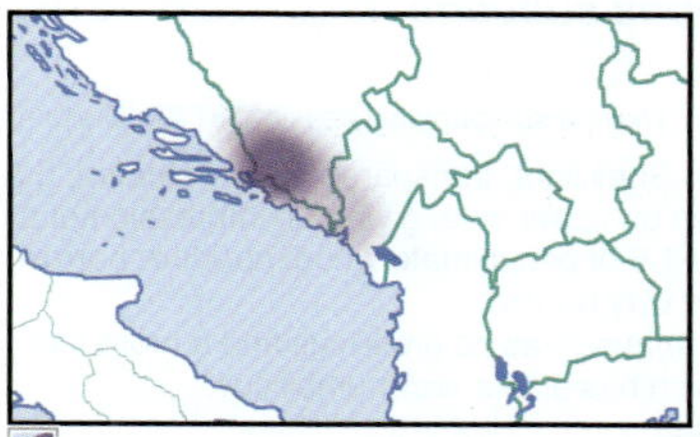

Herzegovina, ? Montenegro

Vitrea sporadica Pintér, 1978

Vitrea sporadica Pintér, 1978 in Pintér 1978: 51

Shell yellowish white, weakly shiny, surface granulated under high magnification, 5-6 whorls, rounded at periphery, upper side flattened, lower side more rounded, periphery in the upper section, last whorl width 2 x of penultimate whorl, columellar margin rounded, basal margin of aperture moderately curved, umbilicus wide, 1/5 of diameter, all whorls visible inside.

1.2-1.4 x 2.8-3.5 mm

Riedel 1992: 33, Triantis et al. 2008: 474.

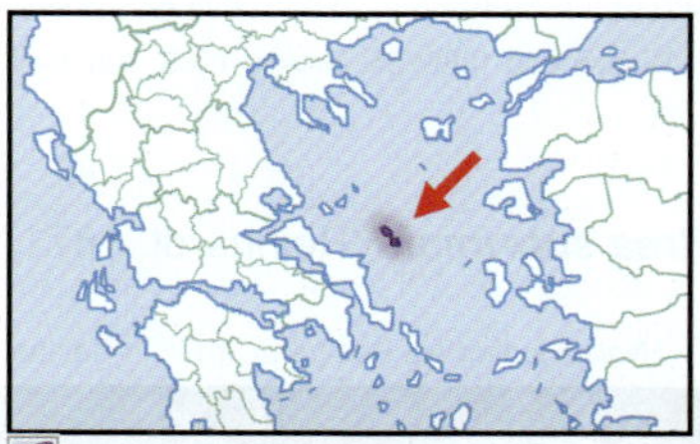

Greece (island group of Skiros)

Vitrea storchi Pintér, 1978

Vitrea storchi Pintér, 1978 in Pintér 1978: 173

Shell very flat, yellowish grey, translucent and shiny, occasionally weakly striated, 4.5 flat and initially slowly increasing whorls, last whorl width 2.5 x of penultimate whorl, umbilicus slowly increasing and finally moderately wide, 1/7-1/6 of shell diameter.

1.3-1.5 x 3.0-3.5 mm

Was found in an ancient quarry among rocks.

Must be extremely rare, only known from one locality, has never been found alive.

Riedel 1992: 33.

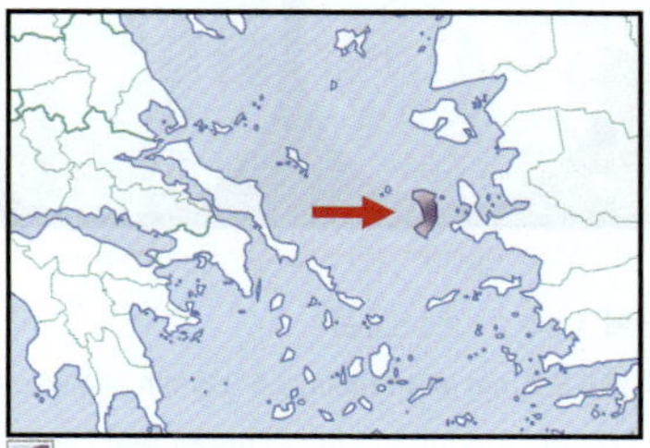

Greece (Chios)

Vitrea striata Norris, Paul & Riedel, 1988

Vitrea striata Norris, Paul & Riedel, 1988 in Norris, Paul & Riedel 1988: 46

Shell glassy, first whorls smooth any shiny, last whorls prominently and irregularly striated, 5-5.5 slowly increasing step-like whorls with moderately impressed suture, upper side nearly flat, lower side rounded or with rather triangular outline, periphery with blunt edge in the upper section, last whorl width 1.5 x of penultimate whorl, umbilicus very narrow, 1/20 of diameter.

1.3-1.5 x 3.2-3.7 mm

Was found at the foot of cliffs.

Beckmann 2007: 79.

Spain (Balearic Islands: Ibiza)

Vitrea sturanyi (Wagner, 1907)

Crystallus sturanyi Wagner, 1907 in A. J. Wagner 1907: 106

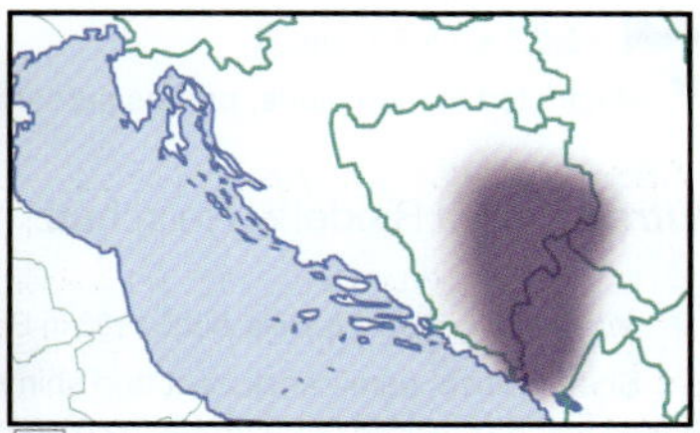

1.7-1.9 x 3.8-4.3 mm

Bosnia: south of Tjentiste near Foča

Montenegro, Bosnia

Shell large, slightly pale greenish translucent, shiny, older shells yellowish, finely striated, more than 5 whorls with deep suture, rounded and not depressed, last whorl not wider than penultimate whorl, umbilicus 1/5-1/6 of shell diameter and deep with almost perpendicular walls.

Pavlović 1912: 27, Tomić 1959: 13, Pintér 1972: 248, Dhora & Welter-Schultes 1996: 143, Irikov & Mollov 2006: 795 (not in Bulgaria).

Vitrea subaii Pintér & Riedel, 1973

Vitrea subaii Pintér & Riedel, 1973 in Pintér & Riedel 1973: 272

1.6-1.9 x 3.5-4 mm

Croatia: Dalmatia, Mljet island

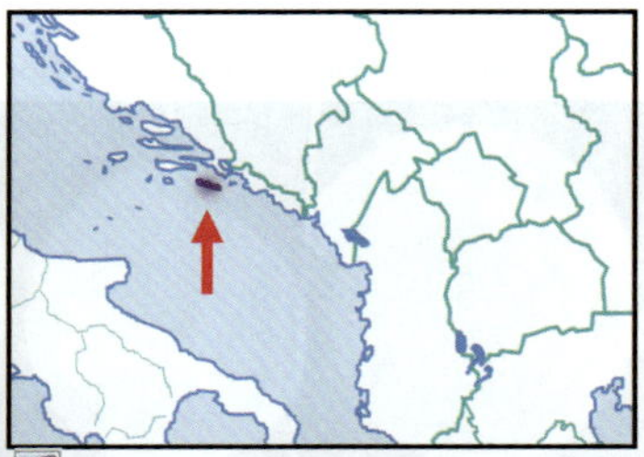

Croatia (Dalmatia: Mljet island)

In rock crevices, under moss and organic litter, abundant where present.

Massimo Prodan

Shell whitish translucent, shiny, finely striated and with very fine spiral lines near the suture, upper and lower side nearly flat, periphery well rounded, 5-5.3 regularly increasing whorls with shallow suture, last whorl width more than 2 x of penultimate whorl, aperture slightly protruding, also at basal side (*Nautilus*-like), umbilicus deep and wide, 1/8-1/7 of shell diameter, all whorls visible inside and rounded.
Differs from *V. illyrica* in its flatter lower side, a wider last whorl and narrower umbilicus, and its *Nautilus*-like aperture.

Differences to nearby populations of *V. illyrica* are less pronounced than the descriptions suggest. More research in the local variations of *V. illyrica* might be useful, it is possible that the form of Mljet is just a local form of *V. illyrica*.

Vitrea subcarinata (Clessin, 1877)

Hyalina subcarinata Clessin, 1877 in Clessin 1877: 129, Taf. 1

Shell light, transparent, finely striated, 5.5-6 whorls with deep suture, last whorl more rounded on lower side than on upper side, giving an impression of a rounded keel in the upper half, last whorl width seen from above 1.3-1.6 of penultimate whorl, aperture narrow, margin seen from lower side forming a weak ~ wave line, umbilicus very narrow.
V. erjaveci has no umbilicus and a more lenticular shape. Whorls more densely coiled than in *V. contracta*, which has also a wider umbilicus.

1.1-1.8 x 2.6-4 mm

Rocky habitats in forests on calcareous substrate, lives in leaf litter. Usually rare except near Petroşani and south of the Mureş river where it is the most frequent *Vitrea* species.

Westerlund 1886: 35, Grossu 1983: 100, Deli & Subai 2011: 24.

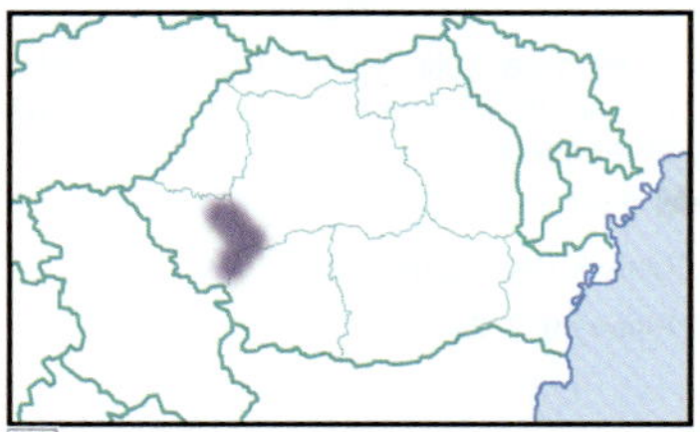

SW Romania

Vitrea subrimata (Reinhardt, 1871)

Hyalina subrimata Reinhardt, 1871 in Reinhardt 1871: 39

1.8-2.2 x 2.5-5 mm

Italy: Sardinia, Nuoro, Laconi

Ø 2.6-2.7 mm at 3 whorls, 3.5-3.8 mm at 4 whorls

Humid mountain forests under leaf litter, stones or in crevices, usually on calcareous substrate, rocks and rock rubble, up to 2600 m, in Bulgaria recorded from 2000 m, in England also in open habitats, between 250 and 600 m.

Vulnerable in Germany.

Damjanov & Likharev 1975: 259, Kerney et al. 1983: 162, Riedel 1992: 34, Riedel 1995: 21, Kerney 1999: 136, Hubenov 2007, Deli & Subai 2011: 16.

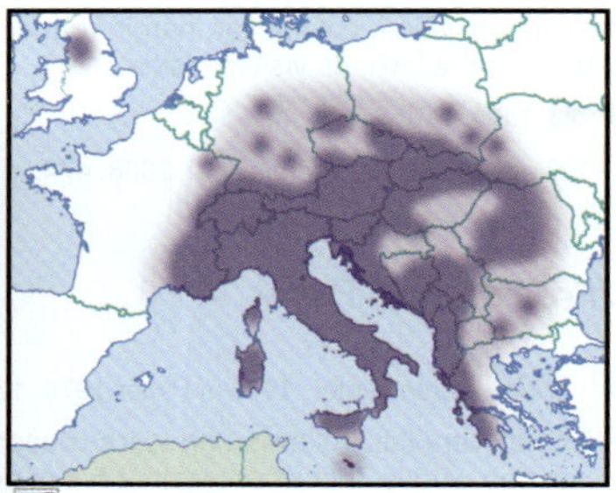

Europe and N Africa

Shell colourless, transparent, 4-5 whorls, last whorl width seen from above 1.2-1.7 x of penultimate whorl, umbilicus very narrow but present (difference to *V. diaphana*), the columellar margin is pointed and slightly protruded (well visible in shell 5) and covers the umbilicus only very weakly. Shell very variable, except the characteristic form of the columellar margin at the umbilicus.

The morphological variation has been associated with subspecies, but the nature of the variation is not yet understood. Two forms may co-occur with and without intermediates, other populations may consist of only one single form.

Vitrea szekeresi Deli & Subai, 2011

Vitrea szekeresi Deli & Subai, 2011 in Deli & Subai 2011: 41

Shell finely striated, 4-5 whorls, last whorl inflated, width seen from above 2.0-2.8 x of penultimate whorl, apertural margin seen from above protruded in Nautilus form, umbilicus narrow and perspectivical.
V. jetschini has a wider umbilicus and more densely coiled whorls, at equal size *V. jetschini* has 0.75-1 whorls more. *V. transsylvanica* has no umbilicus.

1.5-2.5 x 3.5-5.5 mm

Restricted to mountains, prefers subterranean habitats in narrow spaces inside rock rubble.

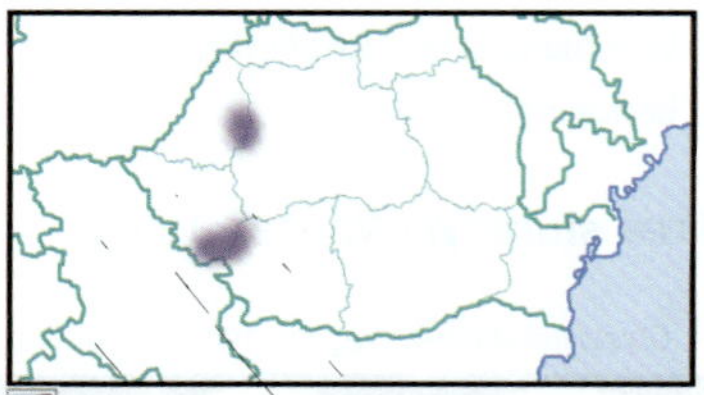

W Romania

Vitrea thasia Riedel & Reischütz, 1983

Vitrea thasia Riedel & Reischütz, 1983 in Reischütz 1983: 137

Shell whitish, semitranslucent and shiny, weakly striated and with distinct and dense spiral lines, upper side almost flat, 4.5-5 slowly and regularly increasing whorls, last whorl width 2 x of penultimate whorl, periphery at last whorl not regularly rounded but upper side more depressed and lower side more rounded, upper side of apertural margin almost horizontal, umbilicus very wide, nearly 1/3 of shell diameter, widened at last whorl.

1.2-1.3 x 2.9-3.3 mm

Was found under large rocks in an olive plantation.

Riedel 1992: 35.

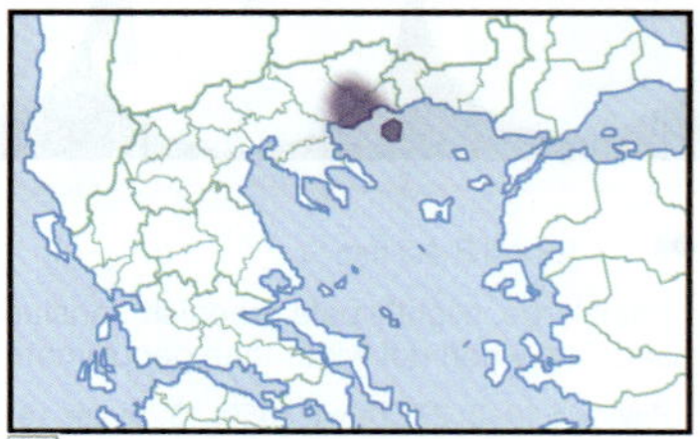

Greece (Makedonía, Thásos)

Vitrea transsylvanica (Clessin, 1877)

Hyalina transsylvanica Clessin, 1877 in Clessin 1877: 133, Taf. 2

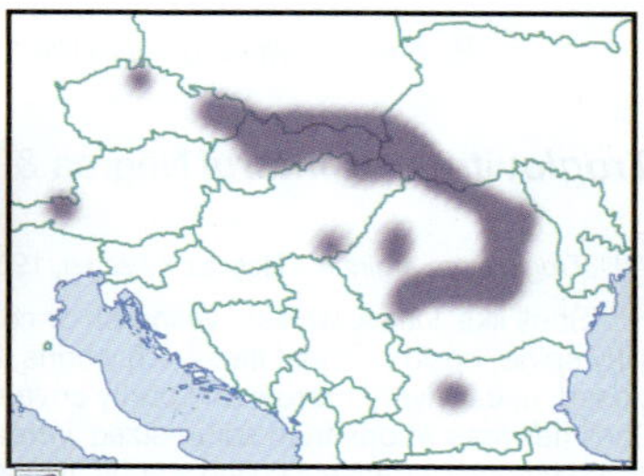

1.3-2.2 x 3.0-4.5 mm

Slovakia: Lesnica Pienicy

Last whorl more inflated than in *V. subrimata* and *V. subcarinata*, width seen from above around 2.3-2.6 x of penultimate whorl, first whorls denser arranged than in *V. subrimata*, aperture seen from above weakly curved (*Nautilus*-like protruded), no umbilicus.
Differs from *V. diaphana* in its wider last whorl and its *Nautilus*-like protruded aperture (in *V. diaphana* straight).

Carpathians

In closed forests, mainly in beech forests.

Negrea & Riedel 1968, Damjanov & Likharev 1975: 258, Grossu 1983: 96, Dedov 1998, Sysoev & Schileyko 2009: 70, Deli & Subai 2011: 3.

Vitrea trolli (Wagner, 1922)

Crystallus trolli Wagner, 1922 in A. J. Wagner 1922: 113, Taf. 5

Shell colourless, translucent, smooth with only faintly visible growth lines, flat, periphery convex, 5.5 whorls, more slowly increasing than in *V. subrimata*, last whorl not much wider than penultimate whorl, umbilicus extremely narrow.

Pintér 1972: 222, Manganelli et al. 1995: 18, Gittenberger & Eikenboom 2006.

1.3-1.5 x 2.9-3.4 mm

Italy: Vicenza, Mt. Zebio, 1700 m

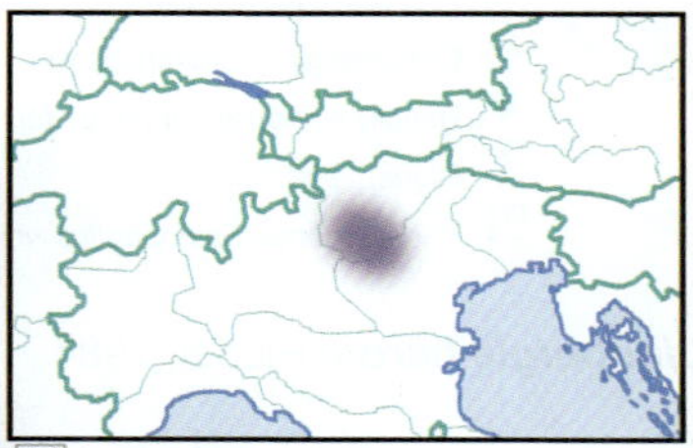

N Italy: Veneto, Trentino
Not in Gargano peninsula.

Maurizio Sosso

Vitrea vereae Irikov, Georgiev & Riedel, 2004

Vitrea vereae Irikov, Georgiev & Riedel, 2004 in Irikov, Georgiev & Riedel 2004: 79

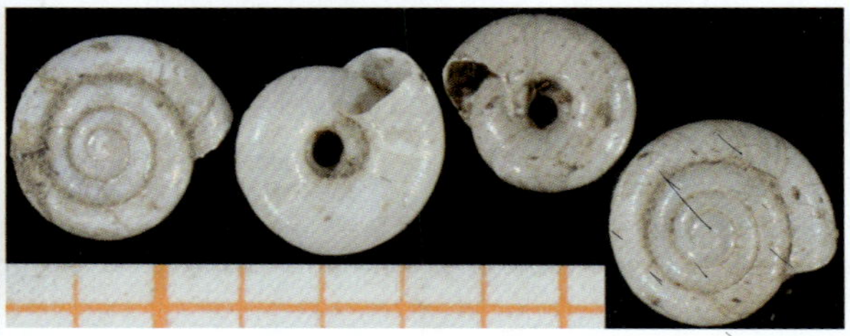

1.0-1.2 x 2.3-2.8 mm

Bulgaria: (1-3) Stara Zagora, (4-7) Osam river deposits near Lovech

(1-3) Jivko Stoilov

Ø 1.3-1.4 mm at 2 whorls, 2.0-2.2 mm (3 whorls), 2.6-2.7 mm (3.5 whorls)

Umbilicus extremely wide, wider than in the other *Vitrea* species. Differs from *Lucilla* and *Hawaiia* in its larger size, and anatomically.

Bulgaria

In moss on rocks, in rock crevices, under stones, also in pine forests.

Rare.

Vitrea zilchi Pintér, 1972

Vitrea zilchi Pintér, 1972 in Pintér 1972: 256

1.4-1.5 x 3.5-3.7 mm

Croatia: Dalmatia, Dubrovnik

Croatia (S Dalmatia)

Not very frequent.

Shell almost flat, can be very flat at lower side, yellowish translucent, almost smooth with irregular growth lines, more than 5 whorls, last whorl almost twice as wide as penultimate whorl, umbilicus wide, showing all whorls inside.

Troglovitrea

Troglovitrea Negrea & Riedel, 1968
in Negrea & Riedel 1968: 209

Type species *Troglovitrea argintarui* Negrea & Riedel, 1968
(original designation)

Shell like *Vitrea*, genitalia like *Oxychilus*.

Troglovitrea argintarui Negrea & Riedel, 1968

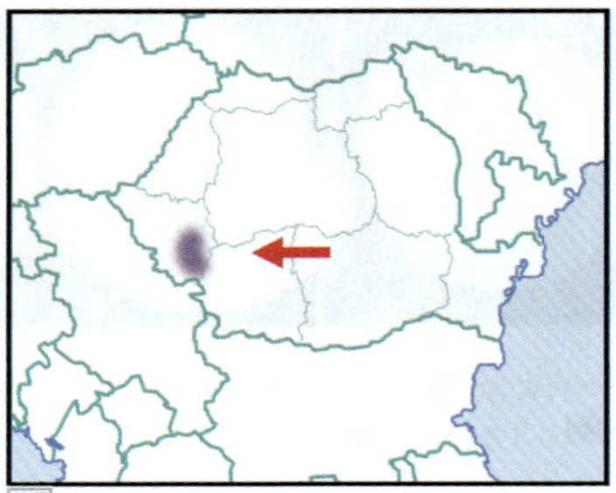

SW Romania

Troglovitrea argintarui Negrea & Riedel, 1968 in Negrea & Riedel 1968: 210

Shell like *Vitrea*, whitish, yellowish or colourless, transparent, smooth with dense growth lines producing a silky shine, upside almost flat, 4-4.5 whorls, last whorl width 1.7-2.7 x of penultimate whorl, umbilicus rarely closed, usually very narrow and partly covered by the slightly protruded columellar margin.
Shell has less whorls than same-sized *Vitrea diaphana* or *V. transsylvanica*. Usually larger than *V. subrimata* which has more densely coiled whorls, its last whorl width is below 1.7 x of the penultimate whorl.

1.8-2.3 x 4-5 mm

In highland valleys in caves and at slopes in more or less narrow spaces inside rock rubble. Seems to feed on disintegrating plant particles or perhaps on guano, obviously not carnivorous.

Grossu 1983: 108, Deli & Subai 2011: 44.

Lindbergia

Lindbergia Riedel, 1959
in Riedel 1959: 110

Type species *Lindbergia spiliaenymphis* Riedel, 1959
(original designation)

Echinophallus Riedel, 1960 in Riedel 1960: 334 *Lindbergia (Echinophallus)*, type species *Lindbergia uminskii* Riedel, 1960 (original designation) not *Echinophallus* Schumacher, 1914 (Cestoda)
Spinophallus Riedel, 1962 in Riedel 1962: 226
Expressly established as a new replacement name for *Echinophallus* Riedel, 1960.

Conchological differences with *Vitrea* species and the *Oxychilus hydatinus* group are unreliable. *Lindbergia* species can only be identified by their anatomy.

Usually subterranean, often found in caves.

Riedel 1992.

The subgenus *Spinophallus* which contains only one species (*L. uminskii* from Bulgaria):

Lindbergia beroni Riedel, 1984

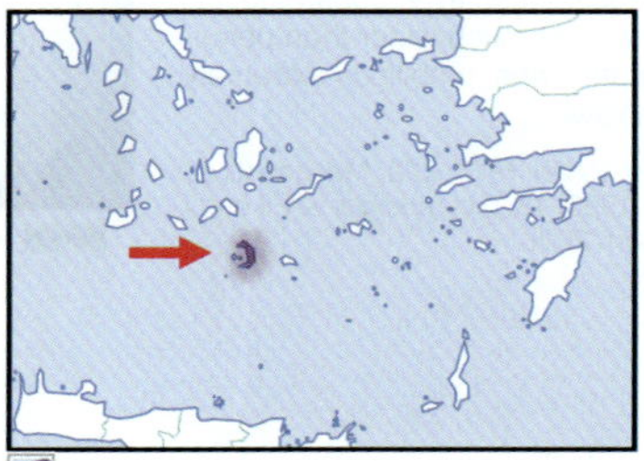

Greece (Santoríni)

Lindbergia beroni Riedel, 1984 in Riedel 1984: 1

Shell *Vitrea*-shaped, almost lentiform, cream and very little translucent, smooth and very shiny, with distinct growth lines, no spiral lines, 5 weakly convex whorls with shallow suture, periphery not rounded but depressed, without edge, upper margin of the aperture not rounded but with edge near the upper suture, also in juveniles, umbilicus wide, regularly increasing, last whorl not much wider, 1/5 of shell diameter.
The animal is cream, its foot undivided.

2.0-2.2 x 4.4-4.8 mm

Endemic species of Thíra/Santoríni, seems to be the only recent endemic species of this island.

Riedel 1992: 40.

Lindbergia garganoensis Gittenberger & Eikenboom, 2006

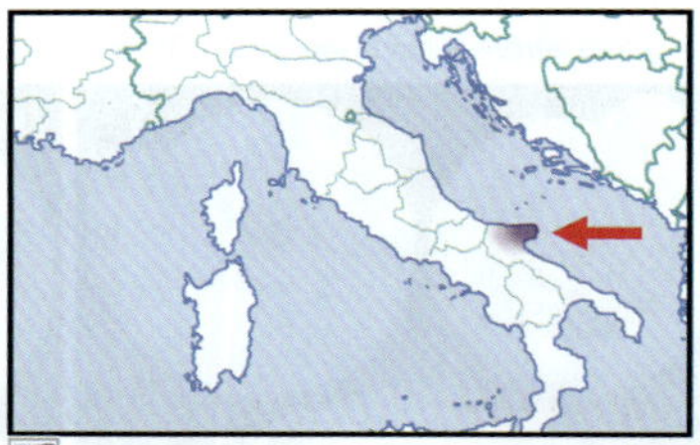

Italy (Gargano peninsula)

Lindbergia garganoensis Gittenberger & Eikenboom, 2006 in Gittenberger & E. 2006: 163

Shell colourless, glossy, irregularly striated, with numerous very regular and fine spiral lines, less distinct at the last whorl, 3.5-4 slightly convex whorls with shallow suture, last whorl width 2.3 x of predecing whorl, umbilicus very narrow but open, less than 1/10 of diameter.

1.0-1.3 x 2.0-2.6 mm

Very scarcely vegetated habitats with limestone rock faces and boulders.

Classification in *Lindbergia* doubtful, the others have all a much wider umbilicus. But *Lindbergia* species in Greece live in likewise scarcely vegetated habitats.

Lindbergia gittenbergeri Pintér & Riedel, 1983

Lindbergia gittenbergeri Pintér & Riedel, 1983 in Pintér & Riedel 1983: 101

Shell yellowish white, translucent and shiny, with very fine spiral lines, 4-4.5 regularly increasing whorls, last whorl width 2 x of penultimate whorl or slightly more, umbilicus moderately wide, 1/6 of shell diameter.

Subterranean.

Riedel 1992: 41.

Anatomy unknown, genus tentative.

2.1-2.5 x 3.8-4.5 mm

Greece: Kérkira, cave near Loutses (juvenile specimens)

Massimo Prodan

NW Greece (island of Kérkira)

Only known from Kérkira, possibly endemic, has only been found in two caves.

Lindbergia karainensis Rähle & Riedel, 1987

Turkey (Vilayet Antalya)

Lindbergia karainensis Rähle & Riedel, 1987 in Rähle & Riedel 1987: 204

Shell like *Vitrea*, smooth and shiny, very weakly striated with very faintly visible spiral lines, 4-4.5 slightly convex whorls with deep suture, regularly rounded at periphery, last whorl width 1.5-2 x of penultimate whorl, umbilicus variable within populations, moderately narrow (0.15-0.22 of shell diameter) and widened at last whorl.

1.4-1.7 x 2.9-3.6 mm

Subterranean, found in caves and cavities.

Riedel 1995: 21, Schütt 2001: 183. Anatomy unknown. Identity of the Kárpathos population uncertain.

Rähle & Riedel classified the species in the genus *Lindbergia* because 1 - first whorls are relatively wide, 2 - shell surface with fine spiral lines, 3 - umbilical region is flattened, 4 - umbilicus width is variable within a population, 5 - frequent occurrence in caves.

Lindbergia orbicularis (Riedel, 1962)

Oxychilus? orbicularis Riedel, 1962 in Riedel 1962: 224

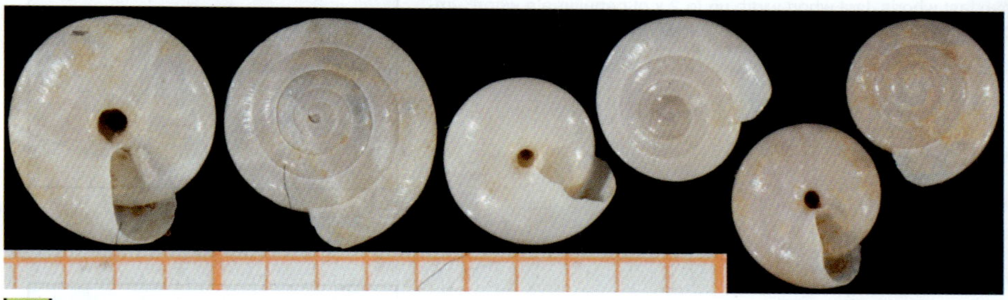

Greece: Crete, Réthimno, near Agios Ioánis

2.7-3 x 5-5.6 mm

Subterranean, a strict cave-dweller, but common.

Riedel 1968, Riedel 1977 (anatomy), Riedel 1992: 38.

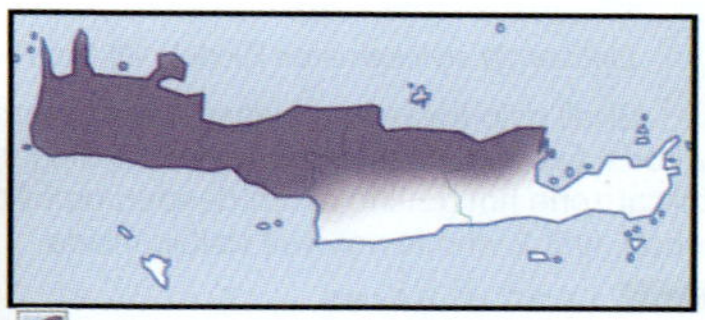

Greece (W and N central Crete)

Shell lentiform but not keeled or angulated, outline regularly rounded at upper side, flatter at lower side, umbilical zone of lower side not getting deeper, in contrast to most other zonitid species, 5-5.5 regularly increasing and weakly convex whorls with shallow suture, last whorl width 1.5 of penultimate whorl, umbilicus with strongly bent margins, narrow and sometimes slightly widened at the last whorl.

Lindbergia pageti Riedel, 1968

Lindbergia pseudoillyrica? pageti Riedel, 1968 in Riedel 1968: 483

Like a small form of *L. pseudoillyrica*, but with 0.3-0.5 more whorls at the same size, umbilicus slightly wider, shell relatively flatter.

1.4-1.7 x 3-3.8 mm

Subterranean, often found in caves, occasionally in rock crevices.

Riedel 1992: 38.

Anatomy unknown, could also belong to *Vitrea*, but less likely. Reliably only known from a single locality (Kolimbári).

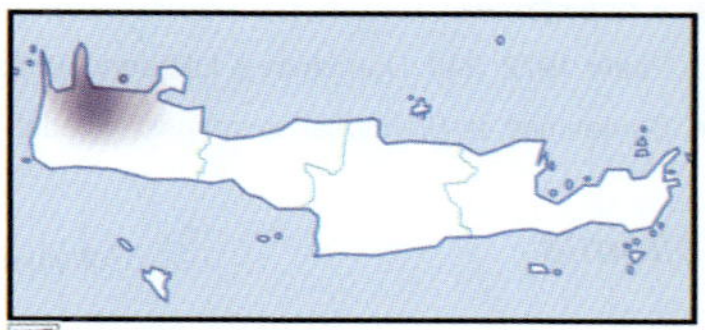

Greece (W Crete)

Lindbergia pinteri Riedel, 1981

Lindbergia pinteri Riedel, 1981 in Riedel 1981: 235

Original description by Riedel, but Riedel quoted a short text attributed to Pintér where the new species was compared with other species. Pintér must be considered as co-author of the description. The name was not attributed to both Pintér and Riedel. In a strict application of Art. 50.1.1 the name in such a case must be attributed to the author of the work.

Shell like *Vitrea*, dirty whitish or cream, weakly translucent in living animals, smooth with weak growth lines, and very fine spiral lines on the first 3 whorls, 4.5-5 weakly convex and slowly and regularly increasing whorls with shallow suture, last whorl width 1.5 x of penultimate whorl, lower side slightly concave, basal and columellar apertural margin slightly thickened, umbilicus moderately wide, 1/5 of diameter, widened at last 1/4 whorl.

2.3-2.6 x 4.4-5 mm

Subterranean, found in a cave.

Anatomy known.

Greece (Ikaría)

Lindbergia pseudoillyrica Riedel, 1960

Lindbergia pseudoillyrica Riedel, 1960 in Riedel 1960: 335

Shell whitish with weak pinkish hue, weakly striated, lower side slightly more flattened than upper side, 4.5-5 weakly convex whorls, regularly increasing, last whorl width 1.5 x of penultimate whorl, umbilical zone not getting deeper, umbilicus moderately wide and deep, 1/7-1/6 of shell diameter, with a blunt edge at its margin.

2-2.5 x 4.5-5.2 mm

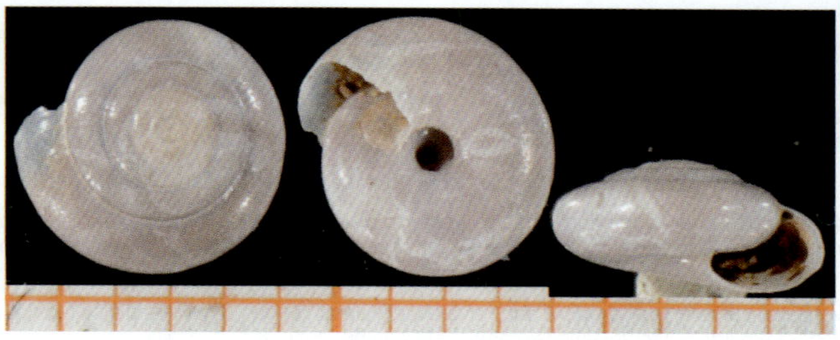

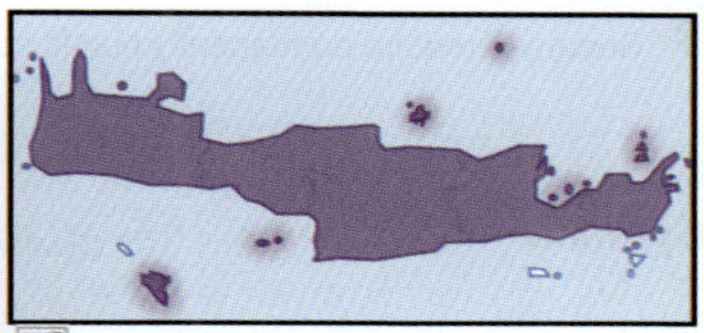

Greece (Crete)

Subterranean, often found in caves.

Riedel 1968: 481 (anatomy), Riedel 1977.

Greece: W Paximádia island S of central Crete

Lindbergia spiliaenymphis Riedel, 1959

Lindbergia spiliaenymphis Riedel, 1959 in Riedel 1959: 110

Shell white like *Vitrea*, weakly shiny, smooth, without microsculpture, slightly translucent at lower side, 3-3.5 weakly convex and slowly increasing whorls, last whorl not much wider than penultimate whorl, umbilicus moderately wide and perspectivical, 1/6 of shell diameter, all whorls visible inside.
Vitrea shells tend to be flatter and have more densely coiled whorls.

1.3-1.5 x 3-3.5 mm

Subterranean.

Most probably strictly endemic of Mount Panion in Attikí, only 3 specimens known (Riedel 1992).

Riedel 1992: 37.

Anatomy known.

Greece (Attikí)

Lindbergia stylokamarae Riedel, 1981

Lindbergia stylokamarae Riedel, 1981 in Riedel 1981: 281

Shell colourless and glassy, transparent and shiny, smooth with some growth lines, 4.2-4.8 weakly convex whorls, first whorls more slowly increasing than last whorls, last whorl width up to 2 x of penultimate whorl, umbilical zone not getting much deeper, umbilicus variable but usually moderately wide and deep, widened at last whorl, first 2 whorls not well visible inside.

1.6-2.3 x 3.5-5 mm

Subterranean, known from a cave.

Only known from the Stilokamara cave of Kásos island, where it is abundant (Riedel 1992).

Riedel 1992: 39.

Anatomy known.

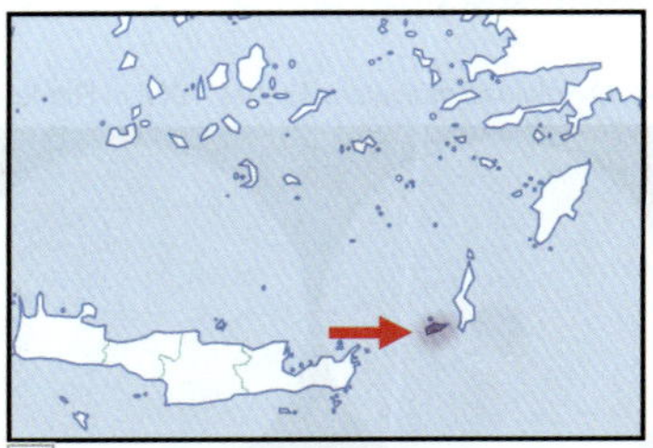

Greece (island of Kásos)

Lindbergia uminskii Riedel, 1960

Lindbergia uminskii Riedel, 1960 in Riedel 1960: 339 (*Lindbergia (Echinophallus) umińskii*)

Shell whitish yellow, shiny and translucent, with dense spiral lines, well visible on all whorls, 4.5-5 weakly convex and slowly and regularly increasing whorls, last whorl width nearly 2 x of penultimate whorl, umbilicus narrow.
Differs from *Vitrea jetschini* in its less translucent and less whitish shell with a spiral surface sculpture.

1.5-1.6 x 2.9-3.3 mm

In caves and valleys, 500-1800 m altitude.

A rare species.

Pintér 1972: 240, Damjanov & Likharev 1975: 266, Dedov 1998, Hubenov 2007.

Sometimes as *Spinophallus uminskii* (Riedel, 1960).

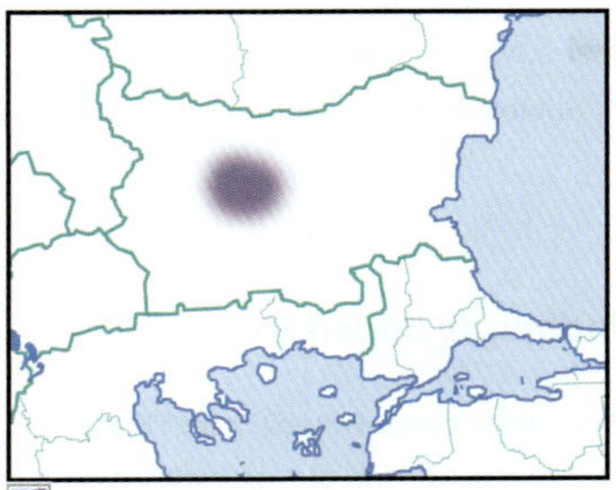

Bulgaria (Stara Planina)

Gollumia

Gollumia Riedel, 1988
in Riedel 1988: 193

Type species *Gollumia pageti* Riedel, 1988
(original designation)
(= currently *Gollumia filocincta* (Hesse, 1915))

Turkey, Cyprus

Schütt 2001 (Stuttg. Beitr.), Schütt 2001: 184.

Only two empty shells of *G. filocincta* were known from S Turkey (Seyhan) until 2001 when Schütt found two slightly different forms in 100-200 km distance (Içel) and in N Cyprus, and classified these as separate species. The new forms differed weakly from the Seyhan shells in their shell size, shape and outline, but although 55 shells were at his disposal, Schütt failed to provide evidence that the new forms fell outside the variation of the known forms that classifying them as separate species would be justified. The main difference given for *G. applanata* was the flatter shell, but Schütt did not present values on the relative shell height of the new form. The height for 20 "adult" shells was given as 1.1-1.8 mm, but the values for the two Seyhan shells were not given.
Keeled hygromiid *Xerocrassa* populations in Crete are known to display much more prominent size and shape variations within much less distance than observed in *Gollumia*. In *Xcr. claudiconus* from Gávdos the average diameter varied between 5 and 13 mm within less than 10 km distance, height ranged between 3 and 7 mm, the height/diameter ratio varied locally between 0.45 and 0.7 (Welter-Schultes 1998: 83). For separating species in *Gollumia* more research would be necessary, to show that the mechanisms behind the local variations in keeled hygromiids in the same climate zone would not be effective in keeled shells of other groups.
The two shells from Cyprus were claimed to differ by a more prominent edge around the umbilicus, a feature of the shell outline not or only weakly observed in the other populations. In the absence of evidence for this feature being associated with reproductive isolation and separate species in other related groups I would also expect spatial variation to account for this observation.

Gollumia filocincta (Hesse, 1915)

Trochula filocincta Hesse, 1915 in Hesse 1915: 61
Gollumia applanata Schütt, 2001 in Schütt 2001 (Stuttg. Beitr.): 4
Schütt cited information provided by A. Riedel in the original description, so both contributed to the description of the species. The name was not attributed to both co-authors. In a strict application of Art. 50.1.1 the name in such a case must be attributed to the author of the work.
Gollumia torumbilicata Schütt, 2001 in Schütt 2001 (Stuttg. Beitr.): 6

Shell flat, pale yellowish, or light grey with brownish hue, thin and weakly translucent, densely striated and with fine and distinct spiral lines, also on lower side, sharply keeled, first whorls relatively wide, the following whorls are equally wide, total 4.5-5 very slowly increasing whorls, last whorl width less than 1.5 x wider than penultimate whorl, not descending near aperture, aperture sharp, umbilicus very wide and deep, 1/3 of diameter.

2.1-2.5 x 4.2-5.5 mm

Seems to require permanently humid soil.

S Turkey (Seyhan, Içel), Cyprus

Gyralina

Gyralina Andreae, 1902
in Andreae 1902: 8 *Hyalinia (Gyralina)*

Type species *Helix circumlineata* Pfeiffer, 1846
(original designation)

Spelaeopatula Wagner, 1922 in A. J. Wagner 1922: 112, type species *Zonitoides candidus* Wagner, 1909 (monotypy)

Shell with usually more or less distinct spiral lines, and often with a characteristic sinus at the upper side of the aperture or a *Nautilus*-like protruded apertural margin. In some species there is no or only a weak spiral micrirelief, no sinus is present and the aperture is normally developed (not protruded) like in other zonitids.

Most species seem to live below the soil surface, they are rare and have usually small ranges. They are difficult to find, live animals are extremely rarely found, usually only by coincidence. Under favourable circumstances many shells can be found at one locaity.

Riedel & Subai 1993 (summary of present knowledge).

Initially two groups were distinguished, *Gyralina* (with spiral microrelief, sinus and protruded aperture) and *Spelaeopatula* (weak spiral microrelief and normaly developed aperture). Until 1980 several intermediate species were discovered so that it became clear that a distinction in two such groups was arbitrary. Riedel & Subai 1993 admitted that it was difficult to subivide the groups, also anatomically, because it was unclear where the exact limits between both groups should be seen, and proposed to regard *Gyralina* and *Spelaeopatula* preliminarily as subgenera.

Gyralina candida (Wagner, 1909)

Without gender agreement: *Gyralina candidus* (Wagner, 1909)

Zonitoides candidus Wagner, 1909 in Wohlberedt 1909: 626
Description and name were exclusively attributed to Wagner.

Shell with 4-5 convex whorls without edge at the periphery, suture normally developed, no spiral relief in the north, moderately dense spiral lines in the south, umbilicus 1/4 of shell diameter, aperture simple (not protruded, without sinus).

2.1-2.5 x 4.2-5.5 mm

Most probably subterranean at limestone rocks.

Riedel 1970: 8, Riedel & Subai 1993: 60.

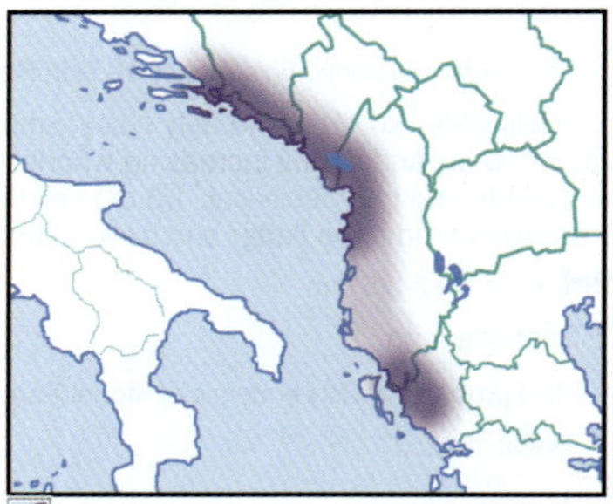

S Dalmatia to NW Greece

Gyralina circumlineata (Pfeiffer, 1846)

Helix circumlineata Pfeiffer, 1846 in L. Pfeiffer 1846 (Symb.): 68

Albania: Mallakastër, near Ballsh

1.3-1.7 x 5-6 mm

Shell flat, with 6-8 fine spiral lines on upper side, finely radially striated on lower side, 5 weakly convex whorls with moderate suture, last whorl not much broader than preceding whorl, upper margin of aperture protruded, umbilicus wide and perspectivical.

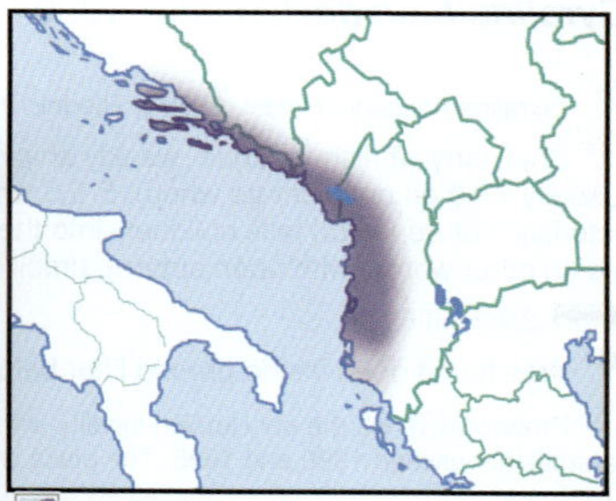

S Dalmatia to S Albania

Westerlund 1886: 45, Riedel & Subai 1993, Dhora & Welter-Schultes 1996: 144, Fehér & Erőss 2009: 11.

Gyralina epeirotica Riedel, 1983

Gyralina epeirotica Riedel, 1983 in Riedel 1983: 17

Shell greyish yellow, not transparent or shiny, with spiral riblets, 4.5 whorls, last whorl not much wider than penultimate whorl, upper side of aperture rounded, not protruded, umbilicus 1/4-1/3 of shell diameter. Shell seems to be variable.
Animal dirty cream, weakly corrugated and slightly transparent.

1.9-2.2 x 5-5.3 mm

Presumably subterranean, but shells were found mainly at the soil surface under stones, not in caves and rock crevices.

Riedel & Subai 1993: 55.

NW Greece (Ipiros)

Gyralina ermonae Gittenberger, 1977

Gyralina ermonae Gittenberger, 1977 in Gittenberger 1977: 53

Shell like *G. tsatsae*, slightly more solid, with a well visible spiral sculpture crossed by radial growth lines, up to more than 4 whorls, aperture simple without sinus (detached section at penultimate whorl), apertural lip thickened, umbilicus 1/3 of diameter.

Up to 2.4 x 5.8 mm

Presumably subterranean, shells were found in a 15 cm small hole at a rock face.

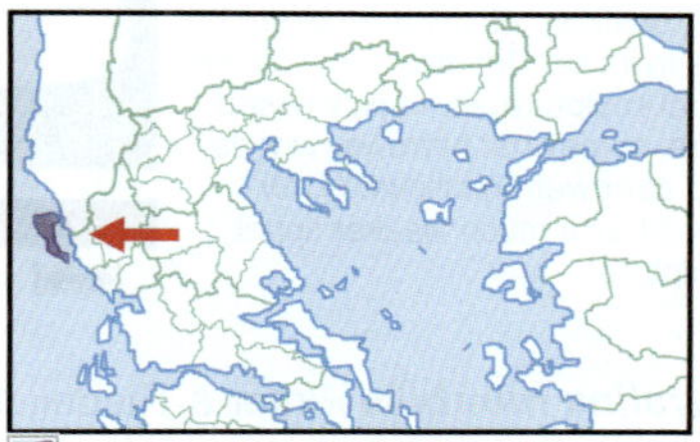

NW Greece (island of Kérkira)

Gyralina formosa Riedel & Subai, 1993

Gyralina formosa Riedel & Subai, 1993 in Riedel & Subai 1993: 54

Shell almost flat on upper side, 4-4.3 whorls, last whorl not much wider than penultimate whorl, with prominent and thickened spiral riblets (4-7 on the penultimate whorl, getting blurry on the last whorl), suture deep, upper section of the aperture *Nautilus*-like protruded, last 1/10 whorl section not connected to the penultimate whorl (also in juvenile shells), umbilicus wide and perspectivical, 1/3 of shell diameter.
G. pageti has less prominent and much denser spiral lines.

1.6-1.9 x 4.1-4.8 mm

Probably subterranean, found in organic litter at limestone rocks.

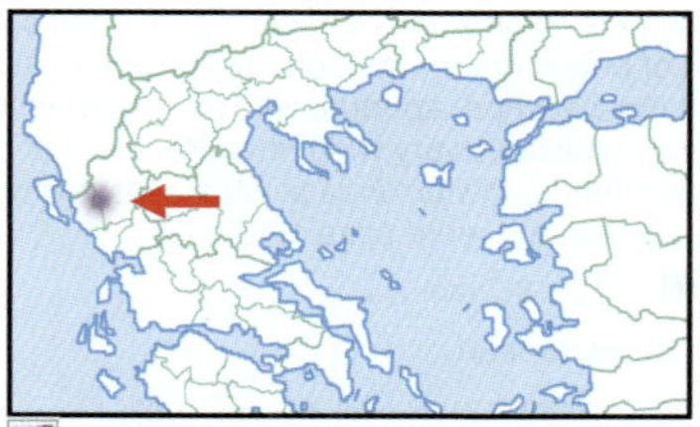

NW Greece (Ipiros)

Gyralina gjirokastrana Riedel & Welter-Schultes, 1996

Gyralina gjirokastrana Riedel & Welter-Schultes, 1996 in Riedel & W. 1996: 18

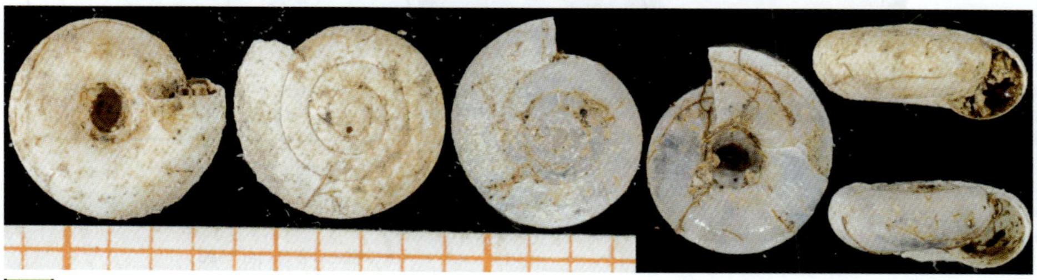

Albania: Gjirokastër

1.7-1.9 x 4.8-5.3 mm

S Albania

Critically Endangered (Dhora 2004).

In rock crevices.

Shell depressed, flat on upper side, dirty whitish, not very shiny, 4.3 whorls, last whorl width 2 x of penultimate whorl, with Nautilus-like growth lines well visible on the last whorl, fine and dense spiral lines (can be weak or partly absent on last whorls), suture deep, aperture *Nautilus*-like protruded but not partly detached from penultimate whorl, umbilicus wide (1/3 of shell diameter) and deeper than in *G. rempei*. *G. rempei* has smooth first whorls without any spiral lines.

Gyralina gyralinaeformis (Riedel & Velkovrh, 1976)

Spelaeopatula gyralinaeformis Riedel & Velkovrh, 1976 in Riedel & Velkovrh 1976: 219

Shell dirty whitish, not transparent, upper side weakly shiny, densely striated, no spiral sculpture, upper side flat, lower side more rounded, 4.5-5 slowly increasing convex whorls with deep suture, last whorl not wider at aperture, umbilicus wide, 1/4 of diameter.
Characteristic for this species is the lack of a spiral sculpture and the deep suture.

2.5-2.7 x 5-5.5 mm

Found in a valley of a moist deciduous forest, in soil litter.

Rep. Makedonija

Gyralina hausdorfi Riedel, 1990

Gyralina hausdorfi Riedel, 1990 in Riedel 1990: 504

Shell dirty whitish, opaque, weakly shiny, distinctly striated, with prominent spiral ridges on the lower whorls (usually 7-10 on penultimate whorl), 5-5.5 whors, first whorls more slowly increasing than last whorls, with characteristic suture (whorl falls obliquely into the suture), last 1/5-1/3 whorl with an elongated sinulus at upper margin (in other words, with open suture), umbilicus very wide, nearly 1/3 of shell diameter.

2.3-3 x 6-8 mm

Was found deep inside ground litter below limestone rocks in a poorly vegetated area with few shrubs.

Probably extinct, the only known locality where this species was known to occur has been completely destroyed by caterpillars between 1990 and 1995. The place was used as a garbage dump afterwards.

Riedel & Welter-Schultes 1996: 18.

Greece (Thessalía)

Gyralina korabensis (Riedel, 1970)

Spelaeopatula? korabensis Riedel, 1970 in Riedel 1970: 9

Shell flat, whitish-creamy, semi-translucent, smooth, not very shiny, weakly striated, 5 whorls with deep suture, last whorl not much wider than penultimate whorl, convex without edge at the periphery, umbilicus 1/3 of shell diameter with all whorls well visible inside. Larger than *G. candida*, suture and umbilicus deeper.

2.8-3.2 x 6.7-7.3 mm

Subterraenan, in caves and dep rock crevices.

Riedel & Subai 1993.

NW Rep. Makedonija, E Albania

Gyralina lunikensis Fehér & Erőss, 2009

Without gender agreement: *Gyralina lunikense* Fehér & Erőss, 2009

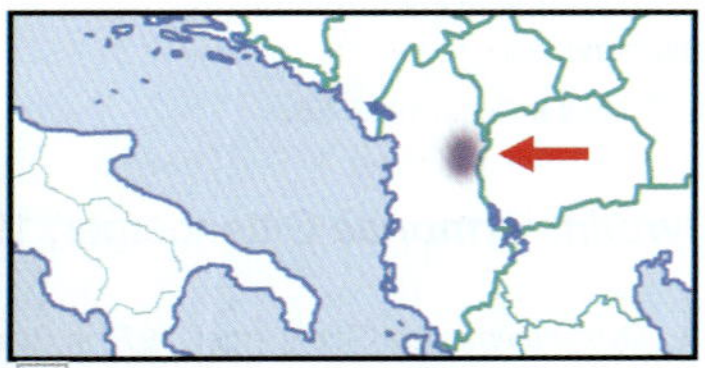

Gyralina lunikense Fehér & Erőss, 2009 in Fehér & Erőss 2009: 16

Name was derived from a locality, *Gyralina* is feminine, should have been *lunikensis*.

Shell like *G. gjirokastrana*, differs in its smaller size and weaker spiral sculpture, its weakly raised spire, somewhat narrower whorls and narrower umbilicus. *G. formosa* has a much weaker spiral sculpture.

2.1-2.2 x 4.4-4.5 mm

Albania: Lunik near Librazhd

Zoltán Fehér

NE Albania

Gyralina mirabilis Pintér & Riedel, 1973

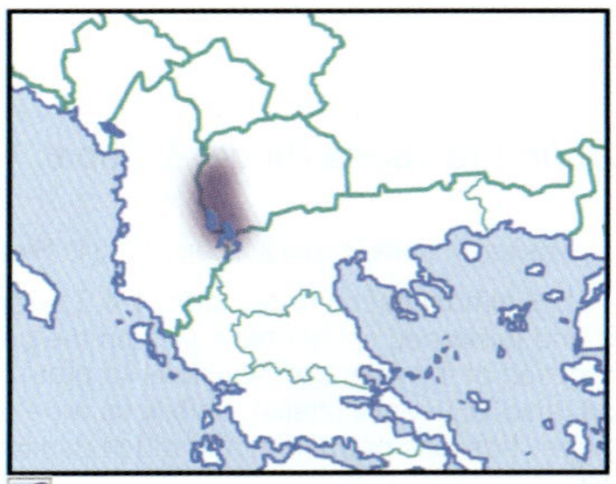

Gyralina mirabilis Pintér & Riedel, 1973 in Pintér & Riedel 1973: 425

Fresh shell white transparent and shiny, flat on upper side, 5 not very convex (at upper side) whorls, first 2-2.5 whorls convex and almost smooth, then suddenly with prominent spiral almost rib-like lines (4-5 on upper side of penultimate whorl, on the last whorl also at the periphery, weaker at lower side), last whorl not wider than penultimate whorl, aperture with sinus (detached in its last section from penultimate whorl, also in subadult shells), umbilicus very wide, more than 1/3 of shell diameter.
Larger than *G. circumlineata*, whorls less convex at upper side, umbilicus wider and less deep, *G. circumlineata* has denser and more spiral lines (7-8).

2-2.5 x 6-7 mm

Subterranean.

Riedel 1985: 31, Riedel & Subai 1993.

W Rep. Makedonija

Gyralina mljetica (Pintér & Riedel, 1973)

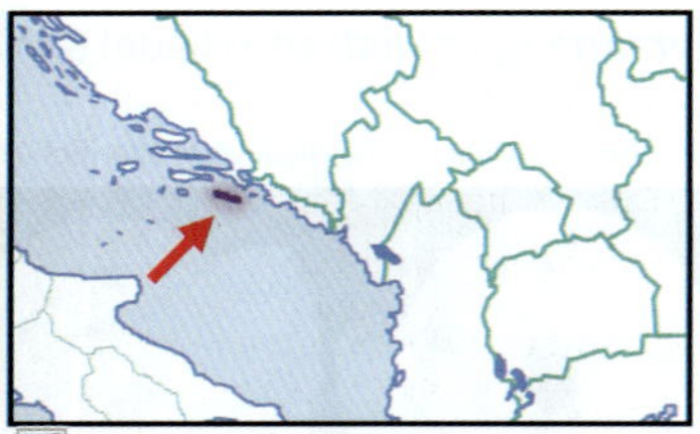

Spelaeopatula mljetica Pintér & Riedel, 1973 in Pintér & Riedel 1973: 271 (*Spelaeopatula? mljetica*)

Shell solid, whitish, not transparent when slightly weathered, smooth and very shiny, with very fine hardly visible spiral lines at apex, upper side almost flat, slightly more than 4 convex whorls, suture deep, last whorl width 1.5 x of penultimate whorl, no edge at periphery, umbilicus wide, deep and perspectivical, slightly more than 1/4 of shell diameter.
Differs from *G. tarabosensis* in its rounded periphery without edge, *G. korabensis* is much larger. Whorls are wider than in *G. candida*, number of whorls at same diameter is smaller, umbilicus deeper and slightly narrower, surface smoother and more shiny.

2.2-2.4 x 4.7-5.3 mm

Croatia: Mljet

Massimo Prodan

Croatia (Dalmatia: Mljet island)

Obviously subterranean, found in rock crevices under dry mosses.

Riedel & Subai 1993.

Gyralina nopcsai Riedel, Fehér & Erőss, 1999

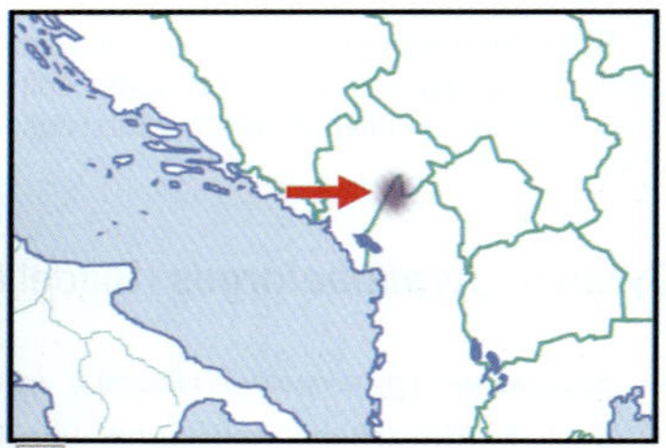

Gyralina nopcsai Riedel, Fehér & Erőss, 1999 in Riedel, Fehér & E. 1999: 241

Shell whitish, weakly shiny, weakly striated or almost smooth without spiral lines, 3.5-4 weakly convex whorls with shallow suture, upper side flat, lower side weakly convex, juvenile shells slightly keeled, last whorl width 1.2-1.3 x of penultimate whorl, apertural margin regularly rounded seen from above, not *Nautilus*-like protruded, without sinus, umbilicus wide, 1/5-1/4 of diameter, all whorls visible inside.

2.6-2.8 x 5.5-6.2 mm

N Albania: mountains at Rraps near Vermosh

Zoltán Fehér

N Albania

Was found in a rocky mountain area with scarce vegetation and steep slopes.

Gyralina pageti Gittenberger, 1988

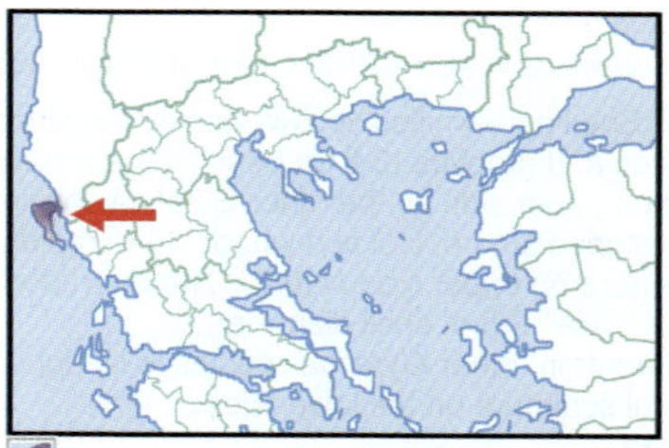

NW Greece (island of Kérkira)

Gyralina pageti Gittenberger, 1988 in Gittenberger 1988: 425

Shell dirty whitish, finely striated and with very fine spiral lines, best visible above the periphery, upper side flat, 4-4.5 whorls with prominently marked suture, apertural margin with long (1/12 of a whorl) sinulus at upper side (provoking the well marked edge at the suture), umbilicus wide, slightly more than 1/4 of shell diameter. Slightly larger than *G. formosa*, with very fine and dense spiral lines.

1.3-1.7 (? estimated, not given anywhere) x 5-5.5 mm

Subterranean.

Riedel & Subai 1993: 55.

Gyralina rempei Gittenberger, 1975

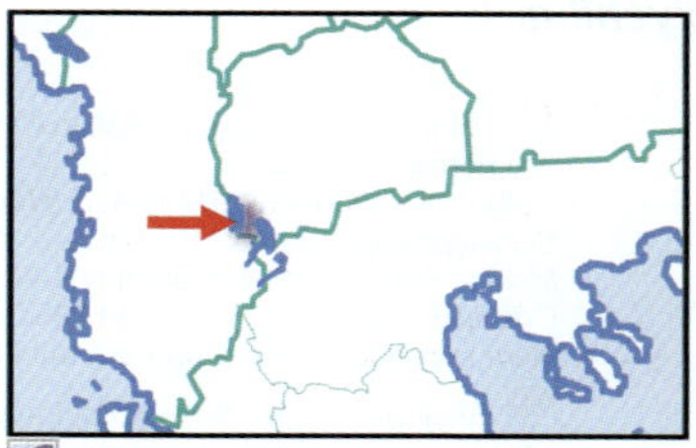

SW Rep. Makedonija

Gyralina rempei Gittenberger, 1975 in Gittenberger 1975: 75

Shell colourless, with 7-10 spiral ridges above the periphery, disappearing on the last whorl, lower side always smooth with weak growth lines, outline flat on upper side, 4-4.5 moderately convex whorls with deep suture, aperture Nautilus-like protruded and usually with elongated sinulus (suture open in last 1/8 whorl), sometimes simple without sinulus, umbilicus wide, 1/3 of shell diameter.
Shell like *G. gjirokastrana*, but with smooth embryonal whorls, no spiral lines present.

1.7-2.3 x 5-6 mm

Subterranean, in the Galičica mountains at 800-1000 m.

Riedel 1985: 34, Riedel & Subai 1993, Riedel & Welter-Schultes 1996.

Gyralina sattmanni Riedel, 1990

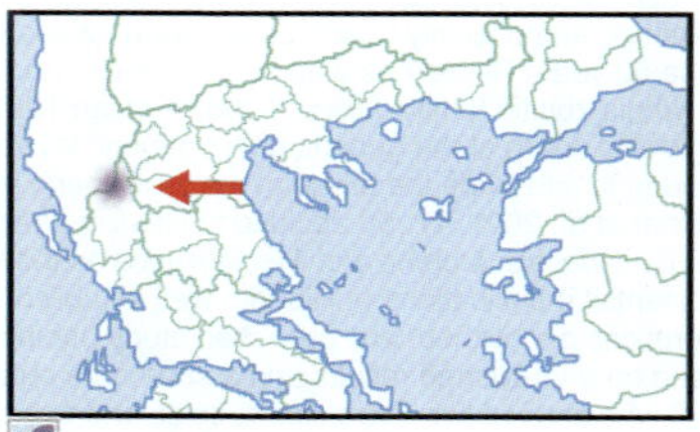

Greece (N Ipiros)

Gyralina sattmanni Riedel, 1990 in Riedel 1990: 497

Shell dirty whitish, slightly transparent at upper side, silky shiny, densely and very obliquely (when see from above) striated, and with occasionally well visible spiral lines (around 8-12 at penultimate whorl), 4.7-5.3 weakly convex and slowly increasing whorls with very deep suture, which has a sharp edge, upper apertural margin *Nautilus*-like protruded and at last 1/6 whorl not connected to suture, umbilicus very wide, nearly 1/3 of shell diameter.

2.2-2.7 x 5-6.2 mm

Subterranean, was found in moist soil samples with fine limestone substrate and plant roots under a meadow at the foot of steep limestones, in mountains at 1200-1400 m.

Gyralina tarabosensis (Riedel, 1970)

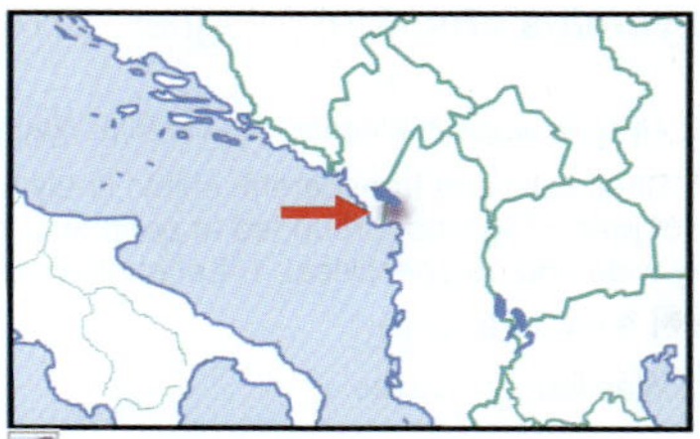

NW Albania

Spelaeopatula? tarabosensis Riedel, 1970 in Riedel 1970: 8

Shell small, lentiform, somehow like Discus rotundatus, dirty whitish, weakly translucent and weakly shiny, first 2 whorls smooth, the others with dense radial growth lines, also at lower side, sometimes also weak spiral lines, up to 5 convex whorls, with edge at periphery, last whorl not much wider than penultimate whorl, lower side of shell more convex in juveniles than in fully grown specimens, umbilicus very wide, more than 1/3 of shell diameter, deep with all whorls well visible inside.
Balcanodiscus frivaldskyanus is two times larger and has a more prominent microrelief.

2.5-2.7 x 5.5-6 mm

Critically Endangered (Dhora 2004). No reason for the threat was given.

Gyralina tsatsae Gittenberger, 1977

Gyralina tsatsae Gittenberger, 1977 in Gittenberger 1977: 52

Shell whitish, transparent, surface with weak radial growth lines and no spiral structure, 4.5 whorls, weakly convex on the upper side, umbilicus slightly less than 1/3 of shell diameter, aperture simple and thin without sinus (detached section) at the penultimate whorl.

Subterranean, known from caves.

Riedel & Subai 1993.

2.0-2.3 x 5-5.5 mm

Greece: Kerkira, Liapades

NW Greece (island of Kérkira)

Gyralina velasensis Riedel & Subai, 1991

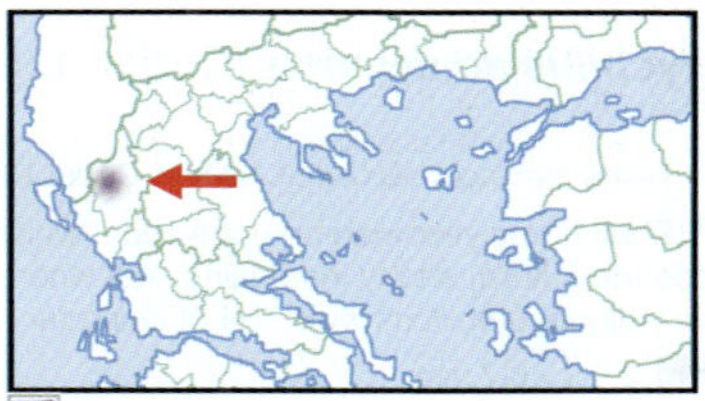

Greece (N Ipiros)

Gyralina velasensis Riedel & Subai, 1991 in Riedel & Subai 1991: 112

Shell with 5.2-5.3 convex whorls, last whorl not much wider than penultimate whorl, suture deep, spiral relief variable and often prominent at the periphery (4-7 riblets), umbilicus relatively narrow, aperture simple (not protruded, without sinus) with short upper side, apertural margin can be thickened.

2.3-2.8 x 5-5.7 mm

Subterranean, found in rock crevices.

Riedel & Subai 1993: 59.

Gyralina velkovrhi Riedel, 1985

S Rep. Makedonija

Gyralina velkovrhi Riedel, 1985 in Riedel 1985: 36

Shell pale whitish yellow, weakly translucent and weakly shiny, with weak growth lines and prominent spiral ridges and grooves (8-9 on penultimate whorl), sometimes even more prominent on lower side, outline not as flat on upper side as the other Gyralina species, 4.5-5 initially slowly increasing and convex whorls with regularly developed suture, last two whorls more rapidly increasing, last whorl only slightly wider than penulimate whorl, aperture on upper side *Nautilus*-like protruded and with sinulus at upper margin, umbilicus very wide and perspectival, 2/5 of shell diameter.
Differs from other *Gyralina* species in its upper side not being flat, its narrower last whorl and its well developed spiral sculpture on the lower side.

2-2.5 x 4.5-5.5 mm

Was found in a mountain gorge composed of volcanic rocks, probably subterranean.

Oxychilus

Oxychilus Fitzinger, 1833 in Fitzinger 1833: 100

Type species *Helix cellaria* Müller, 1774 (subsequent designation by Herrmannsen 1847-1849 [08 Sep 1847]: 183)

Mediterranea Clessin, 1880 in Clessin 1880: 207 *Hyalina (Vitrea (Mediterranea))*, type species *Helix hydatina* Rossmässler, 1838 (subsequent designation by Forcart 1957: 121)
Cellariopsis Wagner, 1914 in A. J. Wagner 1914 (Anz. Wien): 335, type species *Schistophallus deubeli* Wagner, 1914 (original designation)
Schistophallus Wagner, 1914 in A. J. Wagner 1914 (Anz. Wien): 334, type species *Hyalina oscari* Kimakowicz, 1883 (original designation)
Morlina Wagner, 1914 in Sturany & Wagner 1915: 27, type species *Helix glabra* Rossmässler, 1835 (subsequent designation by C. R. Boettger 1930: 580)
Ortizius Forcart, 1957 in Forcart 1957: 125, *Oxychilus (Ortizius)*, type species *Hyalina helvetica* Blum, 1881 (original designation)
More subgenera were proposed, but I leave it with these few here.

Shell small to medium-sized, fragile, nearly flat to low conical. The umbilicus may be narrow to wide, only a few species are without umbilicus. Usually the shell is smooth and glossy, some with fine spiral or reticulate patterns.
Epiphallus inserts subdistally at penis, retractor distally, proximal part of penis with an envelope, vagina elongate, with vaginal gland, a spermatophore is produced. Right ommatophore retractor running between male and female genital organs.

Europe, N Africa, Near East

Researching taxonomy and systematics of this extremely diverse group as we can use it today has been the lifework of the Polish malacologist Adolf Riedel who examined anatomically hundreds of taxa between the 1950s and the late 1990s. In these 50 years Riedel has elaborated a sound and well elaborated classification of the whole zonitid family (Riedel 1998). Forcart 1957: 136 noted that subdividing the *Oxychilus* group at the generic or subgeneric level was arbitrary. Forcart and Riedel decided to subdivide the group at the subgeneric level.
Falkner et al. 2001: 47, 50 decided to classify *Ortizius* as a subgenus of *Oxychilus*, and *Cellariopsis*, *Morlina* and *Mediterranea* as independent genera. These substantial modifications of Riedel's well established system were proposed without references and without any published study behind, but they seem to have been based on unpublished results and a short note published by Giusti & Manganelli 1999 who studied two *Oxychilus* species and expressed some doubts on the classification of the genus. Later the same authors (Giusti & Manganelli 2002: 466) argued that *Oxychilus* could be subdivided into several genera instead of subgenera because no synapomorphies were found, but they expressly decided to maintain the generic classification as established by Riedel and classified *Mediterranea*, *Cellariopsis*, *Conulopolita*, *Morlina*, *Schistophallus* and others as subgenera of *Oxychilus*.
The classification proposed by Falkner, Bank & Proschwitz 2001 was not based on a detailed and published scientific study. It was however used in 2004 by R. Bank who compiled the list of European terrestrial gastropods for the Fauna Europaea database (www.faunaeur.org). Fauna Europaea has often been misinterpreted by users as authorative in terms of current taxonomy and classification. This is a problematic situation in *Oxychilus*, where Riedel's1998 well established system has been forced to compete with a new system that was practically presented without arguments.

Here Riedel's 1998 classification is used.

It is possible that Riedel's system is not in agreement with the eternal truth in science. Certainly some species require more research. But replacing a well elaborated genus-subgenus classification by a new system would require an equally sound, detailed and published scientific study.

Oxychilus absoloni (Wagner, 1914)

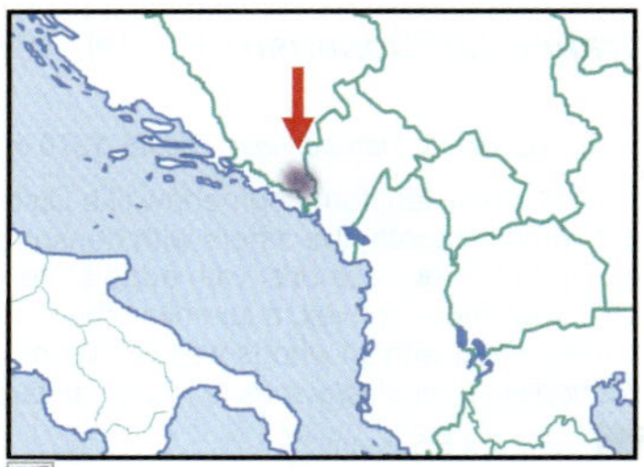

Herzegovina

Hyalinia absoloni Wagner, 1914 in A. J. Wagner 1914 (Sitz.-Ber. Wien): 40

Shell solid and transparent, shiny, finely and distinctly striated, 4-5 weakly convex whorls, last whorl width 2 x of preceding whorl, rounded at periphery, not descending near aperture, apertural margin sharp, umbilicus very wide and perspectival, 1/4 of shell diameter.

3.8-4.3 x 9-12 mm

Was found in caves.

Oxychilus adamii (Westerlund, 1886)

Hyalinia villae var. *adamii* Westerlund, 1886 in Westerlund 1886: 48 (*Hyalinia (Polita) villæ* var. *adamii*)

Shell horny brownish on upper side, whitish with green hue on lower side, indistinctly striated, 5.5-6.5 whorls, suture deep and with sharp edge (this profound suture is characteristic), first 4 whorls slowly increasing, penultimate whorl width about 2 x of preceding whorl, last whorl width about 1.5 x of penultimate whorl, shortly descending near aperture, umbilicus moderately narrow, 1/8 of diameter. Differs from *O. mortilleti* in its more profound suture and narrower umbilicus.
Body greyish blue.

Italy: Lombardia, Como, Canzo
Angelo Vannozzi

5-7 x 14-17 mm

Humid and shady habitats in leaf litter and under stones in forests and under shrubs. In Switzerland in 400-900 m altitude.

Manganelli et al. 1995: 20, Turner et al. 1998: 275, Boschi 2011: 409.

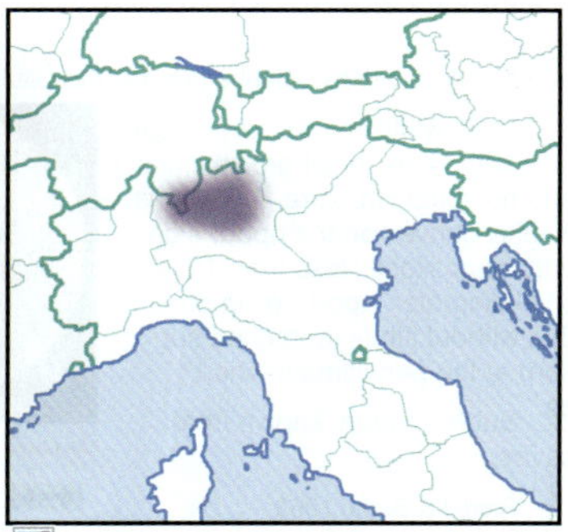

N Italy, S Switzerland

Oxychilus adjaciensis (Caziot, 1904)

France (Corsica)

Hyalinia adjaciensis Caziot, 1904 in Caziot 1904: 34

Shell horny yellowish on upper side, whitish on lower side, fragile, finely striated, 5-6 almost flat whorls with moderate suture, slowly increasing, last whorl initially angulated at periphery, rounded near aperture, aperture very oblique, umbilicus wide, 1/4 of diameter.

2.5-3 x 6-7 mm

Was found in the early 1900s in the city of Ajaccio, in cavities of old stone walls and in gardens.

Oxychilus aegopinoides (Maltzan, 1883)

Helix aegopinoides Maltzan, 1883 in Maltzan 1883: 102

Greece: E Crete, (1) island of Gianisáda, (2-4) near Káto Zákros, (5-7) Akra Síderos

12-13.5 x 21-23 mm

Rocky habitats.

Westerlund 1886: 76, Riedel 1992: 79, Schultes & Wiese 1991.

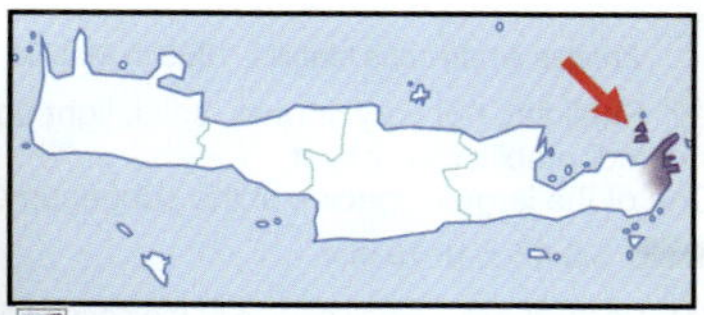

Greece (E Crete)

Shell solid, horny brownish or greenish dark yellow, irregularly striated, 5 whorls with white suture, first whorls sharply keeled and flat on upper side, last whorl usually also keeled but can be rounded in some populations, aperture with white lip inside, umbilicus open and deep.

Oxychilus alicurensis (Benoit, 1857)

Helix alicurensis Benoit, 1857 in Benoit 1875 (Ill. Sist.): 99 (*Helix Alicurensis*)

Shell thin, horny yellowish, striated, 5 slowly increasing whorls, last whorl flatened at lower side and descending deeply towards the umbilicus, umbilicus narrow.

3.7-4.3 x 8-9 mm

Westerlund 1886: 48, Riedel 1973: 16, Giusti 1973: 177, Manganelli et al. 1995: 19.

Italy (Isole Eolie: Alicudi)

Oxychilus alliarius (Miller, 1822)

Without gender agreement: *Oxychilus alliaria* (Miller, 1822)

Helix alliaria Miller, 1822 in Miller 1822: 379

Germany: Schleswig-Holstein near Korügen-Möltenort

3.5-4 x 5-7 mm, diameter usually below 7 mm

0.9-1 mm at 1 whorl, 1.8-2.1 mm (2 wh.), 3.4-3.6 mm (3), 4-4.5 mm (3.5), 5.5-6.5 mm (4 whorls)

Predominantly in swampy habitats of mountain slope forests under moss. Also in humid and shady rocky habitats in creek gorges, in swampy meadows, at water margins and in cultivated areas with moist meadows. In Switzerland up to 1300 m.

Vulnerable in parts of Germany.

Lloyd 1970, Riedel 1992: 77, Giusti & Manganelli 2002, De Oliveira & Altonaga 2010 (Portugal).

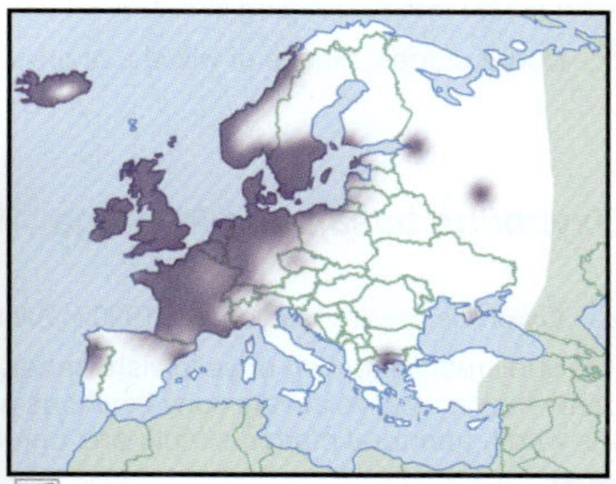

W Europe, Azores, introduced to many other regions

Shell smooth, shiny, reddish brown, umbilicus wide (1/6 of shell diameter), animal blackish blue and smells like garlic when disturbed.

Anatomy: usually lacks blackish mantle edge and has internal ornamentation of proximal penis consisting of not more than four longitudinal pleats, usually straight, sometimes slightly wavy, but never laterally branched or papillate.

Oxychilus altimirai Riedel, 1972

Oxychilus altimirai Riedel, 1972 in Riedel 1972: 126

Shell light amber coloured, translucent, smooth and very shiny, no spiral microsculpture, 4.5-5 weakly convex whorls, last whorl width almost 2 x of penultimate whorl, no edge at periphery, umbilicus narrow, last 2 whorls visible inside, aperture relatively high and rounded, basal margin curved.

4-4.3 x 8-9 mm

Seems to be rare, nowhere abundant.

Pyrenees

Oxychilus amaltheae Riedel & Subai, 1982

Oxychilus amaltheae Riedel & Subai, 1982 in Riedel & Subai 1982: 21

Shell light amber coloured, lower side slightly translucent, older shells dirty yellowish, surface except first whorls with a prominent reticular pattern of spiral lines and radial riblets, 4.5 whorls, last whorl 2 x wider than penultimate whorl, umbilicus very narrow and partly covered by columellar apertural margin.

Subterranean, lives at some 5 m from Dikteon Antron cave entry (1700 m), animals crawl at the foot of the cave walls, also under stones and rocks in this zone, but not deeper inside the dark cave. Presumably carnivorous, seems to feed on arthropods.

3.1-3.5 x 6-7 mm

Probably strictly endemic, very rare, population small and should be protected, no other population besides Dikteon Antron known. Life animals should not be collected, this could lead to the extinction of the species.

Riedel 1992: 91.

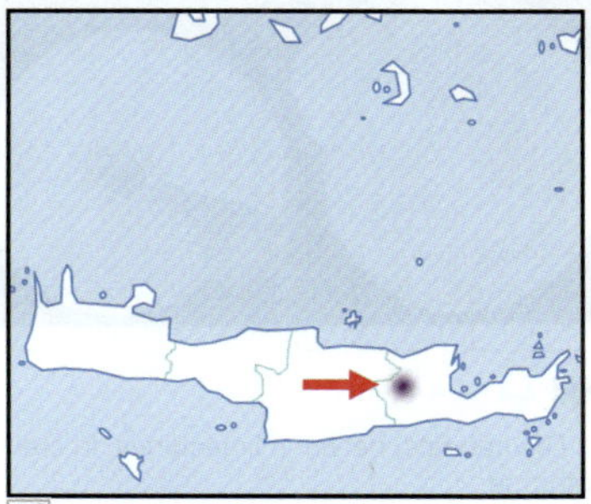

Greece (E central Crete)

Oxychilus amblyopus (Mabille, 1869)

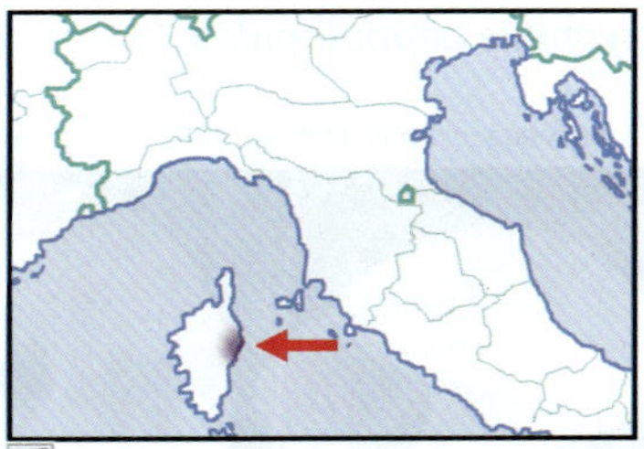

France (E Corsica)

Zonites amblyopus Mabille, 1869 in Mabille 1867-1869: 63

Shell flat, not very convex, solid, light horny coloured, at lower side whitish, almost translucent, shiny, finely striated, umbilicus present.
One of the largest species of the subgenus *Oxychilus*.

5.5-6.5 x 15-19 mm

Restricted to a small portion of the coastal plain at Aléria (Haute-Corse), and seriously threatened by the increasingly intensive agricultural occcupation of the terrain.

Westerlund 1890: 9, Falkner et al. 2002: 123.

Oxychilus anjana Altonaga, 1986

Oxychilus anjana Altonaga, 1986 in Altonaga 1986: 237

Spain: Cantabria, La Cavada S of Santander

N Spain

3.5-4.8 x 7-10 mm

Shell fragile, striated, dotted with numerous point-shaped depressions visible unter microscope, translucent, 4-5 whorls, last whorl not wider than preceding whorl, aperture slightly oblique, umbilicus narrow and deep, 1/10-1/7 of diameter.
Animal pale cream merging into greyish blue towards the tentacles, sole pale cream.

Under stones, pieces of woods, at grass, on river banks with a lot of decomposed material. Also in disturbed habitats among plastic junk.

Oxychilus basajauna Altonaga, 1990

N Spain (Basque Country)

Oxychilus basajauna Altonaga, 1990 in Altonaga 1990: 281

Shell reddish brown or yellowish amber, paler above and whitish around umbilicus, translucent and very shiny, fragile, with relatively convex whorls and shallow suture, up to 5.5 whorls, last whorl width 1.5 x of previous whorl, apertural margin simple and unreflected, umbilicus narrow (1/10 of shell diameter) and deep.

4.8-6.9 x 10-13 mm

In river valleys, under stones and wood on calcareous substrate, requires high humidity, exposed and sheltered habitats.

Oxychilus blauneri (Shuttleworth, 1843)

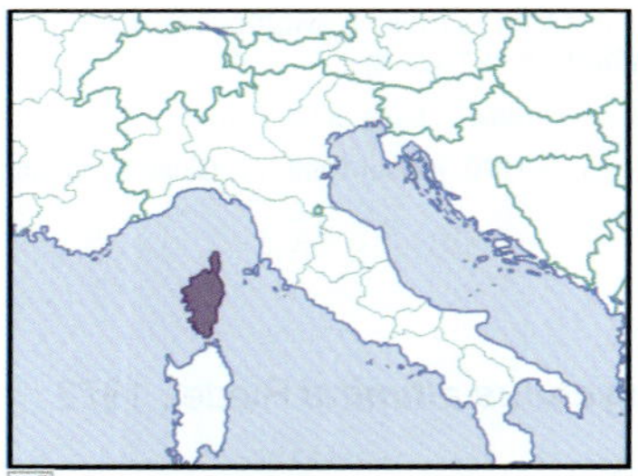

France (Corsica)

Helix cellaria var.? synonym *Helix Blauneri* Shuttleworth, 1843 in Shuttleworth 1843: 13
Mentioned with short description. Not mentioned as a junior synonym of an available name (as demanded by Art. 11.6), but as a synonym for a doubtful subordinate variant for which no name was used. For those who neglect this (as proposed by Kadolsky 2012: 78, this is preliminarily accepted here), the name was made available under Art. 11.6.1 by Requien 1848: 45 who used the name as *Helix Blauneri*, with bibliographical reference to Shuttleworth 1843: 13. For the others, the name must be attributed to Requien 1848.

Shell horny brown, opaque, lower side light greenish, last whorl 2 x as broad as preceding whorl, suture deep, with margin, umbilicus open, columellar margin of the aperture slightly reflected.

6-7 x 12-15 mm

Frequent in humid habitats, under stones and below trees.

Westerlund 1886: 60, Germain 1930: 150.

Oxychilus camelinus (Bourguignat, 1852)

Without gender agreement: *Oxychilus camelina* (Bourguignat, 1852)

Helix camelina Bourguignat, 1852 in Bourguignat 1852: 14

Albania: Shkodër

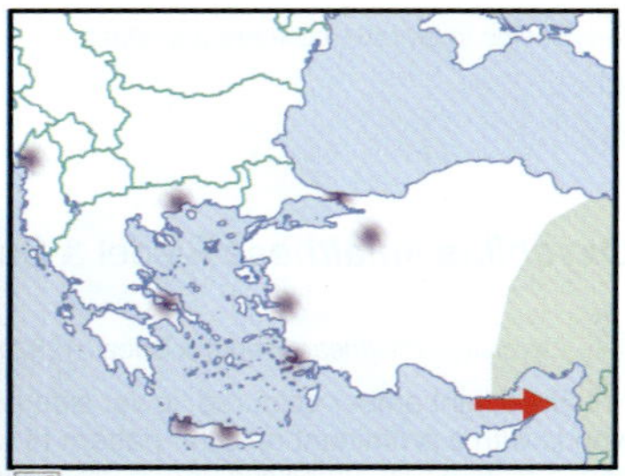

Turkey to Lebanon and Albania
Native probably in the border region Turkey / NW Syria.

Shell light horny coloured, almost flat, with distinct radial growth lines, umbilicus narow, 1/8 of shell diameter, up to almost 6 whorls, slowly increasing, suture deep, last whorl not much wider, flat at lower side, aperture obliquely depressed.

4-4.5 x 9-11 mm

Dry habitats, between limestones, in crevices, under rock rubble. In W Turkey, Greece and Albania as a synanthrope.

Riedel 1992: 82, Dhora & Welter-Schultes 1996: 147, Schütt 2001: 211.

Oxychilus canini (Benoit, 1843)

Helix canini Benoit, 1843 in Benoit 1843: 10 (*Helix Canini*)

Shell pale horny yellowish, lower side lighter, finely striated, 7 weakly convex whorls with weak suture, upper side weakly rounded, last whorl not much broader than preceding whorl, umbilicus narrow but deep.

Dry and open hill slopes covered with dwarf palms and high grasses, under stones and in limestone rock crevices.

5-7 x 10-13 mm, height variable

Italy: Trapani, San Vito lo Capo
Maurizio Sosso

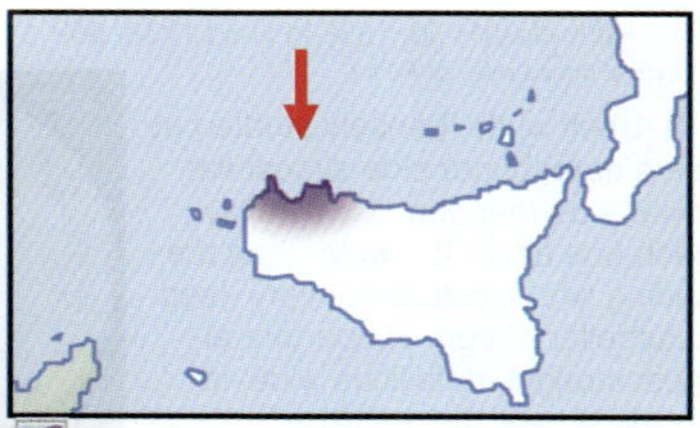

Italy (NW Sicily)

Westerlund 1886: 46 (*Helix testae*), Riedel 1973: 3, Manganelli et al. 1995: 19.

Oxychilus carotii (Paulucci, 1878)

Hyalina carotii Paulucci, 1878 in Paulucci 1878: 26 (*Hyalina Carotii*)

Shell pale brownish yellow, thin, smooth with thin spiral grooves at high magnification, more than 5 whorls, last whorl widened and sometimed downturned, umbilicus narrow (1/12-1/10 of shell diameter), aperture simple.
Animal lemon yellow, posterior end bluish grey.

Mixed forests under thick leaf litter layers, also in subterranean cavities of of dry walls in woods. Animals are carnivorous and feed on any kind of small helicids.

9-13.5 x 20-26 mm, animal 4-5 cm long

Italy: Reggio Calabria, Consolino mountain
Massimo Prodan

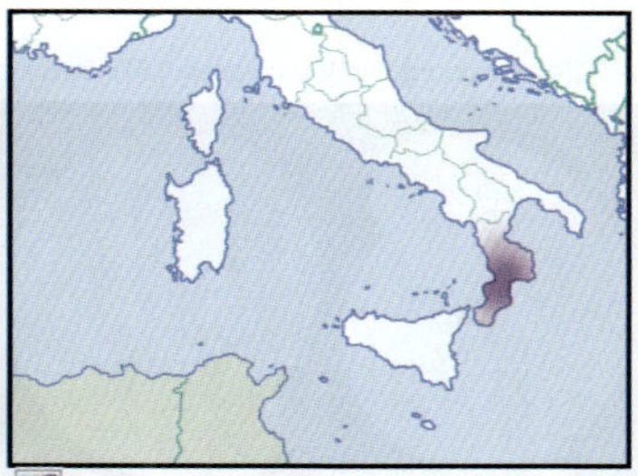

S Italy (Calabria)

Manganelli, Castagnolo & Giusti 1990: 182, Manganelli et al. 1995: 43.

Oxychilus cellarius (Müller, 1774)

Without gender agreement: *Oxychilus cellaria* (Müller, 1774)

Helix cellaria Müller, 1774 in Müller 1774: 28

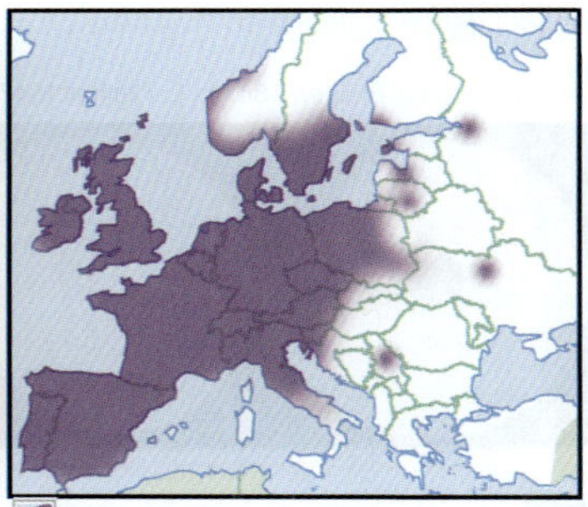

4.5-6 x 7-11 mm, diameter up to 14 mm, usually below 11 mm

Germany: Heikendorf near Kiel

Europe, dispersed worldwide

Ø 0.9-1.1 mm at 1 whorl, 1.8-2.1 mm (2 wh.), 3.4-3.6 mm (3), 4.5-4.9 mm (3.5), 6.0-6.5 mm (4), 7.3-8.1 mm (4.5), 9.3-9.7 mm (5 whorls)

(1) Switzerland: near Schaffhausen, (2) Germany: Baden-Württemberg, near Nördlingen

Shell nearly smooth and shiny, very faintly striated, 5.5-6 regularly increasing whorls, last whorl not inflated, umbilicus moderately deep and open. Last whorl narrower than in *O. draparnaudi*.
Animal usually pale bluish grey, rarely in light colours or darkly pigmented. Penis cylindrical without constrictions (in contrast to *O. draparnaudi*), rows of papillae of penis continuous without interruption (interrupted at the constriction in *O. draparnaudi*).

Humid and shady habitats of any kind, in deciduous forests under leaves, wood or stones, in caves, also in cultivated areas under brick rubbish, in humid cellars and under old humid wood logs, in backyards under garbage. Avoids marshes, but tolerates non-calcareous substrate, can occur in poor heathy British upland habitats, unlike most other shelled gastropods. Frequent in the Scottish mountains, in Switzerland up to 1800 m.

25-45 eggs (diameter 1.3-1.5 mm) are laid between February and October, juveniles hatch after 12-16 days, maturity is reached at the beginning of the following year (Gassies 1849 reported that full shell size was reached after 4 months in S France).

Gassies 1849: 106, Moquin-Tandon 1855: 78, Germain 1930: 147 (life cycles), Giusti & Manganelli 1997, Boschi 2011: 393.

Oxychilus clarus (Held, 1838)

Without gender agreement: *Oxychilus clara* (Held, 1838)

Helix clara Held, 1838 in Held 1838: 902

Original genus was not *Polita*.

Shell white, densely striated at the suture, upper side almost flat, lower side rounded, 4-5 whorls, each one about 2 x wider than preceding whorl, suture not very deep, columellar margin of aperture almost straight, umbilicus wide, increasing near the aperture.
Shell somehow like *Nesovitrea hammonis*, very fine spiral sculpture, like in *Aegopinella*.

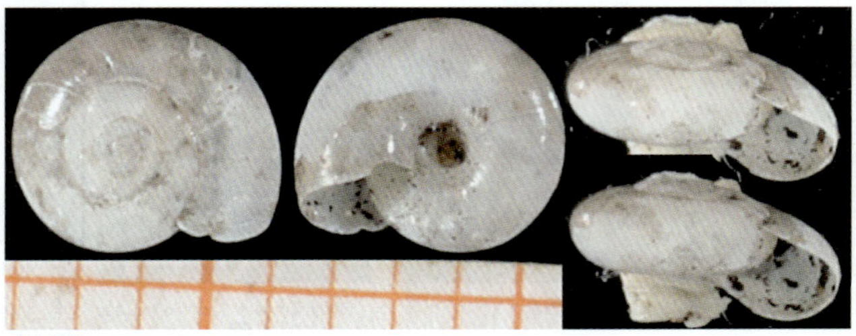

1.6-2 x 4-4.5 mm — Germany: S Bayern, near Mittenwald

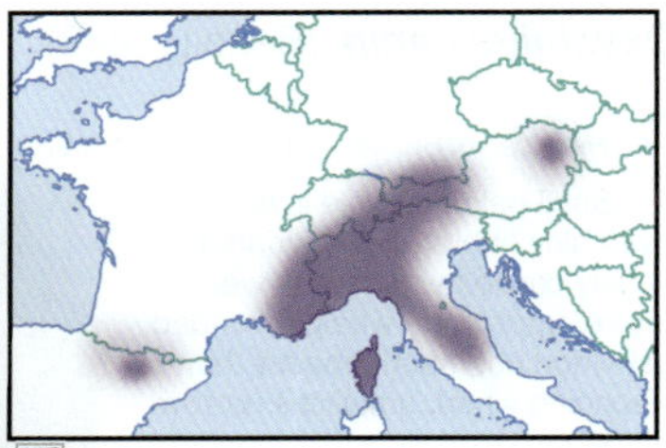

Pyrenees, Alps, N Italy, Corsica

Scattered and very rare, with large gaps between the populations. Critically Endangered in Bayern and Austria, Endangered in Switzerland.

Under stones, between plant roots, in rock crevices, partly subterranean, in woods and on grassy slopes, usually on calcareous substrate, between 400 and 2500 m. In Corsica in mixed woodlands on non-calcareous substrate, 600-1000 m.

Westerlund 1886: 43, Gittenberger, Me. & Ra. 1980: 12, Kerney et al. 1983: 169, Giusti, Hol. & M. 1985, Manganelli & Giusti 1993: 68, Turner et al. 1998: 278.

Oxychilus courquini (Bourguignat, 1870)

Zonites courquini Bourguignat, 1870 in Bourguignat 1870: 93 (*Zonites Courquini*)

2.5-2.8 x 5-6.5 mm — Spain: Alicante, Gata de Gorgos

E Spain

Shell flat, whitish creamy, shiny, almost smooth, whorls regularly increasing, last whorl width 1.5 x of penultimate whorl, umbilicus wide and perspectivical.

Subterranean, lives in caves, but is also found under stones and tree trunks in humid forest habitats, 200-1300 m altitude.

Westerlund 1886: 44, Riedel 1972: 123, Martínez-Ortí & Robles 2003: 102.

Oxychilus cyprius (Pfeiffer, 1847)

Without gender agreement: *Oxychilus cypria* (Pfeiffer, 1847)

Helix cypria Pfeiffer, 1847 in L. Pfeiffer 1847: 12

7-9 x 15-20 mm (10 mm at 5.0 whorls) — Greece: island of Kos, Kos city

Croatia to S Turkey, Cyprus, Near East

Shell light horny brown, lighter on lower side, silky shiny, prominently striated on upper side, at least near the suture, 6-7 weakly convex whorls with relatively deep suture, aperture with whitish layer inside, umbilicus narrow to moderate, becoming wider at last whorl. Can be confounded with *O. deilus*.

Under stones, also in cultivated areas. In natural habitats only in Dalmatia, Makedonia and S Asia Minor.

Westerlund 1886: 73, Falkner 1990: 182, Riedel 1992: 80, Riedel 1995: 45, 49.

Oxychilus deilus (Bourguignat, 1857)

Zonites deilus Bourguignat, 1857 in Bourguignat 1857: 9, Pl. 8

7.0-7.4 x 13-17 mm — N Turkey: near Abana, Black Sea

Diameter of some populations 10-13 mm, others 17-26 mm

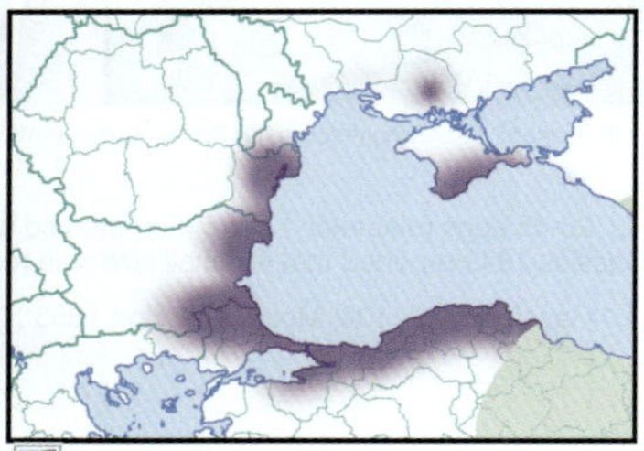

Black Sea coastlands

Shell horny brown, distinctly striated, usually no spiral lines, 5-6 weakly convex whorls, umbilicus moderately narrow, first whorls not visible inside. Conchologically very variable, but rarely within a population. Differences between populations in umbilicus width, height of the shell, width of last whorl, shell colour (mostly yellowish brown, sometimes darker), sometimes worse or better exhibited spiral lines on the upper part of the shell.

Humid habitats under stones, in gorges, river margins, forests and in high grass. Hides during daytime, active in the evening, in the winter as well as in the other seasons. Possibly not carnivorous. Also as a synanthrope. In NW Asia Minor the most frequent zonitid and most frequent land snail at all.

Grossu & Riedel 1958, Riedel 1958: 398, Damjanov & Likharev 1975: 280, Riedel 1995: 43, Schütt 2001: 208, Irikov & Mollov 2006: 798, Sysoev & Schileyko 2009: 126.

Oxychilus denatale (Pfeiffer, 1856)

Helix denatale Pfeiffer, 1856 in L. Pfeiffer 1856: 182 (*Helix De Natale*)

Shell horny yellowish, opaque, rarely dark, spirally striated, 6-7 slowly increasing whorls, coiled somehow like *Helicodonta obvoluta* with narrow first whorls and last whorl substantially increasing and rounded, last whorl width 2-3 x of penultimate whorl, whorls with sharp edge in the lower section of the periphery, aperture narrow, umbilicus moderately wide.

In rocky and stony gorges, at the soil, in leaf litter, lives in shady habitats with some vegetation, the species is quite abundant there.

5-7 x 12-17 mm

S Italy: island of Marettimo

Maurizio Sosso

Westerlund 1886: 47, Riedel 1973: 18, Manganelli et al. 1995: 19.

Italy (Marettimo island)

Oxychilus depressus (Sterki, 1880)

Without gender agreement: *Oxychilus depressa* (Sterki, 1880)

Hyalina depressa Sterki, 1880 in Sterki 1880: 104

3-4.5 x 6.5-8 mm

Germany: Bayern, near Hersbruck

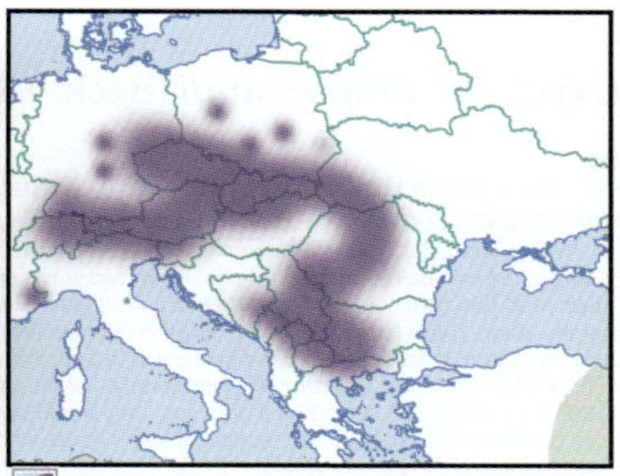

Central Europe to Greece

Shell very flat, white or pale yellowish, shiny, umbilicus very narrow. Glands in the mantle of young animals visible as red spots from outside.

Subterranean, in moderately humid and rocky forests and various other kinds of habitats such as open grassy mountain slopes above the timberline or with shrubs or scattered trees. Humid soils saturated with detritus and other dead organic matter. In S Bulgaria it often falls into caves, but is not considered as a troglobiont. Calcareous substrate is preferred, but silicate substrate is tolerated. In Switzerland and Bulgaria up to 2600 m.

Nowhere abundant, scattered with large gaps between populations. Endangered in Austria and Germany, Vulnerable in Switzerland.

Damjanov & Likharev 1975: 289, Riedel 1992: 98, Turner et al. 1998: 280, Cianfanelli et al. 2005, Irikov & Mollov 2006: 800, Sysoev & Schileyko 2009: 133.

Has recently been classified in the (sub)genera *Riedelius*, *Mediterranea* and *Riedeliconcha*.

Oxychilus deubeli (Wagner, 1914)

Hyalina cellaria var. *orientalis* Clessin 1887 in Clessin, 1887: 69
Not *Hyalina (Vitrea) crystallina* var. *orientalis* Kimakowicz, 1883 in Kimakowicz 1883: 23 (= currently *Vitrea crystallina* (Müller, 1774))
Schistophallus deubeli Wagner, 1914 in A. J. Wagner 1914 (Anz. Wien): 335

4-5 x 9-10 mm

Romania: Bucegi mountains, Sinaia Peles

Penis short, inflated at distal end and distally ending in two appendices, vas deferens in penial proximity thicker than in its distal section.

Grossu 1983: 144, Hausdorf 1993: 44.

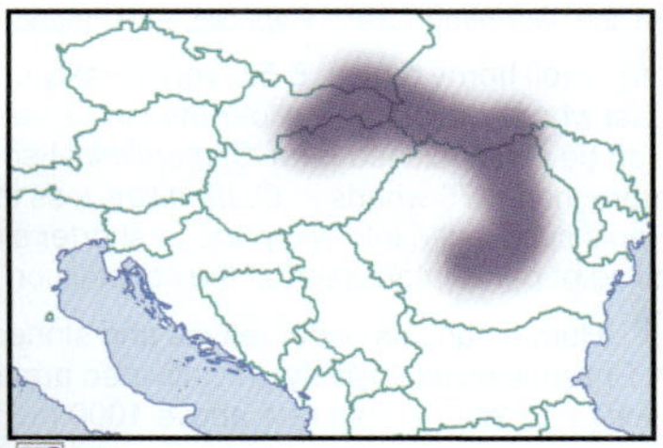

E Carpathians

Shell horny yellowish with greenish hue, lower side more pale whitish, translucent and shiny, with very fine growth lines, 4.5-5 very weakly convex whorls, last whorl width 1.5 x of penultimate whorl, umbilicus narrow, 1/8 of diameter. Shell resembles *O. cellarius*.

Oxychilus diaphanellus (Krynicki, 1836)

Without gender agreement: *Oxychilus diaphanella* (Krynicki, 1836)

Helix diaphanella Krynicki, 1836 in Krynicki 1836: 204
Expressly established as a new replacement name for *Helix diaphana* Krynicki, 1833 in Krynicki 1833: 436 not *Helix diaphana* Poiret, 1801

Shell horny brownish on upper side, lower side whitish, finely striated, 6-6.5 moderately convex whorls with flat suture, very weakly keeled, last whorl width 1.5 x of preceding whorl, umbilicus narrow but deep, becoming wider at last whorl.

3.5-4 x 7-8.5 mm

Ukraine: Crimea, "Peace Valley"

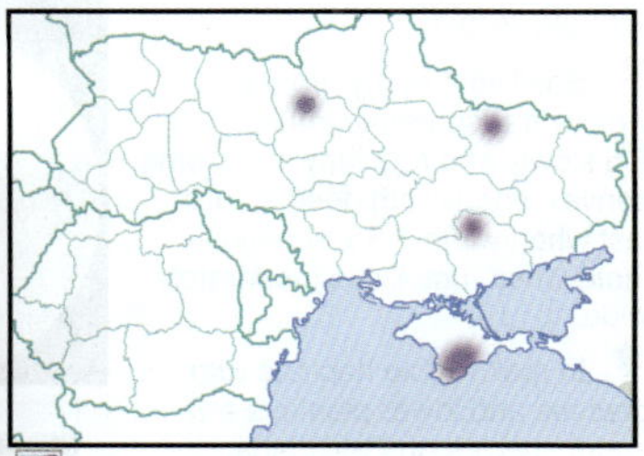

Ukraine (Crimea and some cities)

Westerlund 1886: 49, Riedel 1995: 69, Riedel 1999.

Oxychilus diductus (Westerlund, 1886)

Without gender agreement: *Oxychilus diducta* (Westerlund, 1886)

Hyalinia diducta Westerlund, 1886 in Westerlund 1886: 49

3.5-4 x 9-11 mm

Italy: Lampedusa, La Roccia

Italy (Lampedusa)

Shell densely striated, flat on upper and lower side, 5-6 weakly convex whorls with relatively depp suture, lower side almost flat, last whorl width slightly less than 1.5 x of preceding whorl, with blunt edge at lower part of periphery, aperture horizontal, umbilicus wide and perspectivical.
O. diductus can be identifed by *draparnaudi*-like anatomy and shell depressed, discoidal, not "polygyrate", with umbilicus wide and external surface irregularly striated, opaque and with finely granulated microsculpture.

Westerlund 1886: 49, Manganelli et al. 1995: 19, Manganelli et al. 2002.

Oxychilus draparnaudi (Beck, 1837)

Helix lucida Draparnaud, 1801 in Draparnaud 1801: 96 (not *Helix lucida* Pulteney, 1799 from England)
Helix draparnaudi Beck, 1837 in Beck 1837: 6 (*Helix (Helicella) Draparnaldi*)
ICZN Opinion 336 (1955): spelling emended to *draparnaudi*.
Established with bibliographical references for two subordinate variants "a. gallica" and "b. italica", but not with a distinct description or indication for the main form (the nominal taxon *draparnaudi*). The name *Helix draparnaudi* was made available at this occasion under Art. 12.1 because it had an indication for the taxon (Code Glossary "taxon": a taxon encompasses all included taxa of lower rank). Art. 12.1 does not demand a description (or indication) for the nominal taxon, a description for the taxon is sufficient, this can include subordinate taxa.
Those who do not like to accept this interpretation of the Code are invited to accept that the name was regarded as having been made available by Beck 1837 in Opinions 336 and 1924. If opposed, they would have to look for the next available author or name, and would also need to verify whether Art.11.6.1 can be applied under any interpretation of their choice, since Beck's name could be interpreted as a synonym.
Oxychilus beckmanni Falkner, 2007 in Falkner 2007: 179

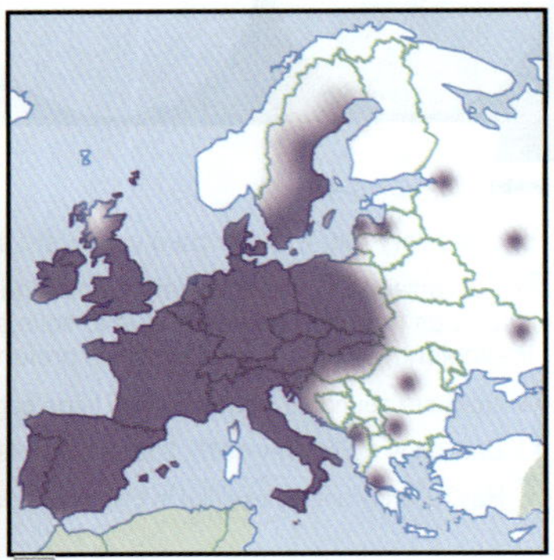

Initially SW and W Europe, today wide-spread, worldwide dispersed

6-7 x 12-14 mm, diameter usually larger than 12 mm

N Germany: Kiel

1.1-1.5 mm at 1 whorl, 2.1-2.6 mm (2 wh.), 3.8-4.5 mm (3), 5.0-6.0 mm (3.5), 6.5-7.8 mm (4), 8.7-10.5 mm (4.5), 11.0-13.5 mm (5 whorls) (values based on shells from Ireland, Germany, Czech Republic, Switzerland, Austria, N Italy and Spain)

Shell horny brown, 5-5.5 very weakly convex whorls, last whorl width 1.7-2.2 x of penultimate whorl, not inflated, umbilicus moderately wide.
Last whorl considerably wider than in *O. cellarius*, diameter more than 1 mm larger at 5 whorls. *O. glaber* has a narrower umbilicus.
Can be distinguished from *O. mortilleti* at shells larger than 10 mm, *O. draparnaudi* has average 0.5 whorl less, there is only little overlap (10 mm shells have more than 4.75 whorls in *O. mortilleti*, less in *O. draparnaudi*). Locally there may be more differences in the elevation of the shell or width of the umbilicus.
Penis subdivided into two parts (a shorter slender proximal and longer broader distal section) separated by a constriction (contrast to *O. cellarius*), rows of papillae of penis interrupted at the constriction (in *O. cellarius* continuous).

Humid habitats under leaves and stones in open and semi-open landscapes. Originally in humid and warm deciduous forests, today mostly (in central and N British Isles exclusively) in cultivated areas in human settlements, gardens, greenhouses, urban waste ground, roadside rubbish, compost heaps. In Switzerland up to 2000 m, but rare above 1000 m.
Feeds on juvenile slugs and snails, earthworms, also cat and dog food.

Reproduction between May and September, animals lay up to 70 eggs (usually 30-40) (diameter 1.5-1.8 mm), every few days some 3-15 eggs, juveniles hatch after 2 weeks and reach maturity after 8-9 months. Egg laying lasts until full shell size is reached, animals die shortly after. Maximum age about 2 years.

Moquin-Tandon 1855: 75 (*Zonites lucidus*), Germain 1930: 149 (*Oxychilus lucidus*), Frömming 1954: 89, Giusti & Manganelli 1997, Boschi 2011: 395.

O. beckmanni from Mallorca is considered as a synonym with regular anatomy (C. R. Altaba, unpublished studies), just with shells looking like *O. cellarius*. The genital morphology is regarded as decisive for the species identity in *Oxychilus*.

Oxychilus egadiensis Riedel, 1973

Oxychilus egadiensis Riedel, 1973 in Riedel 1973: 9

Shell light horny brownish, weakly transparent, striated, 5.5-6.5 slowly and regularly increasing convex whorls with deep suture, last whorl width 1.5 x of penultimate whorl, umbilicus moderately wide.

In rock rubble habitats with grasses and lower plants, a frequent species on Favignana.

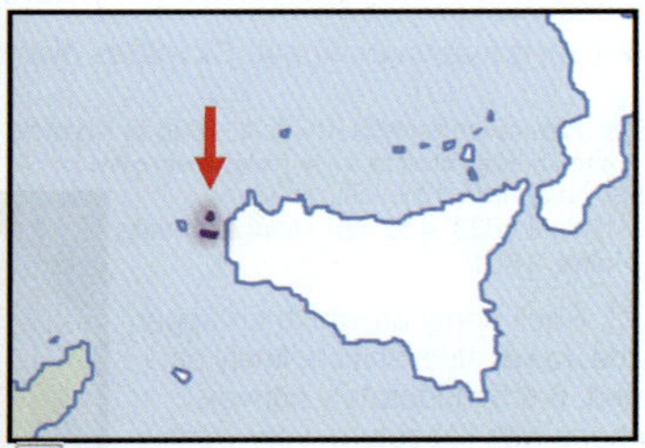

6-7 x 13-15 mm

Italy: Favignana island

Italy (Favignana and Levanzo)

Oxychilus farinesianus (Bourguignat, 1870)

Zonites farinesianus Bourguignat, 1870 in Bourguignat 1870 (Moll. nouv. litig.): 11, Pl. 3, (Rev. Mag. Zool. (2) 22): 18, Pl. 16 (*Zonites Farinesianus*)
I don't know which paper appeared earlier, the texts had identical contents.

Shell horny brownish at upper side, paler or greenish white at lower side, shiny, 6-6.5 weakly convex whorls with weak suture, last whorl only little wider than preceding whorl, umbilicus wide and in form of a funnel (Y). Characterised by its densely coiled whorls, flattened base and the form of the umbilicus.

6.5-7.5 x 14-16 mm

Westerlund 1886: 58, Germain 1930: 149 (synonym of *O. lucidus* = *draparnaudi*), Falkner et al. 2002: 122 (independent species).

This species is weakly justified. Falkner et al. 2002 gave a diagnosis that allows us to distinguish this species from others, and argued that the characters felt outside the regular variability of *O. draparnaudi*. They did not provide results of studies to substantiate this hypothesis. More research is needed.

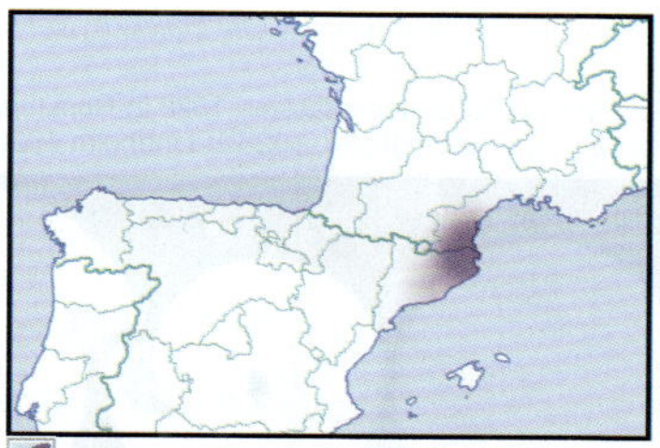

NE Spain, S France

Oxychilus fuscosus (Rossmässler, 1838) dd

Without gender agreement: *Oxychilus fuscosa* (Rossmässler, 1838)

Helix fuscosa Rossmässler, 1838 in Rossmässler 1838-1844 (7-8): 35, Taf. 39

9-16 x 16-33 mm

Italy: Sicily, Monte Pellegrino W Palermo

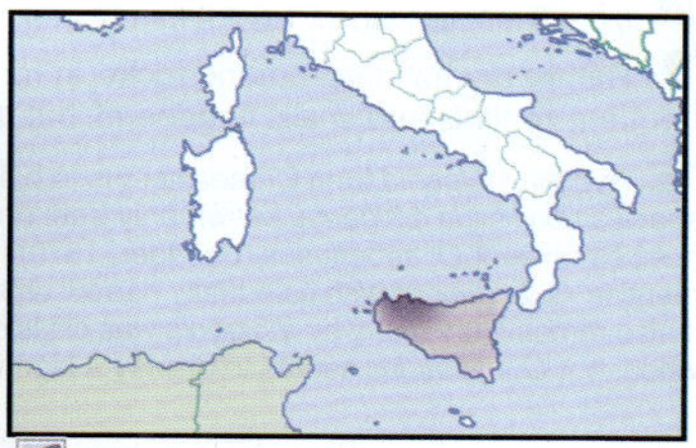

Italy (Sicily)

Shell horny yellow on upper side, greenish or yellowish white on lower side, thin and transparent, on upper side coarsely and densely striated and with fine spiral lines producing a reticulate pattern, lower side smooth, upper side flattened, 6-6.5 weakly convex with weak suture, last whorl width about 3 x of preceding whorl, umbilicus narrow but deep.

Westerlund 1886: 70, Manganelli et al. 1995: 19.

Species seems to need a revision.

Italy: Sicily, Palermo — Agatino Reitano

Oxychilus gardinii Manganelli, Bodon & Giusti, 1991

Oxychilus gardinii Manganelli, Bodon & Giusti, 1991 in Manganelli et al. 1991: 401

Back and tentacles bluish, sides and posterior extremity pale grey, sides of sole pale greish blue. Shell brownish yellow, translucent, depressed, sometimes with a few growth lines, umbilicus quite narrow (1/6-1/8 of shell diameter) and deep. Anatomy: penial complex with very slender proximal penis with variable inner sculpture and widened distal penis.

5.2-6.6 x 12.4-16.5 mm

Hypogean chalky habitats, inside mones in dolomitic areas, but lives also in pine forests and is not confined to limestone.

Lives only in a small restricted area in N Italy, where no other *Oxychilus* species live.

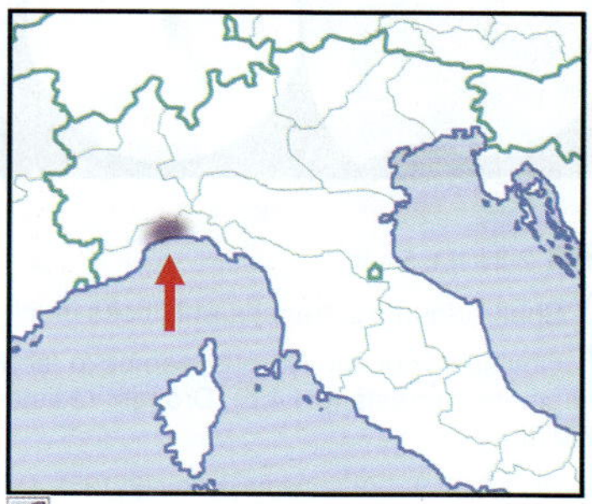

N Italy (Liguria)

Oxychilus glaber (Rossmässler, 1835)

Without gender agreement: *Oxychilus glabra* (Rossmässler, 1835)

Helix glabra Rossmässler, 1835 in Rossmässler 1835-1837 (1): 71
Described by Rossmässler in the text without figure, and attributed to both Studer and Férussac.
not *Helix hispida* var. *glabra* Hartmann, 1821 in Hartmann 1821 (Neue Alpina): 237 - nomen oblitum under Art. 23.9, fixed here:
In accordance with Art. 23.9 *Helix glabra* Rossmässler, 1835 shall be used as "valid" (name in a current classification), *Helix hispida* var. *glabra* Hartmann, 1821 has not been used for a taxon since 1899, *Helix glabra* Rossmässler, 1835 was used as a "valid" name in the following 26 papers: Riedel 1969: 94, Alzona 1971: 129, Damjanov & Likharev 1975: 286, Kerney & Cameron 1979: 126, Kerney et al. 1983: 172, Reischütz 1983: 135, Boato et al. 1984: 300, Jungbluth et al. 1989: 150, Falkner 1990: 182, Riedel 1992: 96, Manganelli et al. 1990: 193, Dhora & Welter-Schultes 1996: 148, Turner et al. 1998: 279, Falkner et al. 2002: 127, Martínez-Ortí & Robles 2003: 108, Wiktor 2004: 178, Ferreri et al. 2005: 68, Irikov & Mollov 2006: 798, Sverlova 2006: 127, Taláván Gómez & T. 2006: 115, Hubenov 2007: 138, Egorov 2008: 71, Uherkovitch et al. 2008, De Oliveira 2009 (Not. Soc. Esp. Mal. 51): 59, Jungbluth & Knorre 2009: 10, Sysoev & Schileyko 2009: 124.

5.5-9 x 12-18 mm, diameter usually up to 14 mm

Albania: Mali Gradecit near Çorovodë

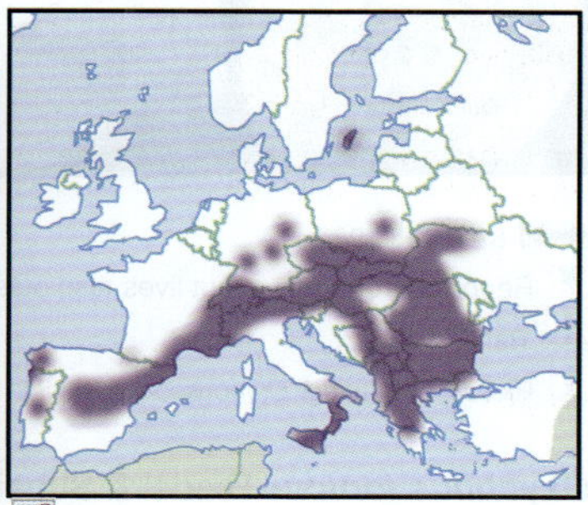

S Europe

France: Ain, Torcieu

5-6 mm at 3 whorls, 8-8.5 mm at 4 whorls, 13-14 mm at 5 whorls

Shell reddish brown or brownish yellow and usually clean, translucent, 5-5.5 whorls, umbilicus deep and very narrow, 1/12 of shell diameter, aperture with a thin and non-reflected peristome.

Humid habitats in leaf litter of deciduous and coniferous forests, below rocks, under stones and old wood logs. In S Bulgaria with a wide ecological amplitude, but prefers humid forest habitats. In Switzerland up to 1800 m, usually between 600 and 1300 m, in Bulgaria up to 1900 m.
Feeds on dead and life parts of plants and juvenile snails of *Succinea*, *Cepaea* and *Arianta*.

Critically Endangered in Germany, Vulnerable in Austria.

Manganelli, Castagnolo & Giusti 1990: 193, Taláván Gómez & T. 2006: 115.

Subgenus *Morlina*.

Cédric Audibert

Oxychilus hydatinus (Rossmässler, 1838)

Without gender agreement: *Oxychilus hydatina* (Rossmässler, 1838)

Helix hydatina Rossmässler, 1838 in Rossmässler 1838-1844 (7-8): 36, Taf. 39
Hyalina iphigenia Lindholm, 1926 in Lindholm 1926: 167

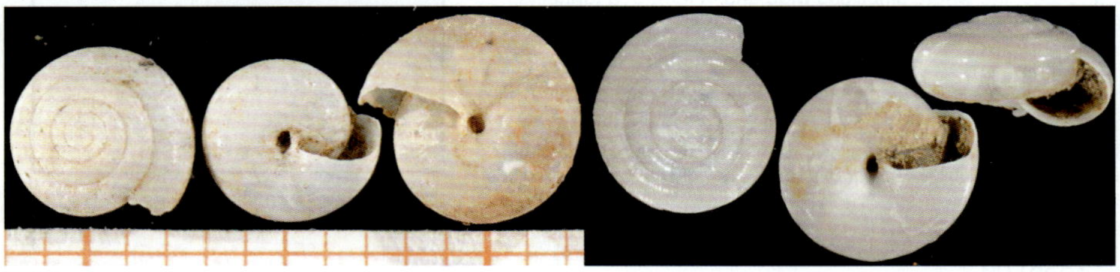

(1-3) W Turkey: near Bodrum, (4-6) Greece: Crete, near Léndas

2.5-3.0 x 5.0-6.5 mm

Ø 0.8-0.9 mm at 1 whorl, 1.5-2.2 mm (2 wh.), 2.4-3.6 mm (3), 3.2-4.4 (3.5), 3.9-5.5 mm (4), 4.7-6 mm (4.5 whorls)

Rather dry habitats, among rocks, in rock crevices, also in caves. In Bulgaria up to 1100 m. One of the most frequent zonitids and snails in general, but usually not very abundant.

Hyalina iphigenia from Crimea seems to belong to this species. Sysoev & Schileyko 2009: 129 listed it as an independent species *Oxychilus iphigenia*, known only from one locality in SW Crimea, and argued that it differed from all other *Oxychilus* species (except *O. subeffusus* (Boettger, 1879) from Caucasus and NE Turkey) in its very narrow umbilicus. They did not mention *O. hydatinus* with its equally narrow umbilicus (which they would probably not have classified in the genus *Oxychilus*, it also remains unclear why *iphigenia* was classified in *Oxychilus*). The Crimean population may differ in the width of the last whorl, but *O. hydatinus* seems to display a high variation in this character.

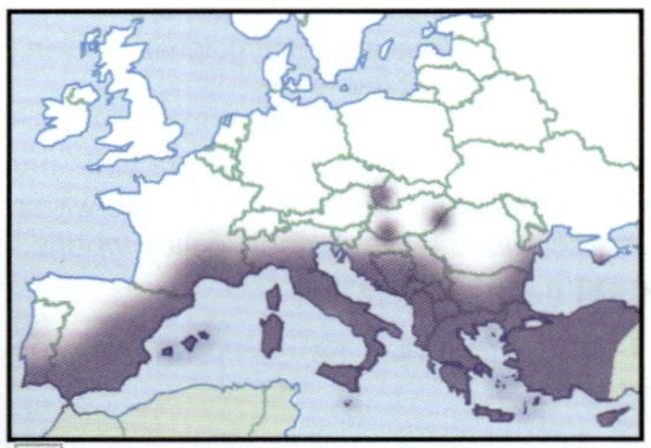

Canary Islands to central Turkey

Shell quite variable, whitish or colourless to pale greenish or with reddish hue, glassy, translucent when fresh, thin, shiny, 3.5-4.5 slowly increasing and weakly convex whorls, whorls rounded at periphery, spire often a little raised, aperture without internal lip, umbilicus very narrow and deep.
Juvenile shells may look like *Vitrea* species, those can be easily distinguished by their more narrowly coiled whorls. *O. hydatinus* has a high degree of variation in the diameter/whorl ratio but has usually at least one whorl less at the same size.

Oxychilus inopinatus (Uličný, 1887)

Without gender agreement: *Oxychilus inopinata* (Uličný, 1887)

Hyalina inopinata Uličný, 1887 in Uličný 1887 [Vesmír, prior to 31 Dec]: 111
Hyalina opinata Clessin, 1887 in Clessin 1887 [31 Dec]: 89
Clessin 1887 attributed the name to Uličný but the description was by Clessin. Uličný had submitted in 1886 or early 1887 a description of *Hyalina inopinata* in Latin/German to Malakozoologische Blätter (see Uličný 1889: 39), but the article appeared only in 1888 (Uličný 1888: 112-114), after the description of *Hyalina opinata* Clessin, 1887. Clessin was editor of Malakozoologische Blätter.

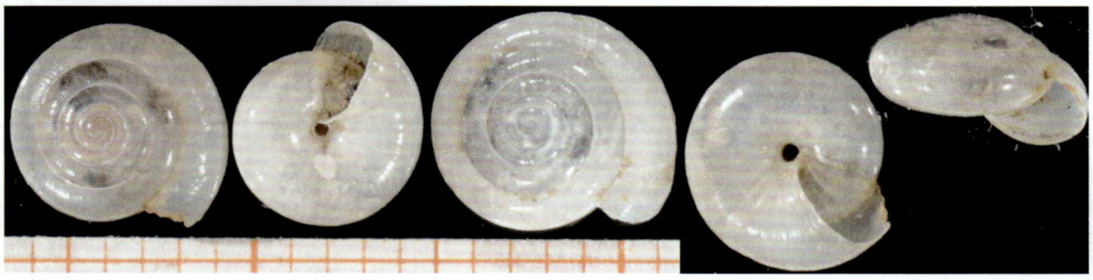

Austria: Steiermark, near Peggau

2-3 x 4-6.5 mm

Shell colourless, transparent, looks much like *Vitrea*, upper side relatively flat, umbilicus narrow but open.

Obviously not carnivorous, seems to feed on disintegrating plant particles. In broad-leaved deciduous forests and pinetree forests (pH 4.5). Drought-resistant, survives considerable temperature fluctuations, in Bulgaria up to 2000 m.

Riedel 1959: 179 (anatomy), Hudec 1961, Damjanov & Likharev 1975: 289, Riedel 1969: 100, Riedel 1992: 92, Štamol & Velkovrh 1995 (Croatia: Cres, 1 of 80 stations), Irikov & Mollov 2006: 800.

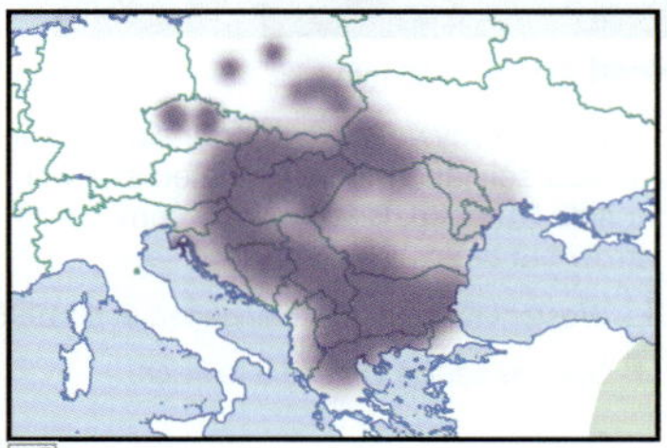

SE Europe

Bulgaria: Sakar Mountain, Sladun
Dilian Georgiev

Oxychilus investigatus Riedel, 1993

Oxychilus investigatus Riedel, 1993 in Riedel & Maassen 1993: 142

Bulgaria: Stara Planina, W Kotel

8-10 x 15-19 mm

Reported from caves, but lives also outside.

Rare in Turkey.

Riedel 1972 (under *O. moussoni*), Riedel 1995: 55, Schütt 2001: 218.

SE Bulgaria to NW Turkey

Conchologically very similar to *O. kobelti* (slightly smaller, last whorl slightly wider), differences in truncus receptaculi and flagellum are more striking, species identification requires anatomical dissection.

Oxychilus ionicus Riedel & Subai, 1978

Oxychilus ionicus Riedel & Subai, 1978 in Riedel & Subai 1978: 79

Shell whitish or almost colourless, thin, translucent but not transparent, smooth or with weak growth lines, very shiny, outline of upper side almost flat, 4.2-4.8 very weakly convex and rapidly increasing whorls with shallow suture, last whorl width 2.5-3 x of penultimate whorl, regularly rounded at periphery, basal and columellar margin of aperture very weakly curved, umbilicus extremely narrow and partly covered by the reflected columellar margin.
Animal colour cream, organs are well visible under the transparent shell.

2-2.5 x 4.5-6 mm

Probably semisubterranean, was found in soil samples under rock rubble and, more rarely, between roots of old olive trees, under leaves and stones.

Riedel & Subai 1982: 25 (anatomy), Riedel 1992: 88, Manganelli et al. 1995: 20.

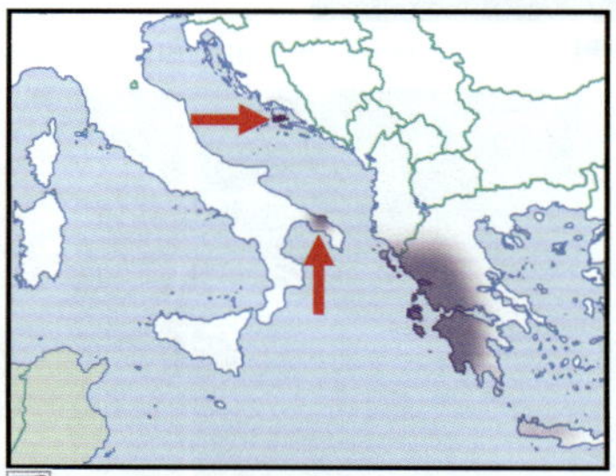

W Greece, Croatia (Dalmatia: island of Hvar), possibly Italy and Crete

Oxychilus jugoslavicus (Urbański, 1939)

Without gender agreement: *Oxychilus jugoslavica* (Urbański, 1939)

Retinella jugoslavica Urbański, 1939 in Urbański 1939: 260, Taf. 9

Shell yellowish white, translucent and shiny but solid, densely striated, less prominently and partly nearly smooth on lower side, disc-like flattened on both sides, 5 slowly and regularly increasing whorls with deep suture, last whorl width more than 2 x of penultimate whorl when seen from above, not descending near aperture, umbilicus very wide and perspectivical, more than 1/3 of shell diameter.

2.4-2.8 x 6-7 mm

Was found at a deep rock crevice in a shady and moderately moist habitat of a deciduous forest, lives probably in subterranean habitats.

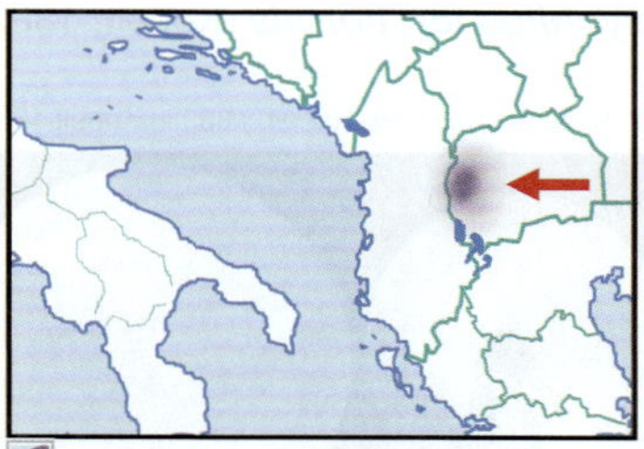

Rep. Makedonija

Oxychilus juliae Riedel, 1990

Oxychilus juliae Riedel, 1990 in Riedel 1990: 523

Shell horny yellowish, sometimes with greenish hue, transparent, smooth and shiny, 4.5-5.5 initially slowly and then more rapidly increasing whorls (number of whorls locally variable), suture shallow and with margin, last whorl width 2-3 x of penultimate whorl (locally variable), umbilicus very narrow, only last whorl visible inside. Animal greyish blue with cream sides and sole, or dark brownish grey with blue foot, seems to vary locally.

6.5-8 x 15-17 mm

In moist habitats of old olive orchards at stone walls covered with mosses and plants, or in caves.

Riedel 1992: 99, Dhora & Welter-Schultes 1996: 149, 188.

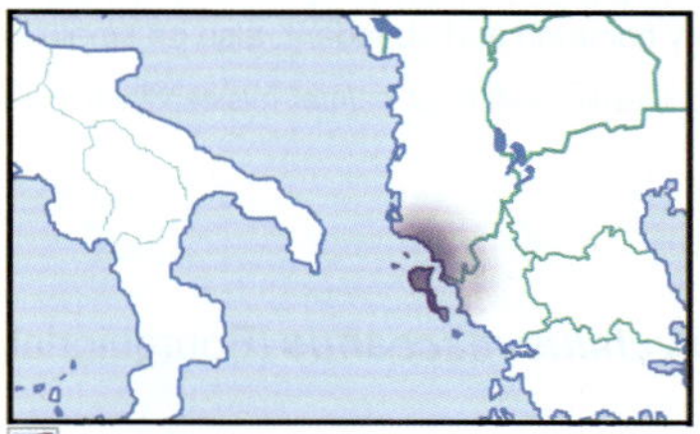

NW Greece, S Albania

Oxychilus kobelti (Lindholm, 1910)

Retinella kobelti Lindholm, 1910 in Lindholm 1910: 223

Turkey: Vilayet Bolu, Abant Gölü

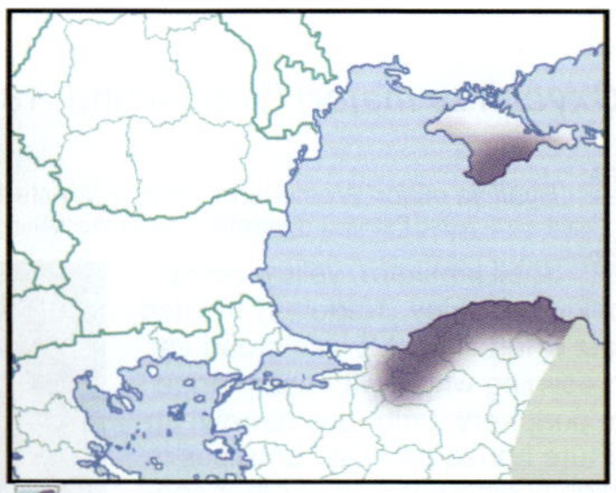

NW Turkey, S Ukraine (Crimea)

10-12 x 19-22 mm

Riedel 1995: 57.

Shell horny reddish on upper side, yellowish white on lower side, striated, with spiral lines on upper side, shiny, 4.5-5.5 slightly convex whorls, last whorl width about 2 x of penultimate whorl, not descending near aperture, umbilicus narrow but deep, not widened at last whorl.

Was found in Ukraine a cool valley in deciduous forest, under moist leaf litter.

Oxychilus lagrecai Giusti, 1973

Oxychilus lagrecai Giusti, 1973 in Giusti 1973: 173

Shell ochrous yellowish, not very shiny, fragile, weakly striated, 5-5.5 slowly and regularly increasing whorls, umbilicus moderately wide, 1/6 of diameter.

Widespread but not frequent, prefers humid and cool habitats under stones, often at walls of ruins.

4.3-5.2 x 10-11 mm

Italy: Lipari island

Maurizio Sosso

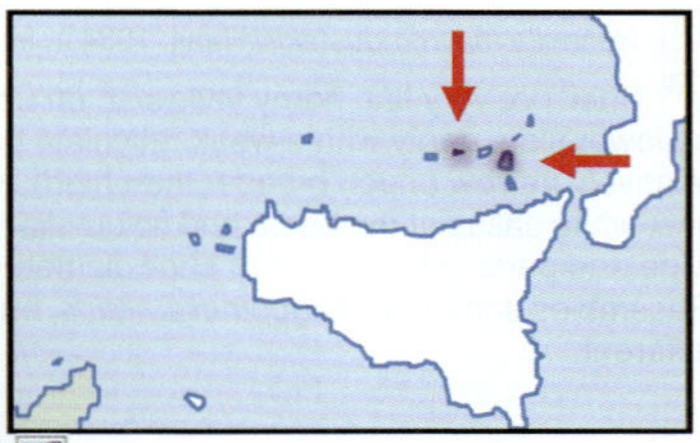

Italy (Isole Eolie: Filicudi, Lipari)

Oxychilus lathyri (Mabille, 1869)

Zonites lathyri Mabille, 1869 in Mabille 1867-1869: 64 (*Zonites Lathyri*)

Shell light horny reddish on upper side, horny whitish on lower side, almost translucent, not very solid, almost regularly striated, first whorls increase slower than last worls, umbilicus present.

5-5.5 x 15-17 mm

Westerlund 1890: 9.

France (Corsica)

Oxychilus lentiformis (Kobelt, 1882)

Hyalinia lentiformis Kobelt, 1882 in Kobelt 1882: 122

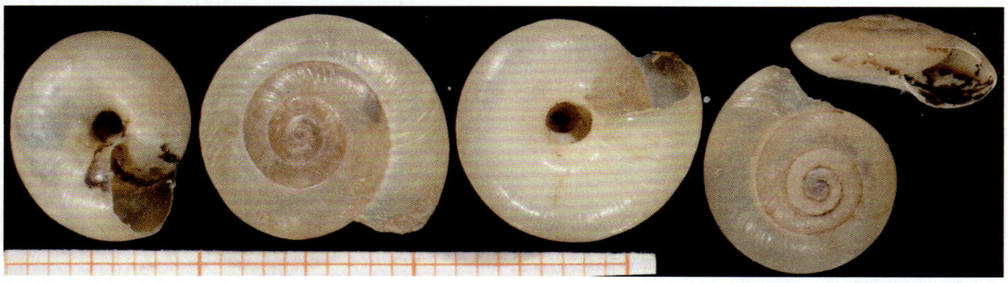

4.5-6 x 10-13 mm

Spain: Balearic islands, E Mallorca

In a broad variety of habitats, in dry zones under stones or in humid habitats in organic litter, up to 700 m. Very common and abundant, also on small islands (Cabrera, Vedranell).

Gasull 1964: 6 (*O. balmei*, misidentified), Riedel 1969, Beckmann 2007: 82.

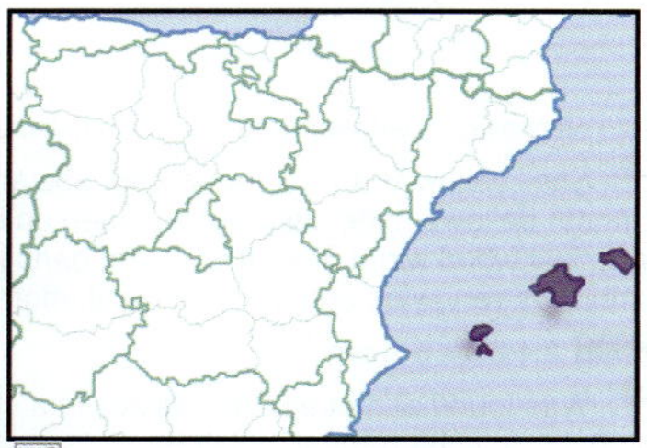

Spain(Balearic islands)

Shell dirty yellow, lighter at lower side, not very translucent, growth lines usually irregular and weak, usually finely and densely spirally striated on upper side, shape very variable, 5-5.5 whorls, with or without edge at periphery, umbilicus wide and perspectivical.

Oxychilus maceanus (Bourguignat, 1869)

Zonites maceanus Bourguignat, 1869 in Bourguignat 1869: 48 (*Zonites Maceanus*)

Shell horny coloured, lower side whitish, shiny, fragile, very weakly striated, with edge at periphery, umbilicus narrow, upper portion of aperture protruding.

4.2-2.7 x 12-14 mm

Westerlund 1886: 63, Germain 1930: 145, Falkner et al. 2002: 123.

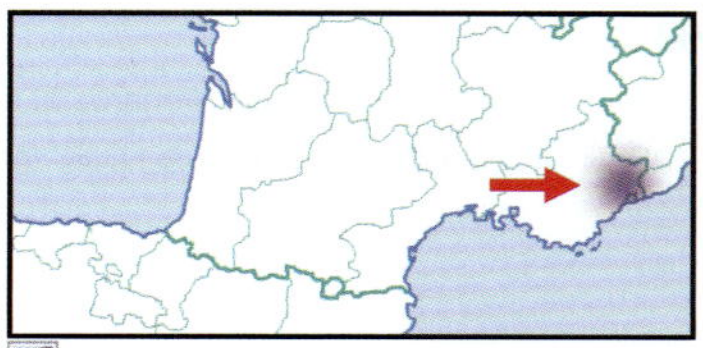

SE France (Alpes Maritimes)

Oxychilus majori (Westerlund, 1886)

Hyalinia majori Westerlund, 1886 in Westerlund 1886 [Apr 1886]: 50 (*Hyalinia (Polita) majori*)

Name attributed to Paulucci, description was by Westerlund. Also described as *Hyalinia (Hyalinia) Majori* in Paulucci 1886 [30 Jun 1886]: 12 Tav. 1.

Shell lentiform, yellow horny coloured, shiny, irregularly striated, last whorl almost 2 x as broad as preceding whorl, with sharp edge at periphery, umbilicus narrow, aperture horizontal. The umbilicus is slightly different than in other species.

Paulucci 1886: 1, Manganelli & Giusti 1993: 70, Manganelli, Favilli & G. 1999.

3-4 x 8-9 mm

Italy: Toscana, Monte Argentario

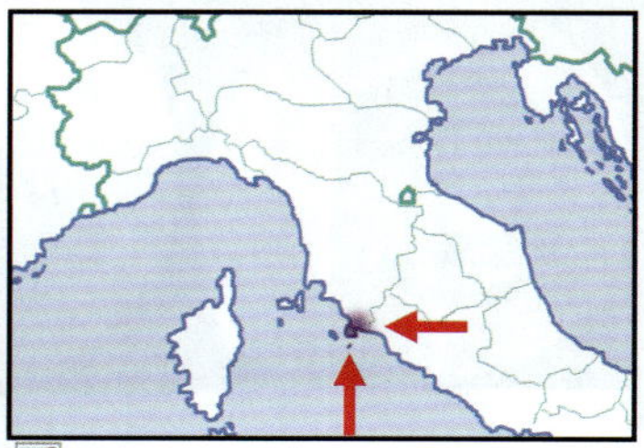

Italy (Toscana)

Angelo Vannozzi

Oxychilus mavromoustakisi (Haas, 1934)

Retinella mavromoustakisi Haas, 1934 in Haas 1934: 16

Shell not very flat, horny yellowish on upper side, whitish on lower side, partly with a weak reticulate pattern, aperture relatively narrow (much broader than high), inside with a distinct white callus at the basal and columellar side, umbilicus wide and perspectivical, 1/5 of shell diameter.
Resembles somehow *Aegopinella nitens*, but the aperture is different.

4.5-5 x 8.5-10 mm

Cyprus

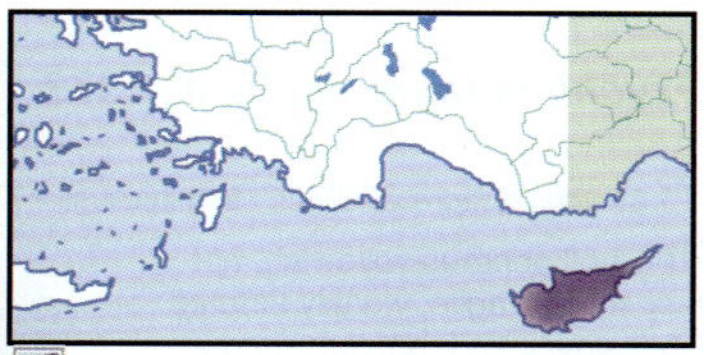

Cyprus

Georgios Konstantinou

Oxychilus mercadali Gasull, 1968

Oxychilus mercadali Gasull, 1968 in Gasull 1968: 143

3-5 x 7-11 mm

Spain: Valencia, Jeresa les Cingles

A trogophilous species, frequent in caves and humid forest habitats, but also found in relatively arid mountain limestone habitats and slopes.

Species with restricted range, usually not very abundant.

Riedel 1972: 133, Martínez-Ortí & Robles 2003: 105.

Spain (Alicante, València)

Shell flatter than other Iberian species, amber coloured, transparent, smooth and very shiny, no spiral sculpture, 5 weakly convex whorls, rounded without edge at periphery, lower side weakly convex, umbilicus wide and perspectivical, aperture not very oblique.

Oxychilus meridionalis (Paulucci, 1881)

Hyalinia meridionalis Paulucci, 1881 in Paulucci 1881: 78

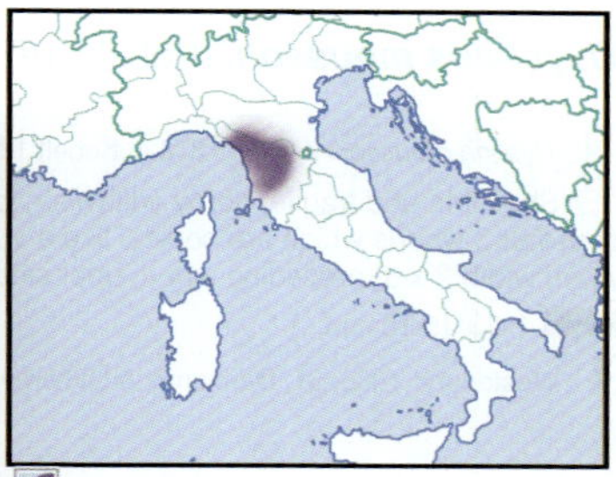

Italy

4-7 x 10-16 mm, shape and size variable, also within populations — Italy: Pisa, Asciano

Westerlund 1886: 61, Manganelli et al. 1995: 19, Manganelli & Giusti 2001 (0.5 whorls must be discounted from their values).

1.1-1.7 mm at 1 whorl, 2-2.7 mm (2 wh.), 3.9-4.7 mm (3), 4.7-6.2 mm (3.5), 7-9 mm (4), 8.5-12 mm (4.5), 11.5-15 mm (5)

Shell thin and fragile, light horny brownish with weak greenish hue, lower side whitish, densely and finely striated, distinctly only at the suture, angulated, 5.5-6 regularly increasing and weakly convex whorls with deep suture and a weak edge at the suture, last whorl width about 2 x of preceding whorl, umbilicus narrow, getting wider near the aperture, 1/7-1/5 of shell diameter.
Penis without clear distinction into proximal and distal parts (no bottleneck). No other differences to *O. draparnaudi*.

Oxychilus minoicus Riedel, 1968

Oxychilus minoicus Riedel, 1968 in Riedel 1968: 503

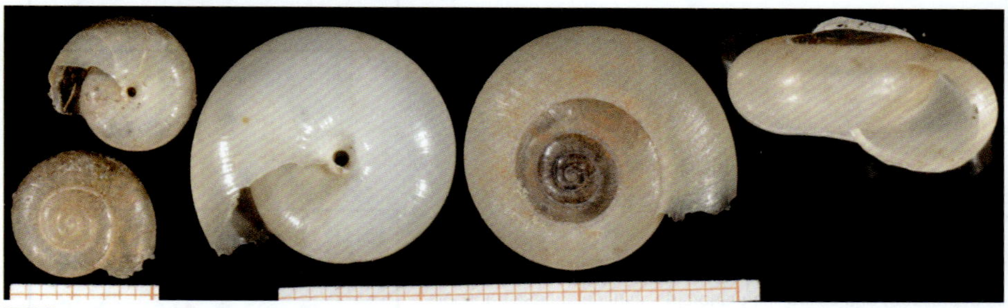

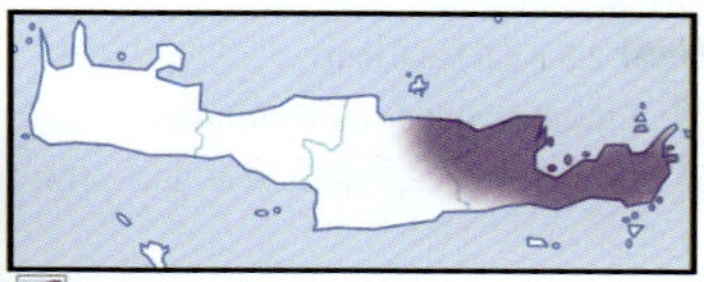

Greece (E Crete)

Shell horny yellow, lower side much lighter, silky shiny, translucent, densely striated and with dense and prominent spiral lines, much weaker on lower side, outline of upper side flattened, 5-5.5 weakly convex and rapidly increasing whorls with shallow suture, last whorl width 2 x of penultimate whorl, umbilicus very narrow, penultimate whorl not well visible inside.

9-12.5 x 16-25 mm — Greece: E Crete, (1-2) near Avrakóndes, (3-5) near Kritsá

Rocky habitats in mountains, requires much humidity, often found in rock crevices or near caves.

Riedel 1992: 95.

Oxychilus montivagus (Kimakowicz, 1890)

Without gender agreement: *Oxychilus montivaga* (Kimakowicz, 1890)

Hyalinia nitidissima var. *montivaga* Kimakowicz, 1890 in Kimakowicz 1890: 27
The word *montivaga* is usually treated as a Latin adjective.

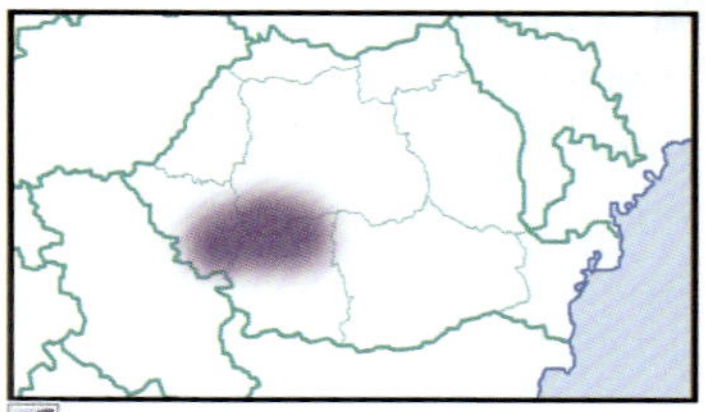

Romania

Shell horny yellow with reddish hue, more whitish on lower side, irregularly striated, 5 convex whorls with shallow suture, last whorl width 2.4-3 x of penultimate whorl, umbilicus very narrow like in *O. glaber*.

4-5 x 9-12 mm

Rocky habitats in mountains, mountain valleys.

Grossu 1983: 140, Riedel 1997: 763.

Oxychilus mortilleti (Pfeiffer, 1859)

Helix mortilleti Pfeiffer, 1859 in L. Pfeiffer 1859: 101 (*Helix Mortilleti*)

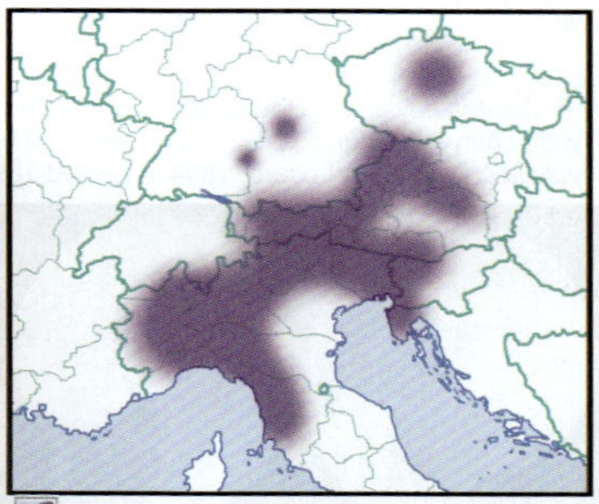

Italy to Czech Republic

5-7 x 10-14 mm, diameter up to 17 mm in the Alps — Switzerland: Ticino, Meride

1.1-1.4 mm at 1 whorl, 1.9-2.5 mm (2 wh.), 3.3-4.2 mm (3), 4.4-5.3 mm (3.5), 5.7-6.8 mm (4), 7.3-9.0 mm (4.5), 9.1-11.3 mm (5), 11.3-13.5 mm (5.5), 14-17 mm (6 whorls) (values based on shells from Meride, Caslano and Manganelli & G. 1998)

Shell horny to reddish brown, 4.5-5 whorls, growing regularly until whorl 3.5, then visibly broader, last whorl width 1.5-2 x of penultimate whorl, umbilicus moderately wide.
Umbilicus more cylindrical than in *O. cellarius*, suture deeper, last whorl broader, diameter more than 1 mm larger at 5 whorls. *O. glaber* has a much narrower umbilicus.
Can be distinguished from *O. draparnaudi* at shells larger than 10 mm, *O. draparnaudi* has average 0.5 whorl less, there is only little overlap (10 mm shells have more than 4.75 whorls in *O. mortilleti*, less in *O. draparnaudi*). Locally there may be more differences in the elevation of the shell or width of the umbilicus.
Penis longer and narrower than in *O. draparnaudi* and *O. cellarius*.

Italy: Lago di Garda, Gargnano

Humid habitats under leaf litter and stones, in open habitats and forests, at old walls, vineyard terraces, in rock crevices. Not restricted to limestone. In Switzerland up to 1600 m. Carnivorous, consumes smaller snails and slugs, it seems that the snail bites and injects poison into the other snail which stops crawling after a short distance.

2-5 eggs (diameter 1.4 mm) are laid under moss, probably between March and May, juveniles grow at a rate of slightly less than 1 mm per month, they seem to become mature at the age of 1 year in early summer at about 9 mm shell diameter, maximum age in captivity 5-6 years.

Frömming 1954: 100, Kerney & Cameron 1979: 124, Falkner 1990: 180, Turner et al. 1998: 274, Manganelli & Giusti 1998 (0.5 whorls must be discounted from their values to come to the values here), Boschi 2011: 397 (used mixed sources in the whorl numbers and lumped these).

Manganelli & G. 1998 and Boschi 2011 saw no differences in the shells between *O. cellarius*, *O. draparnaudi* and *O. mortilleti*. When only few shells are found it is more difficult. When many shells are collected at one locality it is usually possible to produce morphological clusters and to see the differences in the shells between such species.

Oxychilus moussoni (Kobelt, 1878)

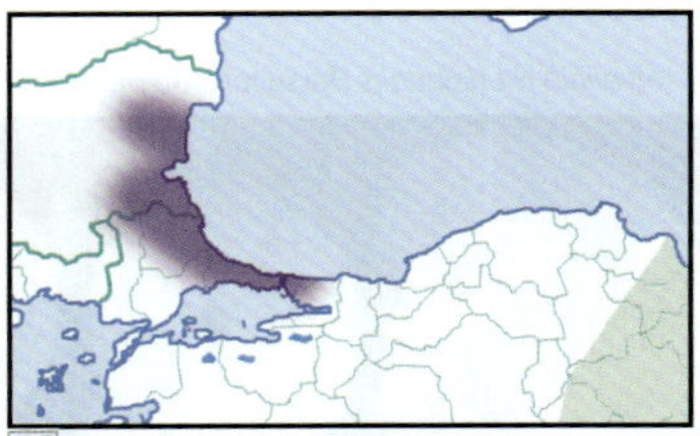

SE Bulgaria to Istanbul region

Hyalina moussoni Kobelt, 1878 in Kobelt 1879: 22, Taf. 156 (*Hyalina Moussoni*)

Shell thin and fragile, horny yellowish, lower side whitish, shiny, very finely striated, upper side flattened except first 3 whorls which are conical like a vulcanic island in the sea, 5-5.5 convex whorls with deep suture, last whorl width 2 x of preceding whorl, uper side more rounded than lower side, umbilicus very narrow.

7-9 x 15-18 mm

Westerlund 1886: 51, Damjanov & Likharev 1975: 284, Riedel 1992: 96, Riedel 1995: 67.

Oxychilus mylonasi Riedel, 1983

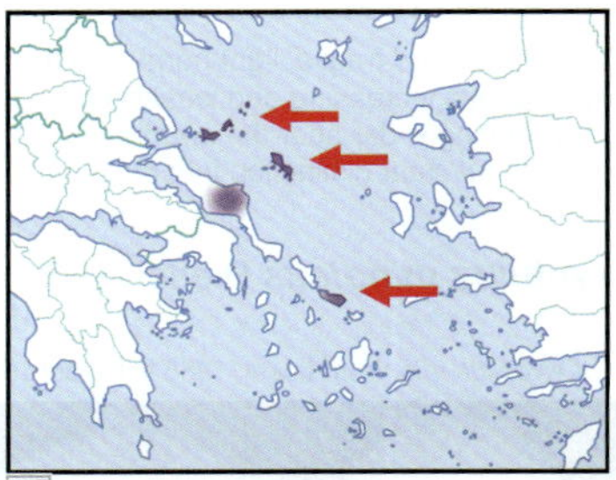

Greece (W Aegean islands)

Oxychilus mylonasi Riedel, 1983 in Riedel 1983: 278

Shell light yellowish, slightly opaque, very shiny, with very weak growth lines, 4-4.5 very weakly convex whorls with shallow suture, first whorls slowly increasing, last whorls more rapidly, last whorl width 2.5 x of penultimate whorl, umbilicus very narrow, inner whorls not visible inside.
Animal pale cream, sole very narrow and subdivided in 3 sections.

2.5-3 x 5.5-6.5 mm

Probably mainly subterranean.

Riedel 1992: 92, Triantis et al. 2008: 474.

Oxychilus nautiliformis (Wagner, 1914)

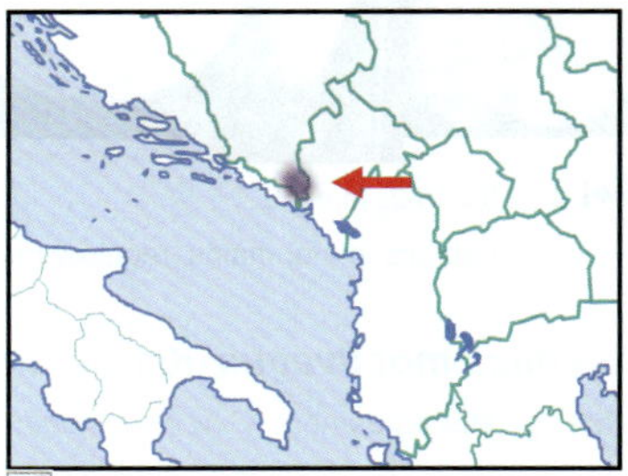

Herzegovina

Hyalinia nautiliformis Wagner, 1914 in A. J. Wagner 1914 (Sitz.-Ber. Wien): 41

Shell outline concave on upper side (apex depressed) and coiled like planorbids, thin, colourless transparent and shiny, with fine S-shaped growth lines, 4 very weakly convex whorls with distinctly deep suture, upper whorls narrow, last whorl width 2.5 x of preceding whorl, not descending near aperture, aperture perpendicular, wider than high, margin thin and sharp, upper side with rounded sinulus at preceding whorl, umbilicus wide, almost 1/3 of shell diameter.
Like juvenile *Zonites*, but without spiral lines and not keeled.

2.8-3.2 x 7-8 mm

Known from caves.

Riedel 1979 (Fragm. Faunist.).

Oxychilus navarricus (Bourguignat, 1870)

Oxychilus helveticus (Blum, 1881)

Zonites navarricus Bourguignat, 1870 in Bourguignat 1870: 20 (*Zonites Navarricus*)
Hyalina helvetica Blum, 1881 in Blum 1881: 141
Hyalinia colliourensis Locard, 1894 in Locard 1894: 52 (*Hyalinia Colliourensis*) (this species?)

England: Yorkshire, Malham

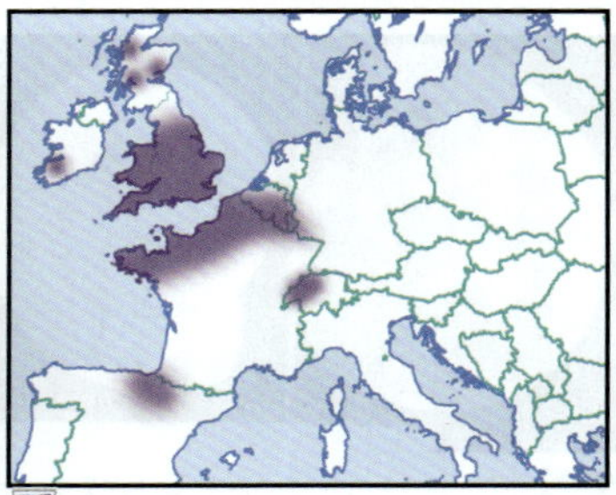

British Isles, W Europe

4.5-6 x 7-13 mm

Shell thin, light horny yellowish, whitish near umbilicus, very shiny, almost smooth, outline almost globulous, 4.5-6 weakly convex whorls with weak suture, last whorl width 1.5 x or less of preceding whorl, well rounded at lower side, umbilicus not very wide, 1/8-1/7 of diameter.
Larger than *O. alliarius*, umbilicus narrower (1/8 of shell diameter), presence of a black spot on the mantle at the aperture.
Anatomy: internal ornamentation of proximal penis consisting of more than seven longitudinal pleats, sometimes straight and distinct, sometimes wavy, slender and connected by lateral projections giving a reticulate appearance.

Humid and shady habitats in forests and between rock rubble. Predominantly deciduous mountain forests under moss, more restricted to calcareous substrate than Oxychilus alliarius. Also in creek gorges, at springs and in caves. In Britain also in old quarries, at roadsides and under human rubbish. In Switzerland up to 1600 m.

Westerlund 1886: 53 (*cantabrica*), 54 (*helvetica*), 56 (*navarrica*), Falkner 1990: 180, Turner et al. 1998: 277, Kerney 1999: 146, Giusti & Manganelli 2002, Falkner et al. 2002: 125. Germain 1930: 150 used *navarricus* for widespread forms in S France and the Pyrenees.

Had been called *Oxychilus helveticus* (Blum, 1881) until 2002. Riedel 1970 (and Riedel 1972, Riedel 1980) classified *helvetica* with *cantabrica*, and Falkner et al. 2002 found that *cantabrica* was a junior synonym of *Zonites navarricus*.
Germain 1930 classified *Hyalinia colliourensis* from the NE Pyrenees as a synonym of *O. lucidus* var. *navarricus* (= currently *O. navarricus* (Bourguignat, 1870), this name was used for widespread forms in S France). Falkner et al. 2002: 125 argued that *colliourensis* was a good species, but gave no constraints or arguments to substantiate this hypothesis. The original information given by Locard 1894 did not provide constraints for an independent species, so the classification proposed by Germain 1930 (as a form that cannot be distinguished with our knowledge from other forms known to occur in S France) represents the current state of published research. A new classification should be based on published results of new scientific studies, and should enable us to distinguish this species from others.

Oxychilus nortoni (Calcara, 1845)

Helix nortoni Calcara, 1845 in Calcara 1845 (Atti Palermo): 42 (*Helix Nortoni*)

Was formerly cited with the date 1843, but a paper published by Calcara 1843 could not be confirmed (Riedel 1973: 24). Probably a manuscript, name was claimed to have been mentioned as *Helix Nortoni*, p. 143. Sherborn 1922 cited the name and the publication, but confirmed not to have seen the journal L'Occhio.

Shell horny yellowish, translucent or opaque, striated, 5-6 slightly convex and slowly and regularly increasing whorls, periphery rounded, last whorl not much wider than penultimate whorl, aperture depressed and oval, margin simple and sharp, umbilicus wide and perspectival, all whorls visible inside.

Riedel 1973: 13, Manganelli et al. 1995: 19.

4-5 x 8-10 mm Italy: Sicily, island of Ustica

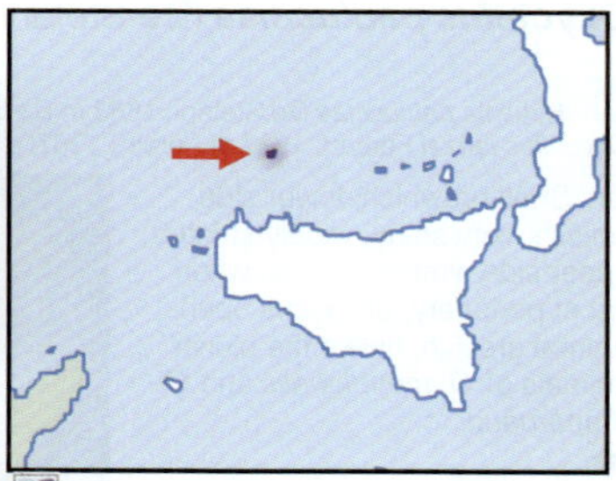

Italy (Ustica island)

Massimo Prodan

Oxychilus obscuratus (Porro, 1841)

Without gender agreement: *Oxychilus obscurata* (Porro, 1841)

Helix obscurata Porro, 1841 in Villa & Villa 1841: 56

Name was attributed to Porro in original source. Author of the description was Porro, not Villa & Villa (see footnote p. 54).

Shell very depressed, weakly convex, greenish or brownish, shiny, finely striated, lower side lighter, 5.5-6 rapidly increasing whorls, last whorl weakly angulated at periphery, umbilicus wide, 1/5 of diameter.

6-7 x 14-18 mm

Under stones, among low herbs and rocks, also in forests.

Westerlund 1886: 60, Germain 1930: 151.

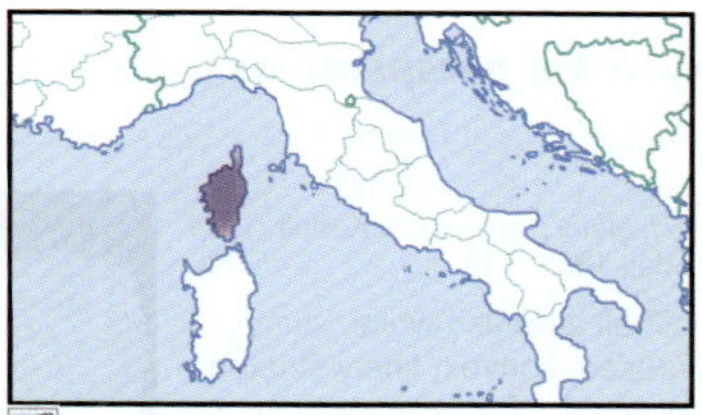

France (Corsica)

Oxychilus oglasicola Giusti, 1968

Oxychilus oglasicola Giusti, 1968 in Giusti 1968: 221

Shell very depressed and strongly keeled.

La Scola 5.5-6.5 x 14-16 mm, in Montecristo slightly larger

On calcareous (La Scola) and granitic (Montecristo) substrate.

Endemic species of Montecristo and La Scola near Pianosa (between Corsica and mainland Italy)

Giusti 1970 (Ann. Genova 78): 73, Giusti 1976: 184, Manganelli et al. 1995: 19, Manganelli, Favilli & G. 1999.

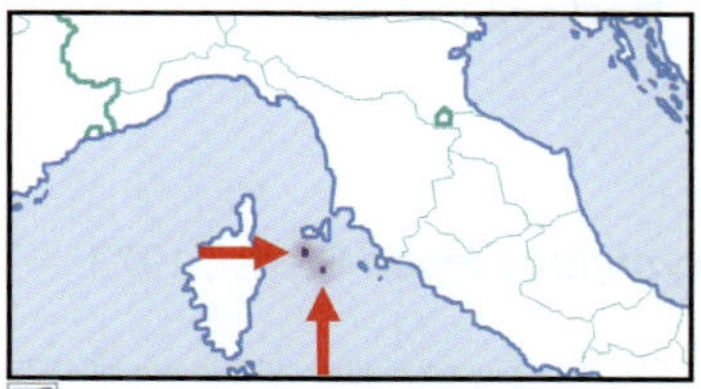

Italy (Montecristo and La Scola)

Oxychilus oppressus (Shuttleworth, 1877)

Without gender agreement: *Oxychilus oppressa* (Studer & Shuttleworth, 1877)

Zonites oppressa Shuttleworth, 1877 in Shuttleworth 1877: 6, Pl. 2 [only verified in the Bern edition] (*Zonites (Retinella) oppressa*)

Name was attributed to Shuttleworth, author of the Latin diagnosis and a short French text (p. 6) was Studer, as explained by Fischer (p. 6). Fischer and Shuttleworth treated *Zonites* as if it was feminine (or forced the specific name to agree with the subgenus *Retinella*). In a strict application of Art. 50.1.1 the name in such a case must be attributed to the author of the work, Shuttleworth. Name also printed on plate, as *Zonites oppressa*.

4.5-5.5 x 10-12 mm Italy: Sardinia, W Tresnuraghes

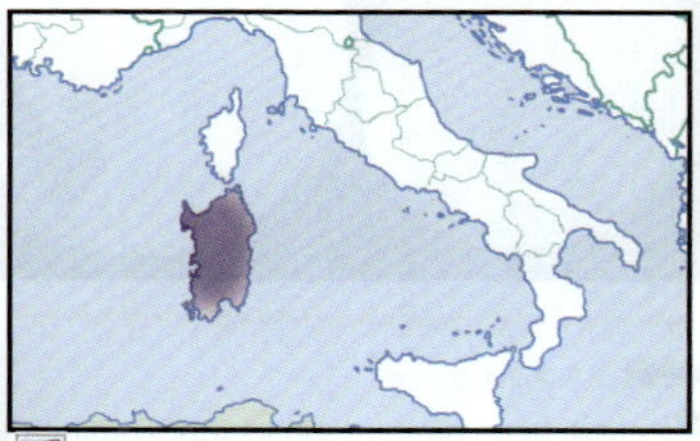

Italy (Sardinia)

Shell light horny yellowish, lower side whitish, finely striated in clusters, very flat with prominent apex, angulated, 5-6 convex whorls with deep suture, last whorl width equalling the two preceding whorls taken together, umbilicus narrow.

Westerlund 1886: 48, Manganelli et al. 1995: 19.

Oxychilus oscari (Kimakowicz, 1883)

Hyalina oscari Kimakowicz, 1883 in Kimakowicz 1883: 19 (*Hyalina (Retinella) Oscari*)

Shell thin, light horny brownish, lower side almost white with greenish hue, shiny, irregularly striated, more distinctly at the suture, 6-6.5 flattened whorls, last whorl width at least 2 x of preceding whorl, umbilicus narrow.

8-9 x 19-23 mm

Westerlund 1886: 68, Grossu 1983: 147.

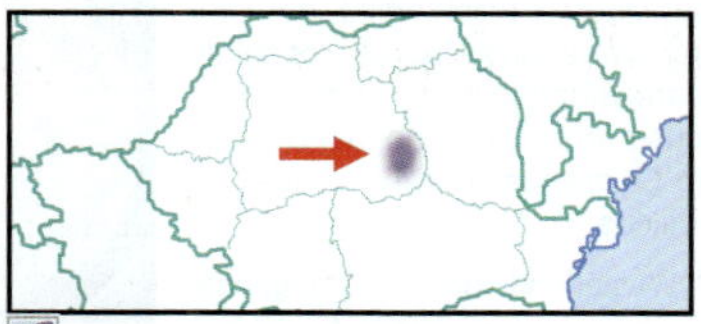

Central Romania

Oxychilus paphlagonicus Riedel, 1993

Oxychilus paphlagonicus Riedel, 1993 in Riedel 1993: 135

Shell flat, yellowish to brownish, upper side not translucent, lower side lighter and more shiny, 6 whorls, last 1.5 whorls faster increasing, last whorl width 2 times of penultimate whorl, whorls not much rounded but with deep suture, aperture wider than high, almost horizontal, umbilicus almost 1/5 of shell diameter, shell surface with distinct reticulate microrelief except on last whorl.

6-7 x 15-18 mm

Under shrubs.

A relatively rare species, but may occasionally occur in high numbers.

Riedel 1995: 50, Schütt 2001: 214.

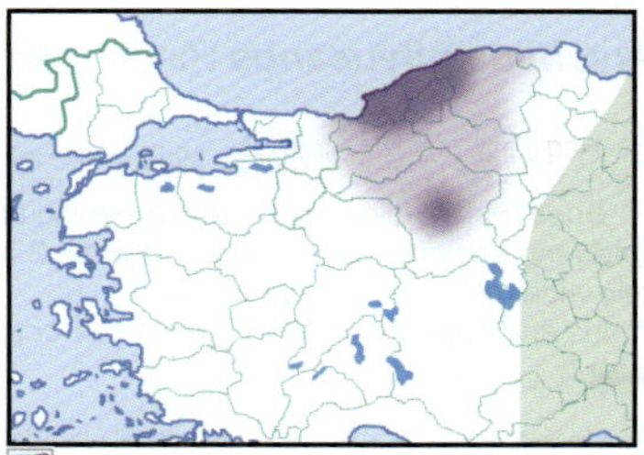

N Turkey

Oxychilus paulucciae (De Stefani, 1883)

Hyalinia paulucciae De Stefani, 1883 in De Stefani 1883-1884: 35 (*Hyalinia Paulucciae*)
Oxychilus lanzai Forcart, 1967 in Forcart 1967: 114

Shell brownish, lower side whitish, very shiny, widely striated, upper side almost flat, last whorl flat at periphery, umbilicus open. Animal greyish, unlike the bluish animals of *O. meridionalis* and *O. draparnaudi*.

In caves.

Westerlund 1886: 63, Manganelli et al. 1995: 19, Manganelli, Cian. et al. 2004.

5.5-6.5 x 14-16 mm

Italy: Toscana, Lucca, Cave of Risvolta
Massimo Prodan

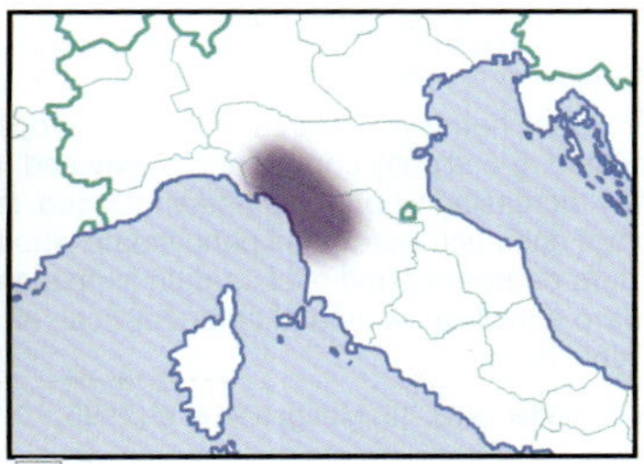

Italy (Alpi Apuane)

Oxychilus perspectivus (Kobelt, 1881)

Without gender agreement: *Oxychilus perspectiva* (Kobelt, 1881)

Hyalinia perspectiva Kobelt, 1881 in Kobelt 1881: 179

Shell greenish white, shiny, depressed, convex, finely striated, every whorl 1.5 x broader than preceding whorl.

Westerlund 1886: 58, Giusti 1973: 159, Manganelli et al. 1995: 19.

2.7-3.2 x 5.5-6.5 mm

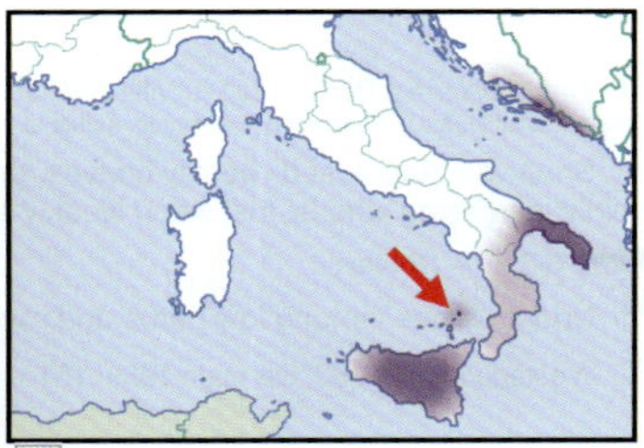

Croatia (Dalmatia), S Italy

Italy: Otranto, Colle della Minerva

Oxychilus pieperi Riedel, 1973

Oxychilus pieperi Riedel, 1973 in Riedel 1973 (Fragm. Faunist. 19): 22

1.5-2 x 4-5 mm

Greece: island of Gávdos S of Crete

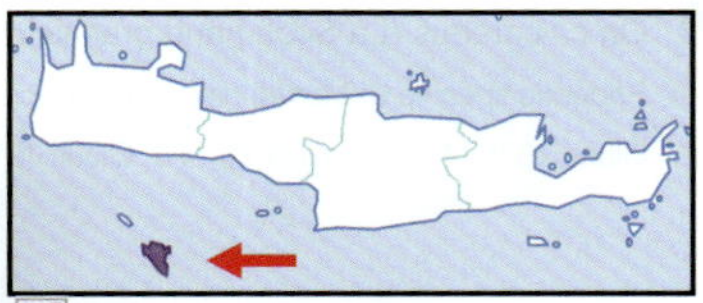

Greece (island of Gávdos)

Shell amber yellow, weakly striated, 3.5-4 very weakly convex whorls, umbilicus very narrow. Juveniles with up to 2 whorls have a glassy and colourless shell, the yellow colour is added with the 3rd whorl.
Animal light with greyish tentacles, the shell of living animals is brightly orange.

Lives in all parts of the island, confined to limestone basements, frequent under *Pinus brutia* in needle layers in open forests, under higher trees also under rotting tree trunks. Rarely in rock crevices and in the rocky habitats in the south of the island, not very frequent on sandy substrate, absent in non-calcareous units.

Riedel 1992: 90, Welter-Schultes 1998: 58.

Oxychilus pilula (Westerlund, 1886)

Hyalinia pilula Westerlund, 1886 in Westerlund 1886: 71 (*Hyalinia (Mesomphix (Retinella)) pilula*)
Date Apr 1886, author was Westerlund alone. Also published as *Hyalinia (Retinella) pilula* in Paulucci 1886 [June 1886]: 14.

Shell small, reddish brown, lower side greenish white, not very shiny, irregularly striated, reticulate pattern with fine spiral lines, umbilicus narrow and conical. Differs from *O. draparnaudi* in its wider umbilicus.

Westerlund 1886: 71, Manganelli et al. 1995: 19, Manganelli, Favilli & G. 1999.

4-5 x 8-9 mm

Italy: Toscana, Livorno, Capraia island

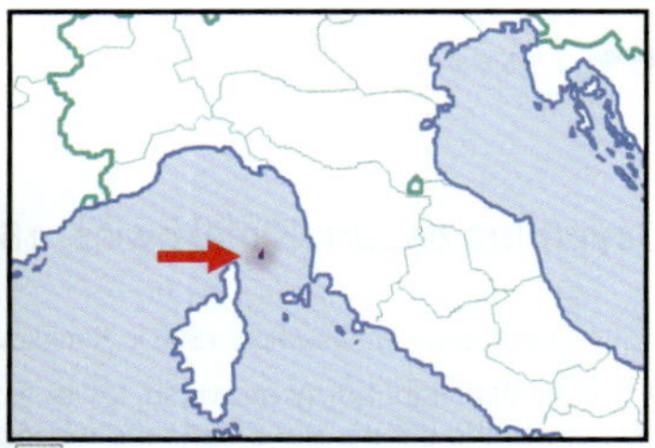

Italy (island of Capraia)

Angelo Vannozzi

Oxychilus pityusanus Riedel, 1969

Oxychilus pityusanus Riedel, 1969 in Riedel 1969: 237

Shell horny yellow, often with greenish hue, slightly translucent, smooth and shiny, 4-4.5 weakly convex and rapidly increasing whorls, last whorl width 2 x of penultimate whorl, aperture not very oblique, umbilicus moderately wide, all whorls visible inside.
Whorls and aperture less high than in *O. cellarius*, umbilicus slightly narrower, suture less deep.

3-4.5 x 6.5-9 mm

Beckmann 2007: 82 (Ibiza and small islands, Formentera).

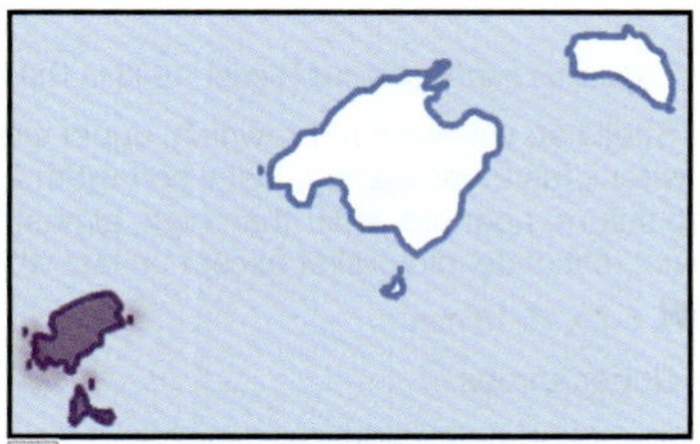

Spain (Ibiza island group)

Oxychilus planellus (Pfeiffer, 1853)

Without gender agreement: *Oxychilus planella* (Pfeiffer, 1853)

Helix planella Pfeiffer, 1853 in L. Pfeiffer 1853 (Z. f. Malak 10): 146

3.5-4 x 7-10 mm

Italy: Sicily, Agrigento (determination uncertain)

Italy (Sicily)

Shell flat, almost disc-like, horny brownish, or dark, distinctly and coarsely striated, weakly shiny, thin, 4-4.5 weakly convex whorls, umbilicus conical, 1/6-1/5 of shell diameter.

Westerlund 1886: 50, Manganelli et al. 1995: 42 (listed as uncertain taxon).

More research is needed. Diagnosis considers uncertainly determined material in Haus der Natur Cismar (see figure).

Oxychilus planorbis (Möllendorff, 1899)

Hyalinia planorbis Möllendorff, 1899 in Möllendorff 1899: 169

Shell horny yellowish, translucent, 4.5 whorls, upper side flatter than lower side, first 3 whorls more slowy increasing than the last 2 whorls, umbilicus narrow.

Riedel 1979 (Fragm. Faunist.), Dhora & Welter-Schultes 1996: 149.

3.6-4 x 9-11 mm

Croatia: Dalmatia, Srebreno near Dubrovnik

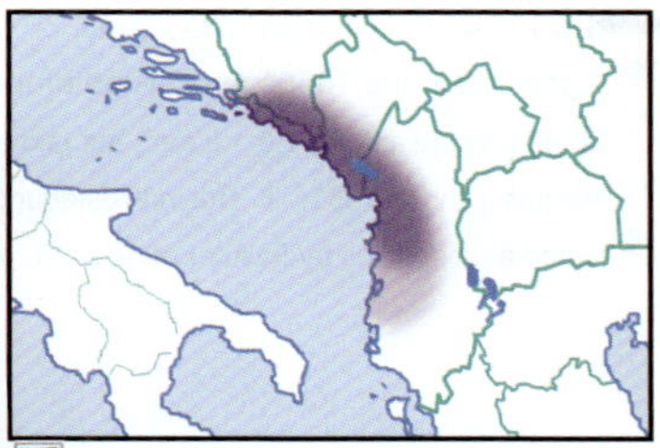

W Herzegovina to Albania

Massimo Prodan

Oxychilus planospiroides Riedel, 1969

Oxychilus planospiroides Riedel, 1969 in Riedel 1969: 127
Expressly established as a new replacement name for
Hyalina planospira Wagner, 1907 in A. J. Wagner 1907: 111 (non Sacco, 1889)

Shell depressed with slightly convex upper side, thin and fragile, finely striated, 4-4.5 weakly convex whorls, last 2 whorls increasing more rapidly than the first whorls, last whorl width less than 2 x of penultimate whorl, umbilicus wide and perspectivical, umbilical suture shallow and not deep as usual in *Oxychilus*.
Shell less depressed than in *O. planorbis*, last whorl less inflated, 0.5 whorls less at same shell diameter, basal apertural margin forms a sharp curve to the columellar margin.

4.2-5 x 11-12 mm

Subterranean, found in caves.

Gittenberger 1976: 278, Riedel 1997 (Genus 8): 764.

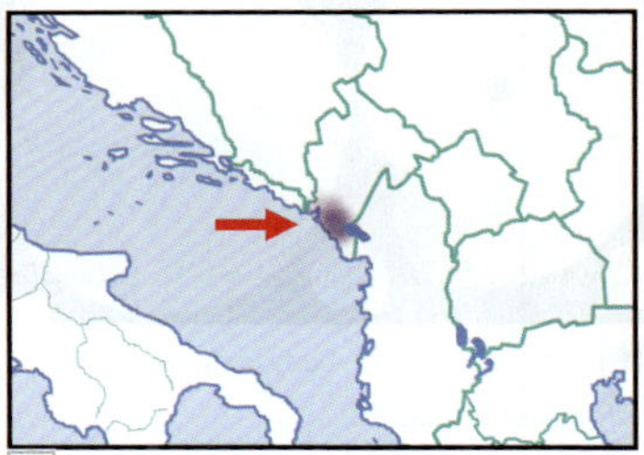

W Montenegro

Oxychilus polygyra (Pollonera, 1885)

Hyalinia polygyra Pollonera, 1885 in Pollonera 1885: 523 (or 681)
Name of Greek origin, not Latinized or a Latin adjective, to remain unchanged under Art. 31.2.3.

Shell horny brownish, shiny, densely striated, 5.5-6 slightly convex whorls with deep suture, slowly increasing, last whorl and aperture not very large, umbilicus moderately narrow, 1/8-1/7 of shell diameter.

4.5-5 x 9-11 mm

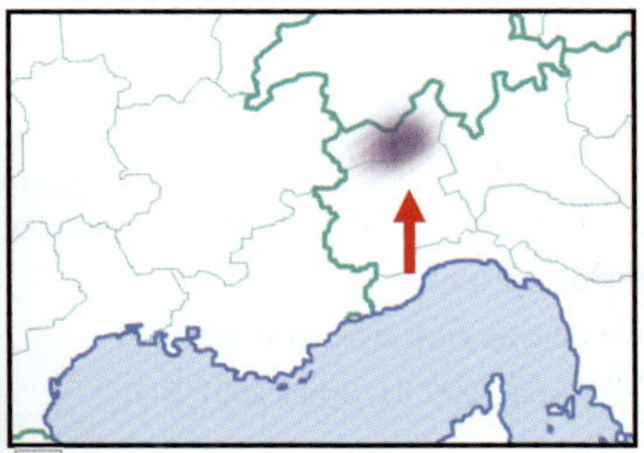

NW Italy

Italy: Biella, Piedicavallo

Oxychilus pygmaeus Riedel, 1983

Oxychilus pygmaeus Riedel, 1983 in Riedel 1983: 280

Shell almost colourless, translucent and very shiny, old shells opaque and dirty cream, very weakly striated, shape like Vitrea, 4.2-4.8 weakly convex and regularly increasing whorls, last whorl width 1.5 x of penultimate whorl, umbilicus narrow, less than 1/10 of shell diameter but all whorls visible inside.
Animal whitish cream, eyes with black pigment, sole subdivided in 3 sections.

1.5-1.7 x 3.2-3.7 mm

Probably subterranean, found at a S exposed hill slope providing open oak forest habitats, snails were found in rock rubble and 20-50 cm deep inside soils with limestones, associated with *Cecilioides* and *Vitrea contracta*.

Riedel 1992: 91.

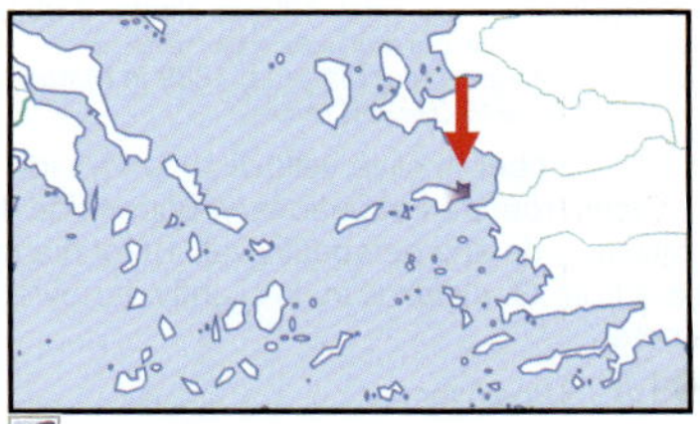

Greece (Sámos)

A strictly endemic species, restricted to the E parts of Sámos.

Oxychilus rateranus (Servain, 1880)

Zonites rateranus Servain, 1880 in Servain 1880: 17

Shell practically not distinguishable from *O. draparnaudi*. Oviduct very long, longer than vagina, flagellum less than 1/5 of penis length.

Rocky limestone habitats in Mediterranean shrublands, in organic soil components deep inside rock crevices and caves.

Riedel 1972: 129, Talaván Gómez & T. 2006: 37.

6-7 x 9-14 mm

S Spain: Sierra Morena, near Aracena

S Spain, possibly also NW Africa

Oxychilus requienii (Moquin-Tandon, 1855)

Zonites lucidus var. *requienii* Moquin-Tandon, 1855 in Moquin-Tandon 1855: 76 (*Zonites lucidus* var. *Requienii*)

Shell transparent, very finely and irregularly striated, 6-7 whorls, shell flat, upper side slightly convex, keeled, aperture simple, umbilicus moderately small.

6-12 x 12-20 mm, animal 30 x 3 mm

Observed in the 1850s as a synanthrope, in holes of old walls and under stones.

30-40 eggs (diameter 1.5-1.8 mm) are laid within 15 days.

Moquin-Tandon 1855: 76. Not in Westerlund 1886 and Germain 1930.

Listed as a species by Falkner et al. 2001: 48. The *Oxychilus* species of Corsica need a revision.

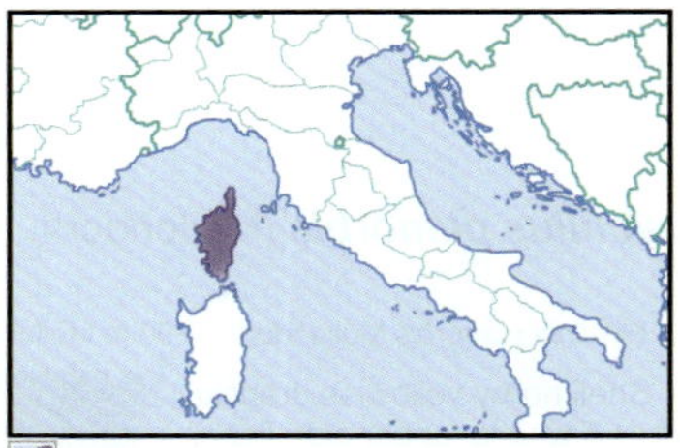

France (Corsica)

Oxychilus samius (Martens, 1889)

Without gender agreement: *Oxychilus samia* (Martens, 1889)

Hyalina nitidissima var. *samia* Martens, 1889 in Martens 1889: 189, Taf. 10 (*Hyalina nitidissima* var. *Samia*)

8-11 x 18-28 mm, body up to 50 mm

Turkey: Davutlar near Kuşadası

Frequent in caves.

Riedel 1958, Falkner 1990: 182, Riedel 1992: 94, Riedel 1995: 59.

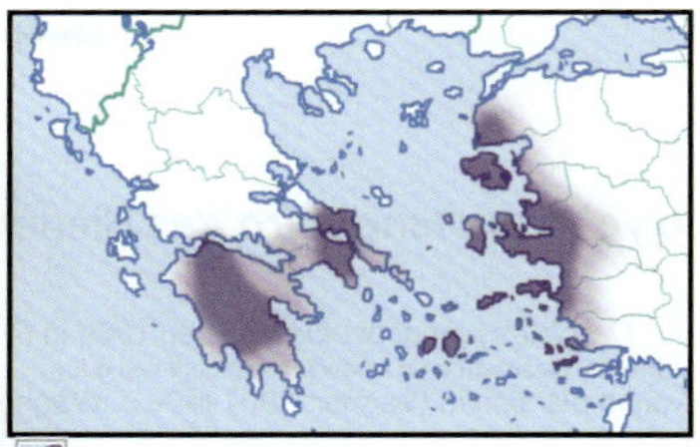

Greece, W Turkey

Shell almost flat, yellowish, very shiny, striated, 4.5-5.5 slightly convex whorls with impressed suture, regularly increasing, last whorl rounded, umbilicus narrow.

Foot at posterior end very narrow and pointed.

Oxychilus samsunensis (Retowski, 1889)

Hyalinia samsunensis Retowski, 1889 in Retowski 1889: 231

Shell yellowish white, translucent, finely striated, shiny, 3.5-4 flat whorls with weakly impressed suture, last whorl width 3 x of penultimate whorl, aperture very oblique with callus at parietal side, upper side of aperture rounded and protruded, nearly not curved at columellar side, umbilicus completely closed. Anatomy unknown.

1.8-2.0 x 3.8-4.2 mm

Only very few shells are known.

Riedel 1995: 54, Schütt 2001: 217.

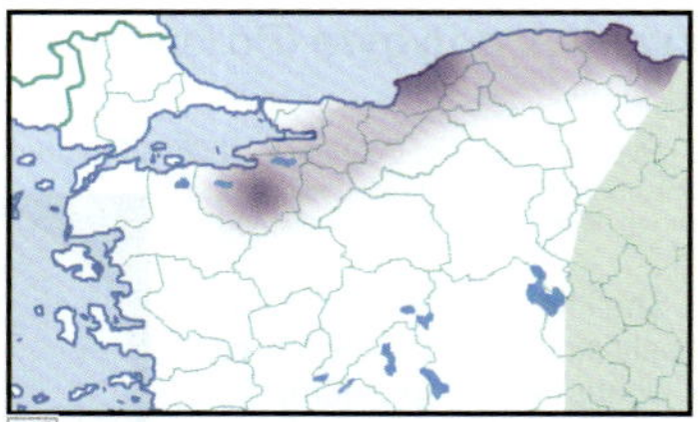

N Turkey

Oxychilus seidli Riedel, 1999

Oxychilus seidli Riedel, 1999 in Riedel 1999: 221
Expressly established as a new replacement name for
Hyalina cypria var. *major* Martens, 1889 in Martens 1889: 189 (*Hyalina Cypria* var. *major*)
Preoccupied by various previous names.

Shell amber to olive yellowish, lower side lighter, shiny, almost smooth, only rarely with a spiral sculpture, 5-5.5 whorls, rounded at periphery, upper side flattened, lower side also relatively flat, aperture almost horizontal, umbilicus wide and perspectivival, 1/6 of diameter.
Differs from *O. syriacus* in its slightly narrower umbilicus and in having 0.5-0.75 whorls less at the same shell diameter.

8-9 x 20-25 mm

Under limestone rocks in shrubland. Predated by small mammals.

Riedel 1999: 221.

Greece (Ikaría and Sámos)

Oxychilus serbicus Riedel, 1969

Oxychilus serbicus Riedel, 1969 in Riedel 1969: 116

Shell horny yellow, smooth, thin, translucent and shiny, lower side lighter and opaque, outline of upper side flat, 4.5-5 rapidly increasing whorls with shallow suture, last whorl width 3 x of penultimate whorl, inflated near aperture, umbilicus extremely narrow.
Shell hard to tell from *O. montivagus*, shell slightly flatter at upper side, last whorl more inflated, upper apertural margin more horizontal, in *O. montivagus* more oblique. Umbilicus width variable in both species.

4.5-6 x 10-15 mm

Between limestone rock rubble and under leaves.

Pavlović 1912: 21 (W Serbia, *montivaga*), Tomić 1959: 9 (*Hyalina nitidissima*), Riedel 1997: 756.

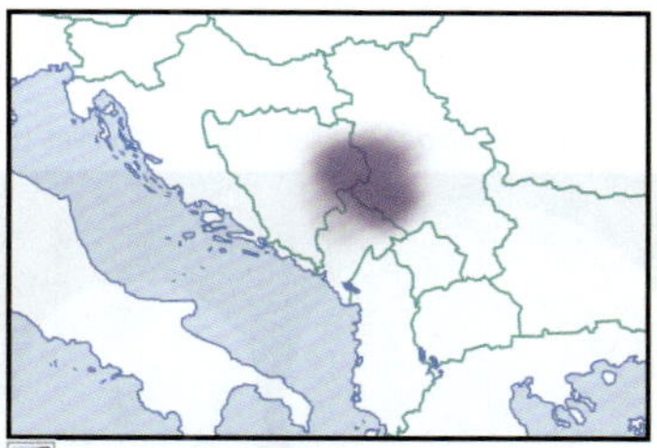

W Serbia, Bosnia

Oxychilus shuttleworthianus (Pini, 1884)

Without gender agreement: *Oxychilus shuttleworthiana* (Pini, 1884)

Hyalinia shuttleworthiana Pini, 1884 in Pini 1884: 399

Shell horny olive or brownish, lower side lighter, very finely striated, 5-6 weakly convex whorls with moderate suture, upper apertural margin protruded, umbilicus moderate.
Like *O. obscuratus*, more distinctly striated on upper side, last whorl narrower.

5.5-6 x 16-18 mm

Westerlund 1886: 60. Not in Germain 1930.

Listed as a species by Falkner et al. 2001: 48. The *Oxychilus* species of Corsica need a revision.

France (Corsica)

Oxychilus spratti (Westerlund, 1892)

Hyalinia spratti Westerlund, 1892 in Westerlund 1892: 44

9-12 x 16-24 mm

Greece: W Crete, near Alikambos

In rock crevices and at the base of rocky walls, often in gorges near steep rocky slopes, in lowlands and mountains up to 2000 m.

Riedel 1968: 500, Riedel 1992: 94.

Greece (W to central Crete)

Shell light horny yellowish, sometimes with greenish hue, lower side lighter, irregularly striated, shiny, outline very flattened but periphery rounded, 4.5-5 slightly convex whorls, last whorl width 2.5-3 x of penultimate whorl (occasionally 2 x), 1/3 of shell diameter (seen from above), aperture not very obblique, umbilicus not very narrow, 1/10-1/9 of shell diameter, widened at last whorl, deep and with all whorls visible inside.

Oxychilus superfluus (Pfeiffer, 1849)

Without gender agreement: *Oxychilus superflua* (Pfeiffer, 1849)

Helix superflua Pfeiffer, 1849 in L. Pfeiffer 1849: 113

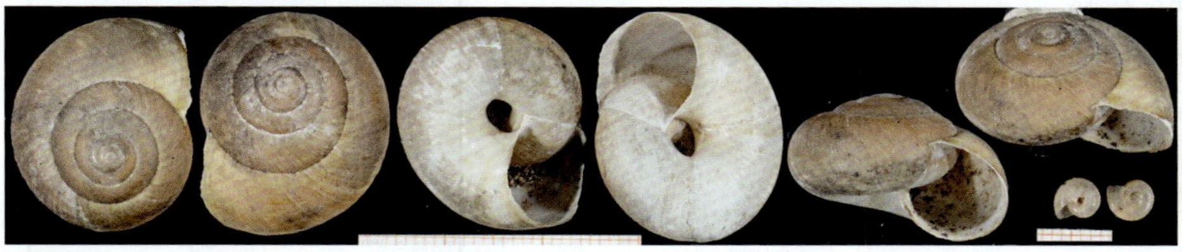

7-14 x 15-30 mm (diameter 10 mm at 4.5 whorls)

Greece: S central Crete, near Agii Déka
(juv. 7-8) Greece: central Crete, Episkopí near Iráklio

Relatively shady habitats under shrubs, among rocks and in rock crevices.

Westerlund 1886: 69, Riedel 1992: 78.

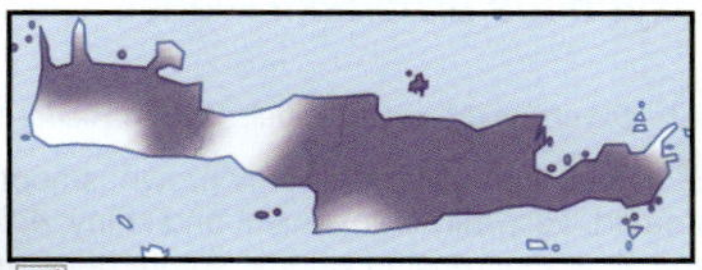

Greece (Crete)

Shell horny brown, sometimes with reddish hue, lower side greenish yellow or whitish, not very shiny, not translucent, densely striated and with a reticulate pattern, 5.5-6.5 moderately convex whorls, rounded at periphery, umbilicus open and perspectivical, 1/4 of diameter.
Differs from *Eopolita protensa* in its regularly rounded and deeper umbilicus, *Eopolita* has a wider last whorl.

Oxychilus tomlini (Smith, 1905)

Vitrea tomlini Smith, 1905 in Smith 1905: 314 (*Vitrea (Polita) Tomlini*)

Shell much depressed, yellowish brown on upper side, paler on lower side, weakly shiny, densely striated and with microscopical spiral structure, 4.5-5 weakly convex whorls, umbilicus wide and perspectivical.

4.5-5.5 x 13-15 mm

Riedel 1992: 77.

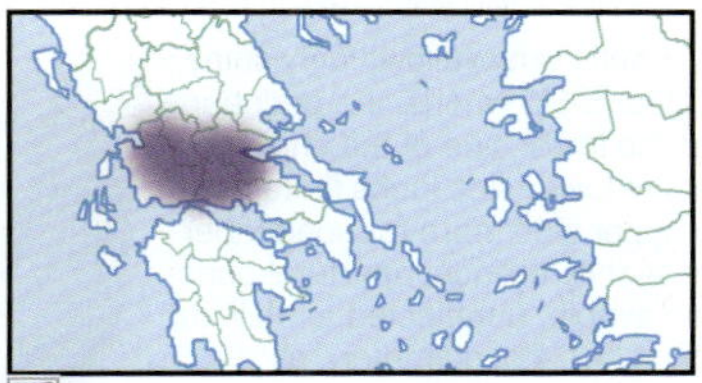

Greece (Stereá Elláda)

Oxychilus translucidus (Mortillet, 1854)

Without gender agreement: *Oxychilus translucida* (Mortillet, 1854)

Helix translucida Mortillet, 1854 in Mortillet 1854: 100

3.5 x 7-8 mm

S Russia: Pyatigorsk

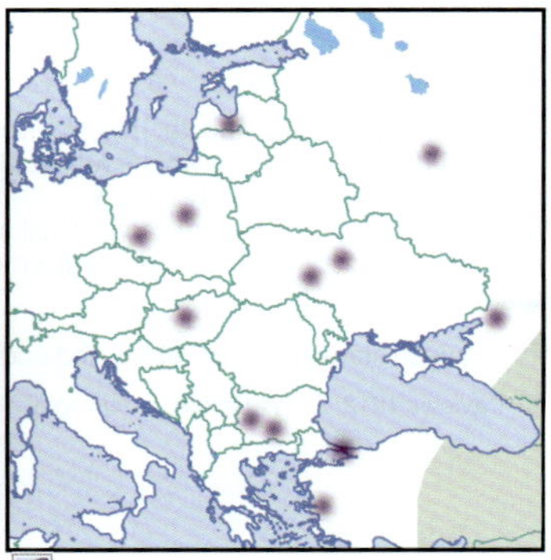

Originally probably E Turkey/N Iran. Introduced to various places in Europe, also to Israel

Shell yellowish, lower side lighter, often with green hue, translucent, smooth, shiny, no spiral microsculpture, umbilicus narrow (1/10 of shell diameter), but initial whorls visible inside.

Synanthrope snail, occurs in parks, botanical gardens and greenhouses, humid habitats in lowlands under shrubs and stones.

Damjanov & Likharev 1975: 278, Riedel 1995: 41, Schütt 2001: 206, Balashov & Gural-Sverlova 2012: 99.

Oxychilus tropidophorus (Mabille, 1869)

Zonites tropidophorus Mabille, 1869 in Mabille 1867-1869: 65
Oxychilus edmundi Falkner, 2008 in Falkner 2008: 136

6-9 x 17-22 mm

France: Corsica, between Bastia and Corte

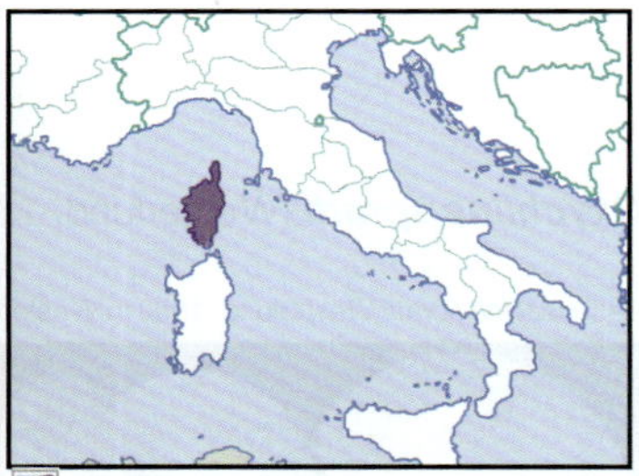

France (Corsica)

Shell light horny brown, adults with reddish hue, striated, shiny, almost flat, 5-5.5 weakly convex whorls with deep suture, first whorls slowly increasing, then rapidly, last whorl bluntly angulated, umbilicus moderate.

Humid and shady habitats, under stones and dead leaves, up to 1500 m. Most frequent *Oxychilus* species of Corsica.

Two snails were live collected in mid-October (Falkner 2008, *edmundi*), one snail laid 18 "large" eggs at dispersed sites in captivity, from which no offspring hatched. Shortly afterwards the single two known animals of the putative new species were killed and dissected. Life span unknown.

Westerlund 1886: 61, Germain 1930: 151.

The taxon *edmundi* is regarded as a synonym of *O. tropidophorus* here. The description of this form was based on only two specimens (more were not found because *Oxychilus* was locally extremely rare), claimed to differ mainly by a narrow bottleneck in the penis. The variation of the width of such bottlenecks could not be studied, the claimed differences were not statistically tested for significance, and no constraints were given to assume absence of genetic interchange between both taxa (for example by showing that in other closely related groups bottlenecks in the penis are associated with reproductive isolation in sympatric species).

Oxychilus urbanskii Riedel, 1963

Oxychilus urbanskii Riedel, 1963 in Riedel 1963: 473 (*Oxychilus (Morlina) urbańskii*)

Shell light yellow to horny brown, smooth or weakly striated, very thin, translucent and shiny, 4.5-5 convex and slowly increasing whorls, suture deep and with margin, last whorl width 1.5 x of penultimate whorl, not inflated near aperture, umbilicus very narrow and deep, partly covered by the reflected columellar margin.
Head, dorsum and foot of the animal blue, sides greyish brown, sole subdivided in 3 longitudinal sections.

S Bulgaria: Sakar mountains, Dabavo village

Atanas Irikov

6-6.7 x 9-12 mm

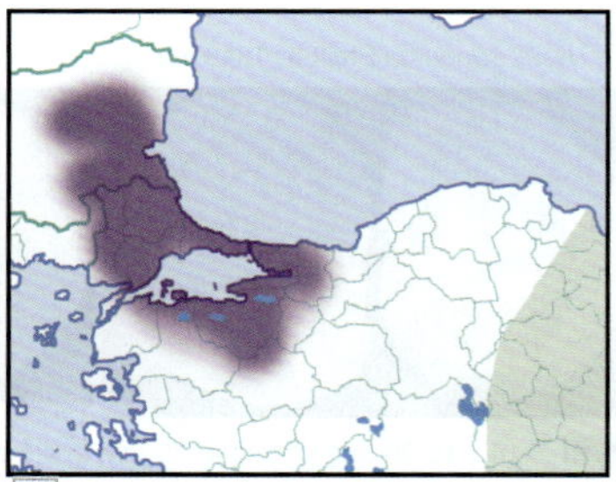

Central Bulgaria to NW Turkey

Was found at slopes in a dark mountain gorge in humid habitats covered with bushes, in rotting leaves and soil crevices, also under rotting wood logs in moist deciduous forests.

Damjanov & Likharev 1975: 287, Schütt 2001: 231, Irikov & Mollov 2006: 800.

Oxychilus uziellii (Issel, 1872)

Zonites uziellii Issel, 1872 in Issel 1872: 60 (*Zonites Uziellii*)

Shell horny yellow, fairly shiny, striated, 5-5.5 whorls, the periphery is at the lower side of the shell, so that the lower side looks flatter and the upper side more rounded, aperture slightly oblique, apertural peristome simple, unreflected, umbilicus wide (1/5-1/4 of diameter), all whorls clearly visible inside.
The animal is blue with conspicuous pits on the upper side of the body, foot pale grey.
Proximal part of penis about 1/3 of distal part of penis.

4-6 x 9.5-12.5 mm

Italy: Pisa, near Chianni

Maurizio Sosso

Ø 1.0-1.4 mm at 1 whorl, 1.5-2.0 mm (2 wh.), 3.2-4.0 mm (3), 4.2-5 mm (3.5), 5.2-6.2 mm (4), 7-9 mm (4.5), 9-11 mm (5), 11-14 mm (5.5 whorls)

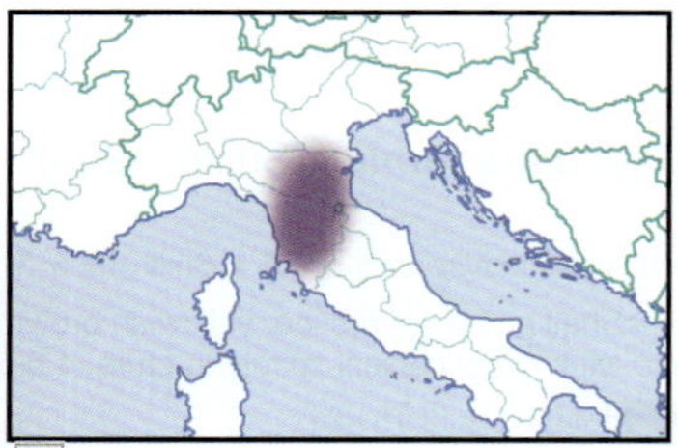

N Italy

Cool and damp habitats in deciduous woods or riverside vegetation, 100-600 m altitude.

Manganelli & Giusti 1993: 70, Manganelli & Giusti 2000.

Oxychilus wiktori Riedel, 1997

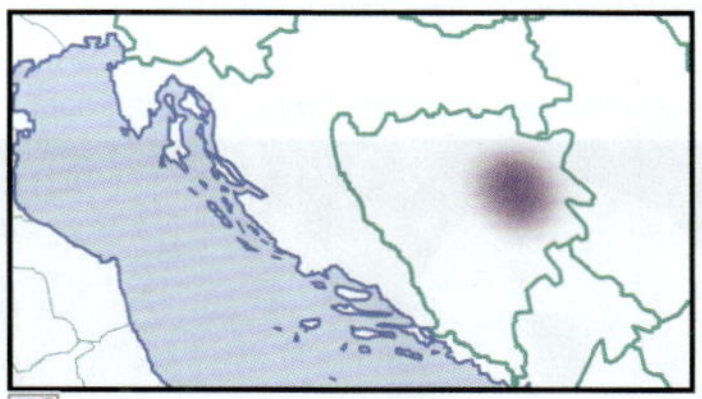

Bosnia

Oxychilus wiktori Riedel, 1997 in Riedel 1997: 757

Shell pale, light yellowish, smooth and very shiny, 4.5-5 very weakly convex whorls with very shallow suture, rapidly increasing, last whorl width 2.7-3 x of penultimate whorl, aperture almost horizontal, umbilicus very narrow, but whorls visible inside, around 1/15 of shell diameter.
Animal light cream, sometimes with light bluish tint on the neck, head and mantle.
Differs from *O. serbicus* in its smaller and lighter shell, slightly more elevated spire and slightly narrower umbilicus. The shell differences do not allow unambiguous separation of the two species.

4.5-5 x 10-12 mm

Eopolita

Eopolita Pollonera, 1916
in Pollonera 1916: 2 *Hyalinia (Eopolita)*

Type species *Zonites aequatus* Mousson, 1854 (monotypy)
(= currently *Eopolita protensa* (Férussac, 1832))

Penis with epiphallus inserting at its distal end, retractor inserting laterally, no terminal flagellum.

E Mediterranean region to Iran

Eopolita protensa (Férussac, 1832)

Helix protensa Férussac, 1832 in A. Férussac & Deshayes 1819-1851: (explication planches) Pl. 82
Original print of explication des planches not verified, second print p. 15.

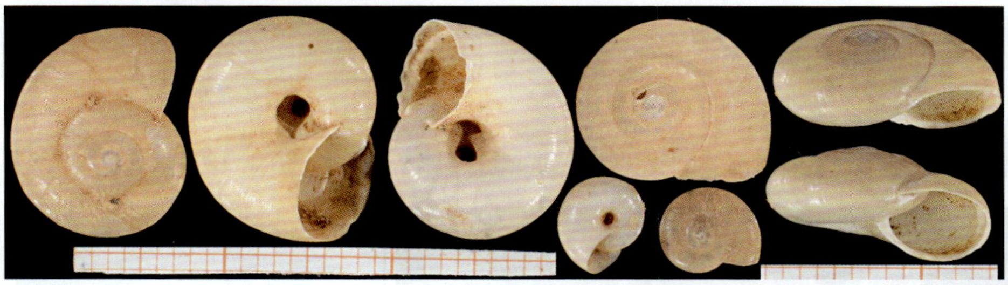

Greece: N Crete, near Gournes

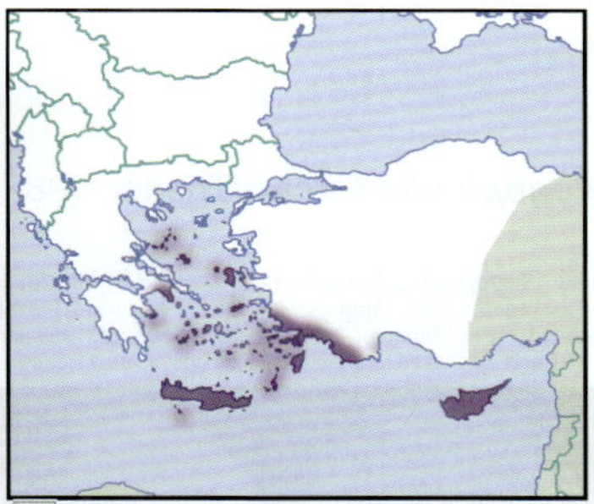

Greece to W Jordan, with gaps

7-12 x x 12-22 mm

Ø 1.0-1.2 mm at 1 whorl, 2.7-3.2 mm (2), 6-7 mm (3), 13-14 mm (4 whorls)

W Turkey: Tuzburgazı near Kuşadası — Aydın Örstan

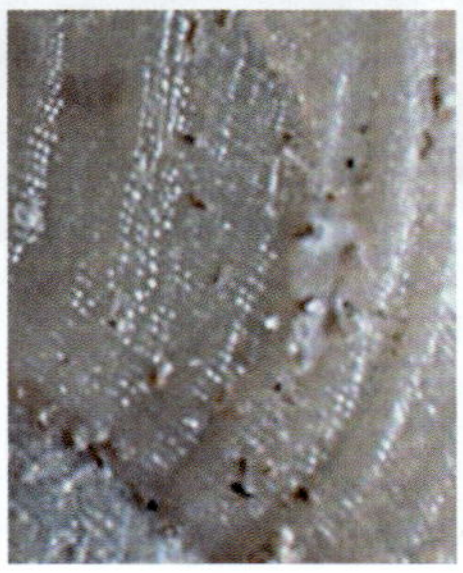

Shell horny brownish or whitish, with 1-3 prominent lines of growth interruptions, and a more or less prominent grid-like microrelief (more prominent in Greece and W Turkey, less in SE Turkey), 4-5 weakly convex whorls, last whorl inflated and predominating, slightly descending near aperture, aperture inside often with prominent tubercles at both sides of the insertions, umbilicus wide, 1/6-1/5 of diameter, all whorls clearly visible inside, highly excentrical with characteristic breaks (difference to *Oxychilus*).
In many regions *E. protensa* displays a very prominent and characteristic grid-like microrelief, so that also fragments of shells can be identified.

Under stones and in rock crevices, often deep in the soil. The species is able to reduce temporarily the diameter of its aperture, an adaptation to resist drought periods. *Eopolita* is carnivorous, feeding on earth worms and snails.

Riedel 1962, Falkner 1990: 180, Riedel 1992: 102, Welter-Schultes 1998: 61, Schütt 2001: 233.

Aegopinella

Aegopinella Lindholm, 1927
in Lindholm 1927: 324 *Aegopina (Aegopinella)*

Type species *Helix pura* Alder, 1830 (original designation)

Intended as a new replacement name for a name *Polita* Held, 1838 (type species *Helix cellaria* Müller, 1774 subsequently designated by Herrmannsen 1847-1849: 313) used in a sense of authors who regarded *Helix pura* Alder, 1830 as its type species. Such a procedure is not covered by the Code (Art. 72.7), Lindholm 1927 established a regular new genus with its type species *Helix pura*.

Shell diameter 4-15 mm, surface striated and with microscopically fine spiral sculpture (difference to *Oxychilus*). Only *Ae. pura*, *ressmanni* and *graziadeii* can reliably be determined conchologically, the others only in regions where no similar species occur (*Ae. nitidula* in northern Europe, *Ae. minor* in Bulgaria).
Penis with epiphallus, penis usually with proximal and distal section, inside with longitudinal folds.

Central and N Europe to Iran

Forcart 1957: 110, Forcart 1959, Riedel 1983: 235.

Aegopinella cisalpina Riedel, 1983

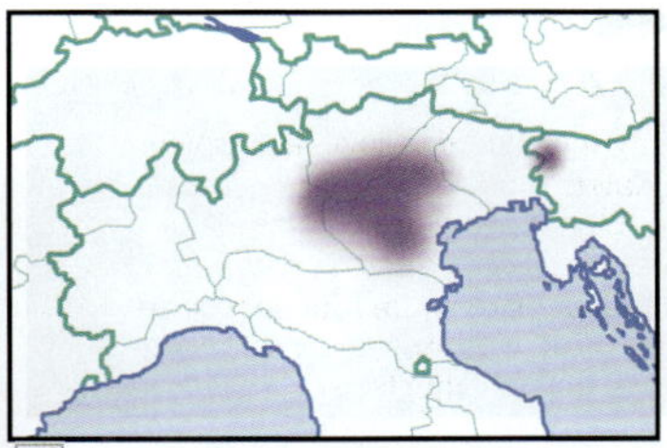

N Italy, W Slovenia

Aegopinella cisalpina Riedel, 1983 in Riedel 1983: 250

Italy: Brescia, Budrio cave — Maurizio Sosso

6-8 x 10-15.5 mm

Shell brownish, striated and with fine spiral lines, but not reticulated, 4-5 whorls, last whorl inflated, width 2 x of penultimate whorl, aperture oblique, umbilicus wide, innermost whorls hardly visible inside.
Shell does not differ substantially from large *A. nitens* and *A. epipedostoma*. The species differ anatomically. *Retinella hiulca* has a slightly narrower penultimate whorl.
Free distal penis section U shaped, long and thick, vas deferens inserts to thinnest distal end of penis, proximal part of oviduct much thicker as distal part.

Riedel 1983: 250, 255, Manganelli et al. 1995: 18, Riedel 1998: 35.

Aegopinella epipedostoma (Fagot, 1879)

Zonites epipedostoma Fagot, 1879 in Fagot 1879: 286

5-7.5 x 10-13 mm

France: Hautes Pyrénées, Lourdes

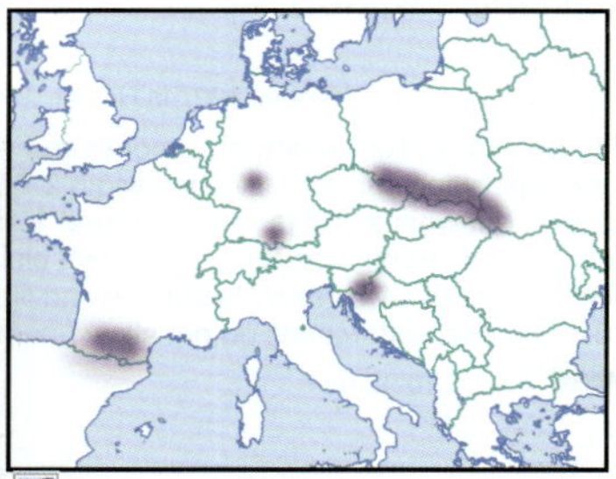

Pyrenees, N Carpathians
Range not well known.

Rare in Poland.

Shell much like *A. nitens*, striated, no or very weak spiral sculpture, 4-4.5 whorls, lower whorls convex, last whorl strongly descending, aperture very oblique, umbilicus wide, all whorls visible inside.
Main difference to *A. nitens* in the aperture.

In Poland in montane forests in and under the litter in humid spoils, cool and humid places near streams.

In Poland eggs are laid deep in moss, juveniles (1.1-1.7 whorls) hatch from July to September, they grow at a rate of 0.5 whorls per month and hibernate at 2.6-3.4 whorls, maturity is reached at 4 whorls in May/June of the next season, some snails hibernate again, life span 2-3 years.

Forcart 1959: 9, 18, Riedel 1983: 243, Kuźnik-Kowalska 2006: 71, Sysoev & Schileyko 2009: 120, Horsák et al. 2010.

Aegopinella forcarti Riedel, 1983

Aegopinella forcarti Riedel, 1983 in Riedel 1983 [Nov 1983]: 239
First mentioned as "*A[egopinella]. forcarti* nov. spec. Riedel mss, Revision in Vorbereitung" by Kerney, Cameron & Jungbluth 1983: 167 [spring 1983], in the synonymy of "*Aegopinella inermis* (sensu Forcart, 1959 non A. J. Wagner, 1907)". The name *A. forcarti* was not used for the taxon by Kerney et al. 1983: 167 and thus failed to meet the conditions of Art. 11.5, so it was not made available at that occasion.

4-5.5 x 8-11 mm

NE Italy: Pordenone

Kerney et al. 183: 167, Manganelli et al. 1995: 18, Štamol & Velkovrh 1995 (Croatia: N Cres).

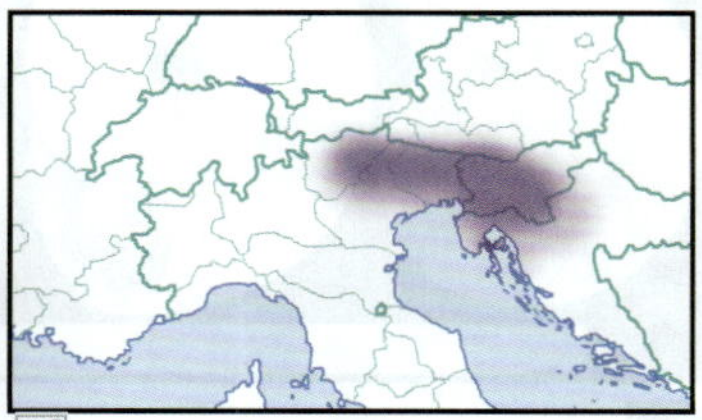

SE Alps to N Croatia

Shell like *A. nitens*, no differences.
Distal penis section at the insertion of proximal penis section (in typical cases) thinner than vas deferens. This is very uncommon. Vas deferens inserts into distal penis section by a perforated inner papilla, can hardly be seen from outside. Vagina short and thick, oviduct long and considerably thinner.

Aegopinella graziadei (Boeckel, 1940)

Retinella graziadei Boeckel, 1940 in Boeckel 1940: 19
Dedicated to Mr. Graziadei, name not spelled *graziadeii* in original source.

4-7 x 8-13 mm

Italy: Lago di Garda, Gargnano

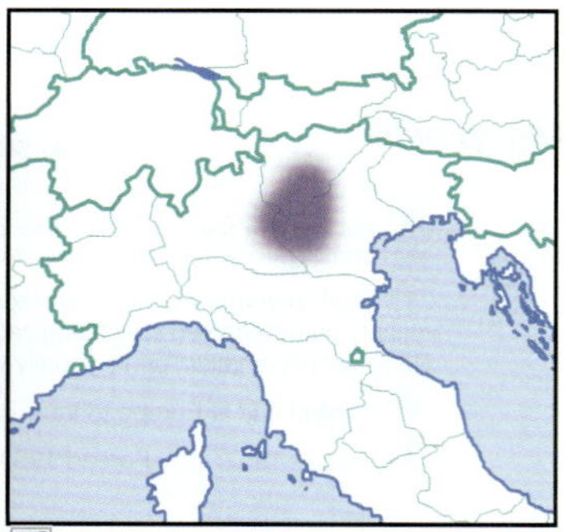

N Italy

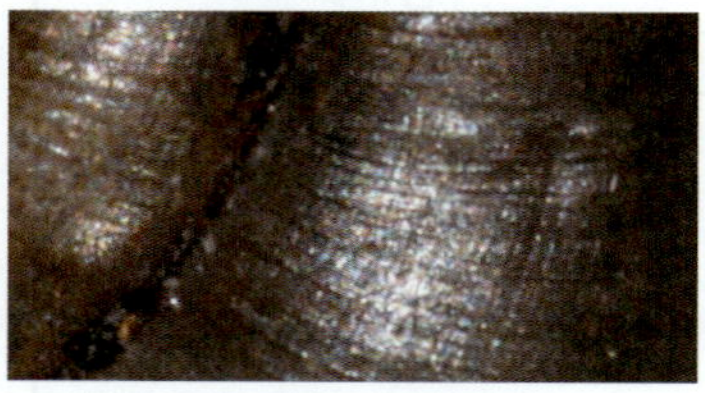

Italy: Verona/Trento, near Brentino Belluno

Shell light horny brownish, translucent, pale shiny, relatively flat, with a very fine but well visible reticulate surface sculpture (difference to *Aeg. cisalpina* and *Retinella hiulca*), 4-5 whorls, last whorl not much wider than penultimate whorl except in its last 1/4 section, descending near aperture, aperture very oblique, umbilicus wide.
Retinella hiulca has a more shiny and more globular shell, the spiral lines are weaker.
In *Aeg. graziadei* the spiral lines do not really form a reticulate pattern (the growth lines are in most views stronger than the spiral lines), but they are slightly more prominent than in *R. hiulca*. Under a certain illumination (like in the insert picture) the surface does appear reticulated.

Forcart 1959: 9, Manganelli et al. 1995: 18.

Aegopinella minor (Stabile, 1864)

Hyalina nitens minor Stabile, 1864 in Stabile 1864: 31 (*Hyalina nitens Minor*)

3.1-3.5 x 6-9 (usually 7-8.5) mm

NE Czech Republic: Olomoucký, Zlaté Hory, Biskupská kupa

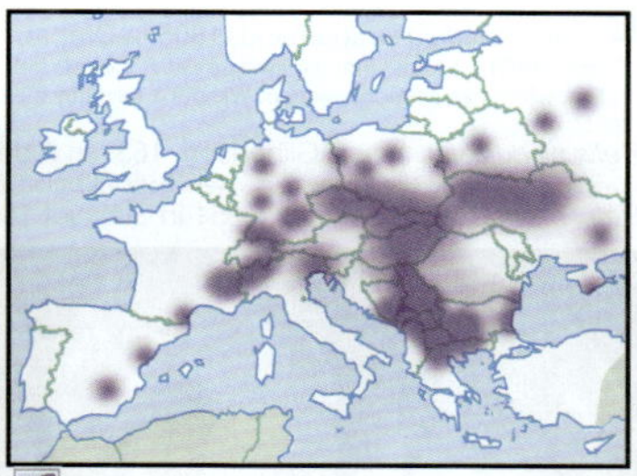

Europe to Caucasus

Endangered in N Switzerland, Niedersachsen and Rheinland-Pfalz, Vulnerable in Germany, probably vulnerable in Austria.

Slovakia: near Banská Bystrica

Jozef Grego

Shell flat, looks somehow like keeled but periphery is rounded, 3.5-4.5 not very convex whorls, last whorl not descending near aperture and width 2 x of penultimate whorl, aperture not very oblique, only little wider than high, umbilicus initially moderately narrow and widened at last whorl, 1/4 of shell diameter. Shell smaller and colour weaker than in *A. nitens*, last whorl less inflated.
Long and thin epiphallus present between shorter and wider penis and the insertion of retractor penis, oviduct very long.

Usually in warm and dry habitats: meadows and shrubs at open mountain slopes, goat pastures, stony and grassy river terraces, karstic limestone rocks and walls, not frequent in forests. In Switzerland up to 2400 m, Bulgaria 1200 m.

Range insufficiently known, published records are certainly distorted by misidentifications. Absence of published records for Austria and Romania suggest that the identity of this form is not yet equally well understood or interpreted in all countries. Serbian records in Tomić 1959 for *A. nitens* seem to refer to this species (Riedel 1983: 238).

Aegopinella nitens (Michaud, 1831)

Helix nitens Michaud, 1831 in Michaud 1831: 44, Pl. 15

4-5.5 x 8-11 mm, in the Alps up to 14 mm

Germany: S Bayern, near Mittenwald

Central Europe

Ø 0.8-0.9 mm at 1 whorl, 1.7-1.8 mm (2), 3.5-3.9 mm (3), 6.3-7 mm (4), 9.2-10 mm (4.5 whorls)

Slovakia: Vel'á Fatra mountains

Jozef Grego

Shell light horny brown, striated and with weak reticular pattern in fresh shells, apex not convex, 4-4.5 weakly convex whorls, last whorl increasing near aperture and flattened, not descending, width about 3 x of penultimate whorl, aperture not very oblique, umbilicus wide.
Animal blue or dark blue, getting lighter down to the sides, upper tentacles can be blackish blue, sole milky white.
Usually slightly larger than *Aeg. nitidula*. Less whorls and last whorl wider than in *Retinella hiulca*.
Anatomy: penis thick and broad, subdivided in two parts, upper end abruptely thinner at insertion of retractor penis.

Under leaf litter and stones in humid forests, in valleys and at slopes, predominantly on calcaeous substrate, particularly in lower and montane altitudes. In Switzerland up to 2300 m, rarely above 1500 m.

30-50 eggs (diameter 1.0-1.6 mm) are laid between June and August (Switzerland: November), in clutches of 6-9 eggs under dead leaves, juveniles hatch after 2 weeks, maturity is reached relatively late, in the second year.

Gassies 1849: 108, Moquin-Tandon 1855: 84, Germain 1930: 155, Frömming 1954: 85, Forcart 1959: 9, Prince 1967, Kerney et al. 1983: 165, Riedel 1983: 237 (in Italy only in Alps), Falkner 1990: 178.

Previous records for Croatia referred to other species, those in Tomić 1959 for Serbia seem to refer to *A. minor*.

Aegopinella nitidula (Draparnaud, 1805)

Helix nitidula Draparnaud, 1805 in Draparnaud 1805: 117

Forcart 1959: 21 found that the syntypes of *H. nitidula* belonged to both *A. nitens* and *Oxychilus draparnaudi*, and proposed to ask the Commission for a ruling, which has never happened. Without a ruling by the Commission the name cannot be used for this species. The next available names seem to be these ones:

Helix nitidula var. *helmii* Gilbertson, 1837 in Alder 1837: 107

Description ("white shell") and name were both attributed to Gilbertson.

Helix margaritacea Schmidt, 1851 in Schmidt 1851: 183

Germany: Schleswig-Holstein, Westensee

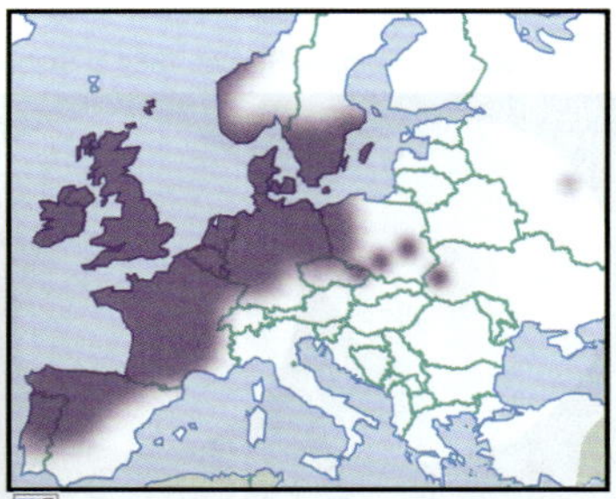

Europe, Azores

4-6 x 6-11 mm, diameter usually 6-8 mm

1.0-1.1 mm at 1 whorl, 2.0-2.3 mm at 2 whorls, 4.5-5 mm at 3 whorls, 7-7.5 mm at 4 whorls

Germany: Bayern, near Walchensee

Under ground litter in a wide variety of sheltered habitats, coniferous and deciduous forests, also in Spain, bases of rocks and walls, moist grasslands, gardens and at roadsides. Tolerates human disturbance. Feeds on small and slow worms, snails and slugs (smaller than *Aegopinella*), also on dead insects and slugs.

Shell amber reddish, rarely withish, milky white near umbilicus, finely striated with fine spiral lines producing a fine reticular pattern (less prominent than in *A. pura*), apex strongly convex (contrast to *A. nitens*), 3.5-4.5 convex and regularly increasing whorls, last whorl not inflated near aperture and not descending, aperture slightly oblique, umbilicus wide. The animal is bluish grey with lighter sole and bluish black upper tentacles. Shell usually smaller than *A. nitens*. Juvenile shells can be distinguished from *A. pura* in their less densely coiled whorls (2.0-2.3 mm at 2 whorls in *A. nitidula*, 1.4-1.7 mm in *A. pura*).

6-9 eggs (diameter 1.1-1.4 mm) are laid at a time in clutches, inside moss.

Frömming 1954: 83, Forcart 1959: 21, Kerney et al. 1983: 166, Riedel 1983: 245, Kerney 1999: 142, Talaván Gómez & T. 2006: 115.

Aegopinella pura (Alder, 1830)

Helix pura Alder, 1830 in Alder 1830: 37

Austria: Tirol, near Landeck

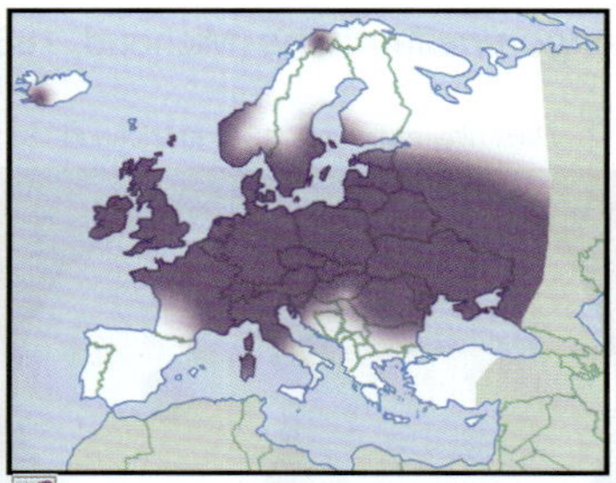

Iceland and Europe to N Iran

2.0-2.7 x 3.5-5.0 mm

0.7-0.8 mm at 1 whorl, 1.6-1.9 mm (2), 3.0-3.3 mm (3 whorls)

Slovakia: Low Tatra National Park

Jozef Grego

Shell colourless or weakly brown, weakly shiny, microsculpture (visible under microscope, 35-40 x) with faint spiral lines crossing irregular radial lines (approximately 1:1), 3.5-4 convex whorls, last whorl increasing and not descending near aperture, umbilicus wide, slightly excentrical. Can easily be identified conchologically.

In litter layers, at the base of vegetation and under dead wood in relatively dry to moderately humid forests, predominantly on calcareous substrate. In Switzerland up to 2300 m, in Bulgaria 2400 m.

Reproduction in spring, adults die after having laid eggs, juveniles reach maturity after 2 years.

Kerney et al. 1983: 164, Riedel 1983: 236, Schwer 1994.

Aegopinella ressmanni (Westerlund, 1883)

Hyalinia nitens var. *ressmanni* Westerlund, 1883 in Westerlund 1883: 56 (*H. nitens* var. *Ressmanni*)

Austria: Kärnten, Gutnitzschlucht

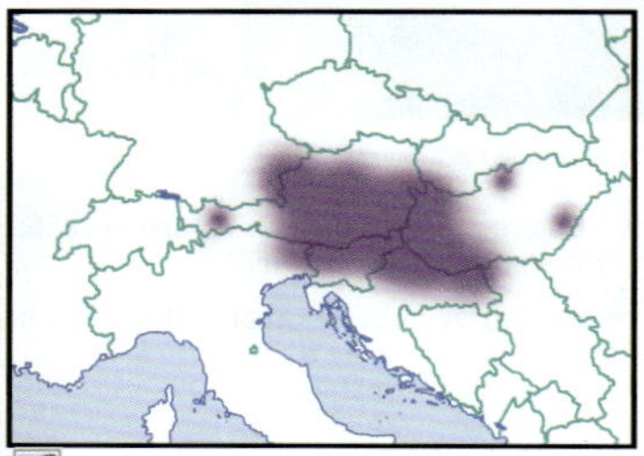

E Alps to Croatia

6-8 x 10-15 mm

Humid forests under leaf litter, in mountains, in Austria 250-1700 m.

Westerlund 1886: 64, Forcart 1959: 9, Stworzewicz 1975, Kerney at al. 1983: 167, Riedel 1983: 236, Manganelli et al. 1995: 18.

Shell larger than other *Aegopinella* species, with relatively high spire, characteristic reticulate surface sculpture consisting of irregular growth lines crossed by spiral lines like in *Aeg. pura*, 4-4.5 convex whorls, last whorl inflated and descending, aperture large and very oblique, umbilicus narrow. First whorls larger than in *A. pura*.

Retinella

Retinella Shuttleworth, 1877 in Shuttleworth 1877: 5, Pl. 2 *Zonites (Retinella)*

Type species usually given as *Helix olivetorum* Gmelin, 1791 (remains to be fixed by the Commission)

[Original source (Leipzig edition) not verified, only the Bern edition from 1878 which I assume had identical contents.]
If the Bern edition had the same content as the Leipzig edition (which appreared earlier), we have the following situation:
The subgenus *Retinella* was established without description but with 4 species included: *Helix fuscosa* Rossmässler, 1838 (= currently *Oxychilus fuscosus*), *Helix balmei* Potiez & Michaud, 1838 (= currently *Pleurodiscus balmei*), *Zonites oppressa* Shuttleworth, 1877 (= currently *Oxychilus oppressus*) and *Zonites duboisi* Mousson, 1863 (from Georgia).
Author of *Retinella* must be Shuttleworth to whom the name was attributed on p. 5, and who was entirely responsible for all information contained on Pl. 2, where the name and those of the 4 species were printed. Fischer (the author of the text) did not give a description and only repeated Shuttleworth's composition of 4 species. The author's name Shuttleworth was indicated on the title page of the work.
On p. 5 Fischer wrote a short text "Shuttleworth en donnant le nom de *Retinella* à un groupe de *Zonites* dont le type est probablement l'*Helix olivetorum* Gmelin, a voulu séparer...", which did not indicate that *olivetorum* was contained at 100 % probability without doubts in this subgenus (Art. 67.2.5). C. R. Boettger 1930: 580 argued that on p. 5 (of the Leipzig edition) Fischer wrote expressly that Shuttleworth gave the name *Retinella* to the group of *Helix olivetorum*. In the Bern edition Fischer's statement on p. 5 implied a speculation and did not indicate total absence of doubts.
In Opinion 335 (1955) the Commission regarded as type *Helix olivetorum*, subsequently designated by Kobelt [1879-1881]: 223. This was certainly not tenable, (1) Kobelt used two alternate specific names (*olivetorum* and *leopoldina*), (2) as "Typus" for two alternate generic names (*Retinella* and *Aegopina*), (3) Kobelt did not use the term "Typus" in the sense of a unique name-bearing type (example p. 206 where 4 specific taxa were regarded as "Typus" for *Daudebardia*), (4) *Helix olivetorum* was not among the originally contained nominal species, see above.
It would be the best solution to regard the type designation as under the plenary powers in Opinion 335.
The next available name seems to be this one:

Lyrodiscus Pilsbry, 1893 in Pilsbry 1894: 48 *Pyramidula (Pyramidula (Lyrodiscus))*, type species *Helix circumsessa* Shuttleworth, 1852 (from Canary islands) (original designation)

Central and W Europe, Canary Islands

Retinella giustii Riedel, 1998

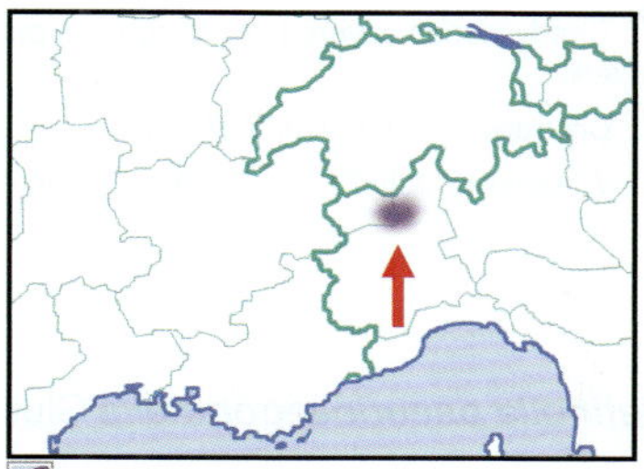

NW Italy

Retinella giustii Riedel, 1998 in Riedel 1998: 6

Shell horny brown or pale yellow, lower side lighter, densely striated, no reticulate micropattern, apex usually eroded, with 4-4.5 moderately convex whorls with deep suture, upper side rounded, almost globulous, lower side convex, last whorl rounded at periphery, width about 2 x of penultimate whorl, but not inflated, umbilicus wide, deep and perspectivical, 1/6-1/4 of diameter.
Animal dark blue, sole lighter.

4.3-4.5 x 7.5-8.5 mm

At slopes and in valleys in deciduous forests, under rock rubble and leaves.

Retinella hiulca (Albers, 1850)

Helix hiulca Albers, 1850 in Albers 1850: p. 68 [footnote 4]
Name *Helix hiulca* was previously mentioned by Rossmässler 1838-1844 (8) [1838]: 35 as a synonym of *Helix nitens* Michaud, 1831 (currently *Aegopinella nitens*). Albers gave a description and a bibliographical reference to Rossmässler 1838 and argued that Rossmässler's classification as a synonym of *Helix nitens* did not agree with his (Albers') specimens.
For those who demand as a requirement for Art. 11.6.1 that both names must have referred to the same taxon without doubts, the name must be attributed to Albers 1850 and can be used here. For the others, to Rossmässler 1838, the types will be those mentioned by Rossmässler, with the effect that the name becomes a junior subjective synonym of *Aegopinella nitens* and cannot be used here.

N Italy: Trento, Lago di Garda, near Torbole

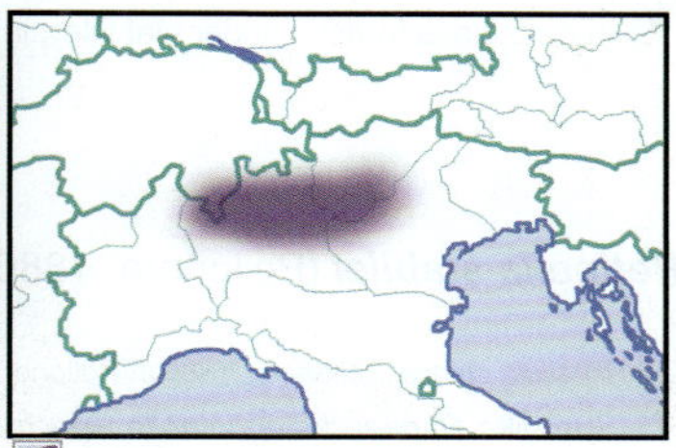

S Alps

6-8 x 10-14 mm

Shell dark horny brownish or with greenish hue, lower side whitish, weakly shiny, moderately solid, indistinctly striated, with very fine spiral lines, 4.5-5 weakly convex whorls, penultimate whorl slightly narrower than in *Aegopinella cisalpina*, last whorl inflated, aperture oblique and not very oval, umbilicus wide, all whorls visible inside.
Aegopinella graziadei has a less shiny shell with slightly better visible spiral lines, its shell is less globular.

In the leaf litter layer of forests, and between rocks in vineyards on calcareous substrate. Also at creek margin vegetation under stones. Not above 1100 m.

Eggs are deposited in May at rotting plants at the soil.

Westerlund 1886: 65, Boeckel 1940: 12 (shell), Forcart 1960: 6, Kerney et al. 1983: 167, Manganelli et al. 1995: 19, Turner et al. 1998: 268.

Retinella incerta (Draparnaud, 1805)

Helix incerta Draparnaud, 1805 in Draparnaud 1805: 109

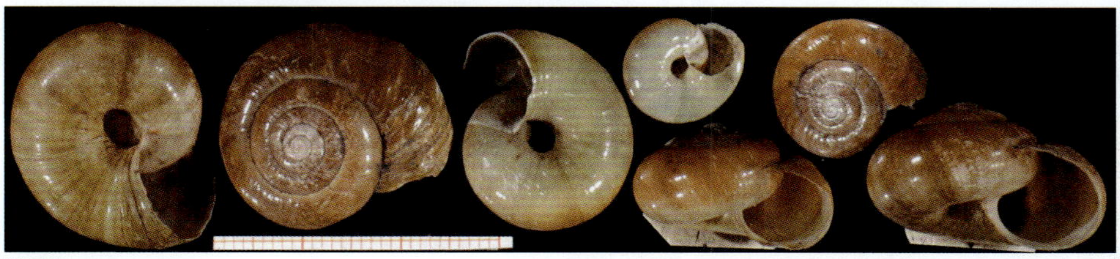

SW France: Pyrénées-Atlantiques, near Bayonne

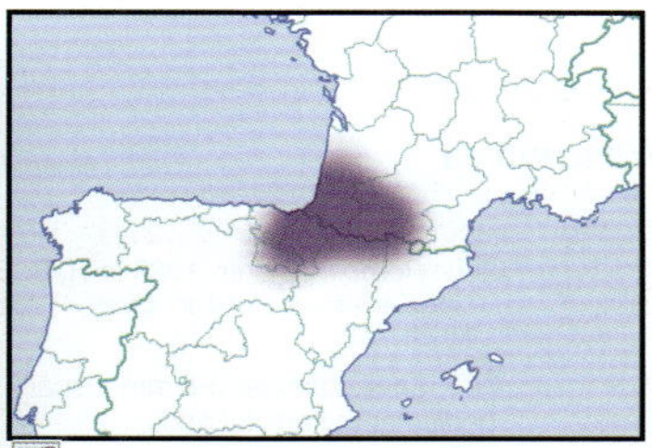

N Spain, SW France

10-16 x 17-27 mm

Shell horny reddish or reddish brown, not very shiny, lower side greenish white and more shiny, thin but solid, 5.5-6.5 slightly convex whorls, last whorl broad but not inflated, rounded at periphery, umbilicus open and much narrower than in *R. olivetorum*.
Animal dark bluish or brownish with blackish tubercles, slightly lighter down to the sides, tentacles greyish black.

Humid habitats in forests under leaves or in holes in the ground near stones.

Copulation was observed in April and in December. 30-40 nearly spherical eggs (diameter 3.5 x 2.5 mm) are laid, juveniles hatch after 20-25 days, maturity is reached after 8-10 months, full growth after more than 2 years.

Dupuy 1848: 266, Moquin-Tandon 1855: 73 (*Zonites olivetorum*), Westerlund 1886: 74, Germain 1930: 154, Forcart 1960: 5.

Retinella olivetorum (Gmelin, 1791)

Helix olivetorum Gmelin, 1791 in Gmelin [1791]: 3639

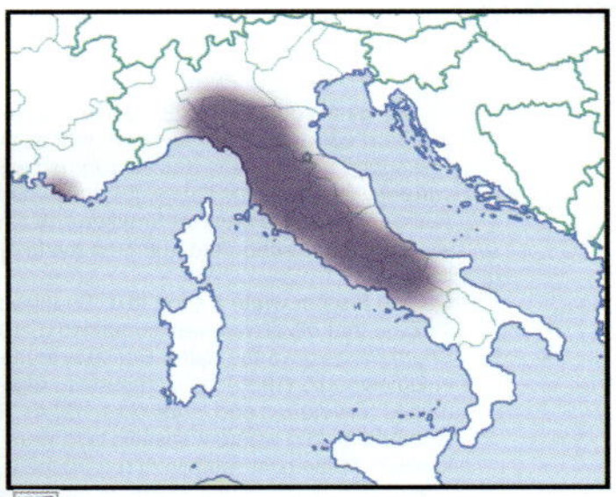

8-17 x 17-30 mm

Italy: near Ronco Scrivia N of Genova

Shady slopes covered with shrubs, appears only during longer periods of rain. Droughts are survived inside the soil.

Copulation was observed in May, life cycle covers probably several years.

Westerlund 1886: 73, Forcart 1960: 3, Manganelli et al. 1995: 19.

Italy

Shell horny brown with greenish hue, finely striated, 5.5-6 weakly convex whorls, last whorl width often 3 x of preceding whorl, rounded at periphery, aperture with white layer inside, umbilicus wide and perspectivical, much wider than in *R. incerta*.

Retinella pseudoaegopinella Giusti, Boato & Bodon, 1986

Retinella pseudoaegopinella Giusti, Boato & Bodon, 1986 in Giusti et al. 1986: 168

Shell horny olive brownish, more rarely whitish, thin and translucent, finely striated and with faint spiral lines producing a finely reticulated microsculpture, protoconch with spiral lines, 3.5-4.5 rapidly increasing convex whorls with moderate suture, last whorl getting wider near aperture, umbilicus wide, 1/4 of shell diameter and slightly excentric.

1.9-2.1 x 4.1-4.5 mm

Italy: Piemonte, Cuneo, Col di Tenda

In deciduous forests, under leaf litter, in shaded habitats, obviously not restricted to calcareous substrate.

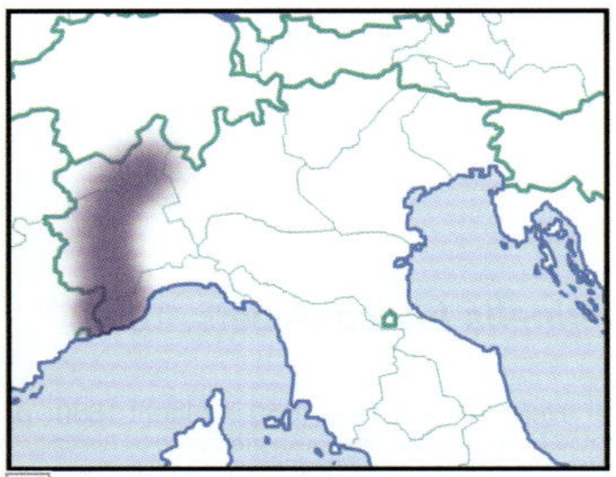

NW Italy

Massimo Prodan

Retinella stabilei (Pollonera, 1886)

Hyalinia stabilei Pollonera, 1886 in Pollonera 1886: 2

Shell like *R. pseudoaegopinella*, species differs only anatomically by absence of a penial flagellum and absence of a glandular covering of the atrium. Shell larger and more globular in *R. incerta*, which differs also by presence of a penial papilla where epiphallus opens into penis.

2.0-2.3 x 4.8-5.3 mm

In leaf litter in moist and shaded habitats, mountain woodlands on non-calcareous soils.

Giusti, Boato & Bodon 1986: 176, Manganelli et al. 1995: 19.

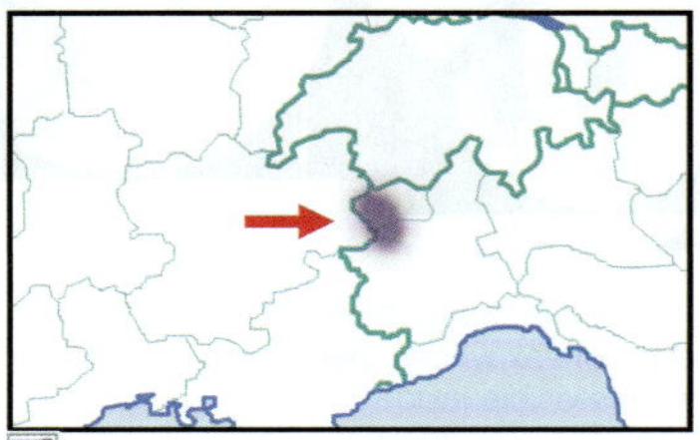

NW Italy

Nesovitrea

Nesovitrea Cooke, 1921 in Cooke 1921: 271

Type species *Helix pauxillus* Gould, 1852 (original designation)
(= currently *Nesovitrea pauxilla* (Gould, 1852) from Hawaii)

Perpolita Baker, 1928 in H. B. Baker 1928: 15 *Retinella (Perpolita)*, type species *Helix hammonis* Strøm, 1765 (original designation)
Type species was based on specimens from Cheboygan County, Michigan, USA. It is likely that the type species was misidentified (Sysoev & Schileyko 2009: 121). A type species should be fixed under Art. 70.3.

Shell diameter 3-8 mm, usually finely striated. Penis without epiphallus, vas deferens inserts directly to penis, penis with longitudinal folds inside with oval bodies at their distal ends.

Northern Hemisphere

The two European species of this group were classified in the genus *Perpolita* by Falkner 1990, Turner et al. 1998: 269, Sysoev & Schileyko 2009 and Horsák et al. 2010, in the genus *Nesovitrea* by Riedel 1980: 85, Kerney et al. 1983, Manganelli et al. 1995, Riedel 1998: 11, Kerney 1999: 139, Falkner et al. 2001, Falkner et al. 2002: 45, Wiktor 2004: 169, Hubenov 2007, Waldén 2007: 168, Boschi 2011: 419. It seems that *Nesovitrea* is currently in front.
Google general search (01.2011, accessed from Germany): in genus *Nesovitrea* 574 (3750), in genus *Perpolita* 255 (1100), Google Scholar (2011): in genus *Nesovitrea* 290, in genus *Perpolita* 108

Nesovitrea hammonis (Strøm, 1765)

Helix hammonis Strøm, 1765 in Strøm 1765: 435, Tab. 6 (*Helix Hammonis*)

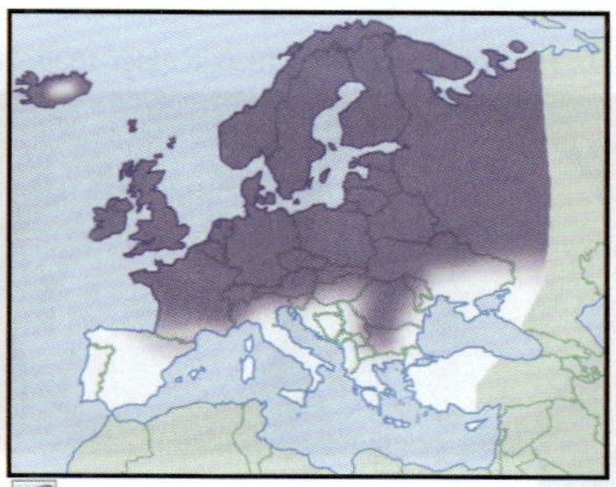

1.9-2.1 x 3.6-4.1 mm, diameter usually below 4 mm

Germany: Flakensee near Berlin

N Eurasia

0.6-0.7 mm at 1 whorl, 1.4-1.7 mm (2 wh.), 3.2-3.7 mm (3 whorls)

Shell usually reddish brown, with regular radial riblets (9-14 riblets/mm), no spiral lines, 3-3.5 whorls, umbilicus open and not very deep, at last whorl slightly excentric. Fresh shells have very faintly visible spiral lines under high magnification, about 10 spiral lines per riblet, 100-150 lines/mm.
Animal slender and blackish, tentacles black, foot narrow, grey with blackish upper sides and black spots on the sides, mantle light grey. *Zonitoides nitidus* is darker and its mantle gland forms an orange spot under the shell's aperture.

In leaf litter of moderately humid forests, often in beech forests with acid soils, sometimes also in open habitats, swampy meadows, reed margins of lakes and sunny dry meadows. Wide ecological amplitude. In Switzerland up to 2400 m, not frequent above 1500 m, in Bulgaria up to 1500 m.

Moquin-Tandon 1855: 86 (*Zonites striatulus*), Germain 1930: 157, Damjanov & Likharev 1975: 269, Kerney et al. 1983: 168, Falkner 1990: 178.

Nesovitrea petronella (Pfeiffer, 1853)

Helix petronella Pfeiffer, 1853 in L. Pfeiffer 1853: 95 (*Helix Petronella*)

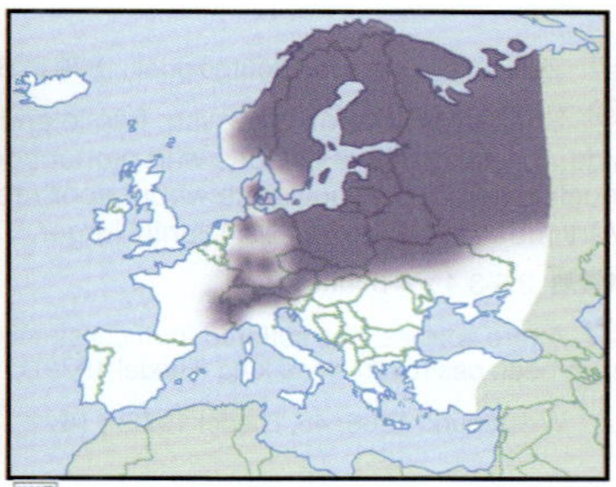

2-2.5 x 4.2-5 mm

Austria: Tirol, near Landeck

Central and N Europe to W Siberia

Shell yellowish white to greenish, not horny brownish, last whorl narrower than in *N. hammonis*, aperture more rounded, umbilicus narrower.

Cooler and more humid habitats than *N. hammonis*, alpine meadows with rock rubble, swamps, in lower altitudes in cool forests in gorges. Tolerates non-calcareous substrate. In Switzerland 1000-2700 m, usually 1600-2100 m.

Endangered in Germany, probably vulnerable in Austria. Frequent only in the north and in the Alps. Sensitive against acid rainfalls in Scandinavia.

Turner et al. 1998: 270.

Zonitoides

Zonitoides Lehmann, 1862 in Lehmann 1862: 111

Type species *Helix nitida* Müller, 1774 (monotypy)

Gender determined to be masculine by the author.

North America, eastern and northern Asia, Europe

Zonitoides arboreus (Say, 1817)

Without gender agreement: *Zonitoides arborea* (Say, 1817)

Helix arborea Say, 1817 in Say 1817: (?) [not verified]

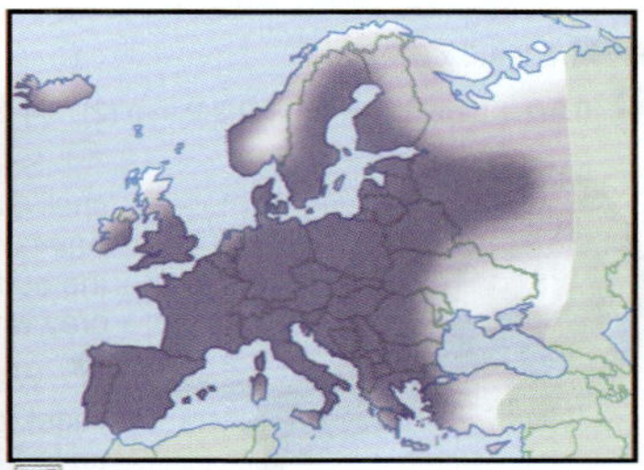

N America (British Columbia to Haiti), dispersed worldwide
Indicated range in Europe based on currently expected approximate distribution.

1.7-3 x 4.5-6 mm

USA: Georgia, Wilkes County, Washington near Athens

Shell pale yellowish brown, translucent, faintly and irregularly striated, 4-4.5 whorls, sometimes very slightly keeled, umbilicus deep, 1/7 of diameter, broadening at the last whorl.

In North America common in woodlands and humid habitats, also in gardens. In Europe in greenhouses and garden centers, from where it is able to escape with plant soil, quite common in outdoor bamboo pots and plantations in SW Germany.

Kerney et al. 1983: 312, Turner et al. 1998: 418, Schmid 2002: 391, Dvořák & Kupka 2007, Beckmann 2007: 80.

Zonitoides excavatus (Alder, 1830)

Without gender agreement: *Zonitoides excavata* (Alder, 1830)

Helix excavata Alder, 1830 in Alder 1830: 38

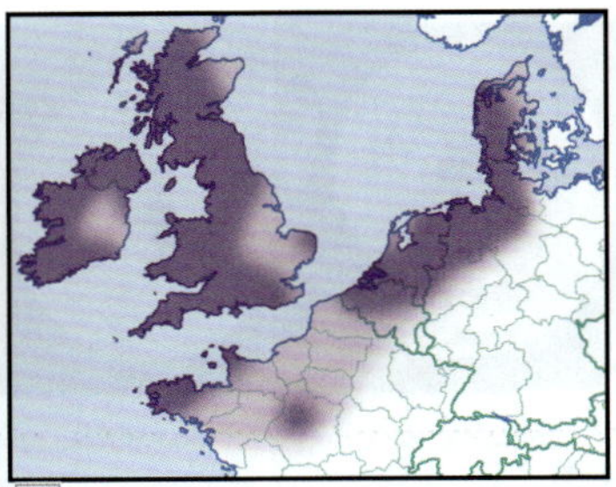

2.8-3.4 x 5.3-6 mm

N Germany: Neuenburger Urwald near Oldenburg

British Isles, Denmark to France

Smaller than *Z. nitidus*, umbilicus extremely wide and perspectivically open (like *Discus rotundatus*), shell weakly brown, slightly transparent, with radial streaks, the animal is dark.

Leaf litter zone and under dead wood in old natural forests, sometimes also in swamps (W Ireland and W Great Britain). Only on non-calcareous soils. Tolerates some degree of human disturbance and replanting, but usually not in forest plantations. In Britain restricted to a few regions, but frequent in the zones where it occurs.

In Britain 2-6 calcified eggs (size 1 mm) are usually laid in batches directly on or in the soil, moss or rotting timber, one batch per day, juveniles with 1.5-2 whorls hatch after 20-35 days, sexual maturity is reached after 4 months (data obtained from laboratory studies).

Endangered in Germany.

Kerney & Cameron 1979: 127, Anderson 1996: 11, Kerney 1999: 147, Kuźnik-Kowalska & Pokryszko 2012.

Zonitoides jaccetanicus (Bourguignat, 1870)

Zonites jaccetanicus Bourguignat, 1870 in Bourguignat 1870: 94, Pl. 16 (*Zonites Jaccetanicus*) Type locality in Barcelona.

Shell evenly horny brownish, fragile and transparent, finely striated on upper side, nearly smooth on lower side, 5 weakly convex whorls with pronounced suture, upper side flattened, lower side convex, last whorl not much wider than penultimate whorl, aperture oblique, margin straight, umbilicus relatively narrow.
Very much like *Z. arboreus*, slightly larger. There are anatomical differences.

2.6-3.4 x 4-5 mm

A rare species, has suffered from construction activities in the growing city of Barcelona. The type locality seems to have been destroyed since long (Riedel 1972).

Westerlund 1886: 45, Riedel 1972: 116.

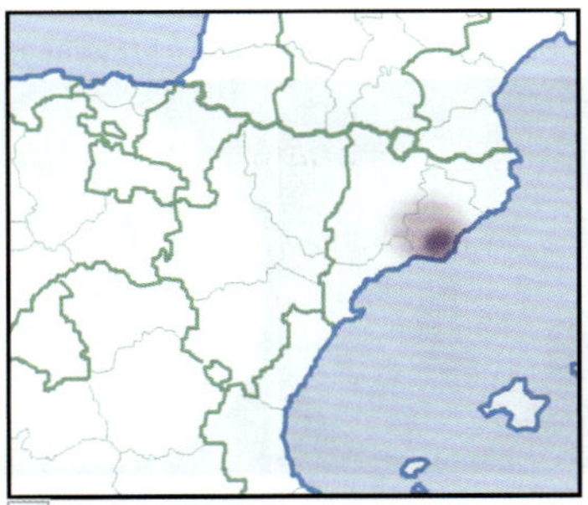

Spain (Barcelona)

Zonitoides nitidus (Müller, 1774)

Without gender agreement: *Zonitoides nitida* (Müller, 1774)

Helix nitida Müller, 1774 in Müller 1774: 32

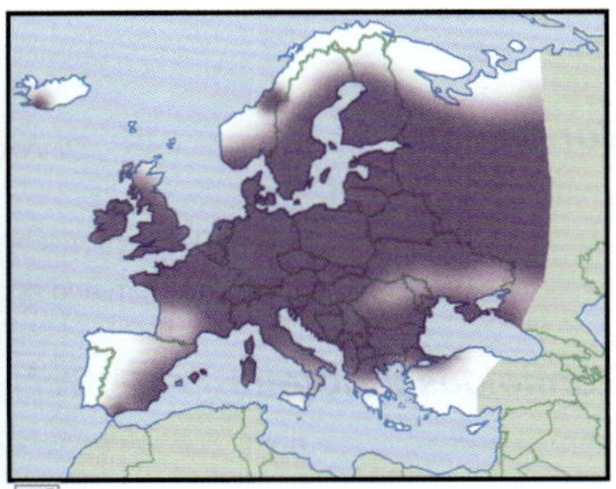

3.5-4.0 x 6-7 mm

Germany: Nordrhein-Westfalen, near Telgte

Holarctic, almost all over Europe

Ø 0.9-1.1 mm at 1 whorl, 1.9-2.2 mm (2), 3.4-3.7 mm (3), 4.3-4.6 mm (3.5 whorls), 5.3-5.7 mm (4 whorls)

Serbia: Vojvodina, near Subotica
József Dószai

Shell reddish brown, distinctly and irregularly striated, 4-4.5 weakly convex whorls, umbilicus large (almost 25 % of shell diameter).
Animal black with a characteristic orange spot (mantle gland visible under the shell's aperture). In *Nesovitrea hammonis* the body is lighter without orange spot, the shell of the living animal is considerably darker in *Z. nitidus*. Juveniles whitish grey with light brown translucent shells.

Wet meadows and river woods, usually near water bodies, swamps and swampy forests, in the zone of emergent vegetation. Man-made habitats such as pools in old quarries are sometimes colonized after a few years. In Switzerland in up to 2100 m altitude.
Herbivorous, snails feed on desintegrating leaves, mushroms, roots and fruit, no dry leaves. Humid leaves are preferred. When consuming soft food such as muchrooms or soft fruits, *Z. nitidus* penetrates perpendicularly inside producing characteristical holes, the entire animal can disappear inside, together with its shell.

In France some 30 eggs were observed to have been laid in May and June (Gassies 1849). In Germany up to three clutches of 2-9 eggs (diameter 1.0-1.6 mm) are laid in all seasons, loose into the soil, with some days or weeks in between. In Poland the offspring hatches from June to October. Juveniles hatch after 21-27 days, they have 1.5 whorls (diameter 1-1.2 mm) after hatch, they start feeding on desintegrating plant remains in the soil. Growth rate is about 0.5-1 whorls/month. After 3 months the shell diameter reaches up to 3 mm under favourable conditions, after 10 months 6 mm, full size after slightly more than one year. Maximum age under laboratory conditions 18 months, in natural habitats in Poland 3 years.

In some regions in Britain the species has declined due to drainage.

Gassies 1849: 107 (eggs France), Frömming 1940 (life cycle), Frömming 1954: 105, Damjanov & Likharev 1975: 299, Kerney et al. 1983: 178, Falkner 1990: 171, Kerney 1999: 148, Kuźnik-Kowalska 2011 (life cycle).

Balcanodiscus

Balcanodiscus Riedel & Urbański, 1964
in Riedel & Urbański 1964: 75 *Paraegopis (Balcanodiscus)*

Type species *Helix frivaldskyana* Rossmässler, 1842 (original designation)

Limestone habitats, mainly subterranean.

Riedel 1992: 55, Erőss, Fe. & P. 2011 (review of species with anatomical diagnoses).

Erőss, Fe. & P. 2011 gave a short review and presented the anatomical differences between all known taxa in this group (not listed here). They gave a statement that shell characters were usable for distinction at species level, but did not explain the extent to which this can be done.
The genitalia of the taxon *stummerorum* were unknown, its shell size did not differ from *B. beroni*, so its status as an independent species can only be speculated. Erőss, Fe. & P. 2011 did not explain why *stummerorum* must be regarded as an independent species and cannot be conspecific with the next known population, *beroni*. More detailed knowledge about local variations in shell shape in the widespread *B. frivaldskyanus* would be a prerequisite to derive constraints from shell characters about the identity of such a form.

Balcanodiscus beroni Riedel, 1995

N Greece (E Makedonía)

Balcanodiscus beroni Riedel, 1995 in Riedel 1995 (Malak. Abh.): 140
Balcanodiscus stummerorum Reischütz, Reischütz & Fischer, 2008 in Reischütz et al. 2008: 11

Shell dirty whitish, slightly translucent, weakly shiny on upper side, lower side more shiny, upper side relatively weakly striated, lower side normally striated (like in *B. frivaldskyanus*), no spiral lines, 5-6 convex whorls, last whorl width 1.7 x of preceding whorl, rounded or very weakly angulated at periphery, lower side concave, umbilicus wide and perspectivical, 1/3 of diameter.
Differs from *B. frivaldskyanus*, *B. cerberus* and *B. gracilis* in its smaller shell, rounded periphery, weaker radial sculpture at the upper side and a concave lower side in the north (in the south in the Pangeon mountains this was not observed).

3.5-4 x 7-9 mm

Known from caves and rock crevices, probably a subterranean species.

The taxon *stummerorum* is here classified as a synonym of *B. beroni* because in the absence of an anatomical record and the main difference to *B. beroni* being only in the shell shape, the taxonomic status of this form as a separate species cannot be established and remains a speculation.

Balcanodiscus carinatus Reischütz, 1983

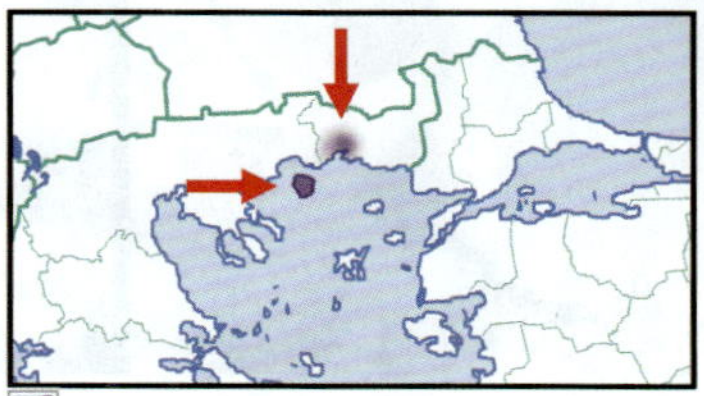

N Greece (Thásos, S Thráki)

Balcanodiscus frivaldskyanus carinatus Reischütz, 1983 in Reischütz 1983: 139

Shell yellowish translucent, keeled, also at last whorl, strongly ribbed and with very fine spiral lines, also at lower side, umbilicus wide.

3.8-4.1 x 9.5-11 mm

In rock crevices and under boulders.

Riedel 1995: 158.

Balcanodiscus cerberus Riedel, 1985

Balcanodiscus cerberus Riedel, 1985 in Riedel 1985 (Malak. Abh.): 142

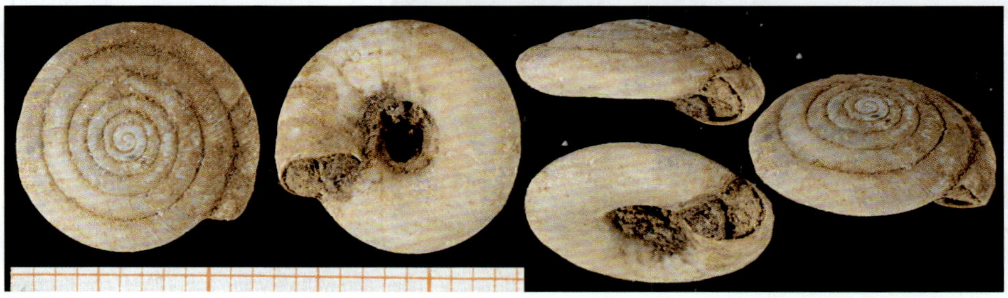

Greece: Rodopi, Maronia cave

N Greece (S Thráki)

Restricted to a small range.

4-5 x 11-14 mm

Shell pale dirty whitish or yellowish, sometimes with a weak greenish hue, weakly shiny, fresh shells semitransparent, finely striated and with well visible spiral lines, much weaker or absent on lower side, juvenile shells keeled, 5.5-6 slowly and regularly increasing whorls, initially more convex and with deeper suture than last 3 whorls, last whorl width 1.1 x of penultimate whorl, with weak edge at periphery (in upper section of whorl height), umbilicus wide and perspectivical, 1/3 of shell diameter.
Shell wider and flatter than *B. frivaldskyanus*, considerably larger than *B. danyi*.
Animal dirty whitish or light brownish, slightly translucent, eyes with very little black pigment.

Most probably subterranean, was found in a cave, more rarely outside the cave under stones and in rock crevices.

Riedel 1988: 101, Riedel 1995: 157.

Balcanodiscus danyii Erőss, Fehér & Páll-Gergely, 2011

N Greece (Rodópi mountains)

Balcanodiscus danyii Erőss, Fehér & Páll-Gergely, 2011 in Erőss et al. 2011: 577

Shell yellowish brown, moderately depressed with angulated periphery, initial whorls keeled, looks like a smoother version of *Discus rotundatus* with curved growth lines, also on lower side, very fine radial lines, 5-5.5 whorls, last whorl width about 1.3 of penultimate whorl, umbilicus wide and very deep, about 1/4 of diameter.
Shell often corroded.
The animal is white or yellowish with short tentacles and black eye spots.
Anatomically closely related with *B. cerberus*, but considerably smaller.

2.3-2.8 x 6-7 mm

Was found in an oak forest among stones, 30-50 cm below soil surface.

Only known from one locality.

Balcanodiscus difficilis Riedel, 1988

Balcanodiscus difficilis Riedel, 1988 in Riedel 1988: 96

Shell dirty beige or greyish, not very translucent, not shiny, prominently ribbed and with fine spiral lines (better visible between the ribs), apex usually corroded, first whorls keeled with keel in upper section of whorl height, 5.5-6.5 weakly convex and very slowly increasing whorls, last whorl bluntly keeled at periphery (in 1/2 of whorl height), umbilicus very wide.
Differs from *B. frivaldskyanus* in its more prominently striated shell sculpture, an in the genital organs.

4.5-5 x 9-11 mm

Occurs together with *B. carinatus*, but more restricted to caves where *B. carinatus* is not found.

Restricted to a small range in the E and central parts of Thásos.

N Greece (Thásos)

Balcanodiscus frivaldskyanus (Rossmässler, 1842)

Without gender agreement: *Balcanodiscus frivaldskyana* (Rossmässler, 1842)

Helix frivaldskyana Rossmässler, 1842 in Rossmässler 1838-1844 (11): 3, Taf. 51

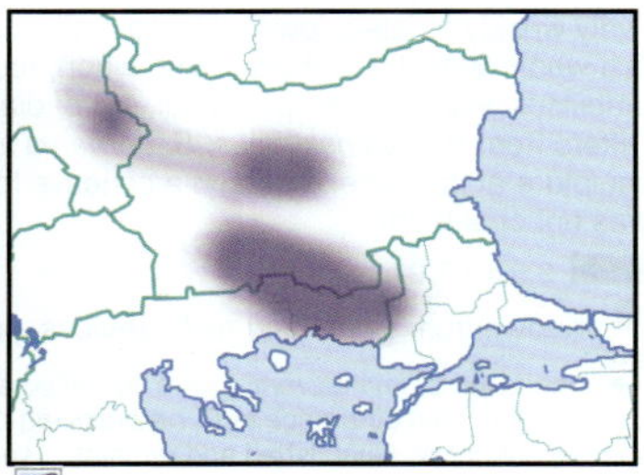

4.0-4.7 x 9-10 mm

Bulgaria: E Rodopi, Dishlik Dere

Bulgaria, NE Greece

Bulgaria: E Rodopi, Oveshari Reserve

Shell more rounded on upper side and flat on lower side, whitish with yellowish hue, finely striated on upper side, more weakly on lower side, 5.5-6 convex whorls with deep suture, periphery rounded or bluntly keeled, umbilicus very wide and perspectivical, 30 % of diameter.

In the W Rodopi mountains most often found in the leaf cover of monodominant beech forests, calciphile, also in caves, in Bulgaria in up to 1100 m.

Westerlund 1889: 13, Damjanov & Likharev 1975: 273, Riedel 1995: 157, Irikov & Mollov 2006: 797.

Dilian Georgiev

Balcanodiscus magnus Reischütz, 1988

Balcanodiscus magnus Reischütz, 1988 in Reischütz 1988: 349

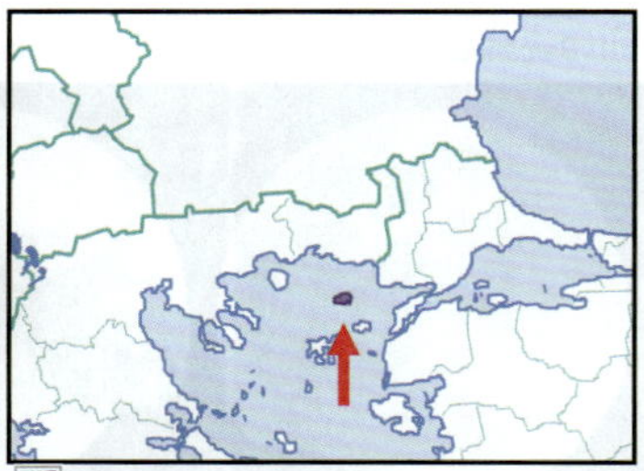

5.2-5.5 x 12-14 mm

Greece: Samothráki, Paleopolis

N Greece (Samothráki)

Shell larger than most other *Balcanodiscus* species, yellowish brown with silky sheen, lentiform, distinctly ribbed and with very fine spiral lines, ribs become weaker at last whorl, also weak at the lower side, first whorls with edge at periphery, vanishing at the last whorl almost entirely, umbilicus wide (1/4 of shell diameter), showing all whorls inside, aperture with sharp margin, attaching directly at the penultimate whorl (no sinus).

Confined to calcareous substrate, lives on the surface in marble habitats, under stones and in leaf litter, locally frequent.

Riedel 1988: 103, Riedel 1995: 157.

Meledella

Meledella Sturany, 1908 in Sturany 1908: 38

Type species *Meledella werneri* Sturany, 1908 (monotypy)

Meledella werneri Sturany, 1908

Meledella werneri Sturany, 1908 in Sturany 1908: 38

Shell yellowish green to pale white, weakly translucent, upper side prominently striated perpendicular to suture, from 2nd whorl onwards, shiny on lower side, 6 not very convex whorls, shell almost flat on upper side, rounded at periphery, umbilicus very wide, all whorls well visible inside with their spiral sculpture, aperture with sharp margin.

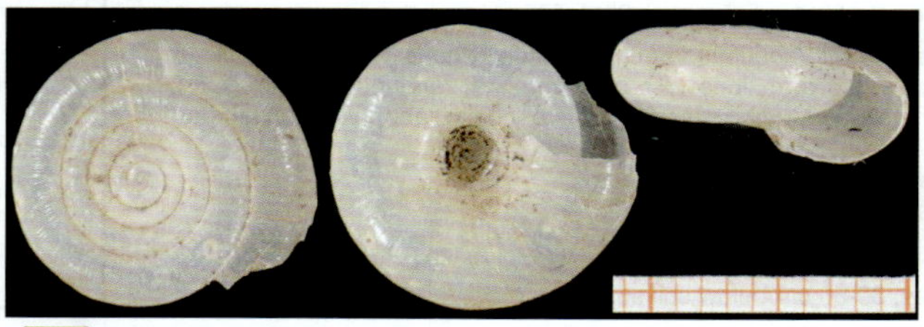

5.5-6.5 x 14-14.5 mm

Croatia: Dalmatia, island of Mljet

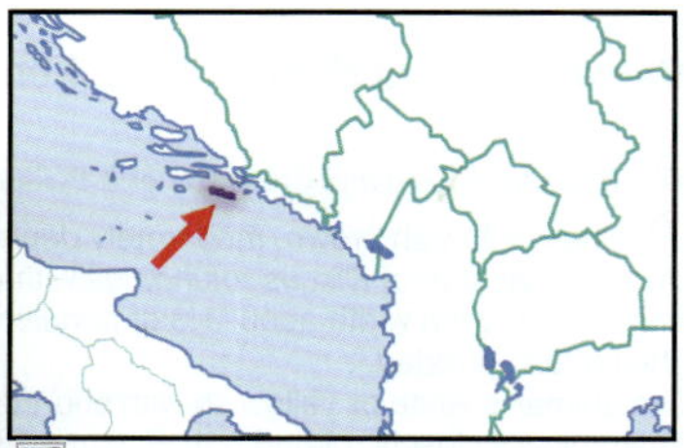

Croatia (island of Mljet)

Subterranean, found in a cave.

Riedel 1960, Riedel 1979 (Fragm. Faunist.): 120.

Aegopis

Aegopis Fitzinger, 1833
in Fitzinger 1833: 99

Type species *Helix verticillus* Lamarck, 1822 (monotypy)

Helix verticillus was the one and only species included by Fitzinger 1833, several other names were cited as synonyms (*Helix lapidicola*, *algiroides*, *oculus*). So monotypy under Art. 68.3.

Shell diameter exceeds 20 mm, embryonal shell 2 whorls and sharply keeled, surface striated and with finer spiral sculpture.
Penis long and not subdivided into sections, proximal part covered by an envelope, to which is attached the vas deferens, penis inside papillated, vas deferens inserts to distal end of penis, no epiphallus, no spermatophores are produced.

Forcart 1957: 112, Bole 1983 (revision).

Aegopis acies (Férussac, 1832)

Helix acies Férussac, 1832 in A. Férussac & Deshayes 1819-1851: (explication planches) Pl. 80
Original print of explication des planches not verified, second print p. 15.

15-21 x 28-34 mm

Croatia: Dalmatia, near Makarska

Under leaves and stones, often in beech forests, mostly in coastal vicinity.

Rossmässler 1835-1837 (3): 3, Westerlund 1886: 81, Kobelt 1902: 8, Falkner 1990: 176, Bole 1983: 144.

Kobelt 1902 suspected continuous gradients from keeled populations in the south (*acies*) to rounded populations in the north (*septentrionalis*), but Bole 1983 suggested that *acies* was an independent species, locally variable but conchologically and anatomically well differentiated.

Croatia (Dalmatia) and Herzegovina

Shell reddish brown, lower side slightly lighter, prominently striated, ribs becoming more prominent downwards, with fine and regular spiral lines producing a granular pattern, 6-7 whorls, first whorls keeled, which is also visible in adult shells, last whorl usually keeled, but can be rounded in some populations (near Sinj in Dalmatia) with an inconspicuous colour band, aperture inside intensively coloured, umbilicus regular or narrow (near Sinj 4-5 mm).

Aegopis compressus (Rossmässler, 1836)

Without gender agreement: *Aegopis compressa* (Rossmässler, 1836)

Helix compressa Rossmässler, 1836 in Rossmässler 1835-1837: (3): 2, Taf. 11

Shell dirty horny brownish, lower side bluish and reddish brown with irregular white and grey radial streaks, 5-6 slightly convex whorls with moderately deep suture, with blunt edge or keel at periphery, upper side often more convex than lower side, umbilicus wide.

11-14 x 26-32 mm

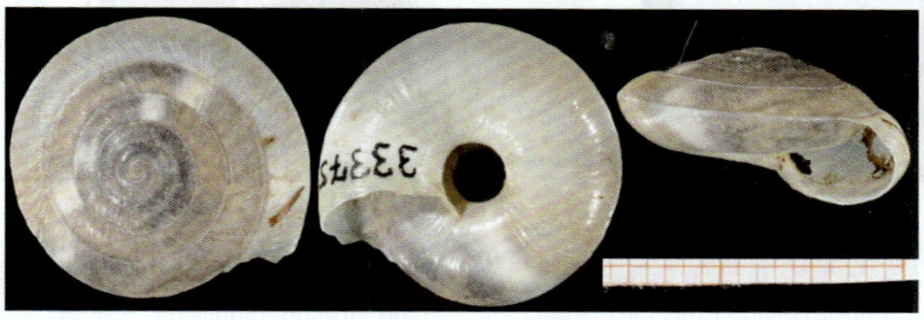

NW Croatia, SW Slovenia

Westerlund 1886: 80, Bole 1983: 142, Štamol & Velkovrh 1995 (Cres island).

Slovenia: Notranjski Snežnik 20 km NNE Rijeka

Aegopis croaticus (Férussac, 1832)

Without gender agreement: *Aegopis croatica* (Férussac, 1832)

Helix croatica Férussac, 1832 in Férussac & Desh. 1819-1851: (explication planches) Pl. 80 (*Helix (Helicella) Croatica*)
Original print of explication des planches not verified, second print p. 15.

12-17 x 26-32 mm, relative height very variable

Croatia: Plitvička Jezera

Ø 19-24 mm at 5 whorls, 23-29 mm at 5.5 whorls, 27-33 mm at 6 whorls

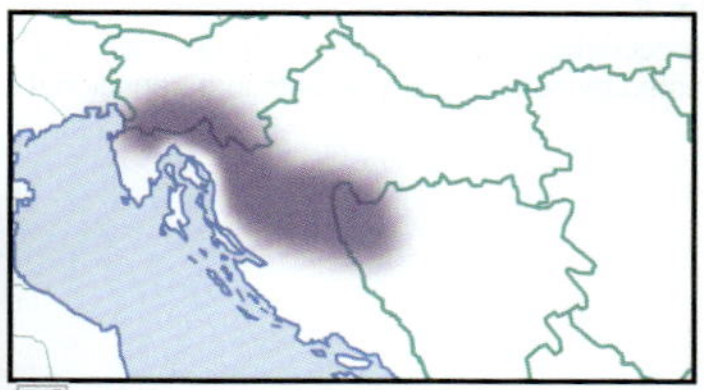

Croatia, S Slovenia

At limestone rocks, in S Slovenia in forests.

Rossmässler 1835-1837 (3): 3, Westerlund 1886: 80, Kobelt 1902: 7 (classified shells of *Aeg. acies* and *septentrionalis* under this name), Bole 1983: 136, Bole & Slapnik 1997: 149 (S Slovenia).

Shell on upper side dirty horny brown with indistinct light spots, densely and irregularly striated, lower side smooth, greenish or yellowish white and shiny, 6-7 flattened whorls with shallow suture, first whorls sharply keeled and flat, last whorl with blunt edge and more convex at lower side, with a white or bluish band along the edge, umbilicus open.
Animal slender, light grey with yellowish hue, dorsal side and tentacles nearly black. Anatomically not very variable.

Aegopis gemonensis (Férussac, 1832)

Helix gemonensis Férussac, 1832 in Férussac & Deshayes 1819-1851: (explication planches) Pl. 80 (*Helix Gemonensis*)
Original print of explication des planches not verified, second print p. 15. Original spelling could have been *Gemonense*, as suggested by Férussac [1821-1822] (Prodrome) folio p. 44.

7-9 x 16-20 mm (16-18 mm at 5 whorls)

Italy: Südtirol, Casotto

Usually in mountain forests under stones, humid habitats.

Westerlund 1886: 82, Kobelt 1902: 4, Bole 1983: 147, Falkner 1990: 176, Manganelli et al. 1995: 18.

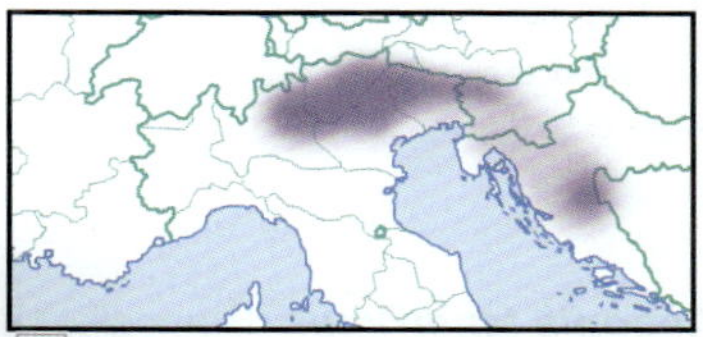

N Italy, W Slovenia, N Croatia

Shell yellowish or greenish, lower side lighter, translucent, densely and finely striated on upper side with spiral lines producing an indistinct reticulate pattern, lower side almost smooth, 5-5.5 almost straight whorls, first whorls keeled, this is also visible at the well-marked suture in adult shells, last whorl usually not keeled but with more or less pronounced edge, aperture inside with strong white lip, umbilicus wide.

Aegopis italicus (Kobelt, 1876)

Without gender agreement: *Aegopis italica* (Kobelt, 1876)

Zonites compressus var. *italica* Kobelt, 1876 in Kobelt 1876: 50, Taf. 111

14-17 x 25-32 mm

Italy: Abruzzo, L'Aquila, near Villalago

Humid habitats in deciduous mountain forests, occasionaly also in open high mountain habitats with sufficient places to provide shelter against desiccation.

Westerlund 1886: 80, Manganelli et al. 1995: 18.

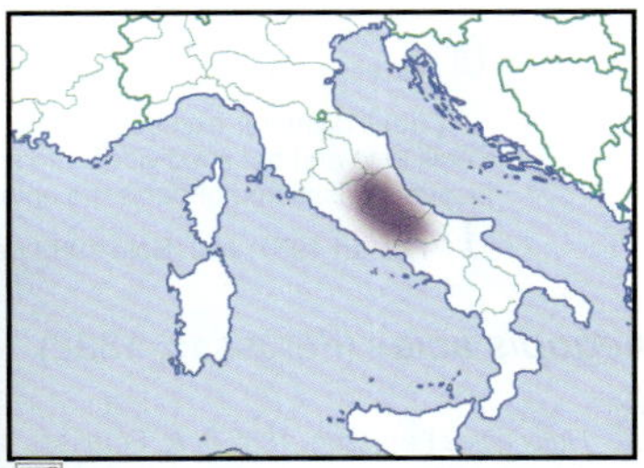

Italy

Shell dirty horny brownish, lower side bluish and reddish brown with irregular white and grey radial streaks, 5-6 slightly convex whorls with moderately deep suture, keeled with well rounded lower side, umbilicus open.
Very much like *A. compressus*, but with higher spire and more rounded lower side at the last whorl.

Aegopis laughofferi Wagner, 1914

Aegopis septentrionalis (Kobelt, 1899)

Zonites croaticus var. *septentrionalis* Kobelt, 1899 in Kobelt 1902: 7, Taf. 244
not *Zonites septentrionalis* Bourguignat, 1870 in Bourguignat 1870 (Rev. Mag. Zool. (2) 22): 17 from France/Switzerland
Aegopis croaticus laughofferi Wagner, 1914 in A. J. Wagner 1914: 335

17-21 x 35-40 mm

Croatia: Dalmatia, near Obrovac

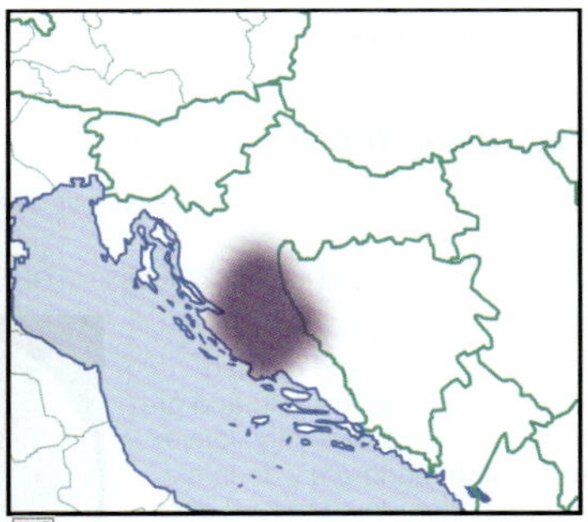

Croatia to N Albania

Shell horny reddish brown with a band at the periphery, regularly striated, upper side not shiny, only very weakly granulated, no spiral sculpture, lower side whitish, smooth and shiny, without any keel or edge at the periphery, 7 weakly convex whorls, suture white, aperture inside brown with keel band shining through, umbilicus wide (diameter more than 6 mm). Juveniles keeled, this is visible on the upper side of adult shells.

Kobelt 1902: 7, Bole 1983: 140 (*septentrionalis*), Dhora & Welter-Schultes 1996: 147.

This species comprises the more widespread form "*septentrionalis*" (this would need a new name, *septentrionalis* is preoccupied) and *laughofferi* from a small range at the SW slopes of the southern Velebit mountains. Both forms are connected by continuous gradients in a transitional zone.

Croatia: Dalmatia, Paklenica

Giuseppe Pocaterra

Aegopis spelaeus Wagner, 1914

Aegopis spelaeus Wagner, 1914 in A. J. Wagner 1914 (Sitz.-Ber. Wien): 39

Shell thin, whitish transparent and shiny, finely and irregularly striated, more weakly at lower side of last whorls, keeled, keel weaker near aperture, 4-5 convex whorls with deep suture, last whorl width 1.5 x of preceding whorl, apertural margin connected at parietal side by a very weak and shiny layer, umbilicus wide and perspectival, 1/6 of shell diameter. Eyes with only very weak pigments.

9-11 x 19-22 mm

Cave snail, troglobiotic.

Rare, not many records.

Štamol et al. 1999.

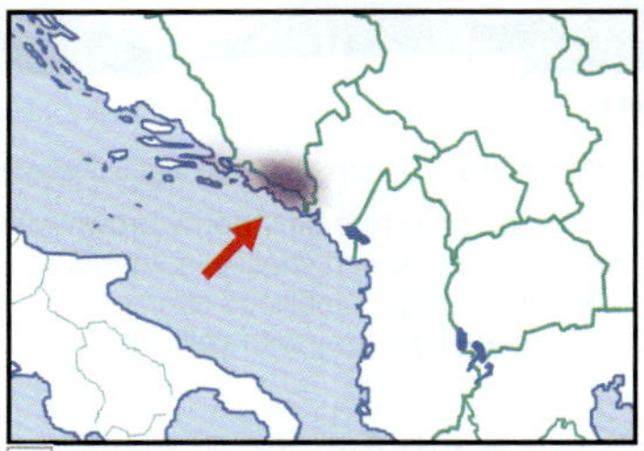

S Dalmatia, Herzegovina

Aegopis tenerrimus (Brancsik, 1890)

Zonites tenerrimus Brancsik, 1890 in Brancsik 1890: 68, Tab. 1

Shell horny or light greenish, thin and translucent, regularly striated, lower side more finely striated, with very fine spiral lines at both sides (difference to *Aeg. gemonensis*), lentiform, sharply keeled, 6 whorls, upper whorls more convex than last whorls, aperture simple, umbilicus open, not very wide, cylindrical.

Deciduous forests.

Kobelt 1902: 4, Pavlović 1912: 17, Tomić 1959: 14.

5-6 x 11-13 mm

Bosnia

Picture: Kobelt 1902: Fig. 1569

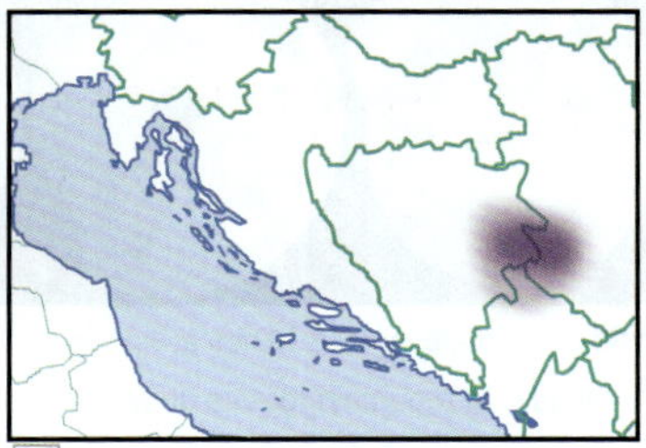

Bosnia, W Serbia

Aegopis verticillus (Lamarck, 1822)

Helix verticillus Lamarck, 1822 in Lamarck 1822: 78
Zonites verticillus var. *corcyrensis* Boettger, 1883 in O. Boettger 1883 (Jahrb. DMG): 315 (*Zonites verticillus* var. *Corcyrensis*)

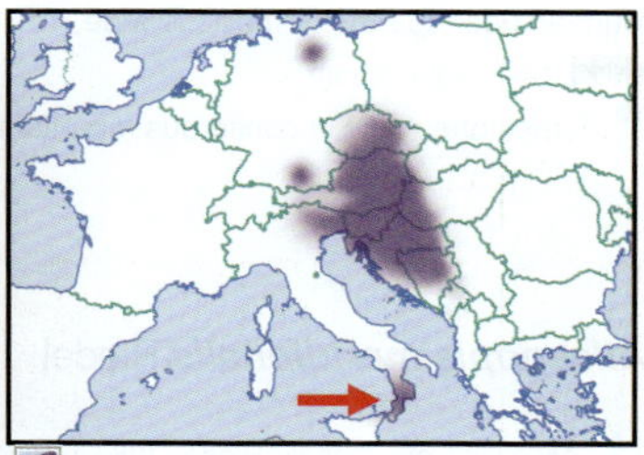

16-17 x 26-32 mm

Austria: Niederösterreich, Trefflingfall

SE Europe, S Italy

20-21 mm at 5 whorls, 24-25 mm at 5.5 whorls, 30-32 mm at 6 whorls

Austria: Niederösterreich, near Kremsmünster Harald Berger

Shell yellowish brown on upper side, greenish yellow around the umbilicus, strongly and regularly striated, a fine spiral scupture produces a granular pattern, on lower side weakly striated, 6-8 distinct radial colour bands, upper whorls keeled, last whorl with an indistinct blunt edge, more convex at lower side.
Animal slender with narrow sole, tentacles and dorsal side nearly black, lighter and with dirty yellowish hue down to the sides.

Under leaves and between stones on humid soils, often in shallow holes at the soil, in mountain forests of lower altitudes. Not only on calcareous substrate. Herbivorous, seems to prefer green parts of meadow plants.

Reproduction in summer, 50-60 eggs (diameter 4 mm) are laid in clutches in humid substrate (soil, leaf litter or rotting wood), hatched juveniles have 1.5 whorls, they stay several weeks inside humid soil or leaf litter substrate and have 3 whorls in late autumn.

Rossmässler 1835-1837 (3): 2, Westerlund 1886: 77, Frömming 1954: 81, Falkner 1990: 176.

Classification of *corcyrensis* is uncertain, this name was based on a single specimen for which the locality "Kérkira" was probably confounded, a similar form was never again found there (Riedel 1992: 51, 119).

Allaegopis

Allaegopis Riedel, 1979
in Riedel 1979: 469 *Aegopis (Allaegopis)*

Type species *Aegopis skanderbegianus* Poliński, 1924
(original designation)

No single perivaginal gland present (encircling the distal end of the vagina), but an appendix-like vaginal gland which is subdivided into two or more parts. Vas deferens inserts distally into penis, no flagellum present.

Riedel 1992, Riedel 1993, Gittenberger 1999.

Allaegopis amphikypellon Riedel, 1982

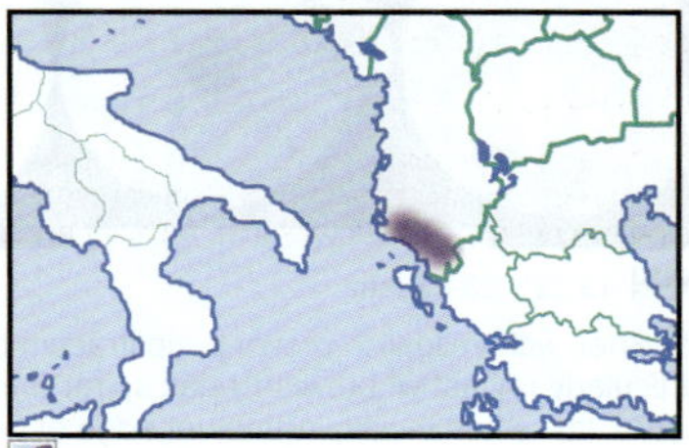

SW Albania

Allaegopis amphikypellon Riedel, 1982 in Riedel 1982: 4

Shell light to dark horny brown, solid, not very translucent, prominently, densely and regularly striated, reticulated by dense spiral lines, same reticulate pattern on lower side, 5.5-6 weakly convex whorls, initially more slowly increasing than later, first whorls keeled, last whorl width 1.5 x of penultimate whorl, more convex and with deeper suture, initally with blunt edge, periphery near aperture rounded, umbilicus moderately wide, 1/5 of shell diameter.

15-17 x 27-35 mm, diameter usually below 31 mm

Allaegopis jonicus (Käufel, 1930)

Aegopis acies jonicus Käufel, 1930 in Käufel 1930: 179

NW Greece

Occurrence in Ipiros questionable (Riedel 1992).

11-14 x 22-27 mm

Greece: Lefkáda, Profítis Ilías mountain

Shell brownish with white keel, prominently and obliquely striated, with weak spiral lines, same surface sculpture on lower side, 5-5.5 slightly convex whorls with prominent and crenulated keel, also at aperture, last whorl width slightly less than 1.5 x of penultimate whorl, umbilicus wide and deep, 1/5 of shell diameter.

In highland habitats at 600-1000 m altitude.

Allaegopis kerketianus Riedel, 1993

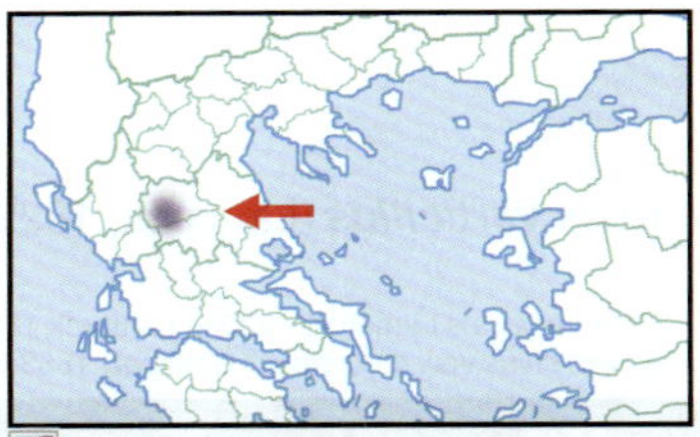

NW Greece (Pindos mountains)

Allaegopis kerketianus Riedel, 1993 in Riedel 1993: 198

Shell light brownish, lower side lighter and with greenish hue, ribbed and with fine spiral lines producing a reticulate pattern on upper side, smoother with weak spiral lines, weakly shiny and slightly transparent on lower side, 5.5-6.5 slowly and regularly increasing whorls with very shallow suture (0.5 whorls more than same sized *A. meridionalis*), juveniles lentiform and strongly keeled, last whorl initially angulated, in large specimens rounded near aperture, umbilicus wide, 1/5-1/4 of diameter.

14-16 x 23-28 mm

Limestone rocks in coniferous mountain forests, 1200-1900 m.

Allaegopis meridionalis Riedel, 1986

Allaegopis meridionalis Riedel, 1986 in Riedel 1986: 87

Greece: Thessalia, Plastíras (shell diameter 19.7 mm, not fully grown)

Panayotis Ovalis

Greece

Shell brownish, lower side lighter, more shiny and with greenish hue, upper side densely reticulated with usually slightly sharper and more regularly arranged ribs than in *All. transiens*, lower side usually smooth (contrast to *All. amphikypellon*), upper side of adults slightly more convex than lower side, 5-5.5 weakly convex whorls (0.5 whorls less than same sized *A. kerketianus*), first whorls keeled, last whorl angulated, aperture almost without edge, umbilicus wide, 1/5-1/4 of shell diameter. Juvenile shells much more flattened on upper side, convex lower side and prominent keel.

13-15 x 26-30 mm

Differences to *All. transiens* and *All. amphikypellon* mainly anatomical: no atrium, distal part of vagina usually thick and with two gland appendices (*All. amphikypellon* has only one).

In limestone rock and rock rubble habitats, often under shrubs and spiny herbs.

Riedel 1992: 49, Riedel 193: 201.

Allaegopis skanderbegianus (Poliński, 1924)

Aegopis skanderbegianus Poliński, 1924 in Poliński 1924: 128

Albania: near Çorovodë

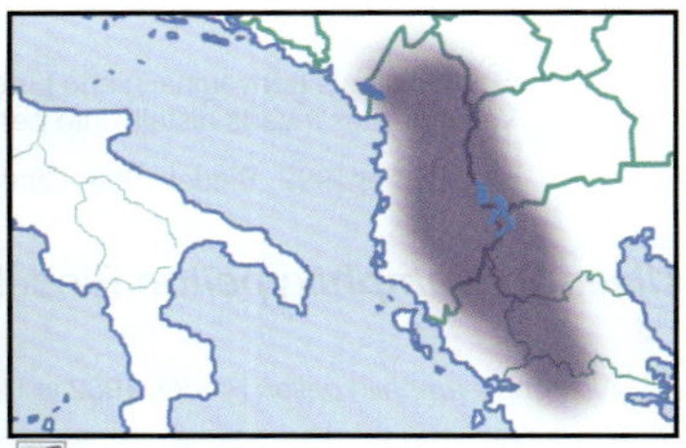

Albania to Greece

13-22 x 23-35 mm

Shell with yellowish brown periostracum covering fresh shells, displaying some concentric radial streaks on lighter background at lower side, finely striated, particularly on last whorl, with weak spiral lines (if at all), shape convex on upper side and flatter on lower side, 5.5-6.5 weakly convex whorls, first whorls keeled, last whorl initially with blunt edge at periphery, near aperture rounded, width 1.3 x of penultimate whorl, aperture large, umbilicus moderately wide and deep, 1/6-1/5 of shell diameter.

Open habitats with vegetational cover.

Allaegopis subariedeli Gittenberger, 1999

NW Greece (Ipiros)

Allaegopis subariedeli Gittenberger, 1999 in Gittenberger 1999: 244

Shell yellowish brown with some radial yellowish streaks and fine growth lines, also fine spiral lines, more than 6 whorls, flat on the upper side, with sharp keel except on the last section of the last whorl, aperture rounded, umbilicus open, 1/5 of diameter.

Vagina with two vaginal glands, penis more than twice as long as vagina and subdivided into three sections, the first one being covered by a penial sheath.

Shell smaller than *A. jonicus*, but not clearly diagnostic, genitalia provide better diagnostic characters.

9-13 x 18.5-24 mm

Allaegopis transiens (Mousson, 1859)

Zonites croaticus var. *transiens* Mousson, 1859 in Mousson 1859: 265

Shell evenly yellowish or horny brown with whitish radial streaks, shiny, densely and irregularly striated, with fine spiral lines, lower side lighter, smooth, no colour band at periphery, 5-6 weakly convex whorls with shallow suture, juveniles keeled, but not visible in adult shells, last whorl with weak edge at the beginning and rounded at aperture, aperture inside with very weak lip, umbilicus wide and perspectival.

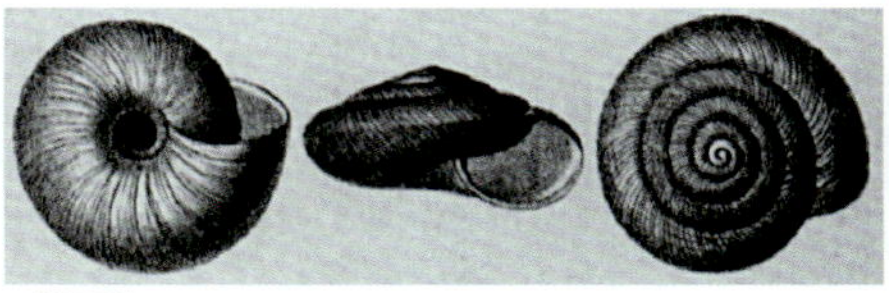

8-14 x 21-28 mm

Greece (?)

Picture: Kobelt 1902: Fig. 1571

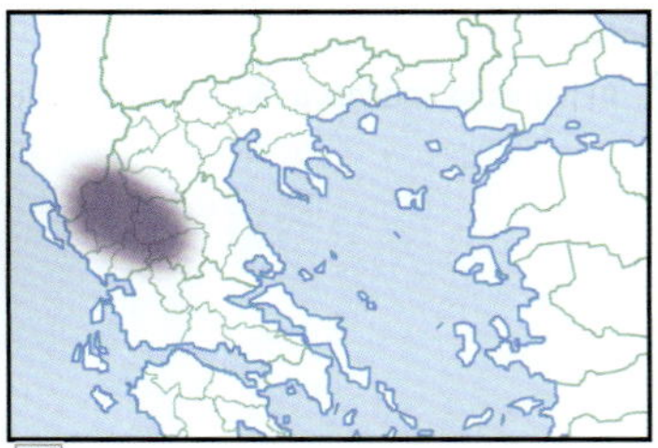

NW Greece (Ipiros), ? S Albania

Westerlund 1886: 80, Kobelt 1902: 5, Riedel 1992: 50, Riedel 1993: 195.

Doraegopis

Doraegopis Riedel, 1982 in Riedel 1982: 20

Type species *Aegopis boeoticus* Riedel, 1980 (original designation)

Presence of one single vaginal gland that is conspicuously stalked and provided with retractor muscles inserting on finger-like glandular processes. Vas deferens inserts laterial into penis, flagellum present.

Riedel 1992, Riedel 1993, Gittenberger 1999.

One additional new species was found in N Pelopónisos (Riedel 1992: 54).

Doraegopis boeoticus (Riedel, 1980)

Aegopis boeoticus Riedel, 1980 in Riedel 1980: 485

Shell light brownish, lighter at lower side, thin, slightly translucent, finely and densely radially and spirally striated, apex shiny and almost smooth, shell keeled at first whorls, with blunt edge at last whorl, umbilicus wide (1/5 of shell diameter), not increasing at last whorl.

7-10 x 15-20 mm

Species was found at mountain slopes covered with herbs and shrubs, rarely near *Pinus* and *Quercus ilex*, also in rock crevices deeply inside humid soil with fine calcareous rocky substrate. Empty shells were also found at rocks and under stones.

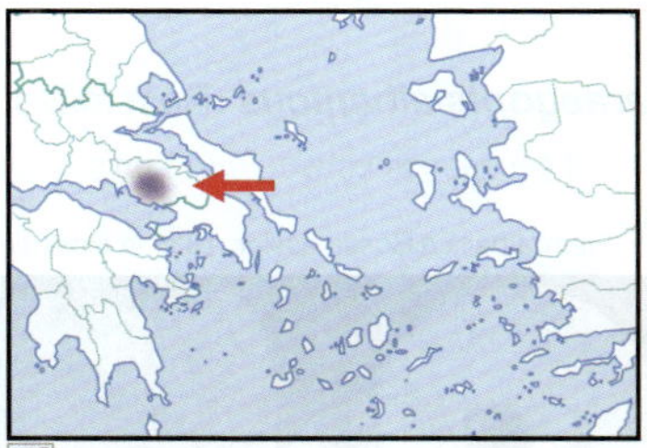

Greece (Viotía)

Doraegopis carinatus Gittenberger, 1999

Doraegopis carinatus Gittenberger, 1999 in Gittenberger 1999: 247

Shell very flat with a strong and sharp keel with indented margin, also at the aperture, colour pale yellowish brown with some radial yellowish streaks, 6 whorls, almost flat on the upper side but concave at both sides of the keel, umbilicus wide (1/4 of shell diameter).

Restricted to very few localities, but may be locally abundant.

5-6.5 x 14-17 mm

Greece: mountain near Ritsona N of Athína

Panayotis Ovalis

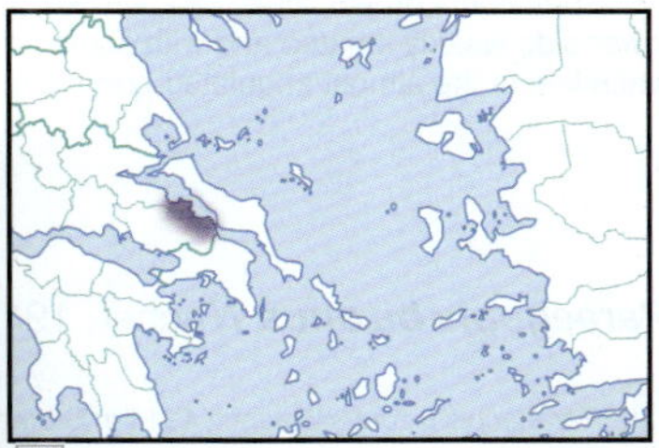

Greece (Viotía)

Doraegopis euboeicus Bank & Menkhorst, 1988

Doraegopis euboeicus Bank & Menkhorst, 1988 in Bank & Menkhorst 1988: 18 147

Shell light horny to light brownish, lower side lighter and more olive greenish, slightly translucent, weakly striated and shiny on upper side, more shiny and smooth on lower side, 6 whorls, last whorl 1.3-1.6 x wider than penultimate whorl, periphery at first whorls with sharp edge (but no keel), blunter at last whorl, umbilicus wide, 1/4 of shell diameter, all whorls visible inside.

7.3-8 x 15.5-18 mm

Shells were found in a quarry.

Sexually mature animals were found at the end of May. In April the genitalia of fully grown individuals were not well developed.

Riedel 1992: 54, Riedel 1993: 212.

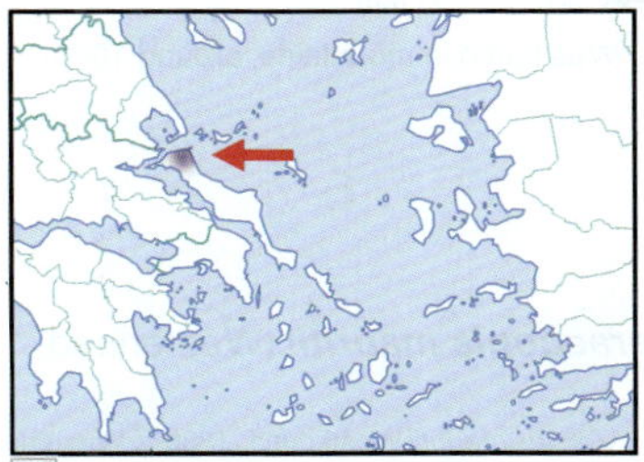

Greece (Evia)

Only known from one locality.

Doraegopis parnonicus Riedel, 1982

Doraegopis parnonicus Riedel, 1982 in Riedel 1982: 21

Shell light horny brownish, apex often whitish, not much translucent, with very pale silky sheen, densely striated with spiral lines forming a distinct reticulate pattern, whorls more convex on upper side than in other species, keeled at first whorls, with blunt edge or almost rounded at last whorl and aperture, umbilicus wide (1/4 of shell diameter) and perspectival, showing all whorls inside.

7.5-8.5 x 15.5-17.5 mm

Lives syntopically with *Zonites oertzeni*.

Greece (Párnonas mountains)

Doraegopis subaii Riedel, 1990

Doraegopis subaii Riedel, 1990 in Riedel 1990: 509

Shell horny yellowish, lower side paler, solid and not translucent, apex shiny, all whorls with dense spiral lines, lower whorls striated provoking a reticulate pattern, slightly less prominent on lower side, 5.5-6 weakly convex and slowly increasing whorls with moderately deep suture, juvenile shells bluntly keeled, last whorl only slightly wider than penultimate whorl, initially with blunt edge at periphery, aperture rounded at periphery, umbilicus wide and perspectival, 1/5-1/4 of diameter, all whorls visible inside.

6-7 x 16.5-18 mm

Was found in a moist cave at a steep limestone rocky slope of a gorge, but the species is probably not a cave snail.

Obviously rare.

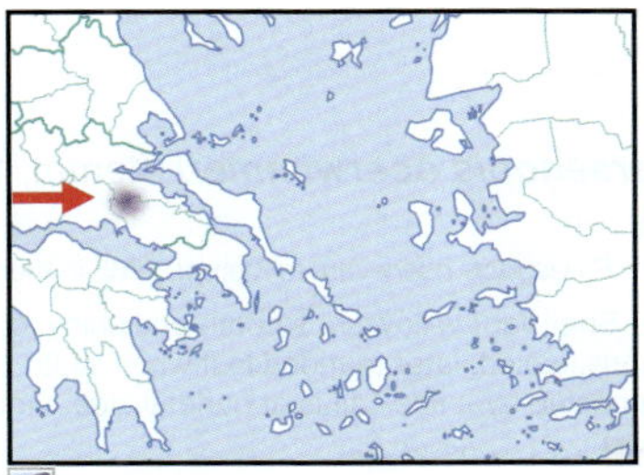

Greece (Parnasós mountains)

Paraegopis

Paraegopis Hesse, 1910
in Hesse 1910: 168 *Zonites (Paraegopis)*

Type species *Helix albanica* Rossmässler, 1836
(original designation)

Shell diameter 17-40 mm, surface granulated, embryonal shell with 2.5 whorls and same surface sculpture as the lower whorls.
Penis thickened in central section, without flagellum but with epiphallus, retractor inserts at distal end of penis. Right ommatophore retractor running between male and female genital organs.

Forcart 1957: 112.

Paraegopis albanicus (Rossmässler, 1836)

Without gender agreement: *Paraegopis albanica* (Rossmässler, 1836)

Helix albanica Rossmässler, 1836 in Rossmässler 1835-1837 (3): 1, Taf. 11

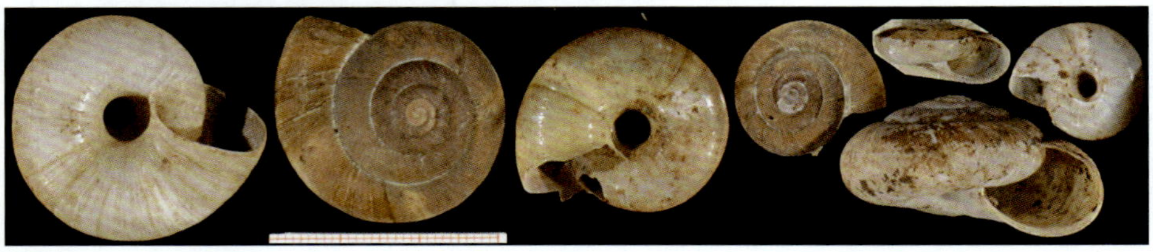

N Albania: near Tamarë

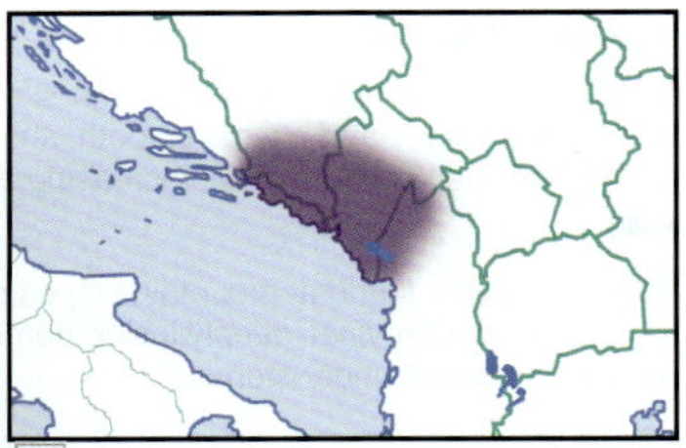

S Bosnia to N Albania

Westerlund 1886: 78, Falkner 1990: 176, Dhora & Welter-Schultes 1996: 146.

13-16 x 32-42 mm

Shell horny brown, lower side whitish, green or brownish yellow, upper side very finely striated and exremely densely spirally or irregularly granulated, lower side weakly striated and with some broad yellowish radial streaks indicating growth interruptions, 5-6 convex whorls with deep suture, last whorl with a white line at the almost angulated periphery, umbilicus very wide.

Paraegopis bizonus Wagner, 1914

Paraegopis bizonus Wagner, 1914 in Sturany & Wagner 1915: 27, Taf. 1, Taf. 3

Shell light horny or dirty yellowish, translucent and thin, weakly and regularly striated but without marked growth interruptions, and with regular spiral lines on the upper side, producing a fine granular structure, lower side almost smooth with weak growth lines, radial and spiral lines visible again in the umbilicus, 4.5-5 weakly convex whorls, juveniles keeled but in adult shells this is not visible at the suture, last whorl can be angulated, apertural margin sharp and straight, umbilicus narrow.
Differs from *P. mauritii* in its smaller size, absence of growth interruptions, more translucent shell with usually rounded periphery and larger umbilicus.

9-11 x 18-23 mm

Was found in mountains, around 1000-1500 m.

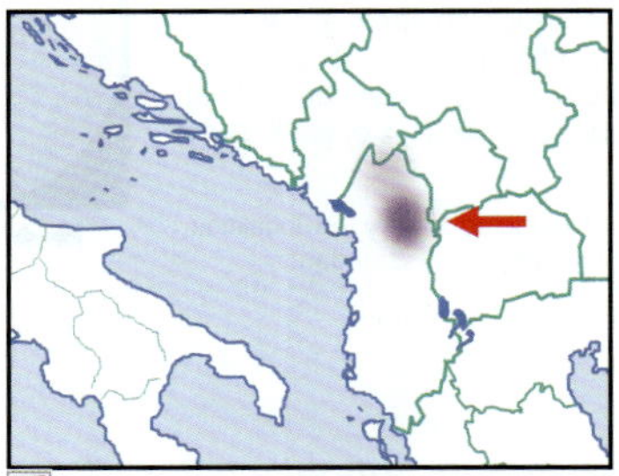

NE Albania

Paraegopis mauritii (Westerlund, 1886)

Zonites mauritii Westerlund, 1886 in Westerlund 1886: 77

Bosnia: Sutjeska National Park, Tientiste near Foča

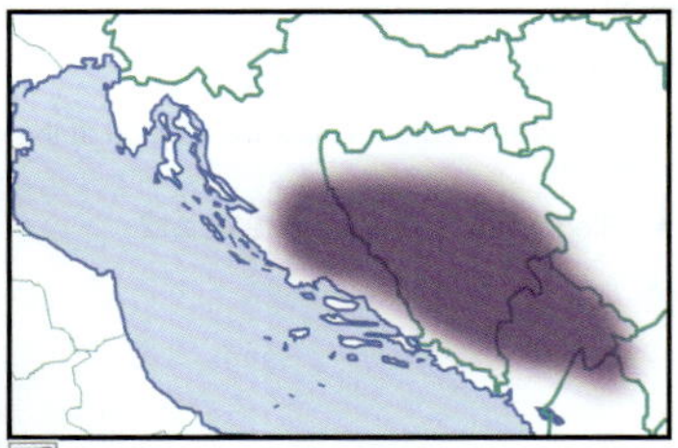

W Croatia to NE Albania

8-14 x 18-28 mm

Shell horny brownish, densely and prominently striated on upper side, and with extremely fine spiral lines, lower side slightly lighter, smooth and shiny, on last whorl with some broad yellow growth interruptions, 5-5.5 weakly convex whorls, first whorls with keel, last whorl with edge at periphery, upper side more rounded than lower side, aperture with weak white lip inside, umbilicus wide and perspectivical. Smaller than *Par. albanicus*.

Rocky habitats in mixed mountain forests, up to at least 1400 m.

Kobelt 1904: 3, Fischer 2003: 5.

Montenegro: Durmitor mountains, near Crno jezero
Cédric Audibert

Paraegopis oberwimmeri Klemm, 1965

Paraegopis oberwimmeri Klemm, 1965 in Klemm 1965: 71

Shell light yellowish, smooth and shiny, obliquely striated and with distinct spiral lines on the upper side, first whorls keeled, later rounded last whorl width 1.5 x of penultimate whorl and not descending, apertural margin connected by a fine callus at parietal side, umbilicus moderate, 1/5 of shell diameter, showing all whorls inside.

10-11 x 27-31 mm

Probably a true cave snail.

Gittenberger 1975: 24.

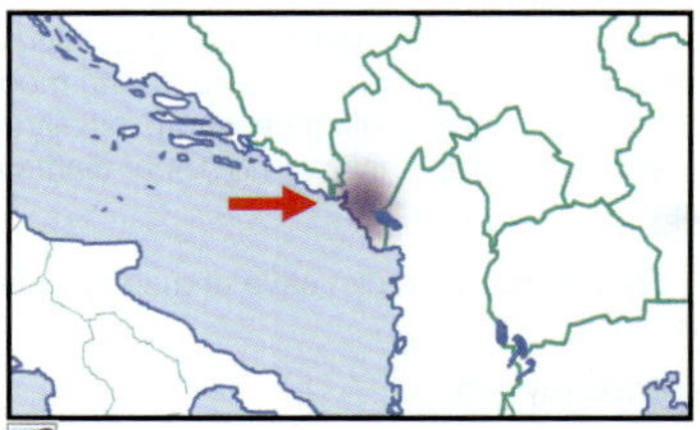

S Montenegro

Paraegopis skipetaricus Wagner, 1914

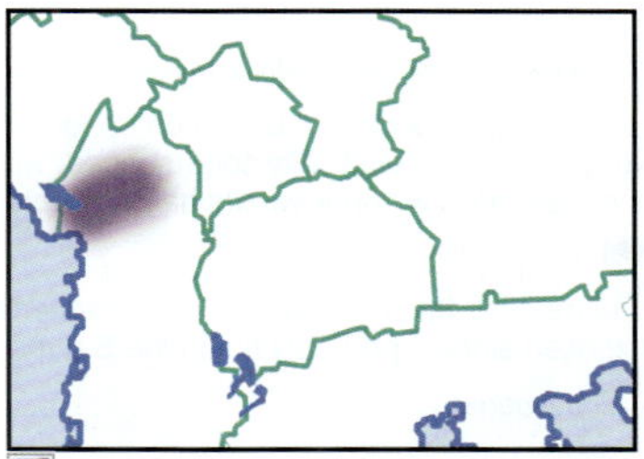

N Albania

Paraegopis skipetaricus Wagner, 1914 in Sturany & Wagner 1915: 26, Taf. 3

Shell keeled except last section which is rounded, horny reddish, translucent on upper side, with some light growth interruptions on the last whorl, radially and spirally granulated, granulation stops abruptly at the keel, lower side yellowish opaque with greenish hue, finely striated, no light band at keel, 5-6 weakly convex whorls with shallow suture, last whorl width 1.3 x of penultimate whorl, umbilicus narrow, 1/7-1/6 of diameter.
Differs from *P. albanicus* in its narrower umbilicus, shells of the same size have 0.5 whorls more (shells with same whorl numbers are smaller), granulated structure stops absuptly at the keel (in *P. albanicus* this extends smoothly to the lower side and becomes gradually weaker).

13-16 x 27-32 mm

Troglaegopis

Troglaegopis Riedel & Radja, 1983
in Riedel & Radja 1983: 259

Type species *Aegopis mosorensis* Kuščer, 1933
(original designation)

Vagina with a long appendix, which works as a stimulatory organ and not as a gland, longer than penis.

Troglaegopis mosorensis (Kuščer, 1933)

Aegopis mosorensis Kuščer, 1933 in Kuščer 1933: 65, Taf. 1

Croatia: Dalmatia, Vranjina Pecina cave

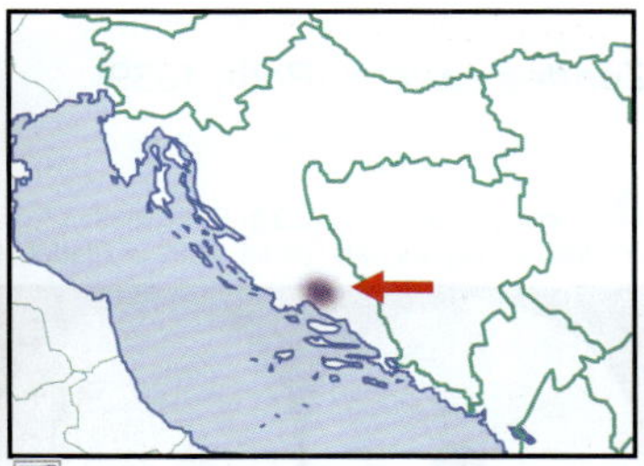

Croatia (Dalmatia)

5-7 x 15-18 mm

Shell dirty whitish or cream, often covered with dust, very densely reticulated, also on lower side, 5-5.5 whorls with shallow suture, last whorl not much wider than penultimate whorl, umbilicus wide, 1/4-1/5 of diameter.
Animal whitish with black eye pigments and blackish ommatophore retractors, body sides slightly transparent, sole subdivided in 3 longitudinal sections.

Only known from caves, probably exclusively subterranean.

Rare, only known from 2 localities (Riedel & Radja 1983).

Zonites

Zonites Montfort, 1810
in Montfort 1810: 282

Type species *Helix algira* Linnæus, 1758
(original designation)

Name was treated as masculine in the original description.

Most species can be considered endangered and should be placed on the Red List (Riedel & Mylonas 1997).

Riedel 1982, Riedel 1985, Riedel 1987, Riedel 1992, Riedel 1995, Kornilios et al. 2009 (phylogeny).

Zonites algirus (Linnæus, 1758)

Without gender agreement: *Zonites algira* (Linnæus, 1758)

Helix algira Linnæus, 1758 in Linnæus 1758: 769

S France: (1-4) Hérault, near Sète, (5) near Avignon

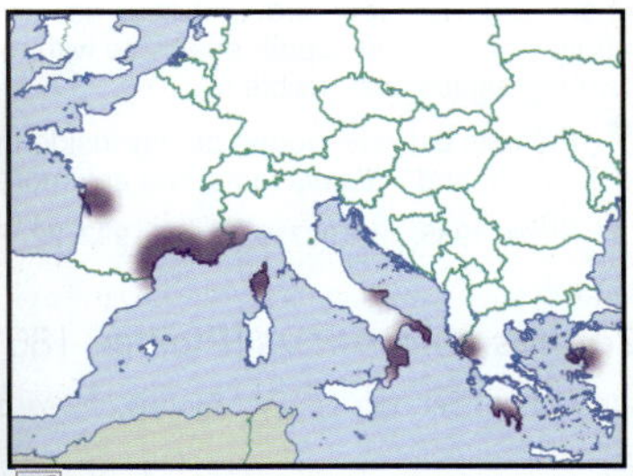

Originally Greece, introduced in historical epochs to France, Italy and elsewhere

15-25 x 35-50 mm, Lésvos up to 42 mm

Shell horny brown or pale yellowish, juveniles almost flat and keeled, last whorls rounded at periphery, umbilicus wide, in Kíthira narrower than in *Z. parnonensis*. Upper shell side granulated, lower side shiny with a coarse reticulate pattern.
Animal grey with bluish hue, lower side blackish.

In S France in cultivated regions, gardens, at margins of arable fields, hides under leaves and in the soil.

In S France eggs are laid in spring, copulation in Greece was made probable for November.

Probably extinct in Liguria (Italy).

Rossmässler 1835-1837 (3): 1, Westerlund 1887: 80, Riedel 1985: 4 (synonyms), Riedel 1992: 61, Manganelli et al. 1995: 18, Örstan 2003 (Istanbul).

France: Provence, near Les Beaux

Zonites anaphiensis Riedel & Mylonas, 1981

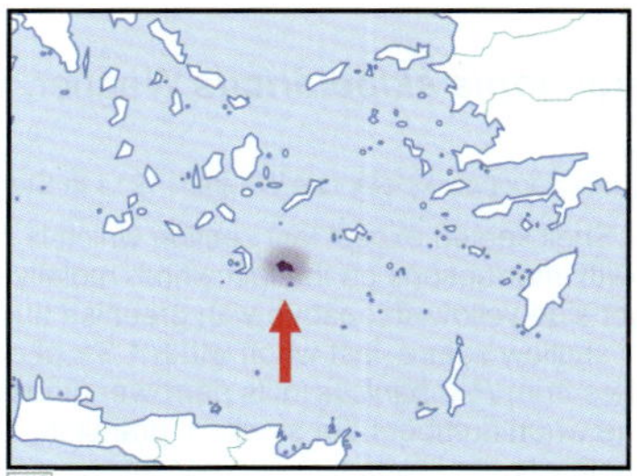

Greece (Anáfi)

Zonites anaphiensis Riedel & Mylonas, 1981 in Riedel & Mylonas 1981: 247

Shell horny reddish or brown on upper side, horny yellowish at lower side, prominently granulated, also on lower side, characteristically cone-shaped, with flat lower side, 5-5.5 whorls, keeled, also at aperture, umbilicus moderately narrow, inner whorls hardly visible inside.

14-18 x 25-32 mm

Limestone habitats covered with Mediterranean shrubland vegetation, particularly under Genista shrubs at N exposed slopes, probably not in the S parts of the island.

Endangered.

Riedel 1985: 38, Riedel 1992: 69.

Zonites astakidae Riedel, 1985

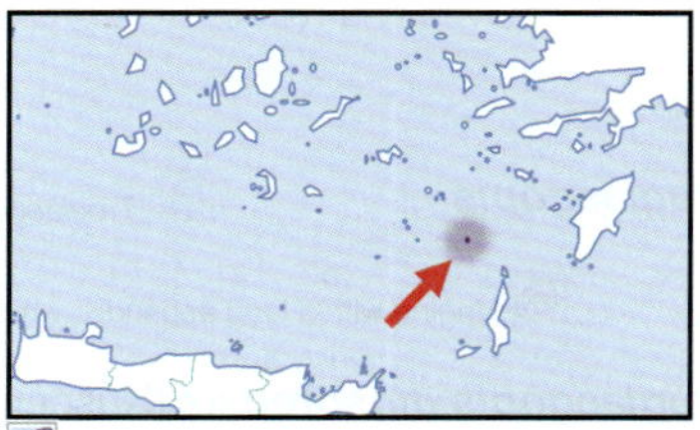

Greece (Astakída near Kárpathos)

Zonites astakidae Riedel, 1985 in Riedel 1985: 51

Shell high, first whorls with sharp keel, disappears between penultimate and last whorl, aperture rounded, umbilicus narrow and deep, upper whorls weakly visible inside. Colour rarely preserved, brownish olice green, spiral granules weak and restricted to last whorls, lower shell side weakly granulated, apex usually smooth.

13-16 x 20-25 mm

Probably endemic for the islet of Astakída 35 km NW of Kárpathos, endangered.

Riedel 1985: 51, Riedel 1992: 72.

Zonites caricus (Roth, 1839)

Without gender agreement: *Zonites carica* (Roth, 1839)

Helix carica Roth, 1839 in Roth 1839: 17, Tab. 1
Zonites beydaglariensis Riedel, 1982 in Riedel 1982: 396

(1-3) Greece: island of Kastelorizo, (4-7) Turkey: Vilayet Antalya, near Kaş

SW Turkey and nearby islands

17-24 x 25-35 mm

Shell locally variable, brownish olive green, lower side slightly lighter and with broad white radial streaks indicating growth interruptions, upper side granulated, lower side smooth, umbilicus 1/6 of shell diameter.
Shells in the east (near Finike) have broader whorls and a convex apex, first whorls with radial lines, those in the west are flatter and possess a distinct keel, also on the last whorls, but very variable within populations, apex is flat or concave, first whorls can be smooth, or have radial lines, which can be interrupted and look like a row of dots, this is locally variable with smooth gradients between the populations. The upper side of the second whorl is usually flat, sometimes concave, sometimes with a crater-like margin, the expression of these features is also locally variable, as well as the internal structure of the penis.

Animals are hidden in the soil, appear only after rain and are not active in September, their general mobility must be low. One of the most frequent land snails between Kaş and Demre.

Riedel 1987: 11, Riedel 1992: 76, Riedel 1995: 24, Neubert et al. 2000, Welter-Schultes 2001 (Journal of Archaeological Science).

Zonites casius Martens, 1889

Zonites casius Martens, 1889 in Martens 1889: 190

Greece: Kásos
Picture: Kobelt 1892: Fig. 764

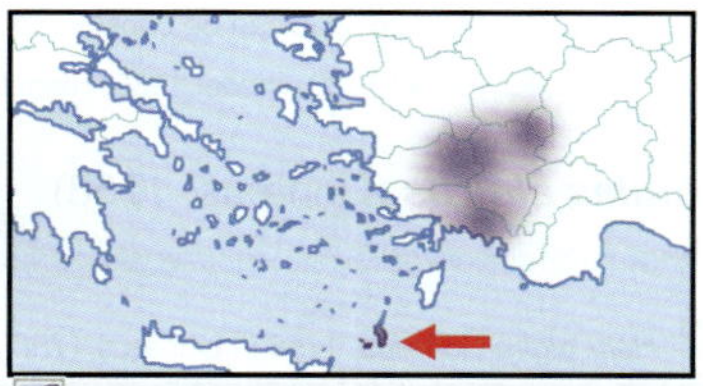

SE Greece (Kásos, Kárpathos), SW Turkey

11-15 x 18-25 mm

Shell not very variable, upper side brownish with yellow streaks, apex reddish brown, lower side lighter, smooth and shiny, periphery with blunt edge at half whorl height, last whorl rounded, umbilicus narrow, upper whorls hardly visible inside. Upper shell side with minute and dense granules, well visible.

Animals estivate digged deeply inside the soil, under rock rubble and *Phlomis* roots. Not any more active in April, rain has no influence. The species is relatively abundant.

Riedel 1985: 46, 1987: 9, Riedel 1992: 72, Riedel 1995: 26.

Zonites chloroticus (Pfeiffer, 1851)

Without gender agreement: *Zonites chlorotica* (Pfeiffer, 1851)

Helix chlorotica Pfeiffer, 1851 in L. Pfeiffer 1851: 127
Date was not 1852, published probably in Sep 1851.

W Turkey: (1-3) Aydın mountains, (4-6) Vilayet Izmir, near Kemalpaşa
(1-3) Aydın Örstan

W Turkey, Sámos

In Chios and Ikaria, probably extinct, not reported after 1900. On Samos rare.

12-32 x 15-40 mm

Shell brownish or dirty greenish yellow with weak silky shine, lower side only slightly lighter, shell usually high, without keel, juveniles with a very blunt edge, suture deep, aperture rounded or slightly flattened, umbilicus narrow (1/7 of shell diameter), upper whorls hardly visible inside. Apex not granulated, with spiral lines and a reticulate pattern, ongoing whorls spiraly granulated, later corrugated in spiral rows. Lower side very different, smooth and shiny, separated by a an edge-like line at the periphery. Whorls on Greek islands less scaly, apex flatter.

Riedel 1987: 4, Riedel 1992: 70, Riedel 1995: 27.

Zonites embolium Fuchs & Käufel, 1936

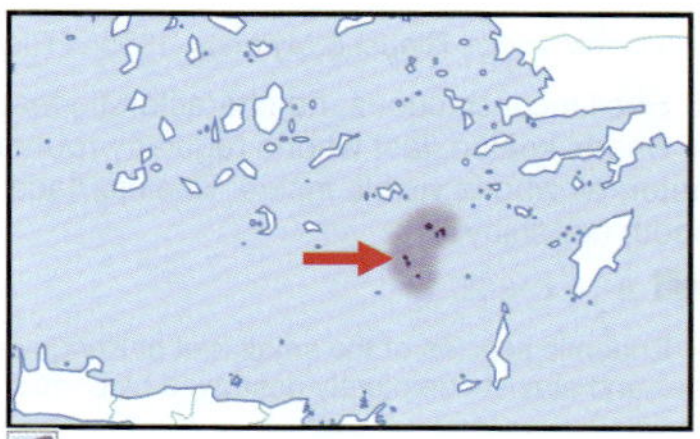

Greece (SE Aegean islands)l

Zonites embolium Fuchs & Käufel, 1936 in Fuchs & Käufel 1936: 612

Shell high, juveniles with edge at periphery, adults rounded, both sides of the shell weakly reticulated. Embryonal whorls smooth or with fine radial lines, not granulated.

13-15 x 25-30 mm

Restricted to some islets between Astipálea and Kárpathos: Dio Adelfi (east), Megalo Zafrano, Karavonisi (Karávi Nisiá), Divounia (= W Unia). Probably extinct on all islets except Megalo Zafrano. Several small islands were used for military purposes and almost completely destroyed by bombs.

Riedel 1985: 50, Riedel 1992: 72.

Zonites euboeicus Kobelt, 1878

Without gender agreement: *Zonites euboeica* Kobelt, 1878

Zonites verticillus var. *euboeica* Kobelt, 1878 in Kobelt 1878: 321 (*Zonites verticillus* var. *eubœica*)

Shell upper side dark brown, lower side light green, finely streaked and granulated, with sharp white keel, at aperture with keel or rounded, umbilicus wide.

Westerlund 1887: 79, Riedel 1985: 30, Riedel 1992: 66.

10-20 x 18-30 mm

Greece: Evia, Niki and Agalis gorge (large shell 36 mm)

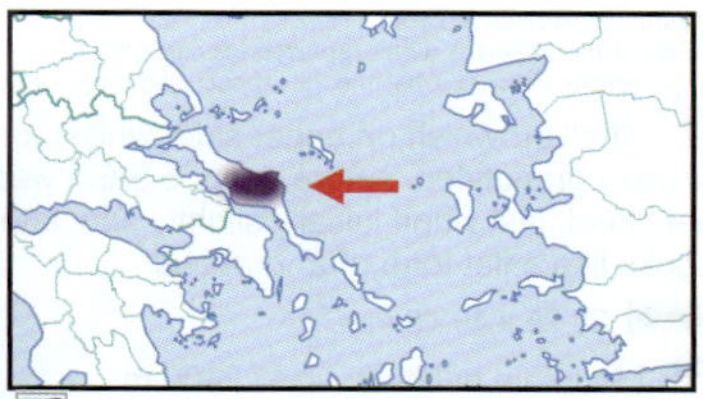

Greece (central Evia)

Robert Reuselaars

Zonites festae Pollonera, 1916

Zonites festai Pollonera, 1916

Zonites festae Pollonera, 1916 in Pollonera 1916: 3 (*Zonites Festæ*)
Name expressly dedicated to Mr. Festa, *festae* not to be corrected to *festai* (Art. 32.5). Art. 31.1 gives no clear rule and has no effective force.

12-20 x 22-34 mm

Turkey: Vilayet Muğla, Gandir near Dalyan

Riedel 1992: 71.

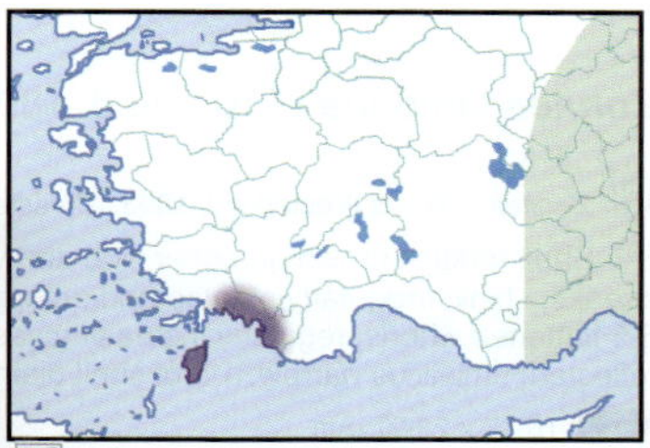

Greece (Ródos), SW Turkey

Last whorl high with blunt edge at periphery, apex flat, aperture rounded, umbilicus narrow, sharp reticular pattern at lower shell side.

Zonites graecus Kobelt, 1876

Without gender agreement: *Zonites graeca* Kobelt, 1876

Zonites verticillus var. *graeca* Kobelt, 1876 in Kobelt 1876: 48, Taf. 110
On the same page was established *Zonites albanicus* var. *graeca* (= currently *Zonites kobelti*). Precedence for *Z. verticillus* var. *graeca* given by First Reviser, if not earlier then herewith (Art. 24.2.3).

Shell brown or lighter, high, in mountains flatter, at 30 mm 5.75-6 whorls in high mountains, 5-5.25 whorls in lower altitudes, last whorls rounded without edge (near Megalopolis with blunt edge), umbilicus narrow, not perspectivical, in mountains wider, near Megalopolis very variable and perspectivical. First whorls flat or concave, almost smooth, next whorls on upper side granulated, lower side smooth and shiny.

15-25 x 25-40 mm, locally variable

Greece: S Pelopónisos, near Gíthio

Greece (Pelopónisos)

Westerlund 1887: 79, Riedel 1985: 8, Riedel 1992: 62.

Could be conspecific with *Z. messenicus*, this is suggested by transitional populations south of Kiparissia.

Zonites humilis Riedel, 1982

Zonites humilis Riedel, 1982 in Riedel 1982: 409

12-20 x 18-30 mm

Turkey: Vilayet Muğla, Iztusu beach rocks

SW Turkey

Shell pale yellow, upper whorls with blunt edge, lower whorls rounded, umbilicus 1/6-1/5 of shell diameter. Embryonal whorls with spiral lines, later continuing as rows of elongated granules.

Riedel 1987: 9, Riedel 1992: 73, Riedel 1995: 29.

Shells of this species were found washed ashore at Kárpathos, Greece.

Zonites invitus Riedel & Mylonas, 1995

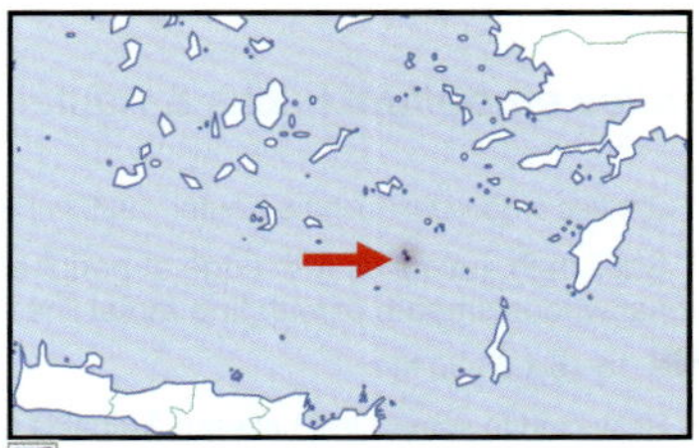

Greece (islet of Kendriko Sofrano)

Zonites invitus Riedel & Mylonas, 1995 in Riedel & Mylonas 1995: 240

Shell inconspicuous, no characterstic features, horny yellow with green hue. All whorls convex, suture deep, first whorls keeled, last whorls regularly rounded, lowr shell side very regularly convex, umbilicus 1/5 of shell diameter, all whorls visible inside. Shell surface structure variable, on lower side weaker, lower side is neither smooth nor shiny.

8-14 x 18-25

Endemic species of the small islet of Kendriko Sofrano. Several small islands in the region were used for military purposes and almost completely destroyed by bombs.

Zonites kobelti Kobelt, 1898

Greece (Pelopónisos)

Zonites kobelti Kobelt, 1898 in Kobelt 1905: 869, Taf. 229

Name was attributed to Böttger, description was not attributed to a person, so under Kobelt's responsibility. In a strict application of Art. 50.1.1 the name in such a case must be attributed to the author of the work. Established as a new name, no new replacement name for *Zonites albanica* var. *graeca* Kobelt, 1876 (as suggested by Riedel 1992: 62) due to the lack of an expressed statement (Art. 72.7) and because nothing indicated why this name should have been replaced.

Even if interpreted as a new replacement name the name must also be attributed to Kobelt because it was Kobelt who would have decided under his own responsibility to substitute the preoccupied name (and to employ this name for the species, not the other name), not Böttger.

Shell brownish on upper side, whitish or with green hue on lower side, striated and on upper side very sharply granulated, suture deep, 5.5-6.5 weakly convex whorls, apex flat and smooth, first whorls without prominent keel, last whorl initially slightly angulated, near aperture rounded, umbilicus wide and perspectivical. Looks like a flat form of *Z. graecus*.

20-25 x 40-45 mm

Mountain habitats, up to 2400 m.

Riedel 1985: 14, Riedel 1992: 63.

Found only sporadically, perhaps only a flat form of *Z. graecus*, and perhaps like this form conspecific with *Z. messenicus*. More research on the local variation of these forms seems to be necessary.

Zonites labiosus Westerlund, 1893

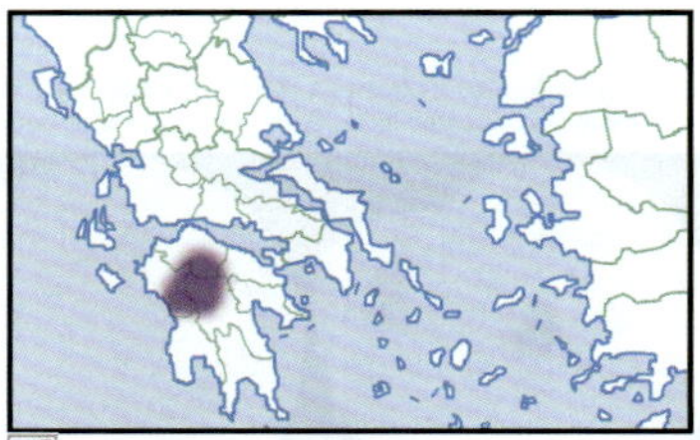

Greece (Pelopónisos)

Zonites labiosus Westerlund, 1893 in Westerlund 1893: 118

Shell reddish brown on upper side and yellowish on lower side, with numerous (11-12) yellowish radial streaks, shiny, irregularly striated and on upper side with very dense spiral lines, forming a granulated pattern, 5-6 flattened whorls, regularly increasing, last whorl initially slightly angulated, upper side of columellar margin reflected, umbilicus narrow, 1/8 of shell diameter.

16-17 x 23-27 mm

Riedel 1992: 64.

Zonites messenicus Zilch, 1965

Greece (Pelopónisos)

Riedel 1985: 14, Riedel 1992: 63.

More research on the local variation is necessary.

Zonites cytherae messenicus Zilch, 1965 in Zilch 1965: 82

Greece: Pelopónisos, Paleo Messini

Shell upper side brown, lower side light greenish with radial streaks and shiny, keeled, last whorl at least with blunt edge. First whorls concave, upper shell side granulated, lower side often also granulated or reticulated.

Typical form near Othoni, shells near Kiparissia slightly smaller and higher, umbilicus narrower, keel of adults not sharp.

15-23 x 25-40 mm

Populations south of Kiparissia may form a transitional gradient to *Z. graecus*, with intermediate characters: shell height and umbilicus like near Kiparissia, edge at periphery weaker but shells longer keeled than in *Z. graecus*, apex flat like in *Z. graecus*, or convex, lower shell side usually reticulate. No great differences in genitalia between the two species, except a characteristically thin, long and curved flagellum.

Zonites nautarum Riedel & Mylonas, 1995

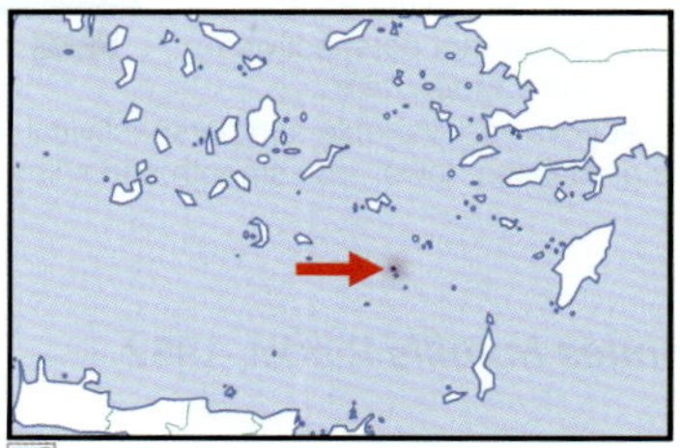

Greece (islet of Megalo Sofrano)

Zonites nautarum Riedel & Mylonas, 1995 in Riedel & Mylonas 1995: 238

Shell dirty horny yellow with slightly lighter lower side, lenticular, with well expressed edge at the peripery, apex almost smooth, umbilicus 1/6 of shell diameter, not enirely open, inner whorls only poorly visible inside. Upper shell side with indistinct and irregular granulation, granulation more distinct and more regular on the lower shell side.

9-14 x 19-27 mm

Endemic species of the small islet of Megalo Sofrano. Several small islands in the region were used for military purposes and almost completely destroyed by bombs.

Zonites nikariae Pfeffer, 1930

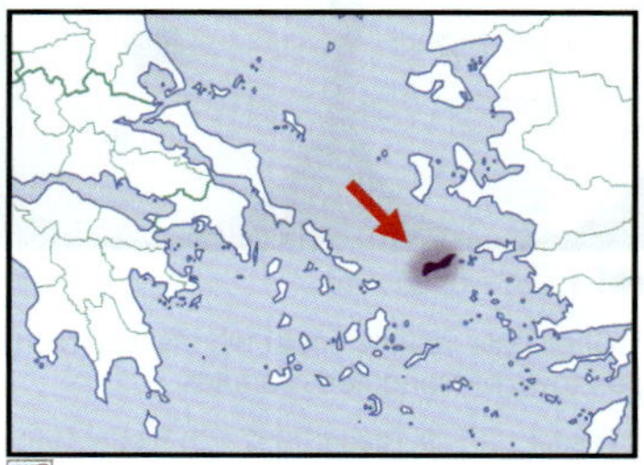

Greece (island of Ikaría)

Zonites nikariae Pfeffer, 1930 in Pfeffer 1930: 488

Shell upper side flatter than lower side, apex blunt, juveniles sharply keeled, adults with sharp edge, periphery near the aperture of fully grown shells rounded. Umbilicus wide, 1/5 of shell diameter, perspectivical. Upper shell side with spiral rows of dense and small granules, blurred on last whorl, on lower side much weaker. Differs from *Z. smyrnensis* by its sharper and higher edge, a more convex lower side, wider umbilicus and smaller granules.

(nowhere given, but probably like similar species)

Endangered, seems to be very rare. Occurrence on the island of Dílos questionable.

Riedel 1985: 43, Riedel 1992: 70.

Zonites nisyrius Riedel & Mylonas, 1997

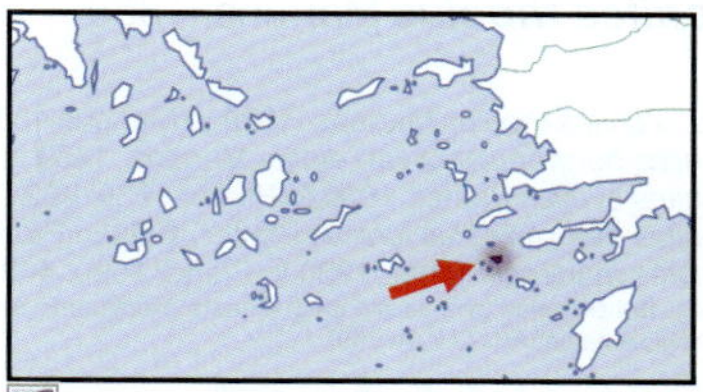

Zonites nisyrius Riedel & Mylonas, 1997 in Riedel & Mylonas 1997: 158

Shell thin and fragile, horny brown periostracum with 1-3 radial streaks in fresh shells, surface densely granulated, 5-5.5 weakly convex whorls, last whorl width 1.5 x of preceding whorl, juveniles keeled, periphery of adults rounded, lower side regularly convex, umbilicus moderately wide, deep and not very perspectivical, slightly covered by the columellar margin.

20-25 x 14-17 mm

Almost certainly an endemic species of the island of Nísiros, endangered.

Greece (island of Nísiros)

Zonites oertzeni Martens, 1889

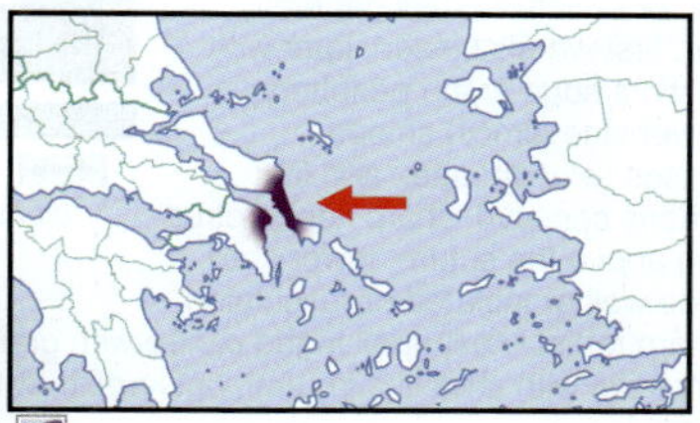

Zonites oertzeni Martens, 1889 in Martens 1889: 173, Taf. 9 (*Zonites örtzeni*)

Shell yellowish, spirally granulated, lower side smooth, upper side flatter than lower side, apex slightly convex, juveniles keeled, following whorls without keel or edge, last whorl with a faintly visible edge, umbilicus wide, all whorls visible inside.
Shells are not different from southern *Z. parnonensis* populations.
Genitalia: flagellum thinner and much longer in southern *Z. parnonensis* populations, oviduct there longer than vagina, receptaculum seminis there very long and slender.

11-18 x 22-35 mm

Greece: Evia (?)
Picture: Kobelt 1892: Fig. 761

Greece (Evia, Attikí)

Copulation possibly in April.

Riedel 1985: 27, Riedel 1992: 66.

Zonites osmanicus Riedel, 1987

Zonites osmanicus Riedel, 1987 in Riedel 1987: 14

Shell pale yellowish with some light radial streaks, lower side with only a weak fatty shine, apex blunt, upper shell side flatter than lower side, with sharp edge at the peripery, first whorls can be keeled, edge at last whorl can be blunt, umbilicus wide, all whorls visible inside. Upper shell side from first whorl onwards with dense spiral rows of small granules, lower side near Denizli distinctly reticulated or granulated, near Pamukkale and Çardak much weaker.
Differs from *Z. humilis* in its lower apertural height and the relief of the apex. *Z. smyrnensis* is larger with narrower umbilicus. *Turcozonites wandae* has coarser granules on the first whorl, its lower shell side is smooth and shiny.

10-14 x 20-27 mm

Riedel 1995: 30.

SW Turkey

Zonites parnonensis Riedel, 1985

Zonites parnonensis Riedel, 1985 in Riedel 1985: 20
Zonites parnonensis fallax Riedel, 1985 in Riedel 1985: 24

Shell yellowish, upper shell side flatter than lower side, apex concave or flat, in the south slightly convex, juveniles keeled, following whorls without keel or edge, last whorl with a faintly visible edge, umbilicus wide, all whorls visible inside, in Kíthira wider than in *Z. algirus*. Shell surface in most populations spirally granulated, lower side smooth. Whorls in the south more narrowly coiled than in the north. In the south shells do not differ from *Z. oertzeni*.
Genitalia: flagellum thinner and much longer in southern *Z. parnonensis* populations, oviduct longer than vagina, receptaculum seminis very long and slender.

Shell diameter up to 40 mm

Riedel 1992: 64.

Two extreme forms *parnonensis* in the north and *fallax* in the south, with extremely broad transitional zones in between. A classification as two subspecies is probably not appropriate in such a situation (no partial barriers to reproduction).

Greece (Pelopónisos, Kíthira)

Zonites pergranulatus Kobelt, 1878

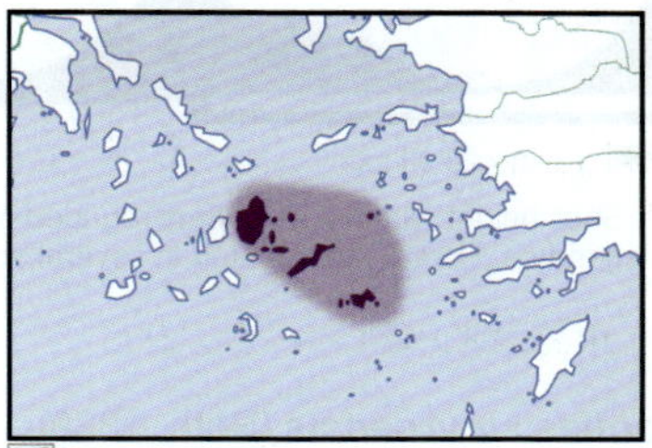

Zonites pergranulatus Kobelt, 1878 in Kobelt 1878: 320

Shell brownish, lower side lighter brownish or olive green, characteristically lenticulate (lower side flatter in Kínaros), with sharp light edge at the periphery (except in Náxos), first whorls with sharp keel, umbilicus narrow (wider in Kínaros), upper whorls hardly visible inside. Spirally granulated on upper shell side, lower side variable, sometimes smooth.

13-17 x 22-30 mm

Copulation possibly in December.

Westerlund 1886: 79, Riedel 1985: 32, Riedel 1992: 66.

Greece (E Kikládes)

Zonites rhodius Martens, 1889

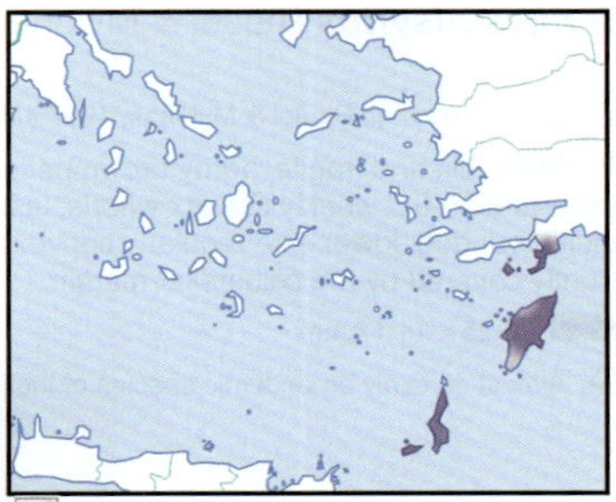

Zonites rhodius Martens, 1889 in Martens 1889: 191, Taf. 3
Zonites pergranulatus var. *elatior* Martens, 1889 in Martens 1889: 190
Zonites symius Pfeffer, 1930 in Pfeffer 1930: 462
Zonites pergranulatus ionicus Pfeffer, 1930 in Pfeffer 1930: 470
Zonites martensi Pfeffer, 1930 in Pfeffer 1930: 440
Regular new name, not (as intended) a new replacement name for *Zonites caricus* as misidentified by (= sensu) Martens 1889 because this was not an available name (Art. 72.7, Art. 49).

12-20 x 25-40 mm, diameter in Ródos up to 30 mm, in Kárpathos, Sími and Turkey up to 42 mm

Greece: Kárpathos, near Othos

Ródos (*rhodius*): upper shell side weakly convex, with fine and well visible granular structure, almost no edge at periphery, umbilicus wide.
Sími and Turkey (*symius*): apex flat or concave, shell generally flatter, first whorls keeled, later with distinct edge at the periphery, lower side almost smooth.
Kásos (*elatior*): apex and first whorls concave, sharp keel protruding above the suture, lower shell side well granulated or reticulated, sharp radial lines, apex in the centre with granules and undulated radial streaks, fossils without sharp keel, only with faintly visible edge. Another form of Kásos (*ionicus*) with shell like in Sími, but keeled, with broader whorls (0.5 whorls more at similar shell diameter in Kásos), lower shell side with much more prominent granules.
Kárpathos (*martensi*): shell high, first whorls keeled and concave like in Kásos, near Volada flat or slightly convex, lower side more convex than upper side, last whorl rounded without edge, umbilicus variable, granulation on postembryonal whorls weak, lower side smooth.

SE Greece, SW Turkey

Copulation probably in April.

Probably shortly before extinction (Riedel 1992, Frank 1997: 13).

Riedel 1985: 55, Riedel 1992: 73, Riedel 1995: 31.

Riedel recognized 4 subspecies *Z. r. rhodius*, *symius*, *elatior* and *martensi*.

Zonites santoriniensis Riedel & Norris, 1987

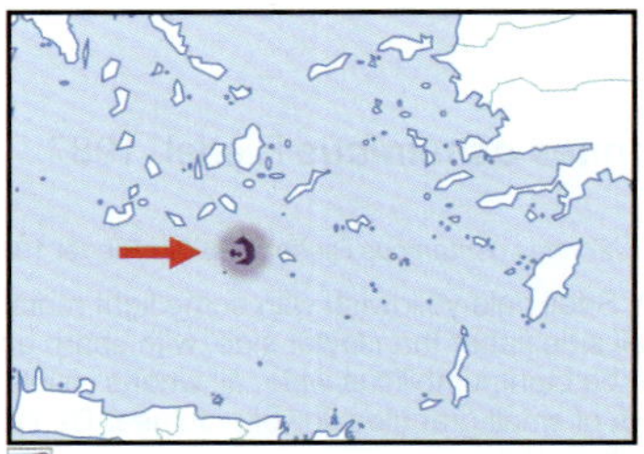

Zonites santoriniensis Riedel & Norris, 1987 in Riedel & Norris 1987: 377

First whorls sharply keeled in the mid-line of the periphery, last whorls rounded, umbilicus wide, 1/6 of shell diameter, all whorls visible inside. Upper shell side spirally granulated, more coarse and less distinctly developed on the last whorl, sculpture on the lower side much weaker.

17-23 x 25-37 mm

Extinct endemic species of Thíra, did probably not survive the volcanic eruption at 1450 BC.

Riedel 1992: 68.

Greece (Santoríni = Thíra)

Zonites sariae Riedel, 1985

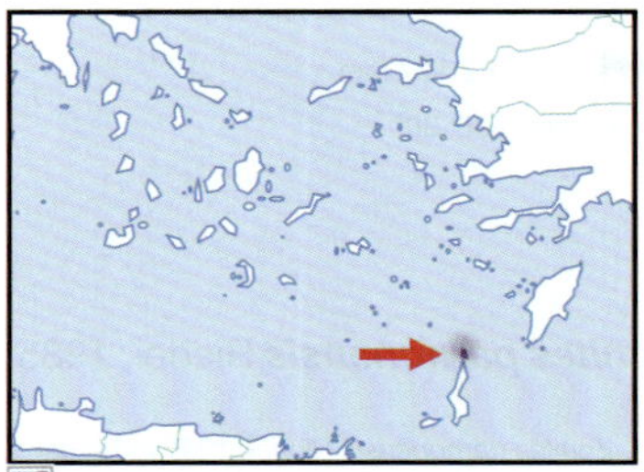

Zonites sariae Riedel, 1985 in Riedel 1985: 52

Shell brown and not shiny on upper side, light olive green on lower side, periphery of first whorls with very blunt edge, last whorl rounded, aperture almost circular, umbilicus wide, inner whorls well visible inside. Apex with spiral scars and corrugated radial lines, later with dense spiral granules or a reticulate pattern, much weaker on lower shell side. First whorls less convex than in *Z. chloroticus*, shells without edge at peripery, but could be conspecific.

13-20 x 22-34 mm

Known only from one locality, rare and occurs probably only on this island.

Riedel 1992: 71.

Greece (island of Sariá)

Zonites siphnicus Fuchs & Käufel, 1936

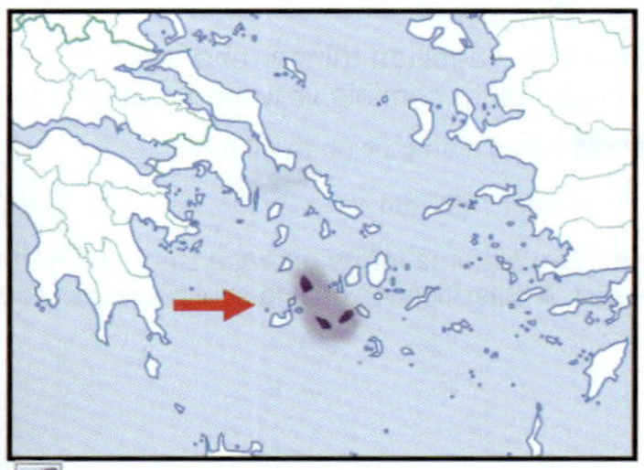

Zonites siphnicus Fuchs & Käufel, 1936 in Fuchs & Käufel 1936: 610

13-16 x 28-33 mm

Greece: Sífnos

Greece (Sífnos, Síkinos, Folégandros)

Probably extinct, only known as subfossils.

Riedel 1985: 43, Riedel 1992: 68.

Shell height variable, periphery rounded or with blunt edge, juveniles not keeled, 5.5-6.5 whorls, umbilicus up to 1/5 of shell diameter, narrower in Folégandros. Radially corrugated in several spiral rows on the first whorls, later a spiral reticulate or granulate pattern, but often blurry.

Zonites smyrnensis (Roth, 1839)

Helix smyrnensis Roth, 1839 in Roth 1839: 16, Tab. 1

Shell yellowish, upper side finely spirally granulated, lower side finely striated forming a reticulated pattern, periphery keeled, also at last whorl, first whorls flat, lower side more convex than upper side.

10-16 x 20-32 mm

Critically Endangered or Extinct in the urbanized area of Izmir.

Westerlund 1886: 78, Riedel 1987: 8, Riedel 1992: 69, Riedel 1995: 33.

W Turkey, Chíos

Turcozonites

Turcozonites Riedel, 1987 in Riedel 1987: 12

Type species *Zonites wandae* Riedel, 1982 (original designation)

S Turkey between Antalya and the Syrian border

Riedel 1995.

Turcozonites anamurensis Neubert & Riedel, 1995

Turcozonites anamurensis Neubert & Riedel, 1995 in Riedel 1995 (Malak. Abh.): 125

Shell yellowish brown on upper side, lower side lighter and sometimes with greenish hue, shiny and slightly translucent, surface sculpture consisting of spirally and densely arranged granules, 5-5.5 slowly and regularly increasing whorls, juveniles keeled, last whorl not much wider than penultimate whorl, initially bluntly keeled, weaker near aperture, umbilicus narrow, 1/7 of diameter.

14-17 x 20-26 mm

S Turkey: Anamur castle, E Antalya

S Turkey

Rocky habitats, at marble rocks and in ruins.

Riedel 1995: 32, Schütt 2001: 195.

Turcozonites megistus (Rolle, 1894)

Zonites megistus Rolle, 1894 in Rolle 1894: 138

Named after and originally reported from Megisti = Kastelorizo island, but the species does not live there.

Shell yellowish brown, with spirally arranged granules, 5.5-6 rapidly increasing and weakly convex or flat whorls, first whorls sharply keeled, last whorl width 1.7-1.8 x of penultimate whorl, initially keeled, near aperture rounded, umbilicus wide and deep, 1/6-1/5 of diameter.

20-30 x 45-55 mm, shape locally variable

Limestone rocks in gorges, under rock rubble and among rocks in densely vegetated zones.

Riedel 1995: 34, Schütt 2001: 198.

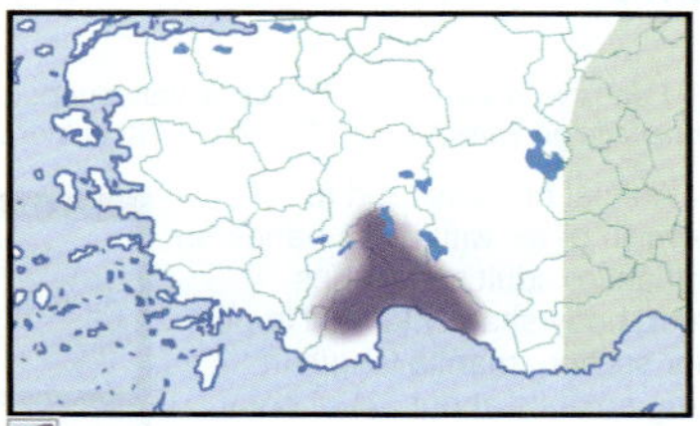

SW Turkey

Turcozonites piratarum (Riedel, 1987)

Zonites piratarum Riedel, 1987 in Riedel 1987: 22

Shell horny brown on upper side and lighter with greenish hue on lower side, thin, lower side shiny, with spirally arranged granules, which can also merge to short spiral lines, lower side smooth, 5.5-6 rapidly increasing weakly convex whorls, last whorl width 1.5 x of penultimate whorl, umbilicus narrow and deep, 1/8-1/7 of shell diameter.

12-22 x 33-39 mm, height variable

Under large boulders and among limestone rocks.

Riedel 1995: 36, Schütt 2001: 199.

S Turkey

Turcozonites wandae (Riedel, 1982)

Zonites wandae Riedel, 1982 in Riedel 1982: 405

12-16 x 22-28 mm

S Turkey: Perge near Antalya

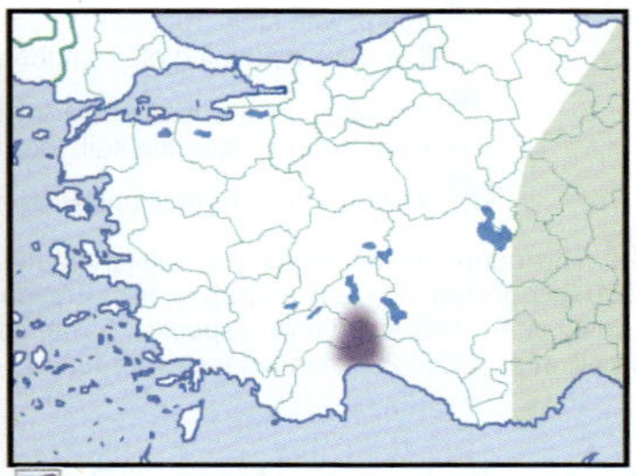

S Turkey

Shell reddish brown, lower side lighter and shiny, with some light radial streaks, solid, upper side densely and prominently granulated, 5-5.5 weakly convex whorls, slowly and regularly increasing, first whorls keeled, later with edge at periphery, also at aperture, last whorl width 1.5 x of penultimate whorl, umbilicus moderate and deep, 1/7-1/6 of shell diameter.

Stony and rocky habitats, not too hot.

Riedel 1995: 37, Schütt 2001: 201.

Parmacella

Parmacella Cuvier, 1804 in Cuvier 1804: 442 *parmacella*

Type species *Parmacella olivieri* Cuvier, 1804 (monotypy)
(= currently *Parmacella olivieri* Cuvier, 1804 from N Africa)

On p. 435 only the French name parmacelle was mentioned, on p. 442 the scientific name *parmacella*.

Animal slug-like with large mantle in central position of the body, foot poited at posterior part, pneumostome in posterior section at right side of mantle. Small shell composed of a very small and more or less coloured dextral protoconch and a large white suboval last 0.5 whorl section.
Juveniles are initially able to retreat entirely within the shell, this ability gets rapidly lost with increasing animal size. The mantle grows over the shell and finally covers it, so that the animal becomes a slug.

Herbivorous and noctural animals, shy and hidden at daytime.

Oval eggs are laid one by one.

Canary Islands, N Africa and Europe to Afghanistan

Germain 1930: 112, Kerney et al. 1983: 294, Wiktor 1983 (Malak. Abh.).

Parmacella ibera Eichwald, 1841

Parmacella olivieri var. *ibera* Eichwald, 1841 in Eichwald 1841: 250

Slug greyish with brownish hue, mantle located in the central part of the body, pneumostome in the posterior part of the mantle, posterior part of body looks like triangulated.
Shell small with circular aperture.

Up to 95 mm long (preserved 65 mm)

Sysoev & Schileyko 2009: 159, Gural-Sverlova et al. 2010, Balashov & Gural-Sverlova 2012: 100.

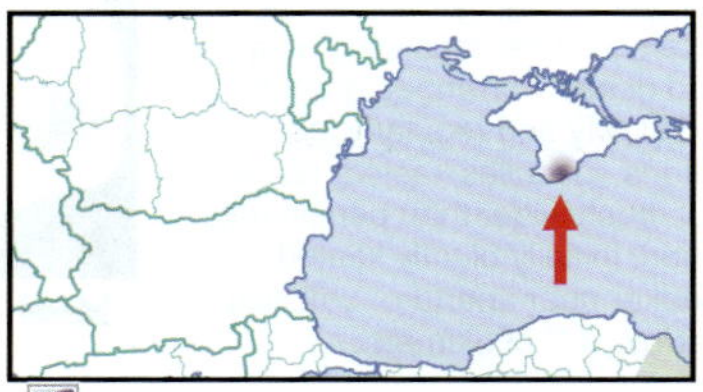

W Kopet Dagh to E Caucasus region, S Crimea

Parmacella valenciennii Webb & Vanbeneden, 1836

Parmacella valenciennii Webb & Vanbeneden, 1836 in Webb & Vanbeneden 1836: 2, Pl. 75 (*Parmacella Valenciennii*)
Parmacella gervaisii Moquin-Tandon, 1850 in Moquin-Tandon 1850: 51 (*Parmacella Gervaisii*)

Animal brownish red to brownish olive, with black bands on the mantle, adult specimens lighter, body sharply keeled in posterior section, mantle length in young animals about 3/4 of body, in adults 40 mm, rounded in anterior and pointed in posterior part. Mucus colourless and sticky. Shell yellow, triangular, fragile and flat, 1.5 whorls, protoconch (first 0.5 whorls) to first whorl width ratio 0.35, anterior part light yellowish with greenish hue, 6 times larger than posterior part.

Portugal: Faro, Algarve
Valter Jacinto

Shell 15-17 x 9-10 x 2-3 (height) mm, animal up to 100 mm long, usually 70-90 x 20 mm

Moquin-Tandon 1855: 34, 37, Nobre 1913: 136, Germain 1930: 112, 114 Nobre 1941: 39, Hutterer 1990: 88.

Moquin-Tandon 1855 and Germain 1930 did not explain why *P. gervaisii* (found in April) could not simply be a juvenile form of *P. valenciennii* (found in May at the same sites).

S Iberian peninsula

Stony open habitats, prefers shady and humid sites between plants and stones. The animals move slowly. High abundance in spring, the animals were so abundant that they were used as pig food in S Portugal. In S France only reported in the 1850s from one site near Arles, was most active in May at rainy nights, population was not found again.

Testacella

Testacella Lamarck, 1801 in Lamarck 1801 [Jan 1801]: 96

Type species undecided and controversial.
Either *Testacella haliotoides* Lamarck, 1801 (monotypy) or *Testacella haliotidea* Draparnaud, 1801 (subsequent monotypy by Draparnaud 1801: 99)

Previously mentioned by Cuvier 1800: Tab. 5, nomen nudum without description or bibliographical reference.
ICZN Opinion 39, Direction 32 regarded names of the systematic tables as available if they were equipped with a bibliographical reference on p. xix of the same work. I do not think that p. xix contains any bibliographical reference at all, I just see general references to the work of various authors, without any published source being cited bibliographically. For the "testacés" Cuvier gave a general remark on p. xx that he took advantage of the work ("travail") of Lamarck. Since Lamarck had published various works on testaceous animals before 1800, the term "travail" cannot be interpreted as a bibliographical reference. The fact that this was a "Direction" published by the Commission does not facilitate the situation.
Lamarck 1801 [Jan 1801]: 96 established *Testacella* with a description and included one species *T. haliotoides* for which no separate description was given, but which was eventually made available under Art. 12.2.6 (combined description of new genus and new species). It was clear from the context and the style of the work that the description was only intended to apply to the genus, not to the species (it was taken as understood that the genera contained more than one species and that under each genus only one species was listed as an example). If regarded as available, *T. haliotoides* would be the type species by monotypy and its syntypes consisted of Maugé's specimens from Tenerife (conspecific with *T. maugei* Férussac, 1819) and Favanne's (1780: Pl. 76 Fig. A1 and A2) specimens, which were conspecific with the species currently known as *T. haliotidea*.
Draparnaud 1801 [July 1801]: 33, 99 used the name in Lamarck's sense and included only one species *T. haliotidea*. The name *haliotoides* was used for a while and distingished from *T. haliotidea* in the early 1800s, for example by Férussac 1807: 41. Gargominy et al. 2011: 358 attributed the name *T. haliotidea* to Lamarck 1801 and regarded *haliotidea* obviously as an incorrect subsequent spelling for it. This interpretation is not applied here.

Slugs with small rudmentary shell at the posterior end of the body, covering the mantle. Two distinct furrows run from the mantle to the anterior end, with small furrows branching off from these.

Usually subterranean, in earthworm burrows and soil crevices, slugs feed on earthworms. Activity periods are short, during cold and dry weather conditions the animals hide inside the soil, they appear at the soil only at nighttime, if at all.

N Africa to W France and Britain

This genus needs certainly a revision (Manganelli et al. 1995: 46). Germain 1930 recognized 10 *Testacella* species in France, 4 of them with large ranges. Kerney et al. 1983 listed only 3 species *T. maugei*, *haliotidea* and *scutulum*. Manganelli et al. 1995: 23 listed 3 species for Italy *T. gestroi*, *haliotidea* and *scutulum*, and regarded *bisulcata* as a synonym of *scutulum*. Kerney 1999: 177 regarded the specific identity of *T. scutulum* as not firmly established, it could also represent a yellow variety of *T. haliotidea*. The unclear taxonomic situation, lack of information how to distinguish species and various different interpretations of species identities are reflected in a blurry distributional record.

Testacella bracciai Nardi & Bodon, 2011

Testacella bracciai Nardi & Bodon, 2011 in Nardi & Bodon 2011: 151

Semislug yellowish white, tentacles light brownish, maximum diameter located in the posterior third, body with two dorsolateral longitudinal grooves, shell is carried at the posterior tenth, mantle very small, pneumostome covered by the shell, posterior end of the animal pointed, sole white and undivided.
Shell with a sharp longitudinal keel (running along a / line when the shell is seen in 0 position with the apex at the upper edge), apex not prominent.
Differs anatomically from *T. gestroi* and *T. scutulum* in its penis having two short cylindrical appendices (or flagella) at its posterior end, vas deferens inserts at the position where the two appendices branch off from penis, penis retractor inserts at the posterior ends of the two appendices. The two appendices are not really separated from each other, their length (around 1 mm) is about 3-4 x of their diameter.

Animal 3-4 cm long (preserved 14-25 x 5-7 mm), shell 2.5-3 x 1.5 x 0.9 (height) mm

Deciduous forests on Mesozoic limestone substrate, lives usually inside the soil or under layers of leaf litter, appears at the surface after heavy rainfalls.

Lives only in a small restricted area in the upper Brembana valley (Val Brembilla, Val Serina).

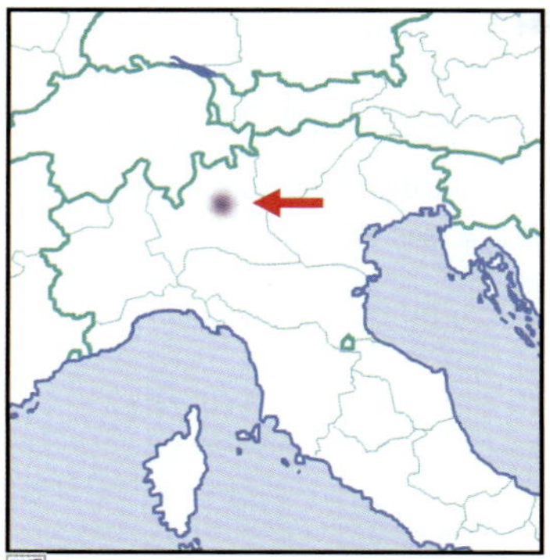

N Italy (Lombardia)

Testacella gestroi Issel, 1873

Italy: Sardinia, Arbus

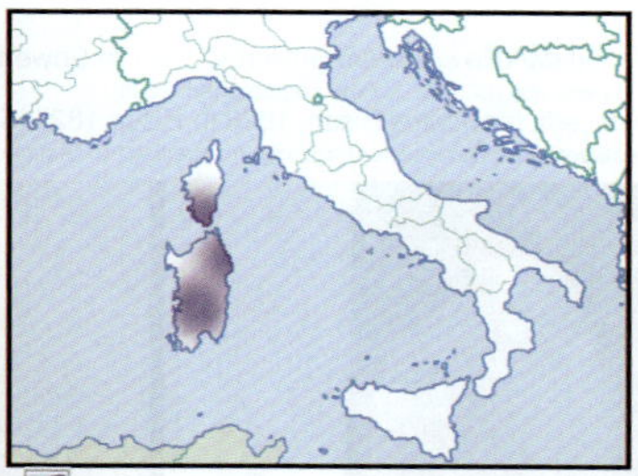

Italy (Sardinia), France (S Corsica)

Francesco Mascia

Testacella gestroi Issel, 1873 in Issel 1873: 277 (*Testacella Gestroi*)

Animal brownish with yellowish grey foot, can be very dark, upper side convex. Lateral grooves not joined near the mantle (_/).
Shell light horny with irregular concentric growth lines, apex small, one whorl.

Subterranean, consumes worms, appears at the surface only at night and in very humid weather conditions.

Giusti 1970: 56-66, Manganelli et al. 1995: 23, 46, Rinaldi 2004.

Animal 30-40 x 15-20 mm (height 8-12 mm), shell 1.4-2 (height) x 5.5-8.5 x 4-5.3 mm

Testacella haliotidea Draparnaud, 1801

Testacella haliotidea Draparnaud, 1801 in Draparnaud 1801 [Jul 1801]: 99
Not *Testacella haliotoides* Lamarck, 1801 [Jan 1801] (= senior synonym of *Testacella maugei* Férussac, 1819, probably not used after 1899). Gargominy et al. 2011 (and Welter-Schultes et al. 2011) suggested that both taxa were the same, which was not true. I am sorry that in our 2011 paper we believed that the authorship given by Gargominy et al. 2011 was correct and referred to the same species, and that we produced confusion.

England: Suffolk, Saxmundham — Roy Anderson

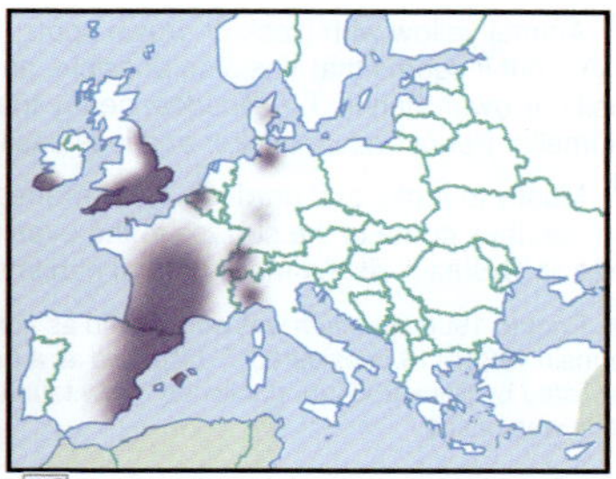

N Africa, W Europe, introduced to Pacific coastlands of northern North America and New Zealand
Range uncertain.

Animal greyish yellow or creamy white, foot sole usually whitish, lateral grooves originate at the front edge of the mantle, they touch the mantle, but are separated from each other (\./).
Shell rudiment triangular and small, at the rear end of the animal, brownish yellow, thin, corrugated, 1.5 whorls.

Herbivorous and noctural animals, shy and hidden at daytime, occurs sometimes in small colonies. Predominantly subterranean, mostly (in Britain almost entirely) in parks and gardens (ornamental and vegetable, in Britain often in traditional kitchen gardens of country houses), arable fields, vineyards, nurseries. In S Ireland also from more natural habitats near the sea.
Consumes earthworms.

Animal 6-12 cm
shell 2 (height) x 6-10 (length) x 4-7 (width) mm

Reproduction in March and April (Gassies 1849) and from August to October (Germain 1930), up to 5 clutches of 8-15 oval eggs (5.5-7 x 4 mm) are laid some days after copulation, juveniles hatch after 10-12 days (Gassies 1849, SW France) or 20-35 days (Germain 1930). Juveniles hatched in early summer reach maturity at the end of autumn.

Gassies 1849: 63 (life cycle), Westerlund 1886: 2, Germain 1930: 117 (life cycle), Nobre 1941: 37, Kerney & Cameron 1979: 173, Kerney et al. 1983: 239, Falkner 1990: 168, Gerber & Heins 1991 (small and stable populations in Elmshorn, N Germany), Manganelli et al. 1995: 23, 46 (in Italy only near Torino), Beckmann 2007: 73 (Mallorca).

Frequently reported populations in W Germany are either extinct (Frankfurt, since around 1970) or have possibly never existed (Heidelberg). Considered as probably introduced in Britain, the few records in the past decades could be due to problems in identification, but probably declining, no recent records from the London area, central and N Ireland, N England and Wales, obviously rare in central England.

Testacella maugei Férussac, 1819

Testacella maugei Férussac, 1819 in A. Férussac & Deshayes 1819-1851 (1): 94 (*Testacellus Maugei*)
Text p. 94 from 04 Dec 1819, Pl. 8 without names on plate or cover from 06 Mar 1819. The name should probably be protected under Art. 23.9.1 against its senior synonym *Testacella haliotoides* Lamarck, 1801.

(1) England: Cornwall, St. Ives, (2-3) Spain: A Coruña, Ferrol
(1) Roy Anderson, (2-3) José Liétor Gallego

N Morocco, Madeira, Azores, Canary Islands, W Europe
Probably introduced to Britain, first records from around 1810.

Extinct in Ireland and central England, at least endangered in S Great Britain except Cornwall, extinct or at least critically endangered in Jersey, Guernsey and Scilly islands.

Animal 6-12 cm
Shell 13-17 x 7-11 x 3.5-3.5 (height) mm

Animal near foot yellowish with greenish or orange hue, on back and sides brownish and darker, with some brown and yellowish spots, shape conical when crawling, anterior part wider than posterior part, posterior part truncated. Points of origin of lateral grooves about 5 mm apart at mantle, not joined (__/). Sole yellowish.
Shell rudiment located on posterior section of dorsum, almost parallel, large, with its apex pointing downwards to the soil and the basal end of the aperture pointing upside, greenish or brownish, 2 whorls, suture keeled.
When disturbed the animal contracts strongly (to ovoid form) and secretes a strong smell and a dense mucus surrounding the whole body.

Subterranean 10-30 cm below the surface, noctural, carnivorous, consumes earthworms. Often in gardens and well-manured ground, in nurseries, in Devon and Cornwall also in more natural habitats among rocks or sea cliffs. If at the soil, then only hidden under dead and humid leaves.

Westerlund 1886: 1, Nobre 1941: 35, Crampton 1975, Kerney 1999: 175, Kittel 2000: 5.

Testacella riedeli Giusti, Manganelli & Schembri, 1995

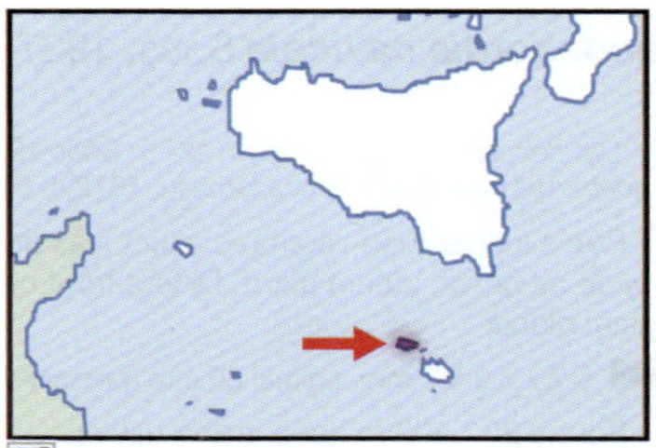

Malta (only Gozo)

Testacella riedeli Giusti, Manganelli & Schembri, 1995 in Giusti et al. 1995: 317

Animal yellow-brown with pale yellow sole, rounded in frontal and pointed in posterior part, dorsoventrally flattened, mantle very small and covered by the shell, tentacles cylindrical, not apically bulbeous. Shell dextral, brown with irregular growth lines, 1.5 whorls, aperture very large, margin simple with thickened columellar margin.

Animal 40 x 18 mm (preserved), shell 8-9 x 5-6 x 2 (height) mm

Not much known. An animal was found on moist sandy substrate under a stone very close to the sea and kept for 6 months in terraria on wet soil where it hid under a stone and fed on earthworms.

I do currently not see constraints in the description supporting the classification of an independent species.

Testacella scutulum Sowerby, 1821

Testacella scutulum Sowerby, 1821 in Sowerby 1821-1824 (part 1): [2], pl. [159] or [26] (*Testacellus Scutulum*)
Date was not 1820 (R. E. Petit, unpublished, 2008).
Testacella bisulcatus Risso, 1826 in Risso 1826: 58 (*Testacellus bisulcatus*) Type locality Alpes-Maritimes (?), SE France.
Testacellus was an incorrect subsequent spelling for *Testacella* Draparnaud 1801.

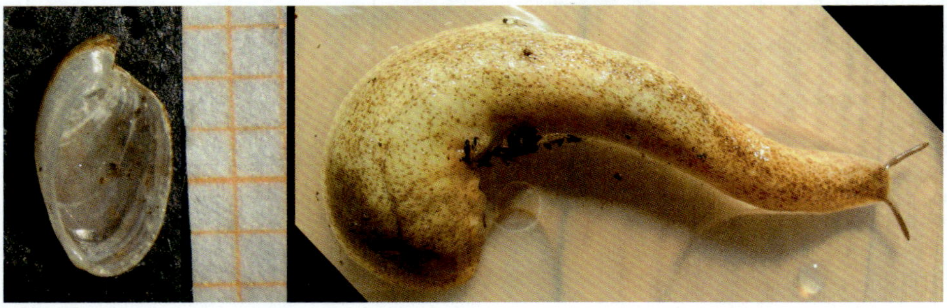

Body 8-12 cm long, shell 6-7 (length) x 3.5-4.5 (width) mm

Italy: Friuli Venezia Giulia, Val Rosandra near Trieste
Massimo Prodan

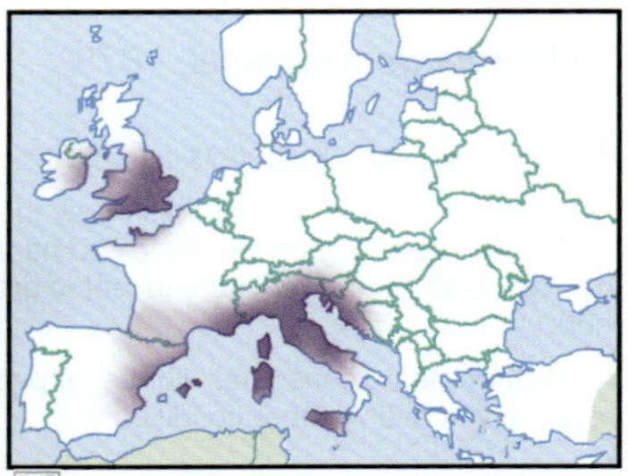

S Europe, British Isles
Range uncertain. Considered as probably introduced to Britain.

In lowland Great Britain and Ireland declining due to the widespread use of pesticides and chemical fertilizers. No recent records from Scotland, N England and N Ireland, possibly rare in Wales, but there are recent records from the London area.

Animal yellow with black or brown spots, foot fringe and sole usually orange, lateral grooves join before reaching their common origin at front edge of mantle (Y), but it seems that V is also possible, and that the grooves can also join below the shell.
Shell narrower than in *T. haliotidea*, nearly triangular, flattened, sometimes concave, outer margin of the aperture sharply truncate and without angle at the columella. Periostracum thicker and less easily eroded than in *T. haliotidea*.

Mostly in parks and gardens. Subterranean, consumes earthworms. During the day the animals are hidden in the soil, perhaps some 10-30 cm below the surface, they come to the surface in the evening. By day it can sometimes be found under flat stones bedded in the earth or sheltering under rubbish near walls. In S Britain also in more natural habitats on cliffs and in woods.

Prior to 1900 *scutulum* was considered as a yellow variety of *T. haliotidea*, this problem is still discussed and not considered as solved (Kerney 1999: 177).
Germain 1930: 119 considered *T. bisulcata* as a frequent species of France, Italy and Algeria. Manganelli et al. 1995: 23 regarded it as a synonym of *scutulum*. *T. bisulcata* was listed by Bank 2004 (unpublished, www.faunaeur.org 01.2011) as an independent species from "French mainland", but it was not explained on which sources this classification was based.

Daudebardia

Daudebardia Hartmann, 1821
in Hartmann 1821 (in Sturm): 41

Type species *Helix rufa* Draparnaud, 1805
(subsequent designation by Herrmannsen 1846 [18-04-1847]: 371)

Animal cannot withdraw into the shell, shell with 0.5-2.5 whorls, last whorl much inflated.
Penis more or less long, usually cylindrical, vas deferens inserting at its apical end, next to retractor, no flagellum or appendix at penis, vagina usually shorter, surrounded by a vaginal gland in its posterior part, spermatheca duct usually short and usually inserting in the zone covered by the vaginal gland, spermatheca spherical or ovoid, oviduct usually short and sometimes thinner than penis and vagina, atrium short.
Differs from *Carpathica* in the absence of an appendix or flagellum at the apical end of the penis, vas deferens and retractor are connected at the apical end of the penis (in *Carpathica* vas deferens is connected subterminally).

Usually in humid and shady habitats.

Central Europe to Near East

Riedel 1978, Grossu 1979 (revision Romania), Grossu 1983: 161 (Romania).

If *Daudebardia transsylvanica* is conspecific with *Carpathica langi*, which is likely if Grossu's 1983 results are tenable (they suggest an uninterrupted gradient between both forms), then there are no differences between *Daudebardia* and *Carpathica*, neither in the shell nor in the anatomy. An independent genus *Carpathica* is unnecessary, it is maintained here only for the purpose of the stability of names.

Grossu 1983: 161 proposed a subgeneric classification of *Daudebardia*: *D.* s. str. (*D. rufa* and *D. brevipes*), *D. (Libania)* containing *cavicola*, *dacica*, *soosi*, *incerta*, *parvula* and *spelaea*, and *D. (Cibinia)* containing *D. transsilvanica*, and an independent genus *Deceballia* for *D. nana*. Since *dacica*, *parvula*, *incerta*, *transsilvanica* and *Carpathica langi* seem to belong to the same species, Grossu's 1983 subgeneric classification is probably not tenable.

Daudebardia brevipes (Draparnaud, 1805)

Helix brevipes Draparnaud, 1805 in Draparnaud 1805: 119, Pl. 8

Shell length 4-4.8 mm, body 17-20 mm

(1-2) Austria: Niederösterreich, near Gaaden, (3-4) Italy: Sicily, Agrigento

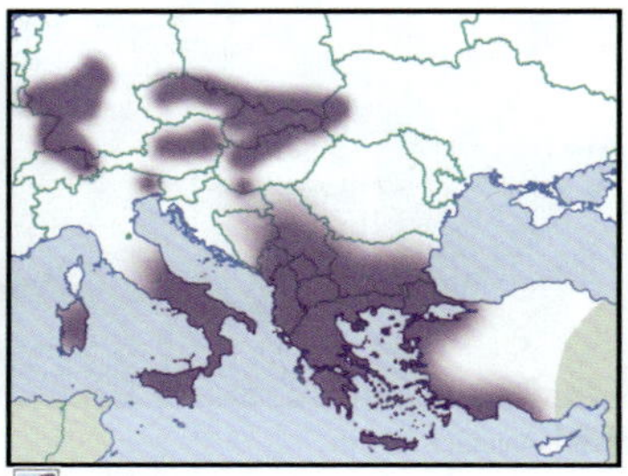

Europe, N Algeria

Endangered in Switzerland, Austria and Germany. Much rarer than *D. rufa*, range more scattered. Threatened by continuous destruction of natural forests.

Shell yellowish, oval, outer margin of last whorl convex rounded, embryonal whorls slightly excentrical, with a small part forming part of the periphery of the fully grown shell, 30 % of the fully grown shell diameter (but not always well visible), the coiled part occupies 30-40 % of the length in a fully grown shell with 2.5 whorls, umbilicus very narrow.
Differs from *D. rufa* in its last whorl not having a straight outline, provoking a more oval shell, the initial whorls are more narrowly coiled and occupy a smaller portion of the shell at equal whorl number, umbilicus narrower.
Adult body not retractable, upper side with dense grey pigments, forming distinct spots at the sides, lower side whitish.

Leaf litter and under stones in natural and original humid mountain and highland forests. Largely subterranean, requires deep soils. In central Europe predominantly in *Alnus* forests. In Switzerland up to 700 m, in Bulgaria 1500 m.
Feeds on earth worms, insect larvas and smaller snails.

Daudebardia cavicola Soós, 1927

Daudebardia cavicola Soós, 1927 in Soós 1927: 168
Daudebardia cavicola ponorica Grossu, 1972 in Grossu 1972: 420

Penis longer than vagina+oviduct, vas deferens inserts apically, no spermatheca, no vaginal gland, vagina as wide as penis, atrium short. Differs from other *Daudebardia* species in the absence of a spermatheca and a vaginal gland.

5-7 x 3.2-4.7 mm, animal 20 mm (preserved)

Found 800-900 m deep inside caves.

Grossu 1983: 183.

D. cavicola seems to be an independent species.

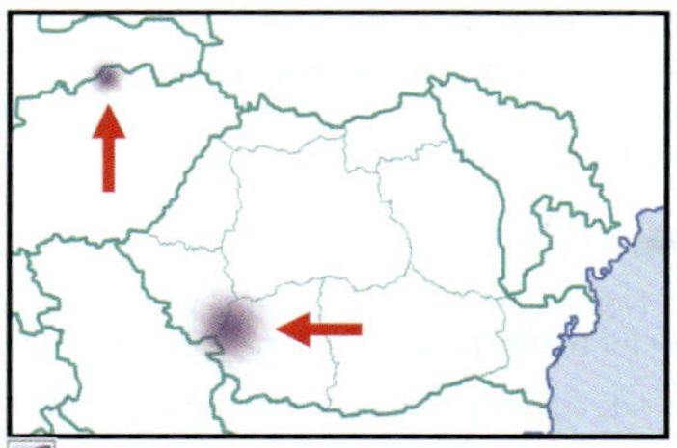

Hungary, Romania

Daudebardia langi (Pfeiffer, 1846)

Italy, Veneto, Treviso, Cansiglio
Massimo Prodan

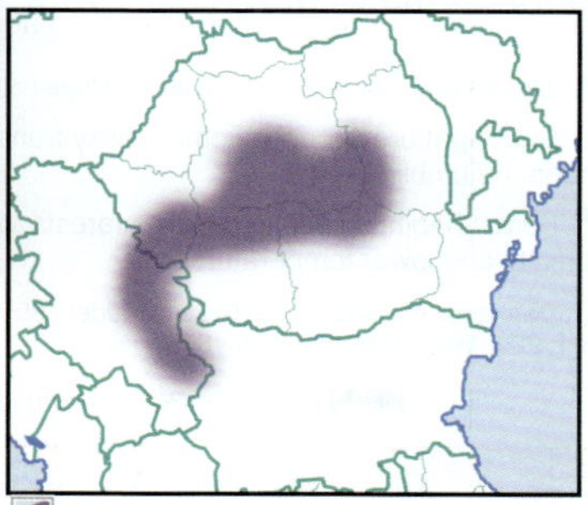

Helicophanta langi Pfeiffer, 1846 in L. Pfeiffer 1846: 81 (*Helicophanta Langi*)
Daudebardia transsilvanica Bielz, 1859 in E. A. Bielz 1859: 214
Original spelling was not *transsylvanica*.
Daudebardia soosi Wagner, 1941 in H. Wagner 1941: 653 (*Daudebardia soósi*)
Daudebardia spelaea Grossu, 1960 in Grossu 1960: 176
Daudebardia dacica Grossu, 1969 in Grossu 1969: 79
Daudebardia parvula Grossu, 1969 in Grossu 1969: 82
Daudebardia dacica incerta Grossu, 1969 in Grossu 1969: 81

1.5-6.5 x 1-4 mm, animal up to 30 mm alive, 7-22 mm preserved

Romania, E Serbia,

Shell yellowish or brownish, shiny, finely striated or with growth lines, with 0.7-1.3 whorls, apex:aperture 1:3-1:5, no umbilicus, membrane at apertural margin at lower side near embryonal whorls smooth without denticles.
Animal usually bluish grey, lighter downwards to the sides, sole margins and sole whitish yellow.
Penis of variable length and usually cylindrical and undivided, tending to be shorter in the west and often with a subdivision penis/epiphallus, vas deferens usually inserting apically/terminally, in the west subterminally (at the sometimes otherwise invisible connection penis/epiphallus), retractor also connecting there, in the west there might be a flagellum (or an appendix) next to the insertion of vas deferens, spermatheca duct inserts to free vagina or in the zone surrounded by the vaginal gland, spermatheca usually ovoid, oviduct usually as wide as penis, atrium short.

Humid and shady mountain habitats, often under leaf litter in deciduous forests.

Westerlund 1886: 8, Pavlović 1912: 17, Wagner 1952: 179, Tomić 1959: 7, Riedel 1978: 165, Grossu 1983: 191, Manganelli et al. 1995: 20, 43 (population in Valsugana, NE Italy, with uncertain identification).

Grossu 1979 found intermediates between *langi* and the *Daudebardia* forms and called them *D. intermedia*.
Grossu's 1983 study suggested a high degree of spatial variation in the anatomy of several parapatric taxa classified as *Daudebardia dacica*, *D. parvula*, *D. soosi*, *D. spelaea*, *D. transsilvanica* and *Carpathica langi*.
Grossu 1983 was aware that his results (presence of continuous gradients and of transitional forms between *Daudebardia* forms and *Carpathica langi*) weakened *Carpathica* as a separate and well-defined genus, and proposed to verify the results with electrophoretical and immunological methods. This however is not necessary, since the classification at generic level is either based on shell or on anatomical characters. Grossu 1983 provided evidence that the separation of *Daudebardia* and *Carpathica* (which was exclusively based on the terminal/subterminal insertion of the vas deferens) was arbitrary.
The logical consequence for a classification based on the current state of knowledge is to classify all these forms in one single species. It is convenient to shift *langi*, hitherto classified in the genus *Carpathica*, to the genus *Daudebardia*, the species is very rarely mentioned (only 23 hits in Google, 10.2009) and all other synonyms have been classified in *Daudebardia*. Using a name *Daudebardia langi* will also imply that *langi* is used in the broad sense outlined here. I usually do not like shifting species into different genera since this provides obstacles for the user to finding species in electronic resources, but in this case it is not otherwise possible to reflect the current state of published research. More research is necessary.

Daudebardia nana (Grossu, 1969)

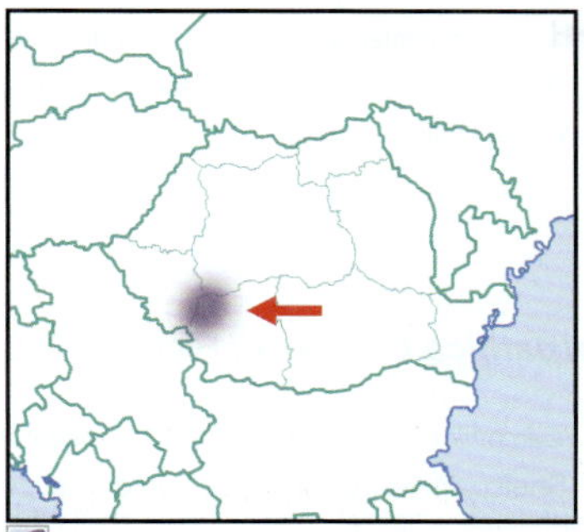

Romania

Deceballia nana Grossu, 1969 in Grossu 1969: 85

Shell light greyish, with a few growth lines, only 0.5 whorls. Animal bluish grey.
Penis as long as vagina+oviduct or longer, at its apical end with one elongate appendix and a very long and thin epiphallus, vas deferens and retractor inserting at its apical end, spermatheca elongate or ovoid without duct, 2/3 of vagina surrounded by vaginal gland, oviduct slightly longer and thinner than free part of vagina, atrium short.
Differs from other *Daudebardia* species in the presence of an appendix and epiphallus, and from *Carpathica* species in that the vas deferens inserts at the apical end of these structures.

1.8 x 1 mm, animal 5.5 x 1.8 mm (preserved)

Only known from caves.

Grossu 1983: 189. Classification follows Riedel 1978: 172.

Seems to be an independent species.

Daudebardia rufa (Draparnaud, 1805)

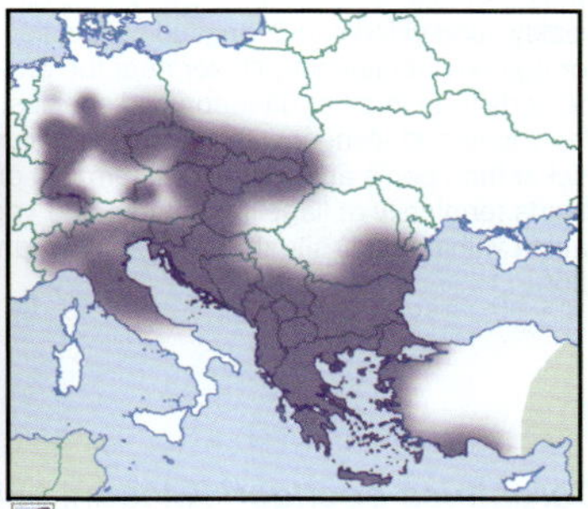

Europe, N Algeria

Helix rufa Draparnaud, 1805 in Draparnaud 1805: 118, Pl. 8

Shell length 4-5.3 mm, body 17-20 mm

(1-2) Austria: Niederösterreich, near Heiligenkreuz, (3-4) N Albania, Malësia, near Tamarë

(2) Czech Republic: Panské Louky nature reserve
(2) Michal Maňas
(1, 3) Slovakia: Slovenská Ľupča
(1, 3) Jozef Grego

Shell horny reddish, white or yellowish, 2.5 whorls, outline of last whorl getting straight from whorl 2.25 onwards (occasionally weakly concave), embryonal whorl surrounded by last whorl, 40 % of fully grown shell (bot not always well visible), the coiled part occupies 40-50 % of the length in a fully grown shell with 2.5 whorls, umbilicus moderately wide.
Adult body not retractable, animal with dense grey pigments on upper side, whitish on lower side.
Differs from *D. brevipes* in its last whorl having a straight outline provoking an oblong instead of an oval shell, the initial whorls are less narrowly coiled and occupy a larger portion of the shell at equal whorl number, umbilicus wider.

Leaf litter and under stones in humid original and natural mountain and highland forests, often near waters. In Switzerland up to 1000 m, in Bulgaria 1500 m. Largely subterranean, avoids rotting microenvironments and acid soils. Feeds on earthworms, insect larvas and smaller snails, which are consumed entirely. Observations suggest that they consume also conspecific animals, at least in captivity. In nature individuals seem to live at some distance to each other.

Life cycle only fragmentarily known (Frömming 1954: 107), reproduction seems to take place from March or April onwards, very young juveniles are present mainly from March to May, larger juveniles in summer and autumn, they seem to hibernate, reproduce in the next year from June onwards and die in autumn. If true, this would indicate a life span of two years.

Endangered in Switzerland, Bayern and Vorarlberg, Vulnerable in Austria and Germany. Scattered and rare in central Europe. The high requirements of natural forests are increasingly rarely met by forests under modern forest managements, the species is threatened by continuous destruction of original forests.

Daudebardia wiktori Riedel, 1967

Daudebardia wiktori Riedel, 1967 in Riedel 1967: 471

Shell light brown, apex lighter, milky transparent, 1.3 whorls, no umbilicus.

Forest habitats, also in pinetree forests, prefers high humidity and lower temperatures.

Damjanov & Likharev 1975: 294, Riedel 1995: 77, Schütt 2001: 239, Irikov & Mollov 2006: 802.

2.6-3.4 x 1.5-2.0 mm, body up to 30 mm

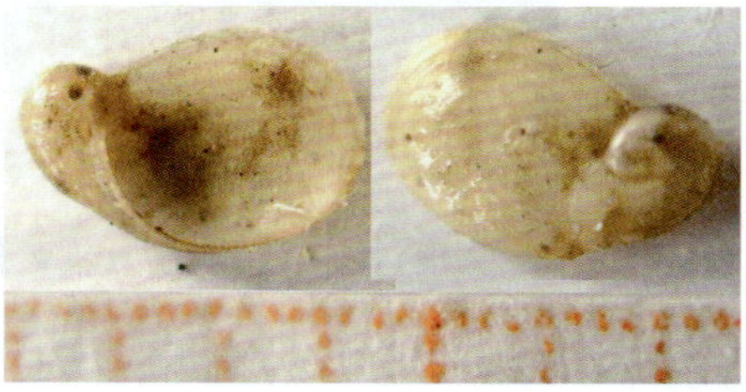

Bulgaria: W Rodopi, near Bachkovo village

Atanas Irikov

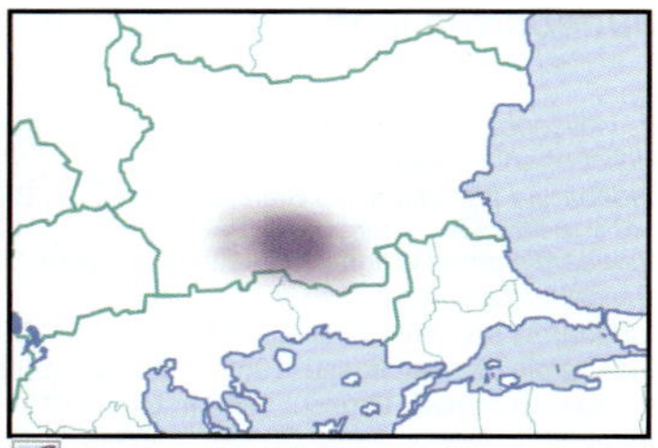

Bulgaria, NE Turkey

Carpathica

Carpathica Wagner, 1895 in A. J. Wagner 1895: 621 *Daudebardia (Carpathica)*

Type species *Daudebardia kimakowiczi* Wagner, 1895 (subsequent designation by Forcart 1950: 110) (= currently *Carpathica calophana* (Westerlund, 1881))

Animal cannot withdraw into the shell, shell with 0.5-2 whorls, last whorl much inflated.
Penis more or less long, usually cylindrical, with apppendix (flagellum) at its terminal end, vas deferens inserting laterally (subterminally), retractor connects at appendix, vagina usually shorter, surrounded by a vaginal gland in its posterior part, spermatheca duct usually short and usually inserting in the zone covered by the vaginal gland, spermatheca spherical or ovoid, oviduct usually short and sometimes thinner than penis and vagina, atrium short.

SE Europe, Turkey

Grossu 1983: 191 (diagnosis sensu Soós 1943, Wagner 1952 and Forcart 1960), Riedel 1991, 1995: 79.

Differs from *Daudebardia* in the presence of an appendix or flagellum at the apical end of the penis, vas deferens inserts subterminally (in *Daudebardia* terminally). This difference does not qualify for a separation from *Daudebardia* at the generic level, since the polytypic and variable species *Daudebardia langi* seems to connect both character states (Grossu 1983). An independent genus *Carpathica* is certainly unnecessary, it is maintained here only for the purpose of the stability of names.

Carpathica amisena (Forcart, 1950)

Pseudolibania amisena Forcart, 1950 in Forcart 1950: 113

Shell oval, embryonal whorl excentrical, oblique, columellar margin of aperture thickened, umbilicus covered.

Shell diameter up to 7.5 mm, body up to 25 mm

River margins, rubbish, under stones, in soils, needs permanently high humidity.

Riedel 1995: 79, Schütt 2001: 244.

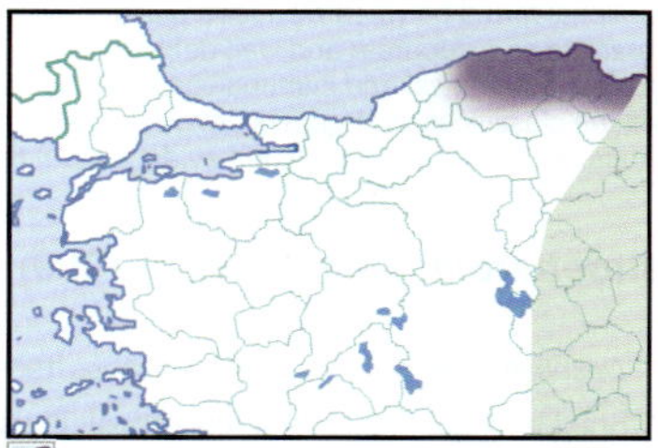

N Turkey

Carpathica bielawskii Riedel, 1963

Carpathica bielawskii Riedel, 1963 in Riedel 1963: 477

Shell pale yellow, translucent and shiny, with distinct growth lines, elliptical, in fully grown shells the embryonal shell is perpendicular to the longer elliptic axis, 1.5 whorls, columellar margin slightly reflected, no umbilicus.
Animal grey, more yellowish near the sole, more corrugated than other daudebardiid species producing rounded areas (not long as in *C. stussineri*), sole dirty yellowish, weakly subdivided into 3 longitudinal sections and also corrugated. Genital pore closer near the anterior end of the body than in other daudebardiid species.
Penis long and slender with terminal flagellum which is thicker than penis and rounded at terminal end, retractor inserts terminally at flagellum, atrium and vagina short, vagina with laterally situated perivaginal gland, oviduct short.

Bulgaria: S Black Sea coast, Kiten village

Atanas Irikov

3.5-5.0 x 2.2-3.4 x 1 (height) mm

In humid deciduous forest habitats under bark or desintegrating wood logs.

Damjanov & Likharev 1975: 296, Riedel 1995: 79.

SE Bulgaria, probably NW Turkey

Carpathica boettgeri (Clessin, 1883)

Daudebardia boettgeri Clessin, 1883 in Clessin 1883: 38, Taf. 2-3 (*Daudebardia Boettgeri*)

Shell diameter up to 5 mm, body up to 20 mm

Ukraine: Crimea, near Simferopol

Shell oval, embryonal whorl 1/5-1/6 of fully grown shell diameter, probably larger than *C. amisena*, no umbilicus. Differs anatomically from *C. amisena*.

Relatively humid habitats under large stones, forests and forest-steppe zones with leaf litter. Feeds on earthworms, insect larva and molluscs, including the own species.

Riedel 1995: 80, Schütt 2001: 242 (in N Turkey probably introduced), Egorov 2008: 64.

Crimea, S Russia, N Turkey

(1-3) Sergey Leonov

Carpathica calophana (Westerlund, 1881)

Daudebardia calophana Westerlund, 1881 in Westerlund 1881: 51
Carpathica denticulata Grossu, 1969 in Grossu 1969: 88

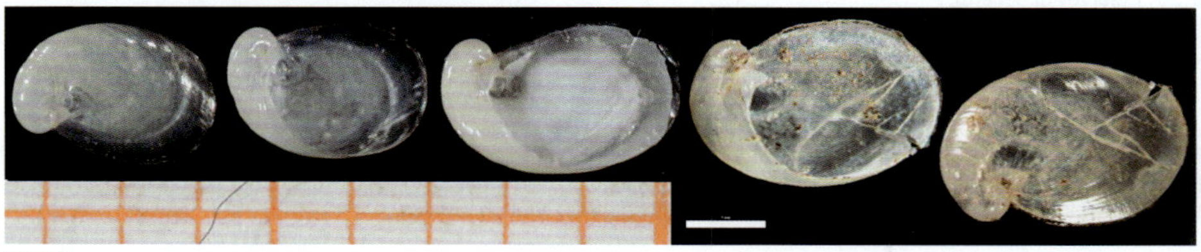

(1-3) Ukraine: Yaremcha, (4-5) Poland: Bieszczady mountains (4-5) Anna Sulikowska-Drozd

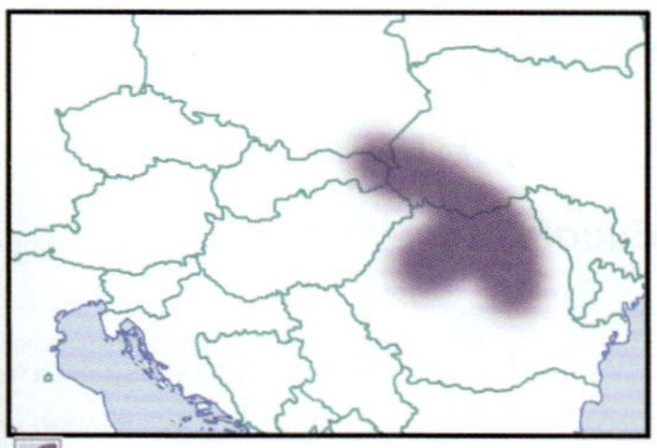

Carpathians

3-5 x 2-3.5 x 1.3 (height) mm, body up to 20 mm (preserved)

Slovakia: Stužica reserve near Nová Sedlica Jozef Grego

Shell yellowish, transparent, weakly striated and fine spiral lines, very shiny, with a white layer covering up to 75 % of the inner side, 0.7-1.0 whorls, apex very small, less than 20 % of shell length, aperture very oblique and oval, upper margin curved, columellar margin long, almost vertical, thickened on upper side (membrane of apertural margin at lower side near embryonal whorls with one peak), lower margin concave, umbilicus covered.
Animal whitish to dark bluish grey, slightly paler down to the sides, with longitudinal structures, sole dirty whitish or with yellowish red hue.
Penis with appendix, vas deferens inserts laterally at appendix, retractor at the terminal end of the appendix, spermatheca duct can insert either in free part of vagina or in the part surrounded by the vaginal gland, oviduct can be slightly thinner than free vagina.

Humid and shady habitats, usually in forests, 200-1900 m altitude.

Westerlund 1886: 9, Grossu 1983: 192, 199.

Only very few specimens of *C. denticulata* were found, obviously there was some variation in the expression of the membrane at the lower side of the shell, and it is possible that the denticles are not a unique specific character. This form is classified here as a synonym of *C. calophana*, the present data are too weak to provide evidence for the presence of an independent local endemic species, and to rule out the more likely hypothesis of a spatially variable shell character. The anatomy of *C. calophana* seems to be quite variable, and *denticulata* seems to fit in this complex. It is also necessary to know more about the life cycle of these animals to determine which ones of the specimens studied by Grossu 1983 were fully adult.

Carpathica cretica (Forcart, 1950)

Pseudolibania cretica Forcart, 1950 in Forcart 1950: 114
Intended as a new replacement name for *Daudebardia saulcyi* (Bourguignat, 1852) sensu Simroth, 1891 and later authors, which was not an available name to be replaced (Art. 72.7), so *P. cretica* was simply a new name.

Shell yellowish brown with milky whitish 1.3 embryonal whorls, translucent and shiny, postembryonal whorl with distinct and irregular growth lines, 1.7-2 whorls, last whorl descending, columellar margin slightly concave, broadly reflected and thickened.
Posterior section of vas deferens enlarged and spirally coiled, inserts laterally to penis, restractor penis inserts to flagellum, vagina covered by a glandular tissue.

Greece: Crete, Asteroúsia near Léndas (juvenile shell)

3.7-4.3 x 1.8-2.2 x 0.9 (height) mm

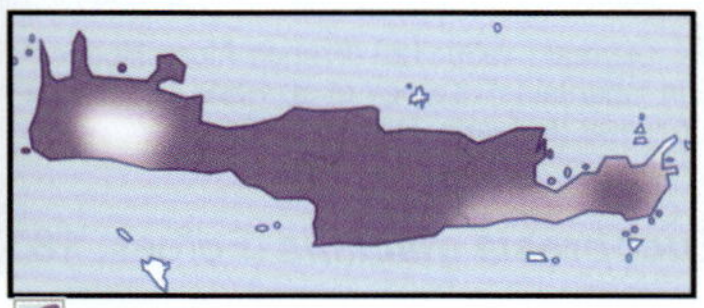

Greece (Crete)

Wagner 1895: 619 (*D. saulcyi*), Riedel 1992: 106.

Carpathica insularis Riedel, 1988

Carpathica insularis Riedel, 1988 in Riedel 1988: 88
Name originally attributed to Riedel & Mylonas, but there was given no statement that besides donation of material Mylonas contributed anything to the description itself. In a strict application of Art. 50.1.1 the name in such a case must be attributed to the author of the work.

Shell yellowish with whitish apex, 1 whorl, with some growth lines after 0.5 whorls.
Shell not distinguishable from *C. cretica*.
Animal brownish with violet hue, often with bronish spots, sides and sole yellowish.
Penis small, with thin terminal flagellum, penis almost entirely in sheeth, retractor apically at flagellum, no epiphallus, vas deferens nearly as wide as epiphallus, vagina short, wider than oviduct, vaginal gland spherical and connecting vagina, oviduct and (thinner) spermatheca duct, spermatheca peach-shaped.
Differs from *C. wirthi* and *C. cretica* in the absence of an epiphallus.

3-4 x 2-3 mm, animal up to 15 mm

Riedel 1992: 107, Bank & Maassen 1998: 63.

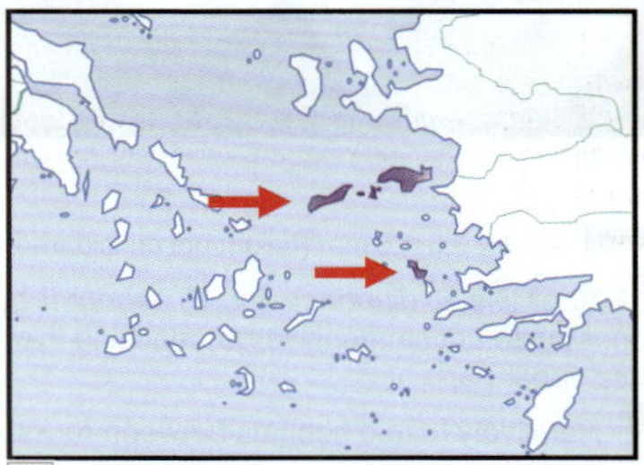

Greece (E Aegean islands)

Carpathica stussineri (Wagner, 1895)

Daudebardia stussineri Wagner, 1895 in A. J. Wagner 1895: 624 (*Daudebardia (Illyrica) Stussineri*)

Shell horny yellowish with milky white and elliptical embryonal shell, shiny and translucent, finely striated with some more prominent growth lines, 1.5 rapidly increasing whorls, suture impressed, descending near aperture, aperture oblique, columellar margin concave, margins connected at parietal side by a weak callus.

Humid habitats, in Bulgaria up to 1400 m.

Pavlović 1912: 17, Tomić 1959: 7, Damjanov & Likharev 1975: 297, Riedel 1967: 478, Dhora & Welter-Schultes 1996: 149, Hubenov 2007.

Bulgaria: Plovdiv Atanas Irikov

3.5-4.2 x 2.2-2.8 x 0.9-1.1 (height) mm

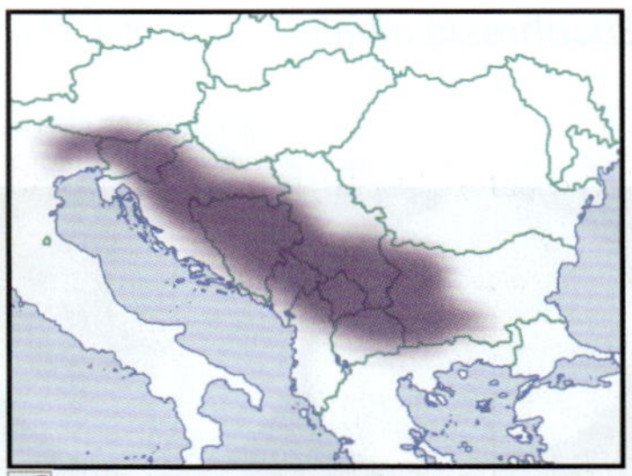

S Alps to Bulgaria

Carpathica wirthi Forcart, 1971

Carpathica wirthi Forcart, 1971 in Forcart 1971: 33

Shell oval, exceeds posterior end of the animal's foot, embryonal whorls oval, adult shell with 1.75 whorls, milky white to light yellow, no umbilicus, columellar margin thickened.
Animal dark bluish grey, foot and mantle surrounded by a narrow yellowish white marginal zone, sole yellowish white.

Bulgaria: W Rodopi, Bachkovo village Atanas Irikov

7.5 x 4.6 x 2.4 (height) mm, embryonal shell diameter 2.3 mm

Calcareous and sandy soils.

Riedel 1992: 107, Riedel 1995: 80, Schütt 2001: 243, Irikov & Mollov 2006: 802 (S Bulgaria?).

Range insufficiently known, Samothráki and Bulgaria uncertain.

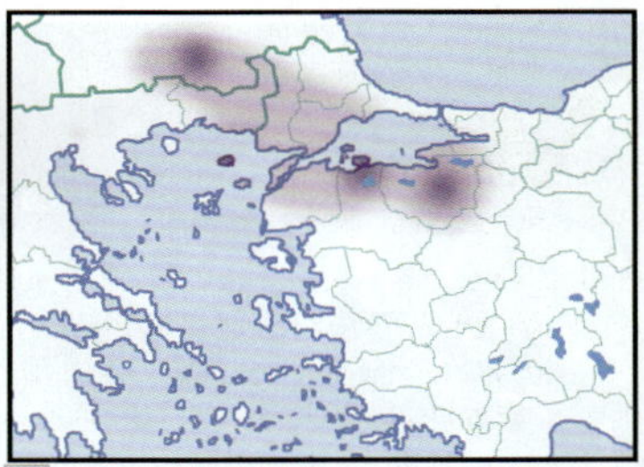

NW Turkey to S Bulgaria

Vitrinidae

Hausdorf 2002, Giusti, Fio. et al. 2011 (survey of genera).

Eucobresia

Eucobresia Baker, 1929
in H. B. Baker 1929: 139 *Vitrina (Vitrina (Eucobresia))*

Type species *Vitrina diaphana* Draparnaud, 1805 (original designation)

Intended as a new replacement name for *Semilimax* sensu Hesse, 1923 non Gray, 1847 or Stabile, 1859. Not a new replacement name because it was not intended to replace an available name (Art. 72.7, Art. 49). So *Eucobresia* was a regular new name. See under *Semilimax* for the synonyms there.

Shell moderately ear-shaped with broad or narrow apertural membrane and large mantle lobe.
No penial appendix, penis joins vagina below the insertion of spermatheca duct, upper part of vagina not strongly swollen.

Europe, northern Asia

Eucobresia diaphana (Draparnaud, 1805)

Vitrina diaphana Draparnaud, 1805 in Draparnaud 1805: 120

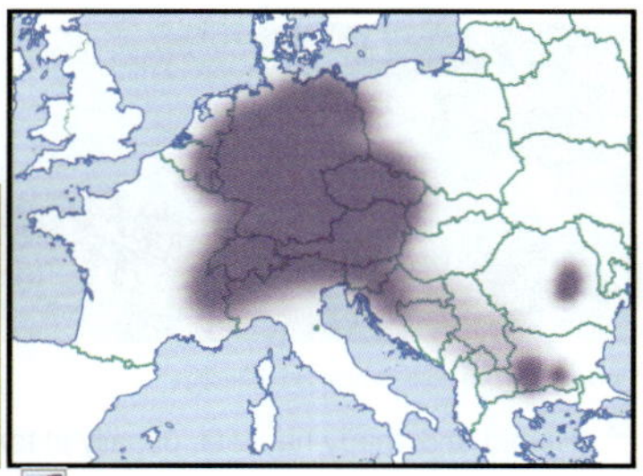

3-4 x 6-7 mm, animal 10-18 mm

Germany: W Allgäu, between Bad Trögen and Weiler

Europe

Shell flat, more than 2 whorls with flat suture, apertural membrane broad (35-50 % of the whorl), columella curved leaving preceding whorls visible, fine whitish streaks.
Animal dark, mantle black, in mountains light grey, cannot withdraw into shell, mantle covers apex entirely.
Penis large, with retractor, atrium very short (in *E. pegorarii* penis is short and atrium longer).

Humid and cool places, valley plains and mountains. In lowlands in shady and humid habitats in forests, in higher altitudes in open habitats, dwarf shrublands and grasslands with sufficient opportunities to hide. In Switzerland and France between 1700 and 2900 m, in Bulgaria up to 2600 m.

Reproduction cycle presumably annual, in higher altitudes reproduction in the late summer. Juveniles hatch after 2 weeks (Alten 1812: 84).

Germain 1930: 129 (French Alps), Hudec & Vasatko 1971: 8 (Bachkovo in W Rodopi mountains at 400 m), Damjanov & Likharev 1975: 250 (Bulgaria: Vitosha, Rila and Pirin mountains), Kerney et al. 1983: 154 (range may extend to Pyrenees), Grossu 1983 (Romania: SE Carpathians), Falkner 1990: 172, Turner et al. 1998: 252, Hubenov 2007: 138 (Bulgaria: Rila and Pirin).

Eucobresia glacialis (Forbes, 1837)

Vitrina glacialis Forbes, 1837 in Forbes 1837: 258

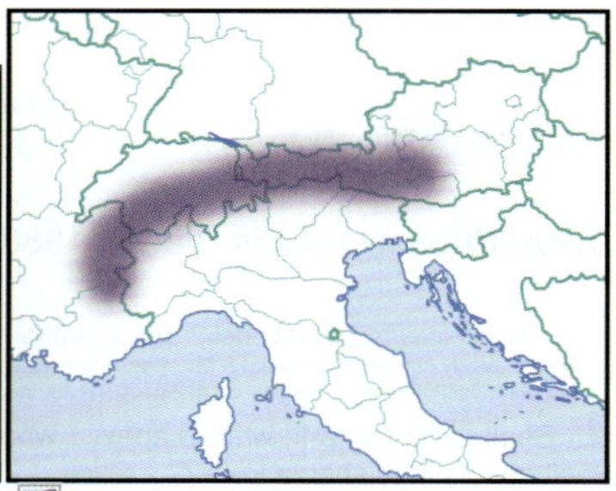

4.5-6 x 3.5-4.5 mm (height not considered)

Switzerland: Diablerets glacier, 2500 m

Alps

Shell like *E. diaphana* and *E. pegorarii*, but apertural margin convex near insertion of preceding whorl. Shells alone are not entirely reliable for determination. Mantle uncommonly large, reaching the base of the tentacles, black mantle appendix covers whorls entirely. Penis broad, conical-pointed, vagina inflated in its upper part.

Relatively humid habitats between rocks and herbal vegetation. Nearly exclusively in high alpine regions above the timberline, 1000-3100 m.

Reproduction cycle covers presumably several years.

Germain 1930: 129, Klemm [1973], Kerney et al. 1983: 157, Manganelli et al. 1995: 18, Turner et al. 1998: 255.

Eucobresia nivalis (Dumont & Mortillet, 1854)

Vitrina nivalis Dumont & Mortillet, 1854 in Dumont & Mortillet 1852-1854: 209 [not verified]
Year could also be 1852, some authors gave p. 299, Germain 1931 gave author Charpentier.

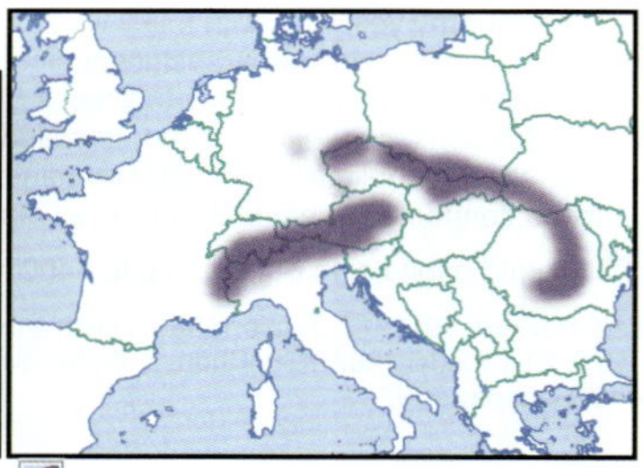

4-6 x 3-4.5 mm (height not considered), animal 14 mm

Switzerland: Kanton Uri, Ursenental

Alps, Carpathians

Slovakia: Slovenský Raj National Park

Jozef Grego

Shell smaller than *E. diaphana*, apertural membrane narrow, no umbilicus-like structure (preceding whorls hardly visible).
Animal grey or dark grey, mantle marbled with greyish spots, mantle appendix relatively small, not covering the apex.

Humid mountain meadow and stony slopes above the timberline, often near snow fields, also in swampy habitats in high mountain forests. Requires strongly stones for shelter against drought and insolation. Always high alpine habitats, up to 2800 m in S Switzerland, mainly between 2000 and 2500 m. Moves quickly, 12-14 mm/s, to reach shelters quickly when weather conditions become dry.

Reproduction in autumn, presumably with annual cycle.

Germain 1930: 136, Kerney et al. 1983: 155, Grossu 1983, Falkner 1990: 172, Turner et al. 1998: 253, Sysoev & Schileyko 2009: 138.

Eucobresia pegorarii (Pollonera, 1884)

Vitrina pegorarii Pollonera, 1884 in Pollonera 1884: 324, 334 (or 414, 424)

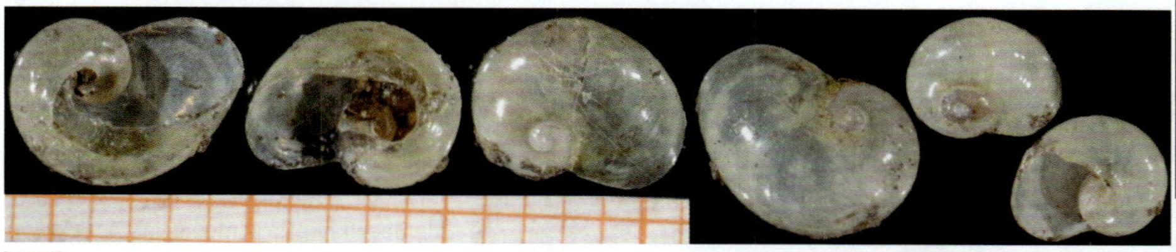

Switzerland: San Bernardino, 2050 m

5-6 x 3.5-4.5 mm (height not considered), animal 14 mm

Humid alpine meadows, stone fields, margins of rock rubble fields, exclusively in high alpine habitats, usually above the timberline, in altitudes of 1200-3200 m.

Reproduction cycle covers several years.

EU Red List: Endangered EN. Seems to be in decline, but the causes and threats are not understood.

Alzona 1971: 133, Klemm [1973], Kerney et al. 1983: 155, Falkner 1990: 172, Manganelli et al. 1995: 17, Turner et al. 1998: 254.

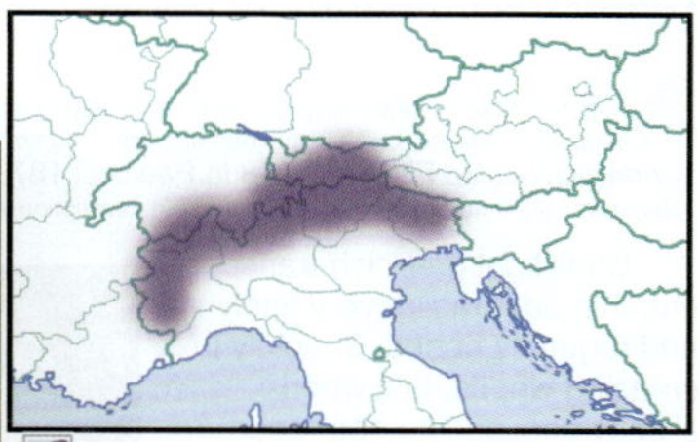

Alps

Shell much like *Phenacolimax glacialis* and *E. nivalis*, differs mainly anatomically, apertural margin more convex than in *Ph. glacialis*. Animal including mantle evenly light to medium grey, mantle appendix black, small, not covering the apex.
Upper part of vagina not inflated, penis rounded and short, not conical.

Oligolimax

Oligolimax Fischer, 1878 in Paulucci 1878: 23 *Vitrina (Oligolimax)*

Type species *Vitrina paulucciae* Fischer, 1878 (original designation)
(= currently *Oligolimax musignani* (Pirajno, 1842))

Author is Fischer in Paulucci 1878, because the whole passage was indicated to have entirely been written by Fischer. Type designated on p. 24.
Gallandia Bourguignat, 1880 in Bourguignat 1880: 3, type species *Vitrina conoidea* Martens, 1874 (original designation)
Hausdorf 1995 found that the type species was misinterpreted and intended to fix as type *Gallandia olympica* Hausdorf, 1995. This was invalid, the type species should be fixed under Art. 70.3.
Sardovitrina Manganelli & Giusti, 2005 in Manganelli & Giusti 2005: 569, type species *Vitrina polloneriana* Fra Piero, 1897 (original designation)

Shell conoid-globose, in which the animal can usually almost entirely withdraw. Penial retractor passes above right ommatophore retractor (in *Phenacolimax* and the many other plutoniine vitrinid species from the Atlantic islands it passes below), no spermatheca retractor, no penial sheath (contrast to *Phenacolimax*).

Manganelli & Giusti 2005 and Giusti, Fio. et al. 2011 proposed a consistent system of genera of vitrinids which is adopted here, except for *Sardovitrina* which is here classified as a subgenus of *Oligolimax*. This seems to have been the classification that was supported by the parsimony analysis of the phylogenetic relationships of vitrinids by Giusti, Fio. et al. 2011: 294-295. The authors did not explain why it was necessary to classify *Sardovitrina* as a separate genus.

Oligolimax annularis (Studer, 1820)

Glischrus annularis Studer, 1820 in Studer 1820: 86 (*Glischrus (Hyalina) annularis*)

(1-4) W Turkey: 10 km S Manisa, (5-6) Greece: Crete, Psiloríti, 2160 m

4.0-6.5 x 2.5-3.5 mm

Moderately humid to fairly dry, open habitats in mountain regions, scarcely overgrown rocks, rock rubble fields, also in gorges and at creek margins. Calcareous substrate is preferred, but silicate substrate also tolerated. In Switzerland up to 2800 m, mainly between 1000 and 2500 m, in Bulgaria up to 2800 m, in Crete 2400 m.

Occurs often in small and restricted populations, far away from each other without interconnections.

Has recently been called *Phenacolimax annularis* and *Gallandia annularis*.
Google hits (01.2011, accessed from Germany): general search engine: in genus *Phenacolimax*: 133, in genus *Oligolimax*: 109, in genus *Gallandia*: 62
Google Scholar: in genus *Phenacolimax*: 32, in genus *Oligolimax*: 15, in genus *Gallandia*: 10

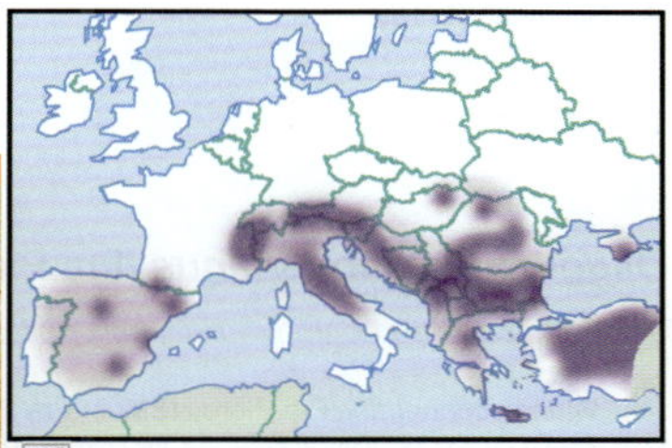

Europe to central Asia
Range fragmentary and insufficiently known.

Shell fragile, higher than in other vitrinids, with characteristic regular growth lines, so that also fragments can be determined.
Animal dark grey, practically no mantle appendix.

Oligolimax apatelus (Soós, 1924)

Vitrinopugio apatelus Soós, 1924 in Soós 1924: 185

Shell greenish yellow, thin and fragile, smooth and translucent, with indistinct growth lines, 1.75 rapidly increasing whorls, last whorl and aperture predominating.

3.5-4 x 7-8 mm

Animals were not perfectly mature in July.

Under stones, at the base of limestone rocks and among leaf litter.

EU Red List: Near Threatened NT. Material from 3 localities has been associated with this species (Mus. Budapest).

wwww.iucnredlist.org (07.2012, Z. Fehér).

Systematic position unclear, no more information has been published since 1924. Sóos 1924 forgot to mention the locality. In 1956 the collection was destroyed by fire. It is assumed to be an endemic species of the Korab mountains.

Albania (Korab mountains)

Oligolimax cephalonicus (Rähle, 1980)

Without gender agreement: *Oligolimax cephalonica* (Rähle, 1980)

Vitrina cephalonica Rähle, 1980 in Rähle 1980: 207

Shell horny brown, smooth, shiny, 3 whorls, last whorl width not more than 3 x of penultimate whorl, umbilicus narrow and half covered by columellar margin of the aperture.
Animal grey to blackish, also at the sole margins, central section of the sole light. Shell appendix of mantle small, no different colour. Animals of higher zones tend to be darker.

Up to 3.2 x 6.1 mm

Was found in a coniferous forest, humid habitats, under bark and desintegrating wood, 1300-1600 m.

Hausdorf 2002: 357 classified this form in *Semilimacella* after anatomical study, without presenting detailed results. Giusti, Fio. et al. 2011: 340 accepted *Semilimacella*, but observed that it was unclear to which genus this species should be assigned, various methods produced different results. More research seems to be necessary to understand this form.

W Greece (Kefaloniá)

Oligolimax musignani (Pirajno, 1842)

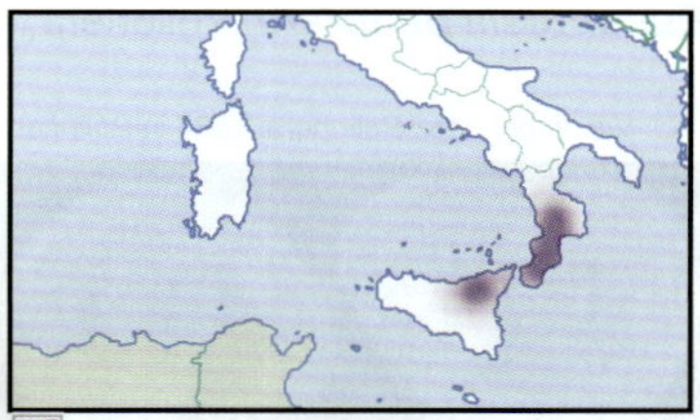

S Italy (Calabria, NE Sicily)

Beech forests in mountains, 1100-1850 m.

EU Red List: Vulnerable VU. Only known from few localities. Threatened by modern forest management.

Manganelli et al. 1995: 18 (*paulucciae*), Manganelli & Giusti 2004.

Vitrina musignani Pirajno, 1842 in Pirajno 1842: 5 [or 157] (*Vitrina Musignani*)
P. 5 in the monograph, p. 157 in the journal article. I do not know which one appeared first.
Vitrina paulucciae Fischer, 1878 in Paulucci 1878: 23 (*Vitrina Paulucciæ*)
Name was attributed to Fischer, the text of the description was also entirely attributed to Fischer.

Shell light greenish transparent, thin and fragile, very finely and regularly ribbed, 3 rapidly increasing whorls, last whorl slightly descending near aperture, umbilicus very narrow.
Shell sculpture resembles some Canarian species, cannot be confounded with other Italian species.
Animal brownish black, mantle with dark spots, withdraws more or less entirely into the shell.

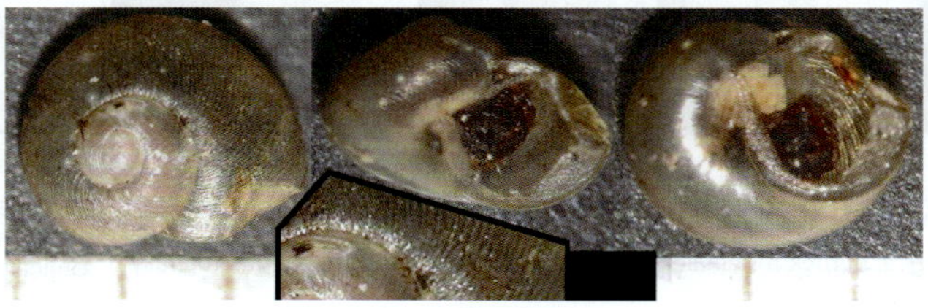

2.1-2.6 x 3.2-3.8 mm (height x diameter)

S Italy: Cosenza, Camigliatello, 1160 m
Angelo Vannozzi

Oligolimax olympicus (Hausdorf, 1995)

Without gender agreement: *Oligolimax olympica* (Hausdorf, 1995)

Gallandia olympica Hausdorf, 1995 in Hausdorf 1995: 69
Intended by Hausdorf 1995 as a new replacement name for *Vitrina conoidea* Martens, 1874 sensu Bourguignat 1880, and taken as the type species for *Gallandia* Bourguignat, 1880 (Bourguignat had misinterpreted *V. conoidea*), this was probably not in accordance with the Code (names established by deliberate misidentifications can only be taken as types for new nominal genera, Art. 11.10). So *G. olympica* was simply a new name with its own types (Art. 72.7, 72.4.4).

Whorls more convex than in *O. annularis*, horny yellowish, with fine ribs. Animal large, mantle appendix covers large parts of the shell.

2.8-4.1 x 3.7-5.4 mm

Humid habitats in mountain forests, exceptionally in dry and open habitats.

Schütt 2001: 166.

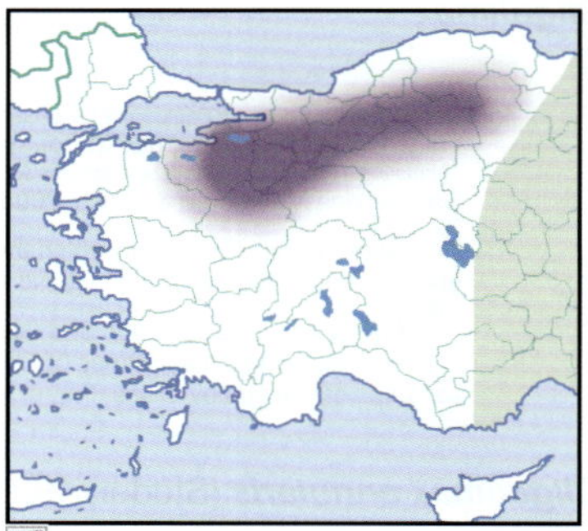

N Turkey

Oligolimax pollonerianus (Fra Piero, 1897)

Without gender agreement: *Oligolimax polloneriana* (Fra Piero, 1897)

Vitrina polloneriana Fra Piero, 1897 in Fra Piero 1897: 70 (*Vitrina Polloneriana*)
The description was attributed to Pollonera, the name was attributed to Fra Piero. In a strict application of Art. 50.1.1 the name in such a case must be attributed to the author of the work.

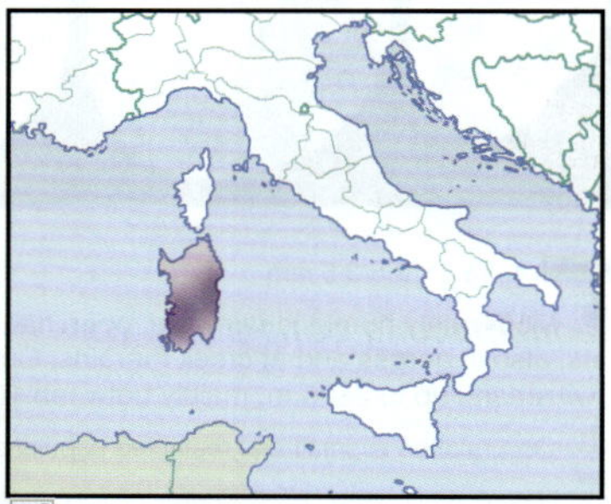

Italy (Sardinia)

Rocky limestone habitats with sparse vegetation, in high mountains and in coastal vicinity.

Shell yellowish translucent, thin and fragile, initially very regularly ribbed and with very small pits, last whorl smooth with weak growth lines, 3 rapidly increasing whorls, apertural diameter more than 1/2 of shell, umbilicus very narrow.
Animal cream with sparse dark spots, posterior section of foot sharply pointed and keeled, body does not withdraw entirely into the shell.

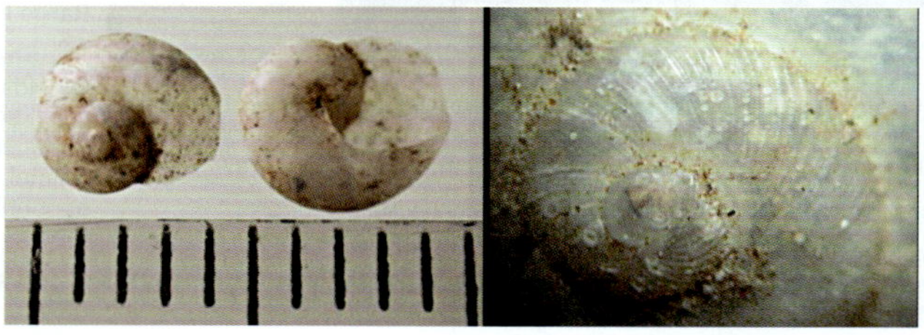

2.2-3 x 4.2-5.8 mm (height x diameter)

Italy: central southern Sardinia
Francesco Mascia

EU Red List: Vulnerable VU. Obviously a rare species, known from only few localities, potentially threatened by habitat alteration.

Was classified in a separate genus *Sardovitrina* by Manganelli & Giusti 2005 and Giusti et al. 2011. This genus was considered most closely related to the other *Oligolimax* species, just with some internal structures in the penial complex that no other species have and certainly allow to distinguish this species from the others. But eventual homologies of these structures do not seem to be well understood in plutoniine vitrinids. It was not explained why the differences should account for a distinction at the generic level in the present state of research, and why a classification as a subgenus would not suffice.

Phenacolimax

Phenacolimax Stabile, 1859 in Stabile 1859: 422

Type species *Helicolimax major* Férussac, 1807 (subsequent designation by Fischer in Paulucci 1878: 24)

Shell globular to moderately ear-shaped with very narrow apertural membrane, mantle lobe small to very large.
No penial appendix, penis joins vagina below the insertion of spermatheca duct, penial sheath present and penial retractor passes below right ommatophore retractor (two differences to *Oligolimax*), upper part of vagina strongly swollen.

Manganelli & Giusti 2005: 576 (differences to *Oligolimax*).

Phenacolimax blanci (Pollonera, 1884)

Vitrina major var. *blanci* Pollonera, 1884 in Pollonera 1884: 337 (or 427)

Shell like *Ph. major*, but colourless, whorls flatter and less globose, mantle appendix partly not adhering to the shell (difference to *Ph. stabilei*). Differs anatomically from *Ph. major* in its thick penis with a sheath in subbasal position, inside with pleats and a papilla in apical vicinity.

Shell diameter 7-10 mm

EU Red List: Vulnerable VU. Threatened by exploitation of habitats for tourism and pasturing.

Boato et al. 1984: 293, Manganelli et al. 1995: 18.

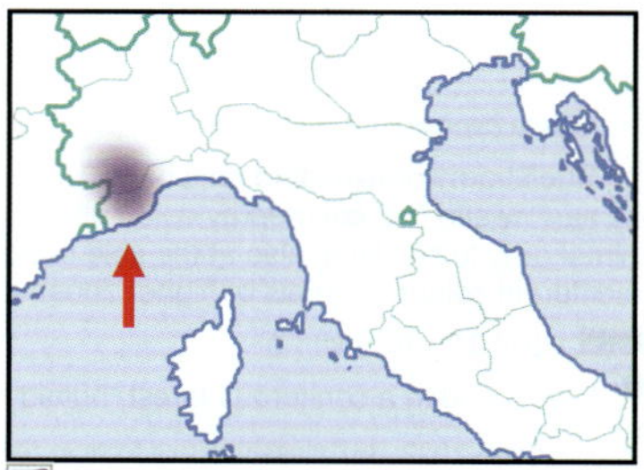

NW Italy

Phenacolimax major (Férussac, 1807)

Helicolimax major Férussac, 1807 in J. Férussac 1807: 43 (*Helico-Limax major*)
I am giving Férussac's single authorship here, as this has been traditionally done. I have not researched which one of the two Férussac's was responsible for this name (or both). It seems to have been overlooked that both Férussacs were co-authors of the 1807 work which should better be cited as "Férussac & Férussac 1807", and that the responsibilities were explained in the introduction.

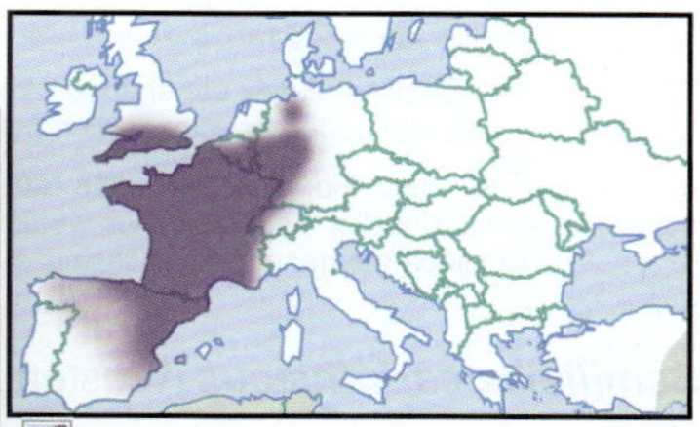

W Europe
Range indicated for Spain uncertain.

5-7 x 4-6 mm (height not considered), animal up to 16 mm   Germany: Sauerland, Hönnetal

Shell flatter than *Vitrina pellucida*, animal larger, gray to black, only juveniles and dehydrated adults can withdraw into the shell. Mantle small with large, meat-coloured appendix which can reach the apex.
Animal yellowish to blackish brown. Oviduct ball-like inflated in its lower part.

In humid forests, also meadows and shrubland on rocky mountain slopes, between rocks, under moist leaf litter and fallen branches, on calcareous and non-calcareous substrate. Requires warm, moderately humid and sheltered habitats. In Britain a useful indicator of primary woodland (but not further south in Europe). Rare above 1000 m, in Switzerland up to 700 m, possibly up to 1700 m in France.

Reproductive season September to November, vagina is exposed during copulation, sperma is exchanged externally. Eggs (diameter 0.3 mm) are laid in clutches of 8-15 eggs and fixed to stones and plants, slightly reddish juveniles hatch after 15-20 days, during growth they become grey, maturity is reached after 8-10 months.

EU Red List: Near Threatened NT. Probably declining overall in S England and S Wales, isolated colonies are at risk from clear felling or replanting with conifers. Endangered in Niedersachsen and Rheinland-Pfalz, Vulnerable in Switzerland and Bayern.

Moquin-Tandon 1855: 49, Germain 1930: 134, Kerney et al. 1983: 156, Jungbluth et al. 1989: 136, Falkner 1990: 174, Vogt et al. 1994: 156, Manganelli et al. 1995: 41 (no confirmed reports for Italy, previous reports referred to *Ph. blanci* or *Ph. stabilei*), Turner et al. 1998: 250, Kerney 1999: 135.

Phenacolimax stabilei (Lessona, 1880)

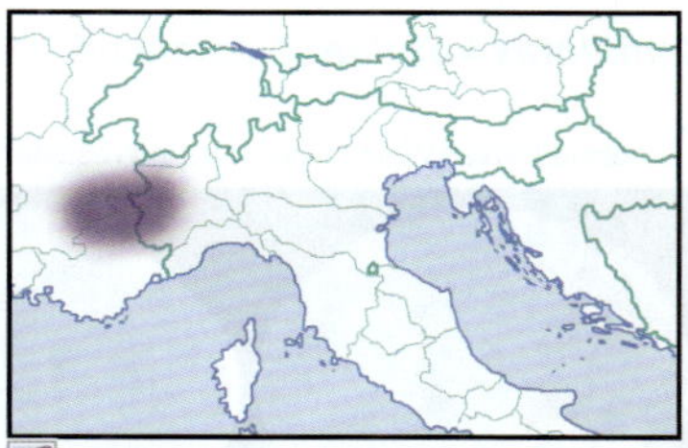

NW Italy

Vitrina major var. *stabilei* Lessona, 1880 in Lessona 1880: 338, Tav. 4 (*Vitrina (Phenacolimax) major* var. *Stabilei*)

Shell flatter than in *Ph. major*, mantle larger, animal colour grey, juveniles light grey, mantle appendix with black margin, can reach the apex.

5-7 x 4-6 mm (height not considered), animal up to 17 mm

In humid mountain forests.

Falkner 1990: 174, Manganelli et al. 1995: 18.

Semilimacella

Semilimacella Soós, 1917 in Soós 1917: 94, 154 *Vitrina (Semilimacella)*

Type species *Vitrina velebitica* Soós, 1917 (monotypy)
(= currently *Semilimacella bonelli* (Targioni Tozzetti, 1873))

Semilimacella bonelli (Targioni Tozzetti, 1873)

Semilimacella reitteri (Boettger, 1880)

Vitrina bonelli Targioni Tozzetti, 1873 in Targioni Tozzetti 1873: 322, Tav. 6 (*Vitrina Bonelli*, *Vitrina Bonellii*)
Name was spelled *Bonellii* on the plate, and *Bonelli* at various occasions in the text (pp. 322, 324, 325). Selection by First Reviser, if not earlier then herewith (Art. 24.2.3).
Vitrina reitteri Boettger, 1880 in O. Boettger 1880 (Ber. Offenb.): 102 (*Vitrina (Phenacolimax) Reitteri*)
Vitrina velebitica Soós, 1917 in Soós 1917: 94, 154

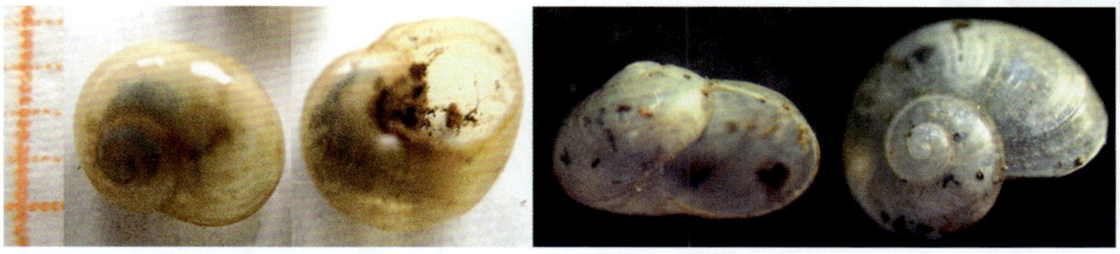

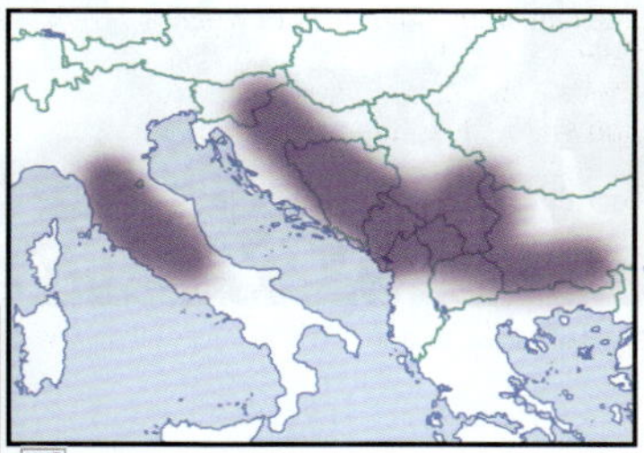

Italy, SE Europe

Shell greenish and ellipsoid, the greenish shell colour is not always visible in living animals.
Animal grey or greenish, tentacles brownish, mantle darker and with spots, mantle appendix does not cover the apex, body can withdraw into the shell entirely. Can hardly be distinguished from *Vitrina pellucida* and others from externally visible characters of the animal and the shell alone.

Shell at 3-3.5 whorls about 4.6 x 9.2 mm   (1-2) Bulgaria: W Rodopi, Chudni Mostove, (3-4) Italy: Apennino Emiliano
(1-2) Atanas Irikov, (3-4) Paolo Paolini

In S Bulgaria found in mountains at 1600 m.

Vulnerable in Albania (Dhora 2004), rare in Bulgaria (Irikov & Erőss 2008).

Ošanova & Pintér 1968 (anatomy), Damjanov & Likharev 1975: 254 (anatomy), Grossu 1983: 82, Irikov & M. 2006: 793.

Systematic position uncertain, possibly two separate species, by some authors classified in genus *Oligolimax*.

Semilimacella carniolica (Boettger, 1884)

Vitrina carniolica Boettger, 1884 in O. Boettger 1884: 184 (*Vitrina (Helicolimax) Carniolica*)

Shell yellowish, transparent and shiny, finely striateds, last whorl occupies slightly more than 50 % of shell diameter, aperture slightly oblique.

Under stones, shady and humid habitats.

Kerney et al. 1983: 150, Grossu 1983: 80, Manganelli et al. 1995: 18, Dhora & Welter-Schultes 1996: 141.

Range and identities of the forms very uncertain.

6-8 x 5-7 mm (height not considered)   Austria: Kärnten, Tscheppaschlucht

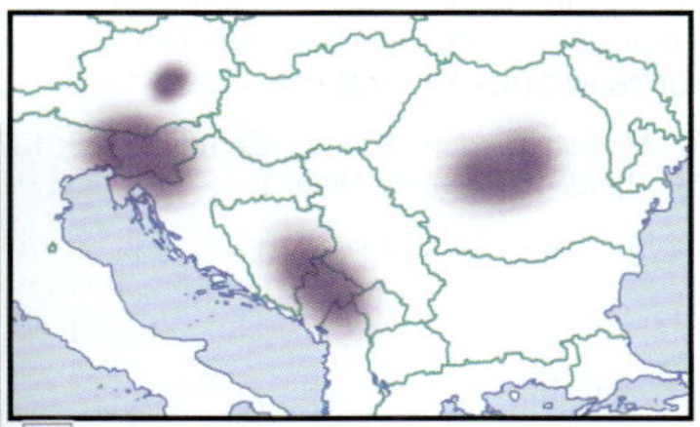

SE Europe

Semilimax

Semilimax Gray, 1847 in Gray 1847: 170 *Vitrina* synonym *Semilimax*

Type species *Vitrina elongata* Draparnaud, 1805 (monotypy, under Art. 67.12.1) (= currrently *Semilimax semilimax* (Férussac, 1802))

Name was made available under Art. 11.6.1 by Baker 1929: 138 who referred bibliographically to Gray 1847. The name has also been attributed to Stabile 1859: 422 who included 7 species (*V. elongata* was among them) and did not refer to a previous use of the name. Fischer in Paulucci 1878: 24 gave as type *Vitrina elongata*.

Shell extremely ear-shaped with broad apertural membrane and very large mantle lobe.
Penis with appendix, joins vagina below the insertion of spermatheca duct.

Giusti, Fio. et al. 2011: 336 classified *kotulae* in a separate genus *Hessemimilax* because they found some anatomical differences. It was not explained why a classification as a subgenus would not suffice in the light of the relationships between the involved taxa currently being sufficiently not well understood to establish separate genera. Their results suggested partly that *kotulae* formed a monophyletic group with other species of *Semilimax*.

Semilimax carinthiacus (Westerlund, 1886)

Without gender agreement: *Semilimax carinthiaca* (Westerlund, 1886)

Vitrina carinthiaca Westerlund, 1886 in Westerlund 1886: 18

Shell much like *Vitrinobrachium breve*, but with more whorls (2 1/8). Inner whorls strongly convex, apex higher. Penis retractor present (unlike in *S. semilimax*).

Mainly in forests, in humid and shady habitats, under stones, in leaf litter and at decaying dead wood.

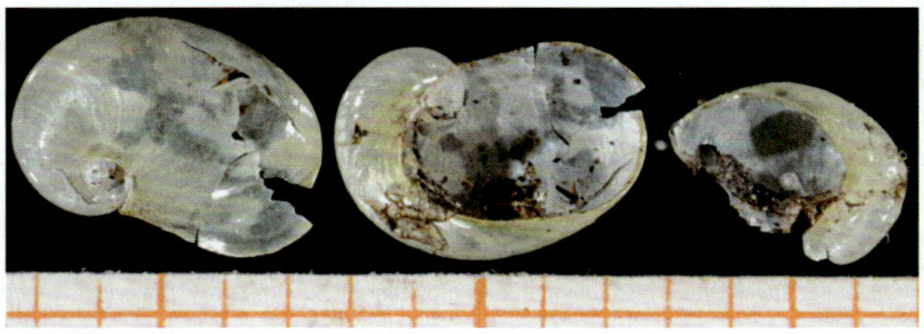

4.5-5 x 3-3.5 mm (height not considered)

Austria: Bärental, 1350 m

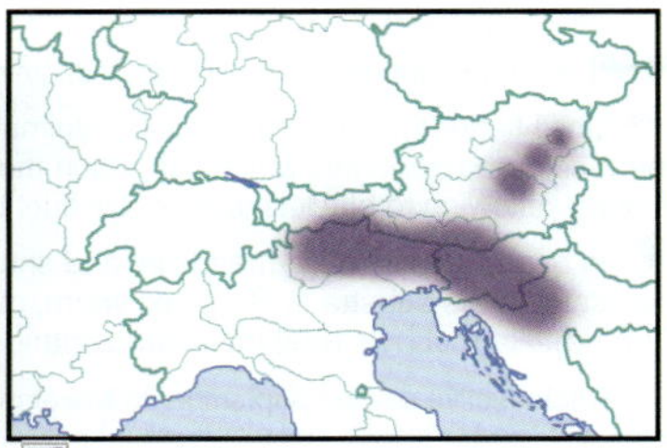

SE Alps

Kerney et al. 1983: 152, Manganelli et al. 1995: 18.

Semilimax kotulae (Westerlund, 1883)

Vitrina kotulae Westerlund, 1883 in Westerlund 1883: 54 (*Vitrina Kotulae*)

4-6 x 3-4.5 mm (height not considered), animal 12-15 mm

Austria: Ötztal, Sölden, 1700 m

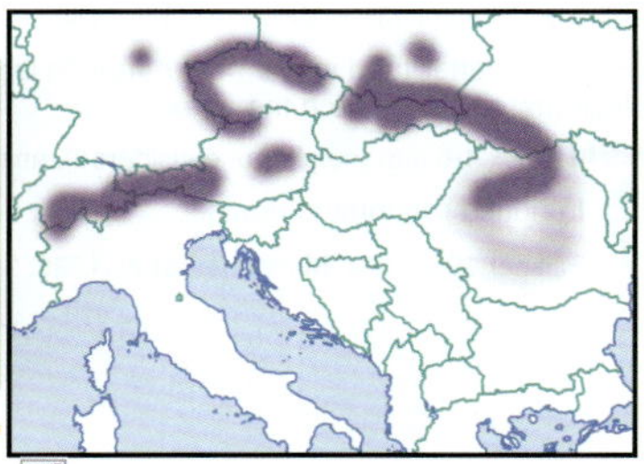

Alps and Carpathians

Slovakia: Poľana, 1300 m — Jozef Grego

Shell much like *S. semilimax*, but with flat whorls, apertural membrane goes further inside in a spiral form, escapes columella and leaves completely visible the upper whorls to the apex. Shell smoother and more shiny than in *S. semilimax*. Mantle lobe covers the apex.

Mainly in forests, cool and humid shady habitats, in moist moss, under stones and in soil litter, only in mountains. Rarely above the timberline in humid dwarf shrubland with mosses and on alpine pastures under stones. In Switzerland between 1300 and 2600 m.

Biology fairly unknown, adult animals were found in August and September.

Endangered in Bayern, Vulnerable in Austria. Everywhere rare and scattered.

Kerney et al. 1983: 153, Grossu 1983, Manganelli et al. 1995: 18, Turner et al. 1998: 257, Wiktor 2004, Sysoev & Schileyko 2009: 138.

Semilimax pyrenaicus (Férussac, 1821)

Without gender agreement: *Semilimax pyrenaica* (Férussac, 1821)

Helicolimax pyrenaica Férussac, 1821 in Férussac [1821-1822]: 25, (expl. planches) p. vj, Pl. 9 (*Helicolimax Pyrenaïca*)
Both text and plate appeared 06 Apr 1821 (Kennard 1942).

Shell like *S. semilimax*, but almost completely flat on upper side, last whorl occupies less than 60 % of shell diameter, finely striated, no umbilicus.
Animal very active, weakly grey, very large, cannot withdraw into the shell, mantle with dark spots, covers the apex entirely.

Humid and shady habitats, between rocks, in Ireland often in moist ground litter in woodland, mostly on non-calcareous soils, also in more open habitats, often in artificially altered or disturbed places, but seems to require absence of grazing animals.

Moquin-Tandon 1855: 48, Germain 1930: 137, Kerney et al. 1983: 153, Kerney 1999: 134.

Northern Ireland: Belvoir forest near Belfast — Roy Anderson

5-6 x 3.5-4.5 mm (height not considered)

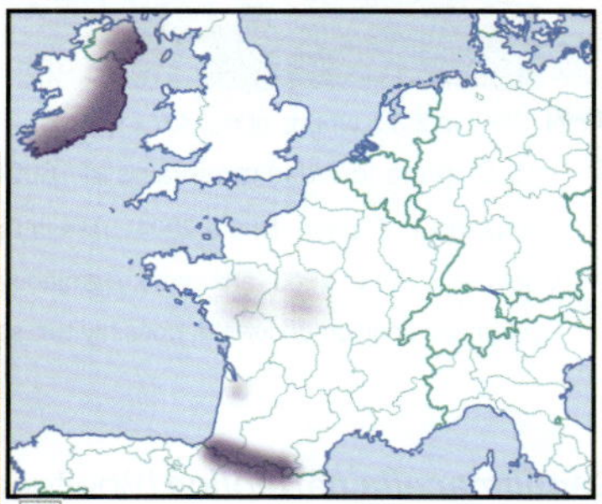

Pyrenees, Ireland
Possibly introduced to Ireland somewhen between 1700 and 1900, but could also be native. Some uncertain records also from W France.

Semilimax semilimax (Férussac, 1802)

Helix semilimax Férussac, 1802 in J. Férussac 1802: 236 (*Helix Semilimax*)

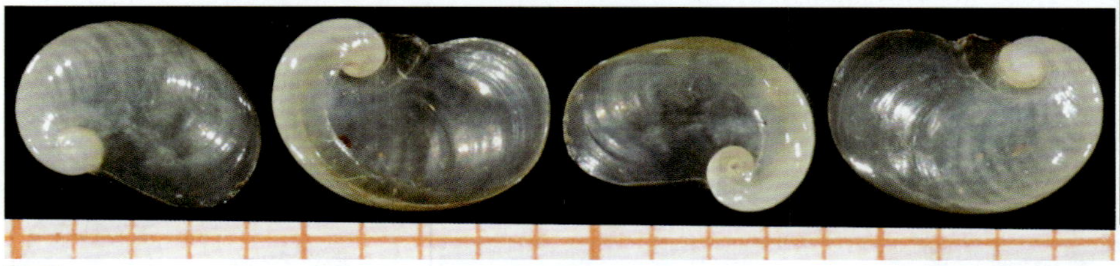

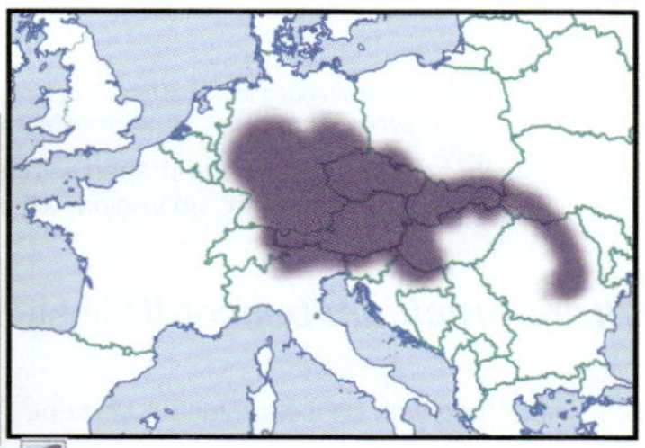

4-5 x 2.5-3 mm (height not considered), animal 12-15 mm

Austria: Osttirol, near Tauernblick, 1300 m

Alps, Carpathians, and highlands

Vulnerable in Switzerland and Germany.

Slovakia: Zvolenská Kotlina   Jozef Grego

Shell with greenish hue, looks reddish with animal inside, very thin and transparent, 2.5 whorls, last whorl occupies more than 60 % of shell diameter, apertural membrane low, no umbilicus.
Animal light to dark grey, mantle lobe covers the apex.

Humid habitats in mountain forests, under leaf litter and stones, prefers creek gorges, but is also found in open habitats with high precipitations above the timberline. In Switzerland up to 1800 m.

Annual life cycle. Reproduction between October and January, 50-90 oval eggs (1-1.5 mm long) are laid in 6-9 clutches between November and March, in humid moss, juveniles hatch after 3-8 weeks, maturity is reached after 4-6 months, most animals die between April and June, age 12-14 months.

Künkel 1933, Frömming 1954: 117, Kerney et al. 1983: 152, Falkner 1990: 172, Manganelli et al. 1995: 18, Turner et al. 1998: 256, Hausdorf 2002, Sysoev & Schileyko 2009: 139.

Vitrina

Vitrina Draparnaud, 1801 in Draparnaud 1801: 33, 98

Type species *Helix pellucida* Müller, 1774 (monotypy)

North America, Greenland, Europe, northern Asia

Shell globular, no apertural membrane, mantle lobe very small.
No penial appendix, no vagina, which means that penis, oviduct and spermatheca duct join at a common point.

Hausdorf 2002, Sysoev & Schileyko 2009: 136, Giusti, Fio. et al. 2011: 356.

One species lives in Greenland and North America and has occasionally been confounded with *V. pellucida* from Iceland and N Scandinavia:
Vitrina angelicae Beck, 1837 in Beck 1837: 1 (*Vitrina angelicæ*)
The abbreviation "fn. gr." in the original source means "Favna Groenlandica" (= Fabricius 1780: 389, where a description for specimens from Greenland was provided). Not easy to recognize, but the name was correctly established (Art. 12.2.1).

Vitrina pellucida (Müller, 1774)

Helix pellucida Müller, 1774 in Müller 1774: 15

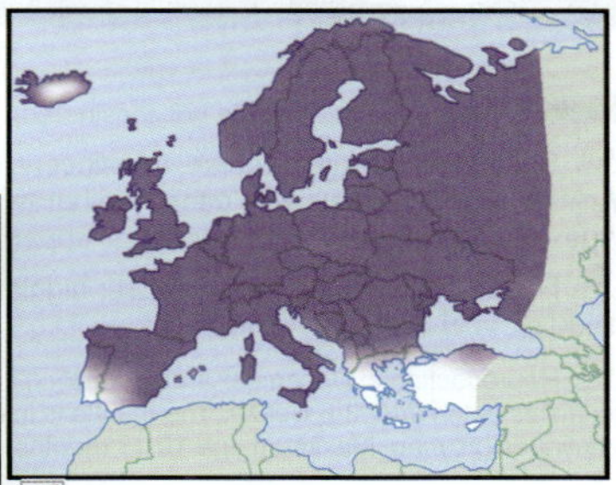

4-6 x 2.5-3.5 mm (height not considered), animal 10 mm

Romania: Transilvania, Cheile Turzii

Europe to central Asia

Slovakia: Low Tatra National Park   Jozef Grego

Shell weakly greenish, smooth and shiny, translucent, 2.5-3 whorls, rounded at periphery, last whorl occupies less than 50 % of the shell diameter (if seen from above), umbilicus open but very narrow.
Animal greyish with darker head and tentacles, can completely withdraw inside the shell, mantle narrow, mantle appendix small and grey or dark grey, does not cover the apex.

Usually in moderately open meadow habitats, in moist or dry habitats in coniferous and deciduous woods, grasslands and waste ground, the dryest habitats of all vitrinids, but with wide ecological amplitude. Occurs also at rocks and walls, in rock rubble of mountain creeks, in the water margin vegetation, on alpine pastures. Sometimes abundant in incompletely vegetated places. Tolerates acid substrate. In Switzerland up to 2700 m, in Bulgaria up to 1800 m.

Annual life cycle, juveniles hatch in spring and become mature in autumn, when they lay eggs (0.9-1.2 mm) and die shortly after. The exact dates seem to depend on regional conditions. In Ireland a winter species, adults lay eggs in spring, juveniles hatch in early summer. In E Germany eggs are laid in October, juveniles hatch in February and March. In Switzerland eggs are laid between September and November, animals can freeze during this process and continue the next day, adults die a few months after egg laying, between November and April/May, depends on the altitude. In S France eggs are laid between July and October.

Gassies 1849: 67, Frömming 1954: 113, Anderson 1996: 10 (life cycles).

Mandahl-Barth 1938: 11, Forcart 1955: 160 and Kerney et al. 1983 classified the Icelandic populations as *V. pellucida*.

Vitrinobrachium

Vitrinobrachium Künkel, 1929 in Künkel 1929: 624

Type species *Helicolimax brevis* Férussac, 1821 (monotypy)

Name intended to "substitute" *Vitrinopugio* Ihering, 1892 for those species where the free atrium appendix had the same function as in *Helicolimax brevis* - thus available under Art. 11.5.1. The other species were not named, so monotypy after Art. 68.3.

Shell strongly ear-shaped with narrow apertural membrane and very large mantle lobe.
Penis with appendix, no vagina, which means that penis, oviduct and spermatheca duct join at a common point.

Vitrinobrachium baccettii Giusti & Mazzini, 1971

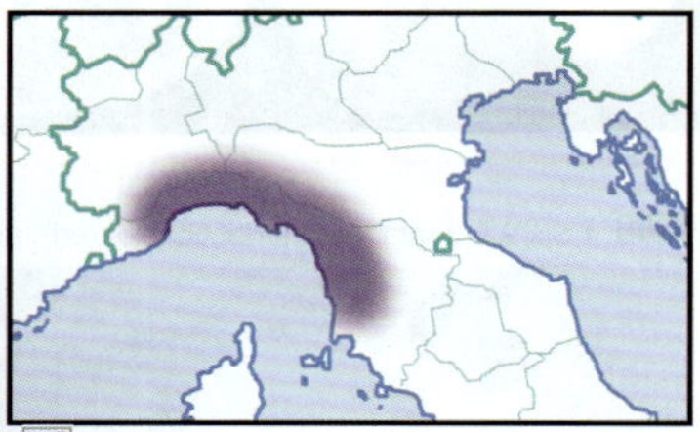

N Italy

Vitrinobrachium baccettii Giusti & Mazzini, 1971 in Giusti & Mazzini 1971: 246

5.5-6.5 x 4-4.5 x 2.1-2.3 (height) mm

Italy: Lucca, near Pizzorno

Maurizio Sosso

Shell horny yellow, transparent, fragile, 1.5-1.8 whorls with weak suture, last whorl predominating, aperture 4.3-4.8 x 3-4 mm, no umbilicus.
Animal grey or dark grey, colour below the shell weakly reddish in posterior third, mantle lobe covers the apex.
Shell differs little from other *Vitrinobrachium* species, but genitalia are different. Unlike *V. breve V. baccettii* does not possess a diverticle containing a copulatory arm (braccio copulatorio), but a special copulatory complex organ attached to the atrium, in which insert penis and receptaculum seminis.

Deciduous and coniferous forests under tree trunks, occasionally not rare in 1000-1300 m, in beach forests not very frequent. Snails tend to escape quickly from direct sunlight.

EU Red List: Near Threatened NT. Threatened by modern forest management (afforestation with conifers, road construction, soil destruction by too large machines). Protected by law in Toscana. This does probably not imply restrictions to forest management activities.

Boato et al. 1984: 292, Manganelli et al. 1995: 18.

Vitrinobrachium breve (Férussac, 1821)

Without gender agreement: *Vitrinobrachium brevis* (Férussac, 1821)

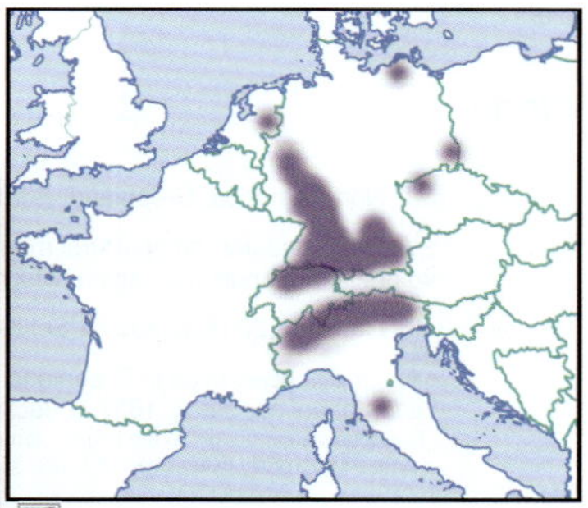

Central Europe, N Italy

Vulnerable in Switzerland.

Helicolimax brevis Férussac, 1821 in A. Férussac [1821-1822]: 21 (folio p. 25), (explication planches) p. vj, Pl. 9
Text and plates are both dated to 06 Apr 1821 (Kennard 1942).

5-6 x 3.5-4.5 mm (height not considered), animal 10-14 mm

Germany: Sauerland, Hönnetal

Shell very weakly greenish, translucent, with slightly less than 2 whorls, last whorl occupies about 60 % of shell diameter, apertural membrane narrow.
Animal large, grey or black, cannot withdraw into shell, mantle covers a large section of the frontal parts of the animal, mantle lobe covers the apex.
Apex not as flat as in *Vb. tridentinum*, upper whorls less densely coiled, last whorl less broad.

Humid deciduous forests in lower highlands and river plains, requires shady habitats with mosses and deep soils on rocky grounds. In S Switzerland up to 1500 m, in N Switzerland 600 m.

Reproduction in December and probably also in summer, clutches of 6-30 oval eggs (1-1.5 mm long) are laid under humid moss, together around 100 eggs per individual, juveniles hatch after 3-8 weeks. The mantle is initially whitish, after one week bluish, one more week later bluish black. Maturity is reached after 3-4 months, but the animal continues growing. Maximum age 10-17 months under artificial conditions.

Frömming 1954: 115, Forcart 1956, Kerney et al. 1983: 150, Falkner 1990: 172, Manganelli et al. 1995: 18, Turner et al. 1998: 259.

Vitrinobrachium tridentinum Forcart, 1956

Vitrinobrachium tridentinum Forcart, 1956 in Forcart 1956: 6

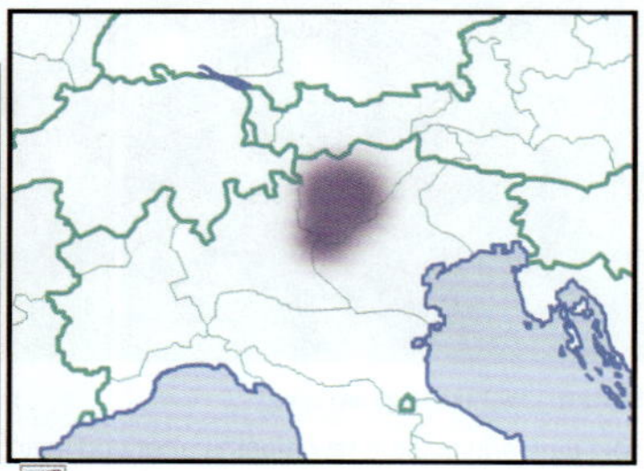

S Alps

2.7 x 5.5-6 x 3.7-4 mm, animal up to 16 mm

Italy: Lago di Garda region, Monte Baldo near Torbole

Shell with slightly more than 2 whorls, apex flat. Inner whorls more densely coiled than in *Vb. breve*, last whorl broader and apex flatter than in *Vb. breve* and *Semilimax carinthiacus*. Animal usually evenly dark grey, looks like *Vb. breve*, mantle larger, covers almost completely the anterior body parts. Receptaculum seminis inserts at the distal section of penis, not at the base as in *Vb. breve*.

Usually in humid mountain forests.

Kerney et al. 1983: 151 (reports for Switzerland were not confirmed by Turner et al. 1998), Falkner 1990: 172, Manganelli et al. 1995: 18 (only Valle del Brenta and Trentino-Alto Adige).

Papilloderma

Papilloderma Wiktor, Martín & Castillejo, 1990 in Wiktor, Martín & Castillejo 1990: 3

Type species *Papilloderma altonagai* Wiktor, Martín & Castillejo, 1990 (original designation)

Gender must be neuter, derma is Greek (= skin).

Slug covered by papillae, mantle short and in mid-section of slug, sole very narrow.

Papilloderma altonagai Wiktor, Martín & Castillejo, 1990

Papilloderma altonagai Wiktor, Martín & Castillejo, 1990 in Wiktor, Martín & Castillejo 1990: 4

Slug cream with darker dorsum, evenly covered with rows of slightly darker papillae, mantle very short and in mid-section of slug, less than 1/10 of body length and 1/3 of width, pneumostome on the right side of mantle, shell below mantle transparent with 0.5 whorls, lower side like *Semilimax*, mid-dorsum line in anterior section consisting of 2 parallel grooves, in posterior section like a keel, tentacles almost black with black eyes, sole cream and very narrow, 1/3 of slug width, no grooves there. Mucus colourless and moderately thick, also during irritation.

Up to 30 x 4 mm (preserved 25 mm long)

Was found on rocky meadows and in the shrub and herb layer in open forests. The slug is extremely difficult to find and its habitat not fully understood. It might live in rock crevices below the soil and appear at the surface only at certain weather conditions.

N Spain (Asturias, Cantabria)

Vulnerable.

Milax

Milax Gray, 1855 in Gray 1855: 162, 174

Type species *Limax gagates* Draparnaud, 1801 (subsequent designation by Connolly 1912: 120)

Gray 1855: [3] wrote that the descriptions of the shells in the work were by L. Pfeiffer, but this did not apply to slugs. Contained originally 9 species, among these were *L. gagates* and *L. sowerbyi*. ICZN Op. 335: type species fixed by Hesse 1926. This must be corrected, Connolly 1912 was earlier.

Slugs up to 80 mm, some species only 20 mm long, with a strongly arched keel.
Diagnostic anatomical characters for species identification: penis and epiphallus shapes, vas deferens sometimes characteristically long and usually opens apically to epiphallus, atrium shape can be characteristical, there is a single atrium accessory gland of specific size and connected to the atrium by one or several ducts, atrium with a specific number of specifically shaped stimulators inside.

Most species are adapted to habitats relatively close to Mediterranean coasts, ground dwellers, not crawling up trees or other plants, can stand long periods of drought, maximum age up to a few years.

Originally Canary Islands and Mediterranean region from Morocco to Tunisia and from S Spain to the Near East and Caucasus, probably introduced to W Europe from Scotland and Ireland to Portugal

Wiktor 1987 (monograph Milacidae), Wiktor 2001 (Fauna Graeciae 8): 23.

Milax aegaeicus Wiktor & Mylonas, 1986

Milax aegaeicus Wiktor & Mylonas, 1986 in Wiktor & Mylonas 1986: 147

Slug blackish grey (preserved, colour of live slugs unrecorded) with unclear darker dots on mantle and back, also near the sole. Two indistinct darker streaks on mantle, pigment accumulations also on posterior body part. Mantle length 30 % of body (preserved state). Keel always light cream, strongly domed, body almost triangular in transverse section. Sole cream with dark lateral zones. Mucus unknown.
Penis in two sections, together shorter than epiphallus, retractor penis bifurcate, inserting first on the boundary penis/epiphallus, the other thinner branch to the spermatheca duct, vas deferens opens apically and symmetrically to epiphallus. Atrium short and hardly visible, with two stimulators inside, atrium accessory gland very uncommon, opening at the base of the large stimulators in a bundle of ducts, then they suddenly widen and turn back to the atrium, where they unite and surround the atrium with a membraneous structure.
The species is anatomically well characterized, externally some slugs resemble *M. gagates. M. gagates* has no colour pattern and no darker spots on the sole. *M. altenai* is smaller and has no dark melanophores on the sole.

Up to 40 x 8.5 mm (preserved)

Wiktor 1987: 186, Wiktor 2001: 24.

Greece (Kikládes)

Milax altenai Forcart, 1972

Milax altenai Forcart, 1972 in Forcart 1972: 108

Slug grey with blackish back, gradually lightening downwards to the sides, slightly darker in the skin grooves, mantle less than 30 % of body length, grey with a characteristic black colour pattern, consisting of a black horse-shoe groove directed backwards and irregular blackish spots on the rest of the mantle. Head almost black. Sole cream.
Penis spherical, epiphallus distinctly thinner and longer, vas deferens opens apically in epiphallus, atrium weakly separated from penis and vagina, atrium accessory gland very big, connected with atrium by 18 thin tubular ducts. Stimulator like a narrow triangle with an oval hollow at its base (the ducts of the accessory gland open there), also wider and shorter than in *M. parvulus*, bent hook-like.
Looks much like *M. parvulus*, differs by its stimulator and the oval hollow at its base.

Up to 30 x 4.5 mm (preserved)

Wiktor 1987: 190, Wiktor 1994: 39.

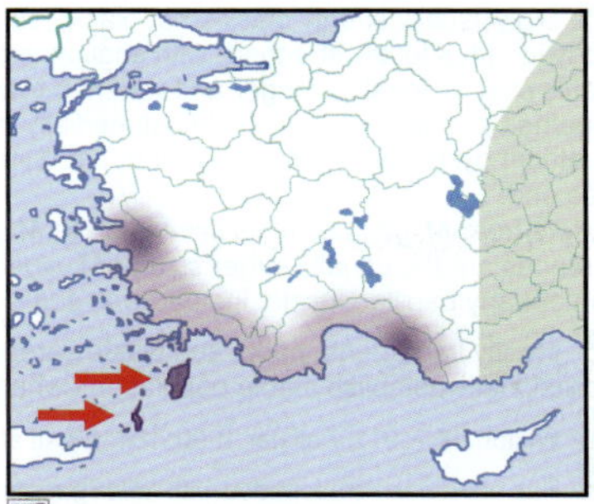

SE Greece, SW Turkey

Milax barypus Bourguignat, 1866

Milax cyprius (Simroth, 1906)

Milax barypus Bourguignat, 1866 [01 Feb 1866] in Bourguignat 1863-1868: 208, Pl. 32
Amalia cypria Simroth, 1906 in Simroth 1906: 90

Slug highly domed with prominently arched keel, colour of preserved slugs olive brownish or yellowish, with grey pattern and darker parts besides horse-shoe groove on mantle and at dorsum next to the keel, keel always lighter, mantle longer than 1/3 of body, narrowed in front, 11-12 grooves between keel and pneumostone.
Epiphallus slightly thinner and longer than penis, vas deferens thin, long and partly coiled, opening apically to epiphallus, oviductus and vagina almost equally wide, atrium is an extension of vagina with a large side pocket, almost as large as penis, atrium accessory gland irregularly shaped with one single duct, slightly ticker than vas deferens, connecting it to the atrium side pocket. Stimulator inside atrium with wide base and a claw-shaped end.

Up to 16 x 4.5 mm (preserved)

Wiktor 1987: 196 synonymized *M. barypus* from N Israel/Palestine with *Amalia cypria*.

Cyprus, Near East

Milax gagates (Draparnaud, 1801)

Limax gagates Draparnaud, 1801 in Draparnaud 1801: 100

Up to 50 mm long (preserved 25-30 x 6-8 mm)

(Shells) Italy: Südtirol, Laag near Neumarkt, (slug): Germany: Helgoland
(slug) Vollrath Wiese

Originally NW Africa and SW Europe, dispersed worldwide

Slug evenly dark grey to black except somewhat lighter sides, no pigment spots. Mantle relatively large (35-40 % of body length), with distinct grooves, keel prominent between mantle and posterior end, skin sculpture weak, 16-17 grooves between keel and pneumostome. Sole with blackish lateral zones and lighter medial zone.
Penis rounded, half as long as epiphallus, epiphallus slightly widened and truncated (as if cut off) at its end, vas deferens short (usually not longer than epiphallus), opens asymmetrically at truncated end of epiphallus, atrium short, not widened, accessory gland consisting of several elongate glands, connected to atrium by some 20 short coiled tubular ducts. Stimulator narrow, conical, slightly flattened, at its end with some papillae.
The stimulator of *M. nigricans* has more papillae, situated at its base and not near its free end.

Mostly in cultivated areas, often in coastal vicinity, also in forests and shrublands, natural meadows, preferably close to water, hides under stones, moist ground litter and in soil cavities. Feeds on fresh herbs, occasionally (W British areas) noxious to gardens and crops (carrots and potatoes).

Copulation in Britain from spring to autumn, during copulation both slugs cling together so closely that no everted genitalia are visible. Self-fertilization is also possible. In France 45-60 eggs (diameter 3 mm) are laid in October. Under laboratory conditions the first eggs are laid 5-15 days after copulation, at one time some 15 eggs are laid, this can be repeated various times, in total not more than 100 eggs are laid by an individual. Juveniles (4 mm long) hatch after 25-40 days, maturity is reached after 18-20 months (4-5 months under laboratory conditions). 15-30 days after egg-laying the slugs die.

Originally from Morocco to Tunisia and S Spain to S France and Balearic Islands, perhaps native in Canary Islands, introduced between Portugal and Galicia, Atlantic coastlands of France to Belgium, British Isles, and almost worldwide (Australia, Tasmania, New Zealand, Japan, Pacific islands, South Africa, Atlantic islands)

In Britain always local except in Cornwall.

Germain 1930: 107, Wiktor 1987: 202, Falkner 1990: 184, Vogt et al. 1994: 177, Manganelli et al. 1995: 20, Turner et al. 1998: 284, Kerney 1999: 149.

Has often been confounded with *Milax nigricans*. Literature records are only reliable if description and drawing of the stimulator was provided. It is also possible that *gagates* and *nigricans* belong to the same species, and that the stimulator displays local variations.

Milax nigricans (Philippi, 1836)

Parmacella nigricans Philippi, 1836 in Philippi 1836: 125, Tav. 8
Schultz was quoted to be responsible for the name and the locality, description was by Philippi alone.

S Europe, Canary Islands, Morocco to Tunisia, often brought abroad

Up to 75 mm long (preserved 65 x 13 mm)

Malta: Msida, Tal-Qroqq David Cilia

Slug usually evenly blackish or black, laterally lighter, sometimes generally lighter to pale bluish grey. Keel well arched. Mantle relatively small (25-30 % of body length if totally streched) with shallow grooves, 14-15 grooves between keel and pneumostome. Juveniles yellowish. Sole dirty cream.
Penis oval, epiphallus longer and widened towards its truncated end (at its end not narrower than penis), vas deferens slightly longer than penis and epiphallus together, opens apically and almost medially to truncated end of epiphallus, vagina and atrium of same diameter, atrium acccessory gland very big, conneted to atrium by 16-18 very thin tubular ducts. Stimulator big, slightly bent, at its free end conical, with several rows of big papillae inside at its base, not near its free end.
Differs from *M. gagates* in its stimulator, in *M. gagates* there are no papillae at its base, only near its free end (this difference works also in juveniles). Other differences are the usually larger size of *M. nigricans*, its smaller mantle, lighter and unicolour sole. *Milax ater* from N Africa has an entirely smooth stimulator without any papillae inside, instead with some processes at its free end.

Mostly in gardens, parks and waste yards, also farmland and greenhouses, clusters under stones, wood and decaying plant material. Synanthropic, may occasionally establish stable populations when brought abroad with vegetables.

Wiktor 1987: 211, Falkner 1990: 184, Giusti et al. 1995: 267, Manganelli et al. 1995: 20, Mienis 2006 (Hoorn, NW Netherlands), Beckmann 2007: 84.

Has often been confounded with *Milax gagates*, only reports with description and figures of stimulators can be regarded as reliable. It is also possible that *gagates* and *nigricans* belong to the same species, and that the stimulator displays local variations with intermediates in large or narrow transitional zones.

Milax ochraceus (Bérenguier, 1900)

Without gender agreement: *Milax ochracea* (Bérenguier, 1900)

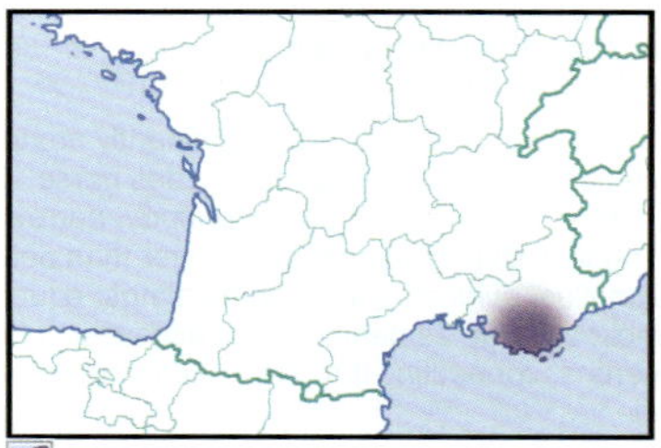

Amalia ochracea Bérenguier, 1900 in Bérenguier 1900: Pl. 1, Pl. 20

Slug evenly yellowish ochrous, never spotted, with grey head and tentacles, pointed in posterior part, dorsal keel prominent, mantle long, finely granulated, respiratory hole in posterior section, with light margin, foot light yellowish. Shell oval and slightly convex at lower side, with central nucleus. Mucus colourless.
Penis cylindrical, short, accessory gland not very long.

Up to 50 mm long, shell 9-10 x 2 mm

Germain 1930: 109. Not mentioned by Wiktor 1987.

More research is necessary to clarify its status, the description of the genitalia is from the 1930s and certainly insufficient.

S France

Milax parvulus Urbański & Wiktor, 1968

Milax parvulus Urbański & Wiktor, 1968 in Urbański & Wiktor 1968: 91
The name was published in the belief that the paper by Wiktor 1968 which was cited as being in press, would have appeared until then, and so the name was given as "*Milax (Milax) parvulus* Wiktor 1968". But Urbański & Wiktor's 1968 paper obviously appeared unexpectedly earlier, or the other article later. The whole paper was written in co-authorship, and there was no statement restricting authorship of parts of the text to one author alone. So both are co-authors of the name (Art. 50.1.1).

Slug bluish ash-grey behind mantle, with spots in grooves forming a reticulate pattern, keel and sole brown-black with orange (orange not in preserved slugs), mantle length 35 % of body (preserved), dark with an irregular reticulate pattern formed by black spots, an irregular black band is present at each side, 11-13 grooves between keel and pneumostome. Head and tentacles blackish. Body broadest at mantle, in the anterior third of the slug, keel prominent, running from the mantle to the posterior end.
Penis and epiphallus like one single hook-bent organ, epiphallus widened at its free end, vas deferens opens in the middle, vagina extremely short, atrium very broad, accessory gland a compact mass, connected to atrium by 9-13 silky-glossy thin ducts. Stimulator very long, thin, coiled in natural posotion and completely smooth.
External appearance and genitalia resemble *M. altenai.*

Up to 25 x 6 mm (preserved)

On limestone rocks, under stones, at shady walls, in open woodland and grasslands. In Central Balkan mountains in up to 1300 m.

Wiktor 1968, Wiktor 1983: 97, Wiktor 1987: 215, Wiktor 2001: 28, Hubenov 2007.

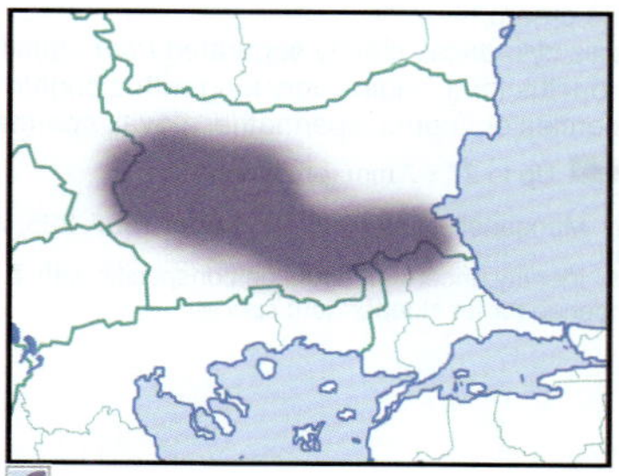

Bulgaria

Milax verrucosus Wiktor, 1969

Milax verrucosus Wiktor, 1969 in Wiktor 1969: 47

Slug evenly white or cream, without spots, largest slugs with bluish or grey dorsum. Tentacles always blackish. Mantle only 15 % of body length, in preserved slugs 20 %, with distinct characteristical and well visible papillae. Keel low, but visible. Mucus colourless and very dense.
M. verrucosus has two special head retractor mucles, unique in *Milax*. Penis not separated from epiphallus, vas deferens not thin and extremely short, atrium not widened, accessory gland is formed very late, when other genital organs are already fully developed, looks forst like a lateral papilla at atrium wall, when fully developed it looks like a bird's wing, connected to atrium by a single thick channel. No stimulator inside atrium.

Up to 35 mm long (preserved 28 x 4 mm)

In short vegetation of valleys, in deciduous and mixed forests, mainly in beech-hornbeam with walnut, also found in caves. Slugs hide under stones, in litter and in the soil, where they dig canals. In Central Balkan mountains in up to 1200 m.

Wiktor 1968, Wiktor 1983: 99, Wiktor 1987: 218, Hubenov 2007.

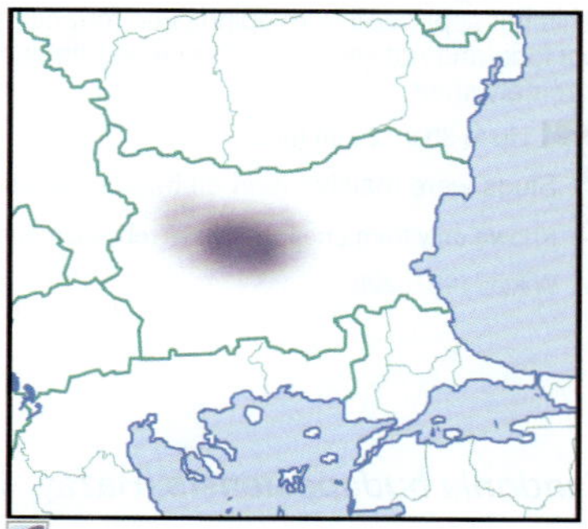

Bulgaria

Tandonia

Tandonia Lessona & Pollonera, 1882 in Lessona & Pollonera 1882: 54 *Amalia (Tandonia)*

Type species *Limax marginatus* Draparnaud, 1805 (original designation)

Aspidoporus Fitzinger, 1833 in Fitzinger 1833: 90, type species *Aspidoporus limax* Fitzinger, 1833 (monotypy)
Aspidoporus limax Fitzinger, 1833 in Fitzinger 1833: 90, no types preserved and no neotype designated. Was considered by Wiktor 2001: 28 as a nomen dubium and senior synonym of *Tandonia.*

Slug body elongate up to 100 mm, usually with almost parallel sides, posterior part abruptly narrowed, keel usually low, restricted to last section of dorsum in some species, reaching mantle margin in others.
Penis usually short, epiphallus long, retractor penis inserting at boundary between penis and epiphallus, vas deferens of specific length opening into posterior end of epiphallus, vagina accessory glands (usually more than one) of specific shapes, connected by ducts to the vagina or to the point where the vagina enters the atrium. Atrium small, tubular, no stimulator inside. Spermatheca of specific size, spermatheca duct longer or shorter than spermatheca.

Europe to N Africa and Asia Minor

Wiktor 1983: 101, Wiktor 1987: 2220, Wiktor 2001: 28.

Some species were described more than 50 years ago without detailed information on their genitalia and were not found afterwards. The types are destroyed, their status is unclear and we cannot say that these are species known to us. This concerns the following taxa:
Milax croaticus Wagner, 1929 in H. Wagner 1929: 333
Amalia dalmatina Simroth, 1900 in Simroth 1900: 106
Tandonia lagostana is not much better known.

Tandonia albanica (Soós, 1924)

Amalia albanica Soós, 1924 in Soós 1924: 191

Slug either black with slightly lighter sides and sole cream (on rocks and rock rubble), or grey-brown-beige with almost black patterns on mantle and back (near rivers and lakes). Mantle 30 % of body length. Keel poorly arched, 12-15 shallow skin grooves between keel and pneumostome. Mucus colourless, milky when irritated.
Penis usually wider than epiphallus, epipallus 1.5 x longer, papilla penis semispherical, smooth or with wrinkles, spermatheca oval, spermatheca duct long and thick-walled, vagina accessory glands lobular. Spermatophore yellowish, at the posterior and not narrowed end with several large hooks, anteriorly with small bifurcate spines, smooth on the rest of the posterior half.
Does not differ externally from *T. macedonica*, only in its spermatophore, its short papilla penis, wider and shorter epiphallus and thick-walled spermatheca duct. Black *T. albanica* can be confounded with *T. melanica*, *T. albanica* differs in its thick-walled conical spermatheca duct, more elongate and larger spermatheca and larger copulatory organs relative to body size.

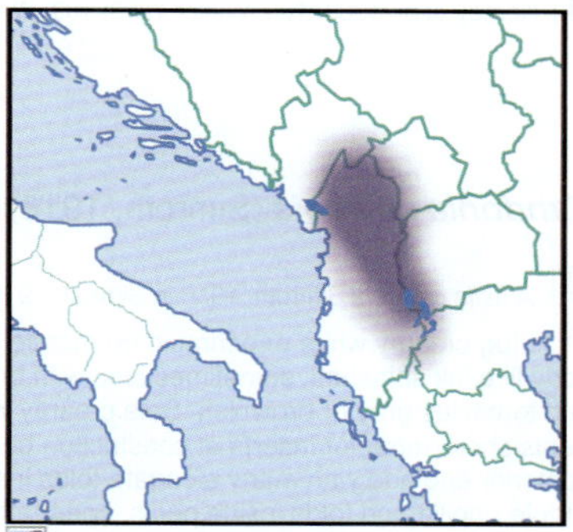

Albania, Rep. Makedonija

Up to 60 mm long (preserved), 30 mm long animals can be mature

Rocky shrublands and rock rubble, very frequent after rain, hides under stones and in rock crevices. Near water bodies beige slugs are more frequent, black forms in more rocky habitats, in mountains and forests, both forms meet in nature and can copulate. In highlands and high mountains, between 600 and 2400 m.

Spermatophores in spermatheca can be found between late September and early November. Slugs are motionless during copulation, usually hidden under stones on warm and wet or rainy days, both partners with arched bodies, adhering closely by the sides, everted atriums are visible like two whitish rolls.

Wiktor 1987: 222.

Tandonia baldensis (Simroth, 1910)

Amalia baldensis Simroth, 1910 in Simroth 1910: 334

Adult slugs blackish with pigments concentrated in body grooves, juveniles evenly black with sides getting lighter downwards. Mantle 30 % of body length, keel well visible, 11 grooves between keel and pneumostome. Sole cream.
Penis cylindrical, clearly separated from equally long epiphallus, papilla inside penis simple, vas deferens and oviductus long, vagina very long with elongate folds inside, accessory glands very small and in the form of small tubercles or fingers, spermatheca oval, spermatheca duct equally long.

Up to 27 x 7 mm (preserved)

Manganelli et al. 1995: 21, Wiktor 1987: 225, Wiktor 2001: 28.

Identity uncertain, could be conspecific with *T. nigra*. The description of the anatomy does not seem to provide strong evidence for an independent species.

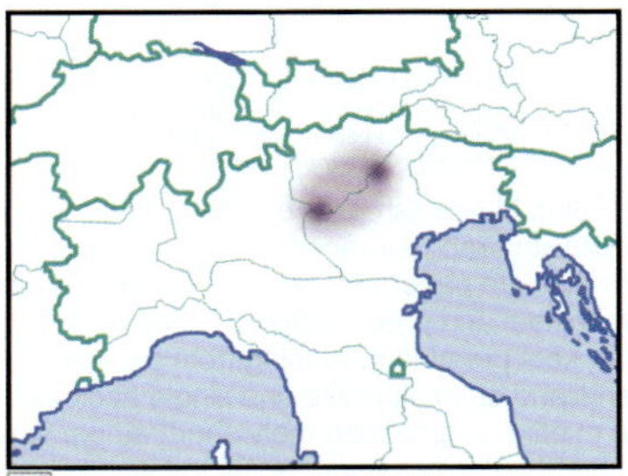

N Italy

Tandonia bosnensis Wiktor, 1986

Tandonia bosnensis Wiktor, 1986 in Wiktor 1986: 162

Slug brownish pink with blackish back and head, also blackish patterns on mantle and sides. Keel pinkish (pink color gets creamy after preservation). Sole cream. Mucus colourless, milky white when irritated.
Penis not separated from epiphallus, retractor penis very thin, vas deferens long, no papilla penis, spermatheca duct long and not separated from elongate spermatheca, together 2.5 x longer than penis+epiphallus, vagina extremely short.

Up to 30 x 12 mm (preserved)

Slugs were mainly found on Fagus leaves at the bottom of low limestone rocks, 1300-1500 m.

Known only from one locality at Trebević mountain near Sarajevo (Wiktor 1987).

Wiktor 1987: 228.

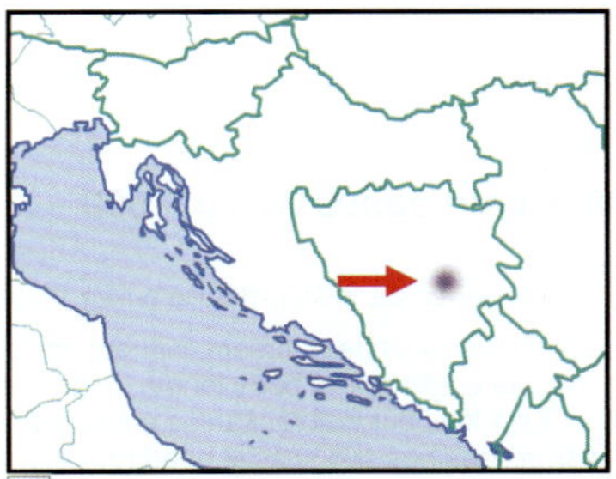

Bosnia (near Sarajevo)

Tandonia budapestensis (Hazay, 1880)

Amalia budapestensis Hazay, 1880 in Hazay 1880: 37

(shells) Italy: Modena, (3) Germany: Rheinland-Pfalz, Soonwald near Mengerschied, (4) Czech Republic: Moravia, Olomouc (4) Michal Maňas

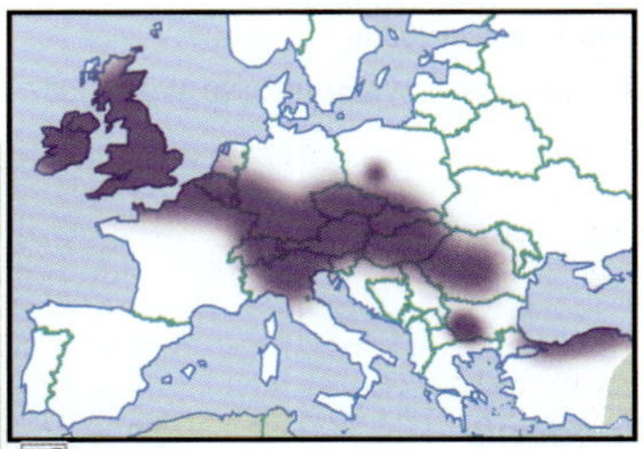

Originally probably SE Alps and N Balkans to Romania, today widespread in Europe

Up to 70 mm long (preserved 30-40 mm)

Slug yellowish grey to brown or dark grey, with numerous black spots so that the slug may appear evenly black-brown, slightly lighter at the sides. In N Italy and Bulgaria often brownish with black colour pigments concentrated mainly in the grooves. Animal very slender, gradually narrowing posteriorly, mantle length less than 1/3 of body, blurry black bands on mantle sides, keel prominent, light, reaching mantle. Head and neck blackish, sole narrow and cream with brown or orange hue, in the south often with slightly darker medial zone. Mucus colourless, yellowish when irritated, thick and sticky.
Penis rounded, wider than and approximately as long as epiphallus, vas deferens opening clearly asymmetrically at the posterior end of epiphallus, small simple papilla inside penis, spermatheca duct usually thick, vagina and atrium short, vagina accessory glands like two lobe-like objects connected to vagina by thin ducts. Spermatophore thin, 16 mm long, covered almost entirely with short spines.

In parks, gardens, ruins and on cultivated fields, lives as a synanthrope and is noxious to crops, in lowland England particularly in potatoes. In natural environments in Britain only where human disturbance is involved. In S Bulgaria usually between 300 and 1000 m, but locally up to 2200 m. Requires humidity and is active at night. Buries also into heavy soils. Widespread and still spreading in the British Isles.

Copulation in Britain from November to January, in Central Europe from April to autumn, slugs may copulate several times in their life. Copulation begins usually at night and may last 15 hours or more, everted genitalia are visible between partners. In Britain juveniles hatch in April or May, maturity is reached next autumn, up to more than 20 eggs are laid at a time. In Central Europe eggs, sub-adults and adults pass the winter.

Kerney et al. 1983: 180, Wiktor 1983: 109, Wiktor 1987: 230, Kerney 1999: 152, Irikov & Mollov 2006: 804, Hubenov 2007.

Tandonia cavicola (Simroth, 1916)

Amalia cavicola Simroth, 1916 in Simroth 1916: 4

Slug creamy white or ochrous, on mantle darker than at the sides, usually with dark spots scattered on the body like ink splashes, sometimes also with brownish reticulate pattern, keel and medial line always light. Head and tentacles grey or brownish. Sole creamy white. Mucus unknown.
Penis short, retractor inserts at constriction between penis and epiphallus, epiphallus 3.5 x longer, broader at posterior end and with many elongate folds inside, vas deferens thin and longer than penis+epiphallus, small simple papilla and folds inside penis, spermatheca duct longer than spermatheca, at its opening to atrium a small papilla is present, vagina accessory glands like small leaves adhering to vagina walls and spermatheca duct, each one connected to boundary of vagina and atrium by a thin duct.
The colour of this slug is probably sufficient for identification.

Up to 70 x 12 mm (preserved)

Not restricted to caves.

Wiktor 1987: 234.

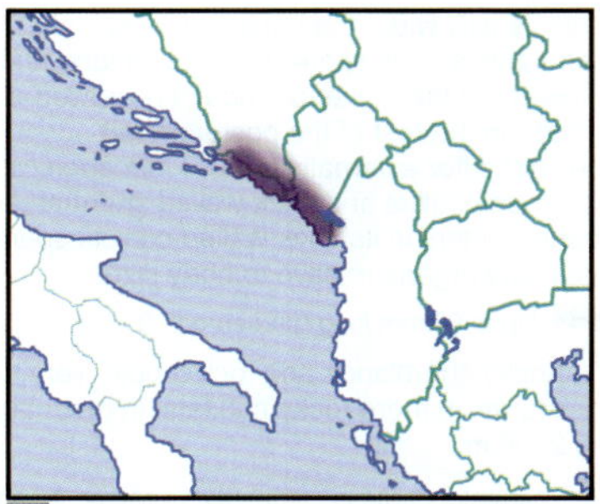

Croatia (Dalmatia) to Montenegro

Tandonia cretica (Simroth, 1885)

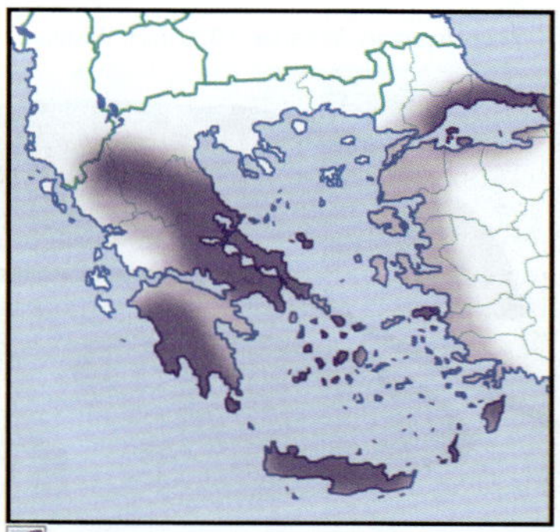

Greece, W Turkey

Amalia cretica Simroth, 1885 in Simroth 1885: 231, Taf. 7, Taf. 10

Head, neck, mantle and dorsum uniformly black or nearly black, usually with a navy-blue tint, sides gradually getting lighter downwards, dirty cream near sole, keel not prominent, but slightly lighter than dorsum, 20-22 grooves between keel and pneumostome. Mantle length 25 % of body. Sole cream-white. Mucus colourless, transparent.
Penis cylindrical with two constrictions, epiphallus with one and slightly shorter, vas deferens thin and long, opening apically into epiphallus, oviductus a short thick tube, atrium and vagina shorter and thicker tubes, accessory glands like short rolls opening with short and thick ducts, spermatheca large, spermatheca duct shorter.

Up to 85 x 15 mm (preserved)

Open habitats, shrubland and open forests, mainly on limestone, often abundant, also as a synanthrope. Shelters under stones. Up to 1700 m. Relatively frequent.

Spermatophores in individuals from October to February.

Wiktor 1987: 236, Wiktor 2001: 29-30.

Tandonia cristata (Kaleniczenko, 1851)

Without gender agreement: *Tandonia cristatus* (Kaleniczenko, 1851)

Krynickillus cristatus Kaleniczenko, 1851 in Kaleniczenko 1851: 225, Tab. 5

(1) Bulgaria: Sredna Gora, Hisar, (2) Ukraine: Crimea, near Yalta (scale bar 10 mm)

(1) Dilian Georgiev, (2) Sergey Leonov

Bulgaria to Crimea and Turkey

Grossu 1983: 226, Wiktor 1983: 115, Bar & Butot 1986, Wiktor 2001: 33, Irikov & Mollov 2006: 804, Sysoev & Schileyko 2009: 160.

Up to 30 mm (preserved 25 x 6 mm)

Slug slender, slightly narrowing posteriorly, in living slugs mantle and dorsum ash-grey-greenish or ash-grey-yellowish, after preservation pale or bluish ash-grey, skin grooves somewhat darker. Keel prominent. Head and tentacles blackish. Mantle length 25 % of body (preserved state). Sole cream. Mucus thick, colourless transparent or yellowish. 8-9, occasionally up to 11 grooves between keel and pneumostome.
Penis broad with a web-like musculus retentor in its anterior section, epiphallus as long as penis but narrower, vas deferens very thin, opening symmetrically at epiphallus end, atrium and vagina short, accessory glands like two wings connected by a thin duct to vagina each, as long as atrium+vagina, spermatheca elongate, unclear where spermatheca duct begins. Spermatophore with a broad end continuosly narrowing to a pointed end, covered with hook-like spines in its middle section.
Differs from *T. samsunensis* mainly in its accessory glands. *T. piriniana* has a narrower penis and epiphallus and a blind process on the vas deferens.

Usually hides under big stones in soil. Most frequently found on heavy loam, in deciduous and mixed forests, among shrubs, lower parts of valleys and their bottoms, gardens and open habitats. Up to 1500 m (central Bulgaria), in S Bulgaria between 500 and 1000 m.

Tandonia ehrmanni (Simroth, 1910)

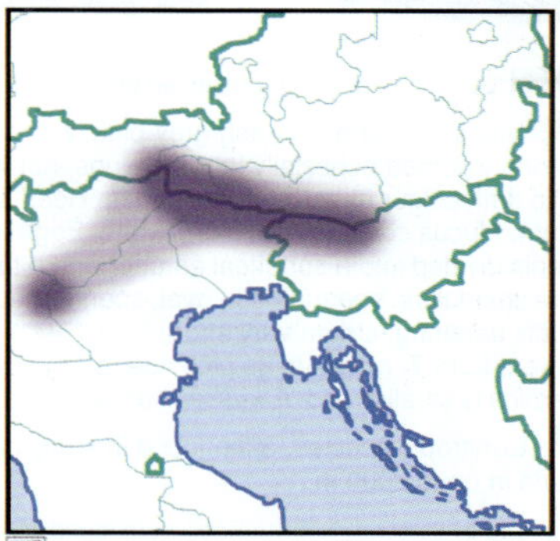

Italian Alps, Austria, Slovenia

Amalia ehrmanni Simroth, 1910 in Simroth 1910: 334 (*Amalia Ehrmanni*)
Amalia kobelti Simroth, 1910 in Simroth 1910: 333 (not *Amalia kobelti* Hesse, 1882)
Milax simrothi Hesse, 1923 in Hesse 1923: 195
Expressly established as a new replacement name for *Amalia kobelti* Simroth, 1910 non Hesse, 1882.

Slug deeply black or blackish, sides and anterior part usually lightened downwards, sole and neck under mantle creamy, mucus colourless.
Penis as long as epiphallus but thicker, vas deferens thin, oviductus a long tube, spermatheca oval, spermatheca duct very short or absent, vagina with two bundles of tubiform accessory glands, atrium with a strong retentor muscle attached.

Up to 40 mm long (preserved 37 mm)

Mountains above timberline, 800-2000 m, in cirques covered with poor gramineous vegetation, also near *Pinus mughus*, under stones.

Wiktor 1987: 246, 299, Wiktor & Milani 1995, Wiktor 1996: 50, Nardi 2011: 15.

Also called *Tandonia simrothi*. I do not see big differences to *T. nigra*.

Tandonia fejervaryi (Wagner, 1929)

Milax fejervaryi Wagner, 1929 in H. Wagner 1929: 331 (*Milax (Subamalia) Fejérváryi*)

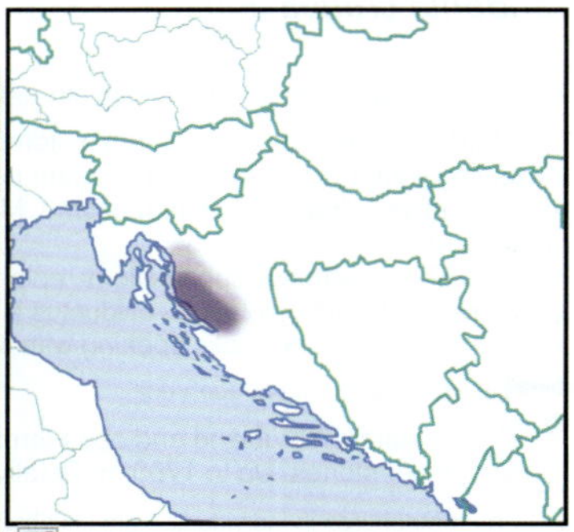

Croatia (Velebit mountains)

Wiktor 1987: 247.

Up to 52 x 12 mm (preserved)

Croatia: Velika Paklenica near Starigrad — Giuseppe Pocaterra

Slug deeply black, also mantle and head, sides sometimes lighter, mantle 30 % of body length. Keel poorly arched, not always reaching mantle, 14 weak grooves between keel and pneumostome. Sole whitish to light brown. Mucus colourless, thick and light bluish opaque when irritated.
Penis as long as epiphallus but thicker, retractor penis very slender and inserts at anterior section of epiphallus, which is unusual, vas deferens thin and short. Spermatheca elongate without visible duct, vagina very long, inside at posterior end with longitudinal grooves, several tubiform accessory glands opening into vagina near atrium, atrium very short with well developed musculus retentor.

Xerothermic habitats, open limestone rocks without much vegetation, hides in rock narrow crevices, low altitudes below 500 m with a climate strongly influenced by the Adriatic Sea. Crawls out in the morning even on dry days.

Tandonia jablanacensis (Wagner, 1930)

Milax jablanacensis Wagner, 1930 in H. Wagner 1930: 46

Slugs dark grey to almost black, getting paler downwards to the sides, mantle dark, paler at the margins, keel week and long, surface with long grooves on the dorsal side. Sole pale yellow with greyish yellow margins. Penis shorter than epiphallus but twice wider, vas deferens possibly not very long, but the other organs very long and slender, spermatheca extremely long without visible spermatheca duct, vagina relatively long with several tubiform accessory glands, atrium tube-shaped.

Up to 50 x 7.5 mm (preserved)

Possibly subterranean, as indicated by the strong elongation of the genitalia.

Wiktor 1987: 251.

The slug has not been found again after 1930.

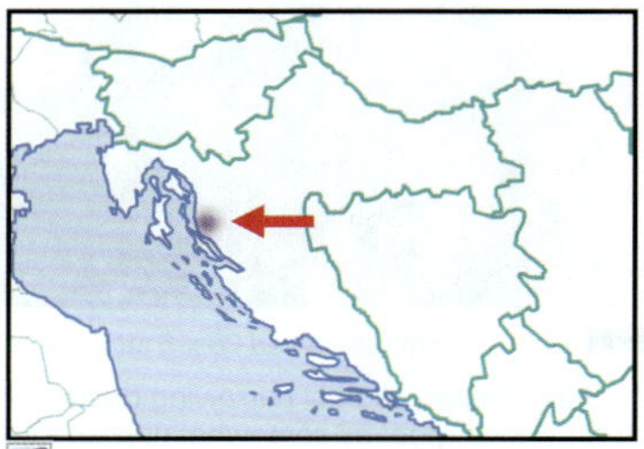

Croatia (Velebit mountains)

Tandonia kusceri (Wagner, 1931)

Milax kusceri Wagner, 1931 in H. Wagner 1931: 200 (*Milax (Milax) Kusceri*)

Up to 100 mm long (preserved 65 x 15 mm)

(1) Ukraine: Crimea, near Simferopol (scale 10 mm), (2) Bulgaria: Stara Zagora

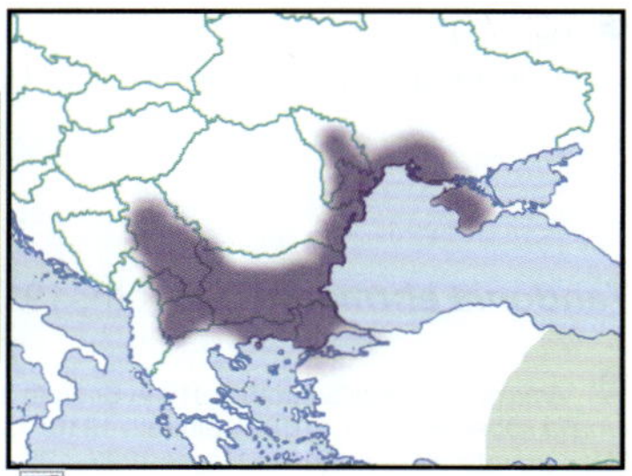

SE Europe

(1) Sergey Leonov, (2) Atanas Irikov

Slugs pink-brown or ash-grey-pink, with small blackish spots with diffuse edges. Keel light, without spots, strongly domed especially in living slugs, not well visible after preservation. Mantle length 25 % of body, characteristic dark colour pattern on the mantle. Head and tentacles blackish. 14-15 wrinkles between keel and pneomostom. Mucus colourless or milky-white. Eggs iridescent, oval, 4 x 4.5 mm.
Penis divided into a spherical anterior and a tubular posterior part, longest epiphallus of all species, 5-8 x of penis length, vas deferens thin and longer than penis+epiphallus, spermatheca oval, spermatheca duct long and equal to penis in diameter, vagina short, with two accessory glands like large lobes of creamy ducts adhering, atrium very short.
Differs from *T. rustica* in its pink hue, a slightly less elongate body, the spermatheca is not apically narrowed and a proportionally much shorter epiphallus. *T. serbica* is smaller and *T. kusceri* has no ink-stains. Vas deferens of *T. serbica* is shorter, delimited from epiphallus, epiphallus shorter, penis larger.

Eurytropic species, often found in stony habitats, most frequent on calcareous substrate. Synanthrope, near shrubs, mixed and deciduous forests. In Bulgaria in up to 2000 m.

Life cycle poorly known. Eggs were found in April, juveniles hatch after 16-20 days under laboratory conditions, slugs of different sizes live simultaneously, so several generations probably overlap.

Grossu 1983: 223, Wiktor 1983: 102, Wiktor 1987: 257, Wiktor 2001: 35, Irikov & Mollov 2006: 803, Sysoev & Schileyko 2009: 160.

Tandonia lagostana (Wagner, 1940)

Without gender agreement: *Tandonia lagostanus* (Wagner, 1940)

Milax lagostanus Wagner, 1940 in H. Wagner 1940: 138
Original spelling was *lagostanus*, derived from Lagosta, not *lagostomus*.

Slug flesh-orange, reddish or beige, gradually paler downwards, brownish-violet with dark pigments on upper side, pigments are concentrated in skin grooves forming a reticulate pattern, mantle length 30-35 % of body, keel light, reaching mantle.
Penis rounded, epiphallus spirally coiled, vas deferens opening almost apically, spermatheca oval with short duct, atrium very short.

Up to 70 x 17 mm (preserved juveniles), adults unknown

Wiktor 1987: 260, Wiktor 1996: 35.

Perhaps conspecific with *T. totevi*, but could also belong to the genus *Milax*. Lives perhaps only on the island of Lastovo, has never been collected after 1940. It would be necessary to study some slugs on Lastovo island.

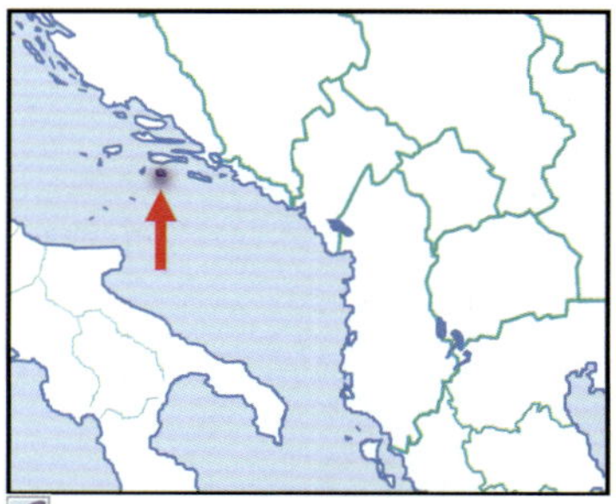

Croatia (Dalmatia: Lastovo island)

Tandonia macedonica (Rähle, 1974)

Without gender agreement: *Tandonia macedonicus* (Rähle, 1974)

Milax macedonicus Rähle, 1974 in Rähle 1974: 51

Slug usually black, sides getting paler downwards, sides below mantle whitish, lateral sections with dark spots, in first row above foot arranged like in a line, on dorsum concentrated in skin grooves forming a reticulate pattern, sometimes slugs are brown with lighter sides. Mantle length 1/3 of body, keel short and restricted to the posterior end of the slug, no colour difference in black slugs, lighter in brown slugs. Sole yellowish white with darker lateral sections. Mucus colourless, milky whitish opaque when irritated.
Penis irregular and wider than epiphallus, inside between penis and epiphallus a very characteristic mushroom-like papilla or stimulator, seen nowhere else in Milacidae, epiphallus with thick walls but with small space inside, vas deferens opening symmetrically, spermatheca oval with a duct longer than penis and epiphallus together, vagina extremely short, oviductus and spermatheca duct being united almost at atrium boundary, accessory glands being 3-4 lobes connected with thin ducts to vagina.

Up to 35 x 10 mm (preserved adults)

Was found in mountains above the timberline up to 1600 m, rocky habitats covered with shrubs, under stones on calcareous substrate.

Copulation possibly in May, nothing more known on its life cycle.

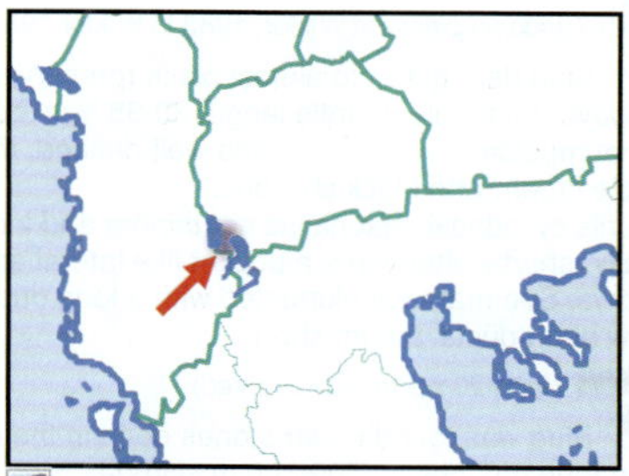

SW Rep. Makedonija

Possibly an endemic species, only known from Lake Ohrid region and Galičica mountains.

Wiktor 1987: 262.

Tandonia melanica Wiktor, 1986

Tandonia melanica Wiktor, 1986 in Wiktor 1986: 157

Slug evenly black, often a violet hue, sides getting somewhat lighter downwards. Keel well visible, 25 % of body length, reaching posterior mantle edge. 15 wrinkles between keel and pneumostome, neck under mantle cream. Sole whitish-cream. Mucus colourless, even when irritated.
Penis subdivided in two rounded parts, epiphallus as a short cone with a wider posterior end, vas deferens opening asymmetrically, spermatheca oval with a tubiform duct, oviductus twisted, vagina very short with large lobular accessory glands surrounding vagina and 2/3 of spermatheca duct and oviductus, atrium very short. Spermatophore yellowish, thin-walled like a shark without flukes, covered with short spines except at its bare narrowed end, more medially near the narrowed end spines are largest and asymmetrically branched. The other end is regularly rounded like an axolotl's or silurifom fish's front.
Resembles much *T. albania*, their ranges overlap. The spermatophore of *T. albanica* is not regularly rounded like a fish at its wider end and has hooked spines there, *T. albanica* also differs in its thick-walled conical spermatheca duct, more elongate and larger spermatheca and larger copulatory organs relative to body size.

Up to 57 x 9.5 mm (preserved)

Occurs on rocks and under rock rubble.

Wiktor 1987: 265, Wiktor 2001: 38.

NW Greece

Tandonia nigra (Pfeiffer, 1894)

Amalia nigra Pfeiffer, 1894 in C. Pfeiffer 1894: 68

Slug black, mantle cream with very dense black spots, tentacles white with black dots, when contracted black, neck and anterior sides dirty cream. Sole dirty cream.
Penis subdivided in two sections, epiphallus shorter and slightly widening at its posterior end, vas deferens opening slightly asymmetrically, oviductus thin, spermatheca elongate with a short duct, vagina almost as long as penis, with various accessory glands like short bundles surrounding the section between oviduct and vagina, atrium tubular, like prolongation of penis.

Up to 35 x 7 mm (preserved)

Known only from mountains peaks and E and W exposed slopes, at 1500-1700 m. Seems to be active at night, presumably herbivorous.

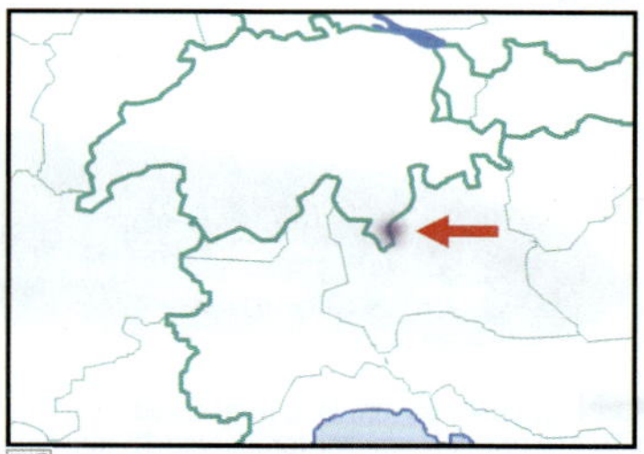

S Switzerland, N Italy

Copulation obviously in July. Adults, semiadults and juveniles were observed simultaneously from May to October, life cycle must at least cover 2 years.

Endemic species with small range, the Swiss habitats are threatened by mass tourism. Critically Endangered in Switzerland.

Wiktor 1987: 269, Turner et al. 1998: 287.

Tandonia pageti (Forcart, 1972)

Milax pageti Forcart, 1972 in Forcart 1972: 106

Slug very large, looks more like *Limax* than like *Tandonia*, reddish-yellow, after preservation blackish-grey, sides and neck dirty cream, keel same colour as dorsum, poorly arched. Mantle 30-35 % of body length, no lateral colour bands, only a very shallow horseshoe-shaped groove, 19-21 grooves between mid-line of dorsum and pneumostome. Sole dirty cream.
Penis wider than epiphallus, epiphallus cylindrical and slightly bent, vas deferens opening laterally in posterior third of epiphallus, not apically, which is unique, spermatheca oval with a wide duct, anteriorly widening, vagina short and twice wider than oviduct, with lobular accessory glands connected with thin ducts, atrium cylindrical and like a prolongation of penis.

Body lenth up to 73 mm, width 11.5 mm, mantle length 24 mm

Wiktor 1987: 271, Wiktor 1994: 43, Wiktor 2001: 41.

Greece (Ródos), SW Turkey

Tandonia pinteri (Wiktor, 1975)

Milax pinteri Wiktor, 1975 in Wiktor 1975: 77

Slug black at dorsum, getting lighter downwards at the sides, sides with scarce black spots, keel light and poorly arched, reaching mantle, mantle with a modern-art blackish and grey colour pattern, mantle length 1/3 of body (preserved state), 11-12 grooves between keel and pneumostome, head and tentacles blackish. Sole dirty cream.
Penis irregular, epiphallus conical and very small, vas deferens thin and long, opening symmetrically, spermatheca oval with a long duct, vagina short with accessory glands looking loke a brain mass connected by several thin ducts around vagina, atrium very short.

Up to 52 x 8 mm long (preserved)

Found in moist and well-shaded sites, at the bottom of mountain valleys, requires limestone substrate, 600-1600 m altitude.

Wiktor 1983: 118, Wiktor 1987: 274, Irikov & Mollov 2006: 804.

S Bulgaria (W Rodopi mountains)

Endemic species in W Rodopi mountains, known only from few localities.

Tandonia piriniana Wiktor, 1983

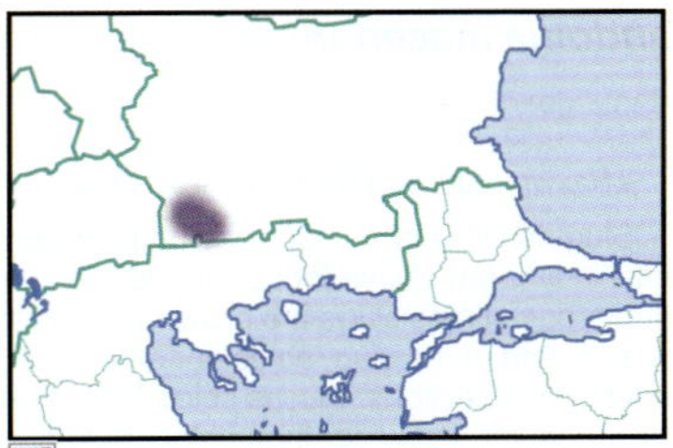

Tandonia piriniana Wiktor, 1983 in Wiktor 1983: 111

Slug dark green to almost black (preserved evenly blackish), with darker skin grooves and a semicircular groove on mantle, mantle length 30-35 % of body (preserved state), 12-13 shallow grooves between keel and pneumostome, keel cream and well-defined, reaching the mantle. Sexual opening anteriorly, in the head region. Sole cream with blackish sides.
Penis cylindrical, epiphallus not thinner and indistinctly separated from penis, vas deferens opening symmetrically, shortly afterwards a pocket-like lateral appendix is present at the vas deferens, a unique character in *Tandonia*. Spermatheca elongated with a long duct, vagina short with small lobular accessory glands opening by few short ducts, atrium short.

Up to 20 x 4 mm (preserved)

Slug was found under stones deep in the ground, among thick shrubs in a dry habitat at the S foothills of the Pirin mountains, 500-800 m altitude.

SW Bulgaria (Pirin mountains)

Rare.

Wiktor 1983: 111, Wiktor 1987: 276, Hubenov 2007.

Tandonia rara Wiktor, 1996

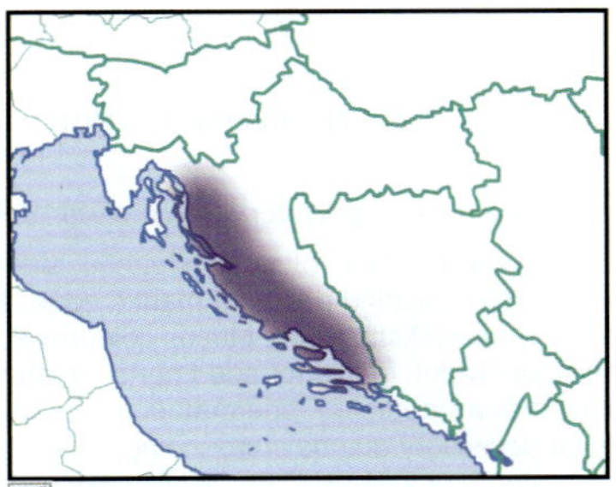

Tandonia rara Wiktor, 1996 in Wiktor 1996: 39

Slug evenly black, keel poorly developed. Mantle 1/3 of body length (preserved state). Sole cream, often with blackish sides.
Penis variable, no retractor penis, epiphallus cylindrical and usually bent, with some constrictions, border to penis unclear, and often covered with papillae from outside, vas deferens thin and long, opens symetrically, spermatheca oval and elongate, duct coiled, together twice as long as penis with epiphallus, vagina and atrium forming a single broad organ, accessory glands as several small unbranched processes, opening at posterior end of vagina surrounding oviduct and spermatheca duct. On the border between oviduct and vagina there are characteristic flat lamellar plates with spines, they form a rosette when everted with atrium.

Up to 45 x 9 mm (preserved)

Limestone rocks at low altitudes, in rock crevices and at bases of rocky walls, up to 600 m.

Croatia

Tandonia retowskii (Böttger, 1882)

Amalia retowskii Böttger, 1882 in O. Böttger 1882 (Nachr.-Bl. DMG): 98 (*Amalia Retowskii*) Type locality Crimea.
Name in original source attributed to Clessin, with a statement by Böttger that the slug had 12-13 furrow folds. Considered as a nomen nudum by Wiktor 1987: 253 and Schütt 2001: 282, but this view is not tenable. A description was given, its quality is irrelevant.
Amalia kalenzkoi Clessin, 1883 in Clessin 1883: 39 (*Amalia Kalenzkoi*) Type locality Crimea, near Yalta.
Name was spelled *Kalenzkoi* on p. 39 and again on p. 51 (no reference to where the name was derived from), was later interpreted as a misspelling for *kaleniczenkoi*. Since the misspelling was not clear in the original source itself, it cannot be considered as an inadvertent error under Art. 32.5.

Ukraine: Crimea, near Simferopol (scale 10 mm) Sergey Leonov

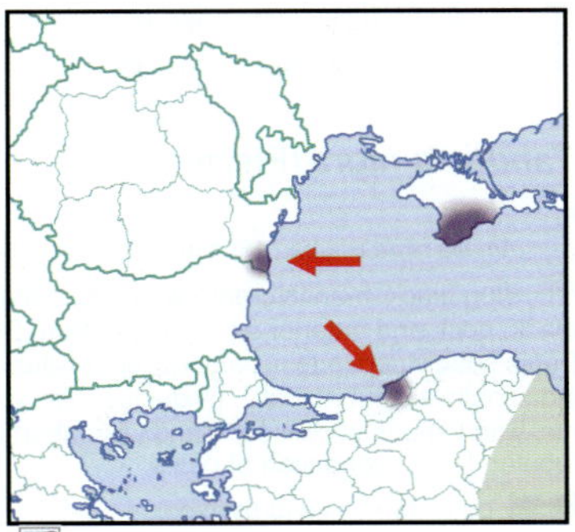

Crimea, W Romania, N Turkey

Up to 40-60 mm long (preserved 27 x 7 mm), in Turkey possibly smaller (preserved 15 mm long)

Slug grey with blackish dorsum, pigments concentrated in skin grooves. Keel distinct, dirty cream without dots, 9-11 grooves between keel and pneumostome. Sole dirty cream, edges can be grey, dark melanophore blots on the sole (only in Crimea). Juveniles whitish or yellowish with bluish tentacles.
Penis as long and as wide as epiphallus, penis wider in anterior section, epiphallus wider in posterior section, vas deferens opening slightly asymmetrically, spermatheca duct long and tubular, clearly limited from potatoe-like spermatheca, vagina short with two lobular accessory glands connected to vagina by thin ducts, atrium widened and unusually large with twisted elongate folds inside.

Wiktor 1987: 253, Wiktor 1994: 41, Schütt 2001: 282, Sysoev & Schileyko 2009: 160.

Was classified as *T. kaleniczenkoi* by Wiktor 1987 (= *T. kalenzkoi*), who regarded *retowskii* as a synonym. So the name must be *T. retowskii*.

Tandonia reuleauxi (Clessin, 1887)

Amalia reuleauxi Clessin, 1887 in Clessin 1887 (Malak. Bl.): 46 (*Amalia Reuleauxi*)

Montenegro: Virpazar Cédric Audibert

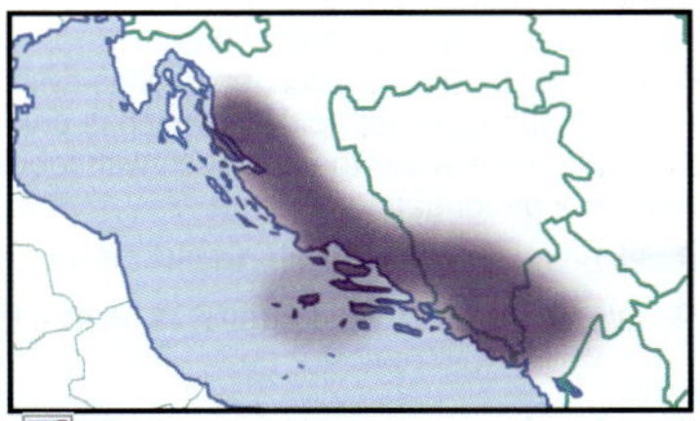

Croatia to Montenegro

Shady habitats in valleys or canyons under trees and herbs, hides under stones and in stone rubble.

Wiktor 1987: 278.

Up to 100 mm long (preserved 55 x 13 mm)

Slug evenly black to dark brown, slightly lighter at the sides, juveniles can be beige, keel prominent and reaching mantle, grooves at surface very shallow, 12-15 between keel and pneumostome, mantle 35 % of body length, with almost circular shallow groove. Sexual opening between pneumostome and mantle front, with a very characteristic swelling. Sole white-grey, sometimes with blackish edges. Mucus colourless, milky or light bluish when irritated.
Penis divided in spherical anterior and elongated posterior section, retractor very thin, epiphallus cylindrical with a spherical posterior end, vas deferens opening asymmetrically, spermatheca oval with a shorter duct, vagina very short with small accessory glands, atrium very short and often everted during preservation, so that vagina accessory glands are in a deep fissure at genital opening.

During copulation slugs are in a curved position (half crescent) towards each other, with space in between and heads near posterior third of partner's body, atria and regions of sexual opening are intensely swollen, so that the everted organs are covered from above, spermatophores of both are transmitted simultaneously, posterior ends can still be in epiphallus of sender while anterior end is already in spermatheca of partner.

Tandonia robici (Simroth, 1885)

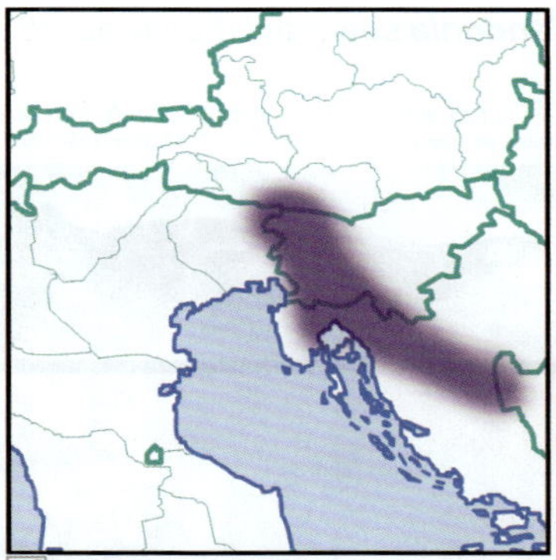

SE Alps to Istria

Amalia robici Simroth, 1885 in Simroth 1885: 230, Taf. 7, Taf. 10 (*Amalia Robici*)

Slug evenly olive-beige, yellowish green or reddish beige (fading after preservation), darker on back, head blackish, tentacles black, keel poorly arched but reaching mantle, mantle 30 % of body length, 8-13 delicate grooves between keel and pneumostome. Sole whitish. Mucus yellow, transparent.
Penis rounded and wider than epiphallus, inside a conspicuous papilla of crater-like or other shape retractor inserting laterally at anterior epiphallus section, epiphallus curved getting narrower towards posterior end, vas deferens thin, opening asymmetrically, spermatheca elongate gradually narrowing to its duct which is longer than penis, vagina short and 4 times wider than oviduct, with various accessory glands like short unbranched processes surrounding posterior vagina end, atrium like a short prolongation of vagina.

Up to 28 x 8.5 mm (preserved)

Often in deciduous forests, among shrubs, usually on limestone substrate, found in litter and under stones, up to 1300 m.

Wiktor 1987: 282, Falkner 1990: 184, Manganelli et al. 1995: 21.

Tandonia rustica (Millet, 1843)

Limax rustica Millet, 1843 in Millet 1843: 1, Pl. 63

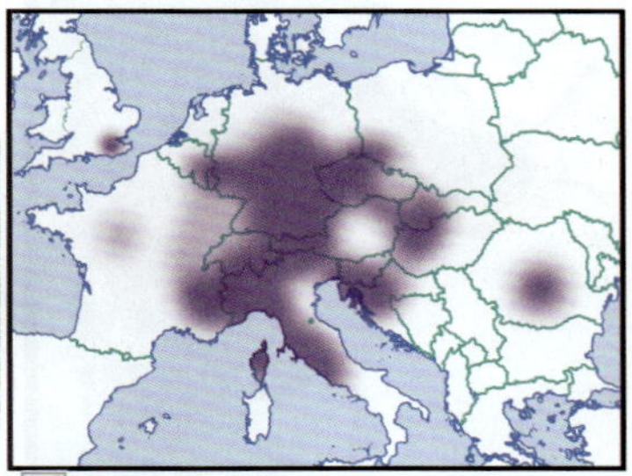

Originally probably S central Europe, today more widespread

Up to 100 mm long (preserved 45 x 12 mm)

(shells) Croatia: Istria, Lovran, (slug, 40 mm long) Germany: Bayern, near Eichstätt

Slug most often whitish, creamy or reddish to yellowish grey with numerous black dots, keel light, mantle 40 % of body length (preserved state), with black streaks at horseshoe groove. Sole cream. Mucus colourless and very thick, milky white and opaque when irritated.
Penis and epiphallus like a single long cylindrical organ, penis with a swelling anteriorly, inside with a richly ornamented papilla, epiphallus obviously longer than penis, vas deferens opening symmetrically, spermatheca elongate with a sharp pointed end, its duct slightly shorter and with a swelling half-way, vagina not much wider than oviduct, accessory glands like compact duct-like canals, surrounding and opening to anterior end of vagina, atrium short.
Lighter than *T. kusceri* and *T. serbica*, with smaller dots giving more contrast, spermatheca with a sharpened end.

Deciduous and mixed forests at mountain slopes with limestone rock rubble, also in open habitats, gardens, shrubs or graveyards, usually on calcareous soils. Needs warm microclimate in Poland, often in ruins of medieval castles. In Switzerland up to 1800 m, but rare above 1200 m.
Feeds predominantly on dead and live parts of plants, mushrooms, but not noxious to crops, consumes occasionally also worms and snails.

3-7 clutches of 10-25 oval eggs (5-6.5 mm long) are laid in two seasons in spring and autumn, in the Netherlands between June and September, inside a digged hole in the soil, juveniles hatch after 40-50 days, or in spring if the eggs were laid in autumn, maturity is reached after 6-9 months. Animals hatched in spring start reproducing in autumn. Life span up to 3 years, slugs can reproduce in 3 subsequent years (in total 200-250 eggs per individual), but they die at the end of the third season.

Populations in Ireland and SE England were found in natural environments (old mixed woodland), so possibly not introduced there, but native and rare. Endangered in Niedersachsen, Vulnerable in Germany, rare in Austria and in some northern parts of Switzerland.

Frömming 1954: 227, Wiktor 1987: 286.

Tandonia samsunensis (Forcart, 1942)

N Turkey

Milax samsunensis Forcart, 1942 in Forcart 1942: 239

Slug reddish with very small dots on back and mantle. Sole reddish with grey dots.
Penis rounded, separated from epiphallus by a strong constriction to which retractor inserts, epiphallus wider than penis, vas deferens opening almost symmetrically, spermatheca large with a relatively short duct, vagina and atrium not really separated from each other, accessory glands like a grape.

Up to 23 x 7 mm (preserved)

Wiktor 1987: 289, Wiktor 1994: 43, Wiktor 2007: 104.

Tandonia serbica (Wagner, 1931)

Without gender agreement: *Tandonia serbicus* (Wagner, 1931)

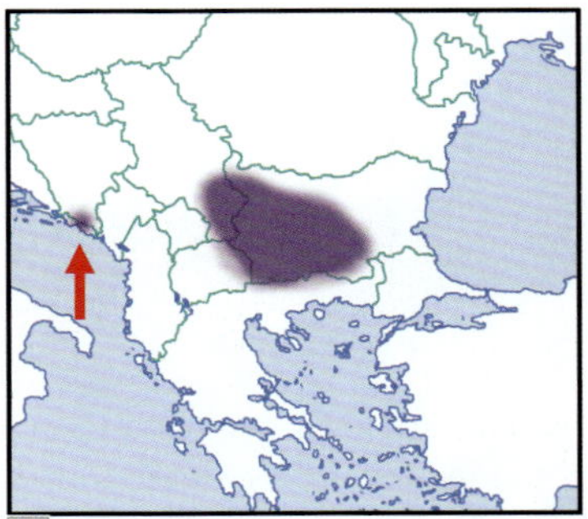

Bulgaria, S Serbia

Milax serbicus Wagner, 1931 in H. Wagner 1931: 63

Slug brownish or dirty cream with blackish reticulate pattern on back, sides and mantle. Black additional spots present on back and sides, diameter 1-2 mm, like ink-stains. Keel pale, mantle darker than the rest, 13 indistinct grooves between keel and pneumostome. Head blackish. Sole dirty cream. Mucus colourless, transparent.
Penis rounded, far wider than epiphallus, epiphallus elongate with pointed end, vas deferens opening symmetrically, spermatheca spherical when empty, with a long tubular duct, vagina very short, accessory glands like adhering lobular formations surrounding vagina, atrium wide and like prolongation of penis.
Resembles *T. kusceri*, with which it occurs often sympatrically, *T. serbica* is smaller and *T. kusceri* has no ink-stains. Vas deferens of *T. serbica* shorter, delimited from epiphallus, epiphallus shorter, penis larger.

Up to 45 x 11 mm (preserved)

Almost exlusively at dry slopes, most often in rock rubble sheltered by bushes, usually on limestone, in Bulgaria in up to 1800 m. Synantrope.

Life cycle very little known. Spermathecas present at least in June and October.

Wiktor 1996: 48, Irikov & Mollov 2006: 803, Hubenov 2007.

Wiktor 1996 pointed out that no spermatophores have ever been observed in *T. kusceri*, and brought up the hypothesis that *T. kusceri* and *T. serbica* could be dimorphic forms of the same species in different stages of sexual activity.

Tandonia sowerbyi (Férussac, 1823)

Limax sowerbyi Férussac, 1823 in A. Férussac & Deshayes 1819-1851 (1): 96 ζ (*Limax Sowerbii*)
Name was corrected to *sowerbyi* in ICZN Opinion 336 (1955).

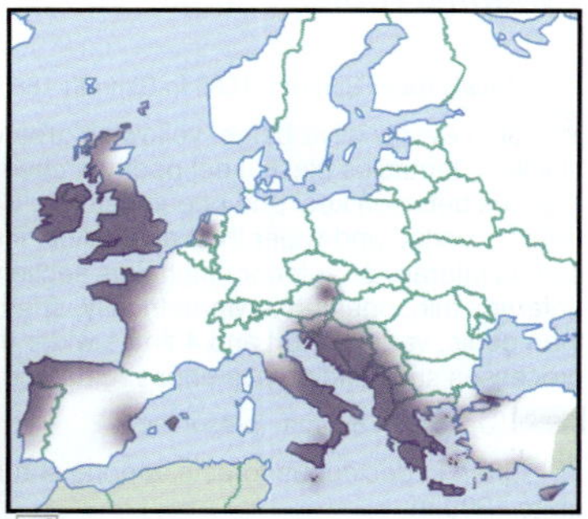

Originally perhaps Greece, today widespread

Up to 60 x 16 mm (preserved)

(1 juv.) Greece: Attiki, Moschato, (2 sole), Croatia: Dalmatia, Brač, (3 big) Cyprus, Pila near Larnaka, (shells) Italy: Calabria, Roccella Jonica

(1) Panayotis Ovalis, (2) Jozef Grego, (3) Georgios Konstantinou

Portugal: Montijo near Lisboa — Emidio Machado

Slug usually brownish-yellow with various dark spots (forming sometimes reticulate patterns), but can also be white, blackish or black. Keel yellow or orange, often without dark pigment, very strongly domed, forming a sharp edge, body in transversal cut almost triangular, a difference to many other species. 16 wrinkles between keel and pneumostome, mantle length 30 % of body (preserved state), on it often one distinct and narrow dark band along a horseshoe-shaped groove. Anterior side often with a narrow dark mid-dorsum line between head and mantle. Sole often orange on edges. Mucus yellowish and thick. Slug very variable.
Penis thinner than thick-walled epiphallus, cylindrical with constrictions, epiphallus with strong retractor and two additional muscles in its anterior part, narrowing at its posterior end, vas deferens thin, twisted, twice longer and opening asymmetrically. Spermatheca long like a bottle, its duct thin and shorter, vagina short with accessory glands consisting of several unbranched lobes connected to vagina by thin ducts, atrium wide.

Lives in a variety of habitats, usually on meadows, grassy slopes, under shrubs and in gardens, frequently as a synanthrope. Hides under stones and pieces of wood, in Britain also under moist ground litter, at the base of plants, in soil cavities and crevices, in Britain nearly always as a synanthrope, can be a pest in gardens. In Greece one of the most frequent slugs, probably native throughout the country. Probably introduced to Britain, where it is frequent, in some regions densely recorded.

Individuals with spermatophores in autumn (former Yugoslavia) or beginning of winter (Britain). 15-20 eggs (diameter 5 mm) are laid in a white envelope.

Hesse 1928: 18, Germain 1930: 110, Wiktor 1983: 119, Wiktor 1987: 300, Wiktor 2001: 42, Kerney 1999: 150, Beckmann 2007: 85.

Tandonia totevi (Wiktor, 1975)

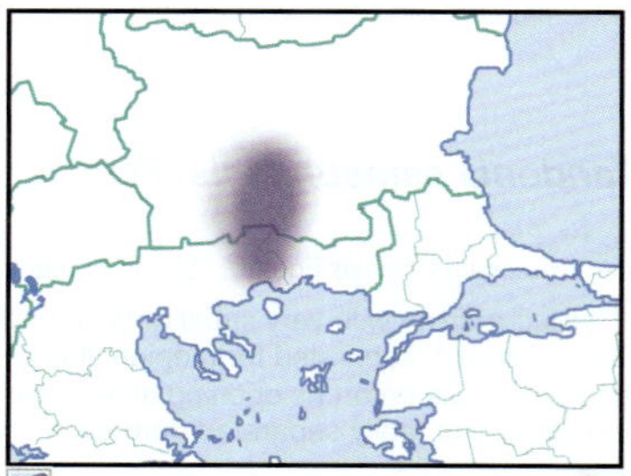

Greece, S Bulgaria

Milax totevi Wiktor, 1975 in Wiktor 1975: 82

Slug very large with short keel (only 5-15 mm, absent in one-third of the animals), light greyish olive to dark olive brown, sides gradually lighter downwards, often with lighter, translucent-like dots, keel and mid-line of dorsum dirty cream. Skin sculpture delicate, 20-21 grooves between mid-dorsum and pneumostome, mantle length 35 % of body (preserved state), horseshoe-shaped groove on mantle very shallow, sometimes barely visibly. Head and tentacles greyish. Sole evenly light yellow to bright orange, this differs locally.
Penis twice as wide as epiphallus but 2-3 x shorter, retractor thin and inserting to anterior epiphallus section, vas deferens opening apically, spermatheca heavily elongate without clearly visible duct, vagina less than half penis length with grape-shaped accessory glands adhering, atrium very short.
Resembles *Limax flavus* both in behaviour and colour.

Up to 180 mm long (preserved 130 mm), in mountains up to 140 mm

Found as a synanthrope in gardens on sandy substrate, near buildings and waste areas, also in subalpine coniferous forests on non-calcareous substrate. Nocturnal. The animals show some social behaviour, sometimes they group together and are not aggressive against conspecifics. Dead conspecifics are eaten.

Reproduction in late autumn and winter, some 10 ovoid eggs (5.5-6 mm) are laid in clutches inside the soil, juveniles hatch after 3-4 weeks, they may reach 60-70 mm length after one year and maturity after 20-26 months, in the second winter. Animals seems to die shortly after reproduction, maximum age 2.5-3 years.

Wiktor 1983: 122, Wiktor 1987: 307, Wiktor 2001: 49, Irikov & Mollov 2006: 805, Schneppat et al. 2011.

Boettgerilla

Boettgerilla Simroth, 1910
in Simroth 1910: 530

Type species *Boettgerilla compressa* Simroth, 1910
(monotypy)
(= currently *Boettgerilla compressa* Simroth, 1910 from N Caucasus)

W Europe to Caucasus and central Asia

Boettgerilla pallens Simroth, 1912

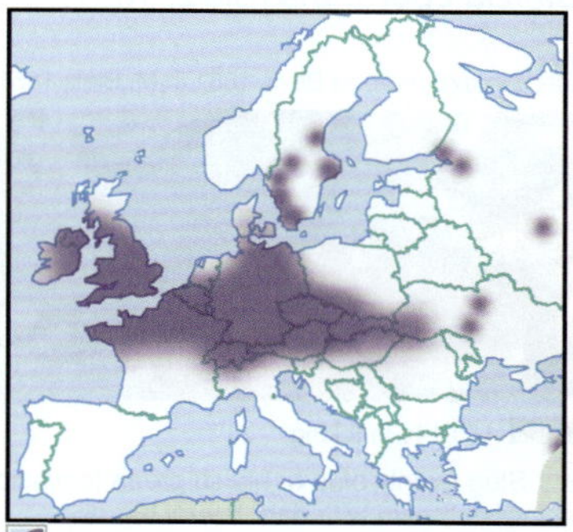

W Europe to Caucasus, introduced to W Canada

Boettgerilla pallens Simroth, 1912 in Simroth 1912: 55, Taf. 3, Taf. 8

Slug very narrow and slender, almost worm-like and sharply keeled, pale greyish or with bluish hue, head and keel slightly darker, mantle pointed at its posterior end and equipped with fine concentrical grooves. Sole pale yellow. Mucus colourless. Juveniles yellowish to dirty white.
Shell small, fragile, 1.5-3 x 0.8-1.5 mm, with median nucleus and growth lines, denser in the posterior part, no thin margin, located below the very terminal section of the pointed end of the mantle.
Much more slender than *Milax gagates*, sole more slender, in *M. gagates* the posterior end of the mantle is rounded, its shell is larger and better visible, its colour more yellowish-brownish.
Penis elongate and broader than vagina, epiphallus and spermatheca, epiphallus swollen and connected to penis laterally near its posterior end, by a long thin duct, which is thinner than vas deferens, vagina as long as penis and almost as narrow as vas deferens, no accessory glands, spermatheca elongate and relatively short.

Humid forests at slopes and in river plains, on calcareous substrate, usually in quite natural habitats, occasionally also in gardens and disturbed habitats. In Switzerland up to 1600 m, usually below 700 m. Subterranean, slug is adapted to moving through earthworm burrows and similar spaces in soil, 2-20 cm (maximum up to 60 cm) below surface.
Feeding is seldom intensive, the slugs usually takes a few bites (earthworm faeces, detritus etc., also arionid eggs) and moves on. Considered as a pest in European greenhouses and gardens, slugs feed on soft plant tissues. Slugs move rapidly away from bright light sources.

Up to 60 x 3 mm

Slovakia: (1-3) Low Tatra National Park, (4-5) Plešivec

Jozef Grego

Mating and egg-laying during late summer and into autumn in Britain (October in Germany), eggs are laid as several 1-6 egg clutches 9-27 cm deep below surface. Adults survive egg-laying but die shortly after. Juveniles hatch after 20-22 days at 17°C (in Germany), between October and December (in Britain). Juveniles turn colour from white to grey in May-June.

Until the 1960s unknown in Europe, but the finds suggest that the species was long time overlooked and misidentified (Turner et al. 1998: 314). Considered as introduced to Great Britain and Ireland, first finds from 1972 in England and 1973 in Ireland (Kerney 1999: 153).

Selenochlamys

Selenochlamys Boettger, 1883
in O. Boettger 1883: 141

Type species *Selenochlamys pallida* Boettger, 1883
(monotypy)
(= currently *Selenochlamys pallida* Boettger, 1883 from Caucasus)

Body elongated, mantle small, broader than long and near the posterior end of the slug.

Subterranean or at least near the soil, slugs are carnivorous and feed on earthworms.

Caucasus to N central Turkey

Schütt 2001: 286, Rowson & Symondson 2008, Sysoev & Schileyko 2009: 157.

Selenochlamys ysbryda Rowson & Symondson, 2008

Selenochlamys ysbryda Rowson & Symondson, 2008 in Rowson & Symondson 2008: 540

Slug very long when crawling in darkness, whitish, tips of tentacles pale brown, mantle very small, near the posterior end, broader than long, pneumostome on right dorsolateral side, two shallow dorsolateral grooves arising from mantle edge and becoming obscure towards head, upper tentacles broad and nearly meeting at the base. Juveniles resemble the adults.
Shell small, situated within mantle, of irregular oval shape, with a nucleus in the upper right section and concentric lines as seen from above, and with a columella-like structure in the upper (northern) section of the lower side.

Was found in a garden habitat in Cardiff. Active at 12-25°C, largely nocturnal. Feeds on earthworms.

Mature animals were found in September and October, colourless spherical eggs (diameter 2 mm) are laid in clutches of around 20 eggs, juveniles were observed to have hatched after 5 weeks in the laboratory.

Wales: Cardiff

Amgueddfa Cymru - National Museum Wales

Up to 110 x 6 mm (undisturbed in darkness), shell 1.8-2.1 mm long

Original range unknown, introduced to Wales

Bielzia

Bielzia Clessin, 1887 in Clessin 1887: 47 *Limax (Bielzia)*

Type species *Limax coerulans* Bielz, 1851 (monotypy)

Bielzia coerulans (Bielz, 1851)

Limax coerulans Bielz, 1851 in M. Bielz 1851: 14

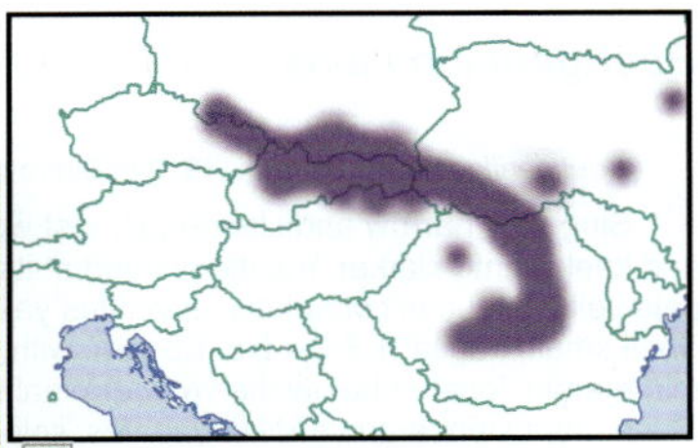

Carpathians and nearby highlands

100-140 mm long

(1) SE Poland, Bieszczady National Park, (2-4, juveniles) Slovakia: near Banská Bystrica — (1) Florian Schwabe, (2-4) Jozef Grego

Slug evenly blue or bluish green (occasionally black) with dark greyish head and tentacles, margins pale yellowish, sole pale yellowish or whitish. Juveniles yellowish brown with dark lateral bands.
Genitalia with 4 branches near the opening: very narrow penis with vas deferens at its end (or seen only as vas deferens), elongate spermatheca, thick atrial appendix (accessory organ for the copulation with retractor) and thick oviduct. Vas deferens and oviduct merge into spermoviduct of variable length, with separated male and female ducts.

Deciduous and coniferous forests in mountains, usually at the bottom, or under dead wood logs.

Annual life cycle. Maturity in June to July, copulation at the soil, 30-80 oval eggs (4-5 mm long) are laid in one clutch or string from July to August, juveniles hatch after 17-25 days, half grown juveniles hibernate, fully grown slugs appear in May, maturity is reached after one year, every individual deposits eggs only once and dies shortly after.

Smolenska 1936 (biology), Frömming 1954: 201, Kerney et al. 1983: 190, Grossu 1983: 304 (anatomy), Falkner 1990: 186, Zhiltsov & Shilejko 2002 (anatomy), Sysoev & Schileyko 2009: 155.

Lehmannia

Lehmannia Heynemann, 1863 in Heynemann 1863: 211

Type species *Limax marginatus* Müller, 1774 (monotypy)

Ambigolimax Pollonera, 1887 in Pollonera 1887: [2] *Agriolimax (Ambigolimax)*, type species *Limax valentianus* Férussac, 1822 (subsequent designation by Wenz & Zilch 1959-1960: 267)
Contained originally *Limax valentianus* and *Limax fulvus* Normand, 1852.

Body narrow, gradually and evenly narrowing towards its rear end. Body length up to 80 mm. Mantle covering less than 1/3 of the body, narrower in the anterior part, surface covered with concentric grooves. Pneumostom postmedial. Keel short. About 20 wrinkles between keel and pneumostom. Colour variable from cream and brownish to black. At least 2 longitudinal stripes on the mantle (except in unicolour forms). Sole is lighter in the middle. Mucus colourless, watery. Lateral teeth of radula S-bent. Marginal teeth on edge of radula with numerous cusps.
Intestine with 3 loops, the first one being the biggest. Last loop very short, with very long rectal caecum reaching posterior body end. Caecum opens into last section of intestine, positioned more anteriorly than posterior bend of the last loop.
Prostata well developed, vas deferens short and thick. Penis short, usually shorter than 2/3 of mantle length, tubular, baggy, or claviform, always crosses right ommatophore retractor. Nodular or flagellate penial gland on posterior part of penis. V-shaped fold inside penis. Spermatheca connected with penis.

Very active slugs. Most feed on lichens, occur on the bark of trees or on rocks. Most species live in mountains.

Europe and N Africa

Wiktor 2001: 55.

15 species known, most of them poorly studied.

Lehmannia brunneri (Wagner, 1931)

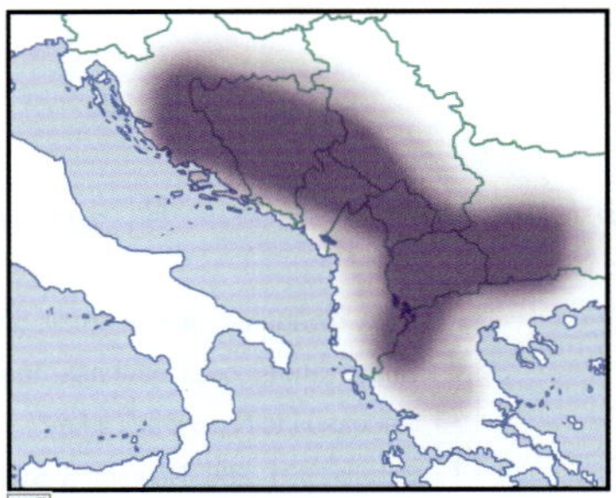

Croatia to N Greece

Agriolimax brunneri Wagner, 1931 in H. Wagner 1931: 197 (*Agriolimax Brunneri*)

Slug usually dark or with a light navy-blue hue, other colours are also possible, up to 5 dark bands on the mantle, mantle length 30 % of body, 19-22 wrinkles between medial line and pneumostome. Sole cream.

Up to 45 x 7 mm

A mountain species preferring upper timberline zones or higher, 900-2650 m. Slug inhabits exclusively rocks, even when trees are nearby, hides in rock crevices, lives on the ground and is difficult to find.

Slug seems to reach maturity in autumn and only few slugs survive the winter, dying in spring. Life span up to 2 years.

Wiktor 1996: 63, Wiktor 2001: 57, Irikov & Mollov 2006: 808.

Lehmannia horezia Grossu & Lupu, 1962

Lehmannia horezia Grossu & Lupu, 1962 in Grossu & Lupu 1962: 196
Lehmannia sarmizegetusae Grossu, 1970 in Grossu 1970: 110

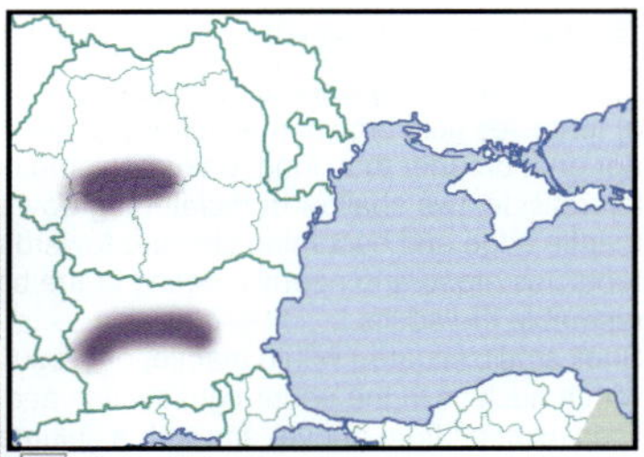

Romania, Bulgaria

Up to 70 mm long (preserved 40 mm)

Bulgaria: Stara Planina, Shipka peak — Dilian Georgiev

Slug slender, grey, lighter and more yellowish downwards to the sides, some slugs with light irregular spots which can form interrupted colour bands ranging from keel to mantle, this can give the impression as if the slug was keeled. Mantle often darker with two connected black bands, looking like a lyra. Sole evenly light.
The form *sarmizegetusae* was described as yellow-brown, darker on dorsum, lightening downwards, mantle with dark bands, head with a dark band, sides of head with longitudinal spots. Penis elongate, penial gland flattened, vas deferens thin, spermatheca duct thick and about as long as oviduct.

At humid weather conditions found on trees and under bark, in river valley woodlands. An agile species. In Bulgaria in up to 2200 m.

Rare in Bulgaria.

Hubenov 2007, Irikov & Erőss 2008.

Grossu 1970 did not include features distinguishing *L. sarmizegetusae* sufficiently from other species known at the time, and did not provide any evidence for an independent species. Irikov & Erőss 2008 considered *L. sarmizegetusae* as a synonym of *L. horezia*.

Lehmannia islandica Forcart, 1966

Lehmannia islandica Forcart, 1966 in Forcart 1966: 234

Slug smaller than *Le. marginata*, penis shorter, thickened in its distal section, with a long and pointed flagellum.

20 mm long (preserved)

The only endemic non-marine mollusc of Iceland. It would be interesting to know more about it, molecular studies.could shed more light on its origin.

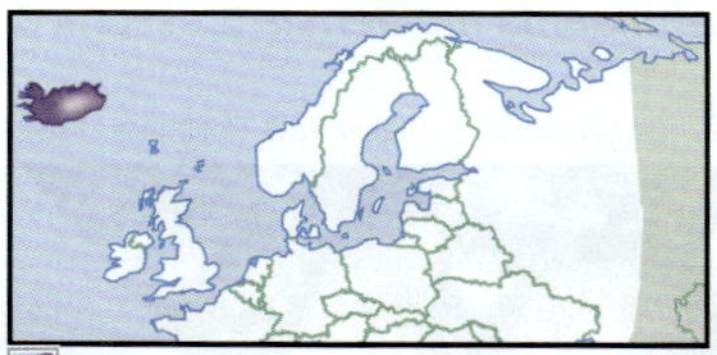

Iceland

Lehmannia jaroslaviae Grossu, 1967

Lehmannia jaroslaviae Grossu, 1967 in Grossu 1967: 120

Slug light grey with irregular white dashes on mantle and dorsum, on mantle forming bands not uniting to a horse-shoe like pattern and not being continued on dorsum. Keel light, reaching mantle, but prominent only on posterior half of dorsum. Sole slender, medial section narrower than lateral sections. Shell 4.5 x 3 mm with thin margins and concentric lines.
Penis long but not evenly cylindrical, vas deferens shorter opening laterally near proximal end of penis, retractor inserts slightly more proximal, posterior end of penis equipped with a well developed flagellum about 1/3 of penis length.

Up to 80 mm long (preserved 35 x 13 mm)

In mixed forests at 700-800 m, on tree trunks, most active in November at rainy and cold weather conditions.

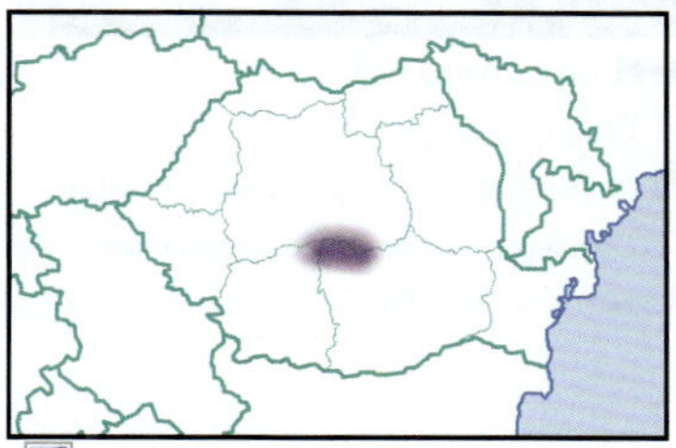

Central Romania

Lehmannia macroflagellata Grossu & Lupu, 1962

Lehmannia macroflagellata Grossu & Lupu, 1962 in Grossu & Lupu 1962: 198

Slug bluish black with numerous black dashes, sometimes with yellowish spots or streaks forming symmetrical colour patterns, median section of dorsum with a light band, keel restricted to the posterior end. Mantle oval, relatively small, usually evenly dark with 5 variably broad light colour bands. Sole has 3 sections, evenly grey, sometimes the lateral zone is darker blue and the median section greyish yellow.

Up to 25 mm long (preserved)

Obviously restricted to high mountains, 1200-1800 m, slugs were found near plants, stones and rocks.

Carpathians (Poland to Romania)

Lehmannia marginata (Müller, 1774)

Without gender agreement: *Lehmannia marginatus* (Müller, 1774)

Limax marginatus Müller, 1774 in Müller 1774: 10
Lehmannia janetscheki Forcart, 1966 in Forcart 1966: 230

Up to 75 mm long

(1-2, green) Slovakia: Veľká Fatra mountains, (3-6) Germany: Bayern, near Walchensee, (shells) Germany: near Kiel

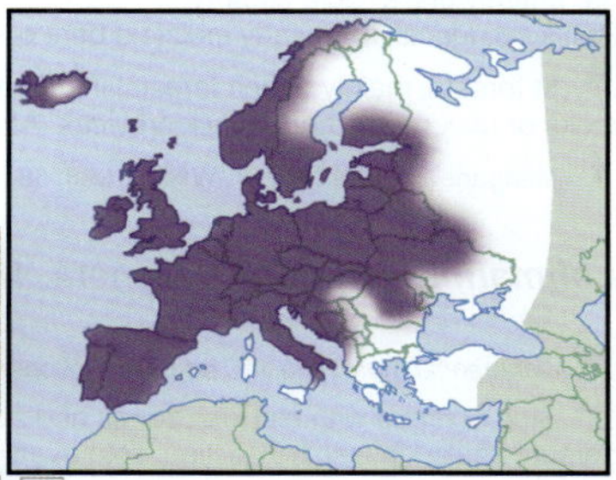

Central, N and W Europe

Slug basically light greyish brown with a light dorsal colour band and black lateral bands (often forming colour spots), posterior end with sharp keel. Juveniles dark reddish brown with a lateral band at each side.
Penis appendix pointed (rounded in *L. valentiana*), spermatheca also slightly pointed and not rounded as in *L. valentiana*.

(1-2) Jozef Grego

Mainly at trees, under the bark of dead timber, climbing up trees during rain. Usually in forests, sometimes also in open habitats, can also survive in gardens if trees and walls are covered by algae. Feeds on algae, lichens and mushrooms, predominantly on algae growing on tree trunks, in coniferous forests also on rocks where they feed on lichens, absent from intensively farmed lowland areas. Does not attack other snails, and consumes dead snails only if no herbs are available. In Switzerland probably up to 2300 m.

Oval eggs (4-5 mm long) are laid inside the soil, not under mosses, in batches of 10-30 eggs, several times per year, in Germany from November to June (France: September to December). Juveniles of 5-10 mm length hatch after 30-120 days, maturity is reached after one year at 50-60 mm size, life span 2.5-3 years.

A frequent species, but absent from intensively farmed lowland areas with few suitable habitats. Threatened by continuous habitat destructions and air pollution damaging epiphytes on tree trunks and walls.

Künkel 1908, Germain 1930: 88, Frömming 1954: 202, Kerney et al. 1983: 186, Falkner 1990: 190, Wiktor 1996: 65 (classified *janetscheki* as a synonym), Wiktor 2001: 62.

Lehmannia medioflagellata Lupu, 1968

Lehmannia medioflagellata Lupu, 1968 in Lupu 1968: 39

Slug reddish brown, lighter down to the sides, with a light and fine medial band and broad and uninterrupted black lateral bands, keel occupies last 1/3 of body, posterior section pointed, mantle perfectly oval with 2 broad lateral black bands (never in lyra form), sole light in subdivided in 3 longitudinal zones.
Penis short (3.5 mm), flagellum inserts laterally (not terminally) at penis, which is unusual.

60-70 mm long (preserved 30 mm)

Found in mountains at 1300 m altitude.

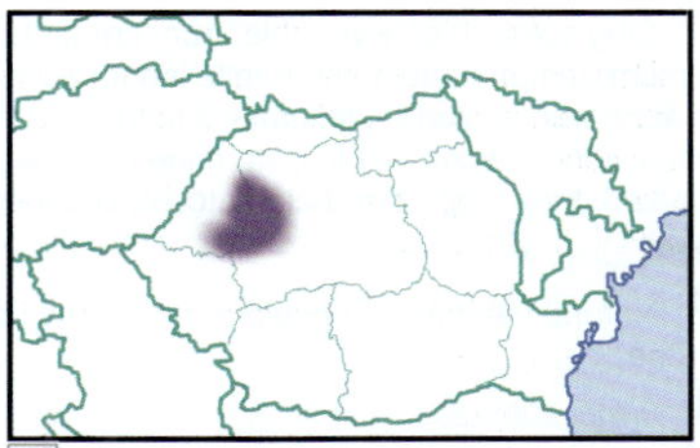

Romania (Apuseni mountains)

Lehmannia melitensis (Lessona & Pollonera, 1882)

Limax melitensis Lessona & Pollonera, 1882 in Lessona & Pollonera 1882: 69

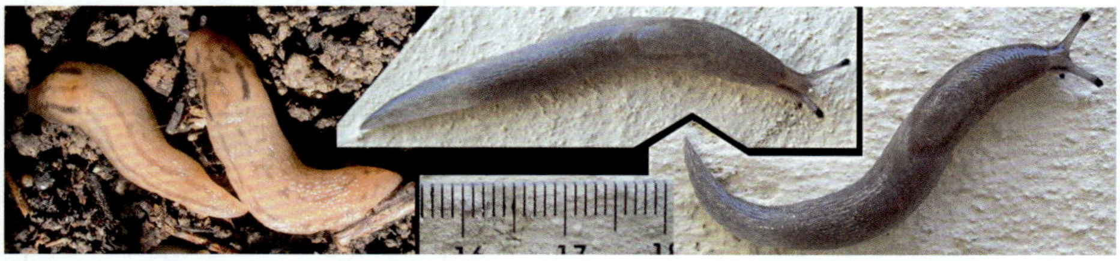

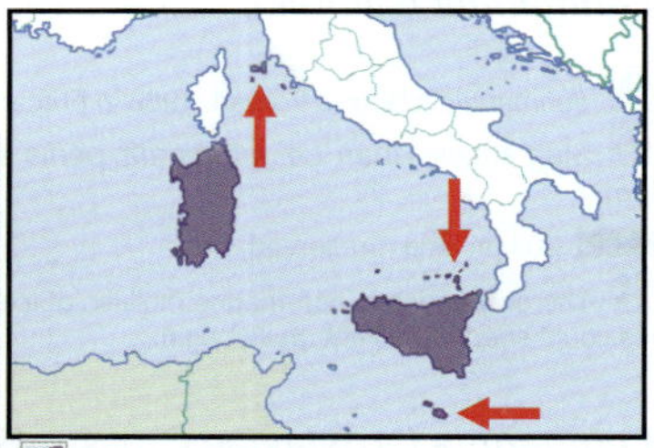

S Italy, Malta

Up to 70 mm long

(1-2) Italy: Sardinia, Cagliari, (3-4) Malta: Mdina bastions (1-2) Francesco Mascia, (3-4) David Cilia

Slug pale greyish with reddish hue, pointed at rear end, mantle granulated, moderately sized and rounded at anterior and porterior ends, no keel.

In moist natural and disturbed habitats.

Copulation was observed in December in the morning, slugs follow each other about 30 min and finally curl up and exchange sperma within 3-5 seconds.

Gerhardt 1940: 560, Giusti et al. 1995: 279, Manganelli et al. 1995: 21.

Lehmannia nyctelia (Bourguignat, 1861)

Without gender agreement: *Lehmannia nyctelius* (Bourguignat, 1861)

Limax nyctelius Bourguignat, 1861 in Bourguignat 1861: 305

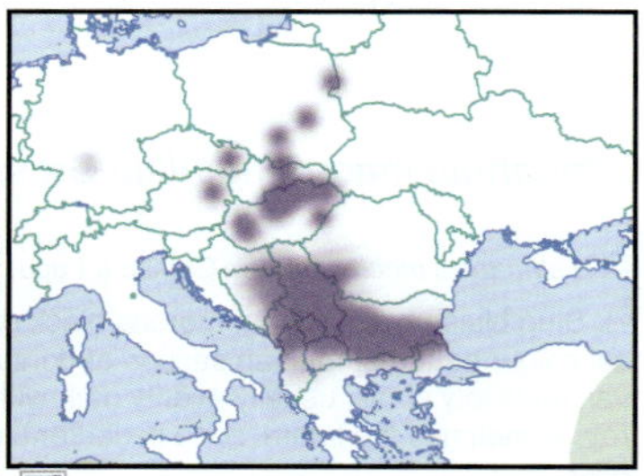

Originally SE Europe
Widely dispersed, range uncertain.

Up to 50 x 9 mm

(1-2) Slovakia: Slovak Karst National Park, (3) Bulgaria: Sredna Gora mountains (1-2) Jozef Grego, (3) Dilian Georgiev

Slug colour variable, most often grey or ash-brown with blackish or dark brown spots and streaks, mantle length 30 % of body. Melanistic mountain forms are entirely black. Sole creamy.
Penis cylindrical, irregularly multifold bent or spirally coiled, posterior part of penis usually dark pigmented. No penial gland.

In forests, mainly beech forests, in mountains also above the timberline. Creeping on trees, hiding under bark, hollows in tree stumps, in mountains under rocks or rock rubble and in rock crevices. After rains the slugs creep out in great numbers, particularly in Bulgaria. In Bulgaria from lowlands up to 2700 m.

Manganelli et al. 1995: 21, Wiktor 1996: 68, Irikov & Mollov 2006: 808, Hubenov 2007.

Lehmannia requienii Pollonera, 1896

Lehmannia marginata var. *requienii* Pollonera, 1896 in Pollonera 1896: 1 (*Le. m.* var. *Requienii*)

Slug ochrous with blackish zones and spots, less transparent and contains less water than *Le. marginata*, dorsum keeled in last 1/3 section, sole evenly pale.

Up to 50 mm long

Probably restricted to mountain habitats in the centre of the island.

Not much known about this form, perhaps an endemic species of Corsica.

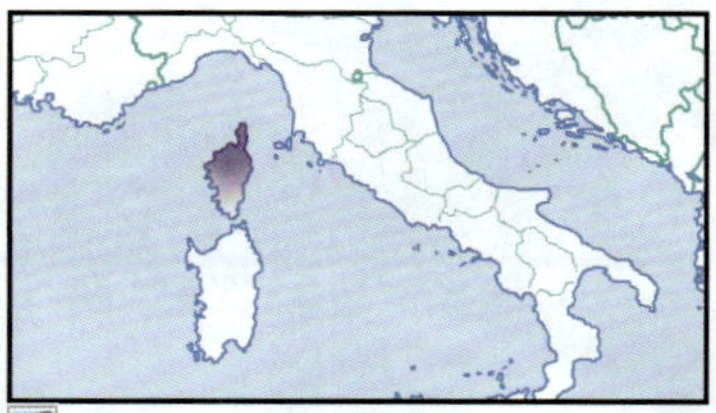

France (Corsica)

Lehmannia rupicola Lessona & Pollonera, 1882

Lehmannia marginata var. *rupicola* Lessona & Pollonera, 1882 in Lessona & P. 1882: 16

Slug basically brown, much like *Le. marginata*, contains less water, light dorsal colour band narrow or absent, lateral bands on mantle shield straighter and with light margins.
Flagellum longer and more slender than in *Le. marginata*, approximately as long as penis, atrium and vas deferens shorter and less slender.

Up to 75 mm long

Mostly in high-altitude mountain forests, between rocks and dwarf shrubs, also in open habitats under stones. Feeds on algae and lichens and requires primary forest-like conditions. In Switzerland between 1000 and 2500 m.

Kerney et al. 1983: 187, Falkner 1990: 190, Manganelli et al. 1995: 21, Turner et al. 1998: 302.

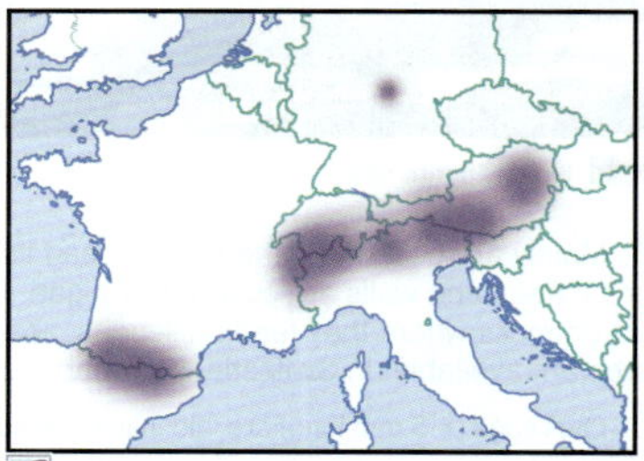

Alps and Pyrenees

Rare in Bayern and Austria.

Lehmannia szigethyae Wiktor, 1975

Lehmannia szigethyae Wiktor, 1975 in Wiktor 1975: 87

Slug colour highly variable, from cream to black all colours are found, 21-22 wrinkles between middle of dorsum and pneumostom, mantle length 33 % of body. Sole also variable. Impossible to recognize the slug on its appearence alone. Ordinarily 3 stripes on the mantle.
Dissection is necessary to determine this species (genitalia see Wiktor 2001). Young individuals cannot be recognized, they might also belong to *Le. brunneri* or to some other species.

Up to 45 x 8 mm

Natural habitats on conglomerate rocks and limestone, less often on tree trunks. The slug is not a synanthrope.

Wiktor 2001: 59.

Range poorly known.

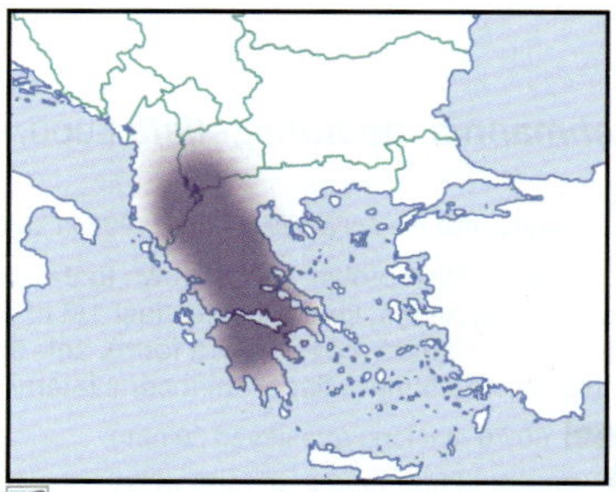

Albania to S Greece

Lehmannia valentiana (Férussac, 1822)

Without gender agreement: *Lehmannia valentianus* (Férussac, 1822)

Limax valentianus Férussac, 1822 in A. Férussac [1821-1822] (prodrome limaces): 21
Publication date for prodrome limaces was 16 Jul 1822. Also mentioned in text of "Histoire" p. 96ε from 27 Sep 1823.

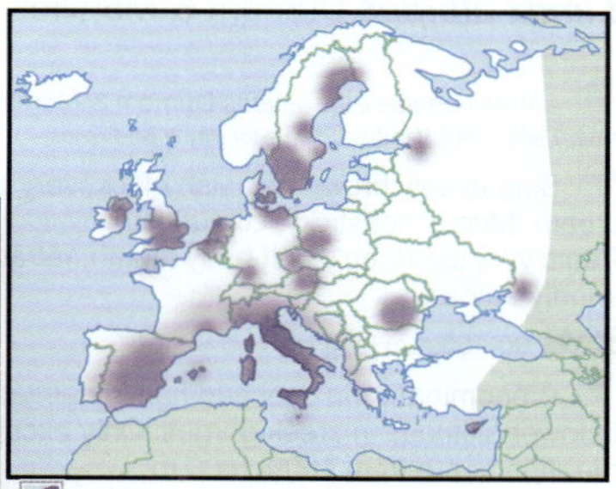

Up to 70 mm long (1) Belgium: Antwerpen, (2-4) USA: Maryland, Germantown, (shell) Germany: Kiel (1-4) Aydın Örstan

Bottom habitats, ground-dwelling, not arboreal and climbing trees like *Le. marginata*. Often in disturbed places, greenhouses, gardens and nurseries, sheltering under wood, stones and moist ground litter. In Italy in anthropogenic habitats, can also be found in salads.

Reproduction was observed between August and October, copulation extremely fast, at any time of the day, slugs curl up on the ground for a short time, sperma is exchanged within less than 3 minutes.

Gerhardt 1940: 562, Kerney et al. 1983: 312, Manganelli et al. 1995: 21, Kerney 1999: 160, Sysoev & Schileyko 2009: 150, www.naturamediterraneo.com (2011), G. Konstantinou (unpublished, Cyprus).

Originally Iberian peninsula, introduced elsewhere

Colour more yellowish than *Le. marginata*, with a dark colour band on each side and one near the mid-line, head slightly darker, keel short, mucus colourless. The two dark longitudinal bands seem to allow recognising this species more easily.
Differs from *Le. marginata* in the shape of the penis appendix (rounded/blunt in *Le. valentiana*, pointed in *Le. marginata*) and the spermatheca (blunt in *Le. valentiana*, pointed in *Le. marginata*).

Lehmannia vrancensis Lupu, 1973

Lehmannia vrancensis Lupu, 1973 in Lupu 1973: 49

Slug light yellowish with black reticulate pattern, with darker appearance near the mid-dorsum line, lighter down to the sides, keel occupies last third of body, mantle with dark colour band in lyra form, sole evenly white with yellowish hue.
Penis long and tubiform, often with lateral incisions, distal end broad and bulbous, flagellum short and pointed (difference to *Le. jaroslaviae* and *Le. nyctelia*).

28-32 mm long (preserved)

E Romania (Vrancea)

Limax

Limax Linnæus, 1758
in Linnæus 1758: 652

Type species *Limax maximus* Linnæus, 1758
(fixed under the plenary powers)

ICZN Opinion 94, and Direction 72: type species designated by Férussac 1819 (Hist. nat. gén. partic. Moll. terrestr. fluviat. 2 (1): 67) - Férussac & Deshayes 1819-1851 should have been meant there, we have this work in Göttingen but I did not find the name mentioned on p. 67 (and Férussac certainly did not designate name-bearing types for genera, this concept came up later in the 1830s). I would preliminarily recommend to recognize *L. maximus* as type species fixed under the plenary powers. Gray 1847: 170 gave as type *Limax rufus* Linnæus, 1758 (= currently *Arion rufus*), so setting this aside and fixing *L. maximus* as type for *Limax* was necessary.
Limacus Lehmann, 1864 in Lehmann 1864: 145, type species *Limacus breckworthianus* Lehmann, 1864 (monotypy) (= currently *Limax flavus* Linnæus, 1758)

Medium or large slugs. Mantle covers about 1/4 of the body. Pneumostom postmedial, keel clearly visible, but often short. Penis long, tubular, convoluted or twisted, less frequently flagellate. Penis and right tentacle retractor crossed. Longitudinal folds inside penis. Vas deferens opens apically or subapically into penis; so the terminal part of the penis can be a blind appendix. Penis without accessory structures.

Almost all species of *Limax (Limax)* occur in forests or in the mountains above timberline. They climb tree trunks and many copulate suspended on a slime thread, others copulate on the substrate.

Mainly SW and Mediterranean Europe, probably also E Asia Minor and N Africa

Together not more than 40 species, but the number is steadily increasing as more information from Italy, Corsica and the S Alps becomes available. It is known that there are definitely more species in the field than in the books. Slugs that are only known by German names or tentative scientific names are not included here. More research is needed to firmly establish their identities.
In a detailed study Nitz et al. 2010 suggested that at least 15 *Limax* species live in Corsica. The taxonomy of *Limax* is still in an initial stage and far from being understood. Many preliminary conclusions are probably of a temporary nature and need to be confirmed and based on a more profound understanding of the diversity observed in the Italian forms. The same study presented almost direct evidence that various *Limax* species in Corsica were reproductively isolated (conclusions derived from accurately studied mating behaviour and penis lengths which strongly suggested intrinsic barriers to reproduction), and that the genetic separation was not visible in the molecular record. This suggested that degrees of molecular divergence in pulmonate molluscs are not directly associated with biological species boundaries.

The system of subgenera and genera associated with *Limax* is under dispute. Authors who follow Wiktor's works (Wiktor 1983: 144, Wiktor 1989: 172, Wiktor 2001: 83) classify ***Limacus*** as a subgenus of *Limax*, authors who follow Forcart 1986 classify *Limacus* as a genus. The arguments were the same. Here Wiktor's system is adopted, who has published more comprehensive works on slugs, with detailed monographs.
Reischütz 1986: 119 argued that classifying *Limacus* as a subgenus would provoke that *Lehmannia* and *Malacolimax* would have to be included to *Limax*. Obviously this was not a binding step, otherwise Wiktor would have considered it later. Falkner (2008: 141) argued that the classification of *Limacus* as a separate genus was profoundly established, and that the classification of *Limax* subgroups should be based on preliminary studies published in abstracts of meetings (Klee et al. 2005, Klee 2006). This view is rejected here. It is usual practice to present the results of such studies more in detail, either in comprehensive monographs or in peer-reviewed journal articles.

Limax aeolianus Giusti, 1973

Limax aeolianus Giusti, 1973 in Giusti 1973: 191

Slug whitish, sometimes yellowish with reddish hue, covered with irregular black spots on the upper and lateral side and mantle, but not on head and neck, and a greyish lateral band-like structure, keel weak and only at posterior 15-18 mm, mantle oval, 17-19 mm long (preserved), posterior margin more pointed, sole whitish or yellowish white, subdivided in 3 longitudinal sections. Shell oval, with nucleus laterally situated.

Shady and humid habitats, under stones and in dense vegetation, also in ruderal environments and under the stones around gardens.

Italy: Eolian islands, Isola di Salina

43-53 x 11-15 mm (preserved) Agatino Reitano

Manganelli et al. 1995: 21.

S Italy (Isole Eolie: Salina, Filicudi)

Endemic species of Eolian islands.

Limax albipes Dumont & Mortillet, 1853

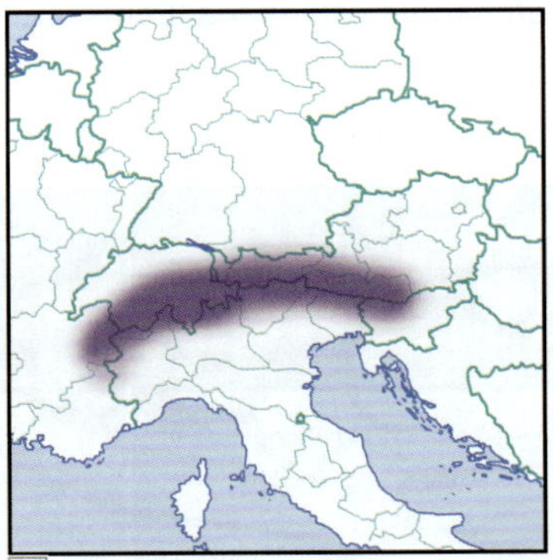

Limax lineatus var. *albipes* Dumont & Mortillet, 1853 in Dumont & Mortillet 1852-1854 (deuxième livraison): 69 [separate 194] [not verified] Type locality in SE France.

Slug evenly brown ot bluish black, very shiny, juveniles with clowdy pattern on the mantle. Sole white or cream. Mucus not sticky, colourless.
Penis very short, only half body length, retractor and vas deferens not opening at posterior end of penis but shortly before.

90-150 mm long

Predominantly in cool and humid habitats with moss on calcareous substrate in natural coniferous forests. In lower altitudes in creek gorges. Only exceptionally above the timberline. In Switzerland up to 2400 m, 85 % of the reports are from 1000-1800 m.
Slugs hide under rotting wood and stone plates, they are mainly active in the late evening, movements are relatively quickly.

Copulation was observed in July, usually at night, there is no fathom of slime as in other *Limax* species, and it takes only 2 minutes.

Gerhardt 1938, Turner et al. 1998: 291, Falkner et al. 2002: 133, Nitz et al. 2009: 289.

Probably not a good species, seems to have been used as a substitute name for various different species. It has not been found recently at its type locality in SE France. Not mentioned by Boschi 2011.

Alps

Relictual distribution, rare and scattered. Critically Endangered in Switzerland, very rare in S Austria (Osttirol, Kärnten).

Limax bielzii Seibert, 1873

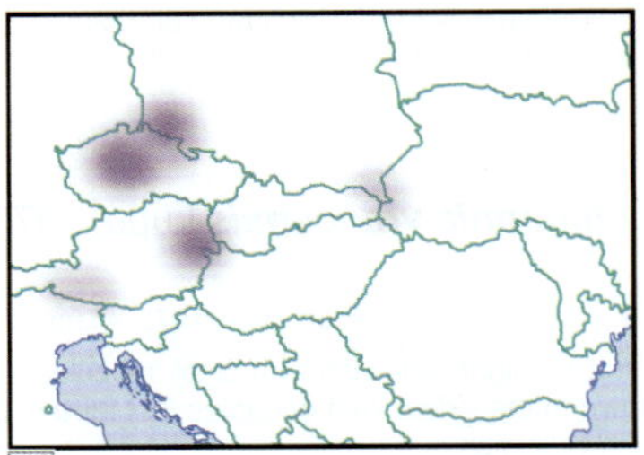

Limax bielzii Seibert, 1873 in Seibert 1873: 195 (*Limax Bielzii*)

Slug pinkish red, sometimes with dark lateral bands, long and slender, posterior end pointed, head and tentacles reddish black to black, mantle dark red to black, 40-44 mm long, rounded in anterior and pointed in posterior part, respiratory hole with black margin, 18-20 grooves between mid line of dorsum and respiratory hole, keel light red, only in posterior section, sole white, medial zone wider than lateral zones, lateral zones grey, margin of sole black. Sole mucus colourless, body mucus red.

Up to 140 x 9-11 mm

Copulation relatively short, takes only 10 minutes. Eggs variable in size and shape, 4-9 mm, are laid in August, in clutches of up to 100 eggs, juveniles hatch after 4 weeks, pale reddish and slightly transparent with red tentacles, some are already pinkish inside the egg, size 20-22 mm, getting reddish grey with darker lateral bands including the mantle after several weeks, mantle gets darker in the following weeks.

Central and E Europe

Gerhardt 1940, Kerney et al. 1983: 184, Manganelli et al. 1995: 21, 44, Wiktor 2004: 189, Cejka et al. 2007: 50 (previous Solvak records referred all to *L. cinereoniger*), Sysoev & Schileyko 2009: 151 (W Ukraine), www.naturamediterraneo.com (2011, Brandstetter, not an Italian species).

Perhaps a form of *L. cinereoniger*, from which it differs in its behaviour. More research is needed.

Limax brandstetteri Falkner, 2008

Limax brandstetteri Falkner, 2008 in Falkner 2008: 134

Slug in high mountain habitats usually black, (in captivity pale brown or grey), mantle short, bluntly angled at posterior side, keel short (1/4 of dorsum) and not different in colour, sole evenly whitish cream. Externally indistinguishable from *L. ianninii*.
Slugs are able to stretch longer than other slugs. Shell 9-10 x 6 mm.
Differs from *L. ianninii* in its more slender penis, presence of a well developed penial coecum and a different structure of a penial wall.

Reproduction probably in summer, slugs lay 25-40 transparent eggs (size 3 mm) into the soil, juveniles hatch after 22-26 days under laboratory conditions, they have a very prominent keel and seem to hibernate as juveniles to reach maturity in the next summer. After the summer they hibernate again, so that in spring both adults and juveniles are present. Maximum age probably two years.

Up to 125 x 15 mm

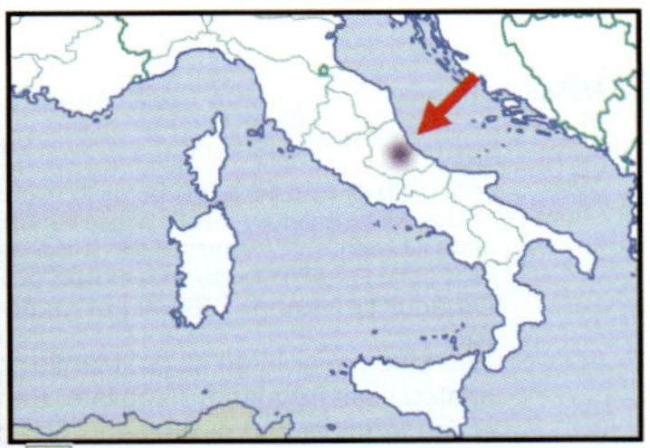

Italy (Abruzzo, Maiella mountains)

On stony and scarcely vegetated alpine meadows, 1750-2100 m. Feeds on *Viola* flowers. Active until December. In spring slugs crawl at the margins of snow fields.

Restricted to alpine zones in the Maiella mountains.

Italy: Abruzzo, Maiella mountains, 1750 m
Clemens M. Brandstetter

Limax canapicianus Pollonera, 1885

Limax canapicianus Pollonera, 1885 in Pollonera 1885: 519 (or 677)

Slug light brown with nearly rounded black dots, mantle rounded at posterior end, respiratory hole in the mid section of mantle, keel short at posterior end and weakly lighter brown than dorsum, sole evenly white.
Genitalia obviously like those of *L. cinereoniger*.

In deciduous forests.

Copulates with the soles attached to a steep wall or tree, penes hanging down.

Manganelli et al. 1995: 21.

Up to 140 mm long

Italy: Piemonte

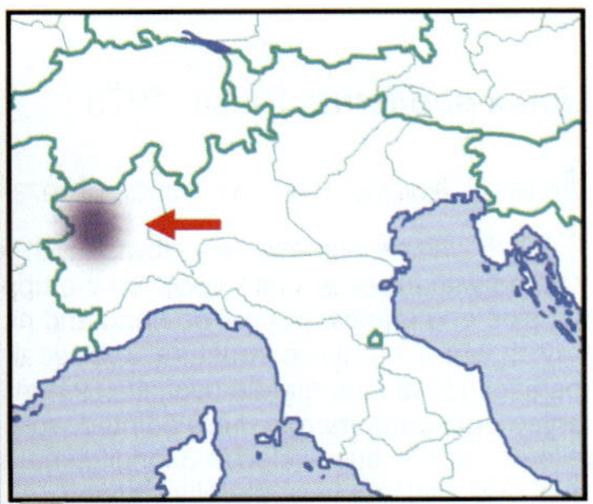

NW Italy (Alpi Graie)

Clemens M. Brandstetter

Limax cephalonicus Simroth, 1886

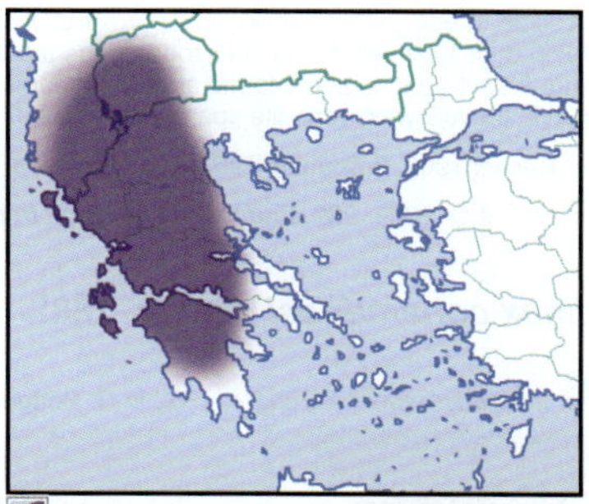

Limax cephalonicus Simroth, 1886 in Simroth 1886: 329, Taf. 11 Type locality in Kefaloniá, Greece.
Limax maximus var. *carbonaria* Boettger, 1885 in O. Boettger 1885: 159 Type locality in Ossa mountains, Greece.
Identity uncertain. Should this turn out to be a synonym of *cephalonicus*, then the species would have to change its name.

Slug mostly black or dark, sometimes with a lighter stripe in the middle of dorsum and keel, many slugs have dark and light spots, forming a reticulate pattern. Skin sculpture fine, 22-25 wrinkles between medial line of dorsum and pneumostom, mantle length 1/3 of body. Keel lighter, poorly arched, short, less than 1/7 of body length. Sole always uniformly cream-white, in juveniles ochre, mucus colourless, but can also be orange.
Penis is specific, slightly longer than mantle, always less than half of the body length. Differs from *L. graecus* only in its penis length.

Up to 125 x 20 mm

A mountain slug, up to 2000 m altitude. On rocks and at walls, sometimes also on tree trunks or under stones. Open habitats and deciduous forests.

Rähle 1981, Wiktor 2001: 68.

Rep. Makedonija to S Greece

Due to the high colour variation many synonyms have been established. Very uncertain is the identity of *L. carbonaria* from the Ossa mountains in Thessalia (Greece), sometimes (Falkner 1990: 190) classified as an independent species *Limax carbonarius*, but the form is too insufficiently known for a definite classification.

Limax ciminensis Pollonera, 1890

Limax corsicus var. *ciminensis* Pollonera, 1890 in Pollonera 1890: [3] (*Limax corsicus* var. *Ciminensis*)

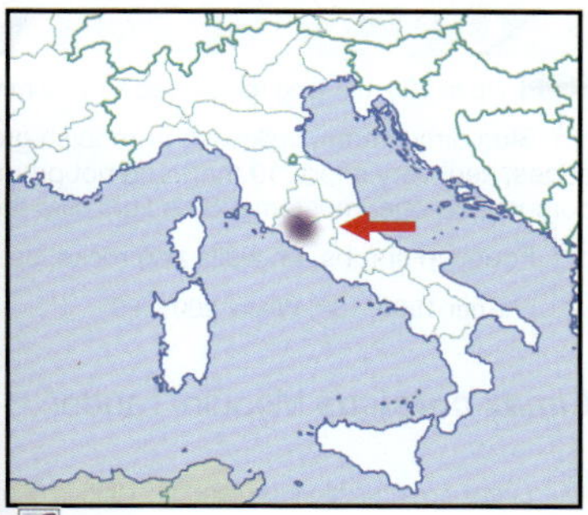

Slug light brownish with orange hue and 4 darker colour bands, keel lighter and slightly more yellowish, extending on posterior 2/3 of dorsum, mantle light brown with some light spots on its anterior part and slightly darker spots on its posterior part, foot fringe reddish brown. Sole whitish with reddish hue in medial zone, lateral zones reddish brown.

In the leaf litter layer and under stones in deciduous forests.

Reproduction was observed in June, copulation attached with the sole to a tree, penes hanging down about 1 m.

Up to 120-130 mm long

Italy: Monte Cimino

Italy (Lazio)

Manganelli et al. 1995: 21.

Clemens M. Brandstetter

Limax cinereoniger Wolf, 1803

Limax cinereoniger Wolf, 1803 in Sturm 1803: 7 (*Limax cinereo-niger*) [not verified]
Not verified, author uncertain, Wolf's name needs to have been mentioned in the original source.
Limax alpinus Férussac, 1822 in A. Férussac [1821-1822] (prodrome limaces): 21
Date Jan 1821 after Falkner et al. 2002: 133 (probably incorrect), 16 Jul 1822 after Bouchet & Rocroi 2005: 304.

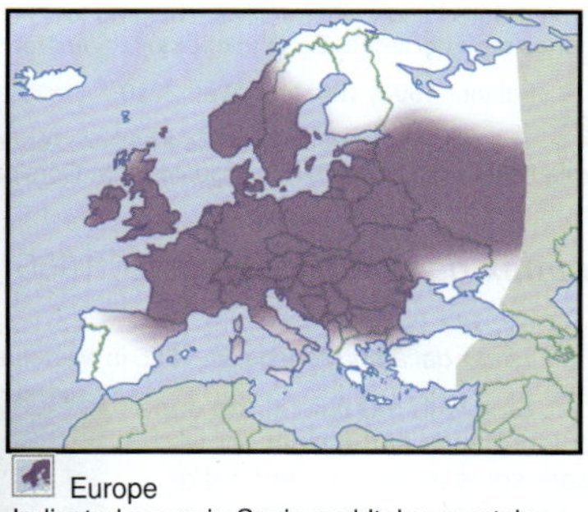

Europe
Indicated range in Spain and Italy uncertain.

Still frequent but declining in Great Britain, threatened by modern forest management. Rare in Ireland, Vulnerable in Spain.

(1) Czech Republic: Moravia, near Těchanov, (3, 6) Germany: Hessen, Hoher Meissner,
(5) Germany: Tambach-Dietharz near Gotha,
(2, 4) Slovakia: Banská Bystrica,
(shells) Germany: Schleswig-Holstein, near Kiel
(inside box) Italy: Sűdtirol, Seiseralm
(1) Michal Maňas, (2, 4) Jozef Grego,
(inside box) Clemens M. Brandstetter

Up to 200 mm long

Slug greyish brown to black, very variable, also tentacles, often with two dark lateral longitudinal bands, tubercles large, keel long and light, extending about two-thirds distance from tail to mantle. Lateral zones of sole grey to black, central zone light and sharply delimited. Mucus colourless. Juveniles brown with indistinct colour pattern and evenly brownish sole.
Penis very long and coiled, epiphallus inserting almost apically, evenly wide throughout.
The characteristic sole pattern is very important to recognise this species.

Natural forests of all types and shrublands, also in pure coniferous forests, crawling on the ground and in the trees. Avoids cultivated forests. Hiding in stumps, decaying tree trunks and under loose bark. Indicates ancient woodland in Britain, even if greatly modified by traditional management. In Romania mainly in the mountains, where the species is common, Britain mostly on non-calcareous soils. In Switzerland up to 2000 m, in Romania and Bulgaria 2500 m. Strongly nocturnal, consumes mainly algae, mushrooms and rotting plant particles. Slugs are also active in winter when temperatures are mild. Only rarely associated with *Limax maximus* in Britain and Romania.

Self-fertilisation is possible. Copulation in early summer (June-July) usually at trees, slugs stick at the substrate with their posterior ends (these are separated, not united at a slime thread), with their penes (approximately body length) twisted screw-like around each other. Individuals lay a total of 400-800 eggs in batches (every few weeks) of 40-250 eggs between July and October, eggs are laid under mosses, not inside the soil. Juveniles hatch after 20-30 days, they are initially 8-9 mm long. Slugs get their dark colour after 1 year, in some forms after 2-3 months, maturity is reached after 1.5-2 years at 160-190 mm length, slugs may survive the first reproduction period and lay eggs again in the following year, life span up to 3 years.

Künkel 1908, Frömming 1954: 194, Grossu & Lupu 1960: 160, Kerney & Cameron 1979: 136, Falkner 1990: 186, Castillejo & Iglesias in Verdú & Galante 2006: 393.

Nitz et al. 2009: 289 designated a neotype from Austria for *Limax alpinus* that was conspecific with *L. cinereoniger* (see under *L. sarnensis*).

Limax conemenosi Böttger, 1882

Limax conemenosi Böttger, 1882 in O. Böttger 1882: 100 (*Limax Conemenosi*)

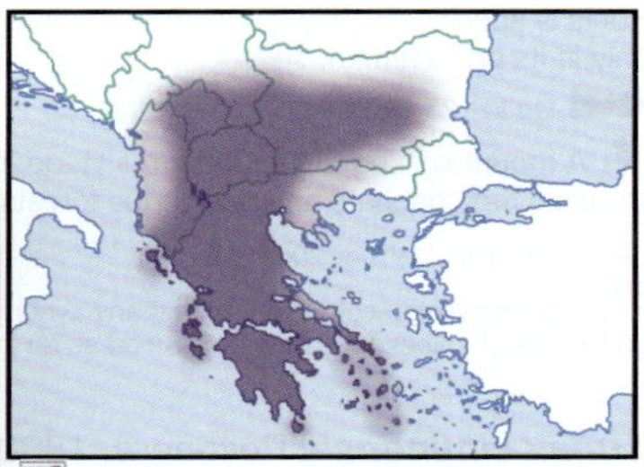

Originally probably Greece, at present also Bulgaria to Albania

Up to 120 mm long (preserved 80 x 12 mm)

(1) Bulgaria: Plovdiv, (2) Pelopónisos, Kitries near Kalamáta — (1) Atanas Irikov, (2) Panayotis Ovalis

Slug dirty cream, yellowish or reddish grey, dorsum rounded and little darker than sides, regularly spotted with numerous 1 mm large black dots with sharp edges, keel very short (10 mm) and not prominent, mantle relatively short (1/4 of body, up to 22 mm), 21-23 fine and flattened wrinkles between medial line of dorsum and pneumostome. Sole light, like the background colour. Mucus colourless.

Found in shrubs, on walls and rocks, usually not crawling up trees. Common as a synanthrope, especially in rural environments. Up to 1300 m in Bulgaria.

Falkner 1990: 188, Wiktor 2001: 69.

Limax corsicus Moquin-Tandon, 1855

Limax corsicus Moquin-Tandon, 1855 in Moquin-Tandon 1855: 26

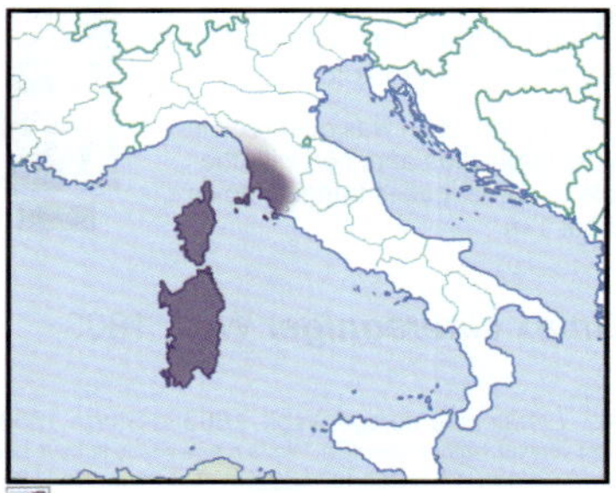

Corsica, Toscana, Sardinia

Slug brownish ochre to dark reddish brown, with light dorsal line, margins lighter, near foot red. Dorsal colour pattern can be blurred so that the animals appear unicolour. Lateral zones of sole red, central zone yellowish to light red.

Up to 150 x 15-23 mm, shell 10 x 5 mm

Humid habitats in deciduous forests and Mediterranean shrubland vegetation.

Copulation takes place mainly during warm and humid weather conditions, slugs crawl to a place at a steep wall or on a tree, they twist around each other and stick at the substrate with a triangular portion of mucus, hanging below it they copulate during 2-3 hours. Penes are everted to about 30 mm, they find each other in the air and twist around each other spirally, the sexual openings approach each other, then the sperma packages are sent down the penes which hang down getting more twisted and longer until reaching up to 60 cm. Sperma is finally fixed at the end of the penes and withdrawn when the penes are shortened. Eggs (globulous, diameter 5 mm) are laid in small packages.

Falkner 1990: 188, Nitz et al. 2010.

The name is here used in the traditional sense, it has generally been applied to *Limax* species between Sardinia and Toscana with brownish to creamy bodies and red sole fields. Nitz et al. 2010 suggested that Corsica may host at least 15 endemic species, but this seems to need further research and confirmation.

Limax dacampi Menegazzi, 1855

Limax dacampi Menegazzi, 1855 in Menegazzi 1855: 63, Tav. 1 (*Limax Dacampi*) [not verified] Type locality Garda.

Slug with red colour and dark colour bands at the sides, or with dark spots at dorsum, but not on mantle, head and tentacles greyish.

Wide ecological amplitude, prefers forests and shrubs in lowlands.

Copulation on a 20-25 cm long slime thread, everted penes reach 30-50 % of body length, duration 1 hour.

Manganelli et al. 1995: 21 (*dacampoi*), Wiktor 1996: 72, 75 (synonym of *L. cinereoniger*), Turner et al. 1998: 295, www.naturamediterraneo.com (2008, copulation at the type locality).

Up to 200 mm long

Italy: Alessandria, Cerreto Ratti — Ettore Balocchi

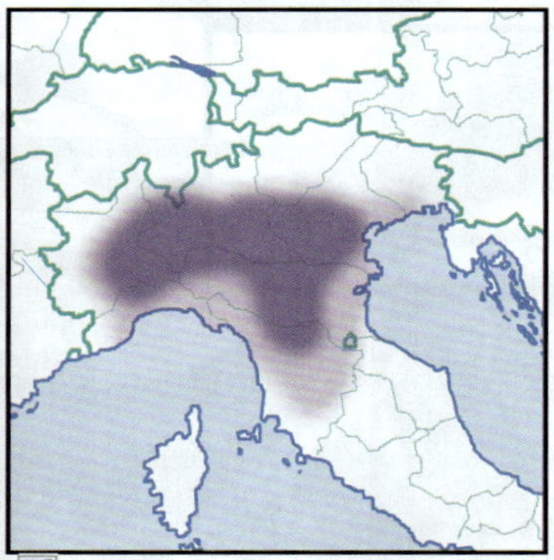

N Italy, S Switzerland (Ticino)

Limax dobrogicus Grossu & Lupu, 1960

Limax dobrogicus Grossu & Lupu, 1960 in Grossu & Lupu 1960: 162

Slug yellowish grey, with deep and regular grooves, tentacles long and bluish, mantle rounded at both sides, keel short and indistinct. Body mucus yellow and sticky, sole mucus colourless.
Slug smaller than *Limax hemmeni*, without distinct spots, posterior margin of mantle rounded (not angular), penis shorter (less than 1/2 of oviduct and spermatheca duct together), vas deferens shorter, no hooked appendix at proximal end of penis.Slug smaller than *L. hemmeni*, without distinct spots, posterior margin of mantle rounded (not angular), penis shorter (less than 1/2 of oviduct and spermatheca duct together), vas deferens shorter, no hooked appendix at proximal end of penis.

Up to 140-150 mm long (preserved 50-60 mm)

Restricted to old natural forsts, under the bark of rotting tree trunks, often in shady and moist crevices of rotting trees. Active only at night, and never at dry weather conditions. The animal is slow, shy and cautious, contracts rapidly after being touched.

A very rare species, restricted to old and natural forests with rotting trees.

Grossu 1983: 254, Rähle 1983: 33, Schütt 2001: 247.

SE Romania (N Dobrogea)

Limax ecarinatus Boettger, 1881

Limax maculatus (Kaleniczenko, 1851) (incorrect name), *Limacus maculatus* (Kaleniczenko, 1851) (correct name)

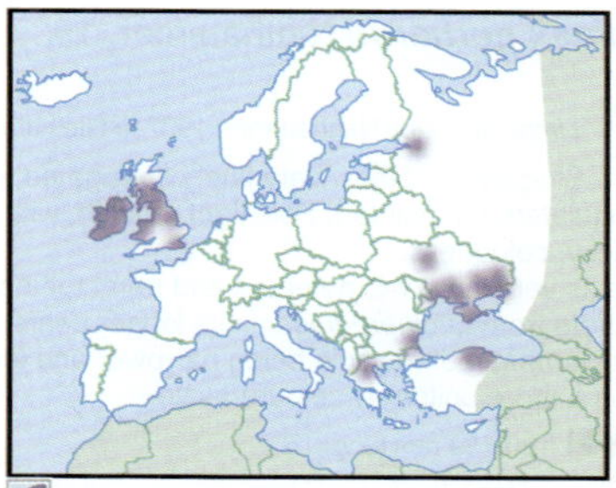

Originally probably Black Sea area, introduced elsewhere

Limax ecarinatus Boettger, 1881 in O. Boettger 1881: 186, Taf. 7
Krynickillus maculatus Kaleniczenko, 1851 in Kaleniczenko 1851: 226, Tab. 6 (not *Limax maculatus* Nunneley, 1837 (= currently *Limax maximus* Linnæus, 1758))
Limax grossui Lupu, 1970 in Lupu 1970: 63
Limax pseudoflavus Evans, 1978 in Evans 1978: 173 [not verified]

Slug colour like *Limax flavus*, but usually darker and more greenish, with darker and lighter irregular spots on body, edges of spots slightly blurred, dots fusing partly with one another, they are more numerous than in *Limax flavus*, and reach lower on body sides, almost near the sole margin. 23 wrinkles between mid line of dorsum and pneumostome, mantle length up to 24 mm. Young slugs can be completely green (the green colour becomes blackish in ethanol). Mucus yellow.

Up to 100 mm long (preserved 72 x 15 mm)

In Crimea and Caucasus in deciduous forests, in N and Central Europe mainly in natural habitats under stones and dead wood, often in forests. In Ireland at roadsides and in plantations, around ruined buildings and crevices of stone walls. In the British Isles often found in the same habitats as *Limax flavus*, in Ukraine in lower abundance than *L. flavus*. In Bulgaria in deciduous forests (hornbeam, *Carpinus betulus*). Shelters in rotting wood. Covers tree-trunks with mucus. Eats various food.

Smaller specimens are often already sexually mature.

Wiktor & Norris 1982, Falkner 1990: 190, Kerney 1999: 157, Wiktor 2001: 85, Chernyshova et al. 2010.

The name combination *Limax maculatus* (Kaleniczenko, 1851) is incorrect, *maculatus* must only be used if the slug is classified in the genus *Limacus*.
Evans 1986 suggested the presence of three different species *L. ecarinatus*, *L. grossui* and *L. pseudoflavus*, but Wiktor 2001: 86 rejected Evans' arguments pointing out that the species was introduced to various places, making it doubtful to define distinct species from these localities. More studies on this slug are necessary.

Limax erythrus Bourguignat, 1864

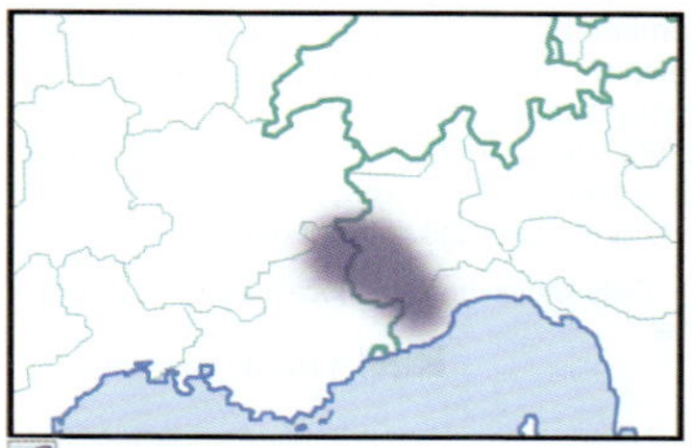

NW Italy, SE France

Limax erythrus Bourguignat, 1864 in Bourguignat 1864 (Mal. Gr-Chartr.): 33, Pl. 2

Slug reddish, slightly lighter downwards at the sides (colours will change in preserved animals), keel restricted to posterior haft section of dorsum behind mantle. Mantle concentrically striated, more intensively red than dorsum, with two darker reddish brown lateral bands, margins of mantle red with orange hue, pneumoston surrounded by a blackish margin. Tentacles brownish. Sole yellowish with violet brownish lateral sides.

Up to 120 mm long

Under stones in natural habitats.

Manganelli et al. 1995: 21.

Limax flavus Linnæus, 1758

Limax flavus Linnæus, 1758 in Linnæus 1758: 652

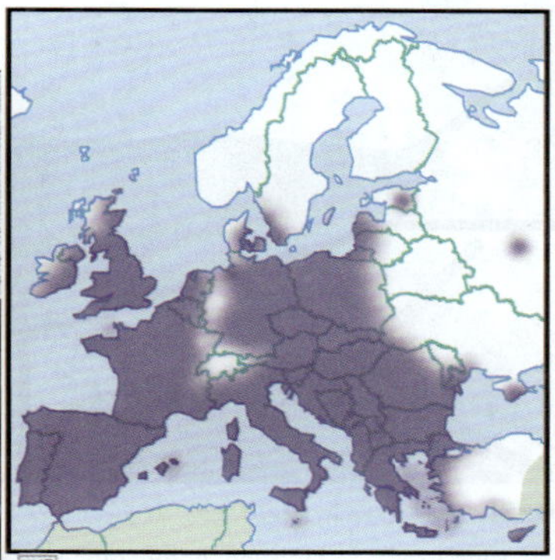

Original range possibly Mediterranean, at present widespread in Europe, Canary Islands, Chile and South Africa

Up to 150 mm long (preserved 80 x 19 mm)

(1) Turkey: Aegean Sea, Bozcaada island, (2 middle) Portugal: Algarve, Castro Marim, (3) Bulgaria: E Rodopi mountains, Dolo Cherkovshte, (4 small) Turkey: Vilayet Erzincan, Kemaliye, (shells) France: Alpes, near Draguignan

(1) Panayotis Ovalis, (2) Valter Jacinto, (3) Dilian Georgiev, (4) Burçin A. Gümüş

Slug grey-greenish-olive, or yellowish brown, covered with lighter or orange-olive spots, lower parts of the sides gradually lighter, at the very bottom the colour is like the light colour of the spots of dorsum. Head and tentacles grey. Dorsum rounded, keel short 10-15 mm, mantle 20 % of body length, 22-23 grooves between mid line of dorsum and pneumostom. Sole cream-olive. Body mucus yellow, sole mucus colourless. Juveniles more greenish.
The light colour spots on the dorsum are often associated with particular wrinkles, or in other words, most wrinkles and the furrows between them are darker, but some wrinkles are light, so the light spots are oval. The keel often consists of 2-3 parallel rows of light wrinkles. On the mantle the light colour spots or dashes are not associated with surface structures.
Penis about 1/6 of body length. Spermatheca duct unites with oviductus and both open together into the atrium as a flattened broadening (an exception in Limacidae, making this slug easy to identify).
Shell oval, white, concentrically striated and slightly concave, size up to 6-8 x 4-5 mm.

Was found in high mountain habitats in Caucasus (1500 m), Rep. Makedonija (1800 m) and Bulgaria (1300 m), in Piemonte (Italy) up to 900 m. One of the most frequent slugs in Greece. In Romania, central and N Europe and Britain almost exclusively as a synanthrope, most often collected near fruit and vegetable stores, compost, wells, gardens, parks, farmyards, caves and moist habitats at old walls, humid cellars and moist kitchens. In Britain not unknown from modern suburban houses, occasionally in woods. Strongly nocturnal, feeds mainly on muchrooms and lichens.

In France, Germany and Britain 40-60 eggs (5-10 x 4-6 mm) are laid in a few clutches (or strings of pearls) between July and December in humid places (they also survive submerged), pale greenish juveniles (10 mm long) hatch after 40-60 days, the colour of adults is taken after few days, maturity is reached after 9-11 months, they are 60-80 mm long after one year, 95-110 mm after 2.5 years. In each season they lay eggs, maximum age 3-4 years. Animals living in humid cellars and moist kitchens are independent from reproductive seasons and may lay eggs at any time in the year, juveniles hatch after 23-35 days, life span 2.5-3 years.

Introduced to Britain prior to 1700, today frequent and widespread in England and Wales. Rare and local in Ireland and Scotland. Threatened in central Europe due to restoring of old buildings and measures of hygene. Critically Endangered in Germany and Switzerland, at least vulnerable in Austria.

Moquin-Tandon 1855: 25, Germain 1930: 98, Nobre 1941: 42, Frömming 1954: 167, Grossu & Lupu 1960: 163, Kerney & Cameron 1979: 137, Liebegott 1986, Falkner 1990: 190, Turner et al. 1998: 299, Kerney 1999: 156, Wiktor 2001: 82, G. Konstantinou (unpublished, 2012, Cyprus).

Also known as *Limacus flavus*.

Limax gerhardti Niethammer, 1937

Limax gerhardti Niethammer, 1937 in Niethammer 1937: 443

Slug evenly black, very finely corrugated, posterior part shortly keeled, posterior end of the mantle smoothly angulated, not pointed. Sole light reddish, with darker reddish margins, margin at the foot black. Mucus not sticky, colourless.
Penis very long, vas deferens and retractor inserting at the distal end of the penis.
Differs from *L. cinereoniger* in its longer penis, a more finely corrugated surface and a shorter keel, and the black margin of the sole being narrower and less clearly delimited against the light central zone. *L. punctulatus* has a more uniformly coloured sole.

130-150 mm long

Copulation from June to October, usually in the evening, first the partners look carefully for a place where their penes can hang down at least 50 cm, it takes 3-4 hours from the beginning to sperma transmission, penis length varies between 15 and 50 cm, after that the partners need another 2-8 hours to get their penes cleaned (by dropping the slime) and separated. Animals copulate repeatedly in the season. Activity and copulation is not much temperature dependent, can also take place at 10-12°C.

Gerhardt 1937: 525, Gerhardt 1940: 568. Not listed by Alzona 1971 and Manganelli et al. 1995.

Italy (Ischia)

Limax giovannellae Falkner & Niederhöfer, 2008

Limax giovannellae Falkner & Niederhöfer, 2008 in Falkner & Niederhöfer 2008: 11

Slug greyish brown with weak light patterns, keel yellowish white in posterior section, in last third sharp, prominent at the posterior end, mantle evenly light brownish, sole evenly white. Penes are everted only slightly during copulation.
Differs from *L. maximus* in its thicker and more massive penis and weak caecum, from *L. veronensis* in its narrower vas deferens without glands.

Only known from two localities, 6 km distant from each other.

115-140 mm long

Italy: Udine, Resia, between Prato and Lischiazze

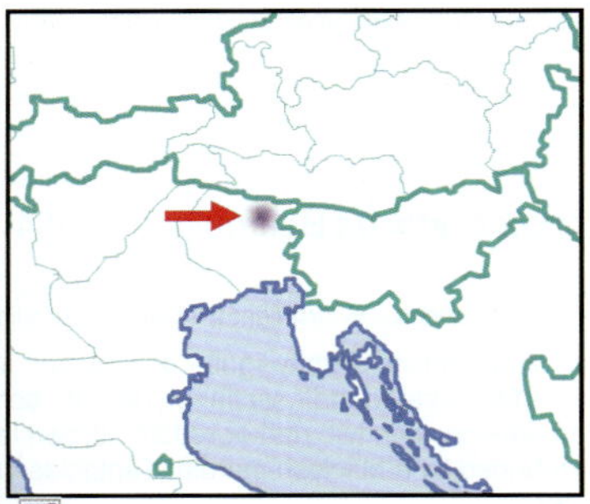

NE Italy (Alpi Giulie)

Clemes M. Brandstetter

Limax giustii Falkner & Nitz, 2010

Limax giustii Falkner & Nitz, 2010 in Nitz et al. 2010: 432

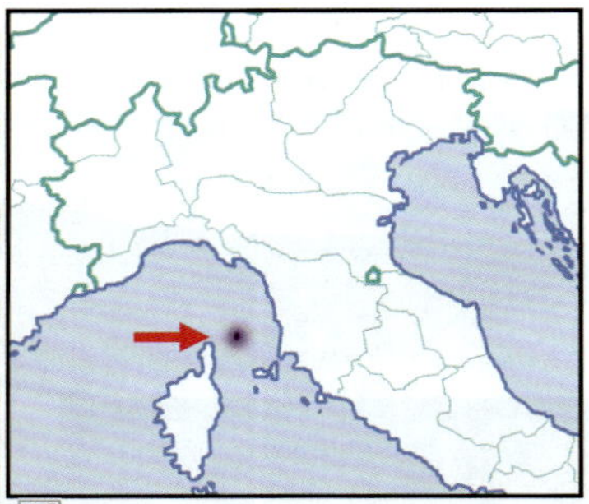

Italy (Capraia island)

Up to 100 mm long

Italy: Capraia island

Paolo Fontanesi

Giusti & Mazzini 1971: 270.

Slug brown to blackish, sometimes with weak or interrupted lateral bands or with spots on the mantle, sole usually weakly reddish, posterior third keeled, keel line sometimes yellowish-white.

Limax graecus Simroth, 1889

Limax graecus Simroth, 1889 in Simroth 1889: 7
Limax macedonicus Hesse, 1928 in Hesse 1928: 10, Taf. 2

Slug colour variable, unicoloured or patterned, dorsum rounded, keel only 10 mm, mantle length 35-40 % of body, 19-26 wrinkles between medial line of dorsum and pneumostom. Sole light cream. Mucus colourless. Penis 2/3 or more of body length, shorter in young individuals, with blind appendix at posterior end.

Deciduous and mixed forests, shrubs, parks etc., rural environments. In Bulgaria in up to 2000 m. Often a synanthrope. Climbs tree trunks, hides under stones and bark of trees. Feeds on lichens.

Bulgaria: Stara Zagora

Up to 95 x 23 mm (preserved)

Atanas Irikov

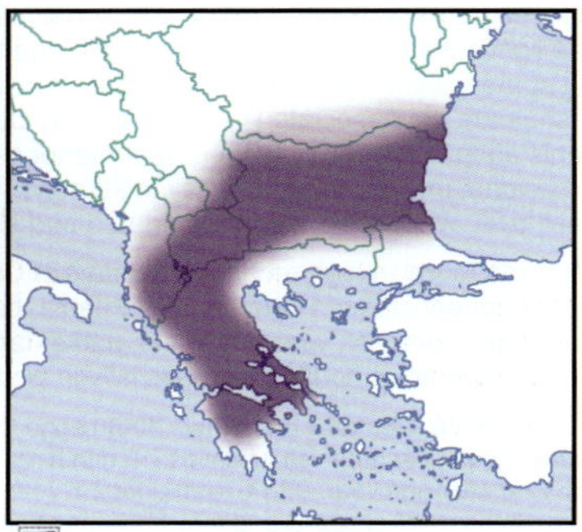

Romania to Greece

Wiktor 2001: 71, Irikov & Mollov 2006: 807 (*Limax macedonicus*), Hubenov 2007.

Has been confounded with other species in the literature. Presence in SE Italy (Lecce) needs confirmation.

Limax hemmeni Rähle, 1983

Limax hemmeni Rähle, 1983 in Rähle 1983: 31

Slug ashy grey with numerous irregular yellowish spots, sole evenly light, mantle length 35 % of body, narrowing posteriorly, not rounded but triangularly elongated, keel short.
Penis short, 3/7 of body length, retractor penis forked into several branches near penis. Vas deferens opens apically into a hooked appendix at the posterior part of the penis. Retractor branches insert there and at different locations of the proximal penis section.

Up to 100 mm (preserved 60-80 mm)

At S exposed slopes under stone plates, 300-1000 m.

Possibly an endemic species of Samos.

Bank & Maassen 1998: 63, Wiktor 2001: 77.

Could be conspecific with *L. dobrogicus*, but our present knowledge indicates that it is a separate species. The degree of variation of both forms should be studied more in detail, also molecular studies could help.

Greece (Sámos)

Limax ianninii Giusti, 1973

Limax ianninii Giusti, 1973 in Giusti 1973: 484

Slug evenly black, keel very weak and restricted to the posterior 16-20 mm, mantle large (23-27 mm long), oval, pointed at posterior end, sole subdivided in 3 longitudinal sections, evenly white or yellowish white.
Penis very short (less than 26 mm, in *L. albipes* up to 60-70 mm).

Meadows in upper mountain zones, under stones and near herbal plants.

Only known from a small restricted range.

Manganelli et al. 1995: 21.

Up to 80 x 14 mm (preserved)

Italy: Monti Reatini Clemens M. Brandstetter

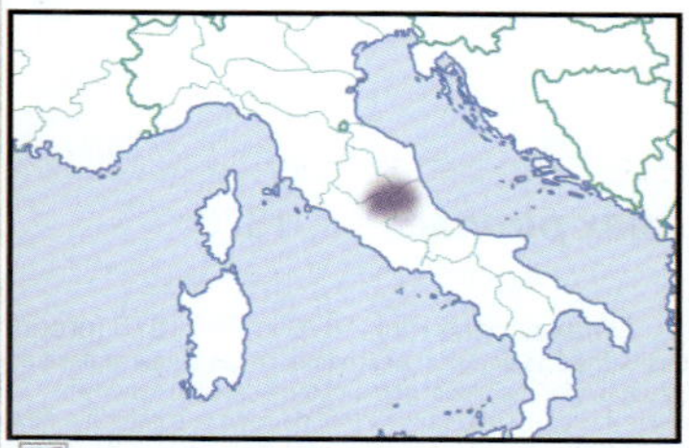

Italy (Monti Reatini, Gran Sasso)

Limax lachensis Bérenguier, 1900

Limax lachensis Bérenguier, 1900 in Bérenguier 1900: Pl. 17, Pl. 22 (*Limax Lachensis*)

Slug reddish, getting pale yellow down to the sides, with black spots on dorsal side, simulating two lateral and one medial lines, mantle rounded in anterior part, finely black spotted, finely concentrically striated, sole evenly yellowish white with orange red margin. Shell subtriangular and shiny, very slightly convex at upper side, concave at lower side, distinctly concentrically striated, nucleus in posterior section, margin translucent. Sole mucus colourless, body mucus reddish orange.
Penis long and massive, spermatheca pear-shaped with short duct.
External apppearence like *Limax maximus*, shell like *Limax cinereoniger*.

120-150 mm long, shell 11-13 x 7-8 mm

In mountain forests, 1300-1600 m.

Germain 1930: 94.

More research is necessary to clarify its classification, the description of the genitalia is from the 1930s and certainly insufficient.

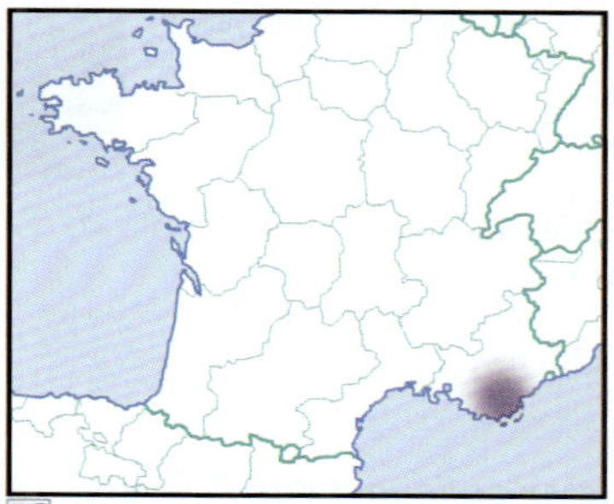

France (Var)

Limax maximus Linnæus, 1758

Limax maximus Linnæus, 1758 in Linnæus 1758: 652

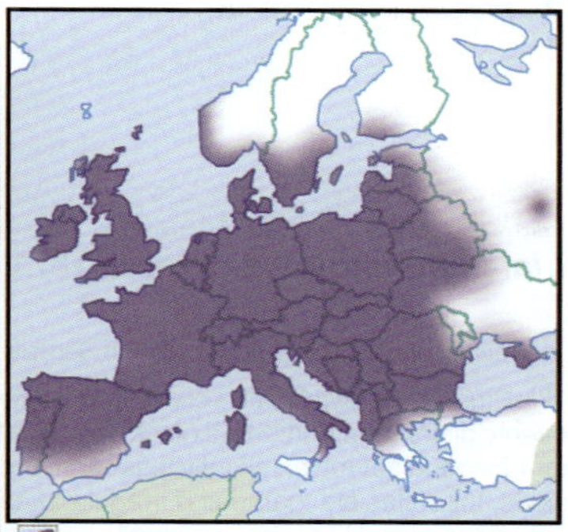

Up to 200 mm long (preserved 130 x 23 mm)

(1) Slovakia: Banská Bystrica, (2) USA: Maryland, Germantown, (3 small) Bulgaria: W Rodopi, Hrabina, (4) Czech Republic: Moravia, Olomouc, (shells) Germany: Kiel
(1) Jozef Grego, (2) Aydın Örstan, (3) Dilian Georgiev, (4 right) Michal Maňas

Originally S and W Europe, perhaps also N Africa, today widespread in Europe, Caucasus and abroad

Slug colour variable, most slugs have large blots, dark mantle spots vary individually, sometimes they are very numerous so that the slug seems black. Spotless slugs are ash-grey-cream or brownish. Keel clearly visible, 1/3 of body length, mantle length 30 % of body, 21-36 wrinkles between medial line of dorsum and pneumostom. Sole uniformly cream-coloured. Mucus colourless.
Penis attains more or less half of the body length.
L. cinereoniger has a longer penis (matching body length in adult slugs), dark lateral zones on the sole and a unicoloured mantle. *L. graecus*, *L. cephalonicus* and *L. conemenosi* have a caecum in the posterior part of the penis, unlike *L. maximus*.

Wide ecological amplitude, inhabits moist sheltered places, often in artificial and disturbed habitats (compost, gardens, cellars, vegetable warehouses, cemeteries, parks), in Romania rarely in habitats distant from human settlements. May damage plant cultivations. Originally probably a woodland species. In Switzerland up to 1900 m, usually below 1600 m, in Bulgaria up to 1500 m. Lives at the soil and hides under pieces of wood, stones, in crevices. Clims tree trunks at night and at wet weather conditions. Omnivorous, consumes other slugs, various invertebrates, mushrooms, lichens, algae, flowers and roots of plants.

Reproduction between July and September, occasionally October. The animals copulate hanging on a slime thread attached to a plant branch, after copulation one animal descends to the bottom and the other ascends the slime thread. 30-60 eggs (4-5.5 mm) are laid, usually attached together at their poles, sometimes simply accumulated. Juveniles hatch after 20-30 days, they reach maturity after one year. Life span 2-3 years.

Moquin-Tandon 1855: 28, Frömming 1954: 183, Grossu & Lupu 1960: 158, Wiktor 1996: 78, Wiktor 2001: 79.

Limax millipunctatus Pini, 1885

Limax millipunctatus Pini, 1885 in Pini 1885: 252

Slug greyish or yellowish with brown-reddish hue on dorsum and mantle, entirely covered with evenly distributed numerous black dots, slightly less regular down to the sides, tubercles oval and not very prominent, keel short and not very prominent, mantle anteriorly rounded and pointed at posterior end, concentrically striated, tentacles finely spotted, respiratory hole without distict margin and slighty posterior of the centre of the mantle, sole evenly white.

115-125 x 10-12 mm (preserved 95 x 10 mm)

Italy: Voltri near Genova
Clemens M. Brandstetter

Differs from *L. punctulatus* in its darker colour, more oval and shorter tubercles, coloured medial line in posterior section, bases of tentacles not approaching each other and lack of a distinct margin of the respiratory hole.

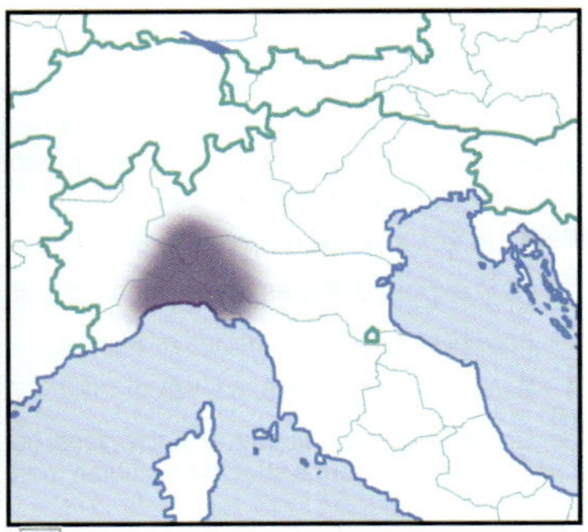

Italy (Lombardia, N Apennines)
Range uncertain.

Manganelli et al. 1995: 21.

Limax pironae Pini, 1876

Limax pironae Pini, 1876 in Pini 1876 [preprint]: 36 (*Limax (Eulimax (Stabilea)) Pironae*)

Established as a regular new name, not as a new replacement name. Falkner & Proschwitz 2009: 173 argued that because Pini 1876 gave 5 synonyms, *L. pironae* must be regarded as a new replacement name, and by default for the oldest name of the attributed synonyms. This procedure was not based on any written rules. There was no expressed statement to qualify for a new replacement name as demanded by Art. 72.7, and it was moreover unclear why Pini decided to establish a new name. It rather seems that Pini 1876 included the synonyms informally without being sure about their identity, also expressed by question marks, for future scientists to verify if those taxa were identical with Pini's species for which he gave a detailed description and figures. The original source gave no indication that *L. pironae* was intended as a new replacement name for either one of the mentioned synonyms, much less for which one.

Slug intensively brown to black, yellowish at the margin of the foot, with sharp keel of the same colour occupying posterior third of the body, mantle very finely and irregularly striated, respiratory hole slightly posterior of the middle of the mantle. Sole evenly whitish with yellow hue.

Under stones and other shelters.

Range small and restricted to massiccio della Grigna in Lombardia.

Manganelli et al. 1995: 21.

Up to 85 x 6 mm

Italy: Lombardia, Esino near Lecco
Clemens M. Brandstetter

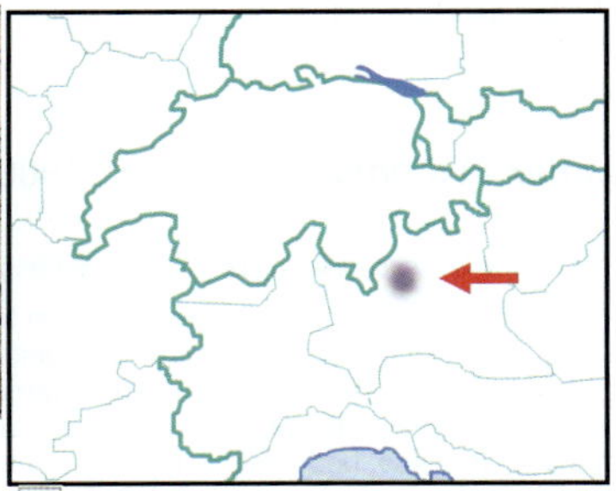

Italy (Lombardia)

Limax polipunctatus Pollonera, 1888

Limax polipunctatus Pollonera, 1888 in Pollonera 1888: 2, Tav. 3

Slug pale brownish with blackish spots on dorsum, less dotted at foot fringe, weakly keeled. Sole evenly white.

Up to 100-120 mm long

Manganelli et al. 1995: 21.

Italy (Alpi Graie, N Apennines)

Limax punctulatus Sordelli, 1871

Limax punctulatus Sordelli, 1871 in Sordelli 1871: 250

Slug light greyish with yellowish hue, covered with evenly distributed numerous black dots (12-14 dots in 3 or 4 longitudinal rows), slightly less regular down to the sides, tubercles elongated and irregular, keel light and extending over 1/2 of dorsum length, mantle with 10-25 small (0.5 mm) black spots, respiratory hole with darker margin and slighty posterior of the centre of the mantle, sole evenly white and subdivided in 3 equally wide longitudinal zones.
Differs from *L. millipunctulatus* in its lighter colour, less oval and longer tubercles, no coloured medial line in posterior section, bases of tentacles approaching each other more, and presence of a distinct margin of the respiratory hole.

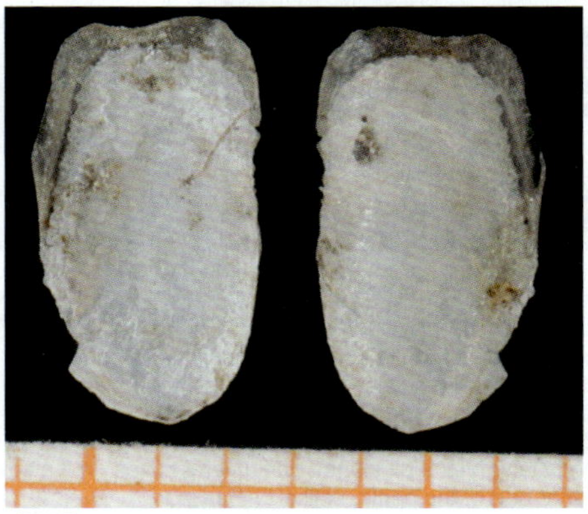

Up to 130 x 14 mm

Italy: Piemont, Novara

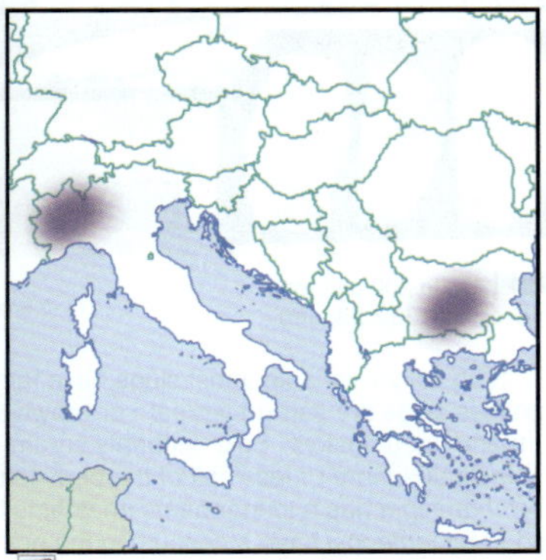

N Italy, Bulgaria (?)
Range perhaps incompletely known.

In forests. In S Bulgaria up to 2100 m.

Pini 1876: 31 [= 93], Pini 1885: 255, Manganelli et al. 1995: 21, Irikov & Mollov 2006: 807.

Limax redii Gerhardt, 1933

Limax redii Gerhardt, 1933 in Gerhardt 1933: 425

Slug creamy or yellowish, sometimes with bluish hue, usually with scattered black dots, less dotted on mantle, head and tentacles grey. Sole unicoloured.
Differs from other species mainly by its suspicuous copulation behaviour: needs to climb up perpendicular rocks or tree trunks of considerable height. Penis longer than in *L. punctulatus*, from which it differs also in its conspicuous proboscid structure during copulation. Copulation takes about 2 hours.

Mixed deciduous forests with dead wood at the soil, on calcareous substrate. In the winter deep inside the soil or between rocks, in the summer usually deep inside leaf litter layers. Best seasons for observation above the soils is in spring after the first heavy rainfalls, or in autumn when the slugs climb tree trunks or perpendicular rocks for copulation. In Switzerland up to 900 m.

Endemic species from a small and restricted area in the S Alps. Vulnerable in Switzerland.

Falkner 1990: 188, Turner et al. 1998: 296.

Up to 150 mm long, exceptionally 230 mm

Italy: near Varese

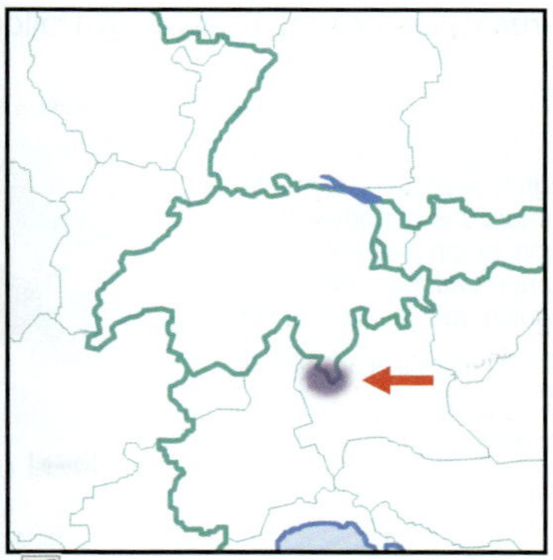

N Italy (Como, Varese), S Switzerland (Ticino, W of Lago di Como)

Carlo Morelli

Limax sarnensis Heim & Nitz, 2009

Limax sarnensis Heim & Nitz, 2009 in Nitz et al. 2009: 284

Slug colour variable, creamy white, brownish or black, often with colour spots or bands, mantle always evenly coloured without patterns (difference to several other species).
Penis 1/3-1/2 of body length, inside with weak transverse folds and one interior cord, vas deferens inserts close to tip, retractor attached at same point as vas deferens. Copulates on slime thread, on trunks 80-180 cm high.

Switzerland: Graubünden, Poschiavo

Up to 200 x 22 mm (preserved 165 x 17 mm)

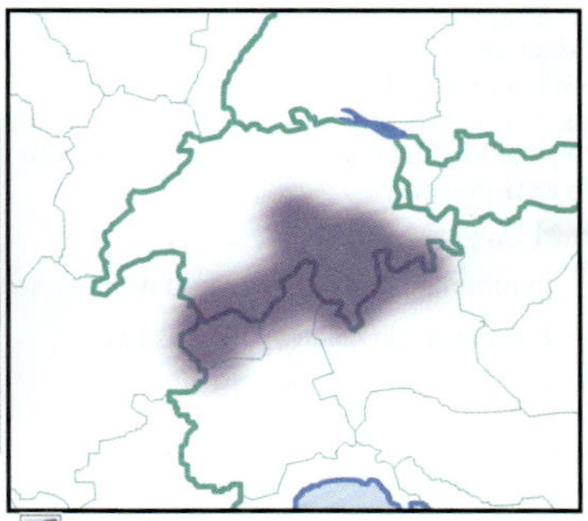

Central and S Switzerland, N Italy

Mountain habitats, usually at 1000-1500 m, also in forests. Locally in high population densities. Active at night, feeds on lichens growing on coniferous trees.

Copulation between May and September, takes place in 80-180 cm height above surface (for example at coniferous tree trunks), posterior ends of both animals are united and sticked on a slime thread (difference to *L. cinereoniger*), penes are everted and reach 50-75 % of body length, duration 2 hours. Life span at least 18 months, perhaps more (C. M. Brandstetter, unpublished data).

Some previous records for *Limax alpinus* Férussac, 1822 referred to *L. sarnensis*, others not. Moquin-Tandon 1855: 27 used *L. alpinus* for an accurately described species from near Grenoble, SE France. This was accepted by Falkner et al. 2002: 133.
The Swiss Alps were accepted as the type locality by Turner et al. 1998: 292 and Falkner et al. 2002: 133. Falkner et al. 2002 had dissected an animal from Ticino that suggested an independent species which they called *L. alpinus*. They did not present detailed results to substantiate their arguments. Turner et al. 1998: 293 classified slugs from Ticino and Graubünden with *alpinus* based on Férussac's information.
By designating a neotype from Ebensee near Linz (Austria) Nitz et al. 2009 converted *Limax alpinus* into a synonym of *L. cinereoniger* Wolf, 1803. Brandstetter 2011 demonstrated that the original locality of *Limax alpinus* should have been near Bern in 700 km distance, and considered the neotype designation as invalid. Nitz et al. 2009: 289 explained that in their subjective perception the type locality was nearby, Studer could have obtained slugs from other persons. Brandstetter 2011 argued that this would contradict Studer's information "dans nos alpes" and that Studer did not correspond with Austrian researchers.
The neotype designation is considered as correct here because the formal requirement in Art. 75.3.6 was met by Nitz et al. 2009. Their careless research has very probably led to incorrect conclusions, but this cannot have influence on the correctness of a type designation. The Code can ask authors to research carefully, but a successful and correct result cannot be required.
Brandstetter 2011 added some additional constraints in the morphological record that Férussac's 1822 name should have referred to the Swiss species, and argued that prior to neotype selection Nitz et al. 2009 neglected critical voices and did not consult other specialists as demanded by Recommendation 75B. Their neotype selection contrasted previous usage by Turner et al. 1998. Nitz et al. 2009 successfully converted *Limax alpinus* into a junior synonym of a different species, and in the same paper established a new name *Limax sarnensis* for the Swiss species.

Limax squamosus Bérenguier, 1900

Limax squamosus Bérenguier, 1900 in Bérenguier 1900: Pl. 3, Pl. 21

Slug yellow with interrupted dark brown lateral bands, keel prominent and yellowish, mantle yellowish with oval black spots, foot whitish with yellowish margin. Shell long and very narrow.
Differs from *L. maximus* in its yellowish borded foot abd narrow shell. Penis cylindrical, vas deferens very long, spermatheca oval.

Up to 150 mm long

In high mountains, where it replaces *Limax maximus*.

Germain 1930: 93.

More research is necessary to clarify its classification, the description of the genitalia is from the 1930s and certainly insufficient.

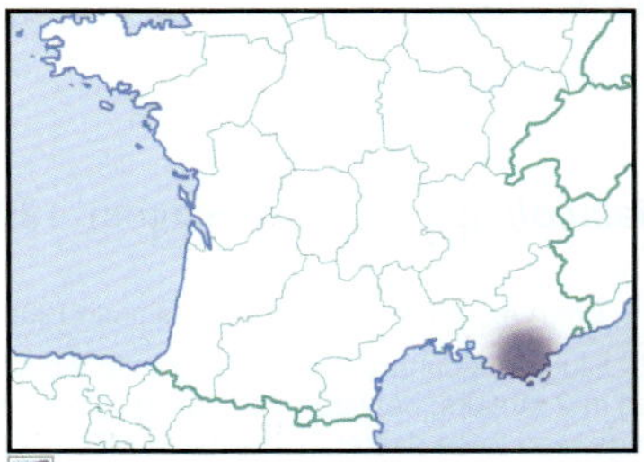

France (Var)

Limax subalpinus Lessona, 1880

Limax subalpinus Lessona, 1880 in Lessona 1880: 332

Slug cream, greyish or greyish brown with usually light keel in the posterior 1/2 of dorsum (occasionally dark), lighter down to the sides, with some lighter or darker dashes or spots, posterior end very steep at both sides, mantle covers 35 % of body length, entirely spotted whitish and black, pointed in posterior part, head and tentacles grey, foot margin and lateral zones of sole bluish black, medial zone of sole cream.

In Switzerland mostly deep inside forest soils, or between rocks, appears at the surface only after long and heavy rainfalls. In Switzerland in up to 1500 m, in Bulgaria 1600 m.

Rare in Bulgaria.

Manganelli et al. 1995: 21, Turner et al. 1998: 294, Irikov & Mollov 2006: 806, Hubenov 2007.

Up to 200 mm long

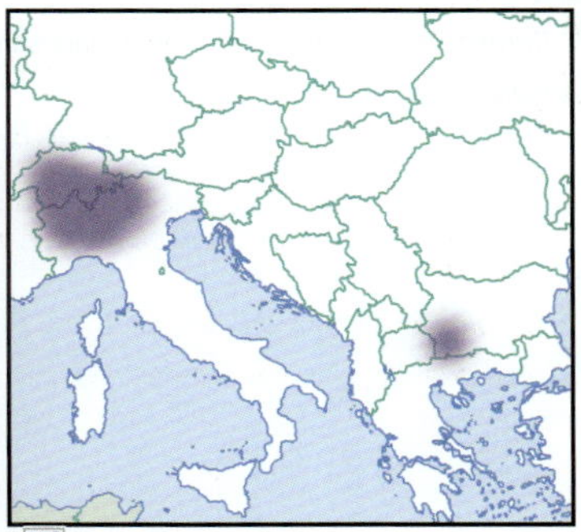

W Alps and S Bulgaria

Italy: Piemonte

Clemens M. Brandstetter

Limax veronensis Lessona & Pollonera, 1882

Limax subalpinus var. *veronensis* Lessona & Pollonera, 1882 in Lessona & Pollonera 1882: 36

Slug olive greyish or with slightly reddish hue, mantle and dorsal zone of body with brown spots or longitudinal dashes, foot margin and marginal zone of sole reddish, medial zone of sole whitish.

Manganelli et al. 1995: 21.

Original description very short and without size. Size probably more or less like *L. maximus*.

Up to 200 mm long

Italy: province of Trento   Clemens M. Brandstetter

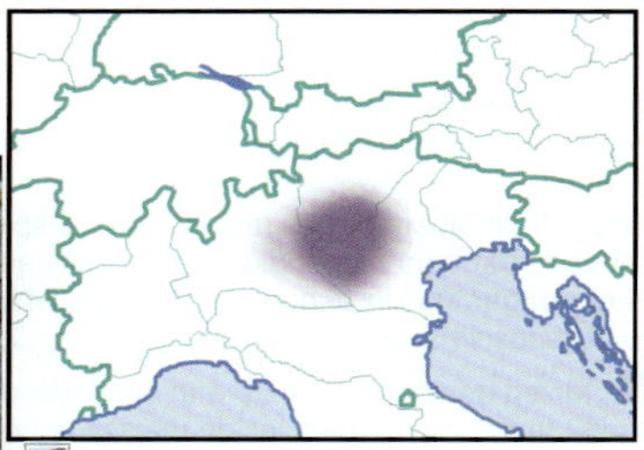

N Italy

Limax wohlberedti Simroth, 1900

Limax wohlberedti Simroth, 1900 in Simroth 1900: 98

Slug evenly deep black, keel sharp and covering 2/3 of dorsum length, up to 20 wrinkles between mid line of dorsum and pneumostome, sole at margin dark grey, mid section of sole slightly lighter.
Genitalia like *L. maximus*, but penis short, 1/6-1/4 of body length (difference to the Greek species *L. cephalonicus*, *L. graecus* and *L. conemenosi*), proximal section wider than distal section, vas deferens inserts immediately next to insertion of retractor, penis with very prominent longitudinal ridge inside, producing many folds in the proximal part.

Up to 180 mm long

In mountain habitats covered with dense vegetation.

Endangered in Albania (Dhora 2004).

Dhora & Welter-Schultes 1996: 153.

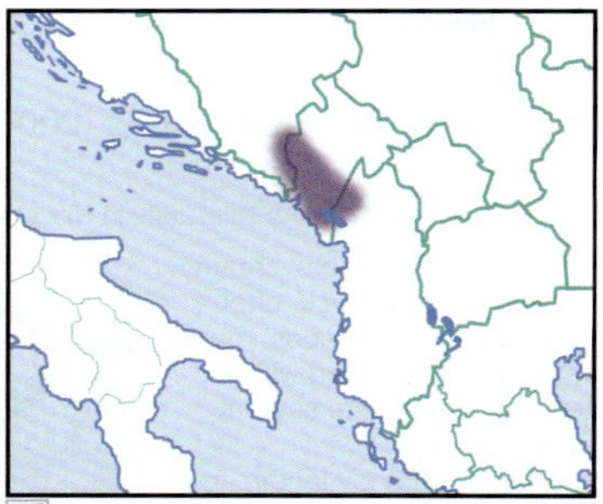

Herzegovina to N Albania

Limax zilchi Grossu & Lupu, 1960

Limax zilchi Grossu & Lupu, 1960 in Grossu & Lupu 1960: 161

Slug yellowish white with grey hue, with two longitudinal rows of elongated blackish brown spots, occasionally only at posterior end, keel not very prominent, mantle large and ellipsoid, concentrically striated, sole unicoloured with 3 distict longitudinal bands.

Up to 150 mm long

Restricted to forest habitats, at the soil, hides under bark and rotting tree trunks and stones, more rarely under leaves. The slug is agile, active at moist weather conditions, not very shy, continues crawling also after having been touched.

Grossu 1983: 253.

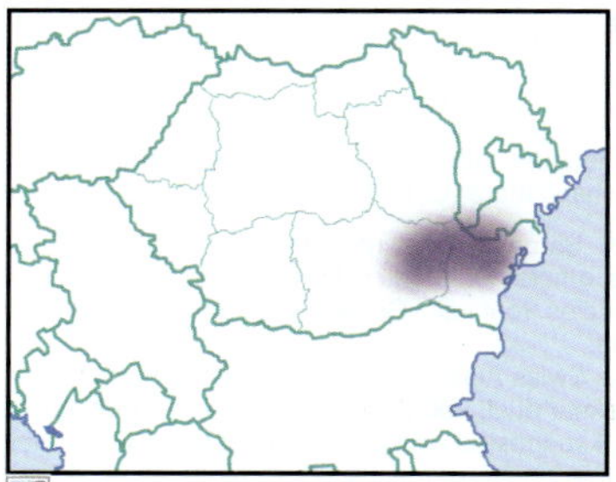

Romania (N Dobrogea)

Malacolimax

Malacolimax Malm, 1868 in Malm 1868: 66

Type species *Limax tenellus* Müller, 1774 (monotypy)

Europe, Canary Islands

Malacolimax mrazeki (Simroth, 1904)

Limax mrazeki Simroth, 1904 in Simroth 1904: 12, 25 (*Limax Mrazeki*, *Limax Mrázeki*)

Slug dark canary-yellow, on mantle 2 blackish streaks, mid-mantle darker, dark pigment also in posterior part in 2 streaks, between the streaks is a light keel, mantle length up to 14 mm, sole whitish, tentacles black. Slugs tend to be darker in the south. Mucus yellow-orange. Penis length exceeds half of the mantle legth and 2.5 x longer than pharynx (difference to *M. tenellus*, where it is shorter that half of the mantle and insificantly longer than the pharynx).

Up to 38 x 8.5 mm (preserved)

Forests, more rarely at sheltered places in open environments.

Wiktor 1996: 58.

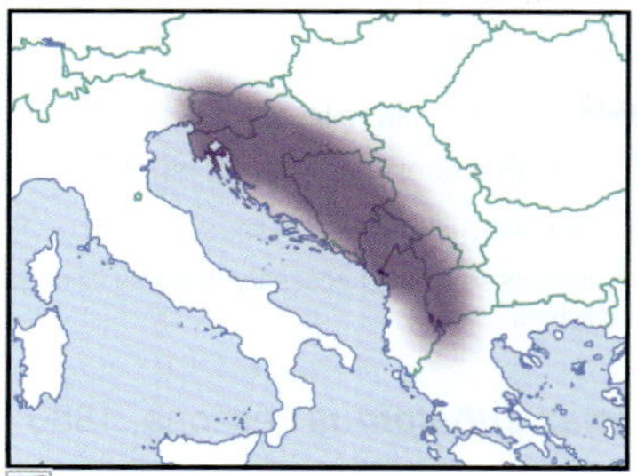

Slovenia to Rep. Makedonija

Malacolimax tenellus (Müller, 1774)

Limax tenellus Müller, 1774 in Müller 1774: 11

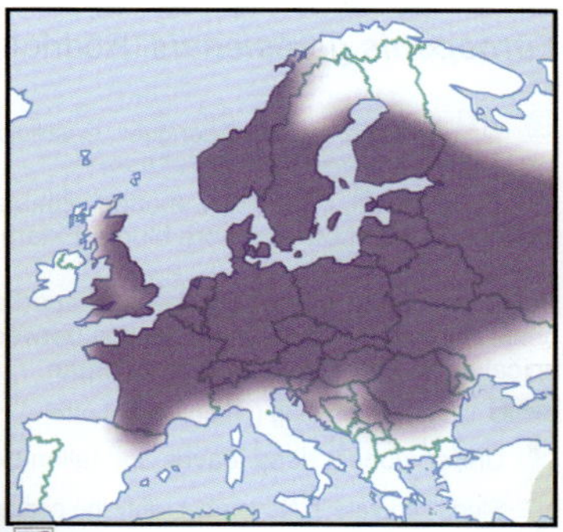

Europe

In Great Britain widespread but local, absent from large regions, and undoubtedly declining, threatened by clear felling and replanting of forests. Vulnerable in some parts of Germany, rare in Bulgaria.

Slug pale to brownish lemon yellow, with very faint darker lateral bands, translucent, head darker, tentacles brown, tubercles small and weak, keel blunt, short and restricted to the posterior end.

Up to 50 mm long

(1) Austria: Tirol, Landeck, (2-4) Slovakia: Veľká Fatra mountains, (shells) Germany: Bayern, near Beilngries (1) Vollrath Wiese, (2-4) Jozef Grego

In old woodlands, both deciduous and coniferous. In the S Pyrenees under stones and tree trunks in very shady habitats. In Britain often in remnant woodland on steep slopes, usually on non-calcareous soils. Tolerates traditional forest management, but not clearing and replanting, unknown from modern plantations. In Switzerland up to 2100 m, usually below 1700 m. Feeds on mushrooms, lichens and algae, at humid weather also taken from smooth tree trunks. Consumes many mushrooms in late summer and autumn.

Reproduction in October and November in France and Germany, 15-35 eggs (diameter 2-4 mm) are laid in batches (every few weeks, 4-6 times) in sheltered and moderately warm places, juveniles hatch after 20-120 days, they are colourless or light yellowish, subterranean and feed mainly on mushroom mycel, animals have full yellow colour after 6 months, maturity is reached after 7-9 months, then they accumulate some dark pigment, life span 12-13 months. Copulation is usual, but self-fertilisation is possible.

Künkel 1908, Germain 1930: 99, Frömming 1954: 199, Kerney & Cameron 1979: 136, Falkner 1990: 190, Irikov & Erőss 2008.

Agriolimacidae: *Furcopenis*, *Krynickillus*, *Megalopelte*, *Mesolimax*, *Deroceras*

Furcopenis

Furcopenis Castillejo & Wiktor, 1983 in Castillejo & Wiktor 1983: 1

Type species *Furcopenis darioi* Castillejo & Wiktor, 1983 (original designation)

Slug up to 30 mm, mantle covers 1/3 of body, skin very thin, creamy to brown, often with irregular spots, externally like some *Deroceras* species.
Penis small, narrow, anteriorly connected with 1-2 accessory organs, each with granular glands on apical ends, no stimulator inside penis, retractor penis bifurcate attaching to penis and accessory organ. Rectum without caecum.

Wiktor 2000: 554, Pesqueira et al. 2003.

Furcopenis circularis Castillejo & Mascato, 1987

Furcopenis circularis Castillejo & Mascato, 1987 in Castillejo & Mascato 1987: 33

Slug pale to dark brown, body covered with fine black dots (no pattern). Mucus colourless and sticky.
Penial gland in the form of 1-3 unforked branches, accessory body besides penis larger than penis, joined with penis in the anterior penis section, on the posterior part of the accessory organ there is a large cap-shaped granular gland opening directly into the penis, without any tubular ducts as present in *F. gallaeciensis*.

Up to 29 mm long

Forest clearings and open habitats with herbal vegetation, also found at road margins, not restricted to calcareous substrate.

Vulnerable.

Wiktor 2000: 555, Castillejo & Iglesias in Verdú & Galante 2006: 343.

NW Spain, N Portugal

Furcopenis darioi Castillejo & Wiktor, 1983

Furcopenis darioi Castillejo & Wiktor, 1983 in Castillejo & Wiktor 1983: 2

Slug evenly chestnut brown, skin very thin, mucus colourless.
Penis small, with two accessory bodies joining penis in anterior penis section, roughly equalling penis in size, both with cap-shaped granulated glands at their apical ends, retractor penis three-forked, attaching to penis and both accessory bodies.

Up to 30 mm long (preserved 20 mm)

In Galicia under leaf litter layers of *Castaneus* trees, in León under stones.

Vulnerable.

Wiktor 2000: 555, Castillejo & Iglesias in Verdú & Galante 2006: 344.

NW Spain, N Portugal

Furcopenis gallaeciensis Castillejo & Wiktor, 1983

Furcopenis gallaeciensis Castillejo & Wiktor, 1983 in Castillejo & Wiktor 1983: 8

Slug dirty cream, slightly darker on back, sometimes with some dark dots, mucus colourless.
Penial gland in the shape of 1-3 unforked processes, penis with one accessory organ, at least as long as penis, with a large granular gland at its end, penis and accessory organ connected by numerous coiled ducts (difference to *F. circularis*), retractor penis forked, attaching to apical penis end and to the membrane enveloping the gland of the accessory body.

Up to 35 mm long (preserved 25 mm)

Slug lives in gardens and open habitats, sometimes as a synanthrope, also forests.

Vulnerable.

Wiktor 2000: 555, Castillejo & Iglesias in Verdú & Galante 2006: 345.

NW Spain

Furcopenis geresiensis (Rodríguez, Castillejo & Outeiro, 1989)

Deroceras geresiensis Rodríguez, Castillejo & Outeiro, 1989 in Rodríguez et al. 1989: 95
Should have been *geresiense*, *Deroceras* is neuter.

Body cream, irrregularly spotted with blackish dots, on mantle yielding irregular reticulate patterns, bedind the mantle the spots are more blurry. Externally indistinguishable from other *Furcopenis* and various *Deroceras* species.
Penis connected with a large heart-shaped accessory organ equipped with glands at the two posterior ends, penis is located in the central section between the two sides of the accessory organ, retractor penis attaching to apical penis end and to accessory organ.

Up to 30 mm long (preserved 24 mm)

Under stones, dead leaves and fallen tree trunks.

Slugs copulate in November, only the accessory organ gets everted.

Vulnerable.

Wiktor 2000: 557, Castillejo & Iglesias in Verdú & Galante 2006: 341.

NW Spain, N Portugal

Krynickillus

Krynickillus Kaleniczenko, 1851 in Kaleniczenko 1851: 220

Type species *Krynickillus melanocephalus* Kaleniczenko, 1851 (subsequent designation by Pilsbry 1922 (The Nautilus 35): 79)

Slugs up to 45 mm, blackish brown or black, slender, somwhat like Limacidae, posterior end strongly narrowed, mantle 1/3 of body, almost half of anterior mantle section can be bent aside, pneumostom surrounded by a porly marked shield, weak keel reaching half the section behind mantle.
Penis oval, no accessory organs, vas deference opens into posterior penis section, retractor penis thin and unfurcate, inserted beside vas deferens opening, spermatheca with long spermatheca duct together usually longer than penis.

Mountain forests, slugs occur in small numbers, hiding under stones and leaf litter.

Black Sea region from Romania to Caspian Sea and N Iran

Wiktor 2000: 569.

Krynickillus hoplites (Simroth, 1899)

Mesolimax hoplites Simroth, 1899 in Simroth 1899: 37

Juvenile slugs dark or black, adults greyish with brownish dorsal part and lighter lateral parts, slender with reticulate surface structure, posterior end keeled and lighter. Mantle not attched to body in its anterior part, only near the pneumostome, with a characteristic furrow near the pneumostome. Shell almost symmetrical, fragile, calcareous.
Anatomy like *K. urbanskii*, the only difference is a hard narrow plate inside the penis of adult slugs.

Up to more than 50 mm

In mountain forests.

Wiktor 1994, Wiktor 2000: 573, Schütt 2001: 272.

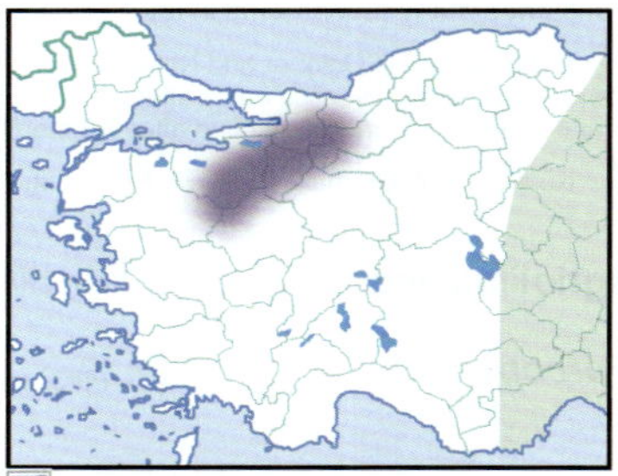

NW Turkey

Krynickillus melanocephalus Kaleniczenko, 1851

Krynickillus melanocephalus Kaleniczenko, 1851 in Kaleniczenko 1851: 221, Tab. 5

Slug dirty whitish to bluish grey, mantle and back darker, head and tentacles black (difference to *K. urbanskii*).
Spermatheca with spermatheca duct roughly as long as penis.

Forests in vicinity of water sources, under leaves and bark, lowlands to subalpine zone.

Wiktor 2000: 570, Schütt 2001: 268, Gural-Sverlova et al. 2009: 59, Sysoev & Schileyko 2009: 145.

Up to 45 mm long

Ukraine: Crimea, near Simferopol

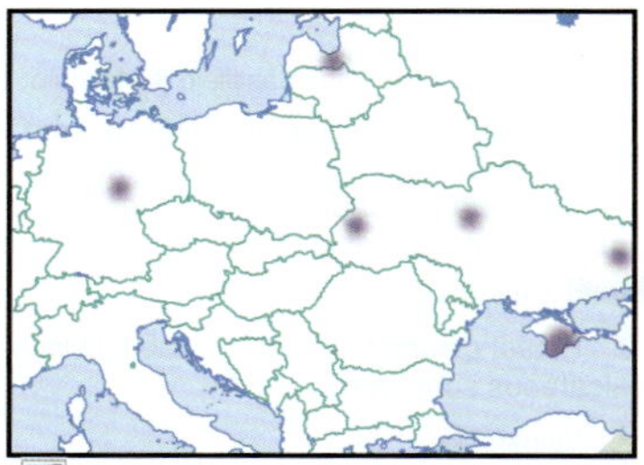

Ukraine to N Iran, introduced elsewhere

Sergey Leonov

Krynickillus urbanskii (Wiktor, 1971)

Deroceras urbanskii Wiktor, 1971 in Wiktor 1971: 264

Up to 45 mm long

Bulgaria: Sredna Gora, NW Stara Zagora

Dilian Georgiev

Slug grey-black, sides getting lighter downwards, head and tentacles darker (but not black), sole creamy, keel reaches 1/3 of section behind mantle. Mucus colourless and watery.
Spermatheca duct longer than in other species, other ducts also very long, copulatory organs in natural position heavily coiled, small spermatheca with long spermatheca duct together twice longer than penis. Inside penis only soft structures (the only difference to *K: hoplites*).

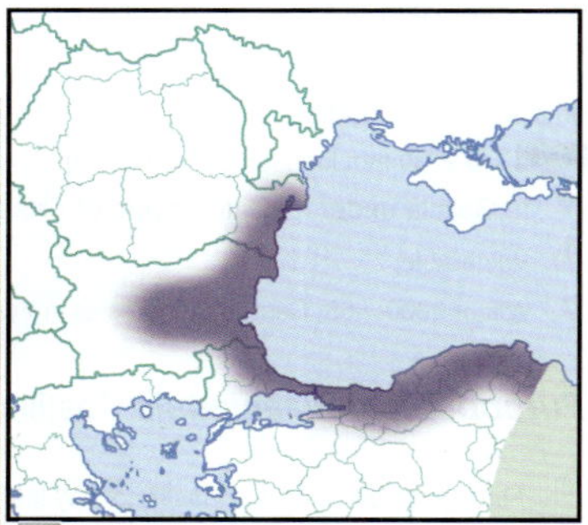

SE Europe, N Turkey

Forest habitats, mainly in beech (*Fagus*) and hornbeam (*Carpinus*), hides in leaf litter or under stones. Often found on mushrooms. In Bulgaria up to 1300 m.

Adults in Bulgaria occur from August to October.

Grossu 1983: 312, Wiktor 2000: 570, Schütt 2001: 269, Hubenov 2007.

Megalopelte

Megalopelte Lindholm, 1914
in Lindholm 1914: 167

Type species *Megalopelte simrothi* Lindholm, 1914
(original designation)

Slugs up to 25 mm long, 2/3 covered by mantle, mantle margins can easily be bent aside from all sides. Penis with broader anterior and irregular and cylindrically elongated posterior section. Retractor penis widened, no other appendages present. Intestine with two loops, no caecum.

Humid deciduous forests.

N Turkey to Georgia and Armenia

Wiktor 2000: 564.

Megalopelte simrothi Lindholm, 1914

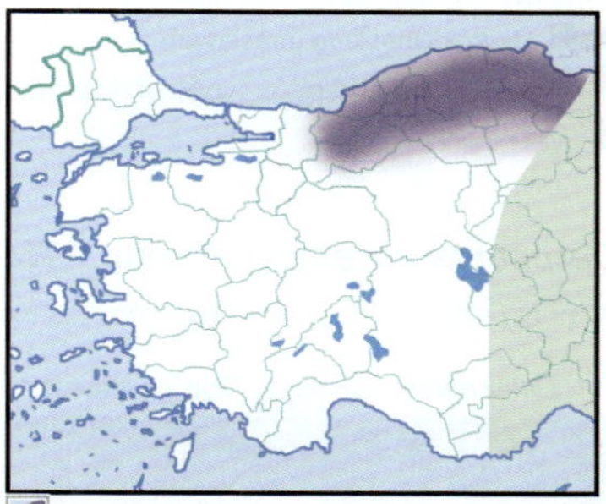

Georgia, N Turkey

Megalopelte simrothi Lindholm, 1914 in Lindholm 1914: 168

Slug brown or dark brown, mantle very large, covers the contracted animal entirely (2/3 of crawling slug), connected to the main body only in its centre. Head, cervix and end of the tail darker. Slug very characteristic and easy to recognize.

20-25 mm long

Humid habitats in deciduous forests under wood and between stones. Lives also in humid synanthropic habitats.

Wiktor 1994: 34, Wiktor 2000: 566, Schütt 2001: 273.

Mesolimax

Mesolimax Pollonera, 1888
in Pollonera 1888: 8

Type species *Mesolimax brauni* Pollonera, 1888
(monotypy)

Medium-sized slugs, up to 53 mm. Mantle less than 1/3 of body length, keel short. Pneumostom slightly postmedial, usually in middle part of mantle. The opening is surrounded by a ring distinguishable form the rest of the mantle. Sole divided in 3 zones, on lateral zones 2 or 3 longitudinal grooves. The central zone has wrinkles forming a V shape. Colour of the slugs brown-grey or brown-reddish to black, light slugs may have dark pigment patterns. Penis most often crossed by the retractor of the right ommatophore. The part of the penis with the vas deferens running parallel to it is surrounded by a thin membrane, it surrounds them both like a pipe (characteristic feature of the genus). Shell symmetrical or nearly, nucleus on longitudinal axis.

Ecology almost unknown. Often abundant, but most slugs seem to be juveniles. Or the adults know better how to escape from the collectors.

SW Turkey and E Aegean islands

Wiktor 2000: 575.

Mesolimax brauni Pollonera, 1888

SW Turkey, E Greece

Mesolimax brauni Pollonera, 1888 in Pollonera 1888: 8, Tav. 2 (*Mesolimax Brauni*)

Body slender, brownish, with darker patterns, keel lighter, lateral zones of the sole subdivded into 3 zones by 2 longitudinal grooves, mantle length up to 8 mm. Some slugs have black back and mantle.
Shell solid, rounded rectangular, with growth lines and asymmetrical nucleus at the lower margin on the left side.
Penis about 1/2 of body length, twisted (in *M. escherichi* the penis is shorter, only 1/3, *M. escherichi* is always black and has 3 elongate grooves on the lateral zones).

Up to 55 x 16-18 mm

Sometimes seems like a synanthrope. Often abundant, but most crawling slugs are juvenile. Adults may live more hidden, they are rare throughout all seasons.

Wiktor 2000: 575, Wiktor 2001: 181, Schütt 2001: 274.

Mesolimax escherichi Simroth, 1899

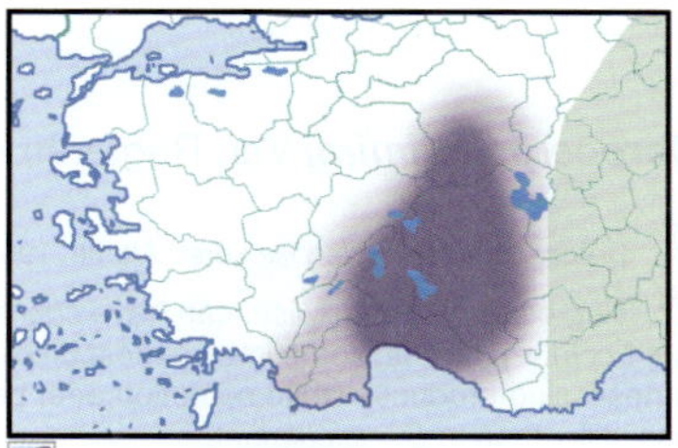

S and central Turkey

Mesolimax escherichi Simroth, 1899 in Simroth 1899: 36 (*Mesolimax Escherichi*)

Quite like *M. brauni*, but with some differences. *M. escherichi* is always black, on lateral part of sole there are 3 grooves, penis is only 1/3 of body length.

Up to 56 x 14 mm, usually 40 x 8 mm (preserved)

Humid lowlands, margins of rivers, creeks and lakes. Most collected slugs are juvenile.

Wiktor 2000: 575, Wiktor 2001: 183, Schütt 2001: 275.

Deroceras

Deroceras Rafinesque, 1820
in Rafinesque 1820: 10 *Limax (Deroceras)*

Type species *Limax gracilis* Rafinesque, 1820
(monotypy)
(= currently *Deroceras reticulatum* (Müller, 1774))

Original description: "It [*Limax gracilis*] may perhaps form a sub-genus *Deroceras*." The subgenus was proposed conditionally (available under Art. 15.1), the species was included without doubts. Available with type species *Limax gracilis*. Gender is neuter (Art. 30.1.2, ends in a Greek word *-ceras*). Rafinesque did not explain the derivation of the name, but even if Dero- is not derived from a Greek word then the compound name still ends in a Greek word. I would favour an interpretation by which such an ending should not be regarded as an arbitrary combination of letters.

Body size up to 45 mm (after preservation slugs become smaller, about 25 mm then), robust, mantle large (sometimes 1/2 of body length). Keel very short. Up to 14 wrinkles between mid line of dorsum and pneumostom. Colour creamy, white, brown, sometimes black. On mantle and dorsum of light forms there are often dark spots. Mucus usually colourless.
Penis always short, variable or characteristical in shape, does not cross right spermatophore retractor, may have appendices and usually a penial gland of variable or characteristical shape. Retractor penis either unforked or in two branches, attached to penis. Inside penis there is a stimulator of different shape, which is often characteristical. Spermatheca duct united with penis. Rectum often with a caecum of characteristical size.

Most species live in open habitats.

Life cycle short, often several generations are present at the same time. Life cycles in species living in colder regions 6-14 months, in the south down to 2 months.

Originally Holarctic N of Sahara, Arabia, India, S China and Central America, except N margins of Siberia and N America, isolated in Ethiopia.

Wiktor 2000 (systematic monograph), Wiktor 2001: 98.

More than 100 species are known. Several early described taxa have traditionally been classified as species, but Wiktor 2000 could not identify them, they are not recognized as species here.

4 subgenera were recognized in the past: *Deroceras*, *Agriolimax*, *Plathystimulus* and *Lytopelte*, but the criteria for such a subdivision were difficult to retain (Wiktor 2000). *Lytopelte* from Asia is at present recognized as an independent genus.

Deroceras adolphi Wiktor, 1998

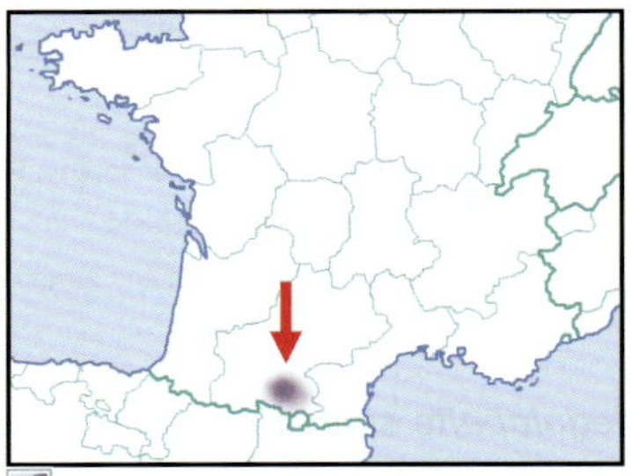

SW France (Ariège)

Deroceras adolphi Wiktor, 1998 in Wiktor 1998: 13

Slug cream beige or whitish, dark spotted, especially on mantle, spots often form a reticulate pattern behind the mantle, irregular blot present above pneumostom, skin thick.
Penis irregulaly shaped, penial gland long, bent, gradually narrowing with smooth surface and no papillae, like a penial extension, retractor penis forked (but only in adults), inserting at penis and at the base of penial gland, stimulator cone-shaped with rounded top. Rectal caecum consisting only of a few folds.

Up to 33 mm long (preserved)

Beech wood and other types of deciduous forest, in rock rubble, under stones, near streams.

Wiktor 2000: 395.

Deroceras agreste (Linnæus, 1758)

Without gender agreement: *Deroceras agrestis* (Linnæus, 1758)

Limax agrestis Linnæus, 1758 in Linnæus 1758: 652

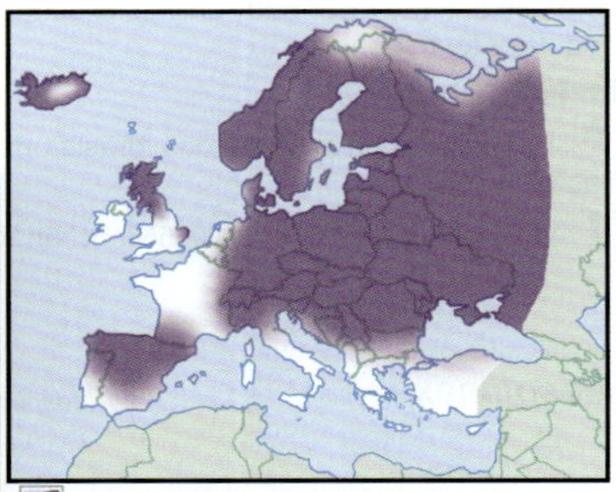

Europe to Siberia

Rare in Central Europe, central and S England. Vulnerable in Bayern and Austria.

Slug yellowish white to light greyish yellow, transparent, no spots. Mantle length 1/3 of body. Mucus colourless, on irritation a milky white secretion appears. Gonad never reaching the posterior end of viscera. Penis oval in juveniles, with lateral constriction in adults, penial gland finger-shaped, never branched, smooth without glandular papillae. Stimulator cone-shaped with narrow base. Caecum on rectum well developed, clearly longer than wide.

Up to 40 mm long

(1) Germany: Schleswig-Holstein, near Grömitz, (2) Germany: Bayern, near Bad Brückenau, (3) Finland: Helsinki, (shells) Germany: Berlin (1) Vollrath Wiese

Humid meadows and river plain woods, in the north also in cultivated areas, in central Europe only, in Britain mostly in natural habitats. Moderately dry alpine meadows and stony alpine pastures are exceptionally also inhabitated. In Scotland in tall herbage with umbellifers at roadsides, or on marshy ground, upland pastures and under stones on hillsides. In Switzerland up to 2500 m, usually below 2000 m, in Bulgaria reported from up to 2600 m.

Reproduction usually after self-fertilization, 10-30 eggs (1.1-2 mm) are laid in clusters in autumn, 3-4 mm long juveniles appear in spring, initially transparent, weakly brown after 3 days, maturity is reached in autumn, life span 1 year.

Frömming 1954: 220, Falkner 1990: 192, Wiktor 1996: 86, Turner et al. 1998: 307, Kerney 1999: 162, Wiktor 2000: 397, Pesqueira et al. 2003, Irikov & Mollov 2006: 810, Hubenov 2007.

Distribution not well-known due to frequent misidentifications. Records for *Agriolimax agrestis* from until the 1940s referred to *D. reticulatum*.

Deroceras altimirai Van Regteren Altena, 1969

Deroceras altimirai Van Regteren Altena, 1969 in Van Regteren Altena 1969: 104

Slug creamy or yellowish, can have characteristic pigment colour patterns, sometimes almost black. Mucus colourless, slug produces white secretion after being irritated, less milky than in *D. reticulatum*.
Penis very similar to *D. levisarcobelum*, posterior end with two pockets, one being the penial gland, stimulator large, like a laterally flattened cone with smooth fingerprint-like ornaments on its surface, rectal caecum well developed.

Mixed forests above 1000 m.

De Winter 1986, Wiktor 2000: 402.

D. levisarcobelum is regarded as a distinct species here, externally similar but without rectal caecum and with a smooth stimulator.

Spain: Barcelona, Montjuïc
Roy Anderson

Up to 45 mm long (preserved 25-38 mm)

NE Spain, Pyrenees

Deroceras astypalaeense Wiktor & Mylonas, 1992

Without gender agreement: *Deroceras astypalaeensis* Wiktor & Mylonas, 1992

Deroceras astypalaeensis Wiktor & Mylonas, 1992 in Wiktor & Mylonas 1992: 1
Should have been *Deroceras astypalaeense*, *Deroceras* is neuter.

Slug dirty cream with a blackish pattern on the head and the neck, skin thin and limp, mantle length up to 7.5 mm.
Penis distended anteriorly, appendix cone-shaped and bent, penial gland consists of a common trunk with 4-7 unbranched proceses, stimulator flat and very large, filling out anterior penis section and also a lateral pocket. Rectal caecum a shallow pocket.

Up to 22 x 5 mm long (preserved)

Wiktor 2000: 402, Wiktor 2001: 146.

Greece (Astipálea island)

Deroceras bakurianum (Simroth, 1912)

Without gender agreement: *Deroceras bakurianus* (Simroth, 1912)

Agriolimax bakurianus Simroth, 1912 in Simroth 1912: 40, Taf. 3, Taf. 7

Slug with creamy backgound colour, densely covered by dark blurry spots so that slugs may look blackish or black except on middle of dorsum and keel, in juveniles the dark spots may form a reticulate pattern. Sole light, in dark individuals with grey lateral zones, often spotted.
Penis oval, penial gland like a thick noodle, retractor penis single (unforked), stimulator flat, huge, asymmetrical at its loose end where a pointed structure is present, bent backwards. Differs from *D. subagreste* in its unforked retractor penis and the shape of the stimulator. *D. subagreste* has usually much less spots on its body.

Up to 23 mm long (preserved)

In humid deciduous forests (oak, beech), also in mixed forests, less frequently in coniferous forests and river valleys.

Wiktor 2000: 405, Balashov & Gural-Sverlova 2012: 101 (possibly also in Crimea).

Transcaucasia from Sochi and Batum to W Dagestan, N Turkey

Deroceras berytense (Bourguignat, 1852)

Without gender agreement: *Deroceras berytensis* (Bourguignat, 1852)

Limax berytensis Bourguignat, 1852 in Bourguignat 1852: 10 (*Limax Berytensis*)
Agriolimax cyprius Simroth, 1906 in Simroth 1906: 86

Slug colour very variable, black or blackish, getting lighter downwards at the sides, or dirty grey with dark colour pigments in skin grooves, mantle with black melanophores, sole uniformly light, in black slugs blackish at lateral zones but medial zone lighter. Mantle can be darker, pneumostom with light margin.
Penis irregularly shaped, penial gland consisting of a common trunk and two undivided branches. Rectal caecum extremely variable from long (Lebanon, Cyprus and Vilayet Konya) to short or almost absent (Cyprus and between Kayseri and Black Sea).

Turkey: Vilayet Erzincan, Kemaliye
Burçin A. Gümüş

Up to 40 mm (preserved 18-32 mm)

The most frequent *Deroceras* species of Cyprus.

Rähle 1984: 48, Wiktor 1994: 16, 2000: 408, Wiktor 2001: 149.

Agriolimax cyprius was considered as a separate species by Rähle 1984. Wiktor 2000 suggested that it was conspecific with *D. berytense*.

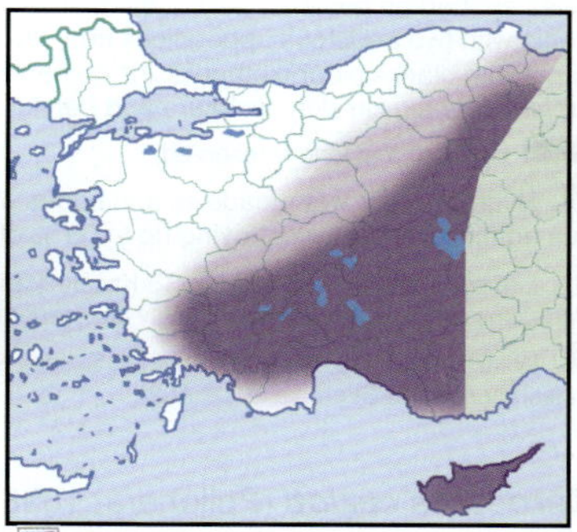

Turkey, Cyprus, Lebanon

Deroceras bisacchianum Bodon, Boato & Giusti, 1982

Deroceras bisacchianum Bodon, Boato & Giusti, 1982 in Bodon, Boato & Giusti 1982: 54

Slug pale cream, with irregular dark-brown spots, not forming a pattern, sole creamy.
Penis elongated, penial gland consisting of a common trunk with 3-8 unbranched arms. No proper stimulator inside penis, a unique character in *Deroceras*. Rectal caecum short.

Up to 33 mm long (preserved)

Lowlands up to 1300 m, in open habitats, woodlands and olive crops.

Only known from a small range in the Ligurian Alps.

Manganelli et al. 1995: 22, Wiktor 2000: 414.

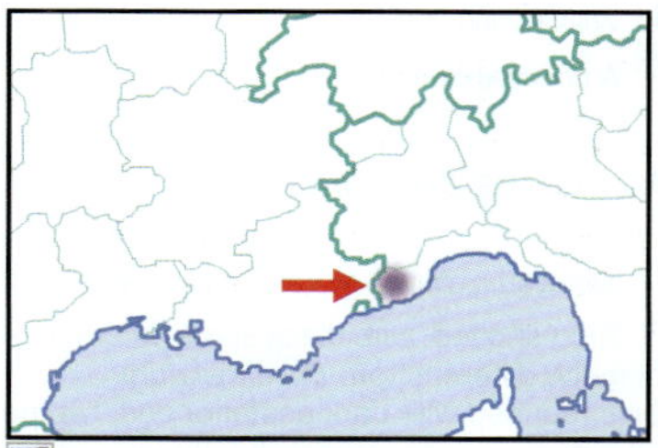

NW Italy (W Liguria)

Deroceras bistimulatum Wiktor, 2000

Deroceras bistimulatum Wiktor, 2000 in Wiktor 2000: 414

Preserved slugs dirty, greyish-cream or brown, slightly darker on the back and on the mantle, skin thin and soft, with small (probably calcareous) crystals which may produce streaks of white colour. Mantle length 35 % of body.
Penis irregularly shaped, with two asymmetrical appendices, penial gland consists of a cluster of 4-8 processes on a short common stem. Stimulator flat, consisting of two parts, looking like two stimulators.

Up to 19.5 x 4.5 mm (preserved)

Only known from one locality at 26 km E of Volos.

Wiktor 2001: 151.

Greece (Thessalía)

Deroceras boeoticum Wiktor, 1984

Deroceras boeoticum Wiktor, 1984 in Wiktor 1984: 152

Body evenly cream, usually without spots, sometimes dark pigments on mantle, dorsum and mantle somewhat darker, accumulations of dark pigments visible there under high magnification, mantle length 40 % of body.
Penis oval in anterior and elongated in posterior part, appendix transverse conical and smaller than in *D. seriphium*, penial gland consisting of 5 long processes with papillae not set on a common shaft, stimulator not flattened. Rectal caecum in the shape of a poorly marked pocket.

Up to 30 x 6 mm (preserved)

Found on rocks under tree shadows, and on meadows. Up to 900 m altitude.

Wiktor 2000: 419, Wiktor 2001: 100.

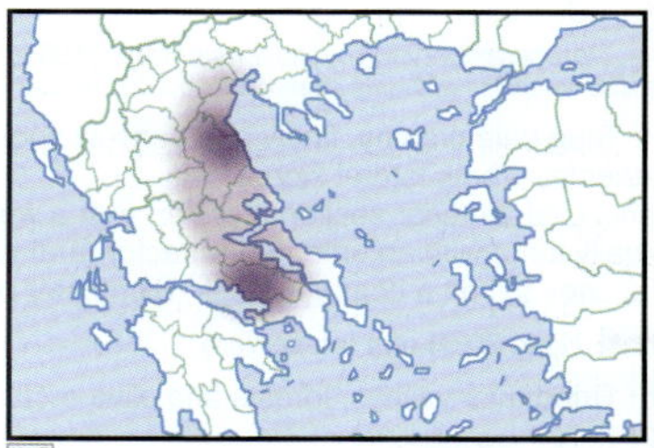

Greece (Viotía, Thessalía)

Deroceras bureschi (Wagner, 1934)

Agriolimax bureschi Wagner, 1934 in H. Wagner 1934: 55 (*Agriolimax (Hydrolimax) Bureschi*)

Slug evenly dark brown or black-spotted, spots may resemble ink-flashes.
Penis more or less oval, with a penial-gland like finger-shaped and smooth process. Stimulator inside penis inconspicuous, on its side an irregular hard calcified yellow-russet plate present, usually round or elongated, with visible growth lines or a distinct "nucleus", its end is entirely smooth and uncrenated, which is considered typical and diagnostically essential.

Up to 30 mm long (preserved)

Beech and hornbeam forests, often in large numbers, reported between lowlands and 2000 m in Bulgaria, hides in leaf litter, under stones and wood.

Adults appear in June, when copulation can be observed.

Wiktor 1983: 187, Wiktor 2000: 536, Irikov & Mollov 2006: 811, Hubenov 2007.

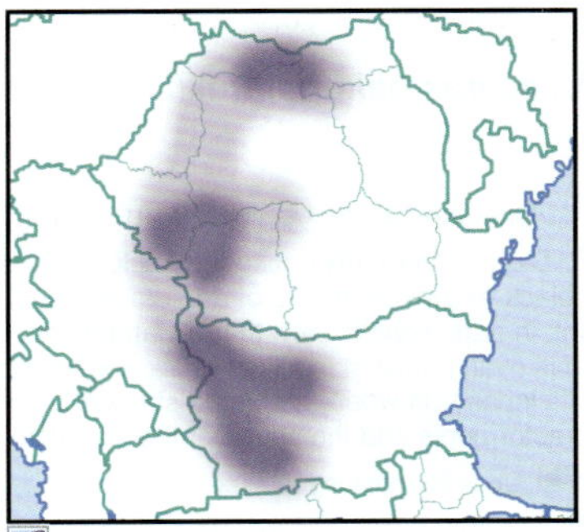

Bulgaria, Romania

Deroceras caucasicum (Simroth, 1901)

Without gender agreement: *Deroceras caucasica* (Simroth, 1901)

Lytopelte caucasica Simroth, 1901 in Simroth 1901: 171 (*Lytopelte (Liolytopelte) caucasica*)

Slug colour variable from whitish to dark brown, head, tentacles and nape blackish or black, juveniles sometimes spotted on posterior part of the mantle, sole pale, mucus colourless.
Penis with two nodulous appendices near its end, vas deference opens between them, posterior fragment of penis with black colour pigments. Inside penis a hard bivalve shell-like plate is present. Differs from *Krynickillus melanocephalus* in skin sculpture and genitalia.

Up to 40 mm long (preserved)

Hygrophilous, wet meadows and edges of water reservoirs (also in semi-deserts). Slug can become a serious agricultural pest, destroying up to 50 % of vegetable crops.

Life cycle varies regionally. In SE Kazakhstan 3 generations per year, half-grown individuals pass the winter. In Causasus and Tajikistan probably only one generation per year, half-grown slugs and eggs hibernate.

Wiktor 2000: 538, Schütt 2001: 270, Sysoev & Schileyko 2009: 142, Tappert 2009, Gural-Sverlova et al. 2009: 57, Balashov & Gural-Sverlova 2012: 102.

Ukraine and N Turkey to N Iran, introduced elsewhere, mainly central and eastern Asia

Deroceras cazioti (Pollonera, 1896)

Agriolimax cazioti Pollonera, 1896 in Pollonera 1896: 3 (*Agriolimax Cazioti*)

Slug blackish brown, getting paler downwards at the sides, sole cream.
Penis elongated, like an S-shaped tube, with hard and strongly muscled walls, partly with silky glittering, no glands in anterior penis part near stimulator.

Up to 39 mm long (preserved 20 mm)

Slugs were found under stones in gardens with trees, forests and meadows, in 400-950 m altitude.

Wiktor 2000: 424.

A poorly known species.

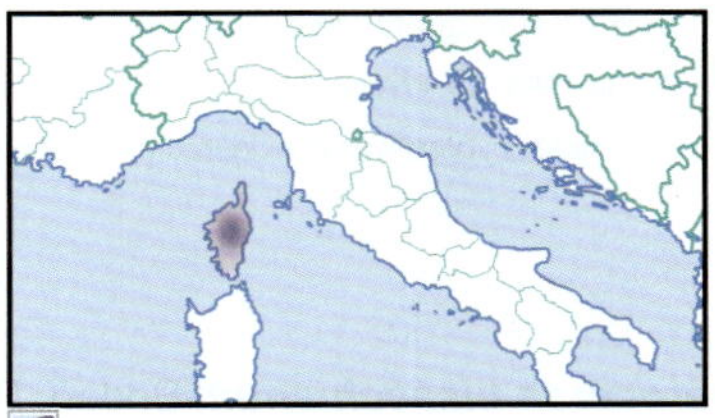

France (Corsica)

Deroceras chevallieri Van Regteren Altena, 1973

Deroceras chevallieri Van Regteren Altena, 1973 in Van Regteren Altena 1973: 25

Slug spotted, with black spots on creamy background, behind mantle forming a reticulate pattern, externally not much different from *D. reticulatum*.
Penis rounded with dark posterior end, two penial glands present consisting of common shafts and 5-10 short and branched processes, retractor unforked, penial stimulator narrow and flat with a blunt apex. Rectal caecum very long.

Up to 20-30 mm long (preserved)

Wiktor 2000: 425.

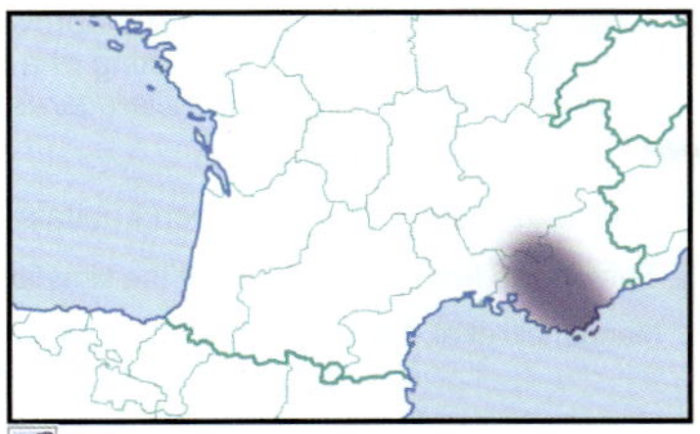

SE France

Deroceras christae Rähle, 1998

Deroceras christae Rähle, 1998 in Rähle 1998: 25

Slug light cream with grey mantle spots, behind mantle dark spots in grooves forming a reticulate pattern, sole pale, darker on lateral zones.
Penial gland as long as penis, in the form of a fern leaf, becoming gradually narrower towards its posterior end. Rectal caecum long.

Up to 40-50 mm (preserved)

Only known from two localities in Vilayet Antalya.

Wiktor 2000: 426, Schütt 2001: 267.

S Turkey

Deroceras chrysorroyatissense Rähle, 1984

Without gender agreement: *Deroceras chrysorroyatissensis* Rähle, 1984

Deroceras chrysorroyatissensis Rähle, 1984 in Rähle 1984: 45
Should have been *chrysorroyatissense*, *Deroceras* is neuter. Code Appendix B 5 was not considered (name is not easily memorable).

Slug pale creamy, finely and densely spotted, distal end shortened, keel fluke-like, like as if pressed together from both lateral sides.
Penis oval, anterior section broader, with a weak furrow between the sections, appendix and penial gland equally long, with an unforked retractor (and vas deferens insertion) in between, penial gland composed of two branches set on a common trunk. Stimulator conical and slightly flattened. Rectum without caecum.

Up to 20-35 mm (preserved)

Dried river valleys, forests and olive orchards, recorded up to 900 m.

Wiktor 2000: 428.

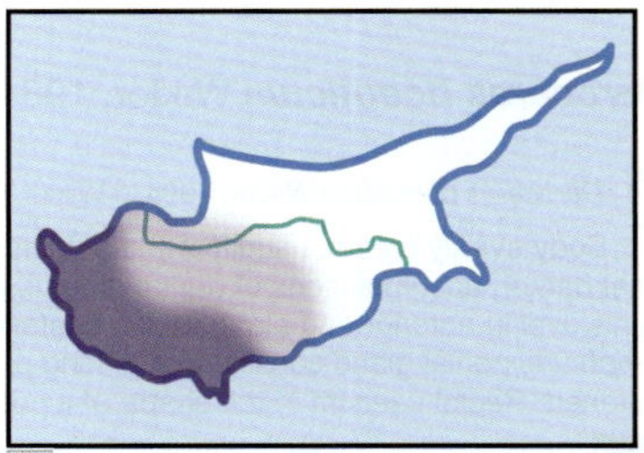

W and SW Cyprus

Could be conspecific with *Deroceras barceum* (Gambetta, 1925) from Cyrenaica, Libya (*Agriolimax barceus* Gambetta, 1925 in Gambetta 1925: 7).

Deroceras cycladicum Wiktor & Mylonas, 1981

Greece (Kiklades)

Deroceras cycladicum Wiktor & Mylonas, 1981 in Wiktor & Mylonas 1981: 190

Slugs creamy with indistinct blackish spots on mantle and dorsum, accumulated in grooves forming an indistinct reticulum. Sometimes dorsum uniformly dark and single dark spots on the rest of the body.
Penis irregularly shaped, with a small appendix and a penial gland usually consisting of two unbranches processes fused at their bases, stimulator strongly flattened. Rectum untypical, no caecum but instead rectum broadens abruptly in its terminal section, like a snake which has eaten a mouse.

Up to 22 mm long (preserved 14 mm)

Nothing published on habitat and ecology, but on the islands slugs live usually in various different habitats.

Wiktor 2000: 430, Wiktor 2001: 154.

Deroceras demirtense Rähle, 1998

Without gender agreement: *Deroceras demirtensis* Rähle, 1998

S Turkey

Deroceras demirtensis Rähle, 1998 in Rähle 1998: 22
Should have been *demirtense*, *Deroceras* is neuter.

Slug reddish brown with dark spots on head and mantle, behind mantle a grey reticulate pattern is formed by grey spots in skin grooves, sole light creamy.
Penis about 1/3 of body length, bent in an S-shape, penial gland composed of 5 branches set on a common trunk, musculus retractor forked, stimulator flat and broad. Caecum on rectum long.

Up to 45 mm long (preserved 33 mm)

Only known from one locality.

Wiktor 2000: 433, Schütt 2001: 267.

Deroceras dewinteri Maassen, 2000

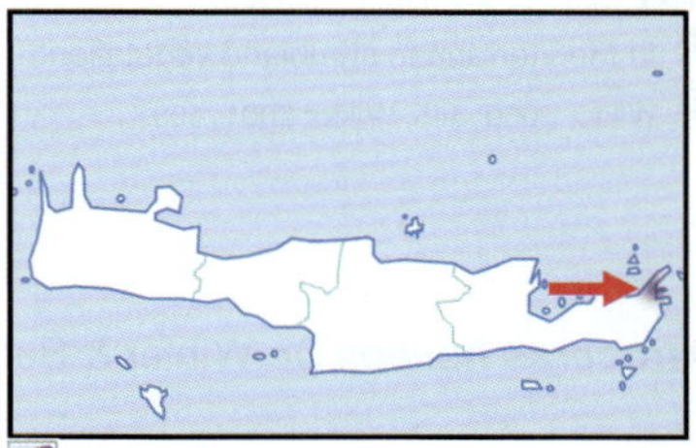

Greece (E Crete)

Deroceras dewinteri Maassen, 2000 in Maassen 2000: 125

Slug blackish without spots, skin thin, grooves hardly discernible, mantle almost 50 % of body length. Externally like *D. oertzeni*.
Penis blackish, oval with constriction in its central section, penial gland with two lobular protuberances and no papillae, posterior section of penis with lateral extension, retractor short and undivided, inserting laterally at penis, stimulator small with an acute tip, spermatheca large and black.

Up to 17 x 5 mm (preserved)

Was found under old wood and rubbish.

Copulation was observed during daytime in January.

Deroceras ercinae De Winter, 1985

N Spain

Deroceras ercinae De Winter, 1985 in De Winter 1985: 69

Slug usually evenly coloured, sometimes there are some dark spots on the mantle.
Penis divided in two parts, posterior part with a small lateral pocket, penial gland like a finger with two rows of big papillae, stimulator conical. No caecum on rectum.

Up to 30 mm long (preserved)

Limestone mountains, in short herbal vegetation, 1000-1200 m altitude.

Vulnerable.

Ondina, Ro. & H. 1994, Wiktor 2000: 435, Pesqueira et al. 2003, Castillejo & Iglesias in Verdú & Galante 2006: 340.

Deroceras famagustense Rähle, 1991

Without gender agreement: *Deroceras famagustensis* Rähle, 1991

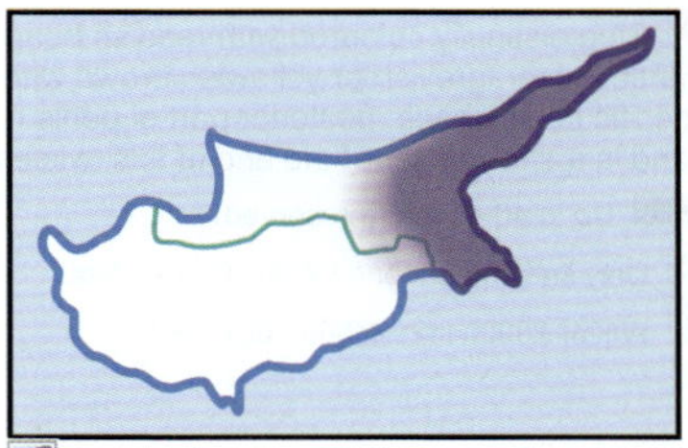

E Cyprus

Deroceras famagustensis Rähle, 1991 in Rähle 1991: 144
Should have been *famagustense*, *Deroceras* is neuter.

Slug light brownish grey, with dark spots in skin grooves of the posterior body section producing a reticulate pattern there.
Penis relatively small, elongated, retractor penis forked, its wider end accreted to the unbranched penial gland, vas deferens surrounds penis half-spirally (in most other *Deroceras* species it runs along penis), no stimulator inside penis, only a long fold along its wall. Caecum on rectum well developed in form of a pocket.

Up to 20 mm long (preserved)

Wiktor 2000: 436.

Deroceras gavdosense Wiktor, Vardinoyannis & Mylonas, 1994

Without gender agreement: *Deroceras gavdosensis* Wiktor, Vardinoyannis & Mylonas, 1994

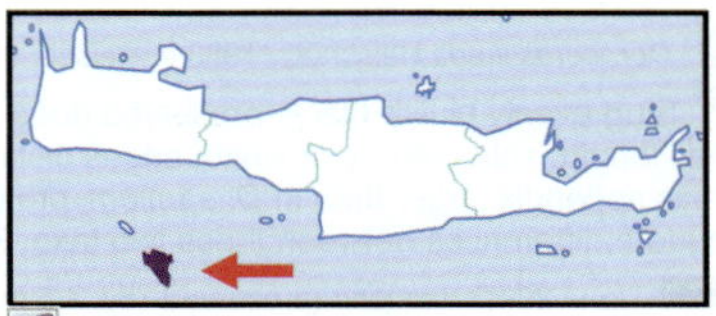

Greece (island of Gávdos)

Greece: island of Gávdos — Pascal Van Herck

Up to 30 x 4 mm (preserved 21 mm long)

Deroceras gavdosensis Wiktor, Vardinoyannis & Mylonas, 1994 in Wiktor et al. 1994: 10
Should have been *gavdosense*, *Deroceras* is neuter.

Slugs under *Pinus* needles evenly reddish brown to brown, in more open habitats front and tentacles light grey to reddish brown, mantle (8.5 mm) lighter and yellowish, posterior part dark brown to grey, sole light grey. In the north of Gávdos (near Ambelos) slugs are darker, almost black.
Penis constricted in its mid section, no appendices, penial gland composed of 3-4 branches set on a common stalk, which joins the penis rather apically, retractor penis and vas deferens opening next to it. Stimulator inside penis conical in shape.

In open pine forests and shrublands, also at margins of cultivated crops. Found mostly under *Pinus brutia* needle layers, but not on sandy substrate.

Life cycle extremely short, apparently only 2 months. Slugs in terraria hatched 30 days after copulation of the parents, laid eggs another 30 days later and lived at least 2 months. The short life cycle is possibly an adaptation to the extremely hot climate.

Welter-Schultes 1998: 62, Wiktor 2000: 438, Wiktor 2001: 102.

Deroceras golcheri Van Regteren Altena, 1962

Deroceras golcheri Van Regteren Altena, 1962 in Van Regteren Altena 1962: 49

Slugs grey to pinkish violet (preserved creamy brownish) with dense brown-black colour spots, on mantle irregularly scattered, elsewhere accumulated in aggregates, also more concentrated in skin grooves, sole evenly creamy.
External appearence does not differ from sympatric *D. panormitanum*.
Penis without appendix in its proximal section (*D. panormitanum* has at least one long appendix = penial caecum), no significant other differences with *D. panormitanum*.

Up to 40 mm (preserved 20-30 mm)

On walls and under stones and wood, in natural and disturbed habitats with trees or shrubs.

Giusti et al. 1995: 288, Manganelli et al. 1995: 22, Wiktor 2000: 443.

Reise et al. 2011 analysed the mating behaviour and found that matings with sympatric *D. panormitanum* were never successful.

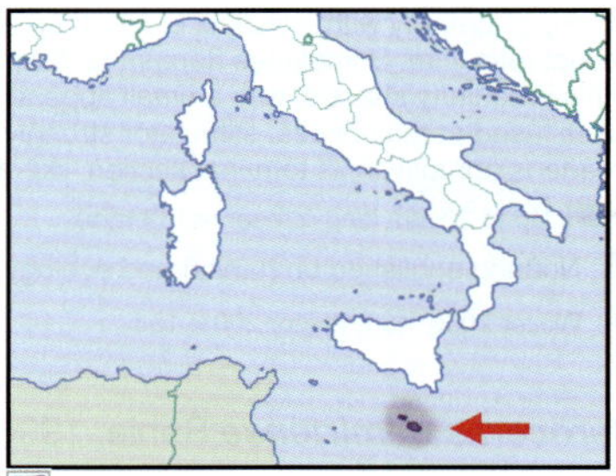

Malta

Deroceras gorgonium Wiktor, Vardinoyannis & Mylonas, 1994

Deroceras gorgonium Wiktor, Vardinoyannis & Mylonas, 1994 in Wiktor et al. 1994: 12

Slug dirty cream with blackish spots on mantle, sides and back, except on the medial part of the back, mantle length up to 50 % of body.
Penial gland at apical end of penis very specific, like a cluster of numerous thin finger-like processes on a short common shaft or directly on the penis. Somehow like *Nautilus* arms.

Up to 38 x 7 mm (preserved)

In Mediterranean shrubland vegetation.

Wiktor 2000: 445, Wiktor 2001: 103.

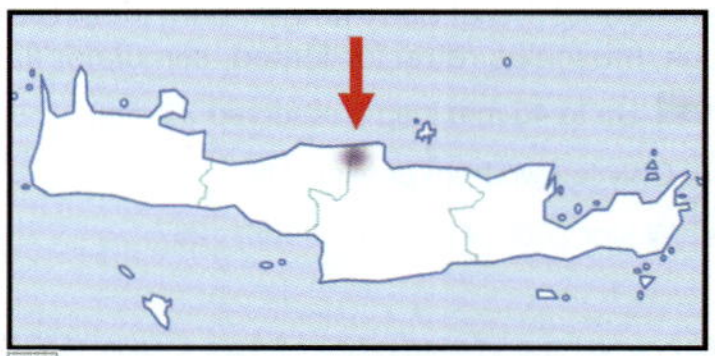

Greece (N central Crete)

Only known from a small restricted range near Síses.

Deroceras halieos De Winter & Butot, 1986

Deroceras halieos De Winter & Butot, 1986 in De Winter & Butot 1986: 35

Slug evenly dark grey, almost black, mantle length 10 mm. Sole laterally greyish, in the middle section dirty cream.
Penis of irregular shape, with a large lateral distension comprising stimulator and 2 curved appendices, on one of the appendices the penial gland is located, consisting of some 4-7 long processes without glandular papillae, stimulator elongatedly tongue-shaped or nearly semi-circular. Caecum on rectum long.

Up to 42 mm long (preserved)

Was found in a *Fagus* forest 200 m above sea level.

Only known from one locality at Athos, Chalkidiki.

Wiktor 2000: 447, Wiktor 2001: 156.

Greece (Chalkidiki: Athos)

Deroceras helicoidale Rähle, 1998

Deroceras helicoidale Rähle, 1998 in Rähle 1998: 22

Slug distinctly coloured with reddish head, mantle with accumulating dark spots on pale background, behind mantle lighter, greyish or yellowish brown with dark pigment in skin grooves, reaching foot margin, sole pale.
Specific penis shape: posterior part of penis runs into a long and spirally snail-like coiled process, the penial gland is a short finger at the end of this twisted penis section. Rectal caecum long.

Up to 40-45 mm (preserved)

Only known from one locality E of Alanya.

Wiktor 2000: 447, Schütt 2001: 267.

S Turkey

Deroceras ikaria Reischütz, 1983

Deroceras ikaria Reischütz, 1983 in Reischütz 1983: 23

Slug evenly bluish-black, preserved deep black, head blackish, sides lighter or almost white, mantle 40 % of body length, sole creamy or white, edges distinctly pigmented black.
Penis appendix larger than in *D. samium*, protuberance on the lateral penis wall larger than in *D. samium*, but no more differences between these two taxa. Penis slightly resembles that of *D. kasium*.

Up to 20-23 mm long (preserved 15 x 4.5 mm)

Possibly an endemic species of Ikaria island.

Wiktor 2000: 450, Wiktor 2001: 159.

Anatomy of genitalia follows Wiktor 2001, the figures by Reischütz 1983 are not precise and were based on immature and ill-preserved specimens. Status uncertain, could be a synonym of *D. samium*. It is necessary to study additional material from Ikaria.

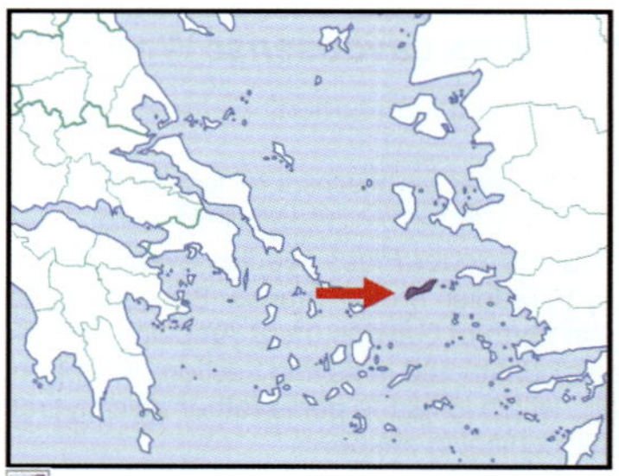

Greece (Ikaría)

Deroceras invadens Reise, Hutchinson, Schunack & Schlitt, 2011

Deroceras invadens Reise, Hutchinson, Schunack & Schlitt, 2011 in Reise et al. 2011: 220

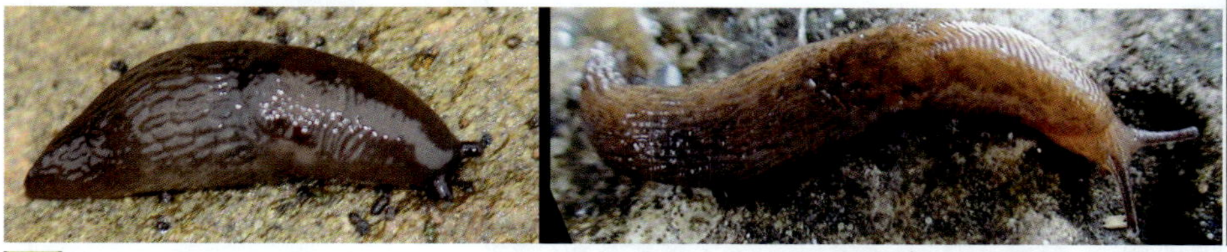

Up to 20 x 4.5 mm (preserved)

(1) Portugal: Algarve, near Castro Marim, (2) Italy: Napoli, Baia
(1) Valter Jacinto, (2) Alessandro Hallgass

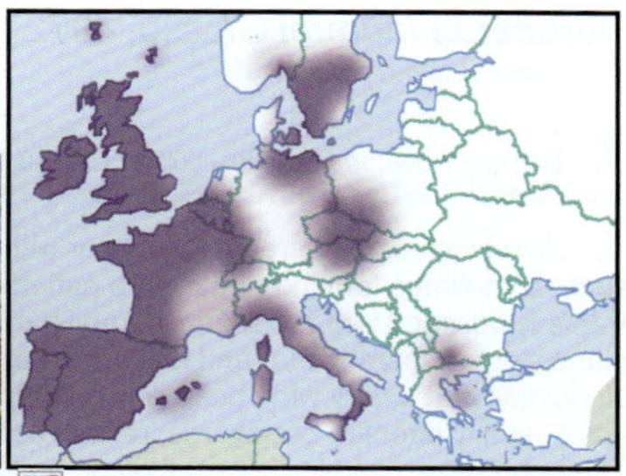

Originally perhaps Italy, at present wide-spread in Europe and abroad

Slug colour variable, usually semitransparent dirty cream or pale brown (preserved almost evenly dark brown), often with a pattern of dark spots, especially on the mantle, some slugs are unspotted, rarely black. Mantle length 40-45 % of body. Sole with distinctly darker coffee-coloured lateral zones and a cream zone in the middle (only the lightest specimens have a uniformly light creamy sole). Mucus colourless and watery.
Penis with a proximal part with two side pockets or appendices (penial lobe and usually slightly longer caecum well developed and bluntly rounded at their ends, the lobe is usually more bent than the caecum), posterior part with cone-shaped stimulator, penial gland consists of 3-7 usually unbranched processes on a common shaft, retractor inserting at proximal penis between the two appendices. Rectal caecum short or absent.
Differs from *D. panormitanum* from Malta/Sicily mainly in its potatoe-shaped penial appendices, in *D. panormitanum* the penial lobe is weaker or invisible after dissection, the penial caecum is pointed with a broad base, getting gradually narrower towards the tip.

Lowlands and low altitudes in mountains, in gardens, parks, semideserts, cultivated land, perferably in shady valleys, near rivers and springs. In Greece and Britain under pieces of wood and stones, in Britain also under cardboard and other litter in bare or sparcely vegetated waste ground. Can be a pest in farmyards and gardens. In Bulgaria up to 1500 m.

Reproduction exclusively by copulation, life cycle short, up to 3 generations per year.

Kerney 1999: 164, Wiktor 2000: 488, Wiktor 2001: 129, Pesqueira et al. 2003, Reise et al. 2011.

Prior to 2011 known as *Deroceras panormitanum*, which was found to represent a Maltese/Sicilian species with limited range.

Deroceras karnaniense Wiktor, 1984

Without gender agreement: *Deroceras karnaniensis* Wiktor, 1984

Deroceras karnaniensis Wiktor, 1984 in Wiktor 1984: 160 (*Deroceras (Plathystimulus) karnaniensis*)
Should have been *karnaniense*, *Deroceras* is neuter.

Slug creamy-brown, with chocolate brown spots, forming a reticulate pattern behind the mantle, mantle 40 % of body length.
Penis elongate, irregular, with a short flattened appendix on the posterior end. Penial gland set apically, roughly in pronlongation of long penis axis, in the shape of 3-4 unbranched processes united at their base. Retractor penis unbranched, inserts at base of appendix. Stimulator long, crest-like. Rectal caecum in the form of a fold on rectum.

Up to 25 x 5.5 mm (preserved)

Only known from one locality near Mesolóngi.

Wiktor 2000: 453, Wiktor 2001: 161.

Greece (Akarnanía)

Deroceras kasium Rähle, 1993

Deroceras kasium Rähle, 1993 in Rähle 1993: 111

Slug unusually small, ochre to pale grey, back and sides covered with dark brown spots in skin grooves, forming a reticulate pattern, mantle darker than the rest of the body, except a lighter plate around pneumostom, evenly pigmented, with darker spotting at the edges, 40 % of body length, sole pale.
Very typical penis and appendix: penis very small, and non-constricted, appendix larger, hook-like bent, with two unilateral constrictions at the surface of the bulging part, subdividing the appendix into three sections, retractor insertion marking probably the border between penis and appendix, penial gland set apically on penis with 5-6 equally long and partly bifurcate processes. Practically no rectal caecum.

Up to perhaps 18 mm (preserved up to 13 mm, adults can have smaller sizes, usually 9 mm)

In Mediterranean shrubland vegetation, finds shelter under stones.

Wiktor 2000: 453, Wiktor 2001: 163.

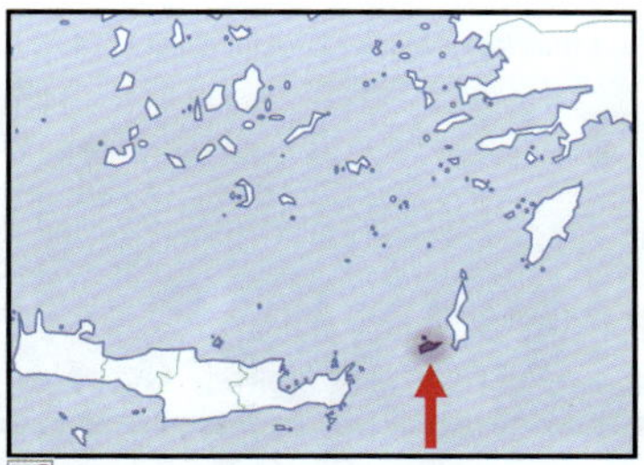

Greece (Kásos)

Deroceras keaense Van Regteren Altena, 1973

Without gender agreement: *Deroceras keaensis* Van Regteren Altena, 1973

Deroceras keaensis Van Regteren Altena, 1973 in Van Regteren Altena 1973: 89
Should have been *keaense*, *Deroceras* is neuter.
Deroceras johannae De Winter & Butot, 1985 in De Winter & Butot 1985: 79

Slug slender and very variable in size, posterior end strongly flattened laterally, frontal third of mantle can be folded backwards, mantle 40 % of body length. Head, mantle and dorsum evenly black or dark grey, sides lighten downwards to dirty cream, sole grey-cream, sometimes lateral zones grey.
Penis elongated, irregularly shaped, with two appendices of variable shape, penial gland with several processes looking loke swollen fingers, stimulator flat, rectangular, with wide base, usually dark pigmented, rectal caecum well developed.

Up to 52 x 12 mm (adults at least 27 mm long) (preserved)

Hides under stones and lives in various kinds of habitats.

Wiktor 2000: 453, Wiktor 2001: 165.

Seems to live parapatrically with *D. samium*, a closely related species. *D. johannae* from Andros and Greek Makedonia is preliminarily regarded as conspecific with *keaense* here, more research is needed, external appearence and rectal caecum are not different, perhaps penis shape, penial gland and stimulator could provide some differences.

Greece

Deroceras klemmi Grossu, 1972
Deroceras lothari Giusti, 1973

Deroceras klemmi Grossu, 1972 in Grossu 1972: 642
Deroceras lothari Giusti, 1973 in Giusti 1973 ("1971"): 491

Slug black or dirty cream, evenly or with rare dark spots sometimes forming a reticulate pattern.
Penis subdivided into two roughly equal parts, penial gland single (unbranched) and long, covered by papillae, stimulator with a very wide base, like a flatttened cone, its apex usually directed backwards, rectal caecum large. The unbranched and large penial gland and the broad and flattened stimulator show relatively little variation.

Germany: Schleswig-Holstein, near Neustadt
Vollrath Wiese

Central Europe

Montane meadows and other open habitats, up to 1500 m.

Up to 36 mm long (preserved)

Manganelli et al. 1995: 22, Turner et al. 1998: 309 (report for near Basel cited from Falkner 1982 and Reischütz 1986), Wiktor 2000: 474, Colling et al. 2008: 125.

Endemic species in Monti Reatini according to Manganelli et al. 1995, widely distributed according to Wiktor 2000 who added *klemmi* to this species as a synonym, but confirmed that F. Giusti did not follow this view. Wiktor's 2000 view is followed here.

Deroceras korthionense De Winter & Butot, 1985

Without gender agreement: *Deroceras korthionensis* De Winter & Butot, 1985

Deroceras korthionensis De Winter & Butot, 1985 in De Winter & Butot 1985: 82
Should have been *korthionense*, *Deroceras* is neuter.

Slug light brown with fine dark spots on mantle and dorsum, sometimes evenly light, mantle 40 % of body length.
Penis consists of 2 parts: the lower contains the stimulator, the apical part is cylindrical, with a conspicuous papilla at the outlet of the vas deferens. Penial gland in the shape of 4-5 processes adjoining with bases. Musculus retractor penis unbranched. Stimulator small, conical. No rectal caecum (contrast to *D. turcicum*).

Up to 22 x 6 mm (preserved)

Wiktor 2000: 458, Wiktor 2001: 105.

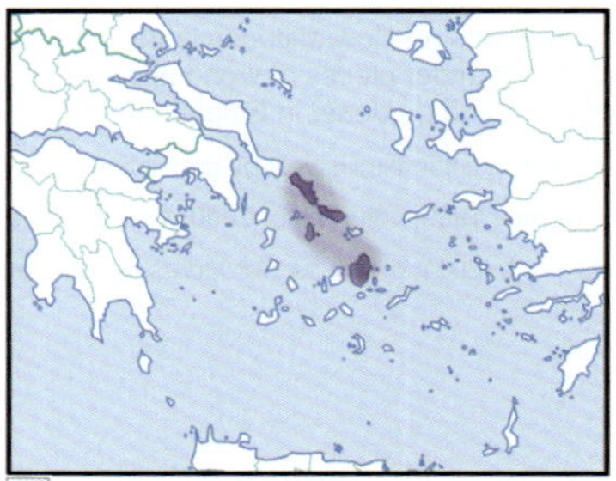

Greece (Kikládes)

Deroceras kythirense Wiktor, Vardinoyannis & Mylonas, 1994

Without gender agreement: *Deroceras kythirensis* Wiktor, Vardinoyannis & Mylonas, 1994

Deroceras kythirensis Wiktor, Vardinoyannis & Mylonas, 1994 in Wiktor et al. 1994: 14
Should have been *kythirense*, *Deroceras* is neuter.

Slug dirty cream, densely covered with dark spots, concentrated mainly in skin grooves, also around pneumostom.
Penis oval, no appendix, penial gland set apically on penis, large with big papillae, sometimes bifurcate, stimulator small and conical. Rectum abruptly wider in its apical section, no typical caecum.

Up to 17 mm long (preserved)

Wiktor 2000: 462.

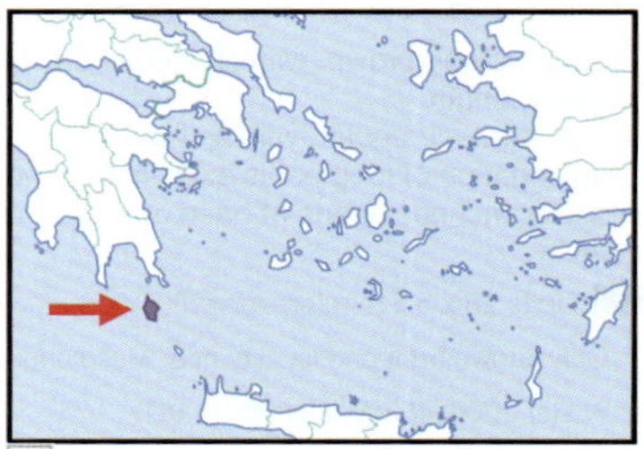

Greece (Kíthira)

Deroceras laeve (Müller, 1774)

Without gender agreement: *Deroceras laevis* (Müller, 1774)

Limax laevis Müller, 1774 in Müller 1774: 1 (*Limax lævis*)

(1-3) Slovakia: Plešivec, (shells) Germany: Schleswig-Holstein, near Selent
(1-3) Jozef Grego

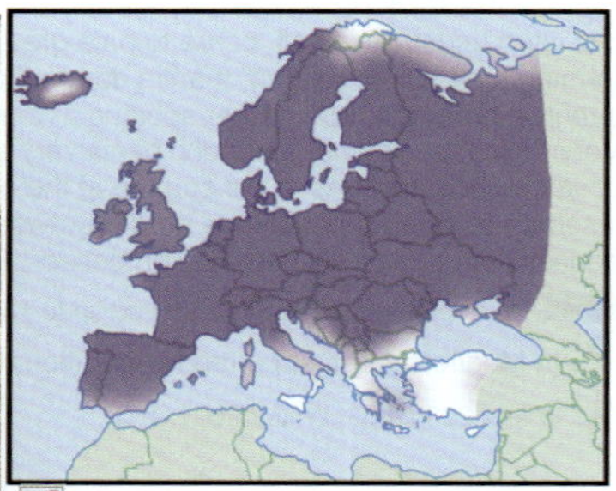

Originally Palearctic, today dispersed worldwide

15-25 mm long (preserved)

Slug brown to dark brown, usually with dark and characteristic but not well visible spots arranged in groups, shape almost cylindrical, posterior end abruptly widened, mantle 50 % of body length (unusually large), wrinkles on skin present (may disappear in preserved slugs). Mucus thin, colourless.
Penis often reduced, elongated if present, without proper penial gland but with two or more tiny glandular papillae and its end, retractor unforked and attached at half penis length, stimulator small, cone-shaped but looks more like a papilla. Tubular oviductus and atrium are unusually long. No rectal caecum.

High ecological tolerance, but needs permanently wet habitats. Usually in lowlands and very humid habitats, swamps, riversides, wetlands, especially alder and oak woods, marshlands and degraded areas, also greenhouses, often near water under wood or detritus. Tolerates subpolar and tropical temperatures. Newly created habitats are often colonized after a few years. In Switzerland up to 1800 m, but usually below 1000 m, in Bulgaria up to 2500 m.
The only land snail that goes deliberately into the water and can survive for days submerged. Can be dispersed by water currents. Eggs also survive submerged, juveniles can hatch under water and then get to the surface. Slugs are agile, crawl quickly. Omnivorous, live and dead plant remains are preferred.

3 days after copulation eggs (1-2.5 mm) are laid singly or in small clusters at intervals of some days, in total 20-30 eggs, 3 mm long reddish juveniles hatch after 20-30 days, maturity is reached after 60-80 days, animals die shortly after the egg laying period, usually at 3-5 months age. Maximum age slightly more than 1 year, but only very few animals reach this age. Life cycle in some regions extremely short, down to less than a month, up to 5 generations in a year, several generations can live at the same time. Frequently there are forms with reduced penis, which reproduce by self-fertilisation.

Threatened by continuous destructions of wet habitats by drainage and construction projects, also roads. On the other hand it can be a serious pest in greenhouses.

Frömming 1954: 206, Kerney et al. 1983: 191, Wiktor 1983, Falkner 1990: 192, Wiktor 2000: 462.

Deroceras lasithionense Wiktor, Vardinoyannis & Mylonas, 1994

Without gender agreement: *Deroceras lasithionensis* Wiktor, Vardinoyannis & Mylonas, 1994

Deroceras lasithionensis Wiktor, Vardinoyannis & Mylonas, 1994 in Wiktor et al. 1994: 16
Should have been *lasithionense*, *Deroceras* is neuter.

Slug often deep black or graphite-black, or slightly lighter with black pigment patterns, gradual lightening down to the body sides, mantle 40 % of body length.
Penis irregularly shaped with small appendix, posterior end of penis with dark blot, penial gland large, 1/2 of penis length, composed of 2-9 lobular processes (flat branches with papillae at both sides) (only in mountains more than 2) on a short common shaft or directly on penis, never on a long common shaft as in *D. rethimnonensis*, retractor penis unforked. Short and often pocket-like rectal caecum.

Up to 47 x 10 mm (preserved)

Slug occurs in maquis and phrygana, pine forests, old cultivations. Most often found in the vicinity of streams.

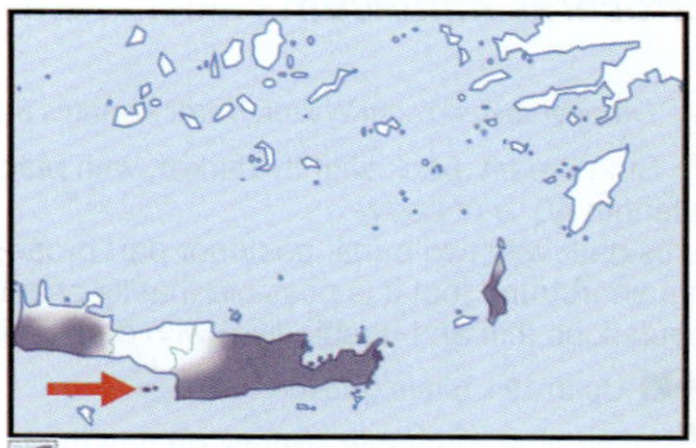

Greece (Crete, Kárpathos)

Wiktor 2000: 465, Wiktor 2001: 110.

Deroceras levisarcobelum De Winter, 1986

Deroceras altimirai levisarcobelum De Winter, 1986 in De Winter 1986: 141

Slug pale brown, yellow-brown or almost orange with dark melanophores sometimes merging into diffuse spots.
Penis very similar to *D. altimirai*, with two pockets at posterior end, one being the penial gland, stimulator large and with smooth surface, not ornamented. Rectal caecum short or absent.

Up to 50 mm long (preserved 35 mm)

Wiktor 2000: 466.

Could be a synonym of *D. altimirai*, the two taxa do not differ markedly in external appearence and penis.

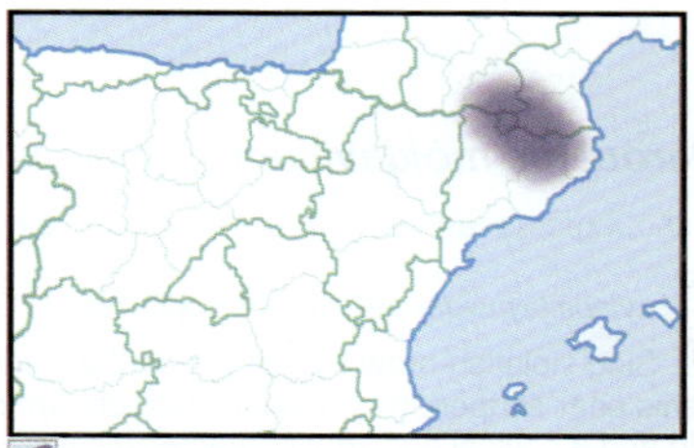

SW France, NE Spain

Deroceras lombricoides (Morelet, 1845)

Limax lombricoides Morelet, 1845 in Morelet 1845: 39, Pl. 3
Deroceras hispaniensis Castillejo & Wiktor, 1983 in Castillejo & Wiktor 1983: 11
Should have been *hispaniense*, *Deroceras* is neuter.

Slug brownish or reddish creamy to blackish, mantle and back darker, sometimes with dark pigment concentrations there, sole creamy. Mucus colourless.
Penis composed of two parts, anterior part spherical and with glandular papillae at one wall, posterior part narrower and usually cylindrical, penial gland consists of 2-5 branches with papillae set on a common shaft. Stimulator-like membraneous and U-formed structure present in anterior part of penis (can easily be damaged during preparation). No distinct rectal caecum, only a pocket-like enlargement.

Various habitats in pine and oak woods, on rocks and in leaf litter.

Adults were found in spring and autumn. The number of laid eggs is low.

Wiktor 2000: 470 (classified *hispaniensis* as a synonym), Pesqueira et al. 2003, Castillejo & Iglesias in Verdú & Galante 2006: 339.

Spain: Galicia, Bastabales

Up to 40 mm long (preserved 23 mm)

Portugal, Spain, Madeira

Vulnerable in Spain. Threatened by forest exploitation, destruction and fire.

Luis Miguel Bugallo Sánchez

Deroceras maasseni Wiktor, 1996

Deroceras maasseni Wiktor, 1996 in Wiktor 1996: 89

Slug evenly black, getting lighter downwards at the sides, sole pale.
Penial gland composed of 3-4 branches, inside penis a flat but fleshy stimulator, its free end broad and like perpendicularly cut, rectal caecum like a pocket.

Up to 21 mm (preserved)

Near rivers in mountains, 650-1150 m altitude.

Only known from a small range in Montenegro.

Wiktor 2000: 474.

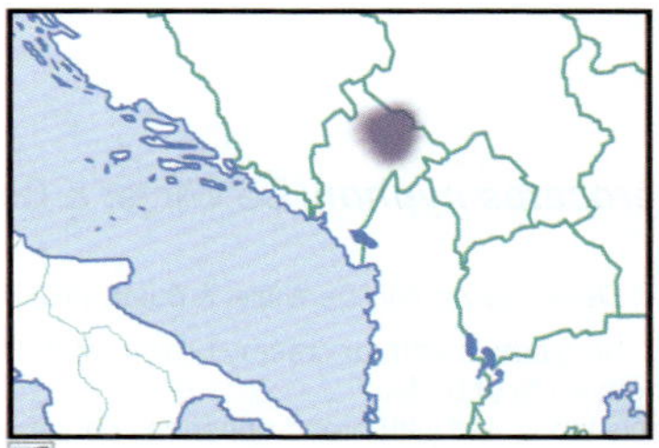

Montenegro

Deroceras malkini Wiktor, 1984

Deroceras malkini Wiktor, 1984 in Wiktor 1984: 151

Slug whitish with few dark spots on mantle and dorsum, mantle length 40-45 % of body.
Penis irregular with swellings, among these are an appendix and an unusually vestigial penial gland varying in shape, stimulator flattened with blunt top. Rectal caecum short but well-marked.

Up to 30 x 7.5 mm (preserved)

Wiktor 2000: 475, Wiktor 2001: 116.

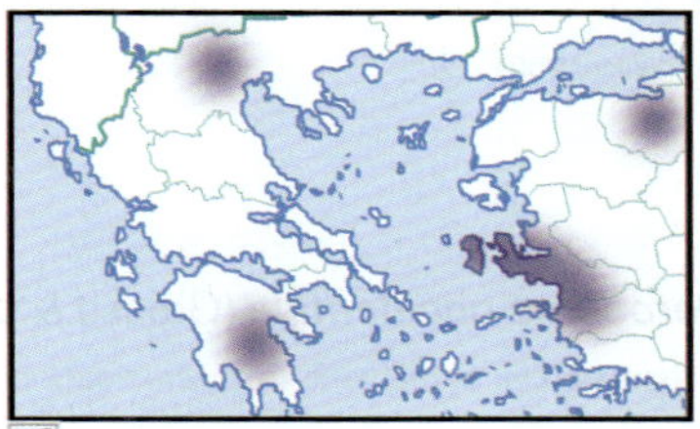

Greece, W Turkey

Deroceras melinum Wiktor & Mylonas, 1981

Deroceras melinum Wiktor & Mylonas, 1981 in Wiktor & Mylonas 1981: 183

Slug cream, dark spots on the mantle, dorsum behind mantle and sides are covered with spots forming a reticulate pattern, no dark pigment on middle of dorsum and keel, mantle length 10 mm.
Penis elongated with different swellings, with a bent appendix, penial gland relatively thick, with papillae, stimulator narrow and flat, tongue-like (its one side sunk). Rectal caecum like a shallow pocket.

Up to 24 mm long (preserved)

Wiktor 2000: 476, Wiktor 2001: 168.

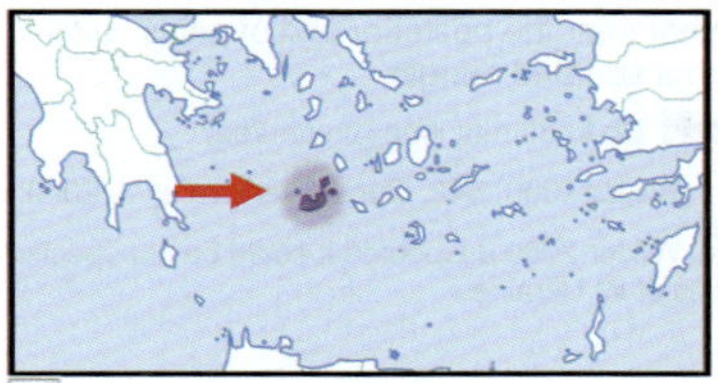

Greece (SW Kikládes)

Deroceras minoicum Wiktor, Vardinoyannis & Mylonas, 1994

Deroceras minoicum Wiktor, Vardinoyannis & Mylonas, 1994 in Wiktor et al. 1994: 25

Slug cream, back slightly darker, with blackish irregular spots, in eastern populations evenly ash-grey, mantle length 40 % of body.
Penis dark with two parts, posterior part broad, penial gland laterally and composed of three rounded processes on a short trunk (but it is possible that its shape is very variable), no papillae, stimulator conical (in eastern populations thin and needle-like) with dark pigments. Practically no rectal caecum.

Up to 19 x 5 mm (preserved)

Lives near mountain springs on a plateau of 1500 m altitude in the Idi mountains, but possibly also some lowland slugs belong to this species.

Wiktor 2000: 476, Wiktor 2001: 118.

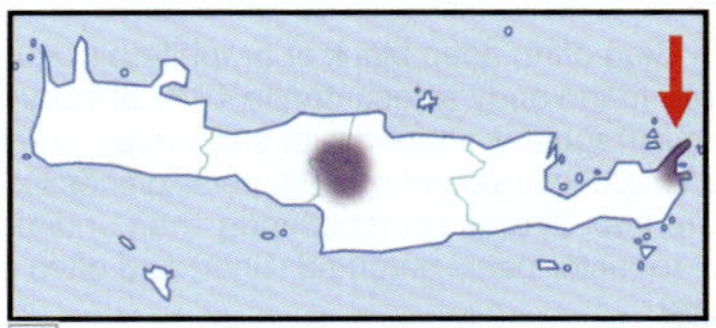

Greece (central and E Crete)
The eastern populations might belong to a different species.

Deroceras moldavicum (Grossu & Lupu, 1961)

Without gender agreement: *Deroceras moldavica* (Grossu & Lupu, 1961)

Lytopelte moldavica Grossu & Lupu, 1961 in Grossu & Lupu 1961: 28

Slug violetish-brown, back darker than sides, preserved specimens paler, mucus thin and colourless.
Penis pear-shaped, no appendix, retractor unforked, attached to posterior part of penis, penis inside with a large fold and a conical and very characteristical stimulator, with a hard glossy transparent or whitish and semicircular plate on its side. Rectum without caecum or any likewise enlargement.

Up to 35 mm long (preserved 23 mm)

Well-shaded and damp places, often at river margins in flood debris, deciduous forests, up to 1000 m altitude.

In Poland maturity is reached in autumn.

Wiktor 2000: 542, Gural-Sverlova et al. 2009: 59, Sysoev & Schileyko 2009: 143.

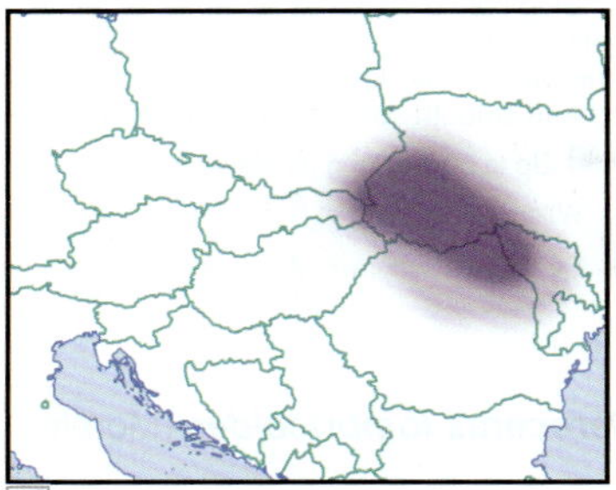

SE Poland to NE Romania

Deroceras neuteboomi Forcart, 1972

Deroceras jaeckeli neuteboomi Forcart, 1972 in Forcart 1972: 112

Slug light meat-coloured, mantle covered with dark grey spots, dorsum behind mantle with fine spots forming a reticulate pattern, no spots in the middle of dorsum, mantle length more than 50 % of body.
Penis elongated or pear-shaped, no appendages, no lateral constriction, retractor penis inserted close to vas deferens outlet, slightly anteriorly from the posterior end of the penis, spermatheca and its duct roughly equal to penis length, stimulator flat with wide base. No caecum on rectum. Glandula albuminalis reaching anteriorly to the mantle margin.

Up to 12 x 2.5 mm (preserved)

Was found at conglomerate rocks nearby streams.

Wiktor 2000: 478, Wiktor 2001: 170.

Greece (Ródos)

Deroceras nyphoni De Winter & Butot, 1986

Deroceras nyphoni De Winter & Butot, 1986 in De Winter & Butot 1986: 40

Slug dirty cream or creamy brown, with diffuse dark spots, no light section in middle of dorsum, mantle length 45 % of body.
Penis elongated with broad appendix, penial gland consisting of up to 9 unbranched processes with papillae, retractor penis two-forked, stimulator cone-shaped with wide base. No typical caecum, only a slight pocket-like distension in rectum.

Up to 31 x 6.5 mm (preserved)

Was found under stones, in open habitats and in olive-yards.

Wiktor 2001: 120.

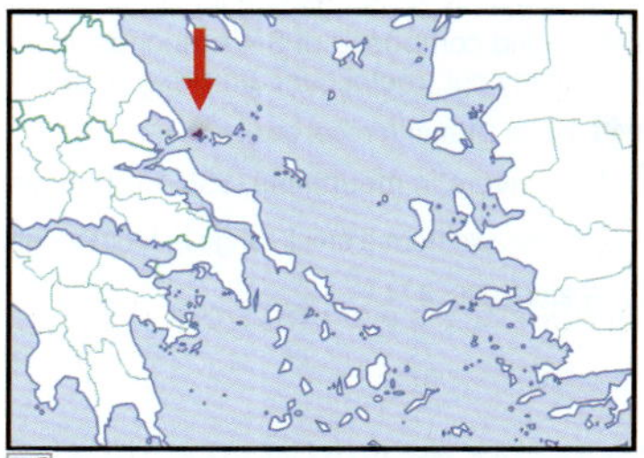

Greece (Skíathos island)

Deroceras occidentale (Grossu & Lupu, 1966)

Without gender agreement: *Deroceras occidentalis* (Grossu & Lupu, 1966)

Lytopelte occidentalis Grossu & Lupu, 1966 in Grossu & Lupu 1966: 25

Slug yellowish white, sometimes with gray pigments on mantle, dorsum with white line in the middle.
Penis egg-shaped without apppendices or penial gland, vas deferens and retractor penis attached closely together near the posterior end of penis, stimulator conical, with a characteristical plate on it, like a flattened cap with a serrated margin.

Up to 30 mm long (preserved)

Wiktor 2000: 543, Gural-Sverlova et al. 2009: 59, Sysoev & Schileyko 2009: 143.

Wiktor 2000 suspected it could be conspecific with *D. bureschi*. Sysoev & Schileyko 2009 reported it from various regions of W Ukraine.

W Ukraine, N Romania

Deroceras oertzeni (Simroth, 1889)

Agriolimax oertzeni Simroth, 1889 in Simroth 1889: 17 (*Agriolimax Oertzeni*)

Up to 43 x 6 mm (preserved) Greece: Samothráki Dilian Georgiev Greece

Slug evenly black or almost black, sides gradually lighter downwards, mantle length 35-40 % of body, posterior end perpendicularly truncated, sole creamy white.
Penis dark-pigmented, of complex shape, elongated with a long twisted appendix with swellings (penis looks like bifurcate), penial gland consisting of 3-8 thin processes sitting directly on penis or on a short common trunk, retractor penis large and forked. Rectal caecum well-developed but of variable length.

Humid habitats, for example near water bodies, not necessarily on calcareous substrate. In the hot summer season the slug seems to be active at night and in the morning, and during the day remains hidden in crevices.

Wiktor 2000: 485, Wiktor 2001: 121. Habitat information provided by D. Georgiev.

Deroceras pageti Grossu, 1972

Deroceras pageti Grossu, 1972 in Grossu 1972: 640

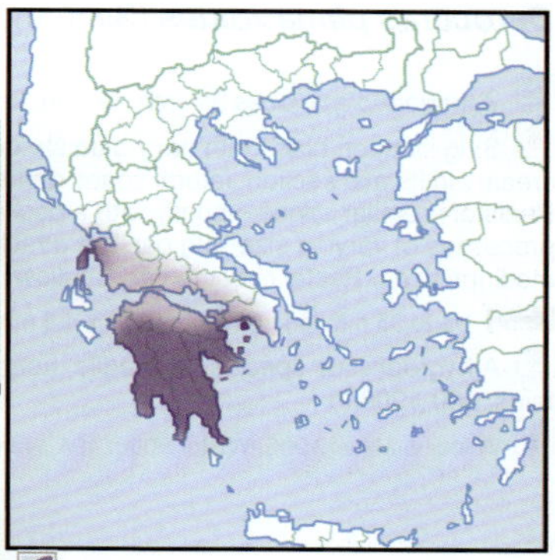

Greece

Up to 32 x 8 mm (preserved) Greece: Salamina island (1+4: adult 25 mm, dark: juvenile 15 mm) Panayotis Ovalis

Slug pinkish or grey and without spots (preserved dirty creamy to blackish), sides whitish, juveniles darker and sometimes with indistinct dark spots, mantle length 35-40 % of body.
Penis in two sections, posterior section shorter and narrower, appendix short and cone-shaped, penial gland apically connected with penis, very long, sometimes even longer than penis, composed of a loing trunk and two even longer processes with papillae and short offshoots, retractor penis unforked, stimulator conical. Vas deferens opens on the top of the appendix, a very typical character. Rectal caecum short.

Usually under stones in ruderal habitats or Mediterranean shrubland vegetation, but occasionally also in garden plantations where they may consume vegetables.

Wiktor 2000: 488, Wiktor 2001: 124, P. Ovalis (unpublished, 2010).

Deroceras panormitanum (Lessona & Pollonera, 1882)

Without gender agreement: *Deroceras panormitanus* (Lessona & Pollonera, 1882)

Agriolimax panormitanus Lessona & Pollonera, 1882 in Lessona & Pollonera 1882: 52
Agriolimax pollonerae Simroth, 1889 in Simroth 1889: 179
Agriolimax caruanae Pollonera, 1891 in Pollonera 1891: 3
Dedicated to Mr. Caruana Gatto, name is not corrected to *caruanai* (Art. 32.5). Art. 31.1 has no effective force.
Deroceras giustianum Wiktor, 1998 in Wiktor 1998: 17

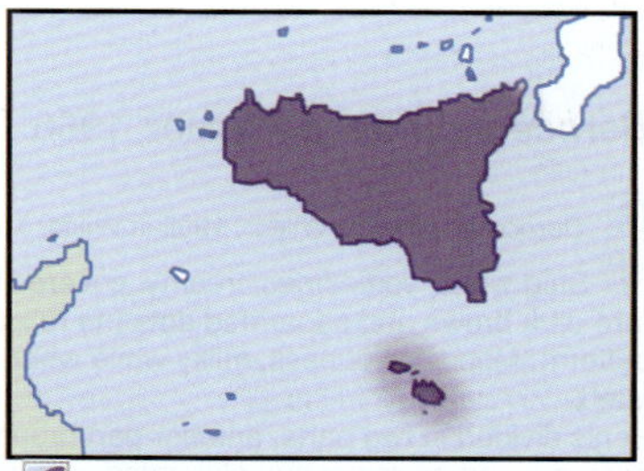

Malta, S Italy (Sicily)

Slug colour variable, usually semitransparent dirty cream or pale brown (preserved almost evenly dark brown), usually with a pattern of dark spots, especially on the mantle, sometimes black. Mantle length 40-45 % of body. Sole with distinctly darker coffee-coloured lateral zones and a cream zone in the middle (only the lightest specimens have a uniformly light creamy sole). Mucus colourless and watery. This description differs only weakly from *D. invadens* and *D. golcheri*.
Penis with a proximal part with two side pockets or appendices (penial lobe of variable length and long and pointed caecum), posterior part with cone-shaped stimulator, penial gland consists of 3-7 usually unbranched processes on a common shaft, retractor inserting at proximal penis between the two appendices. Rectal caecum short or absent.
Differs from *D. invadens* (which is rare in Sicily and absent from Malta) in its weaker penial lobe (usually shorter, can be very difficult to see in dissected specimens, somewhat hidden inside the penis wall, hidden lobes are well visible in everted genitalia) and its long and pointed shape of the penial caecum (it has a broad base and becomes gradually narrower, in *D. invadens* it has the shape of a potatoe). *D. golcheri* in Malta has no penial appendices, only weak lateral swellings.

Up to 20 x 4.5 mm (preserved)

Usage of this name prior to 2011 included a very similar invasive species in Europe and abroad, *Deroceras invadens*. Reise et al. 2011 separated *D. panormitanum* from *D. invadens* and regarded *D. giustianum* as a synonym.

Deroceras parium Wiktor & Mylonas, 1981

Greece (Kikládes)

Deroceras parium Wiktor & Mylonas, 1981 in Wiktor & Mylonas 1981: 187

Slug dirty or brownish cream with numerous darker blurry spots, or evenly brown or blackish with diffuse and merged spots, lowest parts of sides lighter. Sole creamy, in darkest slugs the lateral zones of the sole are dark. Posterior part of body behind mantle getting more slender, last section with a distinct keel.
Penis and penial gland look like one single elongated organ, penial gland with some papillae on the edges, vas deferens opens at base of penial gland, retractor unforked, stimulator very large, filling almost the entire part of the spherical part of the penis, and completely flat. Rectal caecum has the shape of a vestigial enlargement.

Up to 20 x 4 mm (preserved)

Wiktor 2000: 490, Wiktor 2001: 172.

Deroceras parnasium Wiktor, 1984

Greece

Deroceras parnasium Wiktor, 1984 in Wiktor 1984: 159

Slug slender, thin and soft, evenly deep black, margins of sole grey, medial zone of sole whitish. Mantle length 40 % of body (length up to 10 mm), posterior end of dorsum vertically truncated. Slugs from Pilio mountains brown-black, with creamy sole, slugs from Andikithira are blackish.
Penis in S-shape, with a large lateral distension with papillae in its anterior part, surface of last penis section goffered, penial gland consists of 3-5 short processes attached directly to penis, retractor unforked, stimulator conical. Rectal caecum well-developed.

Up to 28 x 5 mm

Lives on calcareous and crystalline substrate and shelters under stones, in up to 1200 m altitude.

Wiktor 2000: 494, Wiktor 2001: 131, Triantis et al. 2008: 474.

Deroceras planarioides (Simroth, 1910)

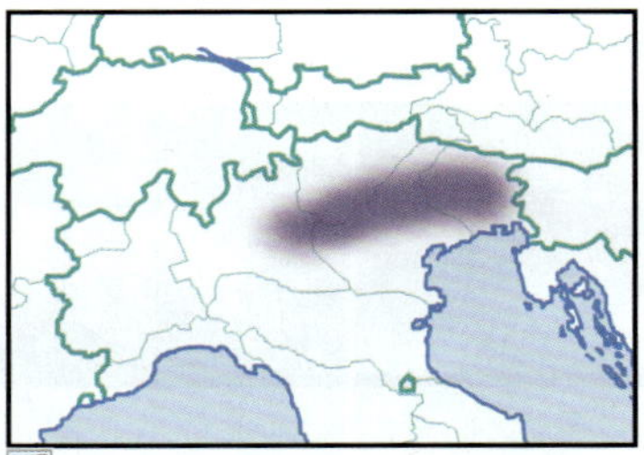
NE Italy

Agriolimax planarioides Simroth, 1910 in Simroth 1910: 326

Slug slender, black with navy-blue shade or dark chocolate brown, sides gradually lighter downwards, sole creamy in its mid section, lateral zones darker. Mucus colourless when undisturbed, otherwise milky.
Penis of irregular shape, usually with a pocked-shaped apppendix, penial gland composed of two unbranched processes of varying size with papillae attached directly to penis (without common shaft), stimulator conical, medium-sized. Rectal caecum very shallow.

Up to 35 mm long (preserved 19 x 4.5 mm)

A hygrophilous species, very agile, near streams, springs and pools, hides under stones, mainly in mountains, 800-1900 m.

Wiktor & Milani 1995, Wiktor 2000: 494, Nardi 2011: 18.

Deroceras ponsonbyi (Hesse, 1884)

Spain (region of Gibraltar)

Limax panormitanus var. *ponsonbyi* Hesse, 1884 in Hesse 1884: 12 (*Limax (Agriolimax) Panormitanus* var. *Ponsonbyi*)

Slug dark grey or dark brown with dark spots on the body, mucus colourless.
Penis anteriorly with an egg-shaped distension with granular glandulous surface, posteriorly elongated, no clearly visible penial gland present, retractor bifurcate, one branch attached to anterior penis distension, which is very unusual, stimulator small and conical. No rectal caecum.

30-35 mm long (preserved 20-30 mm)

Synanthropic, in gardens, damp places, on waste ground.

Vulnerable in Spain (Gómez Moliner et al. 2001).

Wiktor 2000: 495.

Hesse 1884 described a juvenile 14 mm long slug, no types were preserved, no anatomy was discussed, no figure was given. Name was later used for a species from Gibraltar.

Deroceras praecox Wiktor, 1966

Deroceras praecox Wiktor, 1966 in Wiktor 1966: 449

Czech Republic: Moravia, Panské louky nature reserve

Up to 40 mm long (preserved 35 mm) Michal Maňas

S Poland, Czech Republic, Slovakia

Slug white, sometimes creamy, covered by rare dark brown and separated dots (no reticulate pattern). Mucus colourless, milky white when irritated.
Penis divided in two parts, anterior part broad, posterior part narrower and elongated with a helically coiled apppendix at its end, which is usually unilaterally dark-pigmented, penial gland consists of 3-4 long processes set on a common trunk, attached to penis at the base of appendix, retractor's main part inserts laterally at the base of appendix, and has a thinner strand running to the anterior part of the penis, attaching there. Stimulator very flat, fan-shaped. Rectal caecum in the form of a shallow pocket.
Differs from the very similar *D. rodnae* in its spirally coiled appendix and its two-split rectrator penis.

Deciduous and mixed forests, at low plants at mosses, crawls on the undergrowth vegetation and shelters in soil and leaf litter, very rarely under wood. Prefers shady and humid habitats, near springs and streams, occurs in low numbers, up to 1200 m altitude.

Adults occur almost exclusively between April and June, later than in *D. rodnae*.

Kerney et al. 1983: 196, Reise 1995 (mating behaviour), Wiktor 2000: 502, Sysoev & Schileyko 2009: 144 (expected in W Ukraine).

Deroceras pseudopanormitanum Wiktor, 1984

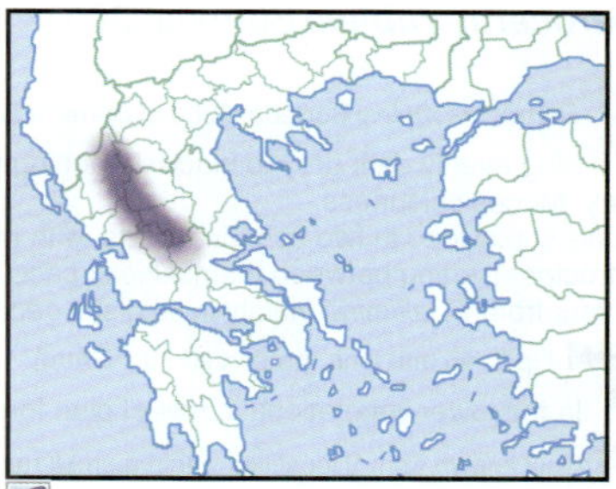

W Greece

Deroceras pseudopanormitanum Wiktor, 1984 in Wiktor 1984: 162

Slug creamy coffee brown with dark spots, in the middle of the back a light, unspotted streak, mantle length 45 % of body, keel short.
Penis broad in anterior and elongated in its posterior section, with a small pocket-shaped appendix, penial gland consists of 3-4 unbranched processes on a common trunk, stimulator in the shape of a wide, completely flat process. No rectal caecum. Differs from *D. panormitanum* in the distinctly flat stimulator.

Up to 18 x 4 mm (preserved)

In mixed forests, of different soils, for example on sandstone.

Seems to be restricted to few localities in W Greece.

Wiktor 2000: 502, Wiktor 2001: 175.

Deroceras rethimnonense De Winter & Butot, 1986

Without gender agreement: *Deroceras rethimnonensis* De Winter & Butot, 1986

S Greece, possibly SW Turkey

Deroceras rethimnonensis De Winter & Butot, 1986 in De Winter & Butot 1986: 41
Should have been *rethimnonense*, *Deroceras* is neuter.

Head and back of preserved specimens evenly blackish or grey, as if faded, getting lighter to dirty cream downwards at the sides, more or less regularly scattered pigment particles only distinguishable under high magnification. Sole cream. Dark slugs have more dark pigment, also on the lateral zones of the sole. Spotted slugs (a 3rd colour form) with unevenly distributed dark pigments, forming accumulations and irregular patterns of spots. Skin thin and soft. Mantle length 35-40 % of body, can also be spotted.
Penis divided in a broad anterior and a narrower often black posterior section with pocket-like distensions, penial gland very characteristical, with a long stem and two long branches distant from penis. Stimulator narrowly conical and usually dark. Rectal caecum long and thin.
Differs from *D. lasithionense* by the penial gland, which in *D. lasithionense* is a very short stalk if present at all. *D. thersites* is similar too, but its skin is thicker and there is a three-branching penial gland.

Up to 40 x 7 mm (preserved)

One of the most common slugs of Crete, rare above 800 m altitude, never found above 1200 m. Lives in practically all kinds of habitats where a slug may occur, also as a synanthrope.

Wiktor 2000: 505, Wiktor 2001: 136, Schütt 2001: 260.

Deroceras reticulatum (Müller, 1774)

Without gender agreement: *Deroceras reticulatus* (Müller, 1774)

Limax reticulatus Müller, 1774in Müller 1774: 10

(1) Slovakia: Starohorské vrchy mountains,
(2) Germany: Bayern, near Murnau,
(3) USA: Colorado, Lakewood, (shells) Germany: near Kiel

(1) Jozef Grego, (3) Joseph Berger

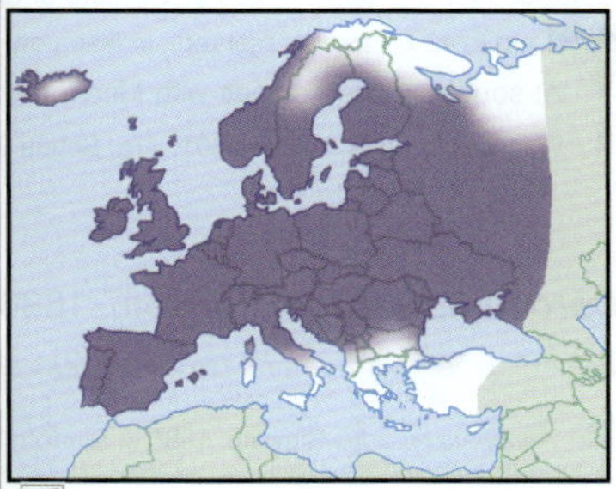

Europe, introduced to North America, Peru, Tasmania, New Zealand, central Asia

Up to 40-60 mm long (preserved 25-30 mm), size variable, depending on habitat

Slug very variable, creamy or light coffee cream, often blackish spotted (slugs with spots may appear blackish), behind the mantle the dark spots form a reticulate pattern, skin thick. Mucus colourless, on irritation milky white. The slug cannot be distinguished from many other *Deroceras* species based only on its external appearance.
Penis fleshy and with a silky sheen, in the shape of an irregular sac, in fully mature specimens divided into 2 parts by a deep lateral constriction. Penial gland of very variable shape, usually a few branches or a single long branch. Stimulator large, conical and narrow. Retractor penis inserted laterally. Vas deferens opens into penis wall facing the external body side (difference to *D. turcicum*). Rectal caecum large.

At present almost exclusively restricted to cultivated areas, usually in open habitats, in meadows, near roadsides, in ruins, gardens and parks, not inside forests. Shelters under stones and ground litter (does not burrow into the soil), active at night. Species is omnivorous, feeding mainly on fresh leaves and fruits or seedings. *D. reticulatum* is a serious pest of agricultural crops and garden cultivations. After several years with continuous moist weather conditions abundance can seriously increase.
In North and central European lowlands, Great Britain and Ireland probably the most widely distributed slug. Scarce and mainly as a synanthrope in N Scandinavia.

Life cycle differs by climatic zone and covers a few months, in Central Europe two generations per year, main reproductive phase in summer and autumn, maximum age about 1 year, slugs die at the first frosts. Usually only eggs hibernate, sometimes also juveniles. Hatched juveniles are brownish. Self-fertilisation is rare.
In SW Poland copulation from June to October, eggs (1.7-2.0 mm) are laid shortly after but can also be laid in winter, in total 300 eggs or more per individual, juveniles appear usually in April and May and reach full length after 40-55 days (in France 80 days were observed, maturity was reached earlier, after 65 days), maximum age in captivity slightly more than 1 year.
In S Finland 10-55 eggs are laid in clusters between August and September, the eggs hibernate, embryonal development takes 20-35 days, juveniles hatch in June and reach maturity in August. Adults die at cold winter temperatures, maximum age in Finland 4 months.

Frömming 1954: 214, Kerney et al. 1983: 195, Falkner 1990: 192, Turner et al. 1998: 308, Kerney 1999: 163, Wiktor 2000: 507, Wiktor 2001: 136, Pesqueira et al. 2003, Martínez-Ortí et al. 2004, Irikov & Mollov 2006: 810.

Was often confounded with *D. turcicum* in Bulgaria, and with *D. lothari*. Records for *Agriolimax agrestis* from prior to the 1940s referred to *D. reticulatum*.

Deroceras roblesi Borredà, 2003

Spain (Castellón)

Deroceras roblesi Borredà, 2003 in Borredà 2003: 78

Slug evenly dark grey to blackish, getting lighter downwards at the sides. Sole light, lateral margins often grey. Mucus colourless.
Penis subdivided in two spherical parts, with two small appendices of different size at its posterior end, unforked retractor inserting between them. Rectal caecum well developed.
Differs from *D. altimirai* mainly in its distal genitalia.

Up to 45 mm long (preserved 33 x 6 mm)

In the lower vegetational cover of pine forests, under stones and tree trunks, 900-1300 m altitude.

Only known from 9 localities within a small range. Threatened by forest destruction and water mismanagement.

Martínez-Ortí & Robles 2003: 113.

Deroceras rodnae Grossu & Lupu, 1965

Deroceras rodnae Grossu & Lupu, 1965 in Grossu & Lupu 1965: 28

Switzerland: Solothurn — Vollrath Wiese

Up to 45 mm long (preserved 32 mm)

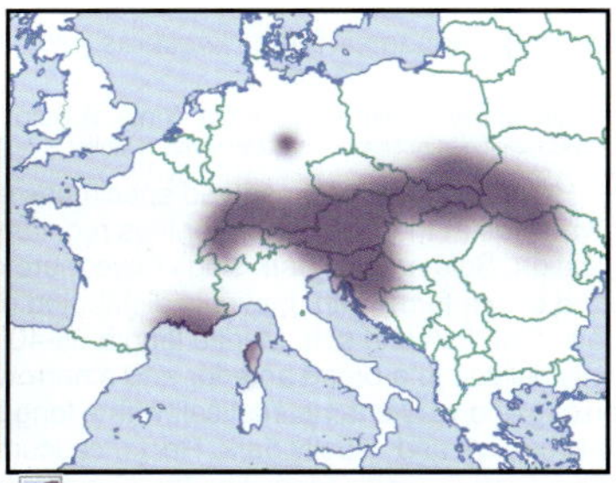

E Carpathians to Croatia, Switzerland, S France, Corsica, probably Spain

Slug light creamy white or dirty cream, more rarely white or straw yellow, almost always with dark or brown dots on the back, except in the middle section, spots can form a reticulate pattern. Blackish colour morphs also occur, against this background entirely black spots are visible. Mucus colourless, but milky when irritated.
Penis shape irregular, posterior section with different swellings, appendix not coiled, penial gland composed of a few short branches on a short common trunk, retractor unforked, stimulator entirely flat. Rectal caecum in the shape of a small pocket, also in slugs from Spain.
Differs from *D. praecox* in its non-coiled appendix, unforked retractor penis (in *D. praecox* retractor has two insertions), more spots on its body, in *D. praecox* they never form a reticulate pattern.

Deciduous and mixed forests, often in dense vegetational cover near springs, on calcareous substrate. In Switzerland up to 2200 m, mainly 700-1300 m.

Reproduction after copulation in central Europe in spring, in warmer regions also in other seasons. Life cycle 1 year, adults in spring, in fall only juveniles.

Falkner 1990: 192, Reise 1995 (mating behaviour), Wiktor 2000: 512.

Deroceras samium Rähle, 1983

W Turkey, E Aegean islands

Deroceras samium Rähle, 1983 in Rähle 1983: 115

Slug evenly black, sides gradually lighter downwards, head blackish. Sole cream, sometimes with blackish lateral zones. Mantle length 40-45 % of body, distinct groove in the posterior part of the mantle, between its margin and pneumostom.
Penis very short, lateral distention in anterior part uncommonly large, occupying 2/3 of the whole organ, appendix in the form of a large pocket, penial gland composed of 10-15 processes joined at the base, vas deferens inserts close to penial gland, stimulator flat and so large that it almost fills up the large anterior distension. Rectal caecum medium-sized.
Closely related to *D. keaense*, but with larger anterior penis section, shorter penis and a large lateral pocket (*D. keaense* has two appendices of variable shape).

Up to 42 x 9 mm (preserved), in Ikaria only up to 22 mm long

At south-exposed slopes with Mediterranean shrubland vegetation, under stones, in coastal vicinity.

Wiktor 2000: 514, Wiktor 2001: 175, Schütt 2001: 265.

Deroceras sardum (Simroth, 1886)

Without gender agreement: *Deroceras sardus* (Simroth, 1886)

Agriolimax sardus Simroth, 1886 in Simroth 1886: 319, Taf. 10
In the plate explanations the name was spelled *Agriolimax sardinus*. Simroth 1900: 101 mentioned *Agriolimax sardus* and fixed this spelling as First Reviser (Art. 24.2.4).

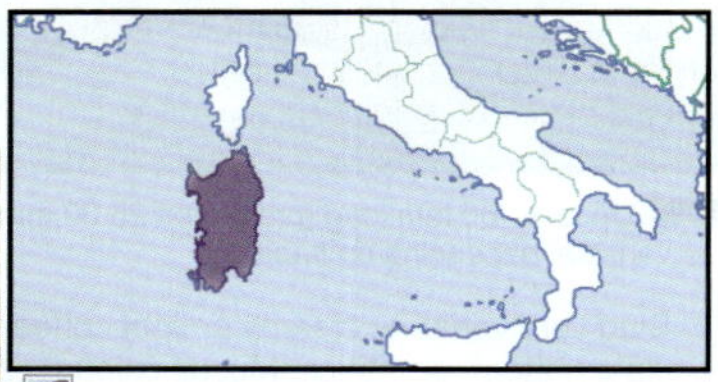

Italy (Sardinia)

Italy: Sardinia, Iglesias (determination uncertain) — Francesco Mascia

10-20 mm long

Slug pale grey, upper sections darker, with dark spots.

Manganelli et al. 1995: 22, Wiktor 2000: 552.

Originally described from a juvenile specimen, type material does probably not exist. The name was used by Manganelli et al. 1995 for an endemic species of Sardinia. The status of the taxon is unclear unless a neotype is fixed.

Deroceras seriphium Wiktor & Mylonas, 1981

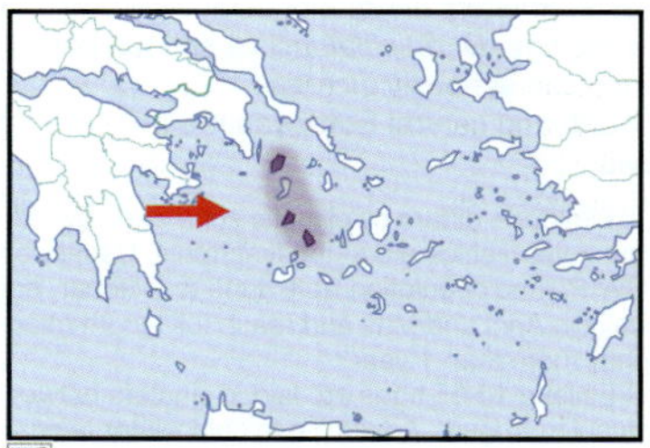

Greece (W Kikládes)

Deroceras seriphium Wiktor & Mylonas, 1981 in Wiktor & Mylonas 1981: 184

Slug creamy, mostly blackish spotted except in the middle section of dorsum. Slugs from Kea nearly evenly creamy with scarce spots on the mantle. Mantle length 40-45 % of body.
Penis with a deep lateral constriction, like a crevice, anterior part oval and larger than posterior part, appendix transversely situated, penial gland composed of a few thick processes, retractor bifurcate, attaching to anterior and posterior penis sections, stimulator flat. Rectal caecum like a shallow pocket.

Up to 27 x 4.5 mm (preserved)

Wiktor 2000: 514, Wiktor 2001: 178.

Deroceras sturanyi (Simroth, 1894)

Agriolimax Sturanyi Simroth, 1894 in Simroth 1894: 393, Taf. 19 (*Agriolimax Sturanyi*)

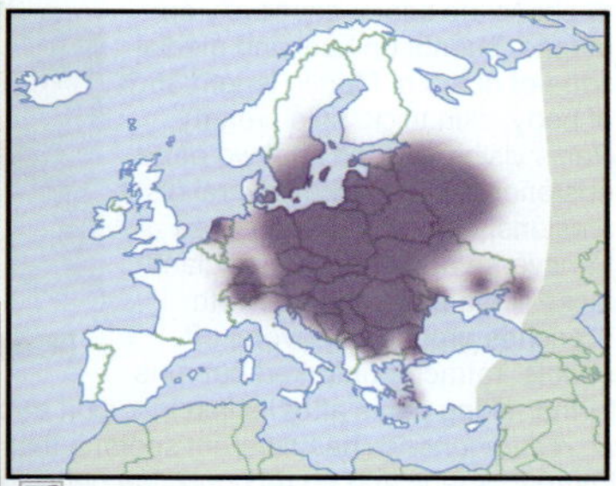

Originally SE Europe, at present central Europe to Kazakhstan

Up to 45 mm long, exceptionally 60 mm (preserved 25 mm)

(1) Bulgaria: NW of Plovdiv, (2-3) S Bulgaria: Trankovo — Dilian Georgiev

Slug body gently rounded, evenly dirty-cream to brown and black (Lake Ohrid region), skin in live slugs partly translucent with internal organs showing through, after preservation light specimens become light cream with crystals of calcium carbonate often visible inside, mantle half of body length. Mucus colourless, watery.
Mesenterium covering internal organs always dark pigmented. Penis hammer-shaped with a short shaft and two variable appendices, at least one is rounded, no properly recognizable penial gland, vas deferens inserts at the base of the larger appendix (which could be the penial gland), retractor unforked, stimulator low and conical. Rectal caecum shallow or pocket-shaped.

Lowlands, most often found in meadows, predominantly near water, in the vicitity of ditches filled with water, near rivers. Also under solitary trees and shrubs, in gardens, cemeteries, rural habitats, is commonly dispersed as a synanthrope. In Bulgaria in up to 900 m. Feeds on green parts of plants and fruits.

Reaches sexual maturity after 2 months, self-fertilization and copulation are both possible, life cycle lasts 1 year and appears to depend on the climate zone. Main reproductive phase is in summer and fall, eggs hibernate in frosty regions.

Kosińska 1980, Falkner 1990: 192, Turner et al. 1998: 304, Wiktor 2000: 516, Wiktor 2001: 139, Irikov & Mollov 2006: 809.

Deroceras subagreste (Simroth, 1892)

Without gender agreement: *Deroceras subagrestis* (Simroth, 1892)

Agriolimax subagrestis Simroth, 1892 in Simroth 1892: 43

Slug brown, dark creamy to coffee, sometimes white except head, tentacles, nape, neck and anterior part of sides which are always evenly black or blackish. Body often covered with dark spots looking splashy. Sole cream. Mucus of juveniles milky when irritated.
Shape of penis and penial gland variable, black blot present on penis between vas deferens insertion and base of penial gland, retractor penis forked, stimulator flat and fan-shaped. Instead of a rectal caecum only a hanging pocket.
Differs from *D. bakurianum* in its black blot on the penis, forked retractor and different stimulator.

Up to 45 mm long (preserved)

Forests, under wood and stones, in grass and under leaves. Also as a synanthrope, in gardens and greenhouses.

Slugs hibernate as juveniles, reach maturity in spring, adults only in early summer. Later in the year only juveniles are found, a part of which reach the male phase of genital development in autumn.

Wiktor 2000: 517, Schütt 2001: 266, Gural-Sverlova et al. 2009: 55, Balashov & Gural-Sverlova 2012: 101.

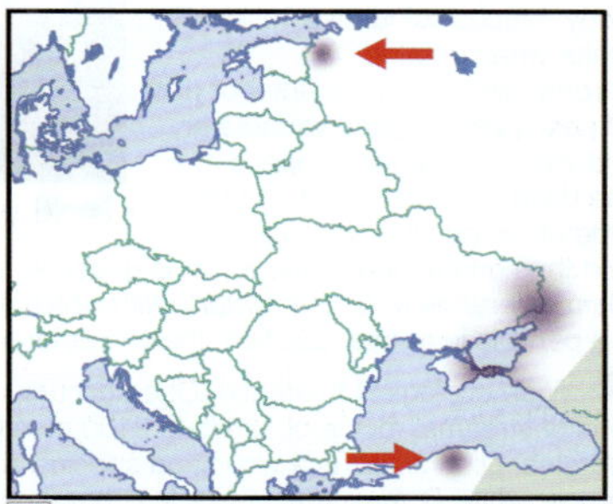

Caucasus and forelands, scattered in N Turkey, introduced to Tashkent and places in Ukraine, Russia and Turkey

Deroceras tarracense Van Regteren Altena, 1969

Deroceras tarracense Van Regteren Altena, 1969 in Van Regteren Altena 1969: 106

Slug brown with grey-brownish or blackish spots. Mucus colourless.
Penis anteriorly with large characteristical appendix of transversely corrugated surface. Penial gland small, undivided, finger-shaped, stimulator in the form of an inconspicuous cylindrical fold pointing to the appendix interior. Rectal caecum well-developed.
Very much like *D. altimirai*, differs in its characteristical appendix and a different stimulator.

Up to 40 mm (preserved 23 mm)

Leaf litter layers of mountain forests, active at night on soil substrate.

Vulnerable. Threatened by urbanisation and road construction activities.

De Winter 1986, Wiktor 2000: 519, Castillejo & Iglesias in Verdú & Galante 2006: 342.

E Spain (Tarragona)

Deroceras tauricum (Simroth, 1901)

Without gender agreement: *Deroceras tauricus* (Simroth, 1901)

Agriolimax tauricus Simroth, 1901 in Simroth 1901: 153

Slug grey-yellowish with brown or blackish spots of varying intensity, concentrated mainly in skin grooves, spots can be very abundant so that slugs may look dark, sole light with lateral zones little or much darker. Lighter slugs are usually larger.
Penis shape variable, with bulges, at the end a hook-like appendix, an additional spherical distension (or bulge) and a single and unbranched penial gland with papillae. Retractor penis unforked, stimulator large and flat with a wide base. No rectal caecum.

Up to 35 mm long (preserved)

In forests and forest-steppe regions, also in parks and gardens.

Reproduction period is from February to April.

Wiktor 2000: 521, Gural-Sverlova et al. 2009: 57.

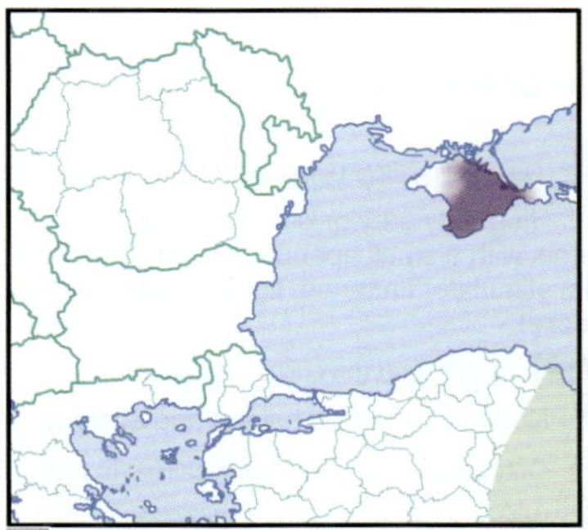

Ukraine (Crimea)

Deroceras thersites (Simroth, 1886)

Agriolimax thersites Simroth, 1886 in Simroth 1886: 317, Taf. 10 (*Agriolimax Thersites*)

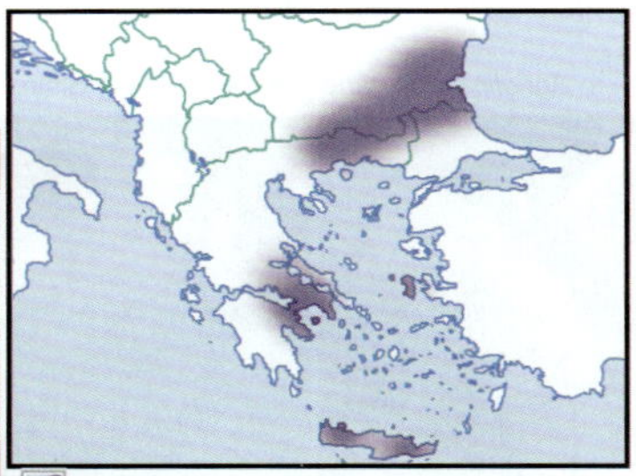

Bulgaria: Rodopi, near Byala Reka river

Dilian Georgiev

Up to 55 mm long (preserved 35 x 11 mm)

Bulgaria, Greece

Always local, in Bulgaria rare.

Slug from light creamy to pale brown, usually with irregular black or chocolate spots, especially on mantle, but not on keel and medial zone of dorsum. Mantle length 35 % of body, skin thick, sole creamy. Penis visibly divided into two parts, posterior sections with several distensions, one looks like a broad transverse appendix, penial gland looks like a branched tree with branches producing several offshoots, retractor unforked but runs further along penis after its first insertion at penis, stimulator large and conical, never pigmented. Rectal caecum long.
D. rethimnonense has different spots, a thinner skin and a very different penial gland. *D. sturanyi* has no spots, a dark pigmented mesenterium covering the internal organs and no recognizable penial gland.

Nothing published on habitat end ecology. The scattered distribution suggests that at least some populations had been artificially dispersed, and that to some extent the slug might be a synanthrope.

Wiktor 1983: 178, Wiktor 2000: 521, Wiktor 2001: 140.

Difficult to establish which ones of the scattered localities represent artificially dispersed populations.

Deroceras turcicum (Simroth, 1894)

Without gender agreement: *Deroceras turcicus* (Simroth, 1894)

Agriolimax turcicus Simroth, 1894 in Simroth 1894: 392, Taf. 19

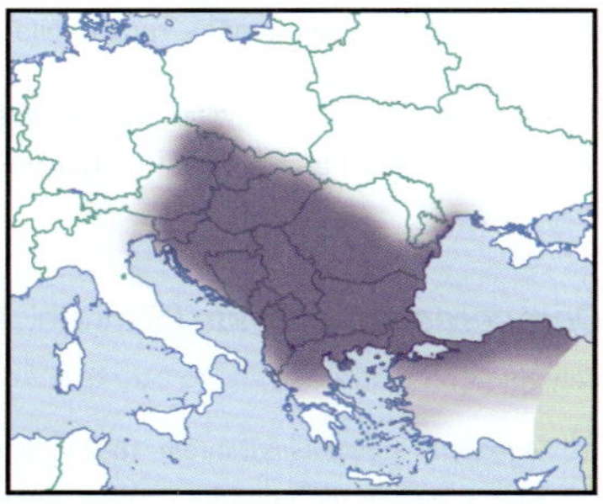

SW Poland to Greece and Turkey

Up to 29 x 5 mm (preserved 20 mm)

S Bulgaria: Stara Zagora, near Bedechka river

Dilian Georgiev

Slug dirty-cream or light cream, covered with dark spots on mantle, behind the mantle the pigment is mainly in skin grooves forming a reticulate pattern. Keel and middle of dorsum without spots. Head usually blackish, sole cream. Mantle length 35 % of body. Mucus colourless, milky white when irritated.
Strong swelling on the anterior part of penis, penial gland variable or reduced, vas deferens opens at the base of penial gland on the side of the penis facing viscera and thus on the body axis, which is unusual, retractor penis unforked, stimulator conical. Rectal caecum well developed.
Main difference to *D. reticulatum*, with which it is often confounded, is the small size and distinctly shorter processes of the penial gland in *D. turcicum*. When the penis is in natural position, the vas deferens in *D. turcicum* opens on its wall facing the body axis, in *D. reticulatum* it is on the external side.

Wide ecological tolerance. Often occurs in masses. Inhabits mainly beech and mixed forests, less often coniferous woodland, mountain meadows, margins of streams, rock rubble. Up to 2100 m in Greece and Bulgaria. Also as a synanthrope. While *D. reticulatum* inhabits open land and is found most at forest margins, *D. turcicum* is a woodland species and is much less found in open habitats. Where both species occur together, *D. reticulatum* is a decided synanthrope, while *D. turcicum* prefers less human-disturbed habitats.

Sexually mature individuals are found in various seasons of the year.

Grossu 1983: 345 (*D. wiktori*), 362 (*D. forcarti*), Falkner 1990: 192, Wiktor 2000: 524, Wiktor 2001: 143, Reise & Hu. 2001, Reise et al. 2005, Irikov & M. 2006: 809, Gural-Sverlova et al. 2009: 55.

Deroceras vascoanum De Winter, 1986

Without gender agreement: *Deroceras vascoana* De Winter, 1986

Deroceras vascoana De Winter, 1986 in De Winter 1986: 143
Should have been *vascoanum*, *Deroceras* is neuter.

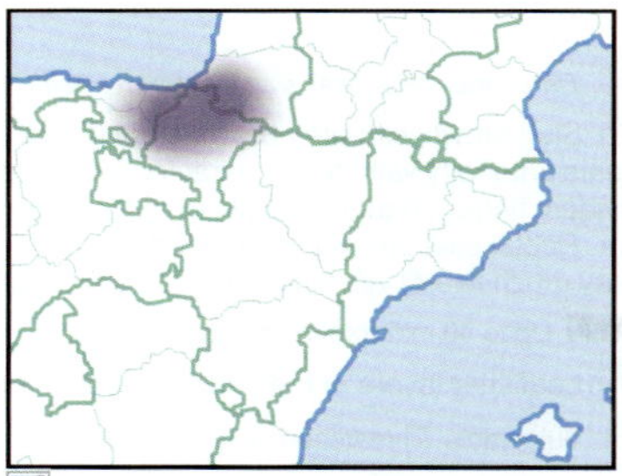

W Pyrenees and Basque country

Slug evenly brown, never with spots. Mucus milky white when irritated.
Penis oval, penial gland composed of several short flattened branches (can have offshoots) attached to penial wall, retractor penis unforked, dark blot at posterior end of penis, stimulator small, like a flattened cone. No rectal caecum.

Up to 30 mm long (preserved)

Little known, was found in beech forests at low altitudes in the Pyrenees.

Wiktor 2000: 532.

Deroceras zilchi Grossu, 1969

Deroceras zilchi Grossu, 1969 in Grossu 1969: 164

Romania (Banat), NW Bulgaria

Slug dirty grey or pale yelowish.
Penis with a small apendix, penial gland composed of 4 usually unbranched processes attached directly to penis, stimulator flattened, ligula-shaped, straited and its free end rounded. Rectal caecum in form of a shallow pocket.

14 mm long (juveniles, preserved)

Nothing known on habitat and ecology.

Only known from very few localities in a poorly studied region, two localities in Bulgaria are known.

Wiktor 2000: 533, Irikov & Erőss 2008: 203.

This seems to be an independent species, but needs to be further studied. The original description was based on juvenile specimens, the drawings freehand and possibly not reliable, the types were deposited in Romania and not available any more for study 30 years later (Wiktor 2000).

Arion

Arion Férussac, 1819
in A. Férussac & Deshayes 1819-1851 (1): 50

Type species *Limax ater* Linnæus, 1758
(subsequent designation by Fleming 1822 (Encycl. Brit. 5): 572 [not verified])

Text p. 50 from 10 July 1819.

Slugs unicoloured, without spots, pneumostome in anterior secton of mantle, no keel, no shell rudiment, only occasionally some $CaCO_3$ crystals, often with lateral colour bands, posterior end rounded, with a groove at the tail, tentacles usually dark. Sole broad and undivided. Anterior section of sole produces mucus on which the slug crawls, the mucus is resorbed at the posterior end and consumed from time to time. Penis generally absent, only epiphallus present, stimulating organ sometimes present in the atrium. During copulation the slugs describe a double C-shaped formation, epiphallus is inserted to atrium (reciprokely).

Generally herbivorous, the slugs consume also dead animals and detritus.

W Eurasia, widely dispersed

The genus *Arion* as a whole seems to require a thorough revision, in which the species are compared with each other in a consistent system. Currently various research groups apply their own systems and names are applied to different concepts, by each researcher or team individually. This makes it difficult to compile consistent range maps. The maps given here must be seen as vague and tentative approaches since they were based on various literature sources.
Several taxa which have recently been regarded as independent species are not accepted as species here, their identity is not understood and our knowledge does not provide evidence for the presence of reproductively isolated units. Based on the present record they can currently not reliably be distinguished from other species. Other species which were recently recognised as such by some experts still require confirmation in independent and less locally restricted studies.

Arion atripunctatus Dumont & Mortillet, 1853 in Dumont & Mortillet 1852-1854: 177 (separate) (*Arion cinctus* var. *atripunctatus*) [not verified]
Described from SE France. Falkner et al. 2002: 140 used it for a form for which they gave a brief distinctive diagnosis (slug larger than *Arion fuscus*, brownish or greyish brown with black dots), based on personal communications by De Winter and Reischütz. No information on the anatomy or anything else on this species has been published, it would be premature to regard it as a well established separate species.

Arion brunneus Lehmann, 1862 in Lehmann 1862: 165
Regarded as an independent species by Falkner 1990: 196, Vogt et al. 1994: 192, Falkner et al. 2002: 141, Colling et al. 2008: 127. Initially described from natural habitats in NE Poland (near Szczecin), but not recently reported from there. The name was later associated with forms from various regions in Europe, but recent studies suggest that the identity of these forms had not been understood, possibly only a colour morph of other species (Jordaens et al. 2010).

Arion euthymeanus Florence, 1886 in Florence 1886: 225
Classified as a synonym of *A. subfuscus* by Wiktor 1996: 7 and Wiktor 2001: 17, as a species by Manganelli et al. 1995: 40, Falkner et al. 2001: 54 and Falkner et al. 2002: 141. The antomy is unknown and no recent studies were published to clarify the identity of this taxon.

Arion simrothi Geyer, 1909 in Geyer 1909: 28 (not *Arion lusitanicus simrothi* Cockerell, 1893 from Azores)
No new records listed by Bürk & Jungbluth 1982: 117, classified as a form of *A. subfuscus* by Backeljau 1989, from which it cannot be distinguished. Jungbluth & Knorre 2009: 17 listed it as a species "data deficient".

Arion vejdovskyi Babor & Košťal, 1893 in Babor & Košťal 1893: 1 (*Arion Vejdovskýi*)
Described from Czech Republic, no types known, not recognized as a species by Wiktor 1996: 16.

Arion anthracius Bourguignat, 1866

Arion anthracius Bourguignat, 1866 in Bourguignat 1863-1868: 178, Pl. 29
Date was 01 Jan 1866.

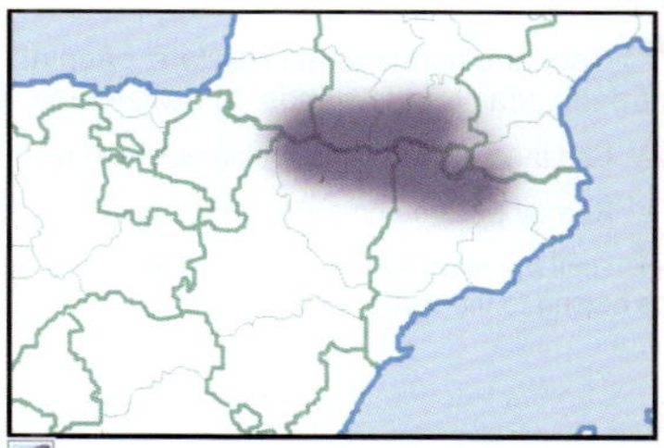

S France (central N Pyrenees)

Slug black with anthracite shine, getting lighter downwards to the sides, with dark lateral bands, sometimes delimited by whitish zones above them, mantle granulated, tentacles black. Sole whitish or yellowish.
Free oviduct with long and narrow proximal and thicker distal section, inserting into proximal section of atrium, no ligula, spermatheca spherical or lenticular, epiphallus and vas deferens equally long (25 mm), insertions of spermatheca duct and epiphallus at atrium have opposite positions, retractor with three branches (one to epiphallus insertion at atrium, one to spermatheca duct and one to central section of free oviduct).

Up to 30-33 mm long (preserved 15-25 mm)

In mixed mountain forests among rocks and accumulations of stones, on various different kinds of soil substrate, 1000-2150 m.

De Winter 1986, Castillejo 1997: 110.

The identity of this slug is uncertain, genitalia not studied in original description, syntypes were preserved but could not be dissected.

Arion ater (Linnæus, 1758)

Limax ater Linnæus, 1758 in Linnæus 1758: 652

Denmark: Bornholm, Ekkodalen

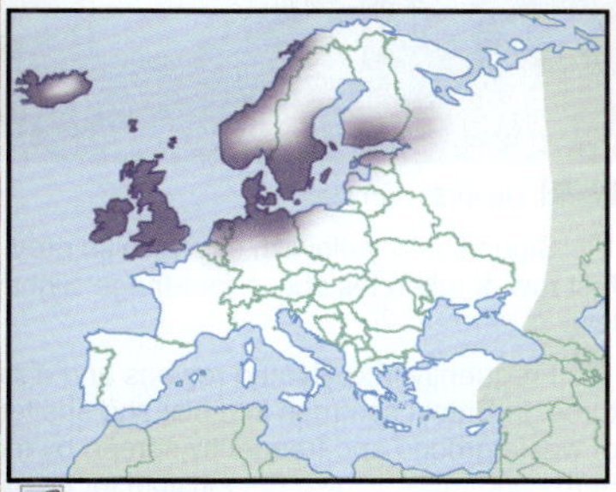

N and NW Europe

Slug deeply black, rarely dark brown, foot-fringe black, tubercles large and elongate. Sole blackish grey. Juveniles ivory whitish with black head, quickly they become greyish, half-grown animals are often black like adults.
A. vulgaris is never black, juveniles are characteristically banded.
Atrium and vagina considerably narrower than spermatheca, oviduct narrow, spermatheca spherical.

Up to 130 mm long

Prefers humid habitats, usually in woods. In Britain in natural habitats of moderately humid habitats, open and in forests, usually in wilder and harsher places than *A. rufus*, in the west and north.

Eggs are laid under mosses, exceptionally also inside the soil, in batches (every 1-3 weeks from August to October) of up to 150 eggs, later in the season only around 20 eggs, juveniles hatch after at least 27 days, this takes longer with cold temperatures, maturity is reached after 9 months, colpulation in early summer, repeated copulations are not necessary, slugs die shortly after the last eggs were laid, life span not more than a year.

Vulnerable in Niedersachsen. In N Germany threatened by the invasive competitor *Arion vulgaris*. In Britain (at least until 1999) one of the most widespread and frequent terrestrial molluscs, records by Kerney 1999 included *Arion rufus*.

Künkel 1908, Kerney & Cameron 1979: 104, Falkner 1990: 194, Kerney 1999: 120.

Sympatric occurrences with *A. rufus* in Denmark, N Germany (Schleswig-Holstein) and S England, in other parts of Britain they seem to interbreed. The taxonomic concept of Falkner 1990 is followed here, perhaps outdated and not in agreement with Castillejo 1997. More research is needed.

Arion baeticus Garrido, Castillejo & Iglesias, 1994

Arion baeticus Garrido, Castillejo & Iglesias, 1994 in Garrido, Castillejo & Iglesias 1994: 37

Slug small to medium sized, dorsum look like uniformly black, but at closer view some are brown or grey with lateral bands, right dark lateral band runs above and distant from pneumostome, sole cream or grenish grey, with a central narrow strip.
Oviduct long, not divided into different portions, epiphallus cylindrical, little longer than vas deferens, gonads and female tracts with dark pigments.

30 mm (preserved)

In humid spots of dry zones.

Adults were observed in spring and autumn.

Vulnerable.

Garrido et al. 1995, Castillejo 1997: 101, Castillejo & Iglesias in Verdú & Galante 2006: 346 (range not well known).

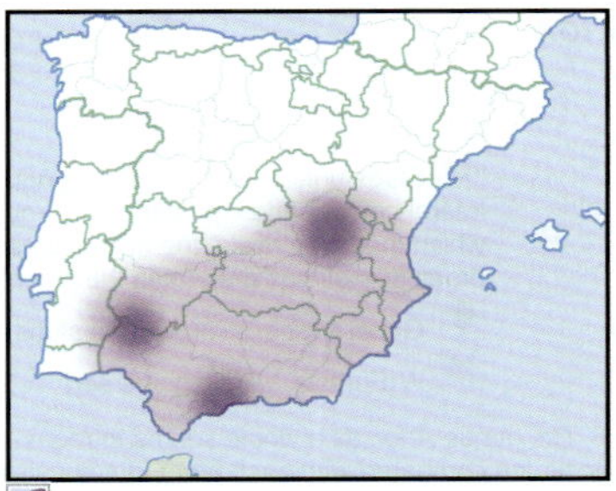

S Spain

Arion circumscriptus Johnston, 1828

Arion circumscriptus Johnston, 1828 in Johnston 1828: 76

Slug dark grey (sometimes with bluish hue) or brown, mantle with dark spots, sides usually lighter near foot-fringe, with one dark longitudinal band on each side, running above pneumostome, dorsum often with a very slight keel, slug bell-shaped in transerse section when contracted. Sole whitish. Mucus colourless. Crawling slugs are slender, when inactive the slugs remain stretched, not contracted like isopods. Juveniles with light grey sides.
Atrium large, epiphallus heavily pigmented and wider than vas deferens, oviduct short and not narrow, spermatheca elongate.

Up to 50 mm long

Ireland

Roy Anderson

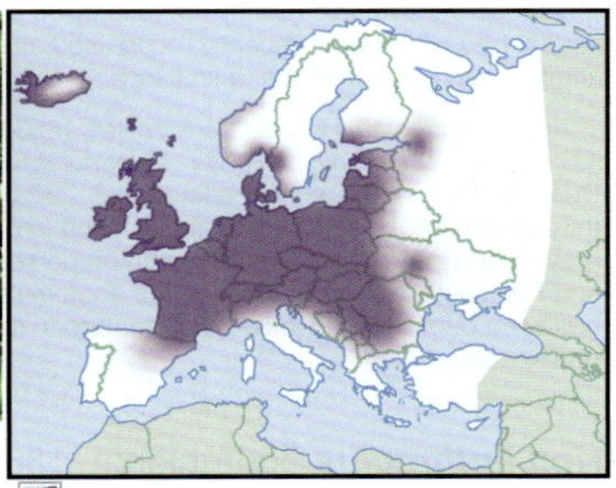

NW and central Europe to N Italy

Cool and humid forest habitats, often at fallen timber, more rarely outside forests and in cultivated land. In Britain also by roadsides and occasionally in gardens, mainly in lowlands. In Switzerland up to 1600 m. Feeds on mushroom mycelia, green plants, flowers and fruits.

Endangered in Austria, Vulnerable in Bayern. Scattered distribution, but common in Britain.

Kerney & Cameron 1979: 105, Falkner 1990: 17, Manganelli et al. 1995: 17, Turner et al. 1998: 320, Kerney 1999: 124, Geenen et al. 2006, Sysoev & Sch. 2009: 161.

Until the mid 1960s *A. circumscriptus* was not distinguished from *A. silvaticus* in Britain. Geenen et al. 2006 suggested that *A. fasciatus*, *A. silvaticus* and *A. circumscriptus* belong to the same species.

Arion distinctus Mabille, 1868

Arion distinctus Mabille, 1868 in Mabille 1867-1869: 42 [not verified]

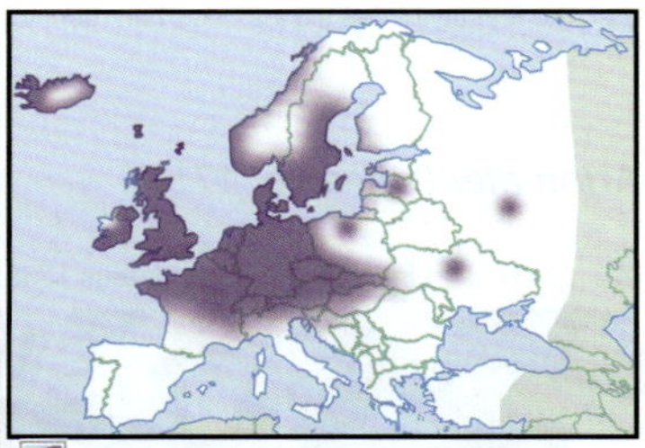

Europe

Up to 50 mm long

(1) Northern Ireland: Belfast Co., Galwally, (2) Czech Republic: Moravia, Olomouc

(1) Roy Anderson, (2) Michal Maňas

Slug usually yellowish or brownish grey, with dark lateral bands running relatively deeply at the sides, pneumostome usually enclosed by right-side band, first row of tubercles above foot-fringe slightly lighter but not as white as in *A. hortensis*, head and tentacles dark bluish grey without reddish hue. Sole light yellow.

Frequently in cultivated regions and disturbed habitats, but also in more or less natural deciduous forests and shrublands, or extensively used alpine pastures. In Britain common in moist and sheltered habitats among herbage and ground litter, often synahtorpic, but also in more natural habitats. A rapid colonist of waste ground and frequently spread by man, can be a pest in gardens and agriculture. In Switzerland up to 2300 m, usually below 1500 m. Feeds on green herbs and fruits, is also responsible for most holes in garden strawberries.

Life cycle 1-1.5 years, maturity is reached after less than 6 months, 200 eggs period can be laid in one period.

Kerney et al. 1983: 140 (*A. hortensis*), Falkner 1990: 198, Manganelli et al. 1995: 17, Turner et al. 1998: 325, Kerney 1999: 130.

Formerly not distinguished from *A. hortensis*, old *A. hortensis* records for Britain refer mostly to *A. distinctus*.

Arion fagophilus De Winter, 1986

Arion fagophilus De Winter, 1986 in De Winter 1986: 147

Slug grey on upper side with white margins and some white pigment granules sparsely dispersed over the body, sometimes with darker lateral bands, the right band passing above the pneumostome. Slug in cross-section somewhat bell-shaped. Tentacles bright reddish. Sole and foot fringe bright yellow orange.
Free oviduct with a large soft part containing a ligula, followed by a muscular shining part like in *A. hortensis*, outlet of epiphallus surrounded by a thin-walled structure, not like the papilla found in other *Arion (Kobeltia)* species.

Up to 40 mm long (preserved 25 x 5.5 mm)

In *Fagus* forests, seems to prefer calcareous substrate, hidden under leaf litter.

Castillejo 1997: 106.

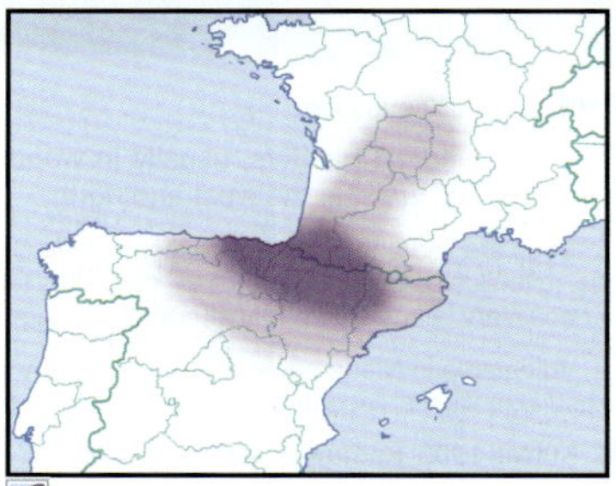

N Spain, SW and central France

Arion fasciatus (Nilsson, 1823)

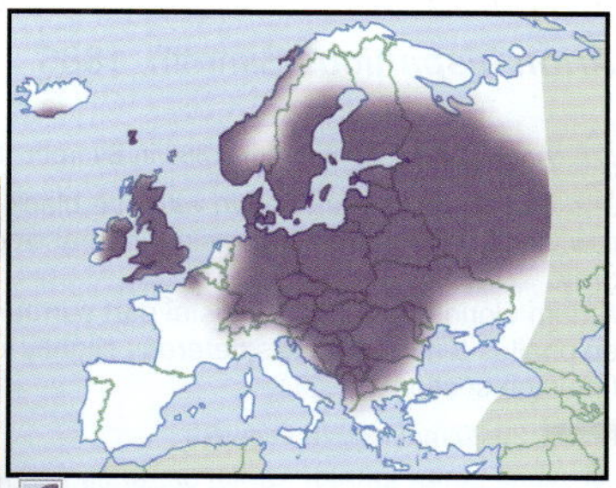

Limax fasciatus Nilsson, 1823 in Nilsson 1822: 3
Protected against its senior homonym *Limax fasciatus* Razoumowsky, 1789 under Art. 23.9 (Falkner et al. 2002: 141), but this was invalid, the name was used by Alzona 1971: 149 and others. The Commission was asked to solve this problem (Case 3569, Proschwitz & Falkner 2012).

Europe

Kerney & Cameron 1979: 105, Falkner 1990: 196, Turner et al. 1998: 321, Kerney 1999: 127.

Up to 50 mm long, exceptionally 70 mm

(1) England: Surrey, South Croydon, (2-3) Germany: Tambach-Dietharz near Gotha (1) Roy Anderson

Slug greyish with yellowish hue to dirty yellowish brown (not purely grey), getting paler downwards to the sides, with dark and usually blurry lateral bands and a yellowish zone below them, right lateral band passes above pneumostome, zone at and above near foot-fringe whitish, dorsum with a very weak keel-like structure, bell-shaped in transverse section when contracted. Sole greyish white. Mucus colourless. Juveniles light greyish brown with dark lateral bands and a more prominently looking light keel between mantle and posterior end.
Differs from *A. circumscriptus* and *A. silvaticus* in its yellow background colour.
Atrium relatively large, epiphallus wider than vas deferens, oviduct very narrow and as long as epiphallus, spermatheca elongate, not spherical.

Seems to prefer open habitats, cultivated areas, frequent in gardens and wastegrounds, but also in shady habitats under trees and shrubs, among herbage and ground litter. In southern Great Britain and Ireland mostly in disturbed habitats, in N Britain more in wild places. In Switzerland up to 1300 m.

Was not recognized as a species in the British Isles before 1962. Geenen et al. 2006 suggested that *A. fasciatus*, *A. silvaticus* and *A. circumscriptus* belong to the same species.

Arion flagellus Collinge, 1893

Arion flagellus Collinge, 1893 in Collinge 1893: 252

Slug light brownish white with dark brownish longitudinal bands in the median line and at the sides, some slugs are dark brown with white sides and black dashes, head bluish white, tentacles getting slightly darker to greyish blue, sole pale yellow.

In a variety of moist sheltered habitats, both wild and disturbed, such as gardens, waste ground, sea cliffs, rocky moorland and deciduous woods.

Castillejo 1992, Castillejo 1997: 82, Kerney 1999: 122.

Up to 60-100 mm long

Northern Ireland: Belfast Roy Anderson

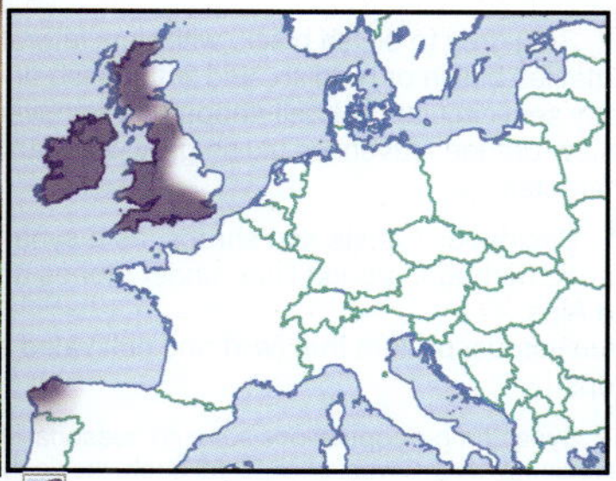

British Isles, NW Spain

Arion franciscoloi Boato, Bodon & Giusti, 1983

Arion franciscoloi Boato, Bodon & Giusti, 1983 in Boato, Bodon & Giusti 1983: 35

Slug grey, sometimes with yellowish or greenish hue, with a darker longitudinal colour band at each side, blurry at lower edge and with a white margin upside, head and tentacles bluish-grey, foot margin yellowish grey with small blackish spots, sole pale yellow. Surface with elongated tubercles surrounded by a fine reticulate pattern of dots, mantle 1/3 of body length, pneumostom on first half with a pale border, caudal mucus gland clearly evident, mucus colourless.
Resembles *A. hortensis*, differences in anatomical characters, characteristic foot-fringe spotting, position of genital aperture with respect to the respiratory orifice (located between right upper tentacle and slit of repiratory orifice).

30-40 mm long, preserved 25-30 mm

Lower montane deciduous woodlands, on acid and calcareous surfaces.

Manganelli et al. 1995: 17.

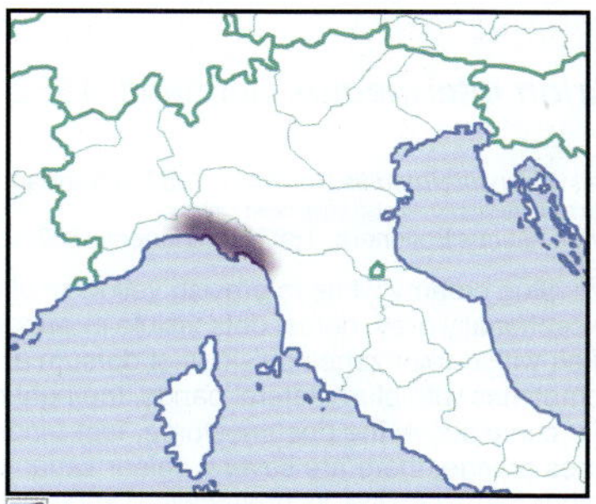

NW Italy (Liguria, Alpi Apuane)

Arion fuligineus Morelet, 1845

Arion fuligineus Morelet, 1845 in Morelet 1845: 30, Pl. 2

Slug brownish or orange with two longitudinal dark bands on dorsum and mantle, mantle covers 1/3 of body length, body below mantle lighter, respiratory hole in anterior section (0.35-0.4), distance between mantle and head short in crawling slug, head and tentacles dark, tentacles short, foot margin narrow and yellow orange in anterior sections, more reddish towards porterior end, posterior end pointed. Sole whitish with orange hue, mucus yellowish white, at the sole with orange hue.

65-70 mm long, maximum 90 mm

Under *Quercus*, *Juniperus* and *Pinus* on granitic soils.

Not much known. One copulation was observed in November.

Castillejo 1997: 80, Verdú & Galante 2006.

Spain (S Galicia), N Portugal
Possibly also in Cantabria.

Vulnerable.

Arion gilvus Torres Mínguez, 1925

Arion gilvus Torres Mínguez, 1925 in Torres Mínguez 1925: 235

Slug yellowish brown on dorsum, getting lighter downwards at the sides, with a central black band and dark lateral bands with a yellowish grey line on their upper side (also on mantle, the right one arching above the pneumostome), foot-fringe yellow without interruptions, sole white or evenly yellowish. Skin tubercles fine. Mucus pale yellow.
Epiphallus, spermatheca duct and free oviduct join the atrium on a single line, epiphallus being in the middle. Epiphallus is longer than vas deferens.

Up to 65 mm long (preserved 40 mm)

Prefers forests, also in *Citrus* cultivations.

Garrido et al. 1995: 112, Castillejo 1997: 93.

Spain (Tarragona)

Arion hispanicus Simroth, 1886

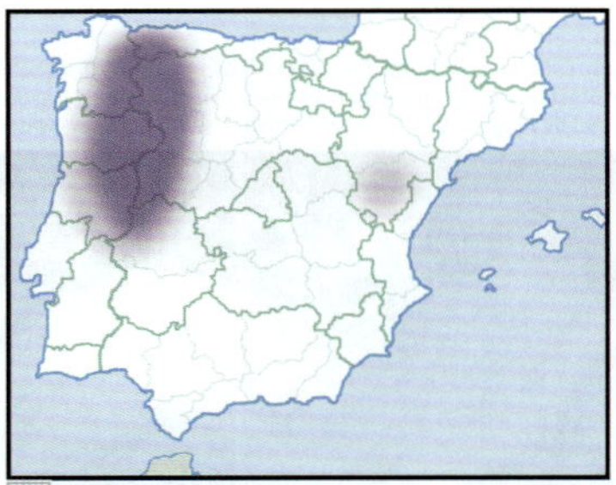

Arion hispanicus Simroth, 1886 in Simroth 1886: 21, Taf. 1

Slug black or dark brown with fine black lines and a longitudinal black band at each side, delimited by a narrow white band on the upper side, lateral sides light brown, head and tentacles grey, sole whitish, yellowish or grey.
Atrium elongated and in its distal part surrounded by a glandulous wall, spermatheca ovoid with short duct, epiphallus 3 mm long, vas deferens slightly longer (4 mm), spermoviduct long and often with dark pigment in its feminine section.

20-30 mm long (preserved)

Was found among herbal vegetation in humid habitats.

Pollonera 1889: 629, Castillejo 1997: 99.

N Portugal, N Spain

Arion hortensis Férussac, 1819

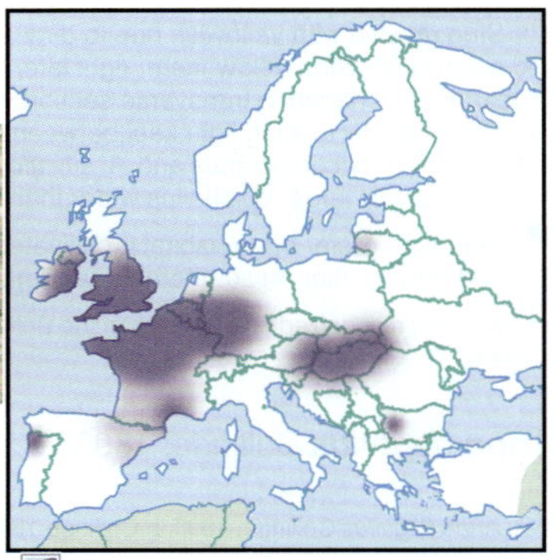

Arion hortensis Férussac, 1819 in A. Férussac & Deshayes 1819-1851 (1): 65
Text p. 65 from 18 Sep 1819, Pl. 12 Fig. 4-6 without name on plate or cover from 6 Mar 1819.

30-40 mm long (preserved 15 mm), rarely 50 mm

W Germany: Hackenheim near Bad Kreuznach — Ira Richling

Slug grey to bluish black, with dark lateral bands (right band running usually entirely above pneumostome), tentacles bluish or reddish, and with bluish or orange tip at the tail, first row of tubercles above foot-fringe white, body semicircular (not bell-shaped) in transverse section when contracted. Sole orange or yellow. Body mucus yellow-orange. Juveniles bluish grey with darker dorsum and yellow sole. The colour of juveniles and adults is characteristic.

British Isles, France, introduced to many regions

Deciduous forests and shrubs, as a synanthrope frequently in cultivated areas (gardens, parks, cultivated crops, greenhouses, cellars, warehouses), among herbage, ground litter, under stones and in soil crevices. Can be a horticultural pest in Britain, in N Britain exclusively as a synanthrope. Up to 2000 m in the Alps.
Omnivorous, prefers fruit (wild and cultivated strawberries), seedlings and young weak plants. Species can be a serious pest in greenhouses and garden crops.

Synanthropic populations have no seasonal life cycle, all stages can be met simultaneously, all stages are able to hibernate. Slugs lay up to 200 eggs (length 2-3.5 mm) in batches (every 1-3 weeks) of 12-80 each, inside the soil, juveniles hatch after 20-40 days, maturity is reached after at least 4-6 months, 6-7 months is usual, life span usually about 9 months, sometimes longer.

Künkel 1908, Kerney & C. 1979: 106, Wiktor 1983: 93, Falkner 1990: 198, Turner et al. 1998: 324, Kerney 1999: 129, Castillejo 1997: 103, Colling et al. 2008: 128.

Old British records for *A. hortensis* refer to *A. distinctus*. Mandahl-Barth 1938 reported the species from Iceland.

Arion intermedius Normand, 1852

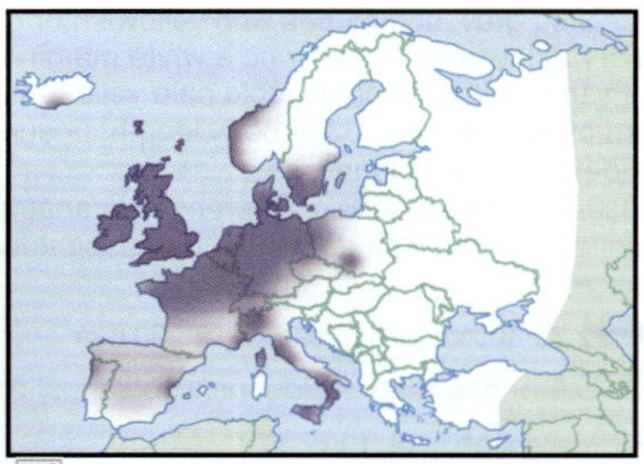

Arion intermedius Normand, 1852 in Normand 1852: 7 [not verified]
Original genus could also have been *Limax*.
Arion alpinus Pollonera, 1887 in Pollonera 1887: 203 (or 305, depends on the copy)

Slug creamy white to greyish yellow or olive brown, exceptionally grey, dorsal dots visible in withdrawn slugs, often with darker zones on sides of dorsum and mantle, sometimes with blurry lateral bands, the right one running in a curve above the pneumostome, foot-fringe sometimes orange, there are several colour variants. Different colour forms can live sympatrically. Tentacles darker and small, pneumostome in the middle or in in 2/5 anterior section of the mantle. Sole yellow. Mucus pale yellow. Shell small and solid, upper side convex and lower side flat. When inactive and not disturbed the slugs are often contracted with spiky tubercles (more frequently observed in semiadults).
Differs from *A. obesoductus* in its short free oviduct.

Northern Ireland: Down County, Hillsborough forest — Roy Anderson

Up to 25-35 mm long (preserved 15 mm)

Europe

In leaf litter layers of humid forests, often on acid substrate, also in river plain woods and at lake margins. In Switzerland up to 900 m, in high altitudes in the Scottish highlands. Seems to feed on mushrooms and dead leaves.

A rare species in central Europe, but widespread and evenly distributed in the British Isles.

Possible exclusively self-fertile. In dry forests slugs estivate with curved dorsum and convex sole in small self-constructed nests made of soil particles carried above the sole. Adults appear in autumn, life cycle 1 year.

Pollonera 1889: 637, Nobre 1941: 70, Kerney & C. 1979: 106, Falkner 1990: 198, Manganelli et al. 1995: 17, 40, Castillejo 1997: 108, Turner et al. 1998: 326, Kerney 1999: 132, Horsák et al. 2010: 26, Manganelli, Bo. & G. 2010.

Prior to 2010 *A. obesoductus* was partly confounded with this species, so that reports of *A. alpinus* referred to *A. obesoductus*.

Arion iratii Garrido, Castillejo & Iglesias, 1995

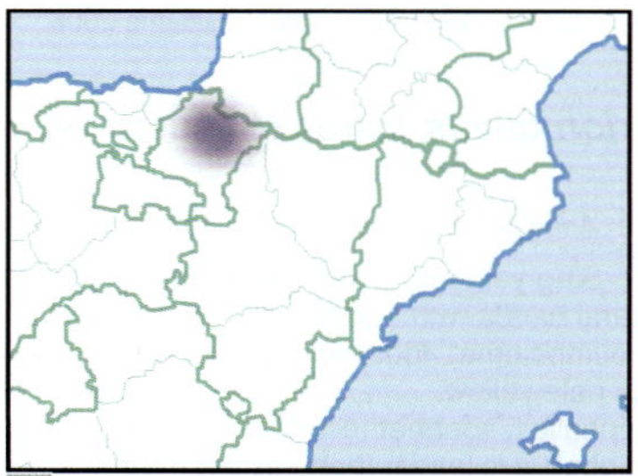

Arion iratii Garrido, Castillejo & Iglesias, 1995 in Garrido, Castillejo & Iglesias 1995: 107

Slug brown or blackish, getting lighter downwards to the sides, with two dark colour bands on the sides and the mantle (right one arching above pneumostome) with a white line above them, colour bands are visible in brown but hardly in black individuals, foot-fringe grey with black interruptions. Skin tubercles unusually thick. Mucus gland white, sole white or yellow.
Epiphallus, spermatheca duct and free oviduct join the atrium on a single line, the epiphallus being in the middle. Epiphallus longer than vas deferens, spermatheca spherical or ovoid.

Up to 45 mm long (preserved)

In leaf litter of beech forests.

Copulation lasts for at least 2 hours, the individuals bend in a C in front of the right side of the partner's head, copulatory organs are everted, two funnel-shaped stimulators are visible.

In the original description the authors missed to explain how to distinguish *A. iratii* from other species, particularly in the genitalia. Castillejo 197: 87 reported the presence of thicker skin tubercles than in the *A. subfuscus* complex.

W Pyrenees (Navarra)

Vulnerable. Only known from one locality. Threatened by potential habitat alterations.

Arion isselii (Lessona & Pollonera, 1882)

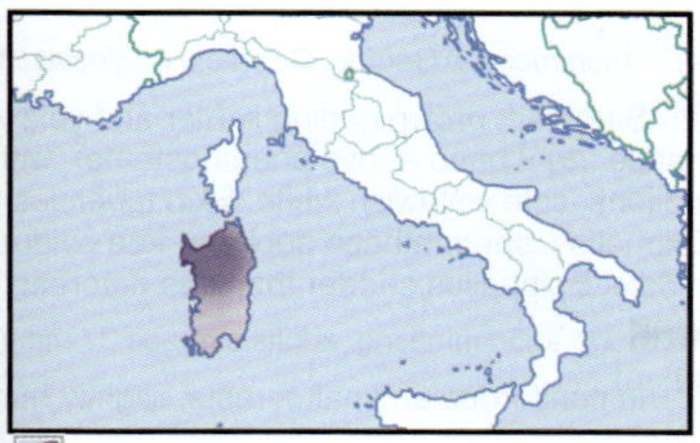

Italy (Sardinia)

Italy: Sardinia (1) Clemens M. Brandstetter, (2) Francesco Mascia

Ariunculus isselii Lessona & Pollonera, 1882 in Lessona & Pollonera 1882: 66

Slug evenly olive blackish, mantle rounded and blackish, genital pore closer to the respiratory hole than to the tentacles (difference to other *Arion* species), margin of foot and margin of sole olive, medial zone of sole whitish.

Moderately humid habitats in deciduous forests at the soil among rocks or stones covered with moss, also in rocky habitats outside forests.

Boato et al. 1983: 41. Up to 110 mm (preserved 35-40 mm)

Arion lizarrustii Garrido, Castillejo & Iglesias, 1995

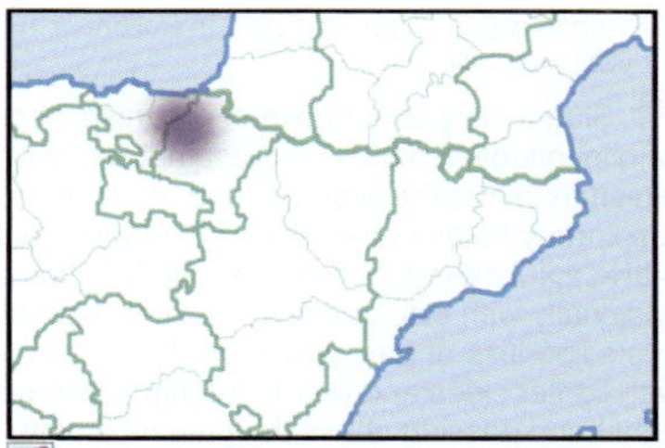

N Spain (Navarra)

Arion lizarrustii Garrido, Castillejo & Iglesias, 1995 in Garrido, Castillejo & Iglesias 1995: 110

Slug dark brown on dorsum with black streaks and dark lateral bands with a yellowish brown line in their upper side, extending to the mantle (the right one arching above the pneuomostome), sides yellowish brown, two rows above foot-fringe white, foot-fringe yellowish, sole cream. Head grey, tentacles dark. Skin tubercles fine. Sole mucus white, body mucus orange.
Epiphallus, spermatheca duct and free oviduct join the atrium on a single line, the epiphallus being in the middle. Epiphallus slightly longer than vas deferens.

Up to 45 mm long, adults average 36 mm (preserved)

Copula is static, copulatory organs are everted in the form of a spherical mass which does not fill up the internal space encircled by the partners (contrast to *A. subfuscus*). At the end of the copula one partner raises the front part of its body over the other slug, before moving away.

In the original description the authors failed to explain explicitly how the species differs from other species, particularly in the genitalia.

Arion lusitanicus Mabille, 1868 (not the invader *A. vulgaris*)

Arion lusitanicus Mabille, 1868 in Mabille 1868 [Apr 1868]: 134 (*Arion Lusitanicus*)
Possibly 1 month earlier published by Mabille 1867-1869 [Mar 1868]: 33-52. I have not verified that source.

Up to 80 mm long

Portugal: near Sintra Gonçalo M. Rosa

Central W Portugal
In the literature only reported for Serra da Arrábida, but possibly also in Serra da Sintra.

Restricted to a small range in Portugal, was not introduced to central Europe.

Slug dark brown with yellowish hue and two light brown bands on dorsum and mantle, the right band passing above the pneumostome, more rarely reddish, tubercles smaller than in *A. ater*, tentacles blackish, margin of foot reddish or yellowish with dark transversal lines reaching into the lateral parts of the sole, sole olive greyish, marginal zones darker, mucus colourless and very sticky in living animals, turning yellow when preserved, at the sole intensively yellow. Juveniles greenish grey with two whitish bands on mantle and posterior part of dorsum, sole whitish or with orange hue.
Atrium spherical and undivided, spermatheca ovoid, spermatheca duct as long as spermatheca or slightly shorter, getting slightly wider near insertion at atrium, epiphallus longer than vas deferens, vagina longer and wider than spermatheca duct.

In the vegetation of natural mountain habitats on calcareous substrate, hides in very narrow holes in the soil (which it enters with its head in front and leaves with its tail in front). Consumes mushrooms in spring and autumn, in the other seasons herbal plants, mosses and lichens.

Copulations were observed in January and February.

Morelet 1845: 29 (forms γ and δ), Mabille 1868: 134, Pollonera 1889: 627, Castillejo 1997: 77, Kerney 1999: 121.

Between 1956 and 1997 the pest slug *Arion vulgaris* which invaded central Europe was misidentified as *A. lusitanicus*. Modern reports for *A. lusitanicus* refer usually to *A. vulgaris* if outdated sources were used. See comments under *A. vulgaris*.

Arion magnus Torres Minguez, 1923

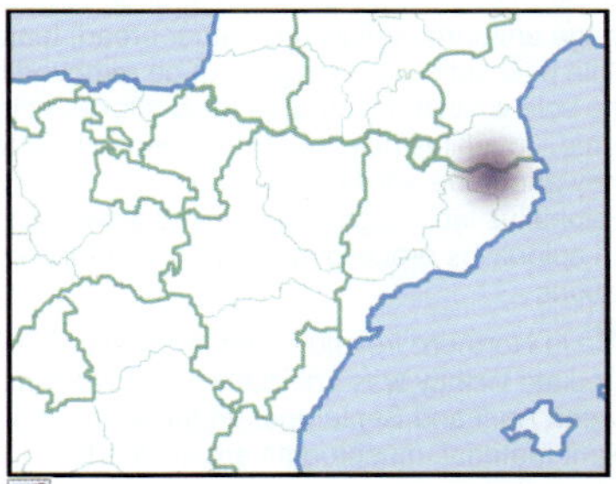

E Pyrenees

Arion magnus Torres Minguez, 1923 in Torres Minguez 1923: 7

Slug large, pale reddish, slightly lighter down to the sides, margin of foot with black transverse lines, tentacles black, mantle oval and rounded at both ends, body mucus pale yellowish, caudal gland mucus reddish-orange.
Genitalia like in *A. vulgaris*, except the atrium which is not spherical but long and slender like in juvenile *A. vulgaris*.

Up to 180-200 x 20-30 mm

Was reported from mountains above 1200 m.

Backeljau 1989: 204.

Poorly known, more information on this form is needed.

Arion molinae Garrido, Castillejo & Iglesias, 1995

Arion molinae Garrido, Castillejo & Iglesias, 1995 in Garrido, Castillejo & Iglesias 1995: 110

Slug dark brown getting lighter and yellowish downwards at the sides, with lateral dark bands on sides and mantle (right band above pneumostome), with a white line above each band, foot-fringe yellowish with grey interruptions, sole yellowish white. Skin tubercles fine. Mucus was forgotten to mention.
Epiphallus, spermatheca duct and free oviduct join the atrium on a single line, the spermatheca duct being in the middle. Epiphallus shorter than vas deferens.

Up to 50 mm long, adults average 37 mm (preserved)

In coniferous and oak forests, slightly humid habitats.

During copula partners place themselves in opposite directions curving in a semicircle, everted copulatory organs have the form of a cylinder, the two individual parts can be distinguished. No rotation during copula. At the end one partner moves away, the other slug licks the secreted mucus.

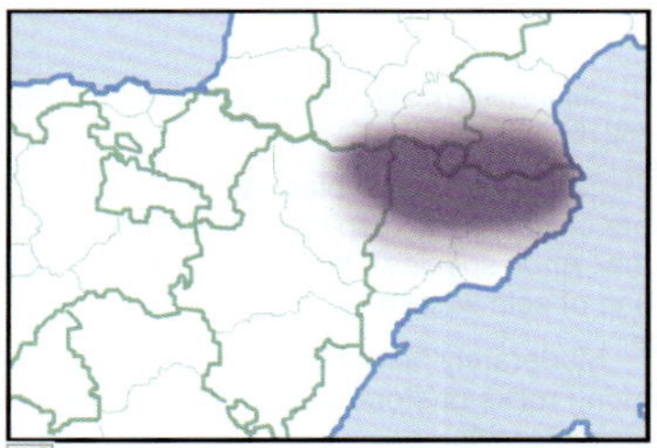

Central and E Pyrenees

In their description the authors missed to explain how to distinguish *A. molinae* from other species, particularly in the genitalia.

Arion nobrei Pollonera, 1889

Arion nobrei Pollonera, 1889 in Pollonera 1889: 404 (or 626) (*Arion Nobrei*)

Slug evenly dark brown or greenish, occasionally with longitudinal bands on dorsum, body slightly more bluish than mantle, posterior end of mantle rounded, respiratory hole in anterior part of mantle, tentacles black, sole margin sometimes slightly lighter and occasionally with black transversal lines, sole black, mucus colourless or pale yellowish. Juveniles dark brown with lighter sides and two dark longitudinal bands, the right band passing above the pneumostome, margins of sole evenly brown, sole light grey.
Differs from *A. ater* in its entirely black sole.

In natural habitats. Frequent in Serra da Estrela.

Castillejo 1997: 74.

90-120 mm long

Portugal, possibly W Spain
Eastern range limit unknown.

Portugal: Serra da Estrela
Gonçalo M. Rosa

Arion obesoductus Reischütz, 1973

Arion obesoductus Reischütz, 1973 in Reischütz 1973: 229

Slug whitish yellow with light greyish brown dorsal band and dark greyish brown lateral bands amtle with small and irregular tubercles, head and tentacles greyish blue, foot and sole orange or yellowish. Mucus yellowish.
Free oviduct long, epiphallus papilla flat.
Differs from juvenile *A. subfuscus* in its yellowish mucus. *A. intermedius* has a short free oviduct, *A. distinctus* a conical epiphallus papilla.

In forests, under the bark of fallen wood, seems to be confined to strictly natural habitats and feeds probably on mushrooms.

Kerney et al. 1983: 142 (*alpinus*), 146, Falkner 1990: 199 Fig. 3, Manganelli, Bo. & G. 2010, Horsák et al. 2010: 26.

Up to 15-25 mm long

Czech Republic: Žákova Hora mountain
Michal Maňas

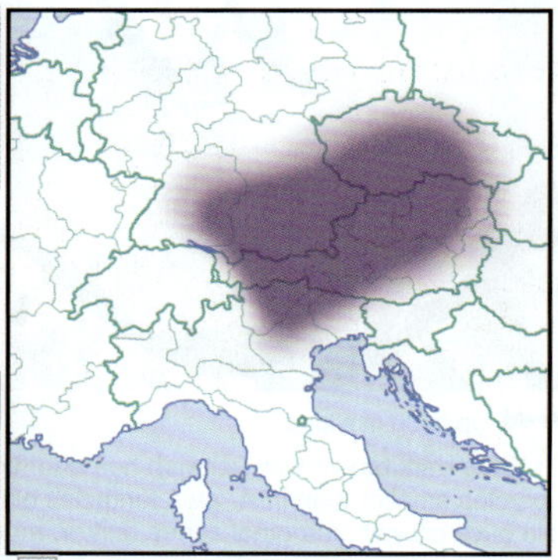

Czech Republic to NE Italy

Was previously called *Arion alpinus* Pollonera, 1887, which was described from Piemonte and Lombardia. It seems that *A. obesoductus* does not live sympatrically with *A. intermedius*. The range of *A. obesoductus* is insufficiently known, Switzerland and NW Italy are also possible. Falkner et al. 2002 argued that specimens from Piemonte belonged to this species, but the arguments were not based on published results of anatomical studies that would have enabled Manganelli et al. 2010 to verify the identity of those specimens.

Arion occultus Anderson, 2004

Arion occultus Anderson, 2004 in Anderson 2004: 343

Slug larger but superficially similar to *A. distinctus*, with broad, high lateral bands, below which the sides are predominantly pale or white, rather than grey. Dorsum with strongly raised and keeled tubercles, foot fringe uniformly yellow and rather broad, tentacles dark with bluish hue. Sole pale yellow or orange, can be removed by rough handling, leaving the sole whitish, the orange mucus takes at least several days to regenerate. Juveniles pale and translucent only a few days after hatch, then they adopt the adult colouration.
Epiphallus lacks pigmentation and a swelling at its distal end, also lacks an epiphallus process at the junction of the epiphallus with the distal oviduct, simple spermatophore present, comprising an oblong, flattened sac lacking serrations.

Northern Ireland: Ballywalter
Roy Anderson

Up to 42 mm long
(preserved 28 mm)

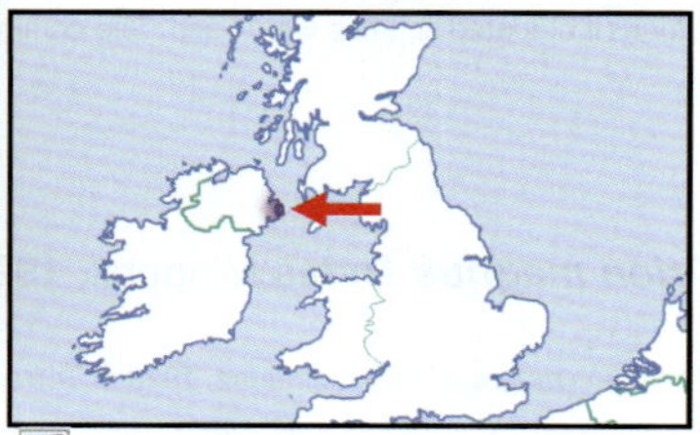

Northern Ireland, probably introduced from somewhere else

In Northern Ireland found in coastal vicinity on sandy substrate planted with woods of *Acer pseudoplatanus*, little or no shrub layer, some herbs, the immediate vicinity was artificially disturbed, wood margins were used as a junkyard for rubbish. Slugs are largely subterranean and appear at the surface between April and September, in the winter they seem to hibernate in self-made chambers (in contrast to *A. hortensis* and *A. distinctus*). Feeds on decaying plant material, mushrooms are avoided.

Breeding in Ireland occurs throughout the year, main periods are in May/June and September/October. Copulation takes one hour, eggs (2.3 x 1.9 mm) are laid on the soil surface in batches between 10 and 20, clutch size 30-50. Juveniles hatch after 4-6 weeks.

Arion owenii Davies, 1979

Northern Ireland: Tyrone County, Termon Glen
Roy Anderson

British Isles, Norway, Hungary

Arion owenii Davies, 1979 in Davies 1979: 126

Slug usually brown with dark colour bands, which can be interrupted (in Cornwall sometimes with few dark dots instead of bands), sides greyish, surface with characteristic very regular and rather sharply ridged tubercles. Tentacles pinkish brown. Sole usually very pale yellow, sometimes yellow or orange.

In moist sheltered habitats under ground litter and in the soil, both in wild and disturbed habitats.

Up to 35 mm long

Maturity is reached in autum, breeding mainly in the winter months. Mating lasts about 45 min, longer than in *A. distinctus* and shorter than in *A. hortensis*.

A local species, frequent in some areas (Cornwall, N Ireland, N England to S Scotland), rare and scattered elsewhere in the British Isles.

Kerney et al. 1983: 141, Kerney 1999: 131, Falkner et al. 2002: 141 (not in Bretagne), Pintér & Suara 2004 (Hungary).

Arion paularensis Wiktor & Parejo, 1989

Central Spain

Arion paularensis Wiktor & Parejo, 1989 in Wiktor & Parejo 1989: 27

Slug brown-blackish with dark lateral bands, middle of dorsum lighter, 13-17 large skin-wrinkles between mid-line of dorsum and pneumostome, sole cream. Mucus yellow.

Up to 27 x 6 mm (preserved)

In mountains under stones, im moss and wet ground depressions of pine forests with thin soil layer on granitic substrate, occasionally in valleys with fewer pine trees.

Slugs found in autumn in the mountains were sexually mature and had probably copulated.

Arion ponsi Quintana Cardona, 2007

Spain (Menorca)

Arion ponsi Quintana Cardona, 2007 in Quintana Cardona 2007: 141

Slug of variable colour between orange and cream, with two longitudinal dark bands on mantle and dorsum, which can be blurry in the posterior section, the right band passes above the respiratory hole, foot-fringe of the same colour and with some dark transversal lines, mantle generally a little lighter, tentacles dark grey, sole cream with greyish hue. Mucus transparent.
Anatomy like in *A. lizarrustii*, *A. gilvus* and *A. iratii*, but epiphallus shorter than vas deferens.

Up to 65 x 9.5 mm

Forest habitats under *Quercus ilex*. Occurs in all parts of Menorca.

Arion rufus (Linnæus, 1758)

Limax rufus Linnæus, 1758 in Linnæus 1758: 652

(1) France: Isère, Réserve naturelle du Grand Lemps, (2 moss) Germany: Rheinland-Pfalz, near Mengerschied, (3 face) unrecorded, (4 black) France: Isère, Grande Charteuse, (5) Denmark: Bornholm, near Hammerhus
(1, 4) Cédric Audibert, (3) Guillaume Brocker

Up to 150 mm long, exceptionally 180 mm

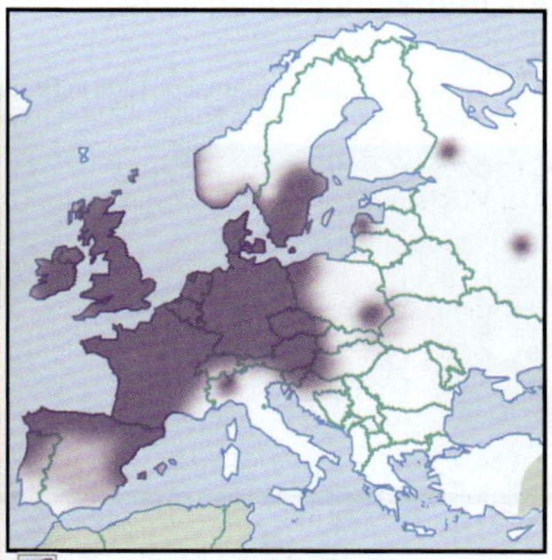

Central and W Europe, N Italy

Slug usually reddish, but also brown, orange or black, foot-fringe nearly always red to yellowish brown, tubercles large and elongate. Sole light grey or cream to reddish. Juveniles pale yellowish to light orange with dark head, tentacles blackish brown, sometimes with pale colour bands. Atrium very large, this is also visible during copulation (white mass considerably larger than in *A. ater* and *A. vulgaris*), approximately same diameter as spermatheca, epiphallus not much wider than vas deferens, oviduct narrow, spermatheca spherical.

Natural humid forests and humid meadows, margins of water bodies, swamps, in central Europe also in cultivated areas until the 1970s when it was replaced by *A. vulgaris*. In Britain mainly in lowlands and man-made habitats in southern Britain and Ireland. Feeds on green and rotting plants, mushrooms, faeces and dead animals. In Spain as a synanthope, feeding on desintegrating and green plants, in many zones a horticultural pest. Up to 1800 m in Alps and Pyrenees.

Life cycle 1 year. Reproduction by copulation, occasionally also by self-fertilization, up to 500 oval eggs (3 x 4 mm) in several clutches of 15-60 eggs are laid between May and September, juveniles hatch after 25-40 days and reach maturity in the next summer after 9-12 months.

Vulnerable in Austria, decreasing in Bayern. Increasingly rare, disappeared nearly completely from cultivated areas. Originally widespread and frequent, but has been replaced by *A. vulgaris* in many regions.

Moquin-Tandon 1855: 10 (life cycle), Falkner 1990: 194.

The taxonomic concepts of Falkner 1990 and Castillejo 1997 (*ater* and *rufus* in Spain/Portugal) are not in agreement with each other, more research is needed. Manganelli, Bo. & G. 2000 reported the species from NW Toscana. Falkner's 1990 concept is applied here.

Arion silvaticus Lohmander, 1937

Without gender agreement: *Arion silvatica* Lohmander, 1937

Arion circumscriptus var. *silvatica* Lohmander, 1937 in Lohmander 1937: 98

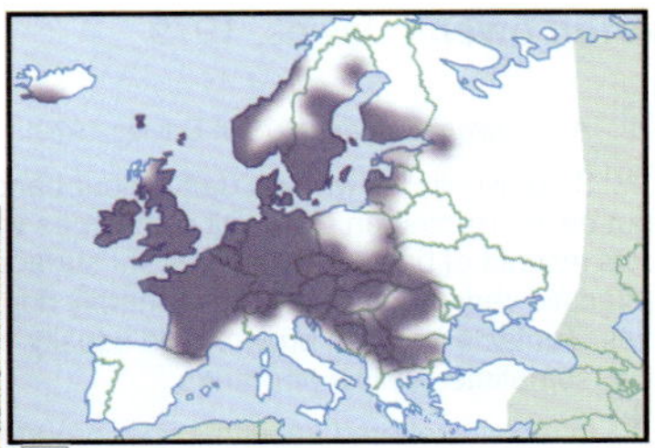

Europe

Up to 40 mm long (preserved 21 x 6 mm) (1-3) Bulgaria: Stara Planina, Shipka peak, (3) Germany: Schleswig-Holstein, near Grömitz (1-3) Dilian Georgiev, (4) Vollrath Wiese

Slug pale grey to greyish brown (in Bulgaria generally lighter than in Central Europe), getting whitish downwards to the sides, with broad and well marked (at least towards the lower side) dark lateral colour bands (overal impression grey and white, not yellowish), often with a keel-like structure. Sole whitish or cream. Mucus transparent or slightly yellowish. When inactive the slugs are often contracted and flat like isopods, with the light lateral zones forming a frame. Juveniles light bluish grey with weak and blurry slightly darker grey lateral bands, and whitish keel-like structure between mantle and posterior end.
Atrium large, broad and flattened, epiphallus wider than vas deferens, oviduct wide, half diameter as atrium, spermatheca elongate and somewhat arrow-shaped.

In leaf litter layers of humid deciduous, mixed and coniferous forests of mountains and highlands, in the north also in moore woodlands, restricted to shady and sheltered habitats. Tolerates non-calcareous substrate. In Bulgaria up to 1600 m, in Switzerland 2400 m. Consumes dead leaves.

Kerney & Cameron 1979: 105, Wiktor 1983: 91, Falkner 1990: 196, Turner et al. 1998: 322, Kerney 1999: 126, Irikov & Mollov 2006: 792.

A. circumscriptus and *A. silvaticus* were not distinguished in Britain until the mid-1960s. Geenen et al. 2006 suggested that *A. fasciatus*, *A. silvaticus* and *A. circumscriptus* belong to the same species.

Arion speziae (Lessona, 1881)

Ariunculus speziae Lessona, 1881 in Lessona 1881: 193, Tav. 4
Expressly dedicated to Mr Spezia, name not to be corrected to *speziai* (Art. 32.5). Art. 31.1 has no effective force.

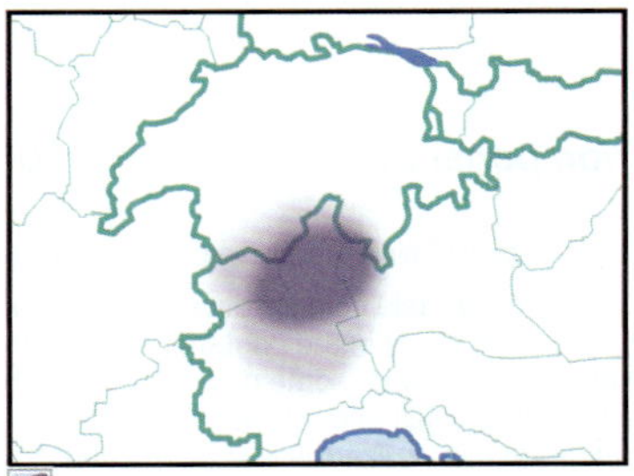

NW Italy (Piemonte)

Slug yellowish brown with dark lateral bands running behind the mantle at each side, posterior end truncated, not pointed, tentacles black, mantle small (1/3 of slug length in preserved specimens) and evenly yellowish brown, respiratory hole in the mid section of mantle and slightly in frontal position, sole not much different from upper body.
Differs from *A. hortensis* and other species in its smaller dimensions, smaller mantle and its dark lateral bands.

20 mm long (preserved)

Under tree trunks in coniferous forests.

Manganelli et al. 1995: 17, Turner et al. 1998: 426.

Arion subfuscus (Draparnaud, 1805)
Arion fuscus (Müller, 1774)

(2 separate and almost cryptic species)

Limax subfuscus Draparnaud, 1805 in Draparnaud 1805: 125, Pl. 9 Type locality in S France.
Limax fuscus Müller, 1774 in Müller, 1774: 11 Type locality in Denmark.

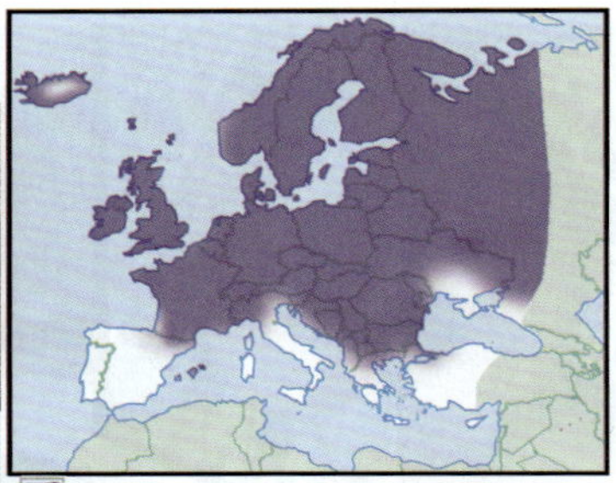

Europe to Caucasus and Ural, introduced to North America
Both species occur sympatrically, or at least in some regions. Reliable records in the literature for *fuscus* from N France to Sweden, for *subfuscus* from S France to N Germany.

(1, 2) W Germany: Buir near Euskirchen
(3, lower left) Bulgaria: near Ravnogor
(4) Czech Republic: Moravia, Panské louky
(1, 2) Ira Richling, (3, lower left) Dilian Georgiev, (4) Michal Maňas,

Up to 75 mm long (preserved 45 mm) (1) *Arion fuscus*, (2, upper right) *Arion subfuscus*, (3, 4) unknown

Slug between orange and brown to dark chocolate, colour varies within populations. Dark lateral bands often in young individuals, after preservation blackish. Dark pigment reaches under the pneumostome from back to front. Sole cream or whitish. Skin tubercles relatively thick. Body mucus bright yellow or orange (can be tested with a finger), in larger specimens sometimes colourless, sole mucus colourless.
Juveniles after hatching are pale brown with blackish tenacles, a few days later they get bright and well contrasting colours with lateral bands uniting at the posterior end, a light zone above the bands, dorsum darker.
Atrium about half diameter of spermatheca, spermatheca spherical, oviduct large and swollen like in *A. vulgaris*, anatomy generally very variable.
A. fuscus: Gonad small and dark, and embedded inside the digestive gland (only visible after removing it).
A. subfuscus: Gonad large and pale, and visible at the periphery of the digestive gland.
Both species differ from other species in its yellow-orange body mucus.

Mainly a woodland slug, inhabiting both deciduous and coniferous forests, also on rocks and meadows, in some regions also frequently in gardens and at roadsides. In Bulgaria up to 2900 m. Slugs ascend trees in moist weather.
Omnivorous, feeding mainly on mushrooms, but consumes also dead and alive parts of flower plants, fruit and vertebrate excrements. Can be noxious to crops cultivating trees for reforestation, since the 1990 also observed as a pest in agricultural crops and gardens.

Life span approximately 1 year, mature individuals occur in summer and autumn. Individuals in central Europe lay up to 50 eggs (diameter 3-4 mm) every 1-3 weeks from June to September, there is a trend to higher egg numbers in the early season, no seasonal preference in the south, the adults still increase their size during this time and become darker. The slugs die a few days after the last eggs were laid. Juveniles hatch after at least 27 days, with cold temperatures up to more than 100 days, in spring when temperatures have reached 10 °C in 5 cm depth. Maturity is reached after at least 4-6 months, but then they do not reach an age of 1 year, usually between 8.5 and 10.5 months, live span under very favourable conditions up to 17 months. 25 mm long snails may also have fully developed genitalia, mature from 35 mm onwards.
Copulation takes at least 2 hours, partners are bent in C shape forming a circle, tentacles are withdrawn, partners do not rotate, everted copulatory organs being a large hyaline ball occupying entirely the circular space between the individuals. Towards the end one partner becomes more active, the organs are withdrawn, the tips of the spermatophores remain visible, the more active partner moves away first.

Künkel 1908, Kerney & Cameron 1979: 104, Wiktor 1983: 86, Garrido et al. 1995, Castillejo 1997: 85, Wiktor 2001: 17, Falkner et al. 2002: 140, Pinceel et al. 2004, Wiktor 2004: 143 (species complex?), Colling et al. 2008: 126, Kozłowski 2008.

Pinceel et al. 2004 studied material from a transect between N France and Sweden and were able to distinguish two forms by molecular methods and gonad morphology, which they identified as two separate species *fuscus* (Belgium to Sweden) and *subfuscus* (France to N Germany). More research is needed: do both species have the same habitat requirements, mating behaviour and life cycle? And where are they distributed? Which species live in the British Isles, which ones in E and SE Europe?

Arion urbiae De Winter, 1986

N Spain

Arion urbiae De Winter, 1986 in De Winter 1986: 152

Slug greyish to blackish, gradually lighter downwards at the sides, sometimes with dirty white colour bands along dorsum and mantle. Sole yellowish white. Mucus colourless, dirty white or very pale yellowish.
Genitalia like *A. subfuscus*, but spermatheca duct wider, epiphallus and oviduct usually with dark pigment, free oviduct with two longitudinal folds, spermoviduct greyish. Spermatophore with a spirally arranged and serrated longitudinal ridge and a long filiform tail.

Up to 60 mm long (preserved 33 mm)

Seems to prefer open habitats with herbal vegetation on calcareous substrate.

Castillejo 1997: 97.

Arion vulgaris Moquin-Tandon, 1855

Arion lusitanicus sensu Van Regteren Altena 1956 non Mabille, 1868

Arion rufus var. *vulgaris* Moquin-Tandon, 1855 in Moquin-Tandon 1855: 10, Pl. 1 Type locality in S and central France.

(1) N Italy: Sondrio, Piuro, Borgonuovo, (2, 4-6) Germany: Göttingen, (3) Germany: Bayern, near Bad Tölz
(1) Clemens M. Brandstetter

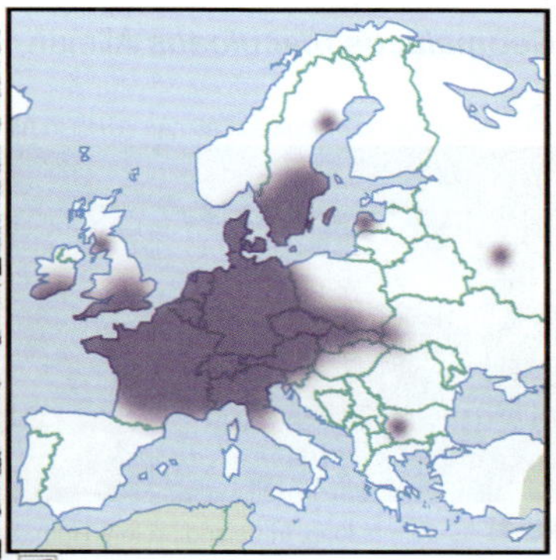

Originally perhaps SW France, today widely dispersed in Europe
In Switzerland since at least 1956, in Britain known since 1954, in W Ukraine since 2011.

80-120 mm long

Slug brownish or reddish brown, or brightly orange. Colouration varies, but not within the same population. Adults differ little from *A. rufus*, in contrast to the brightly coloured and banded juveniles that allow reliable determination of this species.
Atrium small (this is also visible in the copulation, difference to *A. rufus* where the white mass is considerably larger), oviduct large and swollen, same diameter as atrium, spermatheca spherical, its diameter twice of oviduct. The long, anteriorly swollen oviduct with large longitudinal ligula inside is distinctive of *A. vulgaris*.

Cultivated habitats of any kind, also in natural habitats such as river and lake margins, margins of forests, forests in valleys or moderately dry meadows. Serious horticultural pest in large parts of Europe. Dispersal mainly by soil for horticultural purposes containing eggs. In Switzerland up to 1700 m (1998).
Has become the most frequent terrestrial mollusc species in large parts of central Europe. Also spreading in Scandinavia and Britain.

Kerney & Cameron 1979: 104, Falkner 1990: 194, Wiktor 1996: 6, Wiktor 1983: 83, Kerney 1999: 121, Nardi 2011.

Was misidentified as *Arion lusitanicus* since Van Regteren Altena 1956 and prior to Castillejo 1997 and Kerney 1999: 121. Falkner et al. 2002: 139 proposed to officially conserve *lusitanicus* for the invader and to rename to Portuguese endemic, but took no action. Bank et al. 2007: 49 argued that *A. vulgaris* was not used except by Anderson 2005 and Reischütz. Today *A. vulgaris* is widely used. In Jan 2010 there were 3800 Google hits for *A. vulgaris* (and 14000 for *A. lusitanicus*), so *vulgaris* is quickly gaining importance. I such a situation it would make no sense to conserve usage of the name *lusitanicus* for the invader.

The name *vulgaris* is currently not based on types. Its original description gave as type locality northern and central France, a brief description only referring to the colour, and a reference to Linnæus 1758: 652 (description of *A rufus*). Moquin-Tandon 1855 described 11 varieties of *A. rufus* and regarded *vulgaris* as the nominate form (alpha). Much more research is necessary. The French populations of this species need to be studied in detail to finally come up with a sound hypothesis as to which was the original population from which the European invasion originated. Once this is understood, a neotype can be selected for *A. vulgaris* that will fit the original description (which will be easy) and the type locality (only needs to come from France).

Arion wiktori Parejo & Martín, 1990

N Spain

Arion wiktori Parejo & Martín, 1990 in Parejo & Martín 1990: 25

Slug brownish to greyish green with darker brown lateral bands, skin delicate with poorly marked sculpture, sole cream or with orange lateral fields. Mucus transparent, colourless.
Anterior part of genitalia short, free oviduct with folds inside, spermatheca duct with longtudinal folds inside, retractor with two branches attaching to free oviduct and spermatheca duct, one retentor muscle holding the posterior part of the atrium, another one the anterior part of the free oviduct.

Up to 40 x 12 mm (preserved)

In deciduous and coniferous mountain forests, also in ruderal habitats, under stones and bark during daylight, always above 800 m.

Animals were found active between -3 and 14 °C, in autumn and spring.

Geomalacus

Geomalacus Allman, 1843 in Allman 1843: 851

Type species *Geomalacus maculosus* Allman, 1843 (monotypy)

Animal truncated in posterior part, mucus gland only little developed, mantle close to the head, concealing a shell-plate, respiratory hole on the right anterior margin of the mantle, genital pore behind and below the right eye-peduncle. Shell ovate, small and solid.
Wiktor & Parejo 1989 recognized two subgenera: *G. (Geomalacus)* with spotted slugs, long atriopenis (section between epiphallus and atrium) and short spermatheca duct. *G. (Arrudia)* with longitudinal lateral colour bands (as in *Arion*), short atriopenis and long spermatheca duct.

Wiktor & Parejo 1989: 24, Castillejo et al. 1994, Castillejo 1997.

The taxonomic classification proposed by Castillejo 1997 seems to have been accepted by subsequent authors (Verdú & Galante 2006).

Geomalacus anguiformis (Morelet, 1845)

Limax anguiformis Morelet, 1845 in Morelet 1845: 36, Pl. 3

Slug dark grey with brown hue, lighter sides and two light colour bands with dark lower margins running along both sides of the mantle. Mantle elliptical, extending to tentacles in anterior part, tentacles nearly black, thick and not very long. Juveniles dark blue with white spots and four blackish dorsal bands. When crawling, the slug has not always a regularly cylindrical shape, there are depressions and dilatations. Sole light creamy with darker medial zone, with yellowish or greenish hue. Mucus yellow.
Shell elliptical, solid, upper side convex and lower side flat.
Penis long, vas deferens very thin, spermatheca oval with long duct. Differs from *G. oliveirae* in its atriopenis inserting at the prolongation of the atrial long axis (not laterally).

60-70 mm long (preserved 30 mm)

Deciduous forests under bark, often in uncultivated rocky slopes covered by oaks and mossy surfaces. Active at twilight and during rainfalls.

In spring only juveniles, feeding on rock lichens. Adults are found only in autumn, they feed on forest mushrooms.

Nobre 1913: 171, Nobre 1941: 74, Castillejo 1997: 63, Verdú & Galante 2006: 349.

S Portugal, SW Spain

Vulnerable. Threatened by potential habitat destructions.

Geomalacus maculosus Allman, 1843

Geomalacus maculosus Allman, 1843 in Allman 1843: 851

SW Ireland: northern part of Kerry   Roy Anderson

W Europe

55-80 mm long, in Ireland up to 90 mm, Spain up to 120 mm

Slug dark with yellowish hue and lighter sides, and with many light spots from mantle to posterior end, rounded in posterior part, mantle about 1/3 of body length, tentacles not very long, posterior end of body with supra-pedal grooves and a triangular mucus gland which often carries a transparent yellowish ball of mucus. Sole yellowish.Shell rounded oval, solid.

In Spain in deciduous and coniferous forests, frequent also at old granitic walls, in regions of high rainfall and high humidity. In SW Ireland in a restricted area in heather moorland or rough pasture up to 300 m, more rarely in oak woods, mostly on acid sandstone substrate. Feeds on lichens, liverworts, algae, mosses and fungi growing on boulders and trees close to the coasts. In dry weather in Ireland it hides in crevices or under moss. In Spain only active at night, juveniles also at twilight.

From July to October oval translucent eggs (3-4 x 6-8.5 mm) are laid in clusters of 18-30, held together by mucus, colour turns later to yellowish, brown and black, juveniles hatch after 6-8 weeks, with distict lateral bands, maturity is reached after 2 years when length exceeds 25 mm. Life span up to 7 years.

Vulnerable in Spain. Threatened by modification of the habitat and destruction of lichens and moss sources (by intensification of land use or overgrazing) which will lead to a complete local extinction of the species. Protected by Irish Wildlife Act in Ireland. Measures have been taken to save this species by breeding it in captivity.

Nobre 1913: 169, Nobre 1941: 72, Castillejo 1997: 60, Kerney 1999: 119, Gómez Moliner et al. 2001: 202, Verdú & Galante 2006: 351.

Geomalacus moreleti (Hesse, 1884)

Arion moreleti Hesse, 1884 in Hesse 1884: 14 (*Arion (Ariunculus) Moreleti*)
Geomalacus malagensis Wiktor & Norris, 1991 in Wiktor & Norris 1991: 19

Slug pale beige-greenish to beige-grey, sometimes also yellowish-orange, with blackish narrow lateral bands having clearly marked upper edges and blurry lower edges, running from posterior end to frontal section of mantle, above the pneumostome. Slug shape elongate, dorso-ventrally flattened (oval in transversal cut), head very short (only 2-3 mm protruding from under the mantle), tentacles brown and long, mantle length 35 % of body, between medial line of dorsum and pneumostome 26 grooves. Sole dirty cream, transparent (viscera visible in live slugs), medial section narrower than lateral sections and slightly concave.
Animals are very active. Mucus colourless and thin, also when irritated, sometimes with orange hue. Mucus secreted onto the ground while crawling breaks into thin threads and coagulates into a unique spider-web like pattern, not seen in *Arion* species.

Up to 80 x 10.5 mm (preserved 58 mm)

Frequent on calcareous soil, in gardens, shrubland and open forests. Seems to feeds on fruits and other soft parts of plants. Captured slugs were found infected with nematodes.

Sexually mature slugs are found in spring and autumn.

NW Africa, S Spain, Gibraltar

Castillejo 1997: 67 (regarded *moreleti* and *malagensis* as conspecific)

Geomalacus oliveirae Simroth, 1891

Geomalacus oliveirae Simroth, 1891 in Simroth 1891: 359, Tab. 14 (*Geomalacus Oliveirae*)

Slug small, brown, lighter down to the sides, with 4 black colour bands, inner bands are irregularly interrupted, main marginal bands sharply delimited against the light background, sole white. Body mucus yellow. Genital retracor inserts more anteriorly in the median line of dorsum than in the other species.

In Portugal in Mediterranean shrubland and on granitic substrate, also under *Betula*, *Pinus* and *Juniperus* trees, below 1300 m. In Spain under *Pinus* and in olive orchards.

Up to 45 mm long (preserved 20 mm)

Vulnerable. Threatened by conversion of shrublands into suburban industrial areas.

Wiktor & P. 1989: 15 (*anguiformis*) and 23, Castillejo 1997: 65, Gómez Moliner et al. 2001: 200, Verdú & G. 2006: 350.

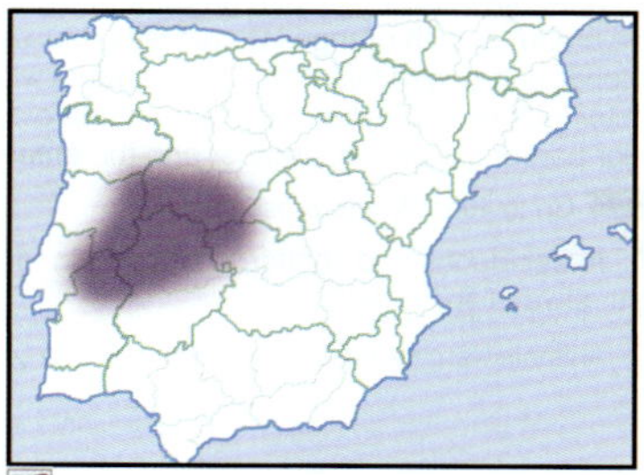

E central Portugal, W Spain

Helicoidea

Genera sorted alphabetically in 10 chains. In the quick identification guide I sorted the hygromiids by region. In the text presentation I sorted them "coarsely" taxonomically.

1 - *Sphincterochila* (quick-id Q61)

2 - Iberian groups (Q61) *Caracollina, Gasullia, Gasulliella, Gittenbergeria, Mastigophallus, Oestophora, Oestophorella, Suboestophora, Trissexodon*
3 - *Fruticicola* (Q62)
4 - *Elona, Norelona* (Q62)
5 - Helicodontidae (Q62): *Atenia, Drepanostoma, Falkneria, Helicodonta, Lindholmiola, Soosia*

6 - Hygromiidae chain 1 (Q63-Q72): *Cochlicella, Ciliata, Montserratina, Plentuisa, Ponentina, Schileykiella, Tyrrheniellina*
7 - Hygromiidae chain 2 (Q63-Q72): *Ashfordia, Euomphalia, Monacha, Prostenomphalia, Trochoidea, Xerocrassa*
8 - Hygromiidae chain 3 (Q63-Q72): all others (more than 30 genera)

9 - Ariantinae (Q73-Q75)
10 - Helicinae (Q76-Q78)

Sphincterochila

Sphincterochila Ancey, 1887
in Ancey 1887: 23

Type species *Helix boissieri* Charpentier, 1847
(subsequent designation by Pilsbry 1894 [1895]: 234)
(= currently *Sphincterochila boissieri* (Charpentier, 1847) from Near East)

Contained originally *Helix filia* Mousson, 1863 and *Helix boissieri*.

Shell heavy thick-walled, spheroidal, with a typical chalky surface, sharply keeled to rounded, surface smooth to heavily granulated. The last whorl always descends under the periphery of the preceding one.
The genitalia are mainly characterised by the presence of a penial stimulatory organ.

Mostly in arid or desert environments, snails can survive more than 1.5 years without food or water.

SW Europe to northern Africa, Near East and northern Arabia

Forcart 1972, Neubert 1998: 416.

Sph. candidissima and *baetica* are regarded as separate species here. This is obviously based on weak evidence, more research is needed.

Sphincterochila baetica (Rossmässler, 1854)

Helix baetica Rossmässler, 1854 in Rossmässler 1854-1859 (13-14): 16, Taf. 65

16-21 x 18-23 mm

Spain: 5 km SW Almería

SE Spain, N Morocco, NW Algeria

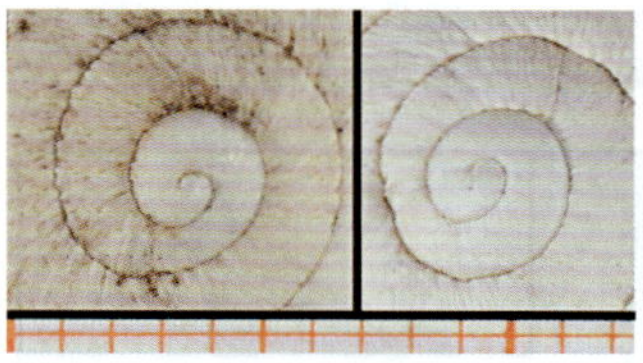

Spain: 5 km SW Almería

Shell greyish white, not shiny, sometimes with indistinct brown spots, upper side corrugated, lower side smooth. Only juvenile shells keeled, last whorl without keel, whorls connect to the previous whorl with a rounded edge slightly above the keel of the previous whorl.
Differs from *Sph. candidissima* mainly in its corrugated shell surface of the first whorls, but some populations of *Sph. candidissima* show the same feature.

Shrublands and semideserts.

Falkner 1990: 200, Martínez-Ortí et al. 2004.

More research to confirm the specific independence of this form seems to be necessary, could be conspecific with *Sph. candidissima*. It is possible that different concepts have been applied in Italy and Spain.

Sphincterochila candidissima (Draparnaud, 1801)

Helix candidissima Draparnaud, 1801 in Draparnaud 1801: 75

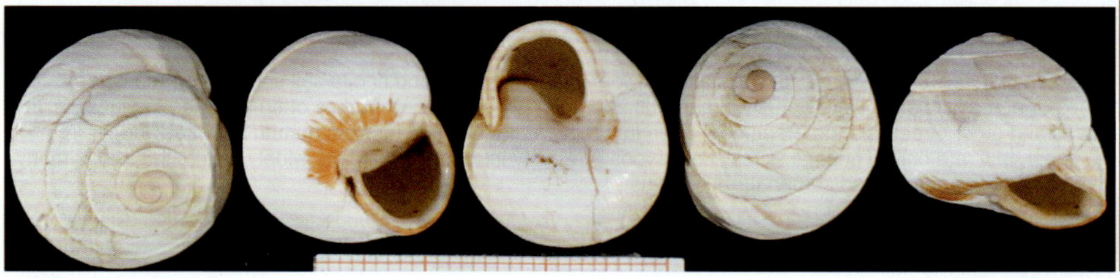

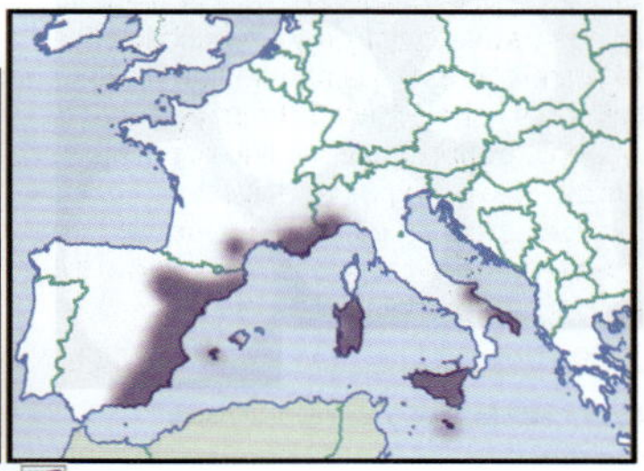

12-20 x 15-22 mm

Malta: Mistra village

S Europe, NW Libya

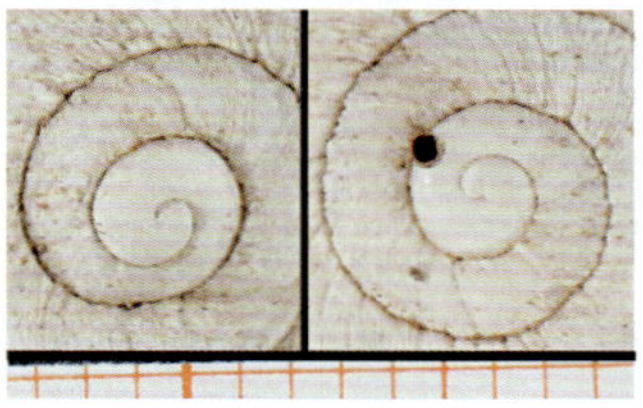

Spain: Almería, near Cala Honda

Shell white, pale or weakly shiny, sometimes with light reddish hue at apex and apertural margin, whorls almost smooth, upper edge of whorls connects usually slightly below the previous whorl's keel, juveniles are strongly keeled. Locally variable, in Sardinia adults are slightly keeled.
Animal greyish brown and very corrugated, tentacles lighter brown.
Differs from *Sph. baetica* mainly in its smoother shell surface, but it is possible that *Sph. candidissima* is locally variable and only one gene pool is present.

Dry shrublands and semideserts. Two peaks of activity, in October and March. In the winter snails bury into the soil to escape from the cold, in the summer they estivate climbing up plants to reduce exposure to heat. In France rarely above 500 m.

Gasull 1964: 17, Kerney et al. 1983: 298, Prieto & Altonaga 1988, Falkner 1990: 200, Manganelli et al. 1995: 27, Beckmann 2007: 89, Moreno-Rueda & C. 2007.

Originally perhaps in Spanish coastlands, introduced to mainland Italy, Mallorca and other sites. In Mallorca only at one locality.

Sphincterochila cariosula (Michaud, 1833)

Helix cariosula Michaud, 1833 in Michaud 1833: 5

Spain: Balearic Islands, Mallorca

11-15 x 16-19 mm

Shrublands and semideserts.

Gasull 1964: 20 (W Mallorca), Falkner 1990: 200, Beckmann 2007: 89.

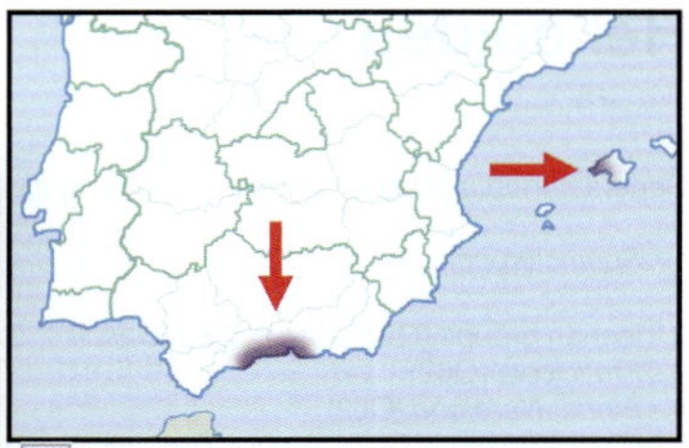

Originally N Algeria, introduced in historical times to S Spain and W Mallorca

Shell whitish, with pressed keel in all stages, whorls at upper side corrugated and rough, below the keel smooth, with knots above the suture, apertural margin sharp and reflected, umbilicus only closed in adult shells.

Sphincterochila insularis (Boettger, 1894)

Leucochroa candidissima var. *insularis* Boettger, 1894 in O. Boettger 1894: 5

Shell white, 4-5 convex whorls with deep suture, last whorl initially angulated. Differs from *Sph. candidissima* in the much more convex whorls with deeper suture.

10-11 x 14-16 mm

Extremely thermophilous, snails estivate on very hot and sun-exposed limestone slabs.

Endemic species of Andikíthira, frequent only at one site.

Gittenberger & Goodfriend 1993: 113, Gittenberger 1993: 530.

Clearly an endemic species, differs from the others also anatomically. The lack in the late Pleistocene 16,000 year-old fossil record is regarded as a coincidence. Closest relatives live in the African coastlands from S Tunisia to E Libya.

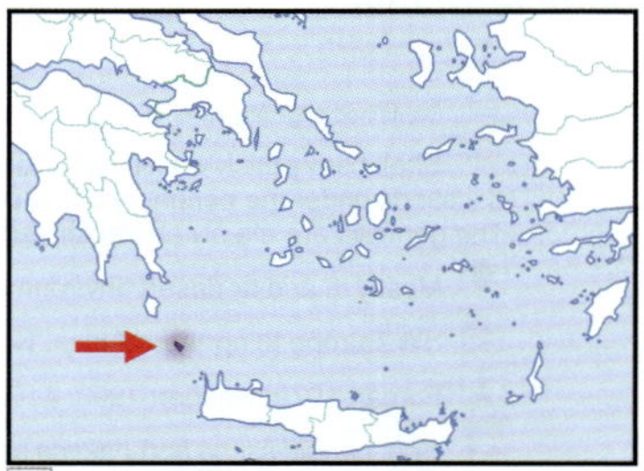

Greece (island of Andikíthira)

Iberian groups *Oestophora-Ponentina-Pyrenaearia*

Caracollina

Caracollina Beck, 1837
in Beck 1837: 28 *Helix (Caracollina)*

Type species *Helix lenticula* Férussac, 1821
(subsequent designation by Herrmannsen 1846 [01 Dec 1846]: 173)

Cabo Verde, Canary Islands, Madeira, Azores, W Mediterranean region

Puente et al. 1998: 53.

Caracollina lenticula (Férussac, 1821)

Caracollina lenticula (Michaud, 1831)

Helix lenticula Férussac, 1821 in A. Férussac [1821-1822]: 37

Established with a description for the subordinate variant α, but not with a distinct description for the main form (nominal taxon). The name *Helix lenticula* was made available at this occasion under Art. 12.1 because it had a description for the taxon (Code Glossary "taxon": a taxon encompasses all included taxa of lower rank). Art. 12.1 does not demand a description for the nominal taxon, a description for the taxon is sufficient, this can include subordinate taxa.

For those who do not like to accept this interpretation of the Code, the next available author is Michaud, 1831 (p. 43, *Helix lenticula*) (they would also need to verify whether Art. 11.6.1 can be applied under any interpretation of their choice, since Férussac's name could be interpreted as a synonym).

Spain: Cataluña, S Ebro estuary, near Rapita

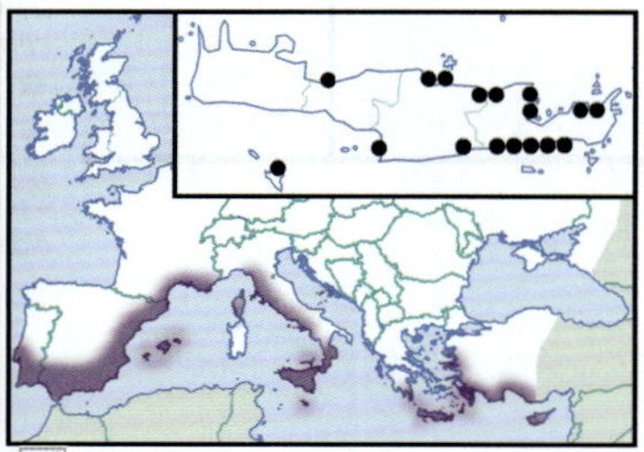

Near East to Cabo Verde

3-4 x 7-9 mm

Spain: Cataluña, S Ebro estuary, near Rapita

Shell flat, lentiform, no teeth, umbilicus wide. A very characteristic shell, also juveniles are easy to recognize with their wide umbilicus. Juvenile shells of *Oestophora barbula* are larger.
Animal light with reddish brown hue, tentacles long and thick with black eyes. When crawling the shell is lifted above the right side of the animal.

In dry habitats under stones, at walls and between plants, under wood and in soil substrate, often behind dunes or at rocky coasts. Occasionally in gardens and greenhouses. In SW Turkey often attached to old walls. In Ródos and other Greek islands (Psérimos) often found in natural habitats far from human activities and far from the coast. In Crete only in coastal vicinity, in Gávdos obviously introduced. In Malta widely distributed and in many different kinds of natural and semi-natural habitats.

Giusti et al. 1995: 463, Frank 1997: 64, Schütt 2001: 442, Vardinoyannis (1994, unpublished thesis), HNC data.

Gasullia

Gasullia Ortiz de Zárate López, 1962
in Ortiz de Zárate López 1962: 101 *Oestophora (Gasullia)*

Type species *Macularia riffensis* Kobelt, 1903
(original designation)
(= currently *Gasullia riffensis* (Kobelt, 1903) from Morocco)

Ortiz de Zárate Rocandio & Ortiz de Zárate López 1961: 177 (footnote) mentioned a new genus *Gasullia*, which included *Oestophora gasulli* and *Macularia riffensis*, but they did not fix a type species. So this was a nomen nudum. Ortiz de Zárate López 1962: 101 mentioned the name *Gasullia* again and this time fixed a type species for it, making the name available at that occasion.

Hatumia Arrébola, Prieto, Puente & Ruiz, 2006 in Arrébola et al. 2006: 126, type species *Macularia riffensis* Kobelt, 1903 (original designation)

Arrébola et al. 2006 cited the type species as *Oestophora (Gasullia) riffensis* Ortiz de Zárate, 1962, but Ortiz de Zárate López 1962 (p. 102) had just used Kobelt's name *Macularia lucasii riffensis* Kobelt, 1903 in Kobelt 1903: 37, Taf. 284 Fig. 1822-1823, author of the description was Kobelt, not Pallary (Kobelt 1903: 37 attributed the name to Pallary, but the description was by Kobelt, Pallary gave him only the shells). *Hatumia* is an objective junior synonym of *Gasullia*.

Shell with rounded or angulated whorls, apertural margin straight, simple and not thickened, apex smooth or with weak spiral crests, teleoconch with punctiform nodules.
Penial papilla present, flagellum short or long, vagina with two divided or undivided mucous glands and with an accessory sac, one small dart sac.

Morocco and S Iberian peninsula

Schileyko 1991 (gave Ortiz de Zárate 1961), Nordsieck 1993 (gave Ortiz de Zárate & Ortiz de Zárate 1961), Puente 1994, Puente 1996 (gave Ortiz de Zárate 1962), Puente et al. 1998: 53, Arrébola et al. 2006: 133 (followed Nordsieck 1993).

Since *Gasullia* and *Hatumia* are objective synonyms, the older name *Gasullia* must be used for the genus. The species *gasulli* is classified in *Suboestophora*.

Gasullia cobosi (Ortiz de Zárate López, 1962)

S Spain

Oestophora cobosi Ortiz de Zárate López, 1962 in Ortiz de Zárate López 1962: 103 (*Oestophora (Gasullia) cobosi*)
Not *Xerocrassa cobosi* (Ortiz de Zárate López, 1962), established as *Helicella cobosi*.

Shell pale brown and slightly transparent, weakly striated with granular surface structure on both sides of the shell, 4-5 convex whorls with deep suture, not keeled but with edge at the periphery, on the last whorl there is a depression between periphery and upper suture so that the upper section of the aperture looks somehow wavy, aperture simple without lip, umbilicus wide and deep, up to 1/3 of shell diameter.

3.4-4.1 x 7-10 mm

Frequently found under rocks and trunks or in rock crevices, in lower altitudes, relatively dry habitats on calcareous surfaces, also in ruderal habitats (disturbed by human influence).

EU Red List: Vulnerable VU. Threatened by human pressure in coastlands, has declined in greenhouse areas.

Arrébola et al. 2006: 128, Talaván Gómez & T. 2006: 30.

Gasullia zapateri (Hidalgo, 1870)

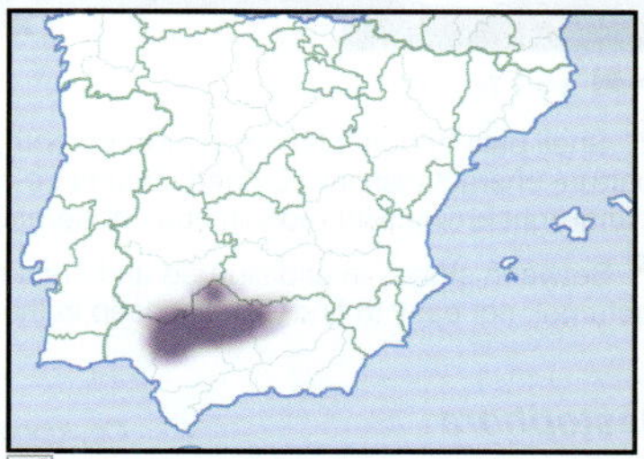

S Spain (Sierra Morena)

Helix zapateri Hidalgo, 1870 in Hidalgo 1870: 299 (*Helix Zapateri*)
Hatumia pseudogasulli Arrébola, Prieto, Puente & Ruiz, 2006 in Arrébola et al. 2006: 129

Shell pale brown, slightly paler on lower side, weakly striated with weak granular surface structure, 4.5-5 convex whorls without keel, first whorl with fine spiral lines, all whorls regularly curved at periphery, aperture simple without lip or thickened margin, umbilicus wide and deep, 1/4 of shell diameter.
Differs from *Suboestophora gasulli* mainly in its smaller size, spiral lines on apex and granular surface structure.

4.5-5.5 x 7-10 mm

Under stones and trunks in humid and shady habitats, in lower sections of Sierra Morena among Cistus shrubs and Quercus trees. Lives sympatrically with *Suboestophora gasulli* in the west of its range.

EU Red List: Near Threatened NT. Some populations are threatened by intensive agriculture and land use.

Arrébola et al. 2006: 129, Martínez-Ortí, Arr. & R. 2008.

Gasulliella

Gasulliella Gittenberger, 1980
in Gittenberger 1980: 206

Type species *Helix simplicula* Morelet, 1845
(original designation)

In the headline of the original description on p. 206 the genus name was misspelled *Gasuliella*, but the species was expressly dedicated to Mr. L. Gasull and the spelling *Gasulliella* appeared various times in the text, correct spelling under Art. 32.5 must be *Gasulliella*. Established without description (Art. 13.1.1), but monotypic in the original source and the species was described.

Shell lenticular, whorls rounded, apex with regular spiral crests, apertural margin simple, umbilicus wide.
Penial papilla present, flagellum short, no accessory (mucus) glands attached to vagina.

NW Africa, SW Iberian peninsula

Puente et al. 1998: 44, 53, Arrébola et al. 2006: 133.

Only one species.

Gasulliella simplicula (Morelet, 1845)

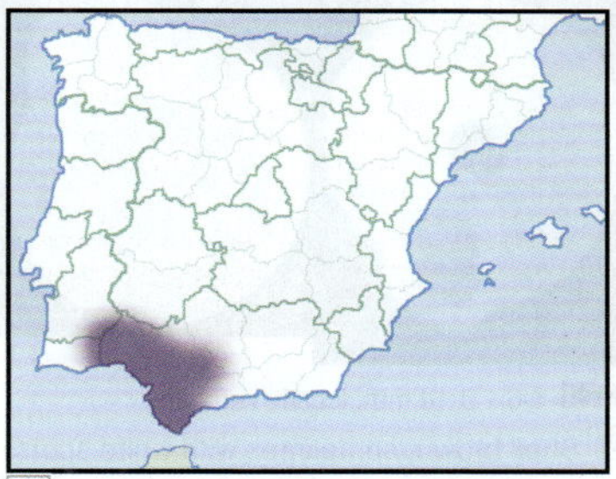

NW Africa, S Portugal, SW Spain

Helix simplicula Morelet, 1845 in Morelet 1845: 56, Pl. 6

Shell pale brown, very weakly striated, 4.5 moderately shouldered whorls, upper side almost flat, lower side more rounded, aperture simple, like in a juvenile *Oestophora*, umbilicus wide, 1/4 of shell diameter.
Differs from juvenile *Oe. barbula* and from zonitids by being much weaker and more irregularly striated.

2.2-3 x 5.5-7 mm

Under stones and leaves, between herbals plants, not frequent but obviously adapted to various kinds of habitats, including disturbed ones such as road and cultivated field margins or pastures. In higher altitudes up to 600 m in more humid and natural habitats.

Gittenberger 1980: 206, Puente et al. 1998: 44.

Gittenbergeria

Gittenbergeria Schileyko, 1991
in Schileyko 1991: 225

Type species *Helix turriplana* Morelet, 1845
(original designation)

Puente 1996: 73.

Gittenbergeria turriplana (Morelet, 1845)

Helix turriplana Morelet, 1845 in Morelet 1845: 59, Pl. 6

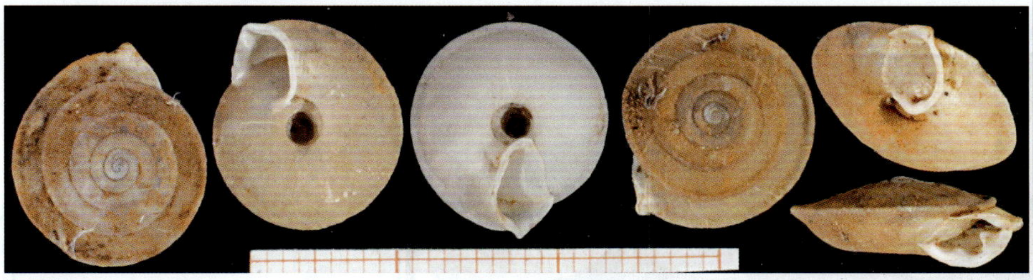

Portugal: Algarve, Tavira

S Portugal (Algarve)

EU Red List: Near Threatened NT. Certainly exaggerated, no real threats known.

5-6 x 12-14 mm

Shell brown with a characterictic microsculpture, depressed with a strong keel, first whorls almost flat, aperture narrow, margin reflected and turned outward, 2 teeth on the lower edge corresponding to 2 conspicuous depressions on the outer shell, umbilicus open and excentric.
Animal almost black with a lighter foot, upper tentacles very long.

Lives in low altitudes on calcareous substrate, close to the coast, frequent at old walls and in old gardens and parks in cities, also in shrublands and cultivated areas.

Nobre 1913: 183, Nobre 1941: 88, Puente 1996: 74, Kittel 2003: 37.

Portugal: Algarve near Faro
Charles Geerts

Mastigophallus

Mastigophallus Hesse, 1918 in Hesse 1918: 103

Type species *Helix rangiana* Michaud, 1831 (original designation)

Puente et al. 1998: 53 (gave a diagnostic genital description).

Mastigophallus rangianus (Michaud, 1831)

Without gender agreement: *Mastigophallus rangiana* (Michaud, 1831)

Helix rangiana Michaud, 1831 in Michaud 1831: 40, Pl. 14 (*Helix Rangiana*)

3-4 x 7-12 mm

Spain: Cataluña, Girona, Port de la Selva

E Pyrenees

EU Red List: Near Threatened NT. Vulnerable in Spain. Threatened by human pressure, mainly by tourism, urbanisation and fire.

Shell pale or horny brown, almost smooth, 6-7 almost flat whorls with strong keel and weakly marked suture, aperture characteristically formed, cannot be confounded, umbilicus 1/8-1/7 of diameter.
Animal light brown with reddish hue, moderately transparent, with very small tubercles, about 15 mm long.

Moquin-Tandon 1855: 111, Puente et al. 1998: 46, Puente et al. in Verdú & Galante 2006: 398.

Between stones in undisturbed and humid habitats of lower altitudes. Also at old walls in vineyards and olive orchards, and in humid soils near creeks. Nocturnal, not easy to find, usually deep in the soil, active in warm and humid nights.

Oestophora

Oestophora Hesse, 1907 in Hesse 1907: 76

Type species *Helix lusitanica* Pfeiffer, 1841 (subsequent designation by Wenz 1923 (18): 469)

Contained originally 6 species mentioned on p. 76 ("Diese südwestlichen Arten" = *Helix maroccana, lusitanica, lenticularis, tarnieri, annai, pechaudi*).

Shell ribbed, with rounded or keeled whorls, apertural margin reflected and thickened.
No flagellum, no penial papilla, one small dart sac, one longer accessory sac separated from vaginal walls, vagina with 3 simple and long acccessory (mucus) glands.

W Iberian peninsula, N Africa, Azores

Puente 1996: 81, Puente et al. 1998: 53, Arrébola et al. 2006.

Oestophora barbella (Servain, 1880)

Helix barbella Servain, 1880 in Servain 1880: 66

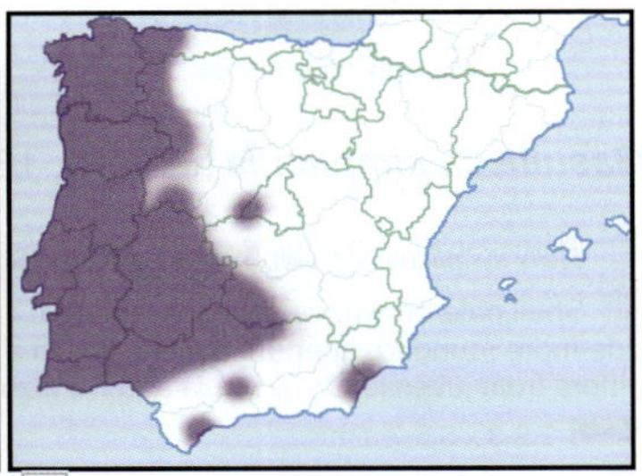

4-6 x 8-14 mm, locally variable

Spain: SW Galicia, Pontevedra, Tui

Azores, Iberian peninsula

Shell brown, sometimes with white spots, very finely and regularly ribbed, lentiform and angulated or moderately keeled, 5-5.5 convex whorls with deep suture, last whorl seen from above not wider than previous whorl, aperture narrow and with two prominent teeth at lower side (palatal tooth stronger than columellar tooth), apertural margin thickened and reflected but without internal lip, umbilicus wide and deep, 1/5-1/4 (19-23 %) of shell diameter. Juveniles without apertural teeth.
Differs from *O. lusitanica* in the presence of teeth in the aperture. *O. barbula* is more prominently keeled at the periphery and has a slightly narrower umbilicus.

Many different kinds of habitats, includes also disturbed places in cities, villages and next to roads and cultivated fields. Often in aggregations of several individuals, when they are found under stones. Mainly in lowlands, not restricted to calcareous substrate, up to 1100 m in Portugal.

Nobre 1913: 182, Nobre 1941: 87, Ortiz de Zárate López 1962: 91, Ondina et al. 1997, Puente 1996: 82, Holyoak & Holyoak 2012.

Reports from before 2012 referred to both *O. barbula* and *O. barbella*.

Oestophora barbula (Rossmässler, 1838)

Helix barbula Rossmässler, 1838 in Rossmässler 1838-1844 (7-8): 11, Taf. 32

4-6 x 8-13 mm, locally variable

Portugal: Sintra

W Portugal

Shell brown, sometimes with white spots, very finely and regularly ribbed, lentiform and keeled, 5-5.5 convex whorls with deep suture, last whorl weakly keeled, seen from above not wider than previous whorl, aperture narrow and with two prominent teeth at lower side (palatal tooth stronger than columellar tooth), apertural margin thickened and reflected but without internal lip, umbilicus wide and deep, 1/6-1/5 (16-20 %) of shell diameter. Juveniles without apertural teeth.
Differs from *O. lusitanica* in the presence of teeth in the aperture. *O. barbella* is more rounded at the periphery and has a slightly wider umbilicus.

Mainly in lowlands in wood and open habitats, near limestone outcrops, often under boulders, seems to be restricted to calcareous substrate, up to 550 m.

Holyoak & Holyoak 2012.

Reports from before 2012 referred to both *O. barbula* and *O. barbella*.

Oestophora calpeana (Morelet, 1854)

Morocco, Gibraltar (introduced?)

Helix calpeana Morelet, 1854 in Morelet 1854: 621 (*Helix Calpeana*)

Shell brown, slightly translucent, ribbed, 6-7 whorls, lentiform with both sides being equally convex, keel sharp, suture not very deep, last whorls almost flat, last whorl not wider than previous whorl, aperture narrow, margin reflected and thickened, umbilicus regular, last three whorls visible inside, 1/6 of shell diameter.

4.9-5.0 x 10.4-11.4 mm

Open habitats covered by dense vegetation, at road margins, under stones and rocks, also in gardens, not exposed to direct sunshine. Frequent where present in Gibraltar.

Individuals are adult at the beginning of spring.

EU Red List: Endangered EN. Threatened by urban pressure.

Ortiz de Zárate López 1962: 90, Puente 1996: 96, Talaván Gómez & T. 2006: 37, Ruiz et al. 2009: 700.

Oestophora dorotheae Hesse, 1930

Morocco, S Spain (?)

Oestophora dorotheae Hesse, 1930 in Hesse 1930: 80
Expressly established as a new replacement name for
Helix lenticularis Morelet, 1853 in Morelet 1853: 140 (not *Helix lenticularis* Alten, 1812 in Alten 1812: 35 from S Germany)

Shell horny reddish, transparent, lighter at lower side, regularly ribbed on upper side, weakly striated on lower side, lentiform, equally convex at both sides, prominently keeled, 7 weakly convex and slowly increasing whorls, aperture in U-shape with whitish lip inside between keel and columellar margin, margin between keel and suture almost horizontal and without lip, umbilicus moderate and perspectival, weakly covered by the reflected columellar margin.
Looks somehow like *Lindholmiola lens*, but more transparent, edge at keel more pointed in the aperture, apertural margin less strongly reflected.

4.5-5.5 x 10-12 mm

At the soil in humid habitats, in plains covered with grasses, also in rocky habitats without much vegetation.

Puente 1996: 94, Puente et al. in Verdú & Galante 2006: 399, Ruiz et al. 2009: 700.

Was once reported from two localities in Spain, but these were probably short-term introductions from Morocco.

Oestophora ebria (Corbella i Alonso, 2004)

S Spain (Málaga)

Suboestophora ebria Corbella i Alonso, 2004 in Corbella i Alonso 2004: 40

Shell evenly pale brown, very regularly ribbed on upper side (60-90 ribs/penultimate whorl), more finely striated on lower side, shiny, upper side almost flat, lower side rounded, 5-6 whorls with deep suture, periphery rounded and located in upper third, last whorl ascending near aperture so that the apertural margin passes over the upper outline of the shell, apertural margin reflected and with white lip inside, strongly S-shaped when seen from a lateral view, and slightly ow-shaped (o = umbilicus) when seen from below (outer edge protruding, ending in a hook when seen from below and above), umbilicus open and deep, 1/6 of shell diameter, all whorls visible inside.

3.5-4.3 x 7-9 mm

In karstic mountain environments, often in rock crevices, but also under oak trees without rocky outcrops, under perennial grass and in the vegetation near riverbanks, 300-1100 m.

EU Red List: Near Threatened NT. Only known from few localities, at some threatened by habitat destruction.

Ruiz, Arrébola & Puente 2009: 698 (studied the anatomy, decisive for the generic classification).

Oestophora granesae Arrébola Burgos, 1998

S Spain

Oestophora granesae Arrébola Burgos, 1998 in Arrébola Burgos 1998: 38

Shell evenly light brownish, slightly translucent, very regularly and finely ribbed, ribs weaker at lower side, 5.5-6 whorls with deep suture, periphery equally convex on upper and lower side, apertural margin reflected, covers slightly the umbilicus, umbilicus deep, 1/8 of diameter.

4.3-4.5 x 8.5-9.0 mm

Was found in open woodland near a small river under stones, at 600-800 m altitude.

EU Red List: Vulnerable VU. Only known from 2 localities in a protected area.

Arrébola & Puente in Verdú & Galante 2006: 400.

Oestophora lusitanica (Pfeiffer, 1841)

Helix lusitanica Pfeiffer, 1841 in L. Pfeiffer 1841: 40 (not *Helix lusitanica* Linné, 1767, a junor synonym of *Helix cornea* Linnæus, 1758, currently *Planorbarius corneus*)
This name can only be used for this species when *Helix lusitanica* Linné, 1767 is suppressed for the principle of homonymy. Perhaps it is not possible to find 25 publications as required by Art. 23.9.1.2. This shall count as a use here. The case should be referred to the Commission under Art. 23.9.5.

Portugal: Sintra

Portugal, NW Spain, introduced to Azores

6-8 x 15-20 mm

Shell densely and regularly ribbed, 5-5.5 whorls, slightly keeled but not on the last whorl, first whorl with fine spiral lines, last whorl width almost 2 x of previous whorl, apertural margin reflected except upper section which is almost not reflected, no apertural teeth, umbilicus open and deep with all whorls visible inside. Differs from *O. barbula* and *O. barbella* in the absence of teeth in the aperture.
Animal blackish with bluish hue on upper side and white on lower side and foot, tentacles long. Epiphragm white and fragile.

In shady and humid habitats, under stones and dead humid leaves, at old stone walls, also in cities.

Nobre 1913: 179, Nobre 1941: 83, Ortiz de Zárate López 1962: 83, Puente 1996: 89.

Oestophora mariae Ruiz, Arrébola & Puente, 2009

Oestophora mariae Ruiz, Arrébola & Puente, 2009 in Ruiz et al. 2009: 697

Shell evenly pale brown, slightly translucent, smooth, depressed with angular periphery, upper and lower side equally convex, 5.5-6 weakly convex whorls, last whorl not descending near aperture, apertural margin thickened and weakly reflected, inside with a weak tooth at the palatal side and a corresponding small external depression, basal margin occasionally with a callus, umbilicus narrow and deep, 1/7 of diameter.

4-5 x 8-10 mm

Mountain habitats under stones and decaying plant litter, found in 970 m altitude.

EU Red List: Vulnerable VU. Only known from one locality in Sierra de Castril mountains. No real threats known.

Spain (E Andalusia)

Oestophora ortizi De Winter & Ripken, 1991

Oestophora ortizi De Winter & Ripken, 1991 in De Winter & Ripken 1991: 3

Shell evenly brown, finely and regularly ribbed on upper side, more weakly on lower side, 5-6 regularly increasing whorls with deep suture, periphery equally convex at upper and lower side, last whorl not descending near aperture, apertural margin not connected at parietal side, slightly thickened and slightly reflected, umbilicus deep, 1/6 of diameter.

Humid and undisturbed habitats on calcareous substrate. Was found at the soil between plants, under stones and near water currents, in open forests with intensive surface vegetation.

EU Red List: Near Threatened NT. No real threats known.

4.3-4.6 x 8-9 mm

S Spain

José Liétor Gallego

S Spain

Puente 1996: 93, Arrébola Burgos 1998.

Oestophora prietoi Ruíz, Arrébola & Puente, 2009

Oestophora prietoi Ruíz, Arrébola & Puente, 2009 in Ruíz, Arrébola & Puente 2009: 695

Shell evenly pale brown, slightly translucent, upper side densely and prominently ribbed, lower side striated, shell shape lenticular with blunt keel at periphery, upper part often flatter than lower part, 6 convex whorls with deep suture, aperture simple without teeth, margin reflected except above, umbilicus deep and open, 1/7-1/6 of diameter.

4-5.5 x 9.5-13 mm

Spain: Jaen, Sierra de Cazorla, La Iruela (D = 11 mm)

Charles Geerts

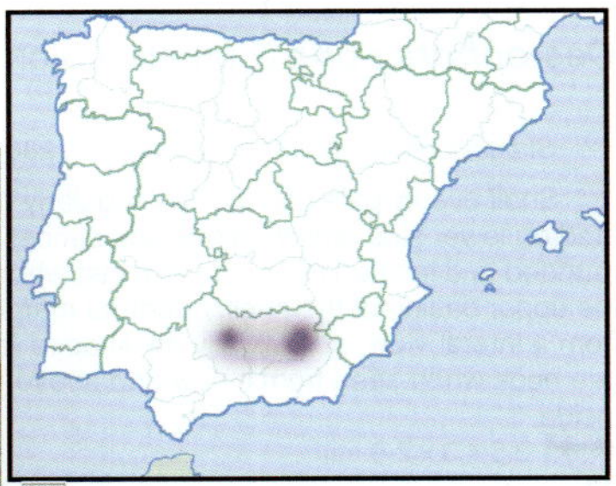

Spain (N Andalusia)

EU Red List: Near Threatened NT. No real threats known.

In inland limestone mountains, mostly in woodlands under dead leaves of *Quercus rotundifolia* and *Pinus halepensis*, also in pastures with dispersed coniferous trees. The animals hide under tree trunks, moss, walls and crevices.

Oestophora silvae Ortiz de Zárate López, 1962

Oestophora silvae Ortiz de Zárate López, 1962 in Ortiz de Zárate López 1962: 84

Shell reddish-brown, finely and regularly ribbed, 5.5 whorls, apertural margin broadly reflected, with a prominent interior callus in its upper section, umbilicus open. Not necessarily smaller than *O. lusitanica*, spiral streaks present but only visible in humid shells. Differs from *O. barbella* in the absence of teeth in the aperture.

4.5-7 x 10-18 mm

Spain: Asturias, near Cangas de Onís (D = 9 mm)

Charles Geerts

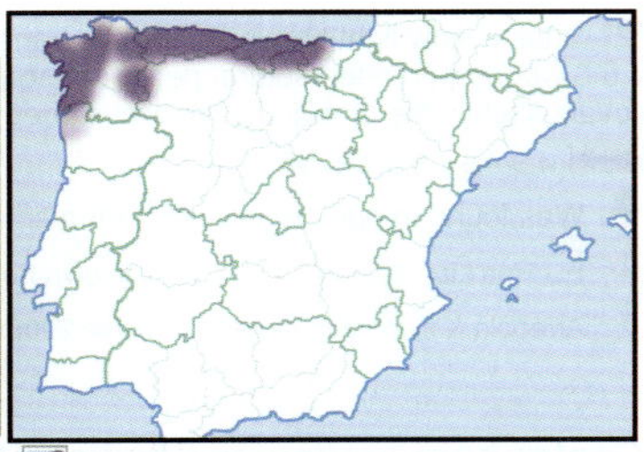

N Spain

Puente 1996: 91, Ondina et al. 1997.

Original description by Ortiz de Zárate López was not exact. Has also been known under the name *Oestophora lusitanica* var. *minor* from Galicia.

Oestophora tarnieri (Morelet, 1854)

Helix tarnieri Morelet, 1854 in Morelet 1854: 623 (*Helix Tarnieri*)

Shell dark brown or reddish yellow and solid, slightly translucent, finely ribbed, 6-7 convex and very regularly increasing whorls with deep suture, periphery weakly keeled, last whorl not much wider than previous whorl, apex with very irregular spiral lines, apertural margin reflected but less in its upper section, umbilicus open and deep, 1/6 of diameter.

5.5-6 x 10.7-12.3 mm, animal up to 50 mm long

Under stones and vegetation, also between desintegrating plant material, usually in relatively humid habitats not directly exposed to sunshine. In lower altitudes 0-200 m in disturbed habitats near roads and villages, in 200-600 m in more natural habitats near water currents and in open forests.

Ortiz de Zárate López 1962: 86, Puente 1996: 96.

S Spain, Morocco

Oestophorella

Oestophorella Pfeffer, 1930
in Pfeffer 1930: 126 (= 276)

Type species *Helix buvinieri* Michaud, 1841 (monotypy)

N Iberian peninsula

Prieto & Altonaga 1988, Puente et al. 1998: 48, 53 (with diagnostic genital description).

Oestophorella buvinieri (Michaud, 1841)

Helix buvinieri Michaud, 1841 in Michaud 1841: 64

5-6 x 8-10 mm

Spain: Caranceja near Villanueva de la Peña

N Spain

Shell light horny or darker brown, finely and regularly ribbed, more weakly at lower side, 6.5-7.5 convex and steadily increasing whorls, last whorl initially weakly angulated, apertural margin white, strongly reflected and conspicuously S-like curved when seem from lateral view, umbilicus covered. Animal bluish black, foot slightly lighter, the shell is carried nearly horizontally.

Between sea level and 1400 m.

Suboestophora

Suboestophora Ortiz de Zárate López, 1962
in Ortiz de Zárate López 1962: 93 *Oestophora (Suboestophora)*

Type species *Helicodonta hispanica* Gude, 1910 (original designation)

Mentioned by Ortiz de Zárate & Ortiz de Zárate 1961: 177, but as a nomen nudum, no type species was fixed there.

Shell ribbed and granulated with rounded or keeled whorls, apex also granulated, apertural margin either simple or with thickened and reflected margin. Penial papilla present, flagellum short, one long accessory sac separated from vagina, at the base with 2 bifurcated accessory (mucus) glands (if *Gasullia gasulli* is included, then the two glands can also be attached to vagina and have 1-6 branches), one dart sac rounded and externally visible.

Puente et al. 1998: 53, Arrébola et al. 2006: 133.

Suboestophora altimirai (Ortiz de Zárate López, 1962)

Oestophora tarraconensis var. *altimirai* Ortiz de Zárate López, 1962 in Ortiz de Zárate López 1962: 99
Name not available under Art. 45.6.3 (use of the term "nov. var.", makes the name infrasubspecific and not available). Expressly dedicated to Carlos Altimira, the original spelling was *altamirai* and should be regarded as an inadvertent error, to be corrected to *altimirai* under Art. 32.5. Name should be attributed to the next author who used this name before 2000 and made it available.

Shell brownish, upper side densely ribbed, less densely on last whorls, with flatter ribs irregularly spaced in between, lower side weakly striated or almost smooth, shell flatter on upper than on lower side, 5.5-6 whorls, slightly keeled but last whorl gets rounded, aperture narrower than in *S. hispanica*, apertural margin broadly reflected on right and lower sides, not affecting umbilicus, with a weak curve on upper side near previous whorl, but no sinus, and with a callus inside on right (palatal) wall, umbilicus regular, wider than in *S. tarraconensis*, 1/5 of shell diameter. Animal whitish grey.

4-5 x 8-11 mm

Spain: Castellón, Sueras (D = 10 mm)

Charles Geerts

E Spain (Castellón)

EU Red List: Vulnerable VU. Only known from few localities, abundant at only one locality. Threatened by habitat destruction.

Pine forests and shrublands, occasionally in caves. Usually 20-40 cm inside rock crevices at places with higher humidity than on the rock faces. Abundant on non-calcareous substrate, 260-640 m.

Martínez-Ortí & Robles 2003: 156, Martínez-Ortí & Puente in Verdú & Galante 2006: 401.

Suboestophora boscae (Hidalgo, 1869)

Helix boscae Hidalgo, 1869 in Hidalgo 1869: 20, Pl. 2 (*Helix Boscæ*)
Oestophora kuiperi Gasull, 1966 in Gasull 1966: 159

5-8 x 8-12 mm

(1) Spain: Alicante, Pego, (2-6) Spain: Valencia, Cullera

E Spain (Alicante, València)

Shell light brownish, irregularly ribbed, less densely on last whorls, almost flat on upper side, convex on lower side, keeled, aperture very wide, margin broadly reflected with a convex curve on the upper side and angular sinus, umbilicus narrow, 1/7 of shell diameter. The form *kuiperi* (2-6) is smaller and less densely ribbed. Animal dark with whitish sole.

Under rocks, in rock crevices and organic litter, also in caves, up to 1100 m. A frequent species.

Ortiz de Zárate López 1962: 95, Robles 1990: 471, Martínez-Ortí & Robles 2002, Martínez-Ortí & Robles 2003: 159.

Suboestophora gasulli (Ortiz de Zárate Rocandio & Ortiz de Zárate López, 1961)

Oestophora gasulli Ortiz de Zárate Rocandio & Ortiz de Zárate López, 1961 in Ortiz de Zárate Rocandio & O. 1961: 178

4.9-5.3 x 10-11 mm Spain: Huelva, La Puente (Castañuelo) Rafael Vázquez Romero

SW Spain (Sierra Morena)

EU Red List: Near Threatened NT. No real threats known.

Arrébola et al. 2006: 121.

Here classified in *Suboestophora*. The genus *Gasullia* cannot be used for this species.

Shell looking like a zonitid with very wide umbilicus, pale brown, slightly paler brown on lower side, translucent, 5 convex whorls with deep suture and not keeled, first 1-2 whorls smooth, the others weakly striated, not granulated, aperture simple without lip or thickened margin, umbilicus wide and deep, 1/4 of shell diameter. Larger than *Gasullia zapateri* and without granulated surface.

Hidden under trunks and rocks in Sierra Morena mountains, in lower altitude habitats in river valleys, also in disturbed habitats, among *Cistus* shrubs and *Quercus suber* (cork oaks), also under pines and eucalyptus. Lives sympatrically with *Gasullia zapateri* in N Sevilla.

Suboestophora hispanica (Gude, 1910)

Helicodonta hispanica Gude, 1910 in Gude 1910: 124 (*Helicodonta Hispanica*)

5-6 x 11-13 mm Spain: Jeresa, Les Congles

E Spain (Alicante, València)

EU Red List: Vulnerable VU. Threatened by fire. Only 5 localities with living animals were found in the past years.

Shell yellowish-brown, slightly translucent, lentiform, densely and regularly ribbed, less densely near the aperture, 5-5.5 whorls, suture not deep, apertural margin curved, reflected and not much thickened, umbilicus deep and relatively narrow, not much more than last whorl visible inside, 1/6-1/5 of shell diameter.

Shrublands and pine forests, under rocks and in rock crevices, also in caves (up to 45 m deep inside), up to 800 m.

Ortiz de Zárate López 1962: 94, Gómez Moliner et al. 2001: 158, Martínez-Ortí & Robles 2003: 162, Martínez-Ortí & Puente in Verdú & Galante 2006: 402

Suboestophora jeresae (Ortiz de Zárate López, 1962)

Oestophora jeresae Ortiz de Zárate López, 1962 in Ortiz de Zárate López 1962: 99

Shell evenly horny brown, fragile, densely ribbed and very strongly keeled, keel with a depressed section on both sides and a toothed band produced by the ribs, and with a sinus at interior side, apertural margin little reflected at upper side, more strongly at lower side, umbilicus narrow, 1/8-1/7 of shell diameter, partly covered by columellar margin of aperture.

4.5-6 x 11-13 mm Spain: València, Jeresa

Mediterranean shrubland, near *Citrus* plantations with slightly higher humidity, under stones, leaves and tree trunks, up to 300 m.

EU Red List: Vulnerable VU. Threatened by fire.

E Spain (València)

Gómez Moliner et al. 2001: 158, Martínez-Ortí & Robles 2003: 165, Martínez-Ortí & Puente in Verdú & Galante 2006: 403.

Suboestophora tarraconensis (Aguilar-Amat, 1935)

Oestophora lusitanica tarraconensis Aguilar-Amat, 1935 in Aguilar-Amat 1935: 259

Shell light, densely ribbed, flatter on upper than on lower side, 5.5 whorls, slightly keeled, aperture narrower than in *S. hispanica*, apertural margin strongly reflected, partly covering the umbilicus, with a weak curve on upper side near previous whorl, but no sinus, and with a callus inside on right wall, umbilicus regular, 1/6-1/5 of shell diameter. Animal whitish grey.

4-5 x 8-11 mm Spain: Castellón, Sueras (D = 11 mm) Charles Geerts

E Spain (Castellón, Tarragona)

In caves where organic substrate from outside is accumulated, also outside caves under stones and dead leaves, between 200 and 800 m.

EU Red List: Near Threatened NT. Theatened by continuous habitat destructions, disappeared in the past years from 17 out of 20 previously known localities. A fire in 1992 in Desierto de las Palmas destroyed local populations.

Ortiz de Zárate López 1962: 97, Martínez-Ortí & Robles 2003: 168, Martínez-Ortí & Puente in Verdú & Galante 2006: 404. Occurrence in Tarragona suggested by recent finds (in Torrelles de Foix) of José Liétor Gallego.

Trissexodon

Trissexodon Pilsbry, 1895
in Pilsbry 1894: 287 *Helicodonta (Trissexodon)*

Type species *Helix constricta* Boubée, 1836
(subsequent designation by Wenz 1923 (18): 457)

Contained originally (p. 288) *Helix constricta* and *H. quadrasi.*

Gittenberger 1968, Prieto & Altonaga 1988, Puente et al. 1998: 51, 53 (with diagnostic genital description).

Trissexodon constrictus (Boubée, 1836)

Without gender agreement: *Trissexodon constricta* (Boubée, 1836)

Helix constricta Boubée, 1836 in Boubée 1836: 220 [not verified]

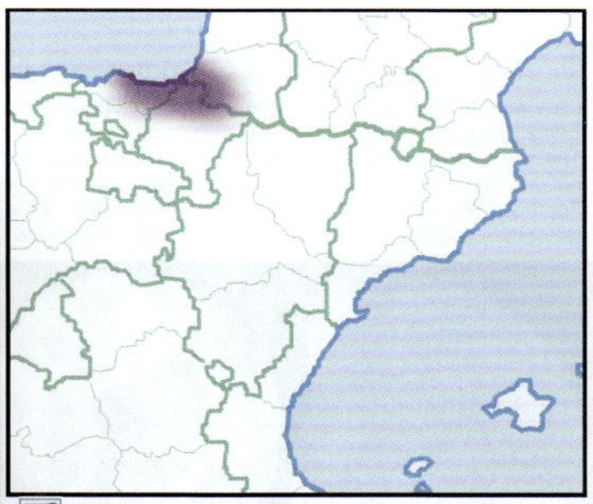

3-4 x 6-8 mm

Spain: San Sebastián

W Pyrenees to Bizkaia

A rare species in Navarra, only at one locality live animals have been observed.

Larraz et al. 2007.

Shell finely and regularly ribbed, strongly depressed on upper side, lower side rounded, 5-6 whorls, aperture with characteristic shape, lip reflected and forming a U when seen from above, umbilicus deep, 1/7 of shell diameter.

Under stones in the soil of humid and shady forests, between moss und under leaves, from sea level to 800 m, exceptionally 1500 m.

Fruticicola

Fruticicola Held, 1838
in Held 1838: 914

Type species *Helix fruticum* Müller, 1774
(subsequent designation by Herrmannsen 1846 [18-04-1847]: 450)

Gender feminine (common gender in Latin): in the Latin descriptive text the name was connected with the Latin adjective "parva" and thus treated as feminine, determining feminine the gender under the French version of Art. 30.1.4.2 (the English version is ambiguous).

Fruticicola fruticum (Müller, 1774)

Helix fruticum Müller, 1774 in Müller 1774: 71

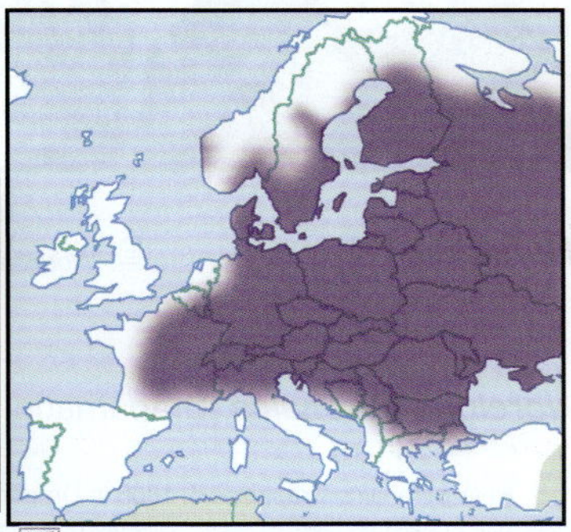

10-20 x 13-25 mm

Austria: Tirol, near Landeck

Europe to Crimea and Caucasus

Germany: Bayern, Barmsee near Mittenwald

Shell yellowish (sometimes reddish brown or even violet), quite transparent, sometimes with colour band at periphery, suture deep, umbilicus open and deep (1/7 of shell diameter). Animal usually light, mantle with dark pigment spots well visible under the shell. *Euomphalia strigella* has a flatter suture, wider umbilicus (1/5), apertural margins approach each other at parietal side.

River floodplains and open shrublands, requires warm temperatures, sensitive against drought, often in tall green vegeration. In Switzerland up to 1800 m, rare above 1500 m, in Bulgaria up to 1800 m. Feeds on humid dead or almost dead leaves of plants, consumes also some herbs and mushrooms.

In N Greece two cohorts exist in the field, reproductive period starts in the beginning of summer, growth increases during spring and autumn. One-year old individuals reach 10 mm shell diameter, 2-year olds 20 mm, maximum diameter of 25 mm is reached after 3 years.
Reproduction in central Europe during humid and warm weather conditions, can be several times between March and August, 10-30 eggs (2.5-3 mm) per clutch, juveniles hatch after 25-50 days, they grow mainly from April to August and reach full size and maturity after 15-22 months. Animals copulate few days after hibernation, probably no eggs are laid in the second year, egg laying starts in the third year, maximum age in captivity 6 years.

Künkel 1928, Frömming 1939, Frömming 1954: 231, Staikou et al. 1990.

Also known as *Bradybaena fruticum* (Müller, 1774). Was introduced to SE England, where it was recorded at various sites 1900-1920, but disappeared again. SW-most record in Haus der Natur Cismar from Les Eyzies-de-Tayac-Sireuil.

Fruticicola schrenkii (Middendorff, 1851)

Helix schrenkii Middendorff, 1851 in Middendorff 1851: 302 (*Helix Schrenkii*)

Shell horny whitish with a narrow and well delimited reddish brown colour band above the suture, slightly translucent, weakly striated, 4.5-5 convex whorls, apex pointed, last whorl not descending, aperture with weak whitish lip inside, umbilicus narrow and slightly covered by the reflected columellar margin. Colour not very variable, sometimes the colour band is absent.
Differs from *Fr. fruticum* in its smaller size, narrower umbilicus and its characteristic colour pattern (which can however also occur in *Fr. fruticum*).

9-14 x 14-18 mm

Sysoev & Schileyko 2009: 177.

Perhaps a synonym of *Fr. fruticum*.

Arkhangelsk region to Altai and Kamchatka

Elona

Elona Adams & Adams, 1855
in Adams & Adams 1858: 211 *Iberus (Elona)*

Type species *Helix quimperiana* Férussac, 1821
(monotypy by Albers 1850: 93)

Adams & Adams gave as synonym "*Sterna*, Albers, not Linn.", thus showing awareness of the homonymy and indicating that *Elona* was meant as a new replacement name for *Sterna* Albers, 1850 in Albers 1850: 93 *Helix (Sterna)* not *Sterna* Linnæus, 1758 (Aves)

Elona quimperiana (Férussac, 1821)

Helix quimperiana Férussac, 1821 in A. Férussac [1821-1822]: 39

Date was 26 May 1821. Established with a bibliographical reference to Férussac & Deshayes 1819-1851: Pl. 76 ("66" printed on the plate in error, previously published, date 05 Aug 1820, on the plate a specimen of *Elona quimperiana* was figured, no specific names were printed on the plate) for the subordinate variant α, but not with a distinct description for the main form (nominal taxon).
The name *Helix quimperiana* was made available at this occasion under Art. 12.1 because it had a description for the taxon (Code Glossary "taxon": a taxon encompasses all included taxa of lower rank). Art. 12.1 does not demand a description for the nominal taxon, a description for the taxon is sufficient, this can include subordinate taxa.
For those who do not like to accept this interpretation of the Code, the next available author is Blainville, 1821 (in Cuvier 1821: 429, *Helix (Helicella) quimperiana*, date 31 Dec 1821 if no other date was made likely in a secondary source that I do not know) (and they would also need to verify whether Art. 11.6.1 can be applied under any interpretation of their choice, since Férussac's name could be interpreted as a synonym).

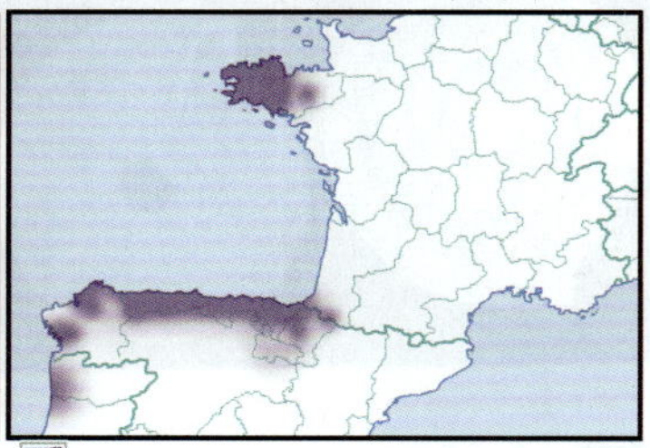

10-14 x 20-30 mm

France: Bretagne, near Lorient

NW France, N Spain, N Portugal

France: Bretagne
Wikipedia User Jymm

Shell with greenish brown periostracum, striated, light radial zones indicate growth interruptions, juveniles with short hairs, flat on upper side with depressed whorls, periphery regularly rounded, last whorl predominating, apertural margin simple, and slightly reflected, inside with white lip, umbilicus open and deep, 1/7 of shell diameter.
Animal greyish, darker on upper side, mantle with irregularly arranged blackish-brown spots.

Humid and shady habitats in woods and shrubland, with herbal vegetation, often near cave entrances, also on moist rocks, in Spain up to 1000 m. Undergoes predation from rodents juveniles also from beetles.

Life span 2-3 years.

In France and Spain strictly protected (including its habitats, best protected snail of France). Populations in SW France do not exist any more.

Kerney et al. 1983: 242, Prieto & Altonaga 1988, Falkner 1990: 200, Kittel 2002: 19, Larraz et al. 2007.

Norelona

Norelona Nordsieck, 1986
in Nordsieck 1986: 118

Type species *Helix pyrenaica* Draparnaud, 1805
(original designation)

Norelona pyrenaica (Draparnaud, 1805)

Helix pyrenaica Draparnaud, 1805 in Draparnaud 1805: 111 (*Helix Pyrenaica*)

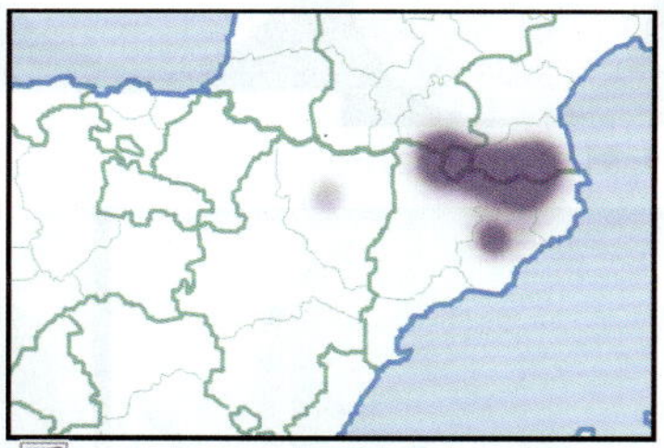

9-11 x 17-21 mm

Spain: (1-3) Girona, Cap de Creus (D = 15 mm), (4-5) Barcelona, Ripol, Rialp
(1-3) Robert Reuselaars, (4-5) José Liétor Gallego

E Pyrenees

Humid habitats at old stone walls, rocks, gardens, also in caves.

Germain 1930: 230, Kerney et al. 1893: 299, Verdú & G. 2006: 354, Guillén & C. 2007.

Shell quite flat, yellowish-brown transparent, no spiral colour bands (unlike *Chilostoma* species), aperture very oblique, with reddish white lip inside, umbilicus narrow and deep, slightly covered by apertural lip.

Vulnerable in Spain, threatened by human pressure in its few habitats. Protected by law in France, eggs and animals must not be collected.

Atenia

Atenia Gittenberger, 1968
in Gittenberger 1968: 169

Type species *Helix quadrasi* Hidalgo, 1885
(original designation)

Gittenberger 1968.

Atenia quadrasi (Hidalgo, 1885)

Helix quadrasi Hidalgo, 1885 in Hidalgo 1885: 193, Pl. 9 (*Helix Quadrasi*)

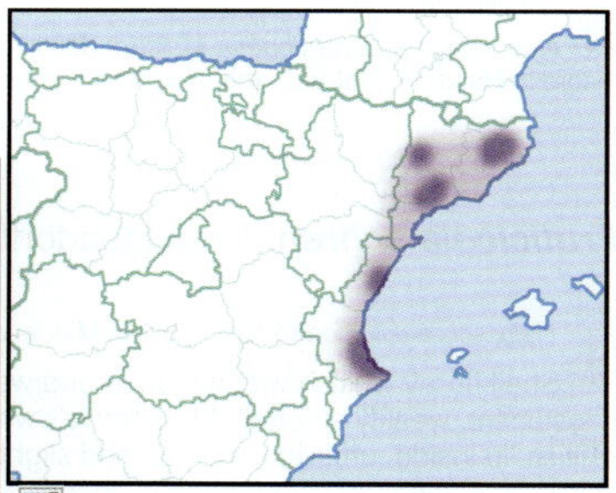

2-2.4 x 4-5 mm (including outer apertural margin)

Spain: Tarragona, La Riga

E Spain

Shady habitats with subterranean access, often associated with *Hedera helix* which helps to provide humid crevices in rocks, also at cave entries, 200-300 m altitude with high levels of microprecipitations carried with winds from the sea.

Vulnerable (Gómez Moliner et al. 2001). Threatened by fire and growing cities.

Shell flat on upper side, fresh shells with hairs and short riblets, periphery regularly rounded, 4-5 whorls, aperture characteristically shaped from upper to lower side, Y- shaped when seen from above, margin reflected, umbilicus very wide and excentrical, more than 1/3 of shell diameter.
The animal is colourless.

Martínez-Ortí & Robles 2003: 120, Martínez-Ortí & Roqué i Pau 2007, Taruella et al. 2011.

Drepanostoma

Drepanostoma Porro, 1836
in Porro 1836: 1, Pl. 71

Type species *Drepanostoma nautiliformis* Porro, 1836
(monotypy)

Drepanostoma nautiliforme Porro, 1836

Without gender agreement: *Drepanostoma nautiliformis* Porro, 1836

Drepanostoma nautiliformis Porro, 1836 in Porro 1836: 1, Pl. 71
Should have been *nautiliforme*, *Drepanostoma* is neuter.

2.5-3 x 5-6 mm

Italy: near Lago Maggiore

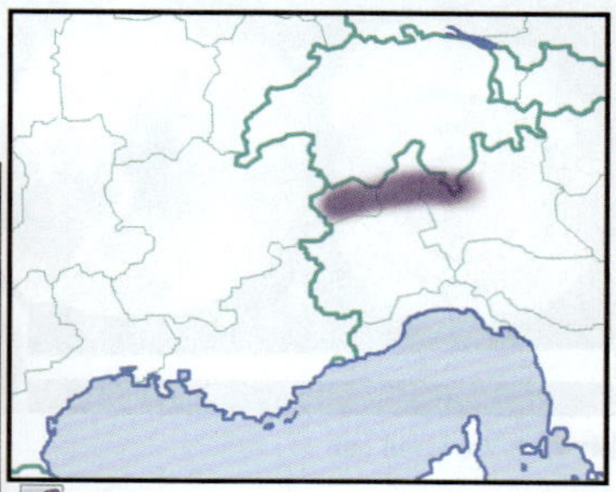

SW Alps

EU Red List: Near Threatened NT. Threatened by modern forest management.

Shell turban-shaped, brown with long hairs, very characteristic, apertural margin protruding and strongly reflected, umbilicus at both sides deep, at lower side wider. Juveniles of below 3 mm white.

In mountain forests, humid and shady habitats, often in small groups, in leaf litter, at old wood, old bark and humid rock rubble. More frequent on calcareous than on silicate substrate. In Switzerland up to 1200 m.

Girod 1968, Falkner 1990: 224, Manganelli et al. 1995: 31, Turner et al. 1998: 329.

Falkneria

Falkneria Nordsieck, 1989
in Nordsieck 1989: 166

Type species *Helix camerani* Lessona, 1880
(original designation)

Falkner 1990: 222.

Falkneria camerani (Lessona, 1880)

Helix camerani Lessona, 1880 in Lessona 1880: 356, Tav. 4

Shell brown, with some hairs, 5 whorls, upper side nearly flat, periphery at last whorl with blunt edge in its upper section, lower side rounded, aperture with white lip and a pointed palatal tooth, margin reflected, umbilicus very wide and perspectivical, all whorls well visible inside, 1/3 of shell diameter.

Mountain forests, mainly above 1000 m, not confined to limestone.

EU Red List: Vulnerable VU. Species with a small and restricted range N of Biella, known from very few localities. The threats are unknown, besides forest road constructions, but a decline has recently been observed in Valle del Cervo.

3.5-4.5 x 9-11 mm

Italy: Biella, Valle del Cervo

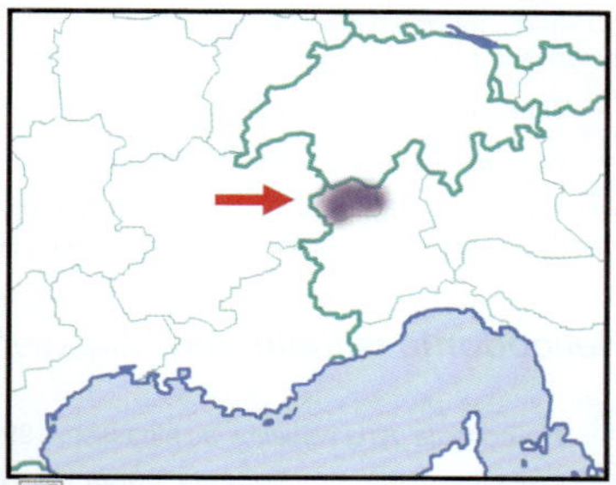

NW Italy

Falkner 1990: 222, www.iucnredlist.org (08.2012).

Paolo Zaccaria
Tiroler Landesmuseen Innsbruck

Helicodonta

Helicodonta Férussac, 1821
in A. Férussac [1821-1822]: 23 (folio p. 27) *Helix (Helicodonta)*

Type species *Helix obvoluta* Müller, 1774
(subsequent designation by Pfeiffer 1848: 545)

Established with short description and without species included, 06 Apr 1821 (livraison 9). First species included in quarto p. 33 and folio p. 37 (26-05-1821), among these was *Helix obvoluta*. Gray 1847: 174 gave as type *Helix gularis*, which was not originally contained. Pfeiffer 1848 (Mon. Helic. 2): 545 cited two names "*Helicodonta* Férussac" and "*Helicodonta* Risso", and gave no type for Férussac's name and *Helix obvoluta* as type for Risso's name. Risso 1826: 65 had only used Férussac's name (Risso 1826 included only *Helix obvoluta*, this was in line with Férussac's 1821 use of the name). So the type designation by Pfeiffer 1848 referred to this name. Wenz 1923 (18): 446 also gave as type *Helix obvoluta*.

Helicodonta angigyra (Rossmässler, 1834)

Helix angigyra Rossmässler, 1834 in Rossmässler [1834]: 4
Original description short and somewhat unintentional: "Dagegen steht unserer Art [*Helix obvoluta* Müller, 1774] die neue *H. angigyra* Z[iegler]. aus Oberitalien sehr nahe, unterscheidet sich jedoch durch ihre mindere Größe und einen Umgang mehr.". The name was made available at this occasion.

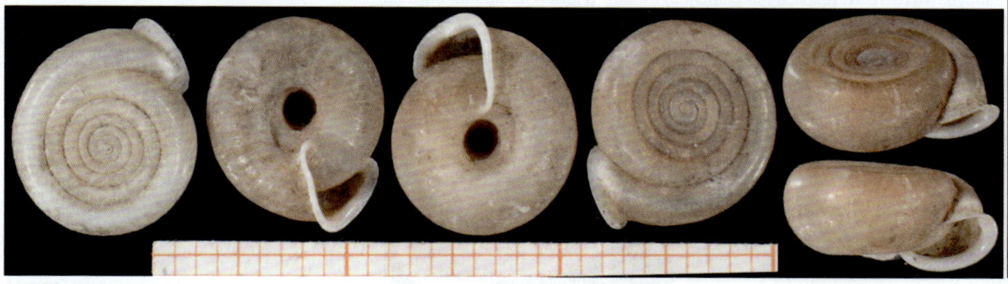

4-5.5 x 9-11 mm

Italy: near Lago di Como, 3 km N Lecco

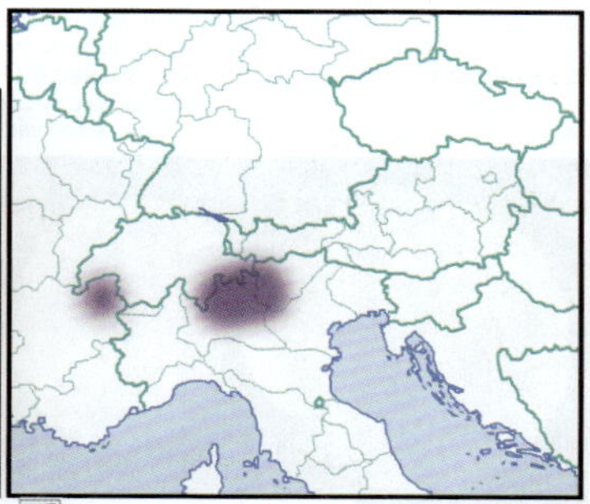

S Alps

Shell smaller than *H. obvoluta*, whorls much narrower, hairs in young animals only 0.1 mm long and very fine, hair scars in adult specimens produce a characteristic pattern, usually no callus in the aperture, umbilicus narrower and slightly more excentric.

Under dense ground litter and in rock rubble at humid places in mountain forests of lower altitudes, also in cultivated land, requires higher temperatures than *H. obvoluta*. In Switzerland up to 1700 m.

Girod 1968.

Helicodonta obvoluta (Müller, 1774)

Helix obvoluta Müller, 1774 in Müller 1774: 27
Helicodonta langhofferi Wagner, 1912 in A. J. Wagner 1912: 249
Sturany & Wagner 1915: 31 claimed a printer's error for *laughofferi*, but on p. 250 the species was expressly dedicated to "Prof. Langhoffer", not Laughoffer. So the author was responsible for the error, and the name *langhofferi* must be maintained.
Helicodonta obvoluta albanica Wagner, 1914 in Sturany & Wagner 1915: 31

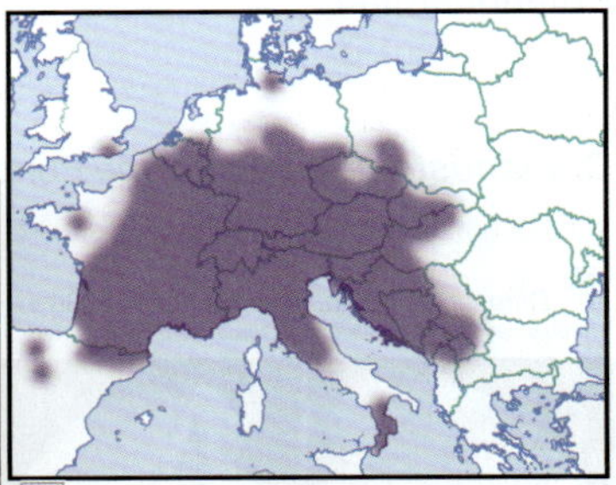

Europe

5-7 x 11-15 mm

Germany: (1-6) "Gambach", (7-8) Thüringen, Hainich forest

Slovakia: Banská Bystrica — Jozef Grego

Shell brown, flat above with apex slightly sunken, hairs usually permanent and up to 1 mm long, 5-6 densely coiled and slightly convex whorls, aperture reflected, with white lip, basal thickening and palatal tooth or callus, umbilicus wide, perspectivical.
Shells from Albania and Montenegro are usually smaller, upper whorls more impressed/concave, whorls more rapidly increasing, hairs 1/3 shorter, aperture more rounded, no teeth at the margin, no corresponding grooves outside, umbilicus slightly wider.

Germany: Thüringen, Hainich forest

In forests under leaf litter, fallen branches and between stones, often in highlands, usually bot not obligatory on calcareous substrate. In the Alps up to 1900 m, rarely above 1500 m. May weakly climb up tree trunks.

In France copulation was observed (Gassies 1849) between May and July, 2-3 days later some 20 eggs were laid, juveniles hatched after 3 weeks with 1.5 whorls, the shell was completed in November. In Poland copulation takes place in the night and lasts 2-3 hours, 15-20 eggs (2.3-2.4 mm diameter) are laid in one clutch in May/June and one in September (egg laying lasts 1-2 days), 4-6 mm deep in rotting timber, producing two generations per season, juveniles (initially with 1.0 whorls) hatch asynchronously (1-4 days difference) after 15-30 days, initial growth rates are at 0.4-0.5 whorls/month, full size is reached after 4-12 months, maturity coincides with the completed lip, life span 2-3 years, maximum number of clutches laid in lifetime is six. At the end of October snails penetrate rotting wood logs from below to hibernate with thick epiphragms, activity restart is in April. Mobility in adults ranges at 4-7 m/month.

Threatened in England by modern forest management (clear felling and replanting), though still common at a number of places.

Gassies 1849: 104, Maltz 2003 (life cycles).

There are currently no constraints for classifying *langhofferi* from Montenegro (?, only known from Dalmatian sea shore deposits and not found living somewhere in the past 100 years) and *albanica* from N Albania as separate species. More research on the spatial variation of shells between Austria and Albania would be useful.

Helicodonta wilhelminae Maassen, 1991

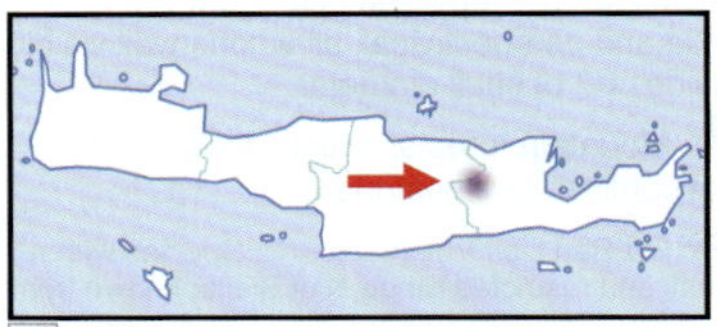

Greece (Crete)

Helicodonta gyria wilhelminae Maassen, 1991 in Maassen 1991: 124

Shell like *Lindholmiola gyria*, upper side weakly striated, 5.5-6 whorls, whorls more rapidly increasing than in *L. gyria*, last whorl width more than 2 x of preceding whorl, slightly descending near aperture, aperture without teeth, umbilicus open, excentrical at last half whorl, 1/4 of diameter.

6-7 x 13.5-16 mm

In deep rock crevices in mountains.

EU Red List: Endangered EN. Only known from a small range, no acute threats known.

Initially classified as a subspecies of "*Helicodonta gyria*" from S Turkey, which later turned out to belong to *Lindholmiola*. Later it was classified as a separate *Helicodonta* species (www.iucnredlist.org , 08.2012). Its relationships to *Hcd. obvoluta* should be clarified, and compared to the spatial variation of this widespread species, also on the molecular level. Seems to represent an old relictual fauna element in Crete.

Lindholmiola

Lindholmiola Hesse, 1931
in Hesse 1931: 50

Type species *Helix lens* Férussac, 1832
(original designation)

This genus presentation was kindly checked by P. Subai.

Lindholmiola barbata (Férussac, 1832)

Helix barbata Férussac, 1821 in A. Férussac [1821-1822]: 37
Established with short descriptions for two subordinate variants α and β, but not with a distinct description or indication for the main form (nominal taxon). The name *Helix barbata* was made available at this occasion under Art. 12.1 because it had an indication for the taxon (Code Glossary "taxon": a taxon encompasses all included taxa of lower rank). Art. 12.1 does not demand a description (or indication) for the nominal taxon, a description for the taxon is sufficient, this can include subordinate taxa.
For those who do not like to accept this interpretation of the Code, the situation is meaningless. ICZN Opinion 1690: ruled that the name was not first available from Férussac 1821, but from Férussac 1832.
Helix barbata Férussac, 1832 in A. Férussac & Deshayes 1819-1851: (explication planches) Pl. 66* (*Helix (Helicigona) barbata*)
Original print of explication des planches reproduced by Gittenberger & Groh 1986, Pl. 66* did not appear on second print p. 13.
ICZN Opinion 1690: as defined by the lectotype designated by Gittenberger & Groh 1986, figured by Férussac 1832 and named as *H. barbata* var. α.

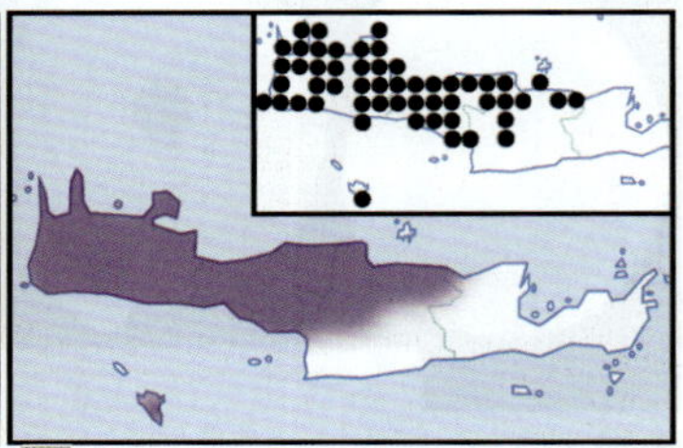

4.5-5.5 x 10-12 mm

Greece: island of Gávdos

Greece (W Crete, Gávdos)

Shell brown, covered with fine hairs, almost flat on upper side, 5-6 whorls, apertural margin white, reflected, covers two thirds of the umbilicus in adult shells (difference to other *Lindholmiola* species), parietal side usually with a whitish layer, umbilicus 1/9-1/6 of diameter, proportionally larger in juveniles. Animal almost black, darker than the shell.

Open forest habitats, under stones, seems to prefer Mesozoic calcareous substrate. Usually less frequent than *Metafruticicola*.

Threatened by destruction of natural forests and forest fires. In Gávdos the area north of Kastrí seems to have been cleared from forest in the antiquity. The species must have gone extinct there and was not able to resettle this habitat after the forest had returned during the past 1000 years.

Maassen 1991: 35, Vardinoyannis (1994, unpublished thesis), Welter-Schultes 1998: 89, unpublished HNC data.

Lindholmiola girva (Frivaldszky, 1835)

Helix girva Frivaldszky, 1835 in Frivaldszky 1835: 274, T. 7
Helix contorta synonym *Helix corcyrensis* Rossmässler, 1838 in Rossmässler 1838-1844 (7-8): 40, Taf. 39
Mentioned with description and figure. Pfeiffer 1848: 415 used the name as *Helix corcyrensis* and gave a bibliographical reference to Rossmässler 1838: 40, making this name available under Art. 11.6.1.

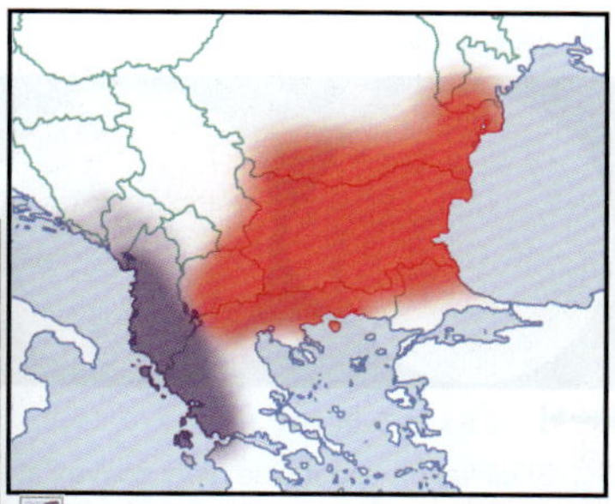

SE Europe
Violet: *corcyrensis*, red: *girva*.

4-7 x 8-14 mm, varies locally

Albania: Shkodër

Shell brown with 6-7 nearly flat whorls, suture deep, hairs are usually permanent, upper side nearly flat, periphery slightly but not prominently keeled in its upper section, lower side rounded, aperture with white reflected margin, covering a small part of the umbilicus.
Form *corcyrensis*: whorls perhaps more convex, upper side perhaps flatter, apex impressed.

Limestones in open habitats with sparce forest or shrub vegetation and relatively low temperatures, also in forest habitats. Survives considerable temperature fluctuations. In Bulgaria up to 2000 m. Frequent in Albania.

Damjanov & Likharev 1975: 341, Shilejko 1978: 117, Falkner 1990: 222, Schütt 1996: 369, Dhora & Welter-Schultes 1996: 157, Irikov & Mollov 2006: 816.

Lindhiolmiola corcyrensis represents a separate species according to the yet unpublished results obtained by P. Subai. All figures represent *corcyrensis*.

Greece: Ipiros, Ditiko Zagori, Aristi — Cédric Audibert

Lindholmiola gyria (Roth, 1839)

Helix gyria Roth, 1839 in Roth 1839: 16, Tab. 1

Shell ribbed on upper side, striated and shiny on lower side, with weak hair scars, upper side flat with apex slightly sunken, periphery regularly convex, aperture rounded, no teeth, margin whitish and reflected, umbilicus wide, not covered by apertural margin.

Forest habitats. In E Crete in deep rock crevices at the W side of mountains.

6-7 x 12-16 mm

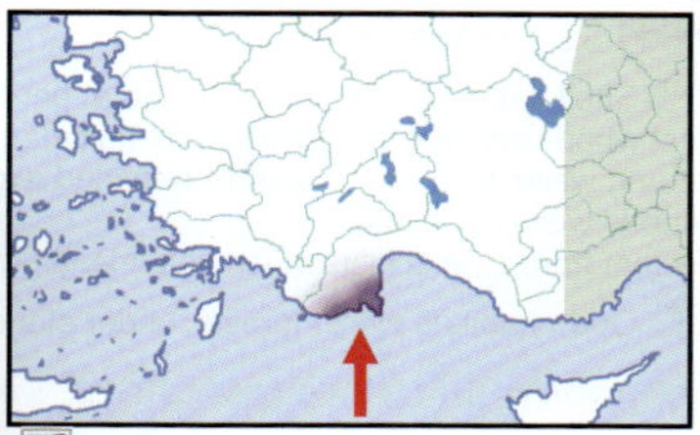

S Turkey

Schütt 2001: 402.

The taxon *wilheminae* from Crete is classified as a separate species in *Helicodonta*.

Turkey: Vilayet Antalya, near Finike

Lindholmiola lens (Férussac, 1832)

Helix lens Férussac, 1832 in Férussac & Deshayes 1819-1851: (explication planches) Pl. 66* (*Helix (Helicigona) Lens*)
Original print of explication des planches reproduced by Gittenberger & Groh 1986, Pl. 66* did not appear in second print, where it appeared as *Helix lens* for Pl. 66.

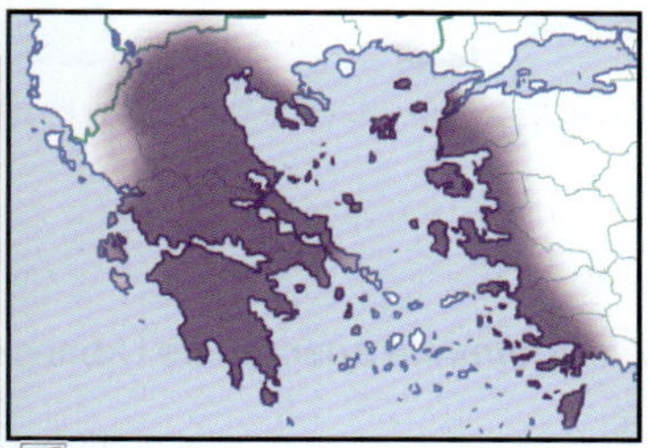

5-6 x 10-13 mm

Greece: island of Psérimos

Greece, W Turkey

Turkey: 80 km S Izmir

Shell brownish, hairs in juveniles present at keel and lower side, in adults not many hairs remain, 6-7 flat (not convex) whorls, upper side usually not flat, periphery with strong (sometimes blunt) keel usually in the upper half of the shell (shell nearly lentiform with lower side usually more rounded), apertural margin white, not much reflected, covers very slightly a small portion of the umbilicus, if at all, umbilicus deep, usually 1/9-1/8 of diameter. Shell quite variable.

Under stones, often in open habitats, also at olive trees.

Partners touch each other for several hours with their tentacles and mouths prior to copulation (in contrast to *L. spectabilis*, which prevents possible cross-breeding in captivity). 10-15 eggs (diameter 2 mm) are laid in autumn, hidden deeply in humid soil substrate, juveniles hatch after 2 weeks. 3 weeks old juveniles have 2-3 whorls.

Falkner 1990: 222, Rähle 1980: 218, Liebegott 1986, Schütt 1996: 368, Gümüş 2010 (W Turkey).

Cédric Audibert

Lindholmiola pirinensis Jaeckel, 1954

Lindholmiola contorta pirinensis Jaeckel, 1954 in Jaeckel 1954: 86

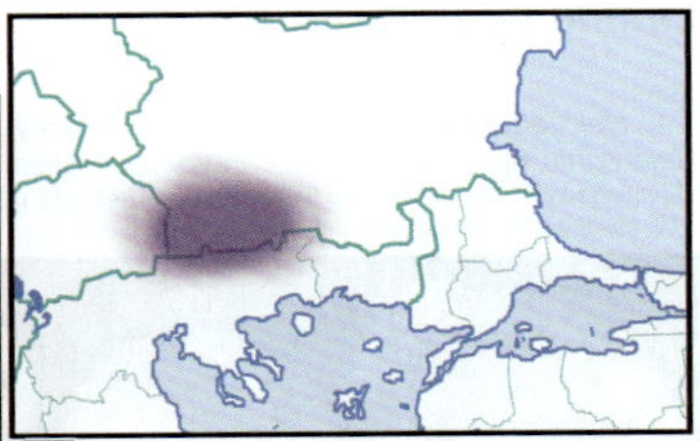

5-7.5 x 11.5-14 mm

Bulgaria: W Rodopi mountains

Shell flat on upper side, very convex at lower side, with only few unstable hairs, with weaker sculpture than other *Lindholmiola* species in the region, aperture larger and higher, reaching more downside, umbilicus 1/6-1/5 of shell diameter, columellar margin of the aperture does not cover partly the umbilicus.

SW Bulgaria, adjacent regions

In mountain habitats, 1500-2000 m.

Damjanov & Likharev 1975: 342, Dedov 2008.

Lindholmiola regisborisi (Wagner, 1928)

Caracollina regisborisi Wagner, 1928 in A. J. Wagner 1928: 364 (*Caracollina regis-borisi*)

5.5-7 x 13-18 mm

NE Greece: Stavroúpoli near Xánthi

NE Greece

Reischütz 1988: 347.

Shell flattened, bluntly angulated at periphery, finely striated and with long hairs except near the umbilicus, 6-6.5 slowly and regularly increasing whorls, last whorl only slightly wider than penultimate whorl, but suddenly wider and suddenly deeply descending near the aperture, aperture very oblique, inside with strong white lip, margins slightly reflected except on upper side, not approaching each other closely at parietal side, umbilicus wide and perspectivical, all whorls well visible inside.

Differs from *L. lens* in its much flatter lentiform shell with more stable hairs, less angulated last whorl, more oblique aperture and wider umbilicus.

Open habitats within forests, under stones.

Lindholmiola reischuetzi Falkner, 1996

Lindholmiola reischuetzi Falkner, 1996 in Falkner 1996 ("1995"): 97

Expressly established as a new replacement name for *Helix lens* var. *insularis* Boettger, 1907 in O. Boettger 1907: 34 (not *Helix serpentina* var. *insularis* Benoit, 1857 in Benoit 1875 (Ill. Sist.): 109)

Shell densely ribbed, 5-6 whorls, moderately strongly keeled, keel at last whorl white, apertural margins connected by a strong white callus at parietal side, umbilicus moderately wide. Smaller than other *Lindholmiola* species, lower number of whorls, parietal callus stronger.

3.7-4.1 x 9-11 mm

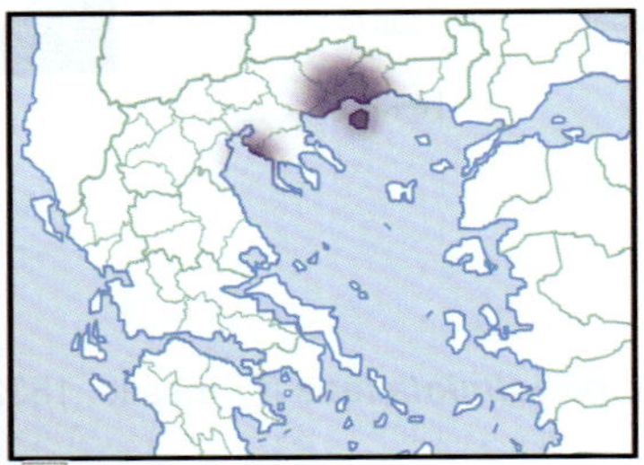

N Greece

Greece: Thásos, Aliki (D = 10 mm)

Robert Reuselaars

Lindholmiola spectabilis Urbański, 1960

Lindholmiola spectabilis Urbański, 1960 [I do not know where this name was established]

Shell very flat, ribbed on upper side, weakly ribbed on lower side, hairs only in juveniles and only at lower side, periphery with strong and white keel in the upper fourth of the whorl, lower side rounded, apertural margin well reflected, its outermost curve does not correspond with the keel which reached the aperture slightly above, umbilicus deep, often slightly excentrical at the last 1/2 whorl, its inside section 1/9 of shell diameter.

4-5 x 10-18 mm

Greece: Makedonia, near Asprovalta

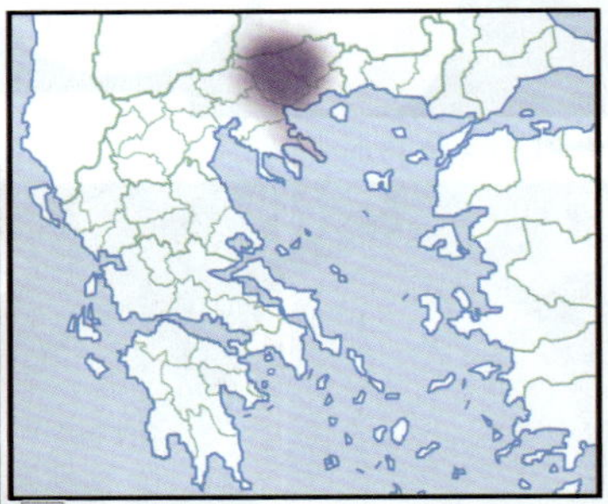

N Greece

Under rocks and dense ground litter, more attached to rocks than *L. lens*.

Falkner 1990: 222.

Needs less time for copulation than *L. lens*, snails do not touch each other with their tentacles prior to copulation.

Soosia

Soosia Hesse, 1918
in Hesse 1918: 104

Type species *Helix diodonta* Férussac, 1832
(original designation)

Soosia diodonta (Férussac, 1832)

Helix diodonta Férussac, 1832 in A. Férussac & Deshayes 1819-1851: (explication planches) Pl. 51A
Original print of explication des planches not verified, second print p. 11.

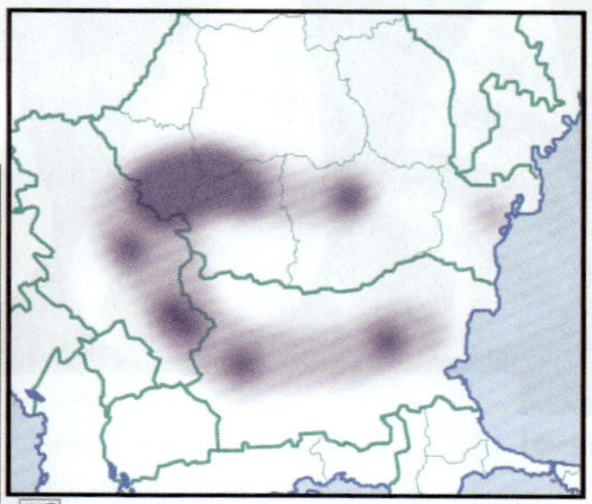

3.5-4 x 10-12 mm

Romania: Caraş-Severin, Băile Herculane

Romania, Serbia, Bulgaria

EU Red List: Near Threatened NT.
Threatened by modern forest management.

Shell extremely flat, like a flying saucer, shiny, 4-4.5 slightly convex whorls, periphery rounded without keel, apertural opening strongly directed downside, margin white and reflected, with two prominent teeth (upper palatal and basal), umbilicus wide and perspectivical, excentrical at last whorl, 1/4 of diameter.

Warm and humid natural forests under bark, dead wood and inside the fibres of humid trunks, the species needs large decaying tree logs.

Pavlović 1912: 33, Tomić 1959: 18, Damjanov & Likharev 1975: 343, Falkner 1990: 200, Georgiev & Stoycheva 2007 (E Bulgaria).

Cochlicella

Cochlicella Férussac, 1821
in A. Férussac [1821-1822]: 24 (folio p. 28) *Helix (Cochlicella)*

Type species *Helix conoidea* Draparnaud, 1801
(subsequent designation by Gray 1847: 173)

Established with short description and without species included, 06 Apr 1821 (livraison 9). First species included on p. 56 folio (26-05-1821) (p. 52 quarto, 13-07-1821).
Prietocella Schileyko & Menkhorst, 1997 in Schileyko & Menkhorst 1997: 54, type species *Helix barbara* Linnæus, 1758 (original designation)

Schileyko & Menkhorst 1997 classified *Coc. barbara* in a separate genus *Prietocella*, but their hypothesis on the phylogenetic relationships would also allow the classify *Cochlicella* and *Prietocella* as subgenera. This is preferred here, in agreement with Alba et al. 2011: 50.

Cochlicella acuta (Müller, 1774)

Helix acuta Müller, 1774 in Müller 1774: 100

9-15 x 4-7 mm

Greece: E Crete, near Ierápetra

Iran to Britain
Insert map for Crete: Vardinoyannis (1994, unpublished thesis), unpublished HNC data.

Shell evenly white or with brown spots, 8-11 whorls, in juvenile specimens with sharp edge.
Shell less brownish, more slender and higher than *C. barbara*, upper whorls slightly more rounded, juveniles with much sharper edge at the periphery of the lowest whorl.
Animal very light yellowish, dorsum with blackish brown pigments, one dark medial line on dorsum, two dark lines (= retractor muscles) from the sides to the upper tentacles.

Grasslands, dunes and sandy habitats, at limestones, open dry habitats in coastal vicinity, often exposed to the sun, but more abundant under shrubs or low trees, introduced also to the interior of some countries. Probably frost-sensitive, in Britain only in frost-free habitats. Habitat requirements less restrictive than *C. barbara* and *C. conoidea*, with which it may live syntopically. *C. barbara* prefers slightly more humid habitats.

Subgenus *Cochlicella*.

Cochlicella barbara (Linnæus, 1758)

Helix barbara Linnæus, 1758 in Linnæus 1758: 773

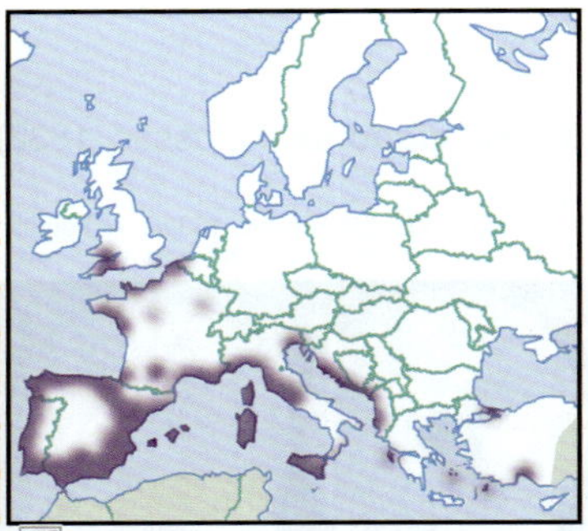

4.5-7 x 7-15 mm, usually 5 x 8-10 mm

Spain: Ebro delta

Mediterranean region to England

Shell whitish, greyish or yellowish with brown spots or bands, 7-8 whorls, umbilicus very narrow but rarely closed. Juveniles slightly keeled. Shells in Portugal more greyish and lighter than in E Spain or S France, where the brown colour is more intensive.
Shell generally more brownish, shorter and less slender than *C. acuta*, first whorls slightly less rounded. Juvenile *C. acuta* have a more sharply angulated periphery at the lowest whorl.
Body very light yellowish, dorsum with blackish brown pigments, one dark medial line on dorsum, two dark lines (= retractor muscles) from the sides to the upper tentacles. The animal is shy and easily irritated, the shell is carried in an almost vertical position.

Sandy and dry habitats in coastal vicinity, dunes with grass cover, on the soil between plants. Probably frost sensitive. In wide regions a very frequent and common species, for example in central and S Portugal. Slightly more humid habitats than *C. acuta*, sometimes syntopically.

Eggs sphercial (1 mm diameter), laid in a whitish and transparent envelope.

Moquin-Tandon 1855: 277, Germain 1930: 317, Nobre 1941: 107, Prieto & Gómez 1988 (N Spain), Falkner 1990: 218, Manganelli et al. 1995: 30, Schütt 1996: 392, Dhora & Welter-Schultes 1996: 156, Ondina et al. 1997, Kerney 1999: 187, Martínez-Ortí, Ap. & R. 2004.

Subgenus *Prietocella*.

Cochlicella conoidea (Draparnaud, 1801)

Helix conoidea Draparnaud, 1801 in Draparnaud 1801: 69 (*Helix conoïdea*)

6-9 x 5-6 mm

S Portugal: Algarve, Ilha de Tavira

Mediterranean region

Shell light greyish or light horny brown, with brown spots and a brown colour band at the base of the last whorl, 4.5-6 whorls, slightly keeled at the base, aperture simple and sharp, umbilicus open and slightly covered by columellar margin of the aperture. Shorter than *C. acuta* and with larger diameter and a slightly more profound suture than *C. barbara*.
Animal yellowish, lighter at the foot, upper tentacles very long, lower tentacles short, posterior end of foot pointed.

Sandy habitats at beaches among coastal vegetation, restricted to localities in coastal vicinity, requires higher temperatures than the other *Cochlicella* species. Lives in large colonies, can locally be very abundant.

Nobre 1941: 111, Falkner 1990: 218, Gittenberger 1991: 126, Manganelli et al. 1995: 30, Schütt 2001: 436, Beckmann 2007: 93.

Subgenus *Cochlicella*.

Ciliella

Ciliella Mousson, 1872
in Mousson 1872: 60 *Helix (Ciliella)*

Type species *Helix ciliata* Hartmann, 1821
(subsequent designation by Pilsbry 1894 [1895]: 276)

Ciliella ciliata (Hartmann, 1821)

Helix ciliata Hartmann, 1821 in Hartmann 1821 (Neue Alpina 1): 240

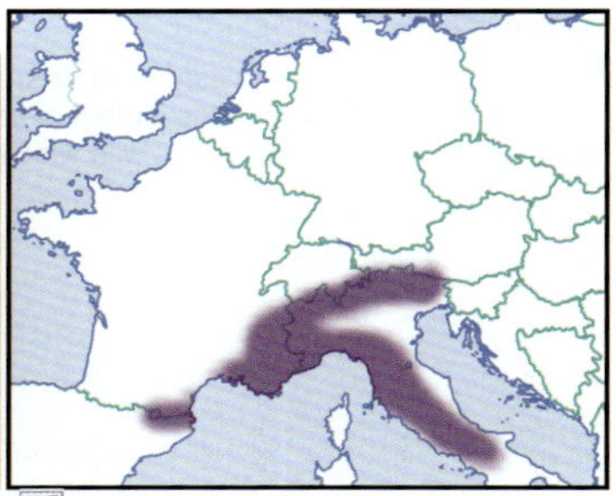

4-6 x 9-12 mm

Italy: Gardone near Lago di Garda

Alps to Toscana, E Pyrenees

Shell brown, striated, relatively flat, low conical above and rounded below, sharply keeled at the periphery, with spines somehow like *Acanthinula aculeata*, apertural margin simple, sharp and reflected, umbilicus narrow but open, slightly covered by the reflected lip.

Grassy slopes, riversides and slopes in mountains, pastures, also in forests, more rarely in gardens and parks. Up to 2100 m in the Alps.

Kerney & Cameron 1979: 197, Falkner 1990: 218.

Ciliellopsis

Ciliellopsis Giusti & Manganelli, 1990 in Giusti & Manganelli 1990: 269

Type species *Ciliellopsis oglasae* Giusti & Manganelli, 1990 (original designation)

Ciliellopsis oglasae Giusti & Manganelli, 1990

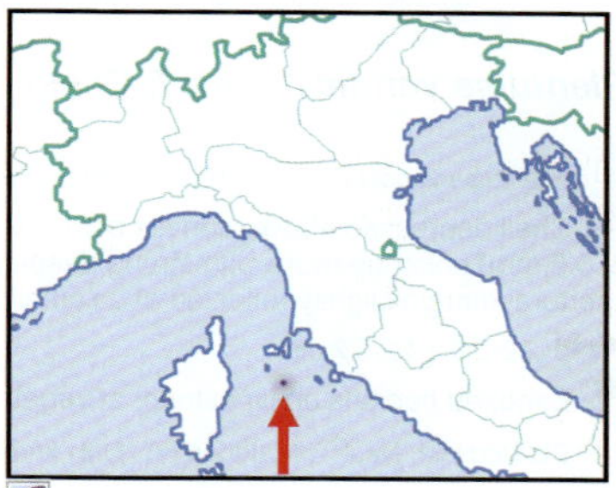

Italy (island of Montecristo)

Ciliellopsis oglasae Giusti & Manganelli, 1990 in Giusti & Manganelli 1990: 270

Shell thin and fragile (due to the granitic substrate), brown, with hairs and dense growth lines, 4 whorls with deep sutures, apertural margin slightly reflected at lower margin, umbilicus wide (1/6-1/5 of shell diameter) and deep.

2.8-3.4 x 5.3-6 mm

Under stones, fallen tree trunks or in leaf litter, in shady habitats with abundant *Ailanthus glandulosa* vegetation, on granitic substrate.

EU Red List: Vulnerable VU. Endemic species of Montecristo island, rare and endangered, the island is a natural reserve. Potentially threatened by overgrazing and fire.

Montserratina

Montserratina Ortiz de Zárate y López, 1946 in Ortiz de Zárate y López, 1946: 342

Type species *Helix bofilliana* Fagot, 1884 (monotypy)

Ortiz de Zárate included two species *H. bofilliana* and with doubts *Helix martorelli* Bourguignat, 1870, the expressed doubts provoke monotypy (Art. 68.3).

Penial walls externally enveloped by a thin muscular sheath, penial retractor muscle very short, epiphallus and true penial papilla present, presence of a muscle connecting the vagina walls to the columellar muscle, vagina without any trace of a dart-sac complex, unbranched digitiform glands consisting of 3 tubular elements.

Giusti & Manganelli 1988: 89, Bank et al. 2001: 99.

Montserratina becasis (Rambur, 1868)

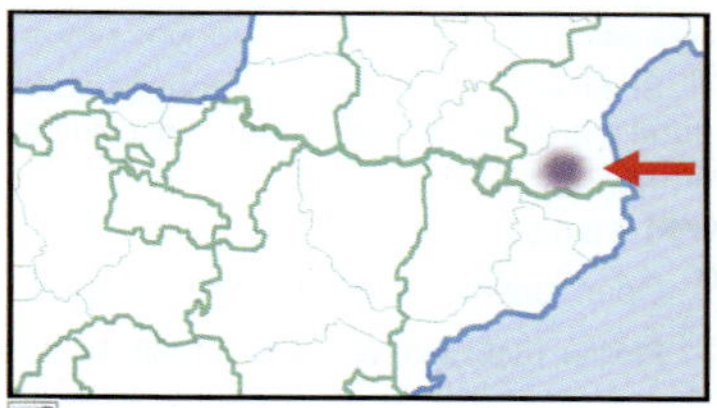

E Pyrenees (?)

Helix becasis Rambur, 1868 in Rambur 1868: 267 (*Helix Becasis*)

Shell thin and transparent, greenish or reddish, irregularly striated, with fine hairs, apex broad, suture deep, apertural margin simple, straight and sharp, umbilicus slightly narrower than in *M. bofilliana*. Differs from *M. martorelli* in its much shorter hairs.

2-2.5 x 4 mm

Humid or wet meadows, under rocks.

EU Red List: Critically Endangered CR. The form has not been found after 1900, either the type locality was given incorrectly, or all habitats have been destroyed by draining of wet meadows for agricultural purposes.

Westerlund 1889: 52, Falkner et al. 2002: 146, Bertrand 2002: 17.

Montserratina bofilliana (Fagot, 1884)

Helix bofilliana Fagot, 1884 in Fagot 1884: 177

Spain: Barcelona, Sant Llorenç del Munt

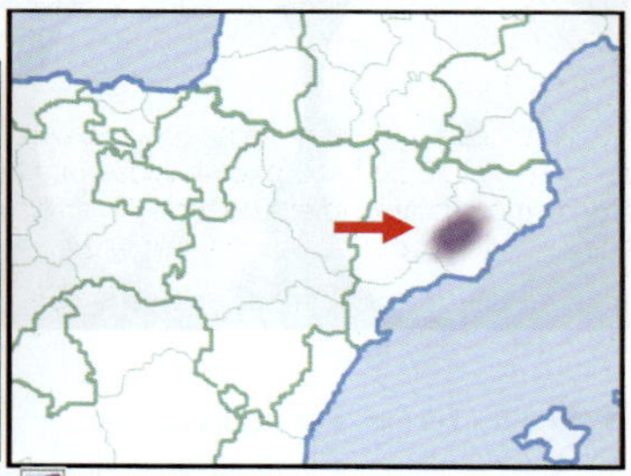

NE Spain (Montserrat)

2.5-3 x 5-6 mm

Shell horny reddish or greenish, irregularly striated, with some hairs, particularly on the last whorl, apex protruding, last whorl depressed and angulated, suture deep, umbilicus wide and regularly coiled, apertural margin simple and reflected at the umbilicus.

Westerlund 1889: 51, Germain 1930: 243, Ortiz de Zárate y López, 1946: 340, Puente et al. in Verdú & G. 2006: 372, www.iucnredlist.org (08.2012).

Cliffs and steep rocks in forests, also present in stony areas.

EU Red List: Near Threatened NT. Only scattered occurrences. Threatened by fire and modern forest management.

Montserratina martorelli (Bourguignat, 1870)

Helix martorelli Bourguignat, 1870 in Bourguignat 1870: 26 (*Helix Martorelli*)

Spain: Cataluña, Santa Maria de Tinestres (?)

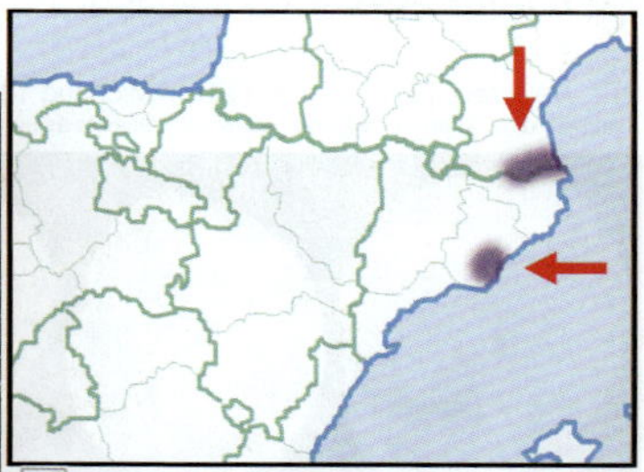

NE Spain, SW France

2.7-3.5 x 4.5-6 mm

Shell dark horny coloured, usually covered with mud, finely striated and with hairs all over the shell, apex protruding, umbilicus narrow, aperture regular and slightly thickened inside, basal margin slightly reflected.

Only scattered occurrences, seems to have disappeared from various localities.

In natural habitats, in France mostly found on acid substrate, occasionally also on calcareous substrate. In Spain in mixed forests and oak woods, also on meadows and pastures at tree trunks, walls or stones.

Westerlund 1889: 51, Germain 1930: 243, Falkner et al. 2002: 146, Bertrand 2002: 17, Verdú & Galante 2006: 373, www.iucnredlist.org (08.2012).

Possibly conspecific with *M. becasis*.

Plentuisa

Plentuisa Puente & Prieto, 1992
in Puente & Prieto 1992: 159

Type species *Plentuisa vendia* Puente & Prieto, 1992
(original designation)

Shell small, brown with hairs, umbilicus deep and wide. Vagina short and thickened, no dart sacs or accessory sacs, with four digitiform mucus glands, flagellum rudimentary. Right ommatophore retractor does not cross genital ducts. Penial nerve originated from right cerebral ganglion.

Plentuisa vendia Puente & Prieto, 1992

N Spain

Plentuisa vendia Puente & Prieto, 1992 in Puente & Prieto 1992: 160

Shell depressed above, convex below, evenly brown, thin, translucent, 4.5 whorls, periostracum with long (> 0.5 mm) hairs, aperture with strong deep-set internal lip, visible from outside as a yellowish or whitish area, apertural margin slightly reflected at columellar edge, umbilicus wide and deep, 1/4 of shell diameter.

2.2-2.8 x 4.2-5.2 mm

Exposed habitats on limestone, at vegetation growing inside rock crevices.

EU Red List: Near Threatened NT. Only known from a small restricted area.

Puente et al. in Verdú & Galante 2006: 374.

Ponentina

Ponentina Hesse, 1921
in Hesse 1921: 60, 62 *Fruticicola (Ponentina)*

Type species *Helix subvirescens* Bellamy, 1839
(original designation, p. 76)

Shell subglobulous with narrow umbilicus, whorls and aperture rounded, aperture with white lip inside.

NW Africa, W Europe

Germain 1930: 245.

Ponentina martigena (Férussac, 1832)

Spain: Málaga, Sierra de las Nieves

José Liétor Gallego

S Spain

Helix martigena Férussac, 1832 in A. Férussac & Deshayes 1819-1851: (explication planches) Pl. 69

Original print of explication des planches not verified, second print p. 13 as *Helix Martigena*. Name had been mentioned by Férussac [1821-1822]: 38 under No. 169 without description or reference.

Shell with yellowish hue, 4 convex whorls, last whorl descending near aperture, apertural margin white and broadened, columellar margin reflected, umbilicus open and narrow.

5.5-6.5 x 7-8 mm

Open forests, grasslands and wastelands, also at roadsides.

Pfeiffer 1848: 365, Westerlund 1889: 61, Arrébola & García 1999.

Ponentina ponentina (Morelet, 1845)

Helix ponentina Morelet, 1845 in Morelet 1845: 65, Pl. 6

Portugal: Lisboa

Portugal

3-5 x 7-9 mm, shell height variable

Shell yellowish with greenish or brownish hue, or brownish with olive green hue, coarsely striated, adults often with short hairs of the same colour as the shell, 4 not very convex whorls, suture deep, aperture with a broad white lip inside, margin white, sharp and reflected, umbilicus open but not very wide. Juveniles usually covered with many hairs, aperture sharp without white lip inside. In the non-calcareous habitats in the north shells are thinner and more often covered with hairs.
Animal greyish yellow, mantle brownish yellow, with white and dark brown pigments. Tentacles not very long.

In shady and sun-exposed habitats, on old walls and among plants at their bases, under stones, on the soil, also on sandy substrate. Approaches very closely to the seaside, at some localities found only a few meters from the shore. In Serra da Estrela it lives also in high altitudes, on mosses covering the rocks. Usually in colonies. A common species in Portugal, but seems to be less common in the south.

Pfeiffer 1848: 132, Westerlund 1889: 61, Nobre 1941: 122.

Ponentina revelata (Michaud, 1831)

Helix revelata Michaud, 1831 in Michaud 1831: 27, Pl. 15
Name did not interfere with *Helix (Helicella) revelata* as mentioned by Férussac 1821-1822: 44 (nomen nudum, no description attached).

France: Bretagne, Finistère, Tréfeuntec near Douarnenez

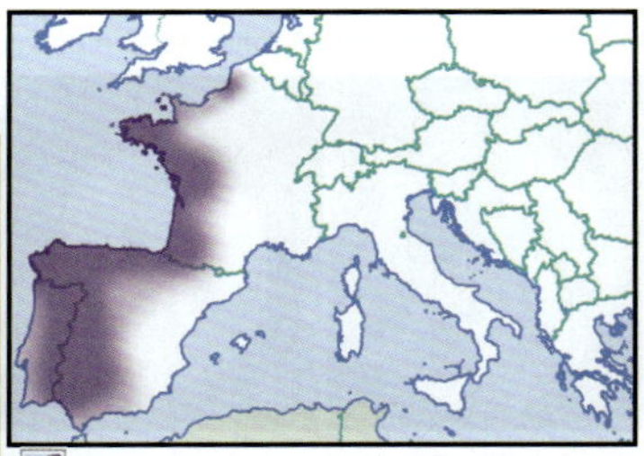

W and N France, N Spain
Range in Spain uncertain, several species may be involved.

3.5-4.5 x 5-7 mm

Shell almost globulous, horny greenish or yellowish, sometimes with two reddish bands, fragile, striated and with short whitish hairs (longer than in *P. subvirescens*), 4.5-5 slightly convex whorls, last whorl inflated, aperture almost perfectly rounded, margins approaching each other very closely, sometimes with whitish surface layer inside, margin slightly broadened, umbilicus narrow.
Differs from *P. subvirescens* in its more horny colour and longer hairs touching each other (Falkner et al. 2002: 147).

Warm and exposed slopes of hills in coastlands under shrubs, among roots, often deep inside the soil.

Moquin-Tandon 1855: 223, Westerlund 1889: 61, Germain 1930: 246 (*montivaga*), Falkner et al. 2002: 147 (Spain to N France).

Has often been misidentified as *subvirescens*. It is possible that *P. revelata* comprises more than one species.

Ponentina subvirescens (Bellamy, 1839)

Helix subvirescens Bellamy, 1839 in Bellamy 1839: 420, Pl. 18 [not verified]

4-6.5 x 5-9 mm

SW England: Pembrokeshire

Shell dull greenish-yellow, weakly striated and with short soft hairs, 4-4.5 convex whorls, aperture large and usually without lip, umbilicus narrow, partly covered by the reflected columellar margin.
Differs from *P. revelata* in its more greenish colour and denser and shorter hairs (Falkner et al. 2002: 147).

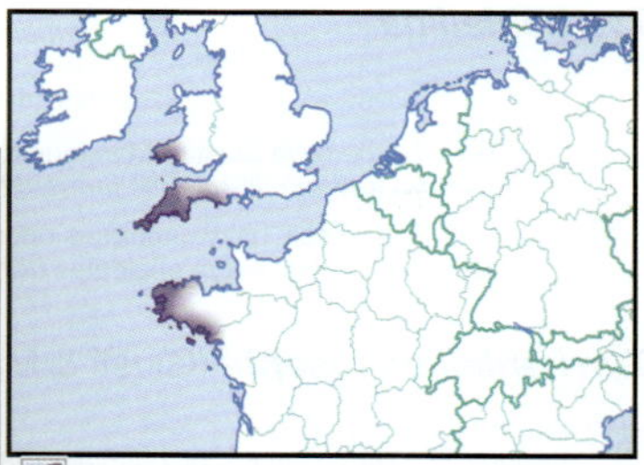

NW France, SW Great Britain

Germain 1930: 245, Kerney & Cameron 1979: 197, Prieto & Gómez 1988, Kerney 1999: 198, Falkner et al. 2002: 147 (Britain and NW France, only Bretagne), Talaván Gómez & T. 2006: 115.

Previous reports from Spain referred to *P. revelata*, usually also those from W France.

At the roots of trees and shrubs, in grasslands and rocky areas, costal flatlands, river plains, roadsides and near villages. Nearly always on well-drained non-calcareous soils, not in humid habitats.

Schileykiella

Schileykiella Manganelli, Sparacio & Giusti, 1989 in Manganelli et al. 1989: 141

Type species *Helix parlatoris* Bivona, 1839 (original designation)

Shell brown, with hairs, 4-5 whorls with deep suture, relatively flat with edge at periphery, apertural margin simple, not thickened, slightly reflected at lower side, umbilicus deep, 1/5-1/4 of shell diameter.
Vagina very short, no dart-sac complex, no digitiform glands, spermatheca duct variable in length, its initial portion flared, spermatheca small and rounded, penis and epiphallus almost equally long, flagellum variable, no typically hygromiid penial papilla, but a papilla-like structure arises from a side of the inner surface of the medial section of the penis. Right ommatophore retractor passes between penis and vagina.

Cianfanelli et al. 2004.

Schileykiella bodoni Cianfanelli, Manganelli & Giusti, 2004

Schileykiella bodoni Cianfanelli, Manganelli & Giusti, 2004 in Cianfanelli et al. 2004: 211

Shell brown, with short hairs, upper side flat, lower side rounded, periphery keeled, about 4 slowly increasing whorls with deep suture, last whorl descending near aperture giving the shell a slightly scalariform shape, margin sharp without lip inside, not connected at parietal side, umblicus very wide, 1/4-1/3 of shell diameter.
Differs from Sch. parlatoris in its shorter hairs, and anatomically.

3.0-3.7 x 7.2-8.6 mm

Leaf litter in open forest habitats and Mediterranean shrubland, usually on calcareous substrate.

EU Red List: Vulnerable VU. Only known from 2 localities on the small island of Marettimo. No acute threats known.

www.iucnredlist.org (08.2012).

Italy (Marettimo island)

Schileykiella parlatoris (Bivona, 1839)

Helix parlatoris Bivona, 1839 in Bivona 1839: 65 (*Helix Parlatoris*) Type locality Monte Cuccio near Palermo and Monte Busambra. Manganelli, Sp. & G. 1989: 144 intended to "restrict" the type locality to Monte Cuccio, this was ineffective since they did not select a name-bearing type.

Shell brown, with (up to 0.25 mm long) hairs, almost flat on upper and rounded on lower side, 4-4.5 whorls, last whorl angled above periphery, slowly descending near aperture, apertural margin simple, not thickened, slightly reflected at lower side, umbilicus deep, 1/5-1/4 of shell diameter. Microsculpture of apex with very fine spiral lines crossed by distinct growth lines. The animal is greyish on head and dorsum, lighter at the sides and foot, upper and lower tentacles grey. Hairs are longer than in *Sch. reinae*, the shell is usually flatter on the upper side.
Penial flagellum very short, slightly shorter than epiphallus, spermatheca duct short (3 mm, 2 x of vagina length).

3.5-4.5 x 6.5-8 mm

Italy: Sicily, Bosco Ficuzza near Palermo

Massimo Prodan

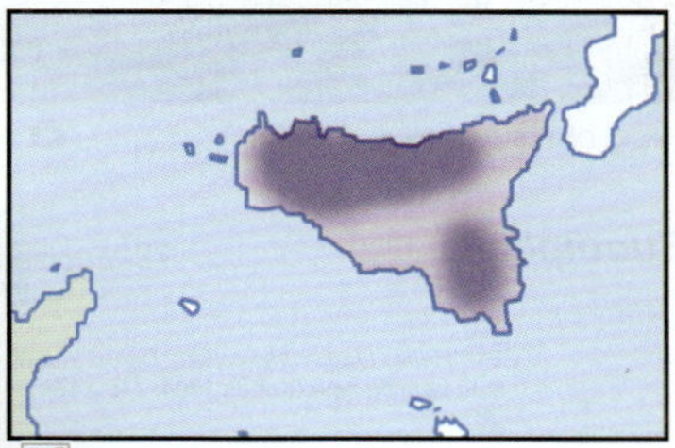

Italy (Sicily)

Open habitats in Mediterranean shrublands or in forests, under stones, on calcareous substrate.

EU Red List: Near Threatened NT. Rare, but no real threats known.

Manganelli, Sp. & G. 1989: 142, Giusti, M. & S. 1995: 506, Manganelli et al. 1995: 29, Cianfanelli et al. 2004: 210.

Schileykiella reinae (Pfeiffer, 1856)

Helix reinae Pfeiffer, 1856 in L. Pfeiffer 1856: 183, Taf. 2 (*Helix Reinae*)

Shell brown, with (up to 0.13 mm long) hairs, almost flat on upper and rounded on lower side, 4-4.5 whorls, last whorl angled above periphery, slowly descending near aperture, apertural margin simple, not thickened, slightly reflected at lower side, umbilicus deep, 1/5-1/4 of shell diameter. Microsculpture of apex with very fine spiral lines crossed by distinct growth lines. Hairs are shorter than in *Sch. parlatoris*, the shell usually not very flat on the upper side.
Penial flagellum about 15 times longer than epiphallus, spermatheca duct very long (25 mm, 20-25 x of vagina length).

4-4.5 x 7.5-8 mm

Italy: Sicily, Monte Pellegrino near Palermo

Massimo Prodan

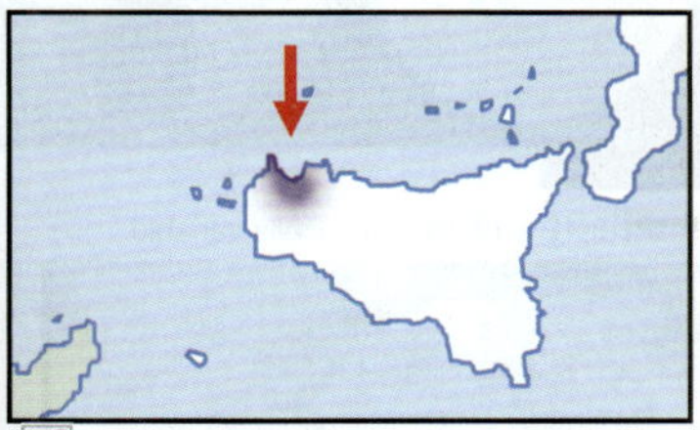

Italy (NW Sicily)

EU Red List: Near Threatened NT. Threatened by human pressure leading to habitat destruction.

Manganelli, S. & G. 1989: 147, Manganelli et al. 1995: 30, Cianfanelli et al. 2004: 211.

Humid habitats in the leaf litter layer in forests, under stones, or in more open Mediterranean shrublands near caves, on calcareous substrate.

Tyrrheniellina

Tyrrheniellina Giusti & Manganelli, 1992 in Giusti & Manganelli 1992: 231
Expressly established as a new replacement name for

Type species *Tyrrheniella josephi* Giusti & Manganelli, 1989 (original designation by Giusti & Manganelli 1989: 25)

Tyrrheniella Giusti & Manganelli, 1989 in Giusti & Manganelli 1989: 25 not *Tyrrheniella* Riedl, 1959 (Plathelminthes)
Original spelling in 1992 was not *Thyrreniellina*, as incorrectly cited by www.faunaeur.org and www.iucnredlist.org (08.2012).

Shell small, hairy, umbilicus open and wide. Vagina short and without dart-sac complex or accessory glands, penis wider and shorter than epiphallus, flagellum short, penial nerve originates from right cerebral ganglion, right ommatophore retractor passes between penis and vagina.

Tyrrheniellina josephi (Giusti & Manganelli, 1989)

Tyrrheniella josephi Giusti & Manganelli, 1989 in Giusti & Manganelli 1989: 27

Shell brown, with 0.1-0.15 mm long hairs on elongated bases, 4.5 convex whorls with deep sutures, last whorl angled at the periphery, aperture simple without internal lip, umbilicus wide, 1/5 of shell diameter.

3.2-3.5 x 5.5-5.8 mm

Under stones in shrubland vegetation or grassy habitats.

EU Red List: Near Threatened NT. No threats known, only known from very few localities in NE Sardinia and the islet of La Praiola near Capraia.

Manganelli et al. 1995: 30, www.naturamediterraneo.com (2007), www.iucnredlist.org (08.2012).

Italy (NE Sardinia, La Praiola)

Ashfordia

Ashfordia Taylor, 1917 in Taylor 1917: 69

Type species *Helix granulata* Alder, 1830 (monotypy)

Ashfordia granulata (Alder, 1830)

Helix granulata as mentioned in Alder 1830: 39 (nomen nudum)
Mentioned without description or reference, only "on plants in woods; rather rare", name in original source attributed to Turton.
Helix globularis Jeffreys, 1833 in Jeffreys 1833: 507 (not *Helix globularis* Philippi, 1836, currently *Marmorana globularis* (Philippi, 1836))
Established with description. If *Helix granulata* cannot be used, this name is available, but will distort *Marmorana* nomenclature.
Helix granulata Alder 1838 in Alder 1838: 341
Published with description, Alder 1838 argued that the name *granulata* was appropriate. Mentioned as *Helix granulata* by Gray in Turton & Gray 1840: 151, with description and figure, and reference to "Alder cat. 107". Possibly Alder published another article in the same journal, or a follow-up, who knows. The notes in Alder 1838: 337 did not support this assumption. The name *granulata* should be protected somehow if it turns out to be unavailable, eventually by the Commission.

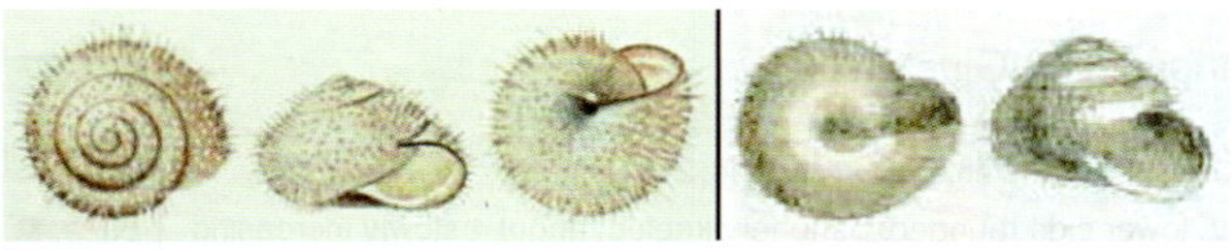

5-7 x 7-9 mm

(1-3) England: Bristol, Ashley Down, (4-5): England
Pictures (1-3): Taylor 1917: Pl. 9, (4-5) Turton & Gray 1840: Pl. 3 Fig. 29

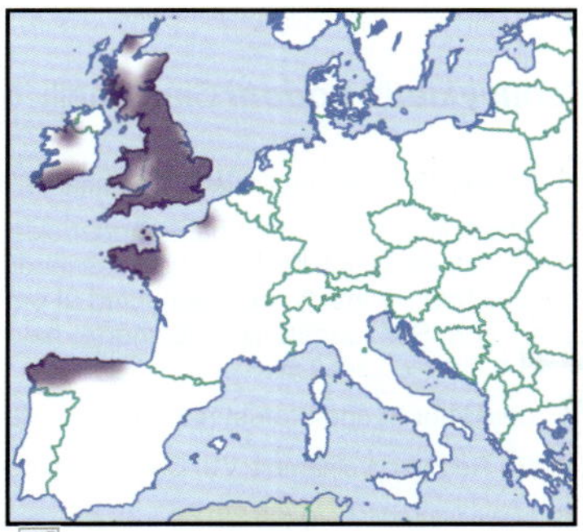

Spain to British Isles

Ireland: Cork, Blarney Castle Woods — Roy Anderson

Shell whitish to pale brown, thin, translucent, with fine straight (not curved) hairs with bulbous bases, 5.5-6 convex whorls with deep suture, aperture simple with thin or without lip, reflected only at columellar side, umbilicus very narrow and partly or almost entirely covered by the reflected columellar margin.
Mantle with black spots.

Herbal layers in humid and shady habitats in woods, marshes and hedges, often near waters (in central and E England almost exclusively), prefers open habitats in Britain, in N Ireland on scrubby pasture.

Has been declining in Ireland.

Taylor 1917: 70, Germain 1930: 269, Kerney & Cameron 1979: 185, Prieto & Gómez 1988, Anderson 1996: 14, Kerney 1999: 190.

Euomphalia

Euomphalia Westerlund, 1889 in Westerlund 1889: 31, 92 *Helix (Fruticicola (Euomphalia))*

Type species *Helix strigella* Draparnaud, 1801 (subsequent designation by Hesse 1931: 19)

Helix strigella was called "Typus" Kobelt 1904, this was not a valid type designation because Kobelt's work contained a disclaimer (Art. 8.3) that he did not use this term in the sense of a unique name-bearing type (Kobelt 1904: 145-147).

Europe to Armenia

Euomphalia strigella (Draparnaud, 1801)

Helix strigella Draparnaud, 1801 in Draparnaud 1801: 84

8-12 x 13-18 mm, locally variable

Austria: Tirol, near Landeck

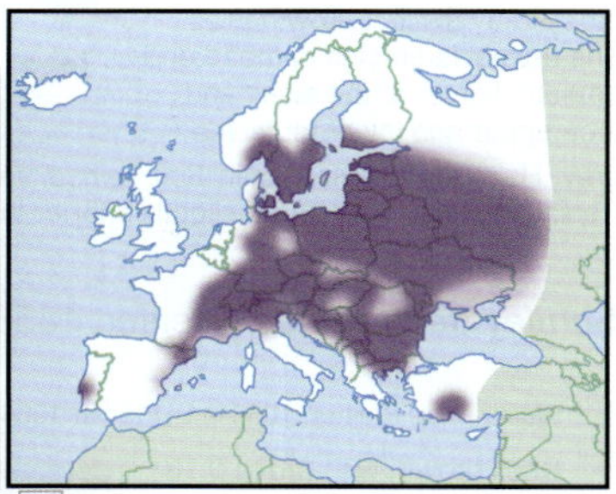

NE Spain to Volga region

Slovakia: Ortaše, near Topľa river — Jozef Grego

Shell yellowish grey to brownish, solid, juveniles with hairs, suture quite flat, umbilicus 1/5 of shell diameter (larger than in *Fruticicola fruticum* and much wider than in *Monacha cantiana*), apertural margins approach each other quite closely at parietal side.
Animal yellowish grey to reddish brown with small black dots visible below the shell, tubercles small, tentacles slightly darker with grey hue, lower tentacles exceptionally long. Moves relatively slowly, is easily irritated and secretes much mucus.

In light forests, under shrubs and on half dry meadows, between leaves, in Bulgaria at humid and shady places, especially in mountain river valleys. In N Switzerland also on semi-dry meadows at sunny slopes. In Switzerland and Bulgaria up to 1600 m.

Critically Endangered in Oberösterreich and Salzburg, Vulnerable in parts of Germany. Threatened by continuous habitat destructions.

Monacha

Monacha Fitzinger, 1833
in Fitzinger 1833: 95

Type species *Helix cartusiana* Müller, 1774
(subsequent designation by Herrmannsen 1847-1849 [17 Jul 1847]: 51)

ICZN Opinion 431: type species designated by Gray 1847 [09 Nov 1847]: 173, incorrect because Herrmannsen's act preceded Gray.
Metatheba Hesse, 1914 in Hesse 1914: 268 *Theba (Metatheba)*, type species *Helix samsunensis* Pfeiffer, 1868 (original designation)
Paratheba Hesse, 1914 in Hesse 1914: 268 *Theba (Paratheba)*, type species *Helix fruticola* Krynicki, 1833 (original designation)

Shell small to medium-sized, depressed to subglobose, somewhat fragile, smooth or with small bristles, may exhibit some faint spiral grooves. Aperture subcircular, often thickened by a lip which may be reddish-coloured. Whorls milky white, an opaque peripheral band may occur.
Genitalia: atrium with 3 branches (1) penis with flagellum, (2) vagina with glandulae mucosae, free oviduct and spermoviduct (and vas deferens connecting penis and spermoviduct), spermatheca duct and spermatheca, and (3) appendicula as a third branch from the atrium next to penis and vagina.

Usually the species of this genus live one year, exceptionally two years.

W Europe and N Africa to Iran and Arabia

Hausdorf 2000.

Metatheba and *Paratheba* are considered as subgenera.
The following two taxa from France are not listed as separate species here, we have currently no published evidence suggesting such a status. We do not know how to distinguish them from others and where they are distributed. Studies on a molecular level could be useful.

Monacha auturica Falkner, 2000 in Falkner 2000: 31
Expressly established as a new replacement name for
Helix avarica Locard, 1893 in Locard 1893: 86 (*Helix Avarica*) (not *Helix avarica* Servain, 1884 in Servain 1884: 352, from Bosnia)
Helix lamalouensis Reynes, 1870 in Reynes 1870: 34

Monacha ascania Hausdorf, 2000

NW Turkey (Vilayet Bursa)

Monacha ascania Hausdorf, 2000 in Hausdorf 2000: 64

Shell light horny brownish, sometimes with an indistinct light band at the periphery, striated and with numerous hair scars, last whorl with spiral lines, 5-5.5 convex whorls, last whorl often initially slightly angulated, near aperture distinctly descending, aperture almost circular and with whitish lip inside, margin sharp, umbilicus very narrow, usually partly covered by the reflected columellar margin.
Shell cannot be distinguished from *M. bithynica*. Flagellum longer, appendicula shorter.

6-7.5 x 8-10 mm

Mountains habitats.

Restricted to a small range north of Iznik Gölü.

Monacha atacis Gittenberger & De Winter, 1985

Monacha atacis Gittenberger & De Winter, 1985 in Gittenberger & De Winter 1985: 200

France: Aude, near Axat

NE Pyrenees

6.6-10.8 x 7.5-15.2 mm

Shell globular, creyish to creamy white, opaque or slightly translucent, irregularly striated and with fine spiral lines, with hair pits on all whorls, 5-6 inflated whorls, aperture with a thickened white internal lip, reflected only near the umbilicus, umbilicus very narrow and partly covered by the reflected apertural margin.
Differs from other *Monacha* species in its hair pits all over the shell.

Calcareous mountains regions of NE Pyrenees, in natural and disturbed habitats, among short vegetation, in meadows, limestone ridges, along rivers and roads. Lives syntopically with *M. cartusiana*. Not in the high mountains.

Monacha badia Hausdorf, 2000

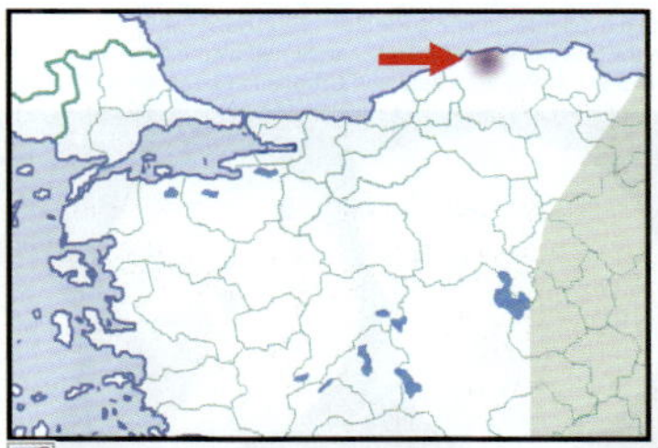

NW Turkey (Vilayet Kastamonu)

Monacha badia Hausdorf, 2000 in Hausdorf 2000: 67

Shell reddish brown, with wrinkled growth lines and indistinct spiral lines and hair scars, 5.5-6 convex whorls, outline conical on upper side, last whorl rounded and slightly descending near aperture, aperture almost circular, with strong whitish lip, umbilicus very narrow, at least half covered by the reflected columellar margin.
Anatomically characterized by a rather long flagellum and appendicula inserting at the base of the vagina.

8-10 x 11-13 mm

Was found in roadside vegetation.

Not very frequent.

Monacha bithynica Hausdorf, 2000

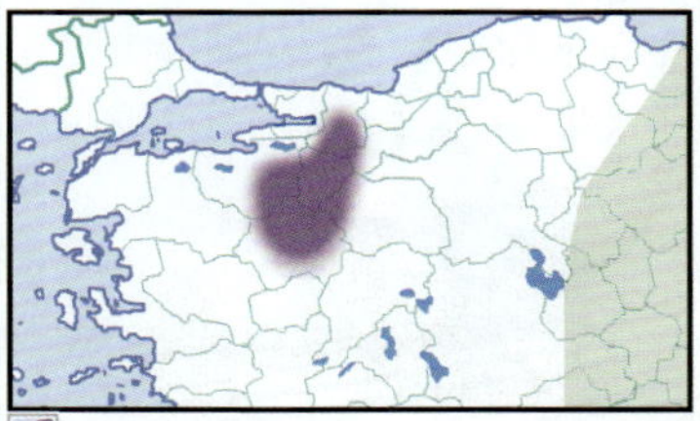

NW Turkey

Monacha bithynica Hausdorf, 2000 in Hausdorf 2000: 67

Shell light horny brownish, sometimes with an indistinct light band at the periphery, striated and with numerous hair scars, last whorl with spiral lines, 5-5.5 convex whorls, last whorl often initially slightly angulated, near aperture distinctly descending, aperture almost circular and with strong whitish lip inside, margin sharp, umbilicus very narrow, usually almost completely covered by the reflected columellar margin.
Shell cannot be distinguished from *M. ascania*. Flagellum shorter, appendicula longer.

6-7.5 x 8-10 mm

Monacha cantiana (Montagu, 1803)

Helix cantiana Montagu, 1803 in Montagu 1803: 422

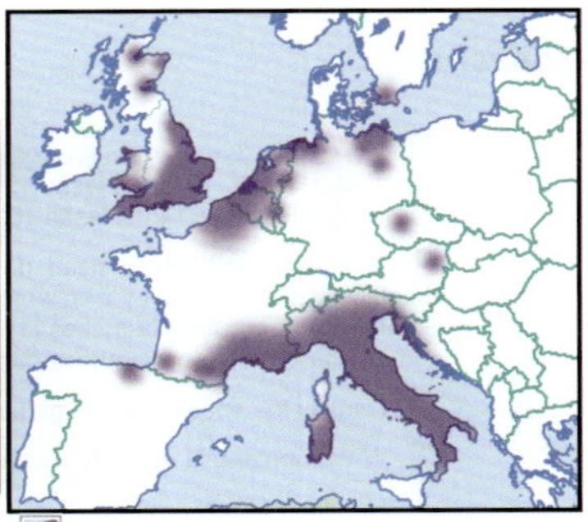

11-14 x 16-20 mm

Germany: Brandenburg, Nauen

Originally Italy and S France, introduced elsewhere

Shell usually creamy white, with hairs on upper whorls, 5-6 slightly convex whorls, aperture with white or reddish internal lip, with much less contrast than in *M. cartusiana*, umbilicus slightly wider than in *M. cartusiana*, 1/11-1/7 of diameter.
Animal light, slightly reddish in frontal section, tubercles well marked and serrated, tentacles greyish brown. The animal moves slowly, is easily irritated, and secretes a colourless mucus.

In herbal layers of hedges, waste ground, shrublands, at roadsides and railways, also in dunes, not in woods, prefers well-drained calcareous soils. Juveniles are often attached high up on the vegetation for long periods.

In N France 60-90 eggs (diameter 1.5 mm) are laid between June and September, juveniles hatch after 14-15 (in England up to 20) days and reach maturity after 1 year, often before the shell is completed, life span 12-18 months.

Moquin-Tandon 1855: 201, Germain 1930: 263, Chatfield 1968, Kerney & C. 1979: 184, Prieto & G. 1988, Falkner 1990: 220, Kerney 1999: 189, Hlaváč & P. 2010.

Introduced to Great Britain with farmers in late Roman times and spread mainly in the medieval period, occupying a compact area covering S and E England. It still continues spreading (isolated sites in Wales, W central England and Scotland). Attempts to introduce the species to Ireland have been unsuccessful.

Monacha carascaloides (Bourguignat, 1855)

Helix carascaloides Bourguignat, 1855 in Bourguignat 1855: 561 (*Helix Carascaloïdes*)

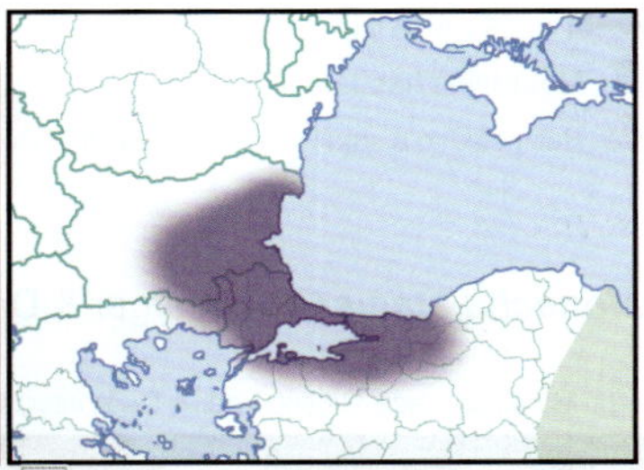

9-13 x 14-19 mm

Bulgaria: Sakar mountains, near Sladun

Bulgaria, NW Turkey

Shell horny brownish with lighter band at the periphery, striated and with scattered hair scars, lower whorls with spiral lines, 4.5-6.5 convex whorls, last whorl rounded or slightly angular, usually distinctly descending near aperture, aperture with strong whitish lip inside, close to margin, umbilicus narrow or very narrow, to a large extent covered by the reflected columellar margin, sometimes closed. Usually larger than *M. solidior*, more depressed and with wider umbilicus.
Differs from *M. solidior* by broader appendicula inserting more distally at the vagina.

In shady habitats, below shrubs in forests, also in dry habitats, up to 600 m in Rodopi, 1000 m in Central Balkan.

Damjanov & Likharev 1975: 364, Hausdorf 2000: 79, Schütt 2001: 418, Irikov & Mollov 2006: 821, Irikov 2008: 787.

Range insufficiently known due to confusions with *M. solidior*, reports for N Greece (Reischütz 1988) need anatomical confirmation.

Monacha cartusiana (Müller, 1774)

Helix cartusiana Müller, 1774 in Müller 1774: 15
Helix olivieri Férussac, 1821 in A. Férussac [1821-1822]:43 (folio p. 47)

Established with descriptions for subordinate variants α, β and γ, but not with a distinct description for the main form (nominal taxon). The name *Helix olivieri* was made available at this occasion under Art. 12.1 because it had a description for the taxon (Code Glossary "taxon": a taxon encompasses all included taxa of lower rank). Art. 12.1 does not demand a description for the nominal taxon, a description for the taxon is sufficient, this can include subordinate taxa.
For those who do not like to accept this interpretation of the Code (and who like to follow Hausdorf 2000: 78 who argued that the name was not made available at this occasion because the types were excluded), the next available author is Pfeiffer 1828 (*Helix Olivieri*, p. 25).
www.faunaeur.org (03.2012) accepted two names: *Helix olivieri* Férussac, 1821 and *Helix olivieri* Rossmässler, 1837 (as a synonym of *M. parumcincta* (Menke, 1828) from Italy and Croatia to Greek islands). Rossmässler 1837 (Icon. 5-6): 37 used only Férussac's name and did not establish a new name.

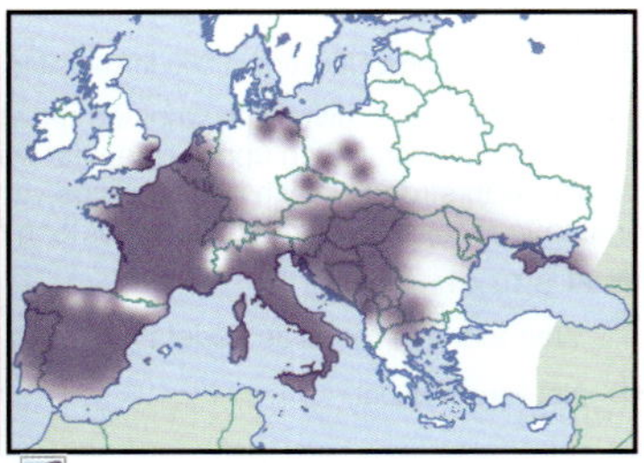

6-10 x 9-18 mm, extremely variable, also in the same population

Albania: Shkodër

Portugal to Ukraine

Croatia: Dalmatia, Rab island

Jozef Grego

Shell horny yellowish, translucent and shiny, aperture with prominent white lip inside, margin reddish, the contrast is clearly visible from outside, umbilicus very narrow, almost covered.
Differs from *M. cantiana* in its usually smaller and more depressed shell, the characteristic red-white colour contrast at the apertural margin, and a slightly narrower umbilicus.
Animal yellow with black pigments, anterior part slightly reddish, mantle with white dots, tentacles yellowish and transparent or darker greyish. Moves slowly, is easily irritated, adheres strongly to any objects, tolerates sunshine, but moves away from cold water.

Open habitats, sunny and dry shrubs, grassy slopes in lowlands, also artificial habitats in gardens, parks, vineyards, road margins and ruins. Tolerates grazing by sheep or cattle in grasslands. Near Porto (Portugal) locally very common in spring, but restricted to only a few habitats. In Switzerland up to 700 m, in Bulgaria and Spain 1000 m.

In France and Poland 50-150 eggs (diameter 1.2-1.3 mm) are laid in clusters (containing 100 eggs) at the end of the summer season, in Hungary 15-20 eggs in September-October, juveniles hatch after 12-15 days and reach maturity after 1 year.
Gassies 1849 reported that eggs in S France are laid in April-May and that maturity is reached at the end of the summer or autumn. Germain 1930 reported that maturity is reached after 12 months. Hazay 1881 reported a strictly annual life cycle in Hungary, adults die after egg laying in autumn.

Vulnerable in N Switzerland, rare in England (most surviving populations must be considered at risk).

Gassies 1849: 89, Moquin-Tandon 1855: 208, Hazay 1881: 82, Germain 1930: 266, Nobre 1941: 119, Frömming 1954: 276, Prieto & Gómez 1988, Turner et al. 1998: 334, Kerney 1999: 188, Lesicki & K. 2007, Sysoev & Schileyko 2009: 214.

Introduced to SE England in prehistoric times by farmers, has been declining seriously since the 1890s due to changes in farming practices. Spreading northwards in Poland. Indicated ranges in S Europe uncertain due to historical misidentifications included in recent reports.

Monacha cemenelea (Risso, 1826)

Theba cemenelea Risso, 1826 in Risso 1826: 75

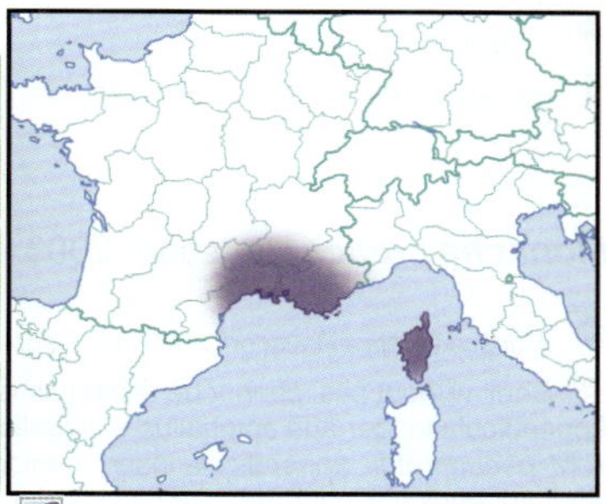

12-13 x 18-21 mm France: Hérault Cédric Audibert S France, Corsica

Shell light or horny reddish, transparent, with fine spiral lines, 5-5.5 convex whorls, last whorl well rounded and very weakly descending near aperture, aperture with light reddish or pinkish lip inside, outside with a yellow margin, margins not approaching each other closely at parietal side, umbilicus narrow and deep.
Shell like *M. cantiana*, broader and with broader aperture, columellar margin less curved, descending obliquely, umbilicus narrower.

Open habitats in shrublands, olive orchards, up to 8-900 m, occasionally higher.

Westerlund 1889: 79, Germain 1930: 265.

Monacha claustralis (Rossmässler, 1834)

Helix carthusianella var. a synonym *Helix claustralis* as mentioned by Menke 1828: 12 (unavailable synonym)
The name *claustralis* was mentioned as a synonym of the variant a, for which no name was used. Since Art. 11.6 demands (for making a synonym available under Art. 11.6.1) that such a name "was treated as a junior synonym of a name then used as valid", *claustralis* cannot be made available from this source. Hausdorf 2000: 81 stated that *Helix claustralis* "was first published in the synonymy of *Helix carthusianella*" (and in the next step looked for the first source that used the name for a taxon). The Code does not say "in the synonymy of", but "for a name used as valid". This was not done by Menke 1828, so the name is unavailable.
Helix carthusianella Varietät *claustralis* Rossmässler, 1834 in Rossmässler [1834]: 5 Type locality Kérkira, Greece.
Established with description for a form from Kérkira, probably seen in the collection in Wien (no types preserved in Senckenberg-Museum Frankfurt SMF, R. Janssen, e-mail, 2009). Taxon was regarded by Rossmässler [1834] as a variety beyond doubts, used for a taxon under Art. 11.5.1. Overlooked by Hausdorf 2000. The lectotype Zoologisches Museum Zürich ZMZ 505571 (Hausdorf 2000: 82) is invalid, type locality is not Préveza, but Kérkira.
Helix subobstructa Bourguignat, 1855 in Bourguignat 1855: 564, Pl. 16
Monacha dissimulans Pintér, 1968 in Pintér 1968: 221
Monacha oshanovae Pintér & Pintér, 1970 in Pintér & Pintér 1970: 96

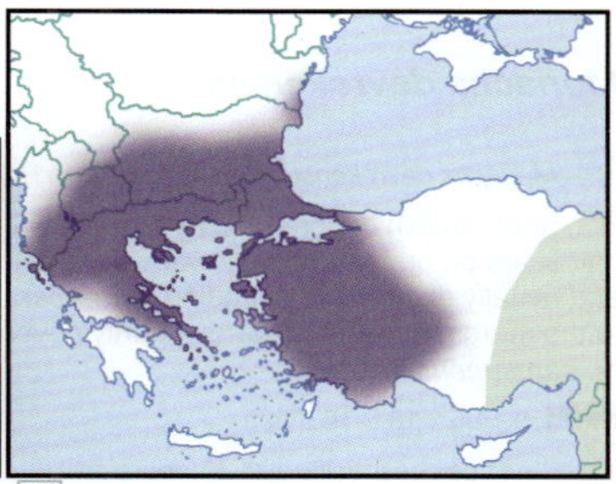

6-11 x 9-18 mm S Albania: near Këlcyrë S Albania to W Turkey

Shell horny yellowish, translucent and shiny, aperture with prominent white lip inside, margin reddish, the contrast is clearly visible from outside, umbilicus very narrow, almost covered.
Belongs conchologically to the *M. cartusiana* complex from which it can only be distinguished anatomically. Differs from *M. solidior* in its lighter shell with less interrupted growth lines an no hair scars. In sympatric occurrences *M. claustralis* shells of the same population the surface can have a hammered sculpture, but not with tiny hair scars or pits.
Differs from *M. cartusiana* in the absence of a lateral bulge of vagina, smaller epiphallus (penis ratio 1-2.3 in *M. claustralis*, 2.3-3.5 in *M. cartusiana*; vagina ratio 0.6-1.8 in *M. claustralis*, 1.5-2.6 in *M. cartusiana*).

Open habitats with herbs and tall grass, sunny and dry places, often as a synanthrope in agricultural lands, meadows and gardens. In S Bulgaria up to 1600 m.

Dhora & Welter-Schultes 1996: 157, Hausdorf 2000: 80 (1 report from Crimea, so possibly also there), Irikov & Mollov 2006: 821, Irikov 2008: 789.

There is no published evidence that *M. oshanovae* from SW Bulgaria represents a separate species, no differences in shell or anatomy were detected by Irikov 2008.

Monacha consona (Rossmässler, 1839)

Helix consona Rossmässler, 1839 in Rossmässler 1838-1844 (9-10): 8, Taf. 43
Helix rizzae Aradas, 1843 in Aradas 1843: [1]

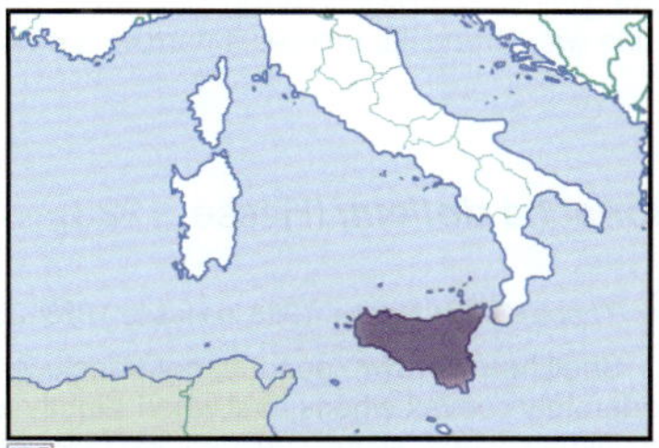

Shell horny yellowish, reddish near aperture, with a weak colour band at the periphery, finely striated, translucent, upper whorls with weak hair scars, 4.5-5.5 whorls, last whorl initially angulated, shortly descending near aperture, aperture with fine whitish or reddish lip inside, margin dark or reddish, sharp, columellar margin ending terminally in a triangular structure, umbilicus narrow and largely covered by the reflected columellar margin.

9-11 x 13-16 mm

Open forests, shrublands and grasslands, also in moderately disturbed habitats.

EU Red List: Near Threatened NT. Threatened by increasing urban pressure in the Palermo region.

Westerlund 1889: 76, Manganelli et al. 1995: 30, 53.

Italy (Sicily)

Giusti 1973: 262 considered *consona* as a synonym of *Monacha gregaria*. It is unclear if this study was based on original types. Reitano et al. 2009: 195 studied the anatomy of a conchologically almost entirely similar E Sicilian form and called it *Monacha rizzae*. Its umbilicus was perhaps slightly better covered.
The *Monacha* species in S Italy need to be revised and better understood. It would be necessary to learn how to distinguish the species, to learn more about their ranges and local variations, for example in the surface sculpture and umbilicus coverage. It is very difficult to understand the identity of a local form if the other forms on the same island were not studied with the same methods and compared with it. It is also possible that *rizzae* represents what has been called *parumcincta* in mainland Italy (Giusti 1973: 268), and that anatomical characters vary over area.

Monacha crenophila (Pfeiffer, 1857)

Helix crenophila Pfeiffer, 1857 in L. Pfeiffer 1857: 162
Expressly established as a new replacement name for
Helix muscicola Bourguignat, 1855 in Bourguignat 1855 [Dec]: 562, Pl. 16 (not *Helix muscicola* Philippi, 1855 [July] from Chile)

Shell horny brownish, irregularly striated, with indistinct spiral lines and short hairs, 4.5-6 convex whorls, last whorl rounded, hardly descending near aperture, aperture almost circular, with strong white lip inside, umbilicus narrow, partly covered by reflected columellar margin.
Differs from *M. ovularis* in its smaller epiphallus:flagellum ratio, longer penis, and shorter apical section of appendicula which is terminally not distinctly enlarged.

4.5-6 x 6-8.5 mm

In forests.

Hausdorf 2000: 84, Schütt 2001: 421.

NW Turkey (Istanbul)

Monacha cretica Hausdorf, 2003

Monacha cretica Hausdorf, 2003 in Hausdorf 2003: 43

Shell without pits, cannot be distinguished from *M. pseudorothii* and pitless forms of *M. rothii*.
Appendicula longer and epiphallus to flagellum ratio lower than in *M. rothii*, epiphallus to vagina ratio higher than in *M. pseudorothii*, appendicula insert at middle section of vagina, glandulae mucosae near proximal end of vagina; appendicula longer and vagina shorter than in *M. maasseni*.

6-8 x 8-11 mm

Greece (W Crete)

Monacha densecostulata (Retowski, 1887)

Helix densecostulata Retowski, 1887 in Retowski 1887: 18

Shell brown with indistinct light band at the periphery, distinctly striated, with very long (up to 1 mm) hairs, last whorls with spiral lines, 4.5-5.5 convex whorls, upper whorls depressed conical, last whorl with weak edge at the periphery, distinctly descending near aperture, aperture elliptical, with strong whitish lip inside, thickened at basal side, umbilicus narrow, partly covered by reflected columellar margin.
Shell more depressed than other *Monacha* species.

5-6 x 8-10 mm

Hausdorf 2000: 106.

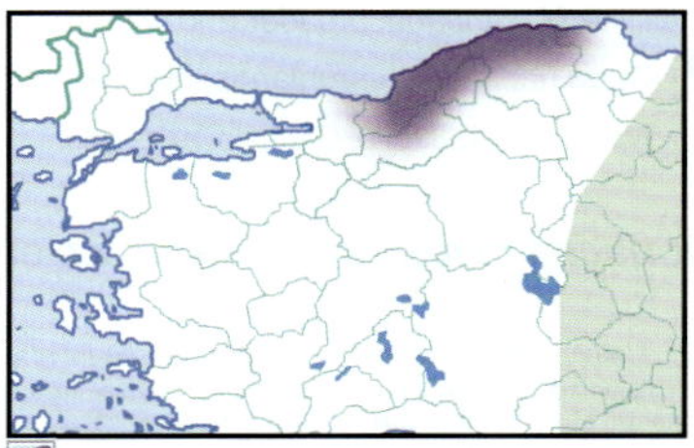

NW Turkey

Monacha devrekensis Hausdorf, 2000

Monacha devrekensis Hausdorf, 2000 in Hausdorf 2000: 106

Shell whitish or light horny yellowish with an indistinct lighter band at the periphery, weakly striated and with hair scars on upper whorls, lower whorls with spiral lines, 5.5-6.5 not very convex whorls, last whorl rounded and usually distinctly descending near aperture, aperture elliptical, with strong whitish internal lip relatively close to the margin, umbilicus narrow, partly covered by the reflected columellar margin. Shell indistinguishable from *M. samsunensis*.

9-11.5 x 13.5-16.5 mm

Could be conspecific with *M. georgievi*.

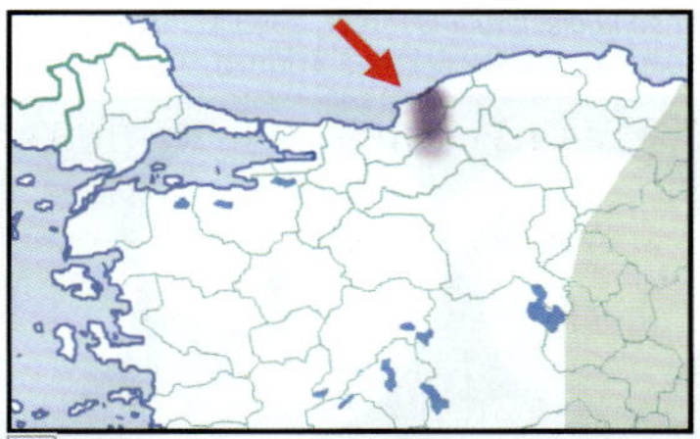

NW Turkey

Monacha dirphica (Martens, 1876)

Helix dirphica Martens, 1876 in Martens 1876: 339, Taf. 12 (*Helix Dirphica*)

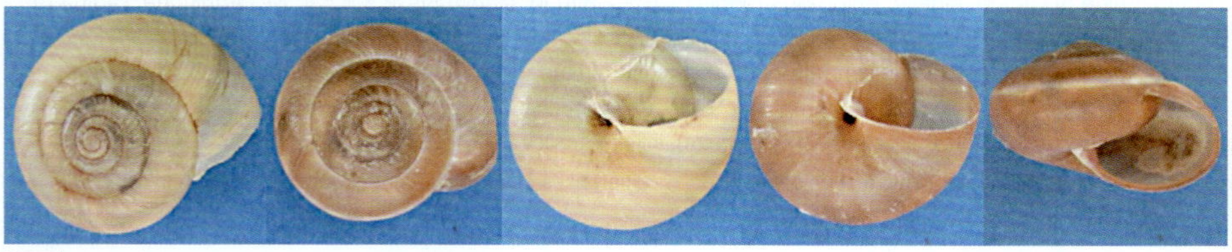

11-13 x 15-19 mm

Greece: Evia, Kimi (D = 15 mm) Robert Reuselaars

Shell pale horny yellow, or evenly reddish or with red spots, with a fine white band at the periphery, often with some yellowish growth interruption streaks, upper whorls regularly, last whorl more irregularly striated, aperture inside with strong lip, umbilicus narrow and not much covered by the reflected columellar margin.

Greece (Evia)

Westerlund 1889: 74.

Monacha dofleini (Hesse, 1928)

Theba dofleini Hesse, 1928 in Hesse 1928: 23, Taf. 1-2

Shell horny brown, opaque, not translucent, regularly and finely striated, at last whorl almost ribbed, 5-6 moderately convex whorls, last whorl slightly angulated at the beginning but then rounded, width almost 2 x of penultimate whorl, slightly descending near aperture, apertural margin straight with brownish lip inside, also visible outside as a reddish structure, margins relatively distant from each other and connected by a thin white layer at the parietal side, umbilicus very narrow and partly covered by the reflected columellar margin.

6-7 x 8.5-11 mm

Obviously a flatland species, known from grasslands, but perhaps also in rocky habitats in highlands.

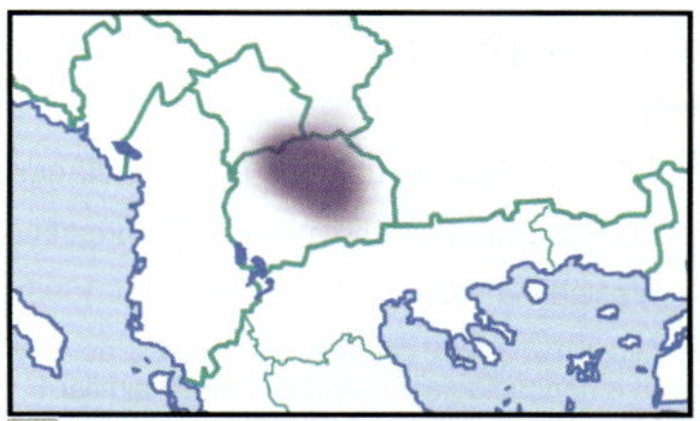

Rep. Makedonija

Monacha emigrata (Westerlund, 1894)

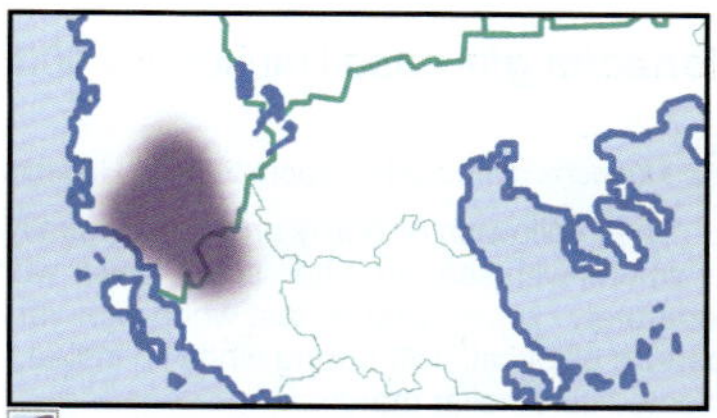

S Albania, NW Greece (Ipiros)

Helix emigrata Westerlund, 1894 in Westerlund 1894: 166

Shell light horny yellowish or whitish, irregularly striated, with weak spiral lines, 5-6 convex whorls, slowly increasing, last whorl initially slightly angulated, descending near aperture, aperture with prominent white lip inside, basal margin slightly reflected, umbilicus narrow and deep, not wider at last whorl, partly coverd by the reflected columellar margin.
Flagellum length 0.5 of penis, base of spermatheca duct not thickened (differences to *M. haussknechti*).

6-7.5 x 9-12 mm

Westerlund 1894: 166, Sattmann & Reischütz 1994: 44, Hausdorf 1996 (Schr. Malak. 9), Dhora & W. 1996: 158.

Monacha euboeica (Kobelt, 1877)

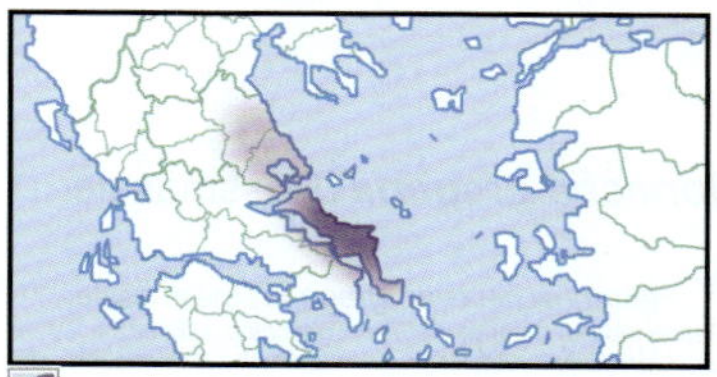

Greece (Evia, possibly Thessalia)

Helix apennina var. *euboeica* Kobelt, 1877 in Kobelt 1877: 24, Taf. 126 (*H. ap.* var. *euböica*)

Shell horny yellowish and shiny, striated, upper whorls with distinct hair scars, 4.5-5.5 weakly convex whorls, last whorl deeply descending near aperture, aperture with narrow lip inside, margins approaching each other closely at parietal side, umbilicus open.

5-6 x 9-12 mm

Westerlund 1889: 74.

Monacha frequens (Mousson, 1859)

Helix frequens Mousson, 1859 in Mousson 1859: 254
Monacha aniliensis Reischütz & Sattmann, 1990 in Reischütz & Sattmann 1990: 262

N Albania: Malësia, near Tamarë

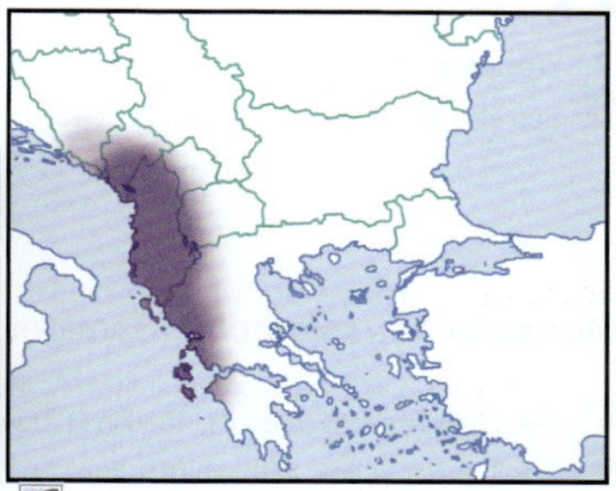

SE Europe

6-12 x 8-15 mm

Shell horny brownish, often weakly reddish near aperture, finely striated or almost smooth, shiny, last whorl with indistinct spiral lines, 5-6 convex whorls, last whorl not angulated, aperture inside light horny coloured, with narrow and white distinct lip, margin straight and often reddish, umbilicus narrow and in adults usually covered.
Differs from *M. cartusiana* and *M. claustralis* in its darker shell, which is usually less shiny, without blurry white colour bands.

Various kinds of open habitats, very frequent in Albania.

Westerlund 1889: 75, Dhora & Welter-Schultes 1996: 158.

M. aniliensis from NW Greece is regarded as a synonym of *M. frequens* here, Reischütz & S. 1990 did not compare their species with forms of the *M. frequens* complex.

Monacha fruticola (Krynicki, 1833)

Helix fruticola Krynicki, 1833 in Krynicki 1833: 429

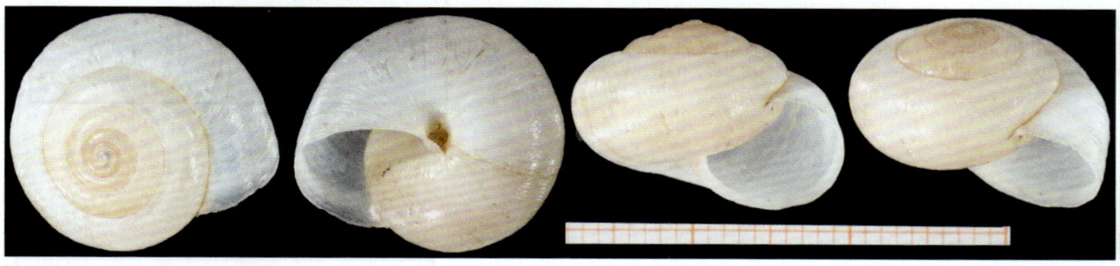

Ukraine: Odessa

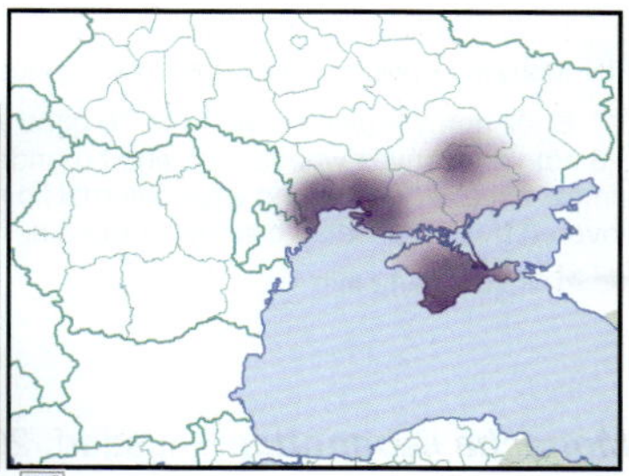

S Ukraine

11-15 x 14-20 mm

Shell light horny yellowish or whitish, with a weak white band at the periphery, upper whorls finely striated, last whorl with dense and weak spiral lines, 5-6 whorls, last whorl descending near aperture, aperture with narrow white lip inside, columellar margin abruptly and broadly reflected, umbilicus very narrow and usually covered.

Westerlund 1889: 76, Shilejko 1978: 282, Sverlova 2006: 135, Egorov 2008: 95, Sysoev & Schileyko 2009: 215.

Monacha galatica Hausdorf, 2000

Central Turkey

Monacha galatica Hausdorf, 2000 in Hausdorf 2000: 109

Shell whitish with indistinct ligther band at the periphery, weakly striated, upper whorls with very short (0.1 mm) hairs bent towards the aperture, lower whorls with spiral lines, 5-5.5 moderately convex whorls, last whorl rounded and usually distinctly descending near aperture, aperture with strong white lip inside, relatively close to the margin, umbilicus narrow or moderately wide, partly but not always much covered by the reflected columellar margin.

5.8-7.6 x 8.4-11.3 mm

Monacha georgievi Páll-Gergely, 2010

Turkey (Vilayet Bolu)

Monacha georgievi Páll-Gergely, 2010 in Páll-Gergely 2010: 91

Shell brownish, sriated and with short (0.1-0.15 mm) hairs, 5-6 convex whorls, last whorl rounded, apertural margin sharp with strong whitish lip inside, umbilicus narrow, partly covered by the reflected columellar margin.

4.5-6.5 x 7-9.5 mm

Mature animals were found in April.

Only known from one locality (near Gerede), where only 4 specimens were collected.

Specific status needs to be confirmed. Description was based on 2 dissected specimens which seemed to differ only in a relatively shorter flagellum from *M. devrekensis*. I am asking myself if this really provides evidence that this form is reproductively isolated from neighbouring *M. devrekensis*. Studies on a molecular level could be helpful.

Monacha glareosa Hausdorf, 2000

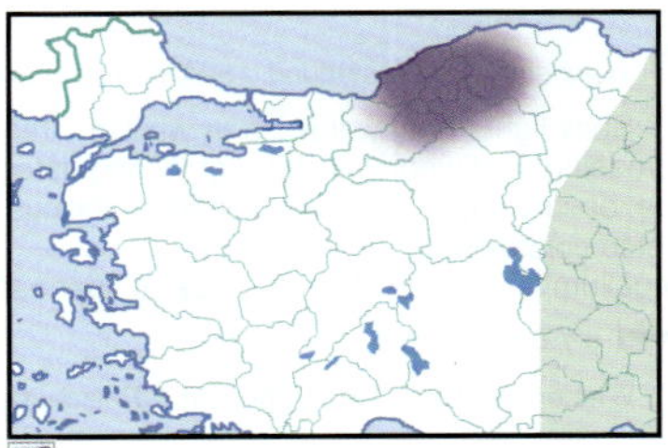

NW Turkey

Monacha glareosa Hausdorf, 2000 in Hausdorf 2000: 71

Shell yellowish or horny brownish, usually with indistinct lighter band at the periphery, distinctly striated, upper whorls with scattered hair scars, lower whorls with gravel-like sculpture and indistinct spiral lines, 5.5-6 moderately convex whorls, last whorl with very blunt edge at the beginning, slightly descending near aperture, aperture elliptical, with strong white lip inside, umbilicus very narrow, partly or almost completely covered by the reflected columellar margin.

9-14 x 13-19 mm

Was found in rocky habitats.

Monacha gregaria (Rossmässler, 1839)

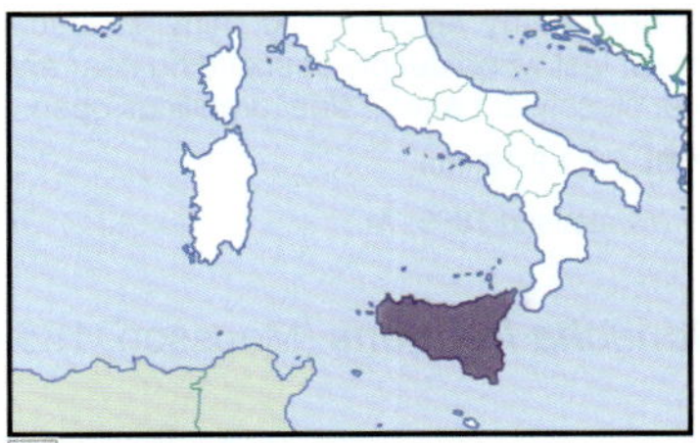

Italy (Sicily)

Helix onychina var. *gregaria* Rossmässler, 1839 in Rossmässler 1838-1844 (9-10): 7, Taf. 43
Mentioned in the chapter of *Helix onychina*, subchapter "Varietäten": "Fig. 569. ist eine kleine ziemlich einfarbige, sonst aber nicht verschiedene Varietät, als *H. gregaria* Z. von Parreyss mitgetheilt." Rossmässler did not make clear that the name *gregaria* was used in his own classification for the variety, but there is no unambiguous evidence of the contrary either.

Shell evenly horny yellowish or reddish, finely striated, 4.5-5.5 whorls, last whorl initially angulated, aperture inside with white lip, outside visible as a red structure, umbilicus usually covered.

7-8 x 10-12 mm

Westerlund 1889: 86, Manganelli et al. 1995: 30, Giusti 1973: 262.

See also *M. consona*, a revision is needed, with detailed information on distribution and habitat.

Monacha haussknechti (Boettger, 1886)

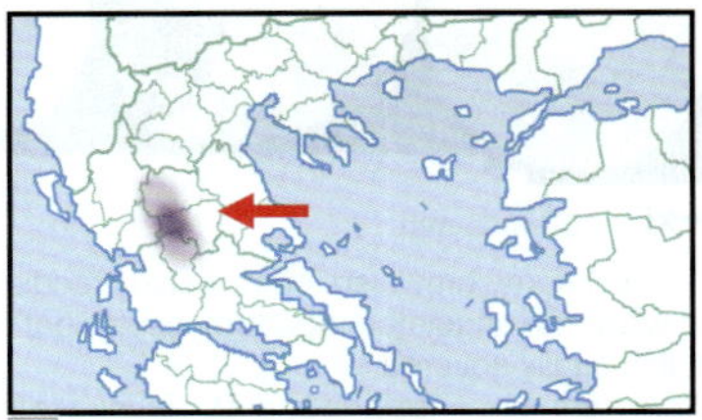

Greece (Thessalía)

Helix haussknechti Boettger, 1886 in O. Boettger 1886: 53, Taf. 2 (*Helix Haussknechti*)

Shell horny yellow with white suture and always with a white band along the periphery, occasionally dark horny brownish, solid, irregularly striated, 5-6 whorls, last whorl initially angulated, weakly descending near aperture, aperture with broad white lip, margins not approaching each other closely at parietal side, columellar margin protruded, umbilicus narrow.
Flagellum as long as penis, base of spermatheca duct thickened (differences to *M. emigrata*).

7-8 x 11-13 mm

Westerlund 1889: 75, Sattmann & Reischütz 1994: 44.

Monacha leucozona Hausdorf, 2000

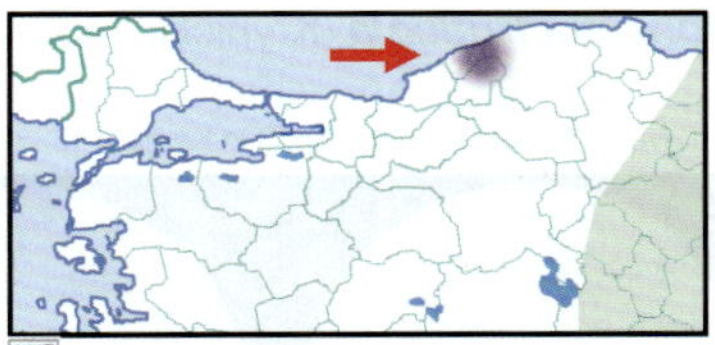

NW Turkey

Monacha leucozona Hausdorf, 2000 in Hausdorf 2000: 72

Shell horny brownish with a white band at the periphery, striated, with hair scars and indistinct spiral lines, 5-6 not very convex whorls, last whorl rounded or slightly angular, slightly descending near aperture, aperture almost circular, with strong white internal lip relatively close to the margin, umbilicus very narrow, at least half covered by the reflected columellar margin.

7-8.5 x 10-12 mm

Monacha liebegottae Hausdorf, 2000

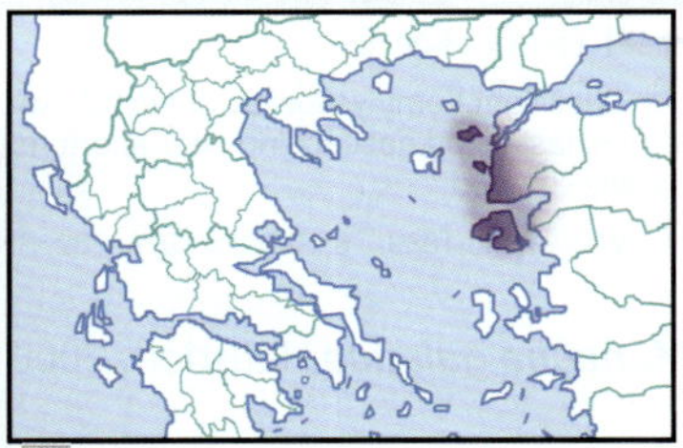

NE Aegean

Monacha liebegottae Hausdorf, 2000 in Hausdorf 2000: 88

Shell horny brownish, striated, very shiny, 5-6 moderately convex whorls, last whorl rounded or slightly angular, descending near aperture, aperture with strong white lip inside, close to the margin, umbilicus narrow, partly (Gökçeada, Bozcaada) or almost completely (Lésvos) covered by the reflected columellar margin.
Shell has one more whorl than *M. claustralis*. Weathered shells are not brownish.

6.5-10 x 9-13.5 mm

NW Turkey: Bozcaada island — Panayotis Ovalis

Monacha maasseni Hausdorf, 2003

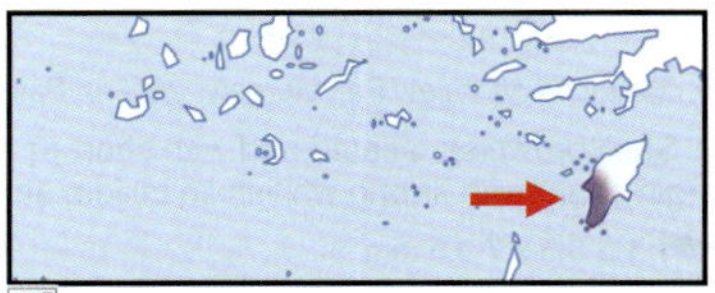

Greece (Ródos)

Monacha maasseni Hausdorf, 2003 in Hausdorf 2003: 43

Shell with pits, cannot be distinguished from *M. rothii* shells with pits.
Epiphallus to flagellum ratio lower than in *M. rothii*, appendicula shorter than in *M. pseudorothii*, inserts at the middle section of the vagina, glandulae mucosae insert near proximal end of vagina, as in *M. rothii*.

6-8 x 9-12 mm

In various kinds of habitats, also in cultivated land, favoured by human activities.

www.iucnredlist.org (08.2012).

Monacha magna Hausdorf, 2000

NW Turkey

Monacha magna Hausdorf, 2000 in Hausdorf 2000: 72

Shell yellowish or horny brownish, sometimes with an indistinct lighter band at the periphery, prominently striated and with distinct spiral lines, 5.5-6.5 moderately convex whorls, last whorl with blunt edge at the beginning, last whorl running below periphery near aperture, aperture elliptical, with strong white lip inside, umbilicus narrow, half covered by the reflected columellar margin.
Differs from *M. glareosa* in its larger shell without gravel-like sculpture and a wider umbilicus.

11.5-15 x 17.5-21 mm

Monacha margarita Hausdorf, 2000

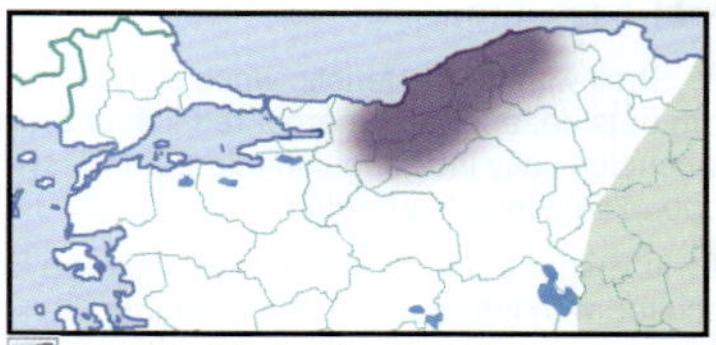

NW Turkey

Monacha margarita Hausdorf, 2000 in Hausdorf 2000: 72

Shell horny brownish, sometimes with indistinct lighter band, densely striated and with distinct spiral lines, 4.5-5.5 convex whorls, last whorl often with a blunt edge at the beginning, usually distinctly descending near aperture, aperture almost circular, with strong whitish lip inside, umbilicus very narrow, slightly or almost completely covered by the reflected columellar margin.

4.5-5.5 x 6-7.5 mm

Monacha messenica (Blanc, 1879)

Helix cantiana var. *messenica* Blanc, 1879 in Westerlund & Blanc 1879: 41, Pl. 2
Name in original publication attributed to Blanc, diagnosis to Westerlund. Under Art. 50.1 the name must be attributed to the co-author of the work who was made responsible for the name.

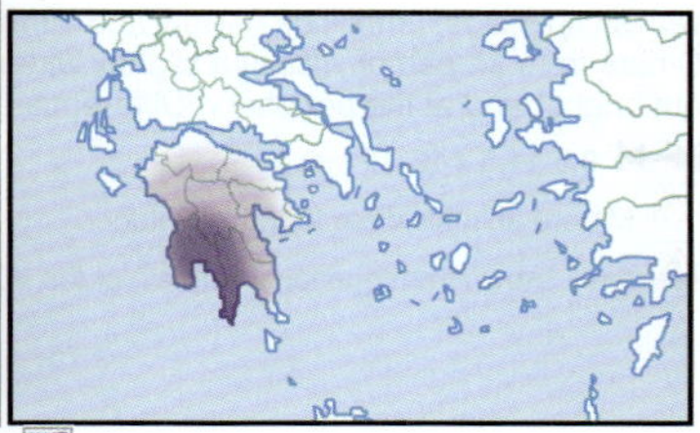

8.5-9.5 x 13-16 mm — Greece: S Pelopónisos, Máni peninsula near southern cape — Greece (Pelopónisos)

Shell bluish grey with olive greenish periostracum, apex pointed, 5-5.5 not very convex whorls, last whorl width more than 2 x of penultimate whorl, descending near aperture, aperture with white lip, margin sharp, umbilicus narrow and partly covered by the reflected columellar margin.

Mediterranean shrubland habitats, also in open forests, not rare.

Westerlund 1889: 74.

Monacha microtricha Jaeckel, 1954

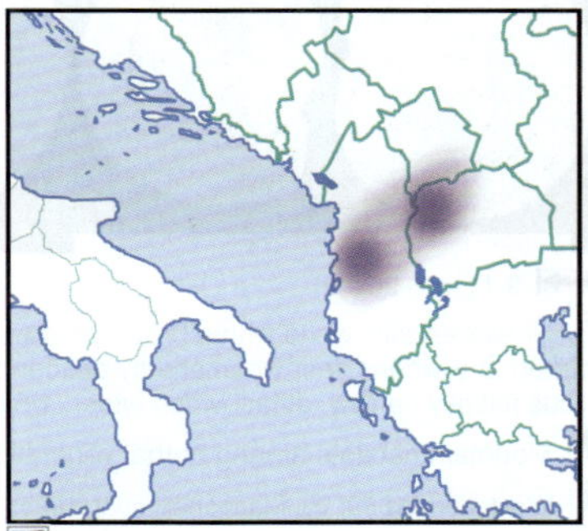

NW Rep. Makedonija, Albania

Monacha microtricha Jaeckel, 1954 in Jaeckel 1954: 84

Shell pale brownish yellow, finely striated with short light and slightly curved hairs arranged in rows (0.2 mm high, 25-30 hairs/mm^2), leaving well visible scars when lost, 5 weakly convex whorls, apertural margin sharp and reddish brown at the inner side, shortly behind is a strong white lip, both are also visible from outside, umbilicus very narrow, partly covered by columellar margin of aperture.
Animal white, head and foot darker, mantle with greyish black dashes, in living snails yellowish streaks are visible below the shell.

4-5.5 x 6-7.5 mm

Was found on relatively humid meadows.

Dhora & Welter-Schultes 1996: 158.

Systematic position unclear.

Monacha obstructa (Pfeiffer, 1842)

Helix obstructa Pfeiffer, 1842 in L. Pfeiffer 1842: 35

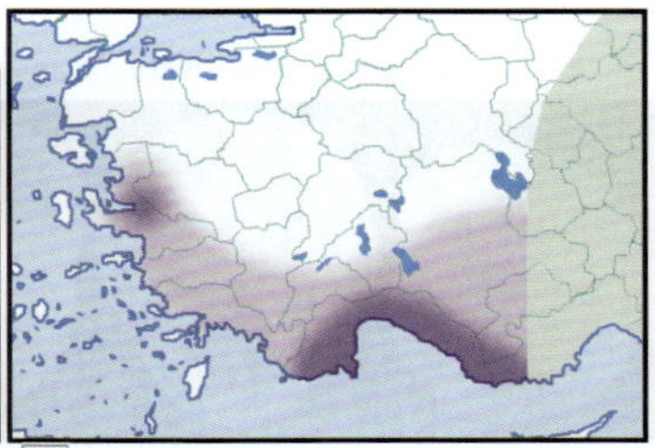

6-8.5 x 9-15 mm — Iraq: Bagdad — Egypt to Iran and S Turkey

Shell whitish opaque or light horny brownish, sculptured with minute riblets and shallow pits (irregular impressions), shiny, 4.5-5 slightly convex whorls, last whorl descending shortly near aperture, aperture with prominent white lip inside, well visible from outside, margin slightly reddish, umbilicus closed but looks like open, the excentrically expanded last whorl forms a pseudo-umbilicus. Shell height and width of the umbilicus variable.

Warm and dry habitats, often on sandy substrate.

Neubert 1998: 422, Hausdorf 2000: 93, Schütt 2001: 425.

Monacha ocellata (Roth, 1839)

Helix olivieri var. *ocellata* Roth, 1839 in Roth 1839: 14

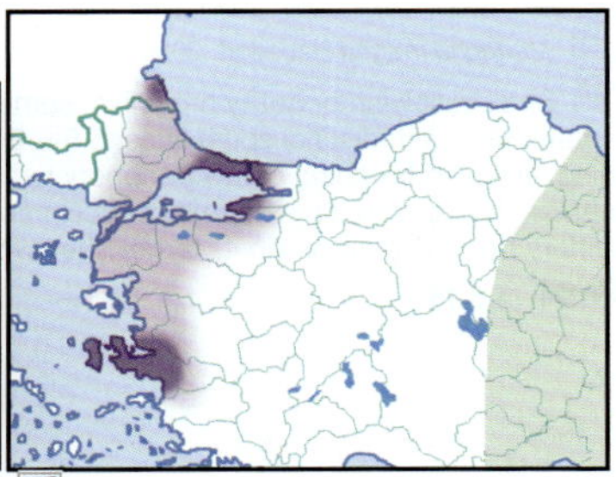

7-12 x 8-16 mm

Turkey: Istanbul

W Turkey to SE Bulgaria

Shell with characteristic white spiral bands below the suture, usually also visible in weathered shells, aperture with prominent white lip, also well visible from outside, apertural margin reddish or brown, umbilicus entirely closed.
Conchologically indistinguishable from *M. syriaca*. Differs by distinctly shorter penis, smaller papilla, s-shaped bend distal epiphallus section, longer flagellum, appendicula inserts at atrium.

In Bulgaria only at one locality near the Black Sea coast, found in dunes with abundant grassy vegetation, snails tend to aggregate in summer, and are associated with introduced *Opuntia* plants.

Hausdorf 2000: 93, Irikov 2008: 792.

Monacha oecali Hausdorf & Páll-Gergely, 2009

Monacha oecali Hausdorf & Páll-Gergely, 2009 in Hausdorf & Páll-Gergely 2009: 15

Shell light horny brownish, irregularly striated with smooth apex, last whorl sometimes with weak spiral sculpture, 5-6 convex whorls, last whorl rounded, slightly descending near aperture, apertural margin sharp, with whitish lip inside, umbilicus narrow and almost completely covered by the reflected columellar margin.
Differs from *M. pamphylica* in a slightly wider umbilicus and larger shell, from *M. ignorata* in a lighter shell without bright band at the periphery. Differences are mainly anatomically.

8.5-11.5 x 13-17 mm

Mature animals were found in October.

Only known from one locality (plain of the Köprülü Çayı creek between Serik and Boğazkent).

Turkey (Vilayet Antalya)

Monacha orsini (Porro, 1841)

Helix orsini Porro, 1841 in Villa & Villa 1841: 54 (*Helix Orsini*)
Name was attributed to Porro in original source, description was by Porro (explained in footnote p. 54). Original spelling was not *orsinii*.
Helix martensiana Tiberi, 1869 in Tiberi 1869: 68 (*Helix Martensiana*)

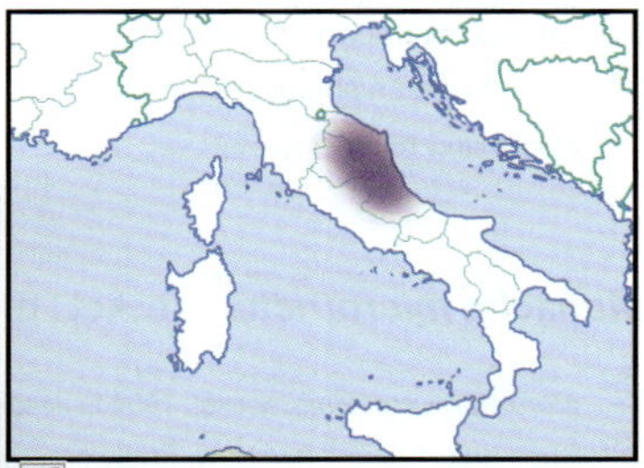

Central Italy

EU Red List: Near Threatened NT.
Threatened by pasturing and land use.

9-11 x 16-20 mm

Italy: L'Aquila, Gran Sasso, Prati di Tivo

Shell basically white with a broad yellowish brown colour band on upper side, lower side evenly yellowish brown, irregularly striated, with spiral lines, apex pointed, 5-6 whorls, last whorl shortly descending, aperture rounded with white lip inside, margin sharp and approaching each other slighty at parietal side, umbilicus initially narrow, at last whorl wider, 1/5 of diameter. Differs from *M. cantiana* in its larger umbilicus.

Mountain habitats among herbal vegetation and grasslands on calcareous substrate.

Westerlund 1889: 95, Manganelli et al. 1995: 30.

Taxonomic status uncertain. *Helix martensiana* has occasionally been recognised as a separate species from central Italy in the recent past. The *Monacha* populations in Italy are not understood and require detailed studies. Calling the species with wider umbilicus *orsini* is only preliminary and tentative.

Monacha ovularis (Bourguignat, 1855)

Helix ovularis Bourguignat, 1855 in Bourguignat 1855: 561, Pl. 16

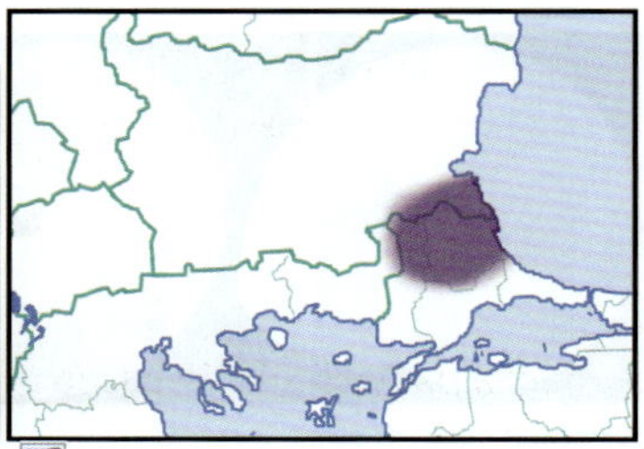

4.5-7 x 6-9 mm

Bulgaria: Sakar mountains, near Sladun

SE Bulgaria, NW Turkey

Shell horny brownish, sometimes with a lighter band at the periphery, striated and with distinct hair scars, 4.5-5 moderately convex whorls, last whorl rounded, hardly descending near aperture, aperture almost circular, with strong white lip inside, umbilicus narrow, partly or almost entirely covered by the reflected columellar margin.

Open grassy habitats with shrubs, also in sandy habitats at the seaside. In the summer the snails are not active in the coastal habitats in drought periods, then they can be found buried 20-40 cm deep into the sandy soil, around succulent plant roots (*Silene euxina*), but they do not build an epiphragm.

EU Red List: Near Threatened NT. Possibly exaggerated, no threats known.

Westerlund 1889: 84, Hausdorf 2000: 96, Irikov 2008: 793.

Monacha pamphylica Hausdorf, 2000

Monacha pamphylica Hausdorf, 2000 in Hausdorf 2000: 96

6-9 x 9.5-13 mm

Turkey: Serik Belek E of Antalya

S Turkey

Shell horny brownish with a blurry whitish band at the periphery, finely striated, lower whorls with irregular impressions, shiny, 4.5-5.5 moderately or weakly convex whorls, last whorl rounded or with blunt edge at the beginning, descending distinctly near aperture, aperture almost circular, with strong whitish lip inside, close to the aperture, umbilicus closed or almost closed.
Differs from the conchologically similar *M. ignorata* in its long flagellum.

Monacha parumcincta (Rossmässler, 1834)

Helix olivieri var. a synonym *Helix parumcincta* Menke, 1828 in Menke 1828: 12 (*Helix Olivieri* synonym *Helix parumcincta*)
Mentioned with a short description ("maior, fasciata"), no types, no locality. Not mentioned as a junior synonym of an available name (as demanded by Art. 11.6), but as a synonym for a subordinate variant for which the letter "a" was used. For those who neglect this (as proposed by Kadolsky 2012: 78, this is preliminarily accepted here), the name was eventually made available under Art. 11.6.1 in a subsequent source (Falkner 1996: 97). For the others, the name must be attributed to Rossmässler 1834.
I did not find a later use of *parumcincta* which referred to Menke 1828.

Helix olivieri var. a synonym *Helix parumcincta* Menke, 1830 in Menke 1830: 21 (*Helix Olivieri* synonym *Helix parumcincta*)
Mentioned with a short description.
Used as *Helix Olivieri* var. *parumcincta* by Mousson 1854: 4 for a form from Kérkira and other localities, with bibliographical reference to Menke 1830.
Not mentioned as a junior synonym of an available name (as demanded by Art. 11.6), but as a synonym for a subordinate variant for which the letter "a" was used. For those who neglect this (as proposed by Kadolsky 2012: 78, this is preliminarily accepted here), the name was made available under Art. 11.6.1 by Mousson 1854: 4 who used the name as *Helix Olivieri* var. *parumcincta* for a form from Kérkira and other localities, with bibliographical reference to Menke 1830. For the others, the name must be attributed to Rossmässler 1834.

Helix parumcincta Rossmässler, 1834 in Rossmässler [1834] (Heft 2): 5 Type locality Kérkira, Greece and Dalmatia, Croatia.
Used as a name for a form from Kérkira and Dalmatia, with a short description, no bibliographical reference to Menke. First use as a name of a taxon.
For those who require a bibliographical reference for making available a synonym under Art. 11.6.1, the name must be attributed to Menke, 1830 made available by Mousson 1854. For the others, to Menke, 1828, made available by Rossmässler 1834. In all cases the name cannot be used for the Italian species because Menke 1828 and Menke 1830 did not "cite" type specimens (the name may have referred to one or more specimens in his collection but he did not cite them in the sense of Art. 72.4.3), so the types would have to be taken from either Rossmässler [1834] or Mousson 1854, and both did not have types from Italy. Those who interpret Art. 72.4.3 in a way that citing specimens is not necessary if a list of species present in a collection is given, could designate a neotype from Italy. Doing so they would confirm that neither Rossmässler 1834 nor Mousson 1854 would have adopted the name, since this would have been outside their initial concept, such a neotype designation would be ineffective.

Shell horny brownish with white bands, rarely whitish, not shiny, striated and slightly corrugated, 6 whorls, last whorl initially angulated, aperture with prominent white lip inside and often a reddish margin, umbilicus usually covered by the reflected columellar margin. Animal light with dark grey tentacles.

7-9 x 11-14 mm

Italy: Friuli-Venezia Giulia, Santa Croce near Trieste

Massimo Prodan

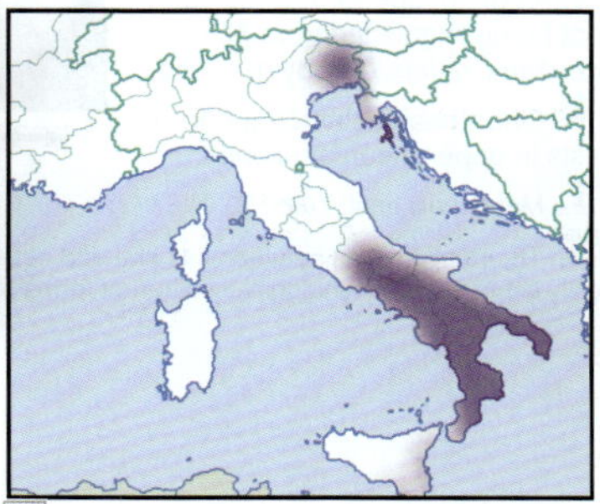

S Italy, perhaps also NE Italy and Sicily

Open habitats at the ground and under stones.

Westerlund 1889: 86 (S Italy to Turkey), Falkner 1990: 220 (S Italy, W and S Balkan peninsula to Aegean islands), Manganelli et al. 1995: 30 (Italy, not Sicily), Cossignani & C. 1995 (Italy), Frank 1997: 42 (several records in Greece, but probably none reliable), Falkner et al. 2002: 50, 144 (Sicily and Toulouse), Triantis et al. 2008: 475 (Skíros, Greece).

M. parumcincta is a very questionable name, its type locality must be in W Greece or Dalmatia. It seems to have been used by various authors for various forms in various regions between France and Turkey. It should preferably be regarded as a junior synonym of *Helix cartusiana* Müller, 1774. I would select a name-bearing type from Rossmässler's collection, from Dalmatia or Montenegro, and to convert it into a junior synonym of *cartusiana*.
Giusti 1973: 268 considered *Helix rizzae* from near Siracula (E Sicily) as a closely related form. See also the comment under *M. consona*.

Monacha phazimonitica Hausdorf, 2000

N Turkey

Monacha phazimonitica Hausdorf, 2000 in Hausdorf 2000: 114

Shell whitish ot light horny brownish, striated and on lower whorls with spiral lines, 5-6 moderately convex whorls, last whorl rounded, slightly descending near aperture, aperture with strong whitish lip inside, close to margin, umbilicus narrow, widely covered by the reflected columellar margin.
Shell indistinguishable from *M. gemina* and forms of *M. samsunensis* without hair scars. *M. phazimonitica* has no penis retractor (contrast to *M. samsunensis*), its flagellum is shorter and penial papilla longer than in *M. gemina*.

9-12 x 14-18 mm

Reported from limestone walls and rocky habitats.

Monacha pseudorothii Hausdorf, 2003

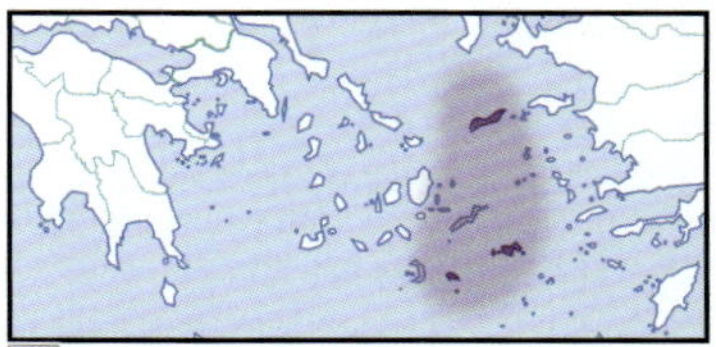

Greece (E Kikládes to Ikaría)

Monacha pseudorothii Hausdorf, 2003 in Hausdorf 2003: 40

Shell without pits, like *M. rothii*.
Lower epiphallus to flagellum ratio than in *M. rothii*, appendicula longer, basally broadening, inserts at distal end of vagina, glandulae mucosae insert at distal section of vagina.

8-10 x 11.5-14.5 mm

Triantis et al. 2008.

Monacha rothii (Pfeiffer, 1841)

Helix rothii Pfeiffer, 1841 in L. Pfeiffer 1841: 218 (*Helix Rothii*)

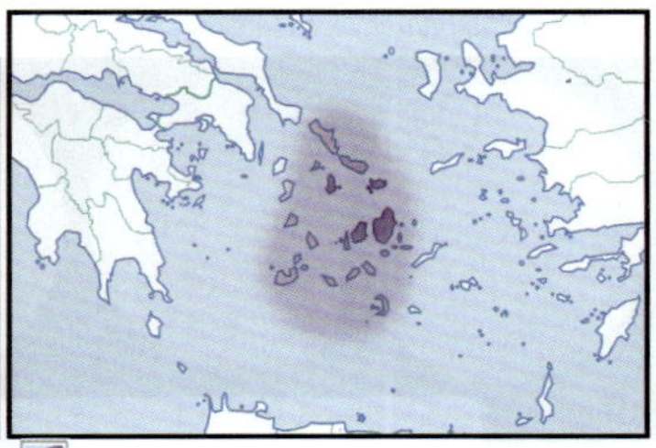

6-11 x 9-14 mm

Greece: island of Sífnos

Greece (W and central Kikládes)

Shell whitish with brown colour band at the periphery, lower side brownish, indistinctly striated, with or without pits, aperture with white lip and brownish margin, umbilicus usually covered by the reflected columellar margin.
Differs from *M. syriaca* and *M. ocellata* in a more closely coiled whitish shell, a distinctly delimited brown band above the periphery, whitish and sharp peristome, umbilicus prick-like or often completely or almost completely covered. Shell almost indistinguishable from *M. pseudorothii*, *M. maasseni* and *M. cretica* (see genitalia differences there).

Hausdorf 2000: 76, Hausdorf 2003.

Has partly been misidentified by various authors. Anatomically confirmed only from Kíthnos, Síros and Náxos, true range perhaps within the area Sérifos-Mílos-Íos-Amorgós-Náxos-Tínos-Sérifos.

Monacha ruffoi Giusti, 1973

Monacha ruffoi Giusti, 1973 in Giusti 1973: 533, Tav. 6

Shell globulous, horny yellowish, finely striated, last whorl predominating, aperture with prominent whitish lip inside, apertural margin reddish brown, umbilicus narrow, partly covered by columellar margin of aperture. Shell cannot be distinguished from *M. cantiana*, differs only by genitalia.

Open habitats and open forests in upper mountain zones.

Manganelli et al. 1995: 30.

The relationships between this one and other local *Monacha* forms need to be clarified, particularly those with anatomically similar and conchologically different forms of lower altitudes.

9.3-11.2 x 14-17.5 mm

Italy: Lazio, Mount Sassetelli near Terminillo

Massimo Prodan

Italy (Monti Reatini)

EU Red List: Vulnerable VU. Only known from few localities in Monti Reatini. Threatened by intensive pasturing and habitat alterations associated with land use in mountains.

Monacha samsunensis (Pfeiffer, 1868)

Helix samsunensis Pfeiffer, 1868 in L. Pfeiffer 1868: 480 (*Helix Samsunensis*)

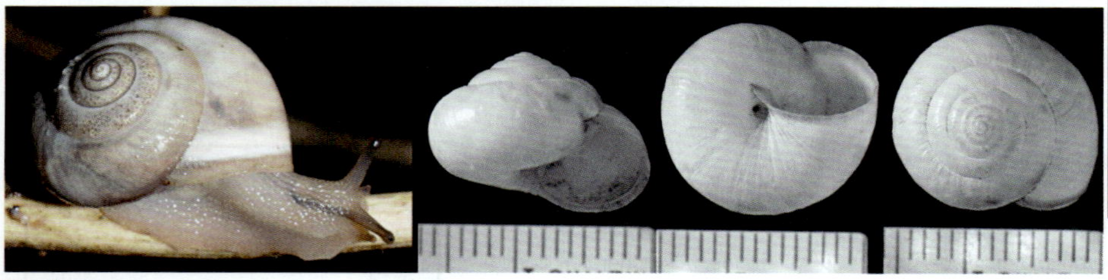

N and central Turkey, possibly also Batumi to Novorossijsk

6.5-13 x 11-19 mm

Turkey: Vilayet Kastamonu

Aydın Örstan

Shell whitish, or light horny brownish with whitish band at the periphery, striated, usually with very short hairs or hair scars at upper whorls, rarely also at lower whorls, lower whorls also with spiral lines, 5-6 moderately convex whorls, last whorl rounded, descending near aperture, aperture elliptical, with strong white lip inside, umbilicus narrow or very narrow, partly or almost completely covered by the reflected columellar margin.
Penis connected to epiphallus by weak muscle bands, penial papilla conical or cylindrical, epiphallus 1.2-2.7 x longer than flagellum, 2.5-6 x longer than penis, spermatheca rounded triangular or oval, two small branched mucus glands insert near proximal end of vagina. Forms without hair scars differ from *M. phazimonitica* in the presence of a penial retractor.

Various habitats in woods or shrublands, also on agricultural crops.

Shilejko 1978: 288, Hausdorf 2000: 114.

Records from Balkan peninsula are probably all based on misidentifications.

Monacha solidior (Mousson, 1863)

Helix rissoana var. *solidior* Mousson, 1863 in Mousson 1863: 369 (*Helix Rissoana* var. *solidior*)

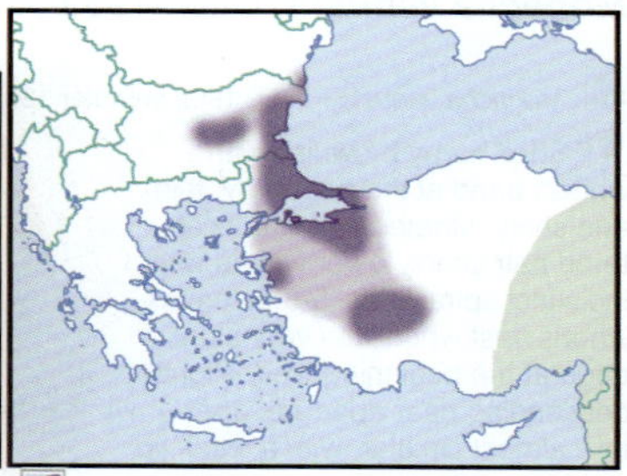

5-8 x 7.5-11 mm (Hausdorf 2000)

Bulgaria: Nikolaevo

SE Romania to W Turkey

Shell horny brownish with lighter band at the periphery, striated and with scattered hair scars, lower whorls with apiral lines, 4.5-6.5 convex whorls, last whorl rounded or slightly angular, usually distinctly descending near aperture, aperture with strong whitish lip inside, close to margin, umbilicus narrow or very narrow, to a large extent covered by the reflected columellar margin, sometimes closed.
Differs from *M. carascaloides* in its narrower appendicula inserting more proximally at the vagina, shell usually smaller and more globular, umbilicus narrower.
Differs from sympatric *M. claustralis* (which can be very similar in its umbilicus and aperture) in its usually darker shell, the presence of hair scars and in some parts of whorls 3 and 4 tiny riblets or interrupted growth lines, generally a slightly more globulous and smaller shell. Small *M. claustralis* shells are lighter and and the growth lines are more regular, in larger *M. claustralis* shells of the same population the surface can have a hammered sculpture, but not with tiny hair scars or pits.

Open habitats and sand dunes, up to 600 m. In coastal dunes the snails prefer sand covered by moss, they aggregate at the soil near plants, or under leaves, where the humidity is higher.

Hausdorf 2000: 100, Irikov 2008: 795 (gave size 10-12 x 13-17 mm, probably incorrect), Haus der Natur Cismar collection material from Mamaia near Constanţa, Romania, collected in 1987.

Monacha stipulifera Hausdorf, 2000

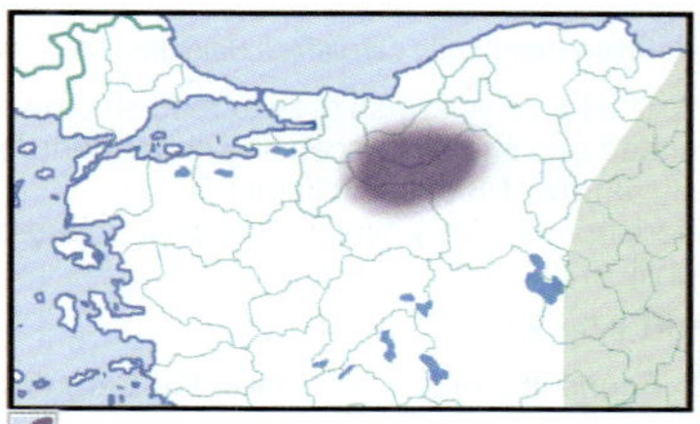

NW Turkey

Monacha stipulifera Hausdorf, 2000 in Hausdorf 2000: 119

Shell light or horny brownish, with an indistinct lighter band at the periphery, striated and with short (0.1-0.2 mm) hairs bent towards aperture, 4-5.5 convex whorls, last whorl rounded or slightly angular, hardly descending near aperture, aperture with strong whitish or brownish lip inside, umbilicus narrow or moderately wide, slightly excentrical at last whorl, usually only slightly covered by the reflected columellar margin.

4-8 x 6-11 mm, locally variable

Reported from coniferous forests.

Monacha subaii Hausdorf, 2000

SW Turkey

Monacha subaii Hausdorf, 2000 in Hausdorf 2000: 103

Shell horny brownish with a distinct white band at the periphery, with wrinkled growth ridges and weak spiral lines, upper whorls with hair scars, 5-5.5 convex whorls, last whorl rounded, slightly descending near aperture, aperture with strong whitish lip inside, umbilicus narrow and partly covered by the reflected columellar margin.
Differs from *M. solidior* in the more distal insertion of the appendicula at the vagina. In *M. subaii* the apical section of the appendicula is 3-4.5 times as long as the basal section, in *M. solidior* only 1-2 times.

6-8 x 8.5-11.5 mm

Was found in mountains at 1200 m altitude.

Only known from one locality.

Monacha syriaca (Ehrenberg, 1831)

Helix syriaca Ehrenberg, 1831 in Ehrenberg 1831: [Gasteropoda p. 7]

6-9 x 8-12 mm

(1-5) Greece: W Crete, Stílos near Vámos, (6-7) Greece: central Crete, near Asítes

S Greece to Syria and Egypt

Shell with characteristic white spiral bands below the suture, usually also visible in weathered shells, aperture with prominent white lip, also well visible from outside, apertural margin reddish or brown, umbilicus entirely closed.
Animal light yellowish with brown tentacles.
Cannot be distinguished from *M. ocellata* by the shell, only anatomically: substantially longer penis with larger papilla, shorter flagellum, appendicula inserts usually at the vagina.

Dry and open habitats of any kind, usually in coastal vicinity, frequently in meadows. In Crete in lowlands, often in agricultural crops.

Frank 1997: 43, Hausdorf 2000: 103, Heller 2009: 327.

Monacha venusta Pintér, 1969

Monacha venusta Pintér, 1969 in Pintér 1969: 93

Shell horny brownish with whitish band at the periphery, thin and shiny, striated and with scattered hair scars, lower whorls with indistinct spiral lines, 5-5.5 convex whorls, last whorl with very blunt edge at the beginning, very slightly descending near aperture, aperture almost circular, with a weak lip inside, umbilicus very narrow, partly or almost completely covered by the reflected columellar margin.

6-9 x 9-15 mm

Bulgaria: S Black Sea coast near Ropotamo river

Atanas Irikov

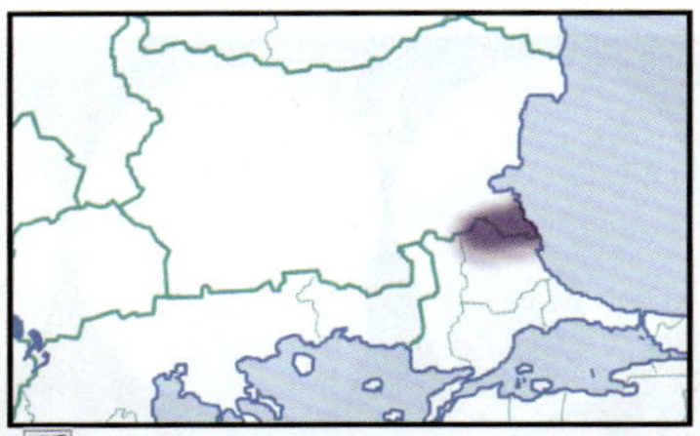

SE Bulgaria, NW Turkey

EU Red List: Near Threatened NT. No real threats known.

Hausdorf 2000: 105, Irikov 2008: 796.

In Bulgaria in coastal vicinity, in open habitats covered by grasses, in the summer under stones, active after rainfall.

Prostenomphalia

Prostenomphalia Baidashnikov, 1985 in Bajdashnikov 1985: 206

Type species *Prostenomphalia carpathica* Baidashnikov, 1985 (original designation)

The name of the author was spelled in Latin Baidashnikov in the original source, this spelling is used here for the nomenclatural purpose. The Russian script name is transcribed as Bajdashnikov in the convention used here, this spelling is used for the bibliographical purpose (it allows unambiguously a correct re-transcription into Russian script).

Prostenomphalia carpathica Baidashnikov, 1985

Prostenomphalia carpathica Baidashnikov, 1985 in Bajdashnikov 1985: 206

Shell pale yellowish grey to reddish brown, promonently striated, 4-5 weakly convex whorls, last whorl rounded and shortly descending near aperture, aperture with whitish lip inside, margin reflected, umbilicus moderately narrow, 1/8 of shell diameter.
Differs from the very similar *Urticicola umbrosa* in its less depressed shell, narrower umbilicus and rounded (not slightly angulated) last whorl.

6.5-8 x 10.5-13 mm

Found in a riverside forest in the Carpathian mountains.

Sysoev & Schileyko 2009: 218.

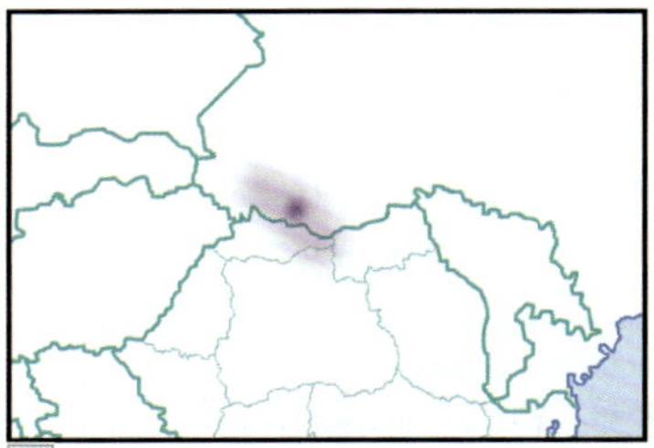

W Ukraine (Carpathians)

Trochoidea

Trochoidea Brown, 1827 in Brown 1827: text for Pl. 41 Fig. 80-81 [not verified]

Type species *Trochus terrestris* Pennant, 1777 (monotypy, but probably misidentified) (= currently *Euconulus fulvus* (Müller, 1774))

Type species by monotypy, Brown gave *Trochoidea terrestre* as of Montagu 1803. *Trochus terrestris* was established by Pennant 1777: 127 Fig. 108, currently considered as a synonym of *Helix fulva* Müller, 1774 (= *Euconulus fulvus*). Montagu 1803: 287 cited Pennant's name *Trochus terrestris* (and did not established a new name), gave a description and several references, among others Lister et al. 1685: Pl. [26] 61 Fig. 58, where a specimen of *Helix elegans* Gmelin, 1791 was figured (Gmelin [1791]: 3642 referred to the same figure). Montagu's 1803 description matched Lister's figure. This suggests that Brown 1827 misidentified the type species and meant *Helix elegans* Gmelin, 1791.
Gude & Woodward 1921: 625 argued that Brown's 1827 figures did not resemble either *Helix fulva* or *Helix elegans*, and argued that *Trochoidea* should disappear from molluscan literature. Malacologists obviously did not follow their proposal.
It is necessary to verify Brown's 1827 original publication. If it provides constraints for a misidentification as suggested above, contrasting Gude & Woodward's 1921 interpretation, then *Helix elegans* Gmelin, 1791 should be fixed as type species for *Trochoidea* under Art. 70.3.

Presence of two symmetrical but rudimentary (not dart-bearing) dart-sacs, usually 4 branched mucus glands around the vagina, large apppendix at the atrium (difference to *Xerocrassa*). Penis is innervated from the right cerebral ganglion.

Hausdorf & Sauer 2009: 404.

Trochoidea carinatoglobosa (Haas, 1934)

Helicella syrensis carinatoglobosa Haas, 1934 in Haas 1934: 18 *Helicella (Jacosta) syrensis carinato-globosa*)

5-9 x 9-13 mm

Cyprus. Akrotiri near Agios Georgios

Charles Geerts

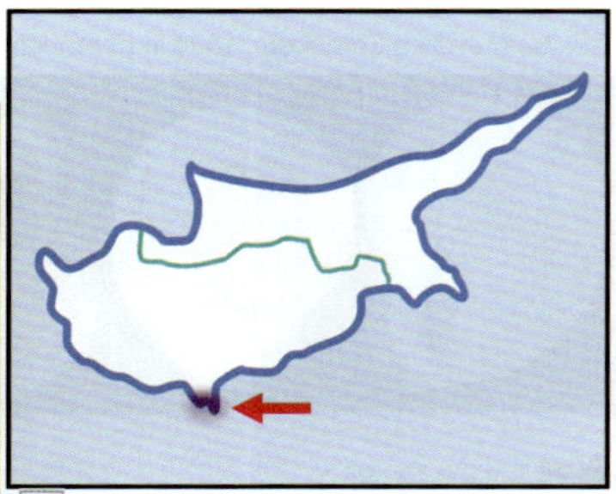

S Cyprus (Akrotíri peninsula)

Open and forest habitats.

Gittenberger 1991: 123.

Shell variable in shape, keeled or regularly rounded, white with a broad brown colour band on upper side, several not always equally broad bands on lower side, solid, slightly translucent, coarsely striated, 4.5-5.5 weakly convex whorls, suture can be scalarid in some specimens, keel weaker at last whorl and may disappear near the aperture, apertural margin with a weak white lip inside, umbilicus open, wider in flatter shells, 1/9-1/6 of shell diameter.

Trochoidea caroni (Deshayes, 1832)

Helix caroni Deshayes, 1832 in Deshayes 1830 [part 2]: 262 (*Helix Caroni*)
Date was 29 Sep 1832 (Sherborn & Woodward 1899: 595, Cowie 1998: 92, "444" should probably read "144").

Italy (Sicily, Capri, Lazio)

Shell brownish yellow, often with dark brown spots, not shiny, ribbed, 8-10 flat whorls, with a sharp and prominentlky indented keel, last whorl not much broader, aperture narrow, with indented keel, with whitish or brownish lip relatively deep inside, margin sharp, umbilicus very narrow.

Calcareous rocks and stones.

Westerlund 1889: 351, Giusti et al. 1995: 423, Manganelli et al. 1995: 30.

7-12 x 5-9 mm

Italy: Sicily, Monte Pellegrino near Palermo

Trochoidea cucullus (Martens, 1875)

Helix cucullus Martens, 1875 in Martens 1875 ["1873"]: 27

5-7 x 9-11 mm

Malta: Mtahleb

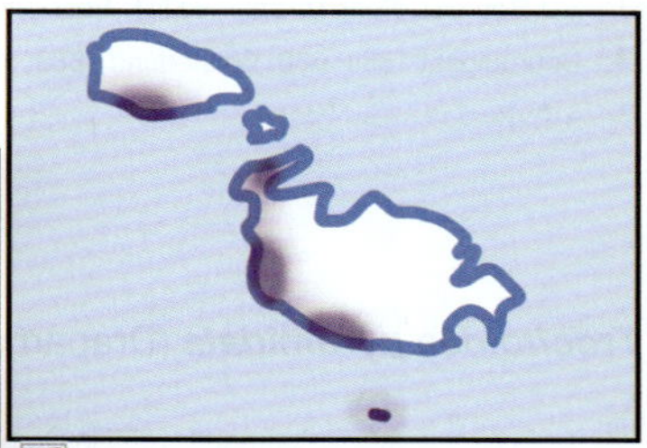

W Malta, S Gozo, Fifla

Open habitats among rocks and shrubs.

Shell greyish white, with quadrangular reddish brown spots near the suture, often also with other interrupted colour bands, finely striated, 4.5-5.5 weakly convex whorls, last whorl initially angulated, rounded at periphery, aperture with strong yellowish white lip inside, umbilicus moderate, 1/6 of diameter. Several geographically isolated and morphologically slightly different populations. Differs from other Maltese species in its slightly larger umbilicus, a slightly more shiny shell and its whorls being more rounded.

Westerlund 1889: 363, Falkner 1990: 216, Beckmann 2003.

Trochoidea cumiae (Calcara, 1847)

Helix cumiae Calcara, 1847 in Calcara 1847: 37 [not verified]

6-7.5 x 7-11 mm

Italy: Lampedusa, Guitgia

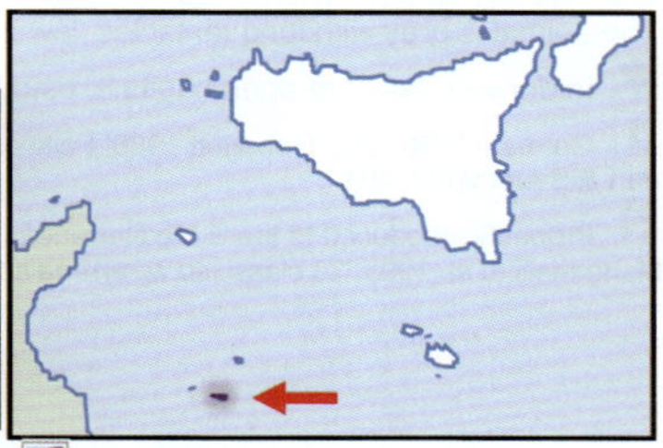

Italy (Lampedusa)
Possibly also in Calabria.

In various kinds of habitats in the island, also in cultivated land.

EU Red List: Near Threatened NT. No acute threats known.

Shell white, often with 1-2 rows of brown spots on the upper and a band on the lower side, sometimes with more brown bands, densely ribbed on upper side, striated on lower side, 5.5-6.5 convex whorls, distinctly keeled, last whorl not wider than penultimate whorl, aperture with white lip inside, umbilicus narrow, sometimes covered by the weakly reflected columellar margin.

Westerlund 1889: 359, Manganelli et al. 1995: 30, Cossignani & Cossignani 1995.

Trochoidea elegans (Gmelin, 1791)

Helix elegans Gmelin, 1791 in Gmelin [1791]: 3642

5-11 x 6-12 mm

Italy: Liguria, Lerici, Maralunga

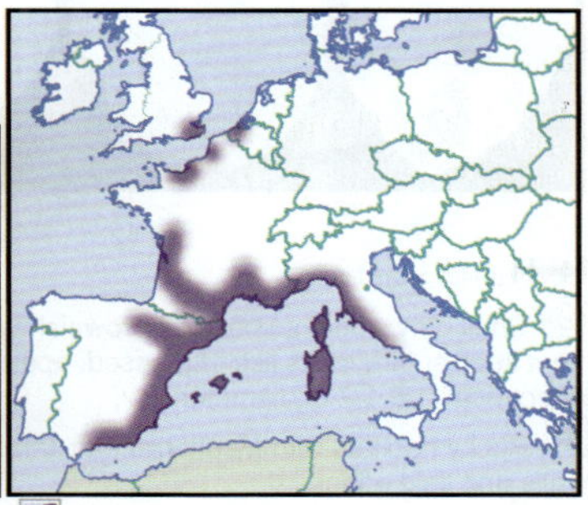

W Europe

Shell greyish white with dark to pale brown colour bands, rarely evenly brown, coarsely ribbed, juveniles with fine hairs, 6-7 sharply keeled, flat and slightly oberlapping whorls, outline conical above, very flattened below, aperture with a notch at the keeled periphery, umbilicus open, moderately narrow, circular and not wider at last whorl.

Dry and exposed habitats, grasslands, dunes, rocky meadows and dry grassy slopes. Often at steep slopes with broken soil cover and limestone rocks showing through.

50-60 eggs are laid in July in small holes in the ground under herbs, juveniles hatch after 3 weeks, maturity is reached after 12 months.

Gassies 1849: 73 (life cycle), Moquin-Tandon 1855: 271, Germain 1930: 314, Kerney et al. 1983: 250, Prieto & Martín 1988, Falkner 1990: 216, Manganelli et al. 1996: 30, Kerney 1999: 183, Beckmann 2007: 94.

Trochoidea liebetruti (Albers, 1852)

Helix liebetruti Albers, 1852 in Albers 1852: 124 (*Helix Liebetruti*)

5-7 x 4-6 mm

SE Cyprus: near Agía Nápa

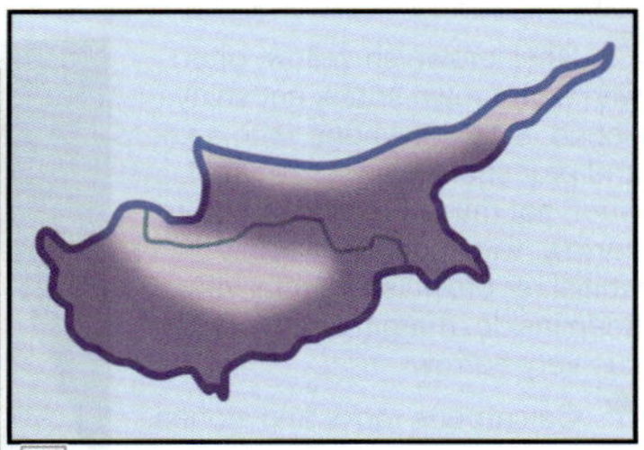

Cyprus

Cyprus: near Lefkosía — Georgios Konstantinou

Shell high spired, like juvenile *Cochlostoma*, white, prominently ribbed, scalarid, 5-6 strongly convex whorls with deep suture, last whorl with distinct and broad keel, with a furrow on the upper side, last whorl not descending, aperture wide with a weak dark lip inside, umbilicus moderately narrow and open.

Open habitats with shrubland vegetation, in coastlands and in the interior parts of the island. Often found sticking in shrubs.

EU Red List: Near Threatened NT. Possibly exaggerated, in the belief that the species was restricted to coastlands.

Bourguignat 1854: 660, Westerlund 1889: 356, Gittenberger 1991: 122.

Trochoidea pyramidata (Draparnaud, 1805)

Helix pyramidata Draparnaud, 1805 in Draparnaud 1805: 80
Helix tarentina Pfeiffer, 1848 in L. Pfeiffer 1848: 442

6-8 x 7-10 mm

Croatia: S Dalmatia, Mlini near Dubrovnik

Mediterranean region

Greece: Kíthnos island — Panayotis Ovalis

Shell conical, evenly white, or with brown bands or spots, weakly shiny, striated (in S Italy more prominently), 4.5-7 weakly convex whorls, last whorl rounded, not keeled, aperture with strong white lip inside, umbilicus narrow. The animal is grey with long tentacles, lower side lighter.

In dry vegetation of open habitats, usually in coastal vicinity.

Germain 1930: 311, Liebegott 1986, Falkner 1990: 216, Gittenberger 1991: 123, Giusti, Ma. & S. 1995 (anatomy), Hausdorf & Sauer 2009: 404.

Probably introduced to the E Mediterranean. Range in S Italy uncertain.
Manganelli et al. 1995: 30 classified ***tarentina*** as a synonym of *pyramidata*. Ferreri et al. 2005: 99 studied *tarentina* forms anatomically and came to the same conclusion.

Trochoidea spratti (Pfeiffer, 1846)

Helix spratti Pfeiffer, 1846 in L. Pfeiffer 1846 (Symb. 3): 69 (*Helix Spratti*)

4-4.5 x 5-11 mm

Malta: near Dingli Cliffs

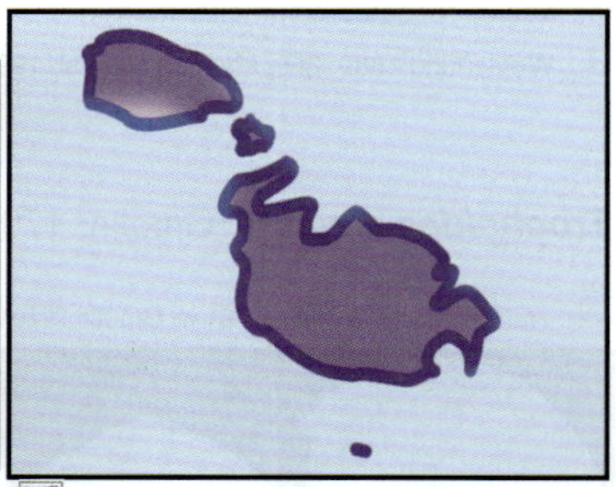

Malta

Westerlund 1889: 324, Falkner 1990: 216, Giusti et al. 1995: 395, Beckmann 2003 (perfect range map).

Shell dirty white, sometimes brownish with darker colour bands, densely ribbed, 5-6 whorls with a very prominent keel, suture looks like impressed, aperture small with weak lip inside, umbilicus moderately wide and perspectival, 1/8-1/7 of diameter.

Rocky habitats with Mediterranean shrub vegetation, under stones, in rock crevices or at the base of plants, also in artificial habitats such as rock faces, walls and field edges.

Trochoidea trochoides (Poiret, 1789)

Helix trochoides Poiret, 1789 in Poiret 1789: 29

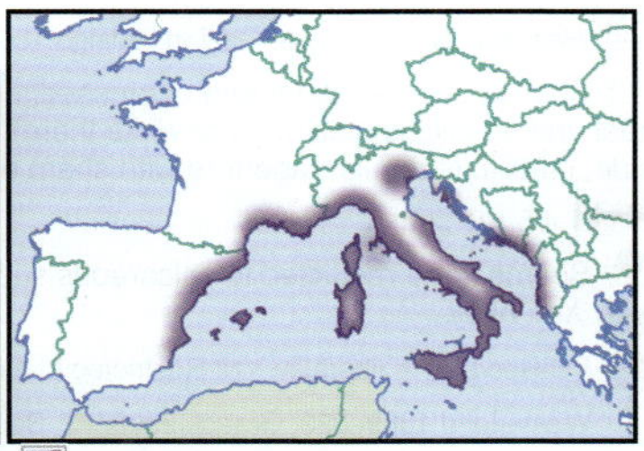

4.5-8 x 4.5-8 mm

(1-3) Albania: near Durrës, (4-6) Croatia: Mlini

W Mediterranean region

Shell opaque whitish or yellowish, sometimes with a brown colour band, on lower side with some partly interrupted narrow bands, striated, 5.5-6 convex whorls, prominently keeled, last whorl convex on upper side, much less convex on lower side, aperture oblique and broad with white lip inside, margin sharp, umbilicus open.
Much like *T. pyramidata*, but whorls more step-like with keel, lower side more flattened, umbilicus narrower.

Dry vegetation in coastal vicinity, usually in dunes, also in ruderal habitats with similar vegetation, up to 900 m.

Vulnerable in Spain, threatened by destruction of dune habitats caused by growing cities and tourism, populations in Alicante and Castellón might already be extinct.

Westerlund 1889: 355, Falkner 1990: 216, Manganelli et al. 1995: 30, Gómez Moliner et al. 2001: 136, Puente et al. in Verdú & Galante 2006.

Xerocrassa

Xerocrassa Monterosato, 1892 in Monterosato 1892: 23 *Helix (Xerocrassa)*

Type species *Helix seetzeni* Pfeiffer, 1847 (monotypy)

Globular, trochiform or nearly flat shells of medium size. In the male genital system, the flagellum is not longer than the epiphallus. No penial appendix. The two dart sacs and the glandulae mucosae are reduced.

SE Europe to Palestine and Libya

Hausdorf & Sauer 2009 (revision Crete).

Xerocrassa amphiconus (Maltzan, 1883)

Helix amphiconus Maltzan, 1883 in Maltzan 1883: 102

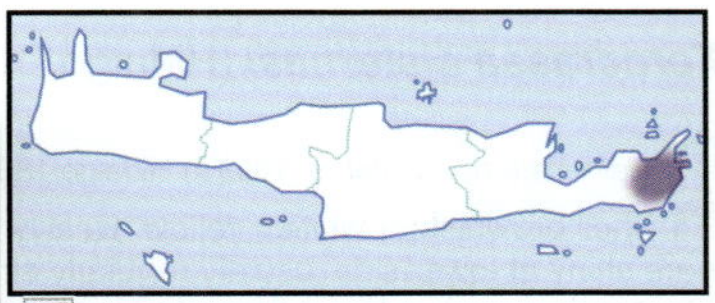

4-6 x 9-12 mm

Greece: E Crete, near Sitía

Greece (E Crete)

Rocky limestone habitats with some vegetation.

Hausdorf & Sauer 2009: 376.

Differences to *Xcr. siderensis* are unclear, could be the same species.

Shell whitish, often with brown spots or interrupted bands, prominently striated, flat, strongly keeled, 4.2-4.7 whorls, often concave with strong keel, last whorl usually not descending near aperture, aperture with white lip inside, umbilicus open but very narrow, 1/30 of diameter, but usually not partly covered by the columellar margin.
Flagellum relatively short (proximal epiphallus : flagellum = 3.3-4.6), penial papilla with long terminally open basal and very short conical apical part, with an open channel.

Xerocrassa arturi (Haas, 1924)

Helicopsis gigaxii arturi Haas, 1924 in Haas 1924: 143, Taf. 8

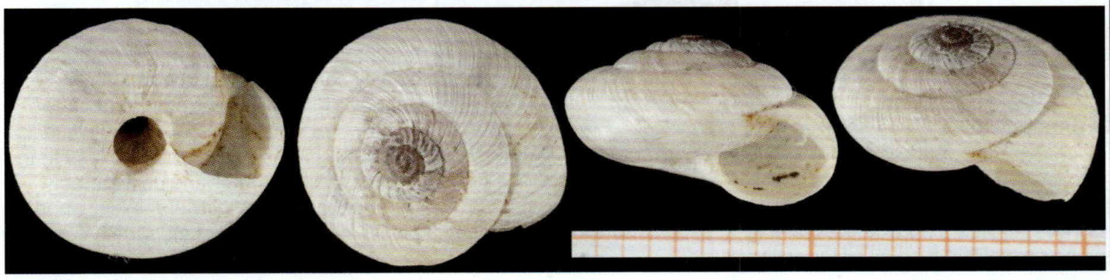

4.5-5.5 x 8-10 mm

Spain: Tarragona, Monte Caro

NE Spain

Shell whitish with dark apex, upper side densely striated, lower side smoother, 5-6 slowly and regularly increasing whorls with shallow suture, last whorl not descending near aperture, umbilicus wide.

Xerocrassa barceloi (Hidalgo, 1878)

Helix barceloi Hidalgo, 1878 in Hidalgo 1878: 230 (*Helix Barceloi*)

Shell yellowish or greyish white, densely ribbed, sometimes brown colour spots on the upper side and fine colour bands at the lower side, sharply keeled, lentiform, upper side flat, lower side convex, keel also visible at suture, last whorl with a crenulated keel, aperture not rounded, inside with a distinct white lip, umbilicus wide.

Westerlund 1889: 331, Martínez-Ortí & Robles 2003: 129.

4-5 x 9-11 mm

Spain: Alicante, Castillo de Santa Barbara

Spain (Alicante, Murcia)

Xerocrassa cardonae (Hidalgo, 1867)

Helix cardonae Hidalgo, 1867 in Hidalgo 1867: 209, Pl. 12 (*Helix Cardonæ*)

Shell keeled, lentiform and depressed, equally convex at both sides, weakly striated, horny brown except keel and umbilicus (yellowish or white there), short and dense hairs, sometimes brown colour bands at the lower side, umbilicus regular, aperture with sharp edges, inside with a weak lip. Differs anatomically from *Xcr. nyeli*.

4-5 x 9-11 mm

Seems to be restricted to calcareous substrate, under stones in pine and oak forests, lives sympatrically with *Xcr. nyeli*.

EU Red List: Vulnerable VU. Restricted to a small site, potentially threatened by human pressure and fire.

Westerlund 1889: 333, Graack 2005: 42, Beckmann 2007: 108.

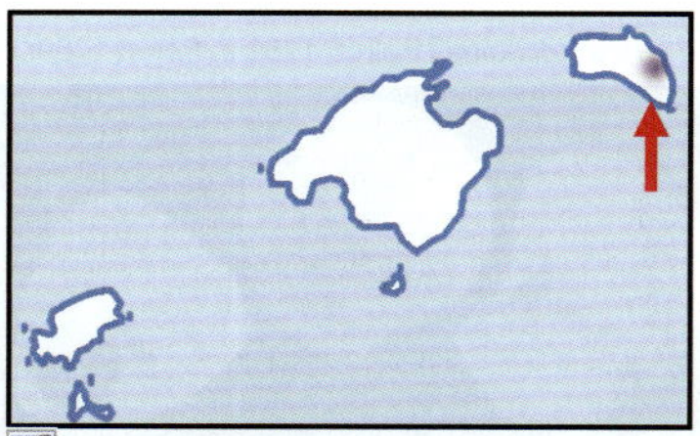

Spain (Menorca)

Xerocrassa caroli (Dohrn & Heynemann, 1862)

Helix caroli Dohrn & Heynemann, 1862 in Dohrn & Heynemann 1862: 107 (*Helix Caroli*)

5.5-6 x 8-9 mm

Spain: Illes Balears, Formentera

Spain (Ibiza island group)

Westerlund 1889: 237, Schröder 1984, Beckmann 2007: 109, Puente et al. in Verdú & Galante 2006: 384 ("Vulnerable").

Shell white with some brown colour bands or (on the upper side) rows of spots, densely striated, umbilicus partly covered, aperture with a white lip. Locally variable.

Dry habitats on the main island and surrounding islets, seems to require unusually high temperatures and little rainfall.

Xerocrassa cisternasi (Hidalgo, 1883)

Helix cisternasi Hidalgo, 1883 in Hidalgo 1883: 56, Pl. 2 (*Helix Cisternasi*)

Shell brownish or whitish yellow, weakly shiny, lighter, smoother and more shiny at lower side, usually regularly striated, umbilicus narrow, aperture with a weak edge at the outer margin, inside with a strong white lip.

In rocky habitats, under rocks or in crevices, also on ocean cliffs.

EU Red List: Near Threatened NT. Restricted to a small range. Main threats are fire and hiking (possibly not a real threat).

Westerlund 1889: 259, Beckmann 2007: 112.

Spain: Ibiza, Portinatx

8-9 x 14-15 mm

Roy Anderson

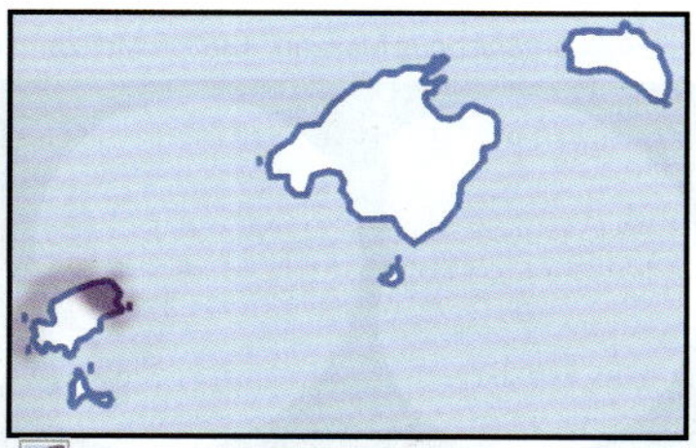

Spain (Ibiza)

Xerocrassa claudia Hausdorf & Welter-Schultes, 1998

Xerocrassa claudia Hausdorf & Welter-Schultes, 1998 in Welter-Schultes 1998: 76

Name and description were attributed to both co-authors in the original source.

5-8 x 9-12.5 mm, varies locally, shells are larger in the south and NW, smaller in the central parts of the island

Greece: NW of Gávdos, S of Crete

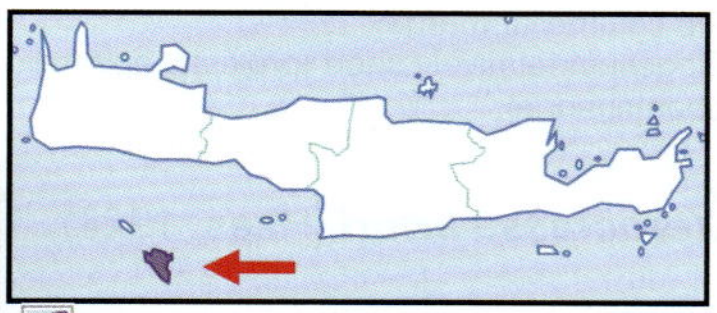

Greece (Gávdos S of Crete)

Shell whitish with brown spots or interrupted bands, prominently striated, 4-5 weakly convex whorls, last whorl initially angulated, near aperture rounded and weakly descending, aperture with white lip inside, umbilicus narrow, slightly less than 1/10 of shell diameter and considerably widened in the last third of the last whorl. Looks like the other small *Xerocrassa* species of Crete. Differs from *X. mesostena*, *X. psiloritana* and *X. subvariegata* in its more flattened shell, slightly faster increasing whorls and slightly wider umbilicus, anatomically in a basal broader flagellum.

Mountain slopes near pine trees, more frequent in natural habitats. In higher altitudes more frequent than *X. claudiconus*. Rare in sandy habitats and mysteriously absent from an area around the village of Vatsianá.

Xerocrassa claudiconus Hausdorf & Welter-Schultes, 1998

Xerocrassa claudiconus Hausdorf & Welter-Schultes, 1998 in Welter-Schultes 1998: 81

Name and description were attributed to both co-authors in the original source.

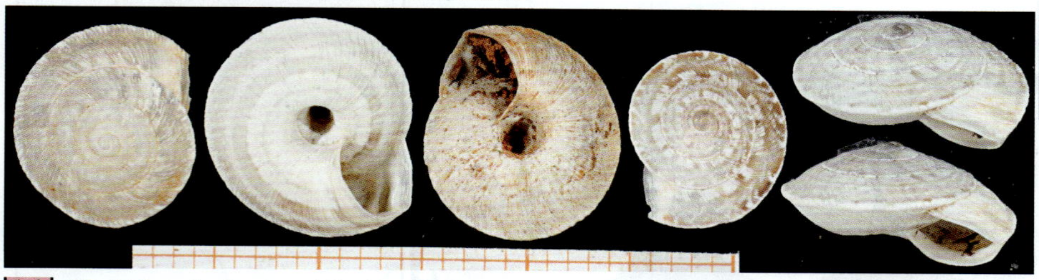

3-7 x 5-13 mm, locally very variable, large in Gavdopoúla and in the S of Gávdos, in the N and centre of Gávdos small

Greece: Gávdos S of Crete, near Kórfos

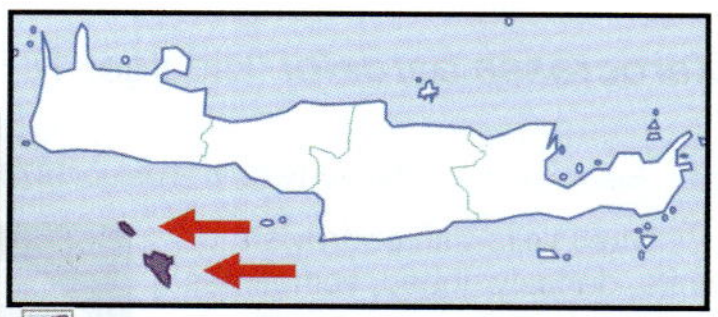

Greece (Gávdos and Gavdopoúla)

Shell whitish with light brown spots, or interrupted bands, prominently striated, strongly keeled, 4-5 flat whorls, last whorl very weakly descending, aperture lemon-shaped, with prominent white lip inside, umbilicus wide and regular, 1/7 of shell diameter.

Nearly all habitats, usually in lower numbers than *X. claudia*. Frequent in open habitats, but also under pine needles and in sandy habitats with vegetational cover. Not necessarily attached to limestone habitats. Also frequent on Gavdopoúla, where *X. claudia* is absent.

Xerocrassa claudinae (Gasull, 1964)

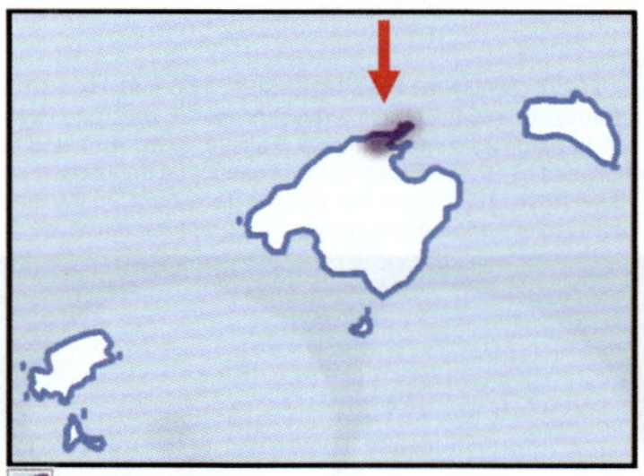

Spain (Mallorca)

Helicella claudinae Gasull, 1964 in Gasull 1964: 89

Shell flat on upper side, lower side rounded, prominently keeled, dirty whitish to brown, occasionally with two brown bands on the last whorl, prominently and irregularly striated, 5-6 regularly increasing whorls, last whorl width 1.5 x of penultimate whorl, aperture characteristically separated from the spire, much more downside than the rest of the shell (I have not seen and understood this), umbilicus wide and perspectivical.

2-6 x 6-11 mm, size increasing from west to east

Open habitats between sea level and 450 m.

Vulnerable. Restricted to a small range in NE Mallorca, seriously threatened by urban pressure and mass tourism.

Classified as a form of *X. moraguesi* by Beckmann 2007, but recognized as an independent species by Spanish authors (Puente et al. in Verdú & Galante 2006: 385, C. R. Altaba, unpublished, Apr 2008).

Xerocrassa cobosi (Ortiz de Zárate López, 1962)

S Spain

Helicella cobosi Ortiz de Zárate López, 1962 in Ortiz de Zárate López 1962: 41, Lám. 1

Shell whitish with horny brown spots, irregularly striated, with short hairs, 3.7-4.3 whorls, last whorl slightly angulated, aperture simple without lip, umbilicus wide, deep and perspectivical, 1/4-1/3 of diameter. Shells are usually dirty, sometimes entirely covered by dust, provoked by the sticky mucus of the animal.

2.2-2.6 x 4.5-5.5 mm

On calcareous substrate, under stones in open habitats exposed to sunshine, in altitudes up to 1600 m.

Talaván Gómez & T. 2006: 31, Ruiz Ruiz et al. [2006]: 108.

Xerocrassa cretica (Férussac, 1821)

Helix cretica Férussac, 1821 in A. Férussac [1821-1822]: 45

Established with a short description for the subordinate variant α, but not with a distinct description for the main form. The name *Helix cretica* was made available at this occasion under Art. 12.1 because it had a description for the taxon (Code Glossary "taxon": a taxon encompasses all included taxa of lower rank). Art. 12.1 does not demand a description for the nominal taxon, a description for the taxon is sufficient, this can include subordinate taxa.

For those who do not like to accept this interpretation of the Code, the next available author is Pfeiffer, 1841 (in L. Pfeiffer 1841: 16, *Helix cretica* with reference to "Fér. 288") (they would also need to verify whether Art. 11.6.1 can be applied under any interpretation of their choice, since Férussac's name could be interpreted as a synonym).

Greece: Gávdos island S of Crete

Greece, W Turkey, W Cyprus

3-15 x 4-23 mm, locally extremely variable

At some locations all shells may remain very small, smooth gradients over few kilometers may lead to populations with extremely big shells. In Koufonísi S of E Crete the diameter of the largest shells in some areas is at 8 mm, at 1 km distance at 15 mm. In Chrisí S of Crete the largest shells in the southeast have a diameter of 7 mm, at 2-3 km distance in the west 18 mm. In Gávdos (local differences between 13 and 20 mm at close distances) and Ródos (10-20 mm at larger distances) this is less well pronounced.

Greece: S central Crete, Pitsídia

Shell basically dirty whitish, often with brown spots, more rarely with uninterrupted narrow bands, weakly striated, 3.5-6 not very convex whorls, first whorls keeled or angulated, last whorl evenly rounded, often with one or several weak lips inside, margin sharp, umbilicus open, circular, deep and moderately wide, 1/7 of shell diameter. In some populations in coastal habitats shells remain very small, with only few keeled whorls, they look like a completely different species.

Juvenile shells are asymmetric, steeply moving from the keel to the umbilicus - the umbilical area looks in small shells as if it was elevated when compared with that of smaller hygromiid species.

Can easily be confounded with *Cernuella virgata*, which has much more frequently uninterrupted brown colour bands, its umbilicus is slightly less regularly circular and not as deep.

The animal is light yellowish or with some reddish hue.

Dry vegetation, often in coastal regions, cultivated areas. In Crete in practically every kind of habitats, from beaches to high mountains, very frequent.

The animal can reach maturity at shell sizes of few mm.

Westerlund 1889: 203, Falkner 1990: 216, Frank 1997: 52, Welter-Schultes 1998: 74, Hausdorf & S. 2009: 380.

The Cretan people used to consume large specimens of this umbilicated species, usually as a snack to drinks, but it has never been sold commercially in large quantities.

Greece: Gávdos island S of Crete Vollrath Wiese

Xerocrassa derogata (Rossmässler, 1854)

Helix derogata Rossmässler, 1854 in Rossmässler 1854-1859 (13-14): 27, Taf. 67
Rossmässler 1854: 27 gave a reference to "*H. derogata* Rossm. in Zeitschr. f. Malak. 1853. p. 145.", which was incorrect (*Helix subrogata* was published there). It is obvious that Rossmässler had based his description of *Helix derogata* on new material (Falkner 1996: 101), and referred to the species described as *Helix subrogata* var. β by Pfeiffer 1853. The name *derogata* is not an emendation (because *subrogata* was not mentioned by Rossmässler 1854). It cannot be doubtlessly interpreted as an incorrect subsequent spelling for *subrogata*. Even if possibly unintended, it was established as a regular new name.
Falkner 1996: 102 argued that Pfeiffer, 1859 must be the author of *Helix derogata*, for first having clarified the synonymy of the involved names (Pfeiffer 1859: 145). This is not tenable. Pfeiffer 1859 clearly used Rossmässler's 1854 name, with a bibliographical reference. A name must always be attributed to the first source where the criteria of availability were met. A deliberate use of a misspelling remains a misspelling.

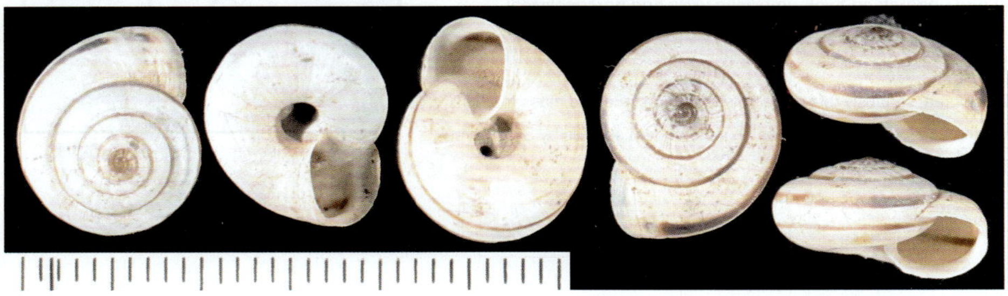

3.5-4.5 x 8-10 mm

Spain: Alicante, Gata de Gorgos

Spain (Alicante, Murcia)

Shell depressed, usually finely striated, white with one well delimited dark brown colour band at the upper side and 2-4 bands at the lower side, periphery well rounded, aperture with a weak white lip inside, umbilicus wide and open (1/5-1/4 of shell diameter).

Mediterranean shrublands and pine and oak forests, under stones and tree trunks, in lowlands a common species, also in cultivated areas, up to 1000 m. Active and abundant at rainy days.

Westerlund 1889: 282, Martínez-Ortí & Robles 2003: 132.

See also *Xcr. subrogata*, a closely related species, or possibly conspecific.
History of names used for this species: Pfeiffer 1853: *Helix subrogata* var. ß. Rossmässler 1854: *H. derogata*. Pfeiffer 1859: *H. derogata*. Westerlund 1889: *H. derogata*. Hesse 1920: *H. subrogata*. Falkner 1996: *X. derogata*.

Xerocrassa diensis (Maltzan, 1883)

Helix diensis Maltzan, 1883 in Maltzan 1883: 104 (*Helix (Candidula?) Diensis*)

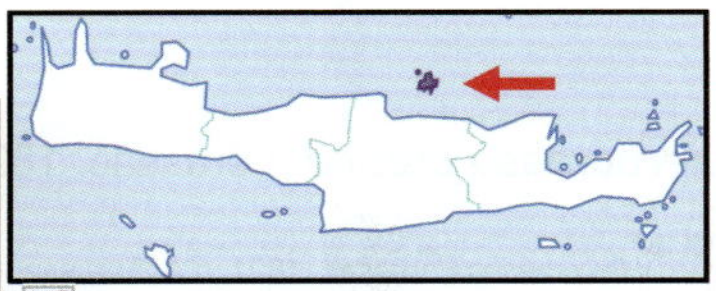

6-8 x 10-13 mm

Greece: Día N of Crete

Greece (island of Día N of Crete)

Rocky limestone habitats with scarce vegetation. A frequent species.

Schultes & Wiese 1990.

Not in Hausdorf & Sauer 2009, this was because this work did not cover the small islands (Hausdorf, e-mail, 2010).

Shell whitish with brown spots or interrupted bands, striated, 4-5 very weakly convex whorls, last whorl initially angulated, near aperture rounded and not descending, aperture with white lip inside, umbilicus very narrow, often slightly excentrical, partly covered by the columellar margin.

Xerocrassa ebusitana (Hidalgo, 1869)

Helix ebusitana Hidalgo, 1869 in Hidalgo 1869: 19, Pl. 2 (*Helix Ebusitana*)

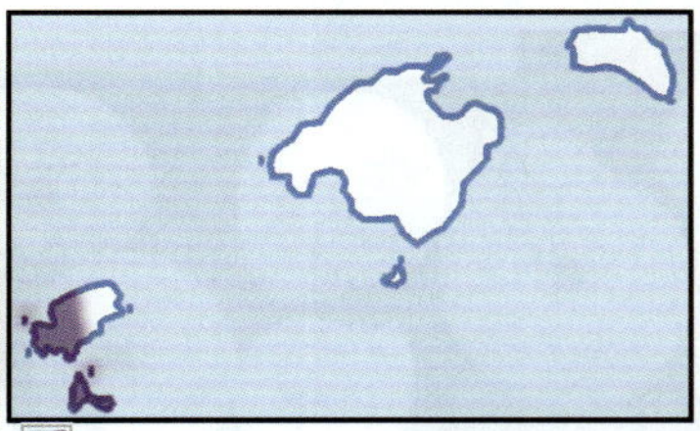

7-8.5 x 14-16 mm

Spain: Illes Balears, (1-3) Formentera, Cap de Barbaria, (4-6) "Bleda Plana"
(1-3) José Liétor Gallego, (4-6) Paolo Zaccaria
(4-6) Tiroler Landesmuseen Innsbruck

Spain (Ibiza island group)

EU Red List: Near Threatened NT. Threatened by fire and habitat loss caused by urban pressure originated by tourism.

Westerlund 1889: 406, Schröder 1984, Puente et al. in Verdú & Galante 2006: 386, Beckmann 2007: 117.

Taxonomy requires more research.

Shell white or horny coloured with brown colour bands or dashes, with brown ziczac patterns in the centre, densely striated, whorls not very convex, near the umbilicus smooth, umbilicus regularly open and deep, aperture almost circular, margin sharp, with a white lip deeply inside, columellar margin may cover a part of the umbilicus. Shell variable, many subspecies were described.

In various kinds of habitats, crop margins, uncultivated fields, rocky beaches, usually with scarce vegetation.

Xerocrassa edmundi Martínez-Ortí, 2006

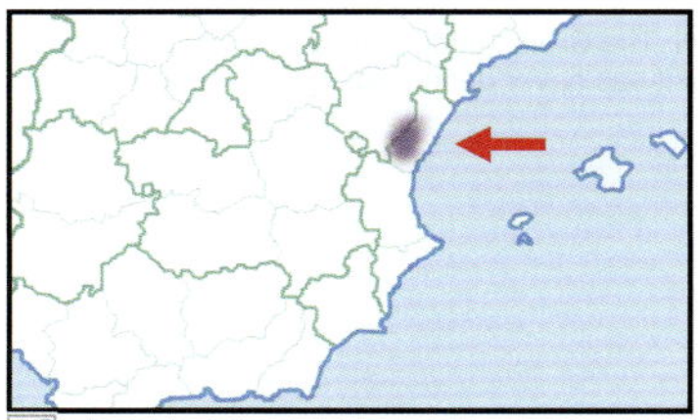

Xerocrassa edmundi Martínez-Ortí, 2006 in Martínez-Ortí 2006: 62

Shell small, light brown, slightly dashed, with short hairs, aperture thin and breaks quickly, peristome interrupted, umbilicus narrow (0.6-1.4 mm).

3.2-4.0 x 6.2-7.1 mm

Under small stones in dry habitats, Mediterranean vegetation with shrubs and pine trees, on non-calcareous substrate, among herbal vegetation in cork oak (*Quercus suber*) forests.

EU Red List: Endangered EN. Restricted to Serra d'Espadà, threatened by fire and human disturbance.

E Spain

Xerocrassa ferreri (Jaeckel, 1952)

Helicella ferreri Jaeckel, 1952 in Jaeckel 1952: 141

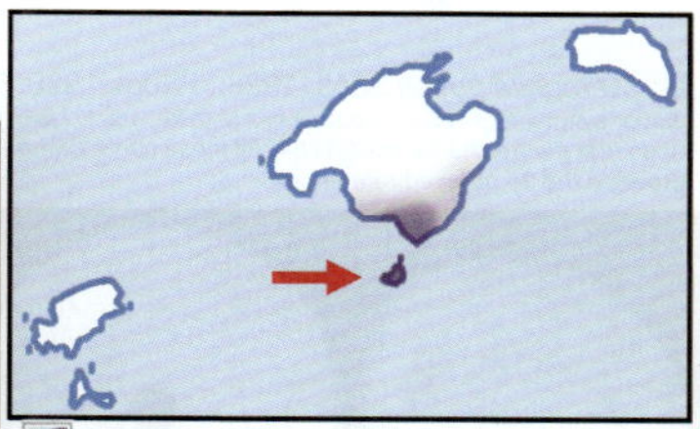

4-6.5 x 7-10 mm

Spain: Mallorca, Cabo Salinas

Spain (S Mallorca, small islands)

In coastal vicinity in pine forests with some bottom vegetation. Lives at some locations syntopically with *Xcr. homeyeri*.

Graack 2005: 6, Beckmann 2007.

Shell light to brownish with two usually interrupted brown bands on upper side, sometimes some more narrower bands at lower side, periphery rounded, 5-5.5 moderately convex whorls, last whorl not or only weakly descending, aperture very oblique with a weak lip inside at the basal margin, margin sharp, slightly reflected only in basal section, umbilicus very narrow.
Very much like *Xcr. frater*, slightly higher and more globulous, last whorl more rounded, no trend to be keeled.

EU Red List: Near Threatened NT. Restricted to a small range, potentially threatened by fire and human pressure such as illegal constructions of residential areas.

Xerocrassa franciscoi Hausdorf & Sauer, 2009

Xerocrassa franciscoi Hausdorf & Sauer, 2009 in Hausdorf & Sauer 2009: 385

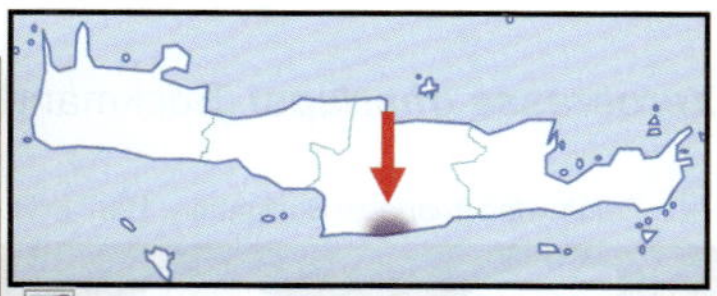

4-6.5 x 9-12 mm

Greece: S central Crete, 8 km E Léndas

Rocky limestone habitats in scarce shrubland vegetation of under pine trees, at southern mountain slopes.

Restricted to a small W-E extending range less than 6 x 2 km in a remote region. No acute threats known.

Greece (S central Crete)

Shell greyish, occasionally with blurry brown spots, densely and regularly ribbed, 4-4.5 whorls, keeled, keel at last whorl prominently protruding, last whorl often descending near aperture, aperture with whitish lip inside, umbilicus wide, about 1/7 of diameter, often widened at last 1/2 whorl and excentric.

Xerocrassa frater (Dohrn & Heynemann, 1862)

Helix frater Dohrn & Heynemann, 1862 in Dohrn & Heynemann 1862: 103

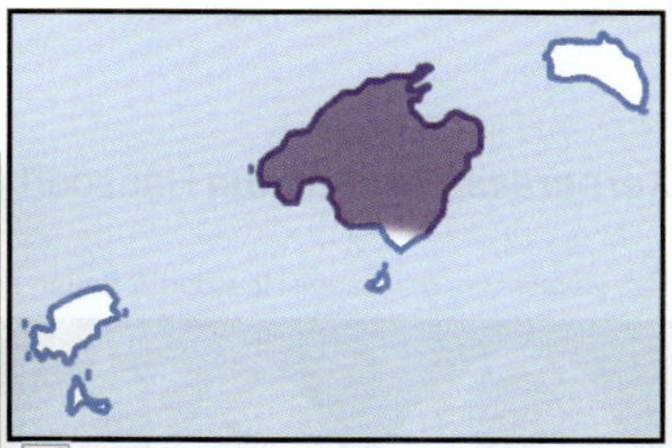

Spain (Mallorca)

3-7 x 7-15 mm

Spain: Mallorca, (1-2) near Porto Cristo, (3-6) Peguera 15 km W Palma
(1-2) Paolo Zaccaria   Tiroler Landesmuseen Innsbruck

Spain: W Mallorca, Calvià, Bendinat
Roy Anderson

Shell whitish or horny brownish, often with more or less interrupted brown colour bands, 5-6 whorls, last whorl not much wider than penultimate whorl, either keeled or at least initially angulated or bluntly keeled, aperture in most populations rounded, with prominent lip inside, umbilicus very variable, either wide or very narrow. Shell locally very variable, from keeled and very flat to highly spired with very narrow umbilicus.
X. ferreri has a regularly rounded periphery and shows no trend to be keeled.

Open habitats in Mediterranean shrubland vegetation, also in coniferous forests, usually under stones and wood.

Westerlund 1889: 329, Graack 2005: 8, Beckmann 2007.

Various subspecies have been described, of questionable taxonomic value. It would be necessary to plot the spatial distribution of various character expressions to see the local trends, this was not done by Graack 2005.
We could be dealing with local gradients of various conchological characters running in various directions over area, in other words, without partial barriers to gene flow along narrow hybrid zones between separate units that would be classified as subspecies.

Xerocrassa geyeri (Soós, 1926)

Xerophila geyeri Soós, 1926 in Soós 1926: 98
Junior synonym of *Helix arceuthophila* Mabille, 1881, *Helix ycaunica* Mabille, 1881, *Helix vicianica* Bourguignat in Locard, 1881, *Helix deana* Berthier, 1884 and *Helix pleurestha* Berthier, 1884. All these names were not used after 1899 and were suppressed under Art. 23.9 by Falkner et al. 2002: 146 who provided the 25 required references.

3.5-5 x 5-8 mm

Germany: near Fulda

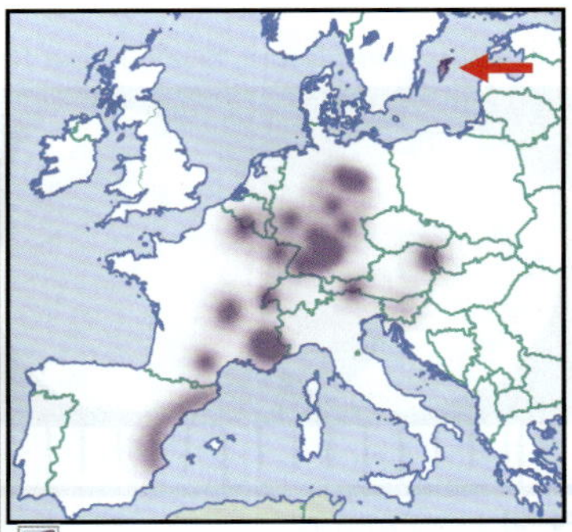

Spain to central Europe
Indicated range in Spain uncertain.

Shell whitish, often with dark and interrupted colour bands, coarsely striated, also at lower side, apex 1.0 mm wide, 4-4.5 weakly convex whorls with shallow suture, relatively flat above but well rounded at periphery, last whorl not descending near aperture, aperture rounded and without or with weak lip inside, umbilicus relatively narrow and sometimes getting slightly wider at last whorl, 1/8-1/7 of shell diameter.
Very much like *Candidula intersecta*, but less flattened above and below and more coarsely ribbed, with slightly deeper suture and without lip inside aperture. Differs from *Helicopsis striata* in a smaller protoconch (diameter 1.5 mm, in *H. striata*), ribs less interrupted, umbilicus slightly wider.
Anatomical difference: in *Xcr. geyeri* 2 rudimentary dart sacs, in *H. striata* 4 active dart sacs.

Dry and open calcareous habitats with short vegetation, rocks or rock rubble, usually in natural and undistubed habitats.

Not very frequent, scattered occurrences, always local. Critically Endangered in Germany, Endangered in Switzerland. Extinct since the early Postglacial period in S England, used to live in grasslands there.

The published record is difficult to interpret because the species has frequently been misidentified.

Xerocrassa gharlapsi (Beckmann, 1987)

Trochoidea gharlapsi Beckmann, 1987 in Beckmann 1987: 26

4.6-4.9 x 12-14 mm

Malta: Ghar Lapsi

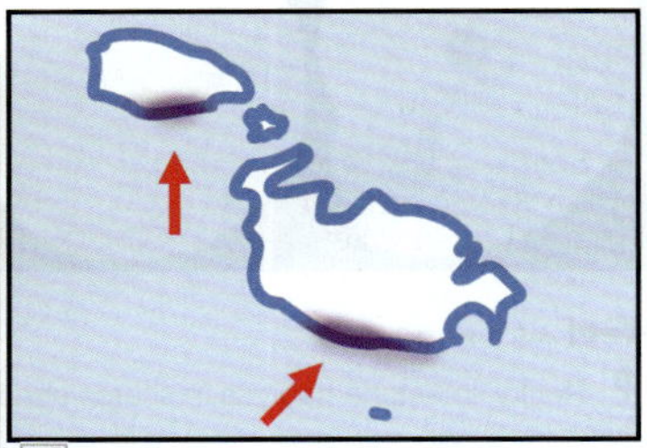

Malta

EU Red List: Vulnerable VU. Restricted to very few localities, threatened by activities related to illegal bird hunting, a major problem in Malta.

Falkner 1990: 216, Giusti, Mang. & S. 1995: 430.

Shell whitish or greyish-yellow with brown bands or spots on the first whorls, prominently striated, more weakly on lower side, shell very flat with prominent keel, umbilicus very wide, 1/4 of diameter.
Animal yellowish-grey, mantle border blackish.

Steep-sided rdum of S coasts of Malta and Gozo, in deep cracks and crevices of limestone rocks in typical Mediterranean shrubland vegetation.

Xerocrassa grabusana Hausdorf & Sauer, 2009

Xerocrassa grabusana Hausdorf & Sauer, 2009 in Hausdorf & Sauer 2009: 386

5-6.5 x 7-10 mm

Greece: W Crete, Kísamos, near Kaliviani

Greece (W Crete)

Rocky limestone habitats in open Mediterranean shrubland vegetation. Restricted to a small range (5 x 15 km), but relatively frequent in this area.

Shell whitish, occasionally with brownish spots or bands, moderately densely ribbed, 4.5-5 convex whorls, keeled, also weakly keeled at last whorl and aperture, last whorl sometimes descending near aperture, aperture with whitish lip inside, umbilicus moderately narrow, 1/10 of shell diameter, weakly obscured by the reflected columellar margin.
Differs from *Xcr. subvariegata* in its more prominent keel and coarser ribs.
Penial papilla with cylindrical basal part and dilated apical part, and subterminal opening.

Xerocrassa grata (Haas, 1924)

Helicopsis murcica grata Haas, 1924 in Haas 1924: 142, Taf. 8

Shell like *X. murcica* but with more whorls, more strongly keeled, last whorl less flattened at lower side.

4-5.5 x 8-11 mm

In mountains, where the species occurs in colonies, not very common.

Taxonomic status needs to be clarified.

NE Spain

Xerocrassa heraklea Hausdorf & Sauer, 2009

Xerocrassa heraklea Hausdorf & Sauer, 2009 in Hausdorf & Sauer 2009: 388

4-5.5 x 5.5-8 mm

Greece: N central Crete, near Limin Chersonisou

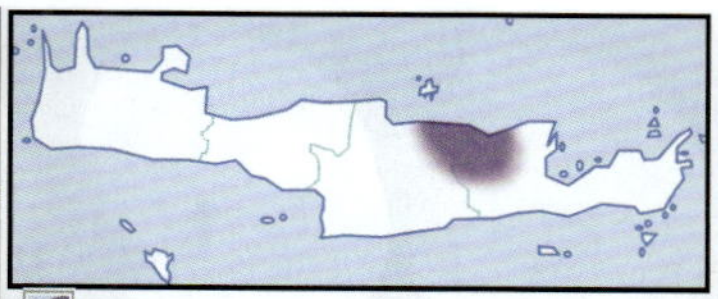

Greece (N central Crete)

Rocky limestone habitats with some vegetation, usually in Mediterranean shrublands.

Shell whitish, occasionally with brown spots or bands, coarsely striated or ribbed, conical, 4.5-5 convex whorls, initially keeled, keel gets weaker or disappears at last whorl, last whorl at least slightly descending near aperture, aperture with whitish lip inside, margin sharp and not reflected, umbilicus narrow or very narrow, often partly covered by the reflected columellar margin.
The shell cannot be distinguished from several other *Xerocrassa* species in Crete, this needs anatomical dissection.

Xerocrassa homeyeri (Dohrn & Heynemann, 1862)

Helix homeyeri Dohrn & Heynemann, 1862 in Dohrn & Heynemann 1862: 103 (*Helix Homeyeri*)

3.5-4 x 8-10 mm

Spain: Mallorca, Cala Mesquida near Capdepera

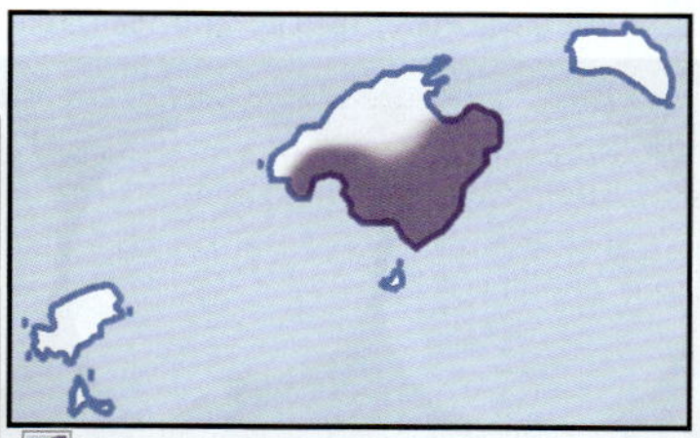

Spain (Mallorca)

Spain: SE Mallorca, Cap Blanc
Roy Anderson

Shell bluish or yellowish white, flat on the upper side, with two rows of brown spots on the upper side and two reddish brown colour bands on the lower side, densely striated, last whorl broader than precedent whorls, with a sharp and serrated keel, lower side irregularly convex, aperture inside with a strong white lip, umbilicus wide and perspectivical, 1/4 of diameter.
Keeled populations of *Xcr. frater* are more coarsely ribbed, their umbilicus is narrower.

In coniferous forests with some bottom vegetation, also in more open habitats in shrubland vegetation.

Westerlund 1889: 329, Graack 2005: 31, Beckmann 2007: 101.

Xerocrassa jimenensis (Puente & Arrébola, 1996)

Trochoidea jimenensis Puente & Arrébola, 1996 in Puente & Arrébola 1996: 58

Shell shape variable (flat to almost globulous), solid or semi transparent, with variable colours or whitish, finely striated, with short hairs (0.24-0.27 mm), lower side convex, umbilicus large (0.9-1.7 mm) and excentrical.
Differences to *X. zaharensis* mainly in the genitalia.

3.5-6 x 6-9 mm

Karstic and ruderal open habitats without much vegetation.

EU Red List: Near Threatened NT. No real threats known.

José Liétor Gallego S Spain: near Jaén

S Spain

Xerocrassa kydonia Hausdorf & Sauer, 2009

Xerocrassa kydonia Hausdorf & Sauer, 2009 in Hausdorf & Sauer 2009: 389

4-5.5 x 5.5-8 mm

Greece: W Crete, Koutsoumados

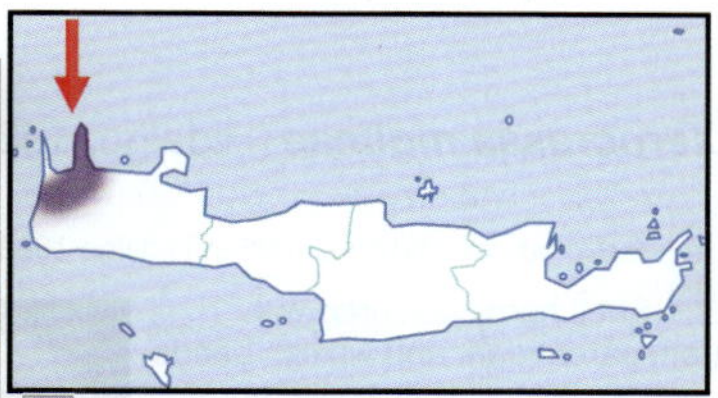

Greece (W Crete)

Rocky habitats in Mediterranean shrubland vegetation, restricted to limestone substrate.

Shell whitish, occasionally with brown spots or bands, coarsely striated or ribbed, conical, 4-5 convex whorls, initially keeled, keel gets weaker or disappears at last whorl, last whorl at least slightly descending near aperture, aperture with whitish lip inside, margin sharp and not reflected, umbilicus narrow or very narrow, often partly covered by the reflected columellar margin. Shell cannot be distinguished from several other *Xerocrassa* species in Crete. Differs from neighbouring *Xcr. mesostena* in its usually smaller and more conical shell with more convex whorls.
Flagellum shorter than in *Xcr. mesostena*, proximal epiphallus : flagellum ratio >= 3.0.

Xerocrassa lasithiensis Hausdorf & Sauer, 2009

Xerocrassa lasithiensis Hausdorf & Sauer, 2009 in Hausdorf & Sauer 2009: 391

4.5-6 x 6.5-8.5 mm

Greece: E central Crete, near Milatos N of Neápoli

Greece (E central Crete)

Rocky limestone habitats in scarce Mediterranean shrubland vegetation, also at old stone walls.

Shell whitish, occasionally with brown spots or bands, coarsely striated or ribbed, conical, 4.5-5 convex whorls, initially keeled, keel gets weaker or disappears at last whorl, last whorl at least slightly descending near aperture, aperture with whitish lip inside, margin sharp and not reflected, umbilicus narrow or very narrow, often partly covered by the reflected columellar margin.
The shell cannot be distinguished from several other *Xerocrassa* species in Crete, needs anatomical dissection.

Xerocrassa meda (Porro, 1840)

Helix meda Porro, 1840 in Porro 1840: 106 (*Helix Meda*) Type locality in Sassari, Sardinia.

6-8 x 9-12 mm

Italy: Sicily, Palermo

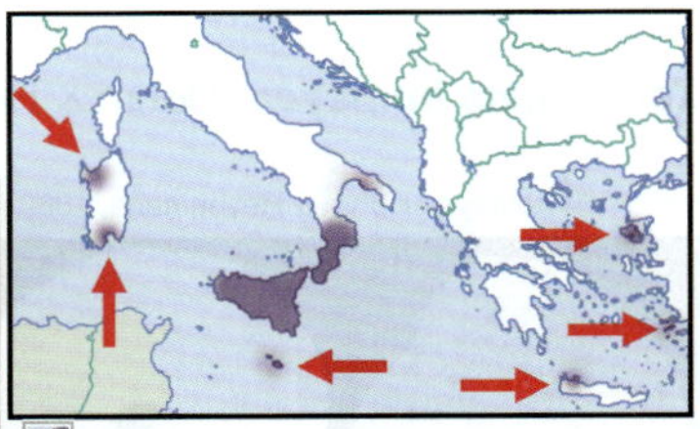

Malta, S Italy, dispersed to Greece

Malta: Santa Venera David Cilia

Shell horny greyish or yellowish white, sometimes with light brownish spots or dashes, finely striated on both sides, lower side less convex than upper side, 4.5-5 slightly convex whorls, last whorl initially angulated, often weakly descending near aperture, aperture inside reddish brown, with white or reddish yellow lip, umbilicus very narrow and partly or entirely covered by the reflected columellar margin.
Differs from *Cernuella* species in its almost closed umbilicus.

Restricted to anthropogenic habitats, only near Palermo in seminatural habitats. Often in gardens climbing shady walls, under low bushes, flowerpots and stones.

Westerlund 1889: 238, Giusti et al. 1995: 426, Manganelli et al. 1995: 30, Hausdorf & Sauer 2009: 393.

Originally perhaps Malta, Sicily and Calabria, introduced to Puglia, Sardinia, Kos, Lesvos, Chaniá in W Crete.

Xerocrassa mesostena (Westerlund, 1879)

Helix mesostena Westerlund, 1879 in Westerlund & Blanc 1879: 71, Pl. 3
Name and Latin diagnosis were attributed to Westerlund alone, the French text was by both co-authors. Under Art. 50.1 the name must be attributed to the co-author of the work who was made responsible for the name.

3-8 x 8-12 mm

Greece: W Crete, near Chrisoskalítisas

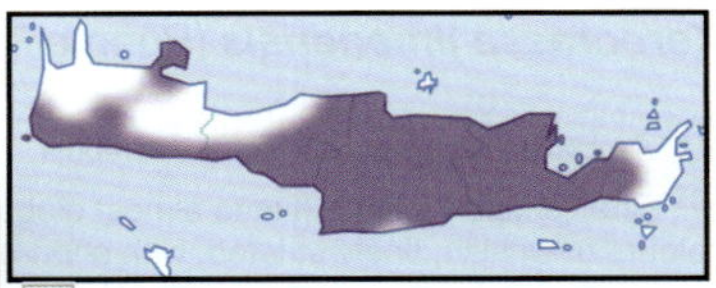

Greece (Crete)

Usually in open and rocky limestone habitats, occurs less frequently on non-calcareous substrate, present but not frequent in forests. Usually not as a synanthrope.

Hausdorf & Sauer 2009: 393.

Shell white with brown spots or more or less interrupted bands, or evenly white, apex often grey or blackish, striated, shape usually globulous, but can also be keeled in some regions, 4-6 (often weakly) convex whorls with slightly impressed suture, last whorl weakly descending near aperture, aperture with prominent white lip inside, umbilicus narrow or wide, open and partly covered by apertural margin, often conspicuously widened at last 1/2 whorl. Shell locally very variable.
Proximal epiphallus : flagellum ratio 0.9–2.3 (rarely up to 2.8), penial papilla with a broader basal and a narrower apical part with a terminal opening.

Xerocrassa molinae (Hidalgo, 1883)

Helix molinae Hidalgo, 1883 in Hidalgo 1883: 59, Pl. 2 (*Helix molinæ*)

Shell horny coloured with transparent dark and whitish bands and dashes (or whitish with horny transparent spots, or brown with white spots), finely striated, last whorl not always rounded at the periphery, umbilicus narrow, aperture inside with a thin lip.

Open and exposed habitats, very frequent under stones. High population density.

Westerlund 1889: 259, Gómez Moliner et al. 2001: 148, Martínez-Ortí & Robles 2003: 135, Puente et al. in Verdú & Galante 2006: 387.

5.5-6 x 10-11 mm

Spain: Columbretes, Illa Grossa

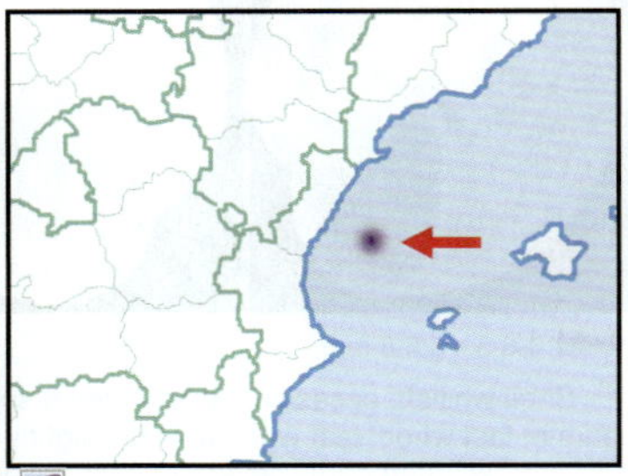

Spain (Illes Columbretes)

EU Red List: Near Threatened NT. Endemic species of Columbretes islands. Since the islands have been protected, the number of populations has increased.

Xerocrassa montserratensis (Hidalgo, 1870)

Helix montserratensis Hidalgo, 1870 in Hidalgo 1870: 298 (*Helix Montserratensis*)

7-8 x 12-14 mm

Spain: (1-2) San Gerónimo near Montserrat, (3-6) Montserrat mountains

Shell whitish to yellowish grey, sometimes with a dark row of dashes below the keel, densely and obliquely striated, keeled, upper side flat, lower side regularly convex, umbilicus narrow, apertural margin sharp, inside with a white lip.

EU Red List: Endangered EN. Only known from few localities, partly in suburban areas, one population has recently been destroyed. The remaining populations are threatened by fire and mountain biking.

NE Spain

In dry and open shrubland, at the stems of shrubs, above 700 m.

Westerlund 1889: 314, Puente et al. in Verdú & Galante 2006: 388.

Xerocrassa muehlfeldtiana (Rossmässler, 1837)

Helix candidula var. *muehlfeldtiana* Rossmässler, 1837 in Rossmässler 1835-1837 (5-6): 26 (*Helix candidula* Varietät *Mühlfeldtiana*)

Mentioned with a short description in the text (without figure) as *H[elix]. Mühlfeldtiana* and classified as "Varietät".

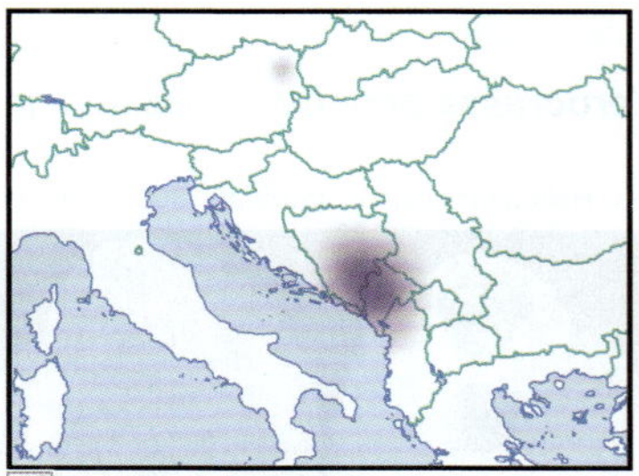

Shell white with 2 or 3 brown colour bands, coarsely striated or ribbed on both sides, fresh shels with short hairs, 4-5 convex whorls with deep suture, last whorl bluntly angulated, much more convex on lower than on upper side, aperture simple, margin sharp, umbilicus narrow.

3-4 x 5-7 mm

Westerlund 1889: 253.

Originally described from specimens collected near Wien, Austria.

Bosnia and Montenegro

Xerocrassa nicosiana (Gittenberger, 1991)

Trochoidea nicosiana Gittenberger, 1991 in Gittenberger 1991: 124

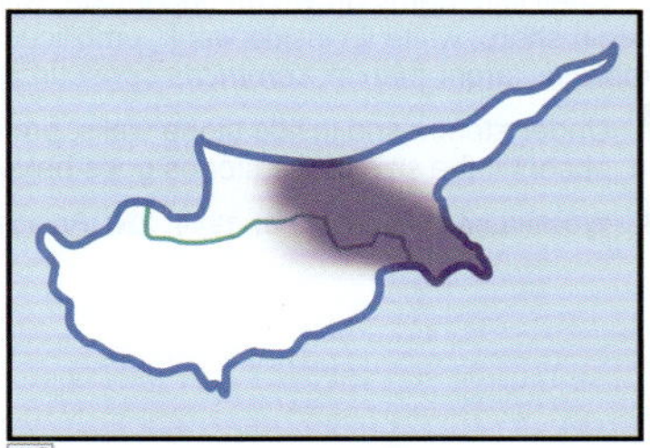

Shell whitish with light horny brownish and spirally arranged colour patterns, densely and irregularly ribbed, 4-6 flat whorls, juveniles angulated or keeled, last whorl often initially angulated, descending near aperture, aperture with prominent lip inside, umbilicus very narrow, 1/10 of shell diameter, often slightly excentric.
Umbilicus narrower and more excentric than in *Xcr. cretica*, slightly lower aperture, thicker internal lip, more glossy shell, slightly higher spire.

9-15 x 11-19 mm

Cyprus

Xerocrassa nyeli (Mittre, 1842)

Helix nyeli Mittre, 1842 in Mittre 1842: 190 (*Helix Nyeli*)

3-5.5 x 7-11.5 mm

Spain: Menorca, Mercadal

Shell whitish, greyish or yellowish, with two rows of brown spots on the upper side, brown colour bands on the lower side, finely striated, last whorl not broader than penultimate whorl, with a crenulated or serrated keel, aperture with a strong white lip inside, margin at the upper side protruded, umbilicus wide and perspectival, 1/8-1/6 of diameter.

Spain (Menorca)

In natural open shrubland habitats and in coniferous forests.

EU Red List: Near Threatened NT. Threatened by habitat fragmentation associated with human pressure.

Westerlund 1889: 329, Graack 2005: 41, Beckmann 2007.

Xerocrassa pallaresica (Fagot, 1886)

Helix pallaresica Fagot, 1886 [not verified]
I did not find more information than that of Westerlund 1889, who gave as source Estudio 1886 p. 4, author Fagot.
Helicella ribasica turolensis Ortiz de Zárate, 1963 in Ortiz de Zárate 1963: 94

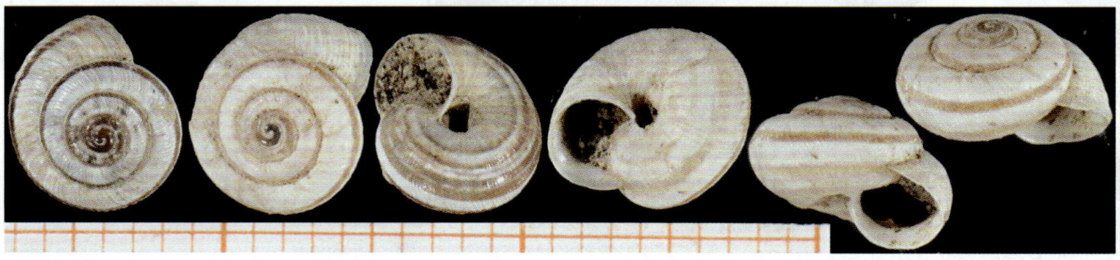

8-10 x 9-11 mm

Spain: Girona, Santa Maria de Besora

NE Spain, S France

Shell yellowish, not shiny, with some grey spots and with dense and oblique white streaks, upper side conoid, lower side convex, umbilicus narrow, aperture circular without lip.

Westerlund 1889: 281, Ortiz de Zárate 1963 (did not mention any diagnostic differences for *turolensis*, only a smaller size), Bech 1988: 107 (central Lleida, dimensions very homogenously 4.5-5.4 x 6.6-8.8 mm, among herbal plants), Bertrand 2011 (France), Alba et al. 2011 (synonym of "*Helicella madritensis*", here in *Xerotricha*).

I have lost track of the taxonomy of the related species. Alba et al. 2011: 51 tried hard to summarize some of the chaos and used a name *Xerocrassa chiae* (Fagot, 1886). But for which species, where distributed? Very difficult to understand the current state of research.

Xerocrassa penchinati (Bourguignat, 1868)

Helix penchinati Bourguignat, 1868 in Bourguignat 1863-1868: 305, Pl. 42 (*Helix Penchinati*)

3-4 x 5-7 mm

Spain: Barcelona, Castellar des Valles

E Spain, Pyrenees

Shell light horny brown on upper side, with white dashes, lower side whitish, can have dark colour bands, regularly striated, 5 convex whorls, apertural margin sharp, white lip inside weak, umbilicus narrow, 1/8-1/7 of shell diameter.
Umbilicus larger than in *Xerotricha conspurcata* and *Microxeromagna armillata*.

Under stones and inside stone walls, preferably in highlands and mountains. In France mostly in heavily eroded shrubland between boulders and rock rubble, absent if the size of the stones goes below 2 cm. Occasionally also on non-calcareous substrate.

Westerlund 1889: 260, Fez 1944: 220, Prieto & Martín 1988, Clanzig & Bertrand 2000, Martínez-Ortí, Ap. & R. 2004, Martínez-Orti 2007.

Xerocrassa poecilodoma (Boettger, 1894)

Helix poecilodoma Boettger, 1894 in O. Boettger 1894: 7

Shell white or yellowish white with some rows of brown spots, solid and opaque, apex black, densely striated, 5-6 slowly increasing convex whorls, last whorl initially angulated, near aperture rounded and slightly descending, aperture with white lip inside, umbilicus moderate, 1/6 of shell diameter.

7.5-9.5 x 11-14 mm

In various more or less open and dry habitats, abundant.

Gittenberger & Goodfriend 1993: 113.

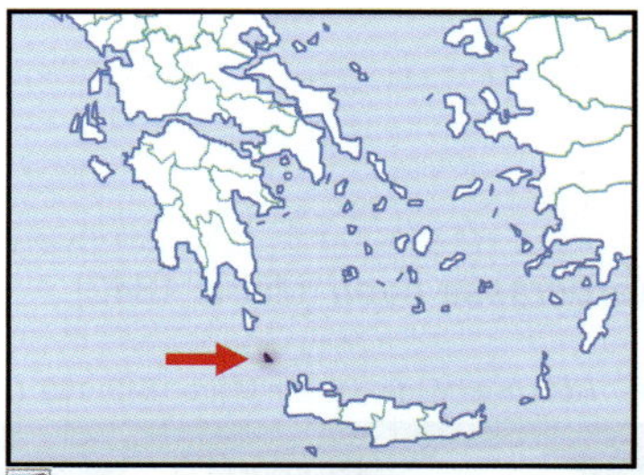

Greece (Andikithira)

Xerocrassa ponsi (Hidalgo, 1878)

Helix ponsi Hidalgo, 1878 in Hidalgo 1878: 232, Pl. 9 (*Helix Ponsi*)
Original type series contained material from Menorca and Cabrera, the Menorcan material belonged to *Xcr. nyeli.*

Shell yellowish brown with broad brown bands and dashes, ribbed, less prominently on lower side, strongly keeled, also near aperture, 4.5 weakly convex whorls, aperture with shiny white lip inside, umbilicus moderate. Differs from *X. ferreri* in its prominent and serrated keel at the periphery.

3.5-4.5 x 9-11 mm

Westerlund 1889: 330, Beckmann 2007: 105. Lives sympatrically with *X. ferreri.*

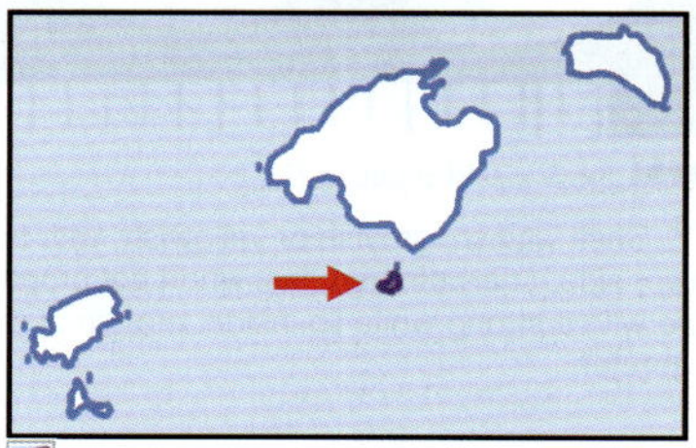

Spain (Cabrera S of Mallorca)

Xerocrassa prietoi (Hidalgo, 1878)

Helix prietoi Hidalgo, 1878 in Hidalgo 1878: 230, Pl. 9 (*Helix Prietoi*)
Helix moraguesi Kobelt, 1883 in Kobelt 1883: 114 (*Helix (Jacosta) Moraguesi*)

3-4 x 7-12 mm — Spain: W Mallorca, Peguera 15 km W Palma

(*moraguesi* form) — Spain: N Mallorca, near Cala de Sant Vicenç

In coniferous forests with some bottom vegetation, in the north also in mountain grasslands under stones.

Westerlund 1889: 330, Graack 2005: 37, 50, Beckmann 2007: 101, 106.

Graack 2005 and previous authors regarded *moraguesi* as an independent species, but it seems this is only an extremely prominently keeled form of *prietoi*, both live parapatrically (C. R. Altaba, unpublished, 2008).

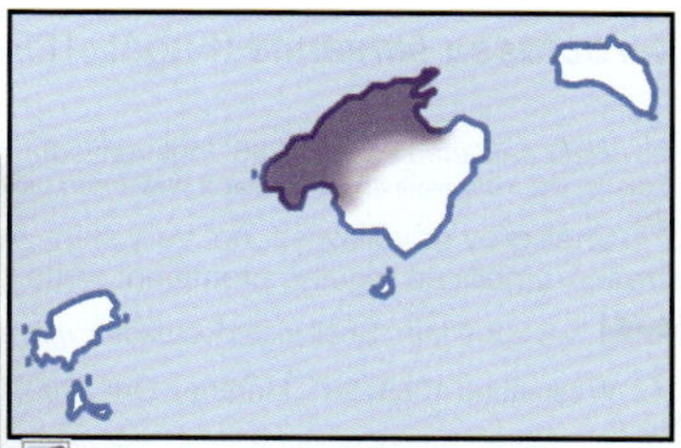

Spain (Mallorca)

Shell brownish white with rows of dark horny spots, upper side flat or slightly convex, densely and irregularly ribbed (ribs blunt or very prominent), usually with a prominent and serrated keel, 5-5.5 whorls, last whorl descending near aperture, umbilicus wide and perspectivical, 1/6-1/3 of shell diameter.
Shell shape, rib and keel prominence locally variable. In the north usually flatter on upper side (*moraguesi*), in the SW less prominently ribbed, more rounded and lower side more convex. In one population near Playa del Muro (extreme east) shells are more lenticular with weaker keel and higher spire, they look like keeled specimens of *Xcr. frater* but transitional zones to typical prietoi were found.

EU Red List: Near Threatened NT. Threatened by continuous habitat destructions and fragmentations related to human pressure.

Xerocrassa rhithymna Hausdorf & Sauer, 2009

Xerocrassa rhithymna Hausdorf & Sauer, 2009 in Hausdorf & Sauer 2009: 399
Name is almost certainly not in accordance with Appendix B 5 of the Code (name is not easily memorable). It is spelled correctly here: RHITHYMNA (this was copied from the original PDF source p. 399). Misspelled *rhitymna* by www.faunaeur.org (08.2012) and *rithymna* by www.iucnredlist.org (08.2012).

4.5-6 x 6.5-8.5 mm — Greece: central Crete, near Moní Asomáton

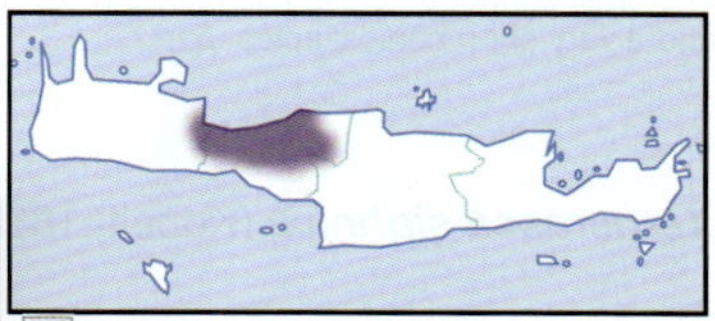

Greece (W central Crete)

Rocky limestone habitats with some vegetation, usually in Mediterranean shrublands.

Attention, the name is frequently misspeled, for example in Fauna Europaea and IUCN Red List web resources, also EOL. Correct is RHITHYMNA.

Shell whitish, occasionally with brown spots or bands, coarsely striated or ribbed, conical, 4.5-5.5 convex whorls, initially keeled, keel gets weaker or disappears at last whorl, last whorl at least slightly descending near aperture, aperture with whitish lip inside, margin sharp and not reflected, umbilicus narrow or very narrow, often partly covered by the reflected columellar margin.
The shell cannot be distinguished from several other *Xerocrassa* species in Crete, needs anatomical dissection.

Xerocrassa ripacurcica (Bofill, 1886)

Helix ripacurcica Bofill, 1886 in Bofill 1886: 151

5-6 x 12-14 mm — Spain: Lleida, Cognost de Terradets

NE Spain (Aragón, Cataluña)

Westerlund 189: 223, Martínez-Ortí & Robles 2003: 138.

Shell horny brown, shiny, relatively flat on upper side, with white spots and corrugated ribs, on lower side weaker striated and whitish, edge present at the periphery, suture deep, umbilicus wide and perspectivical (1/5 of shell diameter), aperture with a weak whitish or violet lip inside.

Pine forests or scarcely vegetated open habitats in mountains and valleys, in rock crevices or under stones and tree trunks, between 450 and 2150 m altitude. At rainy days on the vegetation, tree trunks, stones and rocks.

Xerocrassa roblesi (Martínez-Ortí, 2000)

Trochoidea roblesi Martínez-Ortí, 2000 in Martínez-Ortí 2000: 32 Type locality La Fonteta, Náquera, Valencia (destroyed).

Shell small, brownish with many light dashes, weakly ribbed and covered with small hairs, umbilicus narrow (1.2-1.5 mm), partly covered.

3.7-4.9 x 6.6-7.8 mm

Dry habitats, Mediterranean shrubs or pine forests, under decaying pine leaves and small stones, 260-560 m. Lives syntopically with *Xcr. subrogata*.

EU Red List: Vulnerable VU. Endemic species, threatened by human pressure, growing cities and fire. Only 3 populations with live animals were known in 2003. The type locality was destroyed by fire in 2004.

Martínez-Ortí & Robles 2003: 141, Martínez-Ortí & Puente in Verdú & Galante 2006: 389.

E Spain (Sierra Calderona)

Xerocrassa salvanae (Fagot, 1886)

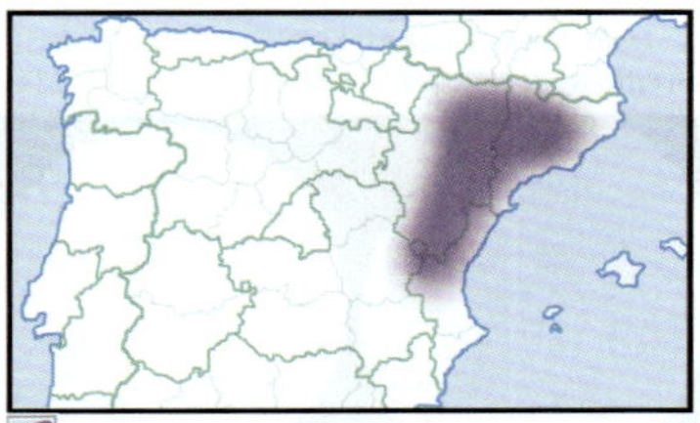

E Spain

Helix salvanae Fagot, 1886 [not verified]
I did not find more information than that of Westerlund 1889, who gave as source Estudio 1886 p. 5, author Fagot.

Shell grey or yellowish, not very shiny, with two blurry colour bands, dense and regular greyish radial streaks, umbilicus regular, prominent white lip inside the aperture.

5-6 x 8-9 mm. Shell size decreases with altitude in the Pyrenees, independently from temperature and rainfall.

Westerlund 1889: 281, Martínez-Ortí 2000, Uribe et al. 2007, Alba et al. 2011: 51 (synonym of "*Helicella madritensis*").

Taxonomic situation chaotic, see the comment under *Xcr. pallaresica*. Alba et al. 2011: 51 used a name *Xerocrassa chiae* (Fagot, 1886). But for which species, where distributed? Very difficult to understand the current state of research.

Xerocrassa siderensis (Maltzan, 1883)

Helix siderensis Maltzan, 1883 in Maltzan 1883: 104 (*Helix (Jacosta) Siderensis*)
Helix euphacodes Maltzan, 1883 in Maltzan 1883: 103 (First Reviser Hausdorf & Sauer 2009: 401)

5-7 x 9-13 mm

(1-2) Greece: E Crete, near Zákros, (3-6) Greece: E Crete, near Akra Síderos

Greece (E Crete)

Rocky limestone habitats in scarce Mediterranean shrubland vegetation.

Hausdorf & Sauer 2009: 401.

Could be conspecific with the parapatrically distributed *Xcr. amphiconus*.

Shell whitish, occasionally with brown spots or bands, striated or ribbed, 4.5-5.5 convex whorls, initially keeled, last whorl initally angulated and usually rounded near aperture, at least slightly descending near aperture, aperture with whitish lip inside, margin sharp and not reflected, umbilicus initially very narrow, often strongly widened at last whorl and not partly covered by the reflected columellar margin. Shell locally very variable.
Differs from *Xcr. amphiconus* exclusively by the rounded or angular but not keeled last whorl.
The shell cannot be distinguished from *Xcr. mesostena*. Proximal epiphallus : flagellum ratio higher (2.7–3.7), penial papilla with a long, terminally open basal and a very short conical apical part.

Xerocrassa siphnica (Kobelt, 1883)

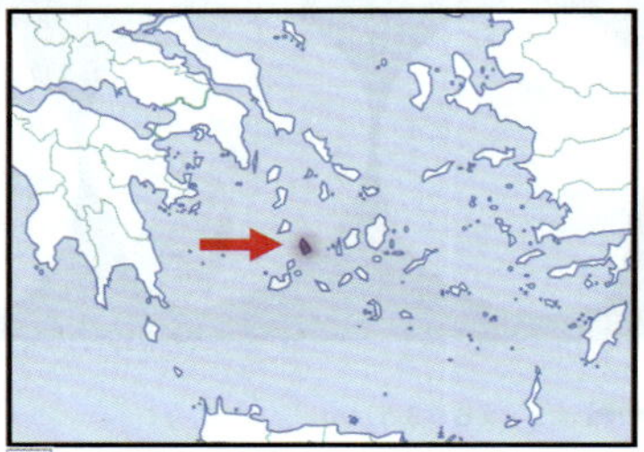

Greece (Sífnos)

Helix siphnica Kobelt, 1883 in Kobelt 1883: 183 (*Helix (Jacosta) Siphnica*)

Shell greyish white, sometimes with dark spots arranged in a band, finely ribbed, shape either globulous or almost lentiform, 4.5 whorls, periphery keeled or rounded, aperture with one or more white lips inside, umbilicus narrow.

6-7 x 11-13 mm

Keeled populations in dry and stony habitats, rounded populations in habitats with more vegetation.

No threats known, but the species seems to adapt to degradation of the forests and desertification of the island by the development of keeled forms.

Westerlund 1889: 333, Mylonas et al. 1995, Hausdorf & Sauer 2009: 375.

Xerocrassa subrogata (Pfeiffer, 1853)

Helix subrogata Pfeiffer, 1853 in L. Pfeiffer 1853: 145
Name attributed to Rossmässler in original source, but author of the description was Pfeiffer.
Pfeiffer recognized two variants α (striated form) and β (weakly striated or smooth form), the main form was implicitly regarded as variant α. This means that only subordinate variants α and β had a description, not the main form as such. Both variants are currently regarded as different species, the form β is currently known as *Xerocrassa derogata* (Rossmässler, 1854), the name *subrogata* is exclusively used for the variant α.
The name *Helix subrogata* was made available at this occasion under Art. 12.1 because it had a description for the taxon (Code Glossary "taxon": a taxon encompasses all included taxa of lower rank). Art. 12.1 does not demand a description for the nominal taxon, a description for the taxon is sufficient, this can include subordinate taxa.
Those who do not like to accept this interpretation of the Code, would need to look for the next available name.
Helix derogata var. *angulata* Rossmässler, 1854 in Rossmässler 1854-1859 (13-14): 27, Taf. 67
Not *Helix angulata* Rackett, 1821 from Canada.
Helix murcica Schmidt, 1854 in Schmidt 1854 [Dec 1854]: 237 (not *Helix lactea* var. *murcica* Rossmässler, 1854 in Rossmässler 1854-1859 (13-14) [Sep 1854]: 13 (= currently *Otala lactea* (Müller, 1774))
Established with bibliographical reference to Rossmässler 1854 (13-14): Fig. 833. Name originally attributed to Guirao. Schmidt proposed to use the name *Helix murcica* for a species, instead of for a variety classified as *Helix derogata* var. *angulata* by Rossmässler.
Falkner 1996: 101 interpreted *H. murcica* Schmidt as a new replacement name for *Helix derogata* var. *angulata* Rossmässler, 1854 non *Helix angulata* Rackett, 1821 from Canada. This is not tenable because there was no expressed statement (Art. 72.7) that (and why) Schmidt intended to substitute Rossmässler's name, and there was no visible intention. Schmidt cited the name *angulata* as used by Rossmässler, next to a second name *murcica*, "determined" for this variety by Guirao. Schmidt considered the name *murcica* as available, and probably thought that Guirao's name *murcica* had precedence. *Helix angulata* Rackett, 1821 was not mentioned.

3.2-3.9 x 8-9 mm

Spain: Alicante, near Orihuela

S Spain

Dry habitats under decaying pine leaves and small stones, among shrubs, in highlands and rocky habitats of lowlands.

Shell slightly depressed, dirty whitish with interrupted dark bands, finely and regularly ribbed, not shiny, 4-5 weakly convex whorls, last whorl keeled or at least initially angulated, descending near aperture, aperture rounded and usually with lip inside, umbilicus initially narrow, widened to 1/6 of diameter at last whorl.

Pfeiffer 1859: 146, Westerlund 1889: 282, Martínez-Ortí 2000, Martínez-Ortí et al. 2004. See also *Xcr. derogata*, a closely related species.

History of names used for this species: Pfeiffer 1853: *Helix subrogata* var. α. Rossmässler 1854: *H. derogata* var. *angulata*. Schmidt 1854: *H. murcica*. Pfeiffer 1859: *H. murcica*. Westerlund 1889: *H. murcica*. Hesse 1920: *H. angulata*. Falkner 1996: *X. subrogata*.

Xerocrassa subvariegata (Maltzan, 1883)

Helix subvariegata Maltzan, 1883 in Maltzan 1883: 105

Greece (W Crete)

Rocky limestone habitats, usually in Mediterranean shrubland vegetation, also in forests and gorges.

5-7 x 7-10 mm   Greece: W Crete, near Chaniá

Hausdorf & Sauer 2009: 403.

Shell whitish, occasionally with brownish spots or bands, moderately densely ribbed, 4-5 convex whorls, initially keeled, last whorl initially weakly angulated and rounded near aperture, sometimes slightly descending, aperture with white lip inside, umbilicus moderately narrow, 1/10-1/7 of shell diameter, slightly excentric, weakly obscured by the reflected columellar margin. Differs from *Xcr. grabusana* in its less prominent keel and denser ribs.
Penial papilla with a cylindrical basal and a dilated apical part with a subterminal opening, proximal epiphallus : flagellum ratio lower than in *Xcr. grabusana*.

Xerocrassa zaharensis (Puente & Arrébola, 1996)

S Spain

Trochoidea zaharensis Puente & Arrébola, 1996 in Puente & Arrébola 1996: 56

Shell slightly depressed, distinctly striated, rounded below, umbilicus wide and excentric, 1/8-1/6 of shell diameter.

5-5.5 x 8-9.5 mm

Under stones in a valley with caroub, olive and fruit trees.

EU Red List: Vulnerable VU. Only known from few localities, not abundant, sensitive to habitat change and fire.

Candidula

Candidula Kobelt, 1871 in Kobelt 1871: 22 *Helix (Xerophila (Candidula))*

Type species *Glischrus candidula* Studer, 1820 (absolute tautonymy)

NW Africa and Europe

One taxon was mentioned as a species by Pfeiffer 1870-1876: 43 and Westerlund 1889: 356 and might be assigned to this genus:
Helix verticillata Pfeiffer, 1871 in L. Pfeiffer 1871: 142 Type locality in Ródos, Greece.
I have not seen any recently published record. Mentioned by www.faunaeur.org (2008, 2011) as *Candidula verticillata*, without any reference, and a distribution record for "Dodecanese islands". Not listed by Frank 1997, which indicates that the species does probably not occur on Ródos. Perhaps the locality was confounded in the 1870s.

Candidula arganica (Servain, 1880)

Helix arganica Servain, 1880 in Servain 1880: 85 (*Helix Arganica*)

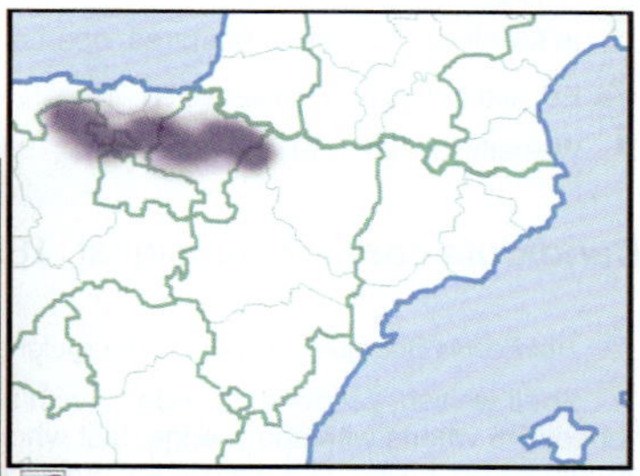

N Spain

2.5-3 x 4.5-5.5 mm   Spain: Navarra, Lecumberri

Westerlund 1889: 277, Ortiz de Zárate y López 1950: 27, Prieto & Martín 1988.

Shell greyish white with horny brown colour bands, shiny, finely striated, fresh shells with short hairs, 4.5-5 slightly convex whorls, last whorl not much wider than precedent whorls, shortly descending near aperture, aperture often quadrangular, with a sharp white lip inside, umbilicus wide and open.
Animal dark grey with white mantle, sometimes with black dots.

In valleys coming down from the Pyrenees southwards.

Candidula belemensis (Servain, 1880)

Portugal, possibly also Galicia

Helix belemensis Servain, 1880 in Servain 1880: 86 (*Helix Belemensis*)

Shell yellowish, very shiny, with a brown colour band, last whorl not much broader than the penultimate whorl, umbilicus narrow, apertural margin sharp with a white lip inside.
Differs from *C. intersecta* in its longer flagellum (1/2 of the length of the epiphallus).

4-5 x 8-9 mm

Threatened by the enormeous land use by eucalyptus plantations resulting in habitat destruction and fragmentation.

Westerlund 1889: 277, Gittenberger 1993: 284, Altaba 1997.

Candidula camporroblensis (Fez, 1944)

Helicella camporroblensis Fez, 1944 in Fez 1944: 214 (*Helicella (Candidula) Camporroblensis*)

Shell evenly dirty whitish, sometimes with a brown colour band above the suture, or spots, lower side with 3-4 blurry brownish bands, shell prominently and coarsely striated, almost ribbed, less prominent at lower side, 4-4.5 convex whorls, last whorl in some specimens slightly keeled, aperture with prominent and white internal lip, umbilicus wide and excentrical, 1/4 of shell diameter.

Was found in 1200 m altitude in highland plains covered with grassland, under stones.

Martínez-Ortí & Robles 2003: 32, Martínez-Ortí 2011: 9.

2.7-3.3 x 4-5 mm   Spain: Valencia, Camporrobles   Paolo Zaccaria   Tiroler Landesmuseen Innsbruck

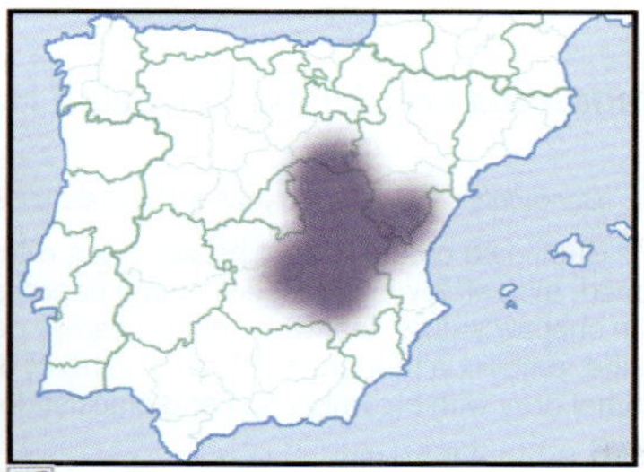

Spain

Candidula castriota Soós, 1924

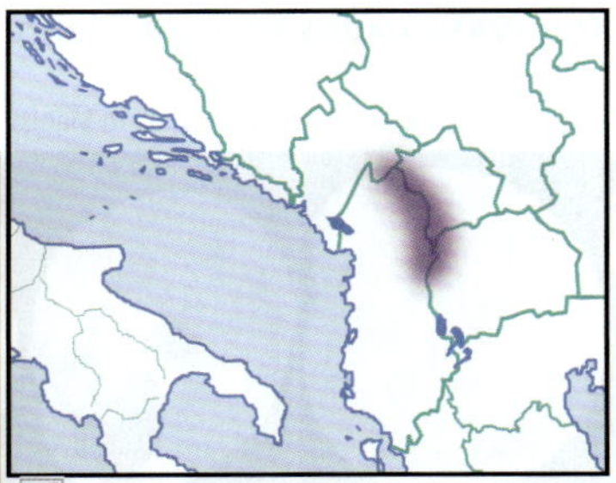

Candidula castriota Soós, 1924 in Soós 1924: 193 (*Candidula Castriota*)

4-5 x 6-7.5 mm — Albania: Mali i Korabit, 2530 m — Zoltán Fehér

NE Albania

EU Red List: Near Threatened NT. Possibly exaggerated, no threats known.

Shell whitish with horny brown apex and brown bands, more intensively brown on upper side, prominently striated, white colour patterns often along growth lines, 4-4.5 convex whorls with deep suture, last whorl not descending, aperture simple and without lip inside, umbilicus very wide, 1/4 of shell diameter.

High mountain limestone habitats in 1400-2500 m altitude, under stones and among leaf litter at the base of rocks.

Hausdorf 1991: 119, Dhora & Welter-Schultes 1996: 155, Fehér & Erőss 2009, www.iucnredlist.org (08.2012).

Information based on material of the Budapest collection. Rarely mentioned after 1924, the original figures of the genitalia suggest that the species belongs to *Candidula*.

Candidula cavannae (Paulucci, 1881)

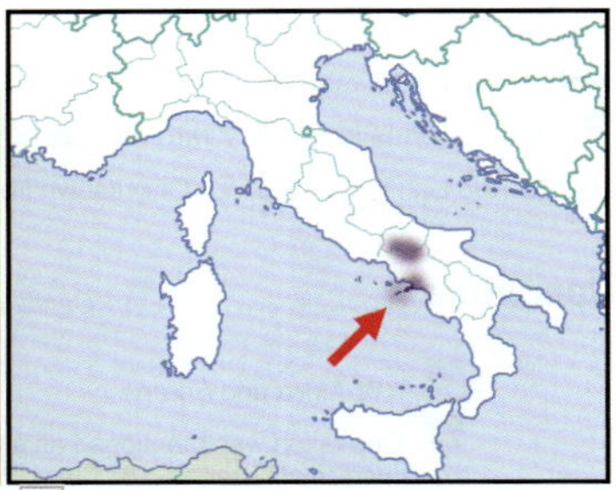

Helix cavannae Paulucci, 1881 in Paulucci 1881: 104, Tav. 2 (*Helix (Xerophila) Cavannæ*)
Expressly dedicated to Mr. Cavanna, name not to be corrected to *cavannai* (Art. 32.5). Art. 31.1 has no effective force.

4-5 x 6-7.5 mm — (1-3) Italy: Napoli, near Vico Equense, (4-8) Italy: L'Aquila, Monte Marsicano, 1950 m — Alessandro Hallgass

Italy (Molise, Campania)

Shell dirty whitish or grey, often with dark spots, densely striated, 5-6 weakly convex whorls with deep suture, last whorl keeled, aperture with weak lip inside, margin sharp, columellar margin slightly curved, umbilicus open.

In Monti del Matese in pastures, open and stony habitats in mountains, up to 2000 m. Near Sorrento and Capri in lower altitudes.

EU Red List: Near Threatened NT. Threatened by habitat destruction associated with increasingly intensive land use, also by tourist activities.

Westerlund 1889: 319, Manganelli et al. 1995: 28, A. Hallgass (e-mail, 2-2008). Range separated: Monti di Matese and Sorrentino peninsula with Capri island.

Candidula codia (Bourguignat, 1859)

S Portugal

Helix codia Bourguignat, 1859 in Bourguignat 1859: 523, Pl. 20

Shell whitish, yellowish or reddish with several interrupted colour bands, usually regularly and coarsely ribbed, 5-6 whorls with sharp edge, last whorl with blunt edge, suddenly descending near aperture, apertural margin sharp, inside with a white lip, umbilicus very narrow and partly covered by columellar margin of the aperture, 1/8-1/6 of diameter.

6-8 x 10-12 mm

Prefers calcareous mountain ridges distant from the coastlands, open and rocky shrublands, at walls, stone accumulations and under limestones.

Westerlund 1889: 258, Gittenberger 1993: 284, Kittel 2002: 5 (São Bartolomeu de Messines to near Moncarapacho).

Candidula corbellai Martínez-Ortí, 2011

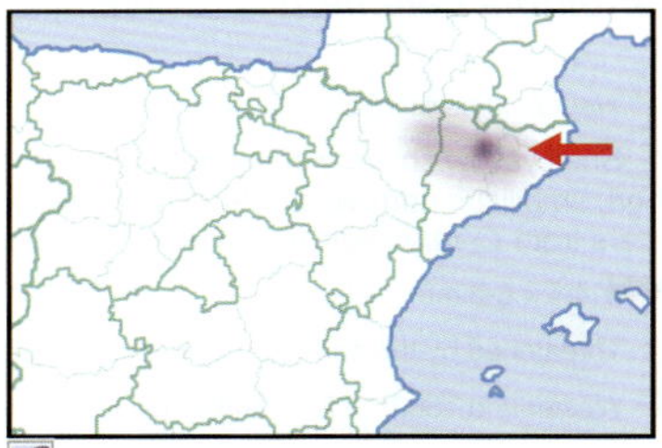

Spain (Lleida)

Candidula corbellai Martínez-Ortí, 2011 in Martínez-Ortí 2011: 2

Shell evenly pale grey, sometimes with a narrow brown band at the periphery, weakly striated, apex brownish, around 5 convex whorls with well-marked suture, last whorl weakly descending near aperture, aperture with white lip inside, also visible externally as a white structure, margin sharp, umbilicus initially narrow, increasing at last whorl, to 1/8-1/7 of shell diameter.
Most closely related to *C. camporroblensis* and *C. rocandioi*, from which it differs anatomically.

3.7-4.7 x 6-8 mm

High mountain grasslands on calcareous substrate, under stones or at the roots of the vegetation.

Only known from one locality, but could be more widespread.

Candidula coudensis Holyoak & Holyoak, 2010

Central Portugal (Leiria)

Only known from a small range less than 1 km in total extent.

Candidula coudensis Holyoak & Holyoak, 2010 in Holyoak & Holyoak 2010: 68

Shell pale cream to light brown, often with diffuse and interrupted brown bands, prominently and regularly ribbed, also on lower side, shell shape depressed with prominent keel at the periphery, 5-5.5 flattened whorls with shallow suture, last whorl not or weakly descending near aperture, apertural margin sharp and with white lip inside, umbilicus moderately wide, 1/7-1/5 of diameter.
Animal grey with brownish hue and blackish foot margins.

9-11 x 4.8-5.3 mm

Was found in an open rocky limestone habitat, among herbs, grasses and low shrubs in a flat and disturbed roadside area, where it lives sympatrically with *C. belemensis*.

Maturity is probably reached in winter, in January 30-70 eggs (diameter 1.0-1.3 mm) are laid in clusters.

Candidula fiorii (Alzona & Alzona Bisacchi, 1938)

Helicella fiorii Alzona & Alzona Bisacchi, 1938 in Alzona & Alzona Bisacchi 1938: 97 [not verified]

5.5-6.5 x 8-11 mm

Italy: Teramo, Gran Sasso, Campo Pericoli, 2350 m

Alessandro Hallgass

Italy (Gran Sasso)

Italy: Gran Sasso, Campo Pericoli, 2350 m

Alessandro Hallgass

Shell dirty whitish with brown interrupted bands, horny brown apex and often a white surface layer, coarsely striated, 4-5 convex whorls with deep suture, last whorl not descending, aperture simple, umbilicus wide, 1/5 of diameter. Differs from sympatrical *Can. spadae* in its slightly smaller dimensions, more intensive colours, blackish umbilicus, absence of an angulated periphery at the beginning of the last whorl, and a darker animal. Shells may superficially look like juvenile *Can. spadae*.

In high mountain habitats under stones and near rocky walls. Sympatrically occurring *Can. spadae* live in herbal vegetation and exposed stony habitats.

EU Red List: Vulnerable VU. Only known from two localities in high altitudes, affected by climate change (dryer conditions).

Manganelli et al. 1995: 28, A. Hallgass in www.naturamediterraneo.com (2007). Needs a revision.

Candidula gigaxii (Pfeiffer, 1847)

Helix caperata var. β synonym *Helix Gigaxii* Pfeiffer, 1847 in L. Pfeiffer 1848: 167

Mentioned with description, name attributed to Charpentier. Not mentioned as a junior synonym of an available name (as demanded by Art. 11.6), but as a synonym for a subordinate variant for which the Greek letter β was used. For those who neglect this (as proposed by Kadolsky 2012: 78, this is preliminarily accepted here), the name was made available under Art. 11.6.1 by Albers 1850: 77 who used the name as "*Helix Gigaxii*", with a bibliographical reference to Pfeiffer 1848: 167 ("No. 430"). For the others, the name must be attributed to Albers 1850.

5.4-8.1 x 7.4-12.2 mm

Germany: Niedersachsen, Alfeld

W Europe

Indicated range in Spain uncertain.

Shell light brown or sandy coloured with pale colour bands, usually flatter and larger than *C. intersecta*, whorls more rapidly increasing, umbilicus inside very narrow, remarkably wider at last whorl.

Near the soil in open habitats, at sunny places, often climbing up vegetation. Frequent as pioneer settlers in badland areas, in Britain mostly on chalk or limestone, also among taller green herbage, but usually not on sandhills.

Critically Endangered in Niedersachsen, Endangered in Germany. Probably introduced to Britain with weeds in the Roman times, today threatened by the retreat of traditional agriculture, seems to have disappeared from Ireland and large parts of Wales and central England.

Kerney et al. 1983: 246, Jungbluth et al. 1989: 190, Falkner 1990: 208, Manganelli et al. 1995: 28, Kerney 1999: 180, Martínez-Ortí, Ap. & R. 2004, Pfenninger et al. 2007.

Candidula grovesiana (Paulucci, 1881)

Helix grovesiana Paulucci, 1881 in Paulucci 1881: 106, Tav. 3 (*Helix (Xerophila) Grovesiana*)

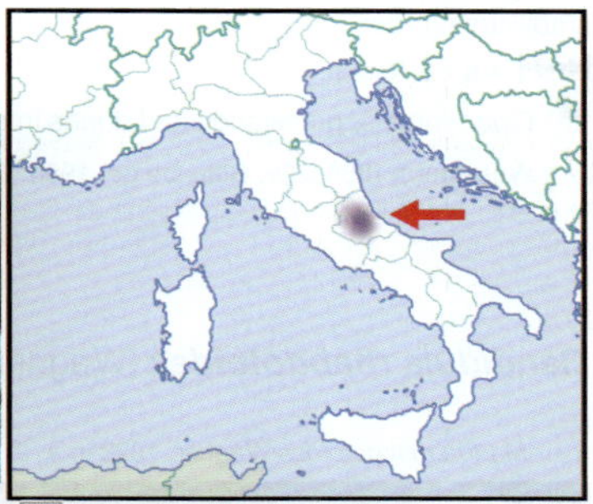

3-7 x 8-10 mm, 3-7 x 9-10 mm,
shell height very variable within a population, populations at lower altitudes always flat

Italy: L'Aquila, Monte Marsicano, 1900 m

Alessandro Hallgass

Italy (Abruzzo)

Shell white, sometimes with horny brown spots, densely and regularly striated, upper side conical, lower side rounded, 5 whorls, last whorl finely keeled with serrated keel (in some populations with rounded periphery), not much wider than penultimate whorl, aperture inside with shiny white lip, margin sharp, umbilicus open, increasing irregularly.

Usually in pastures and grassy mountain plains at 1700-2000 m, but occasionally also in some isolated spots in rocky habitats around 700-900 m where it lives in rock crevices.

EU Red List: Endangered EN. Always local and in isolated occurrences. Populations have declined in the past 20 years with increasingly dryer climatic conditions.

Westerlund 1889: 325, Manganelli et al. 1995: 28, A. Hallgass (e-mail, 01.2012).

Range from Maiella to Abruzzo National Park and border with Lazio, at 1700/1950 m. This complex comprises several forms. One form at Maiella high plain (2600-2750 m), anatomy like *grovesiana*, shell almost without keel. Form of Monte Marsicano (1950-2200 m), shell like in Maiella, but anatomy slightly different from *grovesiana*. The anatomically "typical" *grovesiana* form lives on the same mountain, but separated by a steep wall of 50 m height, the *govesiana* form lives below, the other form above.

Candidula intersecta (Poiret, 1801)

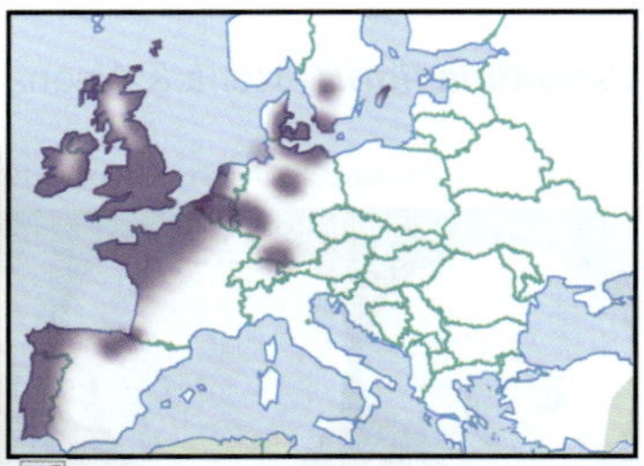

W Europe to Sweden

Eggs (diameter 1 mm) are laid between May and October.

Helix intersecta Poiret, 1801 in Poiret 1801: 81

4-7 x 7-12 mm

Germany: island of Helgoland in the North Sea

Shell whitish or yellowish with brown bands or spots, colour and patterns very variable, overall impression greyish brown to brown, finely striated, 5 convex whorls, aperture simple without lip, margin slightly reflected near umbilicus, umbilicus open and variably wide. Animal yellowish or bluish grey with dark brown pigments, upper tentacles long, lower tentacles very short.

Usually at the soil in dry and open habitats, under stones, at low plants, often in dunes. Occasionally it climbs trees at the edges of woods. Tolerates cultivation, often found in corn stubble.

Moquin-Tandon 1855: 241, Germain 1930: 274, Nobre 1941: 108, Kerney et al. 1983: 246, Prieto & M. 1988, Falkner 1990: 208, Ondina et al. 1997, Kerney 1999: 179.

Candidula lernaea Hausdorf, 1991

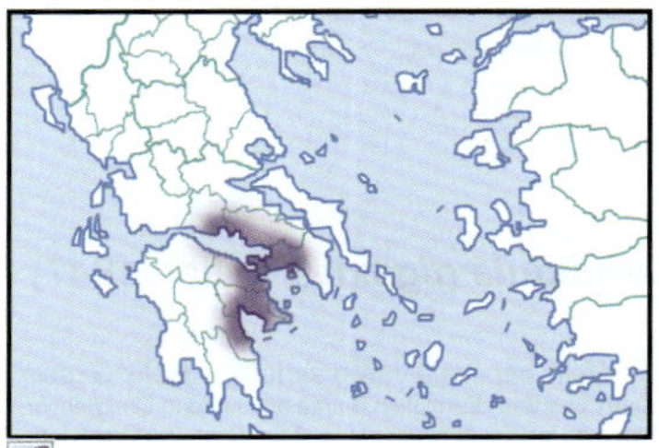

Greece

Candidula lernaea Hausdorf, 1991 in Hausdorf 1991: 124

Shell yellowish white with brown bands and spots, ribbed, less prominently on lower side, 4-5 convex whorls with moderately deep suture, last whorl usually with distict edge at periphery, weaker near aperture, rarely rounded, last whorl sometimes slightly descending near aperture, aperture usually with white lip inside, umbilicus open, 1/7-1/6 of shell diameter, widened at last whorl.
Usually smaller than *Helicopsis gittenbergeri* with same whorl numbers, more prominently striated, last whorl more angulated, aperture with lip inside.

2.5-3.5 x 4.5-6 mm

A lowland species, often in direct coastal vicinity.

Candidula najerensis (Ortiz de Zárate y López, 1950)

N Spain (Navarra, La Rioja)

Helicella najerensis Ortiz de Zárate y López, 1950 in Ortiz de Zárate y López 1950: 29, Lám. 1

Shell horny brown with some dark spots, irregularly ribbed, 4.5 whorls, not keeled but with edge at periphery, aperture with prominent white lip inside, columellar margin slightly covering umbilicus, umbilicus wide, all whorls visible inside.
Animal dark brown, inferior tentacles darker than superior ones, foot lighter with a dark or black line at the margin.

2.6-3.5 x 4.5-6 mm

In valleys, not much above sea level.

Active in the winter season. Produces a solid, white and usually double epiphragm for summer estivation.

Prieto & Martín 1988.

Candidula olisippensis (Servain, 1880)

Portugal

Helix olisippensis Servain, 1880 in Servain 1880: 93 (*Helix Olisippensis*)

Shell yellowish with light spots near the suture, and with a red colour band interrupted by yellow spots, shiny and fragile, indistinctly and finely striated, 4.5-6 weakly convex whorls, last whorl not much wider than preceding whorl, with weak edge, weakly descending near aperture, aperture with white lip inside, margin sharp, umbilicus narrow.

4-6 x 6-10 mm

Open forests, grasslands and dunes in coastal vicinity, under stones, boulders and at tree trunks.

Westerlund 1889: 259, Gittenberger 1993: 285, Altaba 1997, Pfenninger et al. 2003.

Candidula rhabdotoides (Wagner, 1928)

Martha rhabdotoides Wagner, 1928 in A. J. Wagner 1928: 366

5-6 x 6.5-7 mm

Bulgaria: Osam river near Lovech

SE Europe

Mostly in lowlands below 1000 m, but can also live in mountains up to 2000 m.

Damjanov & Likharev 1975: 352, Hausdorf 1991: 120, Fehér & Erőss 2009: 15.

Shell yellowish white with brown dashes, usually prominently ribbed, whorls convex, apertural margin sharp, usually without lip inside, umbilicus 1/7-1/4 of shell diameter, widened at last whorl.
Flagellum length 1/5-1/3 of epiphallus.

Candidula rocandioi (Ortiz de Zárate y López, 1950)

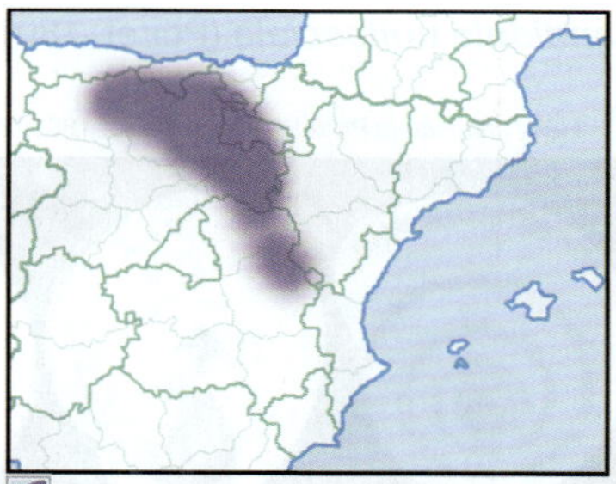

Spain

Helix rocandioi Ortiz de Zárate y López, 1950 in Ortiz de Zárate y López 1950: 34, Lám. 1

Shell solid, distinctly striated, less prominently on lower side, with fine spiral lines on lower side visible under high magnification, 5.5 whorls, aperture with wite lip inside, columellar margin reflected covering partly the umbilicus, umbilicus wide and deep, 1/4 or more of shell diameter, all whorls visible inside.
Animal light brown, also tentacles, foot whitish.

3.5-4 x 5.5-7.5 mm

In the summer under stones, in the autumn on mosses and small grasses, active until mid-December.

Prieto & Martín 1988, Martínez-Ortí 2011: 9.

Candidula rugosiuscula (Michaud, 1831)

S France

Helix rugosiuscula Michaud, 1831 in Michaud 1831: 14 (*Helix Rugosiuscula*)

Shell light greyish with some narrow dark brown bands, prominently striated or ribbed, angulated or weakly keeled, upper side more depressed, 4.5-5.5 weakly convex whorls, aperture oblique, margin with whitish lip inside, umbilicus narrow, penultimate whorl visible inside.

3-5 x 5-8 mm

Scarcely vegetated and dry habitats in areas with most pronounced Mediterranean climate, hot and dry summers and mild winters.

Eggs are laid in June and July under herbs, maturity is reached at the end of the second year.

Gassies 1849: 74, Germain 1930: 277, Pfenninger et al. 2003 (suggested that reproductive isolation to *Cand. unifasciata* was complete), Pfenninger et al. 2007.

Candidula setubalensis (Pfeiffer, 1850)

S Portugal

Helix setubalensis Pfeiffer, 1850 in L. Pfeiffer 1850: 88 (*Helix Setubalensis*)

Shell flat, greyish white, cream or brownish yellow, rarely with some dark colour spots, densely and prominently ribbed, 5-5.5 slowly increasing whorls, periphery with sharp keel, outline nearly lentiform, last whorl suddenly descending, aperture with a sharp edge at the outer side, often with a distinct whitish lip inside, margin thin and sharp, umbilicus wide and perspectivical. Differs from other *Candidula* species in its sharp keel.
The animal is very shy.

3-4 x 7-10 mm

Rocky shrublands and forests, at the bases of walls, between plants and under stones, also on grasses.

EU Red List: Endangered EN. Only in Serra da Arrábida and Serra de São Luis, with small and restricted range.

Westerlund 1889: 331, Nobre 1913: 192, Nobre 1941: 98, Gittenberger 1993: 286, Kittel 2000: 15.

Candidula spadae (Calcara, 1845)

Helix spadae Calcara, 1845 in Calcara 1845 (Cenno sui molluschi): 49, Tav. 3 (*Helix Spadae*)

Italy: Lazio, Monte Teminillo

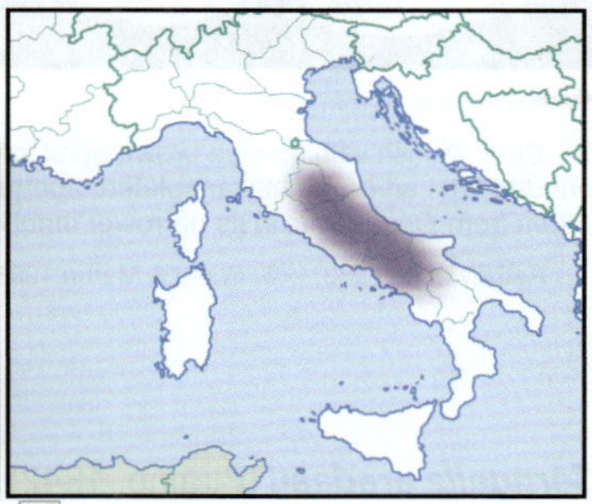

Italy (central Appennino)

6.5-8.5 x 10-16 mm

Shell pale white, sometimes with some narrow light brownish colour bands, almost smooth, 5.5-6 convex whorls, last 2 whorls increasing more rapidly, last whorl initially weakly angulated, not descending near aperture, aperture large, inside white or light yellowish, with white lip inside, upper margin strongly bent, columellar margin widened and slightly reflected, umbilicus initially narrow, at last whorl increasing to 1/4 of shell diameter.

In exposed high mountain habitats among herbs and stones, mostly in rocky grasslands.

EU Red List: Vulnerable VU. Range fragmented, threatened by habitat destruction caused by increased human activities in mountains. Also affected by climate change, populations declined with increasingly dry conditions in the past 20 years.

Westerlund 1889: 344, Manganelli et al. 1995: 28, www.naturamediterraneo.com (2007), www.iucnredlist.org (08.2012).

Candidula syrensis (Pfeiffer, 1846)

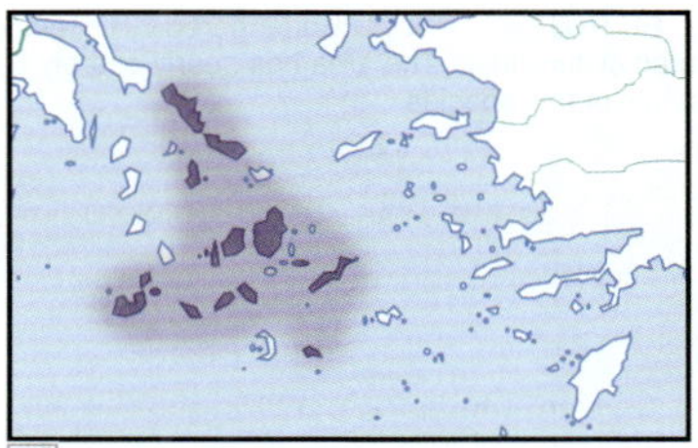

Greece (Aegean islands)

Helix syrensis Pfeiffer, 1846 in L. Pfeiffer 1846 (Symb.): 69 (*Helix Syrensis*)

Shell greyish white with irregular spots and often with a narrow colour band, densely ribbed, 5 very weakly convex whorls, strongly keeled at first whorls, keel visible at suture, last whorl not much wider than penultimate whorl, keel almost disappearing at last whorl, aperture with white lip inside, margin sharp, umbilicus moderately narrow.

4.5-5 x 7-8 mm

Westerlund 1889: 321, Gittenberger 1991: 123.

Seems to be a complex of several species.

Candidula unifasciata (Poiret, 1801)

Helix unifasciata Poiret, 1801 in Poiret 1801: 81

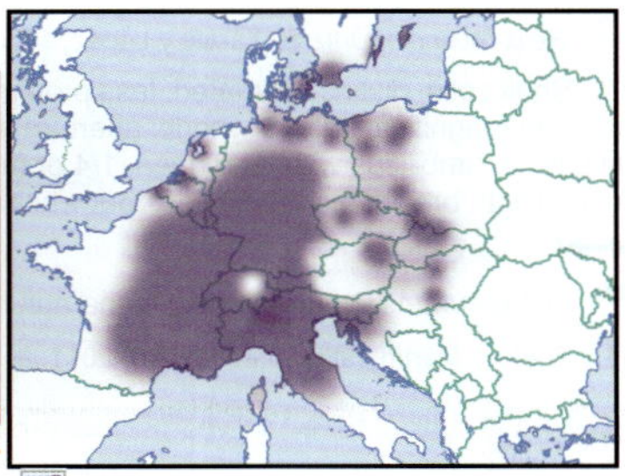

3.5-5 x 5.5-7 mm

Germany: Bayern, near Ansbach (habitat destroyed in 1979)

Europe

Shell white with a dark colour band and some pale bands or spots at lower side, rarely without colour bands, finely and regularly striated, 5-6 slightly convex whorls, last whorl descending, aperture with strong lip inside, lip at basal margin stronger and sometimes like a tooth, umbilicus moderately narrow.

Dry meadows and open lowlands with rocks, S exposed limestone rocks, dry walls (also in gardens and vineyards). In Switzerland up to 2400 m. Animals are also active in the winter season.

Critically Endangered in Austria, Endangered in Germany. Scattered distribution in Switzerland.

Westerlund 1889: 271, Kerney et al. 1983: 245, Réal & R. 1988: 50 (recently reported from Corsica), Falkner 1990: 208, Turner et al. 1998: 353, Pfenninger et al. 2007.

Cernuella

Cernuella Schlüter, 1838 in Schlüter 1838: 6 *Helix (Cernuella)*

Type species *Helix variabilis* Draparnaud, 1801 (subsequent designation by Gude & Woodward 1921: 182) (= currently *Cernuella virgata* (Mendes da Costa, 1778))

Contained originally 16 species.

Shell basically white and without hairs.
Only one dart sac with a large conical papilla inside, and a slightly smaller accessory sac. Penis is innervated from the right pedal ganglion.

Britain to N Africa and Turkey

Fischer 1892: 209, Hausdorf 1988, Manganelli & Giusti 1988, Hausdorf & Sauer 2009: 414.

Cernuella aginnica (Locard, 1882)

Helix aginnica Locard, 1882 in Locard 1882: 341 (*Helix Aginnica*)

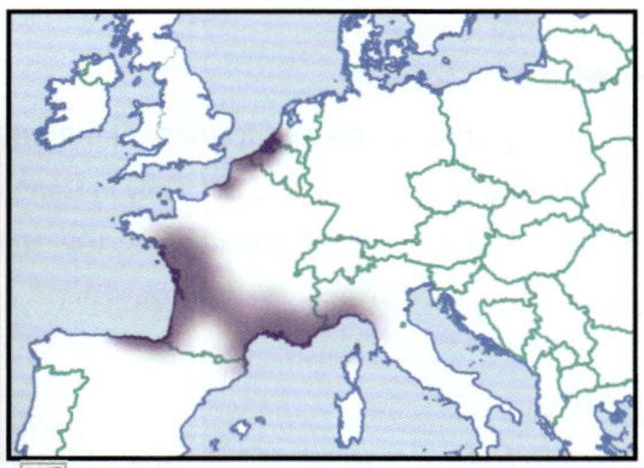

W Europe

6-11 x 8-18 mm

S France: near Marseille

Dry open habitats, including dunes.

Shell whitish with greyish brown apex, often with interrupted colour bands, regularly striated, 5-6 convex whorls with deep suture, last whorl usually rounded but can be slightly angulated, aperture rounded, margin often reddish brown, with whitish lip inside, umbilicus moderately wide.
Differs from *C. neglecta* in its narrower umbilicus, more prominent growth lines, generally less depressed shell shape.

Kerney et al. 1983: 248, Prieto & Martín 1988, Manganelli et al. 1995: 28.

Cernuella aradasii (Pirajno, 1842)

Helix aradasii Pirajno, 1842 in Pirajno 1842: 6 (*Helix Aradasii*)

6-10 x 9-12 mm

Italy: Messina, Punta Faro

José Liétor Gallego

Shell fragile, yellowish white or grey, usually with weak colour bands, very densely and regularly striated, 5-6 weakly convex whorls, last whorl increasing, descending slightly near aperture, aperture with white or reddish lip inside, umbilicus very narrow and partly covered by the reflected columellar margin. Animal light with grey spots on the head and grey tentacles.

EU Red List: Critically Endangered CR. Restricted to a very small locality in a coastal habitat, seriously threatened by human disturbance. The area has nearly entirely been urbanized in the past years. It would be necessary to protect the last habitats of this species.

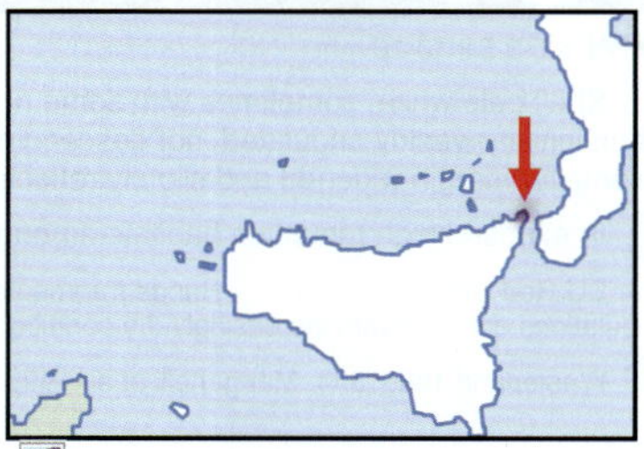

Italy (NE Sicily)

Sandy habitats.

Westerlund 1889: 237 (*Helix subprofuga*), Manganelli et al. 1995: 28.

Cernuella caruanae (Kobelt, 1888)

Helix caruanae Kobelt, 1888 in Kobelt 1888: 119 (*Helix Caruanae*)
Dedicated to Mr. Caruana Gatto, name is not corrected to *caruanai* (Art. 32.5). Art. 31.1 has no effective force.

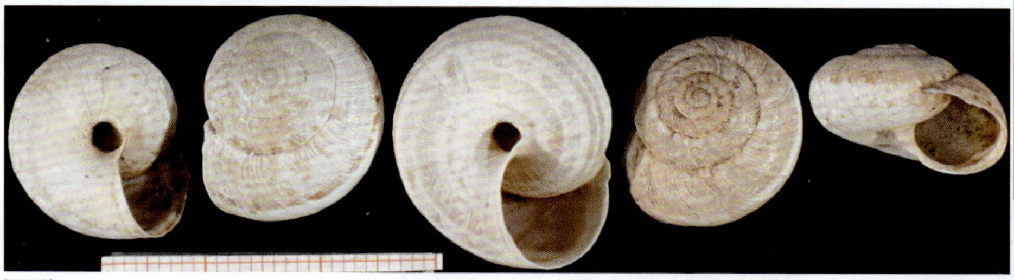

14-15 x 19-23 mm — Malta: Gozo

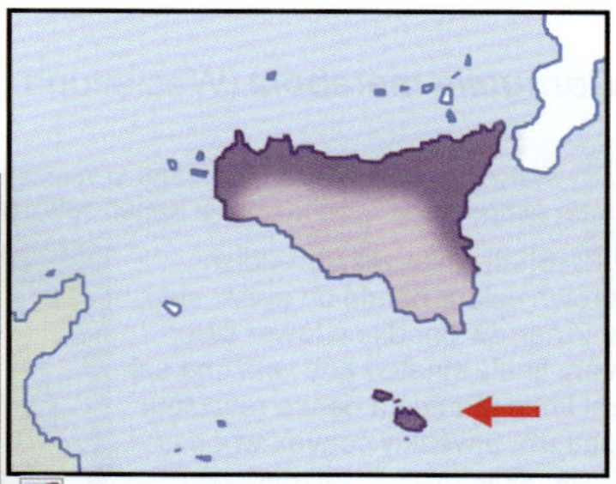

Malta, Italy (Sicily), Libya
In Libya probably introduced.

Westerlund 1889: 202, Giusti et al. 1995: 435, Manganelli et al. 1995: 29, 51 ("Sicily").

Shell thin but solid, yellowish white or grey, with white band at periphery, and a brown band above, and reddish brown spots and bands elsewhere, irregularly striated, also on lower side, 5-5.5 weakly convex whorls, last whorl inflated, descending near aperture, aperture with thin but distinct yellowish lip inside, columellar margin dark, widened and reflected, umbilicus moderately narrow.

Various different kinds of natural and disturbed habitats, on soil, under stones, at walls and rocks, tree branches, shrubs and among low vegetation.

Cernuella cisalpina (Rossmässler, 1837)

Helix striata synonym *Helix cisalpina* Rossmässler, 1837 in Rossmässler 1835-1837 (5-6): 28, Taf. 26
Mentioned with description and figure. Previously mentioned without description as *Helix striata* var. *cisalpina* by De Cristofori & Jan 1832: 4 (nomen nudum). Used by Westerlund 1889: 239 for a species with description and bibliographical reference to Rossmässler Taf. 26 Fig. 354c, thus made available under Art. 11.6.1.
Helix profuga Schmidt, 1854 in Schmidt, 1854: 18

7-8 x 9-12 mm — Italy: Sicilia, Termini Imerese

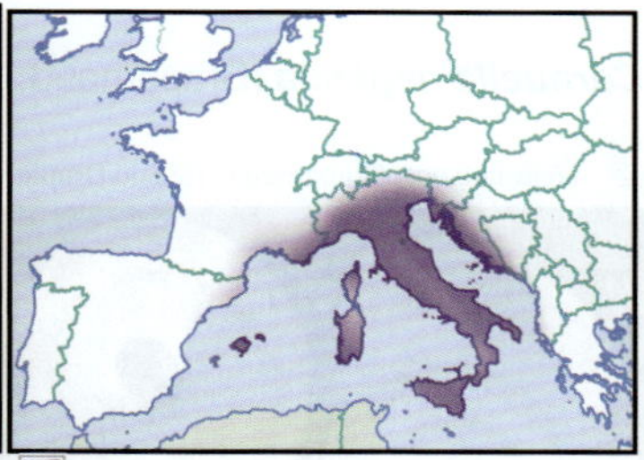

Italy, Balkan coastlands

Shell yellowish to brownish with pale coulour bands, densely and regulately ribbed, aperture with white or reddish lip inside, umbilicus open and of variable width. Locally variable, also at short distances, in terms of colour, shell size, shape and umbilicus width.
No anatomical differences to *C. virgata*, differs in its generally smaller shell with stronger ribs.

Dry soil habitats with little vegetation, also cultivated regions, hides usually under the stones. Habitat and ecological behaviour differs from that of sympatric *C. virgata* which tend to prefer open habitats above the soil, climbing up plants to escape bottom heat.

Falkner 1990: 210, Manganelli et al. 1995: 28, Štamol & Kletečki 2005, Manganelli, Sal. & G. 2005, Beckmann 2007: 121.

Manganelli, Sal. & G. 2005 considered *C. virgata* and *C. cisalpina* from Siena as different species in a molecular study. Hausdorf & Sauer 2009: 415 considered *cisalpina* as a synonym of *C. virgata*, but they gave only very general arguments and a figure of a specimen from Greece which did not fit very well the Italian populations.
C. cisalpina lives sympatrically with *C. virgata* without intermediates at many localities in Italy (www.naturamediterraneo.com, 2007/2012). Populations outside Italy might belong to different species or represent small forms of *C. virgata*.

Cernuella hydruntina (Kobelt, 1883)

Helix hydruntina Kobelt, 1883 in Kobelt 1884: 46, Taf. 17
Original spelling was not *Helix hydrutina*.

10-12 x 16-18 mm — (1-5) Italy: Puglia, Lecce, near Laghi Alimini
(1-2) Giuseppe Pocaterra, (3-5) Alessandro Hallgass

S Italy (S Salento peninsula)

Among herbal vegetation in arid environments.

EU Red List: Near Threatened NT. Possibly exaggerated, lives at many localities.

Westerlund 1889: 178, Manganelli et al. 1995: 28, Ferreri et al. 2005: 91.

Shell depressed, bluish dirty white, near the aperture weakly reddish yellow, weakly striated or almost smooth, aperture oval, margin sharp, inside brownish, with a white lip with thin brown margins, umbilicus wide.

Cannot be distinguished anatomically from *C. virgata*, but the two seem to coexist without intermediate forms.

Cernuella limbata (Philippi, 1836)
Cernuella amanda (Rossmässler, 1838)

Carocolla limbata Philippi, 1836 in Philippi 1836: 137, Tab. 8
Helix amanda Rossmässler, 1838 in Rossmässler 1838-1844 (7-8): 10, Taf. 32
Expressly established as a new replacement name for *Carocolla limbata* Philippi, 1836, classified by Rossmässler in the genus *Helix*, not *Helix limbata* Draparnaud, 1805 (= currently *Hygromia limbata*).
Helix amanda can be used for this species under the provisions of Art. 59.3, as outlined by Manganelli, Fa. & G. 1996: 356, because *amanda* was used in the 1800s when both *limbata* names were classified in *Helix*. However, there are extremely few Google hits for this name (01-2011) and it is rarely mentioned in the literature. I prefer using the oldest name.
In view of the problems caused by Art. 59.3 at other occasions (*Trochulus caelatus, Alopia cyclostoma*), I think this Article should be removed from the Code or substantially modified, and not applied to cases where a name is so rarely mentioned in the literature.

Italy (W Sicily)

Natural habitats on limestone substrate.

EU Red List: Critically Endangered CR. Only known from San Vito lo Capo.

Westerlund 1889: 310, Alzona 1971: 172, Manganelli, Fa. & G. 1996: 350.

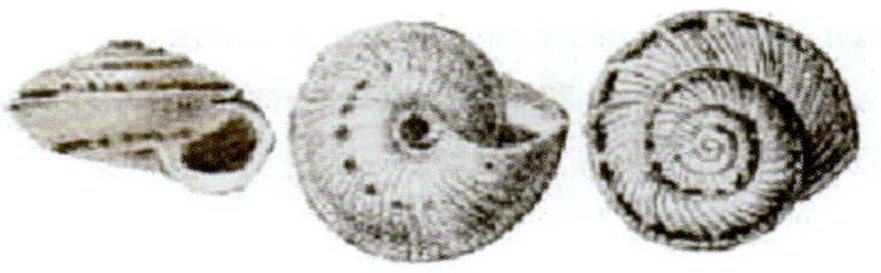

6-7 x 10-12 mm — Italy: Sicily
Picture: Rossmässler 1838-1842 (9-10): Taf. 32, Fig. 449

Shell lentiform, whitish with two fine rows of colour spots, finely striated, 5 sharply keeled and almost straight whorls with shallow suture, keel visible at suture, last whorl increasing, aperture with strong white lip inside, margin sharp, reflected at upper columellar side, umbilicus narrow.

Cernuella metabola (Westerlund, 1889)

Helix metabola Westerlund, 1889 in Westerlund 1889: 236
Helix lampedusae Kobelt, 1890 in Kobelt 1890: 85, Taf. 113 (*Helix Lampedusae*)

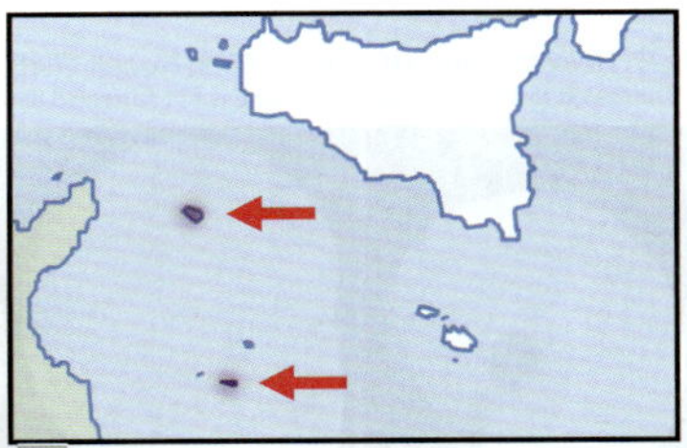

Italy (Lampedusa, Pantelleria)

Mediterranean shubland, grasslands and cultivated land.

EU Red List: Near Threatened NT. Possibly exaggerated, no real threats known.

7-15 x 10-18 mm

Italy: Pantelleria island, La Roccia

Paolo Zaccaria — Tiroler Landesmuseen Innsbruck

Shell white with 3 reddish brown colour bands on upper side, 6-7 narrow bands on lower side, very finely striated and with fine spiral lines forming a dense reticulate pattern, 5 weakly convex whorls, lower side strongly inflated, last whorl rounded near aperture, aperture with white lip inside, margin straight, umbilicus moderately or very narrow. Perhaps locally variable.

Not in Manganelli et al. 1995.

Kobelt 1890: 85 who had two morphologically sightly different rounded shells assumed that Westerlund when describing a shell with initially angulated periphery had juvenile specimens of another species (Falkner et al. in www.iucnredlist.org, 08.2012 suspected that this was *Theba pisana*). *Theba* would contrast Westerlund's description (small apex, 4.5 ("5") whorls and 7.5 mm shell height at 11 mm diameter, *Theba* would have had 15-17 mm at 4.5 whorls, shell height around 9 mm at 11 mm diameter, and a very characteristically big apex that Westerlund would probably have recognised and not described as small). It would be necessary to study Westerlund's type material, the description suggests that his 11 mm large specimen with 4.5 whorls and initally angulated periphery could have been a juvenile shell of Kobelt's *Helix lampedusae*.
The shells deposited in Innsbruck from Pantelleria correspond well with Kobelt's shells in shape, colour patterns and size (diameter 18 mm, 5.5 whorls). The forms of both islands are preliminarily regarded as conspecific here, more research is needed to firmly establish their taxonomic relationships.

Cernuella neglecta (Draparnaud, 1805)

Helix neglecta Draparnaud, 1805 in Draparnaud 1805: 108

6-10 x 9-18 mm

France: Provence, near Montélimar

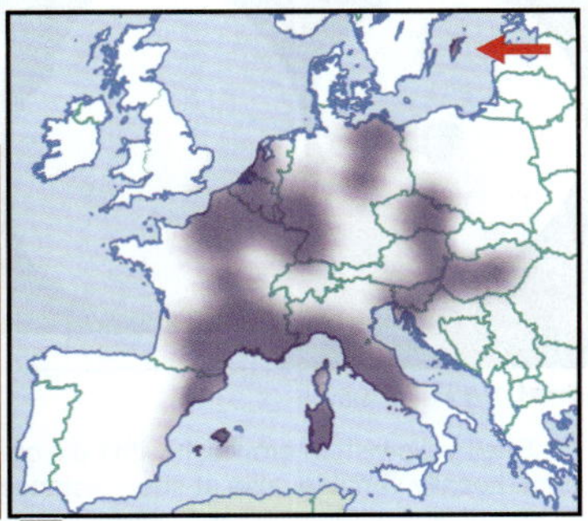

Europe

France: Isère, Réserve naturelle de la Platière — Cédric Audibert

Shell white, usually with brown colour bands, often one slightly above the periphery and several narrower bands below, weakly and irregularly striated, markedly depressed but rounded at periphery, 5-6 weakly convex whorls with shallow suture, aperture with a thin sometimes pinkish lip inside, umbilicus wide, 1/6-1/4 of shell diameter.
Shell somehow like *Xerolenta obvia*, colour bands less dark, aperture inside often reddish brown, lip often with reddish hue, umbilicus narrower. Differs from *C. virgata* and *C. aginnica* in its more depressed and smoother shell, and wider umbilicus.
The animal is dark or greyish with reddish or yellowish hue, moves slowly and is shy.

Rocky dry grasslands and wastelands, gardens, roadsides, climbs tree trunks and herbal plants to estivate.

Moquin-Tandon 1855: 250, Kerney & Cameron 1979: 178, Puente & Prieto 1988, Falkner 1990: 210, Manganelli et al. 1995: 28.

Range scattered and in most regions local. The most serious obstacle for reliable range identification are misidentifications in the published record.

Cernuella rugosa (Lamarck, 1822)

Helix rugosa Lamarck, 1822 in Lamarck 1822: 90

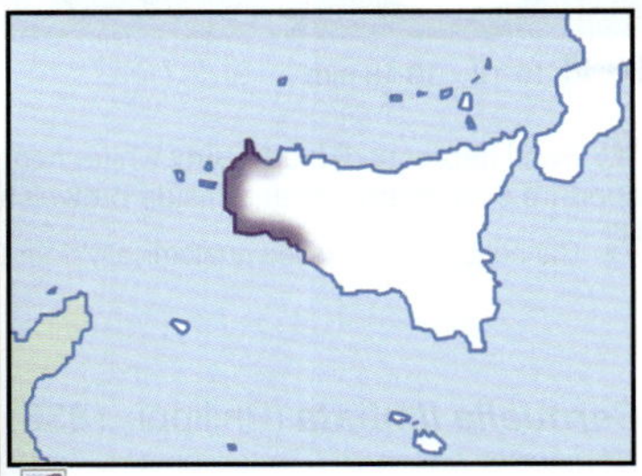

Italy (W Sicily)

5-6.5 x 12-15 mm

Italy: Sicily, (1-3) west of Agrigento, (4-5) Trapani, Monte San Giulio — (1-3) Walter Renda

Shell white, thin and translucent, usually with 3-4 indistinct colour bands on lower side and rarely with a broad band on the upper side or periphery, coarsely ribbed, also on lower side, 5 almost flat worls, periphery weakly keeled, last whorl with a weak edge at the periphery, slightly descending near aperture, aperture with strong light brownish lip inside, umbilicus perspectivical. Locally very variable in terms of ribs and keel prominence and umbilical width.
Body of the animal blackish, darker on upper side.

Dune habitats on calcareous substrate, on the soil among low vegetation and stones.

EU Red List: Endangered EN. Dune habitats are increasingly threatened by human pressure, many populations have disappeared in the past decades.

Westerlund 1889: 320, Manganelli et al. 1995: 28, Manganelli, Fa. & G. 1996: 9.

Cernuella virgata (Mendes da Costa, 1778)

Cochlea virgata Mendes da Costa, 1778 in Mendes da Costa, 1778: 79
Author usually cited as Da Costa, but the name given in the original source was Mendes da Costa.
Helix variegata Adams, 1797 in Adams 1797: 67 (possibly not *Helix variegata* Schröter, 1784)
Helix variabilis Draparnaud, 1801 in Draparnaud 1801: 73
Helix striata var. *jonica* Mousson, 1854 in Mousson 1854: 6

8-15 x 12-23 mm

S Albania: near Përmet

Europe

Gassies 1849: 78, Moquin-Tandon 1855: 262, Germain 1930: 303 (life cycles), Hausdorf & Sauer 2009: 414.

Montenegro: Krute near Ulcinj — Cédric Audibert

Shell whitish or yellowish, sometimes with reddish hue, usually with two brown colour bands on upper side and 3-4 narrower bands on lower side, initially finely ribbed, weakly striated at lower whorls, 4.5-5.5 convex whorls, first whorls keeled, last whorl initially angulated or rounded, aperture rounded, with whitish or reddish lip inside (this does not mean that all semiadult individuals have it), margin not reflected, umbilicus variable but always open, 1/10-1/6 of shell diameter, sometimes slightly excentric. Shell size and colour patterns very variable with local trends, some forms only with one single colour band near the suture, some forms prominently striated (like *cisalpina*). Juveniles are not striated at the lower side. Differs from *Xerocrassa cretica* in its usually more rapidly increasing whorls, irregular ribbing, an often reddish internal lip (easy identification if present), and its often more excentric umbilicus.
Animal light greyish, reddish or yellowish with black dorsum with large tubercles, mantle reddish brown, tentacles greyish transparent and 8 mm long.

Lives in dry habitats, in sandy dunes and on calcareous substrate, on and under herbal plants, frequent in coastal regions, also as a synanthrope at roadsides, cultivated fields and railways. Attaches high on tall plants to escape bottom heat, often in groups, may also ascend vertical walls. Tolerates cultivation, often in harvested fields.
A very common species. Considered as probably introduced with weeds to Britain in the Roman period, where it lives on well-drained calcareous soils, widespread and common in S and E central England, more recently introduced to and only coastal in N Britain and Ireland except the Dublin area, continuously spreading.

In France 30-80 globulous eggs (diameter 1.5 mm) in whitish transparent envelopes are laid at various occasions between September and November, juveniles hatch after 15-20 days, maturity is reached in the second year.

Cernuellopsis

Cernuellopsis Manganelli & Giusti, 1988 in Manganelli & Giusti 1988: 334

Type species *Cernuellopsis ghisottii* Manganelli & Giusti, 1988 (original designation)

Cernuellopsis ghisottii Manganelli & Giusti, 1988

Cernuellopsis ghisottii Manganelli & Giusti, 1988 in Manganelli & Giusti 1988: 336

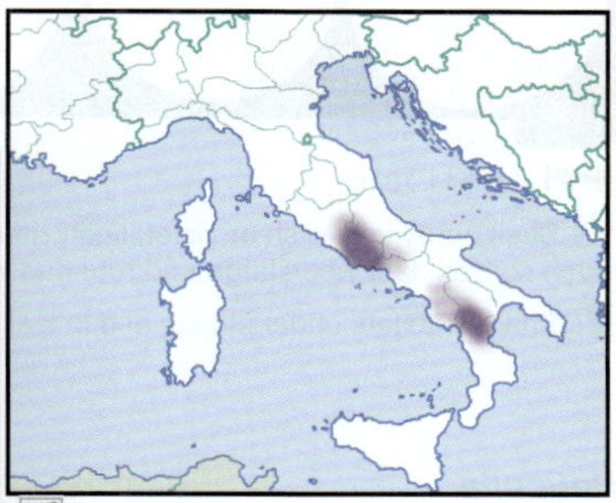

4.5-6.3 x 7.5-10 mm

Italy: Cosenza, Monte Pollino, 1650 m

Italy

Mountain habitats among rocks in small grassland patches, 1200-2200 m.

Shell globose, opaque, white, often with brown bands or spots, apex pinkish or brownish, finely and regularly striated, 4.5-5 slightly convex whorls, last whorl not widened, sometimes slightly angulated, aperture simple with white lip inside, umbilicus moderate, 1/7-1/5 of diameter.
Animal pale grey.

EU Red List: Vulnerable VU. Only known from few localities, range fragmented. Threatened by pasturing and too intensive tourism activities in natural parks.

Cryptosaccus

Cryptosaccus Prieto & Puente, 1994 in Prieto & Puente 1994: 110

Type species *Cryptosaccus asturiensis* Prieto & Puente, 1994 (original designation)

Cryptosaccus asturiensis Prieto & Puente, 1994

Cryptosaccus asturiensis Prieto & Puente, 1994 in Prieto & Puente 1994: 112

Shell evenly light horny cream, weakly translucent, finely striated and densely covered with numerous minute triangular scales, at the peripery under magnification looking like minute hairs or shark teeth, aperture simple, umbilicus narrow (1/8-1/7 of shell diameter) and deep with penultimate whorl visible inside.

4.2-5 x 8.5-10 mm

At exposed vertical limestone walls.

EU Red List: Vulnerable VU. Only known from one locality in the Somiedo valley. This is a remote place to live for a snail, no acute threats known.

Puente et al. in Verdú & Galante 2006: 366.

NW Spain (Asturias)

Cyrnotheba

Cyrnotheba Germain, 1929
in Germain 1929: 282

Type species *Helix corsica* Shuttleworth, 1843
(original designation)

Cyrnotheba corsica (Shuttleworth, 1843)

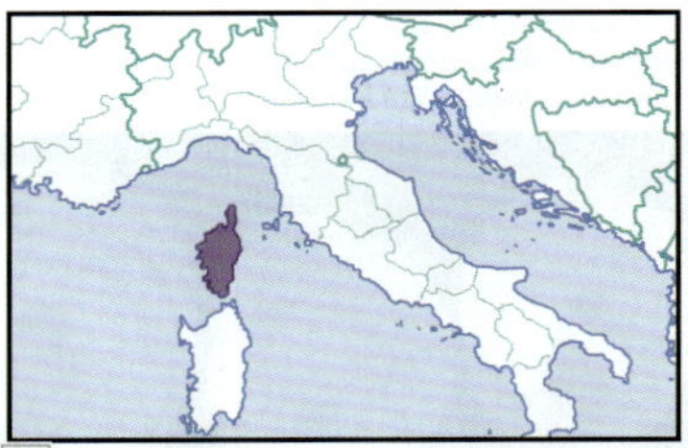

France (Corsica)

Helix corsica Shuttleworth, 1843 in Shuttleworth 1843: 15 (*Helix Corsica*)

Shell thin, horny reddish or yellowish, not shiny, striated and with fine hairs, 5 whorls with indistinct edge at periphery, apertural margin simple and sharp, umbilicus very narrow and partly covered by the reflected columellar margin.

9-10 x 14-16 mm

Under leaf litter in humid places often under trees, in open forests and moderately disturbed habitats.

Not rare, but protected by law in France, eggs and animals must not be collected.

Westerlund 1889: 64, Giusti & Manganelli 1987: 138.

Ganula

Ganula Gittenberger, 1970
in Gittenberger 1970: 64

Type species *Helix lanuginosa* Boissy, 1835
(monotypy)

Name available, type species was fixed in original publication by monotypy (Art. 13.3), not by original designation.

W Mediterranean

Ganula gadirana Muñoz, Almodóvar & Arrébola, 1999

S Spain (Cádiz)

Ganula gadirana Muñoz, Almodóvar & Arrébola, 1999 in Muñoz et al. 1999: 164

Shell evenly light brown, translucent, apex striated and with hair scars, main whorls with 0.4 mm long persistent hairs, 5-5.5 convex whorls, last whorl predominating, 1.5 x wider than preceding whorl, rounded at periphery, aperture with lip inside, umbilicus very narrow, partly covered by the columellar margin.
Shell much like *G. lanuginosa*, the two species differ anatomically.

8-11 x 11-14 mm

Under stones half buried in the ground in Mediterranean shubland vegetation and close to periodical river flows. Some animals are intensively parasited by small nematodes inside the pulmonary cavity.

EU Red List: Vulnerable VU. Flatland habitats near water currents are heavily threatened by habitat destruction.

Arrébola & Puente in Verdú & Galante 2006: 367.

Ganula lanuginosa (Boissy, 1835)

Helix lanuginosa Boissy, 1835 in Boissy 1835: text pp. [1-2], Classe 5 Pl. 69 (*Helix Lanuginosa*)

Spain: Mallorca, Porto Cristo

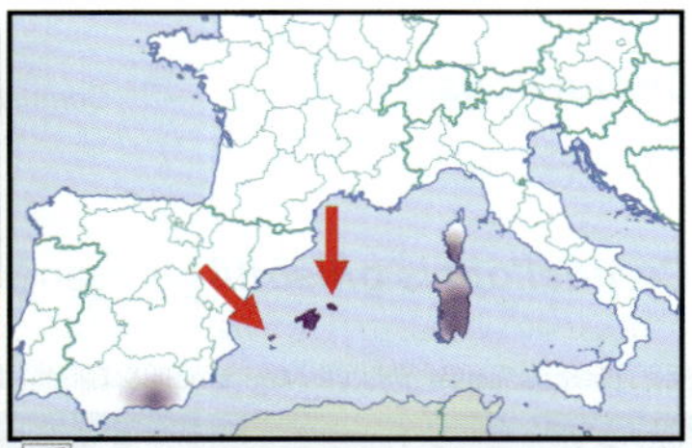

S Europe, N Africa

7-9 x 11-13 mm

Shell horny yellowish or violet, finely striated, irregularly covered with short white persistent hairs, whorls not much convex, last whorl large and rounded with a weak edge, apertural margin sharp and reddish.

Shady habitats under stones and in leaf litter, in forests, wastelands and cultivated land, up to 600 m.

Terver 1839: 22, Pl. 2 Fig. 15-16, Westerlund 1889: 65, Giusti & Manganelli 1984: 86, Beckmann 2007: 123 (introduced to Ibiza).

Helicella

Helicella Férussac, 1821
in A. Férussac [1821-1822]: 24 (folio p. 28) *Helix (Helicella)*

Type species *Helix itala* Linnæus, 1758
(fixed by ICZN under the plenary powers)

ICZN Opinion 431: type species designated under the plenary powers. Name was also mentioned at two occasions (pp. xj, xij) in Férussac & Deshayes 1819-1851, explication des planches, issued at the same date (06 Apr 1821).

This name was used after the 1880s for some decades in a much broader sense, today restricted to a few species.

Helicella bolenensis (Locard, 1882)

Helix bolenensis Locard, 1882 in Locard 1882: 323 (*Helix Bolenensis*)

Germany: Mecklenburg, near Schweriner See

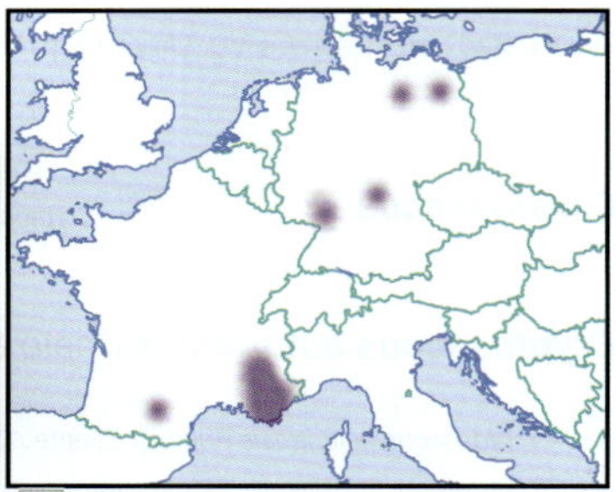

S France, introduced to Germany

6.5-11.5 x 8-14 mm

Shell yellowish or white with horny brownish apex, often with pale and interrupted reddish colour bands at and below the periphery, finely and densely ribbed, 5.5-6 convex whorls with deep suture, last whorl well rounded, aperture with white lip inside, umbilicus moderately wide, up to 1/4 covered by the reflected columellar margin.
Differs from *Candidula intersecta* in its finer ribs, rounded and not shouldered whorls and deeper suture. *Cernuella virgata* has a smoother shell, unbanded shells are rare.

Dry sandy and stony habitats in exposed grasslands, climbs up grasses and herbal plants in drought periods.

Westerlund 1889: 198, Germain 1930: 285, Kerney et al. 1983: 249, Falkner 1990: 210, Vogt et al. 1994: 212.

True range difficult to judge from various sources due to possible misidentifications.

Helicella cistorum (Morelet, 1845)

Helix cistorum Morelet, 1845 in Morelet 1845: 66, Pl. 6

SW Iberian peninsula

7-9.5 x 11-13 mm

(1-2) Portugal: Algarve, Castro Marim, (3-6) Spain: Huelva, La Puente, Castañuelo Aracena (1-2) Valter Jacinto, (3-6) Rafael Vázquez Romero

Shell brownish, lighter and slightly translucent at lower side, with a usually continuous dark brown colour band at periphery, finely and irregularly striated, 5.5-6 whorls, aperture without lip inside, apertural margin sharp and only reflected at columellar side, without covering much of the umbilicus, umbilicus narrow and deep, 1/7-1/6 of shell diameter.
Body and tentacles greyish with large granules or processes near the head, mantle margin and foot white.

In habitats of deserted zones with not much more than some herbal vegetation, tolerates extreme arid conditions, hides under stones and in rock crevices, occurs often in small numbers. Often the only snail living in this kind of habitats, though rare and not easy to find adult specimens. Not rare in lowlands, common in La Sierra Norte in up to 800 m, maximum altitudes 1000 m.

Nobre 1913: 191, Nobre 1941: 97, Ortiz de Zárate y López 1950: 50, Martínez-Ortí, Aparicio & R. 2004.

Helicella iberica (Rambur, 1869)

Helix mirandae Rambur, 1868 in Rambur 1868: 266 (*Helix Mirandæ*)
Not *Helix mirandae* Lowe, 1861 in Lowe 1861: 107 (*Helix (Turricula) Mirandæ*) from Canary Islands.
Helix iberica Rambur, 1869 in Rambur 1869: 254, Pl. 9 (*Helix Iberica*)
Established with expressed statement "*Helix Mirandæ*, Rambur (...) nec Lowe" to indicate a new replacement name for *Helix mirandae* Rambur, 1868 non Lowe, 1861.

Shell light yellow or reddish grey with dark brown colour bands, or rows of dots, on the lower more than on the upper side, finely ribbed, 5-6 convex whorls, last whorl looks uncommonly narrow when seen from above, usually with edge at periphery but rounded near aperture, aperture with white lip relatively deep inside, margins approaching each other, umbilicus initially narrow, at last whorl wider.

5.5-6 x 9-12 mm

Westerlund 1889: 280 (*Helix barcinonensis*), Prieto & Martín 1988.

N Spain

Helicella itala (Linnæus, 1758)

Helix itala Linnæus, 1758 in Linnæus 1758: 772
Helix nubigena Saulcy, 1852 in Saulcy 1852: 439
Helix sabulivaga Mabille, 1881 in Mabille 1881: 125 (*Helix Sabulivaga*)

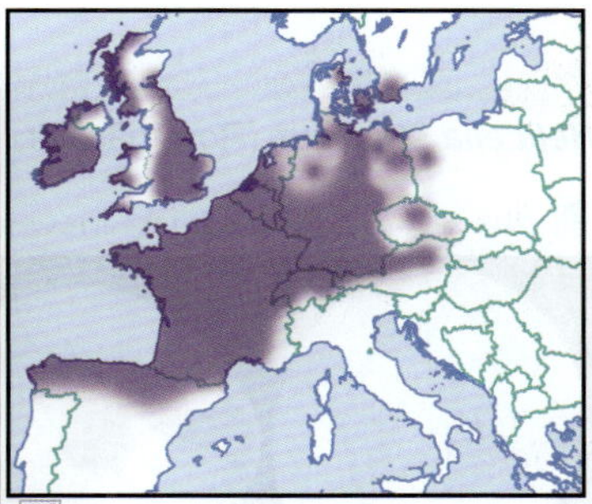

W Europe
Indicated range in Spain uncertain.

5-12 x 9-25 mm (Pyrenees: 6-7 x 10-12 mm, Basque country 3.5-4 x 8-9.5 mm)

N Germany: Lüneburg

Germany: Bodenheim near Mainz
Vera Buhl

Shell white, yellowish or pale pink, usually with pale to dark brown colour bands, weakly and irregularly striated, 5.5-6.5 weakly convex whorls, very depressed above, rounded below, last whorl shortly and sharply descending, aperture usually without internal lip, but can be very slightly reflected, umbilicus very wide and perspectivical, 1/3 of shell diameter. In Pyrenees more globulous, last whorl initially weakly angulated, aperture with white or reddish lip inside, columellar margin light reddish, umbilicus deep and weakly perspectivical (1/4 of shell diameter or more).
Differs from *Xerolenta obvia* in its larger umbilicus and its narrower last whorl descending sharply near the aperture. Upper side of shell flatter than *Cernuella neglecta*, aperture more rounded and umbilicus wider.

Open and exposed habitats, calcareous grasslands, south-exposed slopes covered by grasses and shrubs, also in dunes and sea cliffs. Up to 2000 m in Alps, in Pyrenees 3000 m. Requires more humidity than *Xerolenta obvia*, in Britain it is less tolerant to cultivation than other helicellines. In Pyrenees among *Poa alpina* grasses and dwarf juniper shrubs, active from snowmelt onwards.
Native in the British Isles since the end of the glacial period, declined with forest growth and spread again with human clearances, today common by the sea in N Great Britain and N Ireland, widespread in central and S Ireland.

2 reproduction periods in the year, in spring (January to May) and in autumn (September to November), 1-3 weeks after copulation 18-90 (usually 30-60) eggs (1.2-1.6 mm) are laid in clutches, juveniles hatch after 18-20 days, in late autumn up to 50 days, maturity is reached at the end of the following year. Life span 2-3 years. Adults do not grow any more.

Endangered in Niedersachsen and parts of Austria, Vulnerable in Germany and Ireland. Populations in S and E England have declined strongly since 1900. Mainly threatened by modern agriculture destroying the grassland habitats.

Gassies 1849: 96, Moquin-Tandon 1855: 252, Germain 1930: 286, Frömming 1954: 246, Prince 1967, Kerney et al. 1983: 248, Prieto & Martín 1988, Bech 1988: 107, Ondina et al. 1997.

Falkner et al. 2002: 50, 152 classified *sabulivaga* and *nubigena* (from Pyrenees) as separate species, mainly based on differences in shell shape (*sabulivaga*: shell more depressed than *Hcl. itala*, more densely coiled, umbilicus larger) and longer darts. The identity of these forms as separate species should be confirmed in a more detailed study. It seems that there are altitudinal gradients in the Pyrenees. Shells become continuously more globulous with higher altitude, so that analysing some selected material from a few distant localities, as done by Falkner et al. 2002, would give misleading results (Welter-Schultes et al. 2011, Folia Conchyliologica 12). More research is necessary.

Helicella juglans Gittenberger, 1991

Helicella juglans Gittenberger, 1991 in Gittenberger 1991: 116

3-5 x 5-7.5 mm

Cyprus — Georgios Konstantinou

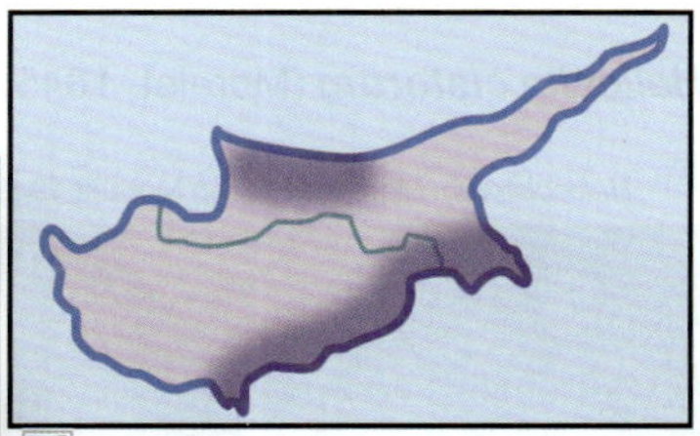

Cyprus

Shrublands, parks and gardens. The species is quite abundant.

Shell whitish with irregular spiral colour patterns, regularly ribbed, 3.5-5 moderately convex and initially regularly increasing whorls, first whorls keeled, last whorl initially sometimes keeled, rounded, slightly enlarged and slightly descending near aperture, aperture with strong lip inside, umbilicus 1/7-1/6 of shell diameter, often slightly excentric.

Helicopsis cypriola is coarser and more irregularly sculptured, and has a deeper suture.

Helicella ordunensis (Kobelt, 1883)

Helix ordunensis Kobelt, 1883 in Kobelt 1883: 115

Shell whitish, densely and regularly striated, with two narrow and sharp brown colour bands at both sides of the periphery, apex small, last whorl almost inflated, aperture with distinct white lip inside, umbilicus narrow and deep.

5-6 x 7-8 mm

Westerlund 1889: 255, Prieto & Martín 1988.

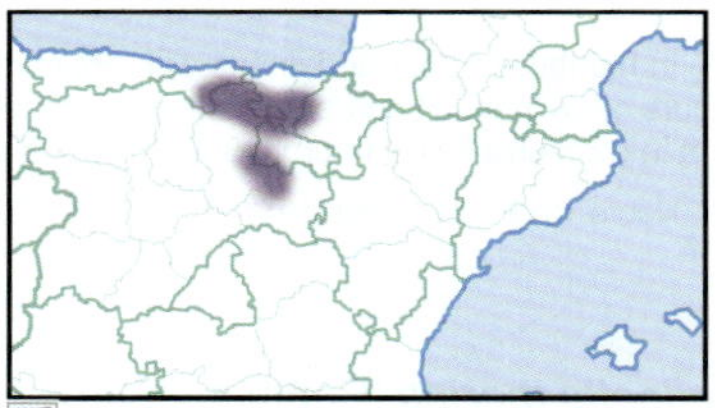

N Spain

Helicella orzai Gittenberger & Manga, 1981

Helicella orzai Gittenberger & Manga, 1981 in Gittenberger & Manga 1981: 121

Shell with variable brown bands on a whitish background, prominently striated, keeled, keel knobbed along the periphery, 4.5-6 moderately convex whorls, aperture simple with U-shaped sinus at the keel, sometimes with a weak white layer inside at the basal side, umbilicus wide, 1/6-1/5 of diameter, first whorl easily visible inside. Flagellum clearly longer than penis.

Under stones and among herbal vegetation in dry and sunny habitats.

3.5-5.5 x 8-12 mm

Spain: Navarra, Huarte-Araquil — Paolo Zaccaria

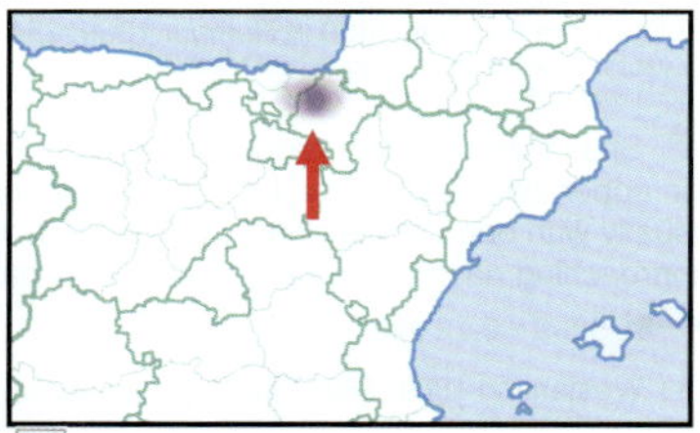

N Spain (Navarra)

EU Red List: Near Threatened NT. Only known from few localities, no real threats known.

Prieto & Martín 1988.

Tiroler Landesmuseen Innsbruck

Helicella stiparum (Rossmässler, 1854)

Helix stiparum Rossmässler, 1854 in Rossmässler 1854-1859 (13-14): 20, Taf. 66 (*Helix Stiparum*)

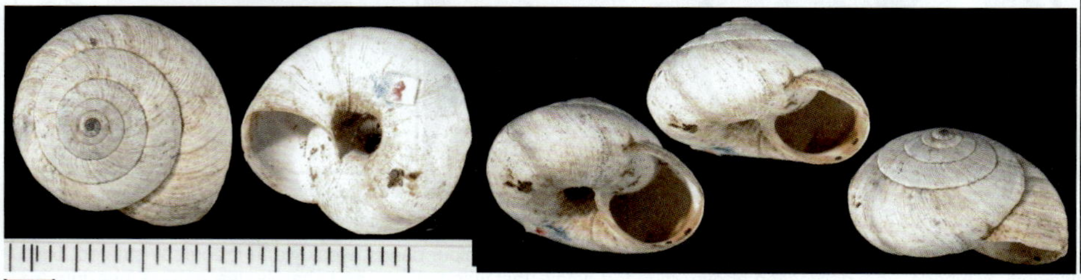

9.5-12.5 x 15.5-20 mm

Spain: Almería

Spain (Almería)

Shell white or with brown or bluish-grey colour bands mainly on upper side, irregularly striated, 5.5-6 whorls, equally convex at both sides, first whorls strongly keeled, later angulated, rounded near aperture, aperture with a lip inside, apertural margin sharp, reflected only slightly at columellar side covering at most a small part of the umbilicus, umbilicus wide, all whorls visible inside. Shell variable from keeled to rounded, the open umbilicus is characteristic. Animal whitish with grey hue, not very granulated, mantle dark grey with yellowish margin.

Dry habitats exposed to sunshine, among medium-sized stones on dry soils and not more than herbal vegetation, in lowlands in coastal vicinity.

EU Red List: Endangered EN. Only known from few localities in fragmented populations of a once wide range. Seriously threatened by intensive argiculture (greenhouses), water mismanagement and direct human pressure by tourism and constructions in coastal lowlands. This species needs serious and direct protection.

Ortiz de Zárate y López 1950: 46, Arrébola 2001: 2, Gómez Moliner et al. 2001: 165, Talaván Gómez & T. 2006: 33.

Classification uncertain.

Helicella striatitala Prieto, 1985

Helicella striatitala Prieto, 1985 in Prieto 1985: 53

Shell white or yellowish brown, with dark brown colour bands, coarsely and regularly ribbed, 4.5-5 moderately convex whorls with well-marked suture, shape usually depressed, almost lenticular, last whorl slightly angulated, descending shortly near aperture, aperture with strong lip inside, umbilicus wide and deep, 1/4-1/3 of diameter, narrower in more globulous shells.

3.6-5.7 x 6-9 mm

In dry Mediterranean shrubland habitats.

Prieto & Martín 1988.

Spain (N Burgos)

Helicella valdeona Gittenberger & Manga, 1977

Helicella valdeona Gittenberger & Manga, 1977 in Gittenberger & Manga 1977: 184

Shell yellowish white with several partly fragmented pale horny brown colour bands, striated but relatively smooth, 5-5.5 whorls, aperture with a weak lip inside, parietal border relatively long, umbilicus 1/6 of shell diameter.

8.5-9.2 x 14.3-15.2 mm

EU Red List: Vulnerable VU. Only known from two localities. No real threats known.

Gittenberger & Carr 2002.

N Spain (NE León)

Helicopsis

Helicopsis Fitzinger, 1833 in Fitzinger 1833: 101

Type species *Helix striata* Müller, 1774 (monotypy)

Europe to Iran

Hausdorf 1996: 65, Gural-Sverlova 2010 (3 new species from Ukraine, not validly published under Art. 8.6, CD-ROMs are not durable media).

Helicopsis cereoflava (Bielz, 1851)

Helix cereoflava Bielz, 1851 in M. Bielz 1851: 56

Shell pale yellow with three interrupted light grey bands, lighter at lower side, solid, not translucent, prominently striated, 4-5 convex whorls, weakly keeled, aperture with weak lip inside, margin sharp, umbilicus open but not very deep.

4.5-5.5 x 8-11 mm

Open ruderal habitats among herbal vegetation.

Hudec 1972, Grossu 1983: 429, Hausdorf 1996: 69.

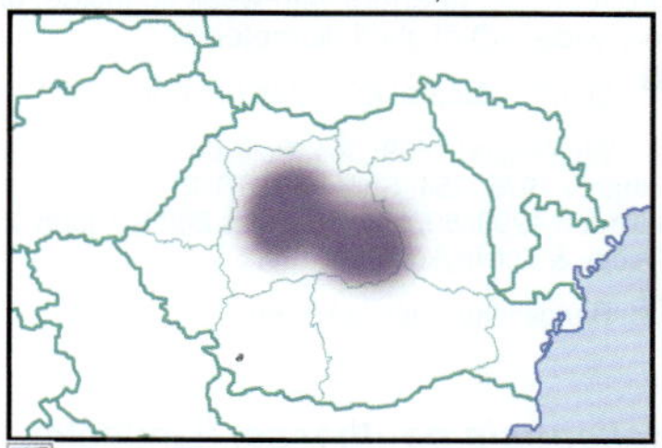

Romania (Transilvania)

Helicopsis cypriola (Westerlund, 1889)

Helix cyparissias var. *cypriola* Westerlund, 1889 in Westerlund 1889: 252

Shell conical, white with spotted upper side and reddish brown colour band above the periphery, lower side with some narrow brown bands, finely striated, 5 convex whorls, last whorl with edge at periphery, aperture without lip, umbilicus open.

Gittenberger 1991: 102, Hausdorf 1996: 69.

3.3-4.3 x 5.0-6.6 mm Cyprus

SW Cyprus

Georgios Konstantinou

Helicopsis arenosa (Krynicki, 1836)
Helicopsis dejecta (Rossmässler, 1838)

Helix arenosa Krynicki, 1836 in Krynicki 1836: 197 Type locality in Crimea.
Helix ericetorum var. *dejecta* Rossmässler, 1838 in Rossmässler 1838-1844 (7-8): 34, Taf. 38 Type locality in Crimea.
Seems to be the first use of the name *dejecta*, attributed to "d. Cr. et J.", description was by Rossmässler alone. Name on plate was given as *H. deiecta*.
Not mentioned by De Cristofori & Jan 1832.
Helix arenosa has precedence.
Sysoev & Schileyko 2009: 189 came to the same conclusion, but preferred to wait until the situation was better understood.
Currently I do not see why *dejecta* should be the name for this species. The name *arenosa* was used after 1899 and cannot be suppressed under Art. 23.9. If nothing else interferes, I would prefer to use *arenosa*.

Shell white with light brown apex, occasionally with one brown band at the periphery, shiny, umbilicus moderately wide, slightly widened at last whorl.
Very much like *Xerolenta obvia*, slightly more globulous, umbilicus slightly narrower.

6-11 x 9-18 mm Ukraine: Crimea, near Simferopol

Sergey Leonov

Westerlund 1889: 339, Hesse 1934, Damjanov & L. 1975: 353, Shilejko 1978: 216, Grossu 1983: 434, Hausdorf 1996: 69, Irikov & M. 2006: 819, Irikov & E. 2008, Egorov 2008: 85, Sysoev & Schileyko 2009: 189, Gural-Sverlova 2012: 25.

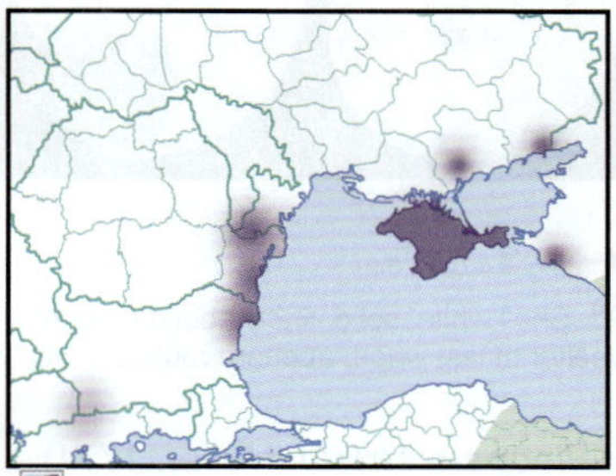

SE Europe

Grasslands and dry slopes.

Rare in Bulgaria.

It is possible that the Bulgarian/Romanian populations represent a different species.

Helicopsis filimargo (Krynicki, 1833)

Helix filimargo Krynicki, 1833 in Krynicki 1833: 435 (*Helix (Helicella) Filimargo*) Type locality in Crimea.

Shell white, rarely with some colour bands, shiny, densely striated, 5 not very convex whorls, initially with edge or keel at periphery, suture with margin, last whorl inflated, rounded near aperture, aperture with shallow white lip very deep inside, umbilicus perspectival.

6.5-9 x 12-17 mm

Under stones and among grasses in rather dry habitats, often on slopes.

Westerlund 1889: 343, Hesse 1934, Shilejko 1978: 217 (Turkey: Zonguldak), Grossu 1983: 435, Hausdorf 1996: 69, Egorov 2008: 86, Sysoev & Schileyko 2009: 190, Gural-Sverlova 2010.

Range insuffiently known and not well understood. Could also be an endemic Crimean species. The E Ukrainian populations belong to different species (their names were not validly published under Art. 8.6 by Gural-Sverlova 2010; CD-ROMs are not durable media).

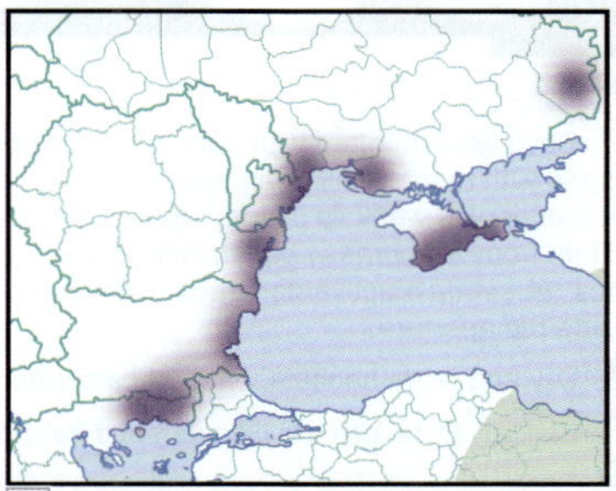

S Ukraine to NE Greece

Helicopsis gittenbergeri Hausdorf, 1990

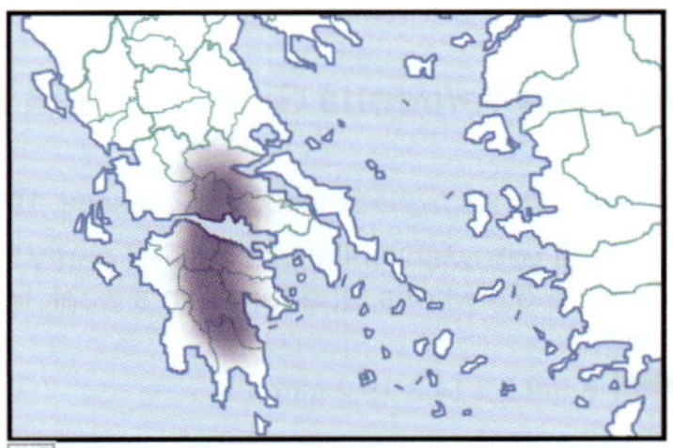

Greece

Helicopsis gittenbergeri Hausdorf, 1990 in Hausdorf 1990: 58

Shell yellowish white with dashy brown colour bands, weakly ribbed, first whorls with weak edge at the periphery, aperture sharp, no white layer at parietal side, umbilicus 1/7-1/4 of shell diameter, initially narrower. Usually larger than *Candidula lernaea* with same whorl numbers, less prominently ribbed, last whorl less angulated, aperture without lip inside.

2.9-4.3 x 4.8-8 mm

In high mountains above 1000 m, alpine meadows and rocky elevations.

Hausdorf 1991: 125, Hausdorf 1996: 69, Gittenberger 1993: 328.

Helicopsis instabilis (Rossmässler, 1838)

Helix instabilis Rossmässler, 1838 in Rossmässler 1838-1844 (7-8): 33, Taf. 38

Shell white or with light brownish hue, rarely with weak and fine yellowish brown bands, prominently and densely striated, 5 weakly convex whorls, last whorl width more than 2 x of penultimate whorl, initial whorls with edge at periphery, last whorl near aperture rounded, aperture with weak lip inside, umbilicus very wide, 1/3 of shell diameter.

Open habitats, tolerates dry conditions.

Westerlund 1889: 344, Poliński 1922, Damjanov & Likharev 1975: 354, Shilejko 1978: 217, Grossu 1983: 431, Hausdorf 1996: 69, Irikov & Erőss 2008, Egorov 2008: 86, Sysoev & Schileyko 2009: 190.

Range insufficiently known.

6.1-7.1 x 11-15 mm

Bulgaria: Galata near Varna

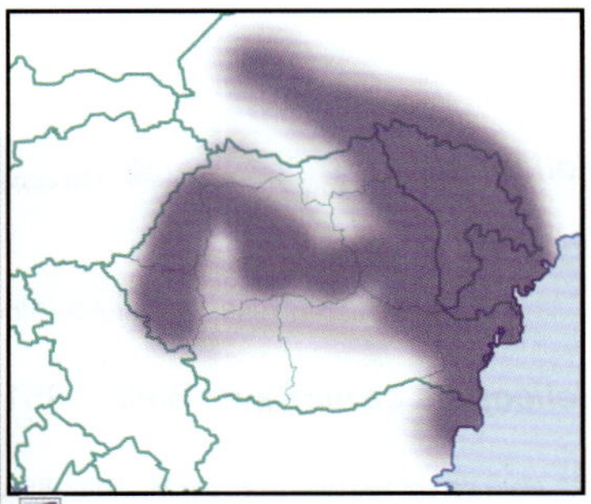

SW Ukraine to Bulgaria

Paolo Zaccaria

Tiroler Landesmuseen Innsbruck

Helicopsis paulhessei (Lindholm, 1936)

S Ukraine (Crimea)

Helicella paulhessei Lindholm, 1936 in Lindholm 1936: 439 (*Helicella paul-hessei*)

Shell white usually with brown colour bands (15 % of the shells in a population are evenly white, another 25 % have pale or weak bands), not shiny, densely and regularly striated, 5-5.5 whorls, aperture without lip, margin sharp and not reflected, umbilicus wide with all whorls visible inside.

7.5-10 x 12.5-15.5 mm

Was found in the park of the village of Gaspra.

Only known from one locality, the park of the village of Gaspra, not recently found again. Either an extinct species, or misidentified and a synonym of *Hcp. retowskii* or *Hcp. dejecta*.

Shilejko 1978: 216, Egorov 2008: 86, Sysoev & Schileyko 2009: 190, Gural-Sverlova 2012.

Helicopsis retowskii (Clessin, 1883)

Helix retowskii Clessin, 1883 in Clessin 1883: 47 (*Helix Retowskii*)

5-8 x 8-13 mm

Ukraine: Crimea, Mnogorichchya near Yalta

Shell white, solid, finely ribbed on both sides, 4-5.5 convex whorls, keeled at periphery, keel crenulated, weaker at last whorl, aperture rounded, with weak white lip deep inside, margin sharp, umbilicus open.

S Ukraine (Crimea)

Dry slopes among grasses.

Westerlund 1889: 323, Shilejko 1978: 215, Egorov 2008: 86, Sysoev & Schileyko 2009: 191, Gural-Sverlova 2012: 17.

Helicopsis striata (Müller, 1774)

Helix striata Müller, 1774 in Müller 1774: 38

4.5-6.5 x 6-9 mm

Sweden: Öland, Möckleby

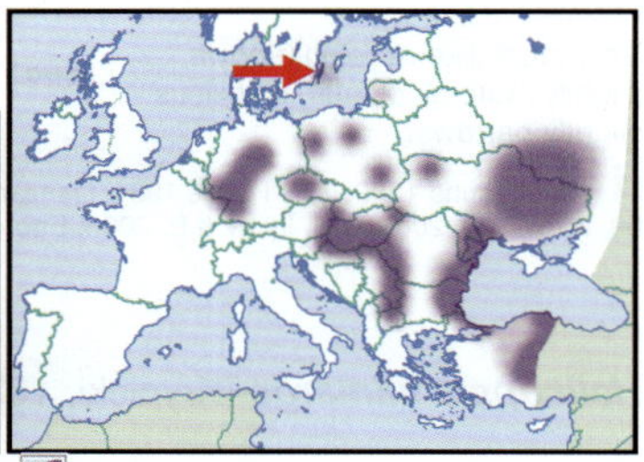

Central and E Europe

Shell whitish, often with dark and interrupted colour bands, coarsely and irregularly striated, also at lower side, apex 1.5 mm wide, 4-4.5 weakly convex whorls with moderately deep suture, relatively flat above but well rounded at periphery, last whorl slightly descending near aperture, aperture rounded and without or with weak lip inside, umbilicus moderatley wide and often getting slightly wider at last whorl, 1/6 of shell diameter.

Differs from *Xerocrassa geyeri* in its larger protoconch (*T. geyeri* 1.0 mm), less interrupted ribs and slightly larger umbilicus. *X. geyeri* has 2 rudimentary dart sacs, *H. striata* 4 active dart sacs. *Candidula intersecta* is more flattened above and below with shallower suture, less coarsely ribbed and has a stronger lip inside the aperture.

Dry and open habitats, often at calcareous rocky places, usually in natural habitats such as stable grasslands but also at margins of cultivated fields.

Critically Endangered in Germany and Poland. Extinct in SE England since the early Postglacial period, disappeared after forest growth, the main threat.

Tomić 1959: 26, Damjanov & Likharev 1975: 351, Shilejko 1978: 214, Kerney et al. 1983: 251, Hausdorf 1990: 64, Hausdorf 1996: 69, Kerney 1999: 185, Schütt 2001: 444, Zajac 2005: 15, Sverlova 2006: 134, Egorov 2008: 86, Sysoev & Schileyko 2009: 191.

Helicopsis subcalcarata (Naegele, 1903)

Helix subcalcarata Naegele, 1903 in Naegele 1903: 173

Shell solid, densely striated, whitish with two keels, lower keel stronger than upper keel, aperture circular, umbilicus wide. Epiphallus : flagellum ratio 2.3-3.0.

4-5 x 7-9 mm

Dry calcareous slopes.

Hausdorf 1990: 60, Schütt 2001: 445.

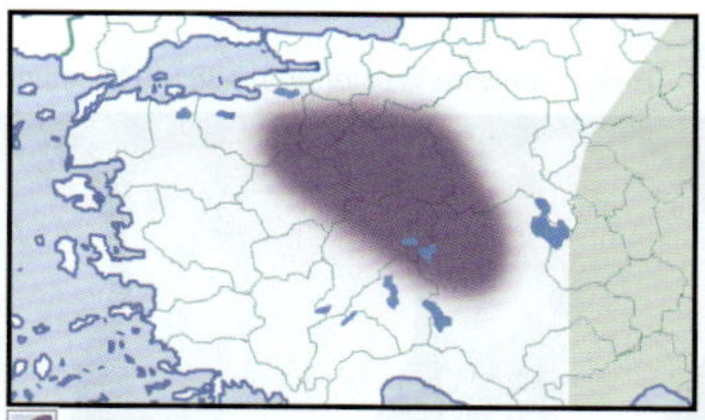

Central and W Turkey

Helicotricha

Helicotricha Giusti, Manganelli & Crisci, 1992 in Giusti et al. 1992: 109

Type species *Helicotricha carusoi* Giusti, Manganelli & Crisci, 1992 (original designation)

Right ommatophore retractor independent of genitalia, penial nerve originating from right cerebral ganglion, distal vagina long, to pairs of dart-sac (small inner and larger outer) stylophores (inner stylophores = accessory sacs) disposed on opposite sites of vagina.

Helicotricha carusoi Giusti, Manganelli & Crisci, 1992

Helicotricha carusoi Giusti, Manganelli & Crisci, 1992 in Giusti, Manganelli & Crisci 1992: 112

Shell opaque brown with white spots, hairy, 4-4.5 convex whorls with deep suture, last whorl slightly angulated, sometimes descending near aperture, aperture simple without internal lip, slightly reflected at columellar margin, umbilicus wide, 1/7-1/5 of diameter. The animal is light greyish transparent.

2.5-3.4 x 4.5-5.4 mm

In various kinds of habitats on small islands, under stones, dry leaves and pieces of wood, often together with *Xerotricha conspurcata*.

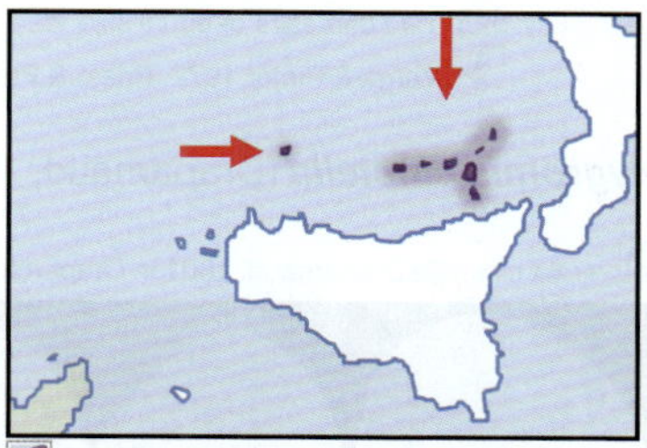

Italy (Eolian islands and ?Ustica)

Hiltrudia

Hiltrudia Nordsieck, 1993 in Nordsieck 1993 (Arch. Moll. 122): 14

Type species *Helix mathildae* Westerlund, 1881 (original designation)

Junction between spermatheca duct and free oviduct is not inflated, flagellum not very long. Surface structure differs from *Metafruticicola*, embryonic shell with distinct spiral sculpture.

Maassen 1995: 9-22.

Hiltrudia globulosa Subai, 2009

Hiltrudia globulosa Subai, 2009 in Subai 2009: 107 (*Hiltrudia (?) globulosa*)

Shell globulous, yellowish to brownish, weakly striated, microsculpture with a fine network of longitudinal scales arranged in radial patterns, 5-6 convex whorls, last whorl only slightly wider than penultimate whorl, slightly descending, apertural margin sharp, weakly reflected, no teeth, umbilicus very narrow, 1/15 of shell diameter and not increasing at lower whorls.
Shell more depressed than in *Hi. kusmici* and *Hi. mathildae*, with narrower umbilicus, scales in *Hi. globulosa* are shorter and more densely arranged.

4-6.5 x 6-8.5 mm

Anatomy unknown and classification in genus *Hiltrudia* uncertain. The status as an independent species was made likely by a biometric analysis.

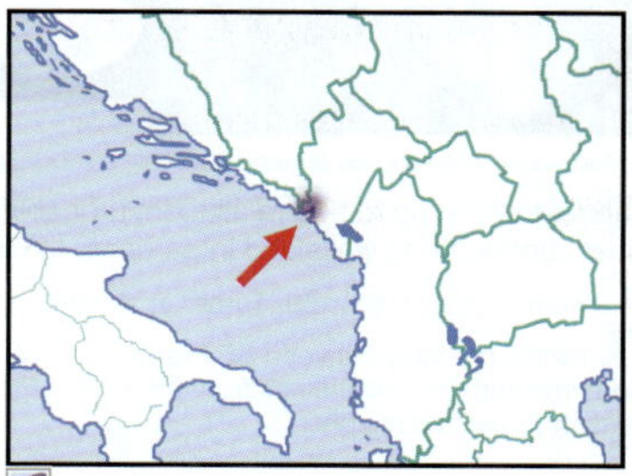

W Montenegro

Only known from 3 localities.

Hiltrudia kusmici (Clessin, 1887)

Helix kusmici Clessin, 1887 in Clessin 1887: 51 (*Helix (Trichia) Kusmici*)

N Albania: Malësia, near Bogë

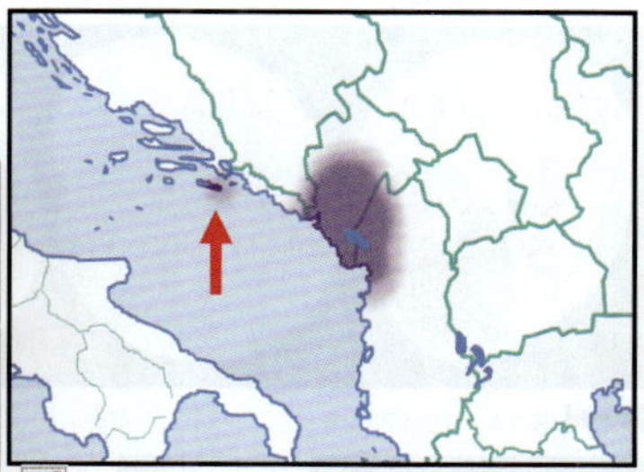

Montenegro, N Albania

Under stones and at the base of limestone rocks among leaf litter.

5-6 x 8-11 mm

Shell thin, pale reddish or browny yellowish, densely and prominently striated, with a dense pattern of tiny riblets (scales), between the scales there is a very fine spiral sculpture, 4.5-5 weakly convex whorls with deep suture, regularly increasing, last whorl slightly more, with weak edge at periphery, aperture inside with thin whitish layer, umbilicus narrow and perspectival, wider near aperture.
Shell differs from *H. mathildae* only in the more prominent scales (in fresh shells) and a less distinct spiral sculpture between.
Vagina very long, slightly longer than penis, epiphallus length 2 x of penis, spermatheca duct at the base broader than oviduct, oviduct and spermatheca duct approximately as long as vagina.

Westerlund 1889: 63, Maassen 1995: 11, Dhora & Welter-Schultes 1996: 157 (as *H. mathildae*, misidentified), Fehér & Erőss 2009: 12, 35.

Hiltrudia mathildae (Westerlund, 1881)

Helix mathildae Westerlund, 1881 in Westerlund 1881: 8

5-6 x 8-11 mm

Croatia: Dalmatia, Brač island

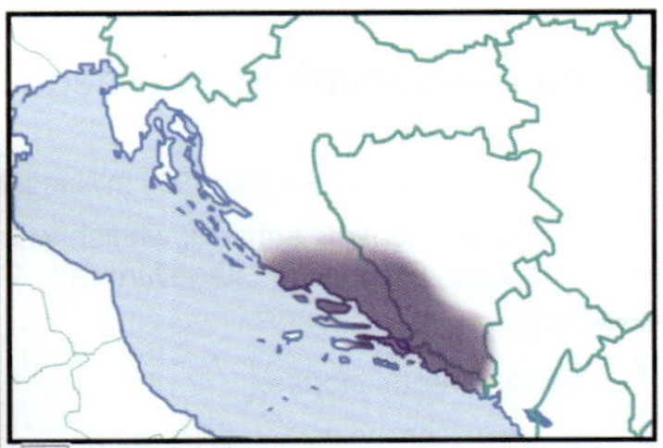

Croatia (Dalmatia), Hercegovina

Westerlund 1889: 64 (synonym of *H. kusmici*), Maassen 1995: 13.

Shell very much like *Hiltrudia kusmici*, perhaps slightly more shiny and with slightly more reddish hue, the scales of the microsculpture are flatter with a more prominent spiral sculpture in between. Weathered shells do not differ from *H. kuscmici*.
Vagina very short, shorter than penis, epiphallus length 2 x of penis, spermatheca duct at the base as broad as oviduct, oviduct and spermatheca duct more than 3 x as long as vagina.

Hygromia

Hygromia Risso, 1826 in Risso 1826: 66

Type species *Helix cinctella* Draparnaud, 1801 (subsequent designation by Herrmannsen 1847 [25-05-1847]: 547)

Two dart sacs of equal size, dart coiled, flagellum short and rudimentary.

Varga & Pintér 1972, Prieto & Puente 1992, Falkner et al. 2002: 153.

Hygromia cinctella (Draparnaud, 1801)

Helix cinctella Draparnaud, 1801 in Draparnaud 1801: 87

6-7 x 10-12 mm

France: Provence, near Montélimar

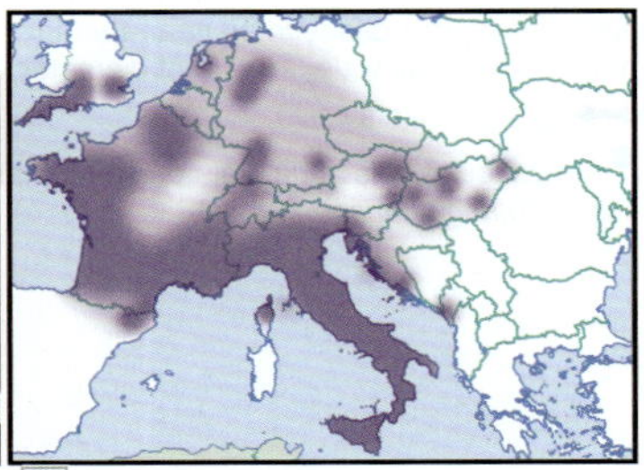

S and central Europe

(1) Switzerland: Rheinfall near Schaffhausen, (2-3) France: near Montélimar

Shell whitish grey to horny brown, slightly translucent, finely and rather regularly striated, 5-6 flattened whorls with shallow suture, last whorl sharply keeled with white edge, aperture simple without lip inside, umbilicus very narrow and almost covered by the reflected columellar margin. In Sicily in a broad variety of colour morphs from green and yellow to reddish, very rarely also with colour bands. The white keel is characteristic. Animal light greyish or with yellowish hue, often with darker greyish or brownish head and tentacles, narrow, moving slowly.

Open habitats, among low shrubs and herbs, often at small water currents or in river valleys. In Italy in cool and humid habitats such as river banks and valleys. In Switzerland almost exclusively in cultivated habitats, in gardens at shrubs and herbs, rarely in natural forests, up to 900 m altitude, probably not native there. In England among grass, *Urtica dioica*, *Hedera helix*, umbellifers and similar plants on road-sides, under stone walls and in gardens. In Germany in disturbed habitats of densely populated regions.

Kerney et al. 1983: 159, Turner et al. 1998: 357, Kerney 1999: 193, Vimpère 1999, Beckmann & Kobialka 2008.

Native probably between SE France and NW Croatia, introduced since the 1930s with cultivated garden plants to N France, Britain and central Europe. First records in Germany and the Netherlands from the 1990s. Probably much more widely distributed than currently known because malacologists rarely look for snails in disturbed and densely populated regions.

Hygromia limbata (Draparnaud, 1805)

Helix limbata Draparnaud, 1805 in Draparnaud 1805: 100
Not *Helix limbata* Mendes da Costa, 1778 (in Mendes da Costa 1778: 63) which was not used after 1899 and suppressed under Art. 23.9.1 by Falkner et al. 2002: 153 who provided the 25 required references.
Helix odeca Bourguignat, 1882 in Locard 1882: 314 Type locality Hendaye (SW France), population today extinct.
Name and entire description were attributed to Bourguignat.

8-14 x 12-17 mm

France: Charente, Angoulème

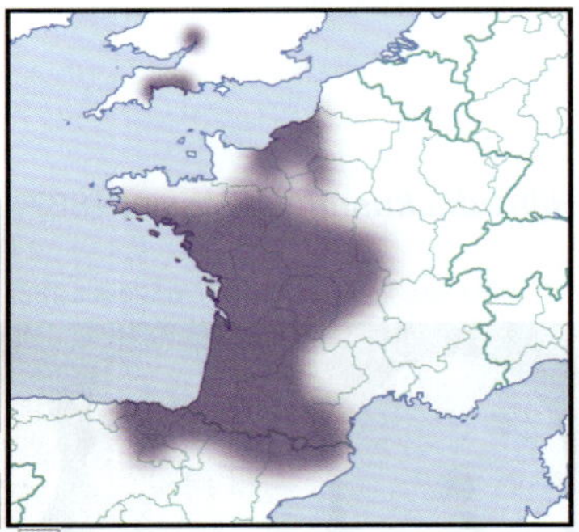

W Europe

Spain: Pyrenees, Espot Giuseppe Pocaterra

Shell creamy white to dark brown, often with white or dark colour bands near the periphery, translucent, striated, 5-6 weakly convex whorls with shallow suture, initial whorls keeled, last whorl with edge at periphery, slightly descending near aperture, aperture with white lip inside, margin reflected, umbilicus very narrow and oblique, partly covered by the reflected columellar margin.
Animal transparent and very pale yellowish, in frontal part with weak reddish hue, tentacles weakly greyish, moves slowly and is shy. Animal colour variable, occasionally reddish or yellowish brown.

Forests, shrubs, hedges, in lower vegetation, up to 2500 m. In England in moist, grassy habitats among herbage and ground litter in gardens and other disturbed places.

30-60 eggs (diameter 1.8 mm) are laid in the summer season in a transparent whitish envelope, juveniles hatch after 15-20 days, maturity is reached in the following year.

Gassies 1849: 88, Moquin-Tandon 1855: 194, Westerlund 1889: 100, Germain 1930: 259, Kerney et al. 1983: 260, Prieto & Puente 1992: 394, Kerney 1999: 194.

Germain 1930 and Prieto & Puente 1992 regarded *odeca* from SW France as a synonym of *limbata*. Falkner et al. 2002: 152 gave a definition for this taxon: type locality near Hendaye at sea level, species characterised by a long dart as found in a specimen at 40 km distance and 1000 m altitude. They argued that different forms of darts (long, short and corkscrew) would provoke reproductive isolation, this was however not based on published results of detailed studies. They used *odeca* for long-dart forms of SW France and Spain. It would be necessary to show where populations with short, long and corkscrew darts actually live, and if they actually occur sympatrically, as suggested by Falkner et al. 2002. Their taxonomic relationships as well as presence or absence of gene flow among populations could be made visible in molecular studies.

Hygromia tassyi (Bourguignat, 1884)

Helix tassyi Bourguignat, 1884 in Bourguignat 1884: 356 (*Helix Tassyi*)
Hygromia gofasi Prieto & Puente, 1992 in Prieto & Puente 1992: 384

Shell evenly pale greenish or horny yellowish and weakly spotted, shiny, locally very fragile like a vitrinid, finely and irregularly striated, occasionally with a whitish band at the periphery which disappears on the last whorl, 5-5.5 weakly or moderately convex and slowly increasing whorls with moderate or weak suture, last whorl rounded or initially angulated, not much wider than penultimate whorl, very weakly descending near aperture, apertural lip whitish, not thickened but reflected, covering completely or largely the umbilicus.

5.5-8 x 8-12 mm, locally variable

Humid and rocky habitats in high mountain valleys.

EU Red List: Vulnerable VU. Restricted to a small range, no acute threats known.

Westerlund 1889: 100, Germain 1930: 260, Prieto & Puente 1992, Bertrand 2000: 37, Falkner et al. 2002: 154 (analysed material from type localities and regarded *gofasi* as a synonym).

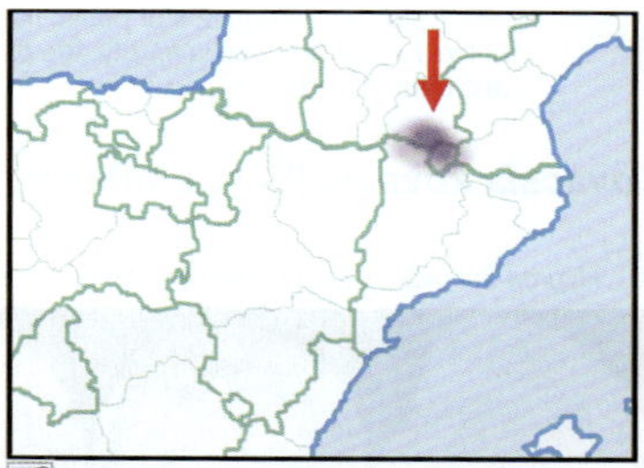

Pyrenees

Ichnusomunda

Ichnusomunda Giusti & Manganelli, 1998 in Giusti & Manganelli 1998: 319

Type species *Ichnusomunda sacchii* Giusti & Manganelli, 1998 (original designation)

Right ommatophore retractor independent of genitalia, penial nerve from right pedal ganglion, left pallial lobe of pallial margin long, vagina with digitiform glands and 0 + 2 dart sac complex, penis with long papilla, epiphallus and flagellum.

Italy (islands of Sardinia and Ustica), possibly also N Africa

Manganelli, Oliv. et al. 2001 tried to find out with which Italian *Cernuella* species *usticensis* and *sacchii* were most closely related, and did not come to clear results. The study left the impression as if the two species were recently introduced to Italian coasts from N Africa, for example by pirates, and belonged to scarcely studied N African groups.

Ichnusomunda sacchii Giusti & Manganelli, 1998

Ichnusomunda sacchii Giusti & Manganelli, 1998 in Giusti & Manganelli 1998: 320

Shell globose, whitish with 2-8 brown interrupted colour bands, 5 convex whorls, also on upper shell side, with deep suture, aperture simple and usually without lip, umbilicus narrow and partly covered by columellar margin of aperture.
Xerosecta contermina is smaller with a slightly straighter outline of the shell and a finely striated sculpture.
Xerosecta dohrni is very similar and differs in its slightly larger size, more distinct colour bands and presence of a lip inside the aperture, also perhaps slightly less convex whorls.

6.5-8.8 x 7.8-10.3 mm

Sandy dunes under shrubs or herbal vergetation, usually hidden in the sand during the day and generally hidden in the dry season.

EU Red List: Critically Endangered CR. A locally frequent but microendemic species only known from one single coastal section of Sardinia with sandy dunes. Threatened by human pressure causing coastland destruction, and pine afforestation.

Italy (SW Sardinia)

Ichnusomunda usticensis (Calcara, 1842)

Helix usticensis Calcara, 1842 in Calcara 1842: 53 (*Helix Usticensis*)

Shell whitish, often with brownish spots or marbled pattern on upper side and with several fine colour bands on lower side, irregularly and finely ribbed, 4.5-5 flat and slowly increasing whorls, with sharp keel at periphery, upper side straight and lower side more rounded, aperture often with strong white lip inside, columellar margin short and descending in a sharp perpendicular angle, widened in upper section, umbilicus narrow. The animal is whitish grey with a black dorsal line.

7-10 x 12-18 mm

Italy: Ustica island
Paolo Zaccaria
Tiroler Landesmuseen Innsbruck

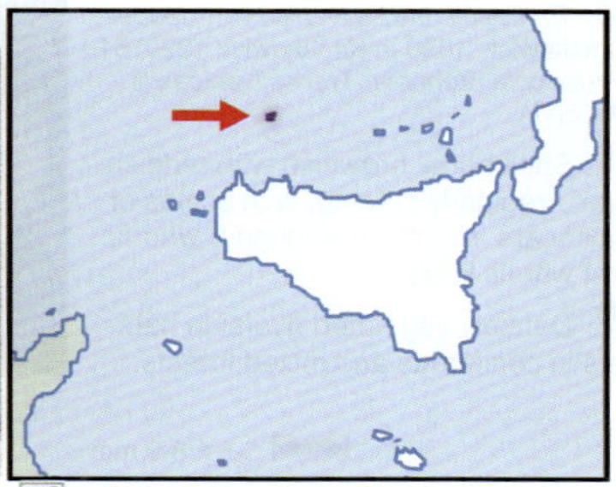

S Italy (Ustica island)

Restricted to anthropogenic and disturbed habitats of Ustica island.

EU Red List: Vulnerable VU. Threatened by coastal land use on the small island.

Calcara 1845 (Atti Pal.): Tav. Fig. 8, Westerlund 1889: 311, Manganelli et al. 1995: 29, Manganelli, Oliv. et al. 2001.

Classification uncertain, relationships with Italian *Cernuella* species unclear. Possibly an endemic species of Ustica, but the anthropogenic occurrence suggests a recent introduction from an unknown locality, possibly located in N Africa.

Ichnusotricha

Ichnusotricha Giusti & Manganelli, 1987 in Giusti & Manganelli 1987: 160

Type species *Ichnusotricha berninii* Giusti & Manganelli, 1987 (original designation)

Ichnusotricha berninii Giusti & Manganelli, 1987

Ichnusotricha berninii Giusti & Manganelli, 1987 in Giusti & Manganelli 1987: 161

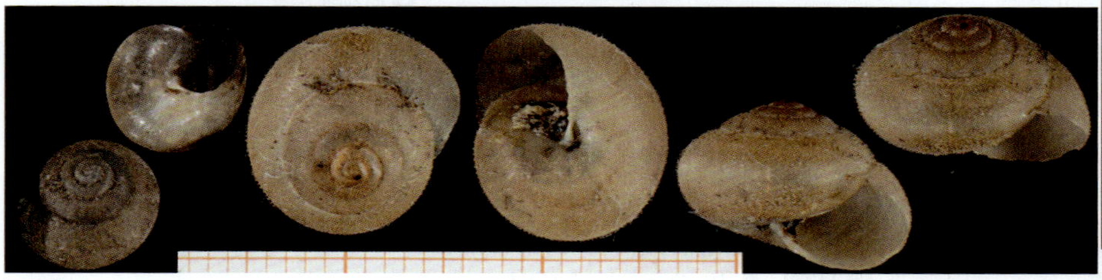

7-10 x 10-11 mm

Italy: Sardinia, Nuoro, Laconi

Shell yellowish brown, with fairly regular transverse rows of short hairs, 4-5 very weakly convex whorls, last whorl can be weakly angulated, aperture simple, not thickened, margin usually not reflected, umbilicus very narrow and almost entirely covered.
Animal pinkish, upper palleal wall with irregular black spots.

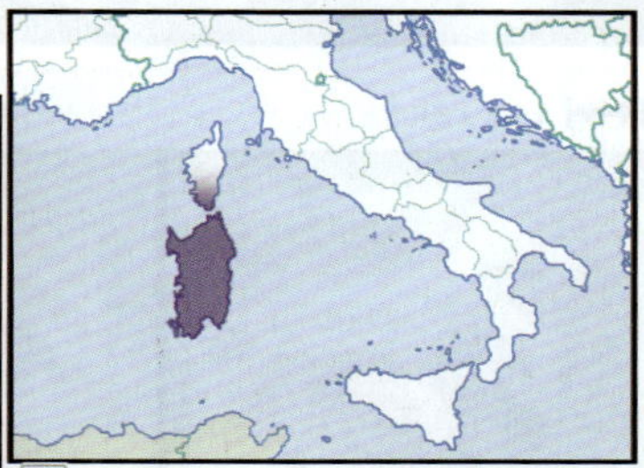

Italy (Sardinia), France (S Corsica)

In leaf litter and among herbal vegetation in moderately open habitats.

Manganelli et al. 1995: 28.

Kovacsia

Kovacsia Nordsieck, 1993
in Nordsieck 1993: 13

Type species *Hygromia kovacsi* Varga & Pintér, 1972
(original designation)

Epiphallus length 2 x of penis, flagellum long, thin and pointed, as long as epiphallus, two roughly equally large dart sacs are fixed symmetrically at the side of the vagina, one sac is empty, the other one has a 0.8 mmn short dart, 4 mucus glands present, having two branches each, spermatheca usually oval with long and often coiled duct.

Kovacsia kovacsi (Varga & Pintér, 1972)

Hygromia kovacsi Varga & Pintér, 1972 in Varga & Pintér 1972: 121

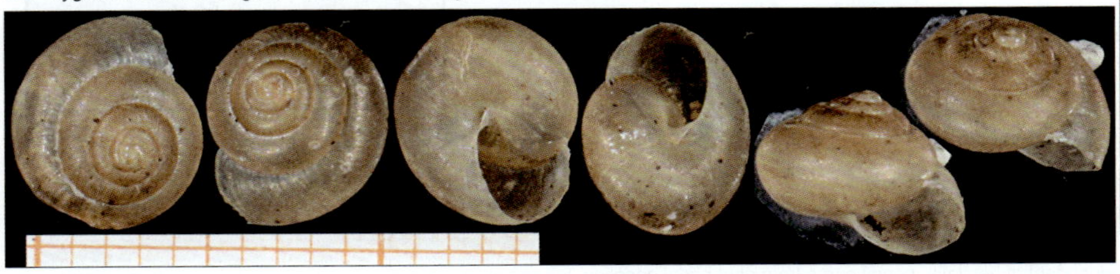

4-5 x 6-7 mm

Hungary: Doboz

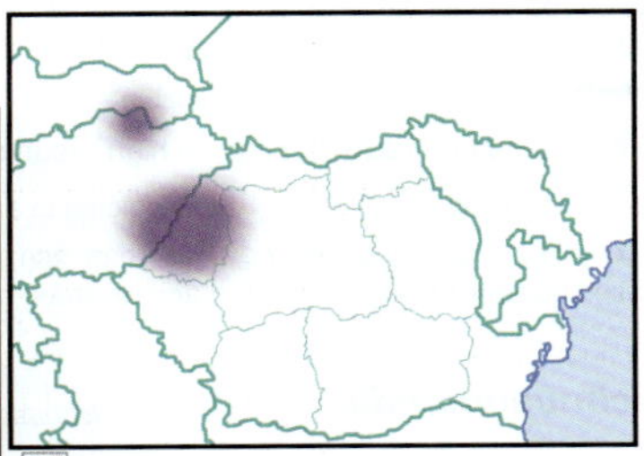

Hungary, Romania

Under leaf litter in deciduous forests.

Shell horny brown, often with a white band at the periphery, slightly translucent, striated, surface structure with very fine, slightly oblique (to the growth lines) and short riblets (0.55-0.09 mm long, 350-500/mm^2 on penultimate whorl, 180-240/mm^2 on last whorl), which become longer with increasing distance from the suture, 4.5-5 whorls, with edge at the periphery but rounded near the aperture, apertural margin sharp and without lip, umbilicus always open and partly covered by the reflected columellar margin of the aperture. Differs from *Lozekia transsilvanica* in its smaller and shorter riblets (120-160 riblets/mm^2 and 0.08-0.25 mm long in *L. transsilvanica*).

Fehér, Va. et al. 2009 (2 papers).

Hungary: Doboz forest
Jozef Grego

Lozekia

Lozekia Hudec, 1970
in Hudec 1970: 41 *Hygromia (Lozekia)*

Type species *Helix transsilvanica* Westerlund, 1876
(original designation)

Two dart sacs of different sizes, dart long and bent, flagellum relatively long.

Varga & Pintér 1972, Grossu 1983: 475, Fehér, Va. et al. 2009 (2 papers).

Lozekia deubeli (Kimakowicz, 1890)

Fruticicola transsilvanica var. *deubeli* Kimakowicz, 1890 in Kimakowicz 1890: 51 (*Fruticicola (Zenobia) Transsilvanica* var. *Deubeli*)

Shell horny brownish with reddish hue, irregularly striated, with scales or hair scars, slightly more than 5 whorls, first whorls keeled.

Densely vegetated riverside habitats in coniferous and mixed forests.

5-6 x 8-9 mm

Romania: Bucegi mountains, W of Predeal

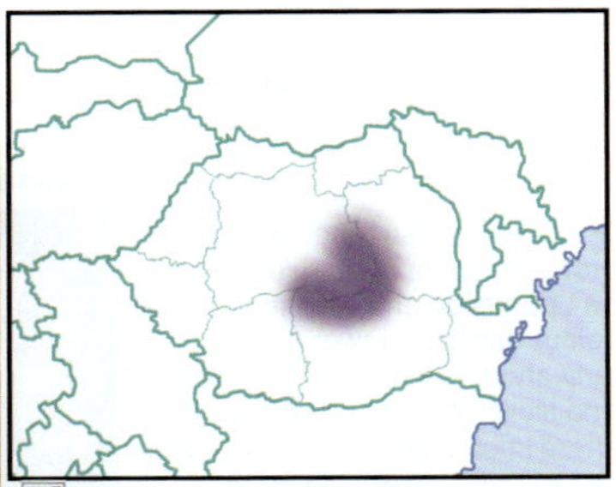

Romania (SE Carpathians)

Zoltán Fehér
Museum Budapest HNHM 96646

Lozekia transsilvanica (Westerlund, 1876)

Helix transsilvanica Westerlund, 1876 in Westerlund 1876-1878: 55
Original spelling was *transsilvanica*.

4.5-6 x 5.5-6.5 mm

(1-4) Slovakia: Šiatorská Bukovinka, near Mt. Karanč, (5) Romania: Transilvania, Sibiu, Dumbrava

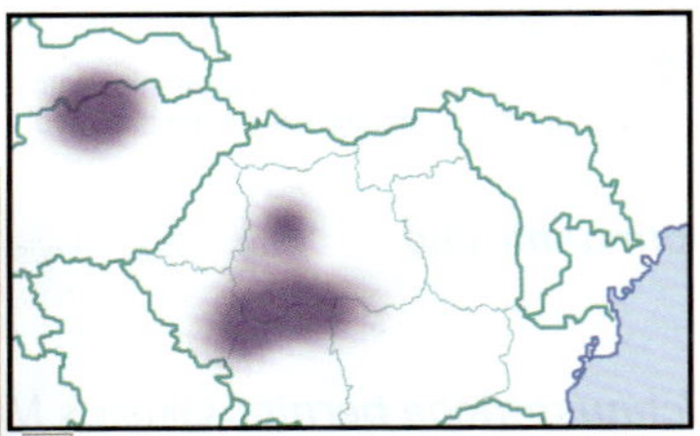

Slovakia, Hungary, Romania

(5) Zoltán Fehér
(5) GNHM Göteborg, Westerlund 488/1 Lectotype

Slovakia: Šiatorská Bukovinka, near Mt. Karanč
Jozef Grego

Shell horny grey, greenish or yellowish brown, extremely finely striated, surface structure with fine and short (0.08-0.25 mm) riblets (120-160/mm^2), 5 weakly convex whorls, rapidly increasing, last whorl inflated and with weak edge at periphery, slowly and straightly descending near aperture, margin straight and thin, umbilicus very narrow and almost covered by the reflected columellar margin.
Differs from *Kovacsia kovacsi* in its larger and longer riblets (*K. kovacsi* 0.06-0.09 mm long, 180-500 riblets/mm^2).

Deciduous forest habitats.

Westerlund 1889: 63, Varga & Pintér 1972, Grossu 1983: 475, Fehér, Va. et al. 2009 (2 papers)..

Mengoana

Mengoana Ortiz de Zárate y López, 1949
in Ortiz de Zárate y López 1949: 291 *Euomphalia (Mengoana)*

Type species *Euomphalia brigantina* Ortiz de Zárate y López, 1949 (monotypy)

Year could be 1951 (this was proposed by Falkner et al. 1996 without source or evidence).

Certainly the most sophisticated establishment of a genus in European molluscs.
Ortiz de Zárate y López 1949 (true date could be 1951) established *Euomphalia (Mengoana)* in soft words ("I think the species [from N Spain] must be included in the genus *Euomphalia*, but forming for it a new subgenus *Mengoana*", "...*Mengoana* for a subgenus where the studied species could be included" (pueda ser incluída)). This could be interpreted as an expression of doubts. I would suggest not to interprete this in a way that the species was doubtfully included (if yes then *Mengoana* would not be available under Art. 67.2.5).

Ortiz de Zárate y López 1949 was aware that *Helix brigantina* as originally established by Da Silva Mengo 1867 could have belonged to *Helix inchoata* Morelet, 1845 (from Portugal), not to the studied species (from Asturias in N Spain) which he had in mind and for which he established the new subgenus *Euomphalia (Mengoana)*.
Kobelt 1879: 37, Taf. 160 Fig. 1626 had assigned the name *Helix brigantina* to the Asturian species, Ortiz de Zárate y López 1949 used Kobelt's interpretation of *H. brigantina* to establish the new subgenus *Mengoana*. Today we know that *Helix brigantina* Da Silva Mengo, 1867 was a junior subjective synonym of *Helix inchoata* Morelet, 1845.
Falkner 1996: 103 proposed to accept *Mengoana* as a correctly established subgenus under Art. 11.10 (deliberate employment of a misidentification as the type species of a new subgenus). I would suggest to accept this proposal.

Ortiz de Zárate y López 1949 expressed some doubts on the correct name *brigantina*, and suspected that the name *Helix jeschaui* Kobelt, 1878 could be the correct name. The doubts concerned only the correct name, not the nominal species from Asturias (Glossary, nominal taxon, this means here the concept of the species), so Art. 67.2.5 (doubtful or conditional inclusion of a species in a genus) does not apply here.
Ortiz de Zárate y López 1949 stated that *Helix brigantina* synonym *Helix Jeschaui* as mentioned by Kobelt 1879 [1878]: 37 (with description) should be used for the Asturian species under the condition that Da Silva Mengo's name referred to another species. This could have made the name *Helix jeschaui* Kobelt, 1878 available under Art. 11.6.1 in combination with 15.1 and 11.5.1.
Strictly taken this name was not made available by Ortiz de Zárate y López 1949: 291 (Art. 11.6.1, 15.1, 11.5 and Glossary "adopt") (in the Glossary "adopt" means use as "the valid name" and lacks an emendation concerning conditional uses, while Art. 11.5 had an extra emendation 11.5.1 that names proposed conditionally should also be available). The others would accept *Helix jeschaui* as available, but then Ortiz de Zárate would have used two available names as valid (*jeschaui* conditionally and *brigantina* really) and *Mengoana* would be left without originally fixed type species (Art. 13.3), and would be unavailable (Art. 68.3 allows only one available name used for the type species by monotypy).
I would suggest to regard *jeschaui* as not having been made available.

Only in this case the type species of *Mengoana* Ortiz de Zárate y López, 1949 should be *Euomphalia brigantina* Ortiz de Zárate y López, 1949 by monotypy.

Mengoana brigantina (Ortiz de Zárate y López, 1949)

Helix brigantina synonym *Helix Jeschaui* Kobelt, 1878 in Kobelt 1879: 37, Taf. 160
Not made available by Ortiz de Zárate y López 1949: 291 because not used as "the valid name" (see above).
Euomphalia brigantina Ortiz de Zárate y López, 1949 in Ortiz de Zárate y López 1949: 291 (*Euomphalia (Mengoana) brigantina*)
Deliberate employment of a misidentification under Art. 11.10, Ortiz de Zárate expressly employed the name *Helix brigantina* in the sense of a misidentification used by Kobelt 1879: 37 and not in the taxonomic sense of Da Silva Mengo 1867: 170 (this was *Helix Brigantina*, junior subjective synonym of *Helix inchoata* Morelet, 1845, currently = *Portugala inchoata* (Morelet, 1845)).

10-11 x 12-15 mm

Spain: Asturias, Picos de Europa, Lago La Ercina

N Spain

Humid habitats in forests, meadows and rocky areas.

Westerlund 1889: 90, Prieto & Gómez 1988, Puente & Prieto 1992: 141, Falkner 1996: 103, Ondina et al. 1997.

Shell evenly horny yellowish, striated, thin, almost translucent, shiny, 5 moderately convex whorls, shape globulous, last whorl only slightly descending near the aperture, apertural margin thin, sharp, without internal lip, umbilicus narrow, partly covered by the reflected columellar margin.

Metafruticicola

Metafruticicola Ihering, 1892
in Ihering 1892: 452 *Helix (Metafruticicola)*

Type species *Helix pellita* Férussac, 1832
(subsequent designation for *Metafruticicola* by Pilsbry 1894 [1895]: 276)

Gender is feminine because the name was derived from *Fruticicola* which is feminine under the French Code ("An sie [*Helix ciliata* and others] schließen sich meines Erachtens diese Formen [*Helix sublecta* Maltzan 1884, *Helix noverca*, *Helix westerlundi*, *Helix pellita*] an, welche Hesse [1884, Jahrbuch DMG 11: 237] als *Pseudocampylaea* vereint, einen bereits von Pfeifer [sic] vergebenen Namen anwendend, den ich in *Metafruticicola* umändere und für die metabelogonen Fruticicolen verwende."). In Latin a noun takes always the gender of the stem.
Expressly established as a new replacement name for
Pseudocampylaea Hesse, 1884 in Hesse 1884: 237 *Helix (Fruticicola (Pseudocampylaea))* (not *Pseudocampylaea* Pfeiffer, 1877, Hygromiidae, currently used for a genus from Madeira)

Albania to Israel, Cyprus

No dart apparatus, junction between spermatheca duct and free oviduct inflated, penis papilla prolongated and with complicate structure, flagellum present and long, protoconch usually sculptured.

Hesse 1931: 28, Maassen 1995: 10, Hausdorf et al. 2004, Welter-Schultes 2007.

The genus needs a revision since long (Schultes & Wiese 1991: 84).

Metafruticicola andria (Martens, 1889)

Helix andria Martens, 1889 in Martens 1889: 181, Taf. 10

Shell horny or greyish yellow, striated and with a dense micropattern of convex riblets, 5-6 slightly convex whorls, last whorl inflated and initially with edge at periphery, descending near aperture, aperture with white or reddish lip inside, margin reflected, umbilicus narrow.
Like *M. schuberti*, but with distinct surface sculpture. The riblets are convex towards the apex, not towards the aperture.

8-11 x 13-15 mm

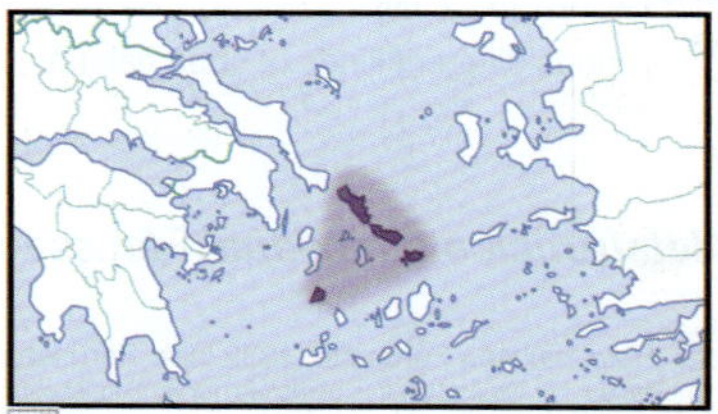

Greece (Kikládes)

Metafruticicola berytensis (Pfeiffer, 1841)

Helix berytensis Pfeiffer, 1841 in L. Pfeiffer 1841 (Symb.): 39
Previously mentioned as *Helix (Helicella) berytensis* by Férussac [1821-1822]: folio p. 47 (nomen nudum, no description).

Shell thin, yellowish brown, lighter near suture and umbilicus, not shiny, finely striated and densely covered with very short hairs, with very fine and dense granular pattern, more prominently striated near suture, 5.5 flat whorls, last whorl broadened, with blunt edge at periphery, suddenly descending near aperture, aperture with thin white lip inside, palatal and basal margins slightly reflected, umbilicus narrow but deep.

10-12 x 14-20 mm

Mountain slopes with low vegetational cover.

Westerlund 1889: 69, Schütt 2001: 459.

N Israel to S Turkey, Cyprus

Metafruticicola coartata Fuchs & Käufel, 1936

Metafruticicola grelloisii coartata Fuchs & Käufel, 1936 in Fuchs & Käufel 1936: 643

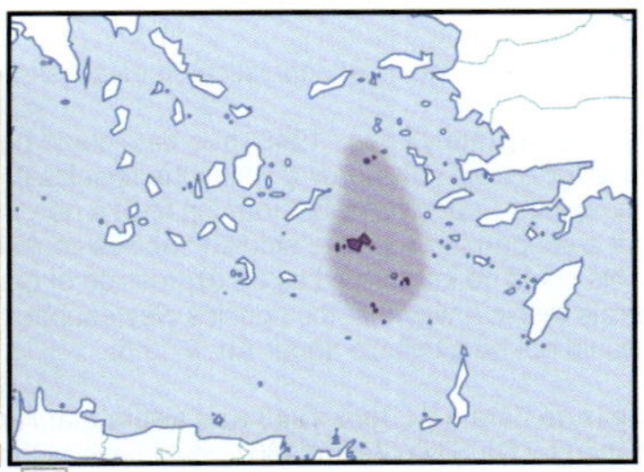

7-10 x 11-16 mm, locally variable

Greece: Kinaros island

Greece (SEAegean islands)

Shell brownish, striated, sculpture locally variable, 5-6 whorls, last whorl strongly descending near aperture, aperture rounded and slightly reflected, margins approaching each other very closely, umbilicus narrow, 1/10 of diameter, often partly covered by the reflected apertural margin.

Metafruticicola dedegoelensis Hausdorf, Gümüş & Yıldırım, 2004

Metafruticicola dedegoelensis Hausdorf, Gümüş & Yıldırım, 2004 in Hausdorf et al. 2004: 168

Shell horny brownish with an indistinct whitish band at the periphery, irregularly and densely ribbed, 4.5-5 weakly convex whorls, last whorl initially bluntly angulated, aperture very oblique, not very closely connected at parietal side, with whitish lip, umbilicus wide, 1/4 of shell diameter.

6-7 x 13-16 mm

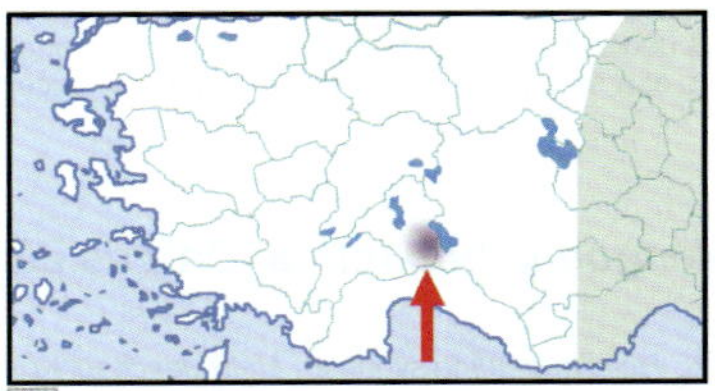

SW Turkey

Metafruticicola kizildagensis Gümüş & Neubert, 2012

Metafruticicola kizildagensis Gümüş & Neubert, 2012 in Gümüş & Neubert 2012: 31

Shell relatively flat, dirty white or pale brownish with a brown colour band at the periphery and a lighter zone below it, aperture nearly circular and reflected, margins distant from each other but connected by a callus at the parietal side, umbilicus open, 1/7 of shell diameter.
Considerably larger than *M. dedegoelensis* and *M. oerstani*.

10-12 x 19-22 mm

A mountain species, found in 1500 m altitude.

Only known from one locality.

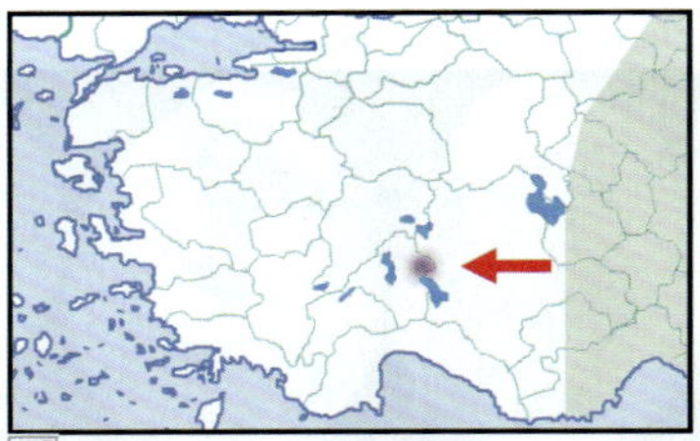

SW Turkey

Metafruticicola lecta (Férussac, 1832)

Helix lecta Férussac, 1832 in A. Férussac & Deshayes 1819-1851: (explication planches) Pl. 69
Original print of explication des planches not verified, second print p. 13
Not *Helix lecta* Rang, 1827 in Rang 1827: 305 (established for a form from Brazil)
Helix lecta Rang, 1827 has probably not been used after 1899. I did not find 25 references for protecting *Helix lecta* Férussac, 1832 under Art. 23.9.1, but I do not know very well the recent Greek literature.
Helix sublecta Maltzan, 1884 in Maltzan 1884: 74

8-13 x 14-20 mm

Greece: E Crete, near Tourloti

Various different kinds of open habitats, requires presence of some vegetation.

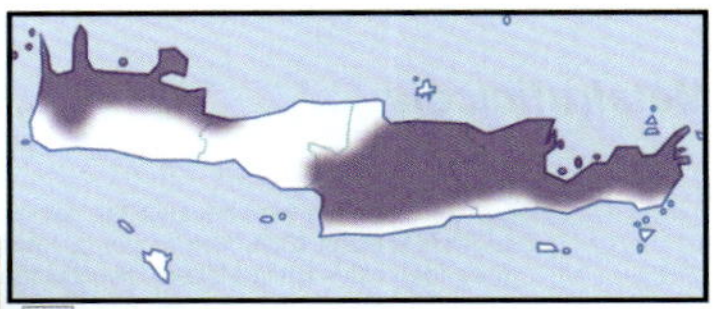

Greece (Crete)

Shell white with 3 narrow brown bands, densely and obliquely striated, 5 very weakly convex whorls, last whorl slightly inflated, descending near aperture, aperture almost circular, with lip inside, margin slightly reflected, approaching each other closely and connected by a whitish layer at parietal side, umbilicus narrow and deep.

Westerlund 1889: 102.

Metafruticicola naxiana (Férussac, 1832)

Helix naxiana Férussac, 1832 in A. Férussac & Deshayes 1819-1851: (explication planches) Pl. 69
Original print of explication des planches not verified, second print p. 13 as *Helix Naxiana*.
Previously mentioned as *Helix naxientia* by Férussac [1821-1822]: folio p. 42 (nomen nudum, without description)

9-13 x 13-20 mm

Greece: N central Crete, E of Iráklio

Shell dark with white streaks and a white band at the peripheral keel, with a reddish band above, striated, 5 not very convex whorls, apertural margin reflected, umbilicus narrow.

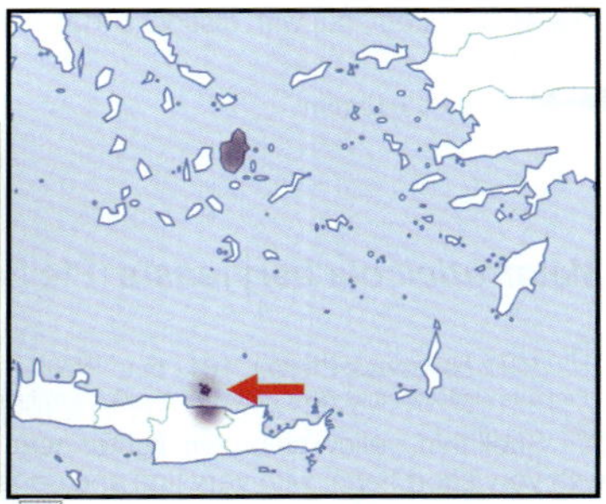

Greece (N Crete, Día, Náxos)

Westerlund 1889: 101, Schultes & Wiese 1990.

Metafruticicola nicosiana (Mousson, 1854)

Helix nicosiana Mousson, 1854 in Mousson 1854: 375
Helix freytagi Maltzan, 1883 in Maltzan 1883: 106 (*Helix Freytagi*)
Helix dictaea Martens, 1889 in Martens 1889: 186, Taf. 3
Metafruticicola nicosiana claudia Bank & Welter-Schultes, 1998 in Welter-Schultes 1998: 69
Name and description were attributed to both co-authors.

5-9 x 10-14 mm

Greece: (1-3, 6) island of Gávdos S of Crete, (4, 5, 7) W Crete, near Argiroúpoli

Greece (Crete to Ródos), Cyprus

Greece: Gávdos S of Crete — Vollrath Wiese

Shell thin, reddish brown with light band at periphery, striated and finely granulated, fresh shells densely covered with hairs, 5-5.5 very slightly convex whorls, initially with edge at periphery, last whorl nearly rounded, aperture with thin lip inside, margin slightly reflected, umbilicus moderate or narrow and partly covered by the reflected columellar margin.
The shell is darker than the animal.

Westerlund 1889: 99, Frank 1997: 87.

In Gávdos mostly below stones or pine needles. Seems to live mostly subterranean. Undergoes predation by *Eopolita protensa* and birds.

Classification of various forms preliminary and uncertain.

Metafruticicola noverca (Pfeiffer, 1853)

Helix noverca Pfeiffer, 1853 in L. Pfeiffer 1853: 648

10-16 x 13-21 mm

Greece: N central Crete, W of Iraklio

Greece (Crete)

Very frequent, lives in various kinds of habitats, usually under stones, all parts of Crete.

Westerlund 1889: 102.

Shell brownish with a whitish band at the periphery, lower side light horny yellowish, densely and regularly obliquely striated, densely granulated, apex prominent, 6 weakly convex whorls, initial whorls with edge at periphery, last whorl rounded, deeply descending near aperture, aperture with white or reddish lip inside, margins approaching each other very closely at parietal side, umbilicus initially very narrow and not deep, wider at last whorl.

Metafruticicola occidentalis Subai, 1999

Metafruticicola occidentalis Subai, 1999 in Subai 1999: 50

8-12 x 11-19.5 mm

Albania: near Librazhd

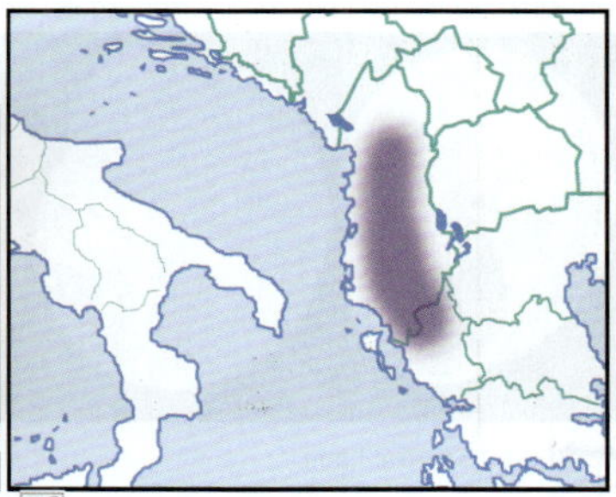

Albania, NW Greece

Shell horny yellow, sometimes with reddish or brownish hue, occasionally with a light or colourless band at the periphery, irregularly striated and very finely granulated, no hairs or hair scars, 5-6 regularly increasing whorls, last whorl width 1.5 x of penultimate whorl, descending near aperture, apertural margin slightly reflected, outside reddish brown, margins approach each other closely (2-4 mm) at parietal side, umbilicus very narrow, widened to 1/15 of shell diameter at last whorl and partly covered by the reflected columellar margin.

Under shrubs, at vegetated slopes, at the soil near rocks under plants or leaves, more rarely under stones.

Metafruticicola oerstani Hausdorf, Gümüş & Yıldırım, 2004

Metafruticicola oerstani Hausdorf, Gümüş & Yıldırım, 2004 in Hausdorf et al. 2004: 170
The name *oerstani* was dedicated to the Turkish malacologist Aydın Örstan, not to be corrected to *orstani* under Art. 32.5 (in contrast to the German ö, the Turkish ö should be reduced to o and not to oe, Art. 32.5.2.1).

Shell light horny yellowish with an indistinct whitish band at the periphery, densely striated and with indistinct spiral lines, and very short (0.1 mm long) hairs, 4.5-5 very weakly convex whorls, last whorl initially bluntly angulated, aperture very oblique, with whitish lip inside, margins not appproaching each other very closely at parietal side, umbilicus wide, 1/5-1/6 of shell diameter.

6-7 x 11-15 mm

SW Turkey

Metafruticicola pellita (Férussac, 1832)

Helix pellita Férussac, 1832 in A. Férussac & Deshayes 1819-1851: (explication planches) Pl. 69
Original print of explication des planches not verified, second print p. 13.
Previously mentioned as *Helix pellita* by Férussac [1821-1822]: folio p. 42 (nomen nudum, without description)

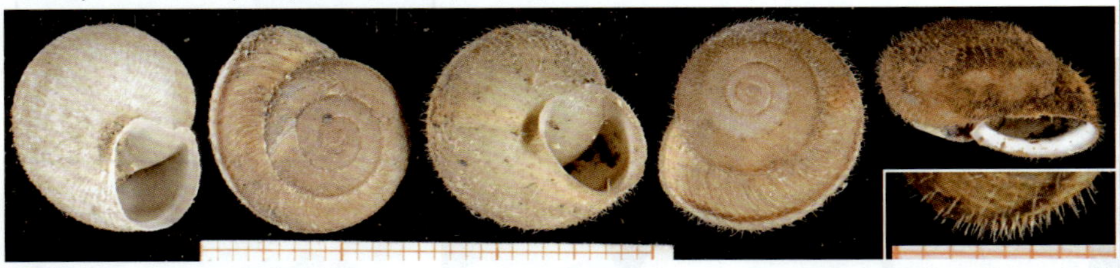

8-12 x 12-16 mm — Greece: E Crete, near Sitia — S Greece

Shell light brownish with weak whitish and reddish brown colour bands, prominently striated and densely covered with long hairs, which can disappear in old shells leaving visible only the hair scars, 4.5-5 initially weakly convex whorls, last whorl rounded and suddenly descending near aperture, aperture almost circular, with strong white lip inside, margins reflected and approaching each other very closely at parietal side, umbilicus narrow and deep.

Limestone habitats.

Westerlund 1889: 101, Vardinoyannis (1994, unpublished thesis), Frank 1997: 91 (Ródos), Schütt 2001: 461 (SW Turkey, uncertain records), unpublished HNC data.

Metafruticicola redtenbacheri (Pfeiffer, 1856)

Helix redtenbacheri Pfeiffer, 1856 in L. Pfeiffer 1856: 176, Taf. 2 (*Helix Redtenbacheri*)
Helix proclivis Martens, 1889 in Martens 1889: 193, Taf. 10

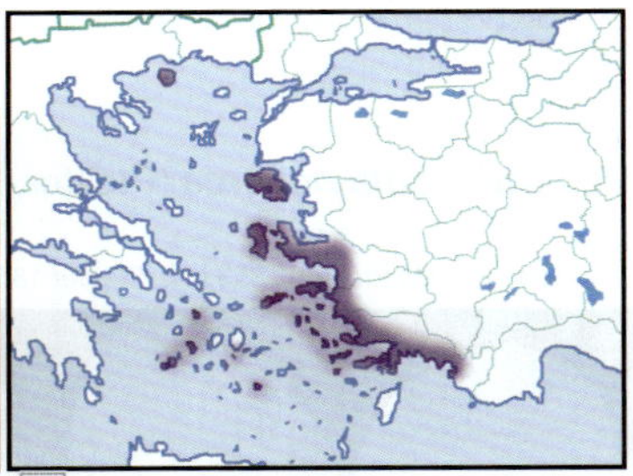

7-10 x 12-17 mm — Turkey: Vilayet Aydın, near Söke — E Greece, W Turkey

Shell pale horny brown, no colour bands, shiny, densely striated and slightly granulated, apex pointed, 4.5-5 weakly convex whorls, initially keeled but last whorl rounded, descending near aperture, aperture almost circular, inside with strong reddish lip, visible outside as a yellowish structure, margins reflected and approaching each other closely, umbilicus moderate and not very deep.

Various different habitats that provide some shelter and vegetational cover, very frequent, found nearly everywhere where terrestrial gastropods are found.

Westerlund 1889: 99, Frank 1997: 101.

Schütt 2001: 452 claimed that *proclivis* was not conspecific with *redtenbacheri*. Such a statement should be combined with published results of detailed analyses. The descriptions of both do not provide evidence for differences.

Metafruticicola schuberti (Roth, 1839)

Helix schuberti Roth, 1839 in Roth 1839: 15, Tab. 1

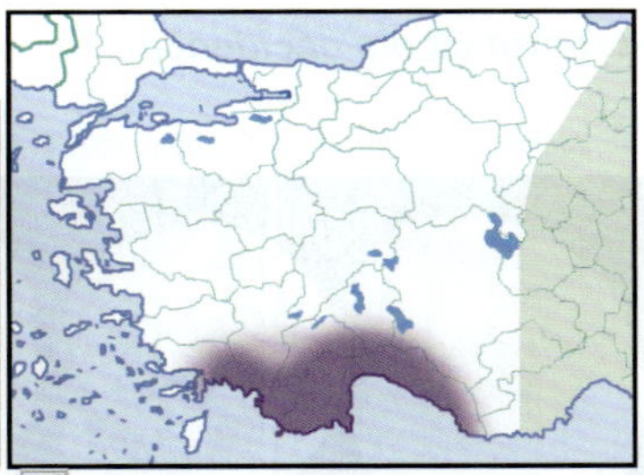

11-15 x 15-22 mm — Turkey: Vilayet Antalya, Demre — SW Turkey
Indicated N and W range limits uncertain.

Shell horny yellowish or with light reddish hue, with whitish band at keeled periphery, shiny, weakly and obliquely striated, with a dense micropattern of hair scars, 5-6 flat whorls, last whorl inflated and with weak edge at periphery, deeply descending near aperture, margins broadly reflected and approaching each other closely at parietal side, connected by a thin whitish layer, umbilicus narrow.

Various kinds of habitats, a widespread and locally very frequent species.

Westerlund 1889: 73, Schütt 1996: 420, Neubert et al. 2000, Welter-Schultes 2001 (J. Arch. Sci.) (described another new species living in the same region but did not name it, for not interfering in R. Bank's ongoing *Metafruticicola* revision).

Metafruticicola zonella (Pfeiffer, 1864)

Helix zonella Pfeiffer, 1864 in L. Pfeiffer 1864: 604

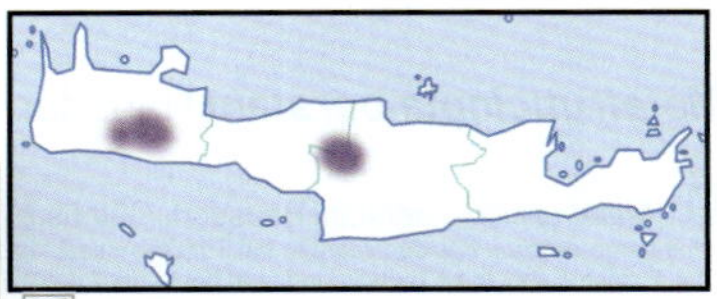

7-9 x 14-17 mm — Greece: central Crete, Psiloríti, 2120 m

Greece (Crete)

Shell dark, often with a light band at periphery, delimited by reddish zones, densely striated, 4.5-5 slightly convex whorls, last whorl rounded, slightly descending near aperture, margin simple, relected only at basal and columellar sides, umbilicus moderately narrow.

In rocky and high mountain habitats, also in scarcely vegetated zones.

Westerlund 1889: 102

Microxeromagna

Microxeromagna Ortiz de Zárate y López, 1950 in Ortiz de Zárate y López 1950: 65

Type species *Helix stolismena* Servain, 1880 (monotypy)

Type species was fixed in original publication by monotypy (Art. 13.3).

Shell medium-sized, with hairs. Presence of one dart-sac without large papilla inside, and one slightly smaller accessory sac. The penis is innervated from the right cerebral ganglion.

Gittenberger 1991: 120, Hausdorf & Sauer 2009: 413.

Microxeromagna lowei (Potiez & Michaud, 1838)

Helix lowei Potiez & Michaud, 1838 in Potiez & Michaud 1838: 91 (*Helix Lowei*)
Helix armillata Lowe, 1852 in Lowe 1852: 113
Helix arrouxi Bourguignat, 1863 in Bourguignat 1863-1868: 44, Pl. 7
Helix vestita Rambur, 1868 in Rambur 1868: 267
Helix stolismena Servain, 1880 in Servain 1880: 78

W Cyprus: Agios Georgios

Originally probably W Mediterranean, historically introduced to Canary Islands and E Mediterranean

3-5 x 5.5-8.5 mm

Shell dirty white with many brownish spots, at lower side more like interrupted narrow colour bands, the entire shell of fresh specimens is densely covered with very short hairs (0.05 mm long), when rubbed off they leave hair scars, 4.5 moderately convex whorls, aperture without lip inside, umbilicus 1/5 of shell diameter. Shells are usually of a relatively uniform size.
Differs from *Xerotricha conspurcata* in its denser and shorter hairs, and a wider umbilicus.
Animal white, only at the margin of the mantle there is a narrow brown spot. Shells of Spanish adults still have hairs in autumn, but in January animals from the previous generation have lost them.

In Spain from lowlands to high altitudes in mountains, elsewhere mainly in synanthropic habitats. One of the most widespread species in the Mediterranean region.

Ortiz de Zárate y López 1950: 81 (*Helicella stolismena*), Hausdorf 1990 (Malak. Abh. 15), Gittenberger 1991: 120, Manganelli et al. 1995: 29, Arrébola 1998, Martínez-Ortí et al. 2004, Hausdorf & Sauer 2009: 414.

This species had many names in the recent past, such as *Microxeromagna vestita*, *Cernuella arrouxi* and *Microxeromagna armillata*.

Monachoides

Monachoides Gude & Woodward, 1921 in Gude & Woodward 1921: 625 *Monachoïdes*

Type species *Helix incarnata* Müller, 1774 (original designation by Gode & Woodward 1921: 179)

Genus ending in -*oides*, masculine by default (Art. 30.1.4.4). Expressly established as a new replacement name for *Monachella* Gode & Woodward, 1921 in Gode & Woodward 1921: 179 not *Monachella* Salvadori, 1874 (Aves)
Originally intended as a new replacement name ("n. nov.") for *Monacha* Fitzinger, 1833 "pars" as used by those authors who would not use it in the sense of the type selection *Helix cartusiana* by Herrmannsen 1847. They added "Non *Monachus* Kaup, 1829 (Aves)". Not a new replacement name for *Monacha* Fitzinger, 1833 because it was not the intention of the authors to replace Fitzinger's name as defined by Herrmannsen 1847, but rather to replace a name *Monacha* in the sense of a subsequent misapplication of this name (the name *Monacha* used in this sense was not an available name in the sense of Art. 49, only available names can be replaced).

N Africa to S Sweden

Monachoides bacescui Grossu, 1979

Monachoides bacescui Grossu, 1979 in Grossu 1979: 22

Shell light horny greyish yellow to reddish brown, pale shiny, rarely with a colour band at the periphery, prominently striated (almost ribbed) and granulated with characteristic scales-like structures, 5.5 convex whorls with deep suture, apertural margin inside with a strong white-reddish lip, also visible from outside as a reddish yellow band, umbilicus very narrow but open, largely covered by the columellar margin of the aperture.
Larger than *M. incarnatus* and *M. vicinus*, which has no umbilicus, and more prominently striated.

8-10 x 14-18 mm

In mountain forests under dead leaves, between grasses and mosses, usually in humid habitats, more rarely found in less humid and sunny habitats. Lives sympatrically with *M. vicinus*.

Grossu 1983: 454.

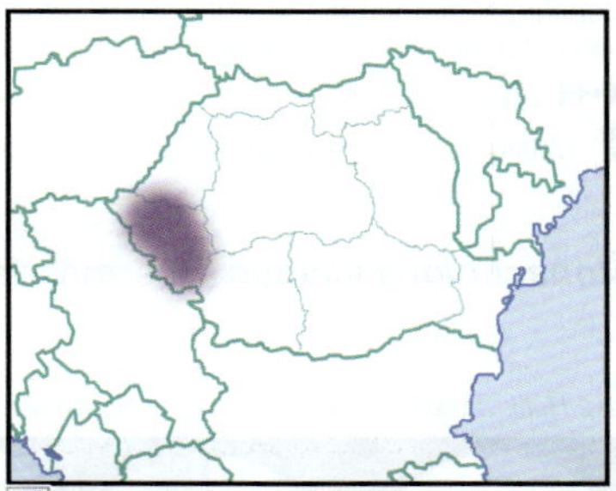

Romania (Banat)

Monachoides fallax (Wagner, 1914)

Monacha fallax Wagner, 1914 in A. J. Wagner 1914 (Anz. Akad. Wiss.): 337 Type locality Trebovič near Sarajevo, Bosnia.

Shell pale brownish, irregularly striated, 4.5-5.5 whorls, periphery weakly angulated or rounded, last whorl descending near aperture, aperture with internal lip, margin reflected at lower palatal, basal and columellar side, umbilicus moderate and deep, 1/6 of diameter.
Penis with short flagellum, mucus glands consisting of 8 tubes.

5-7.5 x 8-13 mm

EU Red List: Vulnerable VU. This was based on the preliminary assumption that the species lived only near Sarajevo.

De Winter & Maassen 1992.

Indicated range uncertain. More research is necessary.

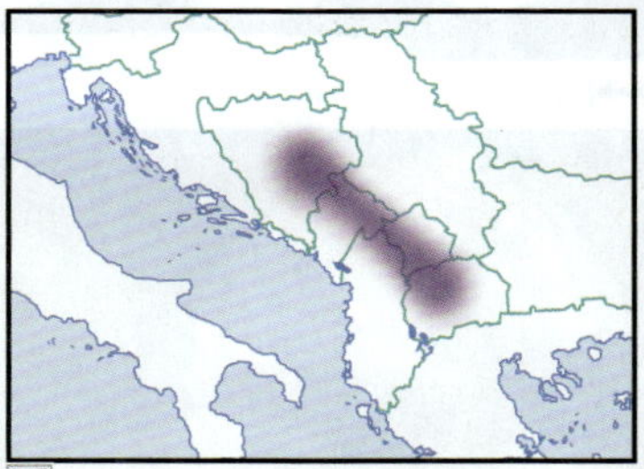

Bosnia to Rep. Makedonija

Monachoides incarnatus (Müller, 1774)

Without gender agreement: *Monachoides incarnata* (Müller, 1774)

Helix incarnata Müller, 1774 in Müller 1774: 63

S Germany: Bodensee lake deposits near Sipplingen

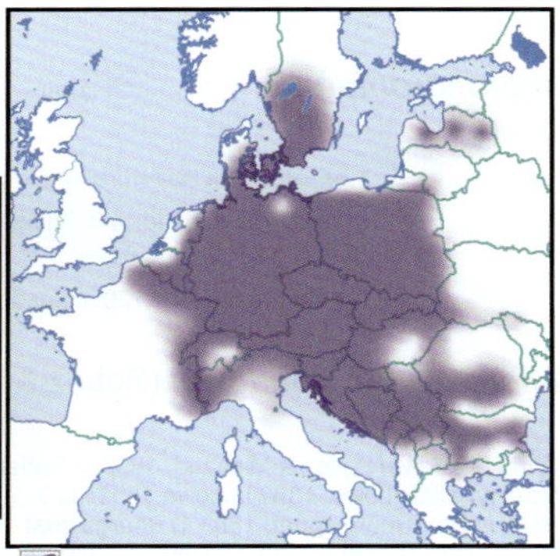

Europe

9-10 x 13-16 mm
(smaller and lighter in more open and dry habitats)

Germany: Bayern, Walchensee

Shell yellowish to reddish brown, dull shiny, with very fine and regular reticulated microsculpture (like very regular small elongate fish scales, 130 scales/mm^2, this allows also to determine shell fragments), apertural margin usually reddish, reflected and with a strong white lip inside (in S Switzerland sometimes with a tooth), umbilicus very narrow but open, less than 1/10 of shell diameter, partly covered by the reflected columellar margin.
Differs from *M. vicinus* in a more depressed upper side, its open umbilicus and finer reticulate surface pattern. *Hygromia limbata*, *Monacha cantiana* and others have no reticulate microsculpture.
The animal is pale reddish with dark greyish anterior part, head and tentacles.

Leaf litter layers of relatively humid forests under shrubs or rock rubble, also outside forests, at the soil, only juveniles tend to climb up plants. Rarely found in humid pinetree forests, gardens and vineyard slopes. In Switzerland and Bulgaria up to 1600 m. Feeds initially on rotting plant material, later also on dead leaves of plants near water bodies. A widespread and frequent species.

Clusters of 20-60 eggs (1.5-2 mm) are mostly laid in late summer, in a hole digged inside the soil, juveniles hatch after 17-24 days, life span 2 years.

Germain 1930: 253, Frömming 1954: 271, Prince 1967, Kerney et al. 1983: 255, Pawłowska-Banasiak 2008.

Monachoides kosovoensis De Winter & Maassen, 1992

Monachoides kosovoensis De Winter & Maassen, 1992 in De Winter & Maassen 1992: 177

Shell pale brown, lighter at the suture, with a light band at the periphery, irregularly striated and finely granulated, 5-6 whorls, periphery weakly angulated or rounded, last whorl weakly descending near aperture, aperture with internal lip, margin reflected at basal and columellar side, umbilicus wide and deep, 1/5 of diameter.

7-9 x 10-14 mm

Forest habitats in mountains.

EU Red List: Near Threatened NT. Only known from one locality, but the potentially inhabitated region in the vicinity cannot be explored due to the danger of land mines.

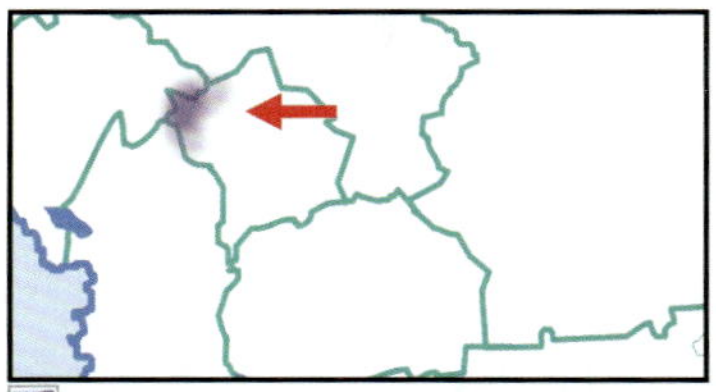

NW Kosovo

Monachoides taraensis De Winter & Maassen, 1992

Monachoides taraensis De Winter & Maassen, 1992 in De Winter & Maassen 1992: 174

Shell pale brownish, lighter at the suture, with a white band at the periphery, irregularly striated and with a weak structure of small pits, 4.7-5.5 whorls, periphery weakly angulated or rounded, last whorl markedly descending near aperture, aperture with internal lip, margin reflected at basal and columellar side, umbilicus narrow and deep, 1/8 of diameter.

7-9 x 12-15 mm

Mostly subterranean among limestone rocks or at their base, and under stones in mountains.

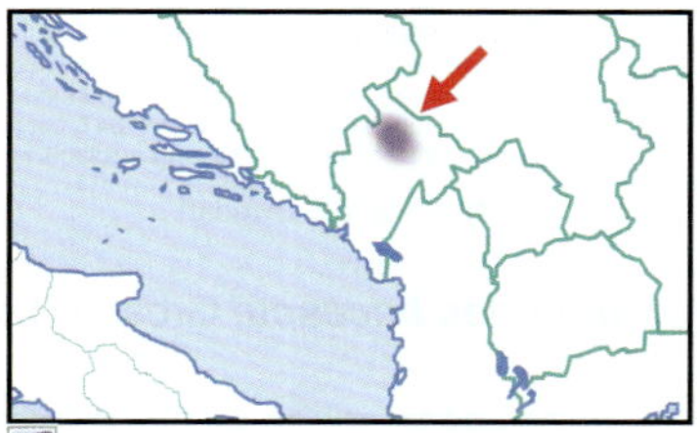

Montenegro

Monachoides vicinus (Rossmässler, 1842)

Without gender agreement: *Monachoides vicina* (Rossmässler, 1842)

Helix vicina Rossmässler, 1842 in Rossmässler 1842 (11): 3, Taf. 51

Slovakia: Low Tatra, Moštenica Kyslá

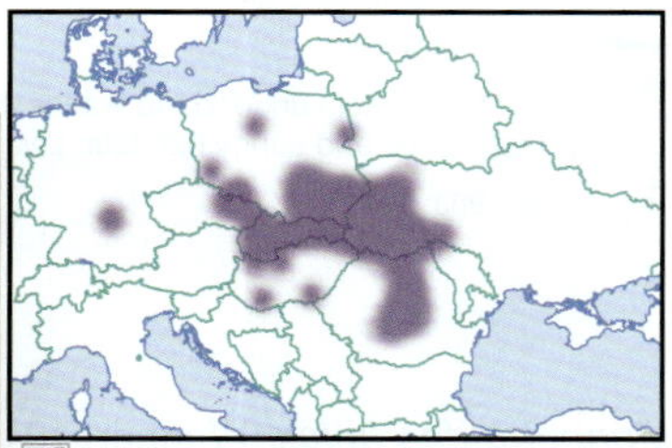

Poland to Romania
Perhaps also in Belarus?

8.5-11.5 x 12-16 mm

Slovakia: Low Tatra, Moštenica Kyslá

Jozef Grego

Shell yellowish to reddish brown, dull shiny, with fine and regular reticulated microsculpture, apertural margin usually reddish, reflected and with a strong white lip inside, umbilicus very narrow and usually entirely covered.
Shell like *M. incarnatus*, but reticulate microsculpture much coarser (35 scales/mm^2 instead of 130), umbilicus almost completely covered by reflected lip.
Differs from *Hygromia limbata* in its more convex whorls, deeper suture, almost closed umbilicus and its reticulate surface microsculpture.

In mountain forests among leaf litter and decaying wood, mostly above 700 m.

Shilejko 1978: 235, Kerney & C. 1979: 187, Grossu 1983: 452, Pawłowska-Banasiak 2008.

Nienhuisiella

Nienhuisiella Giusti & Manganelli, 1987 in Giusti & Manganelli 1987: 167

Type species *Nienhuisiella antonellae* Giusti & Manganelli, 1987 (original designation)

Nienhuisiella antonellae Giusti & Manganelli, 1987

Nienhuisiella antonellae Giusti & Manganelli, 1987 in Giusti & Manganelli 1987: 169

Shell yellow to light brown, with transverse rows of long hairs, which leave scars, 4.5-5 weakly convex whorls, shape moderately lenticular, last whorl variably angulated, aperture simple and not thickened, umbilicus extremely narrow and partly covered by the reflected columellar margin. The hairs often accumulate soil particles in between which serve as camouflage.

6-7.2 x 9-12 mm

Italy: Sardinia, Cagliari, Sant'Antonio Domusnovas
Maurizio Sosso

Differs from other species in its angulated periphery.
Animal pale orange, upper palleal cavity with irregularly shaped black spots.

Leaf litter layers in moderately closed or open *Quercus* forests on limestone or metamorphic substrate, more rarely climbing on rocks and walls, requires presence of leaf litter and moisture.

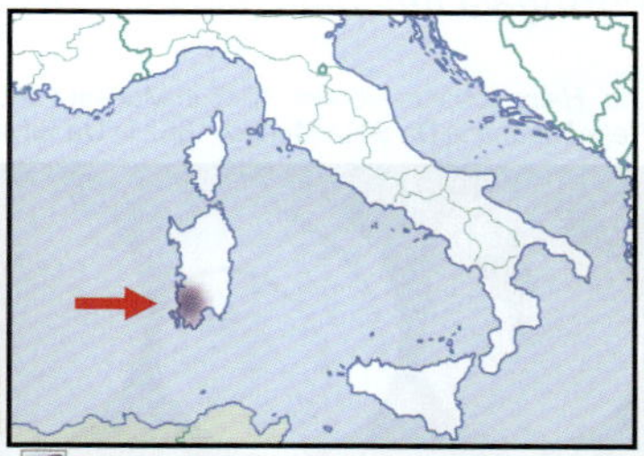

Italy (SW Sardinia)

EU Red List: Vulnerable VU. Threatened by forest fires and too drastically cleaning up of vegetated forest soils in the course of modern forest management.

Perforatella

Perforatella Schlüter, 1838 in Schlüter 1838: 4 *Helix (Perforatella)*

Type species *Helix bidentata* Gmelin, 1791 (monotypy)

Hesse 1931: 16, Egorov 2008: 92 and Sysoev & Schileyko 2009: 209 gave as type "*Trochus bidens* Chemnitz, 1786". *Trochus bidens* was neither mentioned by Schlüter 1838, nor is this name available (published in a rejected work, ICZN direction 1).

NE Europe to Caucasus and Siberia

This genus included until recently more species, among which was *P. incarnata*, here classified as *Monachoides incarnatus*.

Perforatella bidentata (Gmelin, 1791)

Helix bidentata Gmelin, 1791in Gmelin [1791]: 3642

5-7 x 6.5-9 mm

W Slovakia: Striebornica

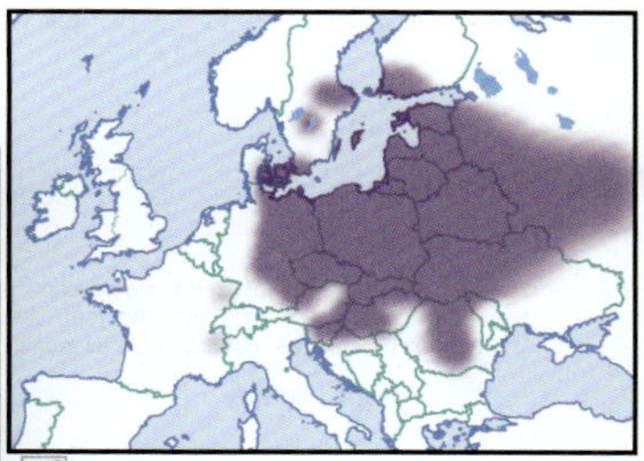

Europe
Range limits in Russia uncertain, in France only fossils.

Shell pale brown, opaque, usually with a whitish band at the periphery, regularly and distinctly striated, 7-8 tightly coiled convex whorls with shallow suture, conical above and more rounded below, last whorl slightly angled, aperture angular, reflected, flattened above and below, with white lip and two teeth at basal side, umbilicus very small and almost entirely covered by the reflected columellar margin.
Differs from *Trochulus edentulus* in its two apertural teeth and its more conical outline of the upper side.

In old woodland in litter and at logs, prefers humid habitats, characteristic in alder forests in river plains. Snails are most active in the night and early morning, mobility is up to 5 m/month, average 2-3 m.

In Poland the reproductive season lasts from May to September, copulation takes place in the evening and lasts 20 min, 4 weeks later 5-15 eggs (1.5 x 1.8 mm) are laid in batches, in shady places under litter or in the uppermost soil layer (in the laboratory also in winter, so this depends on external conditions), juveniles hatch asynchronously after 1-4 weeks (hatching starts in July), they grow at 0.7 whorls/month and hibernate at 3.2-4.1 whorls size, in the next season they reach 5.2-6.1 whorls, maturity is reached either in the second or third season, life span 3 years.

Vulnerable in Germany. Populations disappear when water bodies are dried up.

Frömming 1954: 272, Kerney et al. 1983: 254, Jungbluth & Knorre 2009: 12, Kuźnik-Kowalska & Roksela 2009.

Perforatella dibotrion (Bielz, 1860)

Helix bidens var. *major* synonym *Helix dibotrion* Bielz, 1860 in E. A. Bielz 1860: 160
Mentioned with short description. First used as a name for a taxon by Kimakowicz 1884: 105 as *Helix (Fruticicola "(Petasia)") dibothrion* with bibliographical references to Kimakowicz 1883: 28 (there mentioned as *Helix (Fruticicola) bidens* var. *major* synonym *Helix dibothrion*) and Bielz 1860: 160. Name was made available by Kimakowicz 1884 under Art. 11.6.1. The spellings *dibotrion* and *dibothrion* are demed to be identical under Art. 58.11.

8-9 x 10-13 mm

Poland: Beskid Niski mountains, Puławy Dolne
Agneta Chaniecka

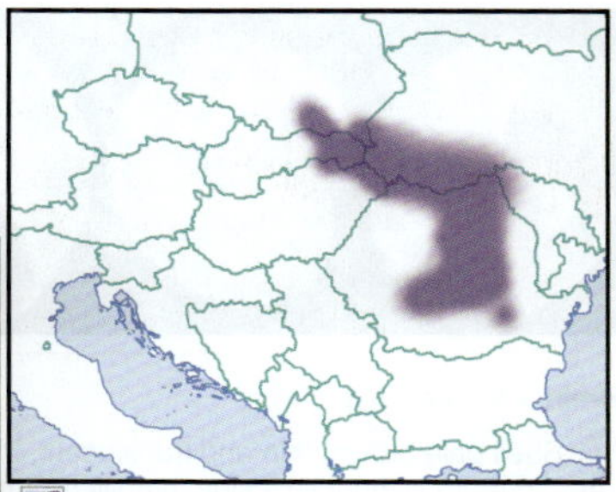

Carpathians, W Ukraine
Range limits in W Ukraine uncertain; presence in Moldova possible, but I have not seen reliable records.

Shell often reddish brown with a broad white band along the periphery, prominently striated, with granulated miscrosculpture, initial whorls with edge at periphery, 6-8 whorls, last whorl rounded, distinctly descending near aperture, aperture with two teeth, margin reddish brown, umbilicus covered by the reflected columelar margin in adults.
Larger than *P. bidentata*.

In highlands, under shrubs in leaf litter.

Westerlund 1889: 32, Grossu 1983: 449, Sysoev & Schileyko 2009: 210.

Slovakia: Slanské Vrchy mountains, Hanušovce
Jozef Grego

Portugala

Portugala Gittenberger, 1980
in Gittenberger 1980: 209

Type species *Helix inchoata* Morelet, 1845
(original designation)

Under certain interpretations of the Code *Portugala* could be a synonym of *Mengoana* Ortiz de Zárate y López, 1949 (see there). *Portugala* was established without description (Art. 13.1.1), but monotypic in the original source and the species was described.

Portugala inchoata (Morelet, 1845)

Helix inchoata Morelet, 1845 in Morelet 1845: 70, Pl. 7
Helix brigantina Da Silva Mengo, 1867 in Da Silva Mengo 1867: 170

15-20 x 17-23 mm

Portugal: Serra de Bom Viagem, N of Figueira da Foz

Portugal, W Spain

Portugal: Algarve, Corte do Gago

Valter Jacinto

Shell yellowish brown with a fine dark reddish brown colour band above the periphery, not reaching the apertural margin in adults, sometimes also with a light band, weakly and irregularly striated, 5.5 whorls, aperture slightly descending in fully grown specimens, margin simple, inside no lip, umbilicus very narrow but open.
Animal light reddish, darker on upper side of the head, upper tentacles long, dark retractor muscles running at both sides to the tentacles, lower tentacles very short, foot lighter than dorsum.

Various habitats including humid parts of valleys, meadows, forests, on alluvial terraces and rud-eral habitats, also open habitats with scarce vegetation and exposed to sunshine, frequently found near rocks, stones and old walls, at northern sides of walls, often near spiny shrubs, in dry habitats deep under rochs and roots.
Very abundant in Portugal N of Lisboa, one of the most characteristic species of Galicia.

Nobre 1913: 208, Nobre 1941: 115, Gittenberger 1980: 209, Ondina et al. 1997, Talaván Gómez & T. 2006: 115.

Pseudotrichia

Pseudotrichia Shilejko, 1970
in Shilejko 1970: 1317

Type species *Helix sericea* synonym *Helix rubiginosa* Rossmässler, 1838
(original designation)

Type species not fixed in the description of the subgenus *Tricheulota (Pseudotrichia)* by Likharev 1949: 417. Likharev 1949 argued that two species (*Tricheulota shadini* Likharev, 1949 and *Helix horripilosella* Heude, 1882) differed from *Tricheulota*, and classified both species in the new subgenus *Pseudotrichia*. Since two species were included, the type species was not fixed in 1949 as demanded by Art. 13.3. *Pseudotrichia* as mentioned by Likharev 1949 was a nomen nudum. Likharev 1955: 180 classified *rubiginosa* in *Zenobiella*.

Europe, Turkey

Pseudotrichia memnonis (Sturany, 1904)

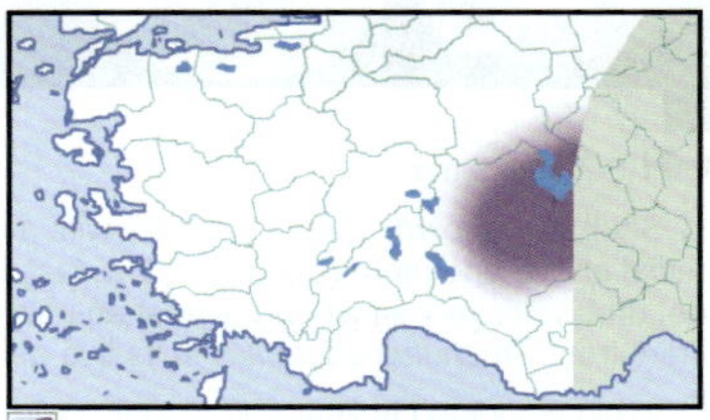

Central Turkey (Konya to Sivas)

Fruticicola memnonis Sturany, 1904 in Sturany 1904: 117

Shell flat, yellowish brown, with short hairs, keel at last whorl only weak if present, umbilicus present, not covered by columellar margin of the aperture.

4-5.5 x 7-10 mm

Nordsieck 1996, Schütt 2001: 443.

Pseudotrichia rubiginosa (Rossmässler, 1838)

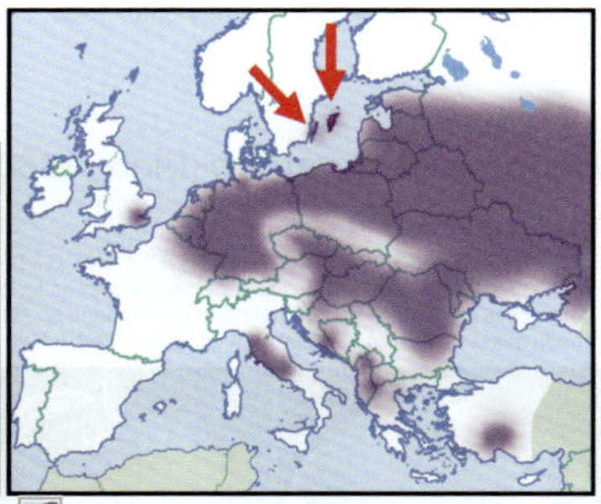

Helix sericea synonym *Helix rubiginosa* Rossmässler, 1838 in Rossmässler 1838-1844 (7-8): 3
Name was mentioned as a synonym for *Helix sericea* s. str. (small form as figured in Fig. 428, "Ziegler nennt die kleinere Form *H. rubiginosa*"). Schmidt 1853: 3 was the first who used the the name for a taxon (*Helix rubiginosa*) and referred bibliographically to Rossmässler 1838: 3 and to Fig. 428, making Rossmässler's synonym available under Art. 11.6.1.
Fruticicola czarnohorica Poliński, 1924 in Poliński 1924: 193

4.5-5 x 6-8 mm

Germany: Schleswig-Holstein, Wedeler Marsch

Europe to N Caucasus and E Asia

Shell pale brown, thin and translucent, irregularly striated, fresh shells with short straight or weakly curved hairs, 4.5-5.5 convex whorls with moderate suture, aperture without lip inside, umbilicus small and partly covered by the reflected columellar margin.
Shell could be confounded with *Trochulus plebeius* and perhaps other *Trochulus* species, which have curved hairs. Differs in the presence of only one single dart (*Trochulus* has two darts).

Meadows and marshes of river plains, swamps in areas periodically flooded by rivers, requires a relatively high degree of insolation and needs periodic inundation. Species hibernates inside the mud, not below wood and grass roots. In SW Bulgaria at few localities between 400 and 900 m of Dobrostan ridge near Plovdiv, and in up to 800 m in Pirin mountains.

2-3 clutches of 2-6 eggs (1.2-1.6 mm) are laid at some 10 mm depth inside the soil, between June and August, whitish juveniles hatch after 15-20 days, they have dark colour after 4 weeks.

Endangered in Germany, Vulnerable in England. Threatened by drainage, flood-control schemes, riverside developments and excessive fertilization. Probably native in England along the Thames but not recognized until 1981 when many populations had already been destroyed.

Schmidt 1853: 3, Steusloff 1949, Frömming 1954: 268, Shilejko 1978: 238, Kerney et al. 1983: 259, Falkner 1990: 214, Schütt 2001: 443.

In the absence of an anatomical record Proćków 2009: 120 suspected that ***czarnohorica*** was a *Trochulus* species. Balashov & Gural-Sverlova 2012: 104 classified it as a synonym of *Ps. rubiginosa*. The status of this form, which is only one from one locality in W Ukraine, should be clarified.

Pseudoxerophila

Pseudoxerophila Westerlund, 1879 in Westerlund & Blanc 1879: 55 *Helix (Pseudoxerophila)*

Type species *Helix bathytera* Blanc, 1879 (subsequent designation by Hausdorf 1988: 20)

Name in original source attributed to Westerlund alone, the descriptive text was not exclusively attributed to Westerlund. Under Art. 50.1 the name must be attributed to the co-author of the work who was made responsible for the name.
Type species not designated by Kobelt 1904: 161 because the work contained a disclaimer (Art. 8.3) that the author did not use the term "Typus" in the sense of a unique name-bearing type to define the identity of a genus (Kobelt 1904: 145-147). Hausdorf 1988: 20 assumed that Kobelt 1904 designated as type validly *Helix bathytera*, type thereby designated under Art. 69.1.1.

Presence of two small symmetrical dart sacs and two accessory sacs, which are attached to a dilatation of the vagina, 4 branched mucus glands around the vagina. Penis is innervated from the right cerebral ganglion.

Hausdorf 1988: 20 (anatomy), Hausdorf & Sauer 2009: 406.

Pseudoxerophila bathytera (Blanc, 1879)

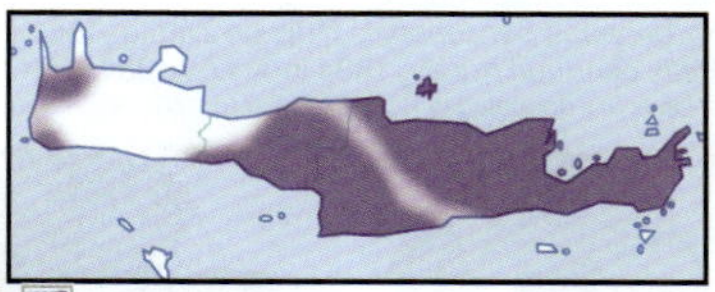

Helix bathytera Blanc, 1879 in Westerlund & Blanc 1879: 55, Pl. 2
Name in original publication attributed to Blanc, diagnosis to Westerlund. Under Art. 50.1 the name must be attributed to the co-author of the work who was made responsible for the name.

7-9 x 11-14 mm — Greece: E Crete, (1) near Chandrás, (2-6) near Sitia

A frequent species, usually in shrublands, lives also on non-calcareous substrate.

Greece (Crete)

Shell white, depressed, occasinally with brown spots or bands, weakly and irregularly striated, 4.5-5 convex whorls, last whorl rounded, rarely initially angulated, umbilicus wide, strongly widened at last 1/2 whorl, to 1/5-1/4 of diameter.
Differs from the other hygromiid species of Crete in its wide and excentrical umbilicus and a long flagellum (proximal epiphallus : flagellum ratio = 0.4–0.7).

Hausdorf & Sauer 2009: 407.

Pseudoxerophila confusa Gittenberger, 1991

Pseudoxerophila confusa Gittenberger, 1991 in Gittenberger 1991 (Zool. Meded. 65): 113

S Cyprus

5-7 x 9-11 mm — S Cyprus — Georgios Konstantinou

Shell whitish with spirally arranged colour patterns, irregularly ribbed, 4-5 slightly convex whorls, al with equally indented sutures, first whorls angulated, last whorl rounded and slightly descending near aperture, aperture with weak lip inside, umbilicus 1/5-1/4 of diameter.
Shell sculpture more irregular than in *Xeropicta krynickii*. Differs from sympatric populations of *Xeropicta ledereri* in its less well rounded aperture and the suture of the first whorls not being shallower than at the last whorl.

Pseudoxerophila oertzeni (Maltzan, 1887)

Helix oertzeni Maltzan, 1887 in Maltzan 1887: 117 (*Helix (Xerophila) Oertzeni*)

4-5 x 7-9 mm — Greece: W Crete, Omalos plain

Greece (W Crete)

Usually in open shrublands.

Hausdorf & Sauer 2009: 409.

Shell flat, white, sometimes with brown colour bands, weakly striated and with fine spiral lines, 4-4.5 convex whorls, blunt edge at the periphery of the first whorls, last whorl rounded near aperture, umbilicus wide and excentrical, widened at last 1/2 whorl. Subfossil shells from near Kournas lake are more strongly keeled.
Differs from *Ps. bathytera* in is much smaller shell and a shorter flagellum (proximal epiphallus : flagellum ratio = 0.8–1.1).

Pyrenaearia

Pyrenaearia Hesse, 1921
in Hesse 1921: 59, 61

Type species *Helix carascalensis* Michaud, 1831
(original designation, p. 77)

Two dart sacs of equal size, dart short and almost straight, flagellum variable between short and very long.

All species occur on limestone rocks, on rock faces and in crevices, below large blocks and in rock rubble.

Elejalde et al. 2009 (detailed molecular study of all known species).

Pyrenaearia cantabrica (Hidalgo, 1873)

Helix cantabrica Hidalgo, 1873 in Hidalgo 1873: 358 (*Helix Cantabrica*)
Helix schaufussii Kobelt, 1876 in Kobelt 1876: 45 (*Helix Schaufussii*)
Pyrenaearia poncebensis Ortiz de Zárate, 1956 in Ortiz de Zárate 1956: 46
Helix oberthuri Ancey, 1884 in Ancey 1884: 159
Name was dedicated to René Oberthur (p. 160), original spelling was not *oberthueri*.

N Spain

4-6 x 10.5-14 mm, in the east 6-8 x 15-17 mm

Spain: Asturias, Oviedo, Santuario de Covadonga

José Liétor Gallego

(*oberthuri* form) Spain: Asturias, Arenas de Cabrales (*schaufussi* form) Spain: Cantabria, Potes, Desfiladero de la Hermida

Shell very flat, horny brownish, transparent but not shiny, finely striated, juveniles with very short hairs (in adults the scars are visible under microscope), in dry habitats adults can also have dense hairs, 4.5-5 whorls, aperture inside horny brownish with a weak whitish lip, margin sharp, umbilicus wide (1/5-1/4 of shell diameter).
Animal with black dorsum and foot, white mantle margin, the rest of the animal is light.

Rocky limestone habitats, at rock faces and in crevices. Individuals in dry habitats under overhanging rocks may have hairs to retain water from fog, the shells at nearby vertical walls have no hairs.

Westerlund 1889: 306, 308, Ortiz de Zárate 1956: 43, 48, Prieto & Gómez 1988, Elejalde et al. 2009.

Ondina et al. 1997 and Elejalde et al. 2009: 70 did not indicate occurrences in Galicia. Ortiz de Zárate 1956: 44 cited findings in Bares, province of La Coruña, reported by Folin 1887: 201. The shells were possibly misidentified.
A morphological distinction between *schaufussi* and *cantabrica* has never been well documented, both taxa clustered closely in molecular studies by Elejalde et al. 2009: 76 who regarded them as synonyms. The same result was obtained for *poncebensis*. Initially *poncebensis* was believed to differ from the others by the presence of hairs, but this seems to be a microhabitat adaptation.
The *oberthuri* form, characterised mainly by a bluish white surface layer and restricted to high mountain habitats, seems to be a high-altitude form from above 1500 m. Also the shell shape displays an altitudinal cline, with larger and flatter shells in the valleys, conical forms in higher altitudes. The molecular record supported this scenario (Elejalde et al. 2009).

Pyrenaearia carascalensis (Michaud, 1831)

Helix carascalensis Michaud, 1831 in Michaud 1831: 29
Previously mentioned as *Helix (Helicella) carascalensis* by Férussac [1821-1822]: 38, nomen nudum, no description added.

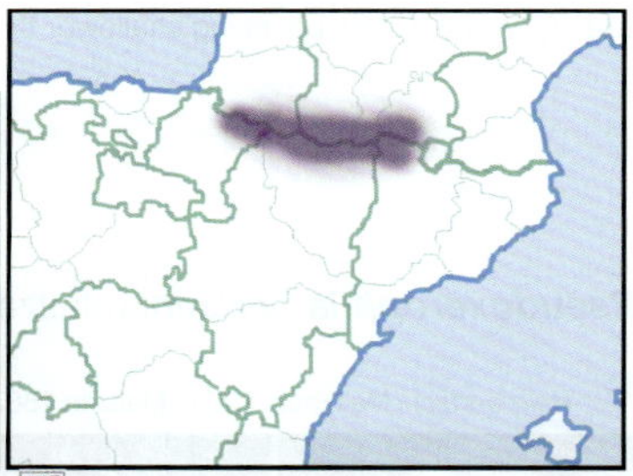

6.5-10 x 10-16 mm

France: Gedre, Cirque de Troumouire, 2100 m

Pyrenees

Shell dirty yellowish with irregular horny coloured rays or colour bands, irregularly striated, not very shiny, 5-5.5 slightly convex whorls with rather shallow suture, aperture slightly broader than high, oblique, margin sharp, inside without or with a weak white lip, columellar margin broadly reflected at the upper side, and prolongated, umbilicus narrow and deep.
The animal is dark.

Open habitats, grasslands and shrubs in high mountains, 700-3000 m, usually above 2000 m.

Westerlund 1889: 307, Ortiz de Zárate 1956: 53, Kerney & Cameron 1979: 190, Kerney et al. 1983: 259, Prieto & Gómez 1988, Elejalde et al. 2009: 73 (could include *cotiellae*, *molae* and *velascoi*, but more detailed studies are needed).

Pyrenaearia carascalopsis (Fagot, 1884)

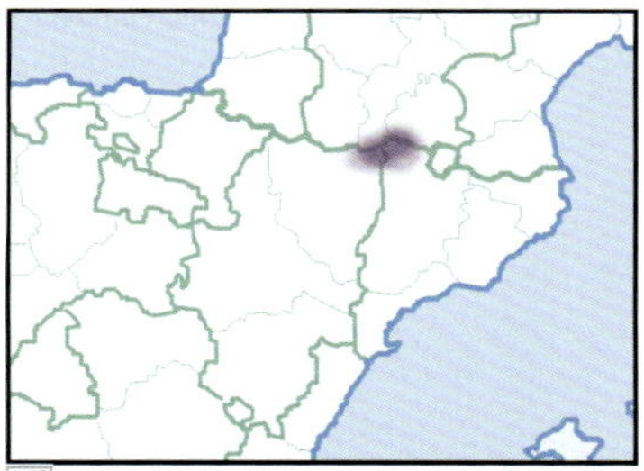

Pyrenees

Helix carascalopsis Fagot, 1884 in Fagot 1884: 178

Shell horny yellowish or grey, irregularly striated and corrugated, shell shape extraordinarly flat, 5-6 weakly convex whorls, aperture oval, margin thin and simple, no lip inside, umbilicus very narrow and almost entirely covered by the reflected columellar margin.

10-12 x 6-7 mm

Under stones in high mountains, was often recorded from 2000-2500 m.

EU Red List: Near Threatened NT. Potentially threatened by human pressure and global warming.

Westerlund 1889: 307, Ortiz de Zárate 1956: 54, Puente et al. in Verdú & Galante 2006: 375, Elejalde et al. 2009: 73 (the molecular record suggested presence of an independent species).

Pyrenaearia cotiellae (Fagot, 1906)

Helix cotiellae Fagot, 1906 in Fagot 1906: 172 (*Helix cotiellæ*)

Shell white, rarely with horny hue, solid, whorls less convex than in other *Pyrenaearia* species, umbilicus very wide.

9-11 x 14-16 mm

At vertical rock walls, limestone and granite, hides in rock crevices or in rock rubble, above 2000 m.

EU Red List: Vulnerable VU. Restricted to a narrow altidudinal zone, threatened by global warming.

Ortiz de Zárate 1956: 52, Elejalde et al. 2009: 73 (probably conspecific with *P. carascalensis*, more detailed studies are needed).

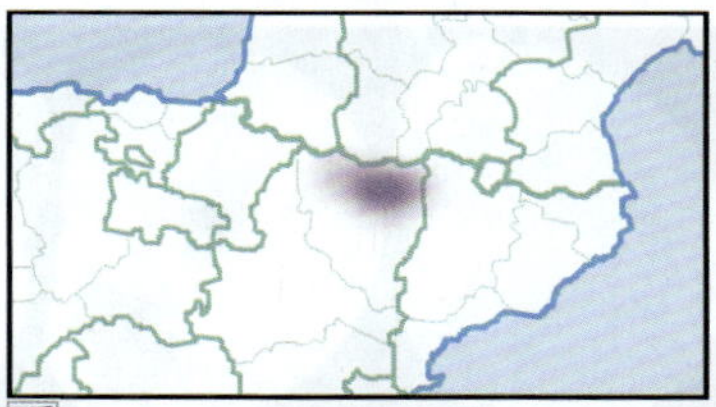

Pyrenees

Pyrenaearia daanidentata Raven, 1988

Pyrenaearia daanidentata Raven, 1988 in Raven 1988: 121

Shell pale grey with irregular dark brownish grey spots, semi-translucent, 4.5-5 convex whorls, last whorl width 1.5 x of penultimate whorl, descending near aperture, aperture inside orange-brown, inside with a prominent and broad white callus with two well developed teeth at the palatal and basal sides, umbilicus wide, 1/5-1/4 of diameter.

4.2-4.6 x 8-9 mm

At vertical limestone rock walls, in rock crevices, under large boulders of a S exposed slope, at 2100 m.

EU Red List: Vulnerable VU. Only known from one locality, threatened by warmer climatic conditions.

Puente et al. in Verdú & Galante 2006: 376, Elejalde et al. 2009.

The molecular record seemed to support the assumption that this was a separate form, perhaps fully or at least partly disconected from genetic interchange with surrounding populations of *P. cantabrica*.

N Spain (Picos de Europa, León)

Pyrenaearia molae Haas, 1924

Pyrenaearia molae Haas, 1924 in Haas 1924: 144

Spain: Tarragona, Mola de Colldejou

Spain (Tarragona)

7-8.5 x 13-16 mm

Shell whitish with horny brownish first 2 whorls and broad radial horny brown streaks, weakly and irregularly striated on upper side, lower side almost smooth but also with horny radial streaks, 5-5.5 whorls, aperture with weak white lip inside, margins not approaching each other closely, umbilicus narrow, 1/7-1/6 of diameter, but all whorls visible inside.

Rocky N exposed mountain slopes in humid and shady habitats, 800-900 m altitude.

EU Red List: Critically Endangered CR. Only known from few sensitive localities which should be protected. The species is affected by changing climatic conditions, increasingly heavy rains may wash the animals down the mountain slopes. More research on the real threats may be helpful.

Ortiz de Zárate 1956: 58, Vilella 1963, Elejalde et al. 2009: 73 (probably conspecific with *P. carascalensis*, more detailed studies are needed).

Pyrenaearia navasi (Fagot, 1907)

Helix navasi Fagot, 1907 in Fagot 1907: 147 (*Helix Navasi*)

Shell evenly dark horny yellowish, very finely striated, 5 whorls, last whorl rounded, aperture without lip inside, umbilicus narrow and partly covered by the reflected columellar margin.
Animal light grey.

10-12 x 5-6.5 mm. Shells at 1200 m are larger and more globose.

One population at 1800-2300 m under stones in alpine zones, another population at 1200 m in forests under tree trunks and in or under leaf litter. Not under rocks at road margins.

EU Red List: Vulnerable VU. Restricted to Sierra de Moncayo. Threatened by climatic changes affecting the vegetation in the high mountain zones, and by moutain hikers descending off the trails directly down the slopes which destroys the vegetational cover on the sensitive N exposed slopes.

Ortiz de Zárate 1956: 51, Prieto & Gómez 1988, Puente et al. in Verdú & Galante 2006: 378, Elejalde et al. 2009: 73 (well supported separate unit in molecular studies, previously recognised subspecies were only ecologically induced shell morphs).

Spain (Sierra de Moncayo)

Pyrenaearia organiaca (Fagot, 1905)

Helix organiaca Fagot, 1905 in Fagot 1905: 123 (*Helix Organiaca*)

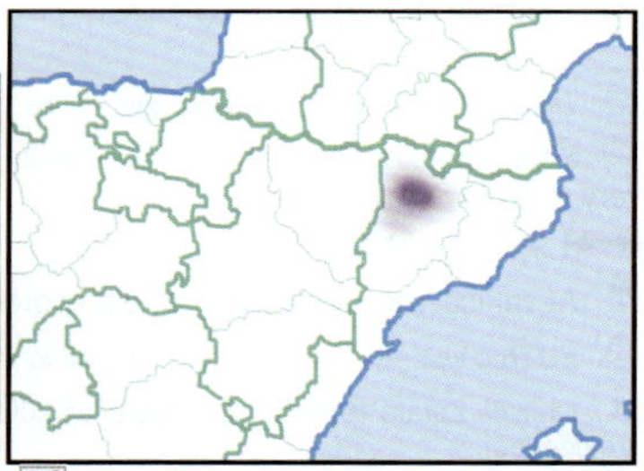

6-9 x 14-17 mm

Spain: Lleida, Congost d'Organya

Pyrenees

Shell horny coloured, not very shiny, with white ribs, slightly less prominent at lower side, depressed on upper side, 5 whorls, last whorl rounded, initially angulated, aperture with a weak white lip inside, margins slightly reflected and approaching each other closely, umbilicus wide, 1/5 of diameter.
Animal dark bluish grey (not white).

Limestone rocks at steep slopes in narrow canyons of lower mountain regions, not in overhanging slopes without access of rain, 500-700 m altitude.

EU Red List: Endangered EN. Restricted to a few canyons of two valleys, only the populations of the Segre valley were found again after 1920. Was once abundant in the 1960s, but the population size has dramatically declined after the suitable habitat at the base of rocks was destroyed by dynamite during road construction. Also seriously threatened by illegal garbage dumping and associated introduced rats.

Ortiz de Zárate 1956: 57, Altimira 1963: 24 (animal colour), Elejalde et al. 2009: 73 (well supported separate unit in molecular studies).

Pyrenaearia parva Ortiz de Zárate, 1956

Pyrenaearia carascalensis var. *parva* Ortiz de Zárate, 1956 in Ortiz de Zárate 1956: 56

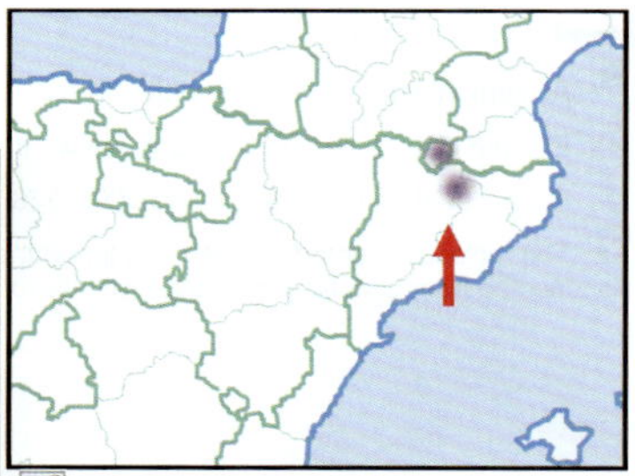

3.8-4.5 x 7.5-9 mm

Spain: Barcelona, Saldes, Pedraforca, 2200 m

Pyrenees

In mountains above 2000 m, mainly unders stones.

Shell dirty whitish with horny brown first 2 whorls, solid, ribbed on upper side, much more weakly striated on lower side, or almost smooth, upper side with horny brown streaks between the ribs, 4.5-5 moderately convex whorls with deep suture, last whorl rounded, but initially angulated, aperture simple with a weak lip inside, margins approaching each other only weakly, umbilicus narrow and deep.

EU Red List: Vulnerable VU. Only known from two disjunct localities. Potentially threatened by the shifting of suitable habitats to higher zones, induced by global warming.

Puente et al. in Verdú & Galante 2006: 380, Elejalde et al. 2009: 73 (well supported species in molecular studies).

Pyrenaearia velascoi (Hidalgo, 1867)

Helix velascoi Hidalgo, 1867 in Hidalgo 1867: 440, Pl. 12 (*Helix Velascoi*)

Shell white with horny streaks and a central white band, strongly and irregularly striated, first whorls with edge, last whorl rounded, suture deep, apex small, aperture extraordinarily broad, with a very weak lip inside, lower side margin reflected, umbilicus narrow.

Limestone rock crevices at N exposed rocks, 1000-1500 m.

Westerlund 1889: 307, Ortiz de Zárate 1956: 49, Prieto & Gómez 1988, Puente et al. in Verdú & Galante 2006: 382, Elejalde et al. 2009: 76.

8-11 x 15-19 mm

Spain: Navarra, Lekumberri, San Miguel de Aralar
José Liétor Gallego

The molecular record suggested no genetic interchange with sympatric *P. cantabrica*; but *velascoi* is probably conspecific with *P. carascalensis*, more detailed studies are needed. The presence of true subspecies was not confirmed.

N Spain (Basque Country)

EU Red List: Vulnerable VU. Restricted and fragmented range in the highest mountains of the zone, potentially threatened by the shifting of suitable climatic conditions above the summits in the course of global warming.

Semifruticicola

Semifruticicola Wagner, 1914
in A. J. Wagner 1914 (Anz. 51): 336

Type species *Semifruticicola serbica* Wagner, 1914
(monotypy)

Gender is feminine because its ending was based on the feminine generic name *Fruticicola* Held, 1838 and orginally combined with the adjectival specific name *serbica* (Art. 30.1.4.2).

Penis with thin and equally long epiphallus, retractor penis inserts between penis and epiphallus, vagina egg-shaped, with two symmetric and unequally shaped dart sacs, containing one dart each, mucus glands consisting of 8 tubes.

Semifruticicola serbica Wagner, 1914

Semifruticicola serbica Wagner, 1914 in A. J. Wagner 1914: 336

Shell white with horny brownish apex and two blurry reddish brown bands, lower band narrower than upper band, granulated and weakly shiny, with prominent growth lines, 4.5-5.5 slightly convex whorls, last whorl slightly inflated, slightly descending near aperture, aperture inside reddish, margin with white lip, umbilicus initially narrow, wider at last whorl and partly covered by the reflected columellar margin.
Shell very much like *Xerocampylaea zelebori*, but surface granulated, growth lines more prominent, bands more blurry, lower band narrower.

7.5-8.5 x 13.5-15 mm

I have not seen a name *Helix costulata* Brancsik, 1897 (non Pfeiffer, 1828 and others), given in this combination by www.faunaeur.org (08.2012), but considering the primary homonymy such a name can probably not be used.

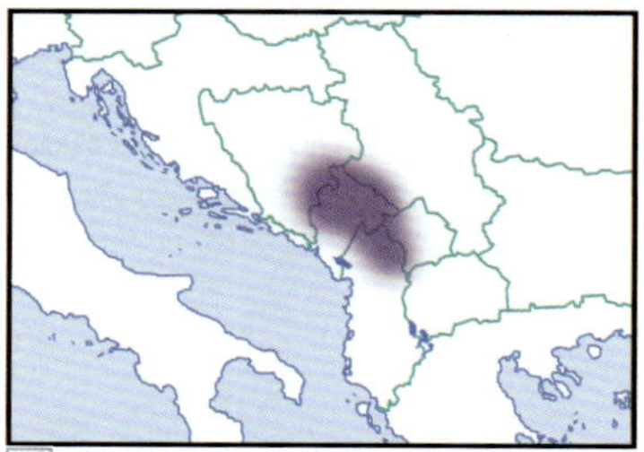

Bosnia to N Albania

Rocky mountain grasslands under stones.

Trochulus

Trochulus Chemnitz, 1786 in Chemnitz 1786: 52

Type species *Helix hispida* Linnæus, 1758 (fixed by ICZN under the plenary powers)

Name made available in ICZN Opinion 2079 (Sep 2004) despite of having been published in a non-binominal work. This was necessary because the frequently used name *Trichia* Hartmann, 1840 was preoccupied, and no other name could be used.

Trichia Hartmann, 1840 in Hartmann 1844: xiii [footnote], type species *Helix hispida* Linnæus, 1758 (subsequent designation by Pfeiffer 1848: 554) (not *Trichia* De Haan, 1839 (Crustacea))

Petasia Beck, 1837 in Beck 1837: 21 *Helix (Petasia)*, type species *Helix trochiformis* Montagu, 1803 (= currently in genus *Euconulus*) (subsequent designation by Gray 1847 [09 Nov 1847]: 173) (not *Petasia* Stephens, 1828 (Lepidoptera))

Contained originally *Helix trochiformis* Montagu, 1803, *Helix edentula* Draparnaud, 1805, *Helix unidens* Rossmässler, 1835, *Helix cobresiana* Alten, 1812, *Helix fulva* Müller, 1774 and *Helix bidens* Rossmässler, 1835, with many synonyms listed under these names, and *Helix fabricii* as a nomen nudum. *Helix unidentata* Draparnaud, 1805 was listed as a synonym of *Helix cobresiana* var. a.
Gray 1847 [09 Nov 1847]: 173 gave as type (for *Petasia*) *Helix trochiformis*. Herrmannsen 1847-1849 [07 Dec 1847]: 240 gave as type (for *Petasia*) *Helix fulva* (= currently *Euconulus fulvus*).

Petasina as mentioned by Beck 1847: 122 *Helix (Petasina)*, availability and type species uncertain and controversial

Contained originally *Helix cobresiana*, *Helix fulva*, *Helix trochiformis*, *Helix fabricii* Möller, 1842, and no synonyms.
Beck 1847 used the name *Petasina* for basically the same species as for *Petasia* in 1837. *Petasina* can either be regarded as an incorrect subsequent spelling for *Petasia*, or as a new name (the Code gives no guide to take a definite decision), but not as a new replacement name (lack of an expressed statement, Art. 72.7) and not as an emendation (nothing qualifies for that under Art. 33.2.1).
Westerlund 1902: 91 gave as type (for *Petasina*) *Helix unidentata* (= currently *Petasina unidentata*). This nominal species was only included in 1837, not in 1847. So Westerlund's act was irrelevant.
Gude & Woodward 1921: 177 argued that *Petasina* (they gave as type *Helix fulva*) was established as a new replacement name for *Petasia* Beck, 1837 non Stephens, 1828 (Lepidoptera). This is not tenable under the Code because Beck 1847 did not publish an expressed statement that the name was established to replace another name (Art. 72.7, 13.1.3), and did not show awareness of the homonymy. For those who regard *Petasina* Beck, 1847 as a new name they fixed at this occasion *Helix fulva* as its type.
The important name *Euconulus* Reinhardt, 1883 is threatened under both interpretations. The Commission should be asked for a ruling (1) to regard *Petasina* as an available name and (2) to fix *Helix unidentata* as its type under the plenary powers.

Edentiella Poliński, 1929 in Poliński 1929: 158 *Fruticicola (Edentiella)*, type species *Helix edentula* Draparnaud, 1805 (original designation)
Plicuteria Shileyko, 1978 in Shileyko 1978 (Malacologia 17): 30, type species *Helix lubomirski* Ślósarski, 1881 (monotypy)

Shell from conical to flattened, striated, with hairs at least in juveniles, 4.5-7 whorls, lip variable, diameter 3.5-14 mm. Longitudinal plicae present on internal vaginal walls, 4 groups of 2 mucous glands each, or glands not grouped.

Usually in humid and shady habitats in forests or river woodlands, some species live in montane or submontane regions, under rotting timber, stones, in moss or leaf litter.

Reproductive biology almost entirely unknown, all species seem to be oviparous.

Proćków 2009 (taxonomic revision).

Proćków's 2009 revision proposed to recognise 21 species. This is accepted here except that *T. suberectus* is regarded as a synonym of *T. striolatus*, that *T. ataxiacus* is accepted as a separate species here, and that *czarnohorica* is regarded as a synonym of *Pseudotrichia rubiginosa* (following the proposal by Balashov & Gural-Sverlova 2012: 104).
The most significant modification of previous systems was certainly the lumping of *hispida*, *sericea* sensu Draparnaud and *plebeium* (as *T. hispidus*). I see this as a tenable step and a logical consequence provoked by the absence of any previously published clear support for a reliable classification of taxa in this complex as separate species. It may be better to wait until the taxonomy is better understood. More research is necessary to understand the nature of these forms, their local morphological variations, their relationships.
Previous authors used 3 genera, but not all authors the same:
Shileyko 1978 (*Trichia*, *Edentiella*, *Plicuteria*), Falkner 1985: 89 (*Trichia*, *Petasina*), Nordsieck 1993 (*Trichia*, *Petasina*, *Plicuteria*).

Trochulus ataxiacus (Fagot, 1883)

Helix ataxiaca Fagot, 1883 in Fagot 1883: 220 (*Helix Ataxiaca*) [not verified]

Shell reddish, not shiny, prominently striated, with some white hairs, 5.5-7 not very convex whorls, slowly and regularly increasing, last whorl not much wider, depressed at both sides and with edge at periphery, aperture with white lip inside, columellar margin strongly reflected, umbilicus straight, cylindrical and deep.
Differs from *T. hispidus* in its larger size, more tightly coiled whorls and more depressed shape.

5-7 x 9-12 mm

In dead wood in *Abies alba* forests on karstic calcareous substrate, 800-1500 m. Locally very frequent at places where wood logs are stored.

EU Red List: Near Threatened NT. Only known from few localities, no threats known.

Westerlund 1889: 39, Falkner et al. 2002: 150, Bertrand 2002: 15.

Falkner et al. 2002 gave a diagnosis and some information to justify its status as a separate species. Proćków 2009 did not have material from Pyrenees.

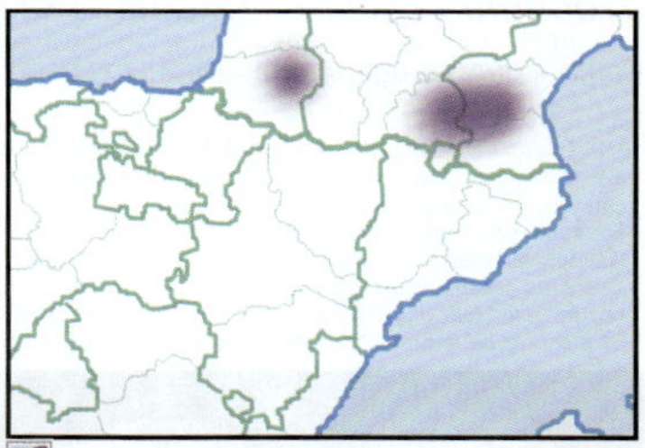

S France

Trochulus bakowskii (Poliński, 1924)

Fruticicola bielzi bakowskii Poliński, 1924 in Poliński 1924: 196 (*Fruticicola bielzi bąkowskii*)

Slovakia: Mt. Pol'ana, 1400 m

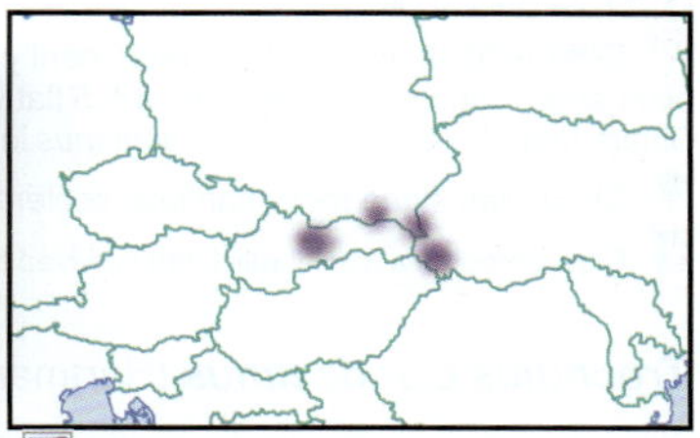

NE Carpathians

3-5 x 5.5-7 mm

Shell brown with a light band at the last whorl, densely striated, with short and weakly curved white hairs, easily lost in adults, scars hardly visible, 5-6 convex and slowly increasing whorls, last whorl not descending, aperture with strong white lip inside, basal margin reflected, umbilicus narrow, less than 1/10 of diameter, partly covered by the reflected columellar margin.
Animal light grey, darker in frontal upper section and lighter down to the sides, tentacles moderately grey.

Slovakia: Pol'ana mountains, main ridge

Jozef Grego

In moist habitats, herb layer of old natural and undisturbed montane forests, or leaf litter, between 700 and 1500 m.

Trochulus biconicus (Eder, 1917)

Without gender agreement: *Trochulus biconica* (Eder, 1917)

Fruticicola biconica Eder, 1917 in Eder 1917: 119

2.5-3.5 x 5-6 mm

Switzerland: Bannalper Schonegg, 2200 m

Shell pale brown, with coarse and widely spaced blunt growth lines above, weaker below, 5-5.5 tightly coiled convex whorls with deep suture, last whorl angled, aperture with thin whitish lip inside, margin slightly reflected, umbilicus narrow, circular and deep.

Grassy limestone rock rubble, on SW exposed mountain slopes with stone plates and scarce vegetation, 2150-2380 m.

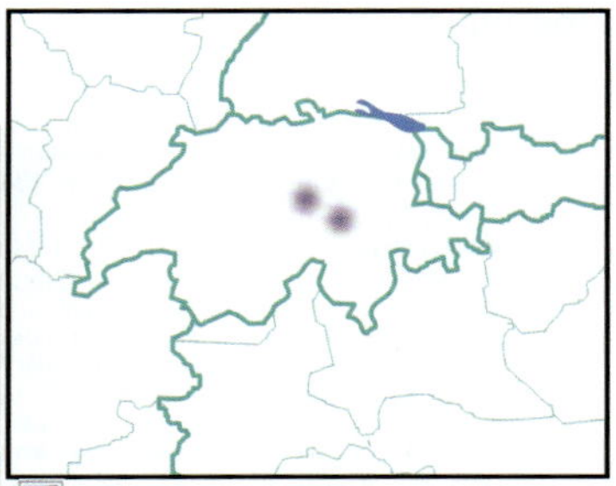

Switzerland (Bannalper Schonegg)

EU Red List: Endangered EN. The sensitive habitat is extensively used as a pasture, also threatened by effects of global warming.

Kerney & Cameron 1979: 196, Turner et al. 1998: 348, Proćków 2009.

Trochulus bielzi (Bielz, 1860)

Helix bielzi Bielz, 1860 in E. A. Bielz 1860: 151 (*Helix Bielzi*)

Established with description. Previously mentioned by Bielz 1859: 223 as *Helix Bielzi* "Ad. Schmidt" without description or bibliographically unambiguous reference (nomen nudum), with a reference to a previous putatively inappropriate use of the name *Helix leucozona* Pfeiffer, 1828. Bielz 1859 referred to unpublished data or to Bielz 1853: 117. *Helix bielzi* was not established as a new replacement name to substitute a name *Helix leucozona* sensu Bielz,1853 non Pfeiffer, 1828, because this was not an available name and there was no expressed statement as required by Art. 72.7. Bielz 1853 had only used Pfeiffers name and misinterpreted it, and had not established a new name (Art. 49). The view expressed by Falkner 1996: 100 (author Bielz 1859, new replacement name) is not tenable.

Shell horny yellow to reddish brown, often with whitish band at the keel, thin and shiny, finely striated and granulated, densely covered with short hairs, easily lost leaving pronounced scars, 5.5-6 regularly increasing convex whorls, initially keeled, last whorl with indistinct edge at periphery, weakly descending near aperture, aperture with prominent white or red lip inside near the margin, margin sightly reflected, umbilicus very narrow, usually entirely covered in adults.

5-7 x 7.5-10 mm

In the moist herbal layer of forests in highlands, 200-700 m, under leaf litter and rotting timber.

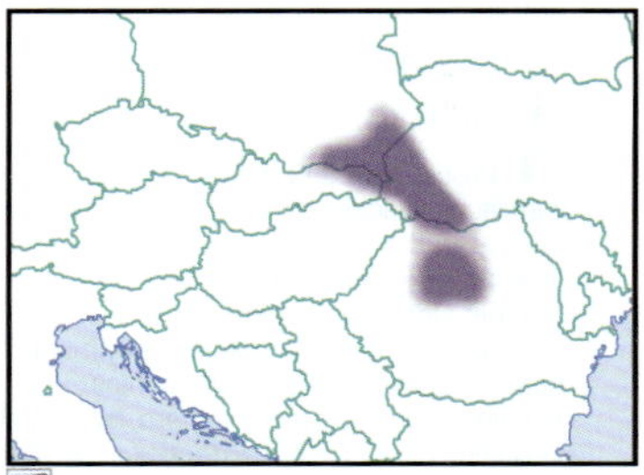

SE Poland to Romania

Bielz 1859: 216, Westerlund 1889: 33, Shilejko 1978: 209, Proćków 2009: 114.

Trochulus caelatus (Studer, 1820)

Without gender agreement: *Trochulus caelata* (Studer, 1820)

Glischrus caelata Studer, 1820 in Studer 1820: 86 (*Glischrus (Helix) cælata*)

Not *Helix caelata* as mentioned by [Vallot], 1801 (regarded here as unpublished under Art. 8.1.1, currently *Candidula intersecta*). *Glischrus caelata* would currently not be a junior homonym of it.
For those who regard [Vallot] 1801 as published work the two available names were temporarily classified in the genus *Helix* in the 1800s. Locard 1880 (Études var. malac. p. 9 [not verified]) established a replacement name *Helix glypta* to escape secondary homonymy. Falkner 1996: 99 considered *Glischrus caelata* as permanently invalid under Art. 59.3.
I do not use *Helix glypta* here, so at the present date (28 Sep 2012) it is not in use as required by Art. 59.3, and *Glischrus caelata* can be used. I think Art. 59.3 should be removed from the Code or substantially modified. It has only caused problems in cases like this one.

3.5-4 x 7-8 mm

Switzerland: Moutier, Bris gorge

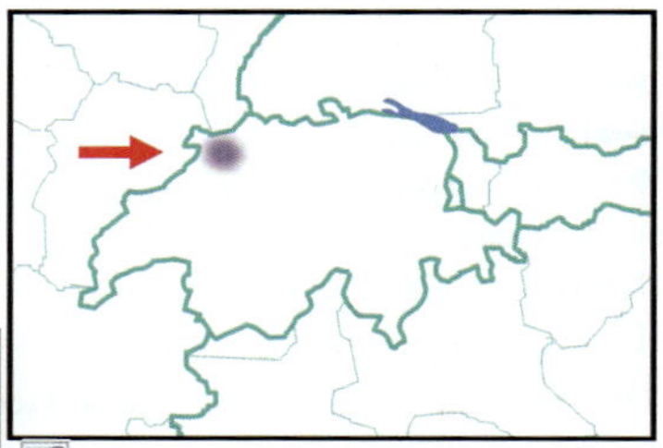

NW Switzerland (near Biel)

EU Red List: Near Threatened NT. Threatened by quarrying taking place in the area, rock climbing and road construction (motorway Biel-Basel).

Kerney & Cameron 1979: 195, Falkner 1990: 204, Turner et al. 1998: 343, Proćków 2009: 116 (considered *coelomphala* as a synonym, here under *T. striolatus*).

Shell reddish brown, shiny, prominently striated, translucent and shiny, hairs only in very young juveniles, shell strongly depressed, nearly disc-like, with blunt edge, convex on lower side, 5-5.5 flat whorls, aperture with white lip, umbilicus deep, open and wide, 1/6-1/5 of diameter. The animal is dark greyish. Differs from *T. striolatus* and *T. montanus* in its very short dart sacs.

Shady limestone rocky habitats, cooler and more shady than for *T. montanus*, at steep rocks, 500-1300 m.

Considered as a glacial relict that survived the glaciations on ice-free tundra areas on the Jura mountains.

Trochulus clandestinus (Hartmann, 1821)

Without gender agreement: *Trochulus clandestina* (Hartmann, 1821)

Helix corrugata var. *clandestina* Hartmann, 1821 in Hartmann 1821 (Neue Alpina 1): 236

5-7 x 9.5-12 mm

Germany: S Baden, near Blumberg

Shell light greyish brown to yellowish grey, sometimes with white band, shiny, adults hairless, last whorl inflated, 5-6 whorls, aperture with weak lip, umbilicus initially narrow, getting wider only at the last whorl, so that only the last whorls are visible inside, 1/6 of diameter.
Differs from *T. striolatus* and *T. montanus* in more rapidly increasing whorls, provoking a lower number of whorls at equal shell size. Flagellum very short in contrast to other *Trochulus* species.

Activity begins in February at 10°C in shadow.

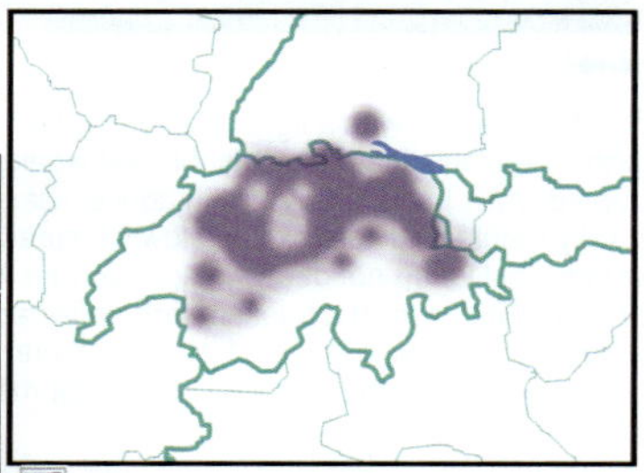

Switzerland, Liechtenstein
Introduced to S Germany (Hegau)

Herbal layers of mountain forests, shrubs near water currents, today frequently in cultivated land, gardens, parks, road margins and in ruderal vegetation. Usually in 400-1000 m altitude, maximum up to 2300 m.

Kerney & Cameron 1979: 195, Turner et al. 1998: 346, Proćków 2009: 118.

Trochulus edentulus (Draparnaud, 1805)

Without gender agreement: *Trochulus edentula* (Draparnaud, 1805)

Helix edentula Draparnaud, 1805 in Draparnaud 1805: 80
Junior synonym of *Helix depilata* Draparnaud, 1801 which was suppressed under Art. 23.9.1 by Falkner et al. 2002: 151 who provided the 25 required references.

4.5-5.5 x 7-8 mm

Germany: S-Baden, Eichberg near Blumberg

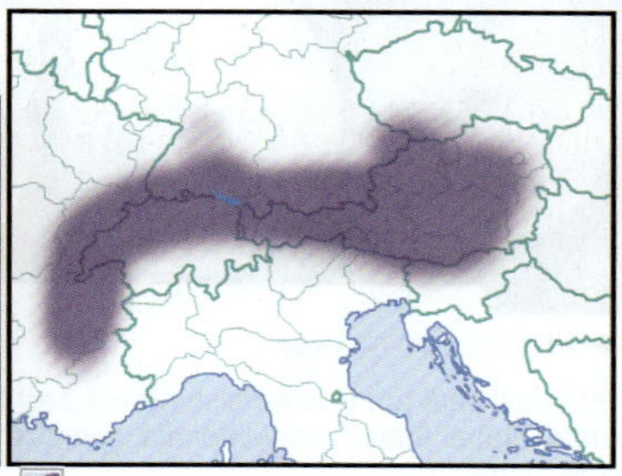

Alps and northern highlands

Austria: Tirol, Karwendel, Enger Grundbach

Shell horny to reddish brown with pale band at the periphery, finely striated, juveniles with dense hairs, often lost in adults, 5-7 tightly coiled whorls, last whorl angled at periphery, shape variable from flat to conical, aperture with white lip inside, slightly thickened at basal side but without teeth, margin slightly reflected at lower side, umbilicus very narrow.
In the west the shells are more cone-like with very narrow (almost covered) umbilicus and only juveniles have hairs, in the east shells are flatter, lower side rounded, umbilicus open or half open (up to 1/10 of diameter), with stronger and more permanent hairs.
Shell usually smaller than *T. unidentatus*, last whorl with more pronounced edge, aperture narrow, no tooth.
Animal light brownish with darker gray upper side of anterior part, head and tentacles.

Among herbs and under leaves and stones in humid mountain forests on calcareous subsrate, usually not above the timberline. Sometimes also in open forests and stony alpine pastures. Between 300 and 2300 m altitude.

Kerney & Cameron 1979: 196, Falkner 1990: 206, Turner et al. 1998: 351, Proćków 2009: 121.

The regional variation in shell morphology has been associated with subspecies, but it has not been established that partial barriers to gene flow are responsible.

Trochulus erjaveci (Brusina, 1870)

Helix erjaveci Brusina, 1870 in Brusina 1870: 26 (*Helix (Fruticicola) Erjaveci*)

6-9 x 10-15 mm

N Albania: Lepushë

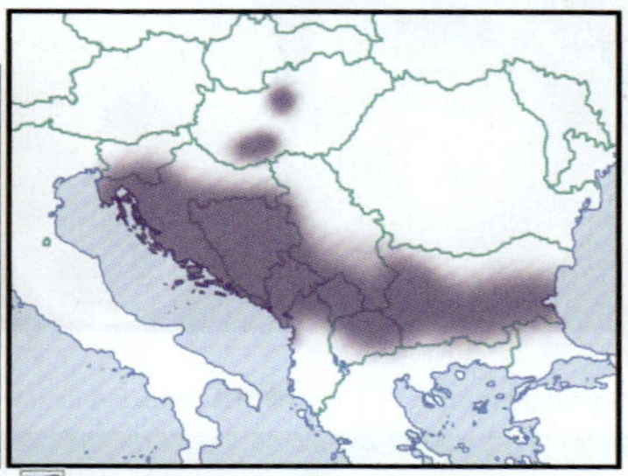

Slovenia to Bulgaria

Shell light yellowish, occasionally with light band at periphery, finely striated, with granulated microsculpture, usually better visible on the lower side, adults without hairs, 5-6 weakly convex whorls with deep suture, initially with edge at periphery, last whorl rounded, descending near aperture, aperture with distinct white lip inside, margins reflected and approaching each other slightly at parietal side, umbilicus deep, initially narrow and distinctly wider at last whorl, 1/6-1/5 of diameter.

Moderately humid habitats, moist and shady forests, rarely in open habitats, usually on shrubs, in Bulgaria between lowlands and 1300 m.

Westerlund 1889: 45, Pavlović 1912: 35 (*sericea, haueri, osoria, floerickei*), Tomić 1959: 19, Damjanov & L. 1975: 370, Dhora & W. 1996: 159, Irikov & M. 2006: 818, Proćków 2009: 124.

Trochulus filicinus (Pfeiffer, 1841)

Without gender agreement: *Trochulus filicina* (Pfeiffer, 1841)

Helix filicina Pfeiffer, 1841 in L. Pfeiffer 1841: 39
Not *Helix (Trichia) filicina* Hartmann, 1841 in Hartmann 1844: 41 from Switzerland (name probably never used for a taxon after 1899). I do not know which name has precedence.

5-8 x 8-11 mm

Austria: Steiermark, Murtal, Ruine Pessau

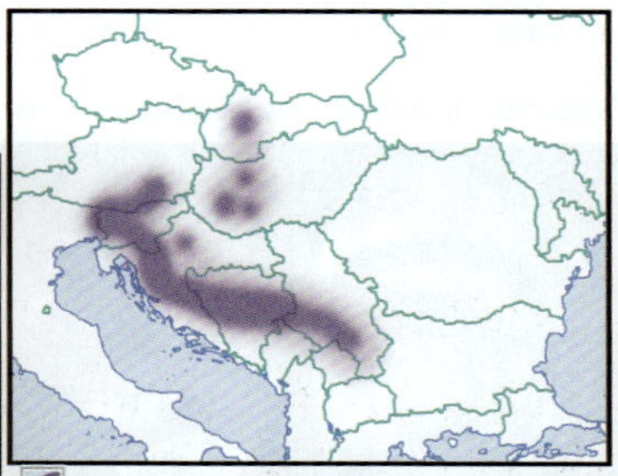

SE Alps to S Serbia, isolated in W Slovakia and W Hungary

Shell horny brownish with whitish band at the keel in the first whorls, transparent but solid, striated, often with soft hairs but not granulated, 5.5-6 convex whorls, regularly increasing, first whorls keeled, last whorl initially with edge at periphery, more rounded near aperture, lower side relatively flat, aperture with lip inside, stronger at basal side, weaker and reddish or whitish at upper side, visible from outside as a yellow structure, aperture flattened at lower side, umbilicus narrow and partly covered by the reflected columellar margin.
Larger than *T. edentulus* and *T. unidentatus*.

Shady habitats in the herbal layer in highlands and lower mountain forests, often in river valleys.

Westerlund 1889: 46, Falkner 1990: 206, Proćków 2009: 125.

The regional variation in shell morphology has been associated with subspecies, but it has not been established that partial barriers to gene flow are responsible.

Trochulus graminicola (Falkner, 1973)

Trichia graminicola Falkner, 1973 in Falkner 1973: 211

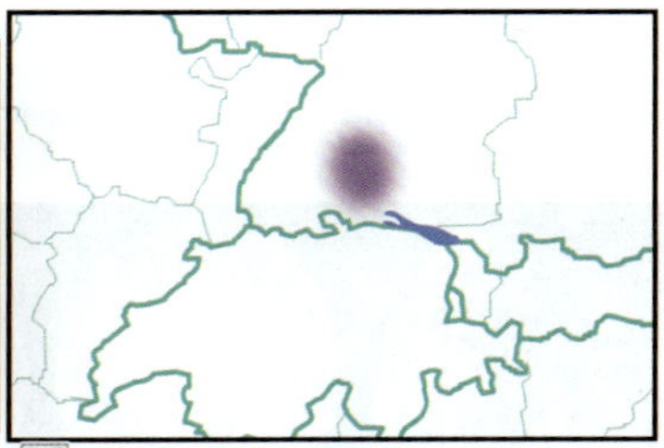

4.5-5.5 x 9.5-11 mm

Germany: S Baden, near Blumberg

S Germany: Schwäbische Alb

Among grasses in rock rubble in light and steep mountain forests, the snails stay firmly attached to grass blades.

EU Red List: Near Threatened NT. Extremely rare, only known from few localities. One of the few endemic molluscs of Germany.

Shell dark horny to reddish brown, slightly translucent, striated, 5-6 slightly convex whorls with moderate suture, flatter than *T. coelomphala*, aperture more oblique, with white lip inside, umbilicus wide (more than 1/5 of shell diameter).
Differs from *T. hispidus* by a flatter shell and a more oblique aperture.
Vaginal neck longer than in the other *Trochulus* species.

Kerney & Cameron 1979: 195, Colling et al. 2008: 131, Proćków 2009: 127.

Trochulus hispidus (Linnæus, 1758)

Without gender agreement: *Trochulus hispida* (Linnæus, 1758)

Helix hispida Linnæus, 1758 in Linnæus 1758: 771
Helix sericea as mentioned by Draparnaud 1801: 85 (subsequent use of *Helix sericea* Müller, 1774, misidentified)

Draparnaud 1801 used and referred to *Helix sericea* Müller, 1774 (in Müller 1774: 62), and obviously misinterpreted it. Rossmässler 1838: 2 attributed the name *sericea* to Draparnaud 1805 who gave a figure. Forcart 1957: 198 proposed to suppress *Helix sericea* Müller, 1774, but a case has never been brought to the ICZN. Forcart 1965: 83 proposed to use the name *Helix plebeia* Draparnaud, 1805 instead of *sericea* sensu Draparnaud 1801. This view was partly followed, for example by the Kerney et al. guides. *Trichia sericea* was applied to various species in the literature, mostly in the sense of Draparnaud, but sometimes (like in the Kerney et al. books) also in the sense of Müller (Falkner et al. 2002: 148). Falkner et al. 2002 did not know what to do with this name (1 - to suppress *Helix sericea* Müller, 1774 and validate *Helix sericea* Draparnaud, 1801, 2 - use the next available synonym for the species Draparnaud 1801 had in mind, 3 - designate a neotype for Müller's 1774 name).

Proćków 2009: 128 classified it as a synonym of *T. hispidus*, which I regard as a proposal to wait until the taxonomy is better understood, certainly the best solution in such a situation. My personal preference would be to follow Falkner et al.'s (2002) second option, but the third option could also be a solution.

Helix plebeium Draparnaud, 1805 in Draparnaud 1805: 105

Comprises forms formerly known as *Trichia hispida*, ***T. sericea*** and ***T. plebeia***. In some regions two distinguishable forms seem to live sympatrically. Here they are referred to as the nominate form and the second form. More research is necessary, also on the molecular level, to learn more about their relationships.

Nominate form

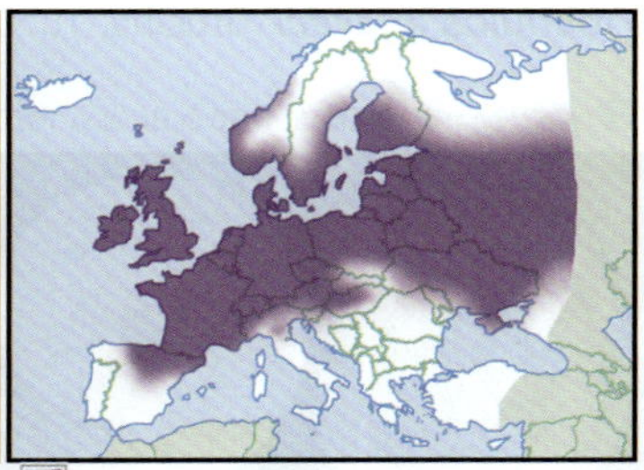

Nominate form. 3-6 x 5-11 mm

Germany: Thüringen, Sondershausen

N Spain to Ural

Shell brown to cream, sometimes with a light band at the periphery, irregularly striated, densely covered with short curved hairs, usually remaining in the umbilicus if worn from the rest of the shell, hairs leave pronounced scars if lost, 5-6 moderately convex whorls, rounded or very slightly keeled at the periphery, aperture with thin white lip inside, umbilicus open and usually wide, 1/8-1/4 of shell diameter.
Shell locally very variable, especially in shell shape, size and size of the umbilicus.
Animal brownish grey with darker anterior part.

(1) Slovakia: Krásnohorská Dlhá Lúka, (2) Austria: Tirol, near Hinterriss
(1) Jozef Grego

Various humid habitats in light forests, shrublands and cultivated areas, often at *Urtica dioica* near water bodies. Sometimes also at rock rubble of alpine pastures, road margins and (N Scotland) in disturbed habitats. In Spain usually in dry places, but absent from very dry sites. Up to 2300 m in Alps and Pyrenees, usually below 1600 m.

Clusters of 15-40 eggs (diameter 1-1.5 mm) are deposited under leaves or at the soil between April and September, juveniles hatch after 18-25 days, maturity is reached at the end of the next year (Moquin-Tandon 1855: 224, Frömming 1954: 260).

Second form (often called ***sericea*** in Germany, Switzerland and Austria, this name cannot be used)

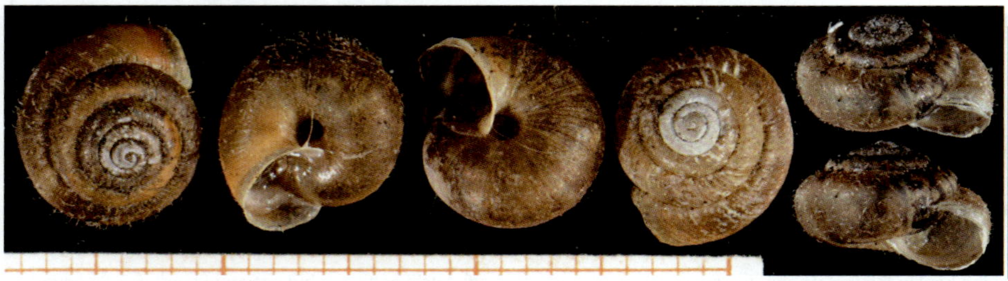

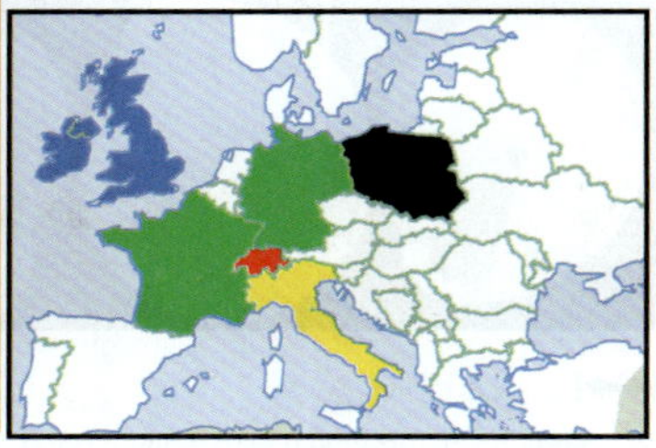

Second form. 4-6 x 5-11 mm

Germany: Hessen, near Dietges

Diversity of *Trochulus* classification in Europe

Shell round with long and curved hairs, whorls more convex and with coarser growth lines than in the nominate form of *T. hispidus*, lip weak, not wave-like, umbilicus narrow. Size and length of hairs locally very variable.

In various habitats, swampy to moderately dry decidious or coniferous forests, also in open forests, shrublands, reed swamps and moist to moderately dry meadows, alpine pastures and old ruins. Usually below 2000 m, occasionally 2400 m.

30-40 eggs (diameter 1 mm) are laid between July and September, deposited on dead leaves, juveniles hatch after 20-25 days, maturity is reached after 12-14 months (Gassies 1849: 91, Moquin-Tandon 1855: 219, Germain 1930: 241, Frömming 1954: 259).

Alps without eastern margin, to Pyrenees, England and W German highlands

Various approaches have recently been proposed to classifiy the forms of this complex. In www.iucnredlist.org (08.2012) some of these were mixed up.
Switzerland: Perrin et al. 1984 found that W Swiss populations classified there as *sericea* and *hispida* separated distinctly in electrophoretical analyses. Turner et al. 1998: 338 accepted this view but admitted that there were many unclear intermediate forms. Dépraz et al. 2009 came to the same result with mitochondrial and nuclear DNA data (apparently they did not read the Perrin et al. 1984 study).
British Isles: Kerney & Cameron 1979, Kerney 1999: 196 and Anderson 2005 accepted *Trichia plebeia* (Anderson 2005: *sericeus*) and *T. hispida* as two separate species in Britain, clearly distinguished anatomically (Naggs 1985 and dfferences in the spermatheca mentioned by Anderson 2005, who reported *sericeus* also from central France). Kerney 1999 noted that more research was necessary.
France and Germany: Falkner et al. 2002: 148, Falkner & Niederhöfer 2008: 131 and Colling et al. 2008: 131 argued that the entire complex comprised some more species in France and Germany, they used *Trichia plebeia* in a different sense than Kerney 1999 for an endemic SW French form and proposed to use *sericea* for another species (in 2008 they were still undecided for which one). The arguments in support of various separate so-called species (which were claimed to be well characterised) were not based on published results of detailed studies. In this sense these arguments were certainly weaker than those of the Swiss and British studies.
Italy: Manganelli et al. 1995: 26, 50 used *plebeia* for a N Italian form and talked about *sericea* sensu Draparnaud 1801 and *plebeia* as if they were two different taxa. Cossignani & C. 1995 used both names *Trichia plebeia* and *sericea* for N Italian forms, without clarifying in which sense.
Poland: Wiktor 2004 used *plebeia* for a SW Polish form. Proćków 2009 classified *sericea, plebeium* and *ataxiaca* as synonyms of *T. hispidus.*

Here the most recent Polish contribution is accepted (Proćków 2009). It is a logical consequence of the absence of published constraints supporting a definite classification of these taxa. In this sense it represents acceptably the current state of our understanding of these forms.

Trochulus leucozonus (Pfeiffer, 1828)

Without gender agreement: *Trochulus leucozona* (Pfeiffer, 1828)

Helix leucozona Pfeiffer, 1828 in C. Pfeiffer 1828: 4
Pfeiffer gave a German translation for the name ("die weissbandirte Schnirkelschnecke"), indicating that the specific name was meant to be used as an adjective.

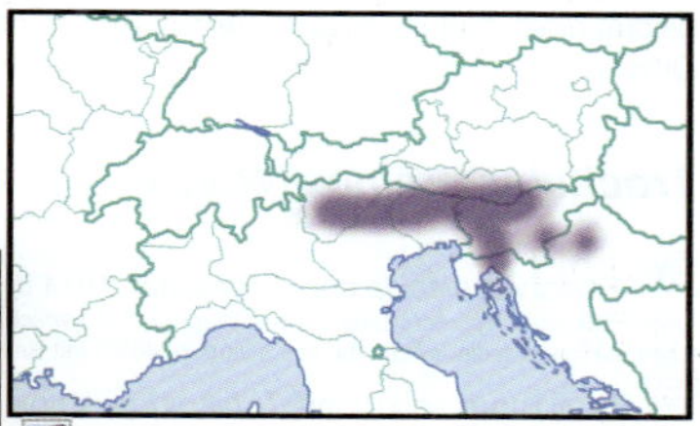

5-7.5 x 8-11.5 mm

Austria: Kärnten, Karawanken, Oberes Bärental

SE Alps

In moist and shady forests, under stones, in leaf litter and in mosses, up to 2200 m.

Westerlund 1889: 33, Proćków 2009: 138.

Shell light horny brownish or pale reddish brown, finely striated, 5.5-6.5 convex whorls, last whorl distinctly wider than penultimate whorl, with blunt edge at periphery, aperture with prominent white lip inside at margin, margin distinctly reflected, visible as a yellow structure from outside, umbilicus usually very narrow, 1/10-1/20 of diameter, sometimes partly covered.

Various subspecies were proposed to reflect the regional variation in shell morphology (Bank 1995).

Trochulus lubomirski (Ślośarski, 1881)

Helix lubomirski Ślośarski, 1881 in Ślośarski 1881: 319 (*Helix (Fruticicola) Lubomirski*)
Original spelling was not *lubomirskii*, the correct spelling *lubomirski* was used in some recent Polish sources.

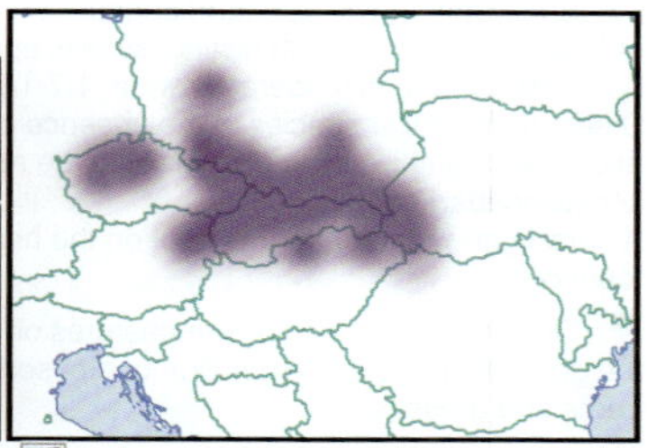

5-7 x 7-10 mm

Slovakia: near Turčianska Blatnica

W Czech Republic to W Ukraine

Shell whitish yellow to greenish yellow, transparent and weakly shiny, few and fine hairs (0.1 mm long), usually lost in adults, 4.5-5 whorls, apex pointed, aperture rounded without or with very weak lip, umbilicus very narrow, less than 1/10 of diameter.

Dense herbal vegetation, under stones in relatively humid forests, sometimes on limestone rocks and meadows, in river plains, highlands and mountains up to 1600 m.

Shilejko 1978: 204, Falkner 1990: 204, Proćków 2009: 140.

Slovakia: Banská Bystrica, Podlavice
Jozef Grego

Trochulus luridus (Pfeiffer, 1828)

Without gender agreement: *Trochulus lurida* (Pfeiffer, 1828)

Helix lurida Pfeiffer, 1828 in C. Pfeiffer 1828: 33

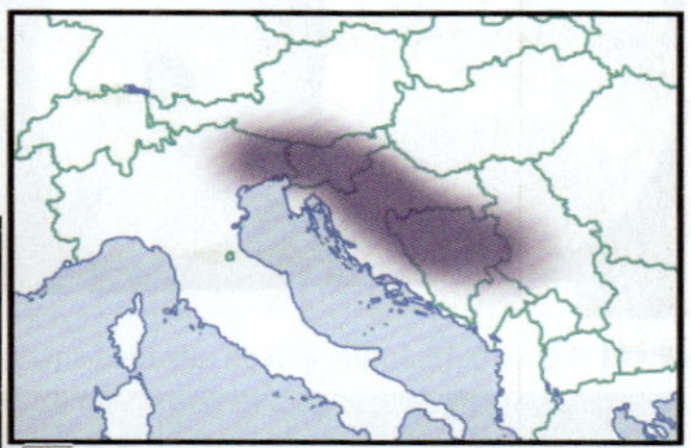

5-8 x 8-12.5 mm

Slovenia: Julian Alps, near Bled

S Austria and N Italy to Bosnia

In moist and shady forests, under leaf litter.

Rossmässler [1834] (Heft 2): 4, Westerlund 1889: 78, Proćków 2009: 141.

Shell horny brownish, finely striated, with dense and short soft hairs, apex pointed, 5.5-6.5 not very convex whorls, irregularly increasing, penultimate whorl width 2 x of preceding whorl, last whorl not much wider, with whitish edge at periphery, aperture with distinct lip inside, visible from outside as a yellow structure, margin sharp and slightly reflected, umbilicus very narrow (less than 1/10 of diameter), partly covered by the reflected columellar margin.

Trochulus montanus (Studer, 1820)

Without gender agreement: *Trochulus montana* (Studer, 1820)

Glischrus montana Studer, 1820 in Studer 1820: 86 (*Glischrus (Helix) montana*)

4.5-7 x 9-12 mm

Switzerland: Kanton Uri, near Realp

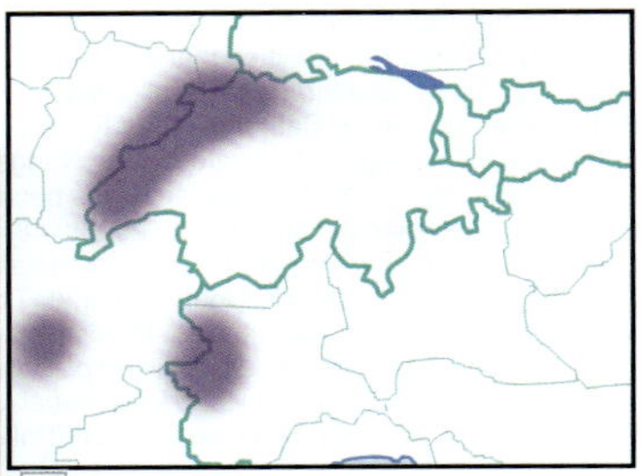

W Alps

Mountain slope forests, shady limestone rocks, deciduous forests in gorges, in 300-1700 m altitude.

Falkner 1973: 215, Falkner 1990: 204, Kerney & Cameron 1979: 194, Turner et al. 1998: 341, Proćków 2009: 143.

Shell light horny to reddish brown, translucent, shiny, finely striated, few hairs only in juveniles, 5-6 convex whorls with moderately deep suture, shell relatively flat but periphery rounded, aperture with white lip inside, umbilicus open and moderately wide, 1/8-1/5 of diameter.
Differs from *T. striolatus* in its longer dart sacs and longer flagellum.

Trochulus oreinos (Wagner, 1914)

Fruticicola hispida oreinos Wagner, 1914 in A. J. Wagner 1915: 484
Not *Fruticicola erjaveci osoria oreinos* Wagner, 1915 which was established on p. 481 in the same work as an infrasubspecific name, not *Fruticicola sericea* forma *oreinos* from the same work p. 484, First Reviser Falkner 1996: 98.

Shell more solid than *T. hispidus*, shiny, coarsely and irregularly striated and finely granulated, with shorter hairs (0.5 mm long) which are lost in 80 % of the shells, last whorl with rounded keel, aperture with strong lip inside, visible as a yellow structure from outside, with a tooth-like structure at the basal margin, umbilicus 1/6 of diameter, slightly larger in the east.

-4 x 5.5-7.5 mm

Rocky habitats of high calcareous mountains, in vegetation-free areas under stones, among sparse alpine grasses, in ground litter of mountain pine shrublands, 1400-2200 m. Frequent within its range.

EU Red List: Near Threatened NT. Threatened by habitat destruction by ski slopes.

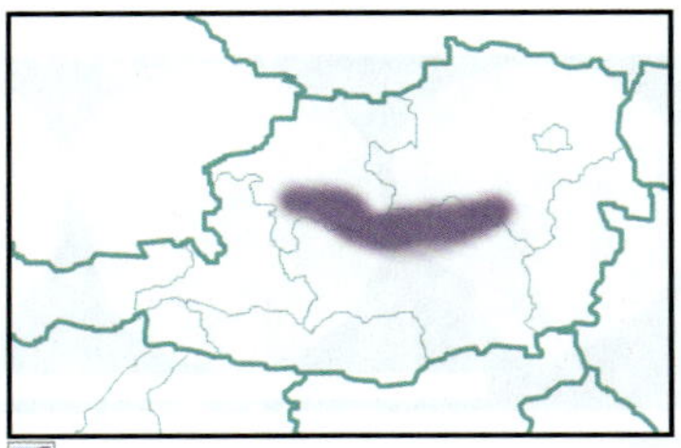

Austria

Klemm 1973: 393, Duda et al. 2011.

Two parapatrically distributed forms (west and east), weakly distinguished morphlogically, separated in molecular analyses, but without constraints to classify them as distinct species. Both forms were not found syntopically with *T. hispidus*, but in close vicinity. Morphological, ecological and molecular analyses suggested that *oreinos* and *hispidus* are different species.

Trochulus piccardi Pfenninger & Pfenninger, 2005

Trochulus piccardi Pfenninger & Pfenninger, 2005 in Pfenninger & Pfenninger 2005: 264

Shell light brown, solid and opaque, without hairs or scars, striated, 4.5-5 whorls with distinct suture, aperture often with white lip inside, umbilicus moderately wide, 1/7-1/6 of diameter.
Differs from similar species in the absence of hairs and hair scars and a wider umbilicus, anatomically in the absence of separate accessory dart sacs.
Animal light, with a dark gray spot on the head behind the ommatophores.

On extensively used alpine pastures on S exposed slopes, on the ground among the vegetation composed of typical alpine herbs and grasses.

Proćków 2009: 146.

Switzerland: Vaud, Château-d'Œx

3.5-4.5 x 6-8 mm

Ira Richling

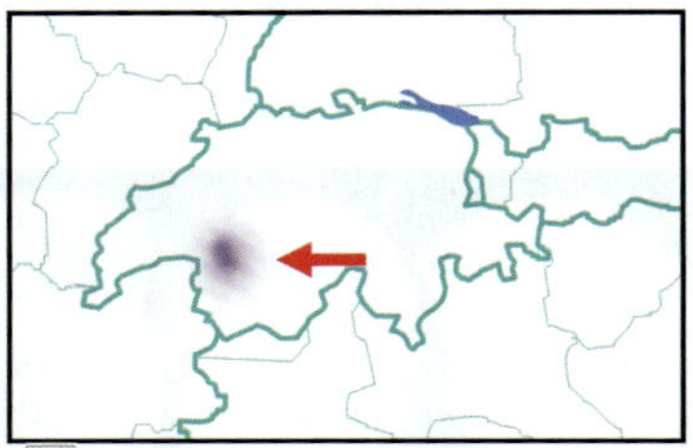

SW Switzerland

Only known from few localities, more research is needed to learn more about the range of this species.

Trochulus striolatus (Pfeiffer, 1828)

Without gender agreement: *Trochulus striolata* (Pfeiffer, 1828).

Helix striolata Pfeiffer, 1828 in C. Pfeiffer 1828: 28
Cochlea rufescens mentioned by Mendes da Costa 1778: 80 was a subsequent use of *Helix rufescens* Pennant, 1777 (currently *Arianta arbustorum* (Linnæus, 1758)), not a new name. Falkner 1990: 204 used it as *Trichia rufescens* (Da Costa, 1778), but this was not tenable.
Helix concinna Jeffreys, 1830 in Jeffreys 1830: 336
Helix suberecta Clessin, 1878 in Clessin 1878: 46 Various type localities in Germany and Austria.
Helix coelomphala Locard, 1888 in Locard 1888: 352 (*Helix cœlomphala*) Various type localities in Germany, France, Switzerland.
Gender unchangeable following Art. 31.2.2 (Falkner & Falkner 2008). Possibly published earlier in "Contr. XII" p. 48, I did not understand what this meant.

6-8 x 10-13.5 mm, in high altitudes smaller

Germany: near Leverkusen

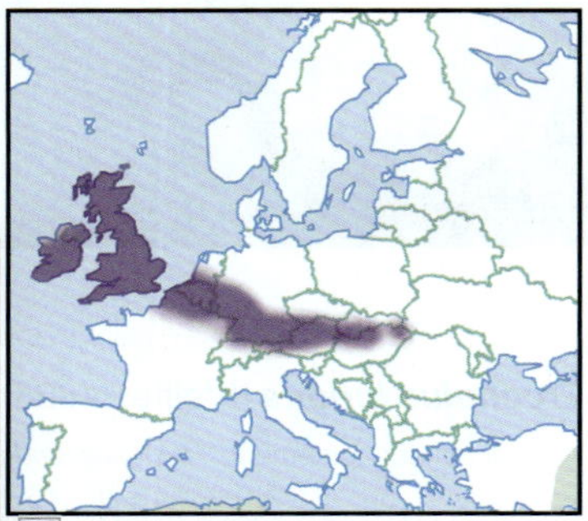

Hungary to British Isles
Native in Britain, but has become much commoner through human activity, still expanding its range in Scotland where it was rare in the 1930s.

Germany: Bayern, Bad Tölz, near Kochel am See, 800 m

Shell dark reddish brown to creamy yellow, usually with a light band at the periphery, adults have usually no hairs, upper side relatively flat, 5.5-6.5 moderately convex whorls with moderate suture, last whorl often angled, aperture with whitish lip inside, margin very slightly reflected, umbilicus open, 1/8-1/5 of shell diameter. Shell size and umbilicus locally very variable. Larger than *T. hispidus*.

Humid and natural shady habitats in forests, shrubs and badlands, road margins, in N France and British Isles among tall grasses and *Urtica dioica*, often artificially dispersed, occasionally a pest in gardens. In N Switzerland between 400 and 800 m, in Austria 150-2000 m.

40-50 eggs (diameter 1.5 mm) are laid between August and October, juveniles hatch after 20-25 days, maturity is reached after 12-14 months.

Germain 1930: 243 (life cycle), Kerney et al. 1983: 263, Falkner 1990: 204 (*T. rufescens*), Turner et al. 1998: 344, Kerney 1999: 195, Egorov 2008: 88 (*Tri. concinna*, probably W Ukraine and W Belarus), Sysoev & Schileyko 2009: 199 (*Tro. concinnus*), Proćków 2009: 146.

Proćków 2009: 153 classified ***T. suberectus*** as a separate species, without arguments supporting such a classification. After having analysed numerous E Swiss specimens Turner et al. 1998: 427 came to the conclusion that the presence of 6 mucus glands was within the variation of *Trichia sericea*, so that the Swiss populations belonged to this species. On the one hand, Proćków 2009 cited this result, and accepted its conclusion, on the other hand the E Swiss populations were again classified with *T. suberectus* (in the map) and the mucus glands in the analysed Austrian populations were not checked to test Turner et al.'s 1998 observations. Another problem consists in the identity of the types, which were not analysed (Turner et al. 1998 considered them as conspecific with *striolata*, this view is certainly acceptable in the absence of evidence of the contrary). Since Proćków 2009 did not weaken Turner et al.'s 1998 conclusions, a classification as a separate species seems badly supported.

Falkner 1990: 202 and Falkner et al. 2001: 57 considered ***coelomphala*** as a separate species without reference to a published study to support this classification, and obviously inconsistently (1990 from Germany and France, 2001 from Germany and Austria, not France). Since no taxonomic arguments were given the identity of the involved units to which this name was referred remained unclear. Proćków 2009 gave no detailed arguments to classify *coelomphala* with *caelatus* (none of the types came from its range). Since the types came probably from various countries (Germany, France, Switzerland) the name can certainly not be used for a species without analysis of the types, and without a proper understanding of the taxonomic diversity in this group. Locard 1888 saw it in the vicinity of *striolata* and others, it is tentatively classified with this species here.

Trochulus monodon (Férussac, 1807)
Trochulus unidentatus (Draparnaud, 1805)

Helix unidentata Draparnaud, 1805 in Draparnaud 1805: 81

Not *Helix unidentata* Holten, 1802 (= currently *Stylodonta unidentata* (Holten, 1802) from Seychelles).

Helix monodon Férussac, 1807 in J. Férussac 1807: 120 (*Helix Monodon*)

I am giving Férussac's single authorship here, as this has been traditionally done. I have not researched which one of the two Férussac's was responsible for this name (or both). It seems to have been overlooked that both Férussacs were co-authors of the 1807 work which should better be cited as "Férussac & Férussac 1807", and that the responsibilities were explained in the introduction.

Mentioned without description, only with a name "*unidentata*" cited from Draparnaud's 1805 work, this work was specified bibliographically at various occasions by Férussac 1807 (p. 85 etc., p. 98, p. 114). Férussac 1807 did not express that or why the name should be replaced, so in the absence of an expressed statement (Art. 72.7) it can be seen as established as a regular new name with Draparnaud 1805 being interpreted as a bibliographical reference in the sense of Art. 12.2.1 (but only because a bibliographic specification was given in Férussac's work).

Bank & Menkhorst 2008: 97 and Colling et al. 2008: 133 remarked that *Helix unidentata* can very probably not be used for the European species. Art. 23.9.5 allows to ask the Commission to solve the case if both were not considered congeneric after 1899. Until 1898 the species from the Seychelles was classified in *Helix*, the first occasion at which it was classified as a separate genus *Stylodonta* was in 1909 (Gerlach 2006: 73). The European species was classified in *Helix* by Kobelt 1906. The arguments which were brought forward by Bank & Menkhorst 2008 were very strong and I do not believe that the usage of the name for the European species can be saved. I do not use *monodon* in the sense of Art. 11.5 here, it is better to wait a little in such a case.

Helix cobresiana Alten, 1812 in Alten 1812: 79 (*Helix Cobresiana*)

Fruticicola unidentata unidentata natio *subtecta* Poliński, 1929 in Poliński, 1929 (Prace Panstw.): 167

Name originally infrasubspecific, used as *Trichia (Petasina) unidentata subtecta* by Klemm 1954: 259 for a subspecies, making it available (Art. 45.6.4).

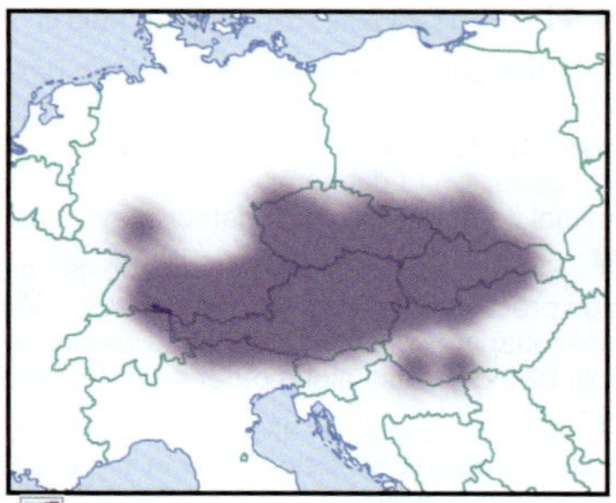

SW Germany to Slovakia

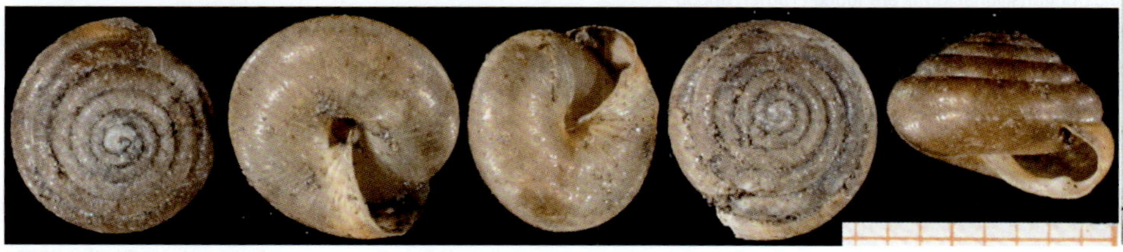

3.5-6 x 5-9 mm

Slovakia: Striebornica

Germany: Bayern, near Walchensee

Shell horny to reddish brown, finely striated, sometimes with a light band at the last whorl, juveniles with dense hairs, adults have usually no hairs any more, 5-7 tightly coiled convex whorls, rounded at periphery, shape somewhat cone-like, aperture with whitish or pinkish lip inside, at basal side with a strong white tooth, umbilicus narrow. Shell locally very variable, in shell size, shape and umbilicus.
T. edentulus has no tooth, *Perforatella bidentata* has two teeth.

Humid to moderately dry mountain forest habitats, also above the timberline, prefers rocky and stony habitats on calcareous substrate. Usually in 500-2000 m, in Switzerland up to 2300 m. Frequent in its core range.

Reproductive season (deposition of eggs) seems to begin in May.

Endangered in Germany. The populations in E Austria (*subtecta*) have seriously declined in the recent past, the species is threatened by modern forest management.

Klemm 1973: 394 (many localities for *subtecta* between Wien and S Steiermark), Falkner 1990: 206, Proćków 2009: 155 (classified *subtecta* as a synonym and found weak support for the presence of various subspecies).

This species was called *Helix cobresiana* between 1840 and 1933 because the homonymy with *unidentata* from Seychelles was known. Ehrmann 1933: 126 used the name *Fruticicola unidentata* (Draparnaud, 1805) for the species, and provoked that subsequent authors also used this name.
The regional shell variation has until recently been associated with some 6 local subspecies. Presence of subspecies in a modern bioscientific sence (based on partial intrinsic barriers to gene flow and narrow hybrid zones) remains to be tested against the alternative explanation of regional morphological gradients and slow but continuous gene flow over area, without partial genetically induced interruptions of gene exchange at narrow hybrid zones.

Trochulus villosulus (Rossmässler, 1838)

Without gender agreement: *Trochulus villosula* (Rossmässler, 1838)

Helix villosa var. *villosula* Rossmässler, 1838 in Rossmässler 1838-1844 (7-8): 1

Mentioned with a short description in the text under *Helix villosa*, subchapter Varietäten, but it is not clear that the name was (conditionally) used for a taxon by Rossmässler under Art. 11.5: "Als solche [= als Varietät] bin ich versucht anzusehen eine viel kleinere, etwas enger genabelte, sonst aber ganz übereinstimmende Form aus Ungarn (...); ich halte diese für *H. villosula* Ziegl., die ich in dem von Ziegler's Sammlung aufgenommenen Verzeichniss (...) finde."

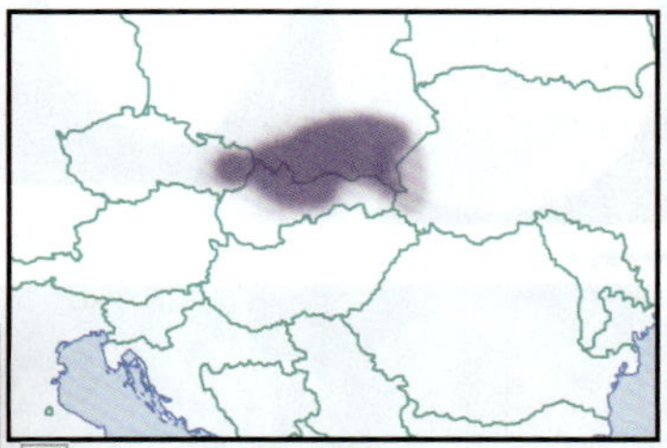

SE Poland, N Slovakia

4.5-6 x 7-9 mm

Czech Republic: Striebornica pri Piestanok

Shell dark brown, thin, coarsely striated, irregularly covered with long yellowish and strongly curved hairs, 5-5.5 convex whorls with deep suture, last whorl with blunt edge at periphery, descending near aperture, aperture with weak lip inside, margin sharp, umbilicus initially moderate, wider at last whorl, 1/6-1/5 of diameter.

Humid habitats in mountains and river valleys, usually in the vegetation, more rarely on the ground among plants and dead leaves, up to 1600 m.

Westerlund 1889: 53, Shilejko 1978: 208, Proćków 2009: 162.

Slovakia: Priechod near Banská Bystrica

Jozef Grego

Trochulus villosus (Draparnaud, 1805)

Without gender agreement: *Trochulus villosa* (Draparnaud, 1805)

Helix villosa Draparnaud, 1805 in Draparnaud 1805: 104
Fruticicola villosa var. *alpicola* Eder, 1921 in Eder 1921: 228
Helix phorochaetia Bourguignat, 1864 in Bourguignat 1864 (Mal. Gr.-Ch.): 52, Pl. 6 (*Helix phorochætia*)

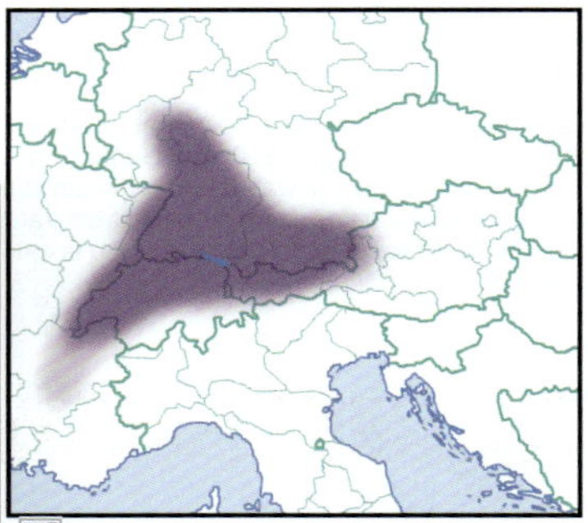

NW and W Alps, SW Germany

6-8 x 11-14 mm, smaller above 1400 m

Germany: Baden-Württemberg, Baar/Schwarzwald

Vulnerable in parts of Germany.

Shell pale greyish yellow to reddish brown, coarsely striated, with (up to 1.5 mm) long hairs, 5.5-6 convex whorls with deep suture, apex blunt, upper side relatively flat but periphery rounded, last whorl slightly descending, aperture rounded with weak lip inside, umbilicus wide and deep, 1/5 of diameter.
Animal light brownish with darker greyish tentacles.

Germany: Bayern, near Walchensee

Shady and humid mountain habitats, usually in forests, requires high humidity, in zones of high precipitations also in open habitats, alpine pastures. Usually in 500-2000 m in the Alps, maximum in Switzerland 2400 m. Follows rivers north of the Alps (Rhein until Mainz, Donau until Dillingen, also along other rivers). Occasionally dispersed to N Germany.

Kerney & Cameron 1979: 195, Falkner 1990: 204, Gittenberger & Neuteboom 1991 (regarded *alpicola* as a separate species), Turner et al. 1998: 347, Proćków 2009: 164 (classified *phorochaetia* and *alpicola* as synonyms).

Eder 1921 suspected that ***alpicola*** was an alpine small and slightly flatter form of *villosus*, and it seems that in montane and subalpine stages all kinds of transitional forms are found (Turner, unpublished).
De Winter 1990 regarded *Trichia* ***phorochaetia*** from SE France as a separate species and proposed to use this name instead of *sericea* or *plebeia* for populations of SW France and Switzerland. The name was accepted by Falkner et al. 2002: 149 (who used it for a French endemic species), but not by Turner et al. 1998: 340 for Switzerland (who classified it with *sericea*), and not by Proćków 2009: 166 who did not consider the species as well characterized and classified it without further arguments with *villosus*, stating that more research was necessary to firmly establish the presence of a separate species.

Urticicola

Urticicola Lindholm, 1927
in Lindholm 1927: 124 *Zenobiella (Urticicola)*

Type species *Helix umbrosa* Pfeiffer, 1828
(original designation)

Name was established as a subgenus and not directly combined with a specific name, so masculine under Art. 30.1.4.2. Manganelli et al. 1995: 52 argued that this name and most names ending in -*cola* were traditionally considered as feminine.

Falkner et al. 2002: 156 intended to present a consistent classification for the *U. glabellus* group (5 species *glabellus*, *isaricus*, *mounierensis*, *moutonii* and *suberinus* and 3 additional subspecies), but failed to provide detailed constraints in the morphological or anatomical record to rule out the presence of spatial gradients to explain the morphological diversity.

Urticicola glabellus (Draparnaud, 1801)

Without gender agreement: *Urticicola glabella* (Draparnaud, 1801)

Helix glabella Draparnaud, 1801 in Draparnaud 1801: 87
Helix telonensis Mittre, 1842 in Mittre 1842: 189 (*Helix Telonensis*)
Helix moutonii Dupuy, 1847 in Dupuy 1847-1852: 95 (*Helix Moutonii*)
First mentioned in a determination key on p. 95 of the work, where the name was correctly established with a description (July 1847). Described in detail on p. 178 (and Pl. 9) which appeared in January 1848.
Helix isarica Locard, 1882 in Locard 1882: 319 (*Helix Isarica*)
Helix suberina Bérenguier, 1882 in Bérenguier 1882: 88
Helix mounierensis Caziot, 1909 in Caziot 1909: 89

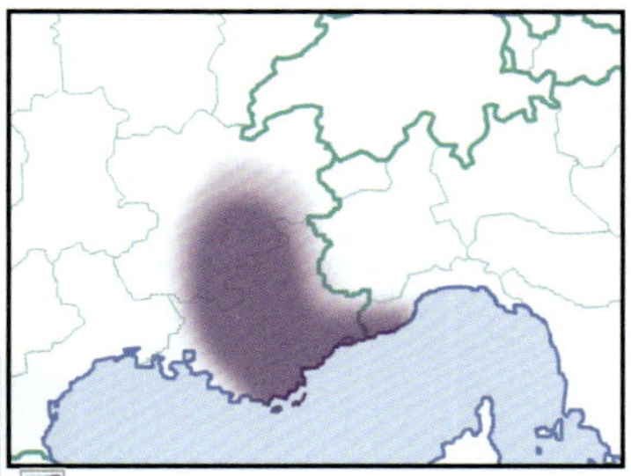

5-8 x 7-14 mm

France: Provence, near Rochefort NW Avignon

France: Vaucluse, near Mt. Vaucluse

SW Alps and promontories

Shell horny whitish to reddish brown, transparent and shiny, finely and irregularly striated, with a microsculpture of cross-shaped scales, 5-6 weakly convex whorls with shallow suture, shell reddish near aperture, aperture usually without lip inside, umblicus locally variable from very narrow to wide.
Shell shape and colour locally very variable, last whorl may be initially angulated.
Animal light or dark grey with weak violetish hue, foot usually lighter, tentacles often darker.

Under stones and plants, open habitats, meadows, tall herbs in grasslands, hedges, usually in calcareous mountains but not restricted to calcareous substrate.

Germain 1930: 252 (*glabella* [= *telonensis, moutoni, suberina*]), Kerney & Cameron 1979: 186 (*glabella*), Boato et al. 1984 (NW Italy, *Perforatella (Monachoides)* sp.), Kerney et al. 1983: 255 (*ventouxiana*, = *glabella*), Falkner 1990: 212 (*glabellus* and *ventouxianus*), Manganelli et al. 1995: 29 (*glabellus*, ?*ventouxianus* and ?*telonensis*), Falkner et al. 2001: 61, Falkner et al. 2002: 156 (*glabellus* [with *telonensis*], *isaricus* [with *ventouxianus*], *mounierensis*, *moutonii* and *suberinus*).

Recent classifications inconsistent and all obviously tentative, based on subjectively judged differences in shell shape, colour and umbilicus width.
The so-called species recognized by Falkner et al. 2002 are parapatrically distributed and locally very variable. Apparently they are not uniform entities.
In such a situation it would be useful to study the spatial distribution of shell parameters and see if we are dealing with local gradients (and slow but continuous gene flow) or environmentally induced characters, or really interrupted gene flow. Anatomical studies could also help. Molecular analyses could shed more light on the degree of genetic isolation of local populations and/or clades.
In the published record I see currently only speculations which would not justify deviations from Germain's (1930) and Kerney & Cameron's (1979) classification.

Urticicola umbrosus (Pfeiffer, 1828)

Without gender agreement: *Urticicola umbrosa* (Pfeiffer, 1828)

Helix umbrosa Pfeiffer, 1828 in C. Pfeiffer 1828: 27

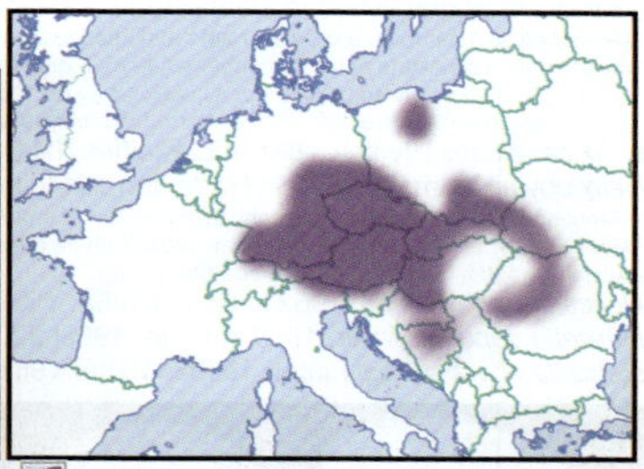

5.5-8 x 10-14 mm

Germany: Bayern, Ansbach

Europe

Slovakia: Bratislava

Jozef Grego

Shell pale yellowish grey to reddish brown, mantle colour pattern often visible below shell, with faint growth lines and a spotted microsculpture, depressed, 4-5 weakly convex whorls, last whorl angulated and strongly descending near aperture, aperture with weak lip inside, umbilicus very wide, deep, eccentric and perspectivical, 1/4-1/3 of shell diameter.
Differs from *Trochulus* species (*T. striolatus*, *T. villosus*) in its spotted microsculpture.

Humid forests, usually in mountains. Seems to feed on rotting plants, fruits and mushrooms.

Frömming 1954: 270, Shilejko 1978: 239, Kerney et al. 1983: 258, Falkner 1990: 212, Sysoev & Schileyko 2009: 211.

Xerocampylaea

Xerocampylaea Kobelt, 1871 in Kobelt 1871: 15 *Helix (Campylaea (Xero-Campylaea))*

Type species *Helix zelebori* Pfeiffer, 1853 (monotypy)

Established conditionally (name proposed under the condition that more similar species would be discovered in the Balkan region) (admissible under Art. 15.1). Name not available under Art. 11.8 (name must be "a word"), 11.9 allows "a compound word" for species-group names, genera have no such provision. But the name is used and I would suggest to continue using it, and propose to insert such a provision for genus-group names in one of the next editions of the Code.

Xerocampylaea zelebori (Pfeiffer, 1853)

7.5-8.5 x 14-16 mm

NW Kosovo: 3 km W Peć

Paolo Zaccaria

Tiroler Landesmuseen Innsbruck

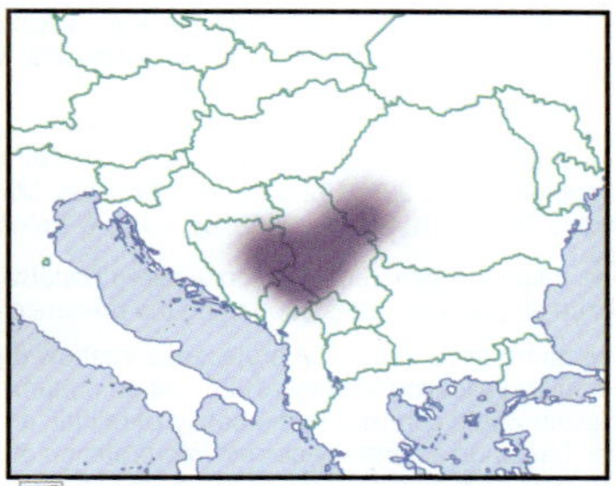

Bosnia to N Kosovo and W Romania

Helix zelebori Pfeiffer, 1853 in L. Pfeiffer 1853 (Z. f. Malak. 10): 186 (*Helix Zelebori*)

Shell white with horny brownish apex, usually with two narrow reddish brown bands, rarely with one band or purely white, shiny and thin but solid, finely striated, 4.5-5.5 slightly convex whorls, last whorl slightly inflated, slightly descending near aperture, aperture inside reddish, margin with white lip, straight and sharp, umbilicus initially narrow, wider at last whorl and partly covered by the reflected columellar margin.

Rocky habitats, usually in crevices or at the base of rocks.

Westerlund 1889: 309, Pavlović 1912: 56, Tomić 1959: 27, Grossu 1983: 465, www.iucnredlist.org (08.2012).

Xerolenta

Xerolenta Monterosato, 1892 in Monterosato 1892: 24 *Helix (Xerolenta)*

Type species *Helix obvia* "Menke, 1828" (subsequent designation by Kobelt 1892: 152)

Contained originally *Helix obvia* and *H. derbentina* (= currently *Xeropicta derbentina* (Krynicki, 1836)).

Differs from *Helicella* in the presence of nerve strings leading from the cerebral ganglion to the penis (like in *Xerotricha*). Differs from *Xerotricha* and *Helicella* in its symmetrically arranged dart apparatus: absence of a thick-walled sheath/container, absence of rudimentary auxiliary dart sacs.

Hausdorf 1988: 22.

Xerolenta macedonica (Hesse, 1928)

Helicella macedonica Hesse, 1928 in Hesse 1928: 20, Taf. 1-2

5-6.5 x 8.5-11.5 mm

Bulgaria: Pirin mountains, 1900 m

Bulgaria, Rep. Makedonija

Shell basically whitish or yellowish with light horny brownish apex, dark brown and partly interrupted colour bands, of which only one continuous band is at the upper side, 4.5-5.5 weakly convex whorls, angulated in juveniles and rounded in adults, last whorl only slightly descending near aperture, apertural margin straight, also at columellar side, thin and without lip inside, umbilicus wide, 1/5 of shell diameter.
Animal blackish grey, slightly lighter at the sides.

Dry grassland and rocky habitats in high mountains around 1900-2200 m.

Damjanov & Likharev 1975: 348, Hubenov 2007 (Pirin mountains), Dedov 2008 (Alibotush mountains, SW Bulgaria).

Xerolenta obvia (Menke, 1828)

Xerolenta candicans (Pfeiffer, 1841)

Helix neglecta Hartmann, 1821 in Hartmann 1821 (Neue Alpina 1): 226 (not *Helix neglecta* Draparnaud, 1805)
Helix obvia as mentioned in Menke 1828: 13 (nomen nudum)
Mentioned as a name for a species, name attributed to Ziegler, without description or bibliographic reference, only a synonym "*H[elix]. neglecta* Hartm[ann]." was added, but no expressed statement that *H. obvia* was meant to be a substitute name. Gittenberger 1975: 5 argued that the name was established as a new replacement name for *Helix neglecta* Hartmann, 1821 non Draparnaud, 1805. This is currently not tenable because of the lack of an expressed statement as required by Art. 72.7. There was no statement to indicate that Menke was aware of the problematic situation of the homonymy (something like "non Drp." would have expressed such an awareness). This can only be inferred implicitly (and speculatively) because at another location in Menke's publication a name *Helix neglecta* Dr[apamaud]. was listed.
Helix candicans Pfeiffer, 1841 in L. Pfeiffer 1841 (Arch. f. Nat. 7): 220
Helix obvia Hartmann 1842 in Hartmann 1844: 148, Tab. 45
Mentioned with description. Hartmann also mentioned that *H. obvia* Parreyss and *H. obvia* Ziegler did not refer to the same taxa, Hartmann used the name in Ziegler's sense and gave also bibliographic references to Menke 1828: 13 and Hartmann in Sturm 1824: 7 (there misinterpreted as *Helix neglecta* Drapamaud, 1805). Hartmann's concept must have included several species because his specimens were from Bohemia, Austria, Croatia, Mantua, Neuwied and Strasbourg.
Helicella pappi Schütt, 1962 in Schütt 1962: 152 (*Helicella (Xerothracia) pappi*)
Helicella depulsa Pintér, 1969 in Pintér 1969: 92 (*Helicella? depulsa*)
Helicella obvia razlogi Pintér, 1969 in Pintér 1969: 91

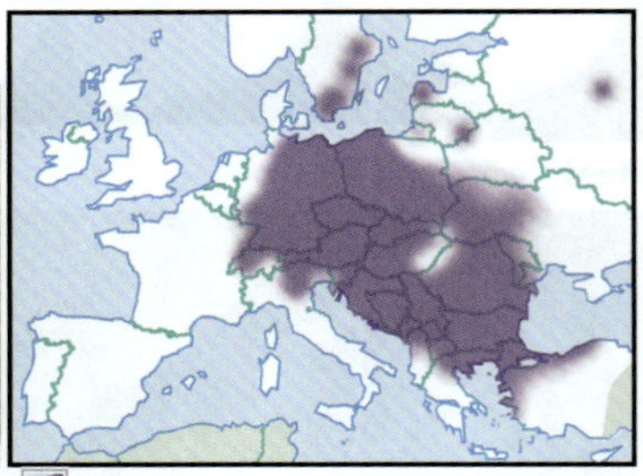

7-10 x 11-20 mm

Austria: Tirol, near Landeck

Europe

Greece: Makedonía: Philippi ruins

Shell usually whitish, often with brown colour bands, mainly at and below the periphery, in most regions weakly striated or almost smooth, 5-6 flattened whorls with shallow suture, very depressed above but with rounded periphery, last whorl not or only very slightly descending near aperture, aperture rounded, often with white lip inside, umbilicus wide, 1/4 of the shell diameter, very excentrically increased at the last whorl. In N Greece shells can also be densely ribbed and more or less sharply keeled with flat upper and rounded lower side, or keeled with both sides rounded.
Differs from *Helicella itala* in its slightly smaller umbilicus and its broader last whorl not distinctly descending at the aperture. Shells from central Albania to W Greece belong to *Xeromunda vulgarissima*, with a slightly broader umbilius.

Open habitats, dry grassy slopes, also dunes, vineyeards, sunny walls of ruins, railway dams, road margins, estivating often in large numbers in the vegetation. Up to 2000 m in the Alps, 1900 m in Bulgaria. Feeds on faded and rotting parts of plants. In summer only active at rainfalls and high humidity, more active in the winter season.

N Italy: Brescia, Gargnano

Two reproduction periods in the year, in spring and autumn. Clutches of 25-40 eggs (1.2-1.5 mm) are laid inside a 15-25 mm deep hole in the soil, which is closed afterwards. Maturity is reached before the shells are fully grown. Life span 2-3 years.

Vulnerable in Germany.

Frömming 1954: 249, Damjanov & Likharev 1975: 347, Gittenberger 1975: 5, Shilejko 1978: 224, Kerney & Cameron 1979: 179, Jungbluth et al. 1989: 192, Falkner 1990: 208, Manganelli et al. 1995: 28, Turner et al. 1998: 356, Sverlova 2006: 133, Irikov & Mollov 2006: 818, Egorov 2008: 85, Irikov & Erőss 2008: 205, Sysoev & Schileyko 2009: 189.

Irikov & Erőss 2008: 205 considered *depulsa*, *razlogi* and *pappi* as conspecific with *obvia*, this view is followed here in the absence of any constraints supporting the presence of separate species. More research is necessary to clarify their relationships.

The name *obvia* should be protected somehow (by the Commission). I would currently not suggest to use another name for this species.
Either by validating Menke's 1828 nomen nudum, or by designating a non-conspecific neotype for *Helix candicans* (for example, one that belongs to *Helicella itala*). The Commission would consider that recently Ukrainian and Russian authors (Sverlova 2006: 133, Sysoev & Schileyko 2009: 189) used *candicans*. In my view this usage of *candicans* was based on a currently correct interpretation of the Code.

Xerolenta spiruloides (Hesse, 1916)

Helicella spiruloides Hesse, 1916 in Hesse 1916: 149
Name was attributed to Wagner, description was partly by Wagner (the shell) and partly by Hesse (the anatomy). In a strict application of Art. 50.1.1 the name in such a case must be attributed to the author of the work.

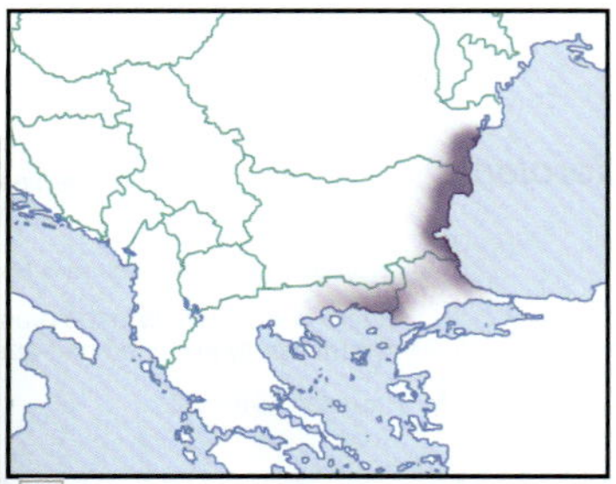

7.5-10 x 11-16 mm

Bulgaria: Pomorie Lake

E Romania to NE Greece
Indicated range in Greece uncertain, no confirmed record from Turkey.

Shell whitish with horny brownish apex and bands, often interrupted, densely striated, 5-5.5 convex whorls with deep suture, spire often higher than in the figured shells here, last whorl slightly descending, aperture without lip inside, umbilicus moderately narrow, 1/6-1/5 of diameter, with deep suture.

Dry habitats exposed to sunshine among grass and herbal plants, often together with halophilous species.

Damjanov & Likharev 1975: 349 (diagnosis), Grossu 1983: 424, www.iucnredlist.org (08.2012).

Xerolenta thasia (Reischütz, 1983)

Helicella rhabdotoides thasia Reischütz, 1983 in Reischütz 1983: 141

Shell yellowish and usually with 4-6 dark bands, of which only one is visible on the upper side, solid, bluntly ribbed, 4.5-5 whorls, last whorl bluntly keeled, very weakly descending near aperture, aperture with prominent lip inside, umbilicus moderately wide and deep.

3.8-4.7 x 6-7 mm

Was found in an olive plantation, attached to dried olive leaves.

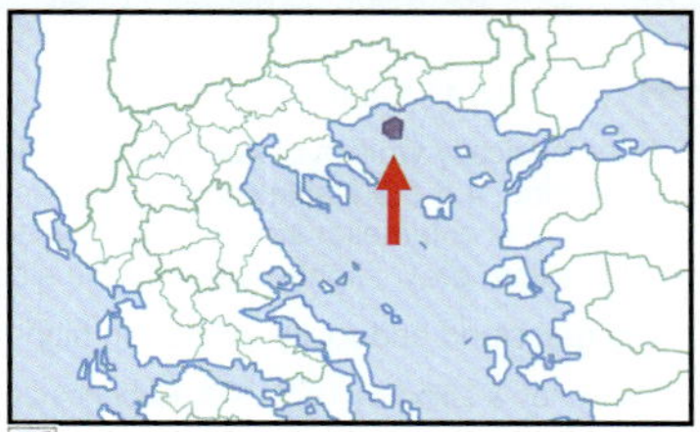

Greece (Thásos)

Xeromunda

Xeromunda Monterosato, 1892
in Monterosato 1892: 25 *Helix (Xeromunda)*

Type species *Helix candiota* Pfeiffer, 1849
(fixed by ICZN under the plenary powers)

Contained originally *Helix turbinata* De Cristofori & Jan, 1832 and *H. candiota*. Kobelt 1892: 152 gave as type *H. turbinata*. Kobelt and Monterosato misidentified the type species (*H. turbinata* is probably conspecific with *Cernuella virgata*) and had *H. candiota* in mind (Hausdorf 1988: 23). ICZN Opinion 1719 from 1993: type species *Helix candiota* fixed under the plenary powers. Today such a case would be solved under Art. 70.3.

Presence of one single dart sac, accessory sac much smaller than dart sac and transformed into a cavity between the vagina wall, the dart sac, and a tissue-layer that envelopes parts of the vagina and the dart sac. Penis is innervated from the right cerebral ganglion.

S Europe, Near East, N Africa

Hausdorf 1988, Manganelli & Giusti 1989, Hausdorf & Sauer 2009: 410.

Xeromunda alticola Hausdorf, 1995

Xeromunda alticola Hausdorf, 1995 in Hausdorf 1995: 25

Shell yellowish white with brown bands, striated, more weakly at lower side, 4-4.5 moderately convex whorls with deep suture, last whorl slightly angulated, rounded and not descending near aperture, aperture without internal lip, umbilicus wide, slightly excentric, increasing at last half whorl, 1/5 of diameter.
Cannot be distinguished from *Helicella gittenbergeri* by shell characters alone.

3.9-4.8 x 6.8-8.2 mm

Greece (N Pelopónisos)

Xeromunda candiota (Pfeiffer, 1849)

Helix turbinata synonym *Helix candiota* Pfeiffer, 1849 in L. Pfeiffer 1846 (Syst. Conch.-Cab. (1) 12 (1)): 255
Mentioned with description.
Helix candiota Mousson, 1854 in Mousson 1854: 10
Mousson 1854: 10 attributed the name to "Friw." and gave a bibliographical reference to "Chemn, Ed. 2. T. 37. f. 17. 18.", making Pfeiffer's name available under Art. 11.6.1 (although Mousson did not refer the name *candiota* directly to this source). Attributed to Pfeiffer 1849 in ICZN Opinion 1719.

Greece: central Crete, near Iráklio

Greece (records from Crete, Pelopónisos and Aegean islands were anatomically confirmed), probably also Turkey, coastlands from Algeria to Syria, and Cyprus

7-11 x 8-12 mm

Shell whitish with greyish spots on upper whorls, last whorl getting brownish with radial streaks, the shell has a characteristic globular shape, 4-5 whorls, last whorl not or weakly descending, aperture with lip inside, inside brownish, umbilicus very narrow and almost but not entirely covered.
Differs from the conchologically similar *Xm. durieui* by a larger dart sac, shorter flagellum, and a smaller basal part of dart apparatus.

Open habitats and cultivated areas. Also as a synanthrope. On the Greek island of Gávdos exclusively in agricultural crops, the species dispeared from arable fields which were not cultivated any more.

Falkner 1990: 206, Hausdorf 1990: 109, Welter-Schultes 1998: 67, Manganelli & Giusti 1989, Hausdorf & Sauer 2009: 411.

Has long been confounded with *Xm. durieui*.

Xeromunda durieui (Pfeiffer, 1848)

Helix durieui Pfeiffer, 1848 in L. Pfeiffer 1848 (Mon. Helic. 1): 441 (*Helix Durieui*)

Shell like *Xm. candiota*, no visible differences. Differs from *Xm. candiota* in its larger dart sac, shorter flagellum, and a smaller basal part of dart apparatus.

7-11 x 8-12 mm

Open habitats and cultivated areas.

Manganelli & Guisti 1989, Hausdorf 1990: 109, Giusti et al. 1995: 452, Manganelli et al. 1995: 28.

In Malta only fossils were found, also in Sicily near Palermo.

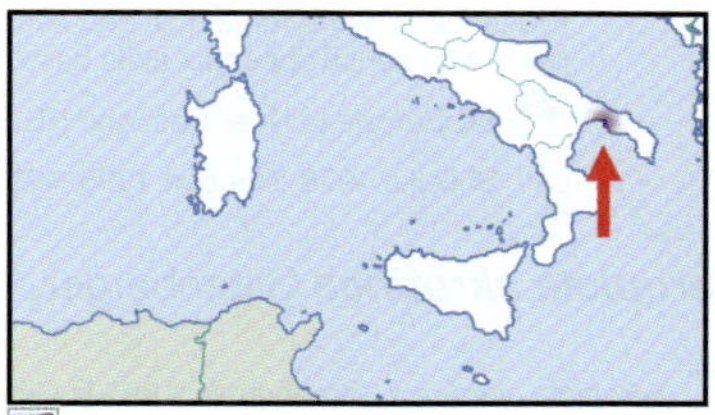

Algeria to Libya, S Italy

Xeromunda peloponnesia Hausdorf, 1990

Xeromunda peloponnesia Hausdorf, 1990 in Hausdorf 1990: 109

Greece: Pelopónisos, Lakonia, Monemvasía

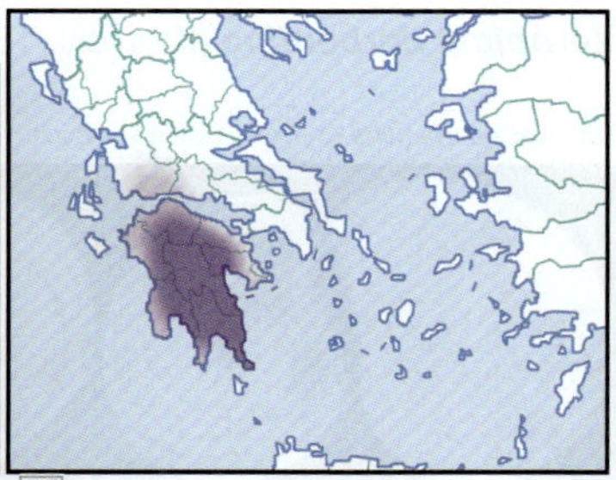

Greece (Peloponisos, ? Etolía)

4.4-6.7 x 7.5-10 mm

Shell yellowish white with brown colour bands or spots, with relatively prominent ribs and blurry spiral lines, 4-5 convex whorls, suture not very deep, last whorl rounded or weakly angulated, no sharp edge on the first whorls, apertural margin sharp, inside with a white lip, umbilicus increases at last whorl, 1/7-1/4 of shell diameter.
Differs from *Cernuella virgata* in its smaller size and its ribs, weak edge on the first whorls (*C. virgata* has a sharp edge) and white colour of the lip.

Xeromunda thessalica Hausdorf, 1990

Xeromunda thessalica Hausdorf, 1990 in Hausdorf 1990: 123

6.8-9.2 x 13-17 mm

Greece: Thessalía, near Vláhos

Greece (Thessalía)

Shell white to yellowish white, sometimes with horny brown colour bands or spots, irregularly striated, sometimes with blurry spiral lines, 4.5-5.5 whorls, rounded, suture not very deep, apertural margin sharp, with white lip inside, umbilicus 1/6-1/4 of shell diameter, initially narrow, at last whorl excentrical. Umbilicus usually more excentrical than in *Xerolenta obvia*, but less than in *Xm. vulgarissima*, which also has flatter whorls.
Has only 2 mucus glands, not 4 like *Xm. vulgarissima*.

Xm. vulgarissima lives in the south, *Xl. obvia* in the north of the range of *Xm. thessalica*.

Xeromunda vulgarissima (Mousson, 1859)

Helix ericetorum var. *vulgarissima* Mousson, 1859 in Mousson 1859: 270

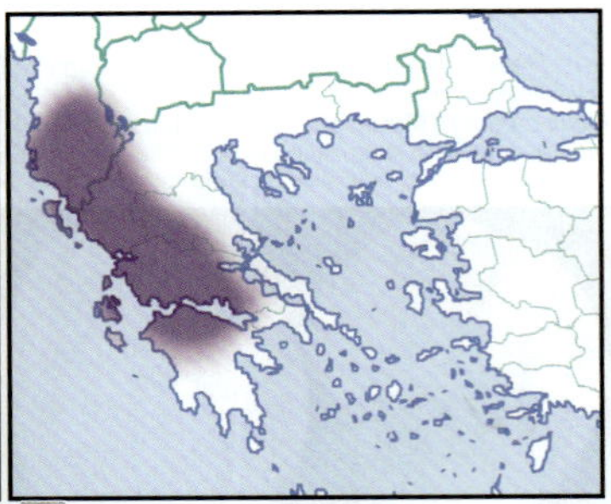

5-9 x 11-18 mm

S Albania: Greshicë near Ballsh

Albania to Greece

Shell flat, white to yellowish white, sometimes with horny brown colour bands or spots, irregularly striated, 4.5-5 whorls, usually rounded at periphery (some populations may be keeled), suture not very deep, apertural margin sharp with a white lip inside, umbilicus wide and very excentrical at last third of last whorl, more than 1/4 of shell diameter.
Differs from *Xerolenta obvia* mainly anatomically (shells of *Xl. obvia* tend to have a less excentrical and narrower umbilicus). Flagellum length 1.3-1.8 times of epiphallus length, penis papilla very long and narrow, dark slightly curved with two lancet wings along half length.

Open habitats.

Hausdorf 1990, Dhora & Welter-Schultes 1996: 160.

Xeropicta

Xeropicta Monterosato, 1892
in Monterosato 1892: 24 *Helix (Xeropicta)*

Type species *Helix krynickii* Krynicki, 1833
(monotypy)

Shell always white with brown spiral bands, medium-sized to small with a subdepressed to subconoidal spire, umbilicus open.
Presence of two symmetrical dart sacs and two long accessory sacs, which are always longer than the dart sacs, 4 branched mucus glands around the vagina, penis with penial appendix at its base. Penis is innervated from the right cerebral ganglion.

SE Europe to Near East and NE Africa

Shilejko 1978, Neubert 1998: 420, Hausdorf & Sauer 2009: 406.

Xeropicta akrotirica Gittenberger, 1991

Xeropicta akrotirica Gittenberger, 1991 in Gittenberger 1991 (Zool. Meded. 65): 107

Shell whitish with variable colour pattern, spirally arranged, coarsely and irregularly ribbed, 4-5 markedly convex whorls, last whorl very weakly descending near aperture, aperture almost perfectly circular when seen in frontal view, with very weak lip inside, umbilicus symmetrical and narrow, 1/9-1/7 of shell diameter.
Differs from other *Xeropicta* species in its globolar shape and very narrow umbilicus.

4.5-7.5 x 6-10 mm

S Cyprus

Xeropicta derbentina (Krynicki, 1836)

Helix derbentina Krynicki, 1836 in Krynicki 1836: 192

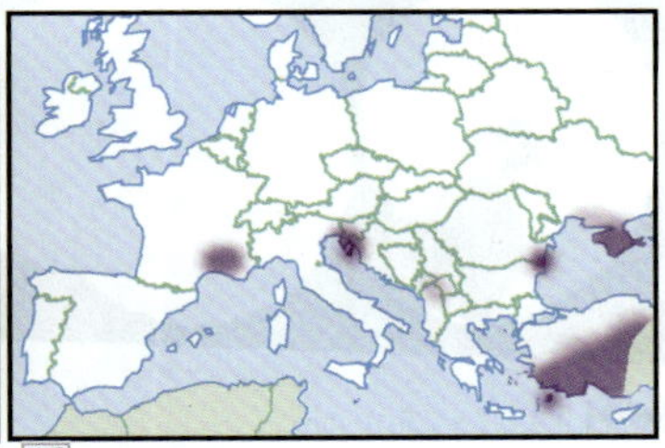

7-12 x 12-20 mm

Turkey: near Kemer

NE Africa, SE Europe to Caucasus

Shell white or with weak reddish brown rows of spots, almost smooth or finely striated, with extremely fine spiral sculpture, 4.5-5.5 convex whorls, rapidly increasing, last whorl slightly descending near aperture, inside with several white lips, umbilicus open and perspectivical.

Seems to be generally absent from humid costal areas and dunes. One of the most frequent species of southern Turkey.

Westerlund 1889: 341, Damjanov & Likharev 1975: 357, Shilejko 1978: 220, Grossu 1983: 437, Schütt 2001: 447, De Mattia 2007, Sysoev & Schileyko 2009: 200.

Xeropicta krynickii (Krynicki, 1833)

Helix krynickii Krynicki, 1833 in Krynicki 1833: 434
Name originally attributed to Andrzejowski, but not the description which was by Krynicki alone.
Helix vestalis Pfeiffer, 1841 in L. Pfeiffer 1841 (Symb.): 40

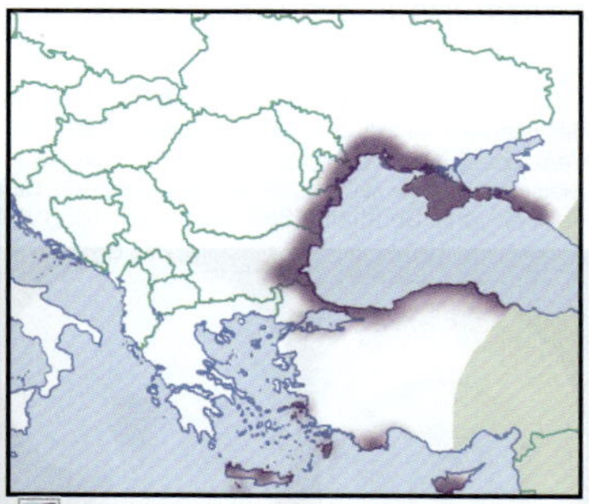

SE Europe to Iran and Arabia

7-11 x 12-17 mm

Greece: island of Gávdos S of Crete

Cyprus — Georgios Konstantinou

Shell white, typically with characteristic dots on the outwards directed side of the suture and several uninterrupted bands or rows of dots on the lower side, irregularly striated and with very fine and distinct spiral sculpture (visible even in weathered specimens), 4-5.5 slightly convex whorls, last whorl rounded, slightly descending near aperture, aperture with weak lip deep inside, margin straight and sharp, umbilicus initially narrow or almost closed, wider and slightly excentric at last whorl.
Flagellum moderately long.

Open habitats and cultivated areas, also forests and sandy habitats. Artificially widely dispersed since antiquity with agricultural products and dunnage, shells of this species were found in a Late Bronze Age (3300 years BP) shipwreck off S Turkey, the ship came from the Syro-Palestinian coast and carried also the Israel/Palestinian endemic species *Xerocrassa langloisiana* (Bourguignat, 1853).

Reaches maturity at an early stage, shells may look like those of juveniles but belong to adult animals.

Westerlund 1889: 192, Damjanov & Likharev 1975: 356, Shilejko 1978: 218, Falkner 1990: 206, Welter-Schultes 1998: 67, Welter-Schultes 2008, Hausdorf & Sauer 2009: 406, Sysoev & Schileyko 2009: 200.

Helix vestalis is considered as a synonym here because there are no significant differences (Hausdorf, unpublished, 2008). Its shell surface is perhaps more shiny with an undulated spiral microsculpture.

Xeropicta ledereri (Pfeiffer, 1856)

Helix ledereri Pfeiffer, 1856 in L. Pfeiffer 1856: 43 (*Helix Ledereri*)

6-9 x 10-14 mm

Cyprus — Georgios Konstantinou

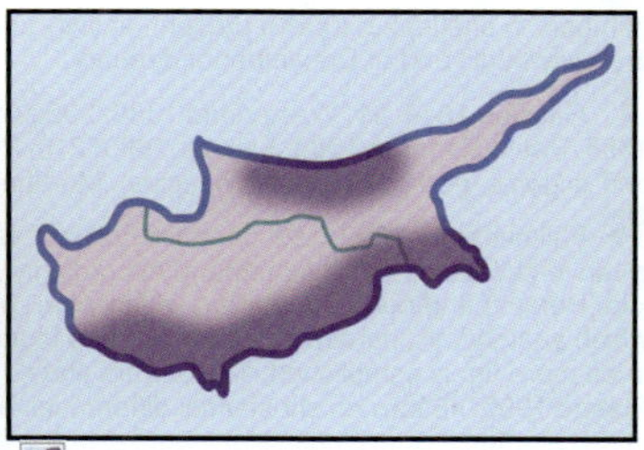

Cyprus

Shell extremely variable, yellowish white, rarely with brown bands at lower side, densely ribbed, 5 slowly increasing whorls with indented suture, upper side depressed and scalarid, juveniles with sharp and crenulated keel, last whorl regularly rounded, slightly descending near aperture, aperture with very weak lip inside, umbilicus initially narrow, regularly increasing to 1/6-1/5 of shell diameter.

Westerlund 1889: 323, Gittenberger 1991: 110.

Xerosecta

Xerosecta Monterosato, 1892
in Monterosato 1892: 21 *Helix (Xerosecta)*

Type species *Helix explanata* Müller, 1774
(monotypy)

Shell flat or globulous, diameter 9-26 mm, 5-6 whorls, umbilicus very variable, aperture oval, not reflected, lip inside white or orange. Penial nerve comes from right cerebral ganglion. Right ommatophoran retractor independient from gential apparatur. Penis globulous, with relatively short papilla, apical (*Polloneriella*) or lateral (*Xerosecta* s. str.) opening, epiphallus larger than penis, flagellum well developed and of variable length, vas deferens rolled above epiphallus, one dart container with a medium large and curved dart.

Western Mediterranean region

Puente 1995.

Xerosecta adolfi (Pfeiffer, 1854)

Helix adolfi Pfeiffer, 1854 in L. Pfeiffer 1854: 264 (*Helix Adolfi*)

11-14 x 20-25 mm

Spain: Almería, Benahadux and El Ejido — José Liétor Gallego

S Spain

Shell whitish with variable brownish or reddish colour patterns, densely and finely striated, shiny, last whorl rounded at periphery, aperture not very oblique, with a thick reddish, brownish or white lip inside (white is less frequent in Almería), more inside eventually 1-2 more lips, umbilicus open and deep.
Adult animal dark or very dark.
Differs from *Xs. cespitum* in being more prominently striated and in the dark colour of the animal.

Open habitats, usually grasslands, dry wastelands with shubs, not in dunes, not in mountains.

Specimens collected in August in Almería were still juvenile, in November sexually mature adults were found.

EU Red List: Endangered EN. Threatened by intensive greenhouse agriculture along with heavy soil and water overexploitation and contamination, mass tourism and urban expansion. In the 1990s 10 % of the suitable habitats were destroyed, in the same period the area occupied by greenhouses increased regionally by 35 %. The increasingly toxic conditions provoke that the species is being replaced by more resistant species such as *Rumina decollata*, *Theba pisana* and *Otala lactea*.

Westerlund 1889: 204, Ortiz de Zárate y López 1950: 71, Puente 1995, Arrébola & Puente in Verdú & Galante 2006: 391, www.iucnredlist.org (08.2012).

Xerosecta cespitum (Draparnaud, 1801)

Helix cespitum Draparnaud, 1801 in Draparnaud 1801: 92
Helix terverii Michaud, 1831 in Michaud 1831: 26, Pl. 14
Helix coespitum var. *introducta* Villa & Villa, 1841 in Villa & Villa 1841: 12
Name was attributed to "Z." (Ziegler) in original source. Description: "major, alba vel fasciata". Authorship for the description was not specified.
Helix arigonis Schmidt, 1853 in Schmidt 1853: 176 (*Helix Arigonis*)

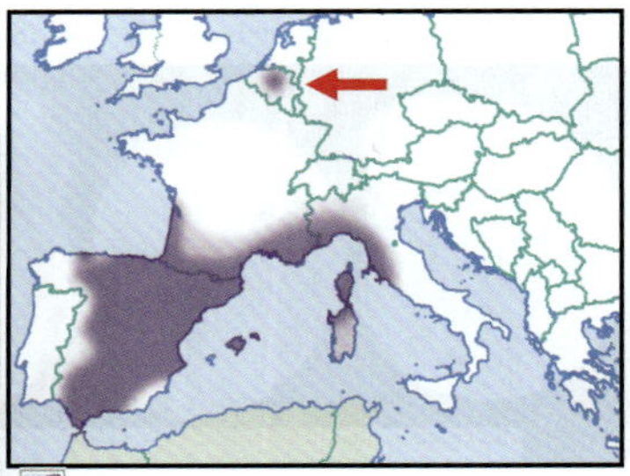

10-13 x 17-22 mm, in Italy 13-15 x 23-27 mm

(*cespitum*) Spain: Isles Balears, Mallorca

(*arigonis*) Spain: near Jaén José Liétor Gallego

SW Europe, Morocco to Tunisia

Frequent in wastelands, margins of roads and arable fields, rock rubble with grasses, mostly near coasts and in valleys, in Italy on sunny dry mountain slopes in rocky highlands. Present but not common in mountains, rarely above 1000 m, in Italy at 500-1500 m, in Spain higher. In Nájera in central Spain the species has been so frequent before 1950 that the snails were collected as food for young ducks.

Shell very variable, basic colour brownish yellow to dirty white with variable colour bands and spots, irregularly striated, also at lower side, last whorl almost smooth, 5-6 not very convex whorls, upper side weakly convex, last whorl only weakly descending near aperture, aperture with weak dirty whitish or reddish lip inside, margin straight and sharp, umbilicus wide, 1/6-1/4 of shell diameter, showing at least the last 2 whorls, in some regions the lower side of washed shells looks like in *Xerolenta obvia*. Also juveniles are striated at the lower side, juveniles are sharply keeled. The *arigonis* forms differ from *Xs. cespitum* in a narrower umbilicus, anatomically in a shorter flagellum.
Juveniles of *Cernuella virgata* are not striated at their lower side.
Animal whitish, margin of mantle can be darker.

Annual life cycle. Mating in central Spain usually in October and November, preferably at days with weak rainfall all day long, in France in August and September. 60-150 eggs (diameter 2 mm) are laid by one individual. At 15°C average temperature hatchlings of Spanish populations (diameter 2 mm with 2 whorls) appear after 22 days, in France 10-15 days were observed. Juveniles in Spain pass the winter, grow in spring to a shell size of 5-8 mm in early summer, when they accumulate to estivate in large numbers in shelters at trees if present, activity and growth continues in late summer until they reach maturity in September or October, in France in summer. Adults in Spain are active until end of December or January.

Gassies 1849: 99, Westerlund 1889: 205 (*arigoi*), 212 (*terverii*), 223 (*cespitum*), Germain 1930: 295, Ortiz de Zárate y López 1950: 67, Kerney et al. 1983: 300, Aparicio 1985, Prieto & Martín 1988, Falkner 1990: 210, Puente 1995: 42 (classified *arigonis* and *terverii* as synonyms), Manganelli et al. 1995: 29 (classified *introducta* as a synyonym and *arigonis* "perhaps" as a subspecies), Martínez-Ortí, Ap. & R. 2004.

Falkner et al. 2002: 159 classified ***terverii*** as a separate species arguing that anatomical dissection showed that it was a distinct species, no detailed arguments were given to substantiate this statement. Accordingly ***introducta*** was classified as a separate species based on unpublished results of anatomical studies of specimens from Poujols (Hérault) and near Montpellier, which were (presumably) compared with specimens from Alpes-Maritimes and the French/Italian border region. It was not explained in which anatomical characters the forms were found to differ, no information on possible local variations of those characters was given. The presence of separate species needs to be more firmly established. The current state of published research does not allow to present them as independent species with different diagnoses and ranges.
Puente 1995: 46 saw no substantial differences between ***arigonis*** and *cespitum* that would justify classifying them as two separate species and to rule out the presence of regional gradients or ecological factors to account for the observed variation.
It seems that more studies in N Africa are needed to come to a more precise understanding, perhaps the species originated there and was introduced long time ago to various parts in Europe independently. This would explain the slightly different morphology observed in European populations.

Xerosecta contermina (Pfeiffer, 1847)

Helix contermina Pfeiffer, 1847 in L. Pfeiffer 1848 (Mon. Helic. 1): 160

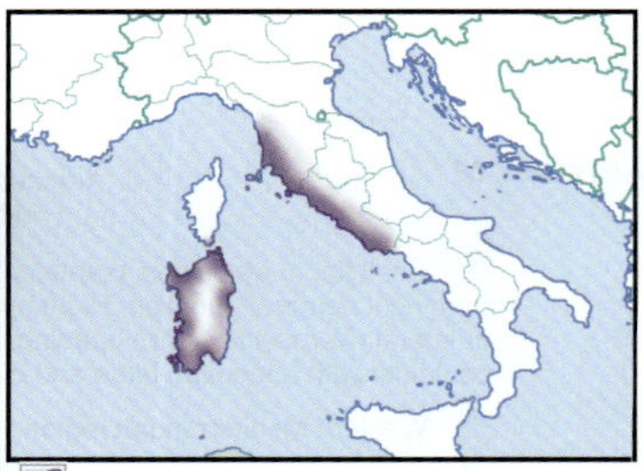

5-7 x 5-7 mm

Italy: (1) Castelporziano near Roma (2-6) Sardinia, Cabras near Oristano

Italy

(1) Alessandro Hallgass

Shell dirty whitish with horny brown spots and horny brown apex, finely and irregularly striated, 5.5-6.5 markedly convex whorls, last whorl not descending near aperture, aperture with weak lip inside, margin sharp, umbilicus narrow and deep.
Differs from *Xsc. dohrni* in its smaller size, slightly straighter outline of the shell and its finely striated sculpture. *Ichnusomunda sacchii* is slightly larger, more whitish with less distinct colour bands and has usually no lip inside the aperture.

Sandy habitats.

Westerlund 1889: 365, Manganelli et al. 1995: 29.

Xerosecta dohrni (Paulucci, 1882)

Helix dohrni Paulucci, 1882 in Paulucci 1882: 252 (*Helix (Helicogena) Dohrni*)

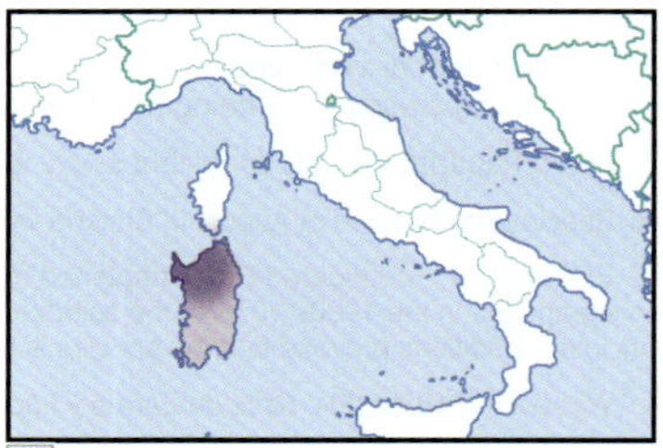

Shell dirty grey with brown spots at the upper suture and a brown band at the lower suture, lower side with fine interrupted brown bands, densely and irregularly striated, 5-5.5 whorls, last two whorls considerably wider, aperture with prominent lip inside, margin sharp, umbilicus narrow and partly or nearly entirely covered by the reflected columellar margin.
Differs from *Xsc. contermina* in its larger size, slightly more convex outline with larger penultimate whorl, and its smoother surface. *Ichnusomunda sacchii* is slightly smaller, more whitish with less distinct colour bands and has usually no lip inside the aperture, also perhaps slightly more convex whorls with deeper suture.

7-10 x 9-11 mm

Among herbal vegetation.

Westerlund 1889: 186, Manganelli et al. 1995: 29. Needs to be revised.

Italy (Sardinia)

Xerosecta explanata (Müller, 1774)

Helix explanata Müller, 1774 in Müller 1774: 26

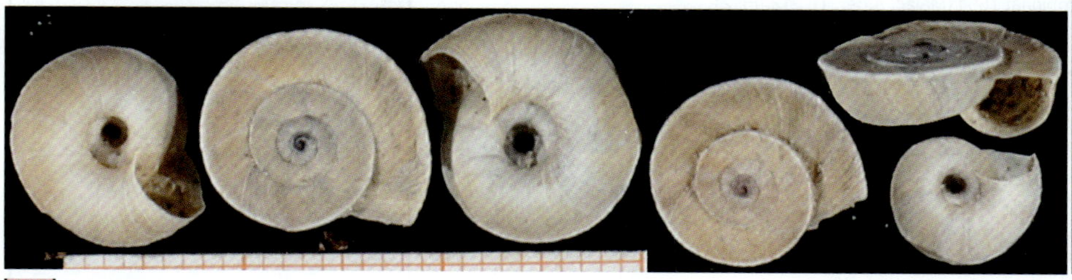

5-6 x 14-17 mm

Spain: Alicante, Dènia

S France, E Spain, Algeria

Shell yellowish white or brownish yellow, often with dark greyish apex, and a white keel at periphery, finely and densely striated, or finely ribbed, 4-5 nearly flat whorls with sharp keel, upper side horizontally flat or slightly concave, lower side rounded, last whorl slighty descending near aperture, aperture with weak lip inside, upper margin amost flat, lower margin rounded, umbilicus wide and perspectivical.

In coastlands on sandy dunes, usually in the vegetation to avoid the heat of the sand, together with *Theba pisana*, *Trochoidea trochoides* and *Cochlicella*.

Vulnerable in Spain. A relictual species, populations are threatened by continuous habitat destructions in coastlands, disappeared already in Alpes Maritimes, Alicante and Castellón, rapidly declining in other areas in Spain.

Westerlund 1889: 325, Testud 1981, Falkner 1990: 210, Puente 1995: 47, Gómez Moliner et al. 2001: 176, Martínez-Ortí & Robles 2003: 153, Martínez-Ortí & Puente in Verdú & Galante 2006: 392, Martínez-Ortí 2010.

Xerosecta giustii Manganelli & Favilli, 1996

Xerosecta giustii Manganelli & Favilli, 1996 in Manganelli & Favilli 1996: 336

Shell evenly grey or whitish grey with dark reddish brown apex, occasionally with weak and thin interrupted brown bands, solid, irregularly striated, 5.5-6 regularly increasing and convex whorls with deep suture, last whorl with 2 x of penultimate whorl, descending near aperture, aperture with prominent white lip inside, margin slightly reflected at lower side, umbilicus open, 1/5-1/4 of diameter.

8-14 x 15-24 mm

Italy: Grosseto, Massa Marítima

José Liétor Gallego

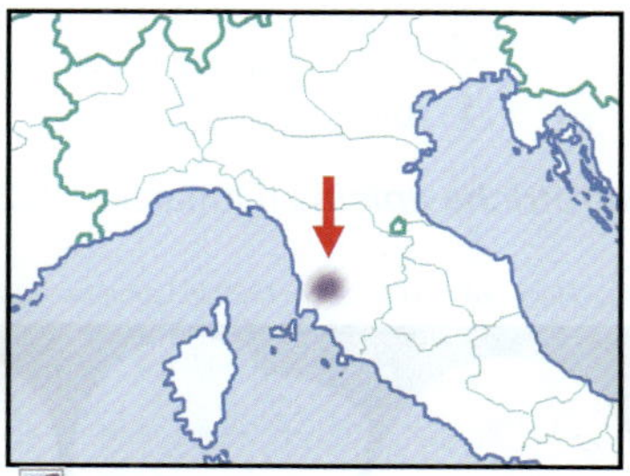

Italy (S Toscana)

Atrium uncommonly large with a large and crest-like structure inside.
Animal yellowish grey with pale grey head and neck, mantle border with five lobes.

www.naturamediterraneo.com (2007).

Open Mediterranean shrubland vegetation, on the soil and under stones, also on branches of bushes and grass stems.

Annual life cycle. Most adults occur in September, the number of individuals differs every year.

EU Red List: Critically Endangered CR. Only known from a very restricted site the size of 100 x 100 m. Threatened by the destruction of large parts of its habitat, overgrazing by goats and destruction of the typical vegetation, only a very small population has survived. A fire could lead to its extinction.

Xerosecta hillyeriana (Paulucci, 1882)

Helix hillyeriana Paulucci, 1882 in Paulucci 1882: 251 (*Helix (Helicogena) Hillyeriana*)

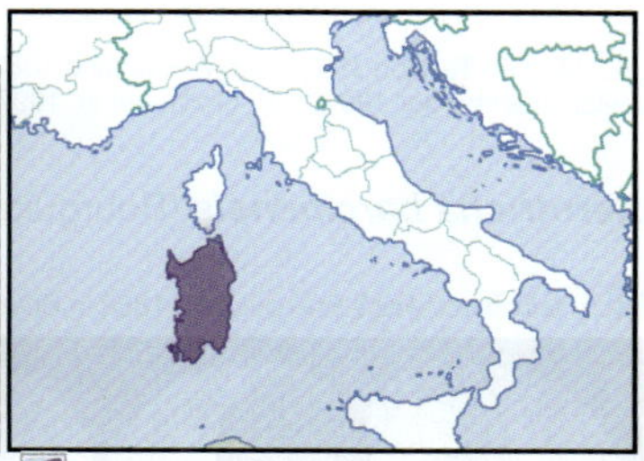

7-10 x 10-13 mm

Italy: Sardinia, Sassari, Alghero, Capo Caccia

Italy (Sardinia)

Shell whitish or yellowish with fine light brownish spots or interrupted bands, thin, irregularly and prominently striated, 4.5-5.5 whorls with protruding apex, regularly increasing, last whorl initially with edge at periphery, rounded and not descending near aperture, aperture with white lip inside, margin sharp and straight, not reflected, umbilicus moderate.

Westerlund 1889: 282, Manganelli et al. 1995: 29. Needs to be revised.

Xerosecta promissa (Westerlund, 1892)

Helix promissa Westerlund, 1892 in Westerlund 1892: 29

Shell brownish with reddish hue, irregularly striated, 4.5-5.5 slightly convex whorls with moderately impressed suture, regularly increasing, last whorl rounded at periphery and slightly descending near aperture, aperture inside reddish brown with 1-2 reddish brown lips inside, umbilicus very small and narrow.

Ruderal habitats, also in cultivated areas, gardens, 50-1200 m.

Puente 1995: 49.

7-8 x 13-15 mm

Spain: Sierra de Cazorla (Jaén) and near Málaga

José Liétor Gallego

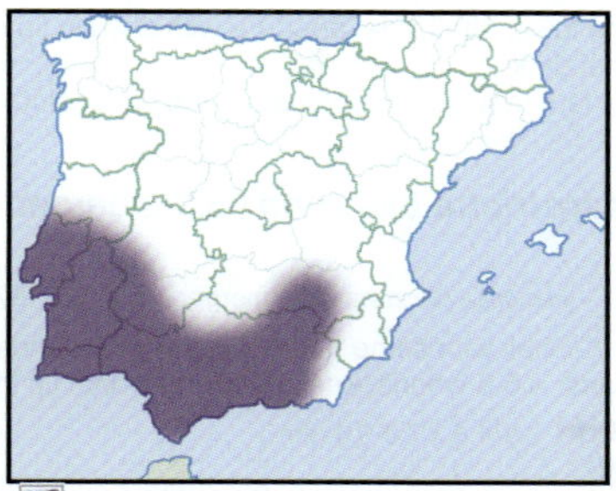

S Spain, S Portugal

Xerosecta reboudiana (Bourguignat, 1863)

Helix reboudiana Bourguignat, 1863 in Bourguignat 1864 (Malac. Algér. 1): 212, Pl. 21

7-8 x 9-12 mm   Spain: Linares (Jaén), Sierra de Segura (Jaén) and El Palo (Málaga)   José Liétor Gallego

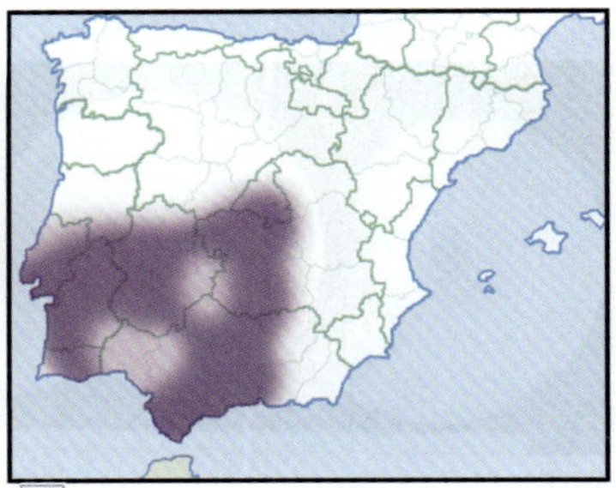

SW Europe, N Algeria

Shell extremely variable, usually whitish with various interrupted dark colour bands, aperture usually with white internal lip, but can also be dark, reddish, there can also be two lips of different colour.

Ruderal habitats, also in mountains (250-600 m, in Granada up to 1300 m). A frequent species.

Ortiz de Zárate y López 1950: 79, Puente 1995: 51. Not mentioned by Nobre 1941.

Xerotricha

Xerotricha Monterosato, 1892 in Monterosato 1892: 23 *Helix (Xerotricha)*

Type species *Helix conspurcata* Draparnaud, 1801 (monotypy)

Shell small to medium-sized (less than 10 mm) and with hairs.
Presence of two symmetrical dart-bearing dart sacs and two small accessory sacs, 4 branched mucus glands around the vagina. Penis is innervated by the right cerebral ganglion.

Canary Islands, N Africa, S Europe

Hausdorf & Sauer 2009: 413.

Xerotricha apicina (Lamarck, 1822)

Helix apicina Lamarck, 1822 in Lamarck 1822: (2) 93

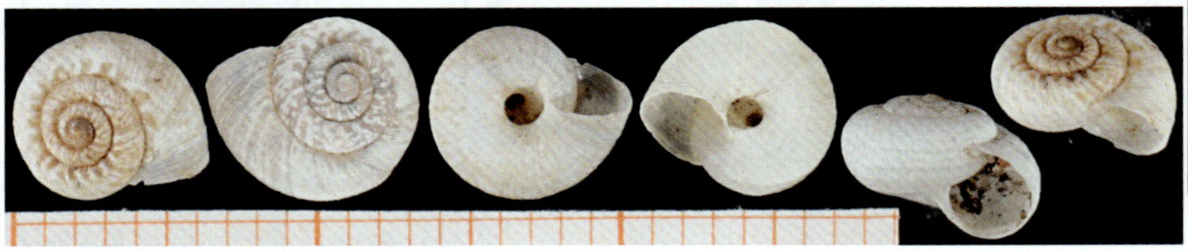

3.5-5 x 5.5-8 mm   Portugal: Algarve, Ilha de Tavira

Cyprus to E Atlantic islands
Originally probably W Mediterranean, historically introduced to the E Mediterranean.

Shell whitish usually with brown spots, concentrated as a series of spots near the suture, 4-5 very convex whorls with deep suture, aperture almost circular, simple without lip inside, margin slightly reflected near umbilicus, umbilicus wide, 1/6-1/4 of shell diameter. Juveniles with short hairs, adults have usually no hairs.
Shell shape variable, shell can also be evenly white, but usually the characteristic series of brown spots near the suture allows to recognize this species easily. Body small, light with dark brown spots on dorsum and head, upper tentacles short and thick, lower tentacles very short, foot larger than body and blackish. The animal moves slowly and is shy, withdraws rapidly at the weakest contact.

In slightly humid and shady habitats, meadows, under leaves and stones, in cultivated areas, along bases of walls. Sometimes in large colonies. A common Mediterranean species.

Eggs (1.5 mm diameter) are laid in packages enveloped by a whitish and stable membrane.

Moquin-Tandon 1855: 232, Westerlund 1889: 252, Nobre 1913: 193, Germain 1930: 184, Nobre 1941: 99, Alzona 1971: 171, Kerney et al. 1983: 250, Prieto & Martín 1988, Gittenberger 1991: 120, Ondina et al. 1997, Hausdorf & Sauer 2009: 413.

Xerotricha barcinensis (Bourguignat, 1868)

Helix barcinensis Bourguignat, 1868 in Bourguignat 1863-1868: 303, Pl. 42 (*Helix Barcinensis*)

5-6 x 10-11 mm   Spain: (1-5) Barcelona, Rellinars, (6-7) Barcelona, Callifo Castielo (?)

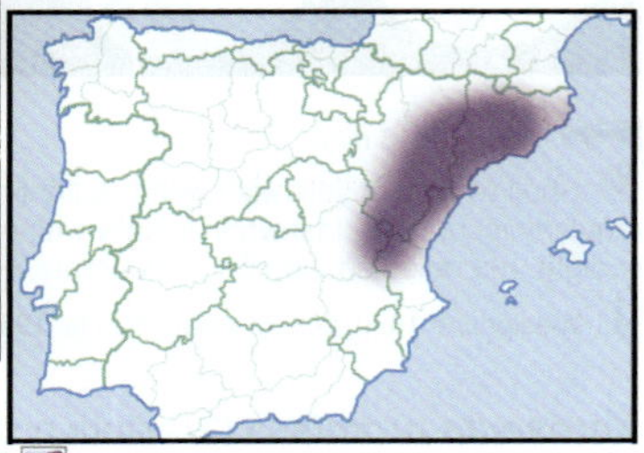

NE Spain

Shell light yellowish or reddish grey with dark brown colour bands, they can be interrupted, more intensive at the lower shell side, aperture with a white lip inside, umbilicus open, 1/8-1/9 of shell diameter, partly covered. Animal dark.
Differs from *Xt. madritensis* in the dark colour of the animal.

Westerlund 1889: 280, Ortiz de Zárate y López 1950: 60.

Xerotricha bierzona (Gittenberger & Manga, 1977)

Helicella bierzona Gittenberger & Manga, 1977 in Gittenberger & Manga 1977: 178

Shell with some fragmented colour bands on light background, prominently ribbed, juveniles with short hairs, 4.5-5 whorls, keeled, aperture with lip inside, umbilicus at least 1/6 of shell diameter.

3.9-4.6 x 7.8-9.4 mm

Rocky shrublands on calcareous substrate, on vertical rock faces or in crevices.

EU Red List: Vulnerable VU. Only known from a very small range, parts of which were probably destroyed by a water reservoir. Threatened by fire and by exaggerated road construction, particularly if vertical rocky slopes are cemented to prevent rocks from falling down onto the road.

Gómez Moliner et al. 2001: 162, Puente et al. in Verdú & Galante 2006: 369, www.iucnredlist.org (08.2012).

N Spain

Xerotricha conspurcata (Draparnaud, 1801)

Helix conspurcata Draparnaud, 1801 in Draparnaud 1801: 93
Helicellea mariae Gasull, 1974 in Gasull 1972: 71

Albania: Shkodër

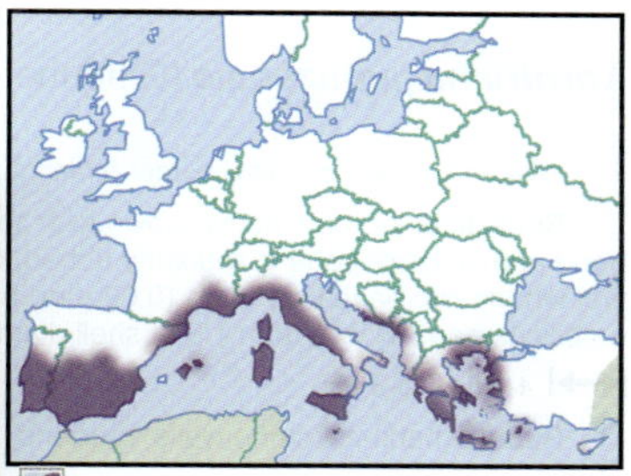

Greece to E Atlantic islands
Originally probably W Mediterranean, historically introduced to the E Mediterranean region.

3.3-4.5 x 4.8-6.8 mm

Italy: Lago di Garda, Gargnano

Shell brownish with irregular light spots, forming patterns on the upper whorls, densely striated and covered with 0.2-0.3 mm long hairs, which are often rare or absent in adults, 4-5 convex whorls, first whorls slightly keeled, aperture simple without lip, margin only slightly reflected near umbilicus, umbilicus 1/7-1/6 of shell diameter. Shell shape not very variable. Its dirty brown general aspect combined with its short hairs allows to recognize this species easily.
Animal small, body light reddish with more or less blackish brown dorsum and head, tentacles not very long, two retractor muscles visible as dark lines. Often with camouflage.
Differs from *Microxeromagna lowei* in its longer hairs and narrower umbilicus.

Under stones and on the soil under dense vegetation, often in wall crevices, sometimes exposed to the sun. Not in high altitudes. Locally very common.

Nobre 1913: 195, 1941: 101, Ortiz de Zárate y López 1950: 82, Hausdorf 1988, Giusti & M. 1989, Falkner 1990: 208, Hausdorf 1990 (Malak. Abh. 15), Gittenberger 1991: 120, Manganelli et al. 1995: 28, Dhora 2004: 146, Talaván Gómez & T. 2006: 114, Martínez-Ortí 2008 (classified *mariae* as a synonym), Hausdorf & Sauer 2009: 413.

Reports for Cyprus referred to other species. The status of *mariae* (from Almería) needs further investigation.

Xerotricha corderoi (Gittenberger & Manga, 1977)

N Spain

Helicella corderoi Gittenberger & Manga, 1977 in Gittenberger & Manga 1977: 179

Shell pale horny brown with whitish spots on upper side, on lower side with colour bands, sometimes also one band above the periphery, irregularly ribbed mainly on upper side, fine spiral lines on lower side, and with hairs on all 4-5 whorls, periphery initially very weakly keeled but rounded near the aperture, aperture with prominent whitish lip or barrier inside, which is also visible from outside, umbilicus wide, 1/5 of shell diameter.
Differs from other species with hairs in its prominent barrier in the aperture.

2.4-3.3 x 4.4-5.6 mm

Xerotricha gasulli (Ortiz de Zárate y López, 1950)

Spain (Andalucía)

Helicella gasulli Ortiz de Zárate y López, 1950 in Ortiz de Zárate y López 1950: 58, Lám. 2

Shell light horny brown with dark brown spots, sometimes with a brown band at the periphery, and with several bands at the lower side, opaque, 4-5 convex whorls, suture deep, aperture with strong white lip inside, margin slightly reflected at columellar side, covering partly the umbilicus, umbilicus narrow and deep, 1/10 of shell diameter, only the last 2 whorls visible inside.
Animal white with dark branched patterns on the mantle.

5.4-6.3 x 6.4-8.1 mm

At margins of cultivated fields and in their vicinity, in lowlands below 150 m. Usually not very abundant.

Adults were found in May.

EU Red List: Endangered EN. Only known from 4 localities. Threatened by intensive agricutural activities and fire.

Arrébola (unpublished, 1995), Arrébola 2001, Gómez Moliner et al. 2001: 163, Arrébola & P. in Verdú & G. 2006: 370.

Xerotricha gonzalezi (Azpeitia Moros, 1925)

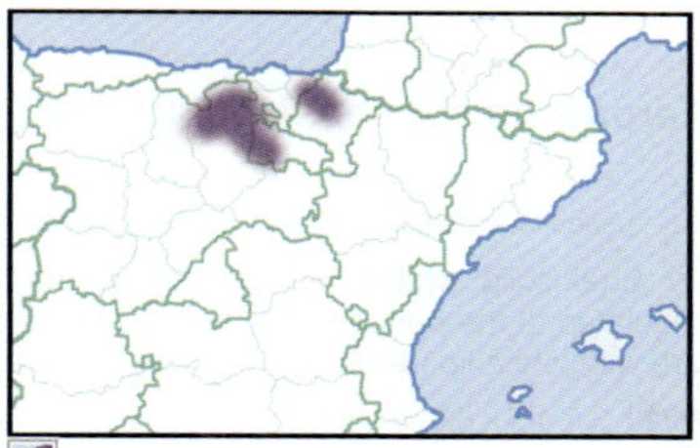

N Spain

Helix gonzalezi Azpeitia Moros, 1925 in Azpeitia Moros 1925: 161

Shell like *Xerocrassa cobosi*, more regularly and less prominently striated, upper side flatter, lower side more rounded, angulated at periphery, umbilicus very wide, up to 1/3 of diameter, whorls well visible inside.

2.2-2.6 x 4.5-5.5 mm

Humid habitats not well exposed to sunshine, often on limestone rock faces.

Prieto & Martín 1988, Gittenberger & Raven 1982, Talaván Gómez & T. 2006: 32.

Xerotricha huidobroi (Azpeitia Moros, 1925)

Helix huidobroi Azpeitia Moros, 1925 in Azpeitia Moros 1925: 32 (*Helix (Jacosta) Huidobroi*)

Spain: Alicante, Torrevieja

Spain

4-5 x 6-7 mm

Shell light, coarsely and irregularly ribbed, 4-4.5 strongly convex whorls, aperture simple, umbilicus 1/6 of shell diameter. Animal white.

Relatively open habitats, exposed to sunshine, Mediterranean shrublands with pine trees, also in cultivated lands, under stones and at the base of vegetation, up to 1500 m.

Animals collected in October in Granada were not completely mature.

EU Red List: Near Threatened NT. Threatened by modern agriculture, urbanisation and tourism. Between 1993 and 2003 live animals were found in only two of 18 known sites in Alicante.

Ortiz de Zárate y López 1950: 56, Martínez-Ortí & Robles 2003: 144.

Xerotricha jamuzensis (Gittenberger & Manga, 1977)

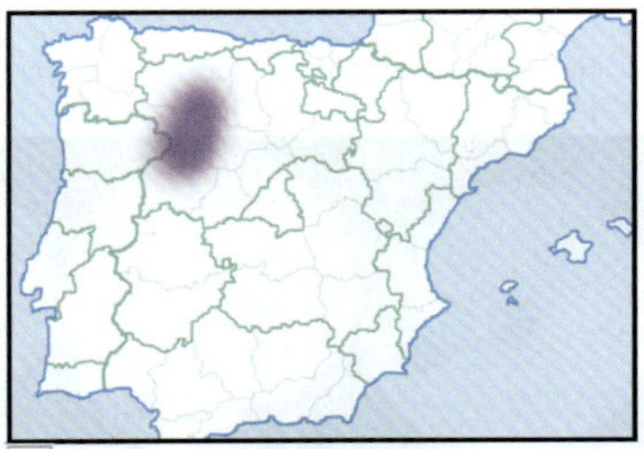

Helicella jamuzensis Gittenberger & Manga, 1977 in Gittenberger & Manga 1977: 181

Shell globular, pale horny brown with some whitish spots and usually a narrow brown fragmented colour band above the periphery, regularly ribbed, covered with periostracum, with a microsculpture of fine spiral lines, first whorls can have very short (0.02 mm) hairs, around 4.5 whorls, aperture simple with or without lip inside, umbilicus narrow, less than 1/6 of shell diameter.

4-5.5 x 6.6-8 mm

Ruderal habitats, stone walls, gardens, also at roadsides, near villages.

Talaván Gómez & T. 2006: 114.

Spain (León to Salamanca)

Xerotricha madritensis (Rambur, 1868)

Helix madritensis Rambur, 1868 in Rambur 1868: 266

Spain: Teruel, Aguaviva de Bergantes

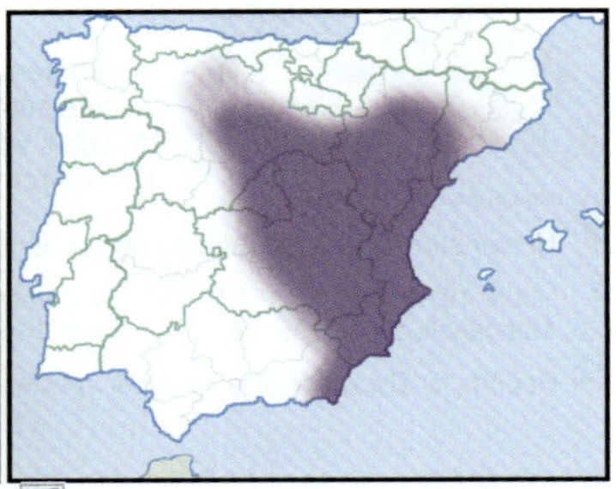

Spain

5-7 x 8-10 mm

Shell white or grey, with two dashy or blurry dark colour bands at the upper side, densely and regularly striated, apertural margins can be united by a callus at the parietal side, umbilicus regular.
Animal whitish with some weak pigments. Differs from *Xt. barcinensis* in the light colour of the animal.

Open and scarely vegetated habitats among stones, exposed to sunshine, usually on calcareous substrate, also on mountain plains.

Westerlund 1889: 281, Ortiz de Zárate y López 1950: 60, Talaván Gómez & T. 2006: 33.

Xerotricha renei (Fagot, 1882)

Helix renei Fagot, 1882 in Fagot 1882: 72 (*Helix Renei*)

France: Hautes-Pyrénées, near Gavarnie

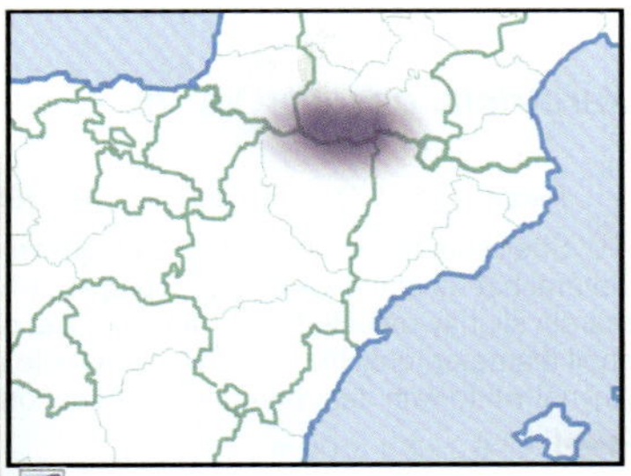

Pyrenees

In high mountains.

Westerlund 1889: 306, Azpeitia Moros 1925: 152, Germain 1930: 282.

Taxonomic status uncertain, requires more research.

4.5-5.5 x 6.5-7.5 mm

Shell dark yellow or brown, no colour bands or dashes, irregularly striated, sometimes with hairs, apex strongly pointed and protruding, last whorls increasing more rapidly than the first whorls, aperture rounded, umbilicus present.

Xerotricha silosensis (Ortiz de Zárate y López, 1950)

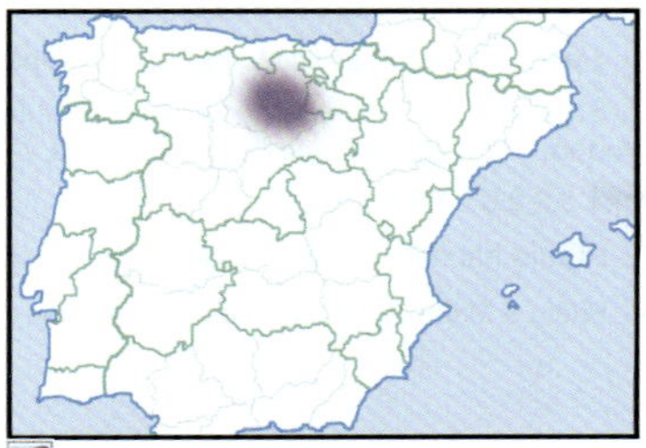

Helicella silosensis Ortiz de Zárate y López, 1950 in Ortiz de Zárate y López 1950: 61, Lám. 2

Shell solid, whitish with dark spots, irregularly and finely striated, apex dark, 5.5 convex whorls, slightly angulated at the periphery except shortly before the aperture, last whorl slightly descending near aperture, aperture with white lip inside, margin only reflected at columellar side, umbilicus 1/5-1/6 of shell diameter.

3.5-4.5 x 6.5-7.5 mm

Forests and open habitats such as pastures in mountains.

Sexually mature adults were found in August.

Prieto & Martín 1988.

N Spain (Burgos)

Xerotricha vatonniana (Bourguignat, 1867)

Helix vatonniana Bourguignat, 1867 in Bourguignat 1863-1868: 254, Pl. 37 (*Helix Vatonniana*)

Shell conical, light yellowish or greyish with light brown spots or fragmented colour bands, iregularly and quite ensely striated, equally on upper and lower side, fresh shells with short hairs, not very densely arranged, 5 convex whorls with deep suture, last whorl not descending near aperture, aperture with internal lip, magrin not reflected except near umbilicus, umbilicus very narrow and partly covered by the apertural margin.

5.3-7.4 x 5.7-7.2 mm

Found at margins of cultivated fields, also at the bottom of olive trees, in coastal vicinity, in Spain not above 50 m altitude.

EU Red List: Vulnerable VU. Seriously threatened by intensive agriculture, chemical products and fire. The species has presumably lost most of its habitats by the agricultural occupation of the land, and is today known from only 9 small and fragmented habitats with low numbers of animals.

N Africa, S Spain, S Portugal

Arrébola (unpublished, 1995), Arrébola 2001, Gómez Moliner et al. 2001: 174, Arrébola & Puente in Verdú & Galante 2006: 390, Cuttelod et al. 2011: 88.

Xerotricha zaratei (Gittenberger & Manga, 1977)

N Spain

Helicella zaratei Gittenberger & Manga, 1977 in Gittenberger & Manga 1977: 185

Shell yellowish white with several fragmented pale horny brown colour bands, regularly ribbed on upper side, striated on lower side, lower side with fine spiral lines, first whorls with short non-persistent hairs (0.05 mm), 4.5 whorls, periphery rounded on last whorl, aperture with internal lip, umbilicus 1/5 of shell diameter, wider than in *Xt. jamuzensis*.

3.6-4.2 x 6.3-7.3 mm

Scattered at limestone outcrops under stones in shrublands and grasslands, up to 1200 m.

EU Red List: Near Threatened NT. Treatened by potential destructions of the fragmented habitats.

Puente et al. in Verdú & Galante 2006: 371.

Xerotricha zujarensis (Ortiz de Zárate y López, 1950)

S Spain

Helicella zujarensis Ortiz de Zárate y López, 1950 in Ortiz de Zárate y López 1950: 63, Lám. 2

Shell solid, white with two zones of dark spots on upper side, and fine and irregular riblets, apex dark, 5 convex whorls, last one very well rounded, aperture simple without lip inside, columellar margin slightly reflected covering part of the umbilicus, umbilicus narrow, 1/7-1/8 of shell diameter, 3/4 of penultimate whorl visible inside.

4.5-5.5 x 6.5-7.5 mm

Open habitats in pine forests, under stones.

Sexually mature adults were found in October.

More research is necessary about the taxonomic status of this form. The few known localities have suffered heavy modification and the species has not been found again after 1975.

Zenobiella

Zenobiella Gude & Woodward, 1921 in Gude & Woodward 1921: 179 *Fruticicola (Fruticicola (Zenobiella))*

Type species *Helix subrufescens* Miller, 1822 (original designation)

Gude & Woodward 1921 intended to establish a new replacement name for *Zenobia* Gray, 1821 (type species *Helix bimarginata* Gray, 1821 by monotypy, = currently *Monacha cartusiana*) used in the sense of authors who regarded *subrufescens* as its type (*Zenobia* used in this sense was not an available name under Art. 49). *Zenobia* was thought to be preoccupied by *Zenobia* Oken, 1815 (Lepidoptera, unavailable name). There was no available name to be replaced, so *Zenobiella* was established as a regular new name with its type designated on p. 179.

Zenobiella subrufescens (Miller, 1822)

Helix subrufescens Miller, 1822 in Miller 1822: 379

France: Orne valley, Le Mesuil Roger

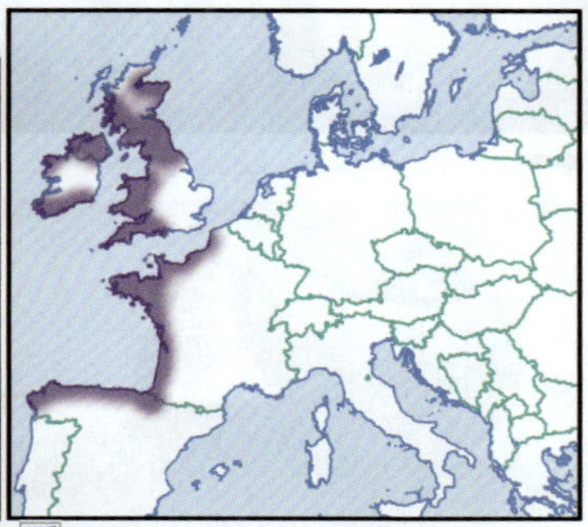

W Europe

4-6 x 6-10 mm

Shell usually pale brown, very finely striated, thin and transparent, 4.5-5 convex whorls with moderate suture, slightly shouldered above periphery, aperture usually without lip inside, umbilicus very narrow, partly covered by the reflected columellar margin.
Differs from *Urticicola glabellus* in its more globolar shell, smaller umbilicus and absence of minute scales on the surface.

Old broad-leaved woodlands, shady and well vegetated valley sides, often in montane habitats. Tolerates non-calcareous soils. In Britain often associated with *Luzula sylvatica*. May climb up tree trunks. Occasionally also in more open but always natural habitats among rocks and sea cliffs, rarely in marshes.

30-50 eggs (diameter 1 mm) are laid between August and October, juveniles hatch after 18-20 days, maturity is reached in the following year.

Vulnerable in Ireland, situation central eastern and SE England not much different. Threatened by continuous destruction of natural habitats in lowlands, particularly by destruction of old woodlands and the forest soil fauna by modern forest management. Many populations in Great Britain and Ireland have been lost since the 1830s. The species is still relatively frequent in Cornwall, Wales, N England, S and SW Scotland and Northern Ireland.

Moquin-Tandon 1855: 212 (life cycle), Germain 1930: 254, Kerney & C. 1979: 185, Kerney et al. 1983: 255, Ondina et al. 1997, Anderson 1996: 14, Kerney 1999: 191.

Ariantinae: which genera should be used?

In the past decades the subfamily Ariantinae (Helicidae) was subdivided into 7-9 genera:
Arianta (6 or 7 species),
Cylindrus (1 species),
Drobacia (2 species),
Vidovicia (1 species),
Isognomostoma (1 or 2 species),
Causa (1 species, or in *Isognomostoma*),
Chilostoma and ***Helicigona*** (more than 115 species; some authors used *Wladislawia* as a separate genus for 2 species in this group).

It has since long been clear that *Helicigona* and *Chilostoma* did certainly not represent monophyletic groups. The question is how to react and produce a taxonomic system in which names can be found quickly in internet resources. Classifying every species in its own genus would be the correctest solution, but probably not very useful because students could not obtain lists of species in a genus, and if genera are overlooked, species are overlooked. I have observed such cases in hygromiids, where species in red lists seem to have been overlooked either because the species have too often changed their genera, or because researchers have overlooked a rarely mentioned genus.
Principally all species in this subfamily could carry the generic name *Arianta*.
A second proposal is to use generic names for small subgroups within the *Chilostoma-Helicigona* complex, such as *Faustina*, *Liburnica* or *Superba*. This has been favoured by P. Subai who has been subsequently revising and defining subgroups in this subfamily. But this enhanced the number of genera enormeously and produced a very much splitted group of European non-marine molluscs:
Ariantopsis, *Campylaea*, *Campylaeopsis*, *Cattania*, *Chilostoma*, *Cingulifera*, *Corneola*, *Delphinatia*, *Dinarica*, *Faustina*, *Helicigona*, *Josephinella*, *Kosicia*, *Liburnica*, *Superba*, *Thiessea*, *Wladislawia* and various new genera, together at least 25 generic names for these species.

The most frequently mentioned species in this complex are *Chilostoma cingulatum* and *Helicigona lapicida* (judging by Google hits, Jan 2010). *Helicigona lapicida* would keep its name (in a then monotypic genus), *Chilostoma cingulatum* would have to change its genus (to *Cingulifera cingulata*).

The third proposal maintains large genera and tolerates their paraphyly. The EU Red List (Cuttelod et al. 2011) used 5 genera *Chilostoma* (71 species), *Helicigona* (1), *Cattania* (6), *Faustina* (4), and *Superba* (5).
My proposal used here sticks more to the traditionally used genera:
Chilostoma (containing *Campylaea*, *Chilostoma*, *Cingulifera*, *Corneola*, *Delphinatia* and *Kosicia*),
Helicigona (containing *lapicida* and all others, from E and SE Europe).
This has the effect that all species north and west of NW Croatia are classified in *Chilostoma*, all south and east of NW Croatia in *Helicigona*. So with 2 generic names students and researchers are able to obtain a complete list of all species of this complex, and none is missed.

The species in this subfamily were kindly checked by P. Subai.

Arianta

Arianta Turton, 1831
in Turton 1831: 35 *Helix* synonym *Arianta*

Type species *Helix arbustorum* Linnæus, 1758
(monotypy by Turton 1831: 35)

Mentioned by Turton 1831: 35 as a synonym of *Helix* in the combination *Arianta Arbustorum* (synonym of *Helix Arbustorum*) with a reference to "Leach Mollusc. p. 86", obviously an unpublished work under Art. 9.3 (Bank, F. & P. 2007: 45). Author is not Leach because the author of a work is alone responsible for introducing synonyms, here Turton (Code 4th ed. Art. 50.7). Spelled *Arantia* in the index p. 143. Type by monotypy because only *Helix arbustorum* was directly associated with *Arianta* (Art. 67.12). Validated under Art. 11.6.1 by Beck 1837: 41 (who used the name as *Helix (Arianta)*, however without bibliographic reference). In ICZN Opinion 335 (from 1955) *Arianta* was placed on the Official List.

Gittenberger, Piel & G. 2004.

Arianta aethyops (Bielz, 1851)

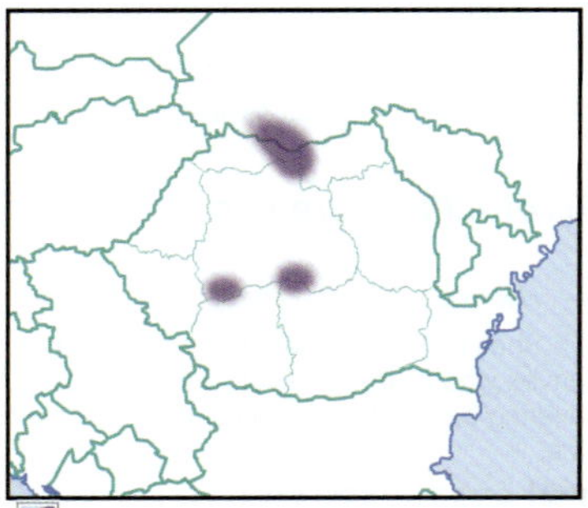

N and central Romania (Făgăraş, Retezat and Rodna mountains), W Ukraine

Helix aethyops Bielz, 1851 in M. Bielz 1851: 55
Original spelling was *aethyops*. E. A. Bielz 1863: 66 spelled the name *aethiops*, an incorrect subsequent spelling.
Campylaea aethiops var. *petrii* Kimakowicz, 1890 in Kimakowicz 1890: 70 (*Campylaea (Arianta) aethiops* var. *Petrii*)

Shell brown to black and shiny, yellowish or greenish near umbilicus, surface slightly corrugated, not very solid, aperture without strong white lip, umbilicus closed.
Animal dark brown or blackish, shiny, coarse elongate granules on dorsal side, sole brown or greyish.
Flagellum longer than in *A. arbustorum*, spermatheca duct shorter, free oviduct length 3 times longer.

9-15 x 17-24 mm

Restricted to altitudes above 1800 m, under stones and among herbs, in wet weather almost everywhere.

Shilejko 1978: 313 (*A. aethiops petrii*), Grossu 1983: 495, Egorov 2008: 81 (*A. petrii*), Sysoev & S. 2009: 164 (*A. petrii*).

Occurs sympatrically with *A. arbustorum* in Făgăraş mountains (Kimakowicz 1890: 71). The northern populations (*petrii*) represent probably a separate species.

Arianta arbustorum (Linnæus, 1758)

Helix arbustorum Linnæus, 1758 in Linnæus 1758: 771

Germany: near Hamburg, Wedeler Marsch

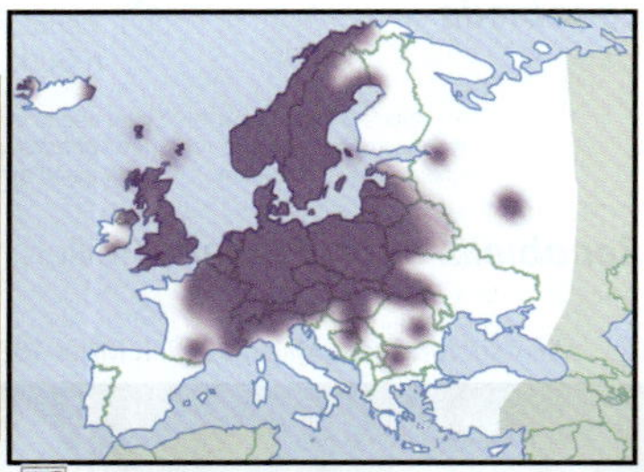

Europe

12-22 x 18-25 mm, locally variable

Germany: S Bayern, near Mittenwald

Shell usually brown with numerous pale yellowish rows of spots and usually with a brown band above the priphery, occasionaly yellowish, reddish or with greenish hue, weakly striated and with fine spiral lines on the upper side, 5-5.5 convex whorls with deep suture, last whorl slightly descending near aperture, aperture with prominent white lip inside, margin reflected, umbilicus entirely covered by the reflected columellar margin.
The animal is usually dark.
Shell shape globular in most present-day populations, but originally believed to have been depressed in the Pleistocene, before lowlands were invaded and shells became globular, re-invading mountain regions except some isolated spots among glaciers.

Forests and open habitats of any kind, requires humidity, lives also in disturbed habitats (not in Ireland where it is restricted to old native woodland). May locally tolerate non-calcareous substrate, in N Scotland also on sandhills. In the Alps up to 2700 m, in Britain 1200 m, in Bulgaria 1500 m. Feeds on green herbs, dead animals and faeces.

Reproduction usually after copulation, but self-fertilization is also possible, in France, Britain and Germany 30-50 eggs (diameter 2.5-3.3 mm) are laid between July and September in a yellowish opaque envelope, juveniles hatch after 15-25 days, maturity is reached with full shell size after 1-2 years, occasionally later, life span around 4 years, maximum age up to 14 years.
If snails hatched more than 50 m distant from each other, they are considered isolated since they would not move more than 25 m (neighbourhood area 32-50 m), usually they move about 7-12 m in a year, mostly along water currents.

Rare in Ireland and Bulgaria.

Moquin-Tandon 1855: 123, Westerlund 1889: 147, Germain 1930: 226, Frömming 1954: 289, Baur & Raboud 1988, Gittenberger 1991 (J. Moll. Stud. 57), Baur 1993.

Several subspecies have been recognized by some authors, their bioscientific value remains to be researched.

Germany: N Bayern, near Bad Brückenau

Arianta chamaeleon (Pfeiffer, 1868)

Helix chamaeleon Pfeiffer, 1868 in L. Pfeiffer 1868 (Malak. Bl. 15): 84

Austria: Kärnten, Karawanken, oberes Bärental

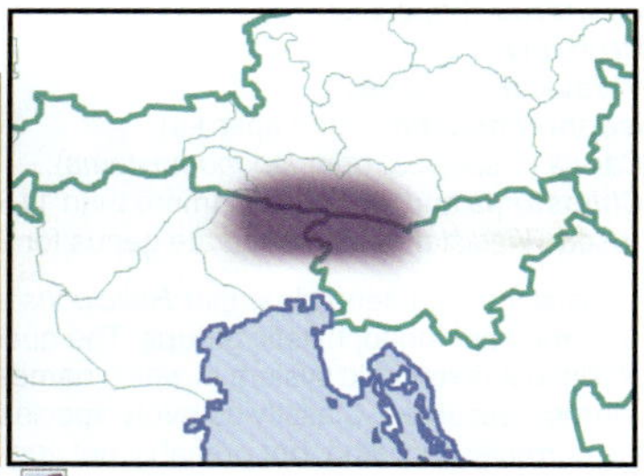

SE Alps

9-11 x 18-22 mm

Shell brownish yellow with brown spots and a distinct dark brown band at the periphery, 5-6 convex whorls, last whorl initially depressed, slightly descending near aperture, aperture often without prominent lip inside, umbilicus moderate and open. The animal is dark brown.

Humid mountain habitats among herbal plants in rock rubble below rocky outcrops, more rarely directly on the rocks, also near small shrubs.

EU Red List: Endangered EN. Threatened by increasing land use by tourism and extensive pasturing in its sensitive mountain habitat. Populations have seriously declined at rates of 40 % in Austria, the same can be inferred for Italy and Slovenia.

Westerlund 1889: 120, Manganelli et al. 1995: 32, Kerney et al. 1983: 272, www.iucnredlist.org (08.2012).

Several subspecies of this locally variable species have been recognized, their bioscientific value remains to be researched.

Arianta hessei (Kimakowicz, 1883)

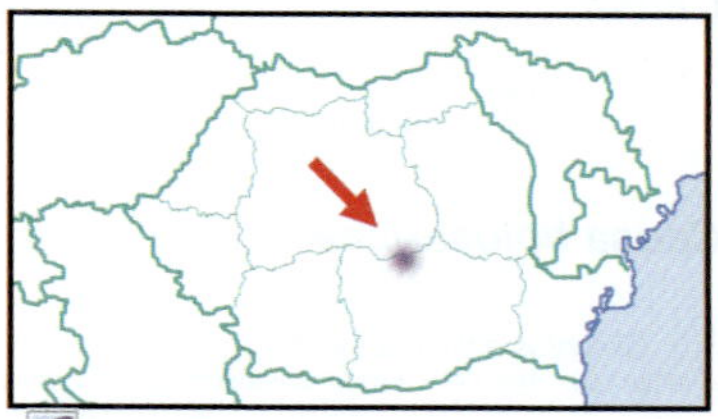

Romania (Bucegi mountains)

Helix schmidti var. *hessei* Kimakowicz, 1883 in Kimakowicz 1883: 33 (*Helix (Campylaea) Schmidti* var. *Hessei*)

Shell relatively flat, spirally striated, last whorl narrower than penultimate whorl, umbilicus open but narrow. Snaller than *Arianta schmidtii*, from which it also differs in its narrower last whorl and narrower aperture.

11-13 x 15-20 mm

Mountain habitats among limestone rocks in a remote and hardly accessible region.

EU Red List: Near Threatened NT. Only known from few localities. No real threats known.

Grossu 1983: 493.

Arianta schmidtii (Rossmässler, 1836)

Helix schmidtii Rossmässler, 1836 in Rossmässler 1835-1837 (3): 5, Taf. 11 (*Helix Schmidtii*)
Name was previously mentioned by Rossmässler 1835 (Icon. (2): p. 2), nomen nudum without description (the only feature assigned to it was "beautiful", insufficient for a description).

9-12 x 17-25 mm

Slovenia: Kaminske Alpe, Grintavec, 2350 m

Shell dirty yellowish grey, more olive brown near aperture, with densely arranged white spots, finely striated with some more or less distinct white spiral lines, 4.5-5.5 convex whorls with deep suture, last whorl depressed but usually rounded, sometimes initially with blunt edge at periphery, slightly descending near aperture, aperture inside brown with prominent white lip, margin reflected, umbilicus moderate and open.
Animal very slender and black.

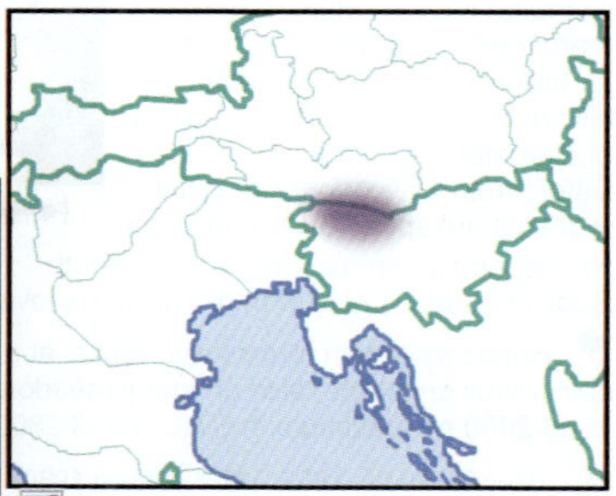

SE Alps

At rocks in grasses and among *Papaver alpinum*.

EU Red List: Near Threatened NT. Acute threats are not known.

Westerlund 1889: 121, Kerney et al. 1983: 273.

Arianta stenzii (Rossmässler, 1835)

Helix arbustorum forma *stenzii* Rossmässler, 1835 in Rossmässler 1835-1837 (1): 57 (*Helix arbustorum* forma *Stenzii*)

Shell like *A. arbustorum*, but prominently striated, with thin and more depressed shell, yellow spots often merging to bands, but without a distinct band at the periphery, umbilicus open.
The animal is brown with reddish or yellowish hue.

Rock rubble at alpine grassland margins.

13-18 x 21-29 mm

Italy: Trento, Passo Rolle, 2000 m

Alessandro Hallgass

Gittenberger, Piel & G. 2004 found no evidence for a separate species, *stenzii* clustered with *A. arbustorum*. This form needs a revision, perhaps it is conspecific with *A. arbustorum*. More detailed local research could shed light on this question, there could exist contact zones with *A. arbustorum*.

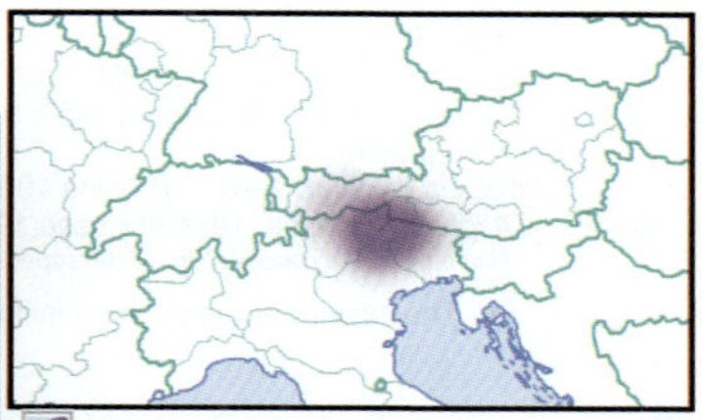

S Alps

Westerlund 1889: 149, Manganelli et al. 1995: 32.

EU Red List: Near Threatened NT. Threatened by increased land use in sensitive mountain habitats, mainly by tourism.

Arianta xatartii (Farines, 1834)

Helix xatartii Farines, 1834 in Farines 1834: 6 (*Helix Xatartii*)

13-15 x 17-20 mm

France: (1-2) Pyrénées-Orientales, Argelès, (3-5) Pyrénées-Orientales

(1-2) Raymond Huet

Shell brown with white spots and a weak and slightly darker colour band at upper side, striated, more finely striated and with radial colour streaks at lower side, 4-5 weakly convex whorls, last whorl suddenly descending near aperture, can be inflated, aperture rounded, margin white and slightly reflected, umbilicus almost covered by the reflected columellar margin.

Westerlund 1889: 148, Germain 1930: 228, Gittenberger, Piel & G. 2004 (found no evidence for a separate species, *xatartii* clustered with *A. arbustorum*), Puente et al. in Verdú & Galante 2006: 360.

Perhaps conspecific with *A. arbustorum*. Range isolated, so no contact zones with *A. arbustorum* can be expected.

E Pyrenees

Humid mountain habitats in summit areas, 1200-2600 m.

EU Red List: Vulnerable VU. Restricted to few localities. Threatened by increased land use, mainly by tourism.

Causa

Causa Shilejko, 1971
in Shilejko 1971: 997

Type species *Helix holosericea* Gmelin, 1791
(original designation)

In original description "monotypic genus with one species", this should be interpreted as original type species designation.

Causa holosericea (Studer, 1820)

Glischrus holosericea Studer, 1820 in Studer 1820: 87 (*Glischrus (Helix) holosericea*) (not *Helix holosericea* Gmelin, 1791 in Gmelin [1791]: 3641)
Forcart 1957 (Mitt. Bern 15): 202 considered Gmelin's name as a replacement name for *Helix obvoluta* Müller, 1774, and proposed to place Gmelin's name on the Official Index and to valididate a name *Glischrus holosericea* Studer, 1820. Gmelin [1791] did not establish this as a new replacement name, and did not mention the name *Helix obvoluta*.
Gmelin [1791] gave a short description and 4 bibliographical references, all referred to the species currently known as *Helicodonta obvoluta* (Müller, 1774). It is not necessary to suppress Gmelin's name because it was established in a different genus, and there are no constraints that Studer referred to Gmelin's name. Studer 1820 mentioned 3 teeth in the aperture, Gmelin [1791] did not mention them and the cited figures showed no teeth.

5-6 x 9-12 mm

Austria: Tirol, Bärental, 1240 m

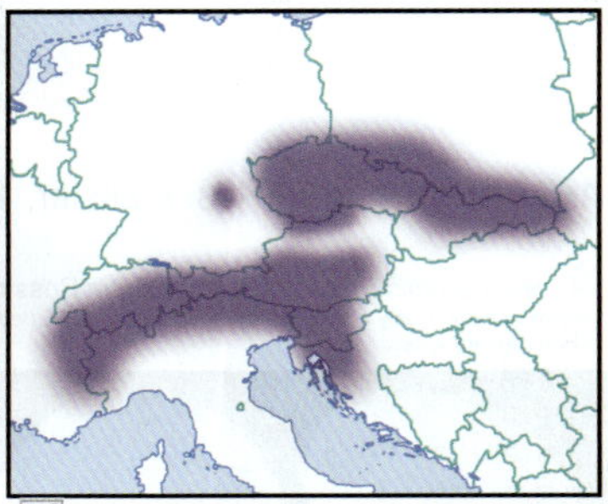

Shell brown and densely covered with hairs, flat above and rounded below, but not keeled, 4-5 weakly convex whorls with well marked suture, last whorl not descending near aperture, apertural margin reflected with 2 or 3 triangular teeth (1-2 basal, 1 palatal), umbilicus very wide and deep, 1/5-1/4 of diameter.
Differs from *Helicodonta* in its flat (not sunken) spire, fewer and wider whorls, and its prominent triangular teeth.
Body moderately grey with bluish hue above, tentacles and mantle darker, foot lighter.

Humid mountain forests with rocks, at and in rotting wood and rock rubble, often in coniferous forests on non-calcareous substrate, also on stony meadow slopes, in clearings from avalanches and at rocks. Usually from 1000-2000 m, maximum in Switzerland 2800 m.

Moquin-Tandon 1855: 117, Germain 1930: 225, Kerney et al. 1983: 280, Falkner 1990: 230.

Alps, W Carpathians

Scattered and in many regions rare.

Usually in genus *Causa*, but sometimes as *Isognomostoma holosericeum*.

Chilostoma

Chilostoma Fitzinger, 1833
in Fitzinger 1833: 98

Type species *Glischrus foetens Studer, 1820*
(subsequent designation by Gray 1847: 172)

Contained originally *Helix cornea* Draparnaud, 1801 (non Linnæus, 1758), *Glischrus zonata* Studer, 1820, *Helix pulchella* Müller, 1774 "and Draparnaud", and *Helix costata* Müller, 1774. Among the synonyms of *Chilostoma zonatum* was listed the nominal species *Glischrus foetens* Studer, 1820. Herrmannsen 1846 [01 Dec 1846]: 225 gave as type both *Helix cornea* and *Helix pulchella*.
Campylaea Beck, 1837 in Beck 1837: 24 *Helix (Campylæa)*, type species *Helix hispana* Linnæus, 1758 [in Linnæus 1758: 772] (subsequent designation by Gray 1847: 172)
Cingulifera Held, 1838 in Held 1838: 911, type species *Glischrus cingulata* Studer, 1820 (subsequent designation by Herrmannsen 1846 [01 Mar 1847]: 235)
Corneola Held, 1838 in Held 1838: 912, type species *Helix cornea* Draparnaud, 1801 [in Draparnaud 1801: 89] (subsequent designation by Herrmannsen 1846 [01 Mar 1847]: 309)
Kosicia Brusina, 1904 in Brusina 1904: 162, type species *Helix intermedia* Pfeiffer, 1828 (original designation)
Delphinatia Hesse, 1931 in Hesse 1931: 60, type species *Helix alpina* Michaud, 1831 (monotypy)
Name proposed conditionally with one species contained, available under Art. 11.5.1 and 13.3.

The system proposed here includes the subgenera *Campylaea*, *Chilostoma*, *Cingulifera*, *Corneola*, *Delphinatia* and *Kosicia*.

Chilostoma achates (Rossmässler, 1835)

Helix foetens var. *achates* Rossmässler, 1835 in Rossmässler 1835-1837 (2) p. 4
Established in the text as a "Varietät" of *Helix foetens*.

7-12 x 17-29 mm

Austria: Osttirol, Leisach

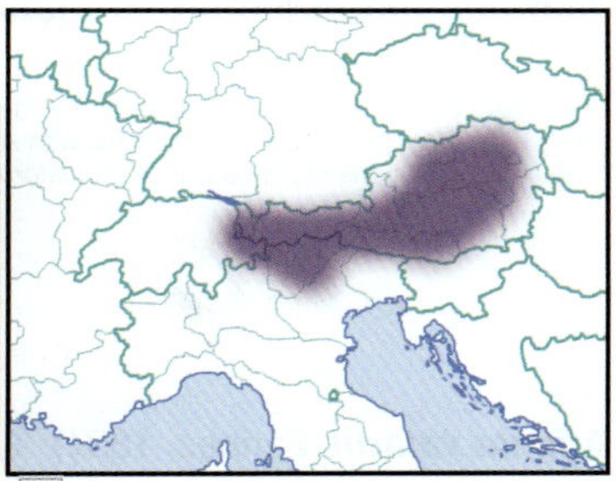

Central and E Alps

Shell very variable, basic colour yellowish white, with a brown colour band and a pale band below the periphery, 5 weakly convex whorls with moderate suture, last whorl deeply descending near aperture, aperture very oblique, margin white and reflected, not connected at parietal side, umbilicus wide, 1/7-1/6 of diameter, slightly excentrical at last whorl so that the reflected lip escapes from covering partly the umbilicus.
Differs from *Ch. zonatum* in its flatter shell, less convex whorls, a more oblique aperture and its colour lacking a green hue. Also *Ch. illyricum* has a less oblique aperture.

Rocks and rock rubble, not confined to limestone. Needs humid and shady rocks, sensitive against sunshine, also hidden at the soil in mountain forests far away from rocky walls, and in valleys. Between 900 and 2300 m.
Active only during humid weather conditions, effectively hidden in periods of drought. Juveniles tend to prefer rocky walls more than adults, which tend to occur below rocky walls in the valleys where they can be found at dead wood. Populations from valley grounds have larger shells.

EU Red List: Near Threatened NT. Populations have seriously decline at rates of 20-50 % in Austria. Threatened by increased land use by tourism and associated reduction of habitat quality, and by modern forest management which tends to alterate habitat conditions and to destroy the soil fauna.

Kerney et al. 1983: 275, Falkner 1990: 228, Manganelli et al. 1995: 31, Turner et al. 1998: 363, Boschi 2011: 571.

Chilostoma. The local variation has been associated with subspecies. Their bioscientific value (the presence of genetically induced partial barriers to reproduction) requires further investigation, local variation could also be caused by slow but continuous gene flow over area.

Chilostoma acrotricha (Fischer, 1877)

Helix desmoulinsii var. *acrotricha* Fischer, 1877 in Fischer 1877: 52 (*Helix Moulinsi* var. *acrotricha*)
The name was attributed to Fischer. Fischer gave a short description and cited a long passage from Dupuy, containing detailed descriptive elements. In a strict application of Art. 50.1.1 the name in such a case must be attributed to the author of the work.
Helix crombezi Locard, 1882 in Locard 1882: 320 (*Helix Crombezi*)
Also mentioned in the same work on p. 91, nomen nudum without description there. The name was attributed to Millière, the entire descriptive text was attributed to Bourguignat. In a strict application of Art. 50.1.1 the name in such a case must be attributed to the author of the work.
Could have been published by Bourguignat 1880 (Descr. Saint-Martin-de-Lantosque): 5 (cited by Alzona 1971: 186), but I could not verify that. Falkner et al. 2002: 162 did not find the name cited there.

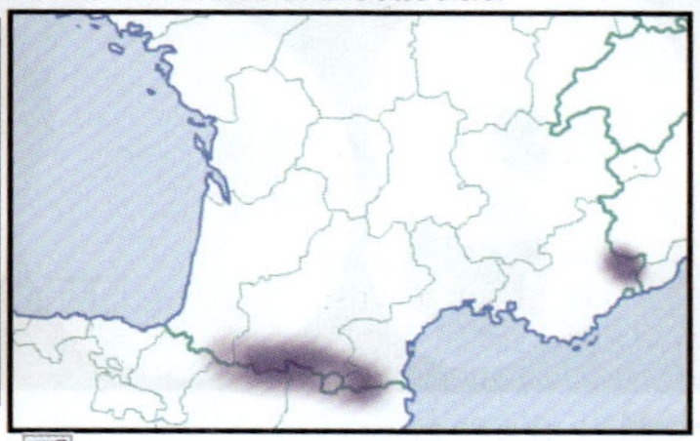

5-8 x 13-17 mm

(1-3) Pyrenees, (4-6) SE France: Parc du Mercantour
Giuseppe Pocaterra

Pyrenees, SW Alps

SE France: Parc du Mercantour
Giuseppe Pocaterra

Shell evenly yellowish red with weak dark bands near the aperture, finely striated, with short hairs arranged in regular rows, 5-5.5 whorls with edge at periphery, upper side nearly flat, lower side convex, last whorl suddenly and deeply descending near aperture, umbilicus narrow.
Looks somehow like a form of *Helicigona lapicida* without keel.
Very much like *Ch. desmoulinsii*, with more rapidly increasing whorls, slightly lighter, smaller and more slender, edge at periphery more pronounced, with regularly arranged short hairs on upper and lower sides, more oblique aperture and narrower umbilicus.

In rock crevices, below walls and stones, in the Alps around 2500 m.

EU Red List: Critically Endangered CR (*crombezi*), Near Threatened NT (*acrotricha*). Very rare. Threats not understood.

Locard 1882: 91, Westerlund 1889: 139 (*Helix desmoulinsii*), 140, Germain 1930: 217 (*desmoulinsii*), Vial & Bertrand 2002: 21, Bertrand 2002: 23.

Subgenus unresolved, perhaps *Corneola*. Forms of Alps (*crombezi*) and Pyrenees (*acrotricha*) have the same shape and microsculpture, they could be the same or separate species. Subspecies have been proposed, but not aligned with bioscientific criteria (genetically induced partial barriers to reproduction). Molecular studies could be helpful.

Chilostoma adelozona (Strobel, 1857)

Helix adelozona Strobel, 1857 in Strobel 1857: 17 (= 245)
Mentioned without description or bibliographical reference, with a reference to "*zonata* Strob. olim.", a previous published or unpublished use of *Glischrus zonata* Studer, 1820 (= currently *Chilostoma zonatum* (Studer, 1820)), and with one subordinate variant for which a new name was used and which had a description (var. *rhaetica*). Not a new replacement name because no available name was intended to be replaced (Art. 72.7). Not established as a synonym under Art. 11.6.1 because no available name was given as a synonym. The name *Helix adelozona* was only made available because it had a subordinate variant (*rhaetica*) for which a description was given, so under Art. 12.1 *adelozona* had a description for the taxon (Code Glossary "taxon": a taxon encompasses all included taxa of lower rank). Art. 12.1 does not demand a description (or indication) for the nominal taxon, a description for the taxon is sufficient, this can include subordinate taxa.
For those who do not like to accept this interpretation of the Code, the next available author is possibly Stabile, 1864. They would have to use the name *rhaetica* Strobel, 1857 for the species. They might also confront debates whether the term "depressior" (= "more flattened [than *adelozona*]) given for *rhaetica* in the 1857 work would constitute or not a description for *adelozona*.
Helix adelozona var. *rhaetica* Strobel, 1857 in Strobel 1857: 17 (= 245)
Helix foetens var. *cisalpina* synonym *Helix adelozona* Stabile, 1864 in Stabile 1864: 51, 53, 54, 55
Mentioned as a synonym with description, without reference to a previous source. Used as *Helix cisalpina* var. *adelozona* by Am Stein 1890: 11 (with a short description by Suter-Näf, and bibliographical reference to Stabile 1864: 54), providing the prerequisites for availability under Art. 11.6.1 at this occasion or earlier.
Helix foetens var. *cisalpina* Stabile, 1864 in Stabile 1864: 51 (not *Helix striata* var. *cisalpina* Rossmässler, 1837, currently *Cernuella virgata* (Mendes da Costa, 1778))
Helix foetens var. *adamii* Kobelt, 1875 in Kobelt 1875: 212 (*Helix foetens* var. *Adamii*)

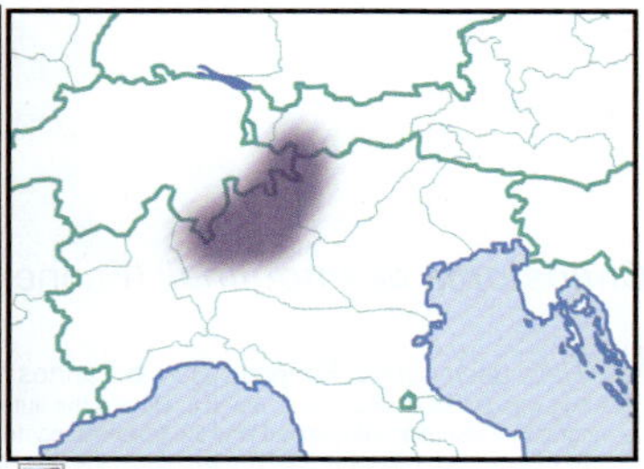

12-14 x 25-29 mm

Austria: Tirol, near Finstermünz

Alps

Shell horny yellowish with a narrow and often indistinct brownish band and a lighter zone below it, finely striated and with fine spiral sculpture, thin and shiny, 4.5-5 weakly convex whorls with deep suture, last whorl width distinctly more than 2 x of penultimate whorl, lower side more convex than upper side, deeply descending near aperture, aperture with a dark band behind the white margin, margins approaching each other closely at parietal side, usually connected by a weak layer, basal and columellar sections reflected, covering sometimes partly the wide (4-4.5 mm) umbilicus.

Predominantly in humid and shady crevices, holes and caves of silicious limestone rocks. Up to 1800 m in Switzerland.

EU Red List: Vulnerable VU. No acute threats known, some habitats in Italy are threatened by quarrying.

Chilostoma. Two subspecies *adelozona* and *rhaetica* were proposed, their relationships and bioscientific value require more research.

Westerlund 1889: 125 (*cisalpina*), Alzona 1971: 195, Manganelli et al. 1995: 31, 54, Turner et al. 1998: 365 (*rhaeticum*, *adelozona*), Boschi 2011: 573.

Chilostoma ambrosi (Strobel, 1852)

Helix ambrosi Strobel, 1852 in Strobel 1851 [Jun 1852]: 59 [not verified, Alzona 1971 spelled *ambrosii*]

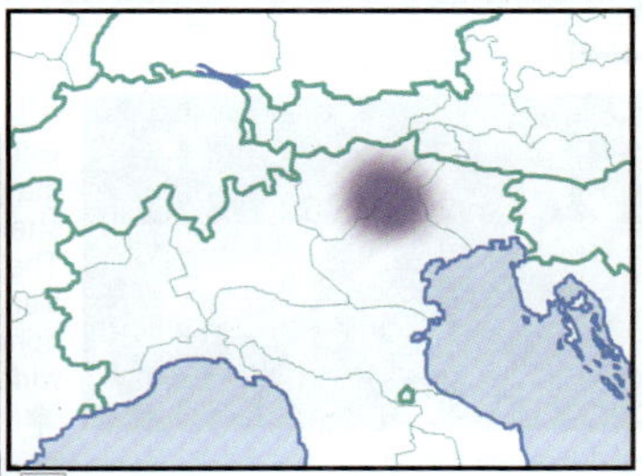

5-6 x 9-12 mm

Italy: Alto Adige/Südtirol, "Val di Oten" (?, perhaps Valdaora)

S Alps

Rocky habitats in mountains.

EU Red List: Near Threatened NT. Range fragmented, but no acute threats known.

Westerlund 1889: 123 (*Helix aemula*), Manganelli et al. 1995: 32.

Kosicia.

Shell whitish with 3 brown bands, the uppermost band always weak, lower side of last whorl yellowish grey, densely striated and with fine spiral lines, thin and shiny, 4-4.5 weakly convex whorls with prominent apex, last whorl suddenly descending near aperture, apertural margin white and with weak lip inside, upper margin straight, lower margin reflected, umbilicus open and moderately narrow.
The animal is dark brown or pale brownish with slightly darker head and upper tentacles.

Chilostoma cingulatum (Studer, 1820)

Without gender agreement: *Chilostoma cingulata* (Studer, 1820)

Glischrus cingulata Studer, 1820 in Studer 1820: 87 (*Glischrus (Helix) cingulata*)

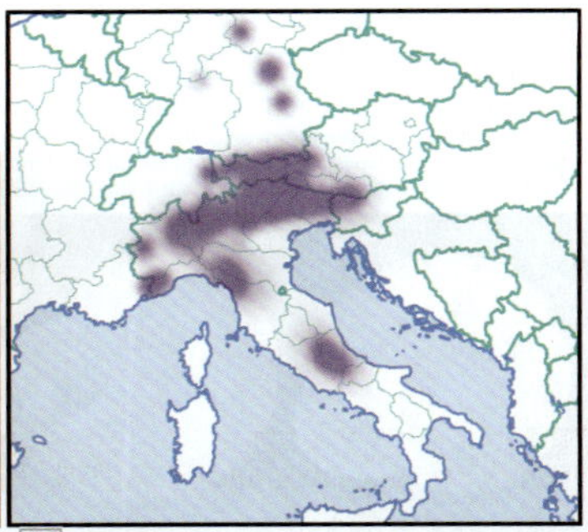

10-12 x 20-27 mm

N Italy: Sirmione near Lago di Garda

Alps, Apennines

Germany: Bayern, near Walchensee

Shell highly variable, whitish with a single brown band just above the periphery, 4.5-5 moderately convex whorls with moderate suture, last whorl descending sharply, aperture very oblique, margin disconnected at parietal side, umbilicus wide, 1/7 of shell diameter.
The animal is usually blackish or brown (can be light grey in forests) with long upper tentacles.

Exclusively on limestone rocks, only present in regions with extended limestone walls, mainly at S exposed slopes and walls. Adults prefer habitats where small spots of grasses or herbs interrupt the smooth rocky walls. In Switzerland in up to 1400 m altitude.

Clutches of 18-100 eggs (2.5-3.3 mm) are laid at intervals of some weeks in the summer season, juveniles hatch after 18-25 days, maturity is reached with full shell size after 1-2 years (1 year activity), life span 4-5 years.

Not abundant where present. Threatened by growing human settlements, many populations disappeared in the past 100 years.

Germain 1930: 221, Klemm 1951, Frömming 1954: 288, Prince 1967, Boato et al. 1984, Kerney et al. 1983: 274, Falkner 1990: 228, Manganelli et al. 1995: 31, Turner et al. 1998: 367.

Cingulifera. More than 30 so-called subspecies have been recognised until recently, of unknown bioscientific value. The populations require more research to learn more about the nature of the morphological variation, particularly if there are genetically induced reasons to account for it, or slow gene flow over area due to the fragmented habitats, without partial barriers to reproduction. Molecular studies could be helpful to show us the relationships, cross-breeding experiments of populations from different distances could allow us to interpret the molecular record and find the appropriate taxonomic conclusions.

Germany: Bayern, near Walchensee

Chilostoma desmoulinsii (Farines, 1834)

Helix desmoulinsii Farines, 1834 in Farines 1834: 5 (*Helix Desmolinsii*)

Spelling was *Desmolinsii* on pp. 5, 6 and 8. On p. 6 the author explained that the name was dedicated to "M. Charles des Moulins". Facing this correct orthography it remains unexplained why the taxnomic name was misspelled (o instead of ou) at 3 occasions. In the other three publications published shortly after in the same year 1834 the spelling of the taxonomic name was consistently *Desmoulinsii*, also in the publication in Bull. Soc. Philomath. Perpignan where the text of the monograph was almost literally copied. This suggests that it was probably an inadvertent error, to be corrected under Art. 32.5.
Falkner et al. 2002: 162 suggested that *Desmoulinsii* was an incorrect subsequent spelling, to be kept by prevailing usage under Art. 33.3.1.

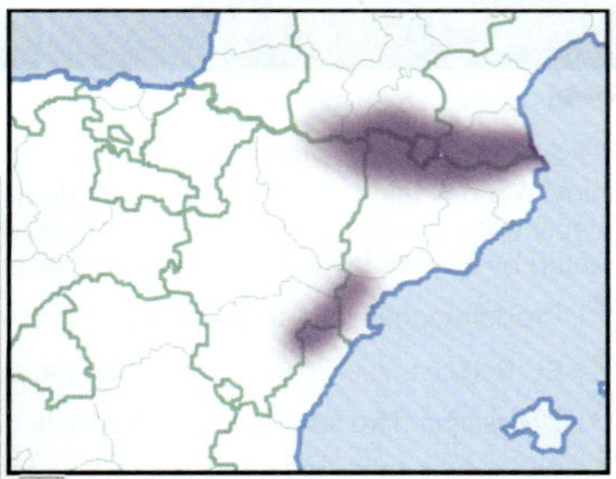

5-7 x 13-16 mm

Andorra: La Vella

Pyrenees, NE Spain

Westerlund 1889: 139, Germain 1930: 217, Kerney et al. 1983: 274, Bertrand 2002:19, Martínez-Ortí & F. 2010.

Spain: Pyrenees, Espot
Giuseppe Pocaterra

Shell yellowish to dark brown, with a faint trace of a dark colour band at the periphery, 5.5-6 weakly convex whorls with shallow suture, clearly angled at periphery, last whorl descending deeply near aperture, aperture very oblique, margin white and reflected, in adults connected by a prominent white layer at parietal side, also visible in weathered shells, umbilicus large, 1/6 of diameter, partly covered by the reflected columellar margin.
The animal is blackish with pale brownish foot margin.
Very much like *Ch. acrotricha*, with less rapidly increasing whorls, slightly darker, larger and less slender, edge at periphery less pronounced, without regularly arranged short hairs on upper and lower sides, less oblique aperture and wider umbilicus.

In rock crevices, between rock rubble, old walls, in mountains, only on calcareous substrate, 600-2800 m altitude.

Corneola. Probably not closely related to *Ch. acrotricha*, but their relationships require further research.

Chilostoma fontenillii (Michaud, 1829)

Helix fontenillii Michaud, 1829 in Michaud 1829: 267
Helix alpina Michaud, 1831 in Michaud 1831: 34, Pl. 14

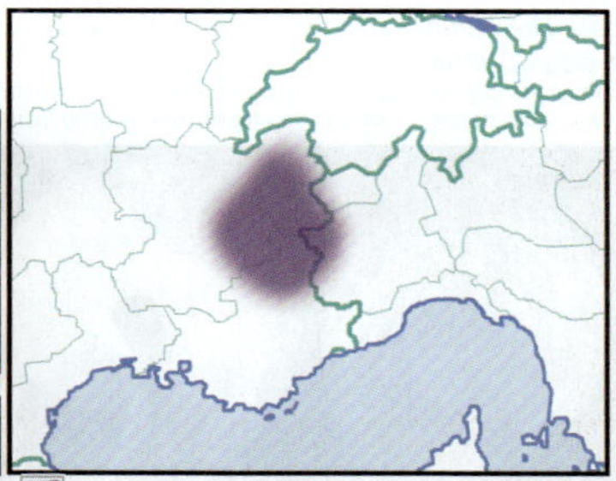

9-12 x 15-22 mm

France: Isère, Col de l'Arc

SW Alps

France: Isère, Grande Charteuse — Cédric Audibert

Shell yellowish white, often with spots and streaks, particularly at lower altitudes, irregularly striated, 5-5.5 slightly convex whorls with moderate suture, slightly keeled or shouldered at the periphery, apertural margin white and reflected, umbilicus wide.
Larger than *Ch. glaciale*, less flattened, more shouldered at the periphery and without spiral colour band.

Mountain habitats between 800 and 2800 m, calcareous meadows, shrubland and rocky areas. Restricted to limestone, usually in humid habitats, but in high altitudes above the timberline also at places exposed to sunshine. Very common where present.

Germain 1930: 218, Boato et al. 1984, Kerney et al. 1983: 176.

Delphinatia. Two subspecies have been recognised, their bioscientific value needs to be verified.

Chilostoma glaciale (Férussac, 1832)

Without gender agreement: *Chilostoma glacialis* (Férussac, 1832)

Helix glacialis Férussac, 1832 in A. Férussac & Deshayes 1819-1851: (explication planches) Pl. 67
Original print of explication des planches not verified, second print p. 13.

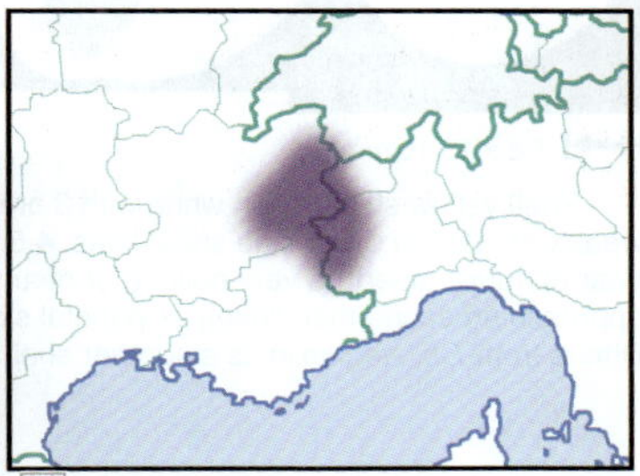

6-8 x 12-16 mm

France: Savoie, Mont Cenis

Shell opaque whitish with brown colour band just above the periphery, coarsely and irregularly striated, with very fine spiral lines, 4.5-5.5 convex whorls with moderate suture, last whorl slightly descending near aperture, aperture not very oblique, margin white and reflected, umbilicus wide, 1/6-1/5 of diameter.
Shell smaller than *Ch. fontenillii*, flatter, less shouldered at the periphery, *Ch. fontenillii* has no spiral colour band.

Rocky shrublands and rock rubble, below cristalline silicate substrate near wild creeks and spots of old snow which have not melted in July-September, usually in 1600-2100 m, up to 2500 m.

SW Alps

Germain 1930: 217, Boato et al. 1984, Kerney et al. 1983: 276, Turner et al. 1998: 368.

Delphinatia.

Chilostoma hirtum (Menke, 1830)

Without gender agreement: *Chilostoma hirta* (Menke, 1830)

Helix hirta Menke, 1830 in Menke 1830: 126
Established with description.

8-10 x 17-21 mm

Croatia: Between Rijeka and Senj

Shell horny greyish with a light brown band in a whitish zone, and 2 other very weak bands, thin, densely covered with short and solid whitish hairs, 4-5 regularly increasing flat whorls with deep suture, upper side nearly flat, lower side rounded, last whorl slightly descending near aperture, aperture with white lip inside, margin slightly reflected, umbilicus moderately narrow.

NW Croatia, S Slovenia

Cool and shady habitats among limestone rocks, in crevices or in leaf litter and decaying wood, in forests and shrublands.

Westerlund 1889: 131, Knipper 1939: 435, Bole & Slapnik 1997: 150.

Campylaea.

Chilostoma illyricum (Stabile, 1864)

Without gender agreement: *Chilostoma illyrica* (Stabile, 1864)

Helix umbilicaris umbilicaris [sic] *illyrica* Stabile, 1864 in Stabile 1864: 62 (*Helix (Campylaea (Corneola)) umbilicaris* typica (α) *Illyrica*)
Forst mentioned as an infrasubspecific name of the nominal subordinate variant. Used for a taxon by Westerlund 1889: 128 (*Helix planospira* var. *illyrica*) with bibliographical reference to Stabile 1864, making he name available under Art. 45.6.4.1.

13-15 x 26-29 mm

Germany: Bayern, Donaustauf near Regensburg

Shell horny brownish with a narrow reddish brown band inside a weak whitish zone, and two additional indistinct bands, finely granulated, 4.5-5 convex whorls with deep suture, last whorl width more than 2 x of penultimate whorl, slightly descending near aperture, margin reflected, umbilicus moderate and slightly covered by the reflected columellar margin.
Differs from *Ch. achates* in its less oblique aperture.

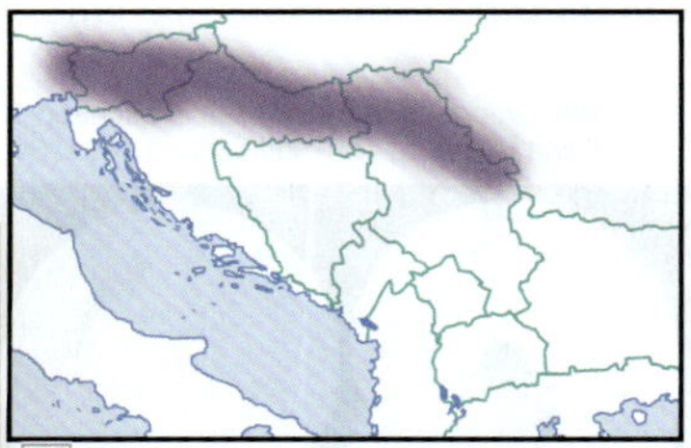

NE Italy, S Austria to S Slovenia
Introduced to near Regensburg, Germany in the 1800s.

In forests, on tree trunks, rocks and rock rubble, also among leaf litter.

Westerlund 1889: 128, Manganelli et al. 1995: 32, Bole & Slapnik 1997: 150.

Campylaea. Possibly conspecific with *Ch. padanum*. Habitat preferences not well known. Has often been confounded with *Ch. planospira*.

Chilostoma intermedium (Pfeiffer, 1828)

Without gender agreement: *Chilostoma intermedia* (Pfeiffer, 1828)

Helix intermedia Pfeiffer, 1828 in C. Pfeiffer 1828: 18

7-8 x 13-16 mm

Italy: Trentino, Valsugana near Tezze

Shell yellowish or bluish white with 3 brown bands, only the central band is usually not interrupted, solid and weakly striated and with fine spiral lines, 4-5 weakly convex whorls, last whorl suddenly and deeply descending near aperture, aperture very oblique or nearly horizontal, apertural margin reflected, more at lower side, margins approaching each other closely at parietal side, umbilicus open.
Differs from *Ch. ziegleri* in its smoother shell surface, regularly rounded periphery and smaller diameter.

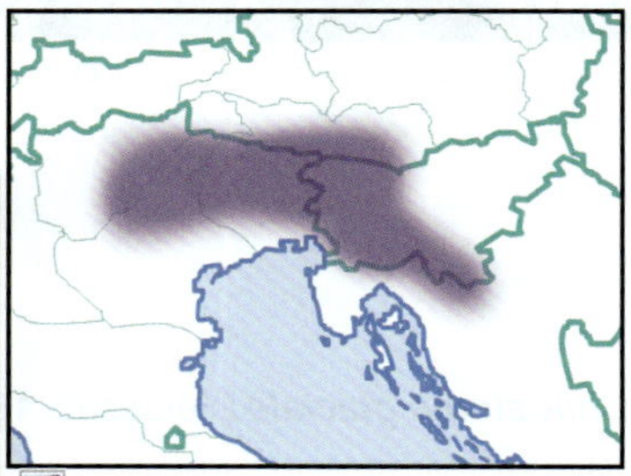

S Alps

Under limestones, in shrubs, at walls. Prefers humid habitats, in Slovenia on rocks and walls covered with moss.

Westerlund 1889: 122, Kerney et al. 1983: 176, Falkner 1990: 228, Bole & Slapnik 1997: 150.

Kosicia.

Chilostoma lefeburianum (Férussac, 1821)

Without gender agreement: *Chilostoma lefeburiana* (Férussac, 1821)

Helix lefeburiana Férussac, 1821 in A. Férussac [1821-1822]: 38
Established with short description.

Shell light horny greyish with a light brown band at the periphery, thin and covered with short yellowish and soft hairs in straight and densely arranged regular rows, 4-4.5 slightly convex whorls with deep suture, upper side nearly flat, lower side rounded, last whorl width more than 2 x of penultimate whorl, deeply descending near aperture, margin reflected and with white lip, umbilicus wide.
Differs from *Helicigona setosa* in its single colour band (not 3), and its apertural margin not being connected at the parietal side.

11-13 x 26-30 mm

Slovenia: Istria, near Trieste

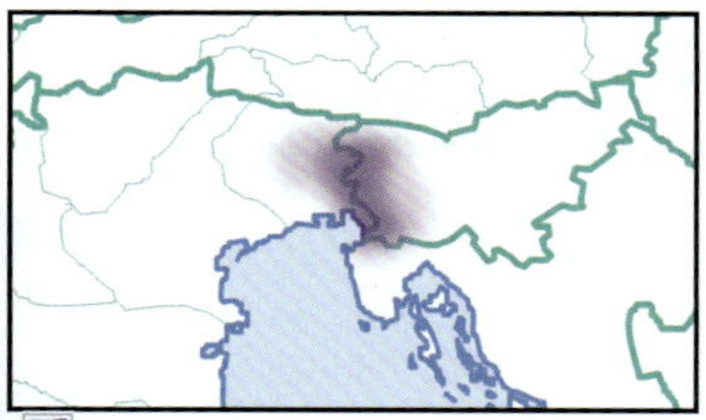

NE Italy, W Slovenia

Humid habitats on calcareous substrate, locally relatively abundant.

Westerlund 1889: 132, Manganelli et al. 1995: 32.

Campylaea.

Chilostoma millieri (Bourguignat, 1880)

Helix millieri Bourguignat, 1880 in Bourguignat 1880: 5 [not verified]

Shell olive greenish with a dark reddish band above the periphery, finely striated, fragile and shiny, 4-5 slightly convex whorls, last whorl initially with a weak edge at the periphery, suddenly and deeply descending near aperture, aperture very oblique, margin yellowish white and slightly thickened, margins approaching each other closely at parietal side, umbilicus wide but not deep.

8-10 x 18-22 mm

SE France: Alpes-Maritimes, Col de Fénestre, 2400 m

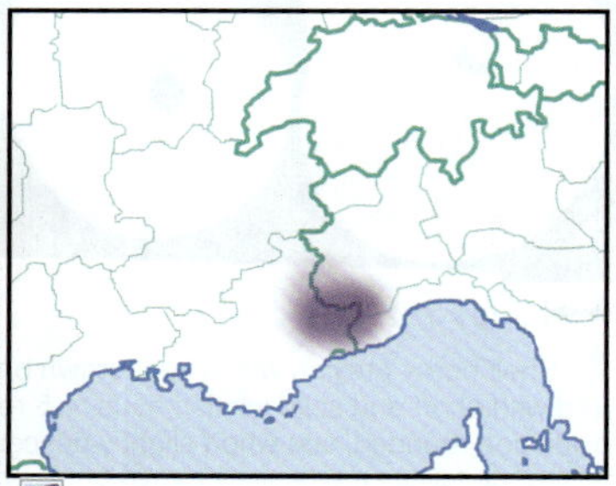

SW Alps

Vittorio Formenti

High mountain slopes under non-calcareous rocks covered by lichens, above 2000 m.

Westerlund 1890: 21, Germain 1930: 221, Alzona 1971: 195, Manganelli et al. 1995: 31. Not in Boato et al. 1984.

Subgenus unresolved, perhaps *Corneola* or *Chilostoma*. Seems to need a revision, possibly an altitudinal ecotype of *Ch. zonatum*.

Chilostoma padanum (Stabile, 1864)

Without gender agreement: *Chilostoma padana* (Stabile, 1864)

Helix umbilicaris padana Stabile, 1864 in Stabile 1864: 55, 62 (*Helix (Campylaea (Corneola)) umbilicaris Padana*)

Shell horny brownish with a narrow reddish brown band in a whitish zone, irregularly striated and with distinct spiral lines, sometimes shiny, 4.5-5 whorls, last whorl width 2 x of penultimate whorl, with convex lower side, descending near aperture, aperture rounded, umbilicus narrow and partly covered by the reflected columellar margin.

11-14 x 22-27 mm — Italy: Piemonte, Cuneo, Calcinero (?), 800 m

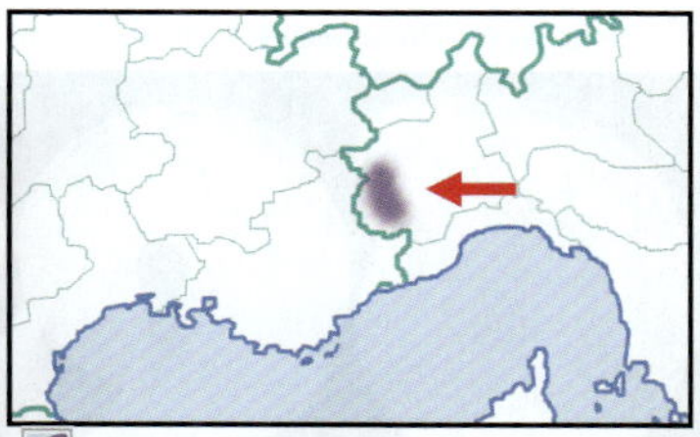

N Italy (Piemonte)

Westerlund 1889: 128, Boato et al. 1984, Manganelli et al. 1995: 32, 55, www.naturamediterraneo.com (2009).

Campylaea. Taxonomic status as a separate species doubtful, needs a revision. Closely related to *Ch. illyricum*.

Chilostoma planospira (Lamarck, 1822)

Helix planospira Lamarck, 1822 in Lamarck 1822: 78

10-14 x 20-33 mm — S Italy: Lazio, Velino valley near Posta

Italy

Shell horny brownish with some light and brownish bands, weakly striated and very finely and densely granulated, juveniles with hairs, adults often with scars, 4-5 weakly convex whorls with deep suture, last whorl width 2 x of penultimate whorl, descending near aperture, margin reflected and inside white, umbilicus wide, deep and regularly increasing.

Humid and shady rocky limestone habitats, frequently near springs, active at night.

Westerlund 1889: 127, Giusti 1973: 274 (Filicudi), Falkner 1990: 226, Manganelli et al. 1995: 32.

Campylaea. Various so-called subspecies of this locally variable species have been recognised, their bioscientific value remains to be verified.

Italy: Lazio, Monti Lepini — Alessandro Hallgass

Chilostoma sadlerianum (Rossmässler, 1838)

Without gender agreement: *Chilostoma sadleriana* (Rossmässler, 1838)

Helix sadleriana Rossmässler, 1838 in Rossmässler 1838-1844 (7-8): 29, Taf. 37 (*Helix Sadleriana*)
Helix planospira var. *istriana* Westerlund, 1889 in Westerlund 1889: 128

Shell horny reddish, often with a brown band at the periphery in a light yellowish zone, densely covered with short whitish hairs, 4.5-5.5 weakly convex and regularly increasing whorls, last whorl slightly and shortly descending near aperture, aperture inside brown, margin white and strongly reflected, outside yellow, umbilicus initially narrow, excentric, widened at last whorl to 1/6-1/5 of diameter, partly covered by the columellar margin.

14-17 x 25-31 mm — Croatia: Istria, Kanfanar, Dvigrad — Massimo Prodan

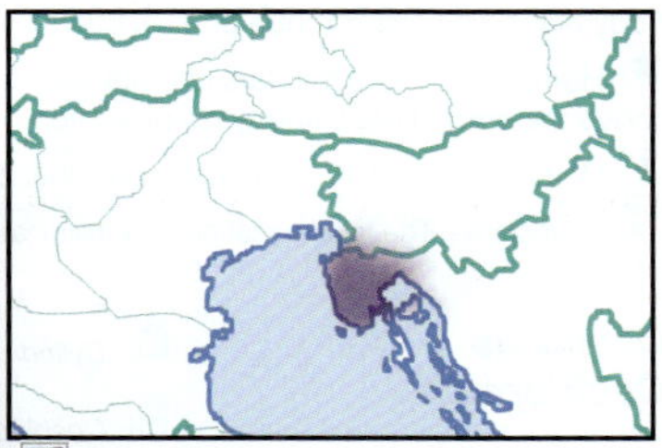

Croatia/Slovenia (Istria)

Westerlund 1889: 128. Classification based on P. Subai (unpublished studies).

Campylaea.

Chilostoma squammatinum (Moquin-Tandon, 1855)

Chilostoma squamatinum (Rossmässler, 1836)

Without gender agreement: *Chilostoma squammatina* (Moquin-Tandon, 1855)

Helix cornea synonym *Helix squamatina* [sic] mentioned by Rossmässler 1835-1837 (2) [1835]: 6
Mentioned with description as a synonym of *Helix cornea* Draparnaud, 1801 (not Linnæus, 1758). Also mentioned as *Helix Squammatina* as a synonym of *Helix cornea* Draparnaud, 1801 by Dupuy 1847-1852 [1848]: 155.
Helix cornea var. *squammatina* Moquin-Tandon, 1855 in Moquin-Tandon 1855: 134
First used as a name for a taxon by Moquin-Tandon 1855 with a short description and a reference to "var. α, Fér., Tabl. Syst., 1822, p. 42", but not to Rossmässler 1835 and not to Dupuy 1848.
For those who require a bibliographical reference for making available a synonym under Art. 11.6.1, the name must be attributed to Moquin-Tandon 1855, and spelled *squammatina*. For the others, to Rossmässler 1835, and spelled *squamatina*.

6-8 x 12-15 mm — Spain: Cataluña, Girona, Vidrà

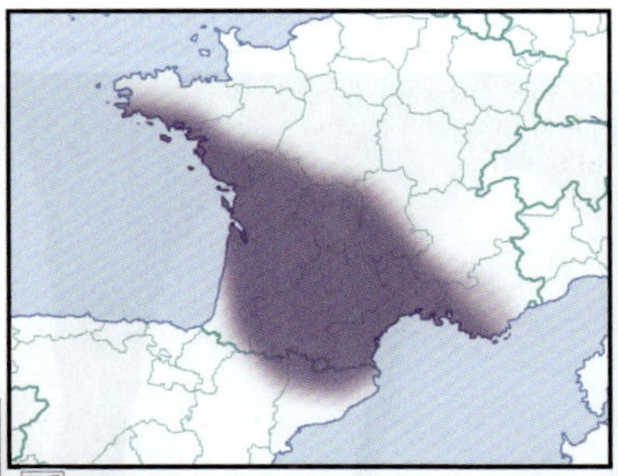

Pyrenees, France

Shell pale to intensively horny brown with reddish brown colour band, 5-6 moderately convex whorls with relatively deep suture, last whorl slightly angled at periphery, descending sharply and deeply near aperture, apertural margin white and reflected, disconnected at parietal side, umbilicus moderately wide, 1/9-1/8 of shell diameter.
Body blackish or greyish brown, tentacles blackish brown. Animal slow, easily irritated, but not very shy.

Corneola.

At rock outcrops in crevices and in rock rubble, also at walls, humid and shady habitats, also in forests. Appears at the surface usually only in wet weather.

25-35 eggs are laid in August and September under stones, maturity is reached in the second year.

Gassies 1849: 100, Moquin-Tandon 1855: 134, Germain 1930: 216, Kerney et al. 1983: 274, Falkner 1990: 228.

Chilostoma ziegleri (Rossmässler, 1836)

Helix ziegleri Rossmässler, 1836 in Rossmässler 1835-1837 (3): 4, Taf. 11 (*Helix Ziegleri*)

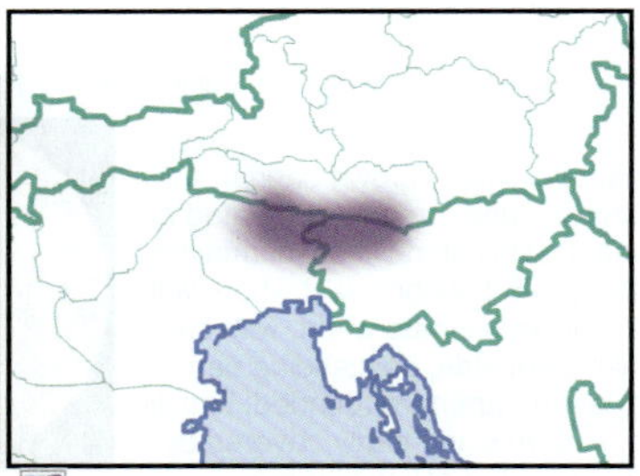

7-9 x 17-21 mm — Austria: Kärnten, near Eisenkappl

SE Alps

Shell light greyish or brownish with 2-3 fine and interrupted brown bands, not shiny, coarsely striated or almost ribbed on upper side, weaker on lower side, 4.5-5 weakly convex whorls with well marked suture, periphery weakly angulated, last whorl very deeply descending near aperture, aperture inside brownish, very oblique and with a weak white lip inside, margins reflected and appproaching each other closely at parietal side, umbilicus open and wide, nearly perspectivical.
Differs from *Ch. intermedium* in its ribbed shell and larger diameter.
Animal slender, light brownish grey, dorsal side nearly black, with a white longitudinal dorsal band, tentacles dark, sole dirty whitish, greyish brown at its margins.

Open and sunny limestone rocks in mountains, above 1000 m.

EU Red List: Near Threatened NT. Rare, but no acute threats known.

Kerney et al. 1983: 276, Falkner 1990: 228.

Kosicia.

Chilostoma zonatum (Studer, 1820)

Without gender agreement: *Chilostoma zonata* (Studer, 1820)

Glischrus zonata Studer, 1820 in Studer 1820: 87 (*Glischrus (Helix) zonata*)
Glischrus foetens Studer, 1820 in Studer 1820: 87 (*Glischrus (Helix) foetens*)
First Reviser Férussac 1821-1822: [quarto] 38 [26 May 1821], selected *zonata*.

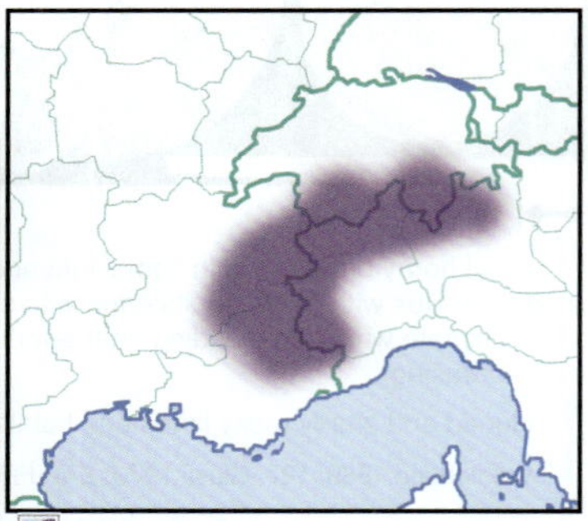

9-13 x 18-26 mm — Italy: Novara, Monte Mottarone near Omegna

W Alps

Shell yellowish to greenish brown, with a brown colour band at the periphery, 4.5-5 moderately convex whorls with moderate suture, last whorl rounded and shortly descending near aperture, aperture not very oblique, margin white and reflected, connected at parietal side by a very thin whitish layer, umbilicus open, 1/10 of diameter, to a small extent covered by the reflected columellar margin.
Differs from *Ch. achates* in its usually larger shell with only one colour band, umbilicus narrower.
Body black with reddish hue or dark brown, sole yellowish grey with greenish hue, tentacles 12 mm long. Animal slow, easily irritated, but indifferent to light.

Humid and rocky habitats in moutains, often at non-calcareous silicate rocks and near water bodies, not at open rock faces. In 800-2900 m, but rarely above the timberline. Undergoes predation by the carabid *Cychrus cordicollis* in high alpine habitats.

Moquin-Tandon 1855: 132, Germain 1930: 220, Boato et al. 1984, Kerney et al. 1983: 275, Falkner 1990: 228, Manganelli et al. 1995: 31, Turner et al. 1998: 362.

Chilostoma. The local variation was associated with 3 so-called subspecies, their bioscientific value needs to be verified.

Cylindrus
Cochlopupa

Cylindrus Fitzinger, 1833 in Fitzinger 1833: 107 — Type species *Pupa obtusa* Draparnaud, 1805 (monotypy)

Cochlopupa Jan, 1830 in Jan 1830: 5 *Helix (Cochlopupa)* — Type species *Pupa obtusa* Draparnaud, 1805 (monotypy)

Cylindrus Fitzinger, 1833 is preoccupied: not *Cylindrus* Batsch, 1789 [in Batsch 1789: 692] (Gastropoda: Conidae, currently used).
Schileyko 2006: 1786 proposed to ask the Commission, in the belief that the correct spelling of the conid genus was *Cylinder* (as used by Montfort 1810: 390). The Batsch 1789 publication was found later (Dubois & Bour 2010: 170). The homonymy has long been known (Kennard 1942: 114). This case seems like the *Trichia* case. If nothing else interferes I would suggest to use *Cochlopupa*.

Cylindrus obtusus (Draparnaud, 1805)
Cochlopupa obtusa (Draparnaud, 1805)

Pupa obtusa Draparnaud, 1805 in Draparnaud 1805: 63
G. obtusa in the original description should be seen as a misprint for *P.* (= *Pupa*) *obtusa*.

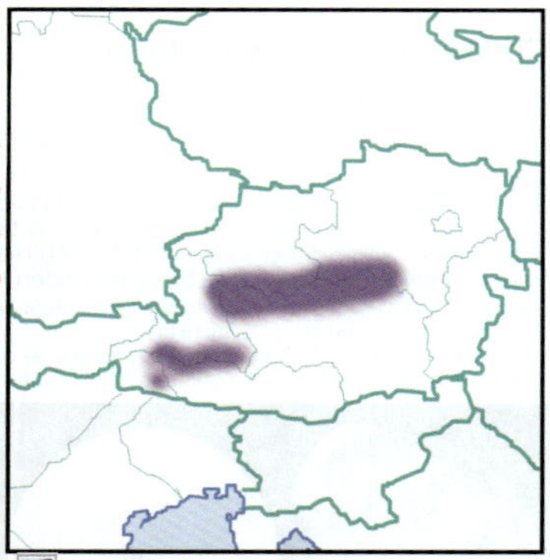

Austrian Alps

Austria: Oberösterreich, Kalkalpen, Hoher Nock
Jan Steger

11-13 x 4-4.5 mm — Austria: Niederösterreich, Ötschergipfel, 1850 m

Shell looks like an orculid, cylindrical, but white and solid, no teeth in the aperture. The animal is big and dark.

Between 900 and 2680 m, in snow valleys and under stones, usually at N exposed slopes, requires humid habitats and limestone, the snails prefer to rest at vegetation and soil, less at rocks. Mean dispersal rate 3 cm/day.

Slow growth rates, snails need several years to become adult.

Klemm [1973], Mildner & Kofler 1990, Falkner 1990: 230, Bisenberger et al. 1999.

Distinct and isolated populations, snails are abundant but the habitats are small.

Drobacia

Drobacia Brusina, 1904
in Brusina 1904: 162

Type species *Helix banatica* Rossmässler, 1838
(original designation)

Drobacia banatica (Rossmässler, 1838)

Helix banatica Rossmässler, 1838 in Rossmässler 1838-1844 (7-8): 14, Taf. 33
Helicigona maeotica Wenz, 1926 in Krejci & Wenz 1926: 60, Taf. 2

12.5-18 x 23-30 mm

Hungary: near Vásárosnamény

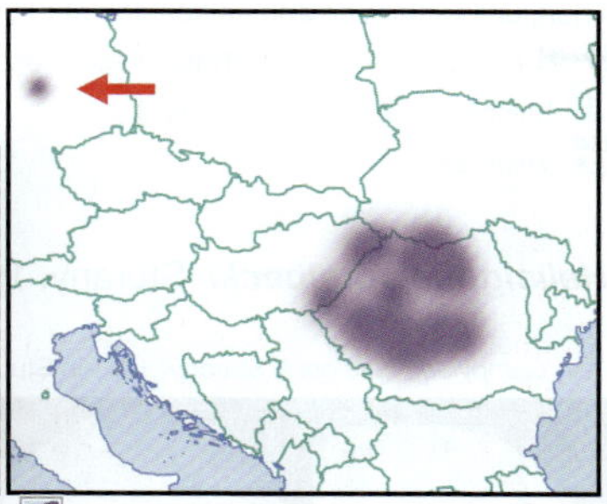

SE Europe

Romania: Valle Butii, Retezat du Mic — Jozef Grego

Shell olive reddish with a darker brown band above the keeled periphery, striated and with dense spiral lines producing a reticulate pattern, 5-5.5 flattened whorls without deep suture, keeled at all whorls, regularly increasing, last whorl weakly descending near aperture, margin rounded (not considering the keel), reflected and with white lip, umbilicus moderately narrow and partly covered by the reflected columellar margin. Shape and size locally variable.
Animal dark greyish or brown.

Westerlund 1889: 116, Lupu 1966, Shilejko 1978: 317, Grossu 1983, Sysoev & Schileyko 2009: 164.

Lengyel & Páll-Gergely 2010: 97 considered the smaller *maeotica* from Stîna de Vale at 1100 m altitude as a synonym. Transitional forms between both parapatrically distributed taxa were found and there were no anatomical differences to justify a status as a separate species *D. maeotica*. Grossu 1983 demonstrated that both taxa differed in their electrophoretical patterns, but this does not necessarily need to be explained by the presence of reproductively isolated units.
The observed transitional forms in the field do not support the presence of two separate species.

Helicigona

Helicigona Férussac, 1821
in A. Férussac [1821-1822]: 23 (folio p. 27), expl. planches viij

Type species *Helix lapicida* Linnæus, 1758
(subsequent designation by Pilsbry 1894 [1895]: 296, probably invalid)

Established with short description and without species included, 06 Apr 1821 (livraison 9), and on explication planches p. viij [06 Apr 1821] with two species included (*Helix listeri* and *Helix pyrostoma* Férussac, 1821). More species were included on p. 36 (26-05-1821), among these was *H. lapicida*. I do not know where *lapicida* was included before that date. Kennard 1942: 115 proposed to regard *H. lapicida* as its type, but demonstrated that this nominal species was not originally contained. The type species *H. lapicida* should probably be fixed under the plenary powers.

Dinarica Kobelt, 1902 in Kobelt 1902: 60 *Campylaea (Eucampylaea (Dinarica))*, type species *Helix dinarica* Bourguignat, 1888 (absolute tautonymy)

Faustina Kobelt, 1904 in Kobelt 1904: 131, 186 *Campylaea (Faustina)*, type species *Helix faustina* Rossmässler, 1835 (absolute tautonymy)
On p. 131 mentioned without description and with *H. faustina* included. On p. 186 *H. faustina* and *H. moellendorffi* were included.

Liburnica Kobelt, 1904 in Kobelt 1904: 76, 131, 154, 185 *Campylaea (Liburnica)*, type species *Helix setosa* Férussac, 1832 (subsequent designation by Boettger & Wenz 1921: 12)
On p. 131 and 154 mentioned without description but with *H. setosa* included (p. 154 "Gruppe der *setosa*"). On p. 185 thirteen species were included. Type not by original designation, Kobelt's work contained a disclaimer (Art. 8.3) (Kobelt 1904: 145-147) that he did use the term "Typus" in the sense of a unique name-bearing type.

Thiessea Kobelt, 1904 in Kobelt 1904: 76, 131, 154, 187 *Campylaea (Thiessea)*, type species *Helix cyclolabris* Deshayes, 1839 (subsequent designation by Boettger & Wenz 1921: 12)
On pp. 76, 130, 154 mentioned without description but with *H. cyclolabris* included. On p. 187, six species were included. Type not by original designation, Kobelt's work contained a disclaimer (Art. 8.3) (Kobelt 1904: 145-147) that he did use the term "Typus" in the sense of a unique name-bearing type.

Cattania Brusina, 1904 in Brusina 1904: 162, type species *Helix trizona* Rossmässler, 1834 (original designation)

Campylaeopsis Sturany & Wagner, 1914 in Sturany & Wagner 1915: 44, type species *Helix moellendorffii* Kobelt, 1871 (subsequent designation by Boettger & Wenz 1921: 12)

Ariantopsis Wagner, 1928 in A. J. Wagner 1928: 377 *Campylaea (Ariantopsis)*, type species *Helicigona pelia* Hesse, 1912 (monotypy)

Wladislawia Wagner, 1928 in A. J. Wagner 1928: 379, type species *Campylaea polinskii* Wagner, 1928 (original designation)

Josephinella Haas, 1936 in Haas 1936: 130 *Campylaea (Josephinella)*, type species *Helix hemonica* Thiesse, 1884 (original designation)

Superba Subai & Fehér, 2006 in Subai & Fehér 2006: 206, type species *Helicigona skipetarica* Subai, 1995 (original designation)

Here used mainly for communicational purposes and not because the genus would be monophyletic. It comprises *lapicida* and all SE European species south of NW Croatia. The EU Red List (Cuttelod et al. 2011) used 5 genera *Chilostoma* (71 species), *Helicigona* (1), *Cattania* (6), *Faustina* (4), and *Superba* (5). Perhaps a peliminary system, the newly inserted species in *Chilostoma* did not have the neuter gender endings.

Helicigona albanica Jaeckel, 1954

Helicigona stenomphala albanica Jaeckel, 1954 in Jaeckel 1954: 87
Not *Campylaea (Dinarica) montenegrina* var. *albanica* Kobelt, 1902 in Kobelt 1902: 69, Taf. 266 (here classified in *Helicigona*, so a senior secondary homonym)
On p. 69 and on Taf. 266 classified in genus *Campylaea*, in the index p. 85 in the same work in genus *Helix*.

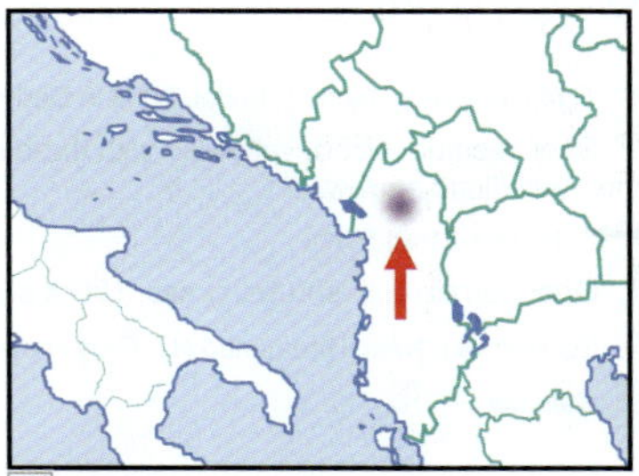

N Albania

Shell evenly light yellowish without colour bands, shiny, without hairs, finely striated, aperture with white layer at the parietal side, margin slightly reflected, last section of palatal side weakly ascending towards the penultimate whorl to which it is attached, umbilicus narrow and partly (50 %) covered by the curved columellar margin of the aperture.

24-26 x 45-49 mm

Subai 1995: 92.

Dinarica.

Helicigona albanograeca Subai, 1995

Helicigona albanograeca Subai, 1995 in Subai 1995: 85

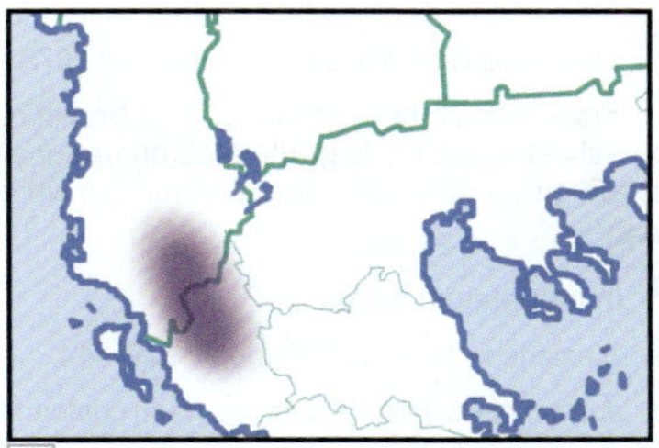

NW Greece, S Albania

Shell dirty white to light horny yellowish, with 3 brown bands, granulated and with fine growth lines, often shiny, 4.5-5.5 regularly increasing whorls, last whorl width up to 2.5 x of penultimate whorl, descending near aperture, apertural margins approach each other closely, occasionally detached, umbilicus deep and perspectivical, initially narrow, considerably widened at last 1/2 whorl, to 1/8-1/7 of diameter.
The animal is evenly light grey, sole cream with slightly darker margins, mantle brownish grey.

11.5-16 x 24.5-34.5 mm

Subai 2002: 13.

Subgenus unresolved.

Helicigona amorgia (Westerlund, 1889)

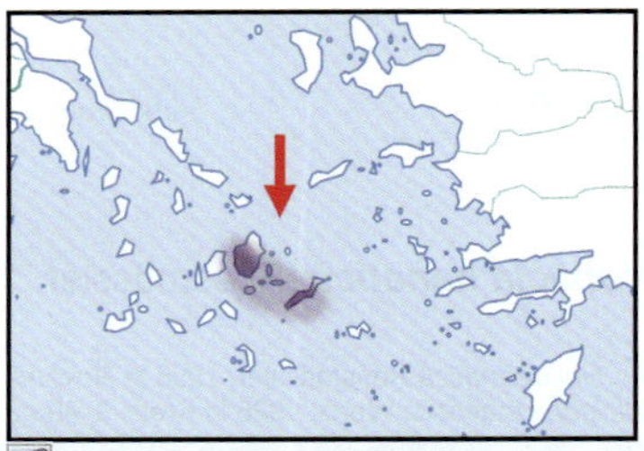

Greece (Amorgós, Náxos)

Helix cyclolabris var. *amorgia* Westerlund, 1889 in Westerlund 1889: 138

Shell flat, horny brownish (in Naxos lighter) with one weak colour band (in Naxos 1 mm wide), granulated with corrugated radial streaks, characteristic hair sculpture with hairs on scars next to depressions, hairs 0.7-1.5 mm long, no hairs above the brown colour band, 1-3 growth interruptions, last whorls width 2 x of penultimate whorl, umbilicus wide and perspectivical (3.5-4.5 mm).

6-11.5 x 13-23 mm (on Náxos larger than on Amorgós)

Knipper 1939: 442, Subai 1996: 4.

Thiessea.

Helicigona apfelbecki (Sturany, 1901)

Campylaea apfelbecki Sturany, 1901 in Sturany 1901: 65

Montenegro: Jeserska Kaar, Durmitor

12-15 x 20-25 mm

Shell basically yellowish brown with coarse white surface layers, sometimes with a weak dark brown colour band on the upper side, striated and with some marked growth interruptions, 4-4.5 weakly convex whorls, last whorl weakly descending, aperture with white lip, inside brownish yellow, margin slightly reflected, sometimes connected at parietal side by a thin layer, umbilicus wide.

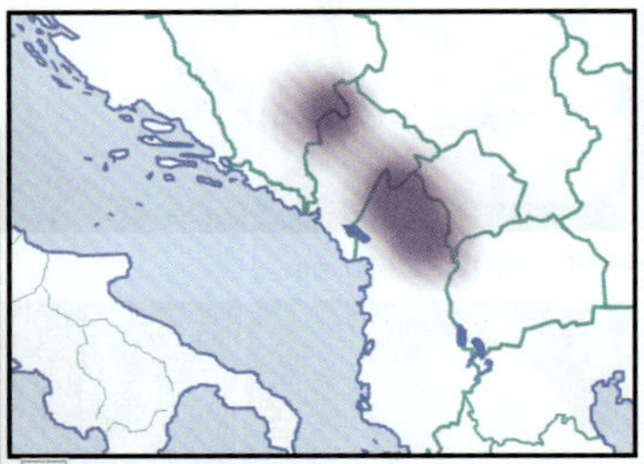

Bosnia, Montenegro, N Albania

Rocky limestone habitats.

Subai 1997: 523.

Subgenus uncertain. Could be two separate species, more research on their relationships and ranges is needed.

Helicigona arcadica (Pfeiffer, 1853)

Helix cyclolabris synonym *Helix arcadica* Pfeiffer, 1853 in L. Pfeiffer 1853 (Mon. Helic. 3): 236 Type locality was given as Tínos.

Mentioned as a synonym of *Helix cyclolabris* Deshayes, 1839, without description or cited types, but with some bibliographical references to earlier descriptions of *Helix cyclolabris*.

Frauenfeld 1867: 777 used the name *Campylaea arcadica* for a species from "Tínos" and gave a description and bibliographical reference to Pfeiffer 1853, making the name available from Pfeiffer 1853 under Art. 11.6.1.

Mousson 1854: 8-9 mentioned *Helix arcadica* as a synonym of *Helix heldreichi* Pfeiffer, 1846 from near Athína, without description for *arcadica* (but yes for *heldreichi*), with locality "Arcadie" for *arcadica*, in the chapter on the results of an expedition to Síros. Frauenfeld 1867 did not refer to Mousson 1854.

Greece: Síros, W of Ano Síros

9.5-12 x 18.5-22.5 mm

Shell horny coloured, depressed, with a brown colour band between two light bands, surface weakly granulated, 1-3 growth interruptions strongly depressed near suture (weakly in *Hcg. melpomene*), hairs yellowish and 0.3-0.5 mm long, last whorl width 1.5-2 x of penultimate whorl (in *Hcg. melpomene* more than 2 x), aperture slightly detached, with sharp margin, umbilicus wide (3.5-5 mm).
Differs anatomically significantly from *Hcg. melpomene*.

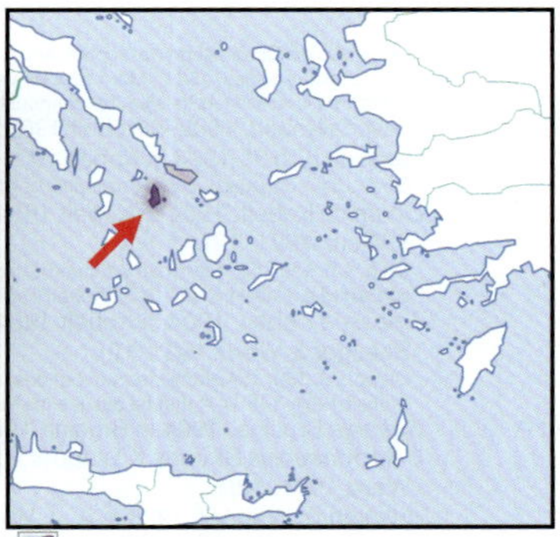

Greece (Síros)

Widespread in various different kinds of habitats on the island.

Subai 1996: 6.

Thiessea. The name *arcadica* is used for an endemic species from Síros, but Frauenfeld's type specimens were reported to have come from Tínos.

Helicigona ardica (Dedov & Subai, 2006)

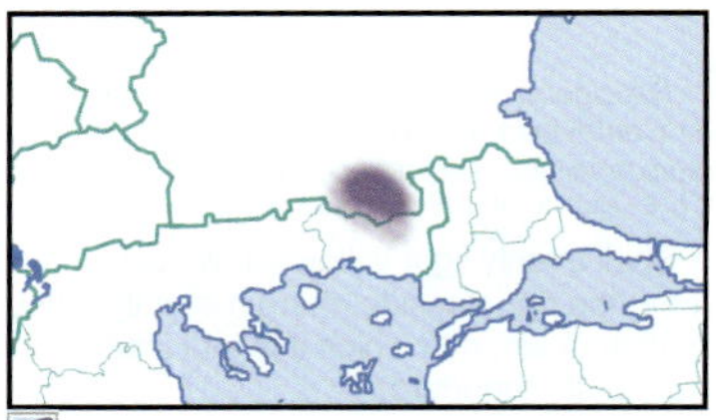

S Bulgaria

Cattania ardica Dedov & Subai, 2006 in Dedov & Subai 2006: 226

Shell medium-sized, yellowish with three brown spiral colour bands, apertural insertions distant from each other, umbilicus narrow.

10-14 x 21-24.5 mm

Open terrain with shrubs, grass, *Urtica* and some trees.

EU Red List: Near Threatened NT. Only known from a small area and few localities close to human settlements.

Cattania.

Helicigona argentellei (Kobelt, 1872)

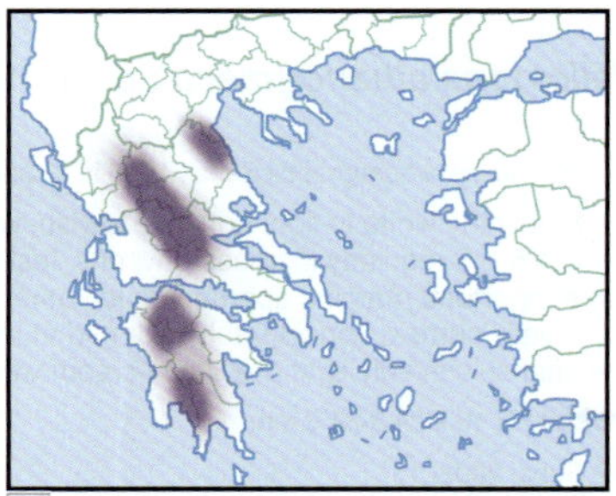

Greece (Pelopónisos)

Helix argentellei Kobelt, 1872 in Kobelt 1872: 49 (*Helix Argentellei*)

Shell reddish horny brown with a brown band above the periphery in a broad light yellow zone, translucent, coarsely striated, finely granulated on upper side, no hairs, 4-5 weakly convex whorls, last whorl suddenly descending near aperture, apertural margin white, triangular at columellar side, umbilicus open, 1/5 of diameter.

9-13 x 18-25 mm

In mountain habitats.

Westerlund 1889: 117, Knipper 1939: 414.

Josephinella. This species has been confounded with other forms in the literature, reports from other Greek regions referred to other species.

Helicigona bacchica (Martens, 1889)

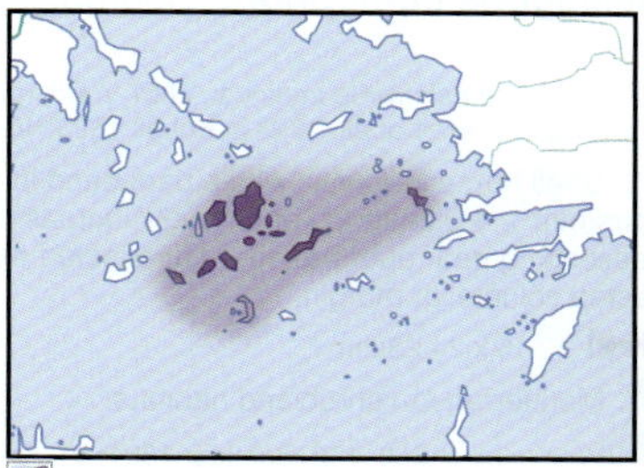
Greece (Kikládes)

Helix cyclolabris var. *bacchica* Martens, 1889 in Martens 1889: 182

Shell depressed, dirty whitish to horny with a brown colour band and two light bands, shiny, finely granulated and irregularly striated (smooth in Síkinos), with weak hair scars on the first whorls, last whorl width at least 2 x of penultimate whorl, umbilicus initially narrow (wider in *Hcg. nympha*), becoming gradually larger, at last whorl 2.5-4.5 mm.
Shells with and without hairs can occur in the same population.
Differences to *Hcg. nympha* mainly anatomically.

8-14.5 x 15-29 mm

Knipper 1939: 446, Subai 1996: 8.

Thiessea. An extinct form of Thira belongs possibly to this species.

Helicigona balcanica (Kobelt, 1875)

Helix trizona var. *balcanica* Kobelt, 1875 in Kobelt 1876: 11
Nomenclature not finally resolved. Was previously mentioned (Pfeiffer 1853: 232) with a brief description as a synonym of *Helix hispana* Linnæus, 1758. Kobelt 1875 added a note that *balcanica* had been classified as a variety of *H. planospira* in Pfeiffer's Monographia Heliceorum, but did not give a precise bibliographical reference. For those who require a bibliographical reference for making available a name under Art. 11.6.1 the name must probably be attributed to Kobelt 1876. For the others, to Pfeiffer 1853 or another previous source where the synonym could have been mentioned.

Bulgaria: Karlukovo

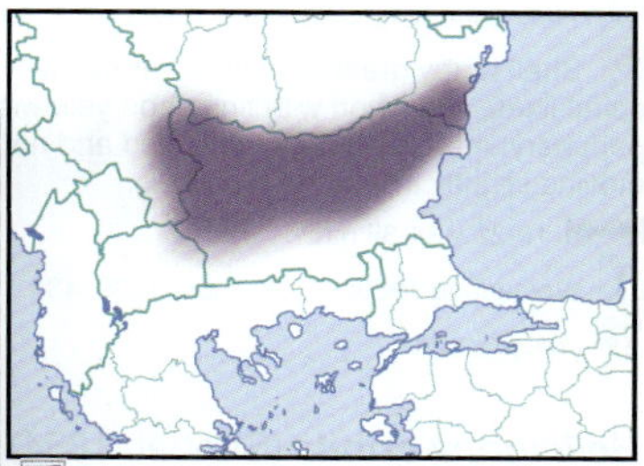
Bulgaria, SE Serbia, S Romania

12.5-16.5 x 22-30 mm

Shell yellowish green, lighter near suture, with blurry bands, prominently striated, 4.5-5 slightly convex whorls, last whorl descending near aperture, aperture with white lip inside, margin thin and broadly reflected, umbilicus not very wide and partly covered by the reflected columellar margin.

Westerlund 1889: 137, Knipper 1939: 399, Damjanov & Likharev 1975: 380, Dedov 1998, Irikov & Mollov 2006: 812.

Cattania.

Helicigona brenskei (Boettger, 1883)

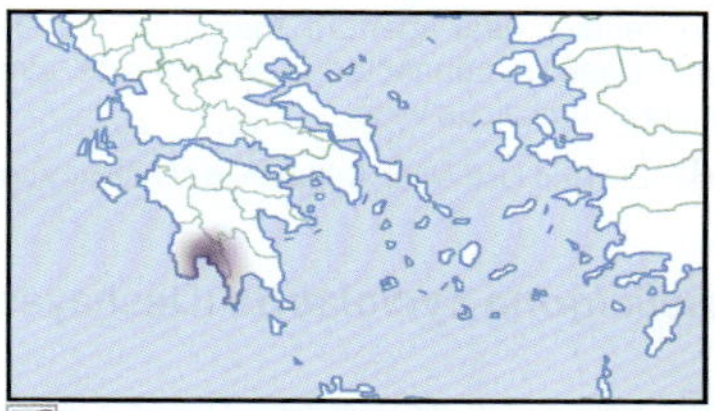
Greece (S Pelopónisos)

Helix brenskei Boettger, 1883 in O. Boettger 1883: 335 (*Helix (Campylaea) Brenskei*)

Shell horny brown with reddish hue, with one brown band and a light yellowish zone below it, prominently striated, densely covered with fine and short, light brownish hairs (16-25 hairs/mm^2), 5-5.5 strongly convex whorls with impressed suture, last whorl wider and descending near aperture, apertural margin whitish and slightly thickened, and distinctly reflected, umbilicus moderate and half covered by the reflected columellar margin.

16-20 x 26-30 mm

Westerlund 1889: 126, Boettger 1891: 84.

Josephinella.

Helicigona byshekensis Knipper, 1941

Helicigona byshekensis Knipper, 1941 in Knipper 1941: 181
Helicigona edlaueri Knipper, 1941 in Knipper 1941: 184

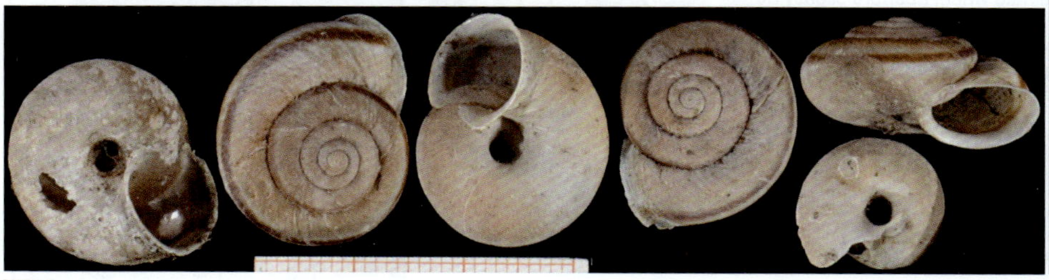
Albania: Mali Gradecit near Çorovodë

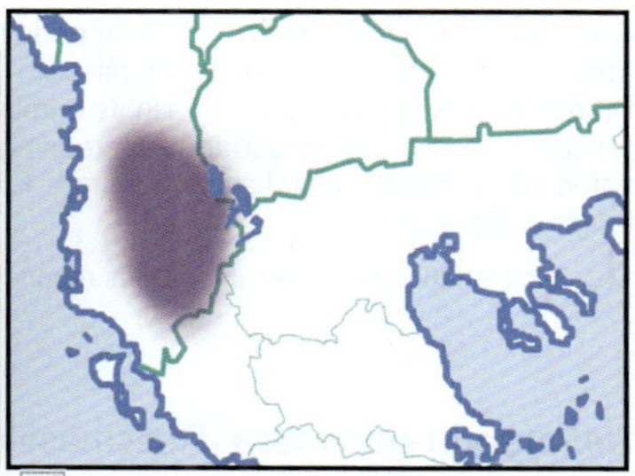
Central Albania to NW Greece

11.5-13.5 x 19-23 mm

Shell evenly light brown with a 2 mm broad brown band above the periphery and occasionally a 3 mm broad light zone below, weakly striated with coarsely arranged short hairs (1.7-2.5/mm^2), 4-4.5 whorls with deep suture, last whorl regularly rounded, gently descending near aperture, apertural margin reflected and without lip inside, margins distant from each other and connected by a whitish callus, umbilicus moderately narrow.

Dhora & Welter-Schultes 1996: 162, Fischer 2003.

Josephinella. *Hcg. edlaueri* seems to differ only weakly in a wider umbilicus and smaller size, either aberrant specimens or (more likely) within the usual variation. Knipper 1941 had only few specimens of these forms.

Helicigona cingulella (Rossmässler, 1837)

Helix cingulella Rossmässler, 1837 in Rossmässler 1835-1837 (5-6): 40, Taf. 27

Shell solid, dirty to yellowish white, pale shiny, with brown color band at periphery, irregular blunt radial streaks, aperture very oblique, umbilicus wide.

Limestone rocks in montane to alpine zones.

Kerney et al. 1983: 278.

Faustina.

Slovakia: Higher Tatra, Vratna valley

6.5-8 x 15-18 mm

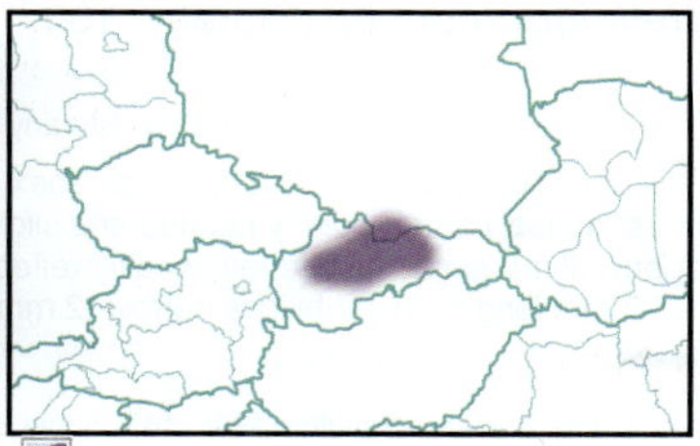
S Poland, Slovakia

Helicigona comephora (Bourguignat, 1857)

Helix comephora Bourguignat, 1857 in Bourguignat 1857: 546, Pl. 16 Type locality Palea Messini, S Pelopónisos.
Campylaea eliaca Kobelt, 1893 in Kobelt 1893: 46 Type locality in NW Pelopónisos.

Shell horny yellowish with a dark band in a whitish zone, thin, with coarsely arranged blackish hairs (4 hairs/mm^2), 4-5 flattened and rapidly increasing whorls, upper side flat, lower side convex, last whorl strongly descending near aperture, margin reflected except upper section, umbilicus narrow, half covered by the reflected columellar margin.

12-16 x 24-28 mm

Mediterranean shrubland habitats.

Westerlund 1889: 132, Boettger 1891: 84, Knipper 1939: 419 (*eliaca*) and 435 (*comephora*, misspelled *comytophora*).

Josephinella. The northern and southern forms are regarded as conspecific here, their descriptions provide no substantial differences. It is unclear why Knipper 1939 did not compare the two forms. More research is needed.

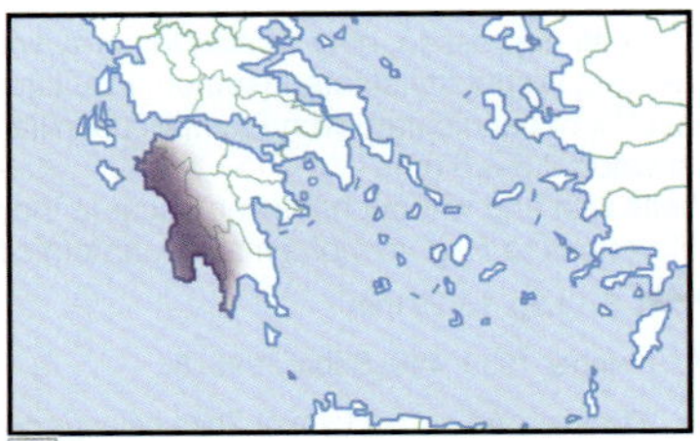

Greece (Pelopónisos)

Helicigona conemenosi (Boettger, 1885)

Helix conemenosi Boettger, 1885 in O. Boettger 1885: 120 (*Helix (Campylaea) Conemenosi*)

Shell horny greenish with a narrow dark brown band, first 2 whorls granulated, the other whorls with fine spiral lines, first whorl with fine short yellowish hairs, last whorl width 2 x of penultimate whorl, well rounded at periphery, margins broadly reflected and not approaching each other closely, umbilicus open, columellar margin moving slightly inside the umbilicus.

19-22 x 34-38 mm

Westerlund 1889: 119, Knipper 1939: 434.

Josephinella.

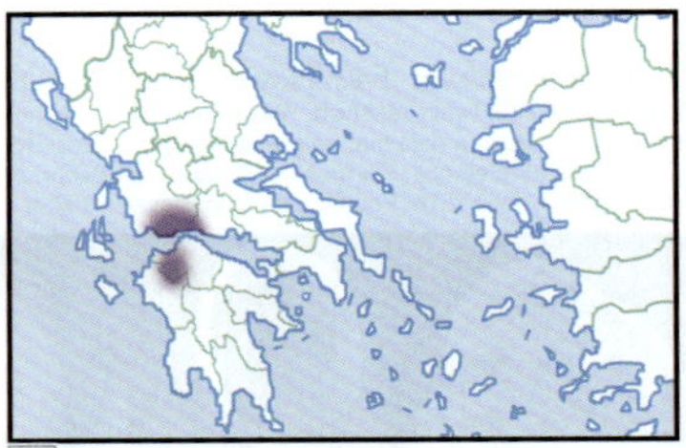

W Central Greece

Helicigona crinita (Strobel, 1854)

Helix crinita Strobel, 1854 in Strobel 1854: 108

Shell dirty horny brownish with 3 indistinct reddish brown bands, thin and solid, finely granulated and coarsely striated on upper side, with long yellowish hairs, 4.5-5 flattened and regularly increasing whorls with deep suture, upper side nearly flat and lower side rounded, last whorl with blunt edge at periphery, suddenly and deeply descending near aperture, margin white, reflected and with weak lip, connected at parietal side, almost detached, umbilicus wide.

12-13.5 x 26-31 mm

Croatia: Dalmatia, Vrlika
Cédric Audibert

EU Red List: Near Threatened NT. Only known from few localities, no real threats known.

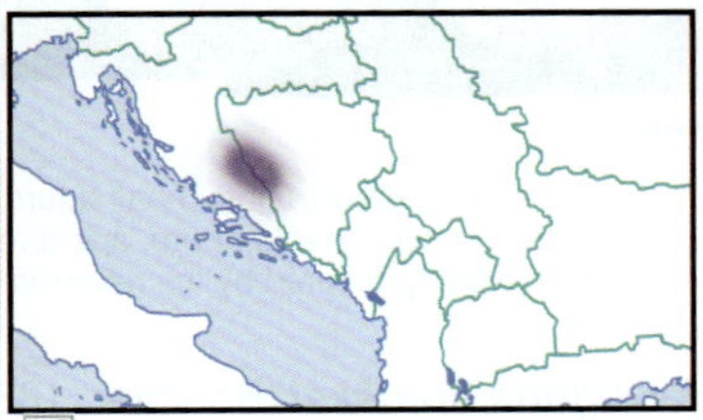

Croatia, Bosnia and Herzegovina

Westerlund 1889: 133, Knipper 1939: 422, Subai 2002: 15.

Liburnica.

Helicigona cyclolabris (Deshayes, 1839)

Helix cyclolabris Deshayes, 1839 in A. Férussac & Deshayes 1819-1851: (2) p. 32, (explication planches) Pl. 68
Original print of explication des planches not verified, second print p. 13 (*Helix cyclolabris*). If the name appeared on the original print of explication des planches, *cyclolabris* should be attributed to Férussac, 1832.
The initial type sample contained 5 species, none of which (including the lectotype) from the region originally given as type locality ("Morée" and islands near Náxos, but should be Vórii Sporádes, possibly Morée was meant to include entire Greece at the time, Subai 1996: 13).

Shell horny brownish with a brown colour band and two indistinct lighter bands, surface finely granulated, on last whorls weakly striated, short fragile hairs (0.1-0.25 mm long) on the first whorls, hair scars on the last whorl not well visible, last whorl width 1.5-2 x of penultimate whorl, aperture slightly detached, umbilicus wide and perspectival.
Differs from *Hcg. sphaeriostoma* in its umbilicus increasing regularly in width (in *sphaeriostoma* more suddenly increasing at last whorl), last whorl less inflated, and anatomically (flagellum length 1.5 x of (penis + epiphallus), dart length 4.6-5.4 mm).

Subai 1996: 10 (revised all previous reports).

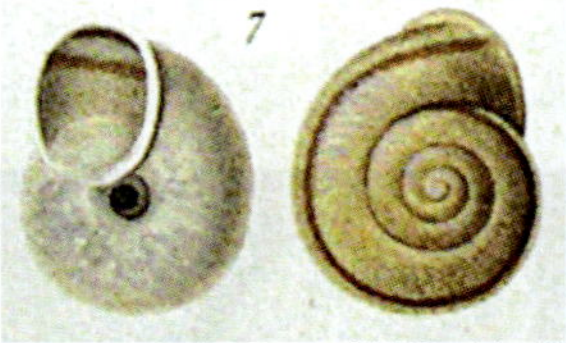

Greece: Vórii Sporádes (?)
Picture: Férussac 1819-1851: Pl. 68

8.5-13 x 17-23 mm

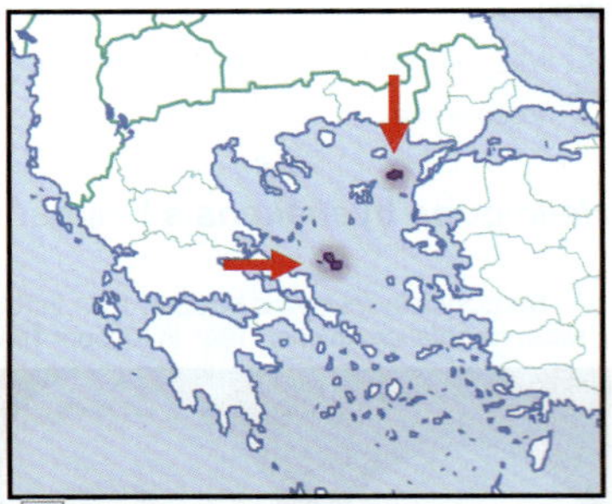

N Aegean islands (Skíros, Gõkçeada)

Thiessea.

Helicigona denudata (Rossmässler, 1836)

Helix denudata Rossmässler, 1836 in Rossmässler 1835-1837 (4): 3, Taf. 16

Shell horny yellow with 3 dark brown bands, striated, solid, 4.5-5 slightly convex whorls, rapidly increasing, last whorl suddenly and deeply descending near aperture, apertural margin reflected and with weak lip, connected at parietal side and detached, with a blunt tooth at basal margin, umbilicus wide.

13-16.5 x 27-32 mm

High mountain habitats, mostly above 1700 m, in beech forests and above the timberline.

EU Red List: Near Threatened NT. No real threats known.

Liburnica.

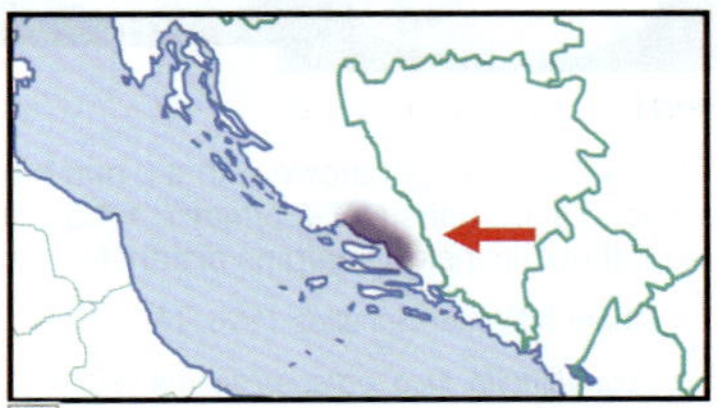

Croatia (Dalmatia)

Westerlund 1889: 136, Knipper 1939: 432, Subai 2002: 16.

Helicigona dochii (Sturany, 1907)

Campylaea dochii Sturany, 1907 in Sturany 1907: 230

Shell light yellowish with brownish apex and three yellowish brown colour bands, solid, densely covered by hairs, densely and irregularly striated and slightly granulated, 4-4.5 whorls, apex and second whorl protruding, apertural margin not thickened, weakly reflected, margins nearly approaching each other at parietal side (with a 1.5-2 mm long gap), umbilicus narrow (2 mm) and perspectival.

7.1-8.7 x 15.5-18.7 mm

Dhora & Welter-Schultes 1996: 162, Subai 2002: 18.

Liburnica.

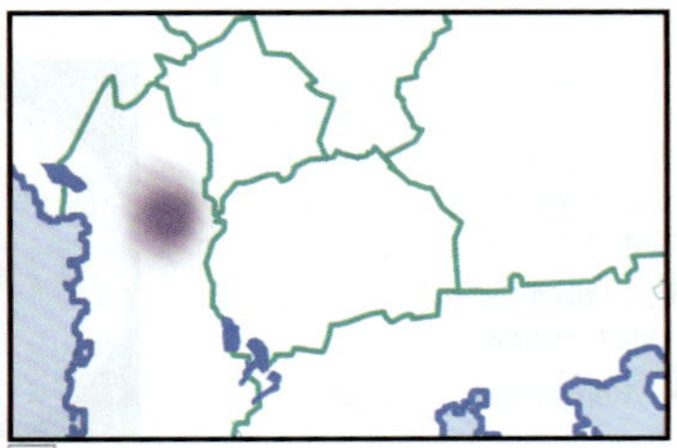

N central Albania

Helicigona dunjana Knipper, 1941

Helicigona dunjana Knipper, 1941 in Knipper 1941: 184

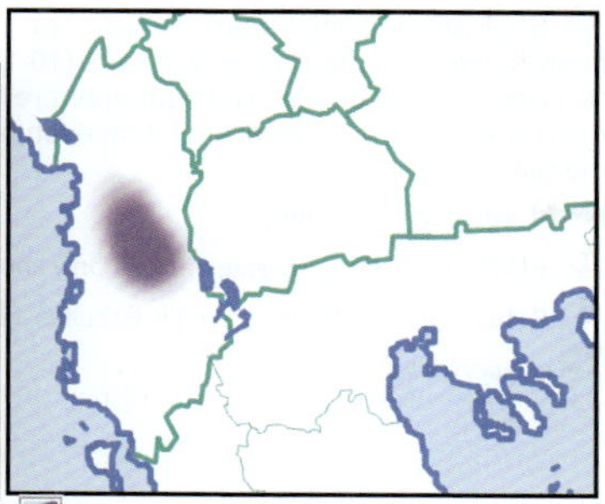

12-15.5 x 23-29 mm

Albania: near Librazhd

Central and S Albania

Shell yellowish grey with 1-3 brown bands, thin, finely and densely striated, with fine hairs (5.5-6.5/mm^2), 4.5 moderately convex whorls with deep suture, last whorl slightly inflated, deeply descending near aperture, apertural margins approaching each other closely and connected at parietal side with a thin callus, umbilicus narrow, slightly widened at last 1/2 whorl.

Subai 1995: 82, Dhora & Welter-Schultes 1996: 163, Subai 2002: 20.

Liburnica. Not in S Albania.

Helicigona euboeae (Frauenfeld, 1867)

Campylaea euboeae Frauenfeld, 1867 in Frauenfeld 1867: 779

Shell light brownish horny with a light brown colour band and two indistinct light bands, irregularly and finely striated, minute hair scars of short and characteristically dense hairs (0.06-0.12 mm long, 50-60 hairs/mm^2), last whorl width 2 x or more of penultimate whorl, aperture slightly detached with a regular 1 mm wide margin, umbilicus wide. Shell not very variable.

11-13.5 x 20-24 mm

EU Red List: Near Threatened NT. In Evia *Hcg. euboeae* is restricted to a small high mountain region, in immediate vicinity (1 km from Káto Stení to Dírfios) lives only *Hcg. sphaeriostoma* which seems to replace slowly *Hcg. euboeae.*

Subai 1996: 13.

Thiessea.

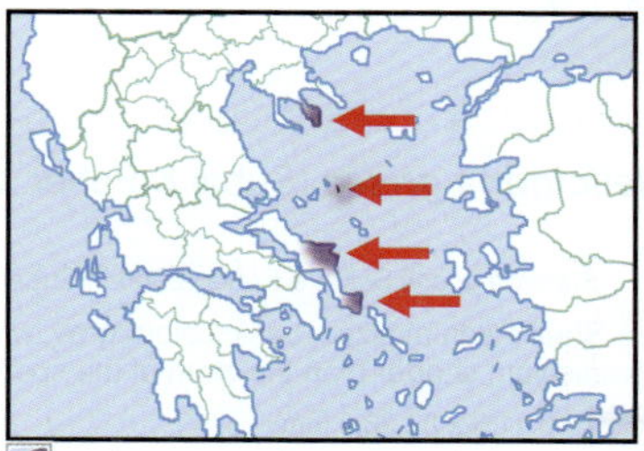

Greece (Evia, Pipéri, Chalkidikí)

Helicigona faueri (Subai, 1990)

Chilostoma faueri Subai, 1990 in Subai 1990: 115

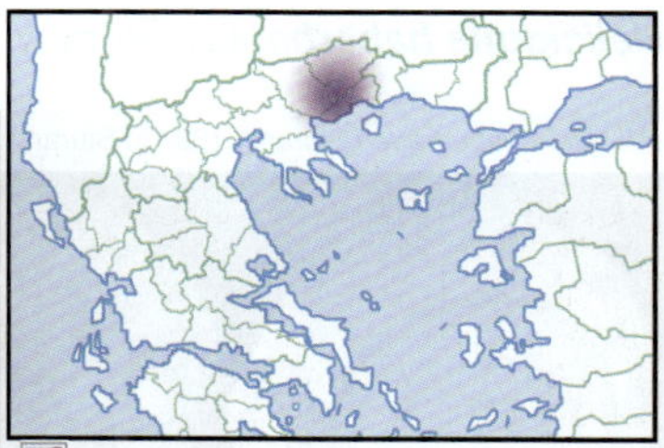

9-15 x 17-26 mm

Greece: Makedonía, Pangeon mountains, 1800 m

Greece (E Makedonía)

Cattania.

Shell depressed, light with light yellowish periostracum, 3 narrow brown colour bands, the central band darker and less blurry than the others, distinctly and densely striated, first whorls covered by hair scars, last whorl width 1.5 x of penultimate whorl, apertural margin sharp and reflected, no white layer at parietal side, umbilicus only a little covered by apertural margin.

Helicigona faustina (Rossmässler, 1835)

Helix faustina Rossmässler, 1835 in Rossmässler 1835-1837 (2): 4, Taf. 6

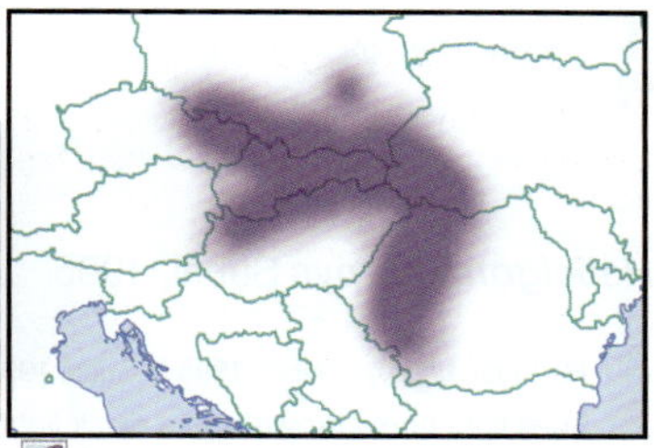

8-11 x 15-20 mm

Poland: Bieszczady mountains, Tworylne

Agneta Chaniecka

S Poland to Romania

Shell light yellow to dark reddish brown, shiny, unicoloured or with broad dark colour band just above periphery, with fine irregular streaks, no spiral line, umbilicus wide or narrow, locally variable.

Forest slopes, humid shady rocks up to subalpine stage, not restricted to limestone areas, not in lowlands.

Shilejko 1978: 309, Falkner 1990: 230, Kerney et al. 1983: 277. Isolated in Lithuania.

Faustina.

Slovakia: Veľká Fatra mountains

Jozef Grego

Helicigona fuchsi Knipper, 1939

Helicigona fuchsi Knipper, 1939 in Knipper 1939: 416

Shell brown, lighter near umbilicus, with a dark brown band above the periphery and a lighter zone below, finely striated, regularly arranged hairs (10-13/mm²), 4.5-5 weakly convex whorls with moderately deep suture, last whorl weakly descending near aperture, aperture not very oblique, with weak lip inside, lip whitish with reddish brown hue, margins distant from each other, umbilicus narrow and partly covered by the reflected columellar margin.

10-12.5 x 15-21 mm

EU Red List: Near Threatened NT. Only known from 2 localities, but might be more widespread. No real threats known.

Subai 1995: 80, Dhora & Welter-Schultes 1996: 163.

Josephinella.

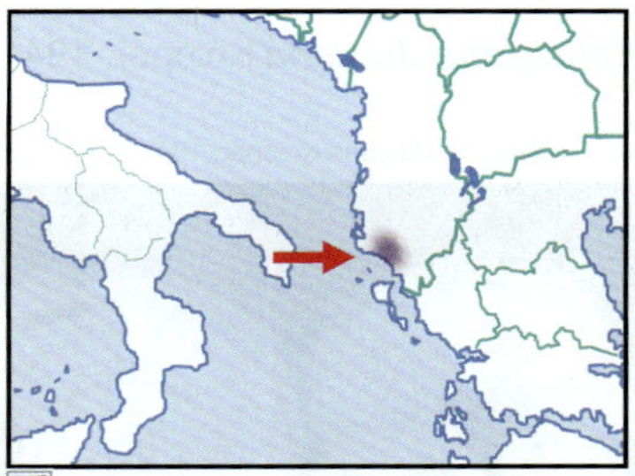

S Albania

Helicigona fuchsiana Knipper, 1939

Helicigona heldreichi fuchsiana Knipper, 1939 in Knipper 1939: 447

Shell horny with a brown colour band and two light bands, finely granulated and irregularly striated with few weak hair scars, last whorl width 1-5.2.5 x of penultimate whorl, aperture larger than in *Hcg. polyhymnia*, umbilicus slightly oblique and perspectivical.

8.5-11 x 16-21 mm

Various different kinds of habitats, mainly found under stones.

Subai 1996: 15, www.iucnredlist.org (08.2012).

Thiessea.

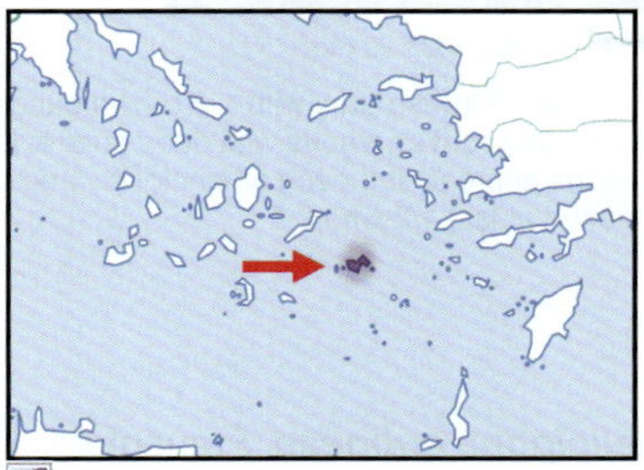

Greece (Astipálea)

Helicigona grisea (Subai & Fehér, 2006)

Superba grisea Subai & Fehér, 2006 in Subai & Fehér 2006: 207

Shell medium sized, basic colour grey, with extraordinarily narrow spiral bands, riblets blunt. Diverticulum shorter than the stem of the spermatheca, flagellum long.

10-11.5 x 22-23.5 mm

Known from a gorge in 1150 m altitude.

EU Red List: Near Threatened NT. Only known from one locality. No acute threats known.

Superba.

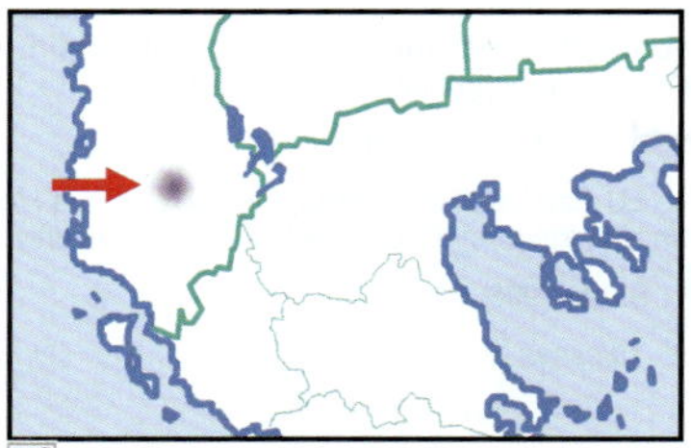

S Albania

Helicigona haberhaueri (Sturany, 1897)

Helix haberhaueri Sturany, 1897 in Sturany 1897: 112

14-18 x 22-27 mm

Bulgaria: W Rodopi, Lepenitza river

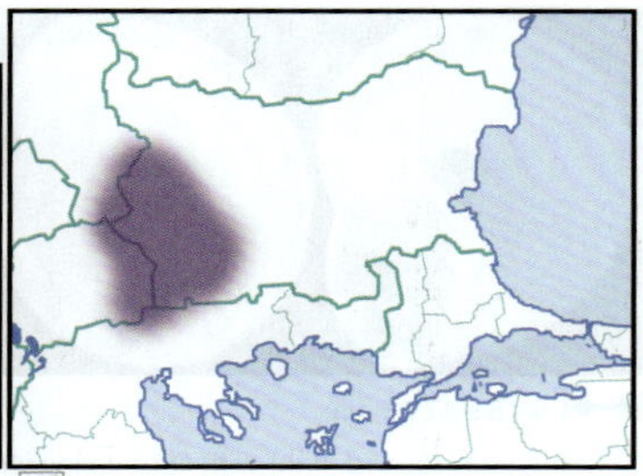

W to central Bulgaria

Shell pale yellowish or darker with one narrow (up to 1 mm) reddish brown band at the periphery and usually above the suture, sometimes with two additional weak bands, with coarsely arranged growth lines on the last 2-3 whorls, 5.5-6 convex whorls, last whorl weakly descending near aperture, apertural margin slightly reflected, connected at parietal side only by a very thin whitish layer, lip at basal margin about 2 mm strong, umbilicus narrow (3 mm), deep and perspectivical, half covered by the reflected columellar margin.

Rocks in broadleaved deciduous mountain forests, rarely in pinetree forests with high humidity. In W Rodopi between 400 and 1700 m, maximum 2000 m.

Knipper 1939: 399, Irikov & Mollov 2006: 812, Hubenov 2007 (Central Balkan, Rila, Pirin).

Cattania.

Helicigona harpya Subai, 1995

Helicigona harpya Subai, 1995 in Subai 1995: 83

Shell light horny brown with a weak light brown band, finely granulated and with short radial riblets, covered with 0.25-0.3 mm long hairs (9-12/mm²), 4-4.5 convex and regularly increasing whorls with deep suture, last whorl width less than 2 x of penultimate whorl, descending near aperture, aperture weakly reflected, margins approach each other with a 3-4 mm wide gap at parietal side, with a very thin whitish layer, umbilicus deep and perspecivical, 1/7-1/6 of diameter.
The animal is dark brown, getting lighter down to the sides, sole dark brown.

8-9 x 18-19 mm

Forest habitats among ground litter and decaying tree trunks.

EU Red List: Critically Endangered CR. Only known from beech forests in Mali i Tomorrit. Threatened by ongoing forest destruction.

Dhora & Welter-Schultes 1996: 164, Subai 2002: 22, www.iucnredlist.org (08.2012).

Liburnica.

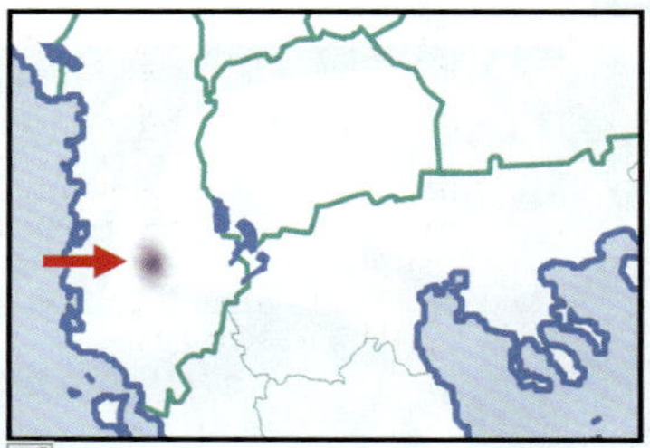

S Albania

Helicigona heldreichi (Pfeiffer, 1846)

Helix heldreichi Pfeiffer, 1846 in L. Pfeiffer 1846: 78 (*Helix Heldreichi*)
Name originally attributed to Shuttleworth, but description was by Pfeiffer alone.

7-11 x 14-22 mm

Greece: Attikí, Athína, Likavitos hill

Shell horny brownish with a brown colour band, two white bands and several hardly visible small brown colour bands, teleoconch with regularly arranged 0.7-0.8 mm long hairs (4-6 hairs/mm^2), granulated and striated, with distinct hair scars below the periostracum, last whorl width 2-3 x of penultimate whorl, aperture almost circular, usually detached, margin strongly reflected outward, umbilicus wide.
Differs from *Hcg. sphaeriostoma* in the surface sculpture and anatomically.

Greece (Attikí)

Seems to be slowly replaced by *Hcg. sphaeriostoma*, which is currently expanding its range.

Knipper 1939: 442, Subai 1996: 17.

Thiessea.

Helicigona hemonica (Thiesse, 1884)

Helix hemonica Thiesse, 1884 in Thiesse 1884: 271

Shell flat, horny coloured with a narrow light brown colour band above the edge, surface in regions of low calcium content finely granulated and coarsely striated, last whorl with blunt edge, 2-3 hairs/mm^2, aperture oblique (45°), margin sharp and reflected, no callus or white layer at parietal wall, umbilicus wide and perspectivical.

7.5-10.5 x 17-23 mm

At rocks and rocky walls, for example at the Meteora monastery, snails hide in deep rock crevices to escape heat, not found on conglomerate substrate.

Westerlund 1890: 140, Pintér & Subai 1980: 173.

Josephinella.

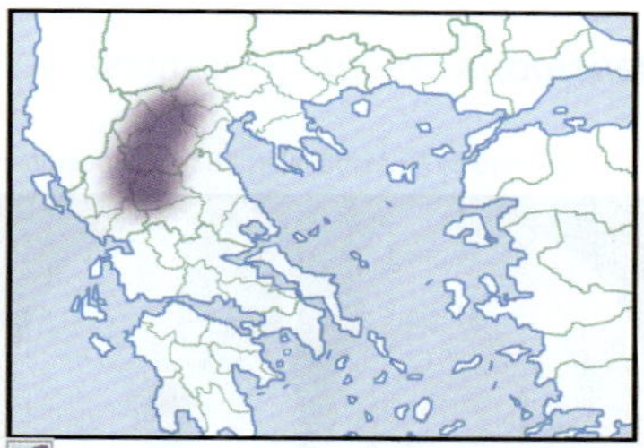

Greece

Helicigona hoffmanni (Rossmässler, 1836)

Helix hoffmanni Rossmässler, 1836 in Rossmässler 1835-1837 (4): 4, Taf. 16 (*Helix Hoffmanni*)
Helix walteri Boettger, 1886 in O. Boettger 1886: 37, Taf. 2 (*Helix (Campylaea) Walteri*)

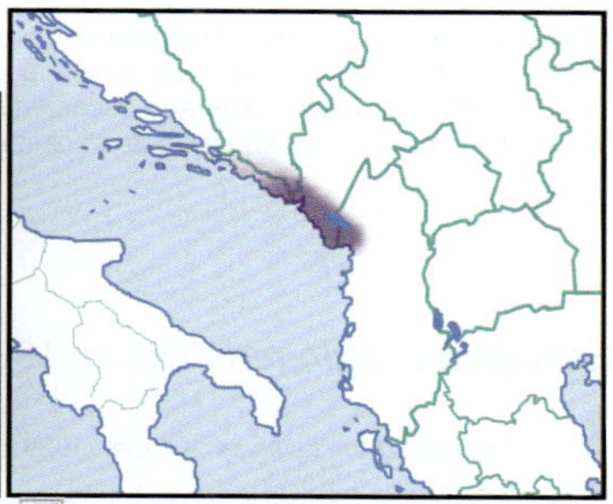

13-21 x 23-35 mm

(*Hcg. h. walteri*)

Albania: Zogaj near Shkodër

S Dalmatia to NW Albania

Shell horny yellowish, usually with 3 narrow reddish brown bands, the two marginal bands often indistinct, thin, finely granulated and striated, with short brown hairs, 4.5-5 convex whorls, last whorl very deeply descending, aperture nearly horizontal, margin reflected and with thin lip, margins approaching each other closely at parietal side, but not connected, columellar margin very broad and with a tooth-like callus, covering the narrow umbilicus partly or entirely.
In the form known as *Hcg. h. walteri* the umbilicus is slightly wider, the 3 bands are equally strongly developed, and the hairs are more densely arranged. We could be dealing with spatial gradients of a locally variable species, or with a well defined uniform subspecies.

Westerlund 1889: 134, Knipper 1939: 424, Dhora & Welter-Schultes 1996: 164, Subai 2002: 23.

Liburnica. Two subspecies have been recognised, they could be separate species. More research seems to be necessary to understand their relationships.

Helicigona hymetti (Mousson, 1854)

Helix hymetti Mousson, 1854 in Mousson 1854: 8 (*Helix Hymetti*)

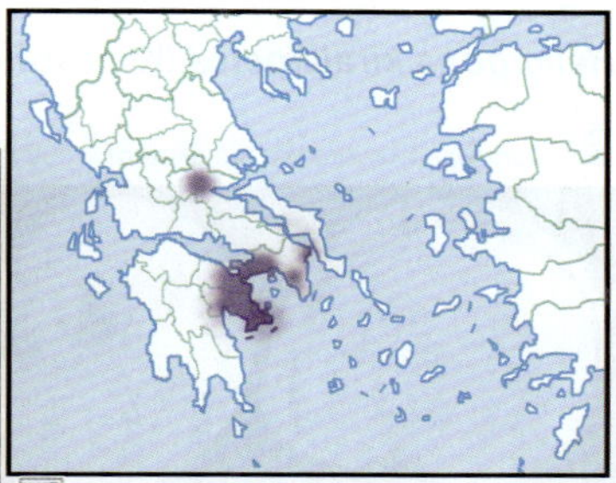

8.5-13 x 17.5-25 mm

Greece: Pelopónisos, Argolis, near Prósimna

Shell light with a brown colour band and two whitish bands, sometimes some additional blurry brown bands are visible, surface smooth, shiny, very finely granulated and striated, sometimes with some hair scars on the first 2 whorls, 4.5-5 whorls with shallow suture, last whorl width 2 x of penultimate whorl, apertural margin sharp and 1 mm wide, often slighty reflected, connected at parietal side by a thin layer or callus, but not detached, umbilicus initially narrow, at last whorl gradually wider (3-5 mm). Shell not very variable.
Differs from *Hcg. sphaeriostoma* in its less "circular" aperture (usually not detached), more depressed shell and less inflated last whorl which descends less sharply. *Hcg. sphaeriostoma* has a very short flagellum.

Greece

In Attikí at present restricted to highland habitats, in Fthiótis in lowland habitats.

Has suffered considerable range reduction in the past centuries, and has been replaced by *Hcg. sphaeriostoma*.

Thiessea. Subai 1996: 18 revised all previous reports which were often based in misinterpretations and misidentifications.

Helicigona inflata (Kobelt, 1876)

Helix trizona var. *inflata* Kobelt, 1876 in Kobelt 1876: 72 (not *Helix inflata* Gmelin, 1791 (= currently ?*Galba truncatula*) and several others).
Name was explicitly mentioned as *Helix trizona* var. *inflata* in the index p. 72. In the text p. 10 (dated May 1875) Kobelt did not unambiguously use the name for a taxon in his classification. Name was not mentioned on Tafel 95.
More research is needed to see if the name can be protected under Art. 23.9.1. This shall count as a use here.

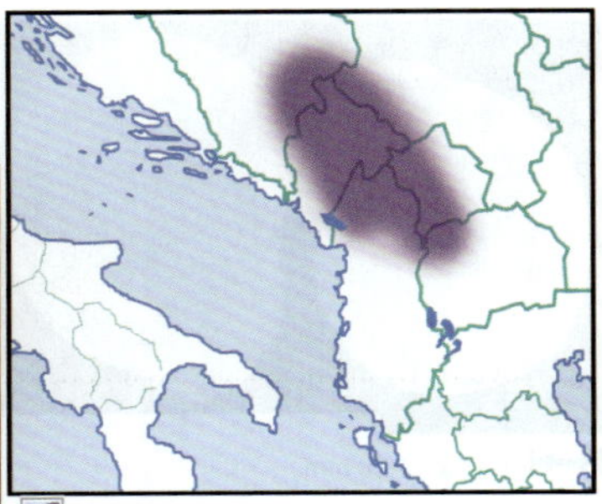

12-15 x 27-32 mm

Crna Gora: between Andrijevica and Čakor pass

Shell greyish or yellowish white with thin yellowish epidermis, which is often entirely preserved, with 3 brown bands, prominently striated and with fine indistinct spiral lines, shiny, 4.5-5.5 convex whorls, last whorl slightly inflated, suddenly and deeply descending near aperture, aperture with white lip at margin, umbilicus narrow and partly covered by the reflected columelar margin, only the last whorl visible inside.
Differs from *Hcg. trizona* in its narrower umbilicus, more convex whorls, more rounded aperture and better preserved epidermis.

NW Makedonija to S Bosnia

Westerlund 1889: 137, Knipper 1939: 399, Subai 1995: 77, Dhora & Welter-Schultes 1996: 164, Dhora 2002: 139.

Cattania.

Helicigona insolita (Rossmässler, 1838)

Helix insolita Rossmässler, 1838 in Rossmässler 1838-1844 (7-8): 31, Taf. 37

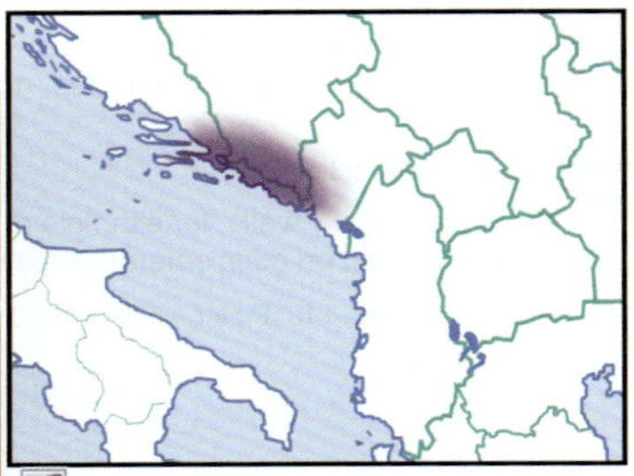

11.5-20 x 22.5-36.5 mm

Croatia: Dalmatia, Mt. Sv. Ilija near Orebic

Shell horny yellow with 3 fine and almost interrupted brown bands, finely striated, very densely granulated, 4-4.5 convex whorls, last whorl inflated and slowly and deeply descending near aperture, aperture very oblique and much wider than high, margin with weak lip and a prominent and broad callus at the basal side, margins approaching each other at parietal side and occasionally connected by a whitish layer, umbilicus narrow and partly covered by the much broadened and reflected columellar margin.

S Dalmatia, S Bosnia, Montenegro

Westerlund 1889: 135, Knipper 1939: 432, Subai 2002: 28.

Liburnica. 3 subspecies were recognised by Subai 2002, their alignment with bioscientific criteria (presence of partial intrinsic barriers to reproduction) requires more research.

Helicigona kattingeri Knipper, 1939

Helicigona trizona kattingeri Knipper, 1939 in Knipper 1939: 394

Fresh shell yellowish to greyish brown, aged shells whitish, usually with 3 dark bands, the middle band being the strongest, weakly striated, 4.5-5 weakly convex whorls, last whorl suddenly and deeply descending near aperture, aperture very oblique, margins reflected and approaching each other closely, giving the aperture an almost circular shape, umbilicus moderately narrow and perspectivical, 1/4 covered by the reflected columellar margin.

11-15 x 22-28 mm

Lowlands and mountains on calcareous substrate, up to 1800 m altitude (R. Reuselaars, 2012, AnimalBase).

Schütt 2001: 465, Hubenov 2007, Irikov & Erőss 2008.

Cattania. Schütt's 2001 report from NW Turkey was probably based on a misidentified name of a locality.

NE Greece, S Bulgaria

Helicigona kiralikoeica (Kimakowicz, 1890)

Campylaea kiralikoeica Kimakowicz, 1890 in Kimakowicz 1890: 54 (*Campylaea (Eucampylaea) Kiralikoeica*)

6-9 x 12.5-17 mm

Romania: S Piatra Craiului, Cheile Seaca, 1050 m

Shell with brown colour bands, also visible inside the aperture, apertural margin not connected at parietal side, umbilicus wide.

Romania (Piatra Craiului)

Mountain habitats under stones and nearby herbs, above 800 m.

Grossu 1983: 507.

Faustina.

Helicigona kleciachi (Pfeiffer, 1871)

Helix kleciachi Pfeiffer, 1871 in L. Pfeiffer 1871: 141 (*Helix Kleciachi*)

Shell horny yellowish with 3 narrow brown bands, densely covered with short yellow hairs, 4.5-5 slightly convex whorls, last whorl rounded and deeply descending near aperture, apertural margin with thin white lip, margins approaching each other at parietal side and connected by a thin whitish layer, basal margin with a broad tooth, umbilicus half covered by the reflected and broadened columellar margin.

12.5-17 x 22-27 mm

Croatia: Dubrovnik-Neretva region, Ston

Massimo Prodan

S Dalmatia, Herzegovina

Rocky limestone habitats.

Westerlund 1889: 134, Subai 2002: 37.

Liburnica.

Helicigona kollari (Pfeiffer, 1856)

Helix kollari Pfeiffer, 1856 in L. Pfeiffer 1856: 180 (*Helix Kollari*)

8.5-12 x 18.5-27 mm

Serbia: Manastir Manasia

Shell horny brownish with a narrow reddish brown band in a broad white zone, lighter near umbilicus, finely striated and initially covered with some hairs, 4.5-5 whorls, upper side more flattened than lower side, last 2 whorls more rapidly increasing than first whorls, last whorl slightly descending near aperture, margin reflected, white and with weak lip, margins aproaching each other, umbilicus wide.

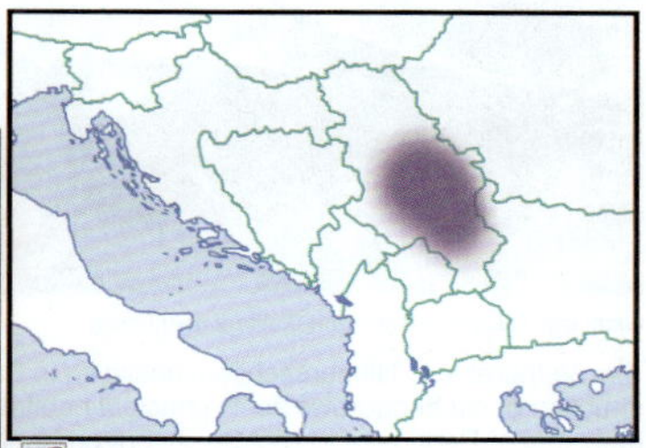

Serbia

At limestone rocks.

Westerlund 1889: 117, Pavlović 1912: 41, Knipper 1939: 436, Tomić 1959: 21, Subai 2002: 41.

Liburnica.

Helicigona korabensis Subai, 1997

Helicigona korabensis Subai, 1997 in Subai 1997: 527

Shell horny yellowish, usually with a blurry brown band, next to two yellowish white zones, occasionally with another blurry brown band, weakly and irregularly striated, 4.5-5.5 regularly increasing whorls, last whorl width at least 2 x of penultimate whorl, slightly descending near aperture, margin weakly reflected, margins not approaching each other closely at parietal side, which has a very thin whitish layer, umbilicus wide and perspectivical, considerably widened at last whorl to 1/5 of diameter.

9-13 x 19-25 mm

Restricted to a small range in a remote region. Not in the EU Red List, probably overlooked, but seems to be rare (Dhora 2004 used the Critically Endangered category for what was considered in the EU Red List as Near Threatened NT).

Subgenus unresolved.

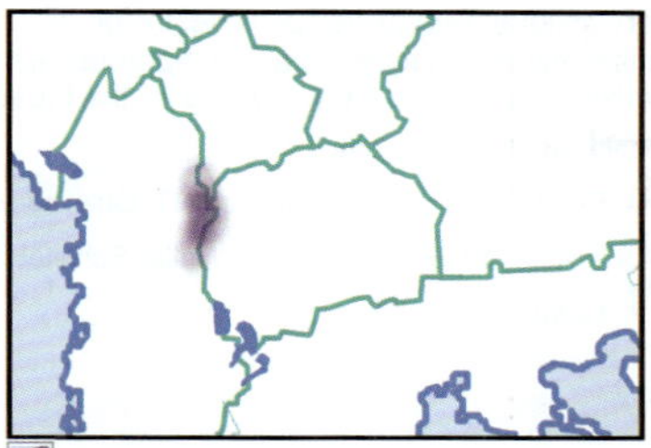

E Albania

Helicigona krueperi (Boettger, 1891)

Helix krueperi Boettger, 1891 in O. Boettger 1891: 84

Shell pale yellowish with greenish-olive hue, opaque and little shiny, with a blurry brown colour band (without lighter zones surrounding it) and moderately densely arranged hairs (9 hairs/mm^2), 5 rapidly increasing convex whorls, last whorl width 2 x of penultimate whorl, slightly descending near aperture, apertural margin reflected, umbilicus narrow.

Hcg. comephora has more coarsely arranged hairs (4/mm^2), *Hcg. brenskei* has denser hairs (16-25/mm^2).

16-18 x 30-34 mm

Knipper 1939: 434.

Josephinella.

Greece (S Pelopónisos)

Helicigona kulmakana (Subai & Fehér, 2006)

Superba kulmakana Subai & Fehér, 2006 in Subai & Fehér 2006: 208

Shell medium sized, three brown spiral colour bands, smooth shell surface with fine growth lines, umbilicus wide.

Mucus glands long, deeply subdivided, diverticulum longer than the stem of the spermatheca.

9-15 x 20-28 mm

Rocky limestone habitats in mountains above 2000 m.

EU Red List: Near Threatened NT. Only known from one locality, habitat in good condition, no threats known.

Superba.

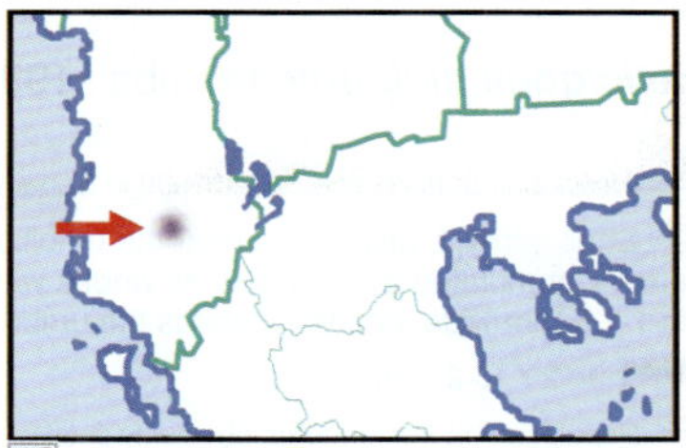

S Albania

Helicigona lapicida (Linnæus, 1758)

Helix lapicida Linnæus, 1758 in Linnæus 1758: 768 (*Helix Lapicida*)

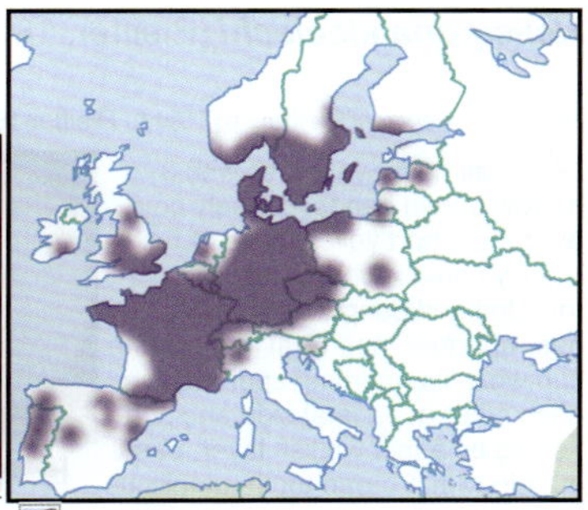

7.5-8.5 x 14-18 mm

Germany: Schleswig-Holstein, near Dänisch-Nienhof

W and central Europe

Germany: Thüringen, Tambach-Dietharz near Gotha

Shell yellowish brown with reddish brown flames, 5 whorls with characteristic keel, apertural margin white, sharp, reflected, connected at parietal side, umbilicus wide, 1/5 of shell diameter. Animal not very shy, dark greyish brown, sometimes yellowish, tentacles long.

Shady limestone rocks, old stone walls, also at tree trunks, often in deciduous highland forests. Needs deep crevices in which to hide. In Switzerland in up to 2000 m altitude. Feeds on algae and lichens at smooth tree trunks during wet weather, in N Germany only during rain, 1-2 hours after rainfall begin, rarely below 10°C. Only 1/1000 of the animals can be found attached to trees during dry weather conditions. Animals in N Germany usually stay at the same tree (*Fagus* and *Carpinus*, more rarely at *Fraxinus*, *Acer campestre* and *Crataegus*, tree trunk diameter usually exceeds 18 cm), maximum distance of dispersal around 5 m.

Some 30 small eggs are laid in May-July. High mortality rates among juveniles during dry weather conditions, lower in adults. Maturity is reached at the end of the second year, maximum age around 2 years, exceptionally 3 years.

Vulnerable in Niedersachsen, declining in Bayern, rare in Ireland. Should be considered at least as vulnerable in E England. Threatened in England by the continuous destruction of old hedgerows as its principal habitat, and by air pollution, the species has disappeared from large parts of E England. In Ireland only one colony is known. Extemely rare in Portugal and NW Spain, but probably native there.

Gassies 1849: 101, Germain 1930: 223, Nobre 1941: 126 (N Portugal), Rolán & Otero-Schmitt 1988: 112 (NW Spain), Prieto & Altonaga 1988, Kerney 1999: 201, Taleván Gómez & T. 2006: 114, Wiese 2009, Rudzīte et al. 2010: 137 (3 isolated localities in Latvia).

Helicigona.

Helicigona maranajensis (Wagner, 1914)

Campylaea maranajensis Wagner, 1914 in A. J. Wagner 1914: 51, Taf. 3

Shell yellowish brown with 3 narrow blackish brown sharply delimited colour bands, striated and with very fine spiral lines, shiny, 4.5-5.5 convex whorls, last whorl inflated, descending strongly near aperture, aperture very oblique, with white lip inside, inside white with colour bands shining through, apertural margin thin and white, slightly reflected, margins approach each other slightly at parietal side and are connected by a thin layer, umbilicus almost entirely covered by the broad columellar margin.

28-33 x 16-20 mm

EU Red List: Near Threatened NT. Only few localities in a weakly explored region known, probably more widespread.

Knipper 1939: 399, Dhora & Welter-Schultes 1996: 165, Dhora 2004, Dhora 2002: 142, Fehér & Erőss 2009.

Cattania.

N Albania

Helicigona matrella (Westerlund, 1898)

Helix matrella Westerlund, 1898 in Westerlund 1898: 161

8-12 x 17-25 mm

W Turkey: NW of Söke near Kuşadası

W Turkey (Izmir, Aydın)

Mountain and highland habitats.

Subai 1996: 22, Schütt 2001: 467, Örstan et al. 2005 (Bodrum peninsula).

Thiessea. The species was found near Manisa and east of Aydın in 2004.

Shell light horny or whitish with a brown colour band and two hardly recognizable light bands, smooth, shiny, densely and irregularly striated, no hairs, last whorl width less than 2 x of penultimate whorl, aperture not detached, insertions relatively distant from each other (3-6 mm), margin sharp and 0.5-1 mm wide, umbilicus slowly increasing (3-5.5 mm).

Helicigona melpomene Subai, 1996

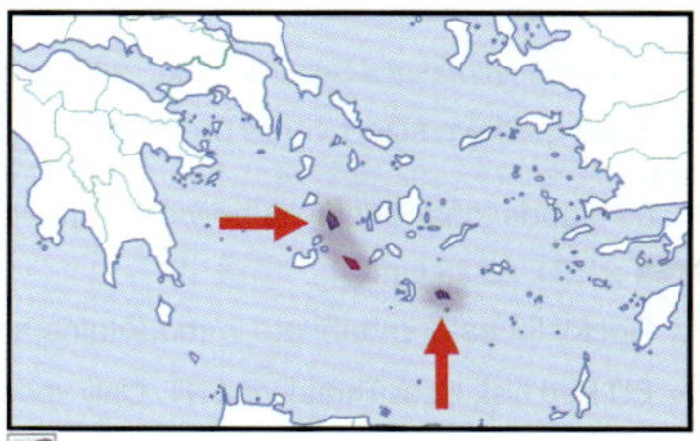

Helicigona melpomene Subai, 1996 in Subai 1996: 24

Shell horny, blurry brown near the umbilicus, with a brown colour band and two blurry light bands, keeled, densely granulated and striated, all whorls with hairs (0.2-0.5 mm long, 3-8 hairs/mm^2), aperture detached, margin 1-1.5 mm wide, pointed towards the umbilicus, umbilicus perspectivical.

9-12 x 18-21 mm

Population on Folégandros obviously extinct.

Thiessea.

Greece (Anáfi, Sífnos)

Helicigona moellendorffii (Kobelt, 1871)

Helix moellendorffii Kobelt, 1871 in Kobelt 1871: 72 (*Helix Möllendorffii*)

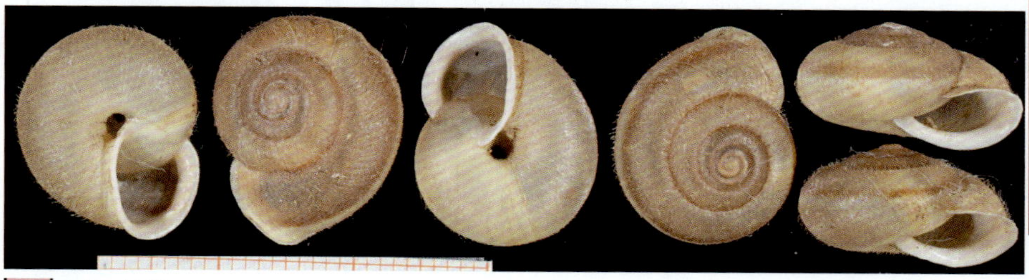

9-11 x 14-18 mm

Bosnia: Mt. Igman near Sarajevo

Shell horny brown with a brown band in a light zone, thin, densely granulated, with long and regularly arranged but widely spaced hairs, 4-5 slightly convex whorls, last whorl slightly increasing and suddenly descending near aperture, aperture oblique, margin broadly reflected and with strong whitish or yellowish lip, umbilicus narrow and partly covered by the columellar margin.

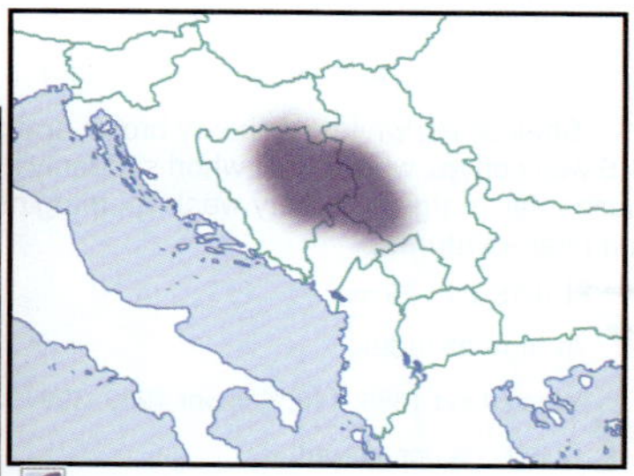

Bosnia to W Serbia

Rocky habitats in forests and shrublands, in crevices and among leaf litter.

Westerlund 1889: 126, Pavlović 1912: 44, Knipper 1939: 419, Tomić 1959: 22.

Campylaeopsis.

Helicigona nympha Subai, 1996

Helicigona nympha Subai, 1996 in Subai 1996: 27

Shell light with an intensively brown colour band and two hardly visible light bands, finely striated and extremely finely granulated, hair scars only on the first whorls and hardly visible, aperture not detached, margin sharp and 1-1.5 mm wide, umbilicus regularly increasing.

10-15 x 19-27 mm

Thiessea.

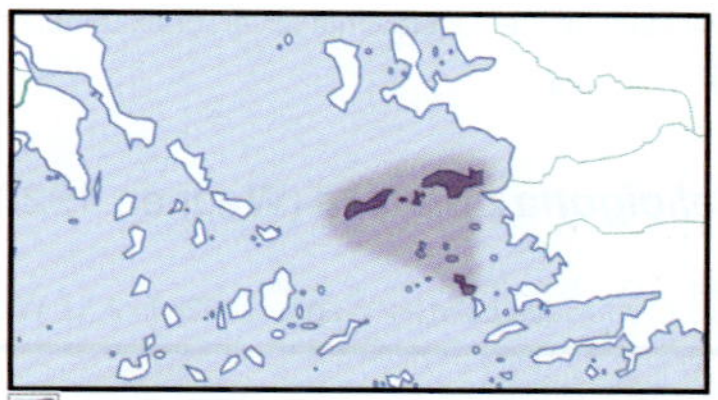

Greece (Sámos, Ikaría, Léros)

Helicigona pelia Hesse, 1912

Helicigona pelia Hesse, 1912 in Hesse 1912: 58

7.5-13 x 11-22 mm

Bulgaria: Rila mountains

Atanas Irikov

Shell dark olive brown with lighter initial whorls, thin and translucent, coarsely striated on last whorl, 5 densely coiled whorls, last whorl slightly descending near aperture, apertural lip weakly reflected, inside with a thin bluish white layer, umbilicus narrow and partly covered by the reflected columellar margin.

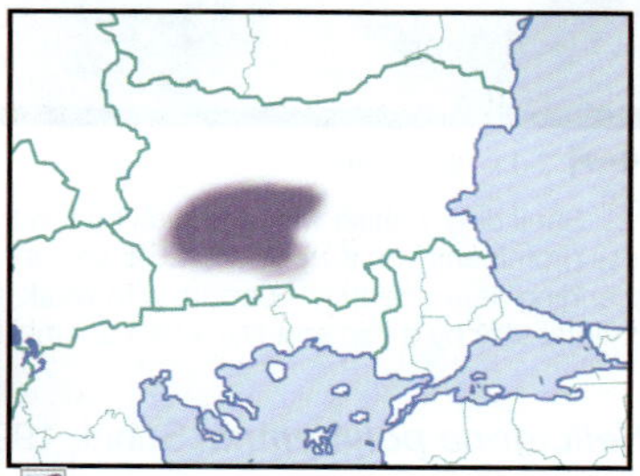

Bulgaria

Moderately humid habitats, reported from between 1000 and 2700 m.

Damjanov & Likharev 1975: 387.

Ariantopsis.

Helicigona pentheri (Sturany, 1901)

Campylaea pentheri Sturany, 1901 in Sturany 1901: 66

Shell yellowish green with 3 brown colour bands, apex yellowish brown and inner whorls with reddish hue, weakly striated, juveniles with short hairs (8 hairs/mm^2), 4-5 weakly convex whorls, last whorl slightly descending near aperture, aperture usually with one basal tooth, margin reflected and white, umbilicus wide and perspectivical, a small part covered by the broadly reflected columellar margin.

11-12.5 x 22.5-25.5 mm

In high altitudes, above 1100 m.

EU Red List: Near Threatened NT. Only known from one location (Plasa mountain near Jablanica) in an inaccurately studied region.

Knipper 1939: 433, Subai 2002: 42.

Liburnica.

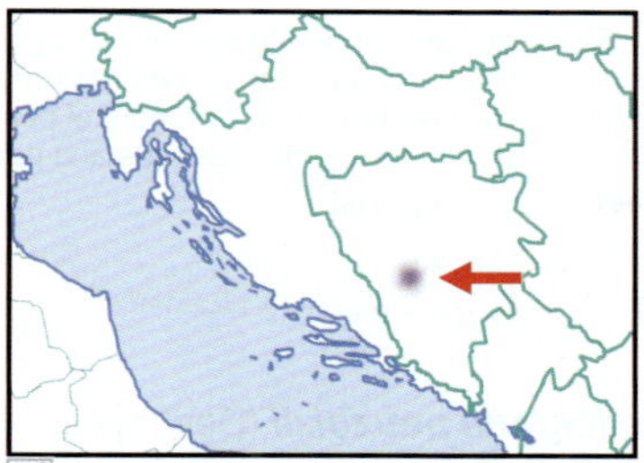

Bosnia-Herzegovina

Helicigona petrovici (Wagner, 1914)

Campylaea maranajensis petrovici Wagner, 1914 in Sturany & Wagner 1915: 51, Taf. 3

Shell yellowish brown with greenish hue and with 3 blackish brown colour bands, the middle band is sharp and the others are blurry, sharply and irregularly striated and with very fine spiral lines, shiny, 4.5-5.5 convex whorls, last whorl inflated, descending strongly near aperture, aperture very oblique, with white lip inside, inside white with colour bands shining through, apertural margin thin and white, slightly reflected, margins approach each other slightly at parietal side and are connected by a thin layer, umbilicus half open.

19-25 x 22-28 mm

EU Red List: Near Threatened NT. Only known from 2 localities, but could be more widespread. No threats known.

Cattania. The species was recently studied anatomically by P. Subai (e-mail, 08.2012).

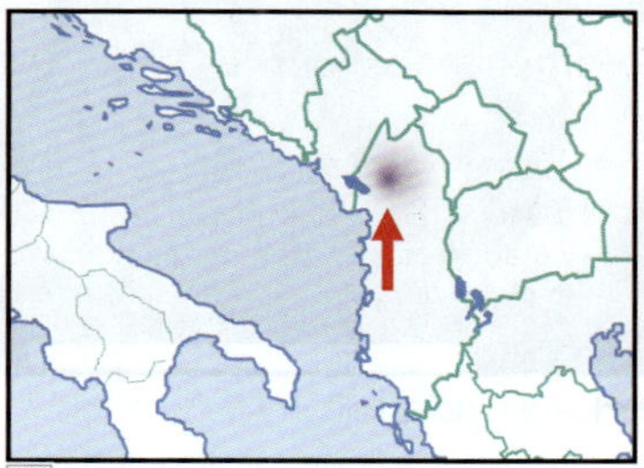

N Albania

Helicigona phocaea (Roth, 1856)

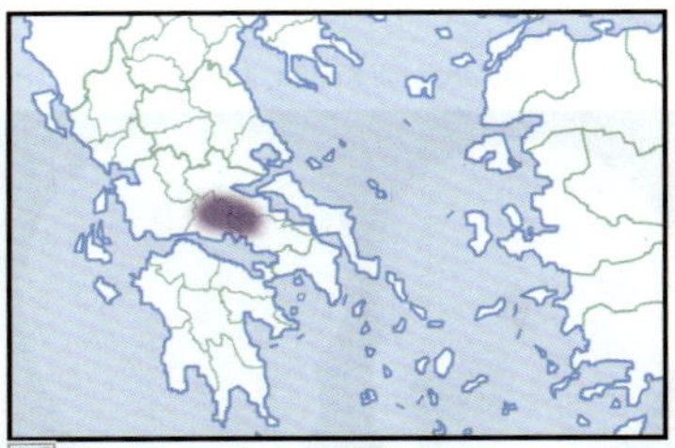

Helix phocaea Roth, 1856 in Roth 1856: 1 (*Helix Phocaea*)

Shell bluish white with horny brown apex, usually with 3 narrow reddish brown bands, irregularly striated, 4-4.5 well convex whorls, last whorl suddenly and substantially descreasing near aperture, aperture looks nearly horizontal, margin with very weak lip, margins approaching each other at parietal side, umbilicus wide, increasing near aperture.

8-15 x 15-25 mm

In high altitudes.

Westerlund 1889: 140, Knipper 1939: 399.

Subgenus unresolved.

Greece (Parnasós mountains)

Helicigona pieperi Subai, 1996

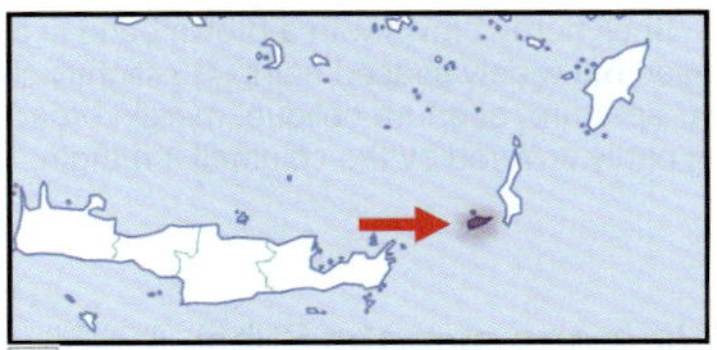

Helicigona pieperi Subai, 1996 in Subai 1996: 28

Shell whitish with sometimes a brown colour band, densely striated, aperture detached, margin sharp and 1.5 mm wide, umbilicus wide and perspectivical.

11-12 x 24-25 mm

EU Red List: Least Concern LC, threat probably underestimated. This species seems to be either extinct due to the destruction of the vegetational cover, or extremely rare (not found by Vardinoyannis 1994).

Thiessea.

Greece (Kásos)

Helicigona polinskii (Wagner, 1928)

Campylaea polinskii Wagner, 1928 in A. J. Wagner 1928: 380 (*Campylaea (Wladislawia) polińskii*)

7-9 x 14.5-18 mm

Bulgaria: Pirin mountains, Vihren peak region

(1-3) Atanas Irikov

Shell dirty whitish with greenish hue, with a well delimited brown band above the periphery, densely striated, less prominently on lower side, 4.5-5 convex and slowly and regularly increasing whorls, last whorl slowly descending near aperture, aperture with weak lip inside, margins not approaching each other closely, connected at parietal side by a thin and shiny callus, umbilicus moderately wide, slightly widened at last whorl.

SW Bulgaria (Pirin mountains)

Moderately humid high mountain habitats, calciphile, 1800-2900 m.

Damjanov & L. 1975: 385, Hubenov 2007.

Wladislawia.

Helicigona polyhymnia Subai, 1996

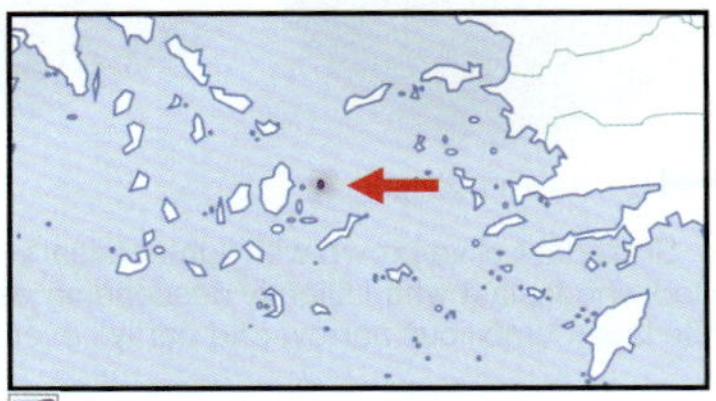

Helicigona polyhymnia Subai, 1996 in Subai 1996: 29

Shell small, light to horny brownish with a brown colour band, two light bands and two blurry narrow brownish bands, granulated and weakly striated with few hairs (0.3 mm long), aperture not detached, connected by a well visible callus, margin less than 1 mm wide, umbilicus initially narrow, at last whorl 3.5 mm wide.

8-9 x 16-18 mm

Thiessea.

Greece (Kikládes: Donoúsa)

Helicigona posthuma Knipper, 1939

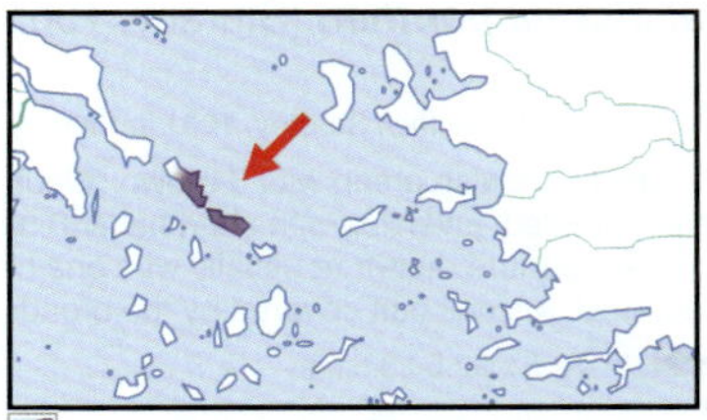

Helicigona posthuma Knipper, 1939 in Knipper 1939: 439

Shell flat, light horny with sometimes a greenish hue, one brown colour band and two yellowish white lighter bands, keeled, finely granulated and irregularly striated, with few hairs (0.25-1.7 mm long, unstable, only 2-3 hairs/mm^2), last whorl width 2 x or more of penultimate whorl, aperture detached, margin sharp and 1 mm wide, pointed near umbilicus, umbilicus wide and perspectivical.

8.5-12 x 18-24 mm

Subai 1996: 30.

Thiessea. Not in Síros.

Greece (Tínos, Andros)

Helicigona pouzolzii (Deshayes, 1832)

Helix pouzolzii Deshayes, 1832 in Deshayes 1830 [part 2]: 233 (*Helix Pouzolzii*)
Date was 29 Sep 1832 (Sherborn & Woodward 1899: 595, Cowie 1998: 92, "444" should probably read "144").

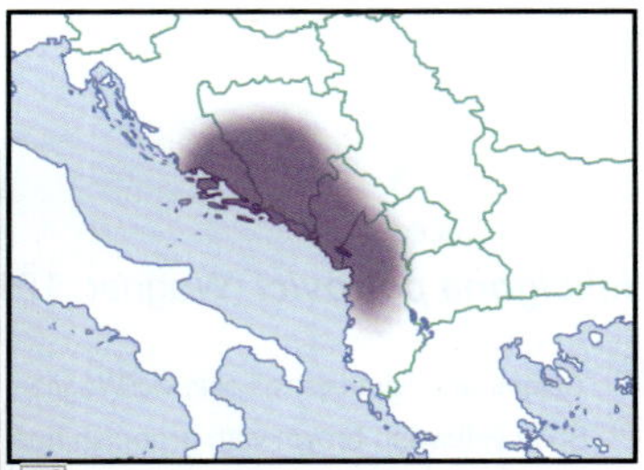

20-33 x 40-60 mm

N Albania: Zogaj near Shkodër

Shell yellowish brown, usually with 3 blackish brown bands, occasionally evenly yellowish, dark or olive greenish without bands, weakly shiny, very solid, usually coarsely striated on upper side and with fine spiral lines, 5-6 weakly convex whorls, last whorl rounded and descending near aperture, margin thickened, basal and columellar margins reflected, umbilicus wide.

Bosnia and Dalmatia to Albania

Vulnerable in Albania (Dhora 2004).

Westerlund 1889: 115, Knipper 1939: 403, Falkner 1990: 230, Dhora & Welter-Schultes 1996: 165.

Dinarica. Some subspecies have been proposed, their alignment with bioscientific criteria (partial barriers to reproduction in their genetics) requires more research.

Helicigona reischuetzi (Subai, 1990)

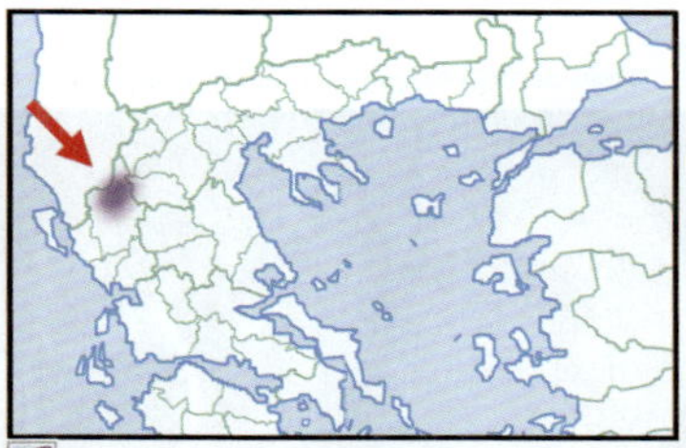

Chilostoma reischuetzi Subai, 1990 in Subai 1990: 244

Shell small, depressed, three brown spiral colour bands, shell surface smooth, umbilicus moderately wide. Penial papilla short, dart slender and long.

7-10 x 16-22 mm

Rocky habitats in river valleys and gorges.

EU Red List: Near Threatened NT. Restricted to a small restricted area. No acute threats known.

Superba.

NE Greece (Ipiros)

Helicigona rossmaessleri (Pfeiffer, 1848)

Helix advena Rossmässler, 1842 in Rossmässler, 1838-1844 (11): 2, Taf. 51 (not *Helix advena* Webb & Berthelot, 1833 from Canary Islands)
Helix rossmaessleri as mentioned by Pfeiffer 1842: 98 (*Helix Rossmässleri*) (nomen nudum)
Name used for a taxon without description or indication as "[Helix] Rossmässleri Pfr. (H. advena Rm.)". Not a new replacement name since it had no expressed statement reflecting the intention and reason to replace this name for nomenclatural reasons, which would qualify for a new replacement name (Art. 72.7, Glossary "expressly"). In this form *advena* was given as a simple synonym.
Helix rossmaessleri Pfeiffer, 1848 in L. Pfeiffer 1848 (Mon. Helic. 1): 360 (*Helix Rossmässleri*)
Established with description and bibliographical reference to Pfeiffer 1842. Art. 11.6.1 cannot come into effect because the name was used in 1842, not mentioned as a junior synonym.

Shell upper side pale yellow with reddish brown colour band above periphery, lower side brownish, shiny transparent, with fine and irregular streaks, umbilicus narrow.

Shady rocks and tree trunks in humid mountain forests, montane and subalpine stages.

Faustina.

6.5-8 x 10-13 mm

Slovakia: Veľká Fatra mountains
Jozef Grego

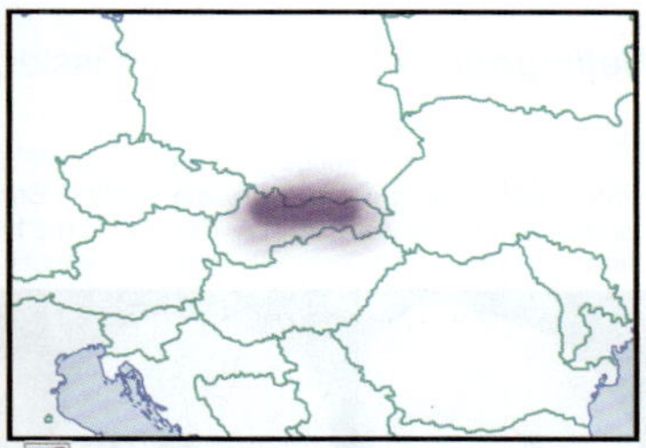

Poland, Slovakia, Romania

Kerney et al. 1983: 278. Misidentified by Grossu 1983: 508.

Helicigona rumelica (Rossmässler, 1838)

Helix rumelica Rossmässler, 1838 in Rossmässler, 1838-1844 (7-8): 30, Taf. 37
Helicigona renschi Knipper, 1939 in Knipper 1939: 453

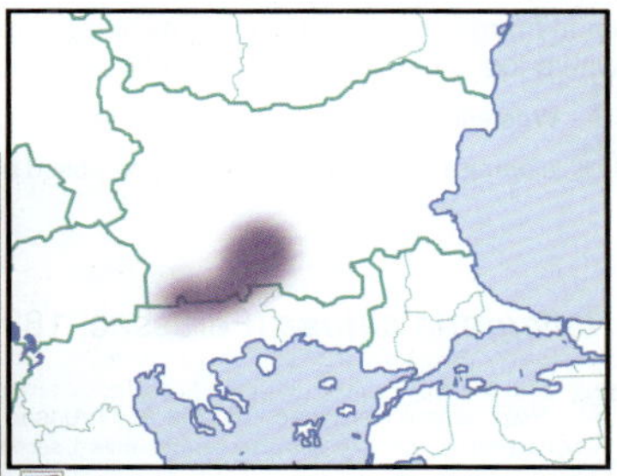

SW Bulgaria

12-14 x 23-27 mm

Bulgaria: (1-2) W Rodopi mountains, Dobrostan ridge near Bachkovo, (3-5) W Rodopi mountains (1-2) Atanas Irikov

Shell greyish or yellowish white with 3 brown bands and light olive remains of an epidermis near the umbilicus and the aperture, 4.5-5.5 slowly increasing convex whorls, last whorl not much broader than penultimate whorl, descending near aperture, aperture very oblique, margins not approaching each other closely, umbilicus narrow and covered by the reflected columellar margin.
Differs from *Hcg. trizona* in its more densely coiled spire and smaller and higher shell, shells of equal diameter have one whorl more than *Hcg. trizona*. Great variation in shell form, colour and size, seems to depend on ecological conditions.

Limestones, deciduous forests, rarely in pinetree forests, open terrains, in up to 2000 m altitude.

Westerlund 1889: 137, Knipper 1939: 399, Damjanov & Likharev 1975: 378, Dhora & Welter-Schultes 1996: 164, Irikov & Mollov 2006: 811.

Cattania. Hcg. renschi is regarded as a synonym here (in agreement with P. Subai e-mail 2010, and www.iucnredlist.org (08.2012)), it was based on two damaged shells which did not differ very much conchologically from *rumelica*, but which were not compared with this species.

Helicigona sattmanni Subai, 1995

Helicigona sattmanni Subai, 1995 in Subai 1995: 90

Shell horny to dark greyish brown with a 1 mm wide brown band next to two lighter yellowish white bands, irregularly striated, not very densely covered with 0.5-0.6 mm long hairs (3-4/mm^2), 4.5-5 regularly increasing whorls with shallow soture, last whorl slightly descending, aperture not extremely oblique, margins not approaching each other closely, parietal side without callus, margin weakly reflected, umbilicus moderately wide.
Animal brown, lighter at the sides, sole dirty whitish, mantle with spots of grey pigment, one large light grey spot is located above the respiratory hole.

8.5-11 x 16.5-22 mm

In deciduous high mountain forests, also in meadows with limestone rocks, 1500-2000 m.

Josephinella.

NW Greece

Helicigona serbica (Kobelt, 1872)

Helix serbica Kobelt, 1872 in Kobelt 1872: 130

16-22 x 24-32 mm

N Albania: near Bogë

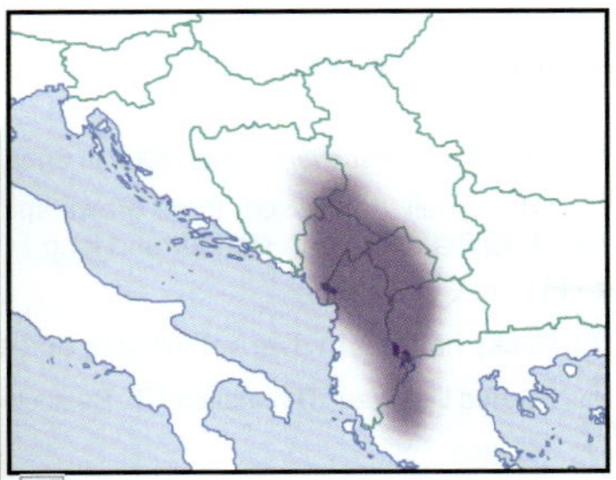

Dalmatia to Rep. Makedonija

Rocky habitats in shrublands and forests, a frequent species in N Albania.

Westerlund 1889: 116, Pavlović 1912: 42 (*pouzolzi*), Knipper 1939: 403, Tomić 1959: 21 (*pouzolzi*), Dhora & W. 1996: 166.

Dinarica.

Shell dark olive or dark brown with two light bands near the periphery, or lighter with 3 dark bands, the central band being more distinct and narrower than the more blurry other two bands, striated and with distinct and very fine spiral lines, 4.5-5.5 slowly increasing whorls with deep suture, last whorl very well rounded and slightly descending near aperture, aperture inside intensively brown with two light bands, umbilicus open and slightly covered by the reflected columellar margin.

Helicigona setigera (Rossmässler, 1836)

Helix setosa var. *setigera* Rossmässler, 1836 in Rossmässler 1835-1837(4): 3
Helix setigera var. *globulosa* Brusina, 1869 in Brusina 1869: 70 (*Helix (Campylæa) setigera* var. *globulosa*) (not *Helix globulosa* Férussac, 1821 in A. Férussac [1821-1822]: 28)
Not established in genus *Campylaea*, Brusina combined the specific name with the subgenus (see pp. 40, 43).

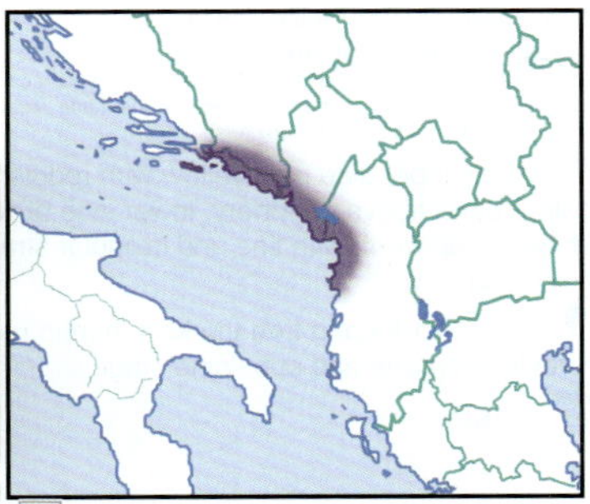

11.5-14 x 21-25 mm, in Mljet 9.5-12 x 17-20 mm

Albania: Mali i Kakariqit near Shkodër

S Dalmatia to Albania

Shell pale horny brown or whitish, usually with one single narrow reddish brown band, thin, finely striated, covered with long and solid yellow hairs, 4-4.5 convex and rapidly increasing whorls, last whorl width 2 x of penultimate whorl, very deeply descending near aperture, aperture nearly horizontal and with white lip at margin, margin thin, margins closely approaching each other and usually connected at parietal side by a whitish layer, but never detached, basal margin with a tooth, umbilicus narrow and usually half covered by the reflected columellar margin.

Westerlund 1889: 134, Knipper 1939: 431, Dhora & Welter-Schultes 1996: 166, Subai 2002: 43.

Liburnica. Two subspecies were proposed by Subai 2002, their bioscientific value remains to be verified.

Helicigona setosa (Férussac, 1832)

Helix setosa Férussac, 1832 in A. Férussac & Deshayes 1819-1851: (explication planches) Pl. 69A
Original print of explication des planches not verified, second print p. 13.

11-15 x 22.5-31.5 mm

Croatia: Tučepi near Makarska

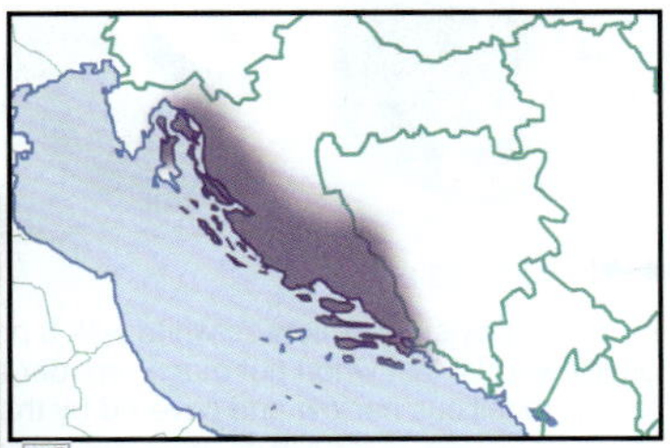

Croatia, W Bosnia

Shell horny yellowish with 3 light brown bands, finely striated, thin and fragile, with long and not very densely arranged stable yellow hairs, 4-4.5 weakly convex whorls with slightly impressed suture, rapidly increasing, last whorl width more than 2 x of penultimate whorl, suddenly and very deeply descending near aperture, margin white and reflected, connected at parietal side and almost detached by a fold, columellar margin often with a broad tooth, umbilicus initially narrow and rapidly widened at last whorl.
Differs from *Chilostoma lefeburianum* in its 3 colour bands (not only one), and its parietal apertural margin being connected and almost detached.

Rocky limestone habitats.

Westerlund 1889: 133, Knipper 1939: 419, Štamol & Velkovrh 1995 (N and central Cres, not Lošinj), Subai 2002: 5.

Liburnica. Two subspecies were recognised, their bioscientific value remains to be verified.

Giuseppe Pocaterra

Croatia: Velika Paklenica near Starigrad

Helicigona skipetarica Subai, 1995

Without gender agreement: *Helicigona skipetaricus* Subai, 1995

Helicigona skipetaricus Subai, 1995 in Subai 1995: 88
Should have been *skipetarica*, *Helicigona* is feminine and *skipetaricus* an adjective.

Shell small to medium sized, basic colour whitish, with three brown colour bands, teleoconch surface with fine radial stripes, umbilicus moderately wide.
Genital organs with grey pigment.

8-11 x 17-24 mm

Mountain habitats in 1330-2240 m.

Restricted to a small range in Mali i Tomorrit. No threats known, the habitat is in good condition.

Superba.

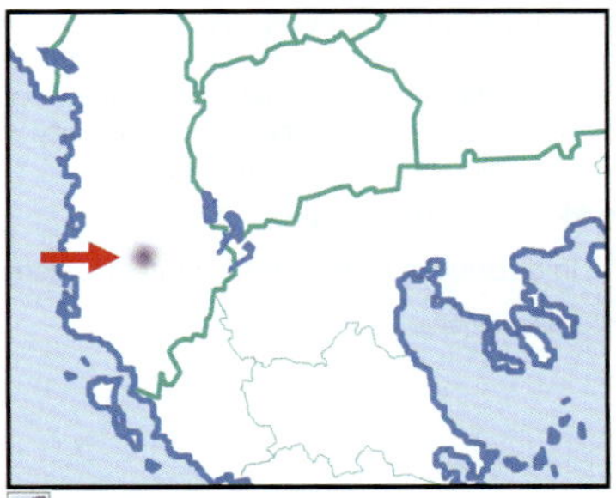

S central Albania

Helicigona sphaeriostoma (Bourguignat, 1857)

Helix sphaeriostoma Bourguignat, 1857 in Bourguignat 1857: 8, Pl. 8 (*Helix sphæriostoma* on p. 8)

7.5-13.5 x 15-27 mm — Greece: Evia, Chalkida

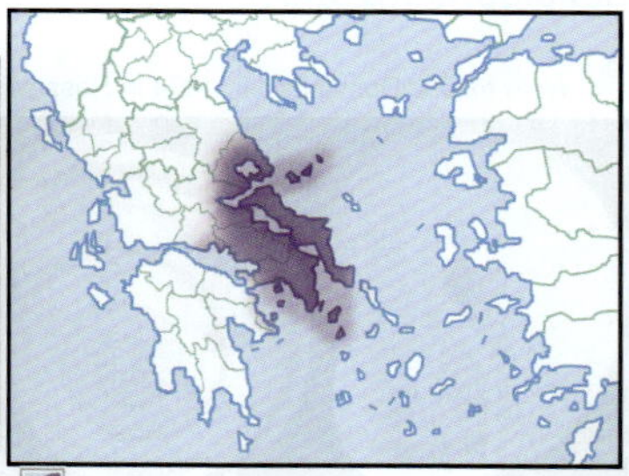

Greece

Shell light horny or dirty whitish with a brown colour band, two whitish bands and two additional blurry brownish bands, finely granulated and irregularly striated, with minute (0.1 mm long) hairs (30-40 hairs/mm^2), last whorl width 2-3 x of penultimate whorl, aperture almost circular, clearly detached, margin 1-2 mm wide and often strongly reflected like a trumpet, umbilicus initially narrow, then suddenly wider at last whorl to reach 3-5 mm width.
Shell relatively variable, hairs are usually hardly visible or absent. Flagellum short, dart length 2.4-2.9 mm. Differs from *Hcg. hymetti* in its usually detached "circular" aperture, less depressed shell and more inflated last whorl which descends more sharply. In *Hcg. cyclolabris* the umbilicus increases more regularly in width, the last whorl is less inflated (slightly less than 2 x of penultimate whorl), there are no hair scars on whorls 3-5, the darts are longer and the flagellum is shorter.

In a large variety of different habitats. Seems to expand its range and replace neighbouring species *Hcg. hymetti*, *Hcg. heldreichi* and *Hcg. euboeae*, possibly supported by human-based dispersal.

Subai 1996: 33 (revised all previous reports).

Thiessea.

Helicigona stenomphala (Menke, 1830)

Helix stenomphala Menke, 1830 in Menke 1830: 126
Established with description.

Shell pale yellowish or brownish yellow, with 3 brown bands, only the central band is distinct and dark, finely striated and with weak spiral lines, 4.5-5.5 convex whorls, last whorl rounded and slowly descending near aperture, margin weakly reflected and with weak white lip, not connected at parietal side, columellar margin broadened and in form of a white fold almost entirely covering the narrow umbilicus.
Spire higher and whorls more inflated than in *Hcg. pouzolzi*, apertural margins more distant from each other, umbilicus narrower.

19-23 x 28-33 mm

Westerlund 1889: 117, Knipper 1939: 406, www.iucnredlist.org (08.2012).

Dinarica. Possibly closely related to *Hcg. albanica* (Jaeckel) from N Albania, more research is needed.

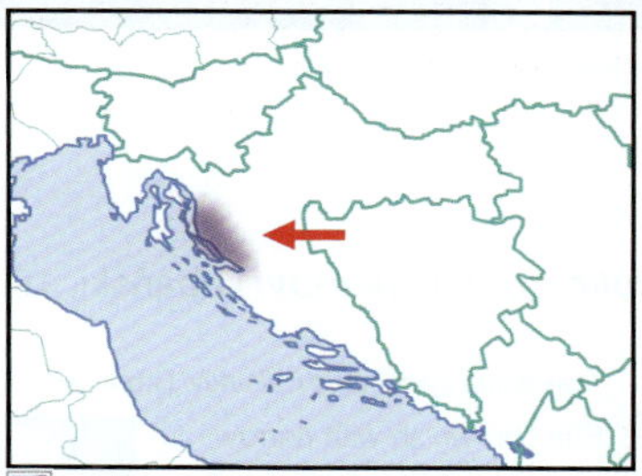

Croatia

Helicigona subaii (Fauer, 1991)

Chilostoma subaii Fauer, 1991 in Fauer 1991: 167

Shell yellowish brown with 3 blurry brown bands at last 1.5 whorls, silky shiny and translucent, striated and finely granulated, upper side of first whorls nearly flat, first whorls strongly keeled, 5-5.5 regularly increasing whorls, last 1.5 whorls with edge at periphery, rounded near aperture, last whorl widened and shortly descending near aperture, margins approach each other closely but are not connected at parietal side, aperture very oblique, oval, with weak white lip inside, umbilicus wide.

11-15 x 24-30 mm — Greece: E Makedonía, between Zarkadiá and Makrichóri (26 mm)
Robert Reuselaars

Greece (E Makedonía)

Stony habitats with little grassy vegetation.

Cattania.

Helicigona subzonata (Mousson, 1859)

Helix subzonata Mousson, 1859 in Mousson 1859: 28

Shell dark horny brown, with a dark brown band and a lighter band running below it, finely striated and very finely granulated, with yellowish red short hairs, 4.5-5 weakly convex whorls, rapidly increasing, each whorl's width 2 x of preceding whorl, last whorl slightly descending near aperture, aperture inside greyish with brown hue, margin lighter, slightly thickened and shortly reflected, umbilicus wide.

10-17 x 20-30 mm

Westerlund 1889: 126, Rähle 1980: 218 (Kefaloniá), Subai 1995: 92, Fehér & E. 2009.

Josephinella. Has been reported from the mainland in W Greece, but these forms belong to undescribed species.

W Greece

Helicigona sztolcmani (Wagner, 1928)

Campylaea sztolcmani Wagner, 1928 in A. J. Wagner 1928: 383

Shell light grey with bluish hue and brownish apex, often with a thin brown band at the periphery of the last whorl, densely and prominently ribbed, 4-4.5 convex and slowly increasing whorls with deep suture, last whorl deeply descending near aperture, often with edge at the periphery, not inflated, aperture inside yellowish brown and with very weak lip, margin slightly reflected at basal and columellar side, margins not approaching each other very closely, umbilicus moderately wide, 1/6 of diameter, widened at last whorl, not covered.

5-6 x 9.5-11.8 mm

High mountain region of Pirin mountains, calciphile, 2000-2700 m.

EU Red List: Near Threatened NT. Only known from one remote and undisturbed locality (the area around Vihren peak), no real threats known.

Damjanov & Likharev 1975: 386, Hubenov 2007.

Wladislawia.

SW Bulgaria

Helicigona trizona (Rossmässler, 1834)

Helix trizona Rossmässler, 1834 in Rossmässler [1834] (2): 4

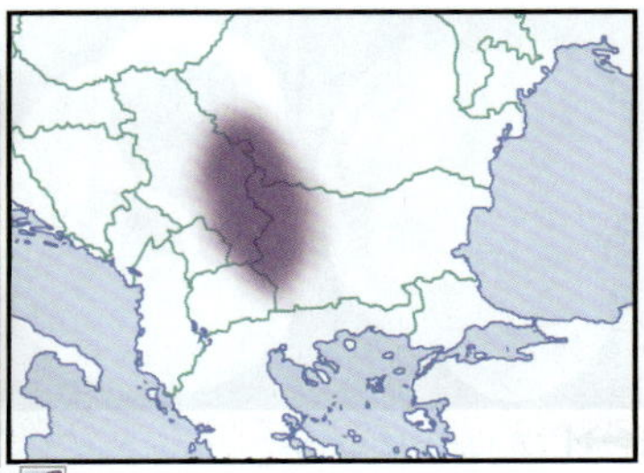

13-18 x 25-32 mm

Romania: Caraş-Severin, Băile Herculane

E Serbia, W Bulgaria, SW Romania

Romania: Danube river near Orşova

Jozef Grego

Shell greyish or yellowish white with 3 brown bands, the central band darker than the others, with remains of a light olive greenish epidermis at the apertural margin and umbilicus, prominently striated and with fine and indistinct spiral lines, 4-5 flattened whorls, upper side nearly flat, lower side convex, last whorl rounded and deeply descending near aperture, margin white and slightly thickened, not connected at parietal side, umbilicus narrow.
The animal is light brownish.

Moderately humid habitats, calciphile, in Bulgaria up to 1700 m. A relatively frequent species within its range.

Westerlund 1889: 136, Pavlović 1912: 45, Knipper 1939: 399, Tomić 1959: 22, Damjanov & Likharev 1975: 380, Falkner 1990: 230, Dedov & Subai 2006.

Cattania. Not in N Albania.

Helicigona valkanovi Urbański, 1960

Helicigona cyclolabris valkanovi Urbański, 1960 in Urbański 1960 (Izv.): 94, Taf. 3

Shell brownish with narrow brown colour band above the periphery and a light zone below, granulated and coarsely striated, more weakly on lower side, fresh shells with hairs, 4-5 weakly convex whorls, last whorl slightly inflated, deeply descending near aperture, apertural margin thin, white and reflected, umbilicus narrow, 1/10 of diameter, slightly covered.

10-12 x 18-22 mm

Greece: Thásos (D = 20 mm)

Robert Reuselaars

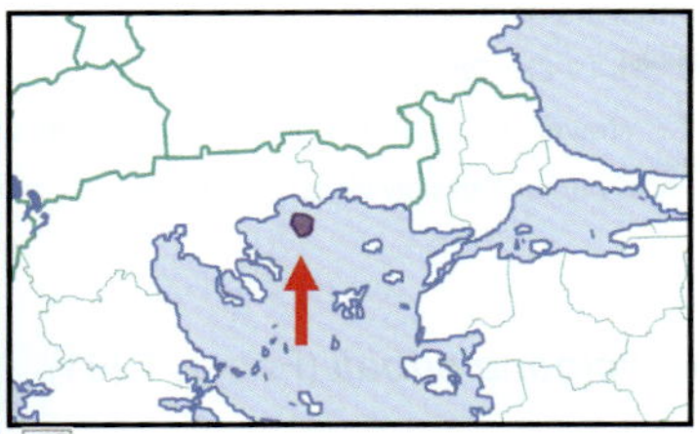

Greece (Thásos)

Thiessea.

Helicigona vikosensis (Subai, 1990)

Chilostoma vikosensis Subai, 1990 in Subai 1990: 245

Shell flat, light horny with brownish-grenish hue, with a brown colour band and two light bands, finely granulated and striated, with few hairs 0.6-0.7 mm long (6-5 hairs/mm^2), last whorl width less than 2 x of penultimate whorl, aperture not detached, insertions 3-6 mm distant, margin 1 mm wide, umbilicus wide and perspectivical (2.5-4 mm).
Mucus glands short, dart slender.

6-13.5 x 16-24 mm

Only known from a small restricted area, the uppermost section of the Vikos gorge.

Subai 1996: 39, Subai & Fehér 2006.

Superba.

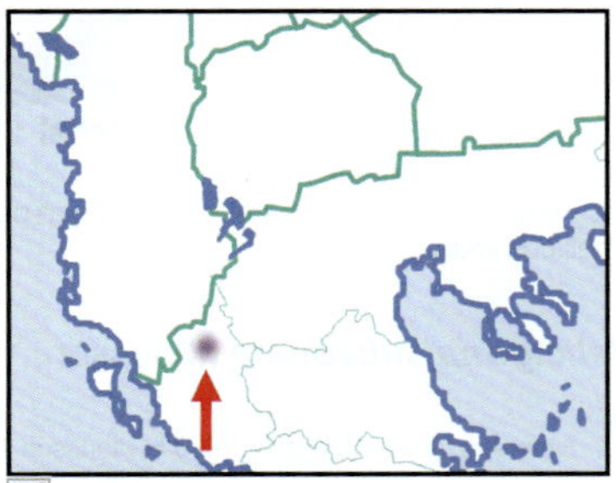

NW Greece

Helicigona zebiana (Sturany, 1907)

Campylaea zebiana Sturany, 1907 in Sturany 1907: 229

Shell yellowish to brown with a brown colour band above the suture, yellowish green on lower side, solid and shiny, finely striated and on upper side with fine spiral lines, 5 weakly convex and slowly increasing whorls, last whorl weakly descending near aperture, aperture with prominent white lip, slightly reflected, distance at parietal side 6 mm, umbilicus open, 4 mm wide, not covered by the columellar margin.

11-14 x 19-25 mm

Albania: Lurë near Liqieni i Zi

Subai 1997: 532.

Subgenus unresolved.

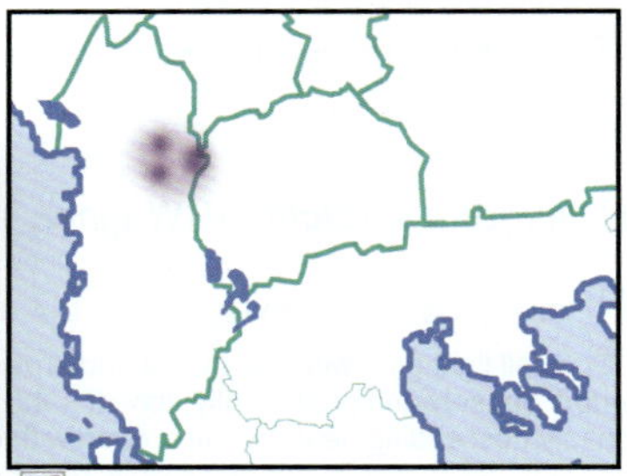

E Albania

EU Red List: Near Threatened NT. Only known from few localities. No threats known.

Isognomostoma

Isognomostoma Fitzinger, 1833
in Fitzinger 1833: 97

Type species *Helix personata* Lamarck, 1792
(monotypy)

Isogn. personatum was treated as the only valid name in this new genus, Fitzinger 1833 cited *Helix isognomostomos* Gmelin, 1791 as a synonym. So monotypy under Art. 68.3.

Isognomostoma isognomostomos (Schröter, 1784)

Helix isognomostomos Schröter, 1784 in Schröter 1784: 194, 699
Treated as a binominal name in the consistently binominal index p. 699. Description on p. 194.

4-7 x 7-11 mm

Germany: Bayern, Franken, Hahnenkamm

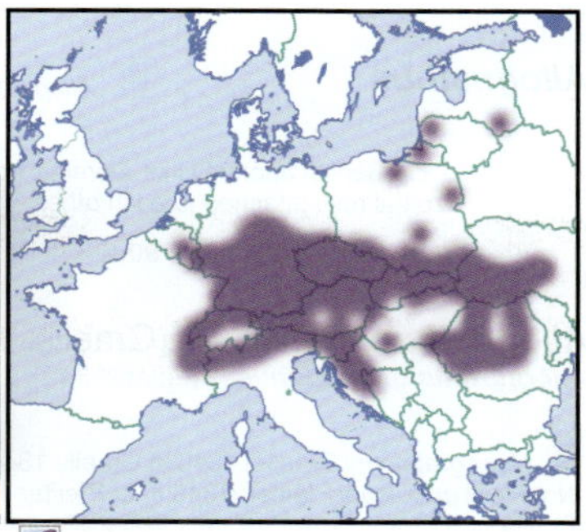

Europe

Germany: Bayern, near Walchensee

Shell brown, smooth with long and slightly curved hairs (can be eroded in old shells), 4-5 convex whorls, rounded at periphery, aperture inside narrow, margin pale brownish and thick, reflected, with a basal and a palatal tooth, parietal side with a prominent white lamella running along the parietal margin of the aperture, but not or only weakly connecting at both ends at the apertural lip, umbilicus in a flattened area and almost entirely covered except a tiny crescentic opening.
Body greyish with brown hue, lighter down to the sides, tentacles grey, up to 5 mm long (not fully extended in the photo). Animals are often shy, but not all.

Under dead wood logs and rock rubble in humid mountain forests, abundant in gorges covered by forests, also in the leaf litter layer. In 300-1800 m altitude.

Vulnerable in parts of Germany. Threatened by modern forest management destroying the soil fauna.

Moquin-Tandon 1855: 118, Germain 1930: 224, Kerney et al. 1983: 279, Grossu 1983: 483, Falkner 1990: 230, Sysoev & Schileyko 2009: 165.

Vidovicia

Vidovicia Brusina, 1904
in Brusina 1904: 162

Type species *Helix lacticina* Rossmässler, 1837
(original designation)
(= currently *Vidovicia caerulans* (Pfeiffer, 1828))

Type species is not *Helix caerulans* Pfeiffer, 1828, as given by Hesse 1931: 89.

Vidovicia caerulans (Pfeiffer, 1828)

Helix caerulans Pfeiffer, 1828 in C. Pfeiffer 1828: 30

8-11 x 15-20 mm

Croatia: near Obrovac

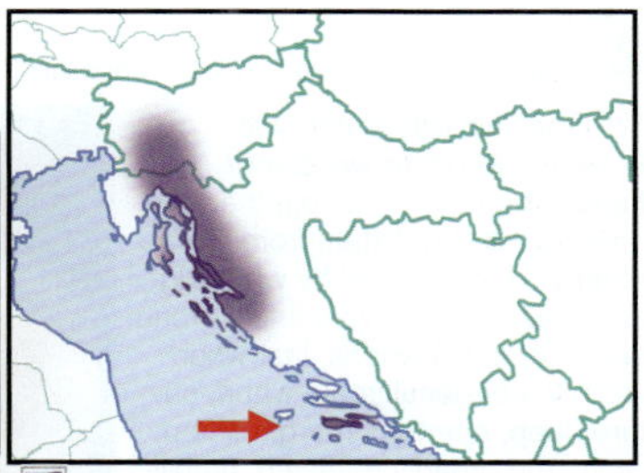

W Croatia, W Slovenia

Shell bluish or violetish grey, not transparent, solid and weakly shiny, 3.5-4.5 weakly convex whorls with protruding apex, last whorl considerably wider, with blunt edge at periphery, lower side more inflated, distinctly descending near aperture, aperture almost circular and connected at parietal side by a strong fold, inside yellow, margin thin, umbilicus narrow and partly covered by the reflected columellar margin.
The animal is grey with reddish hue, broad and relatively large in relation to the shell.

Rocky limestone outcrops, on rock walls and cliffs.

EU Red List: Vulnerable VU. Populations have drastically declined in the past 20 years, at rates of 30 % in the past 10 years. Various subpopulations are now fragmented. The reasons for the observed decline of individuals in intact habitats are not understood. *V. coerulans* is considered a relictual species.

Westerlund 1889: 138, Štamol & Velkovrh 1995 (Cres, at 1 of 80 stations).

There is HNC collection material from Predjamski Grad (W Slovenia), collected in 1974 by S. G. A. Jaeckel.

Croatia: Paklenica National Park

Giuseppe Pocaterra

Helicinae

Allognathus, Assyriella, Caucasotachea, Cepaea, Codringtonia, Eobania, Helix, Iberellus, Iberus, Isaurica, Levantina, Macularia, Marmorana, Otala, Pseudotachea, Tacheocampylaea, Tacheopsis, Theba, Tyrrhenaria, Tyrrheniberus

Allognathus

Allognathus Pilsbry, 1888
in Tryon & Pilsbry 1888: 121, 149 *Helix (Allognathus)*

Type species *Helix grateloupi* Graells, 1846 (original designation)

Genital anatomy like *Iberellus* from which it differs in the radula. Radula with high number of almost similarly shaped and extremely narrow teeth in each horizontal row, attached to each other.

Schileyko 2006: 1790.

Allognathus grateloupi (Graells, 1846)

Allognathus graellsianus (Pfeiffer, 1848)

Helix grateloupi Graells, 1846 in Graells 1846 [after Jan 1846]: p. [24 = Correcciones y Adiciones]
Not *Helix grateloupi* Pfeiffer, 1846 in L. Pfeiffer 1846 (Symb. 3) [after Jul 1846]: 13 (*Helix Grateloupi*)
Established without description and with a bibliographical reference to Grateloup 1840: Pl. 1.
Previously mentioned by Pfeiffer 1842: 99 as *Helix Grateloupii*, a synonym for *Helix rufescens* Grateloup 1840 from Madagascar, not *Helix rufescens* Pennant 1777 (= currently *Arianta arbustorum* (Linnæus, 1758)), without expressed statement indicating a new replacement name (Art.72.7), with a reference to another occasion where the name was claimed to have been used by Pfeiffer 1842, but which was not true. Since the name was not used for a taxon in Pfeiffer's 1842 own classification it can probably not be recognized as a new name. For those who demand a bibliographical reference for making available a synonym under Art. 11.6.1, the name must be attributed to Pfeiffer, 1846, and spelled *grateloupi*; they can use Graells' name for the Spanish species. For the others, to Pfeiffer 1842, and spelled *grateloupii*; they cannot use Graells' name for the Spanish species.
Helix graellsiana Pfeiffer, 1848 in L. Pfeiffer 1848 (Mon. Helic. 1): 437 (*Helix Graellsiana*)

Shell characteristically globulous, almost smooth, whitish with 5 interrupted brown colour bands, darker near the aperture, 4 whorls, last whorl predominating, lip slightly reflected, inside with readdish hue, no umbilicus.
Animal grey.

Rock crevices, also in walls, up to 1000 m.

12-15 x 20-22 mm

Spain: Mallorca, Sierra del Norte

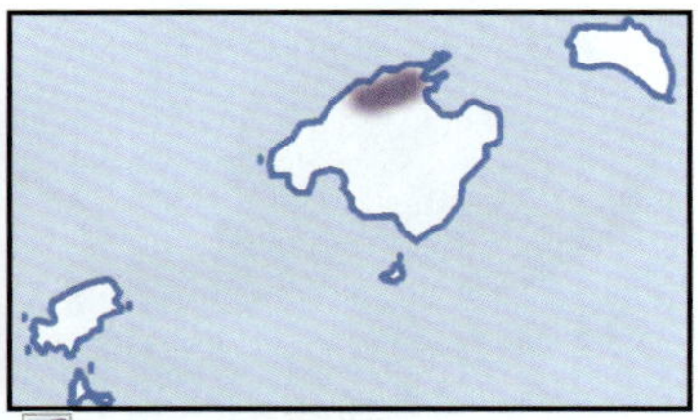

Spain (NW Mallorca)

Vulnerable. Threatened by continuous habitat destructions.

Westerlund 1889: 442, Gasull 1964: 52, Beckmann 2007: 128.

Assyriella

Assyriella Hesse, 1909
in Hesse 1909: 319 *Helix (Assyriella)*

Type species *Helix guttata* Olivier, 1804 (monotypy)

Established with a short description for "die Gruppe der *Lev[antina]. guttata* Ol" without explicit type designation.

S Turkey and Cyprus to Iran

Schütt & Subai 1996 (revision), Schütt 2001: 473.

Assyriella bellardii (Mousson, 1854)

Helix bellardii Mousson, 1854 in Mousson 1854: 379 (*Helix Bellardii*)

Shell whitish or horny to brownish with 5 brown colour bands, 3 bands close together, the other two 3 mm distant from the others, all interrupted by white zigzag lines, smooth, very finely granulated, up to 4.5 whorls, last whorl width 2 x of penultimate whorl, suture deep, apertural margin sharp, white and broadly reflected, umbilicus deep, in many specimens or populations half open, in others closed.

15-21 x 28-38 mm

Cyprus: Nicosia district

Georgios Konstantinou

Cyprus (northern mountain chain)

On limestone rocks.

Assyriella rechingeri (Fuchs & Käufel, 1936)

Levantina rechingeri Fuchs & Käufel, 1936 in Fuchs & Käufel 1936: 658

Shell dirty white with yellowish hue, with 5 brown colour bands, often interrupted by white zigzag lines, but more blurry than in *Levantina spiriplana*, smooth, aperture white with a short and distinct callus at the columellar base, margin sharp, connected by a thin white layer at the parietal side, umbilicus narrow and deep, partly covered by the reflected columellar margin.

15.5-21 x 31-38 mm

Restricted to mountain habitats.

EU Red List: Critically Endangered CR. Rare, usually only subfossil shells are found. It is not known if live animals are still present somewhere, and if yes, by which factors they are threatened.

Pfeiffer & Wächtler 1939, Glaubrecht 1994, Schütt & Subai 1996: 140.

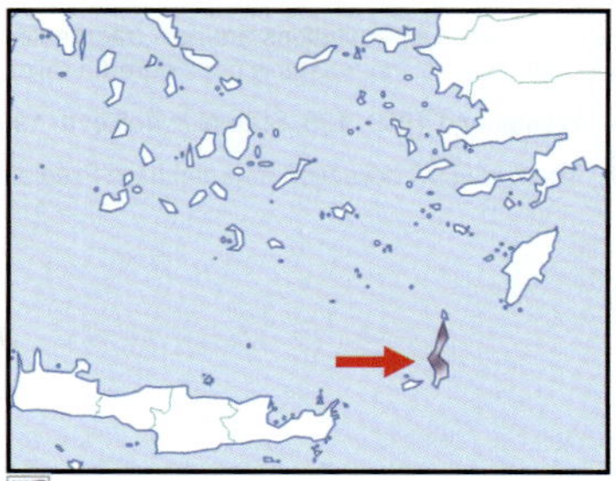

Greece (Kárpathos)

Caucasotachea

Caucasotachea Boettger, 1909 in C. R. Boettger 1909: 10 *Tachea (Caucasotachea)*

Type species *Helix atrolabiata* Krynicki, 1833 (monotypy)

Caucasus and adjacent regions

Neubert & Bank 2006.

Caucasotachea atrolabiata (Krynicki, 1833)

Helix atrolabiata Krynicki, 1833 in Krynicki 1833: 425, Tab. 9 [sic]

Shell whitish with yellowish hue and 2 blackish bands, solid, with prominent growth ridges, 4.5-5.5 regularly increasing and convex whorls, last whorl deeply descending near aperture, apertural margin broadened and dark brown at both sides, margins connected at parietal side by a brown layer, basal margin reflected and with a tooth-like structure, umbilicus entirely covered.

21-36 x 35-42 mm

S Russia: Pyatigorsk

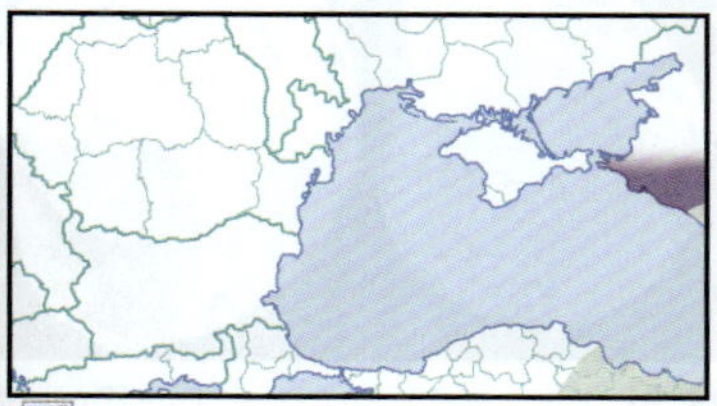

S Russia from Azov Sea to Chechenia, NW Georgia

Westerlund 1889: 443, Shilejko 1978: 332, Neubert & Bank 2006: 107.

Cepaea

Cepaea Held, 1838 in Held 1838: 910

Type species *Helix nemoralis* Linnæus, 1758 (subsequent designation by Herrmannsen 1846: 199)

Cepaea hortensis (Müller, 1774)

Helix hortensis Müller, 1774 in Müller 1774: 52

10-17 x 14-22 mm

Germany: Bayern, Ansbach

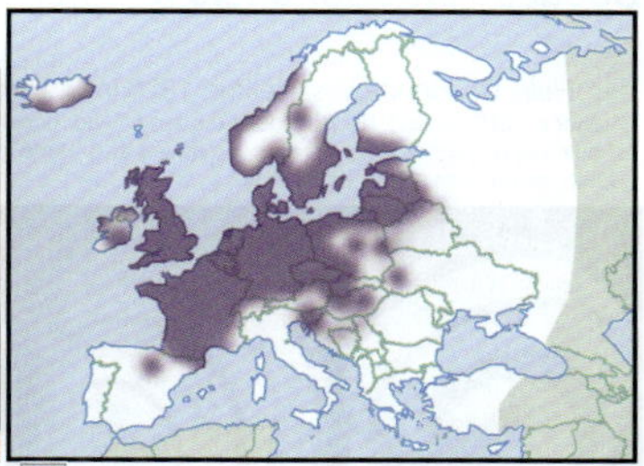

Europe, NE America

Germany: Bayern, near Bad Tölz

Shell reddish, brownish, yellow or whitish, with or without dark brown colour bands, 5-5.5 whorls, apertural lip usually purely white, in dark forests also brown, umbilicus open in juveniles, in adults closed.

C. vindobonensis is less intensively coloured. Differs from *C. nemoralis* usually in its white lip. In areas where lip colour is variable, dissection is necessary. Cross-section of darts in *C. nemoralis* shows a cross with simple blades, *C. hortensis* has bifurcated blades. *C. hortensis* has 4 or more branches of each mucus gland, *C. nemoralis* 3 or less.

Animal light with reddish hue, upper side often slightly darker, tentacles 11 mm long, greyish and weakly translucent.

Habitats very variable, forests, meadows, gardens, near shrubs, dunes. Commonly in more humid and colder habitats than *C. nemoralis* (in N Scotland also in dunes). Up to 2000 m in the Alps, 1000 m in Britain, but rarely above the timberline.

In France 40-80 spherical eggs (2 mm diameter) are laid between May and October (in W France until November), juveniles hatch after 15-20 days, maturity is reached after one year. In Britain snails take 3 years for lip formation, adults can be aged for up to 5 years after lip formation, life span up to 7-8 years, annual survival rate at 50 % (= 3 % in 5 years).

Moquin-Tandon 1855: 167, Germain 1930: 194, Williamson 1979, Kerney et al. 1983: 282, Prieto & Altonaga 1988, Falkner 1990: 240.

Cepaea nemoralis (Linnæus, 1758)

Helix nemoralis Linnæus, 1758 in Linnæus 1758: 773

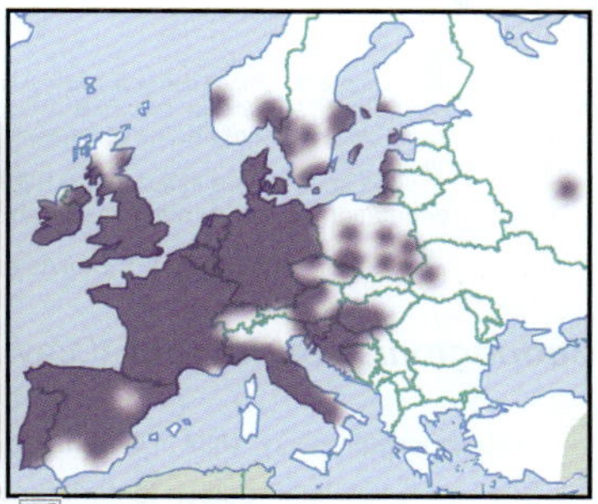

12-22 x 18-25 mm

Germany: Bayern, Ansbach

Europe, dispersed overseas

Switzerland: Berner Oberland, Niesen

Jozef Grego

Shell colour very variable, reddish, brownish, yellow or whitish, with or without dark brown colour bands, apertural lip usually dark brown, rarely white, umbilicus narrow but open in juveniles, closed in adults. For every colour variant names were established in the 1800s, this was later abandoned.

C. vindobonensis is less intensively coloured. Differs from *C. hortensis* usually in its dark lip. In areas where lip colour is variable, dissection is necessary (see under *C. hortensis*).

Body light with reddish or brownish hue, upper side often slightly darker, tentacles darker and 15 mm long.

The animal is moderately slow and not very shy. In Haus der Natur Cismar (N Germany) this species is used to demonstrate snail behaviour to children.

Shrubs and open woods in plains and highlands, dunes, cultivated habitats, gardens and roadsides, locally abundant. Feeds on desintegrating plants, not noxious to crops. Up to 1200 m in the Alps, 1800 m in the Pyrenees, 900 m in Wales, 600 m in Scotland.

Reproduction begins in spring immediately after hibernation, in Britain clutches of 30-50 (in France 40-80) oval eggs (2.3-3.0 mm long) are laid between June and August (in France and Germany May-October, in W France until November), in a hole digged by circulating body movements inside the soil, juveniles hatch after 15-20 days and stay some days inside the nest, maturity is reached after shell is fully grown, in France after 1 year. Life-span up to 7-8 years, annual survival rates at about 50 % (= 3 % in 5 years, older adults suffer higher mortalities).

Moquin-Tandon 1855: 162, Germain 1930: 193, Nobre 1941: 128, Frömming 1954: 314, Prince 1967, Williamson 1979, Kerney et al. 1983: 281, Prieto & A. 1988.

Cepaea sylvatica (Draparnaud, 1801)

Helix sylvatica Draparnaud, 1801 in Draparnaud 1801: 79

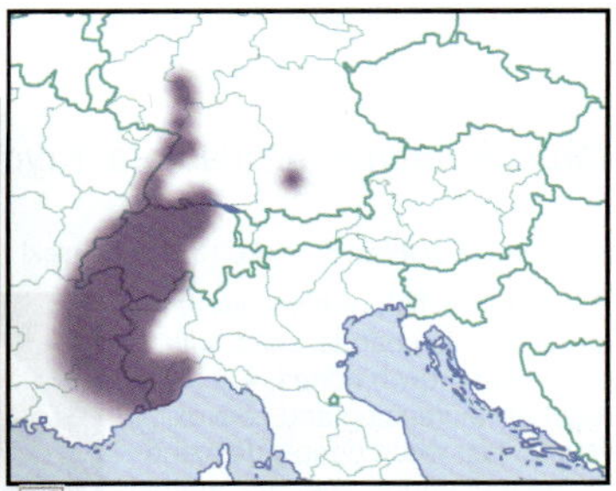

12-16 x 18-25 mm (diameter 16-17 mm above the timberline)

Switzerland: Soloturn, near Aartal, 950 m

W Alps, upper Rhine valley

Shell whitish to yellowish, with up to 5 colour bands, the upper two are usually weak and interrupted, 5-6 whorls, apertural lip usually light reddish, violetish or light brown, lighter near the suture, umbilicus open in juveniles, in adults closed.
Shell flatter than *C. vindobonensis*, less intensively coloured than *C. hortensis* and *C. nemoralis*. In the side view the columellar and upper sides of the aperture are nearly parallel and do not approach each other like in the others.
Body dark brown to nearly black, tentacles dark brown and 9 mm long.

Open mountain forests, meadows near forests, limestone rocks and rock rubble, above the timberline on stony sheep pastures. Almost exclusively on calcareous substrate. Requires high air humidity. Rarely below 500 m altitude, up to 2500 m in Valais. Usually common where present. Secondary habitats have been colonized in river floodplains (Rhein near Karlsruhe, Worms, Mainz, artificially dispersed to Landsberg/Lech).

Moquin-Tandon 1855: 171, Germain 1930: 195, Kerney et al. 1983: 283, Falkner 1990: 240, Vogt et al. 1994: 223, Turner et al. 1998: 375.

Cepaea vindobonensis (Pfeiffer, 1828)

Helix vindobonensis Pfeiffer, 1828 in C. Pfeiffer 1828: 15
Author is not Férussac 1821 because the term "Vindobonensis" there (Férussac [1821-1822] p. 31, folio p. 35, under *Helix sylvatica*, subordinate variant γ) should not be regarded as having been meant as a name (p. 29 terms "Fasciis plurimis" or "Affinis Pomatiæ" were given, if regarded as names the work would not be consistently binominal).

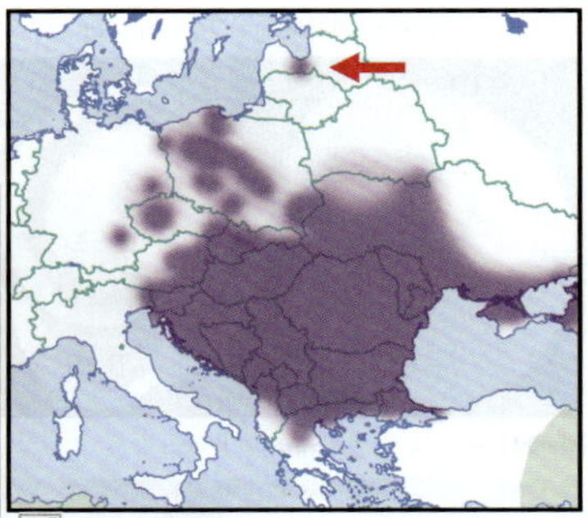

17-21 x 20-25 mm

Austria: Mariabrunn near Wien

E Europe
Ranges in Ukraine and Belarus uncertain.

Austria: Oberösterreich, Plesching — Jan Steger

Shell whitish or dull yellow, with up to 5 colour bands, the upper two are usually weak, 5.5-6 whorls, aperture very oblique, margin brown at the columella to almost white near the suture, umbilicus closed.
Differs from the other *Cepaea* species in its larger size and a more prominently ribbed shell. Colour less intensive than in *C. nemoralis*.

Among shrubs, in open lands and rocky slopes, plains and warm hills, only occasionally in human settlements, in Poland mostly at S exposed and weakly vegetated valley slopes. In S Bulgaria up to 1600 m.
In Hungary probably the most frequent land snail.

Frömming 1954: 308, Damjanov & Likharev 1975: 391, Shilejko 1978: 323, Kerney et al. 1983: 281, Falkner 1990: 240, Manganelli et al. 1995: 33, Schütt 2001: 470, Štamol & Kletečki 2005, Mierzwa 2009, Sysoev & Schileyko 2009: 167.

Codringtonia

Codringtonia Kobelt, 1898
in Kobelt 1898: 208, 306 *Helix (Codringtonia)*

Type species *Helix codringtonii* Gray, 1834
(original designation)

On p. 208 "Gruppe der *Helix codringtonii*" with only one species, this was not a type designation because the term "type" was not involved (Art. 67.5). On p. 306 six species were listed "*C. codringtonii* typica", *C. intusplicata*, *parnassia*, *oetae*, *crassa* and *coracis*. The term "typica" should or could be interpreted in that this species, in contrast to the others, should be regarded as the typical species for the genus *Codringtonia*. The work was not constistent. On p. 322 *codringtonii*, *intusplicata* and *vermiculata* were classified in the genus *Macularia*.

All species are listed in the EU Red List as either Endangered EN or Vulnerable VU. Threatened by global warming and destructive human impact on the vegetation cover (by overgrazing and fire, often in combination) leading to more extreme microclimatic conditions to which the species are not adapted.

Subai 2005 (revision), Giokas et al. 2007. The biology was studied by Hadjiharalambous in 1996, but not published.

Codringtonia codringtonii (Gray, 1834)

Helix codringtonii Gray, 1834 in Gray 1834: 67 (*Helix Codringtonii*)

Shell quite flat, light reddish to dark brownish, colour bands merging, with strong, fathom-like radial lines, umbilicus closed.

23-34 x 38-51 mm

In rock crevices and at large boulders.

Most collection material labelled before 2005 as *C. codringtonii* referred to other species, such as *C. eucineata*.

Greece: SW Pelopónisos

Codringtonia elisabethae Subai, 2005

Cocdrintonia elisabethae Subai, 2005 in Subai 2005: 71

Shell not flat, colour light, with blurry or faintly visible colour bands and weak surface sculpture.

22-31 x 39-48 mm

Greece: Korinthias, Agionori

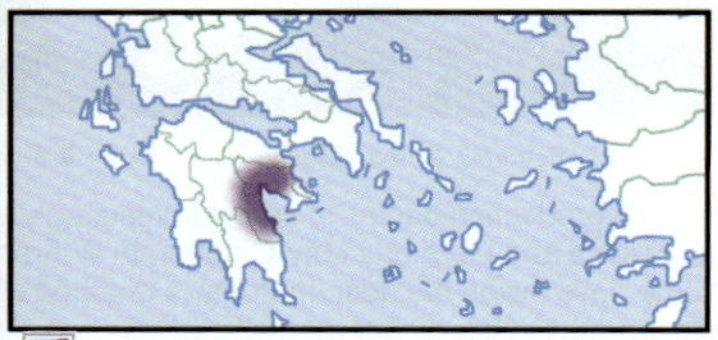

Greece: NE Pelopónisos

Codringtonia eucineta (Bourguignat, 1857)

Helix eucineta Bourguignat, 1857 in Bourguignat 1857: 21, Pl. 9

19-35 x 33-55 mm

Greece: S Pelopónisos

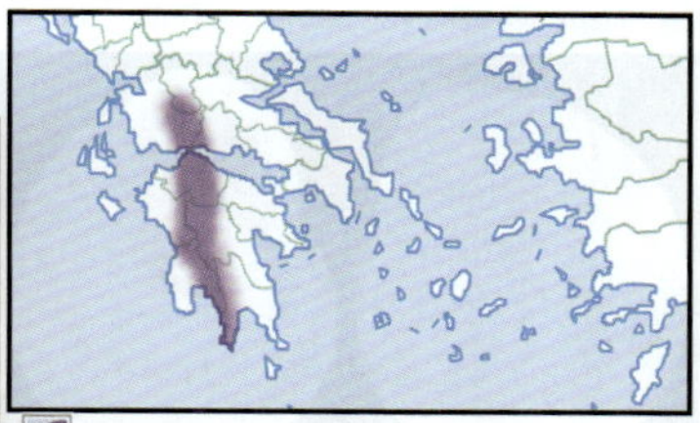

Central Greece to S Pelopónisos

Shell with broad brown colour bands, radial streaks relatively strong, umbilicus covered or nearly covered.

Codringtonia gittenbergeri Subai, 2005

Codringtonia gittenbergeri Subai, 2005 in Subai 2005: 81

25-32 x 43-48 mm

Greece: Pelopónisos, Arkadía, Dafnon gorge (D = 45 mm)
Robert Reuselaars

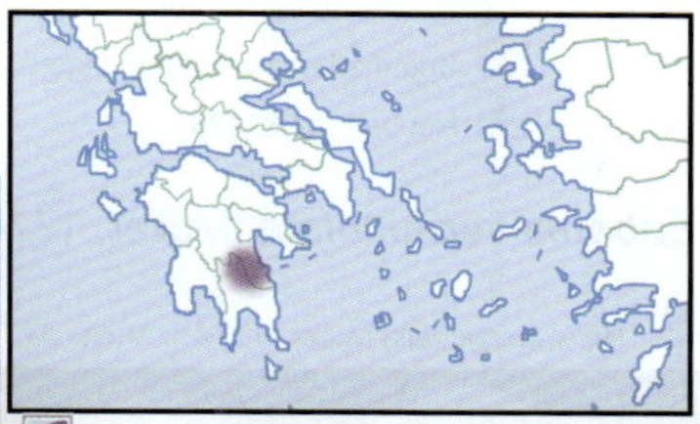

Greece: E Pelopónisos

Shell brown to very dark brown, with dense and fine radial lines, umbilicus closed.

Codringtonia helenae Subai, 2005

Codringtonia helenae Subai, 2005 in Subai 2005: 84

Shell with narrow colour bands and weak surface sculpture, umbilicus closed.

18-27 x 33-47 mm

Limestone rock crevices in open shrublands, coniferous and mixed forests, the species mainly feeds on lichens, also on algae and mosses.

Reproduction in spring, snails mate in April, within 1 month 35-40 eggs are laid in clutches in ground holes, juveniles hatch after 30 days, hatchlings and adults estivate from July to September, in winter snails hibernate from Novermber to March, full shell size and maturity is reached after 4-5 years, adults can breed in various subsequent seasons, life span about 18 years.

Giokas et al. 2007: 270 (life cycle, cited from the unpublished 1996 Ph. D. thesis of Hadjiharalambous).

Greece: central Pelopónisos

Codringtonia intusplicata (Pfeiffer, 1851)

Helix intusplicata Pfeiffer, 1851 in L. Pfeiffer 1851: 26

14-23 x 28-45 mm

Greece: Pelopónisos, Arkadía, Mt. Chelmos

Westerlund 1889: 402, Subai 2005: 87.

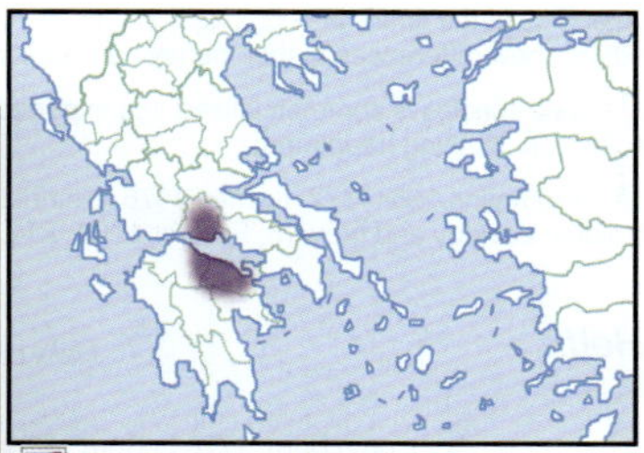

Greece: N Pelopónisos, Fokís

Shell flat, light, with weak colour bands, umbilicus open and covered to only one-third to one-fourth.

Codringtonia neocrassa Zilch, 1952

Helix crassa Pfeiffer, 1850 in L. Pfeiffer 1850: 72 (not *Helix crassa* Mendes da Costa, 1778 from Britain, Planorbidae, currently in genus *Anisus*)

Codringtonia neocrassa Zilch, 1952 in Zilch 1952: 168

Expressly established as a new replacement name for *Helix crassa* Pfeiffer, 1850 non Mendes da Costa, 1778.

Shell solid, flat, with wide blurry and brownish colour bands, and strong, fathom-like radial streaks, umbilicus always open and half covered.

19-26 x 34-48 mm

Greece: Ipiros, between Lias and Lista

S Albania, NW Greece

Codringtonia parnassia (Roth, 1855)

Helix sylvatica var. *parnassia* Roth, 1855 in Roth 1855: 32 (*Helix sylvatica* var. *Parnassia*)

16-27 x 27-44 mm

Greece: Viotia, near Arahova, 1900-2000 m

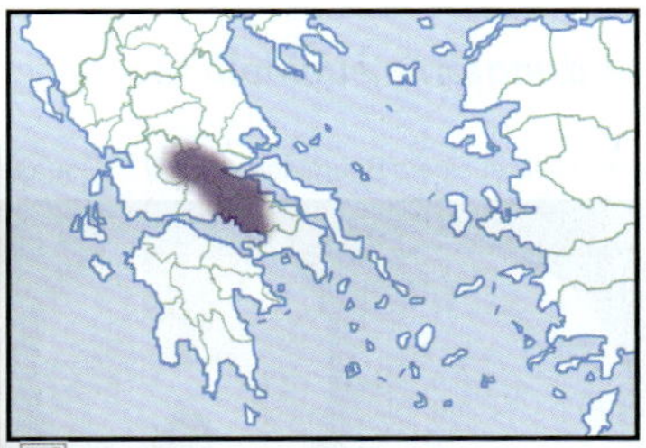

E Central Greece

Shell light, with weak colour bands, umbilicus closed or nearly closed.

Eobania

Eobania Hesse, 1913
in Hesse 1913: 13

Type species *Helix vermiculata* Müller, 1774
(monotypy)

Established without description, but combined with the name *vermiculata* Müller.

Mediterranean region except Morocco and W Algeria

Spermatheca with an elongate diverticulum. One very small dart sac. The flagellum is shorter than penis and epiphallus.

Neubert 1998: 429.

Eobania vermiculata (Müller, 1774)

Helix vermiculata Müller, 1774 in Müller 1774: 20

14-24 x 22-32 mm

Italy: Paestum

In N Greece variation seems to be lower than in S Greece (Gávdos: diameter 24.5-33.5 mm)

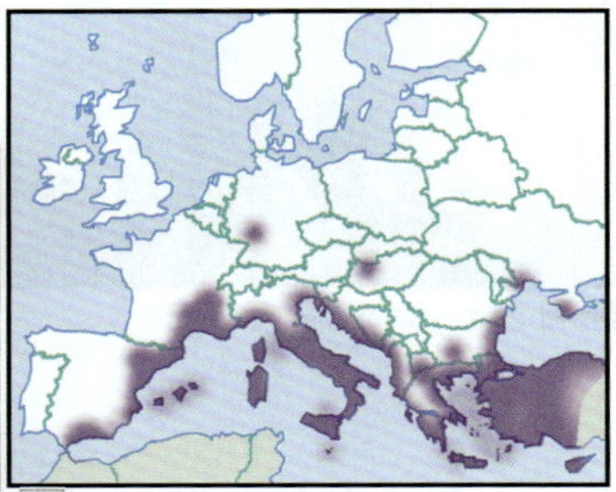

Originally Mediterranean, introduced to many countries abroad
Also in the interior of Turkey, for example in Ankara (A. Örstan, unpublished, 2012)

Croatia: Dalmatia, Rab island — Jozef Grego

Shell colour very variable, whitish to greenish yellow, often with colour bands or spots, lower side frequently with two brown bands and whitish between lowest band and umbilicus, 4-4.5 whorls, last whorl descending abruptly below periphery, apertural margin white, reflected in adult shells, in juveniles only at columellar side, umbilicus narrow and open in juveniles, partly covered by the reflected columellar margin, completely closed in adult shells.
Juveniles differ from *Theba pisana* (with likewise umbilicus) in the larger size of the apex.

In a broad variety of habitats, usually in dry vegetation, mainly in coastal vicinity, also in agricultural crops. Very common in Crete, the species lives on practically every small island in the S Aegean.

In N Greece copulation takes place after the first rainfalls in autumn, around 60-80 eggs (3-4 mm) are laid 20 days later, in a hole digged inside the soil and closed afterwards, juveniles hatch after 13-20 days and grow about 12-13 mm in diameter per year for 2 years (growth is usually restricted to February to June in N Greece, in Crete this period ends already in May). Maturity is reached after 2 years when the diameter reaches 25 mm, the umbilicus becomes closed and the apertural margin becomes reflected. Snails reach 29-30 mm diameter in May/June of the second year in N Greece (in April in Crete), reaching a maximum diameter (33 mm) may take 5 years or more, but mortality increases greatly after 2 years. About 20 % of the snails in a population survive to lay eggs in the 3rd year, 5 % of the snails lay eggs again in the 4th year. The mortality rates decrease with age. The animals hibernate (in N Greece) or estivate (in Crete), but juveniles and adults show differences in their behaviour. Adults dig into the soil and build an epiphragm, while juveniles search protected places under stones or leaves of low plants.

The species is commercialized and exported from Greece to France, which led Lazaridou-Dimitriadou & Kattoulas (1981) to propose restrictions for size and seasons of collection. Nothing happened.

Frömming 1954: 377, Kerney 1976 (distribution Europe), Kerney & Cameron 1979: 203, Lazaridou-Dimitriadou & Kattoulas 1981, 1985, 1986, 1990 (detailed studies on biology, ecology and growth), Falkner 1990: 234, Welter-Schultes 1998: 93 (biology, Gavdos island), Schütt 2001: 469.

Helix

Helix Linnæus, 1758
in Linnæus 1758: 768

Type species *Helix pomatia* Linnæus, 1758
(subsequent designation by Montfort 1810: 231)

Cornu Born, 1778 in Born 1778: 371, type *Cornu copiae* Born, 1778 (monotypy) (= currently *Helix aspersa* Müller, 1774) (name *Cornu* currently under dispute)
Lucena Hartmann, 1821 in Hartmann 1821: 202 ["201"], 208 (footnote), 245, type *Helix tapada* Hartmann, 1821 (monotypy) (= currently *Helix aperta* Born, 1778)
This name has probably not been used since 1899. It was long time regarded to belong to Valloniidae, based on a misunderstanding provoked by Hartmann 1821 (Sturm): 54, where a name *pulchella* was mentioned as a new species without description (nomen nudum), which had nothing to do with *Vallonia pulchella*.
Cantareus Risso, 1826 in Risso 1826: 64, type *Helix naticoides* Draparnaud, 1801 (monotypy) (= currently *Helix aperta* Born, 1778)
Cryptomphalus Charpentier, 1837 in Charpentier 1837: 5, *Helix (Cryptomphalus)*, type *Helix aspersa* Müller, 1774 (subsequent designation by Tryon & Pilsbry 1888: 234)
Erctella Monterosato, 1894 in Monterosato 1894: 168, type *Helix mazzullii* De Cristofori & Jan, 1832 (subsequent designation by Schileyko 2006: 1800)
Contained originally *Helix mazzullii* and *Helix costae* Benoit, 1857. Schileyko 2006 assumed monotypy, type fixed at this occasion under Art. 69.1.1. Colomba et al. 2011 assumed in error that Pilsbry 1895 had designated the type species. Pilsbry 1894 [02 Feb 1895]: 316 wrote that *Erctella* was established "for *H. mazzullii*", this was not a type designation (ambiguous under Art. 67.5.3, the term "type" was not used).

Near East to N Africa and Europe

Helix aspersa and *Helix aperta* are classified in the genus *Helix* here. This has been done until the 1980s. The anatomical differences were known since long. In his widely distributed book, Falkner 1990 used the genus *Cryptomphalus* for *Helix aspersa* and *Cantareus* for *Helix aperta*. This was not based on new evidence, but only a different subjective interpretation of the anatomical record.
Later the genus *Cornu* was proposed to be used for *aspersa*. This name has been under debate until recently. Italian, Russian and Ukrainian have rejected to use it (Manganelli, Salomone & Giusti 2005: 504, Sverlova 2006, Egorov 2008, Sysoev & Schileyko 2009). Currently a Case is pending at the Commission (Cowie 2011). Others have used the genus *Cantareus* for both *Helix aspersa* and *Helix aperta*. No common consensus has been achieved since then.
There is no real need to use a different genus for both species, the names *Helix aspersa* and *Helix aperta* are widely understood, and the main function of names is for communication.
Helix aspersa is quite closely related to the other *Helix* species, although there might be a few species phylogenetically "in between", and which are currently not classified in *Helix*. The genus-species combination *Helix aspersa* is extremely widely used in molecular biology.

Helix species numbers
Taxonomically the genus *Helix* is generally in a bad condition, it certainly needs a profound revision. The identities, delimitations and ranges of several species are uncertain. This unclear taxonomic situation is surprising, since these are the biggest terrestrial snails of Europe and the only ones with major commercial significance as human food.

In SE Europe the last "global" revision was that of Knipper 1939 who proposed to work with 16 species in addition to the 5 undisputed species *Helix aspersa*, *aperta*, *mazzullii*, *pomatia* and *lucorum*. This system seems to be considered as outdated today, but is still used as a base to classify *Helix* in some regions. The general trend is to accept slightly less species in this subgroup.

Manganelli et al. 1995 presented a consistent system for **Italy** which comprised 4 species *cincta*, *delpretiana*, *ligata* and *mileti*.
Grossu 1983 presented 3 species *lutescens*, *secernenda* and *albescens* for **Romania**.
In **Moldova** only *Helix lutescens* was reported by Coadă & Welter-Schultes 2011.
Egorov 2008, Sysoev & Schileyko 2009 and Balashov & Gural-Sverlova 2012 worked with a system of 2 species for **Ukraine** (2 species *albescens*, *lutescens*).
In **Bulgaria** Damjanov & Likharev 1975 recognized 3 species (*vulgaris*, *pomacella*, *figulina*), Irikov & Erőss 2008 splitted *vulgaris* into *albescens* and *philibinensis*, without arguments or references to published studies.
Dhora & Welter-Schultes 1996: 168 and Fehér & Erőss 2009: 35 listed 4 species for **Albania** (*dormitoris*, *oestreichi* with doubts, *secernenda*, *vladika*).

In **Greece** we have no summary of the national fauna, but Mylonas 1985 and others recognised at least *cincta*, *figulina*, *godetiana*, *nucula* and *valentini*. Liebegott 1986 listed *cincta* and *figulina* for the Northern Sporades. Frank 1997 recognised *figulina*, *cincta* and *godetiana* from Ródos, and with doubts another rare form for which she chose the preliminary name *philibinensis*. I do not know recent publications from N Greece with statements concerning *philibinensis*, *secernenda* and *vladika*.
Schütt 2001 tried to give a summary for **Turkey** with 17 species, of which 8 were from the western and central Turkish study area of the present work. In recent studies in W Turkey only *asemnis*, *cincta* and *nucula* were determined by E. Neubert (Neubert et al. 2000, Örstan et al. 2005).

Helix albescens Rossmässler, 1839

Helix obtusata Rossmässler, 1837 in Rossmässler 1835-1837 (5-6): 2
Unclear if this was meant as a name for a taxon in Rossmässler's classification. Not *Helix obtusata* Gmelin, 1791 in Gmelin 1791: 3655 (Achatinoidea or Acavoidea)
Helix albescens Rossmässler, 1839 in Rossmässler 1838-1844 (9-10): 10, Taf. 44
Helix vulgaris Rossmässler, 1839 in Rossmässler 1838-1844 (9-10): 10, Taf. 44
Not *Cochlea vulgaris* Mendes da Costa, 1778 in Mendes da Costa 1778: 72 (= currently *Helix aspersa*)
Helix philibinensis Rossmässler, 1839 in Rossmässler 1838-1844 (9-10): 10 (figured on Taf. 44 Fig. 581 [sic])
On p. 10 in the headline (and in the index p. 45) Rossmässler spelled *Helix Philibinensis* and some lines below on p. 10 said "(Der Name soll wahrscheinlich *Philippinensis* heissen)". Figure caption of Taf. 44 incorrect, name attributed to Fig. 582 and spelled "*H. Philibinenensis*". First Reviser for the three spellings not researched, probably *philibinensis* was selected. If not done yet, then here. Perhaps Rossmässler mentioned it once again in a later work (Art. 24.2.4).
First Reviser of the three names *albescens*, *vulgaris* and *philibinensis* not researched, probably *Helix vulgaris* was selected (Egorov 2008 and Sysoev & Schileyko 2009 used *albescens* because *vulgaris* was preoccupied). If none has been selected, then *albescens* is selected here against both other names. It is convenient to use *albescens* because *vulgaris* can only be used by those who do not classify *Helix aspersa* in the genus *Helix*.

27-36 x 30-38 mm

W Caucasus: Guk mountain

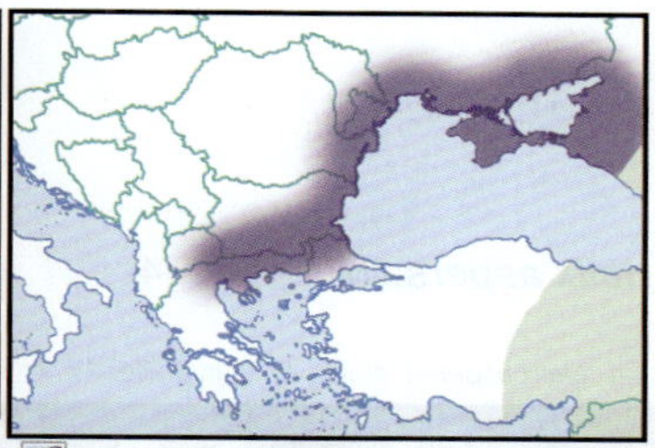

Bulgaria to Caucasus

"N Italy" (= Ukraine ?)
Picture: Rossmässler 1839 (9-10): Taf. 44 Fig. 585-586

Shell dirty yellowish or whitish with 5 brown bands, in Ukraine the 3 middle bands are often interrupted, irregularly striated and finely granulated, 3.5-4 slightly convex whorls, last whorl inflated, descending near aperture, aperture inside in the same colour as outside and with bands, margin blunt and with reddish brown lip inside, often with a brown layer at parietal and umbilical sides, columellar margin reflected, covering entirely the umbilicus.

Open limestone habitats with herbal and shrub vegetation and low humidity, also in forests in leaf litter, in lowlands and mountains. In S Ukraine also in gardens and parks, more rarely at walls and ruins.

Reproduction probably from May to September except in August, sexual activity of the snails could be connected with the moon cycle and rises possibly with increasing moon illumination, clutches of 10-25 eggs (diameter 7-10 mm) are laid in 5-6 cm deep holes of 4-5 cm diameter digged in the soil, the eggs are attached to the walls of the cavity and the entrance is covered by soil, juveniles hatch after 9-15 days under optimal conditions (25°C and 100 % humidity), occasionally after 50-55 days, juveniles may stay inactively in the nest and wait some months.

In Ukraine threatened due to intensive collection provoked by its high commercial importance.

Westerlund 1889: 457 (*obtusata*, 467, Damjanov & Likharev 1975: 396, Shilejko 1978: 341, Grossu 1983: 525 (confirmed that *albescens* has not been found in Romania), Falkner 1990: 248, Zuev et al. 1995 (reproduction), Irikov & Mollov 2006: 813 (considered *philibinensis* and *vulgaris* as synonyms), Hubenov 2007 (used *albescens* for Bulgarian forms, with doubts), Irikov & Erőss 2008 (recognized *albescens* and *philibinensis* as distinct species, without detailed arguments), Egorov 2008: 83 (considered *vulgaris* and *albescens* as synonyms), Sysoev & Schileyko 2009: 168.

Frank 1997: 156 reported one shell from Ródos collected in 1963, possibly not this species.

Helix aperta Born, 1778

Helix aperta Born, 1778 in Born 1778: 399

22-28 x 22-28 mm

Italy: Sicily, Segesta

Mediterranean N Africa, S Europe, introduced to Australia and New Zealand

Greece: Gávdos S of Crete

Vollrath Wiese

Shell with periostracum of green olive colour, last whorl much larger than the others. White convex epiphragma built for estivation. The living snail gives a characteristic noise when disturbed.

Mediterranean shrublands, near cultivated fields, gardens. In Gavdos (Greece) also in woodland spreading on recently abandoned cultivated fields, more rarely in natural habitats. In Crete active after the first rainfalls in October for 3-4 months, estivates buried relatively deep in the soil. Sometimes dispersed with vegetables. Considered as a pest in New Zealand and Australia.

Ovoid eggs (4-5 x 3 mm) are laid in clutches and deposited in moderately deep holes in the soil.

Protected in France, must not be collected for commercial purposes.

Moquin-Tandon 1855: 186, Germain 1930: 187, Falkner 1990: 244, Manganelli et al. 1995: 33, Welter-Schultes 1998: 99.

Originally Italy, introduced in the Roman epoch as food supply to the eastern Mediterranean. There are many different local names for this edible species in Crete, which allow to trace the history of dispersal in combination with the cultural history of the island.

Names: 3 combinations are nomenclaturally correct, but only under certain circumstances.
Helix aperta is always correct, independently from Code interpretations and *Helix aspersa* classification.
Those who do not like to classify this species in *Helix* have various options.
Cantareus apertus is the name for those who like to classify *Helix aperta* and *Helix aspersa* in different genera.
Those who like to classify *Helix aperta* and *Helix aspersa* in the same genus and separate from the genus *Helix*, have two options.
Those who accept *Cornu* as available must say ***Cornu apertum***. For those who regard *Cornu* as unavailable the correct name is ***Cantareus apertus***.
I would recommend to use the genus *Helix* for this species.

Helix asemnis Bourguignat, 1860

Helix asemnis Bourguignat, 1860 in Bourguignat 1860 (Amén malac. 2): 176, Pl. 24

Turkey: Iskenderun

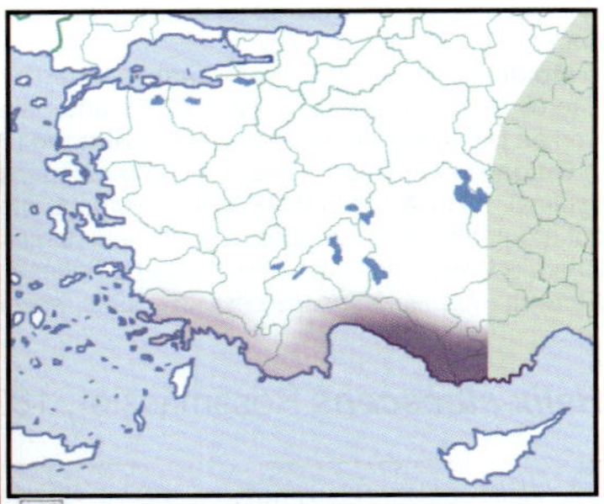

S Turkey to Syria or Lebanon
Range uncertain.

39-45 x 39-45 mm

Westerlund 1889: 454, Schütt 2001: 504.

Shell white or weakly yellowish, usually with two broad colour bands separated by a narrow white zone, striated and with short and irregular spiral lines, 4-4.5 convex whorls, rapidly increasing, last whorl predominating and slowly descending near aperture, aperture widely rounded, inside white or light reddish, margin thick and with a white lip inside, parietal side with a weak white layer, columellar margin reflected, umbilicus covered.

Helix aspersa Müller, 1774

Helix aspersa Müller, 1774 in Müller 1774: 59

(1-5) Spain: Isles Balears, Palma de Mallorca, (6-7, juv.) Greece: Crete, near Pitsídia

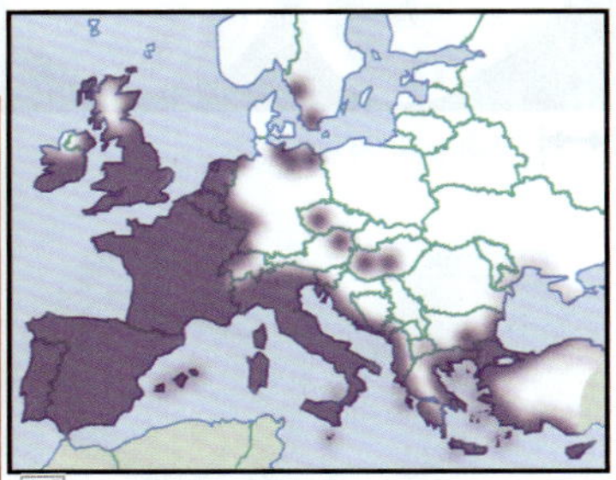

Originally Mediterranean region, today worldwide

30-35 x 32-40 mm

Croatia: Dalmatia, Rab island — Jozef Grego

Fresh shell with brownish golden periostracum and characteristically interrupted brown colour bands, 4-5 whorls, aperture large and characteristically oblique, its margin in adults white and reflected, juveniles with very narrow umbilicus, in adults closed.

Shrubs, light woods, open lands and dunes, often in cultivated areas of any kind. Feeds on herbs and requires cultivated crops or at least presence of garden plants in Greece, can be a troublesome suburban garden pest in Britain. Synanthrope, has been artificially dispersed to many regions. In Switzerland in up to 800 m altitude, in Crete (Greece) up to 1100 m. Tolerates winter frosts, but requires crevices for hibernation in Britain. In N Ireland and N Great Britain mainly coastal. In Portugal the most frequently found land snail species, also in Spain in practically all regions.

In France 50-110 (up to 190) spherical eggs (diameter 4-4.5 mm) agglomerated by a viscose mass are laid 5-8 days after copulation between May and end of October, juveniles hatch after 15-30 days, maturity is reached after 2 years. Life span up to 4 years.

Gassies 1849: 81, Moquin-Tandon 1855: 174, Germain 1930: 185, Nobre 1941: 130, Damjanov & Likharev 1975: 395, Falkner 1990: 244, Manganelli et al. 1995: 33, Ondina et al. 1997, Turner et al. 1998: 376, Kerney 1999: 205, Schütt 2001: 494, Sysoev & Schileyko 2009: 167.

Names: The species has been placed in the genus *Helix* in all sources between 1774 and 1988 and in most bioscientific sources until recently. But in a number of malacological sources since Falkner 1990: 244 the species has been placed in three other genera, depending on the classification in relation to *Helix aperta* and on the accepted interpretation of the ICZN Code for the *Cornu* problem (a Case is currently pending at the Commission, Cowie 2011).

Today the species is known as ***Helix aspersa***, ***Cryptomphalus aspersus***, ***Cornu aspersum*** and ***Cantareus aspersus*** (and also in many incorrect declinations, *Cornu aspersa* and others). In molecular literature widely known as *Helix aspersa*, which is by far the most commonly used name.

Google 08.2009 (02.2011): in *Helix* 85 (88) %, in *Cornu* 9 (7) %, in *Cantareus* 5 (2) %, in *Cryptomphalus* 0.5 (1.3) %, incorrect endings 1 (1.5) %.
Google Scholar 02.2011 "since 2008" ("date ignored"): in *Helix* 88 (96) %, in *Cornu* 7 (2) %, in *Cantareus* 4 (1) %, in *Cryptomphalus* 1 (1) %.

All 4 combinations are nomenclaturally correct, but only under certain circumstances.
Helix aspersa is always correct, independently from Code interpretations and *Helix aperta* classification.
For those who regard *Cornu* as available the name can be ***Cornu aspersum*** (if they do not like to classify it in *Helix*) (the solution proposed by the CLECOM team, Falkner et al. 2001, Bank et al. 2001, Falkner et al. 2002, Bank et al. 2007).
Those who do not regard *Cornu* as available, and do not like to classify the species in *Helix*, can select between ***Cantareus aspersus*** (this is the option for those who classify *Helix aperta* in the same genus as *Helix aspersa*, as done by Italian research teams and others - Manganelli et al. 1995, Giusti et al. 1995, Neubert 1998: 437, Manganelli et al. 2005) and ***Cryptomphalus aspersus*** (for those who like to classify the two species in different genera, as done by Ukrainian and Russian research teams and others (Cossignani & C. 1995: 140, Turner et al. 1998: 376, Sverlova 2006, Schileyko 2006, Egorov 2008, Sysoev & Schileyko 2009).
I would recommend to use the genus *Helix* for this species.

Helix cincta Müller, 1774

Helix cincta Müller, 1774 in Müller 1774: 58
Helix ambigua var. *borealis* Mousson, 1859 in Mousson 1859: 16
Helix cincta var. *anatolica* Kobelt, 1891 in Kobelt 1892: 25, Taf. 127
Pomatia pelagonesica Rolle, 1898 in Rolle 1898: 91 Type locality Pelagonisi, Vórii Sporades.

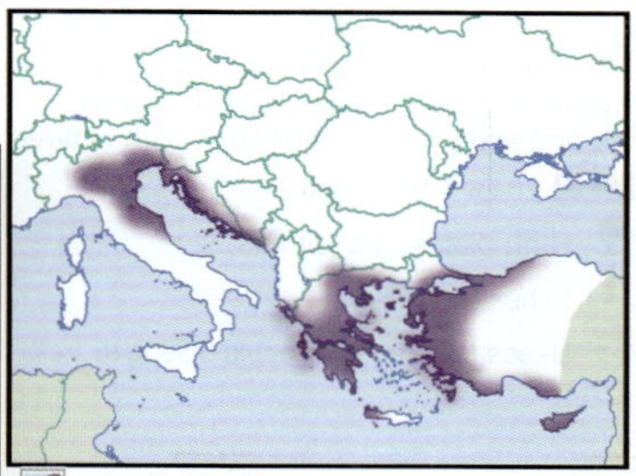

N Italy to Turkey and Lebanon

Should be considered as endangered in Crete.

30-42 x 28-40 mm, size locally variable

W Turkey: near Kuşadası

Shell whitish to yellowish, irregularly striated, often with 5 brown colour bands, the upper 3 bands often melted, in the south the colour bands tend to be weaker, suture fine and deep, aperture higher than wide, inside brown or greyish, lip inside often thickened and brown, columellar margin broadly reflected and covers the umbilicus entirely. Several characters are locally variable.

Shrubby and rocky habitats, often burrowed in the soil. In Crete near Alíkambos the species is burrowed in the soil for most of the year, the snails appear only for a few days in October or November at the first winter rainfalls, then they disappear again in the soil.

Westerlund 1889: 458, Kobelt 1899: 103 (Fig. 1556, regarded *pelagonesica* as a local form of *cincta*), Knipper 1939: 368, Liebegott 1986 (only subfossil shells of *cincta* at the type locality of *pelagonesica*), Falkner 1990: 246, Štamol & V. 1995, Frank 1997: 141, Welter-Schultes 1998: 100, Schütt 2001: 510, Štamol & K. 2005.

Needs a revision. In www.iucnredlist.org (08.2012) it was proposed to split up *cincta* into 3 separate species (*anatolica, borealis, cincta*) that have not been classified as species by Knipper 1939 or in subsequent detailed taxonomic studies, with the remark (*anatolica*) that no anatomical or conchological differences could be found to distinguish those species. In such a form, of course, it is impossible to learn more about these animals. The classification should be based on the published record and only be modified after a scientific revision comprising all species of this group in their entire ranges.

Helix delpretiana Paulucci, 1878

Helix ligata var. *delpretiana* Paulucci, 1878 in Paulucci 1878: 8 (*Helix ligata* var. *Delpretiana*)

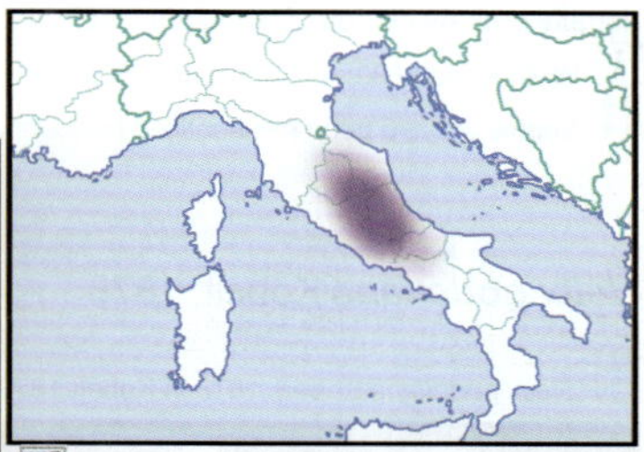

Central Italy

25-35 x 23-33 mm

Italy: L'Aquila, Serra di Celano, 1900 m Alessandro Hallgass

Shell whitish or yellowish with 4 distinct brown colour bands, finely striated, apertural margin inside with broad white lip, columllar margin reflected and broader in its upper section, umbilicus almost or entirely closed.
Differs from sympatrical *Helix ligata* in its lighter background colour. Differs from *Helix mileti* in the presence of a diverticulum of the spermatheca/bursa copulatrix.

In high mountain regions at *Juniperus* shrubs, usually above 1000 m. *H. ligata* is restricted to ground habitats.

Restricted to a small range in the central Apennines.

Westerlund 1889: 466, Alzona 1971: 220, Manganelli et al. 1995: 34, A. Hallgass (e-mail, 2-2008), www.naturamediterraneo.com (2009).

Helix dickhauti (Kobelt, 1903)

Pomatia dickhauti Kobelt, 1903 in Kobelt 1903: 148

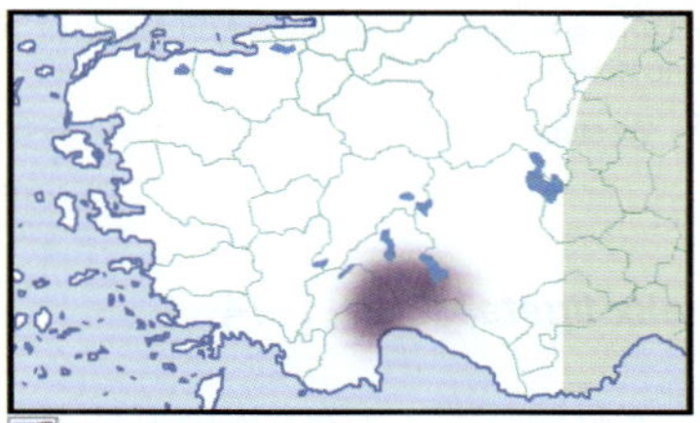

S Turkey

Shell greyish brown with blurry bluish brown bands, fresh shells often with bluish white surface layer, coarsely and irregularly striated, almost ribbed near the aperture, apex broad, smooth and light brown, whorls convex, aperture simple and sharp without callus at parietal side, no lip inside, umbilicus covered in adult shells by the broadly reflected and white columellar margin.

33-41 x 33-41 mm

Exposed habitats on sandy mountain slopes with vegetational cover, usually not in valleys.

Schütt 1976: 66, Schütt 2001: 502.

Helix dormitoris (Kobelt, 1898)

Pomatia dormitoris Kobelt, 1898 in Kobelt 1898 (NachrBl. DMG 30): 163
Pomatia kolaschinensis Kobelt, 1898 in Kobelt 1898 (NachrBl. DMG 30): 164
Selection by First Reviser, if not earlier then herewith (Art. 24.2.3).

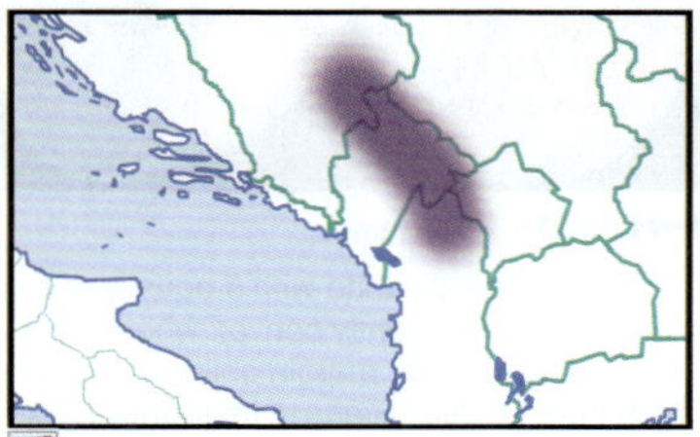

Bosnia to N Albania

Shell light yellowish grey with 5 weak and partly interrupted brown bands, 2nd and 3rd band often merged, solid, coarsely striated, no spiral sculpture, 3.5-4 convex and regularly increasing whorls, last whorl predominating, brownish bands visible inside aperture, margin white and straight, umbilicus closed.
Seems to look somehow like central Italian forms of *Helix ligata*.

31-37 x 31-37 mm

Pavlović 1912: 52 (*kolaschinensis*), Tomić 1959: 24, Urbański 1970.

Helix escherichi Boettger, 1898

Helix escherichi Boettger, 1898 in O. Boettger 1898: 12

Shell small, whitish to yellowish olive brown, nearly without colour bands, irregularly striated, 4 convex whorls with deep suture, aperture inside brownish and shiny, margins connected at parietal side by a weak whitish and shiny layer, umbilicus nearly entirely covered by the reflected columellar margin.

20-24 x 18-22 mm

High plains, grasslands.

Schütt 2001: 496.

Needs to be revised, could belong to *Helix figulina* and related taxa.

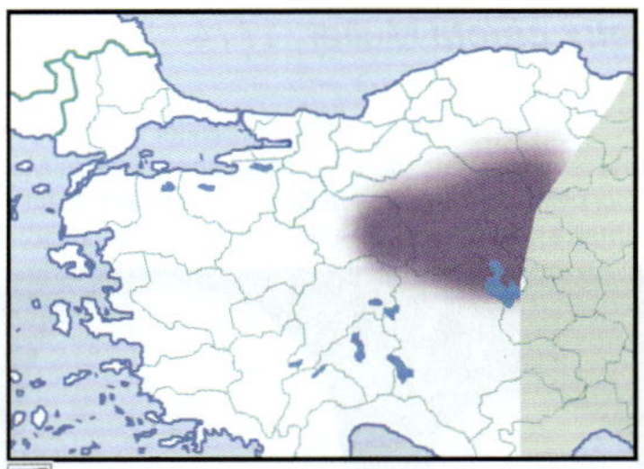

Central Turkey

Helix figulina Rossmässler, 1839

Helix figulina Rossmässler, 1839 in Rossmässler 1838-1844 (9-10): 9, Taf. 44

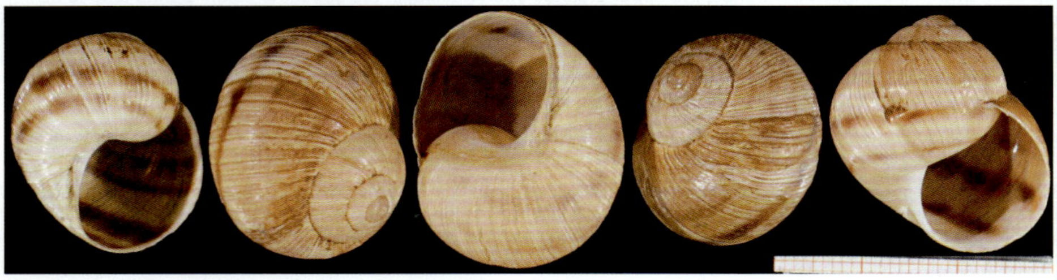

21-27 x 20-26 mm, occasionally 29-31 x 29-30 mm

Greece: Pelopónisos, Argolis, near Prosimna

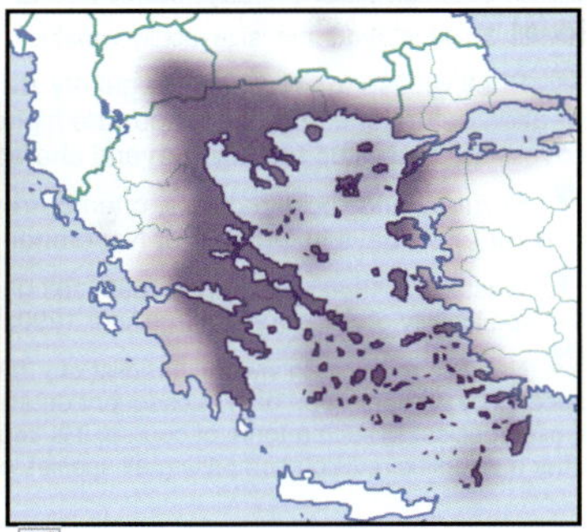

Greece to W Turkey

Shell dirty whitish to yellow brownish, not extremely solid, coarsely and irregularly striated, basically with 5 colour bands, but rarely well visible, aperture vertical, well rounded, colour inside like outside with more intensive colour bands, apertural margin simple with only a weak white lip inside, columellar margin reflected, umbilicus usually entirely closed. Shell highly variable.

Open shrubland habitats, pine woods and dunes, buries deep into the soil. In SW Bulgaria up to at least 700 m.

Westerlund 1889: 454, Falkner 1990: 244, Frank 1997: 136, Schütt 2001: 497, Irikov & Mollov 2006: 815.

Seems to have been misidentified frequently, indicated range uncertain. In Crete, Kíthira and Andikíthira not recent (Gittenberger & Goodfriend 1993).

Helix godetiana Kobelt, 1878

Helix godetiana Kobelt, 1878 in Kobelt 1878: 319 (*Helix (Pomatia) Godetiana*)

35-45 x 32-42 mm

Greece: Náxos, Moutsouna

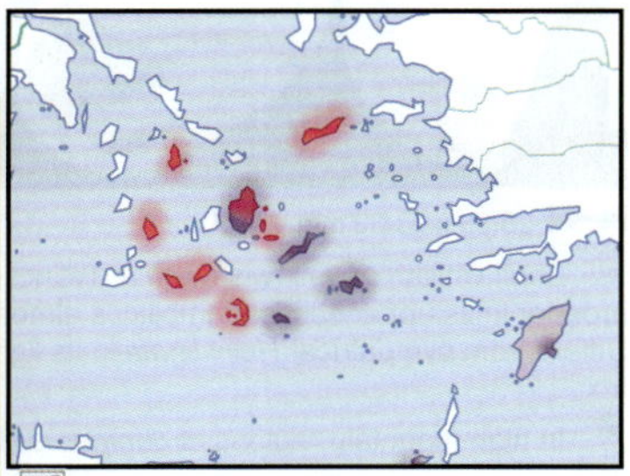

Greece (Kikládes, Ródos, Kásos)
Red: extinct

Shell dark greyish yellow with brown and very broad bands, thin and solid, densely and prominently striated, with very fine spiral sculpture, 3-4 not very convex whorls, last whorl inflated, initially slowly and then more rapidly descending near aperture, aperture inside grey in juveniles and intensively brown in adults, margin thin and with narrow white lip, columellar margin reflected, no umbilicus.

Open Mediterranean shrublands (*Pistacia lentiscus* and *Juniperus phoenicea*) on limestone.

EU Red List: Endangered EN. Protected since 1979 in Kikládes. Threatened by habitat destruction, overgrazing and introduced species *Helix aspersa* and *Helix aperta*, probably also *Eobania*. Recent only in Astipálea, Amorgós, Náxos and Anáfi, possibly extinct in Ródos, extinct in Síros, Sífnos, Kéros, Iraklia, Thíra, Síkinos, Folégandros and Páno Koufonísi.

Westerlund 1889: 451, Gambetta 1929: 83 (anatomy), Mylonas 1985, Frank 1997: 153.

Helix ligata Müller, 1774

Helix ligata Müller, 1774 in Müller 1774: 58

24-37 x 25-40 mm

S Italy

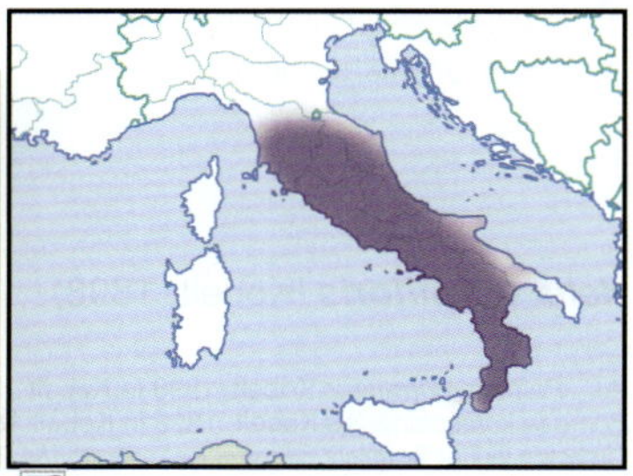

Central and S Italy

Shell yellowish white with 5 equally wide reddish brown bands, striated and with fine spiral lines producing a reticulate pattern, more prominently striated on last whorl, 4-5 slightly convex whorls, last whorl predominating, weakly descending near aperture, aperture inside white with bands, margin inside slightly thickened, columellar margin oblique and almost straight, umbilicus usually slightly open, sometimes completely closed.
Differs from *H. delpretiana* in its slightly darker background colour.

Shrubs and light mountain woods, usually in lower altitudes, sometimes above the timberline, in Abruzzo up to 2000 m where it seems to prefer ground habitats.

Westerlund 1889: 465, Falkner 1990: 246, Manganelli et al. 1995: 34.

Helix lucorum Linnæus, 1758

Helix lucorum Linnæus, 1758 in Linnæus 1758: 773

40-50 x 41-55 mm, diameter up to 60 mm

SE Albania: near Korcë

Europe to Iran
Introduced to various regions, in the 1880s to S France, in the 1990s to Sardinia.

France: Lyon — Cédric Audibert

Shell basically white but almost entirely covered with broad and blurry blackish brown bands with reddish hue, striated and with some spiral lines on the upper whorls, 4.5-5 slightly convex and rapidly increasing whorls, last whorl not much wider, slightly descending near aperture, aperture small, smaller than in other *Helix* species, inside grey with violetish hue and with bands, margin slightly reflected, umbilicus usually covered in adults, but not always completely closed.

Ukraine: Crimea, near Bakhchisaraj — Sergey Leonov

Shrubs, light woods, cultivated areas, at moderate humidity. Active only at night and after rainfalls, at dry weather often burrowed in the soil. In Bulgaria in up to 1200 m. Does not seem to be able to survive in Switzerland.

Germain 1930: 184, Tomić 1959: 22, Alzona 1971: 222, Damjanov & Likharev 1975: 398, Falkner 1990: 246, Dhora & W. 1996: 168, Irikov & Mollov 2006: 813.

Edible snail, commercially collected in many countries for the French market.

Helix lutescens Rossmässler, 1837

Helix lutescens Rossmässler, 1837 in Rossmässler 1835-1837 (5-6): 4, Taf. 21

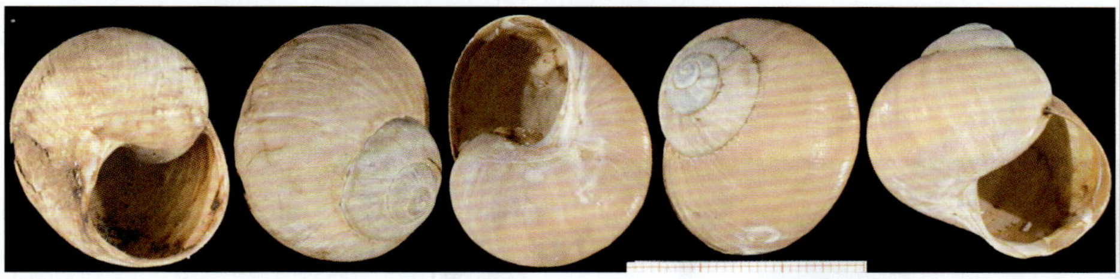

27-33 x 24-31 mm, diameter up to 37 mm

NE Hungary: near Vasarosnameny

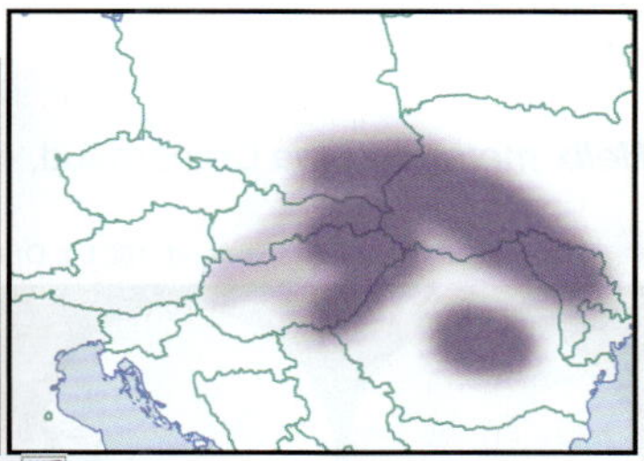

Carpathians and adjacent regions

Slovakia: Slovak Karst mountains, Plešivec — Jozef Grego

Shell usually evenly yellowish white, sometimes with weak colour bands, finely striated, not very shiny, suture well visible, aperture inside whitish, margin with a strong white or reddish lip inside, columellar margin broadly reflected and umbilicus usually completely closed.
Animal greyish with many wrinkles and grooves, and two characteristic longitudinal light streaks long the darker dorsal side.

Meadows, grasslands and shrubs at dry and sunny slopes in lower altitudes. Often in gardens and cemeteries.

Active between afternoon and morning, with a peak in the early morning hours, depending on temperature, humidity and rainfall. Copulation in SE Poland lasts 3 hours, usually in May and early June, two weeks later nests are digged to lay 20-50 eggs (diameter 4-4.5 mm), juveniles hatch after 15 days, maturity is probably reached after 4 years, life span several years. Juveniles of the first and second season prepare shelters to hibernate in mid-September, adults slightly earlier.

Westerlund 1889: 455, Shilejko 1978: 345, Falkner 1990: 246, Koralewska-Batura 1999.

Helix maltzani Kobelt, 1883

Helix maltzani Kobelt, 1883 in Kobelt 1883: 84 (*Helix (Pomatia) Maltzani*)

Shell brown with interrupted bands, coarsely striated, nearly ribbed on last whorl, 4-4.5 weakly convex whorls, last whorl inflated, aperture inside brownish, margins connected at parietal side by a whitish layer which extends around the umbilicus, umbilicus almost or entirely covered by the reflected columellar margin.
Penis long, distal part as long as or longer than proximal part, flagellum longer than penis.

35-50 x 35-50 mm

Forest regions in higher altitudes of mountains north of Izmir.

Schütt 1976: 65, Schütt 2001: 503.

W Turkey

Helix mazzullii De Cristofori & Jan, 1832

Helix mazzullii De Cristofori & Jan, 1832 in De Cristofori & Jan 1832: [mantissa] 1
Helix aspersa var. *insolida* Monterosato, 1892 in Monterosato 1892: 26
Not *Helix insolida* , an incorrect subsequent spelling which was in prevailing usage since the 1840s for *Helix insolita* Rossmässler, 1838 [= currently *Helicigona insolita*], and thus deemed to be the correct name for this species. The Code has a gap and spellings used as correct names under Art. 33.3.1 have no regulation for the homonymy, but it is implicit in the Principle of Homonymy that someone who establishes a species-group name with the same spelling as the correct name of another species, produces a junior homonym of that senior name. Both names were considered as congeneric by Kobelt 1906: 2, so Art. 23.9.5 cannot be applied. An application to the Commission is currently being prepared to legalize the usage of Monterosato's name. Otherwise the next available name *Helix vitincola* would have to be used.
Helix vitincola De Gregorio, 1895 in De Gregorio 1895: 193
Erctella cephalaeditana Giannuzzi-Savelli, Sparacio & Oliva 2012: 109
Name not made available by Giannuzzi-Savelli et al. 1986: 203 who did not use it at a clear rank for a taxon (species? subspecies?), and proposed the name conditionally ("è probabile" = it is probable [that the taxon can be considered as a species or subspecies, under the condition that the stated arguments will once turn out to be tenable]). A name proposed conditionally is not available under Art. 15.1. The name was not made available until 2012.

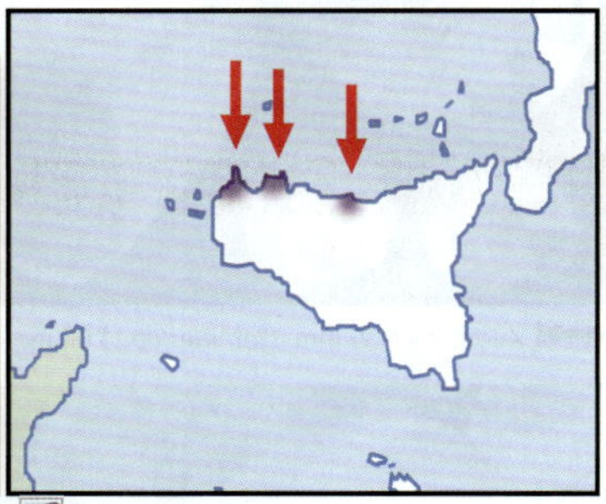

30-40 x 28-35 mm — Italy: Sicily, near Mondello

Italy (NW Sicily)

Shell evenly yellowish or brownish with yellow hue, in some populations with 5 brown colour bands, not spotted, prominently and densely striated, almost ribbed, thin but solid and slightly translucent, apex shiny and very finely striated, 3-4 weakly convex whorls, rapidly increasing, last whorl predominating but not broadened, long and deeply descending near aperture, aperture large and almost vertical, inside white, margins connected at parietal side by a shiny and smooth white layer, no umbilicus. Colour and shape variable, some shells look more like *H. aspersa*, from which they differ in a less oblique aperture (less \ directed) in frontal view. Prominence of surface sculpture increases from west to east, in the eastern population the surface is corrugated. In *H. aspersa* the brown bands are interrupted by yellowish colour spots or dashes, in *H. mazzullii* by more or less corrugated whitish riblets on the shell surface.
The animal is light greyish to yellowish with grey tentacles and a grey band behind the tentacles.

Restricted to limestone cliffs, descend to the soil only at moist weather conditions. The species is able to dug holes in the limestone with its mucus containing H_2CO_3 to dissolve the calcium carbonate of the limestone. The animals live in their holes or tunnels in the limestone walls, and return back home after feeding. Snails feed on living or dried parts of various plants.

EU Red List: Endangered EN. Populations slowly declining due to habitat destructions and commercial collections.

Westerlund 1889: 451, Manganelli et al. 1995: 33, Colomba et al. 2011.

Subgenus *Erctella*. This group is currently being studied, it is possible that we are dealing with 3 separate species.
Colomba et al. 2011 classified the species in a separate genus *Erctella* and splitted *mazzullii* into 3 species with disjunct ranges (from W to E) *insolida*, *mazzullii* and *cephalaeditana*. This splitting into separate species was based on the anatomical, morphological and molecular record. It was however not tested if the differences obtained by these methods corresponded with the presence of intrinsic absolute barriers to reproduction, which would justify a classification at the level of species. Albeit nothing is currently known in the published record about the life cycle and the biology of reproduction of these large-sized animals, such experimental tests in a nearly natural environment would probably not be too difficult, and would permit to align the results of the other methods with the biological species concept.
The taxonomic arguments would have been slightly stronger if various different populations of *Helix aspersa* from various regions of Italy, Europe and abroad had been analysed by the same methods, and the observed genetic distances in this species or species complex could be set in relation to the reproductive behaviour of these animals.

Helix melanostoma Draparnaud, 1801

Helix melanostoma Draparnaud, 1801 in Draparnaud 1801: 77

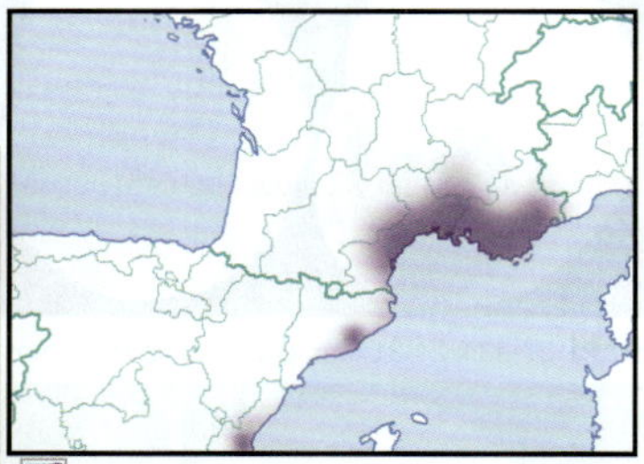

28-36 x 28-33 mm — S France: Provence, near Arles

N Algeria, Tunisia, S France

Must not be collected for commercial purposes in France.

Shell greyish with reddish hue, upper side of last whorl sometimes slightly darker, prominently striated, often less prominently on lower side, 3.5-4 slightly convex whorls, apex small, last whorl inflated, slowly and slightly descending near aperture, aperture inside near margin brownish violet, margin blunt and slightly thickened with a flat lip inside, parietal wall dark brown, often with whitish layers above, columellar margin broadened, umbilicus covered.
Body dark brown or yellowish grey, tentacles light yellowish and 15 mm long.

In France in gardens, olive and wine plantations, hides frequently inside the soil, hibernates with a thick epiphragm.

Eggs numerous, 4 mm diameter and unusually elastic.

Moquin-Tandon 1855: 182, Westerlund 1889: 460, Germain 1930: 186, Kerney et al. 1983: 309, Falkner 1990: 248, Quiñonero Salgado et al. 2010.

Helix mileti Kobelt, 1906

Helix gussoneana var. *mileti* Kobelt, 1906 in Kobelt 1907 (Icon.): 9, Taf. 336

27-33 x 27-35 mm — Italy: Campobasso, Matese, Monte Miletto, 2000 m (type locality)
Alessandro Hallgass

Central Italy

In high mountain regions, under stones, also in alpine zones above the timberline.

Possibly a strictly endemic species.

Manganelli et al. 1995: 34.

Species needs a revision, taxonomic status unclear.

Shell whitish or greyish yellow with five brown colour bands, coarsely striated, aperture simple and sharp, inside light with translucent brownish colour bands, margin white, often connected by a weak whitish and shiny layer at parietal side, umbilicus almost or entirely covered.
Differs from *H. ligata* and *H. delpretiana* in the absence of a diverticulum of the spermatheca/bursa copulatrix (present in *H. delpretiana* and usually present in *H. ligata*).

Helix nucula Mousson, 1854

Helix nucula Mousson, 1854 in Mousson 1854: 367 Type locality in Vilayet Antalya, Turkey.

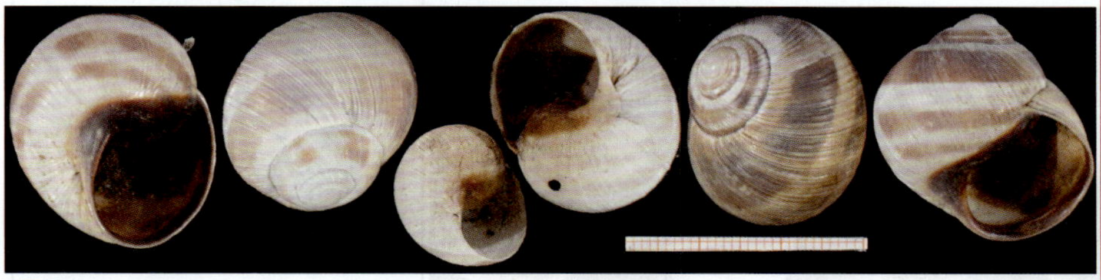

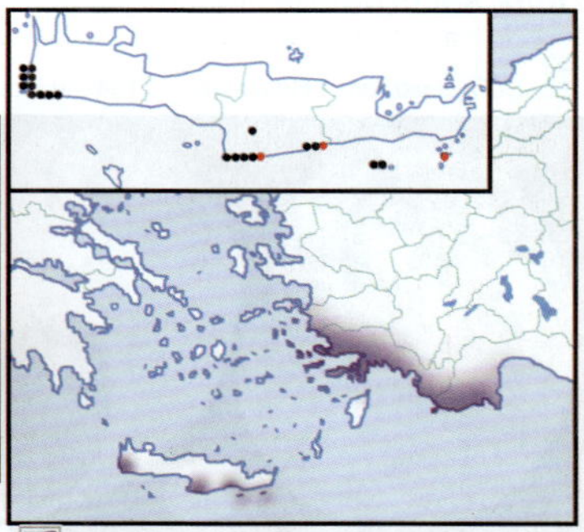

22-28 x 22-28 mm Tunisia: Bizerte

Turkey, Greece, Algeria to Egypt
Insert map: Crete, red: only subfossils

Shell brownish or whitish, usually with 3 or 4 brown bands, locally weaker, densely striated, often with spiral lines giving a reticulate sculpture, 4-4.5 slightly convex whorls, rapidly increasing, last whorl predominating, slowly and slightly descending near aperture, aperture inside white, light bluish or violetish white with broad violetish brown and white bands, margin inside brownish, columellar, umbilical and parietal sides usually intensively dark brown, no umbilicus, not even in juveniles.

In Crete and S Turkey in sandy areas in coastal vicinity, but often also in limestone areas covered by Mediterranean shrubland vegetation. Estivates buried deeply in the soil.

Westerlund 1889: 470, Welter-Schultes 1998: 113-114, Neubert et al. 2000: 113, Örstan et al. 2005: 7.

More research on the identity and distribution of the local forms seems to be necessary. The forms of Crete could belong to a different species. Detailed distribution in Crete based on my own research (HNC collection material), indicated ranges are probably complete.
In Crete the species is consumed, local people from Léndas have tried to rear *H. nucula* near Kolimbári (W of Chaniá) and reported that the species did not survive in northern Crete. The old people in Crete believe that the species was introduced from N Africa, this is also reflected in its local vernacular names (*bárbaros*, *barbarésos*).

Helix pomacella Mousson, 1854

Helix pomacella Mousson, 1854 in Mousson 1854: 365

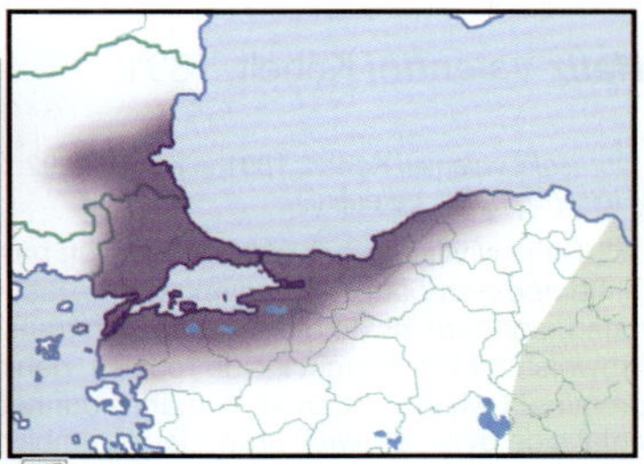

21-26.5 x 22-27.5 mm Bulgaria: near Pomorie Lake

E Bulgaria, NW Turkey

Shell dirty whitish to light horny yellow, finely striated, usually with 5 narrow brown colour bands, aperture rounded, margin reflected, columella usually very oblique, umbilicus closed in adults. Juveniles with open umbilicus.
More globulous than *Helix figulina*, aperture more circular, margin more reflected.

Open habitats with low vegetational cover. In the summer, snails in sandy coastal habitats are buried deep inside the sandy soil, with a well developed epiphragm.

Knipper 1939: 378, Damjanov & Likharev 1975: 403, Schütt 2001: 500.

Helix pomatia Linnæus, 1758

Helix pomatia Linnæus, 1758 in Linnæus 1758: 771 (*Helix Pomatia*)

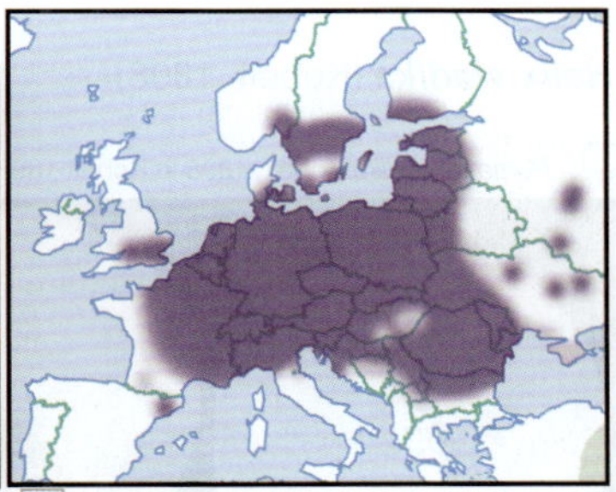

30-45 x 30-50 mm (1-3) W Germany: Heppenheim, (4-7) N Germany: Ostholstein, Fehmarn, near Flügge, (8) France: Alsace, Rouffach

Europe

Germany: Rheinland-Pfalz, Gemünden near Simmern

Shell creamy white to light brownish, often with indistinct brown colour bands, 5-6 whorls, aperture large, margin white and slightly reflected, umbilicus narrow and partly covered by the reflected columellar margin.

In SE Europe in forests and open habitats, gardens, vineyards, especially along rivers, confined to calcareous substrate. In central Europe in open forests and shrubland on calcareous substrate. Prefers high humidity and lower temperatures, needs loose soil to burrow in order to hibernate and lay its eggs. Up to 2100 m in the Alps, usually below 2000 m. In S England restricted to undisturbed grassy or bushy wastelands, usually not in gardens, with a low reproduction rate and low powers of dispersal.

Reproduction in central Europe from end of May to September, 40-80 eggs (diameter 5.5-6.5 mm) are laid in clutches (usually 1 per season and individual), in a hole digged by circulating movements of the body inside the soil, usually after rain, juveniles hatch after 3-4 weeks and may consume their sisters under unfavourable climate conditions, the shell is initially whitish, yellowish brown after 8-14 days, after 1 year the shell has 2.5-4 whorls, in Hungary a first apertural margin is built in the third year. Maturity is reached after 2-5 years, life span up to 20 years, 10 year-old individuals are probably not uncommon in natural populations, maximum 35 years.

Threatened by continuous habitat destructions and drainage, usually less threatened by commercial collections. There were many unsuccessful attempts to establish the species in various parts of England, Scotland and Ireland, it only survived in natural habitats in S England, and is threatened by intensive farming and habitat destruction. In many regions there are restrictions for commercial collections.

Helix secernenda Rossmässler, 1847

Helix secernenda Rossmässler, 1847 in Rossmässler 1847: 163

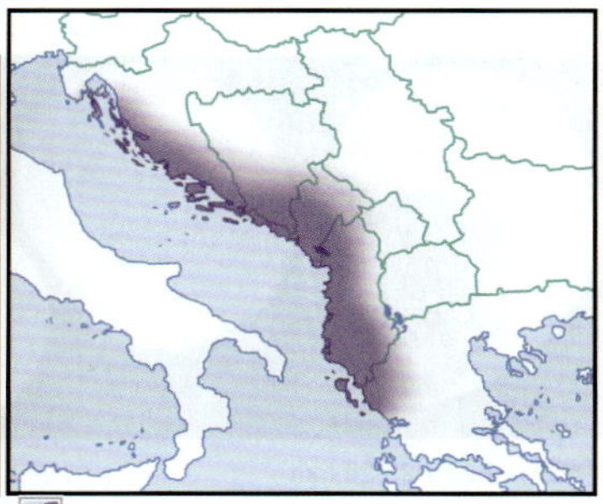

40-45 x 37-45 mm, diameter up to 60 mm

Albania: Zogaj near Shkodër

NW Croatia to NW Greece

Montenegro: Kanjon Mrtvica / Kanjon Morača — Cédric Audibert

Shell large, reddish white with 5 blurry narrow spiral colour bands, the second lowest is quite distant to the others, white near the suture, densely and finely striated, usually without spiral lines, apex large and rounded, aperture inside shiny, reddish white, margin inside brownish, columellar margin also brownish, umbilicus usually completely closed.
Colour bands not always typical, merging bands and other variations are possible.

Rocky limestone habitats in shrublands and forests, also in the vegetation along rivers.

Vulnerable in Albania.

Westerlund 1889: 467, Knipper 1939: 361, Falkner 1990: 248, Štamol & Velkovrh 1995 (N Croatia: Cres, frequent, not Lošinj), Dhora & Welter-Schultes 1996: 169, Dhora 2004: 147.

Two so-called subspecies were propose, their alignment with bioscientific criteria (partial barriers to reproduction) remains to be researched.

Helix valentini Kobelt, 1891

Helix valentini Kobelt, 1891 in Kobelt 1892: 27, Taf. 128 (*Helix (Pomatia) Valentini*)

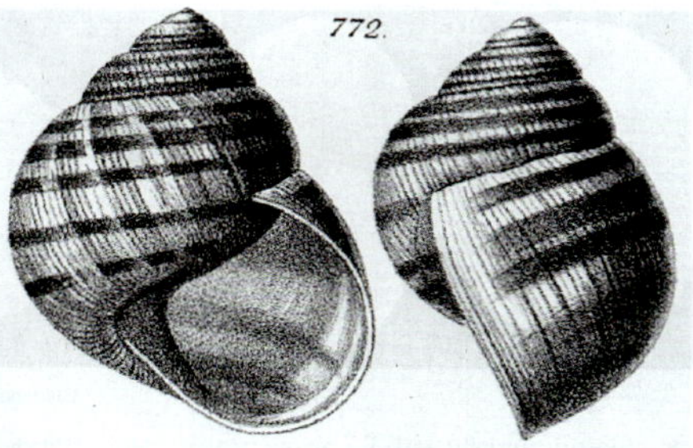

38-46 x 36-44 mm

Greece: Kálimnos
Picture: Kobelt 1792: Taf. 128 Fig. 772

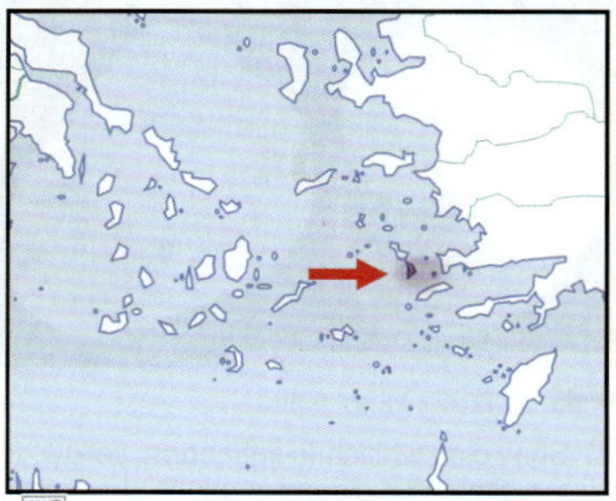

E Greece (Kálimnos, Psérimos)

Shell whitish with five reddish brown bands and white apex, smooth, 4-5 convex whorls, last whorl inflated, aperture inside bluish white with translucent brownish bands, margin inside reddish, behind which a prominent white lip is present, marins connected at parietal side by a yellowish brown layer, umbilicus completely covered.

Shrubland habitats, seems to require a minimum vegetational cover.

EU Red List: Endangered EN. Apparently extinct on Kálimnos after having been commercially exploited.
The strictly endemic species had been sold on the local food market until the 1930s in Kálimnos (Fuchs & Käufel 1936). When the resources on Kálimnos were extinct in the 1930s, the dealers seemed to have exploited the population on Psérimos until the population collapsed there, too. It still lives there, but in low numbers. The species has survived on 5 small uninhabited islands (Telendos, Plati, Nera, Ag. Kiriaki, Safonidi) where it seems to be abundant.
The species was also reported from Kos (Kobelt 1906: 116), but this was not confirmed in the survey by Bank & Neuteboom 1988 who considered the species as extinct.

Knipper 1939: 374, Fuchs & Käufel 1936, Bank & Neuteboom 1988: 56, Frank 1997: 143, Örstan et al. 2005, Triantis et al. 2008: 473, www.iucnredlist.org (08.2012).

Helix vladika (Kobelt, 1898)

Pomatia vladika Kobelt, 1898 in Kobelt 1898: 164

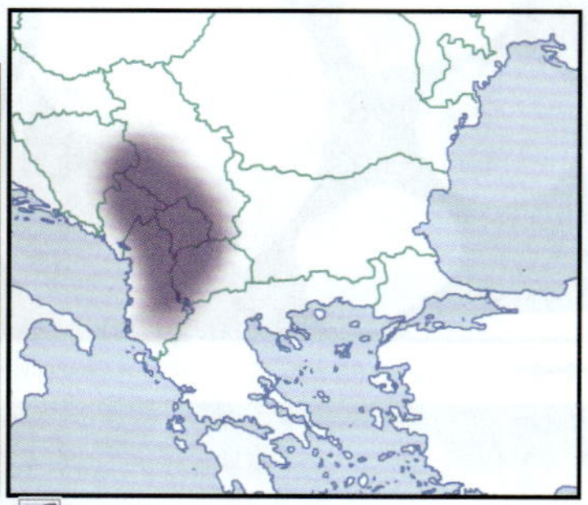

45-55 x 45-55 mm

Montenegro: near Morača monastery

Albania to Serbia

Shell pale yellowish with 5 partly merged weak brownish bands, solid and weakly shiny, 4.5-5 moderately convex and regularly increasing whorls, last whorl weakly descending near aperture, inside light yellowish or with orange hue, colour bands very weakly visible, margin with a strong and often light reddish brown lip, umbilicus usually closed in adults.
Looks like a very big *Helix pomatia*, aperture in relation to shell smaller, shell more solid and more coarsely striated.

Does not crossbreed with *H. pomatia*.

Pavlović 1912: 50, Knipper 1939: 353, Tomić 1959: 23, Falkner 1990: 248, Dhora & Welter-Schultes 1996: 169, Dhora 2004: 147.

Iberellus

Iberellus Hesse, 1908
in Hesse 1908: 131 *Jberellus*

Type species *Helix balearica* Rossmässler, 1838
(absolute tautonymy by Kobelt 1904)

Hesse 1918 used J consistently instead of I as the initial character. A rule should be inserted to the Code that allows to correct that. Expressly established as a new replacement name for
Balearica Kobelt, 1904 in Kobelt 1904: 132, 157, 200 *Iberus (Balearica)*
Contained originally 10 species classified in the genus *Iberus*, among these was *Iberus (Balearica) balearicus*. So this should be absolute tautonomy.

Genital anatomy like *Allognathus*, differs in the radula which is normally developed in *Iberellus*.

Iberellus needs a revision. Sometimes classified as a subgenus of *Allognathus*. The shells are quite variable in various characters, but the nature of this diversity is not yet understood. I found no published constraints for recognising more than one species.

Iberellus hispanicus (Rossmässler, 1838)

Without gender agreement: *Iberellus hispanica* (Rossmässler, 1838)
Iberellus balearicus (Rossmässler, 1838), ***Iberellus pyrenaicus*** (Rossmässler, 1839), ***Iberellus minoricensis*** (Mittre, 1842)

Helix companyonii as mentioned by Companyo 1837: 91, 98 (*Helix Companyonii*), nomen nudum without description or indication
Name was attributed to Aleron in original source, but Aleron was not involved in making the name available. The sentence "cette espèce se rapproche beaucoup de l'*helix serpentina* et *ondulata*" cannot be interpreted as a description "to denote the taxon" (Art. 12.1) (Alonso-Zarazaga 2004, this Code interpretation is accepted here).
Helix hispanica Rossmässler, 1838 in Rossmässler 1838-1844 (7-8) [June 1838]: 15, Taf. 33
Not *Helix hispanica* Potiez & Michaud, 1838 in Potiez & Michaud, 1838 [27 Oct 1838]: 16 (= currently *Otala lactea* (Müller, 1774)).
Helix hispanica synonym *Helix balearica* Rossmässler, 1838 in Rossmässler 1838-1844 (7-8): 15
Also mentioned as a synonym of *Helix hispanica* by Pfeiffer 1848: 273. Used as a name for a species by Pfeiffer 1853: 195, with a bibliographical reference (citing the name *balearica*) to Pfeiffer 1848: 273. I do not know where Rossmässler's 1839 synonym was eventually cited bibliographically and made available under Art. 11.6.1. First Reviser between both names Rossmässler himself at that occasion.
Helix hispanica var. *pyrenaica* Rossmässler, 1839 in Rossmässler 1838-1844 (9-10): 11, Taf. 46
Not *Helix pyrenaica* Draparnaud, 1801 (= currently *Norelona pyrenaica* (Draparnaud, 1805))
Helix hispanica var. *pyrenaica* synonym *Helix Campanyonii* [sic] Rossmässler, 1839 in Rossmässler 1838-1844 (9-10): 11, Taf. 46
Rossmässler (1839: 11) referred to information obtained in a letter from Hermann Eduard Anton from Halle (Germany). Albers 1850: 81 reported that Anton told him to have obtained the snail under the name *campanyoni* from Parreyss and (Albers) suspected that the name referred to a locality Campana at the foot of the Pyrenees near Perpignan. So it is likely that Anton had also written *campanyoni* in the letter to Rossmässler, so Rossmässler did not know the name of the person to which the name was dedicated and the spelling *Campanyonii* was not an inadvertent error under Art. 32.5. Name was subsequently used by Dupuy 1847-1852 (1848): 120 with bibliographical reference to Rossmässler 1839 (however Dupuy spelled *companyonii* and did not quote that Rossmässler had mentioned a synonym), making the name available under Art. 11.6.1. Correct original spelling must be *campanyonii*.
Helix minoricensis Mittre, 1842 in Mittre 1842: 188 (*Helix Minoricensis*)
Helix pythiusensis Bofill i Poch & Aguilar-Amat, 1924 in Bofill i Poch & Aguilar-Amat 1924: 27, Lám. 1
Iberellus tanitianus Forés & Vilella, 1993 in Forés & Vilella 1993: 23

8-10 x 13-15 mm

Spain: Menorca

Spain: Palma de Mallorca

Spain: Ibiza/Eivissa

Spain (Balearic islands, Calaluña)

Shell whitish with 5 interrupted brown bands which tend to create zigzag patterns, finely striated, 4-4.5 weakly convex whorls, last whorl suddenly descending near aperture, aperture inside often reddish brown or red, particularly at parietal and umbilical sides, margin inside with light reddish or white lip, lower margin slightly reflected with a strong callus, umbilicus covered. Smaller than *Eobania vermiculata*.

Open habitats or open forests, in dry and rocky places, at walls and slopes, not much exposed to sunshine, hides under stones.

Westerlund 1889: 410, Beckmann 2007: 128.

Locally variable in shell and possibly also in genitalia. Several separate species or subspecies have been recognised and associated with particular islands.
The name *hispanicus* has not frequently been used for this species. This was mainly because malacologists believed long time that *Helix hispanica* Potiez & Michaud was established in 1835 and had precedence (Falkner et al. 2002: 103). But Rossmässler's name was published earlier (Kadolsky 2012: 83).

Iberus

Iberus Montfort, 1810
in Montfort 1810: 146

Type species *Helix gualtierana* Linnæus, 1758
(original designation)

Aparicio & Ramos 1988, Martínez-Ortí, Ap. & R. 2004: 14, Elejalde at al. 2005, Elejalde at al. 2008.

The number of species recognised in this genus is under debate and has continuously been oscillating in the past decades. It is possible that we are dealing with a single polymorphic species or with several well separated species. The number of species obtained in molecular approaches seems to depend from the number of populations analysed. Elejalde at al. 2008 discovered several so-called "morphologically cryptic lineages" in 40 populations, which means that for profoundly understanding this species complex a very high number of populations would have to be analysed.
It would also be useful to study more closely the situation in contact zones of parapatrically distributed so-called species, to see if absolute or partial barriers to reproduction exist in these zones.

Iberus gualtieranus (Linnæus, 1758)

Without gender agreement: *Iberus gualtierana* (Linnæus, 1758)

Helix gualtierana Linnæus, 1758 in Linnæus 1758: 770 (*Helix Gualtierana*)
Helix alonensis Férussac, 1821 in A. Férussac & Deshayes 1819-1851: Pl. 39 [13 Jul 1821]
Name for the taxon was previously mentioned without description on (explication planches) p. xiv [06 Apr 1821] (*Helix (Helicogena) Alonensis*), also as *Helix (Helicogena) alonensis* Férussac, 1821 in A. Férussac [1821-1822] [25 May 1821]: 31 (folio p. 35) (contained another reference to Pl. 36A which was never issued)
Date was 13 Jul 1821 under Art. 10.1.1.
Established with figures for five subordinate variants α to ε, but not with a distinct description or indication for the main form (nominal taxon). The name *Helix alonensis* was made available at this occasion under Art. 12.1 because it had an indication for the taxon (Code Glossary "taxon": a taxon encompasses all included taxa of lower rank). Art. 12.1 does not demand a description (or indication) for the nominal taxon, a description for the taxon is sufficient, this can include subordinate taxa.
For those who do not like to accept this interpretation of the Code, the next available author is Lamarck 1822 (p. 68, *Helix alonensis*) (they would also need to verify whether Art. 11.6.1 can be applied under any interpretation of their choice, since Férussac's name could be interpreted as a synonym).

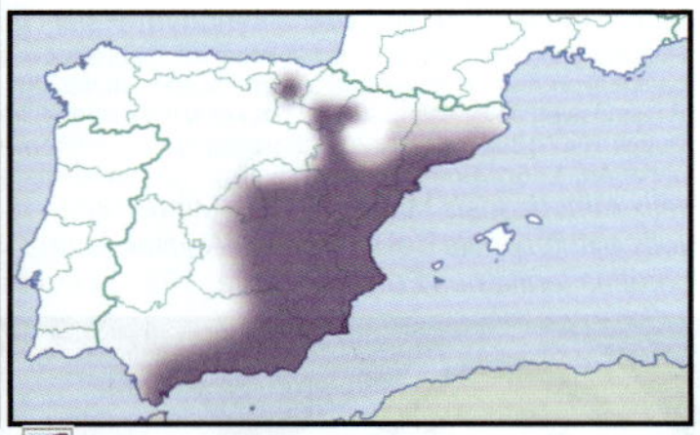

14-22 x 25-50 mm — Spain: Sierra de Gador near Almería — E and central Spain

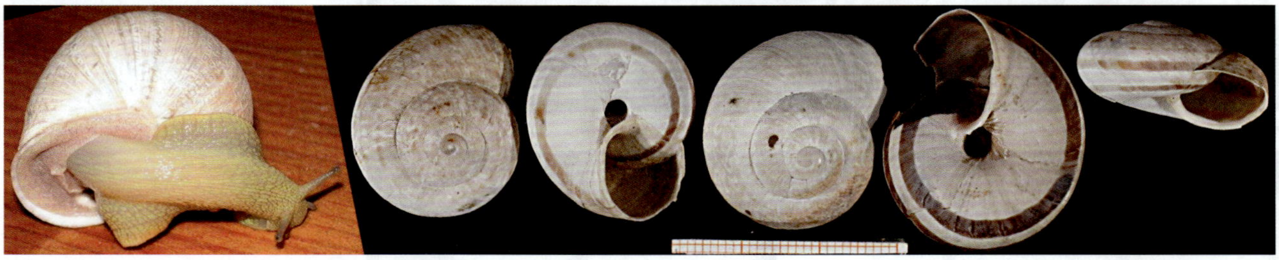

Spain: Sierra Nevada — Charles Geerts — Spain: Sierra de Cazorra, 1500 m

Shell light brownish or whitish with brown bands, moderately granulated and convex whorls (*alonensis* form, with rounded periphery) or prominently reticulated and keeled (*gualtierianus* form, with flat or concave upper whorls), 3.5-4.5 rapidly increasing whorls, last whorl deeply descending near aperture in both morphs, apertural margin white and shiny, reflected except on upper side, umbilicus covered in keeled forms, usually open in rounded forms, occasionally partly or entirely covered.

Rocky open lands not much covered by vegetation, rocky mountain areas. Usually under boulders and in rock crevices. The *alonensis* form is much more abundant than the *gualtierianus* form.

Has been frequently collected and sold on local markets until the 1900s, but after 1950 it has become rare due to human-based pressure. The *gualtierianus* form is considered as Endangered.

Westerlund 1889: 367, Prieto & Altonaga 1988, Falkner 1990: 238.

Elejalde et al. 2005 regarded *alonensis* as a subspecies, but could also be regarded as a form or morph. The *alonensis* form seems to be the most widely distributed form, various other forms (*gualtieranus* being one of them) are locally restricted and usually intermediates with the *alonensis* form are present in the contact zones.

Iberus marmoratus (Férussac, 1821)

Without gender agreement: *Iberus marmorata* (Férussac, 1821)

Helix marmorata Férussac, 1821 in A. Férussac & Deshayes 1819-1851: (explication planches) p. xiv, Pl. 40 (*Helix (Helicogena) marmorata*)
Text p. xiv from 06 Apr 1821, Pl. 40 from 10 Nov 1821, the last date is effective under Art. 10.1.1.

11-14 x 17-25 mm — Spain: Torremolinos near Málaga — S Spain

Shell whitish or yellowish with 4-5 narrow rows of brown spots, lower side lighter, thin and shiny, finely striated and with distinct spiral lines, 3.5-4.5 weakly convex whorls, last whorl wider and deeply descending near aperture, aperture inside white or light reddish with bands, margin sharp and reflected with whitish or light reddish lip inside, no umbilicus.

Rocks covered by vegetation and rocky open shrublands.

Westerlund 1889: 403, Falkner 1990: 238. Not studied by Elejalde et al. 2008.

Isaurica

Isaurica Kobelt, 1901 in Kobelt 1902: 35 *Helix (Isaurica)*

Type species *Helix praecellens* Nägele, 1901 (monotypy)
(= currently *Isaurica praecellens* (Nägele, 1901) from S Turkey

S Turkey

Subai 1994 (revision).

Isaurica lycia (Martens, 1889)

Helix condringtoni var. *lycia* Martens, 1889 in Martens 1889: 182

27-40 x 43-52 mm

Turkey: Vilayet Antalya, (1-2) near Finike, (3-6) near Kaş

Shell large, yellowish grey to dirty white, with 4 broad dark colour bands, can partly be dashed due to radial white ziczac patterns, densely striated, umbilicus closed.

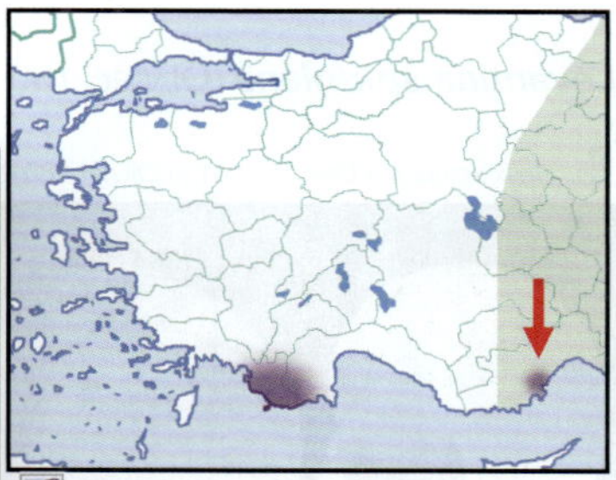

S Turkey, in two areas

Relatively widely distributed, occurs at many localities in natural habitats.

Falkner 1990: 238, Subai 1994: 63, Neubert et al. 2000.

Isaurica pamphylica Subai, 1994

Isaurica pamphylica Subai, 1994 in Subai 1994: 73

Shell dirty to yellowish white with 4 interrupted brown bands, darker near aperture, also visible from inside the aperture, densely striated, 4.5-5 regularly increasing whorls, last whorl width 2 x of penultimate whorl, descending near aperture, apertural margin broadly reflected, umbilicus usually closed.
The animal is grey, lighter near sole.

24-33 x 43-59 mm

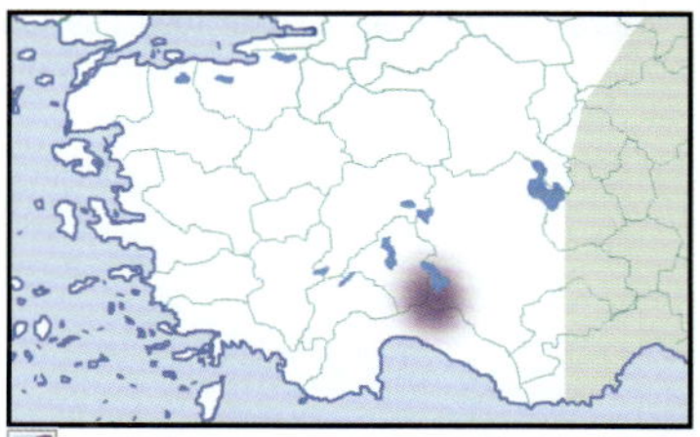

S Turkey

Isaurica riedeli Subai, 1994

Isaurica riedeli Subai, 1994 in Subai 1994: 75

Shell very large, not flat, whitish with 4 broad brown colour bands, the 3rd band is twice as broad as the others, umbilicus entirely covered.

Occurs only in a small restricted area.

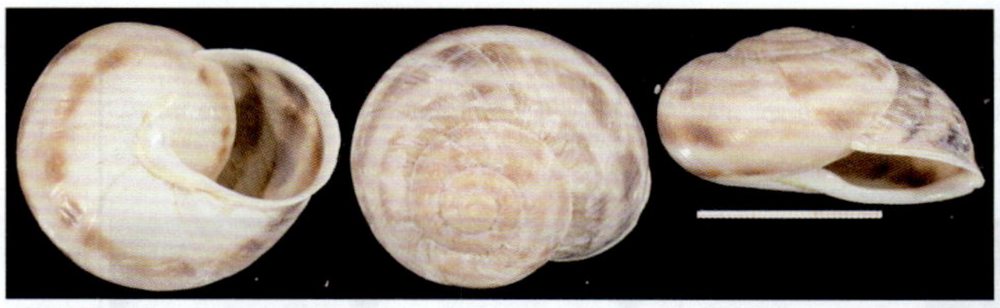

27-36 x 45-54 mm

Turkey: Vilayet Antalya, near Ibradi

S Turkey

Isaurica schuetti Subai, 1994

Isaurica schuetti Subai, 1994 in Subai 1994: 67

20-30 x 38.5-51.5 mm

Turkey: Vilayet Antalya, near Kemer

Shell light with narrow and blurry colour bands, almost smooth or very finely striated, umbilicus closed.

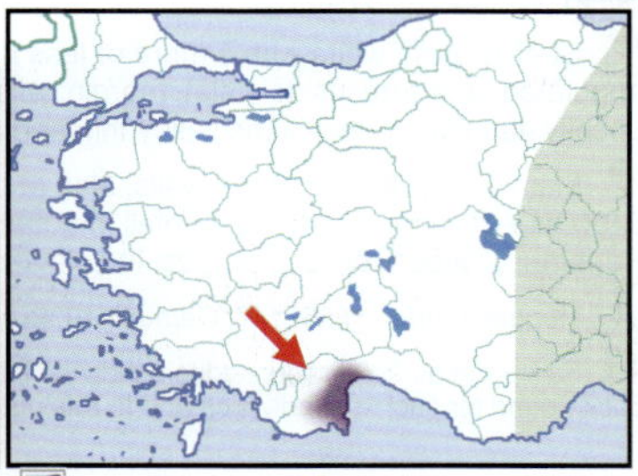

S Turkey

Levantina

Levantina Kobelt, 1871
in Kobelt 1871 (Cat.): 19 *Helix (Pentataenia (Levantina))*

Type species *Helix spiriplana* Olivier, 1801
(subsequent designation by Martens 1873 (Zool. Rec.): 152)

Shell medium-sized to large, with depressed to low conical spire.
Flagellum heavily coiled, glandulae mucosae very small, branching of glandulae is rare.

E Aegean and S Turkey, Syria to Palestine and Jordan, Cyprus

Subai 1994: 59, Neubert 1998: 430.

Levantina spiriplana (Olivier, 1801)

Helix spiriplana Olivier, 1801 in Olivier 1801: 416, Pl. 17

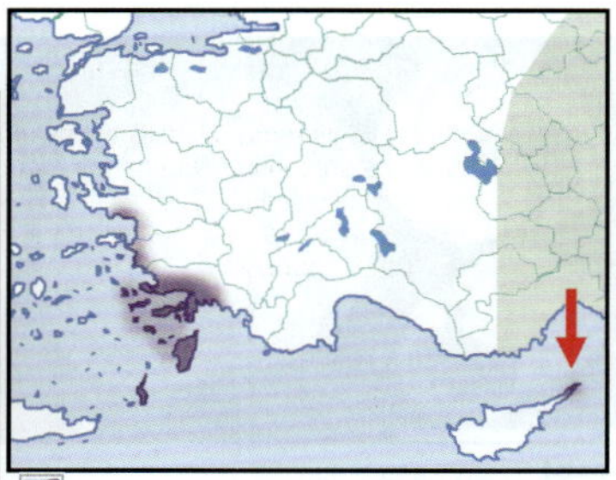

10-20 x 23-40 mm

(1-5) Greece: Ródos, (6-7, juv.) Turkey: near Dalaman

Turkey and SE Greece, E Cyprus, Syria to Palestine/Israel

Shell whitish with horny brown hue, finely granulated on upper side, with 5 interrupted brown colour bands producing a characteristically dashy zigzag colour pattern, the third band exactly on the keel line, juveniles sharply keeled, last whorl can also be weakly keeled until shortly before the aperture, aperture inside weakly reddish, with white reflected lip, connected by a thin callus at the parietal side, umbilicus open in juveniles, in adults half or completely covered.

At limestone rocks, in Karpathos in all kinds of habitats everywhere on the island. Was dispersed by humans in the antiquity, today often at ancient sites.

Westerlund 1889: 391, Falkner 1990: 238, Glaubrecht 1992, Vardinoyannis (unpublished thesis 1994: 119), Schütt 2001: 471, Örstan 2004.

Macularia

Macularia Albers, 1850
in Albers 1850: 80 *Helix (Macularia)*

Type species *Helix niciensis* Férussac, 1821
(subsequent designation by Lowe 1855: 166)

Macularia niciensis (Férussac, 1821)

Helix niciensis Férussac, 1821 in Férussac & D. 1819-1851: (expl. planches) p. xiv, Pl. 39 (*Helix (Helicogena) Niciensis*)
Date was 06 Apr 1821. Also mentioned as *Helix (Helicogena) niciensis* in Prodrome, Férussac [1821-1822]: quarto p. 32 (folio p. 36) from 26 May 1821, with references to Pl. 39A Fig. 1 (date 06 Apr 1821) and Pl. 40 Fig. 9 (date 10 Nov 1821).
Helix saintivesi Kobelt, 1906 in Kobelt 1906: 56, Taf. 328 Type locality Cians valley.

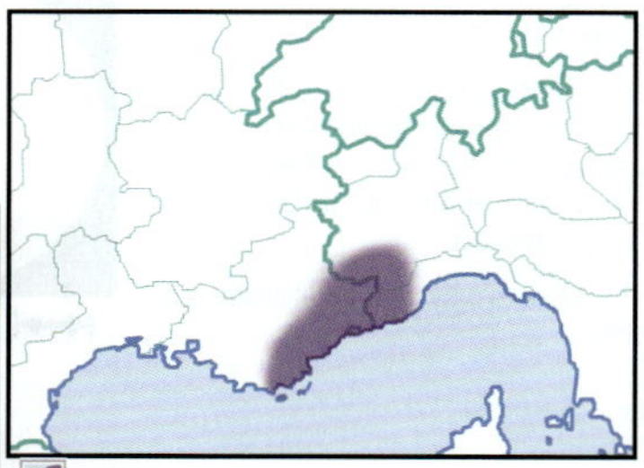

10-15 x 20-26 mm

France: (1-4) Nice, Le Vallon Rouge, (5-7) Alpes-Maritimes, near Pont du Loup

SW Alps

Endangered in Italy. Protected by law in France, eggs and animals must not be collected.

Shell greyish white with 5 more or less interrupted brown bands, sriated, 4.5-5 moderately convex whorls with well marked suture, last whorl initially weakly angulated, aperture inside violet, margin reflected, umbilicus narrow, in adults nearly or entirely covered by the columellar margin.
Locally variable. In the Cians valley thin and almost transparent, apertural margin thin, colour bands also visible inside in the aperture.

Mainly at rocks, old stone walls, also at old olive tree trunks, on surfaces covered by lichens, usually only in limestone regions, also in high altitudes above 2100 m. Active mostly during rainfall in autumn and spring, tends to escape cold and hot temperatures. In the Cians valley at 900-1200 m on constantly humid Permian siliceous rocks in gorges, in colonies of not many indviduals.

Moquin-Tandon 1855: 147, Germain 1930: 199, Boato et al. 1984, Kerney et al. 1983: 307, Falkner 1990: 232, Manganelli et al. 1995: 33, Falkner et al. 2002: 165.

Falkner et al. 2002 proposed to subdivide the complex into 2 species *niciensis* (with 5 subspecies) and *saintivesi*. More research is needed to establish a bioscientific background for such a subdivision and to rule out other options to explain the local variation, such as continuous spatial gradients and environmental factors (Cians valley).

Marmorana

Marmorana Hartmann, 1844
in Hartmann 1844: 210 *Helix (Marmorana)*

Type species *Helix serpentina* Férussac, 1821 (monotypy)

Appeared without description, combined with the name *serpentina*. According to the style of the author it was implicit that *Marmorana* was meant as a subgenus of *Helix* (p. 204).

Murella Pfeiffer, 1877 in Pfeiffer 1877 (Malak. Bl.): 8 *Helix (Iberus (Murella))*, type usually given as *Helix muralis* Müller, 1774 (not verified)

Established without description, with a bibliographical reference to "*Iberus* sect. 1 Alb.", which referred to Albert & Martens 1860 (see p. 5). The generic name was made available under Art. 12.2.5, but it had no originally included species. I have not researched who was the first author to have included species, and which ones. Type not designated by Kobelt 1904 because this work contained a disclaimer.

Fiorentino et al. 2008 and Fiorentino et al. 2010 analysed phylogenetic relationships between *Marmorana* populations and observed local differentiation of parapatrically distributed forms and incongruent patterns between morphology and phylogeny as revealed by the molecular record. The results disagreed with the currently accepted taxonomic classification, in terms of subspecies, species and genera.
The authors found several clades and in 2010 they suggested to classify them in two genera *Murella* (containing at least 2 species in Italy, possibly more in Malta) and *Marmorana* (containing various Italian species). Relationships and behaviour of populations at both sides of contact zones, where shells with intermediate shapes were found, remain to be studied. *Marmorana* species were never found sympatrically. It was also suspected that it could be more appropriate to regard the studied *Marmorana* populations as belonging to one single morphologically variable or polytypic species.
The Fiorentino et al. team treated their preliminary conclusions in a sensitive form and reported possible taxonomic implications (Fiorentino et al. 2010: 35) without presenting new classifications in a too definite frame. This sensitive approach has the effect that confusing new preliminary name usages were avoided, and that we can continue to use the names in their traditional meanings until the taxonomic relationships between the many local populations in Italy are better understood in detailed studies.
Here I am presenting the traditional classification, it would not be useful to interfere in the ongoing process of a revision of the genus. The range maps for Sicily represent approximate ranges of forms for which exactly determined ranges will probably never be established.

Marmorana fuscolabiata (Rossmässler, 1842)

Helix strigata var. *fuscolabiata* Rossmässler, 1842 in Rossmässler 1838-1844 (11). 2, Taf. 51

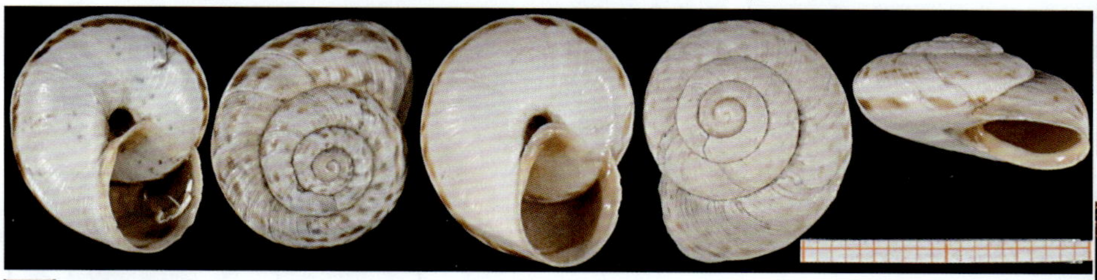

11-13 x 18-25 mm — Italy: Calabria, Monte Tiriolo

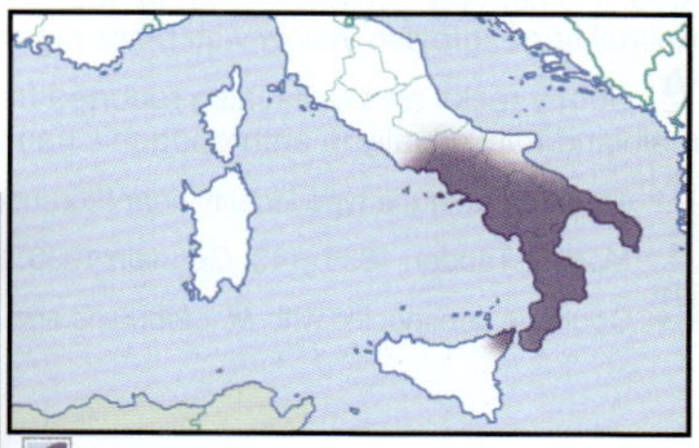

S Italy

Italy: Calabria, Monte Pollino, 2000 m — Francesco Mascia

Shell greyish white, on upper side with white and brown spots produced by several interrupted bands, one interrupted band being visible, ribbed or prominently striated, lower side white and almost smooth, 4-4.5 slightly convex whorls, initally with edge at periphery, last whorl deeply descending near aperture, aperture inside light brown and with lip, margin light brown, umbilicus partly covered by the light brown reflected columellar margin. The animal is grey with bluish or greenish hue.

Westerlund 1889: 384, Manganelli et al. 1995: 32.

Up to 30 so-called subspecies have been proposed to reflect the local morphological variation.

Marmorana globularis (Philippi, 1836)

Helix globularis Philippi, 1836 in Philippi 1836: 127 (not *Helix globularis* Jeffreys, 1833 (= currently *Ashfordia granulata* (Alder, 1830))
Helix granulata was published as a nomen nudum by Alder in 1830 and made available by Alder 1838: 340 - so that Jeffreys' 1833 name would have to be used for that species. The homonymy is not new, it was documented by Sherborn 1922-1931, so Art. 23.9.5 can probably not be applied.

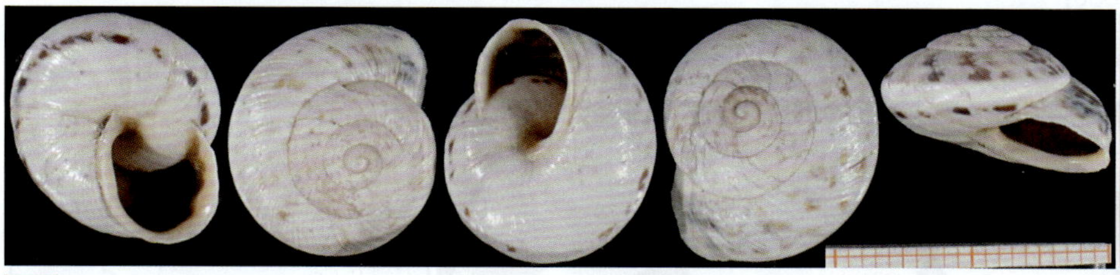

10-14 x 16-21 mm — Italy: Sicily, near Palermo, San Giuseppe Jato

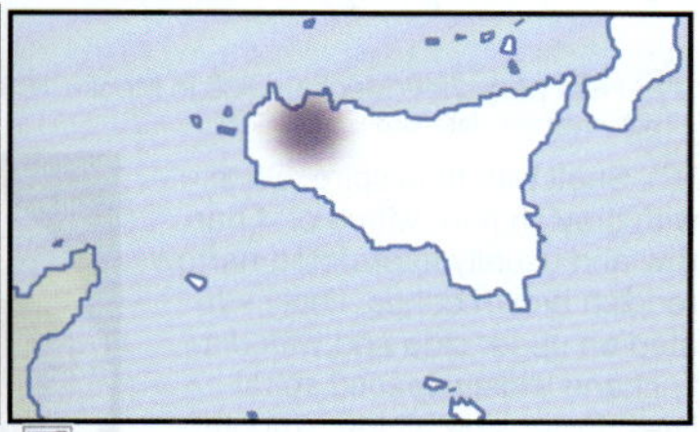

Italy (NW Sicily)

Shell bluish white, with 3 broad rows of reddish brown spots on upper and one on lower side, solid and shiny, very finely striated and densely and distinctly reticulated on upper side, almost smooth on lower side, 4-4.5 convex and regularly increasing whorls, last whorl slowly and deeply descending near aperture, aperture inside greyish white or brownish with 4 usually distinct bands, margin reflected, shiny and with lip, at columellar side often with a callus or straight fold or a tooth, and also with a brown spot on the entirely covered umbilicus.

Westerlund 1889: 375, Manganelli et al. 1995: 32.

M. globularis and *M. platychela* could be conspecific, the populations seem to be locally variable with gradients and intermediates which may connect both forms. Intermediate forms connecting *globularis* and *muralis* are also known. Up to 10 so-called subspecies of *globularis* have been recognised, of unknown bioscientific value.

Marmorana melitensis (Férussac, 1821)

Helix melitensis Férussac, 1821 in A. Férussac & Deshayes 1819-1851: (explication planches) p. xj, Pl. 25 (*Helix (Helicogena) Melitensis*)
Date was 06 Apr 1821, Pl. 25 had been published on 04 Dec 1819 without name printed on it. Also mentioned in Prodrome, Férussac [1821-1822]: quarto p. 31 (folio p. 35) from 26 May 1821.

14-16 x 19-23 mm — Malta: Gozo, Mgarr

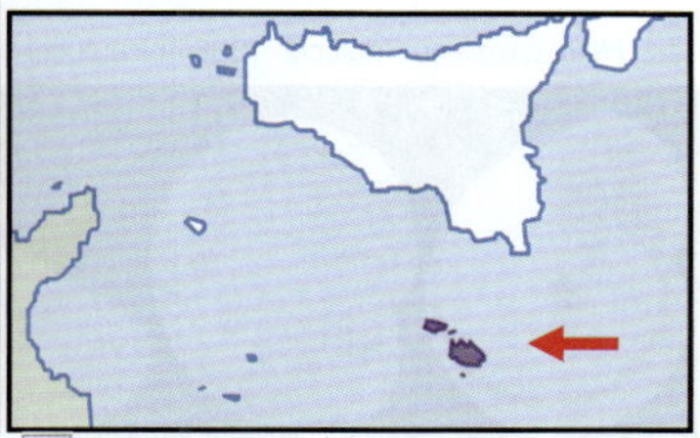

Malta (and Gozo)

Shell usually white, rarely with weak bands or rows of spots, coarsely striated and with coarse malleate or reticulate pattern, 4-4.5 not very convex whorls, initially with weak edge at periphery, last whorl deeply descending near aperture, aperture inside brownish, margin white and reflected except in its upper section, umbilicus in adults entirely covered by the reflected columellar margin, often with a brown spot at the umbilicus.

Usually on vertical or almost vertical limestone rock walls, also in crevices and under tree bark. Restricted to calcareous substrate, does not ascend pine tree, very rarely estivating on cypress trees.

Westerlund 1889: 378, Giusti et al. 1995: 472, David Cilia (unpublished, 2009).

Marmorana muralis (Müller, 1774)

Helix muralis Müller, 1774 in Müller 1774: 14 Type locality near Firenze, Italy.

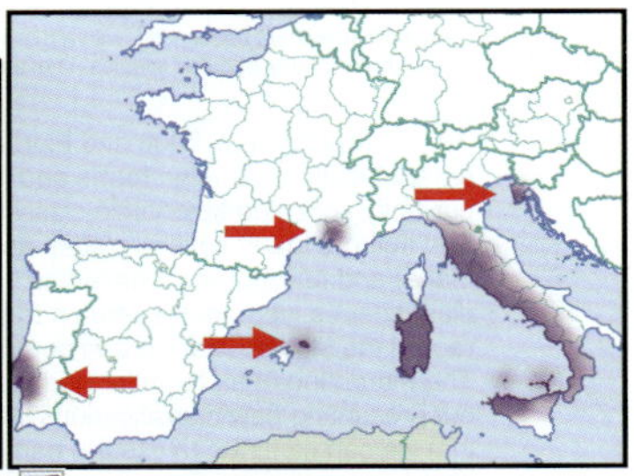

9-12 x 15-20 mm Italy: Sicily, Taormina

Italy (Sicily), dispersed elsewhere

Shell whitish with characteristical reddish brown colour patterns on upper side, lower side more white, weakly striated, some populations prominently striated, 4-5 not very convex whorls, apertural margin sharp and reflected, umbilicus in adults entirely closed.
Animal large, greyish yellow with black pigments on side and posterior part of body, upper tentacles very long and transparent, lower tentacles short.

Sunny rocks and walls, also reported from tree trunks. In France at old walls exposed to sunshine, requires dry conditions and high temperatures. In mainland Italy usually in anthropogenic habitats (usually old walls), occasionally also in natural habitats.

Threatened at the type locality near Florence by the destruction of old walls, same situation in Rome.

Moquin-Tandon 1855: 143, Germain 1930: 209, Nobre 1941: 135, Falkner 1990: 236, Manganelli et al. 1995: 32, Beckmann 2007: 124.

Could be conspecific with *M. globularis* and *M. platychela*.

Marmorana nebrodensis (Pirajno, 1840)

Helix nebrodensis Pirajno, 1840 in Pirajno 1840: 17 (*Helix Nebrodensis*)
Date could be 1842, but I have no source for that date.

Shell evenly whitish or with 2-4 interrupted brown bands, solid, densely striated and with fine and widely spaced spiral lines on the upper side, 4-4.5 regularly increasing and weakly convex whorls, last whorl depressed and rounded, deeply descending near aperture, aperture very oblique, inside yellowish brown with some bands, margin shiny and with white lip, umbilicus usually entirely covered.

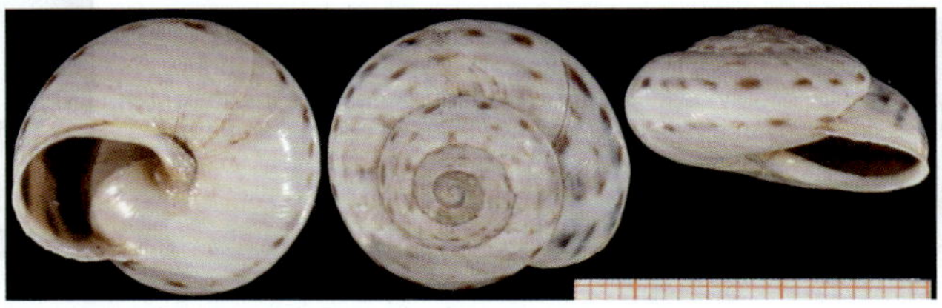

13-18 x 20-26 mm Italy: Sicily, between Rifugio Orest and Piano Battaglia

Italy (N Sicily)

EU Red List: Endangered EN.

Westerlund 1889: 381, Manganelli et al. 1995: 32.

Marmorana platychela (Menke, 1830)

Helix platychela Menke, 1830 in Menke 1830: 125
Established with description.

Shell with high spire, white with greyish hue, with 3 or 4 narrow and usually interrupted narrow reddish brown bands, finely striated on upper side and with distinct and widely spaced spiral lines, lower side almost smooth, 4.5-5 convex and regularly increasing whorls, last whorl wider and rounded, slowly and deeply descending near aperture, aperture very oblique, inside weakly reddish or yellowish white, margin distinctly broadened, reflected and white, upper margin horizontal, columellar margin often with a callus or tooth, umbilicus covered.

12-17 x 17-24 mm Italy: Sicily, Sferracavallo, Monte Gallo

M. globularis and *M. platychela* could be conspecific, the populations seem to be locally variable with gradients and intermediates which may connect both forms. 7 so-called subspecies have been proposed by some authors for *platychela*, of unresolved value under more generally accepted bioscientific criteria (presence of partial barriers to reprocution along narrow hybrid zones).

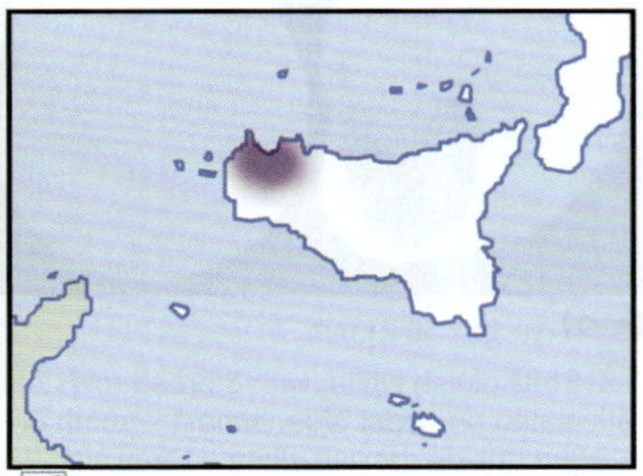

Italy (NW Sicily)

Rocky limestone habitats, in crevices and rock holes built by *Helix mazzullii*.

Westerlund 1889: 376, Manganelli et al. 1995: 32.

Marmorana saxetana (Paulucci, 1886)

Helix saxetana Paulucci, 1886 in Paulucci 1886: 25, Tav. 2

11-15 x 16-23 mm Italy: (1-4) Toscana, Orbetello, (5-6) Toscana, Grosseto, Ansedonia (5-6) Alessandro Hallgass

Shell basically whitish with 4 interrupted dark brown colour bands, shiny, with coarse growth lines, 4-4.5 convex whorls, last whorl deeply descending, aperture almost circular and with lip inside, margin slightly reflected, umbilicus nearly covered and with a brown spot.

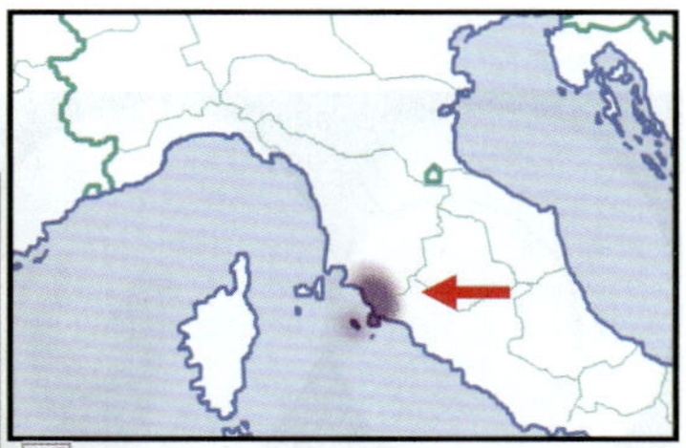

Italy (S Toscana, Giglio island)

Limestone outcrops among Mediterranean vegetation.

EU Red List: Near Threatened NT.

Westerlund 1889: 385, Hesse 1932: 192 (anatomy), Manganelli et al. 1995: 33, 56.

Marmorana scabriuscula (Deshayes, 1832)

Helix scabriuscula Deshayes, 1832 in Deshayes 1830 [part 2]: 258
Date was 29 Sep 1832 (Sherborn & Woodward 1899: 595, Cowie 1998: 92, "444" should probably read "144").

6-12 x 12-25 mm Italy: Sicily, Segesta

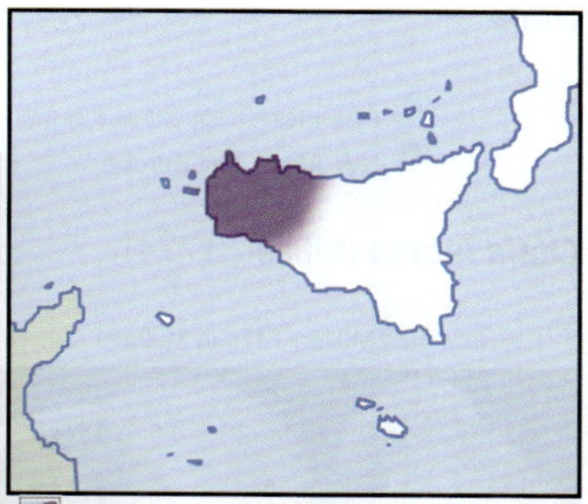

Italy (W Sicily)

Italy: Sicily, Erice — Angelo Vannozzi

Shell brownish grey or brownish white, ribbed or very prominently striated and very finely granulated, 4-4.5 flat whorls with pronounced keel, lentiform or with flat upper and more rounded lower side, last whorl also sharply keeled, suddenly and deeply descending near aperture, aperture inside brown with some bands, margin with white lip, not connected at parietal side, umbilicus partly or entirely covered. Locally variable in terms of shell shape, size and umbilicus.

Rocky limestone habitats, also in cavities of old walls and ruins.

Westerlund 1889: 370, Manganelli et al. 1995: 32.

Could be conspecific with *M. muralis*. Some 14 subspecies have been recognised to reflect the morphological diversity, but such a classification was not confirmed in molecular studies by Fiorentino et al. 2008.

Marmorana serpentina (Férussac, 1821)

Helix serpentina Férussac, 1821 in A. Férussac [1821-1822]: 31 (folio p. 35)
Date was 26 May 1821. Established without description, but with references to Pl. 40 Fig. 7 (ineffective, was published 10 Nov 1821) and to Gualtieri 1742: Tab. 3 Fig. C (effective). Also mentioned in Histoire, Férussac & Deshayes 1819-1851: (explication planches) p. xiv (06 Apr 1821) for Pl. 40 Fig. 7 (ineffective, published 10 Nov 1821).

11-13 x 17-21 mm Italy: W Sardinia, Alghero

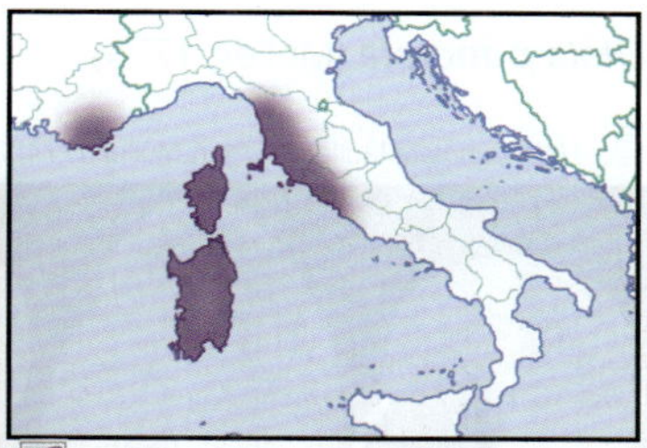

Italy, SE France
Originally Corsica and Sardinia, introduced to mainland sites.

Shell basically cream to weakly brown with 4 brown bands forming a characteristic dark zigzag pattern like in snakes, the 5th and lowest band is often at some distance, 4-4.5 slightly convex whorls, last whorl slightly descending, aperture large, inside whitish with blurry brownish bands, parietal and columellar side intensively brown including the reflected columellar margin, upper and basal margin white and weakly reflected, umbilicus entirely covered. The 5th (lowest) colour band seems to be characteristic. The zigzag pattern is slightly different in *M. muralis*. *M. serpentina* shows often bands of <<< fletches pointing in the direction of the aperture, while the colour streaks in *M. muralis* are often transverse to the suture. Differs from *M. muralis* also anatomically, mainly in the form of the darts.

Stone walls, limestone rocks, olive yards, sometimes crawls up tree trunks at rainy days. In Toscana restricted to ruderal and anthropogenic habitats.

Mating in spring and autumn, with a period of estivation in the summer season, eggs are usually deposited in rock crevices, juveniles hatch after a few weeks. Life cycle at least 3 years.

Moquin-Tandon 1855: 144, Germain 1930: 211, Boato et al. 1984, Kerney et al. 1983: 304, Manganelli et al. 1995: 32, Fiorentino et al. 2009.

Marmorana signata (Férussac, 1821)

Helix signata Férussac, 1821 in A. Férussac & Deshayes 1819-1851: (explication planches) p. xij, Pl. 30 (*Helix (Helicogena) signata*)
Date 06 Apr 1821, for Pl. 30 (name not mentioned on plate) 17 Jun 1820. Also mentioned in Prodrome, Férussac [1821-1822]: 31 (folio p. 35) from 26 May 1821.

11-13 x 16-22 mm Italy: Abruzzo, Gola di Celano

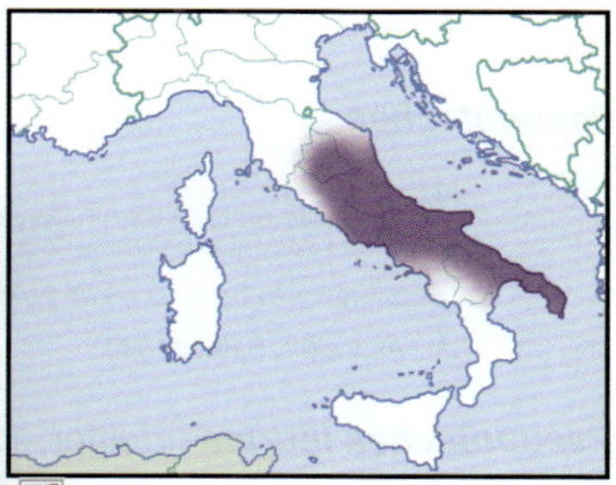

Italy

Italy: Umbria, Gavelli — Giuseppe Pocaterra

Shell white with reddish hue, typically with a broad and intensively brown basal band, upper side sometimes with weak interrupted reddish brown bands, regularly and weakly striated, last whorl with spiral sculpture, or smooth, some populations are prominently ribbed, 4-4.5 convex and regularly increasing whorls, last whorl rounded and descending near aperture, aperture inside light yellowish or white with a brownish band, margin narrow with white lip inside, weakly reflected at basal side, columellar margin covering the umbilicus almost or entirely, with a brown spot.
Locally variable in terms of colour patterns, surface sculpture and umbilicus.

Rocky limestone habitats.

Westerlund 1889: 382, Manganelli et al. 1995: 33, Fiorentino et al. 2010, Nordsieck 2011: 24.

Fiorentino et al. 2010 suggested a trinominal classification (4 subspecies) based on results of DNA studies. Nordsieck 2011 did not accept their classification because not all forms were studied and because they were not consistent with the morphological record. Nordsieck 2011 proposed a different subspecific classification based on morphology.
Fiorentino et al. 2010 argued that *M. signata* could be subdivided into 4 subspecies, but emphasized that more detailed research was needed to verify their distribution and relationships. So this could hardly be understood as a definite proposal of a subspecific classification. The authors made clear that more research was necessary.
The presence of partial intrinsic barriers to reproduction in narrow hybrid zones in the field, which would allow to derive statements on the existence of possible subspecies, still needs to be studied. The genetical background responsible for the morphological variation seems not yet understood. It would be useful to have detailed plots about the spatial distribution of various different morphological characters and to detect and study possible contact zones between morphologically different forms.
Nordsieck's 2011 approach is rejected here because these criteria were not considered.

Otala

Otala Schumacher, 1817
in Schumacher 1817: 58, 191

Type species *Otala atomaria* Schumacher, 1817
(subsequent designation by Pilsbry 1894 [1895]: 323)
(= currently *Otala lactea* (Müller, 1774))

Pilsbry 1894 [1895]: 323 gave as type *Helix lactea* and stated that *atomaria* was a synonym, valid type designation of *O. atomaria* under Art. 69.2.2.

NW Africa, SW Europe

Otala lactea (Müller, 1774)

Helix lactea Müller, 1774 in Müller 1774: 19

SW Europe, Morocco

15-21 x 30-40 mm

(1-5) Spain: Roquetas near Almería, (6-10) Portugal: Algarve, Tavira

Shell colour variable but usually brownish grey with four brown colour bands and many light dots, aperture inside dark brown and shiny, the dark zone extends on the umbilical and parietal sides, apertural margin often with a callus on the columellar side, umbilicus closed in adults and entirely covered by the dark apertural margin. Juveniles with light aperture and narrow circular umbilicus, 1/8 of shell diameter.
Animal light greyish yellow, darker in frontal part, upper tentacles long and greyish, lower tentacles shorter and yellow.

Rocky shrublands and open habitats, locally very abundant.

Nobre 1941: 133, Réal & Réal-Testud 1988: 51 (introduced near Bastia, N Corsica), Falkner 1990: 238, Ruiz Ruiz et al. [2006].

Spain: Almería, Cabo de Gata   Giuseppe Pocaterra

Otala punctata (Müller, 1774)

Helix punctata Müller, 1774 in Müller 1774: 21

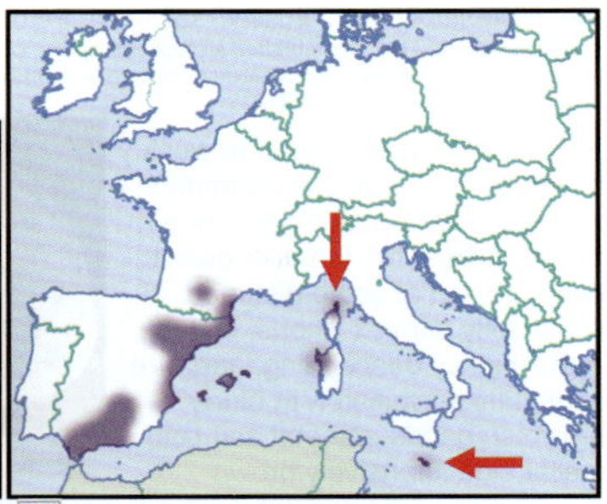

1.6-2.2 x 33-40 mm

Spain: Barcelona, Los Arcos

S Europe, NW Algeria

Shell solid, whitish or greyish brown, with 2-5 indistinct bands interrupted by white spots, 4-5 rapidly and regularly increasing weakly convex whorls with weak suture except at the last whorl, last whorl deeply descending near aperture, aperture inside brown, margin whitish or light brownish, with a callus on the columellar side, umbilicus entirely closed in adults.
Very similar to *O. lactea*, but aperture less intensely coloured, lip usually white. Larger than *Eobania vermiculata*, last whorl more inflated, *Eobania*'s aperture is white.

In cultivated areas of coastal plains, in France in gardens, dry wasteland, vineyards, at rocks and walls, also found among *Urtica dioica* and *Eucalyptus*.

Moquin-Tandon 1855: 157 (introduced from Valencia to Pyrénées-Orientales in 1829, also present in N Corsica), Germain 1930: 189, Bech 1988: 108, Falkner 1990: 236, Clanzig &. Bertrand 2001, Beckmann 2007: 126, Barbara & Schembri 2008, De Mattia & Mascia 2011.

Introduced in the 14th century to Sardinia and in 2003 to Malta.

Pseudotachea

Pseudotachea Boettger, 1909
in C. R. Boettger 1909: 10

Type species *Helix splendida* Draparnaud, 1801
(monotypy)

Established as a genus without description, but with one species included. Type not by original designation. I did not find a statement that the species given as examples for the genera in the list on pp. 9-10 were meant as name-bearing types.

S France to Spain and NW Africa

Aparicio & Ramos 1988.

Pseudotachea liturata (Pfeiffer, 1851)

Helix liturata Pfeiffer, 1851 in L. Pfeiffer 1851: 253 (in original source not spelled *litturata*)

NW Africa, S Spain

11.5-15.5 x 17-23 mm

Morocco: Tetuan

Shell opaque, white, yellowish, reddish or horny brown, often with dark brown colour bands, which can be interrupted, usually 3 on upper and 2 on lower side, finely striated or nearly smooth, 4.5 whorls or slightly more, equally convex at upper and lower side, suture not very deep, last whorl rounded and slowly descending near aperture, aperture with reddish white lip inside, margin weakly reflected, at columellar side and around the umbilicus light reddish, not connected at parietal side, umbilicus closed.
Seems to differ from *P. splendida* in its higher number of whorls.

Between sea level and 600 m altitude, requires a warmer climate and a narrower range of ecological conditions than *P. splendida*.

EU Red List: Near Threatened NT.

Westerlund 1889: 447 (*H. coquandi*), Ruiz Ruiz et al. [2006].

Pseudotachea splendida (Draparnaud, 1801)

Helix splendida Draparnaud, 1801 in Draparnaud 1801: 83

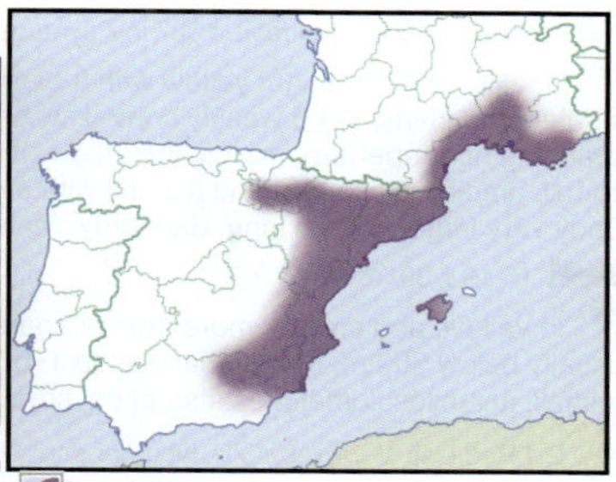

8-14 x 15-24 mm

Spain: Denia

Spain, S France

Shell white with five brown colour bands, sometimes interrupted and some can be absent, upper side lighter than lower side, basic colour at lower side often yellowish or brownish yellow, nearly smooth, shiny, often translucent, 3.5-4 very weakly convex whorls, last whorl gets more rounded near aperture, shortly descending near aperture, aperture inside white with white lip inside, margin and umbilical region occasionally pinkish, umbilicus closed in adults.
Seems to differ from *P. liturata* in its lower number of whorls.
Body yellowish grey with reddish hue on head and foot, tentacles nearly 10 mm long and slightly darker yellowish grey. Animal moves slowly, produces much liquid mucus and carries its shell slightly obliquely.

Dry meadows, shrubs and rocks covered by vegetation, old walls and wine slopes. In France not above 500 m, in E Spain also in 1200 m altitude but not very frequent there.

Moquin-Tandon 1855: 149 (edible), Westerlund 1889: 447, Germain 1930: 197 (not in Corsica), Fez 1944: 213, Gasull 1964: 24, Kerney et al. 1983: 307, Bech 1988: 108, Prieto & Altonaga 1988, Falkner 1990: 238, Bertrand 2000: 39 (SW Andorra).

Tacheocampylaea

Tacheocampylaea Pfeiffer, 1877
in L. Pfeiffer 1877 (Malak. Bl. 24): 8 *Helix (Campylaea (Tacheocampylaea))*

Type species *Helix raspailii* Payraudeau, 1827 (monotypy)

Established without description "für die Verwandten der *H. Raspailf*", no other species mentioned for this group, so monotypy.

Deeply hidden in rock crevices or under stones, in humid and shady natural habitats, active only at night and during rainfall.

Falkner et al. 2002: 165.

Tacheocampylaea acropachia (Mabille, 1880)

Helix acropachia Mabille, 1880 in Mabille 1880: 62 (*Helix Acropachia*)

Shell yellowish with 3 blurry reddish brown bands, apex protruding, 4-4.5 weakly convex whorls, last whorls more rapidly increasing than first whorls, last whorl slowly descending near aperture, margins not much reflected and approaching each other at parietal side, umbilicus covered.

18-22 x 37-44 mm

EU Red List: Endangered EN. Threatened by fire, at least 3 of the few populations were destroyed by fire in 2000.

Westerlund 1889: 109, Germain 1930: 203 (*T. raspailii*), Falkner et al. 2002: 166, www.iucnredlist.org (08.2012).

The local variation has been associated with the presence of various subspecies, of unknown bioscientific value.

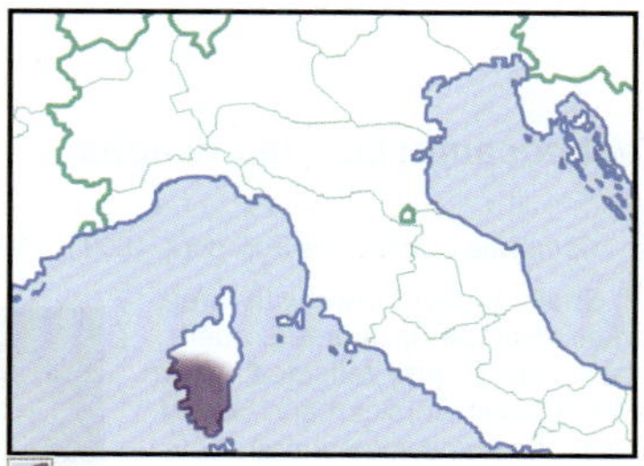

France (Corsica)

Tacheocampylaea carotii (Paulucci, 1882)

Helix carotii Paulucci, 1882 in Paulucci 1882: 203 (*Helix (Tachycampylaea) Carotii*)

16-24 x 22-35 mm

Italy: Sardinia, Cap Monte Santo (?)

Italy (central Sardinia)

Shell opaque, periostracum olive greenish with three brown colour bands at upper side of last whorl, only one at penultimate whorl, weakly striated, 4-4.5 convex whorls with distinct suture, last whorl suddenly descending near aperture, margin connected by a prominent layer at parietal side, umbilicus usually closed.

Westerlund 1889: 111, Manganelli et al. 1995: 33.

Limestone habitats, often deeply inside crevices.

EU Red List: Vulnerable VU. The populations seem to have declined considerably in the past decades, threatened by forest exploitation, fire and urban development.

Tacheocampylaea cyrniaca (Dutailly, 1867)

Helix cyrniaca Dutailly, 1867 in Dutailly 1867: 100 (separate p. 6) (*Helix Cyrniaca*)

Shell light yellowish or reddish olive brown, with 3 blackish brown bands, the central band being very narrow, regularly striated, 3.5-4.5 regularly increasing whorls with deep suture, last whorl inflated producing a concave outline of the upper side of the shell with the first 3 whorls building a cone, rounded and deeply descending near aperture, margin without lip, nearly connected at parietal side, umbilicus initially wide, widened again at last whorl.

12-14 x 26-29 mm

Rock crevices and under stones in mountains, also in shady forests, on granitic substrate, up to 2300 m.

EU Red List: Endangered EN. Threatened by conversion of suitable habitats to pasture land, forest fires, road construction and wild pigs consuming the few remaining snails. About 50 % of the populations have been lost in the past decades.

Westerlund 1889: 111, Germain 1930: 205, Falkner et al. 2002: 165.

The local variation has been associated with the presence of various subspecies, of unknown bioscientific value.

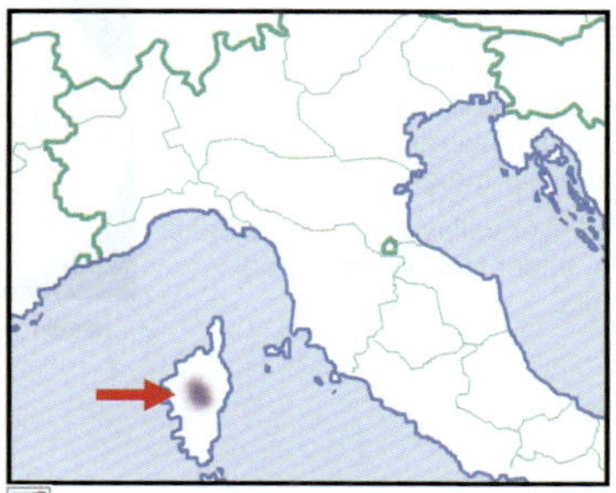

France (Corsica)

Tacheocampylaea raspailii (Payraudeau, 1827)

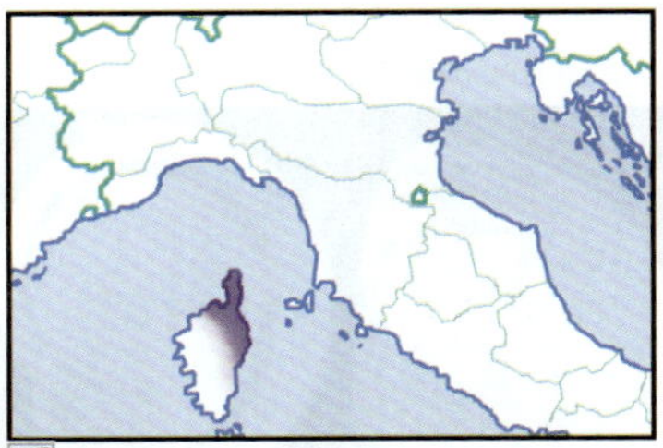

France (Corsica)

Helix raspailii Payraudeau, 1827 in Payraudeau 1826: 102 (*Helix Raspailii*)

Shell brownish or light yellow with 3 blurry reddish brown bands, lowest band disappearing at the last whorl, striated and shiny, 4-4.5 weakly convex whorls, last whorl width less than 2 x of penultimate whorl, sharply descending near aperture, margin light reddish brown and slightly reflected, basal margin with callus, almost horizontal, umbilicus covered and usually with a reddish brown spot.
Body very large, 45 mm long, dark grey, tentacles grey and 22 mm long, longer than in *Chilostoma zonatum*.

15-19 x 30-38 mm

In valleys and creeks, more rarely in gardens, in humid, shady habitats under stones and in rock crevices, usually below 500 m, occasionally up to 1700 m at slopes. Adults are only active in evenings and nights after rainfall, juveniles sometimes also at daytime.

EU Red List: Vulnerable VU. Number and sizes of populations has been declining dramatically in the past 2 decades. Threatened by rapid urban development, massive habitat destruction by new roads, forest destruction and fire, a major problem for the last remaining individuals is wild pigs.

Moquin-Tandon 1855: 152, Westerlund 1889: 109, Germain 1930: 202, Falkner et al. 2002: 165, www.iucnredlist.org (08.2012).

This name was used until recently for all *Tacheocampylaea* species. Previous reports for *raspailii* could have referred to other species.

Tacheocampylaea romagnolii (Dutailly, 1867)

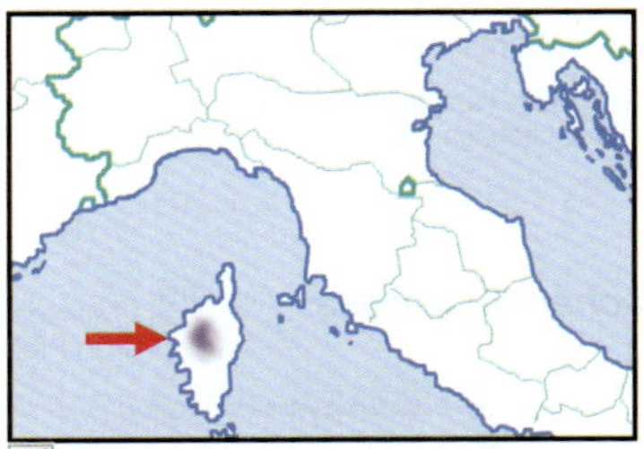

France (Corsica)

Helix romagnolii Dutailly, 1867 in Dutailly 1867: 97 (separate p. 3) (*Helix Romagnolii*)

Shell reddish brown with colour bands, thin, finely striated, covered with short and regularly arranged hairs, 4-4.5 nearly flat whorls, last whorl suddenly descending near aperture, apertural margin yellowish and weakly thickened, margins approaching each other at parietal side, basal and columellar margin reflected, covering entirely the umbilicus.

13-16 x 28-35 mm

EU Red List: Critically Endangered CR. Restricted to a few mountain valleys near Corte. Has been known from some 6 localities, only one population has survived, seriously threatened by human pressure.

Westerlund 1889: 110, Germain 1930: 206, Falkner et al. 2002: 167.

Tacheocampylaea tacheoides (Pollonera, 1909)

Campylaea tacheoides Pollonera, 1909 in Pollonera 1909: 17

Shell reddish-brown with three darker color bands, striated, apertural margin slightly reflected, brown, darker than in Corsican and Sardinian species, umbilicus entirely covered.
Animal blackish grey, some are lighter, tentacles grey with dirty cream eye balls.

At volcanic rocks, only in undisturbed shrubland with agglomerations of stones, or stone walls.

16-20 x 25-32 mm

Italy: Capraia island
Agatino Reitano

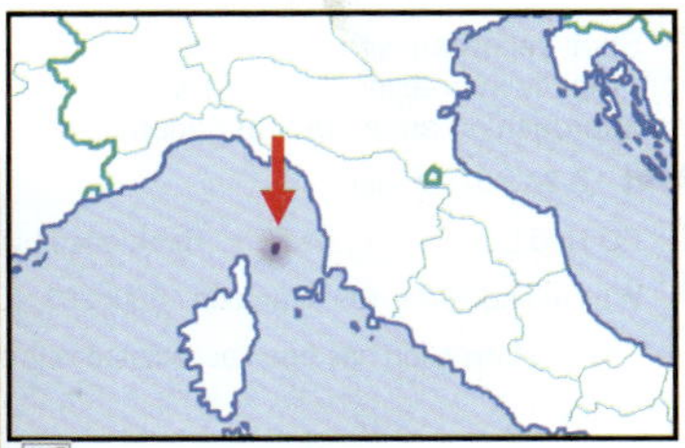

Italy (Capraia island)

Hesse 1932: 190 (anatomy), Giusti 1976: 331, 349, Manganelli et al. 1995: 33, www.iucnredlist.org (08.2012).

EU Red List: Endangered EN. Extremely rare species of Capraia island. Has been kown from 6 sites on the small island, only one of these populations survived. Threatened by fire, predation by rats, overgrazing, and particularly the collecting of live animals for human consumption. Most of the undisturbed shrubland on the island has been modified in the course of history, and the habitats were destroyed.

Tacheopsis

Tacheopsis Boettger, 1909
in C. R. Boettger 1909: 10

Type species *Helix aimophila* Bourguignat, 1859 (monotypy)
(= currently *Tacheopsis nicaeensis* (Férussac, 1821))

Established as a genus without description, but with one species included. Type not by original designation. I did not find a statement that the species given as examples for the genera in the list on pp. 9-10 were meant as name-bearing types.

Turkey

Tacheopsis nicaeensis (Férussac, 1821)

Helix nicaeensis Férussac, 1821 in A. Férussac [1821-1822]: 30, 67 (*Helix (Helicogena) nicæensis*)
Spelled *H. niceensis* on p. 30 (only name and reference to Pl. 28 Fig. 1-2, date of plate was 26 Feb 1820), *H. nicæensis* in corrigenda on p. 67.
Helix aimophila Bourguignat, 1859 in Bourguignat 1859: 522, Pl. 20

Shell solid, white or light reddish, sometimes with two narrow brown colour bands, aperture very oblique, umbilicus closed or open and half covered by the reflected columellar margin.

River valley plains, humid habitats, under shrubs and herbal plants.

20-23 x 25-31 mm

S Turkey: 20 km E Silifke

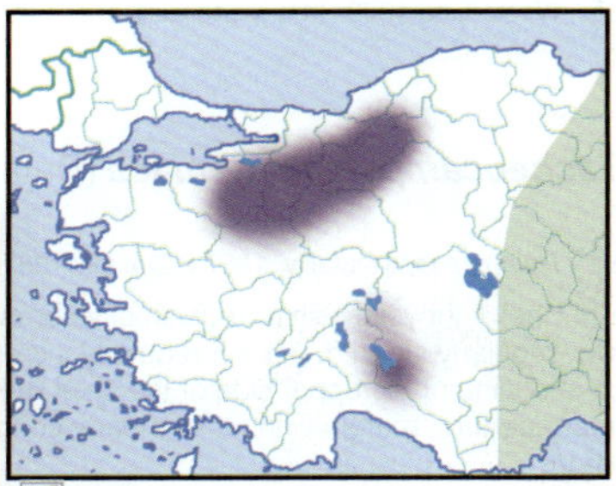

Turkey

Schütt 2001: 493.

Theba

Theba Risso, 1826 in Risso 1826: 73

Type species *Helix pisana* Müller, 1774 (subsequent designation by Gray 1847: 173)

Mediterranean region

Gittenberger & Ripken 1987.

Theba andalusica Gittenberger & Ripken, 1987

Theba andalusica Gittenberger & Ripken, 1987 in Gittenberger & Ripken 1987: 44

12-17 x 15-21 mm

Spain: Cádiz, Tarifa — José Liétor Gallego

Shell usually white or with weak dots or brown colour bands, globulous with very large last whorl, outer lip clearly thickened inside, umbilicus almost completely closed by the columellar lip which has a characteristic notch.

S Spain

Gittenberger & Ripken 1987, Ruiz Ruiz et al. 2006.

Taxomomic position and differences to *Th. pisana* seem to require more research.

Theba pisana (Müller, 1774)

Helix pisana Müller, 1774 in Müller 1774: 60
Helix albella Linnæus, 1758 in Linnæus 1758: 768 (nomen oblitum under Art. 23.9, fixed here)
After Forcart (1965: 255) selected a lectotype for *Helix albella*, this became a senior synonym of *Helix pisana* Müller, 1774. In accordance with Art. 23.9 *Helix pisana* Müller, 1774 shall be used for the taxon, *Helix albella* Linnæus, 1758 has not been used as a name for a taxon since 1899 and shall be regarded as nomen oblitum, *Helix pisana* Müller, 1774 was used as a name for our taxon in the following 28 papers: Hickson 1972, Bar 1978, Kerney & Cameron 1979: 202, Johnson 1981, Lazaridou-Dimitriadou & Daguzan 1981, Cowie 1984 (3 papers), Heller & Gadot 1984, Liebegott 1986, Gittenberger & Ripken 1987: 32, Prieto & Altonaga 1988: 6, Baker 1988, Cowie 1990, Falkner 1990: 234, Manganelli et al. 1995, Dhora & Welter-Schultes 1996, Welter-Schultes & Wiese 1997 (2 papers), Welter-Schultes 1998: 91, Ondina et al. 1997, Kerney 1999: 202, Gómez Moliner et al. 2001, Schütt 2001: 468, Martínez-Ortí et al. 2004, Quintana Cardona 2006, Welter-Schultes 2008: 80, Cuttelod et al. 2012: 82.
Helix arietina Rossmässler, 1846 in Rossmässler 1846: 172

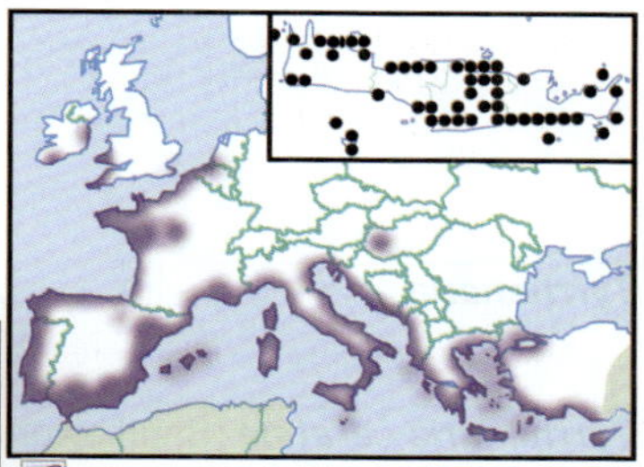

9-20 x 12-25 mm

(1-5) Greece: Zákinthos, Laganas, (6 juv.) N Italy: near Viareggio

Europe, Morocco, Atlantic islands

Ø 1.7-1.9 mm at 1 whorl, 3.7-3.9 mm (2 wh.), 6.8-7.5 mm (3), 11-12 mm (4), 15-17 mm (4.5 wh.)

Montenegro: Ulcinj — C. Audibert

Shell displaying various colour variants, yellow or white with dark colour bands or spots and often a dark bluish grey apex, juveniles sharply keeled, the keel disappears at the last whorl (not in *Th. p. arietina*), aperture often with a light reddish lip inside, margin only reflected at columellar side, umbilicus narrow and half covered. The apex has a characteristic size in the eastern Mediterranean when compared with other species, where there are no other *Theba* species. The umbilicus is also rarely seen in other species, juvenile *Eobania vermiculata* have a considerably larger apex.
Animal very light yellowish with dark colour bands running from the sides to the upper tentacles, tentacles very long.

Usually in coastlands, in or near sandy habitats, in hot climates estivating often directly exposed to the sun, attached to grasses, shrubs or succulent plants. In dunes it can live on nearly bare sand poorly fixed by grasses. One of the most common snails in coastal regions from S Portugal to Greece. Introduced to SW Great Britain and Ireland since at least the 1700s, still spreading along frost-free coastal localities. In the north the snails do not estivate but they climb on plants in dry weather. Does not survive serious winter frosts. Often associated with *Cochlicella acuta* and *Cernuella virgata*, but can live slightly deeper inside pure sandy habitats and is usually more common than *C. acuta*.

In France 40-80 oval eggs (diameter 1.5 mm) are deposited from June to October under stones, between roots or in the soil, hatchling size 2 mm. In Greece maximum shell size is attained after 2 years, maturity is reached at half maximum shell size after 1 year.

Greece: Attiki, Lagonisi beach near Athina — Panayotis Ovalis

Spain: Cádiz, El Portal

Germain 1930: 181, Lazaridou-Dimitriadou & Daguzan 1981.

N Italy: near Viareggio

Theba subdentata (Férussac, 1821)

Helix subdentata Férussac, 1821 in A. Férussac & Deshayes 1819-1851: (explication planches) p. xj, Pl. 27 (*Helix (Helicella) subdentata*)
Date was 06 Apr 1821 (explication planches), plate published 26 Feb 1820. Also mentioned in Prodrome, Férussac [1821-1822]: quarto p. 45 (folio p. 49) from 26 May 1821.

W Morocco, S Spain

Vulnerable in Spain.

12-15 x 16-21 mm — Spain: Almería — Giuseppe Pocaterra

In Almería white with colour bands at last whorl, variably depressed but always keeled, aperture with lip thickened inside, surface with very light reticulate sculpture visible under microscope (regular spiral streaks and broader, more irregular radial ribs, like seersucker texture), umbilicus always closed, the interior of the aperture is pink. In Morocco not keeled.

In Almería in a 2-3 km wide corridor close to the sea in dunes and sand beaches, semiarid climate. Mainly active at night in autumn, in spring and summer the snails are strongly adhered to trunks and bushes, up to more than 1 m above ground, with a very thick epiphragm composed of 5-6 layers of film mucus with a white thin calcareous layer.

Gittenberger & Ripken 1987: 22, Ruiz Ruiz et al. 2006, Moreno & Ramos 2007.

Tyrrhenaria	*Tyrrhenaria* Hesse, 1918 in Hesse 1918: 38 *Helix (Tyrrhenaria)*	Type species *Helix tristis* Pfeiffer, 1845 (original designation)

Tyrrhenaria ceratina (Shuttleworth, 1843)

Helix ceratina Shuttleworth, 1843 in Shuttleworth 1843: 16
Helix tristis Pfeiffer, 1845 in L. Pfeiffer 1845: 66

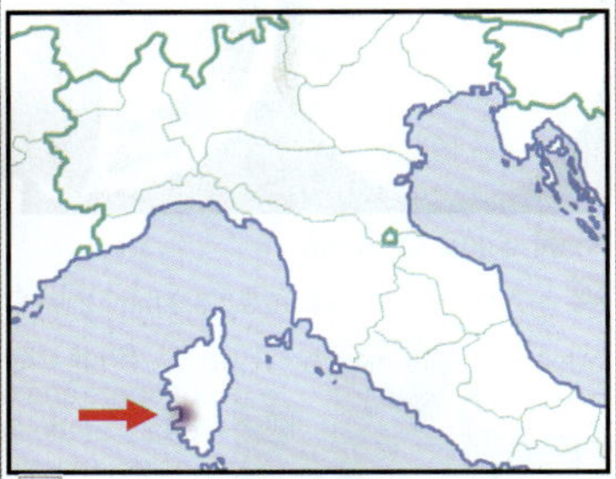

20-25 x 16-25 mm — France: Corsica, Ajaccio, Campo dell'Oro — France (W Corsica)

Shell looks like *Helix*, yellowish brown with 4-5 reddish brown bands, irregularly striated and weakly reticulated, 3.5-4 convex whorls, last whorl slightly descending, aperture inside shiny and with distinct bands, margin straight and sharp, parietal side sometimes with a weak whitish layer, usually no umbilicus.

Sandy and periodically flooded habitats in the coastal plain, in a broken mosaic terrain of isolated shrubs of *Genista salzmannii*, avoids too densely vegetated areas and open areas devoid of *Genista* spots. The snails bury inside the soft and sandy soil (up to 60 cm deep) in periods of heat and appear usually only at night and during rainfall, unable to survive in compact soils.
Herbivorous, consumes primarily leaves, flowers and fruits of *Genista salzmanni*, occasionally also other dune plants, grasses, mosses, lichens and organic material in the sand. Average movement 4 m/yr between isolated spots of vegetation.

Reproduction from end of August to mid-October, up to 20 eggs are laid inside a cavity in the sand, covered by mucus, juveniles hatch after 15-16 days and 2-3 weeks later appear at the surface. Maturity is reached after 2-4 years, life span 6-10 years.

EU Red List: Critically Endangered CR. Protected by law in France, eggs and animals must not be collected (but habitat destruction is allowed). Microendemic species, occurs only in a small area in the suburban area of the growing city of Ajaccio (Ricantu beach in immediate vicinity of the airport Campo dell'Oro).
Must have suffered considerable losses since the Neolithic when the species was present at other sites in Corsica. In the mid-1800s the area near Ajaccio must have covered 50 ha, 90 % of the original area has been lost since then, mainly by the airport, current population size perhaps 5-8000. Seriously threatened by modifications of the supplying water currents to prevent floods, and direct habitat destruction provoked by motorbikes and four-wheel drives. The fragmented habitat needs to be actively protected by local authorities.

Moquin-Tandon 1855: 184, Westerlund 1889: 449, Germain 1930: 188, Gargominy et al. 1999, Charrier et al. 2005.

Tyrrheniberus

Tyrrheniberus Kobelt, 1904 in Kobelt 1904: 132, 157, 199

Type species *Helix sardonia* Martens, 1884 (subsequent designation by Boettger & Wenz 1921: 13)

On p. 132 and 157 mentioned without description but with one species (*Helix sardonia*) included. On p. 199, 8 species were included. On pp. 132 and 157 author was given as "Hesse & Kobelt", on p. 199 "Kobelt & Hesse". The genus was exclusively established on the base of the included species, without description by anyone. The arrangement of the species was Kobelt's work. So Kobelt was the only person who contributed to making the name available. In a strict application of Art. 50.1.1 the name in such a case must be attributed to the author of the work.

Fiorentino et al. 2010 (revision).

Tyrrheniberus ridens (Martens, 1884)

Helix ridens Martens, 1884 in Martens 1884: 197

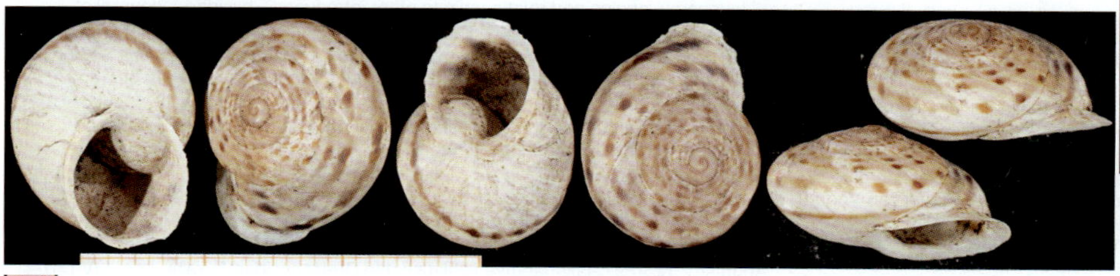

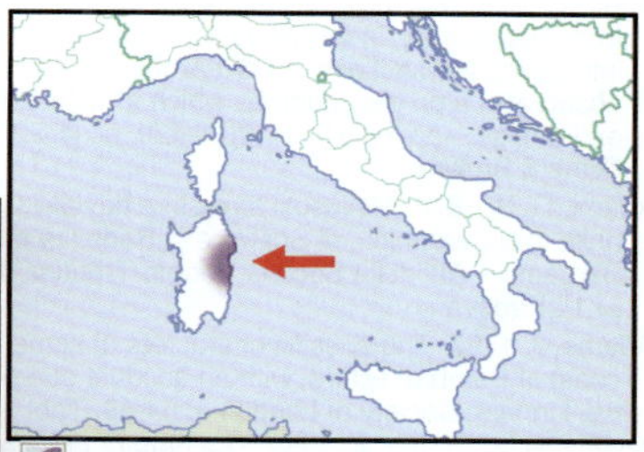

9-12 x 14-18 mm

Italy: Sardinia, Nuoro, Cala Gonone

Italy (central E Sardinia)

Isolated limestone habitats.

Westerlund 1889: 379, Manganelli et al. 1995: 33.

Shell white with reddish hue, with 4 narrow rows of dark or reddish brown spots, finely striated, 3.5-4.5 weakly convex whorls, last whorl broadened and with blunt edge at periphery, suddenly and deeply descending near aperture, aperture almost circular, very oblique and nearly horizontal, inside white or weakly brownish with distinct colour bands, margin thin and white, broadened and reflected at columellar side, covering the umbilicus entirely.
Animal light with two dark longitudinal bands running from the tentacles to the shell, and a dark spot at the head.

EU Red List: Vulnerable VU. Only a few isolated populations in a fragmented range are known. Threatened by increased habitat fragmentation by road construction.

Tyrrheniberus villicus (Paulucci, 1882)

Without gender agreement: *Tyrrheniberus villica* (Paulucci, 1882)

Helix villica Paulucci, 1882 in Paulucci 1882: 233
Helix sardonia Martens, 1884 in Martens 1884: 197

7-13 x 16-21 mm

Italy: Sardinia, Nuoro, Oliena, Monte Corrasi

Italy (central E Sardinia)

Italy: Sardinia, Nuoro, Cala Gonone, Cala Fuili
Giuseppe Pocaterra

Shell white with 4 almost not interrupted brown bands, thin but not translucent, very weakly striated and with dense spiral sculpture, 4-5 weakly convex whorls, last whorl more rapidly increasing than the others, suddenly and deeply descending near aperture, aperture inside white, margin thin, reflected except on upper side, margins approaching each other at parietal side, umbilicus narrow and partly or entirely covered.
The canyon population (*sardonius*) is irregularly ribbed and finely granulated, with a sharp keel.
Differs from *T. ridens* in its larger size. In *T. ridens* the umbilicus is always closed.

Rocky habitats in isolated limestone outcrops in mountains. The *sardonius* population lives inside a canyon.

EU Red List (*T. villicus*): Vulnerable VU. Only known from 4 localities, threatened by habitat fragmentation provoked by road construction.
EU Red List (*T. sardonius*): Critically Endangered CR. Only a single population in a small 2 km extending section in a canyon is known.

Westerlund 1889: 371, 382, Manganelli et al. 1995: 33, www.iucnredlist.org (08.2012).

Italy: Sardinia, Nuoro, Cala Gonone (*sardonius*)

The taxon *sardonius* was regarded as conspecific with *T. villicus* by Fiorentino et al. 2010. Both could not be distinguished in molecular studies. Flat-shelled *T. villicus* and *T. ridens* occur sympatrically. This is a difference to the closely related *Marmorana* species.

:-) This was the last species.
I apologize for the low quality of the typesetting. This book was compiled between 2007 and 28 August 2012 as a Microsoft Word 2003 document and I had to include text and images for each species into a table, to avoid undesired page breaks in the species accounts when I inserted information in previous chapters. The pictures and the corresponding texts should not appear on different pages. I made this visible in the last species *Thyrrheniberus villicus* - why the program decided to show it also one page before in *Tyrrhenaria ceratina* remained its own secret...
The cells of these tables were not really flexible, there were many bugs in the cell width calculation program and it was not comfortable to work with this method. Too many program bugs were the reason why the right margins could not be well aligned (p. 362 is a good example). I was always happy when I obtained something more or less acceptable. Finally the result shown on the screen differed often from that in the outprint.

One of the last steps was a comparison of a nomenclatural compilation of molluscan genera by J.-P. Rocroi (& Ph. Bouchet) with my own files, done in Aug 2012. The result was interesting and also somewhat disppointing. In about 50 % of the 400 generic names the data in both independently elaborated files were incongruent, because either Rocroi or I have made errors (mostly incorrect page numbers, incorrect years, overlooked type designations, more rarely incorrect spellings). In about 30 % of the cases my data turned out to have been incorrect, in 30 % of the cases Rocroi's data were incorrect, in 10 % we both had different incorrect data in the data set for the same genus. These mistakes accumulated in genera established before 1930. The error rate for genera established after 1930 was distinctly below 10 %.
This suggests that in some 5-10 % of the genera established before 1930 we would probably both have the same incorrect data, which of course could not be detected with this method, and are thus still contained in the present book.
This experience demonstrates a little how difficult and sophisticated this matter is, and how limited the perspectives are in such a large animal group for compiling a "List of Available Names in Zoology" (LAN) as outlined in Art. 79 of the Code. It would take several decades and independent cross-checks by at least 3 or 4 skilled scientists, each one would need to invest more than 5 years almost full-time in compiling such a list, before the error rates would reach an acceptably low level.
The comparison was interesting because independently in both compilations we applied basically the same interpretations of the Code.

References

Online links to digitised works, and works that I have forgotten to cite here can be found at AnimalBase, www.animalbase.org.
Bold: only for better finding the *et al.* works and to indicate presence of various authors with the same family name and different surname initials.

Abel, C. 1818. Narrative of a journey in the interior of China, and of a voyage to and from that country, in the years 1816 and 1817; containing an account of the most interesting transactions of Lord Amherst's embassy to the court of Pekin, and observations on the countries which it visited. Illustrated by maps and other engravings. - pp. i-xvi [= 1-16], 1-420, pl. [1-22]. London. (Longman, Hurst, Rees, Orme & Brown).

Adami, G. B. 1876. Molluschi terrestri e fluviatili viventi nella Valle dell'Oglio ossia nelle valli Camonica, di Scalve e di Borlezza spettanti alle provincie di Brescia e Bergamo. - Atti della Società Veneto-Trentina di Scienze Naturali 5 (1): 7-95, Tav. I [= 1]. Padova.

Adams, J. 1797. The specific characters of some minute shells discovered on the coast of Pembrokeshire, with an account of a new marine animal. - Transactions of the Linnean Society of London 3: 64-69, Tab. 13.

Adams, H. & Adams, A. 1858. The genera of recent Mollusca; arranged according to their organization. In three volumes.-Vol. II. - pp. [1-3], 1-661. London. (Van Voorst).
Dates (p. 661): pp. 1-92 (1854), 93-284 (1855), 285-412 (1856), 413-540 (1857), 541-660 (1858).

Adler, M. 1993. Beiträge zur Nomenklatur der europäischen Binnenmollusken, VI. Zur Nomenklatur von *Pisidium pseudosphaerium*. - Heldia 2 (1/2): 53-56, Taf. 8. München.

Agócsy, P. & Pócs, T. 1961. *Alopia soósiana* n. sp. (Mollusca, Gastropoda). - Magyar Nemzeti Múzeum Természettudományi Múzeum Évkönye - Annales Historico-Naturales Musei Nationalis Hungarici 53: 533-536. Budapest.

Aguilar-Amat, J. B. 1935. Observacions malacològiques. XXIV. Segona troballa a Catalunya de *Oestophora lusitanica* Pfr. - Butlletí de la Institució Catalana d'Història Natural (2) 35: 258-259. Barcelona.

Alba, D. M., Tarruella, A., Prats, Ll., Guillén, G. & Corbella, J. 2011. Nova llista actualitzada dels mol·luscos continentals de Catalunya. - Spira 4 (1-2): 39-69. Barcelona.

Albers, J. C. 1850. Die Heliceen, nach natürlicher Verwandtschaft systematisch geordnet. - pp. 1-262, Taf. 1-59. Berlin. (Enslin).

Albers, J. C. 1852. Neue Heliceen. - Zeitschrift für Malakozoologie 9 (8): 124-125.
Date Jul 1852 given on p. 128.

Albers, J. C. & Martens, E. von 1860. Die Heliceen nach natürlicher Verwandtschaft systematisch geordnet. Zweite Ausgabe. - pp. I-XVIII [= 1-18], 1-359. Leipzig. (Engelmann).
Contents mainly by Albers, arrangement of mollusc groups composed by Martens.

Albrecht, C., Hauffe, T., Schreiber, K., Trajanovski, S. & Wilke, T. **2009**. Mollusc biodiversity and endemism in the putative ancient Lake Trichonis (Greece). - Malacologia 51 (2): 357-375.

Albrecht, C., Kuhn, K. & Streit, B. **2007**. A molecular phylogeny of Planorboidea (Gastropoda, Pulmonata): insights from enhanced taxon sampling. - Zoologica Scripta 36 (1): 27-39.

Albrecht, C., Lohfink, D. & Schultheiß, R. **2006**. Dramatic decline and loss of mollusc diversity in long-lived lakes in Greece. - Tentacle 14: 11-13.

Albrecht, C., Trajanovski, S., Kuhn, K., Streit, B. & Wilke, T. **2006**. Rapid evolution of an ancient lake species flock: freshwater limpets (Gastropoda: Ancylidae) in the Balkan Lake Ohrid. - Organisms Diversity & Evolution 6 (4): 294-307.

Albrecht C., Wolff, C., Glöer, P. & Wilke, T. **2008**. Concurrent evolution of ancient sister lakes and sister species: the freshwater gastropod genus *Radix* in lakes Ohrid and Prespa. - Hydrobiologia 615: 157-167.

Alder, J. 1830. A catalogue of the land and fresh-water testaceous Mollusca found in the vicinity of Newcastle upon Tyne, with remarks. - Transactions of the Natural History Society of Northumberland and Durham 1 (1): 26-41. London.

Alder, J. 1837. Notes on the land and fresh water Mollusca of Great Britain, with a revised list of species. - Magazine of Zoology and Botany 2 (8): 101-119. Edinburgh.

Alder, J. 1838. Supplement to a catalogue of the land and fresh-water testaceous Mollusca, found in the vicinity of Newcastle. - Transactions of the Natural History Society of Northumberland and Durham 1 (3): 337-342. Newcastle.

Alexandrowicz, S. W. 2003. *Planorbella duryi* (Wetherby, 1879) from the crater-lake Albano (central Italy). - Folia Malacologica 11 (3/4): 89-93.

Allman, G. J. 1843. On a new genus of terrestrial gasteropod. - The Athenaeum 1843 (829): 851. London.
Date was 16 Sept 1843, printed on p. 852.

Alonso, M. R. 1974. Un nuevo chondrínido de la provincia de Granada: *Chondrina farinesii granatensis*, n. ssp. (Mollusca, Pulmonata, Chondrinidae). - Cuadernos de Ciencias Biológicas 3: 87-90, Lám. I-II [= 1-2]. Granada.

Altaba, C. R. 1997. How many species of *Candidula* (Gastropoda: Hygromiidae) in northern Portugal? - Iberus 15 (1): 1-4.

Altaba, C. R. 2007. À propos de quelques noms de naïades: pourquoi faut-il oublier Potomida et Pseudunio? - MalaCo 4: 148-150.

Altaba, C. R. 2007. Malacofauna Balearica, 1. Dedicada a l'estudi dels mol·luscs de les Illes Balears. - pp. 1-38. Palma de Mallorca.
Cited from a PDF file dated 25 June 2007. In French and Spanish library catalogues registered as electronic resource "CD-ROM" (Oct 2010), not found in other libraries. Probably not published in the sense of Art. 8 and 9.

Alten, J. W. von 1812. Systematische Abhandlung über die Erd- und Flussconchilien welche um Augsburg und der umliegenden Gegend gefunden werden. Als ein Beytrag zur vaterländischen Naturgeschichte. - pp. I-XVI [= 1-16], 1-120, Tab. I-XIV [= 1-14]. Augsburg. (Palm).

Altimira, C. 1963. Notas malacológicas. - Miscelánea Zoológica 1: 15-25. Barcelona.

Altonaga, K. 1986. A new *Oxychilus* (Gastropoda, Stylommatophora, Zonitidae) from N Iberian peninsula. - Iberus 6 (2): 237-244. Barcelona.

Altonaga, K. 1990. A new species from the Iberian peninsula: *Oxychilus (Ortizius?) basajauna* n. sp. (Pulmonata: Zonitidae). - Journal of Conchology 33 (5): 281-289, Pl. 30. London.

Alzona, C. 1971. Malacofauna Italica. Catalogo e bibliografia dei molluschi viventi, terrestri e d'acqua dolce. - Atti della Società Italiana di Scienze Naturali e del Museo Civico di Storia Naturale di Milano 111: 1-433, [1].

Alzona, C. & Alzona Bisacchi, J. 1938. Malacofauna italica. Vol. I. - (da pag. 93 a pag. 128). - pp. 93-128. Genova. (Quinto al Mare). [Cited from original in Museum Leiden Library]

Am Stein, G. 1890. Beiträge zur Mollusken-Fauna Graubünden's, wie solche vom Herbst 1884 bis Herbst 1889 zur Kenntniss gelangt sind. - Jahresbericht der Naturforschenden Gesellschaft Graubündens 33: 3-34.

Ancey, C. F. 1884. Mollusques inédits du système Européen. - Bulletins de la Société Malacologique de France 1: 157-172.
Date Jun 1884 given on p. 172.

Ancey, C. F. 1887. Description of new genera or sub-genera of Helicidæ. (Continued). - Conchologists' Exchange 2 (2): 22-23.

Anderson, R. 1996. Species inventory for Northern Ireland. Land and freshwater Mollusca. - pp. 1-26.
Online publication, possibly unpublished in this form. Recorded in various British libraries with 24 pp. and the date 1997, no library records outside Britain.

Anderson, R. 1996. *Physa gyrina* (Say), a North American freshwater gastropod new to Ireland, with a key to the British Isles Physidae. - Irish Naturalists' Journal 25 (7): 248-253. Belfast.

Anderson, R. 2003. *Physella (Costatella) acuta* (Draparnaud) in Britain and Ireland, its taxonomy, origins and relationship to other introduced Physidae. - Journal of Conchology 38: 7-21. London.

Anderson, R. 2004. *Arion occultus* n. sp, a new slug of subgenus *Kobeltia* Seibert (Pulmonata: Arionidae) from Ireland. - Journal of Conchology 38 (4): 341-353. London.

Anderson, R. 2005. An annotated list of the non-marine Mollusca of Britain and Ireland. - Journal of Conchology 38: 607-637.

Andreae, A. 1883. Ein Beitrag zur Molluskenfauna der Süd-Alpen. - Nachrichtsblatt der Deutschen Malakozoologischen Gesellschaft 15 (9/10): 129-143.

Andreae, A. 1902. Zweiter Beitrag zur Binnenconchylienfauna des Miocäns von Oppeln in Schlesien. - Mitteilungen aus dem Roemer-Museum Hildesheim 18: 1-31.

Andrusov, N. 1897. Iskopaemye i zhivushchie Dreissensidae Evrazii. - pp. I-IV [= 1-4], 1-683, 1-115. S.-Peterburg. (Merkusheva).

Anistratenko, V. V. 1998. Opredelitel' grebnezhabernykh mollyuskov (Gastropoda, Pectinibranchia) fauny Ukrainy. Chast' 2. Presnovodnye i nazemnye. - Vestnik Zoologii Otdel'nj : 67-117. Kiev. [Not cited from original].

Anistratenko, O. Y., Starobogatov, Ya. I. & Anistratenko, V. V. **1999**. Mollusks of the genus *Theodoxus* (Gastropoda, Pectinibranchia, Neritidae) from the Black and the Azov Seas basin. - Vestnik Zoologii 33: 11-19. [Not cited from original].

Anonymous 1847. Atti verbali della sezione di zoologia, anatomia comparata e fisiologia. - Atti della Riunione degli Scienziati Italiani 8: 429-531, Tav. I.II [= 1-2]. Genova.

Anton, H. E. 1839. Verzeichniss der Conchylien welche sich in der Sammlung von Hermann Eduard Anton befinden. - pp. I-XVI [= 1-16], 1-110. Halle. (Anton).
The introcduction on p. IV was signed 3 Aug 1838. A note on p. 110 confirmed that the printing process was finished 9 Oct 1838. On the title page was given the date 1839.
On p. VI the author explained that species printed in capitals were regarded as type species for the corresponding genera.

Aparicio, M. T. 1985. Variabilidad de *Cernuella (Xeromagna) cespitum* (Draparnaud, 1801) en ejemplares españoles e italianos (Pulmonata, Helicidae). - Bollettino Malacologico 21 (10/12): 281-288.

Aparicio, M. T. & Ramos, M. A. 1988. A comparative study of the morphology of the pulmonate snail *Pseudotachea litturata* (Pfeiffer) and other species of *Pseudotachea*, *Iberus* and *Cepaea*. - Journal of Molluscan Studies 54: 287-294.

Araujo, R., Reis, J., Machordom, A., Toledo, C., Madeira, M. J., Gómez, I., Velasco, J. C., Morales, J., Barea, J. M., Ondina, P. & Ayala, I. 2009. Las náyades de la península Ibérica. - Iberus 27 (2): 7-72.

Armbruster, G. 1995. Univariate and multivariate analyses of shell variables within the genus *Cochlicopa* (Gastropoda: Pulmonata: Cochlicopidae). - Journal of Molluscan Studies 61 (2): 225-235.

Armbruster, G. 1997. Evaluations of RAPD markers and allozyme patterns: Evidence for morphological convergence in the morphotype of *Cochlicopa lubricella* (Gastropoda: Pulmonata: Cochlicopidae). - Journal of Molluscan Studies 63 (3): 379-388.

Armbruster, G. F. J. & Bernhard, D. 2000. Taxonomic significance of ribosomal ITS-1 sequence markers in self-fertilizing land snails of *Cochlicopa* (Stylommatophora, Cochlicopidae). - Mitteilungen aus dem Museum für Naturkunde in Berlin/Zoologische Reihe 76 (1): 11-18.

Armbruster, G. & Schlegel, M. 1994. The land snail species of *Cochlicopa* (Gastropoda: Pulmonata: Cochlicopidae): presentation of taxon-specific allozyme patterns, and evidence for a high level of self-fertilization. - Journal of Zoological Systematics and Evolutionary Research 32 (4): 282-296.

Arrébola Burgos, J. R. 1998. On two Andalusian *Oestophora* species: *Oestophora ortizi* De Winter & Ripken, 1991 and *Oestophora granesae* spec. nov. (Gastropoda: Pulmonata). - Journal of Conchology 36 (4): 35-41. London.

Arrébola, J. R. 2001. Nuevos datos sobre dos especies del género *Helicella* Férussac, 1821 en Andalucía: *Helicella stiparum* (Rossmässler, 1854) y *H. gasulli* Ortiz de Zárate, 1950 (Gastropoda, Pulmonata). - Malakos 6: 2-11.

Arrébola, J. R & García, F. J. 1999. Redescripción de *Helix martigena* Fèrussac 1832 como perteneciente al género *Trichia* Hartmann 1840 (Gastropoda: Pulmonata: Hygromiidae). - Iberus 17 (1): 21-30.

Arrébola, J. R. & Gittenberger, E. 1993. New distributional data for *Rupestrella dupotetii* (Terver) (Gastropoda: Chondrinidae) from southern Spain and NW Africa, with notes on allegedly subspecific characters. - Journal of Conchology 31?: 351-355.

Arrébola, J. R. & Gómez, B. J. 1998. Nuevas aportaciones al conocimiento del género *Chondrina* (Gastropoda, Pulmonata) en el sur de la Península Ibérica, incluyendo la descripción de *Chondrina maginensis* spec. nov. - Iberus 16 (2): 109-116.

Arrébola, J. R., Prieto, C. E., Puente, A. I. & Ruiz, A. **2006**. *Hatumia*, a new genus for *Oestophora riffensis* Ortiz de Zárate, 1962, *Oestophora cobosi* Ortiz de Zárate, 1962 and *Hatumia pseudogasulli* n. sp (Pulmonata: Helicoidea: Trissexodontidae). - Journal of Conchology 39 (2): 119-134. London.

Azpeitia Moros, F. 1925. Estudio de algunas especies de moluscos dedicadas al Dr. Hidalgo por diversos autores, y de otras publicadas por dicho Doctor. - Revista de la Real Academia de Ciencias Exactas, Físicas y Naturales 22 (7): 153-186, Lám. 1. Madrid.
11 Feb 1925 was given as the date when the study was presented at the academy session. 1925 was given on the title page of the journal volume.

Azpeitia Moros, F. 1925. Rectificación de nombre para una "*Helix*" española (*H. Huidobroi* Azp.) y revisión de las especies que tienen mayor afinidad con ella. - Trabajos del Museo Nacional de Ciencias Naturales, Serie Zoológica 51: 1-37. Madrid.

Bába, K. 2004. Area-analytical zoogeographic classification of the family Melanopsidae (Gastropoda: Prosobranchia). - Nachrichtenblatt der Ersten Vorarlberger Malakologischen Gesellschaft 12: 30-36.

Babor, J. & Koštal, J. 1893. Příspěvky ku pozni poměrů pohlavních u některých limacidů. - Věstník Královské Společnosti Nauk, Třída mathematicko-přírodovědecká - Sitzungsberichte der Königl.-Böhmischen Gesellschaft der Wissenschaften, mathematisch-naturwissenschaftliche Classe 51 ["1894"]: 1-7. Prag.

Backeljau, T. 1989. The original diagnoses of *Arion simrothi* Künkel and *A. magnus* Torres Minguez (Mollusca, Pulmonata). - Annales de la Société Royale Zoologique de Belgique 119: 199-211. Bruxelles.

Bădărău, A. S. & Szekeres, M. 2001. *Alopia (Kimakowiczia) maciana* n. sp., a Pleistocene relict of the Gilău-Muntele Mare Mts. in Romania (Gastropoda: Clausiliidae: Alopiinae). - Archiv für Molluskenkunde 129 (1/2): 65-68.

Bajdashnikov, A. A. 1990. Obzor mollyuskov rodu *Mentissa* (Gastropoda, Pulmonata). - Zoologicheskij Zhurnal 69 (1): 21-31. Moskva.

Bajdashnikov, A. A. 1990. O vnutrividovykh formakh mollyuskov roda *Mentissa* (Gastropoda, Pulmonata, Clausiliidae). - Zoologicheskij Zhurnal 69 (8): 19-31. Moskva.

Bajdashnikov, A. A. 1985. Novyj dlya nauki nazemnyj legochnyj mollyusk iz vostochnykh Karpat. - Zoologicheskij Zhurnal 64 (2): 206-211.

Baker, **F. C.** 1908. Suggestions for a natural classification of the family Lymnæidae. - Science 27 (703): 942-943. New York.

Baker, **G. H.** 1988. Dispersal of *Theba pisana* (Mollusca: Helicidae). - Journal of Applied Ecology 25: 889-900.

Baker, **H. B.** 1922. The Mollusca collected by the University of Michigan-Walker expedition in southern Vera Cruz, Mexico. I. - Occasional Papers of the Museum of Zoology, University of Michigan 106: 1-92, Pl. I-XVI [= 1-16].

Baker, H. B. 1928. Minute American Zonitidae. - Proceedings of the Academy of Natural Sciences of Philadelphia 80: 1-44, Pl. 1-8.

Baker, H. B. 1929. New southern Appalachian land snails. - The Nautilus 42: 86-93, Pl. 3.

Baker, **H. B.** 1929. Nomenclature in the genus *Vitrina*. - The Nautilus 42 (4): 137-139.

Balashov, I. A. & Gural-Sverlova, N. V. 2011. Nazemnye mollyuski roda *Pyramidula* (Pulmonata, Pyramidulidae) vostochnoj Evropy, tsentral'noj Azii i prilegayushtshikh territorii. - Zoologicheskij Zhurnal 90 (12): 1423-1430.

Balashov, I. & Gural-Sverlova, N. 2012. An annotated checklist of the terrestrial molluscs of Ukraine. - Journal of Conchology 41 (1): 91-109.

Bandel, K. & Riedel, F. 1994. The Late Cretaceous gastropod fauna from Ajka (Bakony mountains, Hungary): a revision. - Annalen des Naturhistorischen Museums in Wien 96A: 1-65.

Bank, R. A. 1985. Eine neue Enide von der griechischen Insel Mytilini (Gastropoda: Pupillacea). - Heldia 1 (2): 41-44, Taf. 6.

Bank, R. A. 1987. Zur rassenmäßigen Gliederung der *Charpentieria (Itala) stenzii* (Rossmässler, 1836) (Gastropoda Pulmonata: Clausiliidae) in den Südalpen. - Basteria 51: 135-140.

Bank, R. A. 1988. Revision der nordostitalienischen Arten und Unterarten der Gattung *Cochlostoma* Jan 1830 (Gastropoda Prosobranchia: Cyclophoridae). - Basteria 52 (4/6): 151-170.

Bank, R. A. 1989. Die Veröffentlichungsdaten der Rossmässler'schen 'Iconographie der Land- und Süßwasser-Mollusken' Europas (1835-1920). - Mitteilungen der Deutschen Malakozoologischen Gesellschaft 44/45: 49-53.

Bank, R. A. 1995. Über *Petasina leucozona* (C. Pfeiffer) und ihre Rassen (Helicoidea, Hygro-miidae). - Heldia 2 (3/4): 75-79. München.

Bank, R. A. 2011. Under threat: The stability of authorships of taxonomic names in malacology. - Mitteilungen der Deutschen Malakozoologischen Gesellschaft 86: 9-12.

Bank, R. A. 2011. Authorships and publication dates in malacology: Some notes on the 2011 French checklist of Welter-Schultes & al. - Mitteilungen der Deutschen Malakozoologischen Gesellschaft 86: 13-24.

Bank, R. A. & Butot, L. J. M. 1990. Eine neue *Turanena*-Art (Gastropoda Pulmonata: Enidae) von der griechischen Insel Samos. - Basteria 54 (4/6): 197-200.

Bank, R. A., Bouchet, P., Falkner, G., Gittenberger, E., Hausdorf, B., Proschwitz, T. von & Ripken, T. **2001**. Supraspecific classification of European non-marina Mollusca (CLECOM Sections I + II). - Heldia 4 (1/2): 77-128.

Bank, R. A., Falkner, G. & Gittenberger, E. **2000**. Nomenclatural notes on a *Cecilioides* species of the Italian and Swiss Alps (Gastropoda, Pulmonata, Ferussaciidae). - Basteria 64: 99-104.

Bank, R. A. & Gittenberger, E. 1985. Notes on Azorean and European *Carychium* species (Gastropoda Basommatophora: Ellobiidae). - Basteria 49 (4/6): 85-100.

Bank, R. A. & Gittenberger, E. 1993. Neither *Rumina truncata*, not *R. gracilis*, but *R. saharica* (Mollusca: Gastropoda Pulmonata: Subulinidae). - Zoologische Mededelingen 67 (38): 525-527. Leiden.

Bank, R. A. & Hovestadt, A. 1991. Notes on Enidae, 3: Revision of the Enidae of Cyprus (Gastropoda: Pupilloidea). - Schriften zur Malakozoologie 4: 1-25.

Bank, R. A. & Maassen, W. J. M. 1998. Beiträge zur Molluskenfauna der Präfektur Samos (Griechenland, östlichen Ägäis). - De Kreukel 34 (4/5): 45-80, Taf. 1-2. Diemen.

Bank, R. A. & Menkhorst, H. P. M. G. 1988. Eine neue *Doraegopis*-Art von Euboea, Griechenland (Gastropoda Pulmonata: Zonitidae). - Basteria 52 (4/6): 147-150.

Bank, R. A. & Menkhorst, H. P. M. G. 1991. Bemerkungen über rezente Arten der Gattung *Pleurodiscus* (Gastropoda Pulmonata: Pupilloidea). - Basteria 55 (1/3): 61-71.

Bank, R. A. & Menkhorst, H. P. M. G. 1992. Notizen zur Familie Enidae, 4. Revision der griechischen Arten der Gattungen *Ena*, *Zebrina*, *Napaeopsis* und *Turanena* (Gastropoda Pulmonata: Pupilloidea). - Basteria 56 (4/6): 105-158.

Bank, R. A. & Menkhorst, H. P. M. G. 1994. Katalog der rezenten Clausiliidae (exkl. Gattung *Albinaria*) der Türkei (Gastropoda, Pulmonata). - Deinsea 1: 85-122. Rotterdam.

Bank, R. A. & Menkhorst, H. P. M. G. 2008. Notes on the nomenclature of some land- and freshwater molluscs of the Seychelles, with consequences for taxa from Africa, Madagascar, India, the Philippines, Jamaica, and Europe. - Basteria 72 (4/6): 93-110.

Bank, R. A. & Neuteboom, W. H. 1988. Zur Molluskenfauna der Dodekanes-Inseln Kos, Kalymnos, Pserimos und Nisyros (Griechenland). - De Kreukel Jubileumnummer: 45-62, Taf. 1-2. Diemen.

Bar, Z. 1978. Variation and natural selection in shell thickness of *Theba pisana* along climatic gradients in Israel. - Journal of Molluscan Studies 44: 322-326.

Bar, Z. & Butot, L. J. M. 1986. The land snails of Chios. - De Kreukel 22 (5): 65-93.

Barbara, N. & Schembri, P. J. 2008. The status of *Otala punctata* (Müller, 1774), a recently established terrestrial gastropod in Malta. - Bollettino Malacologico 44 (5/8): 101-107.

Bargues, M. D., Artigas, P., Jackiewicz, M., Pointier, J. P. & Mas-coma, S. **2005**. Ribosomal DNA ITS-1 sequence analysis of European stagnicoline Lymnaeidae (Gastropoda). - Heldia 6 (1/2): 29-40.

Bargues, M. D., Vigo, M., Horak, P., Dvorak, J., Patzner, R. A., Pointier, J. P., Jackiewicz, M., Meier-Brook, C. & Mas-coma, S. **2001**. European Lymnaeidae (Mollusca: Gastropoda), intermediate hosts of trematodiases, based on nuclear ribosomal DNA ITS-2 sequences. - Infection, Genetics and Evolution 1 (2): 85-107.

Baršiene, J., Ribi, G. & Baršyte, D. 2000. Comparative karyological analysis of five species of *Viviparus* (Gastropoda: Prosobranchia). - Journal of Molluscan Studies 66: 259-271.

Barton, N. H & Hewitt, G. M. 1985. Analysis of hybrid zones. - Annual Review of Ecology and Systematics 16: 113-148.

Baur, A. 1990. Intra- and interspecific influences on age at first reproduction and fecundity in the land snail *Balea perversa*. - Oikos 57: 333-337.

Baur, B. 1989. Growth and reproduction in the minute land snail *Punctum pygmaeum* (Draparnaud). - Journal of Molluscan Studies 55 (3): 383-387.

Baur, B. 1993. Population structure, density, dispersal and neighbourhood size in *Arianta arbustorum* (Linnaeus, 1758) (Pulmonata: Helicidae). - Annalen des Naturhistorischen Museums in Wien 94/95B: 307-321.

Baur, B. & Baur, A. 1995. Habitat-related dispersal in the rock-dwelling land snail *Chondrina clienta*. - Ecography 18: 123-130.

Baur, B. & Raboud, C. 1988. Life history of the land snail *Arianta arbustorum* along an altitudinal gradient. - Journal of Animal Ecology 57: 71-87.

Bech, M. 1988. Contribución al conocimiento de la malacofauna de la Comarca de Solsonés (Lérida). I: Alto Valle de Cardener. - Iberus 8 (1): 101-109.

Beck, H. 1837. Index molluscorum præsentis ævi musei principis augustissimi Christiani Frederici. - pp. 1-124. Hafniæ.
Dates: pp. 1-100 (1837), 101-124 (1838).

Beck, H. 1847. Verzeichniss einer Sammlung von Landconchylien aus den Dänischen Staaten in Europa, bestehend aus 2,058 Individuen, darstellend 158 Arten, die zu 44 Geschlechtern gehören, eingesandt in Folge Allerhöchsten Befehls zur Versammlung Deutscher Naturforscher und Aerzte in Kiel im Jahre 1846 aus dem Königlichen particulären zoologischen Musäum. - Amtlicher Bericht über die Versammlung Deutscher Naturforscher und Ärzte 24: 122-124. Kiel.

Beckmann, K.-H. 1987. Land- und Süßwassermollusken der Maltesischen Inseln. - Heldia 1 (Sonderheft 1): 1-38, Taf. 1-5.

Beckmann, K.-H. 1990. Beiträge zur Kenntnis der Landmolluskenfauna Siziliens mit der Beschreibung von *Muticaria neuteboomi* spec. nov. (Gastropoda Pulmonata: Clausiliidae). - Basteria 54 (1/3): 75-85.

Beckmann, K.-H. 2002. Elemente einer Revision der endemischen Rupestrellen Siziliens. - pp. 49-78, in: Falkner, M., Groh, K. & Speight, M. C. D.: Collectanea malacologica. Festschrift für Gerhard Falkner. -- pp. I-X [= 1-10], 1-547, pl. 1-48.

Beckmann, K.-H. 2003. Beiträge zur Nomenklatur der europäischen Binnenmollusken, XIX. Notizen zu Helicoidea mediterraner Inseln (Pantelleria, Maltesische Inseln). - Heldia 5 (1/2): 33-36. München.

Beckmann, K.-H. 2004. Zur Verbreitung der endemischen nordwestsizilianischen Clausiliidae der Untergattung *Charpentieria (Siciliaria)* mit Beschreibung von zwei neuen Unterarten (Gastropoda: Stylommatophora: Clausiliidae). - Archiv für Molluskenkunde 133 (1/2): 185-191.

Beckmann, K.-H. 2007. Die Land- und Süsswassermollusken der Balearischen Inseln. - pp. 1-255. Hackenheim. (ConchBooks).

Beckmannn, K.-H. & Falkner, G. 2003. Kurze Mitteilungen. Neunachweise von Binnenmollusken auf italienischen Inseln. - Heldia 5 (1/2): 37-40, Taf. 8. München.

Beckmann, K.-H. & Gittenberger, E. 1987. The Clausiliidae (Gastropoda) of the Maltese islands, some additional data. - Journal of Conchology 35: 335-338.

Beckmann, K.-H. & Kobialka, H. 2008. Hygromia *cinctella* (Draparnaud, 1801) auf dem Eroberungszug durch Deutschland (Gastropoda: Hygromiidae). - Club Conchylia Informationen 39 (1/2): 34-41. Ludwigsburg.

Bedulli, D., Cattaneo-Vietti, R., Chemello, R., Ghisotti, F. & Giovine, F. 1995. Fascicolo 15. Gastropoda Opisthobranchia, Divasibranchia, Gymnomorpha. - pp. 1-24, in: Minelli A., Ruffo, S. & La Posta, S.: Checklist delle specie della fauna italiana. Bologna. (Calderini).

Benoit, L. 1843. Ricerche malacologiche. - pp. 1-13, Tav. I-II [= 1-2]. Messina. (La Farfalletta).

Benoit, L. 1875. Illustrazione sistematica critica iconografica de'testacei estramarini della Sicilia ulteriore e delle isole circostanti. - pp. I-XVI [= 1-16], 1-248, Tav. I-IX, XI-XII [= 1-9, 11-12]. Napoli. (Nobile).
Dates (Giusti & Manganelli 1986: 157): part 1 pp. I-XVI, 1-52 (1857 [sic]), part 2 pp. 53-116 (1857), part 3 pp. 117-180 (1859), part 4 pp. 181-248 (1862), In the SUB Göttingen copy <4 ZOOL VII, 1157> dated wrappers were preserved (quaderno I, 1856; quaderno 2, 1857; quaderno III con tavole 5-6, 1859; quaderno 4, 1862), this is not identical with Giusti & Manganelli 1986. From the work itself it is nearly impossible to extract which pages correspond to which quadernos. Title page indicated 1775, not preserved in SUB Göttingen and Smithsonian Institution Libraries, but in Palermo.

Benoit, L. 1876. Catalogo delle conchiglie terrestri e fluviatili della Sicilia e delle isole circostanti. - Bullettino della Società Malacologica Italiana 1 (2) ["1875"]: 129-163. Pisa.
Earliest confirmed date was 09 Jan 1876 (Manganelli & Ci. 2001: 345, Manganelli, Lo. et al. 2008: 31).

Benoit, L. 1881. Nuovo catalogo delle conchiglie terrestri e fluviatili della Sicilia o continuazione alla illustrazione sistematica critica iconografica de'testacei estramarini della Sicilia Ulteriore e delle isole circostanti. - pp. I-VI [= 1-6], 7-176. Messina. (D'Amico).

Beran, L. 2008. A contribution to distribution of genus *Stagnicola* and *Catascopia* (Gastropoda: Lymnaeidae) in the Czech Republic. - Malacologica Bohemoslovaca 7: 70-73.

Beran, L. & Glöer, P. 2006. *Gyraulus chinensis* (Dunker, 1848) – a new greenhouse species for the Czech Republic (Gastropoda: Planorbidae). - Malacologica Bohemoslovaca 5: 25-28.

Beran, L. & Horsák, M. 2007. Distribution of the alien freshwater snail *Ferrissia fragilis* (Tryon, 1863) (Gastropoda: Planorbidae) in the Czech Republic. - Aquatic Invasions 2 (1): 45-54.

Bérenguier, P. 1882. Essai sur la faune malacologique du Var. - pp. 1-106. Draguignan. (Latil).
Preprint, published in 1882, same content as pp. 1-106 of Bérenguier 1884-1885.

Bérenguier, P. 1884-1885. Essai sur la faune malacologique du Var. - Bulletin de la Société d'Etudes Scientifiques et Archéologiques de la Ville de Draguignan 15: 1-118.

Bérenguier, P. 1900. Malacographie de département du Var. Atlas de XXII planches desinées sous la direccion de l'auteur. - Pl. I-XXII [= 1-22]. Draguignan. (Latil).

Bertrand, A. 2000. Les mollusques d'Andorre. Atlas préliminaire de repartition. - Documents Malacologiques 1: 17-39.

Bertrand, A. 2000. *Argna ferrari blanci* (Bourguignat 1874) dans l'Hérault et l'Aveyron (Mollusca: Gastropoda: Argnidae). - Documents Malacologiques 1: 41-44.

Bertrand, A. 2002. *Trichia ataxiaca* (Fagot 1884) en France (Gastropoda: Pulmonata: Hygromiidae). - Documents Malacologiques 3: 15-16.

Bertrand, A. 2002. *Montserratina martorelli* (Bourguignat 1870) en France (Gastropoda: Pulmonata: Hygromiidae). - Documents Malacologiques 3: 17-18.

Bertrand, A. 2002. *Chilostoma desmoulinsii* (Farines 1834) en France (Gastropoda: Pulmonata: Helicidae). - Documents Malacologiques 3: 19-20.

Bertrand, A. 2002. *Chilostoma acrotricha* (P. Fischer 1877) en France (Gastropoda: Pulmonata: Helicidae). - Documents Malacologiques 3: 23-24.

Bertrand, A. 2004. Les Aciculidae des Pyrenées Françaises (Mollusca: Gastropoda). - Documents Malacologiques 4 ["2003"]: 41-46.
Date 22 Jan 2004, not Nov 2003 as stated in the volume (Bichain 2005, MalaCo 1: 22).

Bertrand, A. 2011. *Xerocrassa pallaresica* (Fagot, 1886) en France. -- Folia Conchyliologica 9: 9-10. Lyon.
Unpublished online resource, www.cernuelle.com.

Bielz, **E. A.** 1853. Beitrag zur Kenntniss der siebenbürgischen Land- und Süsswassermollusken. - Verhandlungen und Mittheilungen des Siebenbürgischen Vereins für Naturwissenschaften in Hermannstadt 4 (7): 114-124, 162-165.

Bielz, E. A. 1856. Malakologische Notizen aus Siebenbürgen. - Verhandlungen und Mittheilungen des Siebenbürgischen Vereins für Naturwissenschaften in Hermannstadt 7: 220-228.

Bielz, E. A. 1858. Eine malacologische Excursion in das Burzenland. - Verhandlungen und Mittheilungen des Siebenbürgischen Vereins für Naturwissenschaften in Hermannstadt 9: 142-151.

Bielz, E. A. 1859. Ueber einige neue Arten und Formen der siebenbürgischen Molluskenfauna. - Verhandlungen und Mittheilungen des Siebenbürgischen Vereins für Naturwissenschaften in Hermannstadt 10 (10-11): 212-226.
Bielz, E. A. 1860. Vorarbeiten zu einer Fauna der Land- und Süsswasser-Mollusken Siebenbürgens (Fortsetzung). - Verhandlungen und Mittheilungen des Siebenbürgischen Vereins für Naturwissenschaften in Hermannstadt 11: 149-164.
Bielz, E. A. 1861. Vorarbeiten zu einer Fauna der Land- und Süsswasser-Mollusken Siebenbürgens (Fortsetzung). - Verhandlungen und Mittheilungen des Siebenbürgischen Vereins für Naturwissenschaften in Hermannstadt 12: 63-68, 83-100, 115-116, 131-132, 146-148, 163-164, 176-180, 185-196, 208-227.
Bielz, E. A. 1863. Fauna der Land- und Süsswasser-Mollusken Siebenbürgens. - pp. [1-2], I-VI [= 1-6], 5-206. Hermannstadt. (Steinhaussen).
Bielz, **E. A.** 1867. Fauna der Land- und Süsswasser-Mollusken Siebenbürgens. Zweite Auflage. - pp. I-VIII [= 1-8], 1-216. Hermannstadt. (Filtsch).
Must have appeared before Aug 1867 (Vest 1867: 192).
Bielz, **M.** 1851. Verzeichniss der Land- und Süsswasser-Mollusken Siebenbürgens. - Verhandlungen und Mittheilungen des Siebenbürgischen Vereins für Naturwissenschaften in Hermannstadt 2: 14-16, 55-59, 62-65.
Dates: Jan 1851 (pp. 14-16), Mar 1851 (pp. 55-59), Apr 1851 (pp. 62-65). No author was given on the title page. Author of the preface was C. Fuss, but the article was attributed to M. Bielz on the volume contents page.
Binney, A. 1841. A monograph of the helices inhabiting the United States. - Boston Journal of Natural History 3 (4) ["1840"]: 405-438, Pl. XI, XX-XXV [= 11, 20-25].
On pp. 438 and 465 was printed 17 March 1841, the prints must have appeared later, probably in Nov 1841 (Bank 2011: 19).
Bisenberger, A., Baumgartner, G., Kleewein, D. & Satttmann, H. 1999. Untersuchungen zur Populationsökologie von *Cylindrus obtusus* (Draparnaud, 1805) (Pulmonata, Helicidae). - Annalen des Naturhistorischen Museums in Wien 101B: 453-464.
Bivona, A. 1839. Malacologia. - L'Occhio 9: 65-66. Palermo.
Date 15 May 1839 given on p. 65.
Bivona, A. 1839. Nuovi molluschi terrestri e fluviatili dei dintorni di Palermo ritrovati e descritti. - Giornale di Scienze, Lettere ed Arti per la Sicilia 66: 221-233. [Not cited from original].
Bivona, A. 1839. Nuovi molluschi terrestri e fluviatili dei dintorni di Palermo. - pp. 1-24. Palermo. [Not cited from original].
Could have been an extract of Giornale Letterario, number 198.
Bivona Bernardi, A. 1832. Caratteri d'un nuovo genere di conchiglie della famiglia delle plicacee del signor De Lamarck. - Effemeridi Scientifiche e Letterarie per la Sicilia 2 (2): 58-59. Palermo.
Author was Antonino Bivona Bernardi (p. 55), not Andrea Bivona.
Blanchard, R., Maehrenthal, F. von & Stiles, C. W. 1905. Règles internationales de la nomenclature zoologique adoptées par les Congrès International de Zoologie. International Rules of Zoological Nomenclature. Internationale Regeln der Zoologischen Nomenklatur. - Paris. (Rudeval).
Blum, J. 1881. Schnecken vom Weissenstein bei Solothurn. - Nachrichtsblatt der Deutschen Malakozoologischen Gesellschaft 13 (10): 138-141.
Boato, A. 1988. Microevolution in *Solatopupa* landsnails (Pulmonata Chondrinidae): genetic diversity and founder effects. - Biological Journal of the Linnean Society 34: 327-348.
Boato, A. 1991. Allozyme versus discrete morphologic characters in the phylogenetic analysis of the land snail *Solatopupa* (Pulmonata, Chondrinidae). - Bollettino di Zoologia 58: 345-354.
Boato, A., Bodon, M. & Giusti, F. **1983**. On a new slug from the Northern Apennines (Pulmonata: Arionidae). - Archiv für Molluskenkunde 114 (1/3): 35-44.
Boato, A., Bodon, M. & Giusti, F. **1984**. Molluschi terrestri e d'acqua dolce delle Alpi Liguri. - Lavori della Società Italiana di Biogeografia (Nuova Série) 9: 237-371. Forlì.
Date of publication 30 Dec 1984 given on the title page. List of references contained published articles from 1984 and papers in press indicated for 1985 and 1986.
Bodon, M. 1994. Una nuova *Acicula* Hartmann della Liguria (Prosobranchia: Aciculidae). - Bollettino Malacologico 29 (9/12): 211-222.
Bodon, M., Boato, A. & Giusti, F. **1982**. On the genus *Deroceras* in western Liguria, with description of a new species (Gastropoda: Derocerasidae). - Animalia 9 (1/3): 53-71. Catania.
Date 1982 indicated on title page of the article.
Bodon, M. & Cianfanelli, S. 2008. Una nuova specie di *Platyla* per il sud Italia (Gastropoda: Prosobranchia: Aciculidae). - Bollettino Malacologico 44 (1/4): 27-37.
Bodon, M., Favilli, L., Gianuzzi Savelli, R., Giovine, F., Giusti, F., Manganelli, G., Melone, G. etc. **1995**. Fascicolo 14. Gastropoda Prosobranchia, Heterobranchia Heterostropha. - pp. 1-60, in: Minelli A., Ruffo, S. & La Posta, S.: Checklist delle specie della fauna italiana. - Bologna. (Calderini).
Boeckel, W. 1939. *Cochlostoma*-Arten aus den Karawanken. - Archiv für Molluskenkunde 71: 41-50.
Boeckel, W. 1940. Beitrag zur Systematik alpiner *Retinella*-Arten. - Archiv für Molluskenkunde 72 (1): 10-21.
Boeters, H. D. & Falkner, G. 1998. *Valvata pulchella* S. Studer und *Valvata studeri* n. sp. (Gastropoda, Ectobranchia: Valvatidae). - Heldia 2 (5/6): 113-122, Taf. 14-16.
Boeters, H. D., Gittenberger, E. & Subai, P. **1985**. Eine neue Gattung der Aciculidae (Gastropoda: Prosobranchia) mit zwei neuen Arten. - Basteria 49 (1-3): 59-63.
Boeters, H. D., Gittenberger, E. & Subai, P. **1989**. Die Aciculidae (Mollusca, Gastropoda, Prosobranchia). - Zoologische Verhandelingen 252: 1-234. Leiden.
Boettger, **C. R.** 1909. Ein Beitrag zur Erforschung der europäischen Heliciden. - Nachrichtsblatt der Deutschen Malakozoologischen Gesellschaft 41 (1): 1-19.
Boettger, C. R. 1930. Untersuchungen über die Gewächshausfauna Unter- und Mittelitaliens. - Zeitschrift für Morphologie und Ökologie der Tiere 19 (2/3): 534-590. Berlin.
Boettger, C. R. 1935. Exploration biologique des cavernes de la Belgique et du Limbourg Hollandais. XXIIe contribution: Mollusca. - Zeitschrift für Karst- und Höhlenkunde (Mitteilungen über Höhlen- und Karstforschung) 1935: 49-63. Berlin.
Boettger, C. R. 1939. Bemerkungen über die in Deutschland vorkommenden Bernsteinschnecken (Fam. Succineidae). - Zoologischer Anzeiger 127 (3/4): 49-64. Leipzig.
Boettger, C. R. 1947. Zur Nomenklatur der europäischen Bernsteinschnecken (Fam. Succineidae). - Archiv für Molluskenkunde 76 (4/6): 189-190.
Boettger, C. R. 1949. Eine neue Landschnecke von der Insel Corsica. - Archiv für Molluskenkunde 78 (4/6): 167-168.
Boettger, **C. R.** & Wenz, W. 1921. Zur Systematik der zu den Helicidensubfamilien Campylaeinae und Helicinae gehörigen tertiären Landschnecken. - Archiv für Molluskenkunde 53 (1/2): 6-55.
Boettger, **O.** 1877. Clausilienstudien. - Palaeontographica (Neue Folge) Supplement 3: [1-2], 1-222, Taf. I-IV [= 1-4]. Cassel.
Boettger, O. 1877. Diagnosen neuer Clausilienformen. - Nachrichtsblatt der Deutschen Malakozoologischen Gesellschaft 9 (6): 65-76.
Boettger, O. 1878. Monographie der Clausiliensection *Albinaria* v. Vest. - pp. 39-173, Taf. 145-148, in: Pfeiffer, L.: Novitates Conchologicae 5. Cassel.
Boettger, O. 1878. Neue recente Clausilien. I. - Jahrbücher der Deutschen Malakozoologischen Gesellschaft 5: 33-61, 97-107, Taf. 2-4.
Date 1 Jan 1878 according to a wrapper of Heft 1 preserved in the Smithsonian copy.
Boettger, O. 1878. Neue recente Clausilien. II. - Jahrbücher der Deutschen Malakozoologischen Gesellschaft 5: 291-306, Taf. 10.
Boettger, O. 1879. Neue recente Clausilien. III. - Jahrbücher der Deutschen Malakozoologischen Gesellschaft 6: 101-126, Taf. 2-3.
Boettger, O. 1880. Diagnoses Clausiliarum novarum Graeciae. - Nachrichtsblatt der Deutschen Malakozoologischen Gesellschaft 12 (4): 48-51.
Boettger, O. 1880. Kaukasische Mollusken gesammelt von Herrn Hans Leder, z. Zt. in Tiflis. - Jahrbücher der Deutschen Malakozoologischen Gesellschaft 7: 109-150, Taf. 4.
Boettger, O. 1880. Diagnoses molluscorum novorum ab il. Hans Leder in regione caspia Talysch dicta lectorum. - Jahrbücher der Deutschen Malakozoologischen Gesellschaft 7: 379-383.
Boettger, O. 1881. Sechstes Verzeichnis transkaukasischer, armenischer und nordpersischer Mollusken. - Jahrbücher der Deutschen Malakozoologischen Gesellschaft 8: 167-261, Taf. 7-9.
Böttger, O. 1882. [Nacktschnecken aus Griechenland, den Jonischen Inseln und Epirus.] II. Nacktschnecken aus Epirus und von den Jonischen Inseln. - Nachrichtsblatt der Deutschen Malakozoologischen Gesellschaft 14 (6/7): 96-101.
Author's name not spelled Boettger on the title page.
Boettger, O. 1883. Aufzählung der von den Herren R. Reitter und E. Brenske 1882 in Griechenland und auf den Jonischen Inseln gesammelten Binnenmollusken. - Jahrbücher der Deutschen Malakozoologischen Gesellschaft 10: 313-344.
Böttger, O. 1883. [Malakozoologische Mittheilungen]. II. Binnenconchylien aus Syrien. - Bericht über die Thätigkeit des Offenbacher Vereins für Naturkunde 22/23: 162-176, Taf. 1.
Böttger, O. 1883. Diagnosen neuer Clausilien, gesammelt 1883 auf Creta vom Freiherrn H. v. Maltzan. - Nachrichtsblatt der Deutschen Malakozoologischen Gesellschaft 15 (7/8): 106-113.
Boettger, O. 1883. On new Clausiliæ from the Levant, collected by Vice-Admiral T. Spratt R. N. - Proceedings of the Zoological Society of London 1883: 324-344, Pl. XXXIII-XXXIV [= 33-34].
Date was 10 Apr 1883.
Boettger, O. 1884. Diagnoses specierum Carnioliae, a cl. Jos. Stussiner Labacensi collectarum. - Nachrichtsblatt der Deutschen Malakozoologischen Gesellschaft 16: 184-185.
Boettger, O. 1885. Zur Fauna von Elis und Achaia. - Nachrichtsblatt der Deutschen Malakozoologischen Gesellschaft 17 (7/8): 117-123.
Boettger, O. 1885. On five new species of shells of the genus *Buliminus* from the Levant, collected by Vice-Admiral T. Spratt. - Proceedings of [the Scientific Meetings of] the Zoological Society of London 1885: 23-26.
Boettger, O. 1885. Aufzählung der in Thessalien gesammelten Gastropoden. - Jahrbücher der Deutschen Malakozoologischen Gesellschaft 12: 158-200, Taf. 4.
Boettger, O. 1886. Zur Fauna von Spitza-Sutomore in Süddalmatien. II. - Jahrbücher der Deutschen Malakozoologischen Gesellschaft 13: 34-41, Taf. 2.
Boettger, O. 1886. Aufzählung der in Thessalien gesammelten Schnecken und Muscheln II. - Jahrbücher der Deutschen Malakozoologischen Gesellschaft 13: 50-73, Taf. 2.
Boettger, O. 1888. Ueber einige neue oder bemerkenswerthe Landschnecken aus Griechenland. - Nachrichtsblatt der Deutschen Malakozoologischen Gesellschaft 20 (3/4): 51-58.
Boettger, O. 1889. Die Entwicklung der Pupa-Arten des Mittelrheingebietes in Zeit und Raum. - Jahrbücher des Nassauischen Vereins für Naturkunde 42: 225-327, Taf. VI-VII [= 6-7]. Wiesbaden.
Boettger, O. 1889. Zur Molluskenfauna von Corfu. - Nachrichtsblatt der Deutschen Malakozoologischen Gesellschaft 21 (7/8): 133-138.
Boettger, O. 1889. Verzeichnis der von Herrn E. von Oertzen aus Griechenland und aus Kleinasien mitgebrachten Vertreter der Landschneckengattung *Clausilia* Drp. - Abhandlungen der Senckenbergischen Naturforschenden Gesellschaft 16 (1) ["1890"]: 31-68, Taf. [2]. Frankfurt am Main.
Published in 1889.
Boettger, O. 1891. Weitere Mittheilungen über griechische Mollusken. - Nachrichtsblatt der Deutschen Malakozoologischen Gesellschaft 23 (5/6): 82-91.
Boettger, O. 1892. Schnecken von der Insel Giura, N.-Sporaden, aus dem phokischen Parnass und aus anderen griechischen Gebieten. - Nachrichtsblatt der Deutschen Malakozoologischen Gesellschaft 24 (3/4): 59-66.
Boettger, O. 1894. Die Binnenschnecken der griechischen Inseln Cerigo und Cerigotto. - Nachrichtsblatt der Deutschen Malakozoologischen Gesellschaft 26 (1/2): 1-12.
Boettger, O. 1896. Diagnosen neuer Clausilien. - Nachrichtsblatt der Deutschen Malakozoologischen Gesellschaft 28 (9/10): 124-127.
Boettger, O. 1898. Zwei neue Landschnecken aus Kleinasien. - Nachrichtsblatt der Deutschen Malakozoologischen Gesellschaft 30 (1/2): 12-16.
Boettger, O. 1899. Eine neue Clausilie (*Clausilia blissi*) aus Kleinasien. - Nachrichtsblatt der Deutschen Malakozoologischen Gesellschaft 31 (11/12): 167-169.
Boettger, **O.** 1907. Die ersten Landschnecken von der Insel Thasos. - Nachrichtsblatt der Deutschen Malakozoologischen Gesellschaft 39 (1): 34-40.
Bofill, A. 1886. Contributions à la faune malacologique de la Catalogne. - Bulletins de la Société Malacologique de France 3: 151-164. Paris.
Date July 1886 given on p. 164. Author was spelled Bofill on title page.
Bofill y Poch, A. 1897. Sobre una nueva forma malacológica de la provincia de Gerona (*Nenia subarcuata*). - Boletín de la Real Academia de Ciencias y Artes de Barcelona (3) 1 (18): 364-368.
Date Oct 1897 given on p. 351. Author was spelled Bofill y Poch on title page.
Bofill i Poch, A. & Aguilar-Amat, J. B. de 1924. Malacologia de les Illes Pitiuses. - Treballs del Museu de Ciències Naturals de Barcelona 10 (3): 1-71, Lám. I.II [= 1-2].
Date July 1924 given on the wrapper.
Bogan, A. E., Spamer, E. E. & Hartfield, P. 1992. *Tulotoma* Haldeman 1840 (Gastropoda: Viviparidae): resolution of problems with the designation and nomenclature of the type species. - Malacological Review 25: 13-19.
Boissy, S.-A. de 1835. Hélice. *Helix*. Linnée. H. cotonneuse. *H. lanuginosa*. De Boissy. - Magasin de Zoologie 5: text pp. [1-2], Classe 5 Pl. 69. Paris.
Very difficult to find in the journal, no real pagination. Journal editor is F.-E. Guérin, but author of the unpaginated text pages containing the description is Boissy.
Bole, J. 1960. Novi vrsti iz rodu *Zospeum* Bourg. (Gastropoda). - Biološki Vestnik 7: 61-64. Ljubljana.
Bole, J. 1961. Nove vrste poszemeljskih polžev iz Črne Gore. - II. jugosl. speleol. kongr. Split 1958: 205-207. Zagreb. [Not seen, possibly not a journal but a monographic work, probably not available in the German interlibrary loan system]
Bole, J. 1962. *Congeria kuščeri* sp. n. (Bivalvia, Dreissenidae). - Biološki Vestnik 10: 55-61. Ljubljana.
Bole, J. 1965. Die Vertreter der Gattung *Spelaeodiscus* Brusina 1886 (Gastropoda, Pulmonata) in Jugoslawien. - International Journal of Speleology 1 (3): 349-356, Taf. 76-77. Weinheim.
Bole, J. 1974. Rod *Zospeum* Bourguignat 1856 (Gastropoda, Ellobiidae) v Jugoslaviji. - Razprave [Dissertationes], Slovenska Akademija Znanosti in Umetnosti, Razred za Prirodoslovne in Medicinske Vede, Oddelek za Prirodoslovne Vede 17 (5): 249-291. Ljubljana.
Bole, J. 1983. Rod *Aegopis* Fitzinger 1833 (Gastropoda: Zonitidae) v severozahodni Jugoslaviji. - Razprave [Dissertationes], Slovenska Akademija Znanosti in

Umetnosti [Academia Scientiarum et Artium Slovenica], Razred za Naravoslovne Vede [Classis 4] 24 (3): 1-28 [= 126-154]. Ljubljana.

Bole, J. 1994. Rod *Cochlostoma* Jan 1830 (Gastropoda, Prosobranchia, Cochlostomatidae) v Sloveniji. - Razprave [Dissertationes], Slovenska Akademija Znanosti in Umetnosti [Academia Scientiarum et Artium Slovenica], Razred za Naravoslovne Vede [Classis 4] 35 (11): 187-217. Ljubljana.

Bole, J. 1995. Prispevek k poznavanju rodu *Pagodulina* Clessin 1876 (Gastropoda, Pulmonata: Pagodulinidae) v Sloveniji in na zahodnem Balkanskem Polotoku. - Razprave [Dissertationes], Slovenska Akademija Znanosti in Umetnosti [Academia Scientiarum et Artium Slovenica], Razred za Naravoslovne Vede [Classis 4] 36 (5): 91-106. Ljubljana.

Bole, J. & Slapnik, R. 1997. Molluscs (Gastropoda: Prosobranchia, Pulmonata; Bivalvia) of the Kočevje and Ribnica region (Slovenia). - Razprave [Dissertationes], Slovenska Akademija Znanosti in Umetnosti [Academia Scientiarum et Artium Slovenica], Razred za Naravoslovne Vede [Classis 4] 38 (7): 137-163. Ljubljana.

Bonanno, P. 1684. Recreatio mentis, et ocvli in obseruatione animalium testaceorum curiosis naturæ inspectoribus. Italico sermone primùm proposita. Nunc denuò ab eodem Latinè oblata, centum additis testaceorum iconibus, circa quæ varia problemata proponuntur. - pp. [1-15], 1-270, [1-10], pl. [1-15], Fig. (1) 1-20, Fig. (2) 1-100, Fig. (3) 1-407, pl. [1-5]. Romæ. (Varesij).

Borcherding, J. 1991. The annual reproductive cycle of the freshwater mussel *Dreissena polymorpha* Pallas in lakes. - Oecologia 87: 208-218.

Borredà, V. 2003. *Deroceras roblesi* n. sp., un nuevo agriolimácido (Mollusca, Gastropoda, Pulmonata) del este de la Península Ibérica. - Boletín de la Real Sociedad Española de Historia Natural (Biología) 98 (1-4): 77-83.

Borredà, V. & Martínez-Ortí, A. 2008. Descripción de un nuevo limácido de Menorca (Islas Baleares): *Gigantomilax (Vitrinoides) benjaminus* sp. n. (Gastropoda, Pulmonata). - Animal Biodiversity and Conservation 31 (1): 57-66. Barcelona.

Borredà, V. & Martínez-Ortí, A. 2010. Primera cita confirmada del limácido *Gigantomilax (Vitrinoides) benjaminus* Borredà & Martínez-Ortí, 2008 en la isla de Mallorca. - Noticiario de la Sociedad Española de Malacología 54: 48-49.

Boschi, C. 2011. Die Schneckenfauna der Schweiz. Ein umfassendes Bild- und Bestimmungsbuch. - pp. 1-624. Bern, Stuttgart, Wien. (Haupt).

Bößneck, U. 2000. Eine neue *Abida*-Art aus den südöstlichen Pyrenäen (Gastropoda: Stylommatophora: Chondrinidae). - Malakologische Abhandlungen 20 (1): 5-11. Dresden.

Bößneck, U. 2009. Ein Vorkommen von *Pupilla bigranata* sensu Geyer, 1910 in Ostdeutschland (Gastropoda: Pupillidae). - Mollusca 27 (2): 173-177. Dresden.

Boubée, N. 1836. Hélice nouvelle. [*Helix constricta*]. - L'Echo du Monde Savant et l'Hermes 3 (191): 220. Paris. [Not cited from original].
Date 17 Dec 1836. 2e vol., 2e division (Sci. Nat. Geogr.), No. 50.

Bouchard-Chantereaux, N. R. 1838. Catalogue des mollusques terrestres et fluviatiles observés jusqu'à ce jour à l'état vivant dans le département du Pas-de-Calais. - pp. 1-94, Pl. [1]. Boulogne. (Le Roy-Mabille).
Earliest date 10 Nov 1838, published after Potiez & Michaud 1838 (Kadolsky 2012: 83).

Bouchet, P. & Rocroi, J.-P. 2005. Classification and nomenclator of gastropod famiies. - Malacologia 47 (1-2): 1-397.

Bourguignat, J. R. 1852. Testacea novissima quæ Cl. de Saulcy in itinere per Orientem annis 1850 et 1851, collegit. - pp. [1], 5-31. Lutetiæ. (Baillière).
Date Aug 1852.

Bourguignat, J. R. 1853. Catalogue raisonné des mollusques terrestres et fluviatiles recueillis par M. F. de Saulcy pendant son voyage en Orient. - pp. [1], I-XXVI [= 1-26], 1-96, Pl. 1-4. Paris. (Gide & J. Baudry).

Bourguignat, J.-R. 1854. Aménités malacologiques. - Revue et Magasin de Zoologie pure et appliquée (2) 6: 658-676, Pl. 13-14, 1 table.
Date Dec 1854 given on p. 657.

Bourguignat, J. R. 1855. Aménités malacologiques. - Revue et Magasin de Zoologie pure et appliquée (2) 7: 556-574, Pl. 16. Paris.
Date Dec 1855 given on p. 545.

Bourguignat, J. R. 1856. Aménités malacologiques. - Revue et Magasin de Zoologie pure et appliquée (2) 8: 424-435, Pl. 12. Paris.

Bourguignat, J. R. 1856. Aménités malacologiques. § LI. Du genre *Zospeum*. - Revue et Magasin de Zoologie pure et appliquée (2) 8: 499-516. Paris.

Bourguignat, J. R. 1857. Aménités malacologiques. - Revue et Magasin de Zoologie pure et appliquée (2) 9: 3-21, Pl. 1-4, 8-10. Paris.

Bourguignat, J. R. 1858-1859. Aménités malacologiques. - Revue et Magasin de Zoologie pure et appliquée (2) 10 and 11: vol. 11 pp. 51-59, vol. 10 Pl. 19. Paris.
Falkner et al. 2002, Bank 2011: 16. Text was published in vol. 11 in Feb 1859 (see p. 49 of the volume), Pl. 19 was published in vol. 10 in 1858. The names were quoted on the plate. In the second volume of Aménités malacologiques the same plate was renumbered and published as Pl. 15.

Bourguignat, J. R. 1856. Aménités malacologiques. Tome premier. (Août 1853.-Août 1856.). - pp. [1-2], V-VIII [= 5-8], 1-255, Pl. 1-21. Paris. (Baillière).
1856 indicated on title page, 1 Aug 1856 given on p. VIII. Chapters are entitled "Extrait de la Revue et Magasin de Zoologie" with numbers and years.
p. 1 - No. 8 (1853)
p. 17 - No. 1-2 (1854)
p. 55 - No. 12 (1854) and No. 1 (1855)
p. 107 - No 12 (1855)
p. 131 - No 1 (1856)
p. 151 - No 2 (1856)
p. 165 - No 5, 6 and 7 (1856)
The publications entitled 'Aménités malacologiques' in Revue et Magasin de Zoologie pure et appliquée are different publications, they appeared obviously earlier.

Bourguignat, J. R. 1857. Aménités malacologiques. - Revue et Magasin de Zoologie pure et appliquée (2) 9: 3-21, Pl. 1-4, 8-10. Paris.
Date Jan 1857 printed on volume title page.
§ LII-LXIII [= 52-63].

Bourguignat, J. R. 1857. Aménités malacologiques. - Revue et Magasin de Zoologie pure et appliquée (2) 9: 545-565, Pl. 16-17. Paris.
§ LXV-LXVII [= 65-67].

Bourguignat, J. R. 1859. Aménités malacologiques. - Revue et Magasin de Zoologie pure et appliquée (2) 11: 512-524, Pl. 19-20. Paris.

Bourguignat, J. R. 1860. Aménités malacologiques. Tome second. (Octobre 1856.-Avril 1860.). - pp. [1-2], V-VII [= 5-7], 1-216, Pl. 1-24. Paris. (Baillière).
The publications entitled 'Aménités malacologiques' in Revue et Magasin de Zoologie pure et appliquée are different publications, they appeared obviously earlier.

Bourguignat, J. R. 1861. Des limaces algériennes. - Revue et Magasin de Zoologie pure et appliquée (2) 13: 299-306.

Bourguignat, J. R. 1862. Descriptions des paludinées de l'Algérie, des *Vivipara* d'Europe et de deux espèces nouvelles de la famille des paludinées. - Revue et Magasin de Zoologie pure et appliquée (2) 14: 93-118, Pl. 5. Paris.

Bourguignat, J. R. 1863-1868. Mollusques nouveaux, litigieux ou peu connus. Première centurie (Mars 1863 à Décembre 1868). - pp. [1-2], 1-324, Pl. 1-4, V-XLV [= 5-45]. Paris. (Savy).
Every decade contains 10 species, dates:
1 - pp. 2-22, 01 Mar 1863
2 - pp. 23-49, 01 May 1863
3 - pp. 51-82, 01 Dec 1863
4 - pp. 83-130, 01 May 1864
5 - pp. 131-170, 01 Nov 1865
6 - pp. 171-198, 01 Jan 1866
7 - pp. 199-221, 01 Feb 1866
8 - pp. 223-257, 01 Dec 1867
9 - pp. 259-294, 01 Sep 1868
10 - pp. 295-324, 01 Dec 1868

Bourguignat, J. R. 1864. Malacologie de l'Algérie ou histoire naturelle des animaux mollusques terrestres et fluviatiles recueillis jusqu'à ce jour dans nos possessions du nord de l'Afrique. Tome premier. - pp. [1], I-XII [= 1-12], 1-294, I-XXXII [= 1-32], Pl. I-XXXII [= 1-32], 1 map, 1 table. Paris, Alger. (Challamel Ainé, Bastide).
Dates (Dance 1968) partly 1863, partly 1864.

Bourguignat, J. R. 1864. Malacologie de l'Algérie ou histoire naturelle des animaux mollusques terrestres et fluviatiles recueillis jusqu'à ce jour dans nos possessions du nord de l'Afrique. Tome second. - pp. [1], 1-380, [1-27], 1-4, Pl. I-XXVII [= 1-27]. Paris, Alger. (Challamel Ainé, Bastide).
Date was January 1864.

Bourguignat, J. R. 1864. Malacologie de la Grande-Chartreuse. - pp. [1-2], 1-103, Pl. I-VIII [= 1-8], 8 unpaginated landscape images. Paris. (Savy).
Date was June 1864.

Bourguignat, J.-R. 1869. Description d'espèces nouvelles de mollusques terrestres du Département des Alpes-Maritimes. - Mémoires de la Société des Sciences Naturelles, des Lettres et des Beaux-Arts de Cannes et de l'Arrondissement de Grasse 1: 45-59.

Bourguignat, J. R. 1870. Mollusques nouveaux, litigieux ou peu connus. - Revue et Magasin de Zoologie pure et appliquée et de Séricicultures Comparée (2) 22: 14-30, 87-97, 166-171, Pl. 14-17. Paris.

Bourguignat, J. R. 1870. Mollusques nouveaux, litigieux ou peu connus. Onzième et douzième fascicules. - pp. 1-55, Pl. 1-4. Paris. (Savy).

Bourguignat, J. R. 1876. Histoire des clausilies de France vivantes et fossiles. - Annales des Sciences Naturelles (Série Zoologique) (6) 4: 1-29. Paris.
Date Dec 1876 indicated on p. 5.

Bourguignat, J. R. 1877. Descriptions de deux nouveaux genres algériens, suivies d'une classification des familles et des genres de mollusques terrestres et fluviatiles du système européen. - Bulletin de la Société des Sciences Physiques et Naturelles de Toulouse 3 (1) ["1875-1876"]: 49-101.
Volume must have appeared in early 1877 (pp. 625-632 were published after 1876).

Bourguignat, J. R. 1877. Histoire des clausilies de France vivantes et fossiles. - Annales des Sciences Naturelles (Zoologie) (6) 5 (4): 1-50. Paris.

Bourguignat, J. R. 1880. Description du nouveau genre *Gallandia*. - pp. 1-8. Saint-Germain. (Bardin).

Bourguignat, J. R. 1880. Description de quelques espèces de mollusques terrestres et fluviatiles des environs de Saint-Martin-de-Lantosque (Alpes Maritimes). - pp. 1-8. Cannes. (Vidal). [Not cited from orignal]
Not found in any library catalogue in Europe.

Bourguignat, J. R. 1884. Histoire des mélaniens du système européen. - Annales de Malacologie 2 (1): 1-168. Paris.

Bourguignat, J.-R. 1884. Hélice nouvelle des montagnes de l'Ariège. - Bulletins de la Société Malacologique de France 1: 357-358.
Date Dec 1884 given on p. 358.

Bourguignat, J. R. 1890. Des formes europénnes trocho-hyalinoïdes classées jusqu'à présent sous le nom générique de *Conulus*. - Bulletins de la Société Malacologique de France 7: 325-338, Pl. VIII [= 8]. Paris.

Brancsik, K. 1890. Consignatio systematica specierum in itinere bosnensi anno 1888 per me collectarum, novaque data ad faunam molluscarum Bosniae ac Hercegovinae. - A Trencsén Vármegyei Természettudományi Egylet 11/12: 68-76, Tab. I-II [= 1-2]. Trencsén.

Brandstetter, C. M. 2011. *Limax sarnensis* Heim & Nitz 2009 - ein Synonym zu *Limax alpinus* A. Férussac 1821! - Nachrichtenblatt der Ersten Vorarlberger Malakologischen Gesellschaft 18: 27-32.

Brandt, R. A. 1961. Diagnosen neuer Clausiliiden. - Archiv für Molluskenkunde 90 (1/3): 1-20, Taf. 1-2.

Brandt, R. A. 1962. Über neue und wenig bekannte Clausiliiden. - Archiv für Molluskenkunde 91 (4/6): 127-150, Taf. 4-5.

Braun, **A.** 1847. [Löss bei Krakau und an der Donau; Binnen-Konchylien darin]. - Neues Jahrbuch für Mineralogie, Geognosie, Geologie und Petrefakten-Kunde 1847: 49-54. Stuttgart.

Braun, **M.** 1887. Zur Landmolluskenfauna einiger dalmatinischer Inseln. - Nachrichtsblatt der Deutschen Malakozoologischen Gesellschaft 19 (7/8): 106-111.

Brisson, M. 1759. Observations sur une espèce des limaçon terrestre, dont le sommet de la coquille se trouve cassé, sans que l'animal en souffre. - Mémoires de Mathématique et de Physique, tirés des Registres de l'Académie Royale des Sciences 1759: 99-114, Pl. 3. Paris.

Broderip, W. J. 1836. Descriptions of some species of shells apparently not hitherto recorded. - Proceedings of the Zoological Society of London 4: 43-45.

Brown, D. S., Gracio, M. A. A. & Meier-Brook, C. 1998. The Asian freshwater snail *Gyraulus chinensis* (Dunker, 1848) (Planorbidae) in West Africa and Europe. - Journal of African Zoology 112 (3): 203-213.

Brown, T. 1827. Illustrations of conchology of Great Britain and Ireland drawn from nature. - pp. 1-65, Pl. 1-52. Edinburgh. (Lizars). [Not cited from original].

Bruguière, J. G. 1792. Encyclopédie méthodique. Histoire naturelle des vers. Tome premier. [ABE-CON]. - pp. [1-3], j-xviij [= 1-18], 1-757. Paris. (Panckoucke).
Dates (Sherborn 1902): pp. j-xviij [= 1-18] and 1-344 (1789), 345-757 (1792).

Brumati, L. 1838. Catalogo sistematico delle conchiglie terrestri e fluviatili osservate nel territorio di Monfalcone. - pp. 1-56, [1], Tav. [1]. Gorizia. (Paternolli).

Brusina, S. 1870. Monographie der Gattungen *Emmericia* und *Fossarulus*. - Verhandlungen der kaiserlich-königlichen Zoologisch-Botanischen Gesellschaft in Wien 20: 925-938.

Brusina, S. 1870. Contribution à la malacologie de la Croatie. - pp. 1-40. Zagreb. (Albrecht & Fiedler).

Brusina, S. 1886. Ueber die Mollusken-Fauna Oesterreich-Ungarns. - Mittheilungen des Naturwissenschaftlichen Vereins für Steiermark 22 (1885): 29-56. Graz.

Brusina, S. 1904. Zur Rettung unserer Mollusken-Fauna. - Nachrichtsblatt der Deutschen Malakozoologischen Gesellschaft 36 (4): 157-168.

Bunje, P. M. E. 2007. Fluvial range expansion, allopatry, and parallel evolution in a Danubian snail lineage. - Biological Journal of the Linnean Society 90: 603-617.

Bunje, P. M. E. & Lindberg, D. R. 2007. Lineage divergence of a freshwater snail clade associated with post-Tethys marine basin development. - Molecular Phylogenetics and Evolution 42 (2): 373-387.

Bürk, R. & Jungbluth, J. H. 1982. Erfassung der westpaläarktischen Tiergruppen. Fundortkataster der Bundesrepublik Deutschland. Teil 14: Regionalkataster des Landes Baden-Württemberg: Prodromus zu einem Atlas der Mollusken von Baden-Württemberg. - pp. 1-291. Saarbrücken, Heidelberg.

Bürk, R. & Jungbluth, J. H. 1985. 140 Jahre Molluskenkunde im deutschsprachigen Raum 1844-1984. - pp. 1-348, [1]. Budapest.

Butot, L. J. M. & Welter-Schultes, F. W. 1994. Bibliography of the mollusc fauna of Greece, 1758-1994. - Schriften zur Malakozoologie 7: 1-160.

Buttstedt, L. 2007. Wiederfund einer Restpopulation der Abgeplatteten Teichmuschel (*Pseudanodonta complanata* Rossmässler 1835) für Sachsen-Anhalt. - Mitteilungen der Deutschen Malakozoologischen Gesellschaft 77/78: 11-15.

Calcara, P. 1840. Monografie dei generi Clausilia e Bulimo, coll'aggiunta di alcune nuove specie di conchiglie siciliane esistenti nella collezione della Sig. Teresa Gargotta in Salinas. - Giornale di Scienze, Lettere ed Arti per la Sicilia 76: 1-54, 1 table. Palermo. [Cited from a separate].

Calcara, P. 1842. Descrizione dell'isola di Ustica. - pp. 1-64, 1 map. Palermo.
"Estratto dal Giornale Letterario 229", but obviously not a journal article.

Calcara, P. 1843. Descrizione di alcune nuove specie di conchiglie della Sicilia. - L'Occhio 5: 142-143. [Not cited from original].
Cited by Calcara 1845 and Benoit 1857, not found (neither as a monograph nor the journal) in any library catalogue in Germany and Europe. Sherborn 1922 cited it as an article in L'Occhio but confirmed not to have seen this journal. Riedel 1973 did not find it in mainland Italy and Sicilia. In opac.internetculturale.it (02-2010) recorded as a newspaper (periodico), issued every 2 weeks, probably not for the permanent scientific record (Art. 8.1.1). I consider it as unpublished work.

Calcara, P. 1845. Esposizione dei molluschi terrestri e fluviatili dei dintorni di Palermo. - Atti della Accademia di Scienze e Lettere di Palermo (Nuova Serie) 1 (3): 1-46, Tav. [1].
Date 1845 indicated on the title page of the volume, which is consistent with its content, last page of the article stated "Memoria comunicata in diciembre 1844".

Calcara, P. 1845. Cenno sui molluschi viventi e fossili della Sicilia da servire di supplimento ed insieme di critiche osservazioni all'opera di R. A. Philippi. - pp. 1-65, [1-2], Tav. I-IV [= 1-4]. Palermo. (Reale Stamperia e Libreria).

Calcara, P. 1846. Catalogo dei molluschi terrestri e fluviatili della Sicilia. - pp. 1-4, Tav. [1]. Palermo.
Date 03 Aug 1846 indicated on p. 4, obviously published shortly after. Mentioned by Anonymous 1847: 523.

Calcara, P. 1847. Descrizione dell'isola di Lampedusa. - pp. 1-45. Palermo. (Pagano). [Not cited from original].

Cameron, R. A. D. 2003. Life-cycles, molluscan and botanical associations of *Vertigo angustior* and *Vertigo geyeri* (Gastropoda, Pulmonata: Vertiginidae). - Heldia 5 (Sonderheft 7): 95-110. München.

Cantraine, F. 1835. Les diagnoses ou descriptions succintes de quelques espèces nouvelles de mollusques. - Bulletin de l'Académie Royale des Sciences et Belles-Lettres de Bruxelles 2 (11): 380-401.
Obviously published in Dec 1835. The SUB Göttingen wrapper of vol. 2 quotes 1836.

Cantraine, F. 1840. Malacologie méditerranéenne et littorale, ou description des mollusques qui vivent dans la Méditerranée ou sur le continent de l'Italie, ainsi des coquilles qui se trouvent dans les terrains tertiaires italiens, avec des observations sur leur anatomie, leur mœurs, leur analogie et leur gisement. Ouvrage servant de faune malacologique italienne et de complément à la Conchiologia fossile subapennina de Brocchi. - Nouveaux Mémoires de l'Académie royale des Sciences et des Belles-Lettres de Bruxelles 13 ([7]): 1-173, [1], Pl. I-VI [= 1-6].

Carlsson, R. 2001. Ecology and lifecycle of *Myxas glutinosa* (Müller) in lakes on the Åland islands, southwestern Finland. - Journal of Conchology 37 (2): 105-118.

Carr, R. 2002. Geographical variation of taxa in the genus *Rumina* (Gastropoda: Subulinidae) from the Mediterranean region. - Journal of Conchology 37 (5): 569-577. London.

Caruana Gatto, A. 1892. New *Clausiliæ* from Malta. - The Mediterranean Naturalist 1: 148-149. Malta.

Castillejo, J. 1992. The anatomy of *Arion flagellus* Collinge, 1893, present on the Iberian Peninsula (Gastropoda: Arionidae: terrestria nuda). - The Veliger 35 (2): 146-156.

Castillejo, J. 1997. Las babosas de la familia Arionidae Gray, 1840 en la Península Ibérica e Islas Baleares. Morfología y distribución. (Gastropoda, Pulmonata, terrestria nuda). - Revista, Real Academia Galega de Ciencias 16: 51-118. Santiago de Compostela.

Castillejo, J., Garrido, C. & Iglesias, J. 1994. The slugs of the genus *Geomalacus* Allman, 1843, from the Iberian peninsula (Gastropoda: Pulmonata: Arionidae). - Basteria 58: 15-26.

Castillejo, J. & Mascato, R. 1987. Morphology and anatomy of a new species of *Furcopenis* (Gastropoda Pulmonata Agriolimacidae). - Monitore Zoologico Italiano (Nuova Serie) 21: 33-40. Firenze.

Castillejo, J. & Wiktor, A. 1983. *Furcopenis* gen. n. with its two new species and a new *Deroceras* species from Spain (Gastropoda, Pulmonata, Agriolimacidae). - Malakologische Abhandlungen 9 (1): 1-15.

Caziot, E. 1902. Étude sur la faune des mollusques vivants terrestres et fluviatiles de l'île de Corse. - pp. 1-354, Pl. I-II [= 1-2]. Bastia. (Ollagnier).

Caziot, E. 1904. Complément à l'étude de la faune corse. - Mémoires de la Société Zoologique de France 16 ["1903"]: 33-40. Paris.
Date was 1904 (Falkner et al 2002: 109).

Caziot, E. 1908. Compte rendu d'une excursion malacologique dans la partie supérieure de la vallée de la Roya, et dans le voisinage de la mer, sur la rive droite du Var, près Nice. - Mémoires de la Société Zoologique de France 20: 435-469.

Caziot, E. 1909. Description d'espèces nouvelles de mollusques terrestres et fluviatiles du département des Alpes-Maritimes. - Bulletin de la Société Zoologique de France 34: 87-95, 99-104. Paris.

Caziot, E. 1910. Étude sur les mollusques terrestres et fluviatiles de la principauté de Monaco et du département des Alpes-Maritimes. - pp. [1], 1-559, [1], [1-7], Pl. I-X [=1-10]. Monaco.

Caziot, E. & Margier, E. 1909. Classification proposée pour les espèces de la région paléarctique de la famille des Pupidæ (ancien genre *Pupa*). - Bulletin de la Société Zoologique de France 34: 140-147. Paris.

Čejka, T., Dvořák, L., Horsák, M. & Šteffek, J. 2007. Checklist of molluscs (Mollusca) in the Slovak Republic. - Folia Malacologica 15 (2): 49-58.

Charpentier, J. de 1837. Catalogue des mollusques terrestres et fluviatiles de la Suisse. Formant la seconde partie de la faune Helvétique. - Neue Denkschriften der Allgemeinen Schweizerischen Gesellschaft für die Gesammten Naturwissenschaften - Nouveaux Mémoires de la Société Helvétique des Sciences Naturelles 1 (2): 1-28, Tab. I-II [= 1-2]. Neuchatel.

Charpentier, J. de 1852. Essai d'une classification naturelle des clausilies. - Journal de Conchyliologie 3: 357-408, Pl. XI [= 11]. Paris.

Charrier, M., Chevalier, L., Paradis, G. & Recorbet, B. 2005. Field observations on spatial distribution and diet in the terrestrial snail *Tyrrhenaria ceratina*, an endemic species from Corsica. - Notiziario S.I.M. 23 (5-8): 8. Milano.

Chatfield, J. E. 1968. The life history of the helicid snail *Monacha cantiana* (Montagu), with references also to *M. cartusiana* (Müller). - Proceedings of the Malacological Society of London 38: 233-245.

Chemnitz, J. H. 1786. Neues systematisches Conchylien-Cabinet. Neunten Bandes zwote Abtheilung, enthaltend die ausführliche Beschreibung von den Land- und Flußschnecken, oder von solchen Conchylien, welche nicht im Meere, sondern auf der Erde und in süssen Wassern zu leben pflegen. Mit zwanzig nach der Natur gemalten und durch lebendige Farben erleuchteten Kupfertafeln. - pp. I-XXVI [= 1-26], 1-194, Tab. 117-136. Nürnberg. (Raspe).
Non-binominal work, rejected for nomenclatoral purposes.

Chernyshova, T. M., Garbar, O. V. & Garbar, D. A. 2010. Vydovyj sklad ta poshyrennya pidrodu *Limacus* (Gastropoda, Pulmonata, Limacidae) ta terytoriyi Ukrayiny. - Naukovyj Visnyk Uzhhorod'skoho Universytetu 27: 150-152.

Cianfanelli, S., Manganelli, G. & Giusti, F. **2004**. A new species of *Schileykiella* from Marettimo (Aegadian Islands, Italy) and discussion of relationships of cilielline hygromiids (Gastropoda, Pulmonata, Hygromiidae). - Journal of Conchology 38 (3): 209-230.

Cianfanelli, S., Manganelli, G. & Giusti, F. **2005**. New data on the systematics and distribution of the oxychiline zonitid *Mediterranea depressa* (Sterki, 1880) in Italy (Gastropoda: Pulmonata: Zonitidae). - Journal of Conchology 38: 639-648.

Cianfanelli, S., Talenti, E., Bodon, M. & Manganelli, G. **2000**. Two *Platyla* species from Sardinia (Gastropoda: Prosobranchia: Aciculidae). - Journal of Conchology 37 (1): 61-74.

Cilia, D. P. 2009. On the presence of the alien freshwater gastropod *Ferrissia fragilis* (Tryon, 1863) (Gastropoda: Planorbidae) in the Maltese Islands (central Mediterranean). - Bollettino Malacologico 45: 123-127.

Cioboiu, O. 2006. Diversity of Gastropoda in the Romanian sector of the Danube lower hydrographic basin. - pp. 230-235, in: Austrian Committee Danube Research / IAD: Proceedings 36th International Conference of International Association of Danube Research.

Clanzig, S. & Bertrand, A. 2000. *Trochoidea (Xerocrassa) penchinati* (Bourguignat, 1868) (Gastropoda: Hygromiidae: Geometrinae) un mollusque terrestre de la faune de France peu connu. - Documents Malacologiques 1: 11-15.

Clanzig, S. & Bertrand, A. 2001. *Otala punctata* (O. F. Müller 1774) en France. - Documents Malacologiques 2: 47-48.

Clarke, A. H. 1981. The freshwater molluscs of Canada. - pp. 1-446. Ottawa.

Clench, W. J. & Turner, R. D. 1962. New names introduced by H. A. Pilsbry in the Mollusca and Crustacea. - Special Publication, Academy of Natural Sciences of Philadelphia 4: [1-3], 1-218.

Clessin, S. 1871. Die Mollusken-Fauna der Umgegend von Augsburg. - Bericht des Naturhistorischen Vereins in Augsburg 21: 81-126.

Clessin, S. 1876. Deutsche Excursions-Mollusken-Fauna. - pp. [1-2], 1-581. Nürnberg. (Bauer & Raspe).

Clessin, S. 1877. Die Species der Hyalinen-Gruppe *Vitrea*. - Malakozoologische Blätter 24: 123-134, Taf. I-II [= 1-2].

Clessin, S. 1877. Die Mollusken der Tiefenfauna unserer Alpenseen. - Malakozoologische Blätter 24: 159-185, Taf. III [= 3].

Clessin, S. 1877. Eine neue *Acme*. - Nachrichtsblatt der Deutschen Malakozoologischen Gesellschaft 9 (3): 42-43.

Clessin, S. 1877. Eine neue *Pupa*. - Nachrichtsblatt der Deutschen Malakozoologischen Gesellschaft 9 (4-5): 49-51.

Clessin, S. 1878. Vom Pleistocaen zur Gegenwart. Eine conchyliologische Studie. (Fortsetzung). - Correspondenzblatt des Zoologisch-Mineralogischen Vereines in Regensburg 32: 42-64.

Clessin, S. 1878. Aus meiner Novitäten-Mappe. - Malakozoologische Blätter 25: 122-129, Taf. 5.

Clessin, S. 1879. Die Familie der Cycladeen. In Abbildungen nach der Natur mit Beschreibungen. - Systematisches Conchylien-Cabinet von Martini und Chemnitz 9 (3): 1-283, Taf. 1-46. Nürnberg.
Dates: pp 1-40, Taf. 1-6 (1874), pp 41-48 (1875), pp 49-56 (1876), pp 57-160, Taf. 7-30 (1877), pp 161-200, Taf. 31-42 (1878), pp 201-283, Taf. 43-46 (1879).

Clessin, S. 1879. Zur Molluskenfauna Croatiens. - Nachrichtsblatt der Deutschen Malakozoologischen Gesellschaft 11 (8-9): 116-125.

Clessin, S. 1880. Studien über die Familie der Paludinen. - Malakozoologische Blätter (Neue Folge) 2: 161-196.

Clessin, S. 1880. Die Sectio *Vitrea* des Genus *Hyalina*. - Malakozoologische Blätter (Neue Folge) 2: 204-208.

Clessin, S. 1882. Neue Arten. - Malakozoologische Blätter (Neue Folge) 5: 187-193, Taf. IV [= 4].

Clessin, S. 1882. Ueber *Clausilia silesiaca* A. Schmidt. - Nachrichtsblatt der Deutschen Malakozoologischen Gesellschaft 14 (8-9): 135-137.

Clessin, S. 1883. Anhang zur Molluskenfauna der Krim. - Malakozoologische Blätter (Neue Folge) 6: 37-52, Taf. II-III [= 2-3].

Clessin, S. 1886. Binnenmollusken aus Rumänien. - Malakozoologische Blätter (Neue Folge) 8: 49-56.

Clessin, S. 1887. Beitrag zur Fauna der Binnen-Mollusken Dalmatiens. - Malakozoologische Blätter (Neue Folge) 9: 43-65.

Clessin, S. 1887. Die Molluskenfauna Österreich-Ungarns und der Schweiz. - pp. 1-858. Nürnberg. (Bauer & Raspe).

Clessin, S. 1911. Neue *Acme*-Arten. - Nachrichtsblatt der Deutschen Malakozoologischen Gesellschaft 43 (4): 165-167.

Coadă, V. & Welter-Schultes, F. W. 2011. Checklist of land and freshwater mollusc species in Moldova - Index sistematic al speciilor din Republica Moldova. - Schriften zur Malakozoologie 26: 41-46.

Colling, M., Falkner, G., Groh, K., Jungbluth, J. H., Klemm, M., Niederhöfer, H.-J., Rähle, W. & Schmid, G. 2008. Rote Liste und Artenverzeichnis der Schnecken und Muscheln Baden-Württembergs zweite, neu bearbeitete Fassung. - pp. 1-185. Karlsruhe. (Landesanstalt für Umwelt, Messungen und Naturschutz Baden-Württemberg).

Collinge, W. E. 1893. Description of the anatomy &c. of a new species and variety of *Arion*. - Annals and Magazine of Natural History (6) 12 (70): 252-254, Pl. IX [= 9]. London.

Colomba, M. S., Gregorini, A., Liberto, F., Reitano, A., Giglio, S. & Sparacio, I. **2010**. Molecular analysis of *Muticaria syracusana* and *M. neuteboomi* from southeastern Sicily, Italy (Gastropoda, Pulmonata, Clausiliidae). - Biodiversity Journal 1: 7-14. Palermo.

Colomba, M. S., Gregorini, A., Liberto, F., Reitano, A., Giglio, S. & Sparacio, I. **2011**. Monographic revision of the endemic *Helix mazzullii* De Cristofori & Jan, 1832 complex from Sicily and re-introduction of the genus *Erctella* Monterosato, 1894 (Pulmonata, Stylommatophora, Helicidae). - Zootaxa 3134: 1-42.

Companyo, M. 1837. Rapport de MM. Delocre et Companyo, sur un tableau contenant une collection de mollusques terrestres et fluviatiles du Département des Pyrénées-Orientales, offert à la Société Philomathique par M. Aleron. - Bulletin de la Société Philomathique de Perpignan 3: 85-104.

Companyo, M. & Massot, P. 1845. Description d'une espèce nouvelle de mulette. - Société Agricole, Scientifique et Littéraire des Pyrénées-Orientales 6 ["1843"]: 234-235, Pl. [1].

Conrad, T. A. 1831. Description of fifteen new species of recent, and three of fossil shells, chiefly from the coast of the United States. - Journal of the Academy of Natural Sciences of Philadelphia 6 (2) ["1830"]: 256-268, Pl. XI [= 11]. [Date = Apr 1831, Sherborn 1922].

Conrad, T. A. 1858. Description of a new genus of the family Dreissenidæ. - Proceedings of the Academy of Natural Sciences of Philadelphia 1857: 167.
Date was 1858 (Bank 2011: 22).

Connolly, M. 1912. A revised reference list of South African non-marine Mollusca; with descriptions of new species in the South African Museum. - Annals of the South African Museum 11 (3): 59-306, Pl. II [= 2].

Cooke, C. M. 1921. Notes on Hawaiian Zonitidae and Succineidae. - Occasional Papers Bernice P. Bishop Museum 7 (12): 263-277, Pl. XXIV-XXV [= 24-25]. Honolulu.

Corbella i Alonso, J. 2004. *Suboestophora ebria* sp. nov. (Pulmonata: Trissexodontidae): una nueva especie de gasterópodo de Andalucía. - Spira 1 (4): 39-45. Barcelona.

Cossignani, T. & Cossignani, V. 1995. Atlante delle conchiglie terrestri e dulciacquicole italiane. - pp. 1-208. Ancona. (L'Informatore Piceno).

Cossmann, M. 1889. Gastéropodes. - Annuaire Géologique Universel, Revue de Géologie et Paléontologie 5 (1888): 1079-1109. Paris.

Cossmann, M. 1897. Paléozoologie générale. - Revue Critique de Paléozoologie 1: 129-158.

Cuttelod, A., Seddon, M. & Neubert, E. 2011. European Red List of non-marine molluscs. - pp. i-x [= 1-10], 1-97. Luxembourg. (Publications of the European Union).

Cowie, R. H. 1984. The life cycle and productivity of the land snail *Theba pisana* (Mollusca: Helicidae). - Journal of Animal Ecology 53: 311-325.

Cowie, R. H. 1984. Density, dispersal and neighbourhood size in the land snail *Theba pisana*. - Heredity 52: 391-401.

Cowie, R. H. 1984. Ecogenetics of *Theba pisana* (Pulmonata: Helicidae) at the northern edge of its range. - Malacologia 25: 361-380.

Cowie, R. H. 1990. Climatic selection on body colour in the land snail *Theba pisana* (Pulmonata: Helicidae). - Heredity 65: 123-126.

Cowie, R. H. 1998. Catalog of the nonmarine snails and slugs of the Samoan Islands. - pp. i-viii [= 1-8], 1-122. Honolulu. (Bishop Museum Press).

Cowie, R. H. 2011. Case 3518. *Cornu* Born, 1778 (Mollusca, Gastropoda, Pulmonata, Helicidae): request for a ruling on the availability of the generic name. - Bulletin of Zoological Nomenclature 68 (2): 97-104.

Coxe, W. 1789. Travels in Switzerland, in a series of letters to William Melmoth, Esq. In three volumes. Vol. III. - pp. I-VIII [= 1-8], I-IV [= 1-4], 1-446. London. (Cadell).

Crampton, D. M. 1975. The anatomy and method of functioning of the buccal mass of *Testacella maugei* Férussac. - Journal of Molluscan Studies 41 (3): 549-570.

Crosse, H. 1864. Catalogue des espèces appartenant au genre *Pomatias*, et description d'une espèce nouvelle. - Journal de Conchyliologie 12: 23-33.

Cuvier, **G.** 1798. Tableau élémentaire de l'histoire naturelle des animaux. - pp. j-xvj [= 1-16], 1-710, pl. I-XIV [= 1-14]. Paris. (Baudouin).

Cuvier, G. 1804. Mémoire sur la dolabelle, sur la testacelle et sur un nouveau genre de mollusques à coquille cachée, nommée parmacelle. - Annales du Muséum National d'Histoire Naturelle 5 (30): 435-444, Pl. 29. Paris.

Cuvier, **G.** 1817. Le règne animal distribué d'après son organisation, pour servir de base à l'histoire naturelle des animaux et d'introduction à l'anatomie comparée. Avec figures, dessinées d'après nature. Tome II, contenant les reptiles, les poissons, les mollusques et les annélides. - pp. j-xviij [= 1-18], 1-532. Paris. (Deterville).
On the title page is indicated 1817, but the true date of publication of all 4 volumes of this work is before 2 Dec 1816 (Roux 1976).

Cuvier, **F.** 1821. Dictionnaire des sciences naturelles dans lequel on traite méthodiquement des différens êtres de la nature, considérés soit en eux-mêmes, d'après l'etat actuel de nos connoissances, soit relativement à l'utilité qu'en peuvent retirer la médicine, l'agriculture, le commerce et les arts. Suivie d'une biographie des plus célèbres naturalistes. Ouvrage destiné aux médicins, aux agriculteurs, aux commerçans, aux artistes, aux manufacturiers, et à tous ceux qui ont interêt à connoître les productions de la nature, leurs caractères génériques et spécifiques, leur lieu natal, leurs propriétés et leurs usages. Par plusieurs professeurs du Jardin du Roi, et des principales écoles de Paris. Tome vingtième. [GUA-HEO]. - pp. [1-4], 1-572. Strasbourg, Paris. (Levrault, Le Normant).
Date 1821 given on title page and on a wrapper bound in the Göttingen volume. Author mentioned on p. [4]: "M. F. Cuvier est chargé de la direction générale de l'ouvrage, ...". The authors of the articles (including one author given as M. G. Cuvier) were given on the same page.

Da Costa, E. [= Mendes da Costa, E.] 1778. Historia naturalis testaceorum Britanniæ, or, the British conchology; containing the descriptions and other particulars of natural history of the shells of Great Britain and Ireland: illustrated with figures. In English and French. - Historia naturalis testaceorum Britanniæ, ou, la conchologie Britannique; contenant les descriptions & autres particularités d'histoire naturelle des coquilles de la Grande Bretagne & de l'Irlande: avec figures en taille douce. En anglois & françois. - pp. i-xii [= 1-12], 1-254, i-vii [= 1-7], [1], Pl. I-XVII [= 1-17]. London. (Millan, White, Emsley & Robson).
Name of the author was Emanuel Mendes da Costa, given in this form on the title page. The person was an English naturalist of Portuguese descent, Mendes da Costa was the full Portuguese surname. Zoologists usually cite names of taxa "Da Costa, 1778", possibly in the belief that Mendes was his second first name.

Dall, W. H. 1870. On the genus *Pompholyx* and its allies, with a revision of the Limnæidae of authors. - Annals of the Lyceum of Natural History 9: 333-361. New York.

Dall, W. H. 1898. Tertiary fauna of Florida. - Transactions of the Wagner Free Institute of Philadelphia 3 (4): vii-viii [= 7-8], 571-947.

Dall, W. H. 1903. Notes on the nomenclature of the Pupacea and associated forms. - The Nautilus 17: 114-116.

Damjanov, S. G. & Likharev, I. M. 1975. Fauna na Bâlgarija. 4. Sukhozemni okhljuvi (Gastropoda terrestria). - pp. 1-425, [1]. Sofija. (Bâlgarska Akademija na Naukite).

Damjanov, S. & Pintér, L. 1969. Neue Vitreini aus Bulgarien (Gastropoda: Euthyneura). - Archiv für Molluskenkunde 99 (1/2): 35-40.

Da Silva e Castro, J. 1885. Unionidæ nouveaux du Portugal. - Bulletins de la Société Malacologique de France 2: 277-294. Paris.

Davies, S. M. 1979. Segregates of the *Arion hortensis* complex (Pulmonata: Arionidae), with the description of a new species, *Arion owenii*. - Journal of Conchology 30: 123-127. London.

Dayrat, B. 2009. Review of the current knowledge of the systematics of Onchidiidae (Mollusca: Gastropoda: Pulmonata) with a checklist of nominal species. - Zootaxa 2068: 1-26.

De Betta, E. 1852. Malacologia terrestre e fluviatile della Valle di Non nel Tirolo italiano. Parte I. Molluschi terrestri. - pp. 1-144, Tav. 1. Verona. (Antonelli).

De Betta, E. & Martinati, P. 1855. Catalogo dei molluschi terrestri e fluviatili viventi nelle provincie Venete. - pp. 1-102, [1], Tav. [1]. Verona. (Antonelli).
Date Feb 1855 given on title page, some days after Strobel 1855: 136-141.

De Cristofori, J. & Jan, G. 1832. Catalogus in IV. sectiones divisus rerum naturalium in museo exstantium Josephi De Cristofori et Georgii Jan plurium Acad. Scient. et Societ. Nat. Cur. Sodalium complectens adumbrationem oryctognosiae et geognosiae atque prodrumum faunae et floriae Italiae Superioris. Sectio II. - Pars I. - pp. [1-7], [Conchylia] 1-8, [1], [Mantissa] 1-4, [Excerptum] 1-4, [1], [Conchylia fossilia] 1-16, [1-2]. Parmæ. (Carmignani).

Dedov, I. K. 1998. Annotated check-list of the Bulgarian terrestrial snails (Mollusca, Gastropoda). - Linzer Biologische Beiträge 30 (2): 745-765.

Dedov, I. K. 2008. Terrestrial gastropods (Mollusca, Gastropoda) of the Bulgarian part of the Alibotush Mts. - Malacologica Bohemoslovaca 7: 17-20.

Dedov, I. & Neubert, E. 2002. Contribution to the knowledge of the clausiliid species of the Shar mountains (Republic of Macedonia) with description of new taxa (Pulmonata: Clausiliidae). - Archiv für Molluskenkunde 131 (1/2): 201-209.

Dedov, I. & Neubert, E. 2009. Contribution to the knowledge of the clausiliid fauna from Republic of Macedonia with description of new taxa (Gastropoda: Pulmonata: Clausiliidae). - Archiv für Molluskenkunde 138 (1): 89-101.

Dedov, I. & Subai, P. 2006. Description of a new species of *Cattania* Brusina 1904 from the eastern Rhodopes, Bulgaria (Gastropoda: Pulmonata: Helicidae: Ariantinae). - Archiv für Molluskenkunde 135 (2): 225-230.

Degner, E. 1952. Der Erbgang der Inversion bei *Laciniaria biplicata* Mtg. (Gastr. Pulm.). Nebst Bemerkungen zur Biologie dieser Art. - Mitteilungen aus dem Hamburgischen Zoologischen Museum und Institut 51: 3-61.

Deli, T. 2010. *Agardhiella tunde* spec. nov. (Gastropoda: Pulmonata: Argnidae), a new endemic land snail from Romania. - Journal of Conchology 40 (3): 315-320. London.

Deli, T. & Subai, P. 2011. Revision der *Vitrea*-Arten der Südkarpaten Rumäniens mit Beschreibung einer neuen Art (Gastropoda, Pulmonata, Pristilomatidae). - Contributions to Natural History 19: 1-53. Bern.

Del Prete, R. 1879. Note di conchigliologia apuana. - Bullettino della Società Malacologica Italiana 5 (4/6): 70-91, Tav. 1. Pisa.
Earliest confirmed date was Oct 1879 (Manganelli & Ci. 2001: 345, Manganelli, Lo. et al. 2008: 32).

De Mattia, W. 2004. *Pseudochondrula seductilis* (Rossmässler, 1837) in Italia (Mollusca: Stylommatophora: Enidae). - Bollettino Malacologico 40 (9/12): 104-108.

De Mattia, W. 2007. *Xeropicta derbentina* (Krynicky [sic], 1836) (Gastropoda, Hygromiidae) in Italy and along the Croatian coast, with notes on its systematics and nomenclature. - Basteria 71: 1-12.

De Mattia, W. & Mascia, F. 2011. *Otala punctata* (O. F. Müller, 1774) (Stylommatophora: Helicidae) in Italy. - Iberus 29 (1): 39-46.

De Oliveira, Á. 2007. *Spermodea lamellata* (Jeffreys, 1830) (Pulmonata: Valloniidae, Acanthinulinae): novos dados sobre a sua distribuição em Portugal. - Noticiario de la Sociedad Española de Malacología 47: 29-31.

De Oliveira, A. 2007. *Macrogastra portensis* (Luso da Silva, 1872) (Pulmonata, Clausiliidae): notícia de uma população actual no Noroeste de Portugal. - Iberus 25 (2): 145-147.

De Oliveira, Á. 2009. Materiais para o estudo da malacofauna não-marinha de Portugal. 5. *Cochlicopa lubricella* (Rossmässler, 1834) (Pulmonata, Cochlicopidae). - Noticiario de la Sociedad Española de Malacología 51: 55-58.

De Oliveira, Á. & Altonaga, K. 2010. Materiais para o estudo da malacofauna não-marinha de Portugal. 10. *Oxychilus (Orizius) alliarius* (Miller, 1822) e *Oxychilus (Oxychilus) cellarius* (Müller O. F., 1774) Pulmonata, Oxychilidae). - Noticiario de la Sociedad Española de Malacología 54: 46-47.

De Oliveira, Á., Holyoak, G. A. & Holyoak, D. T. **2010**. Additional records of alien freshwater Mollusca in Portugal (Materiais para o estudo da malacofauna não-marinha de Portugal. 9). - Noticiario de la Sociedad Española de Malacología 54: 41-45.

Dépraz, A., Hausser, J. & Pfenninger, M. 2009. A species delimitation approach in the *Trochulus sericeus/hispidus* complex reveals two cryptic species within a sharp contact zone. - BMC Evolutionary Biology 9: 171.
Electronic only publication, open access without paper outprint, long-term accessibility uncertain.

De Queiroz, K. 2007. Species concepts and species delimitation. - Systematic Biology 56 (6): 879-886.

Deshayes, G. P. 1830. Encyclopédie méthodique. Histoire naturelle des vers. Tome second. - pp. j-vij [= 1-7], 1-256, 1-594. Paris. (Agasse).
Dates (Woodward & Sherborn 1893: 583, Sherborn & Woodward 1899: 595, Cowie 1998: 92): 1830 (part 1 pp. 1-256, part 2 pp. 1-144), Sep 1832 (part 2 pp. 145-594).

Deshayes, G.-P. [1835]. Mollusques. - pp. 81-203, Pl. XVIII-XXVI [= 18-26], in: Geoffroy Saint-Hilaire, I., Geoffroy Saint-Hilaire, E., Deshayes, G.-P., Bibron, G. & Bory de Saint-Vincent, J. B. M. G. ["1832"]: Expédition scientifique de Morée. Section des sciences physiques. Tome III. - 1.re partie. Zoologie. Première section. - Animaux vertrébrés, mollusques et polypiers. -- pp. [1-4], 1-400, Pl. I-XXXVII [= 1-37]. Paris. (Levrault).
Chapter by Deshayes obviously published in 1835, not indicated in the work itself, must have been documented in a secondary source.

Deshayes, G. P. 1848. Exploration scientifique de l'Algérie pendant les années 1840, 1841, 1842 publié par ordre du gouvernement et avec le concours d'une commission académique. Sciences physiques. Zoologie. Histoire naturelle des mollusques. [Tome premier. Mollusques acéphalés. Atlas, 2]. - pp. [1], Pl. LXXVIII-CXLIII [= 78-143]. Paris. (Imprimerie Nationale).
Only plates, no corresponding text volume. Names of taxa were printed on plates.

Des Moulins, C. 1830. Description d'une nouvelle espèce vivante de *Pupa*, du Périgord. - Actes de la Société Linnéenne de Bordeaux 4: 158-163, pl. [1].

De Stefani, C. 1883-1884. Molluschi viventi nelle Alpi Apuane nel Monte Pisano e nell'Appennino adiacente. - Bullettino della Società Malacologica Italiana 9 ["1883"]: 11-253. Pisa.
Earliest confirmed dates were the following (Manganelli & Cianfanelli 2001: 345, Manganelli, Lo. et al. 2008: 33): 15 Jul 1883 (pp. 11-80), 15 Oct 1883 (pp. 81-192), 15 Mar 1884 (pp. 193-253).

De Vico, G., Maio, N. & Castagnolo, L. 2007. Prima segnalazione di *Anodonta (Sinanoonta) woodiana* (Lea, 1834) (Mollusca: Bivalvia: Unionidae) per il Sud Italia. - Notiziario S.I.M. 25 (1/2): 23-25. Milano.

Devriese, H. 1976. *Acicula fusca* (Montagu, 1803) en *Acicula polita* (Hartmann, 1840) uit een Kwartaire laag te Strombeek-Bever (België). - Correspondentieblad van de Nederlandse Malacologische Vereniging 170: 538-539.

De Wilde, J. J., Marquet, R. & Van Goethem, J. L. 1986. Voorlopige atlas van de landslakken van België. - Atlas provisoire des gastéropodes terrestres de la Belgique. - pp. 1-285. Brussel/Bruxelles.

De Winter, A. J. 1985. A new *Deroceras* species from north-western Spain (Gastropoda Pulmonata: Agriolimacidae). - Zoologische Mededelingen 59 (7): 69-77. Leiden.

De Winter, A. J. 1986. Little known and new south-west European slugs (Pulmonata: Agriolimacidae, Arionidae). - Zoologische Mededelingen 60 (10): 135-158. Leiden.

De Winter, A. J. 1990. Little known land snails from the French Alps (Pulmonata). - Basteria 54 (4/6): 227-237.

De Winter, A. J. & Butot, L. J. M. 1985. On some slugs from the isle of Andros, Greece, including the descriptions of two new species (Gastropoda: Pulmonata). - Bulletin. Zoologisch Museum, Universiteit van Amsterdam 10 (12): 77-85.

De Winter, A. J. & Butot, L. J. M. 1986. Three new *Deroceras* species from Greece (Gastropoda, Pulmonata, Agriolimacidae). - Basteria 50 (1/3): 35-44.

De Winter, A. J. & Maassen, W. J. M. 1992. Two new Hygromiidae from Crna Gora and Kosovo (Gastropoda Pulmonata), with a checklist of the hygromiid species described from western Balkan. - Basteria 56: 173-192.

De Winter, A. J. & Ripken, T. E. J. 1991. New land snails (Gastropoda Pulmonata) from the Iberian peninsula. - Basteria 55 (1/3): 3-7.

Dhora, Dh. 2002. Studime mbi molusqet e Shqipërisë. - pp. 1-210. Shkodër. (Camaj-Pipa).

Dhora, Dh. 2004. Lista e Kuqe e ripunuar e molusqeve të Shqipërisë. - pp. 138-148 in: Dhora, Dh.: Mbi molusqet e Shqipërisë. - pp. 1-196. Shkodër. (Camaj-Pipa).

Dhora, Dh. & Welter-Schultes, F. W. 1996. List of species and atlas of the non-marine molluscs of Albania. - Schriften zur Malakozoologie 9: 90-197.

Dieffenbach, E. 1843. Travels in New Zealand; with contributions to the geography, geology, botany, and natural history of that country. In two volumes - Vol. II. - pp. i-iv [= 1-4], 1-396, pl. [1]. London. (Murray).
Author of the fauna chapter (pp. 177-296 "Fauna of New Zealand") is J. E. Gray.

Dillon, R. T., Wethington, A. R., Rhett, J. M. & Smith, T. P. 2002. Populations of the European freshwater pulmonate *Physa acuta* are not reproductively isolated from American *Physa heterostropha* or *Physa integra*. - Invertebrate Biology 121 (3): 226-234.

Diener, C. 1923. Fossilium Catalogus. I: Animalia. Pars 19: Lamellibranchiata triadica. - pp. 1-257, [index 1-2]. Berlin. (Junk).
Date 10 Dec 1923 indicated on the wrapper.

Dohrn, H. & Heynemann, F. D. 1862. Zur Kenntniss der Molluskenfauna der Balearen. - Malakozoologische Blätter 9 (3): 99-111.
Date July 1862 given on p. 97.

Doll, W. 1982. Beobachtungen über Lebensweise und Fortflanzung von *Carychium tridentatum* Risso im Oberrheingebiet (Pulmonata: Ellobiidae). - Archiv für Molluskenkunde 112 (1/6): 1-8.

Douris, V., Giokas, S., Lecanidou, R., Mylonas, M. & Rodakis, G. C. **1998**. Phylogenetic analysis of mitochondrial DNA and morphological characters suggest a need for taxonomic re-evaluation within the Alopiinae (Gastropoda: Clausiliidae). - Journal of Molluscan Studies 64: 81-92.

Douris, V., Giokas, S., Thomaz, D., Lecanidou, R. & Rodakis, G. C. **2007**. Inference of evolutionary patterns of the land snail *Albinaria* in the Aegean archipelago: Is vicariance enough? - Molecular Phylogenetics and Evolution 44: 1224-1236.

Draparnaud, J. P. R. 1801 ("an IX"). Tableau des mollusques terrestres et fluviatiles de la France. - pp. [1-2], 1-116. Montpellier, Paris. (Renaud; Bossange, Masson & Besson).
Date July 1801 (personal communication by R. Bank in Gargominy et al. 2011: 358).

Draparnaud, J.-P.-R. 1805. Histoire naturelle des mollusques terrestres et fluviatiles de la France. Ouvrage posthume. Avec XIII planches. - pp. [1-9], j-viij [= 1-8], 1-134, [Pl. 1-13]. Paris, Montpellier. (Plassan, Renaud).

Drouet, H. 1893. Unionidae de l'Espagne. - Mémoires de l'Académie des Sciences, Arts et Belles-Lettres de Dijon (4) 45: 143-227, Pl. I-II [= 1-2].

Dubois, A. & Bour, R. 2010. The distinction between family-series and class-series nomina in zoological nomenclature, with emphasis on the nomina created by Batsch (1788, 1789) and on the higher nomenclature of turtles. - Bonn Zoological Bulletin 57 (2): 149-171.

Duda, M., Sattmann, H., Haring, E., Bartel, D., Winkler, H., Harl, J. & Kruckenhauser, L. 2011. Genetic differentiation and shell morphology of Trochulus oreinos (Wagner, 1915) and *T. hispidus* (Linnaeus, 1758) (Pulmonata: Hygromiidae) in the northeastern Alps. - Journal of Molluscan Studies 2010 (1): 30-40.

Duméril, A. M. C. 1806. Zoologie analytique, ou méthode naturelle de classification des animaux, rendue plus facile à l'aide de tableaux synoptiques. - pp. i-xxxiii [= 1-33], 1-344. Paris. (Allais).

Dumont, F. & Mortillet, G. 1852-1854. Histoire des mollusques terrestres et d'eau douce vivants et fossiles de la Savoie et du Bassin du Léman. - Bulletin de la Société d'Histoire Naturelle de Savoie 1852: 14-142 [1852], 1-78 [1853], 81-152, 239-248 [1854]. Chambéry. [Only the 1852 part cited from original]

Dundee, D. S. 1986. Notes on the habits and anatomy of the introduced land snails, *Rumina* and *Lamellaxis* (Subulinidae). - The Nautilus 100 (1): 32-37.

Dunker, G. 1848. Diagnoses specierum novarum generis *Planorbis* collectionis Cumingianæ. - Proceedings of the Zoological Society of London 16: 40-43.

Dupuy, D. 1843. Essai sur les mollusques terrestres et fluviatiles et leurs coquilles vivantes et fossiles du Département du Gers. - pp. I-XXXV [= 1-35], 1-140. Auch. (Portes).

Dupuy, D. 1847-1852. Histoire naturelle des mollusques terrestres et d'eau douce qui vivent en France. - pp. I-XXXI [= 1-31], 1-733, [index 1-12], index 735-736, [index 1-4], [errata 1-3], Pl. I-XXXI [= 1-31]. Paris. (Masson).
Dates as indicated on the title page:
1er fasc. pp. I-XXXII, 1-106, Pl. 1-4 (07-1847)
2e fascicule pp. 107-226, Pl. 5-10 (01-1848)
3e fascicule pp. 227-330, Pl. 11-14 (01-1849)
4e fascicule pp. 331-458, Pl. 15-21 (12-1850)
5e fascicule pp. 459-594, Pl. 22-24 (05-1851)
6e fascicule pp. 595-738, Pl. 25-31 (11-1851 to 06-1852)

Dupuy, D. 1849. Catalogus extramarinorum Galliæ testaceorum ordine alphabeticus dispositus, brevioribus specierum nondum descriptarum diagnosibus. - pp. [1-4]. Paris.
Author and date 15 Feb 1849 printed on the last page. Issued during the publication of "Histoire naturelle des mollusques" (Dupuy 1847-1852) and often found bound up with that work.

Dussart, G. B. J. 1979. *Sphaerium corneum* (L.) and *Pisidium* spp. Pfeiffer - the ecology of freshwater bivalve molluscs in relation to water chemistry. - Journal of Molluscan Studies 45 (1): 19-34.

Dutailly, G. 1867. Descriptions de quelques espèces nouvelles du groupe de l'*Helix Raspaili*. - Revue et Magasin de Zoologie pure et appliquée et de Sériciculture Comparée (2) 19: 95-101. Paris.
Date Mar 1867. Seems to have appeared shortly before as a monograph or pre-issued separates in Feb 1867.

Dvořák, L. & Kupka, J. 2007. The first outdoor find of an American snail *Zonitoides arboreus* (Say, 1816) from the Czech Republic. - Malacologica Bohemoslovaca 6: 1-2.

Dybowski, W. 1888. Die Gasteropoden-Fauna des Kaspischen Meeres. Nach der Sammlung des Akademikers Dr. K. E. v. Baer. - Malakozoologische Blätter (Neue Folge) 10: 1-79, Taf. I-III [= 1-3].

Eder, L. 1917. Eine neue Fruticicolenart aus den Schweizeralpen: *Fruticicola biconica* n. sp. - Nachrichtsblatt der Deutschen Malakozoologischen Gesellschaft 49 (3): 119-122.

Eder, L 1921. Zur Gastropodenfauna Unterwaldens (Schweiz). - Archiv für Molluskenkunde 53 (5): 217-237.

Edlinger, K. 1997. Morphological and biometrical researches on Austrian clausiliids. Shell morphology and variability in *Clausilia dubia* Draparnaud, 1805. - Iberus 15 (2): 95-121.

Egorov, R. 2008. Treasure of Russian shells. Supplement 5. Illustrated catalogue of the recent terrestrial molluscs of Russia and adjacent regions. - pp. 1-179, [1]. Moscow.

Egorov, R. & Greke, K. 2003. Treasure of Russian shells, vol. 6. Terrestrial Pectinibranchia. Helicinidae, Cyclophoridae, Diplommatinidae, Cochlostmatidae, Aciculidae, Tateidae, Terrestribythinellidae, Pomatiidae. - pp. 1-43, [1]. (Moscow).

Ehrenberg, C. G. 1831. Symbolae physicae. Animalia Evertebrata exclusis insectis. Series prima cum tabularum decade prima. - pp. [1-135]. Berolini. (Mittler).

Ehrmann, P. 1927. Zur Systematik der Clausiliiden, besonders der ostasiatischen. - Sitzungsberichte der Naturforschenden Gesellschaft zu Leipzig 49-52: 18-59.
Date was Feb 1927 according to a note on the cover and one on the title page of the volume.

Ehrmann, P. 1933. Kreis: Weichtiere, Mollusca. - pp. 1-264, in: Brohmer, P., Ehrmann, P. & Ulmer, G.: Die Tierwelt Mitteleuropas. Leipzig. (Quelle & Meyer).

Eichwald, E. 1829. Zoologia specialis, quam expositis animalibus tum vivis, tum fossilibus potissimum Rossiae in universum, et Poloniae in specie, in usum lectionum publicarum in Universitate caesarea Vilnensi. Pars prior propaedeuticam zoologiae atque specialem Heterozoorum expositionem continens. Cum icone tituli et quinque aliis lithographicis. - pp. I-VI [= 1-6], 1-314, Tab. I-V [= 1-5]. Vilnae. (Zawadzki).

Eichwald, E. 1838. Faunæ Caspii Maris primitiæ. - Bulletin de la Société Impériale des Naturalistes de Moscou 1838: 125-174.

Eichwald, E. 1841. Fauna Caspio-Caucasica nonnullis observationibus novis illustravit. - Nouveaux Mémoires de la Société Impériale des Naturalistes de Moscou 7: I-V [= 1-5], 1-290, [1-2], Tab. I-XL [= 1-40].

Eikenboom, J. C. A. 1993. Enige opmerkingen over Italiaanse landslakken (2). Nieuwe vondsten van *Argna valsabina* (Spinelli, 1851). - De Kreukel 29: 71-74.

Eikenboom, J. 1998. New data on the distribution of *Platyla stussineri* (Boettger, 1884) and *Acicula szigethyannae* Subai, 1977 (Prosobranchia, Aciculidae) in Italy. - Bollettino Malacologico 33 (9-12): 141-142.

Elejalde, M. A., Muñoz, B., Arrébola, J. R. & Gómez-Moliner, B. J. 2005. Phylogenetic relationships of *Iberus gualtieranus* and *I. alonensis* (Gastropoda: Helicidae) based on partial mitochondrial 16S rRNA and COI gene sequences. - Journal of Molluscan Studies 2005: 1-7.

Eleutheriadis, N. & Lazaridou-Dimitriadou, M. 1995. Density and growth of freshwater prosobranch snails (*Bithynia graeca* and *Viviparus contectus*) in relation to water chemistry in Serres, northern Greece. - Journal of Molluscan Studies 61: 347-352.

Eleutheriadis, N. & Lazaridou-Dimitriadou, M. 1995. The life cycle, population dynamics, growth and secondary production of the snail *Viviparus contectus* (Millet) (Gastropoda: Prosobranchia) in the marshes of the river Strymonas, Serres, Macedonia, northern Greece. - Malacologia 37 (1): 41-52.

Eleutheriadis, N. & Lazaridou-Dimitriadou, M. 2001. The life cycle, population dynamics, growth and secondary production of *Bithynia graeca* (Westerlund, 1879) (Gastropoda) in Lake Kerkini, northern Greece. - Journal of Molluscan Studies 67: 319-328.

Elkarmi, A. Z. & Ismail, N. S. 2006. Allometry of the gastropod *Melanopsis praemorsa* (Thiaridae: Prosobranchia) from Azraq Oasis, Jordan. - Pakistan Journal of Biological Sciences 9 (7): 1359-1363. Faisalabad.

Ellis, A. E. 1946. On *Potomida* Swainson. - Proceedings of the Malacological Society of London 27: 105-108, Plate 7.

Erőss, Z. P., Fehér, Z. & Hunyadi, A. **2005**. Invasion of a North American alien, *Planorbella anceps* (Menke, 1830) (Mollusca: Gastropoda: Planorbidae), in the ancient Lake Prespa. - Tentacle 13: 6-7.

Erőss, Z. P., Fehér, Z. & Páll-Gergely, B. **2011**. A new species of *Balcanodiscus* Riedel & Urbański 1964 (Gastropoda: Pulmonata: Zonitidae) from northeastern Greece. - Journal of Conchology 40 (6): 577-581. London.

Erőss, Z., Fehér, Z. & Szekeres, M. **2006**. New taxa of *Montenegrina* O. Boettger, 1877 (Gastropoda: Pulmonata: Clausiliidae) from Albania and the neighbouring regions. - Annalen des Naturhistorischen Museums in Wien 107 B: 181-208.

Eschmeyer, W. M. 1998. Appendix A. Species and genera of fishes and the International Code of Zoological Nomenclature. - pp. 2847-2881, in: Eschmeyer, W. M.: Catalog of Fishes. Center for Biodiversity Research and Information, California Academy of Sciences, Special Publication No.1. -- pp. 1-2905.

Esmark, B. & Hoyer, Z. A. 1886. Die Land- und Süsswassermollusken des arctischen Norwegens. - Malakozoologische Blätter (Neue Folge) 8: 84-123, Taf. 5-6.

Esu, D. 1978. La malacofauna continentale pliopleistocenica della formazione fluvio-lacustre di Nuraghe Su Casteddu (Sardegna orientale) e sue implicazioni paleogeografiche. - Geologica Romana 17: 1-33. Roma.

Evans, H. J. 1978. *Limax pseudoflavus* sp. n. A new species of slug for Ireland. - Irish Naturalist's Journal 19: 173. Belfast. [Not cited from original]

Evans, N. J. 1986. The status of *Limax maculatus* (Kaleniczenko 1851), *Limax grossui* Lupu 1970, and *Limax pseudoflavus* Evans 1978 (Gastropoda, Limacidae). - Proceedings of the Academy of Natural Sciences of Philadelphia 138: 576-588.

Fabricius, O. 1780. Favna Groenlandica, systematice sistens animalia Groenlandiae occidentalis hactenvs indagata, qvoad nomen specificvm, triviale, vernacvlvmque; synonyma avctorvm plvrimvm, descriptionem, locvm, victvm, generationem, mores, vsvm captvramque singvli, provt detegendi occasio fvit, maximaque parte secvndvm proprias observationes. - pp. I-XVI [= 1-16], 1-452, pl. [1]. Hafniae, Lipsiae. (Rothe).

Fagot, P. 1879. Mollusques quaternaires des environs de Toulouse et de Villefranche (Haute-Garonne). - Bulletin de la Société d'Histoire Naturelle de Toulouse 13: 282-304.

Fagot, P. 1882. Mollusques du Pic du Gar (Haute-Garonne). - Bulletin de la Société d'Histoire Naturelle de Toulouse 16: 66-78.

Fagot, P. 1883. Diagnoses d'espèces nouvelles pour la faune française. - Bulletin de la Société d'Histoire Naturelle de Toulouse 17: 207-224. [Not cited from original].

Fagot, P. 1884. Contribution à la faune malacologique de la Catalogne. - Annales de Malacologie 2: 169-194. Paris.

Fagot, P. 1886. Promenades malacologiques dans le sud de la France. - Bulletins de la Société Malacologique de France 3: 165-224.
Date Dec 1886 given on pp. 193 and 224.

Fagot, P. 1887-1888. Contribuciones á la fauna malacológica de Aragón. Catálogo razonado de los moluscos del Valle del Éssera. - Crónica Científica 10 and 11: (10) 345-347, 481-484, (11) 31-39, 103-108, 127-131, 193-198. Barcelona.
Seen in parts. Dates 25 Jan 1888 given on p. 33, 10 Mar 1888 on p. 105, 25 Mar 1888 on p. 129.

Fagot, P. 1891. Histoire malacologique des Pyrénées françaises et espagnoles. - Explorations Pyrénéennes, Ascensions et Recherches Scientifiques, Archéologiques, Historiques (Bull. Société Ramond) 26 (2): 129-143. Toulouse.
Several articles with the same title were published consecutively in the same journal in volumes 25 (3) to 27 (1) (1890-1892), and were bound together as a book in 1892 and issued in Bagnères-de-Bigorre. Article verified, but journal title uncertain.

Fagot, P. 1905. Contributions à la faune malacologique de la Catalogne. - Butlletí de la Institució Catalana d'Història Natural 5 (8, 9): 122-124, 138-143. Barcelona.

Fagot, P. 1906. Mollusca nova provinciæ Aragoniæ. - Boletín de la Sociedad Aragonesa de Ciencias Naturales 5: 171-173. Zaragoza.
Descriptions in Latin, additional comments in French.

Fagot, P. 1907. Contribution à la faune malacologique de la province d'Aragon. - Boletín de la Sociedad Aragonesa de Ciencias Naturales 6: 136-160. Zaragoza.

Falkner, G. 1973. Studien über *Trichia* Hartmann, I. *Trichia (Trichia) graminicola* n. sp. aus Südbaden (Gastropoda: Helicidae). - Archiv für Molluskenkunde 103 (4/6): 209-227.

Falkner, G. 1985. *Lauria (L.) reischuetzi* n. sp. aus Istrien (Gastropoda: Pupillacea). - Heldia 1 (2): 45-46, Taf. 7a.

Falkner, G. 1990. Binnenmollusken. - pp. 112-280, in: Fechter, R, & Falkner, G.: Weichtiere. Europäische Meeres- und Binnenmollusken. Steinbachs Naturführer 10. -- pp. 1-288. München. (Mosaik).

Falkner, G. 1991. Vorschlag für eine Neufassung der Roten Liste der in Bayern vorkommenden Mollusken (Weichtiere). Mit einem revidierten systematischen Verzeichnis der in Bayern nachgewiesenen Molluskenarten. - Schriftenreihe Bayerisches Landesamt für Umweltschutz 97 ["1990"]: 61-112. München.

Falkner, G. 1996. Beiträge zur Nomenklatur der europäischen Binnenmollusken, VIII. Nomenklaturnotizen zu europäischen Hygromiidae (Gastropoda: Stylommatophora). - Heldia 2 (3/4) ["1995"]: 97-107. München.
Date Dec 1995 is printed on the title page of the articles and on the cover of the volume, but must have appeared at the beginning of 1996. On the backside cover is a statement that Heft 5/6 would appear in April 1996 (when it actually appeared it had the date March 1998 printed on it). The invoice sent out with volume 2 Heft 3/4 was dated 15 Feb 1996 (preserved in Haus der Natur Cismar library < Moll J 15>), this can probably be accepted as the true date of publication.

Falkner, G. 2000. Beiträge zur Nomenklatur der europäischen Binnenmollusken, X. Nomenklatur einiger Taxa der Art-Gruppe aus der französischen Fauna (Gastropoda et Bivalvia). - Heldia 3 (1): 27-35, Taf. 4-5.

Falkner, G. 2003. Beiträge zur Nomenklatur der europäischen Binnenmollusken, XVIII. Zur Identität von *Paludina troschelii* Paasch 1842 (Gastropoda: Bithyniidae). - Heldia 5 (1/2): 29-32.

Falkner, G. 2007. *Oxychilus (Oxychilus) beckmanni* n. sp., eine neue Glanzschnecke von der Balearen-Insel Mallorca (Gastropoda: Oxychilidae). - pp. 179-185, in: Beckmann, K.-H.: Die Land- und Süsswassermollusken der Balearischen Inseln. -- pp. 1-255. Hackenheim. (ConchBooks).

Falkner, G. 2008. *Limax (Limax) brandstetteri* n. sp. – ein neuer Hochgebirgsschnegel aus den Abruzzen (Gastropoda: Limacidae). - Stuttgarter Beiträge zur Naturkunde (A, Neue Serie) 1: 133–142.

Falkner, G. 2008. *Oxychilus (Ortizius) edmundi* spec. nov. - a new narrow range endemic from Corsica (Gastropoda, Pulmonata, Oxychilidae). - Basteria 72: 135-141.

Falkner, G., Bank, R. A. & Proschwitz, T. von **2001**. Check-list of the non-marine molluscan species-group taxa of the states of northern, Atlantic and central Europe. - Heldia 4 (1/2): 1-76. München.

Falkner, G. & Falkner, M. 2008. Beiträge zur Nomenklatur der europäischen Binnenmollusken, XXIII. *Trochulus coelomphalus* oder *Trochulus coelomphala*? (Gastropoda: Hygromiidae). - Heldia 5 (4/5): 143-144. München.

Falkner, G. & Niederhöfer, H.-J. 2008. Systematisches Artenverzeichnis. - pp. 62-156, in: Colling, M., Falkner, G., Groh, K., Jungbluth, J. H., Klemm, M., Niederhöfer, H.-J., Rähle, W. & Schmid, G.: Rote Liste und Artenverzeichnis der Schnecken und Muscheln Baden-Württembergs zweite, neu bearbeitete Fassung. -- pp. 1-185. Karlsruhe. (Landesanstalt für Umwelt, Messungen und Naturschutz Baden-Württemberg).

Falkner, G. & Niederhöfer, H.-J. 2008. *Limax (Brachylimax* n. subg.*) giovannellae* n. sp., ein sexualbiologisch definiertes Subgenus und eine neue Art aus den Julischen Alpen (Gastropoda: Limacidae). Mit einer Photodokumentation von Clemes M. Brandstetter. - Club Conchylia Informationen 39 (1/2): 7-19, backside cover. Ludwigsburg.

Falkner, G. & Proschwitz, T. von 2009. Beiträge zur Nomenklatur der europäischen Binnenmollusken, XXV. Nomenclatural remarks on the genus-group names in the synonymy of *Limax* s. str. in Schileyko's treatise on recent terrestrial pulmonate molluscs, part 11 (Gastropoda: Limacidae). - Heldia 5 (6): 171-177.

Falkner, G., Ripken, T. E. J. & Falkner, M. **2002**. Mollusques continentaux de France. Liste de référence annotée et bibliographie. - Collection Patrimoines Naturels 52: [1-2], 1-350, [1-3]. Paris.

Falniowski, A. & Wilke, T. 2001. The genus *Mastoniopsis* (Gastropoda: Rissoidea), intra- and intergeneric phylogenetic relationships. - Journal of Molluscan Studies 67: 483-488.

Faltýnkova, A., Valtonen, E. T. & Karvonen, A. 2008. Spatial and temporal structure of the trematode component community in *Valvata macrostoma* (Gastropoda, Prosobranchia). - Parasitology 135: 1691-1699.

Farines, J. N. 1834. Description de trois espèces nouvelles vivantes du Département des Pyrénées-Orientales. - pp. 1-8, pl. [1]. Perpignan. (Alzine).
Monographic brochure, appeared before the journal articles (Falkner et al. 2002: 249).

Farines, J. N. 1834. Description de trois espèces nouvelles vivantes du Département des Pyrénées Orientales, *Unio Pianensis*, *Helix Desmoulinsiii*, *Helix Xatartii*. - L'Institut [Journal des académies et sociétés scientifiques de la France et de l'étranger] 2 (66): 268-269. Paris.
Date 16 Aug 1834 given on p. 265. No mentioned by Falkner et al. 2002: 249.

Farines, J. N. 1834. Description de trois espèces nouvelles vivantes du Département des Pyrénées orientales; lue à la Société Philomathique de Perpignan. - Annales des Sciences Naturelles (Zoologie) (2) 2: 118-123. Paris.
Date Aug 1834 given on p. 113 Extract of the previously published monograph.

Farines, J. N. 1834. Description de trois espèces nouvelles vivantes du Département des Pyrénées-Orientales. - Bulletin de la Société Philomathique de Perpignan 1 ["1835"]: 59-67, Pl. [1].
Date 03 Dec 1834 given as date of accomplishment of the text on p. 93, volume title page indicated 1835.

Fauer, W. 1978. *Siciliaria ernae* n. sp., eine neue rezente Clausilie aus Italien. - Archiv für Molluskenkunde 108 (4/6): 263-265.

Fauer, W. 1978. Neue Albinarien aus Griechenland. - Archiv für Molluskenkunde 109 (1/3): 53-58.

Fauer, W. 1985. Neue Clausilien aus Griechenland (Pulmonata: Clausiliidae). - Archiv für Molluskenkunde 115 (4/6): 301-303.

Fauer, W. 1991. Eine neue Unterart von *Charpentieria (Itala) stenzii* (Rossmässler 1836) (Gastropoda: Pulmonata Clausiliidae). - Archiv für Molluskenkunde 120 (4/6): 163-166.

Fauer, W. 1991. Eine neue *Chilostoma* aus Griechisch-Ostmakedonien (Gastropoda Pulmonata: Helicidae). - Archiv für Molluskenkunde 120 (4/6): 167-170.

Fauer, W. 1993. Zur Kenntnis der rezenten Clausilienfauna der südlichen Balkanhalbinsel (Gastropoda Stylommatophora: Clausiliidae). - Archiv für Molluskenkunde 122: 49-61.

Favre, J. 1927. Les mollusques post-glaciaires et actuels du bassin de Genève. - Mémoires de la Société de Physique et d'Histoire Naturelle de Genève 40 (3): 171-434, Pl. 14-27.

Fehér, Z. 2000. *Medora macarscarensis* [sic] *brancsiki* subsp. n., a new endemic clausiliid from the Velebit mountains, Croatia (Gastropoda: Stylommatophora: Clausiliidae). - Malakologische Abhandlungen 20 (2): 13-17.

Fehér, Z. 2004. A revision of the genus *Cochlostoma*, subgenus *Titanopoma* (Gastropoda, Caenogastropoda, Cochlostomatidae), in particular the forms occurring in Albania. - Basteria 68: 25-44.

Fehér, Z. & Erőss, Z. 2009. Contribution to the Mollusca fauna of Albania. Results of the field trips of the Hungarian Natural History Museum between 1992 and 2007. - Schriften zur Malakozoologie 25: 3-21.

Fehér, Z. & Erőss, Z. 2009. Checklist of the Albanian mollusc fauna. - Schriften zur Malakozoologie 25: 22-38.

Fehér, Z., Varga, A., Deli, T. & Domokos, T. **2009**. Geographic distribution and genital morphology of the genera *Lozekia* Hudec, 1970 and *Kovacsia* Nordsieck, 1993 (Mollusca, Gastropoda, Hygromiidae). - Zoosystematics and Evolution 85 (1): 151-160.

Fehér, Z., Varga, A., Deli, T. & Domokos, T. **2009**. Phylogeny and phylogeography of the *Lozekia-Kovacsia* species group (Gastropoda: Hygromiidae). - Journal of Zoological Systematics and Evolutionary Research 47 (4): 306-314.

Fehér, Z., Zettler, M. L., Bozsó, M. & Szabó, K. **2009**. An attempt to reveal the systematic relationship between *Theodoxus prevostianus* (C. Pfeiffer, 1828) and *Theodoxus danubialis* (C. Pfeiffer, 1828) (Mollusca, Gastropoda, Neritidae). - Mollusca 27 (2): 95-107. Dresden.

Ferreri, D., Bodon, M. & Manganelli, G. 2005. Molluschi terrestri della provincia di Lecce. - Thalassia Salentina 28: 31-130. Lecce.

Férussac, **A. E. J. P. J. F.** d'Audebard de 1814. Mémoires géologiques sur les terreins formés sous l'eau douce par les débris fossiles des mollusques vivant sur la terre ou dans l'eau non salée. - pp. 1-76. Paris. (Poulet).

Férussac, A. E. J. P. J. F. d'Audebard de [1821-1822]. Tableaux systématiques des animaux mollusques classés en familles naturelles, dans lesquels on a établi la concordance de tous les systèmes; suivis d'un prodrome général pour tous les mollusques terrestres ou fluviatiles, vivants ou fossiles. - pp. j-xlvij [= 1-47], [1], 1-27, 1-110, [1]. Paris, Londres. (Bertrand, Sowerby).
Two editions, one folio and one quarto. Obviously with the same content, but different page numbers. If not otherwise indicated, page numbers for names of taxa refer to the quarto edition. Dates (Kennard 1942, Bouchet & Rocroi 2005: 304):
folio pp 1-32, quarto pp 1-24 = 06-04-1821
fol 33-56 qrt 25-48 = 26-05-1821
fol 57-76 qrt 49-72 = 13-07-1821
fol 77-92 qrt 73-88 = 21-09-1821
fol 93-114 qrt 89-111 = 10-11-1821
i-xxiv = 16-02-1822
xv-xlvii = 13-04-1822
Prodrome limaces "1-28" (incorrect, should be 1-27) = 16-07-1822 (01-1821 after Sherborn 1922 and Falkner et al. 2002: 133, probably incorrect)

Férussac, A. E. J. P. J. F. d'Audebard de 1823. Monographie des espèces vivantes et fossiles du genre mélanopside, *Melanopsis*, et observations géologiques à leur sujet. - Mémoires de la Société d'Histoire Naturelle de Paris 1: 132-164, Pl. VII-VIII [= 7-8].

Férussac, **A. E. J. P. J. F.** d'Audebard de & Deshayes, G.-P. 1819-1851. Histoire naturelle générale et particulière des mollusques terrestres et fluviatiles, tant des espèces que l'on trouve aujourd'hui vivantes, que des dépouilles fossiles de celles qui n'existent plus; classés d'après les caractères essentiels que présentent ces animaux et leurs coquilles. - Tome 1: pp. 1-8, 1-184; Tome 2 (1): pp. 1-402; 2 (2): 1-260, 1-22 , 1-16; Atlas 1: Pl. 1-70.; Atlas 2: pp. 1-166, Pl. 1-5. Paris. (Baillière).
This work is bibliographically extremely difficult. In the period 1819-1832, 28 livraisons appeared, consisting of texts and plates, not necessarily connected with each other. Many plate numbers appeared twice. Covers usually had no important texts or names printed on them. Explanations of plates were published in 1819, 1822 and 1832 (species were printed in small caps in these lists, with references to corresponding numbers in *Prodr.*). When the whole work was finished a final issue of plate explanations appeared in 1851 (specific names were in normal and bold font, authors not bold, no references to *Prodr.*). Usually only the final plate explanations were bound, while the initial plate explanations, different from the final ones, were not conserved by librarians (Gittenberger & Groh 1986).
Dates (Kennard 1942, Bouchet & Rocroi 2005: 304):
Title, préface pp 1-16 (6 Mar 1819)
pp 1-16 (5 Jun 1819)
pp 17-56 (10 Jul 1819)
pp 57-72 (18 Sep 1819)
pp 73-96 (4 Dec 1819)
pp 97-128 (17 Jun 1820)
explanation of plates, pp 1-47 (6 Apr 1821)
explanation of plates 48, 53, 63, 75A, 75B, 113 (2 Nov 1822)
pp 96a-96λ (27 Sep 1823)
explanation of plates of livraisons 22-27 (many plates) (4 Aug 1832)
explanation of plates 8E, 126, 131B, 133, 141, 147 (4 Aug 1832, ?) (cover not conserved in London but should have existed)
Plates issued by Deshayes:
pl. 8F, 10A, 17A, 28A, 29A, 69C, 73B, 83, 84, 106, 107, 129 (1839)
pl. 10B, 62A, 69B, 69D-69H, 69K, 72, 85, 86 (1840)
pl. 37A, 55 [4], 62B, 64A, 69I, 69J, 75C, 87, 89 [2], 90 [3], 127A, 127B (1841).

Férussac, **J. B. L.** d'Audebard de 1802. Ueber eine neue Thierart, welche die Gattungen *Limax* und *Helix* mit einander vereinigt, *Helix Semilimax*. - Der Naturforscher 29: 236-241, Tab. I Fig. A-D. Halle.

Férussac, **J. B. L.** d'Audebard de 1807. Essai d'une méthode conchyliologique appliquée aux mollusques fluviatiles et terrestres d'après la considération de l'animal et de son test, par M. Daudebard de Férussac. Nouvelle édition augmentée d'une synonymie des espèces les plus remarquables, d'une table de concordance systématique de celles qui ont été décrites par Géoffroy, Poiret et Draparnaud, avec Müller et Linné, et terminée par un catalogue d'espèces observées en divers lieux de la France, par J. Daudebard fils. - pp. j-xvj [= 1-16], 1-142. Paris. (Delance).
Bouchet & Rocroi 2005: 304 and previous sources attributed the authorship to Jean Baptiste Louis d'Audebert de Férussac alone. Both Férussac (père) and Férussac fils were indicated on the title page. The authorship of Férussac fils referred only to a part of the work, precisely explained in the introduction. Since both co-authors contributed to the content of the work, and both were mentioned as responsible authors on the title page, the work should be attributed to both co-authors. I noticed this contradiction in Aug 2012 very shortly before printing, so I decided to keep the traditional mode of citing the authorships of this work (exception *Melanopsis*). But this should eventually be fixed in the future.

Fez, S. de 1944. Contribución a la malacología de la provincia de Valencia. - Boletín de la Real Sociedad Española de Historia Natural 42 (3/4): 211-224. Madrid.

Fiorentino, V., Caruso, T., Manganelli, G. & Giusti, F. **2009**. Population dynamics of an urban population of the land snail *Marmorana serpentina* (Gastropoda: Pulmonata). - Malacologia 51 (1): 201-209.

Fiorentino, V., Manganelli, G. & Giusti, F. **2008**. Multiple scale patterns of shell and anatomy variability in land snails: the case of the Sicilian *Marmorana* (Gastropoda: Pulmonata, Helicidae). - Biological Journal of the Linnean Society 93: 359-370.

Fiorentino, V., Salomone, N., Manganelli, G. & Giusti, F. **2008**. Phylogeography and morphological variability in land snails: the Sicilian *Marmorana* (Pulmonata, Helicidae). - Biological Journal of the Linnean Society 94: 809-823.

Fiorentino, V., Salomone, N., Manganelli, G. & Giusti, F. **2010**. Historical biogeography of Tyrrhenian land snails: the *Marmorana-Tyrrheniberus* radiation (Pulmonata, Helicidae). - Molecular Phylogenetics and Evolution 55: 26-37.

Fischer, **P.** 1877. Faune malacologique de la vallée de Cauterets. - Additions et corrections. - Journal de Conchyliologie 25: 49-56.

Fischer, P. 1892. Curiosités bibliographiques. - Le catalogue de la collection Schlüter. - Journal de Conchyliologie 40: 208-212. Paris.

Fischer, P. & Crosse, H. 1878. Etudes sur les mollusques terrestres et fluviatiles du Mexique et du Guatemala. - pp. 1-702, Pl. 1-31, in: Milne Edwards, M.: Recherches Zoologiques pour servir à l'histoire de la faune de l'Amérique Centrale et du Mexique. Septième partie. Tome premier. Paris.
Dates see Cowie 1998: 95.

Fischer, **P.**, Œhlert, P. & Woodward, S. P. **1887**. Manuel de conchyliologie et de paléontologie conchyliologique ou histoire naturelle des mollusques vivants et fossiles suivi d'un appendice sur les brachipodes. Avec 23 planches contenant 600 figures et 1138 gravures dans le texte. - pp. I-XXIV [= 1-24], 1-1369, [1-6], Pl. [A], I-XXIII [= 1-23], map [1]. Paris. (Savy).
Author for the molluscan part (pp. 1-1187) is Fischer.
Dates as given on p. II [= 2]:
pp. 1-112 (21 Sep 1880), 113-192 (16 Mar 1881), 193-304 (28 Jul 1881), 305-416 (05 May 1882), 417-514 (21 Feb 1883), 515-608 (20 Dec 1883), 609-688 (30 Jun 1884), 689-784 (29 Jan 1885), 785-896 (31 Aug 1885), 897-1008 (30 Apr 1886), 1009-1369 (15 Jun 1887).

Fischer, **W.** A. 2003. *Paraegopis mauritii mauritii* (Westerlund 1866) und *Helicigona stenomphala* (Menke 1839) aus dem nördlichen Velebitgebirge bei Oltari, Kroatien. - Nachrichtenblatt der Ersten Vorarlberger Malakologischen Gesellschaft 11: 4-7.

Fischer, W. 2003. *Helicigona byshekensis* Knipper 1941 neu für die griechische Molluskenfauna. - Nachrichtenblatt der Ersten Vorarlberger Malakologischen Gesellschaft 11: 8-9.

Fischer, W., Reischütz, A. & Reischütz, P. L. 2009. *Theodoxus transversalis* (C. Pfeiffer 1828) im Dojransee (Mazedonien). - Nachrichtenblatt der Ersten Vorarlberger Malakologischen Gesellschaft 16: 45-46.

Fish, J. D. & Fish, S. 1981. The early life-cycle stages of *Hydrobia ventrosa* and *Hydrobia neglecta* with observations on Potamopyrgus jenkinsi. - Journal of Molluscan Studies 47: 89-98.

Fitzinger, L. I. 1833. Systematisches Verzeichniß der im Erzherzogthume Oesterreich vorkommenden Weichthiere, als Prodrom einer Fauna derselben. - Beiträge zur Landeskunde Oesterreich's unter der Enns 3: 88-122. Wien.

Flach, K. 1890. Palaeontologische Beiträge. - Verhandlungen der Physikalisch-Medicinischen Gesellschaft zu Würzburg (N. F.) 24 (3): 49-59, Taf. [1].

Flasar, I. 1998. Die Gastropoden Nordwestböhmens und ihre Verbreitung. - Heldia Sonderheft 4: 1-210.

Fleming, J. 1818. Conchology. - pp. 1-79, 1-316, Pl. 54-65, in: Encyclopaedia Britannica. Supplement to the fourth, fifth and sixth editions. Volume 3. [Part 1]. Edinburgh. (Constable). [Not cited from original].

Fleming, J. 1828. A history of British animals, exhibiting the descriptive characters and systematic arrangement of the genera and species of quadrupeds, birds, reptiles, fishes, Mollusca, and Radiata of the United Kingdom; including the indigenous, extirpated, and extinct kinds, together with periodical and occasional visitants. - pp. i-xxxii [= 1-23], 1-565, [1]. Edinburgh. (Bell & Bradfute).

Fleming, J. 1830. Mollusca. - pp. 598-635, in: Brewster, D.: The Edinburgh Encyclopædia. In eighteen volumes. Volume XIV [= 14] [MED-MUS]. -- pp. [1], 1-748, Pl. CCCLXXII-CCCC [= 372-400]. Edinburgh. (Blackwood).
Author's name John Fleming abbreviated as J. F. on p. 635, the abbreviation is decoded in vol. 18 p. xi. Year is given as MDCCCXXX = 1830 on all volumes of the Göttingen copies. Issued Nov 1820 following Kennard & Woodward 1926: 379.

Florence 1886. Espèces nouvelles du Luc (Var). - Bulletins de la Société Malacologique de France 3 (2): 225-230.

Folin, A. G. L. de 1887. Sous les mers. Campagnes d'explorations du "Travailleur" et du "Talisman". - pp. I-XI [= 1-11], 1-340. Paris. (Baillière).

Folin, A. G. L. de & Bérillon, E. 1877. Contribution à la faune malacologique de la région extrême S.-O. de la France. - Bulletin de la Société de Borda à Dax 2 (2): 199-210.

Folin, A. G. L. de & Bérillon, E. 1877. Contributions à la faune malacologique de la région extrème S.-O. de la France. IIIe fascicule. - Bulletin de la Société de Borda à Dax 2: 439-454, Pl. III-IV [= 3-4].

Forbes, E. 1837. On the comparative elevation of Testacea in the Alps. - Magazine of Zoology and Botany 1 (3): 257-259. Edinburgh.

Forbes, E. 1838. On the land and freshwater Mollusca of Algiers and Bougia. - Annals of Natural History 2 (10): 250-255, Pl. XI-XII [= 11-12]. London.

Forcart, L. 1940. Monographie der türkischen Enidae (Moll., Pulm.). - Verhandlungen der Naturforschenden Gesellschaft in Basel 51 (1): 106-263, Taf. 1-3.

Forcart, L. 1942. Zwei neue Schnecken aus der Türkei. - Archiv für Molluskenkunde 74 (5/6): 239-242.

Forcart, L. 1950. Systématique des mollusques en forme de *Daudebaudia* et révision des espèces d'Anatolie et de l'île de Crête [sic]. - Journal de Conchyliologie 90: 107-117, Pl. 1. Paris.

Forcart, L. 1955. Die nordischen Arten der Gattung *Vitrina*. - Archiv für Molluskenkunde 84 (4/5): 155-166, Taf. 12.

Forcart, L. 1956. Die Vitrinidae der Ostalpen. - Archiv für Molluskenkunde 85 (1/3): 1-14, Taf. 1.

Forcart, L. 1957. Taxionomische Revision paläarktischer Zonitinae, I. - Archiv für Molluskenkunde 86 (4/6): 101-136.

Forcart, L. 1957. Ipsa Studeri Conchylia. Professor Samuel Studer (1757-1834), seine Bedeutung als Naturforscher und die von ihm hinterlassene Molluskensammlung. - Mitteilungen der Naturforschenden Gesellschaft in Bern 15: 157-210, Taf. 1-7.

Forcart, L. 1959. Taxionomische Revision paläarktischer Zonitinae, II. Anatomisch untersuchte Arten des Genus *Aegopinella* Lindholm. - Archiv für Molluskenkunde 88 (1/3): 7-34, Taf. 2-4.

Forcart, L. 1960. Taxionomische Revision paläarktischer Zonitinae, III-V. - Archiv für Molluskenkunde 89 (1/3): 1-22, Taf. 1-2.

Forcart, L. 1965. Rezente Land- und Süßwassermollusken der süditalienischen Landschaften Apulien, Basilicata und Calabrien. - Verhandlungen der Naturforschenden Gesellschaft in Basel 78 (1): 59-184.

Forcart, L. 1966. Alpine und nordische Arten der Gattung *Lehmannia* Heynemann (Limacidae). - Archiv für Molluskenkunde 95 (5/6): 225-236.

Forcart, L. 1967. Die systematische Stellung toskanischer *Oxychilus*-Arten und Neubeschreibung von *Oxychilus (Ortizius) lanzai* n. sp. - Archiv für Molluskenkunde 96: 113-123.

Forcart, L. 1970. Die Schalenunterschiede zwischen *Catinella (Quickella) arenaria* (Bouchard-Chantereaux) und *Succinea (Succinella) oblonga* Draparnaud. - Archiv für Molluskenkunde 100 (1/2): 109-111.

Forcart, L. 1971. Revision der Daudebardiinae von Vorderasien (excl. Kaukasusgebiet). - Archiv für Molluskenkunde 101 (1/4): 21-38.

Forcart, L. 1972. Milacidae und Limacidae der Insel Rhodos. - Basteria 36 (2/5): 105-116.

Forcart, L. 1986. *Limacus maculatus* (Kaleniczenko) und *Limacus flavus* (Linnaeus). - Mitteilungen der Deutschen Malakozoologischen Gesellschaft 38: 21-23.

Forés, M. & Vilella, M. 1993. Una nueva especie de *Iberellus* Hesse, 1908 (Pulmonata: Helicidae) en la isla de Eivissa. - Bolletí de la Societat d'Història Natural de les Balears 36: 17-29. Palma de Mallorca.

Forskål, P. 1775. Descriptiones animalium avium, amphibiorum, piscium, insectorum, vermium; quæ in itinere orientali observavit Petrus Forskål. Post mortem auctoris edidit Carsten Niebuhr. Adjuncta est materia medica Kahirina atque tabula maris rubri geographica. - pp. 1-20, I-XXXIV [= 1-34], 1-164, 1 map. Hauniæ. (Möller).

Forster, F. 1840. Animalia mollusca. Mit Anmerkungen von I. v. Voith. - pp. 459-478, in: Fürnrohr, A. E.: Naturhistorische Topographie von Regensburg. In Verbindung mit Forster, Herrich-Schäffer, Koch, v. Schmöger und v. Voith. Dritter Band, die Fauna von Regensburg enthaltend. -- pp. I-XVI [= 1-16], 1-478. Regensburg. (Manz).

Frank, C. 1988. Über Molluskenfunde von den Inseln Chios, Kreta und Zypern sowie vom griechischen Festland (Thessalien). - Berichte des Naturwissenschaftlich-Medizinischen Vereins in Innsbruck 75: 81-90.

Frank, C. 1992. Malakologisches aus dem Ostalpenraum. - Linzer Biologische Beiträge 24 (2): 383-662.

Frank, C. 1997. Die Molluskenfauna der Insel Rhodos, 2. Teil. - Stapfia 48: 1-179. Linz.

Fra Piero [= Arbanasich, P.] 1897. Nota sui limacidi in Sardegna a proposito di una nuova specie di *Vitrina*. - Annuario del Club Alpino Sardo 4: 65-75, Tav. [1]. Cagliari.
Article was published by P. Arbanasich under the pseudonym Fra Piero. The author's desire to publish under a pseudonym is usually respected in zoology.

Frauenfeld, G. 1854. Ueber einen bisher verkannten Laufkäfer, beschrieben von L. Miller; und einen neuen augenlosen Rüsselkäfer, beschrieben von F. Schmidt; ferner einige von Schmidt in Schischka neu entdeckte Höhlenthiere. - Verhandlungen des Zoologisch-Botanischen Vereins in Wien 4 (Abhandlungen): 23-34.

Frauenfeld, G. 1856. Die Gattung *Carychium*. - Sitzungsberichte der Mathematisch-Naturwissenschaftlichen Classe der Kaiserlichen Akademie der Wissenschaften 19 (1-2): 70-93, Taf. [1]. Wien.

Frauenfeld, G. Ritter von 1862. Ueber ein neues Höhlen-*Carychium* (*Zospeum* Brg.) und zwei neue fossile Paludinen. - Verhandlungen der kaiserlich-königlichen Zoologisch-Botanischen Gesellschaft in Wien 12: 969-972.

Frauenfeld, G. Ritter von 1865. Zoologische Miscellen. V. - Verhandlungen der kaiserlich-königlichen Zoologisch-Botanischen Gesellschaft in Wien 15 (Abhandlungen): 525-536, Taf. 8-11.

Frauenfeld, G. Ritter von 1867. Zoologische Miscellen. XII. - Verhandlungen der kaiserlich-königlichen Zoologisch-Botanischen Gesellschaft in Wien 17: 775-784.

Fretter, V. & Graham, A. 1978. The prosobranch molluscs of Britain and Denmark. Part 3 - Neritacea, Viviparacea, Valvatacea, terrestrial and freshwater Littorinacea and Rissoacea. - Journal of Molluscan Studies Supplement 5: 101-152.

Freyer, "C." 1855. Über neu entdeckte Conchylien aus den Geschlechtern *Carychium* und *Pterocera*. - Sitzungsberichte der Mathematisch-Naturwissenschaftlichen Classe der Kaiserlichen Akademie der Wissenschaften 15 (1): 18-23, Taf. [1]. Wien.
On title page "Custos Freyer", no first name given.

Frias Martins, A. de 1996. Anatomy and systematics of the western Atlantic Ellobiidae (Gastropoda: Pulmonata). - Malacologia 37 (2): 163-332.

Fricke, R. 2008. Authorship, availability and validity of fish names described by Peter (Pehr) Simon Forsskål and Johann Christian Fabricius in the 'Descriptiones animalium' by Carsten Niebuhr in 1775 (Pisces). - Stuttgarter Beiträge zur Naturkunde (A Neue Serie) 1: 1-76.

Frivaldszky, I. 1835. Köslések à' Balkány' vidékén tett természettudományi utazásról. - A' Magyar Tudós Társaság' Évkönyvei 2: 235-276, T. I-VII [= 1-7]. Budan.
Name of author is not spelled Frivaldsky in this work.

Frömming, E. 1939. Untersuchungen über die Nahrungsstoffe von *Eulota fruticum* Müller. - Archiv für Molluskenkunde 71 (2/3): 96-100.

Frömming, E. 1940. Kleine Beiträge zur Lebensweise von *Zonitoides nitidus* O. F. Müller. - Archiv für Molluskenkunde 72 (4): 118-123.

Frömming, E. 1954. Biologie der mitteleuropäischen Landgastropoden. - pp. 1-404. Berlin. (Duncker & Humblot).

Froriep, L. F. 1806. C. Dumeril's Analytische Zoologie. Aus dem Französischen mit Zusätzen. - pp. [1], I-VI [1-6], 1-344. Weimar. (Landes-Industrie-Comptoir).

Fuchs, A. & Käufel, F. 1936. Anatomische und systematische Untersuchungen an Land- und Süßwasserschnecken aus Griechenland und von den Inseln des Ägäischen Meeres. - Zeitschrift für Wissenschaftliche Zoologie, Abteilung B, Archiv für Naturgeschichte (Zeitschrift für Systematische Zoologie) (Neue Folge) 5 (4): 541-662. Leipzig.

Gallenstein, M. Ritter von 1848. Systematisches Verzeichniss der in der Provinz Kärnten bisher entdeckten Land- & Süsswasser-Conchylien, mit Angabe der wichtigsten Fundorte, nebst einer kurzen Anleitung für angehende Conchylien-Sammler. - pp. 1-28. Laibach. (Blasnik).
Date seems to be Apr 1848.

Gambetta, L. 1925. Missione zoologica del Dr. E. Festa in Cirenaica. - Bollettino dei Musei di Zoologia ed Anatomia Comparata della R. Università di Torino (Nuova Serie 22) 39 ["1924"]: 1-30.

Gargominy, O., Bouchet, P. & Ripken, T. E. J. **1999**. Fiche 1, *Tyrrhenaria ceratina*. In Muséum National d'Histoire Naturelle [Ed]. 2004 . Inventaire National du Patrimoine Naturel [http://inpn.mnhn.fr] (accessed 02.2011)

Gargominy, O., Prie, V., Bichain, J.-M., Cucherat, X. & Fontaine, B. **2011**. Liste de référence annotée des mollusques continentaux de France. - MalaCo 7: 307-382.
Electronic document (e-only journal).

Garrido, J. A., Arrébola, J. R. & Bertrand, M. **2005**. Extant populations of *Orculella bulgarica* (Hesse, 1915) in Iberia. - Journal of Conchology 38 (6): 653-662.

Garrido, C., Castillejo, J. & Iglesias, J. **1994**. Description of *Arion baeticus* spec. nov. from the Iberian peninsula (Gastropoda: Pulmonata: Arionidae). - Malakologische Abhandlungen 17 (2): 37-45.

Garrido, C., Castillejo, J. & Iglesias, J. **1995**. The *Arion subfuscus* complex in the eastern part of Iberian Peninsula, with redescription of *Arion subfuscus* (Draparnaud 1805) (Gastropoda: Pulmonata: Arionidae). - Archiv für Molluskenkunde 124 (1/2): 103-118.

Gassies, J.-B. 1847. Essai sur le Bulime tronqué. Observations prises depuis l'accouplement jusqu'à l'état adulte; avec l'explication des diverses troncatures de la coquille, suivies de deux planches explicatives du texte. - Actes de la Société Linnéenne de Bordeaux 15: 5-22, Pl. I-II [= 1-2].

Gassies, J.-B. 1849. Tableau méthodique et descriptif des mollusques terrestres et d'eau douce de l'Agenais. - pp. [1], 1-209, [1-4], Pl. 1-4. Paris, Agen. (Ballière).

Gassies, J.-B. 1867. Malacologie terrestre et d'eau douce de la région intra-littorale de l'Aquitaine. - Actes de la Société Linnéenne de Bordeaux (3) 26 (2): 109-136, Pl. I [= 1].

Gasull, L. 1964. Algunos moluscos terrestres y de agua dulce de Baleares. - Boletín de la Sociedad de Historia Natural de Baleares 9 ["1963"]: 3-80. Palma de Mallorca.

Gasull, L. 1964. Descripción de unas nuevas formas del género *Helicella (Xeroplexa)* de Baleares. - Boletín de la Sociedad de Historia Natural de Baleares 9 ["1963"]: 83-92. Palma de Mallorca.

Gasull, L. 1966. Descripción de un nuevo helícido para la fauna valenciana *Oestophora (Suboestophora) kuiperi* nov. sp. - Boletín de la Sociedad de Historia Natural de Baleares 12: 159-160. Palma de Mallorca.

Gasull, L. 1968. Descripción de una nueva especie de zonítido del sudeste Ibérico *Oxychilus mercadali* n. sp. - Boletín de la Sociedad de Historia Natural de Baleares 14: 143-144. Palma de Mallorca.
Date uncertain, could have been published in 1969.

Gasull, L. 1972. Una nueva *Helicella* de la provincia de Almeria, *Helicella (Xerotricha) mariae* n. sp. - Boletín de la Sociedad de Historia Natural de Baleares 17: 71-72. Palma de Mallorca.
Cited in Zoological Record 1975: Moll 37, date 1974 seems to be likely.

Gavira Romero, Ó., Robles Domínguez, E., Becerra Parra, M. & Larraz Azcárate, M. L. 2009. Presencia de *Acicula norrisi* Gittenberger & Boeters, 1977 (Gastropoda: Aciculidae) en España. - Iberus 27 (2): 73-76.

Geenen, S., Jordaens, K. & Backeljau, T. 2006. Molecular systematics of the *Carinarion* complex (Mollusca: Gastropoda: Pulmonata): a taxonomic riddle caused by a mixed breeding system. - Biological Journal of the Linnean Society 89 (4): 589-604.

Gelembiuk, G. W., May, G. E. & Lee, C. E. 2006. Phylogeography and systematics of zebra mussels and related species. - Molecular Ecology 15: 1033-1050.

Geniez, P. & Bertrand, A. 2001. *Abida escudei* sp. nov. (Gastropoda: Chondrinidae) au bassin de l'Aude (France). - Documents Malacologiques 2: 27-29.

Gerber, J. 1996. Revision der Gattung *Vallonia* Risso 1826 (Mollusca: Gastropoda: Valloniidae). - Schriften zur Malakozoologie 8: 1-227.

Gerber, J. 2000. Giusti, F., Manganelli, G. & Schembri, P. J. (1996): The non-marine molluscs of the Maltese Islands. Museo Regionale di Scienze Naturali Torino, Monografie 15: 607 pp., 635 Abb. Torino. - Heldia 3 (1): 42-43.

Gerber, J. & Heins, R. 1991. *Testacella haliotidea* Draparnaud 1801 (Stylommatophora: Testacellidae) seit zehn Jahren im südlichen Schleswig-Holstein. - Schriften zur Malakozoologie 4: 62-64.

Gerhardt, U. 1933. Zur Kopulation der Limaciden. I. Mitteilung. - Zeitschrift für Morphologie und Ökologie der Tiere 27: 401-450. Berlin.

Gerhardt, U. 1937. Weitere Untersuchungen zur Sexualbiologie der Limaciden. - Zeitschrift für Morphologie und Ökologie der Tiere 32 (3): 519-541. Berlin.

Gerhardt, U. 1938. Zur Frage der Sexualbiologie und Artzugehörigkeit von *Limax albipes* Dumont und Mortillet (Limacidae, Pulmonata). - Zeitschrift für Morphologie und Ökologie der Tiere 34 (1): 79-88. Berlin.

Gerhardt, U. 1940. Neue biologische Nacktschneckenstudien. - Zeitschrift für Morphologie und Ökologie der Tiere 36: 557-580. Berlin.

Gerlach, J. 2006. Terrestrial and freshwater Mollusca of the Seychelles islands. - pp. 1-141. Leiden. (Backhuys).

Germain, L. 1911. Mollusques terrestres et fluviatiles de l'Asie Antérieure (2e note). - Bulletin du Muséum National d'Histoire Naturelle 17: 63-67.

Germain, L. 1912. Mollusques terrestres et fluviatiles de l'Asie Antérieure. 5e note. Catalogue des gastéropodes de la Syrie et de la Palèstine. - Bulletin du Muséum National d'Histoire Naturelle 18 (7): 440-452. Paris.

Germain, L. 1929. Les Helicidae de la faune française. - Archives du Muséum d'Histoire Naturelle de Lyon 13: 1-484, Pl. I-XVI [= 1-16].
Date 1929 given on the title page of the volume and on a wrapper preserved in the SUB Göttingen copy.

Germain, L. 1930. Mollusques terrestres et fluviatiles (première partie). - Faune de France 21: 1-477, I-VIII [= 1-8], Pl. I-XIII [= 1-13]. Paris.

Germain, L. 1931. Mollusques terrestres et fluviatiles (deuxième partie). - Faune de France 22: 479-897, IX-XIV [= 9-14], Pl. XIV-XXVI [= 14-26]. Paris.

Girod, A. 1968. Appunti per una biogeografia dei molluschi Lombardi. Distribuzione ed ecologia di *Drepanostoma nautiliforme* Porro, *Helicodonta obvoluta* (Müller), *Helicodonta angigyra* (Rossmässler) (Gastropoda, Pulmonata). - Archiv für Molluskenkunde 98 (3/4): 121-133.

Geyer, D. 1908. Beiträge zur Molluskenfauna Schwabens II. Vallonien. - Jahreshefte des Vereins fuer vaterländische Naturkunde in Württemberg 64: 305-330, Taf. 3-4. Stuttgart.

Geyer, D. 1909. Unsere Land- und Süsswasser-Mollusken. Einführung in die Molluskenfauna Deutschlands. Mit über 500 lithographischen Abbildungen auf 18 Tafeln und Textillustrationen. Nebst einem Anhang über das Sammeln von

Mollusken. Zweite, vollständig neu bearbeitete Auflage. - pp. I-VIII [= 1-8], 1-155, Taf. I-XVIII [= 1-18]. Stuttgart. (Lutz).

Giannuzzi-Savelli, R., Sparacio, I. & Oliva, N. **1986**. I tipi di molluschi terrestri della collezione Pirajno del Museo Mandralisca di Cefalù. - Lavori della Società Italiana di Malacologia 22: 195-208.

Giannuzzi-Savelli, R., Sparacio, I. & Oliva, N. **2012**. Nomenclatural observations on the *Erctella* Monterosato, 1894 species (Gastropoda, Helicidae) from the "Rocca" of Cefalù (Italy, Sicily). - Biodiversity Journal 3 (2): 107-110. Palermo.

Giokas, S., Karkoulis, P., Pafilis, P. & Valakos, E. **2007**. Relictual physiological ecology in the threatened land snail *Codringtonia helenae*: A cause for decline in a changing environment? - Acta Oecologica 32: 269-278.

Giokas, S. & Mylonas, M. 2002. Spatial distribution, density and life history in four *Albinaria* species (Gastropoda, Pulmonata, Clausiliidae). - Malacologia 44 (1): 33-46.

Giokas, S., Mylonas, M. & Sotiropoulos, K. **2000**. Gene flow and differential mortality in a contact zone between two *Albinaria* species (Gastropoda; Clausiliidae). - Biological Journal of the Linnean Society 71: 755-770.

Giokas, S., Thomaz, D., Douris, V., Lecanidou, R. & Rodakis, G. C. **2010**. 5000 years of molecular evolution in a population of the land snail *Albinaria caerulea* transported by humans. - Journal of Molluscan Studies 76: 49-56.

Gittenberger, **A.**, Vrieling, K. & Gittenberger, E. 2001. Restricted gene flow between two alleged subspecies of *Albinaria cretensis* (Gastropoda, Pulmonata, Clausiliidae). - Netherlands Journal of Zoology 51 (1): 71-84.

Gittenberger, **E.** 1967. *Cochlodina laminata* (Montagu) in ihrem südöstlöichen Verbreitungsgebiet. - Archiv für Molluskenkunde 96 (1/2): 25-37.

Gittenberger, E. 1968. Zur Systematik der in die Gattung *Trissexodon* Pilsbry (Helicidae, Helicodontinae) gerechneten Arten. - Zoologische Mededelingen 43 (13): 165-172. Leiden.

Gittenberger, E. 1969. Beiträge zur Kenntnis der Pupillacea I. Die Spelaeodiscinae. - Zoologische Mededelingen 43 (22): 287-306, pl. 1. Leiden.

Gittenberger, E. 1970. Zur systematischen Stellung von *Helix lanuginosa* Boissy, mit Neubeschreibung eines Subgenus. - Boletín de la Sociedad de Historia Natural de Baleares 14: 63-68. Palma de Mallorca.

Gittenberger, E. 1973. Eine *Zospeum*-Art aus den Pyrenäen, *Zospeum bellesi* spec. nov. - Basteria 37 (5/6): 137-140.

Gittenberger, E. 1973. Beiträge zur Kenntnis der Pupillacea III. Chondrininae. - Zoologische Verhandelingen 127: 1-267, Taf. 1-7. Leiden.

Gittenberger, E. 1974. Beiträge zur Kenntnis der Pupillacea IV. Ergänzungen zur Kenntnis der Gattung *Argna*. - Basteria 38: 1-12.

Gittenberger, E. 1975. Zur Synonymie von *Helicella (Helicella) obvia* (Menke, 1828). - Basteria 39 (5-6): 5-6.

Gittenberger, E. 1975. *Gyralina rempei* spec. nov., eine dritte rezente *Gyralina*-Art (Pulmonata, ?Zonitidae). - Basteria 39 (3/6): 75-78.

Gittenberger, E. 1975. Beiträge zur Kenntnis der Pupillacea. V. Die Spelaeodiscinae, erster Nachtrag. - Zoologische Mededelingen 48 (23): 263-277, Taf. 1-3. Leiden.

Gittenberger, E. 1975. Beiträge zur Kenntnis der Pupillacea VI. Die Gattung *Agardhiella* in Jugoslawien. - Zoologische Mededelingen 48 (24): 279-289, Taf. 1. Leiden.

Gittenberger, E. 1976. Vier wenig bekannte troglophile Schneckenarten aus Montenegro. - Zoologische Mededelingen 49 (21): 273-283. Leiden.

Gittenberger, E. 1977. Cave snails from Corfu, Greece. - Comunicacions del Simposium d'Espeleologia (Bioespeleologia) 6: 47-53. Terrassa.

Gittenberger, E. 1977. Beiträge zur Kenntnis der Pupillacea. VII. Über zwei wenig bekannte Valloniidae. - Zoologische Mededelingen 50 (19): 295-301. Leiden.

Gittenberger, E. 1978. Beiträge zur Kenntnis der Pupillacea VIII. Einiges über Orculidae. - Zoologische Verhandelingen 163: 1-44, Taf. 1-4. Leiden.

Gittenberger, E. 1979. Eine neue *Albinaria*-Art (Gastropoda: Clausiliidae) der Insel Kephallinia. - Zoologische Mededelingen 54 (5): 83-85, Taf. 1-2. Leiden.

Gittenberger, E. 1980. Three notes on Iberian terrestrial gastropods. - Zoologische Mededelingen 55 (17): 201-213. Leiden.

Gittenberger, E. 1982. Zur Anatomie von *Clausilia (Clausilia) rugosa* (Draparnaud, 1801). - Basteria 46 (5/6): 132.

Gittenberger, E. 1983. On Iberian Cochlicopidae and the genus *Cryptazeca* (Gastropoda, Pulmonata). - Zoologische Mededelingen 57 (23): 301-320. Leiden.

Gittenberger, E. 1983. Beiträge zur Kenntnis der Pupillacea. IX. Nochmals über Orculidae. - Proceedings of the Koninklijke Nederlandse Akademie van Wetenschappen (C) 86 (3): 325-342. Amsterdam, Oxford, New York.

Gittenberger, E. 1984. Beiträge zur Kenntnis der Pupillacea. X. Eine *Chondrula* der Untergattung *Eubrephulus* auf dem Peloponnes (Mollusca: Gastropoda: Enidae). - Zoologische Mededelingen 58 (20): 329-334. Leiden.

Gittenberger, E. 1984. Vicariantists and dispersalists among the Chondrininae (Gastropoda, Pulmonata). - pp. 56-69, in: Solem, A. & Van Bruggen, A. C.: World wide snails. Biogeographical studies on non-marine Mollusca. -- pp. I-IX [= 1-9], 1-289. Leiden. (Brill/Backhuys).

Gittenberger, E. 1985. Beiträge zur Kenntnis der Pupillacea. XI. *Speleodentorcula beroni* gen. & spec. nov. (Mollusca: Gastropoda: Orculidae) aus einer Höhle in Euboea, Griechenland. - Zoologische Mededelingen 59 (19): 221-228. Leiden.

Gittenberger, E. 1986. Two new species of Enidae (Mollusca: Gastropoda: Pupillacea) from Turkey. - Zoologische Mededelingen 60 (13): 209-216.

Gittenberger, E. 1987. Neue Taxa der sogenannten Gattung *Isabellaria* (Gastropoda Pulmonata: Clausiliidae) vom Peloponnes. - Basteria 51 (1/3): 79-84.

Gittenberger, E. 1988. Über *Virpazaria (V.) pageti* Gittenberger 1969, *Gyralina (G.) pageti* n. sp. und deren Sekundärfundorte (Mollusca: Pulmonata). - Annalen des Naturhistorischen Museums in Wien 90B: 423-426, Taf. 1.

Gittenberger, E. 1991. Comment on the proposed confirmation of a lectotype for *Lindholmiola barbata* (Férussac 1821 or 1832) (Mollusca, Gastropoda). - Bulletin of Zoological Nomenclature 48 (1): 53. London.

Gittenberger, E. 1991. Altitudinal variation and adaptive zones in *Arianta arbustorum*: a new look at a widespread species. - Journal of Molluscan Studies 57 (1): 99-109.

Gittenberger, E. 1991. On Cyprian Helicellinae (Mollusca Gastropoda Pulmonata: Helicidae), making a new start. - Zoologische Mededelingen 65 (7): 99-128. Leiden.

Gittenberger, E. 1991. Two more *Menkia* species (Mollusca: Gastropoda Prosobranchia: Aciculidae). - Zoologische Mededelingen 65: 251-255. Leiden.

Gittenberger, E. 1991. What about non-adaptive radiation? - Biological Journal of the Linnean Society 43: 263-272.

Gittenberger, E. 1993. Digging in the graveyard of synonymy, in research of Portuguese species of *Candidula* Kobelt, 1871 (Mollusca: Gastropoda Pulmonata: Hygromiidae). - Zoologische Mededelingen 67 (17): 283-293.

Gittenberger, E. 1993. On *Idyla pelobsoleta* spec. nov. (Mollusca: Gastropoda Pulmonata: Clausiliidae), and insularity in the molluscan fauna of the highest parts of the Parnon Oros, Peloponnisos, Greece. - Zoologische Mededelingen 67 (20): 321-329. Leiden.

Gittenberger, E. 1993. The southeast European *Hypnophila* species (Mollusca Gastropoda Pulmonata: Cochlicopidae). - Zoologische Mededelingen 67 (37): 517-524. Leiden.

Gittenberger, E. 1993. On the only Greek *Sphincterochila* species, two Libyan close relatives, *Cerigottella* subgen. nov., and an extended diagnosis of the Sphincterochilidae (Mollusca, Gastropoda Pulmonata). - Zoologische Mededelingen 67 (39): 529-535. Leiden.

Gittenberger, E. 1994. Five new *Albinaria* subspecies from the eastern Peloponnese, Greece; with notes on *Isabellaria* s. l. (Gastropoda Pulmonata: Clausiliidae). - Basteria 58: 55-62.

Gittenberger, E. 1996. The westernmost *Turanena* species: *T. katerinae* spec. nov. (Gastropoda Pulmonata: Buliminidae). - Basteria 60 (1-3): 9-11.

Gittenberger, E. 1998. One more *Albinaria* G&N-type species pair from the Peloponnese, once more dictating a revised definition of *Albinaria* and *Isabellaria* (Gastropoda Pulmonata: Clausiliidae). - Basteria 62 (5/6): 263-268.

Gittenberger, E. 1999. Dispersal, vicariance, and partial morphostasis in the evolutionary history of SE European Zonitini (Mollusca, Gastropoda, Pulmonata). - Zoologischer Anzeiger 237: 243-258.

Gittenberger, **E.** 2000. Serrulininae in Greece, there may be more (Gastropoda, Pulmonata, Clausiliidae). - Basteria 64 (4-6): 81-87.

Gittenberger, **E.** 2002. Two limestone 'islands' in central northern Greece, six new clausiliid taxa, three kinds of microarmature (Gastropoda, Pulmonata, Clausiliidae). - Basteria 66 (4-6): 129-138.

Gittenberger, E. 2002. Two more sibling *Chondrina* species, endemic for France (Gastropoda, Pulmonata, Chondrinidae). - pp. 41-47, Colour-Pl. 8b, in: Falkner, M., Groh, K. & Speight, M. C. D.: Collectanea Malacologica. Festschrift für Gerhard Falkner. Hackenheim. (Friedrich-Held-Gesellschaft).

Gittenberger, E. 2004. A surprise in the Pyrénées, another *Renea*-like *Acicula* species (Gastropoda, Caenogastropoda, Aciculidae). - Basteria 68: 86.

Gittenberger, E. 2006. A new locality and synonym for *Abida partioti*. - Basteria 69 (1/3): 20.

Gittenberger, E. 2007. Corrections in *Cochlodina* (Gastropoda, Pulmonata, Clausiliidae). - Basteria 71 (1/3): 77-80.

Gittenberger, E. 2007. The recent Laminiferinae (Gastropoda: Clausiliidae): the reproductive tract and the radula redecribed and illustrated. - Spira 2 (3): 147-155. Barcelona.

Gittenberger, E. 2008. Two sympatric G-type *Albinaria* species, one of which new to science (Gastropoda: Pulmonata: Clausiliidae). - Zoologische Mededelingen 82 (8): 65-68. Leiden.

Gittenberger, E. 2008. A *Napaeopsis* species (Gastropoda: Pulmonata: Enidae) from the Peloponnese, Greece. - Zoologische Mededelingen 82: 271-273. Leiden.

Gittenberger, E. 2010. *Balea lucifuga* Gray, 1824, versus *Balea heydeni* Von Maltzan, 1881. - Basteria 74 (4/6): 110.

Gittenberger, E., Backhuys, W. & Ripken, T. E. J. **1970**. De landslakken van Nederland. - pp. 1-177. Amsterdam. (Koninklijke Nederlandse Natuurhistorische Vereniging).

Gittenberger, E., Backhuys, W. & Ripken, T. E. J. **1984**. De landslakken van Nederland. - pp. 1-184. Amsterdam. (Koninklijke Nederlandse Natuurhistorische Vereniging).

Gittenberger, E. & Bank, R. A. 1996. A new start in *Pyramidula* (Gastropoda Pulmonata: Pyramidulidae). - Basteria 60 (1/3): 71-78.

Gittenberger, E. & Boeters, H. D. 1977. Some remarks on *Acicula* s.s. (Prosobranchia: Rissoacea). - Journal of Conchology 29 (3): 125-128.

Gittenberger, E. & Carr, R. 2002. A second locality for *Helicella (H.) valdeona* (Hygromiidae). - Journal of Conchology 37 (6): 663-666.

Gittenberger, E. & De Winter, A. J. 1985. A Pyrenean *Monacha* species, *Monacha (Metatheba) atacis* spec. nov. (Mollusca: Gastropoda: Pulmonata). - Zoologische Mededelingen 59 (17): 197-207. Leiden.

Gittenberger, E. & Eikenboom, J. C. A. 2006. The genus *Lindbergia* (Gastropoda, Pulmonata, Zonitidae) in Greece and the Promontorio del Gargano in Italy. - Basteria 70 (4/6): 161-163.

Gittenberger, E., Flach, E. C. & Reitsma, N. **1988**. Drei neue *Albinaria*-Arten von Kreta, Nomos Lasithiou (Gastropoda Pulmonata: Clausiliidae). - Basteria 52 (1/3): 111-116.

Gittenberger, E. & Goodfriend, G. A. 1993. Land snails from the Last Glacial Maximum on Andikithira, southern Greece and their paleoclimatic implications. - Journal of Quaternary Science 8 (2): 109-116.

Gittenberger, E. & Groh, K. 1986. Zum Status der Férussac'schen Taxa *Helix lens* und *Helix barbata*. - Archiv für Molluskenkunde 116: 219-223.

Gittenberger, E. & Hausdorf, B. 2004. The *Orculella* species of the South Aegean island arc, a neglected radiation (Gastropoda, Pulmonata, Orculidae). - Basteria 68: 93-124.

Gittenberger, E. & Manga, M. Y. 1977. Some new species of the genus *Helicella* (Pulmonata, Helicidae) from the province León, Spain. - Zoologische Mededelingen 51 (11): 177-189, Pl. 1-2. Leiden.

Gittenberger, E. & Manga, M. Y. 1981. A new *Helicella* (Helicidae, Helicellinae) from Navarra, Spain. - Basteria 45 (4/5): 121-124.

Gittenberger, E. & Menkhorst, H. P. M. G. 1992. Two new *Albinaria* species from Turkey (Gastropoda Pulmonata: Clausiliidae). - Basteria 56 (4/6): 193-196.

Gittenberger, E. & Menkhorst, H. P. M. G. 1993. Die türkischen Enidae: die Gattung *Turanena* Lindholm (Pulmonata: Pupillacea). - Archiv für Molluskenkunde 122: 71-87, Taf. 1.

Gittenberger, E., Menkhorst, H. P. M. G. & Raven, J. G. M. **1980**. New data on four European terrestrial gastropods. - Basteria 44: 11-16.

Gittenberger, E. & Neuteboom, W. H. 1989. The Cyprian *Albinaria* species, keywise descriptions, illustrations and an annotated check-list (Gastropoda Pulmonata: Clausiliidae). - Basteria 4/6: 127-134.

Gittenberger, E. & Neuteboom, W. H. 1991. On *Trichia alpicola* (Eder, 1921) from Switzerland (Mollusca: Gastropoda Pulmonata: Hygromiidae) and the spiral structure on its shell. - Zoologische Mededelingen 65: 247-250. Leiden.

Gittenberger, E., Piel, W. H. & Groenenberg, D. S. J. **2004**. The Pleistocene glaciations and the evolutionary history of the polytypic snail species *Arianta arbustorum* (Gastropoda, Pulmonata, Helicidae). - Molecular Phylogenetics and Evolution 30 (1): 64-73.

Gittenberger, E. & Povel, G. D. E. 1995. Shell growth and decollation in terrestrial gastropods. - The Nautilus 109 (1): 38-40.

Gittenberger, E., Preece, R. C. & Ripken, T. E. J. 2006. *Baley heydeni* von Maltzan, 1881 (Pulmonata: Clausiliidae): an overlooked but widely distributed European species. - Journal of Conchology 39 (2): 145-150.

Gittenberger, E. & Raven, J. G. M. 1982. A new *Helicella* (Helicidae, Helicellinae) from the Cantabrian Mountains, Spain. - Basteria 46: 79-83.

Gittenberger, E. & Ripken, T. E. J. 1987. The genus *Theba* (Mollusca Gastropoda: Helicidae), systematics and distribution. - Zoologische Verhandelingen 241: 3-59. Leiden.

Gittenberger, E. & Ripken, T. E. J. 1998. The resurrection of *Clausilia portensis* Luso da Silva, 1872, as a *Macrogastra (Pseudovestia)* species (Pulmonata: Clausiliidae). - Basteria 62 (3/4): 181-185.

Gittenberger, E. & Ripken, T. E. J. 1993. On *Granaria braunii marcusi* subspec. nov. (Mollusca: Gastropoda Pulmonata: Chondrinidae). - Zoologische Mededelingen 67: 335-339.

Gittenberger, **E.** & Uit de Weerd, D. R. 2006. Reconsidering the generic position of the species once classified in *Carinigera*, *Isabellaria* and *Sericata* (Gastropoda, Pulmonata, Clausiliidae, Alopiinae). - Basteria 70 (1/3): 57-66.

Giusti, F. 1968. Notulae malacologicae - II. Il genere *Oxychilus* nell'Arcipelago Toscano. - Atti della Società Toscana di Scienze Naturali (Memorie B) 75: 218-235.

Giusti, F. 1970. Notulae malacologicae, XII. L'isola di Pianosa e lo scoglio La Scola (Arcipelago Toscano). - Annali del Museo Civico di Storia Naturale di Genova 78: 59-148.

Giusti, F. 1970. Notulae malacologicae. X. *Testacelloides* Wagner e *Testacella (Testacelloides)* gestroi Issel, un buon sottogenere ed una buona specie della Sardegna. - Atti della Società Toscana di Scienze Naturali (Memorie B) 77: 56-66. Pisa.

Giusti, F. 1970. Notulae malacologicae. XI. Alcune nuove specie di molluschi terrestri dell'isola di Sardegna. - Atti della Società Toscana di Scienze Naturali (Memorie B) 77: 67-87.

Giusti, F. 1971. Primo contributo alla revisione del genere *Cochlostoma* e descrizione di *Toffolettia* n. gen. - Atti della Società Italiana di Scienze Naturali e del Museo Civico di Storia Naturale di Milano 112 (3): 353-380.

Giusti, F. 1973. Notulae malacologicae XVI. I molluschi terrestri e di acqua dolce viventi sul massiccio dei Monti Reatini (Appennino Centrale). - Lavori della Società Italiana di Biogeografia (Nuova Serie) 2 ["1971"]: 423-574, Tav. I-VII [= 1-7]. Forlì.
On the title page of the volume 1971 is indicated, but on the last page of the volume p. 848 the "Tipografia Valbonesi - Forlì - 1973" is printed. Accordingly in volume 3 (1972) on the last page a date 15 Dec 1973 is printed at the same location, on the wrapper 1972, but in addition a note "pubblicato nel 25-12-1973" is printed on the wrapper (and on the title page of the volume). Exact references to papers published in 1972 are mentioned in an article on p. 780. There is no evidence that the volume was published in several parts. This makes is likely that volume 2 appeared in 1973.

Giusti, F. 1973. Notulae malacologicae XVIII. I molluschi terrestri e salmastri delle Isole Eolie. - Lavori della Società Italiana di Biogeografia (Nuova Serie) 3 ["1972"]: 113-306, Tav. I-XVI [= 1-16]. Forlì.
Date of publication is 25 Dec 1973.

Giusti, F. 1976. Notulae malacologicae XXIII. I molluschi terrestri, salmastri e di acqua dolce dell'Elba, Giannutri e scogli minori dell'Arcipelago Toscano. Conclusioni generali sul popolamento malacologico dell'Arcipelago Toscano e descrizione di una nove specie (studi sulla riserva naturale dell'isola di Montecristo. IV). - Lavori della Società Italiana di Biogeografia (Nuova Serie) 5: 99-355, Tav. I-XIX [= 1-19]. Bologna.

Giusti, F., Boato, A. & Bodon, M. **1986**. Two small species of the genus *Retinella* from the western Alps (Mollusca: Pulmonata: Zonitidae). - Journal of Conchology 32 (3): 167-179, Pl. 14-17. London.

Giusti, F., Castagnolo, L. & Manganelli, G. **1985**. La fauna malacologica delle faggete Italiane: brevi cenni di ecologia, elenco delle specie e chiavi per il riconoscimento dei generi e delle entità più comuni. - Bollettino Malacologico 21 (5/6): 69-144.

Giusti, F., Fiorentino, V., Benocci, A. & Manganelli, G. **2011**. A survey of vitrinid land snails (Gastropoda: Pulmonata: Limacoidea). - Malacologia 53 (2): 279-363.

Giusti, F., Grappelli, C., Manganelli, G., Fondi, R. & Bullini, L. **1986**. An attempt of natural classification of the genus *Medora* in Italy and Yugoslavia, on the basis of conchological, anatomical and allozymic characters (Pulmonata: Clausiliidae). - Lavori della Società Italiana di Malacologia 22: 259-341.
Title page indicated "pp. 259-362", probably incorrect. Text ends on p. 318, paginated plates end on p. 341.

Giusti, F., Holyoak, D. T. & Manganelli, G. **1985**. *Oxychilus (Ortizius?) clarus* (Held) on Corsica and new data on the systematic position of *Helix hydatina* Rossmässler (Pulmonata: Zonitidae). - Journal of Conchology 32: 17-24, Pl. 1-3. London.

Giusti, F. & Manganelli, G. 1984. Relationships between land evolution and present distribution of terrestrial gastropods in the western Mediterranean area. - pp. 70-92, in: Solem, A. & Van Bruggen, A. C.: World-wide snails. Biogeographical studies on non-marine Mollusca. -- pp. I-IX [= 1-9], 1-289. Leiden. (Brill/Backhuys).

Giusti, F. & Manganelli, G. 1986. Notulae malacologicae, XXXIII. '*Helix' sororcula* Benoit 1859 and its relationships to the genera *Vallonia* Risso and *Planogyra* Morse (Pulmonata: Pupilloidea). - Archiv für Molluskenkunde 116 (4/6): 157-181.

Giusti, F. & Manganelli, G. 1987. Notulae malacologicae, XXXVI. On some Hygromiidae (Gastropoda: Helicoidea) living in Sardinia and in Corsica. (Studies on the Sardinian and Corsican malacofauna VI). - Bollettino Malacologico 23 (5-8): 123-205. Milano.

Giusti, F. & Manganelli, G. 1988. Notulae malacologicae XL. *Montserratina* Ortiz de Zárate López, 1946 (Pulmonata, Hygromiidae): a redescription. - Iberus 8 (1): 89-100.

Giusti, F. & Manganelli, G. 1989. Notulae malacologicae, XLIV. A new Hygromiidae from the Tyrrhenian islands of Capraia and Sardinia with notes on the genera *Xeromicra* and *Xerotricha* (Pulmonata: Helicoidea) (Studies on the Sardinian and Corsican Malacofauna, VIII). - Bollettino Malacologico 25 (1-4): 23-62. Milano.

Giusti, F. & Manganelli, G. 1990. *Ciliellopsis oglasae*, a new hygromiid from Montecristo island (Tuscan Archipelago, Italy) (Pulmonata: Helicoidea). - Journal of Conchology 33 (5): 269-280. London.

Giusti, F. & Manganelli, G. 1992. *Tyrrheniellina* nomen novum pro *Tyrrheniella* Giusti & Manganelli, 1989 (Pulmonata: Hygromiidae). - Bollettino Malacologico 28 (5-12): 231.

Giusti, F. & Manganelli, G. 1997. How to distinguish *Oxychilus cellarius* (Müller, 1774) easily from *Oxychilus draparnaudi* (Beck, 1837) (Gastropoda, Stylommatophora, Zonitidae). - Basteria 61 (1/3): 43-56.

Giusti, F. & Manganelli, G. 1998. *Ichnusomunda sacchii*, a new hygromiid snail from Sardinia island (western Mediterranean): an intriguing case of homoplasy in the anatomical organization (Pulmonata: Hygromiidae). - The Veliger 41 (4): 319-332.

Giusti, F. & Manganelli, G. 1999. Redescription of two problematic Alpine *Oxychilus*: *O. adamii* (Westerlund, 1886) and *O. polygyra* (Pollonera, 1885) (Pulmonata, Zonitidae). - Basteria 63: 27-60.

Giusti, F. & Manganelli, G. 2002. Redescription of two West European *Oxychilus* species: *O. alliarius* (Miller, 1822) and *O. helveticus* (Blum, 1881), and notes on the systematics of *Oxychilus* Fitzinger, 1833 (Gastropoda: Pulmonata: Zonitidae). - Journal of Conchology 37 (5): 455-476.

Giusti, F. & Manganelli, G. 2005. Case 3319. *Helix papillaris* Müller, 1774 (currently *Papillifera papillaris*; Mollusca, Gastropoda): proposed conservation of the specific name. - Bulletin of Zoological Nomenclature 62 (3): 130-133.

Giusti, F., Manganelli, G. & Crisci, J. V. **1989**. A new problematical Hygromiidae from the Aeolian Islands (Italy) (Pulmonata: Helicoidea). - Malacologia 34 (1-2): 107-128.

Giusti, F., Manganelli, G. & Schembri, P. J. **1995**. The non-marine molluscs of the Maltese Islands. - pp. 1-608. Torino.
Date 1995 given in title page, 30 Nov 1995 printed on the last page.

Giusti, F. & Mazzini, M. 1971. Notulae malacologicae XIV. I molluschi delle Alpi Apuane. Elenco delle specie viventi con descrizione di una nuova specie: *Vitrinobrachium baccettii* n. sp. - Lavori della Società Italiana di Biogeografia (Nuova Serie) 1 ["1970"]: 202-335, Tav. I-IX [= 1-9]. Forlì.
Date of print 30 Sep 1971 as indicated on p. [756] of the journal volume.

Giusti, F. & Pezzoli, E. 1984. Notulae Malacologicae, XXIX: Gli Hydrobiidae salmastri delle acque costiere Italiane: primi cenni sulla sistematica del gruppo e sui caratteri distintivi delle singole morfospecie. Atti del Simposio Sistematico dei Prosobranchi del Mediterraneo, Bologna, 24-26 settembre 1982. - Lavori della Società Italiana di Malacologia 21: 117-148.

Glaubrecht, M. 1992. Biogeographie und Systematik der Landschnecke *Levantina spiriplana* (Gastropoda: Helicidae) im Ostmediterran. - Verhandlungen der Deutschen Zoologischen Gesellschaft 85 (1): 297.

Glaubrecht, M. 1993. Mapping the diversity: Geographical distribution of the freshwater snail *Melanopsis* (Gastropoda: ?Cerithioidea: Melanopsidae) with focus on its systematics in the Mediterranean basin. - Mitteilungen aus dem Hamburgischen Zoologischen Museum und Institut 90: 41-97.

Glaubrecht, M. 1994. Systematics and biogeography of *Assyriella rechingeri* (Gastropoda, Pulmonata, Helicidae) on Kárpathos, Greece. - Verhandlungen des Naturwissenschaftlichen Vereins in Hamburg (Neue Folge) 34: 373-384.

Glöer, P. 2002. *Bithynia leachii troschelii* (Paasch 1842) - die östliche Rasse von *B. leachii* (Sheppard 1823) (Gastropoda: Orthogastropoda: Bithyniidae). - Archiv für Molluskenkunde 130 (1/2): 259-265.

Glöer, P. 2002. Mollusca I. Süßwassergastropoden Nord- und Mittelweuropas. Bestimmungsschlüssel, Lebensweise, Verbreitung. 2., neubearbeitete Auflage. - pp. 1-327. Hackenheim. (ConchBooks).

Glöer, P. 2004. Zur Identität von *Bithynia troschelii* (Paasch, 1842) (Gastropoda: Bithyniidae). - Malakologische Abhandlungen 22: 17-22.

Glöer, P. & Bouzid, S. 2008. Redescription of *Planorbis agraulus* Bourguignat, 1864 (Gastropoda: Planorbidae). - Journal of Conchology 39 (5): 487-491. London.

Glöer, P. & Diercking, R. 2010. Atlas der Süßwassermollusken. Rote Liste, Verbreitung, Ökologie, Bestand und Schutz. Hamburg. - pp. 1-180. Hamburg.

Glöer, P. & Falniowski, A. & Pešić, V. 2010. The Bithyniidae of Greece (Gastropoda: Bithyniidae). - Journal of Conchology 40: 179-187.

Glöer, P., Falniowski, A. & Szarowska, M . **2006**. *Bithynia leachii* (Sheppard 1823) and *B. troschelii* (Paasch 1842), two distinct species? - Heldia 6: 49-56. München.

Glöer, P. & Groh, K. 2007. A contribution to the biology and ecology of the threatened species *Anisus vorticulus* (Troschel, 1834) (Gastropoda: Pulmonata: Planorbidae). - Mollusca 25 (1): 33-40. Dresden.

Glöer, P. & Maassen, W. J. M. 2009. Three new species of the family Bithyniidae from Greece (Gastropoda: Bithyniidae). - Mollusca 27 (1): 41-48. Dresden.

Glöer, P. & Meier-Brook, C. 2003. Süsswassermollusken. Ein Bestimmungsschlüssel für die Bundesrepublik Deutschland. 13. neubearbeitete Auflage. - pp. 1-134. Hamburg. (Deutscher Jugendbund für Naturbeobachtung).

Glöer, P. & Meier-Brook, C. 2008. Redescription of *Anisus septemgyratus* (Rossmässler, 1835) and *Anisus leucostoma* (Millet, 1813) (Gastropoda: Planorbidae). - Mollusca 26 (1): 89-94. Dresden.

Glöer, P. & Pešić, V. 2006. On the identity of *Bithynia graeca* Westerlund, 1879 with the description of three new *Pseudobithynia* n. gen. species from Iran and Greece (Gastropoda: Bithyniidae). - Malakologische Abhandlungen 24: 29-36.

Glöer, P. & Pešić, V. 2007. The *Bithynia* species from Skadar Lake (Montenegro) (Gastropoda: Bithyniidae). - Mollusca 25 (1): 7-12. Dresden.

Glöer, P. & Pešić, V. 2008. *Radix skutaris* n. sp., a new species from Montenegro (Gastropoda: Lymnaeidae). - Mollusca 26 (1): 83-88. Dresden.

Glöer, P. & Pešić, V. 2009. *Stagnicola montenegrinus* n. sp., a new species of Montenegro (Gastropoda: Lymnaeidae). - Mollusca 27 (1): 53-56. Dresden.

Glöer, P. & Pešić, V. 2010. The *Planorbis* species of the Balkans with the description of *Planorbis vitojensis* n. sp. (Gastropoda: Planorbidae). – Journal of Conchology 40 (3): 249-257.

Glöer, P. & Vinarski, M. V. 2009. Taxonomical notes on Euro-Siberian freshwater molluscs: 2. Redescription of *Planorbis (Gyraulus) stroemi* Westerlund, 1881 (Mollusca: Gastropoda: Planorbidae). -Journal of Conchology 39 (6): 717-725.

Glöer, P. & Yıldırım, M. Z. 2006. *Stagnicola* records from Turkey with the description of two new species, *Stagnicola tekecus* n. sp. and *S. kayseris* n. sp. (Gastropoda: Lymnaeidae). - Journal of Conchology 39 (1): 85-89.

Glöer, P. & Zettler, M. L. 2005. Kommentierte Artenliste der Süßwassermollusken Deutschlands. - Malakologische Abhandlungen 23: 3-26.

Glöer, P. & Zettler, M. L. 2009. Redescription of *Planorbis moquini* Requien, 1848 (Gastropoda: Planorbidae). - Journal of Conchology 39 (6): 727-732. London.

Gmelin, J. F. [1791]. Caroli a Linné, systema naturae. Tom. I. Pars VI. - pp. 3021-3910. [Lipsiae]. ([G. E. Beer]).

Gofas, S. 1989. Descripción preliminar de dos nuevos *Cochlostoma* (Gastropoda, Prosobranchia) del Cantabrico. - Publicações Ocasionais da Sociedade Portuguesa de Malacologia 14: 43-48, pl. [1-2]. Lisboa.

Gofas, S. 2001. The systematics of Pyrenean and Cantabrian *Cochlostoma* (Gastropoda, Cyclophoroidea) revisited. - Journal of Natural History 35: 1277-1369.

Gofas, S. & Backeljau, T. 1994. *Cochlostoma gigas* spec. nov. (Gastropoda: Cyclophoroidea) de los Pirineos. - Iberus 12 (1): 45-54.

Gómez, B. J. 1990. Description of a new species of the genus *Cryptazeca* from the north of the Iberian Peninsula (Gastropoda: Pulmonata: Cochlicopidae). - Archiv für Molluskenkunde 119 (4/6): 219-225.

Gómez, B. J. 1990. Descripción de una nueva especie troglobia de *Cryptazeca* (Mollusca, Gastropoda). - Bulletin du Museum National d'Histoire Naturelle (Série 4 Section A) 12 (2): 365-374. Paris.

Gómez, B. J & Prieto, C. E. 1983. *Zospeum biscaiense* nov. sp. (Gastropoda, Ellobiidae), otro molusco troglobio para la Península Ibérica. - Speleon 26-27: 7-10. Barcelona.

Gómez Moliner, B., Moreno, D., Rolán, E., Araujo, R. & Álvarez, R. M. **2001**. Protección de moluscos en el Catálogo Nacional de Especies Amenazadas. - Reseñas Malacológicas 11: 1-286. Madrid.

Götting, K.-J. 1963. *Leiostyla schweigeri* n. sp. und *Pseudamnicola lindbergi* Boettger - zwei interessante Gastropoden aus Anatolien. - Archiv für Molluskenkunde 92 (1/2): 31-33.

Gould, A. A. 1841. Report on the Invertebrata of Massachusetts, comprising the Mollusca, Crustacea, Annelida, and Radiata. Published agreeably to an order of the legislature, by the comissioners on the Zoological and Botanical Survey of the state. - pp. i-xiii [= 1-13], 1-373, pl. [1-15]. Cambridge. (Forsom, Wells & Thurston).
No constraints about the date in the work itself (preface is dated to March 1840).

Graack, W. 2005. Die Gattung *Xerocrassa* Monterosato 1892 (Mollusca, Hygromiidae) von Mallorca. - Schriften zur Malakozoologie 22: 1-64.

Graells, M. P. 1846. Catálogo de los moluscos terrestres y de agua dulce observados en España, y descripcion [sic] y notas de algunas especies nuevas ó poco conocidas del mismo pais. - pp. I-VI [= 1-6], 1-23, [1-2], pl. [1]. Madrid, Lima. (Calleja, Ojea & Co.).
Author was spelled Graells on the title page and inside the work.

Graf, D. L. 2010. Funeral for the Nouvelle Ecole *-iana* generic names introduced for freshwater mussels (Mollusca: Bivalvia: Unionoida). - Proceedings of the Academy of Natural Sciences of Philadelphia 159: 1-24.

Graf, D. L. & Cummings, K. S. 2007. Review of the systematics and global diversity of freshwater mussel species (Bivalvia: Unionoidea). - Journal of Molluscan Studies 73: 291-314.

Gray, **J. E.** 1824. On *Balea*. - Zoological Journal 1 ["1825"]: 61-62. London.

Gray, J. E. 1834. [Various undescribed shells]. - Proceedings of the Zoological Society of London 2: 57-68.

Gray, J. E. 1843. Catalogue of the species of Mollusca and their shells, which have hitherto been recorded as found at New Zealand, with the description of some lately discovered species. - pp. 228-265, in: Dieffenbach, E.: Travels in New Zealand; with contributions to the geography, geology, botany, and natural history of that country. In two volumes - Vol. II. -- pp. i-iv [= 1-4], 1-396, pl. [1]. London. (Murray).

Gray, J. E. 1847. A list of the genera of recent Mollusca, their synonyma and types. - Proceedings of the Zoological Society of London 15: 129-219.
Date (Gray 1847: 128) 09 Nov 1847.

Gray, **J. E.** 1850. Nomenclature of molluscous animals and shells in the collection of the British Museum. Part 1. Cyclophoridæ. - pp. 1-68, [1-2]. London.
Gray's name given on p. [3] (preface) suggests authorship for the contents.

Gray, J. E. 1855. Catalogue of Pulmonata or air-breathing Mollusca in the collection of the British Museum. Part I. - pp. [1-3], 1-192, [1-4]. London.
Date was 18 Apr 1855 (Woodward 1903: 261).

Gray, **M. E.** 1850. Figures of molluscous animals, selected from various authors. Etched for the use of students. Vol. IV. Explanation of plates and list of genera. - pp. [1], i-iv [= 1-4], 1-124. London. (Longman).
Author on the title page was given as Maria Emma Gray. The content is usually attributed to J. E. Gray, but he was not the bibliographically relevant author of the work.

Gredler, V. M. 1853. Bemerkungen über einige Conchylien der Gattungen *Pupa* und *Pomatias*. - Programm des Gymnasiums von Bozen 3: 45-52.

Gredler, V. 1856. Tirol's Land- und Süsswasser-Conchylien. I. Die Landconchylien. - Verhandlungen des Zoologisch-Botanischen Vereins in Wien (Abhandlungen) 6: 25-162, Taf. 2, Tabelle [1-2].

Gredler, V. 1869. Nachlese zu Tirol's Land- und Süsswasser-Conchylien. - Verhandlungen der kaiserlich-königlichen Zoologisch-Botanischen Gesellschaft in Wien 19: 909-916.

Gredler, V. 1877. Mittheilungen aus dem Gebiete der Malakozoologie. - Nachrichtsblatt der Deutschen Malakozoologischen Gesellschaft 9 (1): 1-6.
Date was Jan 1877.

Groh, K., Richling, I. & Bößneck, U. 2009. Erstnachweise der Flachen Erbsenmuschel *Pisidium (Cingulipisidium) pseudosphaerium* Favre 1927 in Südwestdeutschland (Bivalvia: Sphaeriidae). - Mitteilungen der Deutschen Malakozoologischen Gesellschaft 82: 40-48.

Grossu, A. V. 1955. Contribuţii la cunoaşterea faunei gasteropodelor din R. P. R. *Hohenwarthia orghidani* n. sp. din familia Ferussaciidae. - Comunicările Academiei Republicii Populare Romîne 5 (11): 1579-1582. Bucureşti.

Grossu, A. V. 1960. Two new species of the genus *Daudebardia* (terrestrial gastropods). - Proceedings of the Malacological Society of London 34 (3): 174-180.

Grossu, A. V. 1967. Două specii noi ale genului *Alopia* in România. - Analele Universităţii Bucureşti, Seria ştiinţele naturii (Biologie) 16: 9-17. Bucureşti.

Grossu, A. V. 1967. Contributions à la connaissance du genre *Lehmannia* Heynemann (Gastropoda, Limacidae). - Journal de Conchyliologie 106 (4): 119-128. Paris.

Grossu, A. V. 1969. Beschreibung einiger neuer taxonomischer Einheiten der Daudebardiidae von Rumänien (Gastropoda, Pulmonata). - Archiv für Molluskenkunde 99 (1/2): 77-89.

Grossu, A. V. 1969. Beschreibung einiger neuer *Deroceras*-Arten (Gastropoda, Limacidae). - Archiv für Molluskenkunde 99 (3/4): 157-170.

Grossu, A. V. 1970. Zwei neue Arten der Gattung *Lehmannia* Heynemann aus Rumänien. - Malakologische Abhandlungen 3 (10): 109-114.

Grossu, A. 1972. Noi contribuţii la cunoaşterea genului *Daudebardia* Hartmann (Gastropoda, Pulmonata) în fauna României. - Studii şi Cercetări de Biologie (Seria Zoologie) 24 (5): 419-422. Bucureşti

Grossu, A. V. 1972. Fünf neue Arten der Gattung *Deroceras* von der Balkanhalbinsel in der Sammlung des Naturhistorischen Museums in Wien (Gastropoda, Limacidae). - Annalen des Naturhistorischen Museums in Wien 76: 639-648.

Grossu, A. V. 1979. Revision of the species of the Daudebardiidae family of Roumania. - Malakologische Abhandlungen 6 (1): 1-41.

Grossu, A. V. 1979. Eine neue Gastropoden-Art: *Monachoides bacescui* sp. n. (Helicidae, Hygromiinae). - Travaux du Muséum d'Histoire Naturelle 'Grigore Antipa' 20 (1): 21-25. Bucureşti.

Grossu, A. V. 1983. Gastropoda Romaniae 4. Ordo Stylommatophora. Suprafam.: Arionacea, Zonitacea, Ariophantacea şi Helicacea. - pp. 1-564. Bucureşti. (Editura Litera).

Grossu, A. V. 1986. Nouvelles espèces de la famille Orculidae (Gastropoda, Pulmonata). - Travaux du Muséum d'Histoire Naturelle 'Grigore Antipa' 28: 7-13. Bucureşti.

Grossu, A. V. 1986. Gastropoda Romaniae 1. I. Caracterele generale, istoricul şi biologia gasteropodelor. II. Subclasa Prosobranchia şi Opisthobranchia. - pp. 1-524. Bucureşti. (Editura Litera).

Grossu, A. V. 1987. Gastropoda Romaniae 2. Subclasa Pulmonata. I Ordo Basommatophora. II Ordo Stylommatophora, Suprafamiliile: Succinacea, Cochlicopacea, Pupillacea. - pp. 1-443, [1]. Bucureşti. (Editura Litera).

Grossu, A. V. & Lupu, D. 1960. Revision der Gattung *Limax* aus Rumänien, nebst Beschreibung einiger neuer Arten. - Archiv für Molluskenkunde 89 (4/6): 157-165.

Grossu, A. V. & Lupu, D. 1961. Die Gattung *Lytopelte* (Limacidae) in den Karpathen. - Archiv für Molluskenkunde 90 (1/3): 27-31.

Grossu, A. V. & Lupu, D. 1962. Zur Kenntnis der Gattung *Lehmannia* Heynemann (Limacidae) und deren große Variabilität, nebst Beschreibung neuer Arten. - Archiv für Molluskenkunde 91 (4/6): 191-201.

Grossu, A. V. & Lupu, D. 1965. Espèces nouvelles du genre *Deroceras* (Gastropoda, Limacidae) en Roumanie. - Travaux du Muséum d'Histoire Naturelle 'Grigore Antipa' 5: 25-31. Bucureşti.

Grossu, A. V. & Lupu, D. 1966. Le genre *Lytopelte* (Limacidae) dans les Carpates occidentales. - Travaux du Muséum d'Histoire Naturelle 'Grigore Antipa' 6: 25-30. Bucureşti.

Grossu, A. V. & Negrea, A. 1968. Révision des espèces du genre *Argna* Cosmann [sic!] (Gasteropoda [sic!], Pulmonata) de Roumanie et description de quelques nouvelles unités taxonomiques. - Travaux du Muséum d'Histoire Naturelle 'Grigore Antipa' 8 (2): 721-734. Bucureşti.

Grossu, A. V & Riedel, A. 1958. *Oxychilus deilus malinowskii* (L. Pfeiffer, 1865) und die verwandten Formen. - Archiv für Molluskenkunde 87 (4/6): 141-148, Taf. 10.

Gualtieri, N. 1742. Index testarvm conchyliorvm qvae adservantvr in mvseo Nicolai Gvaltieri (...) et methodice distribvta exhibentvr. Tabvlis CX. - pp. I-XXIII [= 1-23], I-CX [= 1-110], Tab. I-CX [= 1-110]. Florentinae. (Albizzini).

Gude, G. K. 1910. Description of a new species of *Helicodonta* from Spain. - Proceedings of the Malacological Society of London 9 (2): 124-125.

Gude, G. K. 1911. Note on some preoccupied molluscan generic names and proposed new genera of the family Zonitidae. - Proceedings of the Malacological Society of London 9 (4): 269-273.

Gude, G. K. 1911. Further note on preoccupied molluscan generic names and a proposed new genus of the family Helicidæ. - Proceedings of the Malacological Society of London 9 (6): 361-362.

Gude, G. K. & Woodward, B. B. 1921. On *Helicella*, Férussac. - Proceedings of the Malacological Society of London 14: 174-190.

Gude, G. K. & Woodward, B. B. 1921. Some emendations to their recent paper 'On *Helicella*, Férussac.'. - Annals and Magazine of Natural History (9) 8 (48): 624-626. London.

Guillén, G. & Corbella, J. 2007. Presència de *Norelona pyrenaica* (Draparnaud, 1805) (Gastropoda: Elonidae) al Massís del Montseny (el Vallès Oriental, Catalunya, Espanya). - Spira 3: 189-190.

Guirao, A. 1854. Zwei Novitäten. - Malakozoologische Blätter 1: 32-33.

Gümüş, B. A. 2010. Searching for *Lindholmiola lens* (A. Férussac, 1832) (Mollusca, Gastropoda, Stylommatophora) in western Anatolia. - Triton 21: 31-34.

Gümüş, B. A. & Neubert, E. 2012. New taxa of terrestrial molluscs from Turkey (Gastropoda, Pristilomatidae, Enidae, Hygromiidae, Helicidae). - ZooKeys 171: 17-37.

Gural-Sverlova, N. V. 2010. Obzor nazemnykh mollyuskov roda *Helicopsis* (Hygromiidae) Donetskoj vozvyshennosti i prilegayushtshikh territorij s opisaniem novykh vidov. - Ruthenica 20 (1): 13-26.
Date specified in the PDF was 14 June 2010. The PDF contained a statement that it was published on CD-ROM and sent to 5 public libraries, to comply with Art. 8.6. This procedure was not in agreement with Art. 8.6 because CD-ROMs are not durable media in the sense of the Code, they last only a few years and not several centuries as it is necessary for the purposes of zoological nomenclature. The proposed new names were not made available in this publication.

Gural-Sverlova, N. V. 2012. Predvaritel'nye rezul'taty anatomicheskogo issledovaniya mollyuskov *Helicopsis* (Hygromiidae) Kryma i Prichernomorskoj nizmennosti. - Ruthenica 22 (1): 15-34.

Gural-Sverlova, N. V., Balashov, I. A. & Gural, R. I. **2009**. Sovremennoe rasprostranenie nazemnykh mollyuskov semejstva Agriolimacidae na territorii Ukrainy. - Ruthenica 19 (2): 53-61.

Gural-Sverlova, N. V. & Gural, R. I. 2010. Novye taksony nazemnykh mollyuskov iz rodov Chondrula i Brephulopsis (Enidae) s territorii Ukrainy. - Ruthenica 20 (1): 1-12.
Date specified in the PDF was 14 June 2010. The PDF contained a statement that it was published on CD-ROM and sent to 5 public libraries, to comply with Art. 8.6. This procedure was not in agreement with Art. 8.6 because CD-ROMs are not durable media in the sense of the Code, they last only a few years and not several centuries as it is necessary for the purposes of zoological nomenclature. The proposed new names were not made available in this publication.

Gural-Sverlova, N. V., Martynov, V. V. & Gural, R. I. **2010**. Pervye nakhodki sliznej *Parmacella ibera* i *Deroceras subagreste* (Gastropoda, Pulmonata) v Ukraine. - Vestnik Zoologii 44 (3): 265-269.

Haas, F. 1910. *Pseudunio*, neues Genus für *Unio sinuatus* Lam. - Nachrichtsblatt der Deutschen Malakozoologischen Gesellschaft 42 (4): 181-183.

Haas, F. 1923. Literatur. - Archiv für Molluskenkunde 55 (3): 124-128.

Haas, F. 1924. Beitrag zur Molluskenfauna des unteren Ebrogebietes. - Archiv für Molluskenkunde 56 (4): 137-160, Taf. VIII [= 8].

Haas, F. 1934. Über einige Landschnecken von Zypern. - Senckenbergiana 16 (1): 16-21. Frankfurt am Main.

Haas, F. 1969. Superfamilia Unionacea. - Das Tierreich 88: 1-663.

Hadžišče, S. 1955. Prilog poznavanju Gastropoda Prespanskog i Ohridskog Jezera. - Glasnik Bioloske Sekcije [Periodicum Biologorum], Hrvatsko Prirodoslovno Drustvo (II/B) 7: 174-177. Zagreb.

Hadžišče, S. 1956. III. Beitrag zur Kenntnis der Gastropodenfauna des Ohridsees. Beschreibungen der bis jetzt unbekannten Schnecken und Beispiele der Speciation bei den Gastropoden des Ohridsees. - Zbornik na Rabotite, Khidrobioloshki Zavod Okhrid [Recueil des Traveaux, Station Hydrobiologique Ohrid] 4 (1 (14)): 57-107.
True date Apr 1959 (Radoman 1962: 69).

Haldeman, S. S. 1842-1847. A monograph of the freshwater univalve Mollusca of the United States, including notices of species in other parts of North America. - pp. [9 parts], [40 plates]. Philadelphia.
Dates and titles for each number: Bogan et al. 1992: 14.

Hanley, S. 1855. Ipsa Linnæi Conchylia. The shells of Linnæus, determined from his manuscripts and collection. Also, an exact reprint of the Vermes Testacea of the 'Systema Naturæ' and 'Mantissa'. - pp. [1], 1-556, [1], Pl. I-V [= 1-5]. London. (Williams & Norgate).

Hartmann, J. D. W. 1821. System der Erd- und Flußschnecken der Schweiz. Mit vergleichender Aufzählung aller auch in den benachbarten Ländern, Deutschland, Frankreich und Italien sich vorfindenden Arten. - Neue Alpina 1: 194-268, Taf. I-II [= 1-2]. Winterthur.
Appeared before Sturm's work, as visible in some internal references to Neue Alpina in Sturm's work.

Hartmann, J. D. W. 1821. System der Erd- und Süßwasser Gasteropoden Europa's. In besonderer Hinsicht auf diejenigen Gattungen, welche in Deutschland und der Schweitz angetroffen werden. - pp. 1-60, Tab. I-III [= 1-3], in: Sturm, J.: Deutschlands Fauna in Abbildungen nach der Natur mit Beschreibungen. VI. Abtheilung. Die Würmer. 5. Heft. -- pp. [1-2], 1-60, Tab. I-III [= 1-3]. Nürnberg.
Appeared after Neue Alpina.

Hartmann, J. D. W. 1844. Erd- und Süsswasser-Gasteropoden der Schweiz. Mit Zugabe einiger merkwürdigen exotischen Arten. I. Band. - pp. i-xx [= 1-20], 1-227, Tab. I-XII [= 1-12], I-XII [sic, = 13-24], 25-84. St. Gallen. (Scheitlin & Zollikofer).
Cited from Haus der Natur Cismar library <Moll 1406> and Harvard University, MCZ, Ernst Mayr Library (volumes complete, 1844 on title page).
Appeared in several parts (pp. 37, 50), Tab. I-XII appeared doubly, later the pagination was shifted to Arabic numbers.
Dates (Heppell 1966, Bouchet & Rocroi 2005):
pp. i-xx, 1-36, pl. 1-2 (30-06-1840)
pp. 37-116, pl. 13-36 (1841)
pp. 117-156, pl. 37-60 (1842), pp. 155-156 doubtful (SBPK Berlin copy)
pp. 157-204, pl. 61-72 (1843)
pp. 205-227, pl. 73-84 (1844)
The composition of the Berlin copy (SBPK Berlin <Lq 5935>, volume incomplete, only until p. 154 and plate 60, date 1840 on preliminary title page) suggests that pp. 155-156 were not issued in 1842 but in the next Heft (6, 1843) with the descriptions of the other clausiliid species.
Hartmann used a system of genera and subgenera and usually (but not always and consistently) combined specific names with the subgenera (see the table at the end of the volume). Hartmann called the subgenus "Gattung" and the genus "ältere oder umfassendere Gattung" ("Gattung" translates "genus", but Hartmann used this term incorrectly, both were genus-group names, and two genus-group names arranged in a hierarchy correspond always to genus and subgenus in modern taxonomy).

Hausdorf, B. 1988. Zur Kenntnis der systematischen Beziehungen einiger Taxa der Helicellinae Ihering 1909 (Gastropoda: Hygromiidae). - Archiv für Molluskenkunde 119 (1/3): 9-37.

Hausdorf, B. 1988. Zur Kenntnis von *Orculella templorum* (Benoit 1862) aus Sizilien (Gastropoda: Orculidae). - Archiv für Molluskenkunde 119 (1/3): 77-81.

Hausdorf, B. 1990. Über die Verbreitung von *Microxeromagna armillata* (Lowe 1852) und *Xerotricha conspurcata* (Draparnaud 1801) in Griechenland und der Türkei (Gastropoda Pulmonata: Hygromiidae). - Malakologische Abhandlungen 15 (6): 55-62.

Hausdorf, B. 1990. Zur Kenntnis einiger Arten der Gattung *Leiostyla* Lowe aus NW-Anatolien (Gastropoda: Lauriinae). - Archiv für Molluskenkunde 120 (1/3): 47-56.

Hausdorf, B. 1990. Zur Kenntnis einiger Arten der Gattung *Helicopsis* Fitzinger aus Griechenland und der Türkei (Gastropoda: Hygromiidae). - Archiv für Molluskenkunde 120 (1/3): 57-71, Taf. 1-2.

Hausdorf, B. 1990. Die *Xeromunda*-Arten des griechischen Festlandes (Gastropoda: Hygromiidae). - Archiv für Molluskenkunde 119 (4/6): 107-131, Taf. 1-2.

Hausdorf, B. 1991. Über zwei *Candidula*-Arten von der südlichen Balkanhalbinsel (Gastropoda: Hygromiidae). - Archiv für Molluskenkunde 120 (4/6): 119-129.

Hausdorf, B. 1993. Die Gattung *Thoanteus* Lindholm in Kleinasien (Gastropoda: Buliminidae). - Archiv für Molluskenkunde 122: 89-97.

Hausdorf, B. 1994. Additive typogenesis in *Thoanteus* (Gastropoda: Bulminidae). - Zoological Journal of the Linnean Society 112 (3): 353-361.

Hausdorf, B. 1995. *Xeromunda alticola* n. sp., a species from the alpine region of the Peloponnese, Greece (Gastropoda Pulmonata: Hygromiidae). - Basteria 59: 25-28.

Hausdorf, B. 1995. The Vitrinidae of Turkey, with remarks on the phylogeny of *Gallandia* (Gastropoda: Stylommatophora). - Zoologischer Anzeiger 234: 63-74.
Hausdorf, B. 1996. Revision of *Monacha emigrata* (Westerlund 1984) (Gastropoda: Hygromiidae). - Schriften zur Malakozoologie 9: 1-6.
Hausdorf, B. 1996. Die Orculidae Asiens (Gastropoda: Stylommatophora). - Archiv für Molluskenkunde 125 (1/2): 1-86.
Hausdorf, B. 1996. *Helicopsis aelleni* n. sp. from northern Iran, with remarks on *Helicopsis* Fitzinger 1833 (Gastropoda: Pulmonata: Hygromiidae). - Archiv für Molluskenkunde 126 (1/2): 65-71.
Hausdorf, B. 2000. The genus *Monacha* in Turkey (Gastropoda: Pulmonata: Hygromiidae). - Archiv für Molluskenkunde 128 (1/2): 61-151.
Hausdorf, B. 2001. The genus *Ena* in Turkey, with remarks on its phylogenetic relationships (Gastropoda: Buliminidae). - Journal of Natural History 35: 1627-1638.
Hausdorf, B. 2003. Preliminary revision of the *Monacha (Paratheba) rothii* (L. Pfeiffer, 1841) species complex from the Aegean region (Gastropoda: Hygromiidae). - Journal of Conchology 38 (1): 35-46.
Hausdorf, B. 2006. Is the interspecific variation of body size of land snails correlated with rainfall in Israel and Palestine? - Acta Oecologica 30: 374-379.
Hausdorf, B., Gümüş, B. A. & Yıldırım, M. Z. **2004**. Two new *Metafruticicola* species from the Taurus mountains in Turkey. - Archiv für Molluskenkunde 133 (1/2): 167-171.
Hausdorf, B. & Páll-Gergely, B. 2009. *Monacha oecali* new species from southern Turkey (Gastropoda: Hygromiidae). - Journal of Conchology 40 (1): 15-17.
Hausdorf, B. & Riedel, A. 1996. Wiederbeschreibung von *Vitrea sorella* (Mousson 1863) aus der Türkei (Gastropoda: Zonitoidea). - Archiv für Molluskenkunde 125 (1/2): 113-116.
Hausdorf, B. & Sauer, J. 2009. Revision of the Helicellinae of Crete (Gastropoda: Hygromiidae). - Zoological Journal of the Linnean Society 157: 373-419.
Hauswald, A.-K., Albrecht, C. & Wilke, T. 2008. Testing two contrasting evolutionary patterns in ancient lakes: species flock versus species scatter in valvatid gastropods of Lake Ohrid. - Hydrobiologia 615: 169-179.
Hazay, J. 1880. Die Molluskenfauna von Budapest. - Malakozoologische Blätter (Neue Folge) 3 ["1881"]: 1-69, 160-183, Taf. I-IX [= 1-9].
Was published before 20 Dec 1880 (Falkner et al. 2002: 131).
Hazay, J. 1881. Die Mollusken-Fauna von Budapest mit besonderer Rücksichtnahme auf die embryonalen und biologischen Verhältnisse ihrer Vorkommnisse. - pp. 1-187, I-IX [= 1-9], Taf. I-XV [= 1-15]. Cassel. (Fischer).
Heber, D. & Patzner, R. A. 2009. First report of *Pisidium globulare* Clessin, 1873 (Bivalvia, Sphaeriidae) in the Austrian province of Salzburg. - Mollusca 27 (2): 233-239. Dresden.
Held, F. 1836. Aufzählung der in Bayern lebenden Mollusken. - Isis 1836 (4): 271-282. Leipzig.
Held, F. 1837. Notizen über die Weichthiere Bayerns. - Isis 1837 (4): 303-309. Leipzig.
Held, F. 1838. Notizen über die Weichthiere Bayerns. (Fortsetzung.). - Isis 1837 (12): 902-919. Leipzig.
True date was 1838, in Heft 11 a work published by Zetterstedt in 1838 was cited (Falkner 1991: 100).
Heller, J. 1976. The taxonomy and distribution of the Enidae (Mollusca, Pulmonata) of the Aegean Islands. - Journal of Molluscan Studies 42: 371-393.
Heller, J. 2009. Land snails of the land of Israel. Natural history and field guide. - pp. 1-360. Sofia. (Pensoft).
Heller, J. & Farstay, V. 1989. A field method to separate males and females in the freshwater snail *Melanoides tuberculata*. - Journal of Molluscan Studies 55 (3): 427-429.
Heller, J. & Gadot, M. 1984. Shell polymorphism of *Theba pisana* - the effects of rodent distribution. - Malacologia 25: 349-354.
Heller, J., Mordan, P., Ben-Ami, F. & Sivan, N. **2005**. Conchometrics, systematics and distribution of *Melanopsis* (Mollusca: Gastropoda) in the Levant. - Zoological Journal of the Linnean Society 144: 229-260.
Heller, J., Pimstein, R. & Vaginsky, E. **1991**. Cave-dwelling *Cecilioides genezarethensis* (Pulmonata, Ferussaciidae) from Israel. - Journal of Molluscan Studies 57: 289-293.
Heller, J. & Sivan, N. 1997. Reproductive biology and population dynamics of an ovoviviparous land snail, *Lauria cylindracea* (Pupillidae). - Journal of Zoology 243: 263-280.
Heppell, D. 1966. The dates of publication of J. D. W. Hartmann's 'Erd- und Süsswasser-Gasteropoden'. - Journal of Conchology 26: 84-88.
Herrmannsen, A. N. 1846. Indicis generum malacozoorum primordia. Nomina subgenerum, generum, familiarum, tribuum, ordinum, classium; adjectis auctoribus, temporibus, locis systematicis atque literariis, etymis, synonymis. Praetermittuntur Cirripedia, Tunicata et Rhizopoda. Vol. I. - pp. I-XXVII [= 1-27], 1-637. Cassellis. (Fischer).
A-L. Dates (Herrmannsen 1852: iv, Bouchet & Rocroi 2005): pp. I-XXVII, 1-104 (01-09-1846), 105-232 (01-12-1846), 233-360 (01-03-1847), 361-488 (18-04-1847), 489-616 (25-05-1847), 617-637 (10-07-1847).
Herrmannsen, A. N. 1847-1849. Indicis generum malacozoorum primordia. Nomina subgenerum, generum, familiarum, tribuum, ordinum, classium; adjectis auctoribus, temporibus, locis systematicis atque literariis, etymis, synonymis. Praetermittuntur Cirripedia, Tunicata et Rhizopoda. Vol. II. - pp. XXVIII-XLII [= 28-42], 1-717, [1-2]. Cassellis. (Fischer).
M-Z. Dates (Herrmannsen 1852: iv, Bouchet & Rocroi 2005): pp. 1-104 (17-07-1847), 105-232 (08-09-1847), 233-352 (07-12-1847), 353-492 (18-02-1848), 493-612 (02-1849), 613-717 (03-1849).
Herrmannsen, A. N. 1852. Indicis generum malacozoorum supplementa et corrigenda. - pp. I-V [= 1-5], 1-140. Cassellis. (Fischer).
Date (Herrmannsen 1852: iv, Bouchet & Rocroi 2005: 316) 12-1852.
Hesse, P. 1882. Eine Reise nach Griechenland. - Jahrbücher der Deutschen Malakozoologischen Gesellschaft 9: 283-336, Taf 12.
Hesse, P. 1884. Nacktschnecken von Tanger und Gibraltar. - Malakozoologische Blätter (Neue Folge) 7: 9-17.
Hesse, P. 1884. Beiträge zur Molluskenfauna Griechenlands. III. - Jahrbücher der Deutschen Malakozoologischen Gesellschaft 11: 225-244, Taf. 4-5.
Hesse, P. 1907. Kritische Fragmente. - Nachrichtsblatt der Deutschen Malakozoologischen Gesellschaft 39 (2): 69-77.
Hesse, P. 1908. Kritische Fragmente. [IV.-VI.]. - Nachrichtsblatt der Deutschen Malakozoologischen Gesellschaft 40 (3): 131-141.
Hesse, P. 1909. Die systematische Stellung von *Helix leachii* Fér. und *gyrostoma* Fér. - Zoologische Jahrbücher; Abteilung für Systematik, Geographie und Biologie der Tiere 27 (3): 313-320.
Date was 25 Feb 1909, given on the volume content page.
Hesse, P. 1910. Kritische Fragmente. - Nachrichtsblatt der Deutschen Malakozoologischen Gesellschaft 42 (4): 165-169.
Hesse, P. 1913. Zur Kenntnis der Molluskenfauna von Ostrumelien. II. - Nachrichtsblatt der Deutschen Malakozoologischen Gesellschaft 45 (1): 1-16.
Hesse, P. 1914. Kritische Fragmente. [X.-XII.]. - Nachrichtsblatt der Deutschen Malakozoologischen Gesellschaft 46 (2): 59-64.
Hesse, P. 1914. *Helix frequens* Mousson (Helicidae, Mollusca). - Mitteilungen des Kaukasischen Museums - Izvěstija Kavkazskago Muzeja 6: 253-269, Taf. [1]. Tiflis/Tbilisi.
Hesse, P. 1915. Beschreibungen neuer Arten. - Nachrichtsblatt der Deutschen Malakozoologischen Gesellschaft 47 (2): 58-63.
Hesse, P. 1916. Zur Kenntnis der Molluskenfauna von Ostrumelien. IV. - Nachrichtsblatt der Deutschen Malakozoologischen Gesellschaft 48 (3): 113-122.
Hesse, P. 1916. Kritische Fragmente. - Nachrichtsblatt der Deutschen Malakozoologischen Gesellschaft 48 (3): 122-124.
Hesse, P. 1916. Mollusken von Varna und Umgebung. - Nachrichtsblatt der Deutschen Malakozoologischen Gesellschaft 48 (4): 145-157.
Hesse, P. 1918. Die Subfamilie Helicodontinae. - Nachrichtsblatt der Deutschen Malakozoologischen Gesellschaft 50 (3): 99-110.
Hesse, P. 1918. Kritische Fragmente. - Nachrichtsblatt der Deutschen Malakozoologischen Gesellschaft 50 (1): 34-40.
Hesse, P. 1921. Beiträge zur näheren Kenntnis der Subfamilie Fruticicolinae. - Archiv für Molluskenkunde 53 (1/2): 55-83.
Hesse, P. 1923. Kritische Fragmente. - Archiv für Molluskenkunde 55: 193-198.
Hesse, P. 1928. Die Erscheinungsdaten der Bände XI-XXIII N. F. von Kobelt-Rossmäßler's Iconographie. - Archiv für Molluskenkunde 60: 275.
Hesse, P. 1928. Mazedonische Gastropoden, gesammelt von Herrn Prof. Dr. F. Doflein. - Zoologische Jahrbücher, Abteilung für Allgemeine Zoologie und Physiologie der Tiere 45: 1-30, Taf. 1-2. Jena.
Hesse, P. 1931. Zur Anatomie und Systematik palaearktischer Stylommatophoren. - Zoologica 31 (81): [1-2], 1-118, Taf. 1-16. Stuttgart.
Hesse, P. 1932. Zur genaueren Kenntnis einiger italienischer Heliciden. - Archiv für Molluskenkunde 64: 189-197, Taf. 14.
Hesse, P. 1934. Zur Anatomie und Systematik palaearktischer Stylommatophoren. Zweiter Teil. - Zoologica 33 (85): 1-59, Taf. 1-9. Stuttgart.
Heynemann, F. D. 1863. Einige Mittheilungen der Schneckenzungen mit besonderer Beachtung der Gattung *Limax*. - Malakozoologische Blätter 10 ["1862"]: 200-216, Taf. II-III [= 2-3].
Date Oct 1863 given on pp. 193 and 209.
Hickson, T. G. L. 1972. A possible case of genetic drift in colonies of the land snail *Theba pisana*. - Heredity 29: 177-190.
Hidalgo, J. G. 1867. Diagnose d'une espèce nouvelle d'*Helix*. - Journal de Conchyliologie 15: 209-210, Pl. XII [= 12]. Paris.
Hidalgo, J. G. 1867. Description de deux hélices nouvelles d'Espagne et des îles Baléares. - Journal de Conchyliologie 15: 440-442, Pl. XII [= 12]. Paris.
Hidalgo, J. G. 1869. Descriptions de deux nouvelles espèces d'*Helix* d'Espagne. - Journal de Conchyliologie 17: 19-21, Pl. II [= 2]. Paris.
Hidalgo, J. G. 1870. Description de trois espèces nouvelles d'*Helix* d'Espagne. - Journal de Conchyliologie 18: 298-299. Paris.
Hidalgo, J. G. 1873. Description d'une nouvelle espèce d'*Helix* d'Espagne. - Journal de Conchyliologie 21: 358-359. Paris.
Hidalgo, J. G. 1878. Catalogue des mollusques terrestres des îles Baléares. - Journal de Conchyliologie 26: 215-247, Pl. IX [= 9]. Paris.
Hidalgo, J. G. 1883. Description de deux espèces nouvelles d'*Helix*. - Journal de Conchyliologie 31: 56-58, Pl. II [= 2]. Paris.
Hirc, D. 1881. Die Mollusken-Fauna des liburnischen Karstes. - Verhandlungen der kaiserlich-königlichen Zoologisch-Botanischen Gesellschaft in Wien 30 (Abhandlungen): 519-530.
Hlaváč, J. Č. & Peltanová, A. 2010. First occurrence of the Kentish snail *Monacha cantiana* (Mollusca: Gastropoda: Hygromiidae) in the Czech Republic. - Malacologica Bohemoslovaca 9: 11-15.
Holyoak, **D. T.** 1986. Biological species-limits and systematics of the Clausiliidae (Gastropoda) of the Maltese Islands. - Journal of Conchology 32: 211-220. London.
Holyoak, **D. T.** & Holyoak, G. A. 2012. A taxonomic revision of *Oestophora barbula* (Rossmässler, 1838) and *O. barbella* (Servain, 1880), two Iberian endemic land-snail species (Gastropoda: Trissexodontidae). - Iberus 30 (1): 15-40.
Holyoak, **D. T.** & Holyoak, G. A. 2012. A review of species-limits in some *Cryptazeca* (Gastropoda: Azecidae). - Iberus 30 (1): 91-102.
Holyoak, D. T. & Seddon, M. B. 1985. An undescribed *Acicula* (Gastropoda: Aciculidae) from Portugal and a record of *A. algerensis* in Morocco. - Journal of Conchology 32 (1): 63-66.
Holyoak, **D. T.** & Seddon, M. B. 1985. Non-marine Gastropoda from France, Spain and Italy. - Journal of Conchology 32 (1): 67-69. London.
Holyoak, **G. A.** & Holyoak, D. T. 2009. *Quickella arenaria* (Gastropoda: Succineidae) living in the Cantabrian mountains, NW. Spain. - Noticiario de la Sociedad Española de Malacología 52: 26-27.
Holyoak, **G. A.** & Holyoak, D. T. 2010. A new species of *Candidula* (Gastropoda, Hygromiidae) from central Portugal. - Iberus 28 (1): 67-72.
Horsák, M. & Hájek, M. 2005. Habitat requirements and distribution of *Vertigo geyeri* (Gastropoda: Pulmonata) in western Carpathian rich fens. - Journal of Conchology 38 (6): 683-700.
Horsák, M., Juřičková, L., Beran, L., Čejka, T. & Dvořák, L. **2010**. Komentovaný seznam měkkýšů zjištěných ve volné přírodě České a Slovenské republiky. - Malacologica Bohemoslovaca Suppl. 1: 1-37.
Horsák, M., Schenková, V. & Myšák, J. **2012**. The second site of *Pupilla alpicola* (Charpentier, 1837) and the first recent record of Pupilla pratensis (Clessin, 1871) in Poland. - Folia Malacologica 20 (1): 21-26.
Horsák, M., Škodová, J & Cernohorsky, N. H. **2011**. Ecological and historical determinants of western Carpathian populations of *Pupilla alpicola* (Charpentier, 1837) in relation to its present range and conservation. - Journal of Molluscan Studies 77 (3): 248-254.
Horsák, M., Šteffek, J., Čejka, T., Ložek, V. & Juřičková, L. **2009**. Occurrence of *Lucilla scintilla* (R.T. Lowe, 1852) and *Lucilla singleyana* (Pilsbry, 1890) in the Czech and Slovak Republics – with remarks how to distinguish these two non-native minute snails. - Malacologica Bohemoslovaca 8: 24-27.
Hotopp, K. P., Pearce, T. A., Nekola, J. C. & Schmidt, K. 2010. New land snail (Gastropoda: Pulmonata) distribution records for New York state. - Proceedings of the Academy of Natural Sciences of Philadelphia 159: 25-30.
Hubendick, B. 1951. Recent Lymnaeidae, their variation, morphology, taxonomy, nomenclature, and distribution. - Kungliga Svenska Vetenskapsakademiens Handlingar (4) 3 (1): 1-223, pl. 1-5. Stockholm.
Hubendick, B. 1960. The Ancylidae of Lake Ochrid and their bearing on intralacustrine speciation. - Proceedings of the Zoological Society of London 133 (4): 497-529, Pl. 1-4.
Hubendick, P. & Radoman, P. 1959. Studies on the *Gyraulus* species of Lake Ochrid. Morphology. - Arkiv för Zoologi (2) 12 (16): 223-243. Stockholm.
Hubenov, Z. 2007. Distribution and zoogeographical characteristics of mollusks (Mollusca) from Bulgarian national parks. - Historia Naturalis Bulgarica 18: 127-159.
Hudec, V. 1960. Kritiké hodnocení druhů rodu *Cochlicopa* Risso 1826 (Mollusca) z Československa. - Práce Brněnské Základny Československé Akademie Věd 32 (7 [407]): 277-300, Tab. [1-4]. Brno.
Hudec, V. 1961. Zur Diskussion über die Schnecke *Oxychilus (Riedelius) inopinatus* (Uličný, 1887). - Sbornik Národniho Musea v Praze (B) 16: 97-128, Taf. 8. Praha.
Hudec, V. 1965. Poznámky k anatomii nekterych druhu plzu z Bulharska. - Casopis Národního Muzea. Oddíl Prírodovedny 134 (1): 11-16, Tab. 1-2. Praha.

Hudec, V. 1970. Podnámsky k anatomii některých plžů z Maďarska. - Bemerkungen zur Anatomie einiger Schneckenarten aus Ungarn. - Casopis Národního Muzea, Oddíl Prírodovědný 137 (3/4): 33-43. Praha.
Text basically in Czech, descriptions of taxa were given (exclusively) in German.

Hudec, V. 1972. Was ist eigentlich *Chondrus pupoides* Krynicki 1833? - Archiv für Molluskenkunde 102 (4/6): 211-219.

Hudec, V. 1972. Neue Erkenntnisse über kaukasische Fruticocampylaea-Arten und einige verwandte oder konchyologisch ähnliche Formen (Mollusca: Helicidae). - Malakologische Abhandlungen 3 (18): 251-276.

Huff, S. W., Campbell, D., Gustafson, D. L., Lydeard, C., Altaba, C. R. & Giribet, G. 2004. Investigations into the phylogenetic relationships of freshwater pearl mussels (Bivalvia: Margaritiferidae) based on their molecular data: implications for their taxonomy and biogeography. - Journal of Molluscan Studies 70: 379-388.

Hunyadi, A. & Szekeres, M. 2009. *Tsoukatosia subaii* spec. nov. (Gastropoda: Pulmonata: Clausiliidae), a third representative of a relict genus from Greece. - Schriften zur Malakozoologie 24: 1-4.

Hutterer, R. 1990. Recent and fossil slugs of the genus *Parmacella* in the Canary Islands, with the description of three new species (Pulmonata: Parmacellidae). - Archiv für Molluskenkunde 120 (1/3): 73-93.

ICZN 2004. Opinion 2080 (Case 3233). *Achatina janii* De Betta & Martinati, 1855 (currently *Cecilioides janii*; Mollusca, Gasropoda): proposed conservation of specific name not approved. - Bulletin of Zoological Nomenclature 61 (3): 182-183.

ICZN 2007. Opinion 2176 (Case 3319). *Helix papillaris* Müller, 1774 (currently *Papillifera papillaris*; Mollusca, Gastropoda): specific name not conserved. - Bulletin of Zoological Nomenclature 64 (3): 195.

Ihering, H. von 1892. Morphologie und Systematik des Genitalapparates von *Helix*. - Zeitschrift für Wissenschaftliche Zoologie 54 (1/2; 3): 386-423, 425-520, Taf. 18-19. Leipzig.
The systematic part was in section 3, date 04 Oct 1892.

Iredale, T. 1913. The land Mollusca of the Kermadec islands. - Proceedings of the Malacological Society of London 10 (6): 364-388, Pl. XVIII [= 18].

Irikov, A. A. 2001. K anatomiyata i taksonomiyata ny dva vida ot rod *Vitrea* Fitzinger, 1833 (Gastropoda, Zonitidae). - Nautshny Trudove na PU, Animalia 37 (6): 35-38. Plovdiv.
PU = Plovdiv University.

Irikov, A. 2003. A new species of *Macedonica* O. Boettger 1877 from Bulgaria (Gastropoda: Pulmonata: Clausiliidae). - Archiv für Molluskenkunde 132 (1/2): 143-145.

Irikov, A. 2006. New taxa of Clausiliidae from Bulgaria (Gastropoda: Pulmonata: Clausiliidae). - Archiv für Molluskenkunde 135 (1): 81-89.

Irikov, A. 2008. Genus *Monacha* Fitzinger 1833 in Bulgaria (Gastropoda, Pulmonata, Hygromiidae). - Linzer Biologische Beiträge 40 (1): 785-811.

Irikov, A. & Erőss, Z. 2008. An updated and annotated checklist of Bulgarian terrestrial gastropods (Mollusca: Gastropoda). - Folia Malacologica 16 (4): 197-205.

Irikov, A. A. & Georgiev, D. G. 2008. A review of the *Agardhiella* species (Gastropoda: Pulmonata: Argnidae) in Bulgaria. - Acta Zoologica Bulgarica 60 (2): 135-141.

Irikov A., Georgiev, D. & Riedel, A. **2004**. A new species of the genus *Vitrea* Fitzinger, 1833 from Bulgaria (Gastropoda: Pulmonata: Zonitidae). - Folia Malacologica 12 (2): 79-81.

Irikov, A. & Mollov, I. 2006. Terrestrial gastropods (Mollusca: Gastropoda) of the western Rhodopes (Bulgaria). - pp. 753-832, in: Beron, P.: Biodiversity of Bulgaria. 3. Biodiversity of western Rhodopes (Bulgaria and Greece) I. Sofia.

Issel, A. 1866. Dei molluschi raccolti nella provincia de Pisa. - Memorie della Società Italiana di Scienze Naturali 2 (1): 1-36, (1-2]. Milano

Issel, A. 1873. Di alcuni molluschi raccolti nell'isola di Sardegna dal dott. Gestro. - Annali del Museo Civico di Storia Naturale di Genova 4: 275-281.

Jackiewicz, M. 1959. Badania nad zmiennością i stanowiskiem systematycznym *Galba palustris* O. F. Müll. - Prace Komisji Biologicznej, Poznańskie Towarzystwo Przyjaciół Nauk 19 (3): 1-86, Tab. I-XXV [= 1-25], Mapka 1. Poznań.

Jackiewicz, M. 1979. Materials to Aciculidae fauna of Balcan peninsula (Mollusca, Prosobranchia). - Bulletin de la Société des Amis des Sciences et des Lettres de Poznan (Série D) 18 ["1978"]: 135-138, Tab. 1.

Jackiewicz, M. 1979. *Acicula (Hyalacme) jankowskiana* sp. n. (Gastropoda, Prosobranchia) from east Carpathians. - Bulletin de la Sociéte des Amis des Sciences et des Lettres de Poznan (Série D) 19: 95-98.

Jackiewicz, M. 1980. Some observations on biology of reproduction of *Succinea* Draparnaud (Gastropoda, Pulmonata). - Annales Zoologici 35 (6): 67-73.

Jaeckel, S. sen. 1952. Die Mollusken der spanischen Mittelmeer-Inseln. - Mitteilungen aus dem Zoologischen Museum in Berlin 28: 53-143, Taf. I-IV [= 1-4].

Jaeckel, S. sen. 1954. Zur Systematik und Faunistik der Mollusken der nördlichen Balkanhalbinsel. - Mitteilungen aus dem Zoologischen Museum in Berlin 30: 54-95.

Jakubik, B. 2006. Reproductive pattern of *Viviparus viviparus* (Linnaeus 1758) (Gastropoda, Viviparidae) from littoral aggregations in a through-flow reservoir (Central Poland). - Polish Journal of Ecology 54 (1): 39-55.

Jakubik, B. 2009. Reproduction of the freshwater snail *Viviparus contectus* (Millet, 1813) (Gastropoda: Architaenioglossa: Viviparidae). - Folia Malacologica 17 (4): 223-230.

Jakubik, B. & Lewandowski, K. 2007. Size structure, age, mortality and fecundity in *Viviparus viviparus* (Linnaeus, 1758) (Gastropoda: Architaenioglossa: Viviparidae). - Folia Malacologica 15 (3): 109-117.

Jalžić, B. 1998. The stygobiont bivalve *Congeria kusceri* Bole, 1962 (Bivalvia, Dreissenidae) in Croatia. - Natura Croatica 7 (4): 341-347.

Jalžić, B. 2001. The first finding of a live stygobiont bivalve *Congeria* in the Lika region, Croatia. - Natura Croatica 10 (3): 213-220.

Jan, G. 1830. Scientiae naturalis cultoribus. Conspectus methodicus testaceorum in collectione mea exstantium anno 1830. - pp. 1-8. Parma.

Jeffreys, J. G. 1830. A synopsis on the testaceous pneumonobrancheous Mollusca of Great Britain. - Transactions of the Linnean Society of London 16 (2): 323-392.

Jeffreys, J. G. 1833. A supplement to the "Synopsis of testaceous pneumonobranchous Mollusca of Great Britain". - Transactions of the Linnean Society of London 16: 505-524.

Jenyns, L. 1832. A monograph on the British species of *Cyclas* and *Pisidium*. - Transactions of the Cambridge Philisophical Society 4 (2): 289-311, Pl. 19-21.

Jickeli, C. F. 1882. Diagnosen neuer Conchylien. - Jahrbücher der Deutschen Malakozoologischen Gesellschaft 9: 366-370.

Johnson, M. S. 1981. Effects of migration and habitat choice on shell banding frequencies in *Theba pisana* at a habitat boundary. - Heredity 47: 121-133.

Johnston, G. 1828. A few remarks on the class Mollusca, in Dr Fleming's work of British animals; with descriptions of some new species. - Edinburgh New Philisophical Journal Apr-Sep 1828 (Apr-Jun 1828): 74-81.

Jordaens, K., Pinceel, J., Van Houtte, N., Breugelmans, K. & Backeljau, T. 2010. *Arion transsylvanus* (Mollusca, Pulmonata, Arionidae): rediscovery of a cryptic species. - Zoologica Scripta 39 (4): 343-362.

Jueg, U. 1997. *Pupilla muscorum* (Linnaeus 1758) im NSG 'Klädener Plage' (Mecklenburg-Vorpommern, Landkreis Parchim) – ein Beitrag zur Ökologie, Gehäusemorphologie und Systematik der Art (Gastropoda: Stylommatophora: Pupillidae). - Malakologische Abhandlungen 18 (27): 277-285.

Jueg, U. 2004. Die Verbreitung und Ökologie von *Vertigo moulinsiana* (Dupuy, 1849) in Mecklenburg-Vorpommern (Gastropoda: Stylommatophora: Vertiginidae). - Malakologische Abhandlungen 22: 87-124.

Jungbluth, J. H. 1978. Erfassung der westpaläarktischen Tiergruppen. Fundortkataster der Bundesrepublik Deutschland. Teil 5: Prodromus zu einem Atlas der Mollusken von Hessen. - pp. [1-4], 1-165. Saarbrücken

Jungbluth, J. H. & Knorre, D. von 2008. Trivialnamen der Land- und Süßwassermollusken Deutschlands (Gastropoda et Bivalvia). - Mollusca 26 (1-3): 105-156. Dresden.

Jungbluth, J. H. & Knorre, D. von 2009. Rote Liste der Binnenmollusken (Schnecken (Gastropoda) und Muscheln (Bivalvia)) in Deutschland. - Mitteilungen der Deutschen Malakozoologischen Gesellschaft 81: 1-28.

Jungbluth, J. H., Vogt, D. & Wirth, U. **1989**. Regionalkataster des Landes Niedersachsen. Prodromus zu einem Atlas der Binnen-Mollusken von Niedersachsen. - pp. 1-261. Neckarsteinach.

Kadolsky, D. 2009. *Turbo bidens* Linnaeus 1758 (Gastropoda: Clausiliidae) misidentified for 250 years. - Journal of Conchology 40 (1): 19-30.

Kadolsky, D. 2012. Nomenclatural comments on non-marine molluscs occurring in the British Isles. - Journal of Conchology 41 (1): 65-90.

Kaleniczenko, J. 1851. Description d'un nouveau genre de limaces de la Russie Méridionale. - Bulletin de la Société Impériale des Naturalistes de Moscou 1851 [24] (1): 215-228, Tab. V-VI [= 5-6].

Kaltenbach, H. 1951. Ergebnisse zoologisch-geologischer Sammelreisen in NO-Afrika, 4: *Rumina decollata*. - Archiv für Molluskenkunde 80 (1/3): 27-32, Taf. 2.

Kantor, Yu. I. & Sysoev, A. V. 2005. Katalog mollyuskov Rossii i sopredel'nykh stran. - pp. 1-627. Moskva. (KMK).

Kantor, Yu. I., Vinarski, M. V., Schileyko, A. A. & Sysoev, A. V. **2010**. Catalogue of the continental mollusks of Russia and adjacent territories. Version 2.3.1. - pp. [1-330].
Unpublished internet resource, unpaginated. First version from 2009.

Kaszuba, M. & Stworzewicz, E. 2008. *Hawaiia minuscula* (A. Binney, 1841) - another alien species in Poland (Mollusca: Gastropoda: Zonitidae). - Folia Malacologica 16 (1): 27-30.

Katoh, M. & Ribi, G. 1996. Genetic evidence for natural hybridization and apparent introgression between freshwater snail species (*Viviparus ater* and *V. contectus*). - Journal of Evolutionary Biology 9: 67-82.

Katoh, M. & Ribi, G. 1997. Genetic variation in sympatric and allopatric populations of hybridizing freshwater snail species (*Viviparus ater* and *V. contectus*). - Journal of Molluscan Studies 63: 505-510.

Käufel, F. 1930. X. Teil. Die schalentragenden Land- und Süßwassermollusken. - in: Beier, M. (1930): Zoologische Forschungsreise nach den Jonischen Inseln und dem Peloponnes. - Sitzungsberichte, Akademie der Wissenschaften in Wien, mathematisch-naturwissenschaftliche Klasse, Abteilung I 139 (3/4): 161-188, Taf. I-II [= 1-2].

Keating, W. H. 1824. Narrative of an expedition to the source of St. Peter's River, Lake Winnepeek, Lake of the woods, &c. &c. performed in the Year 1823, by order of the hon. J. C. Calhoun, under the Command of Stephen H. Long. Compiled from the notes of Major Long, Messrs. Say, Keating, and Colhoun. In two volumes. Vol. II. - pp. I-VI [= 1-6], 5-459, Pl. 1-15. Philadelphia. (Carey & Lea).
Chapter on natural history was authored by T. Say, pp. 253-400.

Kemperman, T. C. M. 1992. Systematics and evolutionary history of the *Albinaria* species from the Ionian islands of Kephallinia and Ithaka (Gastropoda Pulmonata: Clausiliidae). - pp. 1-251. Leiden. (Ph. D. thesis). (Backhuys).

Kennard, A. S. 1942. The Histoire and Prodrome of Férussac. Part I. - Proceedings of the Malacological Society of London 25 (1): 12-17.

Kennard, A. S. 1942. The Histoire and Prodrome of Férussac. Part II. - Proceedings of the Malacological Society of London 25 (3): 105-110.

Kennard, A. S. 1942. The Histoire and Prodrome of Férussac. Part III. The divisional names. - Proceedings of the Malacological Society of London 25 (3): 111-118.

Kennard, A. S. & Woodward, B. B. 1923. Note on the nomenclature and systematic arrangement of the Clausiliidae. - Proceedings of the Malacological Society of London 15 (6): 298-308.

Kennard, A. S. & Woodward, B. B. 1924. Nomenclatorial notes relating to British non-marine Mollusca. II. - Proceedings of the Malacological Society of London 16 (1): 7-14.

Kennard, A. S. & Woodward, B. B. 1925. On the *Helix perspectiva* of Megerle von Muhlfeld [sic] versus that of Gray. - The Nautilus 38 (3): 85-86.

Kennard, A. S. & Woodward, B. B. 1926. Synonymy of the British non-marine Mollusca (Recent and Post-Tertiary). - pp. I-XXIV [= 1-24], 1-447. London. (British Museum).

Kennedy, V. S. 2011. Biology of the uncommon dreissenid bivalve *Mytilopsis leucophaeata* (Conrad, 1831) in central Chesapeake Bay. - Journal of Molluscan Studies 77 (2): 154-164.

Kerney, M. P. 1976. Non-marine Mollusca from Faversham, Kent, figured by George Walker in 1784. - Journal of Conchology 29 (2): 75-77.

Kerney, M. 1999. Atlas of the land and freshwater molluscs of Britain and Ireland. - pp. 1-264. Colchester. (Harley).

Kerney, M. P. & Cameron, R. A. D. 1979. A field guide to the landsnails of Britain and North-west Europe. - pp. 288 pp., 24 pl.. London. (Collins).

Kerney, M. P., Cameron, R. A. D. & Jungbluth, J. H. **1983**. Die Landschnecken Nord- und Mitteleuropas. Ein Bestimmungsbuch für Biologen und Naturfreunde. - pp. 384 pp., Taf. 1-24. Hamburg, Berlin. (Parey).
Published in spring 1983.

Kerney, M. P., Preece, R. C. & Turner, C. **1980**. Molluscan and plant biostratigraphy of some Late Devensian and Flandrian deposits in Kent. - Philosophical Transactions of the Royal Society of London (B) 291 (1044): 1-443.

Ketmaier, V., Giusti, F. & Caccone, A. **2006**. Molecular phylogeny and historical biogeography of the land snail genus *Solatopupa* (Pulmonata) in the peri-Tyrrhenian area. - Molecular Phylogenetics and Evolution 39 (2): 439-451.

Ketmaier, V., Manganelli, G., Tiedemann, R. & Giusti, F. **2010**. Peri-Tyrrhenian phylogeography in the land snail *Solatopupa guidoni* (Pulmonata). - Malacologia 52 (1): 81-96.

Killeen, I. J. 2005. Studies on the round-mouthed whorl snail *Vertigo genesii* (Gastropoda: Vertiginidae) in northern England: observations on population dynamics and life history. - Journal of Conchology 38 (6): 701-710.

Killeen, I. J. & McFarland, B. 2004. The distribution and ecology of *Gyraulus acronicus* (Férussac, 1807) (Gastropoda: Planorbidae) in England. - Journal of Conchology 38 (4): 441-488.

Kimakowicz, **M. von** 1883. Beitrag zur Mollusken-Fauna Siebenbürgens. - Verhandlungen und Mittheilungen des Siebenbürgischen Vereins für Naturwissenschaften in Hermannstadt 33: 11-83.

Kimakowicz, M. von 1884. Beitrag zur Mollusken-Fauna Siebenbürgens (Fortsetzung). - Verhandlungen und Mittheilungen des Siebenbürgischen Vereins für Naturwissenschaften in Hermannstadt 34: 57-116.

Kimakowicz, M. von 1890. Beitrag zur Molluskenfauna Siebenbürgens. II. Nachtrag. - Verhandlungen und Mittheilungen des Siebenbürgischen Vereins für Naturwissenschaften in Hermannstadt 40: 1-113.

Kimakowicz, **M. von** 1894. Prodromus zu einer Monographie des *Clausilia*-Subgenus *Alopia* H. et A. Adams. Beitrag zur Mollusken-Fauna Siebenbürgens, III. Nachtrag. -

Verhandlungen und Mittheilungen des Siebenbürgischen Vereins für Naturwissenschaften in Hermannstadt 43: 19-58.

Kimakowicz, **R. von** 1928. *Alopia*-Sammelreise 1927. - Archiv für Molluskenkunde 60 (Taf. III-IV [= 3-4]): 107-126.

Kimakowicz, R. von 1931. *Alopia*-Sammelreise 1930. - Archiv für Molluskenkunde 63: 39-42, Taf. II [= 2].

Kirchner, Ch., Krätzner, R. & Welter-Schultes, F. W. 1997. Flying snails - how far can *Truncatellina* (Pulmonata: Vertiginidae) be blown over the sea? - Journal of Molluscan Studies 63: 479-487.

Kittel, K. 2000. Bemerkenswerte, ungewöhnliche und wenig bekannte Mollusken, IV. Die Braune Rucksackschnecke *Testacella maugei* A. Férussac 1819 und ihre Verbreitung in Portugal (Pulmonata, Testacellidae). - Club Conchylia Informationen 32 (4/6): 5-10. Ludwigsburg.

Kittel, K. 2000. Endemische Landschnecken Portugals und der angrenzenden westlichen Iberischen Halbinsel, 1: *Candidula setubalensis* (L. Pfeiffer 1850) (Pulmonata, Hygromiidae). - Club Conchylia Informationen 32 (4/6): 15-17. Ludwigsburg.

Kittel, K. 2002. Endemische Landschnecken Portugals und der angrenzenden westlichen Iberischen Halbinsel, 2: *Candidula codia* (Bourguignat 1859) (Pulmonata, Hygromiidae). - Club Conchylia Informationen 33 (4/6): 5-7. Ludwigsburg.

Kittel, K. 2002. Bemerkenswerte, ungewöhnliche und wenig bekannte Mollusken, VI. Das Landposthorn *Elona quimperiana* (Blainville 1821) mit einem Nachweis für Portugal (Pulmonata, Elonidae). - Club Conchylia Informationen 34 (1/3): 19-22. Ludwigsburg.

Kittel, K. 2003. Endemische Landschnecken Portugals und der angrenzenden westlichen Iberischen Halbinsel, 3: *Gittenbergeria turriplana* (Morelet 1845) (Pulmonata, Hygromiidae). - Club Conchylia Informationen 34 (4/6): 37-39. Ludwigsburg.

Klee, B. 2005. Preliminary mitochondrial phylogeny of the genus *Limax*. - MalaCo 3: 106.

Klee, B., Falkner, G. & Haszprunar, G. **2005**. Endemic radiations of *Limax* (Gastropoda: Stylommatophora) slugs in Corsica – they came twice. - 78 pp., in: Burckhardt, D.: 8. Jahrestagung der Gesellschaft für biologische Systematik, Basel 13.–16.IX.2005. Abstracts of talks and posters. Basel. (Naturhistorisches Museum).
Organisms, Diversity and Evolution 5, Electronic Supplement 13: 75.

Klemm, W. 1939. Zur rassenmäßigen Gliederung des Genus *Pagodulina* Clessin. - Zeitschrift für Wissenschaftliche Zoologie, Abteilung B, Archiv für Naturgeschichte (Zeitschrift für Systematische Zoologie) (Neue Folge) 8 (2): 198-262, Taf. 1. Leipzig.

Klemm, W. 1943. Schneckenfunde während des Winters 1940-1941 in Frankreich. - Archiv für Molluskenkunde 75 (2/3): 95-106.

Klemm, W. 1951. Ökologische und biologische Beobachtungen an Schnecken, besonders an Felsenschnecken. - Archiv für Molluskenkunde 80 (1/3): 49-56.

Klemm, W. 1954. Klassen Gastropoda und Bivalva [sic]. - pp. 210-280, in: Franz , H.: Die Nordost-Alpen im Spiegel ihrer Landtierwelt. Eine Gebietsmonographie. Der 1. Band. -- pp. [1-5], 1-664. Innsbruck. (Wagner).

Klemm, W. 1962. X. Teil. Die Gehäuseschnecken. - in: Beier, M.: Zoologische Studien in West-Griechenland. - Sitzungsberichte, Österreichische Akademie der Wissenschaften, mathematisch-naturwissenschaftliche Klasse, Abteilung I 171 (6/7): 203-258, Taf. 1-4. Wien.

Klemm, W. 1965. Eine neue Höhlenschnecke, *Paraegopis oberwimmeri* n. sp., aus Montenegro. - Archiv für Molluskenkunde 94 (1/2): 71-73.

Klemm, W. 1967. Über ostalpine Orculae. - Archiv für Molluskenkunde 96: 101-111.

Klemm, W. [1973]. Die Verbreitung der rezenten Land-Gehäuse-Schnecken in Österreich. - Denkschriften der Österreichischen Akademie der Wissenschaften 117: 1-503. Wien, New York.

Knipper, H. 1939. Systematische, anatomische, ökologische und tiergeographische Studien an südosteuropäischen Heliciden (Mollusca, Pulmonata). - Zeitschrift für Wissenschaftliche Zoologie, Abteilung B, Archiv für Naturgeschichte (Zeitschrift für Systematische Zoologie) (Neue Folge) 8 (3/4): 327-517. Leipzig.

Knipper, H. 1941. Nachträge zur südosteuropäischen Helicidenfauna (Moll., Pulm.) I. - Archiv für Molluskenkunde 73 (5/6): 177-186, Taf. 11.

Kobelt, W. 1871. Catalog der im europäischen Faunengebiet lebenden Binnenconchylien. Mit besonderer Berücksichtigung der in Rossmässler's Sammlung enthaltenen Arten. - pp. I-XVI [= 1-16], 1-150. Cassel. (T. Fischer).

Kobelt, W. 1872. Eine neue Campyläe. - Nachrichtsblatt der Deutschen Malakozoologischen Gesellschaft 4 (4): 49.

Kobelt, W. 1872. Neue oder wenig bekannte Campyläen. - Malakozoologische Blätter 19: 125-133, Taf. 4.

Kobelt, W. 1875. Zur Fauna Italiens. - Jahrbücher der Deutschen Malakozoologischen Gesellschaft 2: 192-213.

Kobelt, W. 1876. Iconographie der Land- & Süsswasser-Mollusken mit vorzüglicher Berücksichtigung der europäischen noch nicht abgebildeten Arten von E. A. Rossmässler. IV. Band. Mit dreissig Tafeln. - pp. [1-3], 1-74, Taf. 91-120. Wiesbaden. (Kreidel).
Dates (Hesse 1928, Bank 1989):
Heft 1 (May 1875): pp. [1-3], 1-12, Taf. 91-95
Heft 2-4 (11 Sep 1876): pp. 13-44, Taf. 96-110
Heft 5-6 (14 Dec 1876): pp. 45-72, Taf. 111-120

Kobelt, W. 1877. Über die systematische Anordnung der Heliceen. - Malakozoologische Blätter 24: 1-14.

Kobelt, W. 1877. Iconographie der Land- & Süsswasser-Mollusken mit vorzüglicher Berücksichtigung der europäischen noch nicht abgebildeten Arten von E. A. Rossmässler. V. Band. Mit dreissig Tafeln. - pp. [1-3], 1-129, Taf. 121-150. Wiesbaden. (Kreidel).
Nº 1160-1528. Dates (Hesse 1928, Bank 1989):
Heft 1-3 (28 July 1877): pp. 1-60, Taf. 121-135
Heft 4-6 (20 Dec 1877): pp. 49-158, [1-4], Taf. 136-150

Kobelt, W. 1878. Diagnosen neuer Arten. - Jahrbücher der Deutschen Malakozoologischen Gesellschaft 5: 319-321.

Kobelt, W. 1879. Iconographie der Land- & Süsswasser-Mollusken mit vorzüglicher Berücksichtigung der europäischen noch nicht abgebildeten Arten von E. A. Rossmässler. VI. Band. Mit achtundzwanzig Tafeln. - pp. [1], 1-158, Taf. 151-178. Wiesbaden. (Kreidel).
Nº 1529-1804. No separate title pages for hefts 1/3 and 4/6, so the bibliographic indication year for the work is 1879.
Author for *Clausilia* pp. 52-153 Taf. 167-178 was Boettger, for the other parts Kobelt (see p. [2]).
Dates (Hesse 1928, Bank 1989):
Heft 1-3 (24 July 1878): pp. 1-48, Taf. 151-165;
Heft 4-6 (10 Aug 1879): pp. 49-158, [1-4], Taf. 166-178

Kobelt, W. [1879-1881]. Illustrirtes Conchylienbuch. Zweiter Band. - pp. [1], 145-392, Taf. 51-112. Nürnberg. (Bauer & Raspe).
Dates (handwritten notes on the digitized copy of Smithsonian): pp. 145-264 (1879), 265-312 (1880), 313-392 (1881).

Kobelt, W. 1881. Zwei neue süditalienische Hyalinen. - Nachrichtsblatt der Deutschen Malakozoologischen Gesellschaft 13 (12): 179-180.

Kobelt, W. 1882. Diagnosen neuer Arten. - Jahrbücher der Deutschen Malakozoologischen Gesellschaft 9: 68-72.

Kobelt, W. 1882. Diagnosen neuer Arten. - Nachrichtsblatt der Deutschen Malakozoologischen Gesellschaft 14 (8/9): 121-123.

Kobelt, W. 1883. Diagnosen neuer Arten. - Jahrbücher der Deutschen Malakozoologischen Gesellschaft 10: 84-85.

Kobelt, W. 1883. Diagnosen neuer Arten. - Nachrichtsblatt der Deutschen Malakozoologischen Gesellschaft 15 (7-8): 113-115.

Kobelt, W. 1883. Diagnosen neuer Arten. - Nachrichtsblatt der Deutschen Malakozoologischen Gesellschaft 15: 181-183.

Kobelt, W. 1884. Iconographie der Land- & Süsswasser-Mollusken mit vorzüglicher Berücksichtigung der europäischen noch nicht abgebildeten Arten von E. A. Rossmässler. Neue Folge. Erster Band. - pp. [1-4], 1-72, Taf. 1-30. Wiesbaden. (Kreidel).
No. 1-222. Dates (Hesse 1928, Bank 1989):
Heft 1-2 (31 Oct 1882): pp. 1-32, Taf. 1-10
Heft 3-4 (24 Dec 1883): pp. 33-56, Taf. 11-20
Heft 5-6 (10 Apr 1884): pp. 57-72, [1-4], Taf. 21-30

Kobelt, W. 1886. Iconographie der Land- & Süsswasser-Mollusken mit vorzüglicher Berücksichtigung der europäischen noch nicht abgebildeten Arten von E. A. Rossmässler. Neue Folge. Zweiter Band. - pp. [1-5], 1-56. Wiesbaden. (Kreidel).
No. 223-308. Dates (Hesse 1928, Bank 1989):
Heft 1-2 (25 Oct 1885): pp. 1-24, Taf. 31-40
Heft 3-4 (01 Aug 1886): pp. 25-40, Taf. 41-50
Heft 5-6 (20 Nov 1886): pp. 41-56, [1-5], Taf. 51-60

Kobelt, W. 1888. Diagnose einer neuen Xerophile. - Nachrichtsblatt der Deutschen Malakozoologischen Gesellschaft 20 (7-8): 119.

Kobelt, W. 1888. Iconographie der Land- & Süsswasser-Mollusken mit vorzüglicher Berücksichtigung der europäischen noch nicht abgebildeten Arten von E. A. Rossmässler. Neue Folge. Dritter Band. - pp. 1-60, Taf. 61-90. Wiesbaden. (Kreidel).
No. 309-498. Dates (Hesse 1928, Bank 1989):
Heft 1-2 (15 Apr 1887): pp. 1-12, Taf. 61-70
Heft 3-4 (06 Oct 1887): pp. 13-36, Taf. 71-80
Heft 5-6 (10 Mar 1888): pp. 37-60, Taf. 81-90

Kobelt, W. 1890. Iconographie der Land- & Süsswasser-Mollusken mit vorzüglicher Berücksichtigung der europäischen noch nicht abgebildeten Arten von E. A. Rossmässler. Neue Folge. Vierter Band. - pp. 1-102, Taf. 91-120. Wiesbaden. (Kreidel).
No. 499-716. Dates (Hesse 1928, Bank 1989):
Heft 1-2 (25 Oct 1888): pp. 1-40, Taf. 91-100
Heft 3-4 (07 Nov 1889): pp. 41-80, Taf. 101-110
Heft 5-6 (07 June 1890): pp. 581-102, Taf. 111-120

Kobelt, W. 1892. Iconographie der Land- & Süsswasser-Mollusken mit vorzüglicher Berücksichtigung der europäischen noch nicht abgebildeten Arten von E. A. Rossmässler. Neue Folge. Fünfter Band. - pp. [1], 1-118, Taf. 121-150. Wiesbaden. (Kreidel).
No. 717-960. Dates (Hesse 1928, Bank 1989):
Heft 1-2 (25 Aug 1891): pp. 1-40, Taf. 121-130
Heft 3-4 (20 Dec 1891): pp. 41-80, Taf. 131-140
Heft 5-6 (01 Sep 1892): pp. 81-118, Taf. 141-150

[Kobelt, W.] 1892. Literaturbericht. - Nachrichtsblatt der Deutschen Malakozoologischen Gesellschaft 24 (7/8): 149-152.

Kobelt, W. 1893. Diagnosen neuer griechischer Arten. - Nachrichtsblatt der Deutschen Malakozoologischen Gesellschaft 25 (3/4): 43-46.

Kobelt, W. 1893. Iconographie der Land- & Süsswasser-Mollusken mit vorzüglicher Berücksichtigung der europäischen noch nicht abgebildeten Arten von E. A. Rossmässler. Neue Folge. Sechster Band. - pp. [1], 1-102, Taf. 151-180. Wiesbaden. (Kreidel).
No. 961-1134. Dates (Hesse 1928, Bank 1989):
Heft 1-2 (15 Dec 1892): pp. 1-48, Taf. 151-160
Heft 3-4 (10 Aug 1893): pp. 49-80, Taf. 161-170
Heft 5-6 (02 Dec 1893): pp. 81-102, Taf. 171-180

Kobelt, W. 1896. Iconographie der Land- & Süsswasser-Mollusken mit vorzüglicher Berücksichtigung der europäischen noch nicht abgebildeten Arten von E. A. Rossmässler. Neue Folge. Siebenter Band. - pp. [1], 1-107, Taf. 181-210. Wiesbaden. (Kreidel).
No. 1135-1316. Dates (Hesse 1928, Bank 1989):
Heft 1-2 (10 Aug 1894): pp. 1-40, Taf. 181-190
Heft 3-4 (10 Dec 1894): pp. 41-56, Taf. 191-200
Heft 5-6 (25 Sep 1896): pp. 57-107, Taf. 201-210

Kobelt, W. 1898. Studien zur Zoogeographie. II. Die Fauna der meridionalen Sub-Region. - pp. I-X [= 1-10], 1-368. Wiesbaden. (Kreidel).

Kobelt, W. 1898. Neue *Helix*-Arten aus Montenegro. - Nachrichtsblatt der Deutschen Malakozoologischen Gesellschaft 30 (11/12): 161-165.

Kobelt, W. 1899. Iconographie der Land- & Süsswasser-Mollusken mit vorzüglicher Berücksichtigung der europäischen noch nicht abgebildeten Arten von E. A. Rossmässler. Neue Folge. Achter Band. - pp. [1], 1-111, [1], Taf. 211-240. Wiesbaden. (Kreidel).
No. 1317-1560. Dates (Hesse 1928, Bank 1989):
Heft 1-2 (03 Jan 1898): pp. 1-40, Taf. 211-220
Heft 3-4 (30 Oct 1898): pp. 41-72, Taf. 221-230
Heft 5-6 (20 Mar 1899): pp. 73-111, Taf. 231-240

Kobelt, W. 1902. Das Tierreich. Eine Zusammenstellung und Kennzeichnung der rezenten Tierformen. 16. Lieferung. Mollusca. Cyclophoridae. - pp. I-XXXIX [= 1-39], 1-662, 1 map. Berlin. (Friedländer).

Kobelt, W. 1902. Die Familie Buliminidae. - Systematisches Conchylien-Cabinet von Martini und Chemnitz 1 (13 (2)): 397-1051, Taf. 71-133. Nürnberg.
Dates (Welter-Schultes 1999): pp 397-620, Taf. 71-82 (1899), pp 621-684, Taf. 83-94 (1900), pp 685-836, Taf. 104-123 (1901), pp 837-1051, Taf. 124-133 (1902).

Kobelt, W. 1902. Iconographie der Land- & Süsswasser-Mollusken mit vorzüglicher Berücksichtigung der europäischen noch nicht abgebildeten Arten von E. A. Rossmässler. Neue Folge. Neunter Band. - pp. [1-2], 1-86, Taf. 241-270. Wiesbaden. (Kreidel).
Nº 1561-1740. Dates (Hesse 1928, Bank 1989):
Heft 1-2 (25 Oct 1899): pp. 1-24, Taf. 241-250
Heft 3-4 (26 Nov 1901): pp. 25-56, Taf. 251-260
Heft 5-6 (01 Apr 1902): pp. 57-86, Taf. 261-270

Kobelt, W. 1903. Iconographie der Land- & Süsswasser-Mollusken mit vorzüglicher Berücksichtigung der europäischen noch nicht abgebildeten Arten von E. A. Rossmässler. Neue Folge. Zehnter Band. - pp. [1], 1-77, Taf. 271-300. Wiesbaden. (Kreidel).
Nº 1741-1926. Dates (Hesse 1928, Bank 1989):
Heft 1-2 (29 Jan 1903): pp. 1-32, Taf. 271-280
Heft 3-4 (04 Sep 1903): pp. 33-48, Taf. 281-290;
Heft 5-6 (01 Feb 1904): pp. 49-77, Taf. 291-300

Kobelt, W. 1903. Diagnosen neuer Arten. - Nachrichtsblatt der Deutschen Malakozoologischen Gesellschaft 35 (9/10): 145-151.

Kobelt, W. 1904. Iconographie der Land- & Süsswasser-Mollusken mit vorzüglicher Berücksichtigung der europäischen noch nicht abgebildeten Arten von E. A. Rossmässler. Neue Folge. Elfter (Register-) Band. - pp. I-XII [= 1-12], 1-342, [1], Taf. I-VI [= 1-6]. Wiesbaden. (Kreidel).
Date was Oct 1904 (Bank 1989).
Contains (pp. 129-140) a list of genera and species of Palearctic non-marine molluscs. Each genus is mentioned with one single representative species. These were only meant as examples (Art. 67.5.1). The work contained a disclaimer (Art. 8.3) on pp. 145-148 where Kobelt explained explicitly and in detail that he did not use the term "Typus" in the sense of a name-bearing type for a genus.

Kobelt, W. 1905. Die Schnirkelschnecken nebst den verwandten Gattungen. Fünfter Theil. - Systematisches Conchylien-Cabinet von Martini und Chemnitz 1 (12 (5)): 861-1225, [1], Taf. 229-299, 236a. Nürnberg.
Dates (Welter-Schultes 1999: 163):
Pp 861-884, Taf. 229-234 (1898), pp 885-940, Taf. 235-245 (1899), pp 941-988, Taf. 246-257 (1900), pp 989-1060, Taf. 258-269 (1901), pp 1061-1092, Taf. 270-275 (1902), pp 1093-1116, Taf. 276-281 (1904), pp 1117-1226, Taf. 282-299 (1905).

Kobelt, W. 1906. Iconographie der Land- & Süsswasser-Mollusken mit vorzüglicher Berücksichtigung der europäischen noch nicht abgebildeten Arten von E. A. Rossmässler. Neue Folge. Zwölfter Band. - pp. [1-2], 1-64, Taf. 301-330. Wiesbaden. (Kreidel).
Nº 1927-2067. Dates (Hesse 1928, Bank 1989):
Heft 1-2 (June 1905): pp. 1-24, Taf. 301-310
Heft 3-4 (Jan 1906): pp. 25-40, Taf. 311-320
Heft 5-6 (Aug 1906): pp. 41-64, Taf. 321-330

Kobelt, W. 1906. [Die Familie der Heliceen. Sechste Abtheilung.] Gattung *Helix* (L.) s. str. - Systematisches Conchylien-Cabinet von Martini und Chemnitz 1 (12 (6)): 1-308, Taf. 300-376. Nürnberg.
Dates (Welter-Schultes 1999: 163):
Pp 1-56, Taf. 300-311 (1902), pp 57-120, Taf. 312-322 (1903), pp 121-200, Taf. 323-346 (1904), pp 201-240, Taf. 347-358 (1905), pp 241-308, Taf. 365-376 (1906).

Kobialka, H., Beckmann, K.-H. & Schröder, E. **2006**. Arbeitscheckliste Mollusken NRW, 6. aktualisierte Ausgabe (Stand 15.01.2006). - pp. 1-11. Höxter, Ascheberg-Herbern, Bonn.

Kobialka, H., Schwer, H. & Kappes, H. **2009**. Rote Liste der gefährdeten Schnecken und Muscheln (Mollusca: Gastropoda et Bivalvia) in Nordrhein-Westfalen. 3. Fassung 2009. - Mitteilungen der Deutschen Malakozoologischen Gesellschaft 82: 3-30.

Kofler, A. 1977. Nachtrag zur Faunistik der Weichtiere (Mollusca) Osttirols. - Berichte des Naturwissenschaftlich-Medizinischen Vereins in Innsbruck 64: 67-87.

Kofler, A. & Mildner, P. 2004. Dritter Nachtrag zur Molluskenfauna Osttirols (Mollusca: Gastropoda, Bivalvia). - Berichte des Naturwissenschaftlich-Medizinischen Vereins in Innsbruck 91: 129-155.

Kokshoorn, B. [2009]. Resolving riddles and presenting new puzzles in Chondrinidae phylogenetics. - pp. 1-188. Leiden. (Thesis, Leiden University).
No date of publication given, must have appeared in early 2009. Chapter 8 with "new taxa" is not available for zoological nomenclature (disclaimer p. 115).

Kokshoorn, B. & Gittenberger, E. 2010. Chondrinidae taxonomy revisited: New synonymies, new taxa, and a checklist of species and subspecies (Mollusca: Gastropoda: Pulmonata). - Zootaxa 2539: 1-62.

Kokshoorn, B. & Gittenberger, E. 2012. Phylogeography of the land snail *Abida secale* (Draparnaud, 1801) (Chondrinidae). - Journal of Molluscan Studies 78: 128-133.

Kolouch, L. R. 2003. Suchozemští, sladkovodní a brakičtí měkkýši ostrovů Malty. - Malacologica Bohemoslovaca 2: 43-50.

Korelewska-Batura, E. 1999. *Helix lutescens* Rossmässler, 1837 (Gastropoda: Pulmonata: Helicidae) - its structure, biology and ecology. - Folia Malacologica 7 (4): 197-240.

Korn, K. 1990. Die Mollusken auf dem Hellefelder Kalk zwischen Sundern und Meschede (Sauerland). - Schriften zur Malakozoologie 3: 61-62.

Kornilios, P., Poulakakis, N., Mylonas, M. & Vardinoyannis, K. 2009. The phylogeny and biogeography of the genus *Zonites* Montfort, 1810 (Gastropoda: Pulmonata): preliminary evidence from mitochondrial data. - Journal of Molluscan Studies 75 (3): 109-117.

Korniushin, A. V. 2001. Taxonomic revision of the genus *Sphaerium* sensu lato in the Palaearctic Region, with some notes on the North American species (Bivalvia: Sphaeriidae). - Archiv für Molluskenkunde 129 (1/2): 77-122.

Korniushin, A. V. & Glaubrecht, M. 2002. Phylogenetic analysis based on the morphology of viviparous freshwater clams of the family Sphaeriidae (Mollusca, Bivalvia, Veneroida). - Zoologica Scripta 31: 415-459.

Korniushin, A. V. & Hackenberg, E. 2000. Verwendung konchologischer und anatomischer Merkmale für die Bestimmung mitteleuropäischer Arten der Familie Sphaeriidae (Bivalvia), mit neuem Bestimmungsschlüssel und Diagnosen. - Malakologische Abhandlungen 20 (6): 45-69.

Korniushin, A. V., Janovich, L. N. & Melnichenko, R. K. **2002**. Artenliste der Süßwassermuscheln der Ukraine. Mit Bemerkungen über taxonomischen Status, Verbreitung und Gefährdungskategorien einiger Arten und Formen. - pp. 463-478, in: Falkner, M., Groh, K. & Speight, M. C. D.: Collectanea Malacologica. Festschrift für Gerhard Falkner. -- pp. I-X [= 1-10], 1-547, Pl. 1-48. Hackenheim. (ConchBooks).

Kosińska, M. 1980. The life cycle of *Deroceras sturanyi* (Simroth, 1894) (Pulmonata, Limacidae). - Zoologica Poloniae 28: 113-155.

Kozłowski, J. 2008. On the distribution and food preferences of *Arion subfuscus* (Draparnaud, 1805). - Folia Malacologica 16 (2): 61-67.

Kraszewski, A. 2007. The continuing expansion of *Sinanodonta woodiana* (Lea, 1834) (Bivalvia: Unionidae) in Poland and Europe. - Folia Malacologica 15 (2): 65-69.

Krausse, A. 1917. Eine neue südindische Chalcididen-Gattung. - Archiv für Naturgeschichte, Abteilung A 82 (1) ["1916"]: 93-95. Berlin.
Date was Jan 1917 as indicated on the blue wrapper preserved in the AMNH Library copy.

Krejci, K. & Wenz, W. 1926. Jungtertiäre Landschnecken aus Südrumänien. - Neues Jahrbuch für Mineralogie, Geologie und Paläontologie 55 (B): 53-65, Taf. II [= 2]. Stuttgart.
Stratigraphischer Teil pp. 53-56 by Krejci, Paläontologischer Teil pp. 56-65 and Taf. 2 by Wenz.

Kruglov, N. D. & Soldatenko, E. V. 1997. Reviziya roda *Segmentina* (Planorbidae). - Ruthenica 7 (2): 111-132.

Kruglov, N. D. & Starobogatov, Ya. I. 1983. K morfologii i sistematike Evropejskikh predstavitelej podroda *Peregriana* roda *Lymnaea* (Gastropoda, Pulmonata). - Zoologicheskij Zhurnal 62 (10): 1462-1473.

Kruglov, N. D. & Starobogatov, Ya. I. 1993. Guide to recent molluscs in northern Eurasia. 3. Annotated and illustrated catalogue of species of the family Lymnaeidae (Gastropoda Pulmonata Lymnaeiformes) of Palaearctic and adjacent river drainage areas. Part I. - Ruthenica 3 (1): 65-92.

Kruglov, N. D. & Starobogatov, Ya. I. 1993. Guide to recent molluscs of northern Eurasia. 3. Annotated and illustrated catalogue of species of the family Lymnaeidae (Gastropoda Pulmonata Lymnaeiformes) of Palaearctic and adjacent river drainage areas. Part 2. - Ruthenica 3 (2): 161-180.

Krynicki, J. 1833. Novæ species aut minus cognitæ e chondri, bulimi, peristomæ helicisque generibus præcipue Rossiæ meridionalis. - Bulletin de la Société Impériale des Naturalistes de Moscou 6: 391-436. Moscou [Moskva].
In Russian and Latin.

Krynicki, J. 1836. Helices proprie [sic] dictae hucusque [sic] in limitibus Imperii Rossici observatae. - Bulletin de la Société Impériale des Naturalistes de Moscou 9: 145-214. Moscou [Moskva].

Krynicki, J. 1837. Conchylia tam terrestria, quam fluviatilia et e maribus adjacentibus Imperii Rossici indigena, quæ pro mutua offeruntur historiæ naturalis culturibus commutatione. - Bulletin de la Société Impériale des Naturalistes de Moscou 1837 (2): 50-64.

Książkiewicz, Z. 2009. New localities of *Vertigo moulinsiana* (Dupuy, 1849) (Gastropoda: Pulmonata: Vertiginidae) in northwestern Poland. - Folia Malacologica 17 (1): 219-222.

Kuiper, J. G. J. 1960. Die Pisidien des Ohridsees, Mazedonien, nebst Bemerkungen über die Verbreitung der Pisidien in der Balkanhalbinsel und den Donauländern. - Beaufortia 7: 219-231. Amsterdam.

Kuiper, J. G. J. 1975. Zwei neue boreale *Pisidium*-Arten: *P. hinzi* und *P. waldeni*. - Archiv für Molluskenkunde 106 (1/3): 27-37.

Kuiper, J. G. J. 1981. The distribution of *Pisidium tenuilineatum* Stelfox and *Pisidium annandalei* Prashad in the Mediterranean area. - Basteria 45: 79-84.

Kuiper, J. G. J. 1987. *Pisidium maasseni* n. sp., a new species from Lake Prespa, Jugoslavia (Bivalvia, Sphaeriidae). - Basteria 51 (4/6): 163-165.

Künkel, K. 1908. Vermehrung und Lebensdauer der Nacktschnecken. - Verhandlungen der Deutschen Zoologischen Gesellschaft 18: 153-161. Leipzig.

Künkel, K. 1928. Zur Biologie von *Eulota fruticum* Müller. - Zoologische Jahrbücher, Abteilung für Allgemeine Zoologie und Physiologie der Tiere 45: 317-342.

Künkel, K. 1929. Experimentelle Studie über *Vitrina brevis* Férussac. - Zoologische Jahrbücher, Abteilung für Allgemeine Zoologie und Physiologie der Tiere 46: 575-626, Taf. 5-6. Jena.

Künkel, K. 1933. Vergleichende experimentelle Studie über *Vitrina elongata* Draparnaud und *Vitrina brevis* Férussac. - Zoologische Jahrbücher, Abteilung für Allgemeine Zoologie und Physiologie der Tiere 52: 399-432, Taf. 5-7. Jena.

Kuščer, L. 1928. Primo contributo alla fauna malacologica cavernicola della Venezia Tridentina. *Zospeum globosum* n. sp. - Studi Trentini di Scienze Naturali 9: 185-187. Trento.

Kuščer, L. 1932. Höhlen- und Quellenschnecken aus dem Flußgebiet der Ljubljanica. - Archiv für Molluskenkunde 64 (2): 48-61.

Kuscer, L. 1933. Prispevek k poznavanju podzemskih gastropodov Dalmacije in Hercegovine. - Prirodoslovna Istrazivanja Kraljevine Jugoslavije 18: 59-67, Tab. 1. Zagreb.

Küster, H. C. 1844. Zoologische Notizen. - Isis 1844 (9): 645-656. Leipzig.

Küster, H. C. 1847. Die Schliessschnecken und die verwandten Gattungen (*Clausilia*, *Balea*, *Cylindrella*, *Megaspira*.). In Abbildungen nach der Natur mit Beschreibungen. - Systematisches Conchylien-Cabinet von Martini und Chemnitz 1 (14): 1-355, Taf. 1-38. Nürnberg.
Dates (Welter-Schultes 1999: 165): Taf. 2 (1844), Taf. 1 (1845), pp. 1-32 Taf. 3, 4, 8 (1847), pp. 33-72 Taf. 5, 9, 11 (1850), pp. 73-88 Taf. 10, 12, 16 (1852), pp. 89-150 Taf. 6, 7, 13-15, 19 (1853), Taf. 18, 20 (1854), Taf. 17, 21, 22 (1855), 151-174 Taf. 23-28 (1857), 175-238 Taf. 29-34 (1860), 239-326 Taf. 35-38 (1861), 327-355 (1862).
The plates contained no names printed on them. Sheets (wrappers) had been issued along with the plates, so that subscribers and librarians would know which species were figured, these sheets were not printed for the purpose of providing a permanent scientific record, and do not constitute published work (Art. 8.1.1). They were usually not preserved, there were also no provisions of how to bind these wrappers. The names printed on the wrappers were cited by contemporaneous authors, for example by Pfeiffer 1848 who had received seen wrappers with names of species printed on them before the text passages had been issued. For the dates of publication of names, only the text pages are relevant. This had been overlooked in many publications by H. Nordsieck who assumed that the dates were to be taken from the plates. Probably not a single wrappper of this volume had been preserved anywhere.

Küster, H. C. 1852. Die Gattungen *Paludina*, *Hydrocaena* und *Valvata*. In Abbildungen nach der Natur mit Beschreibungen. - Systematisches Conchylien-Cabinet von Martini und Chemnitz 1 (21): 1-96, Taf. 1-14. Nürnberg.
Dates (Welter-Schultes 1999): pp. 1-56 Taf. 1-8 (1852), pp. 57-96 Taf. 9-14 (1853).

Küster, H. C. 1852. Die Gattungen *Pupa*, *Megaspira*, *Balea* und *Tornatellina*. In Abbildungen nach der Natur mit Beschreibungen. - Systematisches Conchylien-Cabinet von Martini und Chemnitz 1 (15): 1-194, Taf. 1-21. Nürnberg.
Dates (Welter-Schultes 1999): pp. 1-20 Taf. 1-2 (1841), pp. 21-68 Taf. 3, 8-10 (1843), 69-76 Taf 11-12 (1844), 77-88 Taf. 13 (1845), 89-96 Taf. 4, 5, 14 (1847), 97-128 Taf. 6, 15-16 (1850), 129-152 Taf. 7, 17, 18 (1852), Taf. 19-21 (1854), 153-194 (1855).

Küster, H. C. 1856. Nachträge und Berichtigungen zu dem Verzeichnisse der Binnenmollusken Bamberg's. - Bericht ueber das Bestehen und Wirken der Naturforschenden Gesellschaft zu Bamberg 3: 73-78.

Küster, H. C. 1870. Die Binnenmollusken-Fauna von Triest, Istrien, Dalmatien und Montenegro. II. - Bericht der Naturforschenden Gesellschaft zu Bamberg 9: 86-101.

Küster, H. C. 1875. Die Binnenconchylien Dalmatiens mit Zuziehung der Faunen von Triest, Istrien und Montenegro. III. Die Gattung *Clausilia*. - Bericht der Naturforschenden Gesellschaft zu Bamberg 10 (2): 1-132.
Text stops suddenly. Year 1875 indicated on title page. Names usually given with date 1876, I don't know the source for this year.

Küster, H. C., Dunker, W. & Clessin, S. 1886. Die Familie der Limnaeiden enthaltend die Genera *Planorbis*, *Limnaeus*, *Physa* und *Amphipeplea*. In Abbildungen nach der Natur mit Beschreibungen. - Systematisches Conchylien-Cabinet von Martini und Chemnitz 1 (17 (a)): 1-34, 35a, 36a, 35-430, Taf. 1-55. Nürnberg.
Dates (Welter-Schultes 1999): pp 1-8 (1841), Taf. 1 (1843), pp 9-20, Taf. 2-4, 16 (1844), pp 21-28, 35-62, Taf. 5-10 (1850), pp 29-34, 35a, 36a (1878), pp 63-94, Taf. 11-15, 17 (1882), , pp 95-110, Taf. 18-22 (1883), pp 111-222, Taf. 23-39 (1884), pp 223-310, Taf. 40-50 (1885), pp 311-430, Taf. 51-55 (1886).

Küster, H. C. & Pfeiffer, L. 1845-1855. Die Gattungen *Bulimus*, *Partula*, *Achatinella*, *Achatina* und *Azeca*. In Abbildungen nach der Natur mit Beschreibungen. - Systematisches Conchylien-Cabinet von Martini und Chemnitz 1 (13 (1)): III-XIX [= 3-19], 1-395, Taf. 1-70. Nürnberg.
Dates (Welter-Schultes 1999: 164): Taf. 1-2 (1840), pp. 1-32 Taf. 3-7, 11-13 (1841), Taf. 14-15 (1843), pp. 33-48 Taf. 16 (1844), Taf. 17-18 (1845), pp. 49-96, III-XIX (1853?), pp. 97-232 Taf. 30-52 (1854), pp. 233-256 Taf. 53-70 (1855), pp. 257-280 (1856), 281-304 (1857), 305-336 (1860), 337-352 Taf. 25-29 (1863), 353-395 Taf. 37-38 (1865).

Kuźnik-Kowalska, E. 1998. Life cycle and population dynamics of *Alinda biplicata* (Montagu, 1803) (Gastropoda: Pulmonata: Clausiliidae). - Folia Malacologica 6: 33-37.

Kuźnik-Kowalska, E. 1999. Life cycle and population dynamics of *Discus rotundatus* (O. F. Müller, 1774) (Gastropoda: Pulmonata: Endodontidae). - Folia Malacologica 7 (1): 5-17.

Kuźnik-Kowalska, E. 2005. Life cycle and population dynamics of *Discus perspectivus* (Megerle von Mühlfeld, 1818) (Gastropoda: Pulmonata: Endodontidae). - Folia Malacologica 13 (4): 157-168.

Kuźnik-Kowalska, E. 2006. Life cycle and population dynamics of *Discus ruderatus* (Férussac, 1821) (Gastropoda: Pulmonata: Endodontidae). - Folia Malacologica 14 (1): 35-46.

Kuźnik-Kowalska, E. 2006. Age structure and growth rate of *Aegopinella epipedostoma* (Fagot, 1879) (Gastropoda: Pulmonata: Zonitidae). - Folia Malacologica 14 (2): 71-74.

Kuźnik-Kowalska, E. 2011. Age structure and growth rate of Zonitoides nitidus (O. F. Müller, 1774) (Gastropoda: Pulmonata: Gastrodontidae). - Folia Malacologica 19 (1): 51-53.

Kuźnik-Kowalska, E. & Pokryszko, B. M. 2012. Selected aspects of reproductive biology of *Zonitoides excavatus* (Alder, 1830) (Gastropoda: Pulmonata: Gastrodontidae). - Folia Malacologica 20 (1): 57.

Kuźnik-Kowalska, E. & Roksela, A. 2009. Life cycle of *Perforatella bidentata* (Gmelin, 1791) (Gastropoda: Pulmonata: Helicidae). - Folia Malacologica 17 (4): 199-214.

Lacina, A. 2011. Poznámky k ekologii, rozšíření a ochraně vřetence horského (*Pseudofusulus varians*) v České Republice. - Malacologica Bohemoslovaca 10: 18-23.

Lamarck, J. B. P. A. de Monet de 1799. Prodrome d'une nouvelle classification des coquilles, comprenant une rédaction appropriée des caractères génériques, et l'établissement d'un grand nombre de genres nouveaux. - Mémoires de la Société d'Histoire Naturelle de Paris 1 [an VII]: 63-91, 1 tableau. Paris.

Lamarck, J. B. P. A. de Monet de 1801. Systême [sic!] des animaux sans vertèbres, ou tableau général des classes, des ordres et des genres de ces animaux, présentant leurs caractères essentiels et leur distribution, d'après la considération de leurs rapports naturels et de leur organisation, et suivant l'arrangement établi dans les galeries [sic!] du Muséum d'Hist. Naturelle, parmi leurs dépouilles conservées; précédé du discours d'ouverture du Cours de Zoologie, donné dans le Muséum National d'Histoire Naturelle l'an 8 de la République. - pp. j-viij [= 1-8], 1-432, tables [1-6]. Paris.
Date Jan 1801 (personal communication by R. Bank in Gargominy et al. 2011: 358).

Lamarck, J. P. B. A de Monet de 1818. Histoire naturelle des animaux sans vertèbres, présentant les caractères généraux et particuliers de ces animaux, leur distribution, leurs classes, leurs familles, leurs genres, et la citation des principales espèces qui s'y rapportent; précédée d'une introduction offrant la détermination des caractères essentiels de l'animal, sa distinction du végétal et des autres corps naturels, enfin, l'exposition des principes fondamentaux de la zoologie. Tome cinquième. - pp. [1-3], 1-612. Paris. (Deterville, Verdiere).

Lamarck, J. P. B. A de Monet de 1819. Histoire naturelle des animaux sans vertèbres, présentant les caractères généraux et particuliers de ces animaux, leur distribution, leurs classes, leurs familles, leurs genres, et la citation des principales espèces qui s'y rapportent; précédée d'une introduction offrant la détermination des caractères essentiels de l'animal, sa distinction du végétal et des autres corps naturels, enfin, l'exposition des principes fondamentaux de la zoologie. Tome sixième. Ire. partie. - pp. j-vj [= 1-6], 1-343. Paris.

Lamarck, J. P. B. A de Monet de 1822. Histoire naturelle des animaux sans vertèbres, présentant les caractères généraux et particuliers de ces animaux, leur distribution, leurs classes, leurs familles, leurs genres, et la citation des principales espèces qui s'y rapportent; précédée d'une introduction offrant la détermination des caractères essentiels de l'animal, sa distinction du végétal et des autres corps naturels, enfin, l'exposition des principes fondamentaux de la zoologie. Tome sixième. 2me. partie. - pp. [1-3], 1-232. Paris.
Date was Apr 1822.

Larraz, M., Equísoain, J. J., Montilla, E. & Robles, E. 2007. Datos de distribución de *Laminifera pauli* (Mabille, 1865), Elona quimperiana (Férussac, 1821), rissexodon [sic] constrictus (Boubée, 1836) y Helicodonta obvoluta (Müller, 1774) en Navarra. - Noticiario de la Sociedad Española de Malacología 47: 49-54.

Larraz Azcárate, M. L. & Rolán, E. 2003. Distribución de *Lauria sempronii* (Gastropoda, Pupillidae) en la Península Ibérica. - Noticiario de la Sociedad Española de Malacología 39: 69-71.

Lazaridou-Dimitriadou, M. & Daguzan, J. 1981. Etude de l'effet du 'groupement' des individus chez *Theba pisana* (mollusque gastéropode pulmoné stylommatrophore). - Malacologia 20: 195-204.

Lazaridou-Dimitriadou, M. & Kattoulas, M. E. 1981. Contribution à l'étude de la biologie et de la croissance des escargots commercialisés en Grèce: *Eobania vermiculata* (Muller) et *Helix aspersa* Muller. - Haliotis 11: 129-137. Paris.

Lazaridou-Dimitriadou, M. & Kattoulas, M. E. 1985. Contribution à l'étude biologique et écologique d'*Eobania vermiculata* (gastéropode, Helicidae). - Biologia Gallo-hellenica 10: 131-137. Athènes.

Lazaridou-Dimitriadou, M. & Kattoulas, M. E. 1986. Comparaison du cycle biologique et de la croissance d'*Eobania vermiculata* (Müller) dans la nature et dans des conditions expérimentale [sic]. - Bulletin de la Société Zoologique de France 111 (1/2): 99-104. Paris.

Lazaridou-Dimitriadou, M. & Kattoulas, M. E. 1990. Energy flux in a natural population of the land snail *Eobania vermiculata* (Müller) (Gastropoda, Pulmonata, Stylommatophora) in Greece. - Canadian Journal of Zoology 69: 881-891.

Lea, I. 1834. Observations on the naïades; and descriptions of new species of that, and other families. - Transactions of the American Phiolosophical Society (New Series) 5 ["1837"]: 23-119, Pl. I-XVIII [= 1-18]. Philadelphia.

Leach, W. E. 1852. Molluscorum Britanniæ synopsis. A synopsis of the Mollusca of Great Britain, arranged according to their natural affinities and anatomical structure. - pp. i-xvi [= 1-16], 1-376, Pl. I-XIII [= 1-13]. London. (Van Voorst).

Lehmann, R. 1862. Ueber eine neue Heliceen-Gattung. - Malakozoologische Blätter 9: 111-112.
Date July 1862 given on pp. 97 and 113.

Lehmann, R. 1862. Die Nacktschnecken aus der Umgebung Stettins und in Pommern. - Malakozoologische Blätter 9: 156-193, Taf. II-V [= 2-5].
Date July 1862 given on pp. 149 and 161.

Lehmann, R. 1864. Neue Nacktschnecke aus Australien. - Malakozoologische Blätter 11: 145-149.
Date July 1864 given on p. 129.

Lengyel, G. D. & Páll-Gergely, B. 2010. Notes on the landsnail (Gastropoda) and harvestman (Opiliones) fauna of Bihor and Vlădeasa mountains, Romania. - Studii şi Comunicări, Seria Ştiinţele Naturii 10-11: 91-111. Satu Mare.

Lesicki, A. & Koralewska-Batura, E. 2007. *Monacha cartusiana* (O. F. Müller, 1774) (Gastropoda: Pulmonata: Hygromiidae) becomes more frequent in Poland. - Folia Malacologica 15 (4): 181-184.

Lessona, M. 1880. Molluschi viventi del Piemonte. - Memorie della Classe di Scienze Fisiche, Matematiche e Naturali, Accademia Nazionale dei Lincei (3) 7 (277): 317-380, Tav. I-IV [= 1-4]. Roma.

Lessona, M. 1881. Sugli *Arion* del Piemonte. - Atti della Reale Accademia delle Scienze di Torino 16: 185-197, Tav. IV [= 4].

Lessona, M. & Pollonera, C. 1882. Monografia dei limacidi italiani. - pp. 1-82, Tav. I-III [= 1-3]. Torino. (Loescher).
Preprint of the 1884 work, date 1882 given on title page (Kennard & Woodward 1926: 391).

Lessona, M. & Pollonera, C. 1884. Monografia dei limacidi italiani. - Memorie della Reale Accademia delle Scienze di Torino (2) 35: 49-128, Tav. 1-3.

Letourneux, A. & Bourguignat, J.-R. 1887. Prodrome de la malacologie terrestre et fluviatile de la Tunisie. - pp. [1-2], 1-166. Paris. (Imprimerie Nationale).

Liebegott, A. 1986. Die Land- und Süßwassermollusken der Nördlichen Sporaden (Ägäis). - Mitteilungen der Deutschen Malakozoologischen Gesellschaft 39: 1-28.

Lightfoot, J. 1786. An account of some minute British shells, either not duly observed, or totally unnoticed by authors. - Philosophical Transactions of the Royal Society of London 76 (1): 160-170, Tab. I-III [= 1-3].

Likharev, I. M. 1949. Novyj kitajskij element v faune primorskogo kraya. - Doklady Akademii Nauk SSSR (Novaya Seriya) 64 (3): 417-420. Moskva, Leningrad.

Likharev, I. M. 1955. K sistematicheskomu polozheniyu nekotorykh sredneaziatskikh nazemnykh mollyuskov. - Trudy Zoologicheskogo Instituta Akademii Nauk SSSR 21: 179-185.

Likharev, I. M. 1972. Dva novykh vida semejstva Clausiliidae (Gastropda, Pulmonata) iz Bolgarii. - Trudy Zoologicheskogo Instituta Akademii Nauk SSSR 52: 117-120.

Likharev, I. M. & Rammel'mejer, E. S. 1952. Nazemnye mollyuski fauny SSSR. - Opredeliteli po Faune SSSR, Izdavaemye Zoologicheskim Institutom Akademii Nauk SSSR 43: 1-512. Moskva, Leningrad.

Lindholm, W. A. 1908. Materialien zur Molluskenfauena [sic] von Südwestrussland, Polen und der Krim. - Zapiski Novorossijskago Obshchestva Estestvoispytatelej - Mémoires de la Société des Naturalistes de la Nouvelle-Russie 31: 199-232. Odessa.

Lindholm, W. A. 1910. Beschreibung einer neuen *Retinella*-Art aus der Krim. - Abhandlungen der Senckenbergischen Naturforschenden Gesellschaft 32: 221-224. Frankfurt am Main.

Lindholm, W. A. 1914. Miszellen zur Malakozoologie des Russischen Reiches. I-XIII. - Ezhegodnik Zoologicheskago Muzeya Imperatorskoj Akademii Nauk - Annuaire due Musée Zoologique de l'Académie Impériale des Sciences de St.-Pétersbourg 18 (1) ["1913"]: 151-167.
Date was April 1914 as indicated on a separate page bound at the end of the journal volume.

Lindholm, W. A. 1914. Beschreibung einer neuen Nacktschneckengattung aus dem Kaukasusgebiete. - Nachrichtsblatt der Deutschen Malakozoologischen Gesellschaft 46: 167-168.

Lindholm, W. A. 1922. Description of two Bulimini (Gastropoda Pulmonata) from Russian Central Asia. - Ezhegodnik, Zoologicheskago Muzeja Imperatorskoi Akademii Nauk 23: 273-275.

Lindholm, W. A. 1924. A revised systematic list of the genera of the Clausiliidæ, recent and fossil, with their subdivisions, synonymy, and types. - Proceedings of the Malacological Society of London 16 (2): 53-80.

Lindholm, W. A. 1924. K nomenklature nektovykh kaspijskikh gastropod. - Russkij Gidrobiologicheskij Zhurnal 3: 32-34. Saratov.
Author's name Lindgol'm when transcribed from Russian. In German library catalogues found under Russkij gidrobiologičeskij žurnal:

Lindholm, W. A. 1925. Beitrag zur Systematik und Nomenklatur der Familie Enidae (Buliminidae). - Archiv für Molluskenkunde 57: 23-41.

Lindholm, W. A. 1925. Studien an palaearktischen *Vertigo*-Arten. - Archiv für Molluskenkunde 57 (5/6): 241-251.

Lindholm, W. A. 1925. A supplement to the revised systematic list of the genera of the Clausiliidæ. - Proceedings of the Malacological Society of London 16 (6): 261-266.

Lindholm, W. A. 1926. Ein Beitrag zur Molluskenfauna der Krim. - Archiv für Molluskenkunde 58: 161-177.

Lindholm, W. A. 1927. Zur Systematik und Nomenklatur einiger Heliciden und ihrer Verwandten. - Archiv für Molluskenkunde 59 (2): 116-138.

Lindholm, W. A. 1936. Eine neue *Helicella* (Helicidae, Gastropoda) aus der Krim, nebst Notizen zur Oekologie einiger Landschnecken. - Trudy Zoologicheskogo Instituta Akademii Nauk SSSR 3: 439-442.

Link, H. F. 1807. Beschreibung der Naturalien-Sammlung der Universität zu Rostock. Abtheilung 3. - pp. 101-165. Rostock. (Adler).

Linnæus, C. 1758. Systema naturæ per regna tria naturæ, secundum classes, ordines, genera, species, cum characteribus, differentiis, synonymis, locis. Tomus I. Editio decima, reformata. - pp. [1-4], 1-824. Holmiæ. (Salvius).

Linné, C. a 1767. Systema naturæ, Tom. I. Pars II. Editio duodecima reformata. - pp. 533-1327, [1-37]. Holmiæ. (Salvius).
Name was spelled Linné, not Linnæus. Date was 14 June 1767 (Cowie 1998: 101).

Lister, M. 1678. Historiæ animalium Angliæ tres tractatus. Unus de araneis. Alter de cochleis tum terrestribus tum fluviatilibus. Tertius de cochleis marinis. Quibus adjectus est quartus de lapidis ejusdem insulæ ad cochlearum quandam imaginem figuratis. Memoriæ & Rationi. - pp. [1-6], 1-250, [1], Tab. 1-9. Londini. (Martyn).

Lister, M., Lister, S. & Lister, A. 1685. Historiæ sive synopsis methodicæ conchyliorum quorum omnium picturæ, ad vivum delineatæ, exhibētur. Liber primus, qui est de cochleis terrestribus. - pp. [1-6], Pl. [1-42] (Fig. 1-104, Tab. 1-4). Londini.

Lister, M., Lister, S. & Lister, A. 1686. Historiæ [sive synopsis methodicæ] conchyliorum [quorum omnium picturæ, ad vivum delineatæ, exhibētur.] Liber II. Qui est de turbinibus et bivalvibus aquæ dulcis. - pp. [1-2], [3], Pl. 107-160 [= 1-22] (Fig. 1-39, 1-6). Londini.

Locard, A. 1882. Prodrome de la malacologie française. Catalogue général des mollusques vivants de France. Mollusques terrestres, des eaux douces et des eaux saumâtres. - pp. [1-4], I-VI [= 1-6], 1-462. Lyon, Paris. (Georg, Baillière).

Locard, A. 1888. Études critiques sur les Helix du groupe de l'Helix rufescens Pennant (Helix striolata, H. rufescens, H. montana, H. cælata, H. circinata, H. clandestina). - Annales de la Société Linnéenne de Lyon (Novelle Série) 23: 309-370.

Locard, A. 1890. Catalogue des espèces françaises appartenant aux genres *Pseudanodonta* et *Anodonta* connues jusqu'à ce jour. - Annales de la Société Linnéenne de Lyon 36: 49-284.
Appeared shortly before as a monograph.

Locard, A. 1893. Notices conchyliologiques, XXIV. Description de quelques *Helix* nouveaux pour la faune française. - L'Échange 9 (104): 86-87. Lyon.

Locard, A. 1894. Conchyliologie française. Les coquilles terrestres de France. Description des familles, genres et espèces. Aver 515 figures dessinées d'après nature et intercalées dans le texte. - pp. 1-370. Paris. (Baillière).

Lohmander, H. 1937. Über die nordischen Formen von *Arion circumscriptus* Johnston. - Acta Societatis pro Fauna et Flora Fennica 60: 90-112. Helsingfors.

Loosjes, F. E. 1941. Züchtungsversuche mit Clausiliidae. - Basteria 6: 26-36.

Loosjes, F. E. 1955. Eine Clausilien-Ausbeute von Kreta. - Sitzungsberichte, Österreichische Akademie der Wissenschaften, mathematisch-naturwissenschaftliche Klasse, Abteilung I 164 (10): 855-885. Wien.

López, M. A. & Altaba, C. R. 2005. Fish host determination for *Margaritifera auricularia* (Bivalvia: Unionoida): results and implications. - Bollettino Malacologico 41 (9/12): 89-98.

López, M. A., Altaba, C. R., Rouault, T. & Gisbert, E. **2007**. The European sturgeon *Acipenser sturio* is a suitable host for the glochidia of the freshwater pearl mussel *Margaritifera auricularia*. - Journal of Molluscan Studies 73: 207-209.

Lowe, R. T. 1831. Primitiæ faunæ et floræ Maderæ et Portus Sancti; sive species quædam novæ vel hactenus minus rite cognitæ animalium et plantarum in his insulis degentium breviter descriptæ. - Transactions of the Cambridge Philosophical Society 4 (1): 1-70, Pl. 1-6.

Lowe, R. T. 1832. On the genera *Melampus*, *Pedipes* and *Truncatella*: with experiments tending to demonstrate the real nature of the respiratory organs in these Mollusca. - Zoological Journal 5 (19): 280-305, Pl. XIII [= 13]. London.
Date was July 1832.

Lowe, R. T. 1852. Brief diagnostic notices of new Maderan land shells. - Annals and Magazine of Natural History (2) 9 (50): 112-120, 275-279. London.

Lowe, R. T. 1855. Catalogus molluscorum pneumonatorum insularum Maderensium: or a list of all the land and freshwater shells, recent and fossil, of the Madeiran islands: arranged in groups according to their natural affinities; with diagnoses of the groups, and of the new or hitherto imperfectly defined species. - Proceedings of the Zoological Society of London 22 ["1854"]: 161-218.

Lowe, R. T. 1861. Diagnoses of new Canarian land-Mollusca. - Annals and Magazine of Natural History (3) 7: 104-112.

Ložek, V. 1948. Nový plž z rodu *Chondrina* v ČSR. - Věstník Československé Zoologické Společnosti 12: 83-88. Praha.

Lupu, D. 1966. Un fossile vivant de la faune de la Roumanie: *Chilostoma (Drobacia) maeotica* Wenz (Gastropoda - Pulmonata). - Travaux du Muséum d'Histoire Naturelle 'Grigore Antipa' 6: 31-37. Bucureşti.

Lupu, D. 1968. Contributions à la connaissance des limacides (Gastropoda-Pulmonata) des Monts Apuseni. - Travaux du Muséum d'Histoire Naturelle 'Grigore Antipa' 9: 39-44. Bucureşti.

Lupu, D. 1973. Étude anatomique comnparative des espèces *Limax cinereoniger* Wolf 1803 et *Liax* [sic] *maximus* Linné 1758 et description de deux espèces nouvelles du genre *Lehmannia* Heinemann [sic]. - Travaux du Muséum d'Histoire Naturelle 'Grigore Antipa' 13: 47-66. Bucureşti.
Printer's errors in title.

Luso da Silva, A. 1872. Molluscos [sic] terrestres e fluviaes de Portugal. (Continuação). - Jornal de Sciencias Mathematicas Physicas e Naturaes 3: 257-261. Lisboa.

Lvova, A. A. & Stargobogatov, Ya. I. 1982. Novyj vid drejsseny (Bivalvia, Dreissenida [sic!]) iz Okhridskogo ozera. - Zoologicheskij Zhurnal 61 (11): 1749-1752. Moskva.

Maassen, W. J. M. 1981. De Molluskenfauna van het griekse eiland Rhodos. - De Kreukel 17 (3/4): 21-32, pl. 8-15. Diemen.

Maassen, W. J. M. 1989. Eine neue Unterart von *Spelaeoconcha paganettii* Sturany, 1901 (Gastropoda Pulmonata: Enidae) aus Jugoslawien. - Basteria 53: 93-96.

Maassen, W. J. M. 1991. De verspreiding van *Cochlostoma cretense* (Maltzan 1887) en *Poiretia dilatata marginata* (Westerlund 1886) op Kreta. - De Kreukel 27 (3): 35-36, Pl. 1-2. Diemen.

Maassen, W. J. M. 1991. *Helicodonta gyria wilhelminae* nov. subspec. von der griechischen Insel Kreta (Gastropoda Pulmonata: Helicidae). - Basteria 55 (4/6): 123-126.

Maassen, W. J. M. 1995. Die Verbreitung des Genus *Hiltrudia* Nordsieck 1993 im ehemaligen Jugoslawien (Gastropoda: Hygromiidae). - De Kreukel 31: 9-22.

Maassen, W. J. M. 1995. Observations on the genus *Mastus* from Crete (Greece), with descriptions of twelve new species (Gastropoda Pulmonata: Buliminidae). - Basteria 59: 31-64.

Maassen, W. J. M. 1998. *Vitrea meijeri* n. sp. aus Mazedonien (Gastropoda Pulmonata: Zonitidae). - Basteria 62 (5/6): 215-217.

Maassen, W. J. M. 2000. *Deroceras dewinteri* spec. nov. from the island of Crete, Greece (Gastropoda, Pulmonata, Agriolimacidae). - Basteria 64 (4/6): 125-126.

Mabille, J. 1865. Études sur la fauna malacologique de Saint-Jean-de-Luz, de Dinan et de quelques autres points du littoral océanien de la France. - Journal de Conchyliologie 13: 248-265, Pl. XIV [= 14]. Paris.

Mabille, J. 1867-1869. Archives malacologiques. - pp. 1-80. Paris. (Bouchard-Huzard).
Dates March 1868 for fascicule 3 (pp. 33-52) , 01 Feb 1869 for fascicule 4 (pp. 55-71) and 5 (pp. 72-80).

Mabille, J. 1868. Des limaciens européens. - Revue et Magasin de Zoologie pure et appliquée (2) 20: 129-146.
Date was Apr 1868.

Mabille, J. [1871]. Histoire malacologique du Bassin Parisien ou histoire naturelle des animaux mollusques terrestres et fluviatiles qui vivent dans les environs de Paris. Premier fascicule. - pp. 1-128, Pl. I-II [= 1-2]. Paris. (Bouchard-Huzard).
Date given as 03/1870 on p. 128, but wrapper was dated 1871, true date was more likely 05 Sep 1871 (Bibliographie de France 60 [= (2)15] (37): 187, Nr. 3873).

Mabille, J. 1880. Testarum novarum Europæarum diagnoses. - Guide du Naturaliste 2-18: 62-65. Paris.

Mabille, J. 1881. Testarum novarum præsertim Europæarum diagnoses. - Bulletin de la Société Philomathique de Paris (7) 5: 122-130.

Madeira, M. J., Elejalde, M. A., Chueca, L. J. & Gómez-Moliner, B. J. 2010. Phylogenetic position of the genus *Cryptazeca* and the family Azecidae within the system of the Stylommatophora. - Malacologia 52 (1): 163-168.

Mallet, J. 2001. Subspecies, semispecies, superspecies. - pp. 523-526 in: Levin, S. A.: Encyclopedia of biodiversity. Volume 5. R-Z. - i-xxxi [= 1-31], 1-1103. Amsterdam (Elsevier).

Malm, A. W. 1855. Om Svenska landt- och söttvattens mollusker, med särskilt afseende på de arter och former, som förkomma i grannskapet af Christianstad (C) och Göteborg (G). - Götheborgs Kongliga Vetenskaps och Vitterhets Samhälles Handlingar (Ny Tidsföljd) 3: 73-152.

Malm, A. W. 1868. Skandinaviska land-sniglar, limacina, afbildade efter lefvande exemplar och beskrivna. - Göteborgs Kongliga Vetenskaps och Vitterhets Samhälles Handlingar (Ny Tidsföljd) 10 ["1870"]: 26-93, Pl. 1-5.
Date 1870 indicated on title page of volume, date "imlemnad" Nov 1867 indicated on the contents page of the volume, but in the text dates until 05 July 1868 appear at various occasions. Article could have appeared in Nov 1868.

Maltz, T. K. 2003. Life cycle and population dynamics of *Helicodonta obvoluta* (O. F. Müller, 1774) (Gastropoda: Pulmonata: Helicidae). - Folia Malacologica 11 (3/4): 63-88.

Maltz, T. K. & Pokryszko, B. M. 2009. *Macrogastra badia* (C. Pfeiffer, 1828) (Gastropoda: Pulmonata: Clausiliidae) in Zieleniec (Bystrzyckie Mts, central Sudetes) - ecology, conservation status and life history - preliminary data. - Folia Malacologica 17 (2): 53-62.

Maltz, T. K. & Sulikowska-Drozd, A. 2011. Delayed maturation in the genus *Vestia* P. Hesse (Gastropoda: Pulmonata: Clausiliidae): a model for clausiliid lifecycle strategy. - Journal of Molluscan Studies 77 (1): 41-54.

Maltzan, H. von 1881. Description de deux espèces nouvelles. - Journal de Conchyliologie 29 (2): 162-163. Paris.

Maltzan, H. von 1883. Diagnosen neuer cretischer Helices. - Nachrichtsblatt der Deutschen Malakozoologischen Gesellschaft 15 (7/8): 102-106.

Maltzan, H. von 1887. Neue kretenser Landschnecken. - Jahrbücher der Deutschen Malakozoologischen Gesellschaft 14: 117-119.

Mandahl-Barth, G. 1938. The zoology of Iceland. Volume IV, Part 65. Land and freshwater Mollusca. - pp. 1-31. Copenhagen, Reykjavik. (Levin & Munksgaard).

Manganelli, G., Bodon, M., Favilli, L., Castagnolo, L. & Giusti, F. **1998**. Checklist delle specie della fauna d'Italia, molluschi terrestri e d'acqua dolce. Errata ed addenda, 1. - Bollettino Malacologico 33 (9/12): 151-156.

Manganelli, G., Bodon, M., Favilli, L. & Giusti, F. **1995**. Fascicolo 16. Gastropoda Pulmonata. - pp. 1-60, in: Minelli A., Ruffo, S. & La Posta, S.: Checklist delle specie della fauna italiana. Bologna. (Calderini).
Errata and addenda: Manganelli, Bo. et al. 1998, Manganelli, Bo. & G. 2000.

Manganelli, G., Bodon, M. & Giusti, F. **1991**. A new species of *Oxychilus* from the Ligurian Apennines (Italy) (Pulmonata: Zonitidae). - Journal of Molluscan Studies 57 (4): 401-412.

Manganelli, G., Bodon, M. & Giusti, F. **2000**. Checklist delle specie della fauna d'Italia, molluschi terrestri e d'acqua dolce. Errata e addenda, 2. - Bollettino Malacologico 36 (5-8): 125-130.

Manganelli, G., Castagnolo, L. & Giusti, F. **1990**. Notulae malacologicae, XXXIX. Second contribution to the revision of the *Oxychilus*-species living in the Italian Apennine regions: *Hyalina carotii* Paulucci 1878, *Hyalina fragrans* Paulucci 1878 and *Helix ercica* Benoit 1859 (Pulmonata: Zonitidae). - Archiv für Molluskenkunde 119 (4/6): 181-203.

Manganelli, G. & Cianfanelli, S. 2001. The dates of publication of Bullettino malacologico Italiano (1868–1875) and Bullettino della Societá Malacologica Italiana (1875–1899). - Archives of Natural History 28 (3): 337-346.

Manganelli, G., Cianfanelli, S., Brezzi, M. & Favilli, L. **2001**. The distribution and taxonomy of *Vertigo moulinsiana* (Dupuy, 1849) in Italy (Gastropoda: Pulmonata: Vertiginidae). - Journal of Conchology 37 (3): 267-280.

Manganelli, G., Cianfanelli, S., Salomone, N. & Giusti, F. **2004**. Morphological and molecular analysis of the status and relationships of *Oxychilus paulucciae* (De Stefani, 1883) (Gastropoda: Pulmonata: Zonitidae). - Malacologia 46 (1): 19-35.

Manganelli, G. & Favilli, L. 1996. *Xerosecta giustii* a new hygromiid from Tuscany (Italy) close to extinction (Gastropoda, Pulmonata: Helicoidea). - Journal of Conchology 35 (4): 335-355. London.

Manganelli, G., Favilli, L. & Giusti, F. **1996**. The taxonomic status of *Xerotropis* Monterosato, 1892, and redescription of its type species (Pulmonata: Hygromiidae). - The Veliger 39 (1): 1-17.

Manganelli, G., Favilli, L. & Giusti, F. **1996**. The taxonomic status of *Xeroamanda* Monterosato, 1892 (Pulmonata, Hygromiidae). - Malacologia 37 (2): 349-361.

Manganelli, G., Favilli, L. & Giusti, F. **1999**. The *Oxychilus* species endemic to the Tuscan Archipelago: *O. majori* (Paulucci, 1886), *O. oglasicola* Giusti, 1968, and *O. pilula* (Paulucci, 1886) (Pulmonata, Zonitidae). - Bollettino Malacologico 34: 71-86.

Manganelli, G. & Giusti, F. 1988. Notulae malacologicae, XXXVIII [38]. A new Hygromiidae from the Italian Apennines and notes on the genus *Cernuella* and related taxa (Pulmonata: Helicoidea). - Bollettino Malacologico 23 (11/12) ["1987"]: 327-380.
Issue for Nov-Dec 1987, seems to have appeared in 1988.

Manganelli, G. & Giusti, F. 1989. Notulae Malacologicae, XLIII. *Xeromunda* Di Maria Di Monterosato in Italy (Pulmonata: Hygromiidae). - Bollettino Malacologico 25: 1-21. Milano.

Manganelli, G. & Giusti, F. 1993. Notulae Malacologicae, XLIX. Third contribution to the revision of the *Oxychilus*-species living in the Italian Apennine regions: *O. clarus* (Held), *O. majori* (Westerlund) and *O. uziellii* (Issel). - Archiv für Molluskenkunde 121: 67-78.

Manganelli, G. & Giusti, F. 1998. *Oxychilus mortilleti* (Pfeiffer, 1859): a redescription (Pulmonata, Zonitidae). - Basteria 61: 123-143.

Manganelli, G. & Giusti, F. 2000. New data on the distribution of *Oxychilus uziellii* (Issel, 1872) (Pulmonata, Zonitidae) with a redescription of the species. - Bollettino Malacologico 36 (5-8): 83-92.

Manganelli, G. & Giusti, F. 2001. Redescription of *Oxychilus meridionalis* (Paulucci, 1881) (Pulmonata: Zonitidae). - Bollettino Malacologico 37 (9-12): 187-206.

Manganelli, G. & Giusti, F. 2004. Status and relationships of *Vitrina musignani* Pirajno, 1842, and *Vitrina paulucciae* Fischer in Paulucci, 1878 (Gastropoda: Pulmonata: Vitrinidae). - Journal of Conchology 38 (3): 291.

Manganelli, G. & Giusti, F. 2005. The status and relationships of *Vitrina polloneriana* Fra' Piero, 1897, with description of *Sardovitrina* n. gen. (Gastropoda: Pulmonata: Vitrinidae). - Journal of Conchology 38 (5): 567-580.

Manganelli, G., Lori, E., Benocci, A. & Cianfanelli, S. **2008**. Il Bullettino Malacologico Italiano (1868–1875) e il Bullettino della Societá Malacologica Italiana (1875–1899). - Bollettino Malacologico 44 (9/12): 1-52.

Manganelli, G., Salomone, N. & Giusti, F. **2005**. A molecular approach to the phylogenetic relationships of the western palaearctic Helicoidea (Gastropoda: Stylommatophora). - Biological Journal of the Linnean Society 85: 501-512.

Manganelli, G., Sparacio, I & Giusti, F. **1989**. New data on the systematics of two Sicilian land snails, *Helix parlatoris* Bivona, 1839 and *Helix reinae* L. Pfeiffer, 1856 and description of *Schileykiella* n. gen. (Pulmonata: Hygromoidae). - Journal of Conchology 33 (3): 141-156, Pl. 10-16. London.

Martens, **E. von** 1864. Fossile Süsswasser-Conchylien aus Sibirien. - Zeitschrift der Deutschen Geologischen Gesellschaft 16: 345-351. Berlin.

Martens, E. von 1865. Mollusca. - pp. 189-250, in: Günther, A. C. L. G.: The record of zoological literature. 1864. Volume first. -- pp. i-vii [= 1-7], 1-634. London. (Van Voorst).

Martens, E. von 1873. Mollusca. - pp. 113-173, in: Newton, A.: The zoological record for 1871; being volume eighth of the record of zoological literature. -- pp. I-XIV [= 1-14], [1], 1-496. London. (Van Voorst).

Martens, E. von 1873. Ueber Land- und Süsswasser-Conchylien aus dem Peloponnes. - Malakozoologische Blätter 20: 31-50, Taf. 2-3.

Martens, E. von 1875. Sopra alcuni molluschi terrestri di Malta. - Bullettino Malacologico Italiano 6 ["1873"]: 26-29. Pisa.
Appeared not before spring 1874, earliest confirmed date was 14 Apr 1875 (Manganelli & Cianfanelli 2001: 344, Manganelli, Lo. et al. 2008: 30).

Martens, E. von 1876. Einige neue griechische Schnecken. - Jahrbücher der Deutschen Malakozoologischen Gesellschaft 3: 338-343, Taf. 12.

Martens, E. von 1879. Die Gattung *Neritina*. - Systematisches Conchylien-Cabinet von Martini und Chemnitz 2 (10): 1-303, Taf. A, 1-23. Nürnberg.

Martens, E. von 1884. Landschnecken aus Sardinien und aus Südost-Borneo. - Sitzungsberichte der Gesellschaft Naturforschender Freunde zu Berlin 1884 (10): 195-199.

Martens, E. von 1889. Landschnecken aus Lykien. - Sitzungsberichte der Gesellschaft der Naturforschenden Freunde, Berlin 1889 (9): 182-183.

Martens, **E. von** 1889. Griechische Mollusken. Gesammelt von Eberh. von Örtzen. - Archiv für Naturgeschichte 55 (Band 1 Heft 2): 169-240, Taf. 9-11. Berlin.

Martens, **G. von** 1824. Reise nach Venedig. Zweiter Theil. Venedig. Euganeen. Alpen von Belluno. Tirol. Baiern. Naturgeschichtlicher Angang. - pp. I-VI [= 1-6], 1-664. Ulm. (Stettin).

[Martens, **G. von**] 1830. Natur- und Vaterlandskunde. Ueber Würtembergs Fauna. - Correspondenzblatt des Würtembergischen Landwirthschaftlichen Vereins 17 (3): 123-186. Stuttgart, Tübingen.

Martínez-Ortí, A. 2000. Descripción de *Trochoidea (Xerocrassa) roblesi* spec. nov. (Mollusca, Gastropoda, Hygromiidae) de la Comunidad Valenciana (España). - Iberus 18 (2): 31-39.

Martínez-Ortí, A. 2006. *Balea heydeni* Von Maltzan, 1881 (Gastropoda, Clausiliidae) en España: características conquiológicas y distribución. - Noticiario de la Sociedad Española de Malacología 45: 30-37.

Martínez-Ortí, A. 2006. Descripción de un nuevo higrómido ibérico: *Xerocrassa edmundi* spec. nov. (Gastropoda, Pulmonata). - Iberus 24 (2): 61-68.

Martínez-Ortí, A. 2008. Taxonomical clarification of the Iberian endemic *Helicella (Xerotricha) mariae* Gasull, 1972 (Gastropoda, Pulmonata, Hygromiidae). - Basteria 72: 147-154.

Martínez-Ortí, A. 2010. Datos recientes sobre la distribución y el estado de conservación del higrómido *Xerosecta explanata* (O. F. Müller, 1774). - Noticiario de la Sociedad Española de Malacología 53: 45-48.

Martínez-Ortí, A. 2011. A new hygromiid for the Iberian malacofauna: *Candidula corbellai* n. sp. (Gastropoda, Pulmonata). - Animal Biodiversity and Conservation 34 (1): 1-10. Barcelona.

Martínez-Ortí, A., Aparicio, M. T. & Robles, F. **2004**. La malacofauna de la Sierra de Alcaraz (Albacete, España). - Iberus 22 (2): 9-17.

Martínez-Ortí, A., Arantzazu Elejalde, M., Madeira, M. J. & Gomez-Moliner, B. **2008**. Morphological and DNA-based taxonomy of *Tudorella* P. Fischer, 1885 (Caenogastropoda: Pomatiidae). - Journal of Conchology 39 (5): 553-567.

Martínez-Ortí, A., Arrébola, J. R. & Ruiz, A. **2008**. Revisión taxonómica de *Helix zapateri* Hidalgo, 1870 (Pulmonata, Trissexodontidae) y su nuevo estatus en la malacofauna ibérica. - Animal Biodiversity and Conservation 31 (1): 77-81. Barcelona.

Martínez-Ortí, A. & Ferrer, A. P. 2008. Primer hallazgo de *Chilostoma desmoulinsi* (Farines, 1834) y confirmacion de la presencia de *Helicigona lapicida* (Linnaeus, 1758) (Gastropoda, Pulmonata: Helicidae) en la Comunidad Valenciana. - Noticiario de la Sociedad Española de Malacología 50: 35-38.

Martínez-Ortí, A., Gómez-Moliner, B. J. & Prieto, C. E. **2007**. El género *Pyramidula* Fitzinger 1833 (Gastropoda, Pulmonata) en la Península Ibérica. - Iberus 25 (1): 77-87.

Martínez-Ortí, A. & Robles, F. 2002. First anatomic data and taxonomical clarification of *Suboestophora kuiperi* (Gasull, 1966) (Mollusca, Gastropoda: Hygromiidae). - Journal of Conchology 37 (4): 355-362.

Martínez-Ortí, A. & Robles, F. 2003. Moluscos continentales de la Comunidad Valenciana. - pp. 1-261. (Generalitat Valenciana).

Martínez-Ortí, A. & Robles, F. 2005. Los caenogasterópodos terrestres (Mollusca, Orthogastropoda) de la Comunidad Valenciana (España). - Iberus 23 (1): 7-24.

Martínez-Ortí, A. & Roqué i Pau, C. 2007. Nuevas citas, actual y fósil, de *Atenia quadrasi* (Hidalgo, 1885) (Gastropoda, Helicodontidae) en Cataluña. - Noticiario de la Sociedad Española de Malacología 48: 37-39.

Mayr, E. 1942. Systematics and the origin of species, from the viewpoint of a zoologist. - pp. I-XIV [= 1-14], 1-334. New York. (Columbia University Press).

McIvor, A. L. & Alridge, D. C. 2007. The reproductive biology of the depressed river mussel, *Pseudanodonta complanata* (Bivalvia: Unionidae), with implications for its conservation. - Journal of Molluscan Studies 73: 259-266.

Megerle von Mühlfeld, J. K. 1811. Entwurf eines neuen System's der Schalthiergehäuse. - Der Geschellschaft Naturforschender Freunde zu Berlin Magazin für die neuesten Entdeckungen in der gesammten Naturkunde 5 (1): 38-72.

Megerle von Mühlfeld, J. C. 1816. Beschreibung einiger neuen [sic] Conchylien. - Der Gesellschaft Naturforschender Freunde zu Berlin Magazin für die neuesten Entdeckungen in der gesammten Naturkunde 8 (1) ["1818"]: 3-11, Tab. I-II [= 1-2].
The date 1816 was printed on the wrapper of the first part of the journal volume, preserved in the SUB Göttingen copy.

Meier-Brook, C. 1983. Taxonomic studies on *Gyraulus* (Gastropoda: Planorbidae). - Malacologia 24 (1/2): 1-113.

Meier-Brook, C. & Bargues, M. D. 2002. *Catascopia*, a new genus for three nearctic and one palaearctic stagnicoline species (Gastropoda: Lymnaeidae). - Folia Malacologica 10 (2): 83–84.

Meigen, J. W. 1818. Systematische Beschreibung der bekannten europäischen zweiflügeligen Insekten. Erster Theil mit elf Kupfertafeln. - pp. I-XXXVI [= 1-36], 1-324, [1]. Aachen. (Forstmann).

Mendes da Costa, E. 1778. Historia naturalis testaceorum Britanniæ, or, the British conchology; containing the descriptions and other particulars of natural history of the shells of Great Britain and Ireland: illustrated with figures. In English and French. - Historia naturalis testaceorum Britanniæ, ou, la conchologie Britannique; contenant les descriptions & autres particularités d'histoire naturelle des coquilles de la Grande Bretagne & de l'Irlande: avec figures en taille douce. En anglois & françois. - pp. i-xii [= 1-12], 1-254, i-vii [= 1-7], [1], Pl. I-XVII [= 1-17]. London. (Millan, White, Emsley & Robson).
Name of the author was Emanuel Mendes da Costa, given in this form on the title page. The person was an English naturalist of Portuguese descent, Mendes da Costa was the full Portuguese surname. Zoologists usually cite names of taxa "Da Costa, 1778", possibly in the belief that Mendes was his second first name.

Menegazzi, L. 1855. Malacologia veronese. Rapporto letto nella tornata del 14 settembre 1854. - pp. I-XIII [= 1-13], 1-334, Tav. I-II [= 1-2]. Verona. (Vicentini & Franchini). [Not cited from original]
Date was 1855 (De Betta 1870: 8-9 footnote). Recorded in various library catalogues, consistently with the date 1855.

Menke, C. T. 1828. Synopsis methodica molluscorum generum omnium et specierum earum, quae in museo Menkeano adservantur; cum synonymia critica et novarum specierum diagnosibus. - pp. I-XII [= 1-12], 1-91. Pyrmonti. (Gelpke).

Menke, C. T. 1830. Synopsis methodica molluscorum generum omnium et specierum earum, quae in museo Menkeano adservantur; cum synonymia critica et novarum specierum diagnosibus. Editio altera, auctior et emendatior. - pp. I-XVI [= 1-16], 1-168, [1]. Pyrmonti. (Uslar).
Date 12 Apr 1830 is given on p. XVI, must have been printed some time after.

Menke, K. T. 1845. Kritische Übersicht der lebenden *Valvata*-Arten. - Zeitschrift für Malakozoologie 1845 (8): 115-130.
Date Aug 1845 given on p. 115.

Menke, K. T. 1848. Geographische Uebersicht der um die Molluskenfauna Deutschlands verdienten Schriften, Kenner und Sammler. - Zeitschrift für Malakozoologie 54 (4): 49-78.
Date Apr 1848 given on p. 49 (for pp. 49-65), May 1848 on p. 66 (for pp. 66-78).

Mesher, C. & Welter-Schultes, F. W. 2008. Studies on range expansion, predation pressure and insolation in *Albinaria* on the island of Dia (Greece), focused on a recently introduced species (Gastropoda: Clausiliidae). - Species, Phylogeny and Evolution 1 (3): 129-140. Göttingen.

Meyer, K. O. 1955. Naturgeschichte der Strandschnecke *Ovatella myosotis* (Draparnaud). - Archiv für Molluskenkunde 84 (1/3): 1-43.

Michaud, A.-L.-G. 1829. Description de plusieurs espèces nouvelles de coquilles vivantes. - Bulletin d'Histoire Naturelle de la Société Linnéenne de Bordeaux 3: 260-276, pl. [5].

Michaud, A.-L.-G. 1831. Complément de l'histoire naturelle des mollusques terrestres et fluviatiles de la France, de J. P. R. Draparnaud. - pp. j-xv [= 1-15], [1], 1-116, [1-2], 1-12, Pl. XIV-XVI [= 14-16]. Verdun. (Lippmann).

Michaud, A.-L.-G. [1833]. Catalogue des testacés vivans envoyés d'Alger par M. Rozet, capitaine au Corps Royal d'État-Major, au Cabinet d'Histoire Naturelle de Strasbourg. - Mémoires de la Société d'Histoire Naturelle de Strasbourg 1: 1-22.

Michaud, A.-L.-G. 1841. Description d'une nouvelle espèce d'hélice vivante. - Actes de la Société Linnéenne de Bordeaux 12: 64-65.

Middendorff, A. T. von 1851. Mollusken. - pp. [1], 163-464, Taf. VIII-XXX [= 8-30], in: Brandt, F., Erichson, W. F., Fischer, S., Grube, E., Ménétriés, E. & Middendorff, A. T. von: Reise in den äussersten Norden und Osten Sibiriens während der Jahre 1843 und 1844. Band II. Zoologie. Theil 1. Wirbellose Thiere: Annulaten. Echinodermen. Insecten. Krebse. Mollusken. Parasiten. -- pp. 1-516, Taf. I-XXXII [= 1-32]. St. Petersburg.

Mienis, H. K. 2006. De Siciliaanse kielnaaktslak *Milax nigricans* (Philippi, 1936) in Nederland (Gastropoda, Pulmonata, Milacidae). - Basteria Supplement 3: 87-88.

Mienis, H. K. 2008. Does *Rumina saharica* Pallary, 1901 occur in France? - MalaCo 5: 229-230.

Mildner, P. 1997. Die Turmdeckelschnecke *Cochlostoma (Turritus) tergestinum* (Westerlund, 1878) in Kärnten. - Carinthia II 187_107: 225-228. Klagenfurt.

Mildner, P. & Kofler, A. 1990. Zur Verbreitung von *Cylindrus obtusus* (Draparnaud, 1805) in Kärnten und Osttirol (Mollusca, Gastropoda: Helicidae). - Carinthia II 180_100 (2): 455-461. Klagenfurt.

Mildner, P. & Rathmayr, U. 1999. Rote Liste der Weichtiere Kärntens (Mollusca). - Naturschutz in Kärnten 15: 643-662. Klagenfurt.

Miller, J. S. 1822. A list of the freshwater and landshells occurring in the environment of Bristol, with observations. - Annals of Philisophy (New Series) 3 (17): 376-381. London.

Millet, P. A. 1813. Mollusques terrestres et fluviatiles, observés dans le Département de Maine et Loire. - pp. j.xj [= 1-11], 1-82, 1 tableau. Angers. (Pavie).

Millet, P. A. 1843. Description de plusieurs espèces nouvelles de mollusques de la France. - Magasin de Zoologie (2) 5: 1-4, Pl. 63-64. Paris.

Mirolli, M. 1960. Morfologia, biologia e posizione sistematica di *Watsonula wautieri*, n.g., n.s. (Basommatophora, Ancylidae). - Memorie dell'Istituto Italiano di Idrobiologia 12: 121-162.

Mittre, H. 1842. Description de quatre coquilles nouvelles. - Annales des Sciences Naturelles (Zoologie) (2) 18: 188-191. Paris.

Modell, H. 1945. Die Anodontinae, Ortm. emend. (Najad., Mollusca). Eine Studie über die Zusammenhänge von Klimazonen und Entwicklungsgeschichte. (Klimazonentheorie). - Jenaische Zeitschrift für Medizin und Naturwissenschaft 78 (1): 58-100.

Möllendorff, O. von 1873. Beiträge zur Fauna Bosniens. - pp. [1], 1-73, Taf. [1]. Görlitz. (Inauguraldissertation).

Möllendorff, O. von 1873. Zur Molluskenfauna von Serbien. - Malakozoologische Blätter 21: 129-149, Taf. IV [= 4].
Author's name is not spelled Moellendorff on the title page. Date 1873 is likely (there is an announce issued by Theodor Fischer dated Aug 1872, bound at the end of the SUB Göttingen volume).

Möllendorff, O. von 1899. Neue Landschnecken, mitgetheilt von Herrn H. Rolle. - Nachrichtsblatt der Deutschen Malakozoologischen Gesellschaft 31 (9/10): 152-158.
Author's name is not spelled Moellendorff on the title page.

Möllendorff, O. von 1899. Zwei neue Arten aus Montenegro, gesammelt von Otto Wohlberedt, Triebes. - Nachrichtsblatt der Deutschen Malakozoologischen Gesellschaft 31 (11/12): 169-170.
Author's name is not spelled Moellendorff on the title page.

Möller, H. P. C. 1842. Index molluscorum Groenlandiæ. - pp. 1-24. Hafniæ. (Salomon).

Montagu, G. 1803. Testacea Britannica, or natural history of British shells, marine, land, and fresh-water, including the most minute: systematically arranged and embellished with figures. - pp. i-xxxvii [= 1-37], [1-2], 1-606, [1-4], Pl. 1-16. London. (White).

Montagu, G. 1808. Supplement to Testacea Britannica. - pp. [1], i-v [= 1-5], 1-183, Pl. 17-30. London. (White).

Monterosato, T. A. di 1892. Molluschi terrestri delle isole adiacenti alla Sicilia. - Atti della Reale Accademia di Scienze, Lettere e Belle Arti di Palermo (3) 2: 1-33, [1].
On the title page of the volume is indicated "anno 1892" for vol. II, but in the footline is printed 1893 as the date of publication. It is possible that separates were issued earlier. Monterosato's style was somewhat chaotic, but the *Xero-* names on pp. 21-25 were established as subgenera of *Helix*, as becomes evident from p. 16 where *Xerophila* was treated as subgenus of *Helix* in *Helix (Xerophila) aegusae*, while the other specific names were combined with the subgenus.

Monterosato, T. A. di 1894. Conchiglie terrestri viventi e fossili di Monte Pellegrino. - Il Naturalista Siciliano 13 (9): 165-173. Palermo.
Date was June 1894.

Monterosato, T. A. di 1906. Articolo sulle Auriculidae, Assiminidae [sic] e Truncatellidae dei mari d'Europa. - Il Naturalista Siciliano 18: 125-130. Palermo.

Montfort, D. de 1810. Conchyliologie systématique, et classification méthodique des coquilles; offrant leurs figures, leur arrangement générique, leurs descriptions caractéristiques, leurs noms; ainsi que leur synonymie en plusieurs langues. Ouvrage destiné à faciliter l'étude des coquilles, ainsi que leur disposition dans les cabinets d'histoire naturelle. Coquilles univalves, non cloisonnées. Tome second. - pp. [1-3], 1-676. Paris. (Schœll).

Moorkens, E. & Killeen, I. 2004. A survey of the molluscan fauna of a mixed farm in South County Cork, 2004. - Irish Wildlife Manuals 18: 1-55. Dublin.

Moquin-Tandon, A. 1850. Note sur une nouvelle espèce de parmacelle (*Parmacella Gervaisii*), précédée de quelques considérations su ce genre de mollusques. - Mémoires de l'Académie des Sciences, Inscriptions et Belles Lettres de Toulouse (3) 6: 47-56.

Moquin-Tandon, A. 1853. Observations sur trois gastéropodes ovovivipares. - Journal de Conchyliologie 4: 225-227. Paris.

Moquin-Tandon, A. 1855. Histoire naturelle des mollusques terrestres et fluviatiles de la France contenant des études générales sur leur anatomie et leur physiologie et la description particulière des genres, des espèces et des variétés. Tome second. - pp. 1-646, atlas 1-92, Pl. I-LIV [= 1-54]. Paris. (Baillière).
Dates of publication following Falkner et al. 2002: 306, based on 2 sources (Letourneux & Bourguignat 1887 LB and Bibliographie de la France BF), Woodward 1903 WO and Museum Leiden copy ML communicated by E. Gittenberger and cited by Manganelli, Bo., Fa. et al. 1998: 151-152:
Livr. 4 pp. I-III, 1-176, atlas 49-64, Pl. 28-36 (LB 10 Sep 1855) (BF 19 Jan 1856) (WO 19 Jan 1856) (ML 10 Sep 1855)
Livr. 5 pp. 177-368, atlas 65-80, Pl. 37-45 (LB 02 Jan 1856) (BF 19 Jan 1856) (WO 19 Jan 1856) (ML 02 Jan 1856)
Livr. 6 pp. 369-646, atlas I-III, 81-92, Pl. 46-54 (LB 09-04-1856) (BF 26-04-1856) (WO 26 Apr 1856) (ML 09 Apr 1856)
The LB dates seem to be recognized for taxonomical purposes. Falkner et al. 2002 gave no reference to where Letourneux & Bourguignat 1887 published these dates, and did not cite the work in their list of references - I did not find anything in Prodrome malac. Tunisie 1887. Falkner et al. 2002 gave the BF dates in parentheses, suggesting that they preferred the LB dates, but on p. 161 they wrote 1856 for a name published in Moquin-Tandon p. 134.

Mörch, O. A. L. 1864. Fortegnelse over de i Danmark forekommende land- og ferskvandsblöddyr. - Videnskabelige Meddelelser fra den Naturhistoriske Forening i Kjöbenhavn (2) 1863 (17-22): 265-367.

Mordan, P. B. & Woodward, F. R. 1990. *Mytilus anatinus* Linnaeus, 1758 (currently *Anodonta anatina*; Mollusca, Bivalvia): proposed designation of a neotype. - Bulletin of Zoological Nomenclature 47 (2): 110-113.

Morelet, A. 1845. Description des mollusques terrestres et fluviatiles de Portugal. - pp. [1-3], I-VII [= 1-7], 1-116, Pl. I-XIV [= 1-14]. Paris. (Bailliere).

Morelet, A. 1853. Notice sur quelques hélices recueillies dans le Maroc, par M. Coquand. - Journal de Conchyliologie 4: 138-142, Pl. V [= 5]. Paris.

Morelet, A. 1854. Notice sur quelques hélices recueillies dans le midi de l'Espagne et au Maroc, par M. Tarnier. - Revue et Magasin de Zoologie pure et appliquée (2) 6: 614-623. Paris.

Moreno, D. & Ramos, M. A. 2007. New data on *Theba subdentata helicella* (Wood, 1828) (Gastropoda, Helicidae) in Almería (SE Spain). - Iberus 25 (1): 89-113.

Morse, E. S. 1864. Observations on the terrestrial Pulmonifera, including a catalogue of all species of terrestrial and fluviatile Mollusca known to inhabit the state. - Journal of the Portland Society of Natural History 1 (1): 1-63, Pl. I-II [= 1-2], 3-10.

Mortillet, G. 1854. Descriptions de quelques coquilles nouvelles d'Arménie, et considérations malacostatiques. - Mémoires de l'Institut National Genevois 2: 5-15, Tab. 1. Genève.

Morton, B., Velkovrh, F. & Sket, B. **1998**. Biology and anatomy of the 'living fossil' *Congeria kusceri* (Bivalvia: Dreissenidae) from subterranean rivers and caves in the Dinaric karst of the former Yugoslavia. - Journal of Zoology 245 (2): 147-174.

Morton, J. E. 1954. Notes on the ecology and annual cycle of *Carychium tridentatum* at Box Hill. - Proceedings of the Malacological Society of London 31 (2): 30-46.

Mouahid, A., Idaghdour, M., Ghamizi, M. & Mone, H. 1996. Observation of spawn in *Melanopsis praemorsa* (Prosobranchia: Melanopsidae). - Journal of Molluscan Studies 62 (3): 398-402.

Mousson, A. 1847. Bemerkungen ueber die natuerlichen Verhaeltnisse der Thermen von Aix in Savoien. - Neue Denkschriften der Allgemeinen Schweizerischen Gesellschaft für die Gesammten Naturwissenschaften / Nouveaux Mémoires de la Société Helvétique des Sciences Naturelles 8 (2): 1-48, Tab. I-III [= 1-3]. Neuchatel.

Mousson, A. 1854. Coquilles terrestres et fluviatiles, recueillies par M. le Prof. Bellardi dans un voyage en Orient. - Mittheilungen der Naturforschenden Gesellschaft in Zürich 3 (8): 362-402.

Mousson, A. 1854. Coquilles terrestres et fluviatiles recueillies par M. le Prof. Bellardi dans un voyage en Orient. - pp. 1-59, Pl. [1]. Zürich. (Zurcher & Furrer).
Partly identical with the other paper with the same title, which must have been published earlier.

Mousson, A. 1856. Verzeichniss der von Herrn Dr. Schäfli eingekommenen zweiten malacologischen Sendung. - Vierteljahrsschrift der Naturforschenden Gesellschaft in Zürich 1: 395-399.

Mousson, A. 1859. Coquilles terrestres et fluviatiles, recueillies dans l'Orient par M. le Dr. Alex. Schläfli. - Vierteljahrsschrift der Naturforschenden Gesellschaft in Zürich 4: 12-36, 253-297.

Mousson, A. 1863. Coquilles terrestres et fluviatiles, recueillies dans l'Orient par M. le Dr. Alex. Schläfli. [(Suite de la page. 297. 1859.)]. - Vierteljahrsschrift der Naturforschenden Gesellschaft in Zürich 8: 275-320, 368-426.

Mousson, A. 1872. Revision de la faune malacologique des îles Canaries. - Neue Denkschriften der Allgemeinen Schweizerischen Gesellschaft für die Gesammten Naturwissenschaften - Nouveaux Mémoires de la Société Helvétique des Sciences Naturelles 25: 1-176, Taf. I-VI [= 1-6]. Neuchatel.

Mousson, A. 1873. Coquilles recueillies par M. le Dr Sievers dans la Russie méridionale et asiatique. - Journal de Conchyliologie 21: 193-230, Pl. VII-VIII [= 7-8]. Paris.

Mouthon, J. 2007. *Lithoglyphus naticoides* (Pfeiffer) (Gastropoda: Prosobranchia): distribution in France, population dynamics and life cycle in the Saône river at Lyon (France). - Annales de Limnologie - International Journal of Limnology 43 (1): 53-59.

Mouthon, J. 2008. First study of the life cycle of *Pisidium tenuilineatum* Stelfox, 1918 (Bivalvia, Sphaeriidae). - Basteria 72 (1/3): 45-56.

Müller, **O. F.** 1773. Vermivm terrestrium et fluviatilium, seu animalium infusoriorum, helminthicorum et testaceorum, non marinorum, succincta historia. Voluminis Imi pars Ima. - pp. [1-33], 1-135. Havniæ & Lipsiæ. (Heineck & Faber).

Müller, O. F. 1774. Vermivm terrestrium et fluviatilium, seu animalium infusoriorum, helminthicorum, et testaceorum, non marinorum, succincta historia. Volumen alterum. - pp. I-XXVI [= 1-36], 1-214, [1-10]. Havniæ & Lipsiæ. (Heineck & Faber).

Müller, O. F. 1776. Zoologiæ Danicæ prodromus, seu animalium Daniæ et Norvegiæ indigenarum characteres, nomina, et synonyma imprimis popularium. - pp. I-XXXII [= 1-32], 1-274. Havniæ. (Hallager).

Müller, O. F. 1781. Geschichte der Perlen-Blasen. - Der Naturforscher 15: 1-20, Tab. I [= 1]. Halle.

Müller, **R.** 2008. Beitrag zum Vorkommen von *Marstoniopsis scholtzi* (A. Schmidt, 1856) (Hydrobiidae) in Berlin und Brandenburg. - Mollusca 26 (2): 169-174. Dresden.

Muñoz, B., Almodóvar, A. & Arrébola, J. R. 1999. *Ganula gadirana* n. sp., a new Hygromiidae from southern Spain (Pulmonata: Helicoidea). - Malacologia 41 (1): 163-174.

Muus, B. J. 1963. Some Danish Hydrobiidae with the description of a new species, *Hydrobia neglecta*. - Proceedings of the Malacological Society of London 35 (2): 131-138.

Mylonas, M. A. 1982. Meléti páno sti zoogeografía ke ikología ton herséon malakíon ton Kikládon. - pp. 1-236. Didaktorikí Diatriví (unpublished Ph. D. thesis). Athína.

Mylonas, M. 1985. The distribution and ecology of the species *Helix godetiana* Kobelt (Gastropoda: Pulmonata). - Biologia Gallo-hellenica 10 (1/2): 139-142.

Mylonas, M., Botsaris, J., Sourdis, J. & Valakos, E. 1995. On the development, habitat selection and taxonomy of *Helix (Jacosta) siphnica* Kobelt (Gastropoda, Helicellinae). - Zoological Journal of the Linnean Society 115: 347-357.

Myzyk, S. 2002. Life cycle of *Valvata cristata* O. F. Müller, 1774 (Gastropoda: Heterobranchia) in the laboratory. - Folia Malacologica 10 (2): 47-75.

Myzyk, S. 2004. Life cycle of *Valvata macrostoma* Mörch, 1864 (Gastropoda: Heterobranchia) in the laboratory. - Folia Malacologica 12 (3): 110-136.

Myzyk, S. 2005. Egg structure of some vertiginid species (Gastropoda: Pulmonata: Vertiginidae). - Folia Malacologica 13 (4): 169-175.

Myzyk, S. 2007. Life cycle of *Valvata piscinalis* O. F. Müller, 1774 (Gastropoda: Heterobranchia) in the laboratory. - Folia Malacologica 15 (4): 145-174.

Myzyk, S. 2008. Life cycle of *Anisus vorticulus* (Troschel, 1834) (Gastropoda: Pulmonata: Planorbidae) in the laboratory. - Folia Malacologica 16 (4): 207-215.

Myzyk, S. 2011. Contribution to the biology of ten vertiginid species. - Folia Malacologica 19 (2): 55-80.

Naegele, G. 1903. Einiges aus Vorderasien. - Nachrichtsblatt der Deutschen Malakozoologischen Gesellschaft 35 (11/12): 168-177.

Nagel, K.-O., Castagnolo, L., Cencetti, E. & Moro, G. A. 2007. Notes on reproduction, growth and habitat of *Microcondylaea bonellii* (Molluusca: Bivalvia: Unionidae) in the Torrente Versa (Italy). - Mollusca 25 (1): 41-49. Dresden.

Naggs, F. 1985. Some preliminary results of morphometric multivariate analysis of the *Trichia* (Pulmonata: Helicidae) species groups in Britain. - Journal of Natural History 19: 1217-1230.

Najim, A. T. 1961. Further notes on the distribution of molluscs in Iraq with the first record of *Physa*. - Proceedings of the Malacological Society of London 34 (6): 299-301.

Nardi, G. 2011. Nuove segnalazioni di limacce per la malacofauna bresciana (Gastropoda: Arionidae, Milacidae, Limacidae, Agriolimacidae). - Bollettino Malacologico 47 (1): 9-22.

Nardi, G. & Bodon, M. 2011. Una nuova specie di *Testacella* Lamarck, 1801, per l'Italia settentrionale (Gastropoda: Pulmonata: Testacellidae). - Bollettino Malacologico 47 (2): 150-164.

Negrea, A. & Riedel, A. 1968. Eine neue unterirdische Zonitiden-Art und -Gattung (Gastropoda) aus Rumänien. - Annales Zoologici 26 (5): 209-215. Warszawa.

Neubert, E. 1992. Descriptions of new taxa of the Clausiliidae from Turkey (Mollusca: Stylommatophora). - Zoology in the Middle East 7: 65-86. Heidelberg.

Neubert, E. 1998. Annotated checklist of the terrestrial and freshwater molluscs of the Arabian Peninsula with descriptions of new species. - Fauna of Arabia 17: 333-461.

Neubert, E. 1998. New data on the fauna of Clausiliidae of Greece, in particular on *Albinaria* from Attica and the Peloponnese (Gastropoda Pulmonata: Clausiliidae). - Basteria 62: 125-155.

Neubert, E. & Bank, R. A. 2006. Notes on the species of *Caucasotachea* C. Boettger 1909 and *Lindholmia* P. Hesse 1919, with annotations to the Helicidae (Gastropoda: Stylommatophora: Helicidae). - Archiv für Molluskenkunde 135 (1): 101-132.

Neubert, E., Örstan, A. & Welter-Schultes, F. **2000**. The land snails between Kaş and Demre, southwestern Turkey, with special reference to *Albinaria* (Gastropoda, Pulmonata, Clausiliidae). - Basteria 64: 105-123.

Neubert, E. & Riedel, A. 1993. Über einige *Vitrea*-Arten (Gastropoda, Zonitidae) aus der Türkei nebst Beschreibung einer neuen Art. - Basteria 57 (4/6): 155-160.

Nevill, G. 1879. Description of a new species of *Acme* and varieties from the conclomerate beds at Menton. - Annals and Magazine of Natural History (5) 4: 341-342.

Nevill, G. 1880. On the land-shells, extinct and living, of the neighbourhood of Menton (Alpes Maritimes); with descriptions of a new genus and of several new species. - Proceedings of the Zoological Society of London 1880: 94-142, Pl. XIII-XIV [= 13-14].

Niethammer, G. 1937. *Limax gerhardti* n. sp., eine neue Nacktschneckenart aus Italien. - Sitzungsberichte der Gesellschaft Naturforschender Freunde zu Berlin 1936: 443-444.
Date 30 Apr 1937 indicated on p. 470 of the volume.

Nilsson, S. 1822. Historia molluscorum Sveciae terrestrium et fluviatilium breviter delineata. - pp. [1-2], I-XX [= 1-20], 1-124. Lundæ. (Schuboth).
Title page indicated 1822, but must have appeared in early 1823 (Woodward 1924: 23).

Nitz, B., Falkner, G. & Haszprunar, G. **2010**. Inferring multiple Corsican *Limax* (Pulmonata: Limacidae) radiations: A combined approach using morphology and molecules. - pp. 405-435, in: Glaubrecht, M.: Evolution in action. Case studies in adaptive radiation, speciation and the origin of biodiversity. -- pp. i-xxv [= 1-26], 1-586. Heidelberg etc.. (Springer).

Nitz, B., Heim, R., Schneppat, U. E., Hyman, I. & Haszprunar, G. **2009**. Towards a new standard in slug species descriptions: the case of *Limax sarnensis* Heim & Nitz n. sp. (Pulmonata: Limacidae) from the Western Central Alps. - Journal of Molluscan Studies 75 (3): 279-294.

Nobre, A. 1913. Moluscos de Portugal. I. Moluscos terrestres, fluviais e das águas salobras. 2.° fascículo. - Memórias Publicadas pela Sociedade Portuguesa de Sciências Naturais 1: 129-343. Lisboa.

Nobre, A. 1941. Fauna malacológica de Portugal. II. Moluscos terrestres e fluviais. - Memórias e Estudos do Museu Zoológico da Universidade de Coimbra 9 (124): 1-279, Est.I-IV [= 1-4], 1-30.

Nordsieck, H. 1962. Die Chondrinen der Südalpen. - Archiv für Molluskenkunde 91 (1/3): 1-20, Taf. 1.

Nordsieck, H. 1966. Zur Anatomie und Systematik der Clausilien, III. *Clausilia whateliana* und ihre Beziehungen zu den übrigen *Clausilia*-Arten, besonders zum Subgenus *Neostyriaca*. - Archiv für Molluskenkunde 95 (1/2): 19-47.

Nordsieck, H. 1969. Die *Cochlodina*-Arten des westlichen Mittelmeerraumes. - Archiv für Molluskenkunde 99 (1/2): 21-25.

Nordsieck, H. 1969. Zur Anatomie und Systematik der Clausilien, V. Genitalsystem und Systematik des Genus *Cochlodina*. - Archiv für Molluskenkunde 99 (3/4): 107-132.

Nordsieck, H. 1969. Zur Anatomie und Systematik der Clausilien, VII. Dinarische Clausiliidae, I: Das Genus *Delima*. - Archiv für Molluskenkunde 99 (5/6): 267-284.

Nordsieck, H. 1970. Zur Anatomie und Systematik der Clausilien, VIII. Dinarische Clausiliidae, II: Das Genus *Medora*. - Archiv für Molluskenkunde 100 (1/2): 23-75, Taf. 1-6.

Nordsieck, H. 1971. Zur Anatomie und Systematik der Clausilien, IX. Dinarische Clausiliidae, III: Das Genus *Herilla*. - Archiv für Molluskenkunde 101 (1/4): 39-88, Taf. 1-5.

Nordsieck, H. 1972. Zur Anatomie und Systematik der Clausilien, XI. Neue Formen und taxonomische Revision einiger Gruppen der Alopiinae. - Archiv für Molluskenkunde 102 (1/3): 1-51, Taf. 1-5.

Nordsieck, H. 1972. Nomenklatorische und systematische Bemerkungen zu Clausiliidae. - Archiv für Molluskenkunde 102 (1/3): 53-55.

Nordsieck, H. 1973. Zur Anatomie und Systematik der Clausilien, XII. Phaedusinae, I: Phaedusen aus Nepal und ihre systematische Stellung innerhalb der Unterfamilie. - Archiv für Molluskenkunde 103 (1/3): 63-85.

Nordsieck, H. 1973. Zur Anatomie und Systematik der Clausilien, XIII. Neue Balkan-Formen der Mentissoideinae und Baleinae (mit taxonomischer Revision der zugehörigen Gruppen). - Archiv für Molluskenkunde 103 (4/6): 179-208, Taf. 6-7, 7a.

Nordsieck, H. 1974. Zur Anatomie und Systematik der Clausilien, XV. Neue Clausilien der Balkan-Halbinsel (mit taxonomischer Revision einiger Gruppen der Alopiinae und Baleinae). - Archiv für Molluskenkunde 104 (4/6): 123-170.

Nordsieck, H. 1977. Zur Anatomie und Systematik der Clausilien, XVIII. Neue Taxa rezenter Clausilien. - Archiv für Molluskenkunde 108 (1/3): 73-107.

Nordsieck, H. 1977. Zur Anatomie und Systematik der Clausilien, XVII. Taxonomische Revision des Genus *Albinaria* Vest. - Archiv für Molluskenkunde 107 (4/6): 285-307.

Nordsieck, H. 1979. Zur Anatomie und Systematik der Clausilien, XXI. Das System der Clausilien, II: Die rezenten europäischen Clausilien. - Archiv für Molluskenkunde 109 (4/6): 249-275.

Nordsieck, H. 1979. Revision des Genus *Alopia*, I. *Clausilia straminicollis* Charpentier und *C. livida* var. *maxima* A. Schmidt (Gastropoda: Clausiliidae). - Archiv für Molluskenkunde 110 (1/3): 53-61.

Nordsieck, H. 1984. Neue Taxa rezenter europäischer Clausilien, mit Bemerkungen zur Bastardierung bei Clausilien (Gastropoda: Clausiliidae). - Archiv für Molluskenkunde 114 (4/6): 189-211, Taf. 11-12.

Nordsieck, H. 1986. Das System der tertiären Helicoidea Mittel- und Westeuropas (Gastropoda: Stylommatophora). - Heldia 1 (4): 109-120, Taf. 15-17. München.

Nordsieck, H. 1988. Neue Clausilien aus NW-Griechenland. - Annalen des Naturhistorischen Museums in Wien 90B: 197-201, Taf. 1-2.

Nordsieck, H. 1989. *Falkneria* n. gen., eine neue Gattung der Helicodontinae (Gastropoda, Stylommatophora: Hygromiidae). - Heldia 1 (5/6): 165-168, Taf. 25. München.

Nordsieck, H. 1990. Revision der Gattung *Clausilia* Draparnaud, besonders der Arten in SW-Europa (das *Clausilia rugosa*-Problem) (Gastropoda: Stylommatophora: Clausiliidae). - Archiv für Molluskenkunde 119 (4/6): 133-179.

Nordsieck, H. 1993. Das System der paläarktischen Hygromiidae (Gastropoda: Stylommatophora: Helicoidea). - Archiv für Molluskenkunde 122: 1-23.

Nordsieck, H. 1993. Türkische Clausiliidae, I: Neue Arttaxa des Genus *Albinaria* Vest in Süd-Anatolien (Gastropoda: Stylommatophora). - Stuttgarter Beiträge zur Naturkunde (Serie A (Biologie)) 499: 1-31.

Nordsieck, H. 1993. Beiträge zur Nomenklatur der europäischen Binnenmollusken, I. Kritische Anmerkungen und Berichtigungen zur Nomenklatur von Arttaxa der Clausiliidae. - Heldia 2 (1/2): 33-42. München.

Nordsieck, H. 1994. Türkische Clausiliidae, II: Neue Taxa der Unterfamilien Serrulininae und Mentissoideinae in Anatolien (Gastropoda: Stylommatophora). - Stuttgarter Beiträge zur Naturkunde (A (Biologie)) 513: 1-36.

Nordsieck, H. 1996. Beiträge zur Nomenklatur der europäischen Binnenmollusken, VII. Kritische Anmerkungen und Berichtigungen von Arttaxa der Clausiliidae, 2. - Heldia 2 (3/4) ["1995"]: 91-96. München.
Date Dec 1995 is printed on title page of the articles and on the cover of the volume, but must have appeared at the beginning of 1996. On the backside cover was a statement that Heft 5/6 would appear in April 1996 (when it actually appeared it had the date March 1998 printed on it). The invoice sent out with volume 2 Heft 3/4 was dated 15 Feb 1996 (preserved in Haus der Natur Cismar library <Moll J 15>), this is accepted here as the most probable true date of publication.

Nordsieck, H. 1996. Neue Clausilien-Taxa aus Albanien (Gastropoda: Clausiliidae). - Schriften zur Malakozoologie 9: 7-16.

Nordsieck, H. 1999. Annotated check-list of the species of the *Albinaria-Isabellaria* group (Gastropoda: Stylommatophora: Clausiliidae). - Mitteilungen der Deutschen Malakozoologischen Gesellschaft 62/63: 1-21.

Nordsieck, H. 2002. Contributions to the knowledge of the Delimini (Gastropoda: Stylommatophora: Clausiliidae). - Mitteilungen der Deutschen Malakozoologischen Gesellschaft 67: 27-39.

Nordsieck, H. 2002. The subspecies classification of *Clausilia dubia* Draparnaud (Gastropoda: Stylommatophora: Clausiliidae), a critical revision. - Mitteilungen der Deutschen Malakozoologischen Gesellschaft 68: 37-44.

Nordsieck, H. 2002. *Albinaria* taxa (Gastropoda: Pulmonata: Clausiliidae) on Folegandros island (Cyclades, Greece), subspecies or species? - pp. 291-299, in: Falkner, M., Groh, K. & Speight, M. C. D.: Collectanea Malacologica. Festschrift für Gerhard Falkner.

Nordsieck, H. 2003. *Macrogastra mellae* (Stabile) und *M. badia* (C. Pfeiffer), zwei ungenügend bekannte *Macrogastra*-Arten (Gastropoda: Stylommatophora: Clausiliidae). - Mitteilungen der Deutschen Malakozoologischen Gesellschaft 69/70: 61-69.

Nordsieck, H. 2004. Türkische Clausiliidae, III: Neue Arttaxa der Unterfamilien Alopiinae und Mentissoideinae aus Anatolien (Gastropoda: Stylommatophora). - Stuttgarter Beiträge zur Naturkunde (A (Biologie)) 670: 1-28.

Nordsieck, H. 2005. Diversity of the genus *Cochlodina* Férussac in Slovenia and Croatia (Gastropoda: Stylommatophora: Clausiliidae). - Mitteilungen der Deutschen Malakozoologischen Gesellschaft 73/74: 35-40.

Nordsieck, H. 2006. Changes in the systematics and nomenclature of European Clausiliidae (Gastropoda: Stylommatophora). - Mitteilungen der Deutschen Malakozoologischen Gesellschaft 75: 9-13.

Nordsieck, H. 2006. Systematics of the genera *Macrogastra* Hartmann 1841 and *Julica* Nordsieck 1963, with the description of new taxa (Gastropoda: Stylommatophora: Clausiliidae). - Archiv für Molluskenkunde 135 (1): 49-71.

Nordsieck, H. 2007. Worldwide door snails (Clausiliidae), recent and fossil. - pp. 1-214. Hackenheim. (ConchBooks).

Nordsieck, H. 2007. *Cochlodina laminata* (Montagu 1803) - polytypic in genital morphology (Gastropoda: Stylommatophora: Clausiliidae). - Mitteilungen der Deutschen Malakozoologischen Gesellschaft 77/78: 23-25.

Nordsieck, H. 2007. *Balea* Gray 1824 and *Alinda* H. & A. Adams 1855 are separated as genera (Gastropoda: Stylommatophora: Clausiliidae). - Mitteilungen der Deutschen Malakozoologischen Gesellschaft 77/78: 27-30.

Nordsieck, H. 2008. The system of the genus *Alopia* H. & A. Adams 1855 (Gastropoda: Stylommatophora: Clausiliidae). - Mitteilungen der Deutschen Malakozoologischen Gesellschaft 79/80: 7-18.

Nordsieck, H. 2009. Revisory notes on the genus *Montenegrina* O. Boettger 1877 and description of new taxa (Gastropoda: Stylommatophoa: Clausiliidae). - Archiv für Molluskenkunde 138 (1): 71-87.

Nordsieck, H. 2011. Revision of the genus *Leucostigma* A. J. Wagner 1919 (Gastropoda: Stylommatophora: Clausiliidae). - Archiv für Molluskenkunde 140 (1): 123-147.

Nordsieck, H. 2011. Beschreibung einer neuen *Cochlostoma*-Art aus Italien, mit revisorischen Bemerkungen zu den *Cochlostoma*-Arten der Apenninen-Halbinsel (Gastropoda, Caenogastropoda, Architaenioglossa, Cochlostomatidae). - Club Conchylia Informationen 41 (3/4): 13-21.

Nordsieck, H. 2011. Murellen von Mittel- und Süditalien (Gastropoda, Stylommatophora, Helicidae) bestimmen, eine schwierige Aufgabe. - Club Conchylia Informationen 41 (3/4): 22-32.

Normand, N. A. J. 1844. Normand, N.-A.-J. 1844. Notice sur plusieurs nouvelles espèces de cyclades découvertes dans les environs de Valenciennes. - pp. 1-8, pl. [1]. Valenciennes.

Normand, N. A. J. 1852. Description de six limaces nouvelles observées aux environs de Valenciennes. - pp. 1-8. Valenciennes.

Norris, A., Paul, C. R. C. & Riedel, A. 1988. The genus *Vitrea* in Ibiza. - Journal of Conchology 33: 43-48.

Oke, O. C. & Alohan, F. I. 2006. The land snail diversity in a square kilometre of tropical rainforest in Okomu National Park, Edo State, Nigeria. - African Scientist 7 (3): 135-142.

Oke, O. C. & Chokor, J. U 2009. The effect of land use on snail species richness and diversity in the tropical rainforest of south-western Nigeria. - African Scientist 10 (2): 95-108.

Oke, O. C. & Chokor, U. J. 2010. Land snail species richness and diversity in Idanre hills, Ondo State, Nigeria. - African Journal of Ecology 48 (4): 1004-1008.

Økland, F. 1925. Die Verbreitung der Landgastropoden Norwegens. - Skrifter, Norske Videnskaps-Akademi i Oslo, Matematisk-Naturvidenskapelig Klasse 8: I-VIII [= 1-8], 1-168, Taf. I-XII [= 1-12], Karten [1-2].

Økland, J. 1990. Lakes and snails. Environment and Gastropoda in 1,500 Norwegian lakes, ponds and rivers. - pp. 1-516. Oegstgeest. (Backhuys).

Olivier, G. A. 1801. Voyage dans l'Empire Othoman, l'Égypte et la Perse, fait par ordre du gouvernement, pendant les six premières années de la République. Tome premier. - pp. j-xij [= 1-14], [1-2], 1-432, [1], i-vij [= 1-7], Pl. 1-17. Paris. (Agasse).
An 9 given on title page.

Olivier, G. A. 1804. Voyage dans l'Empire Othoman, l'Égypte et la Perse, fait par ordre du gouvernement, pendant les six premières années de la République. Tome second. - pp. [1-2], j-ij [= 1-2], 1-466, [1], [1-7], Pl. 18-33. Paris. (Agasse).
"An 12 - 1804" printed on title page.

Ondina, P., Hermida, J. & Outeiro, A. **1997**. Nuevos datos sobre la distribución de la superfamilia Helicoidea Rafinesque, 1815 (Gastropoda, Pulmonata, Stylommatophora) en el oeste de Galicia. - Iberus 15 (1): 5-22.

Ondina, P., Rodríguez, T. & Hermida, J. **1994**. Sobre la presencia de *Deroceras (Agriolimax) ercinae* De Winter, 1985 (Gastropoda: Pulmonata) en la Península Ibérica. - Iberus 12 (1): 63-65.

Örstan, A. 2001. A preliminary survey of *Albinaria* populations around Kuşadası Bay, Turkey. - Triton 4: 42-44.

Örstan, A. 2003. The rediscovery of *Zonites algirus* in İstanbul, Turkey (Gastropoda: Pulmonata: Zonitidae). - Zoology in the Middle East 29: 75-78. Heidelberg.

Örstan, A. 2003. The first record of *Discus rotundatus* from Turkey. - Triton 7: 27.

Örstan, A. 2004. Possible human-assisted dispersal of *Levantina spiriplana* (Pulmonata: Helicidae) in western Turkey. - Triton 9: 27-28.

Örstan, A. 2009. An observation of the mating of *Chondrus tournefortianus* (Pulmonata: Enidae). - Zoology in the Middle East 48: 117-118.

Örstan, A. 2010. A new record of *Vertigo pusilla* from Turkey. - Triton 21: 27.

Örstan, A., Pearce, T. A. & Welter-Schultes, F. **2005**. Land snail diversity in a threatened limestone district near Istanbul, Turkey. - Animal Biodiversity and Conservation 28 (2): 181-188.

Örstan, A. & Yıldırım, M. Z. 2003. Conchological and genitalic comparisons of *Jaminia loewii* populations in Turkey (Gastropoda: Pulmonata: Enidae). - Zoology in the Middle East 28: 67-76.

Örstan, A. & Yıldırım, M. Z. 2004. *Borlumastus* gen. nov. (Gastropoda, Pulmonata, Enidae), a new land snail genus from Turkey. - Basteria 68 (4/6): 125-129.

Örstan, A., Yıldırım, M. Z., Gümüş, B. A. & Welter-Schultes, F. W. **2005**. The land snails of the Bodrum peninsula, Turkey. - Mitteilungen der Deutschen Malakozoologischen Gesellschaft 73/74: 1-15.

Ortiz de Zárate y López, A. 1946. Observaciones anatómicas y posición sistemática de varios helícidos españoles. II (Moluscos recogidos en Cataluña). - Boletín de la Real Sociedad Española de Historia Natural (Biología) 44: 337-356. Madrid.

Ortiz de Zárate y López, A. 1949. Observaciones sobre la especie conocida con el nombre de *Helix brigantina* Mengo en el Museo de Madrid. - Anales de la Asociación Española para el Progreso de las Ciencias 16 (2): 279-292.
Falkner 1996: 103 argued that the date must be 1951.

Ortiz de Zárate y López, A. 1950. Observaciones anatómicas y posición sistemática de varios helícidos españoles. III. - Boletín de la Real Sociedad Española de Historia Natural (Biología) 48: 21-85, Lám. I-II [= 1-2]. Madrid.

Ortiz de Zárate, A. 1956. Observaciones anatómicas y posición sistemática de varios helícidos españoles. IV. (Género *Pyrenaearia* Hesse 1921). - Boletín de la Real Sociedad Española de Historia Natural (Biología) 54: 35-61. Madrid.

Ortiz de Zárate López, A. 1962. Una especie nueva de *Helicella (Helicella (Xeroplexa) cobosi)*. - Archivos del Instituto de Aclimatación 11: 41-43, Lám. 1.

Ortiz de Zárate López, A. 1962. Observaciones anatómicas y posición sistemática de varios helícidos españoles. V. Género *Oestophora* Hesse, 1907. - Boletín de la Real Sociedad Española de Historia Natural (Biología) 60: 81-104. Madrid.

Ortiz de Zárate, A. 1963. Observaciones anatómicas y posición sistemática de varios helícidos españoles. VI. Anotaciones sobre las especies españolas del género *Helicella*, subgénero *Xeroplexa* (Monterosato) Hesse 1926. - Boletín de la Sociedad de Historia Natural de Baleares 9: 93-100.

Ortiz de Zárate Rocandio, A. & Ortiz de Zárate López, A. 1961. Moluscos terrestres recogidos en la provincia de Huelva. - Boletín de la Real Sociedad Española de Historia Natural (Biología) 59: 169-190. Madrid.

Ošanova, N. & Pintér, L. 1968. Über bulgarische Vitrinidae (Gastropoda, Euthyneura). - Malakologische Abhandlungen 2 (18): 243-247.

Paasch, A. 1842. Beschreibung einer neuen bei Berlin gefundenen *Paludina*. - Archiv für Naturgeschichte 8 (1): 300-301, Taf. 6. Berlin.

Paget, O. E. 1971. Zwei neue Arten der Gattung *Albinaria* Vest 1867, (Moll., Clausiliidae), sowie Neunachweise für die Insel Rhodos. - Annalen des Naturhistorischen Museums in Wien 75: 461-468, Taf. 1-2.

Paget, O. E. 1976. Die Molluskenfauna der Insel Rhodos. 1. Teil. In der Serie: Ergebnisse der von Dr. O. Paget und Dr. E. Kritscher auf Rhodos durchgeführten zoologischen Sammelreisen. - Annalen des Naturhistorischen Museums in Wien 80: 681-780, Taf. 1-7.

Paladilhe, A. 1866. Nouvelles miscellanées malacologiques. - Revue et Magasin de Zoologie (2) 18: 54-70, 89-99, 168-174, Pl. 13. Paris.

Paladilhe, A. 1868. Nouvelles miscellanées malacologiques. Monographie du genre *Acme*. - Revue et Magasin de Zoologie et de Séricіculture Comparée (2) 20: 225-245, 273-283, 321-330, Pl. 13. Paris.

Paladilhe, A. 1874. Monographie du nouveau genre *Peringia* suivie de descriptions d'espèces nouvelles de paludinées françaises. - Annales de Sciences Naturelles (Série Zoologique) 1 (2): 1-38, Pl. 3. Paris.

Páll-Gergely, B. 2009. Revision of the Turkish *Ramusculus* taxa with description of *Ayna* gen. nov. (Gastropoda: Pulmonata: Enidae). - Journal of Conchology 40 (1): 73-77. London.

Páll-Gergely, B. 2010. New and little known land snails from Turkey (Gastropoda: Pulmonata). - Zoology in the Middle East 50: 89-94.

Páll-Gergely, B. 2010. *Montenegrina dofleini sinosi* n. ssp. from north-western Greece (Gastropoda: Pulmonata: Clausiliidae). - Genus 21 (1): 149-153. Warszawa.

Páll-Gergely, B. & Roibu, C.-C. 2011. Land snail fauna of the Humosu Secular forest (Suceava county, Romania); a new locality of *Serrulina serrulata* (L. Pfeiffer, 1847) (Gastropoda: Pulmonata: Clausiliidae). - Malacologica Bohemoslovaca 10: 48-50.

Pallary, P. 1898. Deuxième contribution à l'étude de la faune malacologique du nord-ouest de l'Afrique. Supplément à "La Faune Malacologique du Maroc" de A. Morelet. - Journal de Conchyliologie 46: 49-170, Pl. V-IX [= 5-9]. Paris.
True date was 11 Feb 1899 (Winckworth 1936).

Pallary, P. 1900. Coquilles marines du littoral du département d'Oran. - Journal de Conchyliologie 48: 211-422, Pl. VI-VII [= 6-8]. Paris.

Pallary, P. 1901. Sur les mollusques fossiles terrestres, fluviatiles et saumâtres de l'Algérie. - Mémoires de la Société Géologique de France 9 (22): 1-213, Pl. III-VI [= 3-6]. Paris.

Pallas, P. S. 1771. Reise durch verschiedene Provinzen des Rußischen Reichs. Erster Theil. - pp. [1-12], 1-504, Tab. I-XI [= 1-11], Tab. A-L. St. Petersburg.

Parejo, C. & Martín, R. 1990. *Arion wiktori* sp. n. from the Iberian Peninsula (Gastropoda, Pulmonata: Arionidae). - Malakologische Abhandlungen 15 (3): 25-35.

Parmakelis, A., Spanos, E., Papagiannakis, G., Louis, C. & Mylonas, M. 2003. Mitochondrial DNA phylogeny and morphological diversity in the genus *Mastus* (Beck, 1837): a study in a recent (Holocene) island group (Koufonisi, south-east Crete). - Biological Journal of the Linnean Society 78: 383-399.

Parmakelis, A., Pfenninger, M., Spanos, L., Papagiannakis, G., Louis, C. & Mylonas, M. 2005. Inference of a radiation in *Mastus* (Gastropoda, Pulmonata, Enidae) on the island of Crete. - Evolution 59: 991-1005.

Partsch, P. 1835. Über die sogenannten versteinerten Ziegenklauen aus dem Plattensee in Ungarn, und ein neues, urweltliches Geschlecht zweischaliger Conchylien. - Annalen des Wiener Museums der Naturgeschichte 1: 93-102, Taf. 11-12.

Paulucci, M. 1878. Matériaux pour servir à l'étude de la faune malacologique terrestre et fluviatile de l'Italie et de ses îles. - pp. [1], I-IV [= 1-4], 1-54. Paris. (Savy).

Paulucci, M. 1879. Communicazioni malacologiche. Articolo secondo. Descrizione di alcune nuove specie del genere Pomatias. - Bullettino della Società Malacologica Italiana 5 (1/3): 13-21. Pisa.
Date was Feb 1879 (Paulucci 1879 [Escursione]: 188). Manganelli & Ci. 2001: 345 gave 23 Apr 1879.

Paulucci, M. 1879. Escursione scientifica nella Calabria 1877-78. Fauna malacologica. Specie terrestri e fluviatili. - pp. I-XIX [= 1-19], 1-223, Tav. I-IX [= 1-9]. Firenze. (Arte della Stampa).
1879 was printed on the title page, March 1879 on p. xix (last page of preface). 1880 was indicated on an additional blue wrapper of the SBPK Berlin copy, with names of 3 publishers (Bocca, Friedländer and Savy), possibly a second edition or reprint. The Smithsonian copy does not contain this wrapper. I see currently no evidence that the true date was 1880 and not 1879.

Paulucci, M. 1879. Etude critique sur l'*Helix Balmei*, Potiez & Michaud. - Journal de Conchyliologie 27: 6-15. Paris.

Paulucci, M. 1881. Contribuzioni alla fauna malacologica italiana. Specie raccolte dal Dr. G. Cavanna negli anni 1878, 1879, 1880 con elenco delle conchiglie abruzzesi e descrizione di due nuove *Succinea*. - Bullettino della Società Malacologica Italiana 7 (5/12): 69-180, Tav. I-V [= 1-5]. Pisa.
Date was 01 Oct 1881 (Manganelli & Ci. 2001: 345, Manganelli, Lo. et al. 2008: 33).

Paulucci, M. 1882. Note malacologiche sulla fauna terrestre e fluviale dell'isola di Sardegna. - Bullettino della Società Malacologica Italiana 8 (7/16): 139-381, Tav. I-IX [= 1-9]. Pisa.
Date was 01 Oct 1882 (Manganelli & Cianfanelli 2001: 345, Manganelli, Lo. et al. 2008: 33).

Paulucci, M. 1886. Fauna Italiana. Comunicazioni malacologiche. Articolo nono. Conchiglie terrestri e d'acqua dolce del Monte Argentaro e delle isole circostanti. - Bullettino della Società Malacologica Italiana 12: 5-64, Tav. I-II [= 1-2]. Pisa.
Date was 30 Jul 1886 (Manganelli & Cianfanelli 2001: 345, Manganelli, Lo. et al. 2008: 35), also according to the orange wrapper for fogli 1-4 preserved in the SPBK Berlin <Lq 121> copy.

Pavlović, P. S. 1912. Mekushtsi iz Srbije. I. Sukhozemni puzhevi. - pp. [1], 1-140, Tab. I-II [= 1-2], 1 map. Beograd.

Pawłowska-Banasiak, E. 2008. Shell, genital and colour variation in *Perforatella incarnata* (O. F. Müller, 1774) and *P. vicina* (Rossmässler, 1842) (Gastropoda: Pulmonata: Helicidae). - Folia Malacologica 16 (2): 89-99.

Payraudeau, B.-C. 1826. Catalogue descriptif et méthodique des annelides et des mollusques de l'Ile de Corse; avec huit planches représentant quatre-vingt-huit espèces, dont soixante-huit nouvelles. - pp. [1-3], 1-218, Pl. 1-8. Paris, Genève. (Béchet, Levrault, Paschoud, Treuttel & Wurtz).
Dates (Falkner et al. 2002: 315): pp. 1-7: 12-03-1827; main text: 02/1827, names of taxa are to be cited with year 1827.

Pennant, T. 1777. British Zoology. Vol. IV. Crustacea. Mollusca. Testacea. - pp. [1-3], iii-viii [= 3-8], [1-10], 1-154, [1-2], Pl. I-XCII [= 1-93]. London. (White).

Pérez-Quintero, J. C., Bech Taberner, M. & Huertas Dionisio, L. 2004. Los moluscos de las aguas continentales de la provincia de Huelva (SO España). - Iberus 22 (2): 19-31.

Perrin, N., Honsberger, P. & Pontet, A. 1984. Approche biochimique et morphologique des espèces du genre *Trichia* (Helicidae, Gastropoda) de Suisse occidentale. - Revue Suisse de Zoologie 91 (2): 483-495.

Pesqueira, R., Rodríguez, T. & Outeiro, A. 2003. La familia Agriolimacidae Wagner, 1935 (Gastropoda Pulmonata) en la provincia de Lugo (noroeste de la península Ibérica). - Bollettino Malacologico 39 (1/4): 43-48.

Petit, R. E. 2006. Notes on Sowerby's The genera of recent and fossil shells (1821-1834). - Archives of Natural History 33 (1): 71-89.

Pfeffer, G. 1930. Die Unterfamilie Zonitidae (Moll., Pulm.). - Mitteilungen des Zoologischen Museums Berlin 16: 411-508, Taf. IV-VI [= 4-6].

Pfeffer, G. 1930. Zur Kenntnis tertiärer Landschnecken. - Geologische und Palaeontologische Abhandlungen (Neue Folge) 17 (3) ["1929"]: 1-230 (= 153-380), Taf. I-III [= 1-3], (= XV-XVII [= 15-17]). Jena.
Date 02 Jan 1930, indicated on the contents page of the volume. Two paginations run parallel (headline and footline).

Pfeiffer, **C.** 1821. Naturgeschichte deutscher Land- und Süsswasser-Mollusken. Erste Abtheilung. - pp. I-X [= 1-10], 1-134, [1-2], Taf. I-VIII [= 1-8]. Weimar. (Landes-Industrie-Comptoir).

Pfeiffer, C. 1825. Naturgeschichte deutscher Land- und Süsswasser-Mollusken. Zweite Abtheilung. - pp. I-VIII [= 1-8], 1-40, Taf. I-VIII [= 1-8]. Weimar. (Landes-Industrie-Comptoir).

Pfeiffer, **C.** 1828. Naturgeschichte deutscher Land- und Süsswasser-Mollusken. Dritte Abtheilung. - pp. I-VI [= 1-6], 1-84, Taf. I-VIII [= 1-8]. Weimar. (Landes-Industrie-Comptoir).

Pfeiffer, **C.** 1894. Kleine Reiseergebnisse. - Nachrichtsblatt der Deutschen Malakozoologischen Gesellschaft 26: 68-71.

Pfeiffer, **K. L.** & Wächtler, W. 1939. Über *Codringtonia* und *Isaurica*. - Archiv für Molluskenkunde 71 (2/3): 57-74.

Pfeiffer, **L.** 1841. Beiträge zur Molluskenfauna Deutschlands, insbesondere der österreichischen Staaten. - Archiv für Naturgeschichte 7 (1 (4/5)): 215-230. Berlin.
Possibly published in the first half of the year, around April or May.

Pfeiffer, L. 1841. Symbolae ad historiam heliceorum. - pp. 1-88. Cassellis. (Th. Fischer).

Pfeiffer, L. 1842. Symbolae ad historiam heliceorum. Sectio altera. - pp. 1-147. Cassellis. (Th. Fischer).

Pfeiffer, L. 1845. Descriptions of twenty-two new species of land-shells, belonging to the collection of Mr. H. Cuming. - Proceedings of the Zoological Society of London 13: 63-68.

Pfeiffer, L. 1846. Die gedeckelten Lungenschnecken. (Helicinacea et Cyclostomacea.). In Abbildungen nach der Natur mit Beschreibungen. - Systematisches Conchylien-Cabinet von Martini und Chemnitz 1 (19 (1)): I-IV [= 1-4], 1-228, Taf. A, 1-30. Nürnberg.
Dates (Welter-Schultes 1999): Taf. 4 (1843); pp. 1-24, Taf. 1-3, 5-7 (1846); pp. 25-56, Taf. 8-19 (1847); pp. 57-96, Taf. 20-25 (1848); pp. 97-208, Taf. 26-30 (1849); pp. 209-228 (1850); pp. I-IV (1849?); Taf. A unknown.

Pfeiffer, L. 1846. Symbolae ad historiam heliceorum. Sectio tertia. - pp. 1-100. Cassellis. (Th. Fischer).

Pfeiffer, L. 1846. Die Schnirkelschnecken (Gattung *Helix*). In Abbildungen nach der Natur mit Beschreibungen. - Systematisches Conchylien-Cabinet von Martini und Chemnitz 1 (12 (1)): 1-400, Taf. 1-66. Nürnberg.
Dates (Welter-Schultes 1999): pp 1-32, Taf. 1-4 (1840?), pp 33-48, Taf. 8, 12, 20 (1841), Taf. 21 (1843), Taf. 7 (1845), pp V-XV, 49-56, Taf. 6, 9, 10, 11, 13, 14 (1846), pp 57-120, Taf. 5, 15-19, 22-29, 31, 33, 35, 36 (1847), pp 121-200, Taf. 30, 32, 34, 37-51 (1848), pp 201-288, Taf. 52-57, 59, 61-66 (1849), pp 289-400, Taf. 58, 60, 12* (1850).

Pfeiffer, L. 1847. Descriptions of 38 new species of land-shells, in the collection of Hugh Cuming, Esq. - Proceedings of the Zoological Society of London 14 ["1846"]: 109-116.

Pfeiffer, L. 1847. Diagnosen neuer Heliceen. - Zeitschrift für Malakozoologie 4 (1): 12-16.
Date Jan 1847 given on p. 1.

Pfeiffer, L. 1847. Diagnosen neuer Heliceen. - Zeitschrift für Malakozoologie 4 (5): 65-71.
Date May 1847 given on p. 65.

Pfeiffer, L. 1847. Uebersicht aller bekannten Arten von Cyclostomaceen. - Zeitschrift für Malakozoologie 4 (7): 101-112.
Date July 1847 given on p. 97.

Pfeiffer, L. 1847. Diagnosen neuer von Frivaldsky gesammelter Landschnecken. - Zeitschrift für Malakozoologie 4 (12): 191-192.
Date Dec 1847 given on p. 177.

Pfeiffer, L. 1848. Diagnosen neuer von Frivaldsky gesammelter Landschnecken. (Schluss). - Zeitschrift für Malakozoologie 5 (1): 6-12.
Date Jan 1848 given on p. 1.

Pfeiffer, L. 1848. Monographia heliceorum viventium. Sistens descriptiones systematicas et criticas omnium huius familiae generum et specierum hodie cognitarum. Volumen primum. - pp. I-XXXII [= 1-32], 1-484. Lipsiae. (Brockhaus).
Helix etc.
Dates (Gargominy et al. 2011: 362): pp. 1-160 (before 27 Sep 1847), 161-320 (between 27 Sep and 31 Dec 1847), 321-484, I-XXXI (1848).

Pfeiffer, L. 1848. Monographia heliceorum viventium. Sistens descriptiones systematicas et criticas omnium huius familiae generum et specierum hodie cognitarum. Volumen secundum. - pp. [1-2], 1-594. Lipsiae. (Brockhaus).
Bulimus, Clausilia etc.

Pfeiffer, L. 1849. Nachträge zu L. Pfeiffer Monographia Helicearum [sic!] viventium. (Fortsetzung). - Zeitschrift für Malakozoologie 5 (8) ["1848"]: 113-123.
Date March 1849 given on p. 128.

Pfeiffer, L. 1850. Nachträge zu L. Pfeiffer Monographia Heliceorum. - Zeitschrift für Malakozoologie 6 (7) ["1849"]: 106-112.
Date should be prior to March 1850, but clearly after Nov 1849 (indicated on p. 96).

Pfeiffer, L. 1850. Beschreibungen neuer Landschnecken. - Zeitschrift für Malakozoologie 7 (5): 65-80.
Date Jul 1850 given on p. 80.

Pfeiffer, L. 1850. Beschreibungen neuer Landschnecken. - Zeitschrift für Malakozoologie 7 (6): 81-89.
Date Aug 1850 given on p. 96.

Pfeiffer, L. 1850. Descriptions of thirty new species of *Tornatellina*, *Cylindrella*, and *Clausilia*, from the collection of H. Cuming, Esq. - Proceedings of the Zoological Society of London 17 ["1849"]: 134-141.

Pfeiffer, L. 1851. Description of fifty-four new species of Helicea, from the collection of Hugh Cuming, Esq. - Proceedings of the Zoological Society of London 19: 252-263.

Pfeiffer, L. 1851. Beschreibung neuer Landschnecken. - Zeitschrift für Malakozoologie 8 (1): 25-29.
Date Jun 1851 given on p. 32.

Pfeiffer, L. 1851. Neue Heliceen. - Zeitschrift für Malakozoologie 8 (8): 127-128.
Date after Sep 1851 (given on p. 112) and before 15 Jan 1852 (given on p. 192), most probably Oct 1851.

Pfeiffer, L. 1852. Monographia pneumonopomorum viventium. Sistens descriptiones systematicas et criticas omnium hujus ordinis generum et specierum hodie cognitarum, accedente fossilium enumeratione. - pp. 1-18, 1-439. Cassellis. (Th. Fischer).

Pfeiffer, L. 1853. Diagnosen neuer Heliceen. - Zeitschrift für Malakozoologie 10 (10): 145-151.
Date Nov 1853 given on p. 160. Name of author was not mentioned in the article, but given as L. Pfeiffer in the table of contents issued in Dec 1853.

Pfeiffer, L. 1853. Notiz über serbische und sibirische Schnecken. - Zeitschrift für Malakozoologie 10 (12): 185-189.
Date Dec 1853 given on p. 192.

Pfeiffer, L. 1853. Descriptions of sixty-six new land shells, from the collections of H. Cuming, Esq. - Proceedings of the Zoological Society of London 20 ["1852"]: 56-70.

Pfeiffer, L. 1853. Monographia heliceorum viventium. Sistens descriptiones systematicas et criticas omnium huius familiae generum et specierum hodie cognitarum. Volumen tertium. - pp. I-VIII [= 1-8], 1-711. Lipsiae. (Brockhaus).
Helix, Bulimus, Clausilia etc. Published after May 1853 (Cowie 1998: 106).

Pfeiffer, L. 1854. Nachschrift zu S. 251. - Malakozoologische Blätter 1: 264.
Date Dec 1854 given on p. 257.

Pfeiffer, L. 1854. Descriptions of fifty-seven new species of Helicea, from Mr. Cuming's collection. - Proceedings of the Zoological Society of London 22: 286-298.

Pfeiffer, L. 1855-1856. Versuch einer Anordnung der Heliceen nach natürlichen Gruppen. - Malakozoologische Blätter 2: 112-185.
Date Aug 1855 given on p. 97 (covering pp. 97-112), Nov 1855 on p. 113, Dec 1855 on p. 129, Jan 1856 on pp. 145, 161 and 177.

Pfeiffer, L. 1856. Diagnosen neuer Landschnecken. - Malakozoologische Blätter 3: 43-52.
Dates Apr 1856 for pp. 43-48, Jul 1856 for pp. 49-52.

Pfeiffer, L. 1856. Bericht über weitere Mittheilungen des Herrn Zelebor. - Malakozoologische Blätter 3: 175-186, Taf. 2.
Date Oct 1856 given on pp. 161 and 177.

Pfeiffer, L. 1857. Literatur. - Malakozoologische Blätter 4: 158-167.
Date 21 Jul 1857 given on p. 145.

Pfeiffer, L. 1859. Monographia heliceorum viventium. Sistens descriptiones systematicas et criticas omnium huius familiae generum et specierum hodie cognitarum. Volumen quartum. - pp. I-IX [= 1-9], 1-920. Lipsiae. (Brockhaus).
Helix, Bulimus, Clausilia etc.

Pfeiffer, L. 1861. Diagnosen neu entdeckter Landschnecken. - Malakozoologische Blätter 8: 167-174.
Date Nov 1861 given on p. 161.

Pfeiffer, L. 1862. Diagnosen einiger neuen [sic] Landschnecken. - Malakozoologische Blätter 9: 202-205.
Date Nov 1862 given on p. 193.

Pfeiffer, L. 1863. Beschreibung eines neuen *Pomatias*. - Malakozoologische Blätter 10: 136-137.
Date Aug 1863 given on p. 129.

Pfeiffer, L. 1864. Descriptions of 7 new species of land shells from the collection of H. Cuming, Esq. - Proceedings of the Zoological Society of London 32: 603-605.

Pfeiffer, L. 1865. Die Mollusken der Dobrudscha. - Malakozoologische Blätter 12: 100-105.
Date May 1865 given on p. 97.

Pfeiffer, L. 1866. Beschreibung neuer Clausilien. - Malakozoologische Blätter 13: 146-154.
Date Dec 1866 given on p. 145.

Pfeiffer, L. 1868. Diagnosen neuer Heliceen. - Malakozoologische Blätter 15: 83-85.
Date Aug 1868 given on p. 81.

Pfeiffer, L. 1868. Monographia heliceorum viventium. Sistens descriptiones systematicas et criticas omnium huius familiae generum et specierum hodie cognitarum. Volumen quintum. - pp. I-XII [= 1-12], 1-565. Lipsiae. (Brockhaus).
Helix etc.

Pfeiffer, L. 1868. Monographia heliceorum viventium. Sistens descriptiones systematicas et criticas omnium huius familiae generum et specierum hodie cognitarum. Volumen sextum. - pp. [1-2], 1-598. Lipsiae. (Brockhaus).
Bulimus, Clausilia etc.

Pfeiffer, L. 1870-1876. Novitates Conchologicae. Series prima. Mollusca extramarina. Descriptions et figures de coquilles extramarines nouvelles et peu connues. Tome IV. Avec 29 planches colorées. - pp. [1], 1-171, Taf. CIX-CXXXVII [= 109-137]. Cassel. (Fischer).
Diagnoses in Latin, texts in German and French.

Pfeiffer, L. 1871. Diagnosen neuer Landschnecken. - Malakozoologische Blätter 17 (3) ["1870"]: 141-145.
Date Jan 1871 was quoted on p. 129 of the volume. Date 1870 given by Bürk & Jungbluth 1985: 186, but this referred to a separate article on pp. 93-94 which was published in July 1870.

Pfeiffer, L. 1876. Monographia pneumonopomorum viventium, accedente fossilium enumeratione. Supplementum tertium, monographiae auriculaceorum parte secunda auctum. - pp. 1-479. Cassellis. (Th. Fischer).

Pfeiffer, L. 1876. Monographia heliceorum viventium. Sistens descriptiones systematicas et criticas omnium huius familiae generum et specierum hodie cognitarum. Volumen septimum. - pp. I-X [= 1-10], 1-674. Lipsiae. (Brockhaus).
Helix etc.

Pfeiffer, L. 1877. Ueber die systematische Anordnung der Helicaceen. - Malakozoologische Blätter 24: 1-14.

Pfeiffer, **L.** 1877. Monographia heliceorum viventium. Sistens descriptiones systematicas et criticas omnium huius familiae generum et specierum hodie cognitarum. Volumen octavum. - pp. [1-2], 1-729. Lipsiae. (Brockhaus).
Bulimus, Clausilia etc.

Pfeiffer, L. & Clessin, S. 1881. Nomenclator heliceorum viventium quo continetur nomina omnium hujus familiae generum et specierum hodie cognitarum, disposita ex affinitate naturali. - pp. 1-617. Cassellis. (Th. Fischer).

Pfenninger, M., Cordellier, M & Streit, B. **2006**. Comparing the efficacy of morphologic and DNA-based taxonomy in the freshwater gastropod genus *Radix* (Basommatophora, Pulmonata). - BMC Evolutionary Biology 6 (100): 1-14.

Pfenninger, M., Eppenstein, A. & Magnin, F. **2003**. Evidence for ecological speciation in the sister species *Candidula unifasciata* (Poiret, 1801) and *C. rugosiuscula* (Michaud, 1831) (Helicellinae, Gastropoda). - Biological Journal of the Linnean Society 79: 611-628.

Pfenninger, M., Nowak, C. & Magnin, F. **2007**. Intraspecific range dynamics and niche evolution in *Candidula* land snail species. - Biological Journal of the Linnean Society 90: 303-317.

Pfenninger, M. & Pfenninger, A. 2005. A new *Trochulus* species from Switzerland (Gastropoda: Pulmonata: Hygromiidae). - Archiv für Molluskenkunde 134 (2): 261-269.

Pfenninger, M., Staubach, S., Albrecht, C., Streit, B. & Schwenk, K. **2003**. Ecological and morphological differentiation among cryptic evolutionary lineages in freshwater limpets of the nominal form-group *Ancylus fluviatilis* (O. F. Müller, 1774). - Molecular Ecology 12 (10): 2731-2745.

Pfenninger, M., Véla, E., Jesse. R., Arantzazu Elejalde, M., Liberto, F., Magnin, F. & Martínez-Ortí, A. **2010**. Temporal speciation pattern in the western Mediterranean genus *Tudorella* P. Fischer, 1885 (Gastropoda, Pomatiidae) supports the Tyrrhenian vicariance hypothesis:. - Molecular Phylogenetics and Evolution 54: 427-436.

Philippi, R. A. 1836. Enumeratio molluscorum Siciliæ cum viventium tum in tellure tertiaria fossilium, quae in itinere suo observavit. [Vol. 1]. - pp. I-XIV [= 1-14], 1-267, [1], Tab. I-XII [= 1-12]. Berolini. (Schropp).

Philippi, R. A. 1844. Enumeratio molluscorum Siciliae cum viventium tum in tellure tertiaria fossilium quae in itinere suo observavit. Volumen secundum continens addenda et emendanda, nec non comparationem faunae recentis Siciliae cum faunis aliarum terrarum et com fauna periodi tertiariae. - pp. I-IV [= 1-4], 1-303, [1], Tab. XIII-XXVIII [= 13-28]. Halae. (Anton).
Date should be July or Aug 1844.

Philippi, R. A. 1844. Nachtrag zum zweiten Bande der Enumeratio Molluscorum Siciliae. - Zeitschrift für Malakozoologie [1] (7) ["1845"]: 100-112.
Date should be July or Aug 1844, shortly after the previous paper.

Philippi, R. A. 1845. Abbildungen und Beschreibungen neuer oder wenig gekannter Conchylien, unter Mithülfe mehrerer deutscher Conchyliologen. Erster Band. - pp. [1-11], 1-204, Tab. [1-48]. Cassel. (Th. Fischer).
Plates numbered by chapters. Dates given at the beginnings of the chapters: pp. 1-24 (1842), 25-76 (1843), 77-186 (1844), 187-204 (1845).

Philippi, R. A. 1847. Abbildungen und Beschreibungen neuer oder wenig gekannter Conchylien. Zweiter Band. - pp. [1-9], 1-231, Tab. [1-48]. Cassel. (Th. Fischer).
Plates numbered by chapters. Dates given at the beginnings of the chapters: pp 1-152 (1846), 153-231 (1847).

Piechocki, A. 1982. Life cycle and breeding biology of *Vestia elata* (Rossm.) (Gastropoda, Clausiliidae). - Malacologia 22: 219-223.

Piechocki, A. 1989. The Sphaeriidae of Poland (Bivalvia, Eulamellibranchia). - Annales Zoologici 42 (12): 249-320. Warszawa.

Pieper, H. 1969. Die Clausiliiden (Moll., Gastropoda) der griechischen Insel Karpathos. - Biologia Gallo-hellenica 2: 169-184. Athènes.

Pieper, H. 1984. Die Eniden von Karpathos (Ägäische Inseln) (Pulmonata: Enidae). - Archiv für Molluskenkunde 115 (1/3): 51-52.

Pilsbry, H. A. 1889. New and little-known American molluscs. - Proceedings of the Academy of Natural Sciences of Philadelphia 1889: 81-89.
Date was 14 May 1889 (given on p. [6] of the journal volume, one page behind the title page, indicating that at this date a copy was presented to the meeting of the Academy, that this was a copy of the print run, and that from this moment on the copies were sent to the subscribers)

Pilsbry, H. A. 1894. Manual of Conchology; structural and systematic. With illustrations of the species. Second Series: Pulmonata. Vol. IX [= 9]. (Helicidæ, Vol. 7.). Guide to the study of helices. - pp. 1-366, Pl. 1-71. Philadelphia.
Dates (p. 116-117): pp. 1-48 (16 Nov 1893), 49-112 (19 Mar 1894), 113-160 (27 Jul 1894), i-xlviii and 161-336 (2 Feb 1895), index (Apr 1895).

Pilsbry, H. A. 1907-1908. Manual of Conchology; structural and systematic. With illustrations of the species. Second Series: Pulmonata. Vol. XIX [= 19]. Oleacinidæ. Ferussacidæ. - pp. i-vi [= 1-6], 1-366, Pl. 1-52. Philadelphia.
Dates (p. vi): pp. 1-64 Pl. 1-10 (26 Jun 1907), pp. 65-128 Pl. 11-20 (31 Aug 1907), pp. 129-192 Pl. 21-30 (09 Dec 1907), pp. 193-366 Pl. 31-52, title, contents, introduction (July 1908).

Pilsbry, H. A. 1909-1910. Manual of Conchology; structural and systematic. With illustrations of the species. Second Series: Pulmonata. Vol. XX [= 20]. *Cæcilioides*, *Glessula* and Partulidæ. Index to Vols. XVI-XX [= 16-20]. - pp. i-viii [= 1-8], 1-336, Pl. 1-43. Philadelphia.

Pilsbry, H. A. 1911. On the type of *Congeria*. - The Nautilus 25: 95-96.

Pilsbry, H. A. 1911. Part V. Non-marine Mollusca of Patagonia. - Reports of the Princeton University Expeditions to Patagonia, 1896-1899 3 (2): 513-633.

Pilsbry, H. A. 1916-1918. Manual of Conchology. Second Series: Pulmonata, 24. Pupillidæ (Gastrocoptinæ). - pp. i-xii [= 1-12], 1-380, Pl. 1-49. Philadelphia.
Dates (p. 372): pp. 1-112, Pl. 1-13 (18 Dec 1916), pp. 113-176, Pl. 14-29 (18 July 1917), pp. 177-256, Pl. 30-38 (09 Nov 1917), pp. 257-380, Pl. 39-49 (March 1918).

Pilsbry, H. A. 1922. Observations upon the nomenclature of slugs. - The Nautilus 35 (3): 77-80.

Pilsbry, H. A. 1922. Types of Férussac's subgenera of *Helix*. - The Nautilus 36 (1): 31-32.

Pilsbry, H. A. 1922-1926. Manual of Conchology. Second Series: Pulmonata. Vol. XXVII [= 27]. Pupillidæ (Orculinæ, Pagodulinæ, Acanthinulinæ, etc). - pp. [1], i-iv [= 1-4], 1-369, Pl. 1-32. Philadelphia.
Dates (p. i): pp. 1-80 Pl. 1-5 (29 Aug 1922), pp. 81-128 Pl. 6-11 (13 Jun 1923), pp. 129-176 Pl. 12-18 (16 Jul 1924), pp. 177-369 Pl. 19-32 (March 1926).

Pinceel, J., Jordaens, K., Van Houtte, N., De Winter, A. J. & Backeljau, T. 2004. Molecular and morphological data reveal cryptic taxonomic diversity in the terrestrial slug complex *Arion subfuscus/fuscus* (Mollusca, Pulmonata, Arionidae) in continental north-west Europe. - Biological Journal of the Linnean Society 83: 23-38.

Pini, N. 1876. Molluschi terrestri e d'acqua dolce viventi nel territorio d'Esino. - pp. 1-143, [1-2], Tav. A-B [= 1-2]. Milano.
Preprint of the journal article of the same title which appeared later (Falkner & Proschwitz 2009: 176).

Pini, N. 1877. Molluschi terrestri e d'acqua dolce viventi nel territorio d'Esino. - Bullettino della Società Malacologica Italiana 2 (2) ["1876"]: 67-205, Tav. A-B. Pisa.
Earliest confirmed date was 05 May 1877 (Manganelli & Ci. 2001: 345, Manganelli, Lo. et al. 2008: 31).

Pini, N. 1884. Un po' di luce sulla *Hyalina obscurata* Porro. Studio analitico sintetico. - Atti della Società Italiana di Scienze Naturali 26 (4) ["1883"]: 389-404. Milano.
Wrapper of Fascicolo 4 has printed the date Jan 1884 (preserved in SUB Göttingen, bound before p. 337), title page of the volume has printed 1883.

Pini, N. 1885. Novità malacologiche. - Atti della Società Italiana di Scienze Naturali 27 (3-4) ["1884"]: 368-382, Tav. XII [= 12]. Milano.
Wrapper of Fascicolo 3-4 (preserved in Göttingen) has printed Feb 1885, journal volume title page has printed 1884.

Pini, N. 1886. Nuova forma di *Acme* italiana. - Atti della Società Italiana di Scienze Naturali 29 (4): 521-522. Milano.
Date Dec 1886, printed on wrapper of fascicolo 4 (preserved in SUB Göttingen).

Pintér, I. & Pintér, L. 1970. Mollusken aus Bulgarien. - Malakologische Abhandlungen 3 (8): 81-98. Dresden.

Pintér, L. 1968. Über bulgarische Mollusken. - Malakologische Abhandlungen 2: 209-230. Dresden.

Pintér, L. 1969. Neue Mollusken aus Bulgarien (Gastropoda: Helicidae). - Acta Zoologica Academiae Scientiarum Hungaricae 15 (1/2): 91-96. Budapest.

Pintér, L. 1972. Die Gattung *Vitrea* Fitzinger, 1833 in den Balkanländern (Gastropoda: Zonitidae). - Annales Zoologici 29 (8): 209-315, Taf. 1-10. Warszawa.

Pintér, L. 1977. Zwei neue *Vitrea*-Arten aus Südosteuropa (Gastropoda: Zonitidae). - Acta Zoologica Academiae Scientiarum Hungaricae 23 (1/2): 183-186. Budapest.

Pintér, L. 1978. Eine neue *Vitrea* aus Griechenland (Gastropoda: Zonitidae). - Archiv für Molluskenkunde 109 (1/3): 51-52.

Pintér, L. 1978. Die *Vitrea*-Arten der Ägäischen Inseln (Gastropoda: Zonitidae). - Acta Zoologica Academiae Scientiarum Hungaricae 24 (1/2): 169-176. Budapest.

Pintér, L. 1983. Zwei neue *Vitrea*-Arten (Gastropoda: Zonitidae). - Acta Zoologica Academiae Scientiarum Hungaricae 29 (1/3): 219-222. Budapest.

Pintér, L. & Giusti, F. 1983. Una nuova specie di *Vitrea* delle Prealpi orientali italiane (Mollusca: Gastropoda). - Animalia 8 (1/3) ["1981"]: 5-13. Catania.
True year of publication was obviously 1983, although 1981 was indicated on the title page of the article. Cited in Zoological Record 1983: 136, 500 (with year 1981).

Pintér, L. & Riedel. A. 1973. Zwei neue Zonitiden (Gastropoda) aus Dalmatien. - Bulletin de l'Académie Polonaise des Sciences, Série des Sciences Biologiques 21 (4): 271-273, Taf. [1]. Varsovie.

Pintér, L. & Riedel, A. 1973. Die zweite rezente *Gyralina*-Art (Gastropoda, Zonitidae). - Bulletin de l'Académie Polonaise des Sciences, Série des Sciences Biologiques 21 (6): 425-427, Taf. [1-2]. Varsovie.

Pintér, L. & Riedel, A. 1983. Über eine neue *Lindbergia*-Art von Kérkyra, Griechenland (Gastropoda: Zonitidae). - Zoologische Mededelingen 57 (11): 101-103. Leiden.

Pintér, L. & Suara, R. 2004. Magyarországi puhatestűek katalógusa. Hazai malakológusok gyűjtései alapján. - pp. 1-547. Budapest.

Pintér, L. & Szigethy, A. 1973. Über zwei *Acicula*-Arten aus Jugoslawien (Gastropoda: Prosobranchia). - Archiv für Molluskenkunde 103 (1/3): 97-98.

Pirajno, E. 1840. Catalogo dei molluschi terrestri e fluviatili delle Madonie e luoghi adiacenti. - pp. 1-40, [1]. Palermo. (Oretea).
Author's name on title page Errico Pirajno (Barone di Mandralisca). Date 1840 given on title page and on yellow wrapper bound in the Berlin copy. Date Sep 1840 mentioned on p. 40, possibly the date when the manuscript was finished.

Pirajno, E. 1842. Nota di talune specie di molluschi terrestri e fluviatili di Sicilia. - pp. 1-10. Palermo.
Monograph or separate, title page quoted author Errico Pirajno [Barone di Mandralisca] and year 1842, "Estratto dal Giornale Letterario Num. 230". Date 28 Dec 1841 indicated on p. 10., obviously the date when the manuscript was finished. Giornale di Scienze, Lettere ed Arti per la Sicilia appeared between 1824 and 1842, latest volumes seem to be rarer. Giornale Letterario does not seem to be the name of a journal published in these times.

Pirajno, E. 1842. Nota di talune specie di molluschi terrestri e fluviatili di Sicilia. - Giornale di Scienze, Lettere ed Arti per la Sicilia 74 (230): 156-163. Palermo. [Cited from a separate].
Date Feb 1842 (?).

Pirona, A. 1865. Prospetto dei molluschi terrestri e fluviatili finora raccolti nel Friuli. - Atti del Reale Istituto Veneto di Scienze, Lettere ed Arti (3) 10: 675-708. Venezia.

Poiret, J. L. M. 1789. Voyage en Barbarie, ou lettres écrites de l'ancienne Numidie pendant les années 1785 & 1786, sur la religion, les coutumes & les mœurs des maures & des arabes-bédouins; avec un essai sur l'histoire naturelle de ce pays. Seconde partie. - pp. [1-2], 1-315, [1-4]. Paris. (La Rochelle).

Poiret, J. L. 1801. Coquilles fluviatiles et terrestres observées dans le département de l'Aisne et aux environs de Paris. Prodrome. - pp. i-xi [1-11], 1-119. Paris. (Barrois, Soissons).

Pokryszko, B. M. 1990. The Vertiginidae of Poland (Gastropoda: Pulmonata: Pupilloidea) - a systematic monograph. - Annales Zoologici 43 (8): 133-257. Warszawa.

Pokryszko, B. M. 1992. Life history of *Vertigo pusilla* O. F. Müller, 1774 (Gastropoda, Pulmonata, Vertiginidae). - pp. 247-256, in: Gittenberger, E. & Goud, J.: Proceedings of the Ninth International Malacological Congress Edinburgh, 31 August - 6 September 1986. -- pp. 1-414. Leiden. (Unitas Malacologica).

Pokryszko, B. M. 2001. Observations on seasonal dynamics of age structure and reproduction of *Pupilla muscorum* L. (Gastropoda: Pulmonata: Pupillidae). - Folia Malacologica 9 (1): 45-50.

Pokryszko, B. M. 2003. *Vertigo* of continental Europe - autecology, threats and conservation status (Gastropoda, Pulmonata: Vertiginidae). - Heldia 5 (Sonderheft 7): 13-25. München.

Poli, J. X. 1791. Testacea vtrivsqve Siciliae eorvmqve historia et anatome tabvlis aeneis illvstrata. Tomvs primvs. - pp. [1-6], I-X [= 1-10], 1-90, 1-50, [1], 1-74, i-lxxiii [= 1-73], Tab. I-XVIII [= 1-18]. Parmae.

Poliński, W. 1922. *Auritus hoyeri* n. sp. nebst Beiträgen zur Kenntnis des Subgenus *Titanopoma* A. J. Wagn. - Prace Zoologiczne Polskiego Panstwowego Muzeum Przyrodniczego [Annales Zoologici Musei Polonici Historiae Naturalis] 1 (4): 153-165, Tab. XII [= 12]. Warszawa.

Poliński, W. 1922. Recherches sur l'anatomie et la systématique des xérophilinés de la Pologne et de la Transylvanie. - Prace Zoologiczne Polskiego Panstwowego Muzeum Przyrodniczego [Annales Zoologici Musei Polonici Historiae Naturalis] 1 (4): 166-184, Tab. XII-XIV [= 12-14]. Warszawa.

Poliński, W. 1924. Contributions à l'étude systématique et zoogéographique des mollusques de l'Albanie et des régions limitrophes. - Prace Zoologiczne Polskiego Panstwowego Muzeum Przyrodniczego [Annales Zoologici Musei Polonici Historiae Naturalis] 3 (3/4): 127-150, Tab. IV [= 4]. Warszawa.

Poliński, W. 1924. Anatomisch-systematische und zoogeographische Studien über die Heliciden Polens. - Bulletin international de l'Académie Polonaise des Sciences et des Lettres, Classe des Sciences Mathématiques et Naturelles (Série B) 1924 (2): 131-279, Pl. 6-20. Kraków.

Poliński, W. 1929. Limnoloshka ispitivanja Balkanskog Poluostrva. I. Reliktna fauna gasteropoda Ochridskog Jezera. - Glas Srpska Kraljevske Akademije 137 (65): 129-182. Beograd.

Poliński, W. 1929. Sur certains problèmes du développement morphologique et zoogéographique de la faune des Alpes et des Karpates illustrés par l'étude détaillée des hélicidés du groupe Perforatella auct. - Prace Panstwowego Muzeum Zoologicznego - Annales Musei Zoologici Polonici 7 ["1928"]: 137-229, Tab. XXVI-XXXI [= 24-31]. Warszawa.

Pollonera, C. 1884. Monografia del genere *Vitrina*. - Atti della Reale Accademia delle Scienze di Torino 19 (3): 322-342 (or 412-432), Tav. X [= 10].
Pp. 322-342 in New York Botanical Garden copy, pp. 412-432 in SUB Göttingen copy.

Pollonera, C. 1885. Elenco dei molluschi terrestri viventi in Piemonte. - Atti della Reale Accademia delle Scienze di Torino 20 (5): 517-545 (or 675-703).
Pp. 517-545 in New York Botanical Garden copy, pp. 675-703 in SUB Göttingen copy.

Pollonera, C. 1886. Aggiunte alla malacologia terrestre del Piemonte. - Bollettino dei Musei di Zoologia ed Anatomia Comparata della R. Università di Torino 1 (17): 1-4.
Date was 22 Nov 1886.

Pollonera, C. 1887. Note malacologiche. - Bullettino della Società Malacologica Italiana 12 (9/15) ["1886"]: 204-223, Tav. VI [= 6].
Date was 15 Sep 1887 (Manganelli & Cianfanelli 2001: 345, Manganelli, Lo. et al. 2008: 35), also according to the yellow wrapper for fogli 9-15 preserved and bound in the SPBK Berlin copy.
Article contains 3 chapters: I. Molluschi della valle di Natisone. II. Monografia degli *Sphyradium* italiani. III. Degli *Odontocyclas* italiani.

Pollonera, C. 1887. Specie nuove o mal conosciute di *Arion* europei. - Atti della Reale Accademia delle Scienze di Torino 22 (5): 188-211 (or 290-313), Tav. III [= 3].
Pp. 188-211 in New York Botanical Garden copy, pp. 290-313 in SUB Göttingen copy.

Pollonera, C. 1888. Appunti di malacologia. - Bollettino dei Musei di Zoologia ed Anatomia Comparata della R. Università di Torino 3 (43): 1-10, Tav. II [= 2].
Contains 3 chapters: I. De alcune testacelle raccolte presso Torino. II. Di alcune Testacelle spagnole. III. Un nuovo limacide dell'Asia Minore.

Pollonera, C. 1888. Appunti di malacologia. - Bollettino dei Musei di Zoologia ed Anatomia Comparata della R. Università di Torino 3 (51): 1-4, Tav. III [= 3].
Contains one chapter: IV. Intorno ad alcuni *Limax* italiani. This is not part of the title.

Pollonera, C. 1889. Note malacologiche. - Bullettino della Società Malacologica Italiana 14: 49-64, Tav. II [= 2]. Pisa.
Date was 30 Apr 1889 (Manganelli & Ci. 2001: 346, Manganelli, Lo. et al. 2008: 36).

Pollonera, C. 1889. Nuove contribuzioni allo studio degli *Arion* europei. - Atti della Reale Accademia delle Scienze di Torino 24 (13): 401-418 (or 623-640), Tav. IX [= 9].
Pp. 401-418 in New York Botanical Garden copy, pp. 623-640 in SUB Göttingen copy. Text published in Dispensa 13, Tav. IX possibly in Dispensa 15.

Pollonera, C. 1890. Appunti di malacologia. - Bollettino dei Musei di Zoologia ed Anatomia Comparata della R. Università di Torino 5 (75): [1-4].
Pages not paginated. Date 27 Feb 1890. Contained 2 chapters: V. Un limacide nuovo per l'Italia. VI. Una nuova stazione del *Limax corsicus*.

Pollonera, C. 1891. Appunti di malacologia VII. Intorno ai limacidi di Malta. - Bollettino dei Musei di Zoologia ed Anatomia Comparata della R. Università di Torino 6 (99): 1-4.

Pollonera, C. 1896. Appunti di malacologia. IX-X. - Bollettino dei Musei di Zoologia ed Anatomia Comparata della R. Università di Torino 11 (264): 1-6.

Pollonera, C. 1905. Note malacologiche. - Bollettino dei Musei di Zoologia ed Anatomia Comparata della R. Università di Torino 20 (517): 1-10, Tav. [1].

Pollonera, C. 1909. Note malacologiche. - Bollettino dei Musei di Zoologia ed Anatomia Comparata della R. Università di Torino 24 (608): 1-19, Tav. [1].

Pollonera, C. 1916. Escursioni zoologiche del Dott. Enrico Festa nell'Isola di Rodi. XIII. Molluschi. - Bollettino dei Musei di Zoologia ed Anatomia Comparata della R. Università di Torino 31 (716): 1-9.

Porro, C. 1836. Drépanostome. *Drepanostoma*. - Magasin de Zoologie 6 (Cl. V Pl. 71): 1-2, Pl. 71. Paris.
Date Jan 1836 given on the last page of the article.

Porro, C. 1838. Malacologia terrestre e fluviale della Provincia Comasca. - pp. I-X [= 1-10], 11-132, [1-7], Tav. I-II [= 1-2]. Milano. (Guglielmini e Redaelli).

Porro, C. 1840. Description de deux espèces nouvelles de mollusques d'Europe. - Revue Zoologique par la Société Cuviérienne 3: 106. Paris.

Porter, A. & Ribi, G. 1994. Population genetics of *Viviparus* (Mollusca: Prosobranchia): homogeneity of *V. ater* and apparent introgression into *V. contectus*. - Heredity 73: 170-176.

Potiez, V.-L.-V. & Michaud, A.-L.-G. 1838. Galerie des mollusques, ou catalogue méthodique, descriptif et raisonné des mollusques et coquilles du Muséum de Donai [sic, = Douai]. Tome premier. - pp. I-XXXVI [= 1-36], 1-569, [1-4], altas 1-56, Pl. I-XXXVII [= 1-37]. Paris, Londres. (Baillière).
Kadolsky 2012: 83 gave strong indications that the true date was 1838 (before 27 Oct, published prior to Bouchard-Chantereaux 1838).

Prevost, C. 1821. Sur une nouvelle espèce de mollusque testacé du genre ménalopside [sic]. - Bulletin des Sciences, par la Société Philomathique de Paris 1821: 136-137.

Prieto, C. E. 1985. *Helicella striatitala* sp. nov. (Gastropoda, Helicidae) del N. de la Península Ibérica. - Iberus 5: 53-58.

Prieto, C. E. & Altonaga, K. 1988. Las áreas de distribución de los helicoideos (Gastropoda, Pulmonata) del País Vasco y regiones adyacentes. I. Sphincterochilidae, Elonidae, Helicodontidae y Helicidae. - Iberus 8 (2): 1-8.

Prieto, C. E. & Gómez, B. J. 1988. Las áreas de distribución de los helicoideos (Gastropoda, Pulmonata) del País Vasco y regiones adyacentes. II. Falmilia Hygromiidae (salvo géneros helicelloides). - Iberus 8 (2): 9-14.

Prieto, C. E. & Martín, R. 1988. Las áreas de distribución de los helicoideos (Gastropoda, Pulmonata) del País Vasco y regiones adyacentes. III. Géneros *Helicella, Trochoidea, Helicopsis, Candidula* y *Cernuella*. - Iberus 8 (2): 15-21.

Prieto, C. E., Martín, R., Gómez, B. J. & Larraz, M. **1987**. Nuevos datos sobre *Acanthinula* Beck 1846, *Planogyra* Morse 1864, y *Acicula* Hartmann 1821 (Mollusca, Gastropoda) en la península Ibérica. - Iberus 6 (2) ["1986"]: 257-264.

Prieto, C. E. & Puente, A. I. 1992. El género *Hygromia* Risso, 1826 en la Península Ibérica, con descripción de *Hygromia gofasi* sp. nov., y consideraciones sobre la interpretación funcional del aparato estimulador de Hygromiidae. - Bulletin du Muséum National d'Histoire Naturelle (Série 4 Section A) 14: 383-404. Paris.

Prieto, C. E. & Puente, A. I. 1994. Un nuevo Hygromiinae del norte de la Península Ibérica, *Cryptosaccus asturiensis* n. gen., n. sp. (Pulmonata: Helicoidea: Hygromiidae). - Archiv für Molluskenkunde 123 (1/6): 109-122.

Prince, G. v. 1967. Über Lebensweise, Fortpflanzung und Genitalorgane des terrestrischen Prosobranchiers *Cochlostoma septemspirale*. - Archiv für Molluskenkunde 96 (1/2): 1-18.

Proćków, M. 2009. The genus *Trochulus* Chemnitz, 1786 (Gastropoda: Pulmonata: Hygromiidae) - a taxonomic revision. - Folia Malacologica 17 (3): 101-176.

Proschwitz, T. von 1990. The identity of *Pupa daliaca* Westerlund 1887, with notes on intraspecific variation in and ecology of *Vertigo ronnebyensis* (Westerlund 1871) (Pulmonata: Pupillacea: Vertiginidae). - Archiv für Molluskenkunde 119 (4/6): 241-258.

Proschwitz, T. von 1994. Zoogeographical studies on the land Mollusca of the province of Dalsland (SW. Sweden). - Acta Regiae Societatis Scientiarum et Litterarum Gothoburgensis, Zoologica 15: 1-152. Göteborg.

Proschwitz, T. von 2003. A review of the distribution, habitat selection and conservation status of the species of the genus *Vertigo* in Scandinavia (Denmark, Norway and Sweden) (Gastropoda, Pulmonata: Vertiginidae). - Heldia 5 (Sonderheft 7): 27-50.

Proschwitz, T. von 2004. On the distribution and ecology of *Vertigo substriata* (Jeffreys), *Vertigo modesta arctica* (Wallenberg), *Vertigo lilljeborgi* (Westerlund) and *Vertigo alpestris* Alder in France and on the Iberia Peninsula. - Journal of Conchology 38 (4): 411-420.

Proschwitz, T. von 2007. *Vertigo ultimathule* n. sp., a new whorl-snail from northernmost Sweden (Gastropoda: Pulmonata: Vertiginidae). - Heldia 5 (3): 73-74, Taf. 9.

Proschwitz, T. von 2008. The Chinese giant mussel - *Sinanodonta woodiana* (Lea, 1834) (Bivalvia, Unionidae) - an unwelcome addition to the Swedish fauna. - Basteria 72: 307-311.

Proschwitz, T. von 2010. Three land-snail species new to the Norwegian fauna: *Pupilla pratensis* (Clessin, 1871), *Vertigo ultimathule* von Proschwitz, 2007 and *Balea sarsii* Philippi, 1847 (= *B. heydeni* von Maltzan, 1881). - Fauna Norvegica 30: 13-19.

Proschwitz, T. von & Falkner, G. 2012. Case 3569. *Limax fasciatus* Razoumowsky, 1789 (Linacidae) and *Limax fasciatus* Nilsson, 1823 (currently Arion fasciatus, Arionidae): proposed conservation of both specific names (Gastropoda, Stylommatophora). - Bulletin of Zoological Nomenclature 68 (4): 253-256.

Proschwitz, T. von, Schander, C., Jueg, U. & Thorkildsen, S. 2009. Morphology, ecology and DNA-barcoding distinguish *Pupilla pratensis* (Clessin, 1871) from *Pupilla muscorum* (Linnaeus, 1758) (Pulmonata: Pupillidae). - Journal of Molluscan Studies 75 (4): 315-322.

Protz, A. 1903. Zur Binnenmolluskenfauna der Provinz Ostpreussen. - Nachrichtsblatt der Deutschen Malakozoologischen Gesellschaft 35 (1/2): 1-6.

Prozorova, L. A. 2009. Eshche raz o *Lymnaea ampla* (Hartmann, 1821) i trudnostnakh sistematiki Lymnaeidae (Gastropoda: Pulmonata). - Byulleten Dal'nevostochnogo Malakologicheskogo Obshchestva 139: 47-54. Vladivostok.

Puente, A. I. 1995. El género *Xerosecta* Monterosato, 1892 (Stylommatophora: Helicoidea: Hygromiidae: Hygromiinae) en la Península Ibérica. - Iberus 13 (2): 35-85.

Puente, A. 1996. *Gittenbergeria turriplana* (Morelet 1845): a redescription (Gastropoda: Pulmonata: Hygromiidae: Trissexodontinae). - Archiv für Molluskenkunde 126 (1/2): 73-79.

Puente, A. 1996. El género *Oestophora* Hesse 1907 en la Península Ibérica (Gastropoda: Pulmonata: Hygromiidae: Trissexodontinae). - Archiv für Molluskenkunde 126 (1/2): 81-107.

Puente, A., Altonaga, K., Prieto, C. E. & Ruiz, J. C. **1998**. Los géneros *Gasuliella* Gittenberger 1980, *Mastigophallus* Hesse 1918, *Oestophorella* Pfeffer 1929 y *Trissexodon* Pilsbry 1894 en la Península Ibérica (Gasopoda: Pulmonata: Helicoidea: Hygromiidae: Trissexodontinae). - Archiv für Molluskenkunde 127 (1/2): 43-55.

Puente, A. I. & Arrébola, J. R. 1996. Deux espèces nouvelles de *Trochoidea (Xerocrassa)* du sud de la péninsule Ibérique (Pulmonata, Helicoidea, Hygromiidae). - Bulletin du Muséum National d'Histoire Naturelle (Série 4 Section A) 18 (1-2): 55-67. Paris.

Puente, A. I. & Prieto, C. E. 1988. *Cernuella (Xerocincta) neglecta* (Draparnaud 1805) (Pulmonata: Stylommatophora: Hygromiidae) en la Península Ibérica. - Iberus 8 (2): 31-37.

Puente, A. I. & Prieto, C. E. 1992. *Plentuisa vendia*, a new genus and species from the Picos de Europa (north of the Iberian peninsula) (Gastropoda: Helicoidea: Hygromiidae). - Journal of Conchology 34 (3): 159-168. London.

Pujante Mora, A. M., Tapia Ortega, G. & Martínez López, F. 1998. Los moluscos de los ríos de la Comunidad Valenciana (España). - Iberus 16 (1): 1-19.

Pulteney, R. 1799. Catalogues of the birds, shells, and some of the more rare plants, of Dorsetshire. From the new and enlarged edition of Mr. Hutchins's history of that county. - pp. [1], 1-92. London. (Nichols).

Putnam, F. W. 1863. List of the fishes sent by the museum to different institutions, in exchange for other specimens, with annotations. - Bulletin of the Museum of Comparative Zoology 1: 2-16.

Quiñonero Salgado, S., López Alabau & García Meseguer, A. J. 2010. Nuevas localidades de *Helix lucorum* (Linnaeus, 1758) para la península Ibérica. - Spira 3 (3/4): 193-195. Barcelona.

Quintana Cardona, J. 2006. Mol·luscs terrestres autòctons i introduïts a l'illa de Menorca (Illes Balears, Mediterrània occidental). - Spira 2 (1): 17-26. Barcelona.

Quintana Cardona, J. 2007. Un nuevo molusco terrestre para la fauna balear: *Arion (Mesarion) ponsi* sp. nov. (Gastropoda: Pulmonata: Arionidae). - Spira 2 (3): 139-146. Barcelona.

Quintana Cardona, J. 2010. *Truncatellina beckmanni* sp. nov. (Gastropoda: Pulmonata: Vertiginidae: Truncatellininae), una nueva especie endémica de Menorca (Islas Baleares, Mediterráneo occidental). - Spira 3 (3/4): 149-158. Barcelona.

Radoman, P. 1953. Jedan prilog izuchavanju Ohriskih gasteropoda. - Arhiv Bioloshkih Nauka 5: 55-59. Beograd.

Radoman, P. 1962. Nove okhridske khidrobide (II). - Arkhiv Bioloskikh Nauka [Archives des Sciences Biologiques] 14 (1/2): 69-93. Beograd.

Radoman, P. 1974. Some new gastropod representatives from the brackish waters of the Adriatic and Aegean seasides. - The Veliger 16 (3): 283-288, Fig. 1-2.

Radoman, P. 1977. Hydrobiidae auf der Balkanhalbinsel und in Kleinasien. - Archiv für Molluskenkunde 107 (4/6): 203-223, Taf. 21-23.

Radoman, P. 1979. Nov rod *Pseudoislamia* i novi predstavnici rodova *Islamia* i *Parabythinella*. - Glasnik Prirodnjatshkog Muzeja u Beogradu, Serija B, Bioloshke Nauke [Bulletin du Muséum d'Histoire Naturelle de Belgrade, Série B, Sciences Biologiques] 33 ["1978"]: 23-27. Beograd.

Rafinesque, C. S. 1819. Prodrome de 70 nouveaux genres d'animaux découverts dans 1'intérieur des Etas-Unis d'Amérique, durant 1'année 1818. - Journal de Physique, de Chimie et d'Histoire Naturelle 88: 417-429. Paris.
Date was June 1819 as printed on p. 429.

Rähle, W. 1974. Über einige Milaces vom Galicica-Gebirge (Jugoslawisch-Mazedonien) und Beschreibung einer neuen Art. - Archiv für Molluskenkunde 104 (1/3): 51-56.

Rähle, W. 1980. Land- und Süßwassermollusken von Kephallinia und Zakynthos (Ionische Inseln). - Archiv für Molluskenkunde 110 (4/6): 199-224.

Rähle, W. 1982. *Graecophaedusa sperrlei* n. gen., n. sp.: ein bemerkenswertes Tertiärrelikt aus Griechisch-Mazedonien (Pulmonata: Clausiliidae: Serrulininae). - Archiv für Molluskenkunde 112 (1/6): 45-48.

Rähle, W. 1983. Eine neue *Limax*-Art von der griechischen Insel Samos (Gastropoda, Pulmonata). - Archiv für Molluskenkunde 114 (1/3): 31-33.

Rähle, W. 1983. Eine neue *Deroceras*-Art (Gastropoda, Pulmonata, Agriolimacidae) von der griechischen Insel Samos. - Zoologische Mededelingen 57 (13): 115-120. Leiden.

Rähle, W. 1984. Zur Kenntnis zyprischer *Deroceras*-Arten (Pulmonata: Agriolimacidae). - Archiv für Molluskenkunde 115 (1/3): 45-50.

Rähle, W. 1988. Eine neue *Turanena*-Art aus der südwestlichen Türkei (Gastropoda: Stylommatophora: Enidae). - Archiv für Molluskenkunde 118 (4/6): 175-179.

Rähle, W. 1991. Zur Kenntnis der Nacktschneckenfauna der Insel Zypern (Gastropoda, Pulmonata: Milacidae, Limacidae, Agriolimacidae). - Malakologische Abhandlungen 15 (16): 141-148.

Rähle, W. 1993. *Deroceras kasium* n. sp., eine neue Nacktschnecke von der Insel Kasos (Dodekanes, Griechenland) (Gastropoda: Pulmonata: Agriolimacidae). - Archiv für Molluskenkunde 121 (1/6): 111-115.

Rähle, W. 1998. Drei neue *Deroceras*-Arten aus Südanatolien (Türkei) (Gastropoda: Pulmonata: Agriolimacidae). - Malakologische Abhandlungen 19 (3): 21-28.

Rähle, W. & Riedel, A. 1987. Eine neue, unterirdisch lebende Zonitiden-Art (Gastropoda: Stylommatophora) aus Südwestanatolien, Türkei. - Zoologische Mededelingen 61 (15): 203-207. Leiden.

Rainer, M. 1967. Chromosomenuntersuchungen an Gastropoden (Stylommatophora). - Malacologia 5: 341-373.

Rambur, P. 1868. Diagnoses d'hélices nouvelles. - Journal de Conchyliologie 16: 265-268. Paris.

Rambur, P. 1869. Description de plusieurs hélices inédites, de France et d'Espagne, suivie d'observations et de rectifications concernant deux autres espèces. - Journal de Conchyliologie 17: 252-269, Pl. IX [=9]. Paris.

Rang, P. C. A. L. 1827. Catalogue des espèces des mollusques terrestres et fluviatiles, recueillies par M. Rang, offic. de la marine roy., dans un voyage aux Grandes Indes. - Bulletin des Sciences Naturelles et de Géologie, deuxième Section du Bulletin Universel des Sciences et de l'Industrie 10: 298-307. Paris.

Raven, J. G. M. 1988. *Pyrenaearia daanidentata* spec. nov. (Helicidae), a toothed species from the Cantabrian mountains, Spain. - Basteria 52 (1/3): 121-123.

Raven, J. G. M. 1990. A revision of *Obscurella* Clessin, 1889 (Gastropoda Prosobranchia: Cyclophoridae). - Basteria 54 (1/3): 17-62.

Razoumowsky, G. de 1789. Histoire naturelle du Jorat et de ses environs; et celle des trois lacs de Neufchatel, Morat et Bienne; précédées d'un essai sur le climat, les productions, le commerce, les animaux de la partie du Pays de Vaud ou de la Suisse Romande, qui entre dans le plan de cet ouvrage. Tome premier. - pp. [1-2], I-XVI [= 1-16], 1-322, Pl. 1-3. Lausanne. (Mourer).
Author is Grigorii Kirillovic Razumovskij (1759-1837), spelled Razoumowsky on the title page.

Réal, G. & Réal-Testud, A.-M. 1988. La malacofaune terrestre de l'île de Corse: historique et inventaire actualisé. - Haliotis 18: 43-54. Paris.

Récluz, C. A. 1841. Description de quelques espèces de nérites vivantes. - Revue Zoologique 1841: 337-343. Paris. [Date = Nov 1841 indicated on p. 337]

Reeve, L. A. 1849. Monograph of the genus *Bulimus*. - Conchologia Iconica: or, illustrations of the shells of molluscous animals 5: (Species No. 1-662), Pl. I-LXXXIX [= 1-89]. London.

Reichenbach, H. G. L. 1828. Allgemeine Taschenbibliothek der Naturwissenschaften. 5. Theil. Zoologie oder Naturgeschichte des Thierreichs nach eigenen Ansichten. [Erstes Bändchen]. - Dresden. [Not cited from original]
Libraries seem to have difficulties in citing the title consistently.

Reichenbach, H. G. L. 1836. Allgemeine Taschenbibliothek der Naturwissenschaften. 5. Theil. Zoologie oder Naturgeschichte des Thierreichs nach eigenen Ansichten. Zweites Bändchen. - Zittau, Leipzig. [Not cited from original]
Libraries seem to have difficulties in citing the title consistently.

Reinhardt, O. 1871. Über die in Deutschland vorkommenden Hyalinen aus der *Crystallina*-Gruppe. - Sitzungs-Berichte der Gesellschaft Naturforschender Freunde zu Berlin 2: 39-40.

Reinhardt, O. 1879. Ueber die *Isthmia*-Arten und ihre geographische Verbreitung. - Sitzungs-Berichte der Gesellschaft Naturforschender Freunde zu Berlin 1879 (9): 133-139.

Reinhardt, O. 1880. Über die *Acme*-Arten des Banats und Siebenbürgens. - Sitzungs-Berichte der Gesellschaft Naturforschender Freunde zu Berlin 1880 (2): 45-47.

Reinhardt, O. 1881. Eine Anzahl griechischer Schnecken. - Sitzungsberichte der Gesellschaft der Naturforschenden Freunde, Berlin 1881 (9): 135-137.

Reinhardt, O. 1883. Über die von den Herren Gebrüder Krause auf ihrer Reise gesammelten *Pupa*-, *Hyalina*- und *Vallonia*-Arten. - Sitzungs-Berichte der Gesellschaft Naturforschender Freunde zu Berlin 1883: 37-43.

Reinhardt, O. 1883. Einige von Herrn D. W. Kobelt in Schwanheim a. M. zur Begutachtung übersandte, von Herrn Hungerford gesammelte japanische Hyalinen. - Sitzungs-Berichte der Gesellschaft Naturforschender Freunde zu Berlin 1883: 82-86.

Reinhardt, O. 1916. Einige Bemerkungen über *Pupa minutissima* und Verwandte. - Nachrichtsblatt der Deutschen Malakozoologischen Gesellschaft 48 (4): 158-167.

Reischütz, **A.**, Reischütz, N. & Reischütz, P. L. **2008**. Ein Beitrag zur Kenntnis der Molluskenfauna Albaniens. - Nachrichtenblatt der Ersten Vorarlberger Malakologischen Gesellschaft 15: 37-39. Rankweil.

Reischütz, A., Reischütz, N. & Reischütz, P. L. **2010**. Beiträge zur Kenntnis der Molluskenfauna Albaniens. - Nachrichtenblatt der Ersten Vorarlberger Malakologischen Gesellschaft 17: 29-31. Rankweil.

Reischütz, A., Reischütz, N. & Reischütz, P. L. **2010**. Helleniká pantoía, 29. Neue Fundorte von Serrulininae (Clausiliidae: Gastropoda) in Griechenland und Albanien. - Nachrichtenblatt der Ersten Vorarlberger Malakologischen Gesellschaft 17: 37-39. Rankweil.

Reischütz, A. & Reischütz, P. L. 2003. Helleniká pantoía, 3: Die Schönste von allen. *Tsoukatosia (Agiosspeleikos) christinae* nov. subgen., nov. spec. (Clausiliidae: Pulmonata: Gastropoda). - Nachrichtenblatt der Ersten Vorarlberger Malakologischen Gesellschaft 11: 1-3. Rankweil

Reischütz, A. & Reischütz, P. L. 2003. Helleniká pantoia, 4: Zur Verbreitung von *Pholeoteras zilchi* Subai 1993 (Gastropoda: Prosobranchia) im Louros-Tal (Epirus, Griechenland). - Nachrichtenblatt der Ersten Vorarlberger Malakologischen Gesellschaft 11: 10.

Reischütz, A. & Reischütz, P. L. 2003. Helleniká pantoia, 5: Zur Kenntnis der Molluskenfauna des Limni Trichonida und des Limni Lisimachia (Aitolien/Akarnanien, Griechenland). - Nachrichtenblatt der Ersten Vorarlberger Malakologischen Gesellschaft 10: 28-30.

Reischütz, A. & Reischütz, P. L. 2004. Helleniká pantoia, 9: Zur Verbreitung der Serrulininae (Clausiliidae: Gastropoda) in Griechenland. - Nachrichtenblatt der Ersten Vorarlberger Malakologischen Gesellschaft 12: 5-6.

Reischütz, A. & Reischütz, P. L. 2004. Helleniká pantoia, 10: *Pseudosuccinea columella* (Say 1817) im Nómos Florina (NW-Makedonien, Griechenland). - Nachrichtenblatt der Ersten Vorarlberger Malakologischen Gesellschaft 125: 9.

Reischütz, A. & Reischütz, P. L. 2008. Helleniká pantoia, 12: Bemerkungen zu *Dreissena stankovici* Lvova & Starobogatov 1982 (Bivalvia: Heterodonta: Dreissenidae). - Nachrichtenblatt der Ersten Vorarlberger Malakologischen Gesellschaft 15: 3-4.

Reischütz, A. & Reischütz, P. L. 2009. Ein Beitrag zur Kenntnis der Molluskenfauna von Montenegro, nebst Beschreibung zweier neuer Arten der Gattung *Virpazaria* Gittenberger 1969. - Nachrichtenblatt der Ersten Vorarlberger Malakologischen Gesellschaft 16: 51-60.

Reischütz, A. & Reischütz, P. L. 2009. Helleniká pantoía, 26: Ein Nachweis von *Hawaiia minuscula* (Binney 1840) (Gastropoda: Zonitidae) im Epirus (Griechenland). - Nachrichtenblatt der Ersten Vorarlberger Malakologischen Gesellschaft 16: 41.

Reischütz, A., Reischütz, P. L. & Fischer, W. **2008**. Helleniká pantoía, 13: *Pholeoteras* Sturany 1904 (Gastropoda: Prosobranchia: Cyclophoridae) im Enipeastal in Litochorion (Makedonien, Griechenland). - Nachrichtenblatt der Ersten Vorarlberger Malakologischen Gesellschaft 15: 5-6.

Reischütz, **A.**, Reischütz, P. L. & Fischer, W. **2008**. Helleniká pantoía, 15: Kein Ende anzusehen - Zur Kenntnis der Gattung *Balcanodiscus* Riedel & Urbanski 1964 (Gastropoda: Pulmonata: Zonitidae). - Nachrichtenblatt der Ersten Vorarlberger Malakologischen Gesellschaft 15: 11-13. Rankweil.

Reischütz, **P. L.** 1973. *Arion (Microarion) obesoductus* n. sp., eine neue Nacktschnecke aus Niederösterreich. - Archiv für Molluskenkunde 103: 229-230.

Reischütz, **P. L.** 1983. *Deroceras (Plathystimulus) ikaria* n. sp. von Ikaria, Griechenland (Gastropoda, Pulmonata, Agriolimacidae). - Malakologische Abhandlungen 9 (3): 23-24.

Reischütz, **P. L.** 1983. Ein Beitrag zur Molluskenfauna der Insel Thasos (Griechenland). - Annalen des Naturhistorischen Museums in Wien 85B: 133-146, Taf. 1.

Reischütz, **P. L.** 1986. Die Verbreitung der Nacktschnecken Österreichs (Arionidae, Milacidae, Limacidae, Agriolimacidae, Boettgerillidae). (Supplement 2 des Catalogus Faunae Austriae). - Sitzungsberichte, Österreichische Akademie der Wissenschaften, mathematisch-naturwissenschaftliche Klasse, Abteilun 195 (5): 67-190. Wien.

Reischütz, **P. L.** 1988. Beiträge zur Molluskenfauna Thrakiens und Ostmakedoniens, II. - Annalen des Naturhistorischen Museums in Wien 90B: 341-356, Taf. 1-2.

Reischütz, **P. L.** 2009. Beiträge zur Molluskenfauna Niederösterreichs XXIII (= Phantome der österreichischen Malakofauna, 2). Bemerkungen zur Roten Liste der Weichtiere Österreichs, das *Viviparus*-Problem, sowie eine Artenliste der Süßwasser-Mollusken der Donauauen östlich von Wien. - Nachrichtenblatt der Ersten Vorarlberger Malakologischen Gesellschaft 16: 25-31.

Reischütz, **P. L.** & Sattmann, H. 1990. Beiträge zur Molluskenfauna des Epirus, II. - Annalen des Naturhistorischen Museums in Wien 91B: 253-272, Taf. I-IV [= 1-4].

Reise, H. 1995. Mating behaviour of *Deroceras rodnae* Grossu & Lupu, 1965 and *D. praecox* Wiktor, 1966 (Pulmonata: Agriolimacidae). - Journal of Molluscan Studies 61 (3): 325-330.

Reise, H. & Hutchinson, J. M. C. 2001. Morphological variation in terrestrial slug *Deroceras turcicum* (Simroth, 1894) and a northern extension of its range in Central Europe. - Folia Malacologica 9 (2): 63-71.

Reise, H., Hutchinson, J. M. C., Forsyth, R. G. & Forsyth, T. J. **2005**. First records of the terrestrial slug *Deroceras turcicum* (Simroth, 1894) in Poland. - Folia Malacologica 13 (4): 177-179.

Reise, H., Hutchinson, J. M. C., Schunack, S. & Schlitt, B. **2011**. *Deroceras panormitanum* and congeners from Malta and Sicily, with a redescription of the widespread pest slug as *Deroceras invadens* n. sp. - Folia Malacologica 19 (4): 201-223.

Reitano, A., Liberto, F. & Sparacio, I. **2007**. Nuovi dati su molluschi terrestri e dulciacquicoli di Sicilia. 1° contributo (Gastropoda Prosobranchia Neotaenioglossa; Gastropoda Pulmonata Basommatophora, Stylommatophora). - Il Naturalista Siciliano (4) 31 (3-4): 311-330. Palermo.

Reitano, A., Liberto, F., Sparacio, I. & Giglio, S. **2009**. I molluschi terrestri della R.N.I. "Grotta Palombara" (Melilli, Sicilia sud-orientale) (Gastropoda Architaenioglossa, Neotaenioglossa, Stylommatophora). - Il Naturalista Siciliano (4) 33 (1-2): 177-205. Palermo.

Remigio, E. A. 2002. Molecular phylogenetic relationships in the aquatic snail genus *Lymnaea*, the intermediate host of the causative agent of fascioliasis: insights from broader taxon sampling. - Parasitology Research 88: 687-696.

Renda, W., Bodon, M. & Nardi, G. 2011. On the specific validity of *Rupestrella jaeckeli* Beckmann, 2002 (Gastropoda, Pulmonata, Chondrinidae). - Biodiversity Journal 2 (4): 213-216. Palermo.

Requien, E. 1848. Catalogue des coquilles de l'île de Corse. - pp. I-XII [= 1-12], 13-109. Avignon. (Ainé).
Author's name given as E. R. on p. XII (= Esprit Requien).

Retowski, O. 1883. Am Strande der Krim gefundene, angeschwemmte transcaucasische (?) Binnenconchylien. - Malakozoologische Blätter (Neue Folge) 6: 53-61, Taf. II [= 2].
Explanation of Plate 2 in previous article by Clessin 1883.

Retowski, O. 1886. Am Strande der Krim gefundene angeschwemmte Binnenconchylien. - Malakozoologische Blätter (Neue Folge) 9 (1) ["1887"]: 22-42, Taf. I [= 1].

Retowski, O. 1889. Liste der von mir auf meiner Reise von Konstantinopel nach Batum gesammelten Binnenmollusken. - Bericht der Senckenbergischen Naturforschenden Gesellschaft 1889: 225-265.
Published in Sep 1889 or later, the date indicated on the title page of the volume.

Retzius, A. J. 1788. Dissertatio historico-naturalis sistens nova testaceorum genera. Quam præside D. M. Andr. J. Retzio (...) ad publicum examen defert Laurentius Münter Philipsson. - pp. [1-4], 1-23. Lundæ. (Dissertatio Historico-Naturalis). (Berling).

Reynes, J. 1870. Description d'une hélice nouvelle du Département de l'Hérault. - Annales de Malacologie 1: 34-35. Paris.

Ribi, G. & Oertli, S. 2000. Frequency of interspecific matings and of hybrid offspring in sympatric populations of *Viviparus ater* and *V. contectus* (Mollusca: Prosobranchia). - Biological Journal of the Linnean Society 71: 133-143.

Ribi, G. & Porter, A. H. 1995. Mating between two hybridizing species, *Viviparus ater* and *V. contectus* (Mollusca: Prosobranchia). - Animal Behaviour 49: 1389-1398.

Richardot, M. 1979. The limpet *Ferrissia wautieri*, a model for studies on calcification mechanisms in molluscs: some results. - Malacologia 18: 553-556.

Riedel, A. 1958. Über *Oxychilus samius* (Martens 1889) (Gastropoda: Zonitidae). - Mitteilungen aus dem Zoologischen Museum in Berlin 34: 109-112.

Riedel, A. 1958. Materialien zur Kenntnis der Zonitiden (Gastropoda) des Kaukasus und der Krim. - Annales Zoologici 17 (11): 383-427. Warszawa.

Riedel, A. 1959. Die von Dr. K. Lindberg in Griechenland gesammelten Zonitidae (Gastropoda). - Annales Zoologici 18 (6): 89-117. Warszawa.

Riedel, A. 1959. Materialien zur Kenntnis der paläarktischen Zonitidae (Gastropoda). V-VI. - Annales Zoologici 18 (12): 179-188. Warszawa.

Riedel, A. 1960. Über *Meledella werneri* Sturany 1908 (Gastropoda: Zonitidae). - Fragmenta Balcanica 3 (11): 91-97. Skopje.

Riedel, A. 1960. Die Gattung *Lindbergia* Riedel (Gastropoda, Zonitidae) nebst Angaben über *Vitrea illyrica* (A. J. Wagner). - Annales Zoologici 18 (18): 333-346. Warszawa.

Riedel, A. 1962. Materialien zur Kenntnis der Zonitidae (Gastropoda) des Nahen Ostens, nebst Besprechung der Gattung *Eopolita* Poll. im breiteren geographischen Rahmen. - Annales Zoologici 20 (15): 261-298.

Riedel, A. 1963. Zwei neue Zonitidae (Gastropoda) aus Südostbulgarien. - Annales Zoologici 20 (22): 473-485. Warszawa.

Riedel, A. 1962. Materialien zur Kenntnis der paläarktischen Zonitidae (Gastropoda). VII-VIII. - Annales Zoologici 20 (13): 221-227. Warszawa.

Riedel, A. 1962. *Vitrea argolica* sp. n. aus Griechenland (Gastropoda: Zonitidae). - Bulletin de l'Académie Polonaise des Sciences, Série des Sciences Biologiques 10 (8): 315-317. Varsovie.

Riedel, A. 1963. Zwei neue Zonitidae (Gastropoda) aus Südostbulgarien. - Annales Zoologici 20 (22): 473-485. Warszawa.

Riedel, A. 1967. Daudebardiinae (Gastropoda: Zonitidae) Bulgariens. - Annales Zoologici 24 (8): 463-483. Warszawa.

Riedel, A. 1968. Zonitidae (Gastropoda) Kretas. - Annales Zoologici 25 (13): 473-537, Taf. 1-4. Warszawa.

Riedel, A. 1969. Die Untergattungen *Morlina* A. J. Wagner und *Riedelius* Hudec der Gattung *Oxychilus* Fitzinger (Gastropoda, Zonitidae). - Annales Zoologici 27 (6): 91-131, Taf. 1. Warszawa.

Riedel, A. 1969. Endemische Zonitidae (Gastropoda) der Balearen. - Annales Zoologici 27 (12): 237-248. Warszawa.

Riedel, A. 1970. Zwei neue Zonitidae (Gastropoda) aus Albanien. - Annales Zoologici 28 (2): 7-13. Warszawa.

Riedel, A. 1972. Zur Kenntnis der Zonitidae (Gastropoda) Spaniens. - Annales Zoologici 29 (5): 115-146. Warszawa.

Riedel, A. 1972. Die Untergattung *Schistophallus* A. J. Wagner in Europa und Kleinasien (Gastropoda: Zonitidae). - Annales Zoologici 29 (7): 181-207, Taf. 1-2. Warszawa.

Riedel, A. 1973. Eine Zonitiden-Ausbeute (Gastropoda) von den griechischen Inseln. - Fragmenta Faunistica 19 (2): 21-26. Warszawa.

Riedel, A. 1973. Die Gruppen *Hyalocornea* Monterosato und *Hyalofusca* Monterosato der Gattung *Oxychilus* Fitzinger (Gastropoda, Zonitidae). - Annales Zoologici 30 (1): 1-32. Warszawa.

Riedel, A. 1977. Materialien zur Kenntnis der Zonitidae (Gastropoda). IX-XI. - Annales Zoologici 33 (24): 495-515. Warszawa.

Riedel, A. 1978. Kritische Bemerkungen und Ergänzungen zur Kenntnis der Subfamilie Daudebardiinae (Gastropoda, Zonitidae) mit Verzeichnis aller akzeptierten Arten. - Annales Zoologici 34 (8): 139-206. Warszawa.

Riedel, A. 1979. Materialien zur Kenntnis der paläarktischen Zonitidae (Gastropoda). XII-XIV. - Fragmenta Faunistica 25 (9): 115-125, Taf. 1. Warszawa.

Riedel, A. 1979. Revision von *Aegopis skanderbegianus* Polinski und der verwandten Formen, nebst Aufstellung einer neuen Untergattung (Gastropoda: Zonitidae). - Annales Zoologici 34 (16): 461-473. Warszawa.

Riedel, A. 1980. Genera Zonitidarum. Diagnosen supraspezifischer Taxa der Familie Zonitidae (Gastropoda, Stylommatophora). - pp. 1-197, 2 Taf. Rotterdam. (Backhuys).

Riedel, A. 1980. Eine weitere Zonitiden-Art (Gastropoda) mit vaginalem Anhangsorgan. - Annales Zoologici 35 (27): 485-495. Warszawa.

Riedel, A. 1981. Vitreini (Gastropoda: Zonitidae) von den ägäischen Inseln Chios, Samos und Ikaria. - Annales Zoologici 36 (11): 229-240. Warszawa.

Riedel, A. 1981. Eine neue unterirdische *Lindbergia*-Art von der Insel Kasos, Griechenland (Gastropoda, Zonitidae). - Annales Zoologici 36 (16): 281-287. Warszawa.

Riedel, A. 1982. Die Gattungen *Allaegopis* Riedel und *Doraegopis* gen. n. (Gastropoda, Stylommatophora, Zonitidae). - Malakologische Abhandlungen 8 (1): 1-28.

Riedel, A. 1982. Die Gattung *Zonites* Montfort in Südwest-Kleinasien (Gastropoda, Zonitidae). - Annales Zoologici 36 (21): 391-423. Warszawa.

Riedel, A. 1983. Eine neue *Gyralina*-Art von Épeiros, Griechenland (Gastropoda, Pulmonata, Zonitidae). - Malakologische Abhandlungen 9 (2): 17-22.

Riedel, A. 1983. Über die *Aegopinella*-Arten (Gastropoda: Zonitidae) aus Jugoslawien, Italien und Frankreich. - Annales Zoologici 37 (5): 235-258, Taf. 1-2. Warszawa.
Date was Nov 1983.

Riedel, A. 1983. Manche wenig bekannte und neue *Oxychilus*-Arten aus Griechenland (Gastropoda: Zonitidae). - Annales Zoologici 37 (7): 269-288, Taf. 1. Warszawa.

Riedel, A. 1984. Zwei neue unterirdische Zonitidae aus der Türkei (Gastropoda, Stylommatophora). - Malakologische Abhandlungen 9 (17): 165-170.

Riedel, A. 1984. Eine neue *Lindbergia*-Art von der Insel Thíra, Griechenland (Gastropoda, Stylommatophora, Zonitidae). - Malakologische Abhandlungen 10 (1): 1-4.

Riedel, A. 1985. Zur Kenntnis der Gattung *Balcanodiscus* (Gastropoda, Stylommatophora, Zonitidae). - Malakologische Abhandlungen 10 (14): 139-146.

Riedel, A. 1985. Über die makedonischen *Gyralina*-Arten (Gastropoda, Zonitidae). - Fragmenta Balcanica 12 (4) ["1984"]: 31-43. Skopje.
Title page of the article indicated 15.09.1984, this was probably not the correct date of publication.

Riedel, A. 1985. Revision der Gattung *Zonites* Montfort (Gastropoda: Zonitidae): griechische Arten. - Annales Zoologici 39 (1): 1-67, Taf. 1-9. Warszawa.

Riedel, A. 1986. Eine neue *Allaegopis*-Art aus dem Peloponnes (Gastropoda Stylommatophora: Zonitidae). - Malakologische Abhandlungen 11 (8): 87-92.

Riedel, A. 1987. Revision der Gattung *Zonites* Montfort (Gastropoda: Zonitidae): türkische Arten. Nebst Ergänzungen und Verzeichnis aller *Zonites*-Arten. - Annales Zoologici 41 (1): 1-42, Taf. 1-9. Warszawa.

Riedel, A. 1988. Die Gattung *Carpathica* A. J. Wagner in Griechenland (Gastropoda, Stylommatophora, Daudebardiidae). - Malakologische Abhandlungen 13 (9): 85-92.

Riedel, A. 1988. Zur Kenntnis der Gattung *Balcanodiscus*, II (Gastropoda, Stylommatophora, Zonitidae). - Malakologische Abhandlungen 13 (10): 93-104.

Riedel, A. 1990. Neue und wenig bekannte Zonitidae (Gastropoda) aus Griechenland. - Annales Zoologici 43 (25): 493-534, Taf. 1-2. Warszawa.

Riedel, A. 1992. The Zonitidae (sensu lato) (Gastropoda, Pulmonata) of Greece. - Fauna Graeciae 5: I-VIII, 1-194. Athína/Athens.

Riedel, A. 1993. Eine neue *Oxychilus*-Art aus der Türkei (Gastropoda, Stylommatophora: Zonitidae). - Malakologische Abhandlungen 16 (16): 135-139.

Riedel, A. 1993. Neue Angaben zur Kenntnis der Gattungen *Allaegopis* und *Doraegopis* (Gastropoda, Zonitidae). - Annalen des Naturhistorischen Museums in Wien 94/95B: 193-216, Taf. 1.

Riedel, A. 1995. Wenig bekannte und neue Zonitidae aus der Türkei (Gastropoda, Stylommatophora). - Malakologische Abhandlungen 17: 121-136, Taf. 1-2.

Riedel, A. 1995. Zonitidae sensu lato (Gastropoda, Styommatophora) der Türkei. Übersicht der Arten. - Fragmenta Faunistica 38: 1-86, Warszawa.

Riedel, A. 1997. *Vitrea siveci* Riedel et Velkovrh - neu für Griechenland (Gastropoda: Stylommatophora: Zonitidae). - Malakologische Abhandlungen 18 (24): 261-262.

Riedel, A. 1997. Description of *Oxychilus (Riedelius) wiktori* sp. n., with some notes on other West Balkan species of *Riedelius* (Gastropoda: Pulmonata: Zonitidae). - Genus 8 (3-4): 755-764. Wrocław.

Riedel, A. 1998. Eine neue *Retinella*-Art aus Piemont, Italien (Gastropoda: Stylommatophora: Zonitidae). - Malakologische Abhandlungen 19 (1): 5-11.

Riedel, A. 1998. Genera Zonitidarum: addenda et corrigenda (Gastropoda, Stylommatophora). - pp. 1-91. Warszawa.

Riedel, A. 1999. Zwei neue *Oxychilus*-Arten von Südanatolien und von den Ägäischen Inseln (Gastropoda: Stylommatophora: Zonitidae). - Malakologische Abhandlungen 19 (19): 217-224.

Riedel, A., Fehér, Z. & Erőss, Z. **1999**. Eine neue Zonitiden-Art (Gastropoda: Stylommatophora) aus Nord-Albanien. - Folia Malacologica 7 (4): 241-244.

Riedel, A. & Maassen, W. J. M. 1993. Berichtigung zur Frage von *Hyalina moussoni* Kobelt, 1878 (Gastropoda, Pulmonata: Zonitidae). - Basteria 57 (4/6): 139-146.

Riedel, A. & Mylonas, M. 1980. Eine neue *Vitrea*-Art (Gastropoda, Zonitidae) aus Euböa. - Biologia Gallo-hellenica 9 (1): 135-138. Athènes.

Riedel, A. & Mylonas, M. 1981. Zonitidae (Gastropoda) der griechischen Insel Kea. - Annales Zoologici 36 (6): 117-125, Taf. 1. Warszawa.

Riedel, A. & Mylonas, M. 1981. Eine neue *Zonites*-Art von Kykladen, Griechenland (Gastropoda, Zonitidae). - Annales Zoologici 36 (13): 247-253. Warszawa.

Riedel, A. & Mylonas, M. 1995. The Zonitidae of the south-east Aegean islets (Gastropoda: Pulmonata). - Genus 6 (2): 229-245. Wrocław.

Riedel, A. & Mylonas, M. 1997. Neue Angaben über einige *Zonites*-Arten von den griechischen Inseln, nebst Beschreibung neuer Taxa (Gastropoda: Stylommatophora: Zonitidae). - Malakologische Abhandlungen 18 (14): 153-164.

Riedel, A. & Norris, A. 1987. An undescribed species of *Zonites* from the Island of Santorini, Greece. - Journal of Conchology 32 (6): 377-378, pl. [1]. London.

Riedel, A. & Paul, C. R. C. 1978. Eine neue *Vitrea*-Art aus der balearischen Insel Ibiza. - Boletín de la Sociedad de Historia Natural de Baleares 22: 54-60. Palma de Mallorca.

Riedel, A. & Radja, T. 1983. Systematische Stellung von *Aegopis mosorensis* Kuscer 1933 (Gastropoda: Zonitidae). - Annales Zoologici 37 (6): 259-268. Warszawa.

Riedel, A. & Reischütz, P. L. 1988. Beiträge zur Molluskenfauna Thrakiens und Ostmakedoniens: I. *Vitrea schneideri* n. sp. von Samothrake. - Annalen des Naturhistorischen Museums in Wien 90B: 149-151.

Riedel, A. & Subai, P. 1978. Eine neue *Oxychilus*-Art (Gastropoda: Zonitidae) von den Ionischen Inseln. - Annales Zoologici 34 (4): 79-87. Warszawa.

Riedel, A. & Subai, P. 1982. Beschreibung einer neuen *Oxychilus*-Art aus Kreta nebst neuen Angaben über *Oxychilus ionicus* (Gastropoda: Zonitidae). - Archiv für Molluskenkunde 112 (1/6): 21-26.

Riedel, A. & Subai, P. 1991. Zwei neue Zonitidae aus Griechenland (Gastropoda, Stylommatophora). - Malakologische Abhandlungen 15 (12): 111-114.

Riedel, A. & Subai, P. 1993. Über *Gyralina* und *Spelaeopatula* nebst Beschreibung neuer Taxa (Gastropoda Pulmonata: Zonitidae). - Archiv für Molluskenkunde 121 (1/6): 53-66.

Riedel, A. & Urbański, J. 1964. Systematische Stellung und Angaben über das Vorkommen von *Paraegopis (Balcanodiscus* subg. nov.*) frivaldskyanus* (Rossmässler 1842) (Gastropoda: Zonitidae). - Annales Zoologici 22 (4): 69-79. Warszawa.

Riedel, A. & Velkovrh, F. 1976. Drei neue balkanische Zonitiden (Gastropoda) und neue Funde einiger seltener Arten. - Biološki Vestnik 24 (2): 219-227. Ljubljana.

Riedel, A. & Welter-Schultes, F. W. 1996. Eine neue *Gyralina*-Art aus Albanien (Gastropoda: Zonitidae). - Schriften zur Malakozoologie 9: 17-20.

Rinaldi, A. 2004. New records of *Testacella* (Gastropoda, Pulmonata, Testacellidae) from Abruzzo and Sardinia, Italy. - Bollettino Malacologico 39 (9/12) ["2003"]: 203-206.

Risso, A. 1826. Histoire naturelle des principales productions de l'Europe méridionale et particulièrement de celles des environs de Nice et des Alpes Maritimes. Tome quatrième. - pp. [1-3], j-vij [= 1-7], 1-439, pl. [1-12]. Paris. (Levrault).

Robles, F. 1990. Gasteropodos terrestres de la provincia de Valancia. Revisión bibliográfica y lista de especies. - Iberus 9 (1-2): 467-481.

Robles, F. & Martínez-Ortí, A. 2009. Moluscos continentales de los alrededores de Molina de Aragón (Guadalajara, España), con notas sobre *Orculella bulgarica* (Hesse, 1915) (Gastropoda, Orculidae). - Iberus 27 (2): 99-105.

Rocroi, J.-P., Bichain, J.-M., Gargominy, O. 2007. De *Trichia* à *Trochulus* (Gastropoda, Hygromiidae). - MalaCo 4: 150.
Electronic document, not published in the sense of the ICZN Code.

Rodríguez, T., Castillejo, J. & Outeiro, A. 1989. Morphology and anatomy of a new Iberian species: *Deroceras geresiensis* (Gastropoda: Pulmonata: Agriolimacidae). - The Veliger 32 (1): 95-100.

Rolán, E. 2003. *Leiostyla anglica* (Pulmonata, Pupillidae) en Asturias. - Noticiario de la Sociedad Española de Malacología 39: 67-68.

Rolán, E. 2005. Malacological fauna from the Cape Verde archipelago. Part 1. Polyplacophora and Gastropoda. - pp. 1-455. Hackenheim. (ConchBooks).

Rolán, E. & Otero-Schmitt, J. 1988. Gastrópodos terrestres: nuevas citas para Galicia. - Iberus 8 (1): 111-114.

Rolle, H. 1894. Diagnosen neuer orientalischer Arten. - Nachrichtsblatt der Deutschen Malakozoologischen Gesellschaft 26 (7/8): 136-139.

Rolle, H. 1898. Eine neue *Pomatia*. - Nachrichtsblatt der Deutschen Malakozoologischen Gesellschaft 30 (5/6): 91-92.

Rolle, H. & Kobelt, W. 1895-1897. Iconographie der Land- & Süsswasser-Mollusken mit vorzüglicher Berücksichtigung der europäischen noch nicht abgebildeten Arten von E. A. Rossmässler. Neue Folge. Erster Suppplement-Band. Beiträge zur Molluskenfauna des Orients. - pp. [1-2], 1-72, Taf. 1-28, 1a, 7a. Wiesbaden. (Kreidel).
Main text must have been written by Kobelt alone (p. 59, under *Calaxis rothi* var. *cypria*), all new taxa should be attributed to Kobelt.
Dates (Hesse 1928, Bank 1989):
Heft 1-2 (10 May 1895): pp. 1-32, Taf. 1-6, 8, 10-12
Heft 3-4 (10 Oct 1895): pp. 33-48, Taf. 1a, 7, 7a, 9, 13-18
Heft 5-6 (25 Sep 1896): pp. 49-72, Taf. 19-28.

Roper, P. 2003. ADASTRA 2002. An annual review of wildlife recording in Sussex. - pp. 1-44. Woods Mill, Henfield. (Sussex Biodiversity Record Centre).

Ross, H. C. G. 1984. Catalogue of the land and freshwater Mollusca of the British Isles in the Ulster Museum. - pp. i-vi [= 1-6], 1-160. Belfast. (Ulster Museum).

Rossmässler, E. A. [1834]. Diagnoses conchyliorum terrestrium et fluviatilium. Zugleich zu Fascikeln natürlicher Exemplare. I. Heft. No. 1-20. - pp. 1-10. Dresden, Leipzig. (Arnold).
Date not given in the work, preface is dated Dec 1833. In II. Heft there is a note that I. Heft had not yet appeared in Feb 1834.

Rossmässler, E. A. [1834]. Diagnoses conchyliorum terrestrium et fluviatilium. Zugleich zu Fascikeln natürlicher Exemplare. II. Heft. No. 21-40. - pp. 1-8. Dresden, Leipzig. (Arnold).
Date not given in the work, preface is dated Feb 1834, should have been published some months afterwards.

Rossmässler, E. A. 1835-1837. Iconographie der Land- und Süßwasser-Mollusken, mit vorzüglicher Berücksichtigung der europäischen noch nicht abgebildeten Arten. Erster Band. - Dresden, Leipzig. (Arnold).
Dates (Hesse 1928, Bank 1989):
Heft 1 (Apr 1835): pp. I-VI, 1-132, [1-2], Taf. 1-5 Fig. 1-86
Heft 2 (Aug 1835): pp. [1-2], 1-26, [1-2], Taf. 5-10 Fig. 87-146
Heft 3 (Mar 1836): pp. [1-3], 1-33, Taf. 11-15 Fig. 147-214
Heft 4 (Sep 1836): pp. [1-3], 1-27, Taf. 16-20 Fig. 215-284
Heft 5/6 (July 1837): pp. [1-3], 1-70, Taf. 21-30 Fig. 285-420

Rossmässler, E. A. 1838-1844. Iconographie der Land- und Süßwassermollusken, mit vorzüglicher Berücksichtigung der europäischen noch nicht abgebildeten Arten. II. Band. - Dresden, Leipzig. (Arnold).
Heft 7/8 (June 1838): pp. I-IV, 1-43, [1], Taf. 31-40 Fig. 421-544
Heft 9/10 (Sep 1839): pp. I-IV, 1-46, Taf. 41-50 Fig. 545-681
Heft 11 (July 1842): pp. I-IV, 1-15, [1], Taf. 51-55 Fig. 682-746
Heft 12 (Sep 1844): pp. I-IV, 1-37, [1], Taf. 56-60 Fig. 747-780

Rossmässler, E. A. 1846. Diagnosen einiger neuen [sic] Binnenmollusken. - Zeitschrift für Malakozoologie 3 (11): 172-173.
Date Nov 1846 given on p. 161.

Rossmässler, E. A. 1847. *Helix ligata* Müll. Eine kritische Bemerkung. - Zeitschrift für Malakozoologie 4 (11): 161-164.
Date Nov 1847 given on p. 161.

Rossmässler, E. A. 1854-1859. Iconographie der Land- und Süsswassermollusken Europa's, mit vorzüglicher Berücksichtigung kritischer und noch nicht abgebildeter Arten. III. Band. - Leipzig. (H. Costenoble).
Heft 13/14 (Sep 1854): pp. I-VIII, 1-39, Taf. 61-70 Fig. 781-853
Heft 15/16 (Aug 1856): pp. I-VIII, 1-77, [1], Taf. 71-80 Fig. 854-894
Heft 17/18 (Dec 1859): pp. I-VIII, 1-140, Taf. 81-90 Fig. 895-969

Rossmässler, E. A. 1857. Diagnoses novorum Heliceorum. - Malakozoologische Blätter 4: 38-41.
Date Apr 1857 given on p. 33.

Roth, J. R. 1839. Molluscorum species, quas in itinere per orientem facto comites clariss. Schuberti Doctores M. Erdl et J. R. Roth collegerunt. - pp. I-VIII [= 1-8], 9-26, [1], Tab. I-II [= 1-2]. Monachii. (Dissertatio inauguralis). (C. Wolf).

Roth, J. R. 1855. Spicilegium molluscorum orientalium annis 1852 et 1853 collectorum. - Malakozoologische Blätter 2: 17-58, Tab. I-II [= 1-2].
Date Feb 1855 given on pp. 17, 33 and 49.

Roth, J. R. 1856. Ueber einige griechische Heliceen. - Malakozoologische Blätter 3 ["1857"]: 1-7, Taf. 1.
Date Jan 1856given on p. 1.

Roux, C. 1976. On the dating of the first edition of Cuvier's Règne Animal. - Journal of the Society for the Bibliography of Natural History 8 (1): 31.

Rudzīte, M., Dreijers, E., Ozoliņa-Moll, L., Parele, E., Pilāte, D., Rudzītis, M. & Stalažs, A. 2010. Latvijas gliemji: Sugu noteicējs. A guide to the molluscs of Latvia. - pp. 1-252. Rīga.

Ruiz, A., Arrébola, J. R. & Puente, A. I. 2009. *Oestophora prietoi* n. sp., *O. mariae* n. sp. and *O. ebria* (Corbella, 2004), three *Oestophora* Hesse species (Stylommatophora, Trissexodontidae) from the Iberian peninsula. - Journal of Conchology 39 (6): 693-703. London.

Ruiz Ruiz, A., Cárcaba Pozo, Á., Porras Crevilen, A. I. & Arrébola Burgos, J. R. [2006]. Caracoles terrestres de Andalucía. Guía y manual de identificación. - pp. 1-303. (Fundación Gypaetos).
No date of publication given in the work, but must have appeared in 2006.

Sabrosky, C. W. 1974. Article 50 and questions of authorship. Z.N.(S.) 1925. - Bulletin of Zoological Nomenclature 31 (4): 206-208.

Sacco, F. 1888. Aggiunte alla fauna malacologica estramarina fossile del Piemonte e della Liguria. - Memorie della Accademia delle Scienze di Torino, Classe die Scienze Fisiche, Matematiche e Naturali (2) 39: 61-98, Tav. I-II [= 1-2].

Saint-Simon, A. de 1848. Miscellanées malacologiques. Première décade. - pp. 1-41. Toulouse. (Labouisse-Rochefort).

Saint-Simon, A. de 1867. Mémoire sur les *Pomatias* du Midi de la France. - pp. 1-16. Toulouse.

Saint-Simon, A. de 1869. Descriptions d'espèces nouvelles du genre *Pomatias*, suivies d'un aperçu synonymique sur les espèces de ce genre. - Revue et Magasin de Zoologie pure et appliquée (2) 21: 3-22.

Sajó, I. 1968. Zur Clausiliidenfauna Bulgariens (Gastropoda). - Acta Zoologica Academiae Scientiarum Hungaricae 14 (3/4): 447-454. Budapest.

Sajó, I. E. 1968. Drei neue Clausiliiden aus Griechenland. - Archiv für Molluskenkunde 98 (1/2): 57-60.

Sandberger, C. L. F. von 1870-1875. Die Land- und Süsswasser-Conchylien der Vorwelt. - pp. I-VIII [= 1-8], 1-1000, Taf. I-XXXVI [= 1-36], 1 Tabelle. Wiesbaden. (Kreidel).
Dates were printed on wrappers usually not conserved after binding. Dates were verified by Woodward (1906) in correspondence with the publisher:
pp. 1-96, Taf. 1-12 (1870), pp. 97-160, Taf. 13-20 (1871), pp. 161-256, Taf. 21-32 (1872), pp. 257-352, Taf. 33-36 (1873), pp. I-VIII, 353-1000 (1875)

Sattmann, H. A. & Reischütz, P. L. 1994. Notes on land snails of Epirus (Greece) (Gastropoda: Pulmonata, Prosobranchia). - Annalen des Naturhistorischen Museums in Wien 96B: 39-48.

Saulcy, F. de 1852. Description d'une espèce nouvelle de genre *Helix*. - Journal de Conchyliologie 3: 438-439. Paris.

Say, T 1817. Description of seven species of American fresh water and land shells, not noticed in the systems. - Journal of the Academy of Natural Sciences of Philadelphia 1 (1-2): 13-18.

Say, T. 1817. Conchology. - pp. A-3 - C-6 [= 1-20], pl. [1-4], in: Nicholson, W.: American edition of the British encyclopedia, or, dictionary of arts and sciences comprising an accurate and popular view of the present improved state of human knowledge. First Edition. IV. -- pp. (?). Philadelphia. (Mitchell & Ames).

Say, T. 1821. Descriptions of univalve shells of the United States. - Journal of the Academy of Natural Sciences of Philadelphia 2 (1): 149-179.

Say, T. 1829. Descriptions of some new terrestrial and fluviatile shells of North America. - The Disseminator of Useful Knowledge; containing hints to the youth of the U. States, from the School of Industry 26 (23): 355-356. New Harmony, Indiana.

Scacchi, A. 1833. Osservazioni zoologiche. - pp. 1-27. Napoli.

Scharff, R. F. 1891. The slugs of Ireland. - The Scientific Transactions of the Royal Dublin Society (2) 4: 513-563, Pl. LVI-LVII [= 56-57].

Schenková, V., Horsák, M., Plesková, Z. & Pawlikowski, P. 2012. Habitat preferences and conservation of *Vertigo geyeri* (Gastropoda: Pulmonata) in Slovakia and Poland. - Journal of Molluscan Studies 78: 105-111.

Schileyko - see also Shilejko

Schileyko, A. A. 1991. Taxonomic status, phylogenetic relations and system of the Helicoidea sensu lato (Pulmonata). - Archiv für Molluskenkunde 120 (4/6): 187-236.

Schileyko, A. A. 1998. Treatise on recent terrestrial pulmonate molluscs. - Ruthenica Suppl. 2 (1): 1-127.

Schileyko, A. A. 2000. Treatise on recent terrestrial pulmonate molluscs. Part 5. Clausilliidae [sic]. - Ruthenica Suppl. 2 (5): 565-729.

Schileyko, A. A. 2006. Treatise on recent terrestrial pulmonate molluscs, Part 13. Helicidae, Pleurodontidae, Polygyridae, Ammonitellidae, Oreohelicidae, Thysanophoridae. - Ruthenica Supplement 2 (10): 1765-1906.

Schileyko, A. A. & Menkhorst, H. P. M. G. 1997. Composition and phylogenetic relations of the Cochlicellidae (Gastropoda, Pulmonata). - Ruthenica 7 (1): 51-60.

Schileyko, A. A. & Moisseeva, S. E. 1995. The first sinistral representative of *Turanena* (Gastropoda: Pulmonata: Enidae) and notes on the taxonomy of the genus. - Ruthenica 5 (1): 45-48.

Schilthuizen, M. 1994. Reproductive isolation in parapatric snails of the genus *Albinaria*. - Biological Journal of the Linnean Society 52: 317-324.

Schilthuizen, M. 1994. Differentiation and hybridisation in a polytypic snail. - pp. 1-178. Leiden. (Ph. D. Thesis).

Schilthuizen, M. & Gittenberger, E. 1990. Zur Formenbildung und Verbreitung der *Albinaria*-Arten von Kreta; neue Taxa der 'Gruppe der *candida*' (Gastropoda Pulmonata: Clausiliidae). - Basteria 54 (1/3): 131-142.

Schilthuizen, M., Gutteling, E., Van Moorsel, C. H. M., Welter-Schultes, F. W., Haase, M. & Gittenberger, E. **2004**. Phylogeography of the land snail *Albinaria hippolyti* (Pulmonata: Clausiliidae) from Crete, based on ITS-1 sequences. - Biological Journal of the Linnean Society 83: 317-326.

Schilthuizen, M., Welter-Schultes, F. W. & Wiese, V. **1993**. A revision of the polytypic *Albinaria hippolyti* (Boettger, 1878) from Crete (Gastropoda Pulmonata: Clausiliidae). - Zoologische Mededelingen 67 (9): 137-157. Leiden.

Schlüter, F. 1838. Kurzgefasstes systematisches Verzeichniss meiner Conchyliensammlung nebst Andeutung aller bis jetzt von mir bei Halle gefundenen Land- und Flussconchylien. Zur Erleichterung des Tausches für Freunde der Conchyliologie zusammengestellt. - pp. I-VII [= 1-7], 1-40. Halle. (Gebauer).

Schmid, G. 2002. Der Bambus-Tick oder *Paralaoma servilis*, die Gerippte Punktschnecke, in SW-Deutschland. - pp. 377-403, in: Falkner, M., Groh, K. & Speight, M. C. D.: Collectanea Malacologica. Festschrift für Gerhard Falkner. -- pp. I-X [= 1-10], 1-547, Pl. 1-48. Hackenheim. (ConchBooks).

Schmidt, **A.** 1851. Malakologische Mittheilungen. - Zeitschrift für Malakozoologie 7 (8): 113-120.

Schmidt, A. 1852. Malakologische Mittheilungen. - Zeitschrift für Malakozoologie 8 (12) ["1851"]: 179-192.
Date was 15 Jan 1852 as given on p. 192.

Schmidt, A. 1853. Kritische Bemerkungen über einige Arten von *Clausilia* und *Helix*. - Zeitschrift für die Gesammten Naturwissenschaften 1 (1): 1-10.

Schmidt, A. 1853. Bemerkungen zu dem vorstehenden Aufsatze des Herrn Prof. Rossmässler. - Zeitschrift für Malakozoologie 10 (11): 171-176.
Date Nov 1853 given on p. 176.

Schmidt, A. 1854. Malakologische Mitteilungen. - Malakozoologische Blätter 1 (1-2): 1-25.
Date Dec 1853 given on p. 1 (for pp. 1-16), Feb 1854 on p. 17 (for pp. 17-25).

Schmidt, **A.** 1854. Iconographie der Land- und Süßwasser-Mollusken Europa's, mit vorzüglicher Berücksichtigung kritischer und noch nichl [sic] abgebildeter Arten von E. A. Rossmässler, (...). Mit eingedruckten Holzschnitten und 10 colorirten lithog. Tafeln. III. Band. 1. u. 2. Heft oder 13 u. 14. Heft (Neue Folge). Leipzig, Hermann Costenoble. 1854. - Zeitschrift für Malakozoologie 1 (7): 230-249.
Date Dec 1854 given on pp. 225 and 241, name of author on p. 249.

Schmidt, **A.** 1856. Verzeichniss der Binnenmollusken Norddeutschlands mit kritischen Bemerkungen. - Zeitschrift für die Gesammten Naturwissenschaften 8: 120-169. Berlin.

Schmidt, A. 1857. Die kritischen Gruppen der europäischen Clausilien. - pp. I-VI [= 1-6], 1-63, [1], Taf. I-XI [= 1-11]. Leipzig. (Costenoble).
Almost certainly published in 1856, see Schmidt 1857: 63, Rossmässler 1856: VII-VIII, 71, Nordsieck 1993: 37. Names of taxa are dated to 1856.

Schmidt, **A.** 1868. System der europäischen Clausilien und ihrer nächsten Verwandten. - pp. [1-2], 1-176. Cassel. (Th. Fischer).

Schmidt, **F.** 1855. Beschreibung zweier neuer Höhlenthiere, eines Käfers und einer Schnecke. - Verhandlungen des Zoologisch-Botanischen Vereins in Wien (Abhandlungen) 5: 3-4.

Schneppat, U. E., Georgiev, D. G., Dedov, I. K., Wondrak, G. & Knechtle, F. 2011. *Tandonia totevi* (Wiktor, 1975) (Pulmonata: Milacidae) in Bulgaria and north eastern Greece. Redescription. - Folia Malacologica 19 (4): 225-248.

Schonenberg, D. B. & Gittenberger, E. 2008. The invasive quagga mussel *Dreissena rostriformis bugensis* (Andrusov, 1879) (Bivalvia: Dreissenidae) in the Dutch Harlingvliet, an enclosed freshwater Rhine-Meuse estuary, the westernmost record for Europe. - Basteria 72 (4/6): 345-352.

Schrank, F. von Paula 1803. Favna Boica. Durchgedachte Geschichte der in Baiern einheimischen und zahmen Thiere. Dritten und lezten Bandes zweyte Abtheilung. - pp. [1], 1-372, III-XIX [= 3-19]. Landshut. (Krüll).

Schröder, F. 1984. *Trochoidea (Xerocrassa) ebusitana* (Hidalgo, 1869) und ihre Rassen auf den Pityusen/Spanien. - Veröffentlichungen aus dem Ueberseemuseum Bremen (A) 5: 83-120.

Schröter, J. S. 1784. Einleitung in die Conchylienkenntniß nach Linné. Zweyter Band. Nebst vier Kupfertafeln. - pp. I-VIII [= 1-8], 1-726, Tab. IV-VII [= 4-7]. Halle. (Gebauer).
Consistently binominal (non-binominal names were only used in citations and not used in the consistently binominal second index at the end of the volume). Claimed to be non-binominal by Steiner & Kabat 2001: 441, this was not tenable.

Schubert, G. H. & Wagner, J. A. 1829. Neues Systematisches Conchylien-Cabinet von Martini-Chemnitz, angefangen von Martini und Chemnitz. Zwölfter Band. Erste Abtheilung. - pp. I-XII [= 1-12], 1-196, Tab. 214-237. Nürnberg. (Bauer & Raspe).
Author of the descriptions was Wagner alone, as explained in the preface.

Schultes, W. & Wiese, V. 1990. Die Gattung *Albinaria* auf Kreta: IV. Zur Verbreitung der Landschnecken auf Día. - Schriften zur Malakozoologie 3: 23-47.

Schultes, W. & Wiese, V. 1991. Die Gattung *Albinaria* auf Kreta: VII. Die Landschneckenfauna der Dionisádes-Inseln. - Schriften zur Malakozoologie 4: 76-93.

Schultheiß, R., Albrecht, C., Bößneck, U. & Wilke, T. 2008. The neglected side of speciation in ancient lakes: phylogeography of an inconspicuous mollusc taxon in lakes Ohrid and Prespa. - Hydrobiologia 615: 141-156.

Schumacher, C. F. [1815]. Afhandling over conchyliologiske systemer, og om nogle toskallede conchylier. - Oversigt over det Kongelige Danske Videnskabernes Selskabs Forhandlinger [1813-1815]: 7. København.

Schumacher, C. F. 1817. Essai d'un nouveau système des habitations des vers testacés. Avec XXII planches. - pp. [1-3], 1-287, pl. I-XXII [= 1-22]. Copenhague. (Schultz).

Schütt, H. 1962. Neue Süßwasser-Prosobranchier Griechenlands. - Archiv für Molluskenkunde 91 (4/6): 157-166.

Schütt, H. 1963. Vier bemerkenswerte Höhlenschnecken. - Archiv für Molluskenkunde 92 (5/6): 205-213.

Schütt, H. 1968. Über *Pyrgula annulata* (Linnaeus). - Archiv für Molluskenkunde 98 (1/2): 65-69.

Schütt, H. 1974. Zwei neue reliktäre Süßwassermollusken der Dinariden. - Annalen des Naturhistorischen Museums in Wien 78: 473-480, Taf. 1.

Schütt, H. 1965. Zur Systematik und Ökologie türkischer Süßwasserprosobranchier. - Zoologische Mededelingen 41 (3): 43-72, Pl. 1.

Schütt, H. 1976. Über die *Helix*-Untergattung *Maltzanella* Hesse 1917. - Archiv für Molluskenkunde 107 (1/3): 63-71.

Schütt, H. 1977. Revision der griechischen *Cochlostoma*. - Archiv für Molluskenkunde 108 (1/3): 17-35.

Schütt, H. 1978. Ein Fund von *Cochlostoma* aus der Türkei. - Archiv für Molluskenkunde 109 (1/3): 63-66.

Schütt, H. 1981. Eine unbekannte Rasse von *Cochlostoma achaicum* aus dem Peloponnes (Gastropoda: Cyclophoracea). - Archiv für Molluskenkunde 111 (1/3): 43-44.

Schütt, H. 1982. Die ostmediterrane Verbreitung von *Ancylus fluviatilis*. - Mitteilungen der Deutschen Malakozoologischen Gesellschaft 3 (36): 519-523.

Schütt, H. 1986. The dissemination of *Theodoxus* species in Greece including the islands (Moll. Prosobranchia). - Biologia Gallo-hellenica 12: 283-290. Athènes.

Schütt, H. 1993. Türkische Landschnecken. Vorläufige Zusammenstellung der aus Anatolien bekannt gewordenen gehäusetragenden Landschnecken. - pp. 1-433. Wiesbaden. (Hemmen).

Schütt, H. 1995. Diagnosen türkischer Eniden (Gastropoda: Buliminoidea: Buliminidae). - Malakologische Abhandlungen 17 (13): 161-166.

Schütt, H. 1996. Landschnecken der Türkei. 2., vollständig revidierte und erweiterte Auflage. - pp. [1-2], 1-497. Düsseldorf. (Natur & Wissenschaft).

Schütt, H. 2001. Zwei neue Arten der Gattung *Gollumia* Riedel (Gastropoda: Stylommatophora: Zonitidae). - Stuttgarter Beiträge zur Naturkunde (A (Biologie)) 629: 1-7.

Schütt, H. 2001. Die türkischen Landschnecken 1758-2000. 3., vollständig revidierte und erweiterte Auflage. - pp. 1-549. Solingen. (Natur & Wissenschaft).

Schütt, H. 2005. Turkish land snails 1758-2005. 4th, revised and enlarged edition. - pp. 1-559. Solingen. (Natur & Wissenschaft).

Schütt, H. 2010. Turkish land snails 1758-2005. 5th, revised edition with colour photos. - pp. 1-559. Solingen. (Natur & Wissenschaft).

Schütt, H. & Subai, P. 1996. Revision der Gattung *Assyriella* P. Hesse 1908 (Gastropoda: Pulmonata: Helicidae: Helicinae). - Archiv für Molluskenkunde 125 (1/2): 117-161.

Schwer, H. 1994. Populationsbiologie kleiner laubstreubewohnender Landgastopoden: Lebenszyklus und saisonale Abundanzdynamik am Beispiel von *Aegopinella pura* (Alder 1830). - Mitteilungen der Deutschen Malakozoologischen Gesellschaft 54: 13-14.

Scopoli, J. A. 1777. Introdvctio ad historiam natvralem sistens genera lapidvm, plantarvm, et animalivm hactenvs detecta, caracteribvs essentialibvs donata, in tribvs divisa, svbinde ad leges natvrae. - pp. [1-9], 3-506, [1-34]. Pragæ. (Gerle).

Seddon, M. B., Holyoak, D. T. & Tattersfield, P. 1993. Taxonomy, ecology and distribution of *Leiostyla* (Gastropoda: Pupillidae) in N.W. Europe and N.W. Africa. - Journal of Conchology 34: 299-310.

Seibert, H. 1873. Zur Kenntnis unserer Nacktschnecken. - Malakozoologische Blätter 21: 190-203.

Seidl, F. 1978. Zur Molluskenfauna von Kreta: I. Historische Aufzeichnungen. Aus dem Reisetagebuch von Hofrat Dr. Rudolf Sturany ausgewählt und kommentiert. - Mitteilungen der Zoologischen Gesellschaft in Braunau 3 (5/7): 157-193.

Seidl, F. 1978. Ein Vorkommen von *Acicula (Acicula) fusca* (Montagu) in Österreich. - Mitteilungen der Zoologischen Gesellschaft in Braunau 3 (3/4): 96-97.

Selander, R. K. & Kaufman, D. W. 1973. Self-fertilization and genetic population structure in a colonizing land snail. - Proceedings of the National Academy of Sciences of the United States of America 70 (4): 1186-1190.

Servain, G. 1880. Étude sur les mollusques recueillis en Espagne et en Portugal. - pp. [1-3], 1-172. St. Germain. (Bardin).
Date Aug 1880 indicated on the title page.

Servain, G. 1884. Excursions malacologiques en Bosnie aux environs de Serajewo et aux sources de la Bosna. - Annales de Malacologie 1 (4): 341-380. Paris.

Şeşen, R. & Schütt, H. 2005. The clausilioid genus *Idyla* in Turkey (Gastropoda: Pulmonata: Clausiliidae). - Club Conchylia Informationen 37 (1/2): 3-5.

Sevo, S. 1974. Note sur la repartition en Belgique de trois mollusques gastéropodes terrestres peu connus: *Acme inchoata* (prosobranches, Acmidae), *Abida frumentum* (stylommatophores, Vertiginidae) et *Alexia denticulata* (basommatophores, Ellobiidae). - Malacological Review 7: 1-14.

Sfenthourakis, S. & Legakis, A. 2001. Hotspots of endemic terrestrial invertebrates in southern Greece. - Biodiversity and Conservation 10 (8): 1387-1417.

Shea, M. 1982. First pleurodiscid snail in Australia. - Australian Shell News 40: 6-7.

Sheppard, R. 1823. Descriptions of seven new Bristish land and fresh-water shells, with observations upon many other species, including a list of such as have been found in the county of Suffolk. - Transactions of the Linnean Society of London 14 (1): 148-170.

Sherborn, C. D. 1902. Index animalium sive index nominum quae ab a.d. MDCCLVIII generibus et speciebus animalium imposita sunt. Sectio prima a kalendis Ianuariis, MDCCLVIII usque ad finem Decembris, MDCCC. - pp. i-lix [= 1-59], 1-1195. Cantabrigae. (Cambridge University Press).

Sherborn, C. D. 1922-1931. Index animalium sive index nominum quae ab a.d. MDCCLVIII generibus et speciebus animalium imposita sunt. Sectio secunda a kalendis Ianuariis, MDCCCI usque ad finem Decembris, MDCCCL. - pp. i-cxxxvi [= 1-136], 1-7056, [1]. London. (British Museum, Natural History).
9 separate volumes: A-B (1922), C (1924-1925), D-G (1925-1926), H-L (1927), M-N (1928), O-P (1929), Q-R (1930), S (1930), T-Z (1931), Additions (1832).

Shilejko - see also Schileyko and Shileyko

Shilejko, A. A. 1970. Ob'em, sistema i filogeniya gruppy *Perforatella - Zenobiella - Chilanodon* (Pulmonata, Helicidae). - Zoologicheskij Zhurnal 49 (9): 1306-1321.

Shilejko, A. A. 1971. Srabitel'noanatomicheskoe issledovanie nektovykh Ariantinae (Pulmonata, Helicidae). - Zoologicheskij Zhurnal 50 (7): 990-1001.

Shilejko, A. A. 1978. Fauna SSSR, Molljuski, Tom III, vyp. 6. Nazemnye mollyuski hadsemejctva Helicoidea. - pp. 1-384. Leningrad.

Shilejko, A. A. 1978. Issledovanie tipovykh vidov nekotorykh taksonov rodovoj gruppy v semejstve Buliminidae (= Enidae) (Gastropoda). 3. Vidy Kryma i Kavkaza. Voprosy sistemy semejstva. - Zoologicheskij Zhurnal 57 (6): 834-850. Moskva.

Shilejko, A. A. 1984. Nazemnye mollyuski podotryada Pupillina fauny SSSR (Gastropoda, Pulmonata, Geophila). - Fauna SSSR 3 (3): 1-399. Leningrad.

Shileyko, A. A. 1978. The system of the group *Trichia* s. lat. (Pulmonata: Helicoidea: Hygromiidae). - Malacologia 17 (1): 1-56.

Shubrat, Yu. 2007. Kompleksnyj analis konkhiologichnykh oznak vydiv rodu *Theodoxus* Ukrainy. - Visnyk L'viv Universitetu (Seriya Biologichna) 45: 154-159. [In Ukrainian].

Shuttleworth, R. J. 1843. Über die Land- und Süsswasser-Mollusken von Corsica. - Mittheilungen der Naturforschenden Gesellschaft in Bern 1843 (2/3): 9-21.

Shuttleworth, R. J. 1852. Diagnosen einiger neuen [sic] Mollusken aus den Canarischen Inseln. - Mittheilungen der Naturforschenden Gesellschaft in Bern 1852 (241/242): 137-146.

Shuttleworth, R. J. 1877. Notitiae Malacologicae oder Beiträge zur näheren Kenntnis der Mollusken. 2. Heft. - pp. 1- 16, Pl. 1-15. Leipzig. (Engelmann). [Not cited from original].
Author of the article as given on the title page is Shuttleworth, author of the textual contents is Fischer (see pp. 3-4) who compiled an explanation of Shuttleworth's plates. Text in French. The Bern edition (by Haller) was published in 1878 (Neubert & Gosteli 2003: 10).

Shuttleworth, R. J. 1877. Notitiæ Malacologicæ oder Beiträge zur näheren Kenntnis der Mollusken. II. Heft, enthaltend: I. Monographische Versuche: Paryphanta, Retinella, Mesomphix, Macrocyclis, Patera, Columna, Streptostyla, Rhynchocheila und Trochatella. (Explication des planches inédites de Shuttleworth par le Dr. P. Fischer). - pp. 1-16, Pl. 1-15. Bern. (Stämpfli).
Date of publication is 1878, given incorrectly on title page (Neubert & Gosteli 2003: 10). Had presumably the same content.

Siegismund, H. R. 1982. Life cycle and reproduction of *Hydrobia ventrosa* and *H. neglecta* (Mollusca: Prosobranchia). - Marine Ecology Progress Series 7: 75-82.

Simroth, H. 1885. Versuch einer Naturgeschichte der deutschen Nacktschnecken und ihrer europäischen Verwandten. - Zeitschrift für Wissenschaftliche Zoologie 42 (2): 203-366, Taf. 7-11. Leipzig.

Simroth, H. 1886. Weitere Mittheilungen über palaearktische Nacktschnecken. - Jahrbücher der Deutschen Malakozoologischen Gesellschaft 13: 16-34, Taf. 1.

Simroth, H. 1886. Ueber bekannte und neue palaearktische Nacktschnecken. - Jahrbücher der Deutschen Malakozoologischen Gesellschaft 13: 311-342, Taf. 10-11.

Simroth, H. 1889. Die von Herrn E. v. Oertzen in Griechenland gesammelten Nacktschnecken. - Abhandlungen der Senckenbergischen Naturforschenden Gesellschaft 16: 3-27, Taf. [1].

Simroth, H. 1889. Beiträge zur Kenntnis der Nacktschnecken. - Nachrichtsblatt der Deutschen Malakozoologischen Gesellschaft 21: 177-186.

Simroth, H. 1891. Die Nacktschnecken der portugiesisch-azorischen Fauna in ihrem Verhältnis zu denen der paläarktischen Region überhaupt. - Nova Acta Academiae Caesareae Leopoldino-Carolinae Germanicae Naturae Curiosorum 56: [1], 201-424, Tab. IX-XVIII [= 9-18]. Halle.

Simroth, H. 1892. Ueber die nackten Limaciden und Testacelliden des Kaukasus. - Sitzungsberichte der Naturforschenden Gesellschaft zu Leipzig 17-18: 40-49.

Simroth, H. 1894. Ueber einige von Herrn Dr. Sturany auf der Balkanhalbinsel erbeutete Nacktschnecken. - Annalen des kaiserlich-königlichen Naturhistorischen Hofmuseums 9 (3): 391-394, Taf. 19. Wien.

Simroth, H. 1899. Über einige kleinasiatische Nacktschnecken. - Sitzungsberichte der Naturforschenden Gesellschaft zu Leipzig 24-25 ["1897-1898"]: 35-38.

Simroth, H. 1900. Ueber einige Nacktschnecken von Montenegro und Corsica. - Nachrichtsblatt der Deutschen Malakozoologischen Gesellschaft 32 (5/6, 7/8): 77-85, 97-107. Frankfurt am Main.
Names in 7/8, morphological descriptions in 5/6.

Simroth, H. 1901. Die Nacktschneckenfauna des Russischen Reiches. - pp. I-XI [= 1-9], [1], 1-321, Taf. I-XXVI [= 1-26], Karten I-X [= 1-10]. St. Petersburg.
Date of print Dec 1901 given on p. II.

Simroth, H. 1904. Ueber die von Herrn Dr. Mrázek in Montenegro gesammelten Nacktschnecken unter Hinzunahme verwandten Materiales. - Vestník královské Ceské Spolecnosti Nauk, Trída matematicko-prírodovedecká - Sitzungsberichte der königl. Böhmischen Gesellschaft der Wissenschaften, mathematisch-naturwissenschaftliche Classe 1904 (26): 1-25, Taf. [11]. Praha.

Simroth, H. 1906. Ueber eine Reihe von Nacktschnecken, die Herr Dr. Cecconi auf Cypern und in Palaestina gesammelt hat. - Nachrichtsblatt der Deutschen Malakozoologischen Gesellschaft 38 (1; 2): 17-24, 84-91.

Simroth, H. 1910. Nacktschneckenstudien in den Südalpen. - Abhandlungen der Senckenbergischen Naturforschenden Gesellschaft 32: 275-348, Taf. 23-24.

Simroth, H. 1910. Kaukasische und asiatische Limaciden und Raublungenschnecken. - Ezhegodnik Zoologicheskago Muzeya Imperatorskoj Akademii Nauk - Annuaire du Musée Zoologique de l'Académie Impériale des Sciences de St.-Pétersbourg 15: 499-560, Pl. VI-VIII [= 6-8].

Simroth, H. 1912. Neue Beiträge zur Kenntnis der kaukasischen Nacktschneckenfauna. - Izvěstiya Kavkazskago Muzeya - Mitteilungen des Kaukasischen Museums 6 (1): 1-140. S.-Peterburg.

Simroth, H. 1916. Ueber einige von Herrn Dr. Absolon in der Herzegowina erbeutete höhlenbewohnende Nacktschnecken. - Nachrichtsblatt der Deutschen Malakozoologischen Gesellschaft 48 (1): 1-16.

Skujienė, G 2002. An overview of the data on the terrestrial molluscs in Lithuania. - Folia Malacologica 10 (1): 1-7.

Slapnik, R. 1994. Razširjenost rodu *Zospeum* Bourguignat 1856 (Gastropoda, Pulmonata, Carychiidae) v osamelem krasu vzhodne Slovenije. - Razprave [Dissertationes], Slovenska Akademija Znanosti in Umetnosti [Academia Scientiarum et Artium Slovenica], Razred za Naravoslovne Vede [Classis 4] 35 (13): 297-335. Ljubljana.

Slapnik, R. 2001. Activity and movements of *Zospeum isselianum* Pollonera 1886 (Gastropoda, Pulmonata, Carychiidae) in a cave in the Kaminške-Savinjske Alps (Slovenia). - Natura Croatica 10 (3): 153-162.

Slapnik, R. & Ozimec, R. 2004. Distribution of the genus *Zospeum* Bourguignat 1856 (Gastropoda, Pulmonata, Ellobiidae) in Croatia. - Natura Croatica 13 (2): 115-135.

Ślośarski, A. 1881. Materyjały do fauny malakologicznéj Królestwa Polskiego. - Pamiętnik Fizyograficzny 1: 292-320, Tab. IX [= 9]. Warszawa.

Smith, E. A. 1905. Description of a new species of *Vitrea* from Greece. - Proceedings of the Malacological Society of London 6 (6): 314.

Smolenska, J. 1936. Observations biologiques sur la *Bielzia coerulans* (Bielz). - Annales Musei Zoologici Polonici 11: 241-261. Warszawa.

Soes D. M., Glöer, P. & De Winter, A. J. 2009. *Viviparus acerosus* (Bourguignat, 1862) (Gastropoda: Viviparidae), a new exotic snail species for the Dutch fauna. - Aquatic Invasions 4 (2): 373-375.

Soldatenko, E. V. 2009. On the morphology of two mediterranean species of the genus *Ancylus* (Gastropoda: Pulmnonata [sic]). - Biology Bulletin 36 (4): 397-402.

Soldatenko, E. V. & Starobogatov, Ya. I. 1998. A new species of the genus *Planorbis* (Gastropoda Pulmonata Planorbidae). - Ruthenica 8 (1): 59-63.

Soldatenko, E. V. & Starobogatov, Ya. I. 2000. Genus *Planorbis* (Gastropoda Pulmonata Planorbidae). - Ruthenica 10 (1): 23-35.

Soldatenko, E. V. & Starobogatov, Ya. I. 2004. Rod *Ancylus* Müller, 1774 (Gastropoda Planorbidae). - Ruthenica 14 (1): 37-56.

Soós, L. 1917. Vizsgálatok a magyarországi Pulmonáták rendszertani anatomiája köréből. - Zur systematischen Anatomie der ungarischen Pulmonaten. - Annales Historico-Naturales Musei Nationalis Hungarici 15: 1-165. Budapest.
Main text section (pp. 1-135) in Hungarian, pp. 136-165 in German with slightly less detailed descriptions of all mentioned species.

Soós, L. 1924. Csiki Ernő állattani kutatásai Albániában. - Explorationes zoologicae ab E. Csiki in Albania peractae. XIII. Puhatestűek - Mollusca. - A Magyar Tudományos Akadémia Balkán-kutatásainak Tudományos Eredményei 1: 177-197. Budapest.

Soós, L. 1926. Eine neue Xerophila aus Deutschland, *Xer. geyeri*, und anatomische Bemerkungen über *Xer. barcinonensis* (Bgt.). - Archiv für Molluskenkunde 58 (2): 96-106.

Soós, L. 1927. Adatok a magyarországi barlangok Mollusca-faunájának ismeretéhez. - Állattani Közlemények 24: 163-180, 207-211. Budapest.

Soós, L. 1943. A Kárpát-medence Mollusca-faunája. - pp. I-XXX [= 1-30], 1-478, in: Soós, L.: Magyarország természetrajza, I. Állattani rész.

Sordelli, F. 1871. Anatomia del *Limax Doriae*, Bourg., nei suoi rapporti con altre specie congeneri. - Atti della Società Italiana di Scienze Naturali 13 (3) ["1870"]: 242-253, Tav. 3. Milano.
Wrapper of Fascicolo 3 has printed Jan 1871 (preserved in SUB Göttingen, bound before p. 201), which should be the true date, title page of the volume has printed 1870.

Sosso, M. & Dell'Angelo, B. 2010. Prima segnalazione di *Otina ovata* (Brown, 1827) (Systellomatophora: Otinidae) in Mediterraneo. - Bollettino Malacologico 46: 1-3.

Sowerby, G. B. 1821-1834. The genera of recent and fossil shells, for the use of students in conchology and geology. - pp. [unnumbered], pl. [1-265]. London.
See Petit 2006.

Sowerby, G. B. 1842-1887. Thesaurus Conchyliorum, or monographs of genera of shells. [5 volumes]. - London.

Sparacio, I. 1992. Il *Planorbis moquini* (Requien, 1848) in Sicilia (Gastropoda, Pulmonata). - Il Naturalista Siciliano (4) 16 (1-2): 129-133. Palermo.

Sparacio, I. 1997. La *Ferussacia (Pegea) carnea* (Risso, 1826) dell'isola di Pantelleria (Gastropoda, Pulmonata). - Il Naturalista Siciliano (4) 21 (3-4): 237-241. Palermo.

Spengler, L. 1793. Beskrivelse over et nyt Slægt af de toskallede Konchylier, forhen af mig kaldet Chæna, saa og over det Linnéiske Slægt Mya, hvilket nøiere bestemmes, og inddeles i tvende Slægter. - Skrivter af Naturhistorie-Selskabet 3 (1): 16-69, Tab. II [= 2]. Kiøbenhavn.

Spinelli, G. B. 1851. [Collezione dei molluschi terrestri e fluviatili della provincia Bresciana]. - Commentari dell'Ateneo di Brescia 1851: 122-168, Tav. 1.

Spyra, A. 2008. The septifer form of *Ferrissia wautieri* (Mirolli, 1960) found for the first time in Poland. - Mollusca 26 (1): 95-98. Dresden.

Stabile, J. 1859. Description de quelques coquilles nouvelles ou peu connues. - Revue et Magasin de Zoologie pure et appliquée (2) 11: 419-432, Pl. 15. Paris.

Stabile, G. 1864. Mollusques terrestres vivants du Piémont. - Atti della Società Italiana di Scienze Naturali 7 (1): 1-141, pl. [1-2]. Milano.

Stadnychenko, A. P. & Tarasova-Shubrat, J. V. 2012. The occurrence of *Theodoxus danubialis* (Mollusca: Gastropoda: Neritidae) in Ukraine. - Folia Malacologica 20 (1): 65.

Staikou, A., Lazaridou-Dimitriadou, M. & Pana, E. 1990. The life cycle, population dynamics, growth and secondary production of the edible snail *Bradybaena fruticum* (Müller 1774) (Gastropoda, Pulmonata) in Northern Greece. - Journal of Molluscan Studies 56 (2): 137-146.

Štamol, V. 1995. *Truncatellina velkovrhi* n. sp. and *Truncatellina lussinenis* n. sp., two new species from Croatia (Gastropoda: Pulmonata: Vertiginidae). - Archiv für Molluskenkunde 124: 97-101.

Štamol, V. 2010. A list of the land snails (Mollusca: Gastropoda) of Croatia, with recommendations for their Croatian names. - Natura Croatica 19 (1): 1-76.

Štamol, V., Jalžić, B. & Kletečki, E. **1999**. A contribution to knowledge about the distribution of the troglobiontic snail *Pholeoteras euthrix* Sturany, 1904 (Mollusca, Gastropoda). - Natura Croatica 8 (4): 407-419.

Štamol, V. & Kletečki, E. 2005. Terrestrial snails (Mollusca: Gastropoda terrestria) of Dugi Otok (Croatia). - Natura Croatica 14 (1): 1-28.

Štamol, V. & Kletečki, E. 2009. New finding sites of some interesting species of Croatian terrestrial malacofauna (Mollusca: Gastropoda terrestria). - Natura Croatica 18 (1): 91-112. Zagreb.

Štamol, V. & Mužinić, J. 1998. *Cochlostoma elegans* (Clessin, 1879) on the island of Pag (Mollusca: Gastropoda, Prosobranchia). - Natura Croatica 7 (2): 107-112.

Štamol, V., Riedel, A. & Jalžić, B. **1999**. A contribution to the knowledge of *Aegopis spelaeus* A. J. Wagner, 1914 (Gastropoda: Stylommatophora: Zonitidae). - Malakologische Abhandlungen 19 (21): 233-242.

Štamol, V. & Slapnik, R. 2002. *Delima (Dugiana) edmibrani* n. subgen. and n. sp. from Croatia (Gastropoda: Pulmonata: Clausiliidae). - Archiv für Molluskenkunde 130 (1/2): 239-248.

Štamol, V. & Velkovrh, F. 1995. Contribution to the knowledge of the land snail fauna of the islands of Cres and Lošinj (Croatia) (Gastropoda: Prosobranchia, Basommatophora, Stylommatophora). - Malakologische Abhandlungen 17 (20): 219-237.

Stanković, S. & Radoman, P. 1953. Contribution à la connaissance des gastéropodes ancylides du lac d'Ohrid. - Zbornik na Rabotite, Chidrobioloshki Zavod - Ochrid [Recueil des Traveaux, Station Hydrobiologique - Ohrid] 1 (5): 115-122.

Starobogatov, Ya. I. 1996. Eurasiatic species of the genus *Cochlicopa* (Gastropoda, Pulmonata, Cochlicopidae). - Ruthenica 5 (2): 105-129.

Steenberg, C.-M. 1925. Etudes sur l'anatomie et la systématique des maillots (fam. Pupillidae s. lat.). - Videnskabelige Meddelelser fra Dansk Naturhistorisk Forening i København 80: 1-215, Pl. I-XXIV [= 1-34].

Stefanescu, S. 1896. Etudes sur les terrains tertiaires de Roumanie. Contribution à l'étude des faunes sarmatique, pontique et levantine. - Mémoires de la Société Géologique de la France (Paléontologie) 15: 1-148, pl. 1-11. Paris.

Steiner, G. & Kabat, A. R. 2001. Catalogue of supraspecific taxa of Scaphopoda (Mollusca). - Zoosystema 23 (3): 433-460.

Stelfox, A. W. 1918. The *Pisidium* fauna of the Grand Junction Canal in Herts. and Bucks. - Journal of Conchology 15 (10): 289-304, Pl. 7-9. London.

Stelmashchuk, N. M. 2012. Distribution of Melanopsidae (Mollusca, Gastropoda) in Ukraine. - Folia Malacologica 20 (1): 66.

Stepien, C. A., Morton, B., Dabrowska, K. A., Guarnera, R. A., Radja, T. & Radja, B. **2001**. Genetic diversity and evolutionary relationships of the troglodytic 'living fossil' *Congeria kusceri* (Bivalvia: Dreissenidae). - Molecular Ecology 10: 1873-1879.

Stepien, C. A., Taylor, C. D. & Dabrowska, K. A. **2002**. Genetic variability and phylogeographical patterns of a nonindigenous species invasion: a comparison of exotic vs. native zebra and quagga mussel populations. - Journal of Evolutionary Biology 15: 314-328.

Sterki, V. 1880. *Hyalina depressa* n. sp. - Nachrichtsblatt der Deutschen Malakozoologischen Gesellschaft 12 (10): 104-105.

Sterki, V. 1893. Genus *Vallonia* Risso. (Vol. III, p. 169.). - pp. 247-261, pl. 32, 33, 43, 56, in: Pilsbry, H. A.: Manual of conchology; structural and systematic. With illustrations of the species. Second series: Pulmonata. Vol. VIII [= 8]. Helicidæ, Vol. VI [= 6]. -- pp. 1-314, Pl. 1-58. Philadelphia.

Steusloff, U. 1949. Beiträge zur Lebensgeschichte von *Monacha rubiginosa* und *Vertigo substriata*. (Gastropoda). - Archiv für Molluskenkunde 78 (4/6): 159-166.

Stimpson, W. 1865. Researches upon the Hydrobiinæ and allied forms; chiefly made upon materials in the Museum of the Smithsonian Institution. - Smithsonian Miscellaneous Collections 7 (201): [1-3], 1-59.

Strauch, F. 1977. Die Entwicklung der europäischen Vertreter der Gattung *Carychium* O. F. Müller seit dem Miozän (Mollusca: Basommatophora). (Das

Nordwestdeutsche Tertiärbecken, Beitrag Nr. 18). - Archiv für Molluskenkunde 107 (4/6): 149-193.

Strobel, P. de 1844. Delle conchiglie terrestri dei dintorni di Innsbruck. - Giornale dell'Imperiale Reale Istituto Lombardo di Scienze, Lettere e Arti 9 (26): 301-331.

Strobel, P. 1850. Note malacologiche d'una gita in Valbrembana nel Bergamasco. Per servire ad una topografia generale delle provincie Lombarde. - Giornale dell'Imperiale Reale Istituto Lombardo di Scienze, Lettere e Arti e Biblioteca Italiana (Nuova Serie) 2 (8): 59-70, 114-127, 250-266. Milano.

Strobel, P. 1851. Notizie malacostatiche sul Trentino. - pp. [1-3], 1-117. Pavia. (Fusi).

Strobel, P. 1854. Una nuova *Helix*. - Giornale di Malacologia 2 (7): 108. Pavia.

Strobel, P. 1855. Molluschi terrestri raccolti da Cristoforo Belloti nel 1853 in Dalmazia, con note ed agguinte. (Continuazione). - Giornale di Malacologia 2 (9) ["1854"]: 136-141. Pavia.
Published in February 1855, but a few days prior to De Betta & Martinati 1855.

Strobel, P. 1855. Beitrag zur Mollusken-Fauna von Tirol. Uebersicht der von den Gebrüdern Josef und Peregrin von Strobel in Tirol gesammelten Land-Schnecken, nebst Angabe ihrer Fundorte und ihrer Nord- und Südgrenze gegen das Donau- und Po-Tal. - Verhandlungen des Zoologisch-Botanischen Vereins in Wien 5: 153-176.

Strobel, P. 1857. Essai d'une distribution orographico-géographique des mollusques terrestres dans la Lombardie. - Memorie della Reale Accademia delle Scienze di Torino (2) 18 ["1859"]: 233-280 [= 1-50], Carte 1-2.
Separates (pp. 1-50) were published in 1857, the article (pp. 233-280) was published in 1859.

Strøm, H. 1765. Beskrivelse over Norske insecter, første stykke. - Det Trondhiemske Selskabs Skrifter 3: 376-439, Tab. VI [= 6]. Kiøbenhavn.

Studer, S. 1820. Kurzes Verzeichniss der bis jetzt in unserm Vaterlande entdeckten Conchylien. - Naturwissenschaftlicher Anzeiger der Allgemeinen Schweizerischen Gesellschaft für die Gesammten Naturwissenschaften 3 (11): 83-90, 91-94. Bern.
Dates 1 May 1820 (pp. 83-90) and 1 June 1820 (pp. 91-94): Studer's hierarchical system was somewhat chaotic and inconsistent. The generic hierarchy as interpreted by Forcart (1957: 163) is here marked with [genus], the other names are subgenera.
I. Gasteropodes. 1. [genus *Glischrus*] *Glischrus*. A. Limaces. 1. *Limax*. 2. *Cochlea*. 2. *Hyalina*. 3. *Tapada*. 4. *Helix*. A-F. A. 5. *Bulinus*. B. 6. *Pupa*. C. 7. *Torquilla*. d. A. 8. *Clausilia*. B. 2. [genus *Vertigo*] 9. *Vertigo*. 3. [genus *Carychium*] 10. *Carychium*. 4. [genus *Cyclostoma*] 11. *Cyclostoma*. 5. [genus *Paludina*] 12. *Paludina*. 6. [genus *Valvata*] 13. *Valvata*. 7. [genus *Anisus*] *Anisus*. A. 14. *Planorbis*. B. 15. *Physa*. 8. [genus *Limneus*] 16. *Limneus*. 9. [genus *Ancylus*] 17. *Ancylus*.
II. Acephali. 1. [genus *Cyclas*] 18. *Cyclas*. 2. A. [genus *Unio*] 19. *Unio*. B. [genus *Anodonta*] 20. *Anodonta*.

Stummer, A. 1983. Zur Molluskenfauna des Rudnigbaches im Bereich der Oselitzensäge bei Tröpolach. - Carinthia (2) 173 (93): 123-125. Klagenfurt.

Sturany, R. 1894. Zur Molluskenfauna der europäischen Türkei. - Annalen des kaiserlich-königlichen Naturhistorischen Hofmuseums 9 (3): 369-390, Taf. 18-20. Wien.

Sturany, R. 1897. Ueber die von Dr. H. Rebel in Bulgarien 1896 gesammelten Gehäuseschnecken. - Annalen des kaiserlich-königlichen Naturhistorischen Hofmuseums 12 (2): 111-118, Taf. 3. Wien.

Sturany, R. 1901. Ueber eine neue Höhlenschnecke. - Verhandlungen der kaiserlich-königlichen Zoologisch-Botanischen Gesellschaft in Wien 51: 761-762.

Sturany, R. 1901. Diagnosen neuer Landschnecken aus der Hercegovina. - Annalen des kaiserlich-königlichen Naturhistorischen Hofmuseums 16: [Notizen, = last section] 65-67. Wien.

Sturany, R. 1902. Beitrag zur Kenntnis der kleinasiatischen Molluskenfauna. - Sitzungsberichte, kaiserliche Akademie der Wissenschaften, mathematisch-naturwissenschaftliche Klasse, Abteilung I 111: 123-140, Taf. I-II [= 1-2]. Wien.

Sturany, R. 1904. Kurze Diagnosen neuer Gastropoden. - Anzeiger der kaiserlichen Akademie der Wissenschaften, mathematisch-naturwissenschaftliche Klasse 41 (10): 115-119. Wien.

Sturany, R. 1904. Ueber einige von Herrn G. Paganetti-Hummler entdeckte Höhlenschnecken. - Nachrichtsblatt der Deutschen Malakozoologischen Gesellschaft 36 (3): 103-107.
Date was July 1904, possibly slightly later than the notice in Anz. Akad. Wiss. Wien 41 (10).

Sturany, R. 1907. Kurze Beschreibungen neuer Gastropoden aus der Merdita (Nordalbanien). - Anzeiger der kaiserlichen Akademie der Wissenschaften, mathematisch-naturwissenschaftliche Klasse 44 (12): 229-234. Wien.

Sturany, R. 1908. Die zoologische Reise des naturwissenschaftlichen Vereines nach Dalmatien im April 1906. B. Spezieller Teil. Bearbeitung des gesammelten Materiales. 2. Mollusken. - Mitteilungen des Naturwissenschaftlichen Vereins an der Universität Wien 6 (4-5): 37-43.

Sturany, R. & Wagner, A. J. 1914. Über schalentragende Landmollusken aus Albanien und Nachbargebieten. - Anzeiger der kaiserlichen Akademie der Wissenschaften, mathematisch-naturwissenschaftliche Klasse 51: 190-193. Wien.

Sturany, R. & Wagner, A. J. 1915. Über schalentragende Landmollusken aus Albanien und Nachbargebieten. - Denkschriften der kaiserlichen Akademie der Wissenschaften, mathematisch-naturwissenschaftliche Klasse 91: 19-138, Taf. 1-18, 1 Karte. Wien.
A preprint of this work seems to have appeared in 1914, names of taxa are dated to 1914. The content suggests that the largest part of the text was written by Wagner, including most descriptions of new taxa. I do not know a reliable reference with the exact date when the preprint appeared. I also do not know the precedence between the preprint and the other paper published in 1914 in Anz. Wien 51: 190-193.

Sturm, T. 1803. Deutschlands Fauna in Abbildungen nach der Natur mit Beschreibungen. Bändchen 1. Die Würmer. - Taf. [1-34]. Nürnberg. [Not cited from original]

Stworzewicz, E. 1975. *Aegopinella lozeki* n. sp. and *Aegopinella ressmanni* (West.) (Gastropoda, Zonitidae) from the Early Pleistocene of Poland. - Acta Zoologica Cracoviensia 20 (12): 481-487.

Subai, P. 1976. Über westgriechische *Acicula*-Arten (Gastropoda, Prosobranchia). - Archiv für Molluskenkunde 107 (1/3): 119-122.

Subai, P. 1977. Über italienische *Acicula*-Arten (Gastropoda: Prosobranchia). - Archiv für Molluskenkunde 108 (1/3): 37-43.

Subai, P. 1980. Revision der lebenden Arten der Gattung *Poiretia* (Gastropoda: Oleacinidae). - Archiv für Molluskenkunde 110 (4/6): 151-172, Taf. 11, 11a.

Subai, P. 1990. Beiträge zur Molluskenfauna des Epirus, I. Zwei neue *Chilostoma*-Arten aus Nord-Epirus (Gastropoda: Ariantinae). - Annalen des Naturhistorischen Museums in Wien 91B: 243-247, Taf. 1.

Subai, P. 1990. Beschreibung einer neuen *Chilostoma* aus Griechisch-Makedonien (Gastropoda: Pulmonata: Ariantinae). - Archiv für Molluskenkunde 120 (1/3): 115-118.

Subai, P. 1993. Neue Arten seltener Landschnecken aus Griechenland und aus der Türkei (Gastropoda: Prosobranchia: Cyclophoridae; Pulmonata: Pupillidae, Clausiliidae). - Archiv für Molluskenkunde 122: 63-69.

Subai, P. 1994. Vergleich der mit *Levantina* verwandten großen Heliciden, sowie Revision der Gattung *Isaurica* (Kobelt) (Gastropoda: Helicidae). - Archiv für Molluskenkunde 123: 49-87.

Subai, P. 1995. Beitrag zur Kenntnis albanischer und griechischer Ariantinae (Gastropoda: Helicidae). - Annalen des Naturhistorischen Museums in Wien 97B: 75-94.

Subai, P. 1996. Revision der Ariantinae. 1. Die *Helicigona*-Untergattung *Thiessea* (Gastropoda: Pulmonata: Helicidae). - Archiv für Molluskenkunde 126: 1-49.

Subai, P. 1997. Beitrag zur Kenntnis albanischer und montenegrinischer Ariantinae (Gastropoda: Helicidae). - Annalen des Naturhistorischen Museums in Wien 99 B: 521-534.

Subai, P. 1999. Beschreibung einer neuen Art *Argna thracica*, sowie Angaben über die griechischen Argnidae. - Schriften zur Malakozoologie 13: 43-48.

Subai, P. 1999. Beschreibung einer neuen Art der Gattung *Metafruticicola* (Gastropoda: Hygromiidae) aus NW-Griechenland und Albanien. - Schriften zur Malakozoologie 13: 49-54.

Subai, P. 2002. Revision der Ariantinae. 2. Die *Helicigona*-Untergattung *Liburnica* (Gastropoda: Pulmonata: Helicidae). - Archiv für Molluskenkunde 131 (1/2): 1-65.

Subai, P. 2005. Revision der Gattung *Codringtonia* Kobelt 1898 (Gastropoda: Pulmonata: Helicidae: Helicinae). - Archiv für Molluskenkunde 134 (1): 67-121.

Subai, P. 2008. Revision of the Argninae, 1. The species of *Agardhiella* from the countries bordering the Adriatic Sea (Gastropoda: Pulmonata: Pupillidae). - Archiv für Molluskenkunde 137 (1): 75-103.

Subai, P. 2009. New land snails from Montenegro and Albania (Mollusca: Gastropoda). - Archiv für Molluskenkunde 138 (1): 103-112.

Subai, P. & Dedov. I. 2008. A review of the Bulgarian species of *Aspasita* Westerlund, 1889 (Gastropoda; Pulmonata; Spelaeodiscidae), with description of *A. bulgarica* spec. nov. - Basteria 72: 111-118. Leiden.

Subai, P. & Fehér, Z. 2006. Revision of the Ariantinae, 3. *Superba* n. gen., with the description of three new species (Gastropoda: Pulmonata: Helicidae). - Archiv für Molluskenkunde 135 (2): 205-223.

Subai, P. & Szekeres, M. 1999. *Sciocochlea cryptica* n.sp., a new representative of the *Serrulina* group (Gastropoda: Clausiliidae) from Epirus. - Schriften zur Malakozoologie 13: 9-12.

Sulikowska-Drozd, A. 2005. Distribution and habitat preferences of clausiliids (Gastropoda: Pulmonata: Clausiliidae) in the eastern part of the Polish Carpathians. - Folia Malacologica 13 (2): 49-94. Poznań.

Sulikowska-Drozd, A. 2008. Reproduction and growth of *Vestia elata* (Rossmässler, 1836) (Gastropoda, Clausiliidae) kept under laboratory conditions. - Mollusca 26 (1): 99-104. Dresden.

Sulikowska-Drozd, A. 2009. Egg retention and ovoviviparity in clausiliids of the genus *Vestia* P. Hesse (Gastropoda: Clausiliidae). - Journal of Molluscan Studies 75 (4): 351-359.

Sulikowska-Drozd, A. & Maltz, T. K. 2012. Reproduction of *Balea (Pseudalinda) fallax* (Rossmässler, 1836) (Gastropoda: Pulmonata: Clausiliidae) kept under laboratory conditions. - Folia Malacologica 20 (1): 27-34.

Sulikowska-Drozd, A., Maltz, T. K. & Stachyra, P. 2012. Egg retention in the clausiliid *Balea (Pseudalinda) fallax* (Rossmässler, 1836) from Roztocze (S. E. Poland). - Folia Malacologica 20 (1): 35-38.

Swainson, W. 1840. A treatise on malacology, or shells and shell-fish. - pp. [1-2], vii-viii [= 7-8], 1-419, [1-3]. London. (Longman etc.).

Sysoev, A. & Schileyko, A. 2009. Land snails and slugs of Russia and adjacent countries. - pp. 1-312, Fig. 1-142. Sofia. (Pensoft).

Szekeres, M. I. 1969. Neue Angaben zur Kenntnis der Alopien (Clausiliidae). - Archiv für Molluskenkunde 99 (1/2): 27-33.

Szekeres, M. 1976. New aspects of an *Alopia*-system. - Acta Zoologica Academiae Scientiarum Hungaricae 22: 389-396. Budapest.

Szekeres, M. 2007. Four new subspecies of *Alopia* H. & A. Adams 1855 (Gastropoda, Pulmonata: Clausiliidae). - Schriften zur Malakozoologie 23: 7-18.

Szybiak, K. & Leśniewska, M. 2008. Variability in the sculpture of the shell aperture of *Ruthenica filograna* (Rossmässler, 1836) (Gastropoda: Clausiliidae) in specimens from natural populations and from laboratory breeding. - Journal of Molluscan Studies 74: 183-189.

Talaván Gómez, J. & Talaván Serna, J. 2006. Notas sobre la malacofauna continental de Murcia y Almería. - Spira 2 (1): 27-36.

Talaván Gómez, J. & Talaván Serna, J. 2006. Contribución a la malacofauna terrestre del peñón de Gibraltar. - Spira 2 (1): 37-40.

Talaván Gómez, J. & Talaván Serna, J. 2006. *Cochlostoma (Auritus) fontqueri* (Haas, 1924) (Gastropoda: Architaenioglossa: Cochlostomatidae) en la Comunidad Valenciana. - Spira 2 (1): 65-66.

Talaván Gómez, J. & Talaván Serna, J. 2006. Moluscos testáceos de la Sierra de Francia y regiones adyacentes (Salamanca, España). - Spira 2 (2): 113-117.

Tappert, A. 2009. Die Molluskenfauna von Moskau und der Moskauer Oblast, Russland. - Schriften zur Malakozoologie 24: 5-62.

Targioni Tozzetti, A. 1873. Vertebrati e molluschi osservati o raccolti in una escursione pel Casentoni. - Atti della Società Italiana di Scienze Naturali 15 (4) ["1872"]: 309-372, Tav. 6. Milano.
Wrapper of Fascicolo 4 has printed the date June 1873 (preserved in SUB Göttingen, bound before p. 175), title page of the volume has printed 1872.

Tarruella Ruestes, A. 2006. Presencia de *Abida gittenbergeri* Bössneck, 2000 (Gastropoda, Pupilloidea, Chondrinidae) en la comarca de la Garrotxa, y nuevas citas de distribución en la comarca de l'Alt Empordà (Cataluña, España). - Spira 2 (2): 119-123. Barcelona.

Tarruella, A., Quiñonero Salgado, S. & Quintana Cardona, J. **2011**. Nuevas citas de distribución para Cataluña y la Comunidad Valenciana de *Atenia quadrasi* (Hidalgo, 1885) (Gastropoda: Helicodontidae). - Spira 4 (1/2): 93-96. Barcelona.

Tarruella, A., Quintana, J., Alba, D. M. & Prats, L. **2009**. Primera cita de *Macrogastra (Pyrostoma) attenuata lineolata* (Held, 1836) (Gastropoda: Clausiliidae) para la malacofauna ibérica. - Spira 3 (1/2): 13-25. Barcelona.

Tausch, L. 1886. Über die Fauna der nicht-marinen Ablagerungen der Oberen Kreide des Csingerthales bei Ajka im Bakony (Veszprimer Comitat, Ungarn) und über einige Conchylien der Gosaumergel von Aigen bei Salzburg. - Abhandlungen der Kaiserlich-Königlichen Geologischen Reichsanstalt 12 (1): 1-32, Taf. I-III [= 1-3]. Wien.

Taylor, J. W. 1917. Monograph of the land and freshwater Mollusca of the British Isles. Part 23. - pp. [1-2], 65-112, Pl. VI-IX [= 6-9]. Leeds. (Taylor).
Date 10 Oct 1917.

Temminck, C. J. & Laugier de Chartrouse, M. 1821-1838. Nouveau recueil de planches coloriées d'oiseaux, pour servir de suite et de complement aux planches enluminées de Buffon. - Pl. 1-600 and text, 5 volumes. Paris. (Levrault).

Terrier, A., Castella, E., Falkner, G. & Killeen, I. J. 2006. Species account for *Anisus vorticulus* (Troschel, 1834) (Gastropoda: Planorbidae), a species listed in Annexes II and IV of the Habitats Directive. - Journal of Conchology 39 (2): 193-205. London.

Terver, A.-P. 1839. Catalogue des mollusques terrestres et fluviatiles, observés dans les possessions françaises au nord de l'Afrique. - pp. 1-39, Pl. I-IV [= 1-4]. Paris, Lyon. (Baillière, Crochard, Savy).
Date was June 1839 (Sherborn 1922-1931, see entry *Helix cirtae*).

Testud, A. M. 1981. Repartition en France de l'espèce *Helicella (Jacosta) explanata* (Müller, 1774) (gastéropode pulmoné terrestre). - Haliotis 11: 225-228.

Thake, M. 1985. The biogeography of the Maltese islands, illustrated by the Clausiliidae. - Journal of Biogeography 12: 269-287.

Therriault, T. W., Docker, M. F., Orlova, M. I., Heath, D. D. & MacIsaac, H. J. 2004. Molecular resolution of the family Dreissenidae (Mollusca: Bivalvia) with emphasis on Ponto-Caspian species, including first report of *Mytilopsis leucophaeata* in the Black Sea basin. - Molecular Phylogenetics and Evolution 30: 479-489.

Thiele, J. 1931. Handbuch der systematischen Weichtierkunde. Erster Band. - pp. [1-7], 1-778. Jena. (Fischer).

Thiesse, J. 1884. Nouvelle hélice de Thessalie. - Bulletins de la Société Malacologique de France 1: 271-272. Paris.

Tiberi, D. N. 1869. Note addizionali all'articolo del Sig. Ed. von Martens: Intorno ad alcune conchiglie degli Abruzzi. - Bullettino Malacologico Italiano 2: 33-36, 65-73, 113-123. [Not cited from original]
Dates were May 1869 (pp. 33-36), Jul 1869 (pp. 65-73), Sep 1869 (pp. 113-123) (Manganelli & Cianfanelli 2001: 344).

Tomić, V. 1959. Sbirka retsentikh puzheva P. S. Pavlovića, u Prirodnjatshkom Muzeju u Beogradu. - P. S. Pavlovic's collection of the recent gastropods in the National History Museum in Beograd. - Pozebna Izdanja, Prirodnjatshku Muzej u Beogradu 27: 1-74. Beograd.

Tomlin, J. R. le Brockton 1935. A new *Albinaria* from Cyprus. - Journal of Conchology 20: 164-165. London.

Tomlin, J. R. le Brockton 1943. Gibraltar *Cecilioides*. - Journal of Conchology 22 (1): 23. London.

Torres, J. S. 2010. Nuevos datos sobre gasterópodos terrestres en la provincia de Málaga. - Noticiario de la Sociedad Española de Malacología 54: 38-40.

Torres Mínguez, A. 1923. Notes malacològiques. Un *Arion* nou i a propòsit de la seva descripció, judici critic de la opinió del Dr. Simröth [sic], considerant com una mateixa espècie els *Arion rufus* L. i l'*Arion sulcatus* Morelet. - Butlletí de la Societat de Ciéncies Naturals de Barcelona Club Muntanyenc 3: 7-11.
On the last page of the article was given a date June 1921, obviously the date when the manuscript was submitted. Date of publication was 1923 (Backeljau 1989: 211).

Torres Mínguez, A. 1925. Notas malacolócicas VII: Cuatro nuevos *Arion* ibéricos y dos nuevos limacidos de Guinea (dibujos del autor). - Butlletí de la Institució Catalana d'Història Natural (2 (5)) 8: 228-243. Barcelona.

Tournefort, J. P. de 1717. Relation d'un voyage du Levant, fait par ordre du Roy. Contenant l'histoire ancienne & moderne de plusieurs isles de l'Archipel, de Constantinople, des Côtes de la Mer Noire, de l'Armenie, de la Georgie, des frontieres de Perse & de l'Asie Mineure. Avec les plans des villes & des lieux considerables; le genie, les mœurs, le commerce & la religion des differens peuples qui les habitent; et l'explication des médailles & des monumens antiques. Enrichie de descriptions & de figures d'un grand nombre de plantes rares, de divers animaux, et de plusieurs observations touchant l'histoire naturelle. Tome troisième. - pp. [1-7], 1-404, [1-32], Pl. 1-61. Lyon. (Anisson & Posuel).

Tournouër, R. 1880. Conchyliorum fluviatilium fossilium, in stratis tertiariis superioribus Rumaniæ collectorum, novæ species. - Journal de Conchyliologie 28: 96-100.

Triantis, K. A., Vardinoyannis, K. & Mylonas, M. **2008**. Biogeography, land snails and incomplete data sets: the case of three island groups in the Aegean Sea. - Journal of Natural History 42 (5/8): 467-490.

Triantis, K. A., Pokryszko, B. M., Vardinoyannis, K. & Mylonas, M. **2004**. A new species of *Truncatellina* (Gastropoda: Vertiginidae) from Mount Ossa (= Kissavos) (Greece). - Journal of Conchology 38 (4): 393-397.

Tristram, H. B. 1865. Report on the terrestrial and fluviatile Mollusca of Palestine. - Proceedings of the Zoological Society of London 34/35: 530-545.

Troschel, F. H. 1834. De limnaeaceis seu de gasteropodis pulmonatis quae nostris in aquis vivunt. - pp. [1-5], 1-65, [1]. Berolini. (Dissertatio Zoologica).

Trüb, H. & Ribi, G. 1997. High fecundity of hybrids between the sympatric snail species *Viviparus ater* and *V. contectus* (Gastropoda: Prosobranchia). - Heredity 79: 418-423.

Tryon, G. W. 1885. Manual of Conchology; structural and systematic. Second Series: Pulmonata, 1. Testacellidæ, Oleacinidæ, Streptaxidæ, Helicoidea, Vitrinidæ, Limacidæ, Arionidæ. - pp. 1-364, Pl. 1-60. Philadelphia.

Tryon, G. W. 1863. Descriptions of new species of fresh water Mollusca, belonging to the families Amnicolidæ, Valvatidæ and Limnæidæ; inhabiting California. - Proceedings of the Academy of Natural Sciences of Philadelphia 15: 147-150, Pl. 1.

Tryon, G. W. & Pilsbry, H. A. 1888. Manual of Conchology; structural and systematic. Second Series: Pulmonata, 4. Helicidæ:- Vol. II. - pp. 1-296, Pl. 1-69. Philadelphia.
Author (p. 120) of pp. 1-120 is Tryon, of pp. 120-296 Pilsbry.

Turgeon, D. D., Quinn, J. F. jr., Bogan, A. E., Coan, E. V., Hochberg, F. G., Lyons, W. G. et al. 1998. Common and scientific names of aquatic invertebrates from the United States and Canada: mollusks, 2nd ed. - pp. 1-526. Bethesda. (American Fisheries Society).

Turner, H., Kuiper, J. G. J., Thew, N., Bernasconi, R., Rüetschi, J., Wüthrich, M. & Gosteli, M. 1998. Fauna Helvetica 2. Atlas der Mollusken der Schweiz und Liechtensteins. - pp. 1-527. Neuchâtel.

Turton, W. 1826. Conchological notices. - Zoological Journal 2 (8): 564-567. London.

Turton, W. 1831. A manual of the land and freshwater shells of the British Islands arranged according to the more modern systems of classification; and described from perfect specimens in the author's cabinet: with coloured plates of every species. - pp. i-viii [= 1-8], 1-152, Pl. [1-10]. London. (Longman etc.).
Author was not Leach. Date 1831 indicated on the title page.

Turton, W. & Gray, J. E. 1840. A manual of the land and fresh-water shells of the British Islands, with figures of each of the kinds. A new edition, thoroughly revised and much enlarged. - pp. i-ix [= 1-9], 1-324, Pl. I-XII [= 1-12]. London. (Longman, Orme, Brown, Green & Longmans).
Gray's authorship for all new information inserted to the manual was explained in the preface.

Uherkovich, A., Purger, D. & Csiky, J. 2008. First find of *Pomatias rivularis* (Eichwald, 1829) (Mollusca: Pomatiidae) in Croatia. - Natura Croatica 17 (3): 183-192.

Uit de Weerd, D. R. 2004. Molecular phylogenetic history of eastern Mediterranean Alopiinae, a group of morphologically indeterminate land snails. - pp. [1-6], 3-119. Leiden. (Thesis).

Uličný, J. 1887. Devět nových měkkýšů českých. - Vesmír 16 (10): 111-112. Praha.

Uličný, J. 1888. *Hyalina inopinata* n. sp. - Malakozoologische Blätter (Neue Folge) 10: 112-114.

Uličný, J. 1889. Zweiter Beitrag zur Kenntnis der Molluskenfauna von Mähren. - Verhandlungen des Naturforschenden Vereines in Brünn 27: 37-53.

Umiński, T. & Focht, U. 1979. Population dynamics of some land gastropods in a forest habitat in Poland. - Malacologia 18: 181-184.

Urbański, J. 1939. Über drei neue Schneckenarten aus dem südlichen Teile Jugoslaviens. - Zoologica Poloniae 3 (3): 260-266, Taf. 9. Wrocław, Warszawa.

Urbański, J. 1960. Eine neue *Delima* (Moll., Pulm.) aus Südjugoslawien (Systematische, zoogeographische und ökologische Studien über die Mollusken der Balkan-Halbinsel. III.). - Bulletin de la Société des Amis des Sciences et des Lettres de Poznan (Série D) 1: 51-55.

Urbański, J. 1960. Neue Landschnecken aus Bulgarien (Orculidae u. Pupillidae, Moll. Pulm.) (Systematische, zoogeographische und ökologische Studien über die Mollusken der Balkan-Halbinsel. IV.). - Bulletin de la Société des Amis des Sciences et des Lettres de Poznan (Série D) 1: 57-67.

Urbański, J. 1960. Ein Beitrag zur Kenntnis der Mollusken der Insel Thasos und des Mazedonischen-Thrazischen Küstenlandes. - Izvestija na Zoologitsheskija Institut s Musej 9: 71-105. Sofija.

Urbański, J. 1970. Beiträge zur Kenntnis balkanischer Vertreter des Genus *Helix* L. I (Systematische, zoogeographische und ökologische Studien über die Mollusken der Balkan-Halbinsel. X). - Bulletin de la Société des Amis des Sciences et des Lettres de Poznan (Série D) 11: 63-79, Taf. 1-6.

Urbański, J. 1972. Bemerkungen zur Kenntnis balkanischer Alopiinae (Moll., Pulm.) (Systematische, zoogeographische und ökologische Studien über die Mollusken der Balkan-Halbinsel. XI.). - Bulletin de la Société des Amis des Sciences et des Lettres de Poznan (Série D) 12/13: 251-260, Taf. I-III [= 1-3].

Urbański, J. & Wiktor, A. 1968. Beiträge zur Kenntnis bulgarischer Nacktschnecken (Moll., Pulm.) (Systematische, zoogeographische und ökologische Studien über die Mollusken der Balkan-Halbinsel. VIII). - Bulletin de la Société des Amis des Sciences et des Lettres de Poznan (Série D) 8 ["1967"]: 47-95.

Uribe, F., Hernández, E., Nebot, J., Orozco, A., Bros, V. & Cadevall, J. 2007. Variación del tamaño de la concha en tres especies de caracoles terrestres (Chondrinidae, Hygromiidae) respecto al gradiente altitudinal en los Pireneos. - Iberus 25 (2): 63-71.

[Vallot, J. N.] 1801. Exercice sur l'histoire naturelle. - pp. 1-8. Dijon. (Ecole Centrale du Départment de la Côte-d'Or).
Internal school script, printed for a selected audience, not issued for the purpose of providing a public and scientific record (Art. 8.1.1). Was preserved in the Bibliothèque Municipale de Dijon for historical reasons and many years later in the 1850s discovered in the library. Not recognised as published work by Gofas 2001: 1291, Martínez-Ortí et al. 2007: 79 and Reitano et al. 2007: 314.

Van Beneden, P. J. 1835. Histoire naturelle et anatomique de *Driessena* [sic] *polymorpha*, genre nouveau dans la famille de mylilacées [sic]. - Bulletins de l'Académie Royale des Sciences, des Lettres et des Beaux-Arts de Belgique (1) 2 (1): 25-26.

Van Beneden, P. J. 1835. Histoire naturelle et anatomique du *Dreissena polymorpha*. - Bulletins de l'Académie Royale des Sciences, des Lettres et des Beaux-Arts de Belgique (1) 2 (2): 44-47.

Van Goethem, J. L. 1988. Nouvelle liste commentée des mollusques récents non-marins de Belgique. - Documents de Travail - Studiedocumenten 53: 1-69. Bruxelles.

Van Goethem, J. L. & De Wilde, J. J. 1985. On the taxonomic status of *Deroceras caruanae* (Pollonera 1891) (Gastropoda: Pulmonata: Agriolimacidae). - Archiv für Molluskenkunde 115 (4/6): 305-309.

Van Regteren Altena, C. O. 1936. Remarks on the generic position of *Hydrobia steinii* von Martens and *Paludestrina taylori* E. A. Smith with description of a new genus. - Basteria 1 (4): 64-73.

Van Regteren Altena, C. O. 1956. Notes sur les limaces. 3. Sur la présence en France d'*Arion lusitanicus* Mabille. - Journal de Conchyliologie 95 (4) ["1955"]: 89-99. Paris.

Van Regteren Altena, C. O. 1958. Die anatomischen Merkmale von *Succinea dunkeri* L. Pfeiffer. - Archiv für Molluskenkunde 87 (1/3): 33-36.

Van Regteren Altena, C. O. 1962. Notes sur les limaces, 6-7. - Basteria 26 (3/4): 47-57.

Van Regteren Altena, C. O. 1969. Notes sur les limaces 14. Sur trois espèces de *Deroceras* de la Catalogne dont deux nouvelles. - Journal de Conchyliologie 107: 101-108. Paris.

Van Regteren Altena, C. O. 1973. Note sur les limaces, 19. Description d'une nouvelle espèce du genre *Deroceras* de la France. - Basteria 37 (1/2): 25-28.

Vardinoyannis, K. 1994. Viogeografía ton herséon malakíon sto nótio nisiotikó egeakó tóxo. - pp. [1-13], 1-330, [1-17]. Didaktorikí Diatriví (unpublished Ph. D. thesis). Athens.

Varga, A. & Pintér, L. 1972. Zur Problematik der Gattung *Hygromia* Risso 1826. - Folia Historico-Naturalia Musei Matraensis - A Mátra Múzeum Természetrajzi Közleményei 1: 121-129. Gyöngyös.

Véla, E., Magnin, F., Pavon, D. & Pfenninger, M. 2008. Phylogénie moléculaire et données paléobiogéographiques sur le gastéropode terrestre *Tudorella sulcata* (Draparnaud, 1805) en France et en Algérie orientale. - Geodiversitas 30 (1): 233-246.

Verdcourt, B. 1974. A new species of *Gulella* Pfeiffer known only from European greenhouses (Mollusca: Streptaxidae). - Archiv für Molluskenkunde 104 (4/6): 121-122.

Verdú, J. R. & Galante, E. 2006. Libro Rojo de los invertebrados de España. - pp. 1-411. Madrid. (Dirección General para la Biodiversidad, Ministerio de Medio Ambiente).

Vest, W. von 1859. Ueber die *Clausilia fallax* Rossm. und die ihr zunächst verwandten siebenbürgischen Arten. - Verhandlungen und Mittheilungen des Siebenbürgischen Vereins für Naturwissenschaften in Hermannstadt 10 (12): 259-266.

Vest, W. von 1866. Ueber *Margaritana Bonellii* Fér. (*Alasmodonta compressa* Mke.). - Verhandlungen und Mittheilungen des Siebenbürgischen Vereins für Naturwissenschaften in Hermannstadt 17 (9): 193-201, Taf. I-II [= 1-2].

Vest, W. von 1867. Ueber den Schliessapparat der Clausilien. - Verhandlungen und Mittheilungen des Siebenbürgischen Vereins für Naturwissenschaften in Hermannstadt 18 (1, 7, 8): 5-18, 161-174, 188-196.
Dates: pp. 5-18 (Jan 1867), 161-174 (July 1867), 188-196 (Aug 1867).

Vial, E. & Bertrand, A. 2002. *Chilostoma crombezi* (Locard 1882) en France (Gastropoda: Pulmonata: Helicidae). - Documents Malacologiques 3: 21-22.

Vilella, M. 1963. Notas malacológicas I. Contribución al conocimiento de "*Pyrenaearia molae*" Haas. - Miscelánea Zoológica 1: 27-32. Barcelona.

Villa, A. & Villa, G. ["J."] B. 1841. Dispositio systematica conchyliarum terrestrium et fluviatilium quae adservantur in collectione fratrum Ant. et Jo. Bapt. Villa conspectu abnormitatum novarumque specierum descriptionibus adjectis. - pp. 1-62, [1-2]. Mediolani. (Borroni & Scotti).
According to a footnote on p. 54, the descriptions pp. 54-62 were partly written by Porro.

Vimpère, J. 1999. Nouvelles données sur la repartition d'*Hygromia cinctella* (Gastropoda: Pulmonata: Hygomiidae) (Draparnaud, 1801) en France. - Vertigo 6 (1996): 43-51. Avignon.

Vinarski, M. V. & Glöer, P. 2008. Taxonomic notes on Euro-Siberian freshwater molluscs. 3. *Galba occulta* Jackiewicz, 1959 is a junior synonym of *Limnaea palustris* var. *terebra* Westerlund, 1885. - Mollusca 26 (2): 175-185. Dresden.

Vogt, D., Hey-Reidt, P. & Groh, K. 1994. Erfassung der westpaläarktischen Tiergruppen. Fundortkataster der Bundesrepublik Deutschland. Teil 20: Regionalkataster des Landes Rheinland-Pfalz: Prodromus zu einem Atlas der Mollusken von Rheinland-Pfalz. - pp. [1-3], 1-253, L1-L25. Saarbrücken, Heidelberg.

Wächtler, W. 1927. Zur Biologie der Raublungenschnecke *Poiretia (Glandina) algira* Brug. - Zoologischer Anzeiger 72 (5/8): 191-197.

Wächtler, W. 1929. Anatomie und Biologie der augenlosen Landlungenschnecke *Caecilioides acicula* Müll. - Zeitschrift für Morphologie und Ökologie der Tiere 13 (3/4): 359-462.

Wade, C. M., Mordan, P. B. & Naggs, F. 2006. Evolutionary relationships among the pulmonate land snails and slugs (Pulmonata, Stylommatophora). - Biological Journal of the Linnean Society 87: 593-610.

Wagner, **A. J.** 1895. Die Arten des Genus *Daudebardia* Hartmann in Europa und Westasien. Eine kritische Studie. - Denkschriften der kaiserlichen Akademie der Wissenschaften, mathematisch-naturwissenschaftliche Classe 62: 609-626, Taf. I-V [= 1-5]. Wien.

Wagner, A. J. 1897. Monographie der Gattung *Pomatias* Studer. - Denkschriften der kaiserlichen Akademie der Wissenschaften, mathematisch-naturwissenschaftliche Classe 64: 565-632, Taf. 1-10. Wien.

Wagner, A. 1901. Neue Formen und Fundorte des Genus *Pomatias* Studer. - Annalen des kaiserlich-königlichen Naturhistorischen Hofmuseums 16: [section "Notizen"] 63-65. Wien.

Wagner, A. J. 1906. Neue Formen und Fundorte der Genera *Pomatias* Studer und *Auritus* Westerlund. - Nachrichtsblatt der Deutschen Malakozoologischen Gesellschaft 38 (2; 3): 92-101, 121-140, Taf. 3-4.

Wagner, A. J. 1907. Zur Kenntnis der Molluskenfauna Oesterreichs und Ungarns, sowie der angrenzenden Balkanländer. - Nachrichtsblatt der Deutschen Malakozoologischen Gesellschaft 39 (3): 101-115.

Wagner, A. J. 1910. Neue Arten des Genus *Acme* Hartmann aus Süd-Dalmatien. - Anhandlungen der Senckenbergischen Naturforschenden Gesellschaft 32: 187-188, Taf. 16.

Wagner, A. J. 1912. Beschreibungen neuer Land- und Süßwasserschnecken aus Südösterreich, Kroatien und Bosnien. - Verhandlungen der Zoologisch-botanischen Gesellschaft in Wien 62: 246-260.

Wagner, A. J. 1914. Beiträge zur Anatomie und Systematik der Stylommatophoren aus dem Gebiete der Monarchie und der angrenzenden Balkanländer. - Anzeiger der kaiserlichen Akademie der Wissenschaften, mathematisch-naturwissenschaftliche Klasse 51 (15): 333-338. Wien.

Wagner, A. J. 1914. Höhlenschnecken aus Süddalmatien und der Herzegovina. - Sitzungsberichte, kaiserliche Akademie der Wissenschaften, mathematisch-naturwissenschaftliche Klasse, Abteilung I 123: 33-48. Wien.

Wagner, A. J. 1914. Iconographie der Land- & Süsswasser-Mollusken mit vorzüglicher Berücksichtigung der europäischen noch nicht abgebildeten Arten von E. A. Rossmässler, fortgesetzt von Dr. W. Kobelt. Neue Folge. Einundzwanzigster Band. - pp. 1-65, Taf. 571-600. Wiesbaden. (Kreidel).
Dates (Hesse 1928, Bank 1989):
Heft 1-2 (July 1913): pp. 1-20, Taf. 571-580
Heft 3-4 (Feb 1914): pp. 21-36, Taf. 581-590
Heft 5-6 (Jan 1915): pp. 37-65, Taf. 591-600

Wagner, A. 1915. Beiträge zur Anatomie und Systematik der Stylomatophoren [sic!] aus dem Gebiete der Monarchie und der angrenzenden Balkanländer. - Denkschriften der kaiserlichen Akademie der Wissenschaften, mathematisch-naturwissenschaftliche Klasse 91: 429-498, Taf. 1-24. Wien.
Reprints seem to have appeared in 1914, but possibly not prior to Wagner 1914, Anz. 51: 333-338.

Wagner, A. J. 1919. Zur Anatomie und Systematik der Clausiliiden. - Nachrichtsblatt der Deutschen Malakozoologischen Gesellschaft 51: 49-60, 87-104, 129-147.

Wagner, A. J. 1919. Beschreibungen neuer oder bisher wenig gekannter Clausiliiden (II. Teil). Vorläufige Mitteilung über die von Dr. A. Penther in den Jahren 1914, 1916 und 1918 in Nordalbanien gesammelten Mollusken. - Anzeiger, Akademie der Wissenschaften in Wien, mathematisch-naturwissenschaftliche Klasse 56: 70-73.

Wagner, A. J. 1920. Zur Anatomie und Systematik der Clausiliiden. - Archiv für Molluskenkunde 52: 1-13 ['Nachrichtsblatt DMG'], 67-78, 97-108, 145-158.
The journal changes its name in 1920. Dates: (Heft 1, Jan 1920) pp. 1-13 ['Nachrichtsblatt DMG'], (Heft 2, "1920") 67-78, (Heft 3, "1920") 97-108, (Heft 4, "1920") 145-158. On the orange wrapper of Heft 1 it was informed that the journal was issued quarterly. So the dates would be Apr 1920, Jul 1920 and Oct 1920 for Hefts 2, 3 and 4 respectively.

Wagner, A. J. 1921. Die Molluskensammlung des Polnischen Naturhistorischen Staatsmuseums in Warschau. I. Neue Gruppen und Formen der Subfamilie Alopiinae. - Prace Zoologiczne Polskiego Panstwowego Muzeum Przyrodniczego [Annales Zoologici Musei Polonici Historiae Naturalis] 1 (1): 41-56, Taf. 1-2. Warszawa.

Wagner, A. J. 1922. Beiträge zur Molluskenfauna Zentraleuropas. - Prace Zoologiczne Polskiego Panstwowego Muzeum Przyrodniczego [Annales Zoologici Musei Polonici Historiae Naturalis] 1 (2/3): 112-123, Taf. 5-6. Warszawa.

Wagner, A. J. 1924. Ergänzungen und Erläuterungen zur Systematik der Clausiliiden. II. Neue Formen und Arten des Genus *Albinaria* ex rect. mea. (Schluss). - Prace Zoologiczne Polskiego Panstwowego Muzeum Przyrodniczego [Annales Zoologici Musei Polonici Historiae Naturalis] 2 (2): 9-23. Warszawa.

Wagner, A. J. 1924. Wykaz systematyczny znyanych dotychzas gatunków i form Clausiliidów. III. - Systematisches Verzeichnis der mir heute bekannten Arten und Formen der Clausiliiden. III. - Prace Zoologiczne Polskiego Panstwowego Muzeum Przyrodniczego [Annales Zoologici Musei Polonici Historiae Naturalis] 3 (3/4): 99-126. Warszawa.

Wagner, A. J. 1925. Studien über die Systematik, Stammesgeschichte und geographische Verbreitung des Genus *Delima* (Hartmann) A. J. Wagner. - Prace Zoologiczne Polskiego Panstwowego Muzeum Przyrodniczego [Annales Zoologici Musei Polonici Historiae Naturalis] 4 (1): 1-73, Taf. 1-17. Warszawa.

Wagner, **A. J.** 1928. Studien zur Molluskenfauna der Balkanhalbinsel mit besonderer Berücksichtigung Bulgariens und Thraziens, nebst monographischer Bearbeitung einzelner Gruppen. - Prace Zoologiczne Polskiego Panstwowego Muzeum Przyrodniczego [Annales Zoologici Musei Polonici Historiae Naturalis] 6 (4) ["1927"]: 263-399, Taf. 10-23. Warszawa.
Date 1 Dec 1927 quoted on p. 263, but this was incorrect and 1928 should be used (Poliński 1929, Ann. Mus. Zool. Polonici, 8: 19; Hesse 1933: 224, Falkner et al. 2001: 18, Gittenberger 2006 (Basteria 70): 131).

Wagner, **H.** 1929. Adatok a Milax-nem ismeretéhez. Beiträge zur Kenntnis der Gattung *Milax*. - Annales historico-naturales Musei Nationalis Hungarici (Pars Zoologica) 26: 327-338. Budapest.
Text in both Hungarian and German.

Wagner, H. 1930. Morphologische und anatomische Studien an *Milax*. - Zoologischer Anzeiger 88: 39-57.

Wagner, H. 1931. Die in die Unterfamilie Parmacellinae gehörenden Formen des Naturhistorischen Museums in Wien. (Mit einem Nachtrag.). - Annalen des Naturhistorischen Museums in Wien 46 ["1932-1933"]: 57-76.

Wagner, H. 1931. Diagnosen neuer Limaciden aus dem Naturhistorischen Museum in Wien. - Zoologischer Anzeiger 95 (5/8): 194-202. Leipzig.

Wagner, H. 1934. Die Nacktschnecken des Königlichen Naturhistorischen Museums in Sofia. - Mitteilungen aus den königlichen Naturwissenschaftlichen Instituten in Sofia 7: 51-60. Sofija.

Wagner, H. 1940. Neue Beiträge zur Kenntnis der Nacktschneckenfauna der Balkanhalbinsel, mit besonderer Berücksichtigung der griechischen Arten. - Annales Musei Nationalis Hungarici, Pars Zoologica 33: 137-152. Budapest.

Wagner, H. ["J"]. 1941. Rendszertani tanulmányok magyarországi ragadozó tüdőscsigákon. - Mathematikai és Természettudományi Értesítő 60: 650–662. Budapest.

Wagner, **H.** 1952. Die Raublungenschneckengattungen *Daudebardia*, *Testacella* und *Poiretia*. Eine systematische, zoogeographische, ökologische und entwicklungsgeschichtliche Studie. - pp. 1-259. Budapest. (Akademia Kiado).

Wagner, **M.** 1850. Reise nach Kolchis und nach den deutschen Colonien jenseits des Kaukasus. Mit Beiträgen zur Völkerkunde und Naturgeschichte Transkaukasiens. - pp. I-VIII [= 1-8], 1-341. Leipzig. (Arnold).

Waldén, H. W. 1966. Einige Bemerkungen zum Ergänzungsband zu Ehrmann's "Mollusca", in "Die Tierwelt Mitteleuropas". - Archiv für Molluskenkunde 95 (1/2): 49-68.

Waldén, H. W. 1986. Über Variation, Verbreitung und Ökologie von *Vertigo extima*, einer für Europa neuen und bisher als selbständiges Taxon übersehenen Art (Pulmonata: Pupillacea: Vertiginidae). - Archiv für Molluskenkunde 117 (1/3): 39-59.

Waldén, H. W. 2007. Svensk landmolluskatlas. - pp. 1-271. Stenungsund. (Naturcentrum).

Walderdorff, R. 1864. Systematisches Verzeichniss der im Kreise Cattaro (Süd-Dalmatien) mit Ausnahme der Biela-Gora und in einigen angrenzenden Theilen von Montenegro und türkisch Albanien vorkommenden Land- und Süsswasser-Mollusken. - Verhandlungen der kaiserlich-königlichen Zoologisch-Botanischen Gesellschaft in Wien 14: 503-514.

Wallenberg, C. von 1858. Beschreibung einer neuen *Pupa*. - Malakozoologische Blätter 5: 32.
Date Jan 1858 given on p. 17, prior to the dissertation.

Wallenberg, C. de 1858. De molluscis Lapponiae Lulensis. - pp. 1-48, Tab. [1]. Berolini. (Dissertatio inauguralis). (Schade).
Date was 03 May 1858, mentioned on title page.

Walter, J. E. 1978. Lebenszyklen von *Lymnaea peregra* im Zürichsee (Gastropoda: Basommatophora). - Archiv für Molluskenkunde 108 (4/6): 177-184.

Walther, A. C., Lee, T., Burch, J. B. & Ó Foighil, D. **2006**. Confirmation that the North American ancylid *Ferrissia fragilis* (Tryon, 1863) is a cryptic invader of European and East Asian freshwater ecosystems. - Journal of Molluscan Studies 72 (3): 318-321.

Walther, A. C., Lee, T., Burch, J. B. & Ó Foighil, D. **2006**. *Acroloxus lacustris* is not an ancylid: a case of misidentification involving the cryptic invader *Ferrissia fragilis* (Mollusca: Pulmonata: Hygrophila). - Molecular Phylogenetics and Evolution 39: 271–275.

Webb, P. B. & Vanbeneden, P. J. 1836. Notice sur les mollusques du genre *Parmacella* de Cuvier, et description d'une nouvelle espèce de ce genre. - Magasin de Zoologie 6 (Cl. V Pl. 75 et 76): 1-12, Pl. 75-76. Paris.
In the journal the pages were not consecutively paginated, but by Classes and Plates, the arrangement for the binding was listed on pp. VII-VIII of the journal volume.
Date 1836 was quoted on the first page of the article. Authors' first names were not given on the title page, Pierre Joseph Van Beneden (Belgian name) was spelled Vanbeneden consistently in this work at various occasions.

Weigand, A. M., Götze, M. C. & Jochum, A. **2012**. Outdated but established?! Conchologically driven species delineations in microgastropods (Carychiidae, *Carychium*). - Organisms Diversity & Evolution 12: [1-10].

Weigand, A. M., Jochum, A., Pfenninger, M., Steinke, D. & Klussmann-Kolb, A. **2011**. A new approach to an old conundrum - DNA barcoding sheds new light on phenotypic plasticity and morphological stasis in microsnails (Gastropoda, Pulmonata, Carychiidae). - Molecular Ecology Resources 11: 255-265.

Weiss, K. & Wägele, H. 1998. On the morphology, anatomy and history of three species of *Onchidella* (Gastropoda: Gymnomorpha: Onchidiida). - Archiv für Molluskenkunde 127 (1/2): 69-91.

Welter-Schultes, W. 1992. Notes on the taxonomy of *Albinaria* of Nisos Dia, Crete (Gastropoda: Clausiliidae). - Biologia Gallo-hellenica 19 (1): 55-61.

Welter-Schultes, F. W. 1998. Die Landschnecken der griechischen Insel Gávdos, der südlichsten Insel Europas. - Schriften zur Malakozoologie 12: 1-120.

Welter-Schultes, F. W. 1999. The land snails of the threatened island of Gavdopoula, with description of an endemic *Mastus* species (Gastropoda: Enidae). - Biologia Gallo-hellenica 25 (2): 91-104.

Welter-Schultes, F. W. 1999. Die Gattung *Albinaria* auf Kreta: XI. Die vier *Albinaria*-Arten westlich von Kalí Liménes. - Schriften zur Malakozoologie 13: 66-78.

Welter-Schultes, F. W. 1999. Systematisches Conchylien-Cabinet von Martini und Chemnitz (1837-1920), bibliography of the volumes in Göttingen. - Archives of Natural History 26 (2): 157-203.

Welter-Schultes, F. W. 2000. Spatial analysis of shell parameters suggests low species number of *Mastus* (Gastropoda: Pulmonata: Enidae) in central and eastern Crete. - Folia Malacologica 8 (2): 151-160.

Welter-Schultes, F. W. 2000. The pattern of geographical and altitudinal variation in the land snail *Albinaria idaea* from Crete (Gastropoda: Clausiliidae). - Biological Journal of the Linnean Society 71: 237-250.

Welter-Schultes, F. W. 2000. Approaching the genus *Albinaria* in Crete from an evolutionary point of view (Pulmonata: Clausiliidae). - Schriften zur Malakozoologie 16: 1-208.

Welter-Schultes, F. W. 2001. Land snails from an ancient shipwreck: the need to detect wreck-independent finds in excavation analysis. - Journal of Archaeological Science 28: 19-27.

Welter-Schultes, F. W. 2001. Spatial variations in *Albinaria terebra* land snail morphology in Crete (Pulmonata: Clausiliidae): constraints for older and younger colonizations? - Paleobiology 27 (2): 348-368.

Welter-Schultes, F. W. 2007. The gender of *Metafruticicola* is feminine. - Schriften zur Malakozoologie 23: 87-90.

Welter-Schultes, F. W. 2008. Bronze Age shipwreck snails from Turkey: first direct evidence for oversea carriage of land snails in antiquity. - Journal of Molluscan Studies 74: 79-87.

Welter-Schultes, F. W. 2010. Revision of the genus *Albinaria* in Crete (Greece): presence of geographically variable monotypic and polytypic species. - Archiv für Molluskenkunde 139 (2): 143-245.

Welter-Schultes, F. W. 2011. Authorships of taxonomic names in malacology. - Mitteilungen der Deutschen Malakozoologischen Gesellschaft 85: 35-48.

Welter-Schultes, F. W. 2011. Case 3567. *Bulimus lineatus* Bruguière, 1789 (Gastropoda, Urocoptidae; currently *Macroceramus lineatus*) and *Bulimus lineatus* Draparnaud, 1801 (Gastropoda, Aciculidae; currently *Acicula lineata*): proposed conservation of specific names. - Bulletin of Zoological Nomenclature 68 (4): 250-252.

Welter-Schultes, F. W. 2012. *Turbo bidens* Linnaeus, 1758 (Gastropoda, Clausiliidae): request for setting aside the neotype. - Bulletin of Zoological Nomenclature 69 (2): 85-87.

Welter-Schultes, F. W. & Wiese, V. 1997. Die Landschneckenfauna der Insel Koufonísi (südlich Kreta) mit Nebeninseln. - Schriften zur Malakozoologie 10: 64-79.

Welter-Schultes, F. W. & Wiese, V. 1997. Die Landschneckfauna der Insel Chrisí (südlich Kreta) und der Nebeninsel Mikronísi. - Schriften zur Malakozoologie 10: 80-95.

Wenz, W. 1919. Zur Systematik tertiärer Land- und Süßwasser-Gastropoden. - Nachrichtsblatt der Deutschen Malakozoologischen Gesellschaft 51 (2): 76-79.

Wenz, W. 1923. Fossilium Catalogus. I: Animalia. Pars 17: Gastropoda extramarina tertiaria. - pp. 1-352. Berlin. (Junk).
Date 20 Mar 1923 indicated on the wrapper.

Wenz, W. 1923. Fossilium Catalogus. I: Animalia. Pars 18: Gastropoda extramarina tertiaria. II. - pp. 353-736. Berlin. (Junk).
Date 27 Apr 1923 indicated on the wrapper. Pars 20 see Diener 1923 (Bivalvia).

Wenz, W. 1923. Fossilium Catalogus. I: Animalia. Pars 20: Gastropoda extramarina tertiaria. III. - pp. 737-1068. Berlin. (Junk).
Date 05 Jun 1923 indicated on the wrapper.

Wenz, W. 1923. Fossilium Catalogus. I: Animalia. Pars 21: Gastropoda extramarina tertiaria. IV. - pp. 1069-1420. Berlin. (Junk).
Date 02 Aug 1923 indicated on the wrapper.

Wenz, W. 1923. Fossilium Catalogus. I: Animalia. Pars 22: Gastropoda extramarina tertiaria. V. - pp. 1421-1734. Berlin. (Junk).
Date 09 Sep 1923 indicated on the wrapper.

Wenz, W. 1923. Fossilium Catalogus. I: Animalia. Pars 23: Gastropoda extramarina tertiaria. VI. - pp. 1735-1862. Berlin. (Junk).
Date 20 Nov 1923 indicated on the wrapper.

Wenz, W. & Zilch, A. 1959-1960. Gastropoda. Teil 2. Euthyneura. - pp. I-XII [= 1-12], 1-834. Berlin. (Borntraeger).
Dates (p. XII): 17 Jul 1959 (pp. 1-200), 25 Nov 1959 (pp. 201-400), 30 Mar 1960 (pp. 401-600), 15 Aug 1960 (pp. 601-834, I-XII). Both names Wenz and Zilch appeared on the title page. In the preface (p. V) Zilch explained that the work was based on a manuscript by Wenz which Zilch had subsequently modified.

Westerlund, C. A. 1865. Sveriges land- och sötvatten-mollusker. - pp. [1], 1-142. Lund. (Berling).

Westerlund, C. A. 1868. Catalogus collectionis typicæ molluscorum terrestrium et fluviatilium Sueciæ. - pp. (??). Lund. [Not cited from original]
Data taken from LIBRIS Sweden (http://libris.kb.se, 03.2011). There were two entries for the same title, one from 1867 and one from 1868, both without page numbers. Presumably an unpublished manuscript or collection catalogue, not obtainable when first issued (Art. 8.1.2). Not found in any other library catalogue.

Westerlund, C. A. 1871. Exposé critique des mollusques de terre et d'eau douce de la Suède et de la Norvége. - pp. [1-2], 1-200. Upsal. (Berling).
Seems to have been published prior to Westerlund 1871 Fauna moll. Svec. Norv. Dan.

Westerlund, C. A. 1871. Fauna molluscorum terrestrium et fluviatilium Sveciæ, Norvegiæ et Daniæ. - Sveriges, Norges och Danmarks land- och sötvatten-mollusker. I. Landmolluskerna. - pp. [1-5], 1-296. Stockholm. (Bonnier).
Date 1871 indicated on p. [2]. Text stops suddenly, was continued without visible break in 1873.

Westerlund, C. A. 1872. *Pupa calpica* nov. spec. - Nachrichtsblatt der Deutschen Malakozoologischen Gesellschaft 4 (2): 25-28.

Westerlund, C. A. 1873. Fauna molluscorum terrestrium et fluviatilium Sveciæ, Norvegiæ et Daniæ. - Sveriges, Norges och Danmarks land- och sötvatten-mollusker. II. Sötvattenmollusker. - pp. [1-3], I-V [= 1-5], 297-651. Stockholm. (Bonnier).
Date 1873 indicated on p. [2]. Text starts suddenly, continued from 1871, genus *Carychium*.

Westerlund, C. A. 1876-1878. Fauna Europæa molluscorum extramarinorum. Prodromus. Sistens descriptiones systematicas et criticas omnium generum et specierum horum animalium in Europa viventium et hodie cognitarum. - pp. 1-320. Lundae. (Berlingia).
Fasciculus I pp. 1-160 (1876), Fasciculus II pp. 161-320 (1878, prior to Boettger 1878 in Novitates Conchologicae 5).

Westerlund, C. A. 1877. Sibiriens land- och sötvatten-mollusker. - Kongliga Svenska Vetenskaps-Akademiens Handlingar (Ny Följd) 14 (12): 1-111, Taf. [1]. Stockholm.

Westerlund, C. A. 1878. Diagnosen neuer Mollusken. - Nachrichtsblatt der Deutschen Malakozoologischen Gesellschaft 10: 108-109.

Westerlund, C. A. 1879. Malakozoologische Beiträge. - Jahrbücher der Deutschen Malakozoologischen Gesellschaft 6: 156-168.

Westerlund, C. A. 1881. Kleine kritische Bemerkungen. - Jahrbücher der Deutschen Malakozoologischen Gesellschaft 8: 1-9.

Westerlund, C. A. 1881. Malakologiska bidrag. - Öfversigt af Kongliga Vetenskaps Akademiens Förhandlingar 4: 35-69. Stockholm.

Westerlund, C. A. 1883. Malakologische Miscellen. - Jahrbücher der Deutschen Malakozoologischen Gesellschaft 10: 51-72.

Westerlund, C. A. 1884. Fauna der in der paläarctischen Region (Europa, Kaukasien, Sibirien, Turan, Persien, Kurdistan, Armenien, Mesopotamien, Kleinasien, Syrien, Arabien, Egypten, Tripolis, Tunesien, Algerien und Marocco) lebenden Binnenconchylien. IV. Gen. *Balea* Prid. & *Clausilia* Dr. - pp. 1-212, 1-18, 1-16. Karlskrona. (Länsboktryckeriet).

Westerlund, C. A. 1885. Fauna der in der paläarctischen Region (Europa, Kaukasien...) lebenden Binnenconchylien. V. Fam. Succinidæ, Auriculidæ, Limnæidæ, Cyclostomidæ & Hydrocenidæ. - pp. 1-135, 1-14. Lund. (Håkan Ohlsson).

Westerlund, C. A. 1886. Fauna der in der paläarctischen Region (Europa, Kaukasien...) lebenden Binnenconchylien. I. Fam. Testacellidæ, Glandinidæ, Vitrinidæ & Leucochroidæ. - pp. 1-88, 1-7. Lund. (Håkan Ohlsson). [Date = Apr 1886]

Westerlund, C. A. 1886. Fauna der in der paläarctischen Region (Europa, Kaukasien...) lebenden Binnenconchylien. VI. Fam. Ampullaridæ, Paludinidæ, Hydrobiidæ, Melanidæ, Valvatidæ & Neritidæ. - pp. 1-156, 1-13. Lund. (Håkan Ohlsson).

Westerlund, C. A. 1887. Fauna der in der paläarctischen Region (Europa, Kaukasien...) lebenden Binnenconchylien. III. Gen. *Buliminus*, *Sesteria*, *Pupa*, *Stenogyra* & *Cionella*. - pp. 1-183, 1-15, 1-26. Lund. (Håkan Ohlsson).

Westerlund, C. A. 1889. Fauna der in der paläarctischen Region (Europa, Kaukasien...) lebenden Binnenconchylien. II. Gen. *Helix*. - pp. 1-473, 1-31, 1-8. Berlin. (Friedländer).

Westerlund, C. A. 1890. Fauna der in der paläarctischen Region (Europa, Kaukasien...) lebenden Binnenconchylien. VII. Malacozoa acephala. - pp. 1-319, 1-16, 1-15. Berlin. (Friedländer).

Westerlund, C. A. 1890. Fauna der in der paläarctischen Region (Europa, Kaukasien...) lebenden Binnenconchylien. I. Supplement. - pp. 1-179. Berlin. (Friedländer).

Westerlund, C. A. 1890. Katalog der in der paläarctischen Region lebenden Binnenconchylien. - pp. 1-224, [1], 1-128, 1-8. Karlshamn. (Johansson).

Westerlund, C. A. 1892. Spicilegium malacologicum. Neue Binnenconchylien in der paläarktischen Region. - Verhandlungen der kaiserlich-königlichen Zoologisch-Botanischen Gesellschaft in Wien 42 (Abhandlungen): 25-48.
Date was Apr 1892.

Westerlund, C. A. 1892. Spicilegium malacologicum. Neue Binnen-Conchylien in der paläarktischen Region. III. - Nachrichtsblatt der Deutschen Malakozoologischen Gesellschaft 24 (11/12): 185-201.

Westerlund, C. A. 1893. Specilegium malacologicum. Neue Binnen-Conchylien aus der paläarktischen Region. IV. - Nachrichtsblatt der Deutschen Malakozoologischen Gesellschaft 25 (7/8): 116-132.

Westerlund, C. A. 1894. Specilegium malacologicum. Neue Binnen-Conchylien aus der paläarktischen Region. V. - Nachrichtsblatt der Deutschen Malakozoologischen Gesellschaft 26 (9/10): 163-177, 190-205.

Westerlund, C. A. 1897. Beiträge zur Molluskenfauna Russlands. (Nach den Sammlungen des Zoologischen Museum der Kaiserl. Akademie der Wissenschaften zu St.-Petersburg). - Ezhegodnik. Zoologicheskago Muzeja Imperatorskoi Akademii Nauk [Annuaire du Musée Zoologique de l'Académie Impériale des Sciences de St.-Pétersbourg] 2: 117-143.

Westerlund, C. A. 1898. Novum specilegium malacologicum. Neue Binnenconchylien aus der paläarktischen Region. - Ezhegodnik. Zoologicheskago Muzeja Imperatorskoi Akademii Nauk [Annuaire du Musée Zoologique de l'Académie Impériale des Sciences de St.-Pétersbourg] 3 (2): 155-183. S.-Peterburg.

Westerlund, C. A. 1901. Synopsis molluscorum in regione palaearctica viventium ex typo *Clausilia* Drap. - Zapiski Imperatorskoj Akademii Nauk po fiziko-matematicheskomu Otdeleniyu - Mémoires de l'Académie Impériale des Sciences de St.-Pétersbourg, Classe physico-mathématique (8) 11 (11): [1-2], I-XXXVII [= 1-37], 1-203. S.-Peterburg.

Westerlund, C. A. 1902. Methodus dispositionis conchyliorum extramarinorum in Regione palaearctica viventium, familias, genera, subgenera, et stirpes sistens. - Rad Jugoslavenske Akademije Znanosti i Umjetnosti, Matematičko-Prirodoslovni Razred 151 (32): 82-139. Zagreb.

Westerlund, C. A. 1902. Malacologische Bemerkungen und Beschreibungen. - Nachrichtsblatt der Deutschen Malakozoologischen Gesellschaft 34 (1/2, 3/4): 19-26, 35-47.
Year should be 1902, but there is a date 1 Dec 1902 printed on p. 83 as the date when Westerlund finished the manuscript, so that it is possible that the volume was issued in 1903.

Westerlund, C. A. & Blanc, H. 1879. Aperçu sur la faune malacologique de la Grèce inclus l'Epire et la Thessalie. Coquilles extramarines. - pp. 1-161, [1], Pl. I-IV [= 1-4]. Naples. (Tornese).

Wetherby, A. G. 1879. Notes on some new or little known North American Limnæidæ. - Journal of the Cincinnati Society of Natural History 2 (2): 93-100.

Whitfield, R. P. 1865. Descriptions of new species of Eocene fossils. - American Journal of Conchology 1 (3): 259-268, Pl. 27.

Wiegmann, F. 1901. Beiträge zur Anatomie. - Nachrichtsblatt der Deutschen Malakozoologischen Gesellschaft 33: 8-16.

Wiese, V. 1989. Die Gattung *Albinaria* auf Kreta: II. Clausilien auf der Insel Dia. - Schriften zur Malakozoologie 2: 39-55.

Wiese, V. 1989. Die Gattung *Albinaria* auf Kreta: III. Eine weitere dekollierte *Albinaria*. - Schriften zur Malakozoologie 2: 56-58.

Wiese, V. 1991. Atlas der Land- und Süßwassermollusken in Schleswig-Holstein. - pp. 1-251. Kiel. (Landesamt für Naturschutz und Landschaftspflege Schleswig-Holstein).

Wiese, V. 2009. Notiz zum Verhalten von *Helicigona lapicida* (Linnaeus 1758) in norddeutschen Buchenwäldern. - Schriften zur Malakozoologie 25: 46.

Wiese, V. & Richling, I. 2008. Das Mäuseöhrchen *Myosotella myosotis* (Draparnaud, 1801). Weichtier des Jahres 2008. - Club Conchylia Informationen 39 (3/4): 2-6. Ludwigsburg.

Wiktor, A. 1966. Eine neue Nacktschneckenart (Gastropoda, Limacidae) aus Polen. - Annales Zoologici 23: 449-457. Warszawa.

Wiktor, A. 1968. Eine Nacktschnecke *Milax (Milax) parvulus* Wiktor 1968 aus Bulgarien (Mollusca, Pulmonata). Neubeschreibung. - Bulletin de l'Académie Polonaise des Sciences, Série des Sciences Biologiques 16 (7): 419-422. Varsovie.

Wiktor, A. 1969. *Milax verrucosus* sp. n., eine neue Nacktschnecke aus Bulgarien (Mollusca, Pulmonata). - Bulletin de l'Académie Polonaise des Sciences, Série des Sciences Biologiques 17 (1): 47-49. Varsovie.
Date 1969 printed on title page.

Wiktor, A. 1971. Die von der niederländischen biologischen Expedition in die Türkei in 1959 gesammelten Nacktschnecken (Milacidae und Limacidae, Pulmonata) mit Beschreibung einer neuen *Deroceras*-Art aus dem Balkan-Gebiet und der Türkei. - Zoologische Mededelingen 45 (23): 261-280. Leiden.

Wiktor, A. 1973. Die Nacktschnecken Polens. - Monographiae Fauny Polski 1: 1-180.

Wiktor, A. 1975. New slug species (Pulmonata: Milacidae and Limacidae) from the Balkan Peninsula. - Annales Zoologici 33 (6): 77-91. Warszawa.

Wiktor, A. 1983. Parmacellidae of the Mediterranean Area and Canary Islands (Gastropoda, Pulmonata). - Malakologische Abhandlungen 9 (10): 81-96. Dresden.

Wiktor, A. 1983. The slugs of Bulgaria (Arionidae, Milacidae, Limacidae, Agriolimacidae - Gastropoda Stylommatophora). - Annales Zoologici 37 (3): 71-206. Warszawa.

Wiktor, A. 1984. Six *Deroceras* species from Greece, new for science (Gastropoda, Pulmonata, Agriolimacidae). - Malakologische Abhandlungen 9 (16): 151-164.

Wiktor, A. 1986. New for science Milacidae (Mollusca, Pulmonata). - Malakologische Abhandlungen 11 (13): 155-165.

Wiktor, A. 1987. Milacidae (Gastropoda, Pulmonata) - systematic monograph. - Annales Zoologici 41 (3): 153-319. Warszawa.

Wiktor, A. 1989. Limacoidea et Zonitoidea nuda. Ślimaki pomrowiokształtne (Gastropoda: Stylommatophora). - pp. 1-208. Warszawa.

Wiktor, A. 1994. Contribution to the knowledge of the slugs of Turkey (Gastropoda terrestria nuda). - Archiv für Molluskenkunde 123: 1-47.

Wiktor, A. 1996. The slugs of the former Yugoslavia (Gastropoda terrestria nuda - Arionidae, Milacidae, Limacidae, Agriolimacidae). - Annales Zoologici 46 (1-2): 1-110. Warszawa.

Wiktor, A. 1998. Two new for science *Deroceras* species from France and the Italian island of Sicily (Gastropoda: Pulmonata: Agriolimacidae). - Malakologische Abhandlungen 19: 13-19.

Wiktor, A. 2000. Agriolimacidae (Gastropoda: Pulmonata) - a systematic monograph. - Annales Zoologici 49 (4): 347-590. Warszawa.

Wiktor, A. 2001. The slugs of Greece (Arionidae, Milacidae, Limacidae, Agriolimacidae - Gastropoda, Stylommatophora). - Fauna Graeciae 8: I-VIII, 1-241. Iraklio.

Wiktor, A. 2004. Ślimaki lądowe Polski. - pp. 1-302. Olsztyn. (Mantis).

Wiktor, A. 2007. A check-list of terrestrial slugs of Turkey with some new data and a description of a new species (Gastropoda terrestria nuda). - Folia Malacologica 15 (3): 95-107.

Wiktor, A., Martín, R. & Castillejo, J. **1990**. A new slug family Papillodermidae with description of a new genus and species from Spain (Gastropoda, Pulmonata terrestria nuda). - Malakologische Abhandlungen 15 (1): 1-18. Dresden.

Wiktor, A. & Milani, N. 1995. Contribution to the knowledge of two scarcely known Alpine slugs, *Tandonia simrothi* (Hesse, 1923) and *Deroceras planarioides* (Simroth, 1910) (Gastropoda: Pulmonata: Milacidae et Agriolimacidae). - Malakologische Abhandlungen 17 (12): 151-160.

Wiktor, A. & Mylonas, M. 1981. New *Deroceras* species from Cyclades islands (Greece) (Gastropoda, Pulmonata, Agriolimacidae). - Malakologische Abhandlungen 7 (17): 183-193.

Wiktor, A. & Mylonas, M. 1986. *Milax aegaeicus* n. sp. from Cyclades (Gastropoda, Pulmonata, Milacidae). - Malakologische Abhandlungen 11 (12): 147-153.

Wiktor, A. & Mylonas, M. 1992. *Deroceras astypalaeensis* n. sp. from the Greek island Astipalaia (Dodecanesos) (Gastropoda, Pulmonata: Agriolimacidae). - Malakologische Abhandlungen 16 (1): 1-6.

Wiktor, A. & Norris, A. 1982. The synonymy of *Limax maculatus* (Kaleniczenko 1851) with notes on its European distribution. - Journal of Conchology 31: 75-77.

Wiktor, A. & Norris, A. 1991. *Geomalacus malagensis* sp. n. and some remarks concerning generic level systematics of Arionidae (Gastropoda: Pulmonata). - Journal of Conchology 34 (1): 19-29. London.

Wiktor, A. & Parejo, C. 1989. *Arion (Kobeltia) paularensis* sp. n. from central Spain (Gastropoda, Pulmonata: Arionidae). - Malakologische Abhandlungen 14 (4): 27-33. Dresden.

Wiktor, A., Quintana, J. & Beckmann, K.-H. **2007**. Redescription of "*Limax majoricensis*" (Heynemann 1863) (Gastropoda: Pulmonata: Limacidae) from the Balearic Islands. - pp. 187-197, in: Beckmann, K.-H.: Die Land- und Süsswassermollusken der Balearischen Inseln. -- pp. 1-255. Hackenheim. (ConchBooks).

Wiktor, A., Vardinoyannis, K. & Mylonas, M. **1994**. Slugs of the Greek southern Aegean islands (Mollusca, Gastropoda terrestria nuda: Milacidae, Agriolimacidae et Limacidae). - Malakologische Abhandlungen 17: 1-36.

Williamson, P. 1979. Age determination of juvenile and adult *Cepaea*. - Journal of Molluscan Studies 45 (1): 52-60.

Wilson, E. O. & Brown, W. L. 1953. The subspecies concept and its taxonomic implication. - Systematic Zoology 2: 97-111.

Winckworth, R. 1936. Journal de Conchyliologie: dates of publication. - Journal of Molluscan Studies 22: 153-156.

Wirth, T., Baur, A. & Baur, B. 1997. Mating system and genetic variability in the simultaneously hermaphroditic terrestrial gastropod *Balea perversa* on the Baltic island of Öland, Sweden. - Hereditas 126: 199-209.

Wohlberedt, O. 1907. Kopneni mekushti Tsrne Gore. (Prilozi faune Tsrne Gore). - Glasnik Zemaljskog Muzeja u Bosni i Hercegovini 19 (4): 499-574, Tab. 1-10. Sarajevo.

Wohlberedt, O. 1909. Zur Fauna Montenegros und Nordalbaniens (Mollusken, Käfer, Isopoden, Chilopoden, Diplopoden). - Wissenschaftliche Mitteilungen aus Bosnien und der Herzegowina 11: 585-722, Taf. XLVII-LVII [= 47-57]. Wien.

Woodward, B. B. 1903. Note on the date of publication of J. F. Gray's "Catalogue of Pulmonata.. in the ... British Museum. Pt. I", 12mo, 1855; and of A. Moquin-Tandon's "Histoire naturelle des mollusques terrestres et fluviatles de la France", 2 vols. 8vo, 1855. - Proceedings of the Malacological Society of London 5: 261.

Woodward, B. B. 1906. On the dates of publication of C. L. F. von Sandberger's 'Die Land- und Süsswasser-Conchylien der Vorwelt,' 4to, Wiesbaden (C. W. Kreidel), 1870-1875. - Proceedings of the Malacological Society of London 7: 3-4.

Woodward, B. B. 1924. On the true date of publication of Nilsson's Historia Molluscorum Sveciæ. 8vo. Lundæ, 1822. - Proceedings of the Malacological Society of London 16 (1): 23.

Wüthrich, M. 1984. *Catinella (Quickella) arenaria* (Bouchard-Chantereaux 1837) in den Schweizer Alpen. - Archiv für Molluskenkunde 114 (4/6): 231-241.

Yıldırım, M. Z. 1997. A new subspecies of *Sprattia sowerbyana* O. Boettger 1883 from Eğirdir (Turkey): *Sprattia sowerbyana aksoylari* nov. subsp. (Mollusca, Pulmonata, Clausiliidae). - Turkish Journal of Zoology 21: 219-221.

Yıldırım, M. Z. 1999. Türkiye Prosobranchia (Gastropoda, Mollusca) türleri ve zoocoğrafik yayılışları. 1. Tatlı ve acı sular. - Turkish Journal of Zoology 23: 877-900.

Zajac, K. 2005. Threatened molluscs of Poland. - Tentacle 13: 13-15.

Zettler, M. L., Jueg, U., Menzel-Harloff, H., Göllnitz, U., Petrick, S., Weber, E. & Seemann, R. 2006. Die Land- und Süßwassermollusken Mecklenburg-Vorpommerns. - pp. 1-318. Schwerin. (Obotritendruck).

Zhiltsov, S. S. & Shilejko, A. A. 2002. Stroenie polovogo apparata *Bielzia coerulans* (Gastropoda, Pulmonata) i filogeneticheskie svyazi roda *Bielzia*. - Ruthenica 12 (1): 73-79.

Zilch, A. 1951. Ergebnisse zoologisch-geologischer Sammelreisen H. Kaltenbach's in NO-Afrika, 5: Familie Enidae. - Archiv für Molluskenkunde 80 (1/3): 33-46, Taf. 3-3a.

Zilch, A. 1952. Die Typen und Typoide des Natur-Museums Senckenberg, 8: Mollusca, Helicinae (2). - Archiv für Molluskenkunde 81 (4/6): 135-173, Taf. 5-6.

Zilch, A. 1958. Zur Kenntnis der Gattung *Argna* Cossmann (Pupillidae, Lauriinae). - Archiv für Molluskenkunde 87 (4/6): 149-152, Taf. 11-12.

Zilch, A. 1969. Die Typen und Typoide des Natur-Museums Senckenberg, 43: Mollusca, Pupillacea 2 (Valloniidae, Pleurodiscidae). - Archiv für Molluskenkunde 99 (3/4): 221-245.

Zilch, A. 1965. Die Typen und Typoide des Natur-Museums Senckenberg, 30: Mollusca: Zonitidae: Zonitinae (1). - Archiv für Molluskenkunde 94 (1/2): 75-97.

Zilch, A. 1976. Die Typen und Typoide des Natur-Museums Senckenberg, 56: Mollusca: Aciculidae. - Archiv für Molluskenkunde 107 (1/3): 123-136.

Zilch, A. 1977. Die Typen und Typoide des Natur-Museums Senckenberg, 57: Mollusca: Clausiliidae (4): Alopiinae (2): Alopiini (1). - Archiv für Molluskenkunde 107 (4/6): 309-363, Taf. 25-28.

Zilch, A., Nordsieck, H. & Neubert, E. 2002. Die Typen und Typoide des Natur-Museums Senckenberg, 83: Mollusca: Clausiliidae (7): Alopiinae (5): Delimini (1). - Archiv für Molluskenkunde 130 (1/2): 201-237, Taf. 1-10.

Zimmermann, **F.** 1925. Untersuchungen über die Variabilität einiger Arten des Genus *Carychium* Müller. - Zeitschrift für Induktive Abstammungs- und Vererbungslehre 37: 291-342, Taf. 1-2. Leipzig.

Zimmermann, **S.** 1931. *Orcula fuchsi* n. sp. - Archiv für Molluskenkunde 63: 44-46, Taf. IV [= 4].

Zuev, G. V., Ovcharov, O. P. & Chesalin, M. V. 1995. Reproduction of *Helix albescens* Rossmässler, 1839 (Pulmonata, Helicidae) in captivity. - Ruthenica 5 (1): 49-54.

i Index

This index was compiled by hand (semiautomatically). Some names might have been overlooked, I hope not many. The generic names for species given in this index are meant to facilitate the orientation, they do not imply statements on their current classification. For the same reason synonyms do not necessarily agree in gender with the genus. Usage of names in the sense of Art. 23.9.2 shall not be derived from this index.

Francisco Welter-Schultes

European non-marine molluscs, a guide for species identification

Quick identification plates

Bestimmungsbuch für europäische Land- und Süsswassermollusken

Tafeln zur schnellen Vorbestimmung

Territorial divisions of some European countries

France, Germany, Austria, Italy and Spain see p. 675

If the maps in this atlas contradict the maps in the textbook, those in the textbook are more reliable. The maps in the atlas were manually updated.

Bivalvia

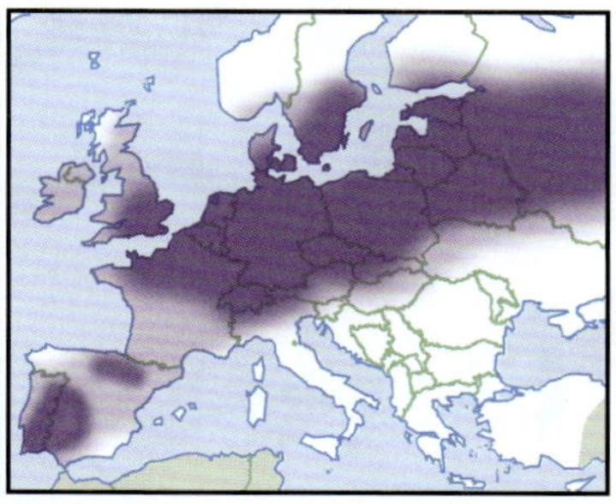

Anodonta anatina

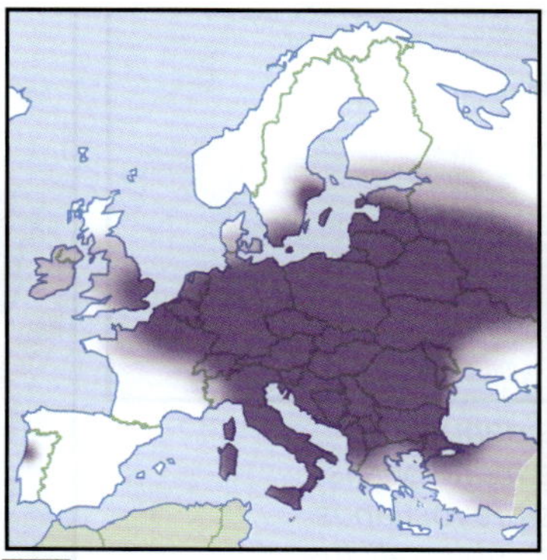

Anodonta cygnea

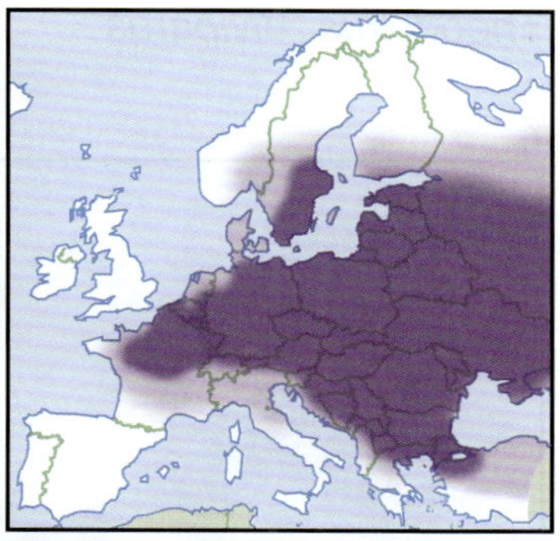

Unio crassus

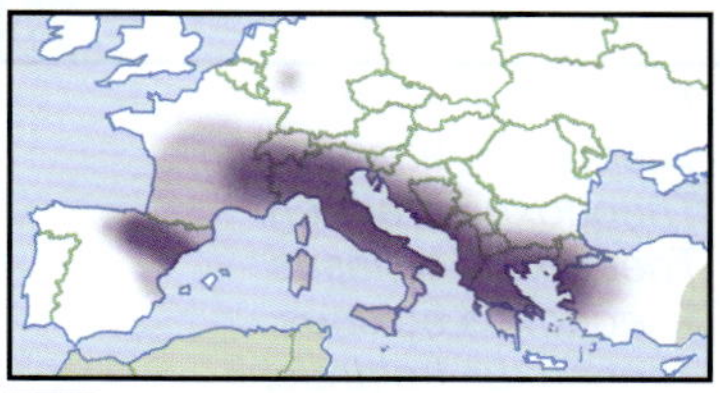

Unio mancus

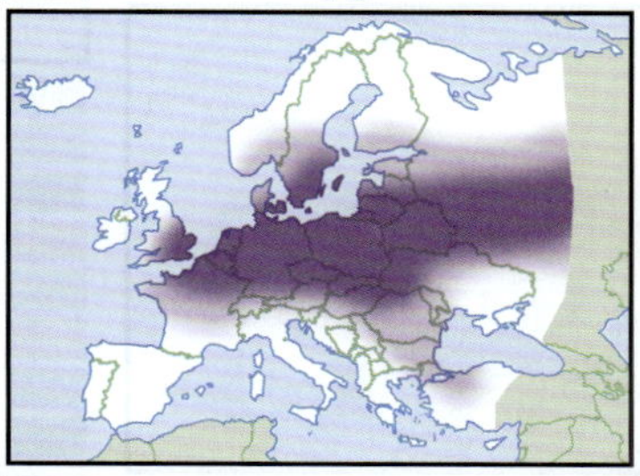

Unio pictorum

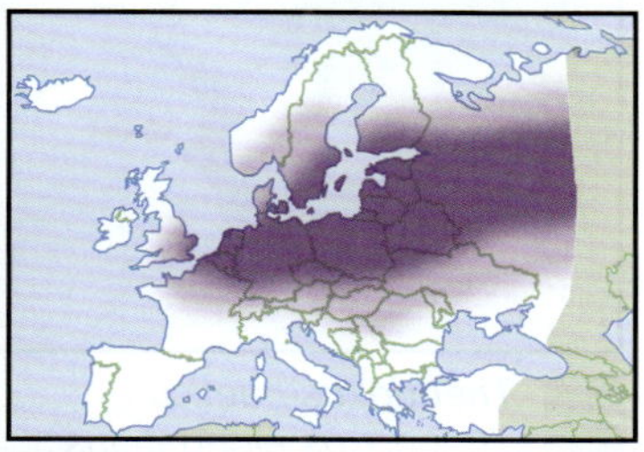

Unio tumidus

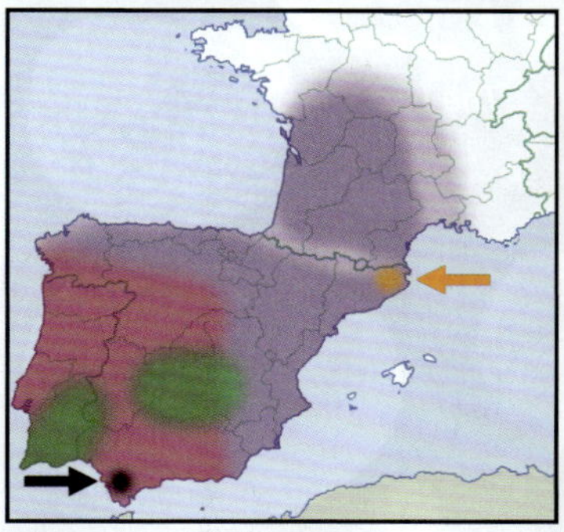

Potomida littoralis
Unio delphinus
Unio gibbus
Unio ravoisieri
Unio tumidiformis

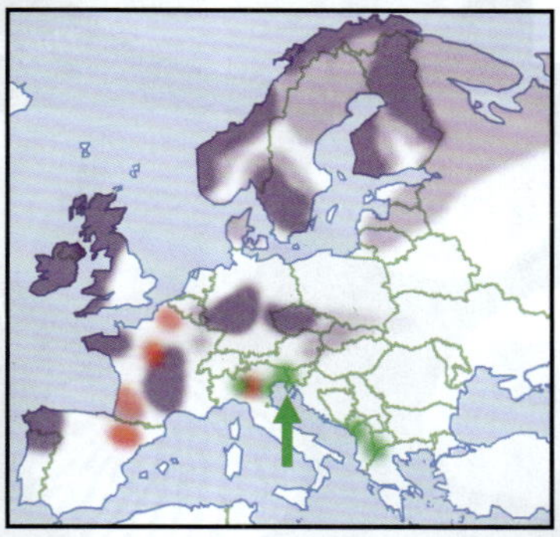

Margaritifera auricularia
Margaritifera margaritifera
Microcondylaea bonellii

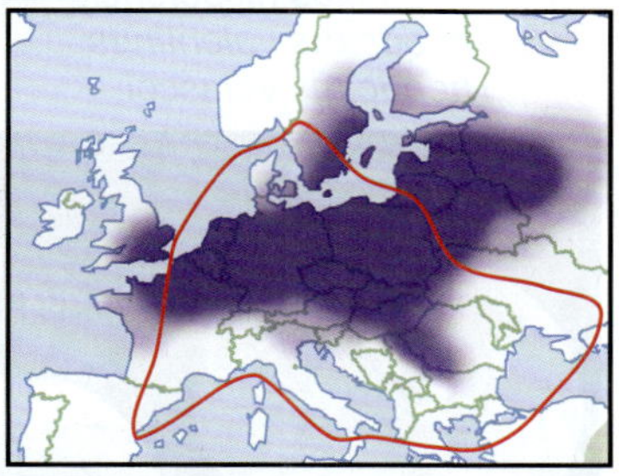

Pseudanodonta complanata
Sinanodonta woodiana

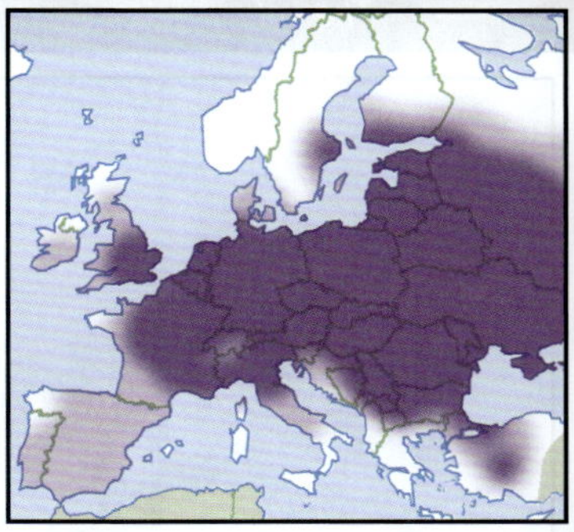

Dreissena polymorpha

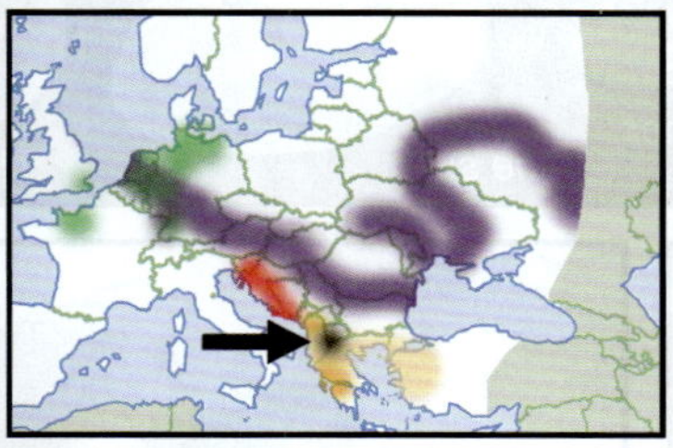

Dreissena blanci
Dreissena rostriformis
Dreissena stankovici
Mytilopsis leucophaeata
Congeria kusceri

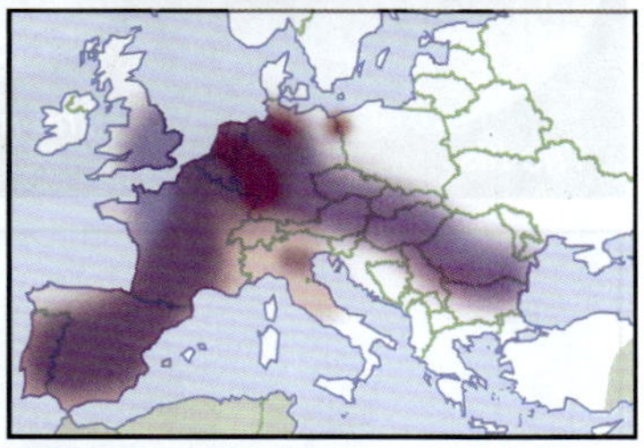

Corbicula fluminea (fl-a)
Corbicula fluminalis (fl-s)

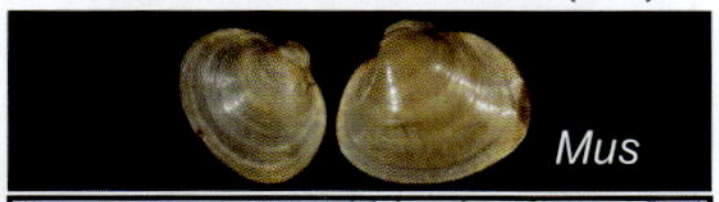

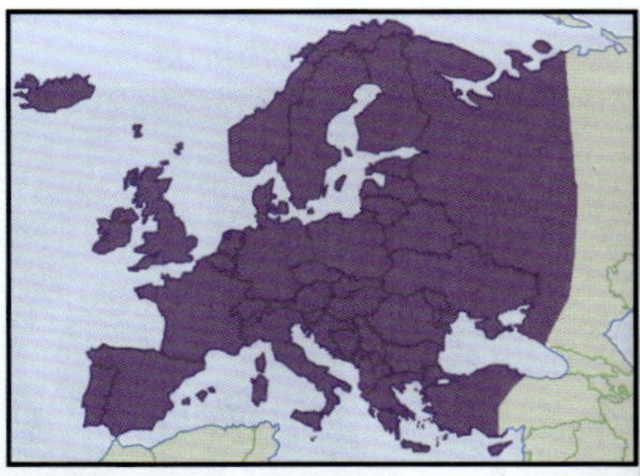

Sphaerium

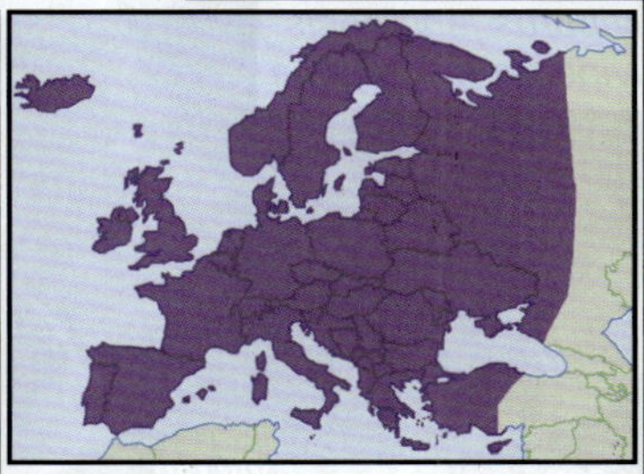

Pisidium

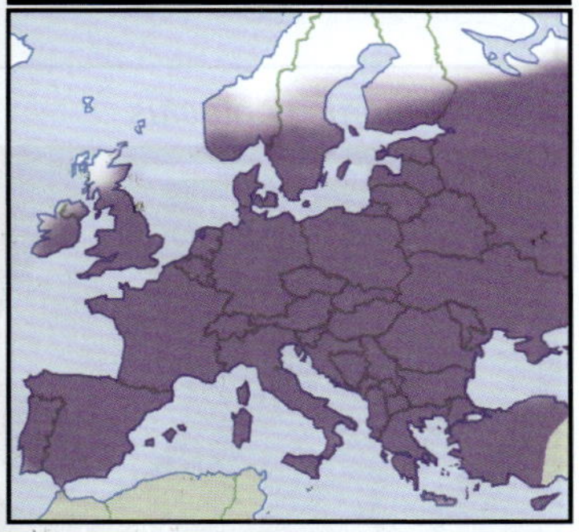

Musculium

Theodoxus, Viviparus

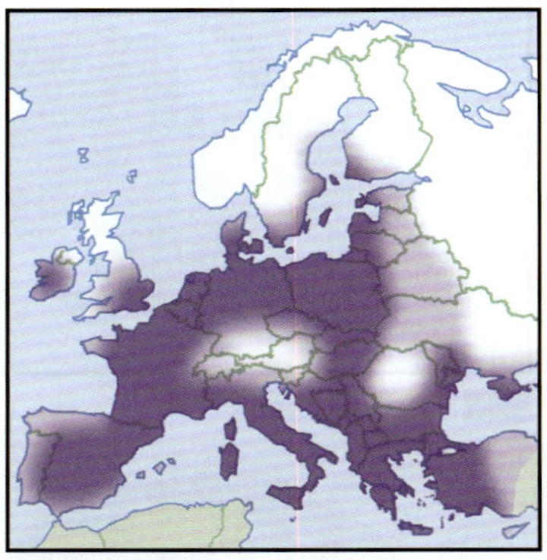

Theodoxus fluviatilis

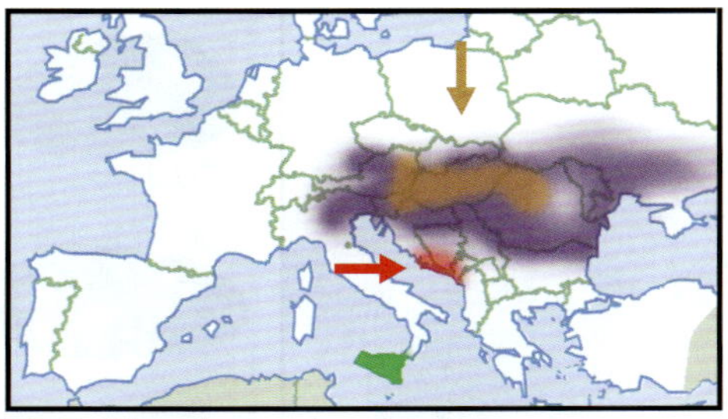

Theodoxus danubialis
Theodoxus meridionalis
Theodoxus prevostianus
Theodoxus subterrelictus

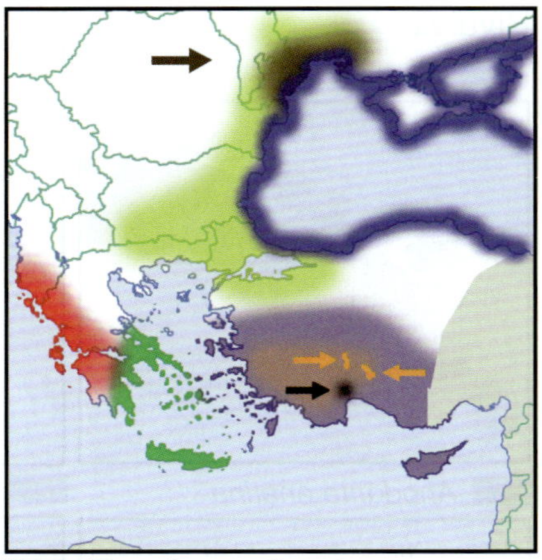

Theodoxus altenai
Theodoxus anatolicus
Theodoxus brauneri
Theodoxus callosus
Theodoxus euxinus
Theodoxus heldreichi
Theodoxus pallasi
Theodoxus saulcyi

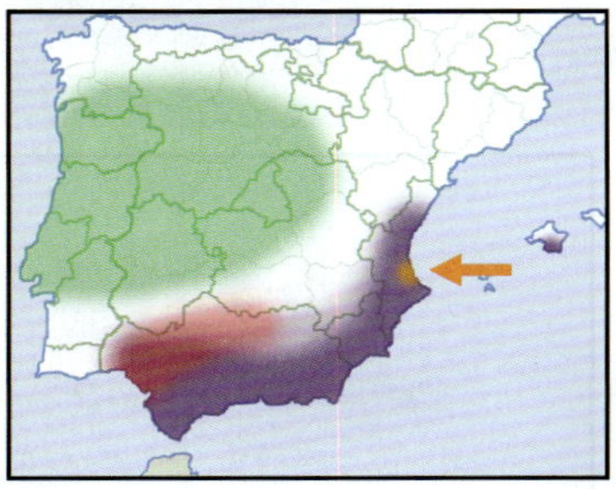

Theodoxus baeticus
Theodoxus elongatulus
Theodoxus hispalensis
Theodoxus valentinus
Theodoxus velascoi

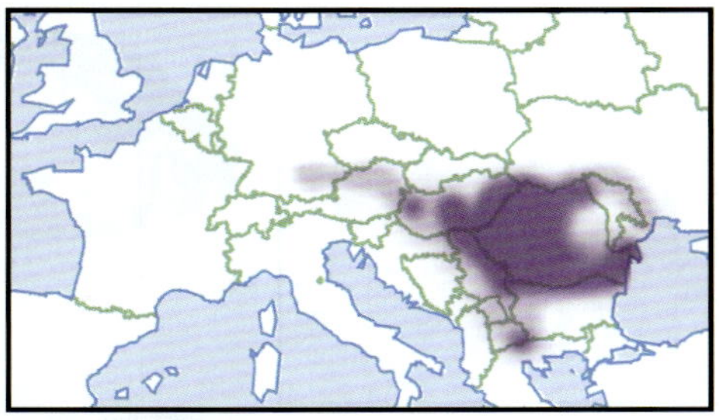

Theodoxus transversalis

The bae R. Vázquez Romero *The elo* Ana Gonçalves *The sub* Z. Fehér *The vel* J. Liétor Gallego

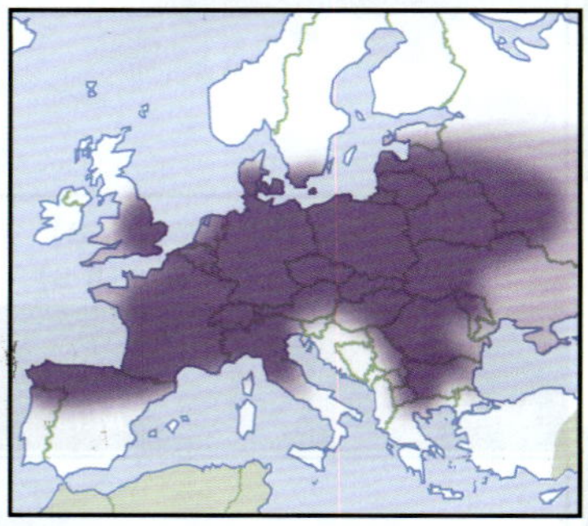

Viviparus contectus

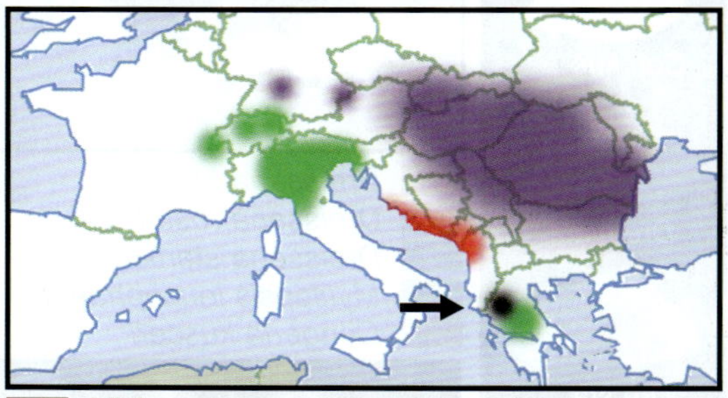

Viviparus acerosus
Viviparus ater
Viviparus janinensis
Viviparus mamillatus

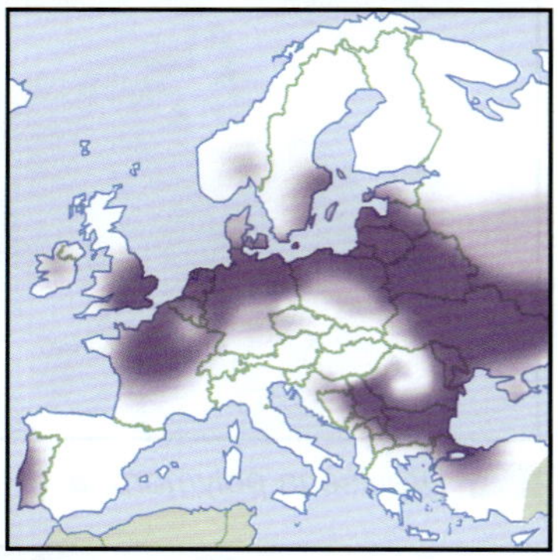

Viviparus viviparus

Cerithioidea

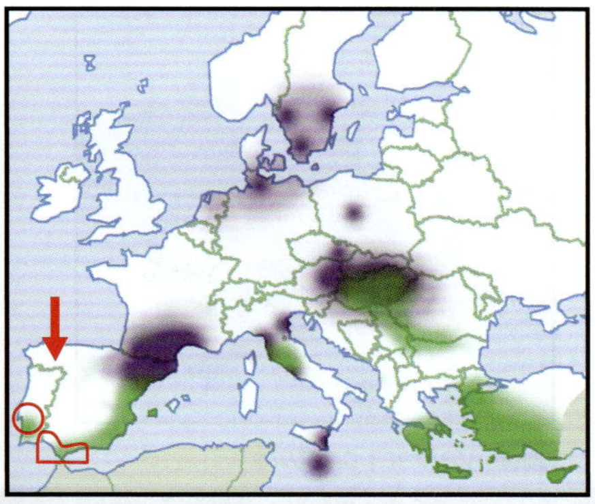

Melanoides tuberculata
Melanopsis cariosa
Melanopsis praemorsa

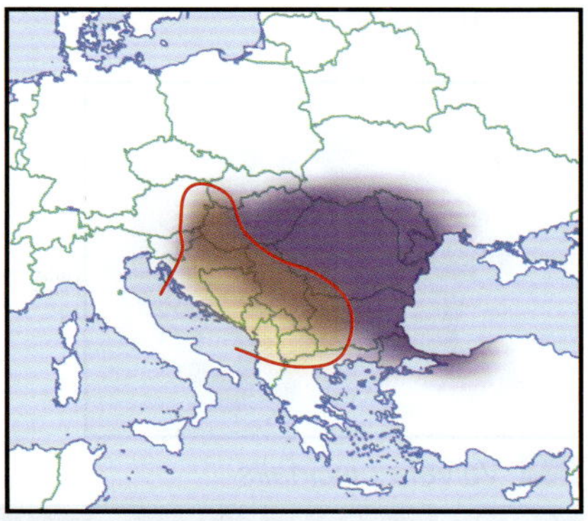

Esperiana esperi
Microcolpia daudebartii
Holandriana holandrii

Hydrobioidea etc.

Only some widespread species are presented here.

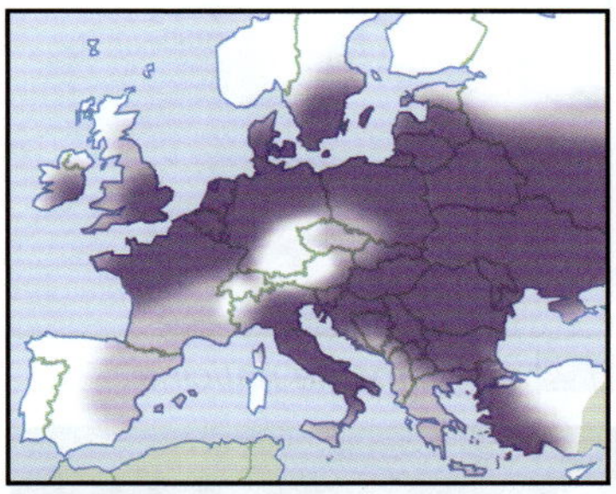

Bithynia leachii

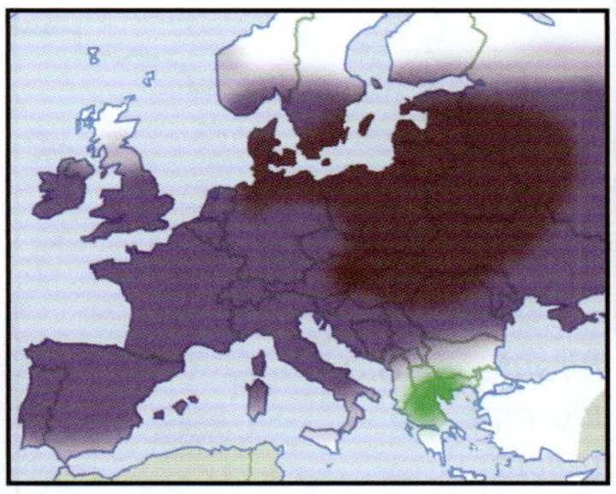

Bithynia tentaculata
Bithynia transsilvanica
Bithynia graeca

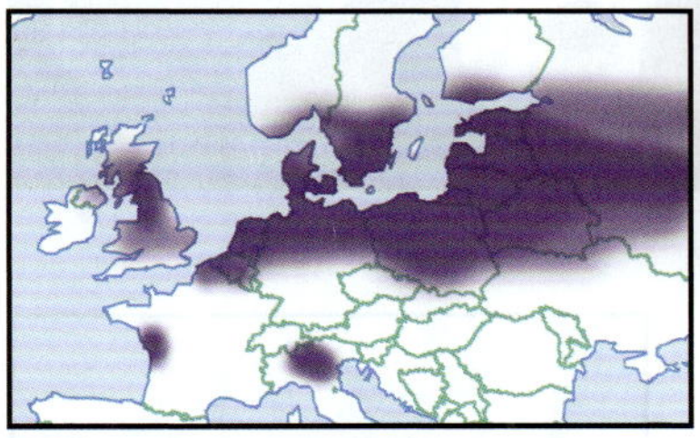

Marstoniopsis insubrica

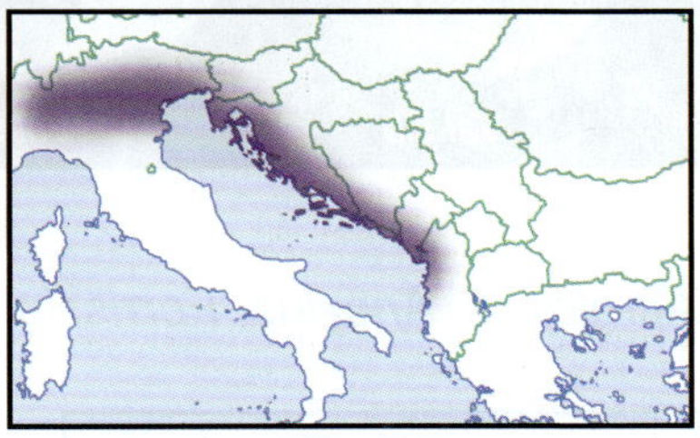

Pyrgula annulata

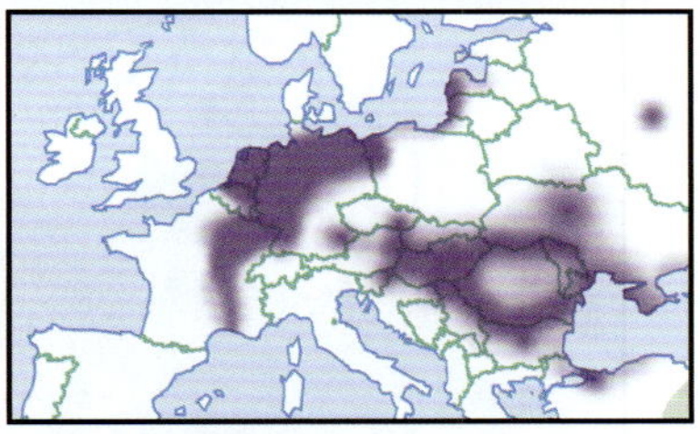

Lithoglyphus naticoides

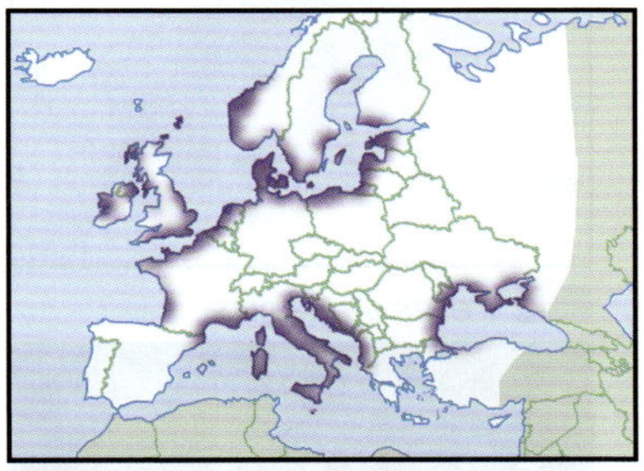

Hydrobia ventrosa

Truncatella subcylindrica

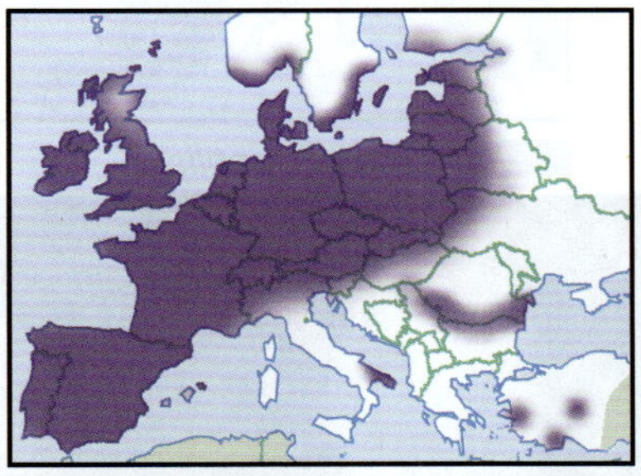

Potamopyrgus antipodarum

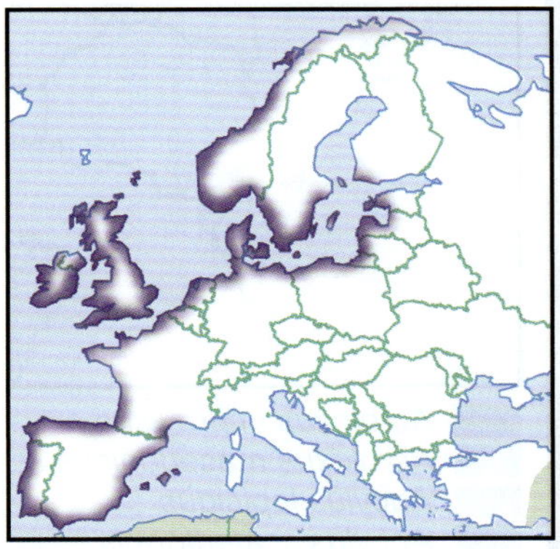

Hydrobia ulvae

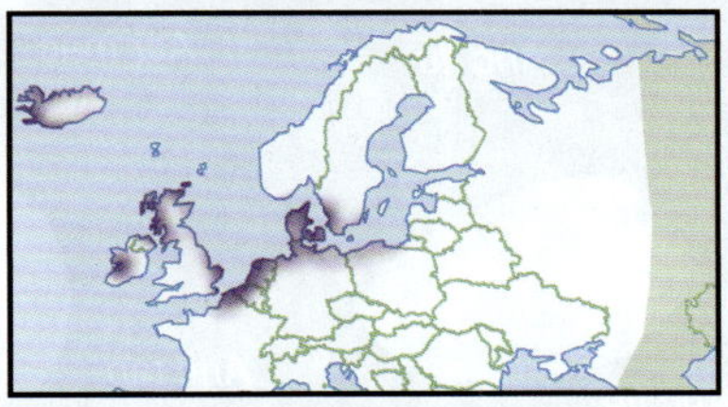

Hydrobia neglecta

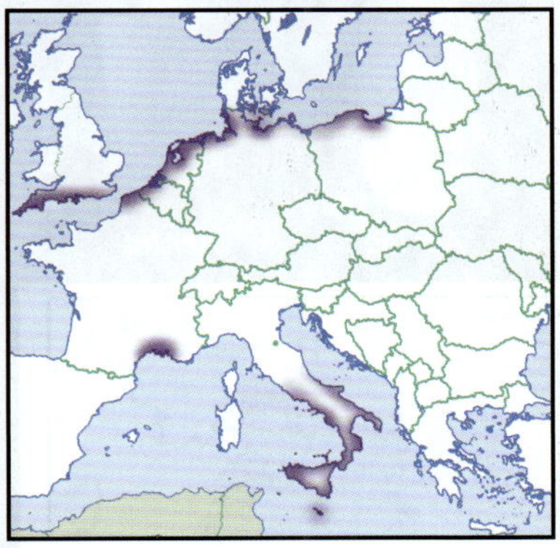

Heleobia stagnorum

Valvatidae

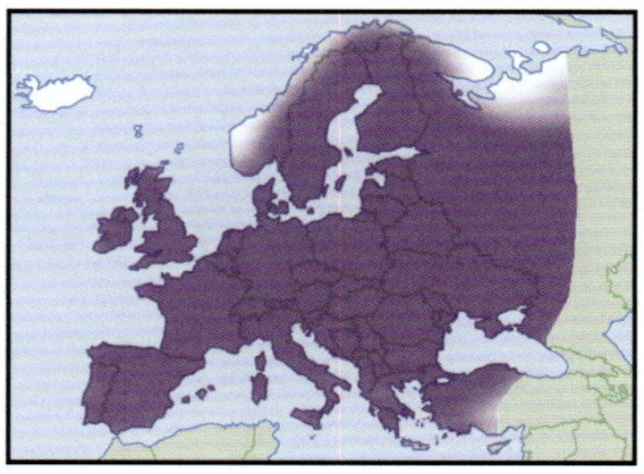

Valvata piscinalis

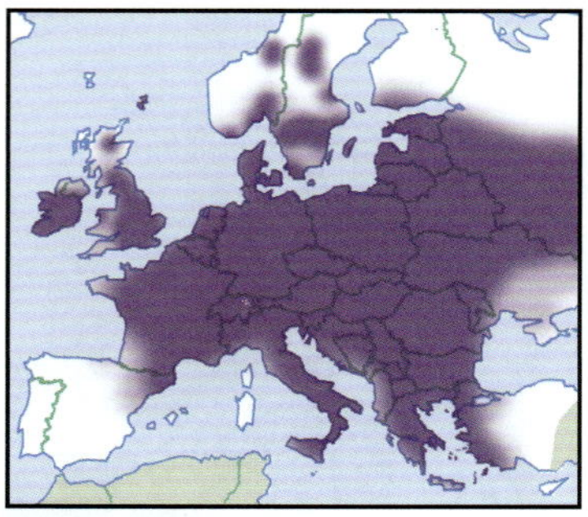

Valvata cristata

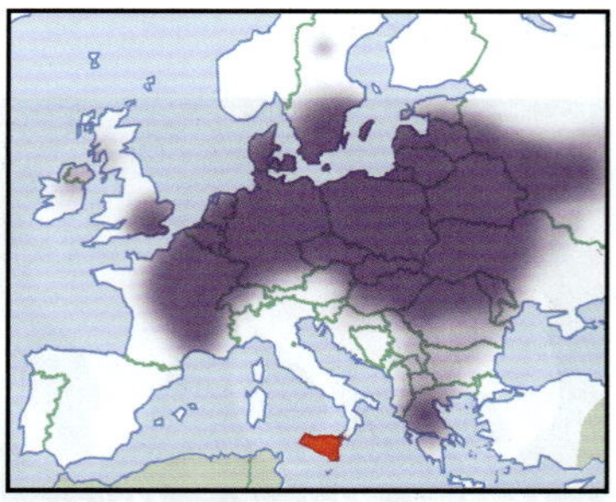

Valvata macrostoma
Valvata saulcyi

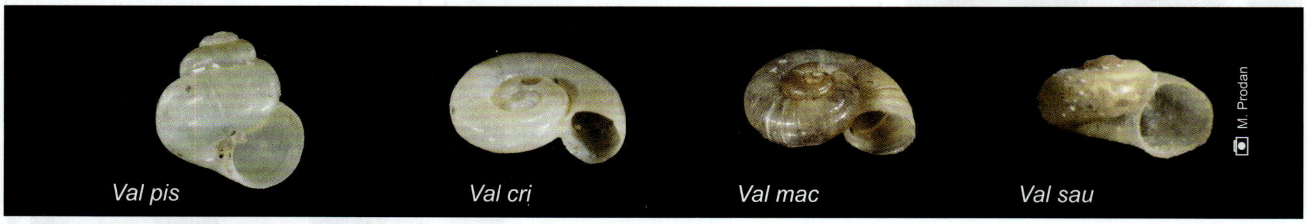

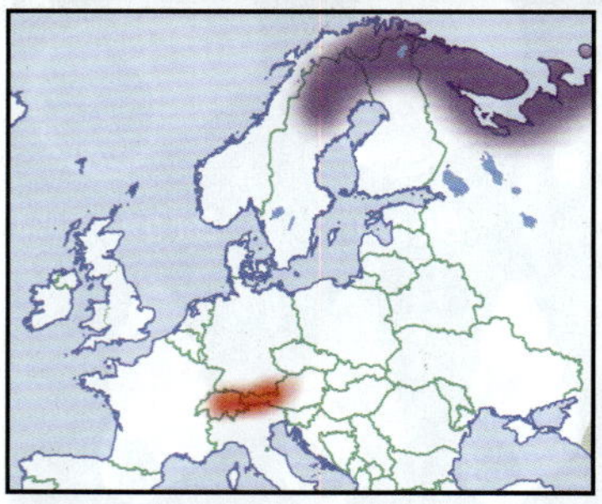

Valvata sibirica
Valvata studeri

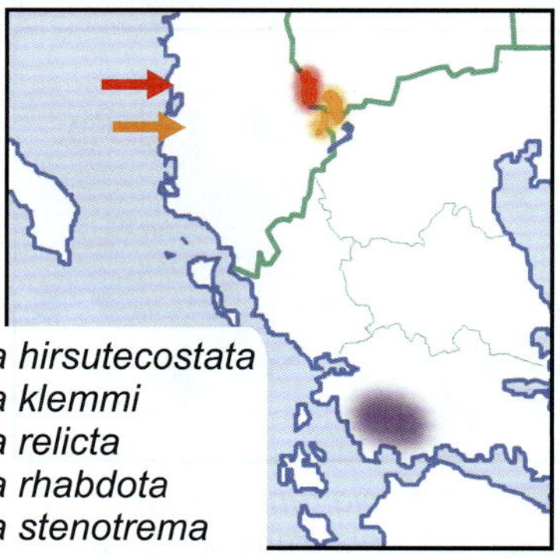

Valvata hirsutecostata
Valvata klemmi
Valvata relicta
Valvata rhabdota
Valvata stenotrema

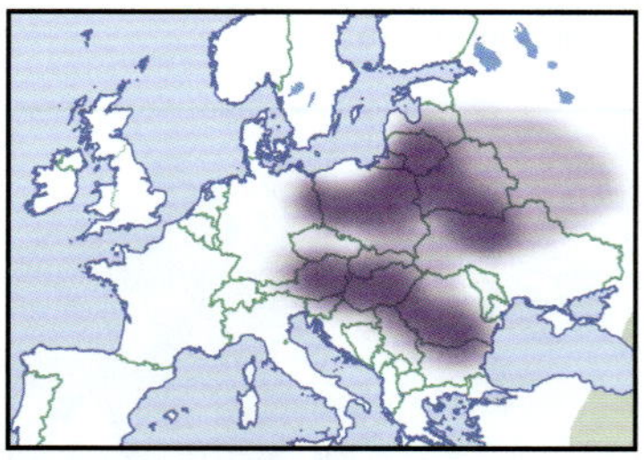

Borysthenia naticina

Acroloxus, Ancylus, Ferrissia

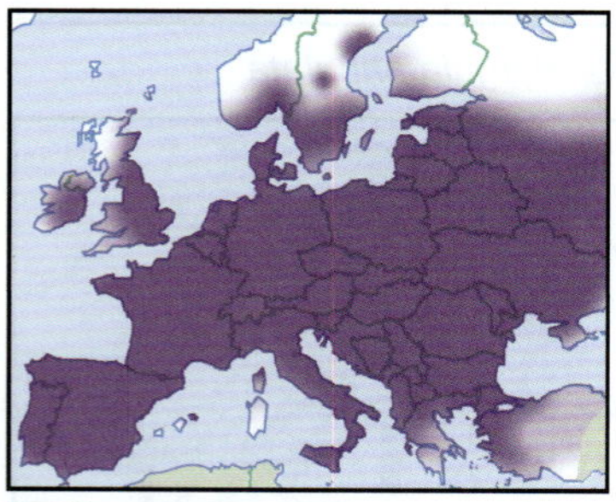

Acroloxus lacustris

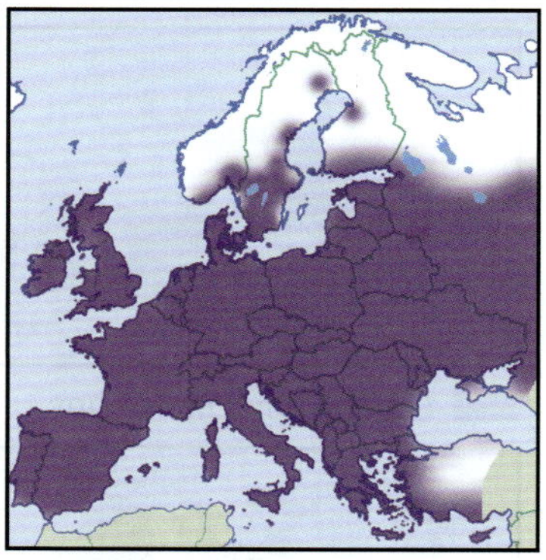

Ancylus fluviatilis

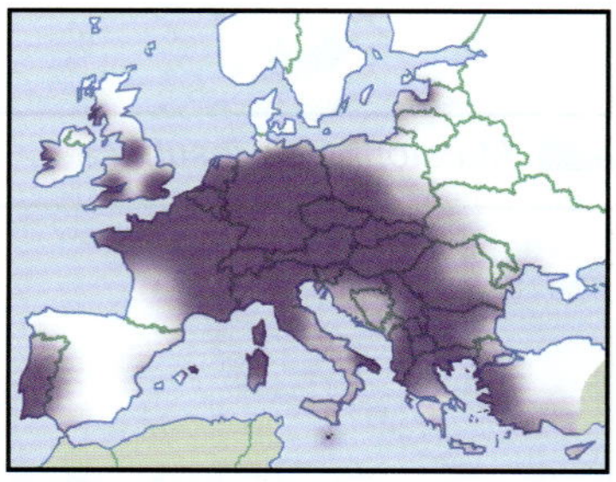

Ferrissia fragilis

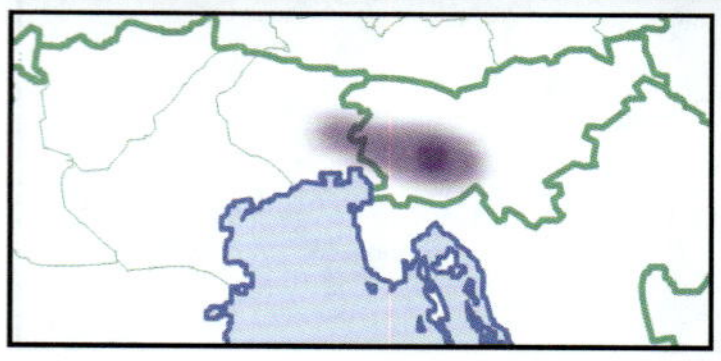

Acroloxus tetensi

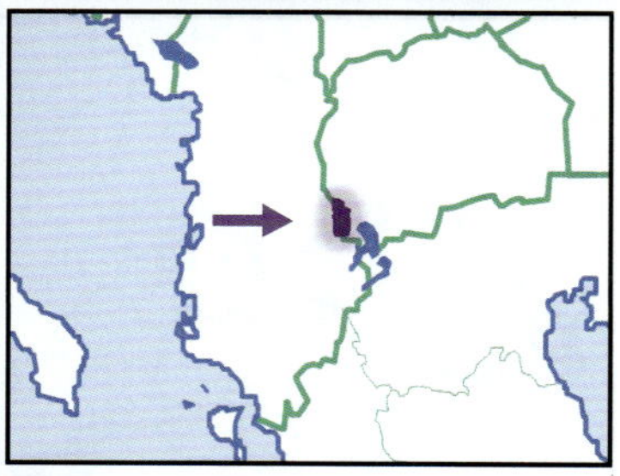

Acroloxus improvisus
Acroloxus macedonicus
Ancylus lapicidus
Ancylus scalariformis
Ancylus tapirulus

Lymnaeidae

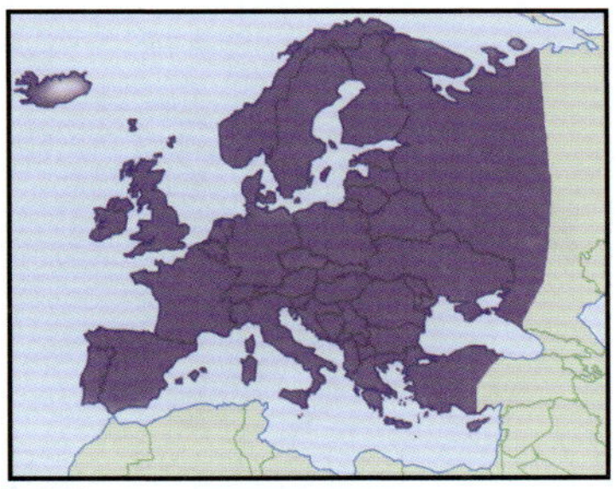

Galba truncatula

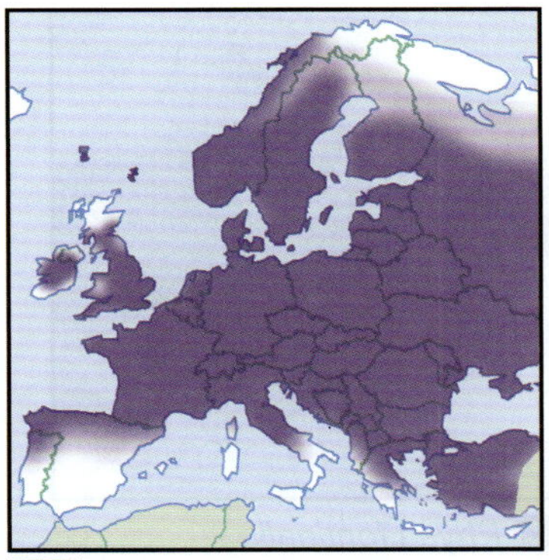

Lymnaea stagnalis

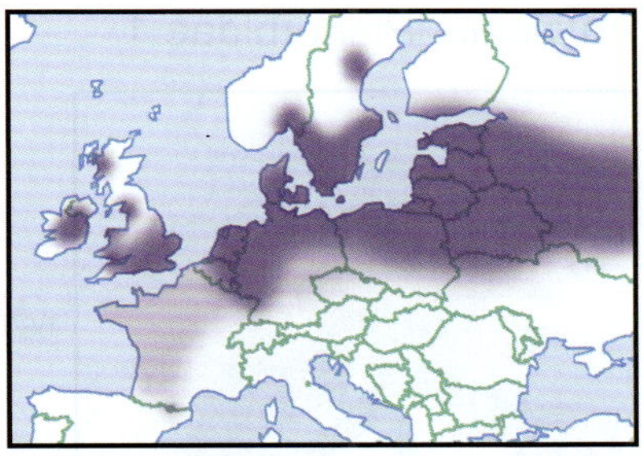

Myxas glutinosa

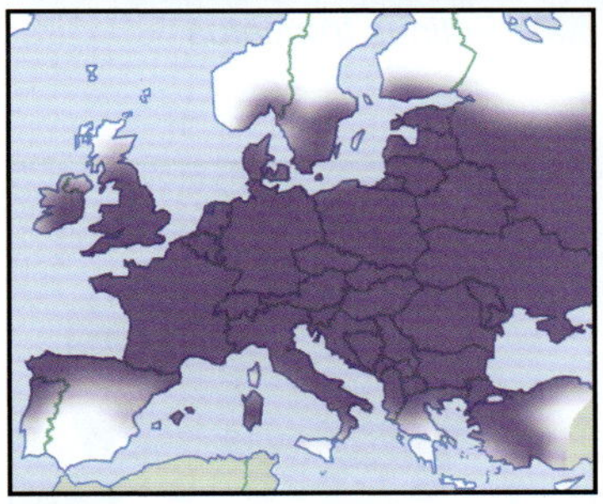

Radix auricularia

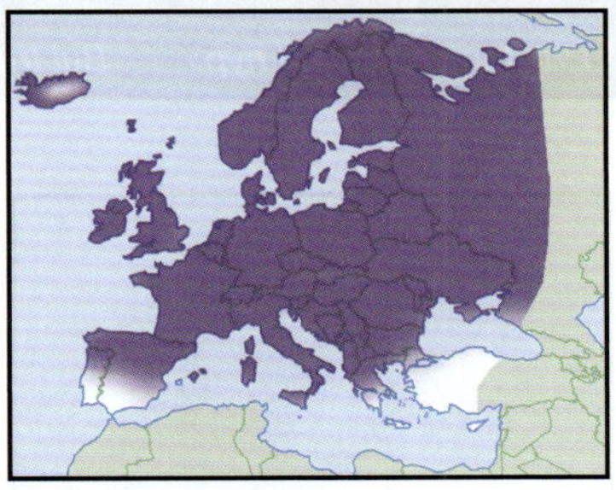

Radix balthica (ovata)

Radix labiata (peregra)

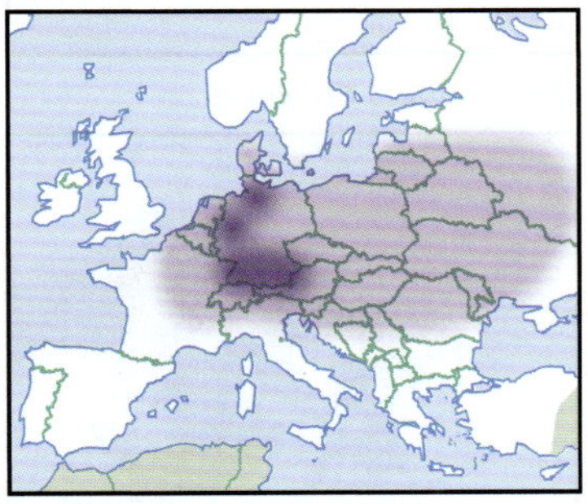

Radix lagotis (ovata)

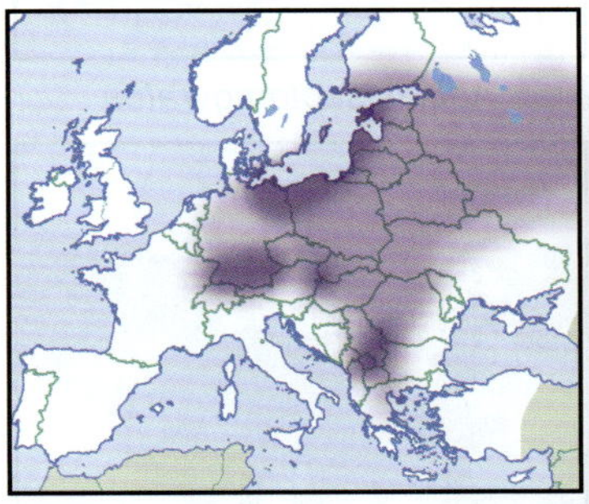

Radix ampla

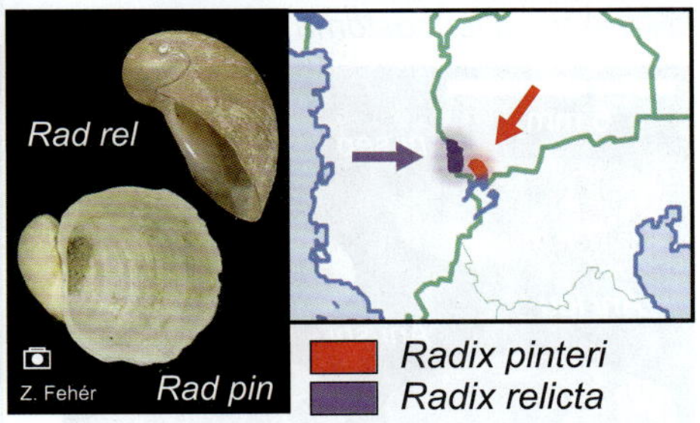

Radix pinteri
Radix relicta

Stagnicola palustris

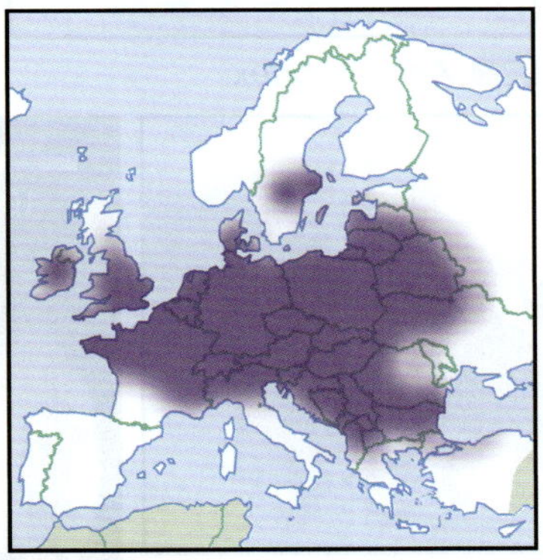

Stagnicola corvus

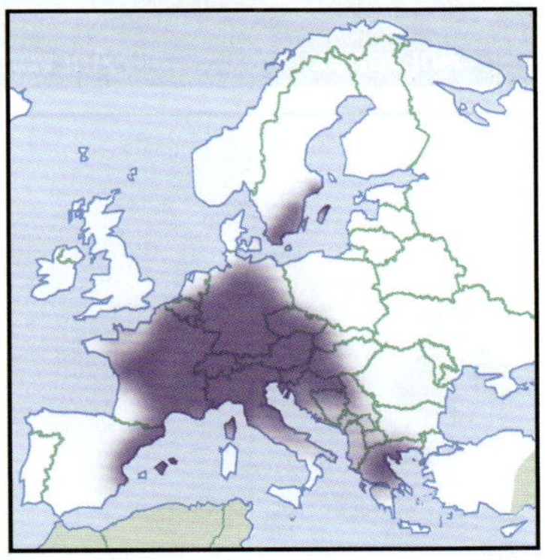

Stagnicola fuscus

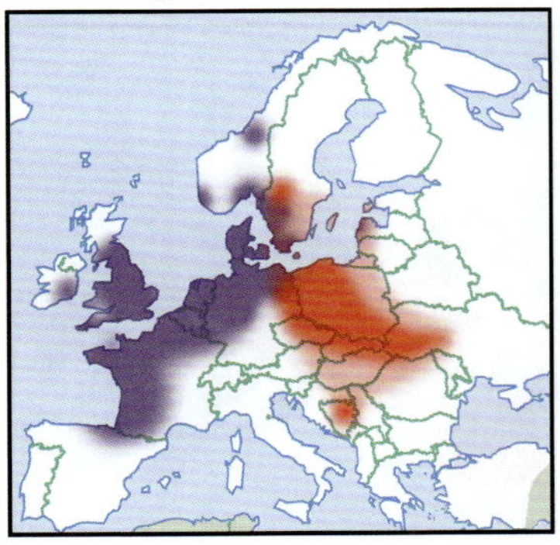

Omphiscola glabra
Stagnicola terebra

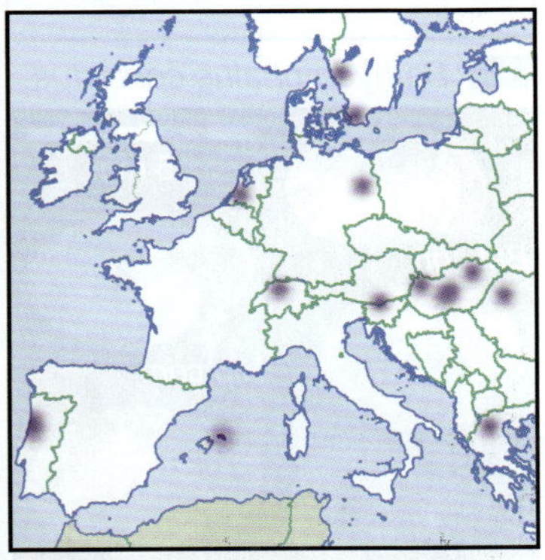

Pseudosuccinea columella

Physidae, Planorbidae 1

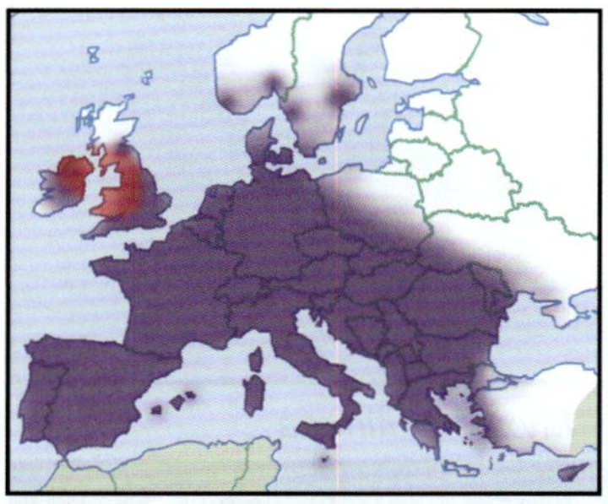

Physa acuta
Physa gyrina

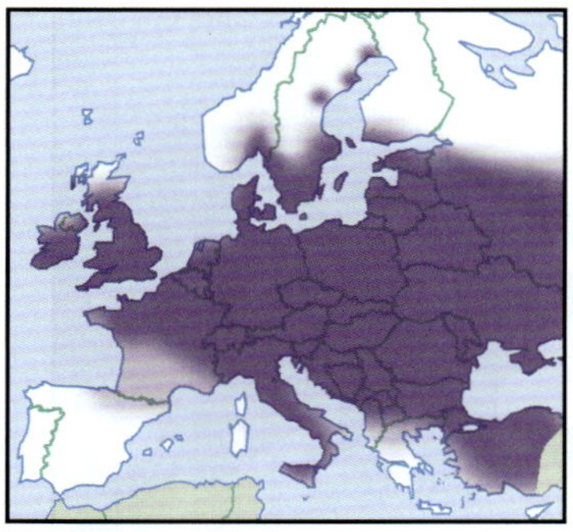

Physa fontinalis

Aplexa hypnorum

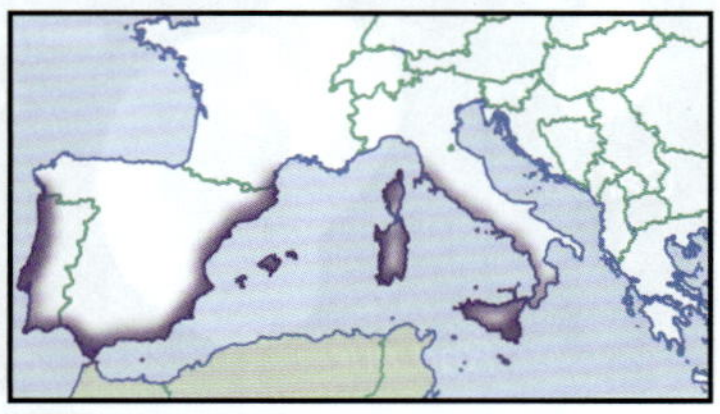

Bulinus truncatus

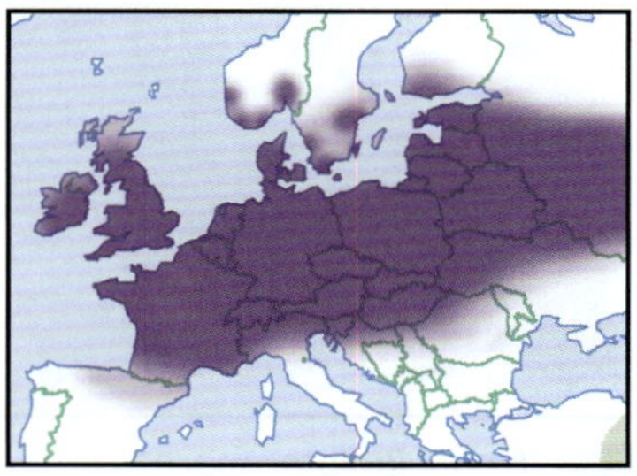

Anisus leucostoma

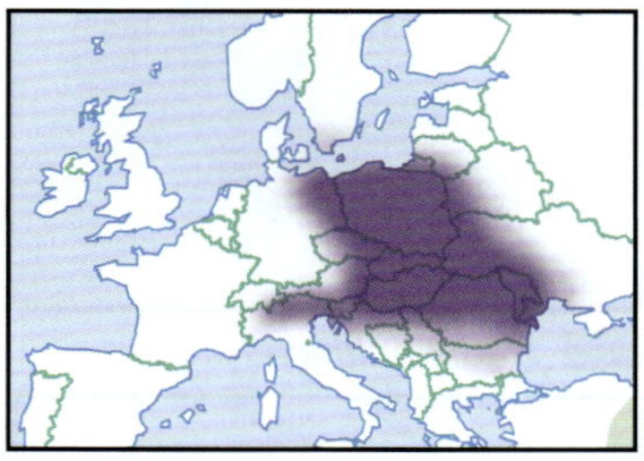

Anisus septemgyratus

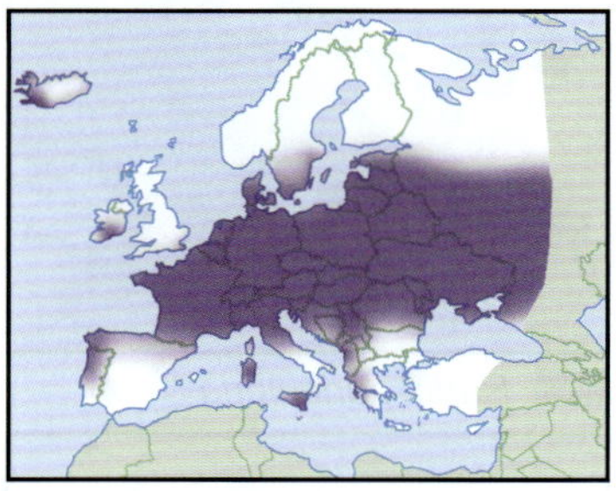

Anisus spirorbis

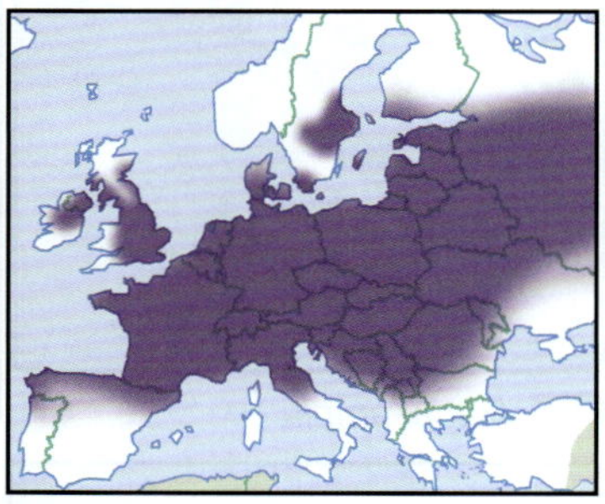

Anisus vortex

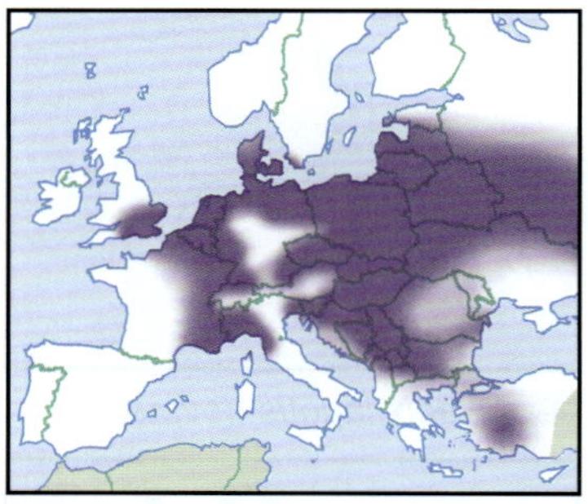

Anisus vorticulus

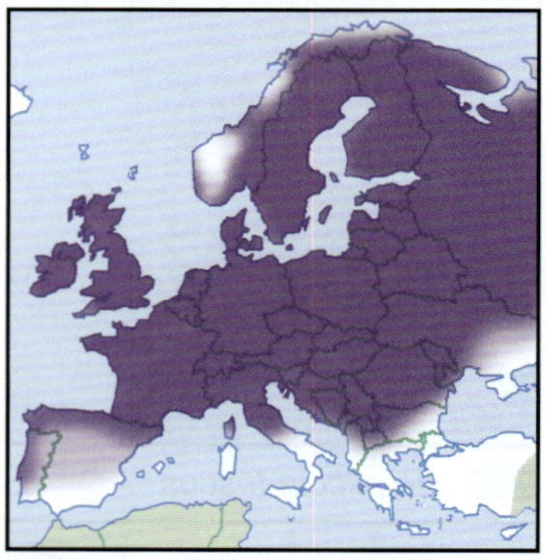

Bathyomphalus contortus

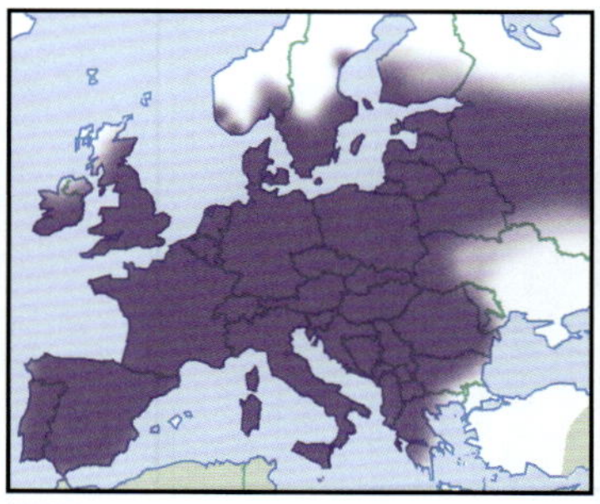

Hippeutis complanatus

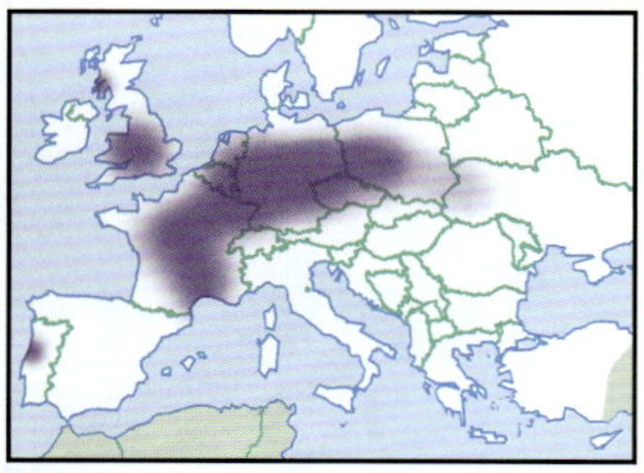

Menetus dilatatus

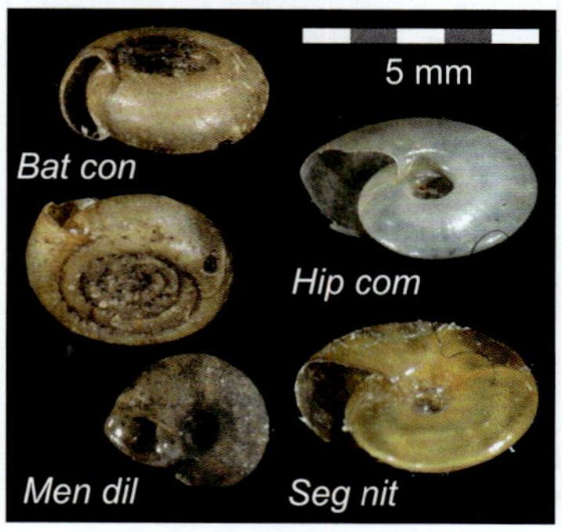

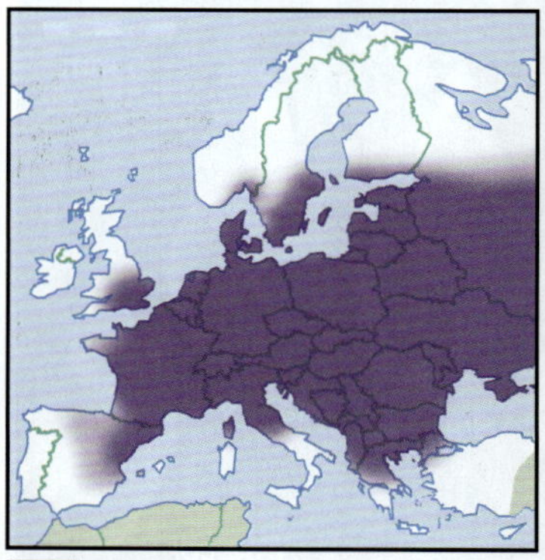

Segmentina nitida

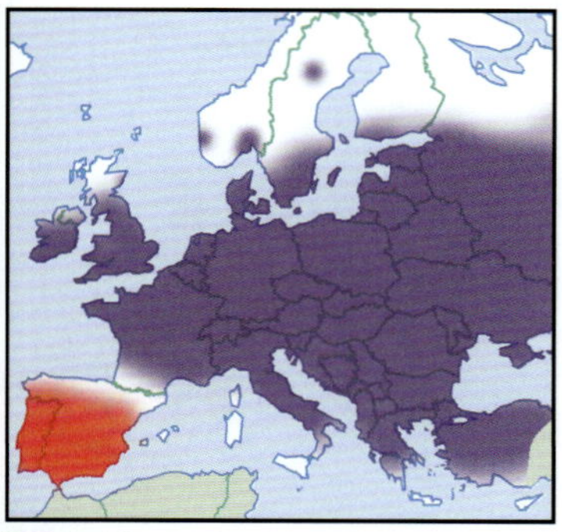

Planorbarius corneus
Planorbarius metidjensis

Planorbidae 2

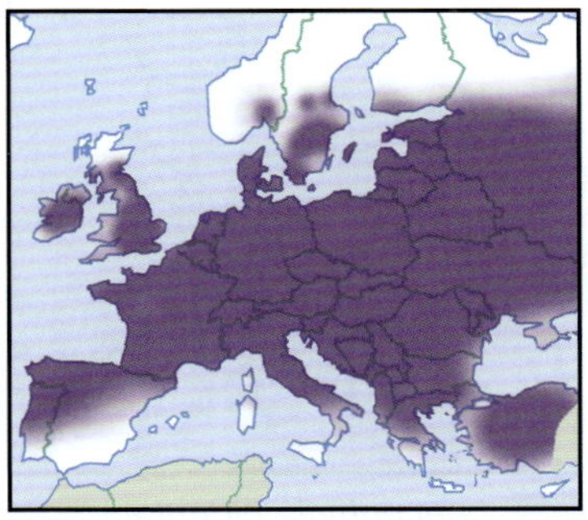

Planorbis carinatus

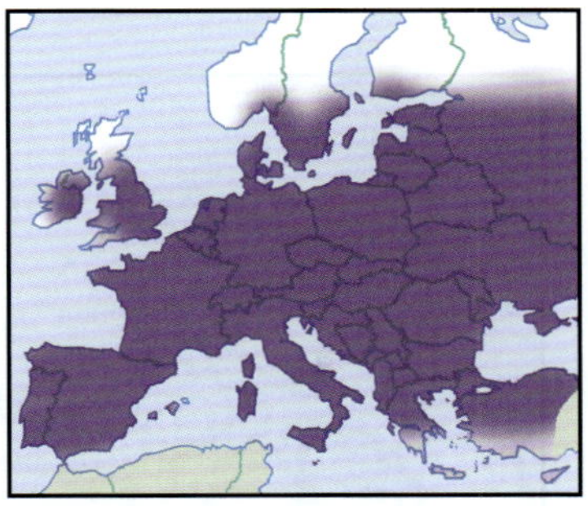

Planorbis planorbis

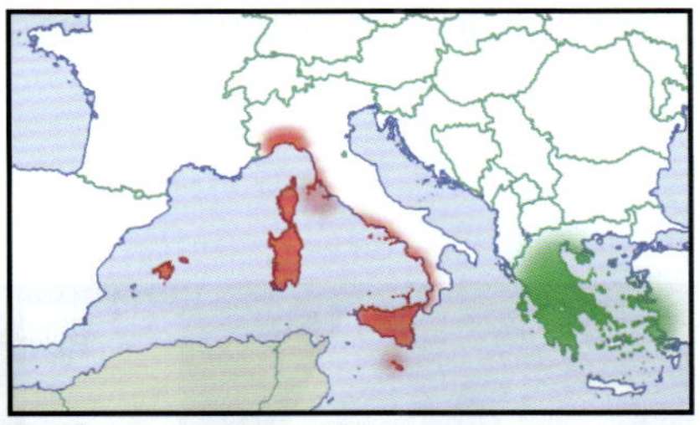

Planorbis atticus
Planorbis moquini

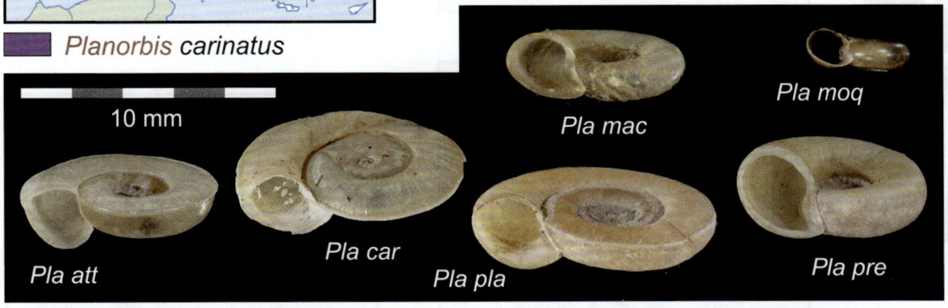

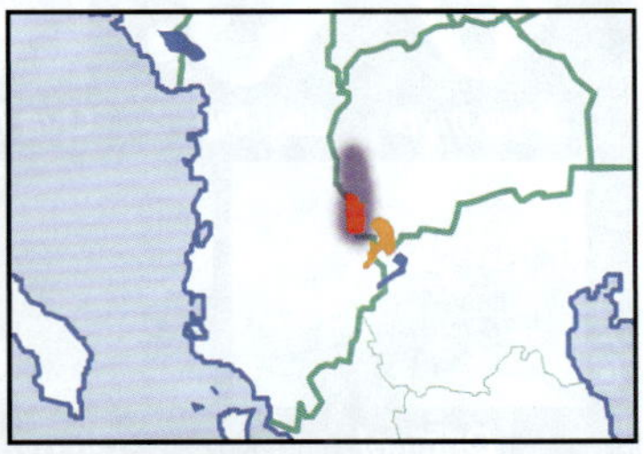

Gyraulus albidus
Gyraulus crenophilus
Gyraulus fontinalis
Gyraulus lychnidicus
Gyraulus stankovici
Gyraulus trapezoides
Planorbis macedonicus
Planorbis presbensis

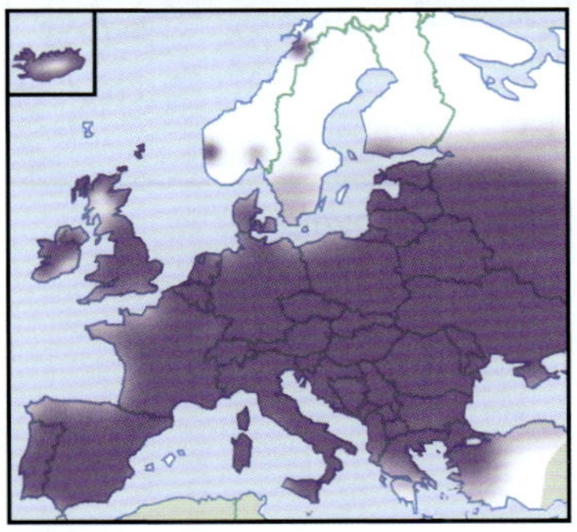

Gyraulus laevis

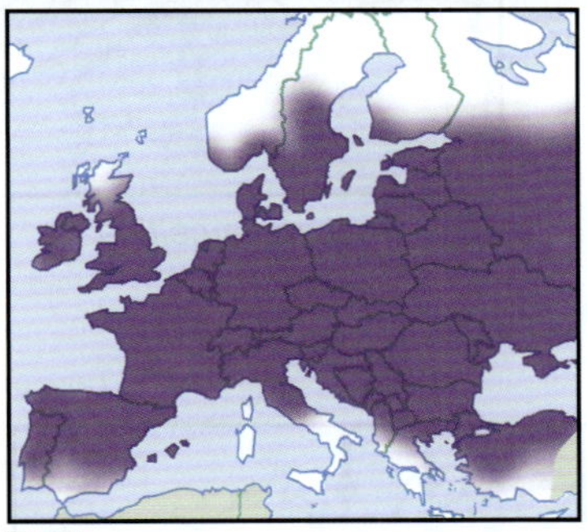

Gyraulus albus

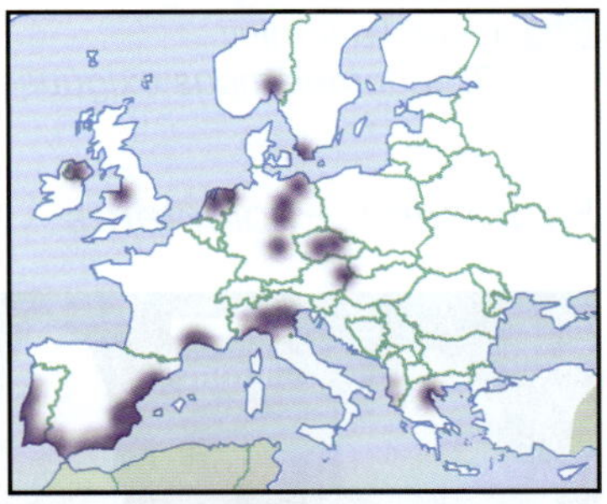

Gyraulus chinensis

Gyraulus crista

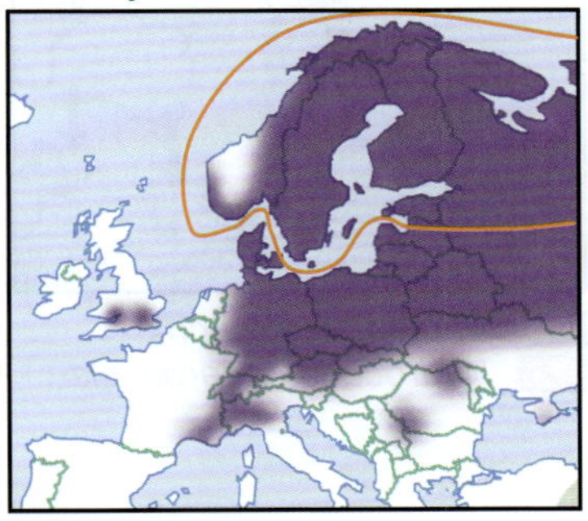

Gyraulus acronicus
Gyraulus stromi

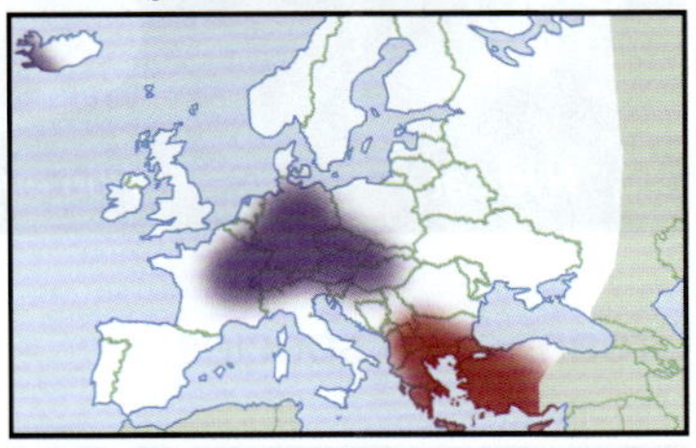

Gyraulus parvus
Gyraulus piscinarum

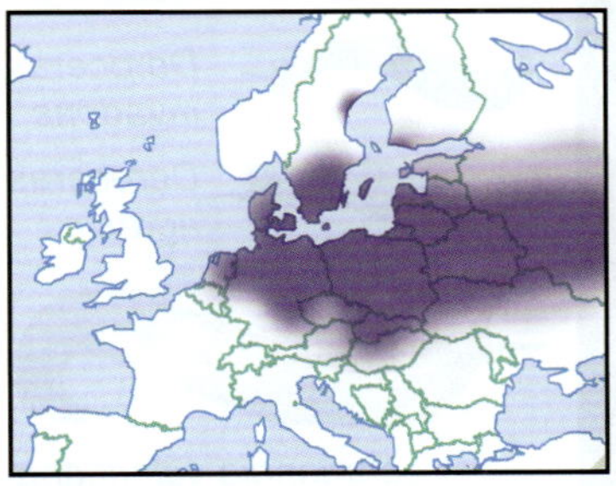

Gyraulus riparius

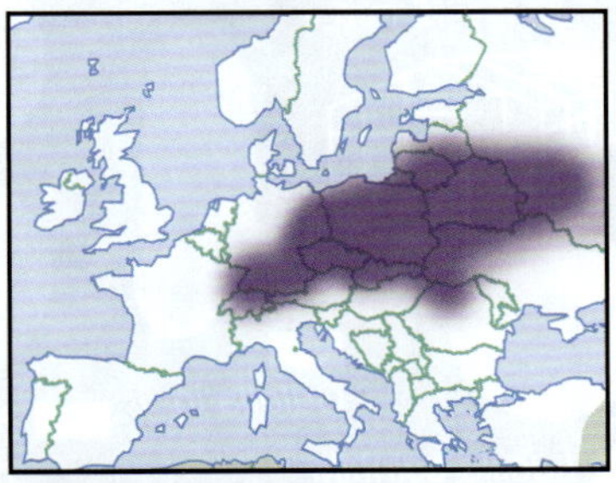

Gyraulus rossmaessleri

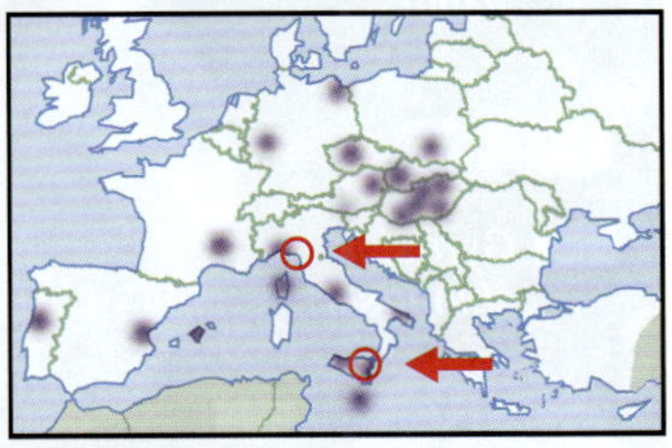

Helisoma anceps
Helisoma duryi

Semi-marine coastland species

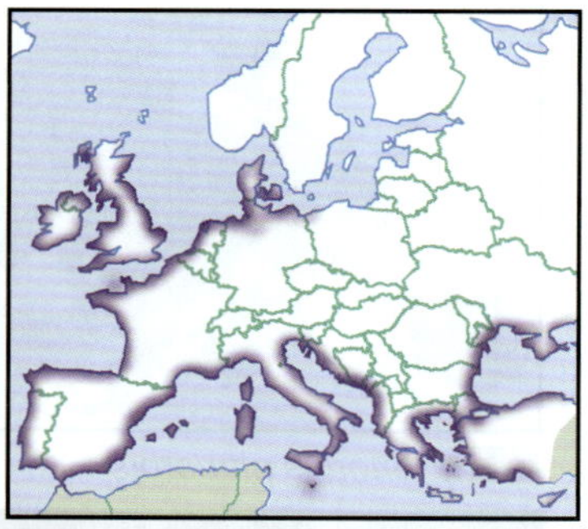

Myosotella myosotis

Leucophytia bidentata

Ovatella firminii
Pseudomelampus exiguus

Onchidella celtica

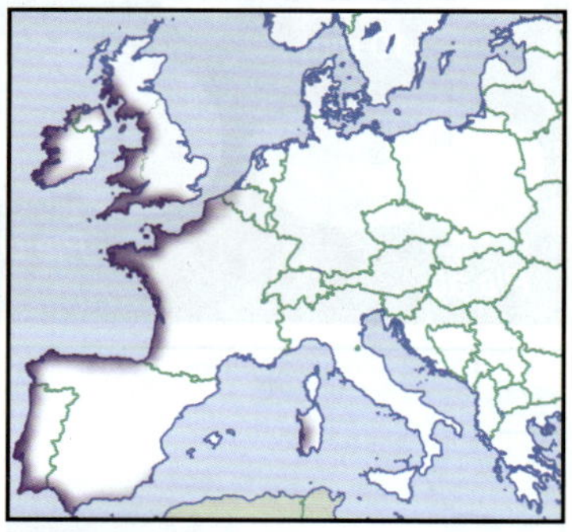

Otina ovata

Hydrocena, *Pholeoteras*

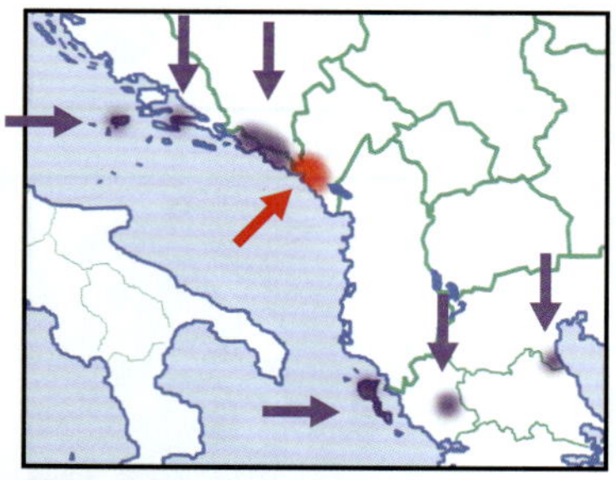

Pholeoteras euthrix
Hydrocena cattaroensis

Greenhouse species

Der agr V. Wiese *Der inv* A. Hallgass *Der lae Boe pal* J. Grego *Mil nig* D. Cilia *Ari hor* R. Anderson

Aciculidae

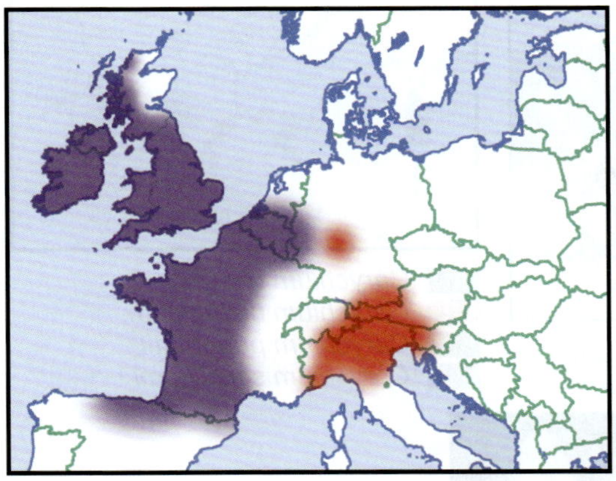

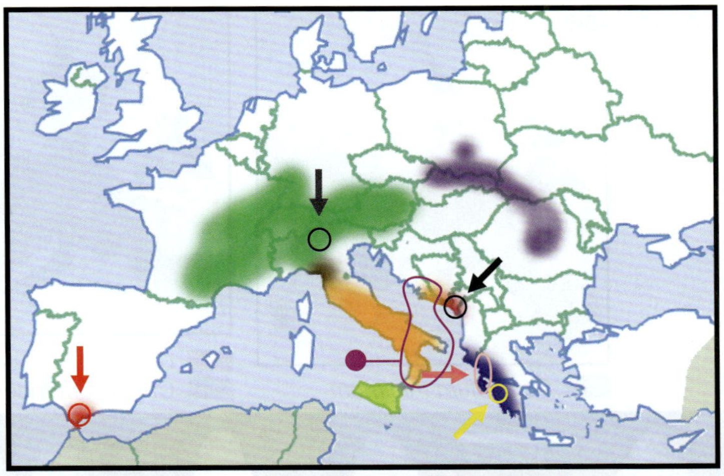

Acicula fusca
Acicula lineolata (lin2)

📷 *Aci cor* M. Prodan *Aci dis* C. Bogi *Aci par* K. Chaniecka *Aci szi* A. Vannozzi *Pla ban Pla per* Z. Fehér *Pla dup* J. Béguinot *Pla mic* M. Sosso

Acicula beneckei
Acicula benoiti
Acicula corcyrensis
Acicula disjuncta
Acicula hausdorfi
Acicula lineata (lin1)
Acicula miaphene
Acicula multilineata
Acicula norrisi
Acicula parcelineata
Acicula szigethyannae
Acicula vezzanii

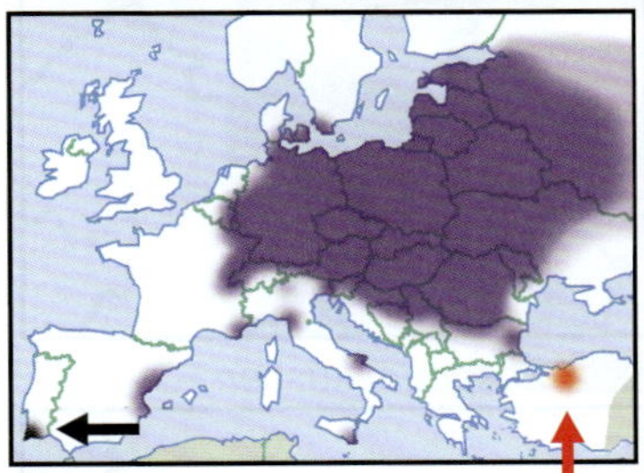

Platyla lusitanica
Platyla polita
Platyla turcica

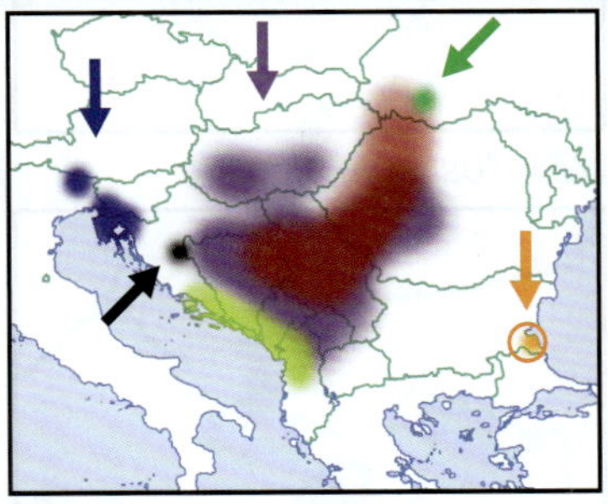

Platyla banatica
Platyla curtii
Platyla jankowskiana
Platyla maasseni
Platyla orthostoma
Platyla perpusilla
Platyla wilhelmi

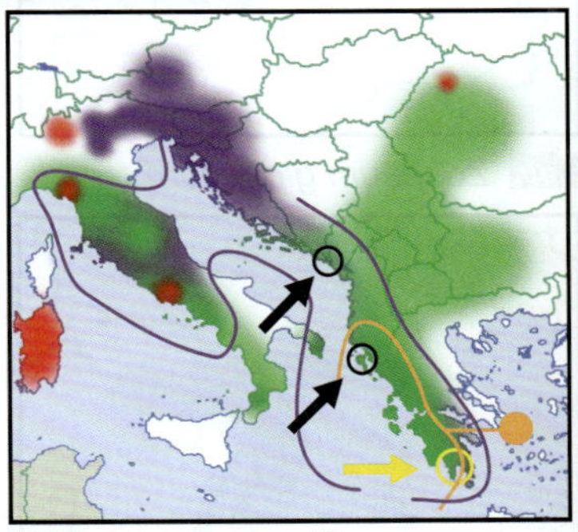

Platyla gracilis
Platyla microspira
Platyla minutissima
Platyla peloponnesica
Platyla pinteri
Platyla similis

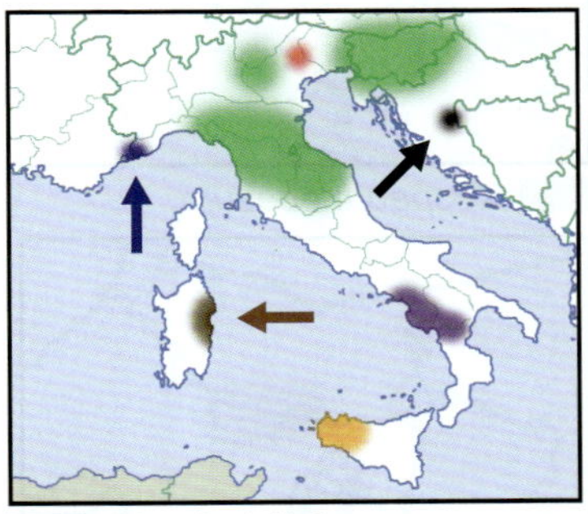

Platyla foliniana
Platyla maasseni
Platyla pezzolii
Platyla sardoa
Platyla stussineri
Platyla subdiaphana
Platyla talentii

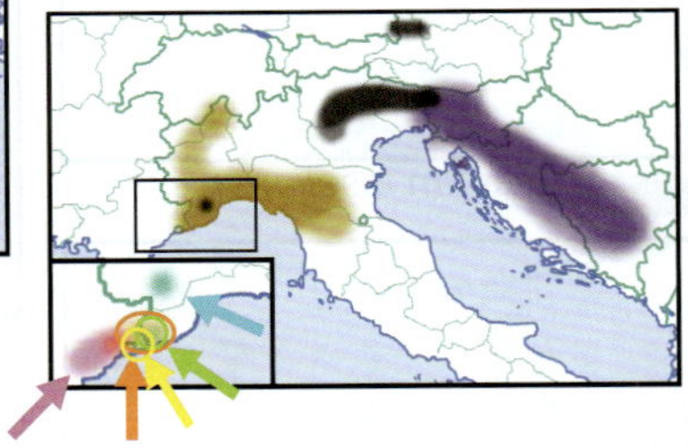

Renea elegantissima
Renea veneta
Renea spectabilis
Renea moutonii
Renea paillona
Renea gormonti
Renea bourguignatiana
Renea gentilei

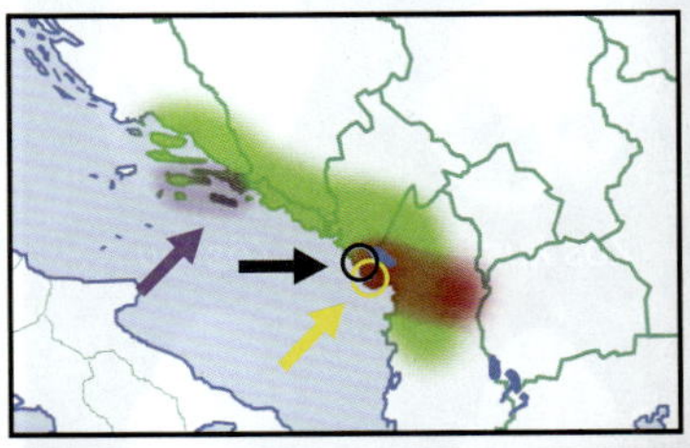

Platyla corpulenta
Platyla elisabethae
Platyla feheri
Platyla procax
Platyla wilhelmi

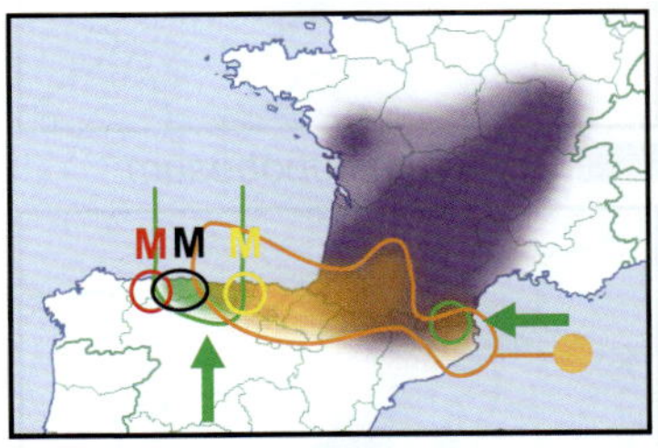

Platyla callostoma
Platyla cryptomena
Platyla dupuyi
M *Menkia rolani*
M *Menkia horsti*
M *Menkia dewinteri*

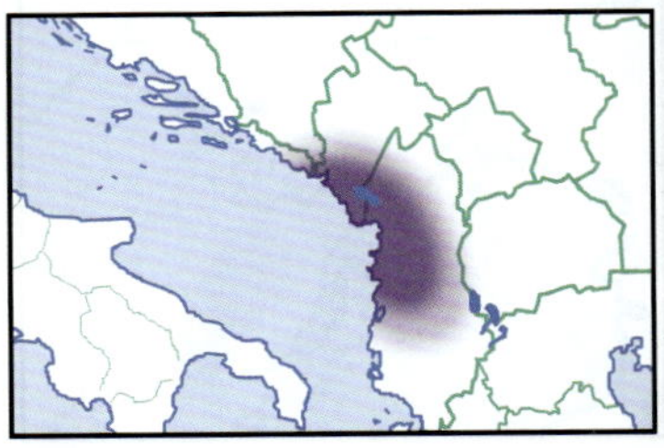

Renea kobelti

Carychiidae

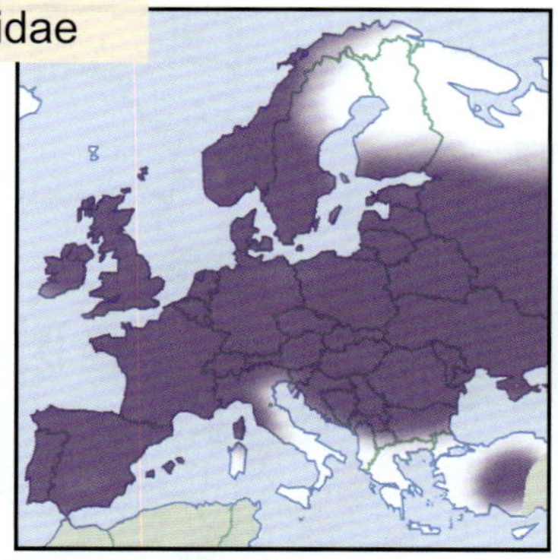

Carychium minimum

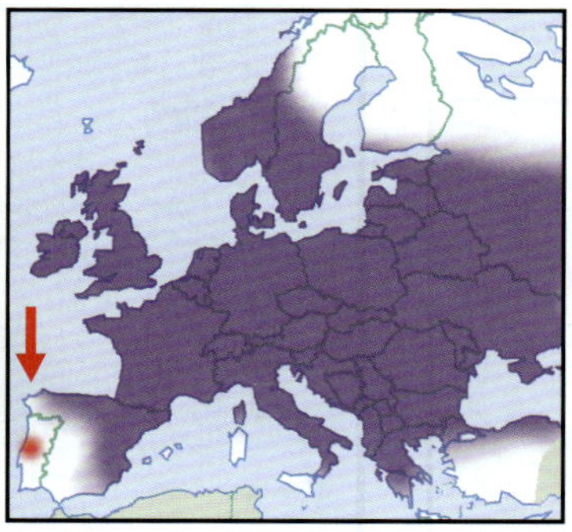

Carychium ibazoricum
Carychium tridentatum

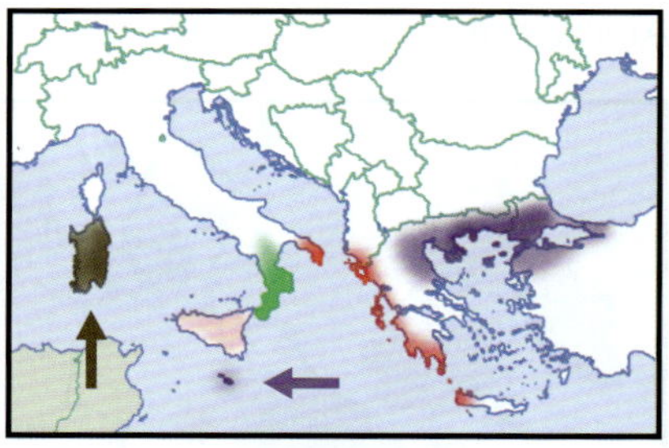

Carychium biondii
Carychium hellenicum
Carychium paganettii
Carychium schlickumi

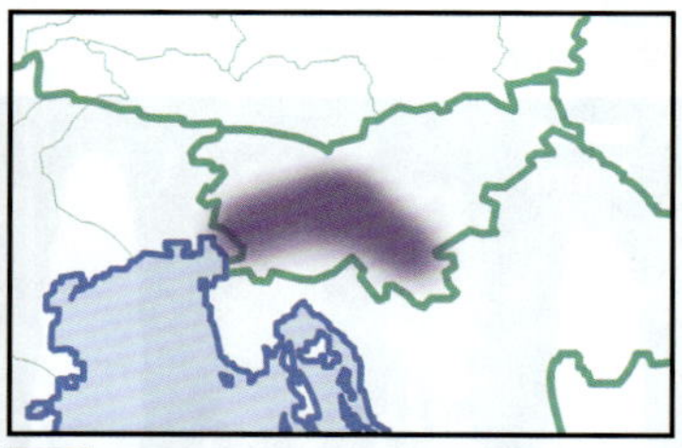

Zospeum lautum

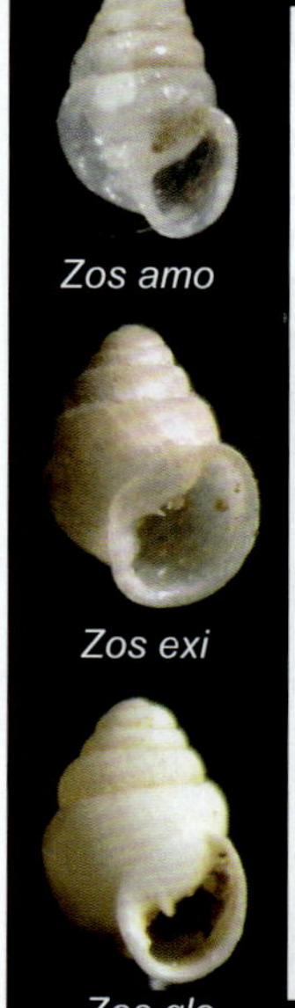

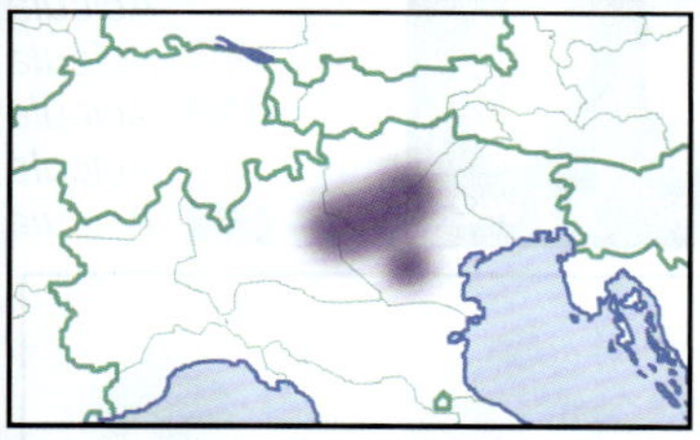

Zospeum globosum

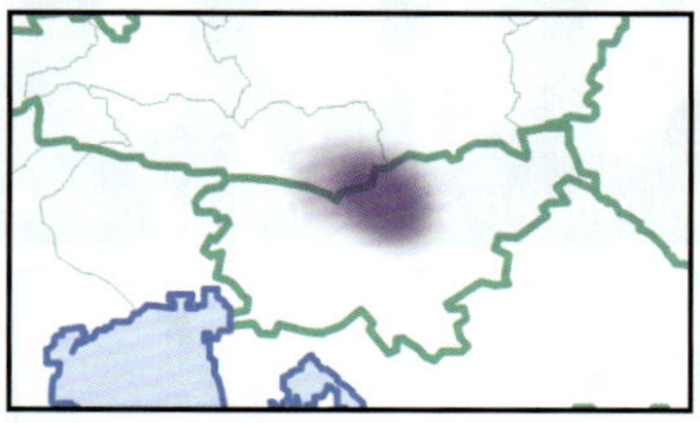

Zospeum alpestre

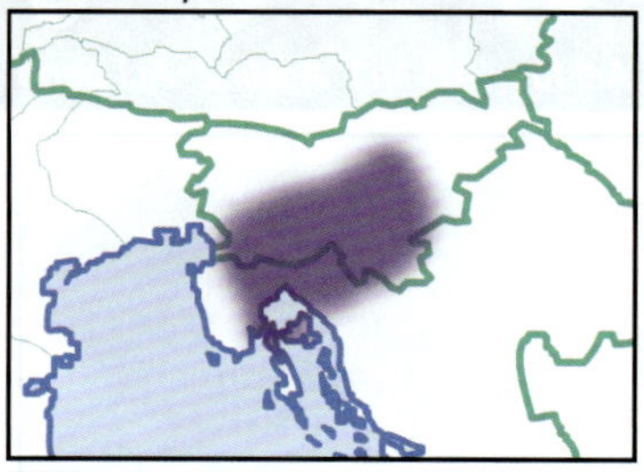

Zospeum kusceri

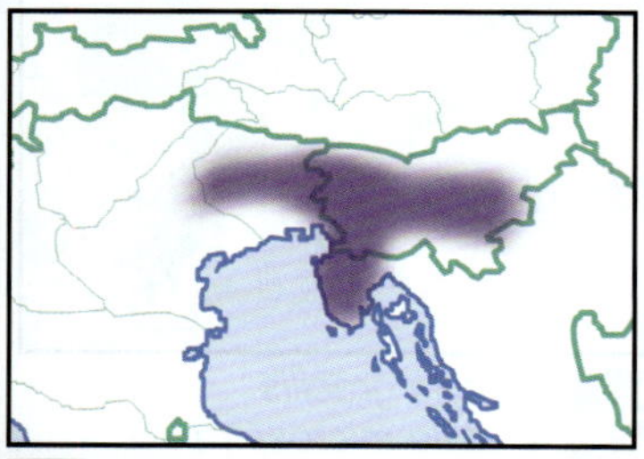

Zospeum spelaeum

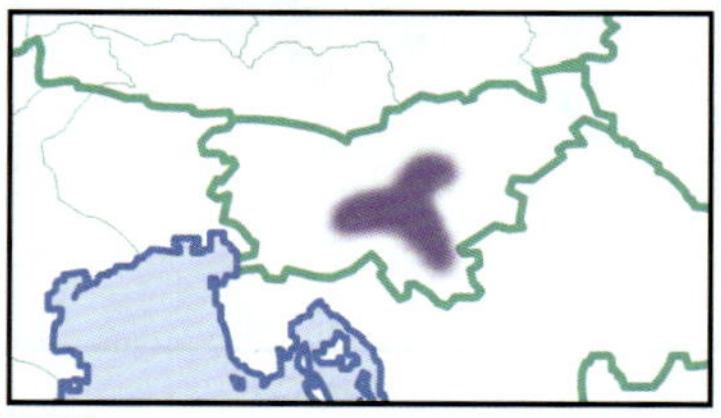

Zospeum frauenfeldi

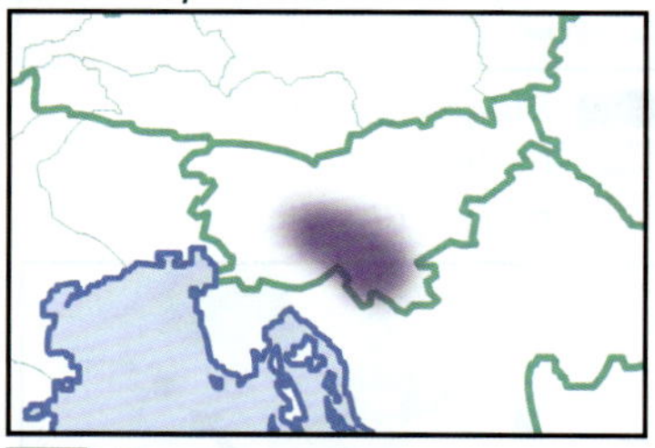

Zospeum obesum

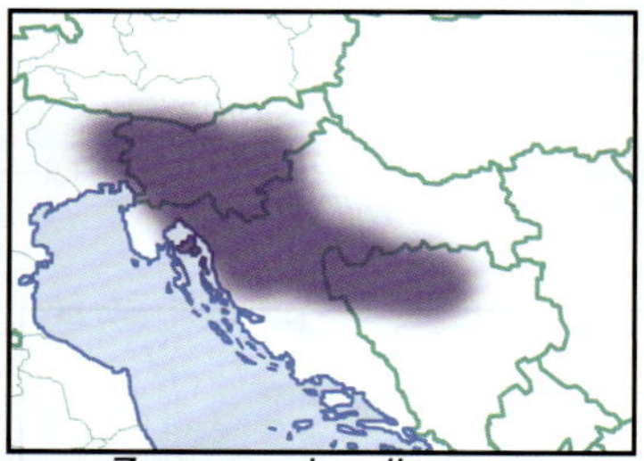

Zospeum isselianum

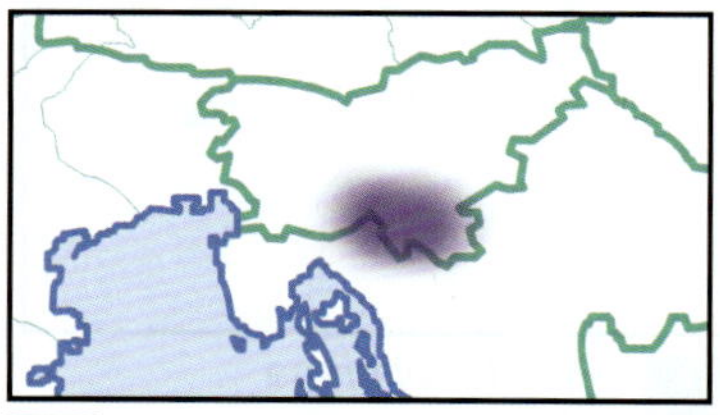

Zospeum exiguum

Zos alp Zos exi Zos glo Zos kus Zos obe Zos sub M. Prodan

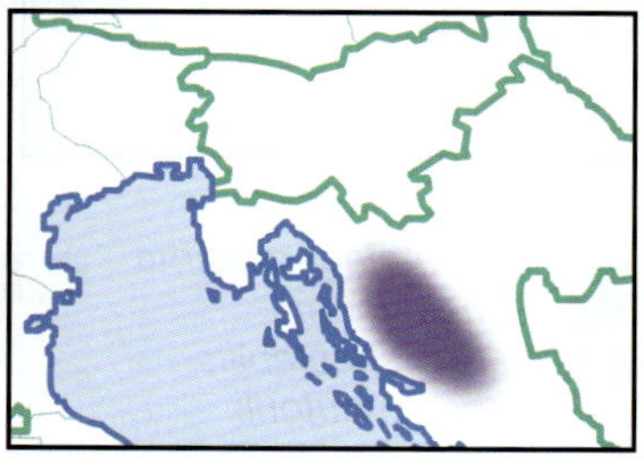

Zospeum pretneri

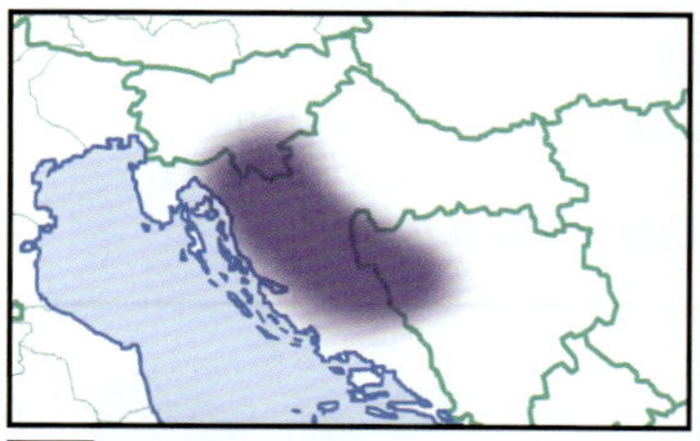

Zospeum subobesum

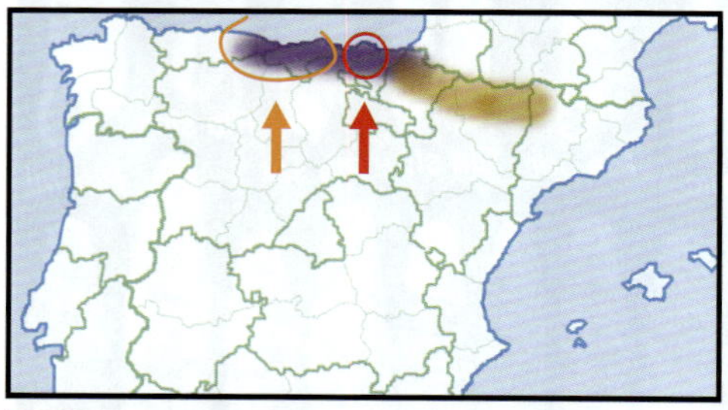

Zospeum bellesi
Zospeum biscaiense
Zospeum schaufussi
Zospeum suarezi

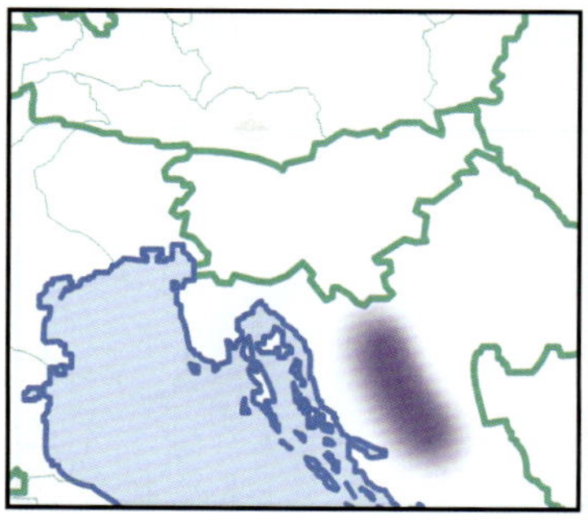

Zospeum likanum

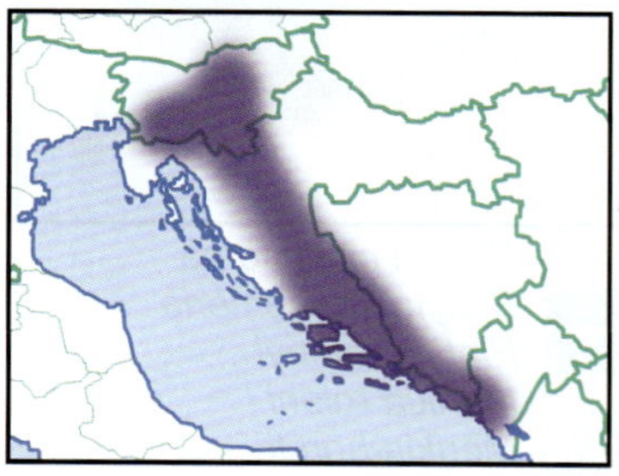

Zospeum amoenum

Cochlostoma 1

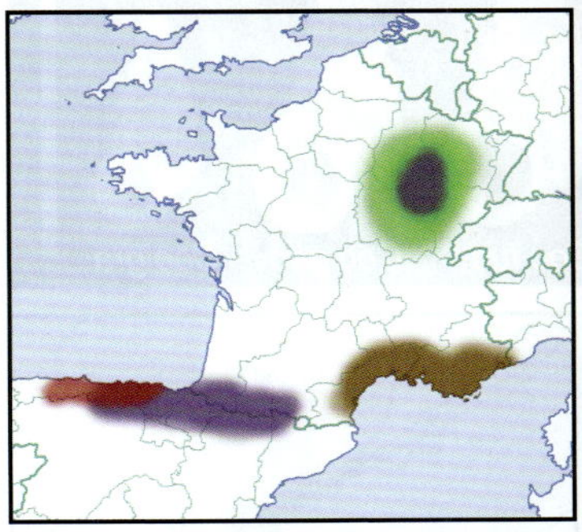

Cochlostoma bicostulatum
Cochlostoma conicum
Cochlostoma obscurum
Cochlostoma patulum

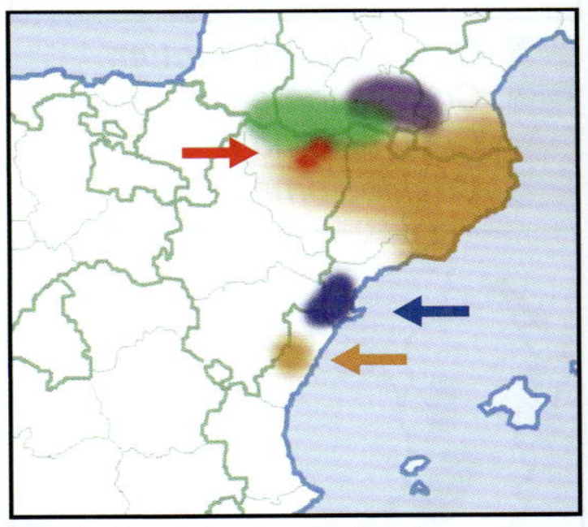

Cochlostoma fontqueri
Cochlostoma gigas
Cochlostoma martorelli
Cochlostoma nouleti
Cochlostoma partioti

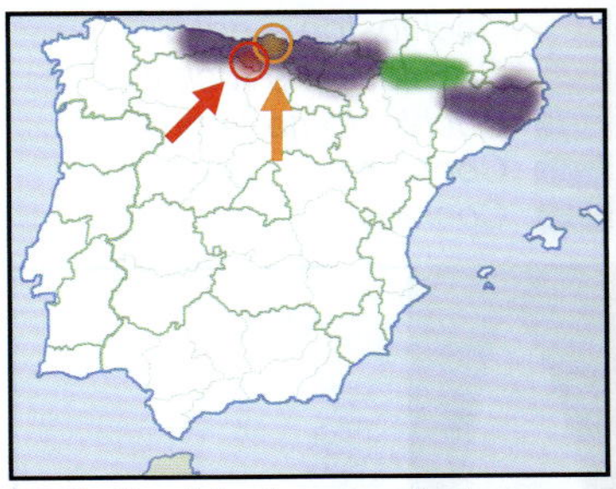

Cochlostoma asturicum
Cochlostoma hidalgoi
Cochlostoma oscitans
Cochlostoma partioti

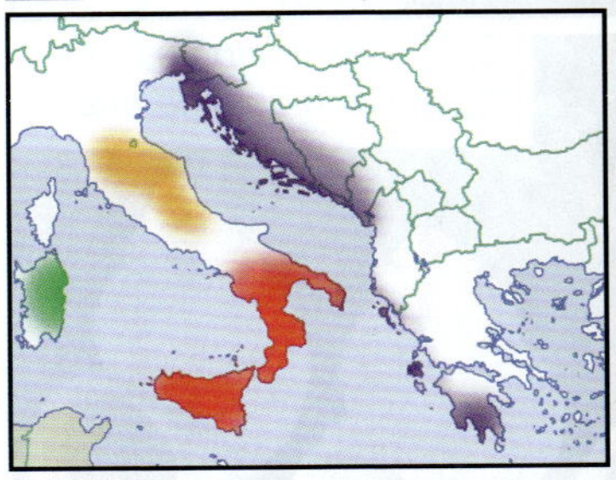

Cochlostoma crosseanum
Cochlostoma sardoum
Cochlostoma scalarinum
Cochlostoma westerlundi

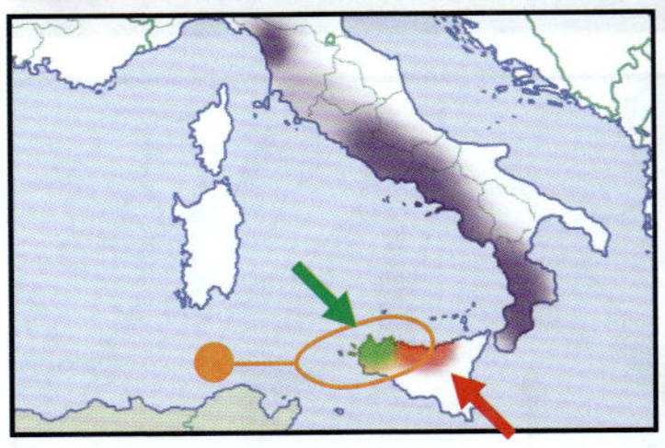

Cochlostoma affine
Cochlostoma alleryanum
Cochlostoma montanum
Cochlostoma paladilhianum

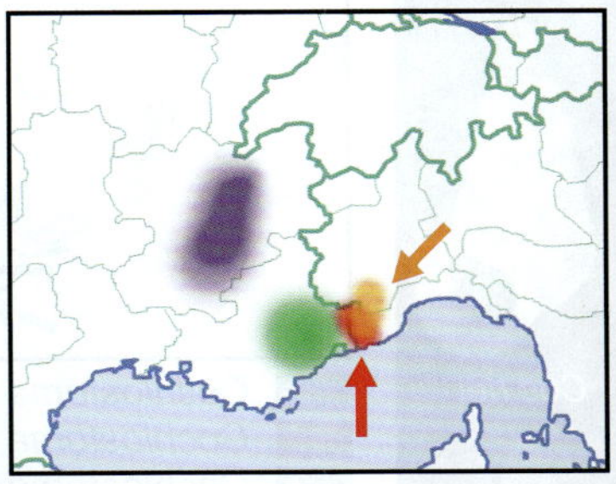

Cochlostoma apricum
Cochlostoma macei
Cochlostoma simrothi
Cochlostoma subalpinum

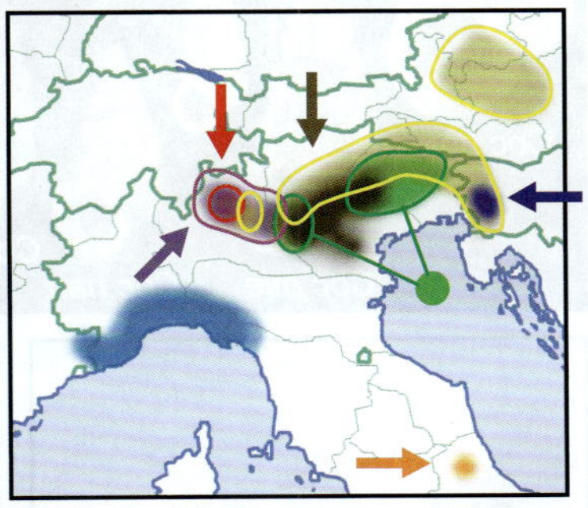

Cochlostoma canestrinii
Cochlostoma henricae
Cochlostoma intermedium
Cochlostoma mariannae
Cochlostoma philippianum
Cochlostoma porro
Cochlostoma villae
Cochlostoma zawinkanum
Striolata striolata

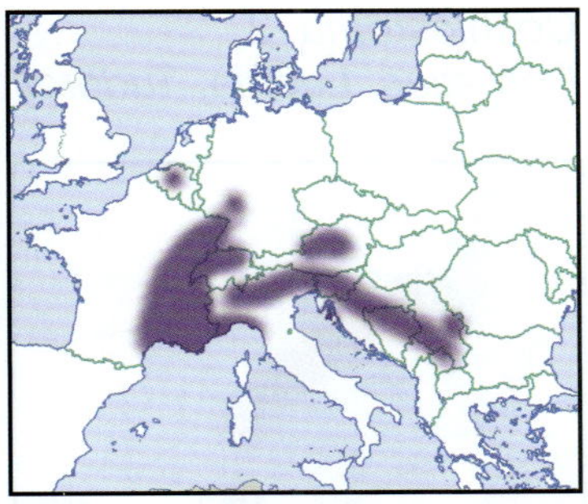

Cochlostoma septemspirale

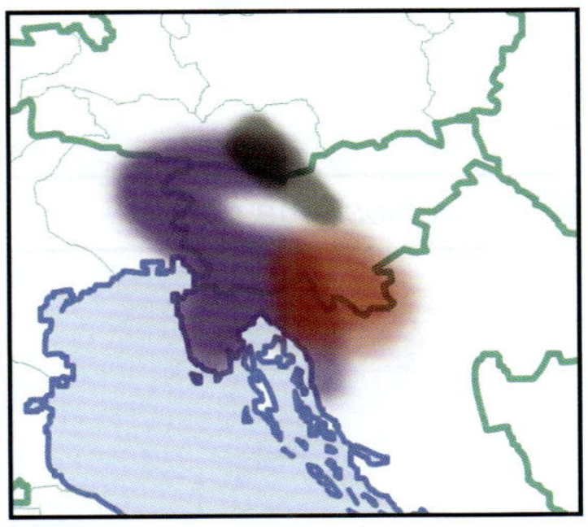

Cochlostoma anomphale
Cochlostoma stossichi
Cochlostoma tergestinum

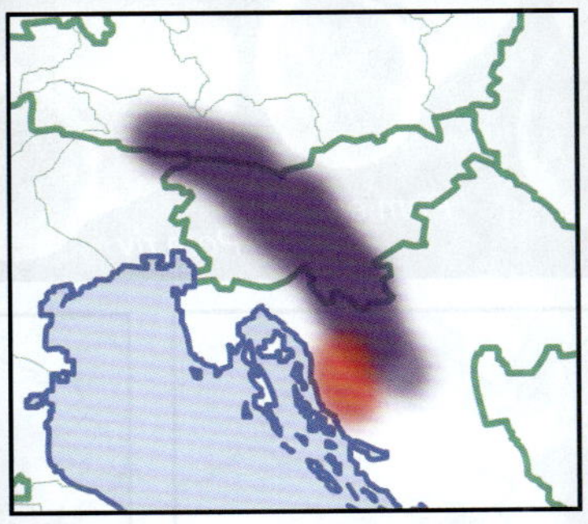

Cochlostoma braueri
Cochlostoma waldemari

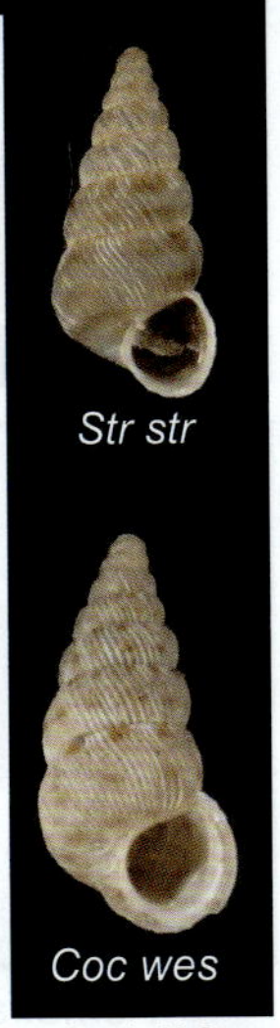

Coc bra Z. Fehér *Coc con* J. Béguinot *Coc phi* E. Zallot *Coc sep* R. Huet *Coc sim* M. Sosso

Cochlostoma 2, *Pomatias, Leonia, Tudorella*

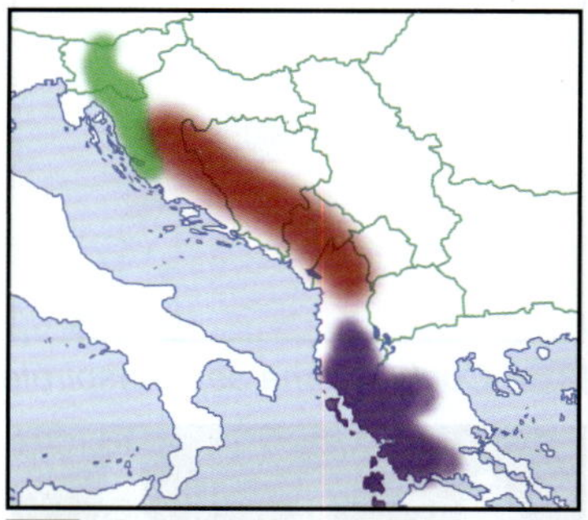

Cochlostoma nanum
Cochlostoma sturanii
Cochlostoma tesselatum

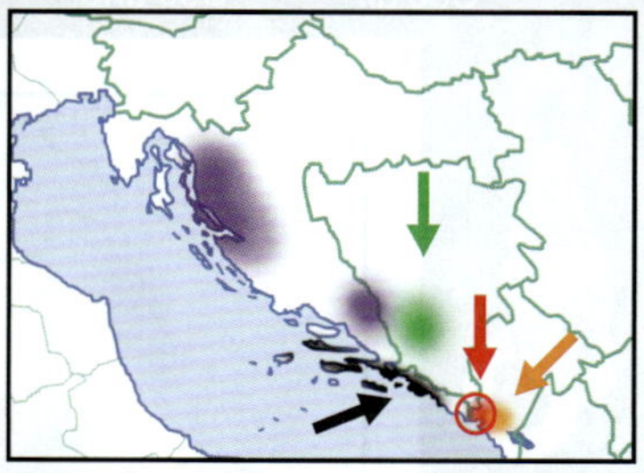

Cochlostoma cinerascens
Cochlostoma dalmatinum
Cochlostoma elegans
Cochlostoma erika
Cochlostoma mostarense

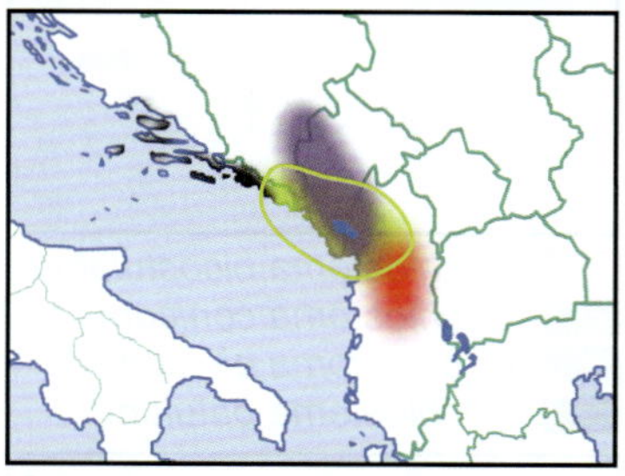

Cochlostoma auritum
Cochlostoma cinerascens
Cochlostoma hoyeri
Cochlostoma roseoli

Cochlostoma gracile

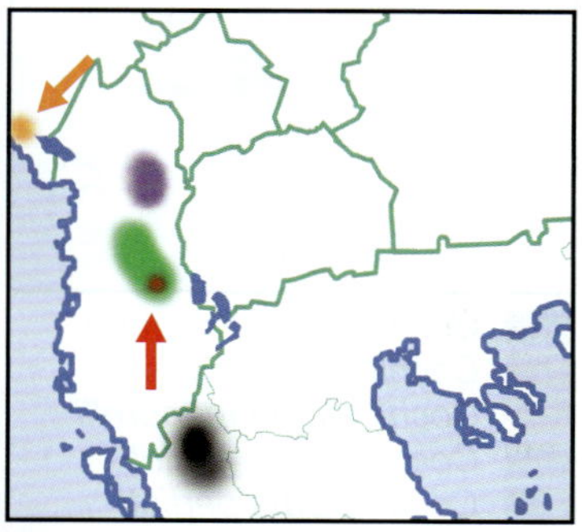

Cochlostoma erika
Cochlostoma fuchsi
Cochlostoma georgi
Cochlostoma pageti
Cochlostoma pinteri

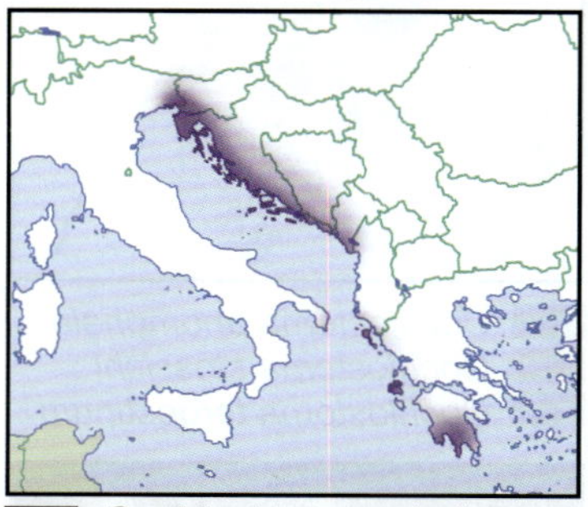

Cochlostoma scalarinum

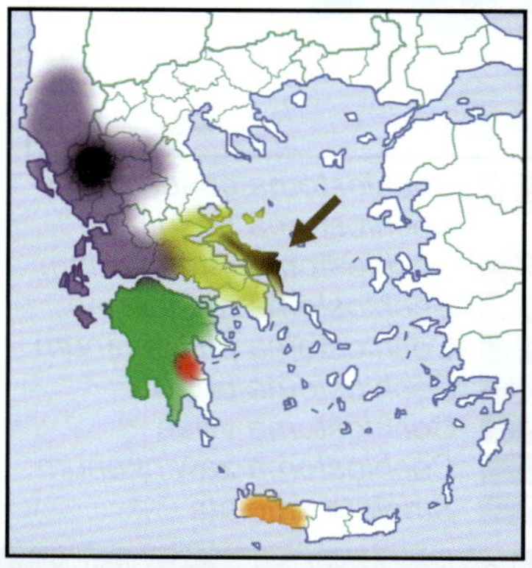

Cochlostoma achaicum
Cochlostoma cretense
Cochlostoma euboicum
Cochlostoma hellenicum
Cochlostoma pageti
Cochlostoma parnonis
Cochlostoma tesselatum

Cochlostoma hueti
Cochlostoma mienisi

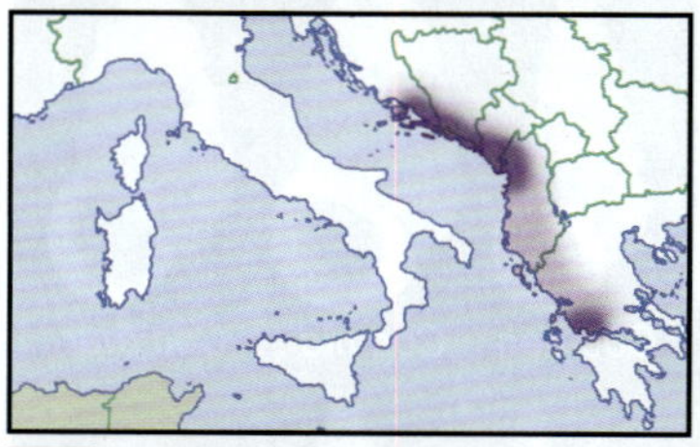

Cochlostoma kleciaki

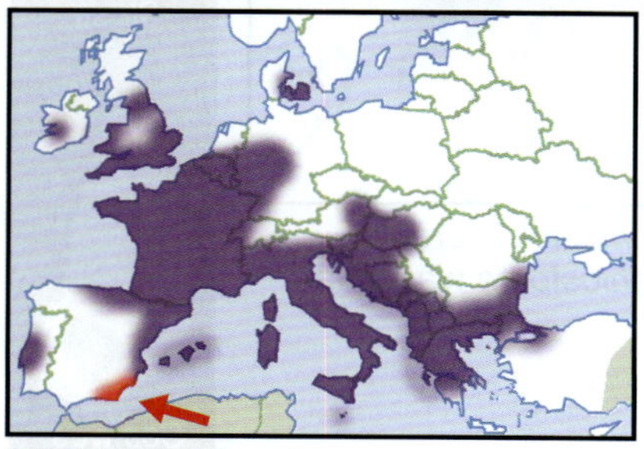

Pomatias elegans
Leonia mammillaris

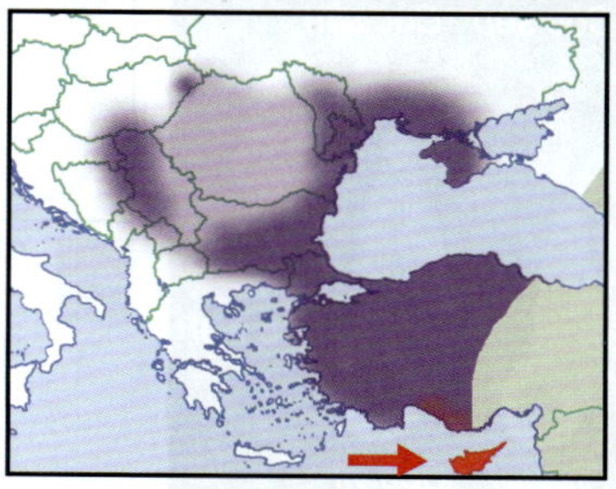

Pomatias glaucus
Pomatias rivularis

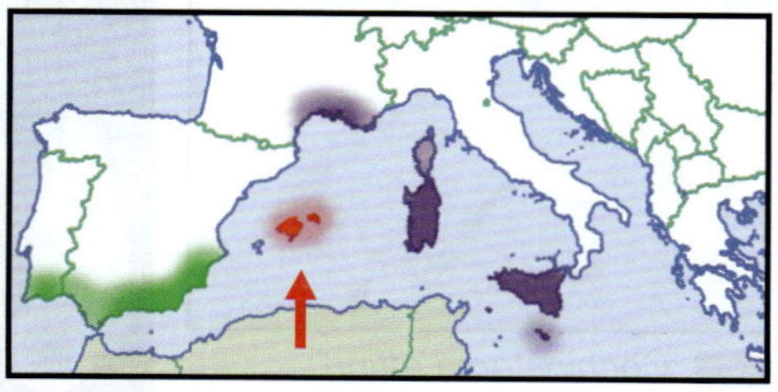

Tudorella ferruginea
Tudorella mauretanica
Tudorella sulcata

Ferussaciidae, Cochlicopidae, *Hypnophila*

📷 *Coc nit* D. Georgiev *Hyp doh* C. Bogi *Hyp emi* C. Bogi *Hyp gir* C. Bogi *Hyp inc* F. Liberto *Hyp pol* M. Prodan

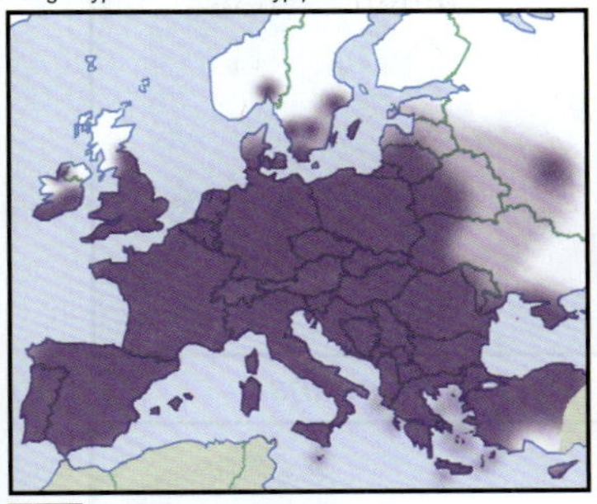
Cecilioides acicula

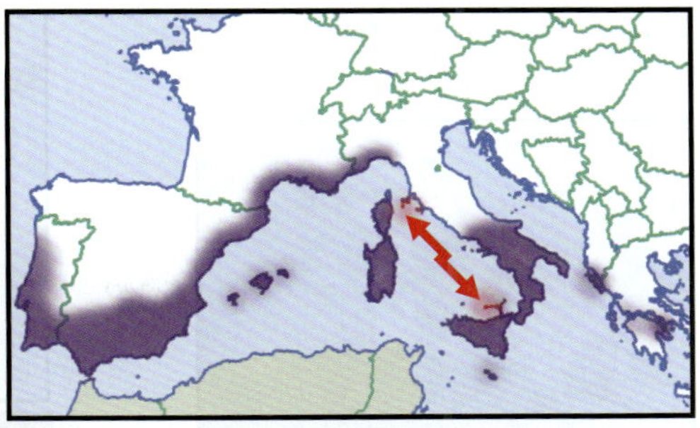
Ferussacia carnea
Ferussacia folliculum

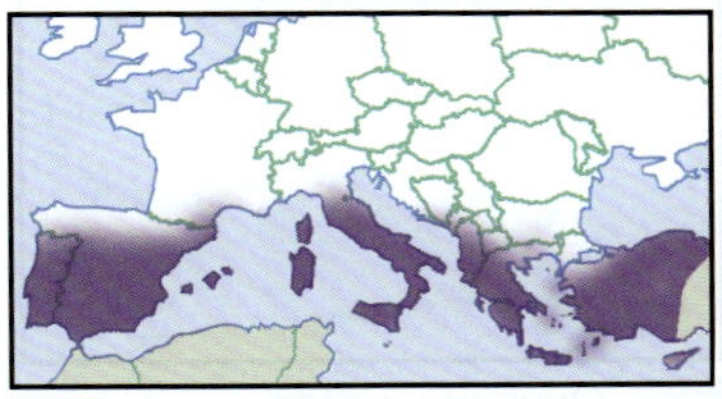
Cecilioides tumulorum

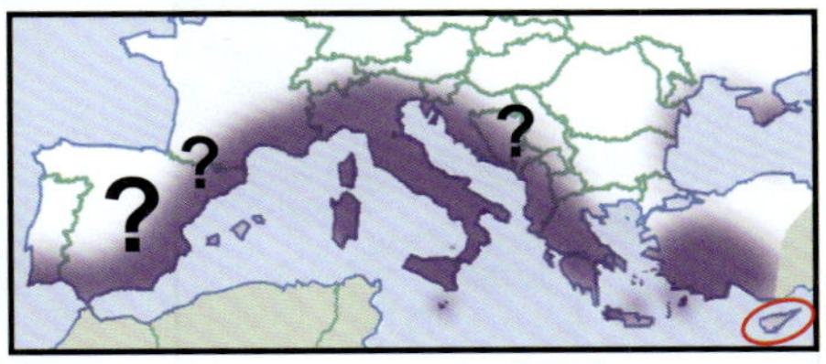

Cecilioides janii
Calaxis hierosolymarum

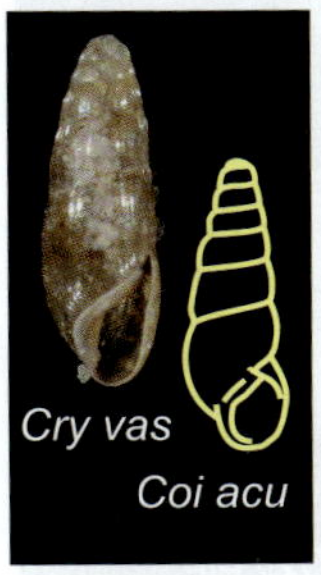

Cryptazeca subcylindrica

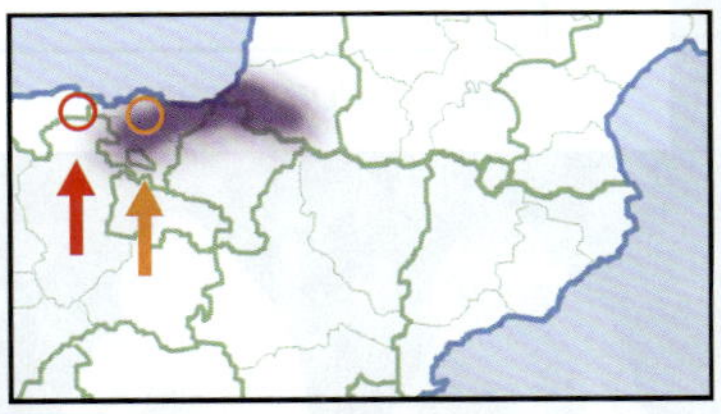
Cryptazeca elongata
Cryptazeca monodonta
Cryptazeca spelaea

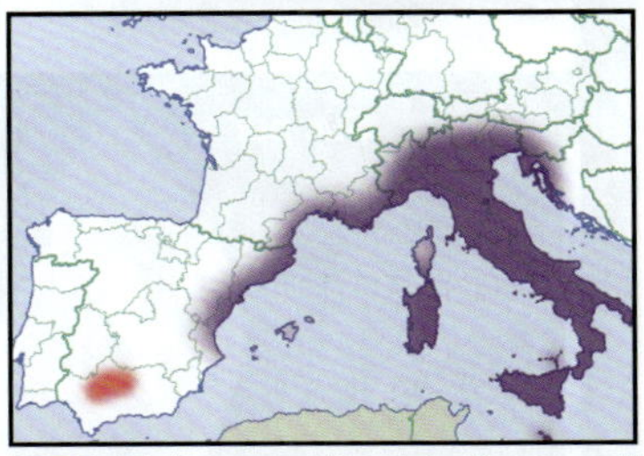
Hohenwartiana hohenwarti
Coilostele acus

Cochlicopa lubrica (lub1)

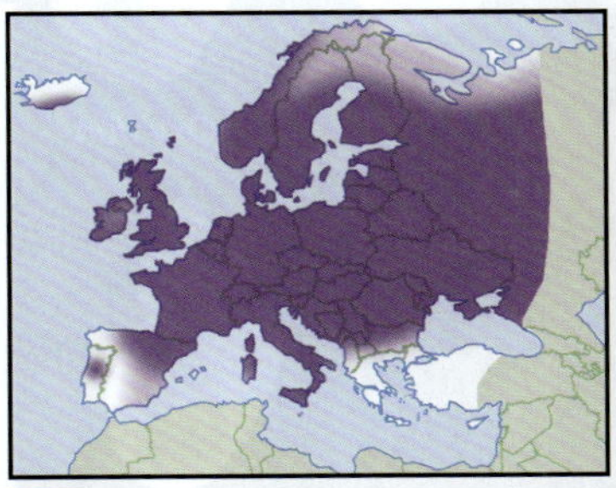
Cochlicopa lubricella (lub2)

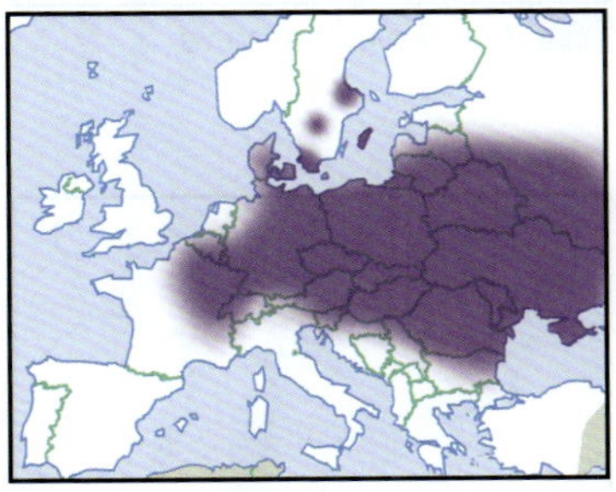
Cochlicopa nitens

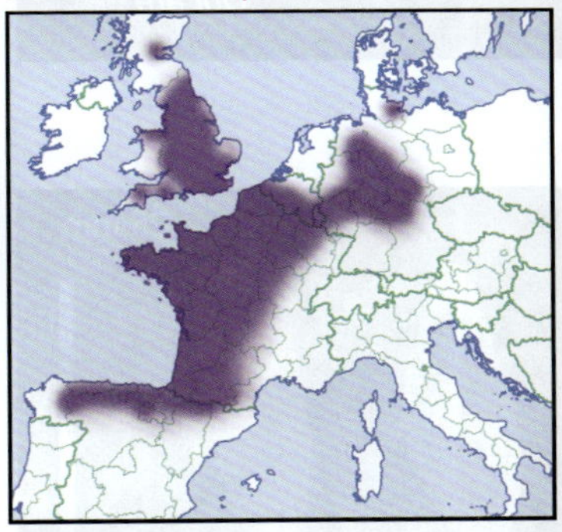
Azeca goodalli

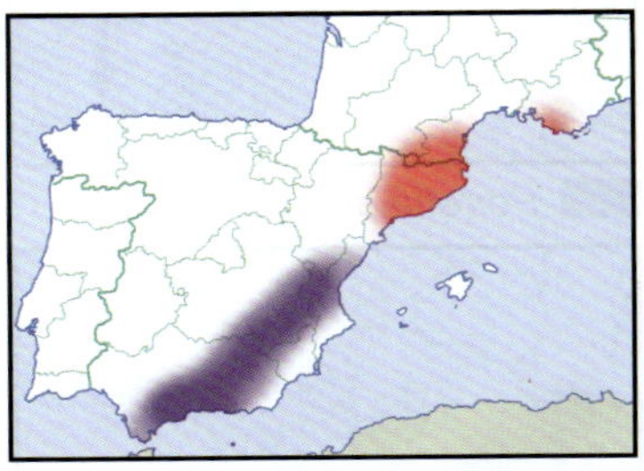
Hypnophila boissii
Hypnophila malagana

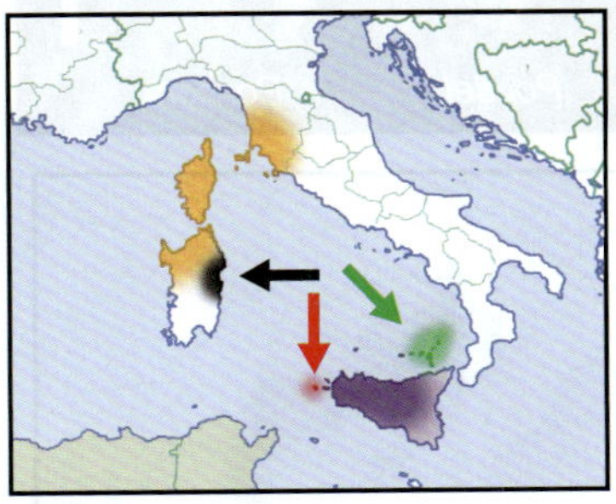
Hypnophila bisacchii
Hypnophila cylindracea
Hypnophila dohrni
Hypnophila emiliana
Hypnophila girottii
Hypnophila incerta

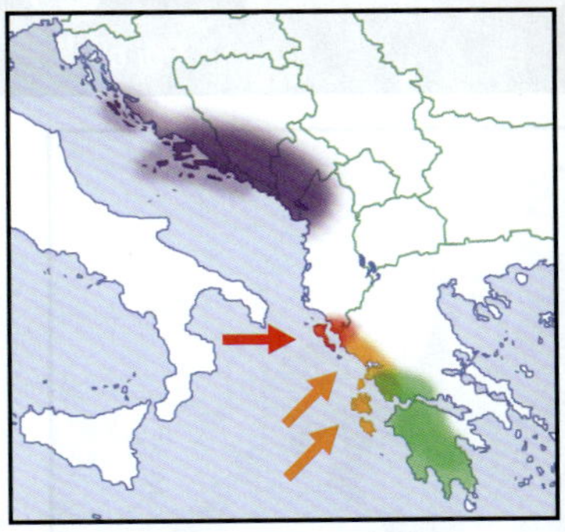
Hypnophila cyclothyra
Hypnophila polita
Hypnophila pupaeformis
Hypnophila zacynthia

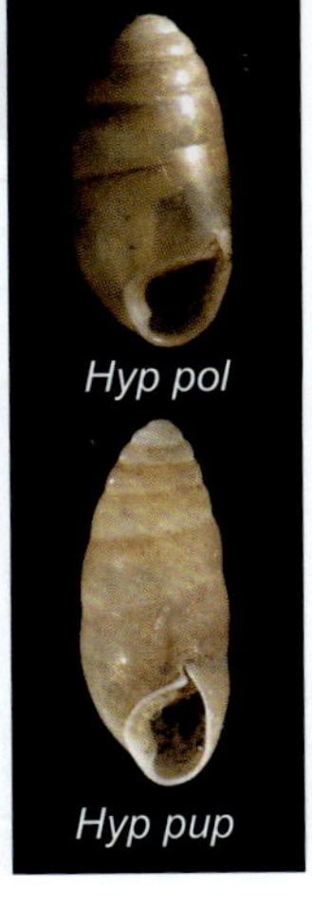

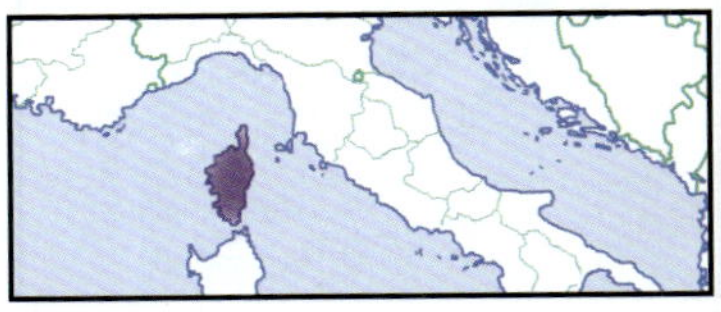
Hypnophila remyi

Succineidae, *Poiretia*

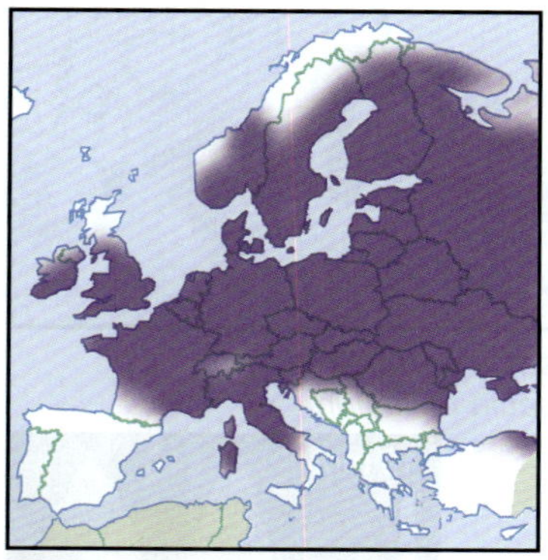

Succinea putris

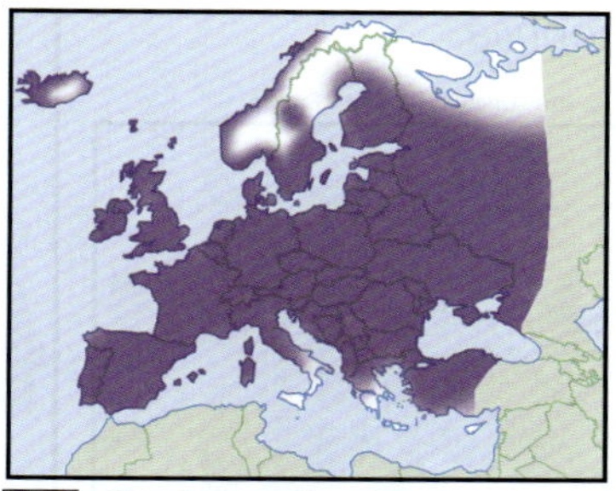

Oxyloma elegans

Oxyloma dunkeri

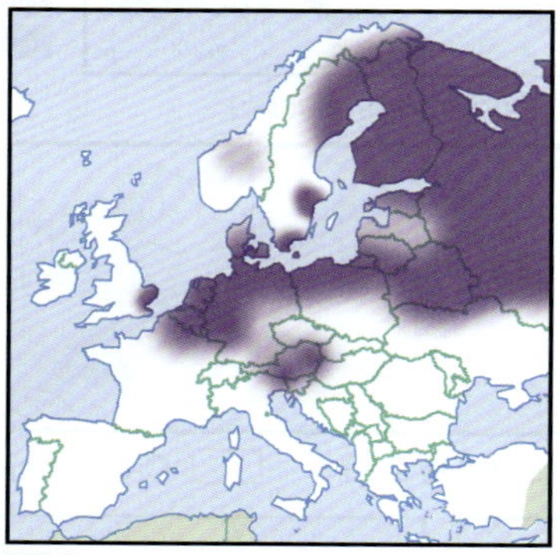

Oxyloma sarsii

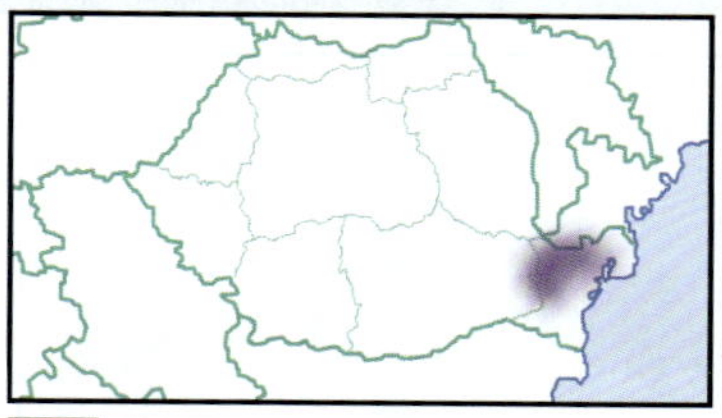

Oxyloma pinteri

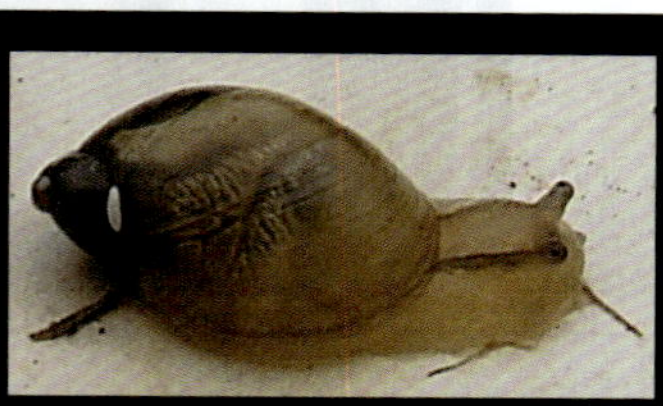

Suc put

10 mm

Oxy ele　Oxy sar　Oxy dun

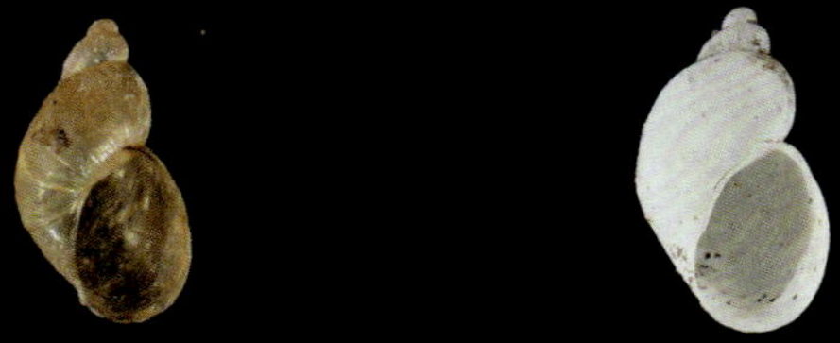

Suc obl　Qui are

Oxy ele

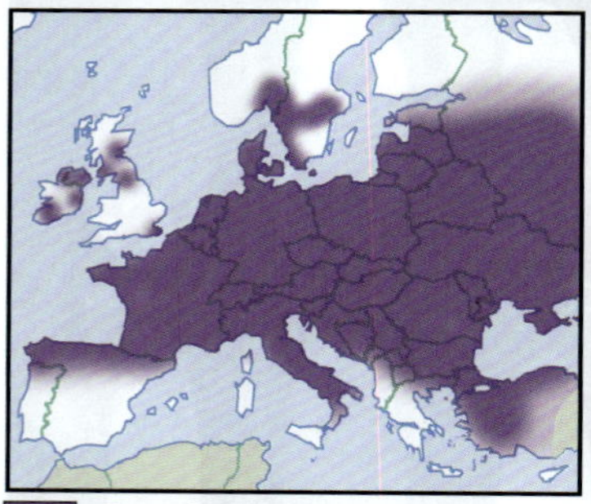

Succinella oblonga

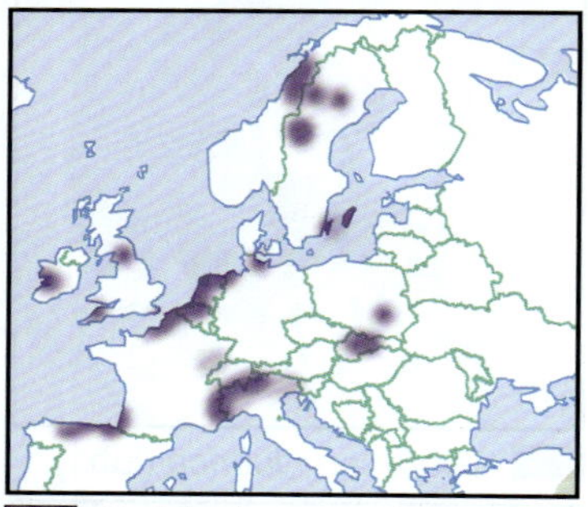

Quickella arenaria

Poi cor

25 mm

Poi cor　Poi del　Poi com

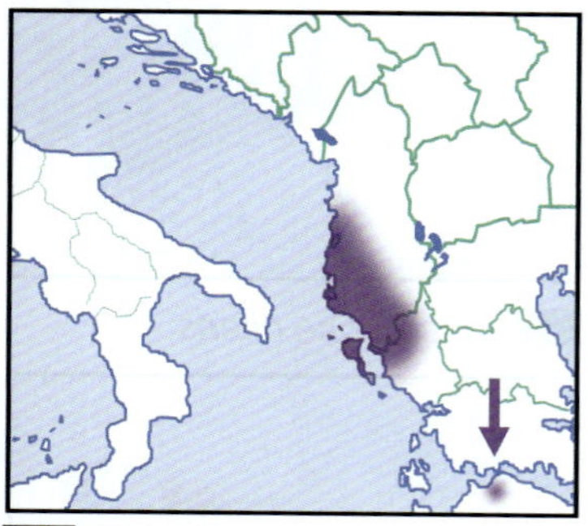

Poiretia delesserti

Poi dil

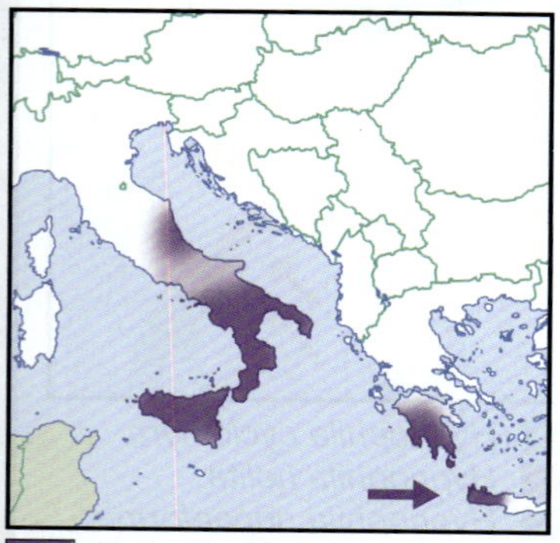

Poiretia dilatata

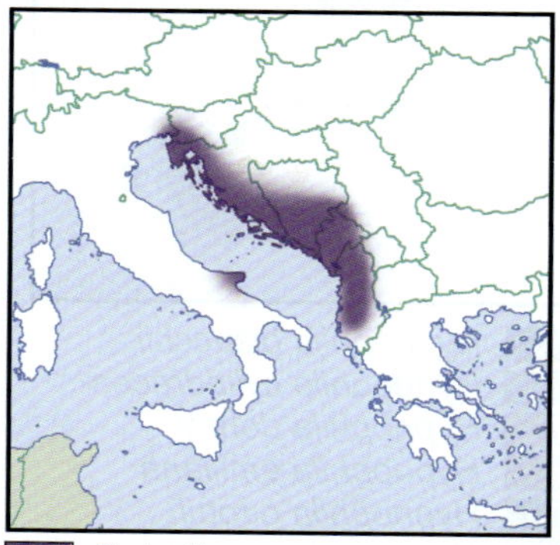

Poiretia cornea

Poiretia compressa

Columella, *Truncatellina*, *Spelaeoconcha*

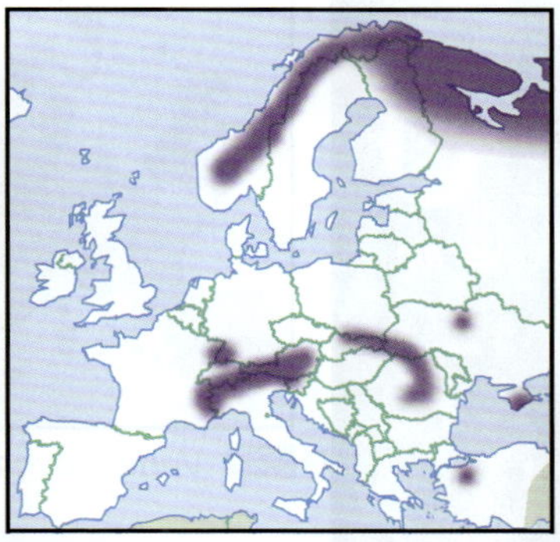

Columella columella

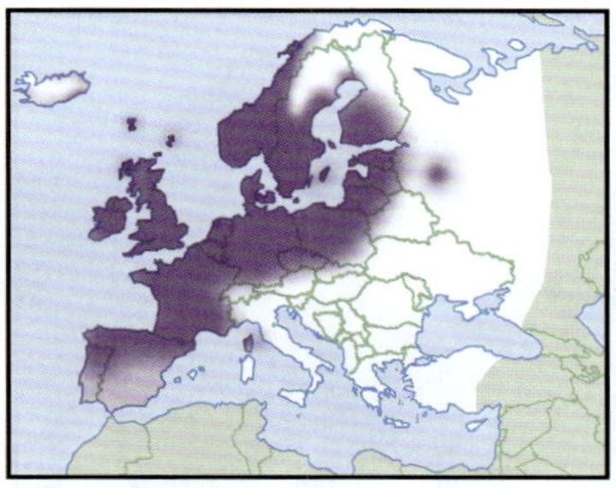

Columella aspera

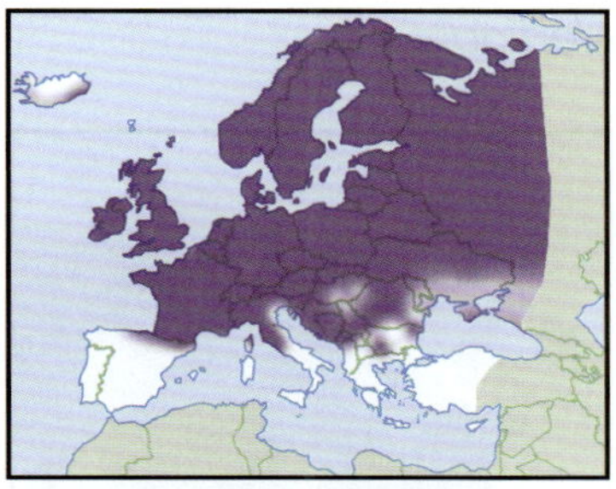

Columella edentula

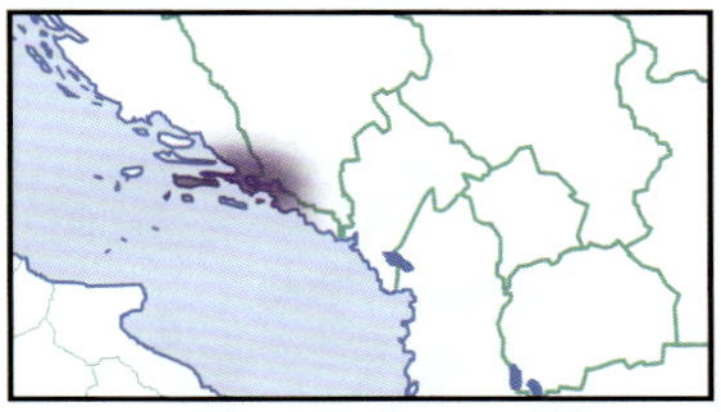

Spelaeoconcha paganettii

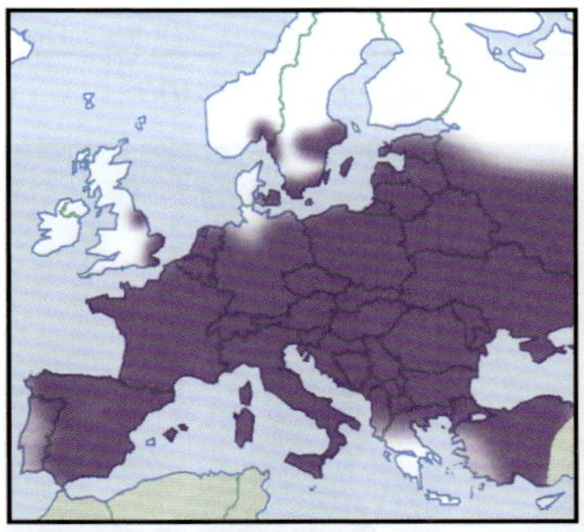

Truncatellina cylindrica

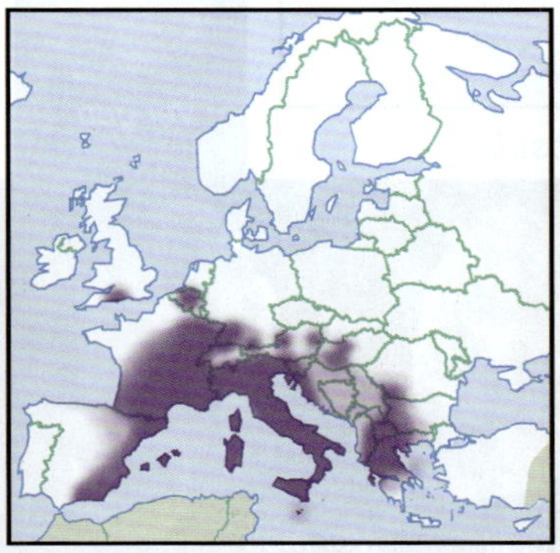

Truncatellina callicratis

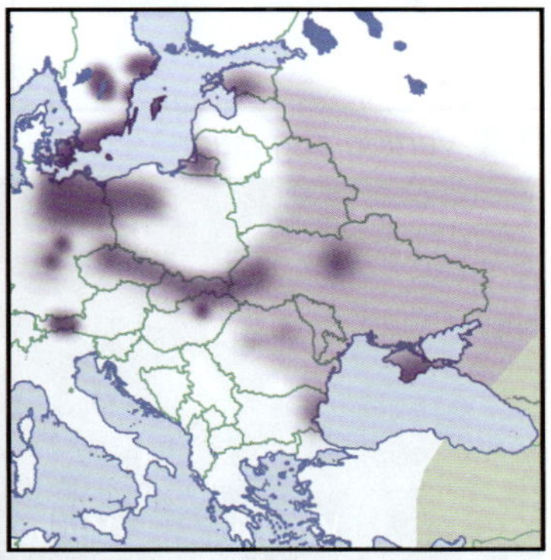

Truncatellina costulata

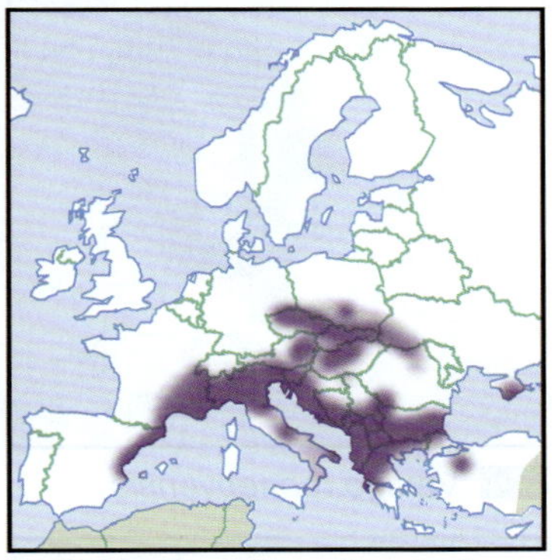

Truncatellina claustralis

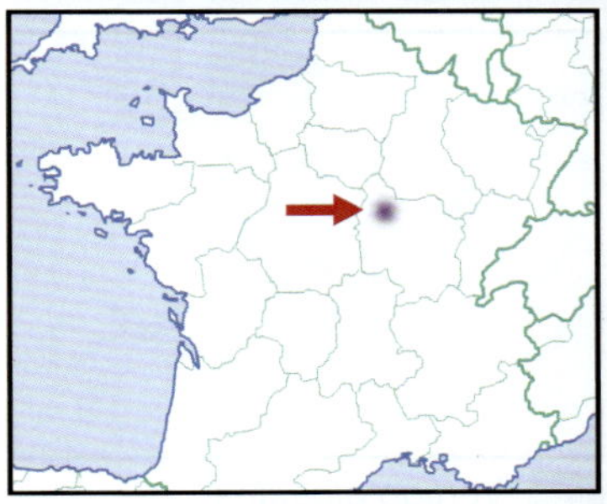

Truncatellina arcyensis

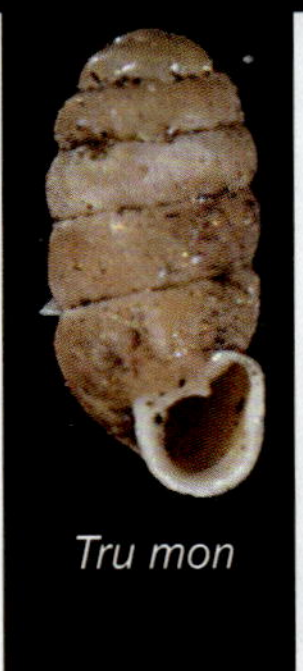

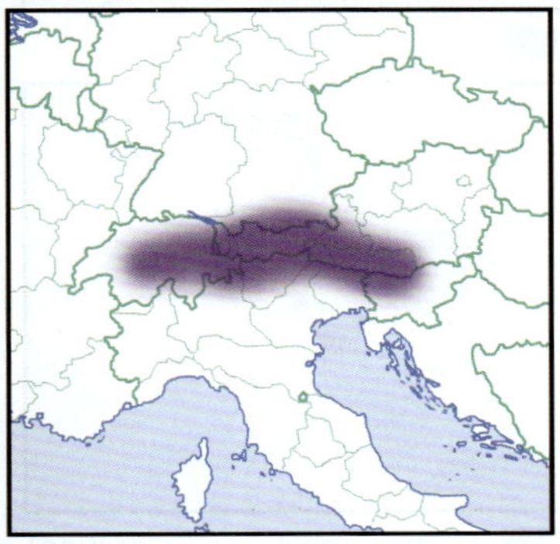

Truncatellina monodon

Truncatellina cameroni

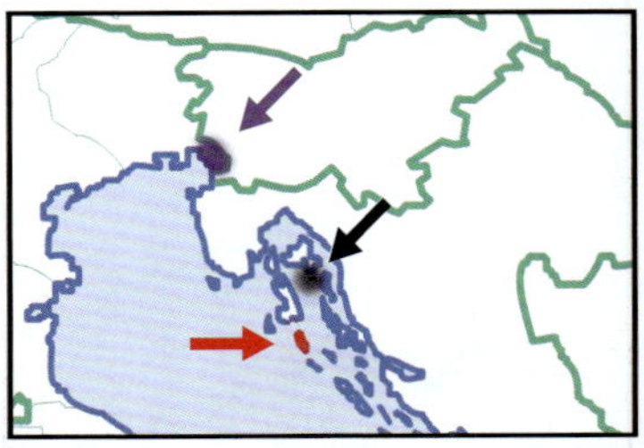

Truncatellina laeviuscula
Truncatellina lussinensis
Truncatellina uniarmata
Truncatellina velkovrhi

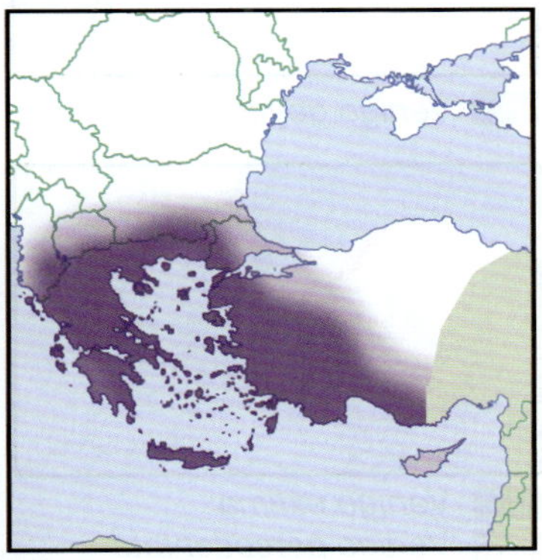

Truncatellina rothi

Vertigo

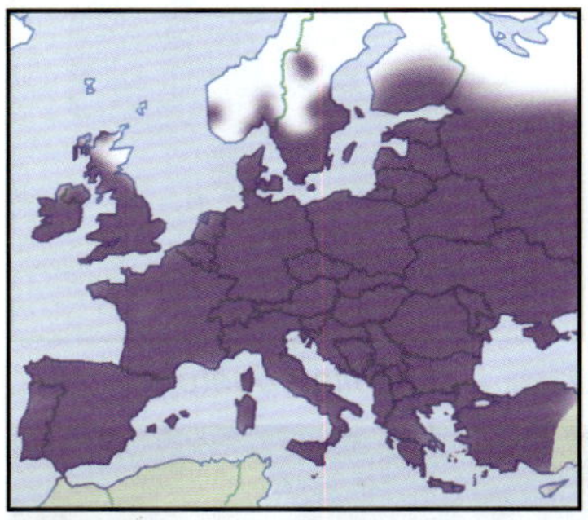

Vertigo antivertigo

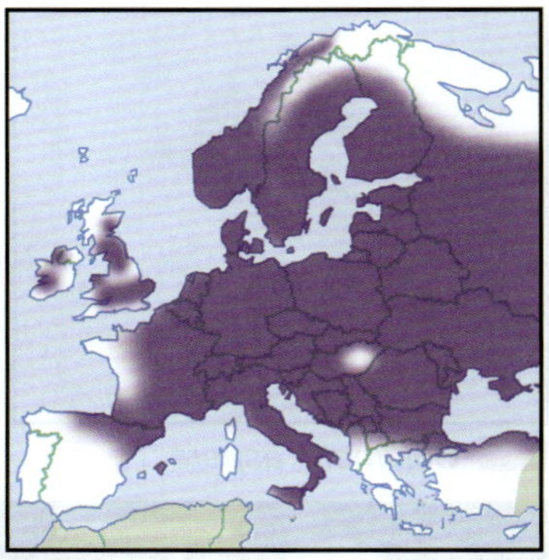

Vertigo pusilla

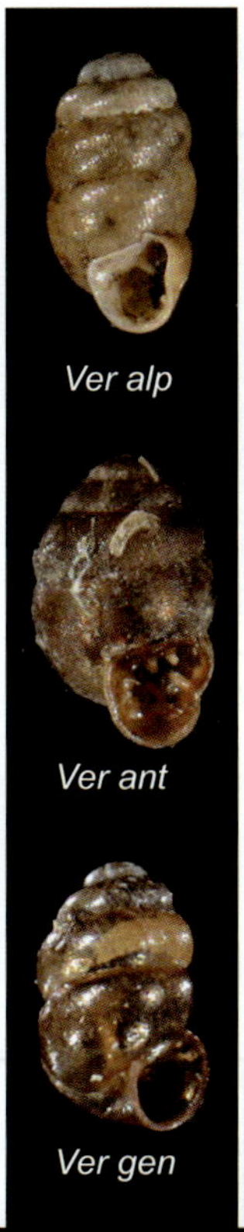

Vertigo pygmaea

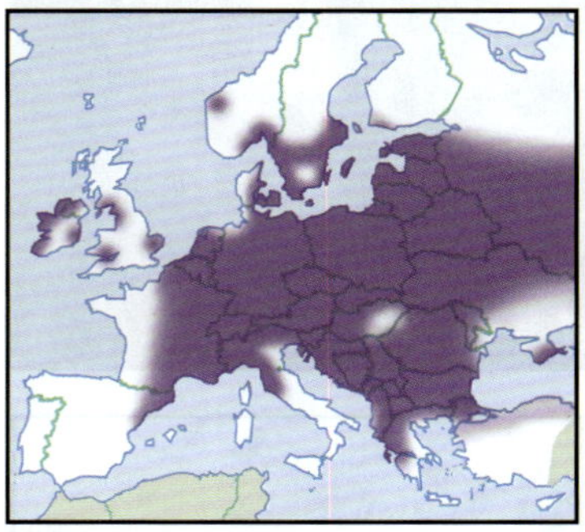

Vertigo angustior

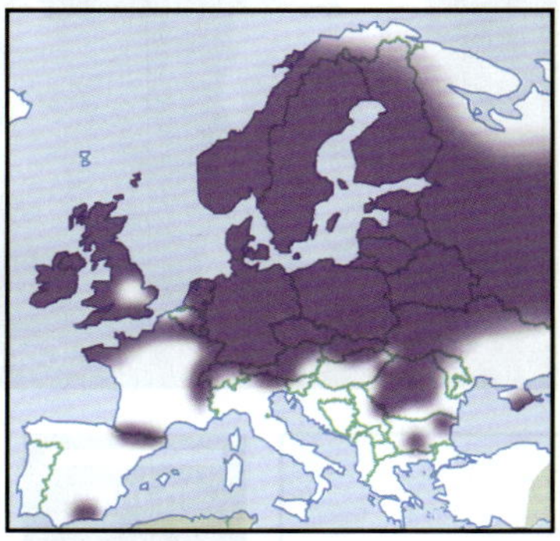

Vertigo substriata

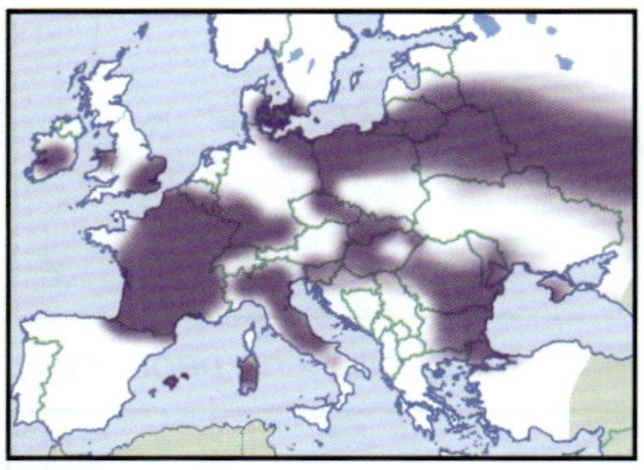

Vertigo moulinsiana

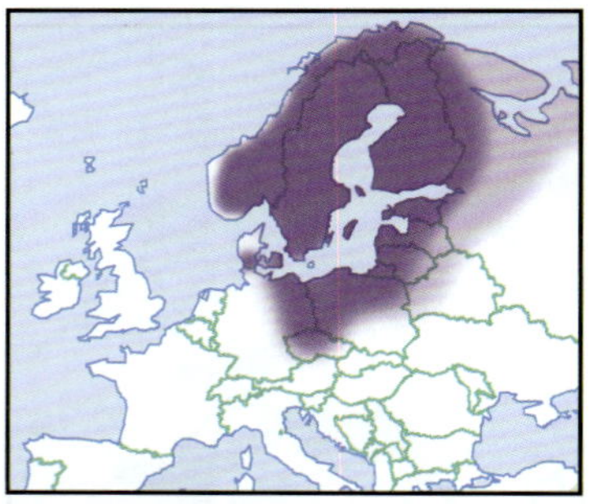

Vertigo ronnebyensis

Vertigo lilljeborgi

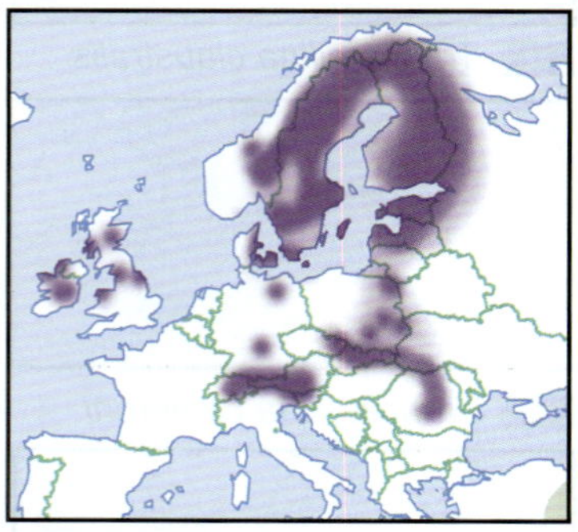

Vertigo geyeri

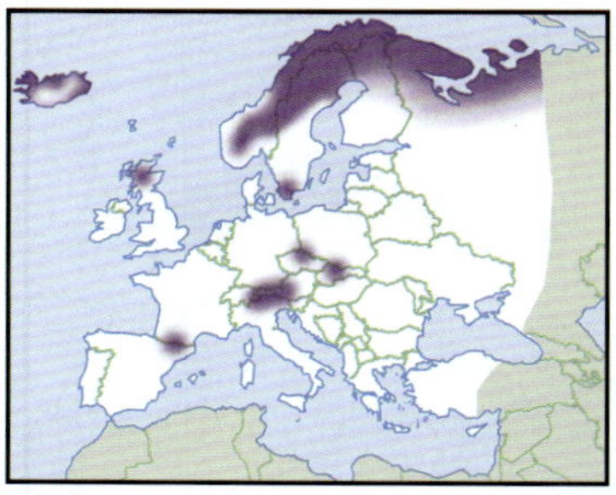

Vertigo modesta (= *arctica*)

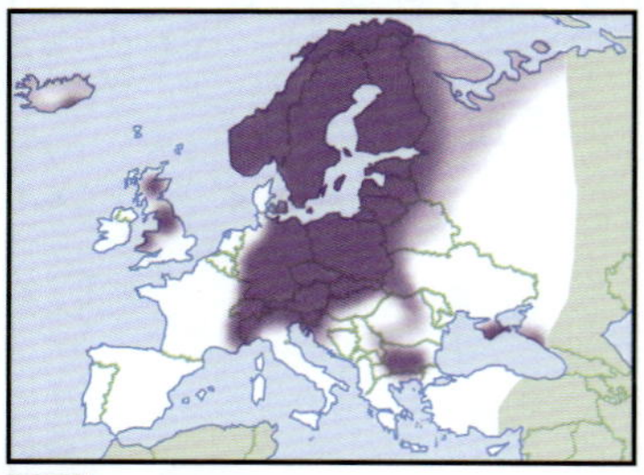

Vertigo alpestris

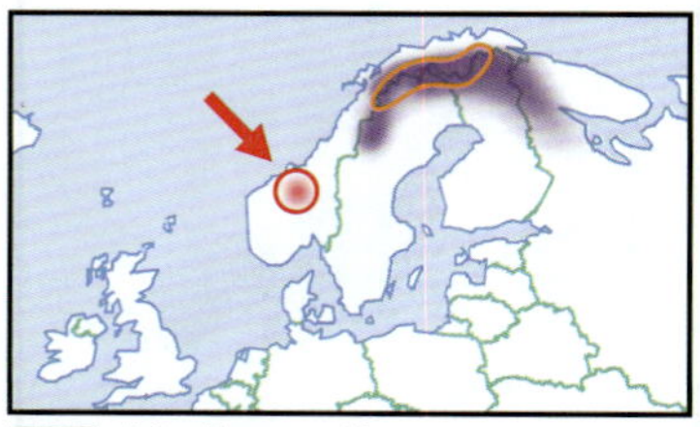

Vertigo extima
Vertigo parcedentata
Vertigo ultimathule

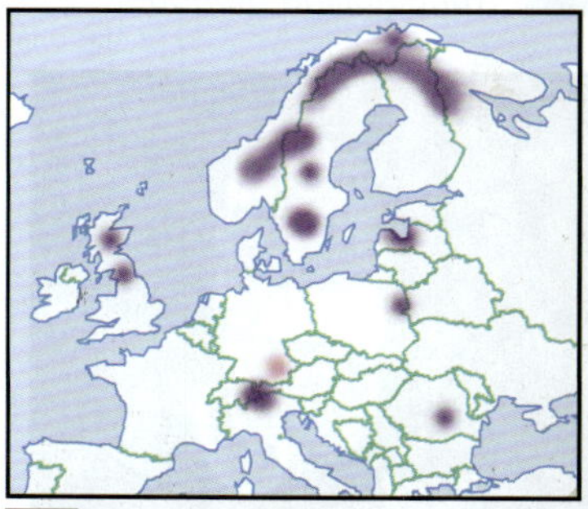

Vertigo genesii

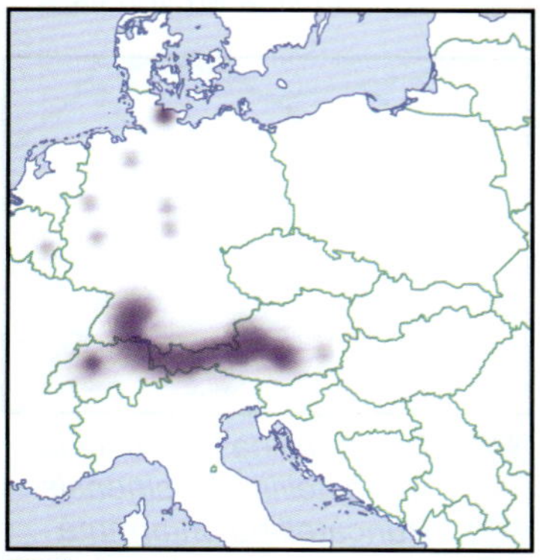

Vertigo heldi

Pupilla, Lauriidae, *Argna*

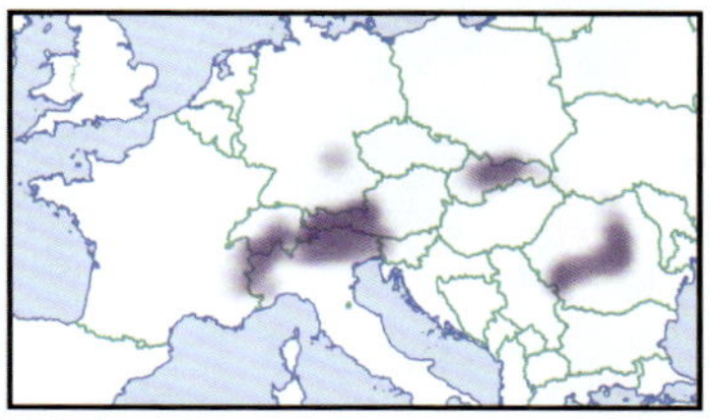

Pupilla alpicola

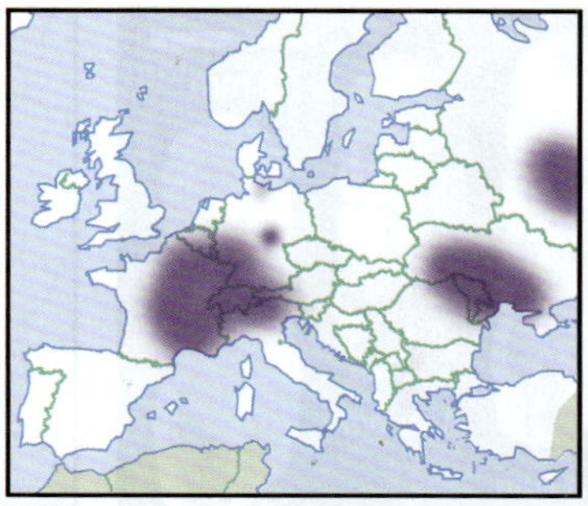

Pupilla bigranata

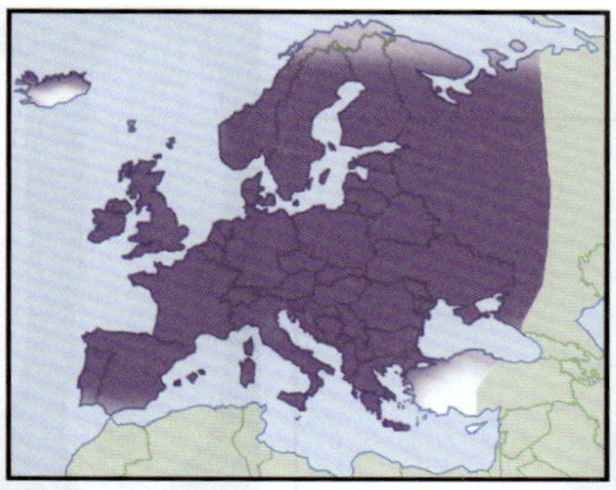

Pupilla muscorum

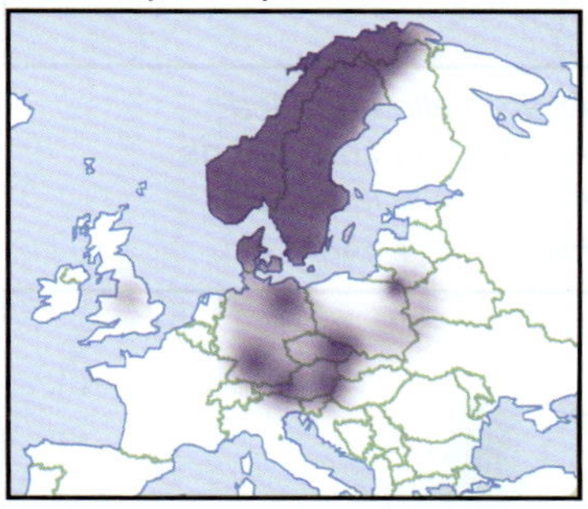

Pupilla pratensis

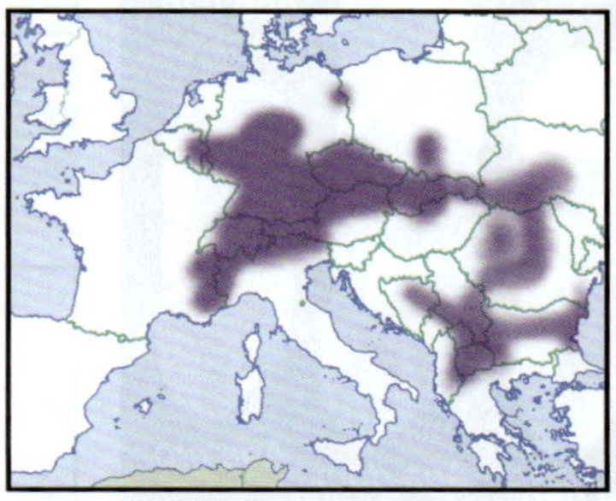

Pupilla sterrii

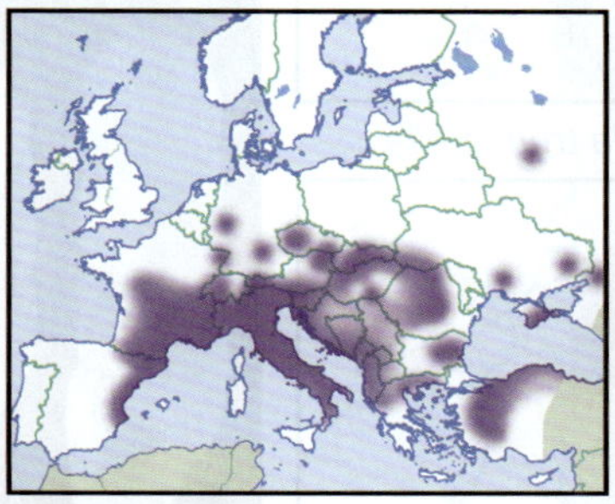

Pupilla triplicata

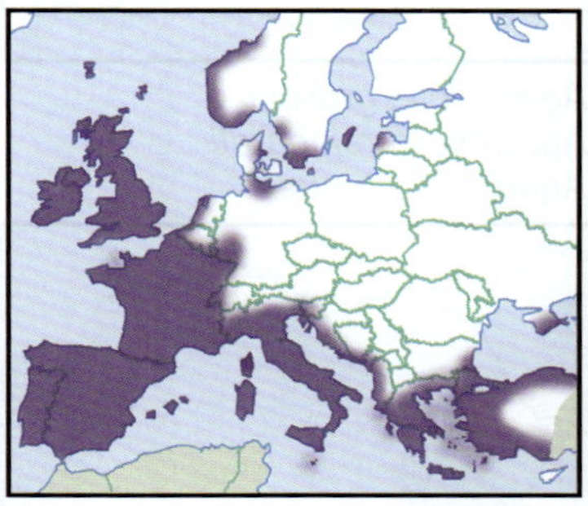

Lauria cylindracea

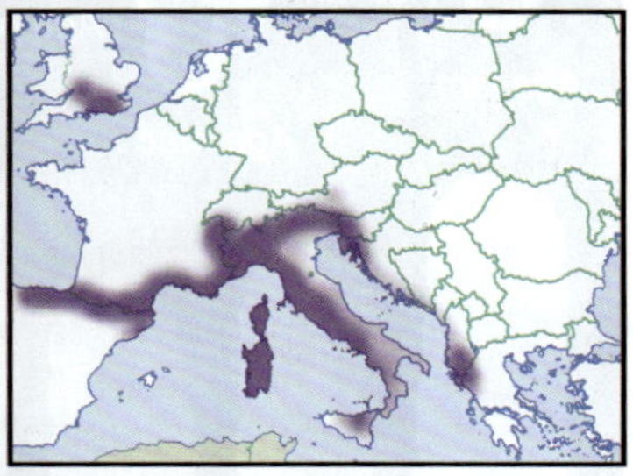

Lauria sempronii

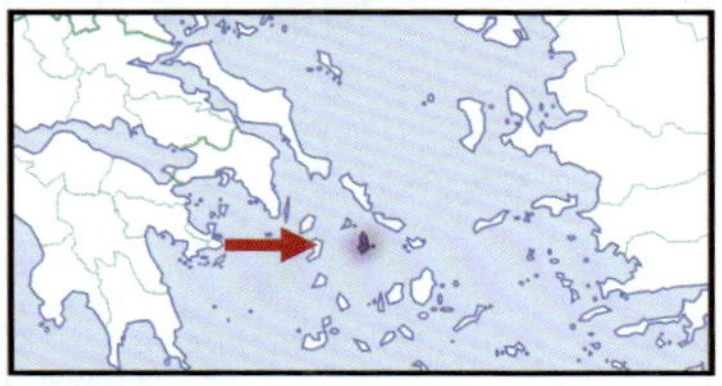

Lauria umbilicus

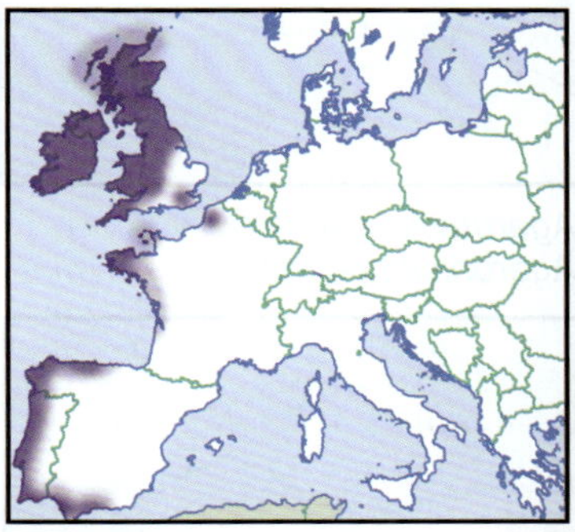

Leiostyla anglica

Lau sem A. Vanozzi Eux sch D. Georgiev
Arg bie J. Grego Arg val C. Bogi

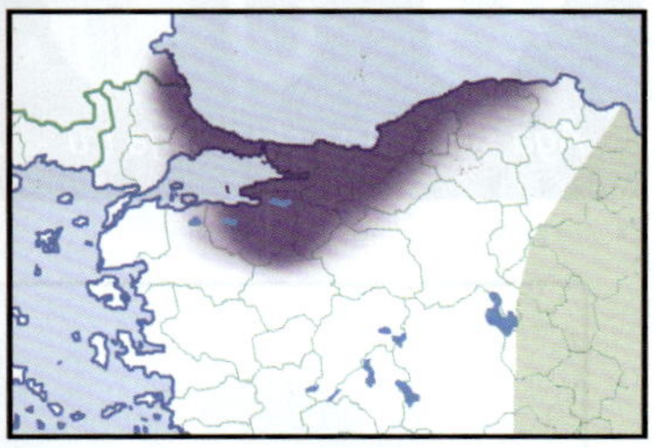

Euxinolauria schweigeri

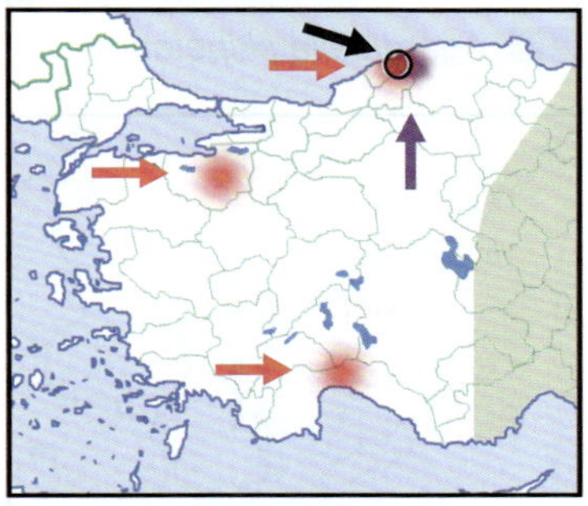

Euxinolauria crassilabris
Euxinolauria paphlagonica
Euxinolauria superba

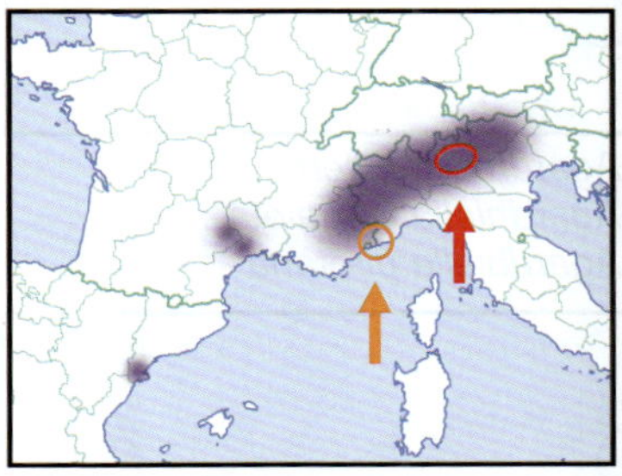

Argna bourguignatiana
Argna ferrari
Argna valsabina

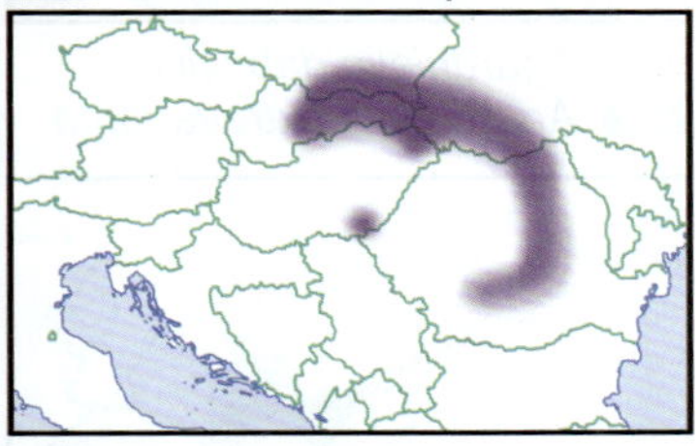

Argna bielzi

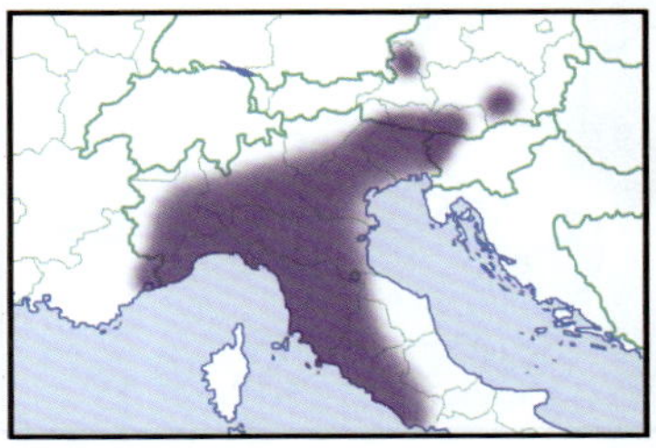

Argna biplicata

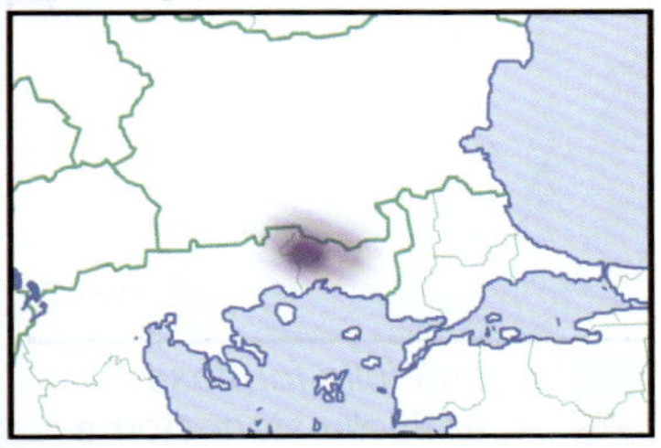

Argna thracica

Agardhiella

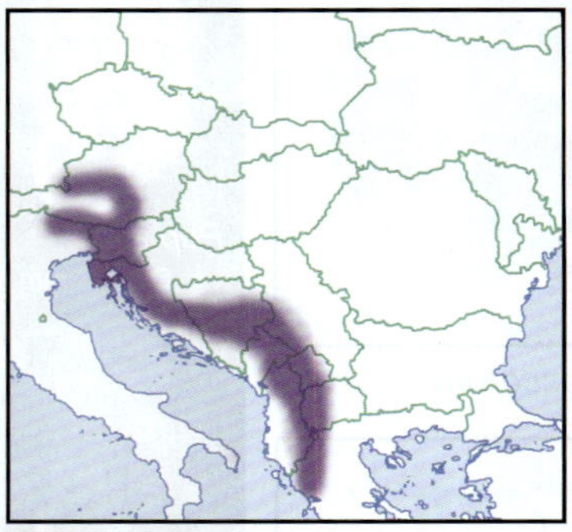

Agardhiella truncatella

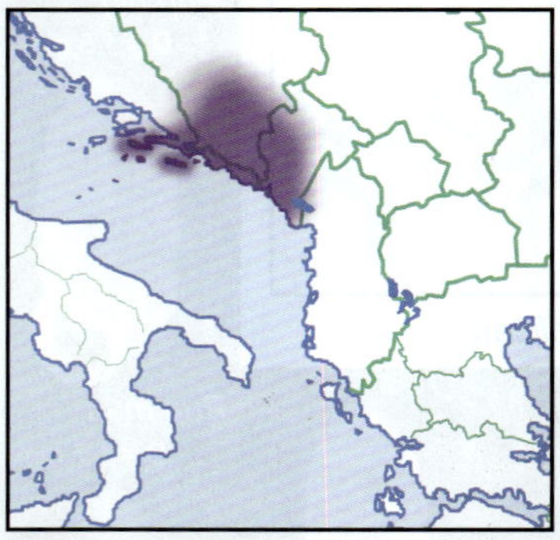

Agardhiella biarmata

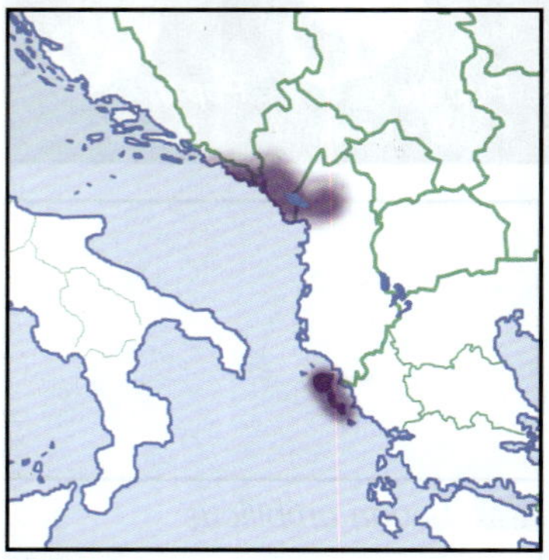

Agardhiella skipetarica

Agardhiella dabovici
Agardhiella extravaganta

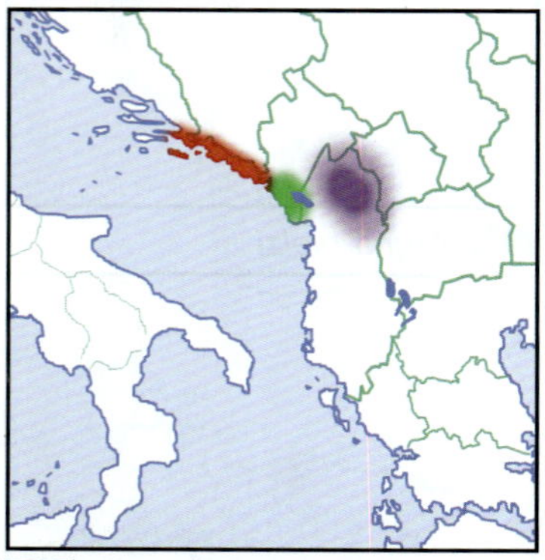

Agardhiella formosa
Agardhiella stenostoma
Agardhiella zoltanorum

Aga bia M. Prodan *Aga ski* M. Prodan

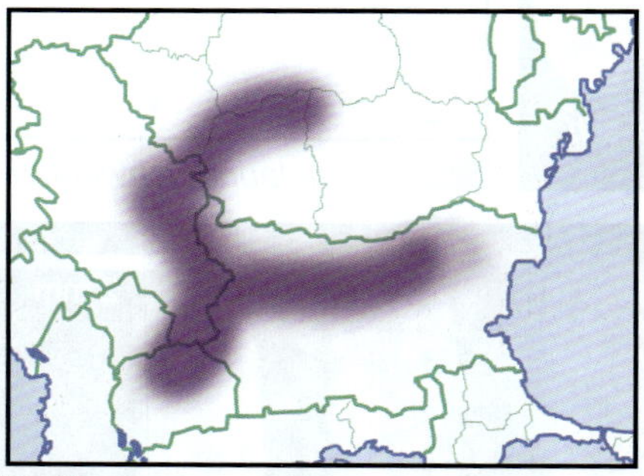

Agardhiella armata

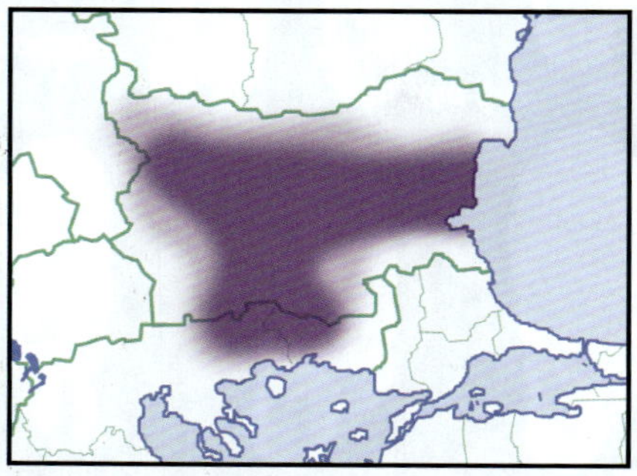

Agardhiella parreyssii

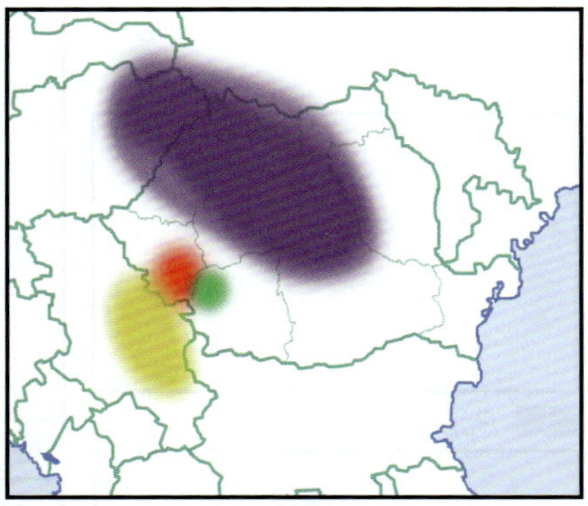

Agardhiella banatica
Agardhiella crassilabris
Agardhiella lamellata
Agardhiella serbica

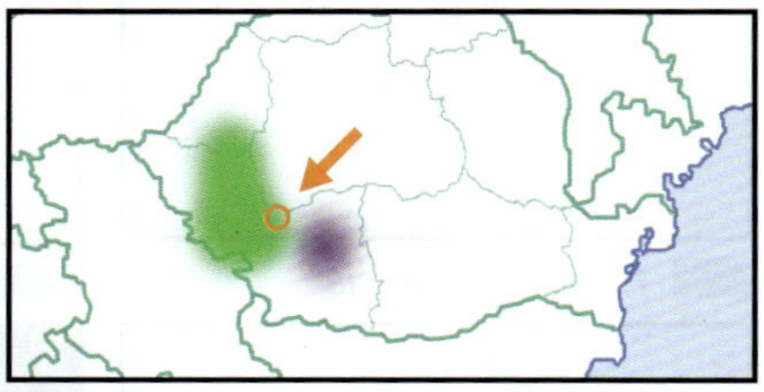

Agardhiella grossui
Agardhiella reinhardti
Agardhiella tunde

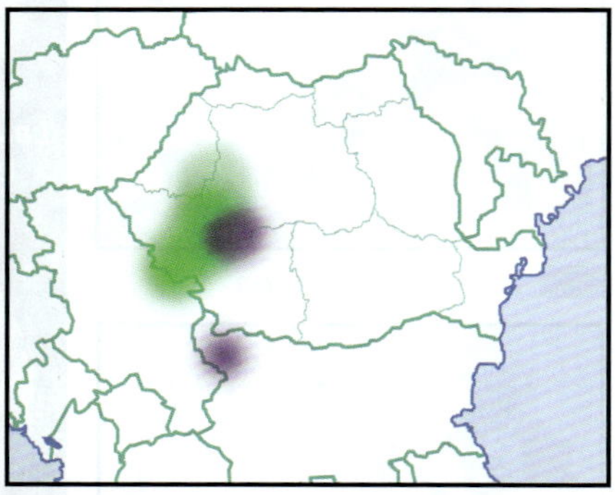

Agardhiella caesa
Agardhiella domokosi

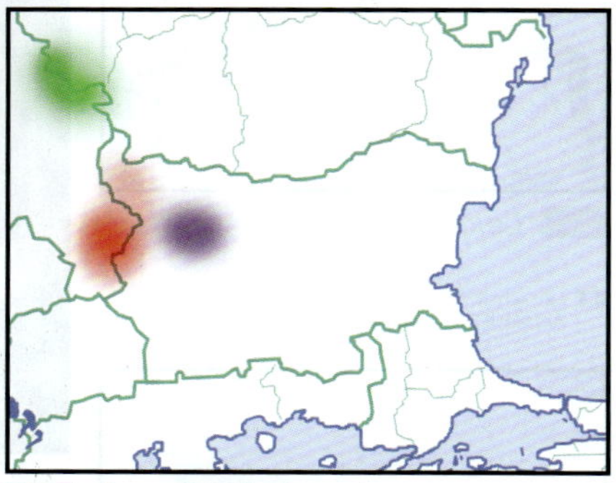

Agardhiella incerta
Agardhiella langaleta
Agardhiella pirotana

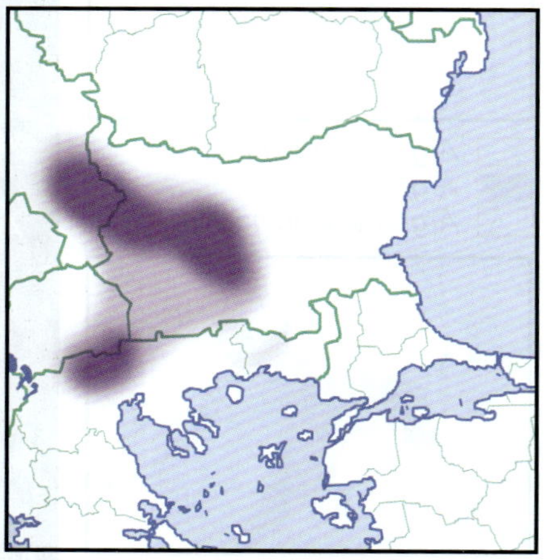

Agardhiella macrodonta

Orculidae 1

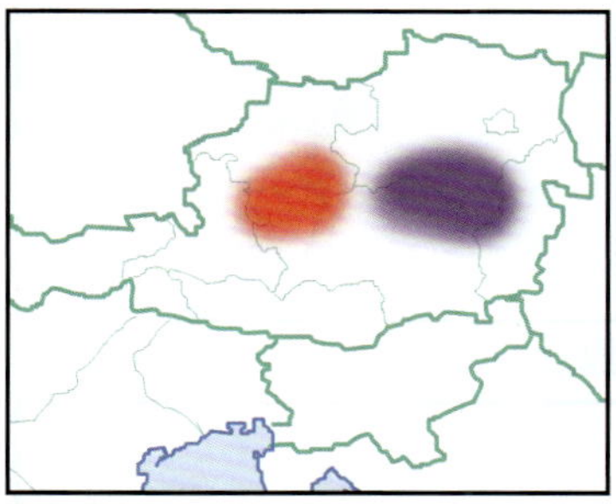

Orcula austriaca
Orcula pseudodolium

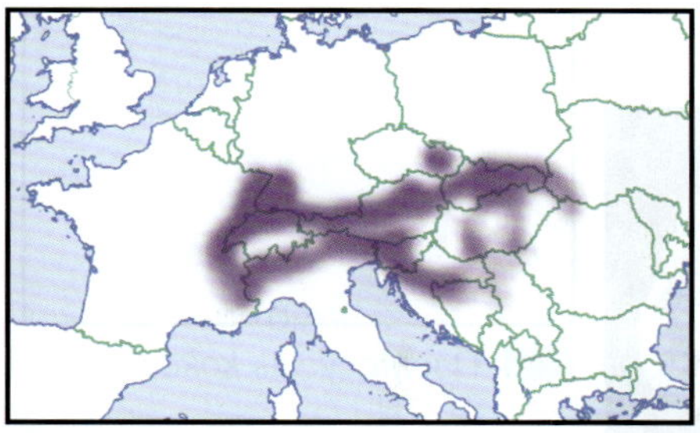

Orcula dolium

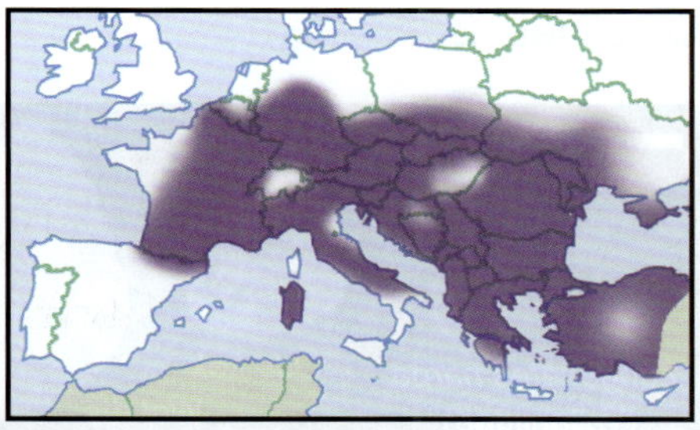

Sphyradium doliolum

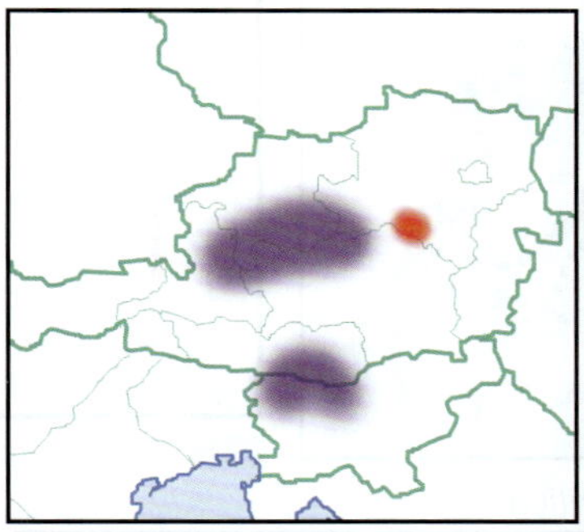

Orcula fuchsi
Orcula tolminensis

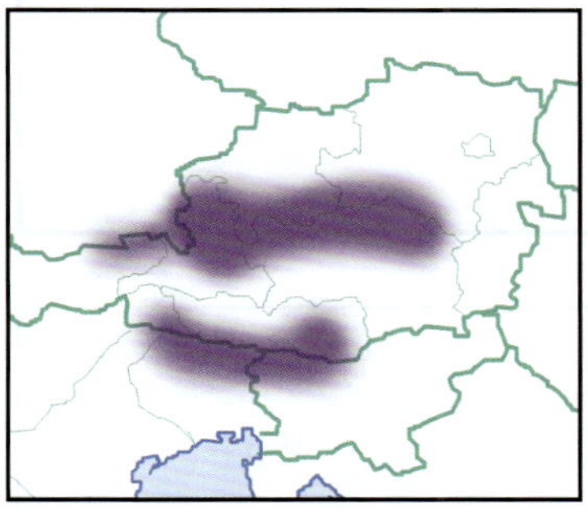

Orcula gularis

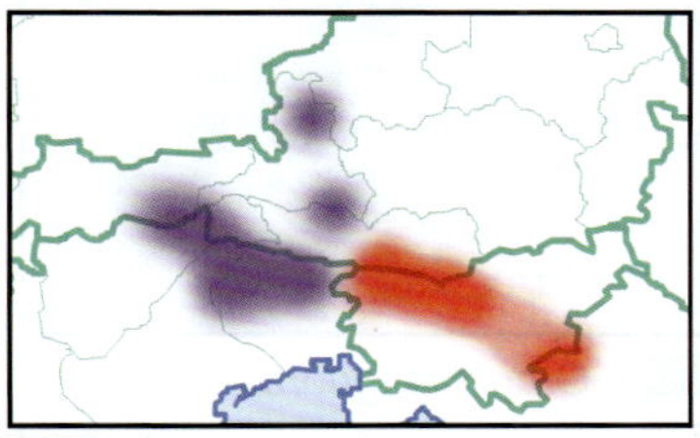

Orcula restituta
Orcula spoliata

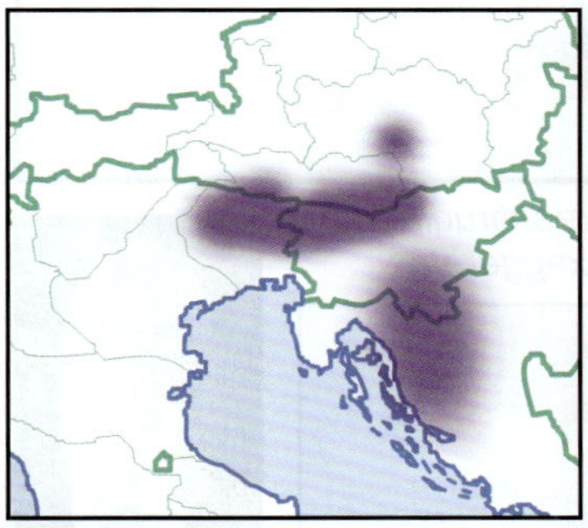

Orcula conica

Orc wag Z. Fehér Orc zil A. Irikov

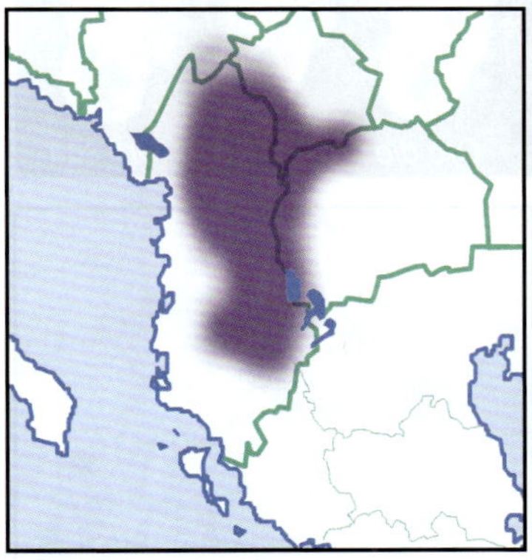

Orcula wagneri

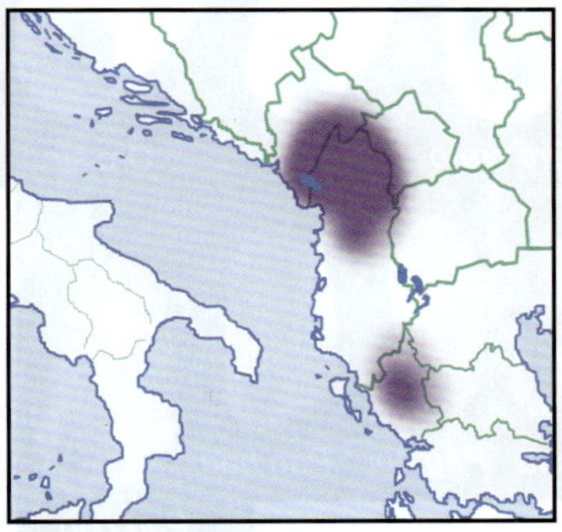

Orcula schmidtii

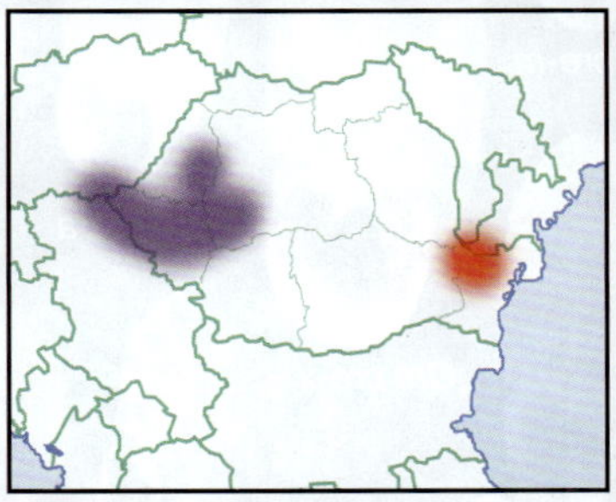

Orcula dobrogica
Orcula jetschini

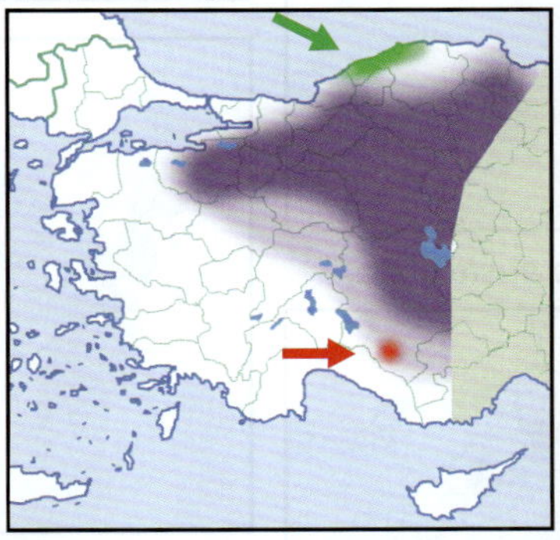

Alvariella multiplicata
Schileykula nordsiecki
Schileykula scyphus

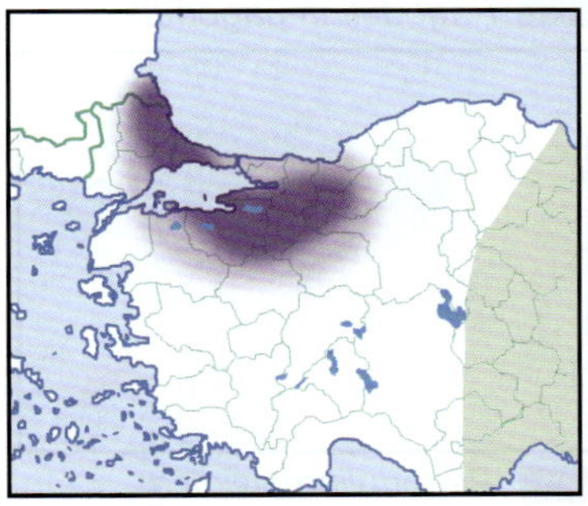

Orcula zilchi

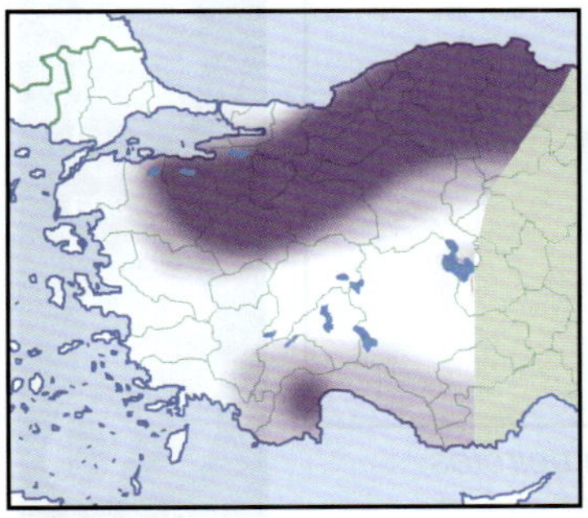

Pilorcula trifilaris

Orculidae 2

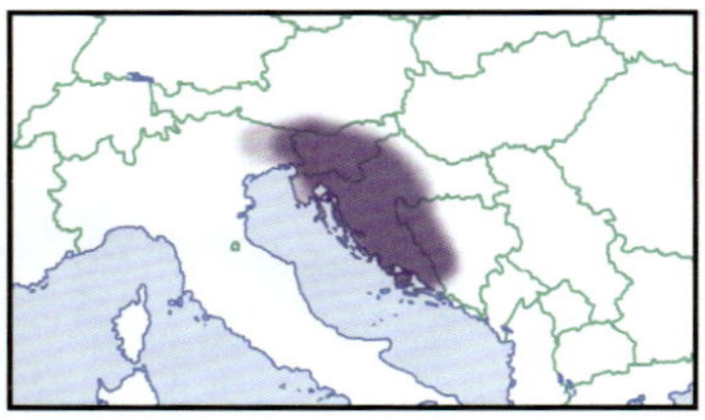

Odontocyclas kokeilii

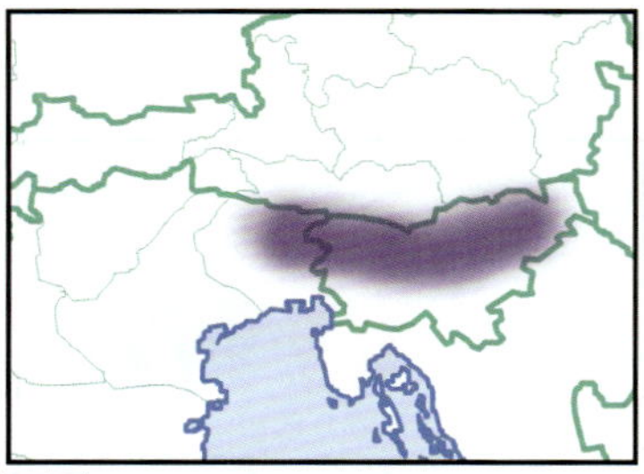

Walklea rossmaessleri

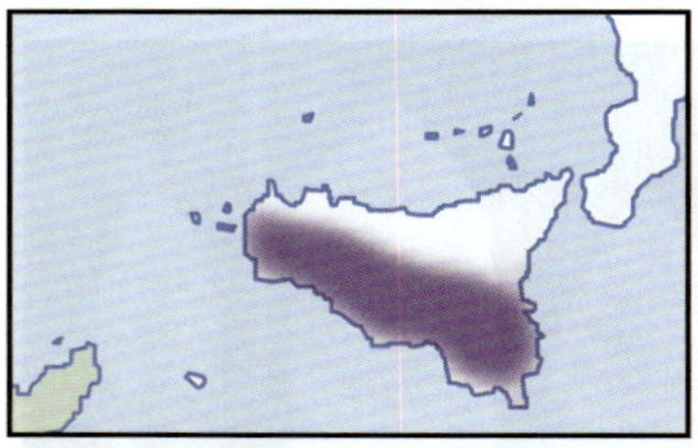

Orculella templorum

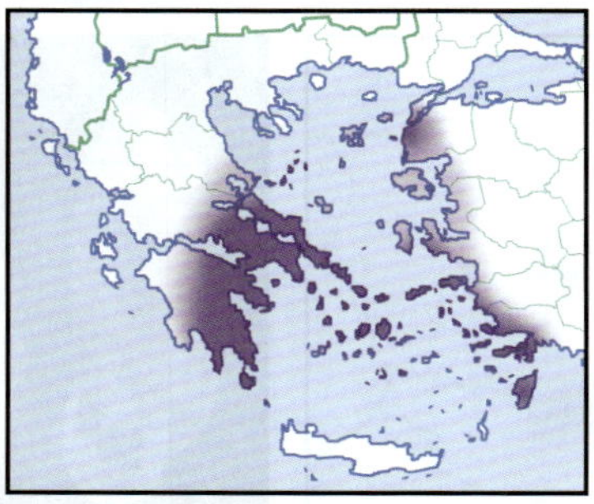

Orculella critica

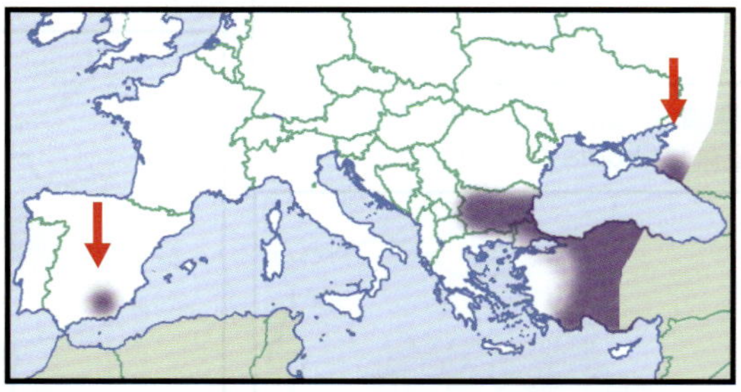

Oculella bulgarica

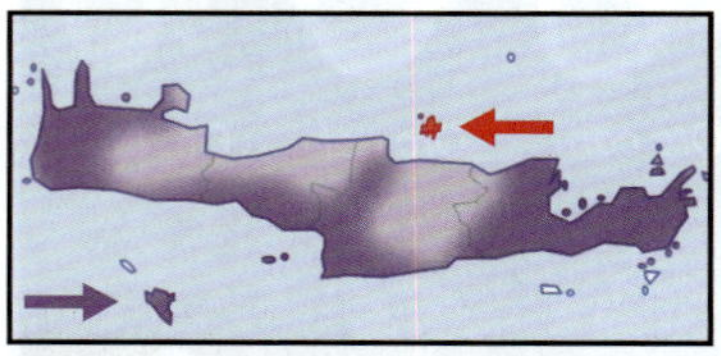

Orculella cretiminuta (cre-m)
Orculella diensis
Orculella franciscoi
Orculella scalaris

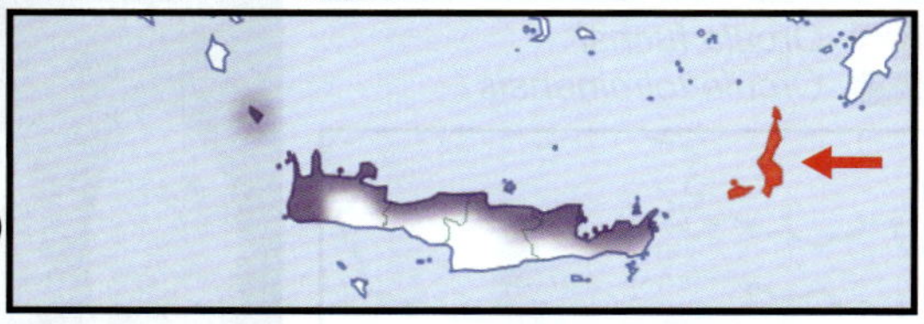

Orculella creantirudis (cre-a)
Orculella exaggerata

Orculella ignorata

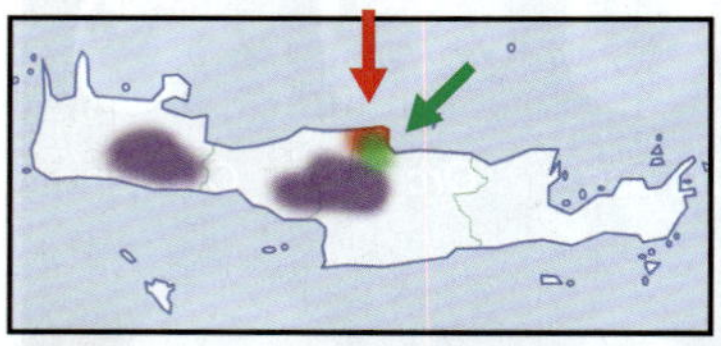

Orculella cretimaxima (cre-x)
Orculella cretioreina (cre-o)
Orculella fodela

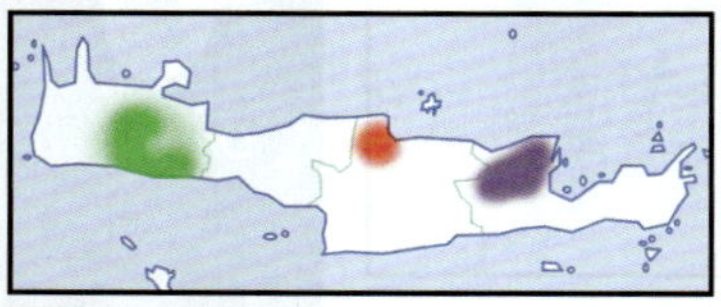

Orculella astirakiensis
Orculella creticostata
Orculella cretilasithi

Orculella sirianocoriensis

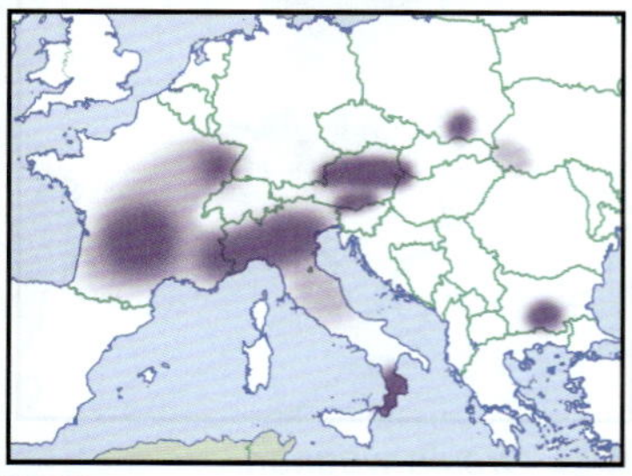

Pagodulina pagodula

Pagodulina austeniana

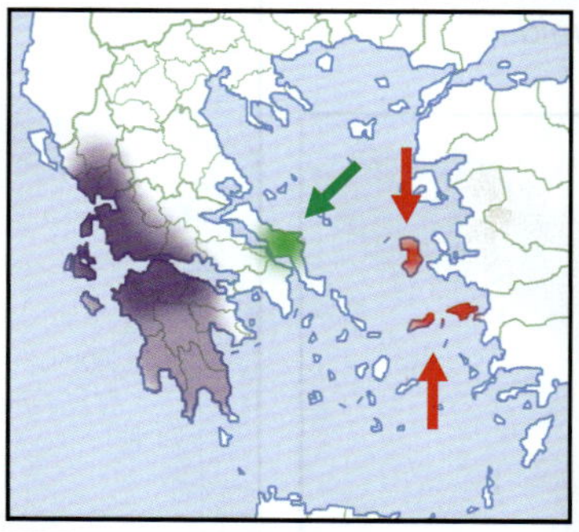

Pagodulina epirotes
Pagodulina hauseri
Speleodentorcula beroni

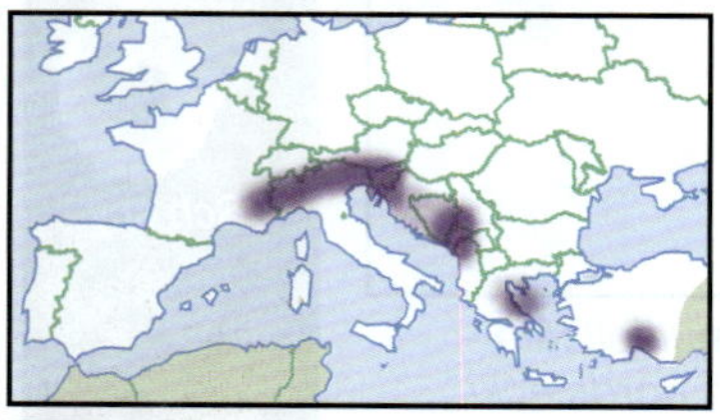

Pagodulina subdola

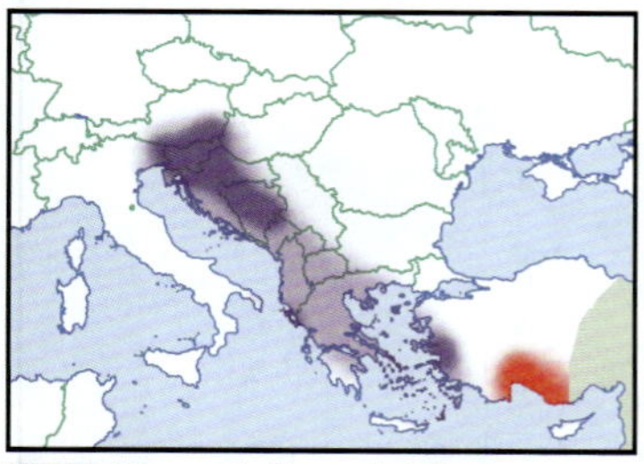

Pagodulina pisidica
Pagodulina sparsa

Chondrinidae 1

Abida, Granaria, Solatopupa

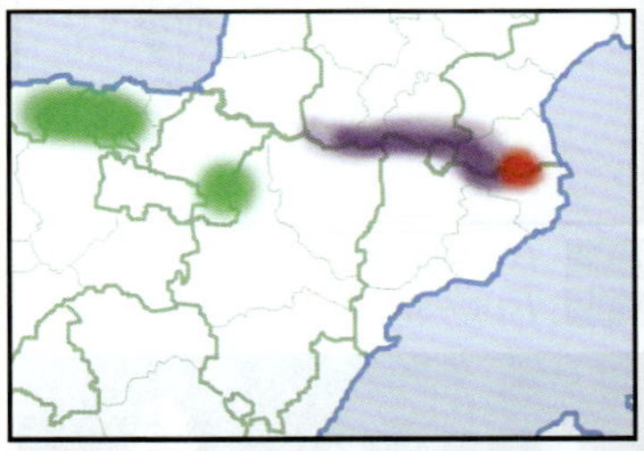

Abida gittenbergeri
Abida occidentalis
Abida vasconica

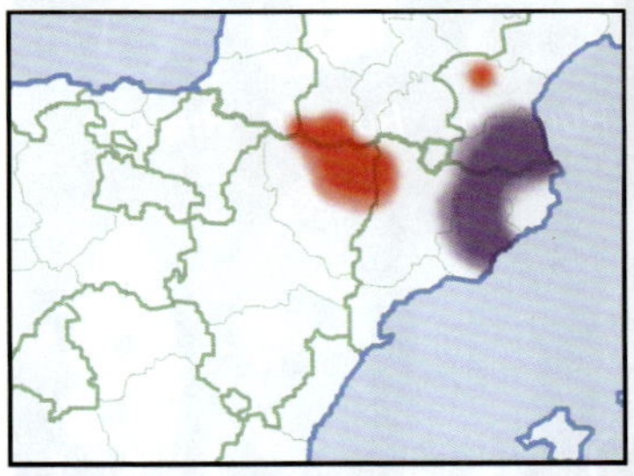

Abida cylindrica
Abida partioti

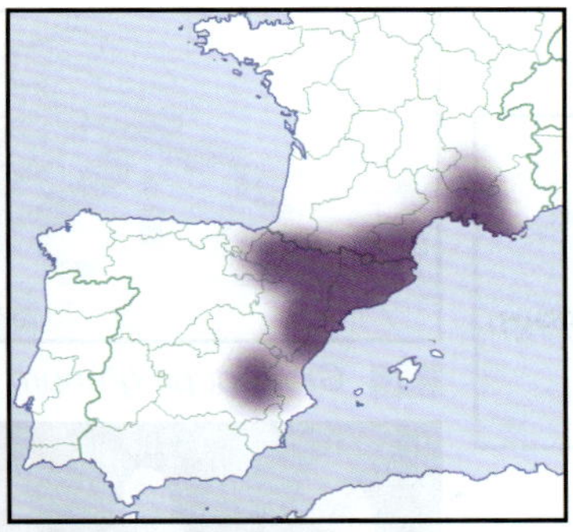

Abida polyodon

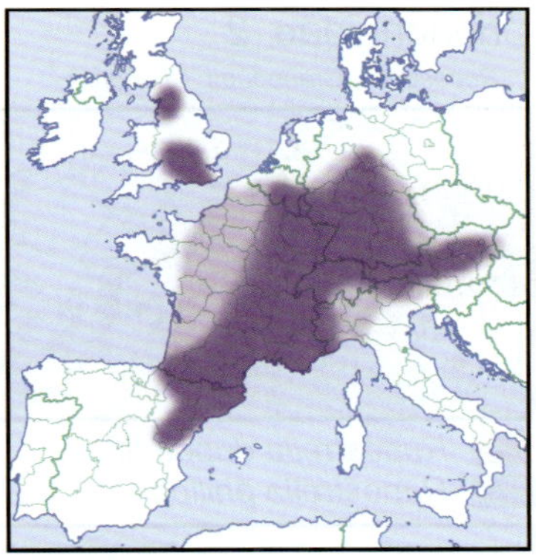

Abida secale

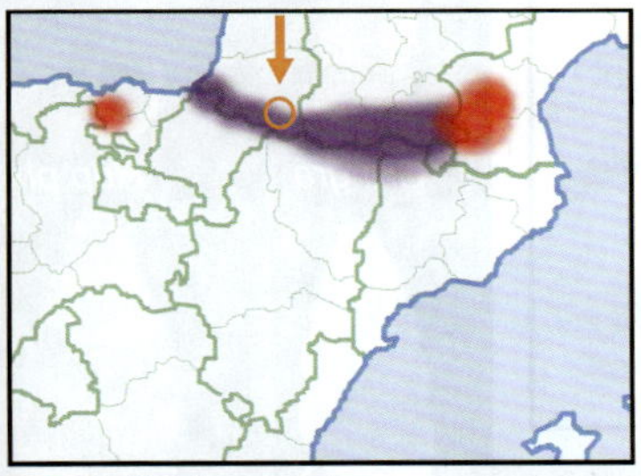

Abida ateni
Abida attenuata
Abida pyrenaearia

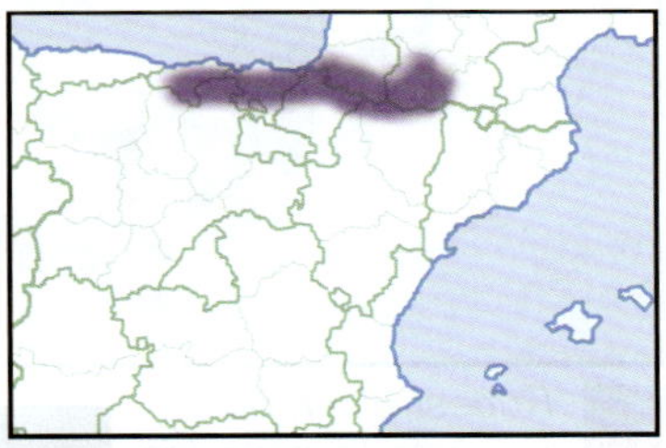

Abida bigerrensis

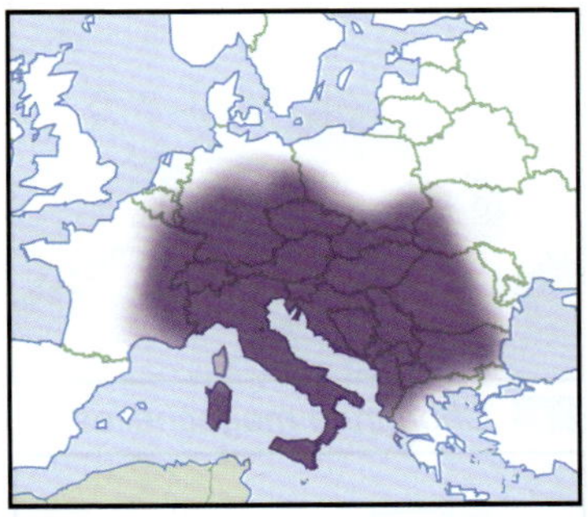

Granaria frumentum

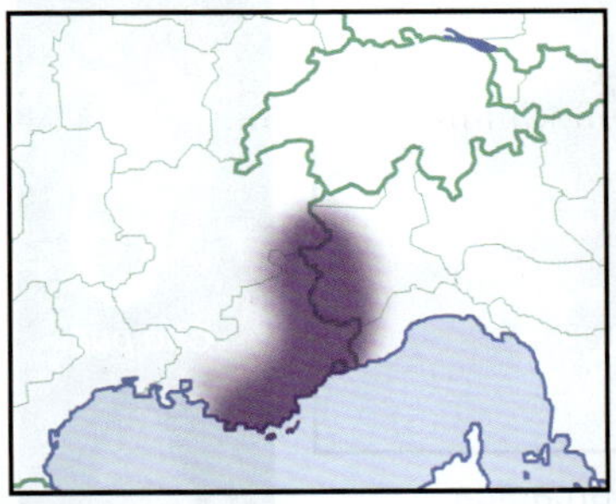

Granaria stabili

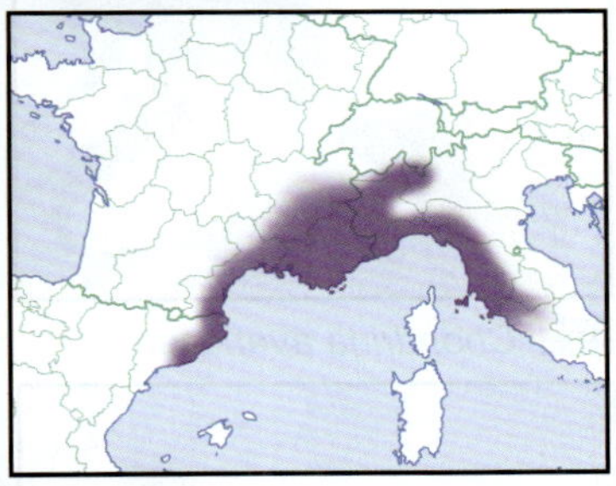

Solatopupa similis

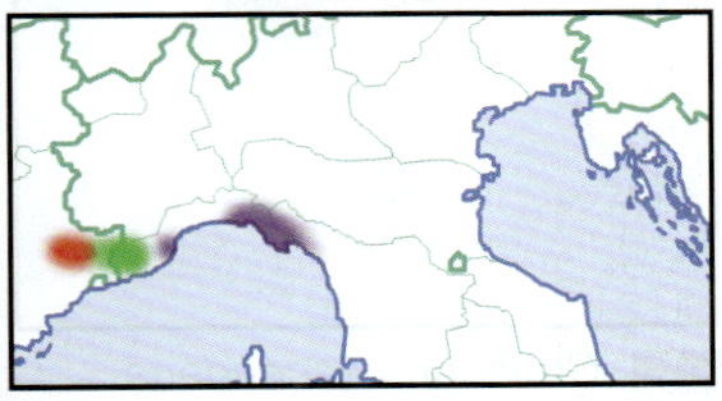

Solatopupa cianensis
Solatopupa pallida
Solatopupa psarolena

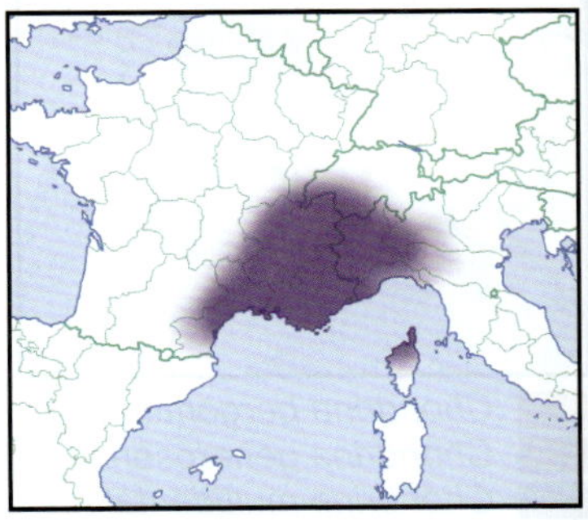

Granaria variabilis

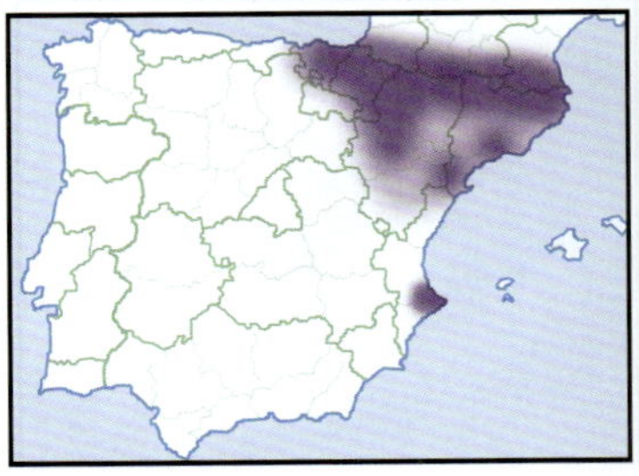

Granaria braunii

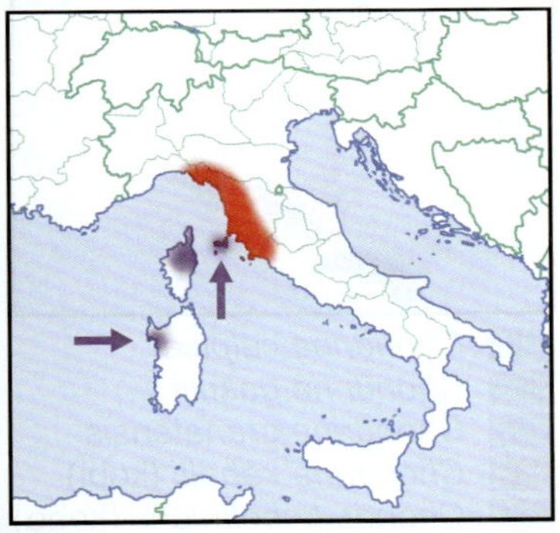

Solatopupa guidoni
Solatopupa juliana

Chondrinidae 2

Rupestrella, Granopupa, Chondrina

Rupestrella dupotetii
Rupestrella philippii

Granopupa granum

Rupestrella homala
Rupestrella occulta

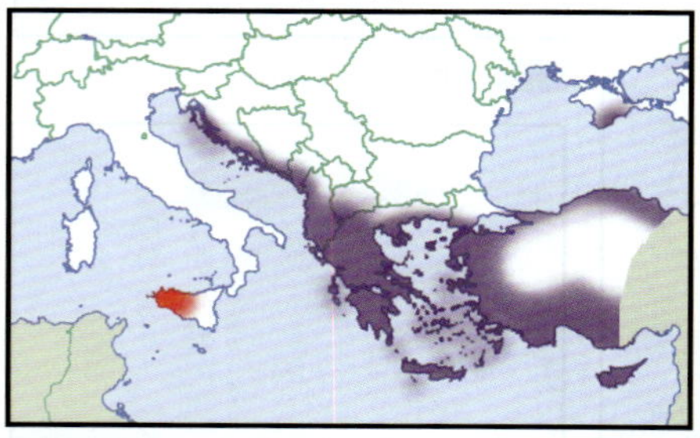

Rupestrella rhodia
Rupestrella rupestris

Rup dup Cho gra J. Liétor Gallego

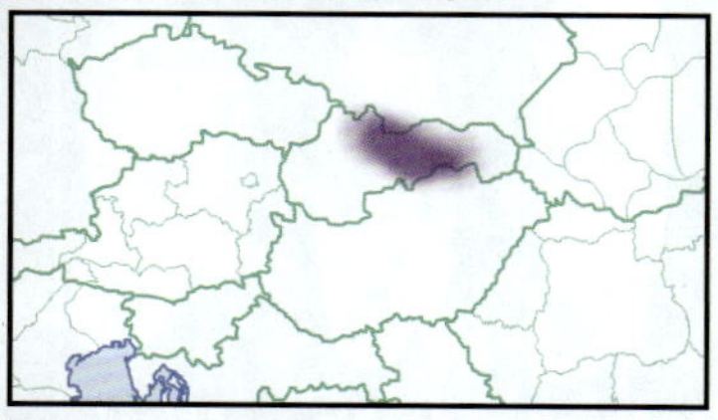

Chondrina tatrica

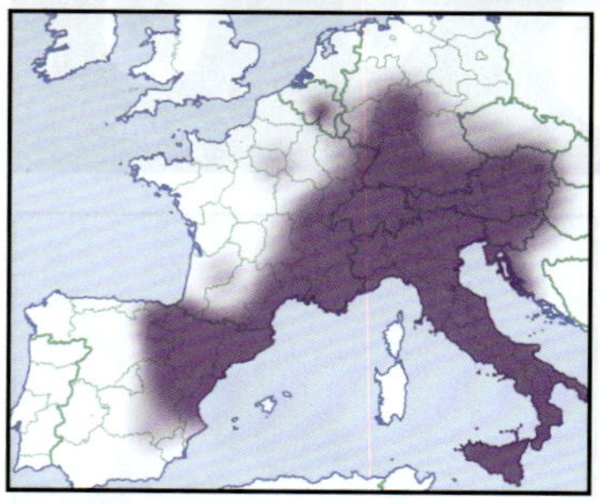

Chondrina avenacea

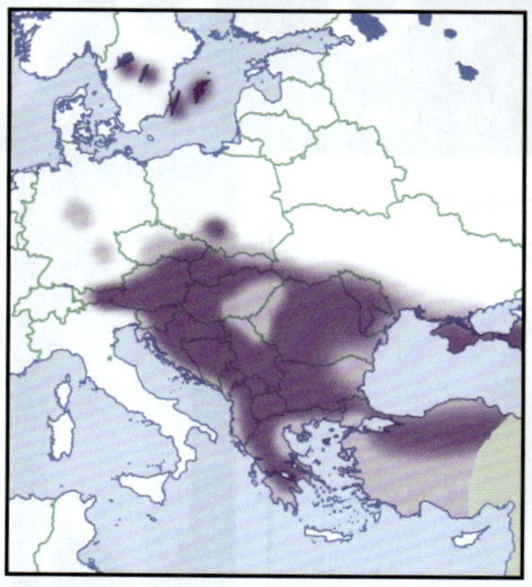

Chondrina arcadica

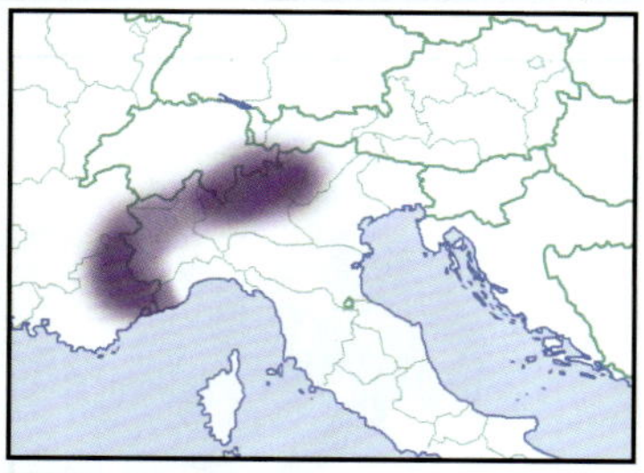

Chondrina megacheilos

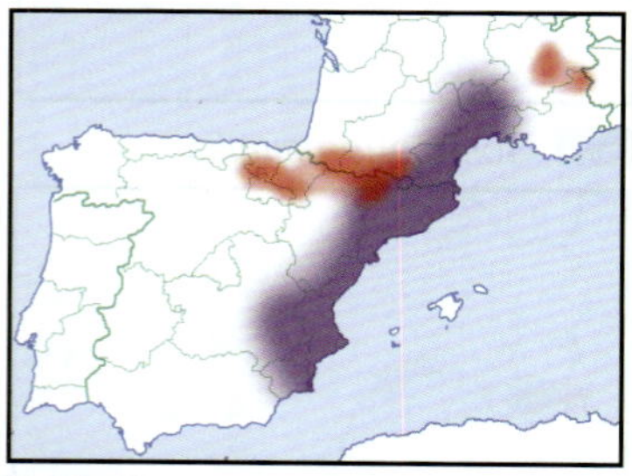

Chondrina ascendens
Chondrina farinesii

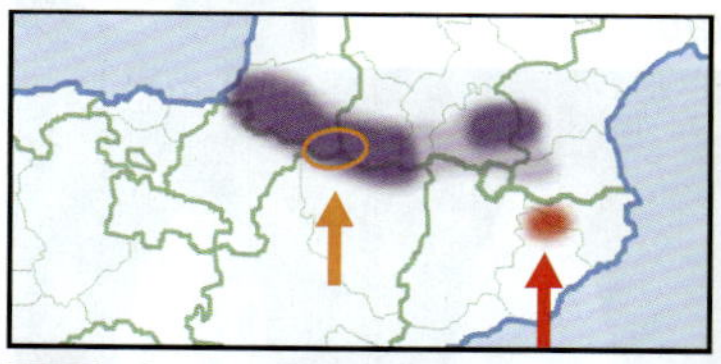

Chondrina altimirai
Chondrina centralis
Chondrina pusilla (= *bigorr.*)

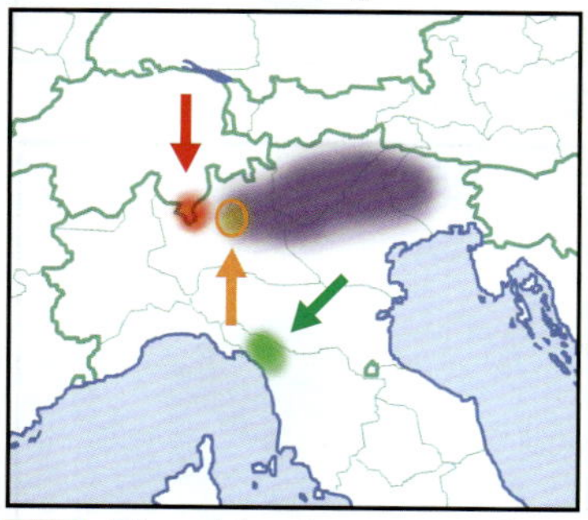

Chondrina bergomensis
Chondrina generosensis
Chondrina multidentata
Chondrina oligodonta

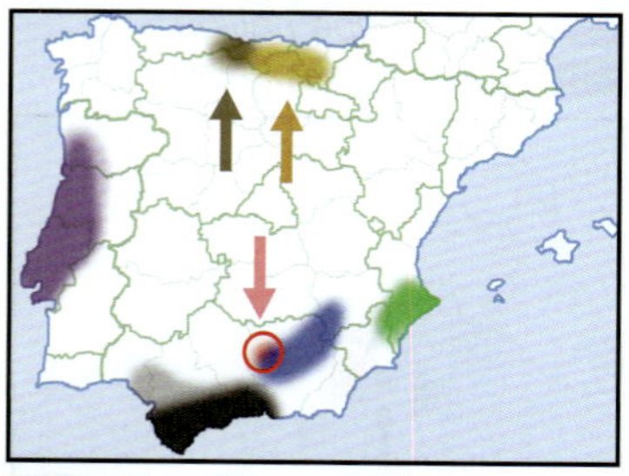

Chondrina calpica
Chondrina gasulli
Chondrina granatensis
Chondrina kobelti (kobi)
Chondr. kobeltoides (kobo)
Chondrina lusitanica
Chondrina maginensis

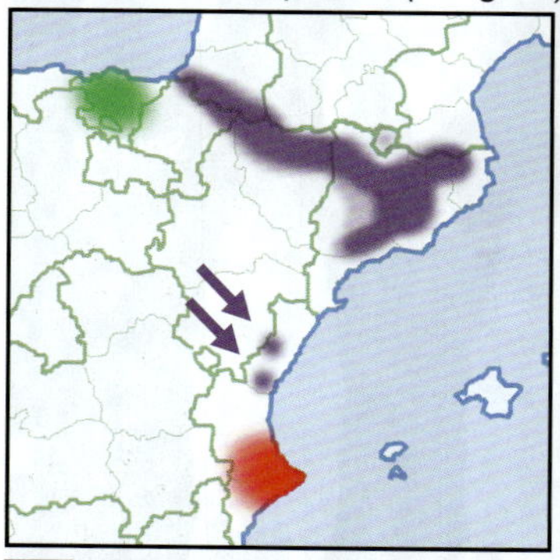

Chondrina arigonis
Chondrina ripkeni
Chondrina tenuimarginata

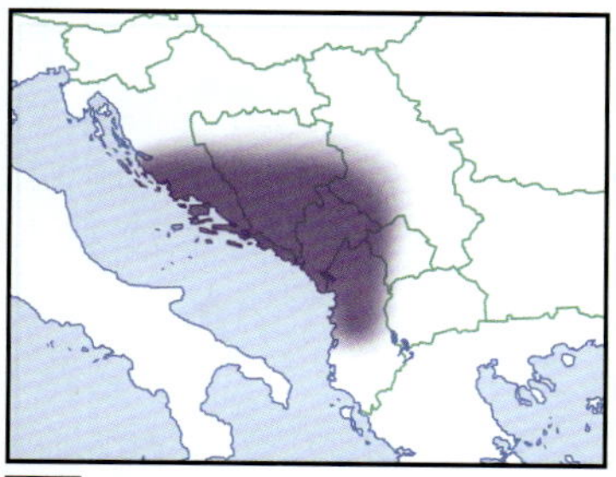

Chondrina spelta

Enidae 1

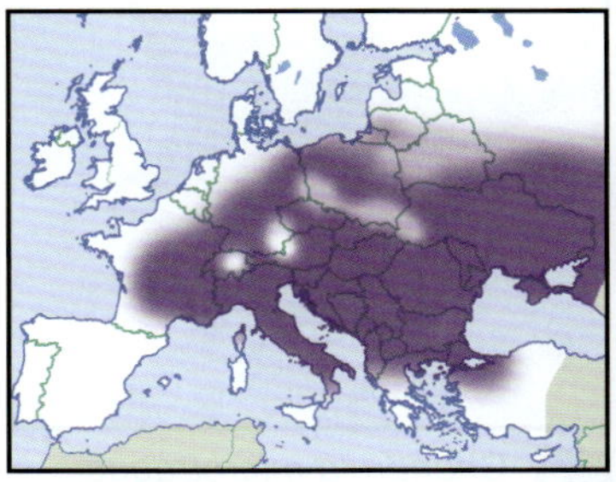

Chondrula tridens

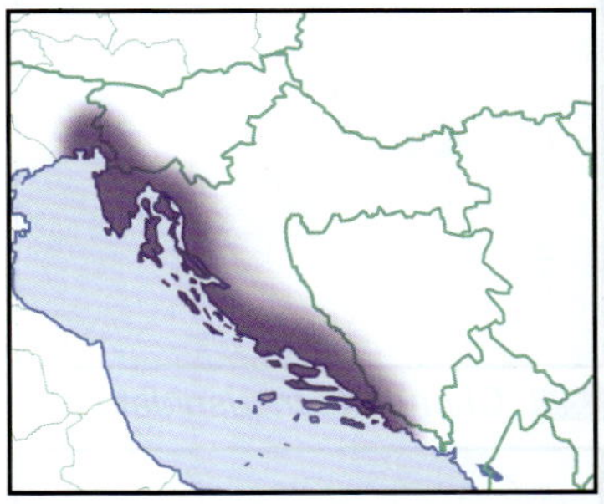

Chondrula quinquedentata

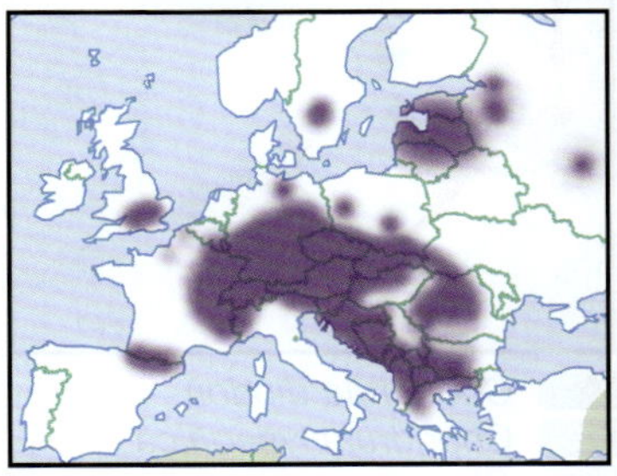

Ena montana

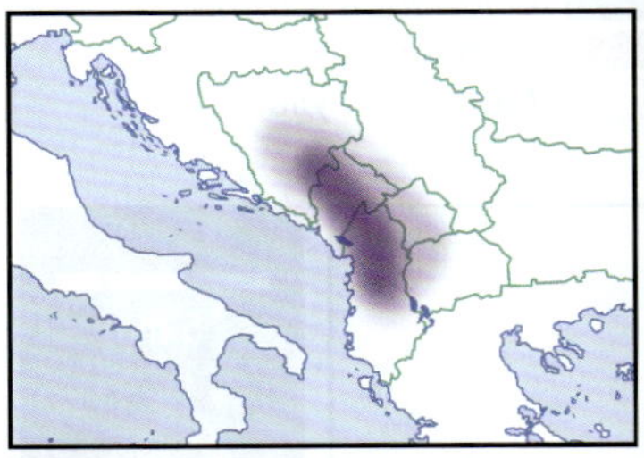

Ena concolor

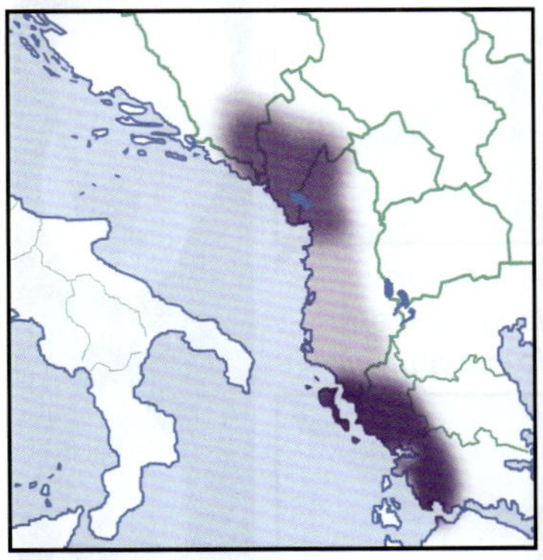

Ena subtilis

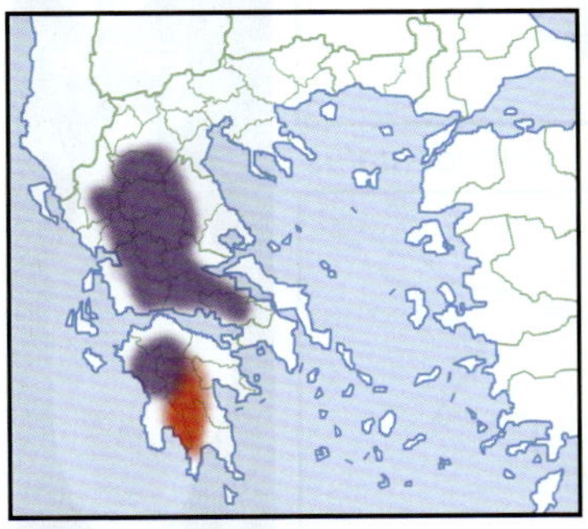

Ena monticola
Ena elongata

Jam thi R. Reuselaars

Ena nogellii
Meijeriella frivaldskyi

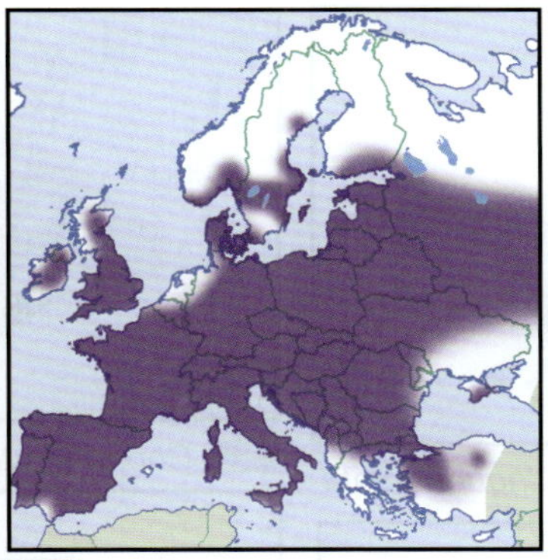

Merdigera obscura

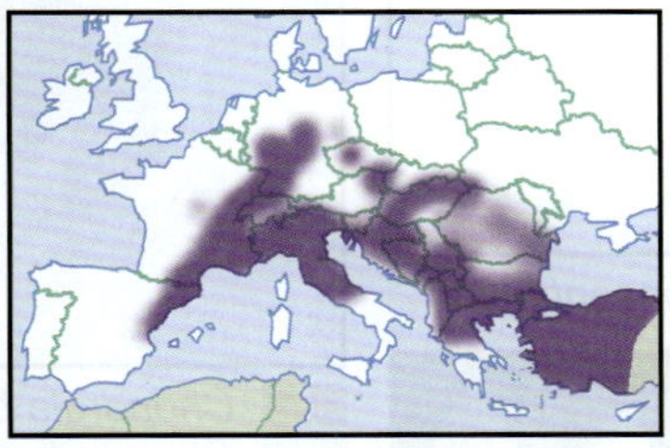

Zebrina detrita

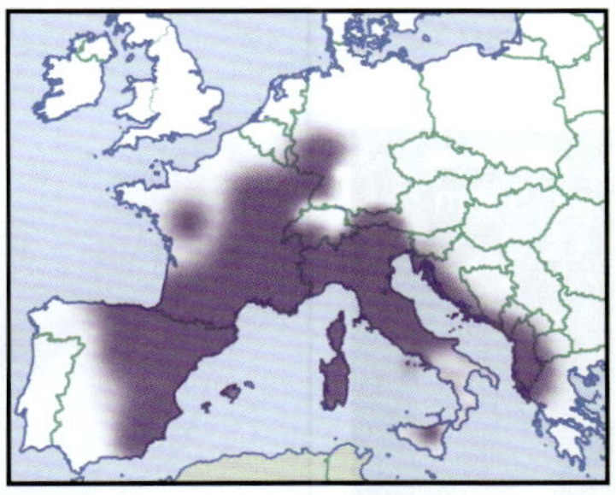

Jaminia quadridens

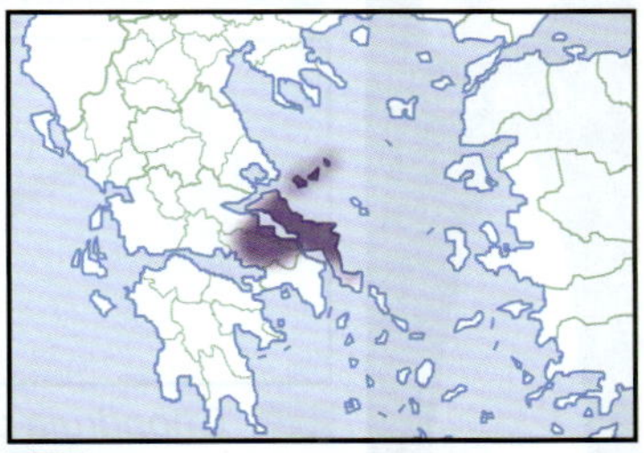

Jaminia thiesseana

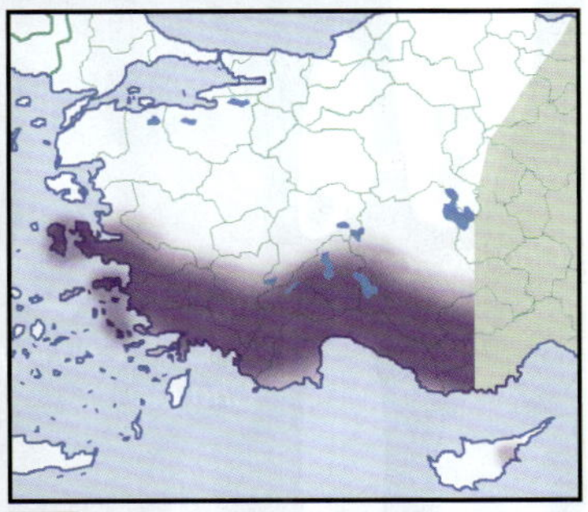

Jaminia loewii

Imparietula seductilis

Brephulopsis fusiformis (= cyl.)

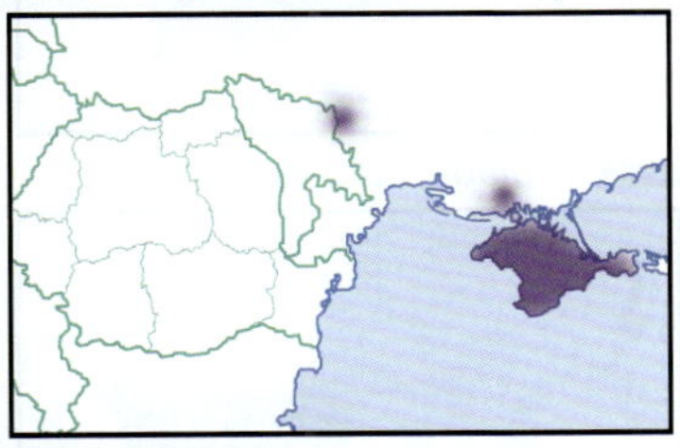

Brephulopsis bidens

Peristoma merduenianum
Peristoma rupestre
Brephulopsis subulata

Enidae 2

Cho lug Z. Fehér *Mei yil* *Cho lyc* A. Örstan *Cho pel* R. Reuselaars *Cho stu* J. Liétor Gallego *Mul squ* Z. Fehér

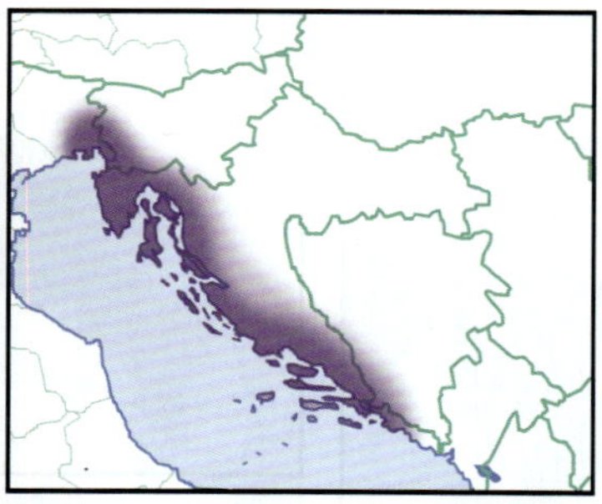

Chondrula quinquedentata

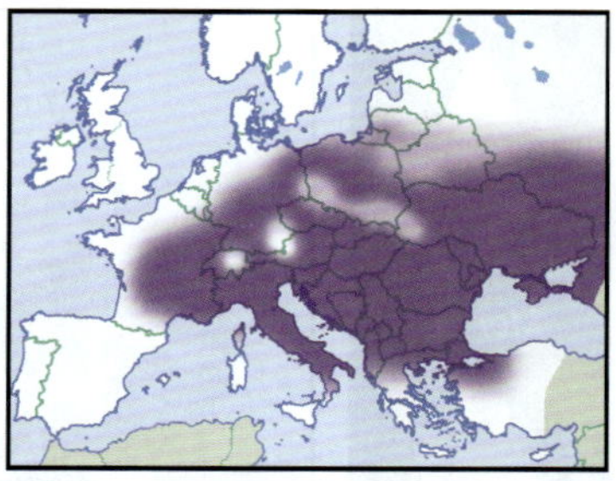

Chondrula tridens

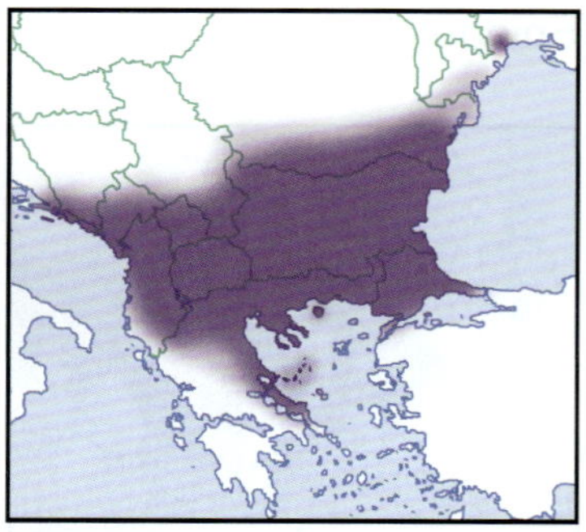

Chondrula microtragus

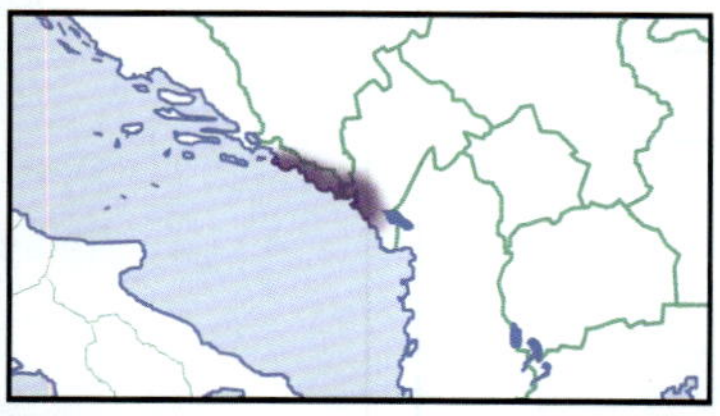

Chondrula consentanea

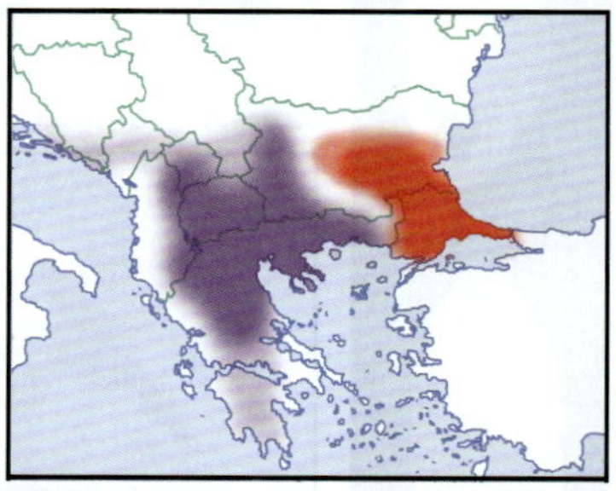

Chondrula bicallosa
Chondrula macedonica

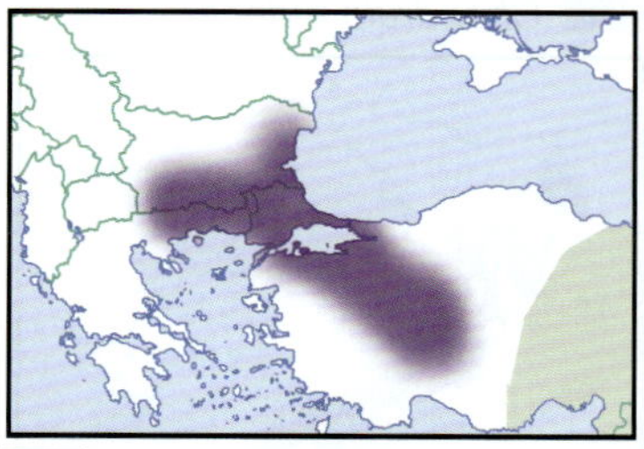

Chondrula tricuspidata

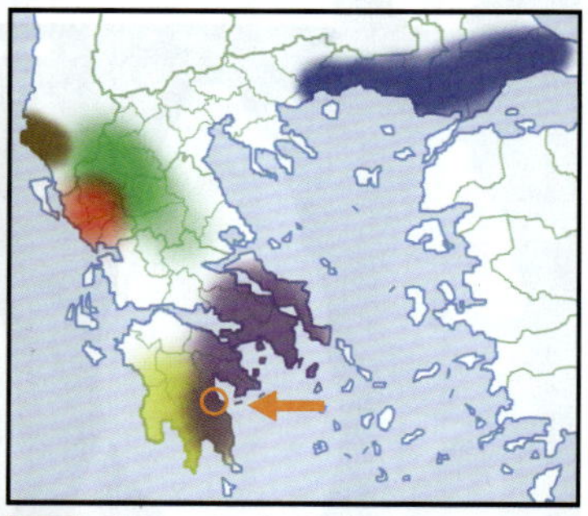

Chondrula beieri
Chondrula bergeri
Chondrula lugorensis
Chondrula munita
Chondrula jaczewskii
Chondrula peloponnesica
Ch. pindica

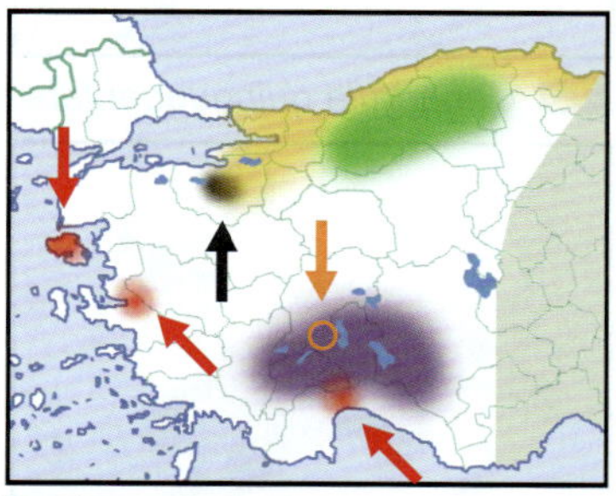

Chondrula diodon
Chondrula lycaonica
Chondrula orientalis
Chondrula sturmii
Chondrula werneri
Meijeriella yildirimi

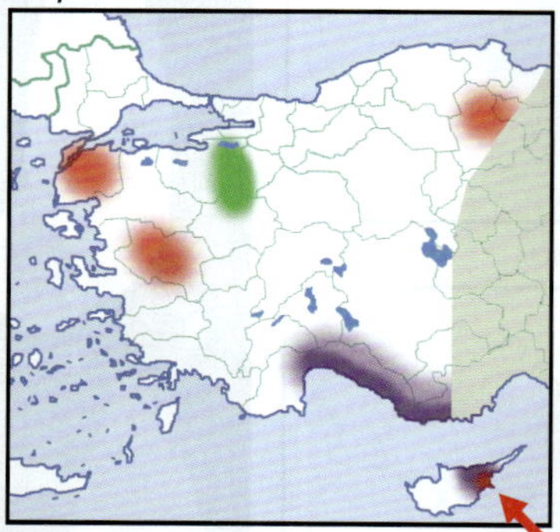

Euchondrus ledereri
Multidentula lamellifera
Multidentula substurmii

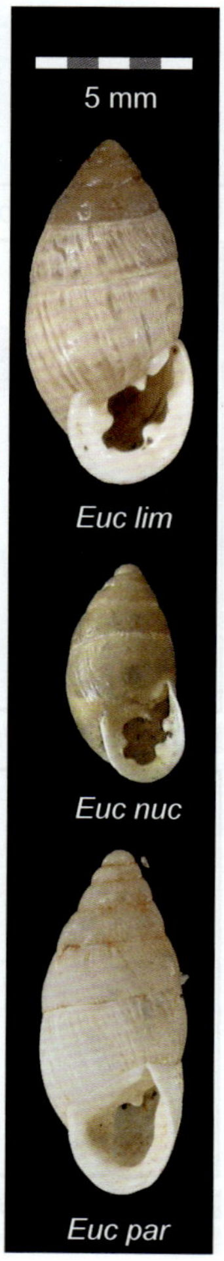

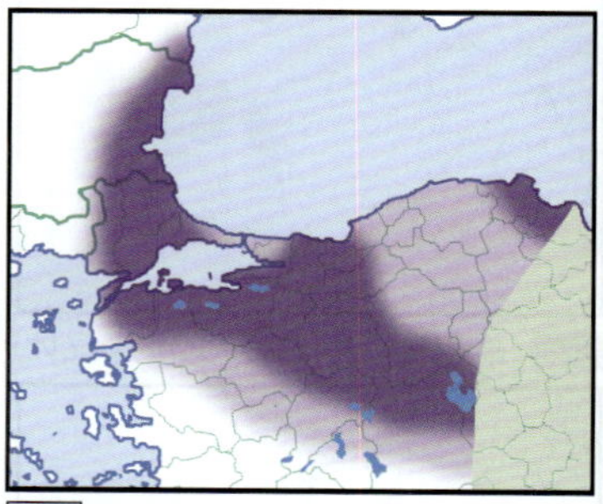

Multidentula ovularis

Multidentula squalina

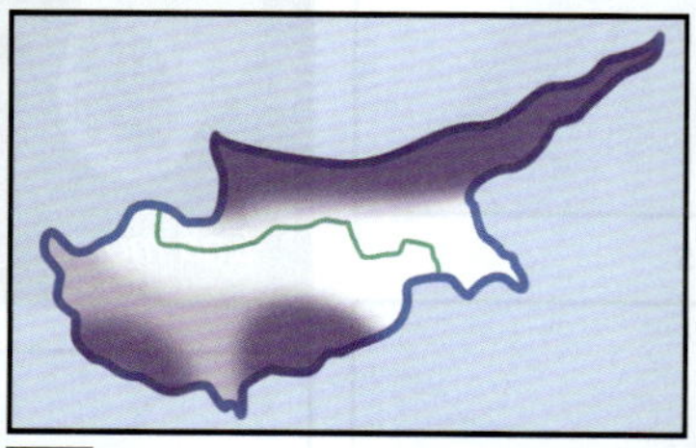

Euchondrus limbodentatus

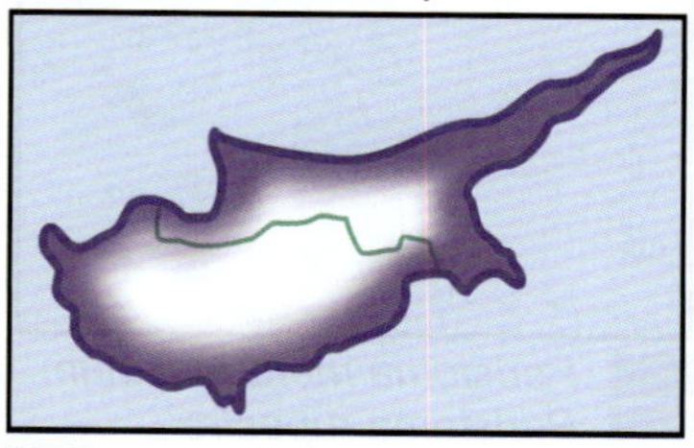

Multidentula stylus

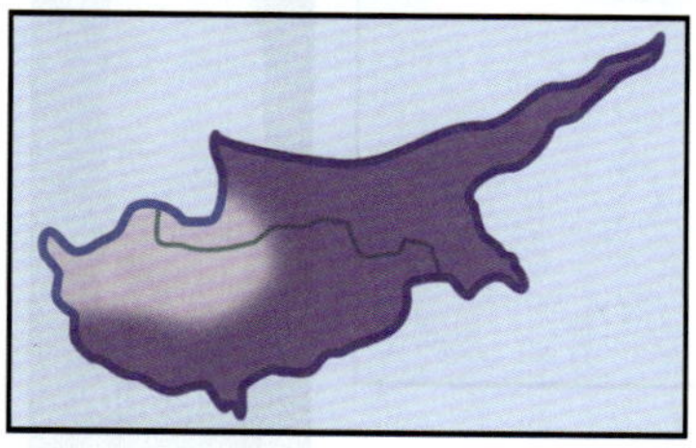

Euchondrus nucifragus

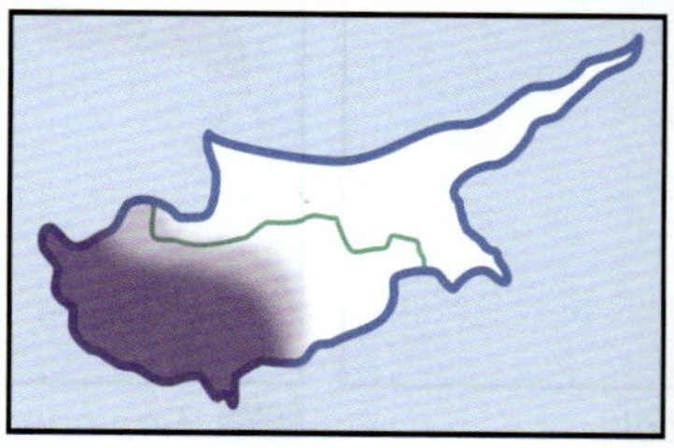

Euchondrus parreyssi

Enidae 3

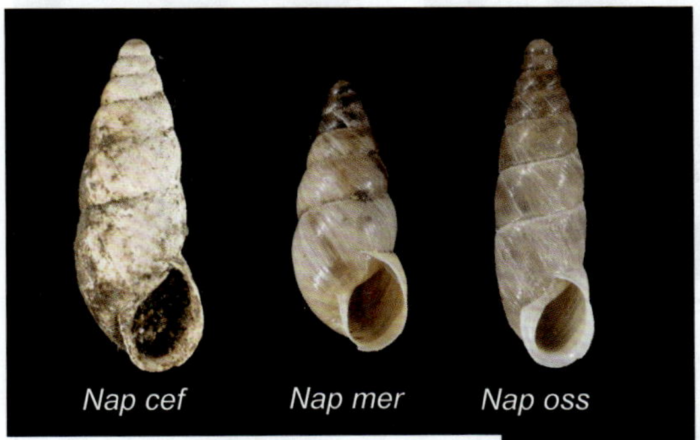

Mei yil A. Örstan Bul car E. Garcia

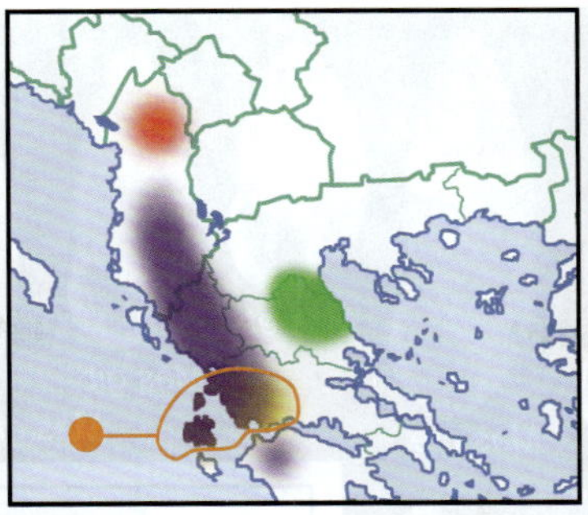

Napaeopsis cefalonica
Napaeopsis merditana
Napaeopsis minima
Napaeopsis ossica

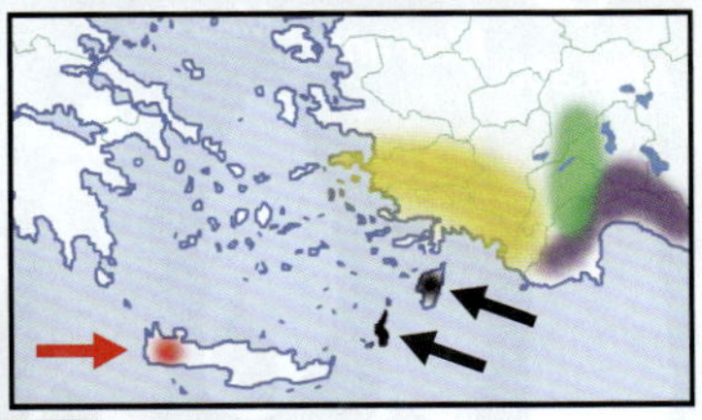

Turanena albrechti
Turanena carpathia
Turanena hemmeni
Turanena katerinae
Turanena tuccari

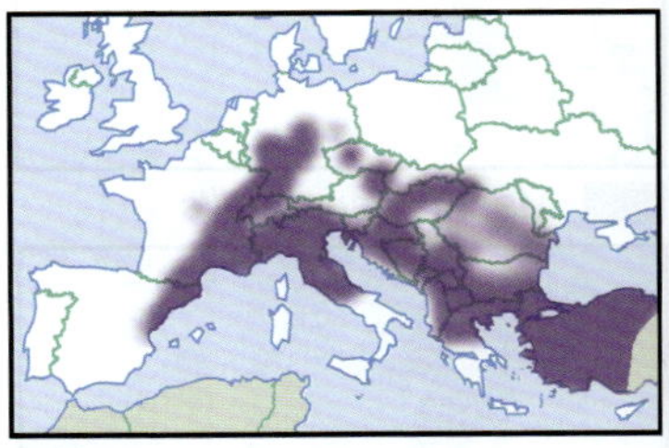

Zebrina detrita

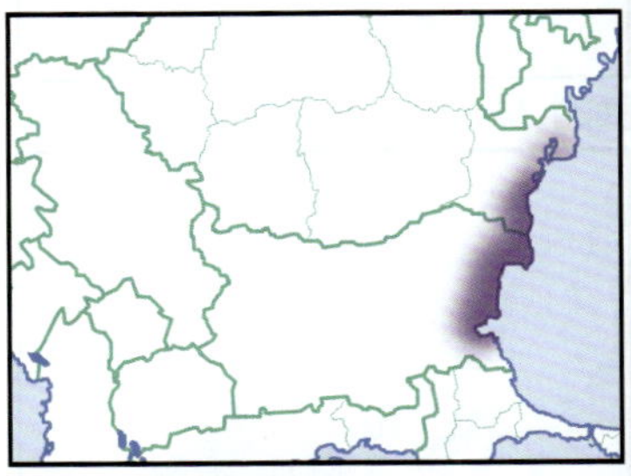

Leucomastus varnensis

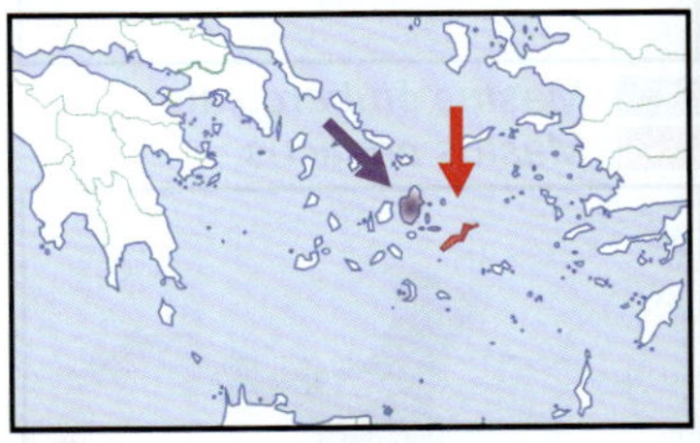

Rhabdoena stokesi
Rhabdoena zasiensis

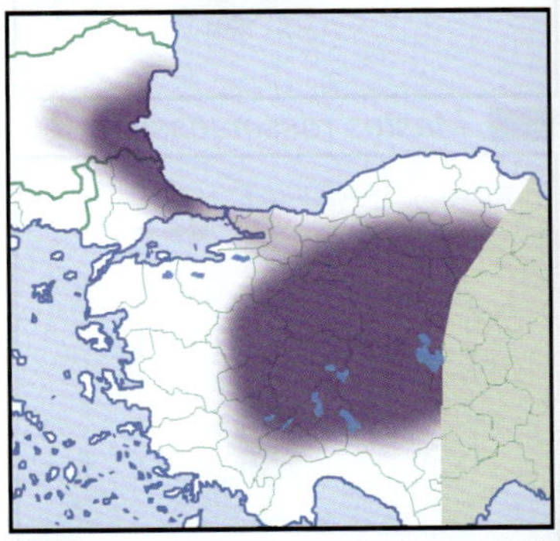

Leucomastus kindermanni

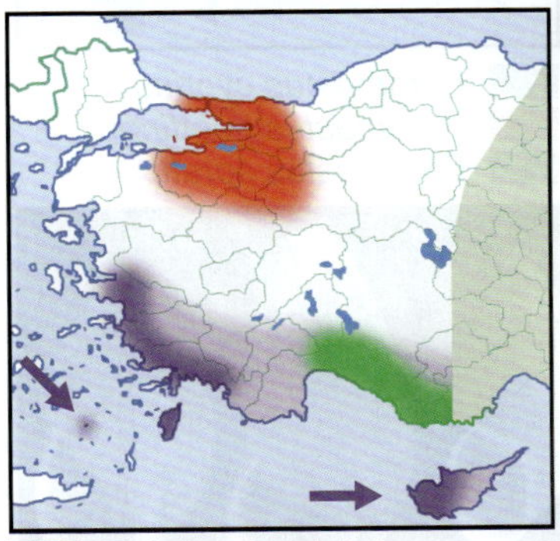

Zebrina fasciolata
Leucomastus dardanus
Leucomastus eburneus

Rhabdoena cosensis
Rhabdoena mirifica

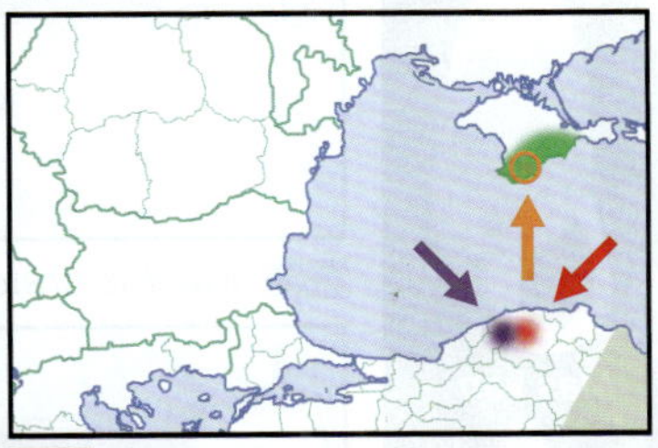

Thoanteus corneus
Thoanteus ferrarii
Thoanteus gibber
Thoanteus zilchi

Chondrus zebrula

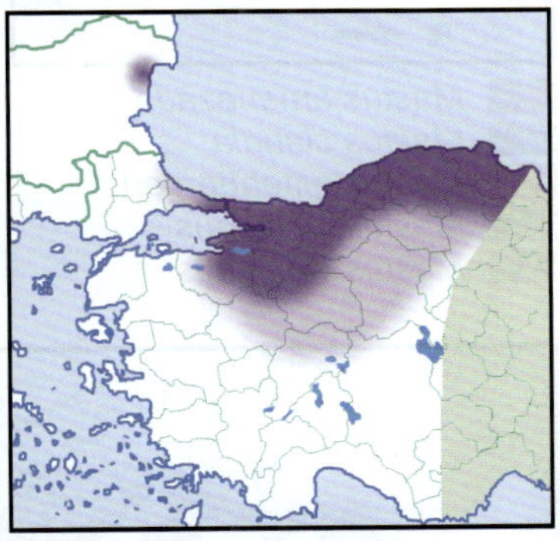

Chondrus tournefortianus

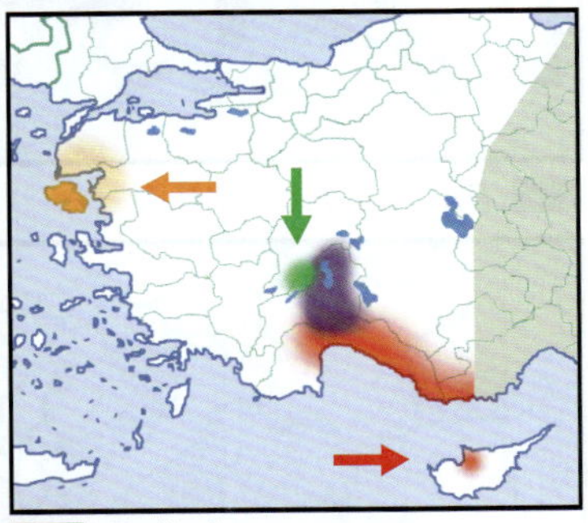

Buliminus carneus
Buliminus lycicus
Meijeriella canaliculata
Meijeriella yildirimi

Enidae 4, *Rumina*

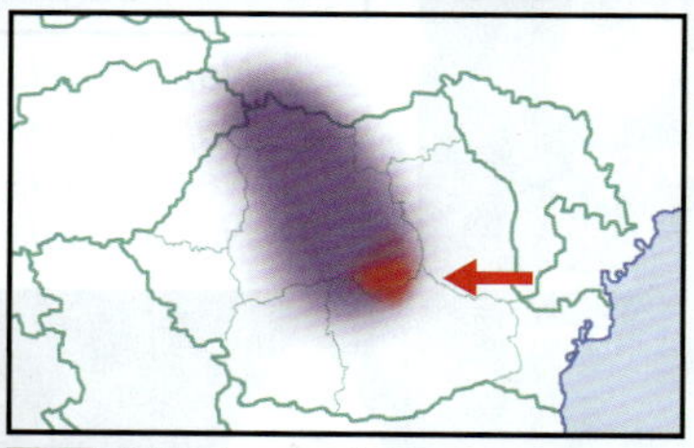

Mastus bielzi
Mastus transsylvanicus

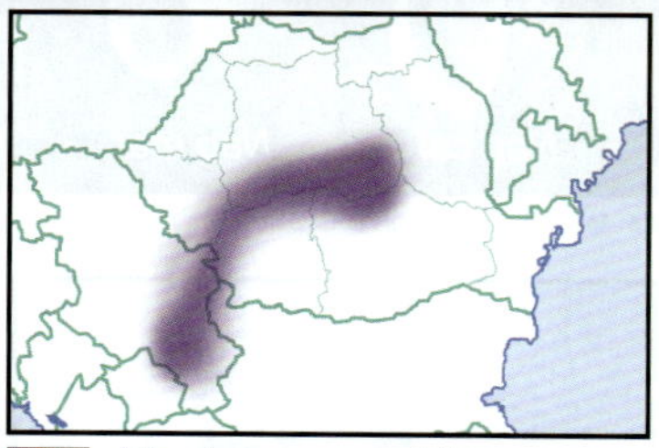

Mastus venerabilis

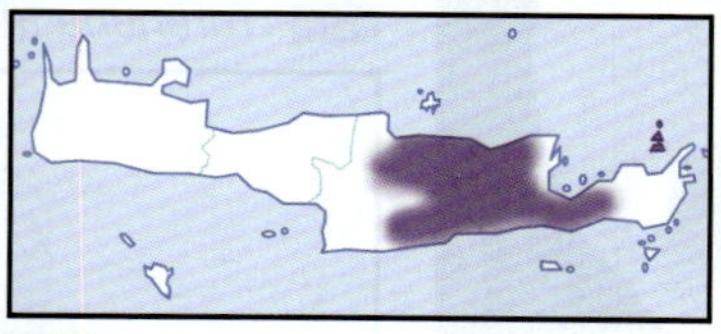

Mastus cretensis

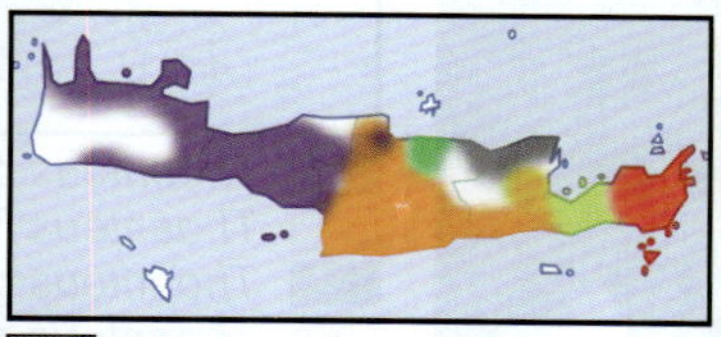

Mastus abundans
Mastus butoti
Mastus ierapetranus
Mastus riedeli
Mastus sitiensis
Mastus sphakiotus

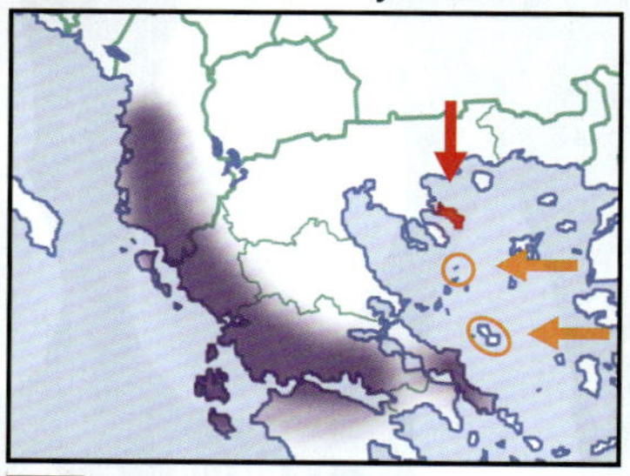

Mastus athensis
Mastus giuricus
Mastus grandis

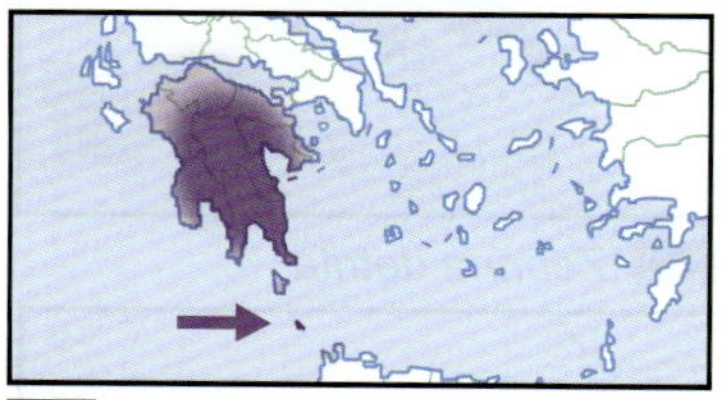

Mastus emarginatus

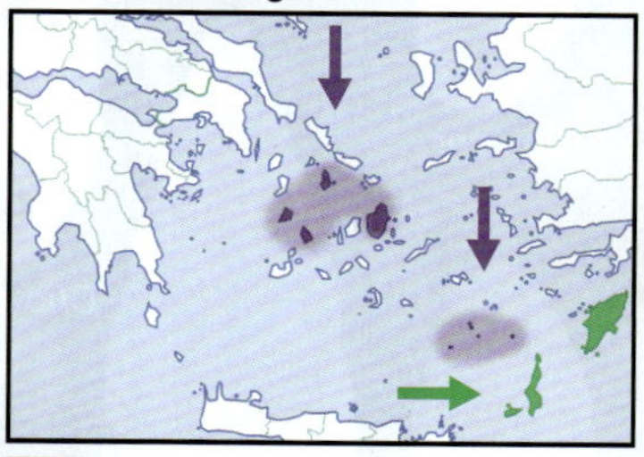

Mastus pusio
Mastus turgidus

Mastus etuberculatus
Mastus ponticus

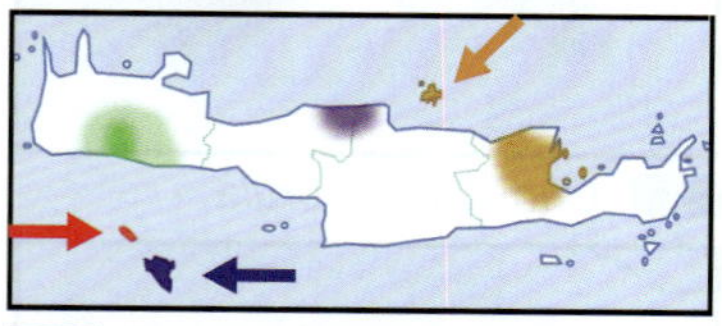

Mastus amenazada
Mastus claudia
Mastus gittenbergeri
Mastus olivaceus
Mastus violaceus

Mastus pupa

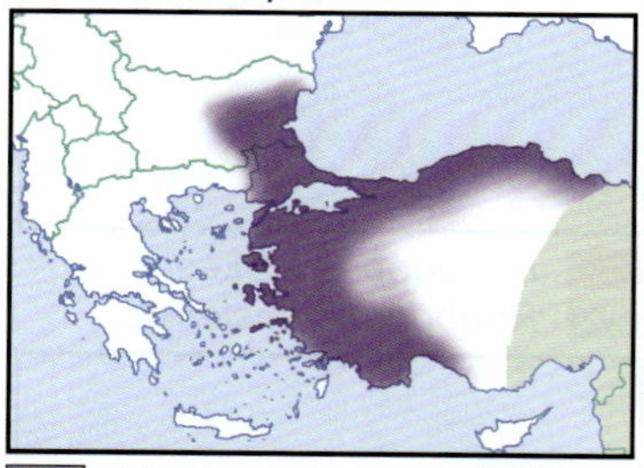

Mastus rossmaessleri

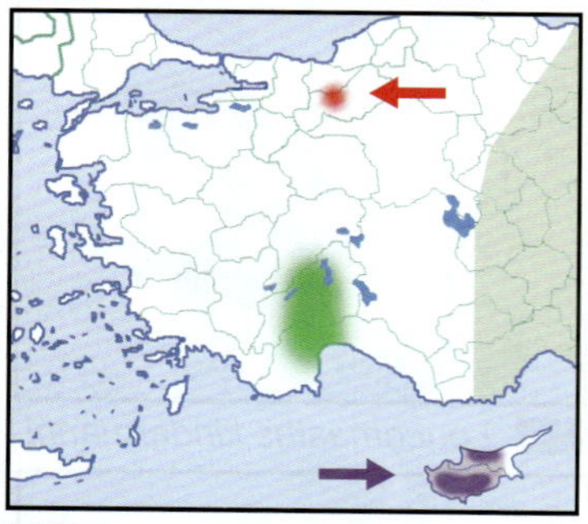

Paramastus cyprius
Paramastus goettingi
Paramastus spratti

Par cyp G. Konstantinou

Rumina decollata

Rumina saharica

Punctum, *Paralaoma*, *Lucilla*, *Hawaiia*, Valloniidae

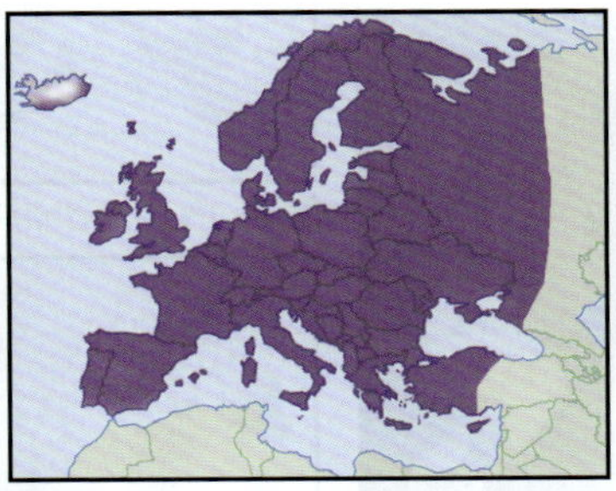
Punctum pygmaeum

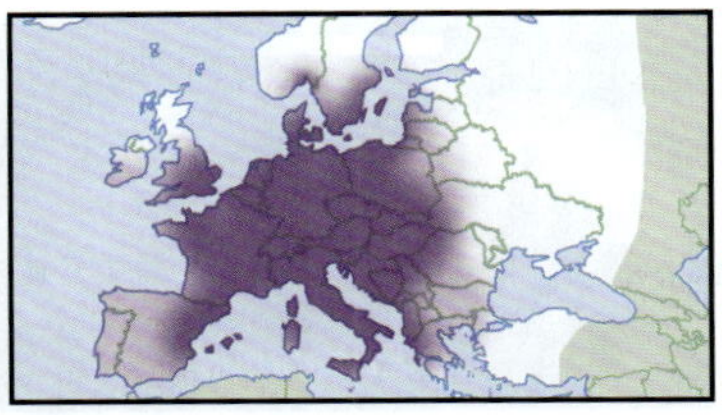
Lucilla singleyana

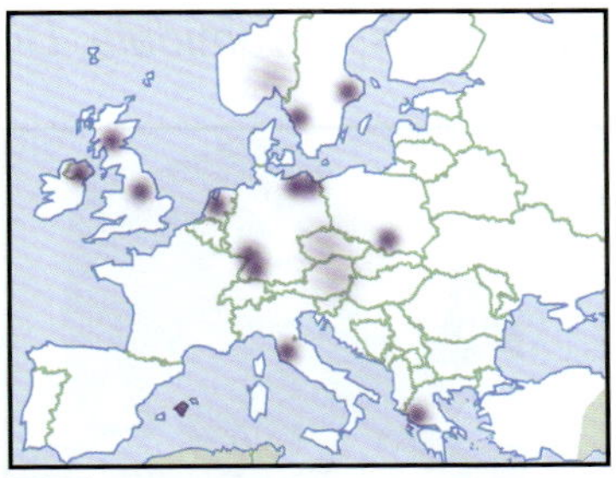
Hawaiia minuscula

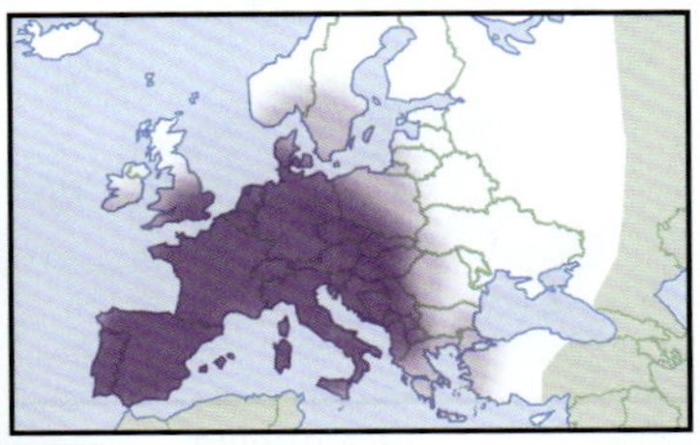
Lucilla scintilla

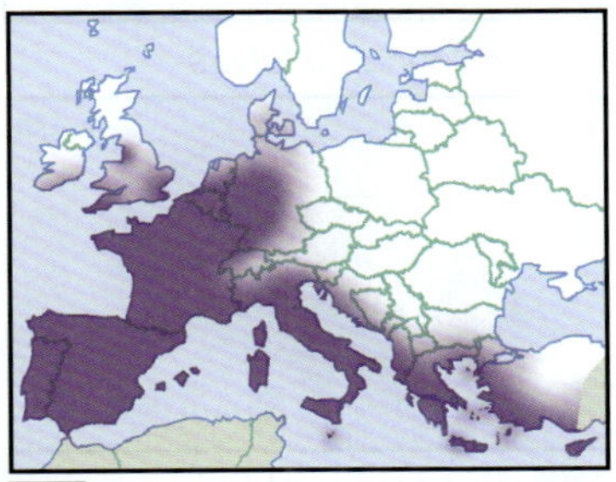
Paralaoma servilis

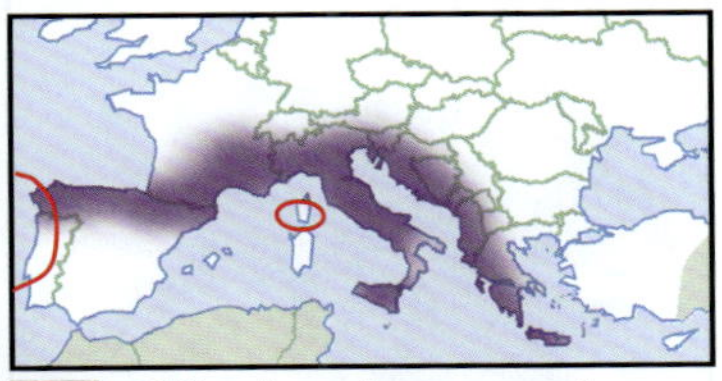
Gittenbergia sororcula
Plagyrona placida

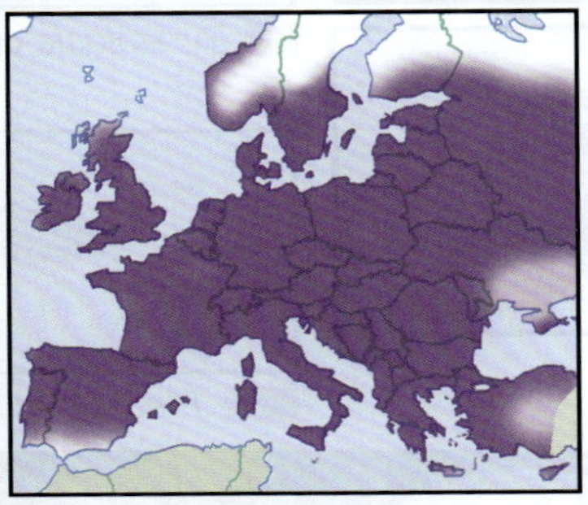
Acanthinula aculeata

Zoogenetes harpa

Spermodea lamellata

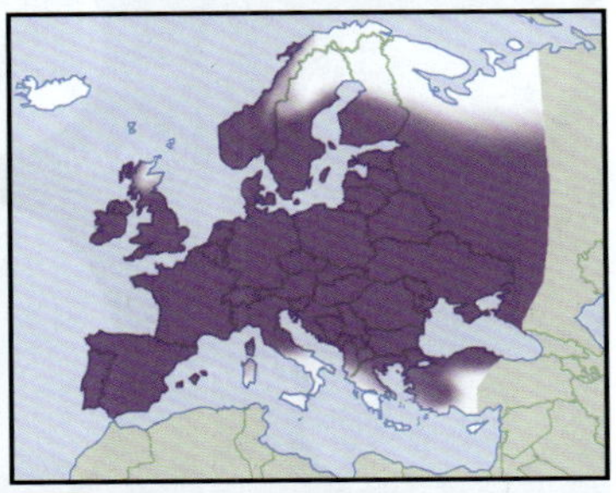
Vallonia costata

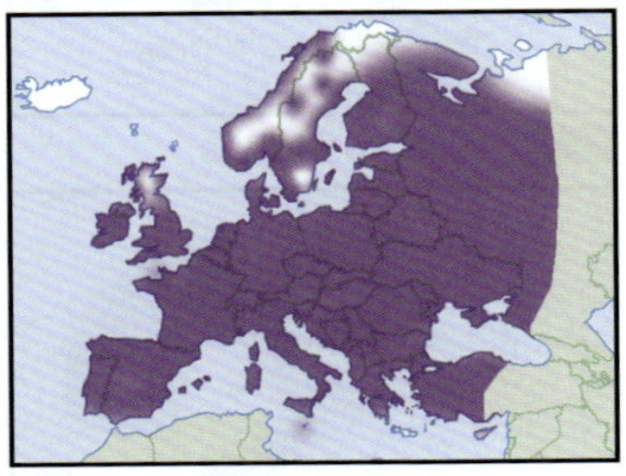
Vallonia pulchella

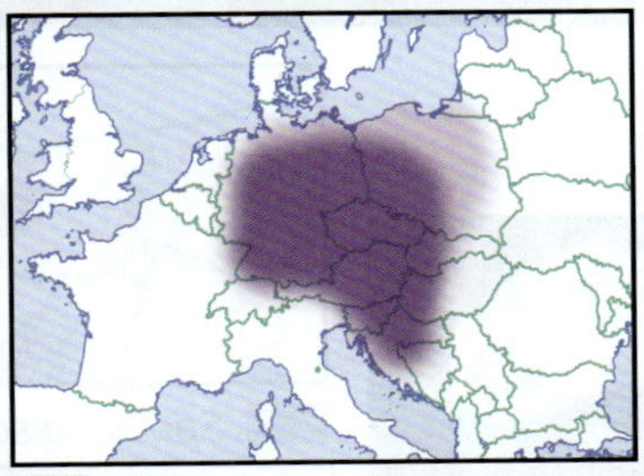
Vallonia declivis

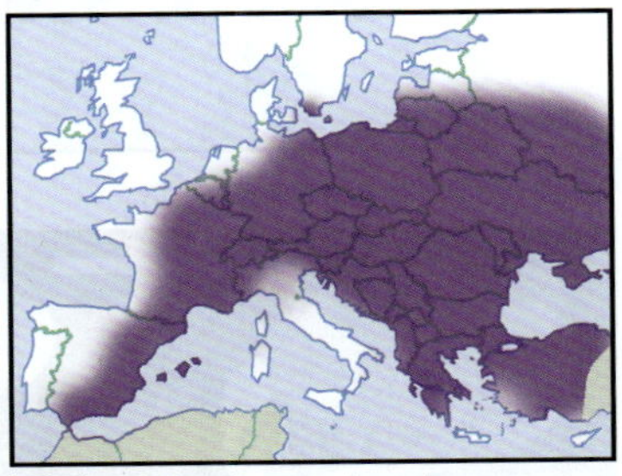
Vallonia enniensis

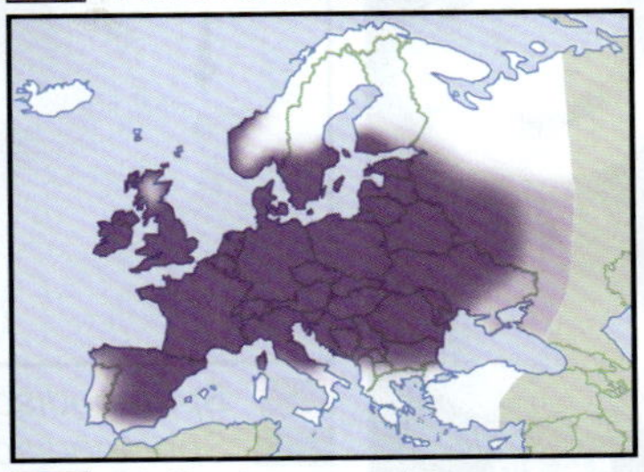
Vallonia excentrica

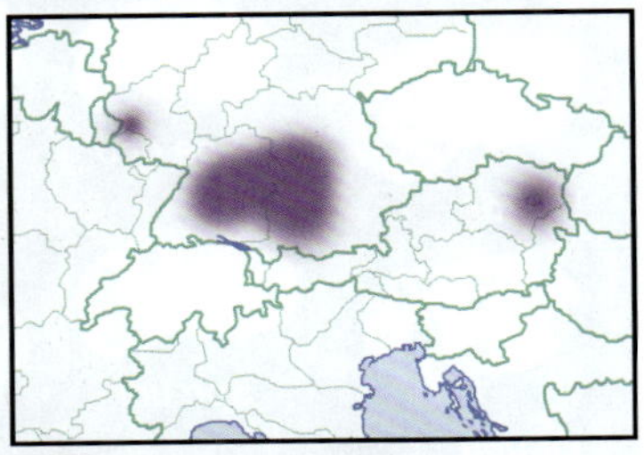
Vallonia suevica

Euconulus, *Pyramidula*, *Discus*, *Pleurodiscus*, Spelaeodiscinae

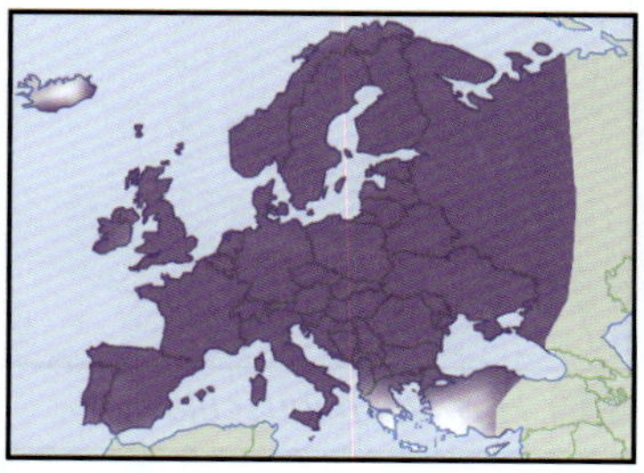

Euconulus fulvus

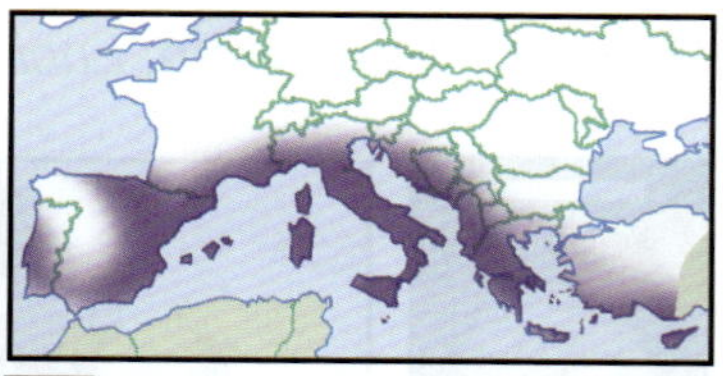

Pyramidula rupestris

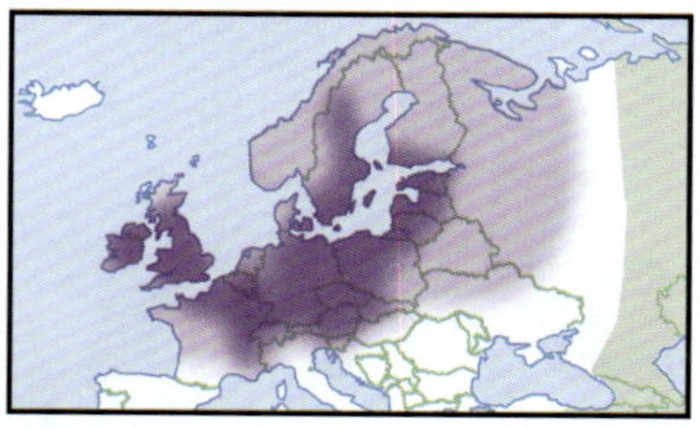

Euconulus alderi

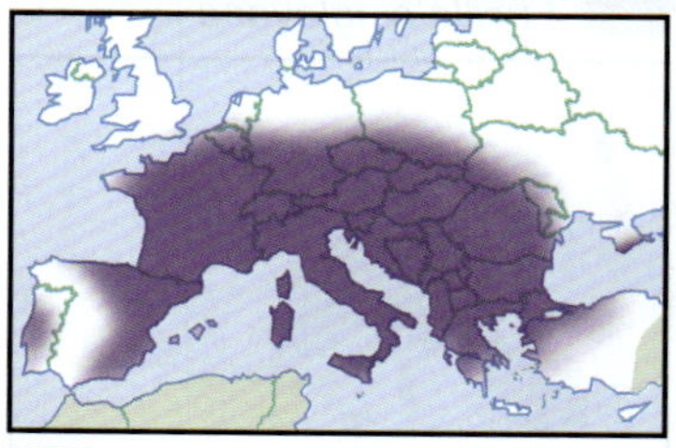

Pyramidula pusilla

Pyramidula jaenensis
Pyramidula umbilicata

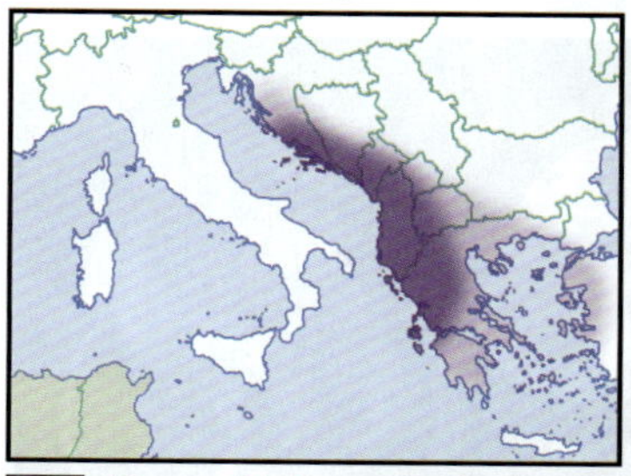

Pyramidula cephalonica

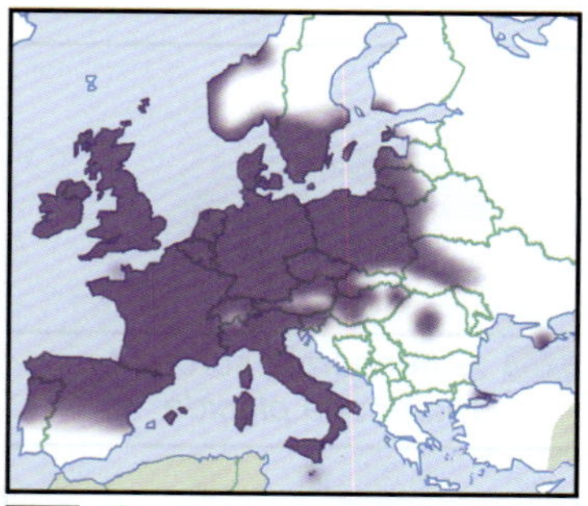

Discus rotundatus

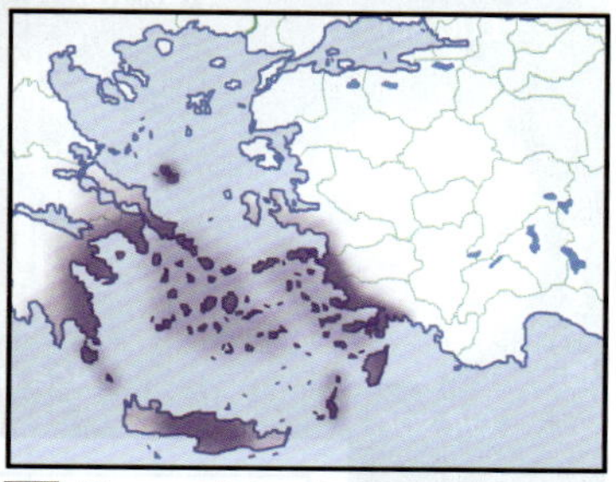

Pyramidula chorismenostoma

Pleurodiscus balmei
Pleurodiscus cyprius

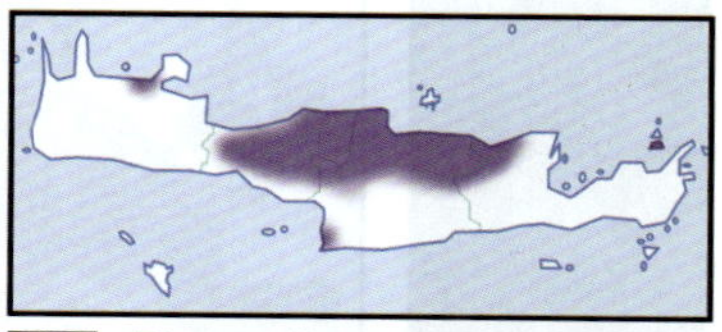

Pleurodiscus sudensis

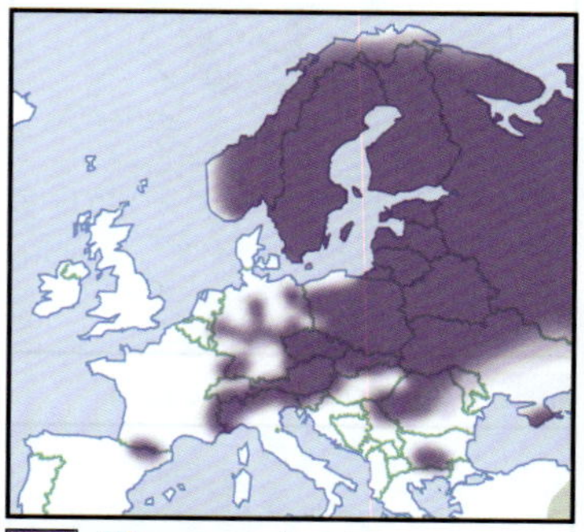

Discus ruderatus

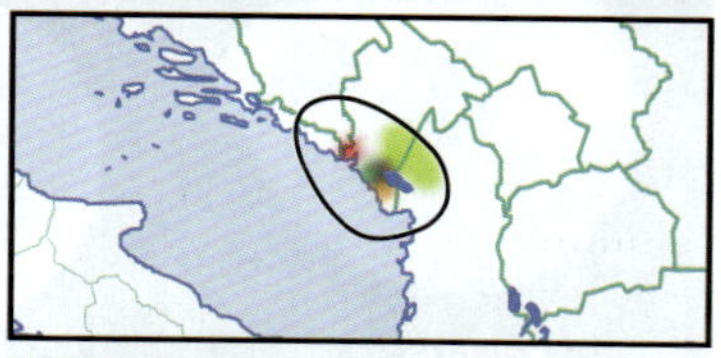

Virpazaria adrianae
Virpazaria aspectulabeatidis
Virpazaria backhuysi
Virpazaria deelemanorum
Virpazaria pageti
Virpazaria ripkeni
Virpazaria stojaspali

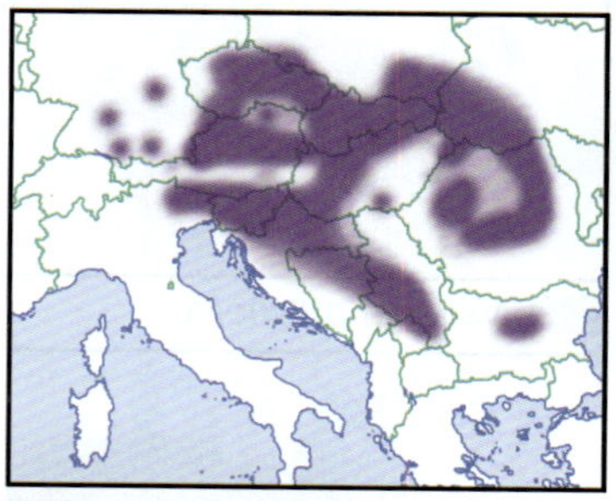

Discus perspectivus

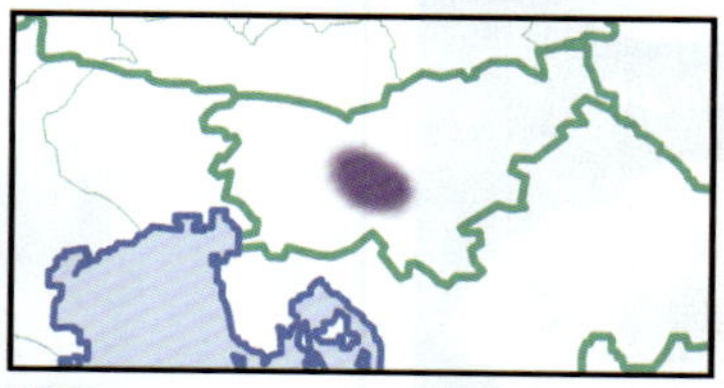

Spelaeodiscus hauffeni

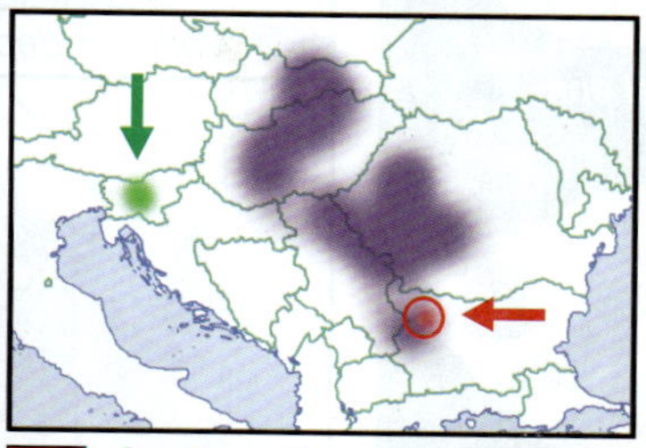

Spelaeodiscus bulgaricus
Spelaeodiscus dejongi
Spelaeodiscus triarius

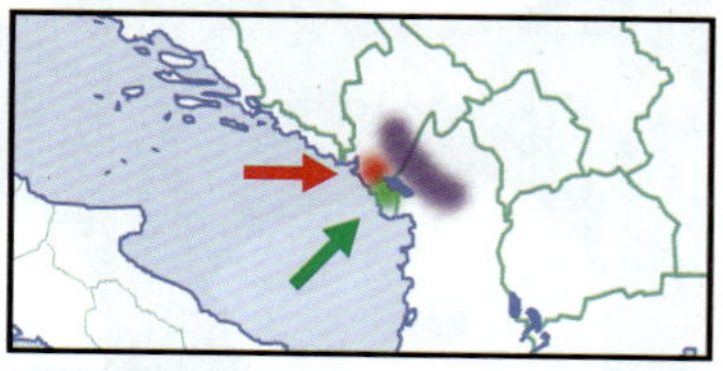

Spelaeodiscus albanicus
Spelaeodiscus obodensis
Spelaeodiscus unidentatus
Klemmia magnicosta
Klemmia sinistrorsa

Clausiliidae 1
Serrulininae, *Agathylla*

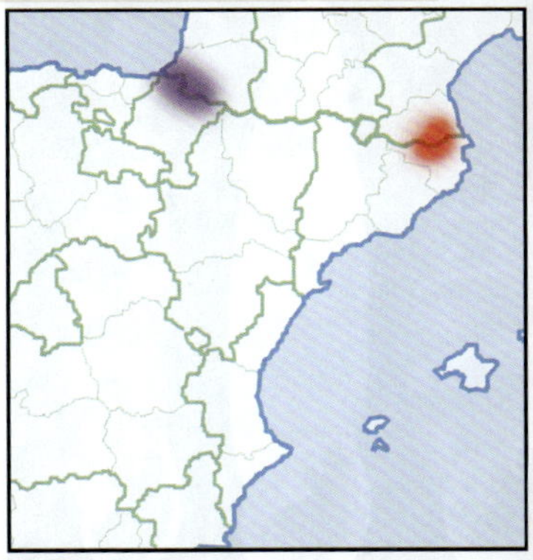

Neniatlanta pauli
Bofilliella subarcuata

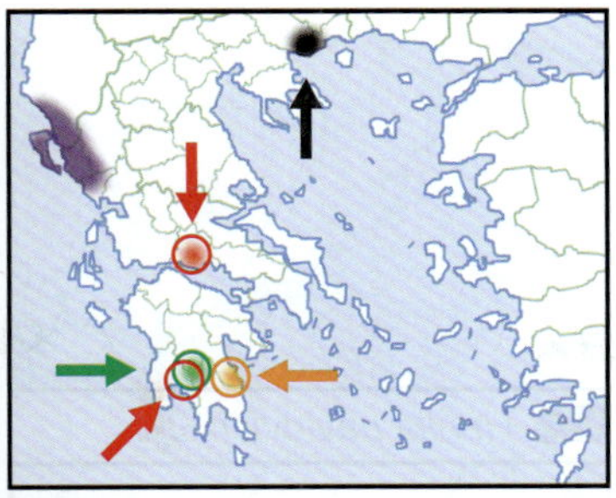

Graecophaedusa sperrlei
Sciocochlea collasi
Tsoukatosia christinae
Tsoukatosia liae
Tsoukatosia subaii

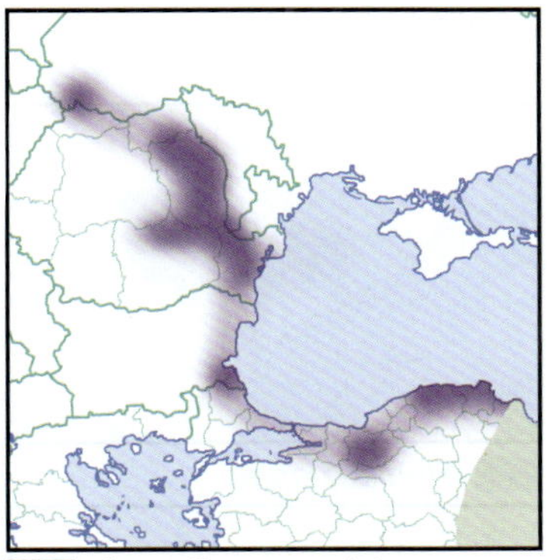

Serrulina serrulata

Dobatia goettingi
Dobatia multidentifera

Agathylla lamellosa

Agathylla narentana

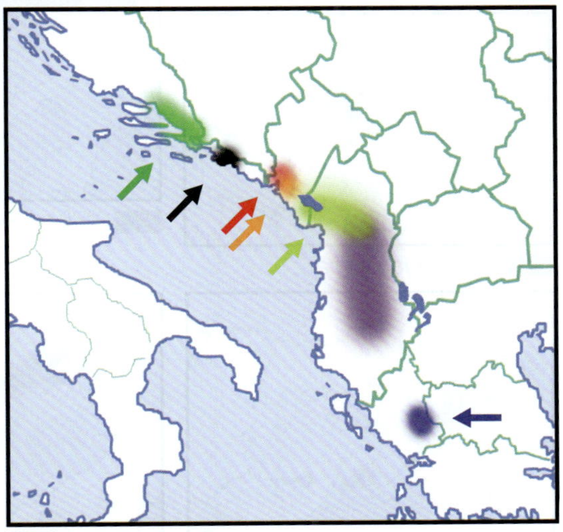

Agathylla biloba
Agathylla exarata
Agathylla goldi
Agathylla merditana
Agathylla neutra
Agathylla regularis
Agathylla sulcosa

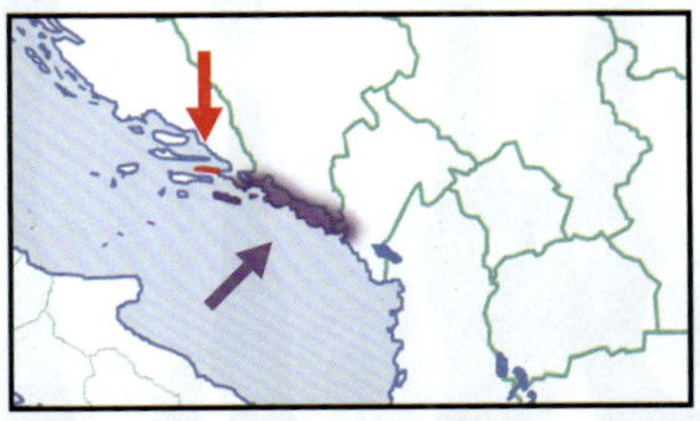

Agathylla formosa
Agathylla viperina

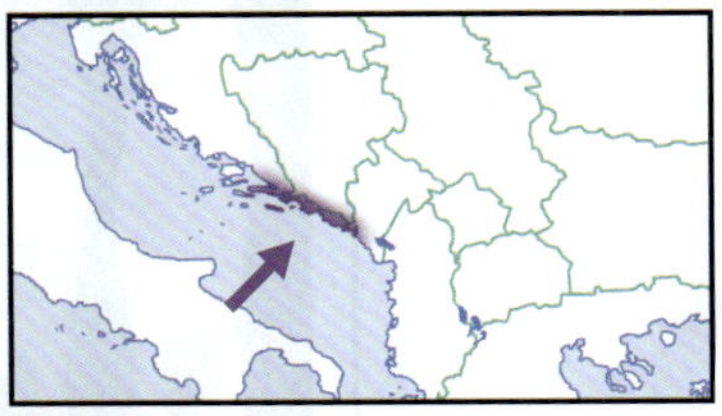

Agathylla abrupta

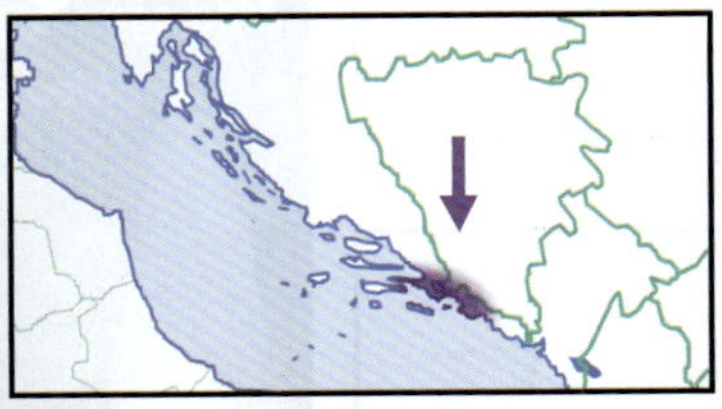

Agathylla strigillata

Clausiliidae 2

Herilla, *Triloba*, *Medora*

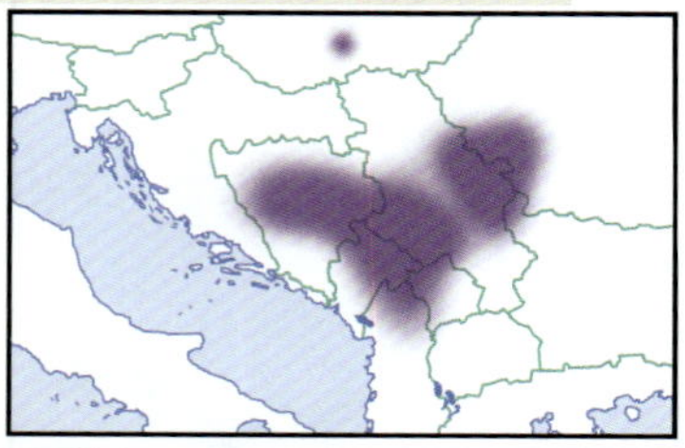

Herilla ziegleri

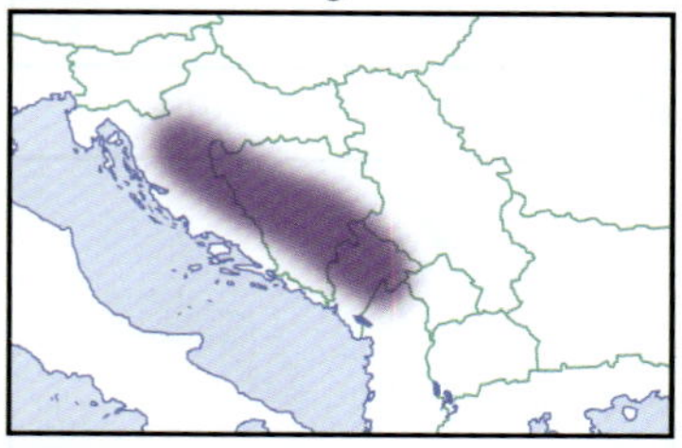

Herilla bosniensis

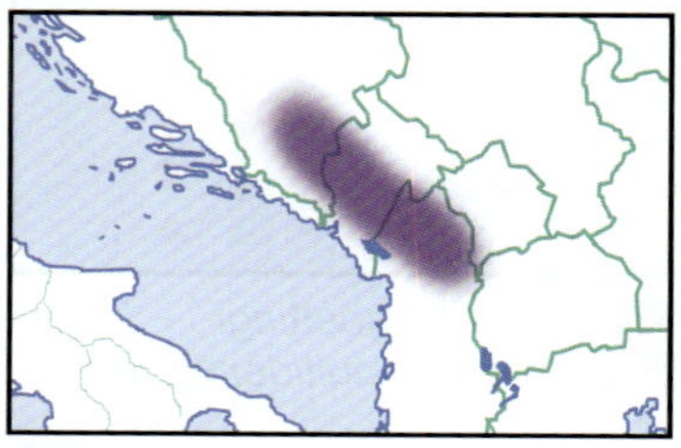

Herilla illyrica

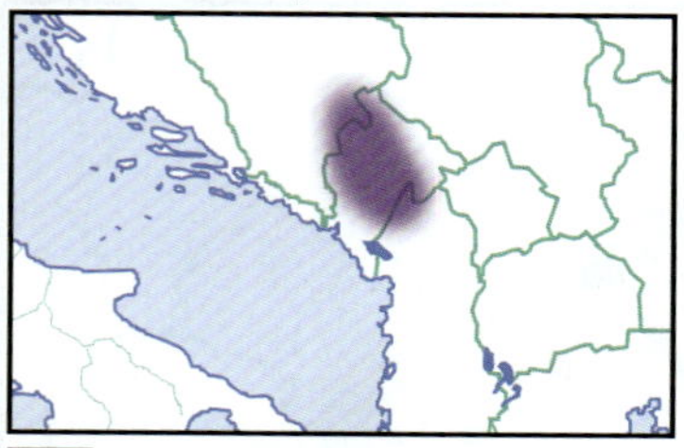

Herilla jabucica

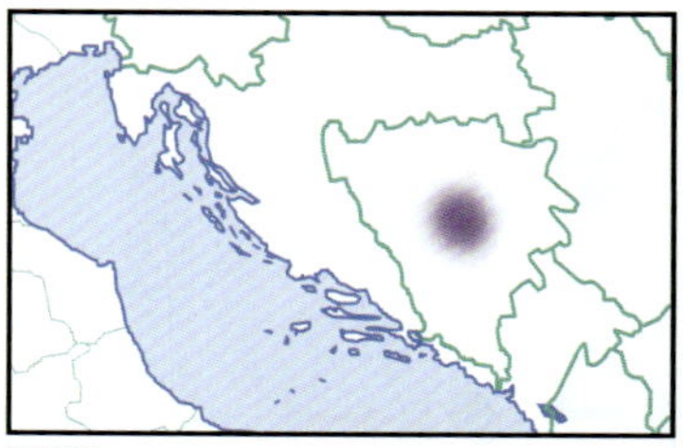

Herilla pavlovici

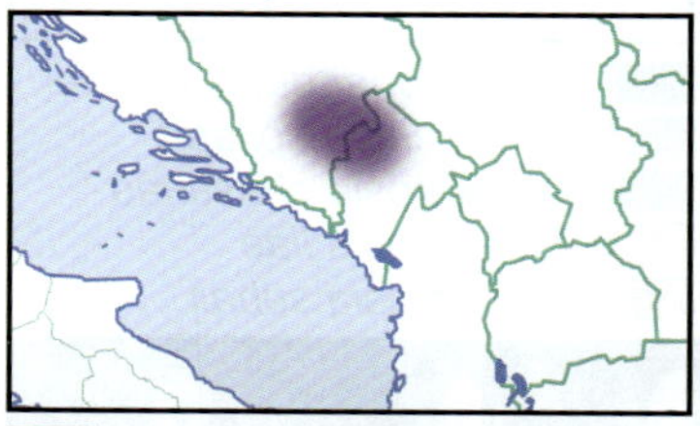

Herilla durmitoris

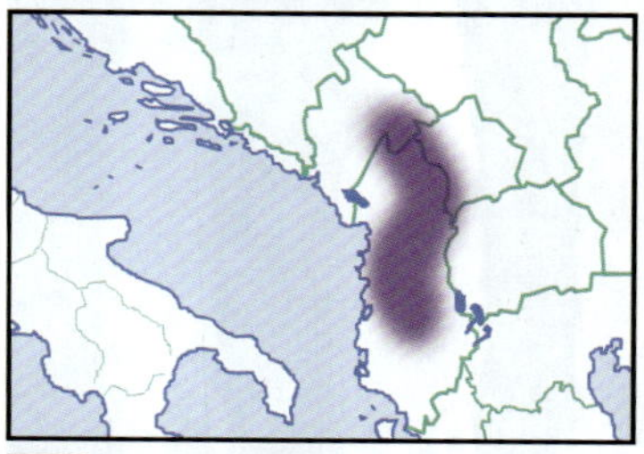

Triloba sandrii

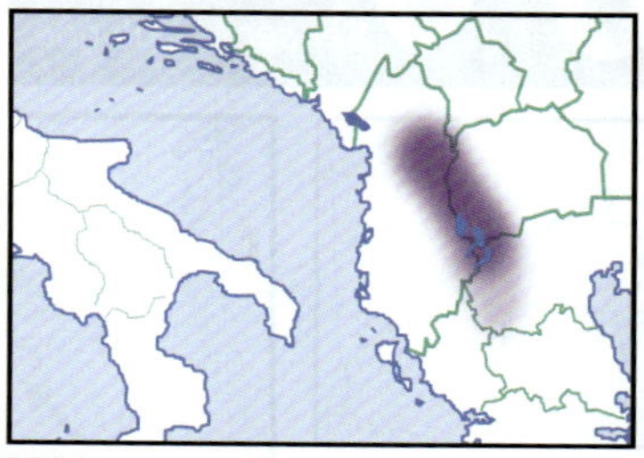

Triloba thaumasia

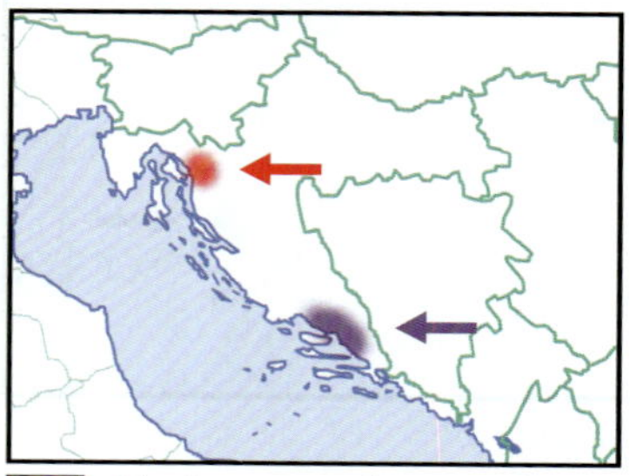

Medora agnata
Medora stenostoma

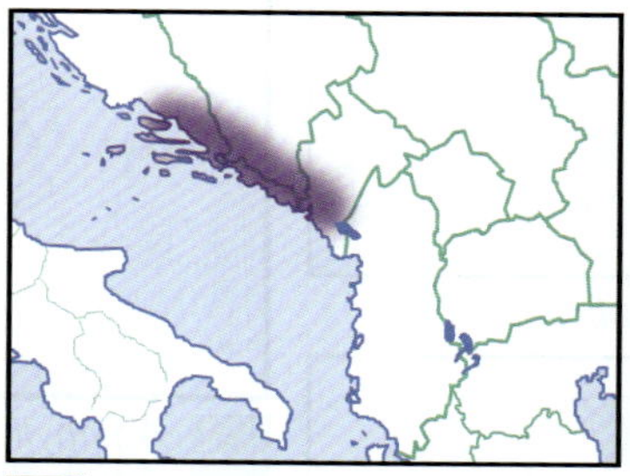

Medora contracta

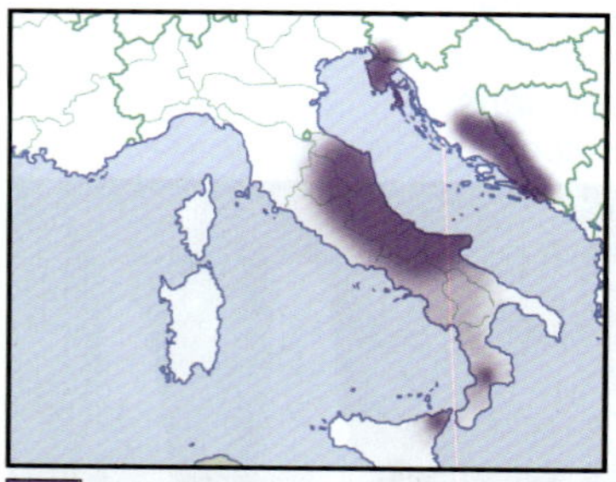

Medora macascarensis

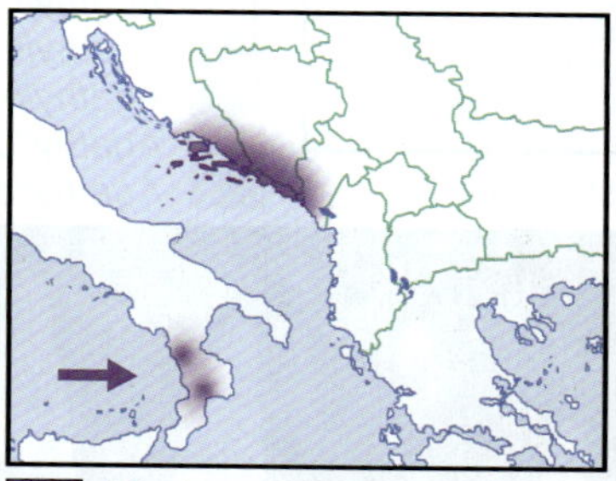

Medora dalmatina

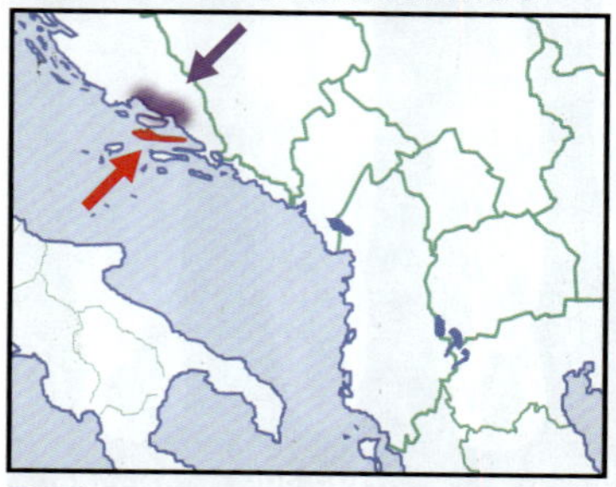

Medora almissana
Medora lesinensis

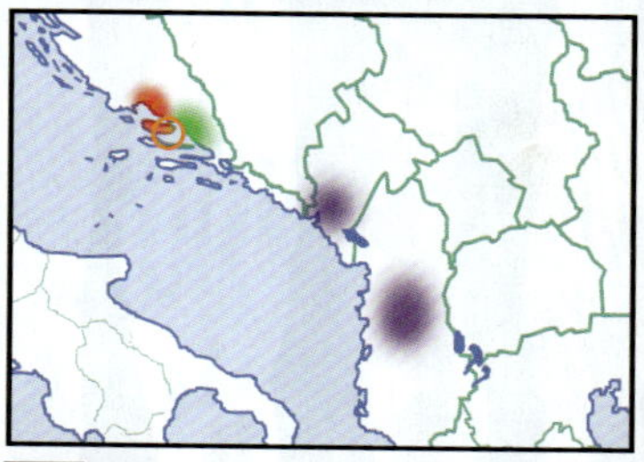

Medora armata
Medora eris
Medora hiltrudae
Medora proxima

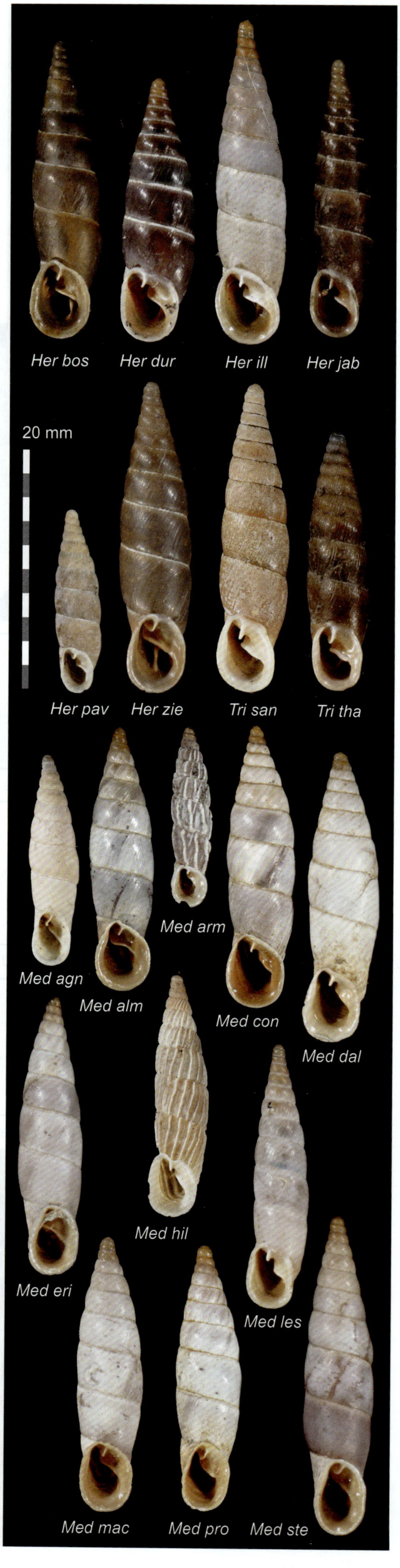

Clausiliidae 3
Alopia, Albinaria group 1

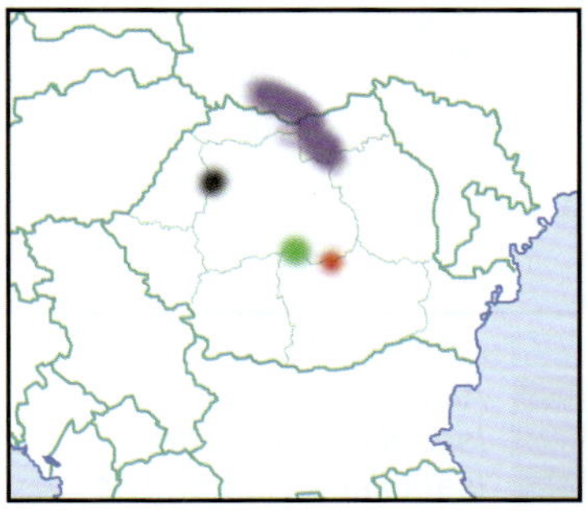

- *Alopia cyclostoma*
- *Alopia glauca*
- *Alopia maciana*
- *Alopia soosiana*

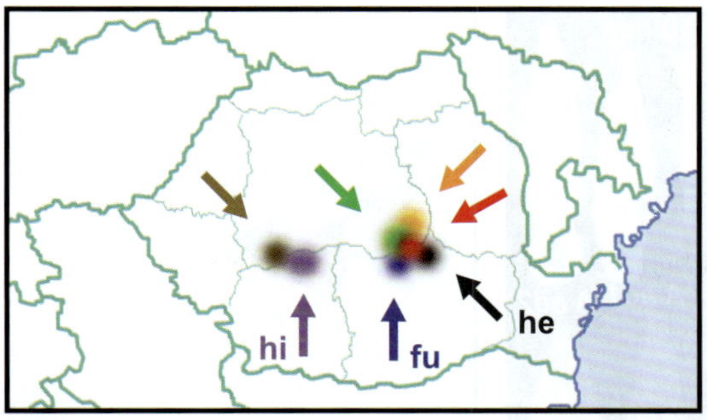

- *Alopia fussi*
- *Alopia helenae*
- *Alopia hildegardae*
- *Alopia meschendorferi*
- *Alopia nefasta*
- *Alopia petrensis*
- *Alopia plumbea*

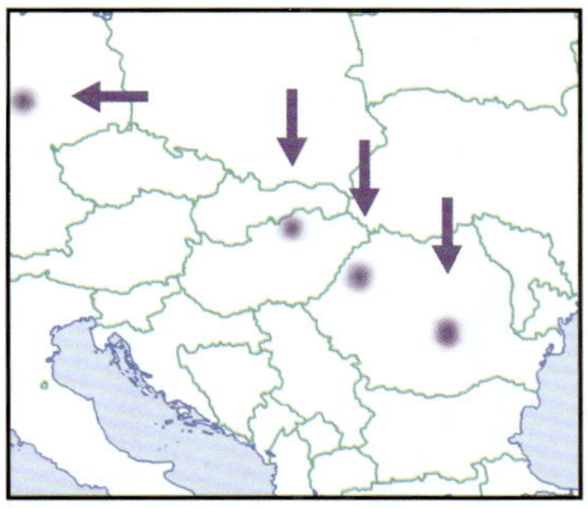

- *Alopia livida*

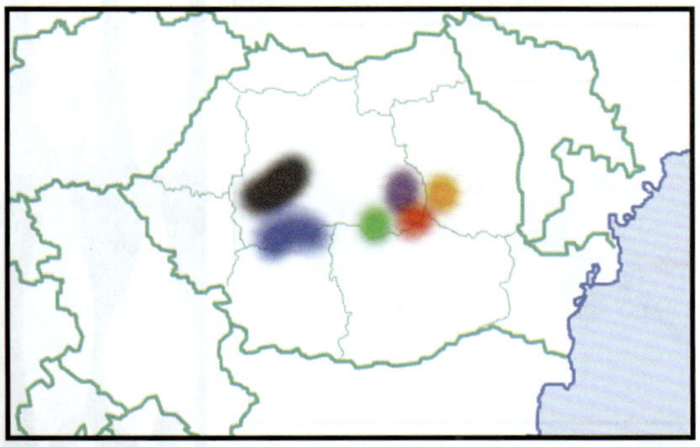

- *Alopia bielzi*
- *Alopia bogatensis*
- *Alopia canescens*
- *Alopia lischkeana*
- *Alopia subcosticollis*
- *Alopia vranceana*

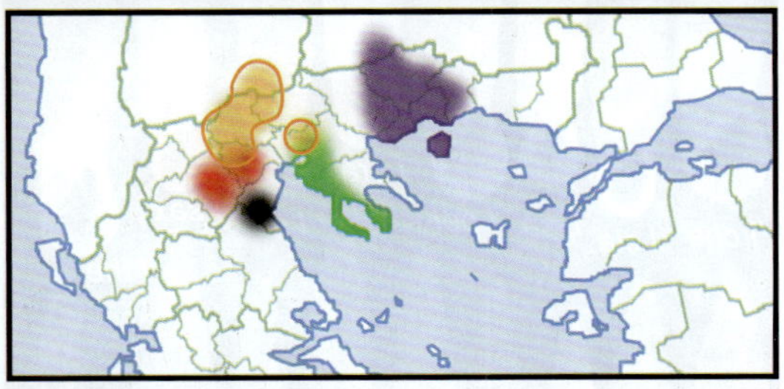

- *Carinigera* buresi
- *Isabellaria* lophauchena
- *Isabellaria* praecipua (prac)
- *Isabellaria* thessalonica
- *Sericata* albicosta

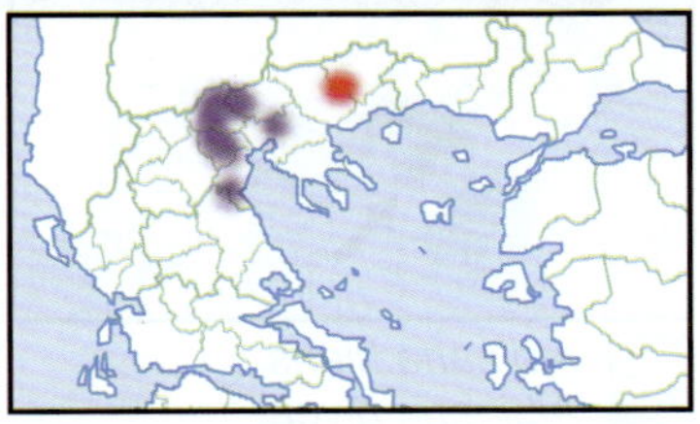

- *Sericata* dextrorsa
- *Carinigera* superba

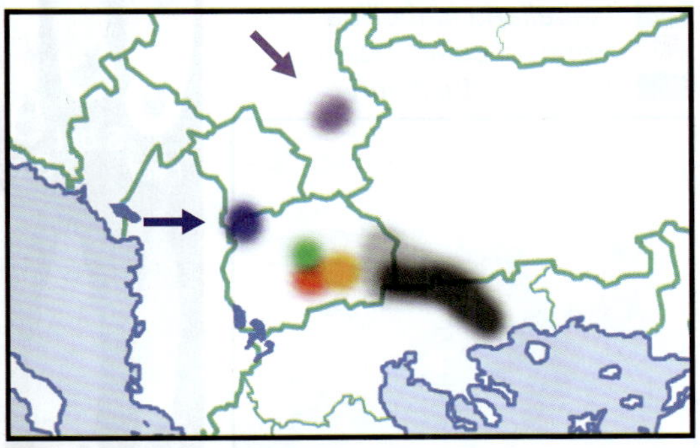

- *Carinigera* drenovoensis
- *Carinigera* eximia
- *Carinigera* octava
- *Carinigera* pellucida
- *Carinigera* schuetti
- *Carinigera* septima

Clausiliidae 4
Albinaria group 2

Albinaria maculosa

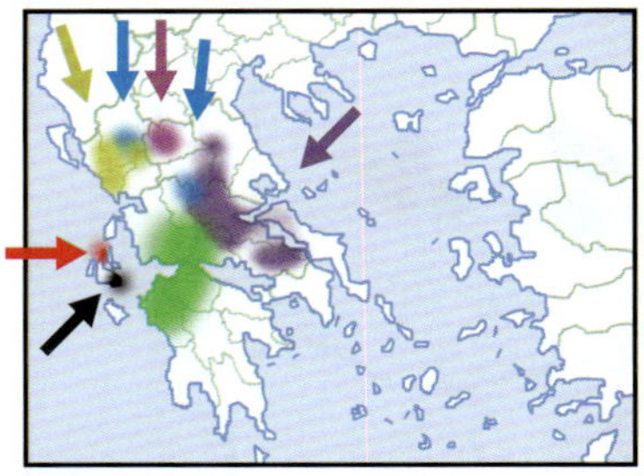

Isabellaria vallata
Isabellaria thermopylarum
Albinaria adrianae (ad1)
Albinaria jonica
Albinaria krueperi
Sericata inchoata
Sericata torifera

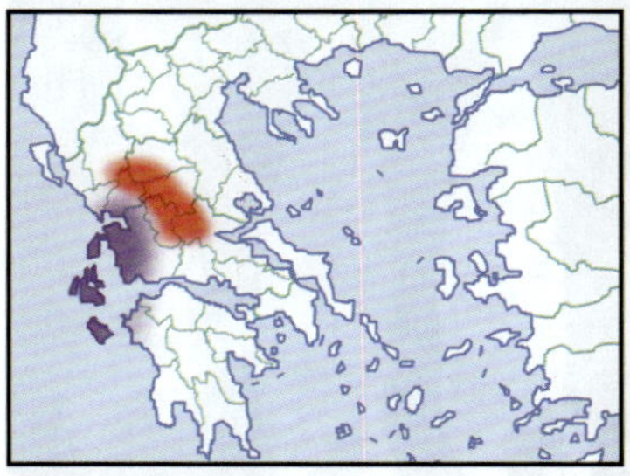

Albinaria contaminata
Carinigera haussknechti

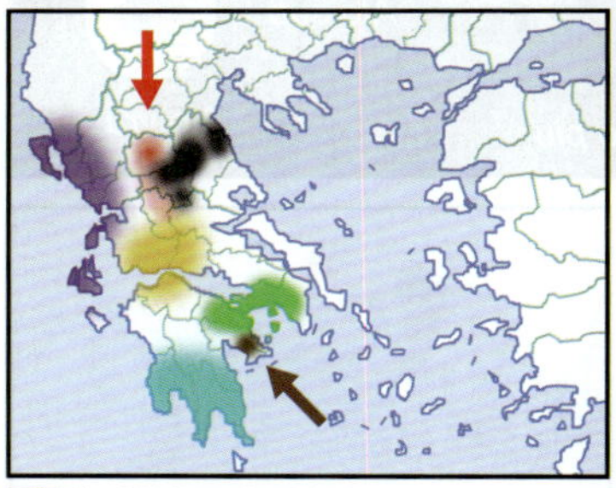

Albinaria hians
Albinaria nivea
Albinaria senilis
Albinaria zilchi
Isabellaria isabellina
Sericata calabacensis
Sericata stussineri

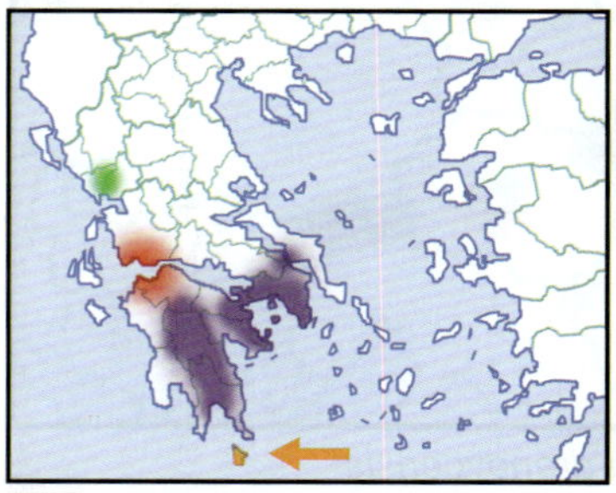

Albinaria compressa
Albinaria cytherae
Albinaria grisea
Albinaria thiesseae
Sericata regina

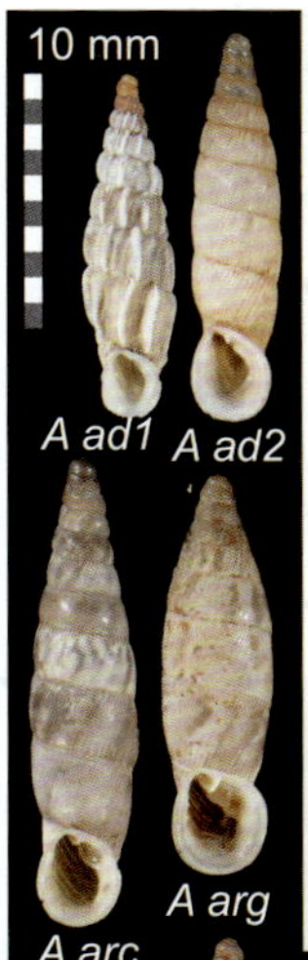

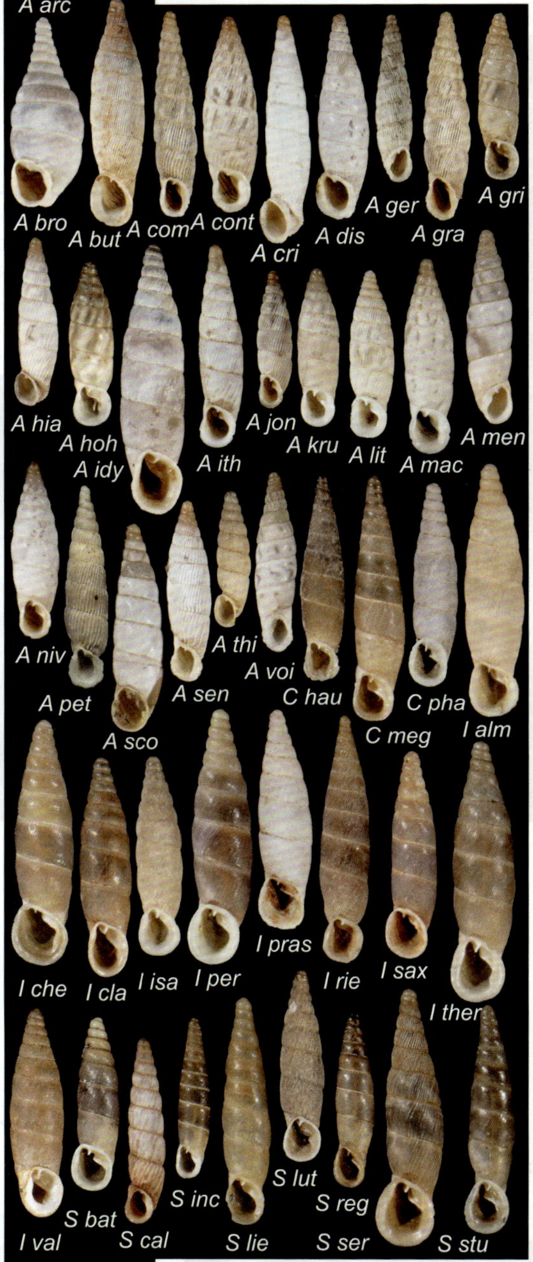

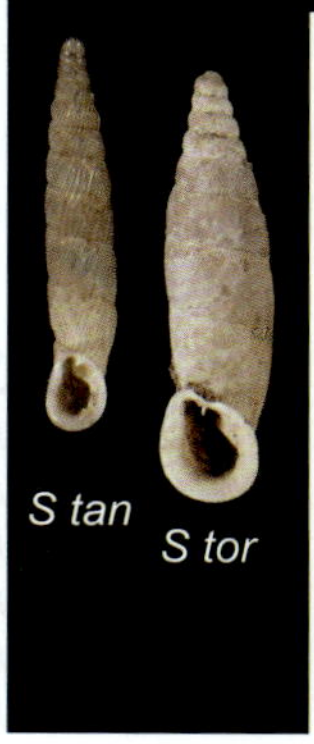

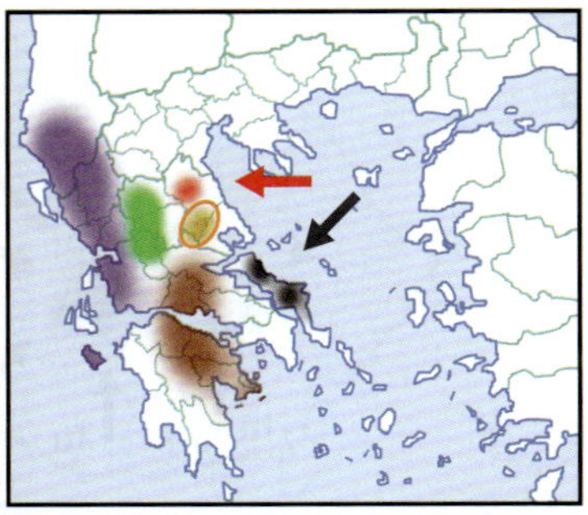

Albinaria arcadica
Albinaria scopulosa
Carinigera megdova
Carinigera pharsalica
Sericata sericata
Sericata tantilla

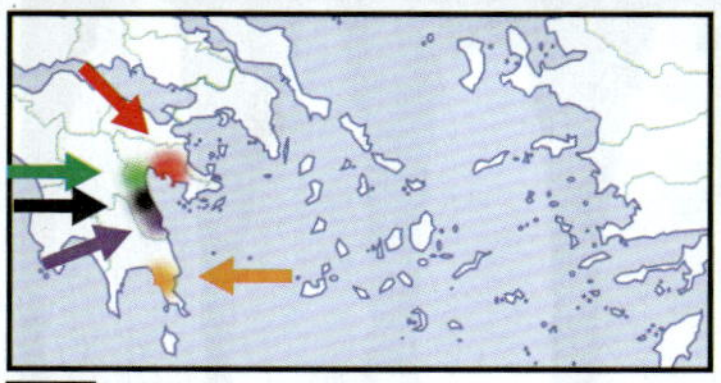

Albinaria adriani (ad2)
Albinaria evelynae
Albinaria hohorsti
Albinaria litoraria
Albinaria mixta
Albinaria solicola

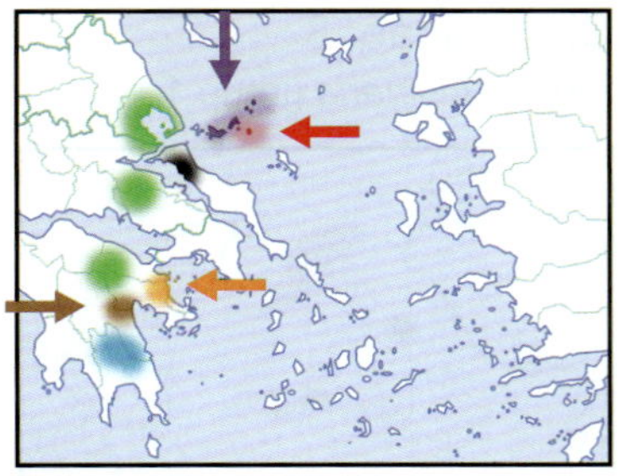

Albinaria butoti
Albinaria petrosa
Isabellaria chelidromia
Isabellaria clandestina
Sericata abyssoclista
Sericata bathyclista
Sericata liebegottae

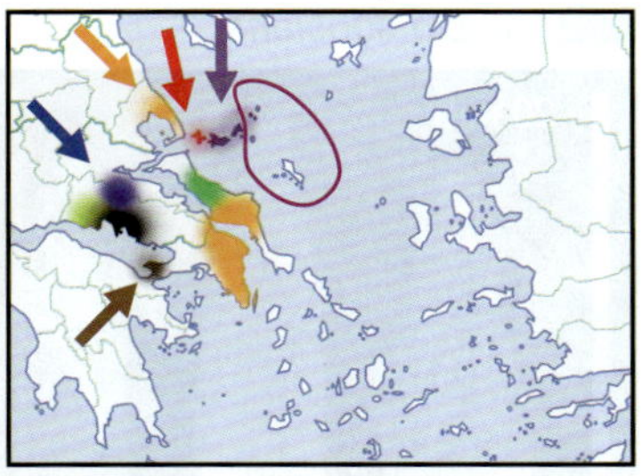

Albinaria cristatella
Isabellaria almae
Isabellaria leucorhaphe
Isabellaria perplana
Isabellaria praestans (pras)
Isabellaria riedeli
Isabellaria saxicola
Sericata lutracana
Sericata parnassia

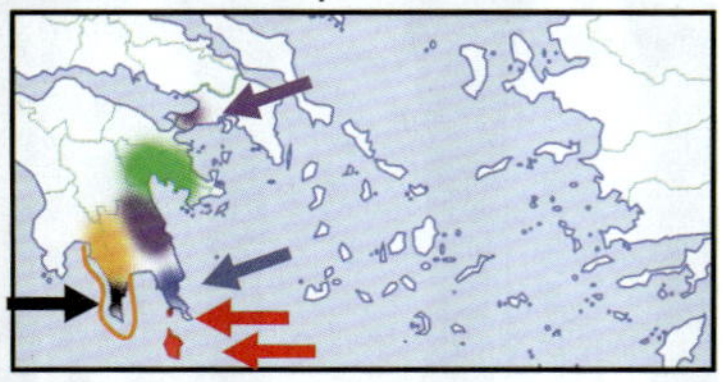

Albinaria argynnis
Albinaria bigibbosula
Albinaria campylauchen
Albinaria grayana
Albinaria menelaus
Albinaria subaii

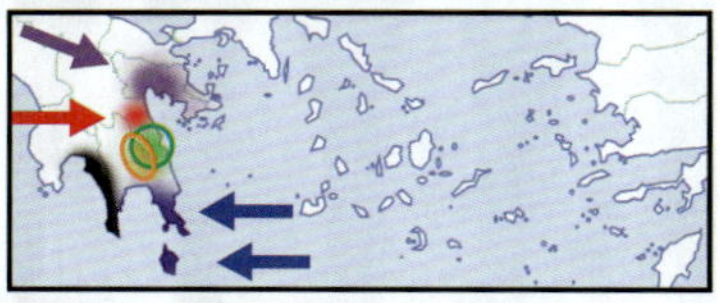

Albinaria confusa
Albinaria edmundi
Albinaria idyllica
Albinaria gerolimena
Albinaria linnei
Albinaria petrosa

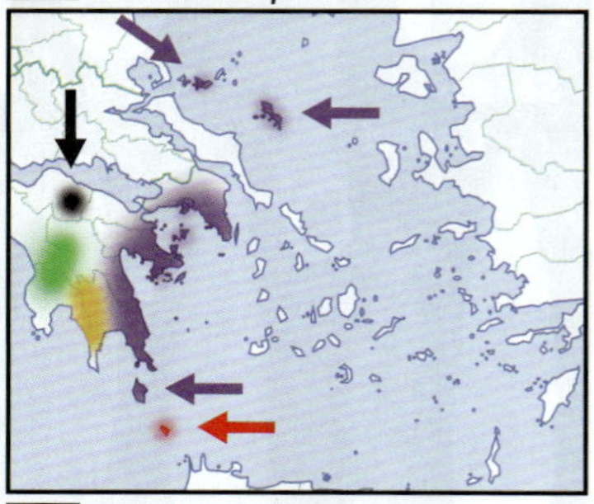

Albinaria broemmei
Albinaria cerigottana
Albinaria discolor
Albinaria ithomensis
Albinaria voithii

Clausiliidae 5
Albinaria group 3

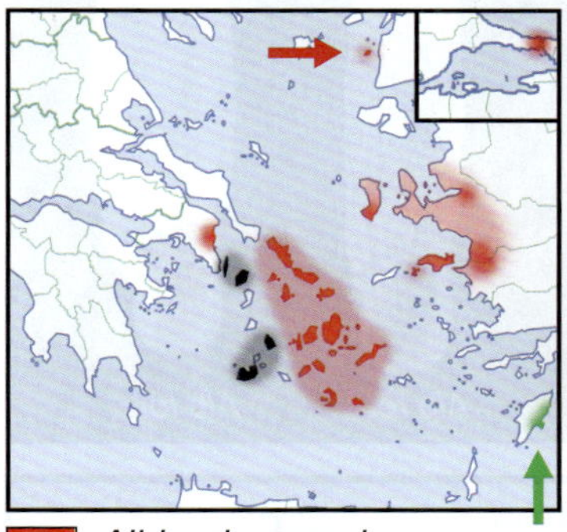

Albinaria caerulea
Albinaria klemmi
Albinaria turrita

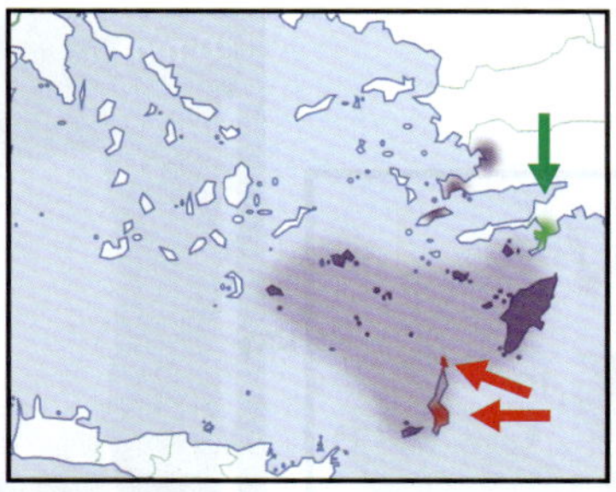

Albinaria basalifera
Albinaria brevicollis
Albinaria proteus

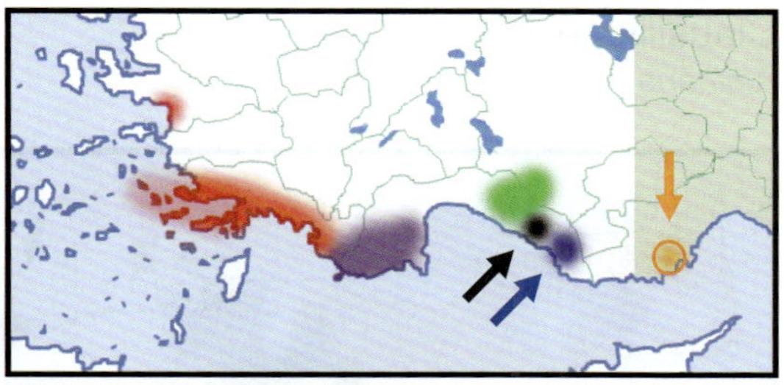

Albinaria anatolica
Albinaria lerosiensis
Albinaria monocristata
Albinaria pellucida
Albinaria schuetti
Albinaria supercarinata

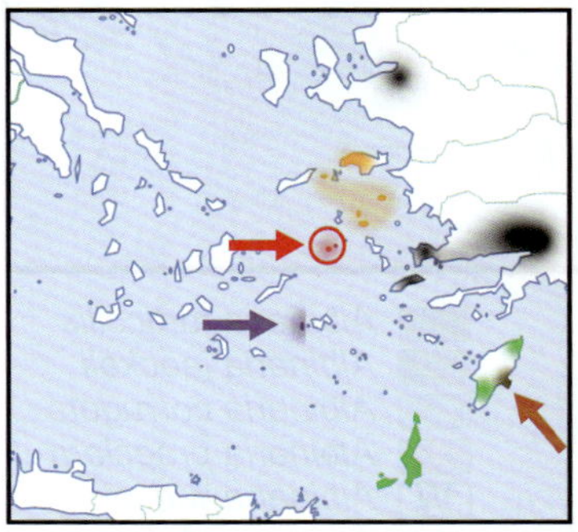

Albinaria freytagi
Albinaria fuchskaeufeli
Albinaria munda
Albinaria olivieri
Albinaria rechingeri
Albinaria wettsteini

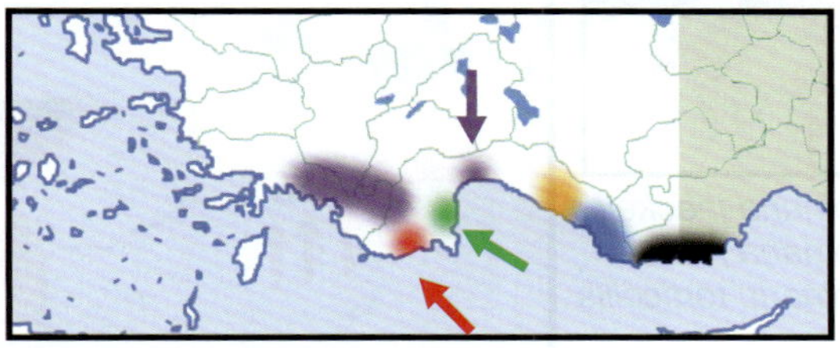

Albinaria forbesiana
Albinaria inauris
Albinaria kemerensis
Albinaria myrensis
Albinaria papillifera
Albinaria percristata

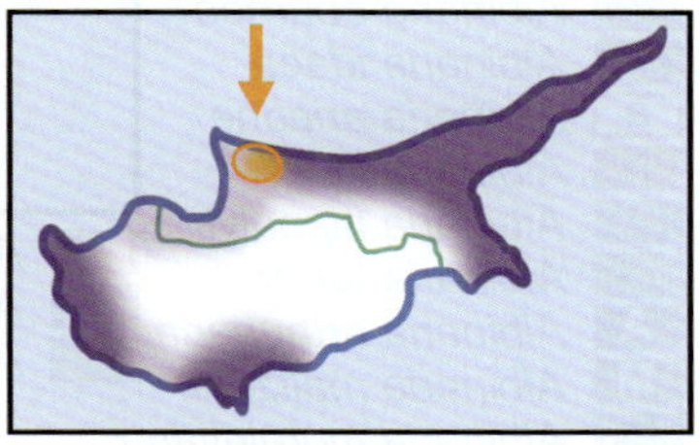

Albinaria mavromoustakisi
Albinaria saxatilis

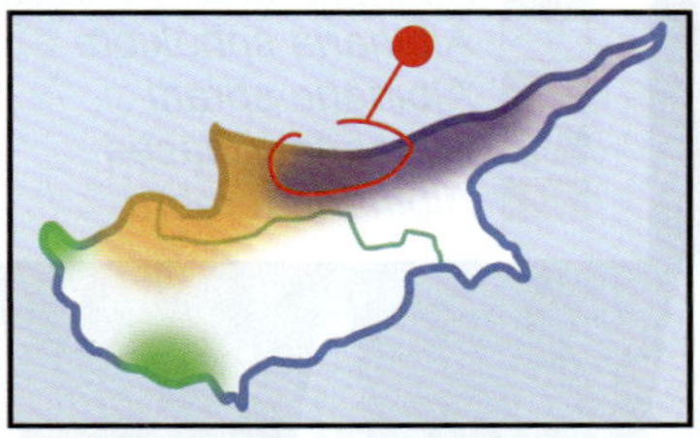

Albinaria alajana
Albinaria greeni
Albinaria rollei
Albinaria virgo

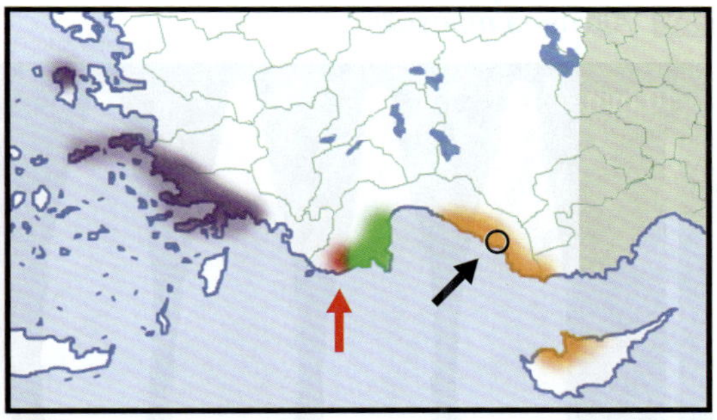

Albinaria alajana
Albinaria ietswaarti
Albinaria latelamellaris
Albinaria lycica
Albinaria puella

Alb for Alb myr B. Gümüş

Clausiliidae 6
Albinaria group 4

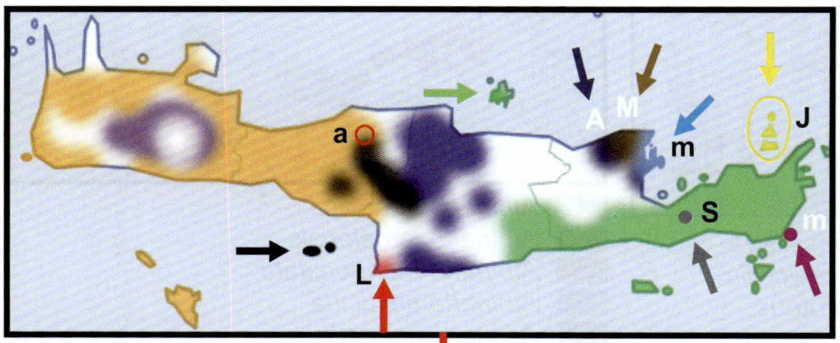

- Albinaria cretensis
- Albinaria eburnea
- Albinaria idaea
- a Albinaria ariadne
- L Albinaria li
- Albinaria hippolyti
- Albinaria teres
- A Albinaria arthuriana
- M Albinaria maltzani
- m Albinaria moreletiana
- S Albinaria sturanyi
- J Albinaria janisadana
- m Albinaria manselli

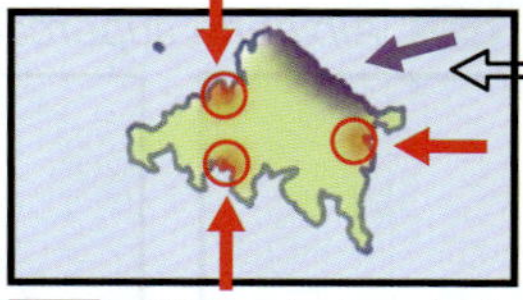

- Albinaria jaeckeli
- Albinaria retusa
- Albinaria torticollis

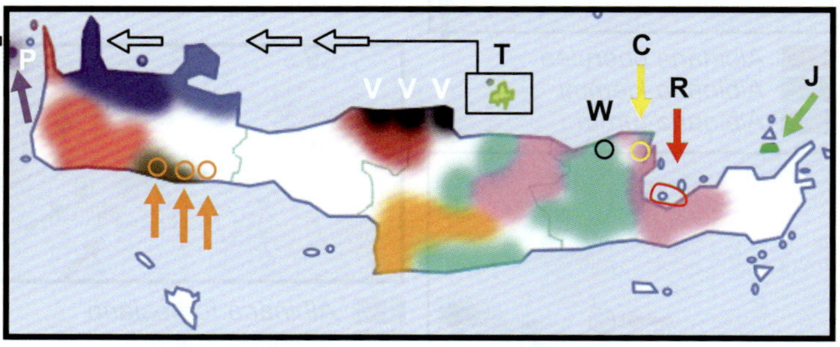

- Albinaria tenuicostata
- P Albinaria pondika
- Albinaria amalthea
- Albinaria candida
- Albinaria sublamellosa
- Albinaria sphakiota
- Albinaria spratti
- V Albinaria violacea
- Albinaria terebra
- T Albinaria torticollis
- Albinaria jaeckeli
- Albinaria corrugata
- Albinaria praeclara
- W Albinaria wiesei
- C Albinaria christae
- R Albinaria rebeli
- J Albinaria janicollis

Welter-Schultes 2010: Fig. 38-41

Clausiliidae 7
Idyla-Elia group

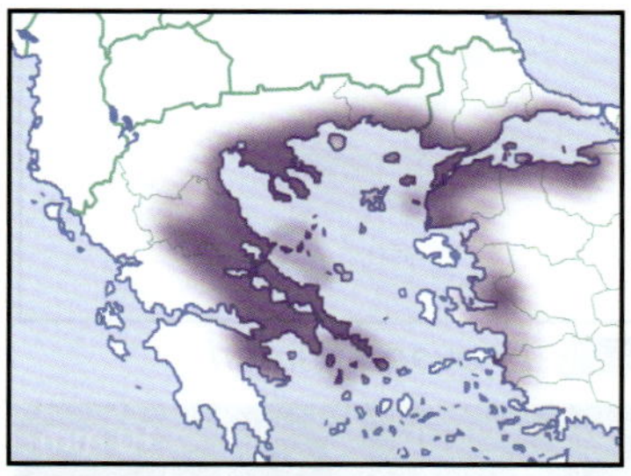

Idyla bicristata

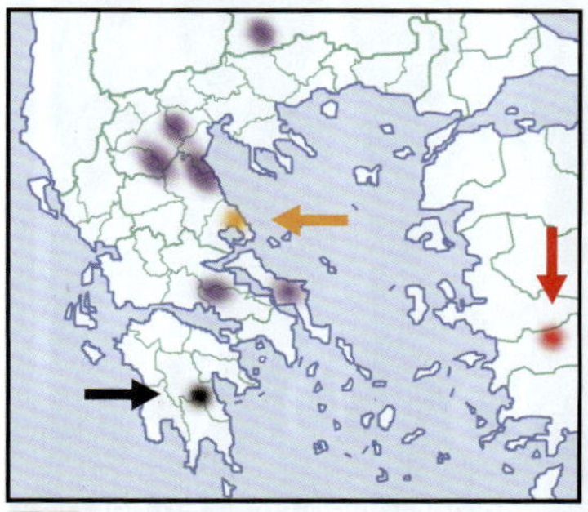

Idyla aydinensis
Idyla castalia
Idyla liebegottae
Idyla pelobsoleta

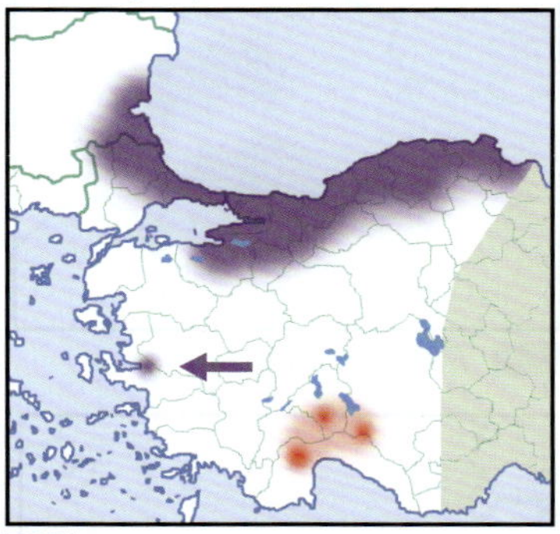

Euxina circumdata
Sprattia beycola

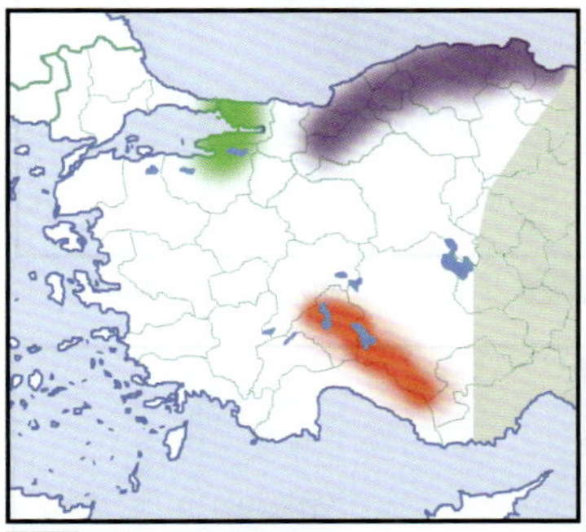

Euxina hetaera
Euxina promta
Sprattia blissi

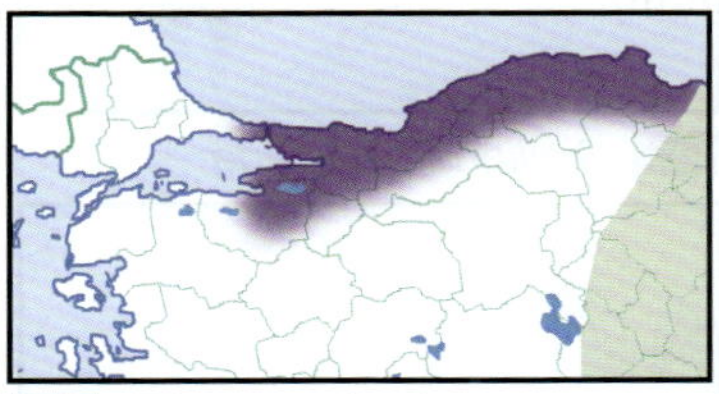

Elia corpulenta

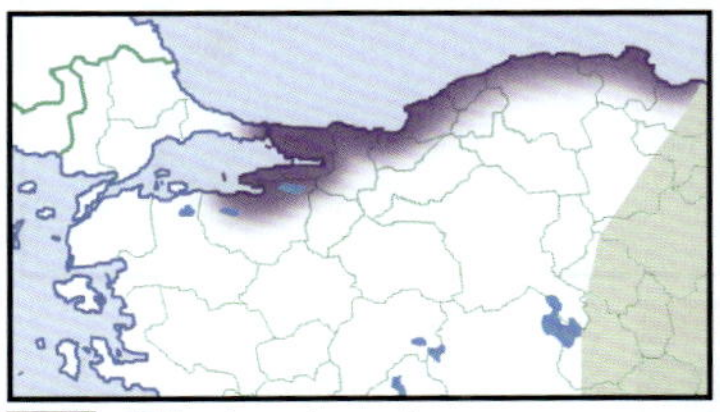

Elia huebneri

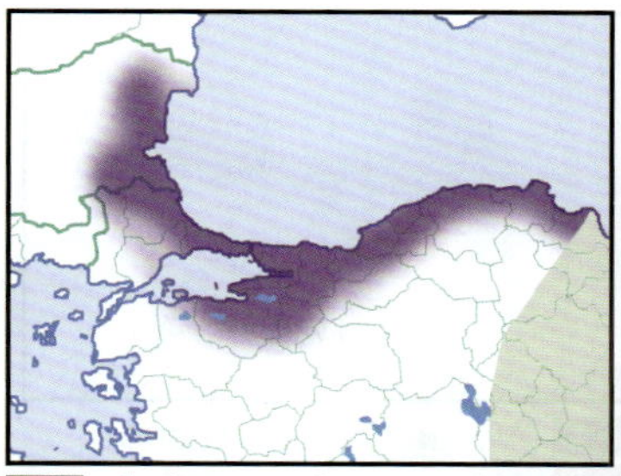

Galeata schwerzenbachii

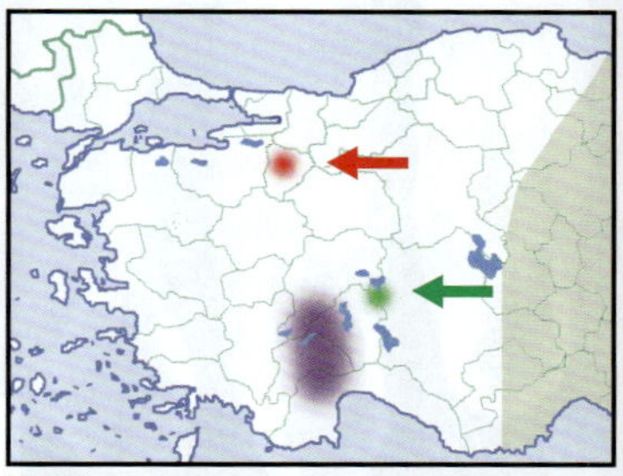

Phrygica euxinaeformis
Phrygica raehlei
Phrygica jelskii

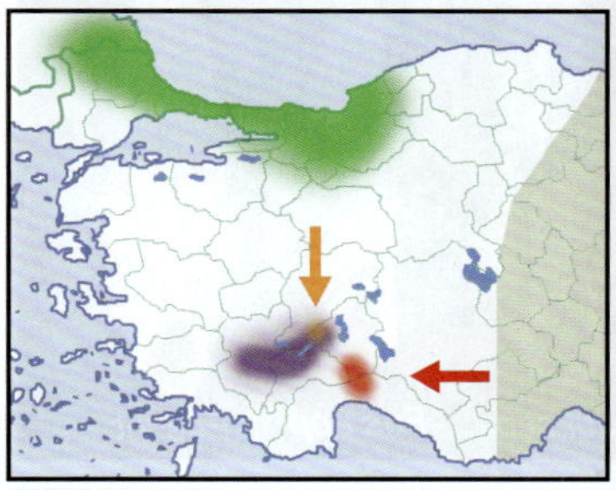

Euxina persica
Phrygica riedeli
Phrygica ilegiensis
Sprattia sillyonensis

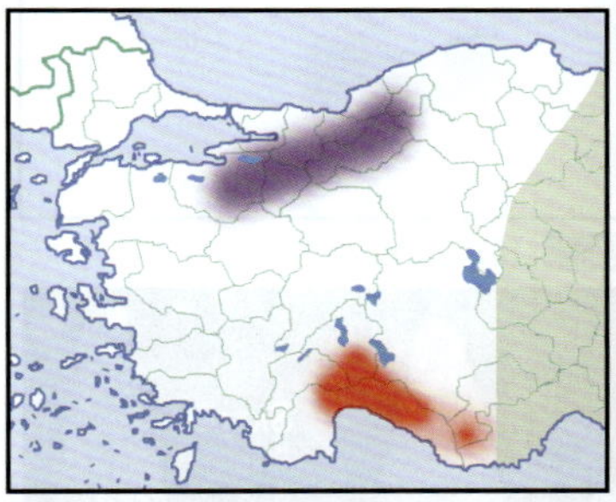

Olympicola olympicus
Sprattia sowerbyana

Elia retowskii

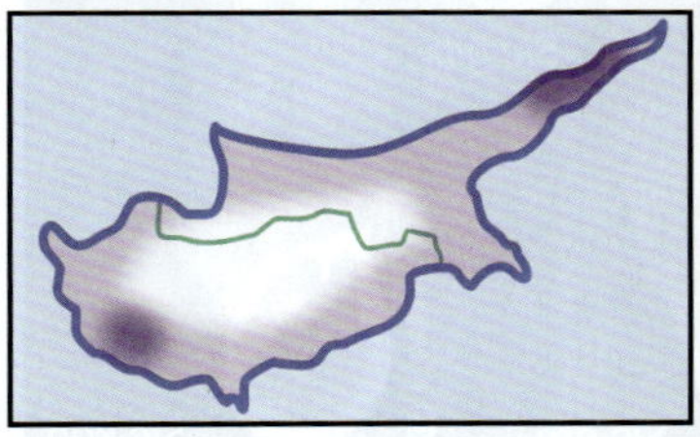

Elia moesta

📷 *Eux het Phr ile* B. Gümüş *Eux per* A. Irikov *Idy ayd* A. Örstan

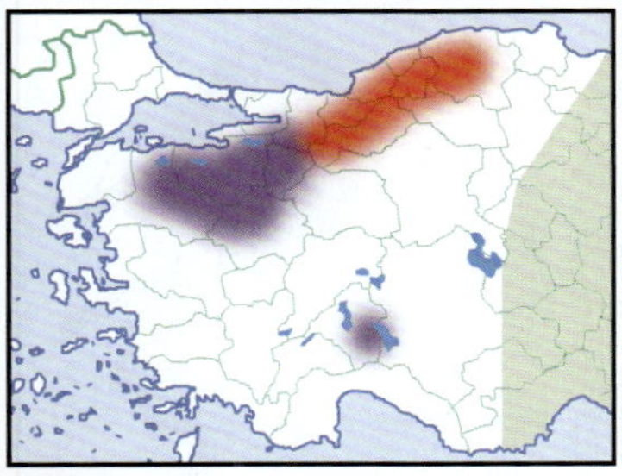

Strumosa abanti
Strumosa strumosa

Clausiliidae 8
all-others group: Alps, Italy

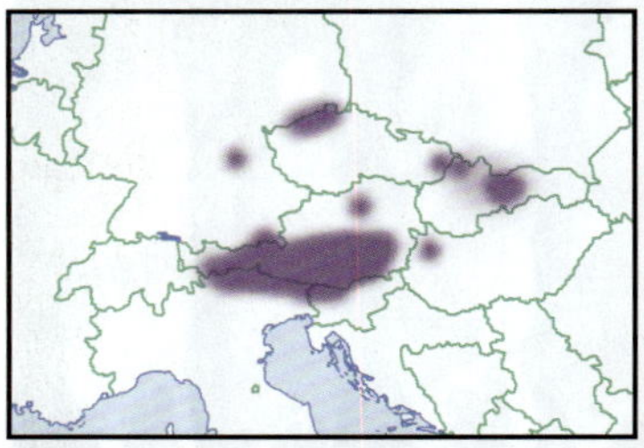

Pseudofusulus varians

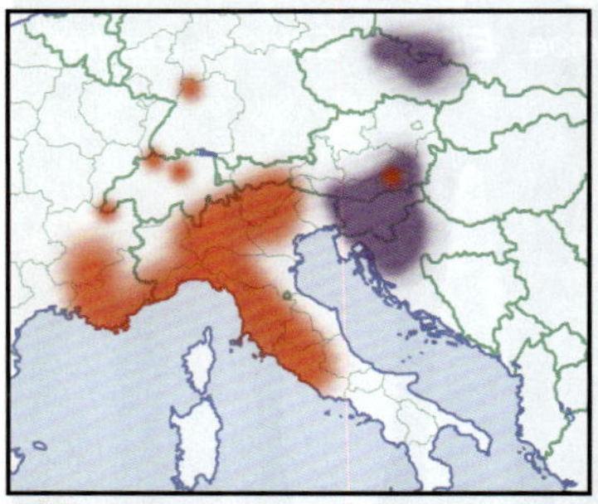

Charpentieria itala
Charpentieria ornata

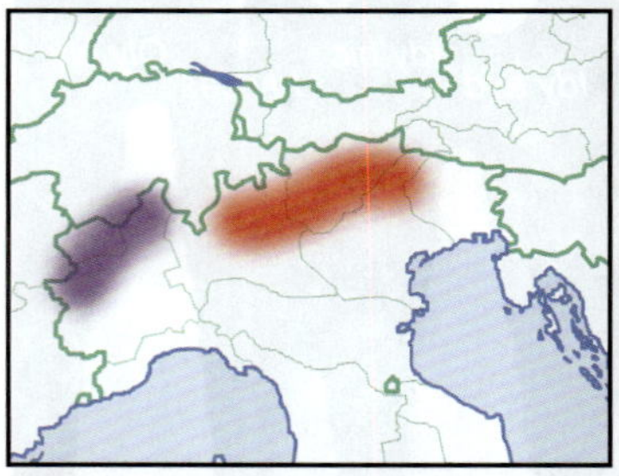

Charpentieria clavata
Charpentieria dyodon

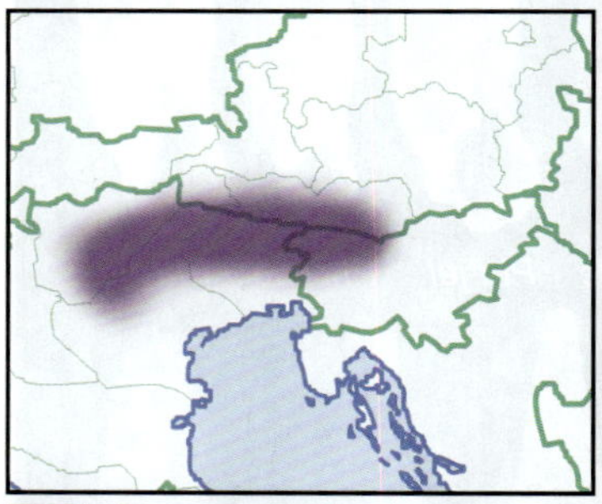

Charpentieria stenzii

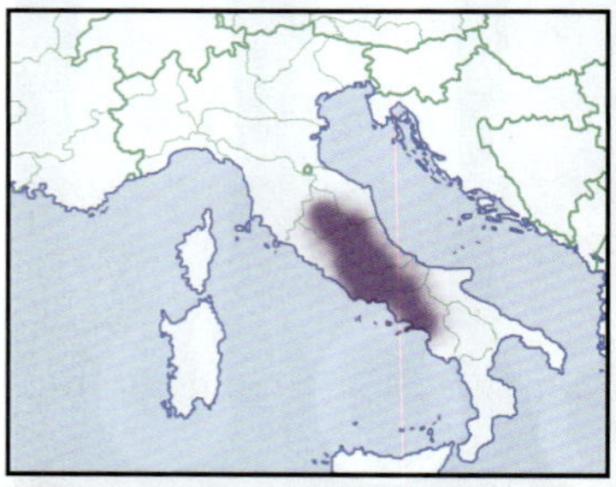

Leucostigma candidescens

Siciliaria gibbula

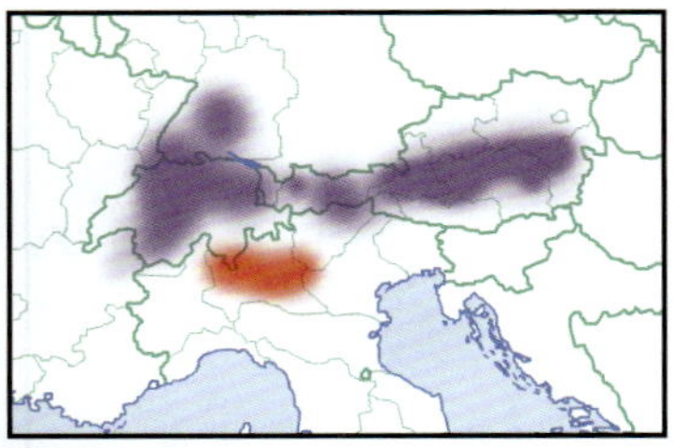

Neostyriaca corynodes
Neostyriaca strobel

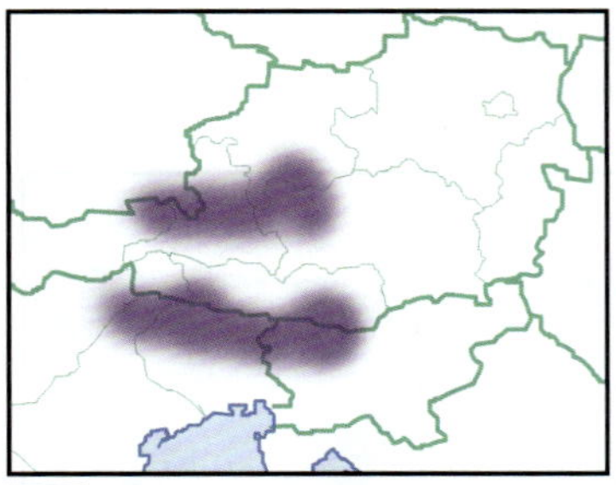

Erjavecia bergeri

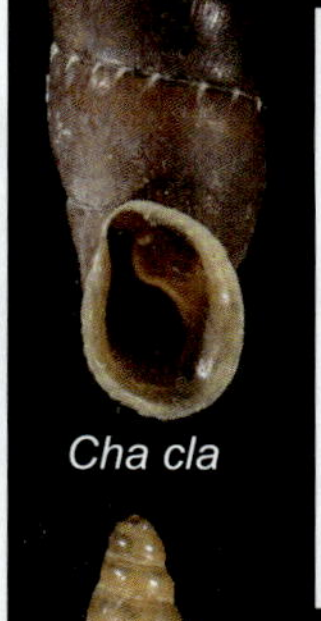

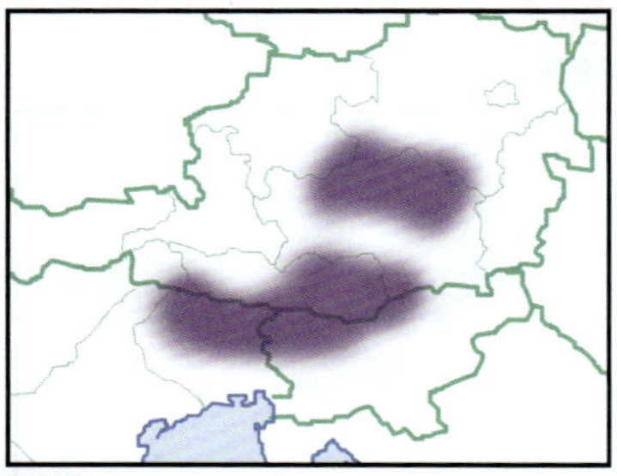

Fusulus interruptus

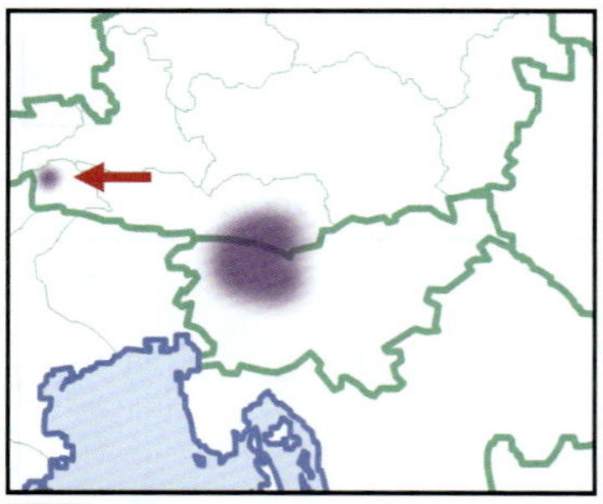

Fusulus approximans

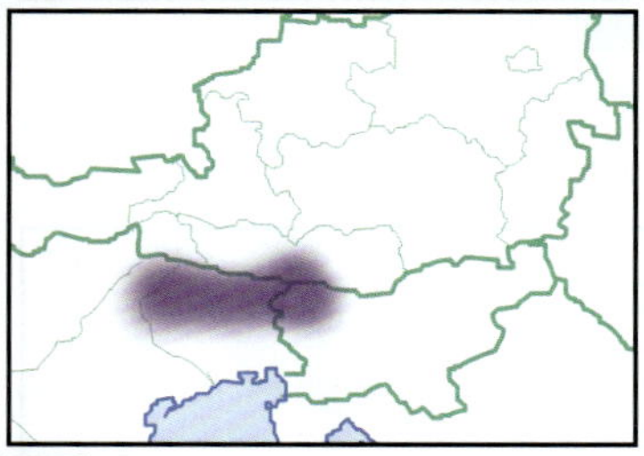

Julica schmidtii

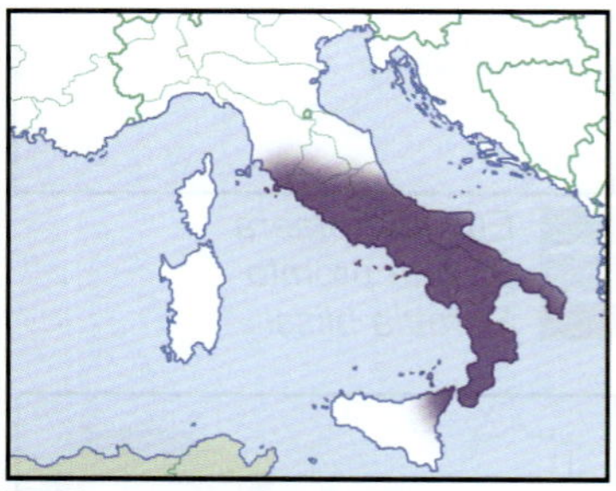

Siciliaria paestana

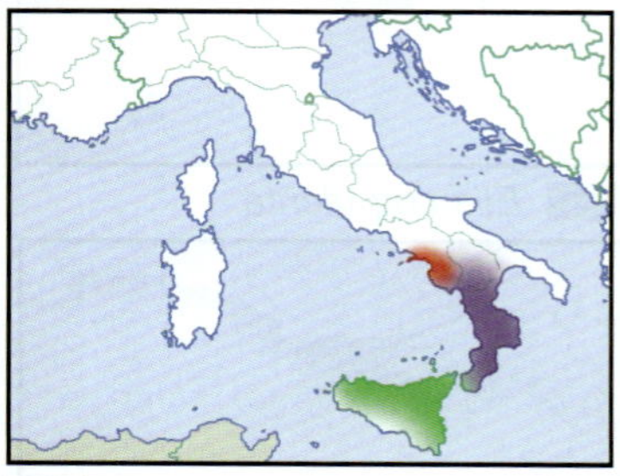

Siciliaria ernae
Siciliaria incerta
Siciliaria kobeltiana

Clausiliidae 9

all-others group: *Lampedusa*, *Muticaria*, *Papillifera*, *Siciliaria*

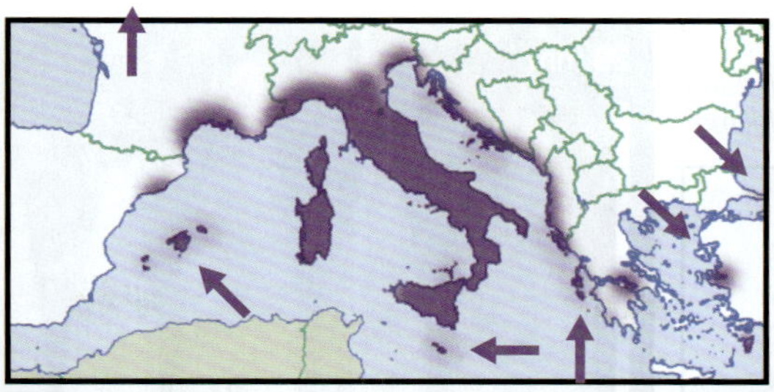

Papillifera bidens

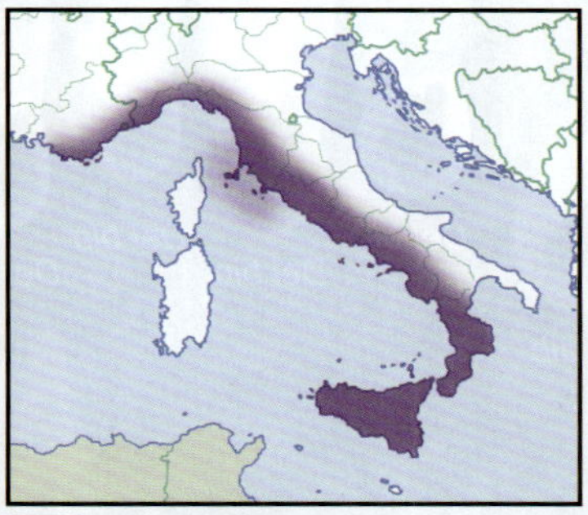

Papillifera solida

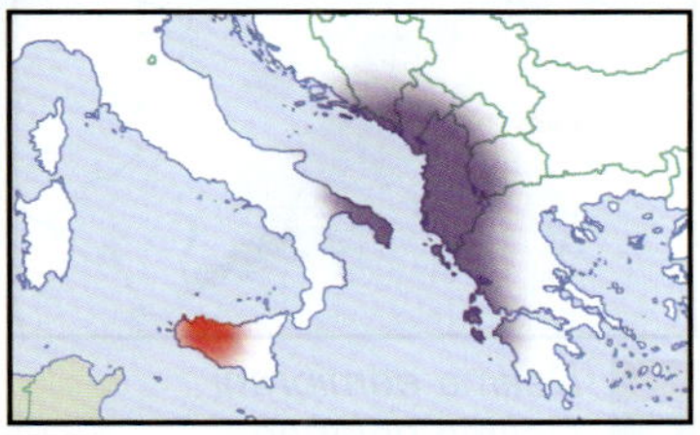

Siciliaria leucophryna
Siciliaria stigmatica

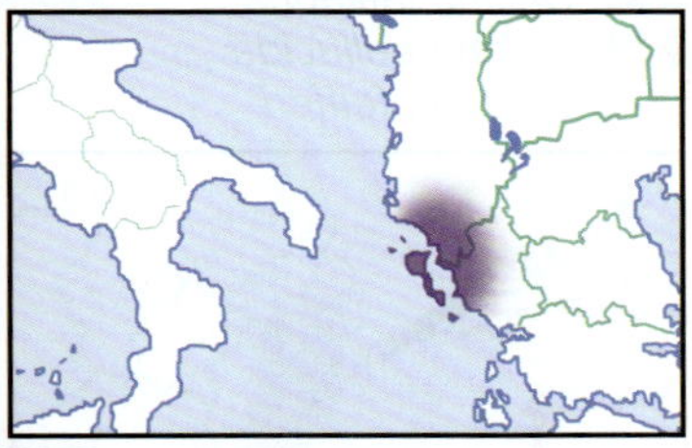

Siciliaria pantocratoris

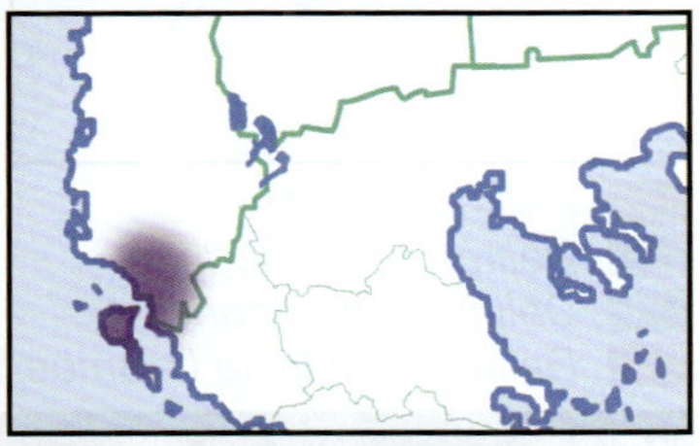

Siciliaria lamellata

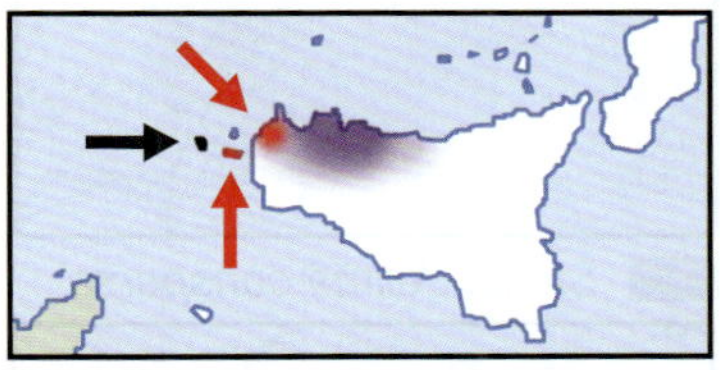

Siciliaria crassicostata
Siciliaria scarificata
Siciliaria septemplicata

Siciliaria calcarae

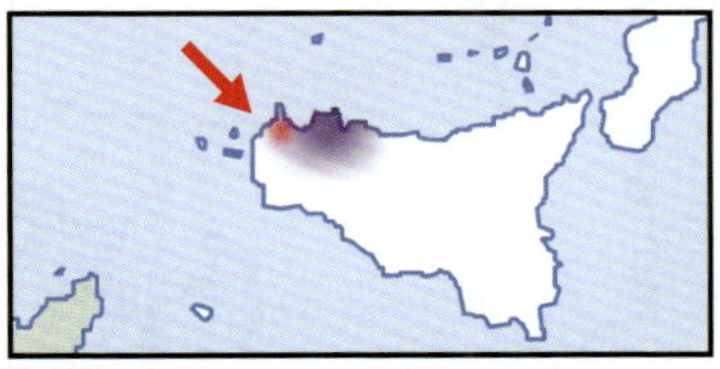

Siciliaria grohmanniana
Siciliaria spezialensis

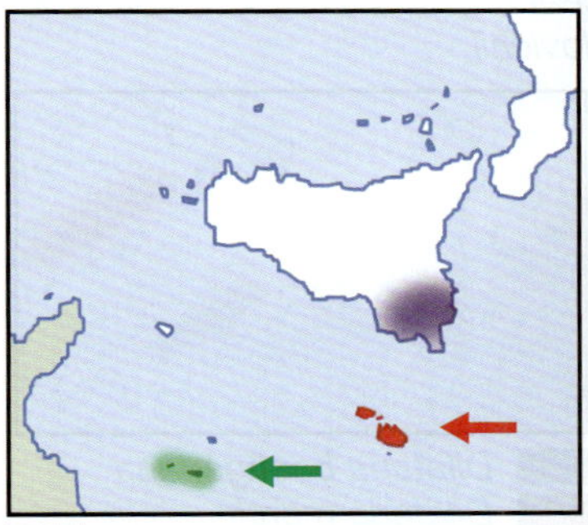

Muticaria macrostoma
Muticaria syracusana
Lampedusa lopadusae

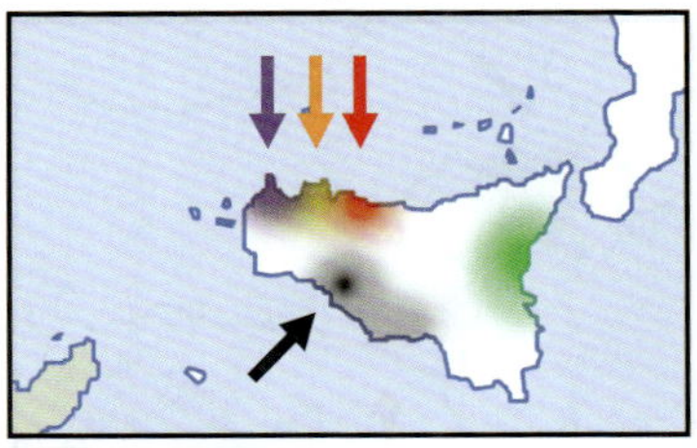

Siciliaria ferrox
Siciliaria nobilis
Siciliaria riberothi
Siciliaria tiberii
Siciliaria vulcanica

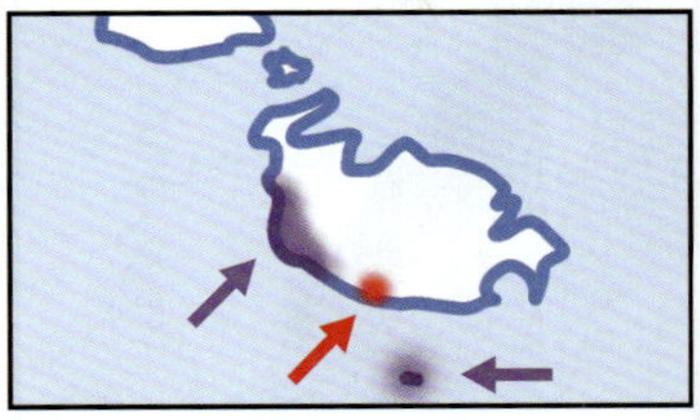

Lampedusa imitatrix
Lampedusa melitensis

Clausiliidae 10

all-others group: *Delima*, *Dilataria*, *Euxinella*, *Protoherilla*, *Strigilodelima*

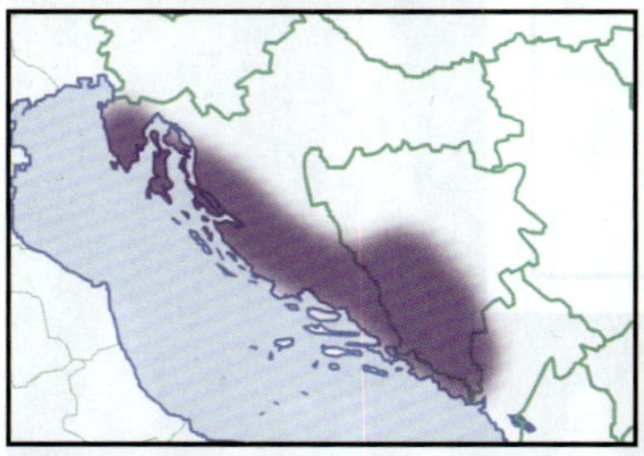

Delima binotata

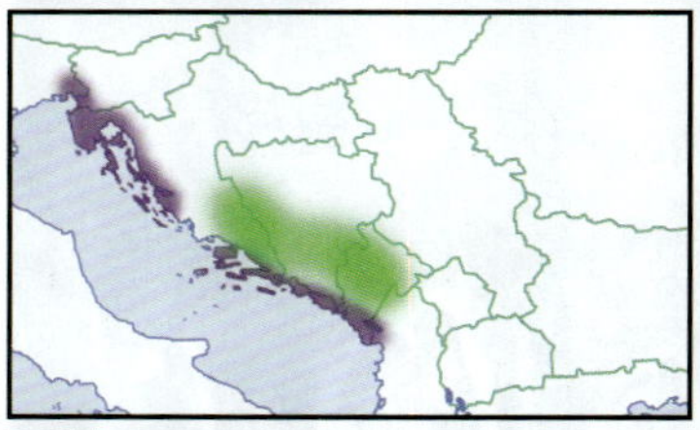

Delima bilabiata
Delima giselae

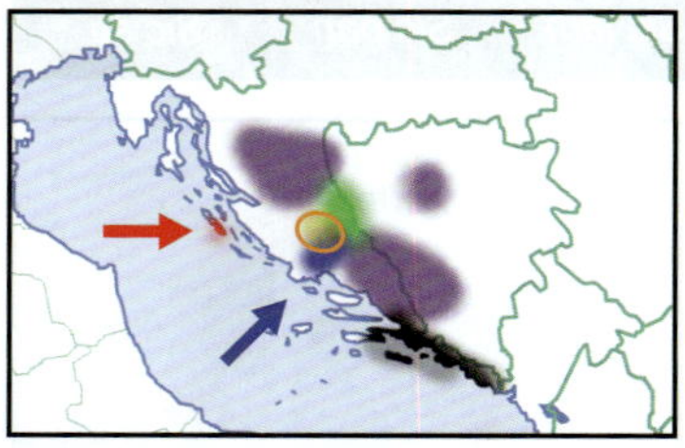

Delima edmibrani
Delima helenae
Delima laevissima
Delima latilabris
Delima pellucida
Delima pfeifferi

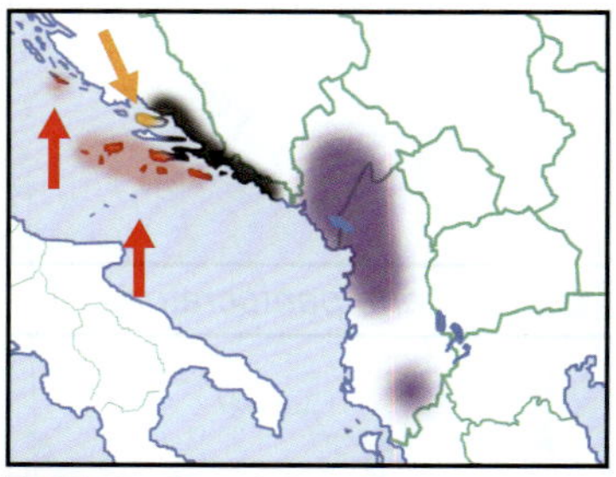

Delima hiltrudis
Delima montenegrina
Delima subcylindrica
Delima substricta (= amoe.)

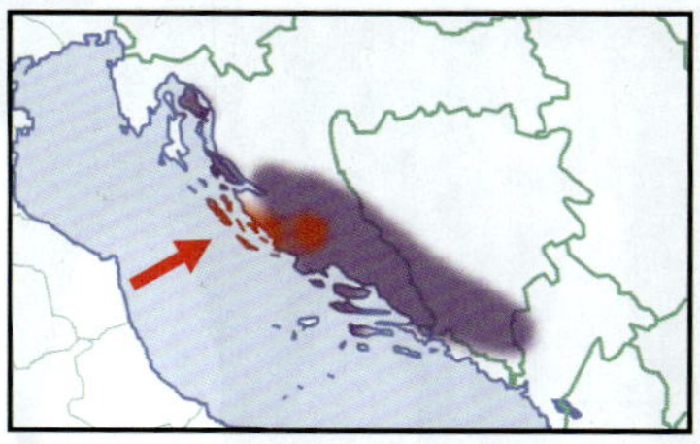

Delima albocincta
Delima semirugata

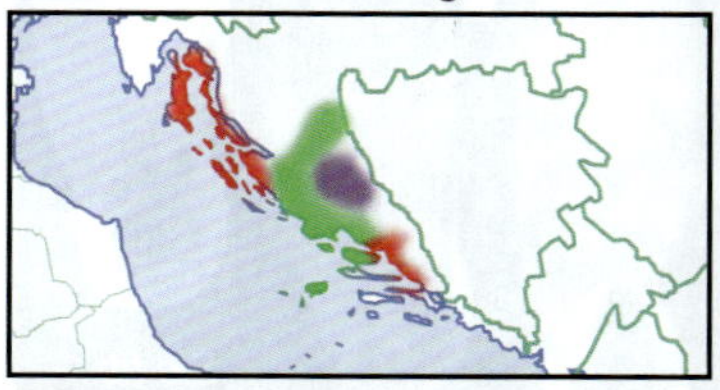

Delima blanda
Del. blanda "conspurcata"
Delima pachystoma

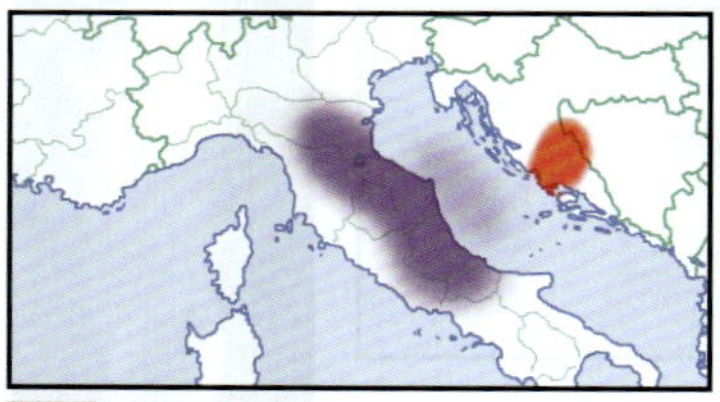

Delima piceata
Delima vidovichii

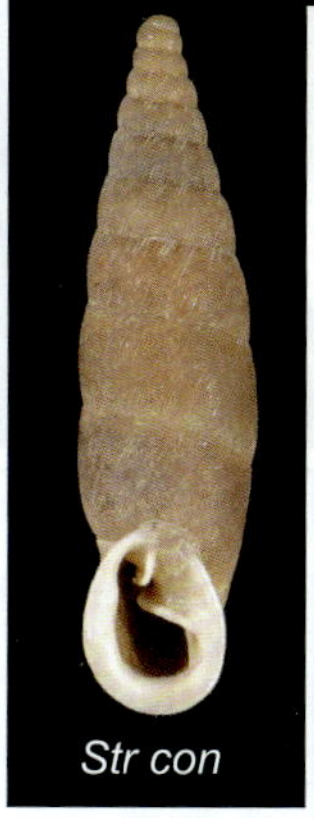

Del hil Del subs
M. Prodan

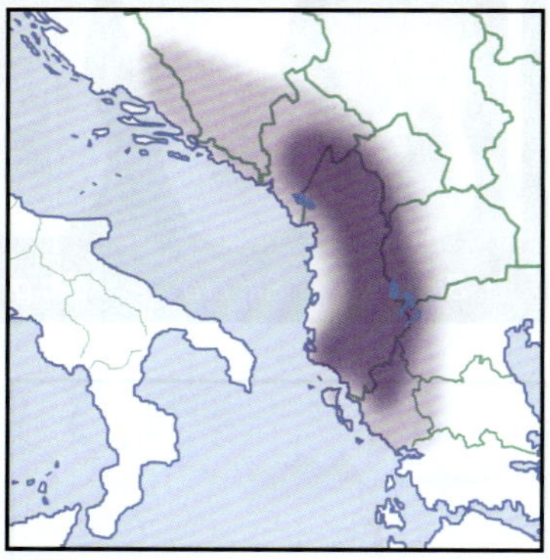

Strigilodelima conspersa

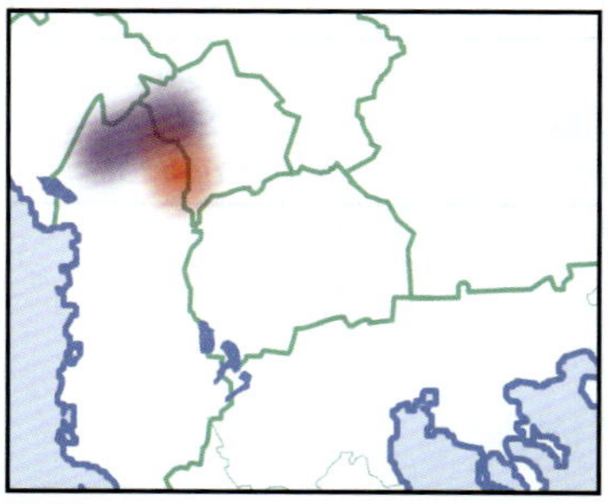

Protoherilla pseudofallax
Strigilodelima pentheri

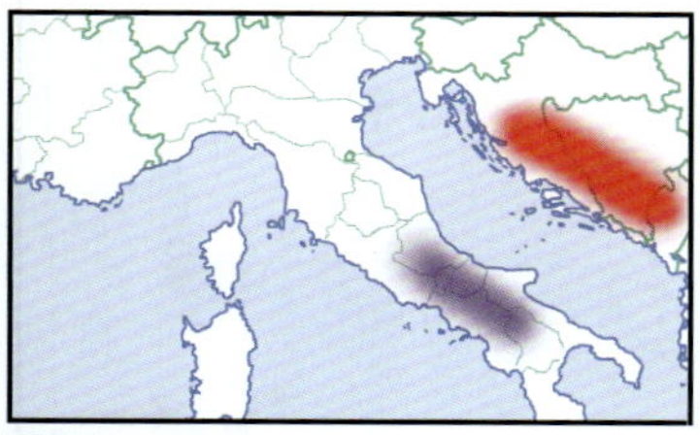

Dilataria boettgeriana
Dilataria marcki

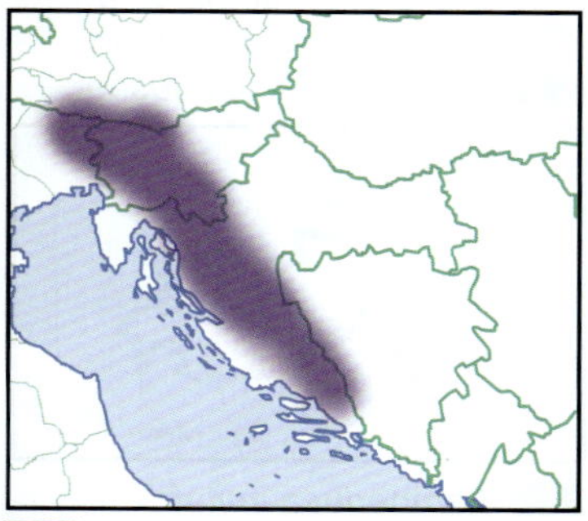

Dilataria succineata

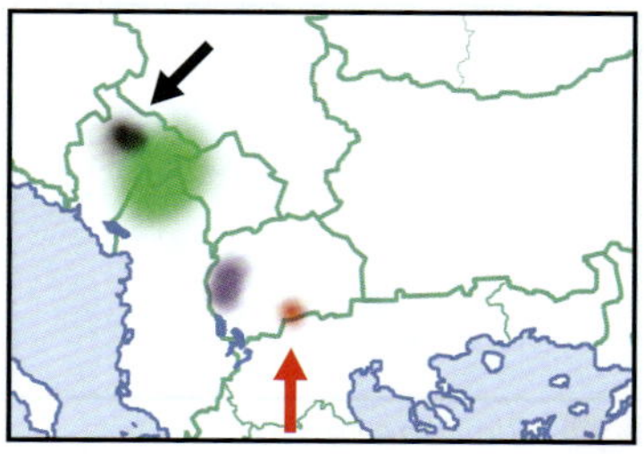

Euxinella radikae
Euxinella subaii
Protoherilla baleiformis
Protoherilla mirabilis

Clausiliidae 11

all-others group: *Montenegrina*, *Macedonica*

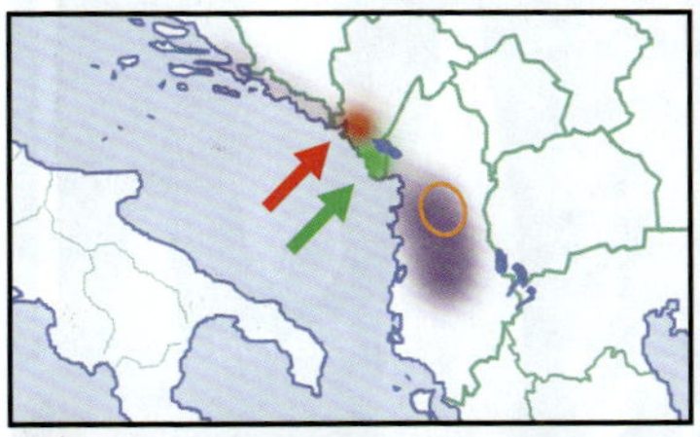

Montenegrina cattaroensis
Montenegrina drimmeri
Montenegrina helvola
Montenegrina umbilicata

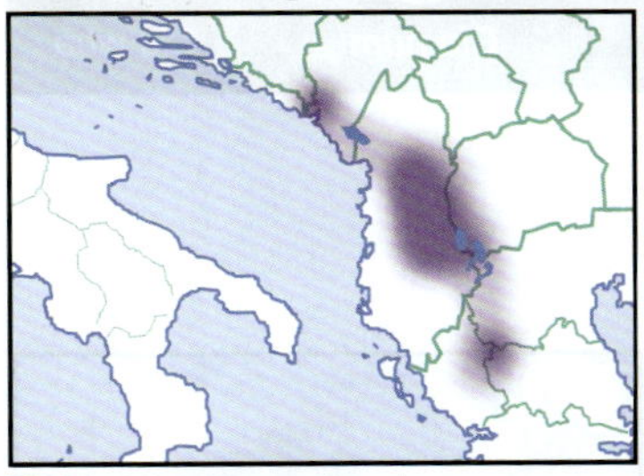

Montenegrina laxa

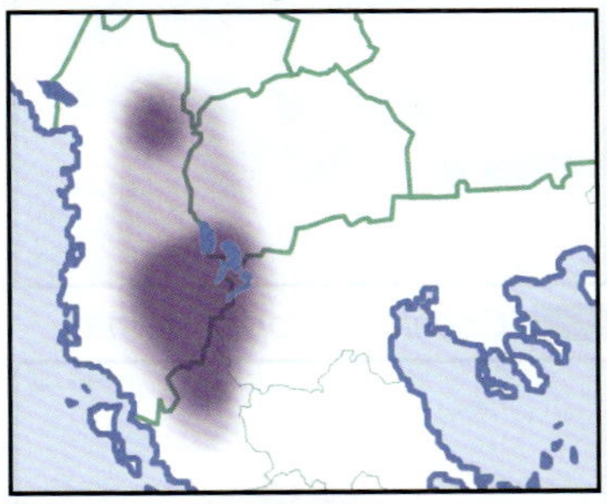

Montenegrina janinensis

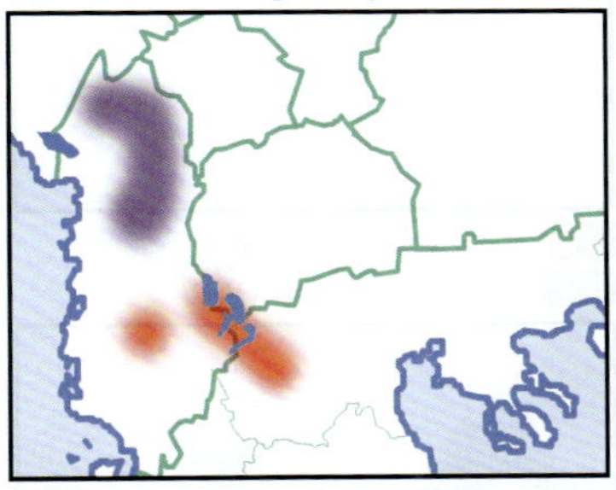

Montenegrina apfelbecki
Montenegrina dofleini

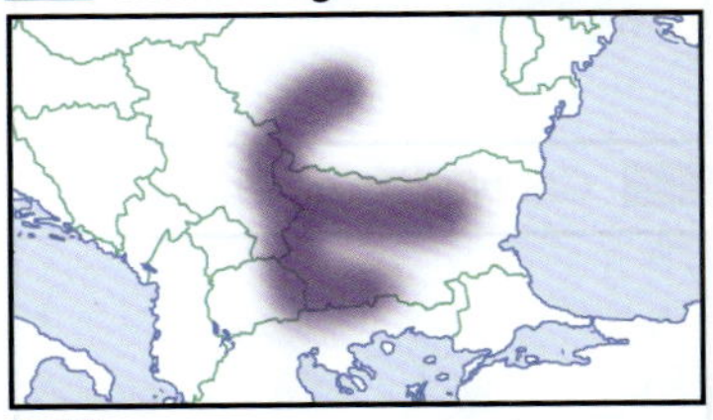

Macedonica marginata

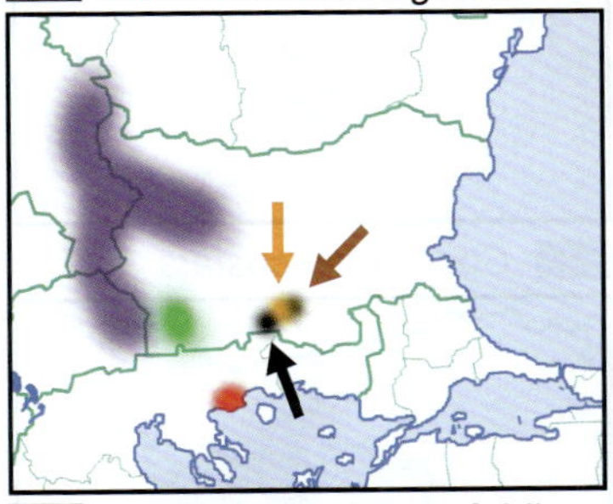

Macedonica frauenfeldi
Macedonica hartmuti
Macedonica martae
Macedonica teodorae
Macedonica ypsilon
Macedonica zilchi

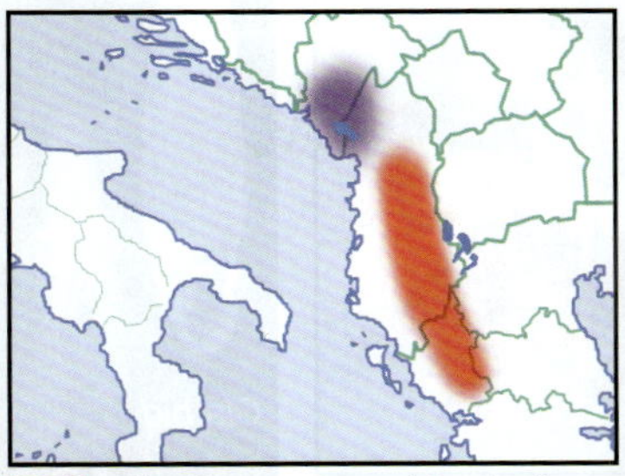

Montenegrina skipetarica
Montenegrina subcristata

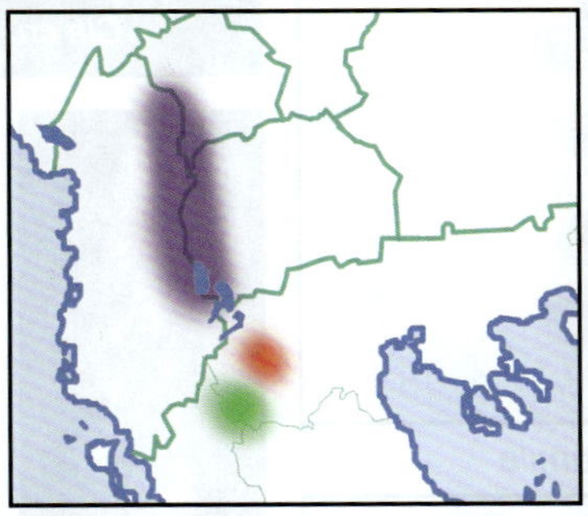

Montenegrina dennisi
Montenegrina hiltrudae
Montenegrina perstriata

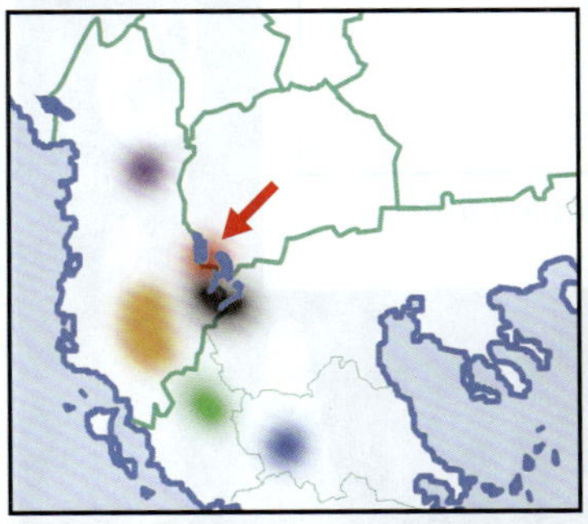

Montenegrina fuchsi
Montenegrina minuscula
Montenegrina rugilabris
Montenegrina sattmanni
Montenegrina stankovici
Montenegrina zilchi

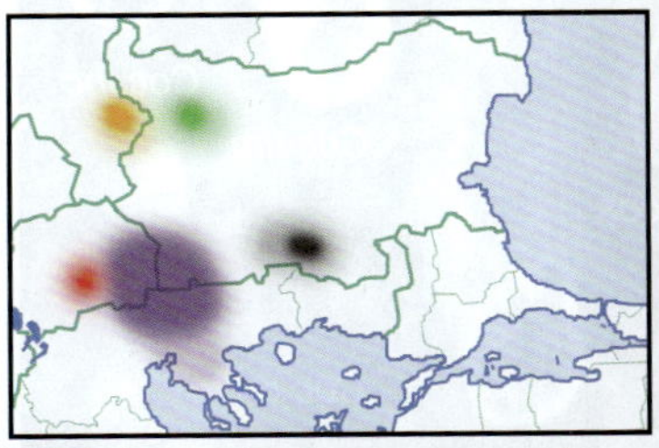

Macedonica brabeneci
Macedonica macedonica
Macedonica slavica
Macedonica pinteri
Macedonica pirotana

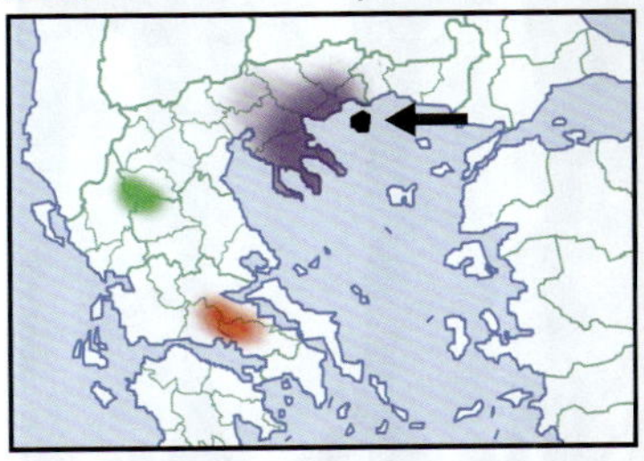

Macedonica guicciardii
Macedonica pindica
Macedonica schatzmayri
Macedonica thasia

Mac fra Mac har Mac mar A. Irikov *Mac teo* J. Stoilov

Clausiliidae 12
all-others group: *Clausilia*, *Cochlodina*

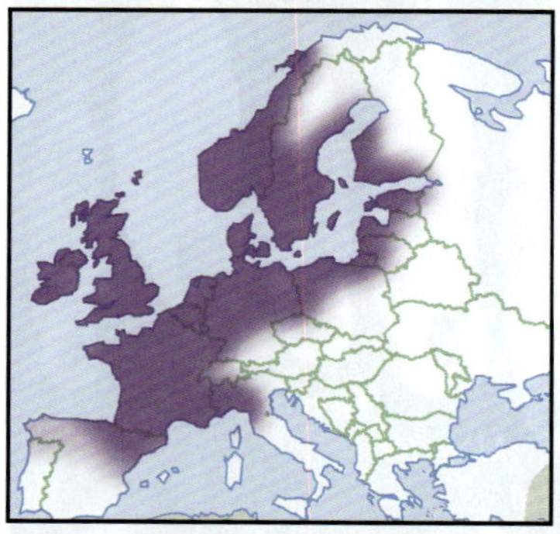

Clausilia bidentata

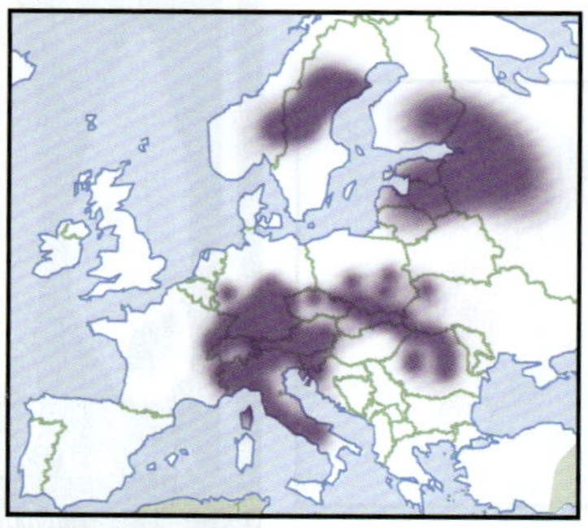

Clausilia cruciata

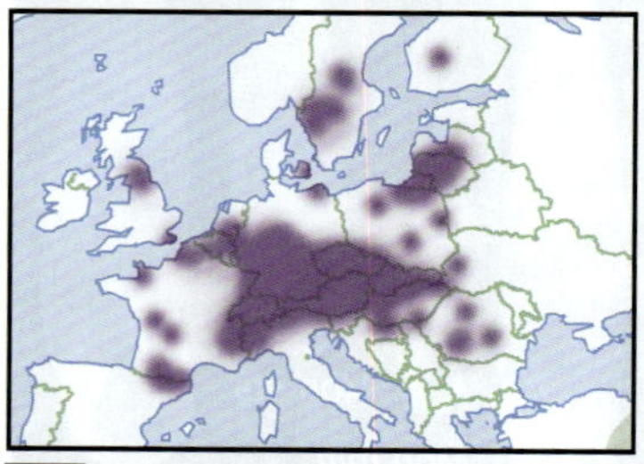

Clausilia dubia

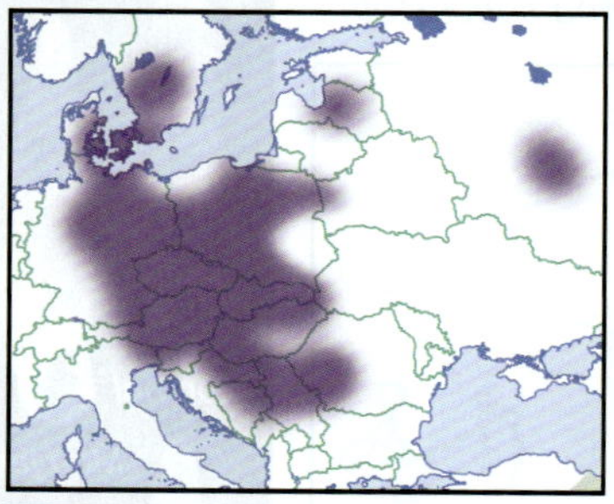

Clausilia pumila

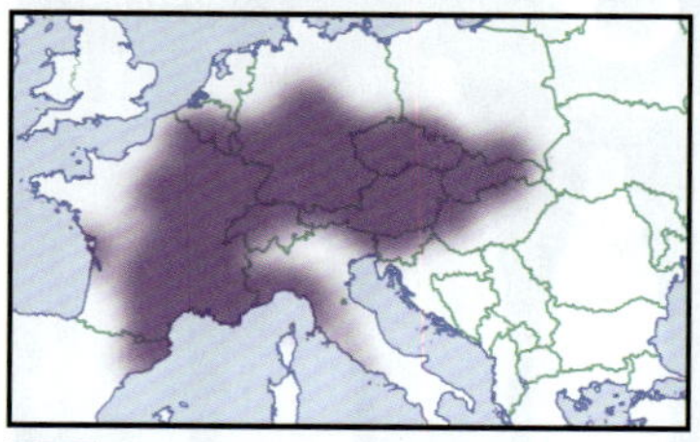

Clausilia rugosa

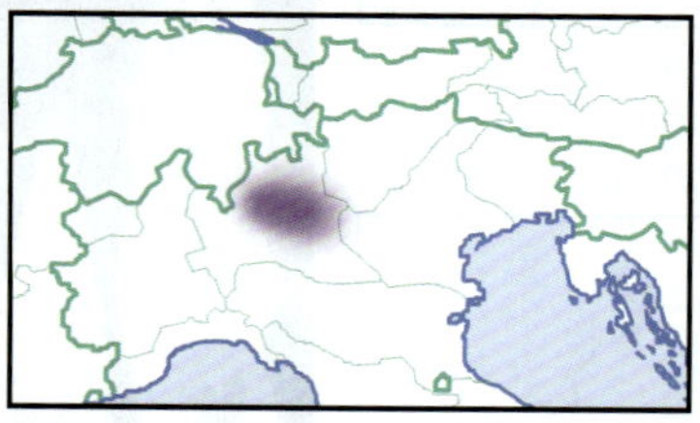

Clausilia whateliana

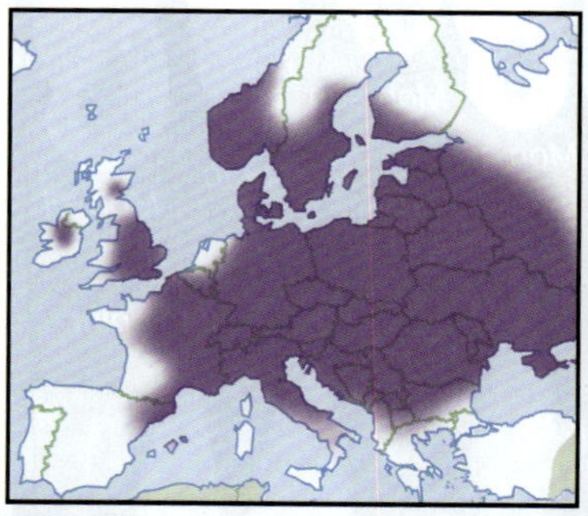

Cochlodina laminata

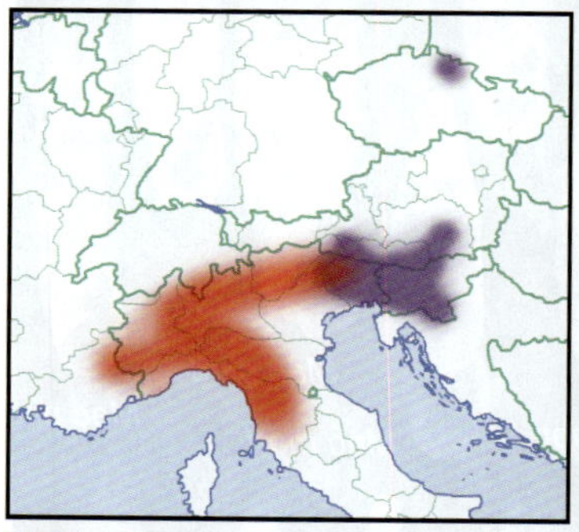

Cochlodina comensis
Cochlodina dubiosa

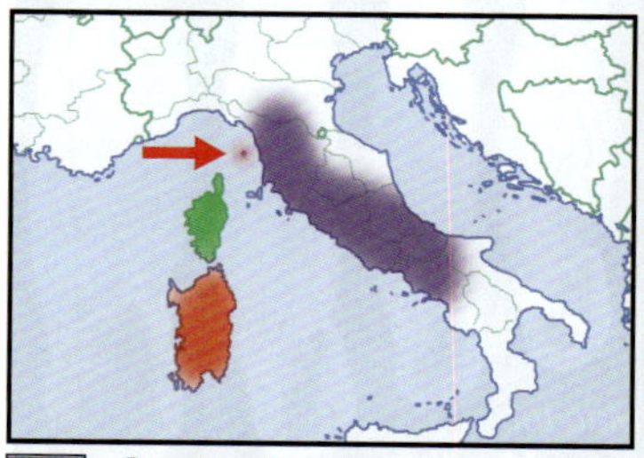

Cochlodina incisa
Cochlodina kuesteri
Cochlodina meisneriana

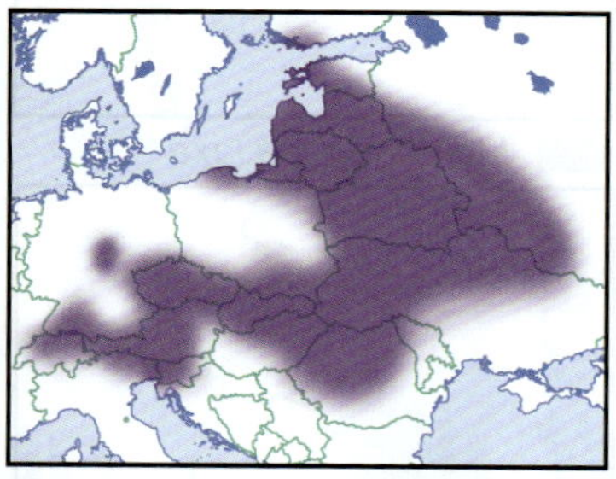

Cochlodina orthostoma

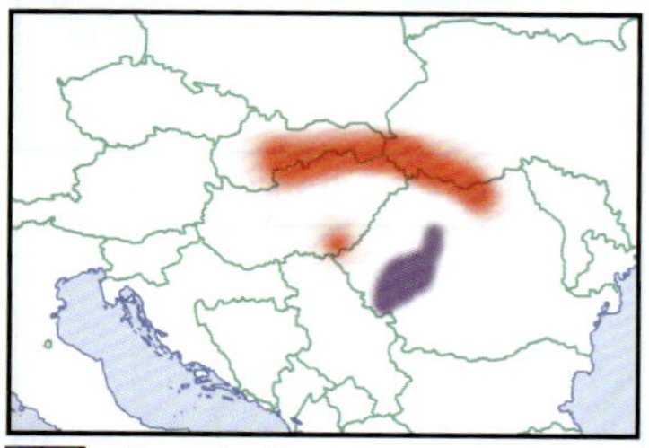

Cochlodina cerata
Cochlodina marisi

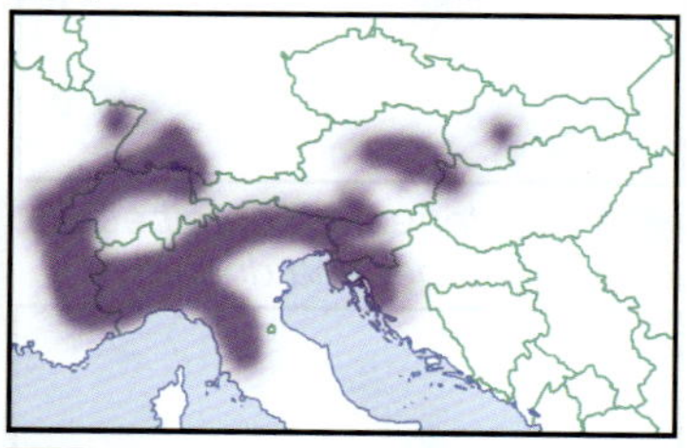

Cochlodina fimbriata

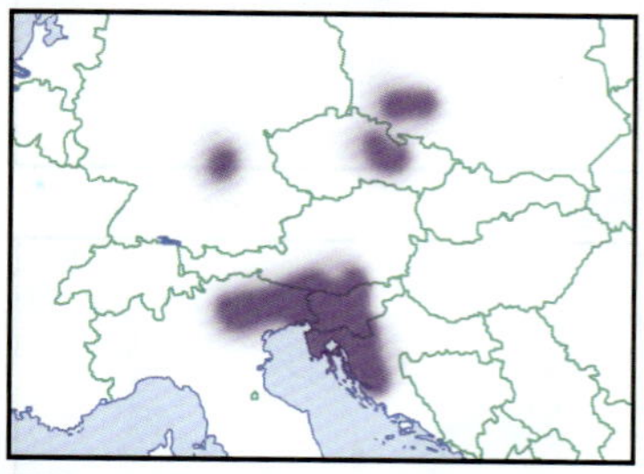

Cochlodina costata

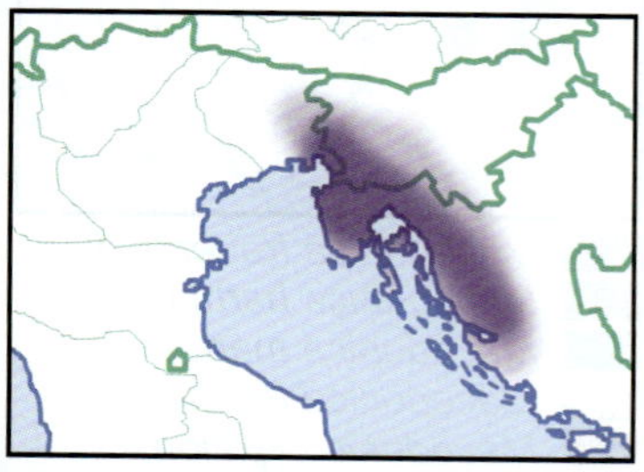

Cochlodina triloba

Clausiliidae 13
all-others group: *Macrogastra*, *Mentissa*, *Micridyla*, *Ruthenica*

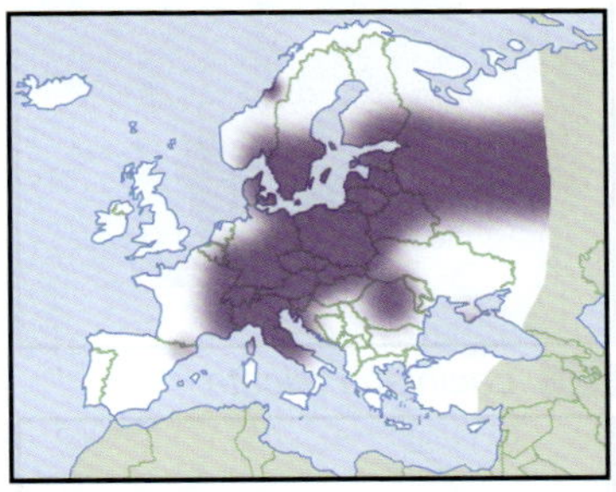

Macrogastra plicatula

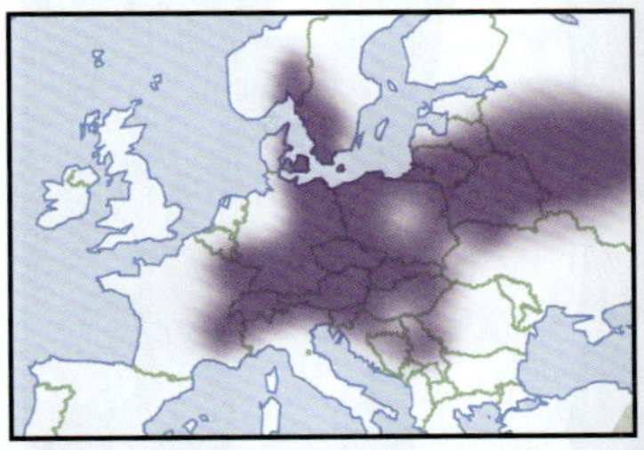

Macrogastra ventricosa

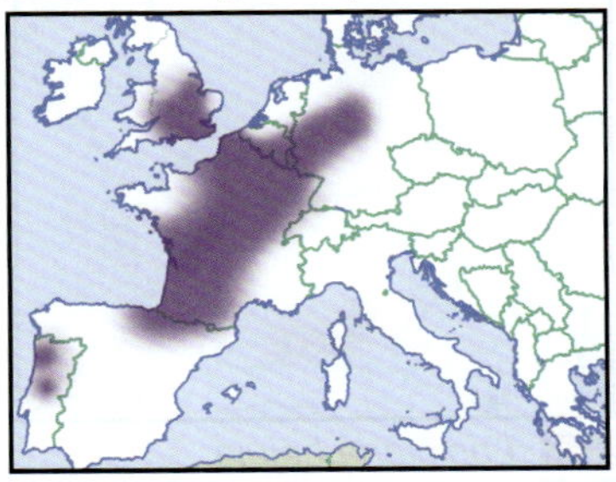

Macrogastra rolphii

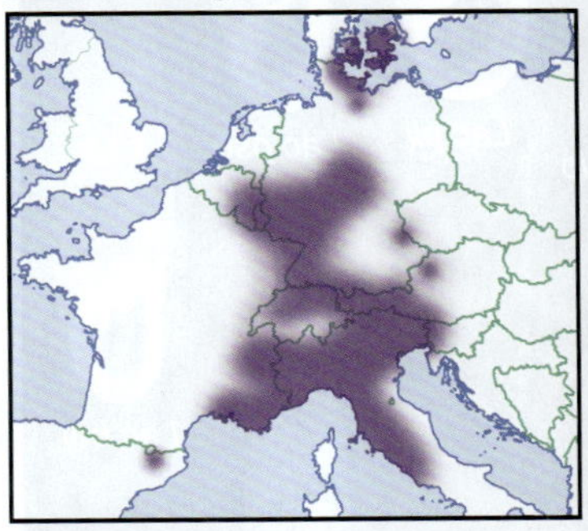

Macrogastra basileensis

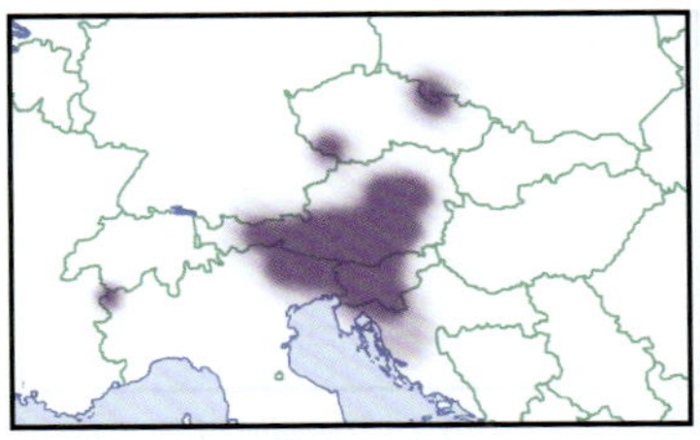

Macrogastra badia

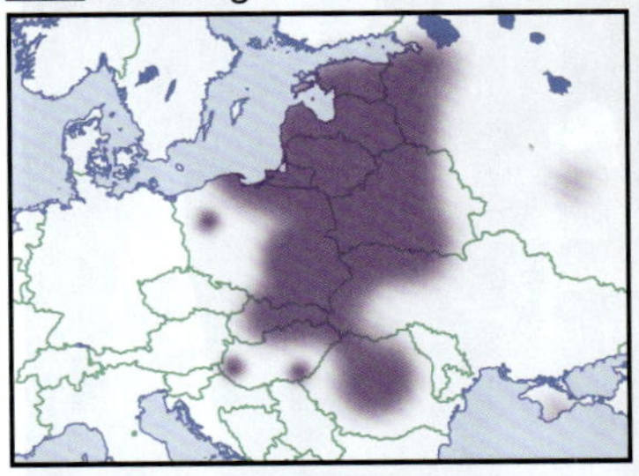

Macrogastra borealis

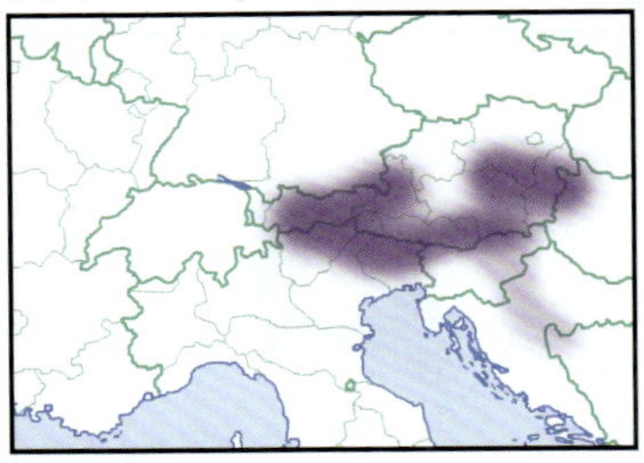

Macrogastra densestriata

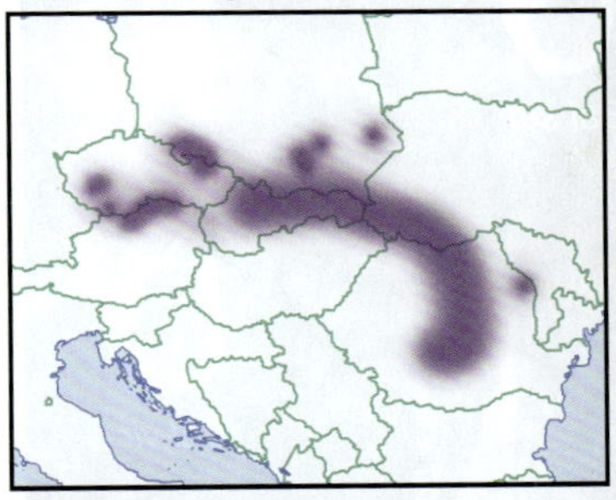

Macrogastra tumida

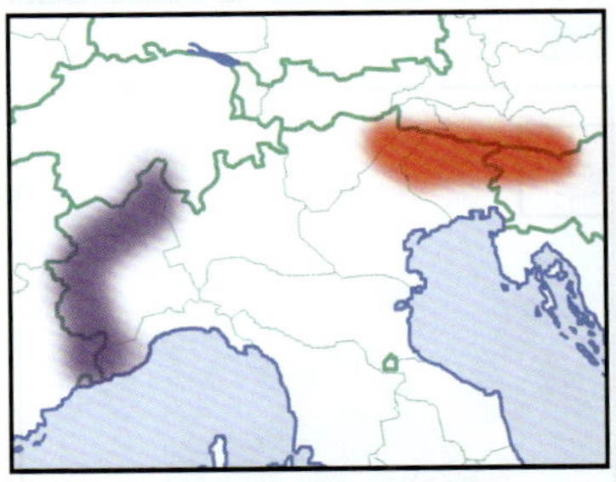

Macrogastra asphaltina
Macrogastra mellae

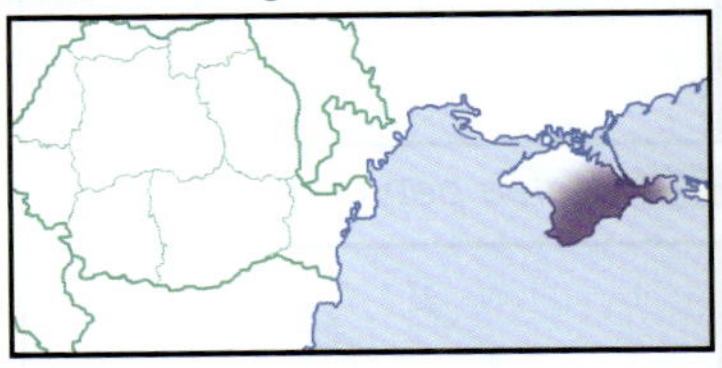

Mentissa canalifera

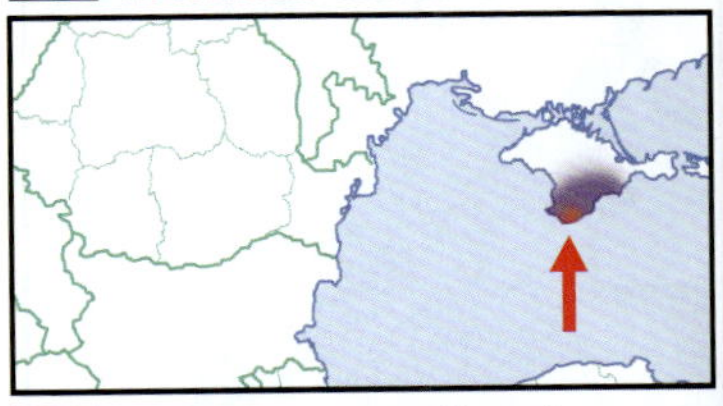

Mentissa gracilicosta
Mentissa velutina

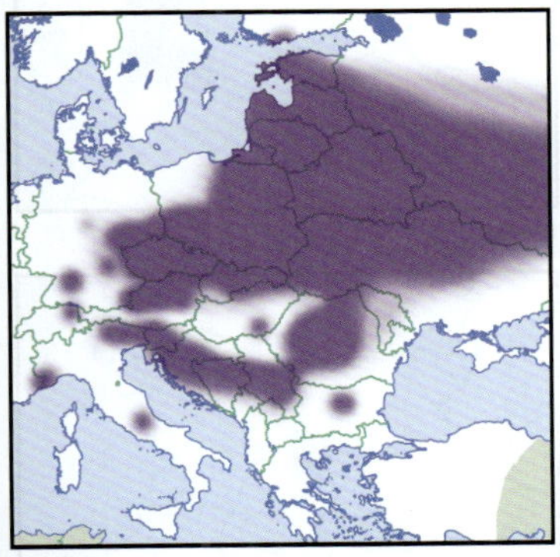

Ruthenica filograna

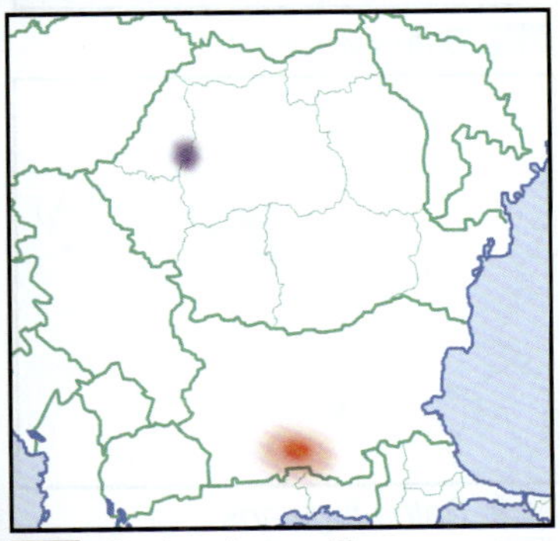

Ruthenica gallinae
Micridyla pinteri

Clausiliidae 14
Balea, Laciniaria

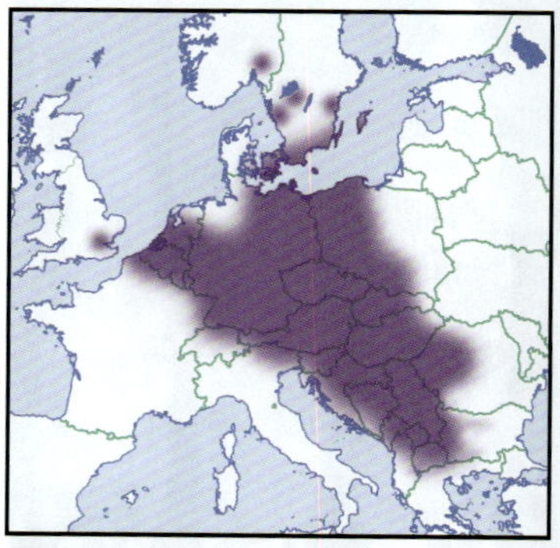

Balea biplicata

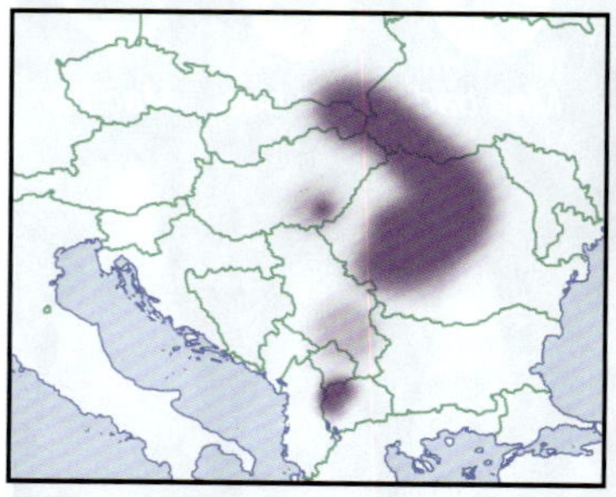

Balea stabilis

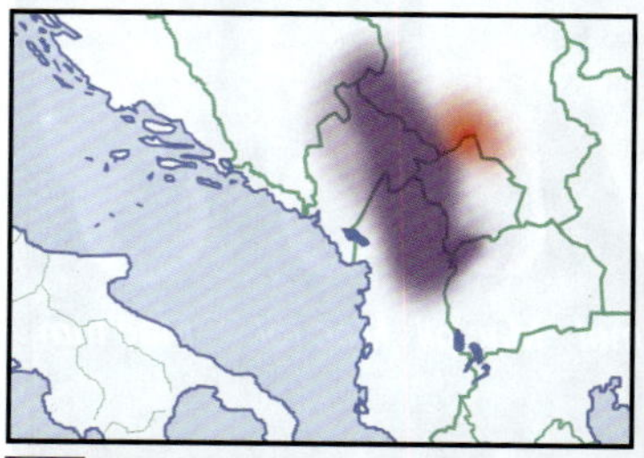

Balea pancici
Balea serbiana

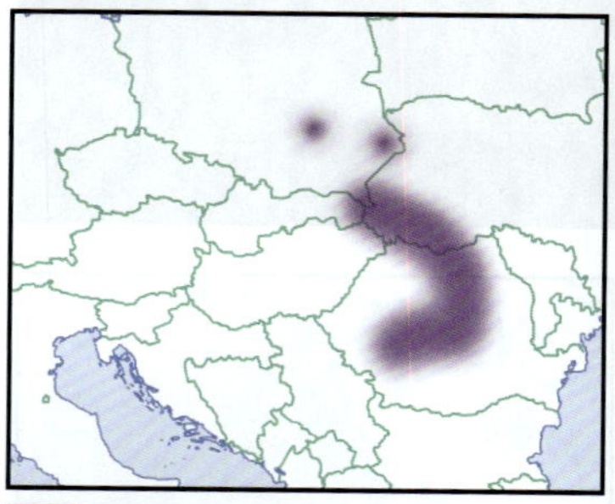

Balea fallax

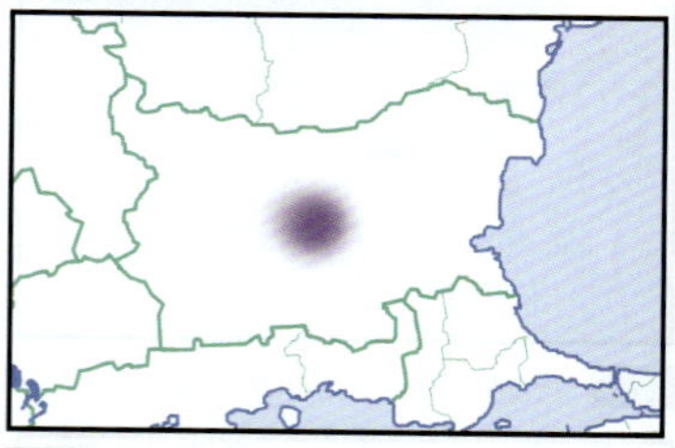

Balea wagneri

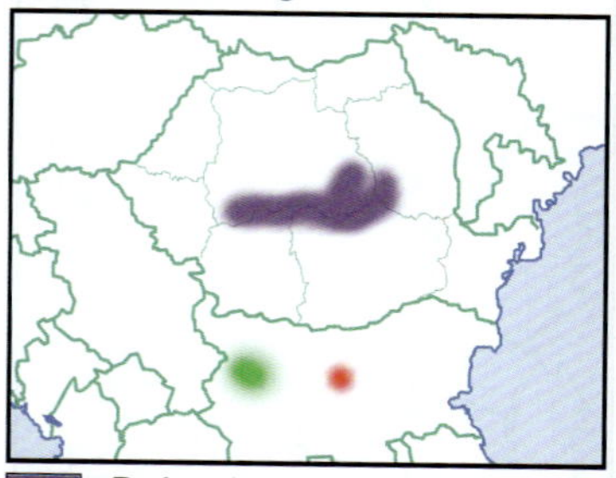

Balea jugularis
Balea kaeufeli
Balea vratzatica

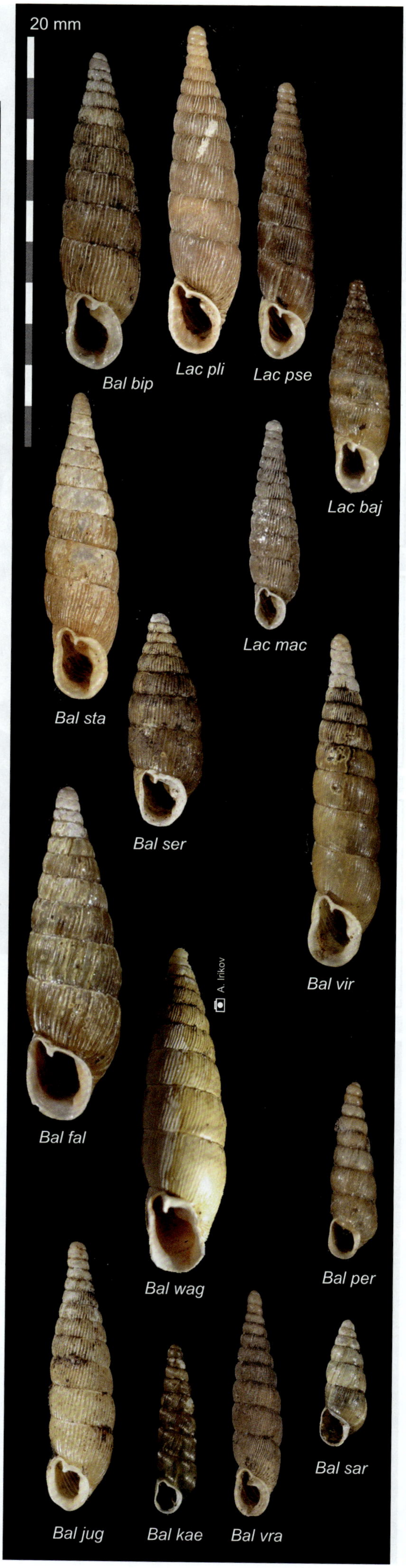

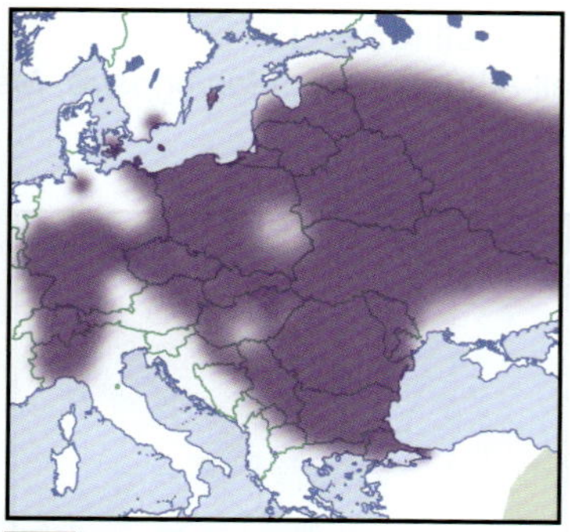

Laciniaria plicata

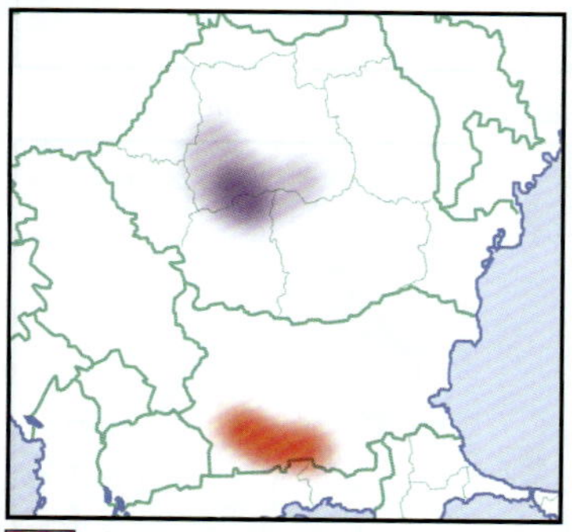

Laciniaria bajula
Laciniaria pseudostabilis

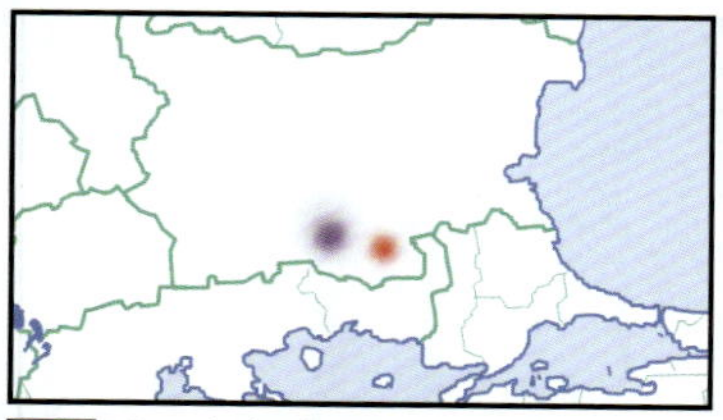

Laciniaria macilenta
Laciniaria potochensis

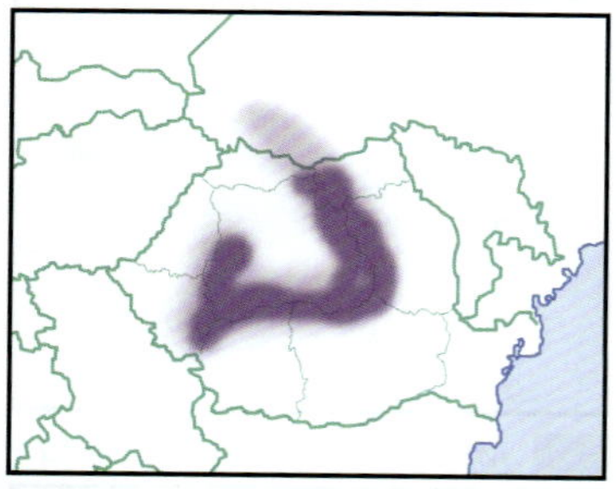

Balea viridana

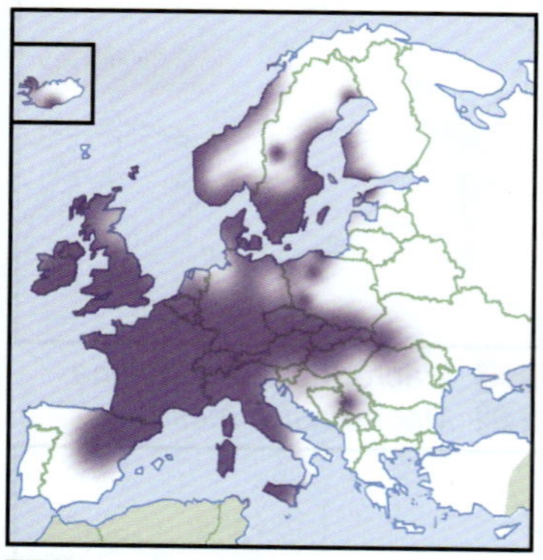

Balea perversa

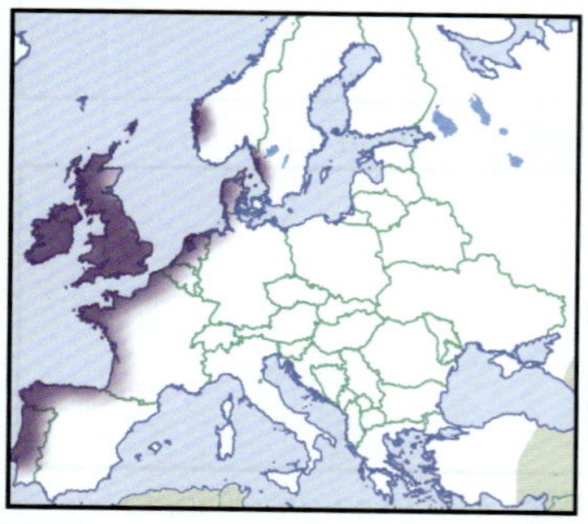

Balea sarsii

Clausiliidae 15
Bulgarica, *Mentissella*, *Vestia*

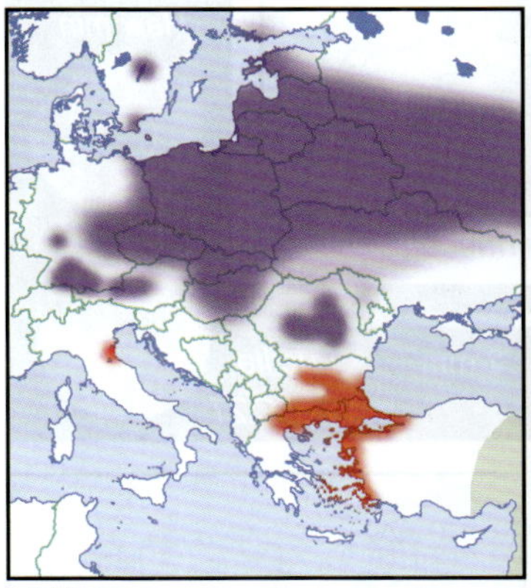

Bulgarica cana
Bulgarica denticulata

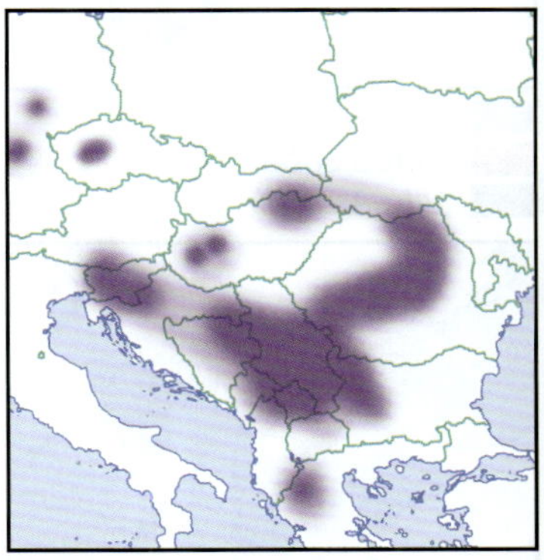

Bulgarica vetusta

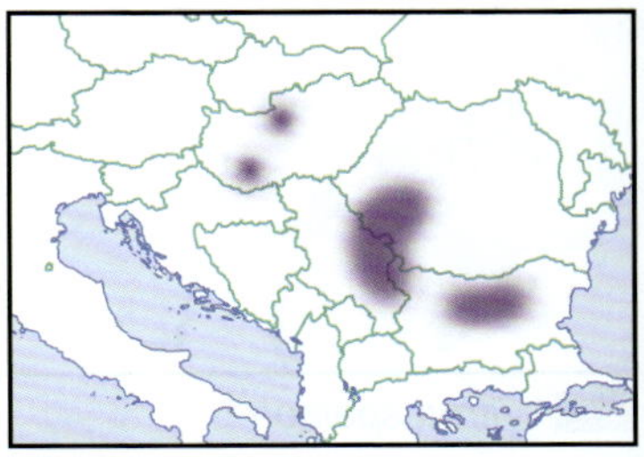

Bulgarica rugicollis

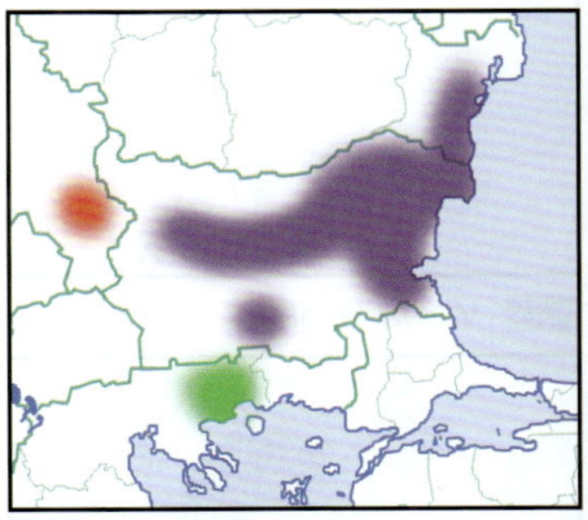

Bulgarica fritillaria
Bulgarica iniucunda
Bulgarica pygmaea

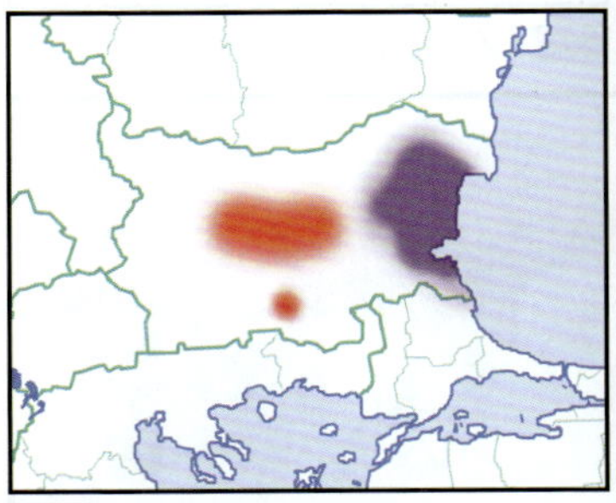

Mentissella rebeli
Bulgarica bulgarica

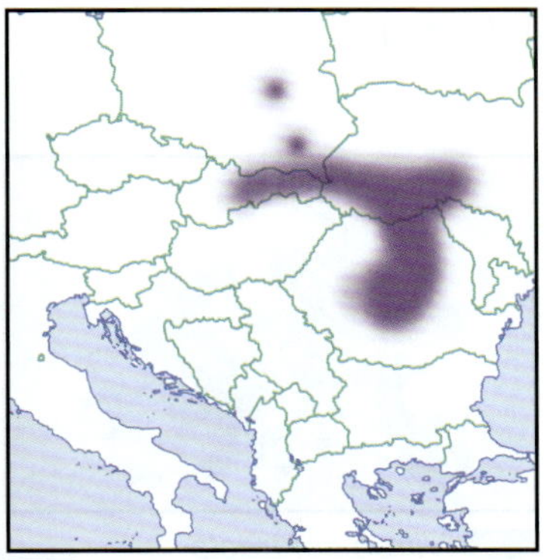

Vestia elata

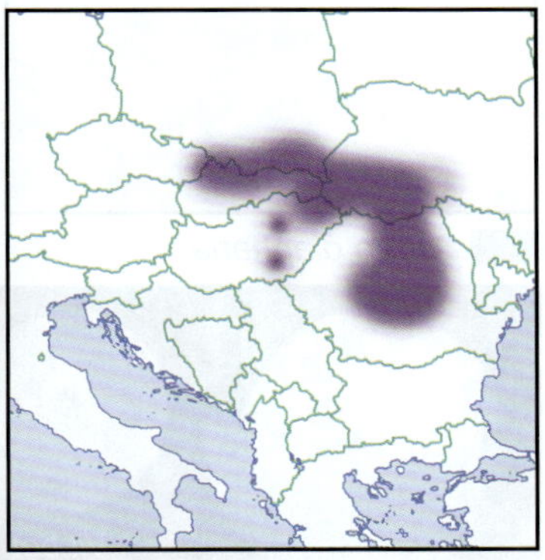

Vestia gulo

Vestia ranojevici

Vestia roschitzi

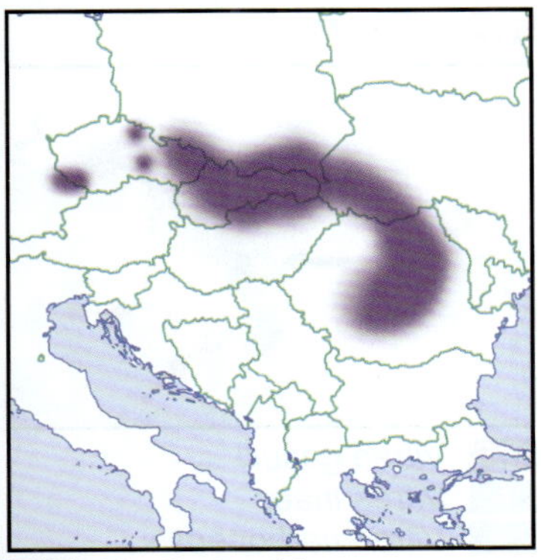

Vestia turgida

Vitrea 1

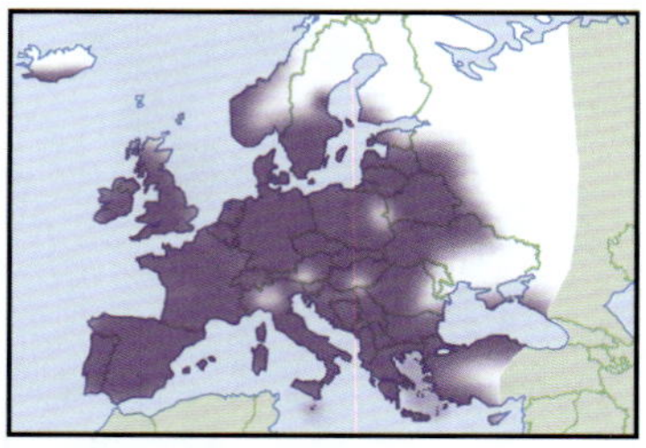

Vitrea contracta

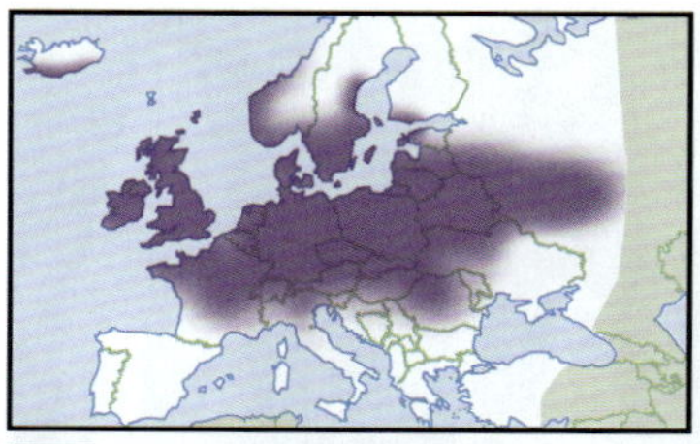

Vitrea crystallina

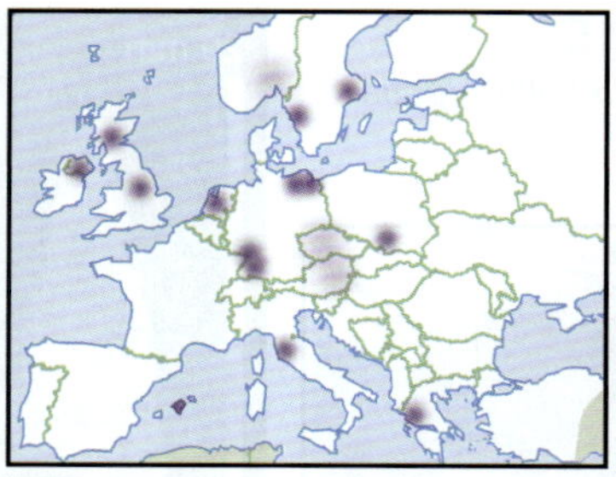

Hawaiia *minuscula*

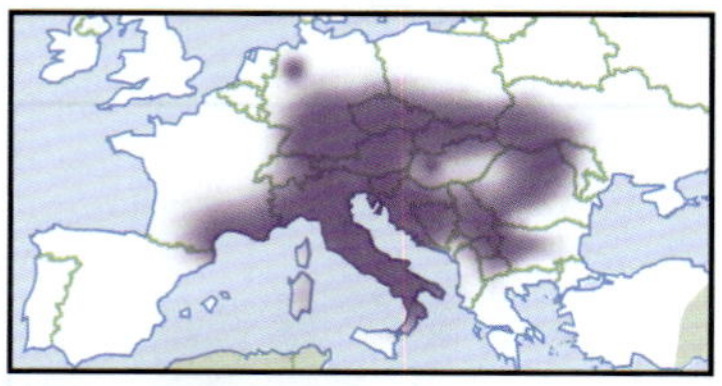

Vitrea diaphana

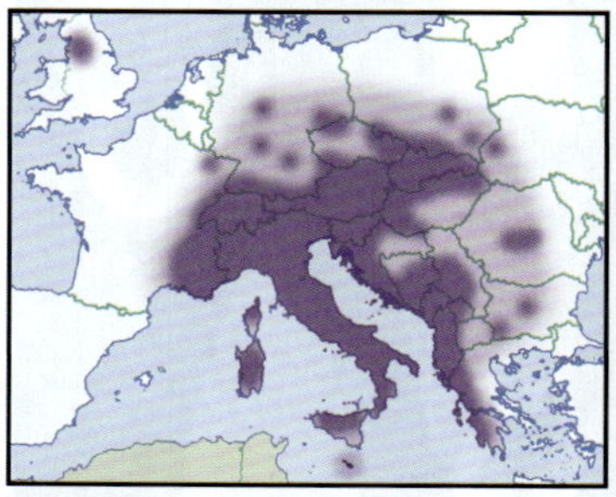

Vitrea subrimata

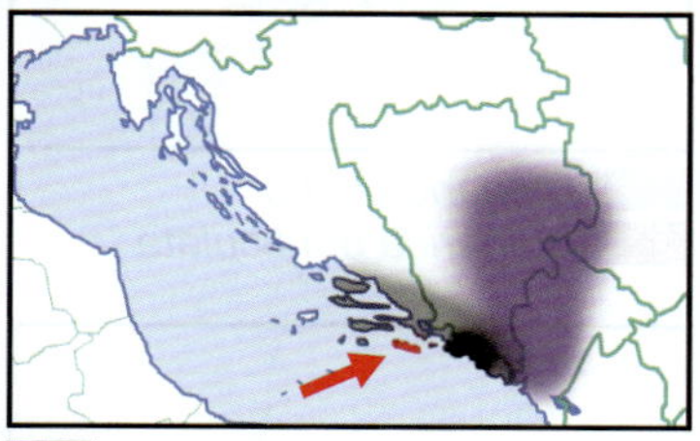

Vitrea sturanyi
Vitrea subaii
Vitrea zilchi

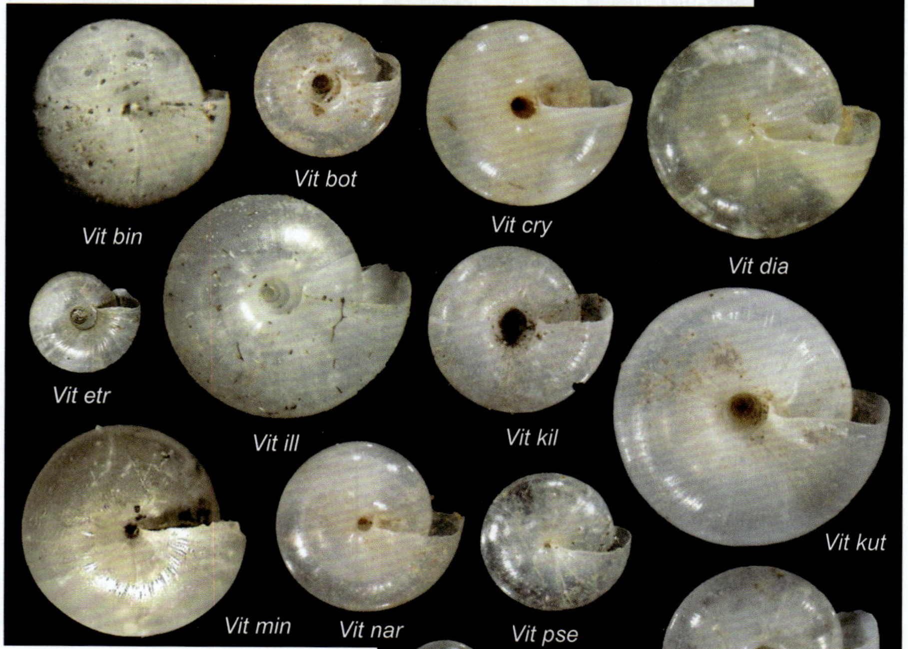

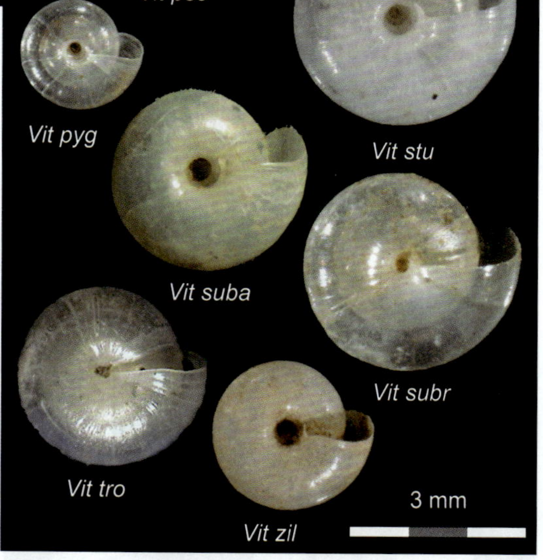

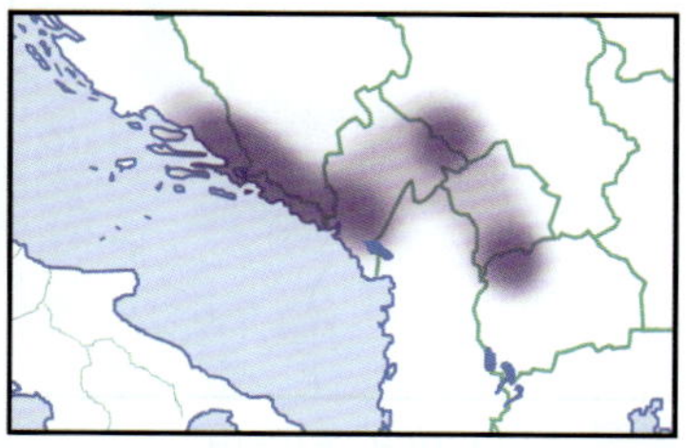

Vitrea kutschigi

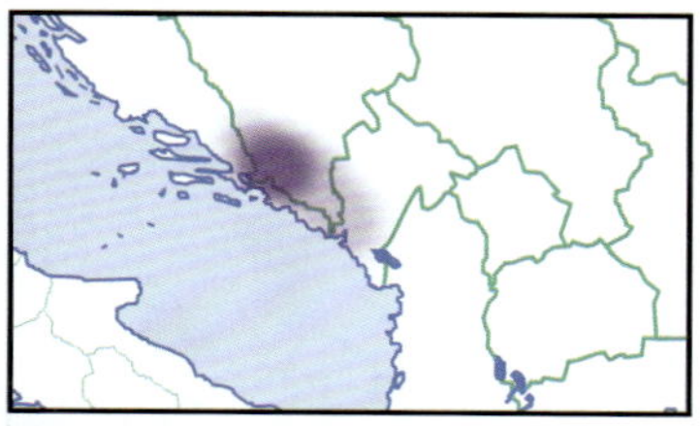

Vitrea spelaea

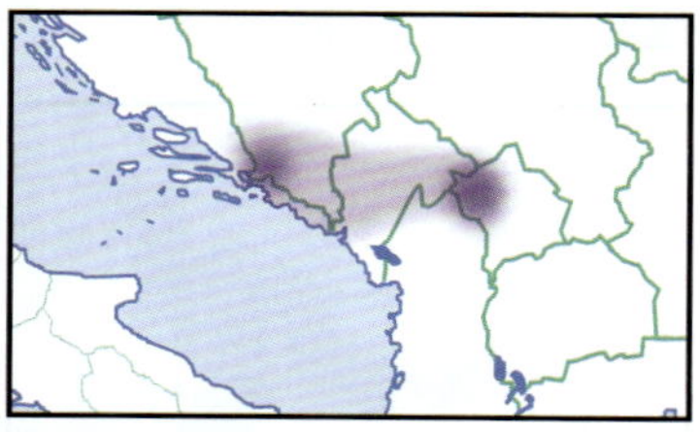

Vitrea kiliasi

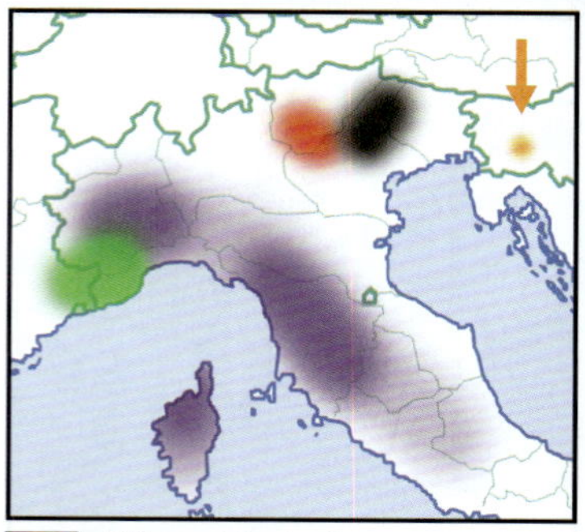

Vitrea binderi
Vitrea etrusca
Vitrea minellii
Vitrea pseudotrolli
Vitrea trolli

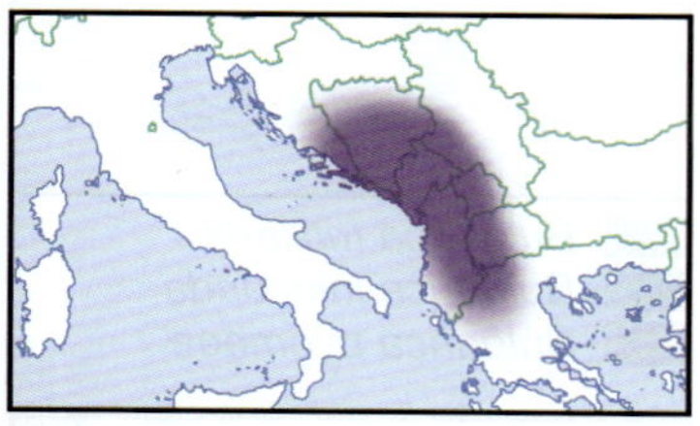

Vitrea illyrica

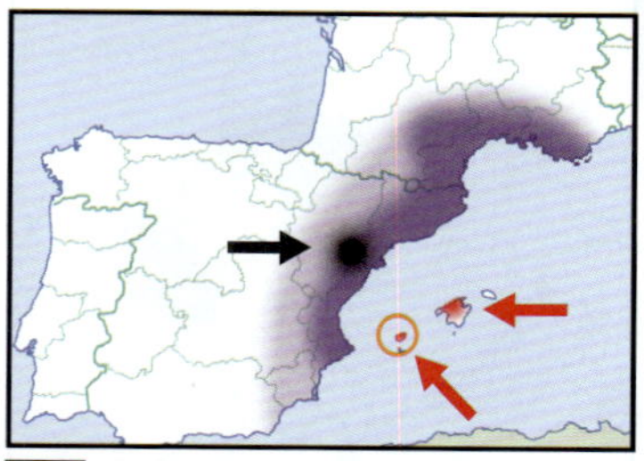

Vitrea gasulli
Vitrea inae
Vitrea narbonensis
Vitrea striata

Vit etr A. Vannozzi *Vit min* *Vit tro* M. Sosso
Vit bin *Vit ill* *Vit suba* M. Prodan

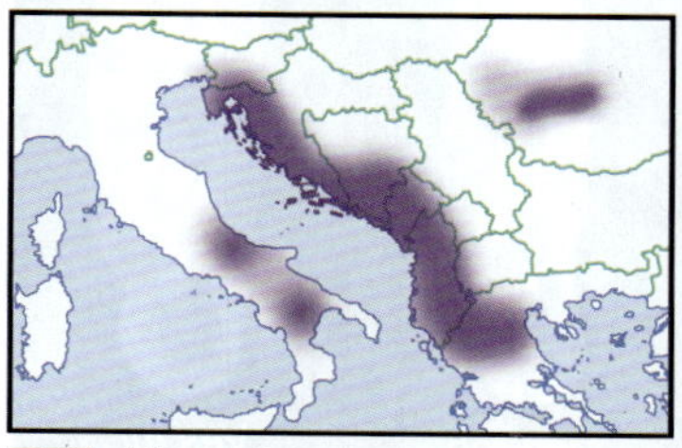

Vitrea botterii

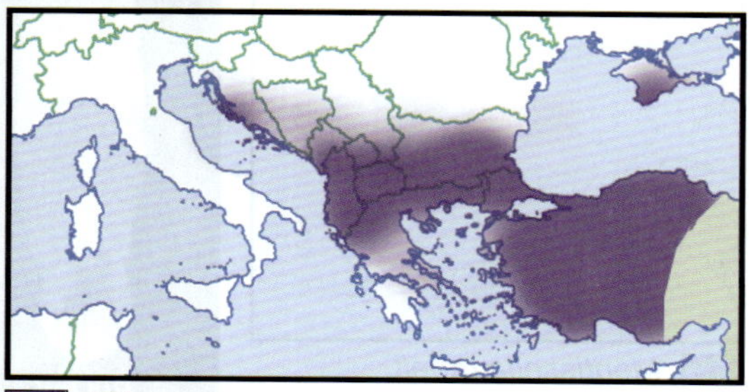

Vitrea pygmaea

Vitrea 2

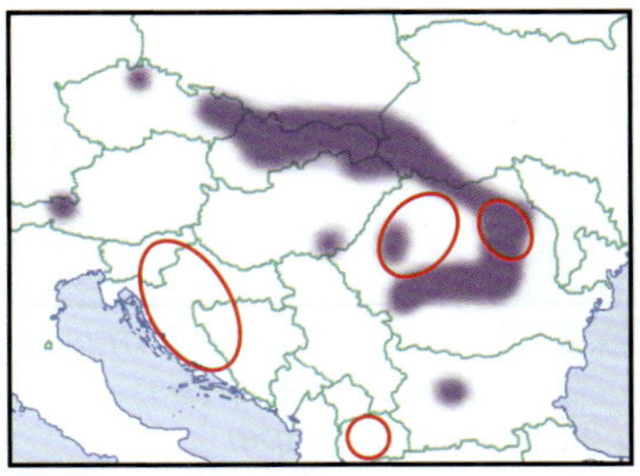

Vitrea erjaveci
Vitrea transsylvanica

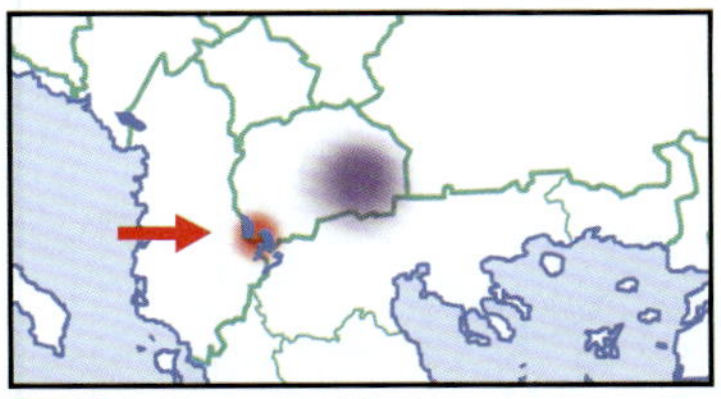

Vitrea meijeri
Vitrea mikuskai

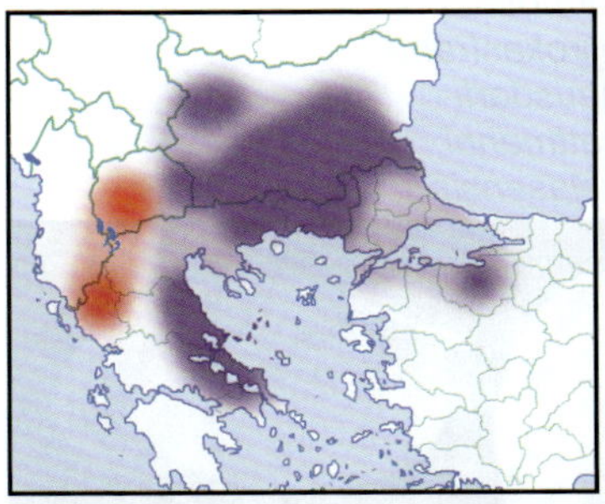

Vitrea neglecta
Vitrea siveci

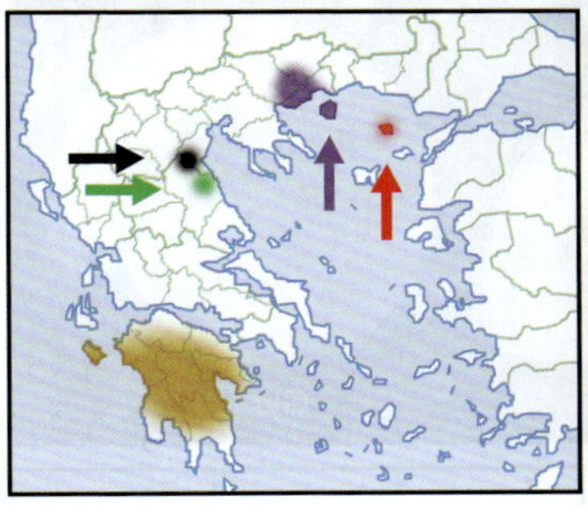

Vitrea argolica
Vitrea olympica
Vitrea ossaea
Vitrea schneideri
Vitrea thasia

Vitrea demiobasensis
Vitrea keaana
Vitrea klemmi
Vitrea matsakisi
Vitrea pageti
Vitrea pieperi
Vitrea pinteri
Vitrea sporadica

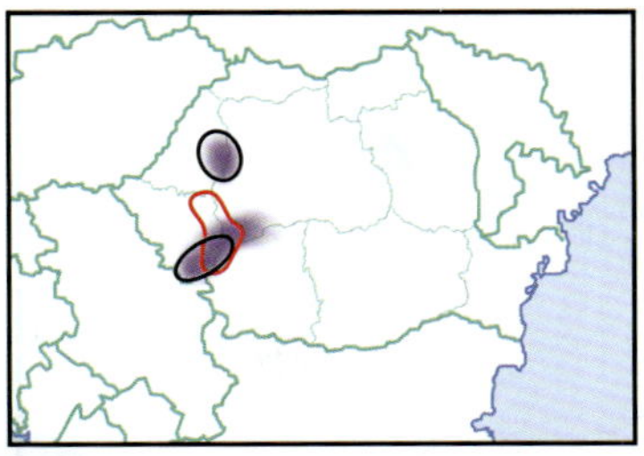

Vitrea jetschini
Vitrea subcarinata
Vitrea szekeresi

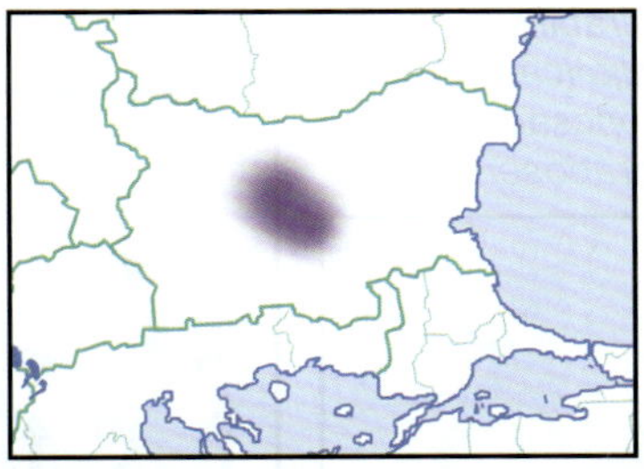

Vitrea vereae

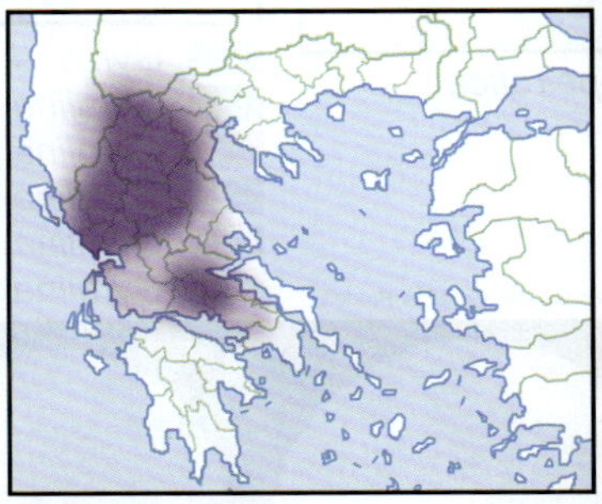

Vitrea selecta

Vitrea ephesina

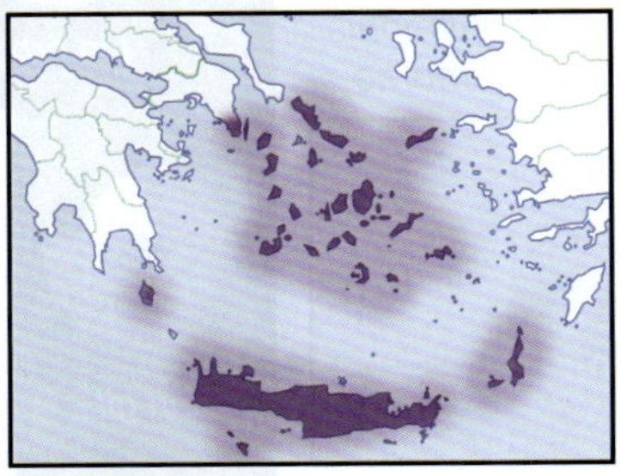

Vitrea clessini

Vitrea nadejdae

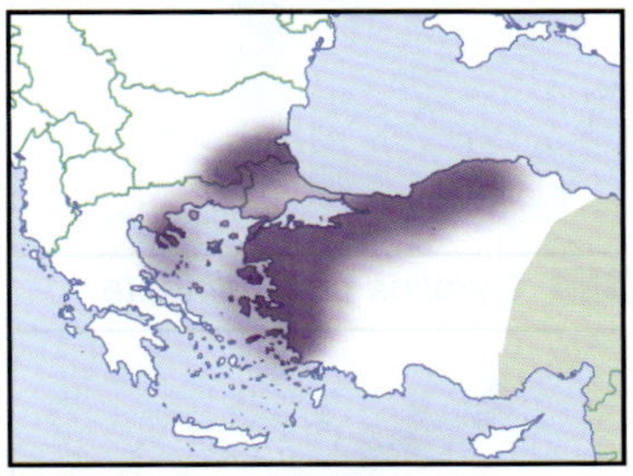

Vitrea riedeli (rie-i)

Vitrea ilgazdaglariensis

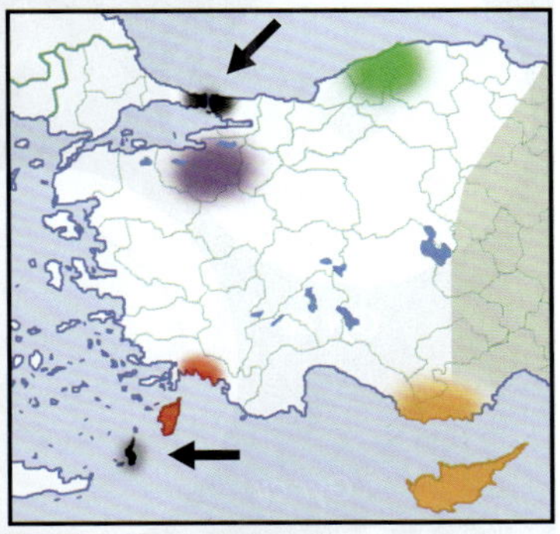

Vitrea angystropha
Vitrea cyprina
Vitrea lodosi
Vitrea riedeliana (rie-a)
Vitrea sorella

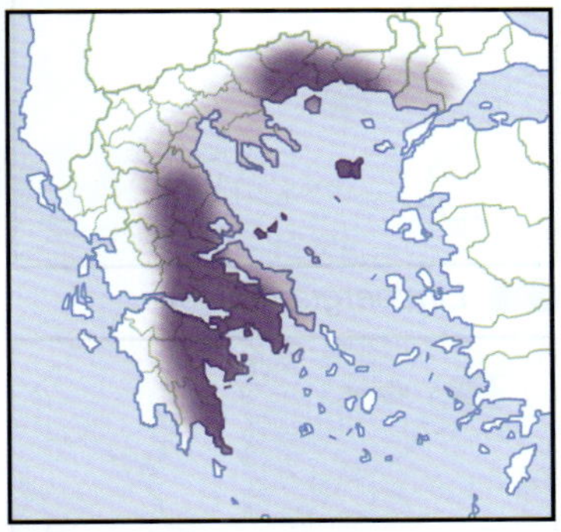

Vitrea schuetti

Vitrea sossellai
Vitrea storchi

Gyralina, *Lindbergia* etc.

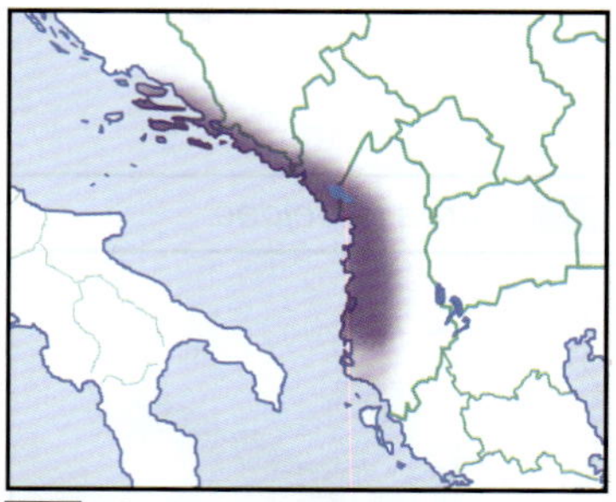

Gyralina circumlineata

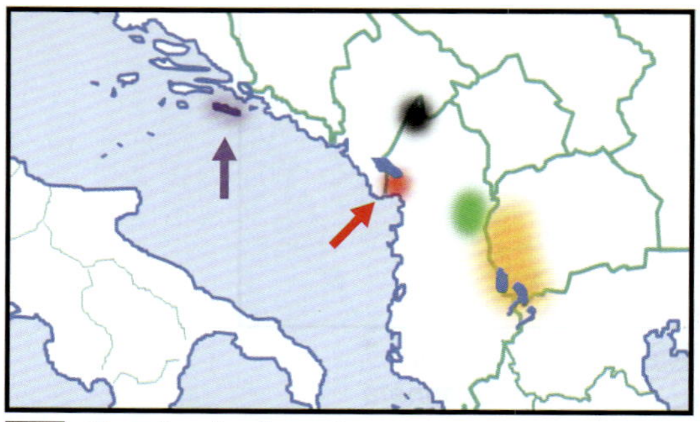

Gyralina lunikensis
Gyralina mirabilis
Gyralina mljetica
Gyralina nopcsai
Gyralina tarabosensis

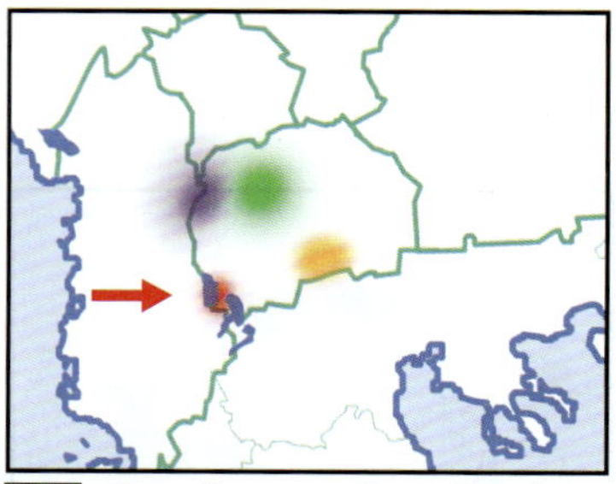

Gyralina gyralinaeformis
Gyralina korabensis
Gyralina rempei
Gyralina velkovrhi

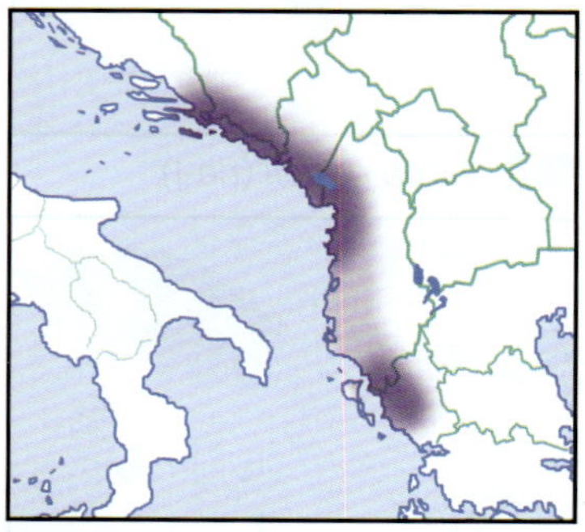

Gyralina candida

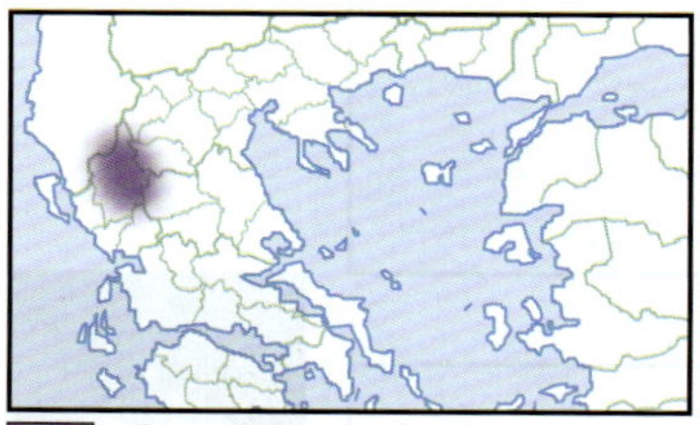

Gyralina epeirotica

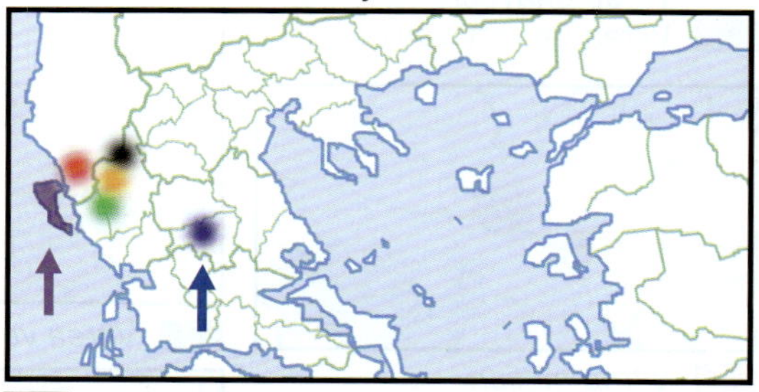

Gyralina ermonae
Gyralina formosa
Gyralina gjirokastrana
Gyralina hausdorfi
Gyralina sattmanni
Gyralina velasensis

Gyr lun Gyr nop Z. Fehér *Gyr mlj Lin git* M. Prodan

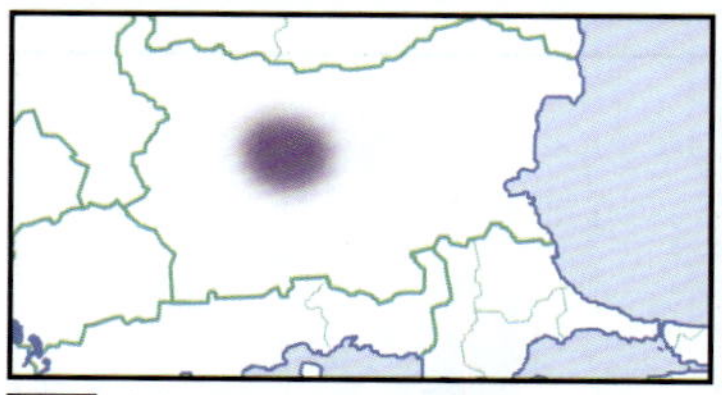

Lindbergia uminskii

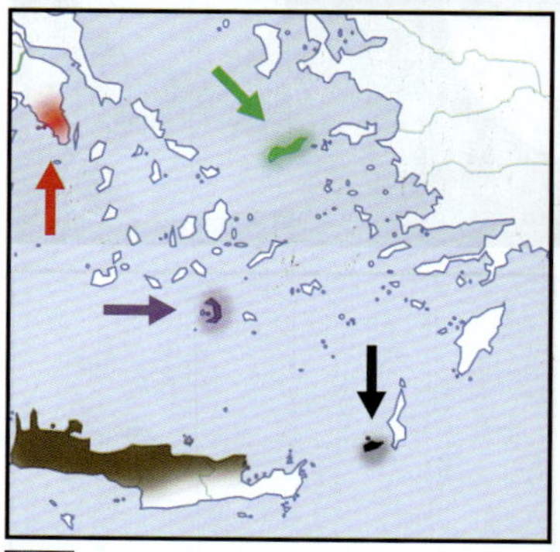

Lindbergia beroni
Lindbergia orbicularis
Lindbergia pinteri
Lindbergia spiliaenymphis
Lindbergia stylokamarae

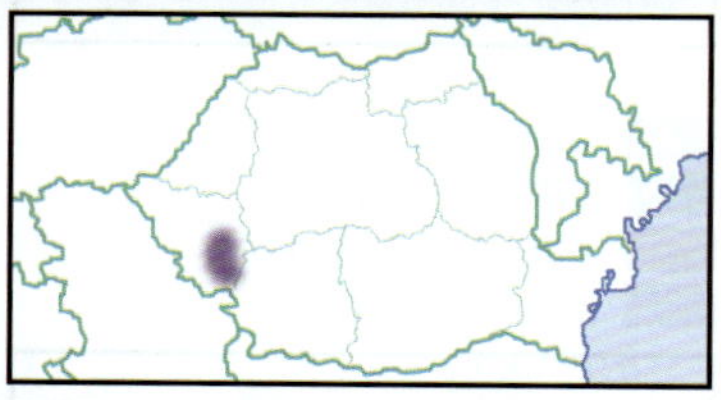

Troglovitrea argintarui

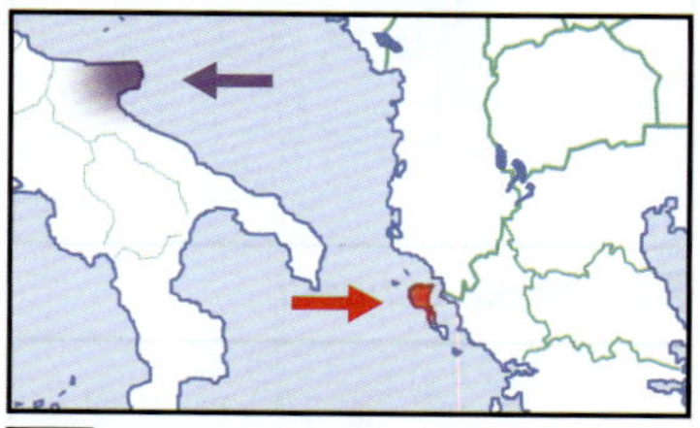

Lindbergia garganoensis
Lindbergia gittenbergeri

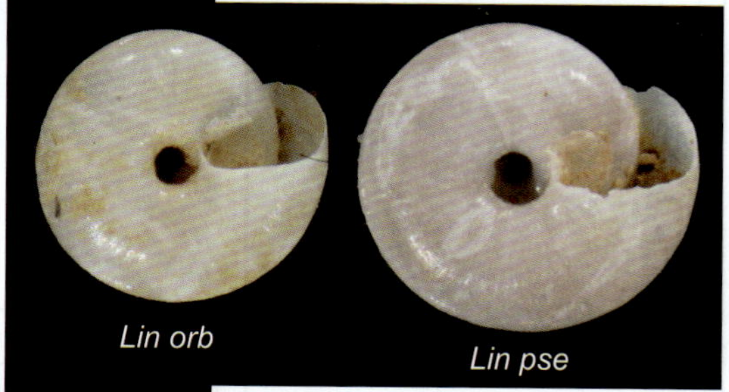

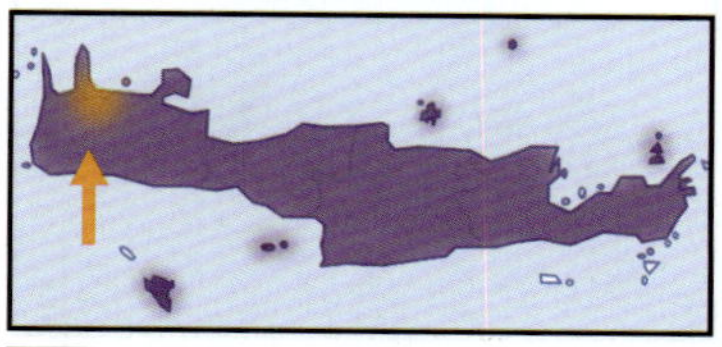

Lindbergia pageti
Lindbergia pseudoillyrica

Lindbergia karainensis

Gollumia filocincta

Oxychilus 1

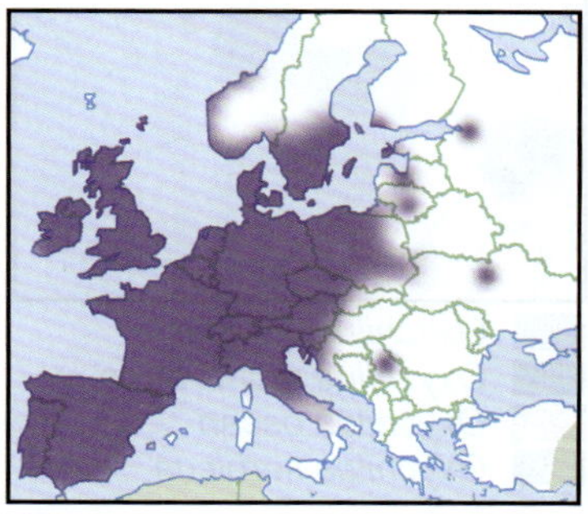

Oxychilus cellarius

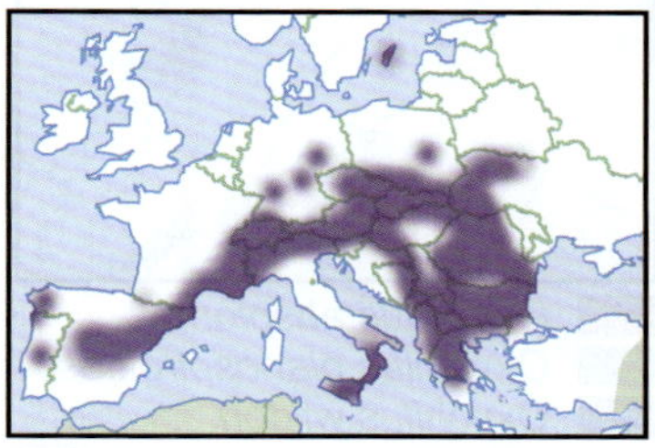

Oxychilus glaber

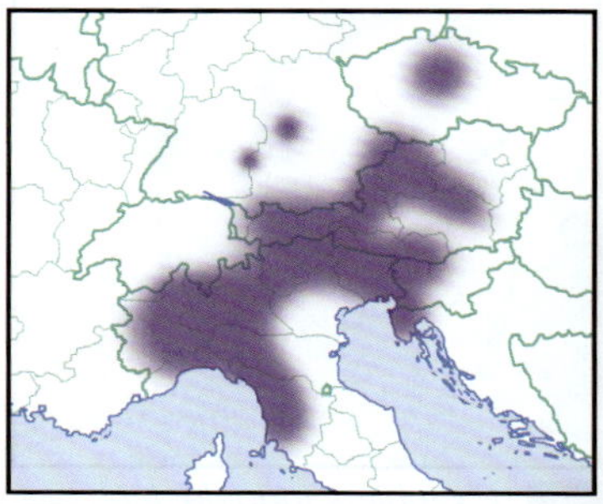

Oxychilus mortilleti

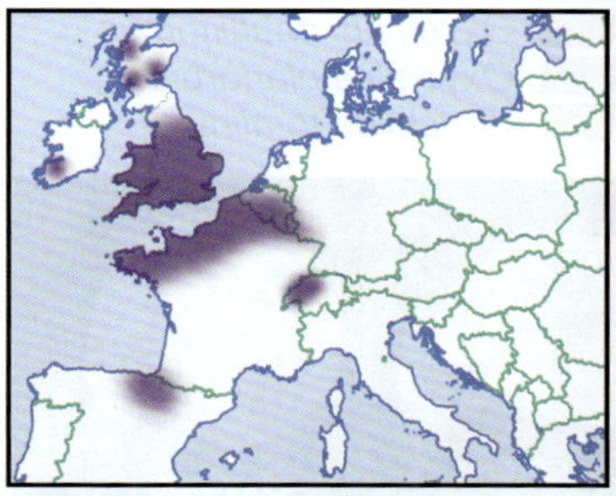

Oxychilus navarricus

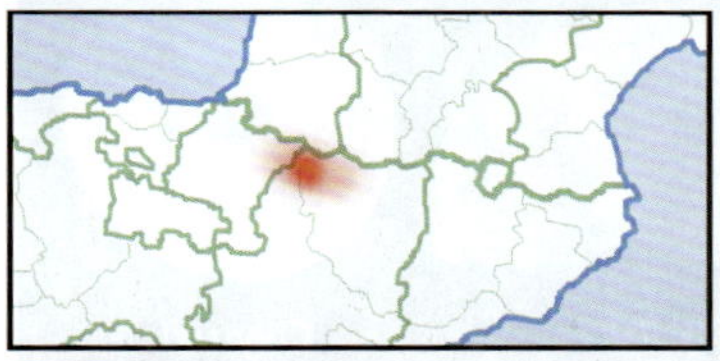

Oxychilus altimirai

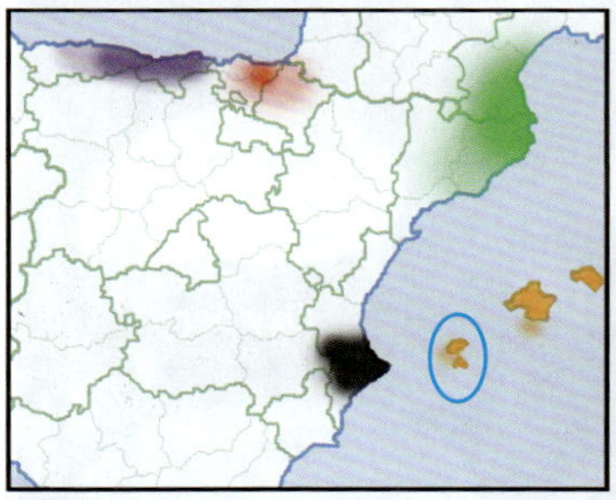

Oxychilus anjana
Oxychilus basajauna
Oxychilus farinesianus
Oxychilus mercadali
Oxychilus lentiformis
Oxychilus pityusanus

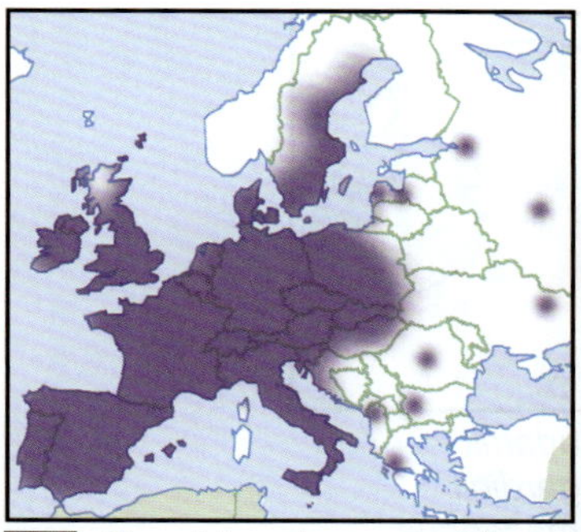

Oxychilus draparnaudi

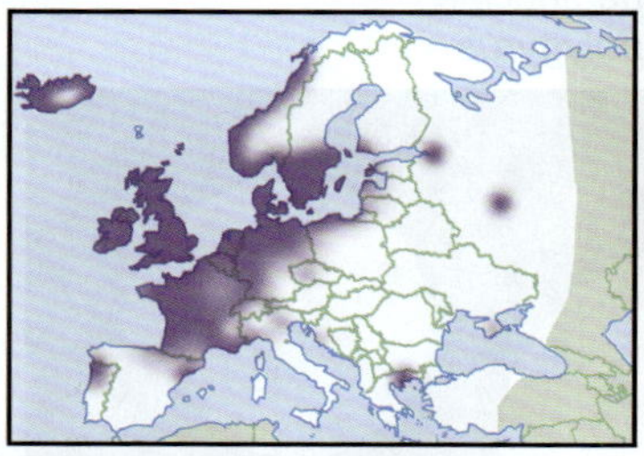

Oxychilus alliarius

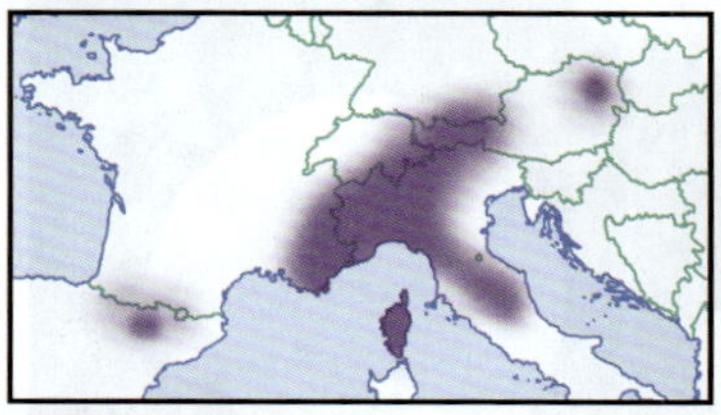

Oxychilus clarus

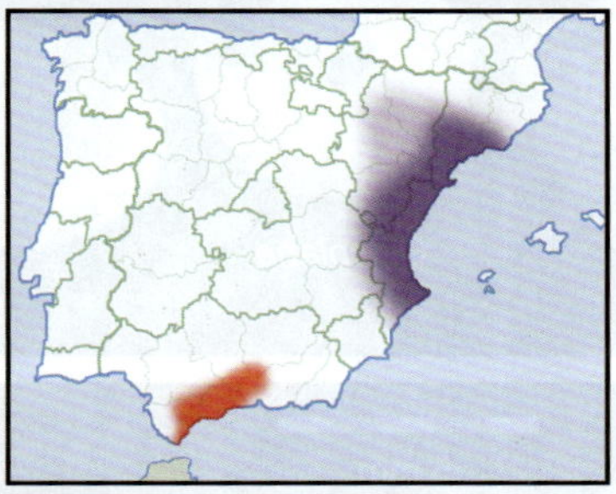

Oxychilus courquini
Oxychilus rateranus

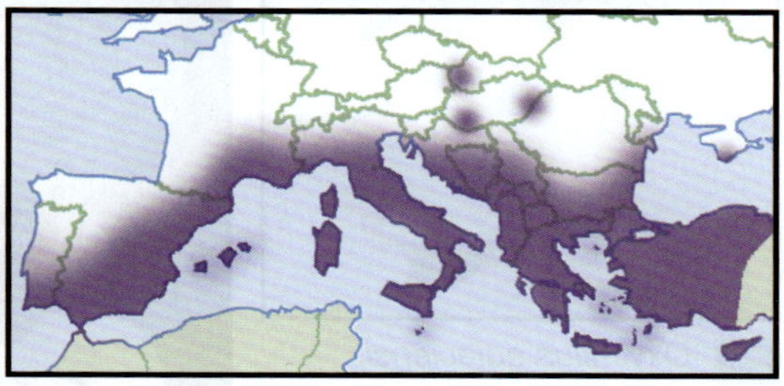

Oxychilus hydatinus

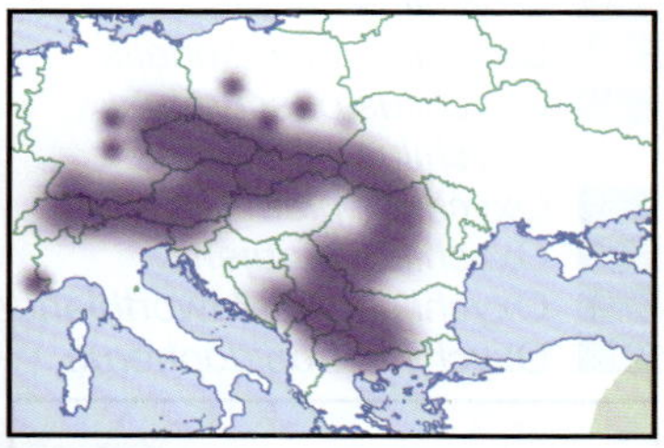

Oxychilus depressus

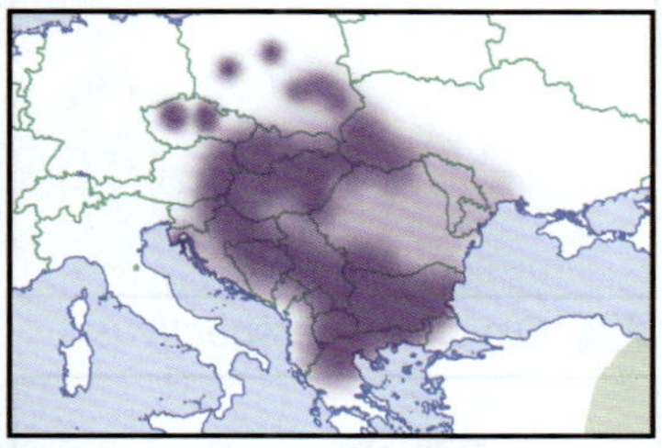

Oxychilus inopinatus

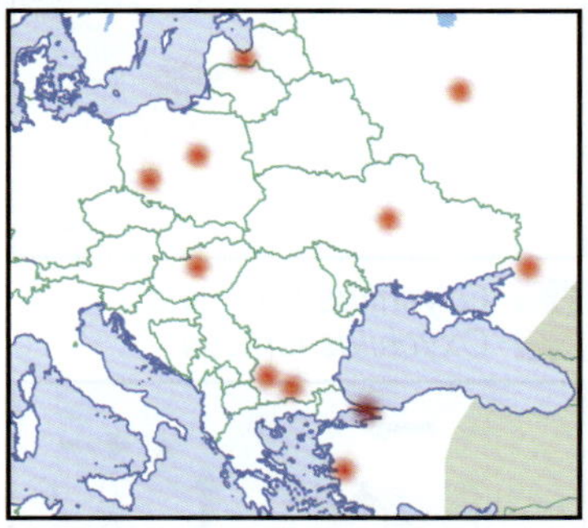

Oxychilus translucidus

Oxychilus 2

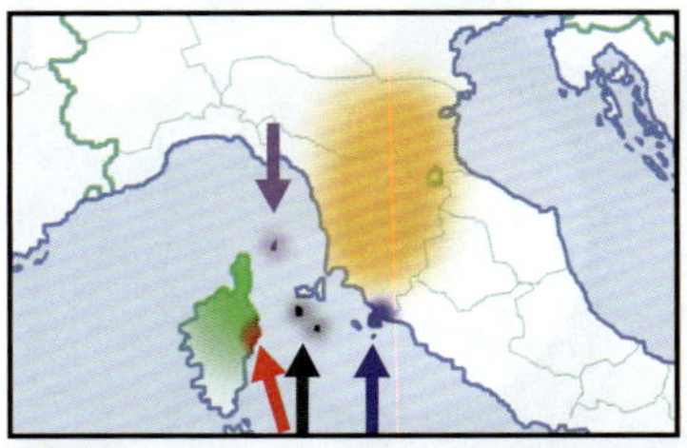

- Oxychilus amblyopus
- Oxychilus lathyri
- Oxychilus majori
- Oxychilus oglasicola
- Oxychilus pilula
- Oxychilus uziellii

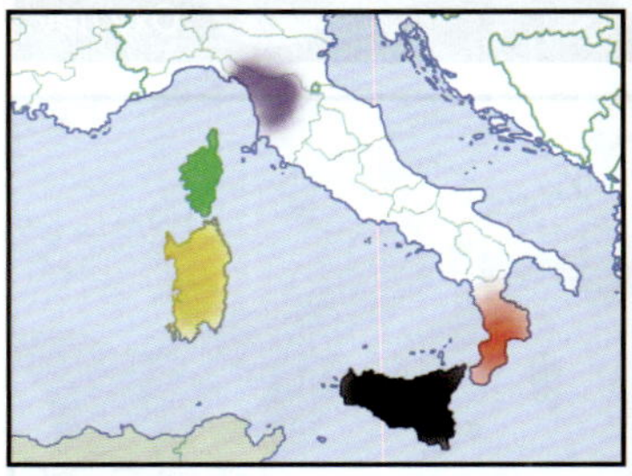

- Oxychilus adjaciensis
- Oxychilus blauneri
- Oxychilus carotii
- Oxychilus meridionalis
- Oxychilus obscuratus
- Oxychilus oppressus
- Oxychilus planellus
- Oxychilus requienii
- Oxychilus shuttleworthianus
- Oxychilus tropidophorus

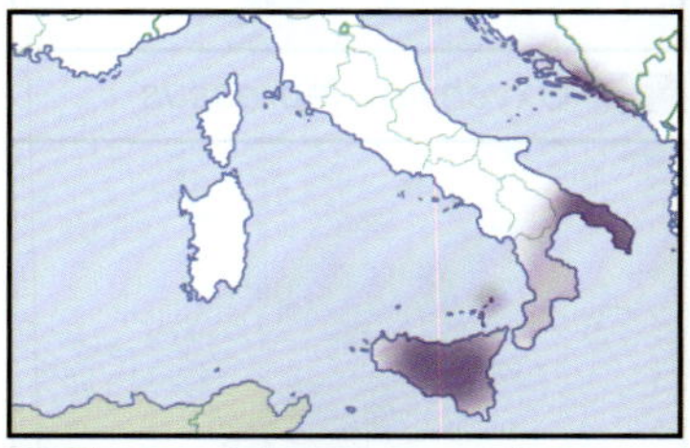

- Oxychilus perspectivus

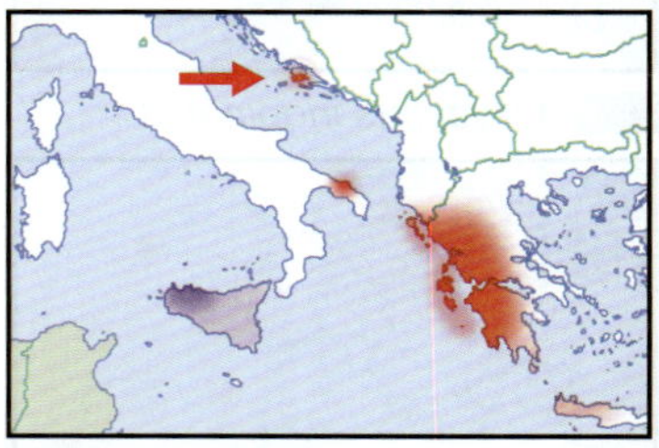

- Oxychilus fuscosus
- Oxychilus ionicus

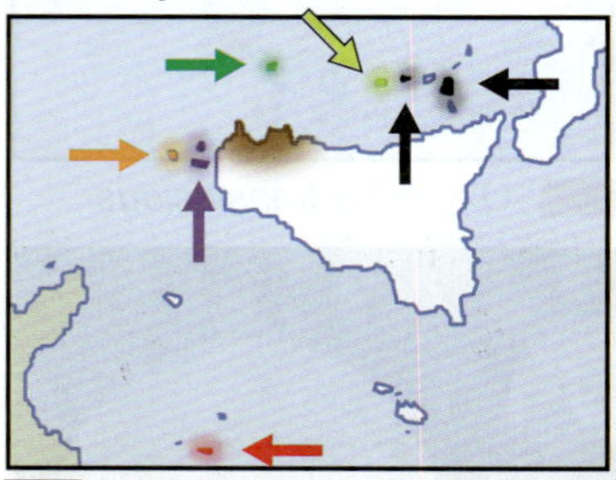

- Oxychilus alicurensis
- Oxychilus canini
- Oxychilus denatale
- Oxychilus diductus
- Oxychilus egadiensis
- Oxychilus lagrecai
- Oxychilus nortoni

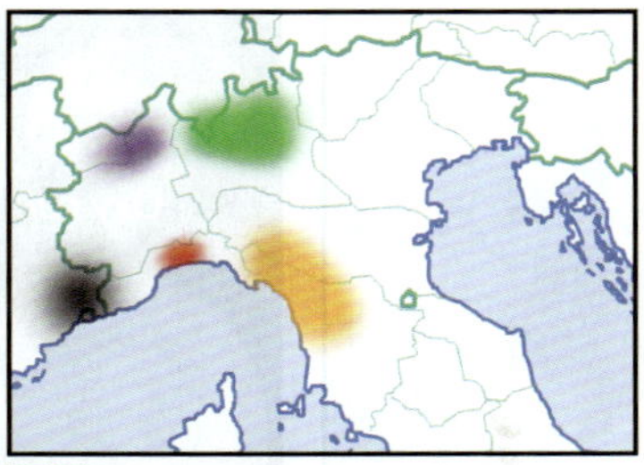

- Oxychilus adamii
- Oxychilus gardinii
- Oxychilus maceanus
- Oxychilus paulucciae
- Oxychilus polygyrus

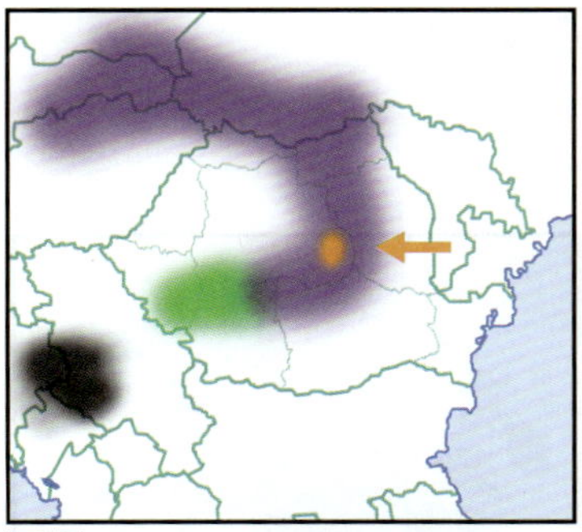

- Oxychilus deubeli
- Oxychilus montivagus
- Oxychilus oscari
- Oxychilus serbicus

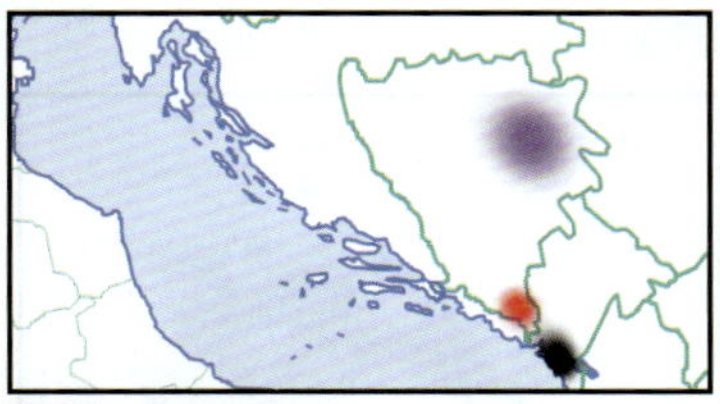

- Oxychilus absoloni
- Oxychilus planospiroides
- Oxychilus wiktori

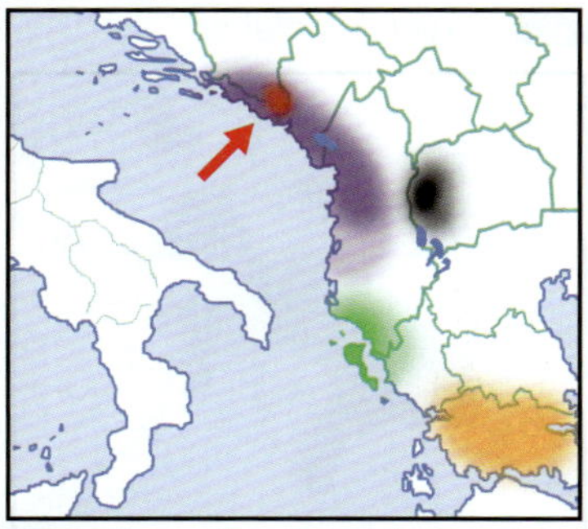

- Oxychilus jugoslavicus
- Oxychilus juliae
- Oxychilus nautiliformis
- Oxychilus planorbis
- Oxychilus tomlini

Oxy can Oxy den Oxy lag Oxy uzi M. Sosso Oxy ada Oxy maj Oxy pil A. Vannozzi Oxy car Oxy nor Oxy pau Oxy planor M. Prodan

Oxychilus 3, *Eopolita*

Eopolita protensa

Oxychilus diaphanellus

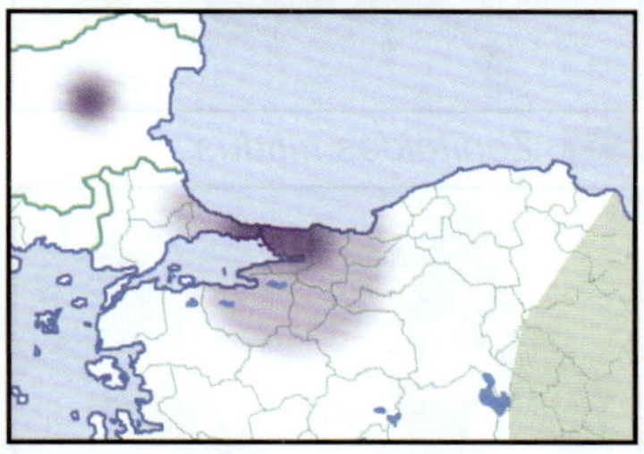

Oxychilus investigatus

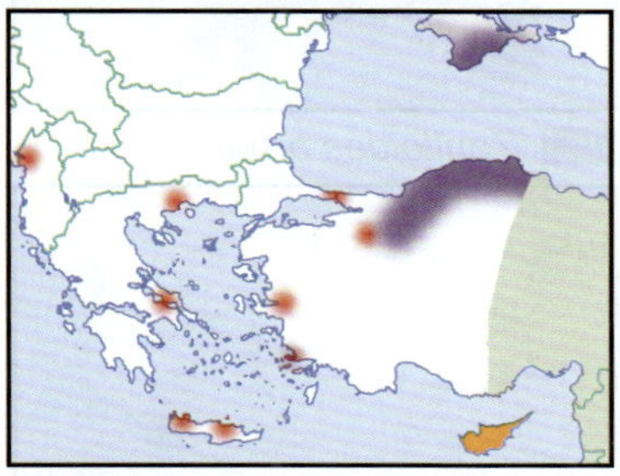

Oxychilus camelinus
Oxychilus kobelti
Oxychilus mavromoustakisi

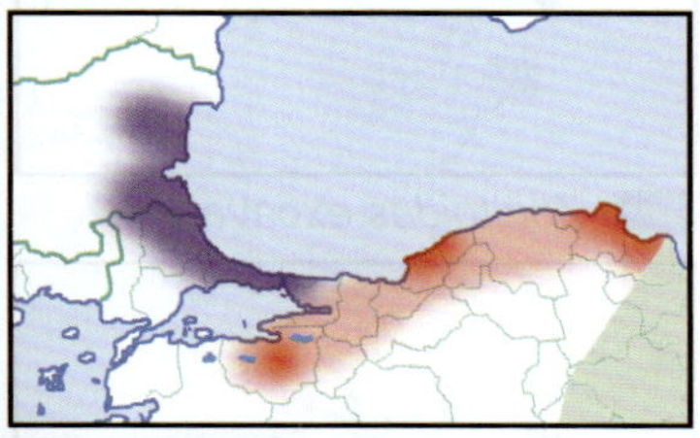

Oxychilus moussoni
Oxychilus samsunensis

Oxy urb A. Irikov
Oxy mav G. Konstantinou

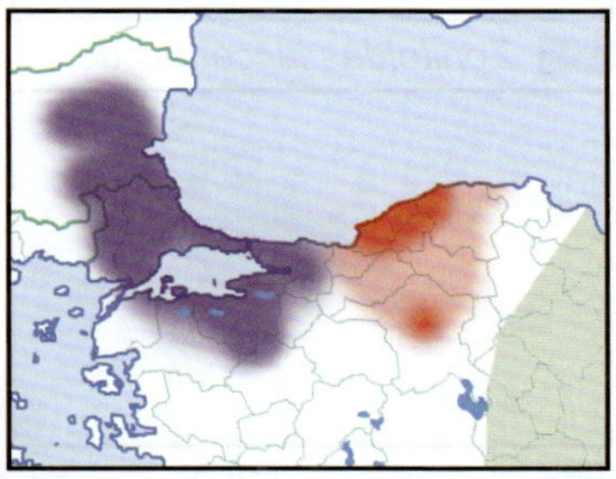

Oxychilus paphlagonicus
Oxychilus urbanskii

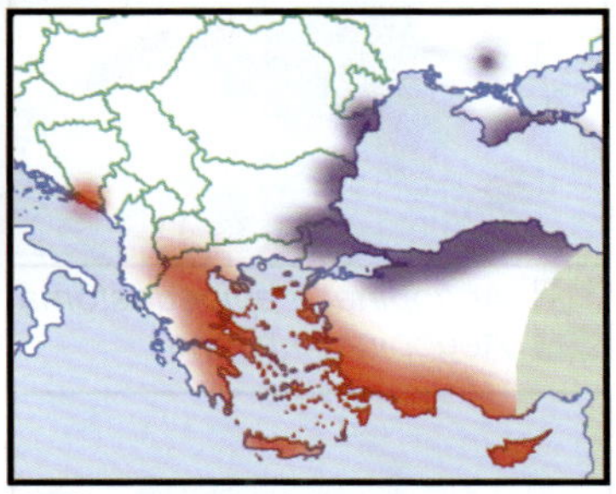

Oxychilus cyprius
Oxychilus deilus

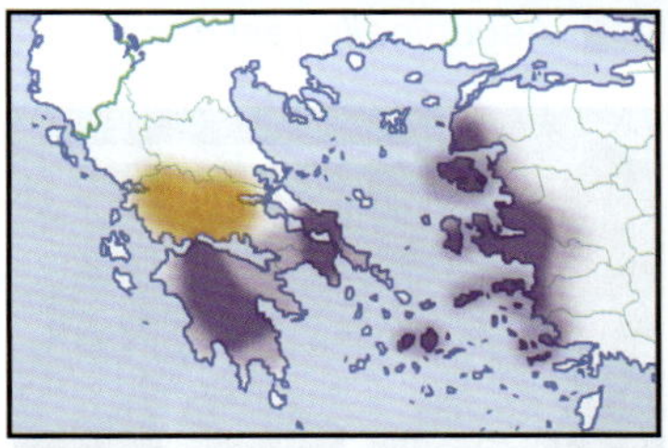

Oxychilus samius
Oxychilus tomlini

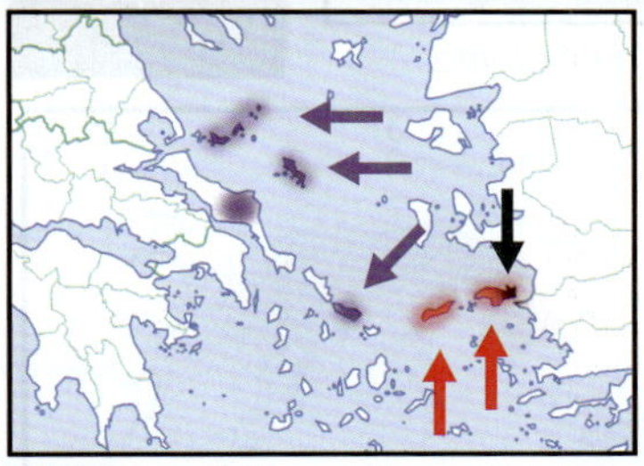

Oxychilus mylonasi
Oxychilus pygmaeus
Oxychilus seidli

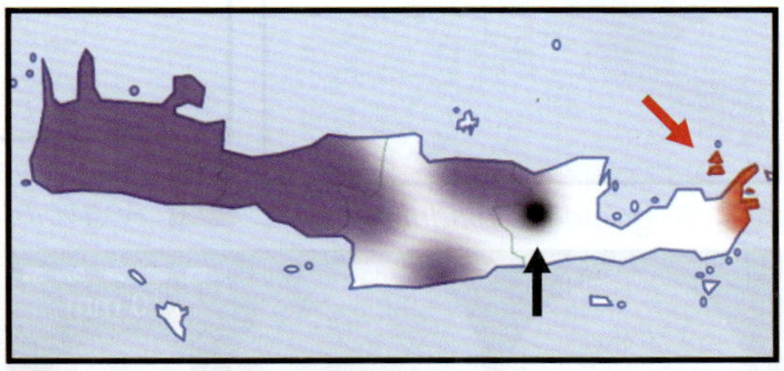

Oxychilus aegopinoides
Oxychilus amaltheae
Oxychilus spratti

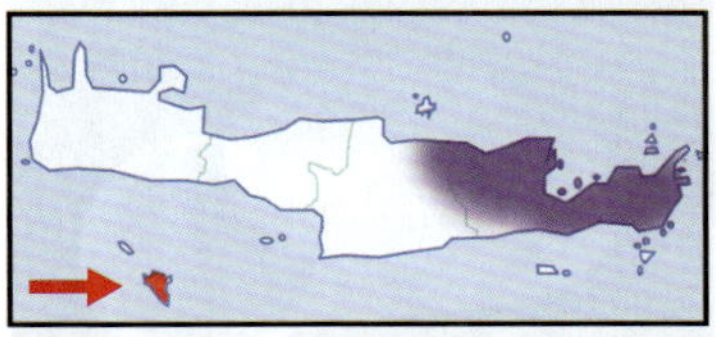

Oxychilus minoicus
Oxychilus pieperi

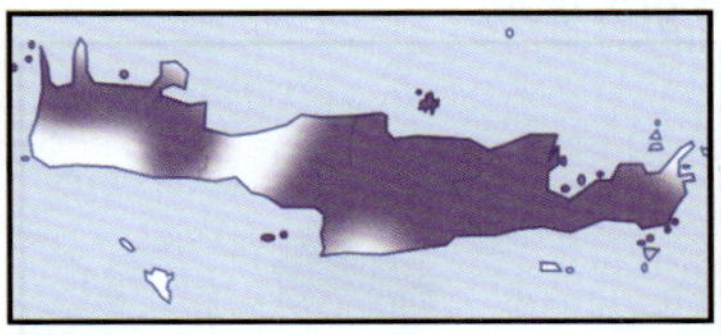

Oxychilus superfluus

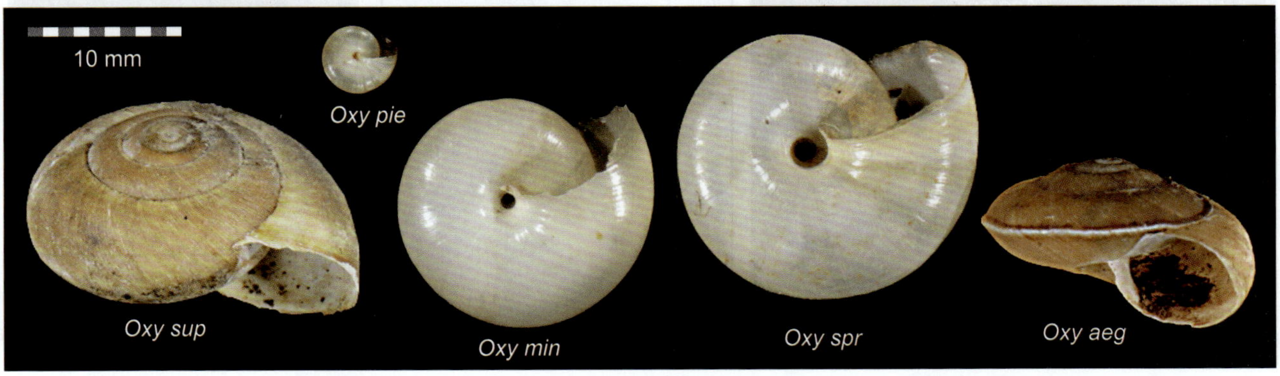

Zonitoides, *Nesovitrea*, *Aegopinella*, *Retinella*

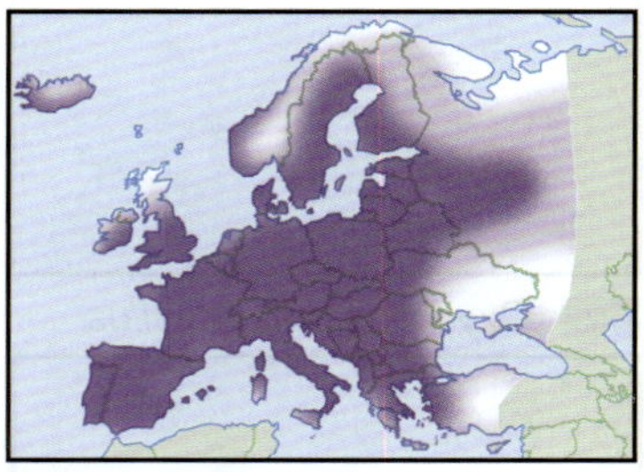

Zonitoides arboreus

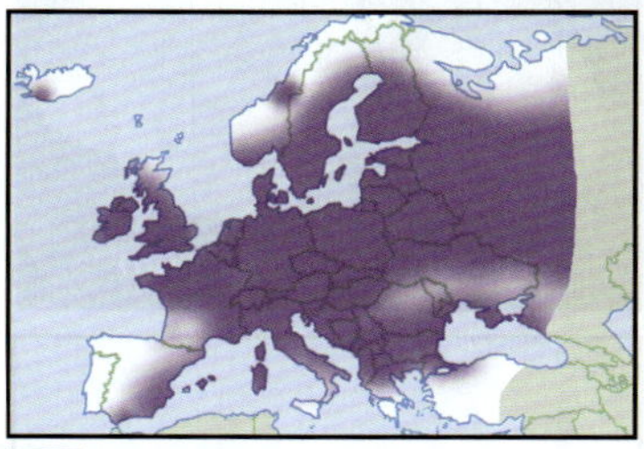

Zonitoides nitidus

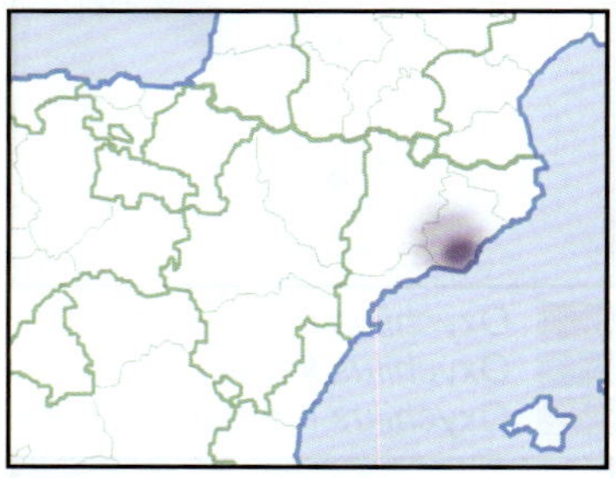

Zonitoides jaccetanicus

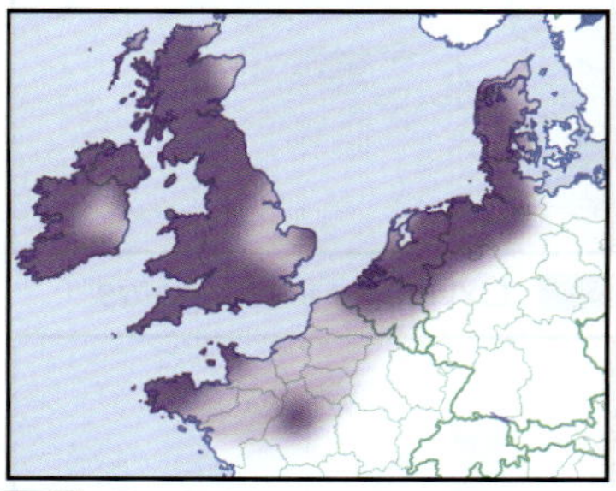

Zonitoides excavatus

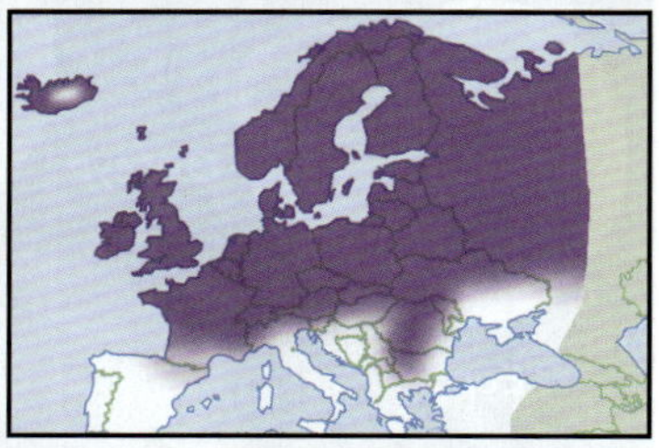

Nesovitrea hammonis

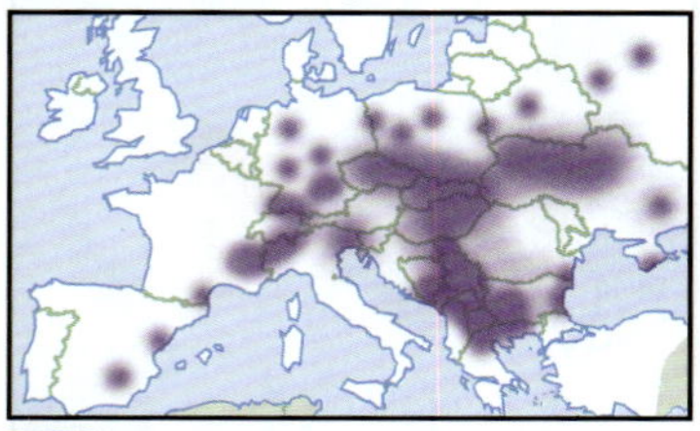

Aegopinella minor

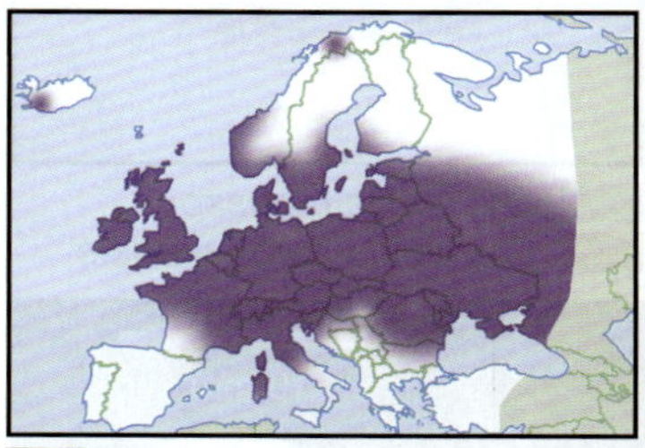

Aegopinella pura

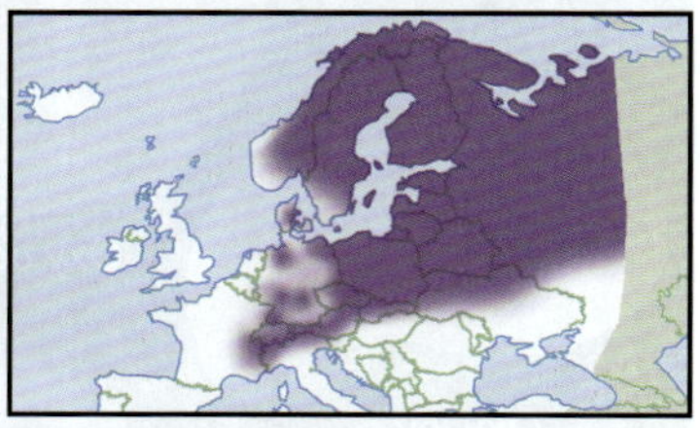

Nesovitrea petronella

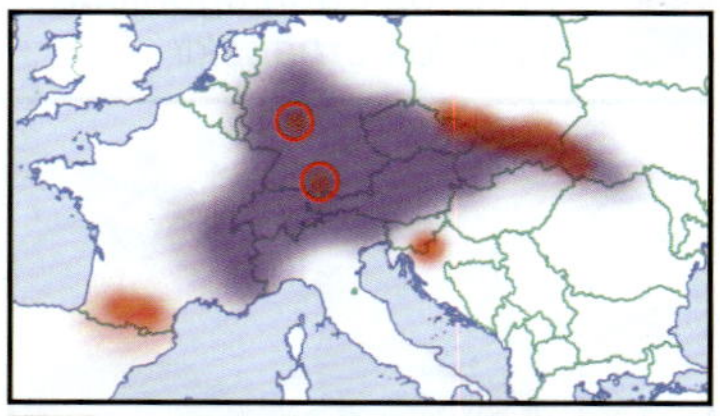

Aegopinella epipedostoma
Aegopinella nitens

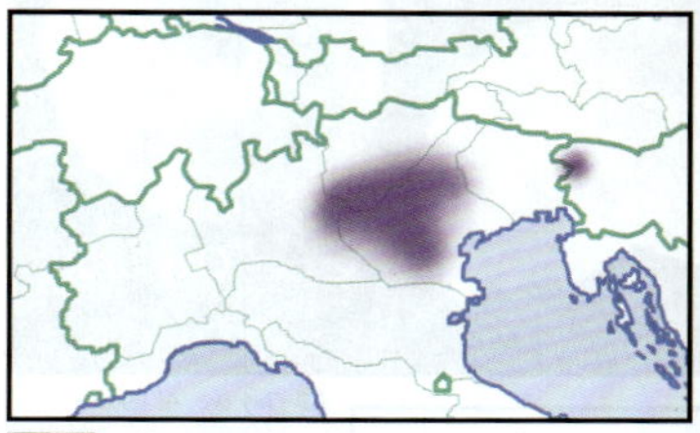

Aegopinella cisalpina

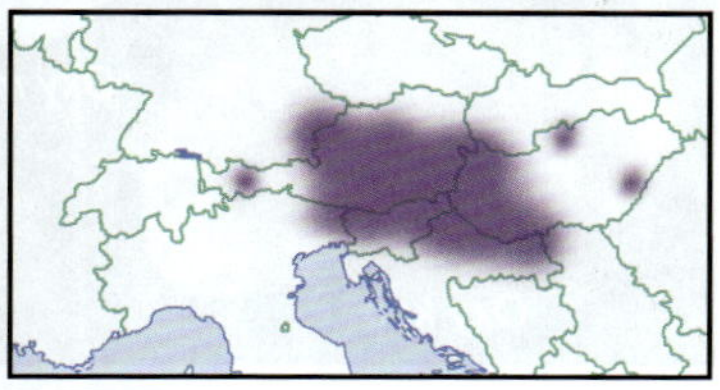

Aegopinella ressmanni

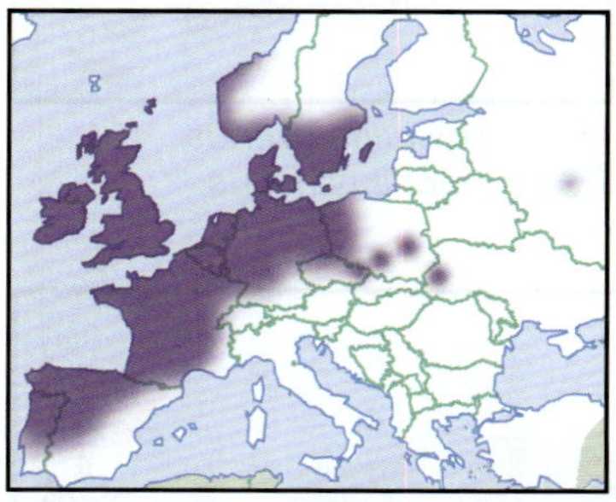

Aegopinella nitidula

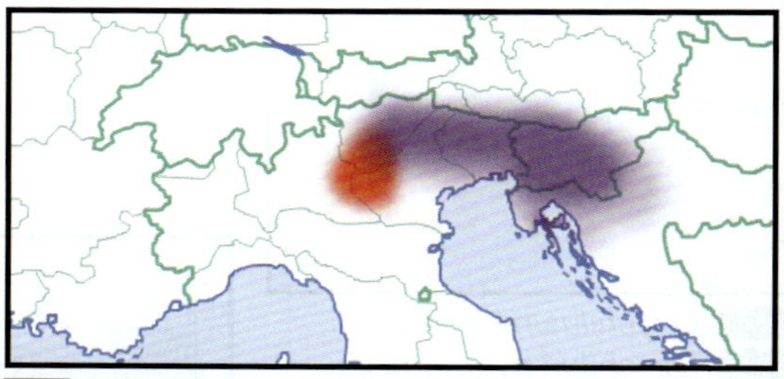

Aegopinella forcarti
Aegopinella graziadei

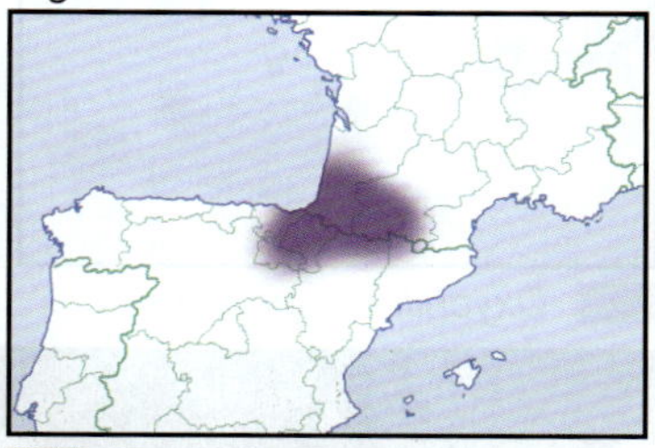

Retinella incerta

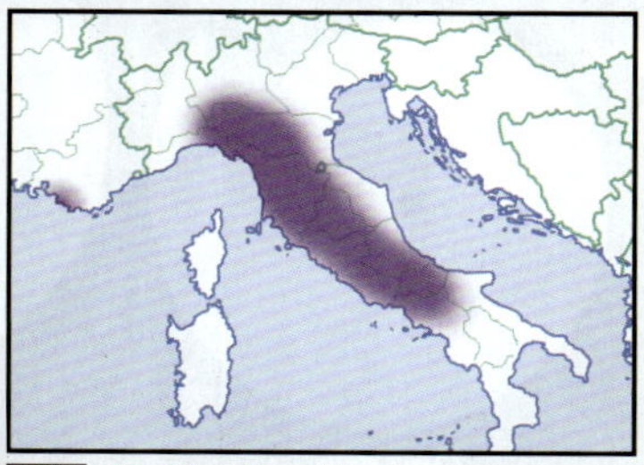

Retinella olivetorum

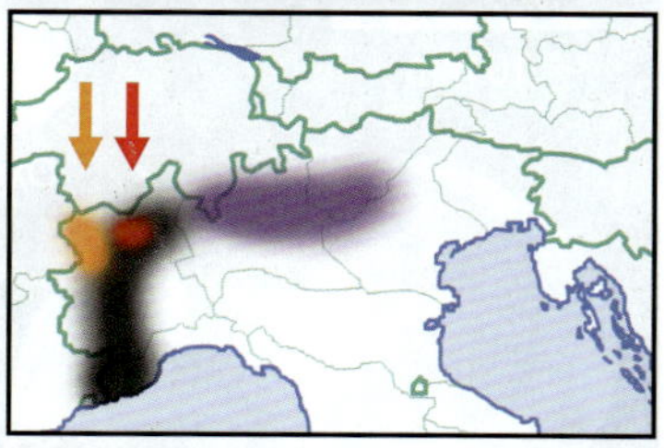

Retinella giustii
Retinella hiulca
Retinella pseudoaegopinella
Retinella stabilei

Aegopis, *Balcanodiscus*, *Allaegopis* etc.

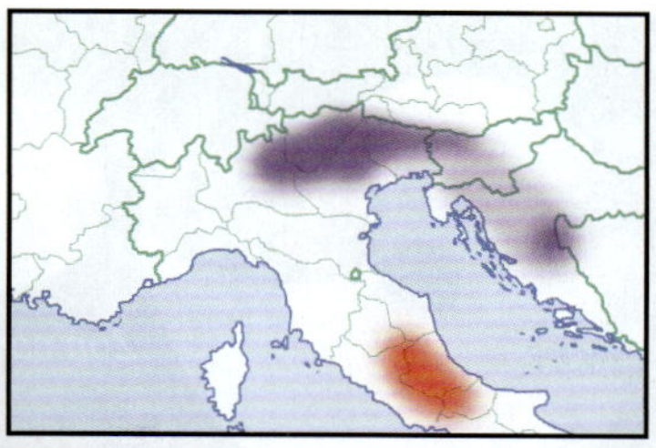

Aegopis gemonensis
Aegopis italicus

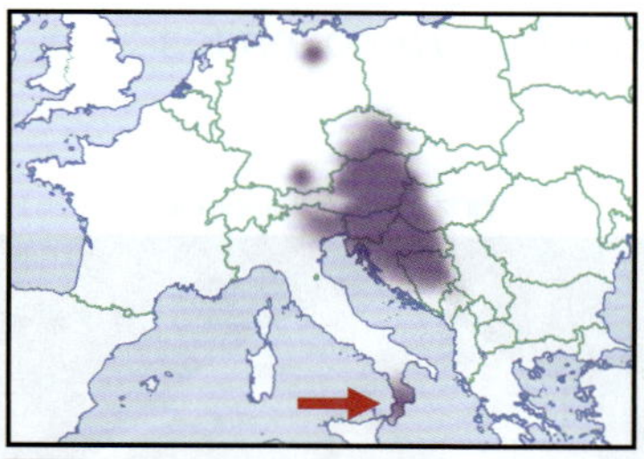

Aegopis verticillus

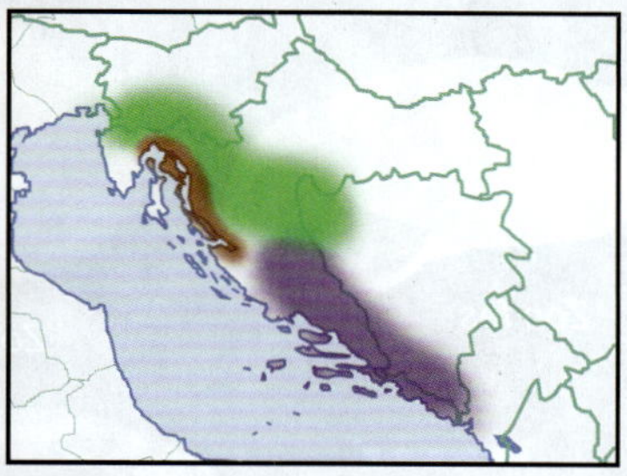

Aegopis acies
Aegopis compressus
Aegopis croaticus

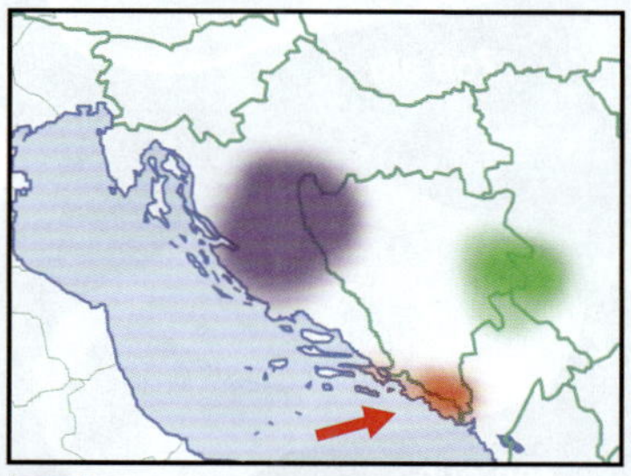

Aegopis laughofferi
Aegopis spelaeus
Aegopis tenerrimus

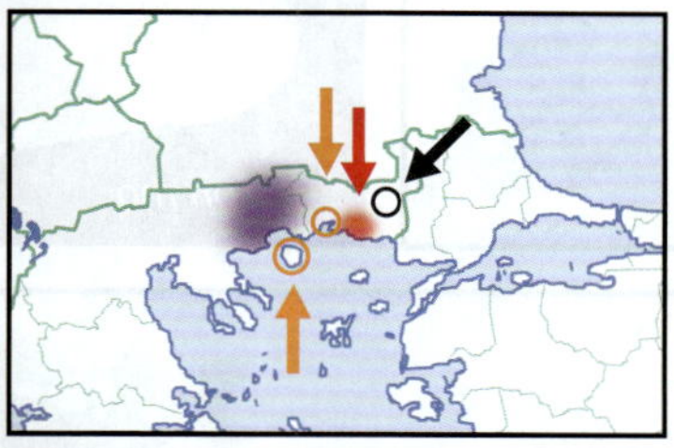

Balcanodiscus beroni
Balcanodiscus carinatus
Balcanodiscus cerberus
Balcanodiscus danii

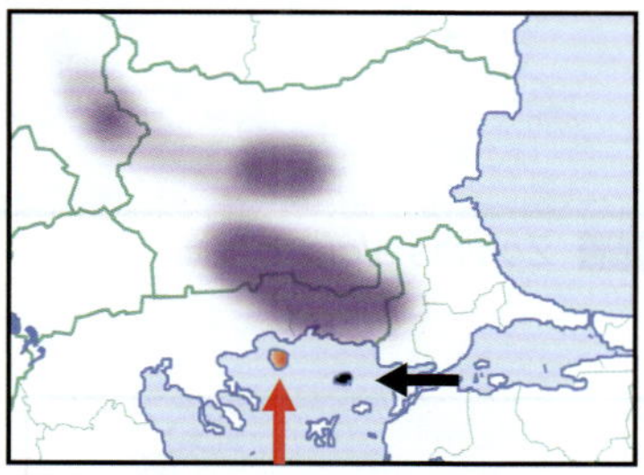

Balcanodiscus difficilis
Balcanodiscus frivaldskyanus
Balcanodiscus magnus

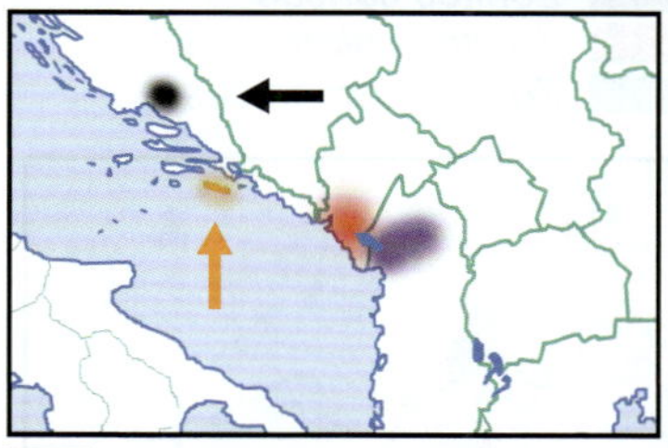

Meledella werneri
Paraegopis oberwimmeri
Paraegopis skipetaricus
Troglaegopis mosorensis

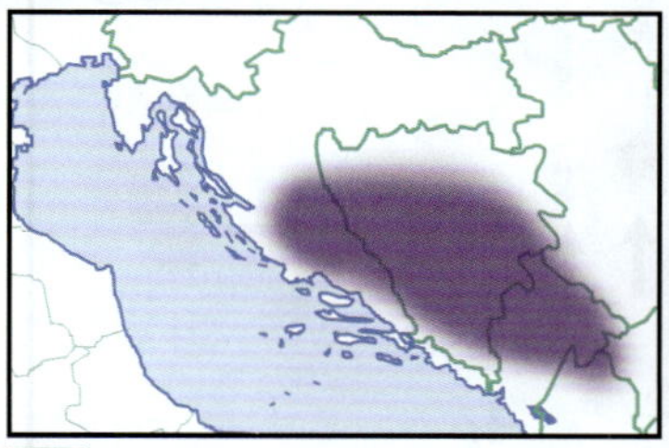

Paraegopis mauritii

All mer Dor car
P. Ovalis

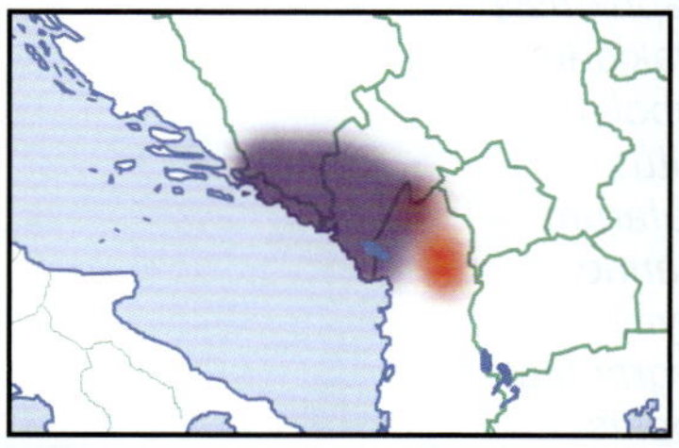

Paraegopis albanicus
Paraegopis bizonus

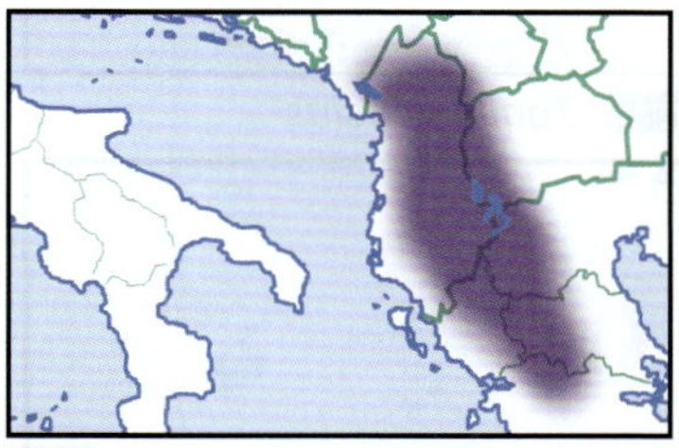

Allaegopis skanderbegianus

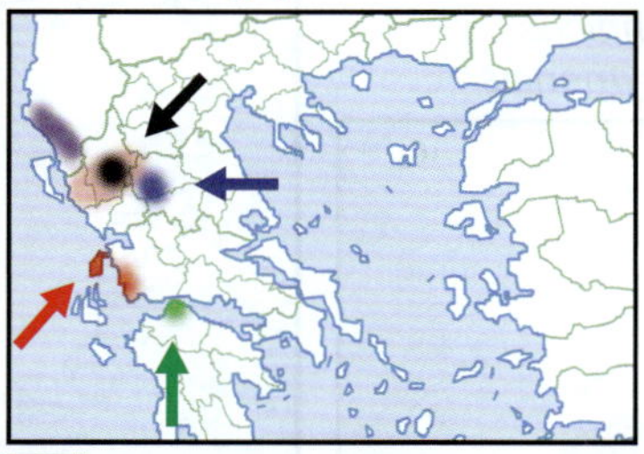

Allaegopis amphikypellon
Allaegopis jonicus
Allaegopis kerketianus
Allaegopis meridionalis
Allaegopis subariedeli

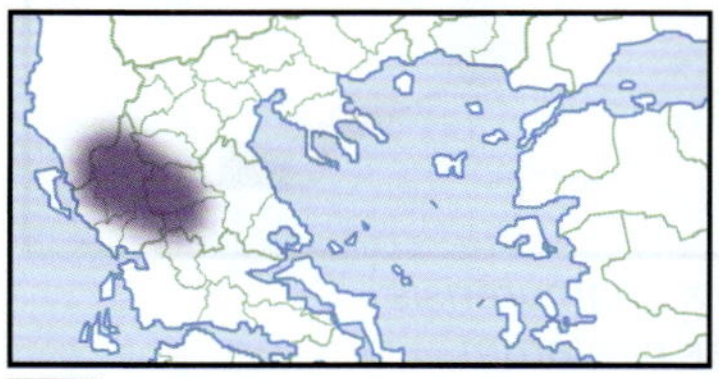

Allaegopis transiens

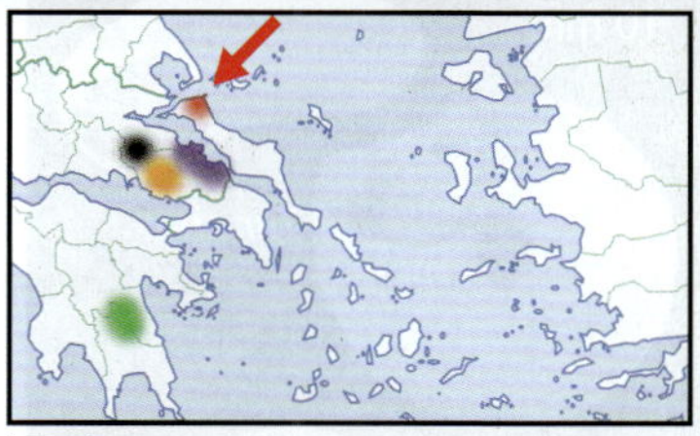

Doraegopis boeoticus
Doraegopis carinatus
Doraegopis euboeicus
Doraegopis parnonicus
Doraegopis subaii

Zonites, *Turcozonites*

Zon chl A. Örstan *Zon eub* R. Reuselaars

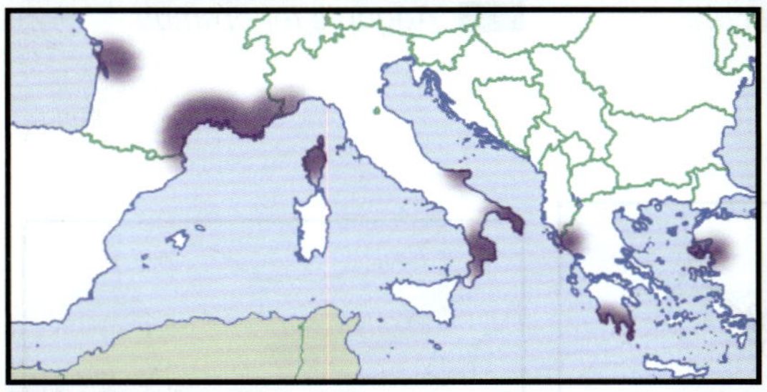

Zonites algirus

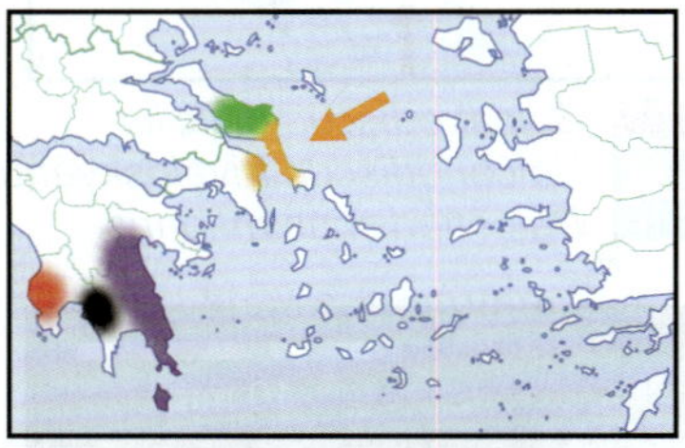

Zonites euboeicus
Zonites kobelti
Zonites messenicus
Zonites oertzeni
Zonites parnonensis

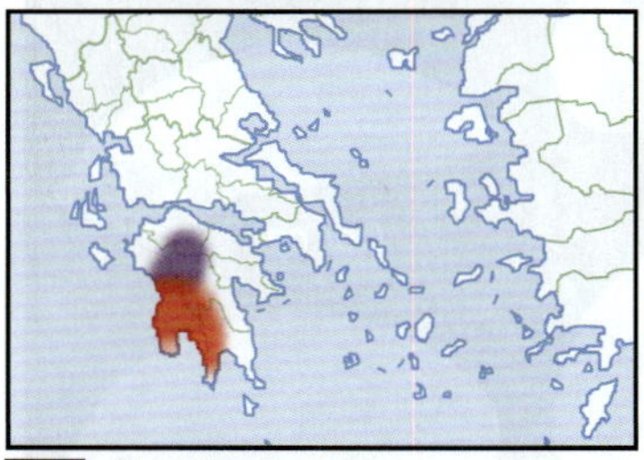

Zonites graecus
Zonites labiosus

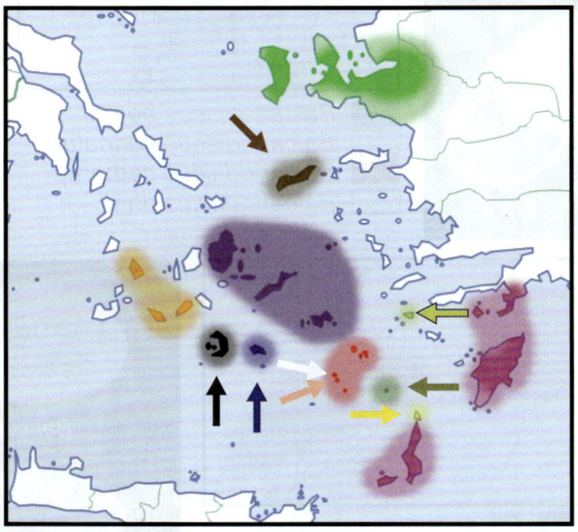

Zonites anaphiensis
Zonites astakidae
Zonites embolium
Zonites invitus
Zonites nautarum
Zonites nikariae
Zonites nisyrius
Zonites pergranulatus
Zonites rhodius
Zonites santorinensis
Zonites sariae
Zonites siphnicus
Zonites smyrnensis

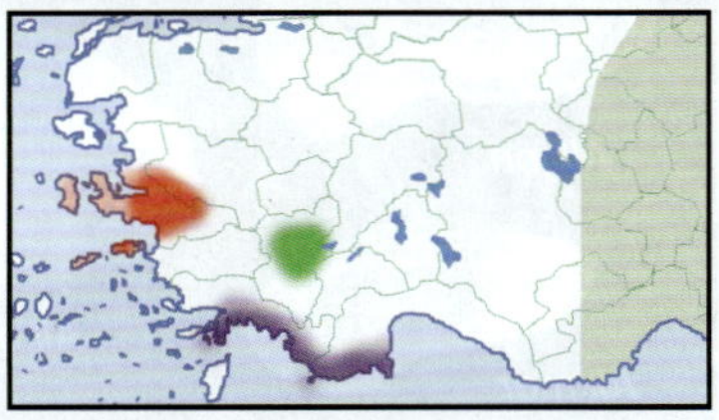

Zonites caricus
Zonites chloroticus
Zonites osmanicus

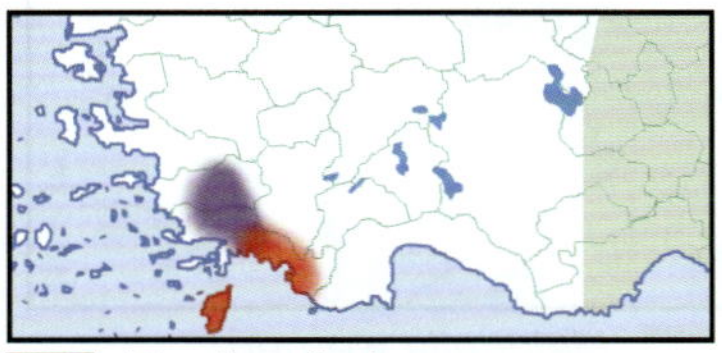

Zonites festae
Zonites humilis

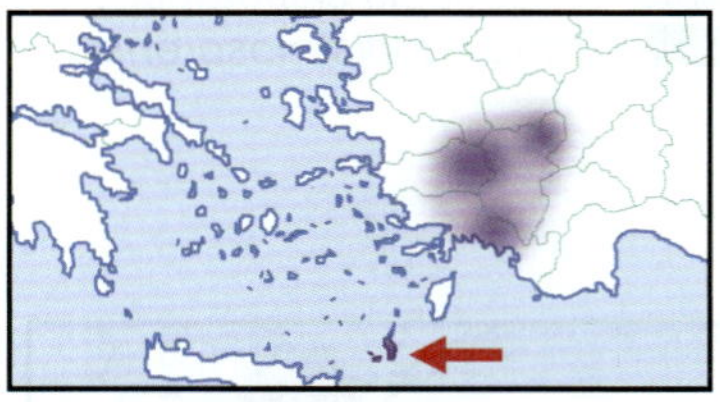

Zonites casius

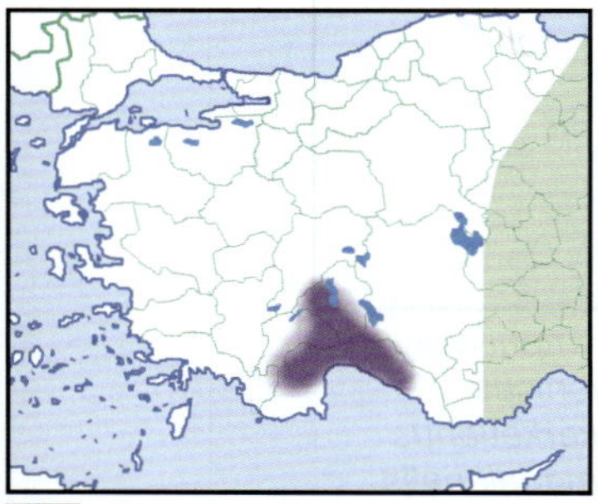

Turcozonites megistus

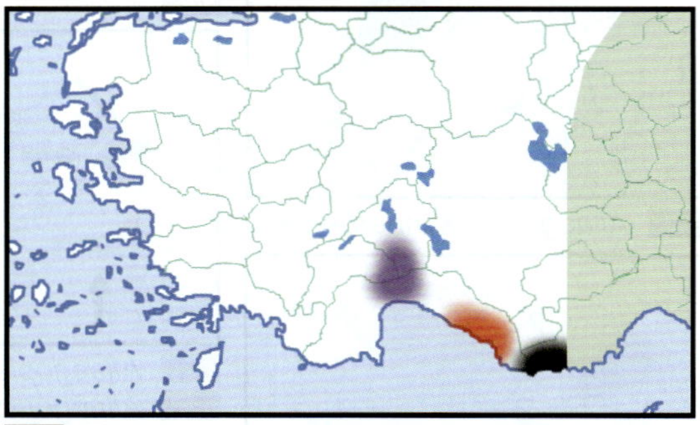

Turcozonites anamurensis
Turcozonites piratarum
Turcozonites wandae

Vitrinidae

Eucobresia (?) from Slovakia — J. Grego

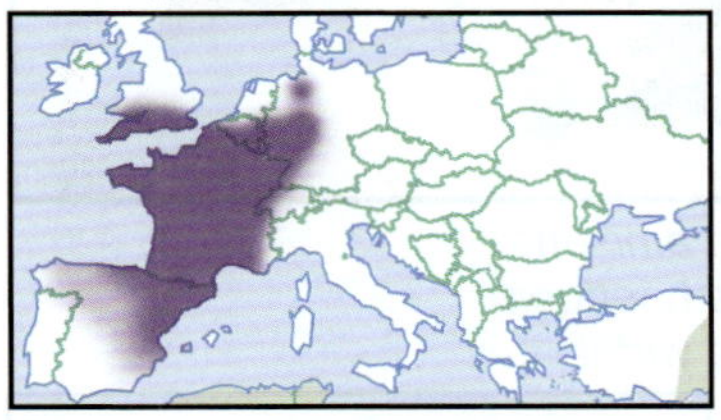

Phenacolimax major

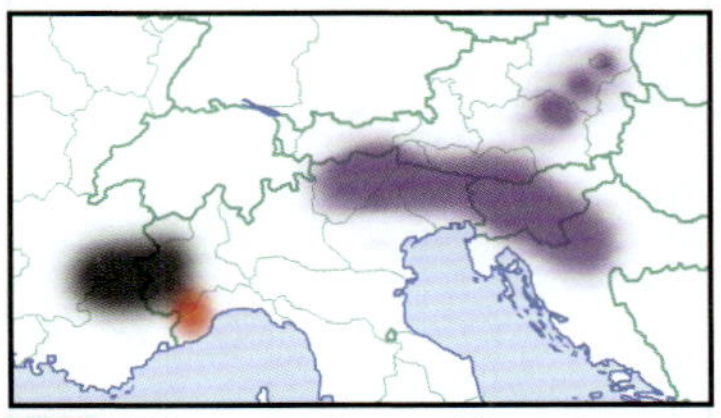

Phenacolimax blanci
Phenacolimax stabilei
Semilimax carinthiacus

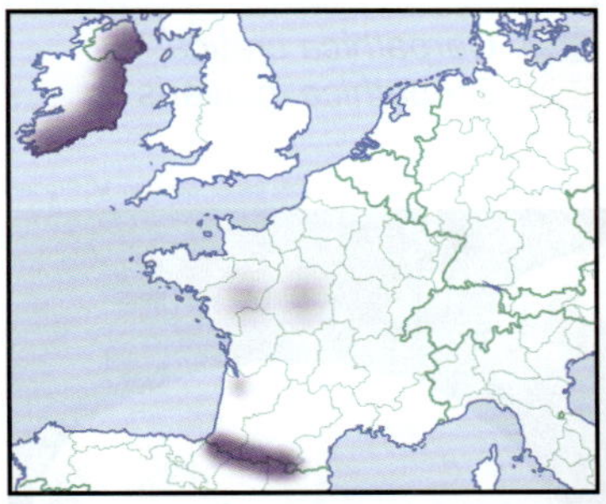

Semilimax pyrenaicus

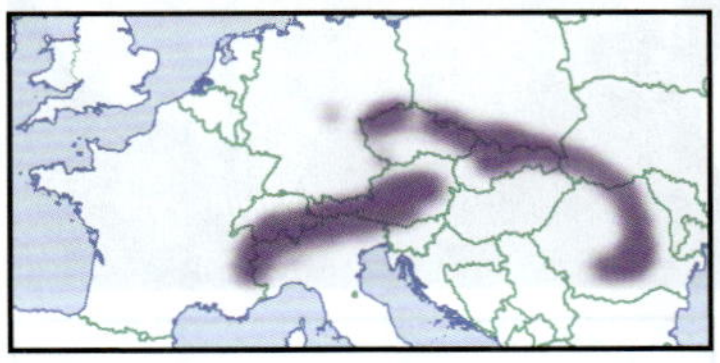

Eucobresia nivalis

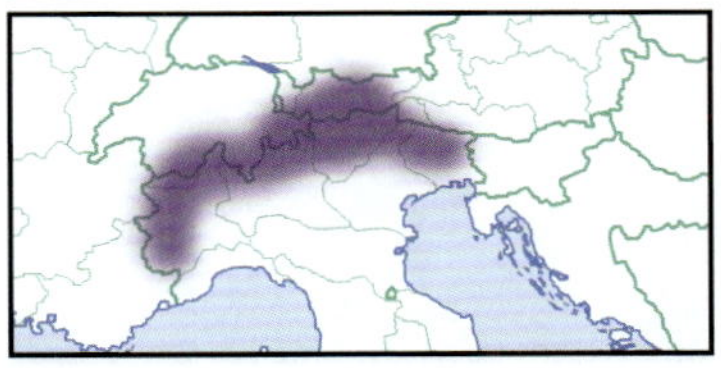

Eucobresia pegorarii

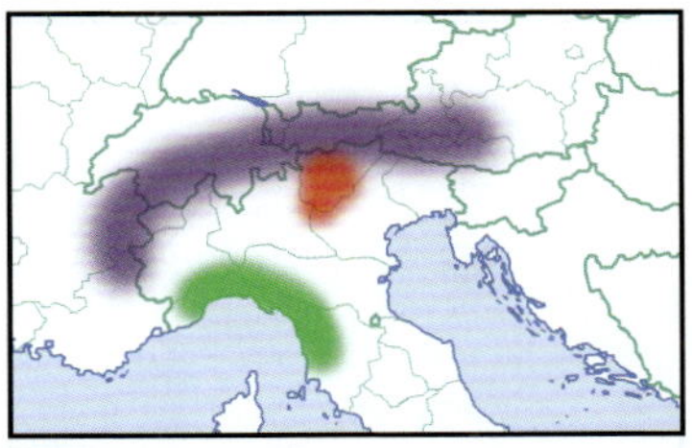

Eucobresia glacialis
Vitrinobrachium baccettii
Vitrinobrachium tridentinum

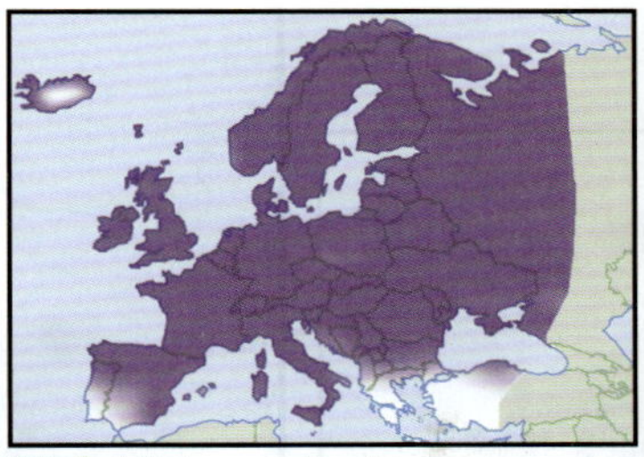

Vitrina pellucida

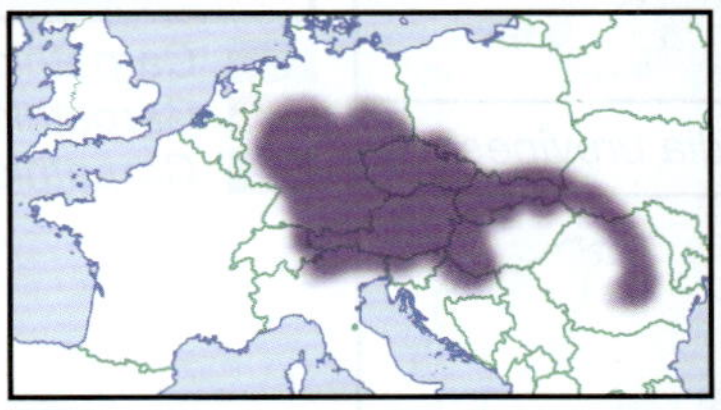

Semilimax semilimax

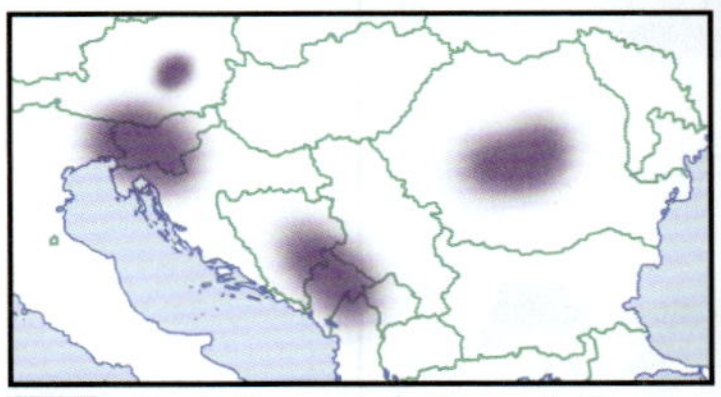

Semilimacella carniolica

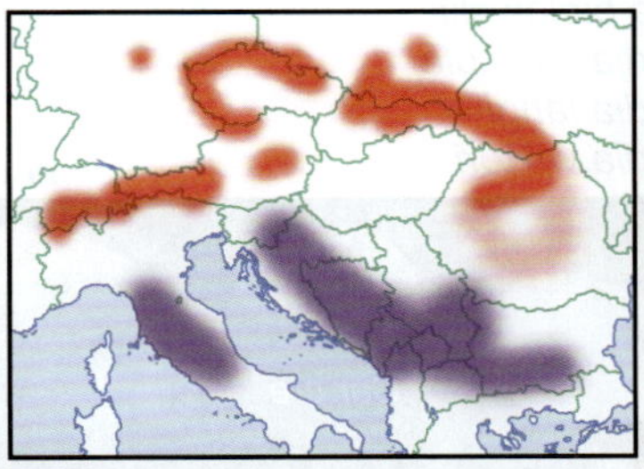

Semilimacella bonelli
Semilimax kotulae

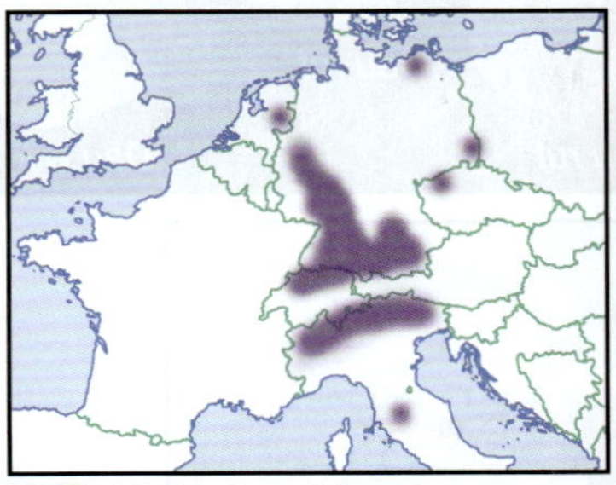

Vitrinobrachium breve

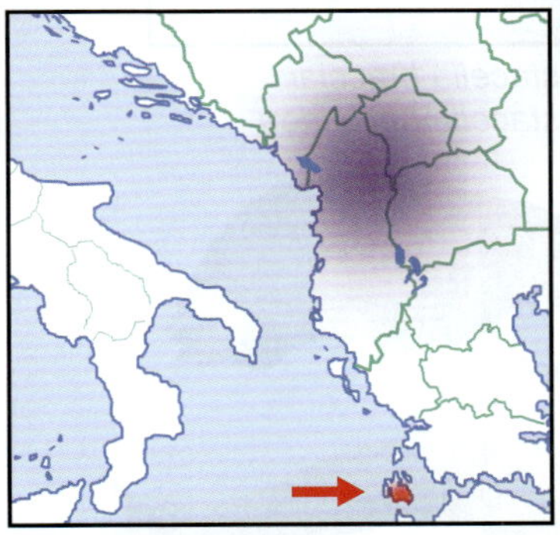

Oligolimax apatelus
Oligolimax cephalonicus

Oli mus A. Vannozzi *Sem bon* Paolo Paolini *Vit bac* M. Sosso

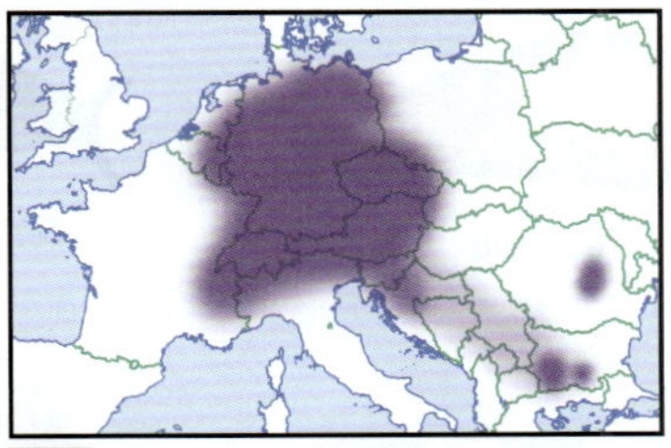

Eucobresia diaphana

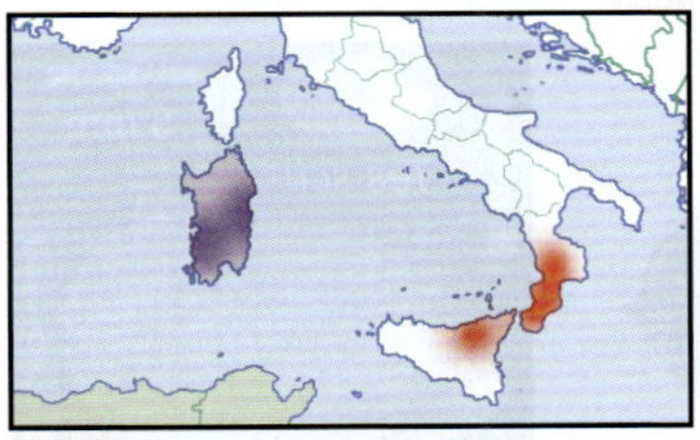

Oligolimax musignani
Oligolimax pollonerianus

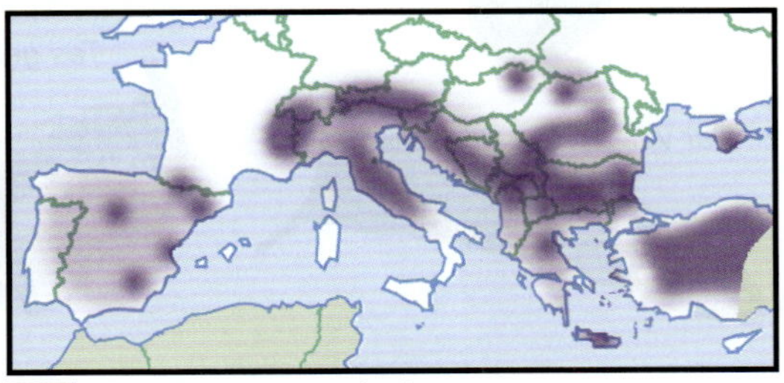

Oligolimax annularis

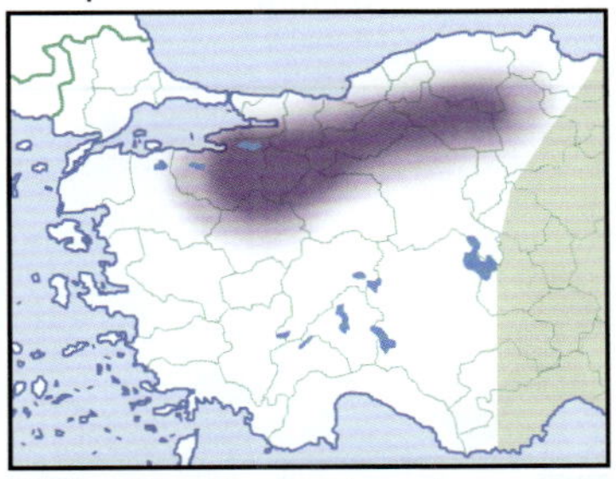

Oligolimax olympicus

Daudebardiidae, *Parmacella*, *Testacella*

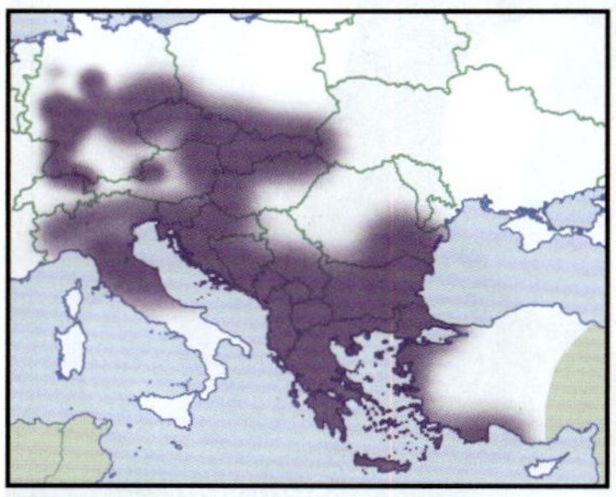

Daudebardia rufa

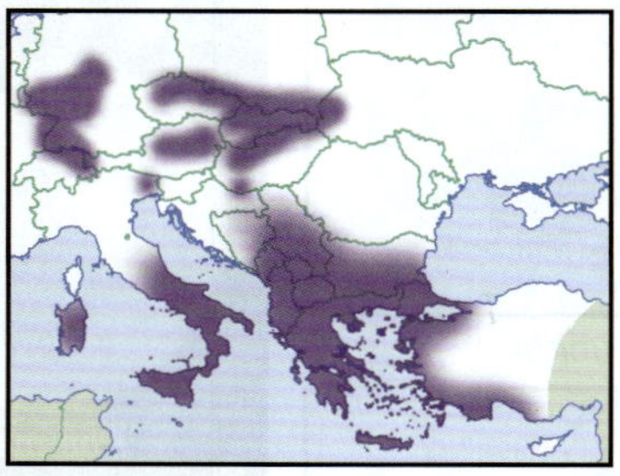

Daudebardia brevipes

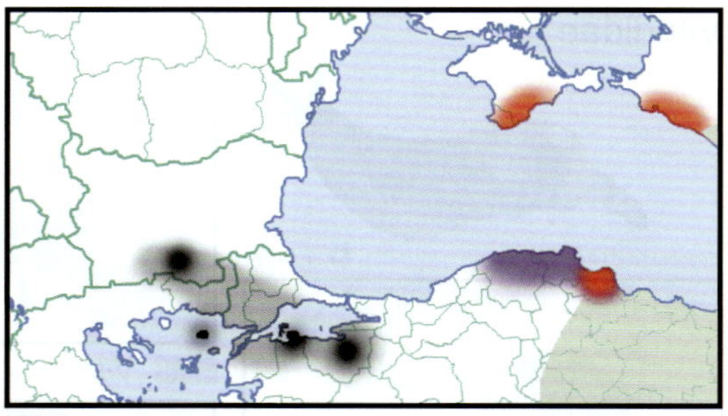

Carpathica amisena
Carpathica boettgeri
Carpathica wirthi

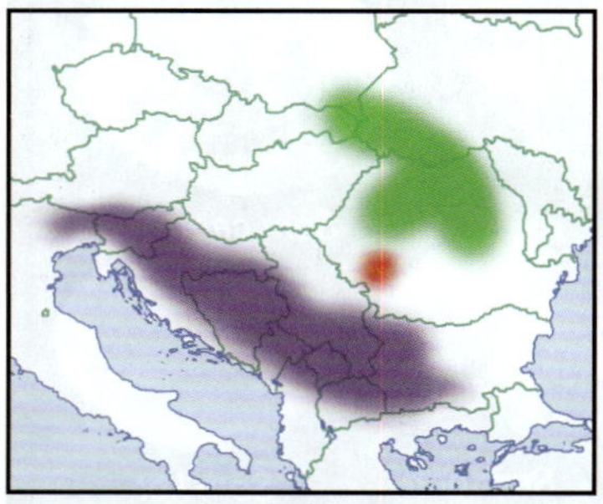

Carpathica calophana
Carpathica stussineri
Daudebardia nana

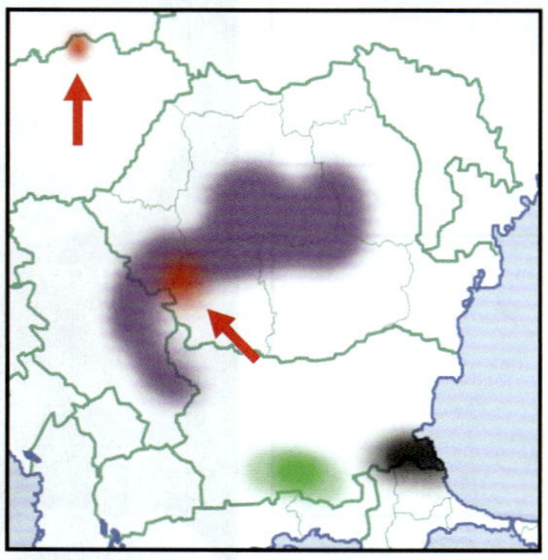

Carpathica bielawskii
Daudebardia cavicola
Daudebardia langi
Daudebardia wiktori

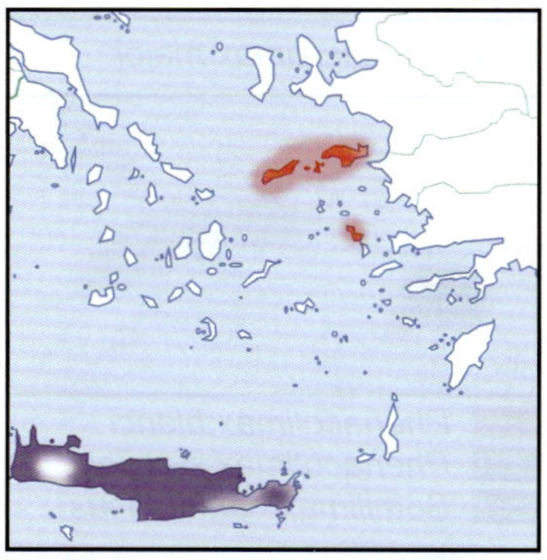

Carpathica cretica
Carpathica insularis

📷 *Dau wik* *Car bie* *Car stu* *Car wir* A. Irikov *Dau lan* M. Prodan

Testacella maugei
Parmacella ibera

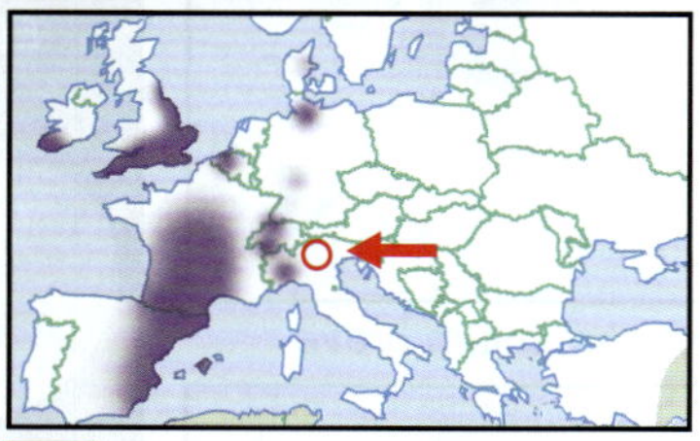

Testacella bracciai
Testacella haliotidea

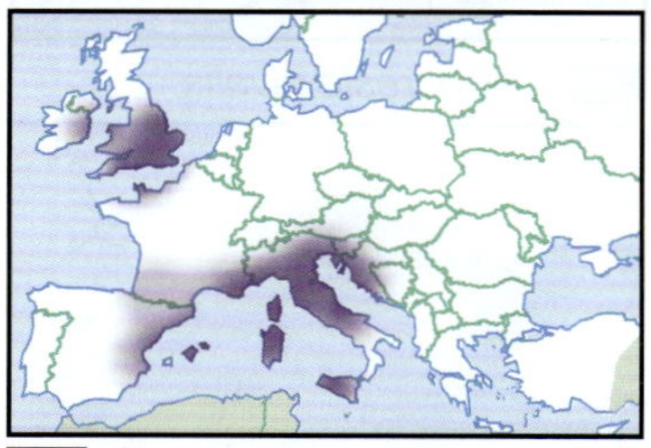

Testacella scutulum

Tes mau

Tes hal

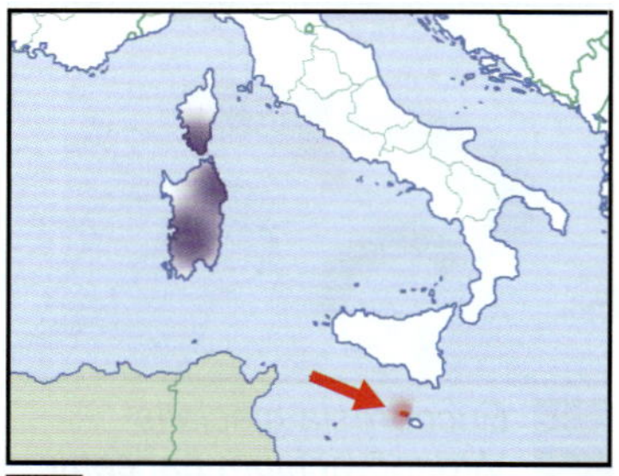

Testacella gestroi
Testacella riedeli

80 mm

Tes scu

Parmacella valenciennii

Par val

Tes ges

📷 *Par val* V. Jacinto *Tes ges* F. Mascia *Tes hal* *Tes mau* R. Anderson *Tes scu* M. Prodan

Milacidae, *Papilloderma*

Milax gagates

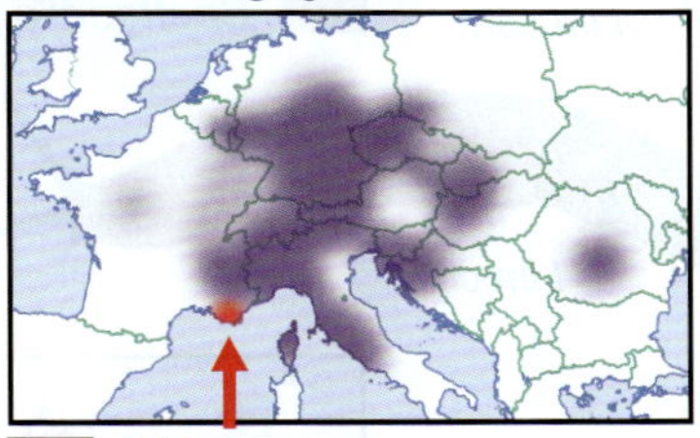

Milax ochraceus
Tandonia rustica

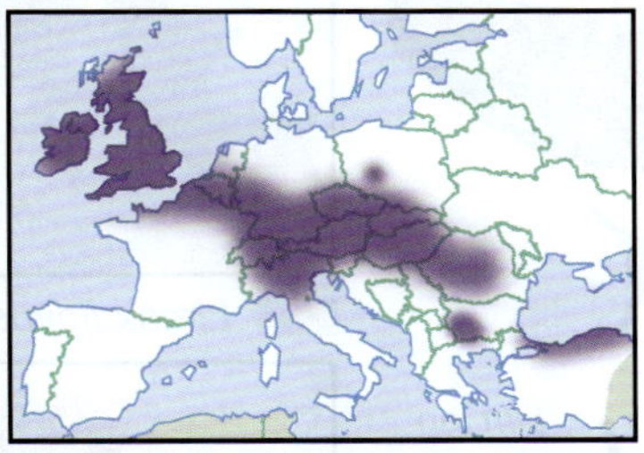

Tandonia budapestensis

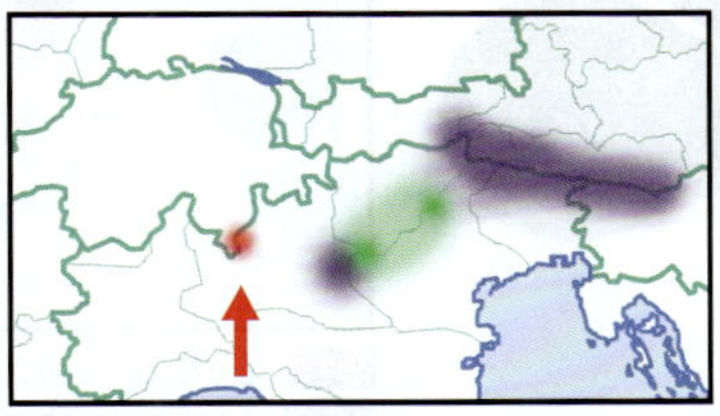

Tandonia baldensis
Tandonia ehrmanni
Tandonia nigra

Milax nigricans

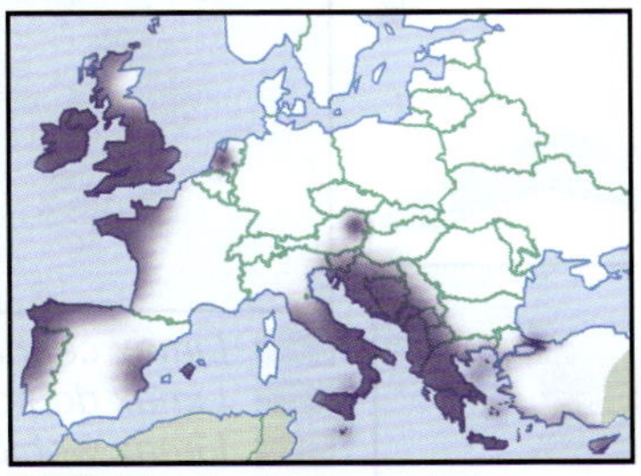

Tandonia sowerbyi

Papilloderma altonagai

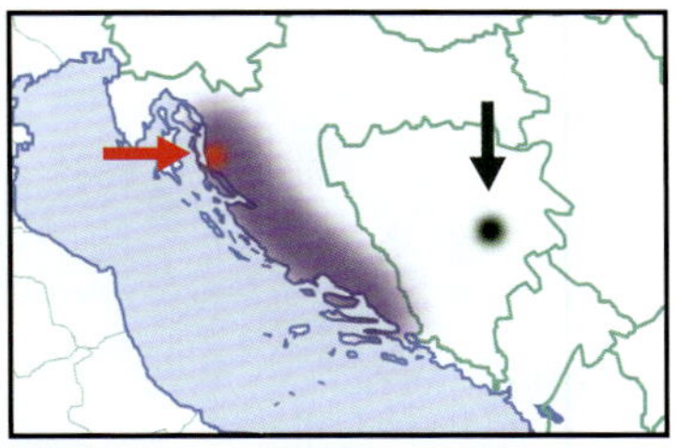

Tandonia bosnensis
Tandonia jablanacensis
Tandonia rara

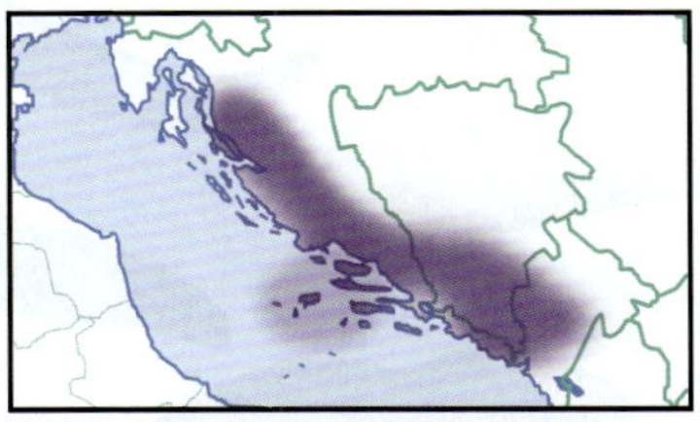

Tandonia reuleauxi

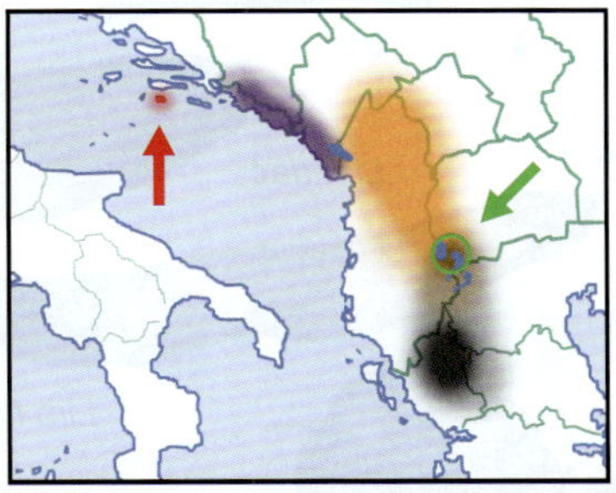

Tandonia albanica
Tandonia cavicola
Tandonia lagostana
Tandonia macedonica
Tandonia melanica

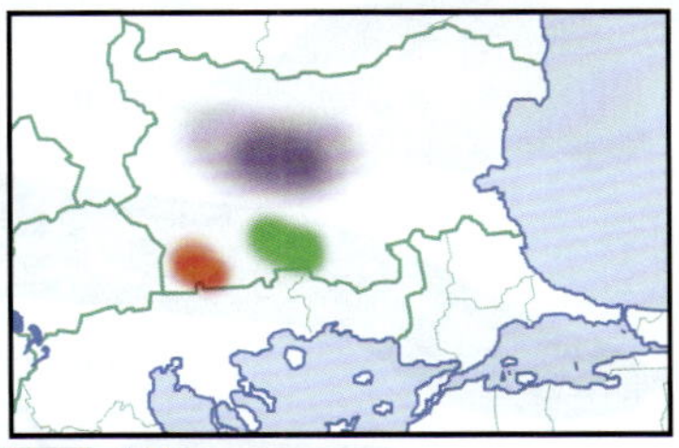

Milax verrucosus
Tandonia pinteri
Tandonia piriniana

Milax aegaeicus
Tandonia pageti
Tandonia totevi

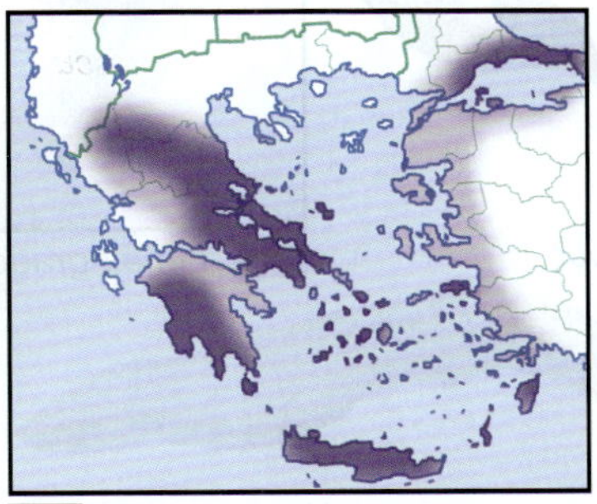

Tandonia cretica

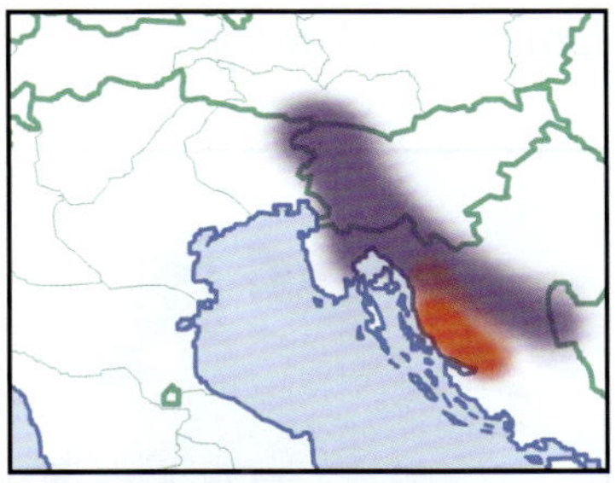

Tandonia fejervaryi
Tandonia robici

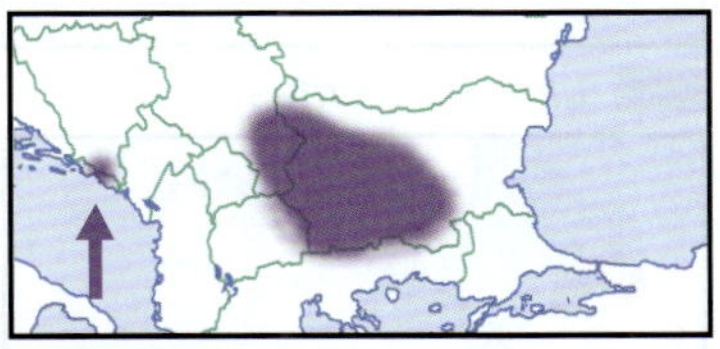

Tandonia serbica

Tandonia kusceri

Mil gag V. Wiese *Mil nig* D. Cilia *Tan bud* M. Mañas *Tan cri Tan kus Tan ret* S. Leonov *Tan fej* G. Pocaterra *Tan reu* C. Audibert *Tan sow* E. Machado, G. Konstantinou

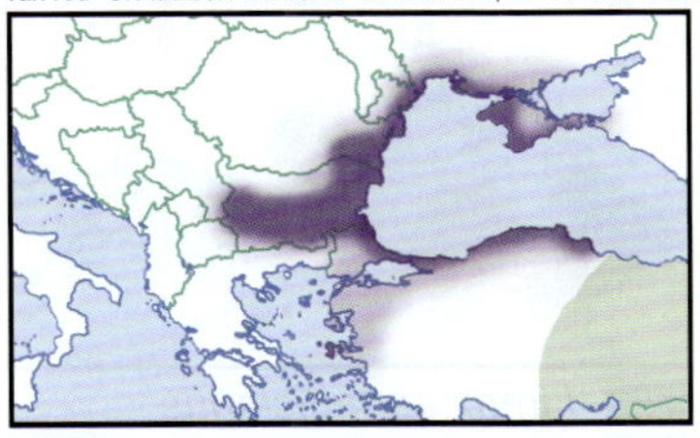

Tandonia cristata

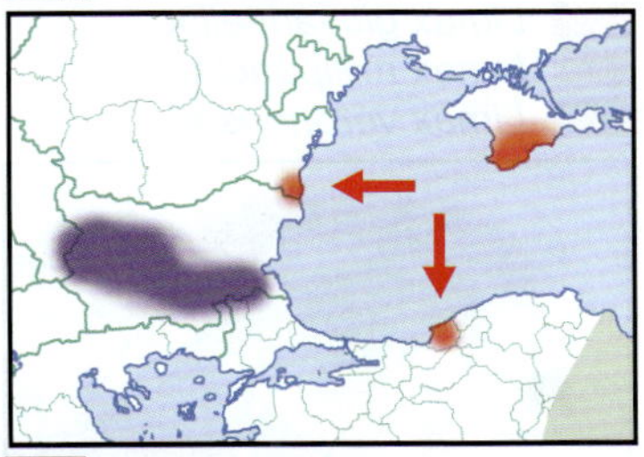

Milax parvulus
Tandonia retowskii

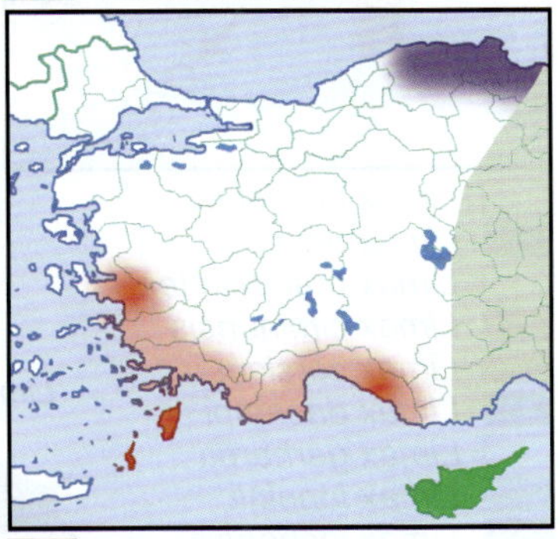

Milax altenai
Milax barypus
Tandonia samsunensis

Limax, *Boettgerilla*

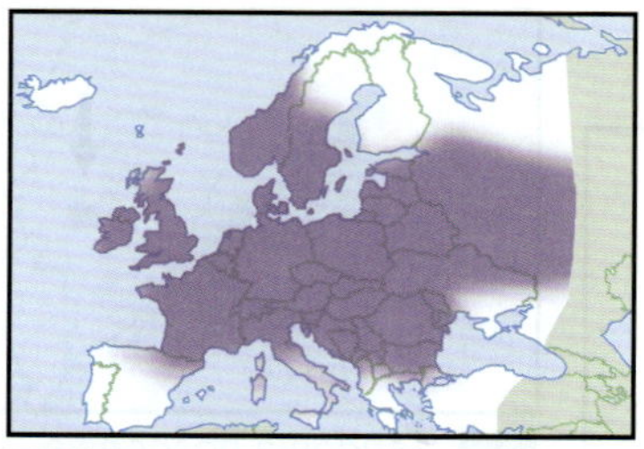

Limax cinereoniger

Boe pal

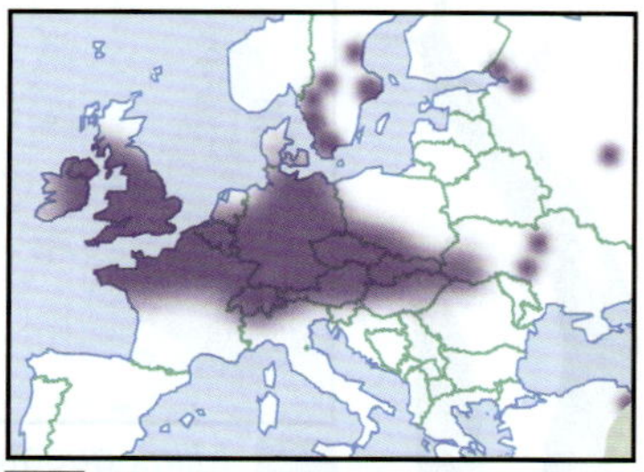

Boettgerilla pallens

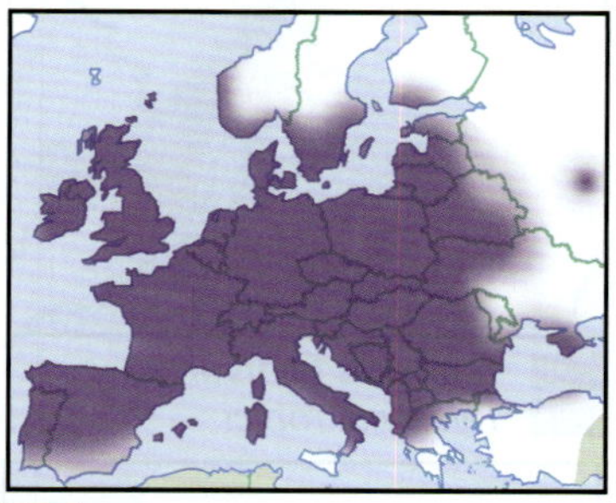

Limax maximus

Lim aeo

Lim can

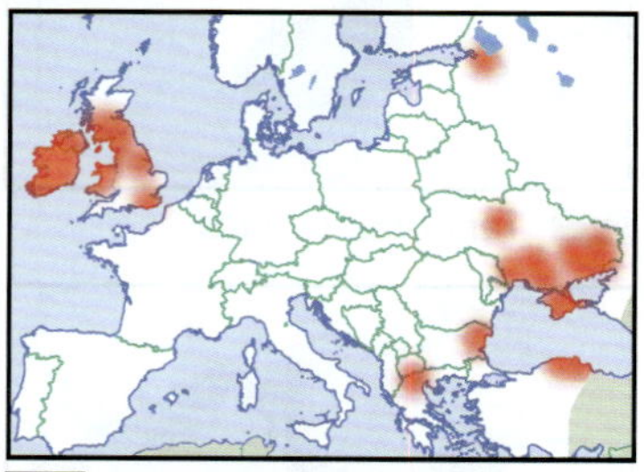

Limax ecarinatus (= *Lmc. mac.*)

Lim cim

Lim cin

Lim con

Lim dac

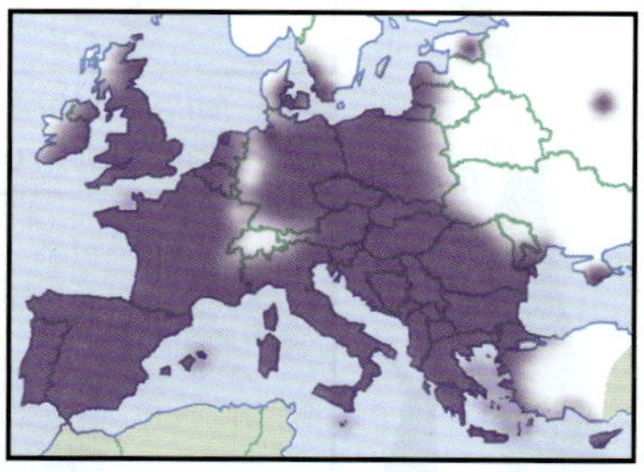

Limax flavus

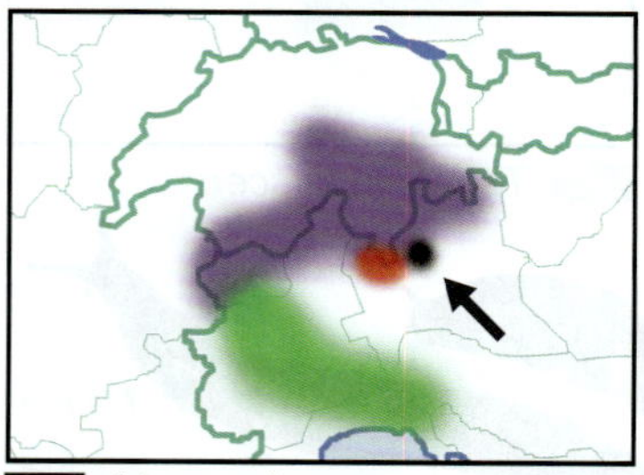

Limax pironae
Limax polipunctatus
Limax redii
Limax sarnensis

Lim fla

Lim gio

Lim ian

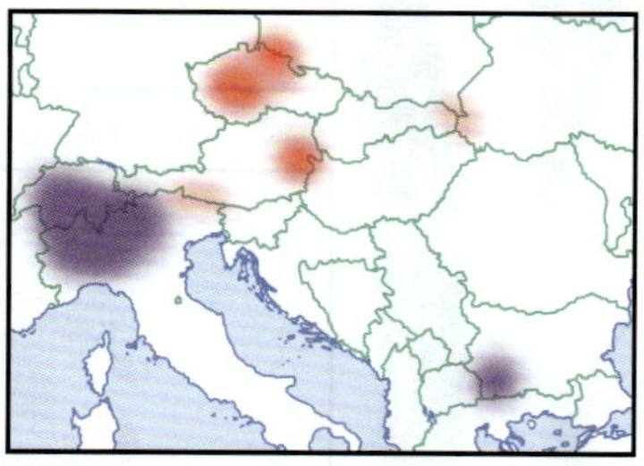

Limax bielzii
Limax subalpinus

Lim gra

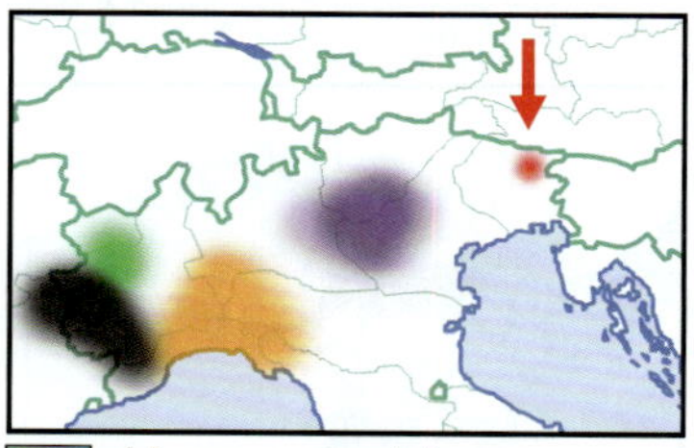

Limax canapicianus
Limax erythrus
Limax giovannellae
Limax millipunctatus
Limax veronensis

Lim max

Limax punctulatus

Lim mil

Lim pir

Lim red

Lim sar

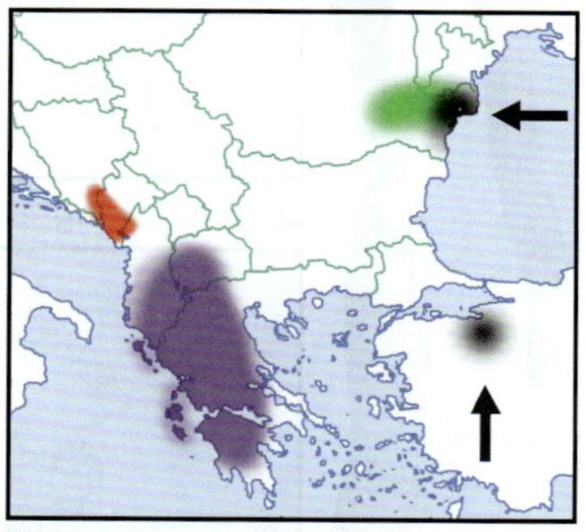

Limax cephalonicus
Limax dobrogicus
Limax wohlberedti
Limax zilchi

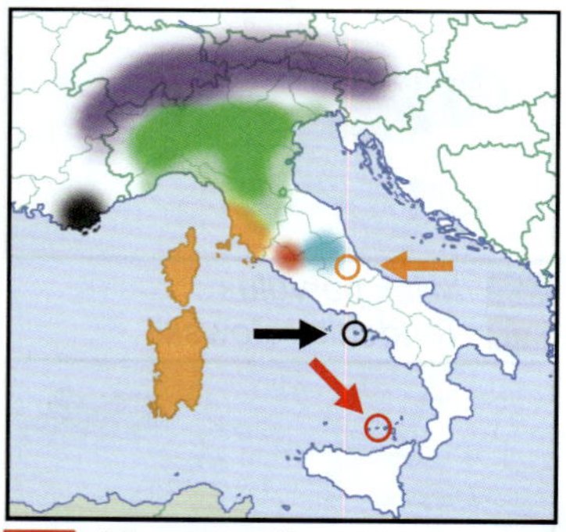

Limax aeolianus
Limax albipes
Limax brandstetteri
Limax ciminensis
Limax corsicus
Limax dacampi
Limax gerhardti
Limax ianninii
Limax lachensis
Limax squamosus

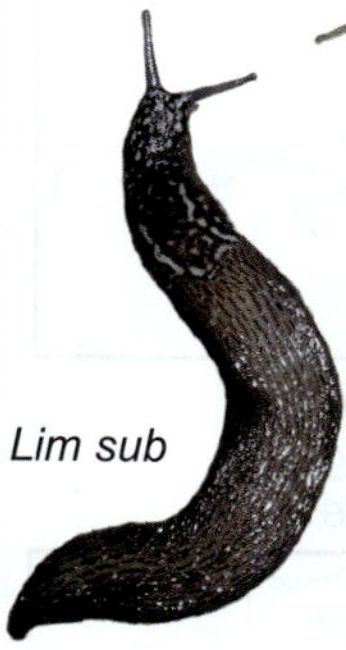

Lim sub

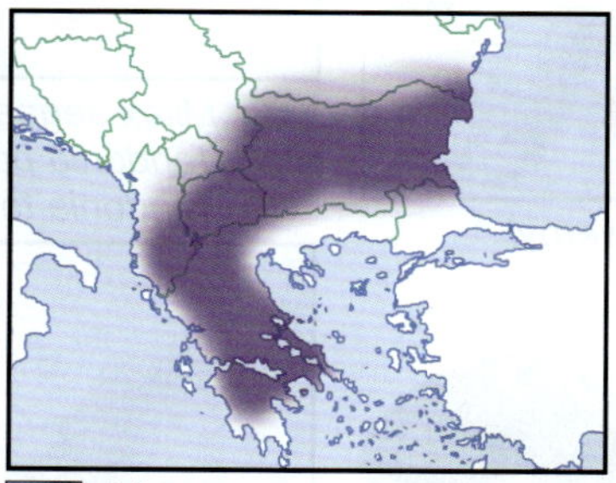

Limax graecus

Lim ver

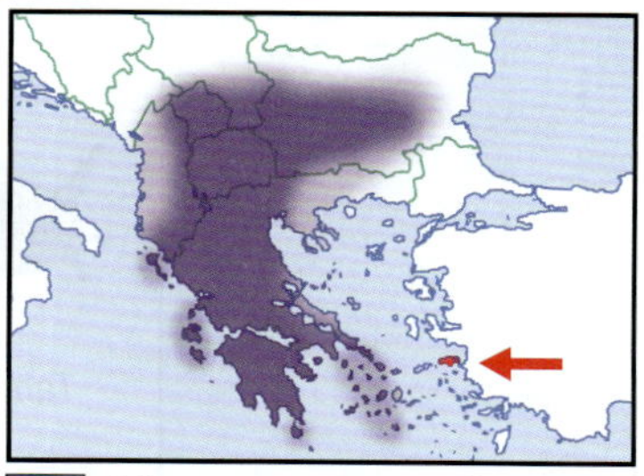

Limax conemenosi
Limax hemmeni

Lim fla D. Georgiev *Lim gra* A. Irikov *Lim red* C. Morelli *Boe pal* *Lim cin* *Lim max* J. Grego *Lim aeo* A. Reitano *Lim can* *Lim cim* *Lim gio* *Lim ian* *Lim mil* *Lim pir* *Lim sub* *Lim ver* C. M. Brandstetter *Lim con* P. Ovalis *Lim dac* E. Balocchi

Lehmannia, Malacolimax, Bielzia, Selenochlamys

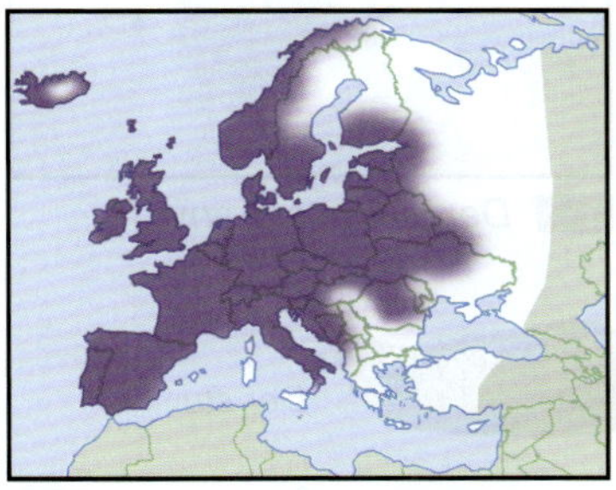

Lehmannia marginata

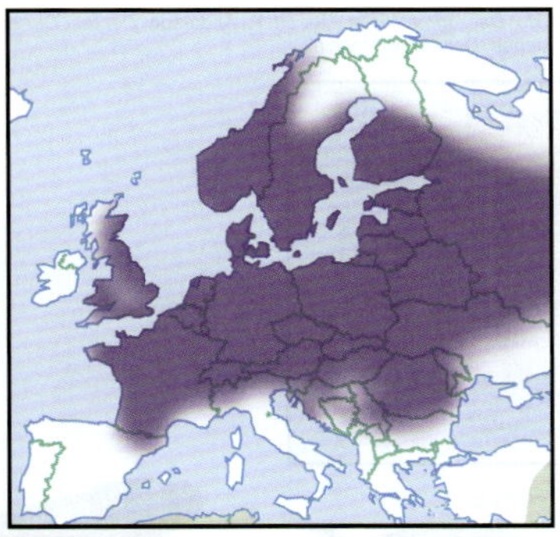

Malacolimax tenellus

Mal ten

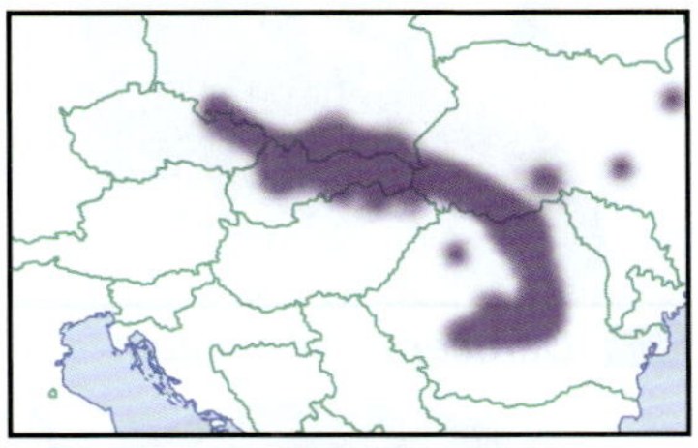

Bielzia coerulans

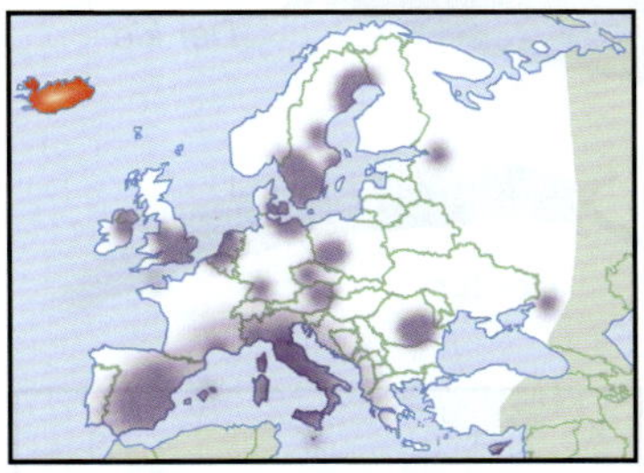

Lehmannia islandica
Lehmannia valentiana

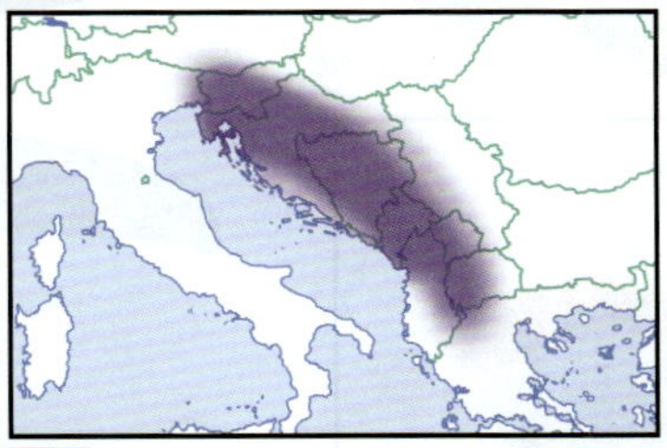

Malacolimax mrazeki

Bie coe

Leh hor

Leh val

Leh mar

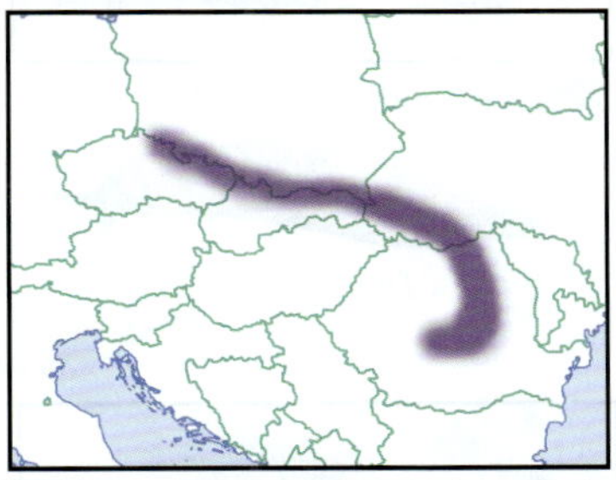

Lehmannia macroflagellata

Sel ysb

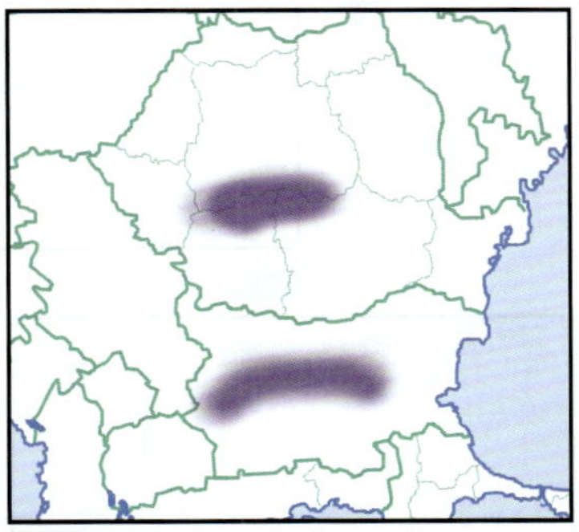

Lehmannia horezia

Selenochlamys ysbryda

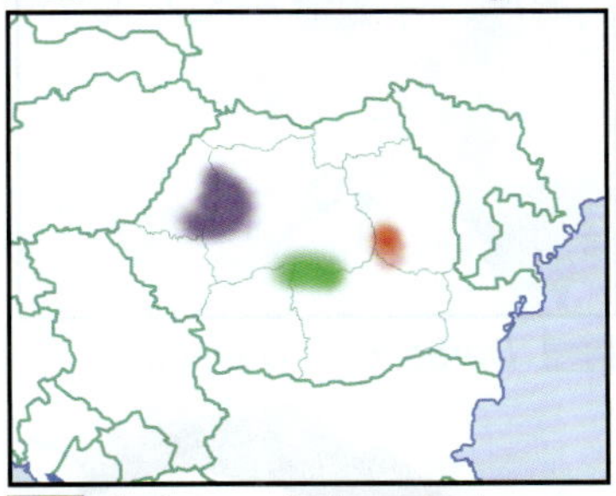

Lehmannia jaroslaviae
Lehmannia medioflagellata
Lehmannia vrancensis

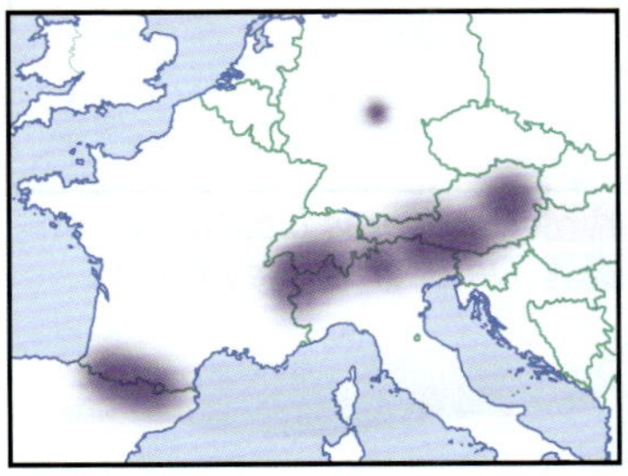

Lehmannia rupicola

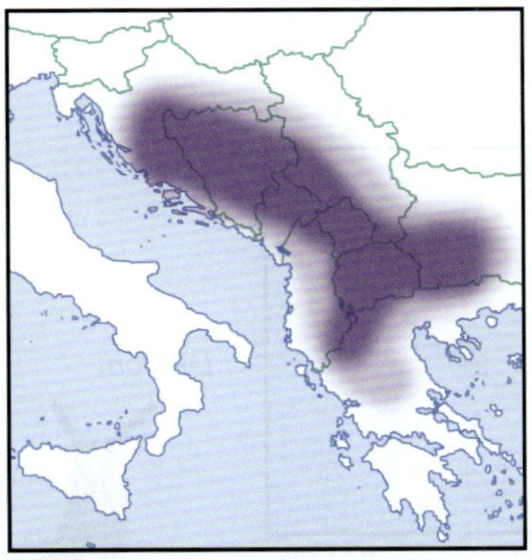

Lehmannia brunneri

Lehmannia nyctelia

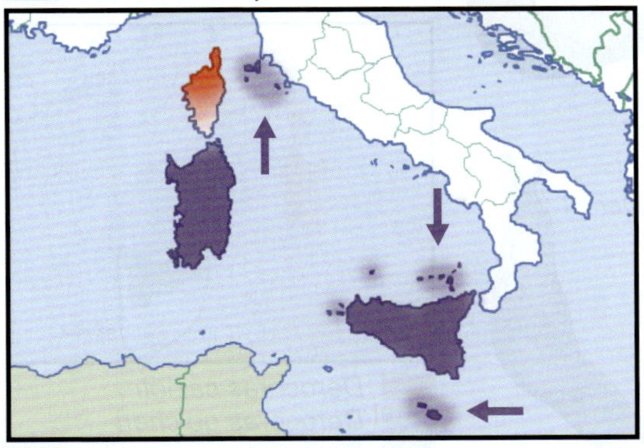

Lehmannia melitensis
Lehmannia requienii

Leh nyc

Leh mel

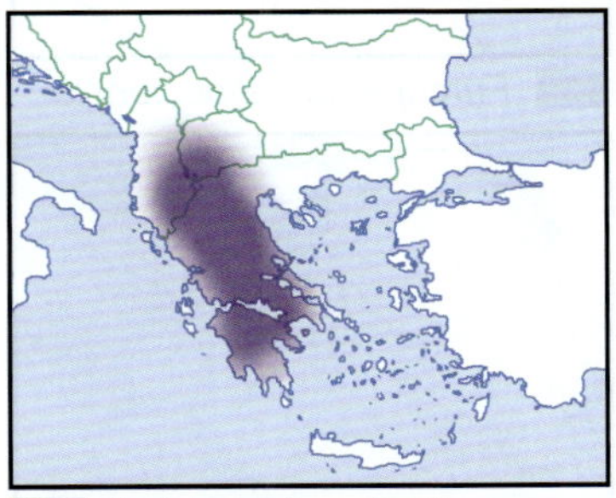

Lehmannia szigethyae

📷 *Bie coe* A. Zelenko, F. Schwabe *Leh nyc Mal ten* J. Grego *Leh hor* D. Georgiev *Leh mel* D. Cilia *Leh val* A. Örstan *Sel ysb* Mus. Wales

Agriolimacidae 1

📷 *Der agr Der kle Der rod* V Wiese *Der alt* R. Anderson *Der lae Der rtc* J. Grego *Der stu Der tur* D. Georgiev *Der pan* A. Hallgass *Der pra* M. Manas *Der sar* F. Mascia

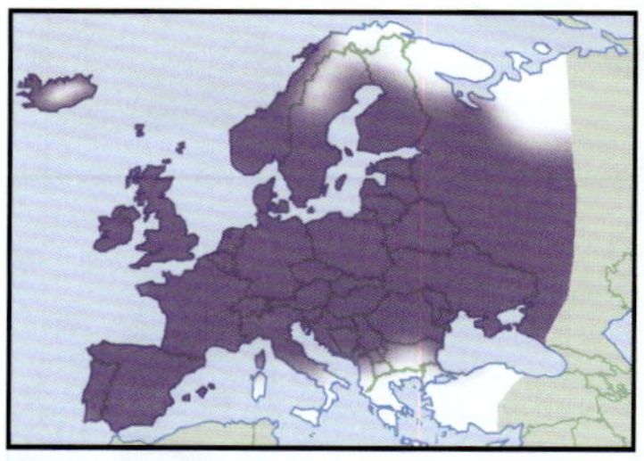

Deroceras reticulatum

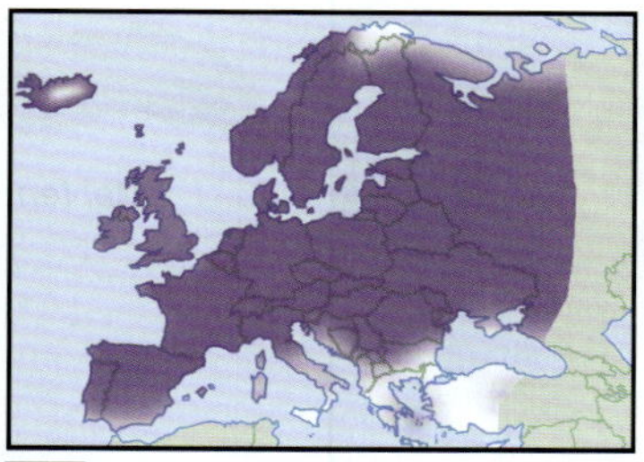

Deroceras laeve

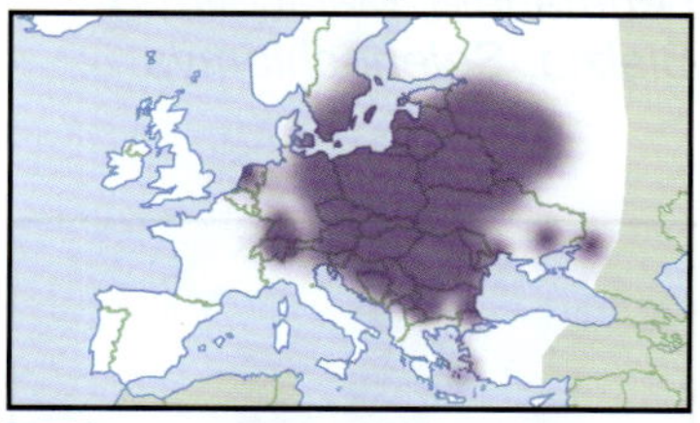

Deroceras sturanyi

Der reti

Der lae

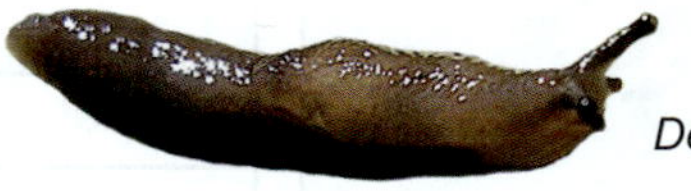

Der stu

Der kle

Der tur

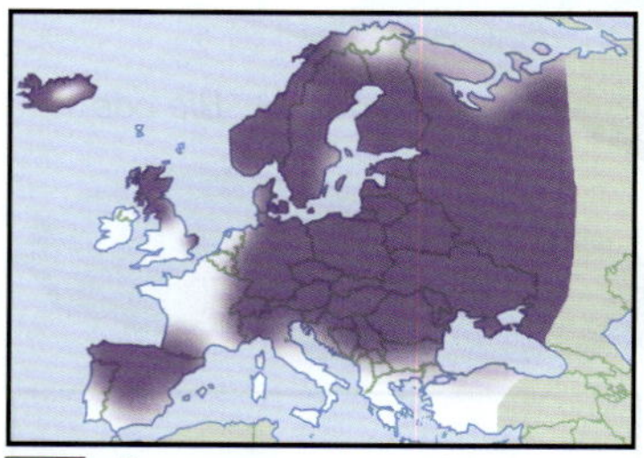

Deroceras agreste

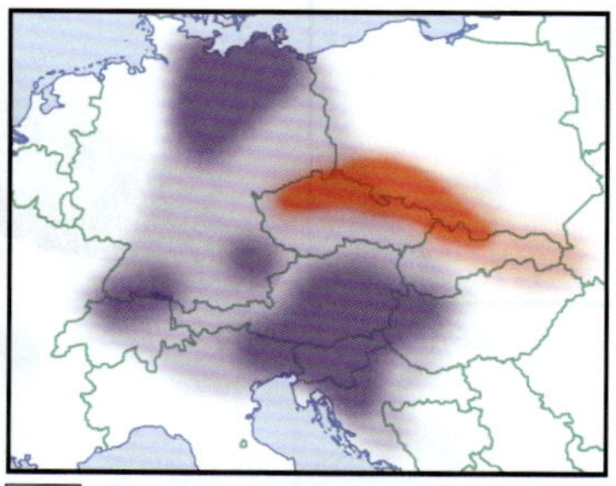

Deroceras klemmi
Deroceras praecox

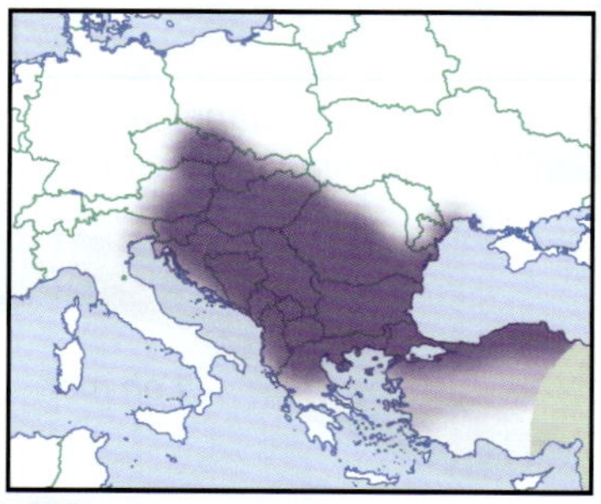

Deroceras turcicum

Der agr

Der pra

Der rod

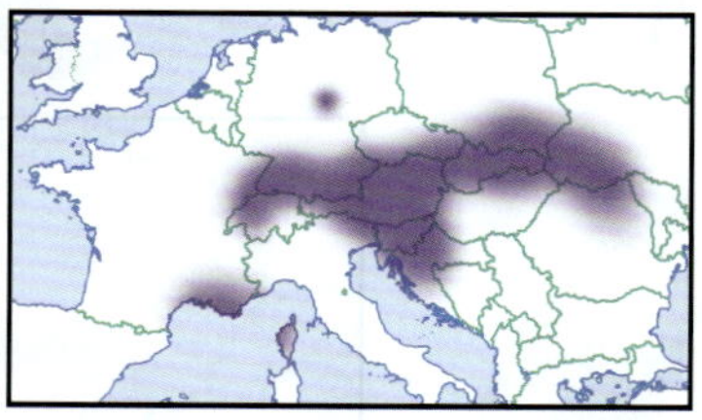

Deroceras rodnae

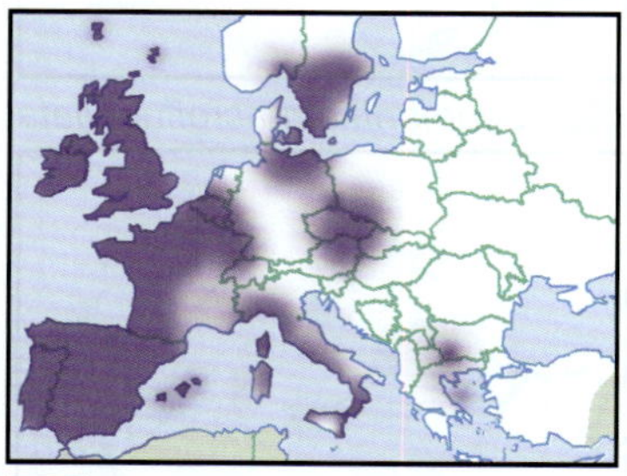

Deroceras invadens

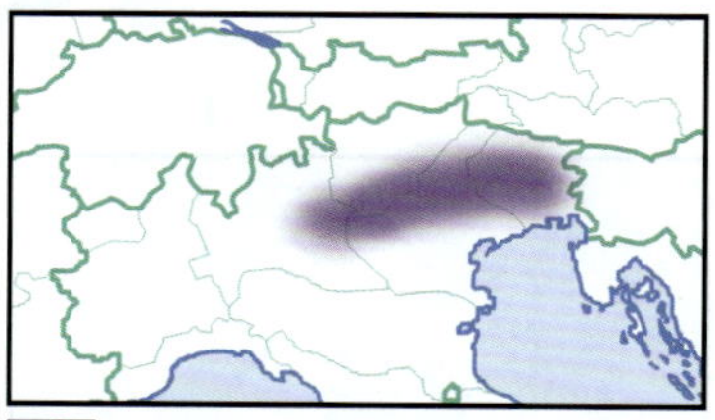

Deroceras planarioides

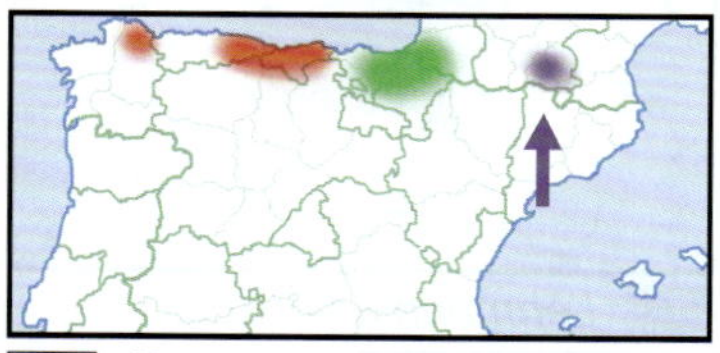

Deroceras adolphi
Deroceras ercinae
Deroceras vascoanum

Der inv

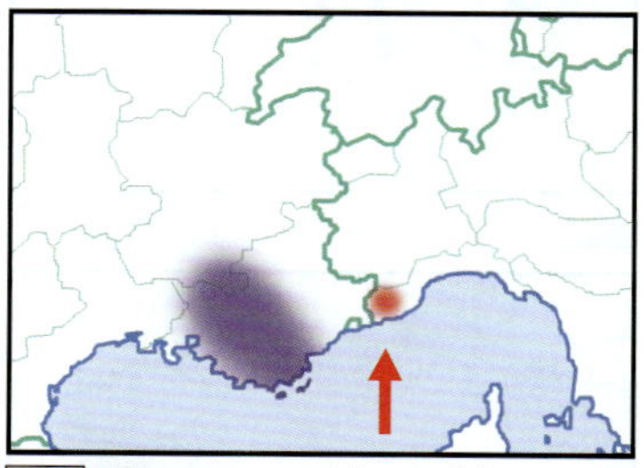

Deroceras bisacchianum
Deroceras chevallieri

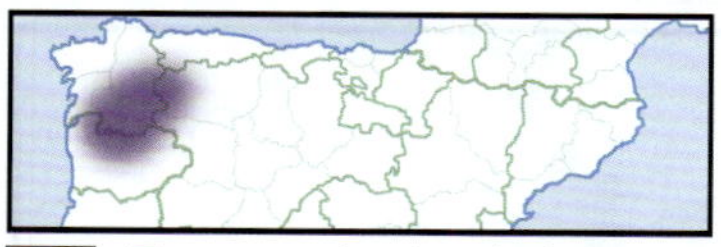

Furcopenis circularis

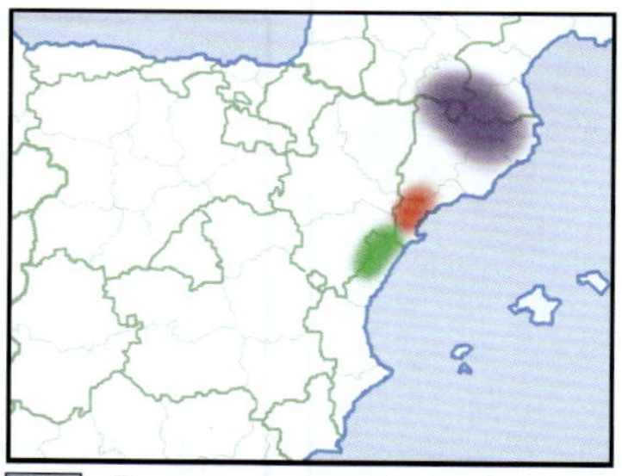

Deroceras levisarcobelum
Deroceras roblesi
Deroceras tarracense

Der lom

Der sar

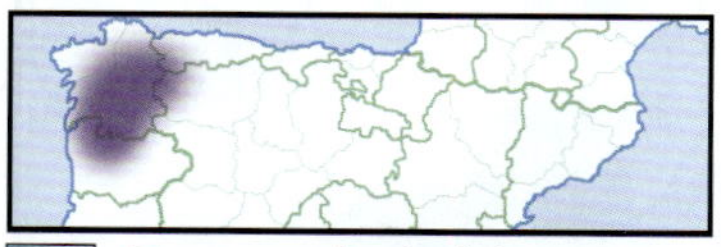

Furcopenis darioi

Furcopenis gallaeciensis

Furcopenis geresiensis

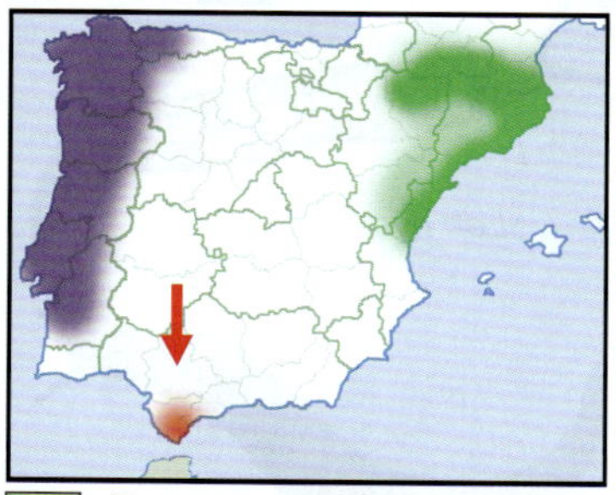

Deroceras altimirai
Deroceras lombrocoides
Deroceras ponsonbyi

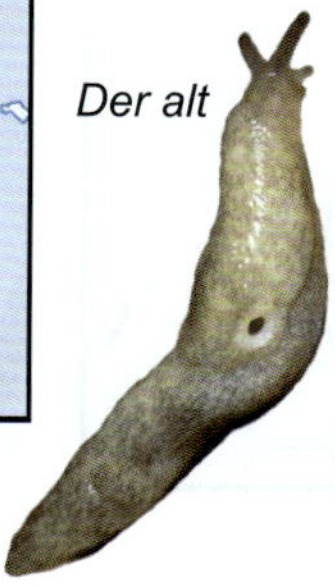

Der alt

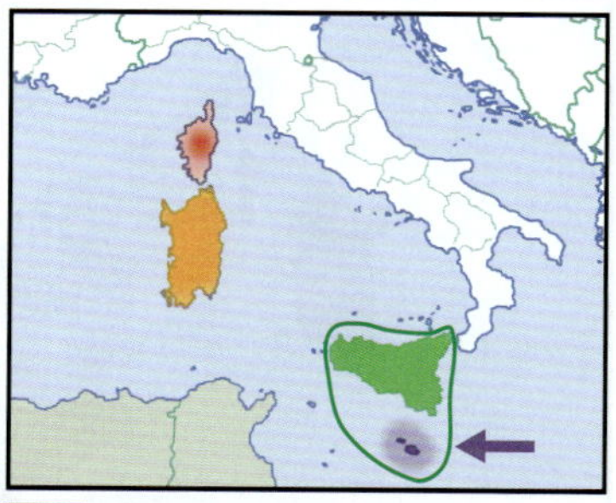

Deroceras cazioti
Deroceras golcheri
Deroceras panormitanum
Deroceras sardum

Agriolimacidae 2

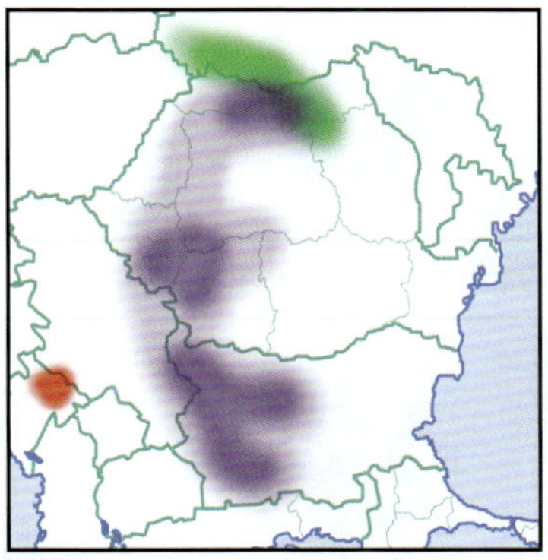

Deroceras bureschi
Deroceras maasseni
Deroceras occidentale

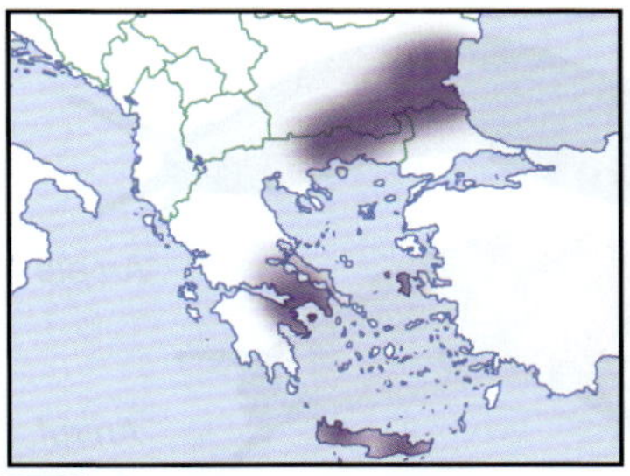

Deroceras thersites

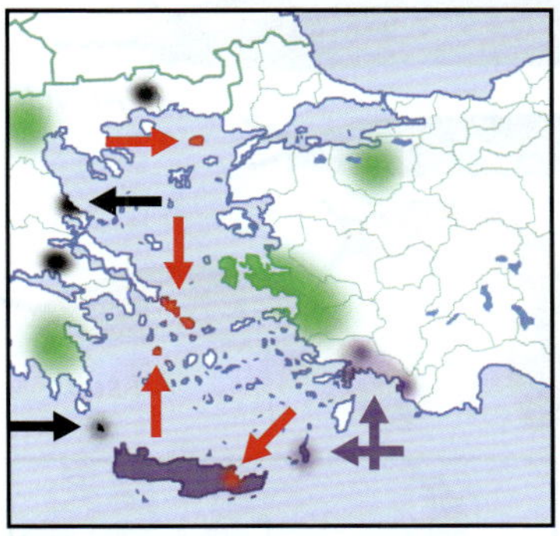

Deroceras malkini
Deroceras oertzeni
Deroceras parnasium
Deroceras rethimnonense

Kry urb

Kry mel

Der the

Der oer

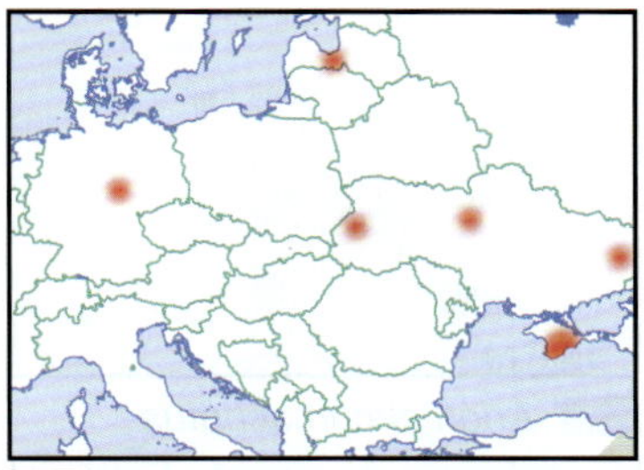

Krynickillus melanocephalus

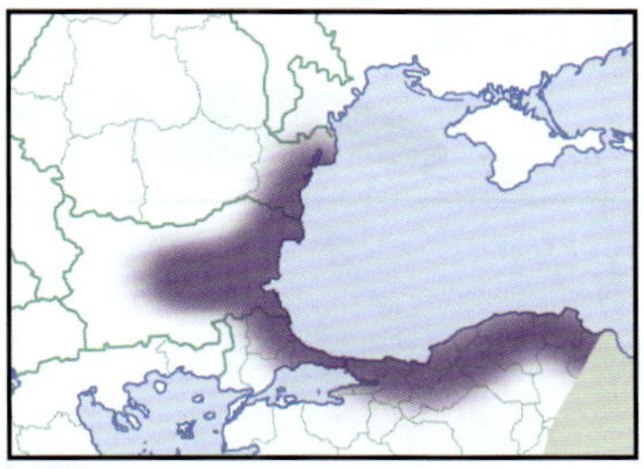

Krynickillus urbanskii

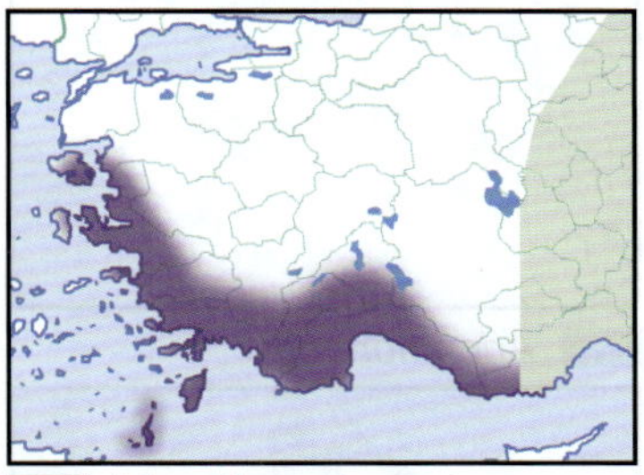

Mesolimax brauni

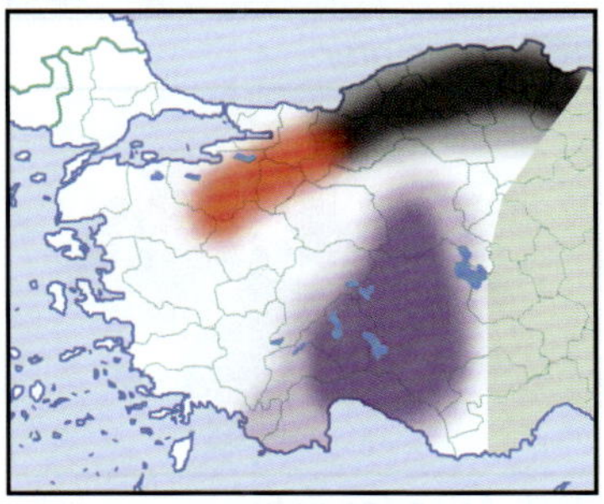

Krynickillus hoplites
Megalopelte simrothi
Mesolimax escherichi

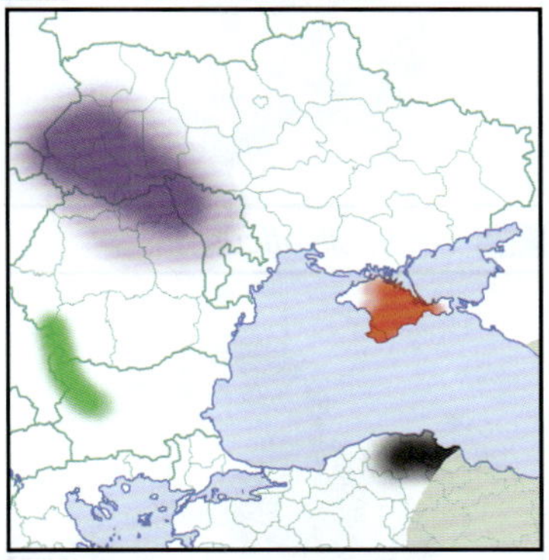

Deroceras bakurianum
Deroceras moldavicum
Deroceras tauricum
Deroceras zilchi

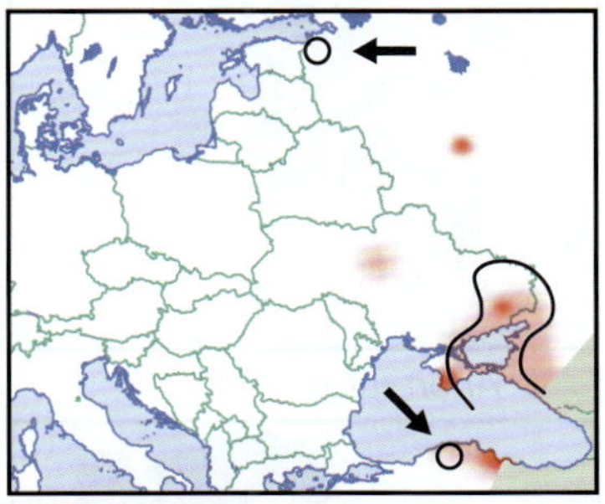

Deroceras caucasicum
Deroceras subagreste

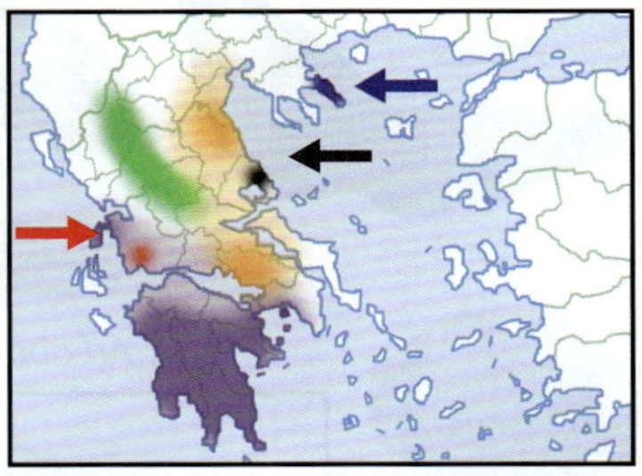

Deroceras bistimulatum
Deroceras boeoticum
Deroceras halieos
Deroceras karnainense
Deroceras pageti
Der. pseudopanormitanum

Der pag

Der ber

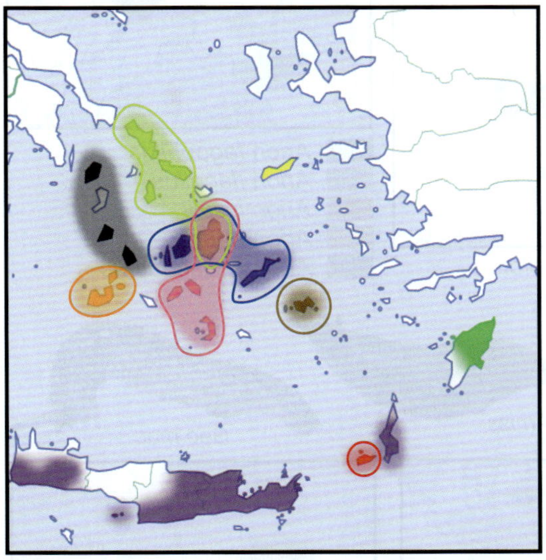

Deroceras astypalaeense
Deroceras cycladicum
Deroceras ikaria
Deroceras kasium
Deroceras korthionense
Deroceras lasithionense
Deroceras melinum
Deroceras neuteboomi
Deroceras parium
Deroceras seriphium

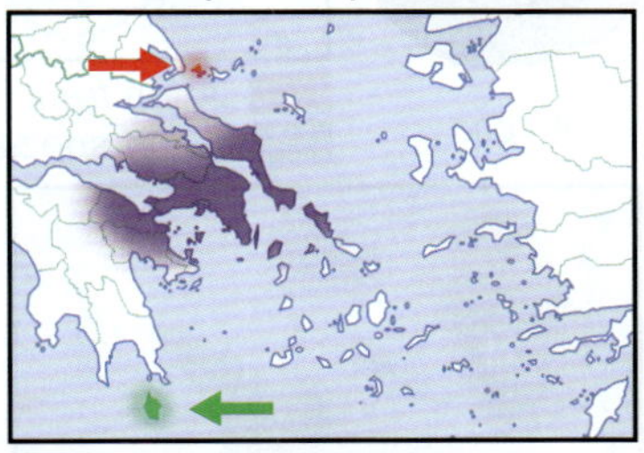

Deroceras keaense
Deroceras kythirense
Deroceras nyphoni

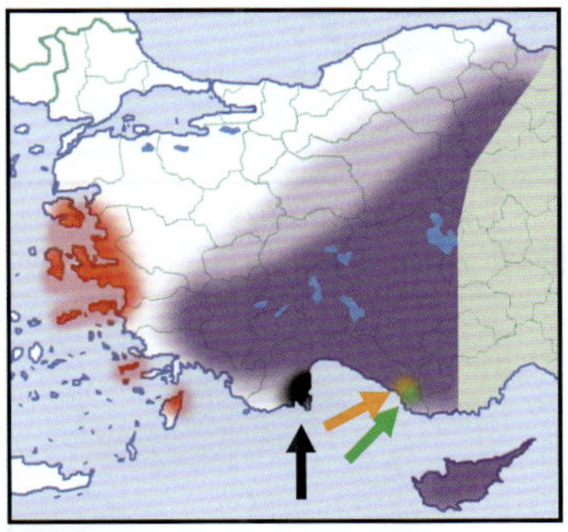

Deroceras berytense
Deroceras christae
Deroceras demirtense
Deroceras helicoidale
Deroceras samium

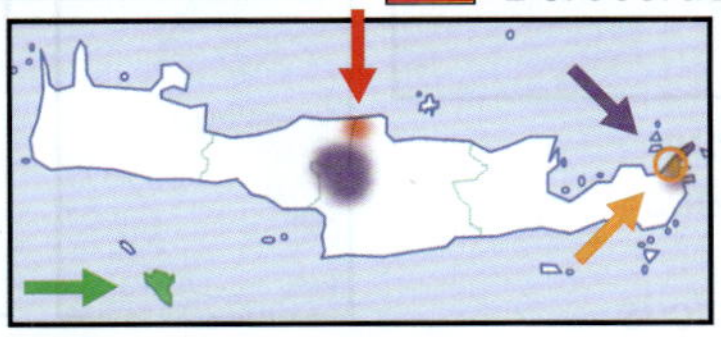

Deroceras gavdosense
Deroceras dewinteri
Deroceras gorgonium
Deroceras monoicum

Der ber B. A. Gümüş
Der gav P. Van Herck
Der oer Der the Kry urb D. Georgiev
Der pag P. Ovalis
Kry mel S. Leonov

Der gav

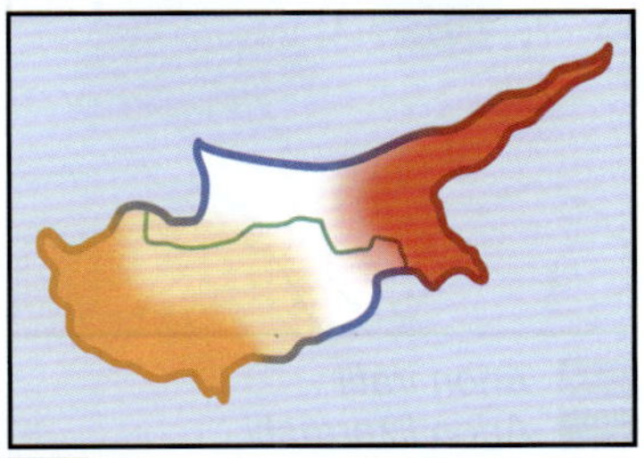

Der. chrysorroyatissense
Deroceras famagustense

Arionidae

Ari sub

Arion ater

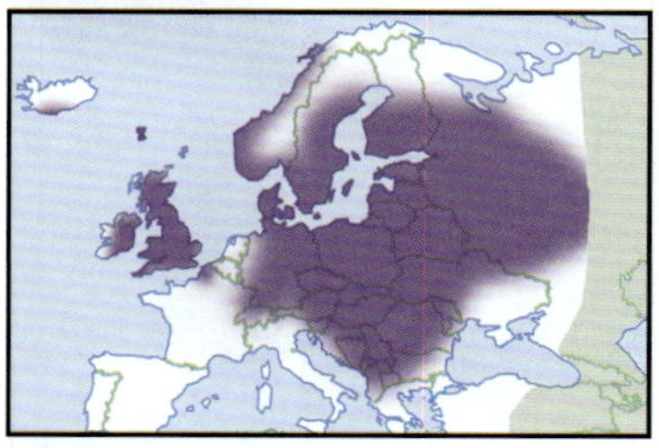
Arion fasciatus

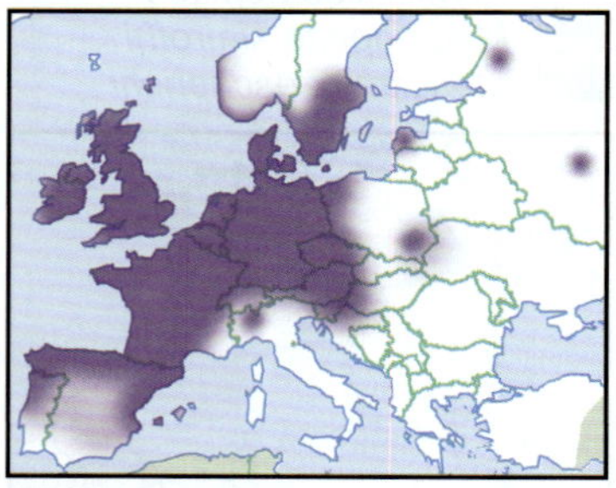
Arion rufus

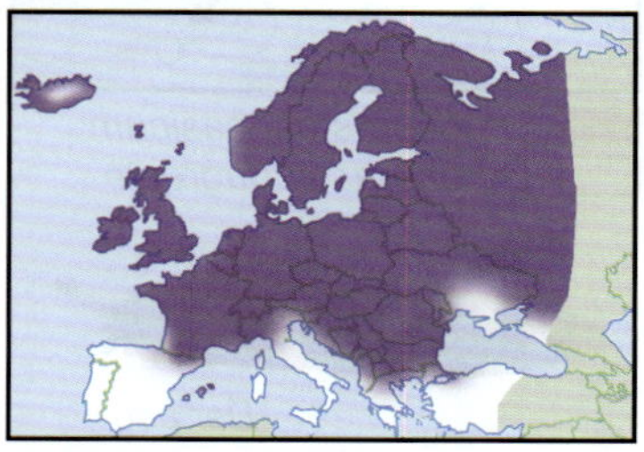
Arion subfuscus, A. fuscus

Arion occultus
Arion owenii

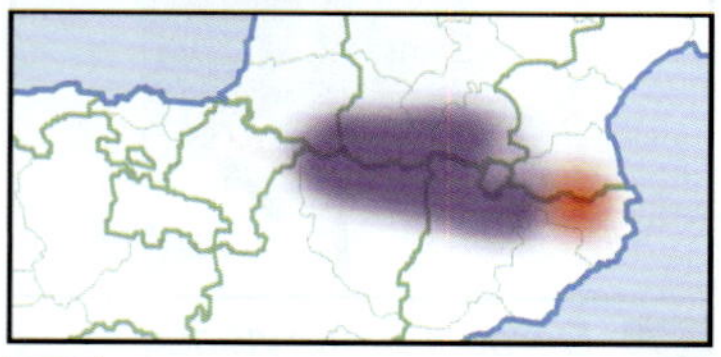
Arion anthracius
Arion magnus

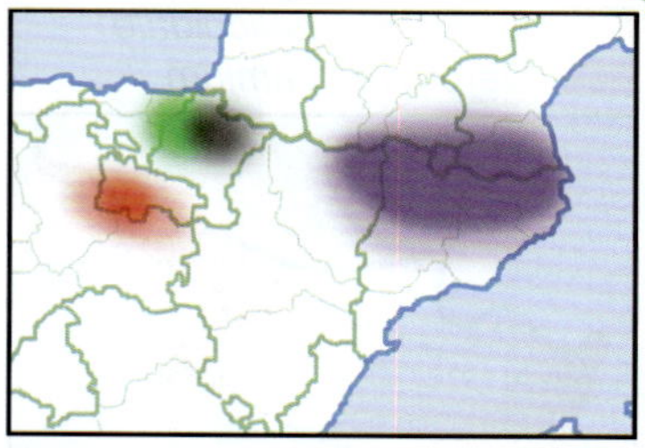
Arion iratii
Arion lizarrustii
Arion molinae
Arion wiktori

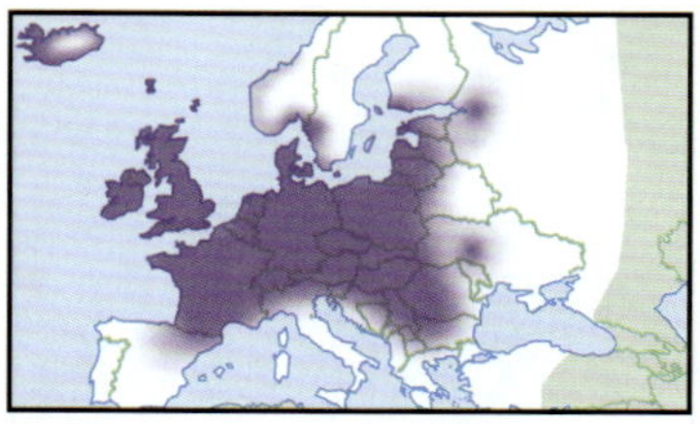
Arion circumscriptus

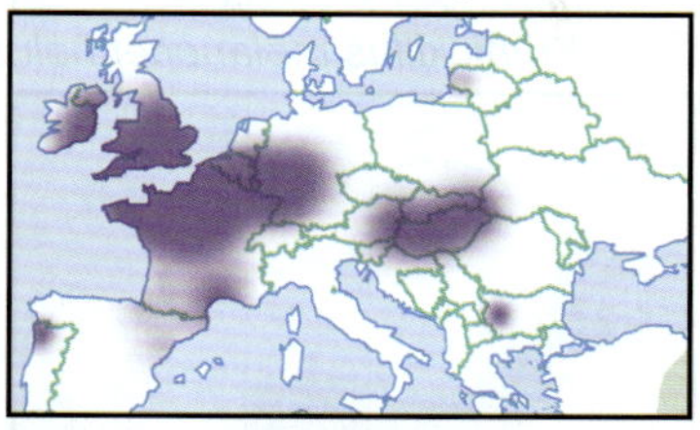
Arion hortensis

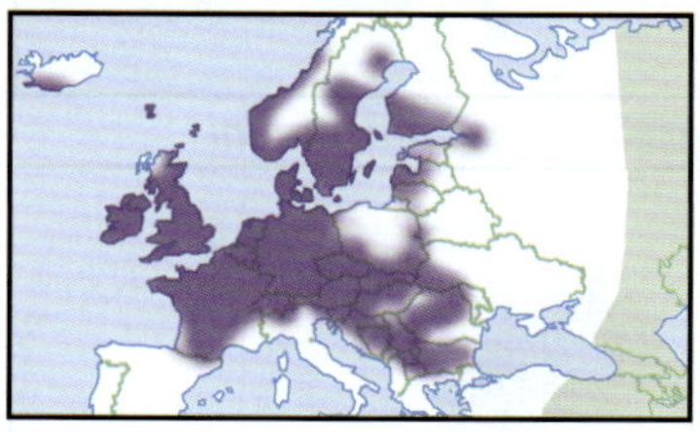
Arion silvaticus

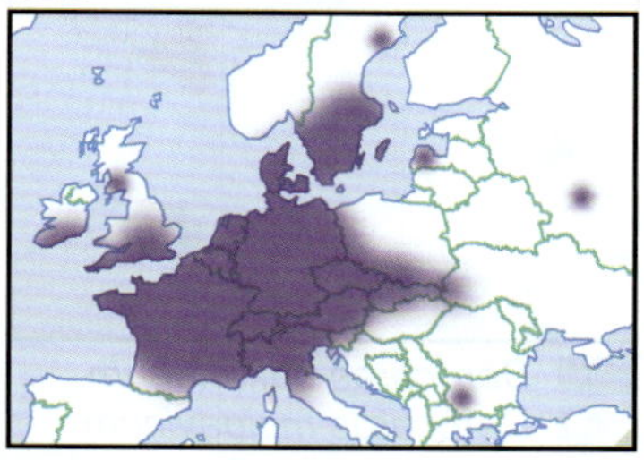
Arion vulgaris

Ari lus

Ari fla

Ari obe

Ari sil

Ari occ

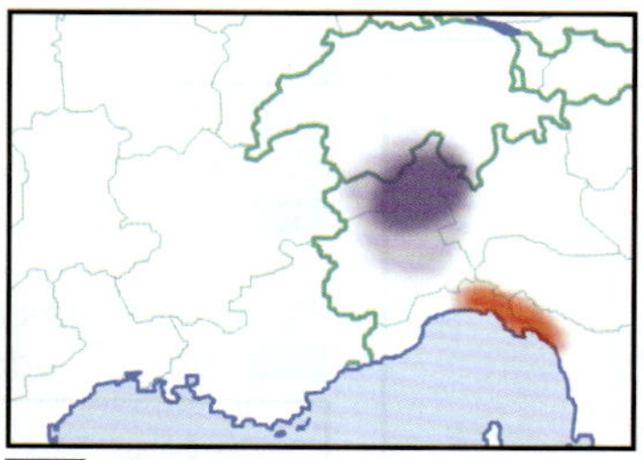
Arion franciscoloi
Arion speziae

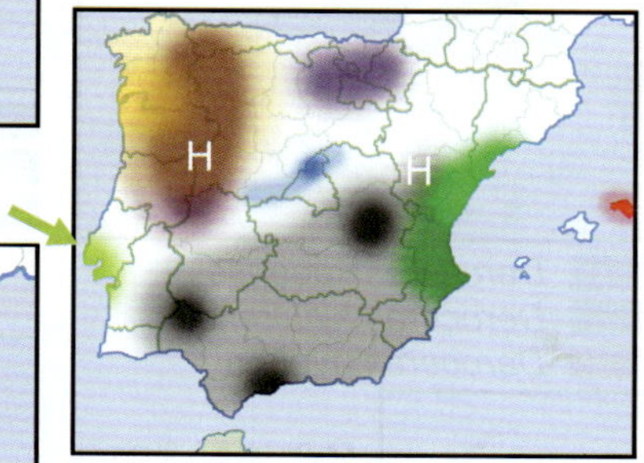

Arion baeticus
Arion fuligineus
Arion gilvus
H Arion hispanicus
Arion lusitanicus
Arion paularensis
Arion ponsi
Arion urbiae

Ari fla Ari occ Ari owe Geo mac R. Anderson Ari fus Ari hor Ari sub I. Richling
Ari dis Ari obe M. Mañas
Ari lus G. M. Rosa
Ari sil D. Georgiev
Ari sub V. Wiese
Ari iss F. Mascia

Ari dis

Ari owe

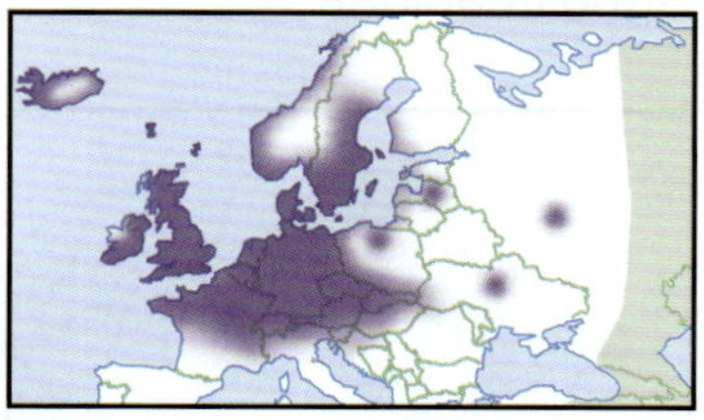
Arion distinctus

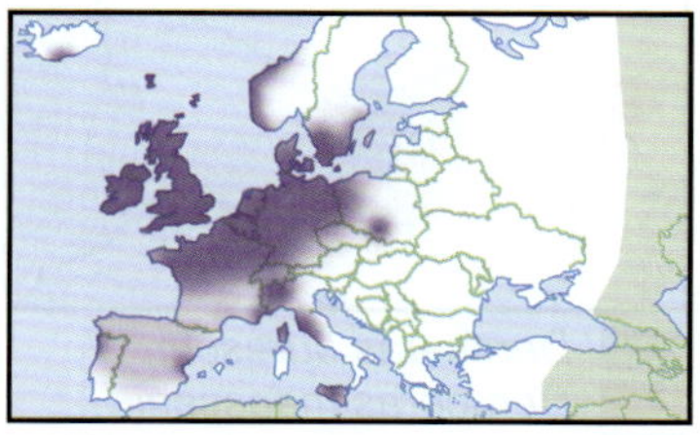
Arion intermedius

Ari ate

Ari ruf

Ari vul

Ari hor

Ari fus

Ari sub

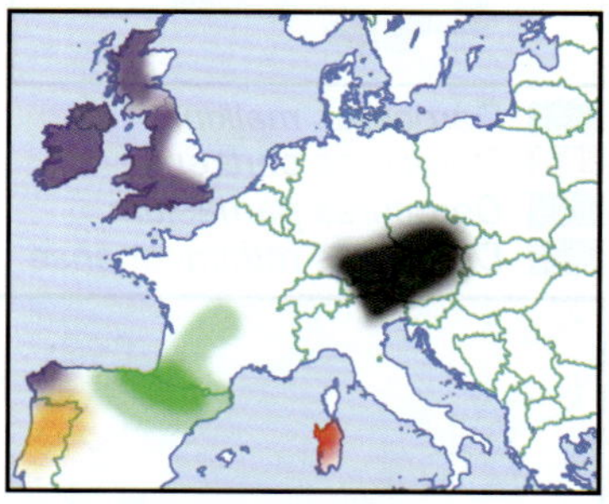
Arion fagophilus
Arion flagellus
Arion isselii
Arion nobrei
Arion obesoductus

Ari iss

Geo mac

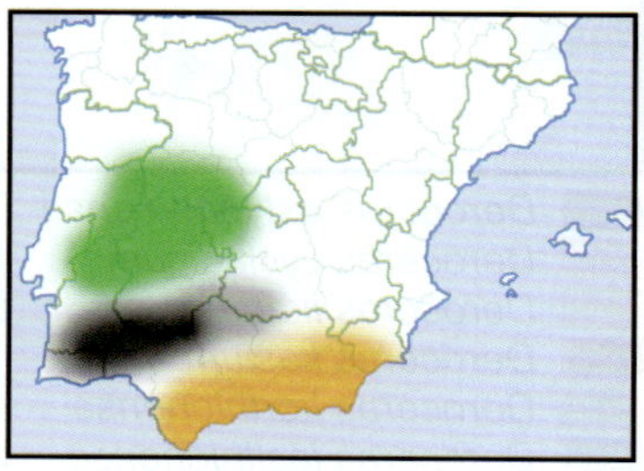
Geomalacus anguiformis
Geomalacus maculosus
Geomalacus moreleti
Geomalacus oliveirae

Sphincterochila, Trissexodontidae

Caracollina, *Gasullia*, *Gasulliella*, *Gittenbergeria*, *Oestophora*, *Oestophorella*, *Suboestophora*, *Mastigophallus*, *Trissexodon*

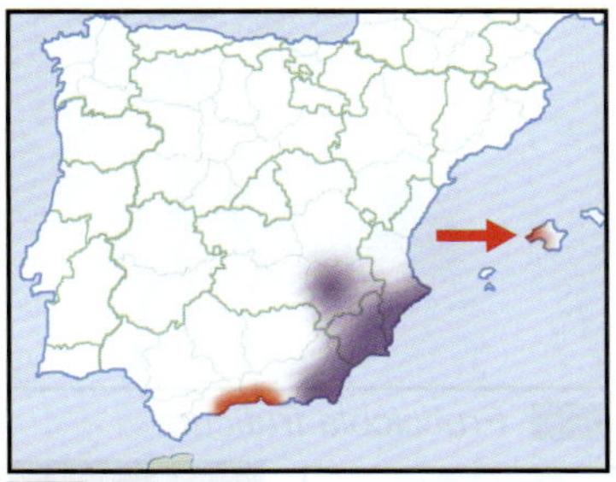

Sphincterochila baetica
Sphincterochila cariosula

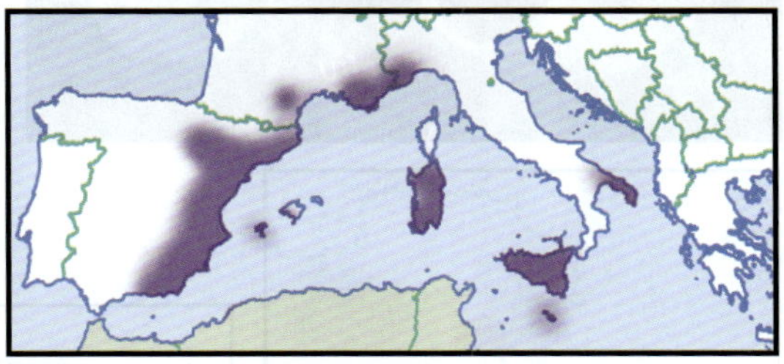

Sphincterochila candidissima

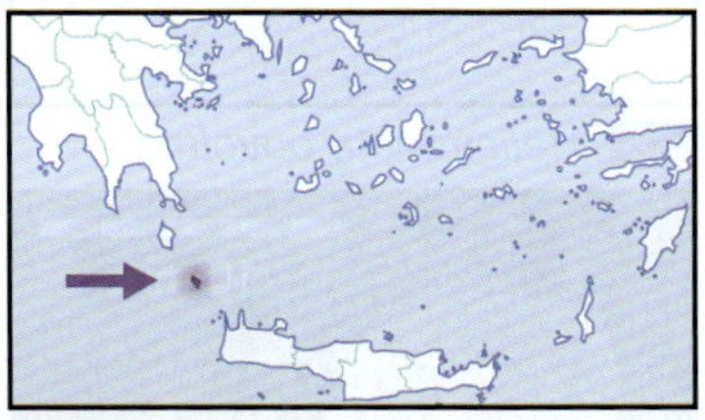

Sphincterochila insularis

Oestophora barbella

Caracollina lenticula

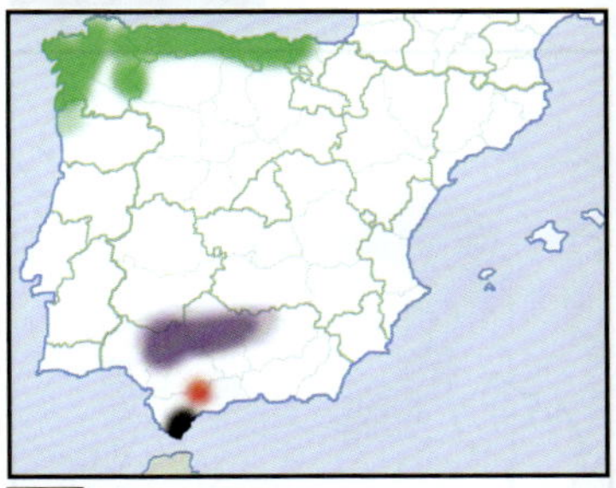

Gasullia zapateri
Oestophora dorotheae
Oestophora granesae
Oestophora silvae

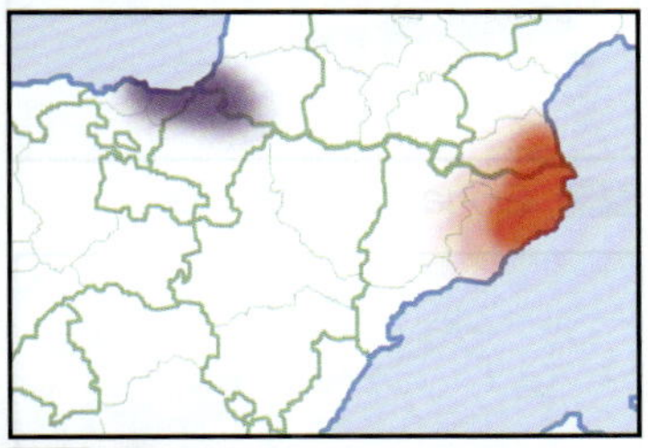

Trissexodon constrictus
Mastigophallus rangianus

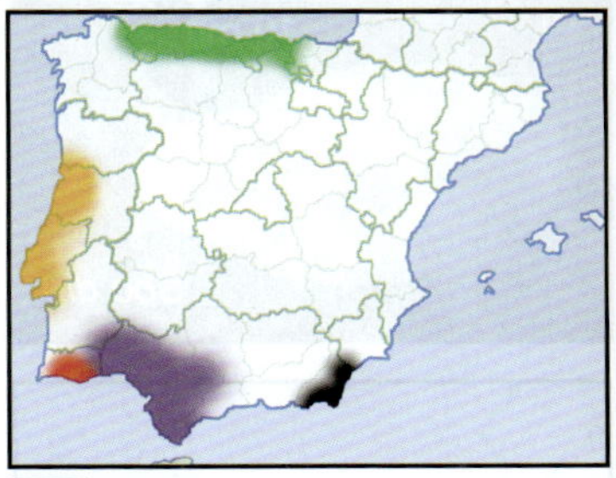

Gasullia cobosi
Gasulliella simplicula
Gittenbergeria turriplana
Oestophora barbula
Oestophorella buvinieri

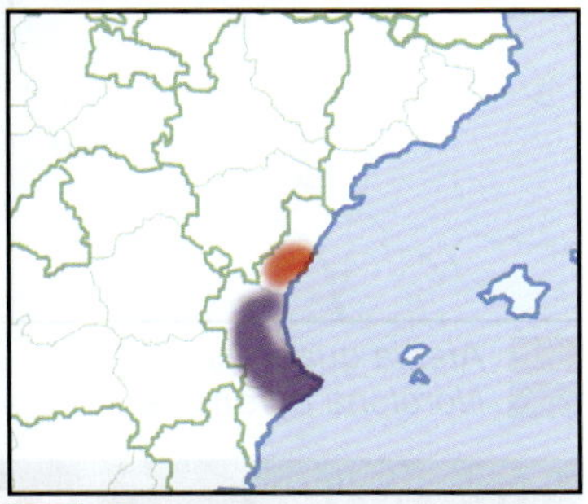

Suboestophora altimirai
Suboestophora boscae

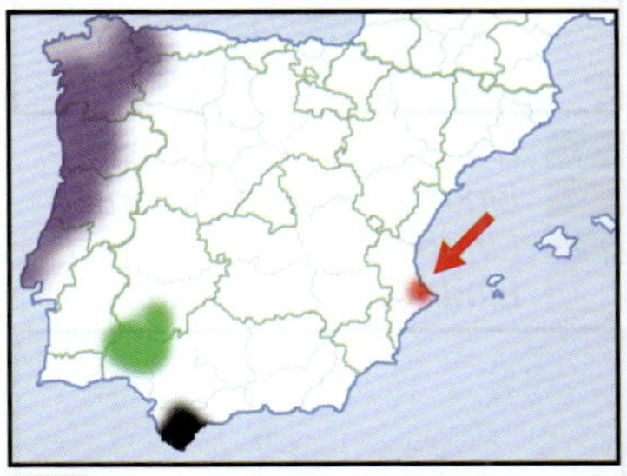

Oestophora lusitanica
Oestophora ortizi
Suboestophora gasulli
Suboestophora jeresae

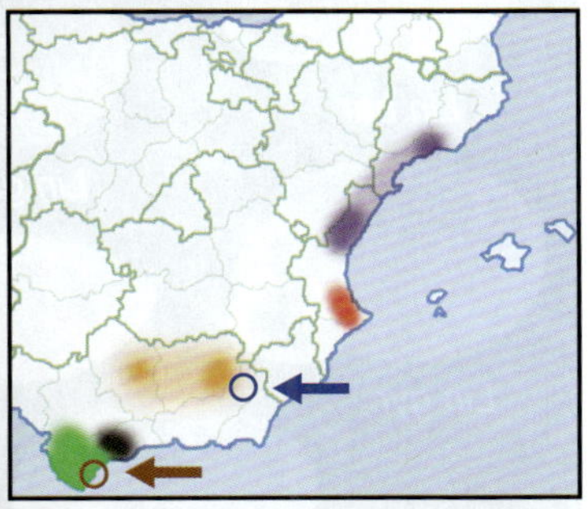

Suboestophora hispanica
Suboestoph. tarraconensis
Oestophora calpeana
Oestophora ebria
Oestophora mariae
Oestophora prietoi
Oestophora tarnieri

Oes ort J. Liétor Gallego *Sub gas* R. Vázquez Romero *Oes sil* *Sub alt* *Sub tar* C. Geerts

Bradybaenidae, Elonidae, Helicodontidae

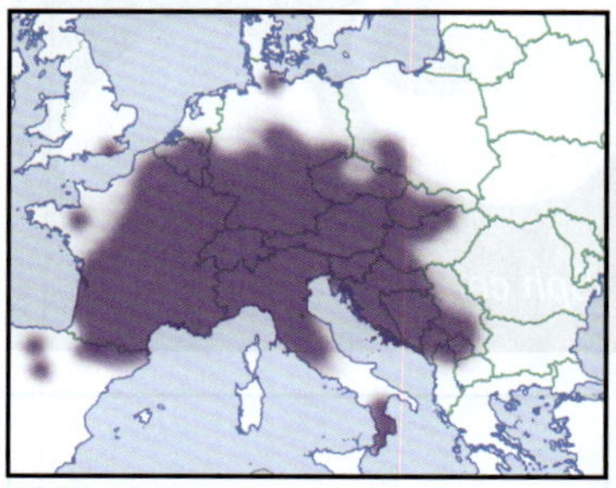

Helicodonta obvoluta

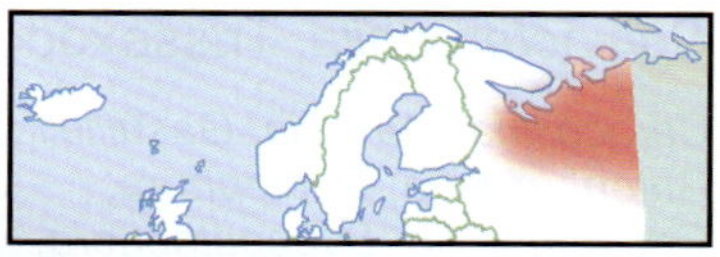

Fruticicola schrenkii

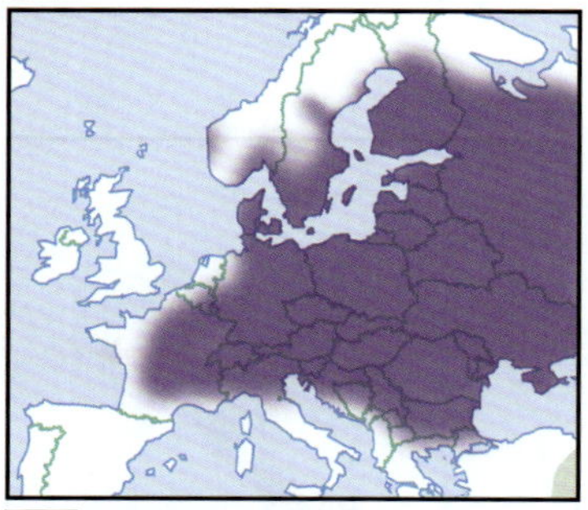

Fruticicola fruticum

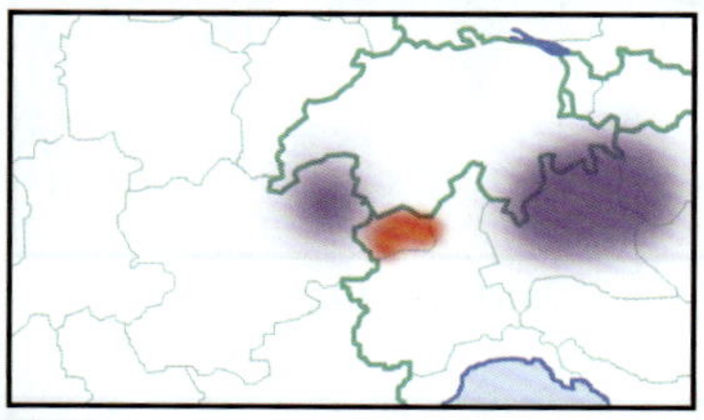

Helicodonta angigyra
Falkneria camerani

Drepanostoma nautiliforme

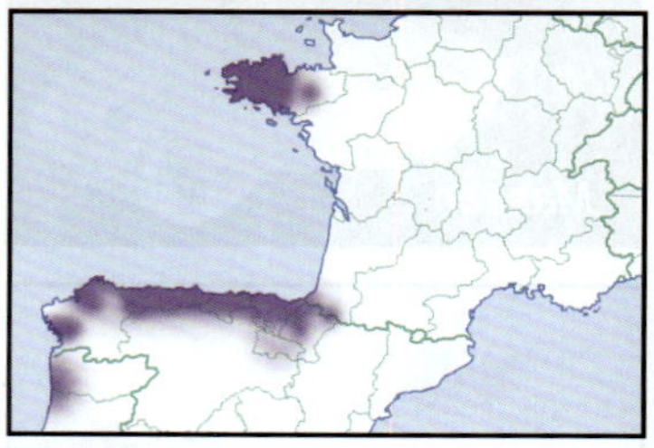

Elona quimperiana

Nor pyr R. Reuselaars

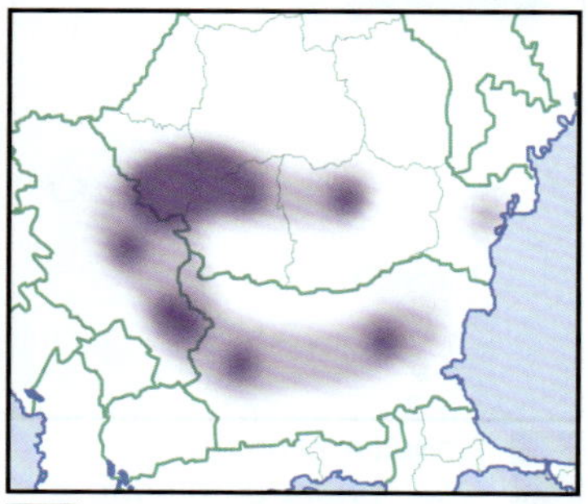

Soosia diodonta

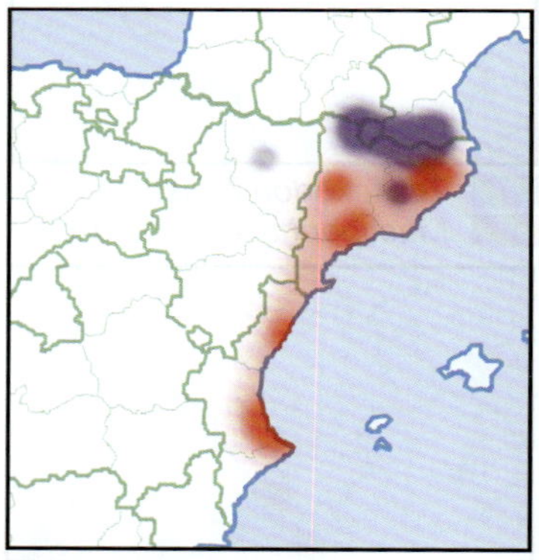

Atenia quadrasi
Norelona pyrenaica

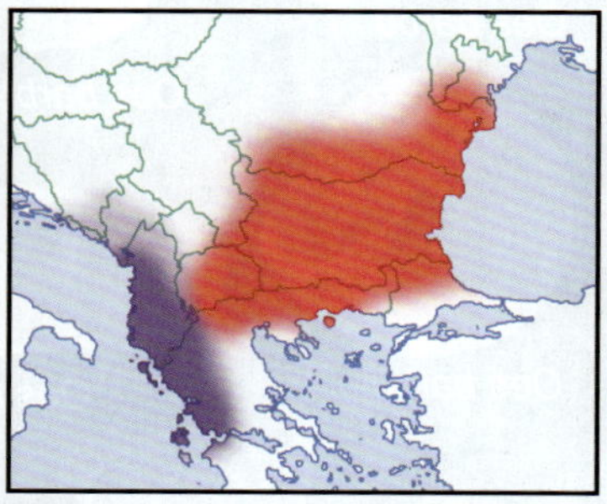

Lindholmiola corcyrensis
Lindholmiola girva

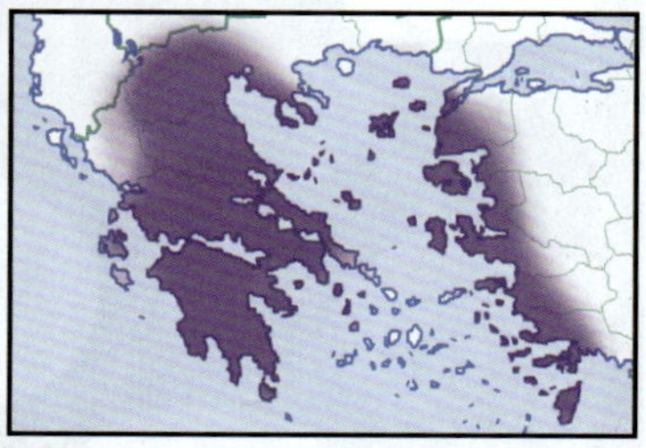

Lindholmiola lens

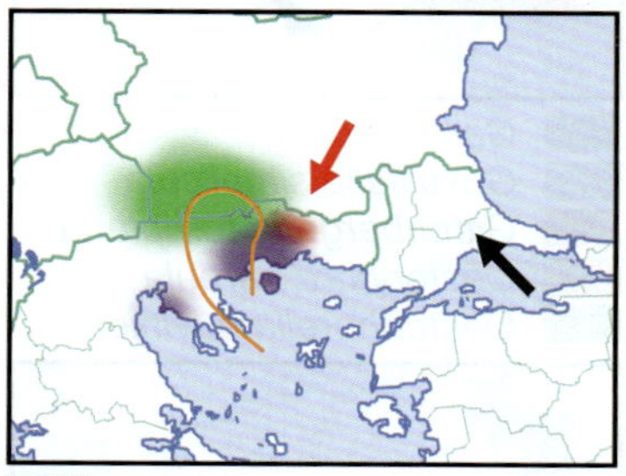

Lindholmiola pirinensis
Lindholmiola regisborisi
Lindholmiola reischuetzi
Lindholmiola spectabilis

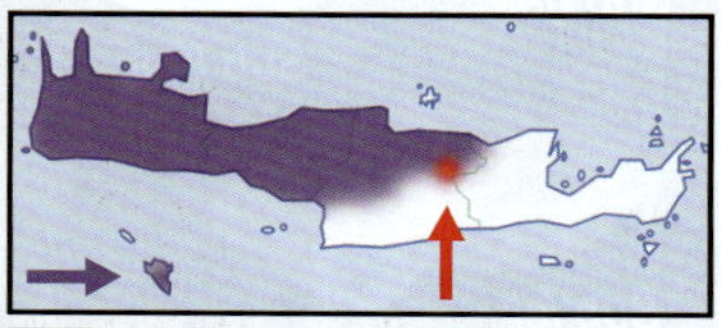

Helicodonta wilhelminae
Lindholmiola barbata

Lindholmiola gyria

Hygromiidae 1 EU
Widespread species

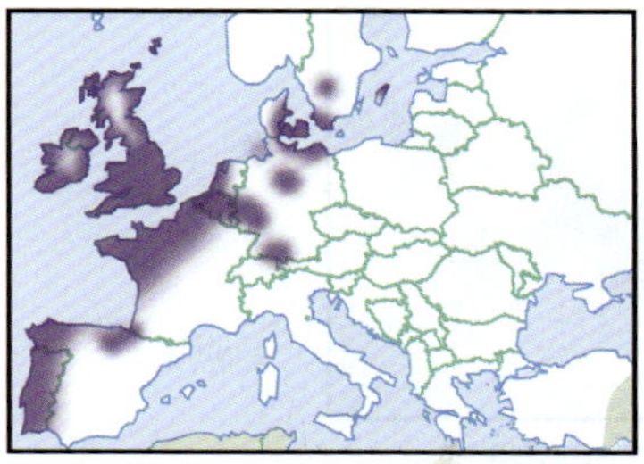

Candidula intersecta

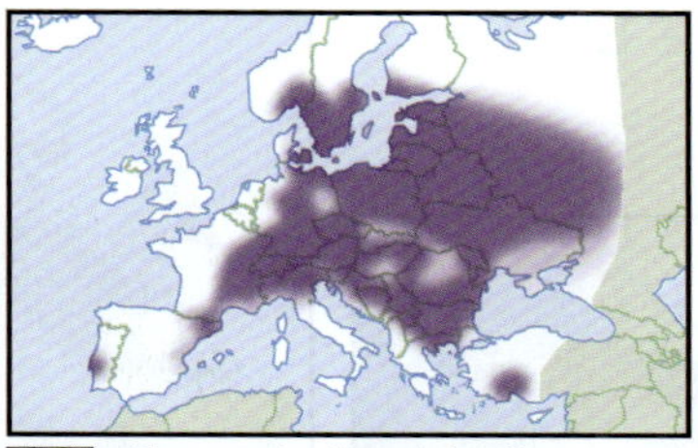

Euomphalia strigella

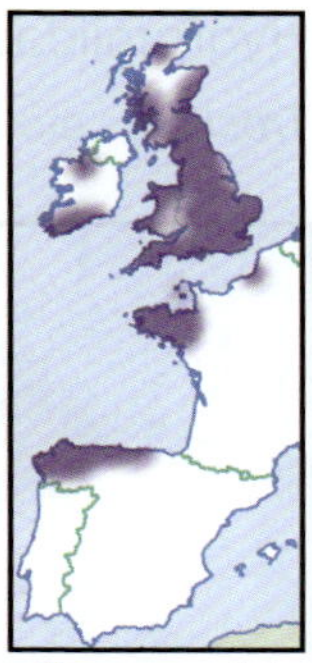

Ashfordia granulata

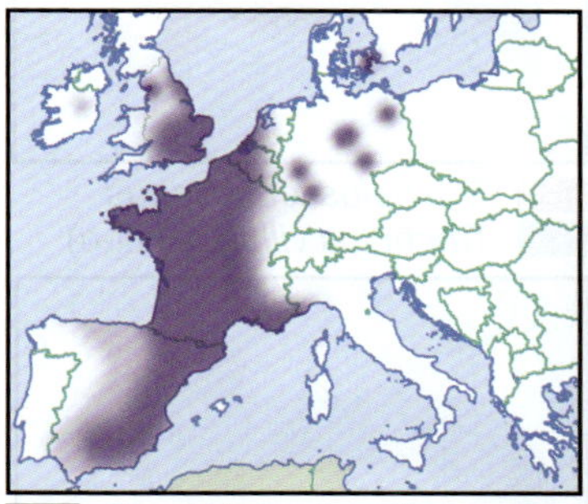

Candidula gigaxii

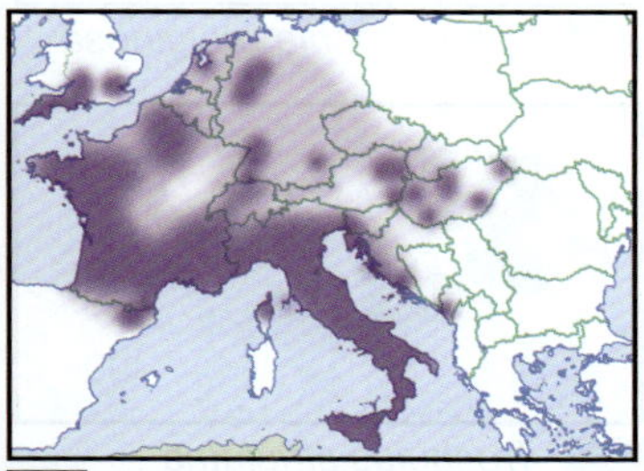

Hygromia cinctella

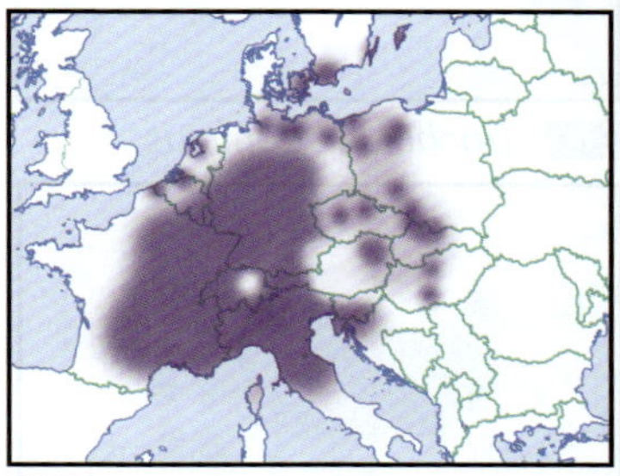

Candidula unifasciata

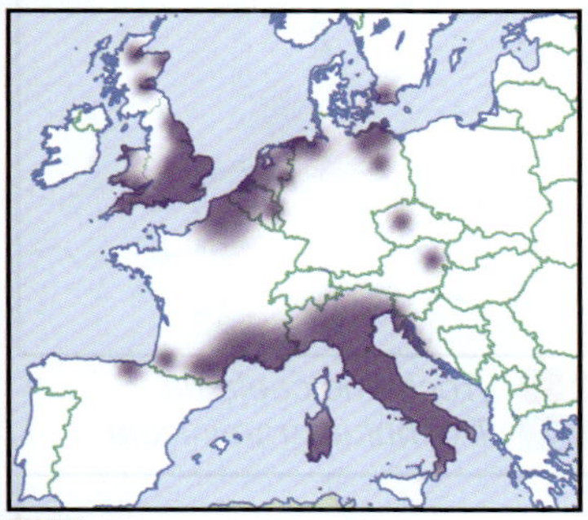

Monacha cantiana

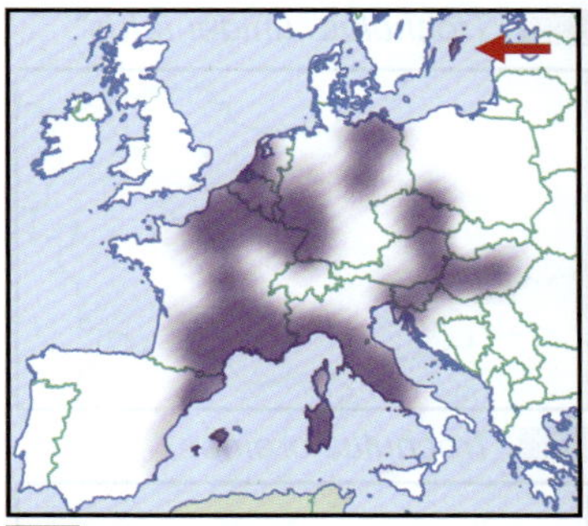

Cernuella neglecta

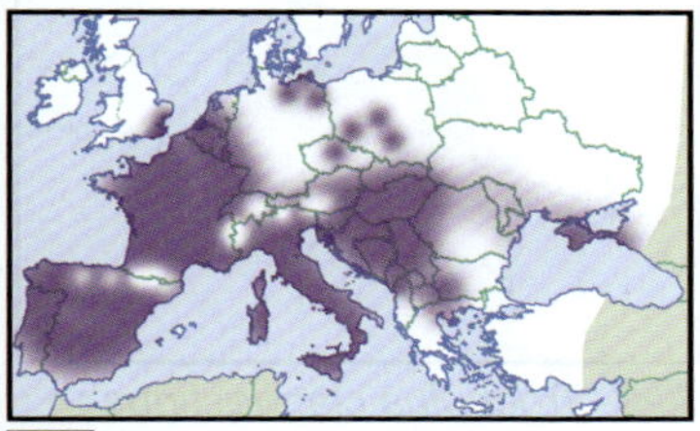

Monacha cartusiana

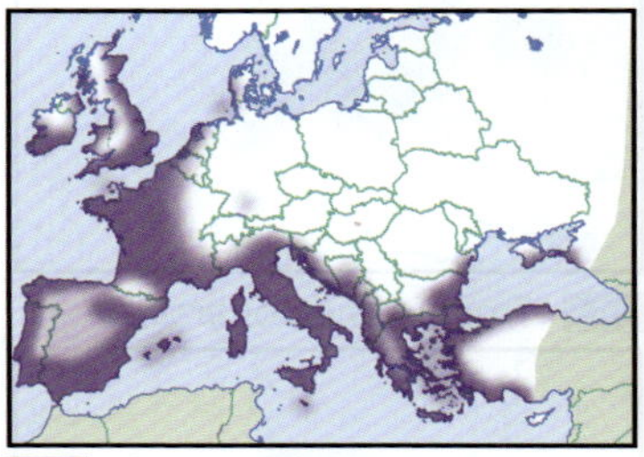

Cernuella virgata

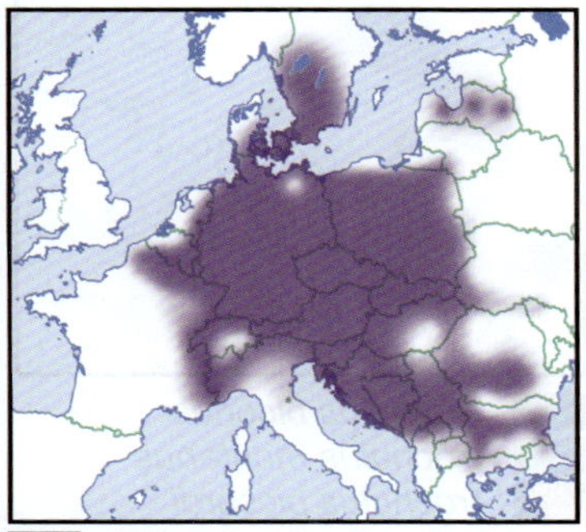

Monachoides incarnatus

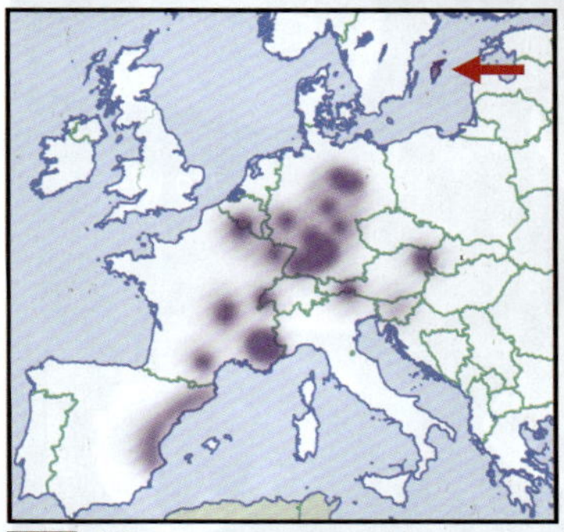

Xerocrassa geyeri

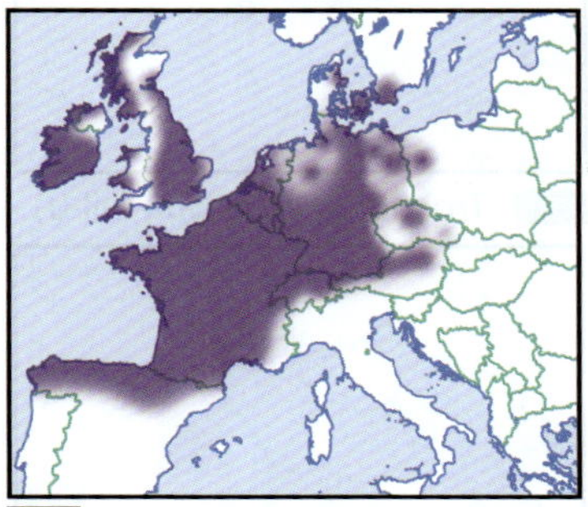

Helicella itala

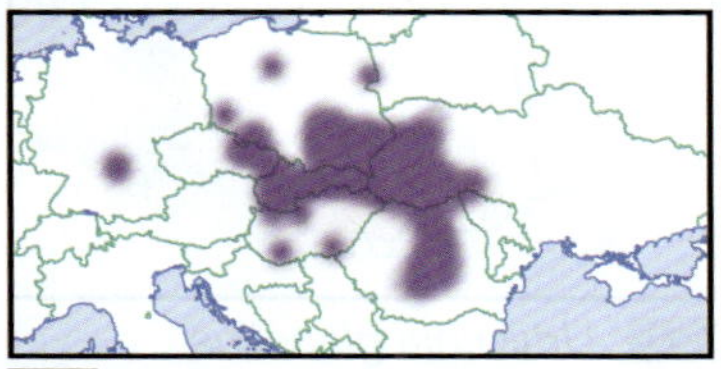

Monachoides vicinus

Zenobiella subrufescens

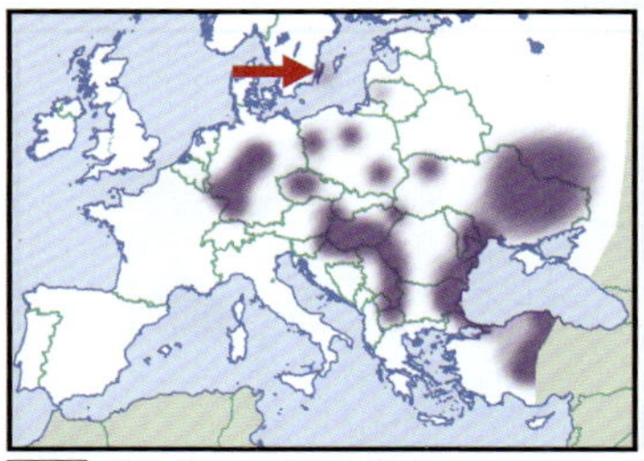

Helicopsis striata

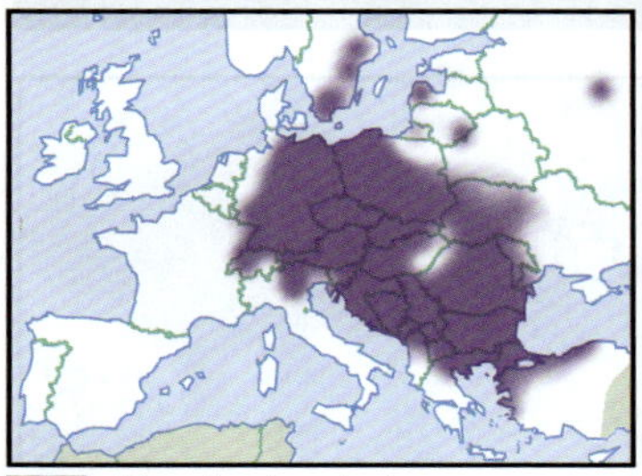

Xerolenta obvia

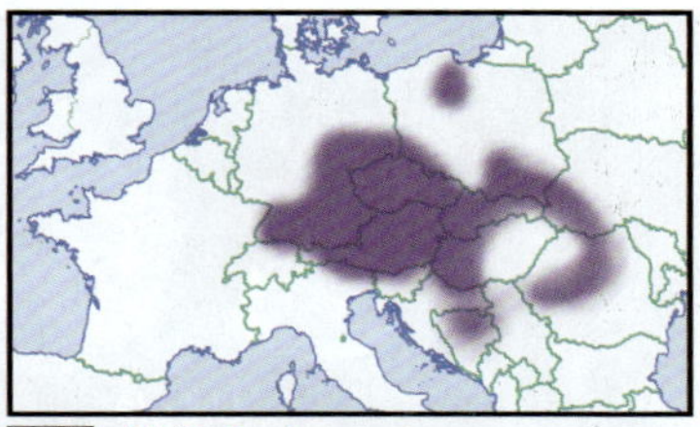

Urticicola umbrosus

Hygromiidae 2 EU

Trochulus etc.

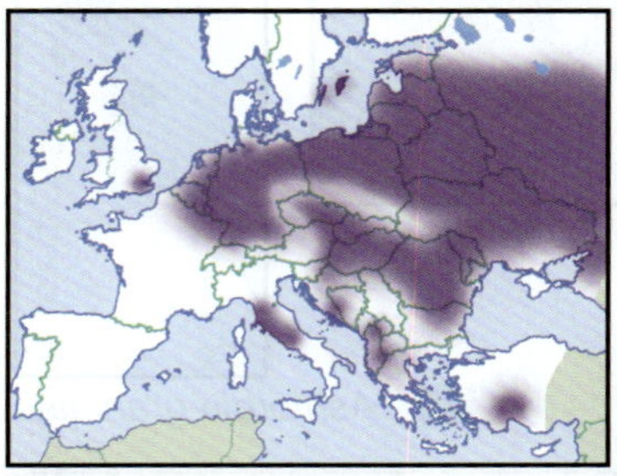

Pseudotrichia rubiginosa

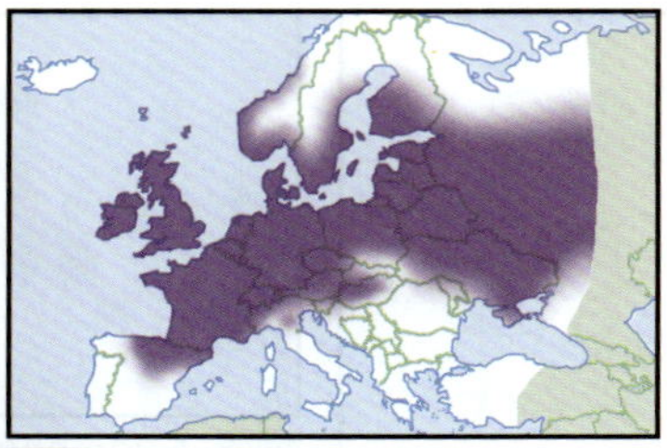

Trochulus hispidus

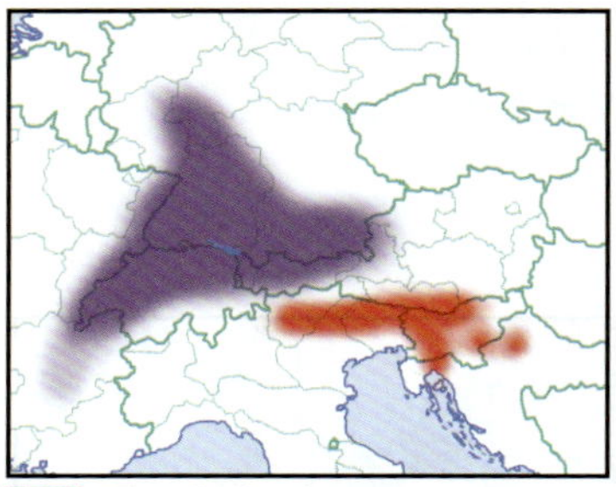

Trochulus leucozonus
Trochulus villosus (vil-o)

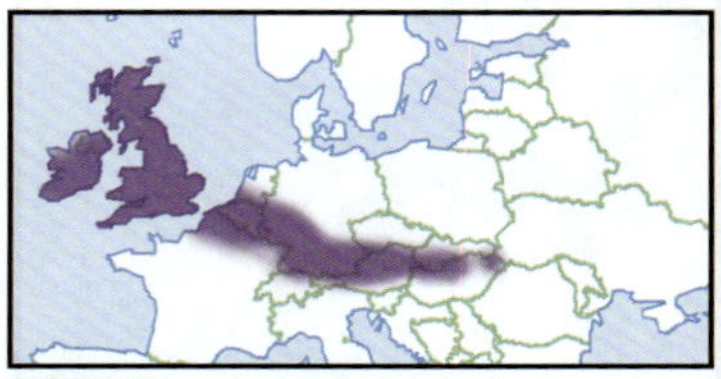

Trochulus striolatus

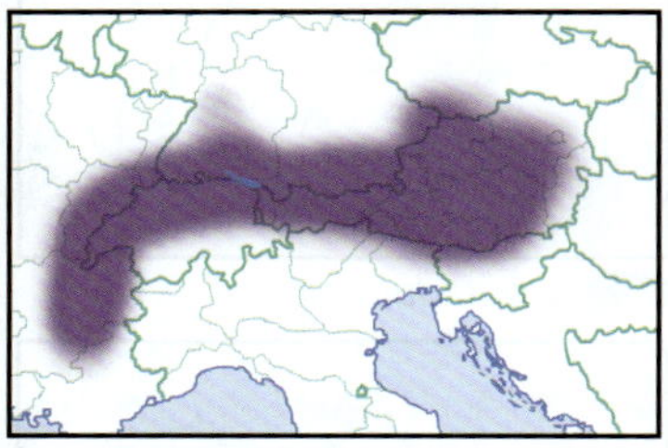

Trochulus edentulus

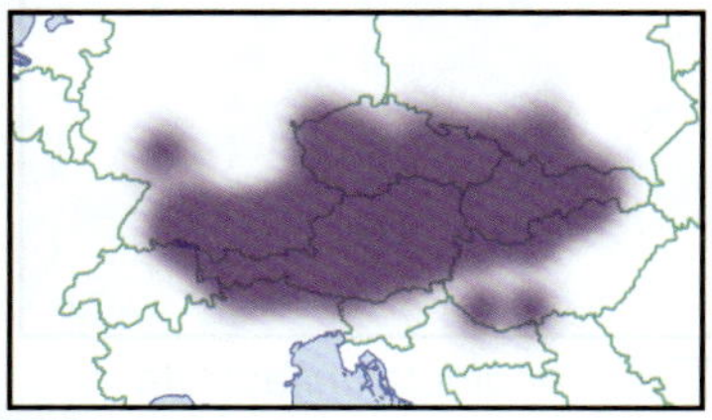

Trochulus "unidentatus"

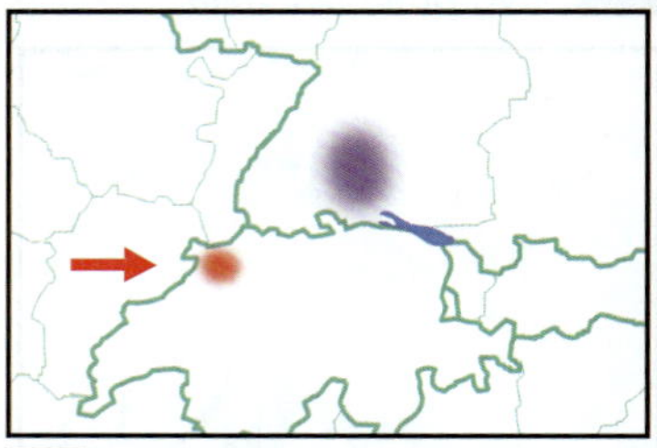

Trochulus caelatus
Trochulus graminicola

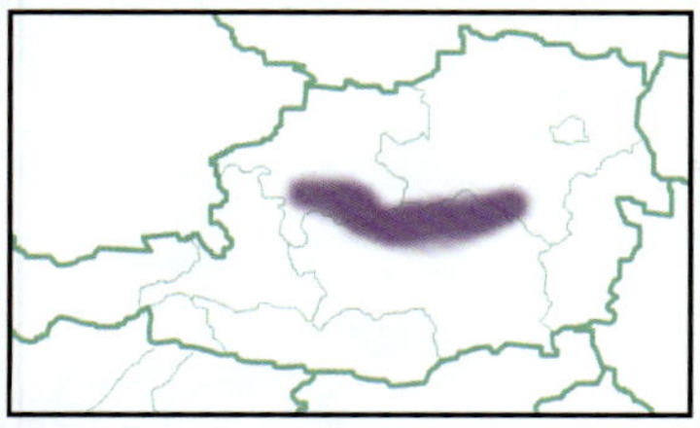

Trochulus oreinos

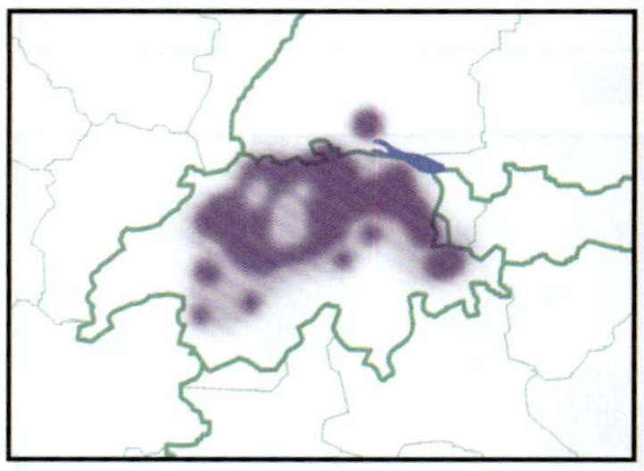

Trochulus clandestinus

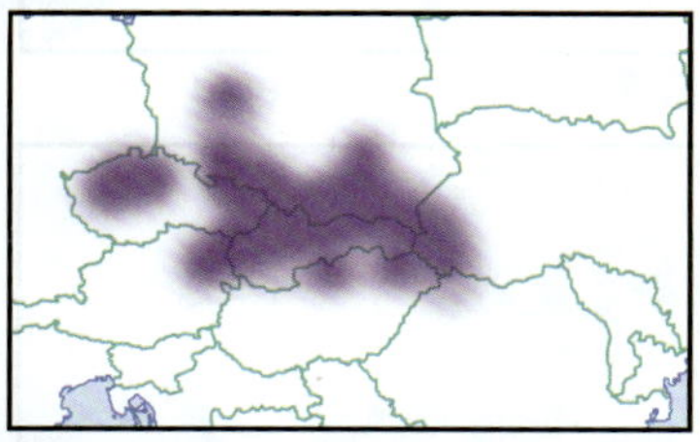

Trochulus lubomirskii

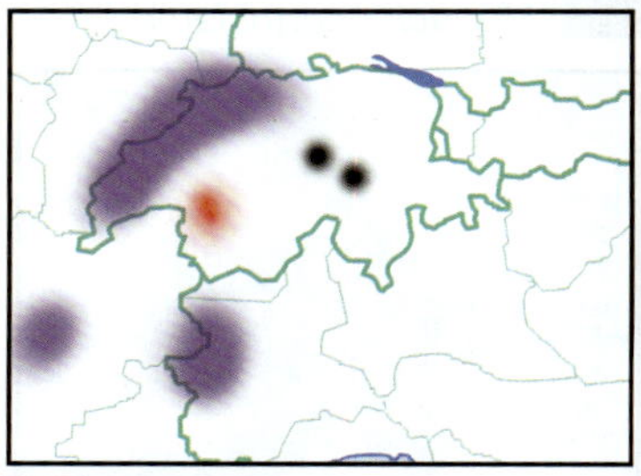

Trochulus biconus
Trochulus montanus
Trochulus piccardi

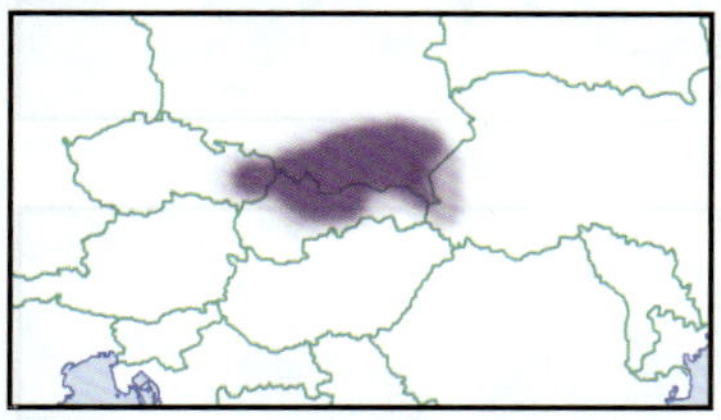

Trochulus villosulus (vil-u)

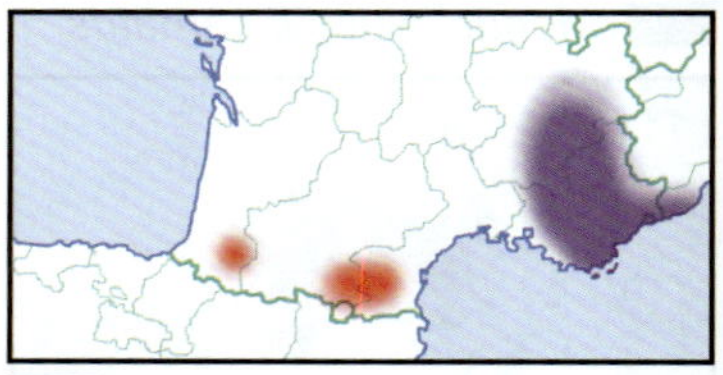

Urticicola glabellus
Trochulus ataxiacus

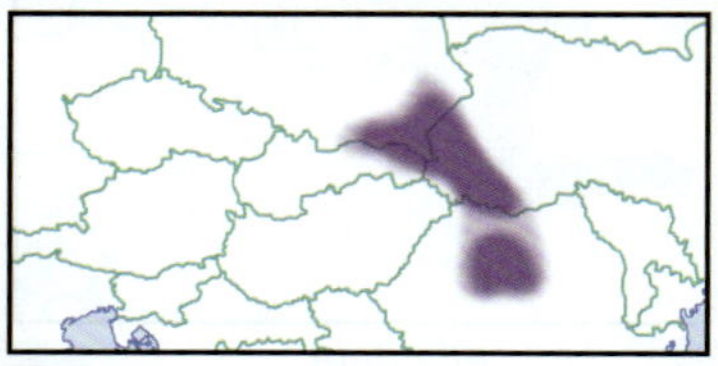

Trochulus bielzi

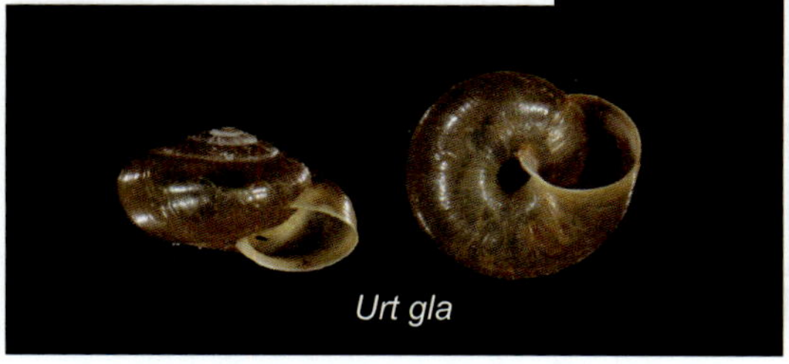

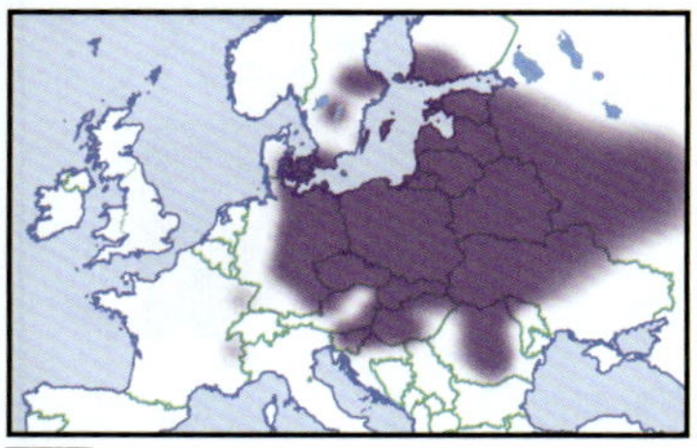

Perforatella bidentata

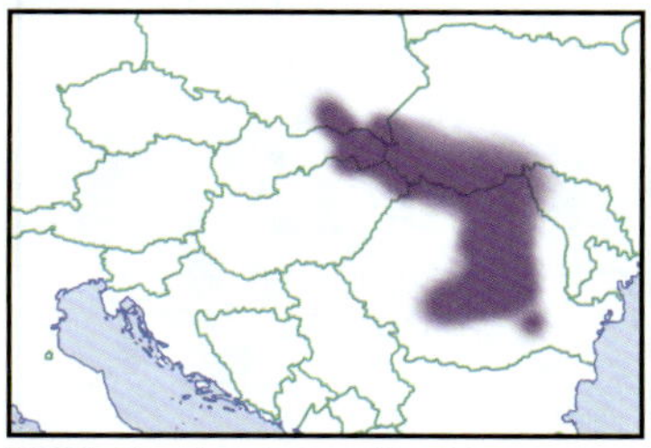

Perforatella dibotrion

Hygromiidae 3 ES

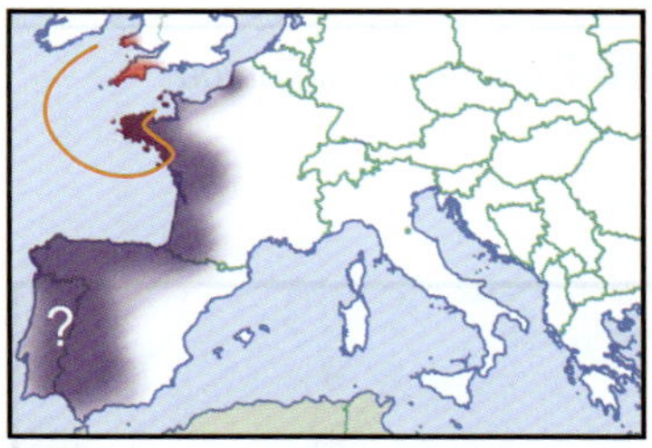

Ponentina revelata
Ponentina subvirescens

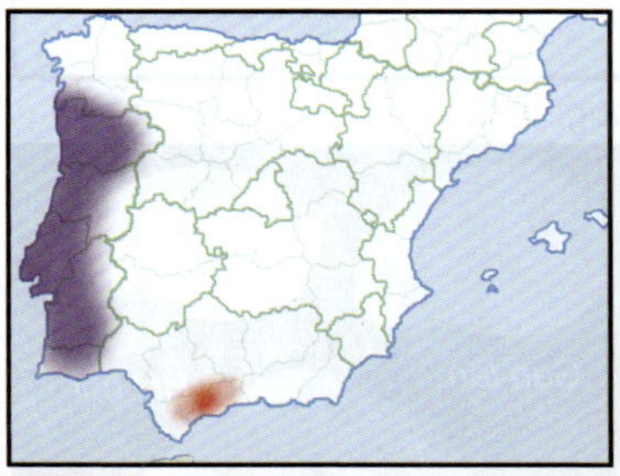

Ponentina martigena
Ponentina ponentina

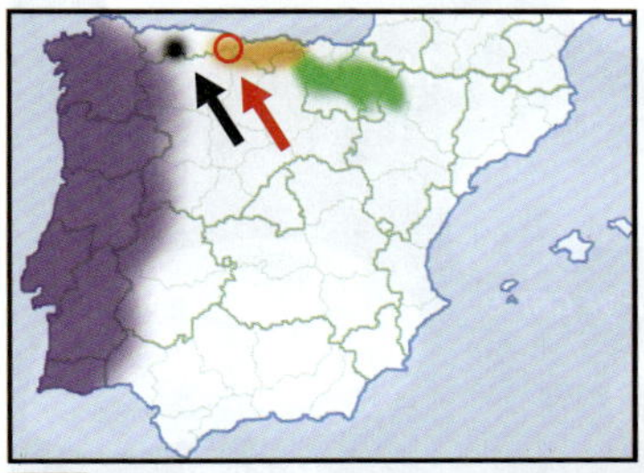

Cryptosaccus asturiensis
Helicella iberica
Mengoana brigantina
Plentuisa vendia
Portugala inchoata

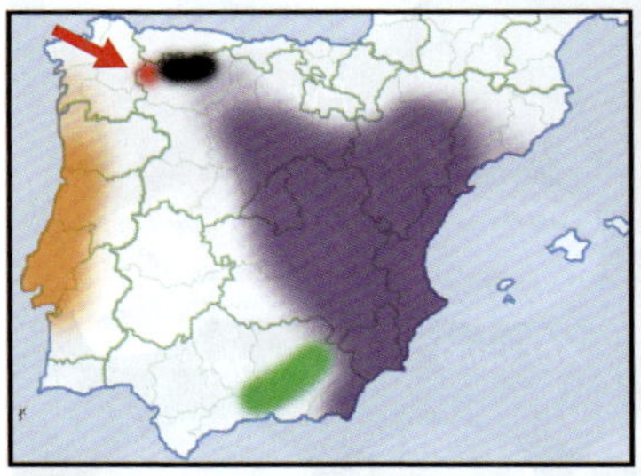

Candidula belemensis
Xerotricha bierzona
Xerotricha corderoi
Xerotricha madritensis
Xerotricha zujarensis

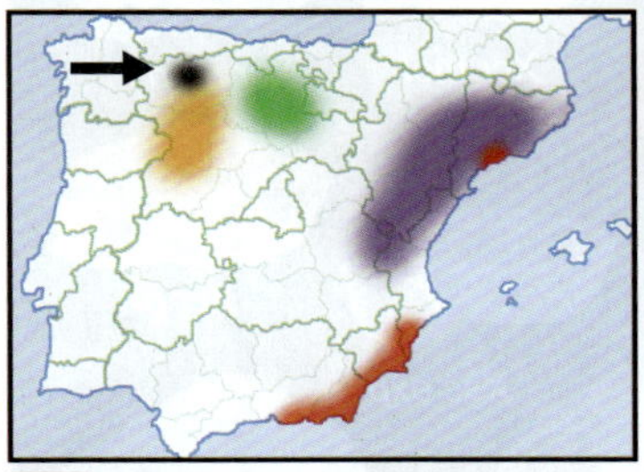

Xerotricha barcinensis
Xerotricha huidobroi
Xerotricha jamuzensis
Xerotricha silosensis
Xerotricha zaratei

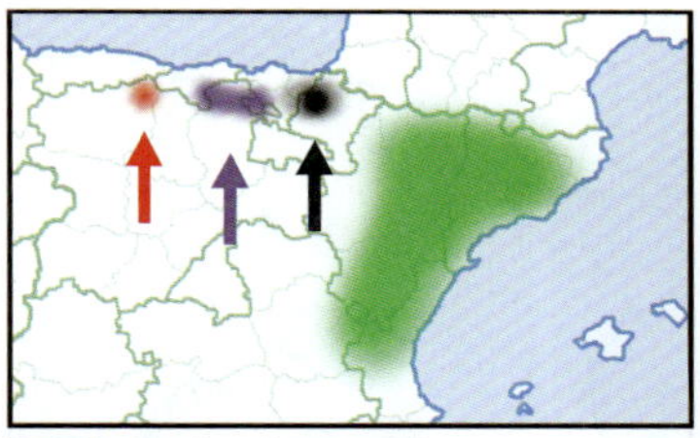

Helicella orzai
Helicella striatitala
Helicella valdeona
Xerocrassa salvanae

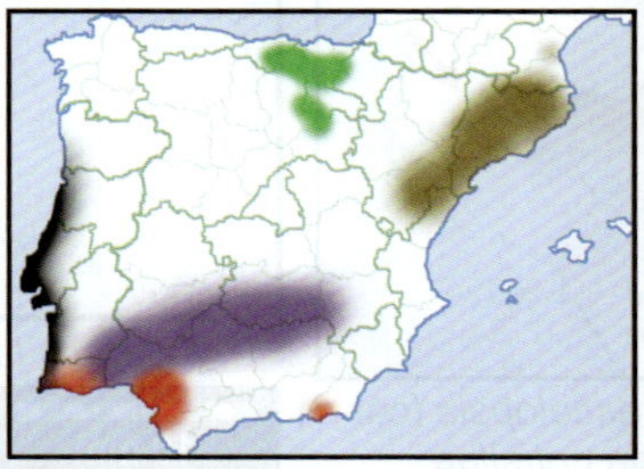

Candidula olisippensis
Helicella cistorum
Helicella ordunensis
Xerocrassa pallaresica
Xerotricha vatonniana

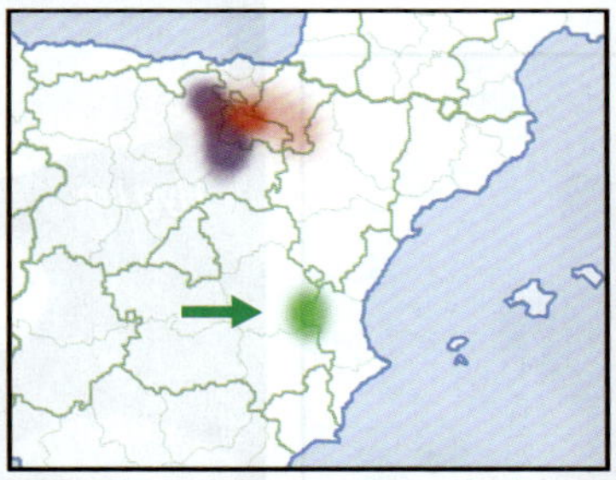

Candidula camporroblensis
Candidula najerensis
Candidula rocandioi

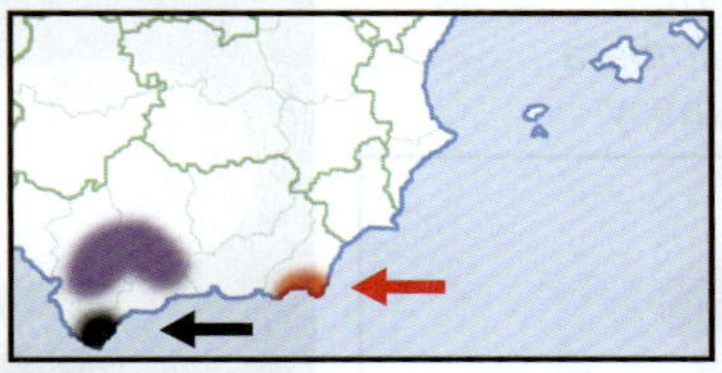

Ganula gadirana
Helicella stiparum
Xerotricha gasulli

Candidula arganica
Candidula codia
Candidula coudensis
Candidula setubalensis
Xerocrassa arturi

Xerocrassa cobosi
Xerocrassa derogata
Xerocrassa zaharensis

Xcr jim Xsc ado Xsc reb J. Liétor Gallego
Hel cis R. Vázquez Romero

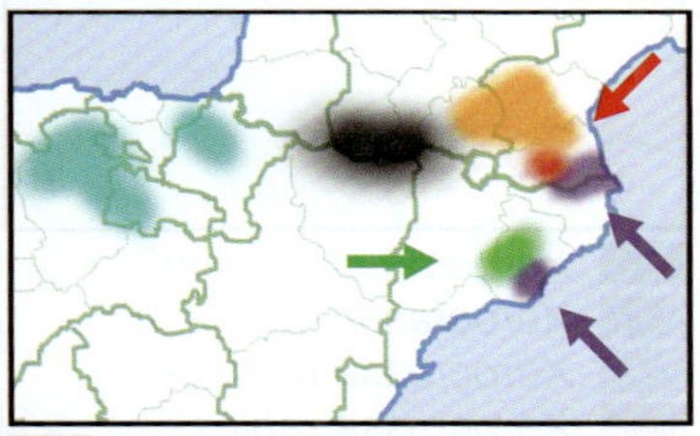

Montserratina becasis
Montserratina bofilliana
Montserratina martorelli
Monacha atacis
Xerotricha gonzalezi
Xerotricha renei

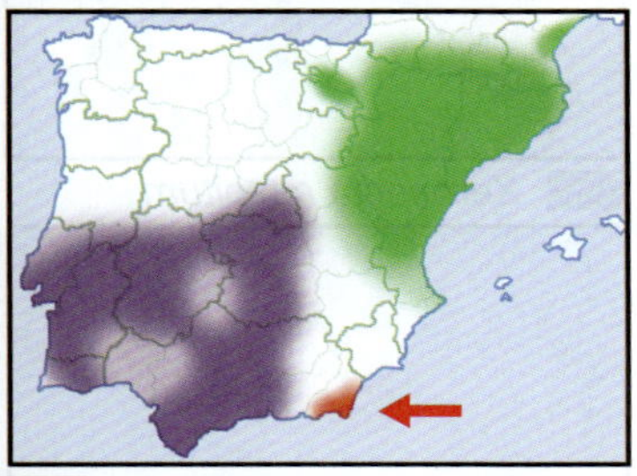

Xerocrassa penchinati
Xerosecta adolfi
Xerosecta reboudiana

Xerocrassa jimenensis
Xerocrassa ripacurcica
Xerocrassa subrogata

Hygromiidae 4 ES-FR

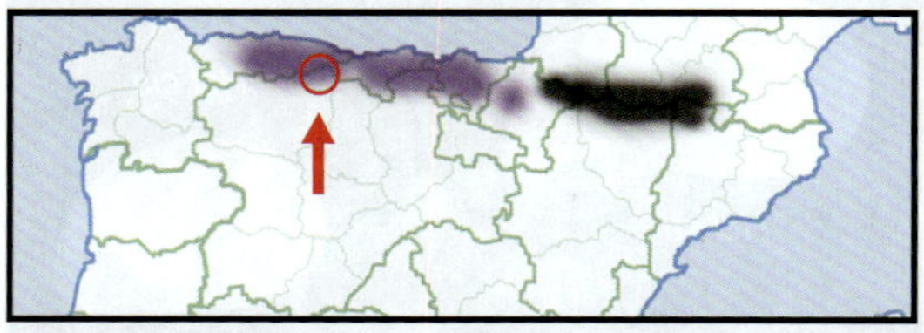

Pyrenaearia cantabrica
Pyren. carascalensis (car-e)
Pyrenaearia daanidentata

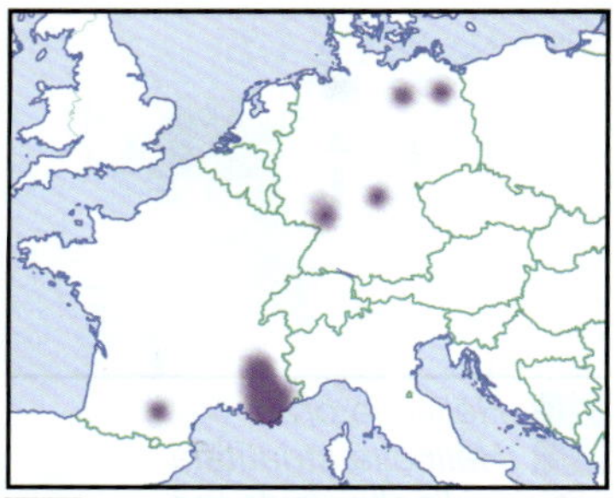

Helicella bolenensis

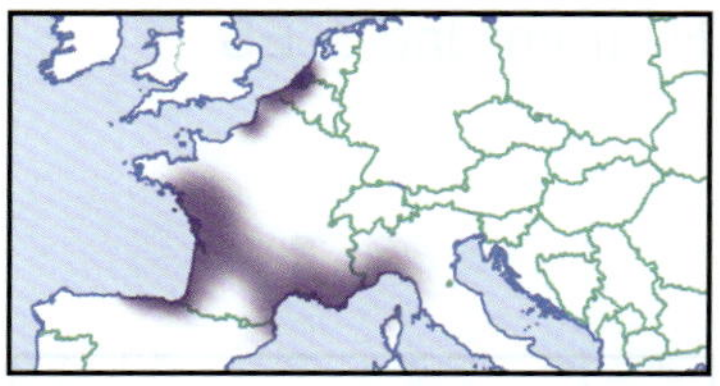

Cernuella aginnica

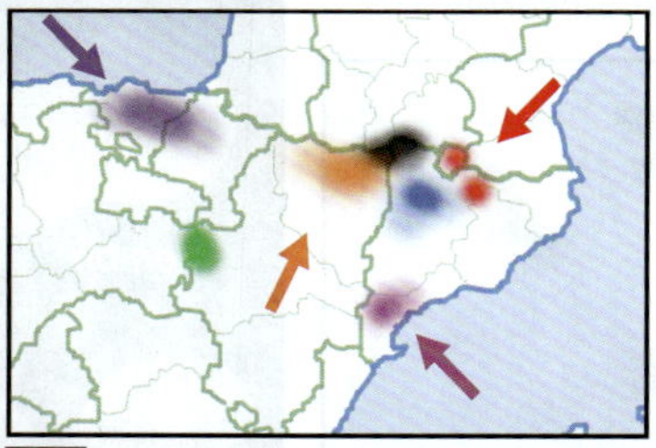

Pyren. carascalopsis (car-o)
Pyrenaearia cotiellae
Pyrenaearia molae
Pyrenaearia navasi
Pyrenaearia organiaca
Pyrenaearia parva
Pyrenaearia velascoi

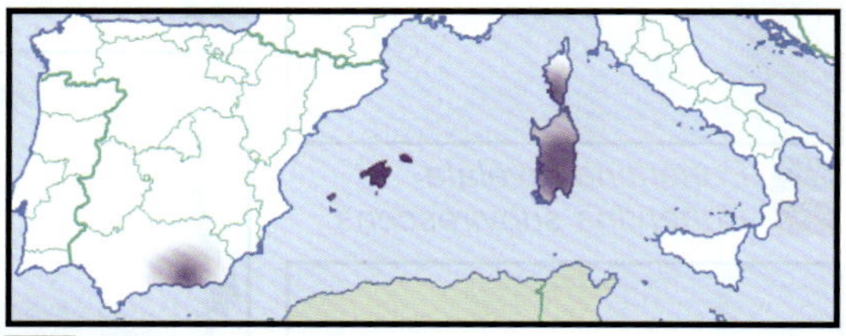

Ganula lanuginosa

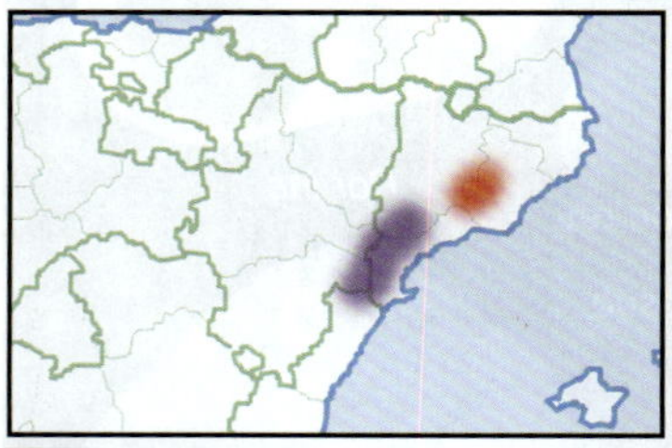

Xerocrassa grata
Xerocrassa montserratensis

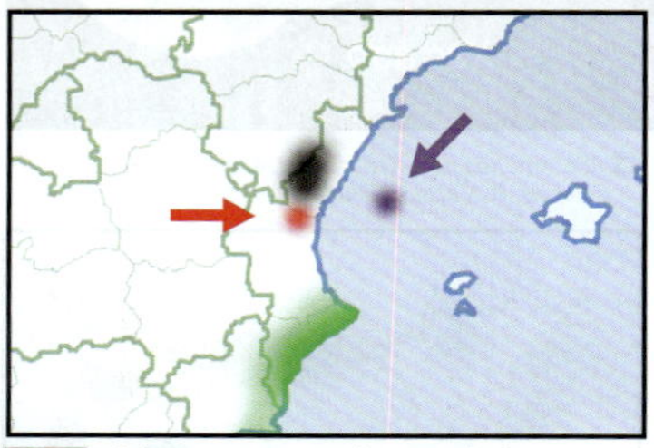

Xerocrassa barceloi
Xerocrassa edmundi
Xerocrassa molinae
Xerocrassa roblesi

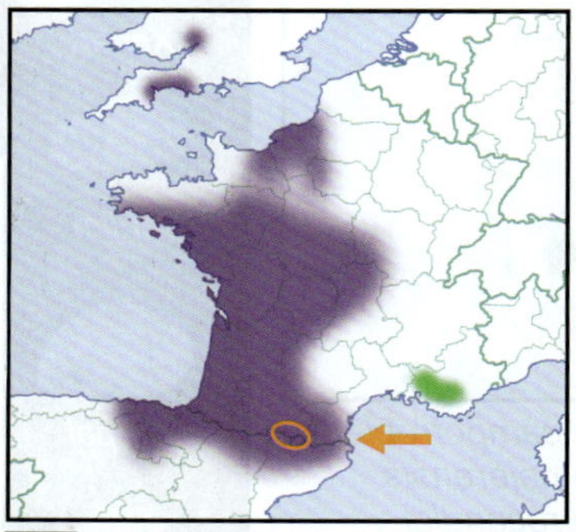

Candidula rugosiuscula
Hygromia limbata
Hygromia tassyi

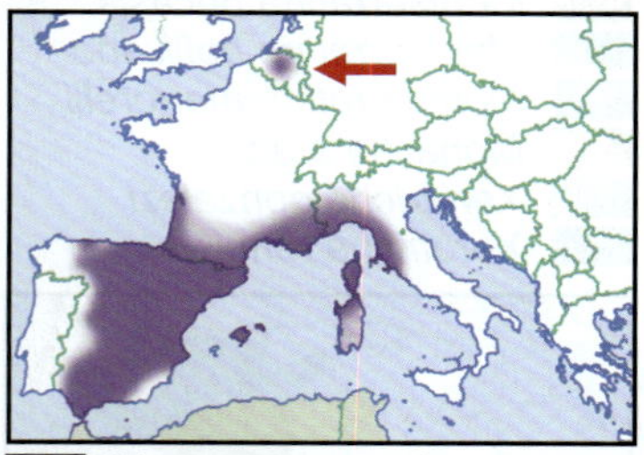

Xerosecta cespitum

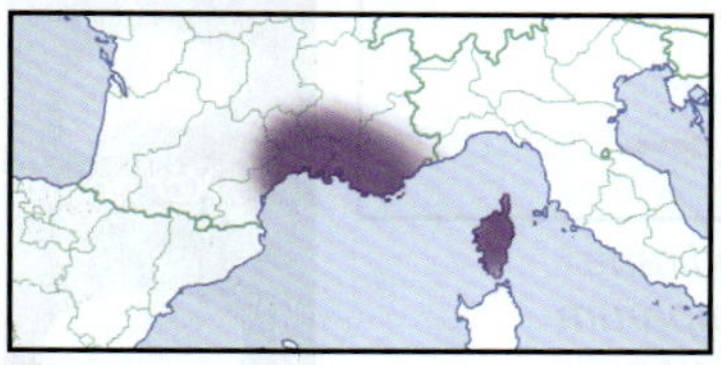

Monacha cemenelea

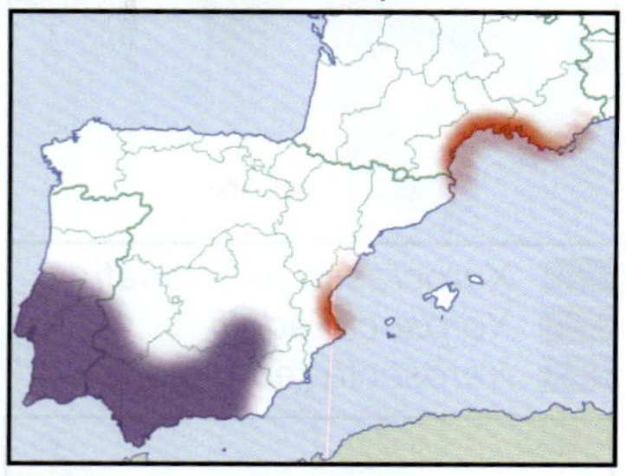

Xerosecta explanata
Xerosecta promissa

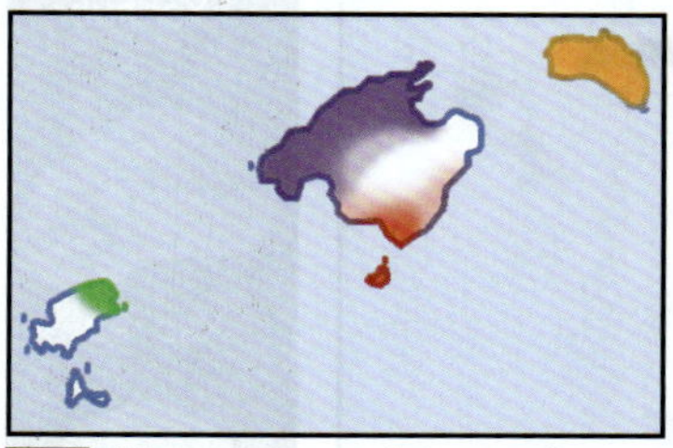

Xerocrassa cisternasi
Xerocrassa ferreri
Xerocrassa nyeli
Xerocrassa prietoi

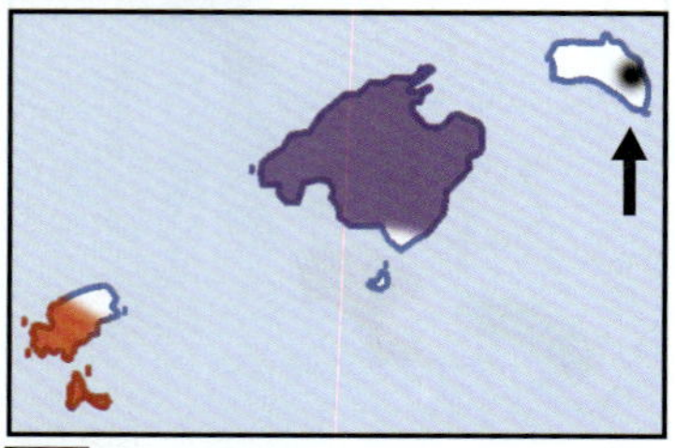

Xerocrassa cardonae
Xerocrassa caroli
Xerocrassa frater

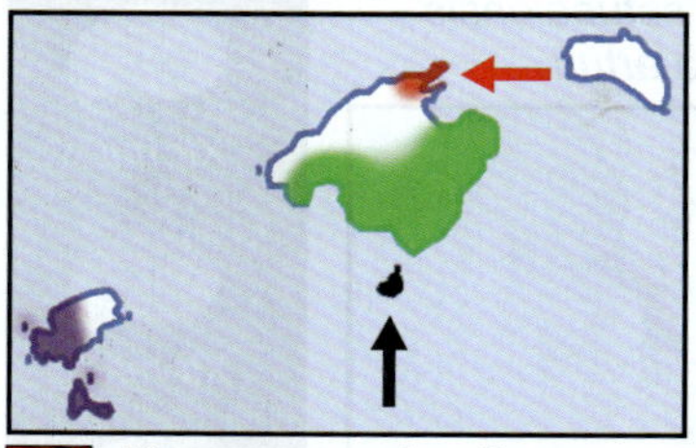

Xerocrassa claudinae
Xerocrassa ebusitana
Xerocrassa homeyeri
Xerocrassa ponsi

Xcr ebu Pyr can Pyr vel J. Liétor Gallego

Hygromiidae 5 S-EU

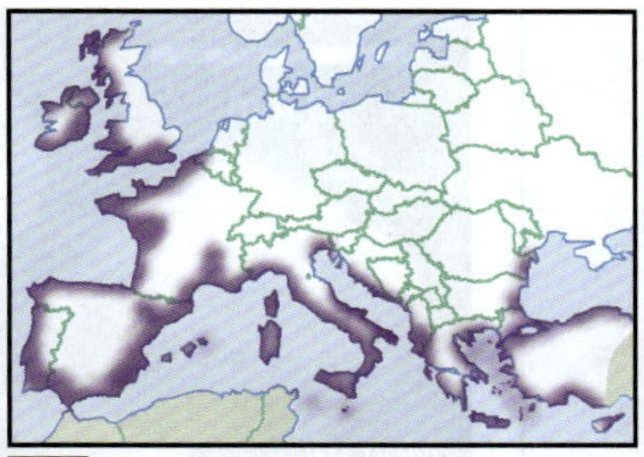

Cochlicella acuta

Cochlicella barbara

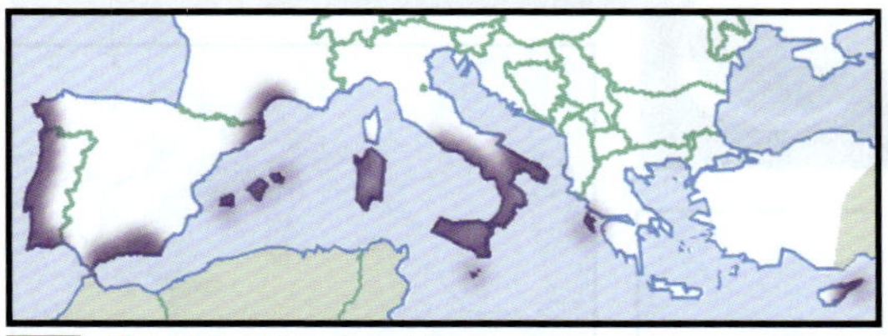

Cochlicella conoidea

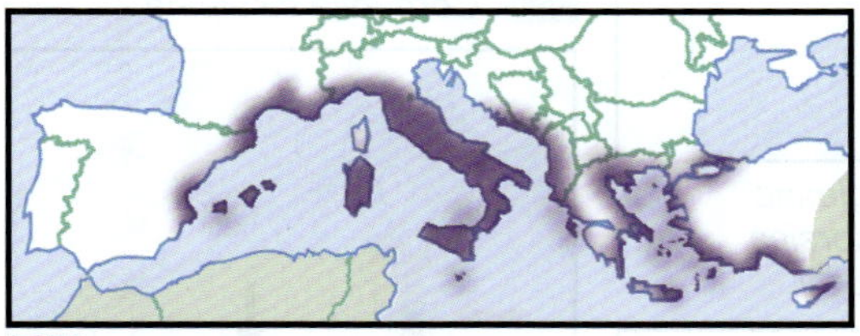

Trochoidea pyramidata

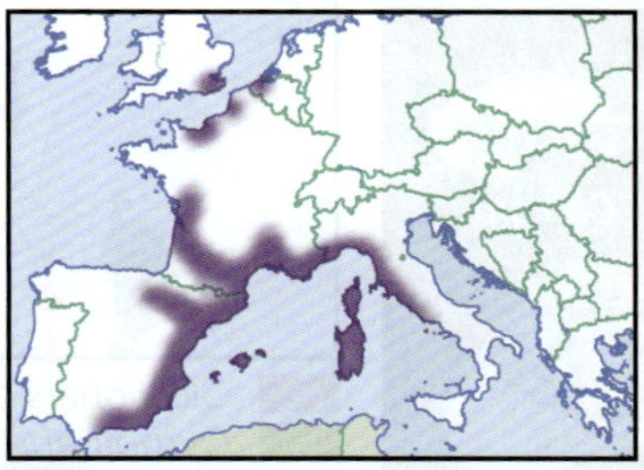

Trochoidea elegans

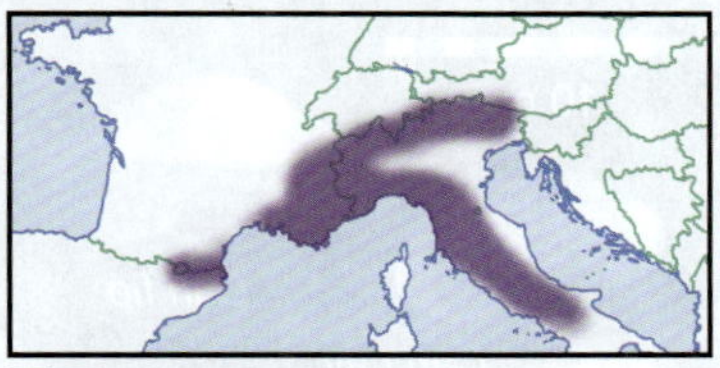

Ciliella ciliata

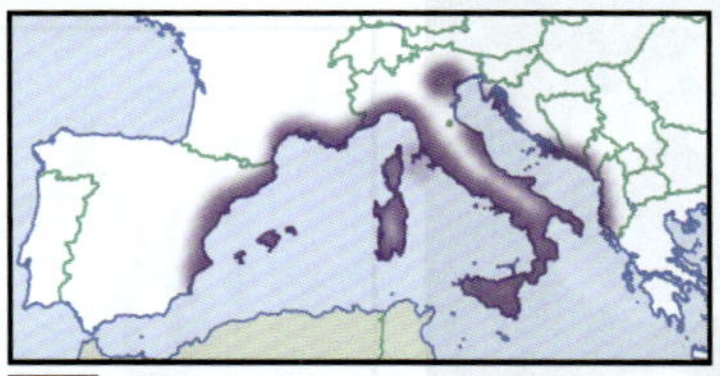

Trochoidea trochoides

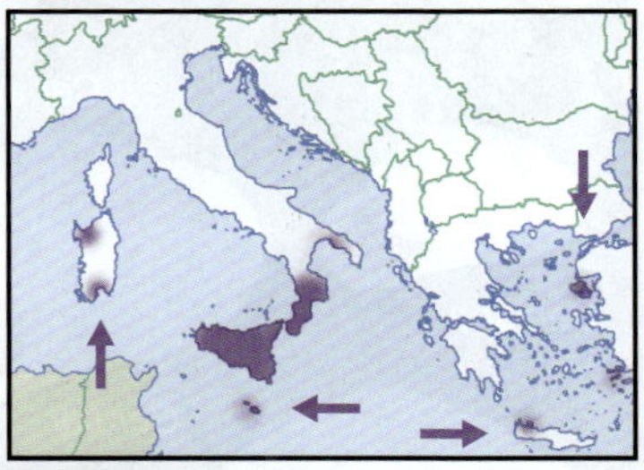

Xerocrassa meda

Xeropicta krynickii

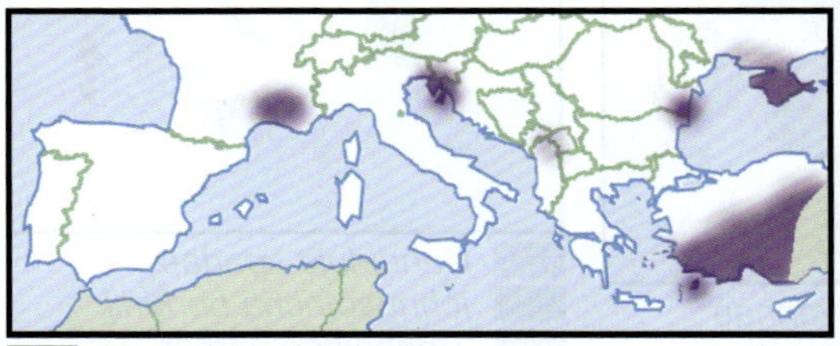

Xeropicta derbentina

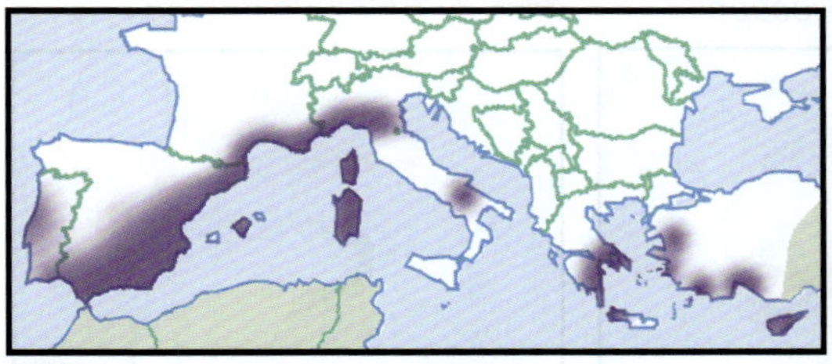

Microxeromagna lowei

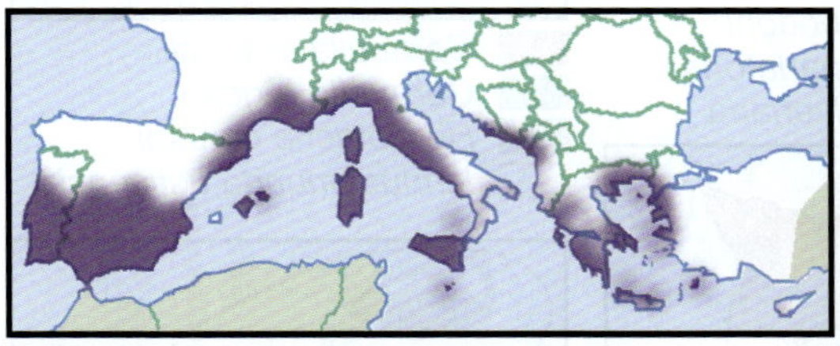

Xerotricha conspurcata

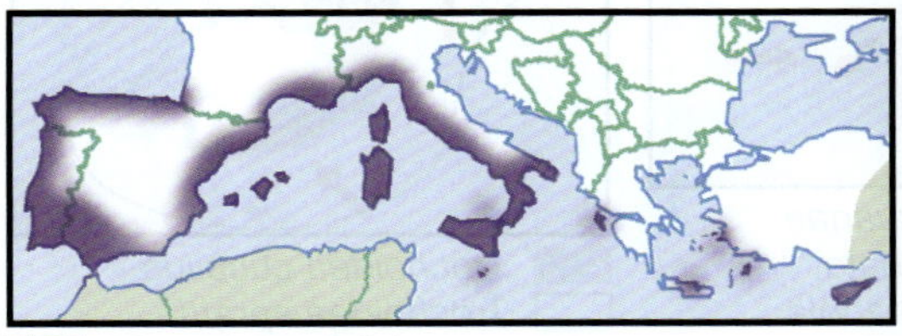

Xerotricha apicina

Hygromiidae 6 IT

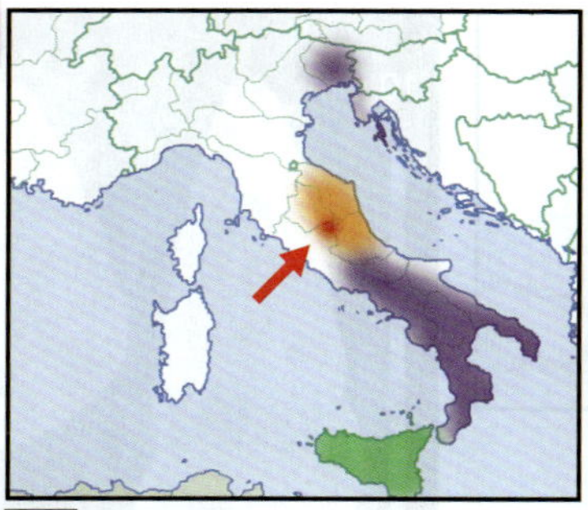

Monacha consona
Monacha gregaria
Monacha orsinii
Monacha "parumcincta"
Monacha ruffoi

Can cav Can fio Cer hyd Xsc con A. Hallgass
Cer ara Xsc giu J. Liétor Gallego
Mon par Mon ruf Sch par
Sch rei M. Prodan Nie ant M. Sosso

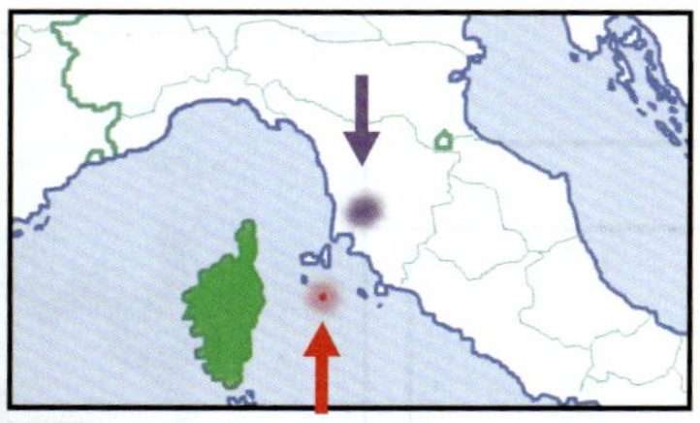

Ciliellopsis oglasae
Cyrnotheba corsica
Xerosecta giustii

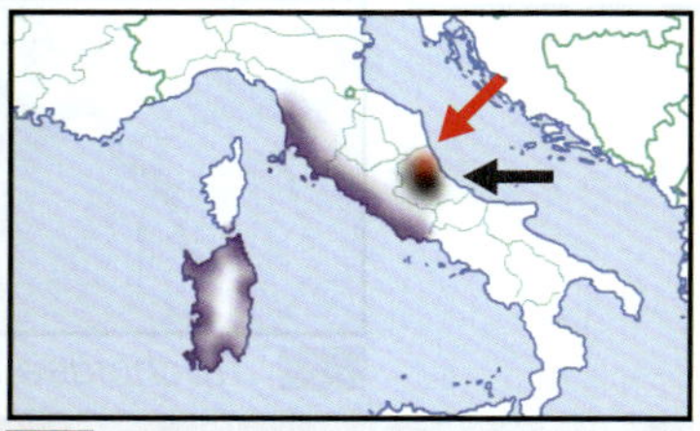

Candidula fiorii
Candidula grovesiana
Xerosecta contermina

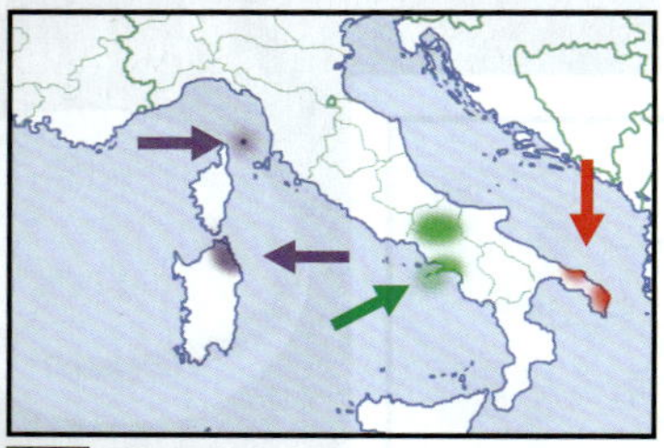

Candidula cavannae
Cernuella hydruntina
Tyrrheniellina josephi

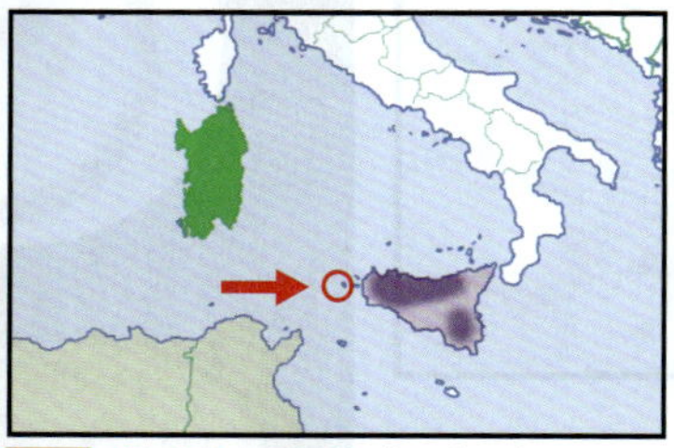

Schileykiella bodoni
Schileykiella parlatoris
Xerosecta hillyeriana

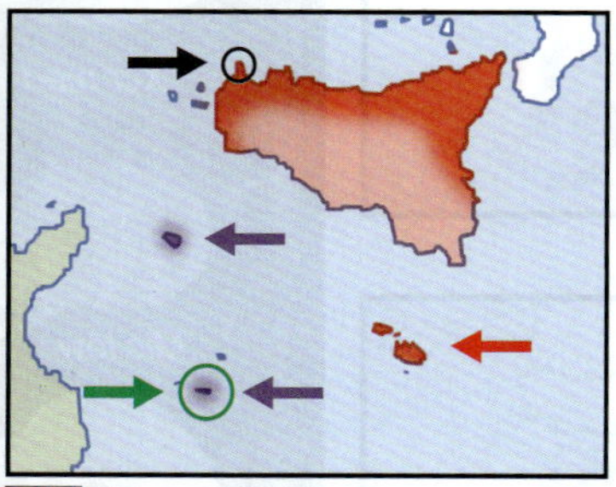

Cernuella caruanae
Cernuella limbata
Cernuella metabola
Trochoidea cumiae

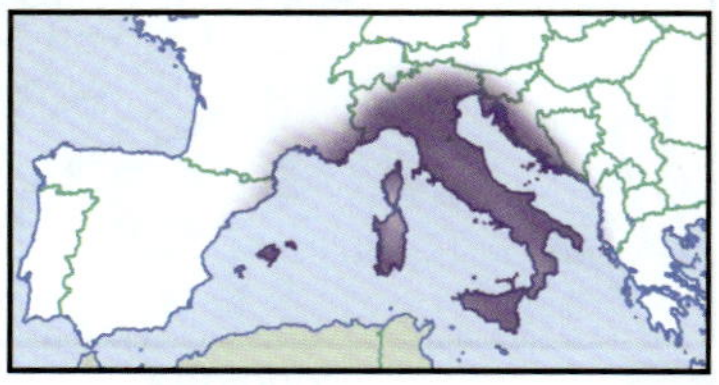

Cernuella cisalpina

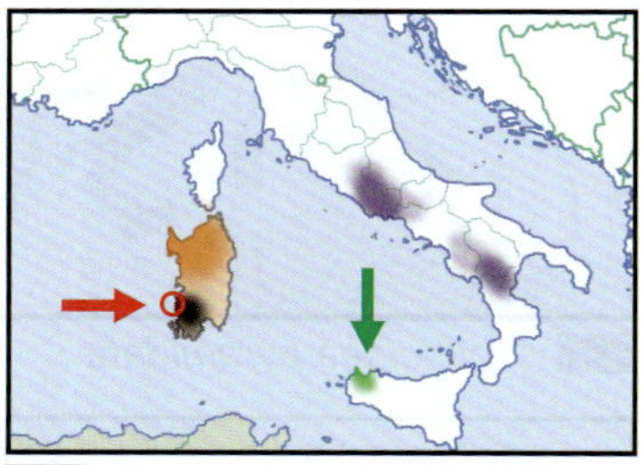

Cernuellopsis ghisottii
Ichnusomunda sacchii
Nienhuisiella antonellae
Schileykiella reinae
Xerosecta dohrni

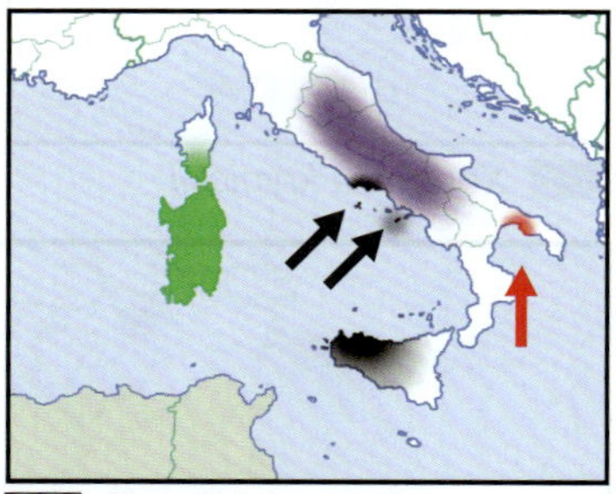

Candidula spadae
Ichnusotricha berninii
Trochoidea caroni
Xeromunda durieui

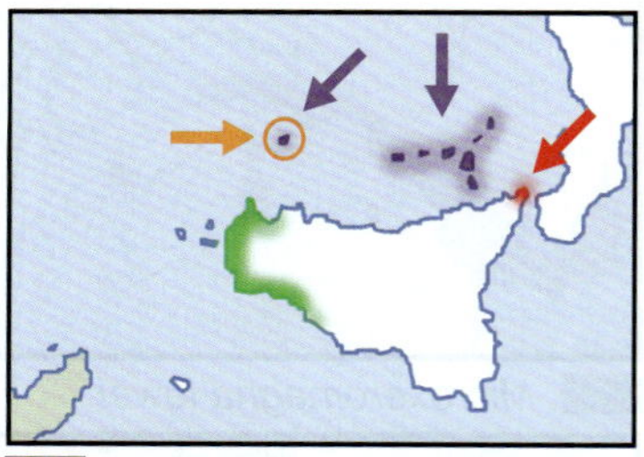

Cernuella aradasii
Cernuella rugosa
Helicotricha carusoi
Ichnusomunda usticensis

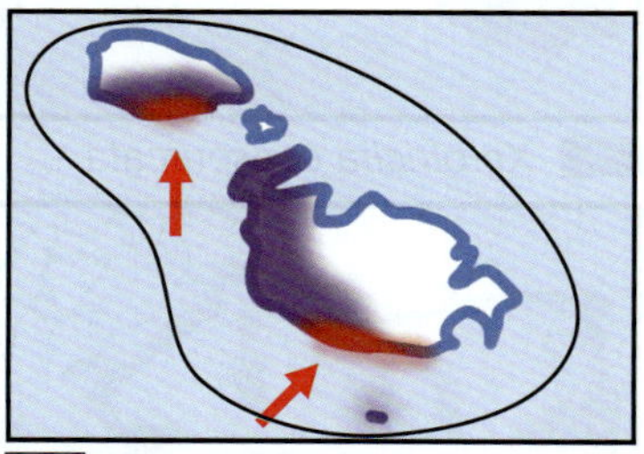

Trochoidea cucullus
Trochoidea spratti
Xerocrassa gharlapsi

Hygromiidae 7 HR-UA

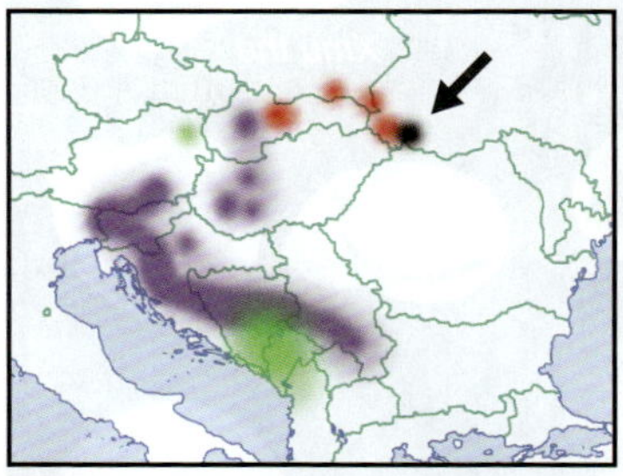

Trochulus bakowskii
Trochulus czarnohoricus
Trochulus filicinus
Xerocrassa muehlfeldtiana

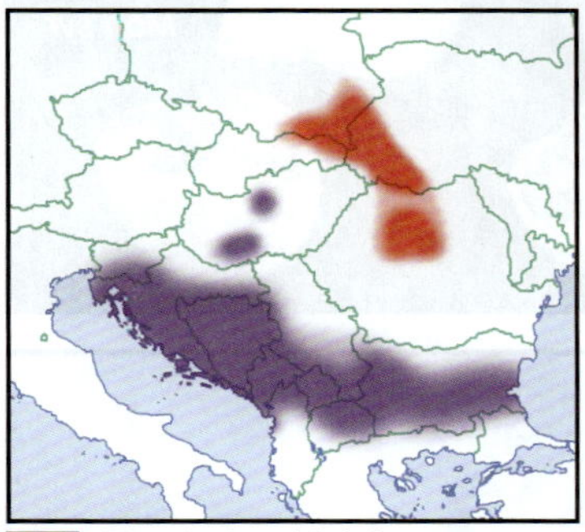

Trochulus bielzi
Trochulus erjaveci

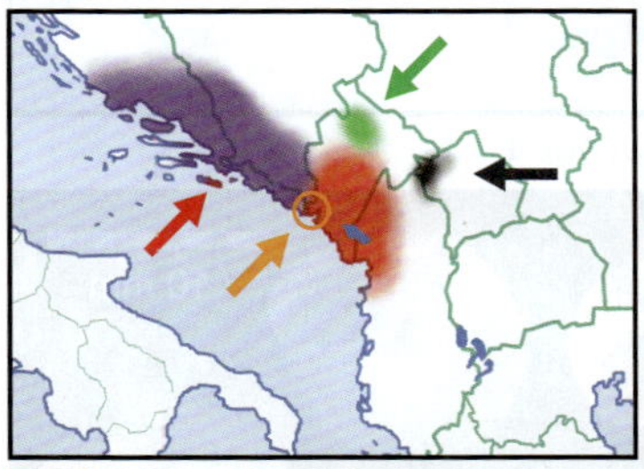

Hiltrudia globulosa
Hiltrudia kusmici
Hiltrudia mathildae
Monachoides kosovoensis
Monachoides taraensis

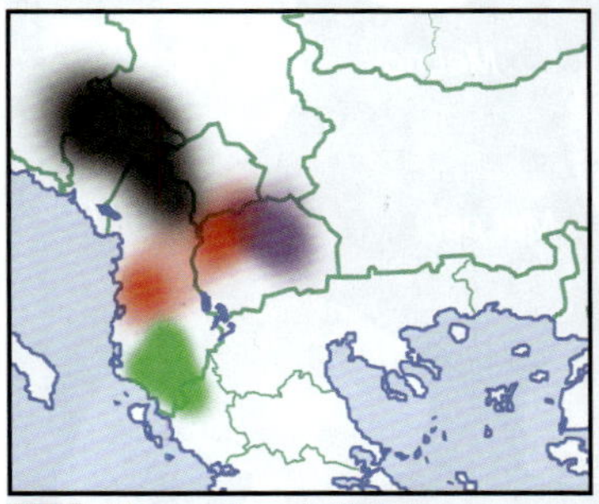

Monacha dofleini
Monacha emigrata
Monacha microtricha
Semifruticicola serbica

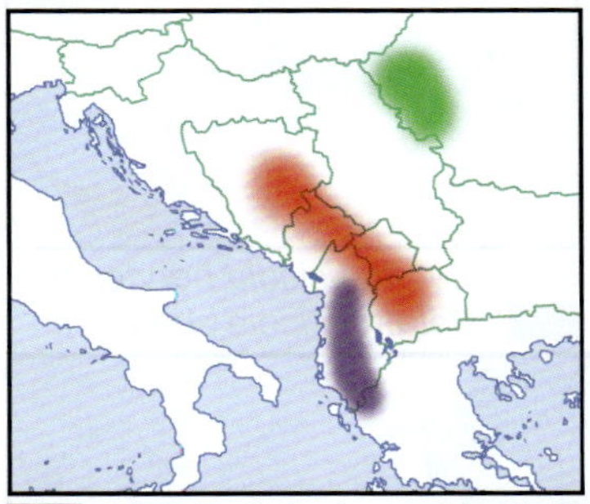

Metafruticicola occidentalis
Monachoides bacescui
Monachoides fallax

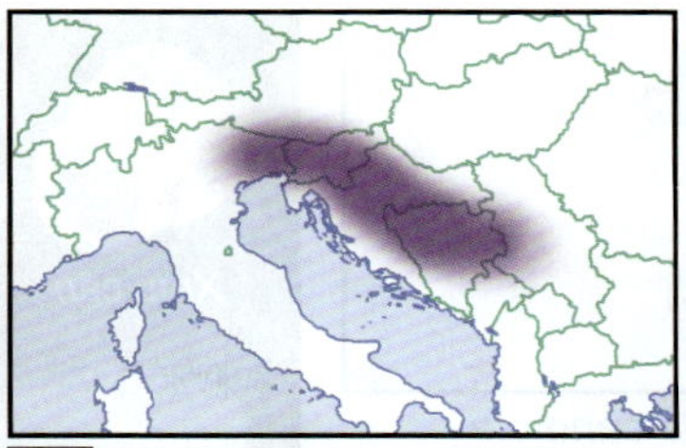

Trochulus luridus

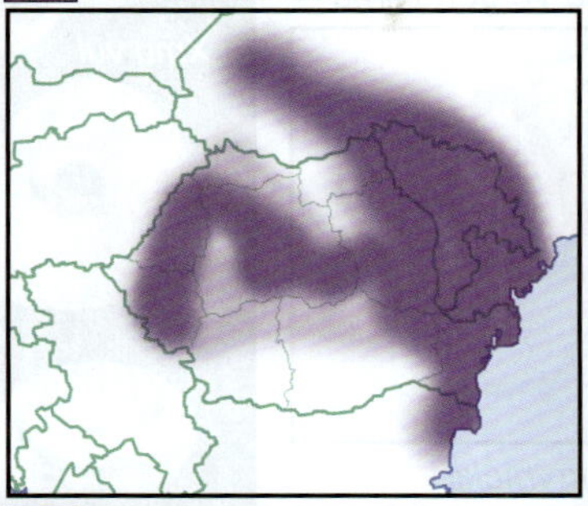

Helicopsis instabilis

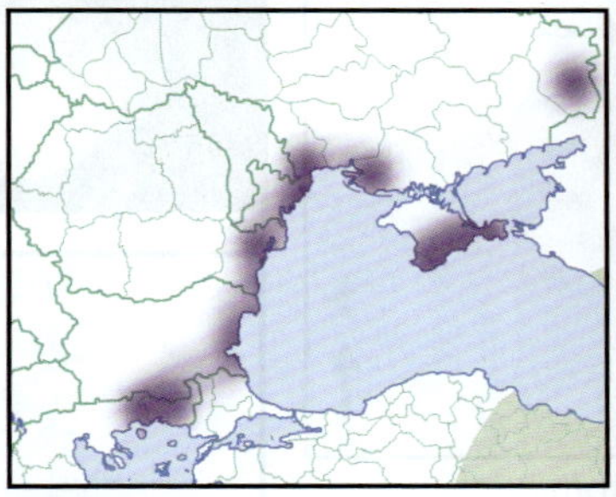

Helicopsis filimargo

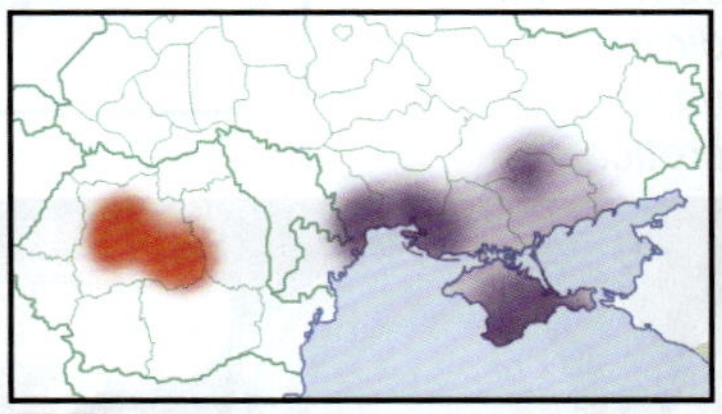

Helicopsis cereoflava
Monacha fruticola

Monacha frequens

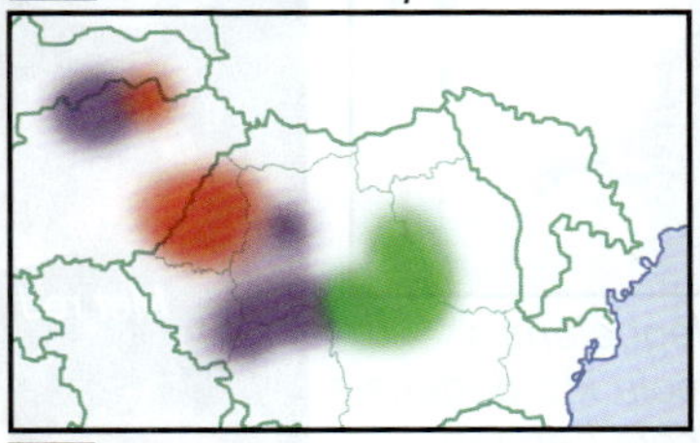

Kovacsia kovacsi
Lozekia deubeli
Lozekia transsilvanica

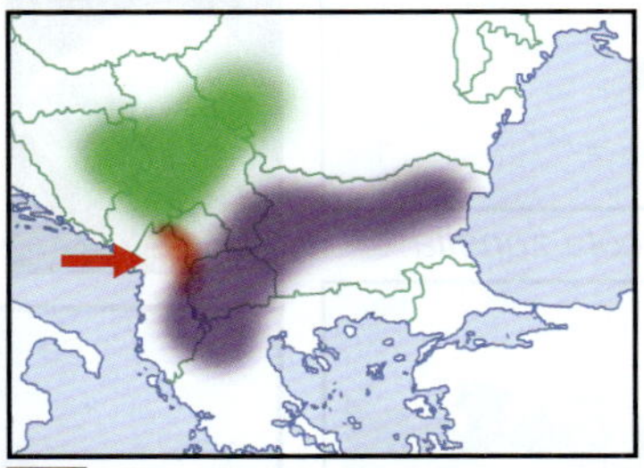

Candidula castriota
Candidula rhabdotoides
Xerocampylaea zelebori

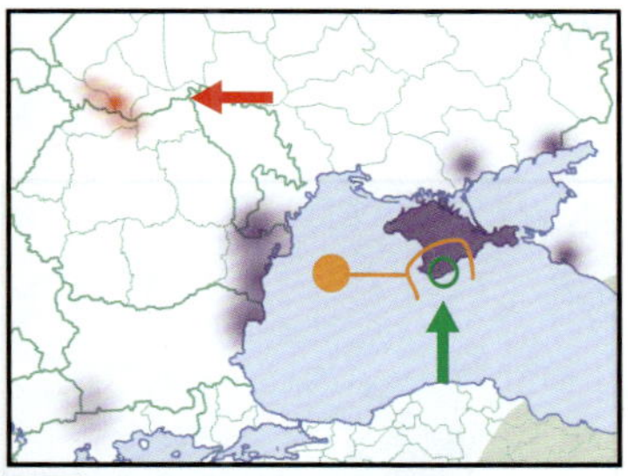

Helicopsis dejecta
Helicopsis retowskii
Helicopsis paulhessei
Prostenomphalia carpathica

Can cas Loz deu Z. Fehér *Hcp dej* S. Leonov

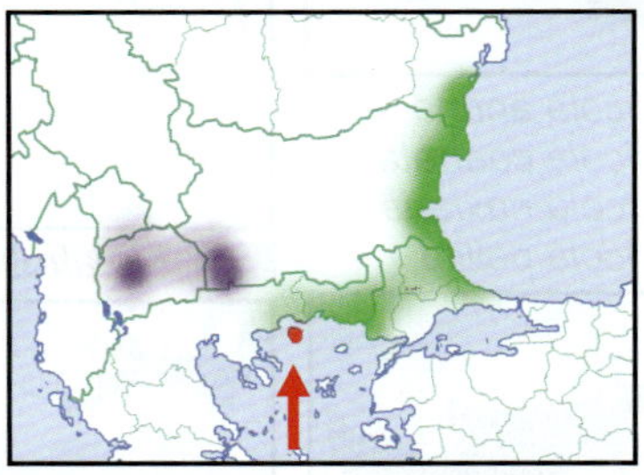

Xerolenta macedonica
Xerolenta spiruloides
Xerolenta thasia

Hygromiidae 8 GR

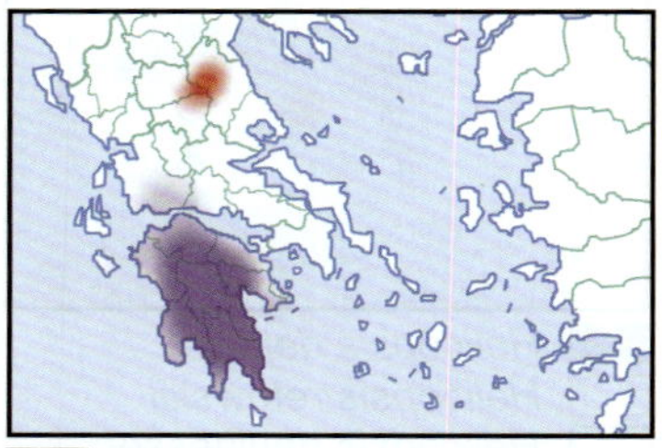

Xeromunda peloponnesia
Xeromunda thessalica

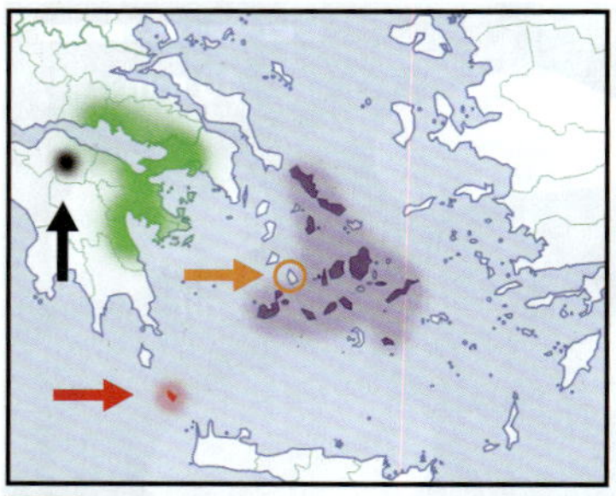

Candidula lernaea
Candidula syrensis
Xeromunda alticola
Xerocrassa poecilodoma
Xerocrassa siphnica

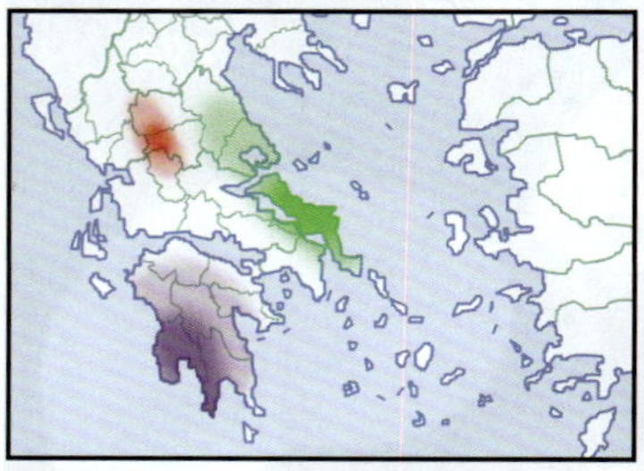

Monacha euboeica
Monacha haussknechti
Monacha messenica

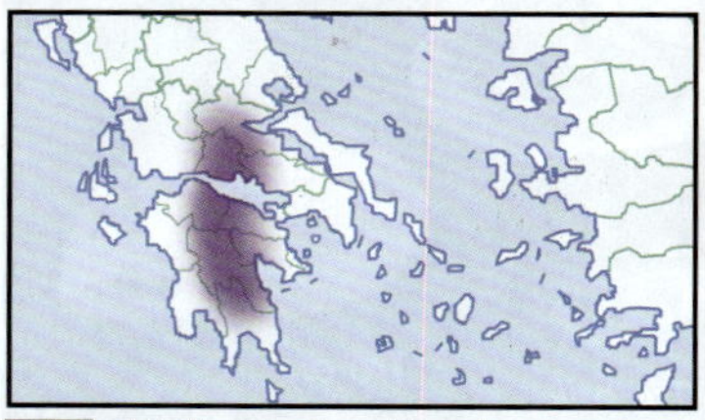

Helicopsis gittenbergeri

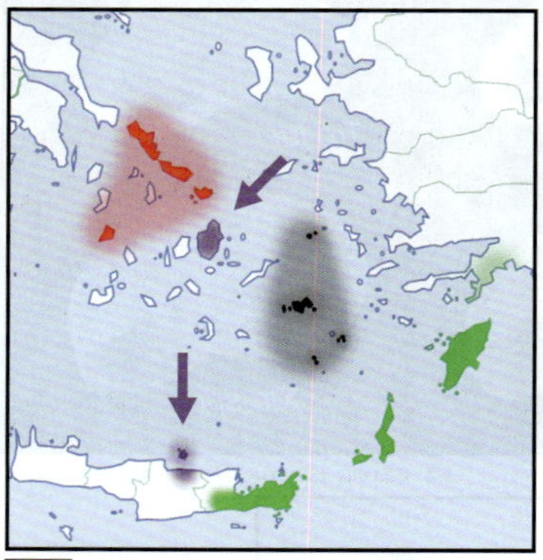

Metafruticicola andria
Metafruticicola coartata
Metafruticicola naxiana
Metafruticicola pellita

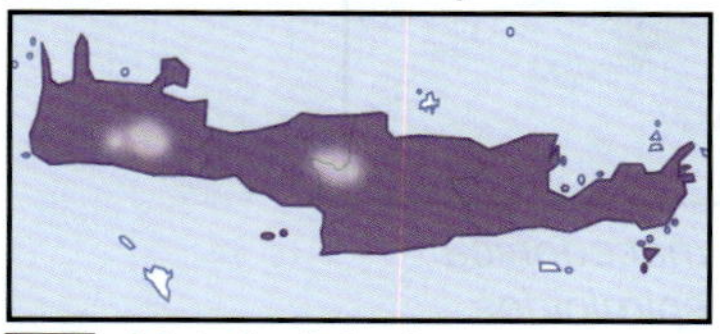

Metafruticicola noverca

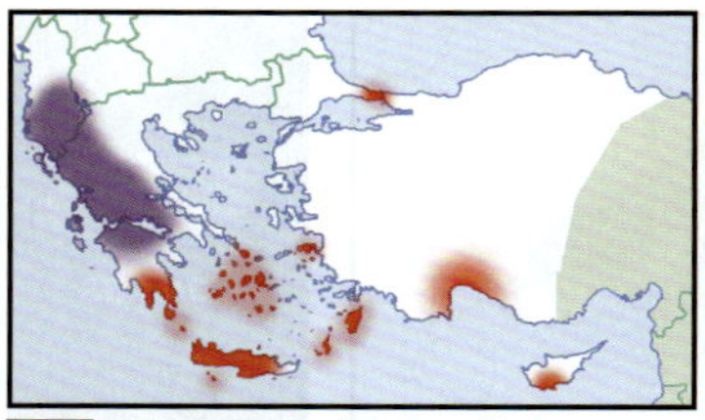

Xeromunda candiota
Xeromunda vulgarissima

Xeropicta krynickii

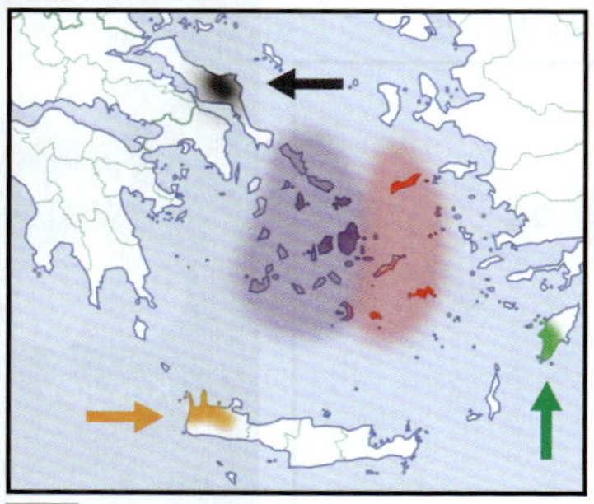

Monacha rothii
Monacha pseudorothii
Monacha dirphica
Monacha maasseni
Monacha cretica

Monacha syriaca

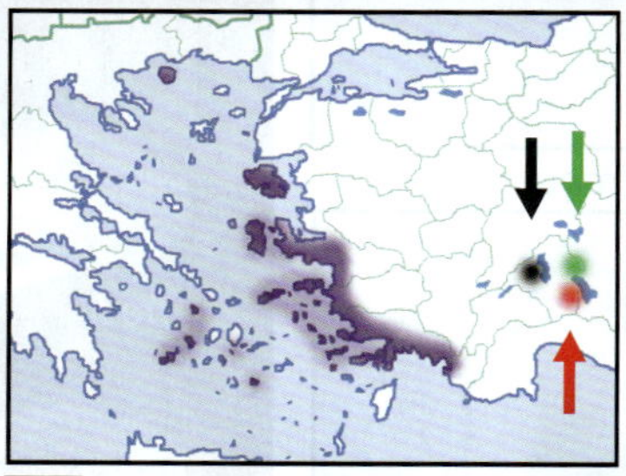

Metaf. dedegoelensis
Metafruticicola kizildagensis
Metafruticicola oerstani
Metaf. redtenbacheri

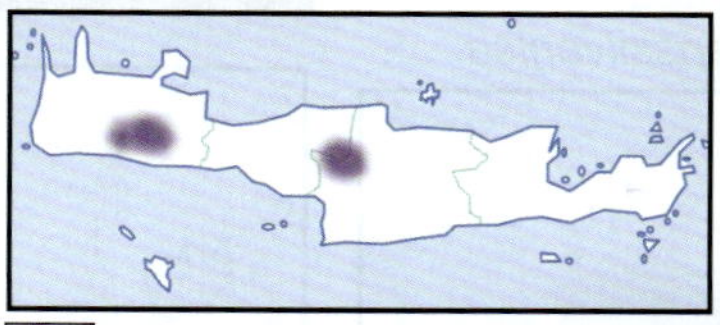

Metafruticicola zonella

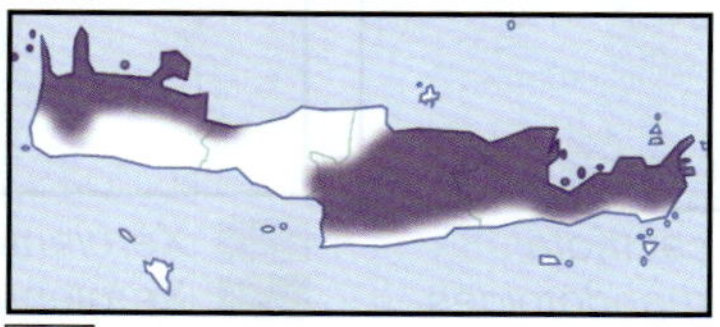

Metafruticicola lecta

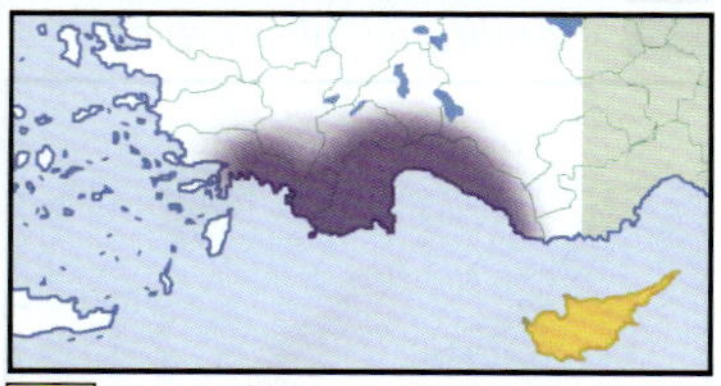

Metafruticicola berytensis
Metafruticicola schuberti

Metafruticicola nicosiana

Hygromiidae 9
GR-CY

Xerocrassa cretica

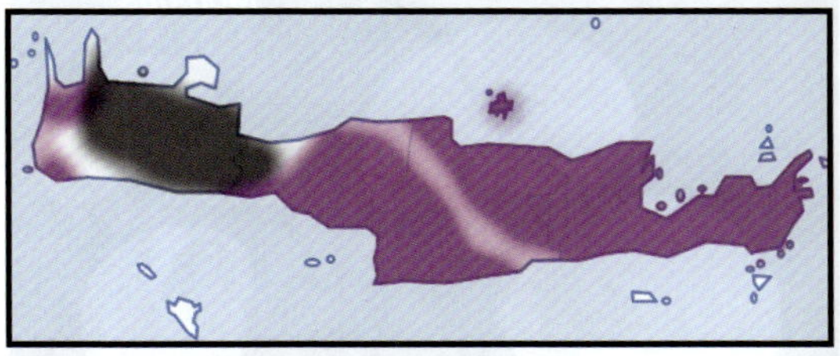

Pseudoxerophila bathytera
Pseudoxerophila oertzeni

Xerocrassa amphiconus
Xerocrassa franciscoi
Xerocrassa grabusana
Xerocrassa heraclea
Xerocrassa subvariegata

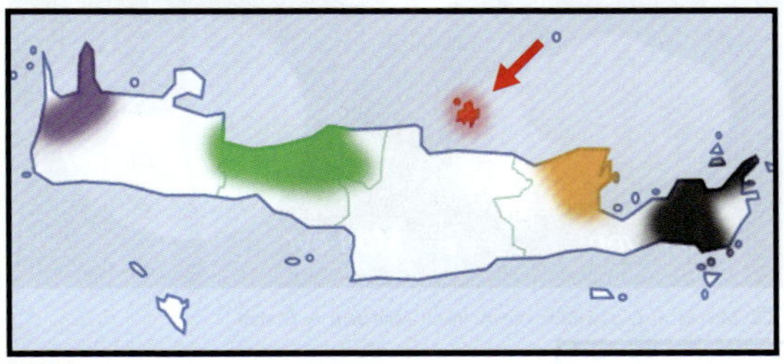

Xerocrassa diensis
Xerocrassa kydonia
Xerocrassa lasithiensis
Xerocrassa rhithymna
Xerocrassa siderensis

Xerocrassa claudia (cla-a)
Xerocrassa claudiconus (cla-c)
Xerocrassa mesostena

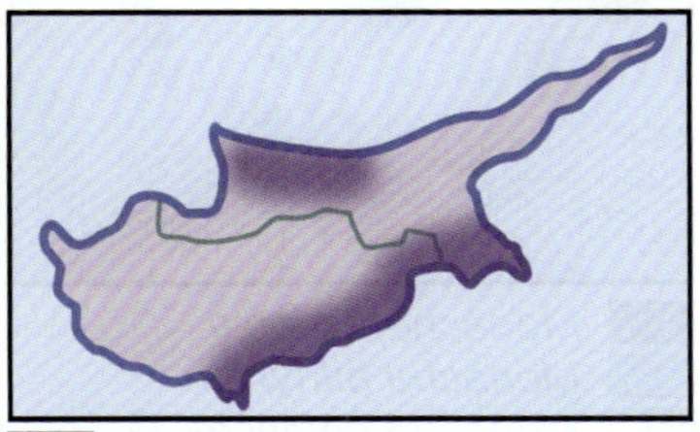

Helicella juglans

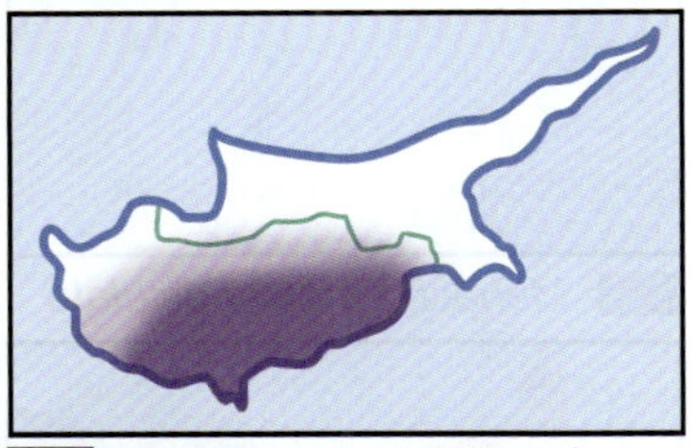

Pseudoxerophila confusa

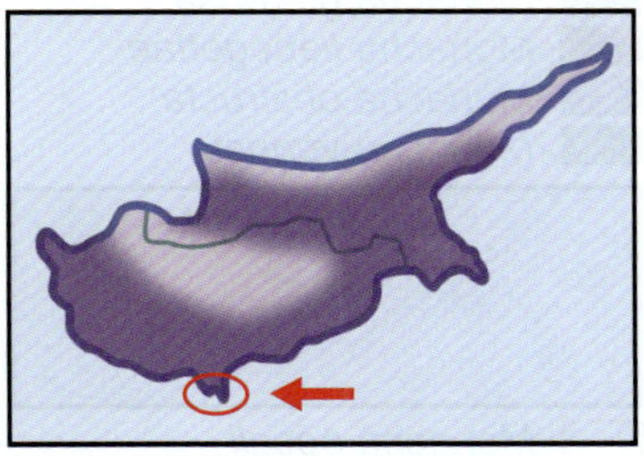

Trochoidea carinatoglobosa
Trochoidea liebetruti

Xeropicta akrotirica

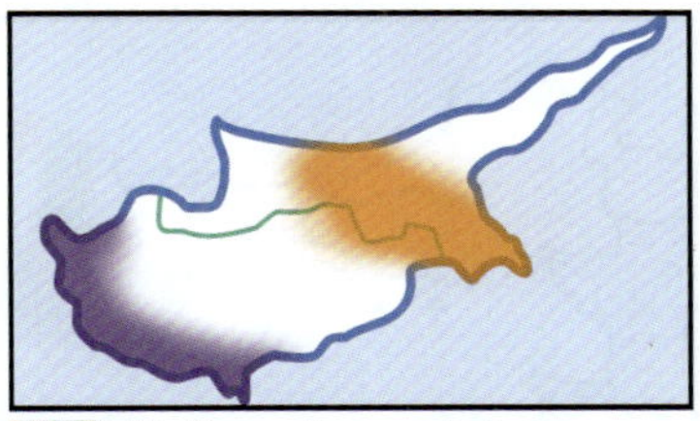

Helicopsis cypriola
Xerocrassa nicosiana

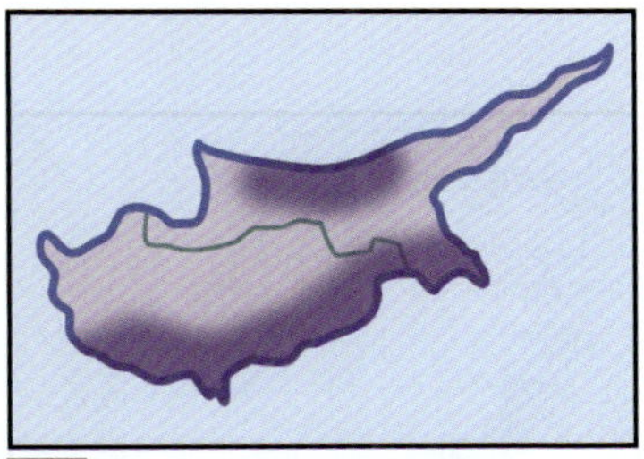

Xeropicta ledereri

Hel jug Pse con Tro car Xpi led G. Konstantinou

Hygromiidae 10 TR

Monacha claustralis

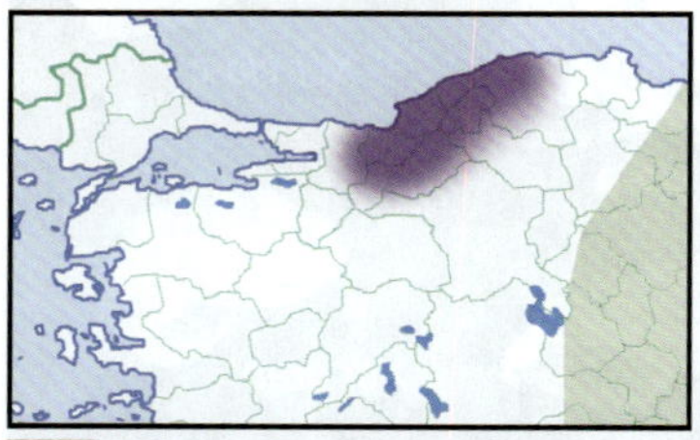

Monacha margarita

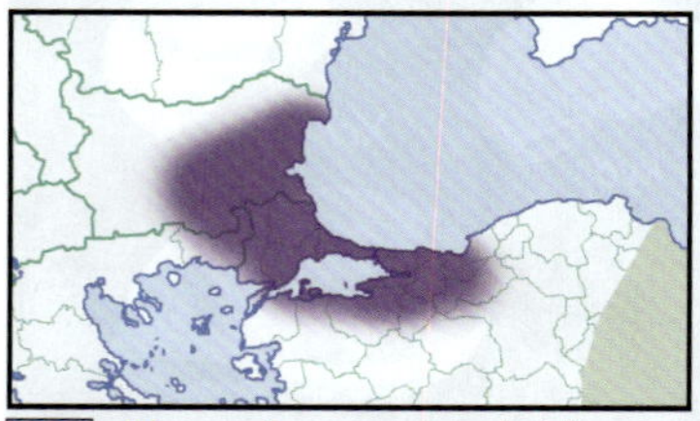

Monacha carascaloides

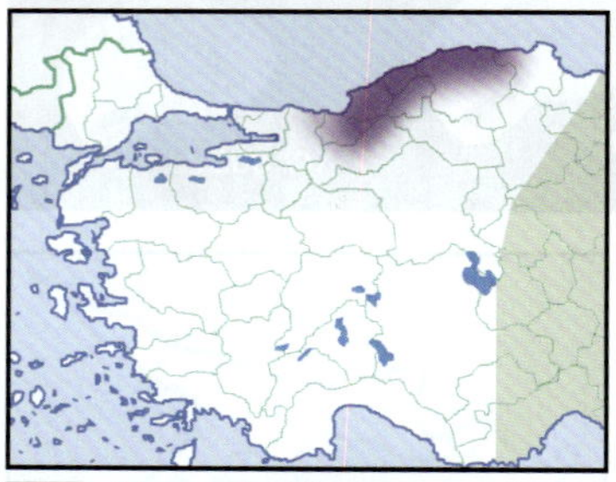

Monacha densecostulata

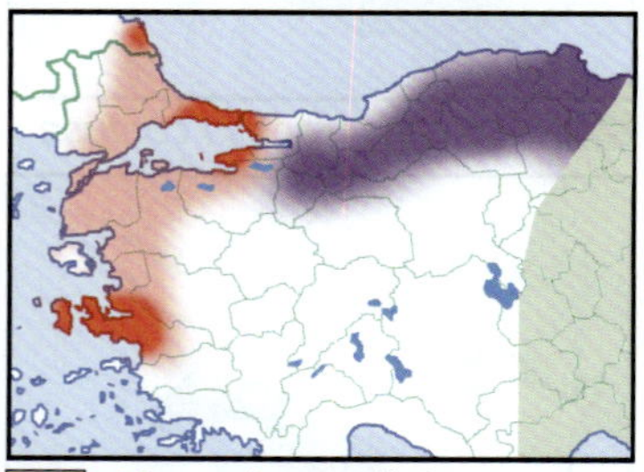

Monacha ocellata
Monacha samsunensis

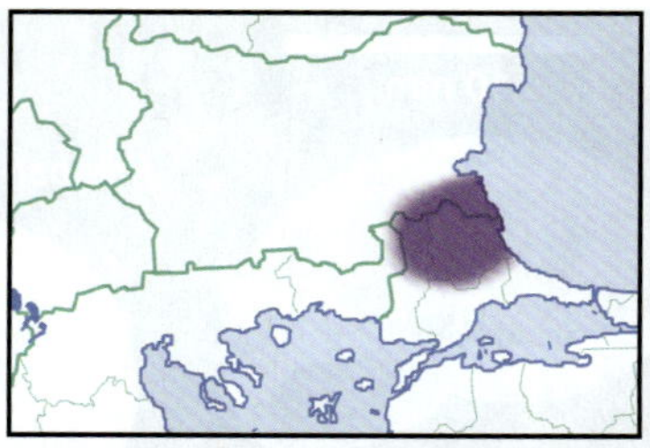

Monacha ovularis

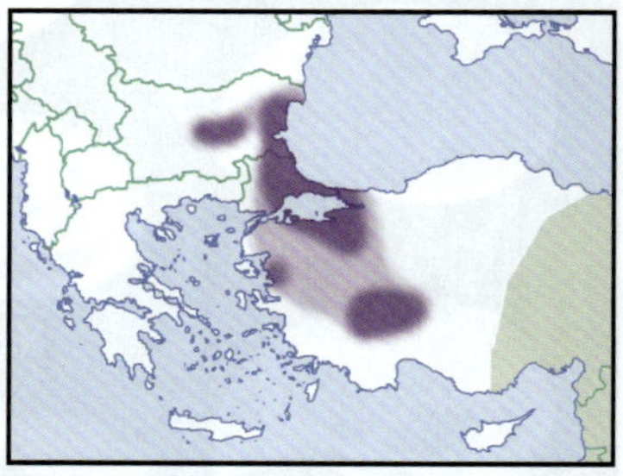

Monacha solidior

Monacha magna

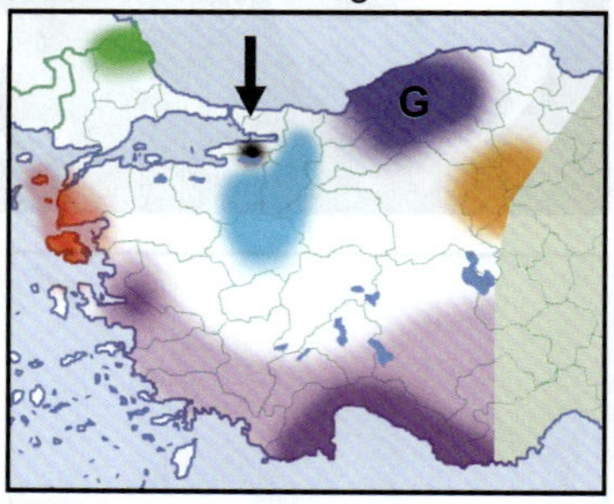

Monacha ascania
Monacha bithynica
Monacha galatica
G *Monacha glareosa*
Monacha liebegottae
Monacha obstructa
Monacha venusta

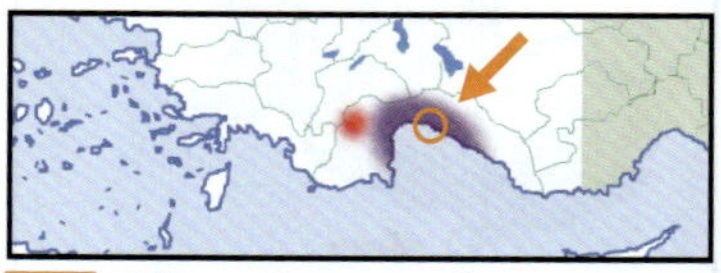

Monacha oecali
Monacha pamphylica
Monacha subaii

Mon lie P. Ovalis *Mon ven* A. Irikov *Mon sam* A. Örstan

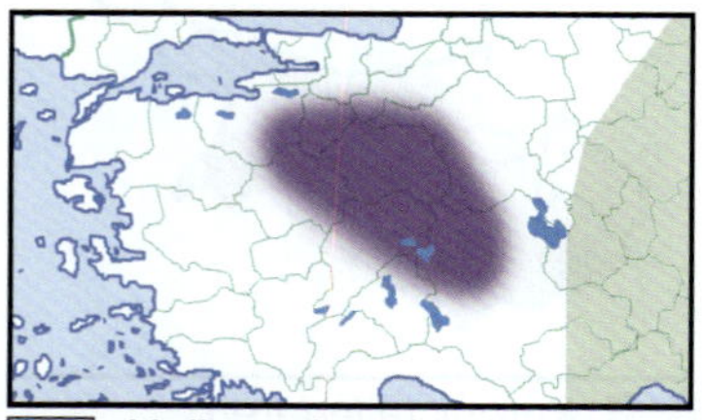

Helicopsis subcalcarata

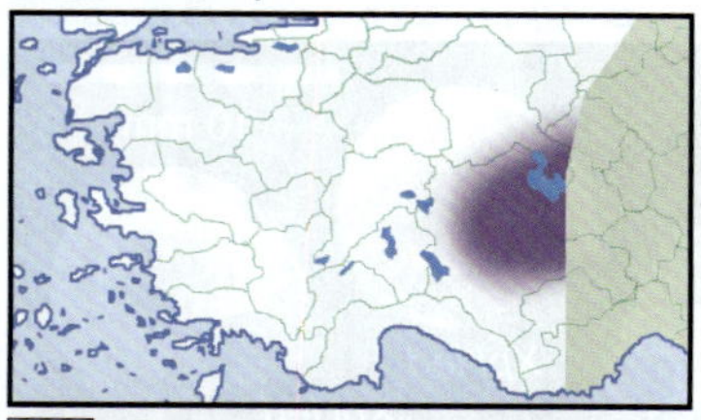

Pseudotrichia memnonis

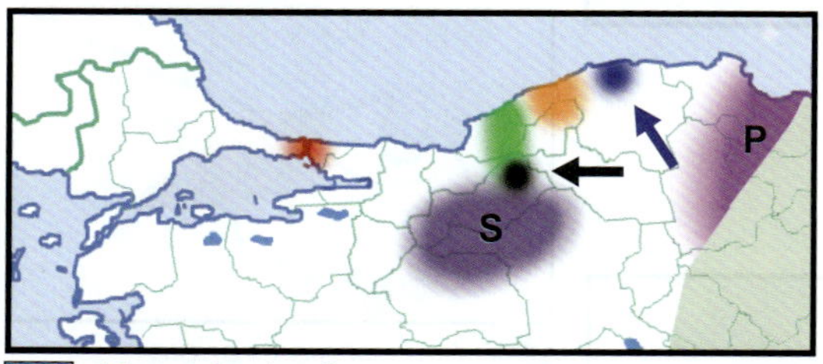

Monacha badia
Monacha crenophila
Monacha devrekensis
Monacha georgievi
Monacha leucozona
P *Monacha phazimonitica*
S *Monacha stipulifera*

Monacha syriaca

Ariantinae 1

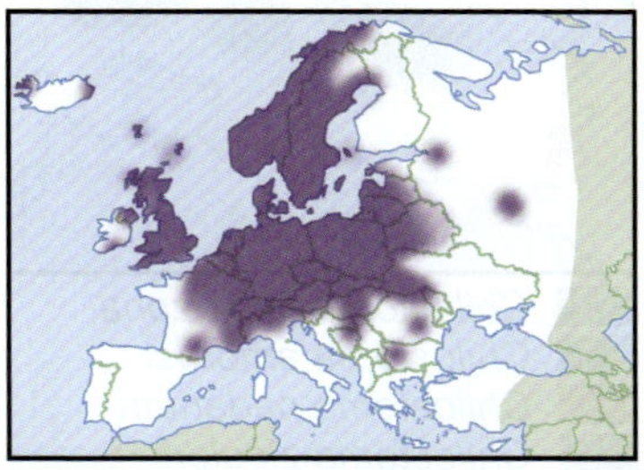

Arianta arbustorum

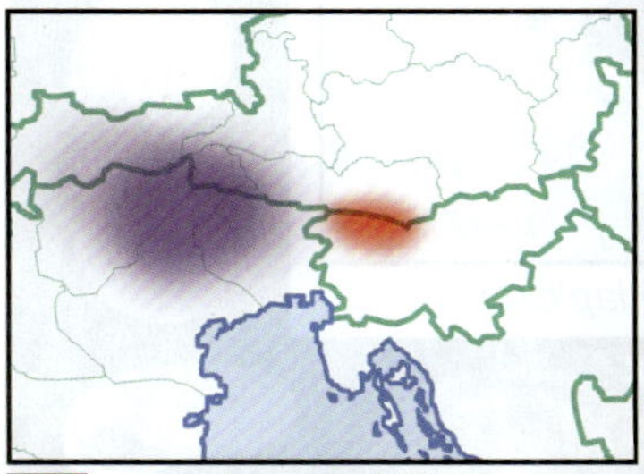

Arianta schmidtii
Arianta stenzii

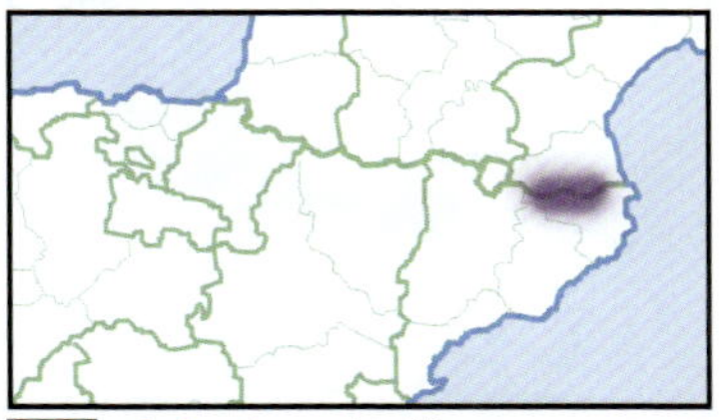

Arianta xatartii

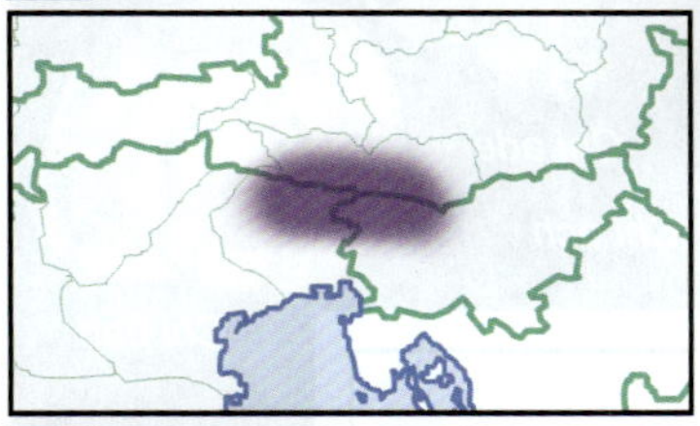

Arianta chamaeleon

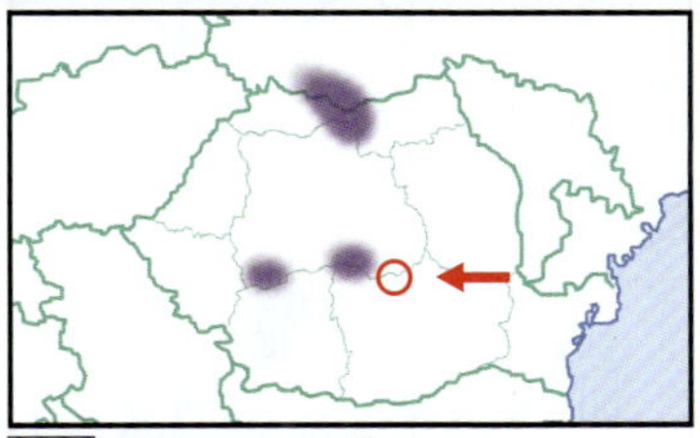

Arianta aethyops
Arianta hessei

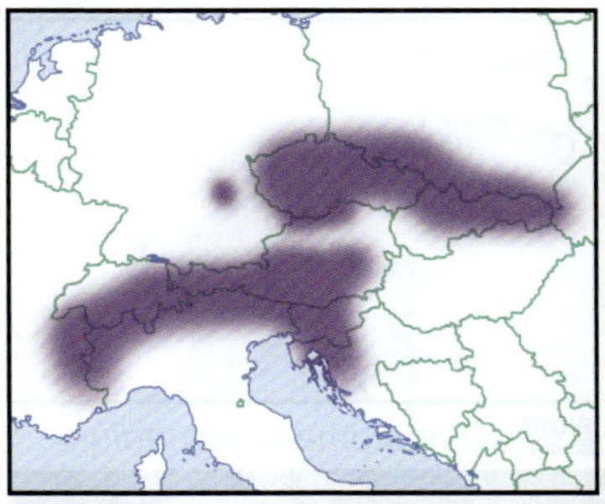

Causa holosericea

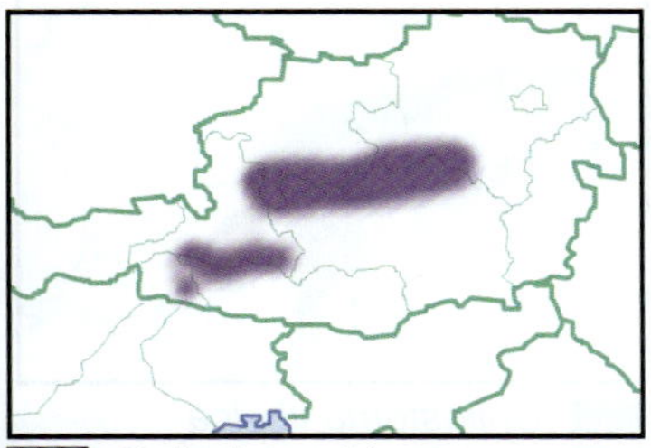

Cylindrus obtusus

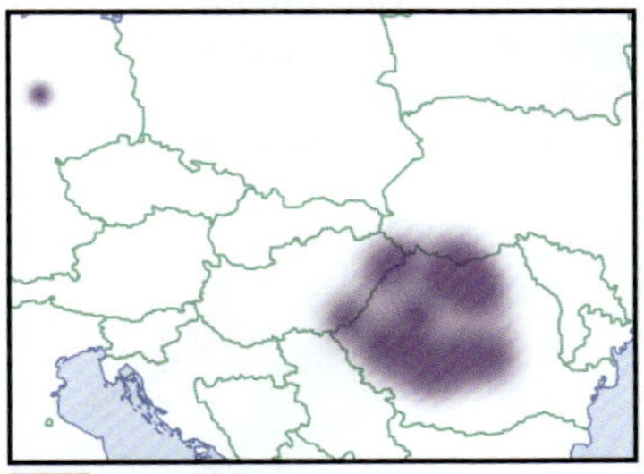

Drobacia banatica

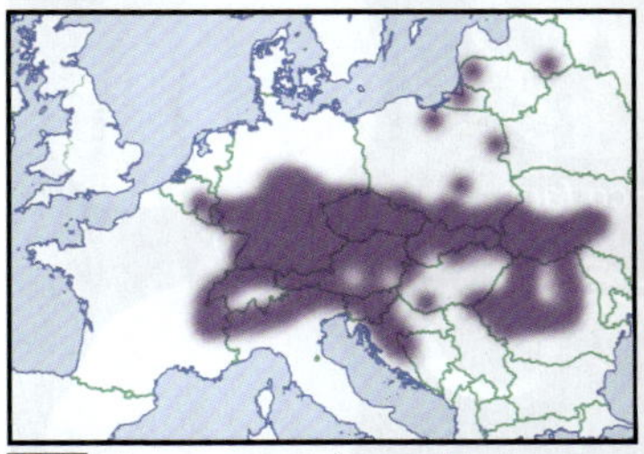

Isognomostoma isognomostomos

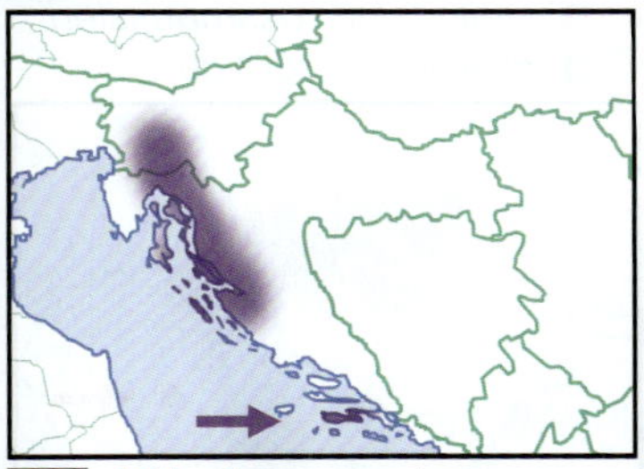

Vidovicia caerulans

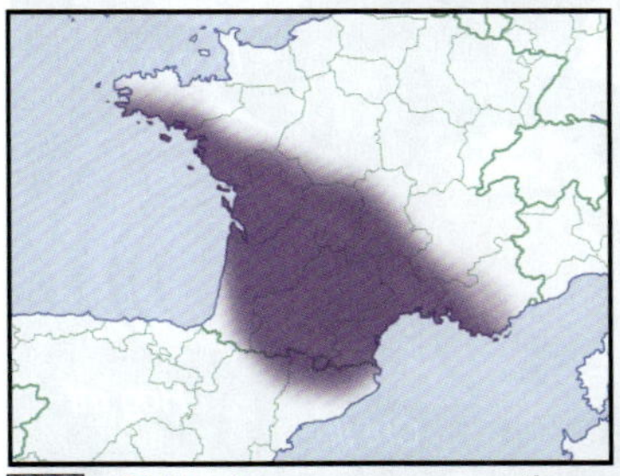

Chilostoma squammatinum

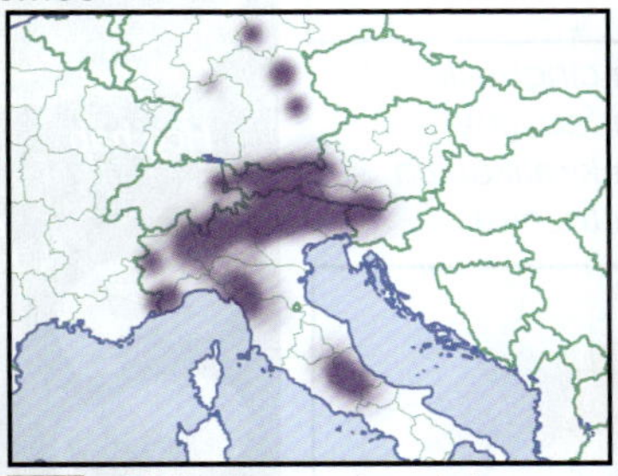

Chilostoma cingulatum

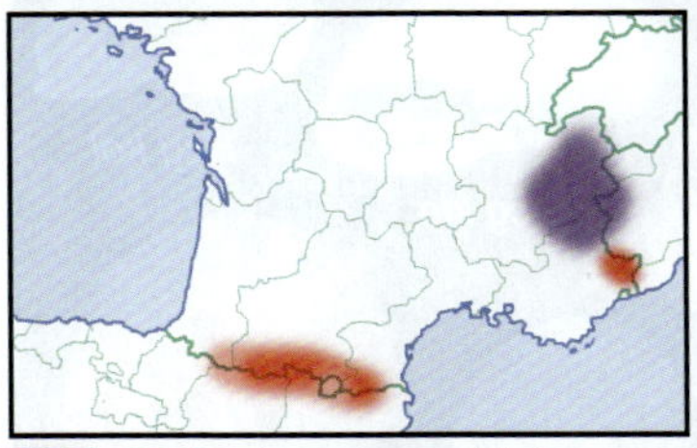

Chilostoma acrotricha
Chilostoma fontenillii

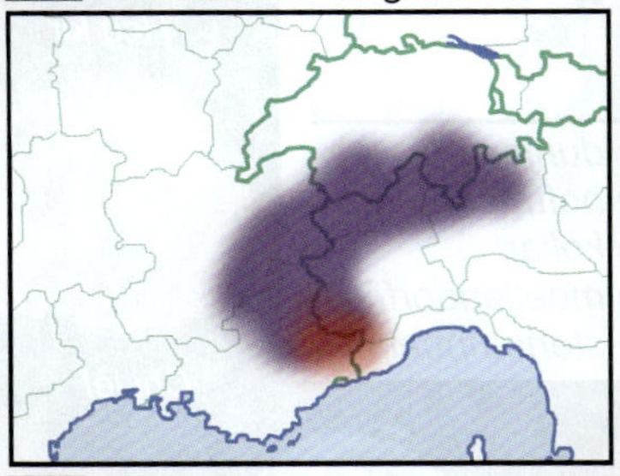

Chilostoma millieri
Chilostoma zonatum

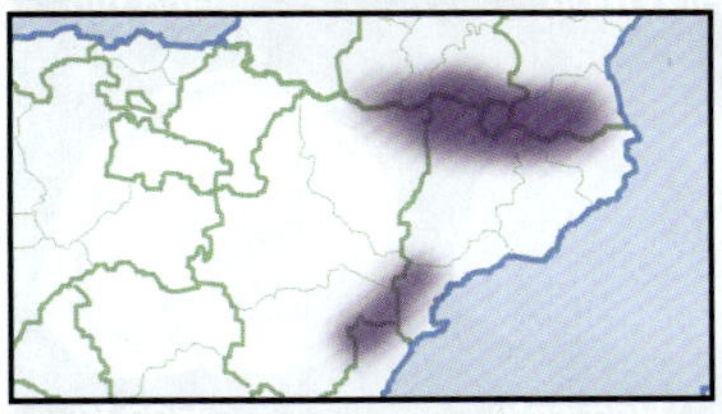

Chilostoma desmoulinsii

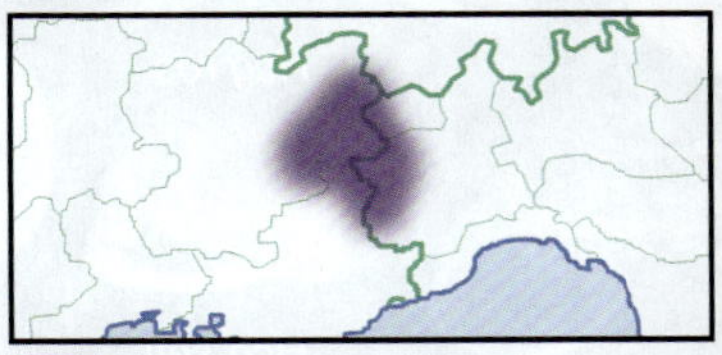

Chilostoma glaciale

Ariantinae 2

Hcg faus A. Chaniecka Hcg kle Hcg sad M. Prodan Hcg lap Boronian Hcg ros J. Grego

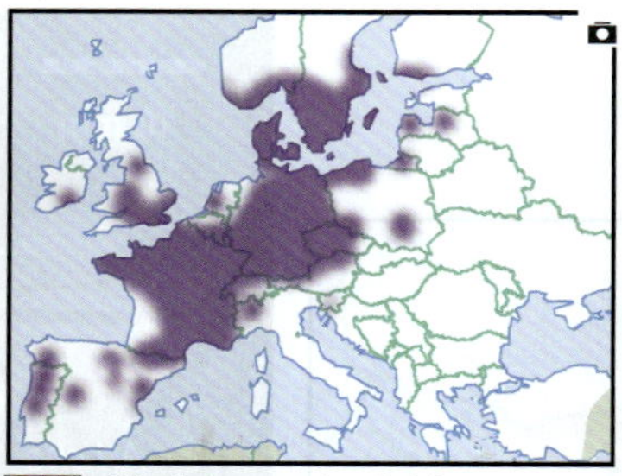

Helicigona lapicida

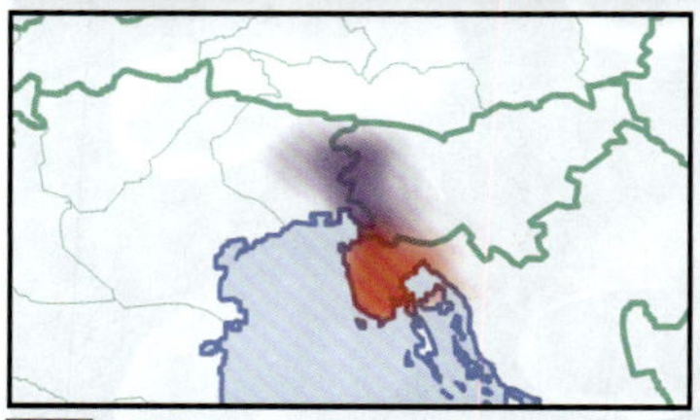

Chilostoma lefeburianum
Chilostoma sadlerianum

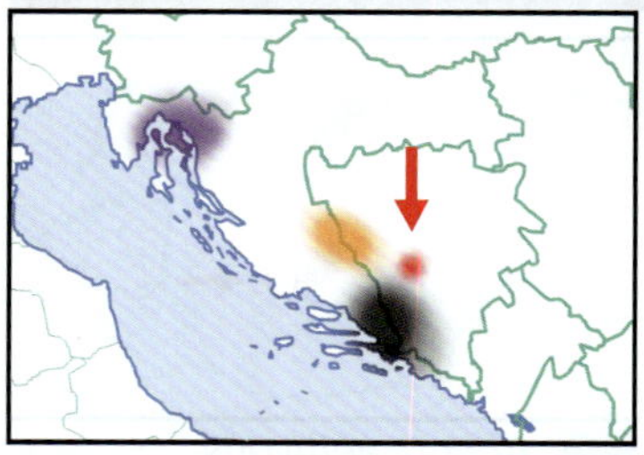

Chilostoma hirtum
Helicigona crinita
Helicigona kleciachi
Helicigona pentheri

Chilostoma planospira
Helicigona setosa

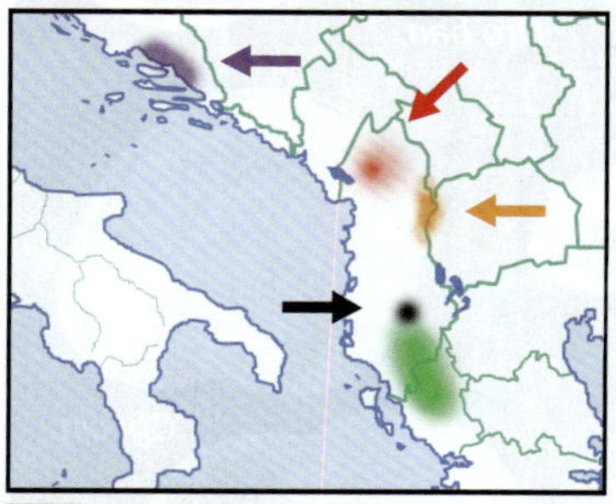

Helicigona albanograeca
Helicigona denudata
Helicigona grisea
Helicigona korabensis
Helicigona petrovici

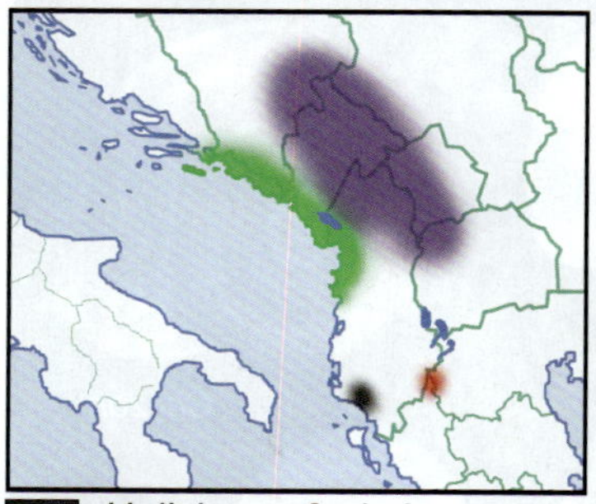

Helicigona fuchsi
Helicigona inflata
Helicigona sattmanni
Helicigona setigera

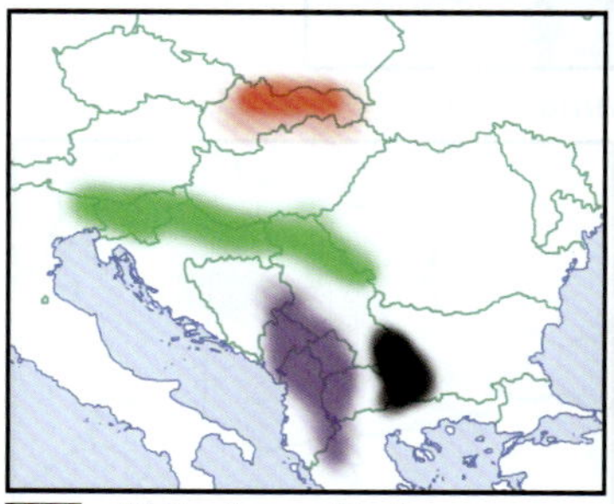

Chilostoma illyrica
Helicigona haberhaueri
Helicigona rossmaessleri
Helicigona serbica

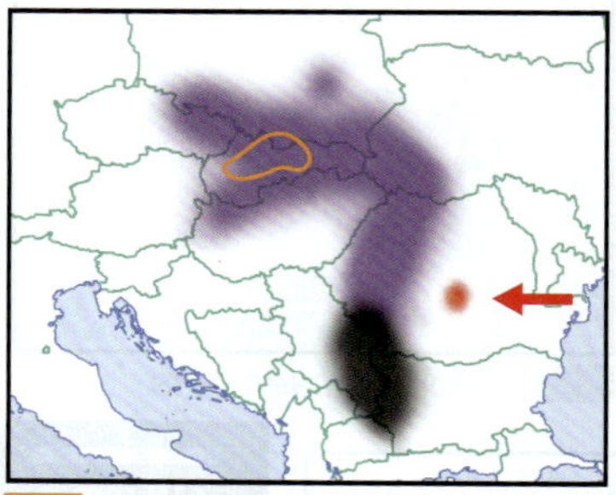

Helicigona cingulella
Helicigona faustina
Helicigona kiralikoeica
Helicigona trizona

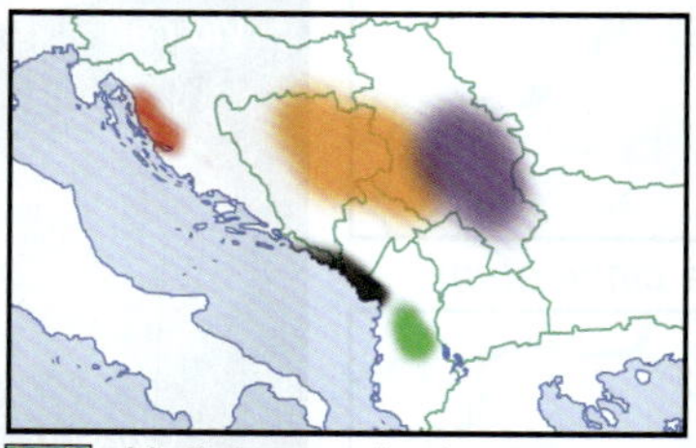

Helicigona dunjana
Helicigona hoffmanni
Helicigona kollari
Helicigona moellendorffii
Helicigona stenomphala

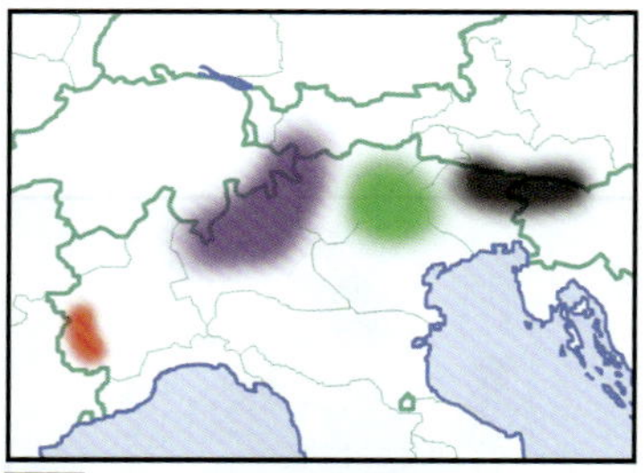

Chilostoma adelozona
Chilostoma ambrosi
Chilostoma padanum
Chilostoma ziegleri

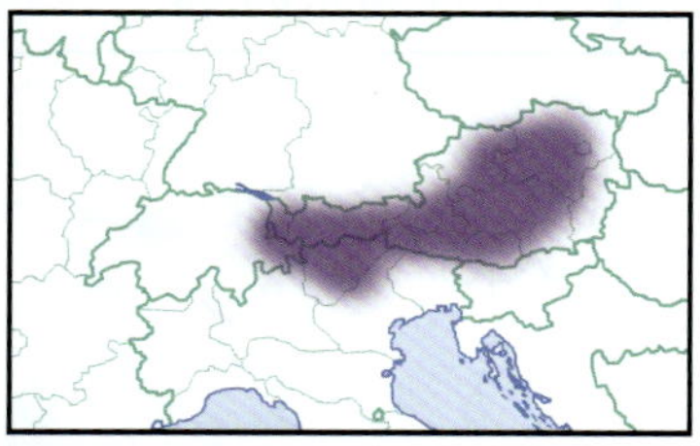

Chilostoma achates

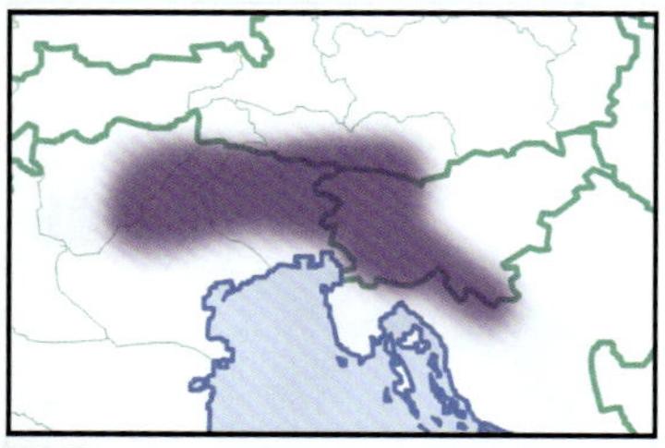

Chilostoma intermedium

Ariantinae 3

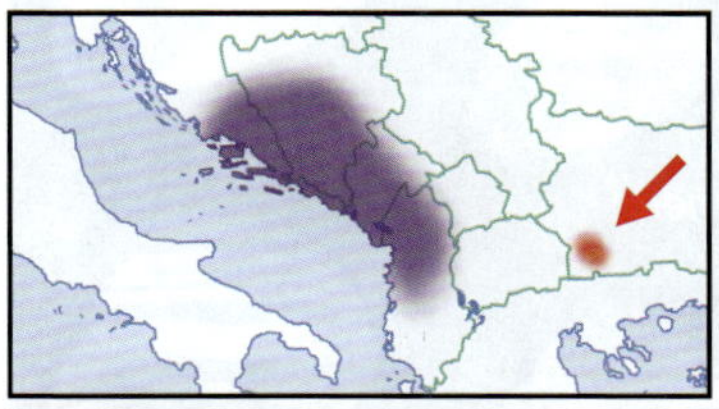

Helicigona polinskii
Helicigona pouzolzii
Helicigona sztolcmani

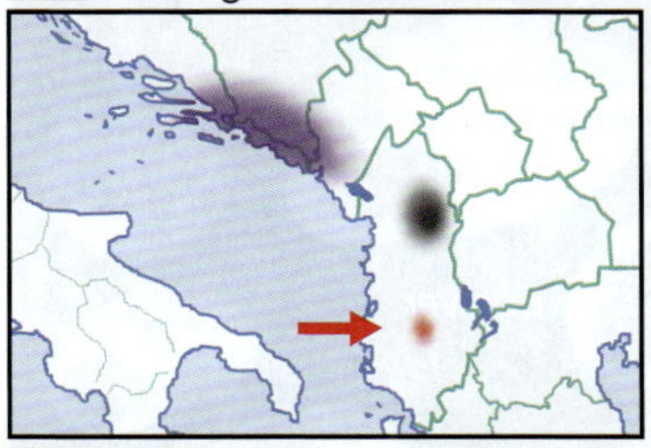

Helicigona dochii
Helicigona harpya
Helicigona insolita

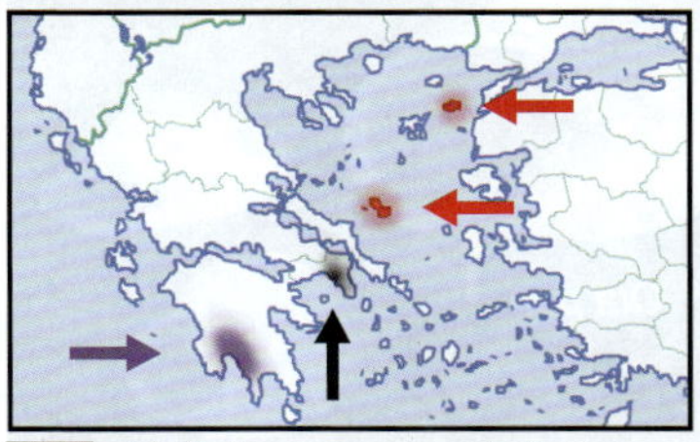

Helicigona argentellei
Helicigona cyclolabris
Helicigona heldreichi

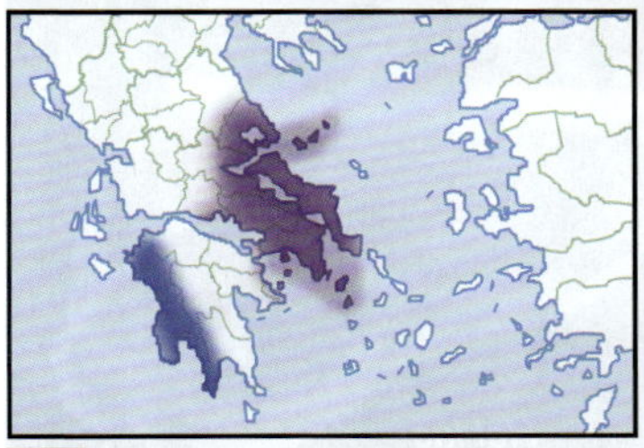

Helicigona comephora
Helicigona sphaeriostoma

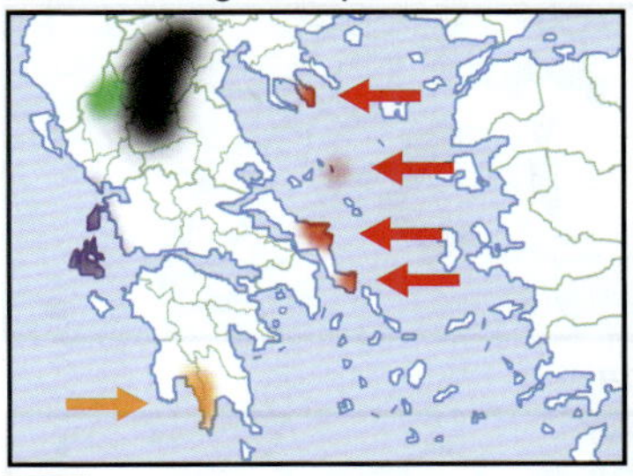

Helicigona euboeae
Helicigona hemonica
Helicigona krueperi
Helicigona reischuetzi
Helicigona subzonata

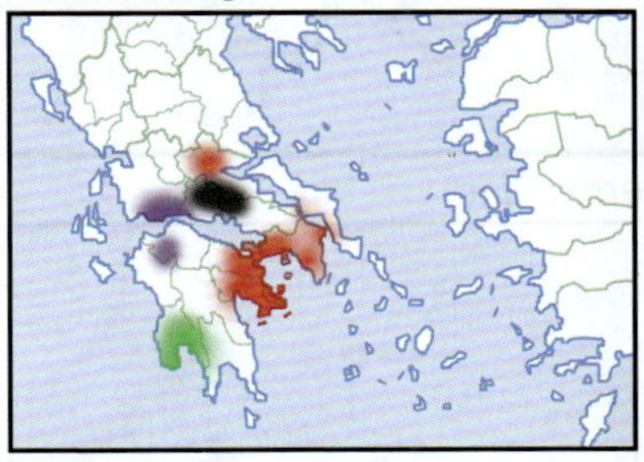

Helicigona brenskei
Helicigona conemenosi
Helicigona hymetti
Helicigona phocaea

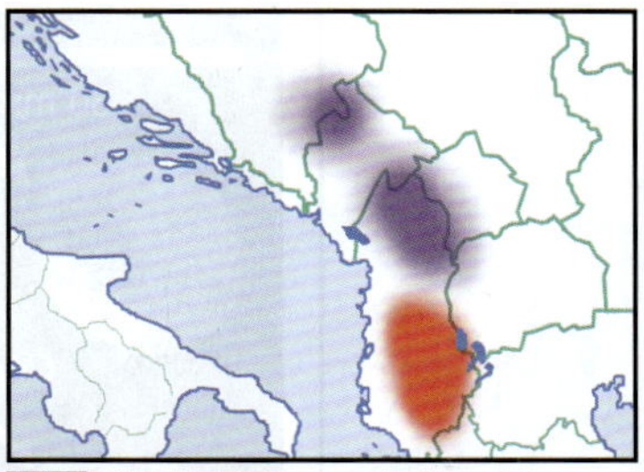

Helicigona apfelbecki
Helicigona byshekensis

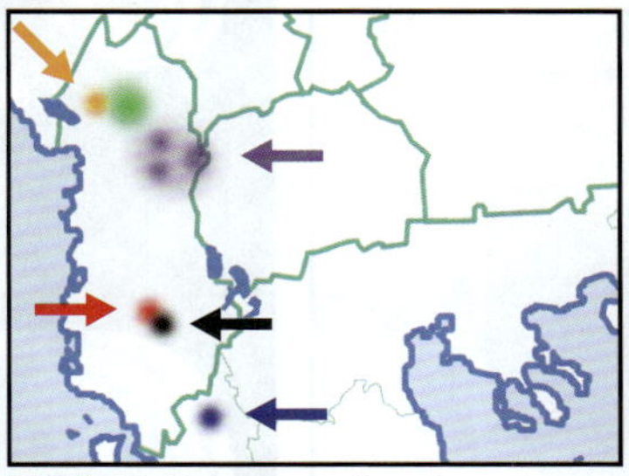

Helicigona albanica
Helicigona kulmakana
Helicigona maranajensis
Helicigona skipetarica
Helicigona vikosensis
Helicigona zebiana

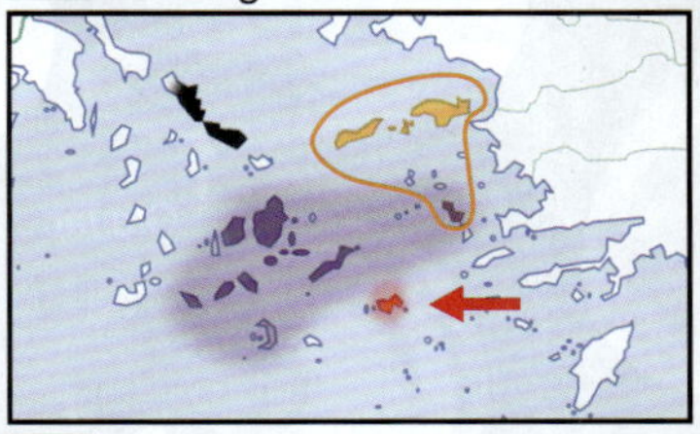

Helicigona bacchica
Helicigona fuchsiana
Helicigona nympha
Helicigona posthuma

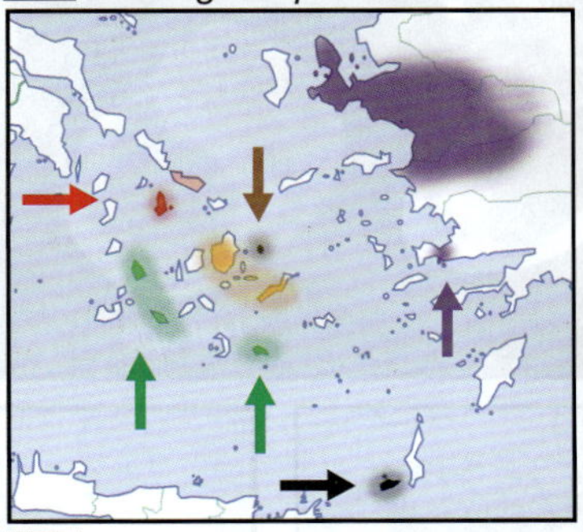

Helicigona amorgia
Helicigona arcadica
Helicigona matrella
Helicigona melpomene
Helicigona pieperi
Helicigona polyhymnia

Hcg mat A. Örstan Hcg pel A. Irikov

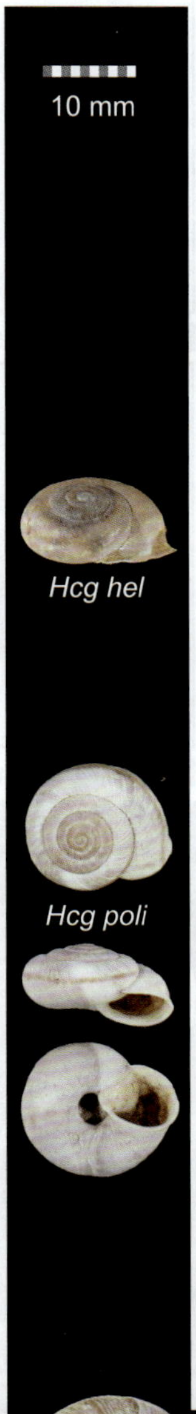

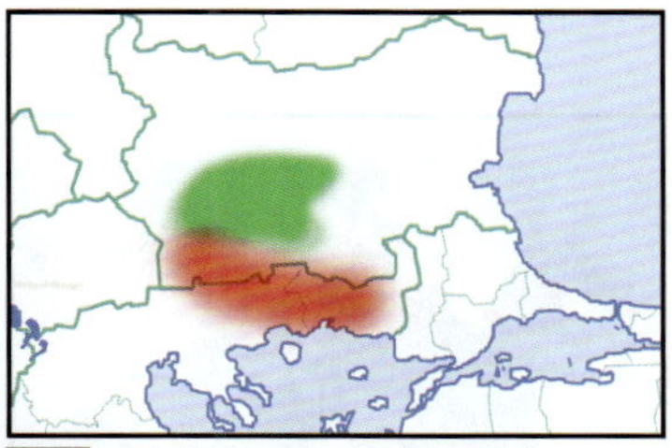

Helicigona kattingeri
Helicigona pelia

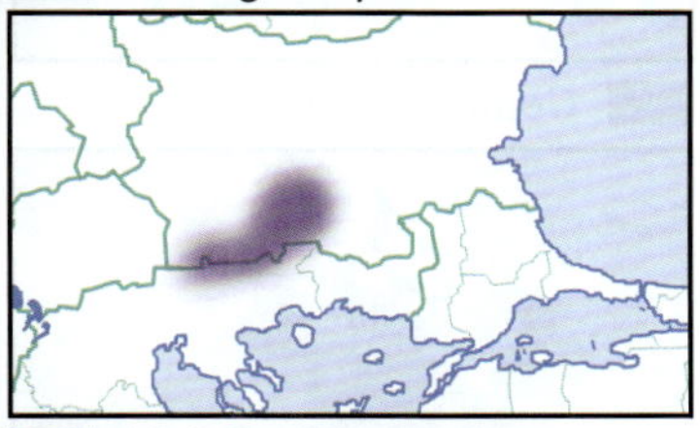

Helicigona rumelica

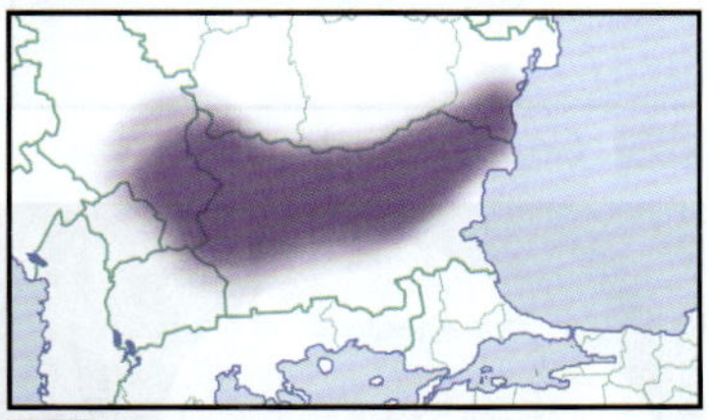

Helicigona balcanica

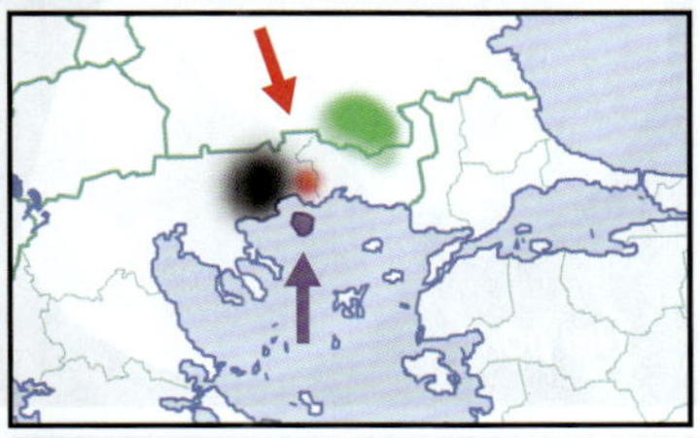

Helicigona ardica
Helicigona faueri
Helicigona subaii
Helicigona valkanovi

Helicinae 1

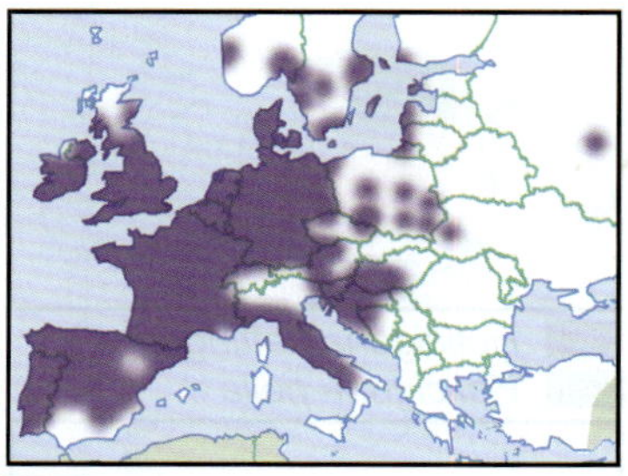

Cepaea nemoralis

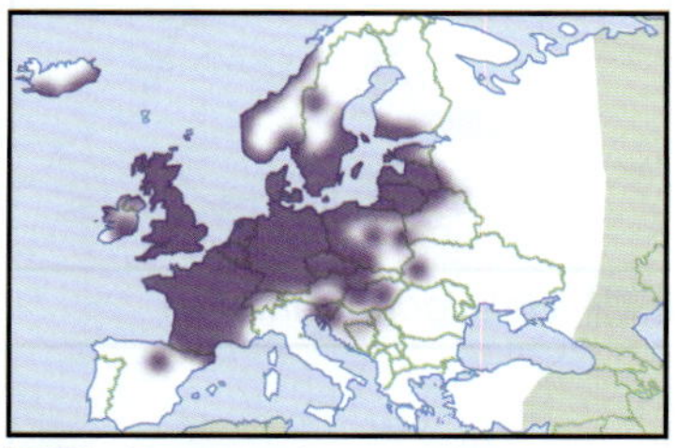

Cepaea hortensis

Cepaea sylvatica

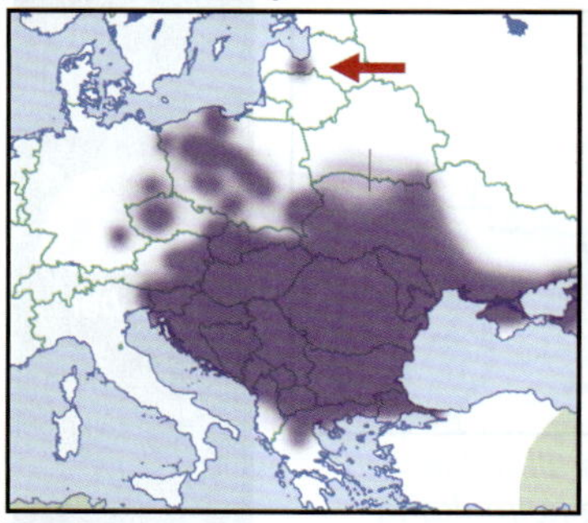

Cepaea vindobonensis

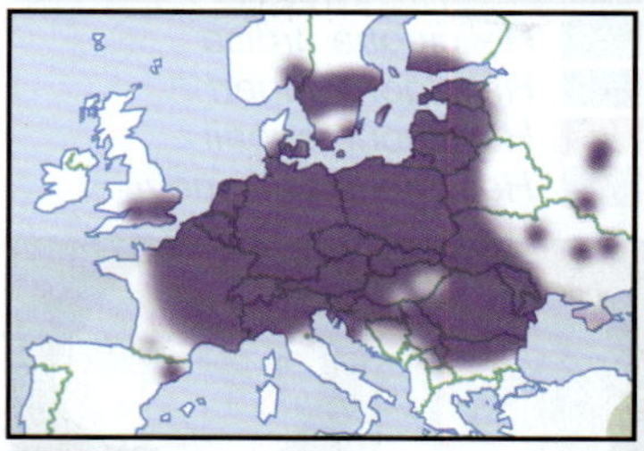

Helix pomatia

Helix cincta

Helix lucorum

Helix aperta

Theba pisana

Helix aspersa

Eobania vermiculata

Otala lactea

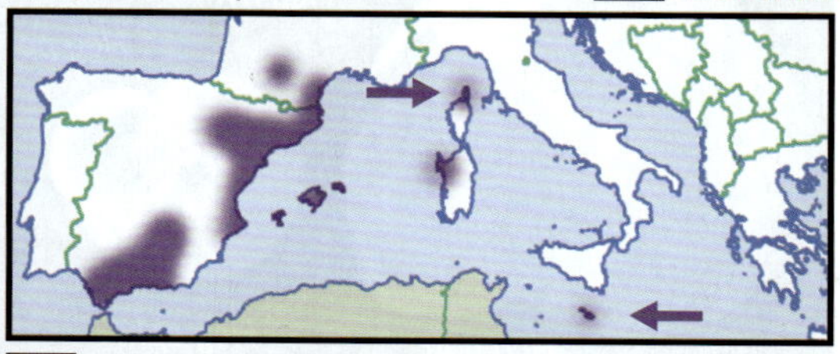

Otala punctata

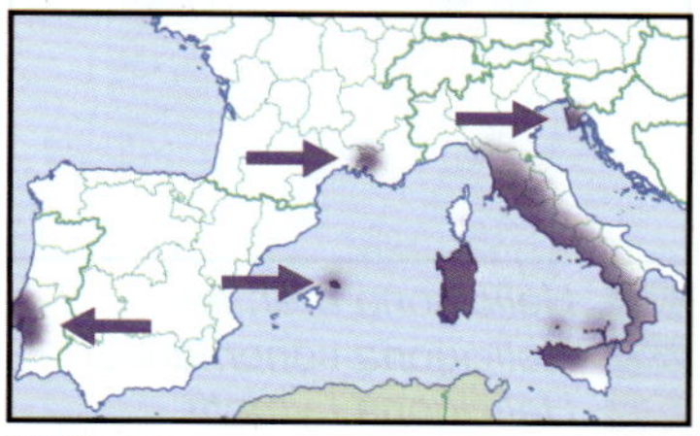

Marmorana muralis

Helicinae 2

Hel del Hel mil Mar sax A. Hallgass
The and J. Liétor Gallego

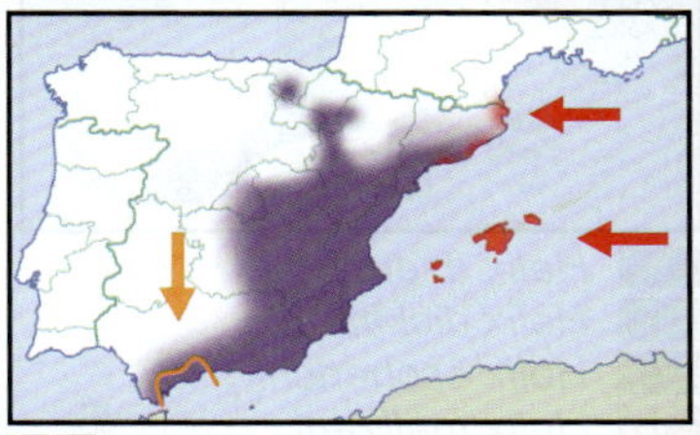

Iberus gualtierianus
Iberus marmoratus
Iberellus hispanicus

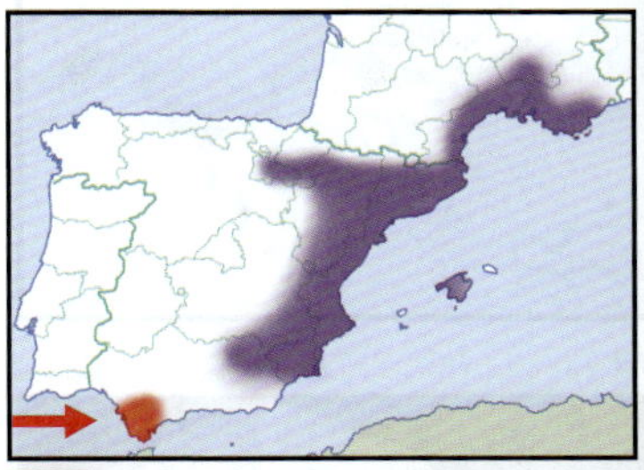

Pseudotachea liturata
Pseudotachea splendida

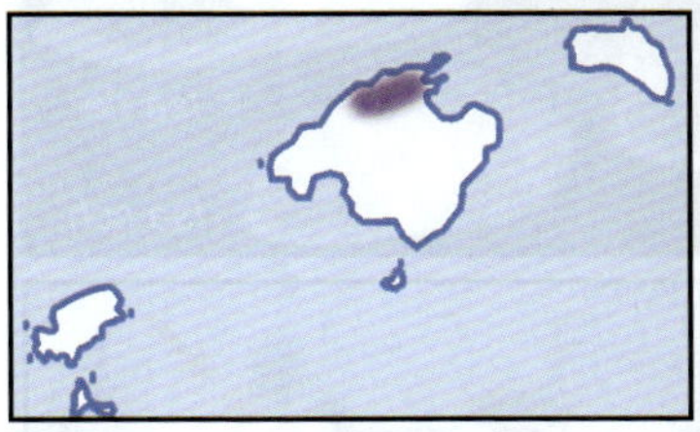

Allognathus grateloupi

Theba andalusica
Theba subdentata

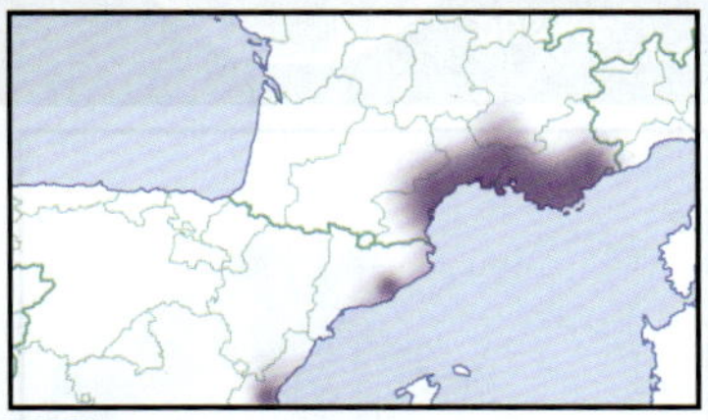

Helix melanostoma

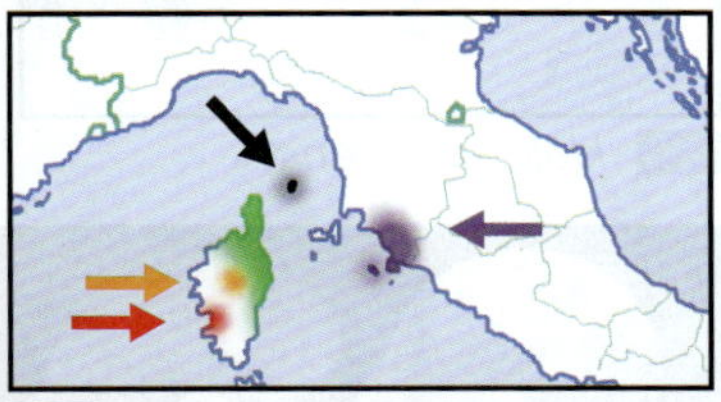

Marmorana saxetana
Tyrrhenaria ceratina
Tacheocampylaea raspailii
Tacheocamp. romagnolii
Tacheocamp. tacheoides

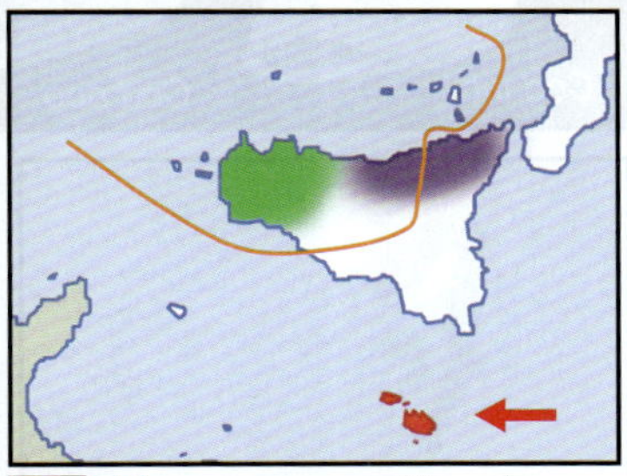

Marmorana melitensis
Marmorana muralis
Marmorana nebrodensis
Marmorana scabriuscula

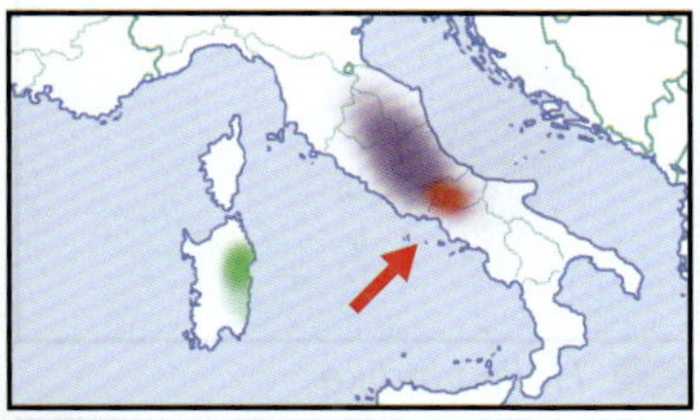

Helix delpretiana
Helix mileti
Tyrrheniberus villicus

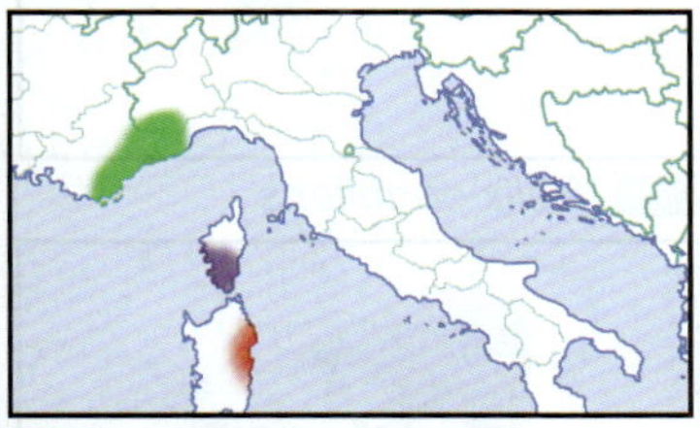

Macularia niciensis
Tacheocampylaea acropachia
Tacheocampylaea carotii

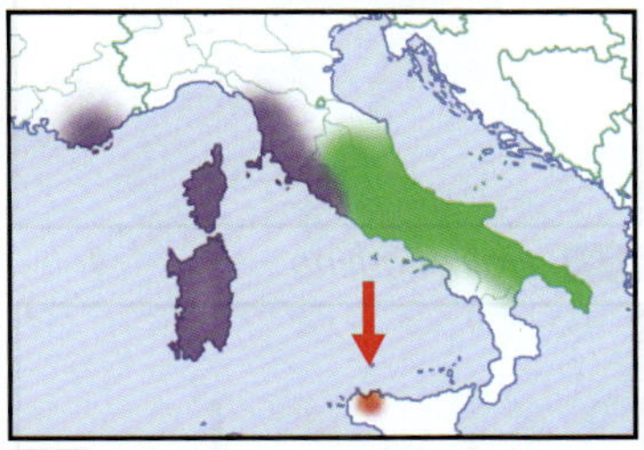

Marmorana globularis
Marmorana serpentina
Marmorana signata

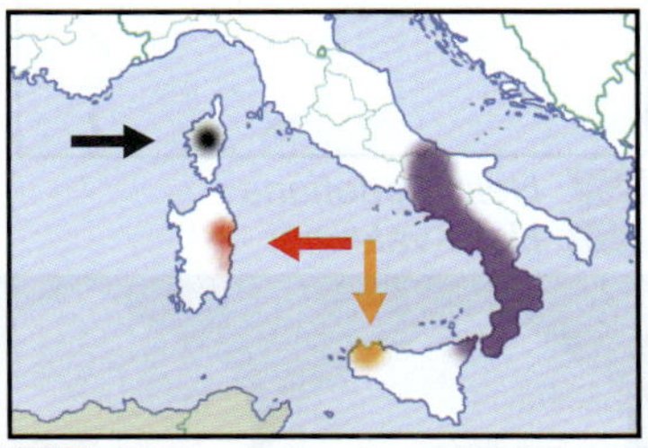

Marmorana fuscolabiata
Marmorana platychela
Tacheocampylaea cyrniaca
Tyrrheniberus sardonius

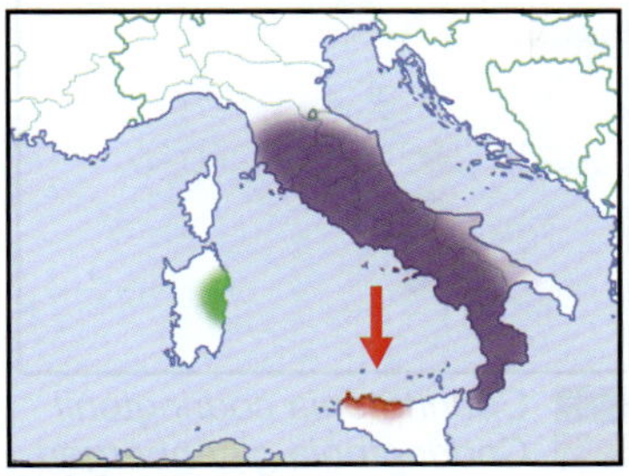

Helix ligata
Helix mazzullii
Tyrrheniberus ridens

Helicinae 3

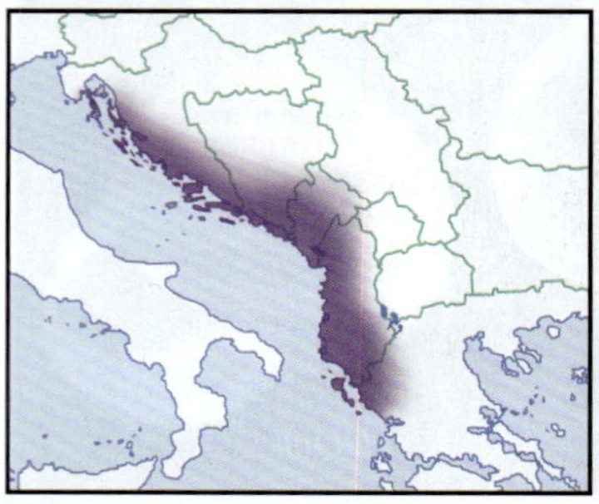

Helix secernenda

Helix vladika

Helix dormitoris

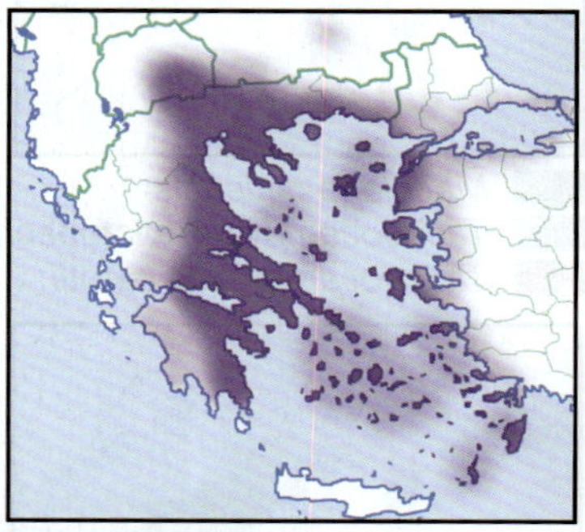

Helix figulina

Tacheopsis nicaeensis
Caucasotachea atrolabiata

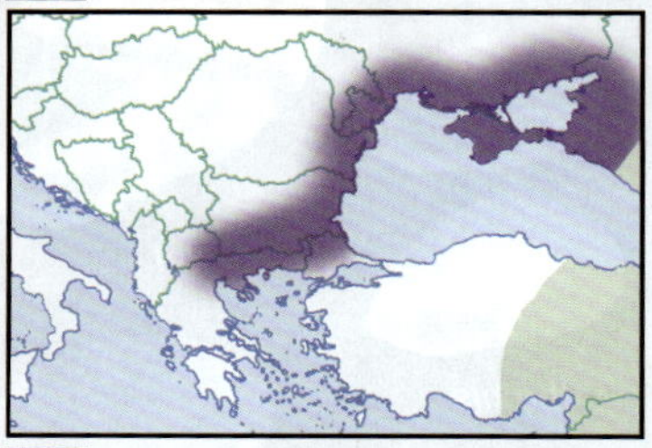

Helix albescens

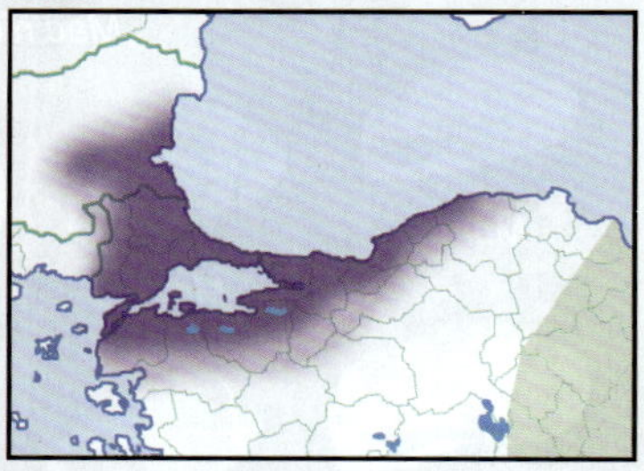

Helix pomacella

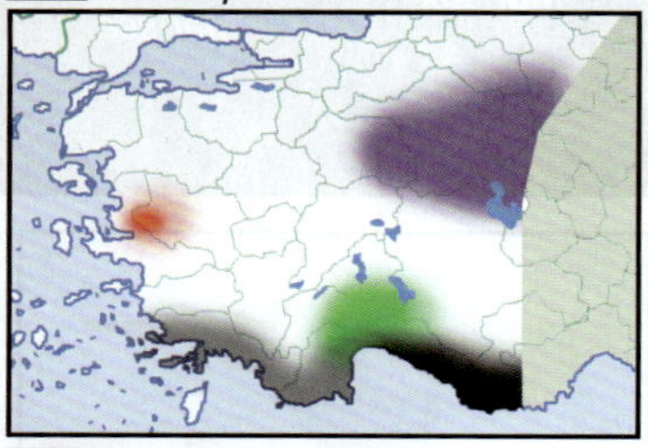

Helix asemnis
Helix dickhauti
Helix escherichi
Helix maltzani

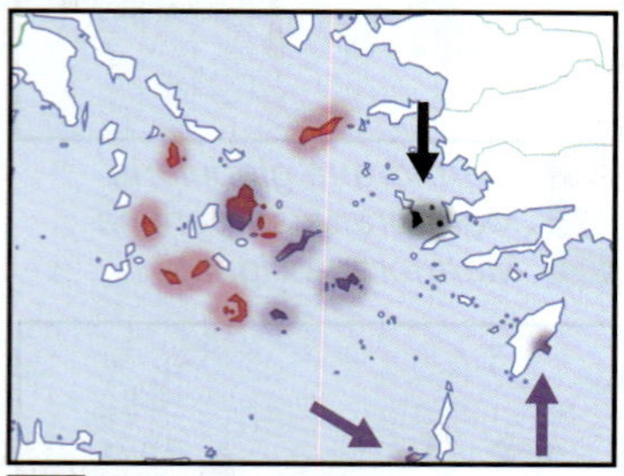

Helix godetiana
Helix valentini

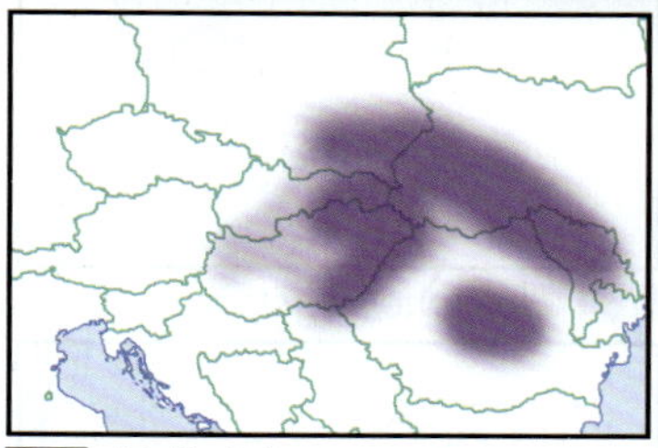

Helix lutescens

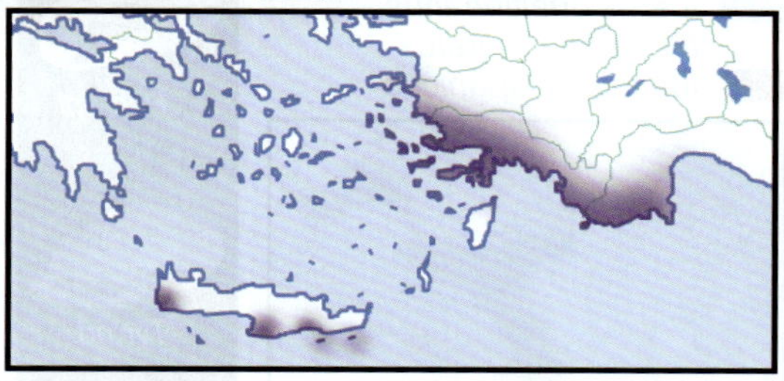

Helix nucula

Ass bel G. Konstantinou

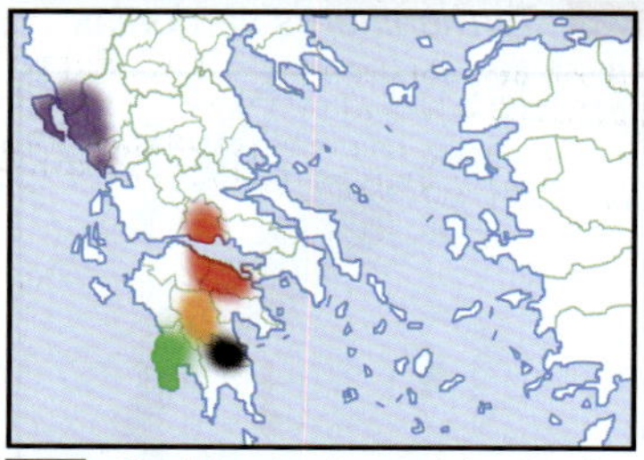

Codringtonia codringtonii
Codringtonia gittenbergeri
Codringtonia helenae
Codringtonia intusplicata
Codringtonia neocrassa

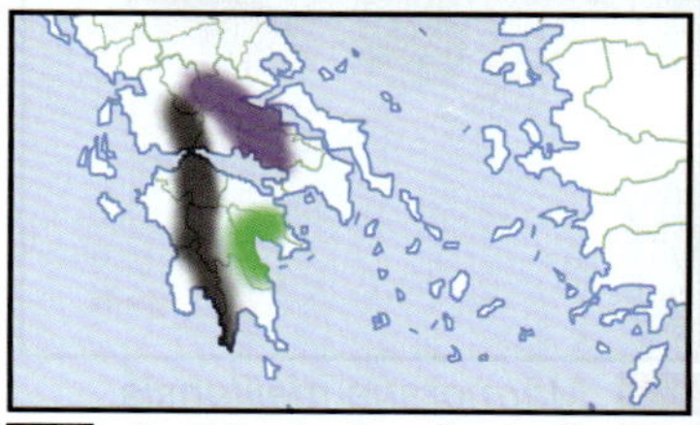

Codringtonia elisabethae
Codringtonia eucineata
Codringtonia parnassia

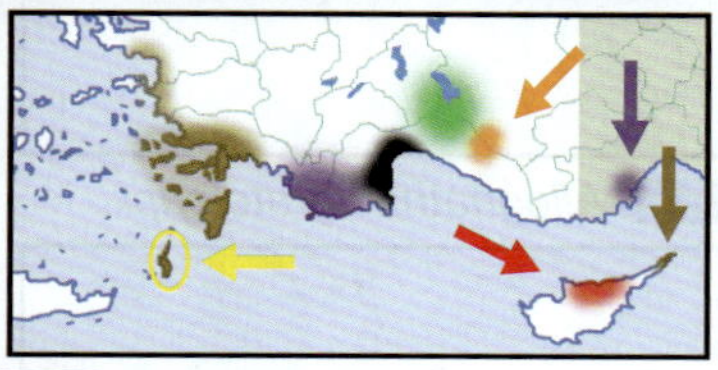

Assyriella bellardii
Assyriella rechingeri
Isaurica lycia
Isaurica pamphylica
Isaurica riedeli
Isaurica schuetti
Levantina spiriplana